Federico Capasso and Dennis Couwenberg (Eds.)

Frontiers in Optics and Photonics

Also of interest

Nanophotonics
Volker Sorger (Editor in Chief)
e-ISSN 2192-8614

Micro-Raman Spectroscopy
Theory and Application
Popp, Mayerhöfer (Eds.), 2020
ISBN 978-3-11-051479-7, e-ISBN 978-3-11-051531-2

Photonic Reservoir Computing
Optical Recurrent Neural Networks
Brunner, Soriano, Van der Sande (Eds.), 2019
ISBN 978-3-11-058200-0, e-ISBN 978-3-11-058211-6

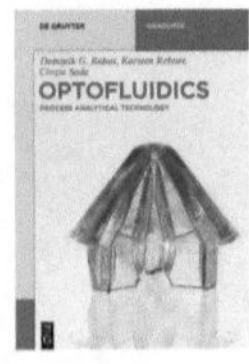

Optofluidics
Process Analytical Technology
Rabus, Sada, Rebner, 2019
ISBN 978-3-11-054614-9, e-ISBN 978-3-11-054615-6

Frontiers in Optics and Photonics

Edited by Federico Capasso and Dennis Couwenberg

DE GRUYTER

Editors
Prof. Federico Capasso
Harvard John A. Paulson
School of Engineering & Applied Sciences
Pierce Hall 205A
29 Oxford Street
Cambridge MA 02138
USA

Drs. Ing. Dennis Couwenberg
Science Wise Publishing
2e Jacob van Campenstraat 89 3hg L
1073 XN Amsterdam
Netherlands

ISBN 978-3-11-070973-5
e-ISBN (PDF) 978-3-11-071068-7
e-ISBN (EPUB) 978-3-11-071070-0

Library of Congress Control Number: 2020952274

Bibliographic information published by the Deutsche Nationalbibliothek
The Deutsche Nationalbibliothek lists this publication in the Deutsche Nationalbibliografie; detailed bibliographic data are available on the Internet at http://dnb.dnb.de.

Cover image: Prof. Marlan Scully
Typesetting: TNQ Technologies Pvt. Ltd.
Printing and binding: CPI books GmbH, Leck

www.degruyter.com

Preface

This book holds a compilation of chapters in many exciting and rapidly developing areas of optics and photonics, a discipline that continues to enable fundamental advances and new technology touching an ever-increasing number of fields and applications.

It includes 10 perspective, 12 review and 41 research related chapters on topics as Quantum Optics, Quantum Computing, QED, Fundamental of Optics, Topological Photonics, Metamaterials, Plasmonics, Spectroscopy, Lasers, LEDs, Fiber Optics, Optical Sensors, Opto-electronics, Integrated Photonics, Flat Optics, Optimization Methods and Biomedical Photonics.

Among the list of scientists that authored these chapters, you will find one by Nobel Laureate Dr. Shuji Nakamura, co-authored with Dr. C. Weisbuch and Dr. Y-R Wu, and Dr. J. S. Speck on "Disorder effects in nitride semiconductors: impact on fundamental and device properties – Optoelectronics and Integrated Photonics".

This book is the result of a COVID-19 benefit issue that was organised by the journal *Nanophotonics* in the Spring of 2020 at the start of the Pandemic. 60+ leading scientists in the field of Optics and Photonics were invited to contribute to this special edition, which has raised a total of Euro 136.800, that will be donated to support the COVID-19 first aid workers and people in need of medical help around the world.

The image of the COVID-19 virus on the cover of this book was kindly offered by Dr. M. Scully and his colleagues; it comes from their chapter titled "A fiber optic–nanophotonic approach to the detection of antibodies and viral particles of COVID-19". In this chapter, they provide a tutorial and a preview on the use of a variation of laser spectroscopic techniques they developed for the rapid detection of anthrax that can also be applied to detect COVID-19.

We like to express our deep gratitude to all the scientists that contributed and supported this benefit. It is inspiring to see how collectively scientists stand up for this humanitarian disaster.

We also like to thank the publisher of Nanophotonics, DeGruyter, who has been so generous to donate to charity all the proceeds from the publication charges of papers published in the *Nanophotonics* issue, as well as the many people at DeGruyter that offered their professional service to publish the journal issue and this book. We also like to thank Tara Dorian for her incredible support and being the backbone of *Nanophotonics*.

We hope you will find this book stimulating.

Federico Capasso, Dennis Couwenberg

https://doi.org/10.1515/9783110710687-201

Contents

Part V: Fundamentals of Optics

Part VI: Optimization Methods

Part VII: Topological Photonics

Part VIII: Quantum Computing, Quantum Optics, and QED

Part X: Metaoptics

Part I: **Optoelectronics and Integrated Photonics**

Claude Weisbuch*, Shuji Nakamura, Yuh-Renn Wu and James S. Speck

Disorder effects in nitride semiconductors: impact on fundamental and device properties

https://doi.org/10.1515/9783110710687-001

Abstract: Semiconductor structures used for fundamental or device applications most often incorporate alloy materials. In "usual" or "common" III–V alloys, based on the InGaAsP or InGaAlAs material systems, the effects of compositional disorder on the electronic properties can be treated in a perturbative approach. This is not the case in the more recent nitride-based GaInAlN alloys, where the potential changes associated with the various atoms induce strong localization effects, which cannot be described perturbatively. Since the early studies of these materials and devices, disorder effects have indeed been identified to play a major role in their properties. Although many studies have been performed on the structural characterization of materials, on intrinsic electronic localization properties, and on the impact of disorder on device operation, there are still many open questions on all these topics. Taking disorder into account also leads to unmanageable problems in simulations. As a prerequisite to address material and device simulations, a critical examination of experiments must be considered to ensure that one measures intrinsic parameters as these materials are difficult to grow with low defect densities. A specific property of nitride semiconductors that can obscure intrinsic properties is the strong spontaneous and piezoelectric fields. We outline in this review the remaining challenges faced when attempting to fully describe nitride-based material systems, taking the examples of LEDs. The objectives of a better understanding of disorder phenomena are to explain the hidden phenomena often forcing one to use ad hoc parameters, or additional poorly defined concepts, to make simulations agree with experiments. Finally, we describe a novel simulation tool based on a mathematical breakthrough to solve the Schrödinger equation in disordered potentials that facilitates 3D simulations that include alloy disorder.

Keywords: III–V alloys; alloy materials; fundamental and device properties; LEDs; nitride semiconductors.

1 Introduction

Modern semiconductor structures and devices combine different elements, compounds and alloys to form heterostructures. Their importance was recognized by the attribution of the 2000 physics Nobel prize to Zhores Alferov and Herbert Krömer who developed the concepts that resulted in the field of bandgap engineering. To realize many heterostructure designs, the material palette includes alloys because pure compounds or elements often do not have the necessary material properties for specific heterostructure designs. This is best seen in Figure 1, which shows the bandgap map of the major semiconductors. With the exception of the GaAs/AlAs material pair, where the compound materials are nearly lattice matched, other pure compound pairs typically have large lattice mismatched. To obtain materials with acceptable defect levels, strain is reduced compared with pure compound associations by using alloys with intermediate lattice constants in heterostructures. Alloys have random substitutions of atoms at well-defined sites within the crystal structure (these alloys do not have positional disorder). Therefore, electrons and holes experience a random potential at a scale of the unit cell.

In usual III–Vs, alloy disorder has nonessential consequences. As the disorder-induced potential fluctuations are weak, band structure properties can be described by the virtual crystal approximation (VCA) where each potentially

*Corresponding author: **Claude Weisbuch**, Materials Department, University of California, Santa Barbara, California 93106-5050, USA; and Laboratoire de Physique de la Matière Condensée, CNRS, Ecole Polytechnique, IP Paris, 91128 Palaiseau, France, E-mail: weisbuch@engineering.ucsb.edu
Shuji Nakamura and James S. Speck, Materials Department, University of California, Santa Barbara, California 93106-5050, USA, E-mail: shuji@engineering.ucsb.edu (S. Nakamura), speck@ucsb.edu (J.S. Speck)
Yuh-Renn Wu, Graduate Institute of Photonics and Optoelectronics and Department of Electrical Engineering, National Taiwan University, Taipei 10617, Taiwan, E-mail: yrwu@cc.ee.ntu.edu.tw

This article has previously been published in the journal Nanophotonics. Please cite as: C. Weisbuch, S. Nakamura, Y.-R. Wu and J. S. Speck "Disorder effects in nitride semiconductors: impact on fundamental and device properties" *Nanophotonics* 2021, 10. DOI: 10.1515/nanoph-2020-0590.

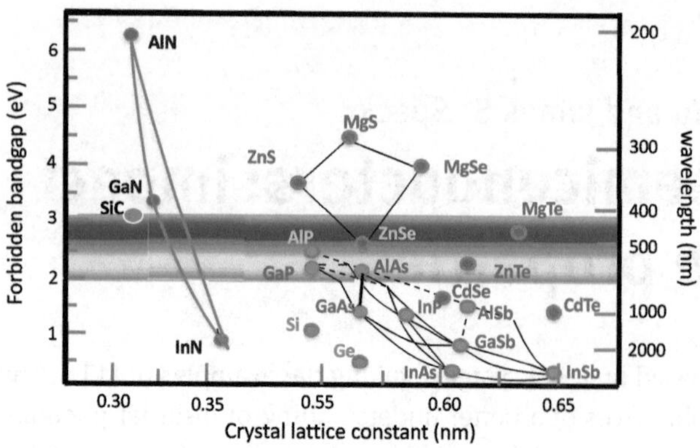

Figure 1: Bandgap map of major semiconductors.

disordered atom site is substituted by an artificial atom interpolating between the properties of the actual constituent atoms (we leave outside this review the very different case of disordered semiconductors where disorder is introduced through dopants, see the study by Shklovskii and Efros [1] for a very thorough analysis). Thus, the crystal is perfectly ordered and carrier quantum states are Bloch waves. To account for disorder, a perturbation is introduced, which is the potential difference between the VCA atom potential and the real atom potential. This allows computation, for example, of the alloy disorder contribution to mobility, as well as the bowing parameter determining the quadratic variation of the bandgap with alloy composition.

In contrast, the random potential due to alloy disorder in nitride semiconductors is so large that it leads to localization properties that cannot be described as perturbations. This significantly impacts in a major way many fundamental properties of the group III nitrides. However, so far, no full microscopic model exists yet to describe optical and transport phenomena as four challenges exist: (1) modeling requires accurate microscopic descriptions of alloy heterostructures, with some open questions on alloy randomness, interface abruptness and composition variation along the growth direction; (2) the high extended and point defect densities existing in nitride materials can add another level of complexity for the description of the physical system. For instance, the large difference between optoelectronic performance of molecular beam epitaxy (MBE)-grown materials compared with metal-organic chemical vapor deposition (MOCVD)-grown ones is still mysterious, although part of the explanation could be due to the presence of Ca impurities in MBE material acting as a killer impurity [2]; another such major effect is the curing of some nonradiative (NR) recombination centers by the growth of superlattices or underlayers before growing the active LED layers [3], recently attributed to the trapping by these structures of surface defects in GaN [4]; in MOCVD materials grown under optimal conditions, NR defects appear in selected layers such as AlGaN electron–blocking layers (EBLs) or in higher In content layers for green LEDs [5–8]; (3) for the electronic quantum description, a number of phenomena need to be further explored such as electron-hole carrier localization and tunneling, Coulomb interactions, … Simulations of basic optical properties requires the computation of numerous energy levels, energy relaxation toward emitting levels, computation of the carrier population, … Simulations of LEDs require the additional computation of transport coefficients taking disorder and localization into account, both for perpendicular transport (I–V characteristics of LEDs, unipolar barrier transport) and in-plane transport. All these tasks require huge computational resources; (iv) finally, for comparisons with simulations, experiments need to determine accurate parameters, avoiding systematic errors.

We discuss disorder effects in nitride alloys in the context of LEDs, as most of these challenges occur there, and their many studies result in a lot of data, as well as in improved materials and devices. In addition, there are fewer carrier transport studies being conducted as alloys are avoided in transistors as their uses in the active channel lead to much reduced mobilities [9]. We emphasize in this article the large number of prerequisites required for a full understanding of LED materials and devices such as the need for intrinsic material parameters, carefully executed experiments and interpretation and powerful computational tools. We also mention alloy-based energy barriers, as they have very weak rectifying properties while they play major roles in devices and need serious design efforts to reach their expected barrier role [10–12].

2 Why do we need to consider disorder effects in nitride LEDs?

There is ample evidence of alloy-induced disorder effects on material and device properties, which can be classified as directly observable either at a *microscopic* scale or at *macroscopic* scale through spatially averaged effects of disorder or indirectly observable where the measured effects are different from those modeled without taking disorder into account. Understanding and modeling the effects of disorder is required if one wishes to reach the physical limits of efficiency in LEDs. At stake is the huge improvement in energy savings still permitted by the physics of nitride LEDs: present day maximum efficiency is ~200 lm/W for white-emitting phosphor-converted LEDs, whereas color mixed LEDs, so far less efficient because of the green gap, could reach 330 lm/W when the full potential of LEDs is reached [13].

3 Experimental evidence of directly observable disorder-induced effects

Optical spectroscopy at the nanoscale should allow direct measurements of alloy disorder-induced energy and localization effects (Figure 2b). There are many observations of compositional variations at the few 100-nm micron length scale through microphotoluminescence [15], near-field scanning optical microscope [16], spatially-resolved cathodoluminescence [17], with a spatial resolution down to ~50 nm. These techniques are used to measure fluctuations that are either intrinsic to alloy disorder or due to details in growth such as QW thickness fluctuations or more involved fluctuations due to specific details of the growth surface [18]. As the scale of intrinsic disorder of alloys is a few nm (Figure 2b), it is not observed through these techniques. Only the recently developed technique of scanning tunneling luminescence (STL) provides spatial resolution at the few-nm scale and allows direct observation of localized states induced by the intrinsic disorder [19].

The indirect impact of nm-scale disorder can however be observed at the *macroscopic* level, through spatially averaged observations. For instance, the increase in emission linewidth compared with pure compounds reflects the variations in local composition, as does the absorption edge broadening [20].

Another macroscopically observed effect of nanoscale disorder is the Stokes shift, that is, the difference between emission and absorption energies. We note that the Stokes shift was originally invoked in the context of molecular fluorescence with well-defined peak emission energies and absorption energies. In semiconductors, both phenomena are broadened by disorder. Thus, taking the Stokes shift as a *quantitative* measurement of disorder is uncertain, at best, as both emission and absorption edges are poorly defined: namely, in QWs, from which most data originate, the absorption edge and emission peak are broadened and shifted by the quantum-confined Stark effect (QCSE) in nitrides because of the electron-hole separation induced by the internal electric fields. Sorting the smaller contribution of disorder is challenging. One could use opposite dopants on both QWs' sides to completely screen the polarization field induced by the spontaneous polarization difference at the InGaN/GaN interface [21]. To avoid the QCSE in QWs altogether and measure intrinsic absorption and emission of InGaN alloys, one should use thick InGaN layers such that the effects of the QCSE can be avoided [22]. In both cases of thin and thick materials, the absorption curve is often approximated by a sigmoid function [23], which for QWs includes both the effect of disorder and the QCSE (Figure 3a). As for emission energy, it is difficult to assign it an accurate value because the *emission peak* results from the history of carrier energy relaxation, and the high-energy *emission edge* is even more dependent on experimental conditions (temperature, carrier density). Therefore, the very large Stokes shifts, up to several hundreds of meV [23, 24], reported might mainly be due to a large QCSE in the case of QWs and large inhomogeneities of materials

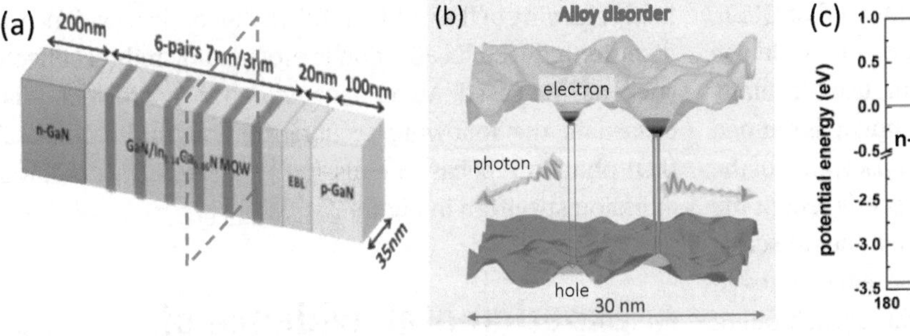

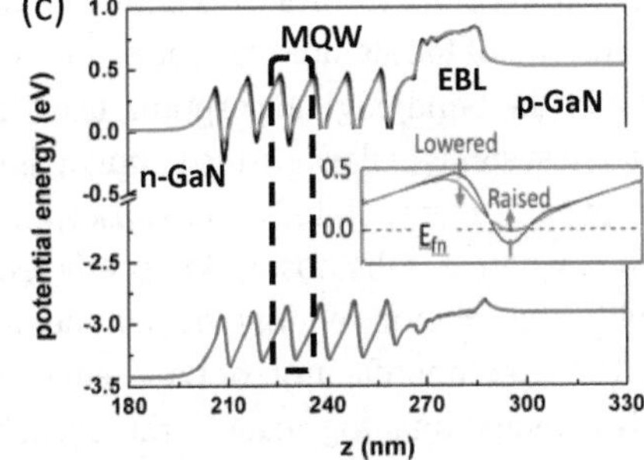

Figure 2: (a) Schematics of the blue-emitting LED to be modeled with 3-nm thick $In_{0.14}Ga_{0.86}N$ quantum wells (QWs) and 7-nm thick GaN barriers; (b) in-plane disordered electron and hole effective energy levels computed with the localization landscape (LL) theory (see below). The emission energy observed at the nanoscale of disorder should reveal the local disordered-localization energy; (c) in-plane averaged, computed band extrema along the growth direction under 2.8 V forward bias. Black curves show the averaged band extrema obtained from the compositional map, whereas the red and green curves show the averaged band extrema computed through the LL theory. The inset shows the effect of quantum disorder, raising and lowering energy extrema due to in-plane quantum confinement and tunneling (from [14]).

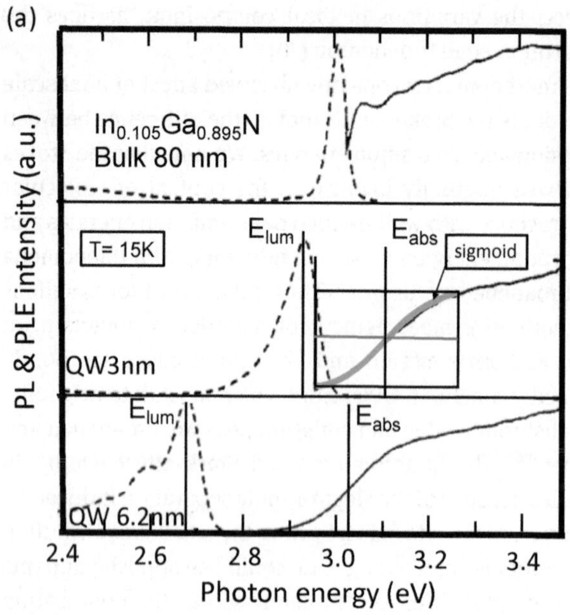

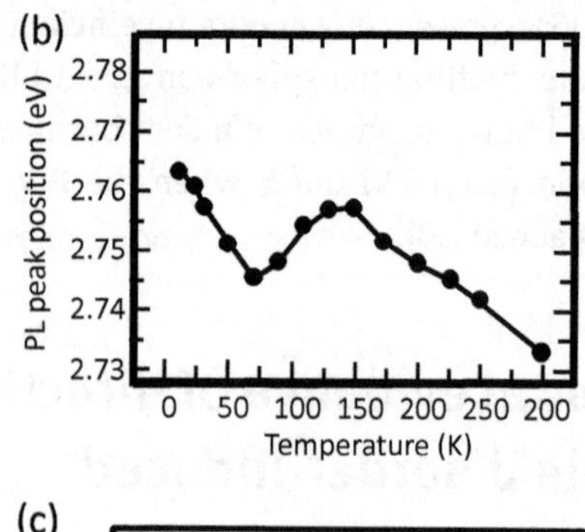

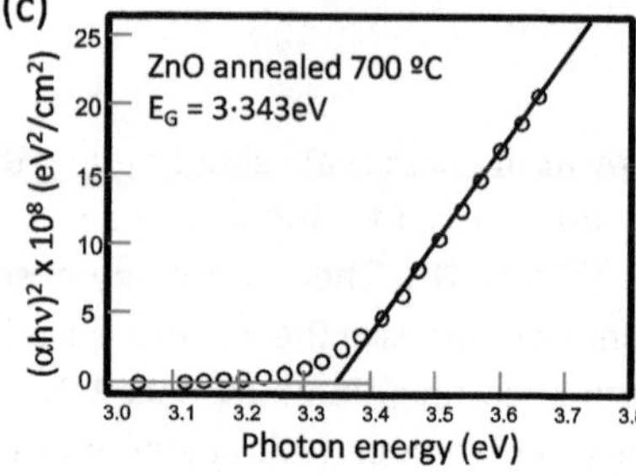

Figure 3: (a) Photoluminescence (PL) and PL excitation (PLE) spectra of InGaN bulk (80-nm thick) and QW (3- and 6.2-nm thick) materials. The bulk material exhibits a small Stokes shift, characteristic of alloy disorder. A large Stoke shift is observed for QWs. Approximating the absorption curve by a sigmoid and the absorption edge E_{abs} as its center leads to a large Stokes shift dependent on the QW thickness (after the study by Berkowicz et al. [26]); (b) example of the S-shape temperature dependence of the peak energy for InGaN-related PL (after the study by Cho et al. [29]); (c) example of a Tauc plot of a ZnO thin film fitting the linear region to evaluate the bandgap at the *x*-axis intercept (after the study by Yang et al. [30]).

as a result of poorly controlled growth (see, e.g., the small Stokes shift of bulk material in Figure 3a compared with the study by O'Donnell et al. [25]). The dominant impact of the QCSE on the Stokes shift in multiple quantum wells (MQWs) is measured in MQW structures *with constant In concentration and variable thickness* (Figure 3a) [26, 27] (for recent measurements on industry-grade materials, see also the study by Nippert [28], p. 32). Applying bias allows to cancel the QCSE when reaching flat-band conditions [31, 32] but should also measure emission under such conditions to extract a value for Stokes shift void of QCSE. In any case, instead of the sigmoid function, one could use the Tauc plot of the band edge absorption, used in many other disordered systems [33] where the bandgap is defined as the intercept of the extrapolated to zero absorption of the quantity $(\alpha h\nu)^2$, where α is the absorption coefficient of the material, with the photon energy $h\nu$ as the abscissa (Figure 3c). The best identification of band edge states in absorption is revealed on a log scale as the Urbach tail of absorption [34], but this is not what is analyzed in the articles stating large Stokes shifts.

A more indirect probe of disorder is provided by the so-called S-shaped curve displayed by the change of the peak emission wavelength with *temperature* (Figure 3b): starting at a low temperature (and increasing the temperature), the decrease in peak energy from 10 to 70 K is interpreted due to improved thermalization between localized states, reaching deeper states, followed at higher temperatures, 70–150 K, by escape from deep localized states into delocalized states emitting at higher energies. At still higher temperatures, the emission energy decreases with the temperature along the materials' bandgap. Another probe of disorder is the blue shift in emission energy *with increased carrier injection* due to band filling of disorder-broadened states, although the major part of the shift in c-plane materials is due to internal field screening of the QCSE in QWs. Other optical evidence of disorder-induced localization is provided by indirect effects such as increased Auger recombination coefficient (discussed in the following sections) or longitudinal-optical (LO) phonon replicas of emission [35] (see, e.g., the lower emission spectrum in Figure 3a).

4 Experimental evidence of indirectly observed disorder-induced effects

Indirectly observed disorder effects are inferred when the modeling of materials and device characteristics cannot describe the measured effects without accounting for disorder.

(i) *In-plane carrier transport properties* are of course sensitive to the in-plane potential fluctuations such as those in Figure 2b as revealed by electron mobilities [9] or carrier diffusion lengths [36–38]. The most direct evaluation of electron localization properties should be the electrical measurements of carrier mobilities. However, the impact of these electrons and hole localization has not yet been theoretically assessed on the in-plane carrier transport properties. The electron mobility measurements of Sohi et al. [9] could be explained by a simple *perturbative model* for electron mobility in InGaN HEMT channels at high carrier densities (2.47×10^{13} cm^{-2}) at room temperature, conditions under which one expects diminished effects of disorder. The mobility decreases from 1340 to 173 cm^2 V^{-1} s^{-1} when the In content varies from 0 to 20%. A surprisingly good agreement with theory is realized by crudely taking an alloy fluctuation potential equal to the conduction band offset between InN and GaN. It remains to be seen whether strong localization would be observed at low temperatures and reduced electron densities.

(ii) *Vertical transport properties* of multilayer LED bipolar structures are also strongly impacted by disorder, but there, disorder improves the transport properties! One would expect that the LED forward voltage V_F (defined as the voltage for a sizeable current, usually 20 or 35 20 A/cm^{-2}) as displayed in *I–V* characteristic curves is such that the energy supplied per electron–hole (*e–h*) pair by the energy source, eV, is equal to the photon energy. This is indeed observed experimentally, at least in blue single QW LEDs, and in usual III–V LEDs. This is due to the electron and hole (*e–h*) quasi-Fermi levels being near the respective *e–h* QW levels, therefore significantly populating them, which leads to a sizeable recombination current. However, for nitride MQW LEDs, where quantum barriers and internal fields hinder well-to-well carrier transport, experiments show excess voltages for longer wavelength LEDs with high indium content in the InGaN QWs. On the theory side, 1D simulations always show larger V_F than experimentally observed, the more so for long-wavelength LEDs. These 1D simulation tools, which obviously cannot include in-plane disorder, need an ad hoc artificial correction of the internal fields to reach reasonable onset values. Although some of the excess voltage can be attributed to large band discontinuities and increased polarization-induced barriers (due to discontinuities of the spontaneous and strain-induced polarization), 1D simulations miss two effects of disorder: first, the in-plane averaged effective barrier height is reduced by compositional fluctuations (Figure 1c); second, percolative carrier transport paths in the disordered potential greatly reduce the effective barrier height carriers (see Figure 8d). The same percolative transport phenomenon is at work in explaining why unipolar energy barriers always display a much reduced resistance to current and even are often ohmic without the rectifying behavior [10–12].

Another often invoked indirect impact of disorder is the high radiative efficiency for blue-violet LEDs [39, 40], seen as a surprise, given that the threading dislocation density is on the order of 10^8–10^9 cm^{-2} when grown on sapphire, values prohibiting efficient recombination in other III–V semiconductors [41]. Some have invoked a lower activity of dislocations in nitrides [42–46]. However, at the beginning of high-efficiency blue LEDs, the preferred explanation was that disorder-induced localization would prevent carriers to reach NR recombination centers [47]. Indeed, the diffusion lengths in active layers, most often alloy QWs, are quite small compared with other materials (although measured values vary substantially [38, 48]), which can make the capture radius of carriers by dislocations smaller than the dislocation separation. The situation is complex to analyze as carriers appear to be thermalized between localized and delocalized states as revealed by the thermal tails of emission in operating LEDs, which indicates a limited effect of localization.

5 The efficiency loss, "droop", at high current densities in nitride LEDs—the role of disorder

A major issue in nitride LED efficiency, at blue and other wavelengths, is the internal quantum efficiency (IQE) droop under high driving current densities. Although peak IQEs reach the 90+% range for blue-violet LEDs, these values are achieved at current densities of a few A/cm^2.

The droop phenomenon is less prevalent in other III–V materials systems as these operate at much lower carrier densities. This originates from the strong QCSE in nitride QWs, which leads to comparatively small radiative recombination coefficient, thus requiring larger carrier densities to emit a given photon flux [31]. Therefore, nonlinear NR effects such as the Auger recombination become more important than in the conventional III–Vs.

The droop phenomenon is often analyzed through the so-called ABC model of the IQE, where A is the Shockley–Read–Hall NR recombination coefficient, B is the bimolecular radiative recombination coefficient and the Cn^3 term is a nonlinear NR recombination term. In this ABC model of droop, the injected current density in the LED J and IQE are given by the following:

$$J = \frac{(An + Bn^2 + Cn^3)ed}{IE} \tag{1}$$

$$\text{IQE} = \frac{Bn^2}{An + Bn^2 + Cn^2} \tag{2}$$

where e is the electron charge, d is the active layer thickness, n is the electron concentration (equal to the hole density p in the nominally undoped active layers), IE is the injection efficiency (fraction of injected e–h pairs that are captured in the active QWs). These two equations are "reasonably" verified in LEDs [49, 50], but with a very wide range of parameters that all lead to excellent fits (Figure 4) (see, e.g., the study by Weisbuch et al. [51] for a discussion of the extraction of the *ABC* parameters from the external quantum efficiency (EQE) curve analysis). In addition, the *ABC* analysis, although very attractive due to its simplicity, depends on a number of important critical approximations: (1) A, B and C are not independent of carrier density. In c-plane–grown nitrides, the internal spontaneous and piezoelectric-related electric fields are progressively screened with increasing injected carrier density, changing accordingly A, B and C; (2) carrier localization effects, such as due to In compositional fluctuations, also change with n; (3) carrier injection is not uniform in the various QWs of MQW LEDs (in particular holes), leading to an inhomogeneous distribution of carriers among QWs, hence of recombination rates between QWs; thus, each QW has its own A, B and C coefficients due to their dependences on carrier densities; (4) density-dependent current crowding at contacts can lead to lateral inhomogeneous carrier distribution, thus A, B and C depend on the position on the chip [52]. For an experimental analysis of variables A, B and C with carrier injection and composition, see the studies by David et al. and David and Hurni [49, 50].

There has been long arguing about the physical mechanism at the origin of the Cn^3 term. At first, the cause for droop was thought to be of extrinsic nature because of decreasing IE [53] or carrier delocalization from localized states with high IQE at increased currents [47]. This changed dramatically when Shen et al. [54] announced that droop was due to an intrinsic effect, the Auger effect, through an analysis by the *ABC* modeling, using carrier lifetime changes with carrier density. This started many efforts to unambiguously determine the origin of droop because the measured Auger coefficient, $1.4–2.0 \times 10^{-30}$ cm^6 s^{-1} [54], was orders of magnitude larger than direct three-body calculations in bulk materials [55]. Then, theorists calculated the effects of indirect phonon-assisted mechanisms or of QW finite thickness [56–59], however, still with smaller values than measured ones, which led to continuing disputing the Auger mechanism with others mechanisms such as diminished IE due to carrier escape [53] or overshoot [60] from active regions (see Figure 6). However, the evidence for carriers escaping the active region relied on misleading interpretations of measurements: forward minority electron current in the LED top p-layer can be mistaken for escape current, whereas it is due to Auger electrons bypassing the EBL [61–63] (it is remarkable that although Vampola et al. [61] concluded that both thermal escape and Auger-assisted electron overflow of electrons could explain his data, most authors citing this article use it as proof of thermal escape of electrons); reverse photocurrent under low bias has been mistaken as an NR recombination channel (for a full discussion of issues in optical measurements, see the study by David et al. [49]).

Then, experimentally, the signature of Auger-generated hot electrons by electron emission in vacuum provided direct proof of the Auger mechanism [64], and theoretically, it became clear compositional fluctuations

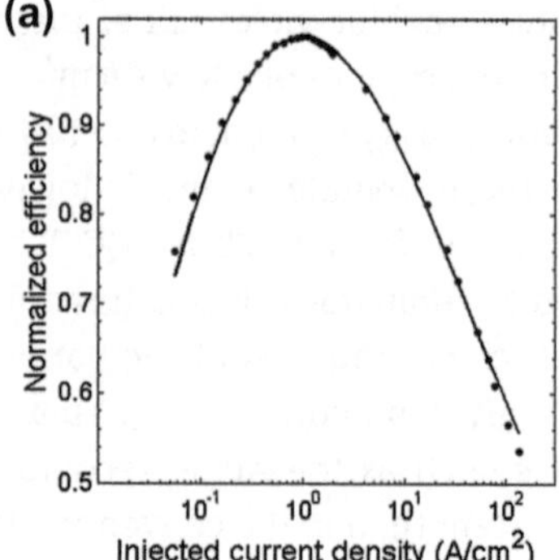

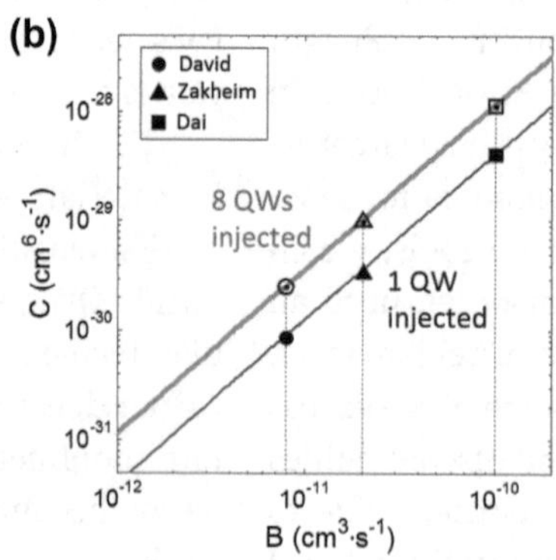

Figure 4: (a) Example of *ABC* fit of EQE vs. injected current in an eight-QW LED. The IQE is 76% as calculated from the position of the peak EQE; (b) calculated C coefficient as a function of B, considering that one or all eight QWs are populated, fitting the IQE curve on the left. Data points correspond to different values of B reported in the literature. Reproduced from the study by Weisbuch et al. [51]; see details there.

would increase the apparent Auger recombination coefficient C [65] (see the following sections).

The connection between disorder and the Auger droop is experimentally well confirmed by the smaller droop or its absence in PL measurements of GaN QWs free from alloy disorder [66, 67].

6 A prerequisite to simulations: the microscopic description of the simulated structures

The material system to be modeled requires the accurate knowledge of both their geometry and composition. These are often limiting factors of simulations due to uncertainty in either structure or physical parameters.

For instance, it is well-known that the transport and optical properties of heterostructures are very sensitive to their detailed geometry. Already in GaAs QWs, interface roughness shows up as a major source of intrinsic geometric disorder leading to inhomogeneous broadening in optical spectroscopy [68]. In turn, interface disorder impacts the design of short wavelength GaAs/AlGaAs lasers [69] and also leads to localization properties at a low temperature displayed by a mobility edge [70]. In the nitride materials system, the fact that the QW material is an alloy leads to a much more complex situation. First, the randomness of the alloy has to be assessed: indium clustering leading to In compositional fluctuations larger than those expected from exact alloy randomness was first invoked to explain the high efficiency of LEDs [25, 71]. Although clustering, indeed observed by transmission electron microscopy (TEM), was later proven as due to TEM measurement artifacts (clustering was created under strong electron irradiation [72]), a recent article suggests the existence of short range clustering, not to be detected in atomic probe tomography (APT) due to the limited detection efficiency of most APT systems [73]. This claim appears however controversial as randomness evaluations in high-efficiency APT systems still conclude to perfect randomness. Second, the geometrical definition of interfaces for QWs, leading, or not, to interface roughness is hard to assess in III nitrides. The composition varies progressively in the growth direction, with the lower interface being abrupt and the upper interface somewhat diffuse, with some In being also present in barriers because of memory effects of growth systems. Attributing interface roughness geometries at nm scale under such conditions is difficult, but evaluations through APT and TEM conclude to interface fluctuations ~5 nm wide and 1–2 monolayer thick [74, 75]. These modified interface geometries help improve simulations of luminescence spectra [76] and of the Urbach tail of the absorption edge [34]. Such measurements are spatially averaged effects of interface disorder and one would wish more direct observations of the interface fluctuations. The STL technique provides spatial resolution at the few nm scale and could probe the interface morphology because of the very localized nature of the probing carrier density [19].

7 Which quantities are to be modeled?—issues in their measurements

For optical properties of LEDs, one needs to simulate the dependence of the emission wavelength on In content, QW thickness and applied bias. The quantities entering equations (1) and (2), namely the A, B and C coefficients and the IE, although their determination require great care as discussed previously, are needed to obtain the IQE. IE is a particularly difficult quantity to measure, and we refer to the detailed discussion in the study by Weisbuch [77]. In addition, great care must be exerted when extracting coefficients from purely optical excitation measurements for use in LED simulations as the internal fields are different from those under electrical excitation. For a description of best practices for such measurements, see the study by David et al. [49].

It is easier to measure the I–V characteristic of an LED, although it also requires some care to extract the potential applied to the junction, taking into account the voltage drops at contacts and through resistance.

We leave out computations of optical properties for the moment as they require to take disorder into account in view of their many displays of disorder effects and first discuss how well LEDs can be simulated without accounting for alloy disorder effects.

8 Simulation of LEDs—without disorder taken into account

Simulations are a very important indication of the understanding LED operation, therefore justifying their use to guide LED designs. For instance, as mentioned previously, some experimental results were invoked to justify carrier escape from the active region and would require mitigation actions by some improved design if proven true.

Simulations of *full LED* structures are usually performed in 1D [78–82] because of the computational challenges to include 3D disorder. LED simulations are based on a known 1D map of the band edges of the devices (eventually computing the QW levels with a 1D Schrödinger equation), solving throughout the device structure drift-diffusion (DD) equations that involve knowing *local* transport parameters of carriers (mobilities, diffusion coefficients, both being related by the Einstein relation $D = (kT/e)\mu$ for homogeneous systems at equilibrium) and recombination times of carriers. Thermal equilibrium is assumed. Local electric potentials and fields are determined by solving Poisson equation, which requires a self-consistency feedback loop to account for the changes in electric potentials and fields caused by carrier injection and propagation (Figure 5a).

Current state-of-the-art 1D models, when used with standard parameters such as band offsets and internal electric fields, fail to describe basic device properties, such as the forward voltage V_F: the simulated blue LED V_F is 3.4 V or more, whereas observed values are well below 3 V (see commercial LED data sheets). To improve the model results so that they fit better experiments, the internal fields are typically adjusted to 50% or lower of their value [83, 84]. There is no clear justification for such an adjustment, which in addition contradict measurements of such fields [31], and measured S-curves and blue shifts. A quantum correction to the DD equations has been proposed, but it is itself adjusted to fit results [85]. Of course, the decreased internal fields decrease the energy barriers for vertical carrier transport and could overestimate carrier leakage out of the active region. However, 1D simulations with such fields adjustments still conclude that there is negligible leakage in typical visible LEDs (see, e.g., Figure 13 in the study by Li et al. [14]).

More advanced modeling approaches than DD models have been implemented to improve the relevance of simulations without disorder. DD equations rely on a semi-classical model where the underlying quantum mechanics is buried in the transport parameters, assuming thermal equilibrium, and separately computing quantized energy levels. The detailed history of carriers from their injection to their recombination from the various localized or delocalized states is hidden in the fact that any state is populated according to equilibrium thermal distributions. Among neglected phenomena are nonequilibrium hot

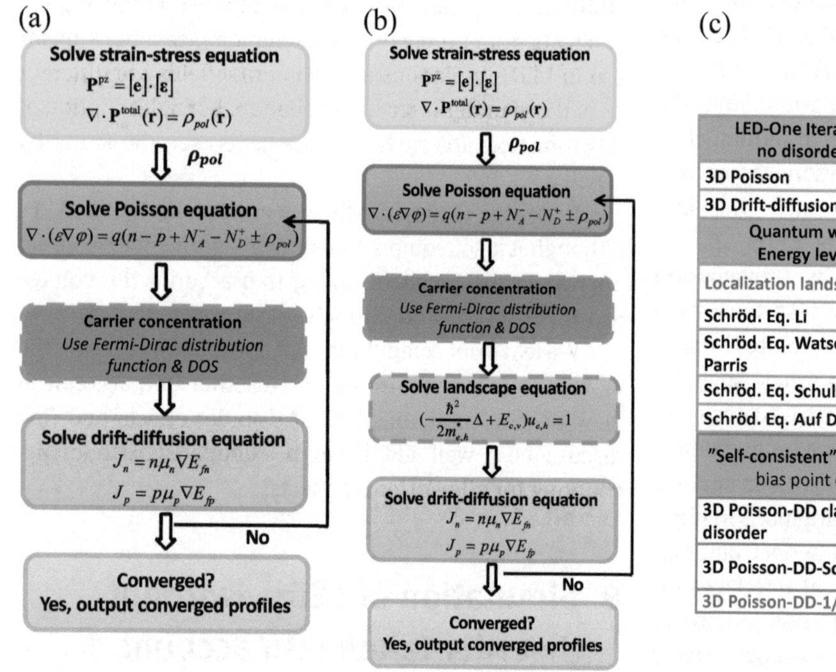

LED-One Iteration no disorder	Node number (matrix size)	Computation time (s)
3D Poisson	428,655	25
3D Drift-diffusion (e &h)	428,655	50
Quantum well Energy level	Node number (matrix size)	Computation time (s)
Localization landscape	428,655	50
Schröd. Eq. Li	428,655	63,650
Schröd. Eq. Watson-Parris	1,500,000	60,000
Schröd. Eq. Schulz	328,000	7,500
Schröd. Eq. Auf Der Maur	100,000	24,000
"Self-consistent" solver for 45 bias point of LED		"Estimated" Total computation time (s)
3D Poisson-DD classical disorder		54,000
3D Poisson-DD-Schrödinger		45,882,000 (531 days!!)
3D Poisson-DD-1/u		90,000 (~1 day)

Figure 5: LED simulation algorithm.
(a) conventional 1D Poisson-DD computations (no disorder). Energy band positions are well known from the layered structure; (b) 3D localization landscape (LL)-Poisson-DD computation involves the computation of the local effective energy levels by the LL theory method (see the following sections) and their input into the Poisson and DD equations. Not shown are similar algorithms for the 3D classical disorder-Poisson-DD computation where the disordered energy maps are used without computing the resulting quantum effects of the disordered potential and the 3D Schrödinger–Poisson-DD computation where the quantum effects of the disordered potential are computed in the effective mass approximation (EMA) or directly by tight-binding (TB) methods; (c) comparison of computation times (after the study by Li et al. [14]).

carriers and their relaxation, quantum tunneling, detailed scattering mechanisms, dynamic effects such as QW capture coefficients, and so forth, as shown in Figure 6. The nonequilibrium green function (NEGF) formalism has been invoked to solve the unrealistic high turn-on voltages of the DD modeling through computation of such phenomena. Indeed, 1D NEGF simulations seem to lead to lower turn-on voltages, in particular for green LEDs. The situation is somewhat ambiguous from the two main publications, as they use structures and parameters quite different from those described in other simulations, which lead to different transport mechanisms. For blue LEDs, Geng et al. [80] computed a correct 2.85 V bias voltage for a 20 A cm^{-2} current density but uses 4.6 nm thick quantum barriers between the QWs. Such thin barriers lead to interwell current being due to tunneling instead of the thermo-ionic emission described in the DD modeling (tunneling being a quantum phenomenon, it is not included in the DD equations). Thus, obviously, this will result in lower voltages for a given current than other simulations that use barrier thicknesses in the 7–10 nm range and where therefore tunneling current would be much suppressed or null. In addition, a 100-meV broadening of QW states, said to originate from the experiment, is used in the computations which of course impacts strongly transport rates. However, such a broadening would be directly observed in precise measurements, while temperature-dependent PL [86] yield an half-width half-maximum (HWHM) of 60 meV at 300 K or Urbach tail measurements yield 30 meV [34]. The use of such smaller broadening parameters would also increase the voltage and lead to diminished agreement with the experiment. For green LED NEGF simulations, Shedbalkar and Witzigmann [87] do not provide detailed information on the I–V characteristics but compute structures at a bias of 3.4 V, far larger than typical experimental V_F for rather thin quantum barriers of 5 nm, and more strikingly requires 30% In concentration to reach green emission. It would clearly be interesting to have NEGF computations for the usual structures and parameters to assess the improvements brought by the NEGF computations.

We note that alloy disorder, so far, is missing from the NEFG simulations. NEGF computations taking alloy disorder into account would be a 4D problem ($\vec{r}$, E), which would require a huge memory and large computer clusters for simulations. Owing to the demand on computing resources of NEGF computations alone, this is so far intractable, even through the more efficient LL theory method.

Another way to treat the hot carrier effects absent in DD simulation is the Monte Carlo method. In the computations by Kivisaari et al., they however do not solve the discrepancy of larger V_F than observed, in particular for green LEDs, with excess V_F from 1 to 2.7 V for 3 or 8 QW LEDs, respectively [88]. Another computation of hot electron effects solves numerically the electron Boltzmann transport equation in the MQW region by assuming a voltage drop of 3 V over the QW stack [89]. With such voltage drop, significant carrier escape out of the active region into the p-contact layer is predicted, arguing that this is a cause for droop, the large excess voltage implied does not represent the V_F of commercial LEDs.

9 Modeling the disordered potential induced by alloy compositional fluctuations

We mentioned that due to the large differences in atom potentials in InGaN, the effects of alloy disorder will be strong and that the usual VCA used for the common III–V alloys is insufficient for simulations of nitrides [90–92].

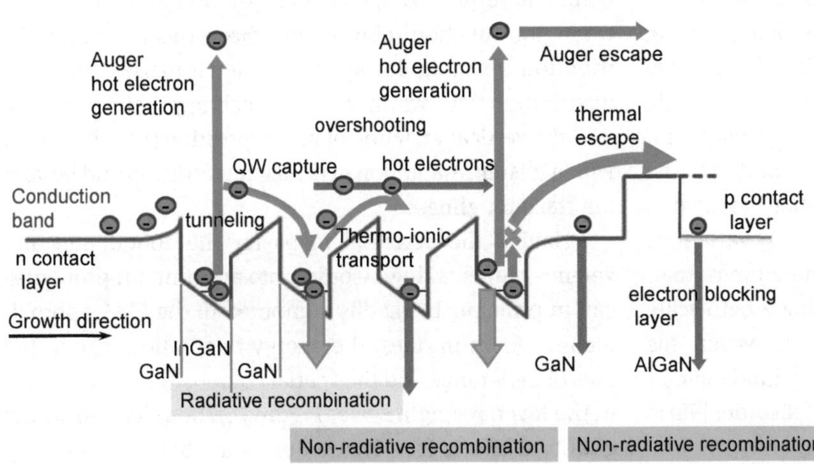

Figure 6: Schematics of the phenomena occurring in LEDs best captured by nonequilibrium computing techniques such as NEGF or Monte Carlo.

One then needs to compute the local energy levels entering the simulation equations.

Some groups compute the energy levels from first principles, usually TB approximation [74, 93] in supercells with random In atom positions that are supposed to represent the alloy at larger scales. The supercell size is clearly a limiting factor (we however note that Tanner et al. used 81,920 atoms in their supercell [94]) and to ensure the reliability of the procedure, many different random supercell configurations are computed, and the final result is the averaged results over the configurations that exhibit sizeable differences (see, e.g., Figure 3 in the study by Tanner et al. [74]).

Many other groups [14, 34, 95, 96] compute energy levels in two steps: they first compute the disordered potential map $V_{e,h}$ (x, y, z) acting on carriers using the EMA, and then compute the energy levels by solving the Schrödinger equation of the carrier envelope wave function. One uses the fact that the potential varies smoothly at the atomic length scale to approximate the local potential as the bandgap of the bulk alloy with the local composition. This is the easiest approach to grasp the impact of compositional disorder.

In this approach, a first input to simulations is the generation of a *compositional map* of the disordered alloy, performed by connecting the discrete quantized atom composition as measured by APT, or from a random number generator, to a continuous variation by weighting each site composition with a continuous Gaussian function (Figure 7a). The standard variation σ is usually taken as twice the lattice constant a (a = 2.833 Å). This choice of weighting is dictated by having a continuous composition that represents a physically meaningful fluctuating compositional map. Too small a σ would lead to maps with very large fluctuations, and too large a σ would average compositional fluctuations. The value $\sigma = 2a$ is commonly used [14, 34, 92, 95]. This approach is questioned as empirical by Di Vito et al. [97] who compare the various approaches of construction of the compositional map. Let us recall here the arguments for this choice given in the study by Li et al. [14]. First, such an averaging length scale 2σ can be extracted from the results of the energy maps and effective energy maps (see, e.g., Figures 7c and 8b): the latter provide the effective potential fluctuations for charge carriers, and, as well as the energy maps, display variations over characteristic lengths of a few nm, quite larger than 2σ, which justifies that the chosen averaging length scale does capture the smaller scale fluctuations. However, this could be seen as a circular argument. A more fundamental argument obtained from a general theory of disorder [98] is given in the study by Li et al. [14].

Knowing the compositional map, one generates maps of energy band extrema $E_c(r)$ and $E_v(r)$ from the bandgap variation given by the bowing parameter, using a 63/37% ratio of conduction band and valence band offsets (Figure 6b). The bowing parameter b describes the quadratic variation of bandgap with alloying as follows:

$$E_g\,(\mathrm{In}_x\mathrm{Ga}_{1-x}\mathrm{N}) = xE_g\,(\mathrm{GaN}) + (1-x)E_g\,(\mathrm{InN}) + bx\,(1-x) \qquad (3)$$

b is also a property of alloy-induced disorder [99] among other sources of bowing [100]. It is clearly an essential parameter for simulations relying on the EMA. Experimentally, it should be determined from the bandgap variation with composition. As discussed previously, the bandgap is an experimentally ill-defined quantity when choosing the PL emission wavelength, the absorption edge, or reflectivity measurements because of their broadening. Therefore, its experimental determination depends on one's analysis of disordered band edges.

For TB computations of supercells, the evaluation of the bowing parameter is unclear. In the most detailed article by Caro et al. [99], the bowing parameter depends on composition, with most of the changes in the conduction band, quite a contrast with the approximation made in EMA computations that split the energy variations 63%/37% in the conduction band and valence band, respectively.

Tanner et al. [94] compared the TB and EMA approaches and found similar results for peak energies and PL FWHM (the differences reported in Table I of the study by Tanner et al. [94] is mainly due to the different QW thicknesses used in the two types of simulations). It should be noted that the reasonably good agreement with experiment assumes QW thickness fluctuations of 2 monolayers [94].

Additional ingredients to generate the 3D band maps are the effects of internal fields due to spontaneous and piezoelectric fields [14, 93, 94, 99] and the vertical potential map. The vertical compositional variation in QWs is added through a function representing the vertical In atom distribution. A simplifying approximation is to use a Gaussian distribution with width $\sigma = 2a$, which approximates quite well the vertical variation of In composition [101], however, with a misrepresentation of the asymmetric top and bottom interface roughness.

Having the potential maps for the conduction and valence bands taking disorder into account, all properties can in principle be readily computed in the EMA approximation. As in nitrides, the energy fluctuations are in the tens of meV range and the spatial compositional variations in the few nm length scale (Figure 7c), both lead to strong quantum effects (see Figure 8c). Thus, before computing

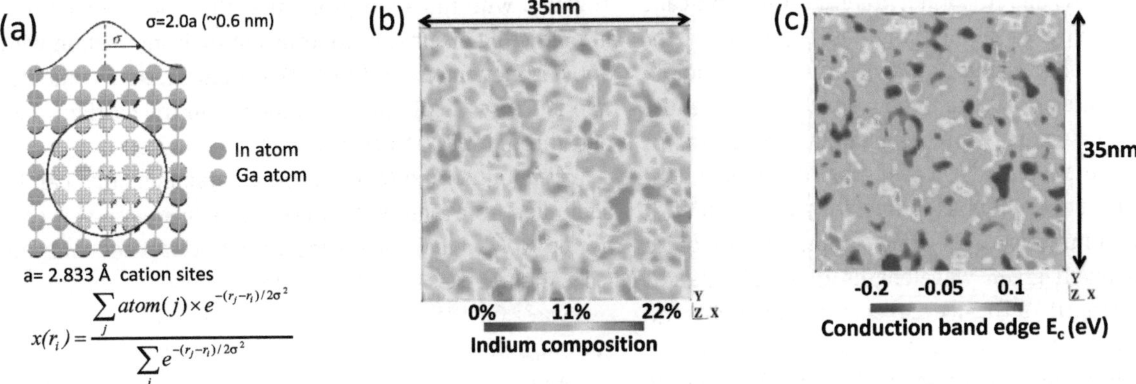

Figure 7: Construction of the potential map seen by electrons; (a) Gaussian averaging of In composition; (b) resulting compositional map in the middle of an $In_{0,14}Ga_{0.86}N$ QW; (c) potential map of conduction electrons. Energy counted from the bottom of the band (from the study by Li et al. [14]).

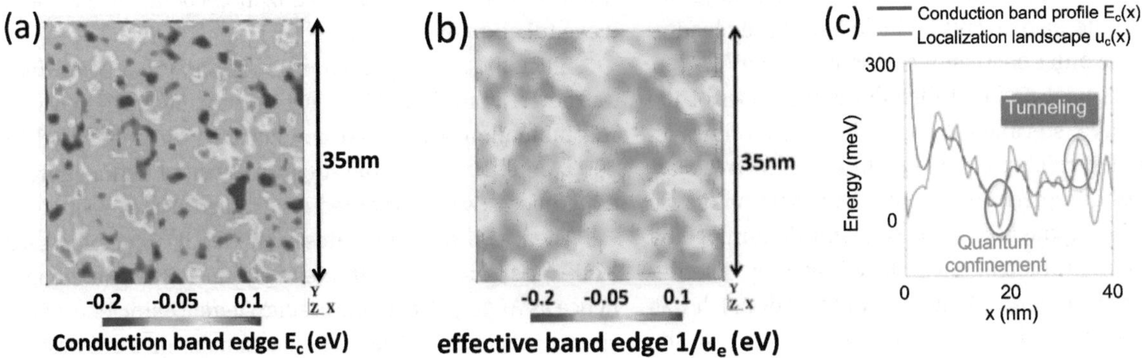

Figure 8: (a) Spatial variation of the conduction band minimum $E_c(r)$ for a random $In_{0,14}Ga_{0.86}N$ alloy and (b) the effective potential map $1/u_e(r)$ computed through the LL theory; (c) a cut across the *x*-direction allows comparisons of the fluctuations of both potentials. One can see that energy fluctuations are quite diminished in the LL theory as it captures two essential ingredients of quantum theory: quantum confinement and tunneling (from the study by Li et al. [14]).

materials and device properties, one needs to compute carrier energy levels quantum mechanically in a disordered material by solving the Schrödinger equation for carriers in the disordered potential.

10 Simulations of optical spectra

There are now quite a few simulations of energy levels, absorption and emission of QWs that include disorder effects (without disorder, predictions are poor as the band tails play an important role in the absorption edge and emission wavelength). The impact of disorder is calculated with energy maps obtained either from first principle TB [92, 93, 97] or from the EMA [14, 92, 93, 96], followed by solving the Schrödinger equations for electrons or holes in their respective disordered potential, or by solving the approximate landscape equations (see the following sections) [14, 34, 96]. The resulting energy levels are used to predict optical properties such as absorption, luminescence, ... These calculations are extremely demanding on computing resources, the number of mesh nodes in finite-element computations exceeding a few $\sim 10^5$ for typical QWs.

An obvious expectation from such simulations is the prediction of absorption and luminescence and of the resulting Stokes shift [32, 95, 102, 103].

To simulate the situation in electrically injected QWs in an LED under bias, a carrier density is numerically injected in a QW structure, with the internal electric fields further computed self-consistently. Another approach relies on the computation, in a tractable 1D self-consistent LED model, of the potentials, the electric fields and the carrier densities, which are then used in 3D Schrödinger computations of optical properties without any further attempt to self-consistency.

Even including alloy disorder, there are still significant discrepancies between simulations and experiments for the dependence of the emission wavelength on In contents and QW thickness: at high In contents beyond blue emission, the simulated emission wavelength differs significantly from the simulated wavelength. For green LEDs, green (525 nm) emission is experimentally obtained with 24% of In for 3-nm-thick QWs while computations require ≈30% In to reach green wavelengths in 1D simulation software, advanced 1D NEGF or 3D computations [87, 91].

To reconciliate measurements and simulations, one can invoke large interface roughness [35] or QW thickness fluctuations. The discrepancy could be alternatively linked to the use of a too small bowing parameter *b*, which would then underestimate the size of energy fluctuations (one usually takes $b = -1.4$ eV, whereas some computations point to much larger values [99]). Electron-hole interactions should also always be included as they lead to redshifts of the order of 50 meV [32] or 30–40 meV [74].

Another optical probe of disorder is the onset of the absorption curve, so-called Urbach tail, which follows an exponential dependence on energy. In InGaN MQW solar-cell structures designed for photo-carrier collection, it was shown that the exponential slope is mainly determined by the compositional fluctuations induced by alloy disorder [34]. Using the LL theory (see the following sections), it was shown that the slope is ~25–35 meV for In content in the range of 11–28%, in reasonable agreement with the ~20 meV of experiment. Direct computations from solving the time-dependent Schrödinger equation [32] found ~20 meV, however, only when taking *e–h* Coulomb interaction into account, vs. 7 meV when ignoring Coulomb interaction.

A significant result of the direct Schrödinger computations is the weaker or nonexistent localization of electrons due to their light effective mass in contrast with the strong localization of holes [74, 95, 97]. As mentioned previously, this is a topic for further investigation of electron transport at low temperatures. For optical properties, this could however be somewhat modified in interband transitions by the *e–h* interaction that could localize electrons because of their interaction with a localized hole [74, 104].

11 Simulations of the *ABC* recombination parameters

The determination of the recombination parameters *A*, *B* and *C* allows to model the IQE, droop, the green gap. Together with transport properties, they play a major role in predicting electrical properties in their interaction with optical phenomena in full LED structures, as recombination is a very dominant phenomenon in the carrier transport through LED structures [105].

Although *B* and *C* are intrinsic parameters that can be computed from materials and structure modeling, *A* is connected to NR mechanisms most often associated with defects. In some cases, the *A* coefficient depends on the vertical *e–h* wave function overlap, therefore on bias and injected current, and should be simulated, as done by David et al. [8].

As *B* and *C* vary so much with structures, good simulations of measured values are needed to further have a good simulation tool. *B* is critical, giving the radiative efficiency. The effect of the internal field separating the electron and holes and the disorder-induced in-plane different electron and hole localizations both affect the *e–h* wave function overlap [32].

Since the first observations of droop, the *C* coefficient has been the subject of intense scrutiny. Even when Auger recombination was rather convincingly demonstrated in 2007, doubts arose as the Auger coefficient deduced from experiments was considered unphysically high [55].

Then, two questions arose: (1) which other processes lead to a carrier loss with a similar carrier-density dependence as Auger; (2) can one predict Auger coefficients of the correct magnitude? The main mechanism invoked besides Auger NR recombination is carrier escape from the active region. Experimental evidence was shown to be disputable [61, 63] and theoretical support further discussed in the LED simulation section shows it to be negligible.

Several effects increase the direct Auger coefficient computed within the VCA, that is, without disorder, $\approx 3.10^{-34}$–10^{-35} cm^6 s^{-1}, either due to phonon- or disorder-assistance which both relax the *k*-conservation rule responsible for the very small direct Auger coefficient in wide-bandgap materials [65]. They however rely on the use of 100–300 meV broadening parameter. Another phenomenon which partially relaxes the *k*-conservation rule is the finite thickness of the QW, which increases *C* [59], while the internal fields decrease *C* due to the *e–h* wave function separations [58]. However, the analysis (without disorder effects) predicts an oscillatory behavior of the Auger rate with QW thickness which has not been observed [59]. From the absence of significant Auger recombination in GaN QWs, beyond the effect of increased bandgap, one concludes that disorder must play a major role in the large Auger coefficient with computed values in the right order of magnitude [65, 106].

A missing item of most *B* and *C* simulations is the *e–h* Coulomb interaction. Although no hydrogen-like relative motion of the e–h pairs is expected because of the diminished e–h overlap due to QCSE and to the localizing potentials that destroy the e-h correlated relative motion [107], the e–h Coulomb interaction leads to an increased optical matrix element (similar to the Sommerfeld factor in unbound hydrogenic states [108]) observed and computed by David et al. [32, 49], with a significant increase of B compared with noninteracting e–h pairs.

12 LED modeling taking disorder into account

As mentioned previously, LED simulations do not include disorder fail to account for a well-measured parameter, the forward voltage V_F. Using the ad hoc diminished internal field yields correct voltages, but other quantities such as the QCSE and wave functions overlaps yield incorrect values.

There is so far only one LED simulation tool taking disorder into account. This is due to the hugely increased required computational power, well beyond that required for optical properties: one needs to add a module for carrier transport to the computation of optical properties. But, and the most important, as one wishes to simulate I–V curves and dependence of the electrical-to-optical conversion on current injection, one will need to compute for a large number of different diode bias voltages and also to run self-consistent loops to account for the modifications of the internal field maps due to injected carriers. A self-consistent loop takes 20 roundtrips to converge on average [14]; thus, one needs at least 500 more time to compute an LED than the optical simulation alone, neglecting the additional time required to solve the DD equations (indeed small, see Figure 5c). Given that computing the optical properties through the Schrödinger equation takes about a day, computing one LED structure would take 500+ days (Figure 5c on computing times).

The tool is based on an approximate solution to the Schrödinger equation in the EMA approximation based on the Filoche–Mayboroda 3D LL theory, which from the original disordered energy map provides an effective potential that allows the use of the standard DD transport equations while accounting for microscopic disorder [109].

The Schrödinger equation is replaced by the landscape equation:

$$\left(-\frac{\hbar^2}{2m^*_{e,h}}\Delta + E_{c,v}\right)u_{e,h} = 1$$

where $m^*_{e,h}$ is the effective mass of the electron/hole, $E_{c,v}$ is the conduction/valence band energy, and $u_{e,h}$ is the landscape function for the electron/hole. The landscape equation is used to predict the energy levels and local density of states in place of the Schrödinger equation [96]. The quantity $1/u_{e,h}$ is interpreted as the effective potential for the charge carrier and in part accounts for their quantum nature and particular behavior, such as in-plane confinement or tunneling due to fluctuating potentials (Figure 8c). Then, $1/u_{e,h}$ is used directly to plot an effective band diagram for a given structure [96]. Compared with the original disordered energy maps $E_{c,v}(r)$, the effective potential increases current at a given bias voltage by smoothing out potential discontinuities in the heterostructure [14].

The use of the landscape equation in lieu of Schrödinger's equation leads to a remarkable gain of 10^3 in computing speed, making 3D self-consistent computations possible based on a finite element method computational approach [14, 110]. The electro-optical behavior of LEDs is simulated by self-consistently solving the Poisson, landscape, and DD equations in 3D structures (Figure 5b). Details of the theory are given in the studies by Li et al. and Wu et al. [14, 110]. Simulations use 100% of the known values for the spontaneous and piezoelectric polarization parameters. As no model exists yet of the electric transport parameters with hopping phenomena (however, strongly suppressed under the carrier densities under LED operation where localized states should be filled), ones uses available DD parameters. In the same vein, one uses A, B and C parameters from the literature, however, choosing their value nearest to those of the disorder-less alloy (see the discussion on the choice of parameters in the study by Li et al. [14]). A better choice might be those determined in the VCA approximation. Simulations of the blue six QW LED of Figure 2a are shown in Figure 9. The strain and polarization fields (spontaneous and piezoelectric) are calculated before entering the self-consistency loop [14].

As can be seen in Figure 9a, the LL computation gives a correct LED forward voltage for blue LEDs without any adjustment of parameters, about 3 V at 20 A/cm^2 vs. an experimental 2.8 V for the better commercial LEDs. The small difference might come from the approximate vertical alloy composition map. The lowering of onset diode voltage compared with simulations without disorder is due to carriers being transported preferentially through regions of lower effective bandgap induced by disorder. The current through the LED structure undergoes complex trajectories, similar to percolation paths in disordered systems, as carriers transport in the perpendicular directions to QW planes through regions where energy barriers are easier to surpass. They will also relax in the QW plane in the domains of lower energies (Figure 9d) [14, 101]. These paths, which can only

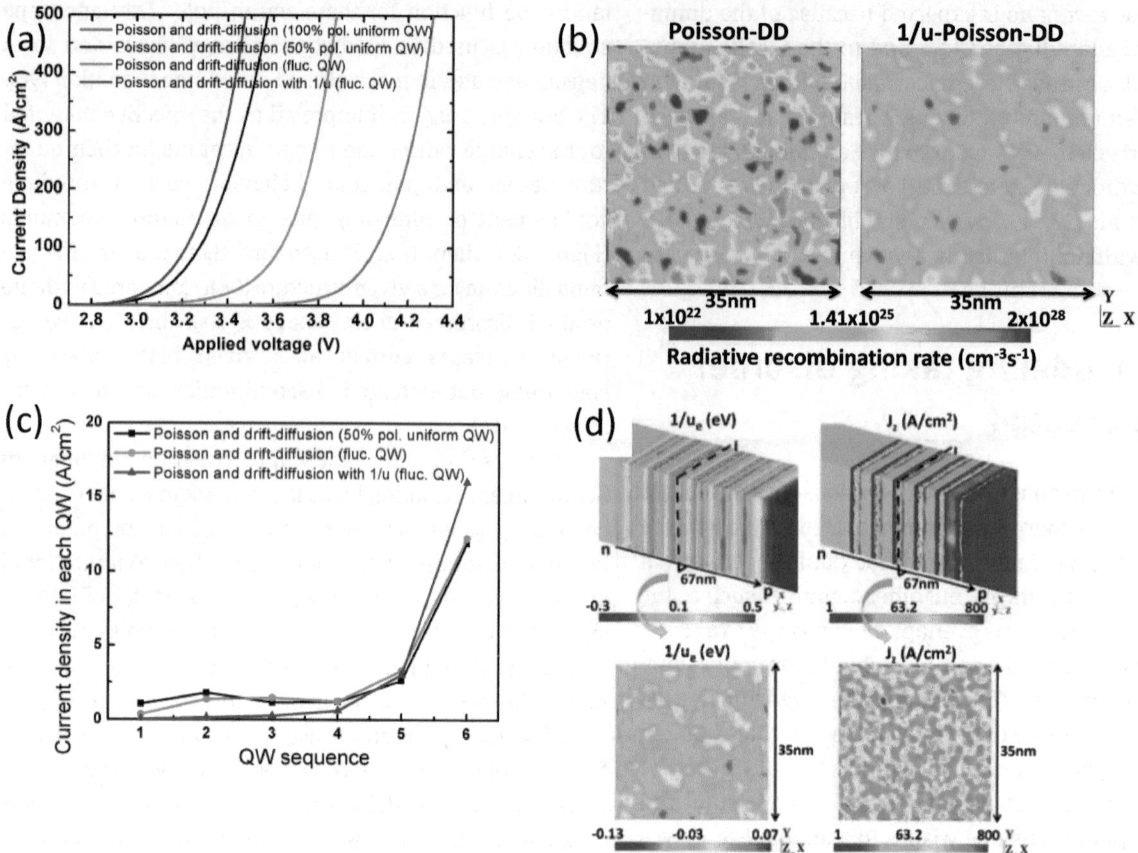

Figure 9: Computations of the six QW LED of Figure 2a.
(a) Computed *I*–*V* curves: dashed black curve: classical Poisson-DD equations assuming full polarization in the LED structure; red curve: Poisson-DD equations with classical treatment of disorder (meaning one uses in the DD equations the computed disordered potential directly obtained from compositional fluctuations without computing the quantum corrections by solving the Schrödinger equation of carriers in the disordered potential); full black curve: no disorder, classical Poisson-DD equations assuming 50% polarization; blue curve: LL theory treatment of disorder with classical Poisson-DD equations; (b) computed self-consistent in-plane electron density in the mid-plane of the third QW with compositional disorder using: (left) the classical disorder, Poisson and DD models; (right) the LL effective potential, Poisson and DD models; (c) recombination currents in each of the 6 QW, displaying the large injection inhomogeneity, which is partly the source of large droop due to the high carrier concentration in the 6th QW (top, *p*-side QW); (d) LL theory implemented in Poisson-DD LED model. Perspective views of the LL function ($1/u_e$) and of the normal component of current (J_z) calculated for an MQW LED in the midplane (*x*–*y* plane) of the third QW. Figures in panels b, c and d are solved by $1/u$-Poisson-DD model, where the LED current density is 20 A cm^{-2} (from the study by Li et al. [14]).

be simulated in 3D computations, lead to these values of V_F in good agreement with measured values for blue LEDs, without having to modify the polarization parameters such as in 1D models. Disorder of course leads to carrier concentration in the high In contents, low energy regions, and one can clearly see the smoothing action of the quantum computation of effective potential (Figure 9b). One also recovers the hole transport mostly limited to the *p* side of MQW LED structures as shown by the concentration of recombination density in the QW nearest to the LED *p*-side (Figure 9c). Among the additional results are the confirmation of (1) unequal injection among the MQWs; (2) the enhancement of Auger-induced droop because of the enhanced carrier concentrations in lower energy regions [14]; and (3) carrier escape being negligible for LED current densities, pointing to the dominant role of Auger NR recombination as the cause for droop through simulation.

The 3D simulations, based on the LL theory, work surprisingly well for blue LEDs given some of the very rough approximations: the absence of vertical QW-to-QW tunneling, DD model of carrier transport instead of accounting for disorder-dominated transport. This is quite certainly due to the fact that the parameters used in the simulations implicitly include the effects of disorder: recombination parameters *A*, *B* and *C*, carrier mobilities, bowing parameter, ...

The LL modeling also provides insight on the green gap. The green gap is due to two factors, the diminished

EQE and the excess voltage ΔV_F needed in longer wavelength MQW LEDs. Let us define ΔV_F. As mentioned previously, the onset voltage V_F is set by the *e–h* recombination energy *hν*. This is indeed the case for blue and shorter wavelength LEDs, but for green and longer wavelength LEDs the excess voltage $\Delta V_F = V_F - V_{ph}$ (where $eV_{ph} = h\nu$) can reach 0.4 V per QW in the sequence of MQWs, which translates into loss of efficiency, the emitted energy per *e–h* pair still being the photon energy. Recent modeling through the LL theory pointed to the excess voltage being due to both increased internal fields and band discontinuities with In concentration, which impede the usual carrier transport from well to well and call for larger bias voltages [111, 112]. This indicates that mitigation of the excess voltage implies diminishing internal energy barriers through doping [21] or alloying, or more radically through injecting carrier laterally to prevent the required cascade of carriers through the multiple barriers in vertically injected MQW LEDs. This has been successfully achieved using *V* defects as lateral injectors [113].

For completeness sake, even this 3D model including disorder fails to capture two important parameters of green LEDs: the computed V_F is too high by about 0.2 V for an SQW LED and by 1 V for a 7 QW LED [111], and the wavelength for a typical experimental In concentration of 24% is too short by about 30 nm compared with the experiment. Several explanations have been invoked for the latter beyond the early-invoked large Stokes shift increasing nonlinearly with In concentration [23], disputable as discussed previously. Other possible causes are large interface roughness or QW thickness fluctuations, the use of a too small bowing parameter to describe the variation of the bandgap with In concentration. Missing items could also the use of a simplified, symmetric vertical In atom distribution and the neglect of Coulomb interactions.

Solutions to resolve the excess voltage contribution to the green gap, other than *V* defects lateral injectors involve using doping of barrier materials do compensate the internal fields [21] or using alloy barriers to diminish the energy barrier to carrier transport throughout the structures. The optimization of such structures, or any other mitigation solution, will require precise full 3D modeling of LEDs.

13 Conclusion

One can see that, while computing optical properties is within reach provided some "educated" guesses on the structure geometries, on carrier thermalization, and so forth, are made, computing full LED device structures is a much more formidable task, given the required computing resources and the lack of detailed transport mechanisms that properly treat alloy disorder effects. While the LL theory markedly reduces the computation time, it still lacks a way to incorporate the out-of-equilibrium phenomena solved by the NEGF technique. Unfortunately, it appears that putting together LL and NEGF computations becomes again a too formidable task to compute. There is clearly a need for new ideas to make the field of LED simulations tractable.

One general remark on the disappointing state of understanding nitride structures is other semiconductor materials such as Ge or Si, or other III–Vs, required long efforts to achieve their full potential. The excellent early results of nitride devices may have hampered the funding resources, and thus the efforts required for a full understanding of this very interesting, specific and rather intriguing materials system. Indeed, with blue LEDs operating at energy conversion efficiency greater than 80% at low current densities, it did not seem obvious that better understanding of structures and device physics would yield clear improvements. We of course believe that it is not so, in view of the need to solve the two major limitations of nitride LEDs, the green gap and the "droop" in efficiency.

Acknowledgments: The authors acknowledge funding from the Solid State Lighting and Energy Electronics Center (SSLEEC) at the University of California, Santa Barbara (UCSB). C.W. thanks Aurélien David for many illuminating discussions and, with Pierre Petroff and Jacques Peretti, for a critical reading of the manuscript.
Author contribution: All the authors have accepted responsibility for the entire content of this submitted manuscript and approved submission.
Research funding: This work was supported by the U.S. Department of Energy under Award No. DE-EE0008204, supported by the National Science Foundation under award # 1839077 (through a subcontract from the University Minnesota), grants from the Simons Foundation (601952, J.S.), (601954, C.W.) and by the Ministry of Science and Technology in Taiwan under Grant No. MOST 108-2628-E-002-010-MY3.
Conflict of interest statement: The authors declare no conflicts of interest regarding this article.

References

[1] B. I. Shklovskii and A. L. Efros, *Electronic Properties of Doped Semiconductors*, Springer Berlin Heidelberg, 1984.
[2] E. C. Young, N. Grandjean, T. E. Mates, and J. S. Speck, "Calcium impurity as a source of non-radiative recombination in (In,Ga)N

layers grown by molecular beam epitaxy," *Appl. Phys. Lett.*, vol. 109, p. 212103, 2016.

[3] C. Haller, J. F. Carlin, G. Jacopin, D. Martin, R. Butté, and N. Grandjean, "Burying non-radiative defects in InGaN underlayer to increase InGaN/GaN quantum well efficiency," *Appl. Phys. Lett.*, vol. 111, p. 262101, 2017.

[4] C. Haller, J. F. Carlin, G. Jacopin, et al., "GaN surface as the source of non-radiative defects in InGaN/GaN quantum wells," *Appl. Phys. Lett.*, vol. 113, pp. 1–5, 2018.

[5] A. C. Espenlaub, D. J. Myers, E. C. Young, S. Marcinkevičius, C. Weisbuch, and J. S. Speck, "Evidence of trap-assisted Auger recombination in low radiative efficiency MBE-grown III-nitride LEDs," *J. Appl. Phys.*, vol. 126, p. 184502, 2019.

[6] D. J. Myers, K. GelŽinytė, A. I. Alhassan, et al., "Direct measurement of hot-carrier generation in a semiconductor barrier heterostructure: identification of the dominant mechanism for thermal droop," *Phys. Rev. B*, vol. 100, p. 125303, 2019.

[7] A. David, N. G. Young, C. A. Hurni, and M. D. Craven, "Quantum efficiency of III-nitride emitters: evidence for defect-assisted nonradiative recombination and its effect on the green gap," *Phys. Rev. Appl.*, vol. 11, p. 031001, 2019.

[8] A. David, G. Young Nathan, C. Lund, and M. D. Craven, "Compensation between radiative and Auger recombinations in III-nitrides: the scaling law of separated-wavefunction recombinations," *Appl. Phys. Lett.*, vol. 115, p. 193502, 2019.

[9] P. Sohi, J. F. Carlin, and N. Grandjean, "Alloy disorder limited mobility of InGaN two-dimensional electron gas," *Appl. Phys. Lett.*, vol. 112, p. 262101, 2018.

[10] D. N. Nath, Z. C. Yang, C. Y. Lee, P. S. Park, Y. R. Wu, and S. Rajan, "Unipolar vertical transport in GaN/AlGaN/GaN heterostructures," *Appl. Phys. Lett.*, vol. 103, p. 022102, 2013.

[11] D. A. Browne, B. Mazumder, Y. R. Wu, and J. S. Speck, "Electron transport in unipolar InGaN/GaN multiple quantum well structures grown by NH3 molecular beam epitaxy," *J. Appl. Phys.*, vol. 117, p. 185703, 2015.

[12] K. S. Qwah, M. Monavarian, G. Lheureux, J. Wang, Y. R. Wu, and J. S. Speck, "Theoretical and experimental investigations of vertical hole transport through unipolar AlGaN structures: impacts of random alloy disorder," *Appl. Phys. Lett.*, vol. 117, p. 022107, 2020.

[13] U. S. Dep. Energy, "2018 Solid-State Lighting R & D Opportunities," 2019, Available at https://www.energy.gov/sites/prod/files/2019/02/f59/edit.ssl_rd-opportunities_jan2019.pdf [accessed: Oct. 31, 2020].

[14] C. K. Li, M. Piccardo, L. S. Lu, et al., "Localization landscape theory of disorder in semiconductors. III. Application to carrier transport and recombination in light emitting diodes," *Phys. Rev. B*, vol. 95, p. 144206, 2017.

[15] S. De, A. Layek, A. Raja, et al., "Two distinct origins of highly localized luminescent centers within InGaN/GaN quantum-well light-emitting diodes," *Adv. Funct. Mater.*, vol. 21, pp. 3828–3835, 2011.

[16] M. Mensi, R. Ivanov, T. K. Uždavinys, et al., "Direct measurement of nanoscale lateral carrier diffusion: toward scanning diffusion microscopy," *ACS Photonics*, vol. 5, pp. 528–534, 2018.

[17] S. Sonderegger, E. Feltin, M. Merano, et al., "High spatial resolution picosecond cathodoluminescence of InGaN quantum wells," *Appl. Phys. Lett.*, vol. 89, p. 232109, 2006.

[18] T. K. Uždavinys, S. Marcinkevičius, M. Mensi, et al., "Impact of surface morphology on the properties of light emission in InGaN epilayers," *Appl. Phys. Express*, vol. 11, p. 051004, 2018.

[19] W. Hahn, J. M. Lentali, P. Polovodov, et al., "Evidence of nanoscale Anderson localization induced by intrinsic compositional disorder in InGaN/GaN quantum wells by scanning tunneling luminescence spectroscopy," *Phys. Rev. B*, vol. 98, p. 045305, 2018.

[20] G. Callsen, R. Butté, and N. Grandjean, "Probing alloy formation using different excitonic species: the particular case of InGaN," *Phys. Rev. X*, vol. 9, p. 031030, 2019.

[21] N. G. Young, R. M. Farrell, S. Oh, et al., "Polarization field screening in thick (0001) InGaN/GaN single quantum well light-emitting diodes," *Appl. Phys. Lett.*, vol. 108, p. 061105, 2016.

[22] S. Srinivasan, F. Bertram, A. Bell, et al., "Low Stokes shift in thick and homogeneous InGaN epilayers," *Appl. Phys. Lett.*, vol. 80, pp. 550–552, 2002.

[23] R. W. Martin, P. G. Middleton, K. P. O'Donnell, and W. Van Der Stricht, "Exciton localization and the Stokes' shift in InGaN epilayers," *Appl. Phys. Lett.*, vol. 74, pp. 263–265, 1999.

[24] M. Meneghini, C. De Santi, A. Tibaldi, et al., "Thermal droop in III-nitride based light-emitting diodes: physical origin and perspectives," *J. Appl. Phys.*, vol. 127, p. 211102, 2020.

[25] K. P. O'Donnell, R. W. Martin, and P. G. Middleton, "Origin of luminescence from InGaN diodes," *Phys. Rev. Lett.*, vol. 82, pp. 237–240, 1999.

[26] E. Berkowicz, D. Gershoni, G. Bahir, A. C. Abare, S. P. DenBaars, and L. A. Coldren, "Optical spectroscopy of InGaN/GaN quantum wells," *Phys. Status Solidi Basic Res.*, vol. 216, pp. 291–300, 1999.

[27] S. F. Chichibu, T. Sota, K. Wada, S. P. DenBaars, and S. Nakamura, "Spectroscopic studies in InGaN quantum wells," *MRS Internet J. Nitride Semicond. Res.*, vol. 4, pp. 93–105, 1999.

[28] F. Nippert, *Non-radiative Loss Mechanisms in InGaN/GaN Multiple Quantum Well Light-Emitting Diodes*, Thesis, Technical University Berlin, 2017.

[29] Y. H. Cho, G. H. Gainer, A. J. Fischer, et al., "'S-shaped' temperature-dependent emission shift and carrier dynamics in InGaN/GaN multiple quantum wells," *Appl. Phys. Lett.*, vol. 73, pp. 1370–1372, 1998.

[30] S. Yang, Y. Liu, Y. Zhang, and D. Mo, "Investigation of annealing-treatment on structural and optical properties of sol–gel-derived zinc oxide thin films," *Bull. Mater. Sci.*, vol. 33, pp. 209–214, 2010.

[31] A. David and M. J. Grundmann, "Influence of polarization fields on carrier lifetime and recombination rates in InGaN-based light-emitting diodes," *Appl. Phys. Lett.*, vol. 97, p. 033501, 2010.

[32] A. David, N. G. Young, and M. D. Craven, "Many-body effects in strongly disordered III-nitride quantum wells: interplay between carrier localization and Coulomb interaction," *Phys. Rev. Appl.*, vol. 12, p. 044059, 2019.

[33] J. Tauc, R. Grigorovici, and A. Vancu, "Optical properties and electronic structure of amorphous germanium," *Phys. Status Solidi*, vol. 15, pp. 627–637, 1966.

[34] M. Piccardo, C. K. Li, Y. R. Wu, et al., "Localization landscape theory of disorder in semiconductors. II. Urbach tails of disordered quantum well layers," *Phys. Rev. B*, vol. 95, p. 144205, 2017.

[35] D. M. Graham, A. Soltani-Vala, P. Dawson, et al., "Optical and microstructural studies of InGaNGaN single-quantum-well structures," *J. Appl. Phys.*, vol. 97, p. 103508, 2005.
[36] J. Danhof, H. M. Solowan, U. T. Schwarz, et al., "Lateral charge carrier diffusion in InGaN quantum wells," *Phys. Status Solidi Basic Res.*, vol. 249, pp. 480–484, 2012.
[37] H. M. Solowan, J. Danhof, and U. T. Schwarz, "Direct observation of charge carrier diffusion and localization in an InGaN multi quantum well," *Jpn. J. Appl. Phys.*, vol. 52, pp. 08JK07-1–08JK07-5, 2013.
[38] K. Kumakura, T. Makimoto, N. Kobayashi, T. Hashizume, T. Fukui, and H. Hasegawa, "Minority carrier diffusion lengths in MOVPE-grown *n*- and *p*-InGaN and performance of AlGaN/InGaN/GaN double heterojunction bipolar transistors," *J. Cryst. Growth*, vol. 298, pp. 787–790, 2007.
[39] S. Nakamura, M. Senoh, N. Iwasa, and S. Nagahama chi, "High-power InGaN single-quantum-well-structure blue and violet light-emitting diodes," *Appl. Phys. Lett.*, vol. 67, pp. 1868–1870, 1995.
[40] S. Nakamura and M. R. Krames, "History of gallium-nitride-based light-emitting diodes for illumination," *Proc. IEEE*, vol. 101, pp. 2211–2220, 2013.
[41] S. D. Lester, F. A. Ponce, M. G. Craford, and D. A. Steigerwald, "High dislocation densities in high efficiency GaN-based light-emitting diodes," *Appl. Phys. Lett.*, vol. 66, p. 1249, 1995.
[42] T. D. Moustakas, "The role of extended defects on the performance of optoelectronic devices in nitride semiconductors," *Phys. Status Solidi Appl. Mater. Sci.*, vol. 210, pp. 169–174, 2013.
[43] M. Meneghini, G. Meneghesso, and E. Zanoni, *Electrical Properties, Reliability Issues, and ESD Robustness of InGaN-Based LEDs. Topics in Applied Physics*, vol. 133, Springer Verlag, 2017, pp. 363–395.
[44] A. Hangleiter, F. Hitzel, C. Netzel, et al., "Suppression of nonradiative recombination by V-shaped pits in GaInN/GaN quantum wells produces a large increase in the light emission efficiency," *Phys. Rev. Lett.*, vol. 95, p. 127402, 2005.
[45] F. Jiang, J. Zhang, L. Xu, et al., "Efficient InGaN-based yellow-light-emitting diodes," *Photonics Res.*, vol. 7, p. 144, 2019.
[46] J. S. Speck and S. J. Rosner, "The role of threading dislocations in the physical properties of GaN and its alloys," *Physica B*, vols. 273–274, pp. 24–32, 1999.
[47] S. F. Chichibu, A. Uedono, T. Onuma, et al., "Origin of defect-insensitive emission probability in In-containing (Al,In,Ga)N alloy semiconductors," *Nat. Mater.*, vol. 5, pp. 810–816, 2006.
[48] J. Danhof, H. M. Solowan, U. T. Schwarz, et al., "Lateral charge carrier diffusion in InGaN quantum wells," *Phys. Status Solidi Basic Res.*, vol. 249, pp. 480–484, 2012.
[49] A. David, N. G. Young, C. Lund, and M. D. Craven, "Review—the physics of recombinations in III-nitride emitters," *ECS J. Solid State Sci. Technol.*, vol. 9, p. 016021, 2020.
[50] A. David and C. A. Y. N. Hurni, "High efficiency group-III nitride light emitting diode," U.S. Patent 10,734,549, 2020.
[51] C. Weisbuch, M. Piccardo, L. Martinelli, J. Iveland, J. Peretti, and J. S. Speck, "The efficiency challenge of nitride light-emitting diodes for lighting," *Phys. Status Solidi Appl. Mater. Sci.*, vol. 212, pp. 899–913, 2015.
[52] Y.-R. W. Chi-Kang Li, "Study on the current spreading effect and light extraction enhancement of," *IEEE Trans. Electron. Devices*, vol. 59, pp. 400–407, 2012.
[53] I. V. Rozhansky and D. A. Zakheim, "Analysis of processes limiting quantum efficiency of AlGaInN LEDs at high pumping," *Phys. Status Solidi Appl. Mater. Sci.*, vol. 204, pp. 227–230, 2007.
[54] Y. C. Shen, G. O. Mueller, S. Watanabe, N. F. Gardner, A. Munkholm, and M. R. Krames, "Auger recombination in InGaN measured by photoluminescence," *Appl. Phys. Lett.*, vol. 91, p. 141101, 2007.
[55] K. A. Bulashevich and S. Y. Karpov, "Is Auger recombination responsible for the efficiency rollover in III-nitride light-emitting diodes?," *Phys. Status Solidi Curr. Top. Solid State Phys.*, vol. 5, pp. 2066–2069, 2008.
[56] E. Kioupakis, P. Rinke, K. T. Delaney, and C. G. Van De Walle, "Indirect Auger recombination as a cause of efficiency droop in nitride light-emitting diodes," *Appl. Phys. Lett.*, vol. 98, p. 161107, 2011.
[57] J. Hader, J. V. Moloney, and S. W. Koch, "Investigation of droop-causing mechanisms in GaN-based devices using fully microscopic many-body theory," *Gall Nitride Mater. Devices VIII*, vol. 8625, p. 86251M, 2013.
[58] R. Vaxenburg, A. Rodina, E. Lifshitz, and A. L. Efros, "The role of polarization fields in Auger-induced efficiency droop in nitride-based light-emitting diodes," *Appl. Phys. Lett.*, vol. 103, p. 221111, 2013.
[59] F. Bertazzi, X. Zhou, M. Goano, G. Ghione, and E. Bellotti, "Auger recombination in InGaN/GaN quantum wells: a full-Brillouin-zone study," *Appl. Phys. Lett.*, vol. 103, p. 081106, 2013.
[60] V. Avrutin, S. Hafiz, F. Zhang, et al., "InGaN light-emitting diodes: efficiency-limiting processes at high injection," *J. Vac. Sci. Technol. A Vac. Surfaces Film*, vol. 31, p. 050809, 2013.
[61] K. J. Vampola, M. Iza, S. Keller, S. P. DenBaars, and S. Nakamura, "Measurement of electron overflow in 450 nm InGaN light-emitting diode structures," *Appl. Phys. Lett.*, vol. 94, pp. 2–5, 2009.
[62] M. Deppner, F. Römer, and B. Witzigmann, "Auger carrier leakage in III-nitride quantum-well light emitting diodes," *Phys. Status Solidi Rapid Res. Lett.*, vol. 6, pp. 418–420, 2012.
[63] A. C. Espenlaub, A. I. Alhassan, S. Nakamura, C. Weisbuch, and J. S. Speck, "Auger-generated hot carrier current in photo-excited forward biased single quantum well blue light emitting diodes," *Appl. Phys. Lett.*, vol. 112, p. 141106, 2018.
[64] J. Iveland, L. Martinelli, J. Peretti, J. S. Speck, and C. Weisbuch, "Direct measurement of auger electrons emitted from a semiconductor light-emitting diode under electrical injection: identification of the dominant mechanism for efficiency droop," *Phys. Rev. Lett.*, vol. 110, p. 177406, 2013.
[65] E. Kioupakis, D. Steiauf, P. Rinke, K. T. Delaney, and C. G. Van De Walle, "First-principles calculations of indirect Auger recombination in nitride semiconductors," *Phys. Rev. B Condens. Matter Mater. Phys.*, vol. 92, p. 035207, 2015.
[66] H. Yoshida, M. Kuwabara, Y. Yamashita, K. Uchiyama, and H. Kan, "Radiative and nonradiative recombination in an ultraviolet GaN/AlGaN multiple-quantum-well laser diode," *Appl. Phys. Lett.*, vol. 96, p. 211122, 2010.
[67] M. Shahmohammadi, W. Liu, G. Rossbach, et al., "Enhancement of Auger recombination induced by carrier localization in InGaN/

GaN quantum wells," *Phys. Rev. B*, vol. 95, pp. 125314-1–125314-10, 2017.
[68] C. Weisbuch, R. Dingle, A. C. Gossard, and W. Wiegmann, "Optical characterization of interface disorder in GaAs-$Ga_{1-x}Al_xAs$ multi-quantum well structures," *Solid State Commun.*, vol. 38, pp. 709–712, 1981.
[69] T. Saku, H. Iwamura, Y. Hirayama, Y. Suzuki, and H. Okamoto, "Room temperature operation of 650 nm Algaas multi-quantum-well laser diode grown by molecular beam epitaxy," *Jpn. J. Appl. Phys.*, vol. 24, pp. L73–L75, 1985.
[70] J. Hegarty and M. D. Sturge, "Studies of exciton localization in quantum-well structures by nonlinear-optical techniques," *J. Opt. Soc. Am. B*, vol. 2, p. 1143, 1985.
[71] S. Chichibu, T. Azuhata, T. Sota, and S. Nakamura, "Luminescences from localized states in InGaN epilayers," *Appl. Phys. Lett.*, vol. 70, pp. 2822–2824, 1997.
[72] T. M. Smeeton, M. J. Kappers, J. S. Barnard, M. E. Vickers, and C. J. Humphreys, "Electron-beam-induced strain within InGaN quantum wells: false indium 'cluster' detection in the transmission electron microscope," *Appl. Phys. Lett.*, vol. 83, pp. 5419–5421, 2003.
[73] A. Di Vito, A. Pecchia, A. Di Carlo, and M. Auf der Maur, "Impact of compositional nonuniformity in (In,Ga)N-based light-emitting diodes," *Phys. Rev. Appl.*, vol. 12, pp. 1–5, 2019.
[74] D. S. P. Tanner, J. M. McMahon, and S. Schulz, "Interface roughness, carrier localization, and wave function overlap in *c*-plane (In,Ga)N/GaN quantum wells: interplay of well width, alloy microstructure, structural in homogeneities, and Coulomb effects," *Phys. Rev. Appl.*, vol. 10, p. 034027, 2018.
[75] R. A. Oliver, S. E. Bennett, T. Zhu, et al., "Microstructural origins of localization in InGaN quantum wells," *J. Phys. D Appl. Phys.*, vol. 43, p. 354003, 2010.
[76] P. Dawson, S. Schulz, R. A. Oliver, M. J. Kappers, and C. J. Humphreys, "The nature of carrier localisation in polar and nonpolar InGaN/GaN quantum wells," *J. Appl. Phys.*, vol. 119, p. 181505, 2016.
[77] C. Weisbuch, "Review—on the search for efficient Solid state light emitters: past, present, future," *ECS J. Solid State Sci. Technol.*, vol. 9, p. 016022, 2020.
[78] S. Y. Karpov, "Modeling of III-nitride light-emitting diodes: progress, problems, and perspectives," *Gallium Nitride Mater. Devices VI*, vol. 7939, p. 79391C, 2011.
[79] M. V. Kisin and H. S. El-Ghoroury, "Inhomogeneous injection in III-nitride light emitters with deep multiple quantum wells," *J. Comput. Electron.*, vol. 14, pp. 432–443, 2015.
[80] J. Geng, P. Sarangapani, K. C. Wang, et al., "Quantitative multi-scale, multi-physics quantum transport modeling of GaN-based light emitting diodes," *Phys. Status Solidi Appl. Mater. Sci.*, vol. 215, pp. 1700662-1–1700662-7, 2018.
[81] Z. M. Simon Li and Z. M. S. Li, "Non-local transport in numerical simulation of GaN LED," *J. Comput. Electron.*, vol. 14, pp. 409–415, 2015.
[82] M. Auf der Maur, "Multiscale approaches for the simulation of InGaN/GaN LEDs," *J. Comput. Electron.*, vol. 14, pp. 398–408, 2015.
[83] Y. K. Kuo, M. C. Tsai, S. H. Yen, T. C. Hsu, and Y. J. Shen, "Effect of P-type last barrier on efficiency droop of blue InGaN light-emitting diodes," *IEEE J. Quantum Electron.*, vol. 119, p. 181505, 2010.
[84] C. Sheng Xia, Z. M. Simon Li, W. Lu, Z. Hua Zhang, Y. Sheng, and L. Wen Cheng, "Droop improvement in blue InGaN/GaN multiple quantum well light-emitting diodes with indium graded last barrier," *Appl. Phys. Lett.*, vol. 99, p. 233501, 2011.
[85] D. A. Zakheim, A. S. Pavluchenko, D. A. Bauman, K. A. Bulashevich, O. V. Khokhlev, and S. Y. Karpov, "Efficiency droop suppression in InGaN-based blue LEDs: experiment and numerical modelling," *Phys. Status Solidi Appl. Mater. Sci.*, vol. 209, pp. 456–460, 2012.
[86] T. Lu, Z. Ma, C. Du, et al., "Temperature-dependent photoluminescence in light-emitting diodes," *Sci. Rep.*, vol. 4, p. 6131, 2014.
[87] A. Shedbalkar and B. Witzigmann, "Non equilibrium Green's function quantum transport for green multi-quantum well nitride light emitting diodes," *Opt. Quantum Electron.*, vol. 50, no. 67, pp. 1–10, 2018.
[88] P. Kivisaari, T. Sadi, J. Oksanen, and J. Tulkki, "Monte Carlo study of non-quasi equilibrium carrier dynamics in III–N LEDs," *Opt. Quantum Electron.*, vol. 48, pp. 1–6, 2016.
[89] V. A. Jhalani, J. J. Zhou, and M. Bernardi, "Ultrafast hot carrier dynamics in GaN and its impact on the efficiency droop," *Nano Lett.*, vol. 17, pp. 5012–5019, 2017.
[90] D. P. Nguyen, N. Regnault, R. Ferreira, and G. Bastard, "Alloy effects in $Ga_{1-x}In_xN$/GaN heterostructures," *Solid State Commun.*, vol. 130, pp. 751–754, 2004.
[91] M. Auf Der Maur, A. Pecchia, G. Penazzi, W. Rodrigues, and A. Di Carlo, "Efficiency drop in green InGaN/GaN light emitting diodes: the role of random alloy fluctuations," *Phys. Rev. Lett.*, vol. 116, pp. 027401-1–027401-5, 2016.
[92] C. M. Jones, C. H. Teng, Q. Yan, P. C. Ku, and E. Kioupakis, "Impact of carrier localization on recombination in InGaN quantum wells and the efficiency of nitride light-emitting diodes: insights from theory and numerical simulations," *Appl. Phys. Lett.*, vol. 111, p. 113501, 2017.
[93] S. Schulz, M. A. Caro, C. Coughlan, and E. P. O'Reilly, "Atomistic analysis of the impact of alloy and well-width fluctuations on the electronic and optical properties of InGaN/GaN quantum wells," *Phys. Rev. B Condens. Matter Mater. Phys.*, vol. 91, p. 035439, 2015.
[94] D. S. P. Tanner, P. Dawson, M. J. Kappers, R. A. Oliver, and S. Schulz, "Polar (In,Ga)N/GaN quantum wells: revisiting the impact of carrier localization on the 'green gap' problem," *Phys. Rev. Appl.*, vol. 13, p. 044068, 2020.
[95] D. Watson-Parris, M. J. Godfrey, P. Dawson, et al., "Carrier localization mechanisms in $In_xGa_{1-x}N$/GaN quantum wells," *Phys. Rev. B Condens. Matter Mater. Phys.*, vol. 83, pp. 115321-1–115321-7, 2011.
[96] M. Filoche, M. Piccardo, Y. R. Wu, C. K. Li, C. Weisbuch, and S. Mayboroda, "Localization landscape theory of disorder in semiconductors. I. Theory and modeling," *Phys. Rev. B*, vol. 95, p. 144204, 2017.
[97] A. Di Vito, A. Pecchia, A. Di Carlo, and M. Auf Der Maur, "Simulating random alloy effects in III-nitride light emitting diodes," *J. Appl. Phys.*, vol. 128, p. 041102, 2020.
[98] S. D. Baranovskii and A. L. Efros, "Band edge smearing in Solid solutions," *Sov. Phys. Semicond.*, vol. 12, pp. 1328–1330, 1978.
[99] M. A. Caro, S. Schulz, and E. P. O'Reilly, "Theory of local electric polarization and its relation to internal strain: impact on polarization potential and electronic properties of group-III

nitrides," *Phys. Rev. B Condens. Matter Mater. Phys.*, vol. 88, p. 214103, 2013.

[100] R. Hill, "Energy-gap variations in semiconductor alloys," *J. Phys. C Solid State Phys.*, vol. 7, pp. 521–526, 1974.

[101] T.-J. Yang, R. Shivaraman, J. S. Speck, and Y.-R. Wu, "The influence of random indium alloy fluctuations in indium gallium nitride quantum wells on the device behavior," *J. Appl. Phys.*, vol. 116, p. 113104, 2014.

[102] J. M. McMahon, D. S. P. Tanner, E. Kioupakis, and S. Schulz, "Atomistic analysis of radiative recombination rate, Stokes shift, and density of states in *c*-plane InGaN/GaN quantum wells," *Appl. Phys. Lett.*, vol. 116, p. 181104, 2020.

[103] A. Di Vito, A. Pecchia, A. Di Carlo, and M. Auf Der Maur, "Characterization of non-uniform InGaN alloys: spatial localization of carriers and optical properties," *Jpn. J. Appl. Phys.*, vol. 58, p. SCCC03-1, 2019.

[104] A. A. Roble, S. K. Patra, F. Massabuau, et al., "Impact of alloy fluctuations and Coulomb effects on the electronic and optical properties of *c*-plane GaN/AlGaN quantum wells," *Sci. Rep.*, vol. 9, p. 18862, 2019.

[105] A. David, C. A. Hurni, N. G. Young, and M. D. Craven, "Electrical properties of III-nitride LEDs: recombination-based injection model and theoretical limits to electrical efficiency and electroluminescent cooling," *Appl. Phys. Lett.*, vol. 109, p. 083501, 2016.

[106] C. M. Jones, C. H. Teng, Q. Yan, P. C. Ku, and E. Kioupakis, "Impact of carrier localization on recombination in InGaN quantum wells and the efficiency of nitride light-emitting diodes: insights from theory and numerical simulations," *Appl. Phys. Lett.*, vol. 111, pp. 1–8, 2017.

[107] E. Hanamura, "Very large optical nonlinearity of semiconductor microcrystallites," *Phys. Rev. B*, vol. 37, pp. 1273–1279, 1988.

[108] R. J. Elliott, "Intensity of optical absorption by excitons," *Phys. Rev.*, vol. 108, pp. 1384–1389, 1957.

[109] M. Filoche and S. Mayboroda, "Universal mechanism for Anderson and weak localization," *Proc. Natl. Acad. Sci. U.S.A.*, vol. 109, pp. 14761–14766, 2012.

[110] C. K. Wu, C. K. Li, and Y. R. Wu, "Percolation transport study in nitride based LED by considering the random alloy fluctuation," *J. Comput. Electron.*, vol. 14, pp. 416–424, 2015.

[111] C. Lynsky, A. I. Alhassan, G. Lheureux, et al., "Barriers to carrier transport in multiple quantum well nitride-based *c*-plane green light emitting diodes," *Phys. Rev. Mater.*, vol. 4, p. 054604, 2020.

[112] G. Lheureux, C. Lynsky, Y.-R. Wu, and J. S. Speck, "A 3D simulation comparison of carrier transport in green and blue *c*-plane multi-quantum well nitride light emitting diodes," J. Appl. Phys., accepted.

[113] F. Jiang, J. Zhang, L. Xu, et al., "Efficient InGaN-based yellow-light-emitting diodes," *Photonics Res.*, vol. 7, p. 144, 2019.

Alexander Raun* and Evelyn Hu

Ultralow threshold blue quantum dot lasers: what's the true recipe for success?

https://doi.org/10.1515/9783110710687-002

Abstract: The family of III-nitride materials has provided a platform for tremendous advances in efficient solid-state lighting sources such as light-emitting diodes and laser diodes. In particular, quantum dot (QD) lasers using the InGaN/GaN material system promise numerous benefits to enhance photonic performance in the blue wavelength regime. Nevertheless, issues of strained growth and difficulties in producing InGaN QDs with uniform composition and size pose daunting challenges in achieving an efficient blue laser. Through a review of two previous studies on InGaN/GaN QD microdisk lasers, we seek to provide a different perspective and approach in better understanding the potential of QD emitters. The lasers studied in this paper contain gain material where QDs are sparsely distributed, comprise a wide distribution of sizes, and are intermixed with "fragmented" quantum well (fQW) material. Despite these circumstances, the use of microdisk cavities, where a few distinct, high-quality modes overlap the gain region, not only produces ultralow lasing thresholds (~6.2 μJ/cm^2) but also allows us to analyze the dynamic competition between QDs and fQWs in determining the final lasing wavelength. These insights can facilitate "modal" optimization of QD lasing and ultimately help to broaden the use of III-nitride QDs in devices.

Keywords: blue semiconductor laser; InGaN/GaN; microdisk cavity; quantum dot laser.

1 Introduction

Microscale light sources have seen impressive advances in sophistication and applicability over the past few decades. The development of increasingly miniaturized light sources has helped realize new technologies across a variety of research fields such as quantum computing, photonic integrated circuits, displays, and biomedicine [1–7]. When research into light-emitting diodes (LEDs) and laser diodes (LDs) began, III–V compound semiconductors (i.e., GaAs: gallium arsenide, InP: indium phosphide, GaN: gallium nitride, etc.) emerged as promising candidate materials owing to their direct bandgaps and high carrier mobilities [8, 9]. Basic research of these devices started with longer wavelength-emitting source materials. In the 1960s, $Ga(As_{1-x}P_x)$, GaAs, and InP diodes were shown to emit stimulated, coherent light in the red and infrared (IR) wavelength regimes [10–12]. As the sophistication of epitaxial growth techniques progressed, quantum heterostructures, beginning with quantum wells (QWs), were studied as enhanced gain materials for LDs. In the 1970s, researchers began demonstrating the use of gallium aluminum arsenide (GaAlAs) QWs in IR LDs [13]. To further improve carrier confinement and material gain, theoretical studies on quantum dots (QDs) began in the early 1980s, and the first GaAs QD laser was shown in 1994 [14–16].

While these discoveries paved the way for adoption of longer wavelength LEDs and LDs in commercial markets, research into high-power UV/blue light sources lagged. This was in part due to challenges associated with growing and doping high-quality, epitaxial layers of gallium nitride (GaN), one of the III–V semiconductors able to emit light in the UV/blue regime [17]. GaN became sought after because of inherent benefits over other semiconductors, including its large bandgap, a large exciton binding energy (allowing it to operate at room temperature more readily), and the ability to emit over the entire visible spectrum through the ternary alloys, $Al_yGa_{1-y}N$ and $In_xGa_{1-x}N$ [18]. However, the lack of lattice-matched growth substrates for GaN and other fabrication difficulties overshadowed these advantages of GaN and paused many research endeavors in GaN-based devices relative to semiconductors like GaAs. Stimulated emission in the UV/blue regime was demonstrated in the 1970s with optically pumped GaN single-crystal needles [19]; however, injection-based GaN LEDs and LDs would not see high-quality production for another

***Corresponding author: Alexander Raun**, Harvard University, 9 Oxford Street, Room 222, Cambridge, MA 02138, USA,
E-mail: raun@g.harvard.edu. https://orcid.org/0000-0001-8832-4197
Evelyn Hu, Harvard University, Cambridge, MA, USA

This article has previously been published in the journal Nanophotonics. Please cite as: A. Raun and E. Hu "Ultralow threshold blue quantum dot lasers: what's the true recipe for success?" *Nanophotonics* 2021, 10. DOI: 10.1515/nanoph-2020-0382.

20 years. It wasn't until the early 1990s that these devices took off with Nakamura et al.'s fabrication of GaN LEDs with GaN buffer layers on top of sapphire substrates [17, 20]. Shortly after, in 1996, using a similar fabrication technique, Nakamura et al. [21] also fabricated the first efficient GaN-based LDs with indium gallium nitride (InGaN) multiple QWs as the gain medium.

Since these developments, InGaN QDs embedded in GaN-based lasers have grown into an exciting area of research to improve gain characteristics and lasing thresholds further. Self-assembled InGaN QDs on GaN have been successfully fabricated through a variety of techniques, including metalorganic chemical vapor phase deposition, molecular beam epitaxy, and modified droplet epitaxy (MDE) [22–24]. However, the same challenges that plagued the growth of high-quality GaN also apply to the controlled formation of InGaN QDs and QWs as gain materials for lasers (as compared to other III–V devices where lattice-matched or nearly lattice-matched substrates are available). Lattice mismatch–induced strain between GaN and sapphire (the most commonly used underlying substrate) can produce threading dislocations which can degrade device performance [25]. Additionally, fabricating the "ideal" array of uniformly sized and spaced InGaN QDs is challenging, and MDE can result in patchy areas of QWs accompanying the QDs [23].

Despite these defects, InGaN QDs have been shown to perform effectively, especially when placed within high-quality optical microcavities. Our research has been at this critical juncture between the two major design specifications of an ultralow threshold, microscale semiconductor laser in the blue wavelength regime: (1) choosing an effective gain material and (2) designing a microscale optical cavity with only a few, high-quality modes. Through fabricating undercut GaN-based microdisk cavities, we have developed an effective "test bed" for exploring the dynamics of low-threshold lasing. Our optically pumped devices contain a unique heterogeneous gain material that includes InGaN QDs and "fragmented" QWs (fQWs), which are a by-product of the MDE method mentioned above. One would expect this nonuniform gain material to be a problem leading to poor device performance and high thresholds. However, through our experiments, we have found the opposite to be true. By first exploring the distinct lasing signature of InGaN QDs and then designing microring cavities, we have observed remarkable behavior, including the consistent dominance of InGaN QDs in the lasing process and modal engineering strategies to push lasing thresholds to still lower values. These insights can help improve understanding of the fundamental lasing dynamics of blue QD lasers, ultimately advancing a wide variety of applications.

2 Gain material

The first consideration in designing a microscale, low-threshold laser is the selection of the gain, or active, material. The gain material acts as an emitter, facilitating exciton recombination to generate photons and promote stimulated light emission [26]. The gain material also determines the laser's general wavelength regime, which can be tailored by using quantum heterostructures. Quantum heterostructures are composed of alternating materials with different bandgaps to produce a potential well with defined energy states where carriers can be captured and thereafter recombine radiatively. The two quantum heterostructures relevant to our research include InGaN QWs and QDs. QWs consist of slabs of InGaN sandwiched between two layers of GaN, leading to a potential energy trap that can confine electron-hole pairs. The slab-like QW structure confines carriers in one dimension, causing it to have a step-like density of states [15]. QDs also trap carriers through potential energy differences between the InGaN dot and the surrounding GaN. However, QDs are shaped like boxes and are much smaller than QWs. QDs confine carriers in three dimensions rather than one, which leads to a variety of advantages over the QW system. The increased confinement restrains carrier diffusion, compared to the situation for QWs, localizing the electron-hole pairs and making QDs less susceptible to defects such as threading dislocations [27]. InGaN QDs also have a higher probability than QWs to produce radiative recombination between carriers because they are less affected by the material's built-in electric field (owing to the polar *c*-axis of the InGaN wurtzite crystal structure), which separates electron and hole wave functions [28]. Finally, unlike QWs, QDs have a delta-like density of states, which in theory leads to higher material gain with a narrowed spectrum [15]. In addition to using QWs and QDs in isolation, there has been research in combining them to capitalize on each of their advantages. This combination can be known as dots-in-well (DWELLs), or quantum well-dots (QWDs), where indium-rich QDs are placed within a QW [29]. QWs have a much higher probability of carrier capture than QDs owing to their larger size, which may allow them to capture electron-hole pairs and then funnel them to QDs, ultimately enhancing the inherent benefits of QDs mentioned above.

In theory, a perfect QD array should provide the best gain characteristics and spontaneous recombination rates and, therefore, the lowest lasing thresholds [15]. This should be especially true when the array is placed in a cavity with high-quality modes; the interaction between the QDs and the cavity modes produce a Purcell

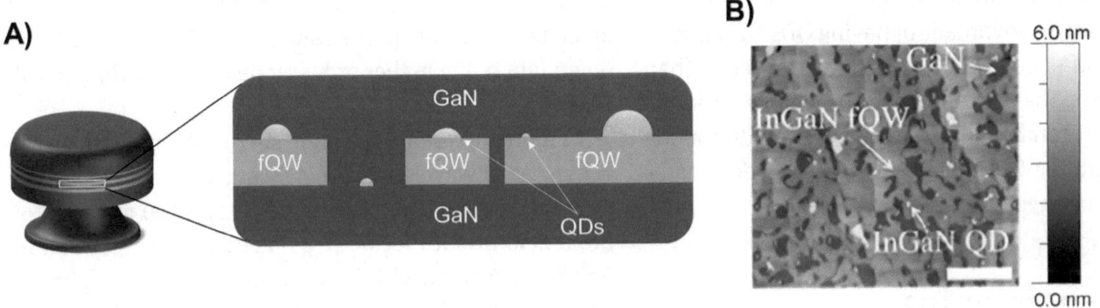

Figure 1: Our microdisk platform.
(A) Schematic of a 1-μm-diameter microdisk laser with three layers of InGaN active material. Inset shows one layer of InGaN fQWs and QDs, showing how QDs form on the fQWs and underlying GaN. (B) An atomic force microscopy (AFM) image of active material, specifying GaN, InGaN fQW, and InGaN QD regions. Taken from the study by Woolf et al. [32]. Scale bar is 500 nm in length, and vertical color scale goes from 0 to 6 nm. QD, quantum dot; fQW, fragmented quantum well; InGaN, indium gallium nitride.

enhancement of the QD emission [30]. However, decreased thresholds in InGaN QD lasers have not been seen experimentally, and InGaN QWs we have studied in undercut microdisk cavities consistently exhibit lower lasing thresholds than samples containing QDs [31]. One reason for this is that producing a "perfect" InGaN QD array with consistent QD size, consistent levels of In, and a sufficiently high density of QDs is quite challenging. While these difficulties are not unique to InGaN QDs (vs. other types of QDs), they can be exacerbated in the InGaN system owing to the material's inherent growth hurdles described above. Additionally, our QD samples are always accompanied with layers of fQWs. Thus, these InGaN QDs are in different potential environments: either directly on top of fQWs or on the underlying GaN. This effect is shown in Figure 1.

The InGaN QDs shown in Figure 1 were provided by Oliver et al. [23] at the University of Cambridge and fabricated via a MDE method. From the results displayed in Figure 1B, we see that owing to the challenges in growth mentioned above, QD formation seems to be randomly distributed over the GaN substrate and fQWs, with a variety of sizes. Additionally, the density of QDs and their individual sizes are quite small compared to the surrounding fQWs. In a given microdisk laser with a diameter of 1 μm and three layers of QDs with an density of 1×10^{10} cm^{-2}, this results in ~240 QDs per laser, with a very small surface area coverage compared to the fQWs. This sparse areal coverage dramatically lowers the probability of QDs to capture excitons compared to the larger fQWs. Therefore, one would expect the fQWs to have a greater contribution to the gain within the active region and to lasing. In fact, because of the only three or so narrow high-quality microcavity modes that overlap the gain region, we do see this when optically pumping our devices at low powers. At input power levels far below the lasing threshold, the large capture cross section of the fQWs results in the gain material predominantly emitting photons into a mode near the material's general background emission spectrum (~460 nm). However, as we increase our pumping power, the dominant emission in the cavity shifts to a shorter wavelength mode that is blue shifted from the center of the general background emission (~430 nm). In fact, this shorter wavelength mode corresponds to the emission wavelength of InGaN QD excitons [32, 33]. The device then ultimately lases at this wavelength associated with the QDs. This distinct signature of QD lasing was shown in a previous work by Woolf et al. [32], and a conceptual representation of this is shown in Figure 2.

Overall, the abrupt shift from the fQW mode to the QD mode with increasing input power underscores the powerful role played by microcavities with only a few, high-quality modes that overlap the gain area. Once

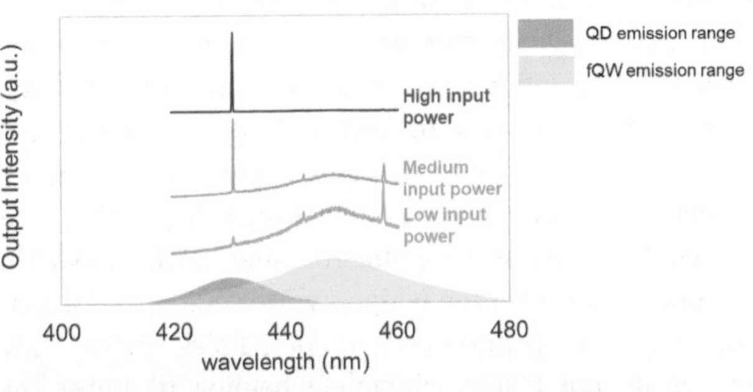

Figure 2: Illustrative example of the dominant mode shifting in a microdisk from the fQW regime to the QD regime with increasing input power. "Low input power" is far below the lasing threshold. "Medium input power" is slightly higher but still below the lasing threshold. "High power" is above the lasing threshold. Modes are indicated by the sharp peaks decorating the background emission. fQW, fragmented quantum well; QD, quantum dot.

enough photons are pumped into the cavity for the QDs to capture, the inherent advantage of having QDs with higher confinement and spontaneous recombination rates (than QWs) takes over. The Purcell effect further increases the QDs' radiative recombination rate through emitter-mode interactions, and the QDs' emission ultimately dominates the luminescence spectrum to achieve lasing.

3 Modal engineering

The research described in the previous section laid the foundation for us to look deeper into the second key component of a microdisk laser: the cavity modes that interact with our emitting material. In a disk-shaped optical cavity, the primary optical modes are whispering gallery modes (WGMs), which are produced by the total internal reflection of light traveling around the circumference of the disk. First-order WGMs are located at the periphery of the disk, while higher order radial modes encroach toward the center. Through finite-difference time-domain (FDTD) simulations, we confirmed that the first-order WGMs exhibit higher quality factors and are therefore better at capturing and storing photons in the microdisk than the higher order modes. Pairing these insights with the knowledge of our heterogenous gain material's unique lasing behavior discussed above, we then studied alterations to our microdisk geometry to see if we could push lasing thresholds lower.

For effective coupling between emitted light and the WGMs, the WGMs should overlap with the gain material spectrally and spatially. As stated in the previous section, our microdisk lasers containing QDs consistently lased via an optical mode with a wavelength of ~430 nm (or centered on the QD emission spectrum). This mode in a 1-micron-diameter disk, as determined through FDTD simulations, is a high-quality first-order mode near the periphery of the disk. Our simulations also reveal the presence of a lossy, higher order mode near the center of the fQW background emission [34]. This motivated us to explore cutting out the center of our microdisks to create microrings. This idea served two purposes. First, since our high-quality modes in the disks occur near the periphery, we could avoid injecting excess energy into the center of the disk where lasing modes do not reside, ultimately helping us lower the threshold. Second, since the low-quality, higher order modes occur closer to the center, by creating a ring, we could selectively remove photons with energies resonant with the lossy higher order modes while preserving photons that overlap with the WGMs near the periphery. This phenomenon is shown through our FDTD simulations in Figure 3, where rings with 200- and 500-nm inner diameters show the preservation of a high-Q WGM and degradation of a higher order mode corresponding to fQW photons.

Figure 3 presents the highest-Q WGM in the rings that occurs at a wavelength of 433 nm and a higher order 449-nm mode that overlaps the fQW spectral region. From the E-field profiles for both geometries, it is clear that the first-order 433-nm WGM appearing near the periphery of the disk is not significantly affected in either of the geometries, and the theoretical Q-factor remains high (~300,000 for each). For the higher order 449-nm mode, the inner circle of the 500-nm ring begins to overlap with the mode itself, which degrades that mode much more drastically than the first-order WGMs, allowing for the selective removal of nonlasing photons.

The idea of removing, or "leaking away" photons from a laser, and yet, still achieving lower threshold lasing, might at first thought seem to be logically inconsistent. In general, it is not possible to delineate the gain region into a "desired" gain material (QDs) and "less desirable" gain

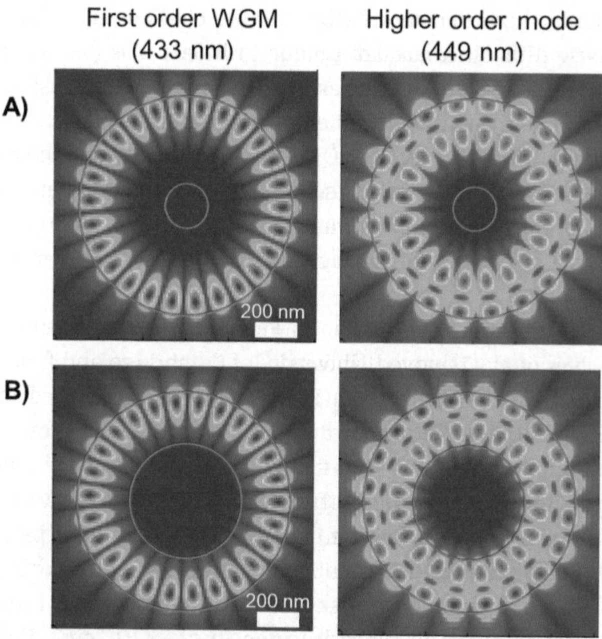

Figure 3: FDTD simulations of microrings.
(A) A microring with a 200-nm inner diameter and 1-μm outer diameter. The concentric red circles superimposed on the E-field profiles represent the inner and outer diameters of the structure. Here, we show a first-order WGM at a wavelength of 433 nm and a higher order mode at a wavelength of 449 nm. (B) A microring with a 500-nm inner diameter and 1-μm outer diameter. Once again, the red circles mark the geometry of the microring, and we present the same modes at 433 and 449 nm shown in part (A). FDTD, finite-difference time-domain; WGM, whispering gallery mode.

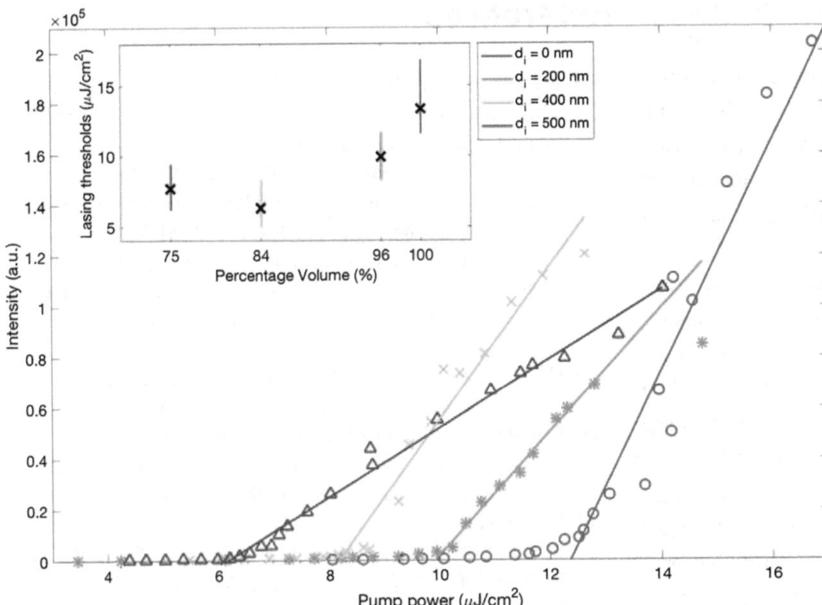

Figure 4: L–O curves for microrings with different inner diameter sizes (all had the same outer diameter length of 1 micron). Taken from the study by Wang et al. [35]. The inset shows average lasing thresholds (denoted by *x*'s) for the different-sized rings, each averaged over eight samples. L–O, Light-in–light-out.

materials (fQWs). Yet, as shown in Figure 2, there are fairly distinct regions separating the shorter wavelength QD emission from the broader, longer wavelength fQW emission. Even with such a heterogeneous gain spectrum, how is it possible to controllably modulate the cavity interaction with the two gain regions, selectively removing fQW photons while ensuring high cavity interaction with QD photons? However, part of the beauty and power of working with microcavity structures is being able to carry out this "modal engineering," as was demonstrated experimentally by Wang et al. [35] and illustrated in Figure 4.

Light-in–light-out (L–O) curves for microring devices with different geometries (different inner diameters or central areas removed) showed dramatically lower lasing thresholds (as low as 6.2 μJ/cm²) than microdisks having the same value for the outer radius [35]. The selected range of inner radii (i.e., 0–500 nm) allowed us to explore the effect of removing the inner area while at the same time ensuring that the largest inner radius (500 nm) would not degrade the high-Q WGMs near the periphery. Moreover, Figure 4 also shows a decrease in the slope efficiency of the L–O curves with an increasing inner diameter for the rings. The lower slope efficiency reflects the greater loss of photons, but the majority of those photons are emitted from the fQWs and do not contribute to lasing, as suggested by our simulations in Figure 3.

As was true for the microdisks, as we pump these microrings at successively higher optical powers, the ~460-nm mode associated with the fQWs dominates at low powers, and as this input power increases, the ~430-nm mode associated with the QDs eventually wins out and ultimately achieves lasing for the device. Once again, even with fewer QDs in the microrings that could be contributing to the lasing modes, QDs still dominate the lasing process, confirming that the QDs' inherent advantages and the cavity's high-quality modes are unaffected in the ring geometry. And these rings produce even lower thresholds, highlighting the advantage of having a heterogeneous gain material emitting in distinct wavelength regimes to selectively remove unwanted photons.

4 Conclusion

We have studied lasing dynamics in blue QD lasers through a unique platform combining undercut GaN microdisk cavities with a heterogeneous gain material composed of InGaN QDs and fQWs. The spectral precision of the WGMs of the microdisk allows us to track the process of electron-hole capture in the gain medium at differing powers. The emitter-cavity interaction depends on both the spatial profiles of the high-quality modes overlapping with the gain material, as well as the resonance in frequencies. Although WGMs overlap the fQW spectral region as well as the QD spectral region, lasing ultimately occurs at the QD wavelengths. The initial advantages of fQWs with a broader capture area, and a larger spectral range, interacting with the WGMs ultimately lose out to the shorter spontaneous emission lifetimes of the QD emitters, more strongly coupling to the WGMs. Thus, QD emission consistently dominates the lasing process in our devices despite a multitude of factors working against it, including low QD carrier capture probabilities, threading dislocations, and inhomogenous broadening, owing to variations in QD size. Additionally, by cutting out the center of our microdisks to create microrings, we were able to remove photons (from the fQWs) that do not contribute to the lasing process, resulting in lower thresholds. This was possible due to our heterogeneous gain material, emitting photons of different wavelengths into engineerable higher and lower quality optical modes.

For further improvement of our devices, several areas can be explored. For example, the contributions to lasing from the different types of QD configurations (i.e., QDs sitting directly on GaN vs. sitting on top of an fQW) could be modeled to enhance our understanding. Do the fQWs help facilitate carrier capture and direct carriers to the QDs (like a DWELL or QWD), therefore, being more advantageous than a QD by itself embedded within the GaN cavity? Advanced cavity geometry studies beyond microdisks could also provide a wealth of knowledge to further capitalize on our gain material's unique emission behavior. Specifically, inverse design of optical cavities has been a burgeoning area of research [36], and we could adapt this design strategy to account for our InGaN fQW/QDs' distinct material properties. While QD placement within cavities is particularly difficult, premapping of InGaN QDs in bulk material and formation of smaller volume cavities around them could be explored to potentially achieve even lower thresholds. Finally, comparisons between optical pumping and electrical injection of our devices could be studied to determine if the trends we have observed with regard to lasing persist.

Ultimately, both studies highlighted in this paper have provided remarkable insight into low-threshold lasing in blue microcavity lasers and the superior qualities of InGaN QDs coupled with high-quality modes. Despite the initial delay and challenges in development of miniaturized commercial blue lasers compared to longer wavelength regimes, we believe our insights have uncovered hidden advantages of particular kinds of InGaN QDs that could be used to improve device performance while the optimization of QD growth itself advances. By studying this dynamic interplay between these two ingredients, namely, the gain material and optical cavity design, it is our hope that our insights can help further inform the quest to find the best recipe for an efficient blue QD laser.

Author contribution: All the authors have accepted responsibility for the entire content of this submitted manuscript and approved submission.
Research funding: None declared.
Conflict of interest statement: The authors declare no conflicts of interest regarding this article.

References

[1] Z. Liu, C. H. Lin, B. R. Hyun, et al., "Micro-light-emitting diodes with quantum dots in display technology," *Light Sci. Appl.*, vol. 9, no. 1, pp. 1–23, Dec. 01, 2020.
[2] N. Martino, S. J. J. Kwok, A. C. Liapis, et al., "Wavelength-encoded laser particles for massively multiplexed cell tagging," *Nat. Photon.*, vol. 13, no. 10, pp. 720–727, Oct. 2019.
[3] J. Van Campenhout, P. Rojo Romeo, P. Regreny, et al., "Electrically pumped InP-based microdisk lasers integrated with a nanophotonic silicon-on-insulator waveguide circuit," *Opt. Express*, vol. 15, no. 11, p. 6744, May 2007.
[4] M. Humar and S. H. Yun, "Intracellular microlasers," *Nat. Photon.*, vol. 9, no. 9, pp. 572–576, Sep. 2015.
[5] A. Imamoglu, D. D. Awschalom, G. Burkard, et al., "Quantum information processing using quantum dot spins and cavity qed," *Phys. Rev. Lett.*, vol. 83, no. 20, pp. 4204–4207, 1999.
[6] S. J. Choi, K. Djordjev, S. J. Choi, and P. D. Dapkus, "Microdisk lasers vertically coupled to output waveguides," *IEEE Photon. Technol. Lett.*, vol. 15, no. 10, pp. 1330–1332, 2003.
[7] Y. Tchoe, K. Chung, K. Lee, et al., "Free-standing and ultrathin inorganic light-emitting diode array," *NPG Asia Mater.*, vol. 11, no. 1, pp. 1–7, Dec. 2019.
[8] A. Nainani, B. R. Bennett, J. Brad Boos, M. G. Ancona, and K. C. Saraswat, "Enhancing hole mobility in III–V semiconductors," *J. Appl. Phys.*, vol. 111, no. 10, p. 103706, May 2012.
[9] M. R. Krames, O. B. Shchekin, R. Mueller-Mach, et al., "Status and future of high-power light-emitting diodes for solid-state lighting," *IEEE/OSA J. Disp. Technol.*, vol. 3, no. 2, pp. 160–175, Jun. 2007.
[10] R. N. Hall, G. E. Fenner, J. D. Kingsley, T. J. Soltys, and R. O. Carlson, "Coherent light emission from GaAs junctions," *Phys. Rev. Lett.*, vol. 9, no. 9, pp. 366–368, Nov. 1962.
[11] K. Weiser and R. S. Levitt, "Stimulated light emission from indium phosphide," *Appl. Phys. Lett.*, vol. 2, no. 9, pp. 178–179, May 1963.
[12] N. Holonyak and S. F. Bevacqua, "Coherent (visible) light emission from $Ga(As_{1-x}P_x)$ junctions," *Appl. Phys. Lett.*, vol. 1, no. 4, pp. 82–83, Dec. 1962.
[13] R. D. Dupuis, P. D. Dapkus, N. Holonyak, E. A. Rezek, and R. Chin, "Room-temperature laser operation of quantum-well $Ga_{(1-x)}Al_xAs$-GaAs laser diodes grown by metalorganic chemical vapor deposition," *Appl. Phys. Lett.*, vol. 32, no. 5, pp. 295–297, Mar. 1978.
[14] Y. Arakawa and H. Sakaki, "Multidimensional quantum well laser and temperature dependence of its threshold current," *Appl. Phys. Lett.*, vol. 40, no. 11, pp. 939–941, Jun. 1982.
[15] M. Asada, Y. Miyamoto, and Y. Suematsu, "Gain and the threshold of three-dimensional quantum-box lasers," *IEEE J. Quantum Electron.*, vol. 22, no. 9, pp. 1915–1921, 1986.
[16] N. Kirstaedter, N. N. Ledentsov, M. Grundmann, et al., "Low threshold, large T_0 injection laser emission from (InGa)As quantum dots," *Electron. Lett.*, vol. 30, no. 17, pp. 1416–1417, Aug. 1994.
[17] G. Fasol, "Room-temperature blue gallium nitride laser diode," *Science*, vol. 272, no. 5269, pp. 1751–1752, 1996.
[18] J. J. Shi and Z. Z. Gan, "Effects of piezoelectricity and spontaneous polarization on localized excitons in self-formed InGaN quantum dots," *J. Appl. Phys.*, vol. 94, no. 1, pp. 407–415, Jul. 2003.

[19] R. Dingle, K. L. Shaklee, R. F. Leheny, and R. B. Zetterstrom, "Stimulated emission and laser action in gallium nitride," *Appl. Phys. Lett.*, vol. 19, no. 1, pp. 5–7, Jul. 1971.

[20] S. Nakamura, T. Mukai, and M. Senoh, "High-power gan p–n junction blue-light-emitting diodes," *Jpn. J. Appl. Phys.*, vol. 30, no. 12A, pp. L1998–L2001, 1991.

[21] S. Nakamura, M. Senoh, S. Ichi Nagahama, et al., "InGaN multi-quantum-well-structure laser diodes with cleaved mirror cavity facets," *Jpn. J. Appl. Phys. Part 2 Lett.*, vol. 35, no. 2B, p. L217, 1996.

[22] C. Adelmann, J. Simon, G. Feuillet, et al., "Self-assembled InGaN quantum dots grown by molecular-beam epitaxy," *Appl. Phys. Lett.*, vol. 76, no. 12, pp. 1570–1572, Mar. 2000 [Online]. Available at: http://aip.scitation.org/doi/10.1063/1.126098 [accessed: Jun. 22, 2020].

[23] R. A. Oliver, G. A. D. Briggs, M. J. Kappers, et al., "InGaN quantum dots grown by metalorganic vapor phase epitaxy employing a post-growth nitrogen anneal," *Appl. Phys. Lett.*, vol. 83, no. 4, pp. 755–757, Jul. 2003.

[24] K. Tachibana, T. Someya, and Y. Arakawa, "Nanometer-scale InGaN self-assembled quantum dots grown by metalorganic chemical vapor deposition," *Appl. Phys. Lett.*, vol. 74, no. 3, pp. 383–385, Jan. 1999.

[25] Y. W. Kim, E. K. Suh, and H. J. Lee, "Dislocation behavior in InGaN/GaN multi-quantum-well structure grown by metalorganic chemical vapor deposition," *Appl. Phys. Lett.*, vol. 80, no. 21, pp. 3949–3951, May 2002.

[26] L. A. Coldren, S. W. Corzine, and M. L. Mašanović, *Diode Lasers and Photonic Integrated Circuits*, Hoboken, NJ, USA, John Wiley & Sons, 2012.

[27] A. Fiore, M. Rossetti, B. Alloing, et al., "Carrier diffusion in low-dimensional semiconductors: a comparison of quantum wells, disordered quantum wells, and quantum dots," *Phys. Rev. B Condens. Matter Mater. Phys.*, vol. 70, no. 20, p. 205311, Nov. 2004.

[28] C. X. Xia and S. Y. Wei, "Built-in electric field effect in wurtzite InGaN/GaN coupled quantum dots," *Phys. Lett. Sect. A Gen. At. Solid State Phys.*, vol. 346, nos. 1–3, pp. 227–231, Oct. 2005.

[29] M. V. Maximov, A. M. Nadtochiy, S. A. Mintairov, et al., "Light emitting devices based on quantum well-dots," *Appl. Sci.*, vol. 10, no. 3, p. 1038. 2020.

[30] G.-H. Ryu, H.-Y. Ryu, and Y.-H. Choi, "Numerical investigation of Purcell enhancement of the internal quantum efficiency of GaN-based green LED structures," *Curr. Opt. Photon.*, vol. 1, no. 6, pp. 626–630, 2017.

[31] A. C. Tamboli, E. D. Haberer, R. Sharma, et al., "Room-temperature continuous-wave lasing in GaNInGaN microdisks," *Nat. Photon.*, vol. 1, no. 1, pp. 61–64, Jan. 2007.

[32] A. Woolf, T. Puchtler, I. Aharonovich, et al., "Distinctive signature of indium gallium nitride quantum dot lasing in microdisk cavities," *Proc. Natl. Acad. Sci. USA*, vol. 111, no. 39, pp. 14042–14046, Sep. 2014.

[33] R. A. Oliver, A. F. Jarjour, R. A. Taylor, et al., "Growth and assessment of InGaN quantum dots in a microcavity: a blue single photon source," *Mater. Sci. Eng. B Solid State Mater. Adv. Technol.*, vol. 147, nos. 2–3, pp. 108–113, Feb. 2008.

[34] D. Wang, *Low Threshold Lasing in Gallium Nitride Based Microcavities*, Doctoral dissertation, Harvard University, Graduate School of Arts & Sciences, Harvard University, 2019.

[35] D. Wang, T. Zhu, R. A. Oliver, and E. L. Hu, "Ultra-low-threshold InGaN/GaN quantum dot micro-ring lasers," *Opt. Lett.*, vol. 43, no. 4, p. 799, Feb. 2018.

[36] S. Molesky, Z. Lin, A. Y. Piggott, et al., "Inverse design in nanophotonics," *Nat. Photon.*, vol. 12, no. 11, pp. 659–670, 2018.

Stephen R. Forrest*

Waiting for Act 2: what lies beyond organic light-emitting diode (OLED) displays for organic electronics?

https://doi.org/10.1515/9783110710687-003

Abstract: Organic light-emitting diode (OLED) displays are now poised to be the dominant mobile display technology and are at the heart of the most attractive televisions and electronic tablets on the market today. But this begs the question: what is the next big opportunity that will be addressed by organic electronics? We attempt to answer this question based on the unique attributes of organic electronic devices: their efficient optical absorption and emission properties, their ability to be deposited on ultrathin foldable, moldable and bendable substrates, the diversity of function due to the limitless palette of organic materials and the low environmental impact of the materials and their means of fabrication. With these unique qualities, organic electronics presents opportunities that range from lighting to solar cells to medical sensing. In this paper, we consider the transformative changes to electronic and photonic technologies that might yet be realized using these unconventional, soft semiconductor thin films.

Keywords: lighting; organic semiconductor; reliability; solar cell; thin film transistor.

1 An introduction to organic electronics

The field of organic electronics, now in its 70th year since the identification of semiconducting properties of violanthrone by Akamatsu and Inokuchi [1], has enjoyed an extended period of discovery of the characteristics of disordered organic materials, ultimately leading to the astonishing success of organic light-emitting diodes (OLEDs). This technology platform has launched a revolution in information displays and lighting, while motivating researchers worldwide to explore a vast variety of new materials with intriguing optical and electronic properties that were never imagined in those early days of discovery [2]. The fundamental discoveries encouraged the small community of researchers to consider if there were any practical outcomes that could be achieved using organic semiconductors. Some of the first devices to exploit these "soft" materials were memories and solar cells. But compared to conventional semiconductor devices (most notably Si), the performance of organic devices was depressingly inferior, and worse, they did not last very long. It was their lack of stability that has given rise to a myth that persists up to the present day: organic devices are inherently unstable. We will return to this issue below.

Formally, an organic material is one that contains a carbon–hydrogen bond. By this definition, fullerenes (e.g., C_{60}), carbon nanotubes and graphene are not organic compounds. But, more practically, it can be considered to be one of a class of carbon-rich compounds. In the context of this paper, the organic materials of interest in electronics and photonics are semiconductors whose energy gaps are typically between 0.75 and 3.5 eV.

The pace of these first, tentative steps in exploiting the unlimited variety of organic materials for optoelectronic applications took an immense leap by the publication of two papers out of Eastman Kodak in 1986 and 1987. The first one, by C. W. Tang, announced the demonstration of an organic solar cell with 1% solar to electrical power conversion efficiency (PCE). The efficiency was not particularly high – that is not what made this demonstration so notable [3]. What was different is that this was a device that mimicked an inorganic p–n junction by combining two different organic semiconductor layers, one an electron donor (D) and the other an acceptor (A), into a bilayer cell. For the first time, the current–voltage characteristics showed nearly ideal rectifying characteristics that up to that time were only found in inorganic junction

***Corresponding author: Stephen R. Forrest**, Departments of Electrical Engineering and Computer Science, Physics, and Materials Science and Engineering, University of Michigan, Ann Arbor, MI 48104, USA, E-mail: stevefor@umich.edu

This article has previously been published in the journal Nanophotonics. Please cite as: S. R. Forrest "Waiting for Act 2: what lies beyond organic light-emitting diode (OLED) displays for organic electronics?" *Nanophotonics* 2021, 10. DOI: 10.1515/nanoph-2020-0322.

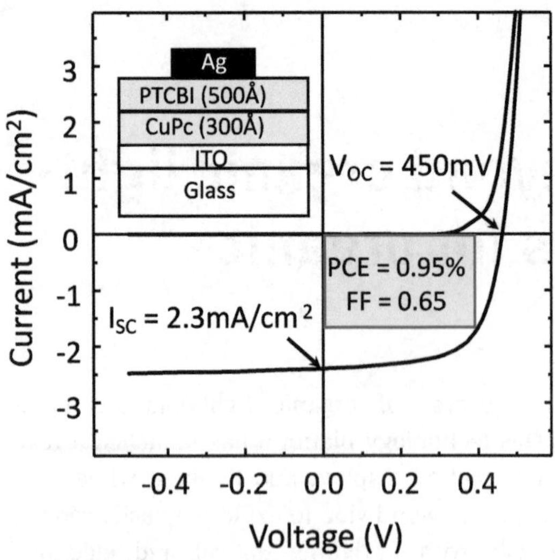

Figure 1: Current–voltage characteristics of the bilayer organic photovoltaic cell shown schematically in the inset [3]. The chemicals used are the acceptor, 3,4,9,10-perylenetetracarboxylic bis-benzimidazole (PTCBI), and the donor, copper phthalocyanine (CuPc). Indium tin oxide (ITO) serves as the transparent anode and Ag as the cathode. The short circuit current (I_{SC}), open circuit voltage (V_{OC}), power conversion efficiency (PCE) and fill factor (FF) under AM2 simulated illumination at 75 mW/cm^2 intensity are indicated. Here, the maximum power generated by the cell is equal to the area in the shaded rectangle and is given by $PCE_{max} = FF \cdot I_{SC} \cdot V_{OC}/P_{sun}$, where P_{sun} is the incident solar power intensity.

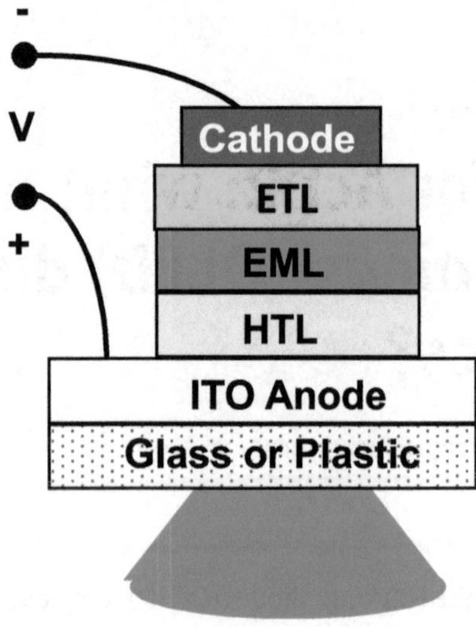

Figure 2: A simplified organic light-emitting diode (OLED) structure indicating the contacts, electron transport layer (ETL), light emission layer (EML) and hole transport layer (HTL). The EML typically comprises a conductive organic host doped at low density with an emissive fluorescent or phosphorescent emitting molecule. The entire device thickness is ~100 nm.

diodes (see Figure 1). From that demonstration forward, all organic solar cells have used the same basic D–A heterojunction (HJ) concept.

The second paper in 1987 was in many ways similar to the solar cell. Again, C. W. Tang, this time in collaboration with Steven van Slyke announced the successful demonstration of a bilayer organic light-emitting diode (OLED) [4]. The OLED, like the solar cell, had a clear rectifying behavior. But most interestingly, it exhibited bright green emission due to exciton recombination on one of the molecules forming the bilayer, namely 8-hydroxyquinoline Aluminum (Alq_3) with an external quantum efficiency of 1%. This was inferior compared to inorganic semiconductors at that time based on GaAs or InP, but the very thin films (~100 nm) comprising the device were grown on a glass substrate. Figure 2 shows a simplified, generic structure of modern OLEDs. Perhaps, if organics could only last long enough, they would be the foundation of a new generation of displays that, at that time, was dominated by cathode ray tubes and the emerging liquid crystal displays (LCDs).

While OLEDs looked promising (their colors could easily be modified across the visible spectrum by implementing only minor modifications to the chemical structures of the fluorescent emitting molecules or fluorophores), their lifetimes and efficiencies still fell short of what was already being achieved by LCDs. This situation changed dramatically with the introduction of electrophosphorescence by Baldo et al. [5, 6] that almost immediately led to OLEDs with 100% internal quantum efficiency [7]. Briefly, molecular excited states, or excitons, fall into two categories based on the spin of the excited electron: singlets and triplets. Singlets have odd symmetry under spin exchange, leading to rapid, fluorescent emission by transitions to the ground state, which also has singlet symmetry. Singlet emission was the basis for the earliest OLEDs. Triplets, on the other hand, have even symmetry and hence are quantum mechanically forbidden to transition to the singlet ground state. However, the selection rule that prevents their relaxation can be perturbed, leading to slow and very inefficient phosphorescence – a process that is not interesting for display applications. Unfortunately, simple statistical arguments show that an injected electron into an organic medium will excite singlet states only 25% of the time, with the other 75% of the electrons wasted on triplet state formation. This constraint is eliminated by the introduction of a heavy metal atom such as Pt or Ir into the luminescent molecule. The atom has a large orbital angular momentum, which when mixed with the electron spin in the organic ligand, results in spin–orbit coupling that leads to violation of the spin selection rule. This is the process of electrophosphorescence that can make every excited molecular state radiative, leading to 100% internal quantum efficiency. With this innovation, the electrophosphorescent OLEDs (PHOLEDs) became the most efficient light emitters known.

Display manufacturers, notably Samsung in the Republic of Korea, took immediate notice of benefits of electrophosphorescence, and the OLED display revolution was off and running. In rapid succession, the Galaxy smartphone series featuring efficient, emissive and surprisingly long lifetime OLED displays was introduced, followed by LG Display producing large, ultrathin and attractive OLED televisions. And now, Apple iPhones and iWatches with OLED displays are flooding the marketplace, bringing OLEDs to a dominant position for information displays supporting a $25 billion panel industry.

The success of OLEDs in the display industry naturally leads us to the question: what is next? Is there another "OLED revolution" waiting just around the corner for organic electronic materials and systems? Before we can answer that question, we must first consider what special attributes are offered by organic semiconductors that are not easily accessed by incumbent and proven inorganic semiconductors. After all, trying to displace an already served and mature market with an upstart technology is generally a fruitless exercise of catch-up, doomed to failure from the start. Listed below are several defining characteristics that point toward application spaces that might best be filled by organic electronics:

1. *Optoelectronics*: The very high absorption coefficients and often 100% emission efficiency of organic semiconductors make them an ideal platform for optoelectronic device applications (e.g., for light emitting displays and illumination, photodiodes and solar cells, etc.). Their electronic properties alone, as exploited in thin film transistors, do not generally provide organic electronics a clear "competitive edge" over conventional thin film semiconductor technologies. But when combined with ultrathin substrates, there are also some

The many faces of organic light-emitting diode (OLED) displays. Beyond smart phones and OLED TVs, the flexibility and scalability of OLEDs has led to a plethora of new form factors and applications. Just a few examples are provided here. Clockwise from top left: Enormous OLED display on exhibit at the ceiling of Seoul-Incheon International Airport. Transparent OLED TV as part of a museum exhibit. Apple iWatch whose volume exceeded that of all Swiss watches sold in 2019. Concept of an OLED illuminated shawl. Rollable OLED (called ROLEDs by LG Display) TVs being deployed to varying heights.

intriguing possibilities for purely electronic technology, as discussed below.

2. *Materials diversity*: The variety of organic molecules that can be developed is limitless. Hence, virtually every application need can be satisfied by engineering molecules that are optimized for a desired function.
3. *Large area*: Given the low cost and abundance of carbon-based materials comprising organic semiconductors, and their ability to be deposited onto substrates of almost arbitrarily large scale, they are ideally suited to applications where large area is a benefit. Displays provide an excellent example. Today, OLED televisions up to 77″ diagonal are on the market. Lighting and solar cells are other examples that benefit from large area.
4. *Flexibility, conformability, foldability*: An unusual feature of organic electronics is that the devices are very thin (typically only a couple of hundred nanometers), employ very flexible, van der Waals bonded thin films and can be deposited onto nearly any flat substrate at low temperature. Hence, organic electronics is easily supported by ultrathin glass, plastic and metal foils. Roll-up and foldable displays are already entering the marketplace owing to this feature. Flexibility also lends itself for their use in "wearable electronics" that can be molded to complex shapes needed for watches, garments and a myriad of other applications. Conformability is also a useful attribute for interior lighting, shaped instrument panels and lighting fixtures such as tail lights in automobiles. Even devices made to conform to the irregular surfaces of living organisms can provide a significant application space for organic electronics.
5. *Environmental friendliness*: Since large area devices are often ubiquitously deployed, it is essential that the technology be nontoxic and easily disposed. Thankfully, organic electronic devices rarely contain materials with significant negative environmental or health impacts. The low deposition and processing temperatures used in organic device fabrication (typically at or only slightly above room temperature) imply a low energy investment and hence a comparatively small environmental impact from their large-scale manufacture.
6. *Low cost*: Production on flexible substrates suggests that organic electronic appliances can be produced in a continuous, high speed, roll-to-roll web process. Organic electronic semiconductors are closely related to inks, paints and dyes used in volume production of newsprint, fabrics, food packaging and a multitude of other common consumer products. These ultrahigh volume production methods are ideally suited to generating the large area devices that are the primary domain of organic devices. Indeed, unless a technology is low cost, there is little reason to believe that it will fill a niche where large area is demanded.

The list of characteristics common to organic electronic devices is undoubtedly longer than the six noted above. Yet, there are few if any thin film electronic technologies that have this collection of attributes that can open the door to many possible applications that remain unaddressed by incumbent materials and systems. With this introduction, we will devote the rest of this paper to answering the question, "What is the next big breakthrough technology that will be served by organic semiconductors?" These materials have set the stage with a brilliant "Act 1: Organic light emitting displays". So what does Act 2 look like, when will it arrive on the stage, and will it hold our attention as well as the opening act?

2 OLEDs for lighting

The value of OLEDs lies in their high efficiency, brilliant colors, flexibility and long operational lifetimes. These are the characteristics needed for all modern lighting fixtures that are now replacing the incandescent light bulb that has been illuminating indoor spaces while wasting an unconscionable amount of energy for over a century. There are advantages and disadvantages to OLED lighting that must be understood before it can become a widespread commodity. Among its disadvantages is that the intensity of an OLED is low compared to conventional LEDs based on InGaN (see Table 1). Thus, to provide sufficient luminosity to light up a room, a large OLED fixture is required, and this increases cost. One common way to increase brightness is to stack individual OLED elements separated by transparent charge generation layers (CGLs), as shown in Figure 3. For example, an electron injected into the OLED element nearest to the cathode draws a hole from the adjacent CGL, forming an exciton that subsequently radiatively recombines. The loss of a hole creates charge imbalance in the CGL, compelling it to emit an electron into the second OLED in the stack. This, in turn, draws a hole from the next lower CGL, creating the second radiative exciton, and so on until the entire stack is once again restored to neutrality. Hence, a single injected charge generates as many photons as there are OLED subelements, resulting in a luminosity that is multiplied times the number of stacked OLEDs. This not only increases luminosity but also increases quantum efficiency well above 100% (but not the power efficiency which is

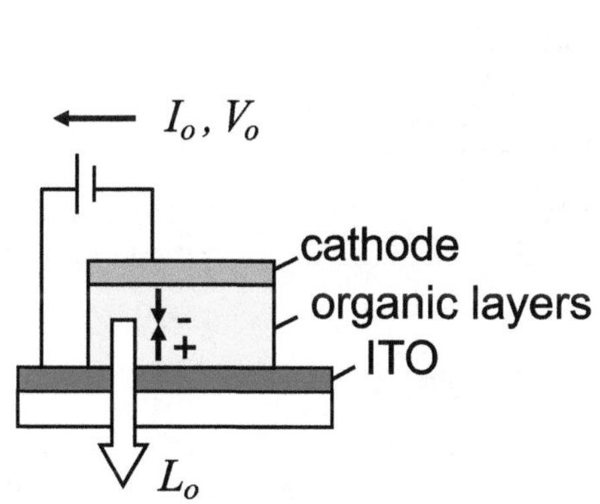

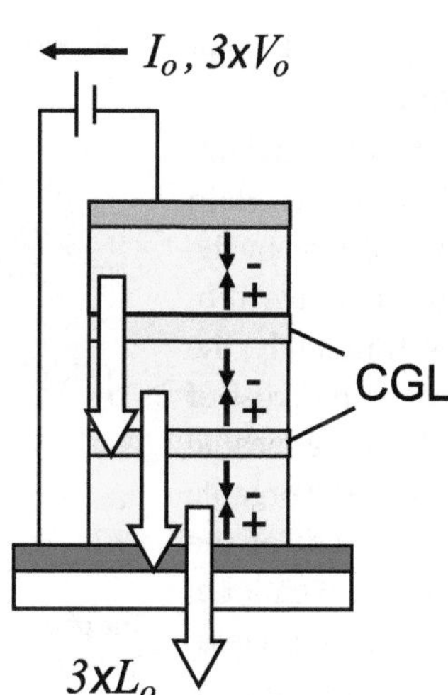

Figure 3: Comparison of a single element organic light-emitting diode (OLED) following the generic design of Figure 2 and a stacked OLED comprising three OLED subelements separated by transparent charge generation layers (CGLs). The CGLs should be nearly optically and electronically lossless. In this case, the current (I_O) and voltage (V_O) required to produce the luminance, L_O, in the single element device are I_O, $3V_O$ to produce $3L_O$ in the stacked device. Note that the lower current required to produce triple the luminosity makes the stacked structure ideal for use in high intensity lighting applications. The CGL conventionally comprises a transparent oxide (e.g., MoO_x) with thin, highly conductive electron and hole injecting films adjacent to the electron transport layer (ETL) and hole transport layer (HTL) of the contacting OLED subelements.

constrained by the law of conservation of energy). To create white light, each subelement can emit in a different zone of the visible spectrum. Alternatively, the red, green and blue emitting molecules can be blended within each subelement (or a combination of subelements), each emitting in an appropriate proportion to provide illumination with the desired color temperature and color rendering index.

The ability to make large white emitting OLED fixtures can mitigate their relatively low brightness. In fact size combined with flexible substrates provides OLED lighting sources with their greatest advantage. An OLED does not require mounting in reflective, light-directing or distributing structures, known as luminaires. To date, all other lighting sources must be mounted in one of these costly fixtures. Yet, the flexible and conformable form factor of an OLED allows it to be shaped to form its own luminaire. For this reason, OLEDs provide architects with possibilities to custom design attractive lighting sources that would be very difficult to achieve using LEDs, fluorescent bulbs or other illuminants. Another attraction of OLEDs is color tunability. By placing separately addressed red, green and blue OLED stripes in a closely spaced, side-by-side arrangement, their white hue, or color temperature, can be tuned according to the time of day, mood or current purpose of the space being illuminated.

Table 1: Comparison of lighting sources.

	Incandescent	Fluorescent	LED	OLED
Efficacy	17 lm/W	100 lm/W	80–90 lm/W – white 65 lm/W – warm white 240 lm/W – lab demo	120 lm/W Lab demos
Color rendering index	100	80–85	80 – white 90 – warm white	Up to 95
Form factor	Heat generating	Long or compact gas filled glass tube	Point source high intensity lamp	Large area thin diffuse source Flexible, transparent
Safety concerns	Very hot	Contains mercury	Very hot in operation	None to date
Lifetime (K h)	1	20	50	40
Dimmable	Yes, but much lower efficacy	Yes, efficiency decreases	Yes, efficiency increases	Yes, efficiency increases
Noise	No	Yes	No	No
Switching lifetime	Poor	Poor	Excellent	Excellent
Color tunable	No	No	Yes	Yes
Cost	$0.50/klm	$1/klm	$3/klm	$50–100/klm

A comparison of performance of OLEDs and other lighting sources is provided in Table 1. With the exception of their high cost, OLEDs fill a niche for highly efficient, pleasant, indirect (soft) interior lighting. But cost is certain to come down in the next few years, primarily driven by the momentum and experience gained in the massive production of large and small screen displays. Ultimately, the cost of OLED lighting will be determined by the cost of encapsulations and substrates that must be impermeable to water and oxygen to prevent degradation of the organic materials. Also, for large area devices, the cost of the organic materials, although low compared to conventional inorganic semiconductors, becomes a factor. Nevertheless, the unique attributes of OLED lighting and the maturity of OLED technology suggest that lighting is poised to become a significant market for organic electronic devices.

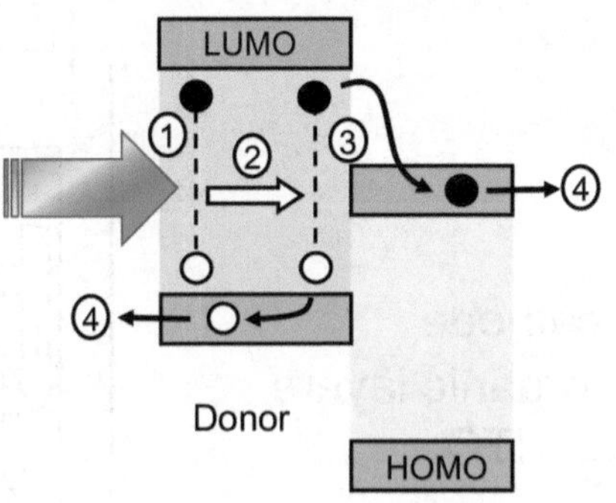

The photogeneration process in organics involves four steps. (1) Light incident (left) is absorbed in either the donor (D) or acceptor (A) layer (here shown only for D absorption), creating an exciton. (2) The exciton diffuses to the D–A heterojunction (HJ) (3) where it encounters an energy step. The electron transfers to the A, dissociating the exciton into a free electron and hole. (4) The electron drifts to the cathode and the hole to the anode. In organics, the conduction band is replaced by the lowest unoccupied molecular orbital level and the valence band by the highest occupied molecular orbital (the LUMO and HOMO, respectively). These are the "frontier orbitals" of the organic molecules themselves. This diagram illustrates a planar HJ. The electron and hole are indicated by filled and open circles, respectively, and the bound exciton state by the dashed lines. Most organic photovoltaics employ a bulk HJ where the D and A regions are intermixed at the nanometer scale to improve the probability that the exciton diffuses to the HJ without first recombining.

3 Organic solar cells

Solar cells would appear to fit all the criteria to which organics are suited. They have large area, require the use of nontoxic materials, they take a relatively small energy investment in their large-scale manufacture and they can conform to whatever surface to which they are attached. The reason they are not widely deployed today is that their efficiency and lifetime, until recently, have been inadequate. Indeed, the increase in efficiency has been slow to materialize. After the 1% efficiency of a bilayer organic photovoltaic (OPV) was demonstrated, no major increase or development emerged until 10 years later with the introduction of the bulk heterojunction (BHJ) solar cell active region [8, 9]. The BHJ is an entangled complex of donor and acceptor materials that eliminate the competition between the relatively long optical absorption length in organics (~100 nm) compared to the diffusion length of excited states (~10 nm) that must find their way to the donor-acceptor junction where they dissociate into a free electron and hole (see Box). Contemporaneous with the demonstration of the BHJ was the introduction of fullerene acceptors in both polymers [9] and small molecule [10] cells to replace the inefficient perylene diimide acceptors of the original bilayer cell. This led to a ten-fold increase in cell efficiency over the course of the next decade. Then, once again, materials innovations led to further efficiency increases by the introduction of thiophene-based "nonfullerene acceptors" [11]. Today, the efficiency of organic solar cells is approximately 17% and will soon reach 20% and beyond. In effect, advances in materials and structure have eliminated the complaint that "organic solar cells are not very efficient".

But what about reliability? A pervasive myth about organic materials and devices is that they lack the capacity for long-term reliability that we expect of our electronic appliances. Table 2 provides a compilation of lifetimes of OLEDs of the type used in displays, giving a hint to what makes some materials and structures more reliable than others. Generally, it is found that red pixels live longer than green pixels, and green live longer than blue. A reasonable conclusion, therefore, is that higher emission energies lead to shorter lifetimes. Indeed, this has been found to be the case. The primary source of molecular degradation is from high energy excitons colliding with other such excitons within the emission layer, and subsequently delivering sufficient energy to a molecular bond to break it [12]. Many strategies have been devised to limit the occurrence of such high energy excited state annihilation reactions, allowing for very long lifetimes of OLEDs now used in billions of displays.

In this same vein, fullerene-based OPVs have demonstrated remarkably long intrinsic lifetimes, extending over thousands of years [15]. But there is a difference between intrinsic lifetime and lifetime in the field where the packaged devices are exposed to the elements, from bright

Table 2: Example lifetimes of OLEDs [2].

OLED type	Chromaticity coordinates	Luminous efficiency (cd/A)	LT95[a] (h)	LT50 (h)
PHOLED[b]				
Deep red	(0.69, 0.31)	17	14,000	250,000
Red	(0.64, 0.36)	30	50,000	900,000
Yellow	(0.44, 0.54)	81	85,000	1,450,000
Green	(0.31, 0.63)	85	18,000	400,000
Light blue	(0.18, 0.42)	50	700	20,000
Fluorescent[c]				
Red	(0.67, 0.33)	11		160,000
Green	(0.29, 0.64)	37		200,000
Blue	(0.14, 0.12)	9.9		11,000
TADF[d]				
Green[e]	(0.34, 0.58)	15	1380[g]	
Light blue[f]	(0.18, 0.34)			40[h]

[a]LTX = time of operation until the luminance drops to X% of its initial value, L_O. For these data, L_O = 1000 cd/m^2, unless otherwise specified.
[b]Source: web sites, Universal Display Corp.
[c]Source: web sites, Idemitsu Kosan.
[d]TADF = thermally assisted delayed fluorescent OLED.
[e]Source: from the study by Tsang et al. [13].
[f]Source: from the study by Cui et al. [14].
[g]LT90.
[h]L_0 = 500 cd/m^2 normalized to L_0 = 1000 cd/m^2 using n = 1.7 acceleration factor.

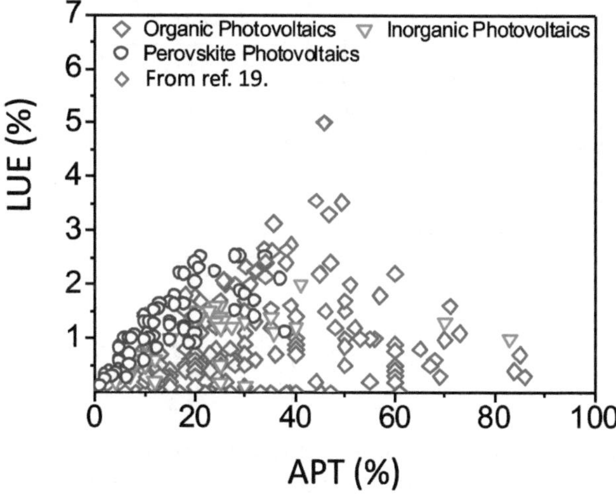

Figure 4: Compilation of selected results for the light utilization efficiency (LUE) versus the power conversion efficiency and the average photopic transparency (APT) for several different solar cell technologies. Here, LUE = PCE × APT. Semitransparent organic photovoltaics (ST-OPVs) are noted by diamond symbols. Data obtained from the studies by Li et al. [17] and Traverse et al. [18]. PCE, power conversion efficiency.

sunlight to rain, snow, ice, hail and dust. Furthermore, while the fullerene-based cells show this extraordinary endurance, the most efficient nonfullerene acceptor cells are much less robust, exhibiting lifetimes of only a couple of years [16]. Undoubtedly there is much work yet to be done to improve the lifetimes of the most efficient cells. Given the vast palette of materials available and that have yet to be synthesized and motivated by the unusual applications addressed by organic thin films, there is little doubt that both high efficiency and long device lifetimes will form the basis of a viable organic solar cell industry in the near future.

It is important to understand the appropriate niche for OPVs that distinguish them from incumbent solar technologies. The widespread generation of commodity power does not provide sufficient motivation for their development given the low costs enjoyed by Si panels. So, what can OPVs offer that Si solar cells cannot? The answer lies in the narrow but intense absorption spectra of organic molecules, allowing for very efficient solar cells that strongly absorb in the near infrared while being semitransparent, and importantly, neutral optical density in the visible. Such a solar module, deposited on a roll of plastic film, can be inserted in the pocket between the two sheets of glass forming a double pane window. Power generating windows, combined with microinverters, can supply considerable energy to buildings if they cover a reasonable fraction of a building surface. Such windows require an appropriate suite of materials to simultaneously provide high efficiency and visible transparency, along with optical coatings that reflect unabsorbed near infrared (NIR) radiation back into the cell for a second pass, while maximizing outcoupling of visible light. A figure of merit that quantifies the performance of semitransparent cells is their light utilization efficiency (LUE), which is the product of the PCE and the average photopic transparency (APT) of the cell. Here, the APT is the perceived transparency that is a convolution of the solar spectrum with the spectral sensitivity of the eye. Figure 4 shows a compilation of LUE vs. APT for thin film solar cells based on a range of technologies. There is little doubt that OPVs have a far more advantageous combination of these parameters than any other technology, including amorphous Si and perovskite solar cells.

An example of a high efficiency, neutral optical density organic solar cell is shown in Figure 5. It combines several solution-processed organic layers, semitransparent anode and cathode contacts, a visible optical outcoupling layer and an antireflection coating. This particular design has an 8.1% PCE with APT = 43.3%, leading to LUE = 3.5% [19]. The calculated thermodynamic efficiency limit for such single junction cells should eventually lead to

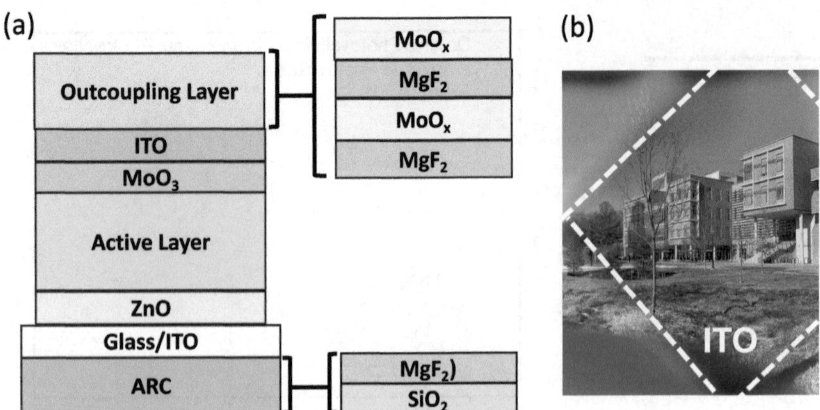

Figure 5: (a) Archetype structure of a high performance semitransparent organic photovoltaic (OPV) cell. The device consists of a glass substrate coated on its distal surface with a bilayer antireflection coating (ARC). The opposite surface comprises an ITO contact, a ZnO nanoparticle buffer layer. The bulk heterojunction active layer consists of a mixture of a solution-deposited nonfullerene acceptor and a polymer donor. This is capped by a MoO_3 electron conducting buffer layer and a second, transparent indium tin oxide (ITO) contact. The device is completed by the deposition of a four-layer outcoupling layer that has a high transmission in the visible while it reflects the near infrared radiation back into the active layer for a second pass at absorption. (b) The solar cell appears nearly transparent with neutral density. In this case, the optical loss in the visible is approximately 50% [19].

LUE > 7%, with even higher efficiencies achieved using multijunction versions.

As in other organic electronic devices, advances will be paced by innovations in materials. Yet the value proposition of ubiquitous solar generation on windows and other building surfaces is substantial. In this era where the production of carbon-free energy is no longer optional on our warming planet, coupled with the very low potential cost of organic devices rapidly produced in a roll-to-roll manufacturing process, it is nearly inevitable that OPVs will form the basis of Act 2 in the historical development of organic electronics.

4 Organic transistors and beyond

Beyond lighting and solar cells, the future opportunities in organic electronics become less clear. Recent advances in organic transistors, however, appear to open up possibilities for some electronic applications that, again, take advantage of the highly flexible form factors of organic thin film devices. The most common organic thin film transistor (OTFT) structure is a lateral geometry, with the organic channel deposited onto the surface of the gate insulator (see Box). The transistor operates in the accumulation mode: charge is drawn in to an otherwise undoped, large energy gap organic semiconductor channel from the source by the gate and drain potentials. As in conventional transistors, the channel current is modulated by the gate potential of ~1 V in optimized devices. The field effect, or channel charge mobilities of organics (<1 cm^2/Vs) are less than or comparable to those of amorphous Si or metal oxide transistors, so mobility alone does not offer a competitive advantage in either transistor gain or bandwidth. In fact, while OTFTs have been the focus of research since their first demonstrations in the mid-1980s [20, 21], it has sometimes been asked if this is a technical solution looking for a problem.

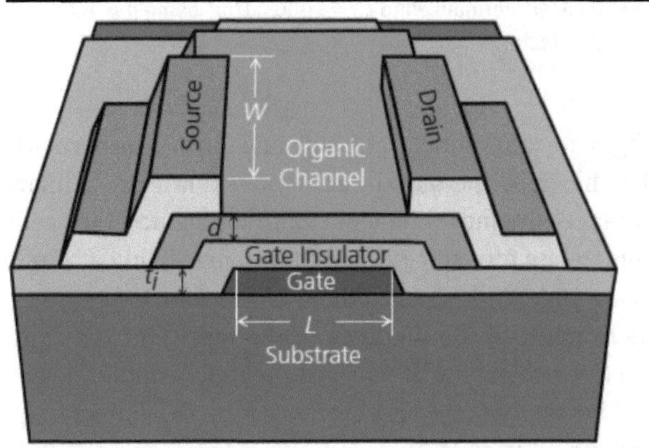

Common configuration of an organic thin film transistor. The source and drain are on the surface of the organic semiconductor channel, and the gate lies underneath the gate insulator. This is known as a bottom gate/top contact organic thin film transistor. Other geometries that switch the layering order of the insulator and gate contact, as well as the source and drain contacts, are also commonly employed.

But there are some interesting application niches for OTFTs that are not easily served by other electronic technologies. One is for selective detection of chemical compounds, agents or threats. The channels can employ organics that can bond or otherwise be altered in the presence of trace (parts per million or billion) concentrations of target chemicals (analytes). This, in turn, can result in a change in the interface charge density, thus shifting

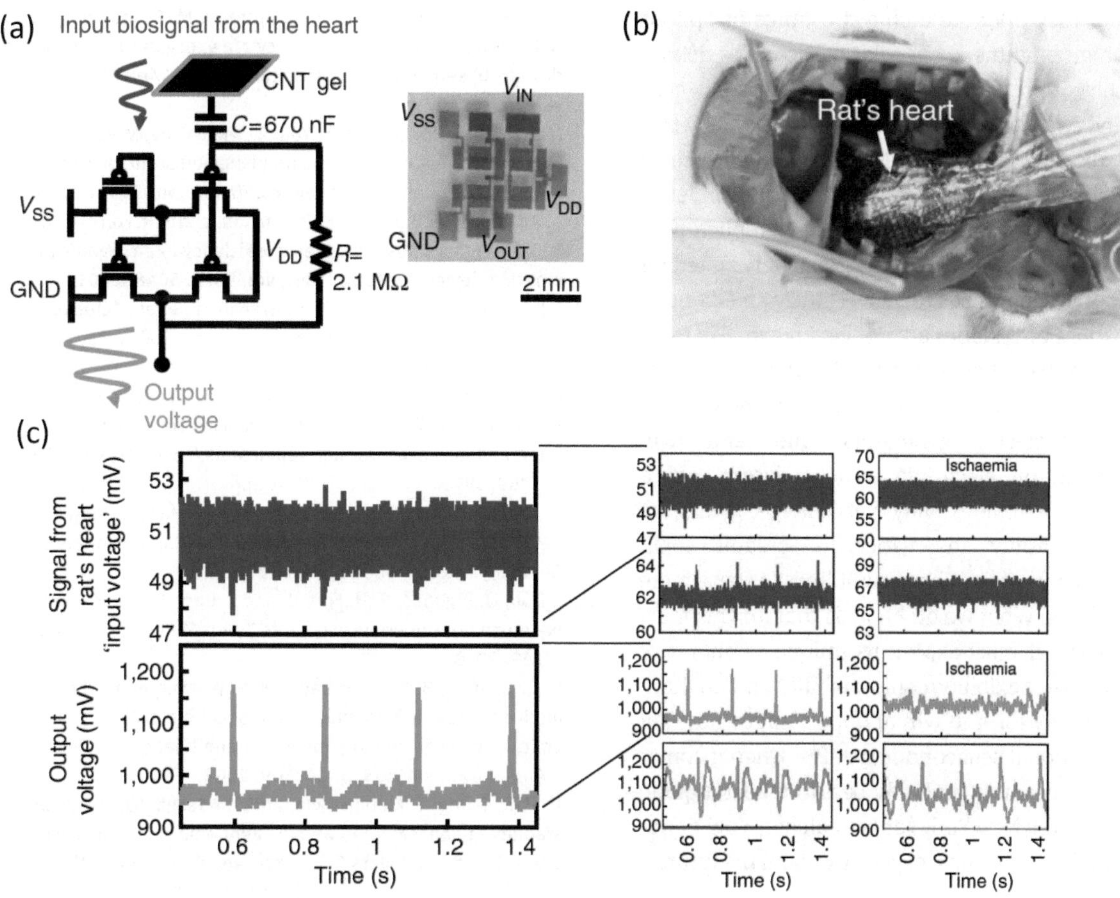

Figure 6: An example of imperceptible organic electronics used to monitor heart function in a rat.
(a) Amplifier circuit used in a biosensing array. Each pixel comprises a carbon nanotube (CNT) gel contact, input capacitor and amplifier circuit. A microscope image of the pixel is shown at right. (b) Photograph of the electrocaridio transducer attached to a rat's heart. (c) Electrocardiograms of the unamplified (blue) and amplified (red) heart impulses for several input conditions, including the ischemic state under myocardial infarction [23].

the transistor threshold voltage [2]. The response can be fast, highly selective, sensitive and reversible, making OTFT chemical sensors an interesting, large and diverse application opportunity.

An even more compelling application for OTFTs is in medical diagnostics. The transistors have been fabricated on plastic substrates that are only ~1 μm thick. Hence, the transistors can easily conform and adhere to irregular living tissues without impacting function. For this reason, organic electronic devices deposited on such ultrathin substrates have been termed "imperceptible electronics", which is yet another form of a wearable display [22]. An elegant example of such a device is the detector/amplifier array in Figure 6 used to monitor heart rhythm by placing the ultrathin electronic circuit in direct contact with the organ [23]. And while the niche for this particular device may be small, medical sensing diagnostics offers an almost limitless opportunity for organic electronics once an initial foothold is established. However, given the relative immaturity of organic electronics for chemical sensing, medical analyses, photodetection and so on, it will be some time before we are likely to see significant penetration by these technologies in the highly diversified but enormous application space of optoelectronic sensors that can be uniquely served by the attributes common to organic devices.

5 Conclusions

In this brief article, I have endeavored to answer the question of what is next in the field of organic electronics that will build on and extend the enormous initial successes of OLED displays? Displays are indeed a hard act to follow. But OLED lighting seems poised to serve interior illumination needs that complement existing inorganic LED lighting sources. I would classify this as Act 1, Scene 2, for organics. It is not easy to predict how large an industry OLED lighting will grow into, given their high cost and relatively low luminosity, but they offer the architect a range of options that existing high brightness, specular LED sources do not. The second big opportunity for organics is in solar cells applied to windows and building facades. It is unlikely that they will ever displace Si (nothing ever does), so OPVs must capture markets not well served or that remain completely unserved by Si. Power windows and building applied photovoltaics seem to be applications that are ideally suited for OPVs. And finally, sensors based on OTFTs are an emerging opportunity whose boundaries are not yet known. Are there opportunities that have not been considered here? Indeed, they are as varied as are organic materials themselves. Memories, thermoelectric generators, one- and two-dimensional quantum electronic devices, lasers and a range of other possibilities may yet emerge given the extraordinary versatility and variety of organic semiconductors [2]. But we do not have sight lines to these more distant possibilities. What we do know is that for this technology to succeed, it must exploit its unique attributes of large area, low cost, flexibility/conformability and environmental compatibility for it to win at opportunities not well served by conventional semiconductors. But when it comes to organic electronics, it is probably best to repeat a quote often attributed to the New York Yankee catcher, Yogi Berra: "It's tough to make predictions, especially about the future".

Author contribution: All the authors have accepted responsibility for the entire content of this submitted manuscript and approved submission.
Research funding: The author thanks the support from the US Office of Naval Research (contract no. N000142012114) US Air Force Office of Scientific Research (contract no. 17RT0908), the National Science Foundation (grant DMR 1905401), and Universal Display Corp.
Conflict of interest statement: The author declares an equity interest in Universal Display Corp. This apparent conflict of interest is under management by University of Michigan's Office of Research.

References

[1] H. Akamatu and H. Inokuchi, "On the electrical conductivity of violanthrone, iso-violanthrone, and pyranthrone," *J. Chem. Phys.*, vol. 18, p. 810, 1950.
[2] S. R. Forrest, *Organic Electronics: Foundations to Applications*, Oxford, UK, Oxford University Press, 2020.
[3] C. W. Tang, "Two-layer organic photovoltaic cell," *Appl. Phys. Lett.*, vol. 48, p. 183, 1986.
[4] C. W. Tang and S. A. VanSlyke, "Organic electroluminescent devices," *Appl. Phys. Lett.*, vol. 51, p. 913, 1987.
[5] M. A. Baldo, S. Lemansky, P. E. Burrows, M. E. Thompson, and S. R. Forrest, "Very high efficiency green organic light emitting devices based on electrophosphorescence," *Appl. Phys. Lett.*, vol. 76, p. 4, 1999.
[6] M. A. Baldo, D. F. O'Brien, Y. You, A. Shoustikov, M. E. Thompson, and S. R. Forrest, "High efficiency phosphorescent emission from organic electroluminescent devices," *Nature*, vol. 395, p. 151, 1998.
[7] C. Adachi, M. A. Baldo, M. E. Thompson, and S. R. Forrest, "Nearly 100% internal phosphorescence efficiency in an organic light emitting device," *J. Appl. Phys.*, vol. 90, p. 5048, 2001.
[8] J. J. M. Halls, C. A. Walsh, N. C. Greenham, et al., "Efficient photodiodes from interpenetrating polymer networks," *Nature*, vol. 376, p. 498, 1995.
[9] G. Yu, J. Gao, J. Hummelen, F. Wudl, and A. J. Heeger, "Polymer photovoltaic cells - enhanced efficiencies via a network of internal donor-acceptor heterojunctions," *Science*, vol. 270, p. 1789, 1995.
[10] P. Peumans and S. R. Forrest, "Very high efficiency double heterostructure copper phthalocyanine/C_{60} photovoltaic cells," *Appl. Phys. Lett.*, vol. 79, p. 126, 2001.
[11] J. Hou, O. Inganäs, R. H. Friend, and F. Gao, "Organic solar cells based on non-fullerene acceptors," *Nat. Mater.*, vol. 17, no. 2, p. 119, 2018.
[12] N. C. Giebink, B. W. D'Andrade, M. S. Weaver, et al., "Intrinsic luminance loss in phosphorescent small-molecule organic light emitting devices due to bimolecular annihilation reactions," *J. Appl. Phys.*, vol. 103, p. 044509, 2008.
[13] D. P.-K. Tsang, T. Matsushima, and C. Adachi, "Operational stability enhancement in organic light-emitting diodes with ultrathin Liq interlayers," *Sci. Rep.*, vol. 6, p. 22463, 2016.
[14] L.-S. Cui, Y.-L. Deng, D. P.-K. Tsang, et al., "Controlling synergistic oxidation processes for efficient and stable blue thermally activated delayed fluorescence devices," *Adv. Mater.*, vol. 28, no. 35, pp. 7620–7625, 2016.
[15] Q. Burlingame, X. Huang, X. Liu, C. Jeong, C. Coburn, and S. R. Forrest, "Intrinsically stable organic solar cells under high intensity illumination," *Nature*, vol. 573, p. 394, 2019.
[16] X. Du, T. Heumueller, W. Gruber, et al., "Efficient polymer solar cells based on non-fullerene acceptors with potential device lifetime approaching 10 years," *Joule*, vol. 3, no. 1, pp. 215–226, 2019.
[17] Y. Li, C. Ji, Y. Qu, et al., "High efficiency semi-transparent organic photovoltaics," *Adv. Mater.*, p. 1903173, 2019.
[18] C. J. Traverse, R. Pandey, M. C. Barr, and R. R. Lunt, "Emergence of highly transparent photovoltaics for distributed applications," *Nat. Energy*, vol. 2, no. 11, pp. 849–860, 2017.
[19] Y. Li, X. Guo, Z. Peng, et al., "Color-neutral, semitransparent organic photovoltaics," *Proc. Natl. Acad. Sci. U. S. A.*, 2020, https://www.pnas.org/cgi/doi/10.1073/pnas.2007799117.
[20] G. Horowitz, D. Fichou, X. Peng, Z. Xu, and F. Garnier, "A field-effect transistor based on conjugated alpha-sexithienyl," *Solid State Commun.*, vol. 72, no. 4, pp. 381–384, 1989.
[21] A. Tsumura, H. Koezuka, and T. Ando, "Polythiophene field-effect transistor: its characteristics and operation mechanism," *Synth. Met.*, vol. 25, no. 1, pp. 11–23, 1988.
[22] T. Someya, S. Bauer, and M. Kaltenbrunner, "Imperceptible organic electronics," *MRS Bull.*, vol. 42, p. 124, 2017.
[23] T. Sekitani, T. Yokota, K. Kuribara, et al., "Ultraflexible organic amplifier with biocompatible gel electrodes," *Nat. Commun.*, vol. 7, p. 11425, 2016.

Bernard C. Kress* and Ishan Chatterjee

Waveguide combiners for mixed reality headsets: a nanophotonics design perspective

https://doi.org/10.1515/9783110710687-004

Abstract: This paper is a review and analysis of the various implementation architectures of diffractive waveguide combiners for augmented reality (AR), mixed reality (MR) headsets, and smart glasses. Extended reality (XR) is another acronym frequently used to refer to all variants across the MR spectrum. Such devices have the potential to revolutionize how we work, communicate, travel, learn, teach, shop, and are entertained. Already, market analysts show very optimistic expectations on return on investment in MR, for both enterprise and consumer applications. Hardware architectures and technologies for AR and MR have made tremendous progress over the past five years, fueled by recent investment hype in start-ups and accelerated mergers and acquisitions by larger corporations. In order to meet such high market expectations, several challenges must be addressed: first, cementing primary use cases for each specific market segment and, second, achieving greater MR performance out of increasingly size-, weight-, cost- and power-constrained hardware. One such crucial component is the optical combiner. Combiners are often considered as critical optical elements in MR headsets, as they are the direct window to both the digital content and the real world for the user's eyes.

Two main pillars defining the MR experience are comfort and immersion. *Comfort* comes in various forms:

- *wearable comfort*—reducing weight and size, pushing back the center of gravity, addressing thermal issues, and so on
- *visual comfort*—providing accurate and natural 3-dimensional cues over a large field of view and a high angular resolution
- *vestibular comfort*—providing stable and realistic virtual overlays that spatially agree with the user's motion
- *social comfort*—allowing for true eye contact, in a socially acceptable form factor.

Immersion can be defined as the multisensory perceptual experience (including audio, display, gestures, haptics) that conveys to the user a sense of realism and envelopment. In order to effectively address both comfort and immersion challenges through improved hardware architectures and software developments, a deep understanding of the specific features and limitations of the human visual perception system is required. We emphasize the need for a human-centric optical design process, which would allow for the most comfortable headset design (wearable, visual, vestibular, and social comfort) without compromising the user's sense of immersion (display, sensing, and interaction). Matching the specifics of the display architecture to the human visual perception system is key to bound the constraints of the hardware allowing for headset development and mass production at reasonable costs, while providing a delightful experience to the end user.

Keywords: augmented reality; diffractive optics; holography; mixed reality; virtual reality; waveguide optics.

***Corresponding author: Bernard C. Kress,** Microsoft Corp. HoloLens Team, 1 Microsoft Way, Redmond, 98052, WA, USA,
E-mail: bernard.kress@gmail.com.
Ishan Chatterjee, Microsoft Corp. HoloLens Team, 1 Microsoft Way, Redmond, 98052, WA, USA

Glossary of Terms, Abbreviations, and Acronyms

We provide this glossary for the reader after the abstract section as these acronyms are used extensively in this review paper.

AR	**Augmented reality, adding virtual content into field of view of reality, can include augmentations created by mixed reality headsets, handhelds, head up displays, smart glasses, camera-projector systems, etc.**
MR	Mixed reality, virtual objects situationalized in 3D in your real space, often interactable
OST-MR	Optical see-through mixed reality, displays are transparent such that real world is viewable optically through the displays

This article has previously been published in the journal Nanophotonics. Please cite as: B. C. Kress and I. Chatterjee "Waveguide combiners for mixed reality headsets: a nanophotonics design perspective" *Nanophotonics* 2021, 10. DOI: 10.1515/nanoph-2020-0410.

(continued)

AR	**Augmented reality, adding virtual content into field of view of reality, can include augmentations created by mixed reality headsets, handhelds, head up displays, smart glasses, camera-projector systems, etc.**
VST-MR	Video see-through mixed reality, virtual reality turned into the mixed reality with camera pass-through of the real-world into the VR environment
XR	Extended reality, a generic term to capture all varieties across MR and AR
VR	Virtual reality, blocks out reality and supplants with virtual objects
Immersion	Sense of realism and development in delivered experience
IMU	Inertial measurement unit consisting of at least an accelerometer, and gyroscope, and often a magnetometer
GPU	Graphical processing unit, parallel architecture suited for graphics render and other matrix operations
HMD	Head-mounted display or helmet-mounted display
HUD	Head up display, refers to see-through display that is often mounted externally (such as above a dashboard) allowing user to see both virtual content and subject of focus (e.g., the road ahead) simultaneously
SLM	Spatial light modulator
LCD	Liquid-crystal display, display technology where electro-sensitive liquid crystal pixels amplitude-modulate light from a global polarized backlight in transmission
LTPS-LCD	Low-temperature polysilicon liquid-crystal display, higher resolution and faster switching speed than amorphous Si LCD
IPS-LCD	In-plane switching liquid-crystal display, liquid-crystal structure twist in-plane of display, allowing for higher viewing angles than twisted nematic (TN) LCDs, used in phones and monitors
HTPS-LCD	High-temperature polysilicon (used for silicon backplanes)
AMOLED	Active-matrix organic light-emitting diode, increased contrast at the cost of lifetime and high brightness, each pixel is its own organic electroluminescent emitter, used commonly in cellphones
mu-OLED, micro-OLED	Micro-organic light-emitting diode, display with emitter size less than 15 μm, used in camera electronic view finders
DLP	Digital light processing, Texas Instrument's colloquially genericized trademark for DMD (digital micromirror device), an array of bi-stable reflective micromirrors, commonly used in projection systems for highly efficiency SLM
LCoS	Liquid crystal on silicon, microdisplay with a switchable liquid-crystal matrix on reflective silicon backplane

(continued)

AR	**Augmented reality, adding virtual content into field of view of reality, can include augmentations created by mixed reality headsets, handhelds, head up displays, smart glasses, camera-projector systems, etc.**
mu-iLED, micro-iLED	Micro inorganic light-emitting diode, actively addressed inorganic LED array with emitter size <50 mu, NTE displays usually require a magnitude lower; can achieve high brightness and contrast, but challenged in maintaining efficiency, multi-color integration and backplane integration
VCSEL	Vertical-cavity surface-emitting laser, laser diode with lower divergence and current threshold than edge-emitting diodes
MEMS	Microelectromechanical system
LBS	Laser beam scanning, type of display where a modulated laser dot is raster scanned across display FOV via system of MEMS mirrors
NTE	Near-to-eye
3DOF	3 degrees of freedom, in the context of tracking usually refers to the rotational axes (pitch, yaw, roll) which can be resolved with only a calibrated IMU
6DOF	6 degrees of freedom, in the context of tracking refers to the rotational and translational axes
CG	Center of gravity, important ergonomic metric in head-worn devices
IPD	Interpupillary distance
PPD	Pixels per degree
HDR	High dynamic range
FOV	Field of view, provided as an angle
Eyebox	The volume that the user's pupil can sit in and view the entire virtual image field-of-view. The box may not be a rectangular prism, but is more often a frustrum
Eye relief	The distance the user's corneal surface is from the display optic surface
UX/UI	User experience/User interface, refers to the design of the experience and applications
VAC	Vergence accommodation conflict, refers to the mismatch experienced when a stereoscopic display's image focal plane does not match the stereo disparity of the virtual image.
Pupil swim	The experience of warp and shift of virtual objects as the user's pupils rove around the eyebox caused by distortion in the projected image across the eyebox
Hard-edge occlusion	The ability for real-world objects to mask virtual content according to the depth the virtual image is in the world
Hologram	Recording of a interference pattern between a reference and a wavefront off a 3D scene… but in AR/VR forums, a virtual stereo image that appears to be positioned in space like a true hologram
ET	Eye tracking
HeT	Head tracking
TIR	Total internal reflection (principle of how light propagates when trapped in a light guide)

(continued)

AR	**Augmented reality, adding virtual content into field of view of reality, can include augmentations created by mixed reality headsets, handhelds, head up displays, smart glasses, camera-projector systems, etc.**
PBS	Polarized beam splitter
EPE	Exit pupil expansion, a technique where a combiner's exit pupil may be replicated in 1D or 2D space allowing for a larger eyebox
LOE	Lightguide optical element
SRG	Surface relief gratings, nanostructure gratings etched into substrate surface, can be blazed, slanted, binary, multilevel, or analog
CGH	Computer-generated hologram, hologram whose wavefront has be calculated computationally rather than recorded in analog
RWG	Resonant waveguide gratings, also known as GMR (guided mode resonant gratings), diffractive, dielectric structures with leaky lateral modes
Metasurface	Surface with nanofabricated, sub-wavelength structures (often high aspect ratio) that can impart arbitrary phase changes in transmission and/or reflection unlocking unique optical functions.
NIL	Nanoimprint lithography
ALD	Atomic layer deposition
MTF	Modulation transfer functions, represents the effect (usually degradation) on spatial frequencies (in resolution and contrast) through an optical element, higher is better
RCWA	Rigorous couple-wave analysis
FMM	Fourier modal method
FDTD	Finite difference time domain
SAW	Surface acoustic wave
AOM	Acousto-optical modulator
EOM	Electro-optical modulator

1 Introduction

Defense has been historically the first application sector for augmented reality (AR) and virtual reality (VR), as far back as the 1950s [1]. Based on these early developments, the first consumer VR/AR boom expanded in the early 1990s and contracted considerably throughout that decade, a poster child of a technology ahead of its time and ahead of its markets [2]. Notably, due to the lack of available consumer display technologies and related sensors, novel optical display concepts were introduced throughout the 1990s [3, 4] that are still considered as state of the art, such as the "Private Eye" smart glass from Reflection Technology (1989) and the "Virtual Boy" from Nintendo (1995)—both based on scanning displays rather than flat-panel displays. Although such display technologies were well ahead of their time [5–7], the lack of consumer-grade inertial measurement unit (IMU) sensors, low-power 3-dimensional (3D)-rendering graphical processing units, and wireless data transfer technologies contributed to the end of this first VR boom. The other reason was the lack of digital content or rather the lack of a clear vision of adapted VR/AR content for enterprise or consumer spaces [8, 9].

The only AR/VR sector that saw sustained efforts and developments throughout the next decade was the defense industry (flight simulation and training, helmet-mounted displays [HMDs] for rotary-wing aircrafts, and head-up displays [HUDs] for fixed-wing aircrafts) [10]. The only effective consumer efforts during 2000–2010 were in the field of automotive HUDs and personal binocular headset video players.

The smartphone technology ecosystem, including the associated display, connectivity, and sensor systems, shaped the emergence of the second VR/AR boom and formed the first building blocks used by early product integrators. Today's engineers, exposed at an early age to ever-present flat-panel display technologies, tend to act as creatures of habit much more than their peers 20 years ago, who had to invent novel immersive display technologies from scratch. We have therefore seen since 2012 the initial implementations of immersive AR/VR HMDs based on readily available smartphone display panels (low-temperature polysilicon liquid-crystal display [LCD], In-plane switching liquid-crystal display, active-matrix organic light-emitting diode) and picoprojector microdisplay panels (High-temperature polysilicon LCD, mu-organic light-emitting diode (OLED), digital light processing (DLP), liquid crystal on silicon (LCoS). (Similarly, the AR/VR industry has been able to leverage the progress made during the smartphone revolution for cheap and reliable sensors as well, such as IMUs and cameras). Currently, HMD display architectures are evolving slowly to more specific technologies, which may be a better fit for immersive requirements than flat panels are, sometimes resembling the display technologies invented throughout the first AR/VR boom two decades earlier (inorganic mu-iLED panels, 1-dimensional [1D] scanned arrays, 2-dimension (2D) laser/vertical-cavity surface-emitting laser [VCSEL] microelectromechanical system [MEMS] scanners, and so on).

Such traditional display technologies will serve as an initial catalyst for what is coming next. The immersive display experience in AR/VR is a paradigm shift from the traditional panel display experiences that have existed for more than half a century, going from cathode ray tube (CRT) TVs to LCD computer monitors and laptop screens, to OLED tablets and smartphones, to LCoS, DLP, and MEMS

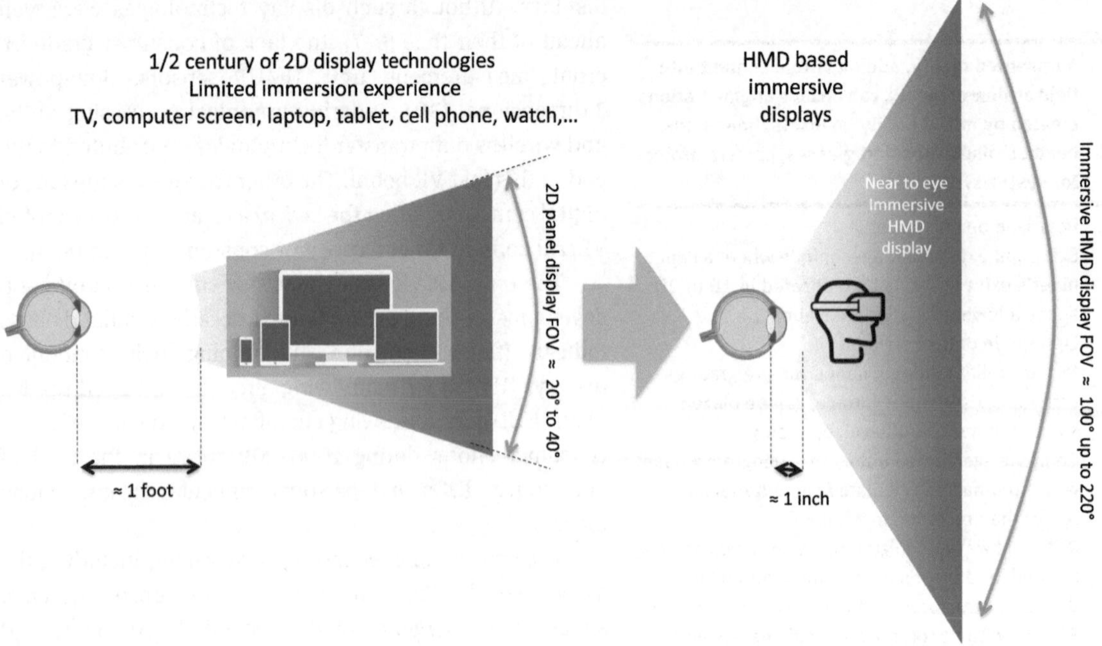

Figure 1: Immersive NTE displays: a paradigm shift in personal information display. NTE, near-to-eye.

scanner digital projectors, and to iLED smartwatches (see Figure 1).

When flat-panel display technologies and architectures (smartphone or microdisplay panels) are used to implement immersive near-to-eye (NTE) display devices, factors such as etendue, static focus, low contrast, and low brightness become severe limitations. Alternative display technologies are required to address the needs of NTE immersive displays to match the specifics of the human visual system.

The emergence of the second VR/AR/smart glasses boom in the early 2010s introduced new naming trends, more inclusive than AR or VR: mixed (or merged) reality (MR), more generally known today as "XR," a generic acronym for "extended reality." The name "smart eyewear" (world locked audio, digital monocular display, and prescription eyewear) tends to replace the initial "smart glass" naming convention.

Figure 2 represents the global MR spectrum continuum, from the real-world experience toward diminished reality (where parts of reality are selectively blocked through hard edge occlusion, such as annoying advertisements while walking or driving through a city, to blinding car headlights while cruising at night on a highway) to AR as in optically see-through MR, to merged reality as in video see-through MR, and to pure virtual worlds (as in VR).

Parallel realities are a new concept that emerged recently with specific optical display hardware, creating from a single-display hardware-specific individual eyeboxes (EBs) with specific information, targeted at multiple viewers detected and tracked through biometrics by sensors around that same display. These dynamic EBs are steered in real time to follow the specific viewers. This is not a wearable display architecture, rather a monitor or transparent window display. In this scenario, different viewers of that same physical display see different information, tuned to their specific interest, depending on their physical location.

2 The emergence of MR as the next computing platform

Smart glasses (also commonly called smart eyewear or digital eyewear) are mainly an extension of prescription eyewear, providing a digital contextual display as an addition to vision prescription correction (see for example Google Glass). This concept is functionally very

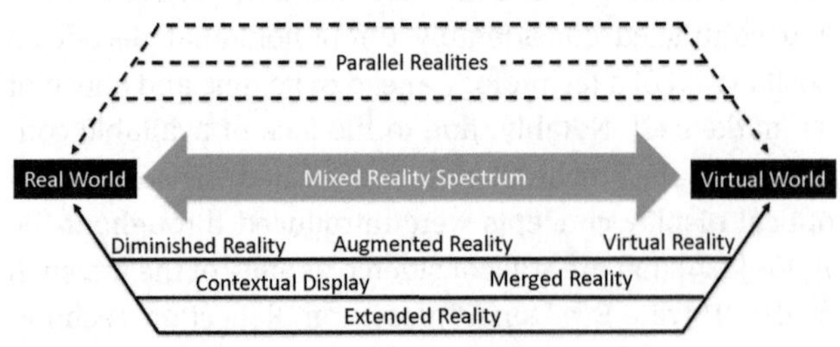

Figure 2: Mixed-reality spectrum continuum.

different from either AR or MR functionality. The typical smart glass field of view remains small (<15°diagonal), is typically monocular, and is often offset from the line of sight. The lack of sensors (apart the IMU) allows for approximate 3 degrees of freedom (3DOF) head tracking, and lack of binocular vision reduces the display to simple, overlaid 2D text and images. Typical 3DOF content is locked relative to the head, while 6 degrees of freedom (6DOF) sensing allows the user to get further and closer to the content.

Monocular displays do not require as much rigidity in the frames as a binocular vision system would (to reduce horizontal and vertical retinal disparity that can produce eye strain). Many smart glass developers also provide prescription correction as a standard feature (e.g., "Focal" by North or Google Glass V2).

The combination of strong connectivity (3G, 4G, WiFi, Bluetooth) and a camera makes it a convincing companion to a smartphone, for contextual display functionality or as a virtual assistant, acting as a global positioning system (GPS)-enabled social network companion. A smart glass does not aim to replace a smartphone, but it intends to contribute as a good addition to it, like a smartwatch.

VR headsets are an extension of simulators and gaming consoles, as shown by major gaming providers such as Sony, Oculus, HTC Vive, and Microsoft Windows MR, with gaming companies such as Valve Corp providing a gaming content ecosystem (Steam VR). The offerings have bifurcated into high-performance personal computer (PC)-tethered headsets (Samsung Odessey, HTC Vive Pro, Oculus Rift) and mobile-first, standalone experiences (Oculus Quest). Pancake optics and hybrid lenses will continue to push the form factor of these devices down.

AR and especially MR systems are poised to become the next computing platform, replacing ailing desktop and laptop hardware, and now even the aging tablet computing hardware. Such systems are mostly untethered for most of them (see HoloLens 1) and require high-end optics for the display engine, combiner optics, and sensors (depth scanner camera, head-tracking cameras to provide 6DOF, accurate eye trackers, and gesture sensors). These are currently the most demanding headsets in terms of hardware, especially optical hardware, and are the basis of this review paper.

Eventually, if technology permits, these three categories will merge into a single hardware concept. This will, however, require improvements in connectivity (5G, WiGig), visual comfort (new display technologies), and wearable comfort (battery life, thermal management, weight/size).

The worldwide sales decline for smartphones and tablets in Q3 2018 was an acute signal for major consumer electronics corporations and VC firms to fund and develop the "next big thing." MR headsets (in all their forms as glasses, goggles or helmets), along with 5G connectivity and subsequent cloud MR services, look like good candidates for many.

2.1 Wearable, visual, vestibular, and social comfort

Comfort, in all four declinations—wearable, visual, vestibular, and social—is key to enabling a large acceptance base of any consumer MR headset candidate architecture. Comfort, especially visual, is a subjective concept. Its impact is therefore difficult to measure or even estimate on a user pool. Careful user testing is required to assess.

Wearable comfort features include the following:

- Untethered headset for best mobility.
- Small size and light weight.
- Thermal management throughout the entire headset (passive or active).
- Skin contact management through pressure points.
- Breathable fabrics to manage sweat and heat.
- Center of gravity (CG) closer to the CG of a human head.

Visual comfort features include the following:

- Large EB to allow for wide interpupillary distance (IPD) coverage. The optics might also come in different stock keeping units (SKUs) for consumers (i.e., small, medium, and large IPDs), but for enterprise, because the headset is shared between employees, it needs to accommodate a wide IPD range.
- Angular resolution close to 20/20 visual acuity (at least 45 pixels per degree [PPD] in the central foveated region), lowered to a few PPD in the peripheral visual region.
- No screen-door effects (large pixel fill factor and high PPD), and no Mura effects.
- High dynamic range through high brightness and high contrast (emissive displays such as MEMS scanners and OLEDs/iLEDs versus nonemissive displays such as LCoS and LCD).
- Ghost images minimized (<1%).
- Unconstrained 200+ degree see-through peripheral vision (especially useful for outdoor activities, defense, and civil engineering).

- Active dimming on visor (uniform shutter or soft-edge dimming).
- Display brightness control (to accommodate various environmental lightning conditions).
- Reduction of any blue remaining ultraviolet (UV) or blue LED light (<415 nm) to limit retinal damage.
- Color accuracy and color uniformity over FOV as well as EB are also important vision comfort keys.

Vestibular comfort features include the following:
- Motion-to-photon latency (time between head movement and display update) below 10 ms (through optimized sensor fusion).
- Spatial stability of holograms in the 3D world across both low and high frequencies.
- User experience/user interface considerations to present content motion do not severely disagrees with a user's sense of motion.

Visual comfort features leveraging eye tracking include the following:
- Vergence–accommodation conflict (VAC) mitigation for close objects located in the foveated cone through vergence tracking from differential eye tracking data (as vergence is the trigger to accommodation).
- Active pupil swim correction for large-FOV optics.
- Active pixel occlusion (hard-edge occlusion) to increase hologram opacity (more realistic).

Social comfort features include the following:
- Unaltered eye view of the HMD wearer, allowing for continuous eye contact and eye expression discernment.
- No world-side image extraction (present in many waveguide combiners).
- Covert multiple-sensor objective cameras pointing to the world (reducing socially unacceptable world spying).

Note: The word "*hologram*" is used extensively by the AR/VR/MR community as referring to "stereo images." For the optical engineer, a hologram is either the volume holographic media (dichromated gelatin [DCG] emulsion, Silver Halide film or Photopolymers films, surface relief element, and so on) that can store phase and/or amplitude information as a phase and/or amplitude modulation or the representation of a true diffracted holographic field, forming an amplitude image, a phase object, or a combination thereof. A hologram in the original sense of the world can thus be also an optical element, such as a grating, a lens, a mirror, a beam shaper, a filter, a spot array generator, and so on. However, throughout this review work, we conform to the new (albeit deformed by the AR/VR/MR community) meaning of the world "hologram" as a stereo image.

2.2 Display immersion

Immersion is the other key to the ultimate MR experience and is not based only on FOV, which is a 2D angular concept; immersive FOV is a 3D concept that includes the z distance from the user's eyes, allowing for arm's length display interaction through VAC mitigation.

Immersive experiences can come in various forms:
- Wide-angle FOV, including peripheral display regions with lower pixels count per degree (resolution) and lower color depth.
- Foveated display that is either fixed/static (foveated rendering) or dynamic (through display steering, mechanically or optically).
- World-locked holograms, hologram occlusion through accurate and fast spatial mapping and hard-edge see-through occlusion.
- World-locked spatial audio.
- Accurate eye/gesture/brain sensing through dedicated sensors.
- Vivid and realistic hologram colors and shading.
- Haptic feedback.

3 Functional optical building blocks of an MR headset

An HMD, and particularly an optically see-through HMD, is a complex system, with at its core various optical subsystems. Once the optical subsystems are defined, such as the choice of the optical engine, the combiner engine and the optical sensors (eye tracking, head tracking, depth scanner, gesture sensors, and so on), all the rest can be engineered around this core.

A typical functional optical building block ecosystem of a MR headset is shown in Figure 3, including display, imaging, and sensing subsystems.

The display engine is where the image is formed and then imaged onwards, forming or not a pupil, and passed through an optical combiner that can include a pupil replication scheme to the eye pupil. Gaze tracking might or might not share optics with the display architecture (which is usually an infinite conjugate system, and eye tracking is usually a finite conjugate system).

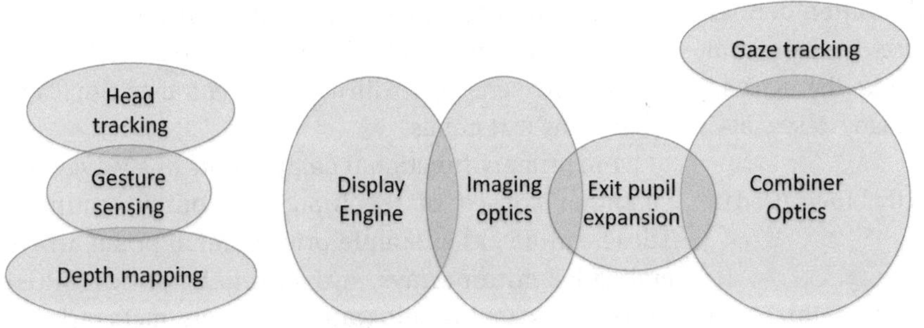

Figure 3: Functional optical building blocks of an MR system.

3.1 Display engine optical architectures

Once the image is formed over a plane, a surface, or through a scanner, there is a need to form an exit pupil, over which the image is either totally or partially collimated and then presented directly to the eye or to an optical combiner (see Figure 4 for the display engine architecture and subsequent waveguide combiner for HoloLens 1 and HoloLens 2). In some cases, an intermediate aerial image can be formed to increase the etendue of the system.

Because the waveguide input pupils for both eyes are located in the upper nasal area in HoloLens 1, several optical elements of the display engine have been shared with both display engines in order to reduce any binocular image misalignments. In the HoloLens 2, this is not the case since the input pupils are centrally located on the waveguide (as the field propagates by total internal reflection [TIR] in both directions in the guides).

Spatially demultiplexed exit pupils (either color or field separated) can be an interesting option, depending on the combiner architecture used (see the Magic Leap One). Imaging optics or relay optics in the display engine are usually free-space optics but in very compact form, including in many cases polarization beam cubes combined with birdbath architectures [12] to fold the optical path in various directions. Reflective/catadioptric optics are also preferred for their reduced achromatic spread.

3.2 Combiner optics and exit pupil expansion

The optical combiner is often the most complex and most costly optical element in the entire MR display architecture: it is the one component seen directly by the user

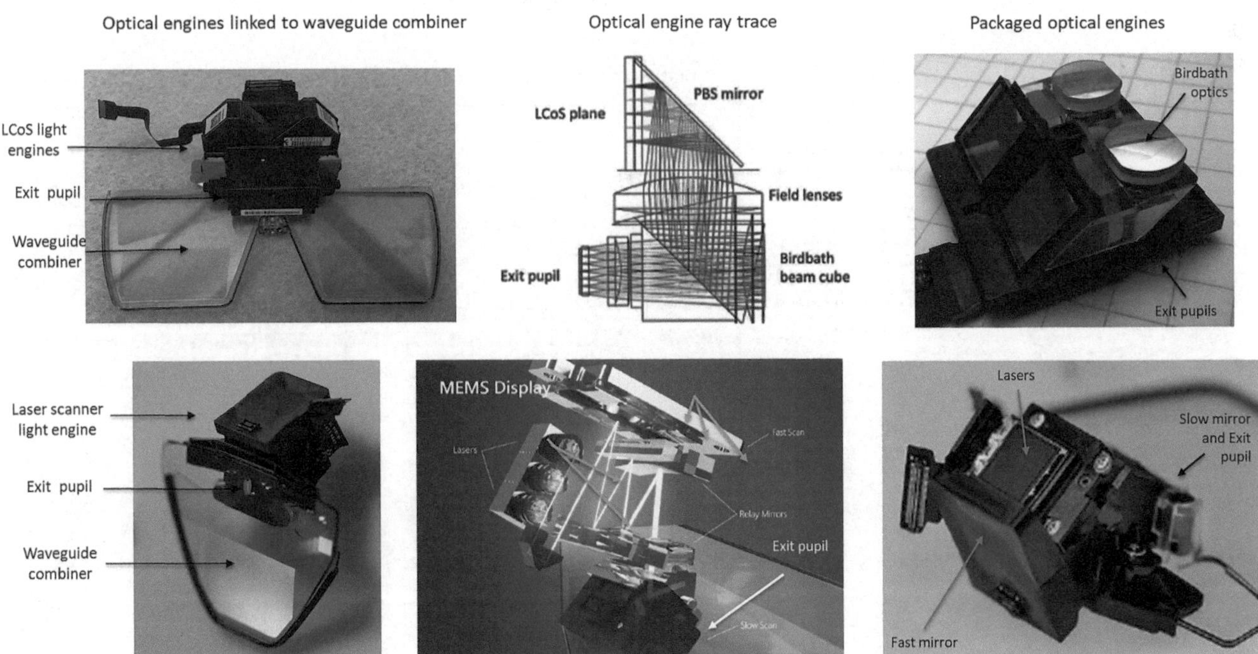

Figure 4: Display engines based on an LCoS imager, as in the HoloLens 1 (top, 2016), and a laser MEMS scanner, as in the HoloLens 2 (bottom, 2019).

and the one seen directly by the world. It often defines the size-and-aspect ratio of the entire headset. It is the critical optical element that reduces the quality of the see-through and the one that defines the EB size (and in many cases, also the FOV).

There are three main types of optical combiners used in most MR/AR/smart glasses today:

- Free-space optical combiners,
- TIR prism optical combiners (and compensators), and
- Waveguide-based optical combiners.

When optimizing an HMD display system, the optical engine must be optimized in concert with the combiner engine. Usually, a team that designs an optical engine without fully understanding the limitations and specifics of a combiner engine designed by another team, and vice versa, can result in a suboptimal system or even a failed optical architecture, no matter how well the individual optical building blocks might be designed.

4 Waveguide combiners

Free-form TIR prism combiners are at the interface between free space and waveguide combiners. When the number of TIR bounces increases, one might refer to them as waveguide combiners. Waveguide combiner architectures are the topic of this review paper.

Waveguide combiners are based on TIR propagation of the entire field in an optical guide, essentially acting as a transparent periscope with a single entrance pupil and often many exit pupils.

The primary functional components of a waveguide combiner consist of the input and output couplers. These can be either simple prisms, microprism arrays, embedded mirror arrays, surface relief gratings (SRGs), thin or thick analog holographic gratings, metasurfaces, or resonant waveguide gratings (RWGs). All of these have their specific advantages and limitations, which will be discussed here. Waveguide combiners have been used historically or tasks very different from AR combiner, such as planar optical interconnections [13] and LCD backlights [14].

Waveguide combiners are an old concept, some of the earliest intellectual property (IP) dates back to 1976 and applied to HUDs. Figure 5(a) shows a patent by Juris Upatnieks dating back 1987, a Latvian/American scientist and one of the pioneers of modern holography [16], implemented in a DCG holographic media. A few years later, 1D EB expansion (1D exit pupil expansion [EPE]) architectures were proposed as well as a variety of alternatives for in-coupler and out-coupler technologies, such as SRG couplers by Thomson CSF (Figure 5(b)). Figure 5(c) shows the original 1991 patent for a waveguide-embedded partial mirror combiner and exit pupil replication. (All of these original waveguide combiner patents have been in the public domain for nearly a decade.)

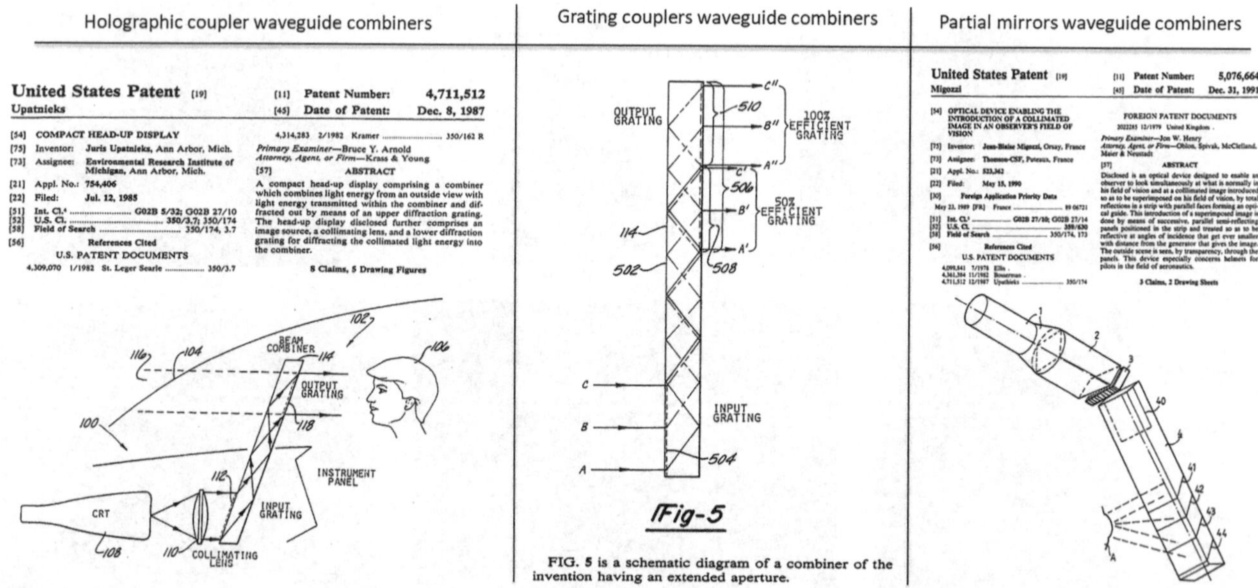

Figure 5: (a) Original waveguide combiner patents including holographic (1987), (b) surface relief grating (1989), and (c) partial mirrors (1991) for HUD and HMD applications.

4.1 Curved waveguide combiners and a single exit pupil

If the FOV is small (<20° diagonally), such as in smart glasses, it might not be necessary to use an exit pupil expansion architecture, which would make the waveguide design much simpler and allow for more degrees of freedom, such as curving the waveguide. Indeed, if there is a single output pupil, the waveguide can imprint optical power onto the TIR field, as is done in the curved waveguide Smart Glass by Zeiss in Germany (developed now with Deutsche Telekom and renamed "Tooz"); see Figure 6.

The other waveguide smart glass shown here (flat waveguide cut as a zero-diopter ophthalmic lens) is the early prototype (1995) from Micro-Optical Corp. in which the extractor is an embedded coated prism.

In the Zeiss "Tooz" smart glass, the exit coupler is an embedded off-axis Fresnel reflector. The FOV as well as the out-coupler is excentered from the line of sight. The FOV remains small (11°) and the thickness of the guide relatively thin (3–4 mm).

Single exit pupils have also been implemented in flat guides, as in the Epson Moverio BT100, BT200, and BT300 (temple-mounted optical engine in a 10-mm-thick guide with curved half-tone extractor in the BT300) or in the Konica Minolta smart glasses, with top-down display injection and a flat RGB panchromatic volume holographic extractor (see Figure 7).

Single exit pupils (no EPE) are well adapted to small-FOV smart glasses. If the FOV gets larger than 20°, especially in a binocular design, 1D or 2D exit pupil replication is required.

Covering a large IPD range (such as a 95 or 98 percentile of the target consumer population, including various facial types) requires a large horizontal EB, typically 10–15 mm. Also, due to fit issues and nose-pad designs, a similar large and vertical EB is also desirable, ranging from 8–12 mm.

4.2 Continuum from flat to curved waveguides and extractor mirrors

One can take the concept of a flat waveguide with a single curved extractor mirror (Epson Moverio BT300) or free-form prism combiner, or a curved waveguide with curved mirror extractor, to the next level by multiplying the mirrors to increase the EB (see the Lumus lightguide optical element [LOE] waveguide combiner) or fracturing metal mirrors into individual pieces (see the Optinvent ORA waveguide combiner or the LetinAR waveguide combiner)

While fracturing the same mirror into individual pieces can increase see through and depth of focus, the use of more mirrors to replicate the pupil is a bit more complicated, especially in a curved waveguide where the two exit pupils need to be spatially demultiplexed to provide a specific mirror curvature to each pupil to correct for image position: this limits the FOV in one direction so that such overlap does not happen.

Figure 8 summarizes some of the possible design configurations with such waveguide mirror architectures. Note that the grating- or holographic-based waveguide combiners are not listed here; they are the subject of the next sections.

Figure 8 shows that many of the waveguide combiner architectures mentioned in this section can be listed in this

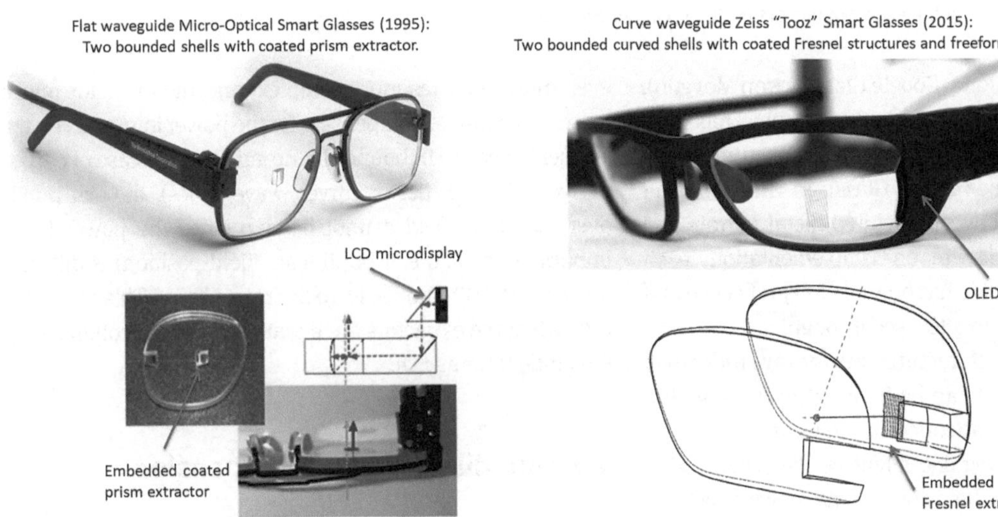

Figure 6: Zeiss "Tooz" smart glass with single exit pupil allowing for curved waveguide.

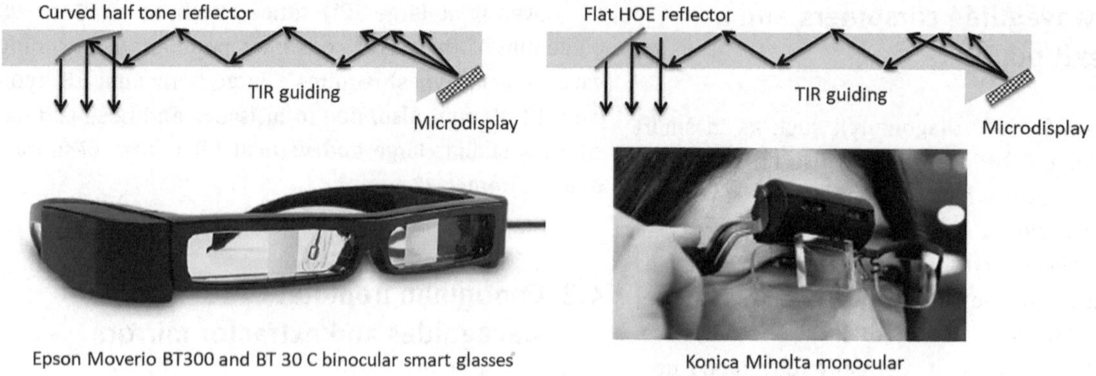

Figure 7: Single-exit-pupil flat waveguide combiners (with curved reflective or flat holographic out-couplers).

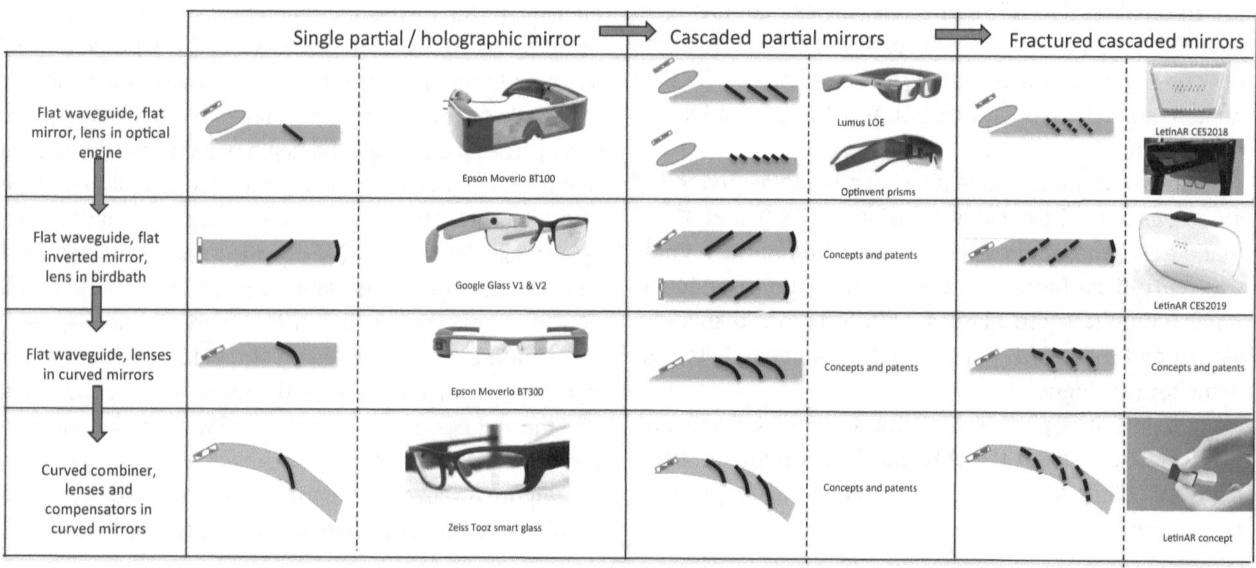

Figure 8: Multiplying or fracturing the extractor mirrors in flat or curved waveguides.

table. Mirrors can be half tone (Google Glass, Epson Moverio), dielectric (Lumus LOE), have volume holographic reflectors (Luminit or Konica Minolta), or the lens can be fractured into a Fresnel element (Zeiss Tooz Smart Glass). In the Optinvent case, we have a hybrid between fractured metal mirrors and cascaded half-tone mirrors. In one implementation, each microprism on the waveguide has one side fully reflective and the other side transparent to allow see through.

In the LetinAR case, all fractured mirrors are reflective, can be flat or curved, and can be inverted to work with a birdbath reflective lens embedded in the guide.

Even though the waveguide might be flat, when using multiple lensed mirrors, the various lens powers will be different since the display is positioned at different distances from these lensed extractors. When the waveguide is curved, everything becomes more complex, and the extractor mirror lenses need also to compensate for the power imprinted on the TIR field at each TIR bounce in the guide. In the case of curved mirrors (either in flat or curved waveguides), the exit pupils over the entire field cannot overlap since the power to be imprinted on each exit pupil (each field position) is different (Moverio BT300 and Zeiss Tooz Smart Glass). This is not the case when the extractors are flat and the field is collimated in the guide (Lumus LOE).

4.3 One-dimensional EB expansion

As the horizontal EB is usually the most critical to accommodate large IPD percentiles, a single-dimensional exit pupil replication might suffice. The first attempts used holographic extractors (Sony Ltd.) [18] with efforts to record RGB holographic extractors as phase-multiplexed volume holograms [19] and also as cascaded half-tone mirror extractors (LOE from Lumus, Israel) or arrays of microprisms (Optinvent, France) [69]. This reduced the 2D footprint of the combiner, which operates only in one direction.

However, to generate a sufficiently large EB in the nonexpanded direction, the input pupil produced by the display engine needs to be quite large in this same direction—larger than the exit pupil in the replicated direction. In many cases, a tall-aspect-ratio input pupil can lead to larger display engines such as in the 1D EPE Lumus LCoS–based engineers. However, a single vertical pupil with natural expansion will provide the best imaging and color uniformity over the EB.

The Lumus LOE has been integrated in various AR glasses at Lumus, as well as in many third-party AR headsets (Daqri, Atheer Labs, Flex, Lenovo). The Lumus LOE can operate in either the vertical direction with the display engine located on the forehead (DK Vision). Lumus is also working on a 2D expansion scheme for its LOE line of combiners (Maximus), with central or lateral input pupils, allowing for a smaller light engine (as the light-engine exit pupil can be symmetric due to 2D expansion). Similarly, the Sony 1D holographic waveguide combiner architecture has been implemented in various products, such as Univet and SwimAR (both using Sony SED 100A waveguide).

4.4 Two-dimensional EB expansion

Two-dimensional EB expansion is desired (or required) when the input pupil cannot be generated by the optical engine over an aspect ratio tall enough to form the 2D EB because of the FOV (etendue limitations) and related size/weight considerations. A 2D EPE is therefore required (see Figure 9).

Various types of 2D EPE replication have been developed: from cascaded X/Y expansion (as in the Digilens, Nokia, Vuzix, HoloLens, and Magic Leap One combiner architectures [21–23]) to combiner 2D expansion [24, 26] (as in the BAE Q-Sight combiner or the WaveOptics Ltd. Phlox 40-degree grating combiner architectures, see Figure 10), to more complex spatially multiplexed gratings (as in the Dispelix combiner).

While holographic recording or holographic volume gratings are usually limited to linear gratings or gratings with slow power (such as off-axis diffractive lenses), SRGs can be either 1D or 2D and either linear or quasi arbitrary in shape. Such structures or structure groups can be optimized by iterative algorithms (topological optimization) rather than designed analytically (WaveOptics computer-generated holograms [CGHs] or Dispelix "mushroom forest" gratings).

Some of these combiners use one guide per color, some use two guides for all three colors, and some use a single guide for RGB; some use glass guides, and others use plastic guides, along with the subsequent compromises one has to make on color uniformity, efficiency, EB, and FOV.

Next, we point out the differences between the various coupler elements and waveguide combiner architecture used in such products. We will also review new coupler technologies that have not yet been applied to enterprise or consumer products. While the basic 2D EPE expansion technique might be straightforward, we will discuss alternative techniques that can allow a larger FOV to be processed by both in-coupler and out-couplers (either as surface gratings or volume holograms). Finally, we will review the mastering and mass replication techniques of such waveguide combiners to allow scaling and consumer cost levels.

4.5 Choosing the right waveguide coupler technology

The coupler element is the key feature of a waveguide combiner. The TIR angle is dictated by the refractive

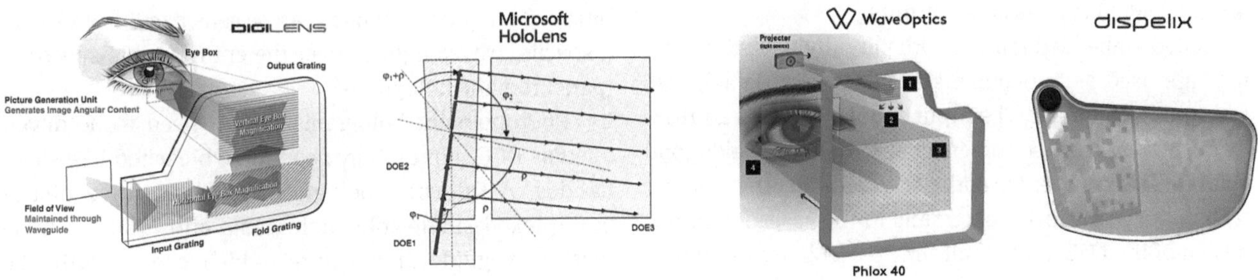

Figure 9: 2D pupil replication architectures in planar optical waveguide combiners.

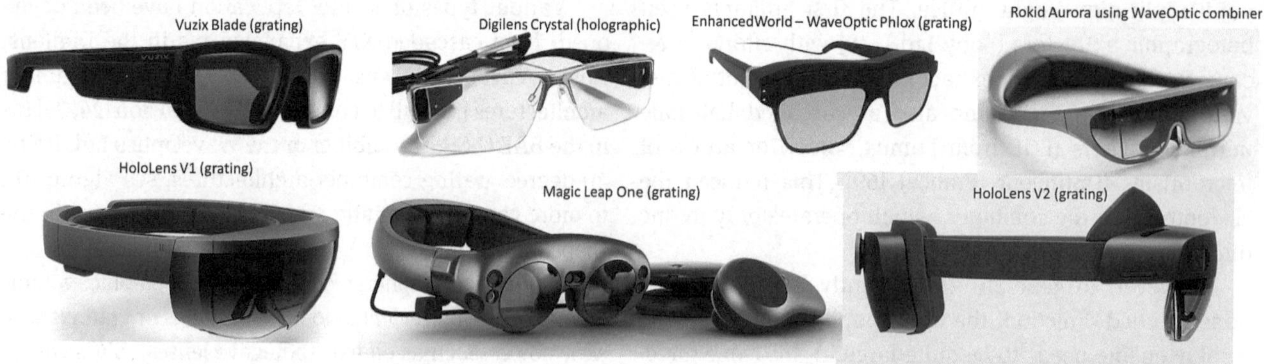

Figure 10: Smart glasses and AR headsets that use 2D EPE diffractive or holographic waveguide combiners.

index of the waveguide, not the refractive index of the coupler nanostructures. Very often, the index of the coupler structure (grating or hologram) prescribes the angular and spectral bandwidth over which this coupler can act, thus impacting the color uniformity over the FOV and EB.

Numerous coupler technologies have been used in industry and academia to implement the in-couplers and out-couplers, and they can be defined either as refractive/reflective or diffractive/holographic coupler elements.

4.5.1 Refractive/reflective couplers elements

4.5.1.1 Macroscopic prism

A prism is the simplest TIR in-coupler and can be very efficient. A prism can be bounded on top of the waveguide, or the waveguide itself can be cut at an angle, to allow normal incident light to enter the waveguide and be guided by TIR (depending on the incoming pupil size). Another way uses a reflective prism on the bottom of the waveguide (metal coated). Using a macroscopic prism as an out-coupler is not impossible, and it requires a compensating prism for see through, with either a reflective coating or a low-index glue line, as done in the Oorym (Israel) light guide combiner concept.

4.5.1.2 Embedded cascaded mirrors

Cascaded embedded mirrors with partially reflective coatings are used as out-couplers in the Lumus (Israel) LOE waveguide combiner. The input coupler remains a prism. As the LOE is composed of reflective surfaces, it yields good color uniformity over the entire FOV. As with other coupler technologies, intrinsic constraints in the cascaded mirror design of the LOE might limit the FOV [26]. See through is very important in AR systems: the Louver effects produced by the cascaded mirrors in earlier versions of LOEs have been reduced recently thanks to better cutting/polishing, coating, and designing.

4.5.1.3 Embedded microprism array

Microprism arrays are used in the Optinvent (France) waveguide as out-couplers [20]. The in-coupler here is again a prism. Such microprism arrays can be surface relief or index matched to produce an unaltered see-through experience. The microprisms can all be coated uniformly with a half-tone mirror layer or can have an alternance of totally reflective and transmissive prism facets, provide a resulting 50% transmission see-through experience. The Optinvent waveguide is the only flat waveguide available today as a plastic guide, thus allowing for a consumer-level cost for the optics. The microprism arrays are injection molded in plastic and bounded on top of the guide.

4.5.2 Diffractive/holographic couplers elements

4.5.2.1 Thin reflective holographic coupler

Transparent volume holograms working in reflection mode —as in DCG, bleached silver halides (Slavic or Ultimate Holography by Yves Gentet), or more recent photopolymers such as Bayfol® photopolymer by Covestro/Bayer, (Germany) [27], and photopolymers by DuPont (US), Polygrama (Brazil), or Dai Nippon (Japan)—have been used to implement in-couplers and out-couplers in waveguide combiners. Such photopolymers can be sensitized to work over a specific wavelength or over the entire visible spectrum (panchromatic holograms).

Photopolymer holograms do not need to be developed as DCG, nor do they need to be bleached like silver halides. A full-color hologram based on three phase-multiplexed single-color holograms allows for a single plate waveguide architecture, which can simplify the combiner and reduce weight, size, and costs while increasing yield (no plate alignment required). However, the efficiency of such full-RGB phase-multiplexed holograms is still quite low when compared to single-color photopolymer holograms.

Also, the limited index swing of photopolymer holograms allows them to work more efficiently in reflection mode than in transmission mode (allowing for better confinement of both the wavelength and angular spectrum bandwidths).

Examples of photopolymer couplers include Sony LMX-001 Waveguides for smart glasses and the TrueLife Optics (UK) process of mastering the hologram in silver halide and replicating it in photopolymer.

Replication of the holographic function in photopolymer through a fixed master has proven to be possible in a roll-to-roll operation by Bayer (Germany). Typical photopolymer holographic media thicknesses range from 16–70 μm, depending on the required angular and spectral bandwidths.

4.5.2.2 Thin transmission holographic coupler

When the index swing of the volume hologram can be increased, the efficiency gets higher and the operation in transmission mode becomes possible. This is the case with Digilens' proprietary holographic polymer-dispersed liquid crystal (H-PDLC) hologram material [28]. Transmission mode requires the hologram to be sandwiched between two plates rather than laminating a layer on top or bottom of the waveguide as with photopolymers, DCG, or silver halides. Digilens' H-PDLC has the largest index swing today and can therefore produce strong coupling efficiency over a thin layer (typically four microns or less). H-PDLC material can be engineered and recorded to work over a wide range of wavelengths to allow full-color operation.

4.5.2.3 Thick holographic coupler

Increasing the index swing can optimize the efficiency and/or angular and spectral bandwidths of the hologram. However, this is difficult to achieve with most available materials and might also produce parasitic effects such as haze. Increasing the thickness of the hologram is another option, especially when sharp angular or spectral bandwidths are desired, such as in telecom spectral and angular filters. This is not the case for an AR combiner, where both spectral and bandwidths need to be wide (to process a wide FOV over a wide spectral band such as LEDs). However, a thicker hologram layer also allows for phase multiplexing over many different holograms, one on top of another, allowing for multiple Bragg conditions to operate in concert to build a wide synthetic spectral and/or angular bandwidth, as modeled by the Kogelnik theory [30]. This is the technique used by Akonia, Inc. (a US start-up in Colorado, formerly InPhase Inc., which was originally funded and focused to produce high-density holographic page data-storage media, ruled by the same basic holographic phase-multiplexing principles [29]).

Thick holographic layers, as thick as 500 μm, work well in transmission and/or reflection modes, but they need to be sandwiched between two glass plates. In some specific operation modes, the light can be guided inside the thick hologram medium, where it is not limited by the TIR angle dictated by the index of the glass plates. As the various hologram bandwidths build the final FOV, one needs to be cautious in developing such phase-multiplexed holograms when using narrow illumination sources such as lasers.

Replication of such thick volume holograms is difficult in roll-to-roll operation, as done with thinner single holograms (Covestro Photopolymers, H-PDLC), and require multiple successive exposures to build the hundreds of phase-multiplexed holograms that compose the final holographic structure. This can however be relatively easy with highly automated recording setups as the ones developed by the now-defunct holographic page data-storage industry (In-Phase Corp., General Electric, and so on).

Note that although the individual holograms acting in slivers of angular and spectral bandwidth spread the incoming spectrum like any other hologram (especially when using LED illumination), the spectral spread over the limited spectral range of the hologram is not wide enough to alter the modulation transfer function (MTF) of the immersive image and thus does not need to be compensated by a symmetric in-coupler and out-coupler as with all other grating or holographic structures. This feature allows this waveguide architecture to be asymmetric, such as having a strong in-coupler as a simple prism: a strong in-coupler is always a challenge for any grating or holographic waveguide combiner architecture, and a macroscopic prism is the best coupler imaginable.

Figure 11 shows both thin and thick volume holograms operating in reflection and/or transmission modes. The top part of the figure shows a typical 1D EPE expander with a single transmission volume hologram sandwiched between two plates. When the field traverses the hologram downward, it is in off/Bragg condition, and when it traverses the volume hologram upward after a TIR reflection, it is in an on/Bragg condition (or close to it), thereby creating a weak (or strong) diffracted beam that breaks the TIR condition.

A hologram sandwiched between plates might look more complex to produce than a reflective or transmission

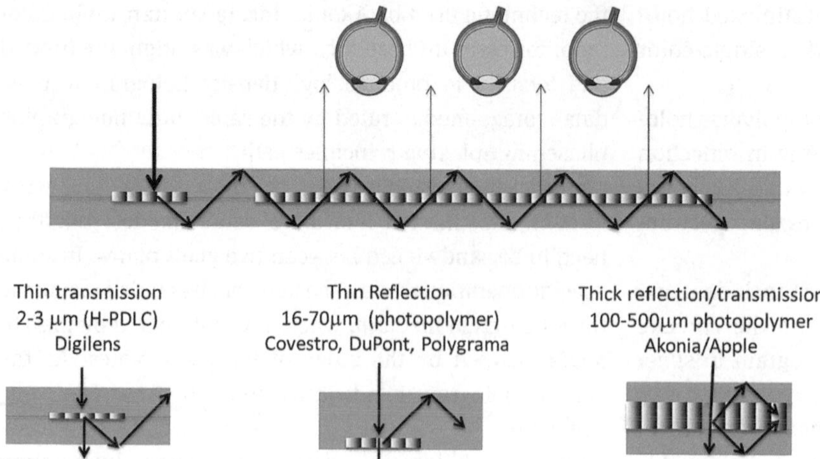

Figure 11: Different types of volume holograms acting as in-couplers and out-couplers in waveguide combiners.

laminated version, but it has the advantage that it can operate in both transmission and reflection modes at the same time (e.g., to increase the pupil replication diversity).

4.5.2.4 Surface-relief grating couplers

Figure 12 reviews the various SRGs used in industry today (blazed, slanted, binary, multilevel, and analog) and how they can be integrated in waveguide combiners as incoupling and outcoupling elements.

Covering a SRG with a reflective metallic surface (see Figure 12) will increase dramatically its efficiency in reflection mode. A transparent grating (no coating) can also work both in transmission and reflection modes, especially as an out-coupler, in which the field has a strong incident angle.

Increasing the number of phase levels from binary to quarternary or even eight or sixteen levels increases its efficiency as predicted by the scalar diffraction theory, for normal incidence. However, for a strong incidence angle and for small periods, this is no longer true. A strong out-coupling can thus be produced in either reflection or transmission mode.

Slanted gratings are very versatile elements, and their spectral and angular bandwidths can be tuned by the slant angles. Front and back slant angles in a same period (or from period to period) can be carefully tuned to achieve the desired angular and spectral operation.

SRGs have been used as a commodity technology since mastering and mass replication techniques technologies were established and made available in the early 1990s

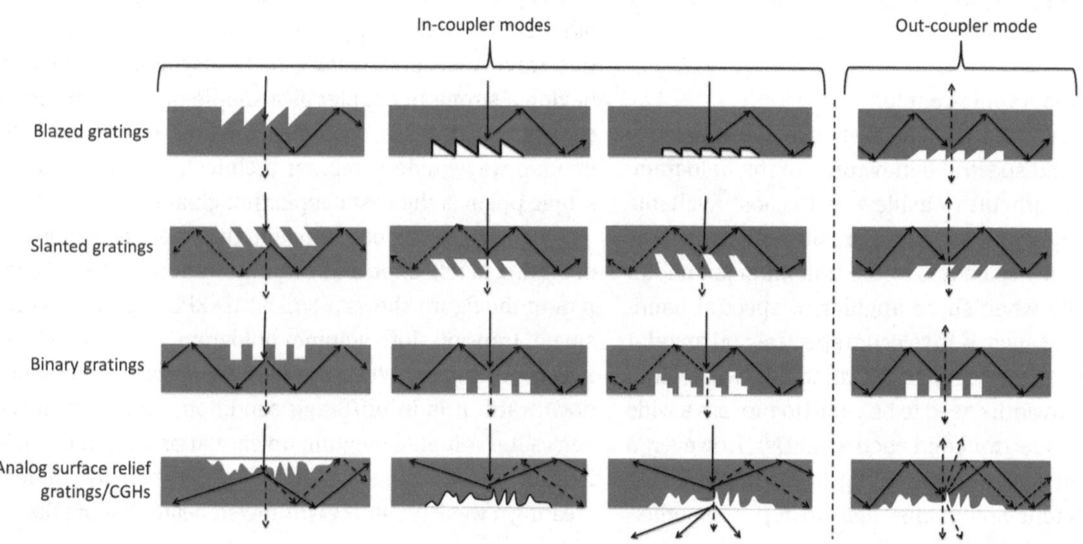

Figure 12: Surface-relief grating types used as waveguide combiner in-couplers and out-couplers. Solid lines indicate reflective coatings on the grating surface, and dashed lines indicate diffracted orders.

[39]. Typical periods for TIR grating couplers in the visible spectrum are below 500 nm, yielding nanostructures of just a few tens of nanometers if multilevel structures are required. This can be achieved by direct e-beam write, i-line (or deep ultra-violet [DUV]) lithography, or even interference lithography (holographic resist exposure) [37]. SRG structures can be replicated in volumes by nano-imprint, a microlithography wafer fabrication technology developed originally for the integrated circuit (IC) industry. Going from wafer-scale fabrication to panel-scale fabrication will reduce costs, allowing for consumer-grade AR and MR products.

Figure 13 and Figure 14 illustrate how some of the SRGs shown in Figure 12 have been applied to the latest waveguide combiners such as the Microsoft HoloLens 1 and Magic Leap One. Multilevel SRGs have been used by companies such as Dispelix Oy, and quasi-analog surface relief CGHs have been used by others, such as WaveOptics Ltd.

Figure 13 shows the waveguide combiner architecture used in the Microsoft HoloLens 1 MR headset (2015). The display engine is located on the opposite side of the EB. The single input pupil carries the entire image over the various colors at infinity (here, only two colors and the central field are depicted for clarity), as in a conventional digital projector architecture. The in-couplers have been chosen to be slanted gratings for their ability to act on a specific spectral range while letting the remaining spectrum unaffected in the zero order, to be processed by the next in-coupler area located on the guide below, and to do this for all three colors. Such uncoated slanted gratings work both in transmission and reflection modes but can be optimized to work more efficiently in a specific mode. The out-couplers here are also slanted gratings, which can be tuned to effectively work over a specific incoming angular range (TIR range) and leave the see-through field quasi unaffected. The part of the see through field that is indeed diffracted by the out-couplers is trapped by TIR and does not make it to the EB. These gratings are modulated in-depth to provide a uniform EB to the user. Note the symmetric in-coupler and out-coupler configuration compensating the spectral spread over the three LED bands.

The redirection gratings are not shown here. Input and output grating slants are set close to 45°, and the redirection grating slants at half this angle. The periods of the gratings are tuned in each guide to produce the right TIR angle for the entire FOV for that specific color (thus the

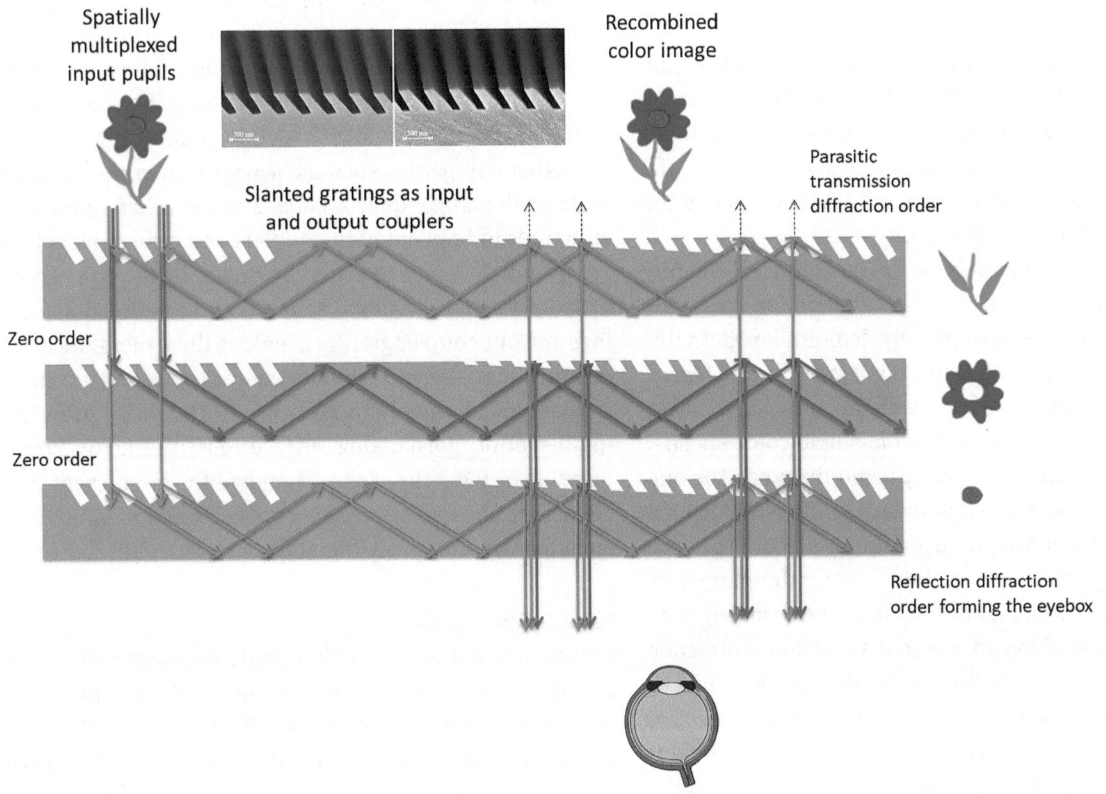

Figure 13: Spatially color-multiplexed input pupils with slanted gratings as in-couplers and out-couplers working in transmission and reflection mode (HoloLens 1 MR headset).

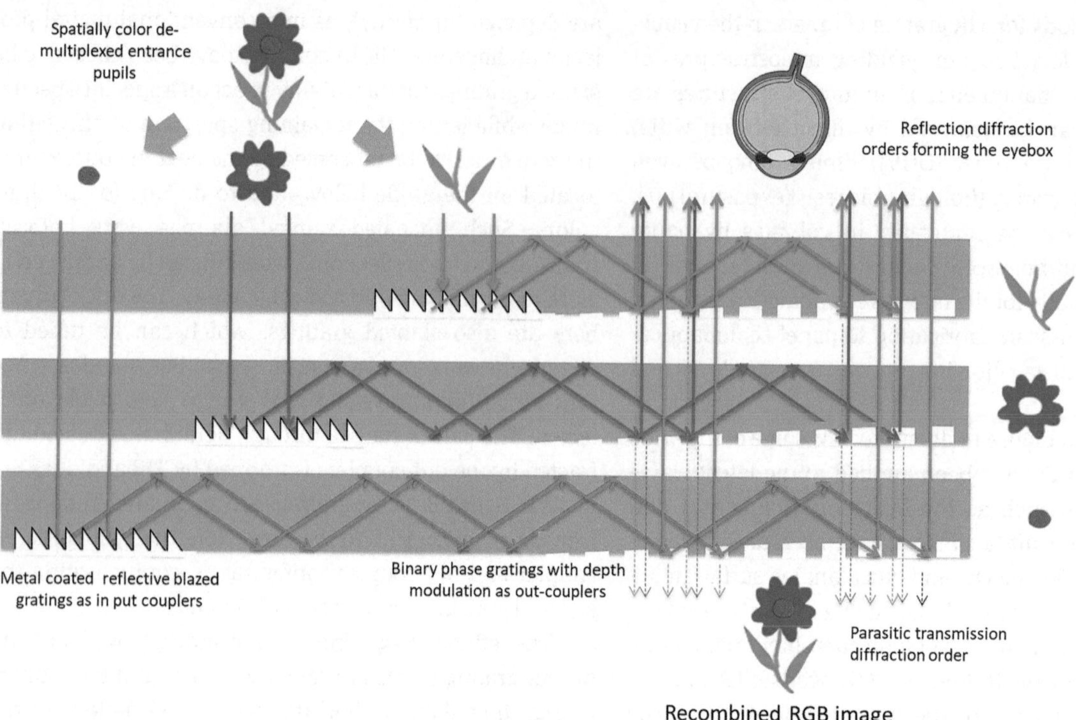

Figure 14: Spatially color-demultiplexed input pupils with 100% reflective blazed gratings as in-couplers and binary phase gratings as out-couplers (Magic Leap One MR headset).

same central diffraction angle in each guide for each RGB LED color band).

Figure 14 depicts the waveguide combiner architecture used in the Magic Leap One MR headset (2018). The display engine is located on the same side as the EB. The input pupils are spatially color demultiplexed, carrying the entire FOV at infinity (here again, only two colors and the central field are depicted for clarity).

Spatial color demultiplexing can be done conveniently with a color sequential LCoS display mode for which the illumination LEDs are also spatially demultiplexed. In this configuration, the input grating couplers are strong blazed gratings, coated with a reflective metal (such as Al). They do not need to work over a specific single-color spectral width since the colors are already demultiplexed. The out-couplers are simple top-down binary gratings, which are also depth modulated to produce a uniform EB for the user. These binary gratings are shallow, acting therefore very little on the see through, but they have much stronger efficiency when working in internal reflection diffraction mode, since the optical path length in this case is longer by a factor of $2n\cos(\alpha)$ than that in transmission mode, (where n is the index of the guide, and α is the angle if there is incidence in the guide). As in the HoloLens 1, most of the see-through field diffracted by the out-couplers is trapped by TIR.

The redirection gratings (not shown here) are also composed of binary top-down structures. The periods of the gratings are tuned in each guide to produce the right TIR angle for the entire FOV for that specific color (same central diffraction angles for each RGB LED color band).

Other companies such as WaveOptics in the UK uses multilevel and/or quasianalog surface relief diffractive structures to implement in-couplers and out-couplers (see Figure 14). This choice is mainly driven by the complexity of the extraction gratings, acting both as redirection gratings and out-coupler gratings, making them therefore more complex than linear or slightly curved (powered) gratings, similar to iteratively optimized CGHs [40]. Allowing multilevel or quasianalog surface relief diffractive structures increases the space bandwidth product of the element to allow more complex optical functionalities to be encoded with relatively high efficiency.

4.5.2.5 RWG couplers

RWGs, also known as guided mode resonant gratings or waveguide-mode resonant gratings [41], are dielectric structures where these resonant diffractive elements benefit from lateral leaky guided modes. A broad range of optical effects are obtained using RWGs such as waveguide coupling, filtering, focusing, field enhancement and nonlinear effects, magneto-optical Kerr effect, or electromagnetically induced transparency. Thanks to their high degree of optical tuning (wavelength, phase, polarization, intensity) and the variety of fabrication processes and materials available, RWGs have been implemented in a broad scope of applications in research and industry. RWGs can therefore also be applied as in-couplers and out-couplers for waveguide gratings.

Figure 15 shows an RWG on top of a lightguide (referred often incorrectly through the popular AR lingo as a "waveguide"), acting as the in-couplers and out-couplers.

Roll-to-roll replication of such grating structures can help bring down overall waveguide combiner costs. The CSEM research center in Switzerland developed the RWG concept back in the 1980s, companies are now actively developing such technologies [90].

4.5.2.6 Metasurface couplers

Metasurfaces are a hot topic in research: they can implement various optical element functionality in an ultraflat form factor by imprinting a specific phase function over the incoming wavefront in reflection or transmission (or both) so that the resulting effect is refractive, reflective, or diffractive or a combination of them. This phase imprint can be done through a traditional optical path difference phase jump or through Pancharatnam–Berry phase gratings/holograms.

Due to their large design space, low track length, and ability to render unconventional optical functions, metalenses could grow out of the laboratory to become an unique item in the engineer's bag of tools. If one can implement in a fabricable metasurface an optical functionality that cannot be implemented by any other known optical element (diffractives, holographics, or Fresnels), it is particularly interesting. For example, having a true achromatic optical element is very desirable not only in imaging but also in many other tasks such as waveguide coupling. Another example is ultralow track length focal stack for IR cameras from Metalenz Corp. Additionally, if one can simplify the fabrication and replication process by using metasurfaces, the design for manufacturing (DFM) can be compelling. However, optical efficiencies, design tools, and large scale fabrication will need to continue to improve find their way into product.

4.5.3 Achromatic coupler technologies

Waveguide combiners could benefit greatly from a true achromatic coupler functionality: incoupling and/or out-coupling RGB FOVs and matching each color FOV to the maximum angular range (FOV) dictated by the waveguide TIR condition. This would reduce the complexity of multiple waveguide stacks for RGB operation.

When it comes to implementing a waveguide coupler as a true achromatic grating coupler, one can either use embedded partial mirror arrays (as in the Lumus LOE combiner), design a complex hybrid refractive/diffractive prism array, or even record phase-multiplexed volume holograms in a single holographic material. However, in the first case, the 2D exit pupil expansion implementation remains complex; in the second case, the microstructures can get very complex and thick; and in the third case, the diffraction efficiency can drop dramatically (as in the Konica Minolta or Sony RGB photopolymers combiners or in the thick Akonia holographic dual photopolymer combiner, now part of Apple, Inc.).

It has been recently demonstrated in literature that metasurfaces can be engineered to provide a true achromatic behavior in a very thin surface with only binary nanostructures [43]. It is easier to fabricate binary nanostructures than complex analog surface relief diffractives, and it is also easier to replicate them by nanoimprint lithography (NIL) or soft lithography and still implement a true analog diffraction function as a lens or a grating. The high index contrast required for such nanostructures can be generated by either direct imprint in high index inorganic spin-on glass or by NIL resist lift-off after an atomic layer deposition process. Direct dry etching of nanostructured remains a costly option for a product.

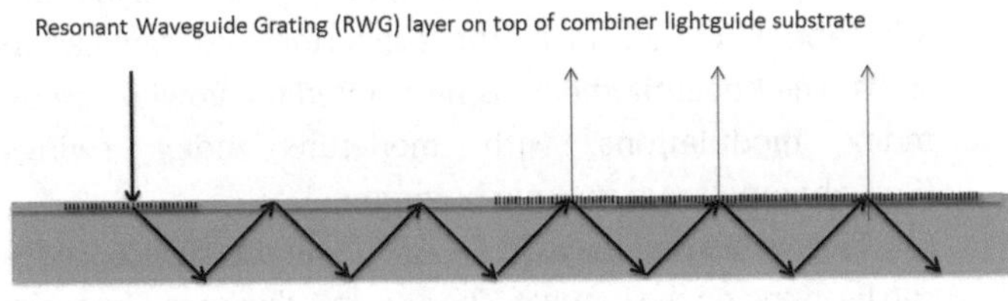

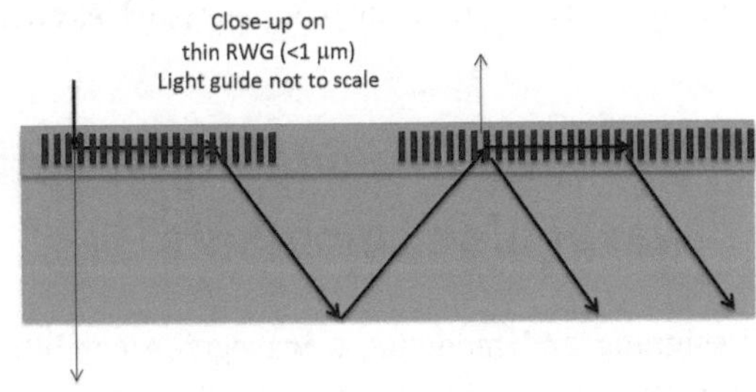

Figure 15: Resonant waveguide gratings as in-couplers and out-couplers on a waveguide combiner.

It is important to remember that metasurfaces or thick volume holograms are not inherently achromatic elements, and never will be. However, when many narrow band diffraction effects are spatially or phase multiplexed in a metasurface or a thick volume hologram, their overall behavior over a much larger spectral bandwidth can effectively lead the viewer to think they are indeed achromatic: although each single hologram or metasurface operation are strongly dispersive, their cascaded contributions may result in a broadband operation which looks achromatic to the human eye (e.g., the remaining dispersion of each individual hologram or metasurface effect affecting a spectral spread that is below human visual acuity—one arcmin or smaller). It is also possible to phase multiplex surface relief holograms to produce achromatic effects but more difficult than with thick volume holograms or thin metasurfaces.

Mirrors are of course perfect achromatic elements and will therefore produce the best polychromatic MTF (such as with Lumus LOE combiners or LetinAR pin mirror waveguides).

4.5.4 Summary of waveguide coupler technologies

Table 1 summarizes the various waveguide coupler technologies reviewed here, along with their specifics and limitations.

Although Table 1 shows a wide variety of optical couplers, most of today's AR/MR/smart glass products are based on only a handful of traditional coupler technologies such as thin volume holograms, slanted SRGs, and embedded half-tone mirrors. The task of the optical designer (or rather the product program manager) is to choose the right balance and the best compromise between coupling efficiency, color uniformity over the EB and FOV, mass production costs, and size/weight.

Figure 16 shows the various coupler elements and waveguide architectures grouped in a single table, including SRG couplers, thin holographic couplers, and thick holographic couplers in three, two, and single flat guides. For geometric waveguide combiners that use embedded mirrors or other reflective/refractive couplers (such as microprisms).

5 Design and modeling of optical waveguide combiners

Designing and modeling a waveguide combiner is very different from designing and modeling a freespace optical combiner. As conventional ray trace in standard optical CAD tools such as Zemax™, CodeV™, Fred™, or TracePro™ are sufficient to design effective free space and even TIR prism combiners and to design waveguide combiners, especially when using diffractive or holographic couplers, a hybrid ray-trace/rigorous electromagnetic diffraction mode is usually necessary.

The modeling efforts is shared between two different tasks:
- Local rigorous EM light interaction with micro- and anno-optics couplers (gratings, holograms, metasurfaces, RWGs, and so on).
- Global architecture design of the waveguide combiner, building up FOV, resolution, color and EB, by the use of more traditional ray trace algorithms.

5.1 Waveguide coupler design, optimization, and modeling

5.1.1 Coupler/light interaction model

Modeling of the angular and spectral Bragg selectivity of volume holograms, thin or thick, in reflection and transmission modes, can be performed with the couple wave theory developed by Kogelnik in 1969 [31, 32].

Similarly, modeling of the efficiency of SRGs can be performed accurately with rigorous coupled-wave analysis (RCWA) [33, 34], especially the Fourier modal method (FMM). The finite difference time domain (FDTD) method—also a rigorous EM nanostructure modeling method—can in many cases be a more accurate modeling technique but also much heavier and more CPU time consuming. However, the FDTD will show all the diffracted fields, the polarization conversions, and the entire complex field, whereas the Kogelnik model and the RCWA will only give efficiency values for particular diffraction orders.

The FDTD can model nonperiodic nanostructures, while RCWA can accurately model quasiperiodic structures. Thus, the FDTD might help with modeling k-vector variations (rolled k-vector) along the grating, slant, depths, and duty cycle variations, as well as random and systematic fabrication errors in the mastering and replication steps. The Kogelnik theory is best suited for slowly varying index modulations with moderate index swings (i.e., photopolymer volume holograms).

Free versions of the RCWA-FMM [35] and FDTD [36] codes can be found on the Internet. The Kogelnik theory can be easily implemented as a straightforward equation set for transmission and reflection modes. Commercial software suites implementing FDTD and RCWA are R-Soft from Synopsys and Lumerical.

Table 1: Benchmark of various waveguide coupler technologies.

Waveguide coupler tech	Operation	Reflective coupling	Transmission coupling	Efficiency modulation	Lensed out-coupler	Spectral dispersion.	Color uniformity	Dynamically tunable	Polarization maintaining	Mass production	Company/Product
Embedded mirrors	Reflective	Yes	No	Complex coatings	No	Minimal	Good	No	Yes	Slicing, coating, polishing.	Lumus Ltd. DK50
Micro-prisms	Reflective	Yes	No	Coatings	No	Minimal	Good	No	Yes	Injection molding	Optinvent SaRL. ORA
Surface relief slanted grating	Diffractive	Yes	Yes	Depth, duty cycle, slant	Yes	Strong	Needs comp.	Possible with LC	No	NIL (wafer, plate)	Microsoft HoloLens, Vuzix Inc, Nokia…
Surface relief blazed grating	Diffractive	Yes	No	Depth	No	Strong	Needs comp.	Possible with LC	No	NIL (wafer, plate)	Magic Leap One,
Surface relief binary grating	Diffractive	Yes	Yes	Depth, duty	Yes	Strong	Needs comp.	Possible with LC	No	NIL (wafer, plate)	Magic Leap One
Multilevel surface relief grating	Diffractive	Yes	Yes	Depth, duty cycle	Yes	Strong	Needs comp.	Possible with LC	Possible, but difficult	NIL (wafer, plate)	WaveOptics Ltd, BAE. Dispelix.
Thin photo-polymer hologram	Diffractive	Yes	Yes	Index swing	Yes, but difficult	Strong	Needs comp.	Possible with shear	No	NIL (wafer, plate)	Sony Ltd, TruelifeOptics Ltd,
H-PDLC volume holographic	Diffractive	No	Yes	Index swing	Yes, but difficult	Strong	OK	Yes (electrical)(	No	Exposure	Digilens Corp. (MonoHUD)
Thick photo-polymer hologram	Diffractive	Yes	Yes	Index swing	Yes, but difficult	Minimal	OK	No	No	Multiple exposure	Akonia Corp (now Apple Inc.)
Resonant waveguide grating	Diffractive	Yes	Yes	Depth. Duty cycle	Yes	Can be mitigated	NA	Possible with LC	Possible	Roll to roll NIL	CSEM/Resonannt screens
Metasurface coupler	Mostly diffractive	Yes	Yes	Various	Yes	Can be mitigated	Needs comp.	Possible with LC	Possible	NIL (wafer, plate)	Metalenz Corp.

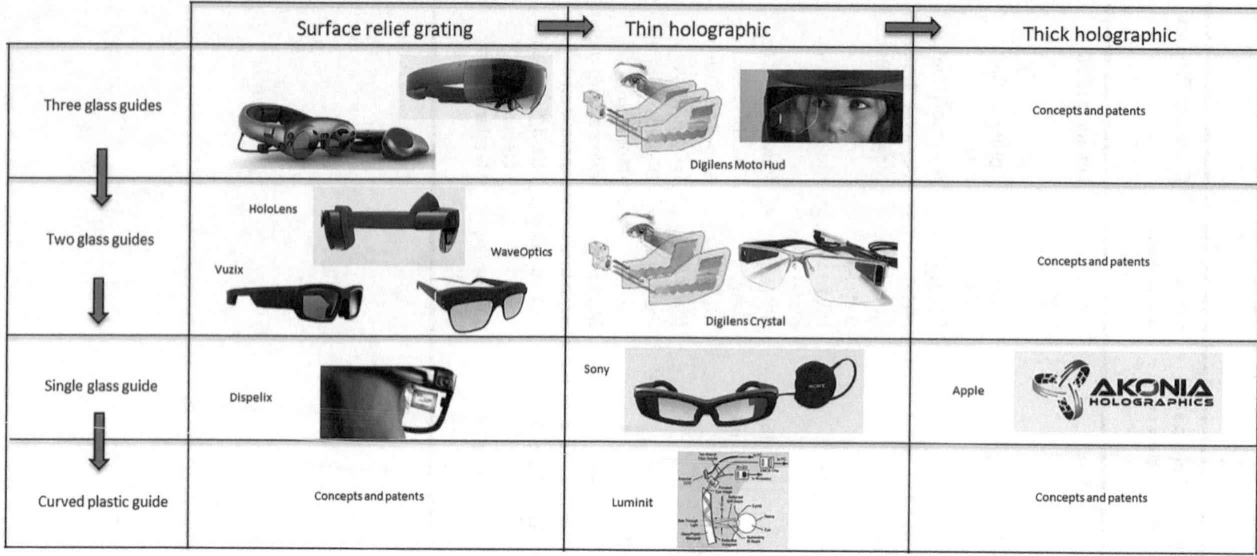

Figure 16: Summary of waveguide combiner architectures with 1D or 2D EPE schemes.

These models predict the efficiency in each order for a single interaction of the light with the coupler element. In order to model the entire waveguide combiner, especially when a pupil replication scheme is used, conventional ray tracing optical design software can be used, such as Zemax, or more specific light-propagation software modules, such as the ones by LightTrans, Germany [37] (see Figure 17 for ray tracing through 2D EPE grating waveguides).

The interaction of the EM field with the coupler regions (surface relief structures or index modulations) modeled through the RCWA or Kogelnik can be implemented via a dynamically linked library (DLL) in conventional optical design software based on ray tracing (e.g., C or Matlab code). As the FDTD numerical algorithm propagates the entire complex field rather than predicting only efficiency values (as in the RCWA or Kogelnik model), it is therefore more difficult to implement as a DLL.

Raytrace optimization of the high-level waveguide combiner architecture with accurate EM light/coupler interactions modeling are both required to design a combiner with good color uniformity over the FOV, a uniform EB over a target area at a desired eye relief, and high efficiency (in one or both polarizations). Inverse propagation from the EB to the optical engine exit pupil is a good way to simplify the optimization process. The design process can also make use of an iterative algorithm to optimize color over the FOV/EB and/or efficiency or even reduce the space of the grating areas by making sure that no light gets lost outside the effective EB.

Waveguide couplers have specific angular and spectral bandwidths that affect both the FOV and the EB uniformity. A typical breakdown of the effects of a 2D EPE waveguide architecture on both spectral and angular bandwidths on the resulting immersive display is shown in Figure 18.

Figure 18 shows that the coupler's spectral and angular bandwidths are critical to the FOV uniformity, especially color uniformity. While embedded mirrors and microprisms have a quasiuniform effect on color and FOV, others do not, such as gratings and holograms. It is therefore interesting to have the flattest and widest spectral and angular bandwidths possible. For volume holograms, this means operating in reflection mode and having a strong index swing (Kogelnik), and for surface gratings, this means a high index (as predicted by the RCWA-FMM or FDTD). The angular bandwidth location can be tuned by the slant angle in both holograms and surface gratings. Multiplexing bandwidths can help to build a larger overall bandwidth, bot spectral, and angular and is used in various implementations today. Such multiplexing can be done in phase, in space, or in time or a combination of the above. Finally, as spectral and angular bandwidths are closely linked, altering the spectral input over the field can have a strong impact on FOV and vice versa.

Polarization and degree of coherence are two other dimensions one should need to investigate especially when lasers or VCSELs are used in the optical engine or if polarization maintaining (or rather polarization conversion) is required. The multiple interactions in the R-E regions can produce multiple miniature Mach–Zehnder interferometers, which might modulate the intensity of the particular fields.

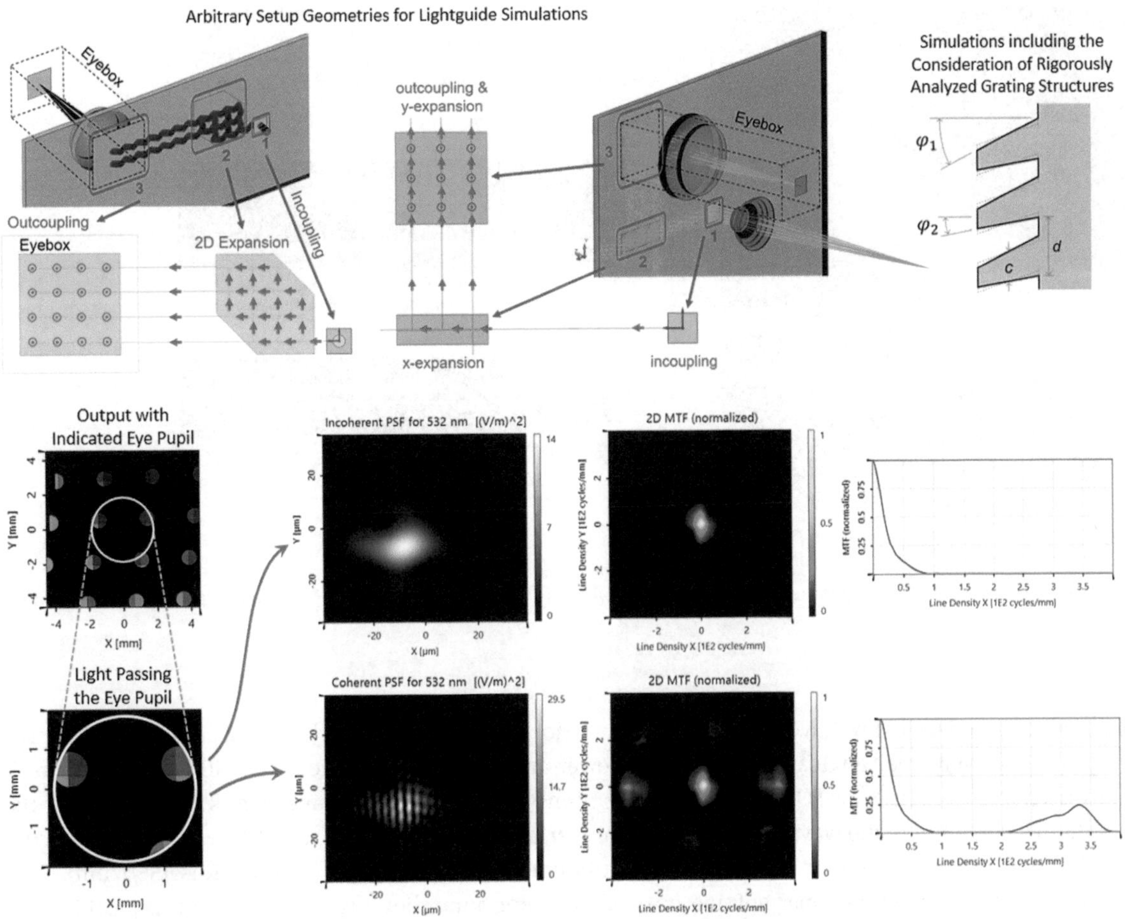

Figure 17: Waveguide grating combiner modeling by LightTrans (Germany) in 2D EPE version.

5.1.2 Increasing FOV by using illumination spectrum

The ultimate task for a holographic or grating coupler is to provide the widest FOV coupling possible, matching the FOV limit dictated by the TIR condition in the underlying waveguide (linked to the refractive index of the waveguide material).

We have seen that volume holographic combiners have been used extensively to provide a decent angular incoupling and outcoupling into the guide. However, most of the

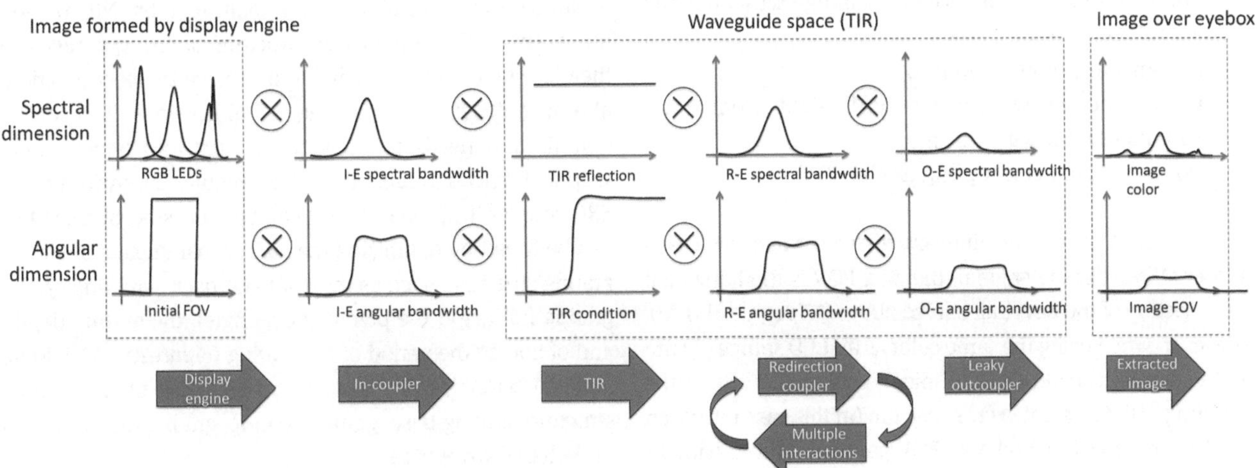

Figure 18: Cascaded effects of the field/coupler interactions on the FOV uniformity.

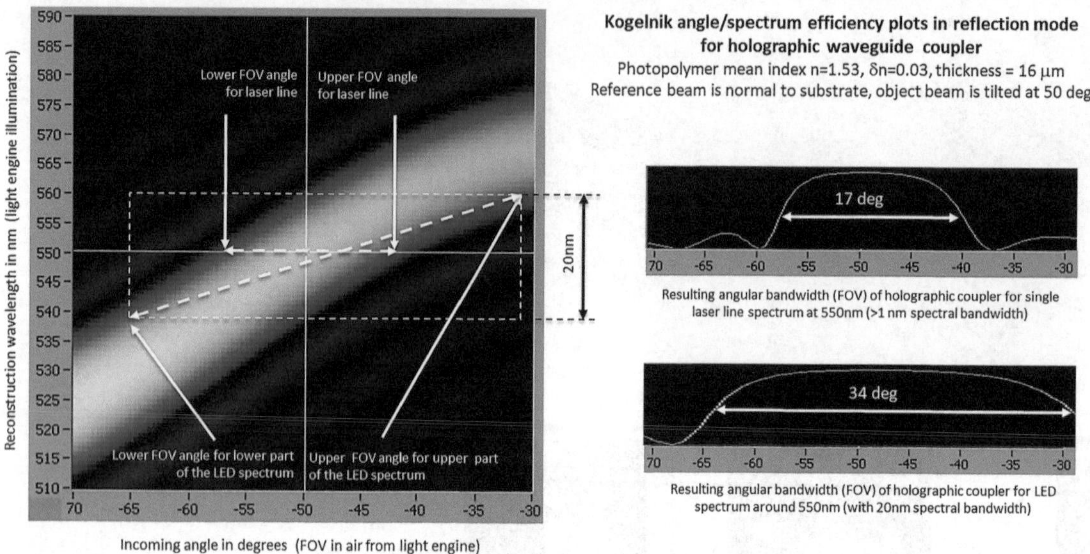

Figure 19: Spectral source bandwidth building larger FOV (angular bandwidth) for photopolymer volume holographic coupler in waveguide combiners.

available holographic materials today have a low index swing and thus yield a relatively small angular bandwidth in the propagation direction. In this case, the FOV bottleneck is the coupler not the TIR condition in the waveguide.

A typical Kogelnik efficiency plot in the angular/spectral space for a reflection photopolymer volume holographic coupler is shown Figure 19 (spectral dimension vertical and angular dimension horizontal).

The hologram specifications and exposure setup in Figure 19 are listed below:

- Mean holographic material index: 1.53,
- Holographic index swing: 0.03,
- Photopolymer thickness: 16 μm,
- Operation mode: reflective,
- Polarization: ("s" but very little change when moving to "p" polarization),
- Design wavelength: 550 nm,
- Reconstruction wavelength: LED light from 540–560 nm (20-nm bandwidth),
- Normal incidence coupling angle: 50° in air.

When using a laser (<1-nm line) as a display source (such as in a laser MEMS display engine), the max FOV is the horizontal cross section of the Kogelnik curved above (17-degree FWHM). However, when using the same color as an LED source (20 nm wide, such as in an LED-lit LCoS micro-display light engine, the resulting FOV is a slanted cross-section (in this case increased to 34-degree FWHM), and a 2× FOV gain is achieved without changing the waveguide index or the holographic coupler, only the illumination's spectral characteristics.

However, this comes at the cost of color uniformity: the lower angles (left side of the FOV) will have more contributions from the shorter wavelengths (540 nm), and the higher angles (right side of the FOV) will have more contributions from the longer wavelengths (560 nm). This slight color nonuniformity over the FOV is typical for volume holographic couplers.

5.1.3 Increasing FOV by optimizing grating coupler parameters

Unlike holographic couplers, which are originated and replicated by holographic interference in a phase change media (see previous section), SRGs are rather originated by traditional IC lithographic techniques and replicated by NIL or soft lithography. The topological structure of the gratings can therefore be optimized digitally to achieve the best functionality in both spectral and angular dimensions. Topological optimization needs to account for DFM and typical lithographic fabrication limitations. The angular bandwidth of an SRG coupler (i.e., the FOV that can be processed by this SRG) can be tuned by optimizing the various parameters of such a grating structure, such as the front and back slant angles, the grating fill factor, the potential coating(s), the grating depth, and of course the period of the grating (Figure20). Additional material variables are the refractive indices of the grating structure, grating base, grating coating, grating top layer, and underlying waveguide.

Figure 20 shows how the SRG grating parameters can be optimized to provide a larger FOV, albeit with a lower

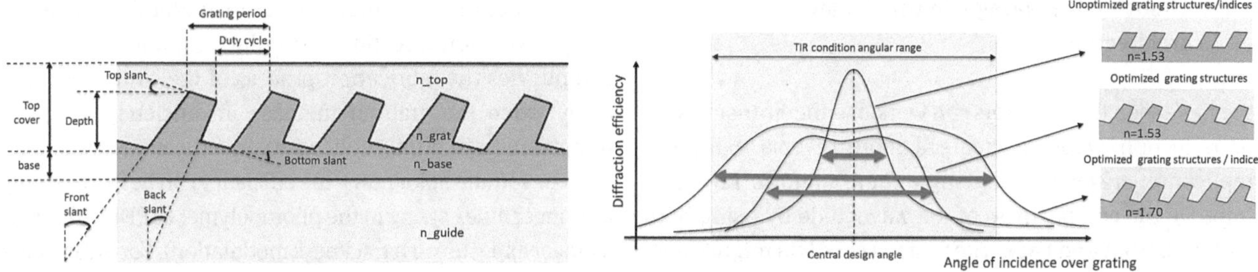

Figure 20: Optimizing the grating parameters to optimize color uniformity over the FOV.

overall efficiency, matching better the available angular bandwidth provided by the TIR condition in the guide. Lower efficiency is okay over the out-couplers since they are tuned in the low-efficiency range to produce a uniform EB (the in-coupler, however, needs to be highly efficient since there is only one grating interaction to couple the entire field into TIR mode).

Calculations of coupling efficiency have been carried out with an RCWA FMM algorithm and topological optimization by a steepest descent algorithm. Note that both unoptimized and optimized gratings have the same grating periods as well as the same central slant angle to position respectively the spectral and the angular bandwidths on identical system design points (with the FOV generated by the display engine and wavelength of the illumination source).

The bottleneck in FOV with the unoptimized grating structure is not the TIR condition (i.e., the index of the waveguide) but rather the grating geometry and the index of the grating. The angular bandwidth of the optimized grating should overlap the angular bandwidth of the waveguide TIR condition for best results over the largest possible FOV. Also, a "top hat" bandwidth makes the color uniformity over the FOV less sensitive to systematic and random fabrication errors in the mastering and the NIL replication of the gratings. Increasing the index of the grating and reducing the back slant while increasing the front slant angle can provide such an improvement.

Additional optimizations over a longer stretch of the grating can include depth modulations, slant modulations (rolling k-vector), or duty cycle modulations to produce an even wider bandwidth over a large, uniform EB.

5.1.4 Using dynamic couplers to increase waveguide combiner functionality

Switchable or tunable TIR couplers can be used to optimize any waveguide combiner architecture, as in

- Increasing the FOV by temporal sub-FOV stitching at double the refresh rate,
- Increasing the brightness at the eye by steering a reduced size EB to the pupil position (thus also increasing the perceived EB size), and
- Increasing the compactness of the waveguide combiner by switching multiple single-color couplers in color sequence in a single guide.

Dynamic couplers can be integrated in various ways: polarization diversity with polarization-dependent couplers (the polarization switching occurring in the optical engine), reconfigurable surface acoustic wave or acousto-optical modulator couplers, electro-optical modulation of buried gratings, switchable SRGs in an LC layer, switchable metasurfaces in an multilayer LC layer, tunable volume holograms (by shearing, pressure, pulling), or switchable H-PDLC, as in Digilens' volume holographic couplers.

5.2 High-level waveguide-combiner design

The previous section discussed ways to model and optimize the performance of individual couplers, in either grating or holographic form. We now go a step further and look at how to design and optimize the overall waveguide combiner architectures.

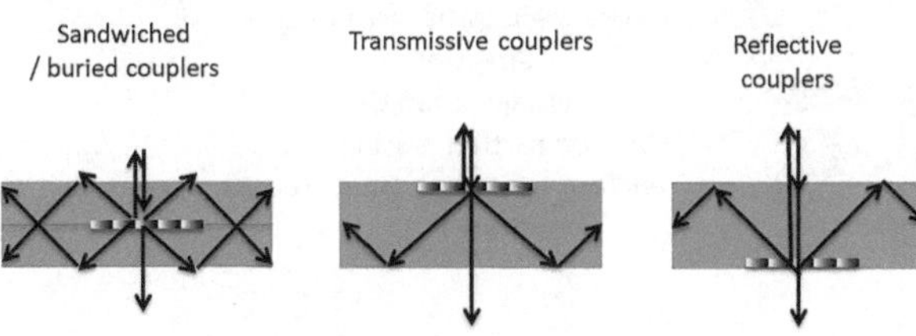

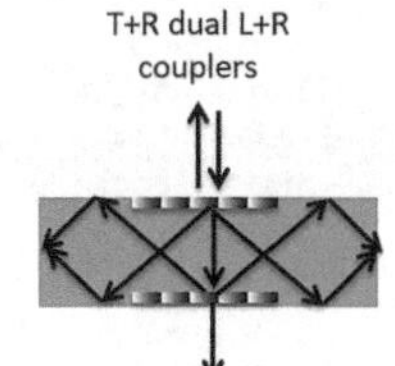

Figure 21: More functional coupler architectures that yield compact and efficient waveguide combiners.

5.2.1 Choosing the waveguide coupler layout architecture

We have seen that couplers can work in either transmission or reflection mode to create a more diverse exit-pupil replication scheme (producing a more uniform EB) or to improve the compactness of the waveguide by using both surfaces, front and back. The various couplers might direct the field in a single direction or in two or more directions, potentially increasing the FOV that can propagate in the waveguide without necessarily increasing its index.

Figure 21shows how the optical designer can expand the functionality of in-couplers or out-couplers, with architectures ranging from bidimensional coupling to dual reflective/transmission operation in the same guide with sandwiched volume holograms or top/bottom grating couplers.

More complex and more functional coupler architectures have specific effects on MTF, efficiency, color uniformity, and FOV. For example, while the index of the guide allows for a larger FOV to propagate, the index of the grating structures in air would increase the spectral and angular bandwidths to process a larger FOV without compromising color uniformity or efficiency. The waviness of the waveguide itself will impact the MTF as random cylindrical powers added to the field. Multiple stacked waveguides might be efficient at processing single colors, but their misalignment will impact the MTF as misaligned color frames. Similarly, hybrid top/bottom couplers will affect the MTF if they are not perfectly aligned (angular alignment within a few arc seconds).

5.2.2 Building a uniform EB

As the TIR field gets depleted when the image gets extracted along the out-coupler region, the extraction efficiency of the out-coupler needs to gradually increase in the propagation direction to produce a uniform EB. This complicates the fabrication process of the couplers, especially when the gradual increase in efficiency needs to happen in both pupil replication directions.

For volume holograms, the efficiency can be increased by a stronger index swing in the photopolymer or PDLC (through a longer exposure or a thickness modulation). For SRGs, there are a few options, as shown in Figure 22. This is true for the redirection grating (R-E) as well as the out-coupler (O-E).

Groove depth and duty cycle modulation can be performed on all type of gratings, binary, multilevel, blazed, and slanted (see Figure 22). Duty cycle modulation has the advantage of modulating only the lateral structures, not the depth, which makes it an easier mastering process. Modulating the depth of the gratings can be done in binary steps (as in the Magic Leap One, Figure 22—right) or in a continuous way (Digilens waveguide combiners).

Grating front- and back-slant angle modulation (in a single grating period or over a larger grating length) can change the angular and spectral bandwidths to modulate efficiency and other aspects of the coupling (angular, spectral, polarization). Periodic modulation of the slant angles is sometimes also called the "rolling *k*-vector" technique and can allow for larger FOV processing due to specific angular bandwidth management over the grating area. Once the master has been fabricated with the correct nanostructure modulation, the NIL replication process of the gratings is the same no matter the complexity of the nanostructures (caution is warranted for slanted gratings where the NIL process must resolve the undercut structures; however, the slanted grating NIL process (with slants up to 50°) has been mastered by many foundries around the world [37]).

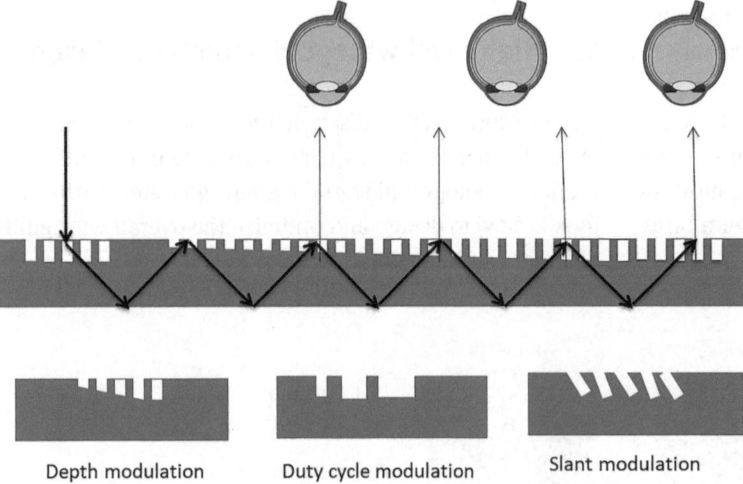

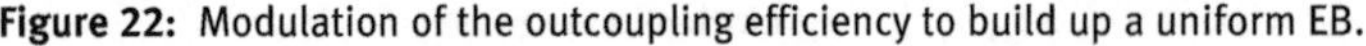

Figure 22: Modulation of the outcoupling efficiency to build up a uniform EB.

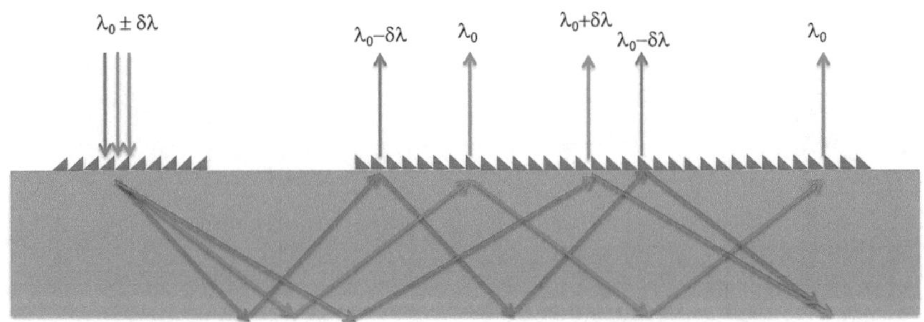

Figure 23: Spectral spread compensation in a symmetric in-coupler/out-coupler waveguide combiner.

5.2.3 Spectral spread compensation in diffractive waveguide combiners

Spectral spread comes to mind as soon as one speaks about gratings or holographic elements. It was the first and is still the main application pool for gratings and holograms: spectroscopy. Spectral spread is especially critical when the display illumination is broadband, such as with LEDs (as in most of the waveguide grating combiner devices today, such as the HoloLens 1, Vuzix, Magic Leap, Digilens, Nokia, and so on), with a notable difference in the HoloLens 2 (laser MEMS display engine). The straightforward technique to compensate for the inevitable spectral spread is to use a symmetric in-coupler/out-coupler configuration in which the gratings or holograms work in opposite direction and thus compensate in the out-coupler any spectral spread impacted in the in-coupler (Figure 23).

Although the spectral spread might be compensated, one can notice in Figure 23 that the individual spectral bands are spatially demultiplexed at the exit ports while multiplexed at the entry port. Strong exit-pupil replication diversity is thus required to smooth out any color non-uniformities generated over the EB.

This symmetric technique might not be used to compensate for spectral spread across different colors (RGB LEDs) but rather for the spread around a single LED color. The spread across colors might stretch the RGB exit pupils too far apart and reduce the FOV over which all RGB colors can propagate by TIR.

The pupil replication diversity can also be increased by introducing a partial reflective layer in the waveguide (by combining two plates with a reflective surface), thus producing a more uniform EB in color and field.

5.2.4 Field spread in waveguide combiners

The different fields propagating by TIR down the guide are also spread out, no matter the coupler technology (mirrors, prisms, gratings, holograms, and so on), see Figure 24.

A uniform FOV (i.e., all fields appearing) can be formed over the EB with a strong exit pupil diversity scheme. This is a concept often misunderstood as in many cases, only one field is represented when schematizing a waveguide combiner. Figure 24 shows the field spread occurring in a diffractive waveguide combiner. The number of replicated fields is also contingent on the size of the human eye pupil. If the ambient light gets bright, i.e., the human eye pupil

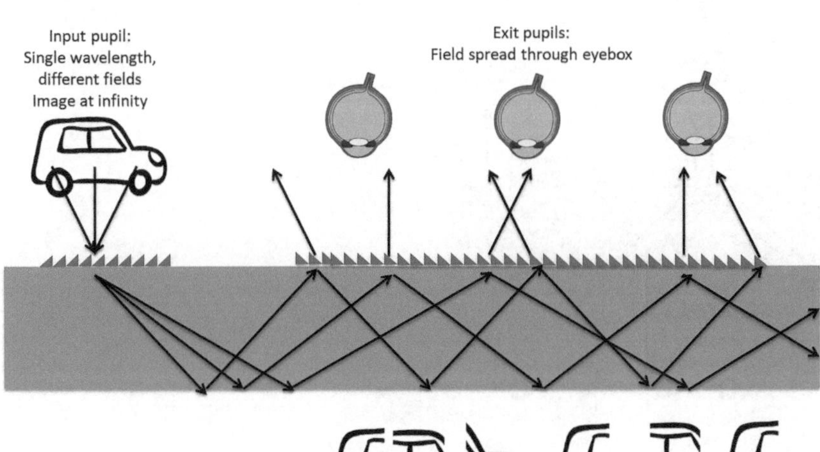

Figure 24: Fractional field spread in a waveguide combiner.

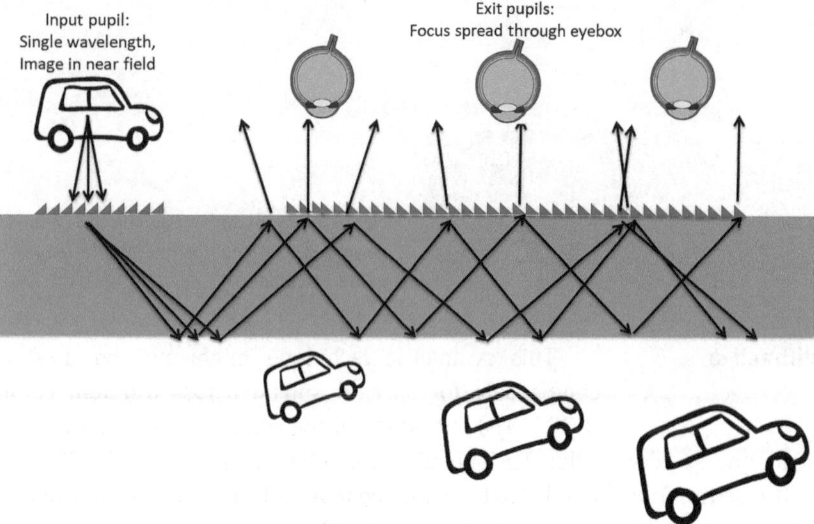

Figure 25: Focus spread in a waveguide combiner with a noncollimated input field.

gets smaller, then only part of the FOV might appear to the user, missing a few fields.

5.2.5 Focus spread in waveguide combiners

When a pupil replication scheme is used in a waveguide combiner, no matter the coupler, the input pupil needs to be formed over a collimated field (image at infinity/far field). If the focus is set to the near field instead of the far field in the display engine, each waveguide exit pupil will produce an image at a slightly different distance, thereby producing a mixed visual experience, overlapping the same image with different focal depths. It is quasi-impossible to compensate for such focus shift over the exit pupils because of both spectral spread and field spread over the exit pupils, as discussed previously. Figure 25 shows such a focus spread over the EB from an input pupil over which the image is formed in the near field.

The image over the input pupil can, however, be located in the near field when no pupil replication scheme is performed in the guide, such as in the Epson Moverio BT300 or in the Zeiss "Tooz" Smart Glass (yielding a small FOV and small EB).

When pupil replication is used in the guide, the virtual image can be set at a closer distance for better visual comfort by using a static (or even tunable) negative lens acting over the entire EB. For an unperturbed see-through experience, such a lens needs to be compensated by its conjugate placed on the world side of the combiner

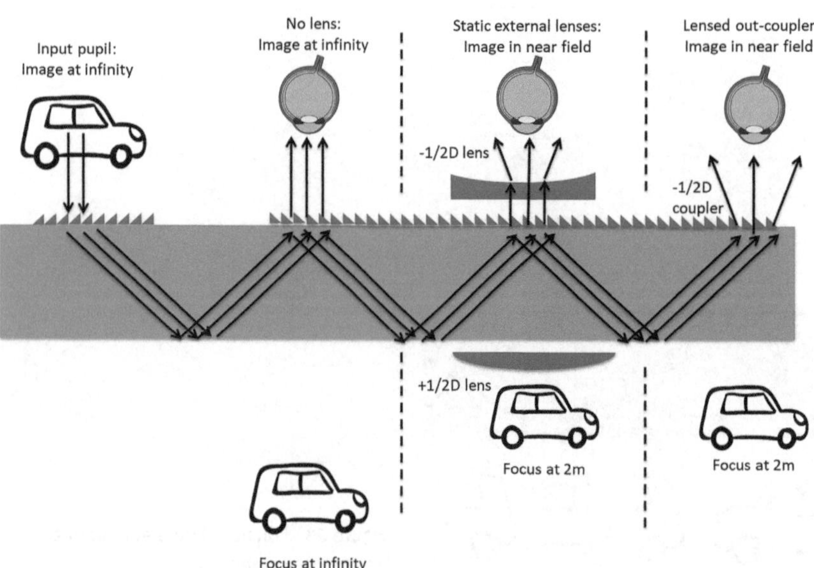

Figure 26: Two out-coupler architectures positioning the virtual image in the near field over all exit pupils.

waveguide. This is the architecture used in the Microsoft HoloLens 1 (2015) [37].

Another, more compact, way would introduce a slight optical power in the O-E, so that this coupler takes the functionality of an off-axis lens (or an off-axis diffractive lens) rather than that of a simple linear grating extractor or linear mirror/prism array. Although this is difficult to implement with a mirror array (as in an LOE), it is fairly easy to implement with a grating or holographic coupler. The grating lens power does not affect the zeroth diffraction order that travels by TIR down the guide but affects only the outcoupled (or diffracted) field. The see-through field is also not affected by such a lensed out-coupler since the see-through field diffracted by such an element would be trapped by TIR and thus not enter the eye pupil of the user.

All three configurations (no lens for image at infinity, static lens with its compensator, and powered O-E grating) are shown in Figure 26. The left part of the EB shows an extracted field with image at infinity (as in the Lumus DK40—2016), the center part shows an extracted field with image at infinity that passes through a negative lens to form a virtual image closer to the user and its counterpart positive lens to compensate for see-through (as in the Microsoft HoloLens 1, 2015), and the right part of the EB shows an extracted field with the image directly located in the near field through a powered grating extractor (as with an off-axis diffractive lens, e.g., the Magic Leap One, 2018).

For example, a half-diopter negative lens power would position the original extracted far field image to a more comfortable 2-m distance, uniformly over the entire EB.

A powered out-coupler grating might reduce the MTF of the image, especially in the direction of the lens offset (direction of TIR propagation), since the input (I-E) and output (O-E) couplers are no more perfectly symmetric (the input coupler being a linear grating in both cases, and the out-coupler an off-axis diffractive lens). Thus, the spectral spread of the image in each color band cannot be compensated perfectly and will produce lateral coromatic aberations (LCA) in the direction of the lens offset. This can be critical when using an LED as an illumination source, but it would affect the MTF much less when using narrower spectral sources, such as lasers or VCSELs.

One of the main problems with such a lensed out-coupler grating configuration when attempting to propagate two colors in the same guide (for example, a two-guide RGB waveguide architecture, as in Figure 34) is the generation of longitudinal chromatic aberrations (due to the focus changing with color since the lens is diffractive). Using a single color per guide and a laser source can greatly simplify the design task.

5.2.6 Propagating full color images in the waveguide combiner over a maximum FOV

We have seen in the previous paragraphs that the spectral spread of grating and holographic couplers can be perfectly compensated with a symmetric in-coupler and out-coupler configuration. This is possible over a single-color band but will considerably reduce the FOV if used over the various color bands (assuming that the couplers will work over these various spectral bands).

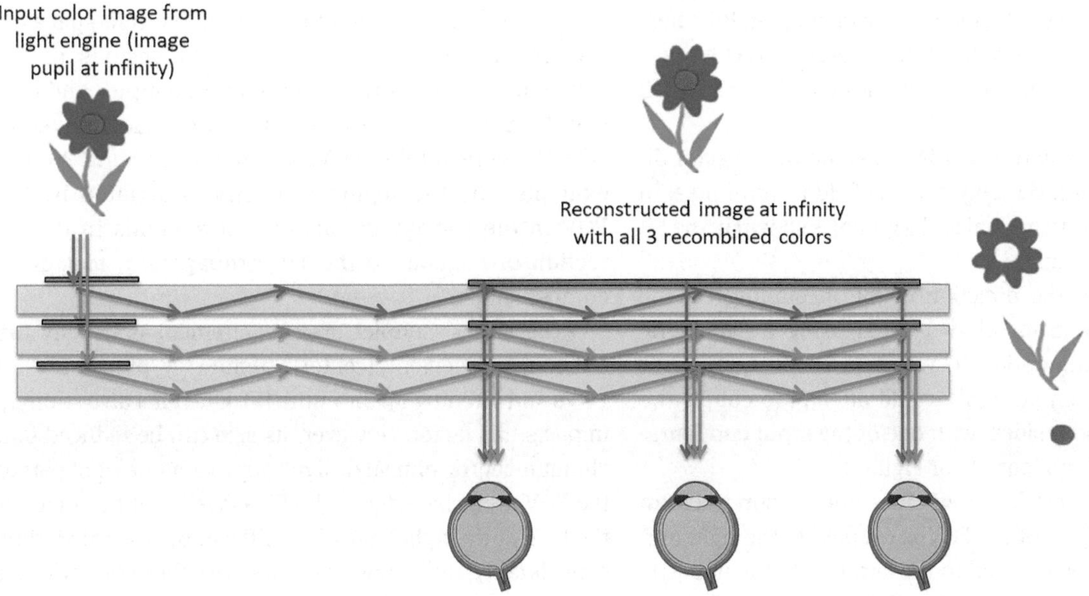

Figure 27: Stacked waveguides combiners that provide the largest FOV TIR propagation over three colors.

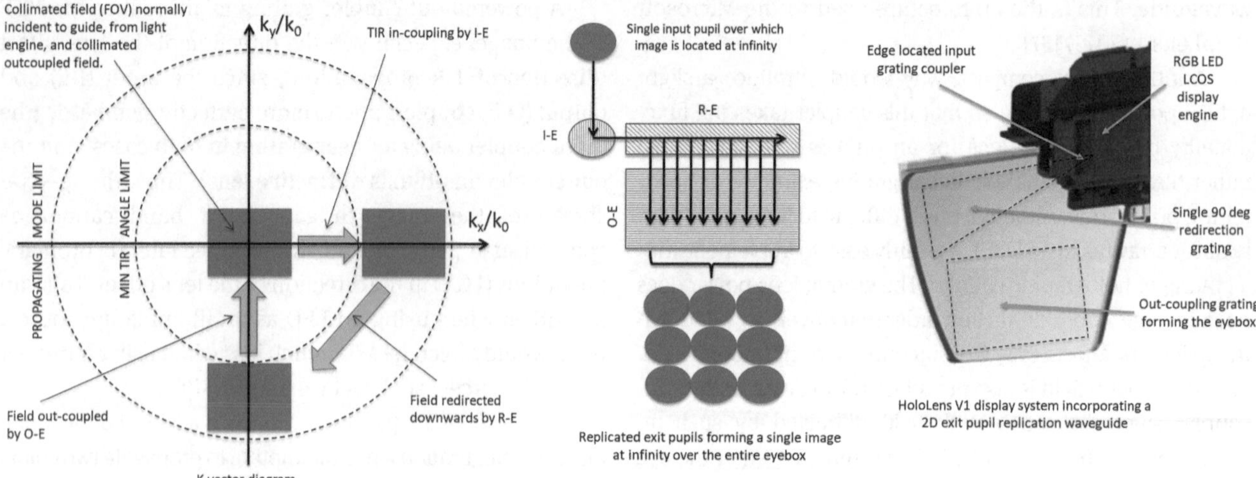

Figure 28: *k*-vector diagram and lateral pupil replication layout for a single guide and single color.

In order to maximize the RGB FOV in a waveguide combiner, one solution is to use stacked guides optimized each for a single-color band, each coupling a maximum FOV by tuning the diffraction angle of the in-couplers and out-couplers accordingly. This is the architecture used in both HoloLens 1 and Magic Leap One (see Figure 27, although the position of the input pupil (light engine) is opposite in both devices.

Air gaps between all plates are required to produce the TIR condition. Such gaps also allow for additional potential filtering in between plates for enhanced performance (such as spectral and polarization filtering).

Figure 28 shows the functional diagram of such a single-color plate as a top view as well as its *k*-vector space depiction. Here again, I-E refers to the in-coupler, R-E refers to the leaky 90-degree redirection element, and O-E refers to the leaky out-coupler that forms the final EB (for 2D pupil replications).

Note that the entire FOV is shown on the *k*-vector diagram (Figure 28), but only a single field (central pixel in the FOV, with entry normal to the guide) is shown in the EB expansion schematic.

The FOV in the direction of the incoupling can be increased by a factor of two when using a symmetric incoupling configuration in which the input grating or hologram (or even prism[s]) would attempt to couple the entire FOV to both sides, with one of the input configurations described in Figure 13 or Figure 14.

As the TIR angular range does not support such an enlarged FOV, part of the FOV is coupled to the right and part of the FOV is coupled to the left. Due to the opposite directions, opposite sides of the FOV travel in each direction. If such TIR fields are then joined back with a single out-coupler, the original FOV can be reconstructed by overlapping both partial FOVs, as in Figure 29.

In the orthogonal direction, the FOV that can be coupled by TIR remains unchanged. This concept can be taken to more than one dimension, but the coupler space on the waveguide can become prohibitive.

5.2.7 Waveguide-coupler lateral geometries

We have reviewed the various coupler technologies that can be used in waveguide combiners, as well as the 2D exit pupil expansion that can be performed in waveguide combiners. Waveguide combiners are desirable since their thickness is not impacted by the FOV, unlike other combiner architectures such as free-space or TIR prisms. However, the lateral dimensions of the waveguide (especially the redirection coupler and out-coupler areas over the waveguide) are closely linked to size of the incoupled FOV, as shown in Figure 30. For example, the R-E region geometry is dictated by the FOV in the waveguide medium: it expands in the direction orthogonal to the TIR propagation, forming a conical shape.

The largest coupler area requirement is usually the out-coupler element (center), aiming at processing all FOVs and building up the entire EB. Eye relief also strongly impacts this factor. However, its size can be reduced in a “human-centric optical design” approach: the right part of the FOV at the left edge of the EB as well as the left part of the FOV at the right edge of the EB can be discarded, thus considerably reducing the size of the O-E without compromising the image over the EB. Note that in Figure 29, the *k*-vector diagram (a) shows the FOV, whereas

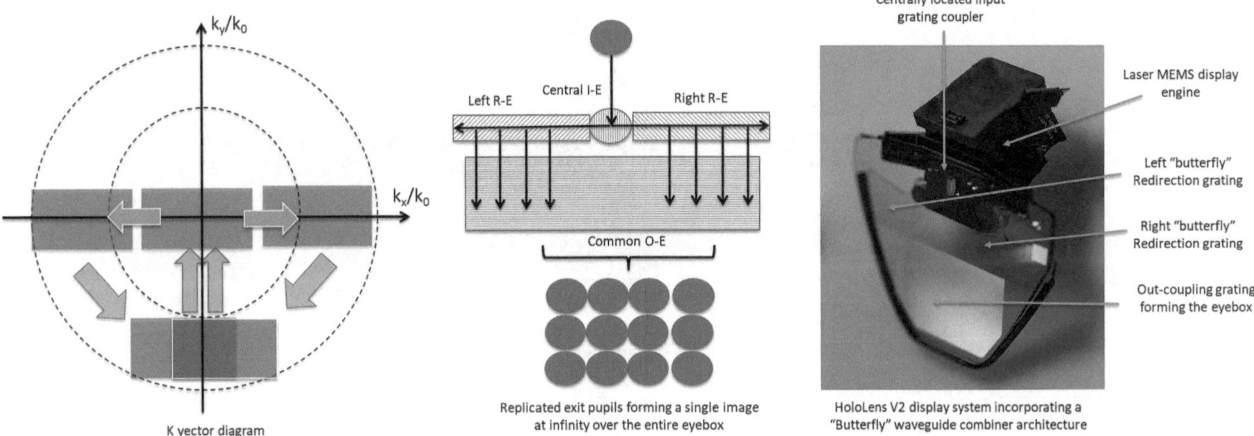

Figure 29: Symmetric incoupling for FOV increase in the direction of incoupling.

the lateral schematics of the waveguide in (b) and (c) show the actual size of the coupler regions.

Reducing the input pupil can help to reduce the overall size and thickness of the combiner. However, the thickness of the guide must be large enough not to allow for a second I-E interaction with the incoming pupil after the first TIR bounce. If there is a second interaction, then by the principle of time reversal, part of the light will be outcoupled and form a partial pupil (partial moon if the input pupil is circular) propagating down the guide instead of the full one. This is more pronounced for the smallest field angle, as depicted in Figure 31.

However, if the polarization of the field is altered after the first TIR reflection at the bottom of the guide, the parasitic outcoupling can be reduced if the I-E is made to be highly polarization sensitive.

Reducing the waveguide thickness can also produce stronger pupil diversity over the EB and thus better EB uniformity. If reducing the guide is not an option (for parasitic outcoupling of the input pupil and also for etendue limitations in the display engine), a semitransparent Fresnel surface can be used inside the guide (as in two guides bounded together), which would reflect only part of the field and leave the other part unperturbed, effectively increasing the exit pupil diversity.

Figure 32 shows how the space of the out-coupler grating is dictated solely by the FOV and the EB. Note that many fields can be canceled at the edges and toward the edges of the EB, as they will not enter the eye pupil (right fields on the left EB edge and left fields on the right EB edge). This can also reduce the size of the redirection grating considerably. This holds true for both EB dimensions.

5.2.8 Reducing the number of plates for RGB display and maintain FOV reach

Reducing the number of plates without altering the color of the image while propagating the maximum FOV allowed by

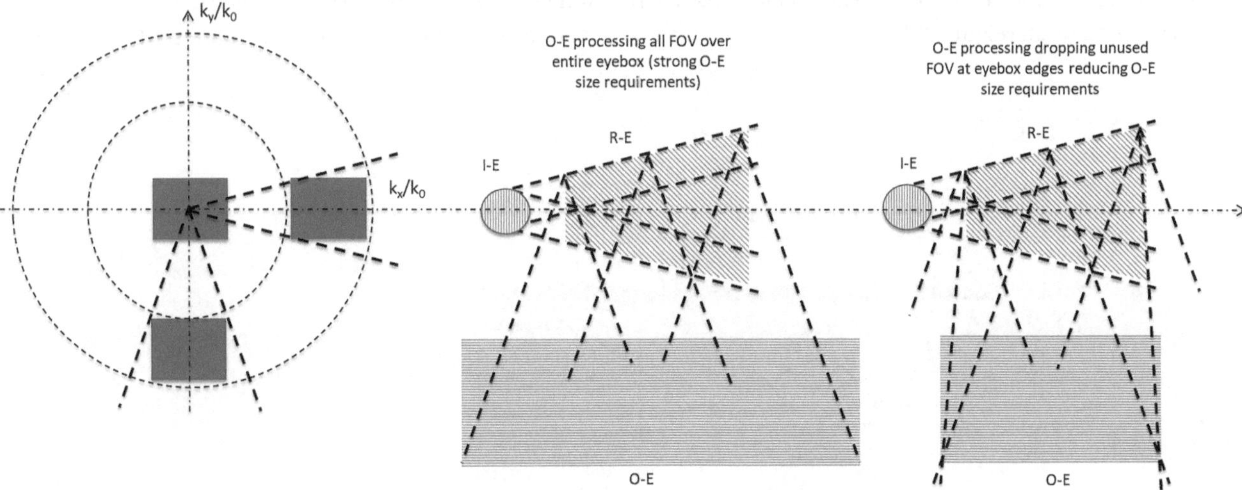

Figure 30: Redirection and out-coupler areas as dictated by the incoupled FOV.

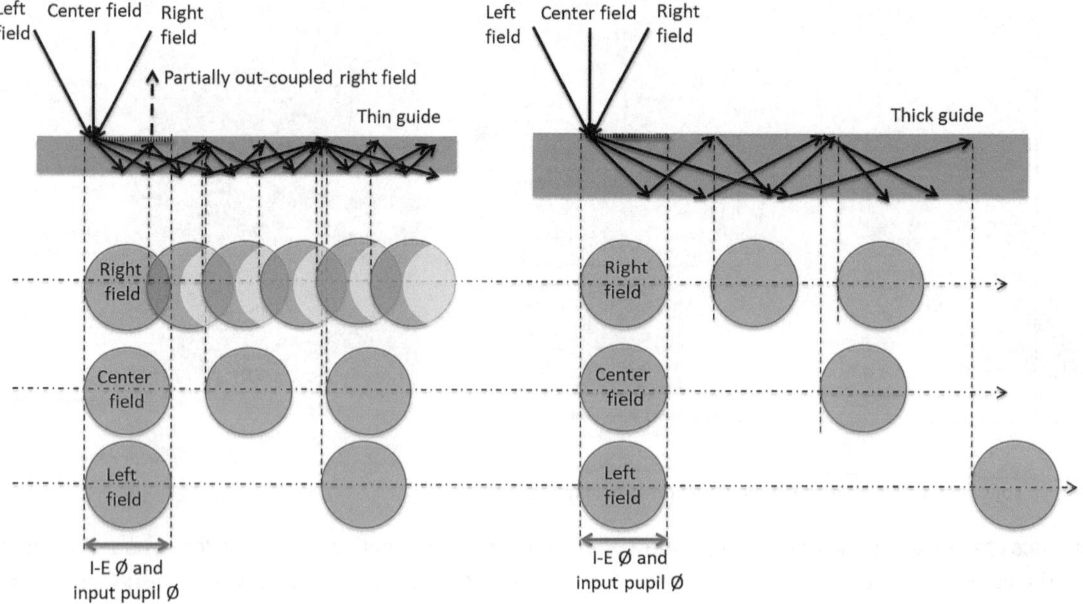

Figure 31: Effects of the input pupil size (and size of the I-E) and thickness of guide on a single field TIR pupil bouncing down the guide.

the index of the guide is a desirable feature since it reduces the weight, size, and complexity of the combiner and make it also less prone to MTF reductions due to guide misalignments. Both lateral and longitudinal angular waveguide misalignments will contribute to a reduction of the MTF built by the display engine. Waveguide surface flatness issues are yet more cause for MTF reduction.

Due to the strong spectral spread of the in-coupler elements (gratings, holograms, RWGs, or metasurfaces), the individual color fields are coupled at higher angles as the wavelength increases, which reduces the overall RGB FOV overlap that can propagate in the guide within the TIR conditions (smallest angle dictated by the TIR condition and largest angle dictated by pupil replication requirements for a uniform EB). This issue is best depicted in the *k*-vector diagram (Figure 33).

A lower spectral spread, such as through a prism in-coupler, would increase the RGB FOV overlap in a single guide, such as in an LOE (embedded partial mirrors out-couplers) from Lumus or in the microprism array couplers from Optinvent.

The left configuration in Figure 33 acts as a hybrid spatial/spectral filter, filtering the left part of the blue FOV, allowing the entire green FOV to be propagated (if the grating coupler periods have been tuned to match the green wavelength), and filtering the right part of the red FOV. The configuration in Figure 33 propagates the entire RGB FOV (assuming the couplers can diffract uniformly over the entire spectrum) at the cost of the FOV extending in the direction of the propagation. However, when considering binocular vision, this limitation could be mitigated by engineering a symmetric color vigneting in each eye (particularly on blue and red), providing a

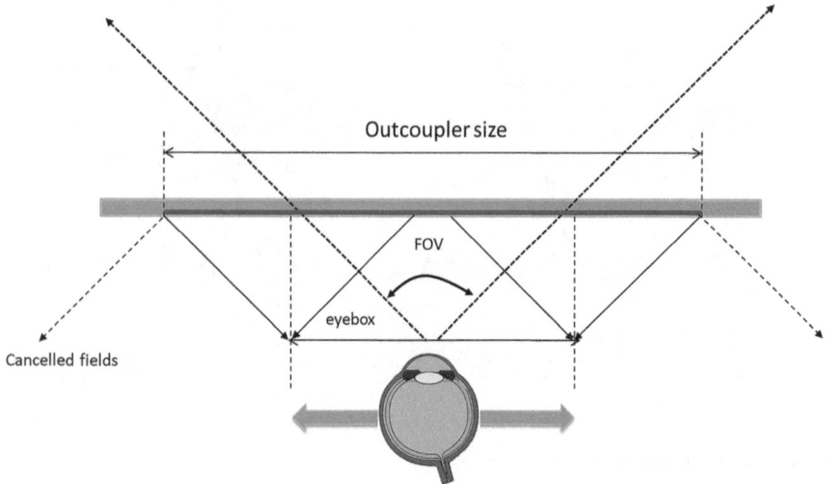

Figure 32: Eyebox and FOV dictate the size of the out-coupler area.

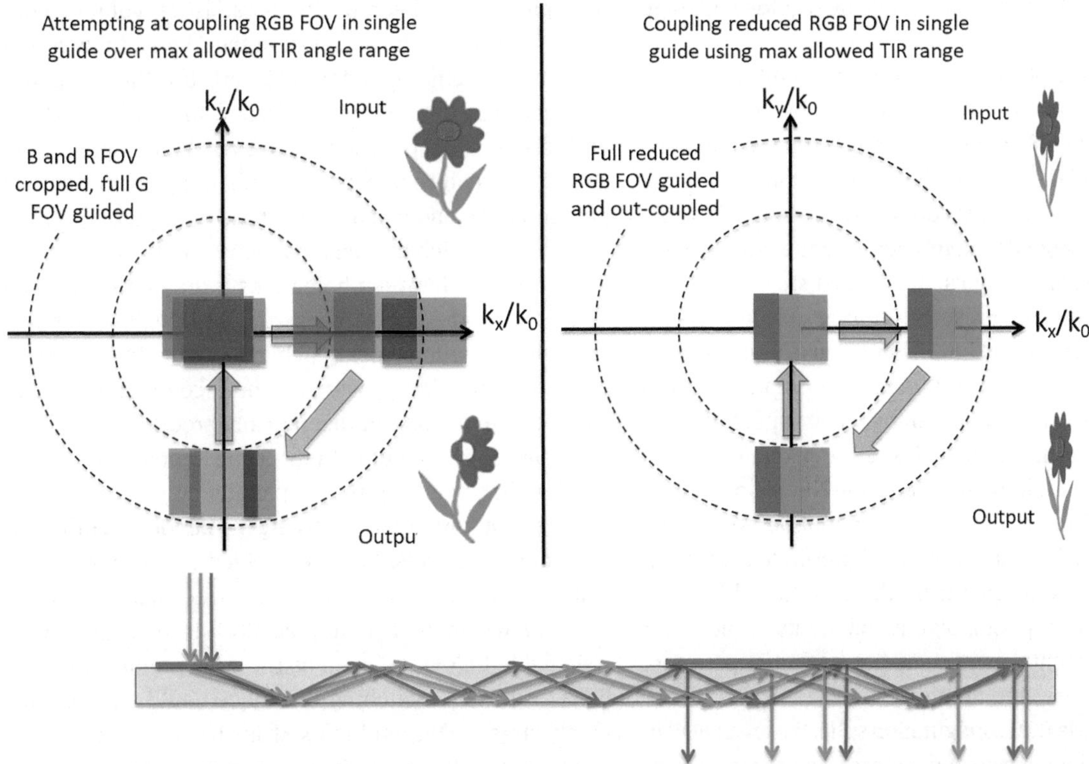

Figure 33: *k*-vector diagram of a single-plate waveguide combiner using (a) RGB FOV coupling over a single-color TIR angular range condition and (b) RGB reduced FOV sharing the same TIR range.

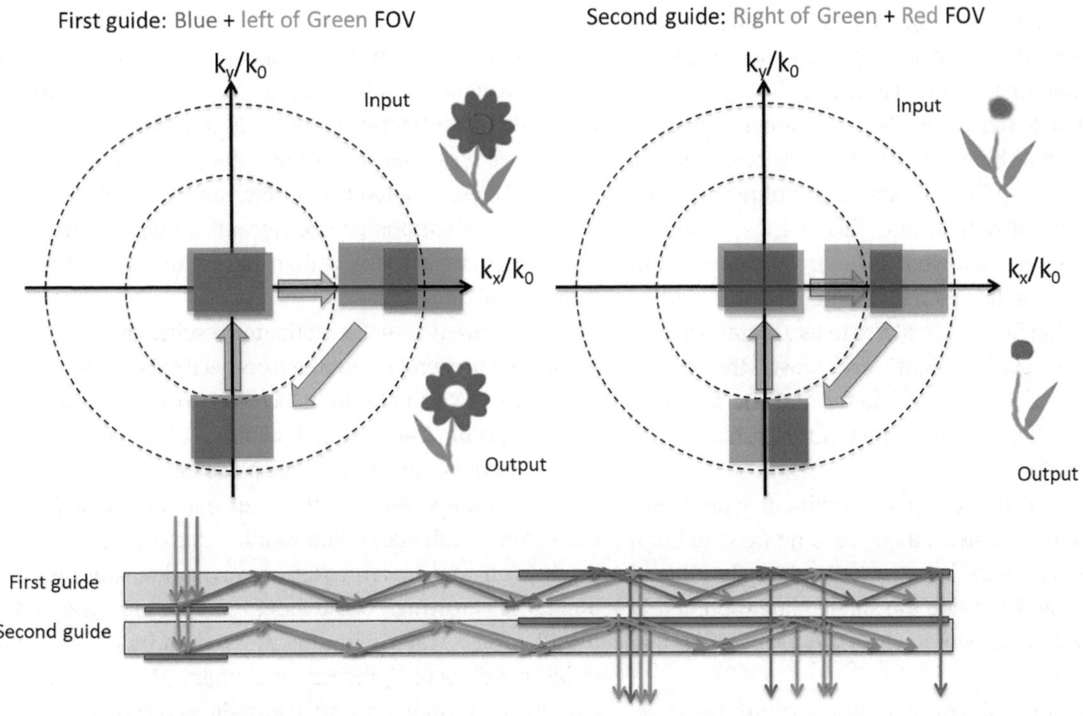

Figure 34: Two-guide RGB waveguide combiner configuration.

uniform stereo color vision in a single RGB guide with high FOV (e.g., Dispelix Oy).

Recently, two plate RGB waveguide combiner architectures have been investigated, reducing by one third the weight and size of traditional three-guide combiners, where the green FOV is shared between the top and bottom layer (see Figure 34. Various companies are using this two-plate RGB waveguide combiner architecture today, including Vuzix, WaveOptics, and Digilens.

However, this requires the grating (or holograms, RWGs, or metasurfaces) to be efficient over a larger spectral band, which implies that SRGs are to be replicated in a higher refractive index, widening their spectral (and angular) bandwidths. High-index grating replication by NIL stretches the traditional wafer-scale NIL resin material science (inclusion of TiO_2 or ZrO_2 nanofiller particles). Nanoimprint at a Gen2 panel size of higher-index inorganic spin-on glass material might be the best fit, which also solves the resin or photopolymer reliability issues over various environmental conditions (temperature, pressure, shear, UV exposure, and humidity).

This two-guide RGB configuration splits the green FOV in two at the in-coupler region and merges them again over the out-coupler region. For good color uniformity over the FOV and the EB, especially in the green field, this technique requires perfect control of the two-guide efficiency balance. Preemphasis compensation of the guide mismatch is possible using the display dynamic range, but this requires precise calibration, reduces the final color depth, and does not solve the stitching region issue where the two fields overlap.

An alternative to the architecture uses the first guide to propagate green and blue FOVs and the second guide to propagate only the red FOV, as green and blue are closer spectrally to each other than red. This change, however, slightly reduces the allowed FOV traveling without vignetting but solves the green FOV stitching problem.

Although going from three plates to two plates brings a small benefit in size, weight, and cost, the added complexity of the color split geometry and the resulting color nonuniformities over the EB might overshadow the initial small benefits.

A single-plate RGB waveguide combiner would provide a much stronger benefit, as there is no need to align multiple guides anymore because everything is aligned lithographically by NIL inside the single plate (potentially also front and back). This would also yield the best possible MTF and the lowest costs.

One single-plate solution is to phase multiplex three different color couplers with three different periods into a single layer and then tune it so that there is no spectral overlap (no color ghost images over the EB). Such phase multiplexing is theoretically possible in volume holograms. This might be achieved in the Akonia (now Apple) thick holographic material (500 μm). If a thinner photopolymer (less than 20 μm) is desired for better reliability and easier mass production, a large holographic index swing is required. Standard photopolymers can be panchromatic and can also be phase multiplexed, but the resulting efficiency remains low, and color cross-contamination between holograms is an additional issue. This is also theoretically possible with SRGs, but it is difficult to simultaneously achieve high efficiency and a high extinction ratio over the three color bands. Metasurfaces and RWGs can theoretically produce such phase-multiplexed layers but with the same limitations.

Another solution is to spatially interleave various grating configurations by varying the periods, depths, and slant angles. This is, however, difficult to achieve practically. Yet another solution to solve the single RGB guide problem would time multiplex RGB gratings through a switchable hologram, such as the ones produced by Digilens Corp. This switching technique could also produce much larger FOVs multiplexed in the time domain and fused in the integration time of the human eye.

6 Conclusion

The aim of this review paper was to capture the state of the art in waveguide combiner optics for AR and MR headsets, especially as diffractive waveguide combiners (surface relief diffractives, volume holographic, and others such as metasurfaces, resonant gratings, and so on). We also reviewed the various geometric waveguides combiner architectures which are rather based on refractive and reflective elements.

We showed that for optimum results, the waveguide grating, display engine, and sensors need to be codesigned as a global system to closely match the optical performances and the specific features and limitations of the human visual system, through human-centric optical design.

The coming years will be an exciting time for MR hardware. A full ecosystem to allow for commodity mass production and lower costs of waveguide grating combiners is growing worldwide, comprising high-index ultraflat glass wafer manufacturers, high-index resin material developers, process equipment developers, NIL equipment developers, and also dedicated software design tools developers allow finally this technology to emerge as a viable option for the upcoming consumer MR and smart glass market.

However, one has to remember that delivering on the promises of the ultimate wearable display hardware is only one aspect of the trial and opportunity ahead for MR, delivering on strong use cases, especially for consumer markets, will be the other critical item to consider.

Author contribution: All the authors have accepted responsibility for the entire content of this submitted manuscript and approved submission.
Research funding: None declared.
Conflict of interest statement: The authors declare no conflicts of interest regarding this article.

References

[1] W. S. Colburn and B. J. Chang, "Holographic combiners for head up displays," Tech Report No. AFAL-TR-77-110, 1977.
[2] J. Jerald, *The VR Book: Human Centered Design for Virtual Reality*, ACM Books, 2016, 978-1-97000-112-9.
[3] W. Barfield, *Fundamentals of Wearable Computers and Augmented Reality*, 2nd ed., CRC Press, Taylor and Francis Group, 2015, 978-1-482243595.
[4] L. Inzerillo, "Augmented reality: past, present and future," in *The Engineering Reality of Virtual Reality*, Vol. 8649, M. Dolinsky and I. E. McDowall, Eds. Proc. of SPIE-IS&T Electronic Imaging, SPIE, 2013.
[5] R. T. Azuma, "A survey of augmented reality," in *Presence, Teleoperators and Virtual Environments*, vol. 6, pp. 355–385, 1997.
[6] O. Cakmakci, J. Rolland, "Head-worn displays: a review," *J. Display. Technol.*, vol. 2, pp. 199–216, 2006.
[7] J. Rolland and O. Cakmakci, "Head-worn displays: the future through new eyes," *Opt. Phot. News*, vol. 20, pp. 20–27, 2009.
[8] D. W. F. Van Krevelen and R. Poelman, "A survey of augmented reality technologies, applications and limitations," *Int. J. Virtual Real.*, vol. 9, pp. 1–20, 2010.
[9] K.-L. Low, A. Ilie, G. Welch, A. Lastra, "Combining head-mounted and projector-based displays for surgical training," *IEEE Virtual Real.*, 2003, Proceedings 10.1109/VR.2003, https://doi.org/10.1109/VR.2003.1191128.
[10] Y. Amitai, A. Friesem, and V. Weiss, "Holographic elements with high efficiency and low aberrations for helmet displays," *Appl. Opt.*, vol. 28, pp. 3405–3416, 1989.
[11] N. Baker, *Mixed Reality Keynote at Hot Chips HC28 – Symposium for High Performance Chips*, Aug. 21-23 2016, www.hotchips.org.
[12] B.Kress and W.Cummins, *Towards the Ultimate Mixed Reality Experience: HoloLens Display Architecture Choices*, SID 2017 Book 1: Session 11: AR/VR Invited Session II.
[13] J. Michael Miller, N. de Beaucoudrey, P. Chavel, J. Turunen, and E. Cambril, "Design and fabrication of binary slanted surface-relief gratings for a planar optical interconnection," *Appl. Opt.*, vol. 36, pp. 5717–5727, 10 August 1997.
[14] J. Kimmel, T. Levola, P. Saarikko, and J. Bergquist, "A novel diffractive backlight concept for mobile displays," *J. SID*, vol. 16, no. 2, 2008.
[15] J. Kimmel and T. Levola, "Diffractive backlight light guide plates in mobile electrowetting display applications," SID 09 Paper 471 Page 2.
[16] J. Liu, N. Zhang, J. Han, et al., "An improved holographic waveguide display system," *Appl. Optic.*, vol. 54, no. 12, pp. 3645–3649, 2015.
[17] T. Yoshida, K. Tokuyama, Y. Takai, et al., "A plastic holographic waveguide combiner for light-weight and highly-transparent augmented reality glasses," *J. SID*, vol. 26, no. 5, 2018.
[18] H. Mukawa, K. Akutsu, I. Matsumura et al., "A full color eyewear display using holographic planar waveguides" 8.4, SID 08 DIGEST 2008.
[19] T. Oku, K. Akutsu, M. Kuwahara et al., "High-luminance see-through eyewear display with novel volume hologram waveguide technology," 15.2, 192 • SID DIGEST 2015.
[20] K. Sarayeddine, P. Benoit, G. Dubroca, and X. Hugel, "Monolithic low-cost plastic light guide for full colour see through personal video glasses," in *ISSN-L 1883-2490/17/1433 ITE and SID (IDW 10)*, 2010, pp. 1433–1435.
[21] T. Levola, "Exit pupil expander with a large field of view based on diffractive optics," *J. Soc. Inf. Disp.*, vol. 17, pp. 659–664, 2009.
[22] T. Levola, "Diffractive optics for virtual reality displays," *J. SID*, vol. 14, no. 5, 2006.
[23] B. Kress, "Diffractive and holographic optics as optical combiners in head mounted displays," in *Proceedings of the 2013 ACM Conference on Pervasive and Ubiquitous Computing – Ubicomp'13*, 2013, pp. 1479–1482.
[24] A. Cameron, "Optical waveguide technology & its application in head mounted displays," in *Head- and Helmet-Mounted Displays XVII; and Display Technologies and Applications for Defense, Security, and Avionics VI*, Vol. 8383, P. L. Marasco, P. R. Havigll, D. D. Desjardins, and K. R. Sarma, Eds., Proc. of SPIE, p. 83830E.
[25] M. Homan, "The use of optical waveguides in Head up Display (HUD) applications," in *Display Technologies and Applications for Defense, Security, and Avionics VII*, Vol. 8736, D. D. Desjardins and K. R. Sarma, Eds, Proc. of SPIE.
[26] D. Cheng, Y. Wang, C. Xu, W. Song, and G. Jin, "Design of an ultra-thin near-eye display with geometrical waveguide and freeform optics," *Opt. Express*, vol. 22, no. 17, pp. 20705–20719, 2014.
[27] D. Jurbergs, F.-K. Bruder, F. Deuber, et al., "New recording materials for the holographic industry," in *Practical Holography XXIII: Materials and Applications*, Vol. 7233, H. I. Bjelkhagen and R. K. Kostuk, Eds. Proc. of SPIE.
[28] www.digilens.com.
[29] K. Curtis and D. Psaltis, "Cross talk in phase coded holographic memories," *J. Opt. Soc. Am. A*, vol. 10, no. 12, December 1993, https://doi.org/10.1364/JOSAA.10.002547.
[30] H. Kogelnik, "Coupled wave theory for thick hologram gratings," *Bell Syst. Tech. J.*, vol. 48, 1969, https://doi.org/10.1002/j.1538-7305.1969.tb01198.x.
[31] M. A. Golub, A. A. Friesem, and L. Eisen, "Bragg properties of efficient surface relief gratings in the resonance domain," *Optic Commun.*, vol. 235, pp. 261–267, 2004.
[32] M. G. Moharam, "Stable implementation of the rigorous coupled wave analysis for surface relief gratings: enhanced transmittance matric approach," *J. Opt. Soc. Am. A*, vol. 12, no. 5, pp. 1077–1086, 1995.

[33] L. Alberto Estepa, C. Neipp, J. Francés, et al., "Corrected coupled-wave theory for non-slanted reflection gratings," in *Physical Optics*, Vol. 8171, D. G. Smith, F. Wyrowski, and A. Erdmann, Eds, Proc. of SPIE.
[34] http://www.kjinnovation.com/.
[35] https://meep.readthedocs.io/en/latest/.
[36] T. Levola and P. Laakkonen, "Replicated slanted gratings with a high refractive index material for in and outcoupling of light," *Opt. Express*, vol. 15, pp. 2067–2074, 2007.
[37] https://www.lighttrans.com/applications/virtual-mixed-reality/waveguide-huds.html.
[38] M. W. Farn, "Binary gratings with increased efficiency," *Appl. Opt.*, vol. 31, no. 22, pp. 4453–4458, 1992.
[39] B. Kress and P. Meyrueis, *Applied Digital Optics: From Micro-optics to Nanophotonics*, 1st ed., John Wiley and Sons Publisher, 2007, -10: 0470022639.
[40] G. Quaranta, G. Basset, O. J. F. Martin, and B. Gallinet, "Steering and filtering white light with resonant waveguide gratings," in *Proc. SPIE 10354, Nanoengineering: Fabrication, Properties, Optics, and Devices XIV*, 2017, p. 1035408.
[41] G. Basset, "Resonant screens focus on the optics of AR," in *Proc. SPIE 10676, Digital Optics for Immersive Displays*, 2018, p. 106760I.
[42] P. Genevet, F. Capasso, F. Aieta, M. Khorasaninejad, and R. Devlin, "Recent advances in planar optics: from plasmonic to dielectric metasurfaces," *Optica*, vol. 4, no. 1, pp. 139–152, 2017.
[43] F. Capasso, "The future and promise of flat optics: a personal perspective," *Nanophotonics*, vol. 7, no. 6, 2018, https://doi.org/10.1515/nanoph-2018-0004.
[44] W. T. Chen, A. Y. Zhu, J. Sisler, et al., "Broadband Achromatic metasurface-refractive optics," *Nano Lett.*, vol. 18, no. 12, pp. 7801–7808, 2018.

Supplementary Material: The online version of this article offers supplementary material (https://doi.org/10.1515/nanoph-2020-0410).

Moritz Merklein*, Birgit Stiller, Khu Vu, Pan Ma, Stephen J. Madden and Benjamin J. Eggleton

On-chip broadband nonreciprocal light storage

https://doi.org/10.1515/9783110710687-005

Abstract: Breaking the symmetry between forward- and backward-propagating optical modes is of fundamental scientific interest and enables crucial functionalities, such as isolators, circulators, and duplex communication systems. Although there has been progress in achieving optical isolation on-chip, integrated broadband nonreciprocal signal processing functionalities that enable transmitting and receiving via the same low-loss planar waveguide, without altering the frequency or mode of the signal, remain elusive. Here, we demonstrate a nonreciprocal delay scheme based on the unidirectional transfer of optical data pulses to acoustic waves in a chip-based integration platform. We experimentally demonstrate that this scheme is not impacted by simultaneously counter-propagating optical signals. Furthermore, we achieve a bandwidth more than an order of magnitude broader than the intrinsic optoacoustic linewidth, linear operation for a wide range of signal powers, and importantly, show that this scheme is wavelength preserving and avoids complicated multimode structures.

Keywords: Brillouin scattering; integrated photonics; nonreciprocity; optical delay.

1 Introduction

Reciprocity is a general concept in optics dictating that a transmission channel does not change, or is symmetric, under the interchange of source and receiver [1–3]. There are, however, different approaches to break this reciprocity, most commonly by utilizing magnetic materials [4–6]. The arguably most common devices based on breaking reciprocity are isolators utilizing magnetic materials. Isolators based on magnetic materials are passive components and have large bandwidth and rejection. Chip integration and material losses, however, are still major challenges despite the great progress made over the recent years.

Another way to achieve nonreciprocal transmission is based on temporal modulation [2, 7]. Although this active method requires some sort of pumping – electrical [8], optical [9], or acoustic [10] – it offers great potential as it does not rely on magnetic materials and hence is more suitable for chip integration with the additional advantage of being reconfigurable. Nonreciprocal elements, such as optical isolators and circulators are crucial building blocks, in particular, for integrated photonic circuits to protect the laser but also to route counterpropagating signals and mitigate back-reflections that can arise from boundaries between multiple elements or are simply caused by Rayleigh backscattering.

A powerful and versatile way to achieve nonreciprocal transmission in a small footprint can be realized by coupling light and mechanical degrees of freedom [11, 12]. This coupling between light and mechanical oscillations is greatly enhanced in resonant structures which lead to demonstrations of chip-scale optomechanical isolators and circulators for optical [13–19] as well as microwave signals [20–23].

Similarly, stimulated Brillouin scattering (SBS), i.e. the resonant coupling of optical waves with propagating acoustic waves in waveguides via optically induced forces is known to be nonreciprocal [9, 24]. The phase-matching

Moritz Merklein and Birgit Stiller have contributed equally to this work.

***Corresponding author: Moritz Merklein,** The University of Sydney Nano Institute (Sydney Nano), The University of Sydney, Sydney, NSW 2006, Australia; and Institute of Photonics and Optical Science (IPOS), School of Physics, The University of Sydney, Sydney, NSW 2006, Australia, E-mail: moritz.merklein@sydney.edu.au. https://orcid.org/0000-0002-5558-2592
Birgit Stiller, The University of Sydney Nano Institute (Sydney Nano), The University of Sydney, Sydney, NSW 2006, Australia; Institute of Photonics and Optical Science (IPOS), School of Physics, The University of Sydney, Sydney, NSW 2006, Australia; and Max-Planck-Institute for the Science of Light, Staudtstr. 2, 91058 Erlangen, Germany
Khu Vu and Stephen J. Madden, Max-Planck-Institute for the Science of Light, Staudtstr. 2, 91058 Erlangen, Germany
Pan Ma, Laser Physics Centre, Research School of Physics and Engineering, Australian National University, Canberra, ACT 2601, Australia
Benjamin J. Eggleton, The University of Sydney Nano Institute (Sydney Nano), The University of Sydney, Sydney, NSW 2006, Australia; and Institute of Photonics and Optical Science (IPOS), School of Physics, The University of Sydney, Sydney, NSW 2006, Australia

This article has previously been published in the journal Nanophotonics. Please cite as: M. Merklein, B. Stiller, K. Vu, P. Ma, S. J. Madden and B. J. Eggleton "On-chip broadband nonreciprocal light storage" *Nanophotonics* 2021, 10. DOI: 10.1515/nanoph-2020-0371.

condition ensures that only pump and probe waves which are counterpropagating (copropagating) couple to the acoustic wave which is mainly longitudinal for backward SBS (transverse for forward Brillouin scattering) [25]. As these acoustic waves carry momentum, Brillouin interactions can be used to induce indirect photonic transitions between different optical modes [9, 24, 26].

Numerous experiments have reported nonreciprocity exploiting Brillouin interactions, including demonstrations in photonic crystal fibers [9], silicon waveguides [27], and fiber-tip resonators [28–30]. Surprisingly, however, so far there was no demonstration of Brillouin-based nonreciprocal schemes harnessing backward SBS. In this approach, the optical wave can couple to a continuum of acoustic modes that provides enormous flexibility, which is particularly important for nonreciprocal signal processing schemes that go beyond providing pure signal isolation. Furthermore, large backward SBS gain can be achieved in small footprint planar integrated circuits [25].

Here, we show a nonreciprocal light storage scheme based on coherent Brillouin coupling of acoustic and optical modes to achieve nonreciprocal delay. We experimentally demonstrate that optical pulses that are traveling simultaneously in the opposite direction, with the same optical frequency and mode, are not impacted nor do they impact the storage process. We show that the bandwidth of the scheme can be broadened beyond the intrinsic acoustic linewidth – in this demonstration by more than one order of magnitude – which was generally thought to be a limiting factor of SBS-based nonreciprocal schemes. Furthermore, the scheme depends linearly on the input data pulse power in the observed range and does neither alter the frequency nor mode of the incoming data – all important requirements for practical nonreciprocal devices.

2 Results

The nonreciprocity is induced by an interaction between two counterpropagating optical modes – here we call them data ω_{data} and write/read $\omega_{\text{w/r}}$ – with a traveling acoustic mode Ω. To enable efficient coupling from optical to acoustic waves, the optical write/read and the data pulses are separated by the frequency of the acoustic wave in the waveguide, known as the Brillouin frequency shift, $\Omega = 2 \cdot n_{\text{eff}} \cdot v_{\text{ac}} \cdot \lambda^{-1}$, where n_{eff} is the effective refractive index, v_{ac} the acoustic sound velocity in the material, and λ the optical pump wavelength. The waveguide used in this demonstration is made out of chalcogenide glass and the corresponding acoustic resonance frequency is $\Omega \approx 7.6$ GHz. The underlying physical mechanisms that couple the two optical and the acoustic waves are electrostriction – the compression of a material in regions of strong light fields and photoelasticity – the change of refractive index with material density [31].

Brillouin interactions can be used to store optical data pulses as acoustic waves on a photonic chip [32]. The optical data pulses are frequency-upshifted from counterpropagating write and read pulses by precisely the Brillouin frequency shift Ω. Hence, the optical data pulses are first depleted by the write pulses and an acoustic wave is generated and are later retrieved using read pulses which deplete the acoustic wave.

This interaction has not only to fulfill energy conservation $\omega_{\text{data}} = \omega_{\text{w/r}} + \Omega$ but also momentum conservation $\mathbf{k}_{\text{data}} = \mathbf{k}_{\text{w/r}} + \mathbf{q}$. As the momentum $\mathbf{q}$ of the traveling acoustic wave in backward SBS is large, approximately twice the magnitude of the momentum vectors of the individual optical modes $|\mathbf{q}| \approx 2 \cdot |\mathbf{k}_{\text{data}}|$ with $|\mathbf{k}_{\text{data}}| \approx |\mathbf{k}_{\text{w/r}}|$, the interaction is only phase-matched in one direction. Note that the phase-matching condition for exciting an acoustic wave with a particular pair of counterpropagating optical pulses can always be fulfilled by matching the detuning to the Brillouin frequency shift in the waveguide as opposed to interdigitated transducer–based approaches that require careful device design and fabrication to ensure phase-matching is fulfilled [10].

That the Brillouin process is only phase-matched for data signals traveling in one direction becomes more evident when looking at the dispersion diagram shown in Figure 1a, where changing the direction of one optical mode leads to a phase mismatch $\Delta\mathbf{q}$. As the interaction takes place over an elongated length given by either the length of the waveguide, for the case of continuous wave (CW) signals, or the length of the optical signal pulse, a small initial phase mismatch builds up to a large mismatch over that interaction length as it was shown for the case of Brillouin interactions that involve multiple optical wavelengths [33]. Even in the case of nanosecond (ns) and subnanosecond pulses, this length scale is in the order of several centimeters.

How this strict phase-matching condition of the opto-acoustic Brillouin interaction can be utilized to achieve unidirectional signal delays is shown in Figure 1b. Optical data pulses ω_{data} that propagate from the left through the waveguide are transferred to acoustic phonons via a Brillouin interaction induced by counterpropagating write pulses ω_{w} and are subsequently retrieved using read pulses ω_{r} [32, 34, 35], whereas optical pulses that travel in the opposite direction are neither effected by the write, the read pulses nor the acoustic wave that stores the original optical pulses ω_{data}. The phase-matching condition

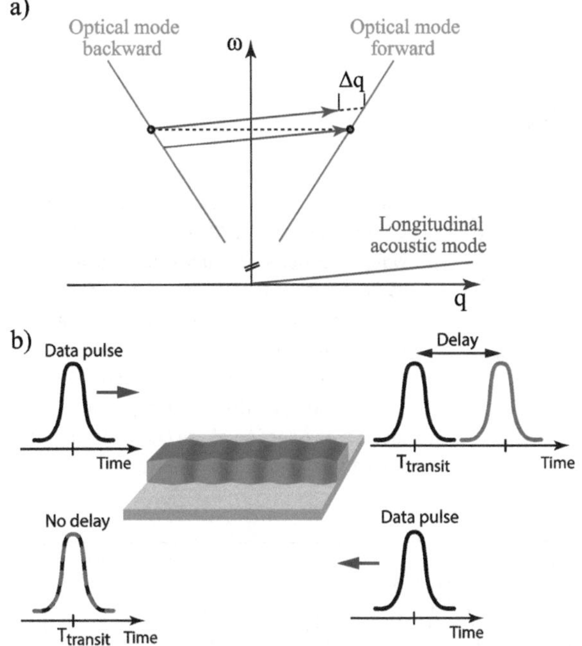

Figure 1: Basic principle and phase-matching diagram. (a) Dispersion diagram for backward Brillouin scattering illustrating the phase-matching condition for data pulses propagating in opposite directions. (b) Optical data pulses that are coupled from the left side into the waveguide are converted to acoustic phonons and experience a delay. Optical data pulses that are coupled simultaneously from the opposite side in the waveguide do not experience any conversion and hence are not delayed.

between the optical modes and the traveling acoustic wave ensures that there is no interaction, even for the extreme case when the counterpropagating optical pulses are in the same optical waveguide mode, at the same frequency and optical power.

Data pulses that are transferred to the acoustic domain accumulate a large delay because of the five orders of magnitude difference in velocity between the acoustic and optical waves. The accumulated delay of the optical signal can hence be approximated by the time difference between the writing and the retrieving operation. Thus, the delay of the optical signal can be continuously tuned within the acoustic lifetime of the acoustic phonon that is given by the material properties of the chalcogenide glass and is in the order of 10 ns [32].

Owing to the phase-matching condition of the Brillouin interaction, the coupling between optical and acoustic wave only occurs for a certain pair of optical write/read and data pulses. Optical data pulses that are simultaneously propagating through the waveguide from the opposite side are neither transferred to the acoustic wave by the write pulses, nor do they interact with the acoustic wave present in the waveguide and hence do also not influence the stored data pulses. The counterpropagating signals could be data pulses that are simultaneously transmitted/received in the opposite direction through the same waveguide or could simply originate from backscattering in the photonic circuit.

2.1 Experimental setup

The simplified experimental setup is shown in Figure 2 (a detailed scheme and description can be found in Section 4). A CW distributed feedback laser is split into two paths, the data and the write/read path. The data signal is shifted by the Brillouin frequency shift Ω of the waveguide. Afterward, the CW signals in both arms are modulated into short pulses using a multichannel arbitrary waveform generator and electro-optic intensity modulators. The data signal is split using a 50/50 coupler. One part of the signal is combined with the write/read arm and coupled to the nonlinear chalcogenide waveguide. The other half passes through an additional 50/50 coupler, to ensure that the path length of both data signals as well as their optical power is the same in both arms, and is coupled to the chalcogenide chip from the opposite side. On both sides of the chip circulators are used to separate input and output. Two narrowband filters (bandwidth ≈ 3 GHz) are used to separate the data signal from the pump signal and fast photodetectors (bandwidth > 10 GHz) and a fast oscilloscope (bandwidth > 10 GHz) are used to detect the data pulses. A cross-section of the chalcogenide waveguides is shown in the inset of Figure 2. A rib structure with a cross-section of $2200 \times 850\ nm^2$ is used to reduce losses caused by sidewall roughness. The chalcogenide glass is surrounded by silica glass to ensure confinement of the optical as well as the acoustic mode [36]. For details on the fabrication of the acousto-optic waveguides, see Section 4. The simulated dispersion of the waveguide for both optical modes (forward and backward) is around −200 ps/nm/km. Although it is larger than in standard single-mode fiber, for the length of the chip (22 cm) and the bandwidth of the pulses (around 4 pm), the signal distortion from dispersion is negligible.

2.2 Nonreciprocal Brillouin light storage

Figure 3a shows the transfer of an optical data pulse ω_{data} to an acoustic wave by a counterpropagating write pulse ω_w (black curve; full-width half-maximum ≈ 1 ns). The depleted optical signal is shown in red (Figure 3a). Around 90% of data pulse depletion could be achieved, whereas a simultaneously counterpropagating data pulse

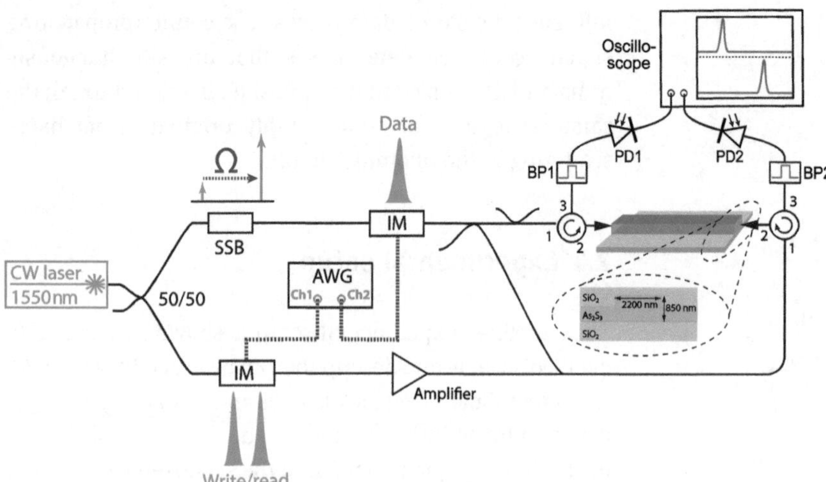

Figure 2: Schematic experimental setup. Continuous wave (CW) laser, continuous wave distributed feedback (DFB) laser; 50/50 fiber coupler; SSB, single-sideband modulator; IM, intensity modulator; AWG, multichannel arbitrary waveform generator; Amplifier, erbium-doped fiber amplifier; BP, bandpass filter; PD, photodetector. Inset: cross-section of the chalcogenide rib waveguide embedded in silica.

(copropagating to the write pulses) is not impacted by the acoustic wave generated in the depletion process (inset Figure 3a). The black curve in the inset shows optical data pulses transmitted in the counterpropagating direction, whereas there is no data transferred to the acoustic wave. The red dashed curve in the inset of Figure 3a shows the counterpropagating optical data pulses while the data traveling in the opposite direction is depleted and transferred to a traveling acoustic wave. Both curves perfectly overlap. In the current experimental implementation, the depletion efficiency of 90% was mainly limited by power constraints and can in principle reach almost full conversion of the optical pulse to the acoustic wave. Figure 3b shows the fast Fourier transform (FFT) of the optical input data pulses demonstrating that the bandwidth of the optical data pulses that can be successfully depleted and transferred to the acoustic domain can greatly exceed the intrinsic Brillouin linewidth of 30 MHz. The dashed lines in Figure 3b indicate the full-width half-maximum of the FFT of the data pulses.

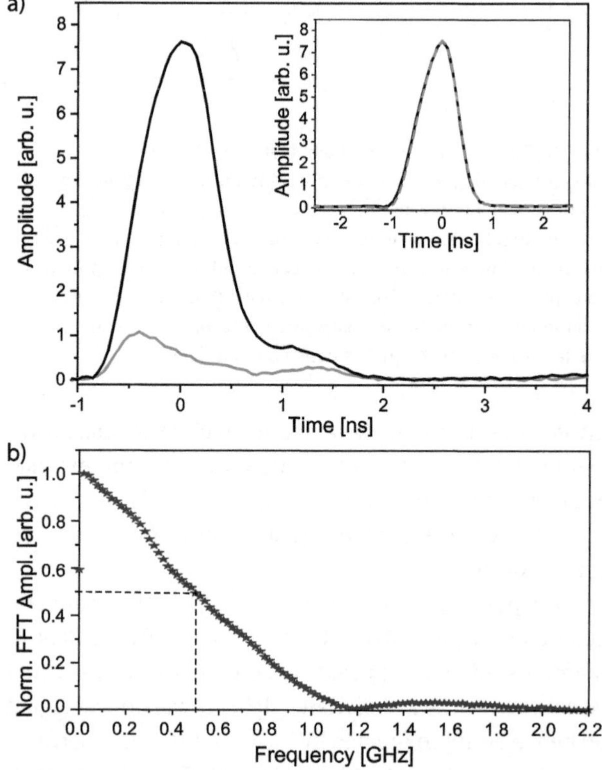

Figure 3: Nonreciprocal pulse depletion.
(a) Conversion of an optical data pulse (black curve) to acoustic phonons (red curve) while a simultaneously counterpropagating data pulse is not impacted (black and dashed red curve in inset). (b) Fast Fourier transform (FFT) of the input data pulse.

This extension of the bandwidth by more than one order of magnitude beyond the intrinsic acoustic linewidth is enabled by the ultrahigh Brillouin gain provided in chalcogenide waveguides which is about two orders of magnitude larger than in standard silica single-mode fiber. The strength of the interaction is usually measured in terms of Brillouin gain G which gives the amplification of the Stokes wave for a given pump power P and is proportional to $\propto \exp(g\, L_{\mathrm{eff}}\, P)$, with $g = g_0 \cdot A_{\mathrm{AO}}^{-1}$ given by the Brillouin gain coefficient g_0 and the acousto-optic area A_{AO}, and L_{eff} being the effective length. If the pump is broadened beyond the 30 MHz intrinsic linewidth of the Brillouin interaction, more overall pump power is required. As in this experiment, the effective length L_{eff} is limited to the interaction length of the counterpropagating short pulses (cm – tens of cm range), a large gain factor g is required to achieve broad bandwidth operation. For our chips, the Brillouin gain coefficient was previously measured to be $g_0 \approx 0.715 \times 10^{-9}$ m/W and the acousto-optic area $A_{\mathrm{AO}} \approx 1.5\ \mu\mathrm{m}^2$ that leads to the approximately two orders of magnitude increase in the Brillouin factor g [37].

The broad bandwidth is one main advantage of the resonator-free waveguide-based Brillouin approach. Although the intrinsic linewidth of the optoacoustic interaction is given by the phonon lifetime for a single frequency CW optical pump the Brillouin response can be broadened by using a broad bandwidth optical pump where the bandwidth of the Brillouin response is then given by the bandwidth of the optical pump itself. The intrinsic Brillouin gain is in this case distributed over the bandwidth and hence the absolute maximum Brillouin gain is reduced accordingly for a given input power.

After demonstrating the nonreciprocal depletion of optical data pulses, we now show that we can retrieve the data back from the acoustic to the optical domain, even in the presence of counterpropagating pulses, and hence unidirectionally delay optical signals relative to the regular transit time of the waveguide. Figure 4a shows experimental measurements of a 1 ns long optical data pulse that is delayed by 4 ns when propagating in one direction in the waveguide, whereas an optical data pulse simultaneously propagating in the opposite direction with the same optical frequency and optical mode is not interacting with the Brillouin storage process or the acoustic wave present in the waveguide (Figure 4). The transit time of the chip is around 1.8 ns and hence our measured delay would equate to an increase in the group index by a factor of about 2.2. As a proof-of-principle demonstration, we only show a pulse delay of 4 ns; however, we note that the delay is given by the arrival time difference between the write and the read pulses and hence can continuously be tuned. As the phonon exponentially decays as e^{-2t/τ_A}, with τ_A being the acoustic lifetime, the readout efficiency decreases for longer storage times. The phonon lifetime in our waveguide structure is mainly limited by the properties of the chalcogenide material. Much longer phonon lifetimes have been shown in acoustic resonators that hence could achieve longer storage times, however, at the expense of the bandwidth [29, 38, 39]. Furthermore, the shape of the pulses is challenging to be maintained in resonator-based acoustic light storage with the spatial extent of the pulses exceeding the circumference of the resonators.

Conversely, in the here-demonstrated delay scheme, the pulse shape is maintained (Figure 4a). The input data pulse and the delayed data pulse in Figure 4a are normalized to visually emphasize that point and show the similarity in the pulse shape as it is a common practice for fiber-based optical pulse delay techniques [35, 40–44]. The readout efficiency of this measurement was around 20% after a delay of 4 ns. Here, the efficiency is slightly below the record of 32% after a delay time of 3.5 ns reported previously [32], which can be accounted to overall power limitations of the experimental setup due to simultaneously counterpropagating signals.

While the data pulses are stored and delayed in one direction, simultaneously counterpropagating data pulses are not delayed (Figure 4b). The black curve in Figure 4b shows an optical pulse when there is no delay applied to the data that travels in the opposite direction, whereas the dashed red curve shows the optical pulse for the case when the counterpropagating channel is delayed via coherent transfer from optical to acoustic and back to the optical domain. Here, the data pulses are not normalized to emphasize that there is neither a change in amplitude nor shape of the pulses. The counterpropagating data pulses do not interact with the acoustic mode present in the waveguide from the delay process of the data pulses propagating in the opposite direction nor do they distort the storage process of these pulses. Hence, we show that the nonreciprocal pulse storage scheme enables full duplex signal processing. It also shows that potential back-reflections which can occur in complex integrated circuits that consist of many discreet components are not distorting the delay process.

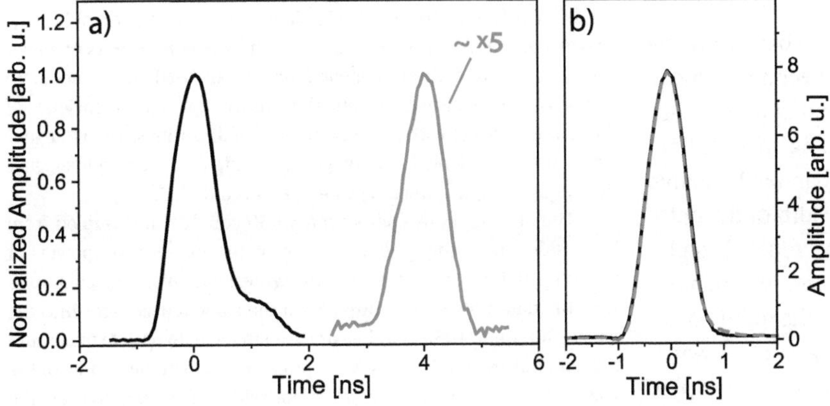

Figure 4: Nonreciprocal light storage. (a) Optical data pulses (black trace) propagating in one direction are delayed by 4 ns (red trace), whereas (b) simultaneously counterpropagating data pulses are not impacted (black trace shows transmitted data without delay applied to the counterpropagating signals, red dotted trace shows transmitted data while counterpropagating data are stored). Note that the readout efficiency of the pulses presented in (a) is around 20% and the data pulses are normalized to visualize the pulse shape before and after the storage process.

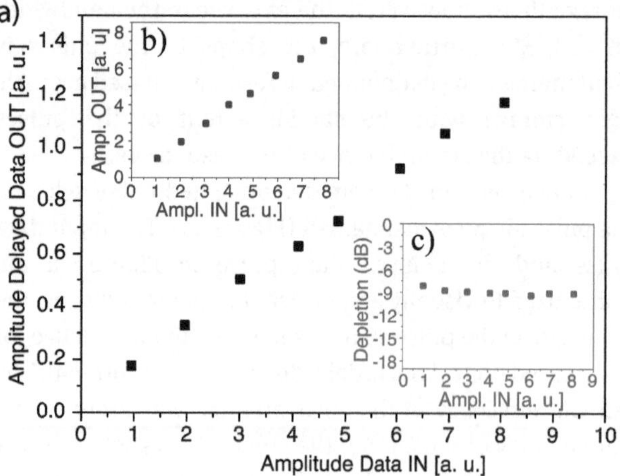

Figure 5: Linearity of nonreciprocal light storage.
(a) Linear amplitude response of the delayed optical data pulses. Inset (b) shows the linearity of the counterpropagating data pulses, whereas inset (c) shows the depletion of the data pulses for different input amplitude levels.

A crucial metric for nonreciprocal devices in general, which is particularly important for optical signal processing applications, is the linearity of the scheme. The linearity ensures that information encoded in the amplitude is maintained during the signal processing operation. Figure 5 shows that the Brillouin-based nonreciprocal delay scheme is linear over a wide range. The coupled power of the input data pulses in Figure 5 is varied by an order of magnitude from −10 to −20 dBm average optical power and we observe a linear relationship between input and output amplitude. We confirm that the same is true for the counterpropagating data channel (Figure 5b). The second inset, Figure 5c, shows that the depletion of the original data pulses which are transferred to the acoustic wave is approximately constant in the measured data power range.

3 Conclusion and outlook

We showed a nonreciprocal delay scheme based on the coherent interaction of photons with traveling acoustic phonons. The phase-matching underlying this process ensures that only optical data pulses traveling in a distinct direction are delayed. We showed that the bandwidth of this scheme is not limited to the intrinsic linewidth of the opto-acoustic interaction, but can, in fact, be much broader approaching the GHz regime. Furthermore, we demonstrated that the scheme depends linearly on the input power and does neither convert the optical mode nor the wavelength of the signal. Hence, it opens a pathway to full duplex signal processing architectures that can greatly reduce size, weight, and power requirements. The delay time is continuously tunable as it is given by the difference in the arrival time of the readout pulse with respect to the write pulse within the phonon lifetime [32]. Recently, however, it was proposed and experimentally shown that the storage time can be extended by refreshing the acoustic phonon with optical pulses [45] overcoming said limitation.

Our demonstration of delaying an optical signal while another optical signal at the same frequency is counter-propagating shows the immunity of the here-presented delay scheme to detrimental back-reflections common in complex integrated photonic circuits that are composed of a multitude of optical elements. In the context of phased array antennas and beam steering elements, inducing nonreciprocal delays could enable new ways of separating transmitted and received signals.

4 Materials and methods

4.1 Experimental setup for nonreciprocal light storage

A layout of the experimental setup is shown in Figure 6.

As a laser source, we use a narrow-linewidth distributed feedback laser (TerraXion NLL) with a wavelength of around 1550 nm. The laser signal is divided into two arms, the data and the write/read arm. The data signal is up-shifted in frequency by the Brillouin frequency shift Ω via a single-sideband modulator. A single laser source is used to avoid relative drift of the data and the write/read arm. Two intensity modulators connected to a multichannel arbitrary waveform generator are used to chop the CW laser signals in both arms into a pulse stream. The write/read pulses are amplified via an erbium-doped fiber amplifier (EDFA) and afterward pass through a nonlinear fiber loop. The nonlinear fiber loop only transmits the write/read pulses and suppresses any background present from the laser or amplifier between the pulses as only the pulses have a high enough intensity to induce a nonlinear phase shift in the fiber loop. Hence, only the pulses are transmitted and the low-intensity background is reflected by the loop. A second EDFA after the loop boosts the signal to peak powers of several Watts. To minimize the effect of white noise, bandpass filters with a bandwidth of around 0.5 nm are implemented after every amplification step by the EDFAs. The one-laser setup, the low-noise EDFA combined with filtering, and the nonlinear loop are all used in the setup to maximize the signal-to-noise ratio that ensures the highest efficiency of the storage process with the least amount of distortions. The write and read pulses are coupled into the photonic chip using lensed fibers and the average coupled on-chip power was around 7 dBm.

The data signal is split with a 50/50 coupler and coupled from both sides into the photonic chip with an average power of around −10 dBm. From one side, the write/read pulses are combined with the data pulses and coupled via the same lensed fiber into the chip. Additional 50/50 couplers are used in the data path to make sure the data pulses and the write/read pulses overlap in the middle of the waveguide. Circulators are used on both sides of the chip to route the transmitted data signal to a two-channel fast oscilloscope. Two

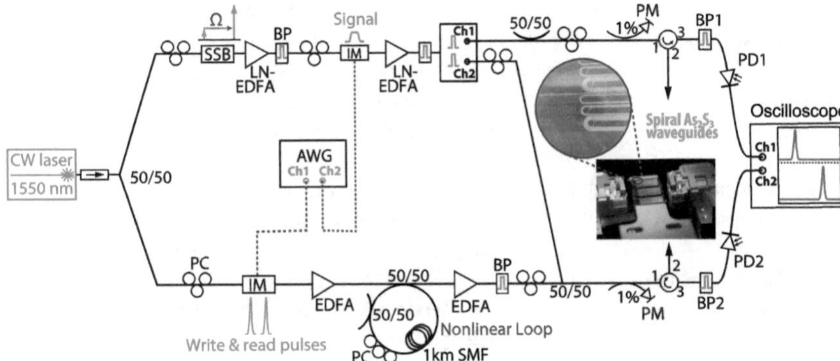

Figure 6: Experimental setup. Continuous wave (CW) laser, continuous-wave laser; 50/50: 50/50 optical fiber coupler; PC, polarization controller; SSB, single-sideband modulator; EDFA, erbium-doped fiber amplifier; LN-EDFA, low-noise EDFA; IM, intensity modulator; BP, bandpass filter; AWG, arbitrary waveform generator; CH 1/2, Channel 1/2; SMF, standard single-mode fiber; PM, power meter; PD, photodetector.

narrowband filters with a bandwidth of 3 GHz are used before the fast photodetectors (bandwidth 12 GHz) to filter residual write/read pulses.

4.2 Storage medium

The storage medium for the nonreciprocal light storage is a chalcogenide rib waveguide [37, 46]. The chalcogenide As_2S_3 thin film of around 850 nm is deposited on a thermal oxide silicon wafer with a variation of the film thickness below 5% [47]. Photolithography is used to pattern the waveguide structures which are etched into the thin film using inductively coupled plasma dry etching with a mixture of CHF_3, O_2, and Ar. The dimensions of the rib structure are 850 nm by 2.2 μm with a 50% etch depth and an overall length of 22 cm. The waveguide is arranged in a spiral to reduce the overall footprint to around 16 mm^2. The bend radius of the spiral is around 200 μm to ensure that there is no additional bending loss introduced for the fundamental optical mode as well as the acoustic mode.

The chalcogenide glass As_2S_3 is sandwiched between a silica substrate and a silica top-cladding to ensure guiding of the acoustic as well as the optical mode [36]. The difference in refractive index of silica $n \approx 1.4$ and the chalcogenide rib waveguides $n \approx 2.4$ guarantees tight confinement of the optical mode, whereas the difference in sound velocity of around 3400 m/s prevents leakage of the acoustic mode into the substrate or cladding which enables strong overlap between the two respective modes. The silica top-cladding is deposited using sputtering.

Light is coupled in and out of the chip using lensed fibers with a roughly 2 μm focus spot size. The coupling loss per facet is around 4 dB. The polarization is adjusted so light is coupled into the fundamental TE mode of the waveguide which is the mode with the lowest loss.

Acknowledgments: This work was supported by the Australian Research Council (ARC) through Laureate Fellowship (FL120100029), Center of Excellence CUDOS (CE110001018), ARC 2020 Discovery Project (DP200101893), ARC Linkage grant (LP170100112), and U.S. Office of Naval Research Global (ONRG) (N62909-18-1-2013). The authors acknowledge the support of the ANFF ACT.
Author contributions: All the authors have accepted responsibility for the entire content of this submitted manuscript and approved submission.
Research funding: This work was supported by the Australian Research Council (ARC) through Laureate Fellowship (FL120100029), Center of Excellence CUDOS (CE110001018), ARC 2020 Discovery Project (DP200101893), ARC Linkage grant (LP170100112), and U.S. Office of Naval Research Global (ONRG) (N62909-18-1-2013).
Conflict of interest statement: The authors declare no conflicts of interest regarding this article.

References

[1] D. Jalas, A. Petrov, M. Eich, et al., "What is – and what is not – an optical isolator," *Nat. Photonics*, vol. 7, pp. 579–582, 2013.
[2] D. L. Sounas and A. Alù, "Non-reciprocal photonics based on time modulation," *Nat. Photonics*, vol. 11, pp. 774–783, 2017.
[3] C. Caloz, A. Andrea, S. Tretyakov, D. Sounas, K. Achouri, and Z. L. Deck-Léger, "Electromagnetic nonreciprocity," *Phys. Rev. Appl.*, vol. 10, p. 1, 2018.
[4] Y. Shoji, T. Mizumoto, H. Yokoi, I. Wei Hsieh, and R. M. Osgood, "Magneto-optical isolator with silicon waveguides fabricated by direct bonding," *Appl. Phys. Lett.*, vol. 92, pp. 2–5, 2008.
[5] L. Bi, J. Hu, P. Jiang, et al., "On-chip optical isolation in monolithically integrated non-reciprocal optical resonators," *Nat. Photonics*, vol. 5, pp. 758–762, 2011.
[6] D. Huang, P. Pintus, C. Zhang, et al., "Dynamically reconfigurable integrated optical circulators," *Optica*, vol. 4, p. 23, 2017.
[7] Z. Yu and S. Fan, "Complete optical isolation created by indirect interband photonic transitions," *Nat. Photonics*, vol. 3, pp. 91–94, 2009.
[8] H. Lira, Z. Yu, S. Fan, and M. Lipson, "Electrically driven nonreciprocity induced by interband photonic transition on a silicon chip," *Phys. Rev. Lett.*, vol. 109, pp. 1–5, 2012.
[9] M. S. Kang, A. Butsch, and P. S. J. Russell, "Reconfigurable light-driven opto-acoustic isolators in photonic crystal fibre," *Nat. Photonics*, vol. 5, pp. 549–553, 2011.
[10] D. B. Sohn, S. Kim, and G. Bahl, "Time-reversal symmetry breaking with acoustic pumping of nanophotonic circuits," *Nat. Photonics*, vol. 12, pp. 91–97, 2018.
[11] E. Verhagen and A. Alù, "Optomechanical nonreciprocity," *Nat. Phys.*, vol. 13, pp. 922–924, 2017.
[12] M. Ali Miri, F. Ruesink, E. Verhagen, and A. Alù, "Optical nonreciprocity based on optomechanical coupling," *Phys. Rev. Appl.*, vol. 7, pp. 1–20, 2017.

[13] S. Manipatruni, J. T. Robinson, and M. Lipson, "Optical nonreciprocity in optomechanical structures," *Phys. Rev. Lett.*, vol. 102, p. 213903, 2009.
[14] M. Hafezi and P. Rabl, "Optomechanically induced non-reciprocity in microring resonators," *Optic Express*, vol. 20, p. 7672, 2012.
[15] X. W. Xu and Y. Li, "Optical nonreciprocity and optomechanical circulator in three-mode optomechanical systems," *Phys. Rev. A Atom. Mol. Opt. Phys.*, vol. 91, pp. 1–8, 2015.
[16] F. Ruesink, M.-A. Miri, A. Alù, and E. Verhagen, "Nonreciprocity and magnetic-free isolation based on optomechanical interactions," *Nat. Commun.*, vol. 7, p. 13662, 2016.
[17] Z. Shen, Y.-L. Zhang, Y. Chen, et al., "Experimental realization of optomechanically induced non-reciprocity," *Nat. Photonics*, vol. 10, pp. 657–661, 2016.
[18] K. Fang, J. Luo, A. Metelmann, et al., "Generalized non-reciprocity in an optomechanical circuit via synthetic magnetism and reservoir engineering," *Nat. Phys.*, vol. 13, pp. 465–471, 2017.
[19] Z. Shen, Y.-L. Zhang, Y. Chen, et al., "Reconfigurable optomechanical circulator and directional amplifier," *Nat. Commun.*, vol. 9, p. 1797, 2018.
[20] N. A. Estep, D. L. Sounas, and A. Alù, "Magnetless microwave circulators based on spatiotemporally modulated rings of coupled resonators," *IEEE Trans. Microw. Theor. Tech.*, vol. 64, pp. 502–518, 2016.
[21] G. A. Peterson, F. Lecocq, K. Cicak, R. W. Simmonds, J. Aumentado, and J. D. Teufel, "Demonstration of efficient nonreciprocity in a microwave optomechanical circuit," *Phys. Rev. X*, vol. 7, p. 031001, 2017.
[22] S. Barzanjeh, M. Wulf, M. Peruzzo, et al., "Mechanical on-chip microwave circulator," *Nat. Commun.*, vol. 8, p. 953, 2017.
[23] N. R. Bernier, L. D. Tóth, A. Koottandavida, et al., "Nonreciprocal reconfigurable microwave optomechanical circuit," *Nat. Commun.*, vol. 8, 2017. https://doi.org/10.1038/s41467-017-00447-1.
[24] X. Huang and S. Fan, "Complete all-optical silica fiber isolator via stimulated Brillouin scattering," *J. Lightwave Technol.*, vol. 29, pp. 2267–2275, 2011.
[25] B. J. Eggleton, C. G. Poulton, P. T. Rakich, M. J. Steel, and G. Bahl, "Brillouin integrated photonics," *Nat. Photonics*, vol. 13, pp. 664–677, 2019.
[26] C. G. Poulton, R. Pant, A. Byrnes, S. Fan, M. J. Steel, and B. J. Eggleton, "Design for broadband on-chip isolator using stimulated Brillouin scattering in dispersion-engineered chalcogenide waveguides," *Opt. Express*, vol. 20, p. 21235, 2012.
[27] E. A. Kittlaus, N. T. Otterstrom, P. Kharel, S. Gertler, and P. T. Rakich, "Non-reciprocal interband Brillouin modulation," *Nat. Photonics*, vol. 12, pp. 613–620, 2018.
[28] J. H. Kim, M. C. Kuzyk, K. Han, H. Wang, and G. Bahl, "Non-reciprocal Brillouin scattering induced transparency," *Nat. Phys.*, vol. 11, pp. 275–280, 2015.
[29] C.-H. Dong, Z. Shen, C.-L. Zou, Y.-L. Zhang, W. Fu, and G.-C. Guo, "Brillouin-scattering-induced transparency and non-reciprocal light storage," *Nat. Commun.*, vol. 6, p. 6193, 2015.
[30] J. H. Kim, S. Kim, and G. Bahl, "Complete linear optical isolation at the microscale with ultralow loss," *Sci. Rep.*, vol. 7, pp. 1–9, 2017.
[31] R. W. Boyd, *Nonlinear Optics*, Acad. Press, 2003.
[32] M. Merklein, B. Stiller, K. Vu, S. J. Madden, and B. J. Eggleton, "A chip-integrated coherent photonic–phononic memory," *Nat. Commun.*, vol. 8, p. 574, 2017.
[33] B. Stiller, M. Merklein, K. Vu, et al., "Cross talk-free coherent multi-wavelength Brillouin interaction," *APL Photonics*, vol. 4, p. 040802, 2019.
[34] Z. Zhu, D. J. Gauthier, and R. W. Boyd, "Stored light in an optical fiber via stimulated Brillouin scattering," *Science*, vol. 318, pp. 1748–50, 2007.
[35] M. Merklein, B. Stiller, and B. J. Eggleton, "Brillouin-based light storage and delay techniques," *J. Opt.*, vol. 20, p. 083003, 2018.
[36] C. G. Poulton, R. Pant, and B. J. Eggleton, "Acoustic confinement and stimulated Brillouin scattering in integrated optical waveguides," *J. Opt. Soc. Am. B*, vol. 30, pp. 2657–2664, 2013.
[37] R. Pant, C. G. Poulton, D.-Y. Choi, et al., "On-chip stimulated Brillouin scattering," *Opt. Express*, vol. 19, pp. 8285–8290, 2011.
[38] V. Fiore, Y. Yang, M. C. Kuzyk, R. Barbour, L. Tian, and H. Wang, "Storing optical information as a mechanical excitation in a silica optomechanical resonator," *Phys. Rev. Lett.*, vol. 107, pp. 1–5, 2011.
[39] V. Fiore, C. Dong, M. C. Kuzyk, and H. Wang, "Optomechanical light storage in a silica microresonator," *Phys. Rev.*, vol. 87, p. 023812, 2013.
[40] Y. Okawachi, M. Bigelow, J. Sharping, et al., "Tunable all-optical delays via Brillouin slow light in an optical fiber," *Phys. Rev. Lett.*, vol. 94, p. 153902, 2005.
[41] K. Y. Song, M. Herráez, and L. Thévenaz, "Observation of pulse delaying and advancement in optical fibers using stimulated Brillouin scattering," *Opt. Express*, vol. 13, pp. 82–88, 2005.
[42] K. Y. Song, K. Lee, and S. B. Lee, "Tunable optical delays based on Brillouin dynamic grating in optical fibers," *Opt Express*, vol. 17, pp. 10344–9, 2009.
[43] S. Preussler, K. Jamshidi, A. Wiatrek, R. Henker, C.-A. Bunge, and T. Schneider, "Quasi-light-storage based on time-frequency coherence," *Opt Express*, vol. 17, pp. 15790–15798, 2009.
[44] S. Chin and L. Thévenaz, "Tunable photonic delay lines in optical fibers," *Laser Photonics Rev.*, vol. 6, pp. 724–738, 2012.
[45] B. Stiller, M. Merklein, C. Wolff, et al., "Coherently refreshing hypersonic phonons for light storage," *Optica*, vol. 7, p. 492, 2020.
[46] S. J. Madden, D.-Y. Choi, D. A. Bulla, et al., "Long, low loss etched As(2)S(3) chalcogenide waveguides for all-optical signal regeneration," *Opt. Express*, vol. 15, pp. 14414–14421, 2007.
[47] A. Zarifi, B. Stiller, M. Merklein, et al., "Highly localized distributed Brillouin scattering response in a photonic integrated circuit," *APL Photonics*, vol. 3, p. 036101, 2018.

David Barton III*, Jack Hu, Jefferson Dixon, Elissa Klopfer, Sahil Dagli, Mark Lawrence* and Jennifer Dionne*

High-*Q* nanophotonics: sculpting wavefronts with slow light

https://doi.org/10.1515/9783110710687-006

Abstract: Densely interconnected, nonlinear, and reconfigurable optical networks represent a route to high-performance optical computing, communications, and sensing technologies. Dielectric nanoantennas are promising building blocks for such architectures since they can precisely control optical diffraction. However, they are traditionally limited in their nonlinear and reconfigurable responses owing to their relatively low-quality factor (*Q*-factor). Here, we highlight new and emerging design strategies to increase the *Q*-factor while maintaining control of optical diffraction, enabling unprecedented spatial and temporal control of light. We describe how multipolar modes and bound states in the continuum increase *Q* and show how these high-*Q* nanoantennas can be cascaded to create almost limitless resonant optical transfer functions. With high-*Q* nanoantennas, new paradigms in reconfigurable wavefront-shaping, low-noise, multiplexed biosensors and quantum transduction are possible.

Keywords: high-*Q*; slow light; wavefront manipulation.

***Corresponding authors: David Barton III and Mark Lawrence,** Department of Materials Science and Engineering, Stanford University, Stanford, CA 94305, USA, E-mail: dbarton@stanford.edu (D. Barton), markl89@stanford.edu (M. Lawrence). https://orcid.org/0000-0002-5409-8512 (D. Barton); and **Jennifer Dionne,** Department of Materials Science and Engineering, Stanford University, Stanford, CA 94305, USA; and Department of Radiology, Stanford University, Stanford, CA 94305, USA, E-mail: jdionne@stanford.edu
Jack Hu, Elissa Klopfer and Sahil Dagli, Department of Materials Science and Engineering, Stanford University, Stanford, CA 94305, USA
Jefferson Dixon, Department of Materials Science and Engineering, Stanford University, Stanford, CA 94305, USA; and Department of Mechanical Engineering, Stanford University, Stanford, CA 94305, USA

1 Introduction

Networks lie at the heart of both natural and engineered systems. The useful information density of a network scales with both the number of elements and, importantly, the number of connections. For optimal optical communications networks, considerable effort has been devoted to miniaturizing photonic components, in order to increase the number of *elements* in the network. Such progress is akin to the semiconductor industry's strides in miniaturizing transistors to increase the power of digital computers. However, unlike electronic chip design, optical engineering is bound by the diffraction limit; plasmonic components have promised to overcome this limit [1–9] but generally suffer from strong absorption. Here, we describe an alternate strategy for scaling photonic networks: increasing the number of *connections* between wavelength-scale components (i.e., components at the diffraction limit), thereby performing highly sophisticated operations. Though the diffraction limit sets a bound on the number of optical elements that can be included in the network, there is no limit on the number of diffracted channels. With this in mind, can we construct highly efficient compound optical devices with nanoantennas? And more importantly, can we dynamically modify their operation in time with electro-optic or nonlinear effects? We believe such photonic networks are possible with high-quality-factor (high-*Q*) nanoantennas and metasurfaces.

Dielectric nanoantennas, or multipolar Mie resonators, represent ideal diffraction-limited optical elements for generating, manipulating, and modulating light waves (Figure 1i). These nanoantennas act as quasi-point sources whose scattering can be understood, much like atoms and molecules, as a superposition of electric and magnetic multipolar modes [10]. The electric and magnetic modes of nanoantennas present a library of available scattering or radiation patterns that can be combined spatially to perform nearly arbitrary transformations to the incident light [11–17]. One drawback of Mie resonators is that light typically only interacts with them over a few hundred femtoseconds. Combined with the usually very weak

This article has previously been published in the journal Nanophotonics. Please cite as: D. Barton III, J. Hu, J. Dixon, E. Klopfer, S. Dagli, M. Lawrence and J. Dionne "High-*Q* nanophotonics: sculpting wavefronts with slow light" *Nanophotonics* 2021, 10. DOI: 10.1515/nanoph-2020-0510.

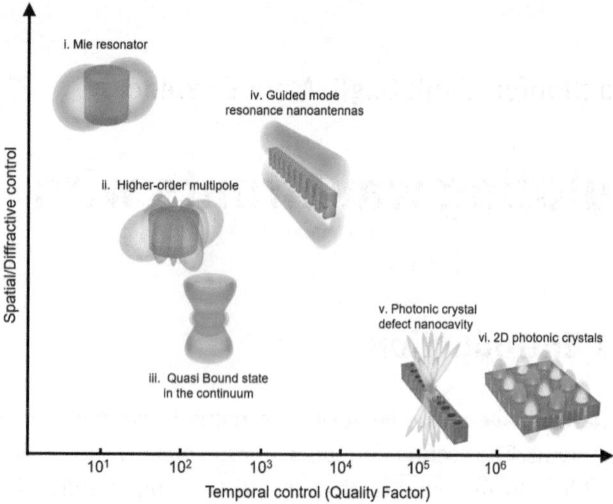

Figure 1: Landscape of dielectric nanophotonic structures for temporal and spatial control of light. From left to right: Conventional Mie resonator used for wavefront shaping in metasurfaces. Higher order multipoles increase the quality factor, at the expense of more complex radiation patterns. Interfering Mie-like and Fabry-Perot–like resonances increases the quality factor compared to dipole resonances, forming a quasi-bound state in the continuum (Q value from the study by Koshelev et al. [31]). Utilizing one translational degree of freedom, coupling to guided mode resonances within a nanoantenna can generate high-Q dipolar-like resonances for wavefront shaping, with a quality factor tuned independently of the metasurface transfer function (Q value taken from the study by Lawrence et al. [30]). Photonic crystal defect nanocavities increase the Q to 10^6 using photonic mirrors but lose the ability to controllably shape far-field radiation (Q from the study by Quan and Loncar [49]). Photonic crystals with two spatial degrees of freedom generate ultrahigh-quality factors whose resonant diffraction is geometrically set (Q from the study by Jin et al. [50]).

light-matter coupling strengths at the material level, this fleeting interaction often limits such devices to passive applications; consequently, generating efficient and active nanoantennas is an active area of research [18–20].

Optical modes appear spectrally as resonances, whose full width at half maximum is inversely proportional to the average lifetime of light within the antenna and local electric field enhancement, important attributes for sensing [21] and nonlinear optics [22, 23]. This lifetime is quantified by the quality factor Q and is composed of various limiting loss mechanisms like radiation loss, material absorption, and fabrication imperfections. The increased lifetime and reduced spectral bandwidth of high-Q resonances are critical to the operation of on-chip photonics. In ring resonators and photonic crystal defect cavities, for example, high-Q modes reduce the required optical or electrical budget for nonlinear optical transformations [24, 25] required for communications technologies. In many ways, these on-chip cavities are the antithesis of Mie scatterers, critically relying on input/output channels being confined to single- or few-mode coupling through a small number of waveguides. By limiting the configurations to 2 or 4 ports and evanescent weak coupling, generating high-quality factors is comparatively easier than in structures that communicate with the entire radiation continuum, such as optical nanoantennas. Therefore, the density of node connections in existing on-chip high-Q cavity designs is extremely limited.

Mie modes are designed with radiation in mind, so their interaction and coupling to free space can be tailored to increase the Q; in parallel, coupling to other antenna modes can be engineered to be as dense as desired. In simple nanophotonic systems, resonant lifetimes tend to increase with multipole order, consistent with a decrease in the angular density of radiation channels and an increased field contribution from large Fourier components (see Figure 1ii). Consequently, the lower order dipolar modes that are commonly used in diffractive metasurface designs necessarily have relatively low quality (Q) factors, of the order ~10. However, there is no physical limitation preventing much higher quality factors. That is to say, highly resonant dipole-like emitters can exist with proper design. After all, nanoantenna modes consist of both radiative (low momentum) and nonradiative (high momentum) waves. Traditionally, these components are highly correlated, but by carefully addressing their relative contributions, the entire parameter space of optical state variables can potentially be manipulated in both space and time, independently. Previous work has already shown that the superposition of two or more multipoles within a subwavelength object can have profound effects on far-field scattering. For example, tailoring a nanoantenna so as to adjust the relative strength and spectral overlap of its electric and magnetic dipole modes produces highly directional scattering, including zero-backscattering and zero-forward scattering conditions [26–28]. Generalizing this notion to device designs composed of point-like sources that can controllably release a very small amount of leakage into interesting spatial distributions will allow for simultaneous control of far-field optics, required for high-density integration, and near-field light localization useful for reconfigurability and sensitive signal readouts. This opportunity requires sculpting the three-dimensional (3D) Fourier map of a nanoantenna in order to fully decouple the directionality and strength of radiation loss.

2 Design methodologies for sculpting free-space light with resonant nanoantennas

Achieving resonant, or high Q, scattering in multipolar nanoantennas is an active area of research [29–31]. Overlapping two or more multipoles (Figure 1iii) whose radiation patterns spatially overlap can lead to strong destructive interference that suppresses far-field scattering and increases the resonance quality factor [31–33]. The moderate enhancement in lifetime over subwavelength volumes may be useful in nonlinear integrated optics. However, current demonstrations of localized quasi-BICs have relied on the combination of distinct spatial modes. The incomplete cancelation of the multipoles not only places bounds on the achievable Q factor, tied to the intrinsic properties of the uncoupled modes, but it will also be very difficult to engineer the corresponding far-field pattern without affecting the Q factor. Additionally, constraining systems to single antennas may reduce the degrees of freedom available for far-field manipulation. The radiation profiles available to multipolar nanoantennas are intrinsically limited by the destructively interfering modes, requiring particular illumination strategies to couple into and out of the modes of interest. Future metasurface-based devices, for example, will rely on judicious design methodologies that allow spatially varying transfer functions composed of many nanoantennas that operate at the same frequency but with variable phase or amplitude.

Rather than relying on the overlap of multiple modes within a single nanoantenna, the resonant lifetime can be increased by capitalizing on the additional degrees of freedom afforded by periodic structures [34, 35]. Periodic systems composed of subwavelength objects provide opportunities for high-quality-factor transmission or diffraction. Two-dimensionally periodic nanostructures whose band structure contains a flat region can induce light localization [36–38], although their stability to structural perturbations is limited. Using degrees of freedom like mode polarization, mode type (electric vs. magnetic) [39], spatial symmetry [40, 41], and others, coupling to radiation can be significantly reduced [42]. Otherwise fully bound modes can leak out to their environment, forming a quasi-bound state. Perhaps the most robust method to date in reducing free-space radiation is in photonic crystal membranes (Figure 1vi), where bound states can be created and destroyed in a deterministic manner and ultrahigh-quality factors have been observed [43]. Here, the interplay between multiple spatial degrees of freedom is required to generate the modes of interest. Accordingly, bound modes such as these exist at particular points in k-space that are not easily moved; their utility in the far field is therefore limited. While these designs can generate ultrahigh-quality factors, they are at the expense of limited momentum-space tunability.

Recent results using bound modes with lower dimensionality or nonlocality have begun to combine resonant optical cavities with wavefront shaping [30, 44]. Here, light trapping can occur along one direction, while the scattering in an orthogonal plane can mimic a dipole requisite for phase-gradient metasurfaces (Figure 1iv). Resonant optical responses within the diffraction plane can be decoupled (that is, separately designed) from the light trapping in this configuration. Exciting avenues of exploration exist with this scheme, spanning tunable beam steering, lensing [45], and nonlinear nanophotonics [46]. However, to date, the design principles exploited require one semi-infinite (or at least multimicron) dimensions for resonant mode coupling [30, 47, 48]. This property intrinsically limits the optical transfer functions to also act one-dimensionally. An open challenge is achieving resonant scattering of point-like sources with complete control over existing wavefronts; quasi-bound or guided states with long lifetimes embedded within a structure capable of arbitrary wavefront shaping in two dimensions would be a transformative development amenable to efficient reconfigurable and tunable devices and will surely be a focus of future research.

Even with new design methodologies, achieving theoretically predicted quality factors can present challenges. Intrinsic material absorption and fabrication-induced scattering losses will naturally reduce the achievable Q. In periodic structures, long-range order, illumination coherence, and device size all contribute to the achievable Q factors. As these modes exist with particular radiation patterns and mode locations in k-space, the careful design of illumination and collection configurations is critical to observe these quasi-bound states. Furthermore, recently developed designs in infinitely two-dimensionally periodic structures such as biperiodic unit cells [51] or other perturbations that couple otherwise dark modes to the radiation continuum all depend on small, perturbative, differences between metasurface elements. The previously derived scaling relations, where the quality factor has an inverse square relationship with the perturbation size [52], require uniformity that is substantially smaller than the perturbation of interest. At optical frequencies, this presents some practical limitations on the devices that can be made. Notions of topologically robust high-Q modes investigated in photonic crystals have shown promise [50], but their translation to lower dimensionality remains an open question. Further developments in robust design strategies will be important for the implementation of these ultrahigh-Q designs in deployable devices.

3 Summary and outlook

The strong resonant interaction and low loss of high-Q nanoantennas make them particularly well suited for application in tunable and nonlinear photonic devices [53, 54]. Individual Mie-resonant structures, for example, can be employed as light sources in a footprint several times smaller than the wavelength of the light being emitted, making them attractive for integration in dense photonic networks and system-on-a-chip sensors [55]. When arranged the form of a metasurface, such resonators may also assist in structuring coherent light, forming "metasurface lasers" [56] that are capable of producing new forms of chiral laser light [57] and accessing subwavelength lasing modes at low pump powers [38, 58]. Access to optical nonlinearities has also been used to form optical switches [59], achieve asymmetric transmission [60], dynamic reciprocity [61], and break Lorentz reciprocity for application in optical isolators and circulators. These Mie-resonant structures can also enhance the nonlinear response of materials with which they are interfaced when the material of interest is placed sufficiently within the near field of the Mie resonance; this property has been especially useful in studying the nonlinear properties of novel two-dimensional materials [62]. It is also possible to interface these structures with a host of other media to lend a strong resonant interaction at any wavelength desirable, from the IR through the UV [63].

In addition to nonlinear wavefront shaping, these resonant antennas can also form the basis of active and reactive optical devices [45, 64, 65]. For example, resonant nanophotonic antennas have allowed for wavelength conversion and modulation [66], quantum state transduction [67], and myriad other technologies for on-chip optical communication and computation. These technologies rely on the additional light-matter interaction and the resonant linewidth to make practical devices. Embedding these antennas into phase-gradient metasurfaces could enable wholly new device designs with complete and tunable control over *free-space* optical wavefronts with an optical or electric field. Furthermore, access to diffraction in a resonant manner could allow for multiplexed sensing modalities, in which an external change in the refractive index or other parameter can impact the diffraction. Optical biosensors relying on high-quality factors have already been demonstrated in transmissive and reflective devices [68], so extending the radiation channels available for sensing promises to increase sensitivity and multiplexing abilities. The local electromagnetic field can also be designed to be sensitive to particular properties of liquid analytes. For example, chiral light-matter interactions can be engineered to have enhanced near-field interactions with observables in the far field [69]. In addition, enhanced sensitivity to vibrational modes of molecules can be used for molecular identification [70]. Resonant nanophotonics that can be arbitrarily multiplexed in energy and momentum will provide increased degrees of freedom for sensitive detection and sensing of materials both in the near field and far field.

Integrated photonics researchers are currently focused on scaling up planar, waveguide-based chips, striving to revolutionize applications spanning light speed matrix multiplication to point-of-care multipathogen medical tests. But, foreshadowed by emerging 3D electronic circuit designs and by progress in artificial neural networks, photonic systems could soon also benefit from less intuitive and highly interconnected network architectures. High-Q nanoscale wavefront shaping structures will undoubtedly play a cornerstone role in bringing this vision to fruition. Developments in device design, material synthesis, nanofabrication, and characterization will need to be combined if the energy scaling requirements for real-world applications are to be met. We envision the intersection of these efforts transforming applications from optical sensing to quantum and classical computing with free-space optical devices.

Author contribution: All the authors have accepted responsibility for the entire content of this submitted manuscript and approved submission.

Research funding: We gratefully acknowledge support from an AFOSR grant (grant no. FA9550-20-1-0120), which supported the work as well as the salaries of D.B, M.L, and J.A.D. We also acknowledge funding from the Moore Foundation, which supported J.H.'s salary through a Moore Inventors Fellowship under grant number 6881. J.D. is supported by a Kodak Stanford Graduate Fellowship. E.K. acknowledges support from a National Science Foundation Graduate Research Fellowship Program under grant no. DGE-1656518. S.D. is supported by the Department of Defense (DOD) through the National Defense Science and Engineering (NDSEG) Fellowship Program.

Conflict of interest statement: The authors declare no conflicts of interest regarding this article.

References

[1] S. Maier and H. Atwater, "Plasmonics: localization and guiding of electromagnetic energy in metal/dielectric structures," *J. Appl. Phys.*, vol. 98, no. 1, p. 011101, 2005.

[2] S. Maier, M. Brongersma, P. Kik, S. Meltzer, A. Requicha, and H. Atwater, "Plasmonics – a route to nanoscale optical devices," *Adv. Mater.*, vol. 13, no. 19, p. 1501–1505, 2001.

[3] J. Schuller, E. Barnard, W. Cai, Y. Jun, J. White, and M. Brongersma, "Plasmonic for extreme light concentration and manipulation," *Nat. Mater.*, vol. 9, p. 193–204, 2010.

[4] N. Engheta, A. Salandrino, and A. Alu, "Circuit elements at optical frequencies: nanoinductors, nanocapacitors, and nanoresistors," *Phys. Rev. Lett.*, vol. 95, no. 9, p. 095504, 2005.

[5] N. Engheta, "Circuits with light at nanoscales: optical nanocircuits inspired by metamaterials," *Science*, vol. 317, no. 5845, pp. 1698–1702, 2007.

[6] S. Bozhevolnyi, V. Volkov, E. Devaux, J. Laluet, and T. Ebbesen, "Channel plasmon subwavelength waveguide components including interferometers and ring resonators," *Nature*, pp. 508–511, 2006, https://doi.org/10.1038/nature04594.

[7] A. Boltasseva, T. Nikolajsen, K. Leosson, K. Kjaer, M. Larsen, and S. Bozhevolnyi, "Integrated optical components utilizing long-range surface plasmon polaritons," *J. Light. Technol*, vol. 23, no. 1, p. 413–422, 2005.

[8] J. Dionne, K. Diest, L. Sweatlock, and H. Atwater, "PlasMOStor: a metal– oxide– Si field effect plasmonic modulator," *Nano Lett.*, pp. 897–902, 2009, https://doi.org/10.1021/nl803868k.

[9] M. Noginov, G. Zhu, A. Belgrave, et al., "Demonstration of a spaser-based nanolaser," *Nature*, pp. 1110–1112, 2009, https://doi.org/10.1038/nature08318.

[10] A. Kuznetsov, A. Miroshnichenko, M. Brongersma, Y. Kivshar, and B. Luk'yanchuk, "Optically resonant dielectric nanostructures," *Science*, vol. 354, no. 6314, p. 846, 2016.

[11] A. C. Overvig, S. Shrestha, S. Malek, et al., "Dielectric metasurfaces for complete and independent control of the optical amplitude and phase," *Light Sci. Appl.*, pp. 1–12, https://doi.org/10.1038/s41377-019-0201-7.

[12] N. Yu, P. Genevet, M. Kats, et al., "Light propagation with phase discontinuities: generalized laws of reflection and refraction," *Science*, pp. 333–337, 2011, https://doi.org/10.1126/science.1210713.

[13] D. Lin, Fan P., Hasman E., and Brongersma M., "Dielectric gradient metasurface optical elements," *Science*, pp. 298–302, 2014, https://doi.org/10.1126/science.1253213.

[14] X. Yin, Ye Z., RhoJ., WangY., and Zhang X., "Photonic spin Hall effect at metasurfaces," *Science*, pp. 1405–1407, 2013, https://doi.org/10.1126/science.1231758.

[15] G. Zheng, Muhlenbernd H., Kenney M., Li G., T. Zentgraf, and S. Zhang, "Metasurface holograms reaching 80% efficiency," *Nat. Nanotechnol.*, pp. 308–312, 2015, https://doi.org/10.1038/nnano.2015.2.

[16] J. Mueller, Rubin N., Devlin R., Groever B., and F. Capasso, "Metasurface polarization optics: independent phase control of arbitrary orthogonal states of polarization," *Phys. Rev. Lett.*, 2017.

[17] Y. Yu, Zhu A., Paniagua-Dominguez R., Luk'yanchuk B., KuznetsovA., "High-transmission dielectric metasurface with 2π phase control at visible wavelengths," *Laser Photon. Rev.* pp. 412–418, 2016, https://doi.org/10.1002/lpor.201500041.

[18] Y. Yao, Shankar R., Kats M., et al., "Electrically tunable metasurface perfect absorbers for ultrathin mid-infrared optical modulators," *Nano Lett.* pp. 6526–6532, 2014, https://doi.org/10.1021/nl503104n.

[19] Y. Huang, Lee H., Sokhoyan R., et al., "Gate-tunable conducting oxide metasurfaces," *Nano Lett.*, pp. 5319–5325, 2016, https://doi.org/10.1021/acs.nanolett.6b00555.

[20] J. Zhang, Wei X., Ruklenko I., Chen H., and Zhu W., "Electrically tunable metasurface with independent frequency and amplitude modulations," *ACS Photonics*, pp. 265–271, 2019, https://doi.org/10.1021/acsphotonics.9b01532.

[21] S. Law, Yu L., Rosenberg A., and Wasserman D., "All-semiconductor plasmonic nanoantennas for infrared sensing," *Nano Lett.*, pp. 4569–4574, 2013, https://doi.org/10.1021/nl402766t.

[22] W. Ye, Zeuner F., Li X., et al., "Spin and wavelength multiplexed nonlinear metasurface holography," *Nat. Commun.*, pp. 1–7, 2016.

[23] N. Nookala, Lee J., Tymchenko M., et al., "Ultrathin gradient nonlinear metasurface with a giant nonlinear response," *Optica*, pp. 283–288, 2016, https://doi.org/10.1364/optica.3.000283.

[24] B. Peng, Ozdemir S., Lei F., et al., "Parity-time-symmetric whispering gallery microcavities," *Nat. Phys.*, pp. 394–398, 2014, https://doi.org/10.1038/nphys2927.

[25] B. Stern, Ji X., Okawachi Y., Gaeta A., and Lipson M., "Battery-operated integrated frequency comb generator," *Nature*, pp. 401–405, 2018, https://doi.org/10.1038/s41586-018-0598-9.

[26] S. Person, Jain M., Lapin Z., Saenz J., Wicks G., and Novotny L., "Demonstration of zero optical backscattering from single nanoparticles," *Nano Lett.*, vol. 13, no. 4, 2013, 1806–1809.

[27] K. Yao and Liu Y., "Controlling electric and magnetic resonances for ultracompact nanoantennas with tunable directionality," *ACS Photonics*, vol. 3, no. 6, pp. 953–963, 2016.

[28] T. Shibanuma, Albella P., and Maier S., "Unidirectional light scattering with high efficiency at optical frequencies based on low-loss dielectric nanoantennas," *Nanoscale*, vol. 8, no. 29, pp. 14184–14192, 2016.

[29] F. Monticone and Alu A., "Embedded photonic eigenvalues in 3D nanostructures," *Phys. Rev. Lett.*, vol. 112, no. 21, p. 213903, 2014.

[30] M. Lawrence, Barton D. III, Dixon J., et al., "High quality factor phase gradient metasurfaces," *Nat. Nanotechnol.*, 2020, https://doi.org/10.1038/s41565-020-0754-x.

[31] K. Koshelev, Kruk S., Melik-Gaykazyan E., et al., "Subwavelength dielectric resonators for nonlinear nanophotonics," *Science*, pp. 288–292, 2020, https://doi.org/10.1126/science.aaz3985.

[32] T. Lepetit and Kante B., "Controlling multipolar radiation with symmetries for electromagnetic bound states in the continuum," *Phys. Rev. B*, 2014 90, no. 24, p. 241103.

[33] C. Wu, Arju N., Kelp G., et al., "Spectrally selective chiral silicon metasurfaces based on infrared Fano resonances," *Nat. Commun.*, pp. 1–9, 2014, https://doi.org/10.1038/ncomms4892.

[34] V. Fedotov, Rose M., Prosvirmin S., Papasimakis N., and Zheludev N., "Sharp trapped-mode resonances in planar metamaterials with a broken structural symmetry," *Phys. Rev. Lett.*, vol. 99, p. 147401, 2007.

[35] V. Savinov and Zheludev N., "High-quality metamaterial dispersive grating on the facet of an optical fiber," *Appl. Phys. Lett.*, 2017.

[36] K. Koshelev, S. Lepeshov, M. Liu, A. Bogdanov, and Y. Kivshar, "Asymmetric metasurfaces with high-Q resonances

governed by bound states in the continuum," *Phys. Rev. Lett.*, vol. 121, no. 19, p. 193903, 2018.

[37] P. Jeong, M. Goldflam, S. Campione, et al., "High quality factor toroidal resonances in dielectric metasurfaces," *ACS Photonics*, pp. 1699–1707, 2020, https://doi.org/10.1021/acsphotonics.0c00179.

[38] S. Ha, Y. Fu, N. Emani, et al., "Directional lasing in resonant semiconductor nanoantenna arrays," *Nat. Nanotechnol.*, pp. 1042–1047, 2018, https://doi.org/10.1038/s41565-018-0245-5.

[39] Y. Yang, I. Kravchenko, D. Briggs, and J. Valentine, "All-dielectric metasurface analogue of electromagnetically induced transparency," *Nat. Commun.*, pp. 1–7, 2014.

[40] A. Overvig, N. Yu, and A. Alu, "Chiral quasi-bound states in the continuum," 2006, ArXiv:2006.05484: 035434.

[41] A. Overvig, S. Malek, M. Carter, S. Shrestha, and N. Yu, "Selection rules for quasibound states in the continuum," *Phys. Rev. B*, 2020, https://doi.org/10.1364/cleo_qels.2020.fm2b.5.

[42] C. Hsu, B. Zhen, A. Stone, J. Joannopoulos, and M. Soljacic, "Bound states in the continuum," *Nat. Rev. Mater.*, pp. 1–3. https://doi.org/10.1038/natrevmats.2016.48.

[43] C. Hsu, B. Zhen, J. Lee, et al., "Observation of trapped light within the radiation continuum," *Nature*, pp. 188–191, 2013, https://doi.org/10.1038/nature12289.

[44] A. Overvig, S. Malek, and N. Yu, "Multifunctional nonlocal metasurfaces," 2020, ArXiv:2005.12332.

[45] E. Klopfer, M. Lawrence, D. Barton, J. Dixon, and J. Dionne, "Dynamic focusing with high quality-factor metalenses," *Nano Lett.*, vol. 20, no. 7, pp. 5127–5132, 2020.

[46] M. Lawrence, D. Barton, and J. Dionne, "Nonreciprocal flat optics with silicon metasurfaces," *Nano Lett.*, vol. 18, pp. 1104–1109, 2018.

[47] S. Kim, K. Kim, and J. Caoon, "Optical bound states in the continuum with nanowire geometric superlattices," *Phys. Rev. Lett.*, vol. 122, no. 18, p. 187402, 2019.

[48] S. Kim and J. Cahoon, "Geometric nanophotonics: light management in single nanowires through morphology," *Acc. Chem. Res.*, pp. 3511–3520, 2019, https://doi.org/10.1021/acs.accounts.9b00515.

[49] Q. Quan and M. Loncar, "Deterministic design of wavelength scale, ultra-high *Q* photonic crystal nanobeam cavities," *Opt. Express*, pp. 18529–18542, 2011, https://doi.org/10.1364/oe.19.018529.

[50] J. Jin, X. Yin, L. Ni, M. Soljacic, B. Zhen, and C. Peng, "Topologically enabled ultrahigh-Q guided resonances robust to out-of-plane scattering," *Nature*, pp. 501–504, 2019, https://doi.org/10.1038/s41586-019-1664-7.

[51] J. Hu, M. Lawrence, and J. Dionne, "High quality factor dielectric metasurfaces for ultraviolet circular dichroism spectroscopy," *ACS Photonics*, pp. 36–42, 2019, https://doi.org/10.1021/acsphotonics.9b01352.

[52] S. Fan, W. Suh, and J. Joannopoulos, "Temporal coupled-mode theory for the Fano resonance in optical resonators," *JOSA A*, pp. 569–572, 2003, https://doi.org/10.1364/josaa.20.000569.

[53] Y. Yang, W. Wang, A. Boulesbaa, et al., "Nonlinear Fano-resonant dielectric metasurfaces," *Nano Lett.*, pp. 7388–7393, 2015, https://doi.org/10.1021/acs.nanolett.5b02802.

[54] H. Liu, C. Guo, G. Vampa, et al., "Enhanced high-harmonic generation from an all-dielectric metasurface," *Nat. Phys.*, vol. 14, no. 10, pp. 1006–1010, 2018.

[55] G. Zograf, D. Ryabov, V. Rutckaia, et al., "Stimulated Raman scattering from mie-resonant subwavelength nanoparticles," *Nano Lett.*, pp. 5578–5791, 2020, https://doi.org/10.1021/acs.nanolett.0c01646.

[56] M. Lawrence and J. Dionne, "Nanoscale nonreciprocity via photon-spin-polarized stimulated Raman scattering," *Nat. Commun.*, pp. 1–8, 2019, https://doi.org/10.1038/s41467-019-11175-z.

[57] H. Sroor, Y. Huang, B. Sephton, et al., "High-purity orbital angular momentum states from a visible metasurface laser," *Nat. Photonics*, pp. 1–6, 2020, https://doi.org/10.1038/s41566-020-0623-z.

[58] A. Kodigala, T. Lepetit, Q. Gu, B. Bahari, Y. Fainman, and B. Kante, "Lasing action from photonic bound states in continuum," *Nature*, pp. 196–199, 2017, https://doi.org/10.1038/nature20799.

[59] M. Shcherbakov, P. Vabishchevich, A. Shorokhov, et al., "Ultrafast all-optical switching with magnetic resonances in nonlinear dielectric nanostructures," *Nano Lett.*, pp. 6985–6990, 2015, https://doi.org/10.1021/acs.nanolett.5b02989.

[60] B. Jin and C. Argyropoulos, "Self-induced passive nonreciprocal transmission by nonlinear bifacial dielectric metasurfaces," *Phys. Rev. Appl.*, p. 13054056, 2020, https://doi.org/10.1364/cleo_at.2020.jw2d.18.

[61] Y. Shi, Z. Yu, and S. Fan, "Limitations of nonlinear optical isolators due to dynamic reciprocity," *Nat. Photonics*, pp. 388–392, 2015, https://doi.org/10.1038/nphoton.2015.79.

[62] N. Bernhardt, K. Koshelev, S. White, et al., "Quasi-BIC resonant enhancement of second-harmonic generation in WS2 monolayers," *Nano Lett.*, pp. 5309–5314, 2020, https://doi.org/10.1021/acs.nanolett.0c01603.

[63] C. Zhang, S. Divitt, Q. Fan, et al., "Low-loss metasurface optics down to the deep ultraviolet region," *Light Sci. Appl.*, pp. 1–10, 2020, https://doi.org/10.1038/s41377-020-0287-y.

[64] S. Li, X. Xu, R. Veetil, V. Valuckas, R. Paniagua-Dominguez, and A. Kuznetsov, "Phase-only transmissive spatial light modulator based on tunable dielectric metasurface," *Science*, pp. 1087–1090, 2019, https://doi.org/10.1126/science.aaw6747.

[65] P. Wu, R. Pala, G. Shirmanesh, et al., "Dynamic beam steering with all-dielectric electro-optic III-V multiple-quantum-well metasurfaces," *Nat. Commun.*, pp. 1–9, 2019, https://doi.org/10.1038/s41467-019-11598-8.

[66] C. Wang, M. Zhang, M. Yu, R. Zhu, H. Hu, and M. Loncar, "Monolithic lithium niobate photonic circuits for Kerr frequency comb generation and modulation," *Nat. Commun.*, pp. 1–6, 2019, https://doi.org/10.1038/s41467-019-08969-6.

[67] J. Bartholomew, J. Rochman, T. Xie, et al., "On-chip coherent microwave-to-optical transduction mediated by ytterbium in YVO4," *Nat. Commun.*, pp. 1–6, 2020, https://doi.org/10.1038/s41467-020-16996-x.

[68] A. Tittl, A. Leitis, M. Liu, et al., "Imaging-based molecular barcoding with pixelated dielectric metasurfaces," *Science*, pp. 1105–1109, 2018, https://doi.org/10.1126/science.aas9768.

[69] K. Yao and Y. Liu, "Enhancing circular dichroism by chiral hotspots in silicon nanocube dimers," *Nanoscale*, vol. 10, no. 18, pp. 8779–8786, 2018.

[70] A. Leitis, Tittle A., Liu M., et al., "Angle-multiplexed all-dielectric metasurfaces for broadband molecular fingerprint retrieval," *Sci. Adv.*, vol. 5, no. 5, pp. 1–8, 2019.

Leonardo Viti, Alisson R. Cadore, Xinxin Yang, Andrei Vorobiev, Jakob E. Muench, Kenji Watanabe, Takashi Taniguchi, Jan Stake, Andrea C. Ferrari and Miriam S. Vitiello*

Thermoelectric graphene photodetectors with sub-nanosecond response times at terahertz frequencies

https://doi.org/10.1515/9783110710687-007

Abstract: Ultrafast and sensitive (noise equivalent power <1 nW $Hz^{-1/2}$) light-detection in the terahertz (THz) frequency range (0.1–10 THz) and at room-temperature is key for applications such as time-resolved THz spectroscopy of gases, complex molecules and cold samples, imaging, metrology, ultra-high-speed data communications, coherent control of quantum systems, quantum optics and for capturing snapshots of ultrafast dynamics, in materials and devices, at the nanoscale. Here, we report room-temperature THz nano-receivers exploiting antenna-coupled graphene field effect transistors integrated with lithographically-patterned high-bandwidth (~100 GHz) chips, operating with a combination of high speed (hundreds ps response time) and high sensitivity (noise equivalent power ≤120 pW $Hz^{-1/2}$) at 3.4 THz. Remarkably, this is achieved with various antenna and transistor architectures (single-gate, dual-gate), whose operation frequency can be extended over the whole 0.1–10 THz range, thus paving the way for the design of ultrafast graphene arrays in the far infrared, opening concrete perspective for targeting the aforementioned applications.

Keywords: 2D materials; nano-detectors; terahertz frequencies.

***Corresponding author: Miriam S. Vitiello,** NEST, Istituto Nanoscienze – CNR and Scuola Normale Superiore, Piazza San Silvestro 12, Pisa, 56127, Italy, E-mail: miriam.vitiello@sns.it. https://orcid.org/0000-0002-4914-0421
Leonardo Viti: NEST, Istituto Nanoscienze – CNR and Scuola Normale Superiore, Piazza San Silvestro 12, Pisa, 56127, Italy, E-mail: leonardo.viti@nano.cnr.it. https://orcid.org/0000-0002-4844-2081
Alisson R. Cadore, Jakob E. Muench and Andrea C. Ferrari: Cambridge Graphene Centre, University of Cambridge, 9, JJ Thomson Avenue, Cambridge, CB3 0FA, UK, E-mail: arc87@eng.cam.ac.uk (A.R. Cadore), jem227@eng.cam.ac.uk (J.E. Muench), acf26@hermes.cam.ac.uk (A.C. Ferrari). https://orcid.org/0000-0003-1081-0915 (A.R. Cadore). https://orcid.org/0000-0002-3124-3385 (J.E. Muench). https://orcid.org/0000-0003-0907-9993 (A.C. Ferrari)
Xinxin Yang, Andrei Vorobiev and Jan Stake: Department of Microtechnology and Nanoscience, Chalmers University of Technology, Gothenburg, SE-41296, Sweden, E-mail: xinxiny@chalmers.se (X. Yang), andrei.vorobiev@chalmers.se (A. Vorobiev), jan.stake@chalmers.se (J. Stake). https://orcid.org/0000-0003-4464-6922 (X. Yang). https://orcid.org/0000-0003-2882-3191 (A. Vorobiev). https://orcid.org/0000-0002-8204-7894 (J. Stake)
Kenji Watanabe and Takashi Taniguchi: National Institute for Materials Science, 1-1 Namiki, Tsukuba, 305-0044, Japan, E-mail: WATANABE.Kenji.AML@nims.go.jp (K. Watanabe), TANIGUCHI.Takashi@nims.go.jp (T. Taniguchi). https://orcid.org/0000-0003-3701-8119 (K. Watanabe). https://orcid.org/0000-0002-1467-3105 (T. Taniguchi)

1 Introduction

Hot-carrier assisted photodetection is an efficient and inherently broadband detection mechanism in single layer graphene (SLG) [1–4]. When a photon is absorbed by the electronic population (either *via* interband or intraband transitions), the photoexcited carriers can relax energy through electron–electron scattering or emission of optical phonons [5, 6], which usually occurs on a time scale of 10–100s fs [5, 6]. However, the electron-to-lattice relaxation via acoustic phonons is slower (1–2 ps) [6], leading to a *quasi*-equilibrium state where the thermal energy is distributed amongst electrons [5, 6] and not shared with the lattice. This produces an intriguing scenario, where the energy is absorbed by a system with an extremely low thermal capacitance ($c_e \sim 2000\ k_B \mu m^{-2}$, k_B is the Boltzmann constant) [7–10], thus leading to the ultrafast (~fs–ps) onset of thermal gradients in SLG-based nanostructures. At terahertz (THz) frequencies this effect is more relevant, since the emission of optical phonons is energetically forbidden [11], thus hindering this additional pathway for energy relaxation. SLG is therefore a promising material for engineering high-speed (~ps response time) opto-electronic THz devices that could benefit from the above mechanism [12].

This article has previously been published in the journal Nanophotonics. Please cite as: L. Viti, A. R. Cadore, X. Yang, A. Vorobiev, J. E. Muench, K. Watanabe, T. Taniguchi, J. Stake, A. C. Ferrari and M. S. Vitiello "Thermoelectric graphene photodetectors with sub-nanosecond response times at terahertz frequencies" *Nanophotonics* 2021, 10. DOI: 10.1515/nanoph-2020-0255.

The detection of THz light is important for applications in imaging [13], tomography [14], security [15, 16], biomedicine [17], and quantum optics [18]. An ideal THz photodetector (PD) should have a low noise equivalent power (NEP < nW $Hz^{-1/2}$), a large dynamic range (ideally >3 decades), have high detection speed (<ns), be broadband (0.1–10 THz), and operate at room temperature (RT). However, current RT THz PDs fail in targeting this combination of sensitivity, speed, and spectral range [19]. Graphene-based THz detectors relying on different physical mechanisms [4] have been widely demonstrated in the last few years [2, 12, 20–28] and include nanodevices exploiting the photovoltaic (PV) [22], the bolometric [23], the photothermoelectric (PTE) [2, 12, 27] and the plasma wave (PW) or Dyakonov–Shur effects, the latter in either its non-resonant [20, 25] or resonant (at low temperatures) [26] configurations. At RT, PTE PDs have proven to be the most sensitive and fast [2, 12, 27], due to the occurrence of photoinduced temperature gradients which alter the electronic thermal distribution on a fast (~100 fs) timescale [5, 6] and to the absence of an applied *dc* current through the SLG channel, which usually increases the noise level (dark current) in alternative physical configurations [23]. PTE detectors are demonstrated to reach response times ~100 ps at 1 THz [12]. The best combination of performance at frequency above 3 THz has been achieved in a thermoelectric RT graphene device [2], showing simultaneously NEP < 100 pW $Hz^{-1/2}$, response time τ ~40 ns (setup-limited), and a three orders of magnitude dynamic range. In this device, an ad hoc dual-gated, H-shaped antenna, having a strongly sub-wavelength gap (100 nm), defines a *p–n* junction, to which the performance improvement is ascribed. More recently, NEP ≤ 160 pW $Hz^{-1/2}$ with response times of 3.3 ns have been also reported in thermoelectric receivers exploiting broadband bow-tie antennas [27].

Here, we undertake the task of boosting the detection performances with respect to that benchmark. We exploit two different architectures: a single-gated hBN/graphene/hBN field effect transistor (GFET) (Figure 1C) and a split-gate hBN/graphene/hBN *p–n* junction (Figure 1D). By deeply investigating the photodetection mechanism, we show that, independently from the geometry, both the architectures operate mainly via the PTE effect. We then evaluate and compare the detection performances, proving that τ can be lowered at the hundreds ps level, without spoiling the detector sensitivity. This is achieved as follows. First, we minimize the absorption area in the GFET channel. This allows maximizing the temperature increase within the electronic thermal distribution, since a smaller absorption area entails a smaller amount of carriers to be heated by the incoming electromagnetic field, and, in turn, a larger temperature increase [2]. Secondly, as a further refinement, we use a novel electrodes design, which features on-chip transmission lines with bandwidth >100 GHz, and readout electronics having bandwidth >1 GHz.

By embedding the hBN/SLG/hBN layered materials heterostructures (LMH) [29, 30] in FET coupled to on-chip planar THz antennas (Figure 1A and B), we demonstrate ultrafast (τ < 1 ns) detection of >3 THz light at RT, with a record combination of speed, NEP and sensitivity, independent on the specific architecture. This is possible owing to the fast (~100 fs) onset of thermal gradients along the SLG channel and the subsequent generation of a PTE photovoltage [1], not dependent on the selected architecture. Thus, encapsulated SLG-based devices coupled to antenna structures can be used for the characterization of high (>10 MHz) repetition rate THz sources and high-speed (<1 ns) and low noise (NEP < 1 nW $Hz^{-1/2}$) THz imaging.

2 Results and discussion

We engineer two photodetector configurations as follows. Sample A is an hBN encapsulated GFET integrated with a planar bow tie antenna, asymmetrically connected to the source (*s*) and top-gate (g_T) electrodes, Figure 1C. Sample B is an hBN encapsulated GFET where two split-gates (g_{TL}, left gate and g_{TR}, right gate, Figure 1D), connected to the two branches of a linear dipole antenna, defining a *p–n* junction at its center [2]. Such antenna geometries are widely used in THz optoelectronics [2, 4, 24, 31] and both enable broadband operation [2, 32].

The hBN encapsulated GFET devices are fabricated as follows. hBN crystals are grown by the temperature-gradient method under high pressures and temperatures [33]. Bulk graphite is sourced from Graphenium. hBN and SLG are individually exfoliated on SiO_2/Si by micromechanical cleavage [34]. Initially, optical contrast [35] is utilized to identify SLG [29, 30]. The transfer technique employs a stamp of polydimethylsiloxane (PDMS) and a film of polycarbonate (PC) mounted on a transparent glass slide for picking up the layered materials and transfer them to the final and undoped SiO_2/Si substrate. The presence and quality of SLG is then confirmed by Raman spectroscopy [36] (see Section 4). The thickness of hBN is determined by atomic force microscope (AFM) and Raman spectroscopy [37, 38]. Combining the results from optical microscopy, Raman spectroscopy and AFM, blister-free areas with full width at half maximum (FWHM) of the 2D peak FWHM(2D) < 18 cm^{-1} are selected for device fabrication.

Following their assembly, we process the heterostructures into antenna-coupled FETs. The GFET channel is

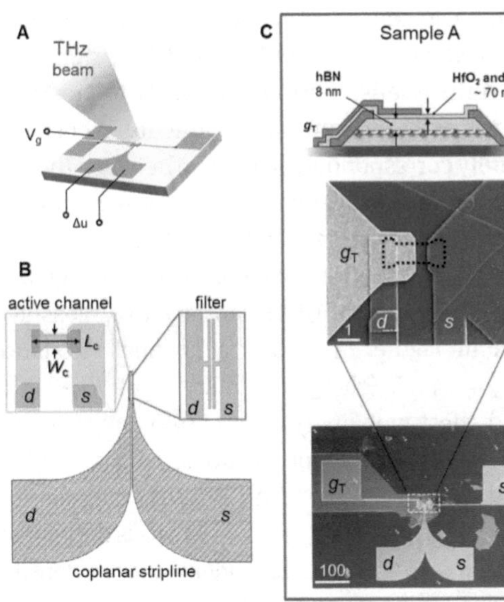

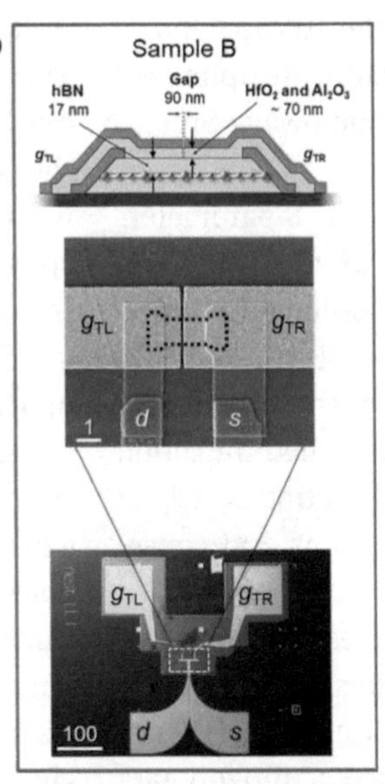

Figure 1: Detector layout. (A) Photodetector schematics: THz radiation is coupled to the GFET by a planar antenna and the photoresponse is recorded as a *dc* photovoltage (Δu) between the *s* and *d* electrodes. (B) On-chip RF components. The *s* and *d* electrodes are shaped in CPS geometry. Inset (left): the shape of the active LMH channel (green area) guarantees a lower contact resistance with respect to a rectangular geometry. The *s* and *d* contacts have a thickness of 45 nm in proximity of the GFET channel (yellow areas) and a thickness of 140 nm far from the GFET channel. Inset (right): planar low-pass filter, with cut-off frequency 300 GHz. (C) Sample A. Top: schematics of the LMH and electrodes layout, highlighting the different layer thicknesses. False color SEM image of the top-gated GFET (center) and optical microscope overview (bottom), where the bow-tie antenna position is marked with a dashed box. (D) Sample B. Top: schematics of the LMH and contacts design. False color SEM image of the GFET showing the split-top-gate geometry with the 90 nm gap (center) and optical microscope overview (bottom), where the position of the planar dipole antenna is marked with a dashed box. All scale bars are in units of micron.

first shaped by electron beam lithography (EBL), followed by dry etching of hBN and SLG [39] in SF_6. The SLG channel geometry is schematically represented in Figure 1: the channel is L_C = 3 μm long and W_C = 0.8 μm wide. The contact regions have lateral extensions. By simple geometrical considerations, it can be demonstrated that these extensions increase the perimeter of the stack, i.e., the length of the edge-contacts, thus reducing the contact resistance by 30%, with respect to more standard rectangular channel geometry. Edge Au/Cr electrodes are defined by standard EBL [39, 40], followed by metallization (40:5 nm) and *lift-off*.

We use, for both samples A and B, bottom hBN flakes of almost identical thickness (h), in order to make the comparison of the device performances consistent and reproducible. It is indeed worth mentioning that, due to the decrease of the electron–hole charge fluctuations at the substrate [41], changes of the bottom hBN layer thickness can significantly affect the FET mobility [29, 42]. In the present case, the flakes thicknesses, retrieved by AFM are: bottom hBN h = 23 nm, top hBN h = 8 nm, for sample A, and bottom hBN h = 25 nm, top hBN h = 17 nm for sample B. The low thickness of the heterostructures (<45 nm) and of the edge-contacts (~45 nm) allows us to use a thinner oxide (70 nm) as encapsulating layer before g_T deposition (Figure 1C and D), thus increasing the effective gate-to-channel capacitance per unit area: $C_g \sim 100$ nF cm^{-2} for both samples. This parameter is important for THz FET detectors [25], since the responsivity (R_v), a figure of merit defined as the ratio between photovoltage (Δu) and impinging optical power, is typically proportional to the sensitivity of the FET conductance to changes in the gate voltage (V_g) [25].

In order to reduce parasitic capacitances, usually detrimental for high-speed (>1 GHz) detection, and simultaneously minimize parasitic losses [43], we design and fabricate a microwave transmission line connected to the *s* and drain (*d*) edge-electrodes based on a coplanar stripline (CPS) geometry [24], Figure 1B. We use this radio frequency (RF) on-chip component because of its simplicity. In contrast to the standard strip-line geometry [44], it does not require a ground plane, and, unlike the coplanar waveguide architecture [44], it consists of only two parallel metallic strips on the substrate top surface. In our devices, the strips are separated by a 2 μm gap, where one conductor (ground electrode, *s*) provides the electrical ground for the other (signal electrode, *d*). This architecture shows an almost perfect transmission below 30 GHz, with S_{21} = 0 dB, S_{11} < −40 dB, whereas at 3.4 THz the transmission is reduced, but not canceled, with S_{21} = −3.5 dB and S_{11} = −25 ÷ −35 dB (details about simulations are given in Supplementary material). The transmission of the THz signal between the antenna-coupled GFET and the contacts can be detrimental for the overall detector performance. This is mainly due to the fact that the antenna modes lose energy (resulting in a decreased resonance quality factor), if the antenna is not isolated from the surrounding circuit. Therefore, our design also includes a low-pass hammer-head filter along the CPS (Figure 1B) [45], with a cutoff frequency $f_{\text{cut-off}} \sim 300$ GHz, which enhances the isolation between antenna and readout circuit. It consists of a capacitive shunt with a lumped capacitance $C_f = 500$ aF. The dimensions of the structure are optimized by time-domain simulations sim (CST Microwave Studio) (see Supplementary material).

The presence of the filter leaves the S-parameters almost unaltered for frequencies <30 GHz: $S_{21} = 0$ dB, $S_{11} < -30$ dB. On the other hand, it modifies the transmission line properties at 3.4 THz: $S_{11} = -4$ dB, $S_{21} \sim -24$ dB. To further increase the signal extraction from the active element, the CPS has an adiabatically matched transition [46] between bonding pads and GFET electrodes, which hinders the formation of spurious reflections and consequent losses.

After this common protocol, samples A and B are processed following different architectures. For sample A, Figure 1C, the lobe of a THz planar bow-tie antenna (110 nm thick) is connected to the s electrode. Then, a thin top-gate oxide bi-layer is placed on the LMH, also covering the s and d contacts: 20 nm HfO_2 deposited via atomic layer deposition (ALD) and 50 nm Al_2O_3 deposited via Ar sputtering. The photodetector is then finalized by the fabrication of g_T, in the shape of the arm of a bow-tie antenna, thus forming a complete bow-tie together with the s electrode. The antenna radius is 21 μm and the gap between antenna arms is 250 nm (Figure 1C). For sample B, Figure 1D, the same oxide bi-layer is deposited before the antenna fabrication. The antenna is here shaped as a linear dipole, with 24 μm arms separated by a gap of 90 nm (Figure 1D, further images are reported in the Supplementary material). The two branches of the antenna also serve as top split-gates for the GFET. The gate voltages (V_{gL}, left gate bias and V_{gR}, right gate bias) can be individually controlled in order to create, at the center of the active channel, a p–n junction whose size is approximately corresponding to the gap between the two split-gates [2, 47]. The gate geometry is therefore nominally the only difference between the two samples.

The devices are then characterized electrically and optically at RT. The two-probe GFET transfer curve, measured for sample A in Figure 2A, shows a channel resistance (R) peak at $V_g = -4.6$ V (charge neutrality point, CNP). The extracted field-effect mobility (μ_{FE}) is 17,000 $cm^2\ V^{-1}\ s^{-1}$ for holes and 19,000 $cm^2\ V^{-1}\ s^{-1}$ for electrons, with a residual carrier density $n_0 \sim 9 \times 10^{11}\ cm^{-2}$. This is fitted using the formula [48] $R = R_0 + (L_C/W_C)\cdot(1/n_{2d}e\mu_{FE})$, where R_0 is the contact resistance and n_{2d} is the gate-dependent charge density, given by [48] $n_{2d} = [n_0^2 + (C_g/e\,(V_g - V_{CNP}))^2]^{1/2}$.

We then test the RT sensitivity using a focused 3.4 THz beam with an average power $P_t = 100$ μW (see Section 4). The intensity distribution on the focal plane (Figure 2D, sample A), displayed through the xy map of Δu, unveils the Airy pattern [49] of the focused beam, showing four concentric rings (maxima) with the central Airy disk. This demonstrates the good signal-to-noise ratio (~1000 at $P_t = 100$ μW) of the proposed device. From the two-dimensional Gaussian fit of the intensity distribution in Figure 2D, we obtain standard deviations $\sigma_x = 95 \pm 1$ μm and $\sigma_y = 87 \pm 1$ μm along the x and y directions, respectively, from which we infer FWHM ~ 303 ± 2 μm

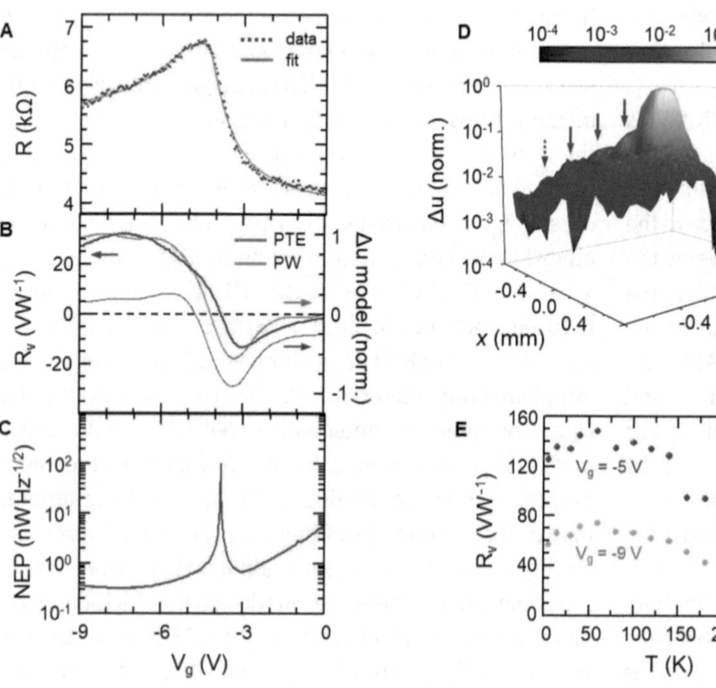

Figure 2: Electrical and optical characteristics of single-gate GFET. (A) Electrical resistance R as a function of V_g at RT in a two-terminal configuration. (B) R_v measured at RT as a function of V_g (left vertical axis), compared with the normalized expected photothermoelectric and overdamped plasma wave photovoltages (right vertical axis). (C) NEP calculated as a function of V_g under the assumption of Johnson–Nyquist dominated noise spectral density [2]. A minimum NEP ~ 350 pW $Hz^{-1/2}$ is obtained for $V_g = -7$ V. (D) Logarithmic plot of the normalized photovoltage on the focal plane, for an average impinging THz power of 100 μW. The four Airy maxima are indicated by blue arrows on the left of the central Airy disk. The red arrow indicates the portion of the focal plane where the beam is blocked by the output window of the cryostat in which the QCL is mounted. The FWHM of the beam is 303 μm. (E) R_v plotted as a function of T measured at $V_g = -5$ V (blue dots) and $V_g = -9$ V (magenta dots).

(see Supplementary material for further details). This is used to estimate the fraction of total power that impinges on the detector $P_a = P_t \cdot (A_\lambda/A_{spot}) = 2.7$ μW, where $A_\lambda = \lambda^2/4 = 1.9 \times 10^{-3}$ mm² is the diffraction limited area (see Supplementary material) and $A_{spot} = \pi \cdot (\text{FWHM}/2)^2 = 72 \times 10^{-3}$ mm² is the beam spot area. Then, by measuring Δu (see Section 4) as a function of V_g and dividing the as-obtained values by P_a, we retrieve the plot of R_v as a function of V_g (Figure 2B). The maximum $R_v = 30$ VW^{-1} is obtained for $V_g = -7$ V and the trend is compatible with a dominant PTE response (see Supplementary material). This is corroborated by the following argument. At $V_{sd} = 0$ V, in a single-gated GFET, connected by identical metallic layers at the *s* and *d* contacts, both the PTE and the non-resonant PW detection mechanisms can in principle be activated [25, 27]. In the geometry of sample A, the PTE photovoltage reads $\Delta u_{PTE} = \Delta T_e \cdot (S_g - S_u)$ [25, 27, 31], where ΔT_e is the THz-induced electronic temperature difference between the (hot) source side of the channel, corresponding to the gap at the center of the bow-tie antenna, and the (cold) drain side (Figure 1C), S_u is the Seebeck coefficient of the ungated region between the *s* and *g* electrodes and S_g is the Seebeck coefficient of the gated LMH channel. By imposing $S_u = S_g$ for $V_g = 0$ V and assuming ΔT_e weakly dependent on V_g [2, 25], we can analytically compute the gate voltage dependence of $\Delta u_{PTE} \propto S_g - S_u$ (see Supplementary material for further details). The same argument applies to the overdamped PW photovoltage [20, 25], $\Delta u_{PW} \propto -\sigma^{-1}(\partial\sigma/\partial V_g)$. The comparison between $\Delta u_{PTE}(V_g)$, $\Delta u_{PW}(V_g)$ and the experimental $R_v(V_g)$ curves (Figure 2B) unveils that the PTE effect well matches with our experimental observation and better reproduces our data with respect to the PW model, which predicts that the maximum response (in absolute value) occurs at $V_g = -3.5$ V and R_v is finite and negative at $V_g = 0$ V, in stark contrast with our measurements, where $R_v \approx 0$ VW^{-1} at $V_g = 0$ V. This conclusion is further supported by the temperature (T) dependent analysis of the responsivity, which unambiguously shed light on the core detection dynamics.

To this purpose we mount the detector in a He flux cryostat and we vary the heat sink T in the 6–260 K range. The measured responsivity (Figure 2E) shows a non-monotonic behavior as a function of T, with a maximum around a crossover temperature $T^\star = 60$ K, in agreement with what observed in other spectral ranges [50]. The origin of such a behavior can be retrieved by the analysis of the electron cooling dynamics in SLG. Δu_{PTE} is proportional to ΔT_e, which, in turn, is proportional to the cooling length $\xi = (k/\gamma c_e)^{1/2}$ [1, 2, 50] (the proportionality holds as long as $\xi < L_C$), where k is the thermal conductivity and γ is the cooling rate. Since both k and c_e scale linearly with T, the functional dependency of the cooling length ξ (and Δu_{PTE}), with respect to T, is the same as $\gamma^{-1/2}$. For $T < T^\star$, $\gamma(T)$ is dominated by acoustic phonon emission and scales as $\sim T^{-1}$, whereas at higher T, the disorder-assisted scattering (supercollision) gives rise to a competing cooling channel which follows the power law $\gamma \sim T$ [50]. The two effects give rise to a crossover temperature ($T^\star$) for which γ is minimum and, consequently, Δu_{PTE} is maximum. We then compare the temperature dependence of R_v at two distinctive gate voltages, $V_g = -5$ V (close to CNP, low carrier density, $n_{2d} \sim 10^{12}$ cm^{-2}) and at $V_g = -9$ V (away from CNP, holes density up to $n_{2d} \sim 4 \times 10^{12}$ cm^{-2}). The non-monotonic behavior is more evident at lower n_{2d}, in qualitative agreement with previous findings on PTE detection [25, 50]. In a non-degenerate electron system, $\Delta u_{PTE}(T)$ is completely determined by ΔT_e, being the Seebeck coefficient weakly dependent from T [25]; conversely, in the degenerate case, S is proportional to T [51] and compensates the decrease of ΔT_e at higher T, resulting in an almost T-independent Δu_{PTE}. For sample A, under the assumption of a noise spectral density (NSD, i.e., noise power per unit bandwidth) dominated by thermal fluctuations [31] (see Supplementary material), we estimate $\text{NEP} = 1/R_v \cdot (4k_BRT)^{1/2}$. The NEP curve as a function of V_g (Figure 2C) shows a minimum NEP ~ 350 pW Hz$^{-1/2}$ at $V_g = -7$ V.

We use a similar approach for the optical and electrical characterization of sample B. Figure 3 plots the device performance as a function of bias applied at the split-gates. By independently varying the two gate voltages, we control the Fermi level (E_F) and, consequently, n_{2d} on each side of the dual-gated SLG junction [2, 47]. The color plot of R with respect to V_{gR} (right gate, horizontal axis) and V_{gL} (left gate, vertical axis) in Figure 3A allows us to extract a hole and electron $\mu_{FE} \sim 19{,}000$ cm² V^{-1} s^{-1} and 15,000 cm² V^{-1} s^{-1}, respectively, with a residual carrier density $n_0 \sim 1 \times 10^{12}$ cm^{-2}.

The independent control of the E_F on each side of the junction allows individual control of the two Seebeck coefficients S_L and S_R [2, 47], which can be used to maximize the photoresponse. THz detection in a graphene *p–n* junction is expected to be dominated by the PTE effect [2]. Δu_{PTE}, measured between the drain and source electrodes, can be written as [52]:

$$\Delta u_{PTE} = \int_d^s \frac{\partial T_e}{\partial x} \cdot S(x)dx = \Delta T_e \cdot (S_L - S_R) \quad (1)$$

where ΔT_e is the electronic temperature increase as a consequence of the absorption of THz radiation at the junction.

Figure 3B is a color map of R_v obtained by continuously changing V_{gR} and V_{gL} in the same ranges of Figure 3A. The

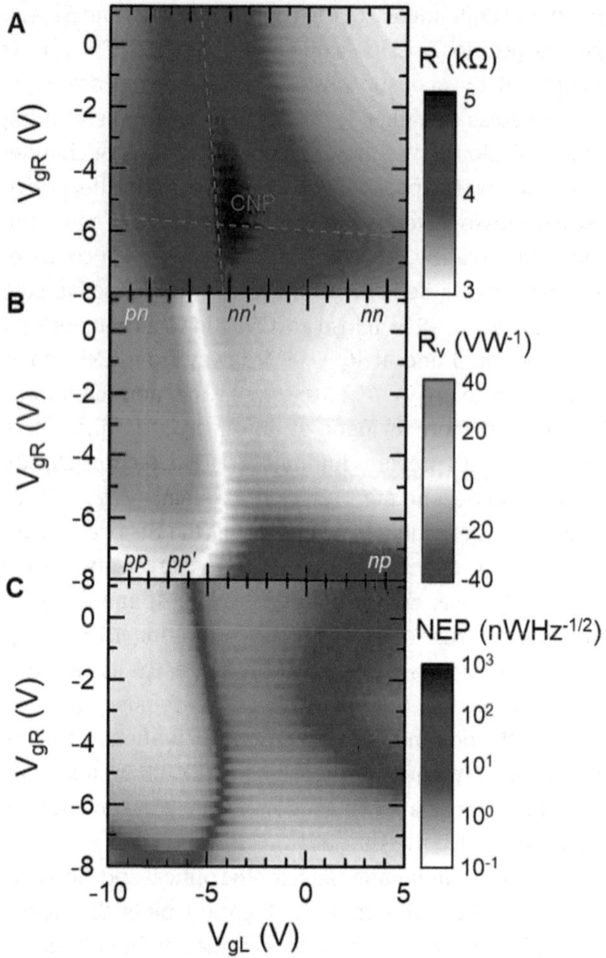

Figure 3: Electrical and optical characteristics of double-gated graphene *p–n* junction. (A) Analysis of electrical transport of GFET: two-terminal *RT* resistance as a function of split-gate biases. The dashed lines indicate the CNP positions for V_{gL} and V_{gR}. (B) Color map of R_v as a function of V_{gL} and V_{gR}. R_v undergoes many sign changes, corresponding to transitions between the different configurations of the *p–n* junction, attainable by polarizing the gates. (C) Two-dimensional plot of the NEP (logarithmic scale) as a function of V_{gL} and V_{gR}. A minimum NEP of 120 pW Hz$^{-1/2}$ is obtained for $V_{gL} = -8$ V and $V_{gR} = -4$ V.

maxima of R_v (~50 V W^{-1}) are obtained when the two local gates have opposite polarity with respect to the CNP, i.e., in *p–n* or *n–p* junction configurations. The resulting *six-fold* pattern in the measured photovoltage is ascribed to the non-monotonic gate voltage dependence of S_L and S_R on each side of the junction, and is a unique fingerprint of a dominant hot-carrier assisted PTE effect in SLG [1, 2, 53]. Therefore, for the *p–n* junction, the room-temperature R_v characterization alone is sufficient to unambiguously unveil the dominant PTE THz detection.

From R and R_v, we can estimate the NEP of sample B, assuming a thermal-noise limited operation. The contour plot of NEP as a function of the two gate voltages (Figure 3C) shows a minimum NEP ~120 pW Hz$^{-1/2}$ at $V_{gL} = -8$ V and $V_{gR} = -4$ V. Sample B is therefore ~3 times more sensitive than sample A. This can be attributed to the larger field enhancement provided by the dual-gate configuration, in particular to the narrow (90 nm) gap between the antenna arms, in agreement with Ref. [2].

To extract the response time and the bandwidth $BW = (2\pi\tau)^{-1}$, we shine light from a pulsed THz quantum cascade laser (QCL, pulse width ~150 ns and repetition rate 333 Hz) and record the signal with a fast oscilloscope (5 GS/s) after a pre-amplification stage (low noise voltage preamplifier, model Femto-DUPVA, bandwidth 1.2 GHz, input impedance 50 Ω).

Figure 4A and B shows the time traces of samples A and B, recorded at zero gate bias with an oscilloscope having a temporal resolution 200 ps. We extract the rise-time τ_{ON} and fall-time τ_{OFF} by using the fitting functions $V_{out} = c_0 + V_{ON} \cdot [1-\exp(-(t-c_1)/\tau_{ON})]$ and $V_{out} = c_2 + V_{OFF} \cdot \exp(-(t - c_3)/\tau_{OFF})$, where c_0, c_1, c_2, c_3 are fitting parameters, and V_{ON} and V_{OFF} are the voltage jumps in the waveforms corresponding to the rising-edge and falling-edge. We find similar results for both devices, with rise-times slightly shorter with respect to fall-times. Sample A shows $\tau_{ON} = 1.3 \pm 0.4$ ns and $\tau_{OFF} = 1.5 \pm 0.6$ ns at $V_g = 0$ V, sample B shows $\tau_{ON} = 890 \pm 150$ ps and $\tau_{OFF} = 1.4$ ns $\pm$ 0.25 ns at $V_{gL} = V_{gR} = 0$ V. These response times are, to the best of our knowledge, the lowest in GFET devices with NEP <1 nW Hz$^{-1/2}$. In terms of *BW*, considering the lower values of τ as limit response time, we obtain $BW = 125 \pm 35$ MHz for sample A and $BW = 180 \pm 30$ MHz for sample B, i.e., 50 times better than in Ref. [2]. The small discrepancy between the latter values can be ascribed to fluctuations in the QCL output power, possibly caused by time jitter (±100 ps [54]) in the electrical circuit employed to drive the laser.

To further validate this assessment, we measure the detector rise-time under different configuration of gate voltages, i.e., at different charge densities and SLG resistances. The response time of a PD is ultimately limited by the *RC* time constant of the circuit [2]. Therefore, if the PD is the key element limiting the detection speed, a change in *R* should directly and proportionally reflect into a change in τ, via $\tau = R \cdot C$. We thus select and investigate three gate voltage configurations, for both devices. The results are shown in the insets of Figure 4A and B.

For sample A, we obtain $\tau_{ON} = 1.3 \pm 0.4$ ns at $V_g = 0$ V ($R = 4.2$ kΩ), $\tau_{ON} = 1.5 \pm 0.6$ ns at $V_g = -5$ V ($R = 6.4$ kΩ) and $\tau_{ON} = 1.4 \pm 0.3$ ns at $V_g = -8$ V ($R = 5.7$ kΩ), showing the lack of a direct proportionality relation between R and τ_{ON}. The same conclusion can be drawn for sample B at $V_{gR} = 0$ V, where $\tau_{ON} = 890 \pm 150$ ns for $V_{gL} = 0$ V ($R = 3.7$ kΩ),

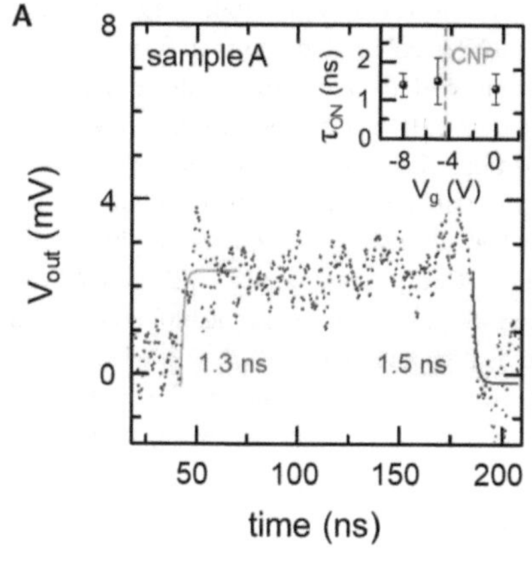

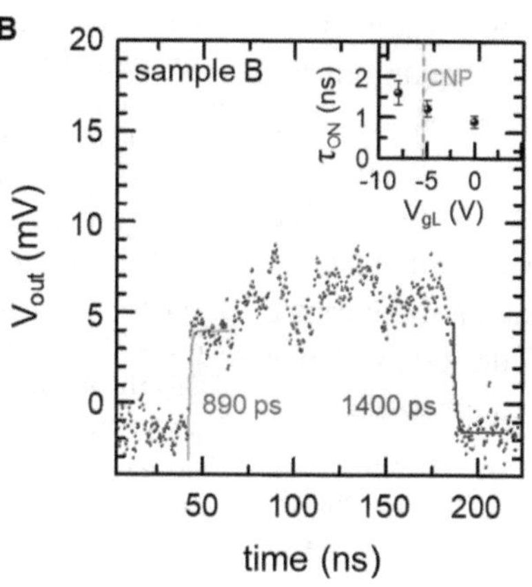

Figure 4: Electrical bandwidth and response time. (A) Photovoltage time-trace under illumination with a 150 ns THz pulse having a peak power of 10 mW, recorded with sample A at $V_g = 0$ V. The time constants $\tau_{ON} = 1.3 \pm 0.4$ ns and $\tau_{OFF} = 1.5 \pm 0.6$ ns are obtained by fitting the data. Inset: variation of τ_{ON} as a function of V_g. The rise-time does not depend on the device resistance. (B) Time trace recorded with sample B at $V_{gL} = V_{gR} = 0$ V, giving $\tau_{ON} = 890 \pm 150$ ps and $\tau_{OFF} = 1400 \pm 250$ ps. Inset: variation of τ_{ON} as a function of V_g. The rise-time does not depend on the device resistance.

$\tau_{ON} = 1.2 \pm 0.2$ ns for $V_{gL} = -5$ V ($R = 4.7$ kΩ), and $\tau_{ON} = 1.6 \pm 0.3$ ns for $V_{gL} = -8$ V ($R = 4.0$ kΩ). This demonstrates that τ is not affected by the SLG resistance in the tested range. This illustrates that the PD itself is not limiting the measured maximum speed, which is instead affected by the switching time of the QCL. A higher intrinsic speed beyond the set-up limited value is in good agreement with reports of high-speed, PTE-based SLG detectors for integrated photonics, with reported 3 dB BW in the tens of GHz [47]. In this work, high-speed performance is enabled by the on-chip architecture, featuring RF electronic components, which mitigates the presence of parasitic capacitances and the undesirable crosstalk between sensing element and outer on-chip components.

Our results show that, up to a bandwidth of 150 MHz, the two proposed architectures are substantially equivalent. Both configurations lead to $\tau \sim$ ns, even though the two geometries are different: in sample A the THz field is distributed along the un-gated portion of the channel (250 nm), whereas in sample B the two symmetric split gates, defining a narrow gap (90 nm), provide a more localized enhancement of the THz field at the center of the SLG channel. The speed limit is, in both cases, lower than that reported in Ref. [2], the switching speed being limited by the onset speed and jitter noise of the employed QCL system. This equivalence is not surprising. As revealed by the low temperature characterization of sample A (Figure 2E), both architectures mainly operate through the same detection mechanism: the PTE effect. This is known to be the dominant mechanism for devices operating through *p–n* junction rectification [1, 2], however it has also been observed in antenna-coupled single-gated architectures [20, 25, 26], where the antenna provided asymmetric THz excitation, essential for the activation of the PTE mechanism. Moreover, our data show that the speed of the two devices does not even depend on the existence of a *p–n* junction, but it only requires that the gates create an imbalance in the Seebeck coefficient along the graphene channel.

3 Conclusions

In summary, the performance achieved at RT on both devices demonstrates that PTE THz detectors, coupled with high-bandwidth on-chip (~100 GHz) and external electronics, detect pulses with sub-ns temporal extension, opening

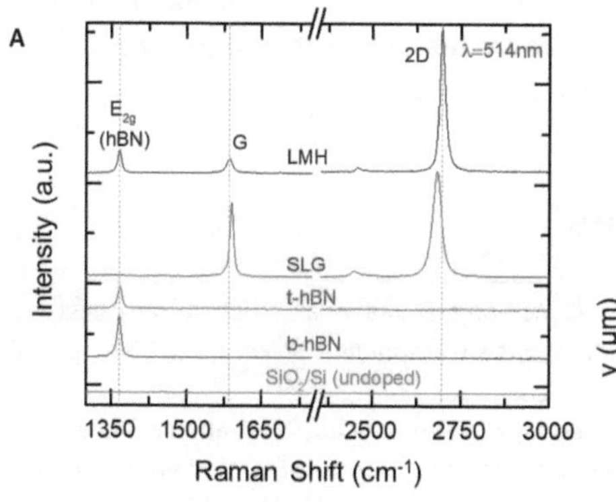

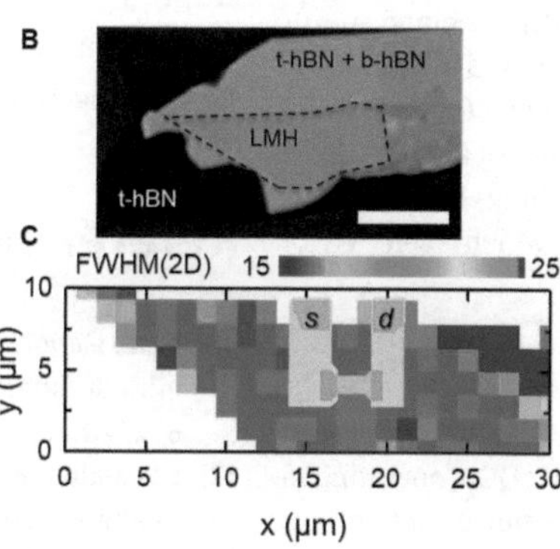

Figure 5: Sample characterization and selection of the device area. (A) Raman spectra before and after LMH assembly measured at 514 nm. The bottom hBN (b-hBN) is shown in blue, the top one (t-hBN) in green, the SLG in purple, and the assembled LMH in black, while the SiO_2/Si substrate in red. The hBN E_{2g}, G, and 2D peaks are highlighted by the dashed gray lines. (B) False color optical image of the LMH. SLG is indicated by a black dashed line. Scale bar is 10 μm. (C) Spatial map of FWHM(2D), indicating the area where the GFET is designed.

unique perspectives for ultrafast applications in a plethora of research field as ultrafast nano-spectroscopy, quantum science, coherent control of quantum nanosystems and high speed communications. Further improvements on the detection performances can be achieved via the on-chip integration of coplanar waveguides and pre-amplification stages. It is worth mentioning that, measuring the intrinsic speed limit of the PTE mechanism in SLG devices, which is expected to be $\tau \sim 10$ ps [2], would require completely avoiding the limitations set by the readout electronics. This could be obtained, for example, by exploiting interferometric techniques, such as pulse autocorrelation measurements [55].

Our results open a route for characterization of high repetition rate THz sources, transient effects in nonlinear optoelectronic devices (e.g., saturable absorbers), time-resolved intracavity-mode dynamics of THz QCL frequency combs and ultimately for high-speed and low noise THz imaging, never pioneered so far.

4 Methods

4.1 Sample characterization

Raman measurements are performed using a Renishaw InVia spectrometer equipped with a 100× objective, 2400 mm^{-1} grating at 514 nm. The power on the sample is <1 mW to avoid any heating and damage. AFM is performed in tapping mode to characterize the topography and thickness of the LMHs using a Bruker Dimension Icon system. Figure 5A plots the spectra of a typical LMH, with 8 and 23 nm thickness top and bottom hBN flakes, while Figure 5B is a false color optical image of the LMH, highlighting the SLG edges. Figure 5A shows that the E_{2g} peak for both bottom and top hBN are ~1366 cm^{-1}, with FWHM(E_{2g}) ~9.3 and 9.7 cm^{-1}, consistent with bulk hBN [37]. Figure 5A plots the SLG G and 2D peaks before and after stacking. Before encapsulation, the 2D and G peaks have FWHM(2D) ~ 27 cm^{-1}, Pos(2D) ~ 2682 cm^{-1}, Pos(G) ~ 1589 cm^{-1}, FWHM(G) ~ 8 cm^{-1}, and the intensity and areas ratio of 2D and G peaks are I(2D)/I(G) ~ 1.4, A(2D)/A(G) ~ 4.6, as expected for SLG with $E_F \geq 250$ meV [56, 57]. No D peak is observed, indicating negligible defects [58]. After LMH assembling, the combined hBN E_{2g} peak is at Pos(E_{2g}) ~ 1366 cm^{-1}, with FWHM(E_{2g}) ~ 9.5 cm^{-1}. For the encapsulated SLG we have Pos(2D) ~ 2697 cm^{-1}, FWHM(2D) ~ 17 cm^{-1}, Pos(G) ~ 1584 cm^{-1}, FWHM(G) ~ 14 cm^{-1}, I(2D)/I(G) ~ 13, and A(2D)/A(G) ~ 12, indicating $E_F \ll 100$ meV [56, 57]. The changes in FWHM(2D) after encapsulation indicates a reduction in the nanometer-scale strain variations within the sample [29, 59]. Figure 5C shows an FHWM(2D) map across a bubble-free LMH sample, exhibiting homogeneous (spread < 1 cm^{-1}) and narrow (~17 cm^{-1}) FWHM(2D), which is selected for the GFET fabrication.

4.2 Optical measurements

In order to test the PD sensitivity, we use a 3.4 THz QCL, operating in pulse mode with a repetition rate of 40 kHz and a pulse width of 1 µs and refrigerated at 30 K by means of a Stirling cryocooler (estimated lattice temperature of the active region 170 K [60]). The divergent beam (divergence angle ~ 30°) is collimated and then focused using two picarin (tsupurica) lenses with focal lengths 50 mm and 30 mm, respectively. The average output power can be continuously varied up to ~1 mW at the PD position. The measurements are performed by keeping the *s* electrode grounded and by extracting the photovoltage signal Δu at the *d* contact. The latter signal is then pre-amplified with a voltage pre-amplifier (FEMTO, input impedance 1 MΩ, gain 40 dB, *BW* 200 MHz) and recorded with a lock-in technique, referenced by a 1333 kHz square wave. Δu is estimated as $2.2\ V_{LI}/\eta$ [31], where V_{LI} is the lock-in signal and η is the voltage preamplifier gain coefficient. The detectors are mounted on a *xyz* stage, allowing automated spatial positioning.

Acknowledgments: We acknowledge funding from the ERC Consolidator Grant SPRINT (681379) and the EU Horizon 2020 research and innovation programme Graphene Flagship under grant agreement No 785219 (GrapheneCore2), ERC grants Hetero2D, GSYNCOR, EPSRC grants EP/L016087/1, EP/K01711X/1, EP/K017144/1. M.S.V. acknowledges partial support from the second half of the Balzan Prize 2016 in applied photonics delivered to Federico Capasso.

Author contribution: L.V and M.S.V. conceived the core idea. A.R.C and A.C.F. prepared the hBN/graphene/hBN heterostrcututes and characterized the quality of the graphene; K.W. and T.T. provided high quality hBN; X.Y., A.V. and J. S. contributed to the design of the microstrip line; L.V. fabricated the sample and performed electrical and optical measurements; L.V. and M.S.V analyzed the data and wrote the manuscript. All authors discussed the results and contributed to the writing of the manuscript. M.S.V. supervised the study.

Research funding: This research was funded by ERC Consolidator Grant SPRINT (681379) and the EU Horizon 2020 research programme Graphene Flagship (785219), ERC grants Hetero2D, GSYNCOR, EPSRC grants EP/L016087/1, EP/K01711X/1, EP/K017144/1.

Data availability: The data that support the plots within this paper and other findings of this study are available from the corresponding authors upon reasonable request.

Conflict of interest statement: The authors declare no competing financial interests.

References

[1] N. M. Gabor, J. C. W. Song, Q. Ma, et al., "Hot carrier-assisted intrinsic photoresponse in graphene," *Science*, vol. 334, pp. 648–652, 2011.

[2] S. Castilla, B. Terrés, M. Autore, et al., "Fast and sensitive terahertz detection using an antenna-integrated graphene pn junction," *Nano Lett.*, vol. 19, pp. 2765–2773, 2019.

[3] K. J. Tielrooij, L. Piatowski, M. Massicotte, et al., "Generation of photovoltage in graphene on a femtosecond timescale through efficient carrier heating," *Nat. Nanotechnol.*, vol. 10, pp. 437–443, 2015.

[4] F. H. L. Koppens, T. Mueller, Ph. Avouris, A. C. Ferrari, M. S. Vitiello, and M. Polini, "Photodetectors based on graphene, other two-dimensional materials and hybrid systems," *Nat. Nanotechnol.*, vol. 9, pp. 780–793, 2014.

[5] A. Tomadin, D. Brida, G. Cerullo, A. C. Ferrari, and M. Polini, "Nonequilibrium dynamics of photoexcited electrons in graphene: Collinear scattering, Auger processes, and the impact of screening," *Phys. Rev. B*, vol. 88, 2013, Art no. 035430.

[6] D. Brida, A. Tomadin, C. Manzoni, et al., "Ultrafast collinear scattering and carrier multiplication in graphene," *Nat. Commun.*, vol. 4, p. 1987, 2013.

[7] K. C. Fong and K. C. Schwab, "Ultrasensitive and wide-bandwidth thermal measurements of graphene at low temperatures," *Phys. Rev. X*, vol. 2, pp. 1–8, 2012.

[8] E. Pop, V. Varshney, and A. K. Roy, "Thermal properties of graphene: Fundamentals and applications," *MRS Bull.*, vol. 37, pp. 1273–1281, 2012.

[9] G. Soavi, G. Wang, H. Rostami, et al., "Broadband, electrically tunable third-harmonic generation in graphene," *Nat. Nanotechnol.*, vol. 13, pp. 583–588, 2018.

[10] G. Soavi, G. Wang, H. Rostami, et al., "Hot electrons modulation of third-harmonic generation in graphene," *ACS Photonics*, vol. 6, pp. 2841–2849, 2019.

[11] M. Lazzeri, S. Piscanec, F. Mauri, A. C. Ferrari, and J. Robertson, "Phonon linewidths and electron-phonon coupling in graphite and nanotubes," *Phys. Rev. B*, vol. 73, 2006, Art no. 155426.

[12] X. Cai, A. B. Sushkov, R. J. Suess, et al., "Sensitive room-temperature terahertz detection via the photothermoelectric effect in graphene," *Nat. Nanotechnol.*, vol. 9, pp. 814–819, 2014.

[13] W. L. Chan, J. Deibel, and D. M. Mittleman, "Imaging with terahertz radiation," *Rep. Prog. Phys.*, vol. 70, pp. 1325–1379, 2007.

[14] J. P. Guillet, B. Recur, L. Frederique, et al., "Review of terahertz tomography techniques," *J. Infrared, Millim. Terahertz Waves*, vol. 35, pp. 382–411, 2014.

[15] L. Kawase, Y. Ogawa, Y. Watanabe, and H. Inoue, "Non-destructive terahertz imaging of illicit drugs using spectral fingerprints," *Optic Express*, vol. 11, p. 2549, 2003.

[16] M. Tonouchi, "Cutting-edge terahertz technology," *Nat. Photon.*, vol. 1, pp. 97–105, 2007.

[17] I. Kašalynas, R. Venckevičius, L. Minkevičius, et al., "Spectroscopic terahertz imaging at room temperature employing microbolometer terahertz sensors and its application to the study of carcinoma tissues," *Sensors*, vol. 16, pp. 1–15, 2016.

[18] S. S. Dhillon, M. S. Vitiello, E. H. Linfield, et al., "The 2017 terahertz science and technology roadmap," *J. Phys. D Appl. Phys.*, vol. 50, 2017, Art no. 043001.

[19] F. Sizov, "Terahertz radiation detectors: the state-of-the-art," *Semicond. Sci. Technol.*, vol. 33, 2018, Art no. 123001.

[20] L. Vicarelli, M. S. Vitiello, D. Coquillat, et al., "Graphene field-effect transistors as room-temperature terahertz detectors," *Nat. Mater.*, vol. 11, pp. 865–871, 2012.

[21] M. Mittendorff, S. Winnerl, J. Kamann, et al., "Ultrafast graphene-based broadband THz detector," *Appl. Phys. Lett.*, vol. 103, 2013, Art no. 021113.

[22] R. Degl'Innocenti, L. Xiao, D. S. Jessop, et al., "Fast room-temperature detection of terahertz quantum cascade lasers with graphene-loaded bow-tie plasmonic antenna arrays," *ACS Photonics*, vol. 3, pp. 1747–1753, 2016.

[23] A. V. Muraviev, S. L. Rumyantsev, G. Liu, A. A. Balandin, W. Knap, and M. S. Shur, "Plasmonic and bolometric terahertz detection by graphene field-effect transistor," *Appl. Phys. Lett.*, vol. 103, 2013, Art no. 181114.

[24] A. A. Generalov, M. A. Andersson, X. Yang, A. Vorobiev, and J. Stake, "A heterodyne graphene FET detector at 400 GHz," in *2017 42nd Int. Conf. Infrared, Millimeter, Terahertz Waves*, 2017, pp. 1–2.

[25] D. A. Bandurin, I. Gayduchenko, I. Cao, et al., "Dual origin of room temperature sub-terahertz photoresponse in graphene field effect transistors," *Appl. Phys. Lett.*, vol. 112, 2018, Art no. 141101.

[26] D. A. Bandurin, D. Svintsov, I. Gayduchenko, et al., "Resonant terahertz detection using graphene plasmons," *Nat. Commun.*, vol. 9, p. 5392, 2018.

[27] L. Viti, D. G. Purdie, A. Lombardo, A. C. Ferrari, and M. S. Vitiello, "HBN-encapsulated, graphene-based, room-temperature terahertz receivers, with high speed and low noise," *Nano Lett.*, vol. 20, pp. 3169–3177, 2020.

[28] G. Auton, D. B. But, J. Zhang, et al., "Terahertz detection and imaging using graphene ballistic rectifiers," *Nano Lett.*, vol. 17, pp. 7015–7020, 2017.

[29] D. G. Purdie, N. M. Pugno, T. Taniguchi, K. Watanabe, A. C. Ferrari, and A. Lombardo, "Cleaning interfaces in layered materials heterostructures," *Nat. Commun.*, vol. 9, p. 5387, 2018.

[30] D. De Fazio, D. G. Purdie, A. K. Ott, et al., "High-mobility, wet-transferred graphene grown by chemical vapor deposition," *ACS Nano*, vol. 13, pp. 8926–8935, 2019.

[31] L. Viti, A. Politano, K. Zhang, and M. S. Vitiello, "Thermoelectric terahertz photodetectors based on selenium-doped black phosphorus flakes," *Nanoscale*, vol. 11, pp. 1995–2002, 2019.

[32] D. Coquillat, J. Marczewski, P. Kopyt, N. Dyakonova, B. Giffard, and W. Knap, "Improvement of terahertz field effect transistor detectors by substrate thinning and radiation losses reduction," *Optic Express*, vol. 24, pp. 272–281, 2016.

[33] K. Watanabe, T. Taniguchi, and H. Kanda, "Direct-bandgap properties and evidence for ultraviolet lasing of hexagonal boron nitride single crystal," *Nat. Mater.*, vol. 3, pp. 404–409, 2004.

[34] K. S. Novoselov, D. Jiang, F. Schedin, et al., "Two-dimensional atomic crystals," *Proc. Natl. Acad. Sci. U. S. A.*, vol. 102, pp. 10451–10453, 2005.

[35] C. Casiraghi, A. Hartschuh, E. Lidorikis, et al., "Rayleigh imaging of graphene and graphene layers," *Nano Lett.*, vol. 7, pp. 2711–2717, 2007.

[36] A. C. Ferrari, J. C. Meyer, V. Scardaci, et al., "Raman spectrum of graphene and graphene layers," *Phys. Rev. Lett.*, vol. 97, 2006, Art no. 187401.

[37] S. Reich, A. C. Ferrari, R. Arenal, A. Loiseau, I. Bello, and J. Robertson, "Resonant Raman scattering in cubic and hexagonal boron nitride," *Phys. Rev. B*, vol. 71, 2005, Art no. 205201.

[38] R. Arenal, A. C. Ferrari, S. Reich, et al., "Raman spectroscopy of single-wall boron nitride nanotubes," *Nano Lett.*, vol. 6, pp. 1812–1816, 2006.

[39] B. S. Jessen, L. Gammelgaard, M. R. Thomsen, et al., "Lithographic band structure engineering of graphene," *Nat. Nanotechnol.*, vol. 14, pp. 340–346, 2019.

[40] L. Wang, I. Meric, P. Y. Huang, et al., "One-dimensional electrical contact to a two-dimensional material," *Science*, vol. 342, pp. 614–617, 2013.

[41] J. Xue, J. Sanchez-Yamanishi, D. Bulmash, et al., "Scanning tunnelling microscopy and spectroscopy of ultra-flat graphene on hexagonal boron nitride," *Nat. Mater.*, vol. 10, 282–285, 2011.

[42] C. R. Dean, A. F. Young, I. Meric, et al., "Boron nitride substrates for high-quality graphene electronics," *Nat. Nanotechnol.*, vol. 5, pp. 722–726, 2010.

[43] P. Kopyt, B. Salski, J. Marczewski, P. Zagrajek, and J. Lusakowski, "Parasitic effects affecting responsivity of sub-THz radiation detector built of a MOSFET," *J. Infrared, Millim. Terahertz Waves*, vol. 36, pp. 1059–1075, 2015.

[44] T. H. Lee, *Planar Microwave Engineering*, Cambridge, Cambridge University Press, 2004.

[45] C. Wang, Y. He, B. Lu, et al., "Robust sub-harmonic mixer at 340 GHz using intrinsic resonances of hammer-head filter and improved diode model," *J. Infrared, Millim. Terahertz Waves*, vol. 38, pp. 1397–1415, 2017.

[46] Z. X. Xu, X. X. Yin, and D. F. Sievenpiper, "Adiabatic mode-matching techniques for coupling between conventional microwave transmission lines and onedimensional impedance interface waveguides," *Phys. Rev. Appl.*, vol. 11, 2019, Art no. 044071.

[47] J. E. Muench, A. Ruocco, M. A. Giambra, et al., "Waveguide-integrated, plasmonic enhanced graphene photodetectors," *Nano Lett.*, vol. 19, pp. 7632–7644, 2019.

[48] S. Kim, J. Nah. I. Jo, et al., "Realization of a high mobility dual-gated graphene field-effect transistor with Al_2O_3 dielectric," *Appl. Phys. Lett.*, vol. 94, 2007, Art no. 062107.

[49] M. Born and E. Wolf, "Elements of the theory of diffraction," in *Principles of Optics*, 7th ed. Cambridge, Cambridge University Press, 2005.

[50] Q. Ma, N. M. Gabor, T. I. Andersen, et al., "Competing channels for hot-electron cooling in graphene," *Phys. Rev. Lett.*, vol. 112, pp. 1–5, 2014.

[51] P. Dollfus, V. H. Nguyen, and J. Saint-Martin, "Thermoelectric effects in graphene nanostructures," *J. Phys. Condens. Matter*, vol. 27, 2015, Art no. 133204.

[52] L. Viti, J. Hu, D. Coquillat, A. Politano, W. Knap, and M. S. Vitiello, "Efficient terahertz detection in black-phosphorus nano-transistors with selective and controllable plasma-wave, bolometric and thermoelectric response," *Sci. Rep.*, vol. 6, p. 20474, 2016.

[53] J. C. W. Song, M. S. Rudner, C. M. Marcus, and L. S. Levitov, "Hot carrier transport and photocurrent response in graphene," *Nano Lett.*, vol. 11, pp. 4688–4692, 2011.

[54] Avtech AVR series, medium to high voltage general purpose pulse generators [Online]. Available at: www.Avtechpulse.Com/Catalog/Avr-1-2-3-4_rev17.Pdf [accessed: Mar., 2020].

[55] A. Lisauskas, K. Ikamas, S. Massabeau, et al., "Field-effect transistors as electrically controllable nonlinear rectifiers for the characterization of terahertz pulses," *APL Photonics*, vol. 3, 2018, Art no. 051705.

[56] A. Das, S. Pisana, B. Chakraborty, et al., "Monitoring dopants by Raman scattering in an electrochemically top-gated graphene transistor," *Nat. Nanotechnol.*, vol. 3, pp. 210–215, 2008.

[57] D. M. Basko, S. Piscanec, and A. C. Ferrari, "Electron-electron interactions and doping dependence of the two-phonon Raman intensity in graphene," *Phys. Rev. B*, vol. 80, pp. 1–10, 2009.

[58] L. G. Cançado, A. Jorio, E. H. Martins Ferreira, et al., "Quantifying defects in graphene via Raman spectroscopy at different excitation energies," *Nano Lett.*, vol. 11, pp. 3190–3196, 2011.

[59] C. Neumann, S. Reichardt, P. Venezuela, et al., "Raman spectroscopy as probe of nanometre-scale strain variations in graphene," *Nat. Commun.*, vol. 6, pp. 1–7, 2015.

[60] M. S. Vitiello and G. Scamarcio, "Measurement of subband electronic temperatures and population inversion in THz quantum-cascade lasers," *Appl. Phys. Lett.*, vol. 86, 2005, Art no. 111115.

Supplementary Material: The online version of this article offers supplementary material (https://doi.org/10.1515/nanoph-2020-0255).

Brian S. Lee, Bumho Kim, Alexandre P. Freitas, Aseema Mohanty, Yibo Zhu, Gaurang R. Bhatt, James Hone and Michal Lipson*

High-performance integrated graphene electro-optic modulator at cryogenic temperature

https://doi.org/10.1515/9783110710687-008

Abstract: High-performance integrated electro-optic modulators operating at low temperature are critical for optical interconnects in cryogenic applications. Existing integrated modulators, however, suffer from reduced modulation efficiency or bandwidth at low temperatures because they rely on tuning mechanisms that degrade with decreasing temperature. Graphene modulators are a promising alternative because graphene's intrinsic carrier mobility increases at low temperature. Here, we demonstrate an integrated graphene-based electro-optic modulator whose 14.7 GHz bandwidth at 4.9 K exceeds the room temperature bandwidth of 12.6 GHz. The bandwidth of the modulator is limited only by high contact resistance, and its intrinsic RC-limited bandwidth is 200 GHz at 4.9 K.

Keywords: 2D materials; cryogenic; graphene; modulator; ring resonator; silicon photonics.

1 Introduction

Integrated electro-optic modulators are essential for high-bandwidth optical links in cryogenic environments, in applications such as delivering control signals and reading out data in solid-state quantum computing [1–6] or inter-satellite optical communications [7, 8]. For example, superconducting nanowire single-photon detectors offer exceptional performance but require optical modulators operating at cryogenic environment to readout signals to room temperature [1, 4]. Superconducting single-flux quantum circuits provide operating frequencies approaching 770 GHz and are expected to provide superior computing performance to conventional electronics at room temperature [6]. Yet, interfacing cryogenic systems to room temperature controllers and processors with electrical connections pose several issues, such as increased thermal load to the cryogenic environment, which potentially degrades cryogenic circuit performance because of increased thermal noise and limited bandwidth due to high frequency losses. Therefore, there is an increasing demand for optical interconnects to interface cryogenic and room temperature systems, and modulators operating at cryogenic temperature are an essential component for establishing such an optical link.

Existing electro-optic modulators, however, suffer from low bandwidth or reduced modulation efficiency at cryogenic temperatures because they rely on tuning mechanisms that degrade with decreasing temperature. Devices based on free carriers such as silicon modulators [9], for example, suffer from carrier freeze-out at low temperature [10]. This may be mitigated with degenerate doping but at the cost of increased insertion loss. Non–carrier-based modulators using Franz–Keldysh or quantum-confined Stark effect [11] or Pockels effect [12] suffer from weak electro-optic strength at low temperature, thus requiring higher drive voltage and increased footprint, which limits the integration density of photonic integrated circuits for cryogenic applications.

Graphene modulators are a promising alternative for low-temperature applications. They rely on tuning graphene's absorption through electrostatic gating [13], a mechanism which does not suffer from degradation at low temperatures [14, 15]. The intrinsic electronic mobility of graphene increases on cooling [16–18], such that the speed, determined by the RC charging time, should not intrinsically degrade at low temperature. However, no work has demonstrated low-temperature operation of

***Corresponding author: Michal Lipson,** Department of Electrical Engineering, Columbia University, New York, NY 10027, USA, E-mail: ml3745@columbia.edu
Brian S. Lee, Aseema Mohanty and Gaurang R. Bhatt, Department of Electrical Engineering, Columbia University, New York, NY 10027, USA. https://orcid.org/0000-0002-8040-3576 (B.S. Lee)
Bumho Kim, Yibo Zhu and James Hone, Department of Mechanical Engineering, Columbia University, New York, NY 10027, USA
Alexandre P. Freitas, School of Electrical and Computer Engineering, University of Campinas, Campinas-SP, 13083-970, Brazil. https://orcid.org/0000-0003-1474-5938

This article has previously been published in the journal Nanophotonics. Please cite as: B. S. Lee, B. Kim, A. P. Freitas, A. Mohanty, Y. Zhu, G. R. Bhatt, J. Hone and M. Lipson "High-performance integrated graphene electro-optic modulator at cryogenic temperature" *Nanophotonics* 2021, 10. DOI: 10.1515/nanoph-2020-0363.

graphene modulators to date. Here, we demonstrate a high-bandwidth graphene electro-optic modulator at 4.9 K. The mobility of the graphene in these devices increases with decreasing temperature, leading to reduced device resistance and better RC-limited bandwidth of graphene modulators. Therefore, graphene enables high-speed integrated electro-absorption modulator that naturally exhibits a high bandwidth at cryogenic temperature.

2 Graphene-silicon nitride electro-absorption ring modulator

The graphene electro-absorption modulator consists of a dual-layer graphene capacitor integrated with a silicon nitride (Si_3N_4) waveguide. The capacitor consists of two graphene sheets (cyan solid lines in Figure 1A) separated by 30 nm alumina (Al_2O_3) gate dielectric (white solid lines in Figure 1A are boundaries of Al_2O_3 gate dielectric, Si_3N_4 waveguide, and metals). The Si_3N_4 waveguide (1300 nm by 330 nm) is designed to ensure significant overlap between the fundamental quasi-transverse electric (quasi-TE) mode and graphene capacitor via evanescent wave as shown in Figure 1A. By applying voltage to the graphene capacitor, we electrostatically gate the graphene sheets and induce Pauli-blocking, that is, reduce optical absorption and mode propagation loss by suppressing interband transitions of carriers in graphene [19]. We describe the fabrication steps of the Si_3N_4 waveguide and dual-layer graphene capacitor in Figure S1 in the *Supplementary material.*

We embed the graphene capacitor/Si_3N_4 waveguide in a ring resonator as shown in Figure 1B to enhance graphene–light interaction while reducing footprint and capacitance for maximum bandwidth. In addition, we use resonator loss modulation at critical coupling [20, 21] to achieve strong modulation even with small voltage swing and small capacitance (about 9 fF, see the optical micrograph inset of Figure 1C for device scale). Figure 1C shows the simulated transmission through the waveguide (blue curve, left axis) and slope of transmission (orange curve, right axis) with respect to resonator round trip loss. The transmission is most sensitive (i.e., largest slope) when the resonator is near critical coupling (i.e., near the dashed vertical line in Figure 1C). We therefore design the resonator–bus coupling gap (around 180 nm) to be near critical coupling to achieve highest sensitivity to changes in graphene absorption and to transition from being critically coupled to over-coupled as graphene absorption is modulated from high to low, respectively.

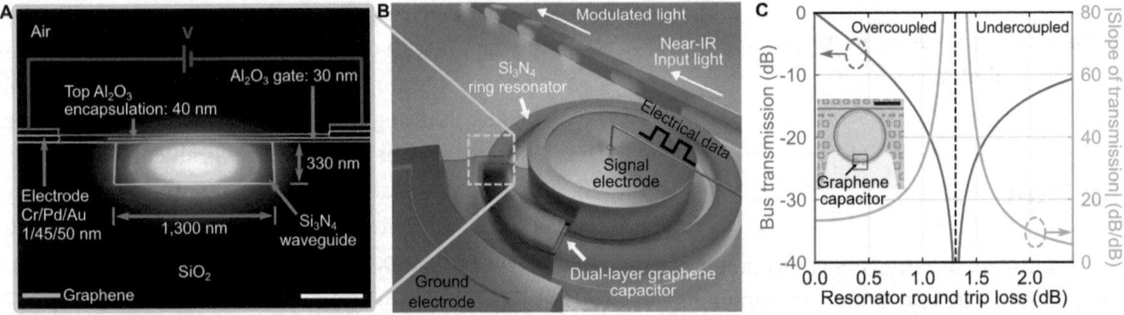

Figure 1: Graphene electro-absorption modulator design and resonator loss modulation at critical coupling.
(A) The graphene modulator consists of a dual-layer graphene capacitor on top of a Si_3N_4 waveguide. The capacitor consists of two graphene sheets (cyan solid lines) separated by a 30 nm Al_2O_3 gate dielectric (white solid lines are boundaries for Al_2O_3 gate dielectric, Si_3N_4 waveguide, and metals). The top graphene sheet is cladded with 40 nm Al_2O_3. The Si_3N_4 waveguide (1300 nm by 330 nm) is designed to ensure significant overlap between the fundamental quasi-TE mode and graphene capacitor via evanescent wave. By applying voltage between the graphene sheets, we electrostatically gate them and induce Pauli-blocking, suppressing interband transitions of carriers in graphene sheets and reducing optical absorption and mode propagation loss. White scale bar, 200 nm. (B) The graphene capacitor is embedded in a ring resonator to enhance graphene–light interaction while reducing footprint and capacitance for maximum bandwidth. (C) Simulated transmission (blue curve, left axis) and slope of transmission (orange curve, right axis) with respect to resonator round trip loss. To achieve strong modulation even with small voltage swing and small capacitance, we modulate resonator loss near critical coupling shown as dashed vertical line. The transmission is most sensitive (i.e., largest slope) when the resonator is near critical coupling. We, therefore, design the resonator–bus coupling gap (around 180 nm) to be near critical coupling. The modulator transitions from being critically coupled to overcoupled as graphene absorption is modulated from high to low, respectively. Inset: An optical micrograph of the fabricated device showing the waveguide and ring resonator, false-colored in blue. The two electrodes for the graphene capacitor are shown in yellow. Boxed region indicates where the graphene capacitor is placed with 5-μm device length around the ring with 40 μm radius. Squares around the device are fill patterns for chemical mechanical planarization. Black scale bar, 40 μm.

3 Results

We show modulation of transmission by more than 7 dB at room temperature by electrostatically gating the graphene and, thus, modulating the resonator round trip loss near critical coupling. In Figure 2A, we show the transmission spectra of the graphene ring modulator at different applied voltages. The transmission is about −25 dB at resonance (around 1586.2 nm) when applied voltage is 0 V, indicating near critical coupling of the ring with high graphene absorption. With applied voltage to the graphene modulator, the resonator–bus coupling condition becomes more overcoupled and the transmission at resonance increases. To measure the change in resonator round trip loss as a function of voltage, we measure the loaded quality factor Q_L from each of the spectra shown in Figure 2A and plot it in Figure 2B. The Q_L increases from about 3500 to 3700 over 9 V, corresponding to resonator round trip loss decreasing from 1.10 to 0.96 dB [22] due to decreasing graphene absorption. We measure Q_L of the ring resonator without graphene sheets to be around 4700 for strongly overcoupled resonator (see Figure S2 in the *Supplementary material*). Therefore, we do not completely achieve Pauli-blocking in our graphene sheets. We attribute this to the lower-than-designed dielectric constant of our atomic layer deposition Al_2O_3 gate, $\varepsilon_{Al_2O_3} = 4.2$ (Figure S3D in the *Supplementary material*). With optimized gate quality, we expect to reach further into the Pauli-blocking regime and achieve higher extinction ratio (see Figure S4 in the *Supplementary material*).

We characterize the frequency response of the graphene modulator at 293 and 4.9 K and measure a 3-dB bandwidth of 12.6 and 14.7 GHz, respectively (Figure 3). We place the graphene modulator in a cryogenic probe station to precisely control the temperature of the chip and perform high-speed electro-optic measurements (see Figure S5 for experimental setup at cryogenic temperature). The bandwidth increases approximately 16% when the device is cooled from room to cryogenic temperature despite being driven with the same d.c. bias (−9 V) and RF power (13.5 dBm, $V_{pp} = 3$ V, see Figure S6 for eye diagrams measured at 293 K in the *Supplementary material*). We normalize the amplitude of each curve to the response at 1 MHz and fit it to a single-pole transfer function $1/(1 + j2\pi f\tau)$, where f is the frequency and τ is the modulator time constant. The 3-dB bandwidth at each temperature is measured from the fitted curve (shown as dashed lines in Figure 3) as $f_{3dB} = 1/(2\pi\tau)$. We do not observe significant change in modulation efficiency at low temperature in agreement with Li et al. who have measured at $T = 45$ K [14]. We suspect that the Pauli-blocking transition becomes negligibly steeper even at cryogenic temperatures due to short lifetimes of excited carriers in graphene on time scales of 100 fs [23, 24].

We show that the increase of graphene modulator's bandwidth at low temperature follows the trend of graphene carrier mobility. We pattern graphene test devices into Hall bar configuration, allowing extraction of the carrier mobility by simultaneous measurement of the four-terminal conductance and carrier density through the Hall effect (see Figure S3 in the *Supplementary material*). As shown in Figure 4A, the mobility increases from approximately 1420 cm^2/V s at 293 K to 1650 cm^2/V s at 4.9 K. This increase reflects the weakening of scattering from phonons

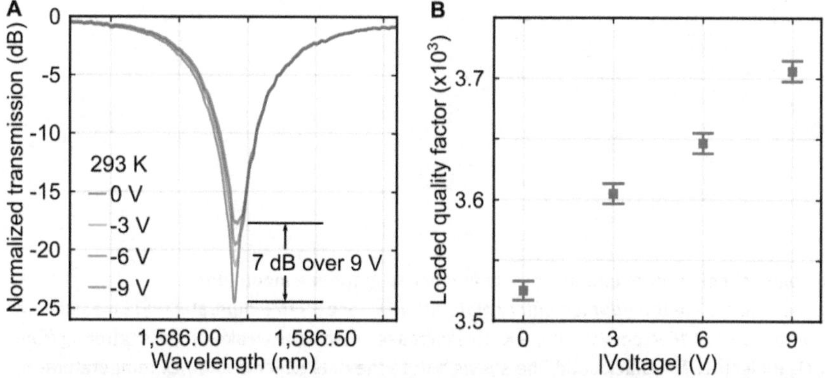

Figure 2: Modulating resonator–bus coupling condition by electrostatically gating the graphene capacitor.
(A) Transmission spectra of the modulator at various d.c. voltages at room temperature. With applied voltage to the graphene modulator, we change the transmission at resonance by more than 7 dB by decreasing graphene absorption and changing the resonator–bus coupling condition. By reducing graphene absorption and resonator round trip loss, we make the resonator more overcoupled as indicated by the increasing transmission at resonance. (B) We measure the loaded quality factor Q_L of the resonator spectra in Figure 2A. The Q_L increases from about 3500 to 3700 over 9 V, corresponding to resonator round trip loss decreasing from 1.10 to 0.96 dB due to decreasing graphene absorption.

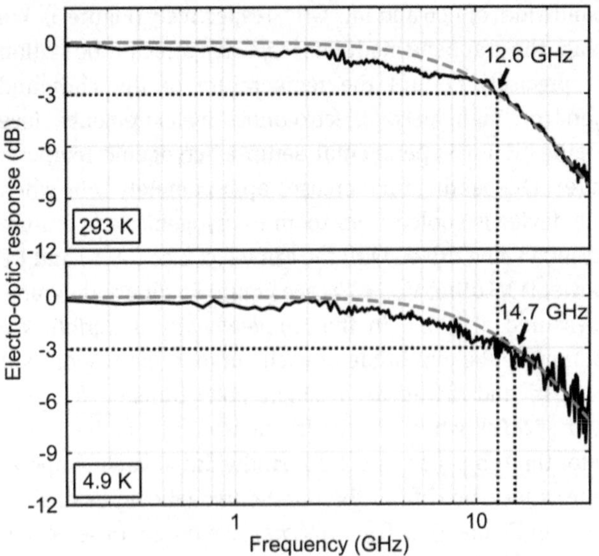

Figure 3: Enhancement of electro-optic bandwidth of graphene modulator at cryogenic temperature.
The measured electro-optic bandwidth at 293 and 4.9 K are 12.6 and 14.7 GHz, respectively. The red dashed lines are single pole fitting to the data. The electro-optic response increases by about 16% when the device is cooled from room to cryogenic temperature. The modulator is driven with a vector network analyzer (see Figure S5 in the *Supplementary material*) with RF power 13.5 dBm (V_{pp} = 3 V) at d.c. bias of −9 V at both temperatures.

in the graphene and the surrounding Al_2O_3 dielectric. The solid line in Figure 4A shows a fit to the data which combines temperature-independent scattering from disorder (such as trapped charge in the Al_2O_3) and temperature-dependent scattering from graphene longitudinal acoustic phonons and Al_2O_3-graphene surface polar phonons [16–18, 25] (see Figure S7 in the *Supplementary material*). To confirm that the device's resistance governs the bandwidth change with temperature, we verify that the capacitance remains constant with temperature. The capacitance changes by less than 5% from 293 to 1.5 K by measuring the change in carrier concentration of the graphene Hall bar at various temperatures (see Figure S8 in the *Supplementary material*).

4 Discussion

Using the measured mobility and two-terminal device resistance, we determine that the graphene modulator bandwidth is currently limited by its high contact resistance of around 2.3 ± 1.2 Ω μm (see Figure S9 in the *Supplementary material*), whereas its high mobility (and resulting low sheet resistance) at cryogenic temperature supports a fundamental bandwidth of 200 GHz. From the RC circuit in Figure 4B, the modulator's total RC-limited bandwidth is,

$$f_{3\text{dB}} = \frac{1}{2\pi(2R_c + 2R_{\text{sh}}g)C}, \qquad (1)$$

where R_c is the contact resistance, R_{sh} is the sheet resistance, g is the gap between the electrodes and capacitor,

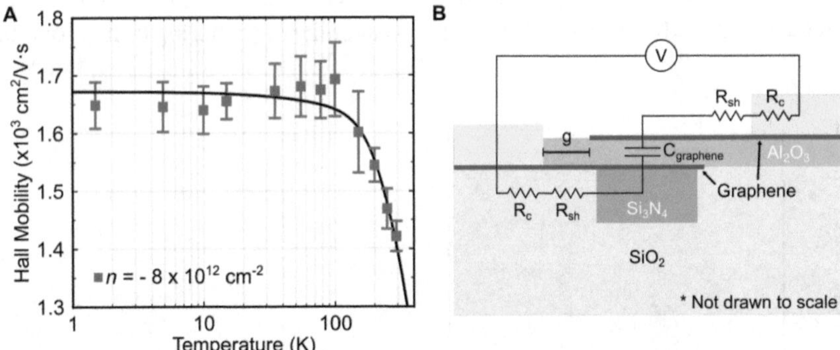

Figure 4: Mobility enhancement of graphene at cryogenic temperature and its correspondence to graphene modulator.
(A) We extract graphene mobility using a Hall bar (see Figure S3 in the *Supplementary material*) with respect to temperature. We measure an increase of mobility from approximately 1420 cm^2/V s at 293 K to 1650 cm^2/V s at 4.9 K. This increase reflects the weakening of scattering from phonons in the graphene and the surrounding Al_2O_3 dielectric. The black solid line shows a fit to the data which combines temperature-independent scattering from disorder (such as trapped charge in the Al_2O_3) and temperature-dependent scattering from graphene longitudinal acoustic phonons and Al_2O_3-graphene surface polar phonons (see Figure S7 in the *Supplementary material*). The measured mobility follows the temperature dependence of the fitted curve well within the measurement error. (B) The equivalent RC circuit of the graphene modulator is overlaid with the device cross section. R_c is the graphene contact resistance, R_{sh} is the graphene sheet resistance, C is the dual-layer graphene capacitor, and g is the gap between the electrode and the capacitor. The enhancement of graphene carrier mobility at cryogenic temperature reduces the sheet resistance of the device, increasing the RC-limited bandwidth.

and C is the capacitance per length. The total bandwidth is governed by contributions from extrinsic (i.e., due to parasitic components) and intrinsic bandwidth:

$$\begin{aligned} 1/f_{3\text{dB}} &= 1/\text{BW}_{\text{ext}} + 1/\text{BW}_{\text{int}} \\ &= 2\pi(2R_cC) + 2\pi(2R_{\text{sh}}gC), \end{aligned} \tag{2}$$

where $\text{BW}_{\text{ext}} = 1/(2\pi[2R_cC])$ is the extrinsic bandwidth governed by parasitic contact resistance and $\text{BW}_{\text{int}} = 1/(2\pi[2R_{\text{sh}}gC])$ is the intrinsic device bandwidth governed by the capacitance and sheet resistance. From Hall bar measurements (see Figure S3 in the *Supplementary material*) and using Equation (1), we extract C and R_{sh} at 4.9 K as 1.85 fF/μm and 470 Ω/sq, respectively, which translates to BW_{int} = 200 GHz. The high contact resistance can be attributed to the ungated contact regions. Unlike electronics, it is challenging to gate graphene contact regions in photonic devices as metal gates could induce unwanted optical absorption, or extending the graphene capacitor sheets to the contacts will parasitically increase the capacitance, hence reduce the device bandwidth. By statically doping these ungated regions, analogous to p or n regions in silicon modulators, we expect to further improve the contact resistance [26, 27] and increase our extrinsic bandwidth. For example, with state-of-the-art graphene contact resistance measured as low as 100 Ω μm [28], we expect to achieve a total bandwidth of >130 GHz at cryogenic temperature with optimized graphene contacts without changing the device's cross section. Note that this high bandwidth requires ensuring that the cavity photon lifetime is not a limiting factor, which can be done by, for example, increasing the graphene capacitor length around the circumference of the ring resonator from 5 to 33 μm (see *Supplementary material*).

5 Conclusion

We have demonstrated high-speed integrated graphene modulator with measured bandwidth of 14.7 GHz with intrinsic bandwidth of 200 GHz at 4.9 K by leveraging graphene carrier mobility improvement at low temperature. In contrast to traditional electro-optic tuning mechanisms that decrease in bandwidth or modulation efficiency at low temperature, the graphene modulator exhibits an increase in electro-optic response from 293 to 4.9 K without trading off voltage or footprint. By optimizing graphene contact resistance and device footprint, we expect the graphene modulator to support bandwidths >130 GHz at cryogenic temperature. In addition, with optimized quality of gate dielectric, the modulation extinction ratio, currently limited by weaker dielectric constant and breakdown field of the Al_2O_3 gate compared with values reported in literature [29], could be further increased (see Figure S4 in the *Supplementary material*). This natural enhancement of electro-optic response at low temperature makes graphene modulators versatile and suitable for high-speed electro-optic applications at cryogenic temperature.

Methods

Please see the *Supplementary material* for detailed description of device fabrication and experimental measurements.

Acknowledgment: The authors would like to thank Dr. Min Sup Choi, Dr. Christopher T. Phare, Dr. Andres Gil-Molina, Dr. Nathan C. Abrams, Ipshita Datta, Min Chul Shin, and Euijae Shim for the fruitful discussions and experimental support. This work was performed in part at the City University of New York Advanced Science Research Center NanoFabrication Facility and in part at the Columbia Nano Initiative (CNI) shared labs at Columbia University in the City of New York.

Author contribution: All the authors have accepted responsibility for the entire content of this submitted manuscript and approved submission.

Research funding: We also gratefully acknowledge support from the Office of Naval Research for award #N00014-16-1-2219, Defense Advanced Research Projects Agency program for award #HR001110720034, National Science Foundation for award #UTA16-000936, National Aeronautics and Space Administration for award #NNX16AD16G, Air Force Office of Scientific Research for award #FA9550-18-1-0379, Air Force Materiel Command for award #FA8650-18-1-7815, and Hypres, Inc. for award #CU15-3759.

Conflict of interest statement: The authors declare no conflicts of interest regarding this article.

References

[1] M. de Cea, E. E. Wollman, A. H. Atabaki, D. J. Gray, M. D. Shaw, and R. J. Ram, "Photonic readout of superconducting nanowire single photon counting detectors," *Sci. Rep.*, vol. 10, p. 9470, 2020.

[2] A. Youssefi, I. Shomroni, Y. J. Joshi, et al., "Cryogenic electro-optic interconnect for superconducting devices," arXiv preprint arXiv: 2004.04705, 2020.

[3] M. Veldhorst, H. G. J. Eenink, C. H. Yang, and A. S. Dzurak, "Silicon CMOS architecture for a spin-based quantum computer," *Nat. Commun.*, vol. 8, no. 1, pp. 1–8, 2017.

[4] J. W. Silverstone, J. Wang, D. Bonneau, et al., "Silicon quantum photonics," in *2016 International Conference on Optical MEMS and Nanophotonics (OMN)*, IEEE, 2016, pp. 1–2.
[5] R. Maurand, X. Jehl, D. Kotekar-Patil, et al., "A CMOS silicon spin qubit," *Nat. Commun.*, vol. 7, no. 1, pp. 1–6, 2016.
[6] D. S. Holmes, A. L. Ripple, and M. A. Manheimer, "Energy-efficient superconducting computing—power budgets and requirements," *IEEE Trans. Appl. Supercond.*, vol. 23, no. 3, pp. 1701610–1701610, 2013.
[7] R. Lange, F. Heine, H. Kämpfner, and R. Meyer, "High data rate optical inter-satellite links," in *Eur. Conf. Opt. Commun.(ECOC 2009)*, Vienna, Austria, Paper, volume 10, 2009.
[8] M. Toyoshima, Y. Takayama, T. Takahashi, et al., "Ground-to-satellite laser communication experiments," *IEEE Aero. Electron. Syst. Mag.*, vol. 23, no. 8, pp. 10–18, 2008.
[9] M. Gehl, C. Long, D. Trotter, et al., "Operation of high-speed silicon photonic micro-disk modulators at cryogenic temperatures," *Optica*, vol. 4, no. 3, pp. 374–382, 2017.
[10] B. Lengeler, "Semiconductor devices suitable for use in cryogenic environments," *Cryogenics*, vol. 14, no. 8, pp. 439–447, 1974.
[11] P. Pintus, Z. Zhang, S. Pinna, et al., "Characterization of heterogeneous InP-on-Si optical modulators operating between 77 K and room temperature," *APL Photonics*, vol. 4, no. 10, p. 100805, 2019.
[12] F. Eltes, G. E. Villarreal-Garcia, D. Caimi, et al., "An integrated optical modulator operating at cryogenic temperatures," *Nat. Mater.*, pp. 1–5, 2020, https://doi.org/10.1038/s41563-020-0725-5.
[13] F. Wang, Y. Zhang, C. Tian, et al., "Gate-variable optical transitions in graphene," *Science*, vol. 320, no. 5873, pp. 206–209, 2008.
[14] Z. Q. Li, E. A. Henriksen, Z. Jiang, et al., "Dirac charge dynamics in graphene by infrared spectroscopy," *Nat. Phys.*, vol. 4, no. 7, pp. 532–535, 2008.
[15] K. I. Bolotin, K. J. Sikes, J. Hone, H. L. Stormer, and P. Kim, "Temperature-dependent transport in suspended graphene," *Phys. Rev. Lett.*, vol. 101, no. 9, p. 096802, 2008.
[16] J. H. Chen, C. Jang, S. Xiao, M. Ishigami, and M. S. Fuhrer, "Intrinsic and extrinsic performance limits of graphene devices on SiO_2," *Nat. Nanotechnol.*, vol. 3, no. 4, p. 206, 2008.
[17] W. Zhu, V. Perebeinos, M. Freitag, and P. Avouris, "Carrier scattering, mobilities, and electrostatic potential in monolayer, bilayer, and trilayer graphene," *Phys. Rev. B*, vol. 80, no. 23, p. 235402, 2009.
[18] S. Fratini and F. Guinea, "Substrate-limited electron dynamics in graphene," *Phys. Rev. B*, vol. 77, no. 19, p. 195415, 2008.
[19] M. Liu, X. Yin, E. Ulin-Avila, et al., "A graphene-based broadband optical modulator," *Nature*, vol. 474, no. 7349, pp. 64–67, 2011.
[20] A. Yariv, "Critical coupling and its control in optical waveguide-ring resonator systems," *IEEE Photonics Technol. Lett.*, vol. 14, no. 4, pp. 483–485, 2002.
[21] C. T. Phare, Y. D. Lee, J. Cardenas, and M. Lipson, "Graphene electro-optic modulator with 30 GHz bandwidth," *Nat. Photonics*, vol. 9, no. 8, pp. 511–514, 2015.
[22] W. Bogaerts, P. De Heyn, T. Van Vaerenbergh, et al., "Silicon microring resonators," *Laser Photonics Rev.*, vol. 6, no. 1, pp. 47–73, 2012.
[23] I. Gierz, J. C. Petersen, M. Mitrano, et al., "Snapshots of non-equilibrium Dirac carrier distributions in graphene," *Nat. Mater.*, vol. 12, no. 12, pp. 1119–1124, 2013.
[24] N. M. R. Peres, T. Stauber, and A. H. C. Neto, "The infrared conductivity of graphene on top of silicon oxide," *EPL (Europhysics Letters)*, vol. 84, no. 3, p. 38002, 2008.
[25] M. V. Fischetti, D. A. Neumayer, and E. A. Cartier, "Effective electron mobility in Si inversion layers in metal–oxide–semiconductor systems with a high-κ insulator: the role of remote phonon scattering," *J. Appl. Phys.*, vol. 90, no. 9, pp. 4587–4608, 2001.
[26] S. Kim, S. Shin, T. Kim, et al., "A reliable and controllable graphene doping method compatible with current CMOS technology and the demonstration of its device applications," *Nanotechnology*, vol. 28, no. 17, p. 175710, 2017.
[27] S. Vaziri, V. Chen, L. Cai, et al., "Ultrahigh doping of Graphene using flame-deposited MoO_3," *IEEE Electron Device Lett*, 2020, https://doi.org/10.1109/LED.2020.3018485.
[28] W. S. Leong, H. Gong, and J. T. L. Thong, "Low-contact-resistance graphene devices with nickel-etched-graphene contacts," *ACS Nano*, vol. 8, no. 1, pp. 994–1001, 2014.
[29] J. Yota, H. Shen, and R. Ramanathan, "Characterization of atomic layer deposition HfO_2, Al_2O_3, and plasma-enhanced chemical vapor deposition Si_3N_4 as metal–insulator–metal capacitor dielectric for GaAs HBT technology," *J. Vac. Sci. Technol. A*, vol. 31, no. 1, p. 01A134, 2013.

Supplementary Material: The online version of this article offers supplementary material (https://doi.org/10.1515/nanoph-2020-0363).

Andrey Sushko*, Kristiaan De Greve, Madeleine Phillips, Bernhard Urbaszek, Andrew Y. Joe, Kenji Watanabe, Takashi Taniguchi, Alexander L. Efros, C. Stephen Hellberg, Hongkun Park, Philip Kim and Mikhail D. Lukin

Asymmetric photoelectric effect: Auger-assisted hot hole photocurrents in transition metal dichalcogenides

https://doi.org/10.1515/9783110710687-009

Abstract: Transition metal dichalcogenide (TMD) semiconductor heterostructures are actively explored as a new platform for quantum optoelectronic systems. Most state of the art devices make use of insulating hexagonal boron nitride (hBN) that acts as a wide-bandgap dielectric encapsulating layer that also provides an atomically smooth and clean interface that is paramount for proper device operation. We report the observation of large, through-hBN photocurrents that are generated upon optical excitation of hBN encapsulated $MoSe_2$ and WSe_2 monolayer devices. We attribute these effects to Auger recombination in the TMDs, in combination with an asymmetric band offset between the TMD and the hBN. We present experimental investigation of these effects and compare our observations with detailed, ab-initio modeling. Our observations have important implications for the design of optoelectronic devices based on encapsulated TMD devices. In systems where precise charge-state control is desired, the out-of-plane current path presents both a challenge and an opportunity for optical doping control. Since the current directly depends on Auger recombination, it can act as a local, direct probe of both the efficiency of the Auger process as well as its dependence on the local density of states in integrated devices.

Keywords: Auger excitation; 2D materials; optoelectronics; transition metal dichalcogenides.

***Corresponding author: Andrey Sushko,** Department of Physics, Harvard University, Cambridge, Massachusetts 02138, USA, E-mail: asushko@g.harvard.edu. https://orcid.org/0000-0002-2756-9753
Kristiaan De Greve, Department of Physics, Harvard University, Cambridge, Massachusetts 02138, USA; and Department of Chemistry and Chemical Biology, Harvard University, Cambridge, Massachusetts 02138, USA; and Currently at Imec, Kapeldreef 75, Leuven, Belgium, E-mail: kristiaan.degreve@gmail.com
Madeleine Phillips, Alexander L. Efros and C. Stephen Hellberg, Naval Research Laboratory (NRL), Washington, DC 20375, USA, E-mail: madeleine.phillips.ctr@nrl.navy.mil (M. Phillips), sasha.efros@nrl.navy.mil (A.L. Efros), steve.hellberg@nrl.navy.mil (C.S. Hellberg)
Bernhard Urbaszek, Université de Toulouse, INSA-CNRS-UPS, LPCNO, 135 Avenue Rangueil, 31077 Toulouse, France, E-mail: urbaszek@insa-toulouse.fr
Andrew Y. Joe, Philip Kim and Mikhail D. Lukin, Department of Physics, Harvard University, Cambridge, Massachusetts 02138, USA, E-mail: andrewjoe@g.harvard.edu (A.Y. Joe), pkim@physics.harvard.edu (P. Kim), lukin@physics.harvard.edu (M.D. Lukin)
Kenji Watanabe, Research Center for Functional Materials, National Institute for Materials Science, 1-1 Namiki, Tsukuba, 305-0044, Japan, E-mail: watanabe.kenji.aml@nims.go.jp. https://orcid.org/0000-0003-3701-8119
Takashi Taniguchi, International Center for Materials Nanoarchitectonics, National Institute for Materials Science, 1-1 Namiki, Tsukuba, 305-0044, Japan, E-mail: TANIGUCHI.Takashi@nims.go.jp
Hongkun Park, Department of Physics, Harvard University, Cambridge, Massachusetts 02138, USA; and Department of Chemistry and Chemical Biology, Harvard University, Cambridge, Massachusetts 02138, USA, E-mail: Hongkun_Park@harvard.edu

1 Introduction

Transition metal dichalcogenides (TMDs) [1] have recently attracted significant interest for their optoelectronic properties [2], which are dominated by strongly bound excitons. As van der Waals (vdW) 2D materials, TMDs can be incorporated into complex, high cleanliness vdW heterostructures tailored to a myriad of possible applications [3]. In particular, such systems can be used to isolate and manipulate electronic and excitonic excitations which allow the creation of engineered, controlled quantum systems [4, 5]. In general, such heterostructures rely on hexagonal boron-nitride (hBN) [6] as an atomically clean dielectric encapsulation layer to separate the active materials from each other, surrounding electrostatic gates,

This article has previously been published in the journal Nanophotonics. Please cite as: A. Sushko, K. De Greve, M. Phillips, B. Urbaszek, A. Y. Joe, K. Watanabe, T. Taniguchi, A. L. Efros, C. S. Hellberg, H. Park, P. Kim and M. D. Lukin "Asymmetric photoelectric effect: Auger-assisted hot hole photocurrents in transition metal dichalcogenides" *Nanophotonics* 2021, 10. DOI: 10.1515/nanoph-2020-0397.

and the environment [7]. Typically, these hBN layers are treated as an inert buffer whose wide 6 eV bandgap [8] allows it to serve as both a physical and electronic barrier between different parts of the heterostructure device. Deviations from this simplified picture are mostly considered in the context of trapped charge defects or dielectric breakdown.

In this article, we describe an experimental observation of a novel optoelectronic effect that challenges this simple physical picture of perfectly insulating hBN encapsulation. Specifically, we report the robust observation of a reversible, photoinduced current that appears across the thick, dielectric hBN layer. The current is observed in dozens of devices with varying geometries and hBN thicknesses, consistent with other recent reports [9, 10]. We report on a systematic doping, wavelength, electric field and thickness dependent study of this effect in two different TMD materials, which allows us to unambiguously point to Auger recombination as the central mechanism involved. Our evidence is multifold. Firstly, we observe photocurrents over a wide range of hBN thicknesses (3–90 nm in our devices) and, secondly, when deconvolved from optical doping (Supplementary Figures 4–6), it is spatially uniform throughout all devices – as verified via spatially scanning the excitation beam and by verification using split gate devices. Such uniformity and thickness independence make alternative explanations such as dielectric breakdown or tunneling via in-gap defect states in the hBN unlikely.

On the other hand, our systematic doping-, field- and wavelength dependence studies, corroborated by theoretical modeling of relevant barrier heights in the two material systems, allow us to extract a Fowler–Nordheim tunneling picture that is activated by an Auger process involving holes and excitons, which differs from previous pictures [9, 11] yet has some similarities with hot-carrier effects as previously reported in graphene-hBN heterostructures [12–14]. Crucially, this picture explains the substantial differences in photocurrent efficiency between the two materials directly from computed band alignments, without postulating significant material-dependent variation in the efficiency of Auger excitation. This process adds an important element to the physics of two-dimensional TMD devices by introducing an optically controlled transport path outside the material. Potentially, it can be leveraged to locally sink unipolar currents from an optically defined "contact" that can be arbitrarily swept over a device structure. In addition, we show that the variation in current generation efficiency can provide insight into the dynamics, transitions, and relaxation pathways of states within the TMD.

2 Photocurrents in encapsulated TMDs

Figure 1a shows the schematic of a typical device structure, consisting in this case of a $MoSe_2$ monolayer encapsulated in hBN over a metal gate electrode. Upon off-resonant optical excitation at 660 nm, a substantial current is measured at the bottom gate, sourced from the $MoSe_2$ via its electrical contact. This photocurrent is substantial in magnitude – up to 6 nA for an excitation power of 15 µW – and appears only in the hole doped regime, as inferred from photoluminescence (PL) emission (Figure 1b).

We first investigate the effect in detail using a dual-gated device structure in which the TMD is grounded through a side contact and the field to the top and bottom gates can be independently varied (Figure 2a). The goal of such a structure is to be able to decouple the *doping* conditions from the *electric field*, and deduce the dependence of the effect on either of them independently. Figure 2b outlines the general band alignment of the TMD/hBN/gate electrode system along with the direction of the observed current. Plotting external quantum efficiency

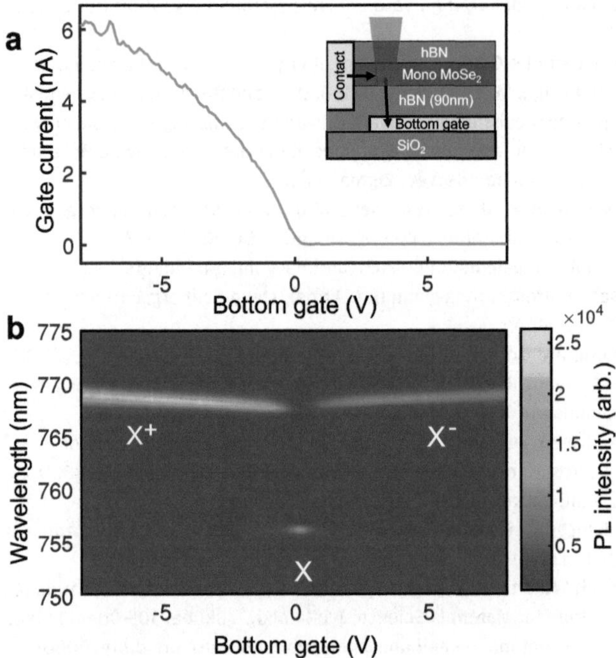

Figure 1: Consistent cross-hexagonal boron nitride (hBN) photocurrent.
a) Gate-voltage dependence of through-hBN photocurrent for the inset device structure (device A) under 15 µW, 660 nm optical excitation. This behavior is spatially uniform and is reproduced in a series of devices of varying hBN thickness. b) Photoluminescent emission from the same device showing presence of photocurrent in the hole doped regime.

(EQE) curves as a function of bottom gate for a $MoSe_2$ device with no top gate (Figure 2c), we first observe linear optical power dependence across three orders of magnitude, consistent with a single-photon excitation process. Here, EQE is given as the number of carriers injected into the gate per photon incident on the device structure. EQE is notably lower for the dual gate $MoSe_2$ device (Figure 2d) due to absorption of incoming photons in the metal top gate. Figure 2d, e shows the top gate current as a function of both gates for $MoSe_2$ and WSe_2, respectively. Cuts along the labeled lines are presented and analyzed in Figure 4. By fixing the potential difference between the TMD and the gate into which current flow is being measured, while varying the potential at the other, we can examine the doping dependence of the photocurrent process at fixed field. At a field of 0.15 V/nm (lines (1) in Figure 2d, e), both $MoSe_2$ and WSe_2 exhibit rapid onset of current upon hole doping, and subsequent saturation with increasing carrier density. Similarly, by incrementing the potential on one gate, in opposition to the other, by a ratio proportional to their relative capacitance, one can maintain constant doping of the sample while varying the field (lines (2) in Figure 2d, e), thereby yielding the pure field dependence, independent of doping effects. While the doping dependences of $MoSe_2$ and WSe_2 appear very similar, the two materials exhibit a qualitative difference in electric field dependence, with the $MoSe_2$ current switching on rapidly at negative field while the WSe_2 current remains negligible until reaching a field of around 0.1 V/nm. Furthermore, even at high field, the EQE of the photocurrent in WSe_2 is an order of magnitude lower than what is observed in $MoSe_2$, a distinction which persists under resonant excitation (see Supplementary Figure 11). The decrease in top gate photocurrent at decreasing bottom gate voltage is attributed to competition between the two gates (see Supplementary Figure 3).

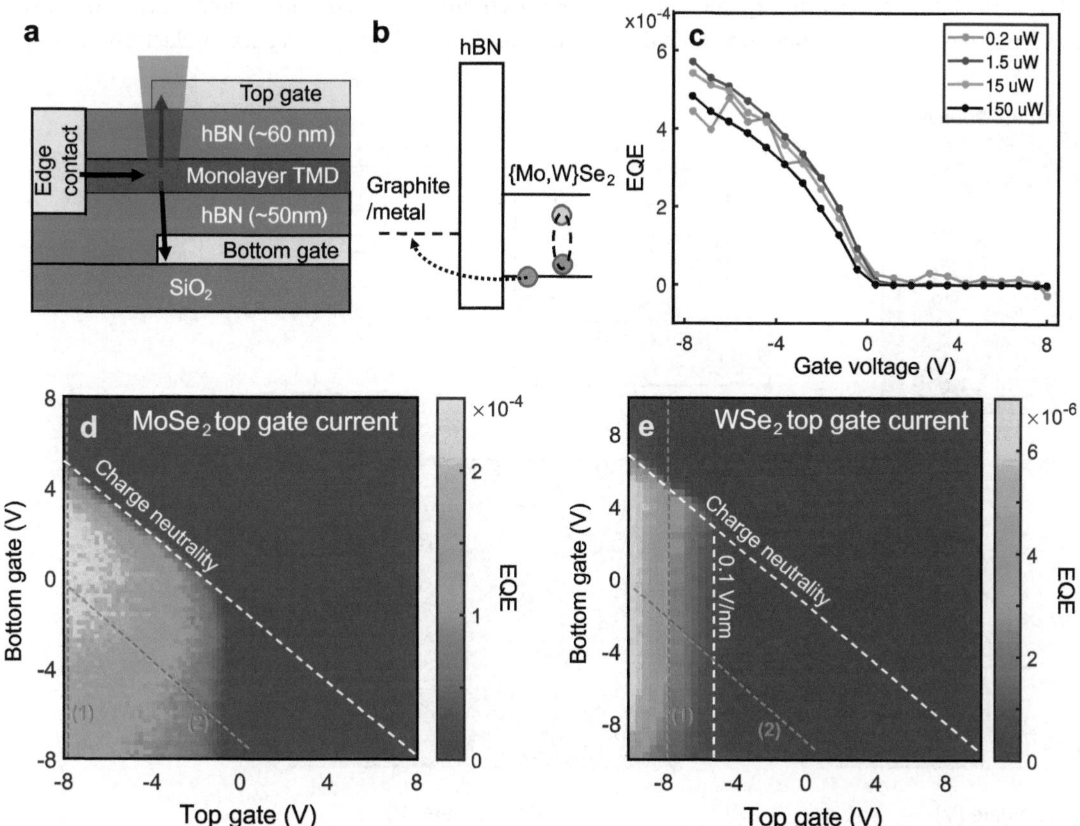

Figure 2: Power, field, and doping dependence in $MoSe_2$ and WSe_2.
a) Schematic of the dual-gated, hexagonal boron nitride (hBN) encapsulated transition metal dichalcogenide (TMD) structure designed to enable independent control of doping and electric field between the TMD and metal gates (device B). b) Band structure schematic for half of the device in (a) illustrating the hole-side photocurrent into one of the top/bottom electrodes when the TMD is hole-doped and electric field oriented toward the electrode. c) Quantum efficiency curves for a single-gated configuration (device A) show linear dependence on optical power. Here, external quantum efficiency (EQE) is the ratio of carriers through the hBN to photons incident on the heterostructure. Dependence of total current on gate conditions for $MoSe_2$ (d) and WSe_2 (e) shows qualitatively distinct characteristics. While both systems require hole doping, WSe_2 also exhibits minimal current below a field of 0.1 V/nm and much lower overall quantum efficiency relative to $MoSe_2$.

Figure 3a shows a spatial map of the photocurrent generated by scanning a diffraction limited excitation spot (see Figure 3b for the sample geometry). A graphite back gate extends under part of a $MoSe_2$ flake, vertically separated from the graphite by 3 nm hBN. This geometry produces a lateral potential step when the gate potential is adjusted relative to the TMD, as schematically shown in Figure 3c. Tuning to the hole side results in enhanced photocurrent generation along the edges of the gate. We attribute this effect to details of the hole-doping mechanism for $MoSe_2$. While the Cr/Au edge contacts are transparent to electrons, they are inefficient at injecting holes into the monolayer. The presence of an in-plane potential step, however, provides an alternative doping mechanism via exciton dissociation upon optical excitation. Indeed, dissociation is energetically favorable for a sufficiently sharp step as long as the in-plane step exceeds the binding energy of the exciton (E_b ~200 meV [15]) as shown in Figure 3c. This process is similar to the well-known, junction-induced exciton dissociation process in organic photodetectors and solar cells, and is common to all semiconductors with tightly bound excitons [16]. The dissociated electron, subsequently, is able to leave via the contact resulting in a net hole photodoping process. As the photocurrent is a sensitive function of the doping in view of the above observations, for sufficient optical power, the steady-state photocurrent will be limited by the rate at which holes can be replenished – which is governed by this very edge-gate photodoping process. This becomes more evident at increasing optical power levels, when other doping mechanisms become comparatively negligible – we refer to Supplementary Figures 4–6 for further details and dual beam, power-dependent measurements confirming this picture. In addition, the doping of the sample can also be inferred from PL emission. As shown in Figure 3d, placing a strong excitation laser at the gate edge while collecting PL from the center results in onset of trion emission whenever the gate potential is greater than E_b from charge neutrality (located at 0.27 V in this device). In contrast, in the absence of edge illumination, the device never accumulates holes or emits any hole-trions (Figure 3e). Interestingly, the onset of electron doping is

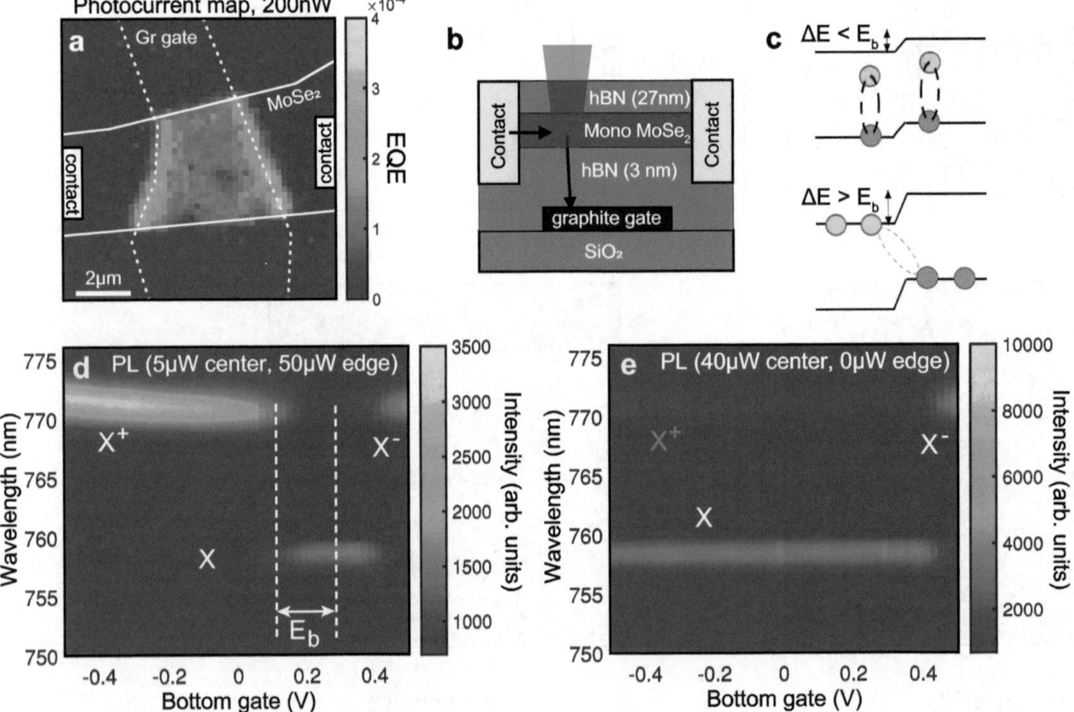

Figure 3: Photodoping via exciton dissociation at lateral potential steps.
a) Spatial photocurrent map of an $MoSe_2$ device (device C) with a local graphite back gate, schematic in (b), for off-resonant excitation at 660 nm. Enhanced current is seen when the excitation laser is located near an edge of the gated region, corresponding to a lateral potential step. c) Schematic of exciton dissociation at a potential step, when the step height exceeds binding energy. Due to the limited ability of Cr/Au edge contacts to inject holes into $MoSe_2$, a hole photocurrent is maintained through neutral-exciton dissociation followed by an electron current into the contacts and a hole current into the gated region. d, e) Photoluminescence (PL) spectra taken from the center of the gated region with and without 660 nm excitation at the gate edge, respectively. The onset of hole doping once the potential step exceeds E_b and lack of trion oscillator strength without edge excitation indicates that photodoping is the primary mechanism for hole-doping the $MoSe_2$ structure.

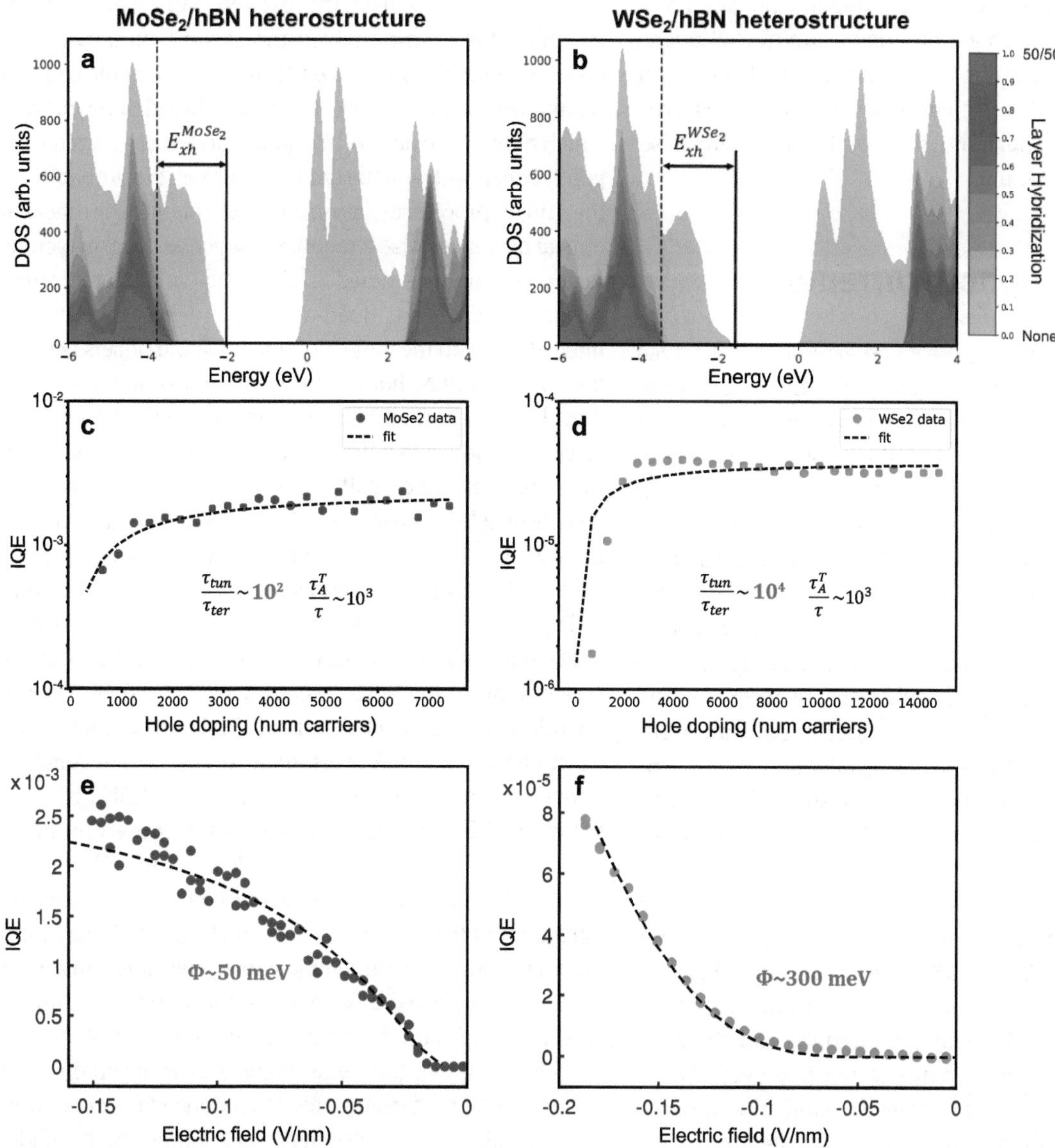

Figure 4: Band structure calculations and kinetics.
Density functional theory (DFT) calculations of hybridization between transition metal dichalcogenide (TMD) and hexagonal boron nitride (hBN) states for $MoSe_2$ (a), and WSe_2 (b) indicate a valence band offset between TMD states and layer-hybridized states on the order of the exciton energy, E_{xh}. c, d) Dual-gate doping dependence of photocurrent from device B fitted to extract hot hole generation rate and relative rates of tunneling and thermalization using a kinetic model. While both materials show a comparable hot hole generation, the lower internal quantum efficiency (IQE) in WSe_2 is explained by substantially slower tunneling relative to thermalization. IQE is obtained from the EQE data in Figure 2, after compensating for absorptivity of the TMD and photon losses in the top gate. Fitting the field-dependence of photocurrent to a Fowler–Nordheim tunneling process gives effective barriers, Φ, of 50 meV in $MoSe_2$ (e) and 300 meV in WSe_2 (f).

unaffected as efficient charge injection can still occur via the contacts, independently of any photodoping effects.

While the former clearly illustrates the role of doping in the process, it does not yet elucidate by which mechanism holes are able to escape the TMD and penetrate or bypass the hBN barrier. For an interface between bulk crystals, a calculation of the band offset along with information about the momentum in the direction perpendicular to the interface would provide the necessary electronic information to understand the transport across the junction. However, for an interface between a monolayer and a bulk crystal, the notion of perpendicular momentum in the monolayer is meaningless in view of the absence of periodicity in this direction. Instead, the relevant picture is one where the entire system is considered as one interface between the monolayer and the bulk. By calculating the properties of this interface, we can then obtain the relevant transport behavior. While the exact values depend on details of the layer interface such as the orientation of the layers, interfacial reconstruction and the exact hBN layer thickness, the trends are clear and explain the observed behavior well – especially, the notable difference between $MoSe_2$ and WSe_2.

3 Physics of photocurrents

To capture the interface physics and explain the observed photocurrents, we start by using first principles methods to compute the density of states (DOS) of a bilayer system that consists of one monolayer of TMD and one monolayer of hBN, after which we color each state according to its layer hybridization (Figure 4a, b). The cyan states are unhybridized, i.e., the state is localized either entirely in the hBN layer or entirely in the TMD layer. The pure cyan states near the band gap in Figure 4a, b should thus be understood as TMD states, as the TMD band gap is much smaller than the hBN band gap. Since the layers form an interface and interact, we now also need to consider hybridization between them, which leads to delocalization across the interface. For example, the magenta states have equal weight in the TMD and hBN layer. The colors in between pure cyan and magenta indicate a state with some weight in each layer, with ratios given according to the scale bar.

To relate these DOS plots to the transmission of charge carriers from the TMD to hBN, we note that hybridized states (denoted by any color other than pure cyan) are effectively delocalized across the interface and therefore represent a pathway for a charge carrier to move between the TMD and the hBN layers. Somewhat similar to the band offset picture in bulk heterojunctions, the distance in energy from the TMD band edge (the cyan states near the gap) to the first hybridized (non-cyan) states can now be considered the effective band offset. In agreement with other studies of $MoSe_2$/hBN systems, the valence band offset is much smaller than the conduction band offset, indicating that holes can much more readily travel from the TMD to the hBN than electrons [9, 10].

These DOS plots elucidate qualitatively why the photocurrent in the $MoSe_2$ system is larger than the photocurrent in the WSe_2 system. They also indicate effective band offsets on the order of the excitonic energy (~1.6–1.7 eV [17, 18], see Figure 4a, b). Since our experiments take place at cryogenic temperatures (6 K), such energies are orders of magnitude higher than the thermal energy in our system. The only process capable of providing such energies is Auger recombination, where the energy of an exciton is transferred nonradiatively to a resident carrier – in our case, a hole. In the presence of free holes, the exciton can undergo Auger recombination, during which the exciton annihilation energy is completely transferred to the hole. The hole can then scatter with a phonon in a very fast broadening process, which allows it to access hybridized states away from the K point. Consistent with the data in Figure 2c, the probability of the Auger process at low excitation intensity is linear in optical power because it requires just one exciton, in contrast to the commonly studied exciton-exciton Auger recombination, which has a quadratic dependence on excitation intensity. Due to the larger conduction band offsets between the TMD and hBN, hot electrons are unable to transfer into the hBN before thermalizing – in line with the observed electron–hole asymmetry in our experiments. In addition, Auger recombination shifts a hole to an energy with a high density of hybridized states in $MoSe_2$, while the same process in WSe_2 leaves the hole at an energy with a much lower density of hybridized states, as indicated by the dashed lines in Figure 4a, b. Thus, Auger recombination is much more likely to result in a hole transmitted to the hBN in the $MoSe_2$ system than in the WSe_2 system. This trend should persist regardless of the exact details of the junction such as the exact hBN thickness. For instance, for the sake of computational efficiency, we model a monolayer of hBN instead of the many different film thicknesses of hBN used in the experiments (see DFT methods for details). Including many layers of hBN should systematically lower the valence band maximum of the hBN in each system [19], which may introduce an additional tunneling barrier the hole must overcome. Yet, a hole in the $MoSe_2$ system would still encounter a smaller tunnel barrier than a hole in the WSe_2 system.

We examine the underlying Auger mechanism in greater detail by deriving the dependence of photocurrent quantum efficiency on doping from a simple kinetic model. In this model, Auger excited holes can either tunnel through an effective barrier (the aforementioned barrier minus the hot hole energy), or thermalize. By properly accounting for charge replenishment and competing (non-Auger) exciton decay paths, we can obtain approximate values for the rates of Auger recombination, hot hole thermalization and tunneling under detailed balance conditions – we refer to the Supplementary materials for details. We can use this model in combination with the deconvolved field- and doping dependence in our devices to extract relevant values. For example, fitting our model to the doping-dependence at fixed field allows us to extract approximate values for the relative times of hot-hole tunneling to thermalization (τ_{tun}/τ_{ter}) and Auger recombination relative to all exciton decay paths (τ_A^T/τ) for each system (Figure 4c, d; Supplementary material for details). We find a comparable τ_A^T/τ of 10^3 for both systems.

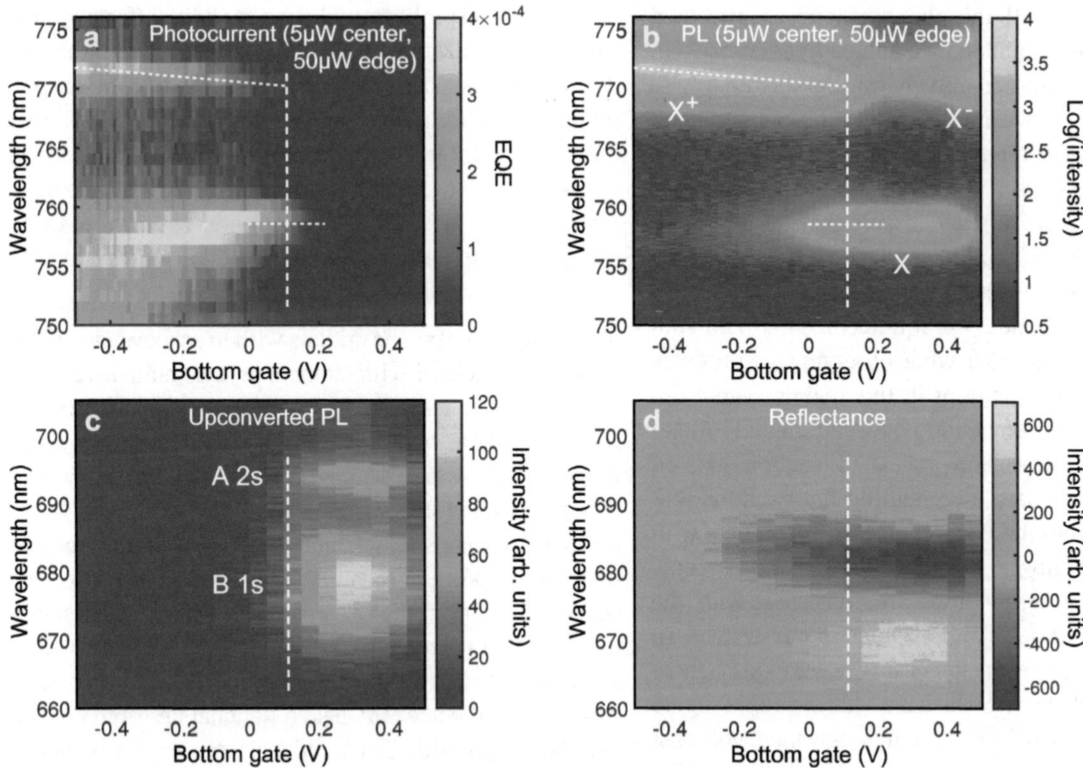

Figure 5: Photocurrent under resonant excitation and competition with exciton upconversion.
a) $MoSe_2$ (device C) photocurrent differential for 50 μW off-resonant gate-edge excitation and 5 μW variable-wavelength center excitation shows current following the exciton and trion resonances seen in photoluminescence (PL) in (b). c) Upconverted photoluminescence spectra taken at 759 nm excitation, documented in literature to arise from Auger excitation of the 1s A exciton, yielding emission of higher Rydberg states, along with the B exciton. d) Reflectance spectra indicating that the B exciton state persists in the hole-doped regime. The onset of hole-doping, however, corresponds to a loss of photoluminescence from the B exciton state and a corresponding onset of photocurrent, suggesting a competition between exciton–exciton Auger and exciton–hole Auger, with the latter dominating in the doped regime.

However, the relative thermalization rate (compared to tunneling) in WSe_2 (10^4) significantly exceeds that of $MoSe_2$ (10^2). This again reflects the higher effective tunnel barrier for WSe_2 as suggested by DFT. For the tunneling process itself, we consider the field dependence of the photocurrent and model it as a Fowler–Nordheim tunneling process (Figure 4e, f). The model reproduces the observed data very well, and allows us to extract effective (net) barrier heights for hot hole tunneling of 50 meV for $MoSe_2$ and 300 meV for WSe_2, consistent with the relative difference seen in DFT calculations.

We next consider the wavelength dependence of the photocurrent under resonant excitation, which clearly confirms the essential role of excitons, consistent with our Auger picture. Figure 5a shows the variation in gate current from the $MoSe_2$ device in Figure 3. We ensure reliable hole doping by photodoping through a strong (50 μW) above-band laser. This value exceeds that of the other rates in our system, and ensures barrier limiting (as opposed to charge-replenishment limited) behavior. When subsequently sweeping a variable wavelength laser (5 μW), we observe a pronounced set of resonances. Comparing those against the photoluminescence emission spectrum (Figure 5b) we observe photocurrent emission coinciding with the exciton and hole-trion resonances – as expected from the perspective of an Auger picture involving excitons and holes. The substantial photocurrent from the neutral exciton in the hole-doped regime, despite low population in PL, suggests an interesting interplay between the hole-exciton scattering mechanisms that create hole-trions (also referred to as attractive polarons [20]) and Auger processes. These observations imply that photocurrent may provide an interesting probe of exciton dynamics, and could be used to shed light on varying decay mechanisms in TMDs as well as novel thermalization physics – as already suggested by our kinetic model.

Finally, we consider similarities between the photocurrent mechanism and a previously documented Auger exciton upconversion process [11]. When a TMD was excited on resonance with the lowest energy 1s exciton, PL emission was observed at higher energy (Rydberg) states [15], including the 2s exciton and B exciton from a higher conduction band. In [11], this phenomenon was attributed to exciton–exciton annihilation, a related Auger process in which one exciton non-radiatively transfers its energy to another, hot exciton. We indeed observe this phenomenon in our $MoSe_2$ devices (Figure 5c), allowing us to examine the relationship between these processes. In reflectance measurements (Figure 5d), we observe that the B exciton state exists in both the neutral and the hole-doped regime. However, in PL, which measures population, we observe a rapid suppression of the B exciton upconverted emission with hole doping (Figure 5c), while observing the presence of a pronounced photocurrent in this regime. These observations suggest a competition between the hole-Auger process and the upconversion process: the exciton–exciton Auger process necessary to create the hot excitons that ultimately relax into the B state appears to compete with the hole-exciton Auger process. From a microscopic perspective, these observations are consistent with the relative densities of holes and excitons in our system. To first order, from a gate capacitance model, we expect a hole density of $\sim 7 \times 10^{12} cm^{-2} V^{-1}$ which exceeds the approximate exciton density of $\sim 10^{10}$ cm^{-2} at only a few megavolt past the onset of hole doping. Assuming somewhat similar exciton–hole and exciton–exciton Auger recombination rates would then indeed suggest a significant suppression of upconversion upon doping due to simple competition between the two processes. More detailed analyses, with independently calibrated carrier and exciton densities, could therefore be used in combination with photocurrent and upconversion measurements to bound the ratio between these respective rates more tightly.

4 Outlook

In conclusion, we have shown that photoexcitation of excitons in hBN-encapsulated TMDs can give rise to a form of "photoelectric effect" for holes that results in a net and substantial current across the nominal hBN dielectric barrier. We attribute this effect due to Auger-generated hot holes being swept through the barrier by the electric field in a tunneling process, which we substantiate by careful field, doping, wavelength and power dependencies. We further support our claims with detailed, ab initio calculations that match well with our observations of a systematically higher effective hole tunnel barrier for WSe_2 as compared to $MoSe_2$. In addition to shedding light on the intrinsic Auger effects in TMDs, which are important to evaluate their device performance as photodetectors and other optoelectronic devices, our studies also demonstrate the spectroscopic potential of photocurrent studies, and provide a novel probe to study non-radiative effects such as carrier thermalization in 2D semiconductors. Intriguingly, if further studies confirm a certain degree of coherence in the photoelectric effect, our findings may open the door for on-chip, integrated probing of the local density of states – in a way similar to advanced spectroscopic techniques such as ARPES, but with greatly reduced complexity and with nanoscale resolution.

Acknowledgments: The authors wish to acknowledge Falko Pientka and Richard Schmidt for many insightful discussions concerning this research. This work was supported by the NSF, CUA, DOE and Vannevar Bush Faculty Fellowship Program. A.S. Acknowledges support from the Fannie and John Hertz Fellowship and Paul and Daisy Soros Fellowships for New Americans. Al.L.E. and C.S.H acknowledge support from the US Office of Naval Research. Al.L.E. also acknowledges support from the Laboratory-University Collaboration Initiative (LUCI) program of the DoD Basic Research Office. B.U. acknowledges a NanoX/NEXT travel grant. This research was performed while M.P. held a National Research Council associateship at NRL. Computational work was supported by a grant of computer time from the DoD High Performance Computing Modernization Program at the U.S. Army Research Laboratory and the U.S. Air Force Research Laboratory Supercomputing Resource Centers (NRLDC04123333). K.W. and T.T. acknowledge support from the Elemental Strategy Initiative conducted by the MEXT, Japan, Grant Number JPMXP0112101001, JSPS KAKENHI Grant Number JP20H00354 and the CREST (JPMJCR15F3), JST.

Author contribution: All the authors have accepted responsibility for the entire content of this submitted manuscript and approved submission.

Research funding: This work was supported by the NSF, CUA, DO and Vannevar Bush Faculty Fellowship Program. A.S. Acknowledges support from the Fannie and John Hertz Fellowship and Paul and Daisy Soros Fellowships for New Americans. Al.L.E. and C.S.H acknowledge support from the US Office of Naval Research. Al.L.E. also acknowledges support from the Laboratory-University Collaboration Initiative (LUCI) program of the DoD Basic Research Office. B.U. acknowledges a NanoX/NEXT travel grant. This research was performed while M.P. held a National Research Council associateship at NRL. Computational work was supported by a grant of computer time from the DoD High Performance Computing Modernization Program at the U.S. Army Research Laboratory and the U.S. Air Force Research Laboratory Supercomputing Resource Centers (NRLDC04123333). K.W. and T.T. acknowledge support from the Elemental Strategy Initiative conducted by the MEXT, Japan, Grant Number JPMXP0112101001,

JSPS KAKENHI Grant Number JP20H00354 and the CREST (JPMJCR15F3), JST.
Conflict of interest statement: The authors declare no conflicts of interest regarding this article.

References

[1] S. Manzeli, D. Ovchinnikov, D. Pasquier, O. V. Yaziev, and A. Kis, "2D transition metal dichalcogenides," *Nat. Rev. Mater.*, vol. 2, p. 17033, 2017.
[2] K. F. Mak and J. Shan, "Photonics and optoelectronics of 2D semiconductor transition metal dichalcogenides," *Nat. Photonics*, vol. 10, p. 216, 2016.
[3] A. K. Geim and I. V. Grigorieva, "Van der Waals heterostructures," *Nature*, vol. 499, p. 419, 2013.
[4] R. Bekenstein, I. Pikovski, H. Pichler, E. Shahmoon, S. F. Yelin, and M. D. Lukin, "Quantum metasurfaces with atom arrays," *Nat. Phys.*, vol. 16, p. 676, 2020.
[5] G. Scuri, Y. Zhou, A. High, et al., "Large excitonic reflectivity of monolayer $MoSe_2$ encapsulated in hexagonal boron nitride," *Phys. Rev. Lett.*, vol. 120, p. 037402, 2018.
[6] C. R. Dean, A. F. Young, I. Meric, et al., "Boron nitride substrates for high-quality graphene electronics," *Nat. Nanotechnol.*, vol. 5, p. 722, 2010.
[7] M. Yankowitz, Q. Ma, P. Jarillo-Herrero, and B. J. LeRoy, "van der Waals heterostructures combining graphene and hexagonal boron nitride," *Nat. Rev. Phys.*, vol. 1, p. 112, 2019.
[8] G. Cassabois, P. Valvin, and B. Gil, "Hexagonal boron nitride is an indirect bandgap semiconductor," *Nat. Photonics*, vol. 10, p. 262, 2016.
[9] E. Linardy, D. Yadav, D. Vella, et al., "Harnessing exciton–exciton annihilation in two-dimensional semiconductors," *Nano Lett.*, vol. 20, p. 1647, 2020.
[10] C. M. E. Chow, H. Yu, J. R. Schaibley, et al., "Monolayer semiconductor Auger detector," *Nano Lett.*, vol. 20, no. 7, pp. 5538–5543, 2020.
[11] B. Han, C. Robert, E. Courtade, et al., "Exciton states in monolayer MoSe2 and MoTe2 probed by upconversion spectroscopy," *Phys. Rev. X*, vol. 8, p. 031073, 2018.
[12] N. M. Gabor, J. C. W. Song, Q. Ma, et al., "Hot carrier–assisted intrinsic photoresponse in graphene," *Science*, vol. 334, p. 648, 2011.
[13] Q. Ma, T. I. Andersen, N. Nair, et al., "Tuning ultrafast electron thermalization pathways in a van der Waals heterostructure," *Nat. Phys.*, vol. 12, p. 455, 2016.
[14] K. J. Tielrooij, J. C. W. Song, S. A. Jensen, et al., "Photoexcitation cascade and multiple hot-carrier generation in graphene," *Nat. Phys.*, vol. 9, p. 248, 2013.
[15] A. Chernikov, T. C. Berkelbach, H. M. Hill, et al., *Phys. Rev. Lett.*, vol. 113, p. 076802, 2014.
[16] M. Massicotte, F. Vialla, P. Schmidt, et al., "Dissociation of two-dimensional excitons in monolayer WSe_2," *Nat. Commun.*, vol. 9, p. 1633, 2018.
[17] S. Shree, M. Semina, C. Robert, et al., "Observation of exciton-phonon coupling in MoSe2 monolayers," *Phys. Rev. B*, vol. 98, p. 035302, 2018.
[18] E. Courtade, M. Semina, M. Manca, et al., "Charged excitons in monolayer WSe2: Experiment and theory," *Phys. Rev. B*, vol. 96, p. 085302, 2017.
[19] D. Wickramaratne, L. Weston, and C. G. Van de Walle, "Monolayer to bulk properties of hexagonal boron nitride," *J. Phys. Chem. C*, vol. 122, p. 25524, 2018.
[20] M. Sidler, P. Back, O. Cotlet, et al., "Fermi polaron-polaritons in charge-tunable atomically thin semiconductors," *Nat. Phys.*, vol. 13, p. 255, 2016.

Supplementary material: The online version of this article offers supplementary material (https://doi.org/10.1515/nanoph-2020-0397).

Harry A. Atwater*

Seeing the light in energy use

https://doi.org/10.1515/9783110710687-010

As photonics researchers work from home, and as we enter the first stages of re-entry into our laboratories, the COVID-19 pandemic has reminded us of two things: 1) meeting global challenges requires globally scalable solutions and 2) infrastructure is important. The current health pandemic, uppermost on everyone's mind at present, is a cautionary reminder of other worldwide challenges—foremost being the need to promptly begin an energy transition to a future steady-state of net-zero carbon energy use and eventually to secure large-scale methods for stabilizing our planet's temperature. What does photonics have to do with this transition? Growth in solar energy is one of the biggest good news stories about photonics science and technology that almost no one seems to be remarking!

Solar energy conversion, which in its current form means the solar photovoltaics industry, has quietly risen to be the world's largest optoelectronics industry—>$150B in 2020 [1, 2]—rivaling in size and turnover the display industry [3] but growing faster, and much larger than, e.g., the solid-state lighting and fiber-telecommunications industries. During 2019–2020, and continuing during the COVID-19 lockdown, renewable energy consumption in the United States passed energy consumption from coal for the first time in 130 years. In this time period, among renewable electricity generation sources in the United States (wind, solar, geothermal, biomass and hydropower), solar photovoltaics grew faster than any other source. Both in the United States and around the world, growth in solar photovoltaic energy systems have been quietly dominating new investments in energy generation. The reason: generating electricity using solar photovoltaics is now less expensive in many locations than using any fossil fuel. You might object and say "What about electricity storage? The sun does not shine at night, and so you will need a way to store electricity". Indeed, this is true. However, recent power purchase agreements for electricity generation systems based on solar-plus-battery-storage are also underbidding competing proposals for electricity generation from fossil fuels [4].

What does the future for solar energy conversion hold, and how can photonics researchers have an impact in this vitally important field? I foresee three areas as exciting opportunities for research, with potentially large impacts on world energy use.

First, we need to increase the efficiency of photovoltaics and accelerate the penetration of photovoltaics in the energy sector. Increasing the efficiency of solar cells touches immediately on fundamental photonics and physics issues, such as the flux balance of absorbed and emitted light and the radiative quantum efficiency of materials. It also harbors opportunities for nanophotonic design in light trapping, in creating antireflection coatings for solar cells—and even in giving solar cells distinct bright colors when, for instance, they are incorporated into building architectural facades.

Second, we can learn to emulate nature's amazing feat of photosynthesis: harvesting solar photons and converting the generated charge carriers into products including fuels, chemicals and polymeric materials. This field of solar fuels blends the primary disciplines of chemistry, physics and engineering, and has many opportunities for photonic design. The photonics challenges include harvesting and exploiting the full solar spectrum and enabling nanophotonic design to create efficient systems that combine semiconductors for light absorption, with often optically opaque catalysts for chemical conversion.

Finally, as our planet warms, one of the biggest challenges for humankind is staying cool. Globally, tropical and equatorial regions are experiencing the fastest growth in population and gross domestic product. Recently, in aggregate worldwide, increases in energy use for cooling have surpassed increases in energy use for heating [5]. The concept of radiative cooling by photonic design, in systems that couple energy from blackbody absorbers at the Earth's surface to the 3 K thermal reservoir in the sky, has rapidly advanced in the research literature over the last few years. If efficient and deployable radiative cooling systems could be developed, this science concept could transform into a technology innovation substantially offsetting increases in global energy use.

As photonics researchers emerge from the COVID-19 lockdown, there is an opportunity to reflect on the future

***Corresponding author: Harry A. Atwater**, California Institute of Technology, Pasadena, CA, USA, E-mail: haa@caltech.edu. https://orcid.org/0000-0001-9435-0201

This article has previously been published in the journal Nanophotonics. Please cite as: H. A. Atwater "Seeing the light in energy use" *Nanophotonics* 2021, 10. DOI: 10.1515/nanoph-2020-0381.

challenges facing our use of energy. There is a great deal to be done, with the promise of a large impact on our world.

Author contribution: The author has accepted responsibility for the entire content of this submitted manuscript and approved submission.

Research funding: This research was funded by U.S. Department of Energy, DE-SC0021266.

Conflict of interest statement: The author declares no conflicts of interest regarding this article.

References

[1] The global Photovoltaics market size is projected to reach US$ 257 billion by 2026, from US$ 231.15 billion in 2020, at a CAGR of 10.4% during 2021–2026; source: https://www.360researchreports.com/global-photovoltaics-market-15911033.

[2] The Solar System market size can also be estimated by multiplying the 115 GW worldwide Solar System production in by the utility-scale photovoltaic system-level cost/Watt of $1.34, yielding a market size estimate of $154B in 2019; source: NREL Q4 2019/Q1 2020 Solar Industry Update, D. Feldman and R. Margolis, May 28, 2020, https://www.nrel.gov/docs/fy20osti/77010.pdf.

[3] The global Display Panel market is valued at $103 billion in 2020 is expected to reach $131 billion by the end of 2026, growing at a CAGR of 3.5% during 2021–2026; source: Global Display Panel Market Research Report 2020, https://www.marketstudyreport.com/reports/global-display-panel-market-research-report-2020?gclid=EAIaIQobChMIu8q96_PT7AIVVR6tBh3SAg-tEAMYASAAEgJ9X_D_BwE.

[4] "Los Angeles OKs a deal for record-cheap solar power and battery storage", Los Angeles Times, September 10th, 2019; https://www.latimes.com/environment/story/2019-09-10/ladwp-votes-on-eland-solar-contract.

[5] "World set to use more energy for cooling than heating", the Guardian, October 26th, 2015; https://www.theguardian.com/environment/2015/oct/26/cold-economy-cop21-global-warming-carbon-emissions.

Part II: **Lasers, Active optical devices and Spectroscopy**

Sara Mikaelsson*, Jan Vogelsang, Chen Guo, Ivan Sytcevich, Anne-Lise Viotti, Fabian Langer, Yu-Chen Cheng, Saikat Nandi, Wenjie Jin, Anna Olofsson, Robin Weissenbilder, Johan Mauritsson, Anne L'Huillier, Mathieu Gisselbrecht and Cord L. Arnold

A high-repetition rate attosecond light source for time-resolved coincidence spectroscopy

https://doi.org/10.1515/9783110710687-011

Abstract: Attosecond pulses, produced through high-order harmonic generation in gases, have been successfully used for observing ultrafast, subfemtosecond electron dynamics in atoms, molecules and solid state systems. Today's typical attosecond sources, however, are often impaired by their low repetition rate and the resulting insufficient statistics, especially when the number of detectable events per shot is limited. This is the case for experiments, where several reaction products must be detected in coincidence, and for surface science applications where space charge effects compromise spectral and spatial resolution. In this work, we present an attosecond light source operating at 200 kHz, which opens up the exploration of phenomena previously inaccessible to attosecond interferometric and spectroscopic techniques. Key to our approach is the combination of a high-repetition rate, few-cycle laser source, a specially designed gas target for efficient high harmonic generation, a passively and actively stabilized pump-probe interferometer and an advanced 3D photoelectron/ion momentum detector. While most experiments in the field of attosecond science so far have been performed with either single attosecond pulses or long trains of pulses, we explore the hitherto mostly overlooked intermediate regime with short trains consisting of only a few attosecond pulses. We also present the first coincidence measurement of single-photon double-ionization of helium with full angular resolution, using an attosecond source. This opens up for future studies of the dynamic evolution of strongly correlated electrons.

Keywords: attosecond science; electron momentum spectroscopy; high-order harmonic generation; ultrafast photonics.

1 Introduction

The advent of attosecond pulses in the beginning of the millennium [1, 2] enabled the study of fundamental light-matter interactions with unprecedented time resolution [3], revealing subfemtosecond electron dynamics in atoms, molecules and solids, such as ionization time delays [4–7], the change of dielectric polarizability [8], and the timescale of electron correlations [9, 10].

Attosecond pulses are generated through high-order harmonic generation (HHG), when intense femtosecond pulses are focused into a generation gas [11]. Close to the peak of each half-cycle, an electron wave packet is born through tunnel ionization. It is subsequently accelerated by the electric field of the driving laser pulse and finally may return to its parent ion and recombine, upon which its excess energy is emitted as an attosecond pulse in the extreme ultraviolet (XUV) to soft-X-ray spectral range [12, 13]. The process repeats itself for every half-cycle of the driving field, resulting in an attosecond pulse train (APT) in the time domain and a comb of odd-order harmonics in the frequency domain. If the emission originates from only one half-cycle, a single attosecond pulse (SAP) is emitted with a continuous frequency spectrum.

Two well-established pump-probe techniques, based on cross-correlating the attosecond pulses with a low frequency field (usually a replica of the generating pulse) while the photoelectron spectrum originating from a detection gas is

***Corresponding author: Sara Mikaelsson**, Department of Physics, Lund University, Professorsgatan 1, 223 63 Lund, Sweden, E-mail: sara.mikaelsson@fysik.lth.se
Jan Vogelsang, Chen Guo, Ivan Sytcevich, Anne-Lise Viotti, Fabian Langer, Yu-Chen Cheng, Saikat Nandi, Anna Olofsson, Robin Weissenbilder, Johan Mauritsson, Anne L'Huillier, Mathieu Gisselbrecht and Cord L. Arnold, Department of Physics, Lund University, Professorsgatan 1, 223 63 Lund, Sweden.
https://orcid.org/0000-0002-9664-6265 (J. Vogelsang).
https://orcid.org/0000-0002-1335-4022 (I. Sytcevich).
https://orcid.org/0000-0003-4249-411X (C.L. Arnold)
Wenjie Jin, ASML Veldhoven, De Run 6501, 5504 DR, Veldhoven, The Netherlands

This article has previously been published in the journal Nanophotonics. Please cite as: S. Mikaelsson, J. Vogelsang, C. Guo, I. Sytcevich, A.-L. Viotti, F. Langer, Y.-C. Cheng, S. Nandi, W. Jin, A. Olofsson, R. Weissenbilder, J. Mauritsson, A. L'Huillier, M. Gisselbrecht and C. L. Arnold "A high-repetition rate attosecond light source for time-resolved coincidence spectroscopy" *Nanophotonics* 2021, 10. DOI: 10.1515/nanoph-2020-0424.

recorded, give access to dynamics on the attosecond time scale. The RABBIT technique (Reconstruction of Attosecond harmonic Beating By Interference of Two-photon transitions) is well suited for the characterization and use of APTs, while the streaking technique is mostly applied with SAPs [1, 2]. The requirements to perform such experiments are challenging in terms of laser sources, HHG and pump-probe interferometric optical setups, as well as photoelectron detectors. Traditionally, mostly chirped pulse amplification, Titanium:Sapphire-based lasers with repetition rates in the low kHz range have been used, rendering experiments that have high demands on statistics or signal-to-noise ratio (SNR) time consuming [9].

Here, we present a high-repetition rate, flexible attosecond light source, particularly designed for the study of gas phase correlated electron dynamics, as well as time-resolved nanoscale imaging. This article both summarizes and extends the previous work [14, 15]. The laser system, located at the Lund High-Power Facility of the Lund Laser Centre, is based on optical parametric chirped pulse amplification (OPCPA), providing sub-6fs long pulses with stabilized carrier-to-envelope phase (CEP) in the near-infrared (IR) with up to 15 μJ pulse energy at a repetition rate of 200 kHz [14]. The 200-fold increase in repetition rate, compared to standard 1 kHz systems, promotes experiments with high demands on statistics. A three-dimensional (3D) momentum spectrometer [16], capable of measuring several correlated photoelectrons/ions in coincidence, fully resolved in momentum and emission direction, has been installed as a permanent experimental end station. The combination of CEP control, where the phase of the electric field is locked to the pulse envelope, and the short pulses, comprising only two cycles of the carrier wavelength, provides control of the characteristics of the generated APTs. For example, we can choose the number of pulses in the train to be equal to two or three pulses, which allows exploring the transition between the traditional streaking and RABBIT regimes. As a proof of concept for a statistically very demanding experiment, we present measurements on single-photon double-ionization of helium, which is an archetype system of strongly correlated pairs of photoelectrons [17–19].

Figure 1 shows a schematic of the experimental setup. The laser pulses enter the first vacuum chamber (A), green, which contains a pump-probe interferometer, a small gas jet for HHG and an XUV spectrometer for diagnostics. The XUV pump and IR probe are then focused by a toroidal mirror (B), yellow, into the sensitive region of the 3D momentum spectrometer (C), red. A refocusing chamber (D), purple, contains a second toroidal mirror for reimaging to the second interaction region, (E), where different end stations for surface science, usually a photoemission electron microscope, can be installed [20–22]. The beamline is designed for simultaneous operation of both end stations.

The paper is structured as follows: The first section introduces the optical setup by briefly discussing the laser source and the XUV-IR pump-probe interferometer, before examining the gas target design for HHG and the control of the emitted APTs and finally, introducing the 3D photoelectron/ion spectrometer. The following section discusses the utilization of our light source for attosecond time-resolved spectroscopy. We close by presenting measurements of the fully differential cross section for double ionization of helium, an experiment that to our best knowledge has not been performed with attosecond pulses before.

2 XUV light source and pump-probe setup

2.1 Laser source characterization

The OPCPA laser that the beamline is operated with is seeded by a Titanium:Sapphire ultrafast oscillator. The seed pulses are amplified in two noncollinear optical parametric amplification stages, pumped by a frequency-doubled, optically synchronized Ytterbium-fiber chirped pulse amplifier. The details of the laser are described elsewhere [14]. Figure 2 shows temporal characterization of the output pulses performed with the dispersion scan (d-scan) technique [23], revealing a pulse duration of 5.8 fs full width at half maximum (FWHM) at a carrier wavelength in the near-infrared of approximately 850 nm.

The CEP stability of the laser was measured out-of-loop with a f:2f interferometer [24], capable of single-shot

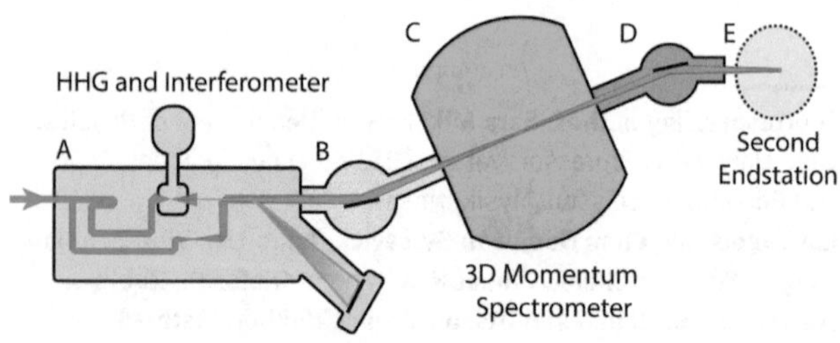

Figure 1: Beamline footprint.
(A) High-order harmonic generation (HHG) and characterization and extreme ultraviolet-infrared (XUV-IR) interferometer. (B) Focusing chamber. (C) First interaction region, three-dimensional (3D) momentum spectrometer. (D) Refocusing chamber. (E) Second interaction region, flexible endstation.

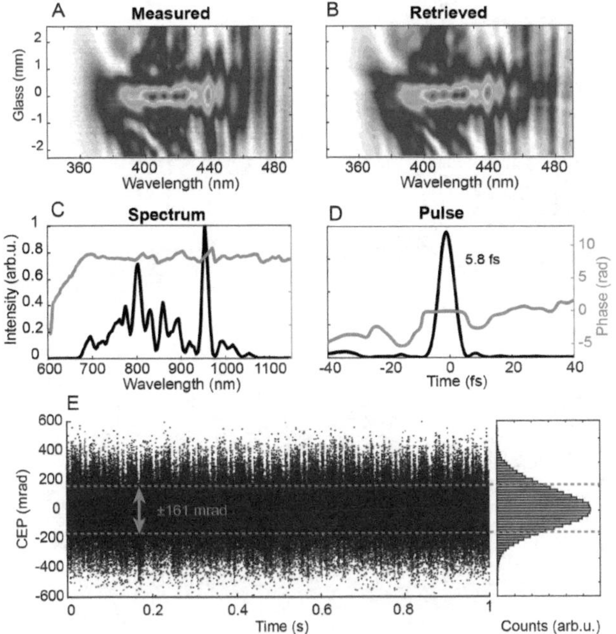

Figure 2: Laser output characterization.
(A) Measured and (B) retrieved d-scan traces; a d-scan trace is a two-dimensional representation of the second harmonic signal (intensity in colors) as a function of wavelength and glass insertion; (C) retrieved spectrum (black) and spectral phase (red), (D) retrieved temporal pulse intensity profile (black) and phase (red), (E) single-shot carrier-to-envelope phase (CEP) measurement, showing the measured CEP as a function of time; the root mean square is indicated in red (see the histogram plot on the right).

acquisition at full repetition rate. This is achieved by recording the spectral interference, encoding the CEP signal, with a high-speed line camera (Basler) with an acquisition rate of >200 kHz. Figure 2E shows measurement results demonstrating a short-term CEP stability with a root mean square of approximately 160 mrad. Single-shot CEP detection is an essential tool to ensure that our light source is stable. Furthermore, the CEP data can be tagged to additionally reduce the noise in photoelectron/ion data. An additional f:2f interferometer is used to actively compensate for long-term drift.

2.2 XUV-IR interferometer

The XUV-IR pump-probe interferometer is designed for high temporal stability. Suppressing instabilities from vibrations and drifts as much as possible is essential for recording data with high SNR [25] over several hours or even days. To promote passive mechanical stability, all optical components of the interferometer are mounted on a single mechanically stable breadboard inside the main vacuum chamber. The breadboard is stiffly mounted to an optical table but vibrationally decoupled from the vacuum chamber in order to isolate from vibrations originating from turbo and roughing pumps. Figure 3A shows the interferometer layout and beam path.

After entering the vacuum chamber, the laser pulses are split into pump- and probe arms of the interferometer by a thin beam splitter optimized for ultrashort laser pulses (Thorlabs UFBS2080), reserving 80% of the power for HHG in the so-called pump arm and 20% for the probe arm. In view of the short pulse duration and broad spectrum of the few-cycle pulses, dispersion management between the two interferometer arms is imperative. It is achieved by a 1 mm thick AR-coated glass plate in the probe arm (compensating the beam splitter) and a pair of fused silica wedges at Brewster angle in the pump arm (for other dispersive elements). A spurious reflex from the glass plate in the probe arm is used to feed a beam pointing stabilization system (TEM-Messtechnik) outside the vacuum chamber.

The delay between the interferometer arms is controlled by a retro reflector in the pump arm, mounted on a linear piezo stage (Piezo Jena Systems) with 80 μm travel. The pump arm is then focused by a 90°, off-axis parabola into the generation gas target for HHG, shown in an insert in Figure 3A and described in more detail in the next section. In the probe arm, a lens is used to mimic the focusing in the pump arm, ensuring that after recombination pump- and probe pulses will focus at the same position. In the pump arm, the generated XUV attosecond pulses are separated from the driving near-IR pulses by different thin (usually 200 nm) metallic filters, transparent to the XUV but opaque to the driving pulses. In order to reduce the thermal load on the filter, an aperture cutting most of the driving pulses is placed before the filter, utilizing the smaller divergence of the attosecond pulses. A holey mirror at 45° is used for recombining the probe- and pump arms of the interferometer (see recombination mirror inset in Figure 3A). The XUV attosecond pulses pass through a 1.6 mm wide hole, chosen such that the XUV beam is barely clipped, while the light from the probe arm is reflected as a ring by the holey surface. Afterwards, a toroidal mirror is used to image the foci of the pump- and probe interferometer arms in 2f:2f configuration into the sensitive region of the 3D photoelectron spectrometer placed downstream. The focal length of the lens in the probe arm of the interferometer is chosen to homogeneously illuminate the toroidal mirror, at 78° angle of incidence with a clear aperture of 125 × 25 and 350 mm focal length, and thus maximize the pulse energy that is refocused.

In order to stabilize the delay for slow thermal drift, we couple a Helium-Neon (He-Ne) laser beam into the

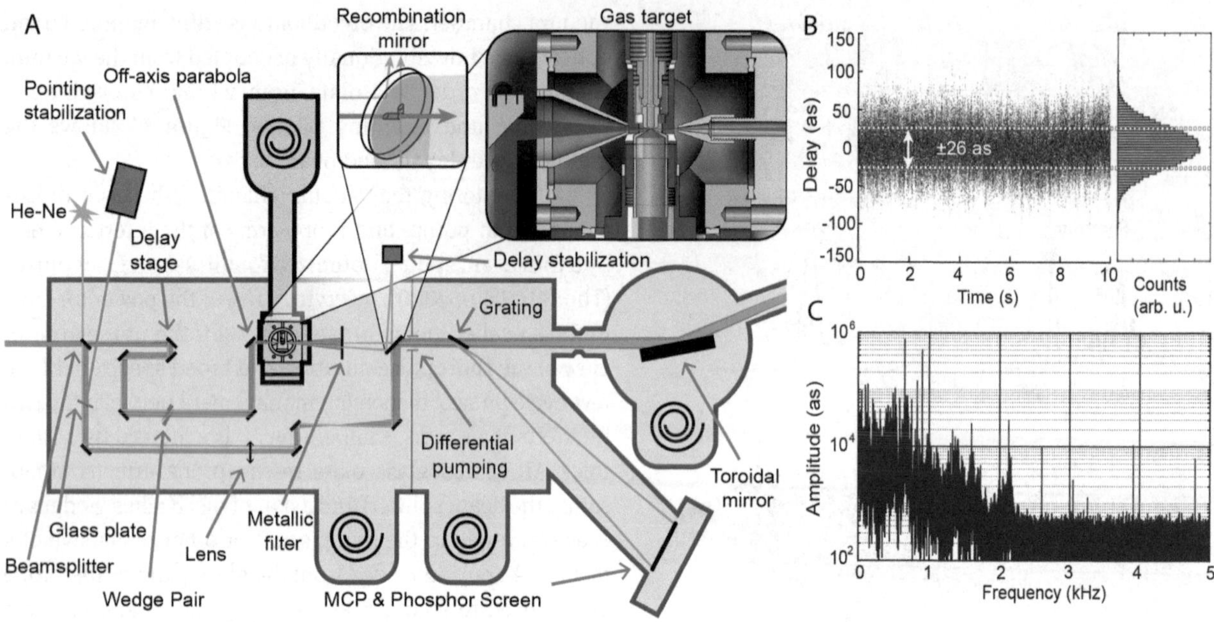

Figure 3: Extreme ultraviolet-infrared (XUV-IR) interferometer.
(A) Interferometer beam path and components. The near-IR beam path is shown in red, the XUV in blue, and a HeNe-laser which can be used for active delay stabilization is shown in green. (B) Stability measurement, showing the measured delay as a function of time; the root mean square is indicated in white (see the histogram plot on the right). (C) Single-sided amplitude spectrum of interferometer stability measurement.

interferometer, shown in green in Figure 3A. This is achieved by illuminating the beamsplitter from the so far unused side. The metallic filter after HHG is suspended on a transparent substrate to transmit the part of the HeNe-laser beam with large divergence, which is then reflected by the back side polished recombination mirror (see recombination mirror inset of Figure 3A). Light from the other arm of the interferometer is transmitted through another hole in the recombination mirror. A spatial interference pattern is observed with a camera and used to record the phase drift in order to feed back to the delay stage. Figure 3B shows the short-term stability of the interferometer, determined by using a fast CCD detector (10 kHz readout). This fast detection enables us to get information on high acoustic frequencies that might contribute to instabilities (Figure 3C). The sharp peaks in Figure 3C can be attributed to the different turbo pumps. A short-term stability of 26 as is achieved.

2.3 Gas target

As a consequence of the high-repetition rate of the few-cycle laser system, the energy of the individual pulses is rather low, compared to the kHz repetition rate Titanium:Sapphire lasers often used for generating high-order harmonics. This has important implications for HHG: First, tight focusing is required to achieve sufficiently high intensity and second, phase-matching and scaling considerations imply a localized, high density generation gas target [26, 27]. The off-axis parabola used to focus the driving pulses into the gas target has a focal length of $f = 5$ cm, resulting in a focal radius around 5 μm and intensities in excess of several 10^{14} W/cm^2 can easily be achieved. Numerical simulations, based on solving the time dependent Schrödinger equation for the nonlinear response with argon as generation gas and the wave equation for propagation [28, 29], suggest a gas density corresponding to 5 bars of pressure (at standard temperature) and a medium length of 40 μm. The gas density should fall off as rapidly as possible outside the interaction region in order to avoid reabsorption of the generated XUV radiation. Such a gas target imposes substantial engineering challenges.

The gas target consists of a nozzle with an exit hole of 42 μm, operated for the case of argon at 12 bars of backing pressure. The small nozzle exit diameter ensures longitudinal (i.e., along the laser propagation direction) and transverse confinement of the interaction region. However, the high mass flow rate, around 4×10^{-3} g/s, of such a nozzle, if used to inject gas directly into the vacuum, would challenge the capacity of the turbo pumps and contaminate the vacuum inside the chamber, resulting in severe reabsorption. Therefore, we use a catcher with a hole of 1 mm diameter mounted less than 200 μm from the injection nozzle and connected to a separate roughing pump. The goal is to catch the majority of the injected gas directly before it expands into the vacuum. The nozzle-catcher configuration is attached to a manual three-dimensional translation stage, for adjusting the position of the gas target with respect to the laser focus without changing the positions of nozzle and catcher with respect to each other. Comparing the pressure inside the vacuum chamber with and without the catcher indicates that the catcher takes away >90% of the injected gas.

In order to further reduce possible contamination of the surrounding vacuum, the gas target is placed in an additional small chamber, pumped by an extra turbo pump (see gas target inset in Figure 3A). The laser enters and exits the cube-shaped chamber through differential pumping holes, placed as closely to the interaction region as possible. As a result, the beam only traverses a distance of approximately 3 mm inside this comparably high pressure environment. During regular operation, when generating in argon (12 bar backing pressure), the pressure inside the cube is around 3×10^{-3} mbar, while the pressure in the surrounding chamber remains at 2.5×10^{-5} mbar. Following the beam path toward the first experimental chamber, the pressure further reduces via differential pumping to 1×10^{-6} mbar in the XUV spectrometer compartment, to 2×10^{-8} mbar in the chamber housing the toroidal mirror and finally to 2×10^{-9} mbar at the 3D photoelectron spectrometer. A pressure below 5×10^{-10} mbar in the second end station is easily reached.

Simulations of the gas density in the interaction region were performed using the STARCCM + compressible flow solver. The results are summarized in Figure 4. The line-out along the direction of the gas flow (Figure 4B) indicates that the density drops rapidly away from the nozzle. Figure 4C shows a line-out transversely to the gas flow at a distance of 10 μm from the nozzle exit (marked with a dashed blue line), which corresponds to the approximate distance at which the laser traverses the gas target during operation. According to the simulations, the gas target has an approximately super-Gaussian shape with a width of 36 μm (FWHM) at this distance to the nozzle with a peak density of 8.5 kg/m^3, corresponding to 4.7 times the density of argon at standard conditions (1.7 kg/m^3). The obtained density and gas medium length are in excellent agreement with the calculated best phase-matching conditions (5 bar, 40 μm). Experimentally, we used the fringe pattern resulting from the interference of the stabilization He-Ne laser pump and probe beams (see Figure 3A) in order to get information on the actual gas density. We measured a phase shift of the fringes, between the cases of active and inactive gas target, equal to 0.65 rad. Using refractive index data for argon [30], a phase shift of 0.625 rad can be calculated from the simulation data in 4C, indicating that the simulation results are in close match to the actual conditions in our gas target.

2.4 High-order harmonic generation

The process of HHG significantly changes from the case of multicycle, long driving pulses, where the attosecond pulses emitted from subsequent half-cycles are nearly identical except for a π-phase shift between them, to the case of few-cycle driving pulses, where attosecond pulses emitted from consecutive half-cycles can be very different from each other. In the latter case, attosecond pulses are only emitted by the most intense half-cycles with an amplitude that is nonlinearly related to the field strength during the half-cycle, the delay between them is not strictly the time between two half-cycles and their phase difference is not exactly π [31]. These properties are also reflected in the spectrum, with spectral peaks which are not

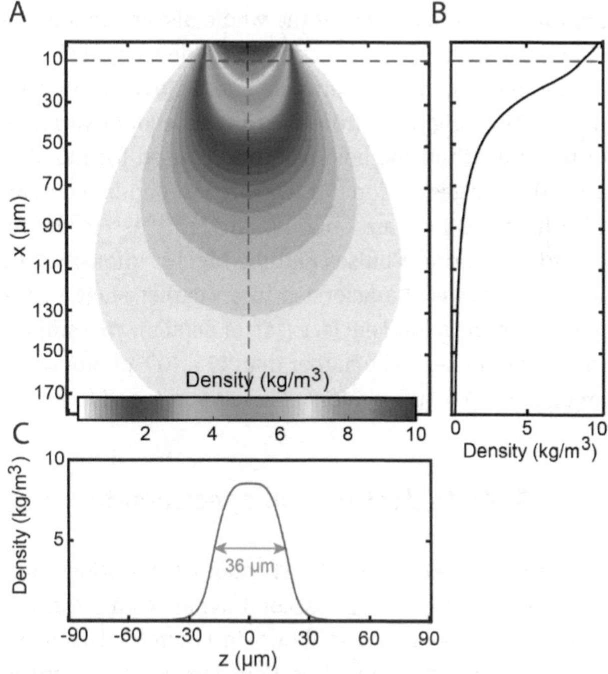

Figure 4: High-order harmonic generation (HHG) gas target simulations for argon as generation gas.
(A) Simulated gas density (12 bar backing pressure) in the interaction region with a 42 μm nozzle centered around $z = 0$. (B) Density at $z = 0$ (black). (C) Line out of the gas density at a distance of 10 μm from the nozzle. In A and B, the dashed blue line indicates the position of the line-out in C. In A, the dashed gray line indicates the position of the line-out in B.

necessarily located at the expected positions for odd-order harmonics.

The electric field of the few-cycle pulse and consequently the characteristics of the HHG pulse train strongly depend on the CEP of the driving pulses, implying that a CEP-stable laser is required for generating a reproducible HHG spectrum and APT [32]. For the two-cycle, sub-6 fs pulses provided by our OPCPA laser, the APTs consist of either two or three strong attosecond pulses, when the CEP of the driving pulse is $\pi/2$ (sine-like pulse) and 0 (cosine-like pulse), respectively [15]. The CEP is controlled by tuning the dispersion of the pulses with a combination of negatively chirped mirrors and motorized BK7 wedges.

To demonstrate the pulse train control achieved in our experiments, we present in Figure 5 the XUV spectrum obtained using neon gas for the generation, as a function of glass insertion. A zirconium filter, cutting all emission below 65 eV, was used to record the spectrum. For large glass insertion (i.e., >2.5 and <–2.5 mm), the pulse is stretched in time to the extent that the peak intensity becomes too low for HHG, while for small amounts of dispersion, the effect is primarily a change of the CEP. As the CEP varies, the positions of the harmonics shift approximately linearly over the whole dispersion and energy range in Figure 5, reflecting a CEP-dependent change of the relative phase between consecutive attosecond pulses. This originates from the CEP-dependent variation of the strength of the field oscillations used for the attosecond pulse generation [31]. The highest photon energies with up to 100 eV are obtained at zero glass insertion, i.e., for the shortest pulses and the highest intensity. The photon flux was characterized in an earlier stage of the light source development (see [14] for details). We estimate the current fluxes to be higher than 15×10^{10} photons/s in argon and 0.7×10^{10} photons/s in neon, respectively.

2.5 3D photoelectron/ion spectrometer

A schematic overview of the 3D photoelectron/ion spectrometer used to study attosecond dynamics in atoms or molecules in gas phase is shown in Figure 6. This spectrometer is based on a revised Coincidences entre ions et electrons localisés (CIEL) design [16], which is conceptually similar to reaction microscope or Cold Target Recoil Ion Momentum Spectroscopy [33, 34]. Momentum imaging instruments of this type have been widely used for the study of photoionization dynamics [35, 36] and can with the help of electric and magnetic fields, collect high-energy electrons over the full solid angle (i.e., 4π collection). The charged particles produced through ionization are accelerated with a weak electric extraction field. In order for the lighter electrons not to escape the spectrometer before they reach the detector, a magnetic field is applied over the whole spectrometer, which confines the electrons to a periodic cyclotron motion with a radius determined by their momentum and direction. By using a position sensitive detector (PSD), which can measure both arrival time and transverse position of the charged particles, and assuming uniform magnetic and electric fields, the full three-dimensional momentum information can be calculated from simple classical equations [16].

If the electrons hit the PSD at an integer multiple of the cyclotron period, the transverse position of these electrons is independent of their initial transverse momentum. The corresponding points in the time-of-flight (ToF) spectrum are called magnetic nodes. At these nodes, the momentum information is no longer unambiguous, leading to loss of data [33, 34]. In the CIEL design, the ToF of all electrons falls between two adjacent magnetic nodes, allowing for 4π-angle resolved measurements without data loss.

The spectrometer was designed to be compact. The dimensions of the extraction region on both the electron and ion detector sides, as well as the length of the drift tube are shown in Figure 6A. The extraction fields are on the order of 5–15 V/cm, and the detected range of electron kinetic energies is defined by adjusting the external magnetic field. Figure 6B introduces the spherical coordinate

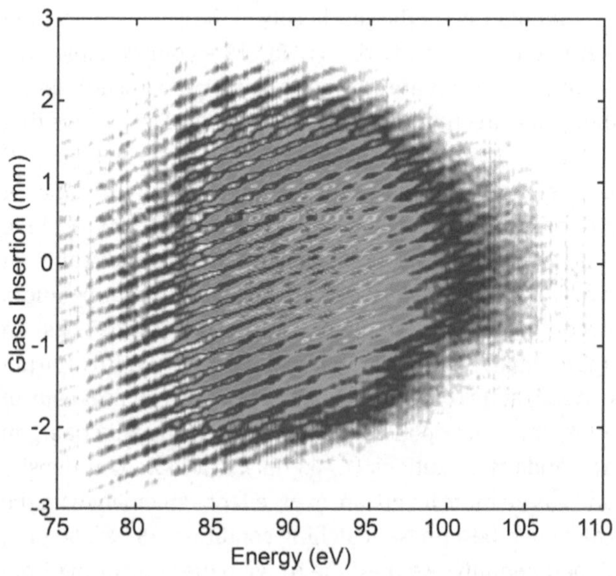

Figure 5: High-order harmonic generation (HHG) dispersion scan. Spectrogram showing the extreme ultraviolet (XUV) spectrum (vertical scale) as a function of BK7 glass insertion for HHG in neon. The spectrum is filtered by a 200 nm thick zirconium thin film.

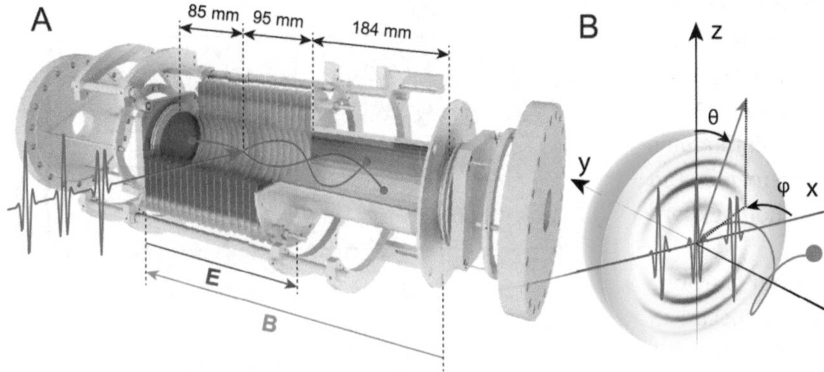

Figure 6: 3D Photoelectron/ion spectrometer (CIEL).
(A) Working principle of the CIEL. (B) Coordinate system used for measurements, along with a 3D representation of the electron momentum distribution, indicating the azimutal angle (φ) and elevation angle (θ) with respect to the plane of the laser polarization (along the z axis).

system used for analysis of measurement data: the elevation θ defines the angle with respect to the light (linear) polarization direction, and the azimutal angle φ determines the position of the projection onto the plane defined by the laser propagation direction (x) and the spectrometer axis (y).

The gas sample is delivered via a long needle with a length-diameter ratio of approximately 1000, resulting in a directional and confined effusive jet [37]. The needle is mounted on a manual three-dimensional translation stage which allows for precise positioning of the gas target with respect to the XUV and IR focus and can be biased to match the potential of the extraction field.

In typical experimental conditions, the photoelectron energy resolution is limited by the light source, i.e., the width of the harmonics and not the spectrometer resolving power. The latter is a complex function of the electron kinetic energy, which also depends on the electric and magnetic fields, source volume, and detector performances [38]. It is typically in the order of a few % for electrons with kinetic energy of 5 eV and above.

The detection system of the spectrometer consists of two commercially available PSDs (RoentDek Handels GmbH) based on multichannel plates and delay line detectors, installed on both sides of the spectrometer, i.e., one to detect ions and the other electrons. The ion anode has a standard design [39], while the electron detector provides unambiguous time and position information for multiple hits [40]. This gives us the possibility to detect one ion in coincidence with several correlated electrons.

3 XUV-IR interferometry in the few attosecond pulse regime

The combination of the CEP-controlled, few-cycle laser source, high-repetition rate, efficient HHG and a highly-stable XUV-IR pump-probe interferometer together with a 3D photoelectron/ion spectrometer that can record several correlated charged particles in coincidence establish excellent experimental conditions for exploring the physics of the interaction of few pulse APTs with atomic and molecular systems.

Recently, we studied single photoionization of helium by a few attosecond pulses (two or three pulses) in combination with a weak IR (dressing) laser field [15]. Instead of using the XUV-IR interferometer described above, the IR laser pulses used for the generation were not eliminated by a metallic filter, but propagated collinearly with the APTs, after attenuation by an aperture to the interaction region of the 3D spectrometer. Both the IR intensity and the XUV-IR delay were therefore fixed. The recorded photoelectron spectra were found to differ considerably depending on whether two or three attosecond pulses were used for the ionization. In the case of two pulses, the spectra were similar to those obtained with XUV-only radiation, except for a shift of the photoelectron peaks induced by the dressing IR field, toward high or low energy depending on the electron emission direction (up or down, relative to the laser polarization direction). In the case of three pulses, additional peaks, so-called sidebands, appeared exactly in between the peaks due to XUV-only absorption. They can be attributed to a two-photon process, where a harmonic photon is absorbed and an additional IR photon is either absorbed or emitted [1]. The reader is referred to our earlier work [15] for details on the experiment and the simulations and for an intuitive interpretation of the results in terms of attosecond time-slit interferences.

In the present work, we use the XUV-IR pump-probe interferometer to record photoelectron spectra in helium as a function of XUV-IR delay in the two attosecond and three attosecond pulse cases. Figure 7 (A, B) present experimental photoelectron spectra at zero delay after integration over the azimutal angle φ as a function of the elevation angle θ, while Figure 7 (C, D) and (E, F) show spectra integrated over 2π solid angle in the down direction as a function of delay for the two pulse and three pulse cases, respectively. (C, D) are experimental results while (E, F) are simulations using the strong-field approximation [15, 41, 42], with an IR intensity of 6×10^{10} W/cm^2. The simulations agree well with the experimental measurements and illustrate the main observable features, without experimental noise and/or irregularities.

Figure 7 (A, B) reproduce and confirm our previous results [15], with energy-shifted photoelectron spectra in the two pulse case and the apparition of sidebands in the three pulse case in the down direction. The energy shift can intuitively be understood by a classical picture, as in streaking [43]. An electron emitted due to ionization by an attosecond pulse with a momentum $\mathbf{p}$ gains or loses momentum from the IR field: $\mathbf{p} \to \mathbf{p} - e\mathbf{A}(t_i)/c$, where $\mathbf{A}(t_i)$ is the vector potential at the time of ionization t_i, e the electron charge and c the speed of light. The 3D momentum distribution becomes therefore asymmetric in the up and down emission directions (Figure 7A). In the three pulse case, the interpretation is more subtle. Here, in the down direction, sidebands appear between the peaks observed in the case of XUV-only radiation. Depending on the delay between the XUV and IR fields, sidebands are observed in only one direction (here the down direction).

In Figure 7 (C, E), the photoelectron peaks are shifted toward lower or higher energies depending on the delay between the APTs and the IR pulses, similarly to attosecond streaking [43, 44]. Unlike a usual streaking trace, where a continuous photoelectron spectrum corresponding to ionization by a SAP is periodically shifted in kinetic energy, here, the spectrum is in addition modulated by interference of the two-electron wave packets created by the two attosecond pulses. The interference structure of the photoelectron spectrum provides spectral resolution, which is missing in ordinary streaking traces obtained with a SAP and may only be obtained by iterative retrieval procedures. The additional spectral resolution provided by the use of two attosecond pulses requires that the spectrometer resolution is not a limiting factor. This is the case in our experiment.

Figure 7 (D, F) show simulated and experimental photoelectron spectra vs. delay for the three pulse case. The results are quite different from those obtained with two attosecond pulses. Instead of modulations of the kinetic energy, sidebands appear at certain delays, similarly to RABBIT spectrograms [45], but with a major difference: The sidebands observed in the present case oscillate with a periodicity equal to the laser period, and not twice the laser period as in RABBIT. Oscillating sidebands are also observed in the up direction, but with a phase shift of π compared to the down direction. The asymmetry in the up and down directions can be attributed to a parity mixing effect and is only observed in the few attosecond pulse case [15].

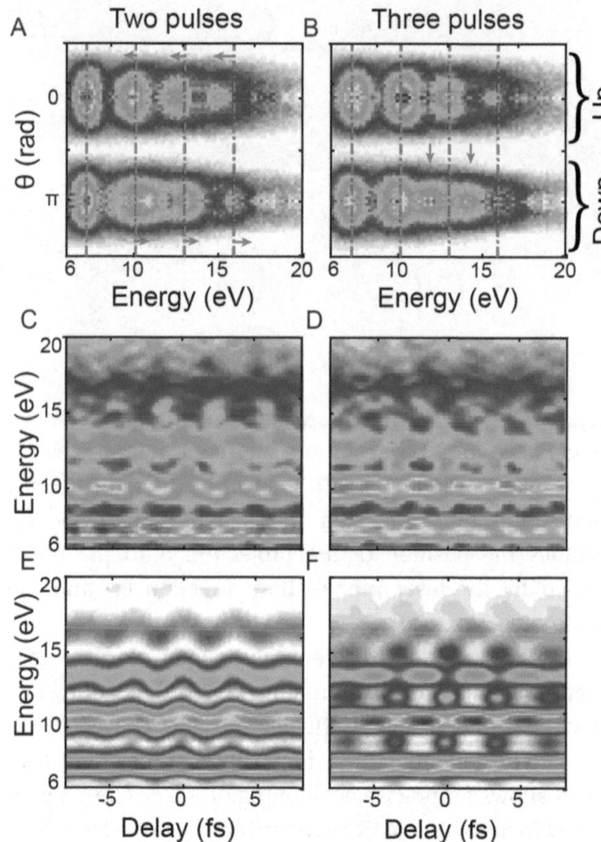

Figure 7: Two-color photoionization of helium. (A, B) Angular-resolved spectrograms. Red dashed lines mark the position of absorption peaks corresponding to odd harmonics in the extreme ultraviolet (XUV) only case. (C, D) Experimental and (E, F) simulated extreme ultraviolet-infrared (XUV-IR) delay traces, by integrating in the down direction. A, C, and E show the two pulse-case (CEP $\pi/2$), and B, D, and F the three pulse case (CEP 0).

The photoionization experiments presented here combining few attosecond pulses and a weak IR field have similarities with streaking (for the two pulse case) and RABBIT (for the three pulse case) experiments, but are also distinctly different, presenting both challenges and new possibilities, especially concerning spectral resolution. In addition, the high-repetition rate enables increased statistics, which is essential for full 3D momentum detection. Advanced detection modes like coincidence become possible. Such a case is discussed in the next section.

4 Single-photon double ionization

Single-photon double ionization in atoms and molecules is one of the most fundamental processes which lead to the emission of correlated photoelectron pairs [17–19]. In the simplest two-electron system, helium, previous works carried out primarily at synchrotron facilities have measured the absolute triply differential cross section (TDCS) for a range of excess energies and angular configurations [46, 47]. The availability of attosecond techniques to probe single-photon double ionization [9] opens up new prospects for measuring the evolution of electron correlation in time.

In order to demonstrate the unique capabilities provided by our high-repetition rate setup, we report the first ever results on single-photon double ionization of helium with full 3D momentum detection using an attosecond light source. Taking benefit of the full imaging capabilities of the CIEL spectrometer, our measurements show that processes hitherto unexplored by attosecond science are now within reach.

Measuring double ionization in helium presents, however, a few challenges for traditional attosecond light sources. Firstly, the threshold for this process at 79 eV is beyond the energies easily achieved with standard HHG in argon using near-IR pulses. This can be circumvented by a change of generation gas, and as shown in Section 2.4, we can reach sufficiently high energies by using neon. Secondly, to perform a kinematically complete experiment we need to detect several charged particles in coincidence. We are thus limited in the final acquisition rate to one-tenth of the repetition rate, to ensure that the event rate is below one event per shot, i.e., to minimize the likelihood of false coincidences. To reduce the number of single ionization events by absorption of low-order harmonics, leading to photoelectrons with similar energies as those due to double ionization by high-order harmonics, we use a zirconium filter which only transmits photon energies at ≈65 eV and above.

Figure 8A shows a typical XUV spectrum generated with neon after this filter (black), along with the cross sections of both the single (red curve) and double (dashed red line) ionization process. In the photon energy region of 79–100 eV, the average ratio of the double versus single ionization cross section is ~1%. With a coincidence rate of 20 kHz, we expect a maximum possible acquisition rate of double ionization events to be 200 Hz. Accounting for the total detection efficiency of typically ~15% for triple coincidence (two electrons and the doubly charged ion), the maximal acquisition rate for a 200 kHz system reduces further to 30 Hz, which is comparable to acquisition rates achievable at synchrotron facilities.

To compensate for the relatively low XUV photon flux in this energy range, e.g., compared to the single photoionization experiments presented in Section 3, the backing pressure of the effusive jet for gas delivery in our spectrometer is increased and the jet is moved very close to the focus. We achieved a final detection rate of ~15 Hz, slightly lower than the nominal 30 Hz, in order to operate the light source with good long-term stability over 40 h. Clearly, without a high laser repetition rate, this experiment would be challenging.

Figure 8B–D show the measured TDCS for equal energy sharing of the two electrons ($E_1 = E_2 = 5 \pm 1$ eV) with emission angles of the first electron of $\theta_1 = 90°$, 60°, and 30° with respect to the polarization axis, and with a total kinetic energy $E_1 + E_2 = 10 \pm 1.5$ eV. The normalization is done according to the procedure outlined in Bräuning et al. [50] using a value of the total cross section of 7.23 kb at 10 eV above threshold [49]. The exact analytical derivation of the TDCS at threshold [51, 46] allows us to further analyze our data. In helium, for a two-electron state with $^1P^o$ symmetry, the differential cross section for equal energy sharing can be written as [46, 51, 52]:

$$\frac{d^3\sigma}{dE_1 d\Omega_1 d\Omega_2} = a_g(E_1, E_2, \theta_{12})(\cos\theta_1 + \cos\theta_2)^2, \quad (1)$$

where θ_1 and θ_2 are the emission angles of the two electrons with respect to the polarization axis (in accordance with the coordinate system defined Figure 6B). The term $(\cos\theta_1 + \cos\theta_2)$ arises from the geometry of the light-matter interaction. A quantum-mechanical description of double photo-ionization with an electron pair in the continuum must obey the Pauli principle and symmetries depending on the total orbital angular momentum, L, the total spin, S, and the parity of the final two-electron state. This leads to a selection rule prohibiting the back-to-back emission, that is observed as a node in the TDCS (see Figure 8B–D). The complex amplitude, a_g, describes the correlation dynamics of the electron pair and only depends on the excess energy and the mutual angle θ_{12}. A Gaussian ansatz provides an excellent parametrization of this amplitude [46, 47]:

$$a_g(E_1, E_2, \theta_{12}) = a \exp\left(-4\ln 2[(\theta_{12} - 180°)/\gamma]^2\right), \quad (2)$$

where a is a scaling factor depending on the cross section and γ is the full width at half maximum correlation factor, which depends on the excess energy. This term affects the opening angle between the two "lobes" seen for example in Figure 8B. A theoretical calculation [53, 54] predicts that the opening angle γ should be around 93°. We find experimentally a value of $\gamma = 90 \pm 3°$, in better agreement than previous reported values of approximately 85° at this excess energy [47, 55].

The results in Figure 8B–D were achieved by integrating over only a small total energy interval for comparison with previous works. However, since the ionizing XUV radiation has a very broad spectrum, this filtering does not give an complete representation of the two-electron wave-packet (EWP) dynamics. To visualize this EWP, we show in Figure 8E the result obtained over the whole energy range, integrating over the azimutal angle and for a 90° emission angle of the first electron, as in Figure 8B. Interestingly, the variation in emission direction and energy of the second electron exhibits the nodal properties (selection rule) that are usually observed in the fully differential cross section. Using the pump-probe capabilities of our setup in a future experiment, the time evolution of the EWP can be studied by observing the change of the final state of the two-electron EWP in the continuum.

5 Conclusion

In this work, we present a compact, high-repetition rate, attosecond light source and demonstrate its capability to perform time-resolved measurements and coincidence experiments. The high-repetition rate, enabled by the partially fiber-based laser architecture, opens up for applying the extraordinary time-resolution promised by attosecond science to new processes and phenomena.

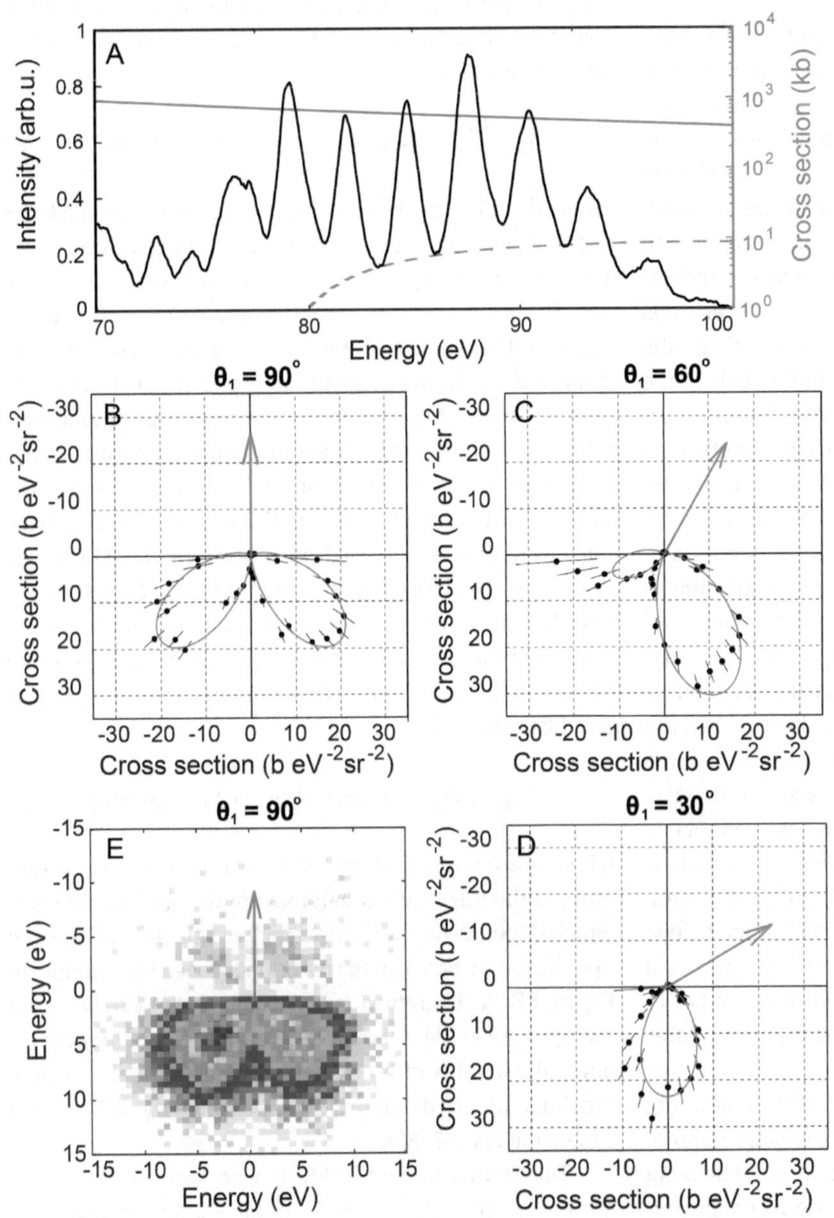

Figure 8: Single-photon double ionization of helium.
(A) Extreme ultraviolet (XUV) spectrum (black) and cross section for single-photon single (solid red) and double (dashed red) ionization of helium [48, 49]. (B–D) Triply Differential Cross Section (TDCS) for second electron (black dots) with equal energy sharing and coplanar emission, with emission angle of the first electron being 90°, 60°, and 30° (red arrow), compared to Eq. (1) (red curve). (E) Distribution of kinetic energies of second electron, integrated over all azimutal angles, when the first electron is emitted at $\theta_1 = 90°$ for coplanar and equal energy sharing emission.

We study single photoionization of helium atoms using APTs consisting of only a few pulses in combination with a weak, delayed, IR field. This new regime opens up new exciting possibilities for control of photoemission through a tailored sequence of pulses. The case of two attosecond pulses, in particular, is promising, as the results can intuitively be understood as in streaking experiments. Compared to streaking with a SAP, it adds spectral resolution due to the interference structure induced by the two attosecond pulses, similar to Ramsey spectroscopy, and lower intensity can be used for the dressing field because of the gained spectral resolution, thus less perturbing the system under study. Generally, the potential of the few pulse attosecond train regime seems to have been widely overlooked by the attosecond science community that traditionally has either worked with SAPs (streaking) or long APTs (RABBIT).

Our setup includes the possibility to add a second end station in order to perform time-resolved surface science experiments (Figure 1). As in coincidence spectroscopy, these experiments strongly benefit from high-repetition rate in order to avoid space charge–related blurring effects in spectroscopy and imaging applications [21]. As the setup is designed for the two end stations to be used simultaneously, the gas phase experiments can serve as a benchmark for simultaneous time-resolved studies on nanostructured surfaces. Our ultimate goal is to investigate and control charge carrier dynamics on the nanoscale with attosecond temporal resolution.

Finally, as a proof of principle, we successfully measure single-photon double ionization in helium by our broadband attosecond XUV source over a complete 4π solid angle. The next experiment will be to add a weak IR laser field in order to study the time evolution of the two-electron wave packet. More generally, our high-repetition rate setup is well adapted for the study of highly correlated many-body processes in the temporal domain.

Acknowledgments: The authors acknowledge support from the Swedish Research Council and the Knut and Alice Wallenberg Foundation. JV acknowledges funding from Marie Skłodowska-Curie Grant Agreement 793604 ATTOPIE.

Author contribution: All the authors have accepted responsibility for the entire content of this submitted manuscript and approved submission.

Research funding: This work was supported by Swedish Research Council and the Knut and Alice Wallenberg Foundation. JV acknowledges funding from Marie Skłodowska-Curie Grant Agreement 793604 ATTOPIE.

Conflict of interest statement: The authors declare no conflicts of interest regarding this article.

References

[1] P. Paul, E. Toma, P. Breger, et al., "Observation of a train of attosecond pulses from high harmonic generation," *Science*, vol. 292, p. 1689, 2001.

[2] M. Hentschel, R. Kienberger, C. Spielmann, et al., "Attosecond metrology," *Nature*, vol. 414, pp. 509–513, 2001.

[3] F. Krausz and M. Ivanov, "Attosecond physics," *Rev. Mod. Phys.*, vol. 81, pp. 163–234, 2009.

[4] A. L. Cavalieri, N. Müller, T. Uphues, et al., "Attosecond spectroscopy in condensed matter," *Nature*, vol. 449, p. 1029, 2007.

[5] M. Schultze, M. Fieß, N. Karpowicz, et al., "Delay in photoemission," *Science*, vol. 328, pp. 1658–1662, 2010.

[6] K. Klünder, J. M. Dahlström, M. Gisselbrecht, et al., "Probing single-photon ionization on the attosecond time scale," *Phys. Rev. Lett.*, vol. 106, p. 143002, 2011.

[7] M. Isinger, R. J. Squibb, D. Busto, et al., "Photoionization in the time and frequency domain," *Science*, vol. 358, pp. 893–896, 2017.

[8] M. Schultze, E. Bothschafter, A. Sommer, et al., "Controlling dielectrics with the electric field of light," *Nature*, vol. 493, pp. 75–78, 2013.

[9] E. P. Månsson, D. Guénot, C. L. Arnold, et al., "Double ionization probed on the attosecond timescale," *Nat. Phys.*, vol. 10, pp. 207–211, 2014.

[10] M. Ossiander, F. Siegrist, V. Shirvanyan, et al., "Attosecond correlation dynamics," *Nat. Phys.*, vol. 13, pp. 280–285, 2017.

[11] M. Ferray, A. L'Huillier, X. Li, L. Lompre, G. Mainfray, and C. Manus, "Multiple-harmonic conversion of 1064 nm radiation in rare gases," *J. Phys. B*, vol. 21, p. L31, 1988.

[12] P. Corkum, "Plasma perspective on strong-field multiphoton ionization," *Phys. Rev. Lett.*, vol. 71, p. 1994, 1993.

[13] J. L. Krause, K. J. Schafer, and K. C. Kulander, "High-order harmonic generation from atoms and ions in the high intensity regime," *Phys. Rev. Lett.*, vol. 68, p. 3535, 1992.

[14] A. Harth, C. Guo, Y.-C. Cheng, et al., "Compact 200 kHz HHG source driven by a few-cycle OPCPA," *J. Opt.*, vol. 20, p. 014007, 2017.

[15] Y.-C. Cheng, S. Mikaelsson, S. Nandi, et al., "Controlling photoionization using attosecond time-slit interferences," *Proc. Natl. Acad. Sci. U. S. A.*, vol. 117, pp. 10727–10732, 2020.

[16] M. Gisselbrecht, A. Huetz, M. Lavollée, T. J. Reddish, and D. P. Seccombe, "Optimization of momentum imaging systems using electric and magnetic fields," *Rev. Sci. Instrum.*, vol. 76, p. 013105, 2005.

[17] N. Chandra and M. Chakraborty, "Entanglement in double photoionization of atoms," *J. Phys. B. At. Mol. Opt. Phys.*, vol. 35, pp. 2219–2238, 2002.

[18] N. Chandra and R. Ghosh, "Generation of tunable entangled states of two electrons and their characterization without entanglement witness," *Phys. Rev. A*, vol. 70, p. 060306, 2004.

[19] D. Akoury, K. Kreidi, T. Jahnke, et al., "The simplest double slit: Interference and entanglement in double photoionization of H2," *Science*, vol. 318, pp. 949–952, 2007.

[20] M. I. Stockman, M. F. Kling, U. Kleineberg, and F. Krausz, "Attosecond nanoplasmonic-field microscope," *Nat. Photonics*, vol. 1, pp. 539–544, 2007.

[21] S. H. Chew, K. Pearce, C. Späth, et al., "Imaging localized surface plasmons by femtosecond to attosecond time-resolved photoelectron emission microscopy – "ATTO-PEEM"," in *Attosecond Nanophysics*, P. Hommelhoff and M. F Kling, Eds., Hoboken, New Jersey, USA, John Wiley & Sons, Ltd, 2014, pp. 325–364.

[22] J.-H. Zhong, J. Vogelsang, J.-M. Yi, et al., "Nonlinear plasmon-exciton coupling enhances sum-frequency generation from a hybrid metal/semiconductor nanostructure," *Nat. Commun.*, vol. 11, pp. 1–10, 2020.

[23] M. Miranda, C. L. Arnold, T. Fordell, et al., "Characterization of broadband few-cycle laser pulses with the d-scan technique," *Opt. Express*, vol. 20, pp. 18732–18743, 2012.

[24] H. R. Telle, G. Steinmeyer, A. E. Dunlop, J. Stenger, D. H. Sutter, and U. Keller, "Carrier-envelope offset phase control: A novel concept for absolute optical frequency measurement and ultrashort pulse generation," *Appl. Phys. B*, vol. 69, p. 327, 1999.

[25] M. Isinger, D. Busto, S. Mikaelsson, et al., "Accuracy and precision of the rabbit technique," *Phil. Trans. Math. Phys. Eng. Sci.*, vol. 377, p. 20170475, 2019.

[26] C. M. Heyl, C. L. Arnold, A. Couairon, and A. L'Huillier, "Introduction to macroscopic power scaling principles for high-order harmonic generation," *J. Phys. B. At. Mol. Opt. Phys.*, vol. 50, p. 013001, 2017.

[27] C. Heyl, H. Coudert-Alteirac, M. Miranda, et al., "Scale-invariant nonlinear optics in gases," *Optica*, vol. 3, pp. 75–81, 2016.

[28] A. L'Huillier, P. Balcou, S. Candel, K. J. Schafer, and K. C. Kulander, "Calculations of high-order harmonic-generation processes in xenon at 1064 nm," *Phys. Rev. A*, vol. 46, p. 2778, 1992.

[29] H. Wikmark, C. Guo, J. Vogelsang, et al., "Spatio-temporal coupling of attosecond pulses," *Proc. Natl. Acad. Sci. U. S. A.*, vol. 116, pp. 4779–4787, 2018.

[30] A. Börzsönyi, Z. Heiner, M. P. Kalashnikov, A. P. Kovács, and K. Osvay, "Dispersion measurement of inert gases and gas mixtures at 800 nm," *Appl. Opt.*, vol. 47, pp. 4856–4863, 2008.

[31] C. Guo, A. Harth, S. Carlström, et al., "Phase control of attosecond pulses in a train," *J. Phys. B*, vol. 51, p. 034006, 2018.

[32] A. Baltuka, T. Udem, M. Uiberacker, et al., "Attosecond control of electronic processes by intense light fields," *Nature*, vol. 421, p. 611, 2003.

[33] R. Dorner, V. Mergel, O. Jagutzki, et al., "Cold target recoil ion momentum spectroscopy: A 'momentum microscope' to view atomic collision dynamics," *Phys. Rep.*, vol. 330, pp. 95–192, 2000.

[34] J. Ullrich, R. Moshammer, A. Dorn, R. Dörner, L. P. H. Schmidt, and H. Schniidt-Böcking, "Recoil-ion and electron momentum spectroscopy: Reaction-microscopes," *Rep. Prog. Phys.*, vol. 66, pp. 1463–1545, 2003.

[35] C. Hogle, X. M. Tong, L. Martin, M. Murnane, H. Kapteyn, and P. Ranitovic, "Attosecond coherent control of single and double photoionization in argon," *Phys. Rev. Lett.*, vol. 115, p. 173004, 2015.

[36] S. Heuser, A. J. Galán, C. Cirelli, et al., "Angular dependence of photoemission time delay in helium," *Phys. Rev. A*, vol. 94, p. 063409, 2016.

[37] W. Steckelmacher, R. Strong, and M. W. Lucas, "A simple atomic or molecular beam as target for ion-atom collision studies," *J. Phys. Appl. Phys.*, vol. 11, pp. 1553–1566, 1978.

[38] Y.-C. Cheng, *Ultrafast Photoionization Dynamics Studied with Coincidence Momentum Imaging Spectrometers*, Lund, Sweden, Lund reports on atomic physics: 555, Division of Atomic Physics, Department of Physics, Lund University, 2019.

[39] O. Jagutzki, V. Mergel, K. Ullmann-Pfleger, et al., "A broad-application microchannel-plate detector system for advanced particle or photon detection tasks: Large area imaging, precise multi-hit timing information and high detection rate," *Nucl. Instrum. Methods Phys. Res. Sect. A Accel. Spectrom. Detect. Assoc. Equip.*, vol. 477, pp. 244–249, 2002.

[40] O. Jagutzki, A. Cerezo, A. Czasch, et al., "Multiple hit readout of a microchannel plate detector with a three-layer delay-line anode," *IEEE Trans. Nucl. Sci.*, vol. 49, pp. 2477–2483, 2002.

[41] M. Lewenstein, P. Balcou, M. Ivanov, A. L'Huillier, and P. Corkum, "Theory of high-order harmonic generation by low-frequency laser fields," *Phys. Rev. A*, vol. 49, p. 2117, 1994.

[42] F. Quéré, Y. Mairesse, and J. Itatani, "Temporal characterization of attosecond XUV fields," *J. Mod. Opt.*, vol. 52, p. 339, 2005.

[43] R. Kienberger, M. Hentschel, M. Uiberacker, et al., "Steering attosecond electron wave packets with light," *Science*, vol. 297, p. 1144, 2002.

[44] R. Pazourek, S. Nagele, and J. Burgdörfer, "Attosecond chronoscopy of photoemission," *Rev. Mod. Phys.*, vol. 87, pp. 765–802, 2015.

[45] R. Lopez-Martens, K. Varjú, P. Johnsson, et al., "Amplitude and phase control of attosecond light pulses," *Phys. Rev. Lett.*, vol. 94, p. 033001, 2005.

[46] J. S. Briggs and V. Schmidt, "Differential cross sections for photo-double-ionization of the helium atom," *J. Phys. B. At. Mol. Opt. Phys.*, vol. 33, pp. R1–R48, 1999.

[47] L. Avaldi and A. Huetz, "Photodouble ionization and the dynamics of electron pairs in the continuum," *J. Phys. B. At. Mol. Opt. Phys.*, vol. 38, pp. S861–S891, 2005.

[48] J. Samson and W. Stolte, "Precision measurements of the total photoionization cross sections of He, Ne, Ar, Kr, and Xe," *J. Electron Spectrosc. Relat. Phenom.*, vol. 123, pp. 265–276, 2002.

[49] J. A. R. Samson, W. C. Stolte, Z.-X. He, J. N. Cutler, Y. Lu, and R. J. Bartlett, "Double photoionization of helium," *Phys. Rev. A*, vol. 57, pp. 1906–1911, 1998.

[50] H. Bräuning, R. Dörner, C. L. Cocke, et al., "Absolute triple differential cross sections for photo-double ionization of helium – experiment and theory," *J. Phys. B. At. Mol. Opt. Phys.*, vol. 31, pp. 5149–5160, 1998.

[51] A. Huetz, P. Selles, D. Waymel, and J. Mazeau, "Wannier theory for double photoionization of noble gases," *J. Phys. B. At. Mol. Opt. Phys.*, vol. 24, pp. 1917–1933, 1991.

[52] G. H. Wannier, "The threshold law for single ionization of atoms or ions by electrons," *Phys. Rev.*, vol. 90, pp. 817–825, 1953.

[53] L. Malegat, P. Selles, and A. Kazansky, "Double photoionization of helium: The hyperspherical $\mathcal{R}$-matrix method with semiclassical outgoing waves," *Phys. Rev. A*, vol. 60, pp. 3667–3676, 1999.

[54] A. Huetz and J. Mazeau, "Double photoionization of helium down to 100 mev above threshold," *Phys. Rev. Lett.*, vol. 85, pp. 530–533, 2000.

[55] R. Dörner, H. Bräuning, J. M. Feagin, et al., "Photo-double-ionization of He: Fully differential and absolute electronic and ionic momentum distributions," *Phys. Rev. A*, vol. 57, pp. 1074–1090, 1998.

Simon Mahler, Yaniv Eliezer, Hasan Yılmaz, Asher A. Friesem, Nir Davidson and Hui Cao*

Fast laser speckle suppression with an intracavity diffuser

https://doi.org/10.1515/9783110710687-012

Abstract: Fast speckle suppression is crucial for time-resolved full-field imaging with laser illumination. Here, we introduce a method to accelerate the spatial decoherence of laser emission, achieving speckle suppression in the nanosecond integration time scale. The method relies on the insertion of an intracavity phase diffuser into a degenerate cavity laser to break the frequency degeneracy of transverse modes and broaden the lasing spectrum. The ultrafast decoherence of laser emission results in the reduction of speckle contrast to 3% in less than 1 ns.

Keywords: optical coherence; optics and lasers; laser dynamics; speckle.

1 Introduction

Conventional lasers have a high degree of spatial coherence, manifesting coherent artifacts and cross-talk. One prominent example is speckle noise, which is detrimental to laser applications such as imaging, display, material processing, photolithography, optical trapping and more [1]. Several techniques have been developed to suppress speckle noise by incoherently integrating many uncorrelated speckle realizations, e.g., by using a moving diffuser or aperture [2–9]. Typically, these methods are effective only at long integration times of a millisecond or longer.

Fast speckle suppression is essential for time-resolved imaging of moving targets or transient phenomena [10–12]. It can be achieved by using multimode lasers with low and tunable spatial coherence [13–19]. The decoherence time of such lasers, critical for fast speckle suppression in short integration times, is determined by the frequency spacing and linewidth of the individual lasing modes, as well as the total width of the emission spectrum $\Delta\Omega$ [20]. Let us consider N transverse modes lasing simultaneously and assume that the linewidth of each individual transverse mode is smaller than the typical frequency spacing $\Delta\omega_t$ of neighboring modes. Only when the integration time τ exceeds $1/\Delta\Omega$, the modal decoherence starts. Once τ exceeds $1/\Delta\omega_t$, the N lasing modes become mutually incoherent and the speckle contrast C is reduced to $1/\sqrt{N}$. Therefore, broadening the laser emission spectrum and increasing the frequency spacing between the transverse modes accelerates speckle suppression, as demonstrated recently with a broad-area semiconductor laser [19].

To reach low spatial coherence, a large number of transverse modes must lase simultaneously. This requires the modes to have a similar loss or quality factor, which can be achieved with a degenerate cavity laser (DCL) [21]. The DCL self-imaging configuration ensures that all transverse modes have an almost identical (degenerate) quality factor. Experimentally, it has been shown that $N \approx 320{,}000$ transverse modes can lase simultaneously and independently in a solid-state DCL [22]. But the transverse modes are also nearly degenerate in frequency, which implies a longer decoherence (integration) time. In the short nanosecond time scale, the longitudinal modes play a critical role in spatial coherence reduction [11]. In particular, the spatiotemporal dynamics of a DCL having M longitudinal modes reduces the speckle contrast to $1/\sqrt{M}$. However, since the number of longitudinal modes is typically far less than the number of transverse modes ($M \ll N$), the speckle contrast reduction at short time scales is limited.

In this work, we develop a simple and robust method for ultrafast speckle suppression. We accelerate the spatial decoherence of a DCL by inserting a phase diffuser

Simon Mahler and Yaniv Eliezer: These authors contributed equally to this work.

***Corresponding author: Hui Cao,** Department of Applied Physics, Yale University, New Haven, Connecticut 06520, USA, E-mail: hui.cao@yale.edu
Simon Mahler, Asher A. Friesem and Nir Davidson, Department of Physics of Complex Systems, Weizmann Institute of Science, Rehovot 761001, Israel. https://orcid.org/0000-0002-9761-445X (S. Mahler)
Yaniv Eliezer and Hasan Yılmaz, Department of Applied Physics, Yale University, New Haven, Connecticut 06520, USA. https://orcid.org/0000-0002-6203-5044 (Y. Eliezer). https://orcid.org/0000-0003-1889-3516 (H. Yılmaz)

This article has previously been published in the journal Nanophotonics. Please cite as: S. Mahler, Y. Eliezer, H. Yılmaz, A. A. Friesem, N. Davidson and H. Cao "Fast laser speckle suppression with an intracavity diffuser" *Nanophotonics* 2021, 10. DOI: 10.1515/nanoph-2020-0390.

(random phase plate) into the cavity. The intracavity phase diffuser lifts the frequency degeneracy of transverse modes and broadens the lasing spectrum. Simultaneously, a large number of transverse modes manage to lase because of their high quality factors. The speckle contrast is reduced to 3% (below human perception level [23]) in less than 1 ns. The lasing threshold is slightly increased (5–10%) with the intracavity phase diffuser, and the output power is reduced by merely 15% over a wide range of pump levels. This work provides a simple and robust method for ultrafast speckle suppression.

2 Degenerate cavity laser configurations

Figure 1 schematically presents several different DCL configurations (details are given in Section 5.1). Figure 1A shows the basic DCL in a self-imaging condition [22]. It comprises of a high-reflectivity flat back mirror, a Nd:YAG gain medium optically pumped by a flash lamp, two spherical lenses of focal lengths f in a $4f$ telescope configuration and an output coupler. We calculate the transverse mode structure (see Section 5.2 for details) and plot the histogram of the frequency differences between the nth order transverse mode ω_n and the fundamental mode ω_0. The difference $\omega_n - \omega_0$ is normalized by the free spectral range (FSR = $\Delta\omega_l$), which is the frequency spacing of longitudinal mode groups. The results shown in the center panel indicate that all the transverse modes in a perfect DCL are exactly degenerate in frequency. The quality factor as a function of the transverse mode index in the right panel exhibits a uniform distribution of high quality factors, indicating that all the transverse modes have an exactly identical (degenerate) quality factor. In this ideal case, despite the fact that many transverse modes are expected to lase, the spectral degeneracy slows down the spatial decoherence. Only when the photodetection integration time exceeds the coherence time given by the inverse of spectral linewidth of individual lasing modes, the degenerate modes become mutually incoherent and the speckle contrast decreases. Note that in practice such an ideal DCL cannot be realized due to the presence of misalignment errors, thermal effects and optical aberrations [24]. Therefore, the transverse lasing modes have slightly different frequencies, which in turn shorten the time of decoherence [22].

In order to accelerate the spatial decoherence, the frequency spacing of the transverse modes has to be increased. Namely, the frequency degeneracy of the modes has to be broken. A conventional method for breaking the frequency degeneracy is detuning the cavity, e.g., translating the output coupler in the longitudinal (z) axis of the cavity, as shown in the configuration of Figure 1B. With a sufficient longitudinal displacement Δz, i.e., $\Delta z = 0.04f$ for our cavity geometry, the frequency spacings of the transverse modes are extended to the entire FSR (center panel). However, the degeneracy in quality factors is also lifted, and many modes suffer a severe quality factor degradation (right panel). Therefore, the number of lasing modes will be significantly reduced, resulting in an effectively higher speckle contrast.

In order to break the frequency degeneracy and increase the frequency spacings of the transverse modes, while minimizing their quality factor degradation, we explore a different approach, where we insert a static intracavity phase diffuser into the DCL, as shown in Figure 1C. The phase diffuser is placed next to the output coupler in order to maintain the self-imaging condition of the cavity. More details are given in Section 5.1. The intracavity phase diffuser is a computer-generated random phase plate made of glass. It introduces an optical phase delay that varies randomly from $-\pi$ to π on a length scale of ≈200 μm (see Section 5.1). The center panel of Figure 1C shows that the transverse modes are spread over the entire FSR of the DCL, increasing the frequency spacings between them. In contrary to the misaligned cavity case, many transverse modes experience minor quality factor degradation. As a result, a large number of transverse modes are expected to lase over a wide spectrum of frequencies, accelerating the speckle suppression process.

3 Ultrafast speckle suppression

To demonstrate the efficiency of our method, we experimentally measure the speckle contrast for integration times in the range of 10^{-10} to 10^{-4} s. The output beam of the DCL is incident onto a thin diffuser placed outside the laser cavity. Then, the speckle intensity is measured by an InGaAs photodiode of 15 GHz bandwidth and an oscilloscope of 4 GHz bandwidth. See Section 5.1 for a detailed description of the experimental setup and the measurement scheme. Figure 2A shows the measured speckle contrast as a function of the photodetector's integration time without and with an intracavity phase diffuser, at the pump power of three times the lasing threshold. By measuring the speckle contrast over many time windows of an equal length, we compute the mean contrast value and estimate the uncertainty that is shown by the shaded area. The lasing pulse is ~100 μs long. To avoid the transient oscillations at the

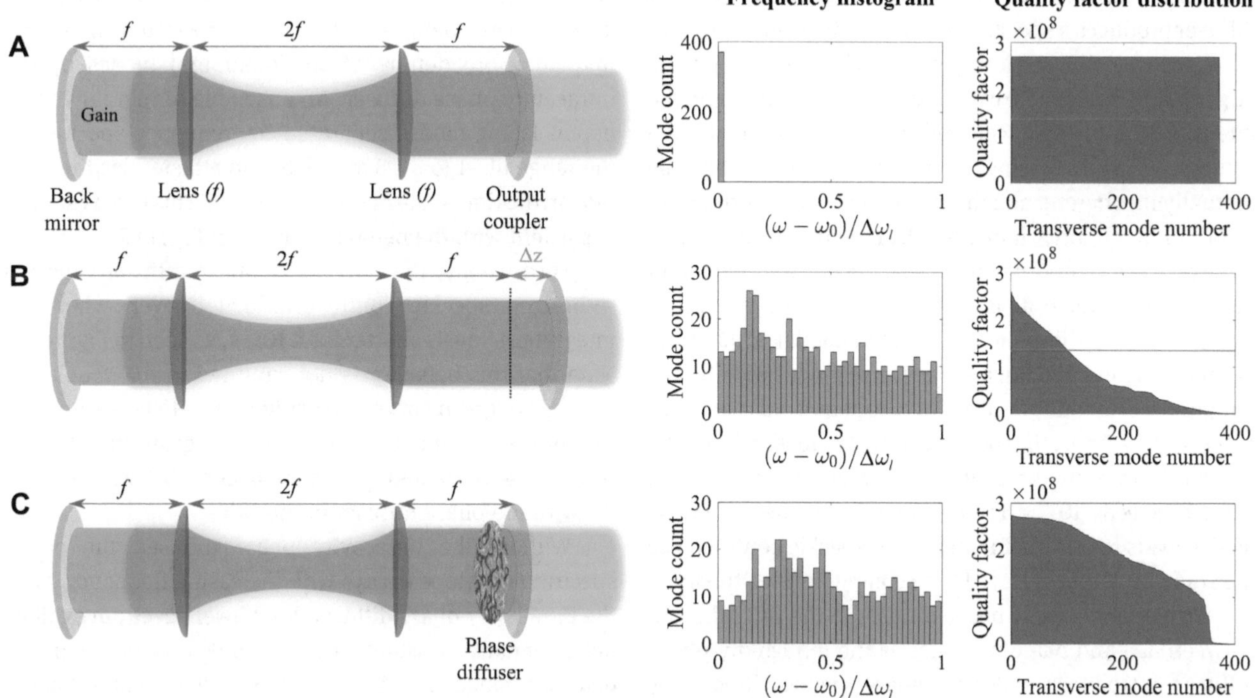

Figure 1: Degenerate cavity laser (DCL) configurations. (A) An ideal DCL with a total length of $4f$. (B) A misaligned DCL. The output coupler is longitudinally translated by Δz along the cavity axis z, with a total cavity length of $4f + \Delta z$. (C) A DCL with an intracavity phase diffuser placed next to the output coupler. The left column contains a sketch of each cavity configuration, the middle column shows the histogram of the frequency differences between the transverse modes and the fundamental one within one longitudinal mode group normalized by the free spectral range: $(\omega - \omega_0)/\Delta\omega_l$ and the right column plots the quality factor versus transverse mode index. The red horizontal line marks half of the maximum quality factor as a reference. The ideal DCL has a large number of high-quality transverse modes, all with the same frequency (both frequency and quality factor degeneracies). The longitudinally misaligned DCL has many modes with different frequencies but also has a relatively small number of transverse modes with high quality factors (no degeneracies). The DCL with a phase diffuser has a relatively large number of high-quality transverse modes with enhanced frequency differences, enabling ultrafast speckle suppression.

beginning of the lasing pulse, we analyze the emission after the laser reaches a quasi steady state. For the effects of lasing transients, see Supplementary material S1. Experimental data with a lower pump power are also presented in Supplementary material S2.

With the intracavity phase diffuser, the speckle contrast at short integration times (between 10^{-10} and 10^{-7} s) is significantly lower than that without the intracavity phase diffuser. Even when the integration time is as short as 10^{-9} s, the speckle contrast is already reduced to 3%. To understand this remarkable result, we numerically calculate the field evolution in a passive cavity with a simplified (1+1)D model. Nonlinear interactions of the lasing modes through the gain medium are neglected (see Section 5.2 for details about the numerical model). The calculated speckle contrast is plotted as a function of integration time τ in Figure 2B. When τ is shorter than the inverse of the emission spectrum width $1/\Delta\Omega$, all lasing modes within $\Delta\Omega$ are mutually coherent with each other.

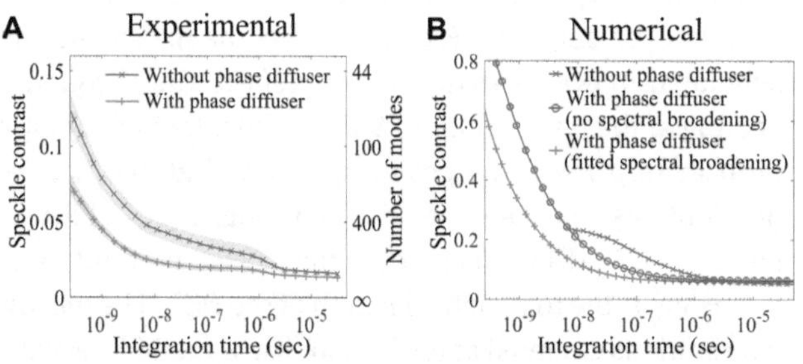

Figure 2: Speckle contrast as a function of the photodetector's integration time measured for the DCL without and with an intracavity phase diffuser. The pump power is roughly three times the lasing threshold ($P \approx 3P_{\text{th}}$). (A) Experimental data. (B) Numerical results. Experimentally, the intracavity phase diffuser introduces a significant reduction in speckle contrast for integration times τ less than 10^{-6} s. Numerically, the reduction of speckle contrast by the intracavity phase diffuser is seen for 10^{-8} s < τ < 10^{-6} s without spectral broadening (green). Further spectral broadening by the intracavity diffuser causes additional speckle reduction in shorter time scales $\tau < 10^{-8}$ s (red).

The interference of their fields scattered by the external diffuser produces a speckle pattern of unity contrast ($C \approx 1$).

Once $\tau > 1/\Delta\Omega$, the lasing modes of frequency spacing larger than $1/\tau$ decohere with respect to each other, and the intensity sum of their scattered light reduces the speckle contrast. With increasing τ, more lasing modes become mutually incoherent, and the speckle contrast continues to drop. In a slightly imperfect DCL without the phase diffuser, the longitudinal mode spacing $\Delta\omega_l$ is much larger than the transverse mode spacing $\Delta\omega_t$. Once τ exceeds $1/\Delta\omega_l \sim 10^{-8}$ s, different longitudinal modal groups are mutually incoherent, but the transverse modes within each longitudinal modal group remain coherent till τ reaches $1/\Delta\omega_t \sim 10^{-6}$ s. Thus, the speckle contrast reduction is greatly slowed down in the time interval between 10^{-8} and 10^{-6} s. Once τ exceeds 10^{-6} s, the decoherence of the transverse modes leads to a further reduction of speckle contrast. See the study by Chriki et al. [11] for a comprehensive theory.

With an intracavity phase diffuser in the DCL, the gap between $\Delta\omega_l$ and $\Delta\omega_t$ diminishes as the intracavity phase diffuser introduces different phase delays (frequency shifts) to individual transverse modes (as depicted in Figure 1C). Meanwhile, the intracavity phase diffuser causes a relatively small reduction in the quality factor of many transverse modes. Thus, a large number of transverse modes can still lase and their frequency detuning accelerates the spatial decoherence. In the time interval of 10^{-8} to 10^{-6} s, the speckle contrast continues to decrease due to the decoherence of the transverse modes within one FSR.

To verify this explanation, we compare the power spectrum of emission intensity of the DCL with the intracavity phase diffuser to that without it. The power spectrum is obtained by Fourier transforming the time intensity signal of the emission. Figure 3 shows the measured and simulated power spectra, which reflect the frequency beating of the lasing modes. Without the intracavity phase diffuser (top row), the power spectrum features narrow distributions peaked at the harmonics of FSR = $c/(2L) \approx 128$ MHz, where c is the speed of light, and $L = 117$ cm is the total optical length of the DCL. The narrow distributions centered at the harmonics of the FSR reveal a slight breaking of frequency degeneracy of the transverse modes, due to the inherent imperfections of the cavity. With the intracavity phase diffuser (bottom row), the power spectrum of emission intensity features many narrow peaks in between the harmonics of the FSR. As the transverse modes move further away from the frequency degeneracy, their frequency differences, which determine their beat frequencies, increase. Nevertheless, the longitudinal mode spacing is unchanged; thus, the peaks at the harmonics of the FSR remain in the power spectrum but appear narrower than that without the intracavity phase diffuser. The changes in the power spectrum indicate a frequency broadening of spatiotemporal modes by the intracavity phase diffuser. An ensemble of mutually incoherent lasing modes separated by frequency spacings in the range of ~1 to ~128 MHz leads to a faster decoherence rate on the time scale of ~10^{-8} to ~10^{-6} s. This observation is consistent with the behavior shown in Figure 2.

Surprisingly, the intracavity phase diffuser causes a significant speckle contrast reduction even when the integration time is shorter than 10^{-8} s, as seen in Figure 2A. Note that this behavior is not captured in the simulation (Figure 2B, green curve). To explain this effect, we analyze the entire experimentally measured power spectra [25]. The results are presented in Figure 4 both (A) without and (B) with the phase diffuser in the DCL.

Without the intracavity phase diffuser, the power spectrum envelope decays with increasing frequency. With the intracavity phase diffuser, the power spectrum exhibits an essentially constant envelope over the entire power detection range of 5 GHz. This difference indicates that the intracavity phase diffuser facilitates lasing in a broader frequency range. With the intracavity phase diffuser, the mutually incoherent lasing modes of frequency spacing well above 1 GHz accelerate the speckle reduction in the subnanosecond time scale. Due to the large number of lasing modes in the DCL, it is extremely difficult to simulate their nonlinear interactions with the gain material. Our numerical model does not account for spatial hole burning and mode

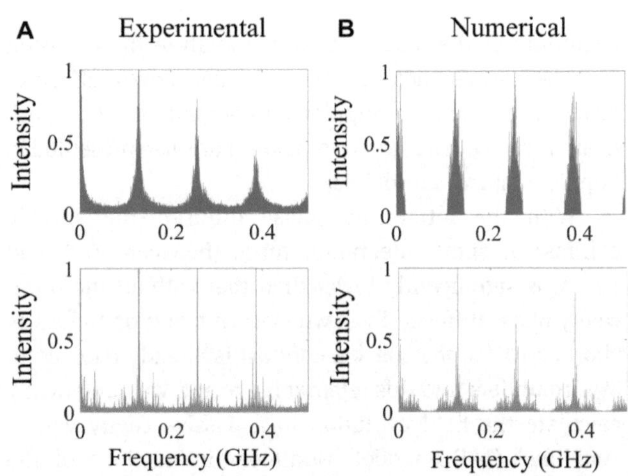

Figure 3: Power spectra of the DCL's emission intensity without (top row) and with (bottom row) the phase diffuser. (A) Experimental data. (B) Numerical results. The intracavity phase diffuser broadens the radiofrequency distribution in each FSR unit, increasing the frequency spacing of the transverse modes and leading to faster spatial decoherence and speckle suppression.

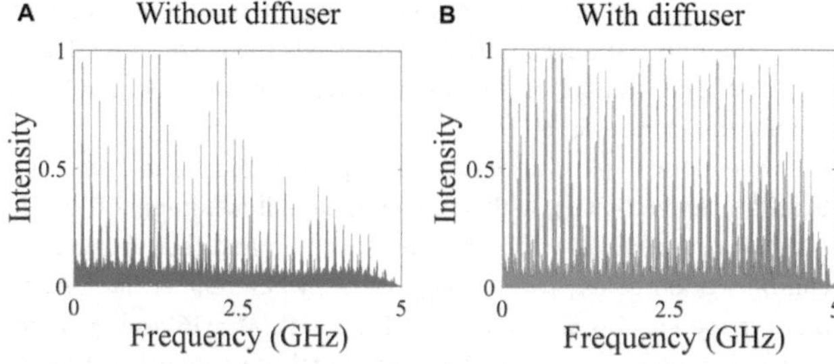

Figure 4: Experimentally measured full-scale power spectrum of emission intensity of the degenerate cavity laser (A) without and (B) with the intracavity phase diffuser. The power spectrum envelope decreases with frequency in (A) and remains nearly constant in (B), indicating that the intracavity phase diffuser enhances lasing in a broader spectral range and accelerates speckle suppression within 1 ns integration time.

competition for gain and thus cannot predict the lasing spectrum broadening induced by the intracavity diffuser. Therefore, the difference between the experimental and the numerical results in the ultrashort time regime is attributed to the absence of nonlinear lasing dynamics in the numerical model. Namely, the intracavity phase diffuser reduces mode competition for gain, allowing modes with wider frequency differences to lase simultaneously.

To complete this observation, we incorporate the diffuser-induced broadening of the lasing spectrum into the numerical model and calculate the spectral contrast as shown by the red curve in Figure 2B. Lasing with more longitudinal modal groups results in a more significant reduction of the speckle contrast at short integration times, in agreement to the experimental data in Figure 2A. This agreement confirms two distinct mechanisms for speckle contrast reduction by the intracavity diffuser. One is the increase of frequency spacing of the transverse modes within each longitudinal modal group, and the other is the broadening of the entire lasing spectrum and an increase in the total number of longitudinal modal groups that can lase. The former mechanism results in speckle reduction in the integration time range of 10^{-8} to 10^{-6} s, while the latter is responsible for speckle reduction in the range of 10^{-10} to 10^{-8} s.

Finally, we measure the total output power of the DCL without and with the intracavity phase diffuser. The lasing threshold is increased by 5–10% after the diffuser is inserted into the DCL. As shown in Section 5.4, the output power is reduced by about 15% over a wide range of pump levels from 1.2 times to 3.3 times the lasing threshold power.

4 Conclusion

In conclusion, we accelerate the spatial decoherence of a degenerate cavity laser (DCL) with an intracavity phase diffuser. In less than 1 ns, the speckle contrast is already reduced to 3%, below the human perception level. Such a light source, together with a time-gated camera, can be used for time-resolved full-field imaging of transient phenomena such as the dynamics of material processing [10] and tracking of moving targets [11, 22]. Our approach is general and will work more efficiently in terms of speckle reduction for compact multimode lasers that have a smaller number of lasing modes and a higher speckle contrast than our DCL. We plan to extend this work by further investigating how the intracavity phase diffuser modifies the nonlinear modal interactions and the spatio-temporal dynamics of a DCL [26].

5 Methods

5.1 Detailed experimental setup

Our experimental setup, shown in Figure 5, consists of two parts: (i) a DCL with a static intracavity phase diffuser and (ii) an imaging system to generate speckle with an external diffuser and to measure speckle contrast [11]. The DCL comprises a flat back mirror with 95% reflectivity, a Nd:YAG crystal rod of 10.9 cm length and 0.95 cm diameter, two spherical lenses of 5.08 cm diameter and f = 25 cm focal length and an output coupler with 80% reflectivity. Adjacent to the output coupler, the phase diffuser is placed inside the cavity.

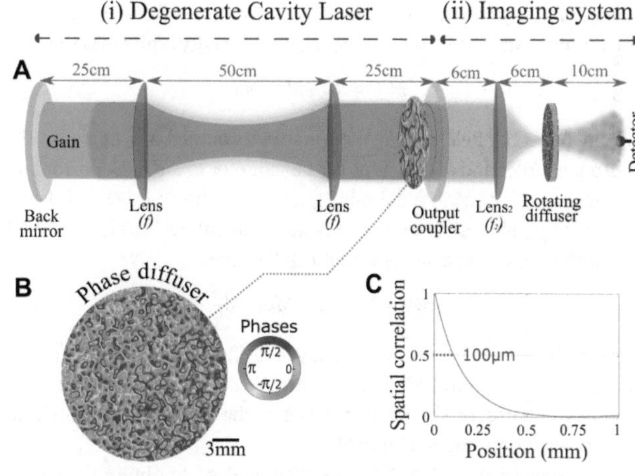

Figure 5: Experimental configuration for the time-resolved speckle intensity measurement. (A) Sketch of (i) a degenerate cavity laser (DCL) with an intracavity phase diffuser and (ii) an imaging system with an external diffuser for generating speckle and a photodetector for measuring the speckle intensity as a function of time. (B) The two-dimensional phase profile of the intracavity phase diffuser, measured by a home-built optical interferometer. (C) Cross section of the two-dimensional autocorrelation function of the phase profile shown in (B), its width gives the typical length scale over which the phase varies.

Lasing occurs at the wavelength of 1064 nm with optical pumping. The output beam is focused by a lens with a diameter of 2.54 cm and a focal length of f_2 = 6 cm onto a thin diffuser with a 10° angular spread of the transmitted light. A photodetector with a 15 GHz bandwidth and 30 μm diameter is placed at a distance of 10 cm from the diffuser and records the scattered light intensity within a single speckle grain in time. We rotate the diffuser and repeat the time intensity trace measurement of a different speckle grain. In total, 100 time intensity traces are recorded.

The intracavity phase diffuser (Figure 5B) is a computer-generated surface relief random phase plate of diameter 5.08 cm and thickness 2.3 mm. The angular spread of the transmitted light is 0.3°. The two-dimensional phase profile across the phase diffuser is measured with a home-built optical interferometer. The phase randomly varies in 16 equal steps between $-\pi$ to π with a uniform probability density. From the measured phase profile, we compute the spatial correlation function, as shown in Figure 5C. Its half width at half maximum is 100 μm. The spatial correlation length is 200 μm, in agreement with the angular spread of the scattered light from the phase diffuser.

In the time-resolved speckle measurement, we use an InGaAs photodiode with 15 GHz bandwidth (Electro-Optics ET-3500). It is connected via a radiofrequency coaxial cable to a Keysight DSO9404A oscilloscope of 4 GHz bandwidth and up to 20 GS/s sampling rate (giga sample per second). The effective bandwidth of our detection system is thus limited to 4 GHz by the oscilloscope.

5.2 Numerical simulation

We simulate continuous wave propagation in a passive degenerate cavity with and without the intracavity phase diffuser. The cavity length and width are identical to those of the DCL in our experiment, except that the cross section is one dimensional in order to shorten the computation time. Without the intracavity phase diffuser, the field evolution matrix of a single round trip in the cavity is given as follows:

$$M^{wo} = M_B \cdot M_\epsilon \cdot M_F \tag{1}$$

where M_F is the field propagation matrix from the back mirror to the output coupler and M_B from the output coupler to the back mirror and M_ϵ represents a small axial misalignment of the DCL [24]. With the intracavity phase diffuser placed next to the output coupler, the field evolution matrix of a single round trip becomes:

$$M^{w} = M_B \cdot M_\epsilon \cdot M_{PD} \cdot M_F, \tag{2}$$

where M_{PD} represents the phase delay of the field induced by the phase diffuser for one round trip in the cavity. To construct M_{PD} in our simulations, we use the spatial distribution of the phase delay taken from the measured profile in Figure 5B.

The matrices M^{wo} and M^{w} are diagonalized to obtain the eigenmodes of the cavity without and with the intracavity phase diffuser. A subset of the eigenmodes has high quality factors (low losses). Hence, they have low lasing threshold and correspond to the lasing modes. The total field in the cavity can be expressed as a sum of these modes:

$$E(x,t) = \sum_{m=-M}^{M} \sum_{n=1}^{N} \alpha_{m,n} \psi_n(x) e^{i[\omega_{m,n}t + \phi_{m,n}(t)]}, \tag{3}$$

where $\alpha_{m,n}$ and $\omega_{m,n}$ denote the amplitude and frequency of a mode, respectively, with a longitudinal index m and a transverse index n and $\psi_n(x)$ represents the transverse field profile for the nth eigenmode. The phase $\phi_{m,n}(t)$ fluctuates randomly in time to simulate the spontaneous emission–induced phase diffusion that leads to spectral broadening [27]. The total number of transverse modes is N, and the number of longitudinal modes is $2M$.

The optical gain spectrum is approximated as a Lorentzian function centered at ω_0 with a full width at half maximum of 32 GHz. All lasing modes are within the gain spectrum and their frequencies can be written as $\omega_{m,n} = \omega_0 + m\Delta\omega_l + \omega_n$, where $\Delta\omega_l$ is the longitudinal mode spacing (FSR), $m = \{-M, \ldots +M\}$, $M = 16$ and ω_n is the transverse mode frequency. The total number of time steps in the simulation of field evolution is 10^6, each step has the duration of 0.1 ns. The power spectrum is calculated by Fourier transforming the time trace of the intensity $|E(x,t)|^2$.

To generate an intensity speckle, we simulate the field propagation from the output coupler of the degenerate cavity to the external diffuser and then from the diffuser to the far field. The field intensity at the far field is used to compute the speckle contrast as a function of the integration time (see Methods 5.3).

5.3 Measurement of speckle contrast

We use the experimental setup in Figure 5A to measure the time-resolved intensity of a single speckle grain behind a diffuser that is placed outside of the DCL. Using the detection device, we record the intensity as a function of time with and without the phase diffuser inside the DCL. The time trace of the intensity is recorded at 100 spatial locations $\vec{r}_i = (x_i, y_i)$, where $i = 1\ldots100$, by rotating the external diffuser by 3.6° for each realization. From the 100 intensity traces, we calculate the speckle contrast C as a function of the integration time τ. First, the total time window T is divided into $J = T/\tau$ intervals. For the jth interval, the intensity is integrated in time: $I_j(\vec{r}_i, \tau) = \int_{j\tau}^{(j+1)\tau} I(\vec{r}_i, t)\,dt$, where $I(\vec{r}_i, t)$ is the time trace of intensity measured at location $\vec{r}_i$. Then, the speckle contrast is calculated for the integration time of τ for the jth interval:

$$C_j(\tau) = \frac{\sigma_j(\tau)}{\mu_j(\tau)}, \tag{4}$$

where $\sigma_j(\tau) = \sqrt{\langle I_j^2(\vec{r}_i, \tau)\rangle_i - \langle I_j(\vec{r}_i, \tau)\rangle_i^2}$ is the standard deviation and $\mu_j(\tau) = \langle I_j(\vec{r}_i, \tau)\rangle_i$ is the mean intensity over $i = 1\ldots100$ spatial locations. Finally, we compute the mean speckle contrast over all time intervals of length τ: $C(\tau) = \langle C_j(\tau)\rangle_j$. The uncertainty of $C(\tau)$ is estimated from the standard deviation: $\sigma_C(\tau) = \sqrt{\langle C_j(\tau)^2\rangle_j - C(\tau)^2}$. Repeating this method, we compute the speckle contrast for different integration times τ in the range from 10^{-10} to 10^{-4} s [11].

5.4 Total output power of the DCL configurations

We experimentally measure the total output power of the DCL without and with the intracavity phase diffuser. As shown in Figure 6A, the total output power with the intracavity phase diffuser is slightly lower than that without it. In Figure 6B, we plot their ratio, which is about 0.85 for all the pump levels. Thus, the intracavity phase diffuser causes a power reduction of about 15%. The lasing threshold is also slightly increased (5–10%) with the intracavity phase diffuser.

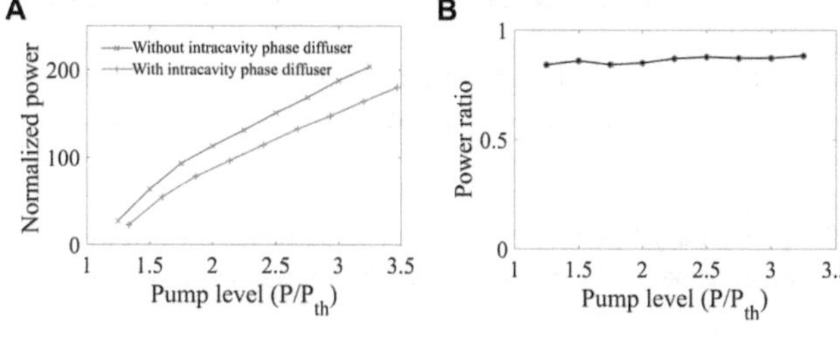

Figure 6: Total output power of the degenerate cavity laser (DCL) with and without the intracavity phase diffuser. (A) The measured output power normalized by the value just above the lasing threshold as a function of the pump power normalized by the threshold value P_{th}. (B) Ratio of the output power from the DCL with the intracavity phase diffuser to that without it. The intracavity phase diffuser reduces the output power by 15% at all pump levels.

Acknowledgments: The authors thank Arnaud Courvoisier and Ronen Chriki for their advice and help in the measurements. This work is partially funded by the US-Israel Binational Science Foundation (BSF) under grant no. 2015509. The work performed at Yale is supported partly by the US Air Force Office of Scientific Research under Grant No. FA 9550-20-1-0129, and the authors acknowledge the computational resources provided by the Yale High Performance Computing Cluster (Yale HPC). The research done at Weizmann is supported by the Israel Science Foundation.

Author contribution: All the authors have accepted responsibility for the entire content of this submitted manuscript and approved submission.

Research funding: This work is partially funded by the US-Israel Binational Science Foundation (BSF) under grant no. 2015509. The work performed at Yale is supported partly by the US Air Force Office of Scientific Research under Grant No. FA 9550-20-1-0129.

Conflict of interest statement: The authors declare no conflicts of interest regarding this article.

References

[1] J. W. Goodman, *Speckle Phenomena in Optics: Theory and Applications*, Englewood, Colorado, Roberts and Company, 2010.

[2] K. V. Chellappan, E. Erden, and H. Urey, "Laser-based displays: a review," *Appl. Opt.*, vol. 49, no. 25, pp. F79–F98, Sep 2010.

[3] S. Lowenthal and D. Joyeux, "Speckle removal by a slowly moving diffuser associated with a motionless diffuser," *J. Opt. Soc. Am.*, vol. 61, no. 7, pp. 847–851, Jul 1971.

[4] T. S. McKechnie, "Reduction of speckle by a moving aperture – first order statistics," *Opt. Commun.*, vol. 13, no. 1, pp. 35–39, 1975.

[5] C. Saloma, S. Kawata, and S. Minami, "Speckle reduction by wavelength and space diversity using a semiconductor laser," *Appl. Opt.*, vol. 29, no. 6, pp. 741–742, Feb 1990.

[6] Y. Kuratomi, K. Sekiya, H. Satoh, et al., "Speckle reduction mechanism in laser rear projection displays using a small moving diffuser," *J. Opt. Soc. Am. A*, vol. 27, no. 8, pp. 1812–1817, Aug 2010.

[7] L. Waller, G. Situ, and J. W. Fleischer, "Phase-space measurement and coherence synthesis of optical beams," *Nat. Photonics*, vol. 6, no. 7, pp. 474–479, 2012.

[8] T.-T.-K. Tran, Ø. Svensen, X. Chen, and M. Nadeem Akram, "Speckle reduction in laser projection displays through angle and wavelength diversity," *Appl. Opt.*, vol. 55, no. 6, pp. 1267–1274, Feb 2016.

[9] M. Nadeem Akram and X. Chen, "Speckle reduction methods in laser-based picture projectors," *Opt. Rev.*, vol. 23, no. 1, pp. 108–120, 2016.

[10] A. Mermillod-Blondin, H. Mentzel, and A. Rosenfeld, "Time-resolved microscopy with random lasers," *Opt. Lett.*, vol. 38, no. 20, pp. 4112–4115, Oct 2013.

[11] C. Ronen, S. Mahler, C. Tradonsky, V. Pal, A. A. Friesem, and N. Davidson, "Spatiotemporal supermodes: rapid reduction of spatial coherence in highly multimode lasers," *Phys. Rev. A*, vol. 98, p. 023812, Aug 2018.

[12] S. Knitter, C. Liu, B. Redding, M. K. Khokha, M. A. Choma, and C. Hui, "Coherence switching of a degenerate vecsel for multimodality imaging," *Optica*, vol. 3, no. 4, pp. 403–406, Apr 2016.

[13] B. Redding, M. A. Choma, and H. Cao, "Spatial coherence of random laser emission," *Opt. Lett.*, vol. 36, no. 17, pp. 3404–3406, 2011.

[14] B. Redding, M. A. Choma, and H. Cao, "Speckle-free laser imaging using random laser illumination," *Nat. Photonics*, vol. 6, no. 6, pp. 355–359, 2012.

[15] M. Nixon, O. Katz, E. Small, et al., "Real-time wavefront shaping through scattering media by all-optical feedback," *Nat. Photonics*, vol. 7, no. 11, pp. 919–924, 2013.

[16] B. Redding, C. Alexander, X. Huang, et al., "Low spatial coherence electrically pumped semiconductor laser for speckle-free full-field imaging," *Proc. Natl. Acad. Sci. U. S. A.*, vol. 112, no. 5, pp. 1304–1309, 2015.

[17] B. H. Hokr, S. Morgan, J. N. Bixler, et al., "A narrow-band speckle-free light source via random Raman lasing," *J. Mod. Opt.*, vol. 63, no. 1, pp. 46–49, 2016.

[18] S. F. Liew, S. Knitter, S. Weiler, et al., "Intracavity frequency-doubled degenerate laser," *Opt. Lett.*, vol. 42, no. 3, pp. 411–414, Feb 2017.

[19] K. Kim, S. Bittner, Y. Zeng, S. F. Liew, Q. Wang, and H. Cao, "Electrically pumped semiconductor laser with low spatial coherence and directional emission," *Appl. Phys. Lett.*, vol. 115, no. 7, p. 071101, 2019.

[20] H. Cao, C. Ronen, S. Bittner, A. A. Friesem, and N. Davidson, "Complex lasers with controllable coherence," *Nat. Rev. Phys.*, vol. 1, no. 2, pp. 156–168, 2019.

[21] J. A. Arnaud, "Degenerate optical cavities," *Appl. Opt.*, vol. 8, no. 1, pp. 189–196, Jan 1969.

[22] M. Nixon, B. Redding, A. A. Friesem, H. Cao, and N. Davidson, "Efficient method for controlling the spatial coherence of a laser," *Opt. Lett.*, vol. 38, no. 19, pp. 3858–3861, Oct 2013.

[23] S. Roelandt, Y. Meuret, C. Gordon, V. Guy, P. Janssens, and H. Thienpont, "Standardized speckle measurement method matched to human speckle perception in laser projection systems," *Opt. Express*, vol. 20, no. 8, pp. 8770–8783, Apr 2012.

[24] J. A. Arnaud, "Degenerate optical cavities. II: Effect of misalignments," *Appl. Opt.*, vol. 8, no. 9, pp. 1909–1917, 1969.
[25] Note that since our detector has an integration time of 0.1 ns, it is not possible to measure the entire beating frequency spectrum (32 GHz bandwidth) of the DCL but a part of it (5 GHz range).
[26] S. Bittner, S. Guazzotti, Y. Zeng, et al., "Suppressing spatiotemporal lasing instabilities with wave-chaotic microcavities," *Science*, vol. 361, no. 6408, pp. 1225–1231, 2018.
[27] D. Rick, *Probability: Theory and Examples*, vol. 49, Cambridge, UK, Cambridge University Press, 2019.

Supplementary Material: The online version of this article offers supplementary material (https://doi.org/10.1515/nanoph-2020-0390).

Giulia Guidetti, Yu Wang and Fiorenzo G. Omenetto*

Active optics with silk

Silk structural changes as enablers of active optical devices

https://doi.org/10.1515/9783110710687-013

Abstract: Optical devices have been traditionally fabricated using materials whose chemical and physical properties are finely tuned to perform a specific, single, and often static function, whereby devices' variability is achieved by design changes. Due to the integration of optical systems in multifunctional platforms, there is an increasing need for intrinsic dynamic behavior, such as devices built with materials whose optical response can be programmed to change by leveraging the material's variability. Here, regenerated silk fibroin is presented as an enabler of devices with active optical response due to the protein's intrinsic properties. Silk's abilities to controllably change conformation, reversibly swell and shrink, and degrade in a programmable way affect the form and the response of the optical structure in which it is molded. Representative silk-based devices whose behavior depends on the silk variability are presented and discussed with a particular focus on structures that display reconfigurable, reversibly tunable and physically transient optical responses. Finally, new research directions are envisioned for silk-based optical materials and devices.

Keywords: biomaterials; optics; reconfigurable; silk; transient; tunable.

Giulia Guidetti and Yu Wang: These authors contributed equally to this work.

***Corresponding author: Fiorenzo G. Omenetto,** Silklab, Department of Biomedical Engineering, Tufts University, Medford, MA 02155, USA; Laboratory for Living Devices, Tufts University, Medford, MA 02155, USA; Department of Physics, Tufts University, Medford, MA 02155, USA; and Department of Electrical and Computer Engineering, Tufts University, Medford, MA 02155, USA, E-mail: fiorenzo.omenetto@tufts.edu. https://orcid.org/0000-0002-0327-853X

Giulia Guidetti and Yu Wang, Silklab, Department of Biomedical Engineering, Tufts University, Medford, MA 02155, USA; and Laboratory for Living Devices, Tufts University, Medford, MA 02155, USA. https://orcid.org/0000-0002-6065-3359 (G. Guidetti). https://orcid.org/0000-0003-0249-4414 (Y. Wang)

1 Introduction

Traditionally, optical devices have been fabricated so that their structure, material and, thus, designed function would either remain unchanged with time, being virtually immutable, or could undergo changes only due to external adaptations of their constructs. This design strategy ensures a practically constant performance as a function of time within the individual components' functional lifetime; additionally, it is based on materials whose properties have been highly engineered to fulfill one specific aim and are, optically, of very high quality [1]. This approach is very advantageous for the fabrication of optical devices that require high specialization and long durability; despite this, optical components built following this design strategy are strongly limited in multifunctionality and often lack inherent physical/chemical sensitivity and specificity to analytes, as they are programmed not to change.

The last few decades have seen a growing interest in materials with dynamic behaviors that are responsive to the environment that surrounds them. In particular, optical structures have been progressively integrated within electronic devices or biomedical platforms (to name a few), causing a shift in the materials and device design specifications. Specifically, new requirements emerged including the need for devices to be reconfigurable, tunable, and resorbable (and, therefore, temporally limited). In addition, the need for disposable devices requires materials with lower production costs and with minimal or no environmental hazards. To produce devices capable of such dynamic responses, it is necessary to change the optical device design strategy in terms of materials used and forms in which they are molded. The advantage of following this design strategy is in having devices which can be programmed to change their response as a function of a set of specific stimuli instead of having to fabricate a dedicated device for each of those stimuli. Regenerated silk protein features a combination of properties that makes it an ideal candidate to build such devices: among the many advantages offered by this biopolymer, the protein's abilities to controllably change conformation, reversibly swell and

This article has previously been published in the journal Nanophotonics. Please cite as: G. Guidetti, Y. Wang and F. G. Omenetto "Active optics with silk" *Nanophotonics* 2021, 10. DOI: 10.1515/nanoph-2020-0358.

shrink, and undergo a programmable degradation are key for integrating an active response in optical platforms.

Here, we present a critical overview of active optical devices that use regenerated silk fibroin as the stimuli-responsive component. We first provide a brief description of the regeneration of silk fibroin, of the manufacturing technologies that allow to transform aqueous silk solutions into multiple silk constructs, and of the silk's structural variability. Then, we discuss recent examples of active silk-based optical devices, focusing on reconfigurable, reversibly tunable, and physically transient optical systems. Finally, we comment on new research possibilities that can be enabled by silk's variable properties.

2 Silk for active optics

The ideal materials for active optical devices should demonstrate modulation across different scales. These range from modifications at the molecular level and at the nanoscale, such as variations in the refractive index and in the formed nanostructure, to macroscopic changes, such as swelling, strain ability, and solubility. These modifications affect the material's form and structural integrity, and, therefore, its overall optical response. Silk, as a naturally-derived biopolymer, can often satisfy those requirements due to its hierarchical structure, controllable transition between its polymorphic structures [2], reversible volume swelling and shrinking, and programmable biodegradability [3].

Silk can be extracted from the cocoons of the *Bombyx mori* lepidoptera, which are spun by the caterpillars during their metamorphosis [4–6]. Natural silk fibers are fabricated by dry spinning of a concentrated silk protein aqueous solution into an insoluble and hierarchically arranged bundle of silk fibroin microfibers surrounded by a sericin gluey layer [4,7]. Such insoluble silk fibers can be retransformed into aqueous solutions that can then be processed using multiscale fabrication technologies to give a variety of forms generating diverse optical responses. Silk cocoons are first degummed through a boiling process, then dissolved in a concentrated salt solution (usually LiBr) and, finally, transformed in an aqueous solution purified by dialysis [8] (Figure 1A). The resulting regenerated silk fibroin solution displays a combination of properties which converge into making silk fibroin an ideal building block for optical and photonic applications; these include nanoscale processability, water-based processing, ease of functionalization, and the capacity to form mechanically robust, highly optically transparent, thermally stable, and multifunctional materials.

Silk fibroin solutions can be processed using multiscale, water-based fabrication techniques to obtain a wide variety of multidimensional structures, including films, hydrogels, fibers, sponges, and solid blocks [10]. Remarkably, silk fibroin's relatively high thermal stability [11] and solubility in water allows it to be processed using a broad variety of already optimized multiscale fabrication techniques for optical and electronic devices without the need of dedicated fabrication processes [12]. Among the broad variety of biomanufacturing techniques available for silk processing, the ones mostly used for the fabrication of silk-based optical devices include electron beam lithography [13, 14], soft lithography [10, 15], photolithography [16], nanoimprinting [17, 18], inkjet printing [19, 20] direct ink writing [21], direct transfer [22], spin coating, and self-assembly [9, 20, 23]. Notably, as some of these techniques require energy transfer to the material being processed and can control its sol–gel–solid transition that can activate crosslinking [10, 13, 14, 17, 18], they can implicitly impart conformational changes to the protein, therefore modulating its water solubility during the fabrication of the material, without the need for post-treatments.

One of the key traits that distinguishes biopolymers from other materials is their variability. As a structural protein, silk fibroin undergoes conformational transition among β-sheets, random coils, and helices upon exposure to external stimuli, such as water vapor, methanol, and deep UV light (Figure 1B) [14, 24, 25]. The interactions between water/methanol molecules and the polar groups of silk fibroin chains can affect the hydrogen bonding between protein chains, thus leading to the conversion of a water-soluble material (amorphous silk) into a water-insoluble format (crystalline silk) by inducing the formation of β-sheets [24, 26]. By contrast, UV irradiation induces peptide chain scission and photodegradation of the silk fibroin initially at the weaker C–N bonds, giving rise to a considerable decrease in the protein's degree of crystallinity [27]. In practice, the higher the β-sheets content is, the higher the crystallinity and hence the lower the water solubility and the longer the functional lifetime of the silk structure. The final protein's conformation is crucial in determining both the silk material macroscopic physical properties and, especially, its modulation possibilities. The essence of silk's polymorphic transitions lies in the molecular rearrangement at the nanoscale, offering the possibility to create reconfigurable and programmable silk-based optical devices. At the same time, as a biodegradable material, the structural integrity of silk-based materials can be modulated when exposed to aqueous solutions or implanted *in vivo*. The dissolution and/or the biodegradation rates can be programmed to range from minutes to years by controlling the

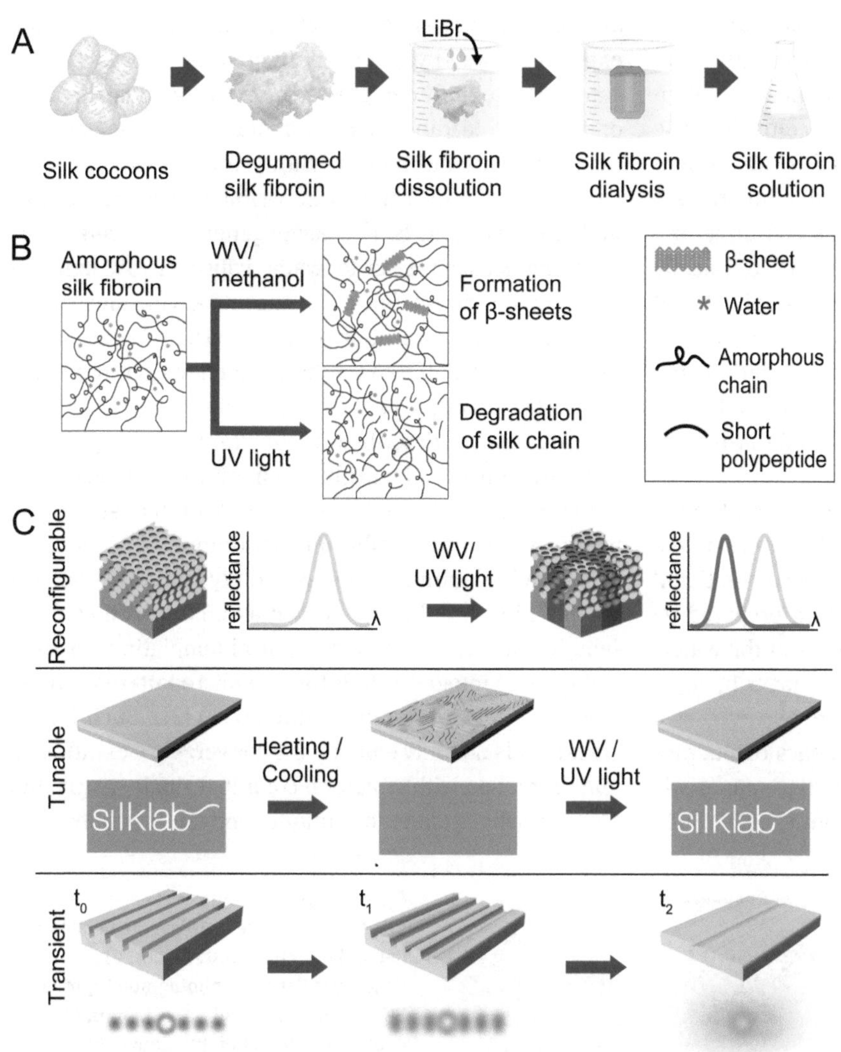

Figure 1: (A) Schematic representation of the silk regeneration process. (B) Schematic representation of amorphous silk fibroin conformational changes induced by water vapor (WV), methanol, and UV light, corresponding, respectively, to the formation of *β*-sheets and the degradation of the silk chains. Adapted from Ref. [9]. Copyright 2017, WILEY-VCH Verlag GmbH & Co. (C) Representative strategies to obtain reconfigurable, reversibly tunable, and temporally transient silk-based optical structures. Schematic representation of the structure (from top to bottom, silk inverse opal, silk-based wrinkled surface, and silk diffractive optical element) and of the corresponding optical response (from top to bottom, shift of the bandgap position, concealing and revealing of information, and gradual degradation of the diffracted pattern).

abovementioned conformational transitions, allowing for the emergence of physical transient optical devices that can be dissolved in a customized way.

By using the above-described fabrication techniques, silk fibroin solutions can be shaped and programmed to display diverse optical responses. Notably, due to its nanoscale conformability resolution [28], silk can be molded into structures that can display optical functions including lensing and diffractive properties [29, 30], photonic response [9, 13, 20, 31], waveguiding [21], lasing [32], fluorescent response [33], and nonlinear optical behaviors [34–36]. Devices based on these optical functions can be programmed to display an active response by dedicated modifications that ensure silk's transformation abilities. As discussed in detail in the next sections, the main active responses that can be embedded in silk are reconfiguration, reversible tunability, and physical transient behavior (Figure 1C).

3 Active optical devices

3.1 Reconfigurable optics

Silk's polymorphic transitions have been recently harnessed to design silk-based three-dimensional (3D) photonic lattices which can undergo non-reversible conformational changes that affect the construct's optical response. The combination of protein self-assembly with colloidal assembly has been utilized to transform silk aqueous solutions into 3D photonic crystals [9, 23]. The latter is a facile, scalable, and cost-effective manufacturing technique that is widely adopted for the bottom-up fabrication of 3D photonic crystals [37, 38]. The resulting colloidal crystals can be further used as templates for the formation of inverted photonic lattices, which are obtained by template removal through solvent etching. Silk solutions can readily infiltrate such templates and self-assemble into a freestanding silk film through control over the dynamics of water evaporation. The fabricated silk inverse opals (SIOs) present bright structural colors because of the diffraction of incident light induced by the periodicity of the nanostructures. SIOs can be either amorphous or crystalline depending on the protein's conformation. Since the protein chains in crystalline SIOs are physically crosslinked, this type of photonic lattice is robust and difficult to tune once the structure is fabricated [23]. On the contrary, amorphous SIOs possess a photonic lattice that can be easily modulated under external stimuli, allowing for the fabrication of materials with variable structural color.

Large-scale, highly-ordered, and reconfigurable SIOs have been recently fabricated by using polystyrene sphere multilayers as templates [9]. Instead of using conventional deposition techniques, such templates were prepared by scooping transfer of the floating monolayers at the water/air interface and by stacking them layer-by-layer (Figure 2A). Uniform structural colors were observed as a result of the favorable material characteristics of silk protein, including robust mechanical properties, nanoscale processability, and conformability (Figure 2B). Such amorphous SIOs are stable at ambient conditions; upon exposure to water vapor or UV light, SIOs' photonic lattice can be controllably compressed along the vertical direction of the film, leading to reconfiguration of the photonic bandgap. This mechanism allows for the generation of structural color over the entire visible spectrum. Multiple multicolor patterns can be easily generated by selectively exposing masked SIOs to water vapor or UV light for different durations (Figure 2C). The underlying mechanism that enables the fabrication of patterned SIOs is the stimuli-induced controllable conformational changes through the molecular rearrangement of silk fibroin.

The above described reconfigurable 3D photonic nanostructures can be further coupled with 2D microscale patterns to fabricate hierarchical photonic structures, where the reconfigurability allows additional functionality by providing extra spectral selectivity. In particular, the combination of silk protein self-assembly, colloidal assembly, and top-down topographical templating has been exploited to integrate 2D diffractive micropatterns with 3D inverse colloidal crystals (Figure 2D–F) [31]. Such hierarchical opals not only embody the convergence of diffusion and diffraction with structural color in a single matrix, but also present a suite of unique optical functions that

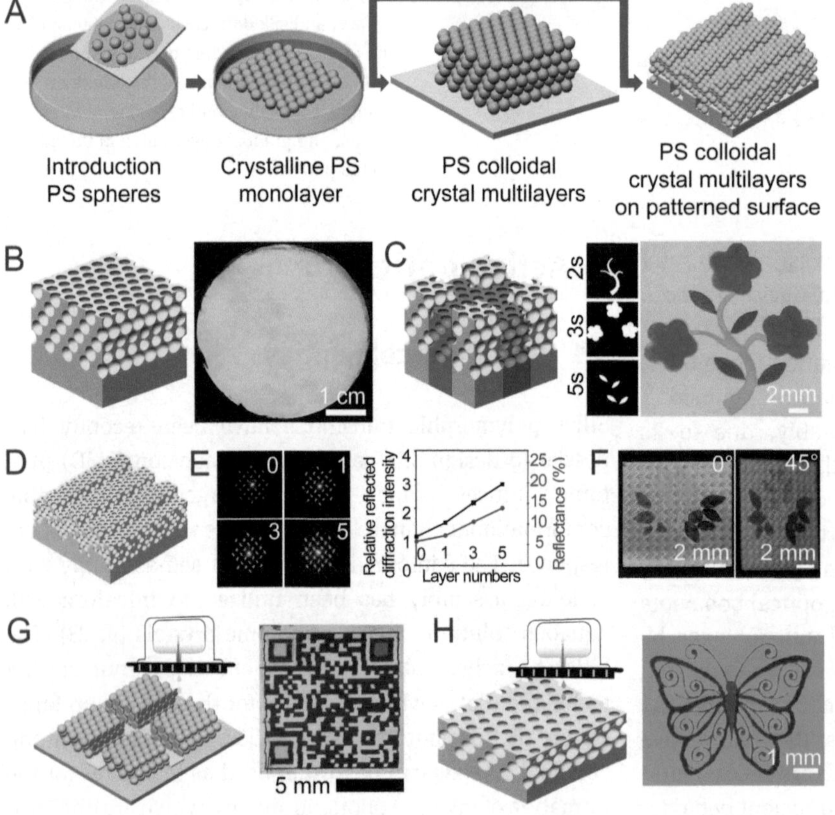

Figure 2: (A) Fabrication schematic of polystyrene sphere multilayer templates on flat and patterned surfaces. (B, C) Schematic (left) and photograph (right) of SIO (B) and patterned SIO (C). (A, B, C) Reproduced from Ref. [9]. Copyright 2017, WILEY-VCH Verlag GmbH & Co. (D) Schematic of hierarchical inverse opal assembled on a diffraction grating; (E) reflected diffraction patterns (left) of hierarchical opals assembled on pattern generators with different layers of inverse colloidal crystals (from 0 to 5) and corresponding relative diffraction and absolute reflection intensity as a function of layers number (right). (F) Photographs of hierarchical opal with a multicolor tree pattern observed at 0° and 45°. (D, E, F) Reproduced from Ref. [31]. Copyright 2018, WILEY-VCH Verlag GmbH & Co. (G) Fabrication schematic (left) of patterned polystyrene multilayer template by depositing ethyl acetate on the structure using inkjet printing and photograph (right) of a multispectral SIO QR code. (H) Fabrication schematic (left) of patterned SIO by depositing MeOH/water mixture on SIO using inkjet printing and photograph (right) of SIO with a butterfly pattern. (G, H) Reproduced from Ref. [20]. Copyright 2019, WILEY-VCH Verlag GmbH & Co.

leverage the interplay of the individual responses. On the one hand, the existence of a 3D nanoscale photonic lattice modulates the diffraction of the 2D optical element by tuning its photonic bandgap; for instance, this can be achieved by controlling the number of layers and the lattice constant of the assembled colloidal crystal or by reconfiguring the photonic lattice by water vapor or UV light treatment. The variation of the reflected (or transmitted) diffraction intensity is consistent with the change of the reflection (or transmission) intensity because of the photonic nanostructure modulation (Figure 2E). On the other hand, the 2D microscale topography affects the overall color appearance of the structures, resulting in uniform structural color over a broad range of viewing angles. Such feature, combined with the reconfigurability of the nanoscale lattice and with the design of multicolor patterns, allows these structures to be used for wide-angle multicolor pattern displays with spatially dependent iridescence (Figure 2F).

The ability to reconfigure photonic lattices provides additional levels of functionality to these engineered photonic crystals, which become structures that not only display multispectral responses but can also be used for information encoding. This has been recently demonstrated by creating multispectral QR codes through the combination of inkjet printing and water vapor treatment (Figure 2G, H) [20]. Inkjet printing was used to modify preassembled photonic crystal lattices to create patterned templates (e.g., a QR code pattern) for silk infiltration and subsequent fabrication of patterned inverse opals. In particular, ethyl acetate was chosen as the ink to selectively remove the photonic lattice in the printed areas, therefore, creating a structure which displayed inverse opal lattices only in the non-printed regions. The resulting SIO QR code was further programmed by exposing specific lattice-bearing regions (colored squares in Figure 2G micrograph) to water vapor for different times, leading to a multispectral QR code with double encryption layers, namely, (i) the scanned message from the QR code, and (ii) a three-digit key defined by the RGB intensity of the reconfigured colors. A reconfigurable SIO itself is an ideal optical platform for pattern writing, which can be achieved by direct integration with inkjet printing and by using a solvent that can trigger the protein's conformational transition (Figure 2H). Different volumetric changes of the silk matrix, and thus distinct differences in the lattice constant and in the reflected colors, can be easily generated by controlling the ink composition or volume. For example, light-blue and dark-blue colors were obtained, respectively, by inkjet printing MeOH/water mixtures with water ratios of 3 and 18% on a 10-layer SIO substrate. This strategy provides great opportunities for the fabrication of arbitrary, complex, and multispectral photonic patterns by manipulating the protein's initial conformation, the lattice's configuration, and the ink's formulation.

In addition to forming homogeneous inverse opal structures, silk-based reconfigurable photonic crystal superlattices can also be fabricated by stacking layers of 3D nanolattices with different periodicity [39]. Multispectral structural colors can be easily achieved by reconfiguring the lattices of the formed heterostructures by water vapor exposure.

3.2 Reversibly tunable optics

Silk protein materials are also capable of reversibly responding to external stimuli, allowing for the fabrication of dynamic tunable optical devices. Crystalline silk offers an ideal platform for the generation of reversible stimuli-responsiveness, due to its molecular chains being anchored through physical crosslinking. The general strategy for implementing dynamic tunable optical function relies primarily on the reversible volume change of the silk matrix in response to various environmental fluctuations such as humidity, aqueous solvents, mechanical strain, and temperature. Among these, humidity is the most widely utilized stimulus given its ease and speed in inducing swelling of the silk matrix. Silk's high sensitivity to humidity makes silk-based optical devices suitable for sensing applications as demonstrated by the several silk-based optical platforms used to generate dynamic tunable responses to humidity, such as inverse opals [40], thin-film and multilayer interference structures [41, 42], optical fibers [43], and metamaterials [44, 45]. As an example, silk-coated terahertz metamaterials can be programmed to display a dynamic response to humidity changes (Figure 3A), whereby the frequency of the devices is gradually blue-shifted upon humidity increase due to the variation of the dielectric constant of the silk's film induced by the swelling [45]. Another type of reversible responsiveness that relies on the physical swelling of silk matrix is achieved through aqueous solvent infiltration; in this case, the water content in the solvent determines the degree of swelling of the silk constituent within the optical device and thus its optical response. Examples of such tunable devices include metal–insulator–metal resonators using silk protein as the insulator layer [46, 47], and silk plasmonic devices [48]. Mechanically induced tunable optics are based on the continuous and reversible variation of optical structures upon changes in the tensile strain applied to them. In this scenario, silk hydrogels provide an

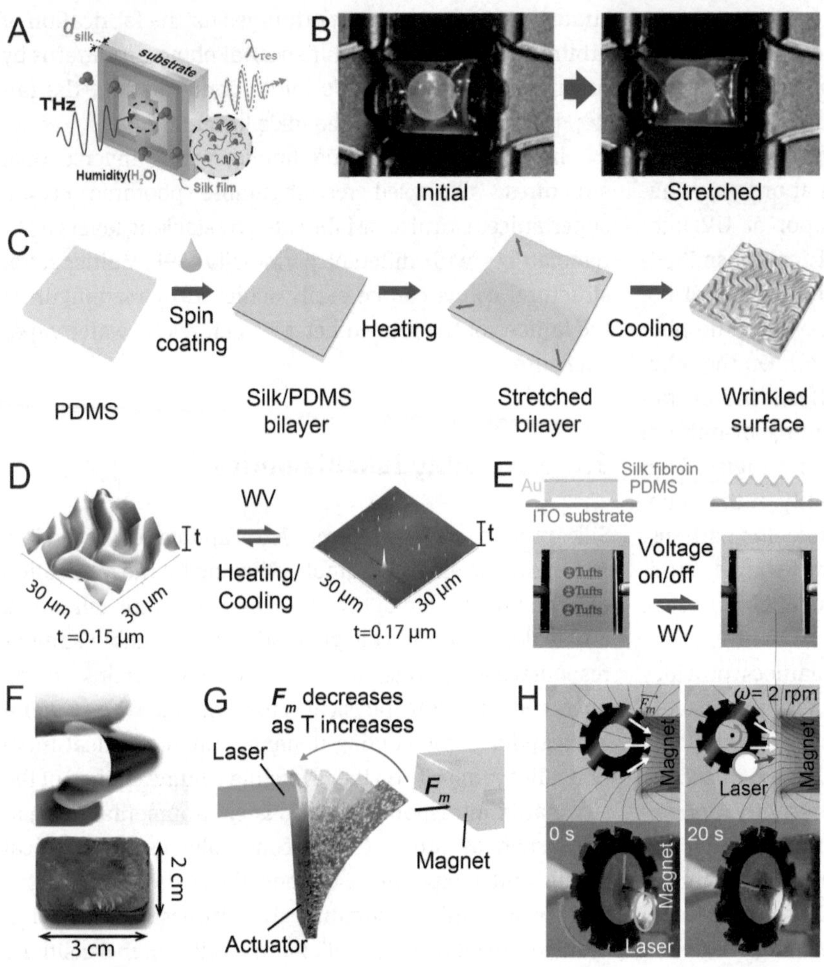

Figure 3: (A) Schematic of humidity sensors based on silk-coated terahertz metamaterials. Reproduced from Ref. [45]. Copyright 2018, OSA publishing, CC BY. (B) Photographs of silk hydrogel inverse opal before (left) and after (right) stretching. Reproduced from Ref. [49]. Copyright 2017, National Academy of Sciences. (C) Fabrication schematic of a silk-based dynamic wrinkle pattern. (D) AFM images showing the reversible transition between wrinkled and wrinkle-free state of the silk-PDMS bilayer film. (E) Schematics (top) and corresponding photographs (bottom) of wrinkle patterns showing the switching between transparent and opaque by respectively exposure to WV and voltage application. (C, D, E) Reproduced from Ref. [51]. Copyright 2019, National Academy of Sciences. (F) Photographs of a magnetic silk film (top) and a silk sponge (bottom) with 50 wt% CrO_2 relative to silk fibroin. (G) Schematic of the actuation mechanism of a cantilever actuator based on heat-induced demagnetization and photothermal effect. (H) Schematics (top) and photographs (bottom) of a Curie engine rotating at 2 rpm under fixed local light illumination and immobile magnet. (F, G, H) Reproduced from Ref. [52]. Copyright 2018, National Academy of Sciences.

ideal material format to transform a mechanical stimulus into an optical response. A typical example is deformable silk hydrogel inverse opals (Figure 3B) [49], in which the elastic deformation of the hydrogel induces a change in the photonic lattice constant of the SIO resulting in reversible changes in the reflected structural color. Further, the large thermal expansion coefficient of silk allows for the development of thermally responsive optical devices. For instance, silk fibroin whispering gallery microresonators showed a significant resonant wavelength shift as a function of temperature, making them suitable for high-precision thermal sensors [50].

Dynamic, tunable optical functions can also be obtained through the combination of two independent stimuli-responsive behaviors. As a demonstration, the polymorphic transition of silk fibroin has been combined with its thermal responsiveness to develop reversible and dynamic wrinkling micropatterns [51]. These wrinkling systems consist of a bilayer structure composed of a stiff silk thin layer and a soft PDMS substrate. The mechanical mismatch between these two materials induces the formation of a wrinkled surface after application of thermal stimuli (Figure 3C). This wrinkle pattern accumulates compressive stress after its formation and can be tuned or erased through molecular rearrangement of silk with either water vapor, methanol vapor, or exposure to UV light (Figure 3D); these treatments cause the continuous disturbance of the localized stress field and the subsequent release of the compressive stress within the bilayer system. The wrinkles' dynamics depend on the initial conformation of the silk film, showing that wrinkle tuning is governed by the interplay between silk protein chains and external stimuli. The switchable wrinkle morphology (Figure 3E), together with the versatility of silk protein, enables applications such as wearable optics, information coding, and thermal management.

Furthermore, reversible stimuli-responsiveness can be achieved by doping silk solutions with optically active components, adding functionality to mechanically robust and biocompatible platforms. One demonstration of this strategy is the fabrication of magnetic composites by incorporating chromium dioxide (CrO_2) in multiple silk formats such as films and sponges (Figure 3F) [52]. The low Curie temperature of CrO_2 along with its high photothermal conversion efficiency allows for the generation of a tunable magnetic response as a function of light, leading to light-controlled and dynamically tunable motions activated by fixed light source and magnetic field (Figure 3G, H). Devices fabricated from these composites show multiple complex locomotions, such as oscillation, gripping, and heliotactic movements, along with continuous rotation (Figure 3H). In addition to solution mixing, optical functionality can also be embedded in silk-based materials by chemical modification of the silk protein by taking advantage of its abundant surface chemistry. For instance, the incorporation of azobenzene moieties into silk structures creates a photo-responsive biomaterial (Azosilk) [53–55], whose optical absorption can be tuned by leveraging the reversible photo-switching between the *trans* and *cis* geometric isomers of azobenzene molecules. Optically switchable diffractive gratings have been developed by integrating an Azosilk grating with an elastomeric substrate, where the diffraction properties can be reversibly modulated by switching the light on and off [55].

3.3 Physically transient optics

Physically transient devices are programmed to either display an on/off signal or a modulation of their response as a function of time. Silk has been recently demonstrated to be a compelling material for the fabrication of both bioelectronic [56–58] and optically transient devices [15, 33, 59, 60] mostly due the ability to modulate its degradation. By properly tuning the cross-linking degree of the silk construct, its functional lifetime can be precisely programmed into the material so that it retains its structural integrity, and therefore, its form and optical functions, for a precise amount of time (from minutes to years) behaving as an intrinsically temporally transient device [24, 25]. The correlation between the silk structural degradation and the device's optical response variation enables the fabrication of devices which can display an intrinsic, self-reporting, and continuous monitoring system. Silk-based transient devices find applications in drug delivery, resorbable implants and information security.

Silk's high cytocompatibility [61, 62], and ability to stabilize labile compounds [63] have been exploited to use silk as a platform for drug-delivery devices. Therefore, silk-based transient optical devices can be particularly useful for the fabrication of comprehensive biomedical devices which simultaneously feature drug stabilization, controlled drug release and optical feedback (indirect drug release monitoring). The optical signature of these devices depends on their structural integrity and, if present, on the dopant type and loading [30]. Typically, silk fibroin is first mixed with therapeutics [19], and then molded into an optical element, such as microprism arrays [3], diffraction optical elements (DOEs) [15, 64], or fluorescent fibers [33], and finally annealed through water/methanol treatments to tune its degradation time. The progressive proteolytic cleaving of silk promotes the drug release while affecting the device's form and, therefore, its optical response (Figure 4A). In particular, the amount of drug released can be correlated to changes in reflectivity for prism arrays [3] and to variations of the photoluminescent intensity and spectral position for fluorescent silk fibers [33]. For monochromatic DOEs, the signal-to-noise ratio (SNR) of the diffracted signal is used, instead, for the optical feedback: the simultaneous dissolution of the device and the release of the drug induce physical and chemical deterioration of the DOE (Figure 4B), leading to a progressive disappearance of the diffracted pattern and, therefore, to a decrease of the SNR [15, 18] (Figure 4C). The SNR variation can be correlated to the amount of drug released and, therefore, used as a self-monitoring system. Multiple drug release can also be monitored at the same time by embedding each one in a different component of multichromatic DOEs [64]. Depending on the silk conformation state, these devices can be programmed not only to lose their optical imprint within a specific timeframe but also to fully dissolve. This kind of transient devices provides monitoring of the therapeutic release while eliminating the need for retrieval of the drug-support system after completion of their functional lifetimes.

Another simple yet elegant application of transient silk DOEs is for information security, in which the variation of the refractive index contrast is used to create transient patterns. Silk DOEs can display three operation modes: information concealment, information reappearance, and information destruction (Figure 4D–F). The information is first encoded by designing the structure of the DOE, it can then be revealed by illuminating the DOE with the specific wavelength it has been optimized for, and it can be further destructed by degrading the microstructure of the DOE through silk dissolution. Information concealment is achieved by using a combination of silk and another material with similar refractive index and by exploiting the variation of the refractive index contrast. For instance, a DOE pattern fabricated on glass ($n_{glass} \sim 1.52$) can be clearly visualized when exposed to air and illuminated with a laser beam; its message can then be temporarily hidden by

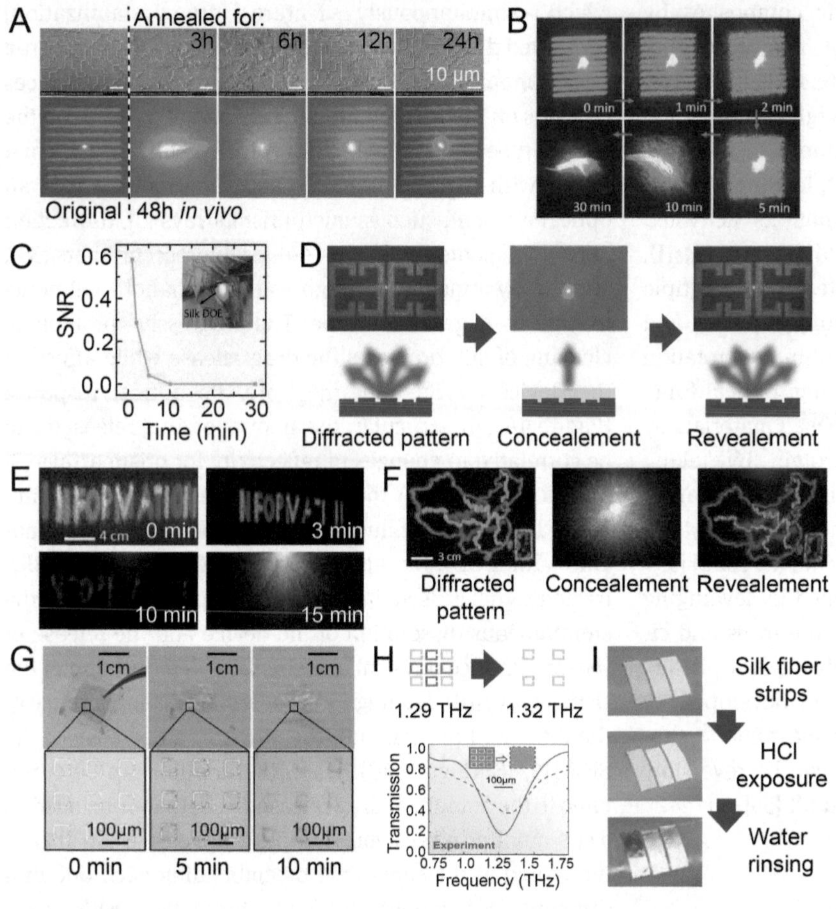

Figure 4: (A) Effect of the annealing time (3, 6, 12, and 24 h) on silk DOEs structure degradation (top row, SEM images) and on the corresponding optical response (bottom row, diffracted pattern) as quantified after 48 h *in vivo*. (B) Reflected diffraction patterns of a drug-doped silk DOE at different times showing the progressive degradation of the diffracted signal. (C) Variation of the surface-to-noise ratio (SNR) of the DOE as a function of time. Inset: picture of the DOE implanted on the skin of a rat on an infection site. (D) Diffracted pattern (top) and schematics of the structure (bottom) demonstrating information concealment and reappearance obtained by combining a glass DOE (green) with a silk coating (yellow). (A, B, C, D) Adapted from Ref. [15]. Copyright 2017, WILEY-VCH Verlag GmbH & Co. (E) Silk multi-colored DOE with multiple crystallinity degrees: the blue DOE starts degrading after 3 min, the red DOE after 10 min and the blue DOE after 15 min leading to gradual destruction of the information. (F) Glass multi-colored DOE showing information concealment and reveal due to, respectively, the application and the removal of a silk coating. (E, F) Adapted from Ref. [64]. Copyright 2018, Shanghai Institute of Microsystem and Information Technology (SIMIT), Chinese Academy of Sciences (CAS). Published by WILEY-VCH Verlag GmbH & Co. (G) Multi-metal metamaterials on silk substrates immersed in deionized water for 0, 5, and 10 min. (H) Schematic of multi-metal metamaterials on silk substrates showing the variation of the split ring resonator unit (top) and of the corresponding resonance (bottom) as a function of metals solubility. (G, H) Reproduced from Ref. [65]. Copyright 2020, WILEY-VCH Verlag GmbH & Co. (I) Doped fluorescent silk nanofiber strips after exposure to HCl vapors and after water rinsing. Adapted from Ref. [33]. Copyright 2017, The Authors.

covering it with a silk coating ($n_{silk} \sim 1.54$) that creates optical continuity by strongly decreasing the refractive index contrast and, therefore, it nullifies the DOE effect (Figure 4D); the pattern can, then, be further revealed by removal of the silk layer, which can, again, be intrinsically programmed by tuning the silk crystallinity [15].

Multichromatic silk-based DOEs enable multilevel encryption by containing multiple DOEs, each one optimized to work at a specific wavelength (Figure 4E, F) [64]. Notably, the degradation rate and sequence can be independently tuned for each monochromatic DOE, therefore enabling individual functional lifetimes for each diffracting pattern. This vanishing DOE system can store both temporally and chromatic dependent information. Multichromatic DOEs can also be used for quantitively monitoring chemical reactions by embedding solid reagents in silk DOEs and by immersing them in a liquid reagent to trigger the reaction; the monitoring of the variation of the SNR and of the optical density of the DOEs as the reaction progresses allows to obtain quantitative information on the formed products, provided prior calibration [64]. In addition, transient silk-based distributed feedback (DFB) lasers can be fabricated by repeated application and dissolution of a dye-doped silk solution on diffraction gratings [60].

Based on the same operation principle of controlled silk degradation, metamaterial devices able to work as transient, implantable and drug-delivery systems have been fabricated by patterning split ring resonators made of electrically conductive, biocompatible, and biodegradable metals on flexible silk substrates [65]. Both the metamaterial and the silk substrate can be programmed to independently affect the time-dependent resonance frequency response, respectively, by controlling the initial conformation of the silk matrix, as previously discussed (Figure 4G), and by using metals with different dissolution rates (Figure 4H). The ability to selectively remove structural components of the actual resonating structures (for instance achieved by using metals with different solubilities) allows for further complexity in the overall metamaterial response and in the structure's degradation path. Notably, small changes in the structure of the metamaterial induce significant variations in the inductance and capacitance, therefore, inducing a shift in the resonant response and making these devices suitable for high precision monitoring. Though these devices are currently operating in the terahertz regime, they could be fabricated to display an optical response in the visible range useful, for instance, for *in vivo* sensing and for information encryption.

The low cost of silk-based transient optical devices makes them ideal for the fabrication of, for instance, disposable acid vapor sensors. Fluorescent electrospun silk fibers doped with sodium fluorescein have been, indeed, reported as high precision, disposable, low cost and conformable sensors for highly toxic acid vapors, such as HCl (Figure 4I) [33]. Fluorescent silk fiber patches can detect HCl in a broad range of concentrations (5–300 ppm) inducing a deterioration in the photoluminescent signal and a variation of its spectral position upon exposure to the vapor. Due to silk's biocompatibility and tunability, it is possible to apply these sensors directly on the skin and to program them to vanish upon water rinsing without creating environmental hazards [33].

4 Conclusion and outlook

Regenerated silk fibroin has become a compelling material platform for the fabrication of active optical systems that are able to sense and transform diverse external stimuli into specific optical signals. This technological use of such an ancient material has been enabled by the protein's intrinsic properties, which include controllable conformational transitions, reversible physical swelling and shrinking, and programmable biodegradation in response to external physical and chemical perturbations. The past decade has witnessed a continuous growth in the development of silk-based stimuli-responsive optical devices, including reconfigurable photonic crystals, reversibly tunable optical structures, and physically transient devices. These stimuli-responsive optical systems, together with the biological traits of silk, enable the fabrication of wearable, disposable, and implantable devices with multiple functionalities, and thus drive application opportunities toward, but not limited to, bio- and environmental sensing, smart displays, information encoding, and drug delivery systems with self-monitoring capabilities.

While silk has shown great potential for the development of active optics, many scientific and technological challenges still need to be overcome for silk-based devices to fulfill their full potential. For instance, in spite of the opportunities enabled by the conformational transitions of silk, some critical questions remain unanswered: (i) how to fabricate "structurally patterned" silk matrices with high resolution over large areas; (ii) how to spatially and precisely control the protein's conformation in thick structures (e.g., thicker than a few tens of micrometers). In addition, for silk-based bio-/environmental sensors, their sensitivity, response rate, and selectivity must be further improved to meet the specific requirements of practical sensing applications. Furthermore, it is still challenging to achieve precise control over the degradation process of silk materials given the difficulty in precisely controlling the degree of crystallinity, film thickness, and the molecular weight of the protein. Other compelling critical concerns include the inherent batch-to-batch variability of silk solutions, the scale-up production and processing, the devices' integration, long-term durability, conformance, and cost; these have not yet been adequately addressed and thus significantly impede the use of silk-based active optical devices for commercial applications. These challenges can be solved by a comprehensive understanding of the "structure–property–process–function" relationships of silk protein.

The use of a versatile biopolymer material format (silk protein) as a stimuli-responsive platform for active optical devices provides a sustainable alternative to traditional inorganic materials, semiconductors, and non-biodegra dable synthetic polymers. Although a lot of progress has been made in silk-based dynamic optical devices, several directions should be payed attention to and be explored in the future. First, the current studies have mainly focused on the fabrication of active optical systems with well-defined single stimuli-responsive behaviors, and thus single output signals. Devices accommodating multiple dynamic behaviors provide emerging opportunities toward intelligent optics. The combinations of protein self-assembly, top-down transformation technologies, and silk's ease of functionalization are promising avenues for the development of these devices. Another interesting direction is the creation of optically functionalized and stimuli-responsive silk platforms, such as optomechanical, photoluminescent, and chromogenic materials, through either chemical modification of the silk protein or genetic engineering to insert specific optical functional domains into the molecular sequences of silk. In addition, it is worthwhile to exploit silk-based active optical materials with different formats, scales, and geometries. Promising strategies for this research direction involve the integration of protein self-assembly with other manufacturing techniques including 3D/4D printing, microfluidic technology, and origami/kirigami. Furthermore, silk provides an excellent stimuli-responsive material platform to interface optics with the biological world due to its comprehensive mechanical, biological, and optical properties. Current focuses on this territory are primarily on silk-based transient optical devices by leveraging the degradation process of silk matrix. In the future, efforts need to be also devoted to developing non-transient, active silk bio-optical devices for chronic epidermal and *in vivo* applications. The key to this purpose is to extend the lifetime of silk matrix at ambient conditions for skin-mounted devices and in wet media for silk implants through, for example, accurately controlling the structural organization from the nano- to the macroscale. Last but not least, nonlinear optical properties observed in natural silk fibers can be an interesting output signals for the future development of silk-based dynamic optics for sensing applications. Finally, interdisciplinary collaborations constitute a further step toward the goal of devising the next-generation of silk-based sustainable, active, and intelligent optical devices that can subtly, adaptively, and multiply respond to environmental changes in a wide variety of application fields.

Acknowledgments: The authors would like to acknowledge support from the Office of Naval Research (grant N00014-19-1-2399).
Author contribution: All the authors have accepted responsibility for the entire content of this submitted manuscript and approved submission.
Research funding: The authors would like to acknowledge support from the Office of Naval Research (grant N00014-19-1-2399).
Conflict of interest statement: The authors declare no conflicts of interest regarding this article.

References

[1] K. A. Arpin, A. Mihi, H. T. Johnson, et al., "Multidimensional architectures for functional optical devices," *Adv. Mater.*, vol. 22, pp. 1084–1101, 2010.
[2] T. Asakura, Y. Sato, and A. Aoki, "Stretching-induced conformational transition of the crystalline and noncrystalline domains of 13C-labeled bombyx mori silk fibroin monitored by solid state NMR," *Macromolecules*, vol. 48, pp. 5761–5769, 2015.
[3] H. Tao, J. M. Kainerstorfer, S. M. Siebert, et al., "Implantable, multifunctional, bioresorbable optics," *Proc. Natl. Acad. Sci. U. S. A.*, vol. 109, pp. 19584–19589, 2012.
[4] H. J. Jin and D. L. Kaplan, "Mechanism of silk processing in insects and spiders," *Nature*, vol. 424, pp. 1057–1061, 2003.
[5] R. F. P. Pereira, M. M. Silva, and V. De Zea Bermudez, "*Bombyx mori* silk fibers: an outstanding family of materials," *Macromol. Mater. Eng.*, vol. 300, pp. 1171–1198, 2015.
[6] F. Chen, D. Porter, and F. Vollrath, "Structure and physical properties of silkworm cocoons," *J. R. Soc. Interface*, vol. 9, pp. 2299–2308, 2012.
[7] C. Dicko, J. M. Kenney, and F. Vollrath, "β-Silks: enhancing and controlling aggregation," *Adv. Protein Chem.*, vol. 73, pp. 17–53, 2006.
[8] D. N. Rockwood, R. C. Preda, T. Yücel, X. Wang, M. L. Lovett, and D. L. Kaplan, "Materials fabrication from *Bombyx mori* silk fibroin," *Nat. Protoc.*, vol. 6, pp. 1612–1631, 2011.
[9] Y. Wang, D. Aurelio, W. Li, et al., "Modulation of multiscale 3D lattices through conformational control: painting silk inverse opals with water and light," *Adv. Mater.*, vol. 29, pp. 1–9, 2017.
[10] B. Marelli, N. Patel, T. Duggan, et al., "Programming function into mechanical forms by directed assembly of silk bulk materials," *Proc. Natl. Acad. Sci. U. S. A.*, vol. 114, pp. 451–456, 2017.
[11] K. Yazawa, K. Ishida, H. Masunaga, T. Hikima, and K. Numata, "Influence of water content on the β-sheet formation, thermal stability, water removal, and mechanical properties of silk materials," *Biomacromolecules*, vol. 17, pp. 1057–1066, 2016.
[12] Z. Zhou, S. Zhang, Y. Cao, B. Marelli, X. Xia, and T. H. Tao, "Engineering the future of silk materials through advanced manufacturing," *Adv. Mater.*, vol. 30, p. 1706983, 2018.
[13] S. Kim, B. Marelli, M. A. Brenckle, et al., "All-water-based electron-beam lithography using silk as a resist," *Nat. Nanotechnol.*, vol. 9, pp. 306–310, 2014.
[14] N. Qin, S. Zhang, J. Jiang, et al., "Nanoscale probing of electron-regulated structural transitions in silk proteins by near-field IR imaging and nano-spectroscopy," *Nat. Commun.*, vol. 7, 2016. https://doi.org/10.1038/ncomms13079.
[15] Z. Zhou, Z. Shi, X. Cai, et al., "The use of functionalized silk fibroin films as a platform for optical diffraction-based sensing applications," *Adv. Mater.*, vol. 29, p. 1605471, 2017.
[16] R. K. Pal, N. E. Kurland, C. Wang, S. C. Kundu, and V. K. Yadavalli, "Biopatterning of silk proteins for soft micro-optics," *ACS Appl. Mater. Interfaces*, vol. 7, pp. 8809–8816, 2015.
[17] M. A. Brenckle, H. Tao, S. Kim, M. Paquette, D. L. Kaplan, and F. G. Omenetto, "Protein-protein nanoimprinting of silk fibroin films," *Adv. Mater.*, vol. 25, pp. 2409–2414, 2013.
[18] J. P. Mondia, J. J. Amsden, D. Lin, L. D. Negro, D. L. Kaplan, and F. G. Omenetto, "Rapid nanoimprinting of doped silk films for enhanced fluorescent emission," *Adv. Mater.*, vol. 22, pp. 4596–4599, 2010.
[19] H. Tao, B. Marelli, M. Yang, et al., "Inkjet printing of regenerated silk fibroin: from printable forms to printable functions," *Adv. Mater.*, vol. 27, pp. 4273–4279, 2015.
[20] W. Li, Y. Wang, M. Li, L. P. Garbarini, and F. G. Omenetto, "Inkjet printing of patterned, multispectral, and biocompatible photonic crystals," *Adv. Mater.*, vol. 31, p. 1901036, 2019.
[21] S. T. Parker, P. Domachuk, J. Amsden, et al., "Biocompatible silk printed optical waveguides," *Adv. Mater.*, vol. 21, pp. 2411–2415, 2009.
[22] D. Lin, H. Tao, J. Trevino, et al., "Direct transfer of subwavelength plasmonic nanostructures on bioactive silk films," *Adv. Mater.*, vol. 24, pp. 6088–6093, 2012.

[23] S. Kim, A. N. Mitropoulos, J. D. Spitzberg, H. Tao, D. L. Kaplan, and F. G. Omenetto, “Silk inverse opals,” *Nat. Photonics*, vol. 6, pp. 818–823, 2012.
[24] X. Hu, K. Shmelev, L. Sun, et al., “Regulation of silk material structure by temperature-controlled water vapor annealing,” *Biomacromolecules*, vol. 12, pp. 1686–1696, 2011.
[25] Q. Lu, X. Hu, X. Wang, et al., “Water-insoluble silk films with silk I structure,” *Acta Biomater.*, vol. 6, pp. 1380–1387, 2010.
[26] A. Sagnella, A. Pistone, S. Bonetti, et al., “Effect of different fabrication methods on the chemo-physical properties of silk fibroin films and on their interaction with neural cells,” *RSC Adv.*, vol. 6, pp. 9304–9314, 2016.
[27] J. Shao, J. Zheng, J. Liu, and C. M. Carr, “Fourier transform Raman and Fourier transform infrared spectroscopy studies of silk fibroin,” *J. Appl. Polym. Sci.*, vol. 96, pp. 1999–2004, 2005.
[28] H. Perry, A. Gopinath, D. L. Kaplan, L. D. Negro, and F. G. Omenetto, “Nano- and micropatterning of optically transparent, mechanically robust, biocompatible silk fibroin films,” *Adv. Mater.*, vol. 20, pp. 3070–3072, 2008.
[29] B. D. Lawrence, M. Cronin-Golomb, I. Georgakoudi, D. L. Kaplan, and F. G. Omenetto, “Bioactive silk protein biomaterial systems for optical devices,” *Biomacromolecules*, vol. 9, pp. 1214–1220, 2008.
[30] P. Domachuk, H. Perry, J. J. Amsden, D. L. Kaplan, and F. G. Omenetto, “Bioactive “self-sensing” optical systems,” *Appl. Phys. Lett.*, vol. 95, p. 253702, 2009.
[31] Y. Wang, W. Li, M. Li, et al., “Biomaterial-based “structured opals” with programmable combination of diffractive optical elements and photonic bandgap effects,” *Adv. Mater.*, vol. 31, 2019. https://doi.org/10.1002/adma.201970030.
[32] S. Caixeiro, M. Gaio, B. Marelli, F. G. Omenetto, and R. Sapienza, “Silk-based biocompatible random lasing,” *Adv. Opt. Mater.*, vol. 4, pp. 998–1003, 2016.
[33] K. Min, S. Kim, C. G. Kim, and S. Kim, “Colored and fluorescent nanofibrous silk as a physically transient chemosensor and vitamin deliverer,” *Sci. Rep.*, vol. 7, 2017. https://doi.org/10.1038/s41598-017-05842-8.
[34] W. L. Rice, S. Firdous, S. Gupta, et al., “Non-invasive characterization of structure and morphology of silk fibroin biomaterials using non-linear microscopy,” *Biomaterials*, vol. 29, pp. 2015–2024, 2008.
[35] S. Kujala, A. Mannila, L. Karvonen, K. Kieu, and Z. Sun, “Natural silk as a photonics component: a study on its light guiding and nonlinear optical properties,” *Sci. Rep.*, vol. 6, 2016. https://doi.org/10.1038/srep22358.
[36] Y. Zhao, K. T. T. Hien, G. Mizutani, and H. N. Rutt, “Second-order nonlinear optical microscopy of spider silk,” *Appl. Phys. B Lasers Opt.*, vol. 123, p. 188, 2017.
[37] N. Vogel, M. Retsch, C. A. Fustin, A. Del Campo, and U. Jonas, “Advances in colloidal assembly: the design of structure and hierarchy in two and three dimensions,” *Chem. Rev.*, vol. 115, pp. 6265–6311, 2015.
[38] K. R. Phillips, G. T. England, S. Sunny, et al., “A colloidoscope of colloid-based porous materials and their uses,” *Chem. Soc. Rev.*, vol. 45, pp. 281–322, 2016.
[39] Y. Wang, M. Li, E. Colusso, W. Li, and F. G. Omenetto, “Designing the iridescences of biopolymers by assembly of photonic crystal superlattices,” *Adv. Opt. Mater.*, vol. 6, pp. 1–7, 2018.
[40] Y. Y. Diao, X. Y. Liu, G. W. Toh, L. Shi, and J. Zi, “Multiple structural coloring of silk-fibroin photonic crystals and humidity-responsive color sensing,” *Adv. Funct. Mater.*, vol. 23, pp. 5373–5380, 2013.
[41] Q. Li, N. Qi, Y. Peng, et al., “Sub-micron silk fibroin film with high humidity sensibility through color changing,” *RSC Adv.*, vol. 7, no. 29, pp. 17889–17897, 2017.
[42] E. Colusso, G. Perotto, Y. Wang, M. Sturaro, F. Omenetto, and A. Martucci, “Bioinspired stimuli-responsive multilayer film made of silk-titanate nanocomposites,” *J. Mater. Chem. C*, vol. 5, pp. 3924–3931, 2017.
[43] K. Hey Tow, D. M. Chow, F. Vollrath, I. Dicaire, T. Gheysens, and L. Thevenaz, “Exploring the use of native spider silk as an optical fiber for chemical sensing,” *J. Light. Technol.*, vol. 36, pp. 1138–1144, 2018.
[44] H. Tao, J. J. Amsden, A. C. Strikwerda, et al., “Metamaterial silk composites at terahertz frequencies,” *Adv. Mater.*, vol. 22, pp. 3527–3531, 2010.
[45] H. S. Kim, S. H. Cha, B. Roy, S. Kim, and Y. H. Ahn, “Humidity sensing using THz metamaterial with silk protein fibroin,” *Opt. Express*, vol. 26, p. 33575, 2018.
[46] H. Kwon and S. Kim, “Chemically tunable, biocompatible, and cost-effective metal-insulator-metal resonators using silk protein and ultrathin silver films,” *ACS Photonics*, vol. 2, pp. 1675–1680, 2015.
[47] S. Arif, M. Umar, and S. Kim, “Interacting metal-insulator-metal resonator by nanoporous silver and silk protein nanomembranes and its water-sensing application,” *ACS Omega*, vol. 4, pp. 9010–9016, 2019.
[48] M. Lee, H. Jeon, and S. Kim, “A highly tunable and fully biocompatible silk nanoplasmonic optical sensor,” *Nano Lett.*, vol. 15, p. 49, 2015.
[49] K. Min, S. Kim, and S. Kim, “Deformable and conformal silk hydrogel inverse opal,” *Proc. Natl. Acad. Sci. U. S. A.*, vol. 114, pp. 6185–6190, 2017.
[50] L. Xu, X. Jiang, G. Zhao, et al., “High-Q silk fibroin whispering gallery microresonator,” *Opt. Express*, vol. 24, p. 20825, 2016.
[51] Y. Wang, B. J. Kim, B. Peng, et al., “Controlling silk fibroin conformation for dynamic, responsive, multifunctional, micropatterned surfaces,” *Proc. Natl. Acad. Sci. U. S. A.*, vol. 116, pp. 21361–21368, 2019.
[52] M. Li, Y. Wang, A. Chen, et al., “Flexible magnetic composites for light-controlled actuation and interfaces,” *Proc. Natl. Acad. Sci. U. S. A.*, vol. 115, pp. 8119–8124, 2018.
[53] M. Cronin-Golomb, A. R. Murphy, J. P. Mondia, D. L. Kaplan, and F. G. Omenetto, “Optically induced birefringence and holography in silk,” *J. Polym. Sci. B*, vol. 50, pp. 257–262, 2011.
[54] M. J. Landry, M. B. Applegate, O. S. Bushuyev, et al., “Photo-induced structural modification of silk gels containing azobenzene side groups,” *Soft Matter*, vol. 13, pp. 2903–2906, 2017.
[55] G. Palermo, L. Barberi, G. Perotto, et al., “Conformal silk-azobenzene composite for optically switchable diffractive structures,” *ACS Appl. Mater. Interfaces*, vol. 9, pp. 30951–30957, 2017.
[56] L. Sun, Z. Zhou, N. Qin, and T. H. Tao, “Transient multi-mode silk memory devices,” in *Proceedings of the IEEE International Conference on Micro Electro Mechanical Systems (MEMS)*, vol. 2019-Janua, Institute of Electrical and Electronics Engineers Inc., 2019, pp. 519–521.
[57] M. A. Brenckle, H. Cheng, S. Hwang, et al., “Modulated degradation of transient electronic devices through multilayer

silk fibroin pockets," *ACS Appl. Mater. Interfaces*, vol. 7, no. 36, pp. 19870–19875, 2015.

[58] S. W. Hwang, H. Tao, D. H. Kim, et al., "A physically transient form of silicon electronics," *Science*, vol. 337, pp. 1640–1644, 2012.

[59] S. Toffanin, S. Kim, S. Cavallini, et al., "Low-threshold blue lasing from silk fibroin thin films," *Appl. Phys. Lett.*, vol. 101, p. 91110, 2012.

[60] H. Jung, K. Min, H. Jeon, and S. Kim, "Physically transient distributed feedback laser using optically activated silk bio-ink," *Adv. Opt. Mater.*, vol. 4, pp. 1738–1743, 2016.

[61] C. Vepari and D. L. Kaplan, "Silk as a biomaterial," *Prog. Polym. Sci.*, vol. 32, pp. 991–1007, 2007.

[62] A. R. Murphy and D. L. Kaplan, "Biomedical applications of chemically-modified silk fibroin," *J. Mater. Chem.*, vol. 19, pp. 6443–6450, 2009.

[63] J. A. Kluge, A. B. Li, B. T. Kahn, D. S. Michaud, F. G. Omenetto, and D. L. Kaplan, "Silk-based blood stabilization for diagnostics," *Proc. Natl. Acad. Sci. U. S. A.*, vol. 113, pp. 5892–5897, 2016.

[64] X. Cai, Z. Zhou, and T. H. Tao, "Programmable vanishing multifunctional optics," *Adv. Sci.*, vol. 6, p. 1801746, 2019.

[65] L. Sun, Z. Zhou, J. Zhong, et al., "Implantable, degradable, therapeutic terahertz metamaterial devices," *Small*, vol. 16, p. 2000294, 2020.

Suruj S. Deka*, Sizhu Jiang, Si Hui Pan and Yeshaiahu Fainman

Nanolaser arrays: toward application-driven dense integration

https://doi.org/10.1515/9783110710687-014

Abstract: The past two decades have seen widespread efforts being directed toward the development of nanoscale lasers. A plethora of studies on single such emitters have helped demonstrate their advantageous characteristics such as ultrasmall footprints, low power consumption, and room-temperature operation. Leveraging knowledge about single nanolasers, the next phase of nanolaser technology will be geared toward scaling up design to form arrays for important applications. In this review, we discuss recent progress on the development of such array architectures of nanolasers. We focus on valuable attributes and phenomena realized due to unique array designs that may help enable real-world, practical applications. Arrays consisting of exactly two nanolasers are first introduced since they can serve as a building block toward comprehending the behavior of larger lattices. These larger-sized lattices can be distinguished depending on whether or not their constituent elements are coupled to one another in some form. While uncoupled arrays are suitable for applications such as imaging, biosensing, and even cryptography, coupling in arrays allows control over many aspects of the emission behavior such as beam directionality, mode switching, and orbital angular momentum. We conclude by discussing some important future directions involving nanolaser arrays.

Keywords: applications; arrays; coupling; dynamics; integration; nanolasers.

***Corresponding author: Suruj S. Deka,** Department of Electrical and Computer Engineering, University of California at San Diego, La Jolla, CA, 92093-0407, USA, E-mail: sdeka@eng.ucsd.edu.
Sizhu Jiang, Si Hui Pan and Yeshaiahu Fainman, Department of Electrical and Computer Engineering, University of California at San Diego, La Jolla, CA, 92093-0407, USA

1 Introduction

In the quest for attaining Moore's law-type scaling for photonics, miniaturization of components for dense integration is imperative [1]. One of these essential nanophotonic components for future photonic integrated circuits (PICs) is a chip-scale light source. To this end, the better part of the past two decades has seen the development of nanoscale lasers that offer salient advantages for dense integration such as ultracompact footprints, low thresholds, and room-temperature operation [2–4]. These nanolasers have been demonstrated based on a myriad of cavity designs and physical mechanisms some of which include photonic crystal nanolasers [5–8], metallo-dielectric nanolasers [9–15], coaxial-metal nanolasers [16, 17], and plasmonic lasers or spasers [18–22]. The applications of such ultrasmall lasers are not just limited to on-chip communications as their unique characteristics allow them to be used for biological sensing [7, 23, 24], super-resolution imaging [25, 26], as well as optical interconnects [27].

In most practical applications, however, a single nanolaser would never operate in isolation but rather in tandem with multiple such devices. Doing so can enable novel applications that would otherwise be infeasible to achieve using an isolated laser. Therefore, the next logical step in the evolution of nanolaser technology is to realize large-scale, dense arrays of nanolasers by leveraging the knowledge gained so far from the plethora of studies on single such emitters.

In this review, we focus on the progress made toward realizing nanolaser arrays. In Section 2, we elaborate on the interaction between only two nanolaser devices. Such a dual nanolaser system can be considered as a unit cell that can assist in understanding larger scale arrays. Section 3 discusses the distinct types of arrays demonstrated thus far and the unique applications they can help enable. Finally, we conclude and present possible research directions yet to be explored on this front in Section 4.

This article has previously been published in the journal Nanophotonics. Please cite as: S. S. Deka, S. Jiang, S. H. Pan and Y. Fainman "Nanolaser arrays: toward application-driven dense integration" *Nanophotonics* 2021, 10. DOI: 10.1515/nanoph-2020-0372.

2 Interaction between two nanolasers

Perusing the recent literature on array architectures of nanolasers reveals that based on array size, most studies can be segmented into two categories. If N is used to denote the number of individual elements (or nanolasers) in the array, the categories correspond to the cases of $N = 2$ and $N > 2$. Prior to reviewing what occurs in larger arrays, it is helpful to first consider an array of only two resonant nanocavities (i.e., $N = 2$). The primary reason for this is that a dual nanolaser system can be regarded as a type of unit cell or building block that can help explain the behaviors observed in larger lattices. Since lasers are inherently nonlinear systems, the interactions between them can produce rich and complex dynamics. The linewidth enhancement factor for semiconductor materials used in most gain media further complicates the physics involved [28]. In this scenario, limiting the number of nanolasers to two simplifies the associated rate equations and can be a stepping-stone toward understanding array behaviors. This section reviews a few aspects of the interesting physics that can result from the interaction of two nanolasers.

2.1 Creation of supermodes

The coupling between two lasers can be introduced in various near-field and far-field manners including evanescent coupling [29–37], leaky wave coupling [38], and mutual injection [39]. Among them, evanescent coupling induced between two resonant cavities in close proximity is the most common and feasible way to build densely integrated nanolaser arrays and has already been extensively investigated in microscale cavities [29–34]. This type of coupling occurs when increased interaction between the evanescent electromagnetic fields of the two individual resonators gives rise to a characteristic splitting of the observed modes in both frequency and loss [29, 32–34]. The bifurcation is the consequence of the creation of bonding and antibonding supermodes which are dissimilar in both their optical losses and frequency.

With advances in fabrication technology, the corresponding higher precision has facilitated moving from the microscale to the nanoscale. Consequentially, similar coupling behavior has now been reported in coupled nanolaser cavities. Deka et al. demonstrate such coupling in a dual nanolaser system comprising two metallo-dielectric nanocavities shown in the schematic in Figure 1A [35]. In their design, the distance between the two resonators, represented by d in the schematic, is varied, while the modes supported by the system are recorded. The authors first consider two extreme cases of when the cavities are spaced far apart (d = 90 nm) and when their dielectric shields are in contact (d = 0 nm). For each case, the electric field intensity is calculated, as portrayed in a two-dimensional side and top cross section in Figure 1B. When designed far apart, the TE_{011} modes supported by the two equally sized nanocavities are independent and identical (Figure 1B, left). This results from the metal between the resonators damping any evanescent fields arising from the cavities, thus inhibiting coupling [35]. If d = 0 nm, however, the increased evanescent coupling between the cavities leads to the formation of two new modes – the antibonding supermode (Figure 1B, middle) and the bonding supermode (Figure 1B, right). The former exhibits strong confinement of the electromagnetic field to the individual gain media while the latter demonstrates an electric field maximum in the central region between the two gains due to constructive interference.

Furthermore, by varying d in discrete steps and calculating the eigenmode wavelength (λ) and quality factor (Q) of the cavity modes for each distance, the authors report a split in the two parameters as d is reduced (Figure 1C, top and middle). Since the cavities are purposed for lasing, the gain threshold, g_{th}, of each mode is also calculated (Figure 1C, bottom). Owing to a larger overlap with the dissipative metal, the bonding supermode experiences a higher loss and is therefore less likely to lase as evidenced by its higher threshold [35]. The creation of these dissimilar supermodes due to evanescent coupling has also been reported in other material systems such as photonic crystals. Figure 1D depicts a schematic of two coupled photonic crystal nanolasers taken from the work of Hamel et al. [36]. When pumped equally, their system supports bonding and antibonding supermodes with the latter becoming the dominant lasing mode at higher pump powers (Figure 1E). Therefore, the formation of bonding and antibonding supermodes is one of the most basic phenomena to occur due to increased evanescent coupling between nanolasers.

It is important to note, however, that besides bonding and antibonding modes, various other types of supermodes can also be observed in coupled nanolaser systems. Parto et al. [40] demonstrate that depending on the size of the metallic nanodisks used in their study, either ferromagnetic (FM) or antiferromagnetic (AF) coupling is exhibited between the constituent elements of the lattice. These FM and AF couplings arise due to the interaction between the vectorial electromagnetic modes supported by the nanocavities which exchange spin Hamiltonians like in

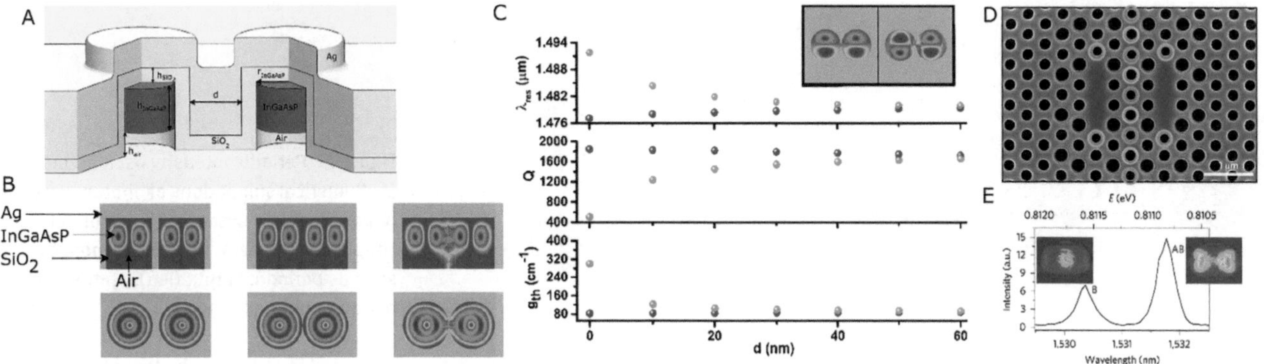

Figure 1: Supermode creation in dual nanolaser systems.
(A) Two coupled metallo-dielectric nanolasers, where the distance between cavities is represented by *d*. (B) Electric field intensity profiles across the side (top row) and top (bottom row) cross sections of the nanolaser system shown in (A). When $d = 90$ nm (left), system supports two identical and independent TE_{011} modes. When $d = 0$ nm, two new modes are supported by the system – antibonding (middle) and bonding supermodes (right). (C) Eigenmode wavelengths (λ), quality factors (Q), and gain thresholds (g_{th}) for the two modes supported by the dual nanolaser system at varying *d*. Inset: electric field distributions of the antibonding (left) and bonding (right) supermodes. (A), (B) and (C) reprinted from a study by Deka et al. [35] with permission. (D) Two coupled photonic crystal nanolasers and (E) the bonding (B, left peak) and antibonding (AB, right peak) modes supported in this design. (D) and (E) adapted and reprinted from a study by Hamel et al. [36] with permission.

magnetic materials [40, 41]. These recent results can help pave the path toward achieving ultracompact, on-chip photonic platforms that implement spin Hamiltonians for solving complex optimization problems [41].

2.2 Analysis of nonlinear dynamics

Besides supermodes creation, the coupling between two lasers can also produce rich nonlinear dynamics. Specifically, for the case of coupled nanolasers, a multitude of dynamical regimes including stable phase locking, periodic intensity oscillations, as well as chaotic fluctuations can be achieved that can help enable a wide variety of applications. For instance, coupled nanolasers operating in a stable phase locking manner are essential for building high power laser arrays [42, 43], whereas periodic oscillations may be used for on-chip modulation and radio frequency (RF) wave generation [44]. Recently, chaotic synchronization of nanolasers which can prove useful in quantum experiments for random number generation and secure key exchange has also garnered some attention [45]. Consequentially, analyzing the diverse dynamical regimes at play for two coupled nanolasers is of critical importance.

Investigation of dynamics usually involves solving the coupled rate equations and analyzing the dependence of the dynamical regimes observed on essential parameters. These can include, among other parameters, injection strength, bias current, and the intercavity distance between constituent elements. This analytical method can be universally applied regardless of the form of coupling at play albeit with some modifications according to the cavity geometries used. For the case of mutual injection, Han et al. numerically explore the dynamics of coupled nanolasers by considering the influence of the spontaneous emission factor β, the Purcell factor F, and the linewidth enhancement factor α for varying optical injection strengths and intercavity distances [39]. Tuning the injection strength alone, the authors can observe periodic oscillations and both stationary and nonstationary periodic doubling as illustrated in Figure 2A. More importantly, an enhanced stability of the coupled system with high values of F is reported, thus underlining the advantage of nanoscale lasers due to their pronounced Purcell factors [39].

Besides stable operation, the high-frequency periodic oscillations exhibited by a dual nanolaser system are also of interest for any applications requiring high speeds. Adams et al. [46] theoretically predict such periodic oscillations to occur outside the stability region for two laterally coupled nanowire lasers. The frequency of these oscillations as a function of both the separation between the nanolasers and the pumping rate is shown in Figure 2B. It can be clearly observed from the plot that ultrahigh frequencies on the order of at least 100 GHz can be reached when the separation is reduced to ~300 nm or so. Moreover, the oscillation frequency increases for larger pumping rates.

The studies used as examples in this subsection highlight the importance of investigating the dynamics of coupled nanolasers. Identifying the stable, oscillatory, and

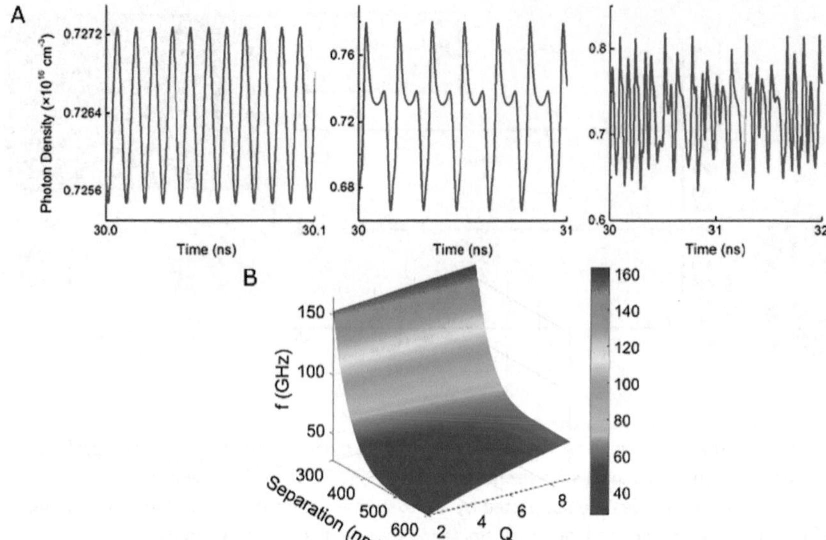

Figure 2: Periodic intensity oscillations. (A) Numerical simulations of photon density of one laser in a system of two mutually coupled nanolasers. Unstable dynamics such as periodic output (left), stationary period doubling (middle) or nonstationary period doubling (right) can be observed. Reprinted from a study by Han et al. [39] with permission. (B) Periodic oscillations for two evanescently coupled nanowires as a function of separation between lasers and pumping rate (*Q*). Oscillations can reach over 100 GHz. Reprinted from a study by Adams et al. [46] with permission.

chaotic regimes as a function of varying control parameters allows the coupled system to exhibit the desired behavior for the application demanded. Theoretical analyses can also shed light on bifurcations that can arise due to coupling. Such bifurcations are especially significant as they can lead to switching of optical modes as elaborated on in the next section.

2.3 Mode selection and switching

What is especially advantageous for nanoscale systems compared to their micron-sized counterparts, is their increased sensitivity to small perturbations induced by nuanced changes in the geometry or in the pumping scheme. These perturbations can result in alterations to the supermodes – such as the bonding and antibonding ones described in Section 2.1 – that are supported by the systems. Haddadi et al. [47] show in their study how varying the size of the middle row of holes in their coupled photonic crystal nanolaser structure can enable them to control the coupling behavior. In other words, by engineering the barrier between the two cavities (Figure 3A), both the amount of wavelength splitting as well as the excited state mode can be selected.

Figure 3B depicts the absolute value of the wavelength difference – $|\Delta\lambda|$ – between the bonding and antibonding supermodes as the barrier hole radius is varied. The distinct colors represent different radii of the nanocavities themselves. Through this graphic it can be inferred that the coupling strength can be tuned via the barrier size yielding either an increase or decrease in $|\Delta\lambda|$. Additionally, the authors report that the branches on either side of the dip in $|\Delta\lambda|$ correspond to a flip in the parity of the bonding mode. To be more accurate, to the left of the dip, the bonding mode becomes the excited state, whereas to the right, it remains the ground state mode [47]. Finally, increasing the size of the cavity results in a monotonic shift of the point where the parity flip of the bonding mode occurs. Efforts to select the ground state mode have also been reported in other studies, albeit with microcavities [48, 49].

Although static geometry tuning can determine the ground state supermode, alterations in the pumping scheme can lead to bifurcations which in turn, can allow for real-time switching between the coupled cavity modes. In their work, Marconi et al. [50] demonstrate that as they increase the optical pumping to their coupled photonic crystal nanolaser system, a mode-switching behavior is observed from the blue-detuned bonding mode to the red-detuned antibonding mode. Figure 3C portrays this switch for both simulated (top) and experimental (bottom) data. The mechanism behind the switching is attributed to the asymmetric stimulated light scattering induced by carrier oscillations which manifests itself through a Hopf bifurcation [50]. Such mode-beating oscillations and their corresponding decay rate were characterized through the statistical intensity experiments in the authors' more recent work [51].

Finally, in addition to switching between coupled supermodes that have dissimilar stability conditions, a spontaneous transition between coexisting eigenmodes is also possible, as demonstrated in quantum well-based photonic crystal nanolasers by Hamel et al. [36]. The light-out vs. light-in (*LL*) curve of the coupled system shown in Figure 3D illustrates that a pitchfork bifurcation occurs at a pump power $P = 1.33P_{th}$, where P_{th} is the lasing threshold. At this bifurcation point, the antibonding mode loses its

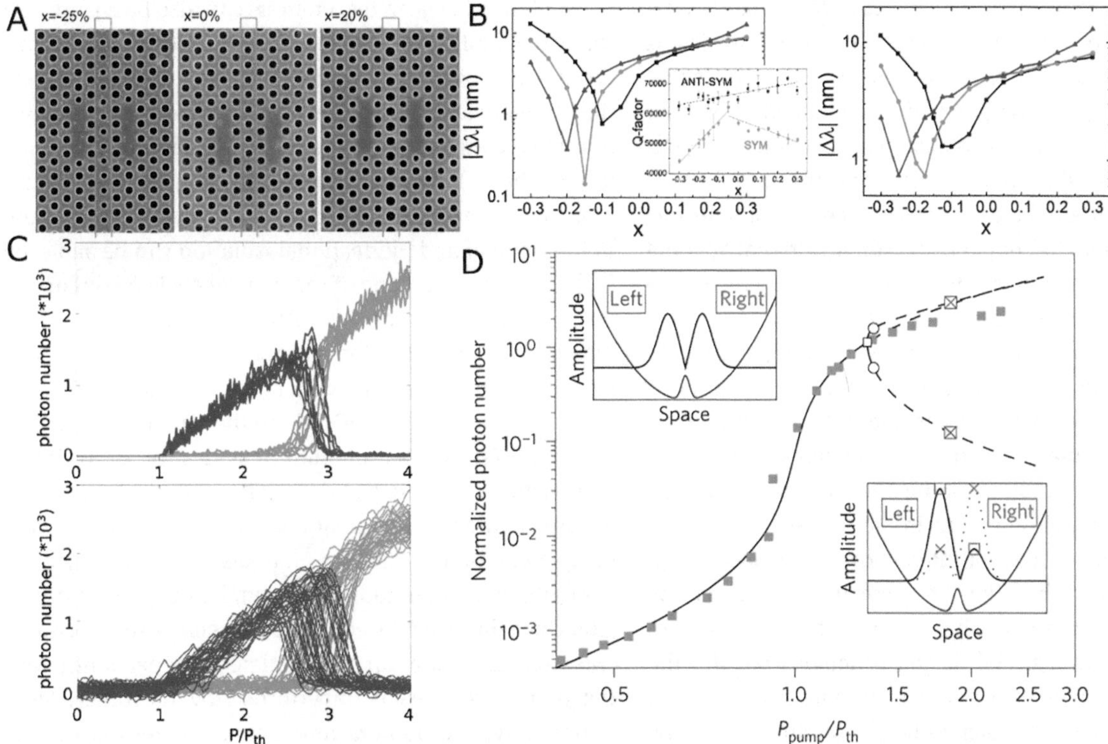

Figure 3: Mode selection and switching.
(A) Barrier engineering in photonic crystal nanolasers by changing size of the middle row of holes. For the case of $x = 0\%$, the radius of the highlighted row of holes is ~115 nm. (B) Absolute value of wavelength difference between B and AB modes – $|\Delta\lambda|$ – vs. size of the middle row of holes. Simulations (left) and experiment (right). Different colors represent different cavity sizes. The dip corresponds to the crossing or transition point where AB mode goes from being the excited state (right of dip) to becoming the ground state (left of dip); the opposite is true of the B mode. (A) and (B) adapted and reprinted from a study by Haddadi et al. [47] with permission. (C) Simulated (top) and experimental (bottom) results showing switching from B (blue) to AB (red) mode due to increased pump powers in two coupled photonic crystal nanolasers. Adapted from a study by Marconi et al. [50] with permission. (D) *LL* curve of two coupled photonic crystal nanolasers (green filled squares-experimental results; black solid line-numerical simulation) showing pitchfork bifurcation that results in two coexisting and stable, broken parity states (right inset). Left inset: Stable solution before bifurcation. Adapted from a study by Hamel et al. [36] with permission.

stability and two stable and coexisting solutions emerge. The first solution represents a high electric field amplitude in the left cavity and low amplitude in the right cavity [36]. The second solution is a mirror opposite of the first. For this spontaneous symmetry breaking to occur, the coupled nanolasers have to be designed with the exact parity and strength of the coupling coefficient needed to support these special states. The selective switching between the two stable states can then be triggered by pumping one of the cavities with a short pulse. Depending on the amplitude in the cavity prior to the incoming pulse, the state can either be retained or switched [36].

As evident from these results, the ability to engineer coupled modes as well as switch between them is of interest not only for the underlying physics involved but also for the potential applications. For instance, measuring the second order correlation function at zero delay – $g^2(0)$ – yields a value greater than 2 for a bonding supermode indicating the superthermal nature of light emitted from this mode [52]. For the antibonding mode, this value approaches 1 confirming lasing behavior. Therefore, systems where the mode selection is enabled are ideal test beds for important studies concerning quantum and nonlinear optics, out-of-equilibrium thermodynamics, quantum correlation devices, and engineering of superthermal sources [36, 50, 52]. Mode-switching devices with the ability to affect the transition between optical modes depending on the current state, as was explained for the broken parity states, can be said to exhibit memory and can therefore, be potentially used as ultracompact logic gates or optical flip flops [36].

2.4 Cross talk isolation

Although coupling between two nanolasers can yield complex dynamical phenomena rich in physics, certain applications might require the independent functioning of the emitters even when placed in close proximity. Biological sensing and imaging are two such areas, where maintaining isolation between the emission of the two cavities is imperative (discussed later in Section 3.1). This implies that each nanocavity can be pumped and modulated without influencing the behavior of its neighbor. For photonic crystal cavities mentioned previously, barrier engineering – where the barrier between the cavities is altered – is an effective method to inhibit the coupling [34, 46]. An alternative route taken to prevent coupling is to detune one cavity relative to the other via thermal or carrier effects [53].

In cavities that share a common cladding, such techniques are difficult to implement since any alterations made to one cavity will be transmitted to its neighbor via the shared cladding. For the case of metallo-dielectric nanolasers, the metallic coating serves as the common cladding for the two resonators. In such a scenario, a viable way to curb the evanescent interaction of the electromagnetic fields is to introduce a resonance mismatch by designing one cavity to be slightly larger than its neighbor. By doing so, Deka et al. [35] are able to demonstrate that the degree of splitting in the eigenmode frequency and optical loss – a characteristic signature of coupling – can be minimized. The case of equal-sized resonators was presented earlier (Figure 1C), where the pronounced splitting in the λ, Q, and g_{th} confirmed the effects of coupling. In order to mitigate this coupling, Deka et al. [35] altered the radius of one cavity to be 5% larger than its neighbor. This size mismatch yields splitting that is now hardly discernible, as depicted in Figure 4A, for the same parameters demonstrated in Figure 1C. This implies that despite the continual reduction in d as before, the dissimilar resonances now lead to inhibited coupling.

Another geometry-based technique that can be implemented to mitigate coupling in metallo-dielectric nanolasers is to increase the cavity radii in order to exploit the higher-order modes supported by larger cavities. Such modes exhibit stronger optical confinement to the gain media. As a result, less of the modes leaks out to evanescently interact with one another which is the exact phenomena that usually leads to the creation of coupled supermodes. This form of strong confinement of higher-order modes has been previously reported in microcavities [32, 54]. Pan et al. [55] demonstrate this behavior for the same dual metallo-dielectric nanolaser system first presented in a study by Deka et al. [35]. Figure 4B depicts their results, where the difference between the Q-factors – ΔQ – for the coupled mode for two different intercavity distances (d = 100 and 0 nm) are plotted as a function of the cavity radius. It can be clearly inferred from this plot that as the radii of the nanocavities are increased, the split in their losses diminishes due to reduced coupling.

Isolation between two emitters can also be ensured for forms of coupling besides just the evanescent interactions discussed above, such as for mutual injection coupling. In their numerical analysis, Han et al. consider the case of two mutually coupled but independently modulated nanolasers where, by tuning parameters such as the bias current, injection strength, and modulation depth, both unidirectional and bidirectional isolation can be achieved [56]. For instance, if one of the nanolasers is modulated at 50 GHz while the other at 20 GHz, for certain ranges of the bias current and injection strength, the dynamics of one laser is affected by the other but not vice versa. This result is encapsulated in Figure 4C, where the Fast Fourier Transforms (FFTs) of the photon densities of both lasers demonstrate unidirectional isolation. If the goal is to ensure bidirectional isolation, eliminating cross talk completely in the process, the bias current, injection strength, and modulation depth can be tuned accordingly [56]. Once this is achieved, Figure 4D shows that the same two lasers as before can be modulated independently with the dynamics of each unperturbed by the other. More importantly, the authors report that the zero cross talk regime can only be observed in coupled systems with large β and Purcell enhancement factors, which implies the suitability of nanolasers (which demonstrate intrinsic high-β and F) for use in PICs.

The studies reviewed in Section 2 emphasize the value of a two-nanolaser system as a test bed to further understanding of complex physics such as quantum and nonlinear optics, phase-locking, and cross talk isolation. For instance, when trying to implement independent control over each emitter, it is much easier to grasp the complex nonlinear dynamics involved by limiting the system size to just two. Next, we will discuss nanolaser arrays and how some of the ideas presented in this section can be scaled up and implemented in large-scale lattices.

3 Nanolaser arrays

Scaling up design to allow individual light sources to work in conjunction in an array format falls into the natural roadmap of the nanolaser technology [1, 2]. It is important to emphasize here, however, that not all nanolaser arrays rely on the same operating principles. The main distinction to be made is between uncoupled and coupled arrays. Uncoupled arrays comprise individual nanolasers that function independently and do not interact with their nearest neighbors. Some of the techniques to ensure isolation for two nanolaser systems reviewed in Section 2.4 can now be extended to ensure zero cross talk for dense

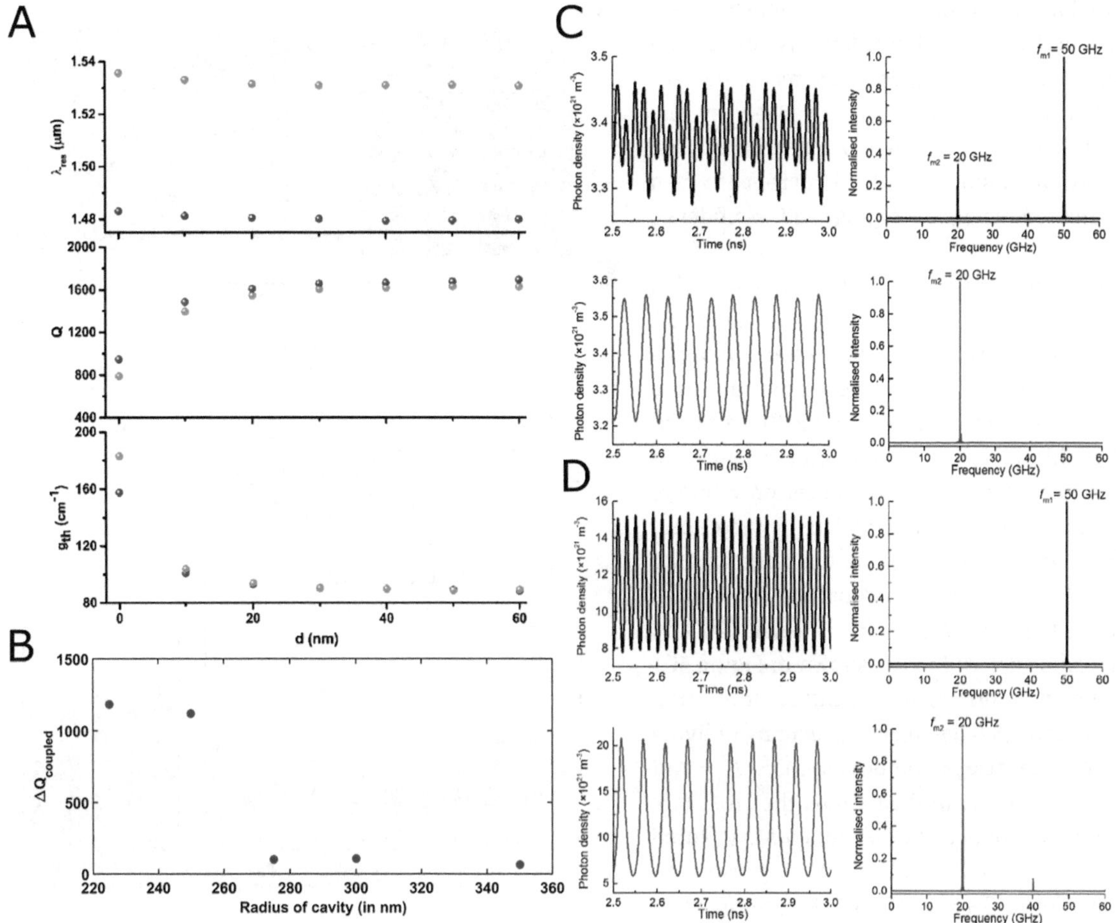

Figure 4: Cross talk isolation techniques.
(A) Eigenmode wavelengths (λ), quality factors (Q), and gain thresholds (g_{th}) for the two modes supported in two unequal sized metallo-dielectric cavities, where one is 5% larger in radius. Unlike in Figure 1C, no pronounced splitting is observed due to reduced coupling. Adapted from a study by Deka et al. [35] with permission. (B) Difference in Q for the coupled mode – ΔQcoupled – for $d = 100$ nm and $d = 0$ nm as a function of radii of the cavities. Larger sized cavities impede coupling. Reprinted from a study by Deka et al. [55] with permission. (C) Photon densities (left) and their Fast Fourier Transforms (FFTs, right) for two mutually coupled nanolasers in unidirectional isolation. Laser 1 (top) is affected by dynamics of laser 2 (bottom) but not vice versa. (D) Same as (C) but now nanolasers are in bidirectional isolation regime. Each laser is unperturbed by dynamics of its neighbor. (C) and (D) reprinted from a study by Han et al. [56] with permission.

lattices. In direct contrast, coupled arrays, as the name suggests, rely on coupling of some form involving the constituent resonators comprising the system. This may be exhibited either in terms of evanescent coupling between the nanocavities or excitonic state-plasmonic surface lattice resonance coupling. Owing to their clear disparity, coupled and uncoupled arrays are each appropriate for different types of applications.

3.1 Uncoupled arrays

Coupling, especially strong coupling, makes the task of distinguishing between the individual resonators more complex since the two (or multiple) emitters can now be viewed as a new, larger system [35]. Therefore, for applications that rely on the individual state or output of each element in the lattice, uncoupled arrays are most suitable. A majority of sensing and imaging applications fall into this category as they rely on recording the wavelength shift of each nanolaser due to changes in the refractive index environment.

One such study authored by Hachuda et al. [57] demonstrates the detection of protein in the form of streptavidin (SA) by using a 16-element 2D photonic crystal nanolaser array. The nanolaser geometry and array design, depicted in Figure 5A, show how a nanoslot is incorporated into the design. These nanoslots help with the localization of the optical mode which is especially significant, given that the measurements are performed in water to help with thermal stability [57, 58]. Although all the nanolasers are designed to be identical, the independence is maintained by optically pumping each one separately to record the red-shift in emission wavelength – $\Delta\lambda$ – caused by the adsorption of protein in each individual nanolaser. The overall wavelength alterations for all 16 devices are then statistically evaluated using averaging and confidence intervals.

Figure 5B portrays the results for $\Delta\lambda$ for each element in the array when impure solutions containing the protein SA are exposed to the nanolaser array. With water serving as the control (blue), it can be clearly observed that the array can distinguish between a sample containing SA (red) and those without it (black) based on the shift in emission wavelengths. In this experiment, bovine serum albumin (BSA) is treated as a contaminant to which the target protein, SA, is attached [57]. By averaging the results for all nanolasers in the array and increasing the amount of contaminant BSA, the limit of the array's sensitivity and selectivity can also be determined as shown in Figure 5C.

Based on the same principle of using $\Delta\lambda$ from individual lasers, Abe et al. [59] demonstrate imaging of living cells using uncoupled arrays comprising 21 × 21 = 441 photonic crystal nanolasers. In their study, the cross talk isolation between individual array constituents is ensured by designing an offset in radii for all neighboring lasers, an idea reviewed earlier in Section 2.4 [35]. Figure 6A illustrates that it is possible to optically pump the entire array yet maintain independent operation due to the radii mismatch engineered in the design. To perform imaging, the target cell was deposited on top of the nanolaser array, and the subsequent shift in the emission wavelength was recorded for each array element. The $\Delta\lambda$ image is created by measuring the reference λ for each nanolaser and then mapping the $\Delta\lambda$ at each laser's position in the array.

The results – illustrated in Figure 6B – not only provide an accurate albeit rough image of the cell but also demonstrate time evolution since the detection is continuous. Additionally, by employing nanoslots in their design, a Δn image is created which suppresses the noise and calibrates nonuniformity to yield a more accurate capture than its $\Delta\lambda$-based counterpart [59]. Figure 6C describes the ability of a Δn image to track the movement of a cell until it is desorbed, which in this case is shown to take upwards of 10 h.

Besides sensing and imaging, uncoupled lasing arrays can also be purposed to address other complex problems. By creating organic molecule-based laser arrays and using them in conjunction with distinct organic solutions, Feng et al. [60] demonstrate the possibility of creating nondeterministic cryptographic primitives. The randomness in

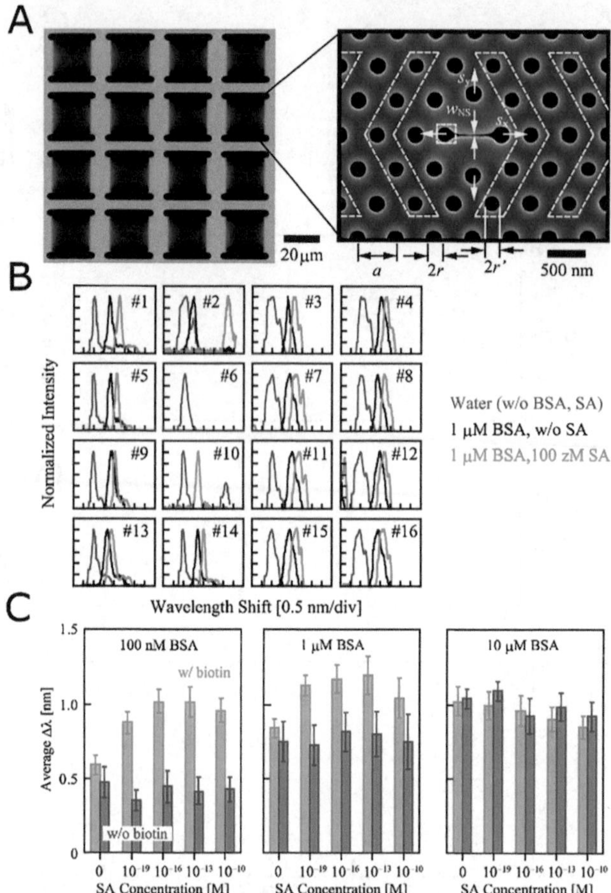

Figure 5: Biological sensing of proteins.
(A) Schematic of 16-element photonic crystal uncoupled array (left) and Scanning Electron Microscope (SEM) image of nanoslot-incorporated single laser (right). (B) Normalized intensity showing wavelength shift for all 16 elements under varying solutions. Water serves as the control. (C) Average wavelength shift calculated from all lasers – $\Delta\lambda$ – plotted for increasing amount of contaminant bovine serum albumin (BSA). Adapted and reprinted from a study by Hachuda et al. [57] with permission.

the size distribution of the individual nanolasers is caused by the stochastic manner in which the organic solution forms capillary bridges around the array elements. The varying array types formed with four distinct organic solutions is shown in Figure 7A. Owing to multiple vibrational sublevels, the organic molecule used in this study is capable of exhibiting dual wavelength lasing at either 660 or 720 nm or both depending on the length of the cavity. Figure 7B describes the emission behavior as a function of the cavity length for arrays created with the different solutions. Clearly, four distinct emission states can be observed depending on the stochastic size distribution of the nanolasers – (1) no lasing, (2) lasing at 660 nm only, (3) lasing at both 660 and 720 nm, and (4) lasing at 720 nm

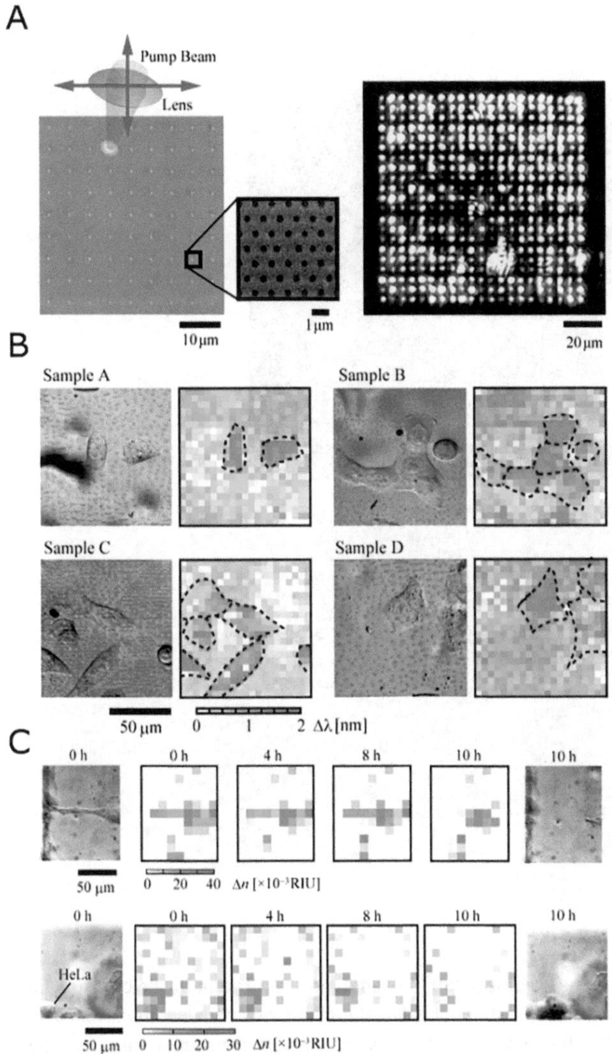

Figure 6: Imaging living cells.
(A) SEM image of uncoupled nanolaser array (left) and near-field emission of the array (right). (B) $\Delta\lambda$ image (right panels) of different samples of cells (optical micrographs; left panels) and (C) Δn image (center panels) tracking movement of a single cell (optical micrographs; left and right panels). Reprinted from a study by Abe et al. [59] with permission.

only. These states can be represented as either quaternary bits ('0' for no lasing, '1' for lasing at 660 nm etc.) or double binary bits ('00' for no lasing, '01' for lasing at 660 nm, etc.). In order to generate a cryptographic sequence from the arrays, each nanolaser is pumped separately with the pump then subsequently scanned to get the emission spectra from all other devices [60]. This technique, portrayed in Figure 7C, yields different encoding bits for each nanolaser (depending on its cavity length) which can be used to generate cryptographic bit sequences like in Figure 7D. In fact, encoding as double binary bits makes it possible to generate up to 2048 binary bits.

The studies elaborated above are only a fraction of the multiple works demonstrating how uncoupled arrays can achieve unique applications. Whether used for refractive index sensing, imaging or developing next generation, all-photonic cryptographic primitives, maintaining the independence of each nanolaser in the array is of vital importance for these applications [59–61]. Additionally, nanolaser arrays offer distinct advantages for some of these applications such as higher sensitivity compared to Raman-based sensors as well as a label-free imaging method [59, 61]. The use of such arrays can also be extended to telecommunications, lab-on-a-chip applications, spectroscopy, and parallel detection [62–64]. Finally, integration of up to 11,664 nanolasers has already been demonstrated in a photonic crystal uncoupled array, underlining the feasibility of achieving even higher on-chip packing density in the future [65].

3.2 Coupled arrays

Contrary to independent nanolaser operation in uncoupled arrays, coupling can give rise to different types of nonlinear dynamics and in general, increases the complexity of the physical mechanisms involved in the process. It is well worth investing effort to understand these underlying phenomena however, as coupling allows much greater control on the emission properties than is possible with uncoupled operation. Coupling in arrays has been demonstrated in a variety of manners including bound state in-continuum mode coupling [66, 67], interferential coupling [68], transverse-mode coupling [69], surface plasmon–based coupling [70–78], and evanescent coupling [40, 79–82]. We will focus this review mainly on the latter two forms since a majority of the literature on nanolaser arrays was found to rely primarily on these two mechanisms.

In order to present an idea of the breadth of functionalities enabled due to coupling, in this section, we review some of the relevant alterations to the emission along with the associated studies demonstrating the principle. It is important to note here that plasmon-based coupling is usually reported to affect properties of the emission such as directionality and wavelength. For most implementations of coupled arrays based on this physics, a lattice of metal nanoparticles creates the localized surface plasmon (LSP) resonance while some form of liquid dye medium, in which the lattice is immersed, serves as the gain or exciton states (ESs). This type of hybrid resonance is referred to as the ES-LSP. Although some aspects of the lasing phenomena are yet to be fully understood, the general consensus is that

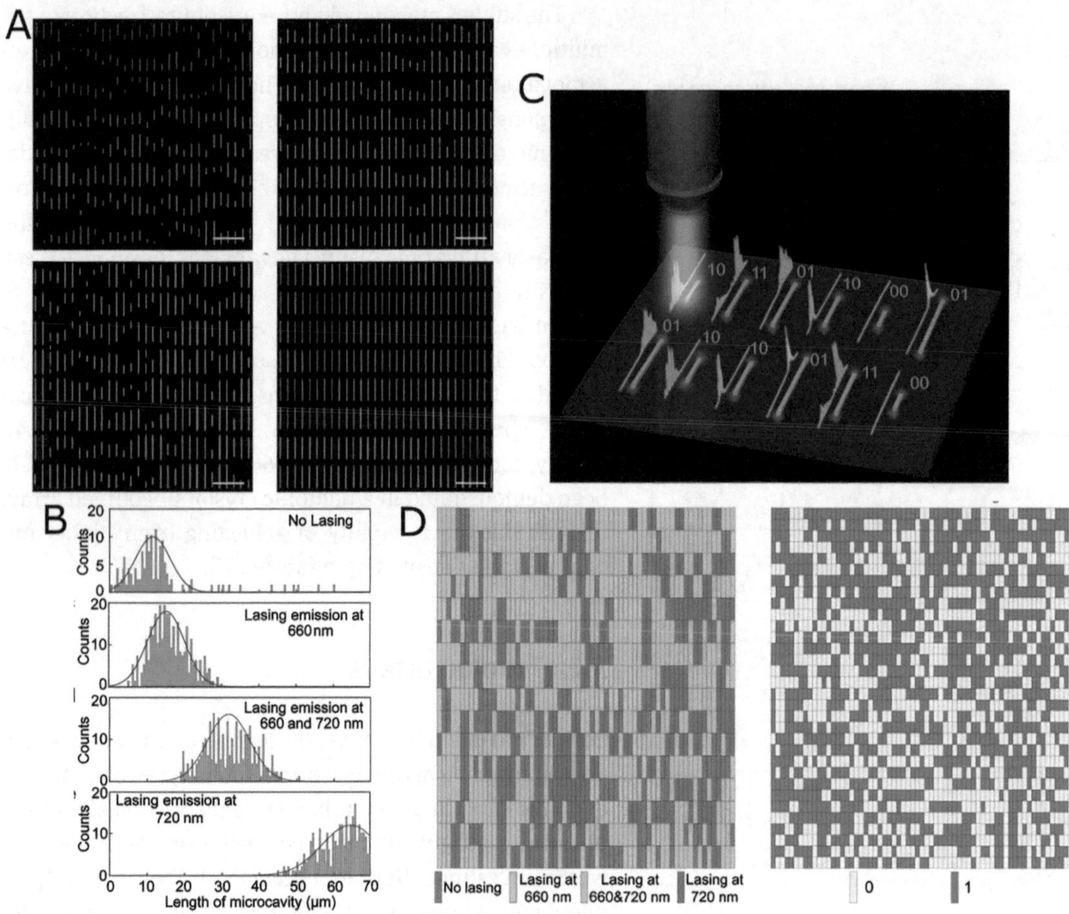

Figure 7: Optical cryptography.
(A) Organic nanolaser arrays with stochastic size distribution dependent on the organic solution used during fabrication. Scale bar: 100 μm. (B) Distribution of cavity lengths for four distinct emission states. The distribution illustrates how the length of the nanolaser determines the emission state. (C) Scanning of each individual nanolaser with a pump pulse. Depending on the emission state, the spectral information can be encoded as a bit. (D) Cryptographic sequence generated by treating emission states as quaternary bits (left) and double binary bits (right). Adapted and reprinted from a study by Feng et al. [60] with permission.

it occurs due to the excited-state molecules being stimulated to transfer energy to the lattice plasmons of the same frequency, phase, and polarization [72, 83]. In other words, the localized near-fields of the plasmonic particles comprising the array can stimulate the gain regions surrounding the particles to emit stimulated light at a wavelength that matches that of the lattice plasmon mode. With stimulated emission from the dye gain and a distributed cavity-like resonance provided by the lattice plasmon, lasing action can thus be obtained. It is important to note here that the reason for the tight confinement of light in these 2D metallic nanoparticle arrays is due to the strong interaction between the LSP resonances of individual particles and the far-field diffractive modes that satisfy the Bragg conditions of the array [84].

In comparison, evanescently coupled arrays can yield high powers and even generate states with orbital angular momentum (OAM) through the interplay between the lattice geometry and the modes of the individual lasers. Unlike for ES-LSP coupling where the resonance structure and gain are disparate media, for evanescent interactions, the active medium (usually comprised III-V semiconductors) is not external to the resonant structure; instead, it is a part of the cavity that supports the electromagnetic mode.

3.2.1 Beam directionality

Ability to control the direction of emission is of notable significance for wireless communications and nanoscale biosensors among other applications. A manner in which the angle of maximum emission from a nanolaser array can be modified involves altering the angle and/or polarization of the input optical pump. Zhou et al. [72] demonstrate this functionality by employing Au and Ag nanoparticle arrays immersed in a polymer gain comprising polyurethane and IR-140 dye gain. The schematic of the nanoparticle array is reproduced in Figure 8A, where the glass substrate and coverslip sandwich the lattice and gain layers. When the authors tune the pump angle to either be parallel to or 45° to the lattice direction, the far-field lasing beam patterns from the array are noticeably distinct as illustrated in Figure 8B. Depending on the angle of the pump, certain nanoparticles exhibit increased localizations of the electromagnetic mode, which in turn, affects the overall direction of the beam emitted from the array. It is also observed that the Ag array performed significantly better than its Au-based counterpart owing to reduced optical losses of the former [72].

Based on a similar though not identical material system as the work above, Meng et al. [73] also demonstrate highly directional lasing from their coupled spaser array. Their design consists of an Ag film with nanometer-sized holes, as opposed to nanoparticles, which is covered with an organic dye-polymer gain as shown in Figure 8C. Instead of altering the pump incidence angle, however, the detector itself is rotated to measure the directionality of the array output emission. Measuring the emission at varying detector angles along both the horizontal and vertical directions yields the results in Figure 8D. The narrow width of the measured emission as a function of the detector angle confirms the coherent, directional nature of the output beam. Lasing from this array is attributed to a surface plasmon polariton Bloch wave which also relies on some amount of feedback from plasmonic mode coupling between the Ag holes. The presence of the feedback is confirmed by the absence of lasing when an aperiodic lattice is used instead of a periodic one [71].

3.2.2 Tunable emission wavelength

Control over the emission wavelength of nanoscale light sources is desirable for dense wavelength division multiplexing (WDM) applications at a chip-scale level [85]. Other potential uses of tuning can be in lidar and imaging/sensing systems. In addition to achieving directionality of beam emission, it is also possible to alter the wavelength of output light from coupled arrays. Incidence angle and polarization of the pump can play a determining role in this respect as well. In one study, Knudson et al. [74] create a rhombohedral Al nanoparticle array which is then immersed in dye gain like the works referenced to in the previous section. Depending on the in-plane pump polarization, the ES-LSP–based mechanism leads to the array emitting at either 513 and 570 nm or both. This tuning of the output light from the array is described via Figure 9A and B which show the experimental streak camera images and numerically simulated spectra of the structure used in the study, respectively. The authors also elaborate on how

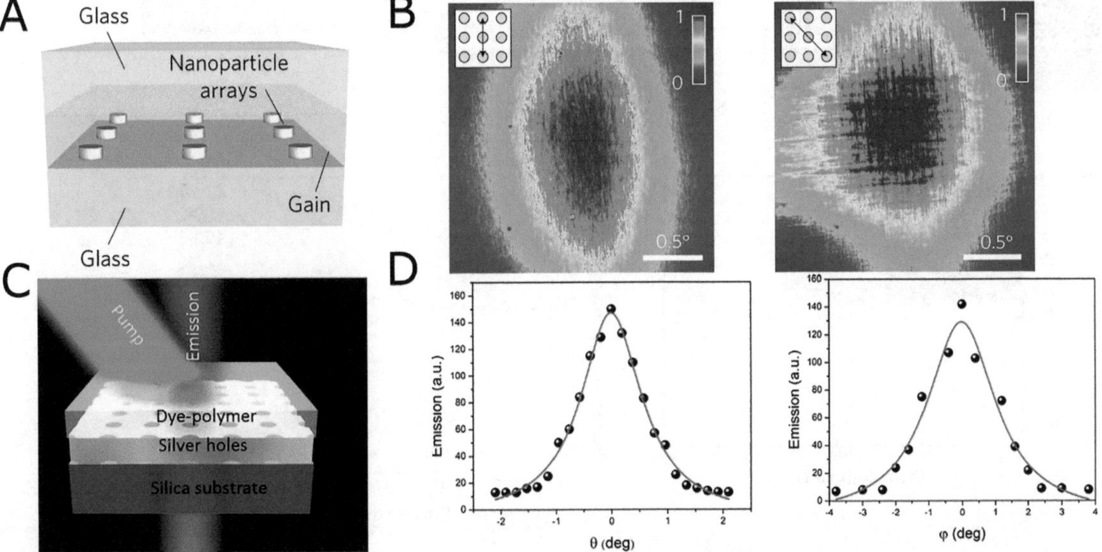

Figure 8: Directional emission.
(A) Schematic of metal nanoparticles embedded in dye gain and sandwiched between two glass slides. (B) Far-field emission patterns of the nanolaser array in (A) for pump incidence angle parallel to (left) and 45° (right) to the lattice direction. (A) and (B) adapted and reprinted from a study by Zhou et al. [72] with permission. (C) Schematic of Ag film with nanoholes immersed in dye-polymer gain. (D) Emission intensity from the coupled nanolaser array in (C) as the detector angle is varied along the horizontal (left) and vertical (right) directions. (C) and (D) adapted and reprinted from a study by Meng et al. [73] with permission.

selection of the nanoparticle shape comprising the array can determine the location of the electric field enhancement (i.e., the plasmon hotspots) [74]. In fact, other studies have demonstrated that unique shapes such as bowties can also exhibit ES-LSP–based nanocavity array lasing along with wavelength tuning [75]. Moreover, the Purcell factor is significantly enhanced due to the bowtie design, which in turn drives down the threshold allowing for room temperature lasing [75]. Finally, van Beijnum et al. [76] use a related phenomenon – surface plasmon associated lasing – to report that depending on the angle at which the emission from their array (comprising Au holes and InGaAs gain) is measured, the wavelength recorded varies.

Besides angle-resolved experiments to select the peak emission wavelength, altering the temperature is another technique that can be leveraged to attain the same goal. This may be accomplished via two main mechanisms: the first involves modifying the stoichiometry of the gain material based on thermal annealing. The second pertains to the temperature-induced red-shift of the bandgap commonly referred to as the Varshni shift [86]. Huang et al. [77] demonstrate a wavelength-tunable device based on the former principle by combining an Au/SiO_2 grating resonance with lead halide perovskite gain material. By thermally annealing their structure in a CH_3NH_3Br environment, the hybrid plasmonic mode in which the coupled nanolaser array operates in is observed to be blue-shifted in emission wavelength as shown in Figure 9C. More importantly, this modification of the wavelength is a reversible change and the original peak wavelength can be recovered after the annealing process [77].

Similar to thermal annealing-based alterations, tuning predicated on the Varshni shift is also caused by temperature acting as the catalyzing factor. However, the latter method differs in that it does not require any specific chemical environment to be implemented and the alteration in the emission wavelength is always a red-shift irrespective of the material. The pseudowedge plasmonic nanolaser array presented by Chou et al. [78] exhibits such a red-shift in the emission wavelength based on operating temperature. Their structure, consisting of a ZnO nanowire placed on an Ag grating (Figure 9D), forms an

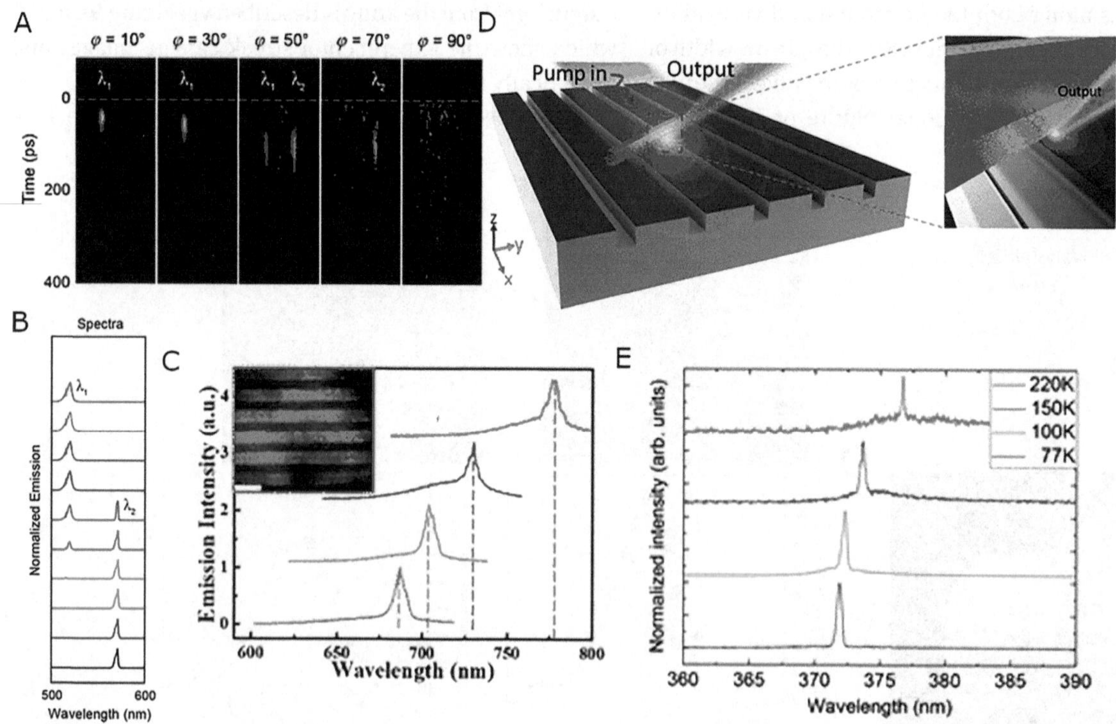

Figure 9: Tuning peak wavelength of emission.
(A) Experimental streak camera images of nanolaser array as the in-plane pump polarization is altered. (B) Numerical simulation of the spectra for varying pump polarization. (A) and (B) Adapted and reprinted from a study by Knudson et al. [74] with permission. (C) Spectra of nanolaser array as sample is annealed. A blue-shift in the peak wavelength occurs; Inset: Fluorescent microscope image of nanolaser array. Reprinted from a study by Huang et al. [77] with permission. (D) Schematic of pseudowedge nanolaser array with ZnO nanowire placed on top of Ag grating. (E) Varshni red-shift of emission wavelength as ambient temperature is increased. (D) and (E) adapted and reprinted from a study by Chou et al. [78].

unconventional array from the intersection points of the nanowire and the grating notches. This array displays single-mode lasing albeit subject to the manner in which the nanowire is positioned. By increasing the operating temperature from 77 to 220 K, the array undergoes a clear spectral red-shift as evidenced in Figure 9E due to the bandgap alterations in the ZnO caused by increased temperature.

3.2.3 Single and multimode switching

The studies on coupled arrays mentioned thus far portray instances of single-mode lasing. However, some specific cases such as multimode fiber–based WDM sources and on-chip multiplexing in photonic devices may also benefit from multimodal operation instead [87]. To meet the need in these niche areas, some studies such as the one authored by Wang et al. [70] have created ES-LSP nanolasing arrays with the capability to switch between the more common single-mode operation and a multimodal one. The researchers are able to do so by designing two distinct types of lattices – a single lattice where the individual nanoparticles collectively contribute to the resonance and a superlattice, where several single lattices combine to give rise to multiple band edge states. This contrast is displayed in Figure 10A, where the lasing emission from a single lattice (left, bottom) is seen to be single mode (right, bottom) while that from the superlattice (left, top) is observed to be multimodal in nature (right, top).

While single-mode and multimode operation can be demonstrated on individual lattices, an altered superlattice as illustrated in Figure 10B is designed to combine the two functionalities. This new design can be viewed as a single lattice from one direction and a superlattice from another. By doing so, both single and multimode emission are achieved from the same sample based on the polarization and direction of the input pump (Figure 10B). Specifically, if the pump direction is perpendicular to the lattice, the array operates in the single-mode lasing regime, whereas when the pump is parallel to the lattice direction, multimodal lasing is observed. In addition, the wavelength of emission can be tuned by altering either the size of the Au nanoparticles or the concentration of the dye gain comprising the nanostructure [70].

By choosing materials with unique properties for the coupled array, the applications for these devices can be extended to an even wider range of platforms. For instance, by employing a FM material like Ni to create nanodisk arrays in conjunction with dye gain, Pourjamal et al. [71] demonstrate the possibility of overcoming inherent losses in magnetoplasmonic systems. Their experimental results portrayed in Figure 10C underline the array's capability to switch between single and multimode lasing by modifying the particle periodicities in both the x and y directions. The authors claim that these arrays can potentially be used in the emerging field of topological photonics [71].

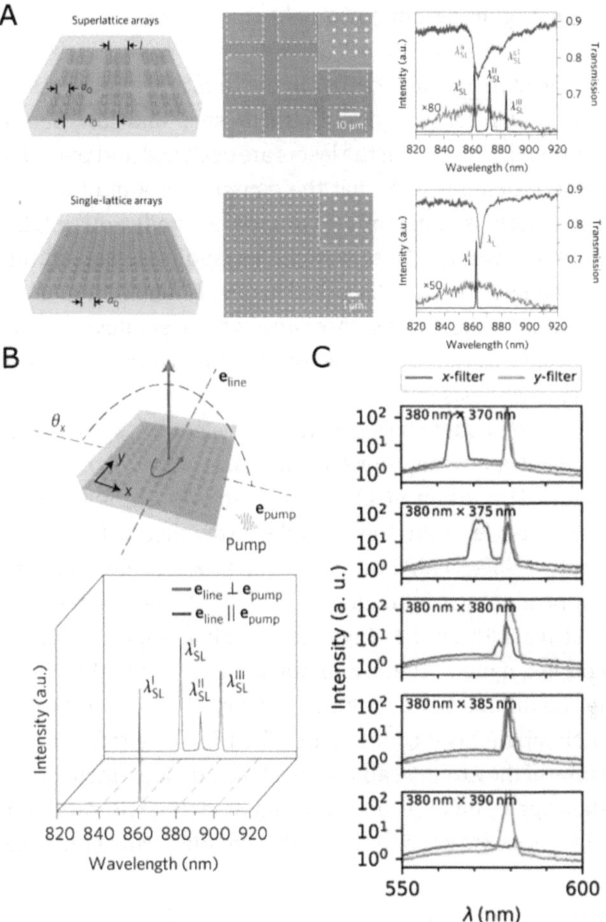

Figure 10: Switching between single and multimodal operation. (A) Schematic of superlattice and single lattice arrays (left), their SEM images (middle), and their corresponding spectra (right). The superlattice exhibits multimodal lasing (top, right), whereas the single lattice demonstrates single mode lasing (bottom, left) as evidenced by the lasing spectra shown as black solid curves in both figures. (B) Schematic of altered superlattice (top) and its output lasing spectra (bottom). If the pump direction is perpendicular to the lattice, the array operates in single mode lasing regime, whereas when the pump is parallel to the lattice direction, multimodal lasing is observed. (A) and (B) adapted and reprinted from a study by Wang et al. [70] with permission. (C) Experimental emission spectra for Ni nanodisk array for varying lattice periodicities. The presence of both *x* and *y* polarized modes confirms multimode lasing for some periodicities. Adapted and reprinted from a study by Pourjamal et al. [71] with permission.

3.2.4 Higher output power

Owing to their compact size, nanolasers inherently possess low power consumption characteristics and low output power [3]. However, if the lasers are designed and placed in close proximity such that the emission of a multitude of such lasers is coherently combined, a significantly higher output power can be obtained. The studies demonstrating this concept that are discussed in this section underline the great potential for nanolaser arrays to be employed in far-field applications, such as optical interconnects [2] and beam synthesis [88].

In their study based on a $N = 81$ element evanescently coupled photonic crystal nanolaser array, shown in Figure 11A, Altug et al. [79] are able to observe higher output powers with their coupled array than with a single emitter. Specifically, the maximum power achieved by the coupled array is found to be about ~100 times higher than that reached by the single cavity. More importantly, the coupled nanolaser array demonstrates a ~20-fold increase in the differential quantum efficiency (DQE) compared to their single laser counterpart. The DQE here refers to the slope of the *LL* curve above threshold and is extracted from the experimental results depicted in Figure 11B. Additionally, numerical analysis of the coupled rate equations reveals that with increasing *N*, both DQE and the maximum output power achievable show a corresponding increase (Figure 11C).

Increased output power has also been observed with other material systems such as metal-clad nanolaser arrays as reported by Hayenga et al. [80]. In this investigation, the overall power emitted by a seven-element metallic nanodisk array arranged in a hexagonal pattern (Figure 11D) is measured. The electromagnetic mode pattern supported by this design is first simulated and is shown in Figure 11D. Then, upon measurement, the output intensity of this array is found to be 35 times higher than that of a single nanodisk. Additionally, the array's slope efficiency is five times that of the single nanolaser. These results, encapsulated in Figure 11E, emphasize the ability to coherently combine the emission of multiple nanolasers to yield higher powers.

3.2.5 Orbital angular momentum

In addition to obtaining higher power, nanolasers may also be engineered to produce unique properties such as vortex beams with OAM in the far-field. Operating in such a state requires careful consideration of the lattice size, shape, and even type of nanolaser comprising the array. Hayenga

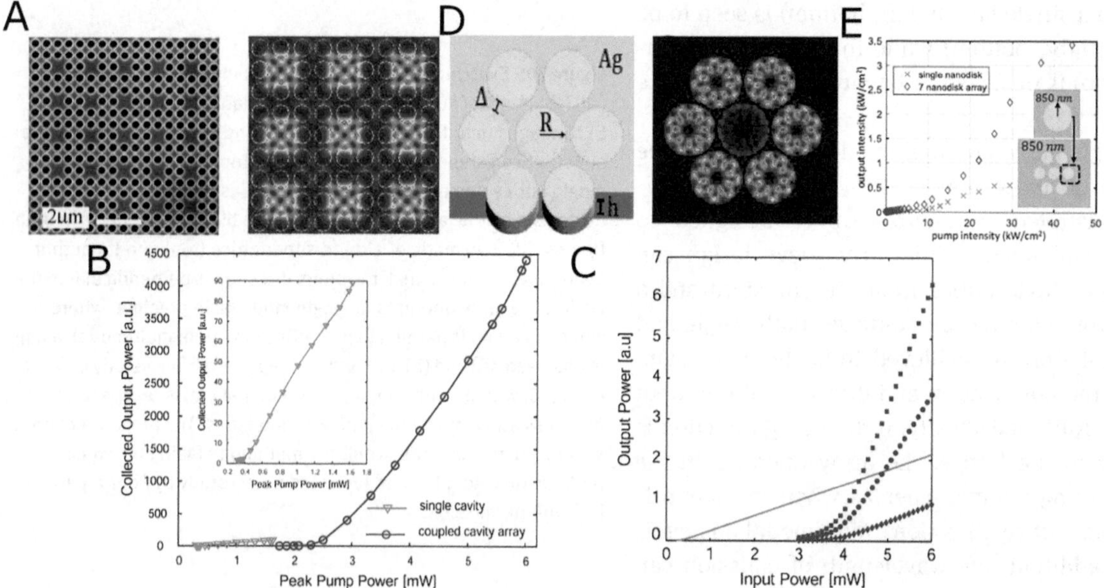

Figure 11: Higher output power.
(A) SEM image of photonic crystal nanolaser array (left) and simulation of the modes supported by the system (right). (B) Output power of coupled cavity array shown in (A) compared to that of a single cavity. Inset: Magnified version of curve for single cavity. (C) Numerical simulations of coupled rate equations comparing output power for single cavity (red) vs. that for a coupled cavity array with $N = 10$ (diamond), $N = 40$ (circle), and $N = 70$ (square). (A), (B), and (C) adapted and reprinted from a study by Altug[79] with permission. (D) Schematic of seven hexagonally designed metal-coated nanodisk lasers (left) and simulation of the mode structure supported (right). (E) Output intensity of one nanodisk vs. array as pump intensity is increased. Slope efficiency and power of array are much greater than that of single emitter. (D) and (E) adapted and reprinted from a study by Hayenga et al. [80] with permission.

et al. [81] demonstrate that by altering the types of evanescently coupled metallic nanolasers, it is possible to segment arrays into those that exclusively output vortex beams but do not carry OAM and those that display both characteristics. Specifically, an array of 500 nm diameter coaxial nanolasers exhibits the former, regardless of the array size, while a nanodisk array with 850 nm radius produces the latter as portrayed in Figure 12A and B, respectively. The dissimilarity in the two designs is explained by the interplay between the geometrical shape of the lattice and the whispering gallery modes supported by individual nanolasers. Whereas rotation is essential for higher order modes supported by the nanodisks to reduce the overlap with the lossy metal, it has no significant effect on the TE_{01} modes supported by the comparatively smaller-sized coaxial nanolasers. Furthermore, the topological charge associated with the beams carrying OAM can be tuned according to the number of lasers in the array as depicted in Figure 12C.

It is also possible to multiplex OAM beams with high topological charges using integrated microscale lasers as demonstrated by Bahari et al. [89] in a recent study. This is accomplished by designing circular boundaries between topologically distinct photonic crystal structures as shown in Figure 13A. In this Scanning Electron Microscope (SEM) image, rings 1 and 3 are composed of a photonic crystal with a nontrivial bandgap obtained by bonding InGaAsP multiple quantum wells on yttrium iron garnet (YIG). In contrast, ring 2 comprises a trivial bandgap photonic crystal. The dissimilarities in these concentric resonators gives rise to orthogonal OAM beams of alternating chirality [89]. In other words, the sign of the topological charge alternates as one moves from the innermost to the outermost ring. The chirality of the beams can also be reversed by applying an external magnetic field. Figure 13B demonstrates the far-field intensity patterns of OAM beams arising from each individual laser and also from the multiplexed array. The topological charges associated with rings 1, 2, and 3 are $|l_1| = 100$, $|l_2| = 156$, and $|l_3| = 276$, respectively. The observation of interference fringes in both theory and experiment for ring 2 is characteristic of beams carrying OAM. Although there is no coupling between the individual lasers in this array, it nevertheless presents a tantalizing possibility of dense integration of any arbitrary number of lasers for multiplexed OAM generation [89]. Generating such beams on a more compact platform based on nanolaser arrays can be a promising direction for future research.

Finally, another manner in which the properties of the beam emission such as OAM and directionality can be controlled is by choosing whether the topological nanolasers operate in a bulk or edge state. Using semiconductor nanodisk arrays, Shao et al. [90] demonstrate single-mode lasing from a bulk state by relying on band-inversion–

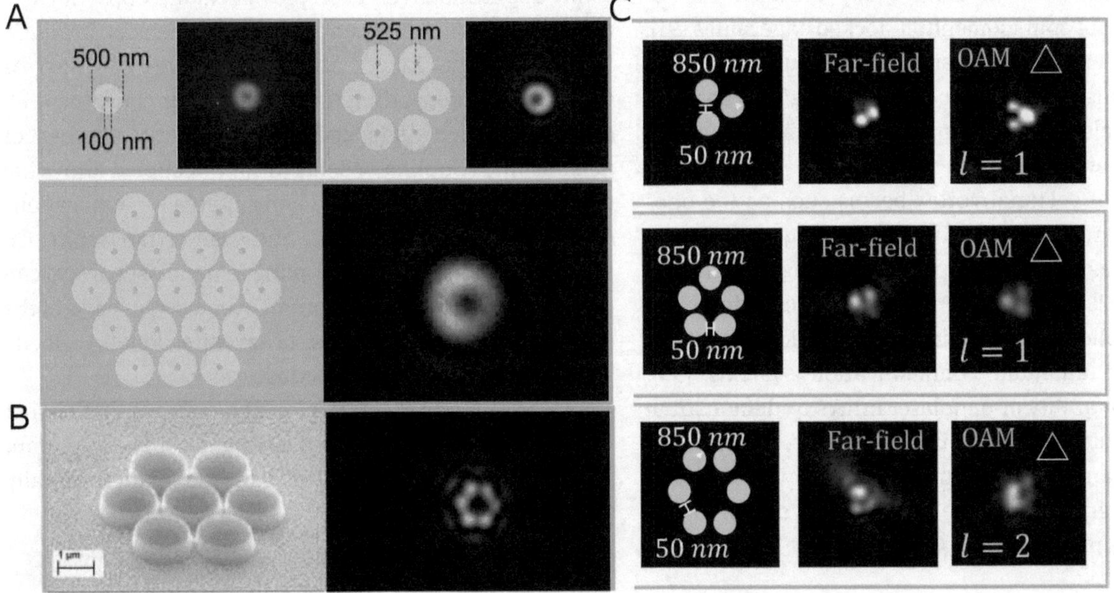

Figure 12: Vortex beams and orbital angular momentum.
(A) Schematics and experimental far-field mode structure for coaxial metal nanolasers of different array sizes. Vortex beam with no angular momentum is observed regardless of array size. (B) SEM image (left) and far-field mode structure of seven nanodisk array (right). A vortex beam with angular momentum is confirmed. (C) Varying nanodisk array sizes (left), their corresponding far-field patterns (middle), and their respective topological charges (right). (A), (B), and (C) adapted and reprinted from a study by Hayenga et al. [81] with permission.

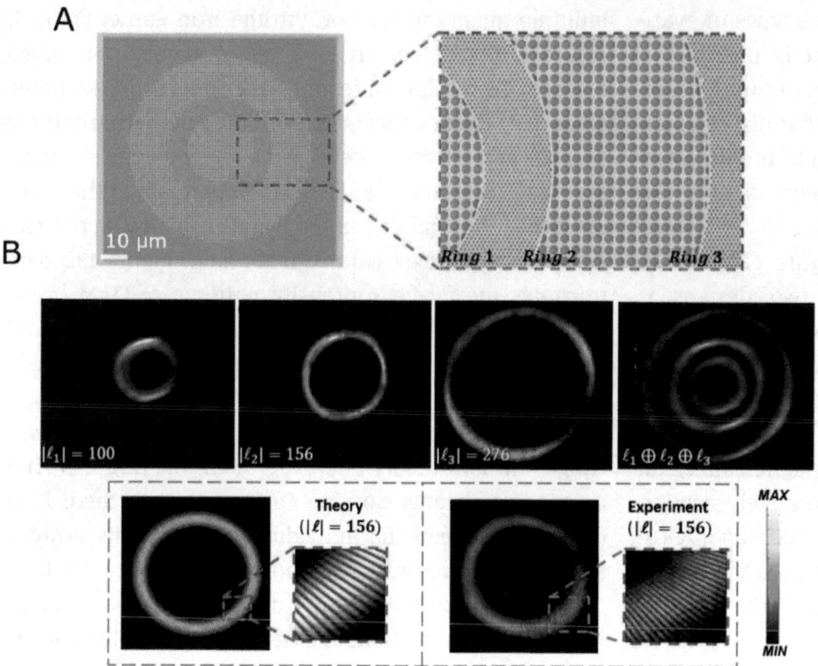

Figure 13: Multiplexed orbital angular momentum (OAM) beams.
(A) SEM image of three concentric ring lasers composed of two different photonic crystal structures. Inset: rings 1 and 3 are composed of a photonic crystal with a nontrivial bandgap obtained by bonding InGaAsP multiple quantum wells on yttrium iron garnet (YIG). In contrast, ring 2 comprises a trivial bandgap photonic crystal and cylindrical air-holes. (B) Measured far-field intensity of the rings showing their individual OAM beams (top row, first three images from left) and the multiplexed OAM beam formed (top row, far right image) when all rings are pumped simultaneously. The interference pattern observed in the far-field emission from ring 2 matches well with theory and confirms the OAM carried in the beams. (A) and (B) adapted and reprinted from a study by Bahari et al. [89] with permission.

induced reflection between trivial and topological photonic crystal cavities which exhibit opposite parities. As a result, although this bulk state mode does not carry OAM ($l = 0$), the emission is highly directional in the axis vertical to the cavity plane with divergence angles less than 6° and side-mode suppression ratios of over 36 dB. Based on a similar material system but with a slightly altered cavity design, authors from the same group are also able to observe lasing of spin-momentum–locked edge states [91]. In addition to vertical emission, the output beam from these cavities is observed to carry a topological charge of $l = -2$, while also allowing for higher side-mode suppression ratios of over 42 dB.

Creation of vectorial vortex beams such as the ones mentioned in the above studies can be of great value in areas such as imaging, optical trapping, and laser machining [92]. At the same time, beams carrying OAM can find applications in micromanipulation and both classical and quantum communication systems [93]. Therefore, the ability of nanolaser arrays to demonstrate useful attributes such as in-phase and out-of-phase supermodes and vortex beams with and without OAM makes them an ideal device platform for catering to a plethora of applications [40, 82].

4 Conclusion and future outlook

To conclude, recent progress on the development of array architectures of nanolasers is reviewed in this article. The focus was on valuable attributes realized due to unique array designs and the underlying physics that may help enable real-world applications such as biological sensing, imaging, and on-chip communications. A distinction was made between an array size of just two nanolasers and larger arrays. The former can not only serve as a testbed to understand the fundamental physics and enable interesting applications (for example optical flip flops) but also as a building block that can aid in comprehension of larger-lattice behavior. Larger arrays themselves can also be distinguished depending on whether their constituent elements function independently from one another or demonstrate coupling of some form. Uncoupled arrays are more intuitive to understand and suitable for applications such as imaging, biosensing, and even cryptography. On the other hand, although coupling in nanolaser arrays can create complex dynamics, control over many aspects of the emission behavior such as beam directionality, mode switching, and OAM are afforded.

In fact, it is in the area of coupled arrays that further research efforts are required owing to some important applications that still need to be experimentally demonstrated.

4.1 Phase-locked laser arrays

As mentioned previously in Section 2, the coupling between nanolasers can yield a plethora of dynamical regimes such as periodic intensity oscillations, chaotic fluctuations, and stable phase-locking. Among these, the regime of stable phase-locking holds great potential as it can help in the realization of phase-locked laser arrays with demonstrably higher power. Though the ability of evanescently coupled nanolasers to generate higher power has already been experimentally demonstrated, the studies mainly focused on how the accumulated power and slope efficiency of the arrays exceeded those of single emitters [79, 80]. Ideally, if all the constituent lasers are locked in-phase, their electric field amplitudes will interfere constructively, and the resulting far-field power density can be much greater than without in-phase locking. Specifically, for an array of N identical nanolasers with identical output intensities, the far-field power density that is measured will be N^2 times greater than that of a single nanolaser. Note here that the greater power density that is afforded is simply due to the coherent superposition of field amplitudes of the emitted waves, which means that conservation of energy is not violated in any manner.

Achieving in-phase stable locking, however, is not trivial as the stability regions are narrow and sensitive to the parameters in the rate equations, especially so for semiconductor lasers [42]. In a past work, Shahin et al. [94] demonstrated this increased sensitivity to the tuning parameters by considering the case of three free-running semiconductor lasers coupled to each other via optoelectronic feedback. The researchers observed that distinct dynamical regimes of operation can be achieved with their system including winner-takes-all (WTA), winner-shares-all (WSA), and winnerless competitions [94]. To gauge the stability of these different regimes, the coupled rate equations from a study by Shahin et al. [94] are numerically solved for the steady-state behaviors while varying the feedback coupling strength (ξ_{32}) as well as the detuning (Ω_{12}) between two of the lasers. This allows us to create a bifurcation map (shown in Figure 14A) for the same system considered in a study by Shahin et al. [94]. In this diagram, the different dynamical regimes are represented by color codes such as black for chaos, red for WTA competition, yellow for WSA competition, and white for winnerless competition. It is evident from the diagram, that the network of three coupled semiconductor lasers has numerous small regions of stability and for the slightest variation in the input parameters, the system transitions from one state to the other. This makes the system extremely unstable, thus rendering it challenging to put it to practical use.

Nanolasers inherently possess certain attributes such as high-β's and Purcell factors that can help alleviate this issue of instability as theorized in some works [42, 43, 95]. To demonstrate the effects such parameters can have on

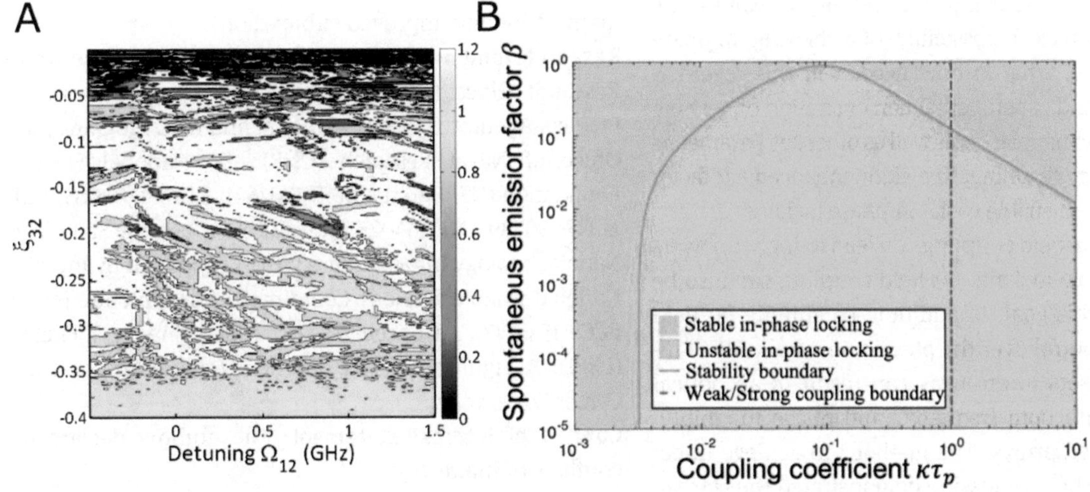

Figure 14: Stability maps.
(A) Calculated bifurcation diagram of three optoelectronically coupled free running semiconductor lasers as a function of detuning Ω_{12} and the coupling coefficient ξ_{32}. Varying system dynamics are observed such as chaos (magnitude of 0 in the color bar with black color), winner-takes-all (WTA) competition (magnitude of 0.5 with red color), winner-shares-all (WSA) competition (magnitude of 1 with yellow color), and winnerless competition (magnitude of 1.2 with white color). (B) Stability regions for two laterally coupled metallo-dielectric nanolasers as a function of coupling coefficient κ and spontaneous emission factor β. The dash-dotted purple line demarcates the weak coupling ($\kappa\tau_p < 1$) and strong coupling ($\kappa\tau_p > 1$) regions.

the stability, the coupled laser rate equations for two laterally coupled metallo-dielectric nanolasers are numerically solved. These equations are shown below:

$$\frac{dE_{1,2}}{dt} = \frac{1}{2}(1+i\alpha)\left(\Gamma G(N_{1,2}) - \frac{1}{\tau_p}\right)E_{1,2} + iw_{1,2}E_{1,2}$$
$$\text{\&doublehyphen; 29pt} + \frac{\Gamma F_p \beta N_{1,2}}{2\tau_{\mathrm{rad}}|E_{1,2}|^2}E_{1,2} + i\kappa E_{2,1}, \quad (1)$$

$$\frac{dN_{1,2}}{dt} = P - \frac{N_{1,2}}{\tau_{nr}} - \frac{(F_p\beta + 1 - \beta)N_{1,2}}{\tau_{\mathrm{rad}}}$$
$$\text{\&doublehyphen; 17pt} - G(N_{1,2})|E_{1,2}|^2, \quad (2)$$

where $E_{1,2}$ are the electric field amplitudes, $N_{1,2}$ are the carrier densities, $w_{1,2}$ are the intrinsic resonance frequencies, α is the phase-amplitude coupling (Henry) factor for a quantum well laser, $G(N)$ is the carrier density-dependent gain function, τ_p is photon lifetime, Γ is the confinement factor, F_p is Purcell enhancement factor, τ_{rad} and τ_{nr} are the radiative and nonradiative recombination lifetimes of the carriers, respectively, and P is the pump rate. The subscripts denote the two cavities that are coupled to each other.

By solving for the steady state solutions of equations (1) and (2), Figure 14B is created which illustrates how the stability region is affected by both β as well as the coupling coefficient κ. It can be inferred from the plot that although varying κ yields a more complex effect, higher values of β contribute to increased possibility of achieving in-phase locking. Therefore, what is now needed in this regard is further analysis and experimental demonstration of how parameters unique to nanolasers as well as other key parameters (such as frequency detuning, linewidth enhancement factor and pump rate) can enable easier in-phase locking.

Whereas near-field coupling can lead to higher powers via in-phase stable locking, far-field coupling can also be used to realize this goal. In addition, combining far-field coupling with control over the phase and amplitude of individual nanolasers in an array can result in an optical phased array with both frequency and phase tunability, akin to RF phased arrays. The methods to achieve directionality in array beam emission demonstrated thus far are limited to altering the incidence angle/polarization of the pump. In contrast, a true optical phased array of nanolasers can offer multiple degrees of freedom and much more nuanced control of the directionality since each emitter in the array can be individually tuned in both frequency and phase. Such far-field coupling with quantum cascade laser arrays has already been demonstrated [96]. The challenge that lies ahead is to achieve the same for subwavelength nanolaser arrays and preferably, with current injection.

4.2 Ultrashort pulse generation

When discussing nanolasers, an often-overlooked topic is ultrashort pulse generation. Typically, such pulses are created using mode-locking techniques, which can include both active and passive mode-locking. Due to the need for external design elements such as saturable absorbers or electro-optic modulators, achieving mode-locking with nanoscale lasers faces impediments. However, Gongora et al. [97] demonstrate in a recent work that it is in fact possible to mode-lock nanolasers in an array without external design elements by relying on the nonradiative nature of anapole states. Using this technique, the authors numerically demonstrate an ultrashort pulse down to 95 fs generated from an array whose constituent nanolasers are spaced apart evenly in frequency from one another to mirror mode-spacing in traditional mode-locking theory. It is also observed that the position of the nanolasers can alter the duration of the pulse generated [97]. Despite this result, experimental results of mode-locked nanolasers for ultrashort pulse generation are yet to be demonstrated. Doing so can unlock a wealth of applications in areas requiring short optical pulses such as LiDAR, optical regeneration, nonlinear optics, and optical sampling [98].

Author contribution: All the authors have accepted responsibility for the entire content of this submitted manuscript and approved submission.
Research funding: This work was supported by the Army Research Office (ARO), the Defense Advanced Research Projects Agency (DARPA) DSO NLM and NAC Programs, the Office of Naval Research (ONR), the National Science Foundation (NSF) grants DMR-1707641, CBET-1704085, NSF ECCS-180789, CCF-1640227, NSF ECCS-190184, the San Diego Nanotechnology Infrastructure (SDNI) supported by the NSF National Nanotechnology Coordinated Infrastructure (grant ECCS-1542148), Advanced Research Projects Agency—Energy (LEED: A Lightwave Energy-Efficient Datacenter), and the Cymer Corporation.
Conflict of interest statement: The authors declare no conflicts of interest.

References

[1] M. Smit, J. J. Van der Tol, and M. Hill, "Moore's law in photonics," *Laser Photonics Rev.*, vol. 6, pp. 1–3, 2012.
[2] R. M. Ma and R. F. Oulton, "Applications of nanolasers," *Nat. Nanotechnol.*, vol. 14, pp. 12–22, 2019.
[3] M. T. Hill and M. C. Gather, "Advances in small lasers," *Nat. Photonics*, vol. 8, p. 908, 2014.

[4] Q. Gu, J. S. Smalley, M. P. Nezhad, et al., "Subwavelength semiconductor lasers for dense chip-scale integration," *Adv. Opt. Photonics*, vol. 6, pp. 1–56, 2014.
[5] H. Altug, D. Englund, and J. Vučković, "Ultrafast photonic crystal nanocavity laser," *Nat. Phys.*, vol. 2, pp. 484–488, 2006.
[6] K. Nozaki, S. Kita, and T. Baba, "Room temperature continuous wave operation and controlled spontaneous emission in ultrasmall photonic crystal nanolaser," *Opt. Express*, vol. 15, pp. 7506–7514, 2007.
[7] S. H. Kim, J. H. Choi, S. K. Lee, et al., "Optofluidic integration of a photonic crystal nanolaser," *Opt. Express*, vol. 16, pp. 6515–6527, 2008.
[8] S. Matsuo, A. Shinya, T. Kakitsuka, et al., "High-speed ultracompact buried heterostructure photonic-crystal laser with 13 fJ of energy consumed per bit transmitted," *Nat. Photonics*, vol. 4, p. 648, 2010.
[9] M. T. Hill, Y. S. Oei, B. Smalbrugge, et al., "Lasing in metallic-coated nanocavities," *Nat. Photonics*, vol. 1, p. 589, 2007.
[10] M. P. Nezhad, A. Simic, O. Bondarenko, et al., "Room-temperature subwavelength metallo-dielectric lasers," *Nat. Photonics*, vol. 4, p. 395, 2010.
[11] K. Ding, Z. C. Liu, L. J. Yin, et al., "Room-temperature continuous wave lasing in deep-subwavelength metallic cavities under electrical injection," *Phys. Rev. B*, vol. 85, p. 041301, 2012.
[12] K. Ding, M. T. Hill, Z. C. Liu, L. J. Yin, P. J. Van Veldhoven, and C. Z. Ning, "Record performance of electrical injection sub-wavelength metallic-cavity semiconductor lasers at room temperature," *Opt. Express*, vol. 21, pp. 4728–4733, 2013.
[13] Q. Gu, J. Shane, F. Vallini, et al., "Amorphous Al_2O_3 shield for thermal management in electrically pumped metallo-dielectric nanolasers," *IEEE J. Quant. Electron.*, vol. 50, pp. 499–509, 2014.
[14] S. H. Pan, Q. Gu, A. El Amili, F. Vallini, and Y. Fainman, "Dynamic hysteresis in a coherent high-β nanolaser," *Optica*, vol. 3, pp. 1260–1265, 2016.
[15] C. Y. Fang, S. H. Pan, F. Vallini, et al., "Lasing action in low-resistance nanolasers based on tunnel junctions," *Opt. Lett.*, vol. 44, pp. 3669–3672, 2019.
[16] M. Khajavikhan, A. Simic, M. Katz, et al., "Thresholdless nanoscale coaxial lasers," *Nature*, vol. 482, pp. 204–207, 2012.
[17] W. E. Hayenga, H. Garcia-Gracia, H. Hodaei, et al., "Second-order coherence properties of metallic nanolasers," *Optica*, vol. 3, pp. 1187–1193, 2016.
[18] M. A. Noginov, G. Zhu, A. M. Belgrave, et al., "Demonstration of a spaser-based nanolaser," *Nature*, vol. 460, pp. 1110–1112, 2009.
[19] R. F. Oulton, V. J. Sorger, T. Zentgraf, et al., "Plasmon lasers at deep subwavelength scale," *Nature*, vol. 461, pp. 629–632, 2009.
[20] Y. J. Lu, J. Kim, H. Y. Chen, et al., "Plasmonic nanolaser using epitaxially grown silver film," *Science*, vol. 337, pp. 450–453, 2012.
[21] Y. J. Lu, C. Y. Wang, J. Kim, et al., "All-color plasmonic nanolasers with ultralow thresholds: autotuning mechanism for single-mode lasing," *Nano Lett.*, vol. 14, pp. 4381–4388, 2014.
[22] Y. H. Chou, Y. M. Wu, K. B. Hong, et al., "High-operation-temperature plasmonic nanolasers on single-crystalline aluminum," *Nano Lett.*, vol. 16, pp. 3179–3186, 2016.
[23] M. I. Stockman, "Nanoplasmonic sensing and detection," *Science*, vol. 348, pp. 287–288, 2015.
[24] E. I. Galanzha, R. Weingold, D. A. Nedosekin, et al., "Spaser as a biological probe," *Nat. Commun.*, vol. 8, p. 15528, 2017.
[25] J. A. Schuller, E. S. Barnard, W. Cai, Y. C. Jun, J. S. White, and M. L. Brongersma, "Plasmonics for extreme light concentration and manipulation," *Nat. Mater.*, vol. 9, pp. 193–204, 2010.
[26] S. Cho, M. Humar, N. Martino, and S. H. Yun, "Laser particle stimulated emission microscopy," *Phys. Rev. Lett.*, vol. 117, p. 193902, 2016.
[27] D. A. Miller, "Device requirements for optical interconnects to silicon chips," *Proc. IEEE*, vol. 97, pp. 1166–1185, 2009.
[28] C. Henry, "Theory of the linewidth of semiconductor lasers," *IEEE J. Quant. Electron.*, vol. 18, pp. 259–264, 1982.
[29] A. Nakagawa, S. Ishii, and T. Baba, "Photonic molecule laser composed of GaInAsP microdisks," *Appl. Phys. Lett.*, vol. 86, p. 041112, 2005.
[30] S. V. Boriskina, "Spectrally engineered photonic molecules as optical sensors with enhanced sensitivity: a proposal and numerical analysis," *JOSA B*, vol. 23, pp. 1565–1573, 2006.
[31] S. V. Boriskina, "Coupling of whispering-gallery modes in size-mismatched microdisk photonic molecules," *Opt. Lett.*, vol. 32, pp. 1557–1559, 2007.
[32] E. I. Smotrova, A. I. Nosich, T. M. Benson, and P. Sewell, "Optical coupling of whispering-gallery modes of two identical microdisks and its effect on photonic molecule lasing," *IEEE J. Sel. Top. Quant. Electron.*, vol. 12, pp. 78–85, 2006.
[33] S. Ishii, A. Nakagawa, and T. Baba, "Modal characteristics and bistability in twin microdisk photonic molecule lasers," *IEEE J. Sel. Top. Quant. Electron.*, vol. 12, pp. 71–77, 2006.
[34] K. A. Atlasov, K. F. Karlsson, A. Rudra, B. Dwir, and E. Kapon, "Wavelength and loss splitting in directly coupled photonic-crystal defect microcavities," *Opt. Express*, vol. 16, pp. 16255–16264, 2008.
[35] S. S. Deka, S. H. Pan, Q. Gu, Y. Fainman, and A. El Amili, "Coupling in a dual metallo-dielectric nanolaser system," *Opt. Lett.*, vol. 42, pp. 4760–4763, 2017.
[36] P. Hamel, S. Haddadi, F. Raineri, et al., "Spontaneous mirror-symmetry breaking in coupled photonic-crystal nanolasers," *Nat. Photonics*, vol. 9, pp. 311–315, 2015.
[37] A. M. Yacomotti, S. Haddadi, and S. Barbay, "Self-pulsing nanocavity laser," *Phys. Rev. A*, vol. 87, 2013, https://doi.org/10.1103/physreva.87.041804.
[38] M. J. Adams, N. Li, B. R. Cemlyn, H. Susanto, and I. D. Henning, "Effects of detuning, gain-guiding, and index antiguiding on the dynamics of two laterally coupled semiconductor lasers," *Phys. Rev. A*, vol. 95, p. 053869, 2017.
[39] H. Han and K. A. Shore, "Dynamics and stability of mutually coupled nano-lasers," *IEEE J. Quant. Electron.*, vol. 52, pp. 1–6, 2016.
[40] M. Parto, W. Hayenga, A. Marandi, D. N. Christodoulides, and M. Khajavikhan, "Realizing spin Hamiltonians in nanoscale active photonic lattices," *Nat. Mater.*, vol. 19, pp. 725–731, 2020.
[41] M. Parto, W. E. Hayenga, A. Marandi, D. N. Christodoulides, and M. Khajavikhan, "Nanolaser-based emulators of spin Hamiltonians," *Nanophotonics*, vol. 9, pp. 4193–4198, 2020.
[42] H. G. Winful and S. S. Wang, "Stability of phase locking in coupled semiconductor laser arrays," *Appl. Phys. Lett.*, vol. 53, pp. 1894–1896, 1988.
[43] S. S. Wang and H. G. Winful, "Dynamics of phase-locked semiconductor laser arrays," *Appl. Phys. Lett.*, vol. 52, pp. 1774–1776, 1988.

[44] S. Ji, Y. Hong, P. S. Spencer, J. Benedikt, and I. Davies, "Broad tunable photonic microwave generation based on period-one dynamics of optical injection vertical-cavity surface-emitting lasers," *Opt. Express*, vol. 25, pp. 19863–19871, 2017.

[45] X. Porte, M. C. Soriano, D. Brunner, and I. Fischer, "Bidirectional private key exchange using delay-coupled semiconductor lasers," *Opt. Lett.*, vol. 41, pp. 2871–2874, 2016.

[46] M. J. Adams, D. Jevtics, M. J. Strain, I. D. Henning, and A. Hurtado, "High-frequency dynamics of evanescently-coupled nanowire lasers," *Sci. Rep.*, vol. 9, pp. 1–7, 2019.

[47] S. Haddadi, P. Hamel, G. Beaudoin, et al., "Photonic molecules: tailoring the coupling strength and sign," *Opt. Express*, vol. 22, pp. 12359–12368, 2014.

[48] N. Caselli, F. Intonti, F. Riboli, et al., "Antibonding ground state in photonic crystal molecules," *Phys. Rev. B*, vol. 86, p. 035133, 2012.

[49] N. Caselli, F. Intonti, F. Riboli, and M. Gurioli, "Engineering the mode parity of the ground state in photonic crystal molecules," *Opt. Express*, vol. 22, pp. 4953–4959, 2014.

[50] M. Marconi, J. Javaloyes, F. Raineri, J. A. Levenson, and A. M. Yacomotti, "Asymmetric mode scattering in strongly coupled photonic crystal nanolasers," *Opt. Lett.*, vol. 41, pp. 5628–5631, 2016.

[51] M. Marconi, F. Raineri, A. Levenson, et al., "Mesoscopic limit cycles in coupled nanolasers," *Phys. Rev. Lett.*, vol. 124, p. 213602, 2020.

[52] M. Marconi, J. Javaloyes, P. Hamel, F. Raineri, A. Levenson, and A. M. Yacomotti, "Far-from-equilibrium route to superthermal light in bimodal nanolasers," *Phys. Rev. X*, vol. 8, p. 011013, 2018.

[53] D. O'Brien, M. D. Settle, T. Karle, A. Michaeli, M. Salib, and T. F. Krauss, "Coupled photonic crystal heterostructure nanocavities," *Opt. Express*, vol. 15, pp. 1228–1233, 2007.

[54] Q. Wang, H. Zhao, X. Du, W. Zhang, M. Qiu, and Q. Li, "Hybrid photonic-plasmonic molecule based on metal/Si disks," *Opt. Express*, vol. 21, pp. 11037–11047, 2013.

[55] S. H. Pan, S. S. Deka, A. El Amili, Q. Gu, and Y. Fainman, "Nanolasers: second-order intensity correlation, direct modulation and electromagnetic isolation in array architectures," *Prog. Quant. Electron.*, vol. 59, pp. 1–8, 2018.

[56] H. Han and K. A. Shore, "Zero crosstalk regime direct modulation of mutually coupled nanolasers," *IEEE Photonics J.*, vol. 9, pp. 1–2, 2017.

[57] S. Hachuda, S. Otsuka, S. Kita, et al., "Selective detection of sub-atto-molar streptavidin in 10^{13}-fold impure sample using photonic crystal nanolaser sensors," *Opt. Express*, vol. 21, pp. 12815–12821, 2013.

[58] S. Kita, K. Nozaki, S. Hachuda, et al., "Photonic crystal point-shift nanolasers with and without nanoslots—design, fabrication, lasing, and sensing characteristics," *IEEE J. Sel. Top. Quant. Electron.*, vol. 17, pp. 1632–1647, 2011.

[59] H. Abe, M. Narimatsu, T. Watanabe, et al., "Living-cell imaging using a photonic crystal nanolaser array," *Opt. Express*, vol. 23, pp. 17056–17066, 2015.

[60] J. Feng, W. Wen, X. Wei, et al., "Random organic nanolaser arrays for cryptographic primitives," *Adv. Mater.*, vol. 31, p. 1807880, 2019.

[61] S. Kita, K. Nozaki, and T. Baba, "Refractive index sensing utilizing a cw photonic crystal nanolaser and its array configuration," *Opt. Express*, vol. 16, pp. 8174–8180, 2008.

[62] J. J. Wu, H. Gao, R. Lai, et al., "Near-infrared organic single-crystal nanolaser arrays activated by excited-state intramolecular proton transfer," *Matter*, vol. 2, pp. 1233–1243, 2020.

[63] H. Kim, W. J. Lee, A. C. Farrell, et al., "Monolithic InGaAs nanowire array lasers on silicon-on-insulator operating at room temperature," *Nano Lett.*, vol. 17, pp. 3465–3470, 2017.

[64] H. Yan, R. He, J. Johnson, M. Law, R. J. Saykally, and P. Yang, "Dendritic nanowire ultraviolet laser array," *J. Am. Chem. Soc.*, vol. 125, pp. 4728–4729, 2003.

[65] T. Watanabe, H. Abe, Y. Nishijima, and T. Baba, "Array integration of thousands of photonic crystal nanolasers," *Appl. Phys. Lett.*, vol. 104, p. 121108, 2014.

[66] S. T. Ha, Y. H. Fu, N. K. Emani, et al., "Directional lasing in resonant semiconductor nanoantenna arrays," *Nat. Nanotechnol.*, vol. 13, pp. 1042–1047, 2018.

[67] A. Kodigala, T. Lepetit, Q. Gu, B. Bahari, Y. Fainman, and B. Kanté, "Lasing action from photonic bound states in continuum," *Nature*, vol. 541, p. 196, 2017.

[68] T. J. Lin, H. L. Chen, Y. F. Chen, and S. Cheng, "Room-temperature nanolaser from CdSe nanotubes embedded in anodic aluminum oxide nanocavity arrays," *Appl. Phys. Lett.*, vol. 93, p. 223903, 2008.

[69] K. Wang, Z. Gu, S. Liu, et al., "High-density and uniform lead halide perovskite nanolaser array on silicon," *J. Phys. Chem. Lett.*, vol. 7, pp. 2549–2555, 2016.

[70] D. Wang, A. Yang, W. Wang, et al., "Band-edge engineering for controlled multi-modal nanolasing in plasmonic superlattices," *Nat. Nanotechnol.*, vol. 12, p. 889, 2017.

[71] S. Pourjamal, T. K. Hakala, M. Nečada, et al., "Lasing in Ni nanodisk arrays," *ACS Nano*, vol. 13, pp. 5686–5692, 2019.

[72] W. Zhou, M. Dridi, J. Y. Suh, et al., "Lasing action in strongly coupled plasmonic nanocavity arrays," *Nat. Nanotechnol.*, vol. 8, p. 506, 2013.

[73] X. Meng, J. Liu, A. V. Kildishev, and V. M. Shalaev, "Highly directional spaser array for the red wavelength region," *Laser Photonics Rev.*, vol. 8, pp. 896–903, 2014.

[74] M. P. Knudson, R. Li, D. Wang, W. Wang, R. D. Schaller, and T. W. Odom, "Polarization-dependent lasing behavior from low-symmetry nanocavity arrays," *ACS Nano*, vol. 13, pp. 7435–7441, 2019.

[75] J. Y. Suh, C. H. Kim, W. Zhou, et al., "Plasmonic bowtie nanolaser arrays," *Nano Lett.*, vol. 12, pp. 5769–5774, 2012.

[76] F. van Beijnum, P. J. van Veldhoven, E. J. Geluk, M. J. de Dood, W. Gert, and M. P. van Exter, "Surface plasmon lasing observed in metal hole arrays," *Phys. Rev. Lett.*, vol. 110, p. 206802, 2013.

[77] C. Huang, W. Sun, Y. Fan, et al., "Formation of lead halide perovskite based plasmonic nanolasers and nanolaser arrays by tailoring the substrate," *ACS Nano*, vol. 12, pp. 3865–3874, 2018.

[78] Y. H. Chou, K. B. Hong, C. T. Chang, et al., "Ultracompact pseudowedge plasmonic lasers and laser arrays," *Nano Lett.*, vol. 18, pp. 747–753, 2018.

[79] H. Altug and J. Vučković, "Photonic crystal nanocavity array laser," *Opt. Express*, vol. 13, pp. 8819–8828, 2005.

[80] W. E. Hayenga, M. Parto, H. Hodaei, P. LiKamWa, D. N. Christodoulides, and M. Khajavikhan, "Coupled metallic nanolaser arrays," in *Conference on Lasers and Electro-Optics (CLEO)*, San Jose Optical Society of America (OSA), 2017, pp. 1–1, https://doi.org/10.1364/cleo_qels.2017.ftu3h.7.

[81] W. E. Hayenga, M. Parto, E. S. Cristobal, D. N. Christodoulides, and M. Khajavikhan, "Direct generation of structured light in

metallic nanolaser arrays," in *Conf. on Lasers and Electro-Optics (CLEO)* San Jose, Optical Society of America (OSA), 2018, pp. 1–2.

[82] M. Parto, W. Hayenga, D. N. Christodoulides, and M. Khajavikhan, "Mode-dependent coupling and vectorial optical vortices in metallic nanolaser arrays," in *Conf. on Lasers and Electro-Optics (CLEO)*, San Jose, Optical Society of America (OSA), 2019, FM1D-2.

[83] A. F. Koenderink, "Plasmon nanocavity array lasers: cooperating over losses and competing for gain," *ACS Nano*, vol. 13, pp. 7377–7382, 2019.

[84] M. B. Ross, C. A. Mirkin, and G. C. Schatz, "Optical properties of one-, two-, and three-dimensional arrays of plasmonic nanostructures," *J. Phys. Chem. C*, vol. 120, pp. 816–830, 2016.

[85] G. Cossu, A. M. Khalid, P. Choudhury, R. Corsini, and E. Ciaramella, "3.4 Gbit/s visible optical wireless transmission based on RGB LED," *Opt. Express*, vol. 20, pp. B501–B506, 2012.

[86] P. Xu, J. Gong, X. Guo, et al., "Fast lasing wavelength tuning in single nanowires," *Adv. Opt. Mater.*, vol. 7, p. 1900797, 2019.

[87] J. Cheng, C. L. Shieh, X. D. Huang, et al., "Efficient long wavelength AlGaInAs vertical-cavity surface-emitting lasers for coarse WDM applications over multimode fibre," *Electron. Lett.*, vol. 40, pp. 1184–1185, 2004.

[88] M. Lorke, T. Suhr, N. Gregersen, and J. Mørk, "Theory of nanolaser devices: rate equation analysis versus microscopic theory," *Phys. Rev. B*, vol. 87, no. 20, p. 205310, 28 May 2013.

[89] B. Bahari, L. Y. Hsu, S. H. Pan, et al., "Topological lasers generating and multiplexing topological light," arXiv preprint 1904.11873, 2019.

[90] Z. K. Shao, H. Z. Chen, S. Wang, et al., "A high-performance topological bulk laser based on band-inversion-induced reflection," *Nat. Nanotechnol.*, vol. 15, pp. 67–72, 2020.

[91] Z. Q. Yang, Z. K. Shao, H. Z. Chen, X. R. Mao, and R. M. Ma, "Spin-momentum-locked edge mode for topological vortex lasing," *Phys. Rev. Lett.*, vol. 125, p. 013903, 2020.

[92] Q. Zhan, "Cylindrical vector beams: from mathematical concepts to applications," *Adv. Opt. Photonics*, vol. 1, pp. 1–57, 2009.

[93] A. M. Yao and M. J. Padgett, "Orbital angular momentum: origins, behavior and applications," *Adv. Opt. Photonics*, vol. 3, pp. 161–204, 2011.

[94] S. Shahin, F. Vallini, F. Monifi, M. Rabinovich, and Y. Fainman, "Heteroclinic dynamics of coupled semiconductor lasers with optoelectronic feedback," *Opt. Lett.*, vol. 41, pp. 5238–5241, 2016.

[95] T. Suhr, P. T. Kristensen, and J. Mørk, "Phase-locking regimes of photonic crystal nanocavity laser arrays," *Appl. Phys. Lett.*, vol. 99, p. 251104, 2011.

[96] T. Y. Kao, J. L. Reno, and Q. Hu, "Phase-locked laser arrays through global antenna mutual coupling," *Nat. Photonics*, vol. 10, p. 541, 2016.

[97] J. S. Gongora, A. E. Miroshnichenko, Y. S. Kivshar, and A. Fratalocchi, "Anapole nanolasers for mode-locking and ultrafast pulse generation," *Nat. Commun.*, vol. 8, pp. 1–9, 2017.

[98] L. A. Coldren, S. W. Corzine, and M. L. Mashanovitch, *Diode Lasers and Photonic Integrated Circuits*, New Jersey: John Wiley & Sons, 2012.

Sergej Markmann*, Martin Franckié, Shovon Pal, David Stark, Mattias Beck, Manfred Fiebig, Giacomo Scalari and Jérôme Faist

Two-dimensional spectroscopy on a THz quantum cascade structure

https://doi.org/10.1515/9783110710687-015

Abstract: Understanding and controlling the nonlinear optical properties and coherent quantum evolution of complex multilevel systems out of equilibrium is essential for the new semiconductor device generation. In this work, we investigate the nonlinear system properties of an unbiased quantum cascade structure by performing two-dimensional THz spectroscopy. We study the time-resolved coherent quantum evolution after it is driven far from equilibrium by strong THz pulses and demonstrate the existence of multiple nonlinear signals originating from the engineered subbands and find the lifetimes of those states to be in the order of 4–8 ps. Moreover, we observe a coherent population exchange among the first four intersubband levels during the relaxation, which have been confirmed with our simulation. We model the experimental results with a time-resolved density matrix based on the master equation in Lindblad form, including both coherent and incoherent transitions between all density matrix elements. This allows us to replicate qualitatively the experimental observations and provides access to their microscopic origin.

Keywords: 2D spectroscopy; nonlinear spectroscopy; quantum cascade laser; quantum wells; semiconductors; THz; time-resolved spectroscopy.

***Corresponding author: Sergej Markmann,** Institute of Quantum Electronics, ETH Zurich, 8093 Zurich, Switzerland, E-mail: msergej@ethz.ch. https://orcid.org/0000-0002-1508-4126
Martin Franckié, David Stark, Mattias Beck, Giacomo Scalari and Jérôme Faist, Institute of Quantum Electronics, ETH Zurich, 8093 Zurich, Switzerland. https://orcid.org/0000-0003-2162-2274 (M. Franckié). https://orcid.org/0000-0002-9850-4914 (D. Stark). https://orcid.org/0000-0002-0260-5797 (M. Beck). https://orcid.org/0000-0003-4028-803X (G. Scalari). https://orcid.org/0000-0003-4429-7988 (J. Faist)
Shovon Pal and Manfred Fiebig, Department of Materials, ETH Zurich, 8093 Zurich, Switzerland. https://orcid.org/0000-0002-1222-9852 (S. Pal). https://orcid.org/0000-0003-4998-7179 (M. Fiebig)

1 Introduction

The terahertz (THz) spectral region has a long standing interest for a wide range of applications such as imaging [1], telecommunication [2, 3], spectroscopy [4], as well as thickness measurements [5]. It is also indispensable in current fundamental research, e.g., in the study of superconductivity [6, 7], quasiparticles such as magnons [8], heavy fermions [9] and in dynamically induced phase transitions [10]. However, a lack of compact, powerful and efficient THz radiation sources has hindered a wide-spread deployment of such applications. A promising candidate is the THz quantum cascade laser (QCL) [11], a unipolar device based on intersubband transitions within the conduction band of an electrically pumped semiconductor heterostructure. The state-of-the-art devices have emission bandwidths of 1–5 THz [12, 13], Watt-level power on a single mode [14] and operate up to 210 K in pulsed mode [15]. Recent developments in the field of optical sensing by means of frequency combs in the THz [16] and mid-infrared [17] spectral regions has gained a lot of attraction. This is because such a spectroscopic technique (dual-comb spectroscopy [18]) provides high resolution and fast acquisition speeds without moving optical or opto-mechanical parts. One important aspect for spectroscopy applications is to increase the spectral coverage of the frequency combs and to control their noise properties. This requires understanding of the frequency comb formation mechanism. Currently, there are two proposed mechanisms which have different origin. The first relies on four-wave mixing (FWM) which arises from the third-order susceptibility ($\chi^{(3)}$) of the gain material itself [17, 19]. The second one is associated with spatial hole burning (SHB) in a Fabry–Perot (FP) cavity. SHB is essentially a dynamical gain grating formed by gain saturation within the laser cavity, which enables mode coupling and can be seen as an additional, effective $\chi^{(3)}$ nonlinearity within the cavity. Recent models [20–22], that take SHB into account, reproduced spontaneous comb formation with the experimentally observed [23] linear frequency chirp in mid-infrared QCLs. However, in a ring cavity laser, SHB can be

This article has previously been published in the journal Nanophotonics. Please cite as: S. Markmann, M. Franckié, S. Pal, D. Stark, M. Beck, M. Fiebig, G. Scalari and J. Faist "Two-dimensional spectroscopy on a THz quantum cascade structure" *Nanophotonics* 2021, 10. DOI: 10.1515/nanoph-2020-0369.

completely eliminated, and hence only FWM from the gain medium is present and responsible for comb formation [24]. The interplay of SHB and FWM makes comb formation in FP cavities a highly nonlinear phenomenon and offer a rich playground for cavity and laser structure designs. Since both mechanisms rely on $\chi^{(3)}$ processes, it is essential to investigate the fundamental nonlinear material properties. In addition, devices such as quantum cascade detectors [25], saturable absorbers [26], and nonlinear conversion media [27] with giant nonlinearities rely on the same operation principle of intersubband transitions in multi-quantum-well structures.

To our knowledge, no investigation of nonlinear properties has been carried out on quantum cascade structures (QCSs) at THz frequencies with two-dimensional spectroscopy, although single and double quantum wells has been studied extensively [28–31]. Since several systems rely on cascaded intersubband transitions, it is of vital importance to analyze and explore the nonlinear optical properties and quantum-coherent evolution in these type of structures, since modelocking of QCLs due to self-induced transparency has been predicted [32].

The focus of our investigation is a THz QCS and its nonlinear intersubband properties, as well as its coherent quantum evolution. We perform coherent collinear two-dimensional THz (2D-THz) spectroscopy on a passive QCS with the aim to determine the ultrafast nonlinear response and relaxation processes originating from the intersubband states. The measurements in this work are performed on an unbiased QCS because this structure has also the potential to be used as a saturable absorber, which can be monolithically integrated into the QCL cavity. Since this multi-quantum-well system is far from a trivial system and involves many coupled electron states, an advanced modeling is required to fully understand and interpret the experimental results. Therefore, we go beyond the commonly employed models based on classical [33] or few-level Maxwell–Bloch equations with phenomenological scattering and dephasing rates [26, 30, 34, 35]. We model our experiment using a time-resolved density matrix model formalism which takes into account all the relevant intersubband states, as well as the relevant scattering mechanisms (except electron–electron interactions) treated with a master equation in Lindblad form [36]. This allows for *ab initio* dynamical simulations of the quantum evolution of the full system density matrix, containing all interactions between coherences and populations mediated by the various scattering mechanisms. Explicitly including the time-dependent electric field pulses on a semiclassical level, allows for calculating the nonlinear optical response nonperturbatively.

2 Materials and methods

2.1 Sample and measurements

In order to gain a detailed insight into the rich nonlinear features of a multilevel intersubband system, we perform 2D-THz spectroscopy on an unbiased THz QCL structure [37] which is designed to lase at 1.2 THz. Additional reason to perform the 2D spectroscopy on an unbiased structure is to keep the system complexity and validate the developed model. Figure 1(a) shows the schematic of the experiment (performed at 10 K), where two phase-locked THz pulses, A and B, are generated by optical rectification in ZnTe and GaP, respectively. These pulses arrive at the sample with a relative time delay τ. The investigated active region is 10 μm in height, 800 μm long, 400 μm wide and is grown on a 500 μm thick *n*-doped GaAs substrate. We lap the substrate down to 190 μm thickness in order to increase the efficiency of the light–matter interaction, polish the substrate to optical quality and coat it with 200 gold for better wave guiding. In this way, the active region and 190-μm-thick substrate are embedded in to a 200-μm-thick single-plasmon wave guide. The optical rectification in the two crystals is realized with 120 fs infrared pulses originating from an amplified Ti:Sapphire laser operating at 1 kHz repetition rate with the center wavelength of 800 nm. In order to obey the intersubband selection rules, the THz pulses are polarized along the growth direction of the QCS. The transmitted THz pulses, which carry the imprint of the induced nonlinear sample response, are detected by a phase-resolved linear detector based on free-space electro-optic (EO) sampling, hence measuring the electric field as a function of detection time t. Figure 1(b) shows the net electric field of two incident THz pulses in air separated by a delay time τ = 2.5 ps. The corresponding spectrum of a single excitation pulse is shown in Figure 1(c) and covers a bandwidth of ~2.5 THz. In order to separate the nonlinear sample response, which arises due to the presence of the combined pulses from the single excitation pulses, a sequence of measurements is performed for each delay τ where the sample interacts with the electric field of pulse A (E_A) and pulse B (E_B) separately, as well as with the two pulses superimposed (E_{AB}). The nonlinear system response is extracted as

$$E_{NL}(t,\tau) = E_{AB}(t,\tau) - E_A(t,\tau) - E_B(t). \quad (1)$$

This results in a two-dimensional (2D) map of E_{NL} (t, τ) as shown in Figure 1(d). Since the measurement is performed in a collinear geometry, all nonlinear signals are simultaneously present in the same measured 2D map. Different orders of the nonlinear sample response can be separated by Fourier transforming into the frequency domain, along the delay axis τ and sampling axis t. The corresponding 2D frequency map as a function of the excitation frequency ν_τ and detection frequency ν_t is shown in Figure 2(a). The collinear geometry of the experiment makes the frequency vectors of the two pulses correspond to vectors with specific directions in 2D frequency space [38, 39]; the frequency vector ν_B of pulse B is parallel to the detection frequency axis, whereas the frequency vector ν_A of pulse A is along the diagonal. This is due to the fact that pulse A is moving with respect to the pulse B (which is kept fixed in time), and thus the wave fronts are tilted by 45° with regards to the sampling axis. Thus, the location of the nonlinear signals in the 2D frequency map can be expressed as a linear combination of frequency vectors ν_A and ν_B, whose lengths are given by the frequencies of the two pulses. The experimental results in Figure 2(a) show multiple nonlinear resonances, marked by circles and labeled as PP_1, PP_2 and PP_3. These correspond to

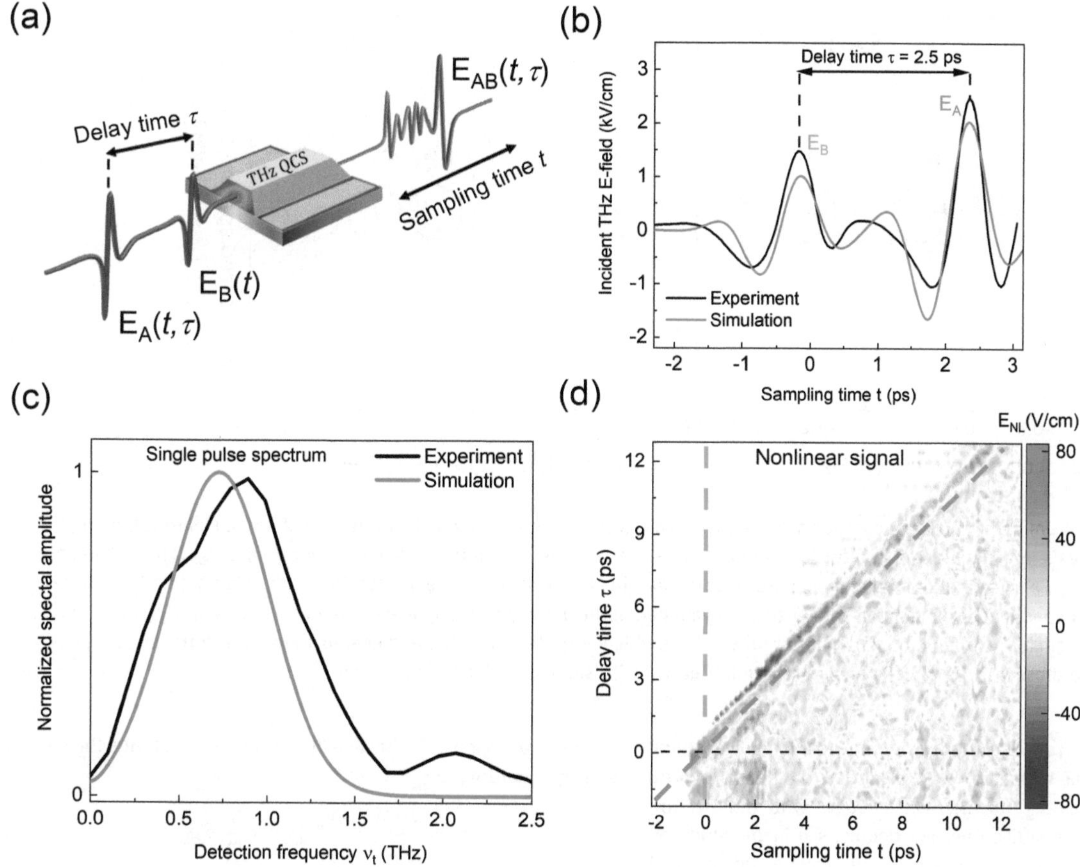

Figure 1: (a) Schematic of the experimental setup showing a collinear pulse geometry used to perform a 2D-THz spectroscopy on quantum cascade structure (QCS). (b) THz electric filed amplitude of pulses A and B separated by a time delay $\tau = 2.5$ ps before coupling in to the quantum cascade wave guide. Black curve represent the experimental pulses while the red curve shows the pulses used in the simulation. (c) Spectra of a single experimental and simulation pulse. (d) Nonlinear signal extracted from the pulse sequence measurements: $E_{NL}(t,\tau) = E_{AB}(t,\tau) - E_A(t,\tau) - E_B(t)$. The green and orange dashed lines indicate the position of the driving fields A and B, respectively. The dashed horizontal line is a guide to the eye for $\tau = 0$ ps.

pump-probe (PP) signals and are generated by the following interaction sequence: Two field interactions of pulse A, creating a nonequilibrium level occupation, followed by a time-delayed single field interaction of pulse B, emitting the nonlinear signal. The nonlinear signals are then detected with EO sampling. Since the detected signal is generated by three light–matter interactions, the origin of this signal is due to the strong $\chi^{(3)}$ of the electronic intersubband system, and the location in the 2D frequency map is given by the frequency vectors $\nu_{AB} = \nu_A - \nu_A + \nu_B$. We note that any higher odd order could contribute to this signal, but for the experimental field strength of ~3 kV/cm we estimate the $\chi^{(5)}$ contribution to be at least three orders of magnitude smaller. The signals along the diagonal in Figure 2(a) are also PP signals, but where the order of pulses A and B is reversed. In this way, there are two interactions with pulse B and one with pulse A. These PP signals are located along the diagonal given by $\nu_{BA} = \nu_B - \nu_B + \nu_A$. Since the electric field strength of pulse A is larger than that of pulse B, the PP signals are smeared out along the diagonal due to the strong system perturbation caused by pulse A. The elongated shape of the B–B–A pulse sequence makes the PP signals almost impossible to distinguish individually. Hence, in the following analysis, we will focus only on the PP signals at zero excitation frequency resulting from A–A–B pulse sequences.

2.2 Simulation model

For simulating the optical response of the structure, we employ a time-resolved density matrix model extended from a study by Kiršanskas et al. [36] to include arbitrary pulse shapes, mean-field potential, as well as a phenomenological electron–electron dephasing time [40]. The elements ρ_{ij} of the full density matrix are evolved according to the Liouville–von-Neumann equation in Lindblad form

$$\dot{\rho}_{ij} = \frac{1}{i\hbar}[H_S,\rho]_{ij} + \sum_{\alpha,kl}\left(L^{\alpha}_{ik}\rho_{kl}L^{\alpha}_{kj} - \frac{1}{2}L^{\alpha}_{ik}L^{\alpha}_{kl}\rho_{lj} - \frac{1}{2}\rho_{ik}L^{\alpha}_{kl}L^{\alpha}_{lj}\right) \equiv \frac{1}{i\hbar}[H_S,\rho]_{ij} + \sum_{\alpha,kl} R^{\alpha}_{ijkl}\rho_{kl} \quad (2)$$

where $H_S = H_0 + H_{EM}(t)$ denotes the Hamiltonian of the intersubband system (H_0) including electromagnetic field ($E(t)$)

$$H_{EM}(t) = -|e|\hat{z}E(t) \quad (3)$$

in Lorenz gauge, L^{α}_{ij} denote Lindblad jump operators as defined in [36], and the equivalent scattering tensors R^{α}_{ijkl} contain couplings between the intersubband system and multiple baths labeled by α, computed

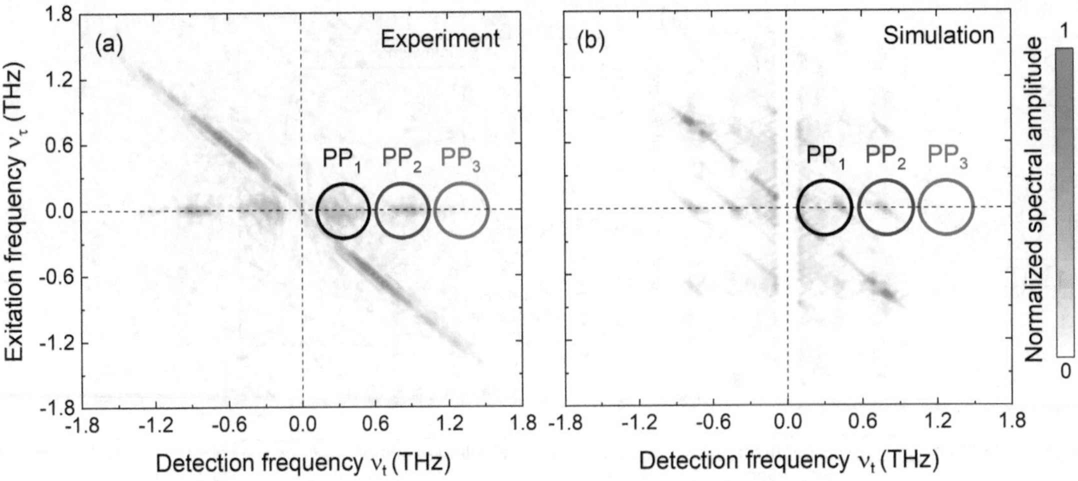

Figure 2: (a) 2D spectrum of the quantum cascade structure (QCS) obtained by performing a 2D Fourier transform of Figure 1(d) along the sampling and delay axis. Signals along the detection frequency axis v_t at zero excitation frequency are pump-probe ($A_{pump}B_{probe}$) signals. Pump-probe ($B_{pump}A_{probe}$) signals along the diagonal are smeared out due to the strong system saturation with pulse A. (b) Simulated E_{NL} induced by the time-dependent nonlinear current density j_{NL} evaluated in the density matrix model. All relevant states of QCS are taken into account. Color-coded circles in (a) and (b) mark pump-probe signals which originates from the engineered states. The corresponding energy levels contributing to these $A_{pump}B_{probe}$ signals are shown in Figure 3 with similarly color-coded arrows.

within Fermi's golden rule. Due to the computational load of resolving explicitly the in-plane momenta of the electrons, the density matrix indices and the rates R^{α}_{ijkl} are thermal averages with the electron temperature set to the lattice one. For details, see in the study by Kiršanskas et al. [36]. In our simulations, we include longitudinal optical (LO) phonon scattering and pure dephasing implemented as in the study by Gordon and Majer [40], as well as scattering with ionized impurities and interface roughness. The coherent system evolution is governed by the first term H_S, which includes the interaction with the electric field pulses. Since the model is solved in the time domain, the nonlinear response is included to infinite order, and it is thus able to capture any nonlinear effects that may occur in the experiment. The second term R^{α}_{ijkl} is responsible for decoherence and dephasing. In addition, this full treatment of scattering on Lindblad form (without the secular approximation), allows the baths to generate coherences as well, allowing for a full treatment of coherent effects such as scattering assisted tunneling, important when the levels are closely spaced as in the present structure. As a basis, we use the Wannier–Stark states of the unbiased structure (a negligible bias of 0.1 meV/period is necessary to break the periodic symmetry of the structure). Due to the computational cost of calculating the scattering tensors R^{α}_{ijkl}, these are calculated without time-dependent fields, and assumed to remain constant. This will make a quantitative difference to the exact scattering rates in the system while the pulses act but will not affect the relaxation behavior once the pulses have subsided or affect significantly the coherent evolution imposed by the pulse, since its duration is short compared to the subband lifetimes. The first simulation step is to reach the self-consistent equilibrium state by iterating the Poisson equation and the evolution of Eq. (2) without external fields. This results in the basis states and potential profile (including the mean-field from the electrons and ionized dopants) shown in Figure 3. The electromagnetic field pulses (parametrized to be similar to the experimental ones and shown in Figure 1(b)) are then applied with varying delay according to the experimental parameters. This results in a response of the density matrix $\rho_{ij}(t)$, and thus the current density according to

$$\langle j \rangle = \frac{Ne\langle \hat{p} \rangle}{m^*} = \frac{Ne}{m^*}\mathrm{Tr}\{\rho\hat{p}\} = \frac{Ne}{m^*}\sum_{kl}\rho_{kl}p_{lk} \tag{4}$$

where N is the 3D electron density, m^* the effective mass and $\hat{p}$ is the momentum operator. Since $p_{ii} = 0$, the current is given solely by the off-diagonal elements of the density matrix, i.e., the coherences. In order to allow for a direct comparison to the experiment, which measures the electric field emitted from the structure rather than the induced polarization current, the induced electric field is evaluated in the frequency domain as

$$E(\omega) = -\mathrm{i}\frac{\mu_0 j(\omega)}{\omega}, \tag{5}$$

where μ_0 denotes the vacuum permeability. The nonlinear contribution to the electric field $E_{NL}(\omega)$ can be evaluated via the Fourier transform of

$$j_{NL}(t,\tau) = j_{AB}(t,\tau) - j_A(t-\tau) - j_B(t) + j_{bg}(t) \tag{6}$$

where $j_{AB}(t,\tau)$ is the current response of both pulse B (with fixed delay) and A (with delay τ), and $j_{A/B}(t)$ are the single-pulse current responses. Here, we also subtract each of the current contributions on the right hand side with the background current $j_{bg}(t)$, which removes any small transients and DC currents coming from the small added bias and the finite number of periods included in the simulation.

By varying the delay between the pulses over the same range as used in the experiment, we obtain a similar 2D map of $j_{NL}(v_\tau, v_t)$, which is Fourier transformed to $E_{NL}(v_\tau, v_t)$ using Eq. (5). As seen in Figure 2(b), the simulation replicate qualitatively all experimental features. Both experiment and simulation show multiple resonances along the detection frequency axis v_t at zero excitation frequency and a strong smeared-out PP signals along the diagonal. We verified, with simulations, that by reducing the A pulse intensity the PP signals

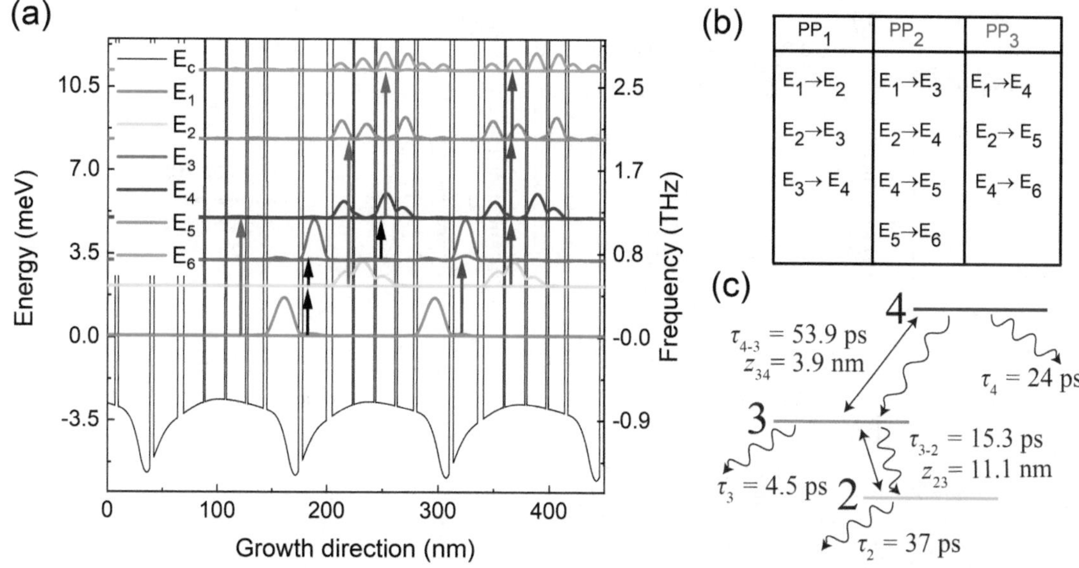

Figure 3: (a) Calculated band diagram of the quantum cascade structure at zero bias that shows the conduction band profile E_C and the six relevant energy states which contribute to the nonlinear signals. (b) The transitions which give rise to the same pump-probe signal. (c) Three of four levels (ladder-type system) which have a dominant contribution to the PP_1 signal and are responsible for nontrivial relaxation decay after the pump drives the system far from equilibrium.

along the diagonal are more localized in the 2D frequency space. By increasing the barrier AlAs fraction by ~3% (from 0.10 to 0.103) from its nominal value, which results in a change in band offset which is small compared to the discrepancy between literature values [41, 42], a good agreement for the absolute frequency positions of the PP resonances PP_1, PP_2 and PP_3 is found with respect to the experiment. This confirms the computed band structure shown in Figure 3, and allows us to identify the transitions responsible for each PP peak as indicated by the arrows in Figure 3.

3 Results

The origin of the PP signals at detection frequencies of 0.2–0.5, 0.7–0.9 and 1.2–1.5 THz can be deduced from the energy band diagram of the QCS shown in Figure 3(a). Due to the low doping (2.6×10^{16} cm^{-3}), only the lowest three energy levels are significantly populated at 10 K (in equilibrium). Three groups of intersubband transitions with energies close to the different PP signals are identified and marked in Figure 3(a) as black, blue, and green arrows, respectively. For the sake of clarity we have tabulated these transitions in Figure 3(b). Since the calculated band diagram at zero bias is highly sensitive to a number of factors, such as the mean-field potential, barrier height, and effective masses, the slight frequency deviation of a few tens of GHz of the PP signals between the experiment and the simulation is expected. The spectral amplitude of the pulses has a minimum at the frequency of the PP_3 signal, as shown in Figure 1(c). The higher energy states (E_4, E_5 and E_6) which contributes to this signal are insignificantly populated. Therefore the PP_3 signal strength is one order of magnitude lower than the other PP signal strengths, even though the oscillator strengths in the PP_3 transitions are in general higher.

Next, we study the time-resolved dynamics of individual nonlinear PP signals. This is realized by multiplying the 2D spectrum with a 2D Gaussian filter and performing the inverse Fourier transforms of the individual frequency components back into the time domain. The 2D Gaussian filter has a spectral bandwidth with full width half maximum as shown in Figure 2 (color-coded circles). The results are shown in Figure 4 (a), (b) and (c). As multiple states contribute to the same PP signal, it is difficult to extract the incoherent state life time of individual energy levels directly from the time domain. We therefore compute a spectrogram by performing a Fourier transform along the sampling time axis t for each delay time τ. The corresponding spectrograms are shown in Figure 4 (a_i), (b_i) and (c_i), which describe the time evolution of the frequency-resolved (nonlinear) state populations of the respective PP signals. In order to extract the population decay time of individual PP signals, cuts in the spectrograms along the delay time axis τ (dashed lines in Figure 4 (a_i), (b_i), and (c_i)) are extracted and shown in Figure 4 (a_{ii}), (b_{ii}), and (c_{ii}). The spectrally resolved PP signals show a general decay trend with an additional nontrivial oscillatory behavior. The overall decay times of the PP_1, PP_2 and PP_3 signals are around 4, 8 and 6 ps, respectively. Note that, since several

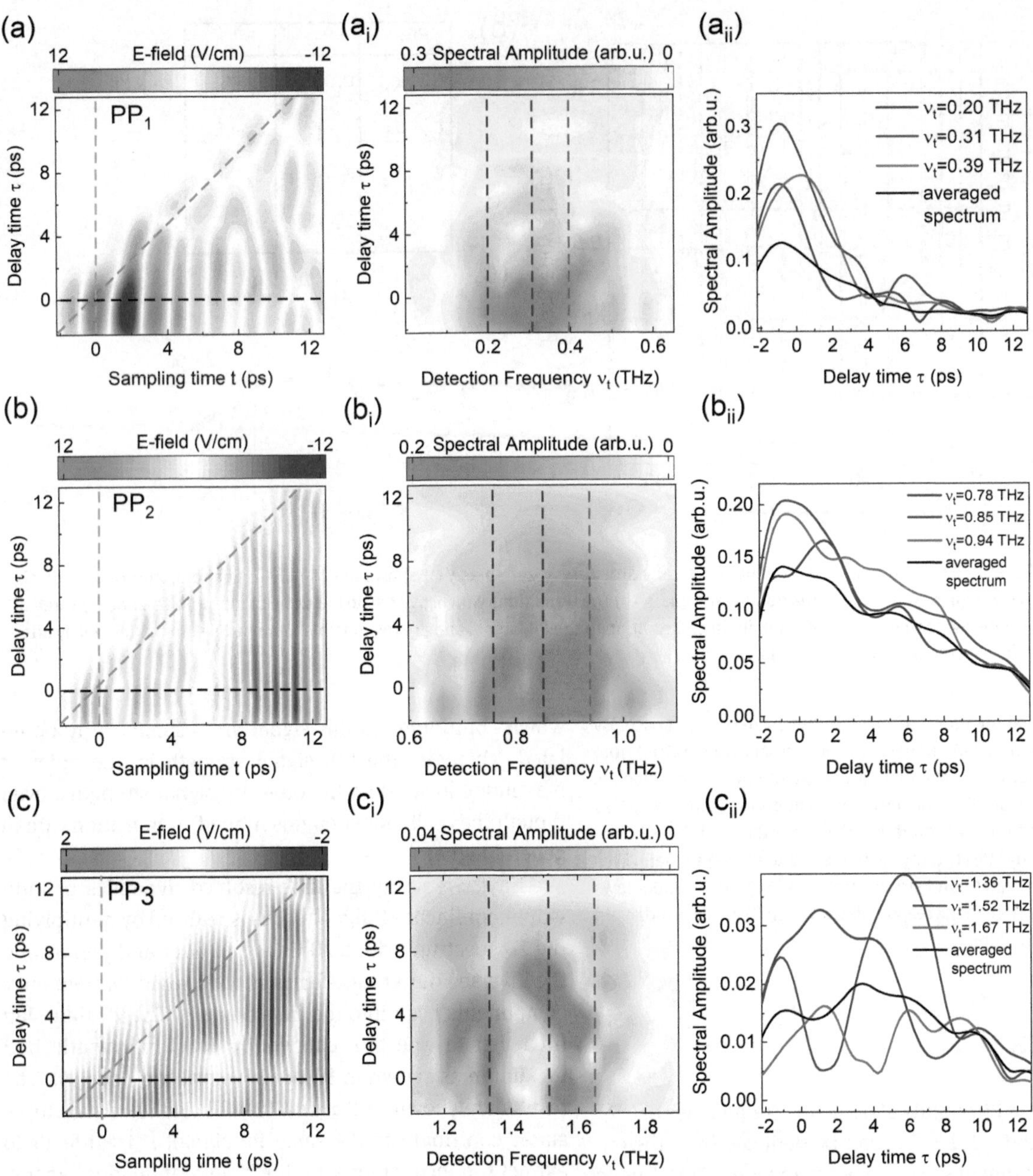

Figure 4: (a), (b) and (c) are inverse Fourier transforms of pump-probe signals PP_1, PP_2 and PP_3 from Figure 2(a) and show the time evolution of each nonlinear signal. Since multiple states contribute to the same pump-probe signal, it is more convenient to study the system dynamics by looking at the time-resolved spectrum along the system evolution axis. This is realized by calculation of the Fourier transform of (a), (b) and (c) along the sampling time axis t for every time delay τ. In this way calculated spectrograms are shown in (a_i), (b_i) and (c_i). Extracted cuts at fixed detection frequencies along delay time τ display spectral time evolution and are shown in (a_{ii}), (b_{ii}) and (c_{ii}). They are also indicated with color-coded dashed lines in (a_i), (b_i) and (c_i). The black curve in (a_{ii}),(b_{ii}) and (c_{ii}) shows the average detection frequency evolution spectrum.

transitions contribute to each PP signal, this method does not provide decay times of individual transitions. Rather, it provides the average rate of decay of the nonlinear response through all "incoherent channels" with transition energies corresponding to the respective PP signal frequency. By incoherent channel, we mean essentially the component of the population decay which is independent of the delay time τ (thus showing no significant frequency components at the filtered $\nu_\tau = 0$). For clarity, the average spectral decay is computed and is visible as a black curve in Figure 4 (a_{ii}), (b_{ii}), and (c_{ii}). Additionally, we have verified the presence of coherent oscillations by computing the Fourier transform of average signals in Figure 4 (a_{ii}), (b_{ii}) and (c_{ii}). The corresponding oscillations appear at frequencies of 0.4, 0.33 and 0.22 THz for PP_1, PP_2 and PP_3, respectively.

In contrast to previous studies on similar, but simpler, quantum well samples, our system shows a population decay as well as non-trivial oscillation (see Figure 4). As already mentioned, Raab et al. [26] investigated a multi-quantum-well system which contains only two levels. Such a system will possess only two PP signals ($A_{\text{pump}}B_{\text{probe}}$ and $B_{\text{pump}}A_{\text{probe}}$). In that two-level system, the expected monotonic, exponential decay of the exited state population was observed. Systems with more than two levels, such as that studied by Kuehn et al. [31], can possess several PP signals at zero excitation frequency, which can be separated in frequency domain provided the energy level separation is larger than the frequency resolution and the level broadening. In these cases, several transitions also contribute to the same PP signals. However, due to the fast dephasing times in the structure in a study by Kuehn et al. [31], with level separations above the LO phonon energy, pulsations in population differences, such as those we observe in Figure 4, could not be seen. We attribute the observed nontrivial oscillatory decay to the coherent carrier exchange among the states. This is enabled by the existence of long coherence times as well as several populated intersubband states. Such a coherent population exchange is driven by the coherent evolution of coherences sharing common state, which are excited by the strong THz pulses. The fact that we can observe it experimentally means that it is associated with an electrical dipole, which requires a non-negligible dipole matrix element between the involved subbands. The existence of such nontrivial incoherent decay is supported by our simulations. Figure 5 show the same analysis on computed 2D spectrum (PP signals from Figure 2(b)). We can identify qualitatively oscillations of the average spectrum in Figure 4 (a_{ii}), (b_{ii}) and (c_{ii}) as observed in the experiment. A detailed discussion on observed coherent oscillations in experiment and theory is elaborated in the following section.

4 Discussion

In order to find a microscopic explanation to the non-monotonic decay patterns seen in Figure 4 (a_{ii}–c_{ii}) and Figure 5 (a_{ii}–c_{ii}), we have investigated the nonlinear contributions to the density matrix elements contributing to the respective PP peaks in Figure 2(b). The nonlinear density matrix of the system provides access to the full time evolution of the nonlinear system. The experimental and computed signals at zero excitation frequency include only the indirect contributions to the decay, i.e. not involving the coherences attributed to the respective PP signals. By including the signals around zero excitation frequency, coherent oscillation effects can be taken into account, which influence the overall population decay. In our model, the nonlinear contribution to the diagonal density matrix elements, which describe the populations, show an oscillating behavior after the excitation pulse has acted. Furthermore, the population oscillations in levels 2 and 3 are out of phase, which indicates a coherent population exchange between the levels. This phenomenon is known from atomic physics, where coherent light fields interact with an atomic system, commonly with three levels. Such a three level system can be either of Λ, V or ladder type [43]. Depending on the system parameters and the system drive, a quantum interference between the excitation pathways will control the optical response of the system and can result in effects such as electromagnetically induced transparency or coherent population trapping (transfer). In our case, the first subband plays a negligible role for the coherences of the system. Thus, the first three excited states are almost decoupled from the ground state and forms a ladder-type system where the population can be coherently exchanged among the levels, as schematically shown in Figure 3(c). Thus, since all PP peaks involve transitions with at least one of levels 2 and 3, they all show similar oscillation patterns in the delay time. The exact behavior also depends on the scattering rates, which are underestimated in the simulation model since electron–electron scattering is neglected. We have checked the calculations by adding an extra pure dephasing channel [40], which does not alter the general trends as it does not increase the transition rates between populations.

Besides the microscopic origin of the observed coherent oscillations, we would like to discuss whether the usage of the single-plasmon wave guide influences the THz pulses (dispersion and nonlinear generation along the entire wave guide) and hence the experimental results. We have computed the losses in the wave guide which are around 44 cm^{-1} and mainly are due to the highly doped substrate. Such relatively high loss results in rapid absorption of the in-coupled THz pulses. In this way, the measured nonlinear signal originates from the first ~200 μm of the wave guide. From the measured transmission, we can observe a slight broadening of the THz pulse which can be explained by the group velocity dispersion along the wave guide.

Naturally, the simulation model does not include wave guide dispersion or loss, which is equivalent to assuming a thin wave guide slab. However, since the

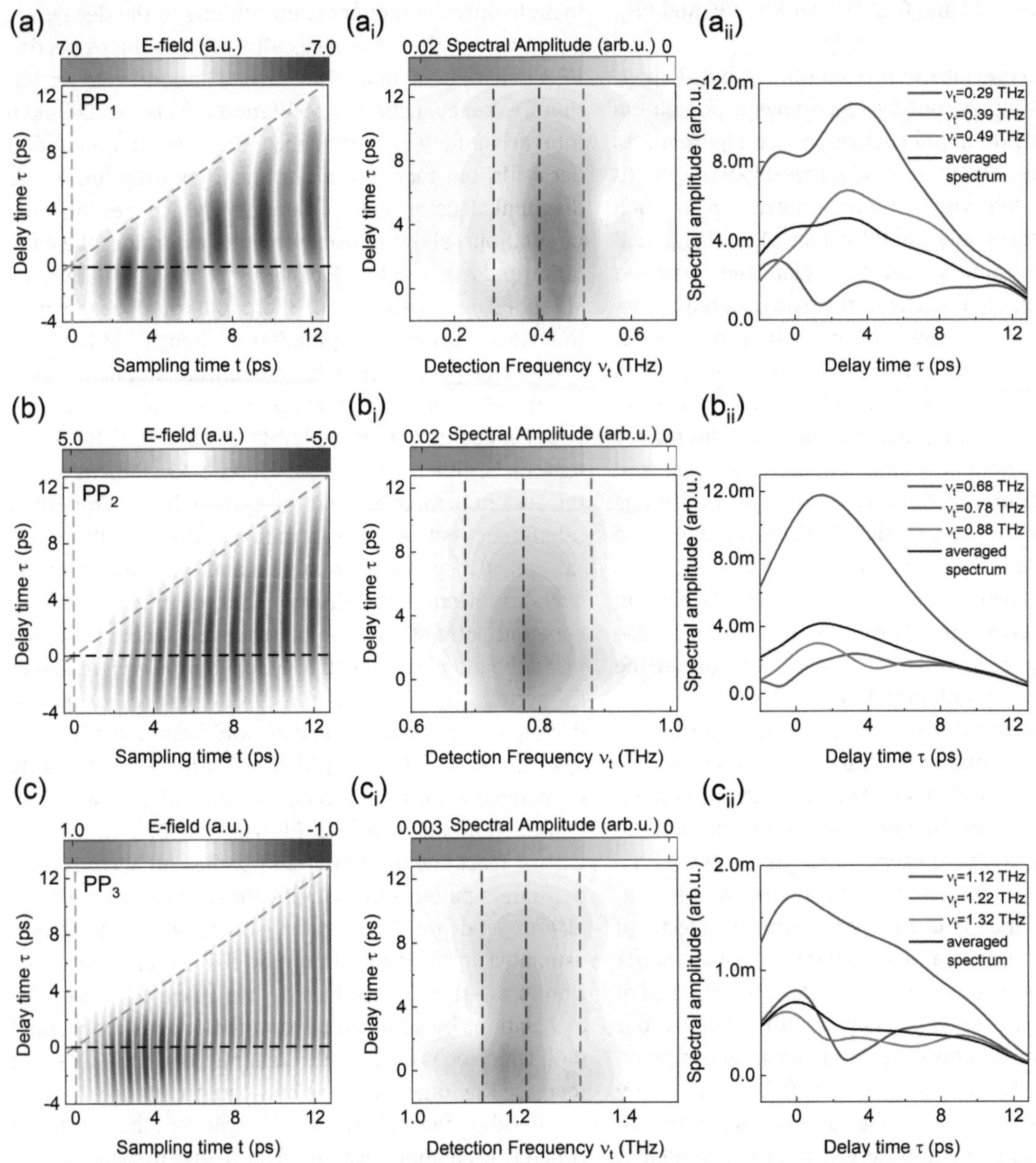

Figure 5: (a), (b) and (c) are inverse Fourier transforms of pump-probe signals PP_1, PP_2 and PP_3 from Figure 2(b) and show the time evolution of each nonlinear signal. Since multiple states contribute to the same pump-probe signal, it is more convenient to study the system dynamics by looking at the time-resolved spectrum along the system evolution axis. This is realized by calculation of the Fourier transform of (a), (b) and (c) along the sampling time axis t for every time delay τ. In this way calculated spectrograms are shown in (a_i), (b_i) and (c_i). Extracted cuts at fixed detection frequencies along delay time τ display spectral time evolution and are shown in (a_{ii}), (b_{ii}) and (c_{ii}). They are also indicated with color-coded dashed lines in (a_i), (b_i) and (c_i).The black curve in (a_{ii}), (b_{ii}) and (c_{ii}) shows the average detection frequency evolution spectrum.

nonlinear signal mainly originates from the facet of the wave guide as discussed above, the qualitative comparison between the experiment and simulation is still valid. Thus, we do not expect that including a more realistic wave guide, by e.g. performing several simulations with decreasing THz pulse amplitudes and obtaining an average 2D map, would alter the quality of the observed phenomenon (coherent oscillation).

5 Conclusion

In conclusion, we have conducted the first experimental and theoretical 2D-THz spectroscopic study on a QCS at THz frequencies. We observe multiple nonlinear (PP) signals originating from the engineered intersubband states, from which we determine the average incoherent intersubband lifetimes. Moreover, we observe nontrivial system dynamics attributed to field-induced coherences, the so-called coherent population exchange. We find incoherent lifetimes to be several times smaller than the typical gain recovery times in THz QCLs [44–46], as well as saturation times to be shorter than the QCL gain saturation time. Under these circumstances, one can consider such a structure as a monolithically integrated saturable absorber into a QCL cavity in order to achieve a passive mode locking. However, this is complicated due to the observed nontrivial dynamics, as the interaction between all density matrix elements must be considered. Further investigation of this QCS with respect to bias, field strength and temperature dependence is needed in order to explore its full potential as saturable absorber, ultrafast pulse shaper, frequency converter, as well as its FWM efficiency.

Acknowledgment: The authors acknowledge the ETH Zurich cluster Euler for performing our numerical simulation.
Author contribution: The sample was grown by M.B. Sample processing was done by S.M. and D.S. S.M. and S.P. performed the 2D THz measurements and S.P. designed the experimental scheme. M. Franckié performed the theoretical simulations and developed the simulation model. S.M., M. Franckié and S.P. analyzed the data. G. S., M. F., and J. F. supervised the work. All authors contributed to the manuscript writing.
Research funding: The authors gratefully acknowledge financial support from the Qombs Project funded by the European Union's Horizon 2020 research and innovation program under Grant Agreement No. 820419. The authors also thank European Union for research and innovation program Horizon 2020 under Grant No. 766719-FLASH Project and the ERC Grant CHIC (No. 724344). M. Fiebig and S. Pal would also like to acknowledge the financial support by the Swiss National Science Foundation (SNSF) via project No. 200021_178825, NCCR MUST via PSP 1-003448-051, NCCR ETH FAST 3 via PSP 1-003448-054. S. Pal further acknowledges the support by ETH Career Seed Grant SEED-17 18-1.
Conflict of interest statement: The authors declare no conflicts of interest regarding this article.

References

[1] P. Dean, A. Valavanis, J. Keeley, et al., "Terahertz imaging using quantum cascade lasers—a review of systems and applications," *J. Phys. D Appl. Phys.*, vol. 47, p. 374008, 2014.

[2] Z. Chen, Z. Y. Tan, Y. J. Han, et al., "Wireless communication demonstration at 4.1 THz using quantum cascade laser and quantum well photodetector," *Electron. Lett.*, vol. 47, p. 1002, 2011.

[3] J. R. Freeman, L. Ponnampalam, H. Shams, et al., "Injection locking of a terahertz quantum cascade laser to a telecommunications wavelength frequency comb," *Optica*, vol. 4, p. 1059, 2017.

[4] H.-W. Hübers, H. Richter, and M. Wienold, "High-resolution terahertz spectroscopy with quantum-cascade lasers," *J. Appl. Phys.*, vol. 125, p. 151401, 2019.

[5] T. D. Nguyen, J. D. R. Valera, and A. J. Moore, "Optical thickness measurement with multi-wavelength THz interferometry," *Optic Laser. Eng.*, vol. 61, p. 19, 2014.

[6] X. Yang, C. Vaswani, C. Sundahl, et al., "Terahertz-light quantum tuning of a metastable emergent phase hidden by superconductivity," *Nat. Mater.*, vol. 17, p. 586, 2018.

[7] X. Yang, C. Vaswani, C. Sundahl, et al., "Lightwave-driven gapless superconductivity and forbidden quantum beats by terahertz symmetry breaking," *Nat. Photonics*, vol. 13, p. 707, 2019.

[8] J. Lu, X. Li, H. Y. Hwang, et al., "Coherent two-dimensional terahertz magnetic resonance spectroscopy of collective spin waves," *Phys. Rev. Lett.*, vol. 118, p. 207204, 2017.

[9] C. Wetli, S. Pal, J. Kroha, et al., "Time-resolved collapse and revival of the Kondo state near a quantum phase transition," *Nat. Phys.*, vol. 14, p. 1103, 2018.

[10] X. Li, T. Qiu, J. Zhang, et al., "Terahertz field-induced ferroelectricity in quantum paraelectric $SrTiO_3$," *Science*, vol. 364, p. 1079, 2019.

[11] R. Köhler, A. Tredicucci, F. Beltram, et al., "Terahertz semiconductor-heterostructure laser," *Nature*, vol. 417, p. 156, 2002.

[12] G. Liang, T. Liu, and Q. J. Wang, "Recent developments of terahertz quantum cascade lasers," *IEEE J. Sel. Top. Quant. Electron.*, vol. 23, p. 1, 2017.

[13] M. Rösch, M. Beck, M. J. Süess, et al., "Heterogeneous terahertz quantum cascade lasers exceeding 1.9 THz spectral bandwidth and featuring dual comb operation," *Nanophotonics*, vol. 7, p. 237, 2018.

[14] C. A. Curwen, J. L. Reno, and B. S. Williams, "Broadband continuous single-mode tuning of a short-cavity quantum-cascade VECSEL," *Nat. Photonics*, pp. 1–5, 2019. https://doi.org/10.1038/s41566-019-0518-z.

[15] L. Bosco, M. Franckié, G. Scalari, M. Beck, A. Wacker, and J. Faist, "Thermoelectrically cooled THz quantum cascade laser operating up to 210 K," *Appl. Phys. Lett.*, vol. 115, p. 010601, 2019.

[16] Y. Yang, D. Burghoff, D. J. Hayton, J. R. Gao, J. L. Reno, and Q. Hu, "Terahertz multiheterodyne spectroscopy using laser frequency combs," *Optica*, vol. 3, p. 499, 2016.

[17] A. Hugi, G. Villares, S. Blaser, H. C. Liu, and J. Faist, "Mid-infrared frequency comb based on a quantum cascade laser," *Nature*, vol. 492, p. 229, 2012.

[18] I. Coddington, N. Newbury, and W. Swann, "Dual-comb spectroscopy," *Optica*, vol. 3, p. 414, 2016.

[19] P. Friedli, H. Sigg, B. Hinkov, et al., "Four-wave mixing in a quantum cascade laser amplifier," *Appl. Phys. Lett.*, vol. 102, p. 222104, 2013.

[20] D. Burghoff, "Frequency-modulated combs as phase solitons," arXiv preprint 2020. https://arxiv.org/abs/2006.12397.

[21] C. Silvestri, L. L. Columbo, M. Brambilla, and M. Gioannini, "Coherent multi-mode dynamics in a quantum cascade laser: Amplitude- and frequency-modulated optical frequency combs," *Opt. Express*, vol. 28, p. 23846, 2020.

[22] N. Opačak, and B. Schwarz, "Theory of frequency-modulated combs in lasers with spatial hole burning, dispersion, and kerr nonlinearity," *Phys. Rev. Lett.*, vol. 123, p. 243902, 2019.

[23] M. Singleton, P. Jouy, M. Beck, and J. Faist, "Evidence of linear chirp in mid-infrared quantum cascade lasers," *Optica*, vol. 5, p. 948, 2018.

[24] B. Meng, M. Singleton, M. Shahmohammadi, et al., "Mid-infrared frequency comb from a ring quantum cascade laser," *Optica*, vol. 7, p. 162, 2020.

[25] D. Hofstetter, F. R. Giorgetta, E. Baumann, Q. Yang, C. Manz, and K. Köhler, "Mid-infrared quantum cascade detectors for applications in spectroscopy and pyrometry," *Appl. Phys. B*, vol. 100, p. 313, 2010.

[26] J. Raab, C. Lange, J. L. Boland, et al., "Ultrafast two-dimensional field spectroscopy of terahertz intersubband saturable absorbers," *Opt. Express*, vol. 27, p. 2248, 2019.

[27] E. Rosencher and P. Bois, "Model system for optical nonlinearities: asymmetric quantum wells," *Phys. Rev. B*, vol. 44, p. 11315, 1991.

[28] M. S. Sherwin, K. Craig, B. Galdrikian, et al., "Nonlinear quantum dynamics in semiconductor quantum wells," *Phys. Nonlinear Phenom.*, vol. 83, p. 229, 1995.

[29] T. Müller, W. Parz, G. Strasser, and K. Unterrainer, "Pulse-induced quantum interference of intersubband transitions in coupled quantum wells," *Appl. Phys. Lett.*, vol. 84, p. 64, 2003.

[30] W. Kuehn, K. Reimann, M. Woerner, and T. Elsaesser, "Phase-resolved two-dimensional spectroscopy based on collinear *n*-wave mixing in the ultrafast time domain," *J. Chem. Phys.*, vol. 130, p. 164503, 2009.

[31] W. Kuehn, K. Reimann, M. Woerner, T. Elsaesser, and R. Hey, "Two-dimensional terahertz correlation spectra of electronic excitations in semiconductor quantum wells," *J. Phys. Chem. B*, vol. 115, p. 5448, 2011.

[32] C. R. Menyuk and M. A. Talukder, "Self-induced transparency modelocking of quantum cascade lasers," *Phys. Rev. Lett.*, vol. 102, p. 023903, 2009.

[33] F. Xiang, K. Wang, Z. Yang, J. Liu, and S. Wang, "A direct method to calculate second-order two-dimensional terahertz spectroscopy in frequency-domain based on classical theory," *Front. Optoelectron.*, vol. 11, p. 413, 2018.

[34] M. Woerner, W. Kuehn, P. Bowlan, K. Reimann, and T. Elsaesser, "Ultrafast two-dimensional terahertz spectroscopy of elementary excitations in solids," *New J. Phys.*, vol. 15, p. 025039, 2013.

[35] M. S. Sherwin, K. Craig, B. Galdrikian, et al., "Nonlinear quantum dynamics in semiconductor quantum wells," *Phys. Nonlinear Phenom.*, vol. 83, p. 229, 1995.

[36] G. Kiršanskas, M. Franckié, and A. Wacker, "Phenomenological position and energy resolving Lindblad approach to quantum kinetics," *Phys. Rev. B*, vol. 97, p. 035432, 2018.

[37] C. Walther, G. Scalari, J. Faist, H. Beere, and D. Ritchie, "Low frequency terahertz quantum cascade laser operating from 1.6 to 1.8 THz," *Appl. Phys. Lett.*, vol. 89, p. 231121, 2006.

[38] J. Lu, X. Li, Y. Zhang, et al., in *Multidimensional Time-Resolved Spectroscopy*, T. Buckup and J. Léonard, Eds., Topics in Current Chemistry Collections, Cham, Springer International Publishing, 2019, pp. 275–320.

[39] T. Elsaesser, *Concepts and Applications of Nonlinear Terahertz Spectroscopy*, IOP Publishing, 2019.

[40] A. Gordon and D. Majer, "Coherent transport in semiconductor heterostructures: a phenomenological approach," *Phys. Rev. B*, vol. 80, p. 195317, 2009.

[41] I. Vurgaftman, J. R. Meyer, and L. R. Ram-Mohan,"Band parameters for III-V compound semiconductors and their alloys," *J. Appl. Phys.*, vol. 89, p. 5815, 2001.

[42] W. Yi, V. Narayanamurti, H. Lu, M. A. Scarpulla, and A. C. Gossard, "Probing semiconductor band structures and heterojunction interface properties with ballistic carrier emission: GaAs/Al_xGa_{1-x} As as a model system," *Phys. Rev. B*, vol. 81, p. 235325, 2010.

[43] Z. Ficek and S. Swain, *Quantum Interference and Coherence: Theory and Experiments, Springer Series in Optical Sciences*, New York, Springer, 2005.

[44] S. Markmann, H. Nong, S. Pal, et al., "Two-dimensional coherent spectroscopy of a THz quantum cascade laser: observation of multiple harmonics," *Opt. Express*, vol. 25, p. 21753, 2017.

[45] D. R. Bacon, J. R. Freeman, R. A. Mohandas, et al., "Gain recovery time in a terahertz quantum cascade laser," *Appl. Phys. Lett.*, vol. 108, p. 081104, 2016.

[46] C. G. Derntl, G. Scalari, D. Bachmann, et al., "Gain dynamics in a heterogeneous terahertz quantum cascade laser," *Appl. Phys. Lett.*, vol. 113, p. 181102, 2018.

Alessandra Di Gaspare, Leonardo Viti, Harvey E. Beere, David D. Ritchie and Miriam S. Vitiello*

Homogeneous quantum cascade lasers operating as terahertz frequency combs over their entire operational regime

https://doi.org/10.1515/9783110710687-016

Abstract: We report a homogeneous quantum cascade laser (QCL) emitting at terahertz (THz) frequencies, with a total spectral emission of about 0.6 THz, centered around 3.3 THz, a current density dynamic range J_{dr} = 1.53, and a continuous wave output power of 7 mW. The analysis of the intermode beatnote unveils that the devised laser operates as an optical frequency comb (FC) synthesizer over the whole laser operational regime, with up to 36 optically active laser modes delivering ~200 μW of optical power per optical mode, a power level unreached so far in any THz QCL FC. A stable and narrow single beatnote, reaching a minimum linewidth of about 500 Hz, is observed over a current density range of 240 A/cm^2 and even across the negative differential resistance region. We further prove that the QCL FC can be injection locked with moderate radio frequency power at the intermode beatnote frequency, covering a locking range of 1.2 MHz. The demonstration of stable FC operation, in a QCL, over the full current density dynamic range, and without any external dispersion compensation mechanism, makes our proposed homogenous THz QCL an ideal tool for metrological applications requiring mode-hop electrical tunability and a tight control of the frequency and phase jitter.

Alessandra Di Gaspare and Leonardo Viti contributed equally to this work

***Corresponding author: Miriam S. Vitiello,** NEST, CNR – Istituto Nanoscienze and Scuola Normale Superiore, Piazza San Silvestro 12, 56127, Pisa, Italy, E-mail: miriam.vitiello@sns.it.
Alessandra Di Gaspare and Leonardo Viti, NEST, CNR – Istituto Nanoscienze and Scuola Normale Superiore, Piazza San Silvestro 12, 56127, Pisa, Italy
Harvey E. Beere and David D. Ritchie, Cavendish Laboratory, University of Cambridge, Cambridge CB3 0HE, UK

Keywords: frequency comb; injection locking; terahertz; quantum cascade laser.

1 Introduction

The terahertz (THz) region of the electromagnetic spectrum, loosely defined in the frequency range between 0.1 and 10 THz [1], attracted a renewed attention, in recent years, for targeting far-infrared applications requiring a tight control of the frequency and phase jitter of the laser modes, such as high-resolution and high-sensitivity spectroscopy [2, 3], telecommunications [4, 5], and quantum metrology [6], amongst many others.

Although relevant models for gigahertz–terahertz generation exploiting nonlinearity in semiconductor superlattices have been recently proposed [7], electrically pumped quantum cascade lasers (QCLs) are the most efficient on-chip sources of optical frequency comb (FC) synthesizers at THz frequencies [8–11], thanks to the wide frequency coverage [10, 12] and the inherently high optical power levels [13, 14], which allow continuous-wave (CW) emitting powers per comb tooth in the 3 μW [12] to 60 μW [8] range and the high spectral purity (intrinsic linewidths of ~100 Hz) [15]. Such a combination of performances makes THz QCL FCs the most suitable choice for the aforementioned applications.

In a THz QCL, the inherently high optical nonlinearity of the quantum engineered gain medium allows locking in phase the laser modes, passively [8, 16]. Specifically, the resonant third-order active material susceptibility inherently induces self-phase locking through the four-wave mixing (FWM) process. FWM tends to homogenize the mode spacing and consequently promotes the spontaneous proliferation of phase-locked equispaced optical modes.

However, to achieve a stable comb regime, the total chromatic dispersion has to be minimized over the laser operational bandwidth [8] so that the phase mismatch from the material, the laser waveguide, and the active region

This article has previously been published in the journal Nanophotonics. Please cite as: A. Di Gaspare, L. Viti, H. E. Beere, D. D. Ritchie and M. S. Vitiello "Homogeneous quantum cascade lasers operating as terahertz frequency combs over their entire operational regime" *Nanophotonics* 2021, 10. DOI: 10.1515/nanoph-2020-0378.

gain approaches zero in the QCL cavity. This has been achieved either by integrating a dispersion compensator in a homogeneous QCL active region [8] providing a negative group velocity dispersion (GVD) to cancel the positive cavity dispersion or by gain medium engineering of heterogeneous multistacked active regions (ARs) [10, 17, 18] or of individual homogeneous ARs [19], in both cases designed to have a relatively flat gain top, within which the intrinsic dispersion is small enough to preserve comb formation. Heterogeneous designs are ideal for achieving a broad spectral coverage, but they also come with some inherent disadvantages, such as the design-related difficulties associated with a proper matching of the threshold currents between the different AR modules, which can prevent the simultaneous mode proliferation over the full operation range of the QCL. On the other hand, homogenous ARs are usually engineered with a narrower gain profile but are in principle easier to be injection locked in the laser operation regime in which dispersion is not compensated.

Although a broad spectral coverage (0.7–1.1 THz) [8, 10, 12, 19] has been demonstrated using both a dispersion compensator [8] or through heterogeneous [10, 12] or homogeneous [8, 19] gain medium engineering, both approaches are ineffective in handling the bias-dependent cavity dispersion, meaning that spontaneous comb operation has been demonstrated to occur only over a limited driving current dynamic range, usually <20–29% of the QCL operational regime [8–10, 12, 19]. To date, external approaches such as coupled *dc*-biased cavities [20] or Gires–Tournois interferometers [21] proved to be the only way to increase the operational dynamic range of THz FC synthesizers, although with poor spectral coverage [20], optical power outputs [20], or with only a moderate increase of the current dynamic range [21].

In this work, we quantum engineer and devise a broadband homogeneous QCL FC covering a bandwidth of 0.6 THz (3.05–3.65 THz), with a dynamic range $J_{dr} = J_{max}/J_{th} = 1.53$ (J_{th} is the threshold current density and J_{max} the maximum current density value) and a maximum CW output power of 7 mW. Remarkably, the collected electrical intermode beatnote map reveals spontaneous FC operation over the entire operational range of the laser, classifying the proposed device as a unique frequency-tunable metrological tool across the far-infrared.

2 Results and discussion

The active region design of the homogeneous gain medium is a slightly modified version [22, 23] of the four quantum well bound-to-continuum structures described in Ref. [24]. The highly diagonal laser transition ensures gain recovery times much larger (>35 ps) [25] than those usually achieved in THz QCLs (5–10 ps), together with shorter upper state lifetimes, above threshold. This is an ideal condition for amplitude modulations up to tens of GHz, ideal for FC operation [19]. Indeed, a GHz modulation envelope, possibly synchronized to the THz field in the time domain, could also help stabilizing the laser in a fashion reminiscent of active mode-locking, supporting the formation of an amplitude-modulated comb. The GaAs/AlGaAs hetoerostructure used in this work is grown by molecular beam epitaxy (MBE) on a semi-insulating GaAs substrate. The final gain medium includes a sequence of 160 periods of a four-well quantum cascade design centered at 3.3 THz. The 10-μm thick final structure consists of a 250 nm undoped GaAs buffer layer, an undoped 250 nm $Al_{0.5}Ga_{0.5}As$ etch-stop layer, a 700 nm Si-doped (2×10^{18} cm^{-3}) GaAs layer, the AR (doped at 3×10^{16} cm^{-3}), and a 80 nm heavily Si-doped (5×10^{18} cm^{-3}) GaAs top contact layer.

Device fabrication is based on a standard metal–metal processing technique that relies on Au–Au thermocompression wafer bonding of the MBE sample onto a highly doped GaAs substrate. A Cr/Au (10 nm/150 nm) top contact is lithographically patterned on the top laser surface, leaving uncoated the two sides of the ridge along two lateral stripes, whose width ranges between 3 and 5 μm, depending on the ridge width [10]. Laser bars are then realized via deep inductively coupled plasma reactive ion etching (ICP-RIE) with vertical sidewalls to allow uniform current injection and ridge widths in the 50–90 μm range. A 5-nm thick Ni layer is then deposited on the uncoated top lateral stripes to define lossy side absorbers needed to suppress undesired high-order lateral modes. Laser bars 50–70 μm wide and 2.75 mm long were then In soldered on a copper plate, wire bonded, and mounted on the cold head of a helium flow cryostat.

The voltage–current density (*V–J*) and the light–current density (*L–J*) characteristics, measured while driving the QCL in CW, as a function of the heat sink temperature (T_H), show laser action up to a maximum heat sink temperature $T_H = 88$ K. Figure 1A shows the *V–J–L* collected in the $T_H = 15$–75 K interval. At $T_H = 27$ K, the CW threshold current density is J_{th} = (428 A cm^{-2}), the maximum current density $J_{max} = 653$ A cm^{-2}, the operational dynamic range is $J_{dr} = J_{max}/J_{th} = 1.53$, and the maximum peak optical power is 7 mW.

CW Fourier transform infrared spectra (FTIR), acquired under vacuum (Bruker vertex 80) with a 0.075 cm^{-1} spectral resolution (Figure 1B–I), while progressively increasing the driving current, show that the laser is initially emitting a single frequency mode (~3.3 THz) at threshold

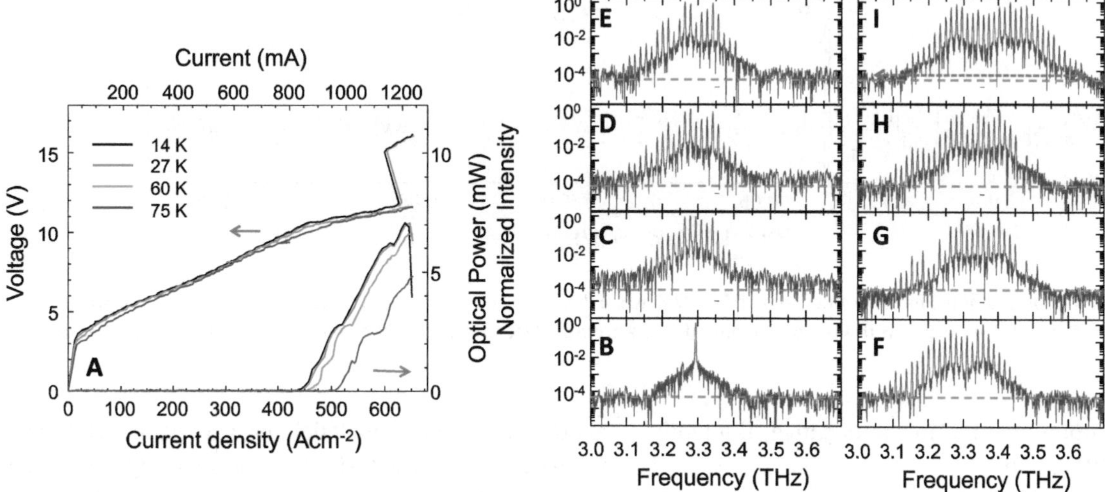

Figure 1: (A) Voltage–current density (V–J) and light–current density (L–J) characteristics, measured in continuous wave, as a function of the heat sink temperature in the 15–75 K range. (B–I) Fourier transform infrared spectra measured at T_H = 27 K, in rapid scan mode, under vacuum with a 0.075 cm^{-1} spectral resolution, while driving the quantum cascade laser (QCL) in continuous wave at (B) 445 A cm^{-2}, (C) 465 A cm^{-2}, (D) 470 A cm^{-2}, (E) 480 A cm^{-2}, (F) 512 A cm^{-2}, (G) 550 A cm^{-2}, (H) 583 A cm^{-2}, and (I) 645 A cm^{-2}. The red dashed lines indicate, roughly, the noise floor of the measurements. The green arrow in panel I marks the laser bandwidth.

(J = 445 A cm^{-2}, Figure 1B) and then turns multimode with a progressively richer (Figure 1C–I) sequence of equidistant optical modes, spaced by the cavity round-trip frequency. The overall spectral coverage reaches 600 GHz (Figure 1I), and differently from the very first demonstration [8], does not show any spectral dip; at a current density of 645 A cm^{-2} (Figure 1I), corresponding to 7 mW peak optical power, 36 optically active laser modes are retrieved. Under this condition, the CW optical power per optical mode is ~200 μW, significantly larger than that of any previous THz QCL FC, demonstrated so far [8–12, 19].

To investigate the coherence properties of the devised THz QCL, we then trace the intermode beatnote radio frequency (RF) as a function of the driving current (Figure 2A). The QCL is driven in CW by a low-noise power supply (Wavelength Electronics QCL2000 LAB), and the RF signal is recorded using a bias tee (Tektronix AM60434) connecting the QCL and an RF spectrum analyzer (Rohde and Schwarz FSW43). Remarkably, the intermode beatnote map shows a single narrow beatnote over the entire dynamic range of the QCL. An idividual beatnote appears at a current density just above threshold and persists,

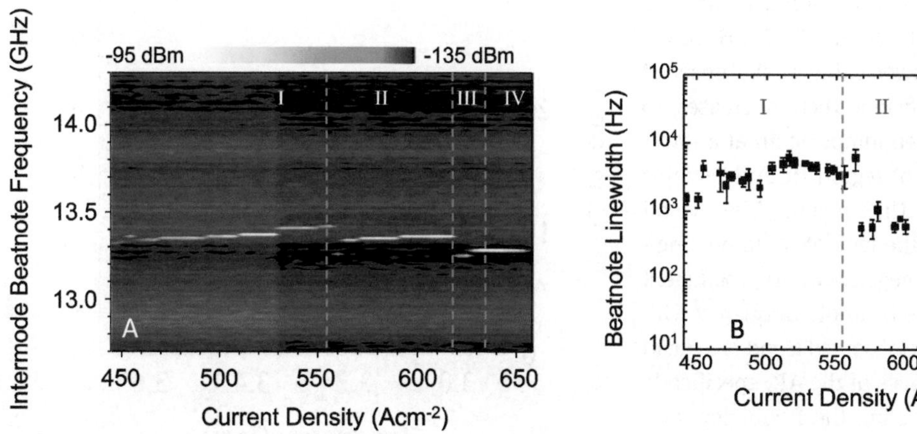

Figure 2: (A) Intermode beatnote map as a function of the driving current density measured at 27 K in a 2.75-mm-long, 70-μm-wide laser bar, operating in continuous wave (CW). The beatnote signal is extracted from the laser bias line using a bias tee and is recorded with a radio frequency (RF) spectrum analyzer (Rohde & Schwarz FSW; resolution bandwidth [RBW]: 500 Hz, video bandwidth [VBW]: 500 Hz, root mean square (RMS) acquisition mode). The vertical dashed lines identify four relevant transport regimes: (I) from onset for conduction to band alignment ; II (band alignment); III (peak optical power); IV (NDR). NDR, negative differential resistance.

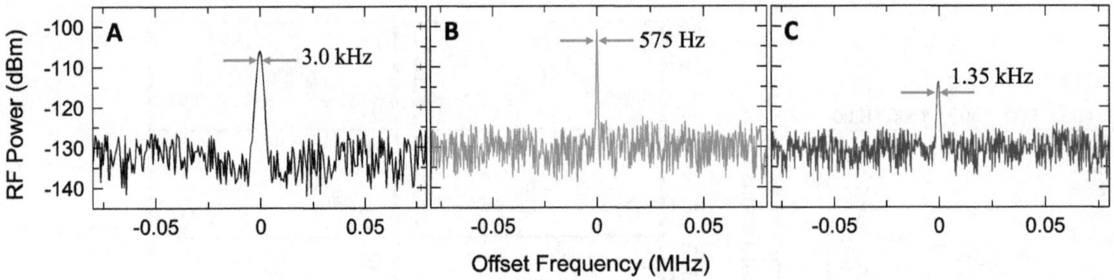

Figure 3: (A–C) Intermode beatnote linewidth measured at $T_H = 27$ K, while driving the quantum cascade laser (QCL) in continuous wave at (A) 480 A cm^{-2}, (B) 570 A cm^{-2}, (C) 645 A cm^{-2} with a radio frequency (RF) spectrum analyzer (resolution bandwidth [RBW]: 500 Hz, video bandwidth [VBW]: 500 Hz).

continuously, upon entering in the negative differential resistance (NDR) region.

The analysis of the beatnote linewithds (LWs) (Figure 2B) shows that the LW is initially varying in the range of ~2–5 kHz (Figure 3A) for J up to 558 A cm^{-2} (region I in the beatnote map in Figure 2A and B), then it becomes narrower (~500 Hz, Figure 3B) for current densities up to 616 A cm^{-2}, (region II). This is followed by a limited current density region, around the peak optical power (J = 632 A cm^{-2}) (region III), in which the LW increases progressively from 500 Hz up to 2 kHz. Interestingly, when the QCL is then driven beyond the roll-off current (region IV), the beatnote is still narrow (2–7 kHz, Figure 3C), but weaker (~15 dBm), due to the electric field instabilities, associated with the bias fluctuations, occurring in this regime. Surprisingly, the laser does not show any high phase noise regime [8–12, 19, 21].

The analysis of the electrical frequency tuning of the intermode beatnote reflects the different dynamics along the distinctive operational bias regimes. Indeed, by varying the driving current at a fixed operating temperature T_H = 27 K, the beatnote initially blue shifts across region I, spanning a 70 MHz range with a tuning coefficient of +0.47 MHz mA^{-1}, then the beatnote jumps at a lower value at the onset of region II and the tuning coefficient then decreases to +0.2 MHz mA^{-1}. The beatnote then jumps again at a lower value (13.315 GHz) at the onset of region III and remains almost constant over the entire NDR regime. The overall tuning behavior clearly reflects the lack of Joule heating-related effects that would led to a negative tuning coefficient with a monotonic decrease of the beatnote frequency with respect to the driving current. The observed trend is instead ascribed to the intracavity dynamics of the AR, specifically to the chromatic dispersion affecting the frequency, and thus the spacing of the cavity modes owing to the variation in the effective refractive index of the gain medium.

To get a deeper insight on the intracavity dynamics, we perform numerical simulations of the group delay dispersion (GDD). The dispersion profile is retrieved including the contributions from the material, the waveguide and the gain of the QCL [9, 21]. The first two terms are computed considering a Drude–Lorentz model for the frequency-dependent refractive index of the material; we then compute the QCL gain from the experimental emission spectra and evaluate the refractive index deviation as a consequence of the gain by applying the Kramers–Kronig equations. Finally, the dispersion provided by the whole structure is computed from the second derivative of the phase. Figure 4 shows the individual GDD contributions, calculated at J = 645 A cm^{-2}, i.e. in the high voltage regime, in which the laser spectrum is particularly rich of optical modes.

In the spectral region where the QCL shows laser action, the waveguide shows a low dispersive refractive index; therefore, the waveguide-related contribution to the dispersion is negligible (<10^5 fs^2) with respect to the other terms. The gain profile (Figure 4), centered at 3.3 THz is extracted from the emission spectra of Figure 1I and it is then normalized

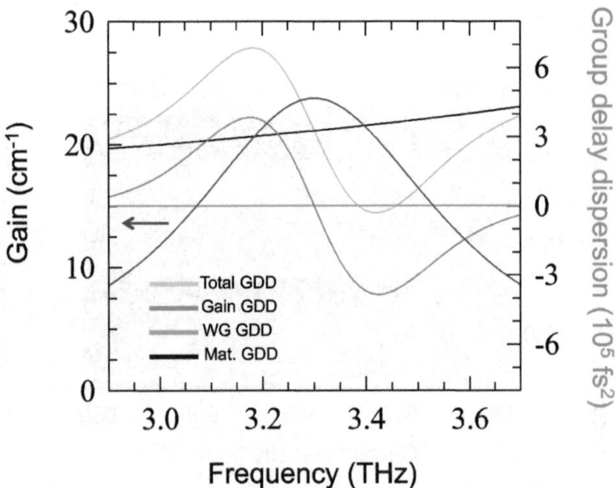

Figure 4: Simulations of the group delay dispersion including contributions from the material (GaAs), waveguide and gain, performed at J = 645 A cm^{-2}. The estimated gain curve (blue) is plotted on the left vertical axis of the graph and marked with an arrow.

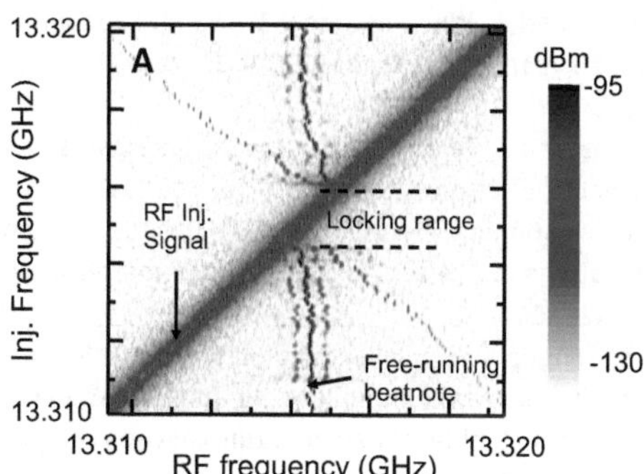

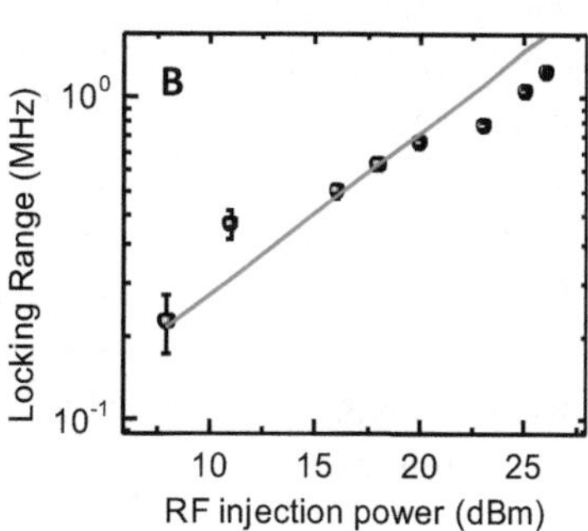

Figure 5: (A) Injection locking map for an injected radio frequency (RF) power of +25 dBm. The map is recorded by sweeping the RF injection signal (diagonal line across the panel) toward the free running beatnote and then locking it. The quantum cascade laser (QCL) is 2.7 mm long and 70 μm wide. (B) Locking range extracted from the QCL injection locking at different RF injection power (black squares), following a 0.5-slope dependence in log–log scale (red line).

considering the cavity length, the light intensity ratio between the center of the bandwidth and the peripheral lasing modes and adding up the total losses from the waveguide (~7.3 cm^{-1}), the material (~3.0 cm^{-1}) [9] and the QCL facets (~1.6 cm^{-1}). The total GDD shows a visible oscillating behavior at the center of the laser gain bandwidth; in the operational range of the QCL, its value ranges between 3×10^5 fs^2 and 2.5×10^5 fs^2, reaching a maximum of 6.5×10^5 fs^2 at 3.2 THz and approaches zero over a range of about 200 GHz. On average, even in such a high-voltage regime, the GDD is equivalent to the values reported in heterogeneous FCs, just above threshold, i.e. in the regime in which the QCL clearly behaves as a comb [9, 26].

We then investigate the frequency locking characteristics of the laser structure at the onset of the NDR range. Injection locking maps are acquired by using an RF synthesizer (Rohde and Schwarz SMA100B) connected to the laser driver line, for the simultaneous supply of both the CW *dc* bias and the RF injection signal over a bias tee. The RF and *dc* biases are fed to the QCL through 18 GHz-cutoff sub-miniature type A (SMA) cables and connectors up to the cryostat input ports. The simultaneous monitoring of the injection signal and of the beatnote was performed positioning a RF antenna in close proximity (~3 cm) of the QCL and recording the antenna signal through the spectrum analyzer.

The injection-locking map, acquired with a RF injection power of 25 dBm while the QCL is driven at the onset of the NDR region (J = 635 A cm^{-2}), is reported in Figure 5A. Under these conditions, the emission spectrum of the QCL, under injection locking, comprises more than 36 optically active modes. The collected map discloses the typical injection-locking behavior, with the initial pulling and then locking of the beatnote to the injection signal, with the simultaneous appearance of multiple sidebands.

In Figure 5B, we report the locking range as a function of the injection power. The plot shows a good agreement with the 0.5-slope dependence in log–log scale foreseen in the Adler's equation [27] particularly in the lowest power region. We observed a maximum locking range of 1.2 MHz with an RF power of 25 dBm. The locking range values are slightly lower than that recently reported for homogeneus QCL FCs [19, 28], around the peak optical power. Conversely, the injection locking power is significantly higher than the typical values reported to date [19, 28, 29], while the actual power injected to the QCL is much lower. This can be understood by considering the total RF losses from cables, connectors, Au pads and bonding wires; by directly measuring the RF power transmitted to the laser chip, we estimate a total RF attenuation of ~35 dB. This latter value is extracted from a direct measurement of the RF power transmitted to the cold finger.

3 Conclusions

In conclusion, we demonstrate a THz QCL based on a homogenous active region design with an emission bandwidth of 600 GHz, a CW optical output power of 7 mW, operating as a FC synthesizer over the entire laser operation range, including the NDR regime. The electrical intermode beatnote map unveils stable comb operation over the entire dynamic range J_{max}/J_{th} = 1.53, with a single narrow beatnote reaching minimum linewidth = 500 Hz under free running operation. The laser shows more than 36 optically active, equally spaced, modes delivering 200 μW of optical power per comb tooth, the largest value reported to date in any THz QCL FC. We furthermore prove the RF injection locking capability of the devised laser bars in a regime conventionally characterized by electric field instabilities. The achieved results provide a concrete route for the development of miniaturized homogeneous chip-scale FC spectroscopic setups for addressing metrological-grade applications in the far-infrared, addressing absorption line strengths comparable or even stronger than fundamental, mid-infrared vibration transitions, but with much narrower Doppler-limited LWs, ruled by inverse linear relationship with the wavelength [3].

Acknowledgments: The authors acknowledge financial support from the ERC Project 681379 (SPRINT) and the EU union project MIR-BOSE (737017). The authors acknowledge useful discussions with Valentino Pistore.
Author contribution: All the authors have accepted responsibility for the entire content of this submitted manuscript and approved submission.
Research funding: The authors acknowledge financial support from the ERC Project 681379 (SPRINT) and the EU union project MIR-BOSE (737017).
Conflict of interest statement: The authors declare no conflicts of interest regarding this article.

References

[1] M. Tonouchi, "Cutting-edge terahertz technology," *Nat. Photonics*, vol. 1, pp. 97–105, 2007.
[2] S. Bartalini, L. Consolino, P. Cancio, et al., "Frequency-comb-assisted terahertz quantum cascade laser spectroscopy," *Phys. Rev. X*, vol. 4, p. 021006, 2014.
[3] L. Consolino, M. Nafa, M. De Regis, et al., "Quantum cascade laser based hybrid dual comb spectrometer," *Commun. Phys.*, vol. 3, p. 69, 2020.
[4] S. Koenig, D. Lopez-Diaz, J. Antes, et al., "Wireless sub-THz communication system with high data rate," *Nat. Photonics*, vol. 7, pp. 977–981, 2013.
[5] A. J. Seeds, H. Shams, M. J. Fice, and C. C. Renaud, "TeraHertz photonics for wireless communications," *J. Lightwave Technol.*, vol. 33, pp. 579–587, 2015.
[6] L. Consolino, F. Cappelli, M. Siciliani de Cumis, and P. De Natale, "QCL-based frequency metrology from the mid-infrared to the THz range: a review," *Nanophotonics*, vol. 8, pp. 181–204, 2018.
[7] A. Apostolakis and M. F. Pereira, "Superlattice nonlinearities for gigahertz–terahertz generation in harmonic multipliers," *Nanophotonics*, pp. 20200155-1–20200155-12, 2020. https://doi.org/10.1515/nanoph-2020-0155.
[8] D. Burghoff, T.-Y. Kao, N. Han, et al., "Terahertz laser frequency combs," *Nat. Photonics*, vol. 8, pp. 462–467, 2014.
[9] M. Rösch, G. Scalari, M. Beck, and J. Faist, "Octave-spanning semiconductor laser," *Nat. Photonics*, vol. 9, pp. 42–47, 2015.
[10] K. Garrasi, F. P. Mezzapesa, L. Salemi, et al., "High dynamic range, heterogeneous, terahertz quantum cascade lasers featuring thermally tunable frequency comb operation over a broad current range," *ACS Photonics*, vol. 6, pp. 73–78, 2019.
[11] Q. Lu, F. Wang, D. Wu, S. Slivken, and M. Razeghi, "Room temperature terahertz semiconductor frequency comb," *Nat. Commun.*, vol. 10, p. 2403, 2019.
[12] M. Rösch, M. Beck, M. J. Süess, et al., "Heterogeneous terahertz quantum cascade lasers exceeding 1.9 THz spectral bandwidth and featuring dual comb operation," *Nanophotonics*, vol. 7, pp. 237–242, 2018.
[13] M. S. Vitiello, G. Scalari, B. Williams, and P. De Natale, "Quantum cascade lasers: 20 years of challenges," *Opt. Express*, vol. 23, pp. 5167–5182, 2015.
[14] L. Li, L. Chen, J. Zhu, et al., "Terahertz quantum cascade lasers with >1 W output powers," *Electron. Lett.*, vol. 50, pp. 309–311, 2014.
[15] M. S. Vitiello, L. Consolino, S. Bartalini, et al., "Quantum-limited frequency fluctuations in a terahertz laser," *Nat. Photonics*, vol. 6, pp. 525–528, 2012.
[16] Q. Lu, D. Wu, S. Sengupta, S. Slivken, and M. Razeghi, "Room temperature continuous wave, monolithic tunable THz sources based on highly efficient mid-infrared quantum cascade lasers," *Sci. Rep.*, vol. 6, p. 23595, 2016.
[17] L. Li, K. Garrasi, I. Kundu, et al., "Broadband heterogeneous terahertz frequency quantum cascade laser," *Electron. Lett.*, vol. 54, pp. 1229–1231, 2018.
[18] D. Bachmann, M. Rösch, C. Deutsch, et al., "Spectral gain profile of a multi-stack terahertz quantum cascade laser," *Appl. Phys. Lett.*, vol. 105, p. 181118, 2014.
[19] A. Forrer, M. Franckié, D. Stark, et al., "Photon-driven broadband emission and frequency comb RF injection locking in THz quantum cascade lasers," *ACS Photonics*, vol. 7, pp. 784–791, 2020.
[20] Y. Yang, D. Burghoff, J. Reno, and Q. Hu, "Achieving comb formation over the entire lasing range of quantum cascade lasers," *Opt. Lett.*, vol. 42, pp. 3888–3891, 2017.
[21] F. P. Mezzapesa, V. Pistore, K. Garrasi, et al., "Tunable and compact dispersion compensation of broadband THz quantum cascade laser frequency combs," *Opt. Express*, vol. 27, pp. 20231–20240, 2019.
[22] S. Biasco, H. E. Beere, D. A. Ritchie, et al., "Frequency-tunable continuous-wave random lasers at terahertz frequencies," *Light Sci. Appl.*, vol. 8, p. 43, 2019.
[23] S. Biasco, A. Ciavatti, L. Li, et al., "Highly efficient surface-emitting semiconductor lasers exploiting quasi-crystalline distributed feedback photonic patterns," *Light Sci. Appl.*, vol. 9, p. 54, 2020.
[24] M. Amanti, G. Scalari, R. Terazzi, et al., "Bound-to-continuum terahertz quantum cascade laser with a single-quantum-well phonon extraction/injection stage," *New J. Phys.*, vol. 11, p. 125022, 2009.
[25] C. Derntl, G. Scalari, D. Bachmann et al., "Gain dynamics in a heterogeneous terahertz quantum cascade laser," *Appl. Phys. Lett.*, vol. 113, p. 181102, 2018.
[26] L. Consolino, M. Nafa, F. Cappelli et al., "Fully phase-stabilized quantum cascade laser frequency comb," *Nat. Commun.*, vol. 10, pp. 1–7, 2019.
[27] R. Adler, "A study of locking phenomena in oscillators," *Proc. IEEE*, vol. 61, pp. 1380–1385, 1973.
[28] J. Hillbrand, A. M. Andrews, H. Detz, G. Strasser, and B. Schwarz, "Coherent injection locking of quantum cascade laser frequency combs," *Nat. Photonics*, vol. 13, pp. 101–104, 2019.
[29] P. Gellie, S. Barbieri, J.-F. Lampin, et al., "Injection-locking of terahertz quantum cascade lasers up to 35 GHz using RF amplitude modulation," *Opt. Express*, vol. 18, pp. 20799–20816, 2010.

Miriam S. Vitiello*, Luigi Consolino, Massimo Inguscio and Paolo De Natale

Toward new frontiers for terahertz quantum cascade laser frequency combs

https://doi.org/10.1515/9783110710687-017

Abstract: Broadband, quantum-engineered, quantum cascade lasers (QCLs) are the most powerful chip-scale sources of optical frequency combs (FCs) across the mid-infrared and the terahertz (THz) frequency range. The inherently short intersubband upper state lifetime spontaneously allows mode proliferation, with large quantum efficiencies, as a result of the intracavity four-wave mixing. QCLs can be easily integrated with external elements or engineered for intracavity embedding of nonlinear optical components and can inherently operate as quantum detectors, providing an intriguing technological platform for on-chip quantum investigations at the nanoscale. The research field of THz FCs is extremely vibrant and promises major impacts in several application domains crossing dual-comb spectroscopy, hyperspectral imaging, time-domain nanoimaging, quantum science and technology, metrology and nonlinear optics in a miniaturized and compact architecture. Here, we discuss the fundamental physical properties and the technological performances of THz QCL FCs, highlighting the future perspectives of this frontier research field.

Keywords: frequency combs; quantum cascade lasers; terahertz.

1 Introduction

Frequency comb (FC) synthesizers [1] revolutionized the field of optical metrology and spectroscopy, emerging as one of the main photonic tools of the third millennium. The peculiar characteristic, which makes a FC a unique optical device, is the multifrequency coherent state of its emission, which is composed, in the frequency domain, by a series of evenly spaced optical modes, all locked in phase. They inherently generate a precise frequency ruler, capable of providing a direct link between optical and microwave/radio frequencies. Due to the tight phase relation among all optical modes, FC emission can be described by two key parameters: the carrier frequency offset and the spectral frequency spacing [2, 3].

Traditionally generated by frequency-stabilized and controlled femtosecond mode-locked lasers, FCs quickly conquered the visible and near-infrared (IR) domain, where different generation techniques were successfully demonstrated, tested and commercialized, such as Ti:Sa [4], fiber-based lasers [5], optical microresonators [6], upconversion of low frequency sources [7] or down-conversion of high-frequency combs [8]. High optical powers, outstanding long-term stability, broad spectral coverage and frequency tunability have been demonstrated so far [9]. Despite such impressive performances, most of these systems have a relatively large footprint, even in a fiber-based configuration.

In the mid-infrared domain, FCs are conventionally generated via difference frequency generation (DFG) [10–12] by continuous-wave (CW) or pulsed lasers and have been widely exploited for high-resolution sensing of molecular fingerprints. Mid-IR FCs also include optically pumped microresonators [13] or electrically pumped sources as interband cascade lasers (ICL–FCs) [14] and quantum cascade lasers (QCLs) (QCL–FCs) [15–18]. A common characteristic of these latter sources is the high third-order (Kerr) nonlinearity, which allows comb generation via degenerate and nondegenerate four-wave mixing (FWM) [15, 19] (Figure 1).

At THz frequencies, supercontinuum generation, recently recognized as FC emission [20], has been conventionally obtained by optical rectification of pulsed lasers. A few microwatts power output has been achieved in a configuration relying on bulky mode-locked lasers. This limits a broader use of these sources in application fields like

***Corresponding author: Miriam S. Vitiello** NEST, CNR - Istituto Nanoscienze and Scuola Normale Superiore, Piazza San Silvestro 12, 56127, Pisa, Italy, E-mail: miriam.vitiello@sns.it. https://orcid.org/0000-0002-4914-0421
Luigi Consolino and Paolo De Natale, CNR-Istituto Nazionale di Ottica and LENS, Via N. Carrara 1, 50019, Sesto Fiorentino, FI, Italy
Massimo Inguscio, CNR-Istituto Nazionale di Ottica and LENS, Via N. Carrara 1, 50019, Sesto Fiorentino, FI, Italy; and Department of Engineering, Campus Bio-Medico University of Rome, 00128 Rome, Italy

This article has previously been published in the journal Nanophotonics. Please cite as: M. S. Vitiello, L. Consolino, M. Inguscio and P. De Natale "Toward new frontiers for terahertz quantum cascade laser frequency combs" *Nanophotonics* 2021, 10. DOI: 10.1515/nanoph-2020-0429.

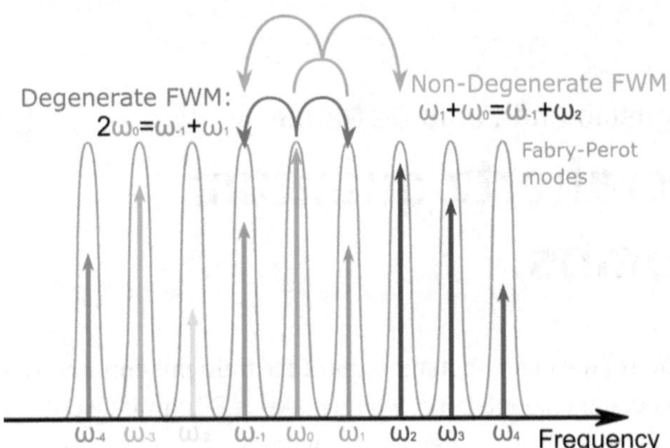

Figure 1: Schematic representation of frequency comb formation mechanisms through degenerate and nondegenerate four-wave mixing (FWM).

optical communications, in-situ molecular spectroscopy, gas sensing, homeland security, quantum optics, where chip-scale configurations are highly desirable.

The advent of miniaturized THz QCL frequency combs [21–23] revolutionized the field, opening new avenues in unexplored domains. In a THz QCL, FC operation occurs spontaneously owing to intracavity third-order nonlinear optical processes, stemming from a large nonlinear $\chi^{(3)}$ susceptibility (7×10^{16} $(m/V)^2$) [24] in the GaAs/AlGaAs active material. Specifically, the resonant third-order nonlinearity inherently induces self-phase locking by FWM. FWM tends to homogenize the mode spacing and, consequently, acts as the main mode proliferation and comb generation mechanism in a free running QCL.

In this opinionated article, we will review QCL FCs emitting at THz frequencies, discussing their performances, present techniques to assess their coherence properties, applications, and with a final outlook on future perspectives of this rapidly progressing research field.

2 Quantum cascade laser frequency combs at THz frequencies: architectures and performances

Electrically pumped QCLs, in either standard [21–23] or DFG configurations [25], are the most compact on-chip sources of FCs at THz frequencies. QCLs have indeed some peculiar features that make them unique for devising FCs: the broad optical bandwidth (OB) (up to an octave) [22], the inherently high optical power levels (hundred mW in CW, W- in pulsed mode) [26–28] and the high spectral purity (intrinsic linewidths of ~100 Hz) [29]. Such a combination of specs recently allowed designing THz QCL FCs covering an OB ≤ 1.2 THz [22, 23] delivering CW optical powers of 4–8 mW [23, 30] with a THz power/comb tooth in the 3 μW [31] to 6 μW [21] range in standard double-metal configurations and hundred nW in the DFG configuration [25].

THz QCL FCs can be quantum engineered with either homogeneous [21, 32] or heterogeneous designs [22, 23, 31]. The latter configuration relies on multistacked active regions (ARs) that ideally should display common threshold currents [23] and that are conventionally exploited to target a broad spectral coverage; conversely, homogenous ARs usually display much narrower gain profiles (≤600 GHz) [21, 32, 33], but are in principle easier to be injection locked in the laser operation regime in which dispersion is not compensated.

In a free-running multimode QCL, the modes are not uniformly spaced owing to chromatic dispersion. As result of the frequency-dependent refractive index, the laser free spectral range is index dependent, producing an unevenly spaced spectrum. The interplay between FWM and group velocity dispersion (GVD) in the laser cavity therefore has a major role in determining the QCL comb degradation mechanism. Figure 2a and b show the gain profile and the corresponding group delay dispersion (GDD) in the case of a hererogeneous (Figure 2a) and homogeneous (Figure 2b) THz QCL, respectively.

Gain medium engineering inherently leads to a flat top gain implemented only at the desired bias point, meaning that the nature of that gain media itself entangles the dispersion dynamics at other biases. As a result, FC operation usually occurs over less than 25% of the lasing bias range [21–23, 31].

Handling such a bias-dependent dispersion compensation, at small far-infrared photon energies (10 meV), is a very challenging task. Present attempts include passive or active schemes. In the first case, the laser bar can be shaped with an integrated dispersion compensator [21] providing a negative GVD to cancel the positive cavity dispersion but allows achieving FC operation over only 29% of the QCL operational regime. Alternative configurations include tunable Gires–Tournois interferometers (GTIs), comprising a movable gold mirror backcoupled with the QCL (Figure 3a) defining an external cavity, which leads to a further 15% increase of the regime where the THz QCL operates as a FC [34], as a result of a partial compensation of the total GDD.

Active approaches rely on the use of external cavities, as coupled *dc*-biased sections [35] or GTIs [36], allowing either a full coverage of the bias range, at the price of a

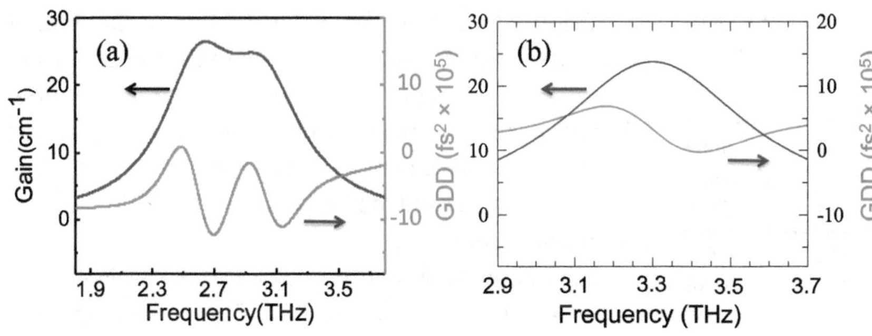

Figure 2: Gain of the QCL (black), plotted as a function of frequency, together with the corresponding group delay dispersion (red), in the case of a (a) heterogeneous (Reproduced from the study by Garrasi et al. [23]) or (b) homogeneous THz QCL. QCLs, quantum cascade lasers.

reduced OB (average OB = 400 GHz) [34] and negligible (μW) optical power outputs [35] or only at a specific operational point [36].

A summary of present-day performances of THz QCL FCs is shown in Table 1.

Extending FC operation in THz QCLs over most of the laser operational range and particularly at the operation point (J_{max}) of high power emission, with a broad (>1 THz) spectral coverage and without any spectral gap, is a very demanding task. Novel architectures exploiting more exotic device technologies and/or physical phenomena have been very recently proposed. One option is to use single-layer graphene grating–gated modulators acting as passive dispersion compensators. After proper integration with THz QCLs, the QCL emission can be amplitude modulated. This scheme enables FC operation over 35% of the laser operational range [33]. Another option relies on GTIs comprising an ultrafast (fs switching time) intersubband polaritonic mirror [37] backcoupled with a THz QCL. The THz electric field–induced bleaching of the intersubband polariton causes a reflectivity change in the polaritonic mirror. Once properly matched with the QCL gain bandwidth, compensation of GDD over a wide spectral portion of the QCL emission (35% of the laser operational range in the FC regime) is enabled. Finally, solution processed THz graphene saturable absorbers [38] can be embedded on the back facet of a THz QCL FC; the fast intraband-driven graphene saturable absorption can then reshape the intracavity mode dynamics leading to FC operation over 55% of the QCL driving regime [30].

3 Characterization techniques

The total complex field of a FC can be written as follows:

$$E = \sum_n E_n e^{i\left[2\pi\left(f_s n + f_o\right)t + \phi_n\right]}$$

where E_n is the amplitude of the nth comb mode, f_s is the mode spacing, f_o is the frequency offset and φ_n is the Fourier phase of the nth comb mode. Ultimately, the most important characteristic of a FC, which makes it different from a standard multimode laser, is the fixed (i.e., constant over time) phase relation between the emitted optical

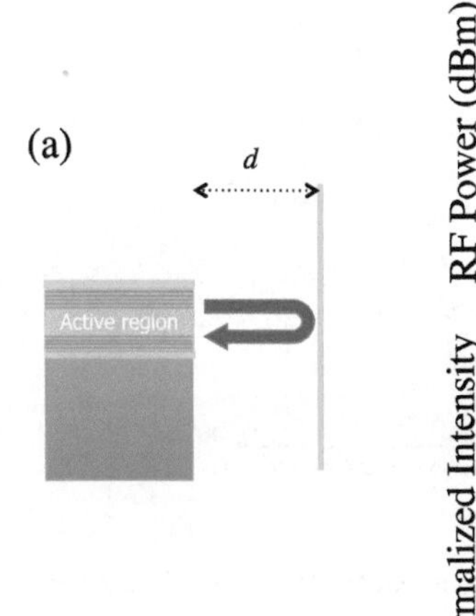

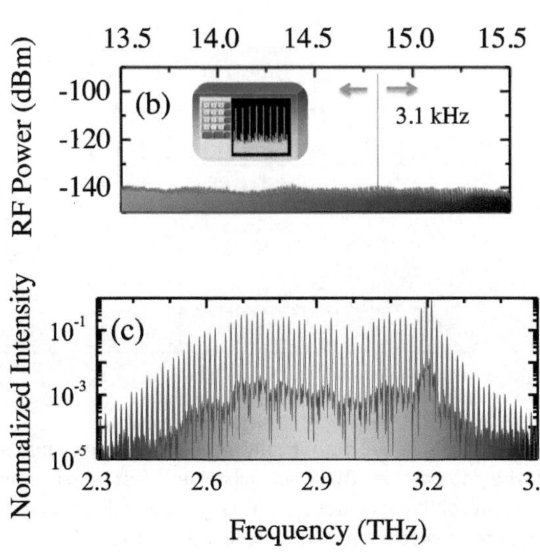

Figure 3: (a) Chip-scale Gires–Tournois interferometer (GTI): Part of the light emitted from the QCL back facet is reflected back into the waveguide mode. (b) Intermode beatnote frequency measured in a heterogenorus THz quantum cascade laser while driving it in continuous wave (CW) at a fixed current density $J = 0.67\,J_{max}$ and at a heat sink temperature $T_H = 18$ K. The IBN linewidth is 3.1 kHz. (c) Fourier-transform infrared spectra collected under vacuum in rapid-scan mode, with a 0.075 cm^{-1} resolution, while driving the QCL at 18 K in CW and at $J = 0.67\,J_{max.}$ QCL, quantum cascade laser; IBN, intermodal beatnote.

Table 1: Summary of the actual performances of the THz QCL comb in terms of power per mode, number of modes, single IBN regime and optical bandwidth in this latter regime.

References	Power per mode	# modes	Single IBN regime (% operational range)	OB of single IBN regime (THz)
Ref. [21]	60 uW	~62	29%	0.6
Ref. [22]	1.2 uW	~34	20%	0.58
Ref. [23]	15 uW	~50	16%	1.05
Ref. [31]	1.6 uW	~50	20%	1.1
Ref. [32]	(Not available)	~24	29%	1
Ref. [34]	20 uW	~52	23%	1.12
Ref. [35]	(Not available)	~22	100%	0.4
Ref. [30]	90 uW	~90	55%	1.2
Ref. [33]	42 uW	~98	35%	1

*: with active modulation. QCL, quantum cascade laser; OB, optical bandwidth; IBN, intermodal beatnote.

modes that translates into fixed Fourier phases for all the emitted modes. In a traditional passive mode-locked comb source, the presence of sharp pulses corresponds to a linear phase relation among the modes [39], but this specific trend is not guaranteed under different circumstances.

A preliminary analysis of FC operation in a THz QCL is usually done through a characterization of the driving current-dependent intermodal beatnote (IBN). The latter is easily extracted either by means of a bias tee in the driving current line or using a resonating antenna placed closed to the device. The presence of a single and sharp (linewidths in the kHz range) IBN may indicate that at least some of the modes emitted by the device are fairly well equally spaced in frequency, thus representing a good precursor for a genuine comb operation. However, such a technique lacks to provide information about the stability of the Fourier phases that can be extracted via alternative optical techniques. Figure 3b and c show the IBN and the Fourier infrared spectrum retrieved while driving a heterogeneous THz QCL at a current density value $J = 0.67\,J_{\max}$ being $J_{\max}$ the maximum current density value at which the QCL shows laser action.

One option to assess the phases is the shifted wave interference Fourier-transform (SWIFT) spectroscopy [40, 41] that gives access to the phase domain by measuring the phase difference between adjacent FC modes. Such a procedure allows retrieving the phase relation of continuous portions of the FC spectrum by a cumulative sum on the phases. In SWIFT spectroscopy, the need for a mechanical scan does not allow for a simultaneous analysis of the phases but, provided that all the measurement frequency chain is properly calibrated, this technique can assess if the phases are stable during the measurement time.

A more recent technique relies on extracting the modal phases of the THz QCL comb through a Fourier analysis of the comb emission (FACE) [39, 42]. Relying on a multi-heterodyne detection scheme (see Figure 4), this technique is capable of real-time tracing the phases of the modes emitted by the FC, thus working even if the comb presents spectral gaps. A phase-stable reference comb overlapping with the sample comb under analysis is required for FACE taking to a more complicated setup but providing a more general and accurate assessment of FC operation.

Phase measurements on THz QCLs to assess a genuine FC operation regime, performed with both SWIFT spectroscopy and FACE, proved that FWM establishes a tight phase relation among the modes emitted by the analyzed devices, resulting in stable Fourier modal phases [39–42].

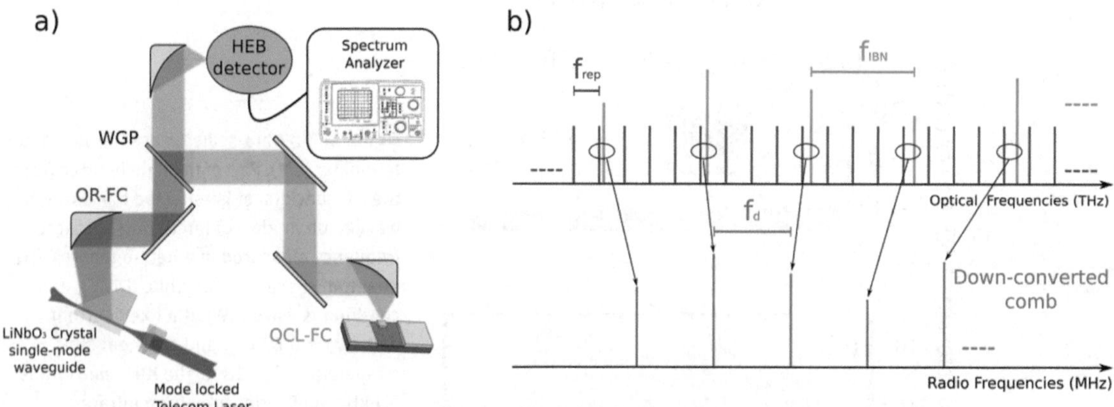

Figure 4: (a) Representation of the multiheterodyne detection scheme: the optically rectified (OR) and quantum cascade laser (QCL) frequency comb (FC) beams are superimposed by means of a wire grip polarizer (WGP) and mixed on a fast hot-electron bolometer (HEB) detector. (b) Sketch of the downconversion process: the two combs, with different modes spacing (f_{rep}: repetition rate of the OR FC and f_{IBN}, frequency of the QCL's intermodal beatnote), close to an integer ratio, are downconverted into a radio frequency comb.

The retrieved phase relations among the modes are never trivial (i.e., not linear, like in pulsed operation combs), leading to a frequency and amplitude modulated laser emission, different from that obtained by passive mode-locking pulsed operation. Nevertheless, these measurements confirm that FWM-based mode-locking is ensured at least in a limited operation range, usually very close to the laser threshold, and that QCL-based FCs are definitely well suited for most metrological-grade applications.

3.1 Applications and perspectives

The assessment of stable FC operation in a THz QCL enabled the quest for novel applications, aiming to exploit the large optical power per comb tooth (with respect to optically rectified [OR] THz combs) and intrinsic spectral purity of these devices.

One of the most straightforward applications of FCs is certainly dual-comb spectroscopy (DCS), which allows retrieving Fourier spectra without the need of moving interferometric components, therefore resulting in a fast and spectrally resolved acquisition procedure. However, while QCL–FC–based DCS setups have been successfully adopted in the mid-IR [43, 44], the THz region is lagging behind. After the first attempts of a proof-of-principle acquisition of an etalon signal simulating a molecular absorption [45], or of a low resolution spectrum of ammonia gas, dual-comb THz setups have been developed for pioneering hyperspectral imaging [46], self-detection schemes [47] and time resolved spectroscopy on molecular gas mixtures [48]. At the same time, a hybrid dual-comb spectrometer based on a OR– and a QCL–FC has been recently demonstrated [49], merging the sensitivity allowed by a 4 mW [23] THz QCL, with the accuracy of the absolute frequency scale, provided by the referenced OR–FC. Such a technique leads to a state-of-the-art accuracy, of the order of 10^{-8}, in the determination of molecular transition line centers.

However, it is worth mentioning that although, in the described experiments, THz QCLs are driven by low-noise current drivers and their operational temperature is stabilized, they are always free running. This means that their full metrological-grade potential, proven by the SWIFT and FACE measurements, has never been exploited, so far, for spectroscopy. Indeed, this can only be done by tight phase referencing both the QCL-FC modes spacing and the frequency offset to a precise frequency standard. These two degrees of freedom usually require two different orthogonal actuators to be stabilized, while, to date, the only available fast actuator, acting simultaneously on spacing and offset, is the QCL driving current. Recently, these difficulties have been overcome by exploiting the driving current actuator in two very different frequency ranges [42] (Figure 5). This technique allows full phase stabilization of a QCL–FC, the emission linewidth of each comb mode being narrowed down to ~2 Hz in 1 s, and a metrological-grade tuning of their individual modes frequencies.

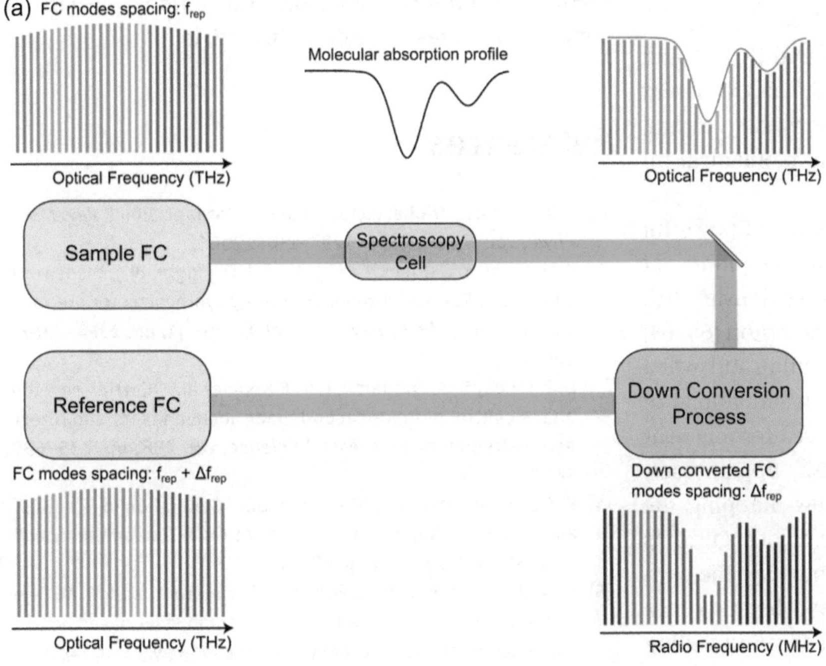

Figure 5: (a) Schematic representation of a generic dual-comb spectroscopy setup. A sample frequency comb (FC) with repetition rate f_{rep} interrogates a molecular sample inside a spectroscopy cell. The molecular information is encoded in the sample FC and is downconverted to radio frequencies, thanks to multiheterodyne mixing with a reference FC with slightly different repetition rate $f_{rep} + \Delta f_{rep}$.

Currently, the use of visible light as an additional actuator on QCL–FC degrees of freedom is under investigation and has already taken to demonstration of phase stabilization of the QCL comb mode spacing [50].

Full control of FCs, i.e., of linewidth and phase of teeth and spacing between them, as well as of their shape, that means getting control on the overall number and on the intensity pattern of the manifold [51] is the key challenge to widen the application spectrum of FCs. In particular, phase control and a significant increase of the intensity of each tooth (e.g., hundreds μW power range), together with the assessment of a stable FC operation over the entire operational range of the QCL, will make it easier to access the nonlinear interaction regime [52]. Saturation of gas-phase molecular lines is a straightforward example of application that could be accompanied by cavity-enhanced setups, although the cavity finesse reported to date, at THz frequencies, is very limited [53, 54]. Nonetheless, progress in this direction could take to THz development of much more sophisticated spectroscopic setups, similar to those working, till now, not beyond the mid-IR [55]. They could be used, e.g., to interrogate molecules prepared in unprecedented thermodynamic states, like ultracold molecules obtained by buffer-gas cooling schemes [56]. Such perspectives could fully exploit the parallelism between mid-IR, that is, the well-assessed "fingerprint region", and the THz frequency range that could have an even higher potential, due to the similar molecular line strengths of polar molecules but with up to two orders of magnitude narrower molecular linewidths, similarly improving line discrimination [57, 58].

Chip-scale THz QCL FCs can allow developing new instruments, as compact time-domain THz spectrometers, mimicking previous experiments in the mid-IR [59] and pioneering new application areas as multicolor coherent THz tomography, with major implications for cultural heritage, quality and process control and biomedical inspections [60].

Also, near-field microscopy could benefit of powerful QCL FCs for pioneering fundamental investigations at the nanoscale, as inspecting charge carrier density [61], plasmon–polariton [62, 63] and phonon–polariton [64, 65] modes with an unprecedented spatial resolution and over a broad bandwidth, overcoming the present limitations of time domain spectroscopy combined with scatterning near field optical microscopy (TDS s-SNOM) apparatuses. FC-based near-field microscopy can allow mapping the spatial variation and the bias dependence of local currents induced by light illumination, therefore tracking the photocarrier transport and the electronic band bending in a plethora of electronic and photonic nanodevices, unveiling fundamental physical phenomena in semiconductors, superconductors or more exotic bidimensional materials as graphene, phosphorene or transition metal dichalcogenides.

Quantum technologies represent another key area of development for THz QCL FCs. Indeed, QCLs are "quantum lasers" by design, with a relatively simple quantum well–based architecture, which is now the subject of analog simulation by Bose–Einstein condensate platforms to fully exploit their quantum properties that cannot be tailored by classical designs [66]. This could be a really disruptive area of application for THz FCs, including novel quantum sensors, quantum imaging devices and, in principle, q-bits made by entangled teeth for photonic-based quantum computation. Of course, most of these quantum applications need preliminary demonstration of subshot-noise operation of THz QCLs, hitherto impossible to achieve due to the technological gap of THz components, as compared to other spectral regions, as well as to the strong background thermal noise typical at these frequencies, generally requiring cryogenic detectors.

Acknowledgments: The authors acknowledge financial support from the ERC Project 681379 (SPRINT) and FET Flagship on Quantum Technologies Project 820419 (QOMBS).
Author contribution: All the authors have accepted responsibility for the entire content of this submitted manuscript and approved submission.
Research funding: The authors acknowledge financial support from the ERC Project 681379 (SPRINT) and FET Flagship on Quantum Technologies Project 820419 (QOMBS).
Conflict of interest statement: The authors declare no conflicts of interest regarding this article.

References

[1] T. W. Hänsch, "Nobel lecture: passion for precision," *Rev. Mod. Phys.*, vol. 78, no. 4, pp. 1297–1309, 2006.
[2] R. Holzwarth, T. Udem, T. W. Hänsch, J. C. Knight, W. J. Wadsworth, and P. S. J. Russell, "Optical frequency synthesizer for precision spectroscopy," *Phys. Rev. Lett.*, vol. 85, no. 11, pp. 2264–2267, 2000.
[3] D. J. Jones, S. A. Diddams, J. K. Ranka, et al., "Carrier-envelope phase control of femtosecond mode-locked lasers and direct optical frequency synthesis," *Science*, vol. 288, pp. 635–639, 2000.
[4] R. Ell, U. Morgner, F. X. Kärtner, et al., "Generation of 5-fs pulses and octave-spanning spectra directly from a Ti:sapphire laser," *Opt. Lett.*, vol. 26, no. 6, p. 373, 2001.
[5] I. Hartl, G. Imeshev, M. E. Fermann, C. Langrock, and M. M. Fejer, "Integrated self-referenced frequency-comb laser based on a combination of fiber and waveguide technology," *Opt. Express*, vol. 13, no. 17, p. 6490, 2005.

[6] I. S. Grudinin, A. B. Matsko, A. A. Savchenkov, D. Strekalov, V. S. Ilchenko, and L. Maleki, "Ultra high Q crystalline microcavities," *Opt. Commun.*, vol. 265, no. 1, pp. 33–38, 2006.
[7] K.-L. Yeh, M. C. Hoffmann, J. Hebling, and K. A. Nelson, "Generation of 10 mJ ultrashort terahertz pulses by optical rectification," *Appl. Phys. Lett.*, vol. 90, p. 171121, 2007.
[8] J. C. Pearson, B. J. Drouin, A. Maestrini, et al., "Demonstration of a room temperature 2.48–2.75 THz coherent spectroscopy source," *Rev. Sci. Instrum.*, vol. 82, p. 093105, 2011.
[9] T. Fortier and E. Baumann, "20 Years of developments in optical frequency comb technology and applications," *Commun. Phys.*, vol. 2, no. 1, p. 153, 2019.
[10] P. Maddaloni, P. Malara, G. Gagliardi, and P. De Natale, "Mid-infrared fiber-based optical comb," *New J. Phys.*, vol. 8, p. 262, 2006.
[11] I. Galli, F. Cappelli, P. Cancio, et al., "High-coherence mid-infrared frequency comb," *Opt. Express*, vol. 21, no. 23, p. 28877, 2013.
[12] A. Gambetta, N. Coluccelli, M. Cassinerio, et al., "Milliwatt-level frequency combs in the 8–14 μm range via difference frequency generation from an Er:fiber oscillator," *Opt. Lett.*, vol. 38, no. 7, p. 1155, 2013.
[13] T. J. Kippenberg, R. Holzwarth, and S. A. Diddams, "Microresonator-based optical frequency combs," *Science* (80-,) vol. 332, no. 6029, pp. 555–559, 2011.
[14] B. Schwarz, J. Hillbrand, M. Beiser, et al., "Monolithic frequency comb platform based on interband cascade lasers and detectors," *Optica*, vol. 6, no. 7, p. 890, 2019.
[15] A. Hugi, G. Villares, S. Blaser, H. C. Liu, and J. Faist, "Mid-infrared frequency comb based on a quantum cascade laser," *Nature*, vol. 492, no. 7428, pp. 229–233, 2012.
[16] C. Y. Wang, L. Kuznetsova, V. M. Gkortsas, et al., "Mode-locked pulses from mid-infrared quantum cascade lasers," *Opt. Express*, vol. 17, no. 15, p. 12929, 2009.
[17] A. K. Wójcik, P. Malara, R. Blanchard, T. S. Mansuripur, F. Capasso, and A. Belyanin, "Generation of picosecond pulses and frequency combs in actively mode locked external ring cavity quantum cascade lasers," *Appl. Phys. Lett.*, vol. 103, no. 23, p. 231102, 2013.
[18] M. Piccardo, B. Schwarz, D. Kazakov, et al., "Frequency combs induced by phase turbulence," *Nature*, vol. 582, no. 7812, pp. 360–364, 2020.
[19] J. Khurgin, Y. Dikmelik, A. Hugi, and J. Faist, "Coherent frequency combs produced by self frequency modulation in quantum cascade lasers," *Appl. Phys. Lett.*, vol. 104, p. 081118, 2014.
[20] S. Bartalini, L. Consolino, P. Cancio, et al., "Frequency-comb-assisted terahertz quantum cascade laser spectroscopy," *Phys. Rev. X*, vol. 4, p. 021006, 2014.
[21] D. Burghoff, T.-Y. Y. Kao, N. Han, et al., "Terahertz laser frequency combs," *Nat. Photonics*, vol. 8, no. 6, pp. 462–467, 2014.
[22] M. Rösch, G. Scalari, M. Beck, and J. Faist, "Octave-spanning semiconductor laser," *Nat. Photonics*, vol. 9, no. 1, pp. 42–47, 2014.
[23] K. Garrasi, F. P. Mezzapesa, L. Salemi, et al., "High dynamic range, heterogeneous, terahertz quantum cascade lasers featuring thermally tunable frequency comb operation over a broad current range," *ACS Photonics*, vol. 6, no. 1, pp. 73–78, 2019.
[24] P. Cavalié, J. Freeman, K. Maussang, et al., "High order sideband generation in terahertz quantum cascade lasers," *Appl. Phys. Lett.*, vol. 102, p. 221101, 2013.
[25] Q. Lu, F. Wang, D. Wu, S. Slivken, and M. Razeghi, "Room temperature terahertz semiconductor frequency comb," *Nat. Commun.*, vol. 10, p. 2403, 2019.
[26] L. H. Li, K. Garrasi, I. Kundu, et al., "Broadband heterogeneous terahertz frequency quantum cascade laser," *Electron. Lett.*, vol. 54, no. 21, pp. 1229–1231, 2018.
[27] M. Brandstetter, C. Deutsch, M. Krall, et al., "High power terahertz quantum cascade lasers with symmetric wafer bonded active regions," *Appl. Phys. Lett.*, vol. 103, no. 17, p. 171113, 2013.
[28] M. Wienold, B. Röben, L. Schrottke, et al., "High-temperature, continuous-wave operation of terahertz quantum-cascade lasers with metal-metal waveguides and third-order distributed feedback," *Opt. Express*, vol. 22, pp. 3334–3348, 2014.
[29] M. S. Vitiello, L. Consolino, S. Bartalini, et al., "Quantum limited fluctuations in a THz lasers," *Nat. Photonics*, vol. 6, pp. 525–528, 2012.
[30] F. P. Mezzapesa, K. Garrasi, J. Schmidt, et al., submitted (2020).
[31] M. Rösch, M. Beck, M. J. Süess, et al., "Heterogeneous terahertz quantum cascade lasers exceeding 1.9 THz spectral bandwidth and featuring dual comb operation," *Nanophotonics*, vol. 7, no. 1, pp. 237–242, 2018.
[32] A. Forrer, M. Franckié, D. Stark, et al., "Photon-driven broadband emission and frequency comb RF injection locking in THz quantum cascade lasers," *ACS Photonics*, vol. 7, no. 3, pp. 784–791, 2020.
[33] A. Di Gaspare, E. A. A. Pogna, L. Salemi et al., submitted (2020).
[34] F. P. Mezzapesa, V. Pistore, K. Garrasi, et al., "Tunable and compact dispersion compensation of broadband THz quantum cascade laser frequency combs," *Opt. Express*, vol. 27, no. 15, pp. 20231–20240, 2019.
[35] Y. Yang, D. Burghoff, J. Reno, and Q. Hu, "Achieving comb formation over the entire lasing range of quantum cascade lasers," *Opt. Lett.*, vol. 42, no. 19, pp. 3888–3891, 2017.
[36] F. Wang, H. Nong, T. Fobbe, et al., "Short terahertz pulse generation from a dispersion compensated modelocked semiconductor laser," *Laser Photonics Rev.*, vol. 11, no. 4, pp. 1–9, 2017.
[37] J. Raab, F. P. Mezzapesa, L. Viti, et al., "Ultrafast terahertz saturable absorbers using tailored intersubband polaritons," *Nat. Commun.*, vol. 11, pp. 1–8, 2020.
[38] V. Bianchi, T. Carey, L. Viti, et al., "Terahertz saturable absorbers from liquid phase exfoliation of graphite," *Nat. Commun.*, vol. 8, p. 15763, 2017.
[39] F. Cappelli, L. Consolino, G. Campo, et al., "Retrieval of phase relation and emission profile of quantum cascade laser frequency combs," *Nat. Photonics*, vol. 13, pp. 562–568, 2019.
[40] D. Burghoff, Y. Yang, D. J. Hayton, J.-R. Gao, J. L. Reno, and Q. Hu, "Evaluating the coherence and time-domain profile of quantum cascade laser frequency combs," *Opt. Express*, vol. 23, no. 2, p. 1190, 2015.
[41] Z. Han, D. Ren, and D. Burghoff, "Sensitivity of SWIFT spectroscopy," *Opt. Express*, vol. 28, no. 5, p. 6002, 2020.
[42] L. Consolino, M. Nafa, F. Cappelli, et al., "Fully phase-stabilized quantum cascade laser frequency comb," *Nat. Commun.*, vol. 10, no. 1, 2019. https://doi.org/10.1038/s41467-019-10913-7.
[43] G. Villares, A. Hugi, S. Blaser, and J. Faist, "Dual-comb spectroscopy based on quantum-cascade-laser frequency combs," *Nat. Commun.*, vol. 5, p. 5192, 2014.
[44] M. Gianella, A. Nataraj, B. Tuzson, et al., "High-resolution and gapless dual comb spectroscopy with current-tuned quantum cascade lasers," *Opt. Express*, vol. 28, no. 5, p. 6197, 2020.

[45] Y. Yang, D. Burghoff, D. J. Hayton, J.-R. Gao, J. L. Reno, and Q. Hu, "Terahertz multiheterodyne spectroscopy using laser frequency combs," *Optica*, vol. 3, no. 5, pp. 499–502, 2016.

[46] L. A. Sterczewski, J. Westberg, Y. Yang, et al., "Terahertz hyperspectral imaging with dual chip-scale combs," *Optica*, vol. 6, no. 6, p. 766, 2019.

[47] H. Li, Z. Li, W. Wan, et al., "Toward compact and real-time terahertz dual-comb spectroscopy employing a self-detection scheme," *ACS Photonics*, vol. 7, no. 1, pp. 49–56, 2020.

[48] L. A. Sterczewski, J. Westberg, Y. Yang, et al., "Terahertz spectroscopy of gas mixtures with dual quantum cascade laser frequency combs," *ACS Photonics*, vol. 7, no. 5, pp. 1082–1087, 2020.

[49] L. Consolino, M. Nafa, M. De Regis, et al., "Quantum cascade laser based hybrid dual comb spectrometer," *Commun. Phys.*, vol. 3, p. 69, 2020.

[50] L. Consolino, A. Campa, M. De Regis, et al., Controlling and Phase-Locking a THz Quantum Cascade Laser Frequency Comb by Small Optical Frequency Tuning, 2020. Arxiv arXiv: 2006.

[51] G. Campo, A. Leshem, F. Cappelli, et al., "Shaping the spectrum of a down-converted mid-infrared frequency comb," *J. Opt. Soc. Am. B*, vol. 34, no. 11, p. 2287, 2017.

[52] M. Wienold, T. Alam, L. Schrottke, H. T. Grahn, and H.-W. Hübers, "Doppler-free spectroscopy with a terahertz quantum-cascade laser," *Opt. Express*, vol. 26, pp. 6692–6699, 2018.

[53] A. Campa, L. Consolino, M. Ravaro, et al., "High-Q resonant cavities for terahertz quantum cascade lasers," *Opt. Express*, vol. 23, no. 3, p. 3751, 2015.

[54] L. Consolino, A. Campa, D. Mazzotti, M. S. Vitiello, P. De Natale, and S. Bartalini, "Bow-tie cavity for terahertz radiation," *Photonics*, vol. 6, no. 1, p. 1, 2018.

[55] G. Giusfredi, S. Bartalini, S. Borri, et al., "Saturated-absorption cavity ring-down spectroscopy," *Phys. Rev. Lett.*, vol. 104, no. 11, 2010. https://doi.org/10.1103/physrevlett.104.110801.

[56] V. Di Sarno, R. Aiello, M. De Rosa, et al., "Lamb-dip spectroscopy of buffer-gas-cooled molecules," *Optica*, vol. 6, no. 4, p. 436, 2019.

[57] L. Consolino, F. Cappelli, M. Siciliani de Cumis, and P. De Natale, "QCL-based frequency metrology from the mid-infrared to the THz range: a review," *Nanophotonics*, vol. 8, pp. 181–204, 2018.

[58] L. Consolino, S. Bartalini, P. De Natale, J. Infrared, and Millimeter, "Terahertz frequency metrology for spectroscopic applications: a review," *Terahertz Waves*, vol. 38, no. 11, pp. 1289–1315, 2017.

[59] F. Keilmann, C. Gohle, and R. Holzwarth, "Time-domain mid-infrared frequency- comb spectrometer," *Opt. Lett.*, vol. 29, pp. 1542–1544, 2004.

[60] A. W. M. Lee, T.-Y. Kao, D. Burghoff, Q. Hu, and J. L. Reno, "Terahertz tomography using quantum-cascade lasers," *Opt. Lett.*, vol. 37, pp. 217–219, 2012.

[61] A. J. Huber, F. Keilmann, J. Wittborn, J. Aizpurua, and R. Hillenbrand, "Terahertz near-field nanoscopy of mobile carriers in single semiconductor nanodevices," *Nano Lett.*, vol. 8, pp. 3766–3770, 2008.

[62] J. Chen, M. Badioli, P. Alonso-González, et al., "Optical nano-imaging of gate-tunable graphene plasmons," *Nature*, vol. 487, pp. 77–81, 2012.

[63] Z. Fei, A. S. Rodin, G. O. Andreev, et al., "Gate-tuning of graphene plasmons revealed by infrared nano-imaging," *Nature*, vol. 486, pp. 82–85, 2012.

[64] P. Li, M. Lewin, A. V. Kretinin, et al., "Hyperbolic phonon-polaritons in boron nitride for near-field optical imaging and focusing," *Nat. Commun.*, vol. 6, p. 7507, 2015.

[65] T. Low, A. Chaves, J. D. Caldwell, et al., "Polaritons in layered two-dimensional materials," *Nat. Mater.*, vol. 16, p. 182, 2017.

[66] F. Scazza, et al., submitted.

Franco Prati*, Massimo Brambilla, Marco Piccardo, Lorenzo Luigi Columbo, Carlo Silvestri, Mariangela Gioannini, Alessandra Gatti, Luigi A. Lugiato and Federico Capasso

Soliton dynamics of ring quantum cascade lasers with injected signal

https://doi.org/10.1515/9783110710687-018

Abstract: Nonlinear interactions in many physical systems lead to symmetry breaking phenomena in which an initial spatially homogeneous stationary solution becomes modulated. Modulation instabilities have been widely studied since the 1960s in different branches of nonlinear physics. In optics, they may result in the formation of optical solitons, localized structures that maintain their shape as they propagate, which have been investigated in systems ranging from optical fibres to passive microresonators. Recently, a generalized version of the Lugiato–Lefever equation predicted their existence in ring quantum cascade lasers with an external driving field, a configuration that enables the bistability mechanism at the basis of the formation of optical solitons. Here, we consider this driven emitter and extensively study the structures emerging therein. The most promising regimes for localized structure formation are assessed by means of a linear stability analysis of the homogeneous stationary solution (or continuous-wave solution). In particular, we show the existence of phase solitons – chiral structures excited by phase jumps in the cavity – and cavity solitons. The latter can be deterministically excited by means of writing pulses and manipulated by the application of intensity gradients, making them promising as frequency combs (in the spectral domain) or reconfigurable bit sequences that can encode information inside the ring cavity.

Keywords: frequency combs; quantum cascade lasers; solitons.

1 Introduction

Optical frequency combs [1, 2] have revolutionized the field of optics and optoelectronics, both from the fundamental and from the application standpoint. The realization, nearly 13 years ago, of frequency combs in high-Q monolithic microresonators filled with Kerr media [3] raised an enormous attention because of its potential for miniaturization and chip-scale photonic integration, and stimulated a great deal of activities [4–6]. More recently, another system relevant for photonic integration, namely the quantum cascade laser (QCL) emerged as a source of frequency combs [7–12]. Close to its lasing threshold, the dynamics of a QCL is governed by a cubic nonlinearity similar to the one of Kerr cavities. On the other hand, unlike Kerr cavities which are passive, in the QCL the medium is active, i.e. with a population inversion.

Frequency combs in active and passive systems have been so far studied in distinct frameworks, and only recently a definite connection emerged [13, 14]. In particular, in the study by Columbo et al. [14], the treatment of frequency combs in passive and active systems was unified by formulating a generalized version of the Lugiato–Lefever equation (LLE) [15, 16]. For passive systems, this generalized model reduces to the LLE of Lugiato and Lefever [15], largely used to describe frequency combs in Kerr ring microresonators [4]. In the active case, under the approximations of fast material dynamics and near-threshold

***Corresponding author: Franco Prati,** Dipartimento di Scienza e Alta Tecnologia, Università dell'Insubria, Via Valleggio 11, 22100 Como, Italy, E-mail: franco.prati@uninsubria.it.
Massimo Brambilla, Dipartimento di Fisica Interateneo and CNR-IFN, Università e Politecnico di Bari, Via Amendola 173, 70123 Bari, Italy
Marco Piccardo, Center for Nano Science and Technology, Fondazione Istituto Italiano di Tecnologia, Via Giovanni Pascoli 70, 20133 Milano, Italy; Harvard John A. Paulson School of Engineering and Applied Sciences, Harvard University, Cambridge, MA, USA
Lorenzo Luigi Columbo, Carlo Silvestri and Mariangela Gioannini, Dipartimento di Elettronica e Telecomunicazioni, Politecnico di Torino, Corso Duca degli Abruzzi 24, 10129 Torino, Italy. https://orcid.org/0000-0002-6566-9763 (L.L. Columbo)
Alessandra Gatti, Dipartimento di Scienza e Alta Tecnologia, Università dell'Insubria, Via Valleggio 11, 22100 Como, Italy; Istituto di Fotonica e Nanotecnologie, IFN-CNR Piazza Leonardo da Vinci 32, Milano, Italy
Luigi A. Lugiato, Dipartimento di Scienza e Alta Tecnologia, Università dell'Insubria, Via Valleggio 11, 22100 Como, Italy
Federico Capasso, Harvard John A. Paulson School of Engineering and Applied Sciences, Harvard University, Cambridge, MA, USA

This article has previously been published in the journal Nanophotonics. Please cite as: F. Prati, M. Brambilla, M. Piccardo, L. L. Columbo, C. Silvestri, M. Gioannini, A. Gatti, L. A. Lugiato and F. Capasso "Soliton dynamics of ring quantum cascade lasers with injected signal" *Nanophotonics* 2021, 10. DOI: 10.1515/nanoph-2020-0409.

operation, it becomes the complex Ginzburg–Landau equation (CGLE) derived in the study by Piccardo et al. [13] for a free-running ring QCL, on the one hand, and the cubic equation formulated in the study by Lugiato et al. [16] for a two-level ring laser, on the other.

A relevant point is that while in the passive case frequency combs need an external field providing the necessary energy intake, in active systems like the QCL frequency combs are generated in a free-running setup [7, 9–13]. The generalized LLE also contemplates a novel configuration, to our knowledge never analysed before, in which a ring QCL is driven by an external coherent field (QCL with injected signal). This setup is of special interest because it introduces two new control parameters, namely the intensity and the frequency of the injected signal, which creates conditions favourable for the generation and control of temporal solitons [17], relevant for new applications in integrated comb technology such as metrology and spectroscopy. Indeed a first analysis of this model [14] demonstrated the emergence of temporal solitons in ring QCL with injected signal, and the possibility of addressing them by means of external pulses.

In the present work, we perform a more extensive analysis, which provides novel insights in the rich dynamical scenery of the injected ring QCL. The linear stability analysis of the S-shaped homogeneous stationary solution (HSS) allows establishing the conditions for the stability of the lower and upper branch by associating the bifurcation point of the lower branch with a plane wave Hopf instability and that of the upper branch with a modulational instability. We show that when the upper branch is stable and the lower branch is unstable, the system supports localized structures characterized by phase jumps equal to (a multiple of) 2π, named phase solitons (PSs), whereas in the opposite situation, when the upper branch is unstable and the lower branch is stable, the localized structures are bright solitons on a homogeneous background, similar to those observed in Kerr microresonators [18]. Focussing on cavity solitons (CSs), we study the basic properties, fundamental for their exploitation as elements for optical information encoding. Namely, we show independent switch-on of CSs, by means of suitable address pulses superimposed to the constant driving field. We determine the optimal amplitude and duration of such pulses for the creation of a single soliton and the minimum distance at which pairs of solitons can be created independently. The possibility of controlling the solitons by means of appropriate gradients in the driving field is verified, thus assessing the CSs as plastic information units which can be deterministically drifted/relocated across the cavity field profile.

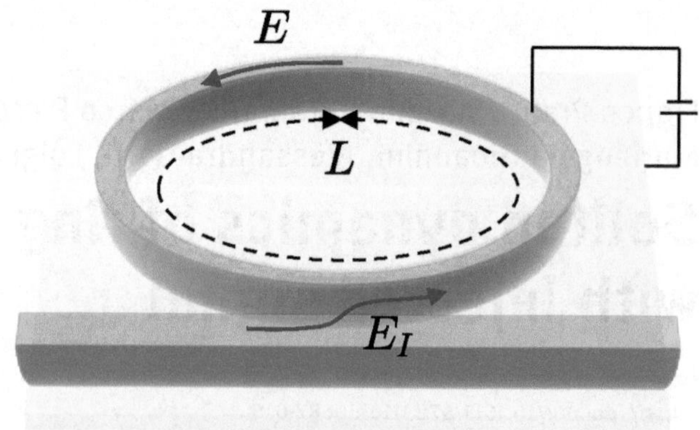

Figure 1: Schematic of a ring quantum cascade laser (QCL) under electrical bias with an injected optical signal. The ring cavity has a length L and the intracavity field is E. The external field E_I is injected into the QCL from a straight waveguide.

2 The model

A possible realization of the system analysed in this work is schematically shown in Figure 1: we consider a ring QCL in a ridge-waveguide geometry, similar to the one used in the study by Piccardo et al. [13], but coupled to a straight waveguide by which a coherent field can be injected into the ring with round-trip cavity length L.

We showed in the Supplementary material of the study by Columbo et al. [14] that a driven QCL can be suitably described by the generalized LLE

$$\tau_p \frac{\partial E}{\partial t} = E_I - (1 + i\theta_0)E + (1 - i\Delta)\left(\mu - |E|^2\right)E + (d_R + id_I)\frac{\partial^2 E}{\partial z^2}, \tag{1}$$

where t is the time variable and z the longitudinal coordinate along the ring cavity in a reference frame moving at the phase velocity $\tilde{c} = c/n_h$, n_h being the refractive index of the host material. E and E_I are the envelopes of the intracavity electric field and of the external field injected into the cavity, respectively, normalized as in the study by Columbo et al. [14]. When referred to a QCL close to its lasing threshold, the parameters appearing in the equation are as follows:

- τ_p is the damping time of the cavity field (typical values are some tens of ps);
- μ is the pump parameter, such that the laser threshold is at $\mu = \mu_{thr} = 1$;
- $\Delta = \alpha + \beta$, where α is the so-called linewidth enhancement factor (LEF) [19] and β is the Kerr nonlinear coefficient of the host medium, which in a QCL is normally small;
- $\theta_0 = (\omega_c - \omega_0)\tau_p - \mu\beta$, is a detuning parameter, where ω_c is the cavity frequency closest to the reference

frequency ω_0 of the injected field. The term $\mu\beta$ arises from the specific form chosen for the generalized LLE in [14], but it does not contribute to the equation because it cancels out with another identical term;
- $d_R = (\tilde{c}\tau_d)^2/(1+\alpha^2)$ represents a diffusion term, where τ_d is the dipole dephasing time. $\sqrt{d_R}$ has the dimension of a length, and defines the spatial length scale. By using typical values for a QCL [13] ($\tau_d = 60$ fs, $n_h = 3.3$, $\alpha \simeq 1.9$), we get $\sqrt{d_R} \simeq 2.5\,\mu$m.
- $d_I = d_R(\alpha + \zeta)$ represents a second order dispersion term, where $\zeta = -(1+\alpha^2)\tilde{c}\tau_p k''/(2\tau_d^2)$, and k'' is the group velocity dispersion (GVD) coefficient of the host medium. Notice that $\zeta > 0$ in the case of anomalous dispersion. Assuming $\tau_p = 50$ ps and $k'' = -300$ fs^{2}mm^{-1} [14] we have $\zeta \simeq 0.9$. Notice however that the GVD can be engineered to take much smaller values, such that $\zeta \simeq 0$.

We note that in the study by Columbo et al. [14], the third term at the r.h.s. of the equation was written as $\mu(1-i\Delta)(1-|E|^2)E$ rather than $(1-i\Delta)(\mu-|E|^2)E$, but the two forms are equivalent in the limit of a laser very close to threshold where $\mu \approx 1$ and $|E|^2 \ll 1$ is on the same order as $|\mu-1|$.

Next, we introduce a small (but not necessarily infinitesimal) parameter $r = \mu - \mu_{thr}$ measuring the distance from the laser threshold. To have finite quantities appearing in the equation, and to minimize the number of parameters, we introduce the following scaling:

$$\tau = t|r|/\tau_p, \quad \eta = z\sqrt{|r|/d_R}, \quad F = E/\sqrt{|r|}, \quad F_I = E_I/|r|^{3/2}, \tag{2}$$

Eq. (1) then takes the form of a forced complex Ginzburg–Landau equation [20]

$$\frac{\partial F(\tau,\eta)}{\partial\tau} = F_I + y(1-i\theta)F - (1-i\Delta)|F|^2F + (1+iG)\frac{\partial^2 F}{\partial\eta^2}, \tag{3}$$

where

$$y = r/|r|, \quad \theta = \left[(\omega_c-\omega_0)\tau_p + \alpha\right]/r + \alpha, \quad G = d_I/d_R = \alpha + \zeta. \tag{4}$$

In this work, we shall focus on the above threshold case, and fix $y = 1$.

An important point to remark is that, because F_I and F need to be on the same order of magnitude, a consequence of the scaling (2) is that E_I is smaller than E by a factor $|r| \ll 1$; this implies that the system can be operated with an injected field of small intensity, which is particularly convenient when the laser cavity is ring-shaped. A second remark concerns the relevant temporal and spatial scales of variation of the intracavity field. These are established by $\tau_p/|r|$ (time) and $\sqrt{d_R/|r|}$ (longitudinal coordinate) and depend on the distance from threshold, getting larger and larger as threshold is approached, as typical of phase transitions.

Besides the ring geometry considered here, a connection between QCLs and the LLE was also established recently in the case of Fabry–Perot devices [21].

3 Homogeneous stationary solution and its stability

As typical of forced cubic equations, the HSS of Eq. (3) may show a bistable behaviour. This in turn creates conditions favourable to the emergence of localized structures, as e.g. a modulation instability appearing in the upper branch in the presence of a stable lower branch [22]. The choice of parameters for the simulations presented in Section 4 will thus be guided by the results of the stability analysis presented in this section.

By introducing the quantities $Y = F_I^2$ and $X = |F|^2$, proportional to the input and output intensity, respectively, the homogeneous and stationary solution of Eq. (3) has the form

$$Y = X\left[(1-X)^2 + (\theta - \Delta X)^2\right]. \tag{5}$$

When plotting X versus Y, the curve is S-shaped provided that

$$1 + \Delta\theta > \sqrt{3}|\Delta - \theta|. \tag{6}$$

An example of such stationary curve is shown in Figure 2 (see also Figure 5). The blue symbols in this figure are

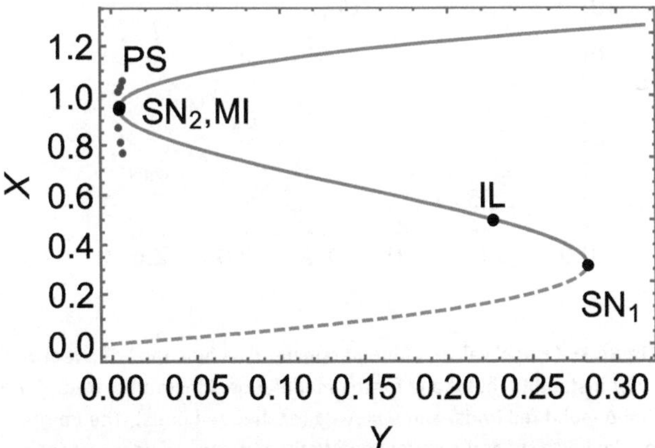

Figure 2: Stationary homogeneous solution of Eq. (3), where the solid and dashed blue lines denote stable and unstable configurations. The blue symbols correspond to a phase soliton branch from numerical simulations of Eq. (3) with $\eta_{max} = 200$ (the symbols indicate the maximum and minimum intensity). $\Delta = G = 1.1$, $\theta = 1$ (point **a** of Figure 3).

related to the PSs that will be discussed in the following Section 4. SN_1 and SN_2 denote the lower and upper turning points of the S-shaped curve (the origin of the naming is that in the single-mode limit these points are related to a saddle-node bifurcation). The grey portion of the curve between the points, having a negative slope, is not accessible. If $\theta = \Delta$ the upper turning point SN_2 touches the X axis at $X = 1$.

3.1 Hopf instability

In Figure 2, IL denotes the so-called injection locking point, below which a Hopf instability takes place. Irrespective of the other parameters, the solution (5) is temporally unstable for $X < X_{IL} = 0.5$. The presence of such an injection locking point is characteristic of a laser above threshold when a slightly detuned external field is injected. Then, only for a sufficiently high injection amplitude the laser locks to the external field [23]. In class A lasers, as the QCL, the injection locking occurs at an output intensity $|E|^2 = \sqrt{\mu_{\text{thr}}}(\sqrt{\mu} - \sqrt{\mu_{\text{thr}}})$ [see Eq. (25.3) of [24]]. For our scaled quantities $\mu_{\text{thr}} = 1$, $\mu = 1 + r$, and $X = |E|^2/r$ (see Eq. (2)), so that $X_{IL} = (\sqrt{1+r} - 1)/r \to 0.5$ in the limit $r \ll 1$.

In the bistable case, it is important to determine the position of the injection point with respect to the turning points, in order to assess the existence of a stable portion of the lower branch of the stationary curve. To this end, there exist two critical values of the parameter θ, that we denote by $\theta_{IL1}(\Delta)$ and $\theta_{IL2}(\Delta)$, given by Eq. (S6) and Eq. (S7) of the Supplementary material, respectively, and plotted by the blue curves in Figure 3, such that:

- For $\theta < \theta_{IL1}(\Delta)$ the injection locking point is always above the lower turning point SN_1. In this case, the whole lower branch of the S-shaped stationary curve is unstable, as in the example of Figure 2.
- For $\theta_{IL1}(\Delta) < \theta < \theta_{IL2}(\Delta)$, the injection locking point is below SN_1, but still in the bistable part of the curve, so that the lower branch of the curve has an unstable and a stable portion between the two turning points, as in Figure S2.
- For $\theta > \theta_{IL2}(\Delta)$, the injection point is at the left of the upper turning point SN_2, and the Hopf instability does not affect the bistability region. In this case, the whole lower branch between the turning points is stable.

In the example of Figure 5, the lower branch of the curve is almost entirely stable, because θ is very close to θ_{IL2}.

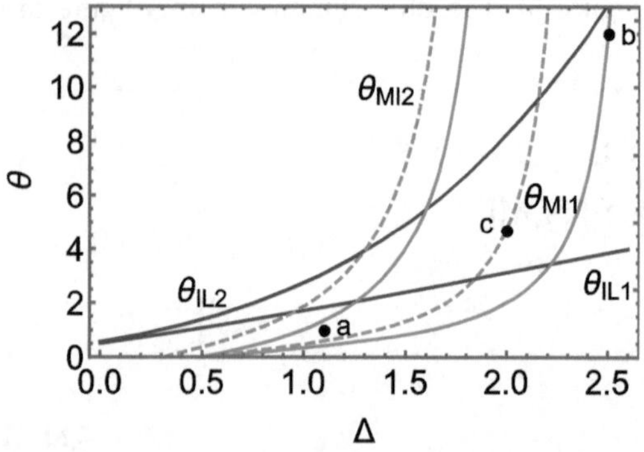

Figure 3: Critical values of the parameter θ as functions of Δ for the Hopf instability (blue) and for the modulational instability (red), for $G = \Delta$ (solid red lines) and $G = \Delta + 1$ (dashed red lines). The points marked as **a**, **b**, and **c** correspond to the parameters used in the numerical simulations, with $G = \Delta$ for **a** and **b**, and $G = \Delta + 1$ for **c**. In the region below the blue curves (point **a**), the lower branch is unstable because it is below the injection locking point. In the region between them (points **b** and **c**), the lower branch of the S has a stable portion. In the region between the red curves, the bifurcation point MI of the modulational instability is placed in the upper branch between the two turning points.

3.2 Modulational instability

The upper branch of the stationary curve is affected by a modulational instability from the left turning point SN_2 up to the bifurcation point MI. Again, it is important to determine the position of the MI point with respect to the turning points.

The position of the MI point on the stationary curve now depends on the whole triplet of parameters Δ, G and θ. As for the Hopf instability, we can introduce two critical values of θ, that we denote by $\theta_{MI1}(\Delta, G)$ and $\theta_{MI2}(\Delta, G)$. Their explicit expressions are given by Eq. (S10) and Eq. (S8), respectively, whereas examples are provided by the red curves in Figure 3.

- For $\theta < \theta_{MI1}(\Delta, G)$, the whole upper branch between the two turning points is modulationally unstable.
- For $\theta_{MI1}(\Delta, G) < \theta < \theta_{MI2}(\Delta, G)$, the MI point is in the upper branch of the curve between the turning points, so that only a portion of the upper branch is modulationally unstable.
- For $\theta = \theta_{MI2}(\Delta, G)$, the modulational instability disappears, while it appears again at higher values of θ.

Figure 3 shows the critical values of θ as functions of Δ both for the Hopf instability (blue lines) and for the MI (red lines). In the figure are also shown the points which will be numerically studied in the rest of the article. Point **a** provides the ideal conditions for the emergence of the PSs that will be discussed in Section 4.1; point **b** has instead the proper conditions for the emergence of the CSs with stable background that will be analysed in Section 4.2, whereas for the parameters of point **c**, the background of the CSs can be stable or unstable depending on the amplitude of the driving field. That case has been already presented in the study by Columbo et al. [14]; here, we provide further details and results in Section 5 and in the Supplementary material.

4 Optical solitons

In this section, we present results of numerical simulations of Eq. (3). The scaled cavity coordinate runs from 0 to $\eta_{\max} = L\sqrt{|r|/d_R}$, where L is the real cavity length. In the figures, the cavity coordinate is shown as $\eta/\eta_{\max} = z/L$. The temporal coordinate is shown as the scaled time $\tau = tr/\tau_p$ (Eq. (2)). The connection with physical quantities, as the cavity roundtrip time, then depends on the distance from threshold. By assuming, e.g. a loss coefficient of the cavity $\simeq 0.7$ (including both the external coupling and distributed losses) [14], the photon lifetime is $\tau_p \simeq L/(0.7\tilde{c})$, so that one scaled time unit corresponds to $\sim 1.43/r$ roundtrips, that is 14.3 roundtrips for e.g. a laser 10% above threshold.

4.1 Phase solitons

In the first simulation, we consider point **a** of Figure 3, by setting $\Delta = G = 1.1$, a value which is realistic for a QCL (LEF = 1.1 and negligible GVD), and $\theta = 1$. This value of θ is smaller than θ_{IL1}, so that the lower branch is entirely unstable, whereas it is close to θ_{MI2}, and the point MI is very close to SN_2, so that the upper branch is almost entirely stable. For input intensities close enough to SN_2, these conditions are favourable for observing excitable pulses in the single mode limit, and localized structures associated with a phase kink in the multimode regime [25, 26]. The latter can be excited taking as initial condition the stable state of the upper branch to which a phase kink is superimposed, i.e. a phase profile along z with a sharp jump equal to $2l\pi$, where l is an integer number. In this way, the boundary condition on the phase imposed by the injected field is still obeyed but, as light propagates along the cavity, the phasor associated with the complex electric field rotates l times around the origin. Because the dynamics of the phase is coupled to that of the amplitude, the length of this phasor varies accordingly, letting the field intensity profile to exhibit a local modulation in correspondence with the phase jump. This kind of structure is called phase soliton (PS) because its dynamics is dominated by the phase, and it is chiral in nature, having positive or negative chiral charge depending on the sign of the integer l. Here we consider only the PS with chiral charge equal to 1.

With the parameters of Figure 2, PSs are stable in the interval $0.0043 \le Y \le 0.0058$. Figure 4(a,b) illustrates the PS with $Y = 0.005$. The upper plot in panel (a) shows the space-time evolution of the emitted intensity. Note that the PS trace is slanted to the left, which means that the PS travels along the cavity at a speed slightly smaller than $\tilde{c}$. The PS accumulates a delay of one roundtrip time τ_r in about 600 time units. Assuming as aforementioned that 1 unit in τ corresponds to $1.43/r$ roundtrip times, the speed of the PS is $V_{PS} = \tilde{c}/[1 + r/(1.43 \times 600)] \simeq \tilde{c}\,(1 - 1.17 \times 10^{-4})$ for a laser 10% above threshold. The bottom plot shows the field optical spectrum at the end of simulation, which for a PS is an asymmetric frequency comb. The first two plots from the top in panel b show the intensity and phase profiles along the cavity at the last round trip. The phase displays a negative jump of 2π in correspondence with the PS. A modulation of the intensity, consisting in a maximum followed by a minimum, is associated with the phase jump. The bottom graph in panel b shows the trajectory described by the tip of the electric field phasor in the complex plane. The trajectory is close to a circle, drawn in green in the bottom panel of Figure 4(b), indicating that the dynamics is an almost pure phase dynamics, but nevertheless the length of the phasor is not constant and this causes the modulation of the intensity.

In this figure, we show the PS with positive chiral charge. The PS with negative chiral charge is also stable. For that PS, the trace in the space-time plot is slanted to the right (the PS travels faster than $\tilde{c}$), in the intensity profile the minimum precedes the maximum, the phase jump is positive, the spectrum is identical to the previous one but with $n \to -n$, and the electric field vector rotates in the complex plane in the opposite direction.

PSs in a forced CGLE, whose existence is made possible by the presence of an unstable focus close to the origin of the complex plane, were predicted in the study by Chaté, Pikovsky and Rudzick [25] and observed in the study by Gustave et al. [26] in a driven semiconductor laser. In the study by Gustave et al. [26], however, the active medium was an interband semiconductor with a slow recovery time of gain (~1 ns) and just one sign of the chirality [27, 28] was observed because of the inertia of the medium. This feature cannot be captured by the forced CGLE alone, which in fact was coupled with a dynamical equation for the gain in previous studies [26–28]. In the aforementioned simulations, PSs are found stable with both signs of the chirality in a driven QCL, based on our model where, although, the adiabatic elimination of all material variables is a

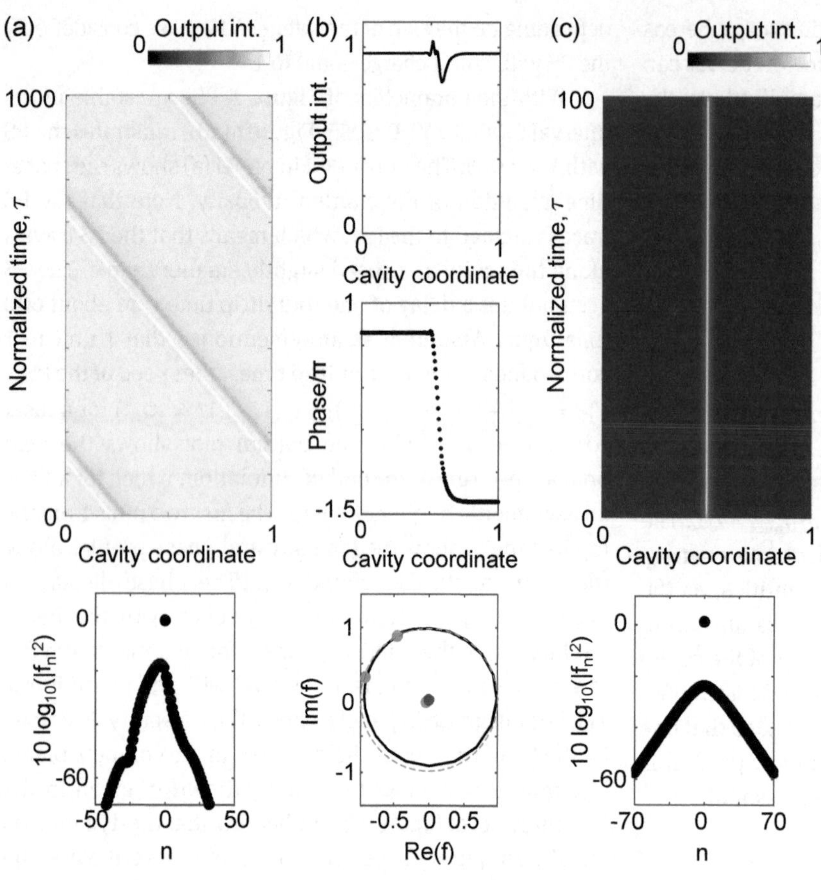

Figure 4: (a) Phase soliton for the same parameters of Figure 2 and $Y = 0.005$. The upper plot is the space-time evolution of the intensity, whereas the bottom plot is the optical spectrum at the last roundtrip, n being the index of the empty cavity mode of complex amplitude f_n. (b) From top to bottom: field intensity along the cavity at the end of simulation; corresponding phase profile; trajectory described by the tip of the electric field phasor in the complex plane, where the red symbols are the three stationary states corresponding to $Y = 0.005$, whereas the blue one shows the origin. (c) Cavity soliton for the same parameters of Figure 5 and $Y = 100$. The upper and bottom panels show the space-time evolution and the optical spectrum at the last roundtrip, respectively.

background assumption. To substantiate a prediction for stable solitons of either charge in such a device, a more complete analysis based on a full set of effective semiconductor Maxwell–Bloch equations [14] with the appropriate medium temporal timescales will be needed.

4.2 Cavity solitons

In the second simulation, we consider point **b** of Figure 3, by setting $\Delta = G = 2.5$ and $\theta = 12$. Such a value of Δ (and hence of the LEF) is probably a bit higher than the typical one for a QCL [13], but it allows obtaining in the system stable stationary temporal solitons, also called cavity solitons (CSs), for a large interval of the input intensity. For these values of the parameters, θ is close to both θ_{IL2} and θ_{MI1}, which means that in the region between the two turning points, the lower branch is almost entirely stable and the upper branch is almost entirely unstable. These are the best conditions to find stable stationary CSs [24, 29].

Figure 5 indeed shows branches of Turing rolls and CSs as defined in the study by Columbo et al. [14] emerging in the upper branch of the stationary homogeneous solution. If we move on the upper branch from the right to the left crossing the bifurcation point MI, we observe the onset of Turing rolls, whose maximum and minimum intensity are shown by the red symbols in Figure 5. At $Y = 82$, the Turing pattern becomes unstable and one or more CSs emerge from it. The blue symbols indicate the maximum intensity of the CSs along their branch. Figure 4(c) shows a CS at steady state for $Y = 100$.

In Figure 3, the curves θ_{IL1} and θ_{MI1} for $G = \Delta$ (solid line) intersect at a point whose abscissa can be calculated analytically and it is $\Delta = (4 + \sqrt{7})/3 \simeq 2.125$. For smaller Δ, if Y_{MI} coincides or is close to Y_{SN_1} in such a way that the upper branch is modulationally unstable between the two turning points, the lower branch is entirely unstable and no stable stationary CSs exist. The picture changes if we allow G to be different from, and in particular larger than, Δ, i.e. if $G = \Delta + \zeta$ with $\zeta > 0$, which means that GVD is not negligible and dispersion is anomalous. In Figure 3, the two dashed red curves have been obtained setting $G = \Delta + 1$. They are displaced to the left with respect to the solid curves obtained with $G = \Delta$ and now the curve θ_{IL1} intersects the curve θ_{MI1} at $\Delta \simeq 1.843$. This means that if we set for instance $\Delta = 2$ and $\theta = 4.7$ (point **c** of Figure 3), we have $Y_{MI} \simeq Y_{SN1}$ and yet a

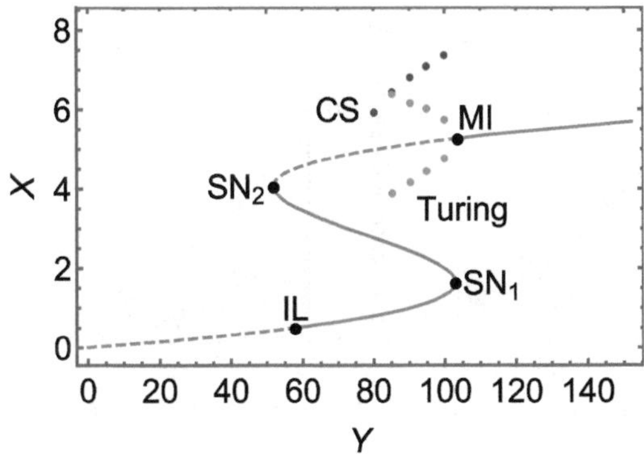

Figure 5: Stationary homogeneous solution (solid and dashed lines), cavity soliton branch (blue symbols), and Turing pattern branch (red symbols) for Eq. (3). $\Delta = 2.5$, $G = 2.5$, $\theta = 12$ (point **b** of Figure 3), and $\eta_{max} = 100$.

part of the lower branch from MI to SN_1 is stable so that stable stationary CSs may exist. For what concerns Turing patterns and CSs this set of parameters was extensively studied by Columbo et al. [14] and the main results are recalled in the Supplementary material (Figure S2). Here instead, in Section 5, we focus on the process of writing the CSs, their interaction and on how they can be manipulated by means of spatial modulations in the external field or in the bias current.

5 Cavity soliton encoding and interaction

CSs exhibit a remarkable potential for application in two ways. In the temporal domain, we can regard them as self-confined intensity pulses which, as shown by Columbo et al. [14], can not only be driven by an external continuous-wave field but also excited in arbitrary positions inside the laser cavity by the injection of optical pulses. A collection of independent CSs can thus be encoded in the cavity and associated with a bit sequence; as the CSs collection impinges on the output mirror each roundtrip, a bit train is emitted through the output facet. The ring thus acts as a buffer, which can be swiftly reset and rewritten and wherein the solitons can also undergo manipulation by means of external field tailoring, as it will be treated in the following.

In the spectral domain, CSs are associated with the onset of an optical frequency comb (see Figure 2(g) in the study by Columbo et al. [14]) so that the possibility to turn multiple CSs on and to control their relative distance provides a unique way to modify in real time the spectrum shape and the frequency sets appearing in the emitted spectrum.

5.1 Switching

An especially interesting feature for any multistable, localized structure such as the CS is the possibility to be deterministically excited at any location of the extended system by means of some external control parameter. In the case of an active optical system (a laser or an amplifier), most often this “encoding channel” is a coherent external field or an incoherent pump (a bias current, in the specific case of a semiconductor laser). In our injected laser layout, we add to the homogeneous component E_I in Eq. (3) suitably tailored pulses, by intensity, duration, and delay (in case multiple CSs switching is required).

The first issue addressed here is thus the relation between the pulse magnitude and the duration of the injection to observe the formation of a stable CS. To this purpose, we modified the input field term in numerically solving Eq. (3); we adopted the form

$$\overline{F}_I(\tau,\eta) = F_I + \epsilon\,\mathrm{sech}\left[(\eta-\eta_0)w/L\right]^8 \prod(\tau/T). \quad (7)$$

where $\prod(x)$ is the rectangular function, equal to 1 for $0<x<1$, and to 0 elsewhere, T is the pulse duration, η_0 is the position of the pulse inside the cavity, w/L is its width scaled to the cavity length and ϵ is the magnitude of the pulse. The $\mathrm{sech}(x)^8$ choice ensures a steep raise/drop of the pulse. In our simulations, the pulse width was taken equal to the CS full width half maximum and its location at $\eta_0/\eta_{max} = 0.5$ was never changed.

We considered two representative cases corresponding to the parametric regime of Figure S2 in the Supplementary material: the first at $Y = 7$ where the CS sits on a stable homogeneous background and the second at $Y = 6.6$ where the background is irregular but the CS is still stable although subjected to jitters (see also [14]). In the first case, when the pulse is too weak or the injection time too short, an intensity peak locally appears at the injection location but it dies rapidly away as the system returns to the HSS, emitting what could conceivably be dispersive waves [30] that ring off on a long timescale, compared with the frustrated CS decay rate (see bottom plot of Figure 6(a)). When the required magnitude and duration are met, a CS appears and rapidly becomes the stationary structure already studied by Columbo et al. [14] (Figure 6(b)). The peak intensity evolution vs. time shown in the bottom plot of Figure 6(b) evidences a latency around $X = 1.5$ as if realizing a metastable structure of intermediate intensity before reaching the CS intensity at $X = 3.7$. We remark that this is consistent with the studies performed on transverse 2D CSs [31] and pointed to the existence of an unstable CS branch bifurcating subcritically from the MI threshold which acts as a separatrix between the HSS branch and the stable CS branch coexisting with it. In this viewframe, a localized pulse causes the local field to grow from the HSS value; if the injection is large enough and it lasts long enough to draw the local field beyond the separatrix, the system locally reaches the CS branch and the structure is formed at regime. While a proper proof that this scenery occurs in our injected laser would require analyses of the stable and unstable parts of the CS branch, and of the CS eigenspectrum, we can suggest that the latency on the intermediate state, may flag the persistence of the system state around the attraction basin of the unstable CS before being finally attracted by the absolute stable solution. Note that in Figure 6(b), the plateau reached by the frustrated CS lies below $X = 1.5$ which seems to separate simulations of frustrated switch-on from successful ones. As the pulse duration grows larger than the critical value reported

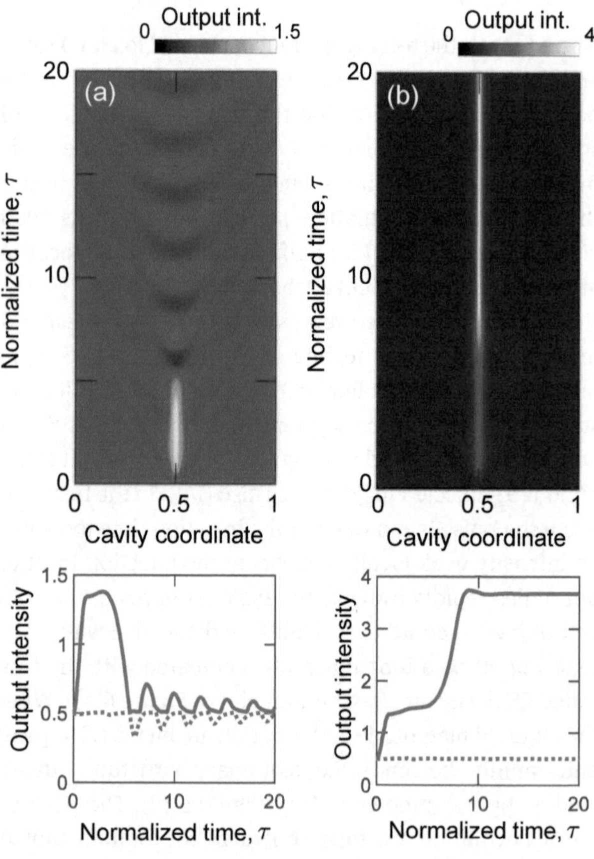

Figure 6: $Y = 7$. (a) Frustrated cavity soliton (CS) switching. The pulse width and duration are $\epsilon = 0.5$ and $T = 0.88$, respectively. The local field maximum does not evolve to a stable CS and rings off in rippling waves. (b) CS switch-on. Here the pulse width and duration are $\epsilon = 0.5$ and $T = 0.89$ where T is just above the critical switch-on value. Space-time diagrams (upper panels) and output intensity vs time (bottom panels) are reported in both cases.

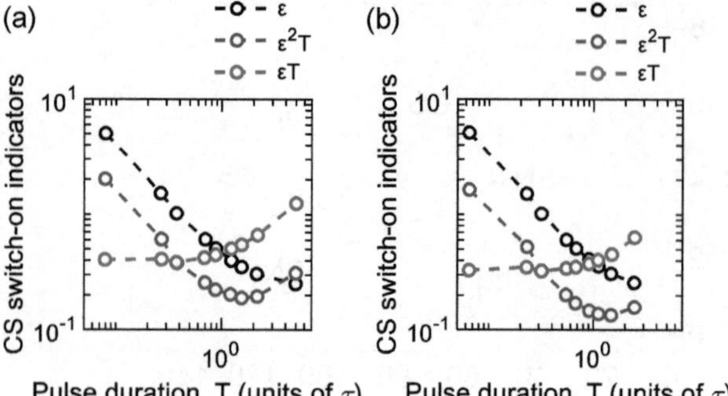

Figure 7: The critical values of a few indicators which correspond to successful cavity soliton (CS) switch-on for (a) $Y = 7$; (b) $Y = 6.6$.

previously ($T = 0.89$), the CS transient, i.e. the CS switch-up time, becomes shorter and settles at about 3. For a resonator with τ_p on the order of some tens of ps and 10% above threshold, it means that the switching time is on the order of some ns.

Systematic simulations in both study cases $Y = 7$ and $Y = 6.6$ for various pulse amplitude ϵ and duration T allow estimating the critical values for the pair and plot the relative curve. In agreement with Brambilla, Lugiato and Stefani [32], the product ϵT is approximately constant, as compared with e.g. the quantity $\epsilon^2 T$, proportional to energy as Figure 7 shows for both cases inspected, revealing small differences between the two, meaning that the switch-on process is not strongly influenced by the background behaviour. The major departures from constant ϵT occur for small injection values, when longer and longer injection times are required. In fact, for $\epsilon < 0.25$ no CS could be switched on, regardless the injection time. This suggests that the CS switch-on is a coherent phenomenon, similar to the onset of the self-induced-transparency solitons of the Sine–Gordon equation [24]. Finally, we observed that when the pulse intensity increases well beyond the critical value, the switch-on time decreases and tends to settle to about one roundtrip time.

5.2 Interaction

To determine the bit density which a cavity can sustain by allocating a certain number of CSs, it is necessary to estimate their interaction distance, defined here as the minimum distance two CSs can be excited at and reach a regime where no CS merging, or other forms of permanent pattern deformation, can be observed. We remark that, together with the CS transient time and the reset time required to start CS encoding anew, this figure contributes to the determination of the bit flux sustained by the device.

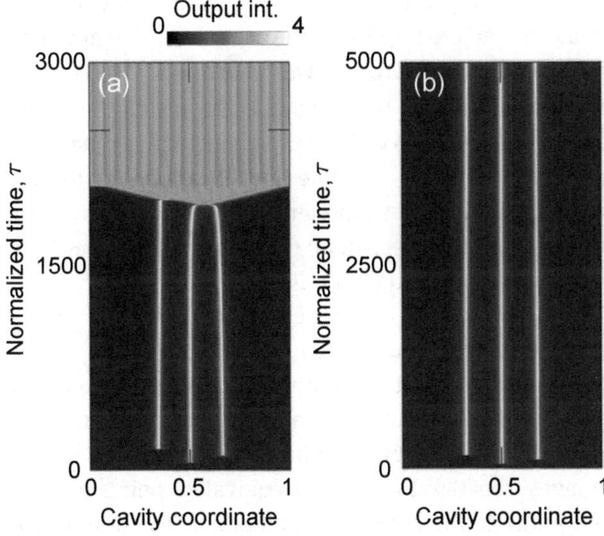

Figure 8: $Y = 7$. (a) A pulse is injected too close ($D = 0.16$) to an existing cavity soliton (CS); the emission profile is destabilized towards rolls. (b) The distance between the CS and two adjacent pulses is now $D = 0.17$, three independent CSs emerge at regime.

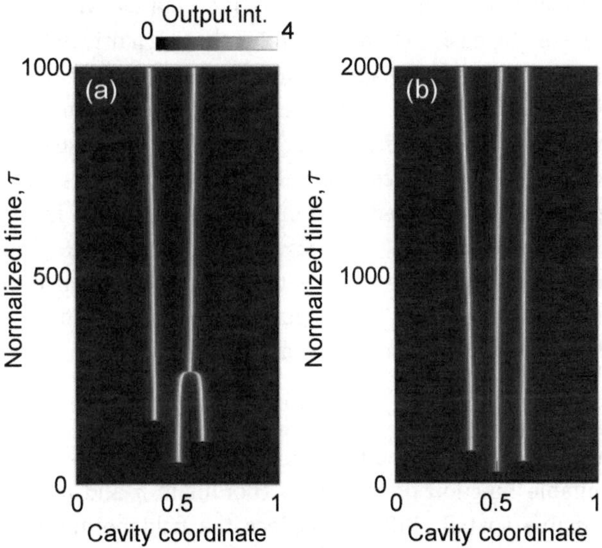

Figure 9: (a) Cavity solitons (CSs) merge into a single CS for $Y = 6.6$. (b) For $Y = 6.4$ CSs do not merge at reported minimum distances, but they push/pull each other modifying their separation.

We first considered the case $Y = 7$ where the HSS is stable and the system is close to the lower turning point. The initial state is precisely the HSS, then three short pulses of amplitude $\epsilon = 0.3$ are added to Y in sequence after $\tau = 50$, 100 and 150 at cavity locations $\eta_0/\eta_{\max} = 0.5$, $0.5 + D$ and $0.5 - D$, respectively. As it turns out, the system is quite sensitive to the perturbation of an existing CS, caused by the turn-on process of another one, so that when CSs are excited too closely, they interact attractively and the merger causes the emission profile to switch to the rolls (Figure 8(a)). For $D \geq 0.17$ instead the CSs are correctly written and reach a regime (Figure 8(b)).

We point out that the CS in presence of a stable homogeneous background does not exhibit oscillating tails, see Figure 2c in the study by Columbo et al. [14]. This excludes the possibility that bounded states can be formed among CS in this instance due to tail interaction [33, 34]. Although long-range CS interaction was predicted in CS without tails [35, 36] or much beyond the tail ringing range [37], a thorough analysis of CS interaction mechanisms exceeds the scopes of the present work and will be studied in the future.

Reducing Y brings the system farther from the turning point and makes it less sensitive to CS interactions. We may interpret in this sense the evidence that, when CSs are excited below the critical distance, they merge but no rolls switching occurs (Figure 9(a)). Also accordingly, we found that for $Y = 6.8$, 6.7, 6.6 interaction, distances to obtain independent CSs are $D = 0.16$, 0.14, 0.13, respectively. We could observe that for $Y < 6.8$, the background fluctuations grow more and more relevant and they seem to 'convey' perturbations across CS locations. This means that CSs do not merge at reported minimum distances, but they push/pull each other modifying their separation as in Figure 9(b), and more markedly as Y decreases. No such pushing occurs for e.g. $Y = 7$.

It is interesting that in this regime, where the CS background is unstable, the profile of the CS tails cannot be estimated altogether because it is drowned in the background fluctuations. Thus, again, interactions cannot be interpreted in terms of tail interaction. Similar CS behaviour (attraction/repulsion) in planar devices with irregular background emission was reported by Rahmani Anbardan et al. [38].

We remark that, in agreement with what was determined for transverse 2D CSs [32], the critical distances are on the order of the spatial modulation of the Turing rolls emerging from the MI which was estimated at $\lambda_{roll} = 0.117$ [14].

5.3 CS manipulation

It is well known that dissipative structures are sensitive to gradients appearing in control parameters, which set them in motion allowing the information they carry to be transported and redistributed (see the review [39] and references quoted therein). Indeed, other mechanisms can set CS in motion, such as thermal effects [40], higher-order dispersion and inhomogeneities [41, 42], coherent optical feedback [43] or delayed response [44], but we will focus here on the basic external control paths that a driven laser provides, namely, we apply gradients in the input field and in the pump current. Experiments with 2D, transverse CSs measured soliton drift speeds at 470 m/s in broad-area vertical cavity surface emitting lasers [45] when a field gradient was provided. This property is confirmed in the case of our devices, where we could observe CS drift longitudinally inside the cavity, under the influence of input field gradients. To this purpose, the input profile the external field F_I was modified as follows:

$$\overline{F}_I(\tau,\eta) = F_I[1 + \epsilon\mathcal{M}(\eta - \eta_s)], \tag{8}$$

where ϵ measures the field modulation strength, $\mathcal{M}$ is a suitable function of the cavity coordinate η and η_s is a possible spatial shift, to displace the field profile with respect to the cavity centre. Note that the value of ϵ must be small enough to ensure that, throughout the whole cavity, the field never exceeds the ranges where the CS is stable.

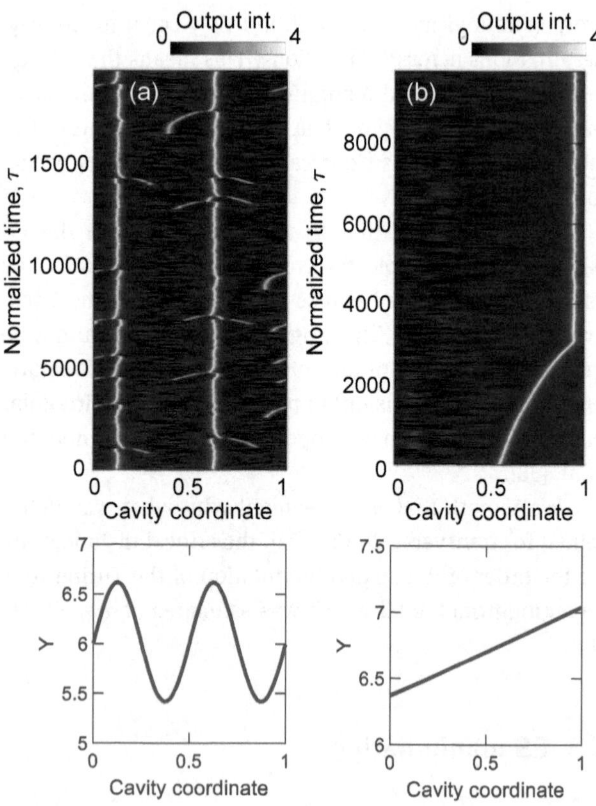

Figure 10: (a) Upper panel: cavity solitons (CSs) are spontaneously excited inside the cavity and follow the positive gradient of the field profile, being trapped at regime on the local maxima. Lower panel: intensity profile of the input field across the cavity. ($Y = 6.0$) (b) Upper panel: a CS follows the linear gradient with non-uniform velocity. Lower panel: intensity profile of the input field across the cavity ($Y = 6.7$).

We initially aimed to confirm that CS follow positive gradients and are trapped in the maxima of the input field profile, to this purpose we chose $\epsilon = 0.05$, $\mathcal{M} = \sin(4\pi\eta/\eta_{\max})$ so to have two maxima inside the cavity, and we selected $Y = 6$ which corresponds to a regime where the lower branch is unstable and CS are spontaneously created due to the spatio-temporal fluctuations of the background. As expected, all CS are attracted towards the maxima, as shown in Figure 10(a).

To estimate the drift speed of the CS a more regular field modulation was chosen by taking $\mathcal{M} = \eta - \eta_s$. The simulations reported in Figure 10(b) show that the CS follows the gradient and stops in the proximity of the cavity edge, where the input field abruptly changes by ϵ when η changes from 0 to 1 (which are equivalent points in a ring cavity due to the boundary conditions). The output field shows a negative hump on the left side of the cavity whose negative gradient, felt by the CS across the ring boundary, may balance the positive input field gradient, as shown in Figure 11. Surprisingly, the soliton does not move at constant speed as opposed to what was found for 2D CSs in semiconductor devices, where their velocity was shown proportional to the neutral mode corresponding to translation symmetries, which in turn is associated to the field gradient [31]. The reason for this discrepancy will be the object of future investigations.

While a precise evaluation of the CS speed is beyond the scopes of this work, by varying the parameter ϵ and shifting the gradient by adopting different values of η_s we could estimate the average CS speed $V_{\text{CS}} = \frac{\Delta\eta}{\Delta\tau}$ on the order

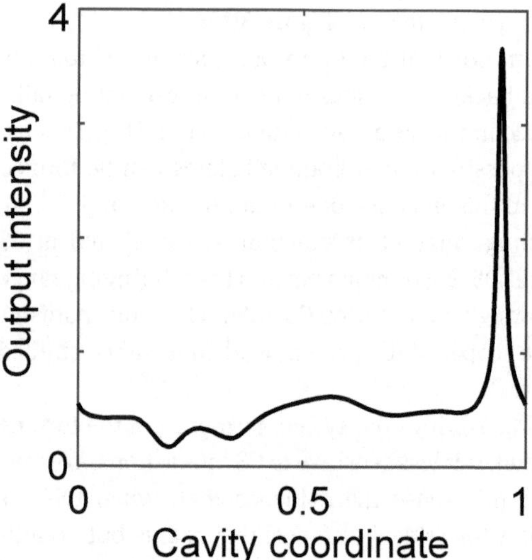

Figure 11: The output intensity profile relative to the simulation of Figure 10(b) shows that a negative hump is present on the left of the cavity, producing an effective negative gradient.

of $0.5 \times 10^{-3}\,\epsilon$. Considering a 3-mm cavity and the unit of time τ equal to 0.5 ns, this translates to $V_{CS} \approx 3 \times 10^3\,\epsilon$ m/s.

While coupling a modulated input field into the ring cavity might pose technical problems, assuming a modulated pump profile is less challenging, given recent RF techniques capable of modulating a microresonator at the beatnote frequency [46]. We wished to assess the behaviour of a CS when a weak modulation of the pump appearing in Eq. (1) is modelled by assuming $\mu = \mu_0 + \mu_1(\eta)$. By following the scaling leading to Eq. (3), one finds the form

$$\frac{\partial F(\tau,\eta)}{\partial \tau} = F_I + \gamma[1+\xi(\eta) - i\theta - i\Delta\xi(\eta)]F - (1-i\Delta)|F|^2 F + (1+iG)\frac{\partial^2 F}{\partial \eta^2}, \quad (9)$$

with $\xi(\eta) = \mu_1(\eta)/(\mu_0 - 1)$. We simulated such a spatial dependent term in our model selecting $\xi(\eta) = \xi_0 \sin(2\pi\eta/\eta_{\max} - \pi)$ and we could verify that the CS shifts at constant speed and is not pinned in modulation extrema as Figure 12 shows. In addition, reversing the sign of the modulated pump component changes the sign of the CS velocity leaving its speed unchanged. Although speed evaluation is left for future work, we checked that speed may be varied with the modulation amplitude ξ_0 and possibly by choosing different modulation shapes.

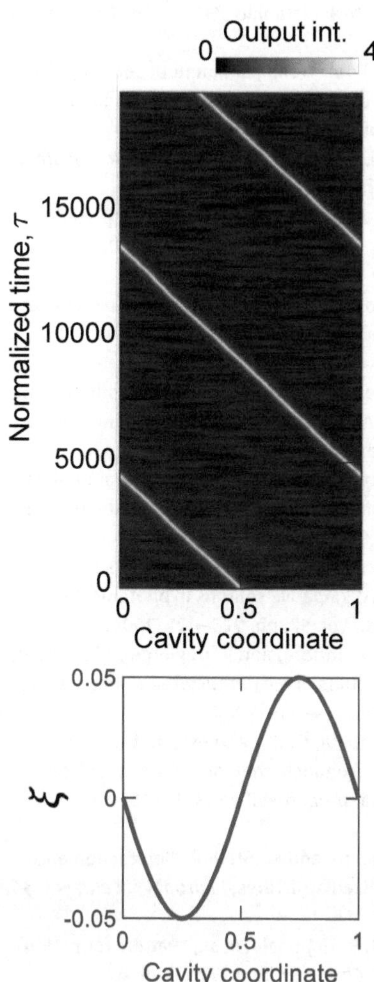

Figure 12: A cavity soliton (CS) drifts continuously along the cavity in presence of pump modulation, plotted in the lower panel. Here $Y = 6.7$ and $\xi_0 = 0.05$.

We remark that, in accordance with our preliminary evidence, a pump modulation will cause the CS to change its position inside the cavity, i.e. to change its propagation speed, and thus will influence the roundtrip time, although this change is quite modest. A regular change of the modulation, e.g. an AC modulation of the pump, may instead continuously change the lag between successive passages of the CS at the exit mirror, inducing a bit FM and possibly changing the associated comb composition.

6 Conclusion and perspectives

In this work, we presented the study of the HSS (or continuous wave) instabilities and the localized pattern dynamics occurring in a unidirectional ring QCL, driven by an external field. By exploiting a previously assessed model based on a generalized LLE, we could identify the character of the HSS emission curve as well as the oscillatory and modulational instabilities which affect it and determine the dynamical scenery the system will exhibit.

Our analysis allows considering the influence of all the main features affecting the semiconductor laser dynamics, such as the LEF, the forcing amplitude and detuning, the bias and, possibly, the GVD. The analyses allowed us to establish credible and promising candidates for the observation of localized structures of different classes, which exhibit remarkable interest for applications, relative to both optical comb formation and ultrashort pulse generation.

Extended simulation streams allowed characterizing different emission regimes, such as Turing rolls, but most importantly, evidencing the formation of two distinct classes of optical solitons. The first one, the PS, is associated to a chiral charge and connects the QCL to other classes of previously studied semiconductor lasers. The second, the cavity soliton, is a well-known pivotal element in optical comb formation (e.g. in Kerr microresonator) and information encoding, which we can now confirm in this new class of forefront, chip-scale and ultrafast lasers.

We confirm CS viability for applications, by showing independent addressing as well as all their salient features such as pair interaction and controlled drift. Further research will certainly lead us to address the possibility to exploit suitably assembled collections of solitary structures to tailor the shape of the associated spectral comb. We will also investigate in this respect the effects of shaping and modulating in time the external field and the bias current.

Finally the outcomes of this work will be a valuable guideline to assess operating regimes and layout configurations for ongoing experimental activities.

Author contribution: All the authors have accepted responsibility for the entire content of this submitted manuscript and approved submission.
Research funding: None declared.
Conflict of interest statement: The authors declare no conflicts of interest regarding this article.

References

[1] T. Udem, R. Holzwarth, and T. W. Hänsch, "Optical frequency metrology," *Nature*, vol. 416, pp. 233–237, 2002.
[2] D. J. Jones, S. A. Diddams, J. K. Ranka, et al., "Carrier-envelope phase control of femtosecond mode-locked lasers and direct optical frequency synthesis," *Science*, vol. 288, p. 635, 2000.
[3] P. Del'Haye, A. Schliesser, O. Arcizet, T. Wilken, R. Holzwarth, and T. J. Kippenberg, "Optical frequency comb generation from a monolithic microresonator," *Nature*, vol. 450, pp. 1214–1217, 2007.
[4] T. J. Kippenberg, R. Holzwarth, and S. A. Diddams, "Microresonator-based optical frequency combs," *Science*, vol. 332, 2011, https://doi.org/10.1126/science.1193968.
[5] Chembo, and Y. K. Kerr, "Optical frequency combs: theory, applications and perspectives," *Nanophotonics*, vol. 5, pp. 214–230, 2016.
[6] L. A. Lugiato, F. Prati, M. L. Gorodetsky, and T. J. Kippenberg, "From the Lugiato-Lefever equation to microresonator-based soliton Kerr frequency combs," *Philos. T. R. Soc. A*, vol. 376, p. 20180113, 2018.
[7] A. Hugi, G. Villares, S. Blaser, H. Liu, and J. Faist, "Mid-infrared frequency comb based on a quantum cascade laser," *Nature*, vol. 492, pp. 229–233, 2012.
[8] D. Burghoff, T.-Y. Kao, N. Han, et al., "Terahertz laser frequency combs," *Nat. Photonics*, vol. 8, pp. 462–467, 2014.
[9] Q. Lu, M. Razeghi, S. Slivken, et al., "High power frequency comb based on mid-infrared quantum cascade laser at λ ~9μm," *Appl. Phys. Lett.*, vol. 106, p. 051105, 2015.
[10] J. Faist, G. Villares, G. Scalari, et al., "Quantum cascade laser frequency combs," *Nanophotonics*, vol. 5, pp. 272–291, 2016.
[11] T. Mansuripur, C. Vernet, P. Chevalier, et al., "Single-mode instability in standing-wave lasers: the quantum cascade laser as a self-pumped parametric oscillator," *Phys. Rev. A*, vol. 94, p. 063807, 2016.
[12] L. Columbo, S. Barbieri, C. Sirtori, and M. Brambilla, "Dynamics of a broad-band quantum cascade laser: from chaos to coherent dynamics and mode-locking," *Opt. Express*, vol. 26, pp. 2829–2847, 2018.
[13] M. Piccardo, B. Schwarz, D. Kazakov, et al., "Frequency combs induced by phase turbulence," *Nature*, vol. 582, pp. 360–364, 2020.
[14] L. Columbo, M. Piccardo, F. Prati, et al., "Unifying frequency combs in active and passive cavities: temporal solitons in externally-driven ring lasers," arXiv:20 07.07533 [physics.optics].
[15] L. A. Lugiato, and R. Lefever, "Spatial dissipative structures in passive optical systems," *Phys. Rev. Lett.*, vol. 58, pp. 2209–2211, 1987.
[16] L. A. Lugiato, C. Oldano, and L. M. Narducci, "Cooperative frequency locking and stationary spatial structures in lasers," *J. Opt. Soc. Am. B*, vol. 5, pp. 879–888, 1988.
[17] N. Akhmediev and A. Ankiewicz, *Dissipative Solitons. Lecture Notes in Physics*, Berlin, Heidelberg, Springer, 2005.
[18] T. J. Kippenberg, A. L. Gaeta, M. Lipson, and M. L. Gorodetsky, "Dissipative Kerr solitons in optical microresonators," *Science*, vol. 361, 2018, https://doi.org/10.1126/science.aan8083.
[19] C. Henry, "Theory of the linewidth of semiconductor lasers," *IEEE J. Quantum Electron.*, vol. 18, pp. 259–264, 1982.
[20] I. Aranson, and L. Kramer, "The world of the complex Ginzburg-Landau equation," *Rev. Mod. Phys.*, vol. 74, pp. 99–143, 2002.
[21] D. Burghoff, "Frequency-modulated combs as phase solitons," arXiv:2006.12397, 2020.
[22] M. Tlidi, P. Mandel, and R. Lefever, "Localized structures and localized patterns in optical bistability," *Phys. Rev. Lett.*, vol. 73, pp. 640–643, 1994.
[23] S. Wieczorek, B. Krauskopf, T. Simpson, and D. Lenstra, "The dynamical complexity of optically injected semiconductor lasers," *Phys. Rep.*, vol. 416, pp. 1–128, 2005.
[24] L. Lugiato, F. Prati, and M. Brambilla, *Nonlinear Optical Systems*, Cambridge, UK, Cambridge University Press, 2015.
[25] H. Chaté, A. Pikovsky, and O. Rudzick, "Forcing oscillatory media: phase kinks vs. synchronization," *Physica D*, vol. 131, pp. 17–30, 1999.
[26] F. Gustave, L. Columbo, G. Tissoni, et al., "Dissipative phase solitons in semiconductor lasers," *Phys. Rev. Lett.*, vol. 115, p. 043902, 2015.
[27] F. Gustave, L. Columbo, G. Tissoni, et al., "Phase solitons and domain dynamics in an optically injected semiconductor laser," *Phys. Rev. A*, vol. 93, p. 063824, 2016.
[28] F. Gustave, L. Columbo, G. Tissoni, et al., "Formation of phase soliton complexes in an optically injected semiconductor laser," *Eur. Phys. J. D*, vol. 71, p. 154, 2017.
[29] T. Ackemann, W. Firth, and G. Oppo, "Fundamentals and applications of spatial dissipative solitons in photonic devices," *Adv. At. Mol. Opt. Phys.*, vol. 57, pp. 323–421, 2009.
[30] I. Cristiani, R. Tediosi, L. Tartara, and V. Degiorgio, "Dispersive wave generation by solitons in microstructured optical fibers," *Opt. Express*, vol. 12, pp. 124–135, 2004.
[31] T. Maggipinto, M. Brambilla, G. K. Harkness, and W. J. Firth, "Cavity solitons in semiconductor microresonators: existence, stability, and dynamical properties," *Phys. Rev. E*, vol. 62, pp. 8726–8739, 2000.
[32] M. Brambilla, L. A. Lugiato, and M. Stefani, "Interaction and control of optical localized structures," *Europhys. Lett.*, vol. 34, pp. 109–114, 1996.
[33] A. Scroggie, W. Firth, G. S. McDonald, et al., "Pattern formation in a passive Kerr cavity," *Chaos Solitons Fract.*, vol. 4, pp. 1323–1354, 1994.
[34] A. Vladimirov, J. McSloy, D. Skryabin, and W. Firth, "Two-dimensional clusters of solitary structures in driven optical cavities," *Phys. Rev. E*, vol. 65, p. 11, 2002.

[35] H. Vahed, R. Kheradmand, H. Tajalli, et al., "Phase-mediated long-range interactions of cavity solitons in a semiconductor laser with a saturable absorber," *Phys. Rev. A*, vol. 84, pp. 063814-1–063814-6, 2011.
[36] S. Rahmani Anbardan, C. Rimoldi, R. Kheradmand, G. Tissoni, and F. Prati, "Exponentially decaying interaction potential of cavity solitons," *Phys. Rev. E*, vol. 97, p. 032208, 2018.
[37] Y. Wang, F. Leo, J. Fatome, et al., "Universal mechanism for the binding of temporal cavity solitons," *Optica*, vol. 4, pp. 855–863, 2017.
[38] S. Rahmani Anbardan, C. Rimoldi, R. Kheradmand, G. Tissoni, and F. Prati, "Interaction of cavity solitons on an unstable background," *Phys. Rev. E*, vol. 101, p. 042210, 2020.
[39] W. J. Firth, in *Soliton-driven Photonics, NATO Science Series (Series II: Mathematics, Physics and Chemistry)*, A. D. Boardman and A. P. Sukhorukov, Eds., Dordrecht, Springer, 2001.
[40] A. Scroggie, J. McSloy, and W. Firth, "Self-propelled cavity solitons in semiconductor microcavities," *Phys. Rev. E*, vol. 66, p. 036607, 2002.
[41] P. Parra-Rivas, D. Gomila, M. Matías, P. Colet, and L. Gelens, "Effects of inhomogeneities and drift on the dynamics of temporal solitons in fiber cavities and microresonators," *Opt. Express*, vol. 22, pp. 30943–30954, 2014.
[42] M. Liu, L. Wang, Q. Sun, et al., "Influences of high-order dispersion on temporal and spectral properties of microcavity solitons," *Opt. Express*, vol. 26, pp. 16477–16487, 2018.
[43] M. Tlidi, and K. Panajotov, "Two-dimensional dissipative rogue waves due to timedelayed feedback in cavity nonlinear optics," *Chaos*, vol. 27, p. 013119, 2017.
[44] M. G. Clerc, S. Coulibaly, and M. Tlidi, "Time-delayed nonlocal response inducing traveling temporal localized structures," *Phys. Rev. Res.*, vol. 2, p. 013024, 2020.
[45] X. Hachair, S. Barland, L. Furfaro, et al., "Cavity solitons in broad-area vertical-cavity surface-emitting lasers below threshold," *Phys. Rev. A*, vol. 69, no. 4, p. 043817, 2004.
[46] H. Li, P. Laffaille, D. Gacemi, et al., "Dynamics of ultra-broadband terahertz quantum cascade lasers for comb operation," *Opt. Express*, vol. 23, pp. 33270–33294, 2015.

Supplementary Material: The online version of this article offers supplementary material (https://doi.org/10.1515/nanoph-2020-0409).

Part III: **Fiber Optics and Optical Communications**

Zelin Ma and Siddharth Ramachandran*

Propagation stability in optical fibers: role of path memory and angular momentum

https://doi.org/10.1515/9783110710687-019

Abstract: With growing interest in the spatial dimension of light, multimode fibers, which support eigenmodes with unique spatial and polarization attributes, have experienced resurgent attention. Exploiting this spatial diversity often requires robust modes during propagation, which, in realistic fibers, experience perturbations such as bends and path redirections. By isolating the effects of different perturbations an optical fiber experiences, we study the fundamental characteristics that distinguish the propagation stability of different spatial modes. Fiber perturbations can be cast in terms of the angular momentum they impart on light. Hence, the angular momentum content of eigenmodes (including their polarization states) plays a crucial role in how different modes are affected by fiber perturbations. We show that, accounting for common fiber-deployment conditions, including the more subtle effect of light's path memory arising from geometric Pancharatnam–Berry phases, circularly polarized orbital angular momentum modes are the most stable eigenbasis for light propagation in suitably designed fibers. Aided by this stability, we show a controllable, wavelength-agnostic means of tailoring light's phase due to its geometric phase arising from path memory effects. We expect that these findings will help inform the optimal modal basis to use in the variety of applications that envisage using higher-order modes of optical fibers.

Keywords: higher-order modes; multimode fibers; orbital angular momentum; Pancharatnam–Berry phase; vector beams; propagation stability.

***Corresponding author: Siddharth Ramachandran**, Boston University, Boston, MA, USA, E-mail: sidr@bu.edu. https://orcid.org/0000-0001-9356-6377
Zelin Ma, Boston University, Boston, MA, USA. https://orcid.org/0000-0002-2801-4004

1 Introduction

Multimode fibers (MMFs) and their spatially diverse higher-order modes (HOMs) have experienced alternating levels of interest ever since the invention of optical fibers. Although one of the first applications of light propagation, for image transport with flexible optical fiber waveguides, utilized MMFs [1], the development of single-mode fibers (SMFs) quickly diverted attention away from MMFs. One important reason was that any realistic deployment of optical fibers includes perturbations, such as bends, twists, 3D paths as well as thermal, mechanical, and environmental fluctuations. While modes in a perfectly straight, circular fiber are theoretically orthogonal, perturbations typically cause coupling between them, leading to potential loss of signal purity or information content. Subsequent advances in light guidance, including microstructuring [2], photonic bandgaps [3], or antiresonant structures [4], primarily focused on means to strip out [5–7] HOMs to effectively realize single-mode guidance. In fact, even an SMF is two-moded, accounting for the two orthogonal polarization eigenmodes. As such, bend- or geometry-induced fiber birefringence [8] can cause polarization-mode dispersion in classical communications links [9] and loss of entanglement preservation in quantum links [10].

The advent of improved signal conditioning and reception technologies over the last decade has, however, refocused investigations on MMFs in which modes mix. Because this mixing predominantly represents unitary transformations, multi-in multi-out digital signal processing can disentangle mode mixing in the electronic domain [11], resulting in scaling the capacity of telecommunications links [12]. Analysis of the speckle patterns out of MMFs enables spectrometry [13]. On the other hand, adaptively controlling the speckle pattern at the input or output enables imaging with MMFs [14, 15]. Linear mixing, when combined with nonlinear coupling, leads to effects such as multimode solitons [16, 17], nondissipative beam cleanup [18] and geometric parametric instabilities [19], among a host of multimode nonlinear effects not seen in SMFs.

This article has previously been published in the journal Nanophotonics. Please cite as: Z. Ma and S. Ramachandran "Propagation stability in optical fibers: role of path memory and angular momentum" *Nanophotonics* 2021, 10. DOI: 10.1515/nanoph-2020-0404.

The aforementioned benefits of a mixed ensemble of modes notwithstanding the ability to excite and propagate specific HOMs remain especially desirable. As is evident from any elementary solution of eigenmodes in waveguides, each HOM has a characteristic phase and group velocity, group-velocity dispersion, modal area (A_{eff}) [20, 21]. For instance, HOMs can be tailored to have large normal [22] or anomalous dispersion [23], with applications in dispersion control for telecom links [24, 25] and ultrashort pulses [26]. One of the first demonstrations of nonlinear wave mixing in optical fibers involved intermodal phase matching between HOMs [27], the diversity of HOMs yielding enhanced degrees of freedom to achieve momentum conservation [28, 29]. Raman [30] or Brilioüin [31] scattering have shown to be strongly dependent on, and hence be tailorable by, the mode(s) in which light propagates. These concepts have received increased attention for applications such as third-harmonic generation [32], extending supercontinuum generation [33], power-scalable source engineering [30, 34] or new forms of quantum sources [35]. The inherently large A_{eff} of HOMs has led to ultralarge A_{eff}, low-nonlinearity flexible fibers [36], with applications in fiber lasers [37]. More recently, fiber modes with orbital angular momentum (OAM) have been shown to yield an additional degree of freedom with which to control the nonlinear interactions of light in fibers [38–40]. Fiber sources operating in HOMs are also interesting for applications where a non–Gaussian-shaped emission is desired, such as in nanoscale microscopy [41, 42] and laser machining [43]. Finally, perhaps most significantly over the last few years, there has been an emerging realization that individual modes, especially those carrying OAM, can enable signal propagation with low or limited mode mixing [44], as a means of increasing the capacity of classical communications networks [45–48] or for enhancing the security of quantum links [49]. All these applications have two critical requirements: (1) the ability to accurately control mode transformations with, for instance, fiber gratings [50], diffractive optics [51], Pancharatnam–Berry optical elements (PBOE) [52], spatial light modulators [53], or metasurfaces [54]; and (2) crucially, the need for linearly, stably propagating desired modes in fibers.

Here, we address the latter issue – the propagation stability of optical fiber modes. Any realistic analysis of an optical fiber must necessarily consider the perturbations it encounters over the long lengths of signal transmission in facilitates. The key question is, in the presence of perturbations, how do otherwise theoretically orthogonal modes of a cylindrically symmetric fiber mix with each other? This is a very complex problem for long-haul networks where propagation over 100s–1000s of kilometers of fiber encounters a wide array of stochastically varying perturbations. As a result, many realistic models for such mode coupling are phenomenological in nature [55, 56]. For shorter lengths, spanning a few meters to kilometers (lengths representing scales of fiber usage in most applications, such as fiber lasers and amplifiers, data-center links, nonlinear devices or sensors), this is a more tractable problem that can yield some first-principles insight. We show that certain classes of spatial modes are more stable and propagation-tolerant than others. Somewhat counterintuitively, we also show that this stability depends even on the bases of modes used – that is, one set of modes can be more stable than modes represented in a mathematically equivalent but rotated basis. Fundamentally, we show that, accounting for typical perturbations an optical fiber encounters, the circularly polarized OAM eigenbasis represents the most stable set of modes for light transmission.

2 Mode classifications and fiber perturbations

2.1 Mode classifications

We start with a brief description of the eigenmodes of a cylindrically symmetric, step-index profile fiber. The 2D cross-section allows modes to be classified by two orthogonal polarization distributions, as well as a radial order index m, and an azimuthal order index L. Equation (1) shows the field distributions for the class of $L = 0$ modes ($\mathrm{HE}_{1,m}$) in two polarization bases:

$$\boldsymbol{E}_{0,m} = e^{i\beta_{0,m}z} \cdot \left\{ \begin{bmatrix} \mathrm{HE}^{x}_{1,m} \\ \mathrm{HE}^{y}_{1,m} \end{bmatrix} \text{ or } \begin{bmatrix} \mathrm{HE}^{+}_{1,m} \\ \mathrm{HE}^{-}_{1,m} \end{bmatrix} \right\} \tag{1}$$

where

$$\begin{bmatrix} \mathrm{HE}^{x}_{1,m} \\ \mathrm{HE}^{y}_{1,m} \end{bmatrix} = F_{0,m}(r) \begin{bmatrix} \widehat{x} \\ \widehat{y} \end{bmatrix}, \quad \begin{bmatrix} \mathrm{HE}^{+}_{1,m} \\ \mathrm{HE}^{-}_{1,m} \end{bmatrix} = F_{0,m}(r) \begin{bmatrix} \widehat{\sigma}^{+} \\ \widehat{\sigma}^{-} \end{bmatrix}$$

where $\widehat{\sigma}^{\pm} = \widehat{x} \pm i\widehat{y}$, representing left or right handed circular polarization, respectively; $F_{0,\mathrm{m}}(r)$ represents the radial profile of the field, which is typically a piecewise linear combination of Bessel functions for most profiles; β is the projection of the wave vector $\vec{k}$ in the propagation direction (usually referenced by the mode's effective index,

$n_{\text{eff}} = \lambda\beta/2\pi$, where λ is the free-space wavelength of light). Paying attention to its subscripts, we see that β is distinct for each radial order m, but is identical for either polarization eigenmodes, represented in either bases. Modes designated as $HE_{L+1,m}$ (or $EH_{L-1,m}$, relevant for HOMs, described later) signify that fields of fiber modes are not strictly transverse to their propagation direction, but are, instead, a hybrid of electric and magnetic fields. This hybridization is typically negligible in weakly guiding fibers, where a scalar approximation leads to the commonly encountered linearly polarized (LP) modes, possessing a uniform, linear spatial polarization distribution. Similarly, we can have circularly polarized (CP) modes, denoting spatially uniform circular polarization distributions. For $L = 0$ modes in most weakly guided fibers, the scalar modes ($LP_{0,m}$ and $CP_{0,m}$) are almost identical to the vector modes $HE_{1,m}$ (with higher index contrast, this approximation breaks down, but, to first order, the field distributions remain the same, with only a small modification to β):

$$\breve{\boldsymbol{E}}_{0,m} = e^{i\breve{\beta}_{0,m}z} \cdot \left\{ \begin{bmatrix} LP^{x}_{0,m} \\ LP^{y}_{0,m} \end{bmatrix} \text{ or } \begin{bmatrix} CP^{+}_{0,m} \\ CP^{-}_{0,m} \end{bmatrix} \right\} \tag{2}$$

where

$$\begin{bmatrix} LP^{x}_{0,m} \\ LP^{y}_{0,m} \end{bmatrix} = F_{0,m}(r)\begin{bmatrix} \hat{x} \\ \hat{y} \end{bmatrix}, \quad \begin{bmatrix} CP^{+}_{0,m} \\ CP^{-}_{0,m} \end{bmatrix} = F_{0,m}(r)\begin{bmatrix} \hat{\sigma}^{+} \\ \hat{\sigma}^{-} \end{bmatrix}$$

where the accent ˘ denotes that the quantities have been calculated under the scalar approximation.

Figure 1 shows the intensity profiles of two representative modes, with $L = 0$, and $m = 1$ and 3, respectively, with $(m - 1)$ signifying the number of intensity nulls in the radial direction. Only the LP and CP modes are shown because the vector and scalar modes are almost identical for $L = 0$. The field with $L = 0$ and $m = 1$ is the well-known fundamental mode of SMFs. Linear combinations of the modes in any basis yield modes in another basis, as illustrated by the lines along with the $+i$ and $-i$ signs connecting modes of the different bases. Such linear combinations represent coordinate rotation among mutually unbiased bases (MUB), often used to transmit quantum information. Generally, the fundamental mode of SMFs and the entire class of $L = 0$ modes are twofold degenerate (in polarization) and any arbitrary polarization state of these modes propagates similarly in a fiber.

The situation is more complex for $L \neq 0$ modes. We leave aside the case of $|L| = 1$, which has its own peculiar behavior but which has been well-studied in the past [57]. Equation (3) shows the field distributions in two representations (assuming $L > 1$) similar to those used in Eq. (1):

$$\boldsymbol{E}_{L,m} = \left\{ \begin{bmatrix} HE^{\text{even}}_{L+1,m} e^{i\beta'_{L,m}z} \\ HE^{\text{odd}}_{L+1,m} e^{i\beta'_{L,m}z} \\ EH^{\text{even}}_{L-1,m} e^{i\beta''_{L,m}z} \\ EH^{\text{odd}}_{L-1,1} e^{i\beta''_{L,m}z} \end{bmatrix} or \begin{bmatrix} OAM^{+}_{+L} e^{i\beta'_{L,m}z} \\ OAM^{-}_{-L} e^{i\beta'_{L,m}z} \\ OAM^{-}_{+L} e^{i\beta''_{L,m}z} \\ OAM^{+}_{-L} e^{i\beta''_{L,m}z} \end{bmatrix} \right\} \tag{3}$$

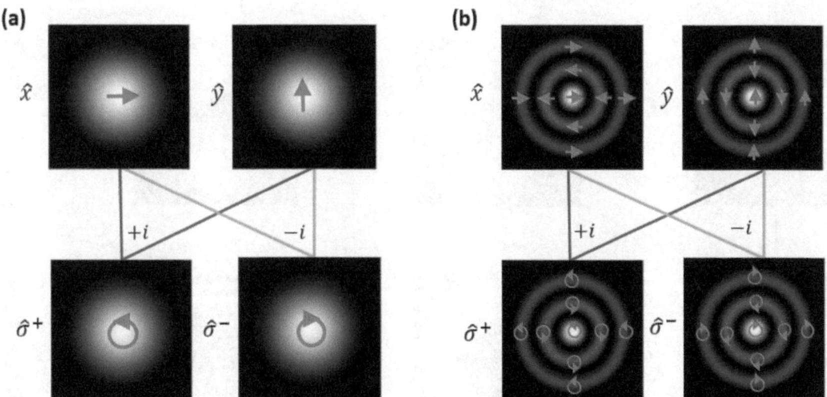

Figure 1: Field representations for $L = 0$ modes in a circular, step-index-guided fiber. Gray scale images show intensity distributions and red arrows indicate polarization state, either linear (straight arrows in top row) or circular (circular arrows in bottom row). The relationship between linear and circular polarizations are shown for the (a) $m = 1$ and (b) $m = 3$ modes. The colored lines indicate that the circular polarization modes can be represented as linear combinations of the two orthgonal linear polarization modes. The $+i$ or $-i$ terms represent a $\pi/2$ or $-\pi/2$ phase shift in the linear combinations. Conversely, linear polarization modes can be decomposed into two orthogonal circular polarization modes as well. The position where the arrowheads is shown within the circles representing circular polarizations are intentionally distinct, indicating a phase shift. Circular polarization representations where the arrowhead is on the top portion of respective circles is π phase-shifted from those in which they are on the bottom of the circle. This is a common feature of $LP_{0,m}$ or $CP_{0,m}$ modes – that each adjacent intensity ring of the mode accumulates a π phase shift and hence their fields are flipped.

where

$$\begin{bmatrix} \mathrm{HE}^{\mathrm{even}}_{L+1,m} \\ \mathrm{HE}^{\mathrm{odd}}_{L+1,m} \\ \mathrm{EH}^{\mathrm{even}}_{L-1,m} \\ \mathrm{EH}^{\mathrm{odd}}_{L-1,m} \end{bmatrix} = F_{L,m}(r) \begin{bmatrix} \hat{x}\cos(L\varphi) - \hat{y}\sin(L\varphi) \\ \hat{x}\sin(L\varphi) + \hat{y}\cos(L\varphi) \\ \hat{x}\cos(L\varphi) + \hat{y}\sin(L\varphi) \\ \hat{x}\sin(L\varphi) - \hat{y}\cos(L\varphi) \end{bmatrix},$$

$$\begin{bmatrix} \mathrm{OAM}^{+}_{+L} \\ \mathrm{OAM}^{-}_{-L} \\ \mathrm{OAM}^{-}_{+L} \\ \mathrm{OAM}^{+}_{-L} \end{bmatrix} = F_{L,m}(r) \begin{bmatrix} \hat{\sigma}^{+}\exp(iL\varphi) \\ \hat{\sigma}^{-}\exp(-iL\varphi) \\ \hat{\sigma}^{-}\exp(iL\varphi) \\ \hat{\sigma}^{+}\exp(-iL\varphi) \end{bmatrix}$$

where $F_{L,m}(r)$ represents the radial profile of the fields, and φ is the azimuthal coordinate. In analogy with Eq. (1) (or Eq. (2)), the field is either linearly or circularly polarized at any specific transverse position in either representation. However, in contrast with the $L = 0$ modes, in one of the representations (HE/EH, also commonly called vector modes), the polarization distributions are spatially nonuniform. The uniformly circularly polarized mode basis is denoted as the OAM basis because the similarity of these fiber modes with free-space beams carrying OAM on account of the helical phase $e^{iL\varphi}$ of the electric field [58]. These vector and OAM modes are illustrated in Figure 2, where the lines along with the $+i$ and $-i$ signs show the linear combinations that rotate modes from one basis to another. There are notable differences compared with the $L = 0$ modes. First, for each value of $|L|$ and m, there are four, instead of two, modes. Next, the propagation constants for the exact modes are not all degenerate, but instead depend on the internal symmetries of the modes. In the OAM basis, the modes are pairwise degenerate, but have different βs depending on whether signs of $\hat{\sigma}$ and L are the same (Spin–Orbit aligned – SOa – modes, with $\beta'_{L,m}$ shown in Eq. (3)) or if they are opposite (Spin–Orbit anti-aligned – SOaa – modes, with $\beta''_{L,m}$ shown in Eq. (3)). Correspondingly, in the vector basis, the β of the HE and EH modes differ. As in the case of the $L = 0$ modes shown in Figure 1, the SOa OAM modes and HE vector modes form an MUB and are linear combinations of one another, whereas the SOaa OAM modes and EH vector modes form a *separate* MUB.

Another, in fact, better known representation for the $L > 1$ modes is the LP designation, equivalent, as in the case of the $L = 0$ modes shown in Eq. (2), to the CP designation.

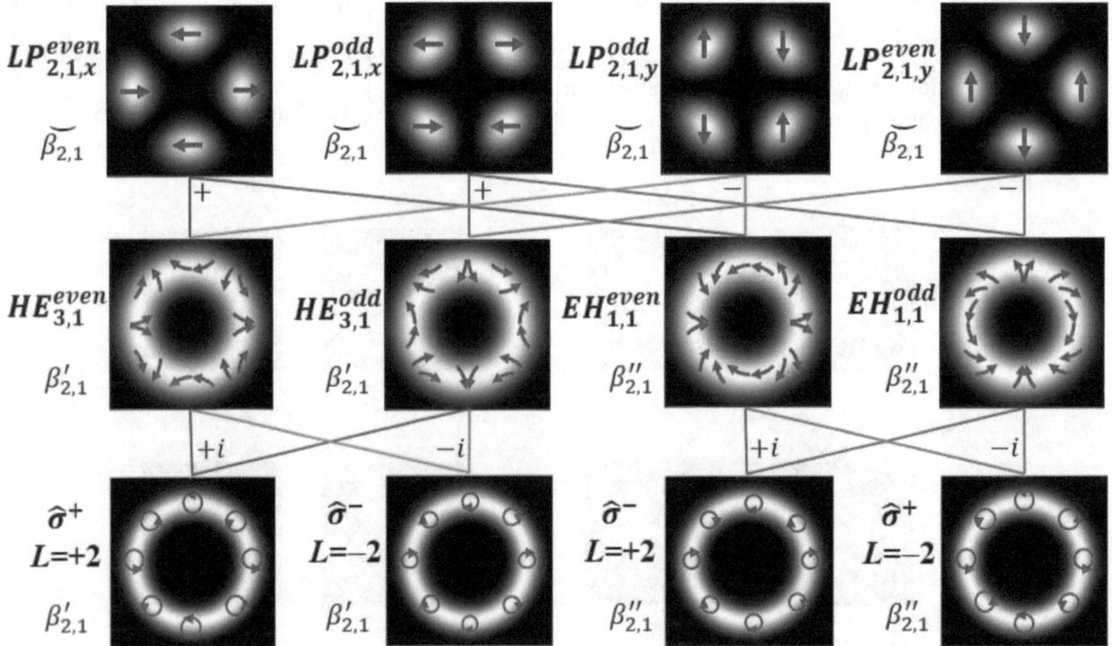

Figure 2: Intensity and polarization patterns of linearly polarized (LP) (top row), vector (middle row), and orbital angular momentum (OAM) modes (bottom row) with azimuthal index $|L| = 2$ and radial index $m = 1$. Colored lines show linear combinations between groups. The $+i$ or $-i$ terms represent a $\pi/2$ or $-\pi/2$ phase shift in the linear combinations. As described in the caption of Figure 1, azimuthal shifts of the arrows on the OAM modes indicate an azimuthal phase shift. The propagation constant is $\beta'_{2,1}$ for Spin–Orbit aligned (SOa) modes (and the corresponding HE modes), and $\beta''_{2,1}$ for SOaa modes (and the corresponding EH modes). The propagation constant of the LP designation is $\breve{\beta}_{0,m}$, which is an average of $\beta'_{2,1}$ and $\beta''_{2,1}$.

Equation (4) shows the field distributions of this scalar designation in two representations similar to those used in Eq. (2):

$$\undergroup{E}_{L,m} = e^{i\undergroup{\beta}_{L,m}z} \cdot \left\{ \begin{bmatrix} \mathrm{LP}^{\mathrm{even}}_{L,m,x} \\ \mathrm{LP}^{\mathrm{odd}}_{L,m,x} \\ \mathrm{LP}^{\mathrm{odd}}_{L,m,y} \\ \mathrm{LP}^{\mathrm{even}}_{L,m,y} \end{bmatrix} \text{ or } \begin{bmatrix} \mathrm{CP}^{\mathrm{even}}_{L,m,+} \\ \mathrm{CP}^{\mathrm{odd}}_{L,m,+} \\ \mathrm{CP}^{\mathrm{odd}}_{L,m,-} \\ \mathrm{CP}^{\mathrm{even}}_{L,m-} \end{bmatrix} \right\} \quad (4)$$

where

$$\begin{bmatrix} \mathrm{LP}^{\mathrm{even}}_{L,m,x} \\ \mathrm{LP}^{\mathrm{odd}}_{L,m,x} \\ \mathrm{LP}^{\mathrm{odd}}_{L,m,y} \\ \mathrm{LP}^{\mathrm{even}}_{L,m,y} \end{bmatrix} = F_{L,m}(r) \begin{bmatrix} \hat{x}\cos(L\varphi) \\ \hat{x}\sin(L\varphi) \\ \hat{y}\sin(L\varphi) \\ \hat{y}\cos(L\varphi) \end{bmatrix},$$

$$\begin{bmatrix} \mathrm{CP}^{\mathrm{even}}_{L,m,+} \\ \mathrm{CP}^{\mathrm{odd}}_{L,m,+} \\ \mathrm{CP}^{\mathrm{odd}}_{L,m,-} \\ \mathrm{CP}^{\mathrm{even}}_{L,m,-} \end{bmatrix} = F_{L,m}(r) \begin{bmatrix} \hat{\sigma}^{+}\cos(L\varphi) \\ \hat{\sigma}^{+}\sin(L\varphi) \\ \hat{\sigma}^{-}\sin(L\varphi) \\ \hat{\sigma}^{-}\cos(L\varphi) \end{bmatrix}$$

These intensity and polarization pattern of these LP designations (the CP designation is ignored here as their relationship with LP counterparts is the same as that in the $L = 0$ cases) are illustrated in the first row of Figure 2. Again, lines along with the + and – signs show how they can also be represented as a linear combination of vector (or OAM) modes. As evident from Figure 2, in contrast to the exact solutions (vector/OAM modes), the LP basis for $|L| > 1$ modes is actually a mixture of two OAM or vector modes of *different* βs. This has consequences for mode stability, as elaborated in Section 3.

2.2 Fiber perturbations

Figure 3 schematically shows the n_{eff} for select modes with indices L and m in select index-guided fiber designs. The coupling efficiency η, between these modes is given by [59]:

$$\eta \propto \int\int\int \overline{F_{L,m}(r)} \cdot \overline{\overline{P_{\mathrm{pert}}(r,\varphi,z)}} \cdot \overline{F^{*}_{L',m'}(r)} \cdot e^{i(\beta-\beta')z} \cdot e^{i(L-L')\varphi} \cdot dA \cdot dz \quad (5)$$

where the perturbation term P_{pert} is a matrix, accounting for the fact that the fields here are vector instead of scalar quantities. The most common perturbation $\overline{\overline{P_{\mathrm{pert}}}}$, in a fiber is a bend, which induces birefringence (i.e., it has an off-diagonal matrix element that mixes the orthogonal components of $\overline{F_{L,m}}$ and $\overline{F_{L',m'}}$), and imparts OAM (i.e., it has a matrix element of the form $e^{i\Delta L\varphi}$ that spans all ΔL, although $|\Delta L| = 1$ is often the strongest component). Note that, although this integral was written with fields described in the OAM basis shown in Eq. (3), similar behavior may be expected of the integral written in other bases, because, after all, they are rotated MUBs of one another. Much physical intuition can be gained from inspecting the form of this integral.

Coupling is expected to be highest for degenerate modes, that is, when $\beta = \beta'$. Referring back to Eq. (2) and Figure 1, this clarifies why any bend causes polarization mixing between the two degenerate $L = 0$ modes, including SMF. Coupling between symmetric and antisymmetric modes, with $|\Delta L| = 1$, also appears to be easy with bends, and circumventing this has involved several instances of fiber designs or the use of modes where the effective-index separation (β–β') between desired and relevant undesired modes is exacerbated. The amount these modes should be separated depends on specific experimental conditions and modes of interest, but as a general rule, $\overline{\overline{P_{\mathrm{pert}}}}$, for a bent, flexible fiber comprises a z-dependent matrix element $e^{i2\pi z/L_c}$, where L_c is a correlation length representing characteristic beat lengths for the perturbations. Typically, $L_c \sim$ 1 mm to 1 cm, and hence it is easy to see that Δn_{eff} between modes of ~10^{-4} to 10^{-3} typically reduces the integral in Eq. (5), hence minimizing coupling. Early dispersion compensation efforts [60, 61] involved "W" fiber designs that separated the n_{eff} of the desired $\mathrm{LP}_{0,2}$ mode from the $\mathrm{LP}_{2,1}$ and $\mathrm{LP}_{1,2}$ modes (Δn_{eff} denoted as red arrows in Figure 3) by $\Delta n_{\mathrm{eff}} > 10^{-3}$. Simple step index fibers have a naturally mode-separating feature, where Δn_{eff} between a desired $\mathrm{LP}_{0,m}$ and the undesired $\mathrm{LP}_{1,m}$ or $\mathrm{LP}_{1,m-1}$ modes monotonically increases with radial order m [36] (green arrows in Figure 3). This feature has been used for scaling the A_{eff} of fiber modes [62], and stable modes with $A_{\mathrm{eff}} \sim 6000\ \mu\mathrm{m}^2$ and yet $\Delta n_{\mathrm{eff}} > 5 \times 10^{-4}$ have now been demonstrated [63] for fiber laser applications. Ring-core designs [64] perform two functions that enable stable OAM mode propagation [65, 66]. Their high index steps [57, 67, 68] help exacerbate the mode separations between SOa and SOaa modes (or, equivalently, HE and EH modes) to the order of $\Delta n_{eff} > 5 \times 10^{-5}$ (blue arrow in Figure 3), and the thin ring helps minimize the number of other radial order modes [69] that might be accidentally degenerate (and hence inadvertently mix) with the desired OAM modes. This design methodology has analogies with spin–orbit coupling for electrons in confined potentials [70], and has enabled device length (>10 m) fiber propagation of 24

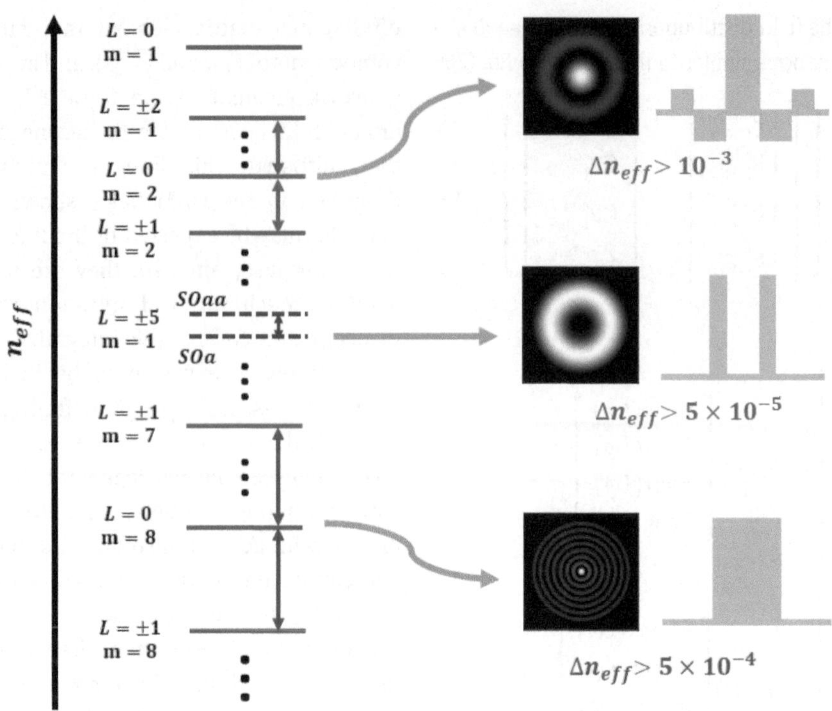

Figure 3: n_{eff} distributions for select modes with indices L and m. For visual clarity, not illustrated are polarization degeneracies of any of the modes or the n_{eff} degeneracies of any except for the $|L| = 5$ mode. Orange arrows leading to representative mode images for corresponding fiber designs (gray shaded features) describe mode separations (quantified by Δn_{eff}) for a select class of modes in their respective fibers. The n_{eff} of the $\text{LP}_{0,2}$ mode can be separated from $\text{LP}_{2,1}$ and $\text{LP}_{1,2}$ modes (red arrows) by using the class of "W" shape fiber designs, enabling its stable propagation for dispersion control designs. The n_{eff} splitting of $|L| = 5$ OAM modes (blue arrow) can be at least 5×10^{-5} in ring-core fiber designs, such that both $|L| = 5$ SOa and SOaa modes can propagate stably, yielding OAM mode stability in fibers. The n_{eff} splitting between $\text{LP}_{0,8}$ and its nearest neighboring $\text{LP}_{1,m}$ modes can be larger than 5×10^{-4} in simple step-index fibers (green arrows), enabling stable guidance of modes with large A_{eff}.

modes [69], to date. Equation (5) and the form of $\overline{\overline{P_{\text{pert}}}}$ point to an interesting mode stability criterion for OAM modes with high L indices. $\overline{\overline{P_{\text{pert}}}}$ from a bent fiber causes birefringence, and this is especially efficient in mixing two degenerate states of orthogonal polarizations – this was discussed in the context of $L = 0$ modes, which described why $\text{LP}_{0,m}$ modes seldom maintain polarization in a bent fiber. Because the cylindrical symmetry of an optical fiber implies that all modes have, at least, twofold (polarization) degeneracy (see Eqs. (1)–(4)), one expects this mixing to be commonly encountered, as in SMF or $\text{LP}_{0,m}$ modes. For low $|L|$ modes, this has indeed been observed [71]. But as $|L|$ of OAM modes increases, note that coupling between their degenerate counterparts additionally involves changing OAM from $+L$ to $-L$, that is, by $|\Delta L| = |2L|$. Because $\overline{\overline{P_{\text{pert}}}}$ primarily induces $|\Delta L| = 1$ coupling, mixing between degenerate OAM modes reduces with $|L|$. Hence, even degenerate states of high $|L|$ OAM modes in suitably designed fibers do not mix [72]. That is, select modes can be polarization maintaining even in strictly circular fibers.

The (bend) perturbations considered thus far assumed in-plane redirection of light. Although it is possible to generalize Eq. (5) to consider more complex perturbations, considerable insight is obtained by independently considering the perturbation associated with a slow, adiabatic redirection of light in 3D space (out of plane). After all, this is a rather common perturbation encountered with a flexible fiber. Equation (5) suggests that, in the absence of other perturbations (bend-induced birefringence or angular momentum exchange, considered previously), such a slow, adiabatic change would have no effect. However, in fact, it does. The phase added by geometrical transformations is distinct from the more common propagating phase associated with βz of a beam of light. The discovery and exposition of this geometric phase, radically different from the propagating dynamic phase, dates back to the seminal report by S. Pancharatnam in 1956 [73]. It took ~30 years for its significance to be appreciated, awaiting the generalization of this concept in quantum mechanics by M. Berry [74]. One important manifestation of this concept is the spin-redirection phase demonstrated by Tomita et al. [75]. A carefully constructed experiment with SMF showed that a fiber, configured to traverse a 3D route in space, acquired phase that was dependent only on the solid angle subtended by the fiber path in momentum (wavevector) space. The sign of this phase depends on the handedness of the circular polarization of a photon. Each degenerate mode in the $CP_{0,1}$ basis (see Eq. (2)) acquires a geometric phase $\varnothing_g$ of sign opposite to that of its polarization ($\hat{\sigma}^{\pm}$). This concept is intimately related to the idea that light carrying circular polarization denotes photons carrying spin angular momentum (SAM) [76], and that a 3D path of light imparts extrinsic angular momentum to it. Thus, an $LP_{0,1}$ mode, which is the linear combination of two orthogonal $CP_{0,1}$ modes, rotates under such geometric perturbations. Fundamentally distinct from the conventional dynamic phase (which includes, birefringence, angular momentum exchange, etc., discussed earlier) with dependence on propagation length, the geometric phase stores "memory" of the evolution (like geometry of the pathway) of a lightwave [77, 78]. The ray trajectory can be continuously deformed into any shape without changing the geometric phase as long as the solid angle remains unchanged, pointing to the topological nature of the effect. The composite effect of (bend-induced) birefringence as well as geometric perturbations on SMFs is illustrated in Figure 4. The first section of the fiber illustrates only a 3D path (geometric transformation – the fact that the fiber is lifted out of plane is schematically illustrated by a shadow it subtends, in-plane). This adds phase in the $CP_{0,1}$ basis, hence a single $CP_{0,1}$ mode merely acquires a phase. By contrast, the $LP_{0,1}$ mode rotates in polarization orientation. After that, the second section of the fiber illustrates a conventional bend that induces birefringence, which serves to convert both the $CP_{0,1}$ as well as $LP_{0,1}$ modes into modes with arbitrary elliptical polarization states.

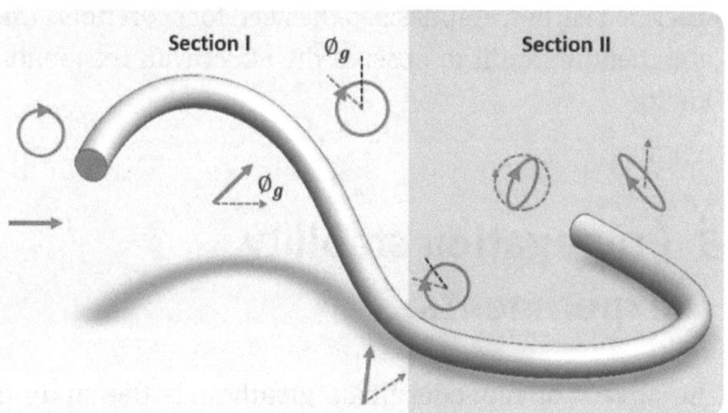

Figure 4: The effect of nonplanar and birefringent perturbations on the polarization of light launched into single-mode fibers (SMFs), illustrated as a flexed gray cylinder. Solid red arrows represent linear polarization states of light in the fiber at different positions along the propagation direction, with the dashed arrows denoting the state it possessed just before propagating to that position. Similarly, arrows on red circles represent circular polarization states and ellipses denote arbitrary elliptically polarized states. Mode transformations described below assume that light enters the fiber at the upper left end. The first section (I) represents an out-of-plane path (schematically illustrated by a shadow it subtends in plane) that is of large bending radii and hence free from fiber birefringence. The second section (II) represents an in-plane path that has strong birefringence. For a fundamental $CP_{0,1}$ mode with right-hand circular polarization, the nonplanar path imparts an extra phase $\varnothing_g$ due to geometric effects arising from Pancharatnam–Berry phases, but otherwise does not perturb the polarization state. This extra phase is illustrated by an azimuthal shift of the arrowhead. A mode with left circular polarization would behave similarly but accumulate an opposite phase $\varnothing_g$. By contrast, a fundamental $LP_{0,1}$ mode with horizontal polarization, being a linear combination the two orthogonal circular polarizations, is rotated by an angle $\varnothing_g$ along the nonplanar path. The subsequent bend of section II induces birefringence, and hence transforms any $LP_{0,1}$ or $CP_{0,1}$ mode into arbitrary elliptically polarized states, with ellipticity and handedness controlled by the strength of the bend-induced birefringence.

The preceding analysis was restricted to the fundamental mode of SMF, that is, the $L = 0$, $m = 1$ mode (though similar behavior is expected for higher m $LP_{0,m}$/ $CP_{0,m}$ modes). In these modes, the only contribution to angular momentum arises from the polarization ($\hat{\sigma}^{\pm}$). This concept is extendable to beams carrying OAM in addition to SAM, and the resultant geometric phase is given by [79]:

$$\varnothing_g(C) = -(\sigma + L)\Omega(C) \tag{6}$$

where C represents the path contour, $\Omega(C)$ represents the solid angle subtended by this path in momentum space, and σ represents the handedness of circular polarization or amount of SAM, taking values of ± 1 for light with $\hat{\sigma}^{\pm}$, and all other quantities have been previously defined. As is evident, geometric phase is enhanced for OAM modes, and depends on the total angular momentum (TAM) of a photon. Observing this effect has historically been obscured by the fact that fiber modes experience all aforementioned dynamic and geometric perturbations simultaneously. One report [80], describing the strength of a so-called optical Magnus effect, showed that the speckle pattern out of a multimode fiber rotates when changing the sign of circular polarization, with the effect being proportional to the solid angle subtended by the fiber coil in the momentum space. A fiber in which low $|L|$ OAM or vector modes were excited showed [81] the rotation of polarizations pattern to be explicitly dependent on mode order L. Unfortunately, concurrent birefringent and angular momentum coupling implied that the experimentally observed rotation did not match the theory well because the SOa and SOaa modes also coupled due to bends and birefringence. In fact, the first experiment [75] with SMFs, described earlier, emphasized the need for short fibers and large bending radii to observe the effect with reasonable fidelity.

3 Propagation stability experiments

The advent of ring-core fibers greatly aids the study of propagation effects in perturbed fibers because of the ability to isolate the effects of the disparate perturbations described in section 2. As described earlier, the ring-core fiber minimizes coupling within the mode group – that is, coupling between the SOa and SOaa pairs of modes (see Figure 3), and the angular momentum conservation effect forbids coupling between degenerate states for high enough $|L|$.

In this section, we describe experiments probing the propagation stability of different fiber modes. We do not consider the $L = 0$ LP or CP modes because the fundamental SMF mode has been well-studied and higher $LP_{0,m}$/ $CP_{0,m}$ modes essentially behave similarly in fibers where they are sufficiently isolated. We also limit ourselves to the study of $|L| > 1$ modes but with $m = 1$ because higher radial orders (higher m) modes also essentially behave similarly. We first consider modes in the LP basis, shown in the top row of Figure 2(a). Note that each LP mode is a linear combination of two vector or OAM modes of distinct propagation constants β' and β'', respectively – indeed, this is the origin of the average $\breve{\beta}$ defined under this scalar approximation $\left(\breve{\beta} \sim \frac{\beta' + \beta''}{2}\right)$. Hence, these modes are fundamentally unstable, in that their spatial patterns and polarizations rotate upon propagation, even in a strictly straight, perturbation-free fiber. This is evident from inspecting the two distinct LP mode images in Figure 2(a) arising from the addition or subtraction of nondegenerate vector or OAM modes – the nondegeneracy necessarily means that the phase difference between them oscillates with propagation. As such, LP modes of $L \neq 0$ are *never* stable in a fiber, and their use necessarily requires pre- or post-processing, optically or electronically, for information recovery.

Figure 5(a) shows the experimental setup [82] used for studying the mixing of two degenerate OAM or vector modes, in a 4-m long ring-core fiber [72] supporting stable propagation (i.e., without SOa–SOaa mixing) of high-$|L|$ modes ($|L| = 5, 6, 7$). The incoming Gaussian beam at 1550 nm from an external cavity laser (ECL) is converted into the desired OAM or vector mode using a PBOE called q-plate [52, 83]. A q-plate with topological charge q can project circular polarization onto OAM modes of order $|L| = |2q|$, with the spin–orbit alignment dependent on the sign of q (Eq. (7)).

$$A\hat{\sigma}^+ + B\hat{\sigma}^- \xrightarrow{q} A\hat{\sigma}^- e^{i2q\varphi} + B\hat{\sigma}^+ e^{-i2q\varphi} \tag{7}$$

where the arrow denotes the transformation induced by the q-plate and A and B are mode amplitudes.

In the following representative experiments, we used $q = 7/2$, which causes Gaussian beams of two circular polarizations $\hat{\sigma}^+$ and $\hat{\sigma}^-$ to be converted into two degenerate SOaa OAM modes of $L = +7\hat{\sigma}^-$ and $L = -7\hat{\sigma}^+$, respectively. The purity of OAM modes excited in the fiber is confirmed to be greater than 15 dB via spatial interferometry [71]. The output of the ring-core fiber is then converted back to a fundamental Gaussian-shaped free-space modes using an identical q-plate. Thereafter, with appropriate polarization optics, the two orthogonal polarization components of the output beam can be spatially separated and projected onto a camera. Therefore, the power ratio between polarization bins on the camera represents the mixing ratio of two degenerate OAM modes in the fiber. By launching a pure mode into the fiber, its stability within the fiber can be deduced by measuring the relative power scattered into its degenerate counterpart. The combination of q-plate and wave plates can not only generate two degenerate OAM modes but also any of their linear combinations. Because linear polarizations are linear combinations of circular polarizations and vector modes are linear combinations of OAM modes, Gaussian beams with two orthogonal linear polarizations can be mapped onto two degenerate vector modes ($A = \pm B$ in Eq. (7)). Reciprocally, measuring the power ratio in the two linearly polarized bins yields the mode-mixing ratio between the two degenerate vector modes. By switching the polarization between circular and linear using quarter-wave plates (or their lack, thereof), we are able to switch between the OAM and vector modal bases with ease while maintaining all other experimental (perturbative) conditions. Representative experimental results on propagation stability in the presence of in-plane and out-of-plane bends are shown for the $|L| = 7$ SOaa OAM modes and the corresponding mathematically rotated basis of $EH_{6,1}$ odd and even modes.

The plots in Figure 5(c) and (d) show the measured power fluctuations between two degenerate OAM and vector modes, respectively, as the fiber is bent, in plane, as illustrated in Figure 5(b). When the input is an OAM mode ($L = -7\hat{\sigma}^+$), a negligible amount of power (~−12 dB) is scattered to its degenerate counterpart. This is consistent with earlier observations that high $|L|$ OAM modes are stable, even between degenerate modes, in ring-core fibers [72]. For a vector mode input ($EH_{6,1}^{even}$), the power of parasitic degenerate mode ($EH_{6,1}^{odd}$) remains mostly low, at

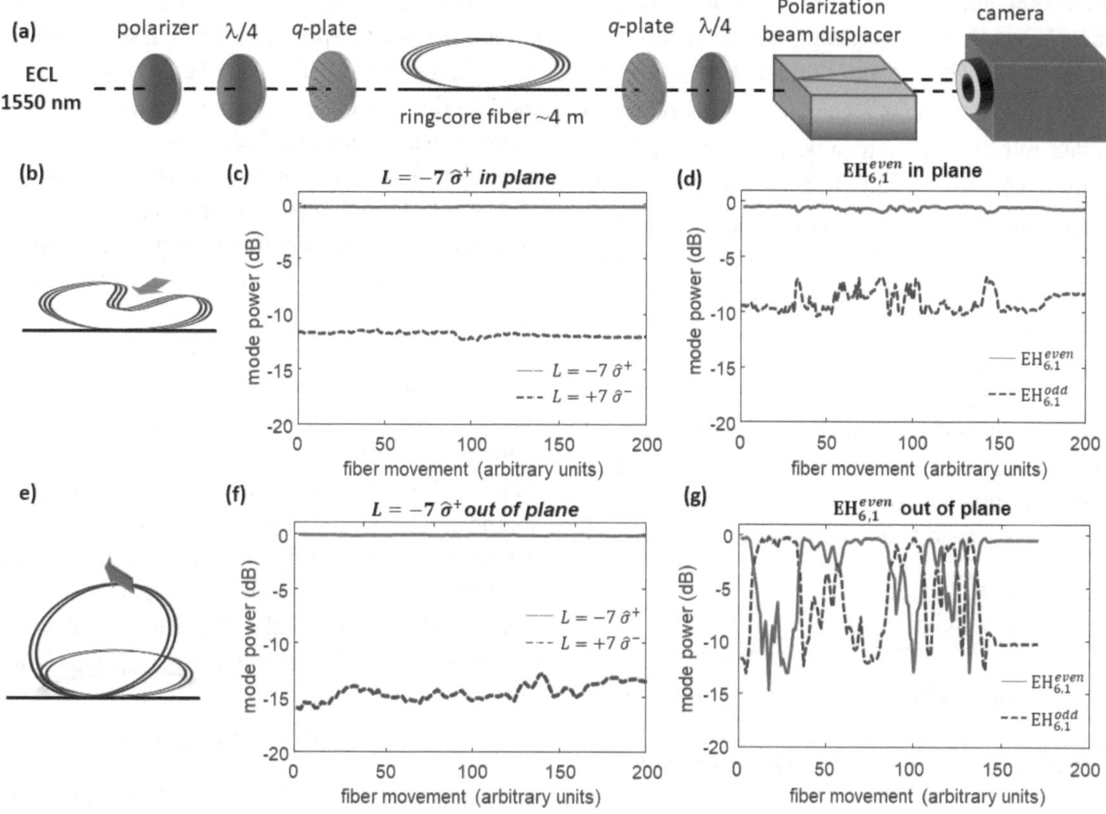

Figure 5: (a) Experimental setup used for studying the mixing ratio of two degenerate orbital angular momentum (OAM) or vector modes. For the OAM basis, the quarter-wave plate ($\lambda/4$) is rotated to an angle such that the fast axis is 45° with respect to the axis of polarizer to generate circularly polarized light, which is converted to OAM modes by the q-plate. In contrast, for the vector basis, the fast axis of the quarter-wave plate is aligned with the axis of the polarizer, such that linearly polarized light can be projected on to corresponding vector modes; A reciprocal setup at the fiber output converts the modes back to Gaussian beams, with the power in each polarization bin being proportional to the power of the individual degenerate (OAM or vector) modes at the fiber output. For the vector mode measurement, the output quarter-wave plate is removed. When measuring OAM mode stability, the input was a ~15-dB pure $L = -7\hat{\sigma}^+$ mode, whereas, during the vector mode stability measurements, the input was a ~10-dB pure $\mathrm{EH}_{6,1}^{\mathrm{even}}$ mode. (b) Schematic of fiber loops being bent in-plane to varying degrees, during the measurement. Part of the fiber loops (radius ~12 cm) are gradually bent into radius ~4 cm, and then return back to the original; (c) Plot of relative power in the two degenerate OAM modes ($L = \mp 7\hat{\sigma}^{\pm}$) under the application of perturbations as shown in (b); (d) Corresponding plot of relative power in the two degenerate $\mathrm{EH}_{6,1}$ modes for in-plane perturbations; (e) Schematic of fiber partly (2 out of 4 fiber loops) being lifted out of plane to different heights during measurements. The plane of lifted fiber is moved to the plane perpendicular to the original and then moved back; (f) Plot of relative power in the two degenerate OAM modes ($L = \mp 7\hat{\sigma}^{\pm}$) as the fiber is moved out of plane, as shown in (e); (g) Corresponding plot of relative power in the two degenerate EH_{61} modes for out-of-plane perturbations. OAM modes remain stable to bend as well as 3D perturbations, while the vector modes are completely mixed by 3D path redirections of the fiber.

the −10 dB level. Some power jumps to around −7 dB are evident. This is due the experimental inability of a strictly in-plane perturbation. Later, we will describe the origin of this discrepancy, but a higher level summary of these two experiments is that OAM and vector modes remain robust to degenerate mode coupling in the presence of in-plane bend perturbations. Given that OAM mode stability was already known [72], this was to be reasonably expected, given that its mathematically equivalent counterpart, the vector modes, would also possess similar stability. The plots in Figure 5(f) and (g) show the mode mixing between two degenerate OAM modes and vector modes as the fiber is moved out of plane, as illustrated in Figure 5(e). Again, for the OAM mode input ($L = -7\hat{\sigma}^+$), the power of parasitic degenerate mode remains at a very low level (~14 dB). By contrast, when the input is the $\mathrm{EH}_{61}^{\mathrm{even}}$ mode, the two degenerate modes completely mix with each other with 3D fiber perturbations.

This curious result, of two mathematically identical sets of modes behaving differently under 3D perturbations is a manifestation of the geometric phase discussed in section 2. An OAM mode traversing a nonplanar path (modified mode represented as $\widetilde{\mathrm{OAM}}$) obtains an extra phase factor compared with the input (Eq. (8)). As two degenerate OAM modes have opposite sign of L and σ, the geometric phases they accumulate, as per Eq. (6), have opposite signs. The vector modes under such a perturbation ($\widetilde{EH}_{6,1}^{even}$) remain a linear combination of perturbed OAM modes, but they are now projected onto two degenerate vector modes $EH_{6,1}^{even}$ and $EH_{6,1}^{odd}$, as shown in Eq. (9). Hence, out-of-plane geometric perturbations fundamentally lead to mode mixing in the vector basis but not in the OAM basis. Note that this result follows previous experiments on geometric phases [78, 79], but here, realistic lengths of fibers could be used, by contrast, because the ring-core fiber design and use of high $|L|$ modes helped avoid the competing effects of mode coupling due to bends and birefringence. Although the length of fiber used in this experiment was only 4 m, OAM stability in ring core fibers has been observed up to 13.4 km propagation lengths [84].

$$\begin{cases} \widetilde{OAM}_{+L}^{-} = OAM_{+L}^{-} e^{i\varnothing_g} \\ \widetilde{OAM}_{-L}^{+} = OAM_{-L}^{+} e^{-i\varnothing_g} \end{cases} \tag{8}$$

$$\begin{cases} \widetilde{EH}_{L-1,1}^{even} = \frac{1}{2}\left(\widetilde{OAM}_{+L}^{-} + \widetilde{OAM}_{-L}^{+}\right) = \cos\varnothing_g \cdot EH_{L-1,1}^{even} - \sin\varnothing_g \cdot EH_{L-1,1}^{odd} \\ \widetilde{EH}_{L-1,1}^{odd} = \frac{1}{2i}\left(\widetilde{OAM}_{+L}^{-} - \widetilde{OAM}_{-L}^{+}\right) = \sin\varnothing_g \cdot EH_{L-1,1}^{even} + \cos\varnothing_g \cdot EH_{L-1,1}^{odd} \end{cases} \tag{9}$$

4 Geometric phase control

To quantitatively study the effect of geometric phase on high $|L|$ modes, we configure the fiber into a uniform helix shown in Figure 6(a). Input and output OAM and vector modes are shown schematically, for visual clarity. Although both feature a donut-shaped intensity profile, the illustrations here show spiral patterns for OAM modes, obtained when an OAM mode is interfered with an expanded Gaussian (with the number and orientation of parastiches denoting L and its sign, respectively). Likewise vector modes are schematically illustrated by their projection patterns, obtained when they are imaged through a polarizer (with the number of "beads" denoting $|2L|$). Figure 6(b) shows that the k-vector of light in a helical path encloses a solid angle, $\mathbf{\Omega}$. As the fiber helix is compressed, the solid angle increases accordingly, resulting in an extra geometric phase of the beam traveling in the fiber, without changing the path length of the light (and hence its dynamic phase). This solid angle is related to the period of the helix $\mathbf{\Lambda}$, by $\Omega = 2\pi(1 - \Lambda/l)$, where l represents the length of fiber in one loop [75]. This helical arrangement is realized by loosely inserting the fiber into a Teflon tube with 1-mm inner diameter to minimize any torsion, stress, or stretching during winding. This Teflon tube is then adhered to a metal spring, with which solid angle $\mathbf{\Omega}$ can be controllably varied. This level of care is not needed for the OAM mode, which is stable, but is required for the vector mode, which, as shown in Figure 5(g) is sensitive to 3D fiber movement. Part of a 3.4-m-long segment of a ring-core fiber of length is wound into a uniform helix of 6.5 loops (light propagating in and out of fibers are in opposite directions). The length of each loop l is 16.3 cm. Note that the k-vector of the mode in the fiber, with magnitude β (its propagation constant), is well approximated to be parallel to the axis of the fiber under the weakly guiding approximation. Therefore, the solid angle subtended by the k-vector is approximately equal to the solid angle spanned by the fiber's path, which follows that of the metal spring. As the spring is compressed, the pitch period $\mathbf{\Lambda}$ decreases from 2.3 to 0.2 cm, and the solid angle correspondingly increases from 1.72π to 1.96π. The total geometric phase acquired for a mode in this setup is equal to the geometric phase acquired in one loop multiplied by the number of loops N, as shown in Eq. (10):

$$\varnothing_g(C) = -N(\sigma + L)\Omega(C) \tag{10}$$

For an OAM mode with $L = -7\hat{\sigma}^{+}$, the extra geometric phase results in a counterclockwise rotation of the beam. This rotation angle is equal to the additional phase divided by OAM order L, or $\Theta = \varnothing_g(C)/L$. Similarly, an OAM mode with $L = +7\hat{\sigma}^{-}$ would obtain an extra geometric phase of the same amount but with opposite sign. As the signs of both OAM order and geometric phase are flipped, the beam rotation would have the same magnitude and direction. Therefore, the vector modes (EH_{61} odd/even), which are the combinations of the two degenerate OAM modes, rotate counterclockwise with the same angle (shown schematically on the output patterns of Figure 6(a)). However, this rotation would lead to power oscillation between the odd and even degenerate vector modes because they are not rotationally invariant.

Figure 6(c) shows the measured power fluctuations between even and odd modes as the spring is gradually compressed, for an input comprising a pure $EH_{6,1}^{even}$ mode. As with the previous experiment of Figure 5(g), the two degenerate vector modes mix completely. The main difference is that the oscillation is now periodic and systematic because the geometric phase is accumulated monotonically, in a controlled fashion, with the helical arrangement. Based on the rotation angle of the vector modes, the accumulated geometric phase of the OAM modes can be calculated. As shown in Figure 6(d), the geometric phase shows a linear relationship with the solid

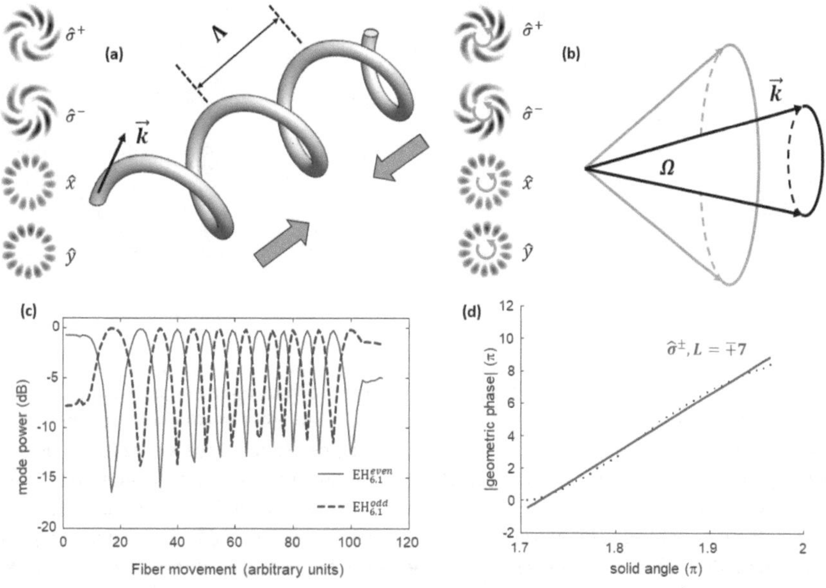

Figure 6: Systematic control of Pancharatnam–Berry phases in optical fibers.
(a) An orbital angular momentum (OAM) mode supporting ring-core fiber, inserted in a loose Teflon tube, is attached to a flexible spring to configure a helical with variable period Λ. A combination of two degenerate OAM modes ($L = -7\hat{\sigma}^+$ and $L = +7\hat{\sigma}^-$) is launched into the fiber. For visual clarity, the OAM modes are illustrated with spiral patterns that arise from their interference with an expanded Gaussian beam. The corresponding projection of this superposition state into $\hat{x}$ and $\hat{y}$ polarization-bins yields petal patterns, illustrated at the bottom left hand side of (a). The mode illustrations on the right side of the helical arrangement show the corresponding modes at the output of the fiber, all of which are rotated counterclockwise due to the geometrical transformation; (b) Geometric illustration of the solid angle **Ω** enclosed by the k-vector of the light path in this helical arrangement. The illustration in orange depicts the higher **Ω** obtained from compressing a spring from its original state, depicted in the black illustration; (c) Measured power fluctuations between even and odd modes for EH_{61}, using the reciprocal mode transmission setup of Figure 5(a), when the spring was compressed (pitch period Λ decreases from 2.3 to 0.2 cm). The uneven periodicity results from the uneven speed with which the spring was compressed; (d) Geometric phase, measured from vector mode power ratios, versus solid angle, for two degenerate OAM modes $L = -7\hat{\sigma}^+$ and $L = +7\hat{\sigma}^-$ as the helical spring is compressed. The red line is a linear fit of the experimental data (solid circles). Near-linear relationship shows exclusive influence of fiber path on geometric phase, and hence relative mode amplitudes.

angle, which matches the theoretical prediction (Eq. (10)) of a linear relationship between these parameters. It clearly shows that image rotation, and hence mixing, of vector modes is linearly proportional to the solid angle enclosed by the k-vector of light.

We repeat this experiment on five other pairs of modes that are stable in this ring-core fiber. As shown in Figure 7(a), the geometric phase shows a linear relationship with the solid angle in all cases. As evident, image rotation, and hence vector mode instability, increases as the TAM (equal to $L + \sigma$) of participating modes increases. Figure 7(b) shows that the slopes for each pair of OAM modes is linearly proportional to the TAM of the corresponding OAM modes. The magnitude of this slope (i.e., slope of the slopes vs. TAM), which is, effectively the number of loops N (per Eq. (10)), is 6.2, which is close to the expected value of $N = 6.5$. The lack of a better match may be due to the fact that input mode purity was only 10 dB, but even so, this confirms that the perturbations experienced by these modes predominantly arise from the experimentally induced geometric, and not inadvertent bend or birefringence, perturbations

While the controlled experiments helped rigorously verify the influence of different kinds of perturbations on an optical fiber and, especially, their influence on different modes, the results of Figure 7(a) also point to applications toward a novel type of shape sensor, with sensitivity controlled by the OAM content of light in a fiber. One key distinction from other types of interferometric sensors that depend on the conventional dynamic phase of light is that this depends only on geometry. As mentioned earlier, dynamic phase arises from $e^{i\beta z}$, which is strongly wavelength dependent. By contrast, there is no wavelength dependence in geometric phase, meaning that its sensitivity does not depend on the bandwidth of the source, facilitating the use of low-cost sources even in high-sensitivity applications. Likewise, the lack of dynamic phase dependence also makes such sensors robust to ambient perturbations,

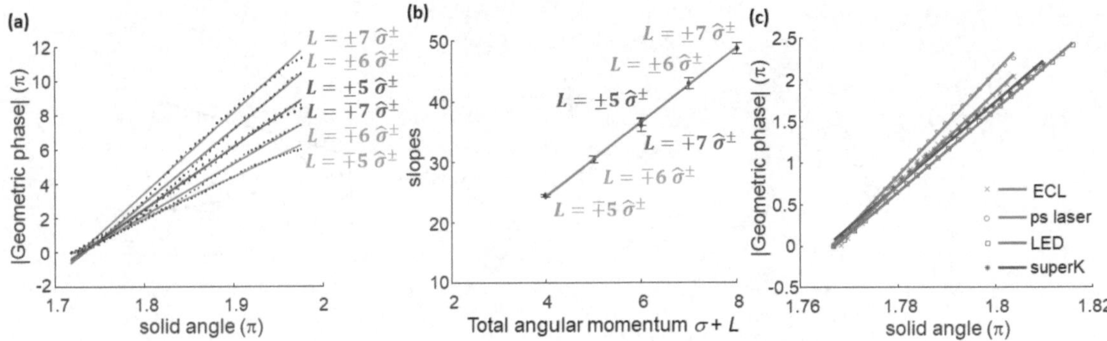

Figure 7: (a) Geometric phase versus fiber path solid angle for six distinct pairs of orbital angular momentum (OAM) modes. The colored lines are linear fits of the experimental data (black solid circles). All modes show a clear linear relationship; (b) The slope of each trace (for each mode) shown in (a) versus the TAM = $L + \sigma$ of the respective modes. The high degree of linearity as well as the slope of this line match well with theory that accounts only for geometric effects; (c) Geometric phase versus solid angle for two degenerate OAM modes $L = \pm 7\hat{\sigma}^{\pm}$ (as shown earlier in (a)), using multiple light sources. External cavity laser (ECL) is a narrow linewidth (100 kHz) source at 1550 nm; ps laser is a picosecond laser at 1550 nm with ~0.5 nm bandwidth; the LED has a bandwidth ~35 nm around 1525 nm; superK represents a supercontinuum source with 3 dB bandwidth up to ~250 nm at around 1475 nm. Similar slopes for all these sources demonstrates weak dependence of wavelength on geometric phase.

such as temperature or pressure induced changes of the refractive index of the fiber.

To demonstrate this independence to wavelength, we conduct the same helix experiment with $|L|$ = 7 SOa modes ($L = \pm 7\hat{\sigma}^{\pm}$) using light sources of different bandwidths. Figure 7(c) shows the measured geometric phase as a function of solid angle, just as in Figures 6(d) and 7(a), when using sources of varying bandwidths. ECL denotes a narrow-linewidth (100 kHz) source at 1550 nm; "ps laser" is a picosecond laser at 1550 nm with ~0.5 nm bandwidth; the LED has a bandwidth ~35 nm around 1525 nm; and "superK" represents a supercontinuum source with 3-dB bandwidth of ~250 nm centered at ~1475 nm. The spatial interferometry method used to previously guarantee mode purity does not work with broadband sources (because it utilizes *dynamic* phase!). Hence, the purity of OAM modes is adjusted to be higher than 15 dB using the ECL, as before, and then the light source is carefully switch to other broadband sources without disturbing alignment, expecting minimal changes in mode purity. The geometric phase shows a linear dependence on solid angle regardless of the bandwidth of light source, as shown in Figure 7(c). However, the slopes obtained with the broadband sources differ from that obtained with the ECL by up to 12%. The mismatch probably arises from the lack of our ability to maintain high-purity excitation with the broadband sources, a problem easily solved in the future with the plethora of emerging mode-conversion technologies for OAM fiber modes [85]. Nevertheless, the results point to a novel means of developing low-cost shape sensors that are insensitive to environmental perturbations such as temperature, pressure, mechanical vibrations and bends, while maintaining high sensitivity.

5 Discussion, summary, and conclusions

In summary, the ability to excite and propagate specific individual or a subset of HOMs in an optical fiber enables a variety of applications ranging from scaling information capacity and enabling new nonlinear interactions, to new forms of sensors and photonic devices. The key in several of these applications is the ability to exploit specific, distinct characteristics of HOMs, such as enabling unique nonlinear coupling pathways, yielding large A_{eff} or tailorable dispersion, and realizing large unmixed states of information carriers. In these applications, finding the subset of modes that propagate with high linear stability is of paramount importance. This linear stability is intimately connected to properties of the modes themselves (their angular momentum content, and even the mathematical basis used to describe them) as well as the form of perturbations a fiber encounters. Considering the two most common fiber perturbations – bends, which induce birefringence as well as OAM transfer, and light's path memory, manifested in the 3D trace that light follows – we arrive at the following conclusions related to modal stability for HOMs (illustrated in Figure 8). The commonly used LP modes are actually linear combinations of eigenmodes of dissimilar phase velocities, and thus they are not

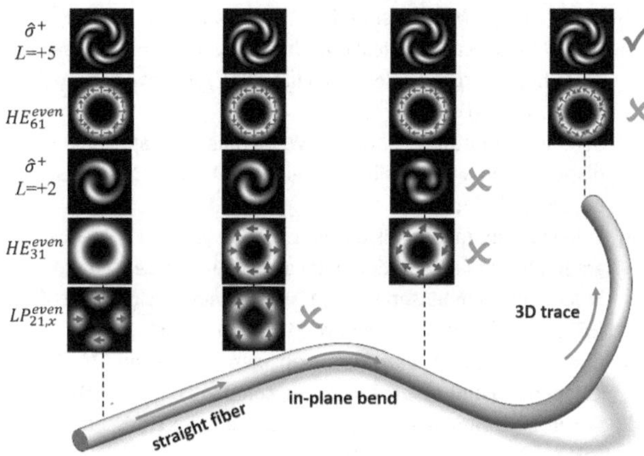

Figure 8: Summary of the propagation stability of optical fiber modes of different classes when the fiber is deployed with commonly encountered perturbations. The illustration depicts modes launched at the left end of a perfectly circular fiber, and all modes are schematically illustrated at four positions (black dashed lines) along the fiber propagation axis. From left to right: The input comprises pure modes in all the classes; the second position represents propagation through a straight fiber without any perturbations; this is followed by a position after propagation through a fiber that is bent only in plane; and, the final position represents propagation through a fiber that experiences an out-of-plane (3D) redirection as well. The modes from bottom to top represent LP modes ($LP_{2,1}$), low order vector modes ($EH_{3,1}^{even}$), low-order OAM modes ($L = 2\hat{\sigma}^+$), high-order vector modes ($EH_{6,1}^{even}$), and high-order OAM modes ($L = 5\hat{\sigma}^+$). After propagating through the straight fiber section, LP modes mix between orthogonal modes of the same class, whereas all the other modes remain stable. As such, LP modes, designated at this point with an orange cross, are not illustrated across subsequent perturbations, having failed to propagate through the most elementary arrangement. In-plane fiber bends easily couple a low-order vector ($EH_{3,1}^{even}$) or orbital angular momentum (OAM) ($L = 2\hat{\sigma}^+$) mode with its degenerate counterpart, while it does not impart enough angular momentum to couple a higher-order OAM mode $L = 5\hat{\sigma}^+$ or the corresponding vector mode EH_{61}^{even} to their respective degenerate counterparts. Again, therefore, no further depiction of low L vector or OAM modes is illustrated (as indicated by the orange cross). After propagating through the 3D trace, a higher-order vector mode $EH_{6.1}^{even}$ mixes completely with its degenerate counterpart, whereas the corresponding OAM mode $L = 5\hat{\sigma}^+$ merely acquire a common geometric phase and remains remarkably stable. Hence, across all perturbations, only the high $|L|$ OAM mode survives without coupling, to its degenerate or nondegenerate counterparts.

translationally invariant even in a perfect, straight fiber. On the other hand, the vector and OAM modes, as two mathematically equivalent bases for mode representations, remain stable in an unperturbed fiber. However, when their modal index $|L|$ is low, they mix completely with their degenerate counterpart in a fiber that is bent (in-plane) because of birefringent coupling that couples polarizations in SMFs too. High-$|L|$ vector and OAM modes are, by contrast, stable even across (in-plane) bent fibers because of inherent OAM conservation rules. Finally, when a fiber is not only bent, but also lifted out of plane, even high-$|L|$ vector modes become unstable, in that their polarization distributions rotate, because of the Pancharatnam–Berry phase that light accumulates in 3D paths. By contrast, a high-$|L|$ OAM mode remains remarkably stable, except for accumulating a common phase. Hence, as mode propagation is studied across a range of perturbations, starting from none (straight fiber) to bends, to, finally, 3D paths, modes of the same L and m indices, but represented in different mathematical bases are not, somewhat counterintuitively, identical. Considering all these perturbations, OAM modes of sufficiently high $|L|$ are the most stable eigenmodes of a circularly symmetric optical fiber. A few important clarifications are in order: this stability is observable only once a fiber is designed such that n_{eff} splittings between pertinent nondegenerate modes is maximized, and this analysis ignores very long (>> km) length propagation, where higher order effects of bends and twists may play a role. In such cases, one would expect modes of any class to mix, although the fundamental nature of the effects described here suggest that even in conditions where all modes mix, the circularly polarized OAM modes will likely be more robust compared with others. In addition, the OAM eigenbasis yields a stable platform in which to exploit path-memory effects arising from geometric transformations (Pancharatnam–Berry phases), which while studied extensively in free space, may now lead to new opportunities for wavelength-agnostic or wavelength-insensitive phase control with fibers. We expect that these findings will help inform the optimal modal basis to use in the variety of applications that envisage using HOMs of optical fibers.

Acknowledgments: The authors would like to thank Dr. P. Kristensen for manufacturing the ring-core fibers used in these experiments, and Drs. P.G. Kwiat, P. Gregg and G. Prabhakar for insightful discussions.

Author contribution: All the authors have accepted responsibility for the entire content of this submitted manuscript and approved submission.

Research funding: This work is supported, in part, by the Vannevar Bush Faculty Fellowship (N00014-19-1-2632), Brookhaven National Labs (Contract: 354281), Office of Naval Research MURI program (N00014-20-1-2450) and the National Science Foundation (ECCS-1610190).

Conflict of interest statement: The authors declare no conflicts of interest regarding this article.

References

[1] H. H. Hopkins and N. S. Kapany, “A flexible fibre scope, using static scanning,” *Nature*, vol. 173, pp. 39–41, 1954.

[2] T. A. Birks, J. C. Knight, and P. J. St. Russell, “Endlessly single-mode photonic crystal fiber,” *Opt. Lett.*, vol. 22, p. 961, 1997.

[3] R. F. Cregan, B. J. Mangan, J. C. Knight, et al., “Singlemode photonic band gap guidance of light in air,” *Science*, vol. 285, pp. 1537–1539, 1999.

[4] F. Benabid, J. C. Knight, G. Antonopoulos, and P. J. St. Russell, “Stimulated Raman scattering in hydrogen-filled hollow-core photonic crystal fiber,” *Science*, vol. 298, pp. 399–402, 2002.

[5] P. Koplow, L. Goldberg, R. Moeller, and D. Klinder, “Singlemode operation of a coiled multimode fibre amplifier,” *Opt. Lett.*, vol. 25, p. 442, 2000.

[6] W. Wong, X. Peng, J. McLaughlin, and L. Dong, “Breaking the limit of maximum effective area for robust single-mode propagation in optical fibers,” *Opt. Lett.*, vol. 30, pp. 2855–2857, 2005.

[7] J. M. Fini, J. W. Nicholson, B. Mangan, et al., “Polarization maintaining single-mode low-loss hollow-core fibres,” *Nat. Commun.*, vol. 5, p. 5085, 2014.

[8] R. Ulrich, R. C. Rashleigh, and W. Eickhoff, “Bending-induced birefringence in single-mode fibers,” *Opt. Lett.*, vol. 5, pp. 273–275, 1980.

[9] P. K. A. Wai and C. R. Menyuk, “Polarization mode dispersion, decorrelation, and diffusion in optical fibers with randomly varying birefringence,” *J. Lightwave Technol.*, vol. 14, pp. 148–157, 1996.

[10] M. Brodsky, K. E. George, C. Antonelli, and M. Shtaif, “Loss of polarization entanglement in a fiber-optic system with polarization mode dispersion in one optical path,” *Opt. Lett.*, vol. 36, pp. 43–45, 2011.

[11] H. R. Stuart, “Dispersive multiplexing in multimode optical fiber,” *Science*, vol. 289, pp. 281–283, 2000.

[12] R. Ryf, N. K. Fontaine, S. Wittek, et al., “High-spectral-efficiency mode-multiplexed transmission over graded-index multimode fiber,” in *ECOC’18*, 2018, Th3B.1.

[13] B. Redding, S. M. Popoff, and H. Cao, “All-fiber spectrometer based on speckle pattern reconstruction,” *Opt. Express*, vol. 21, pp. 6584–6600, 2013.

[14] T. Cižmár and K. Dholakia, “Exploiting multimode waveguides for pure fibre-based imaging,” *Nat. Commun.* vol. 3, p. 1027, 2012.

[15] N. Borhani, E. Kakkava, C. Moser, and D. Psaltis, “Learning to see through multimode fibers,” *Optica*, vol. 5, pp. 960–966, 2018.

[16] A. Hasegawa, “Self-confinement of multimode optical pulse in a glass fiber,” *Opt. Lett.*, vol. 5, pp. 416–417, 1980.

[17] L. G. Wright, D. N. Christodoulides, and F. W. Wise, “Controllable spatiotemporal nonlinear effects in multimode fibres,” *Nat. Photonics*, vol. 9, pp. 306–310, 2015.

[18] K. Krupa, A. Tonello, B. M. Shalaby, et al., “Spatial beam self-cleaning in multimode fibres,” *Nat. Photonics*, vol. 11, pp. 237–241, 2017.

[19] K. Krupa, A. Tonello, A. Barthélémy, et al., “Observation of geometric parametric instability induced by the periodic spatial self-imaging of multimode waves,” *Phys. Rev. Lett.*, vol. 116, p. 183901, 2016.

[20] A. W. Snyder and J. D. Love, *Optical Waveguide Theory*, London, U.K., Chapman and Hall, 1983.

[21] A. Ghatak and W. J. Thyagarajan, *Introduction to Fiber Optics*, Cambridge, U.K., Cambridge University Press, 1998.

[22] C. D. Poole, J. M. Weisenfeld, D. J. DiGiovanni, and A. M. Vengsarkar, “Optical fiber-based dispersion compensation using higher order modes near cutoff,” *J. Lightwave Technol.*, vol. 12, pp. 1746–1758, 1994.

[23] S. Ramachandran, S. Ghalmi, J. W. Nicholson, et al., “Anomalous dispersion in a solid, silica-based fiber,” *Opt. Lett.*, vol. 31, pp. 2532–2534, 2006.

[24] S. Ramachandran, B. Mikkelsen, L. C. Cowsar, et al., “All-fiber grating-based higher order mode dispersion compensator for broad-band compensation and 1000-km transmission at 40 Gb/s,” *IEEE Photonics Technol. Lett.*, vol. 13, pp. 632–634, 2001.

[25] A. H. Gnauck, L. D. Garrett, Y. Danziger, U. Levy, and M. Tur, “Dispersion and dispersion-slope compensation of NZDSF over the entire C band using higher-order-mode fibre,” *Electron. Lett.*, vol. 36, pp. 1946–1947, 2000.

[26] S. Ramachandran, M. F. Yan, J. Jasapara, et al., “High-energy (nanojoule) femtosecond pulse delivery with record dispersion higher-order mode fiber,” *Opt. Lett.*, vol. 30, pp. 3225–3227, 2005.

[27] R. H. Stolen, J. E. Bjorkholm, and A. Ashkin, “Phase-matched three-wave mixing in silica fiber optical waveguides,” *Appl. Phys. Lett.*, vol. 24, no. 7, pp. 308–310, 1974.

[28] J. Demas, P. Steinvurzel, B. Tai, L. Rishøj, Y. Chen, and S. Ramachandran, “Intermodal nonlinear mixing with Bessel beams in optical fiber,” *Optica*, vol. 2, pp. 14–17, 2015.

[29] J. Demas, L. Rishøj, X. Liu, G. Prabhakar, and S. Ramachandran, “Intermodal group-velocity engineering for broadband nonlinear optics,” *Photonics Res.*, vol. 7, pp. 1–7, 2019.

[30] L. Rishøj, B. Tai, P. Kristensen, and S. Ramachandran, “Soliton self-mode conversion: revisiting Raman scattering of ultrashort pulses,” *Optica*. vol. 6, pp. 304–308, 2019.

[31] P. J. St. Russell, R. Culverhouse, and F. Farahi, “Experimental observation of forward stimulated Brillouin scattering in dual-mode single core fiber,” *Electron. Lett.*, vol. 26, pp. 1195–1196, 1990.

[32] F. G. Omenetto, A. J. Taylor, M. D. Moores, et al., “Simultaneous generation of spectrally distinct third harmonics in a photonic crystal fiber,” *Opt. Lett.*, vol. 26, pp. 1158–1160, 2001.

[33] A. Efimov, A. J. Taylor, F. G. Omenetto, J. C. Knight, W. J. Wadsworth, and P. J. St. Russell, “Nonlinear generation of very high-order UV modes in microstructured fibers,” *Opt. Express*, vol. 11, pp. 910–918, 2003.

[34] J. Demas, G. Prabhakar, T. He, and S. Ramachandran, “Wavelength-agile high-power sources via four-wave mixing in higher-order fiber modes,” *Opt. Express*, vol. 25, pp. 7455–7464, 2017.

[35] D. Cruz-Delgado, R. Ramirez-Alarcon, E. Ortiz-Ricardo, et al., “Fiber-based photon-pair source capable of hybrid entanglement in frequency and transverse mode, controllably scalable to higher dimensions,” *Sci. Rep.*, vol. 6, p. 27377, 2016.

[36] S. Ramachandran, J. W. Nicholson, S. Ghalmi, et al., “Light propagation with ultra-large modal areas in optical fibers,” *Opt. Lett.*, vol. 31, pp. 1797–9, 2006.

[37] K. S. Abedin, R. Ahmad, A. M. DeSantolo, and D. J. DiGiovanni, “Reconversion of higher-order-mode (HOM) output from cladding-pumped hybrid Yb:HOM fiber amplifier,” *Opt. Express*, vol. 27, pp. 8585–8595, 2019.

[38] K. Rottwitt, J. G. Koefoed, K. Ingerslev, and P. Kristensen, “Intermodal Raman amplification of OAM fiber modes,” *APL Photonics*, vol. 4, p. 030802, 2019.

[39] X. Liu, E. N. Christensen, K. Rottwitt, and S. Ramachandran, “Nonlinear four-wave mixing with enhanced diversity and

selectivity via spin and orbital angular momentum conservation," *APL Photonics*, vol. 5, p. 010802, 2020.

[40] S. Zhu, S. Pachava, S. Pidishety, Y. Feng, B. Srinivasan, and J. Nilsson, "Raman amplification of charge-15 orbital angular momentum mode in a large core step index fiber," in *CLEO*, 2020, SM1P.2.

[41] L. Yan, P. Kristensen, and S. Ramachandran, "Vortex fibers for STED microscopy," *APL Photonics*, vol. 4, p. 022903, 2019.

[42] B. M. Heffernan, S. A. Meyer, D. Restrepo, M. E. Siemens, E. A. Gibson, and J. T. Gopinath, "A fiber-coupled stimulated emission depletion microscope for bend-insensitive through-fiber imaging," *Sci. Rep.*, vol. 9, p. 11137, 2019.

[43] M. Fridman, G. Machavariani, N. Davidson, and A. A. Friesem, "Fiber lasers generating radially and azimuthally polarized light," *Appl. Phys. Lett.*, vol. 93, p. 191104, 2008.

[44] N. Bozinovic, Y. Yue, Y. Ren, et al., "Terabit-scale orbital angular momentum mode division multiplexing in fibers," *Science*, vol. 340, pp. 1545–1548, 2013.

[45] J. Liu, L. Zhu, A. Wang, et al., "All-fiber pre- and post-data exchange in km-scale fiber-based twisted lights multiplexing," *Opt. Lett.*, vol. 41, pp. 3896–3899, 2016.

[46] R. M. Nejad, K. Allahverdyan, P. Vaity, et al., "Mode division multiplexing using orbital angular momentum modes over 1.4 km ring core fiber," *J. Lightwave Technol.*, vol. 34, pp. 4252–4258, 2016.

[47] G. Zhu, Z. Hu, X. Wu, et al., "Scalable mode division multiplexed transmission over a 10-km ring-core fiber using high-order orbital angular momentum modes," *Opt. Express*, vol. 26, pp. 594–604, 2018.

[48] K. Ingerslev, P. Gregg, M. Galili, et al., "12 mode, WDM, MIMO-free orbital angular momentum transmission," *Opt. Express*, vol. 26, pp. 20225–20232, 2018.

[49] D. Cozzolino, D. Bacco, B. Da Lio, et al., "Orbital angular momentum states enabling fiber-based high-dimensional quantum communication," *Phys. Rev. Appl.*, vol. 11, p. 064058, 2019.

[50] S. Ramachandran, Z. Wang, and M. F. Yan, "Bandwidth control of long-period grating-based mode-converters in few-mode fibers," *Opt. Lett.*, vol. 27, pp. 698–700, 2002.

[51] M. Tur, D. Menashe, Y. Japha, and Y. Danziger, "High-order mode based dispersion compensating modules using spatial mode conversion," *J. Opt. Fiber Commun. Rep.*, vol. 5, pp. 249–311, 2007.

[52] L. Marrucci, E. Karimi, S. Slussarenko, et al., "Spin-to-orbital conversion of the angular momentum of light and its classical and quantum applications," *J. Opt.*, vol. 13, p. 064001, 2011.

[53] A. Astorino, J. Glückstad, and K. Rottwitt, "Fiber mode excitation using phase-only spatial light modulation: guideline on free-space path design and lossless optimization," *AIP Adv.*, vol. 8, p. 095111, 2018.

[54] Y. Zhao, J. Zhang, J. Du, and J. Wang, "Meta-facet fiber for twisting ultra-broadband light with high phase purity," *Appl. Phys. Lett.*, vol. 113, p. 061103, 2018.

[55] R. Olshansky, "Mode coupling effects in graded-index optical fibers," *Appl. Opt.*, vol. 14, pp. 935–945, 1975.

[56] S. Savović and A. Djordjevich, "Solution of mode coupling in step index optical fibers by Fokker–Planck equation and Langevin equation," *Appl. Opt.*, vol. 41, pp. 2826–2830, 2002.

[57] S. Ramachandran and P. Kristensen, "Optical vortices in fiber," *Nanophotonics Berlin*, vol. 2, pp. 455–74, 2013.

[58] A. M. Yao and M. J. Padgett, "Orbital angular momentum: origins, behaviour and applications," *Adv. Opt. Photonics*, vol. 3, pp. 161–204, 2011.

[59] A. Bjarklev, "Microdeformation losses of single-mode fibers with step-index profiles," *J. Lightwave Technol.*, vol. 4, pp. 341–346, 1986.

[60] S. Ramachandran and M. F. Yan, "Static and tunable dispersion management with higher order mode fibers," in *Fiber Based Dispersion Compensation*, S. Ramachandran, Ed., New York, Springer, 2007.

[61] M. Tur, D. Menashe, Y. Japha, and Y. Danziger, "High-order mode based dispersion compensating modules using spatial mode conversion," *J. Opt. Fiber Commun. Rep.*, vol. 5, pp. 249–311, 2007.

[62] S. Ramachandran, J. M. Fini, M. Mermelstein, J. W. Nicholson, S. Ghalmi, and M. F. Yan, "Ultra-large effective-area, higher-order mode fibers: a new strategy for high-power lasers. Invited Paper," *Laser Photonics Rev.*, vol. 2, pp. 429–48, 2008.

[63] J. W. Nicholson, J. M. Fini, A. M. DeSantolo, et al., "Scaling the effective area of higher-order-mode erbium-doped fiber amplifiers," *Opt. Express*, vol. 20, pp. 24575–24584, 2012.

[64] S. Ramachandran, S. Golowich, M. F. Yan, et al., "Lifting polarization degeneracy of modes by fiber design: a platform for polarization-insensitive microbend fiber gratings," *Opt. Lett.*, vol. 30, pp. 2864–2866, 2005.

[65] S. Ramachandran, P. Kristensen, and M. F. Yan, "Generation and propagation of radially polarized beams in optical fibers," *Opt. Lett.*, vol. 34, pp. 2525–2527, 2009.

[66] B. Ung, P. Vaity, L. Wang, Y. Messaddeq, L. A. Rusch, and S. LaRochelle, "Few-mode fiber with inverse parabolic graded-index profile for transmission of OAM-carrying modes," *Opt. Express*, vol. 22, pp. 18044–18055, 2014.

[67] C. Brunet, B. Ung, P. A. Belanger, Y. Messaddeq, S. LaRochelle, and L. A. Rusch, "Vector mode analysis of ring-core fibers: design tools for spatial division multiplexing," *J. Lightwave Technol.*, vol. 32, pp. 4046–4057, 2014.

[68] S. Ramachandran, P. Gregg, P. Kristensen, and S. E. Golowich, "On the scalability of ring fiber designs for OAM multiplexing," *Opt. Express*, vol. 23, pp. 3721–3730, 2015.

[69] P. Gregg, P. Kristensen, A. Rubano, S. Golowich, L. Marrucci, and S. Ramachandran, "Enhanced spin orbit interaction of light in highly confining optical fibers for mode division multiplexing," *Nat. Commun.*, vol. 10, p. 4707, 2019.

[70] D. L. P. Vitullo, C. C. Leary, P. Gregg, et al., "Observation of interaction of spin and intrinsic orbital angular momentum of light," *Phys. Rev. Lett.*, vol. 118, p. 083601, 2017.

[71] N Bozinovic, S Golowich, P Kristensen, and S Ramachandran, "Control of orbital angular momentum of light with optical fibers," *Opt. Lett.*, vol. 37, p. 2451, 2012.

[72] P. Gregg, P. Kristensen, and S. Ramachandran, "Conservation of orbital angular momentum in air-core optical fibers," *Optica*, vol. 2, p. 267, 2015.

[73] S. Pancharatnam, "Generalized theory of interference and its applications. Part I. Coherent pencils," *Proc. Indiana Acad. Sci.*, vol. A44, p. 247, 1956.

[74] M. V. Berry, "Quantal phase factors accompanying adiabatic changes," *Proc. R. Soc. Lond. A*. vol. 392, pp. 45–57, 1984.

[75] A. Tomita and R. Y. Chiao, "Observation of Berry's topological phase by use of an optical fiber," *Phys. Rev. Lett.*, vol. 57, pp. 937–940, 1986.
[76] R. A. Beth, "Mechanical detection and measurement of the angular momentum of light," *Phys. Rev.*, vol. 50, pp. 115–25, 1936.
[77] M. Berry, "Anticipations of the geometric phase," *Phys. Today*, vol. 43, pp. 34–40, 1990.
[78] J. Anandan, "The geometric phase," *Nature*, vol. 360, pp. 307–313, 1992.
[79] K. Y. Bliokh, "Geometrical optics of beams with vortices: berry phase and orbital angular momentum Hall effect," *Phys. Rev. Lett.*, vol. 97, p. 043901, 2006.
[80] S. Abdulkareem and N. Kundikova, "Joint effect of polarization and the propagation path of a light beam on its intrinsic structure," *Opt. Express*, vol. 24, p. 19157, 2016.
[81] X. Huang, S. Gao, B. Huang, and W. Liu, "Demonstration of spin and extrinsic orbital-angular-momentum interaction using a few-mode optical fiber," *Phys. Rev. A*, vol. 97, p. 033845, 2018.
[82] Z. Ma, G. Prabhakar, P. Gregg, and S. Ramachandran, "Robustness of OAM fiber modes to geometric perturbations," in *Conference on Lasers and Electro-Optics*, 2018.
[83] P. Gregg, M. Mirhosseini, A. Rubano, et al., "Q-plates as higher order polarization controllers for orbital angular momentum modes of fiber," *Opt. Lett.*, vol. 40, p. 1729, 2015.
[84] P. Gregg, P. Kristensen, and S. Ramachandran, "13.4 km OAM state propagation by recirculating fiber loop," *Opt. Express*, vol. 24, pp. 18938–18947, 2016.
[85] X. Wang, Z. Nie, Y. Liang, J. Wang, T. Li, and B. Jia, "Recent advances on optical vortex generation," *Nanophotonics*, vol. 7, pp. 1533–1556, 2018.

Alan E. Willner* and Cong Liu

Perspective on using multiple orbital-angular-momentum beams for enhanced capacity in free-space optical communication links

https://doi.org/10.1515/9783110710687-020

Abstract: Structured light has gained much interest in increasing communications capacity through the simultaneous transmission of multiple orthogonal beams. This paper gives a perspective on the current state of the art and future challenges, especially with regards to the use of multiple orbital angular momentum modes for system performance enhancement.

Keywords: free-space links; optical communications; optical fibers; orbital angular momentum; space division multiplexing; structured light.

1 Introduction

In 1992, Allen et al. [1] reported that orbital angular momentum (OAM) can be carried by an optical vortex beam. This beam has unique spatial structure, such that the amplitude has a ring-like doughnut profile and the phasefront "twists" in a helical fashion as it propagates. The number of 2π phase changes in the azimuthal direction is the OAM mode order, and beams with different OAM values can be orthogonal to each other. Such structured beams are a subset of the Laguerre–Gaussian (LG_{lp}) modal basis set, which has two modal indices: (1) l represents the number of 2π phase shifts in the azimuthal direction and the size of the ring grows with l; and (2) p+1 represents the number of concentric amplitude rings (see Figure 1) [2]. This orthogonality enables multiple independent optical beams to be multiplexed, spatially copropagate, and be demultiplexed – all with minimal inherent cross talk [3–5].

This orthogonality is of crucial benefit for a communications engineer. It implies that multiple independent data-carrying optical beams can be multiplexed and simultaneously transmitted in either free-space or fiber, thereby multiplying the system data capacity by the total number of beams (see Figure 2). Moreover, since all the beams are in the same frequency band, the system spectral efficiency (i.e., bits/s/Hz) is also increased. These multiplexed orthogonal OAM beams are a form of mode-division multiplexing (MDM), a subset of space-division multiplexing [4–7].

MDM has similarities to wavelength-division multiplexing (WDM), in which multiple independent data-carrying optical beams of different wavelengths can be multiplexed and simultaneously transmitted. WDM revolutionized optical communication systems and is ubiquitously deployed worldwide. Importantly, MDM is generally compatible with and can complement WDM, such that each of many wavelengths can contain many orthogonal structured beams and thus dramatically increase data capacity [8].

The field of OAM-based optical communications: (i) is considered young and rich with scientific and technical challenges, (ii) holds promise for technological advances and applications, and (iii) has produced much research worldwide. Excitingly, the number of publications per year that deal with OAM for communications has grown significantly over the past several years (see Figure 3). Capacities, distances, and number of data channels have all increased [9, 10], and approaches for mitigating degrading effects have produced encouraging results.

A key question remains as to how this young field may develop over the next decade. It is in this spirit that this article is aimed, taking an educated guess as to the subjective and relative merits of different aspects of this field. Specifically, these opinions try to address promising aspects that might be interesting to explore.

As far as context, this article will address a series of short subtopics and present our reasoned opinions. Moreover, due to the nature of this article, references will be

***Corresponding author: Alan E. Willner,** University of Southern California, Los Angeles, CA, USA, E-mail: willner@usc.edu
Cong Liu, University of Southern California, Los Angeles, CA, USA. https://orcid.org/0000-0003-0089-0763

This article has previously been published in the journal Nanophotonics. Please cite as: A. E. Willner and C. Liu "Perspective on using multiple orbital-angular-momentum beams for enhanced capacity in free-space optical communication links" *Nanophotonics* 2021, 10. DOI: 10.1515/nanoph-2020-0435.

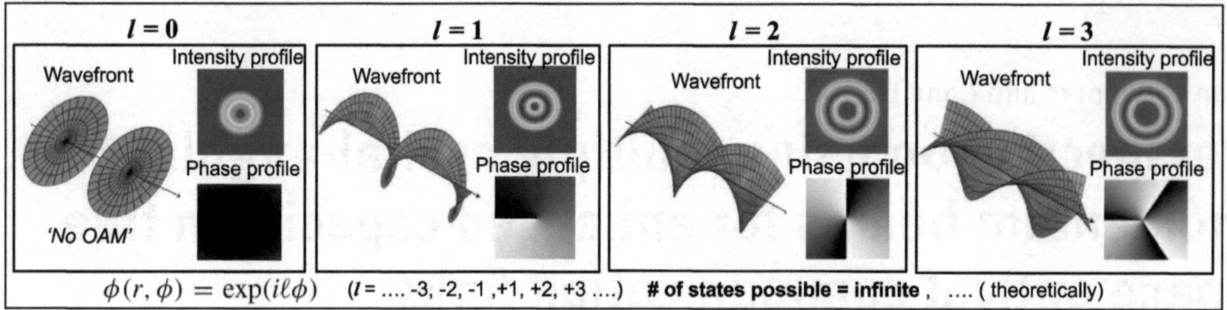

Figure 1: The wavefronts, intensity profiles, and phase profiles of orbital angular momentum (OAM) modes $l = 0, 1, 2$, and 3. The OAM mode with a nonzero order has a donut shape intensity profile and helical phasefront. The size of the ring in the intensity profile grows with l. We note that $p+1$ represents the number of concentric amplitude rings and $p=0$ is shown.

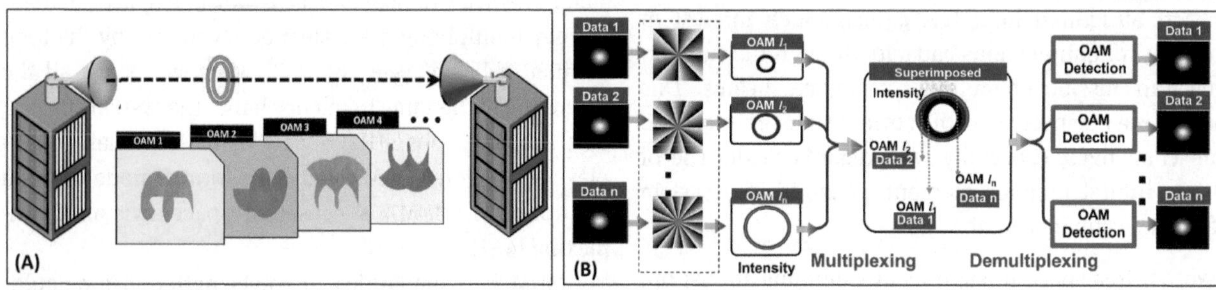

Figure 2: Concept of orbital-angular-momentum (OAM)–multiplexed free-space optical (FSO) links.
(A) Multiple OAM beams are coaxially transmitted through free space. (B) Each orthogonal OAM beam carries an independent data stream.

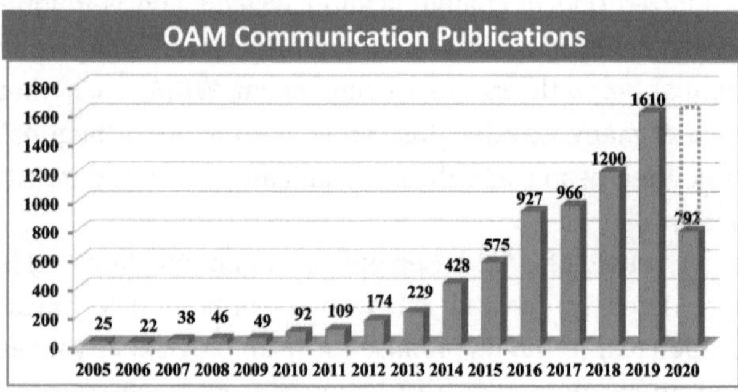

Figure 3: The orbital angular momentum (OAM) communications related publications yearly statistics (until July 2nd, 2020 from Google Scholar, provided by Guodong Xie) [11].

provided for specific background, but the intent is to give a perspective rather than detail. For a basic treatment of OAM-multiplexed communications, the reader is welcome to read A.E. Willner, "Communication With a Twist", *IEEE Spectrum* [12]. Finally and for the sake of readability, the article will generally assume OAM-multiplexed "*free-space classical*" optical communications as the basic default system. Separate subsections will be dedicated to topics that deviate from this system, such as for quantum communications (Section 7) or optical fiber transmission (Section 9).

2 Mitigation of modal coupling and channel crosstalk

A key issue in almost any MDM communication system is dealing with intermodal power coupling and deleterious inter-data-channel crosstalk. There are many causes of modal coupling and crosstalk, including the following for free-space OAM-multiplexed optical communication links:

(a) *Turbulence*: Atmospheric turbulence can cause a phase differential at different cross-sectional locations of a propagating beam. Given this phase change distribution in a changing environment, power can couple from the intended mode into other modes dynamically (e.g., perhaps on the order of milliseconds) [13, 14].

(b) *Misalignment*: Misalignment between the transmitter and receiver means that the receiver aperture is not coaxial with the incoming OAM beams. In order to operate an OAM-multiplexed link, one needs to know the mode that is being transmitted. A receiver aperture that captures power around the center of the beam will recover the full azimuthal phase change and know which l mode was transmitted. However, a limited-size receiver aperture that is off-axis will not recover the full phase change and inadvertently "think" that some power resides in other modes [15].

(c) *Divergence*: Free-space beams of higher OAM orders diverge faster than lower-order beams, thus making it difficult to fully capture the higher-order OAM beam at a limited-sized receiver aperture. Power loss obviously occurs if the beam power is not fully captured, but even modal coupling can occur due to the truncation of the beam's radial profile. This truncation can result in power being coupled to higher-order *p* modes [16, 17].

There are several approaches to potentially mitigate coupling and crosstalk in free-space OAM-multiplexed systems, including (see Figure 4):

(i) *Electrical digital signal processing (DSP)*: Crosstalk due to modal coupling has many similarities to crosstalk that occurs in multiple-transmitter-multiple-receiver (i.e., multiple-input multiple-output, MIMO) radio systems [18]. Multiple optical modes are similar to parallel radio frequency (RF) beams that experience crosstalk. Similar to electronic DSP that can undo much of the crosstalk in MIMO RF systems, these DSP approaches could also be used for mitigating OAM modal crosstalk [19].

(ii) *Adaptive optics*: Adaptive optics, such as by using digital micromirrors, spatial light modulators (SLMs) or multi-plane-light-converters (MPLCs), can mitigate modal crosstalk [20–22]. For example, if atmospheric turbulence causes a certain phase distortion on an optical beam, an SLM at the receiver can induce an inverse phase function to partially undo the effects of turbulence [21]. Typically, there could be a feedback loop, such that a data or probe beam is being monitored for dynamic changes and the new phase function is fed to an SLM.

(iii) *Modifying transmitted beams*: The modal structure of the transmitted beams themselves can be modified. In this approach: (a) the medium is probed by taking power measurements and determining the system modal coupling and channel crosstalk matrix, and (b) transmitting each beam with a combination of modes that represent the "inverse matrix", such that the received data channels would have little crosstalk [23].

Currently, there is an increasing array of potential methods. As with most issues, cost and complexity will play a key role in determining which, if any, mitigating approach should be used.

It should also be mentioned that: (i) modal coupling "tends" to be higher to the adjacent modes, and (ii) separating data channels with a larger modal differential can help in alleviating the problem [14, 24, 25]. Of course, larger modal separation leads to larger beam divergence, so a trade-off analysis is usually recommended.

3 Free-space links

As compared to RF links, optics in general can provide: (a) more bandwidth and higher data capacity due to the higher carrier wave frequency, and (b) better beam directionality and lower divergence, thus making eavesdropping more difficult [15]. When incorporating MDM using OAM multiplexing, such optical links can potentially achieve capacity enhancement and increased difficulty to eavesdropping. This lower probability of intercept stems from the issue that any misalignment causes intermodal coupling, such that it is extremely difficult for an off-axis eavesdropper to recover the signals, and even an on-axis eavesdropper would need to know the modal properties in order to recover the data, again fairly difficult. In addition, these free-space applications share some common desirable characteristics, including: (1) low size, weight and power (SWaP), which can be alleviated by advances in integrated OAM devices [26]; and (2) accurate pointing,

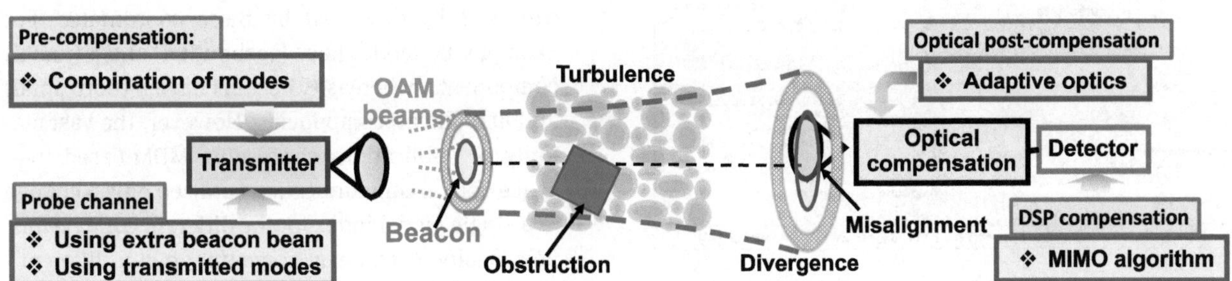

Figure 4: Various crosstalk compensation approaches in orbital-angular-momentum (OAM)–multiplexed links.

acquisition and tracking (PAT) systems, which helps limit modal coupling and crosstalk [15].

These advantages have generated interest in free-space MDM communications in the following scenarios:

(i) *Atmosphere*: OAM multiplexing can potentially benefit communication to: (a) unmanned aerial vehicles, for which distances may be relatively short range and a key challenge is to miniaturize the optical hardware, and (b) airplanes and other flying platforms, for which distances may require turbulence compensation and highly accurate pointing/tracking [27–29] (see Figure 5).

(ii) *Underwater*: Blue–green light has relatively low absorption in water, thereby potentially enabling high-capacity links over ~100 m [30, 31] (see Figure 6). Note that radio waves simply do not propagate well underwater, and common underwater acoustic links have a very low bit rate. For underwater OAM links, challenges include loss, turbidity, scattering, currents, and turbulence. An interesting challenge is transmitting from above water to below the water, such that the structured optical beam would pass through inhomogeneous media surrounding the interface, including nonuniform aerosols above water, the dynamically changing geometry of the air–water interface, and bubbles/surf below the surface.

(iii) *Satellites*: OAM multiplexing may have interesting advantages for up–down links to satellites. However, cross-links that are ultralong might necessitate extremely large apertures due to the increased beam divergence of higher-order modes [32, 33] (see Figure 5).

4 "Why use OAM? Should both modal indices be used?"

(a) *Why use OAM?:* Although there has been significant interest in OAM as a modal basis set for MDM communications, what is the rationale for choosing OAM over other types of modes? On a fundamental level, MDM requires that you can efficiently combine and separate different modes, so almost any complete orthogonal basis set could work. Indeed, many different types of modes were demonstrated in free-space and fiber, including Hermite–Gaussian (HG), LG, and linearly polarized (LP) modes [4–6, 34–37]. In discussions with Robert Boyd and Miles Padgett [38], two practical issues seemed to emerge as to reasons that one "might" prefer OAM modes (as a subset of LG modes) to other modal basis sets:

   (i) OAM modes are round, and free-space optical components are readily available in round form.

   (ii) It is important to maintain interchannel orthogonality and minimize crosstalk. This can be accomplished by fully capturing the specific parameter that defines the modal orthogonality. For a case in which different channels can be defined by different OAM l values, the channel and mode can be fully determined by azimuthally capturing a full 360° circle no matter the size of the round aperture [39].

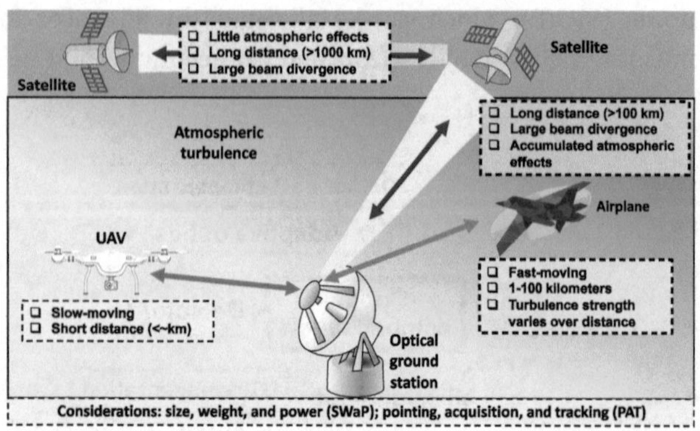

Figure 5: Orbital-angular-momentum (OAM)–multiplexed free-space optical airborne and satellite communications.

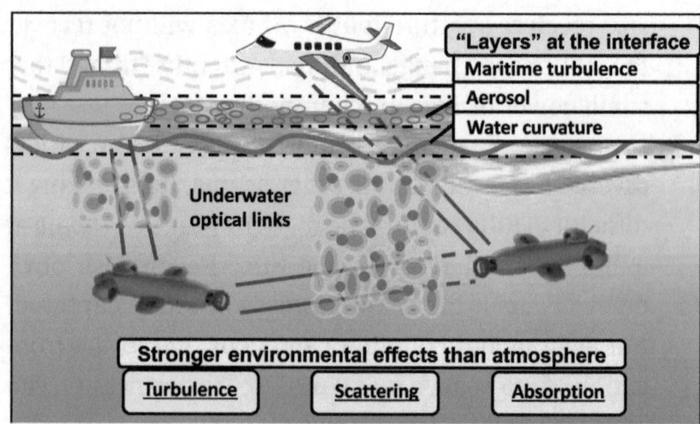

Figure 6: Challenges of different scenarios in underwater free-space optical communications.

(b) *Should both modal indices be used?*: Structured beams that are from a modal basis set can generally be described by two modal indices, such that the beam can be fully described by these coordinates. For example, LG modes have l (azimuthal) and p (radial) components, whereas HG beams have n (horizontal) and m (vertical) components. However, the vast majority of publications concerning MDM-based free-space optical communications utilized only a change in a single modal index for the different OAM beams. Specifically, each beam commonly had a different l value but the same p=0 value [4, 5, 8]. This one-dimensional system can accommodate many orthogonal beams, but a system designer could also use the other beam modal index in order to possibly achieve a larger two-dimensional set of data channels. This two-dimensional approach was shown experimentally for LG and HG beams [6, 34]. It is important to note that a significant challenge is the sufficient capture of the beam at the receiver aperture to ensure accurate phase recovery and orthogonality along both indices [34].

5 Photonic integration and component ecosystem

Back in the 1980s, many WDM optical communications experiments were performed on large optical tables using expensive devices that were often either not meant for communications or one-off, custom-built components. The development of cost-effective, integrated devices was deemed important for WDM to be deployed widely.

The same could be said about OAM-multiplexed optical communications. Many systems experiments were performed on large optical tables using devices that were not originally meant for MDM optical communications. For the future of mode-multiplexing to thrive, R&D in integrated devices would seem to be of signaficants. Indeed, we have been keen advocates of photonic integrated circuits for OAM-based optical communications, as can be seen in these quotes from our prior papers:

(a) *Nature Photonics, Interview* 2012 [40]: "Schemes for the generation, multiplexing and demultiplexing of OAM beams using superior SLMs or integrated devices would help to improve the maximum number of available OAM beams."
(b) *Advances in Optics and Photonics* 2015 [7]: "As was the case for many previous advances in optical communications, the future of OAM deployment would greatly benefit from advances in the enabling devices and subsystems (e.g., transmitters, (de) multiplexers, and receivers). Particularly with regard to integration, this represents significant opportunity to reduce cost and size and to also increase performance."

System development would benefit from a full ecosystem of devices, including the above-mentioned components as well as: (i) amplifiers that uniformly provide gain to different modes, and (ii) waveguides that efficiently guide OAM modes with little modal coupling.

Key desirable features for these integrated devices include [26]: low insertion loss, high amplifier gain, uniform performance for different modes, high modal purity, low modal coupling and intermodal crosstalk, high efficiency for mode conversion, high dynamic range, small size, large wavelength range, and accommodation of high numbers of modes. Other functions that could be advantageous include: (i) fast tunability and reconfigurability covering a range of OAM modes, and (ii) integration of an OAM communication system-on-a-chip that incorporates a full transceiver.

Finally, experiments commonly use SLMs to tailor the beam structure, but commercially available SLMs are generally bulky, expensive, and slow. Our favorite wishlist device would be the creation of a "super" SLM that has low cost, small footprint, large dynamic range in amplitude and phase, wide spectral range, high modal purity, fast tunability (the faster the better, to even encode data bits), and high resolution [41, 42].

6 Novel beams

The excitement in this field originated by the ability to utilize orthogonal structured optical beams. However, there is much work in the fields of optics and photonics on several types of novel variations of optical beams (e.g., Airy and Bessel types), with more being explored at an exciting pace.

Over the next several years, it would not be surprising if novel beams are used to minimize certain system degrading effects. There have been initial results for some of these concepts, but a partial "wish list" for novel beams could be beams:

(a) that are more resilient to modal coupling caused by turbulence and turbidity;
(b) that have limited divergence in free space;
(c) that are resilient to partial obstruction, such that their phase structure can "self-heal" (e.g., Bessel-type beams);
(d) whose phase structure can readily be recovered even if the transmitter and receiver are misaligned.

7 Quantum communications

Another important advantage of OAM orthogonality is that one can use OAM mode order as a data encoding scheme [43–46]. For example in the case of a quantum communication system, an individual photon can carry one of the many different OAM values; this is similar to digital data taking on one of many different amplitude values. A binary data symbol (i.e., one data bit) has two values of “0” and “1”, whereas an M-ary symbol may have many more possible values ranging from “0” to “M–1”. The number of data bits per unit time would be $\log_2 M$. If each photon can be encoded with a specific OAM value from M possibilities, the photon efficiency in bits/photon can be increased. This has the potential to be quite useful for quantum communication systems which are typically photon “starved” and of which qubits commonly can be encoded on one of only two orthogonal polarization states [46] (see Figure 7).

A larger alphabet for each qubit is, in general, highly desirable for enhancing system performance. However, there is much research needed to overcome the challenges in fielding an OAM-encoded quantum communication system, such as: (i) mitigating coupling among orthogonal states, and (ii) developing transmitters that can be tuned rapidly to encode each photon on one of many modes.

8 Different frequency ranges

Separate from using optical beams, free-space communication links can take advantage of mode multiplexing in many other carrier-wave-frequency ranges to increase system capacity. For example, OAM can be manifest in many types of electromagnetic and mechanical waves (see Figure 8), and interesting reports have explored the use of OAM in millimeter, acoustic, and THz waves [47–55].

From a system designer’s perspective, there tends to be a trade-off in different frequency ranges:

(i) *Divergence*: Lower frequencies have much higher beam divergence, exacerbating the problem of collecting enough of the beam to recover the data channels.
(ii) *Interaction with Matter*: Lower frequencies tend to have much lower interaction with matter, such that radio waves are less affected by atmospheric-turbulence-induced modal coupling than optical waves.

There are exciting developments in the millimeter-wave application space, for which industrial labs are increasingly engaging in R&D in order to significantly increase the potential capacity of fronthaul and backhaul links [47, 52–55]. Advances in this area include the use of RF antenna arrays that are fabricated on printed-circuit boards [54]. For example, a multiantenna element ring can emit a millimeter-wave OAM beam by selectively exciting different antenna elements with a differential phase delay [55]. Moreover, multiple concentric rings can be fabricated, thereby emitting a larger number of multiplexed OAM beams [52, 53].

9 Optical fiber transmission

MDM can be achieved in both free-space and fiber, with much of the transmitter and receiver technology being similar. However, the channel medium is different, which gives rise to the following distinctions:

(a) There is no beam divergence in light-guiding fiber.
(b) Fiber has various kinds of inhomogeneities, and coupling can occur among modes within a specific mode group or between mode groups, thereby creating deleterious interchannel crosstalk [36, 56–59]; typically, intramodal group crosstalk is higher than intermodal group crosstalk.

The excitement around using MDM for capacity increase originally occurred primarily in the fiber transmission world, especially in research laboratories [35, 36, 60, 61]. There was much important work using LP modes as the modal set in fiber. However, since there was significant modal crosstalk when propagating through conventional-central-core few-mode fiber, MIMO-like DSP was used with impressive results to mitigate crosstalk [35, 60].

OAM has also been used as the modal basis set for fiber transmission, both for central-core and ring-core few-mode fibers [4, 36, 37, 62]. Importantly, the modal coupling itself can be reduced in the optical domain by utilizing specialty fiber that makes the propagation constants of different modes quite different, thus reducing intermodal coupling.

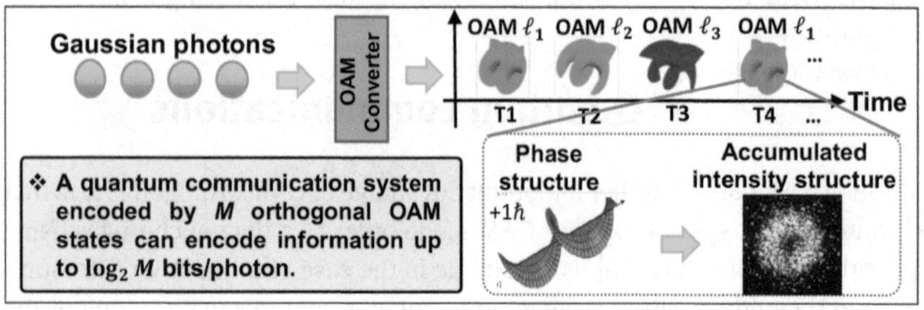

Figure 7: Concept of orbital-angular-momentum (OAM)–based quantum data encoding. Within each symbol period, a Gaussian photon is converted to one of the M OAM states, resulting in information encoding of up to $\log_2 M$ bit/photon. The accumulated intensity structure image is recorded using a single-photon sensitivity, low-noise–intensified charge-coupled device camera [46].

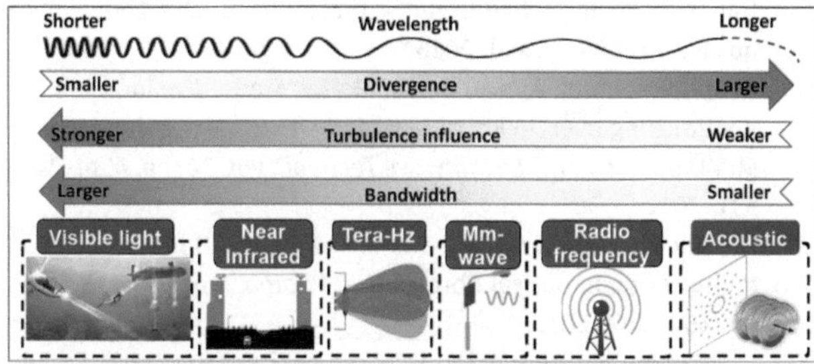

Figure 8: Orbital angular momentum (OAM) applications in different frequencies for communications.

Such fibers include ring-core and elliptical-core fibers [4, 36, 62], and 10's of modes with low crosstalk have been demonstrated. These specialty fibers have produced exciting results, but they are structurally different than conventional fiber and thus require a little more resolve in order for them to be widely adopted.

10 Summary

Will OAM be widely deployed in communication systems? Not clear. However, our opinion is that the R&D community is producing excellent advances that, with all likelihood, will be valuable in some important aspects that use structured light.

Author contribution: All the authors have accepted responsibility for the entire content of this submitted manuscript and approved submission.

Research funding: We acknowledge the generous supports from Vannevar Bush Faculty Fellowship sponsored by the Basic Research Office of the Assistant Secretary of Defense (ASD) for Research and Engineering (R&E) and funded by the Office of Naval Research (ONR) (N00014-16-1-2813); Defense Security Cooperation Agency (DSCA 4441006051); Air Force Research Laboratory (FA8650-20-C-1105); National Science Foundation (NSF) (ECCS-1509965); and Office of Naval Research through a MURI grant (N00014-20-1-2558).

Conflict of interest statement: The authors declare no conflicts of interest regarding this article.

References

[1] L. Allen, M. W. Beijersbergen, R. J. Spreeuw, and J. P. Woerdman, "Orbital angular momentum of light and the transformation of Laguerre–Gaussian laser modes," *Phys. Rev. A*, vol. 45, no. 11, p. 8185, 1992.

[2] A. M. Yao and M. J. Padgett, "Orbital angular momentum: origins, behavior and applications," *Adv. Opt. Photonics*, vol. 3, no. 2, pp. 161–204, 2011.

[3] G. Gibson, J. Courtial, M. J. Padgett, et al., "Free-space information transfer using light beams carrying orbital angular momentum," *Optics Express*, vol. 12, no. 22, pp. 5448–5456, 2004.

[4] N. Bozinovic, Y. Yue, Y. Ren, et al., "Terabit-scale orbital angular momentum mode division multiplexing in fibers," *Science*, vol. 340, no. 6140, pp. 1545–1548, 2013.

[5] J. Wang, J. Y. Yang, I. M. Fazal, et al., "Terabit free-space data transmission employing orbital angular momentum multiplexing," *Nat. Photon.*, vol. 6, no. 7, pp. 488–496, 2012.

[6] K. Pang, H. Song, Z. Zhao, et al., "400-Gbit/s QPSK free-space optical communication link based on four-fold multiplexing of Hermite–Gaussian or Laguerre–Gaussian modes by varying both modal indices," *Opt. Lett.*, vol. 43, no. 16, pp. 3889–3892, 2018.

[7] A. E. Willner, H. Huang, Y. Yan, et al., "Optical communications using orbital angular momentum beams," *Adv. Opt. Photonics*, vol. 7, no. 1, pp. 66–106, 2015.

[8] H. Huang, G. Xie, Y. Yan, et al., "100 Tbit/s free-space data link enabled by three-dimensional multiplexing of orbital angular momentum, polarization, and wavelength," *Opt. Lett.*, vol. 39, no. 2, pp. 197–200, 2014.

[9] M. Krenn, J. Handsteiner, M. Fink, et al., "Twisted light transmission over 143 km," *Proc. Natl. Acad. Sci. U. S. A.*, vol. 113, no. 48, pp. 13648–13653, 2016.

[10] J. Wang, S. Li, M. Luo, et al., "N-dimentional multiplexing link with 1.036-Pbit/s transmission capacity and 112.6-bit/s/Hz spectral efficiency using OFDM-8QAM signals over 368 WDM pol-muxed 26 OAM modes." in *The European Conference on Optical Communication (ECOC)*. IEEE, 2014, pp. 1–3.

[11] G. Xie, Private Communication, 2020.

[12] A. E. Willner, "Communication with a twist," *IEEE Spectrum*, vol. 53, no. 8, pp. 34–39, 2016.

[13] L. C. Andrews and R. L. Phillips, *Laser Beam Propagation through Random Media*, SPIE, 2005. https://doi.org/10.1117/3.626196.

[14] Y. Ren, H. Huang, G. Xie, et al., "Atmospheric turbulence effects on the performance of a free space optical link employing orbital angular momentum multiplexing," *Opt. Lett.*, vol. 38, no. 20, pp. 4062–4065, 2013.

[15] G. Xie, L. Li, Y. Ren, et al., "Performance metrics and design considerations for a free-space optical orbital-angular-momentum–multiplexed communication link," *Optica*, vol. 2, no. 4, pp. 357–365, 2015.

[16] K. Pang, H. Song, X. Su, et al., "Simultaneous orthogonalizing and shaping of multiple LG beams to mitigate crosstalk and power loss by transmitting each of four data channels on multiple modes in a 400-Gbit/s free-space link," in *Optical Fiber Communication Conference (OFC), OSA*, 2020, pp. W1G-2.

[17] Z. Xin, Z. Yaqin, R. Guanghui, et al., "Influence of finite apertures on orthogonality and completeness of Laguerre-Gaussian beams," *IEEE Access*, vol. 6, pp. 8742–8754, 2018.

[18] P. J. Winzer and G. J. Foschini, "MIMO capacities and outage probabilities in spatially multiplexed optical transport systems," *Opt. Express*, vol. 19, no. 17, pp. 16680–16696, 2011.

[19] H. Huang, Y. Cao, G. Xie, et al., "Crosstalk mitigation in a free-space orbital angular momentum multiplexed communication

link using 4 × 4 MIMO equalization," *Opt. Lett.*, vol. 39, no. 15, pp. 4360–4363, 2014.

[20] G. A. Tyler, "Adaptive optics compensation for propagation through deep turbulence: a study of some interesting approaches," *Opt. Eng.*, vol. 52, no. 2, p. 021011, 2012.

[21] Y. Ren, G. Xie, H. Huang, et al., "Adaptive-optics-based simultaneous pre- and post-turbulence compensation of multiple orbital-angular-momentum beams in a bidirectional free-space optical link," *Optica*, vol. 1, no. 6, pp. 376–382, 2014.

[22] N. K. Fontaine, R. Ryf, H. Chen, D. T. Neilson, K. Kim, and J. Carpenter, "Laguerre–Gaussian mode sorter," *Nat. Commun.*, vol. 10, no. 1, pp. 1–7, 2019.

[23] H. Song, H. Song, R. Zhang, et al., "Experimental mitigation of atmospheric turbulence effect using pre-signal combining for uni- and bi-directional free-space optical links with two 100-Gbit/s OAM-multiplexed channels," *J. Lightwave Technol.*, vol. 38, no. 1, pp. 82–89, 2020.

[24] J. A. Anguita, M. A. Neifeld, and B. V. Vasic, "Turbulence-induced channel crosstalk in an orbital angular momentum-multiplexed free-space optical link," *Appl. Opt.*, vol. 47, no. 13, pp. 2414–2429, 2008.

[25] N. Chandrasekaran and J. H. Shapiro, "Photon information efficient communication through atmospheric turbulence—part I: channel model and propagation statistics," *J. Lightwave Technol.*, vol. 32, no. 6, pp. 1075–1087, 2014.

[26] A. E. Willner, "Advances in components and integrated devices for OAM-based systems," Invited Tutorial in Conference on Lasers and Electro-Optics (CLEO), 2018, OSA, pp. Stu3B.1.

[27] S. S. Muhammad, T. Plank, E. Leitgeb, et al., "Challenges in establishing free space optical communications between flying vehicles," in *6th Int'l Symposium on Communication Systems, Networks and Digital Signal Processing*, IEEE, 2008, pp. 82–86.

[28] A. Kaadan, H. Refai, and P. Lopresti, "Spherical FSO receivers for UAV communication: geometric coverage models," *IEEE Trans. Aero. Electron. Syst.*, vol. 52, no. 5, pp. 2157–2167, 2016.

[29] L. Li, R. Zhang, Z. Zhao, et al., "High-capacity free-space optical communications between a ground transmitter and a ground receiver via a UAV using multiplexing of multiple orbital-angular-momentum beams," *Sci. Rep.*, vol. 7, no. 1, pp. 1–2, 2017.

[30] J. Baghdady, K. Miller, K. Morgan, et al., "Multi-gigabit/s underwater optical communication link using orbital angular momentum multiplexing," *Opt. Express*, vol. 24, no. 9, pp. 9794–9805, 2016.

[31] Y. Ren, L. Li, Z. Wang, et al., "Orbital angular momentum-based space division multiplexing for high-capacity underwater optical communications," *Sci. Rep.*, vol. 6, p. 33306, 2016.

[32] R. Fields, C. Lunde, R. Wong, et al., "NFIRE-to-TerraSAR-X laser communication results: satellite pointing, disturbances, and other attributes consistent with successful performance," in *Sensors and Systems for Space Applications III*, vol. 7330, International Society for Optics and Photonics, 2009, p. 73300Q.

[33] F. Heine, G. Mühlnikel, H. Zech, S. Philipp-May, and R. Meyer, "The European Data Relay System, high speed laser based data links," in *2014 7th Advanced Satellite Multimedia Systems Conference and the 13th Signal Processing for Space Communications Workshop (ASMS/SPSC)*, IEEE, 2014, pp. 284–286.

[34] G. Xie, Y. Ren, Y. Yan, et al., "Experimental demonstration of a 200-Gbit/s free-space optical link by multiplexing Laguerre–Gaussian beams with different radial indices," *Opt. Lett.*, vol. 41, no. 15, pp. 3447–3450, 2016.

[35] R. Ryf, S. Randel, A. H. Gnauck, et al., "Mode-division multiplexing over 96 km of few-mode fiber using coherent 6 × 6 MIMO processing," *J. Lightwave Technol.*, vol. 30, no. 4, pp. 521–531, 2011.

[36] D. J. Richardson, J. M. Fini, and L. E. Nelson, "Space-division multiplexing in optical fibres," *Nat. Photon.*, vol. 7, no. 5, pp. 354–362, 2013.

[37] B. Ndagano, R. Brüning, M. McLaren, M. Duparré, and A. Forbes, "Fiber propagation of vector modes," *Opt. Express*, vol. 23, no. 13, pp. 17330–17336, 2015.

[38] R. W. Boyd and M. J. Padgett, Private Communication, 2020.

[39] S. Restuccia, D. Giovannini, G. Gibson, and M. J. Padgett, "Comparing the information capacity of Laguerre–Gaussian and Hermite–Gaussian modal sets in a finite-aperture system," *Opt. Express*, vol. 24, pp. 27127–27136, 2016.

[40] O. Graydon, "Interview: a new twist for communications," *Nat. Photon.*, vol. 6, p. 498, 2012.

[41] J. Lin, P. Genevet, M. A. Kats, N. Antoniou, and F. Capasso, "Nanostructured holograms for broadband manipulation of vector beams," *Nano Lett.*, vol. 13, no. 9, pp. 4269–4274, 2013.

[42] V. Liu, D. A. B. Miller, and S. Fan, "Ultra-compact photonic crystal waveguide spatial mode converter and its connection to the optical diode effect," *Opt. Express*, vol. 20, pp. 28388–28397, 2012.

[43] M. Mafu, A. Dudley, S. Goyal, et al., "Higher-dimensional orbital-angular-momentum-based quantum key distribution with mutually unbiased bases," *Phys. Rev. A*, vol. 88, no. 3, p. 032305, 2013.

[44] M. Mirhosseini, O. S. Magaña-Loaiza, M. N. O'Sullivan, et al., "High-dimensional quantum cryptography with twisted light," *New J. Phys.*, vol. 17, no. 3, p. 033033, 2015.

[45] C. Liu, K. Pang, Z. Zhao, et al., "Single-end adaptive optics compensation for emulated turbulence in a bi-directional 10-Mbit/s per channel free-space quantum communication link using OAM encoding," *Research*, vol. 2019, p. 8326701, 2019.

[46] M. Erhard, R. Fickler, M. Krenn, and A. Zeilinger, "Twisted photons: new quantum perspectives in high dimensions," *Light Sci. Appl.*, vol. 7, no. 3, p. 17146, 2018.

[47] Y. Yan, G. Xie, M. P. Lavery, et al., "High-capacity millimetre-wave communications with orbital angular momentum multiplexing," *Nat. Commun.*, vol. 5, no. 1, pp. 1–9, 2014.

[48] C. Shi, M. Dubois, Y. Wang, and X. Zhang, "High-speed acoustic communication by multiplexing orbital angular momentum," *Proc. Natl. Acad. Sci. U. S. A.*, vol. 114, no. 28, pp. 7250–7253, 2017.

[49] Z. Zhao, R. Zhang, H. Song, et al., "Fundamental system-degrading effects in THz communications using multiple OAM beams with turbulence," in *2020 IEEE International Conference on Communications (ICC)*, IEEE, 2020, pp. WC24-4.

[50] X. Wei, L. Zhu, Z. Zhang, K. Wang, J. Liu, and J. Wang, "Orbit angular momentum multiplexing in 0.1-THz free-space communication via 3D printed spiral phase plates," in *2014 Conference on Lasers and Electro-Optics (CLEO)-Laser Science to Photonic Applications*, IEEE, 2014, pp. STu2F.2.

[51] C. Liu, X. Wei, L. Niu, K. Wang, Z. Yang, and J. Liu, "Discrimination of orbital angular momentum modes of the terahertz vortex beam using a diffractive mode transformer," *Opt. Express*, vol. 24, no. 12, pp. 12534–12541, 2016.

[52] H. Sasaki, D. Lee, H. Fukumoto, et al., "Experiment on over-100-Gbps wireless transmission with OAM-MIMO multiplexing system in 28-GHz band," in *IEEE Global Communications Conference*, IEEE, 2018, pp. 1–6.
[53] H. Sasaki, Y. Yagi, T. Yamada, and D. Lee, "Field experimental demonstration on OAM-MIMO wireless transmission on 28 GHz band," in *2019 IEEE Globecom Workshops (GC Workshops)*, IEEE, 2019, pp. 1–4.
[54] Z. Zhao, Y. Yan, L. Li, et al., "A dual-channel 60 GHz communications link using patch antenna arrays to generate data-carrying orbital-angular-momentum beams," in *IEEE International Conference on Communications (ICC)*, IEEE, 2016, pp. 1–6.
[55] G. Xie, Z. Zhao, Y. Yan, et al., "Demonstration of tunable steering and multiplexing of two 28 GHz data carrying orbital angular momentum beams using antenna array," *Sci. Rep.*, vol. 6, p. 37078, 2016.
[56] K. I. Kitayama and N. P. Diamantopoulos, "Few-mode optical fibers: original motivation and recent progress," *IEEE Commun. Mag.*, vol. 55, no. 8, pp. 163–169, 2017.
[57] P. J. Winzer, "Making spatial multiplexing a reality," *Nat. Photon.*, vol. 8, no. 5, pp. 345–348, 2014.
[58] R. Zhang, H. Song, H. Song, et al., "Utilizing adaptive optics to mitigate intra-modal-group power coupling of graded-index few-mode fiber in a 200-Gbit/s mode-division-multiplexed link," *Opt. Letters*, vol. 45, no. 13, pp. 3577–3580, 2020.
[59] J. Carpenter, B. C. Thomsen, T. D. Wilkinson, "Degenerate mode-group division multiplexing." *J. Lightwave Technol.*, vol. 30, no. 24, pp. 3946–3952, 2012.
[60] S. Randel, R. Ryf, A. Sierra, et al., "6×56-Gb/s mode-division multiplexed transmission over 33-km few-mode fiber enabled by 6×6 MIMO equalization," *Opt. Express*, vol. 19, no. 17, pp. 16697–16707, 2011.
[61] S. Bae, Y. Jung, B. G. Kim, and Y. C. Chung, "Compensation of mode crosstalk in MDM system using digital optical phase conjugation," *IEEE Photon. Technol. Lett.*, vol. 31, no. 10, pp. 739–742, 2019.
[62] C. Brunet, P. Vaity, Y. Messaddeq, S. LaRochelle, and L. A. Rusch, "Design, fabrication and validation of an OAM fiber supporting 36 states," *Opt. Express*, vol. 22, no. 21, pp. 26117–26127, 2014.

Part IV: **Biomedical Photonics**

Navid Rajil, Alexei Sokolov, Zhenhuan Yi, Garry Adams, Girish Agarwal, Vsevolod Belousov, Robert Brick, Kimberly Chapin, Jeffrey Cirillo, Volker Deckert, Sahar Delfan, Shahriar Esmaeili, Alma Fernández-González, Edward Fry, Zehua Han, Philip Hemmer, George Kattawar, Moochan Kim, Ming-Che Lee, Chao-Yang Lu, Jon Mogford, Benjamin Neuman, Jian-Wei Pan, Tao Peng, Vincent Poor, Steven Scully, Yanhua Shih, Szymon Suckewer, Anatoly Svidzinsky, Aart Verhoef, Dawei Wang, Kai Wang, Lan Yang, Aleksei Zheltikov, Shiyao Zhu, Suhail Zubairy and Marlan Scully*

A fiber optic–nanophotonic approach to the detection of antibodies and viral particles of COVID-19

https://doi.org/10.1515/9783110710687-021

Abstract: Dr. Deborah Birx, the White House Coronavirus Task Force coordinator, told NBC News on "Meet the Press" that "[T]he U.S. needs a 'breakthrough' in coronavirus testing to help screen Americans and get a more accurate picture of the virus' spread." We have been involved with biopathogen detection since the 2001 anthrax attacks and were the first to detect anthrax in real-time. A variation on the laser spectroscopic techniques we developed for the rapid detection of anthrax can be applied to detect the Severe Acute Respiratory Syndrome-Corona Virus-2 (SARS-CoV-2 virus). In addition to detecting a single virus, this technique allows us to read its surface protein structure. In particular, we have been conducting research based on a variety of quantum optical approaches aimed at improving our ability to detect Corona Virus Disease-2019 (COVID-19) viral infection.

The list of coauthors represents only a fraction of the team working on the real-time detection of pathogens from anthrax to SARS-CoV-2. They are the team members most involved in the current COVID-19 research and we thank coeditors Federico Capasso and Dennis Couwenberg for the invitation to contribute to this volume honoring the COVID-19 workers.

Navid Rajil, Alexei Sokolov and **Zhenhuan Yi**: contributed equally to this work.

***Corresponding author: Marlan Scully,** Texas A&M University, College Station, TX 77843, USA; Baylor University, Waco, TX 76798, USA; and Princeton University, Princeton, NJ 08544, USA,
E-mail: scully@tamu.edu

Navid Rajil, Alexei Sokolov, Zhenhuan Yi, Garry Adams, Girish Agarwal, Robert Brick, Kimberly Chapin, Jeffrey Cirillo, Sahar Delfan, Shahriar Esmaeili, Alma Fernández-González, Edward Fry, Zehua Han, George Kattawar, Moochan Kim, Ming-Che Lee, Tao Peng, Anatoly Svidzinsky, Aart Verhoef, Kai Wang and Suhail Zubairy, Texas A&M University, College Station, TX 77843, USA. https://orcid.org/0000-0002-9220-4080 (M. Kim)

Vsevolod Belousov, Shemyakin–Ovchinnikov Institute of Bioorganic Chemistry, Russian Academy of Sciences, Moscow 117997, Russia; Pirogov Russian National Research Medical University, Moscow 117997, Russia; and Federal Center of Brain Research and Neurotechnologies of the Federal Medical Biological Agency, Moscow 117997, Russia

Volker Deckert, Texas A&M University, College Station, TX 77843, USA; Leibniz Institute of Photonic Technology, 07745 Jena, Germany; and Friedrich Schiller University, 07743 Jena, Germany

Philip Hemmer, Texas A&M University, College Station, TX 77843, USA; and Zavoisky Physical-Technical Institute, 420029 Kazan, Russia

Chao-Yang Lu and Jian-Wei Pan, University of Science and Technology of China, Hefei, Anhui, 230026, P. R. China

Jon Mogford, Texas A&M University System, College Station, TX 77840, USA

Benjamin Neuman, Texas A&M University, College Station, TX 77843, USA; and Texas A&M University, Texarkana, TX 75503, USA

Vincent Poor, Princeton University, Princeton, NJ 08544, USA

Steven Scully, Collins Aerospace, Richardson, TX 75082, USA

Yanhua Shih, University of Maryland, Baltimore County, 1000 Hilltop Circle, Baltimore, MD 21250, USA

Szymon Suckewer, Texas A&M University, College Station, TX 77843, USA; and Princeton University, Princeton, NJ 08544, USA

Dawei Wang and Shiyao Zhu, Zhejiang University, 38 Zheda Rd, Hangzhou, 310027, P. R. China

Lan Yang, Washington University, St. Louis, MO 63130, USA

Aleksei Zheltikov, Texas A&M University, College Station, TX 77843, USA; International Laser Center, Moscow State University, Moscow 119992, Russia; and Russian Quantum Center, Skolkovo, Moscow Region, 143025, Russia

This article has previously been published in the journal Nanophotonics. Please cite as: N. Rajil, A. Sokolov, Z. Yi, G. Adams, G. Agarwal, V. Belousov, R. Brick, K. Chapin, J. Cirillo, V. Deckert, S. Delfan, S. Esmaeili, A. Fernández-González, E. Fry, Z. Han, P. Hemmer, G. Kattawar, M. Kim, M.-C. Lee, C.-Y. Lu, J. Mogford, B. Neuman, J.-W. Pan, T. Peng, V. Poor, S. Scully, Y. Shih, S. Suckewer, A. Svidzinsky, A. Verhoef, D. Wang, K. Wang, L. Yang, A. Zheltikov, S. Zhu, S. Zubairy and M. Scully "A fiber optic–nanophotonic approach to the detection of antibodies and viral particles of COVID-19" *Nanophotonics* 2021, 10. DOI: 10.1515/nanoph-2020-0357.

Indeed, the detection of a small concentration of antibodies, after an infection has passed, is a challenging problem. Likewise, the early detection of disease, even before a detectible antibody population has been established, is very important. Our team is researching both aspects of this problem. The paper is written to stimulate the interest of both physical and biological scientists in this important problem. It is thus written as a combination of tutorial (review) and future work (preview). We join Prof. Federico Capasso and Editor Dennis Couwenberg in expressing our appreciation to all those working so heroically on all aspects of the COVID-19 problem. And we thank Drs. Capasso and Couwenberg for their invitation to write this paper.

Keywords: detection of SAR-CoV-2 virus; hollow-core fibers; laser spectroscopic technique; nanophotonics.

1 Introduction

As the anthrax attacks and the present pandemic demonstrate, improved strategies to detect viral [1] and bacterial pathogens [2] are urgently needed. Recognizing the importance of the problem, our team has been researching a variety of quantum optical techniques aimed at improving the detection of the Corona Virus Disease-2019 (COVID-19) viral infection.

Much, but not all, of our present work follows on our earlier research using the laser spectroscopic technique called Coherent anti-Stokes Raman scattering (CARS). In particular, here, we utilize innovative fiber optic platforms for antibody and virus detection based on a multichannel fiber sensor such as that sketched in Figure 2a–c.

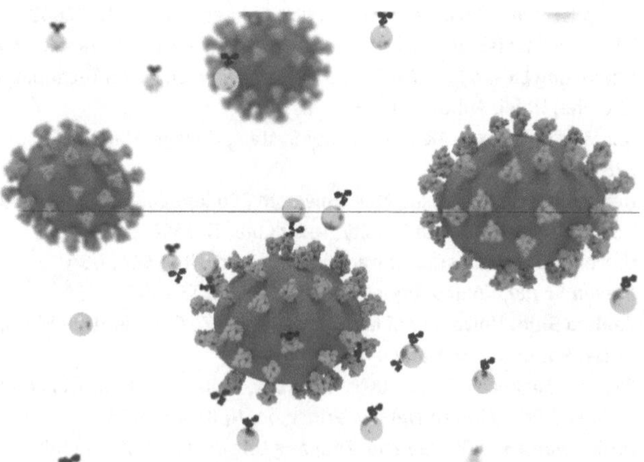

Figure 1: The virion (green) with red spikes, gold nanoparticles (gold) and the antibodies against Severe Acute Respiratory Syndrome-Corona Virus-2 (SARS-CoV-2) (blue). The virion is approximately 100 nm in diameter, the gold nanoparticles are about 40 nm in diameter, and the antibodies are about 10 nm in diameter, respectively.

The present paper focuses on new approaches, building on existing work in modern optics and biochemistry, and is (hopefully) written in a tutorial fashion in hopes of drawing a broader participation from both physical and biological scientists to this important problem.

Figure 1 shows an illustration of virion attachment to antibodies produced by host immune systems. In the present paper, we: (1) develop a coherent laser technique, tailored to the detection of antibodies in the blood and (2) engineer a new approach to detect the presence of Severe Acute Respiratory Syndrome-Corona Virus-2 (SARS-CoV-2 virus), the causative agent of COVID-19, via a fiber-based reverse-transcription polymerase chain reaction (RT-PCR) test.

Central to the broad functionality of the present approach is an optical platform such as shown in Figure 2a–c. In general, we intend to combine microfluidic, optical-interrogation, and PCR-/RT-PCR–amplified nucleic acid identification capabilities. An analyte, such as a blood sample, can flow through the air holes that run along the fiber length [3] and thus form a periodic (Figure 2b) or a spatially chirped (Figure 2c) photonic lattice within a cylindrical rod of host glass. The geometry of this lattice along with its pitch, the diameter of air holes (from 1 to 50 μm [4]), and the glass-analyte index step define the structure of field modes supported by the fiber [5, 6]. When coupled into one of its guided modes, an optical field tends to peak within high-index glass channels (with a notable exception of hollow-core photonic crystal fibers. Its evanescent tails, however, can interact with the analyte (Figure 2a), driving its optical response through a linear [3] or nonlinear [7, 8] optical process, thus providing an optical readout for the detection of viruses and other microorganisms in the analyte. With suitable optimization of the fiber structure, the optical readout can be dramatically enhanced due to improved spatial overlap between the optical field and the analyte (Figrue 2a), improving, sometimes by several orders of magnitudes, the sensitivity of antibody and/or nucleic acid identification and, hence as is discussed below, coronavirus detection.

2 Biophotonic SARS-CoV-2 detection schemes

2.1 Fiber-optical antibody tests

It is our intention that the present approach will result in an improved antibody test, with increased sensitivity and reduced false counts. We envision that, due to the improved characteristics at a relatively low production

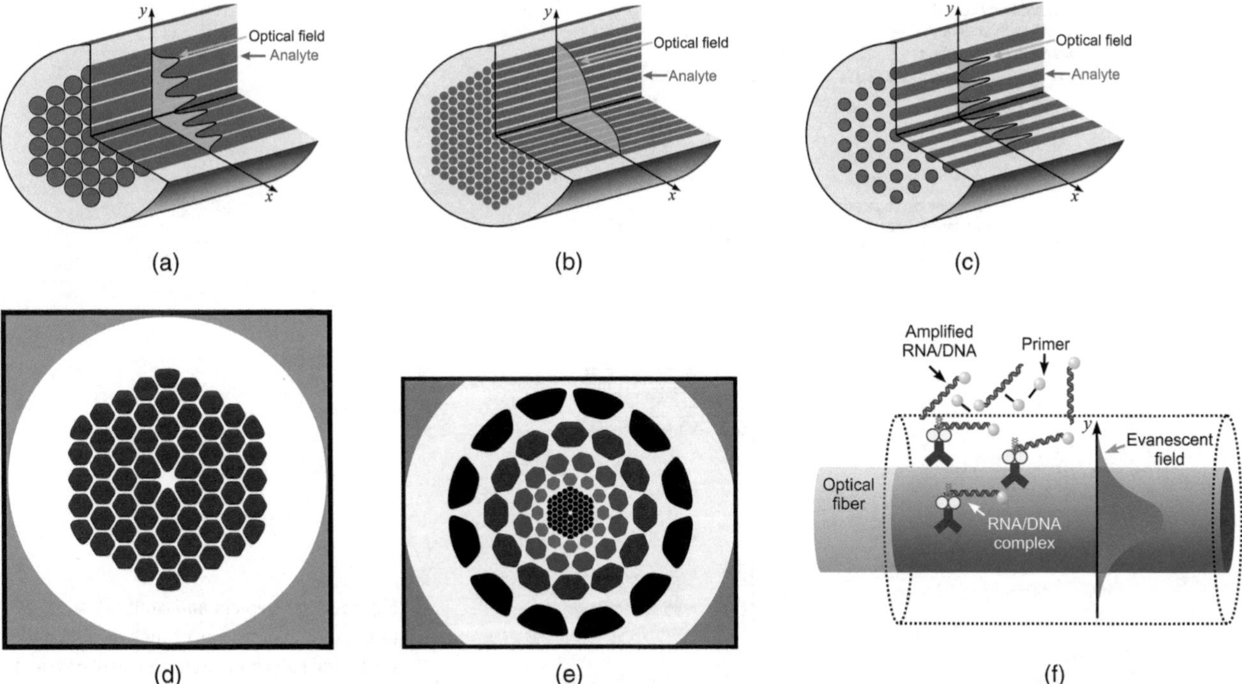

Figure 2: Schematics of the multichannel fiber sensor (a–c) and optical fiber structures (d, e) designed for the sensor. Shown in (f) is a fiber optic fluorometric sensor for the detection of polymerase chain reaction (PCR)–amplified DNA [26].

cost, these tests will become available to a broad population, not limited only to suspected COVID-19 cases. In pursuing this, we build upon the existing ideas, based on the human IgG and IgM antibody property to bind to SARS-CoV-2–specific proteins, and on the availability of antigens capable of binding the IgG and IgM antibodies. We extend prior work by using fluorescent markers together with quantum optical techniques such as higher order, e.g., Brown-Twiss, correlations, etc. In general, we employ optical detection configurations that will make sensitive quantitative measurements with relatively low levels of antibodies. The proposed "fiber" optical scheme is sketched below in Figure 3.

In particular, we shall illustrate our approach by concentrating on IgM and IgG antibodies against SARS-CoV-2 in blood as in Figure 3. There, we see blood mixed with buffer solution containing fluorescent nanoparticles (FNPs) such as nitrogen vacancy nanodiamonds, upconversion nanoparticles or semiconductor quantum dots (QDs), which have been coated with an antigen protein designed from the surface spike protein of SARS-CoV-2. In other words, the QDs, thus coated, look like a SARS-CoV-2 virus to the antibodies and are conjugated (bound) to the antibody as in Figure 3a.

The next step involves forcing the antibody-QD fluid through the tube or fiber optics waveguide, where it binds with the secondary antibodies (anti-antibody) that are themselves "stuck" to the (properly coated, Section 3) walls. The QDs are now fixed in place and, when driven by an external laser as in Figure 3, can constitute a lasing configuration. Thus, when mirrors are included, the system will "lase" when the antibody particle number (and therefore the antibody count) increases beyond a certain number.

The present scheme provides another way to gather and utilize data. For example, the steady-state photon statistical distribution is given by

$$P(n) = \left(\frac{\mathscr{A}}{\mathscr{B}}\right)! \left(\frac{\mathscr{A}^2}{\mathscr{B}\mathscr{C}}\right)^n \frac{1}{\left(n + \frac{\mathscr{A}}{\mathscr{B}}\right)} \frac{1}{F\left(1; \frac{\mathscr{A}}{\mathscr{B}} + 1; \frac{\mathscr{A}^2}{\mathscr{B}\mathscr{C}}\right)}, \quad (1)$$

where F is the confluent hypergeometric function, $\mathscr{A} = 2r\left(\frac{g}{\gamma}\right)^2$ is the gain that is directly proportional to the antibody count and $\mathscr{C}$ is governed by the cavity quality factor, and $\mathscr{B} = \frac{4g^2}{\gamma^2}\mathscr{A}$ where g is the atom-field coupling and γ is the radiative decay rate (11.2 of Ref. [9]).

From Eq. (1), the average number of photons in the cavity is given by

$$\langle n \rangle \cong \frac{\mathscr{A}}{\mathscr{C}}\left(\frac{\mathscr{A} - \mathscr{C}}{\mathscr{B}}\right), \quad (2a)$$

and the variance is found from

$$\langle n^2 \rangle - \langle n \rangle^2 = \frac{\mathscr{A}^2}{\mathscr{B}\mathscr{C}}. \quad (2b)$$

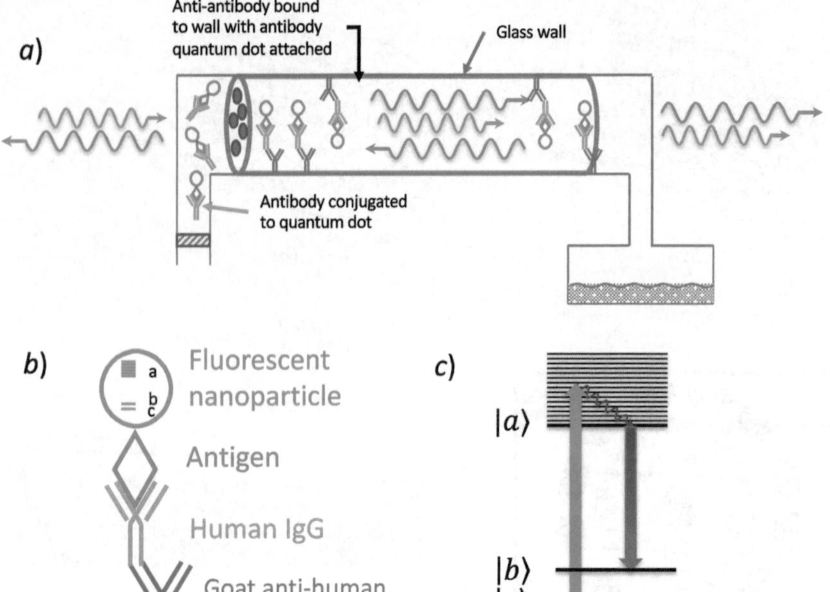

Figure 3: (a) Depicts anti-antibody + antibody-quantum pairs stuck to wall, with the bound pair configuration shown in (b) and the Raman level structure shown in (c).

The fiber bundle configuration can be capped with Bragg reflectors which have a high reflectivity resulting in a cavity loss rate of $\mathscr{C} = (c/l)(1 - R)$, where R is the reflection coefficient. For a "high Q" configuration with $l \approx 1$ cm the loss rate $\mathscr{C} \equiv \nu/Q \approx 10^6 - 10^7\ \mathrm{s}^{-1}$, where ν is the frequency of light. Using the fact that the coupling constant $g = \frac{\wp}{\hbar}\sqrt{\frac{\hbar\nu}{\epsilon_0 V}}$, where $V = \pi d^2 l$ and radiative spontaneous decay rate $\gamma_r = \wp^2\nu^3/3\pi\hbar\epsilon_0 c^3$, we may write $\mathscr{A}$ as

$$\mathscr{A} = 2r\left(\frac{g}{\gamma_t}\right)^2 = \left\{\frac{3}{2\pi^2}\right\}\frac{c}{l}N\frac{\lambda^2}{d^2}\frac{\gamma_r}{\gamma_t}, \tag{3}$$

with antibody number N, wavelength of photons λ, fiber hole radius d, and effective decay rate γ_t. At threshold where $\mathscr{A} = \mathscr{C}$, we find the antibody number

$$N = \left(\frac{2\pi^2}{3}\right)\frac{d^2}{\lambda^2}\frac{\gamma_t}{\gamma_r}(1 - R). \tag{4}$$

On the other hand, if we have no mirrors then the photon flux at $z = l$ will be given by

$$\frac{\mathrm{d}\langle n\rangle}{\mathrm{d}z} = \frac{1}{c}\mathscr{A}\langle n\rangle = \left(\frac{3}{2\pi^2}\right)\frac{N}{l}\frac{\lambda^2}{d^2}\frac{\gamma_r}{\gamma_t}\langle n\rangle$$

so that

$$\langle n(l)\rangle = \left(\exp\frac{\mathscr{A}}{\mathscr{C}}l\right)\langle n(0)\rangle \simeq 1 + \left(\frac{3}{2\pi^2}\right)N\frac{\lambda^2}{d^2}\frac{\gamma_r}{\gamma_t}. \tag{5}$$

So to detect one photon beyond the incident photon number of $\langle n(0)\rangle = 1$, we have the necessary antibody number

$$N = \left(\frac{2\pi^2}{3}\right)\frac{d^2}{\lambda^2}\frac{\gamma_t}{\gamma_r}. \tag{6}$$

The effect of adding "mirror" shows that (4) is smaller than (6) by the factor (1–R). That is to say the treated fiber bundle is more sensitive when the Bragg mirrors are in phase.

2.2 Using a cavity Quantum Electrodynamics (QED) platform for enhancing detection for antibodies via fluorescence

Biomarkers like green fluorescent proteins are extensively used for detecting tagged biomolecules and antibodies. Here, we suggest a new platform, where fluorescence detection can be enhanced especially for situations where either the fluorescent yield is low or the density of biomolecules is low. We make use of the well-known Purcell effect from the field of cavity QED, a field that was honored by the award of the Nobel Prize to S. Haroche. The new platform is sketched in Figure 4, with the test sample in a capillary or hollow-core fiber as indicated in the figure, i.e., antigen with human IgG (Figure 3b). Antibodies tagged to biomarkers are detected by fluorescence, which is enhanced as the system is contained inside a high quality cavity. Enhancement in fluorescence could be many times that in the absence of the cavity [10]. The result depends on the quality factor Q of the cavity. The number of photons per second detected outside the cavity is $(\frac{2g^2}{\kappa})$ times the number of excitations inside the cavity and hence, in technical terms, the enhancement factor is $\frac{3\lambda^3 Q}{4\pi^2 V}$, where V is the volume of the cavity and λ is the wavelength of transition. In other words, fluorescence probability in the direction of the cavity axis (thin red arrow labeled as detection Figure 4) is given by the factor $(\frac{3\lambda^3 Q}{4\pi^2 V})\gamma_r$. For N active molecules in the cavity the net fluorescence photon flux I along the cavity axis will be

$$I = \left(\frac{3\lambda^3 QN}{4\pi^2 V}\gamma_r\right) \tag{7}$$

and thus

$$N = \frac{4\pi^2 V}{3\lambda^3 \gamma_r Q} I. \tag{8}$$

Note that the quality factor, Q, of the cavity is inversely proportional to the cavity leakage, which is proportional to $(1–R)$, where R is the reflection coefficient of the mirror. It is therefore counter intuitive (but true) that the signal is enhanced. The enhancement of the detected signal originates from the Purcell effect. The measurement of the fluorescence using the scheme of Figure 4 can be highly beneficial for the detection of a small number of IgG molecules attached to antigen-conjugated FNP. In addition, we can improve the detection of IgG by reducing the volume of the cavity and, thus, micron size cavities would be recommended.

Spherical dielectric microresonators supporting whispering gallery modes (WGMs) (see Figure 5) could be an important variation of the above platform. WGMs are resonances of a wave field that are confined inside a cavity with smooth edges. They correspond to standing waves in the cavity. Electromagnetic WGMs with ultra-high Q factors, low mode volumes, and small resonators supporting them are the objective (see Appendix B). For example, a quality factor of $Q = 8 \times 10^9$ has been measured at $\lambda = 633$ nm in fused silica [11, 12] and $Q = 3 \times 10^{11}$ at $\lambda = 1.5\,\mu$m for crystalline CaF_2 [13].

Since the evanescent field of WGMs protrudes outside the resonator volume, such modes are affected by the environment in which the resonator is placed. The environment shifts the resonant frequencies. Because whispering gallery resonators can have extremely large Q-factors (small linewidth) a shift in their resonant frequencies is easily measured. This means that they act as very sensitive sensors which can be affected by chemical (or bio) composition of their surroundings.

Figure 5 shows a possible scheme for measuring the presence of viruses or antibodies (with single particle resolution [14]) that bind to the surface of a glass microsphere. Coupling is achieved by a tapered optical fiber, and a tunable near Infrared laser that sweeps across different wavelengths to determine the resonance frequencies of the resonator (on-resonance transmission of the laser beam through the optical fiber drops), which directly depend on the number of bound antibodies.

The use of light transmission is a very sensitive tool for the detection of antibodies and can work even without the use of labels [15–20]. This technique has the capability of detecting a single nanoparticle of about 20 nm as demonstrated by Lan Yang et al. [15, 21]. The transmitted spectra could be considerably shifted due to the interaction with the WGM's. The shifts will be proportional to the number of active molecules on the surface of microresonator.

2.3 Photonic PCR for SARS-CoV-2 detection

In this subsection, we propose a photonic platform for real-time polymerase chain reaction (qPCR) and reverse-

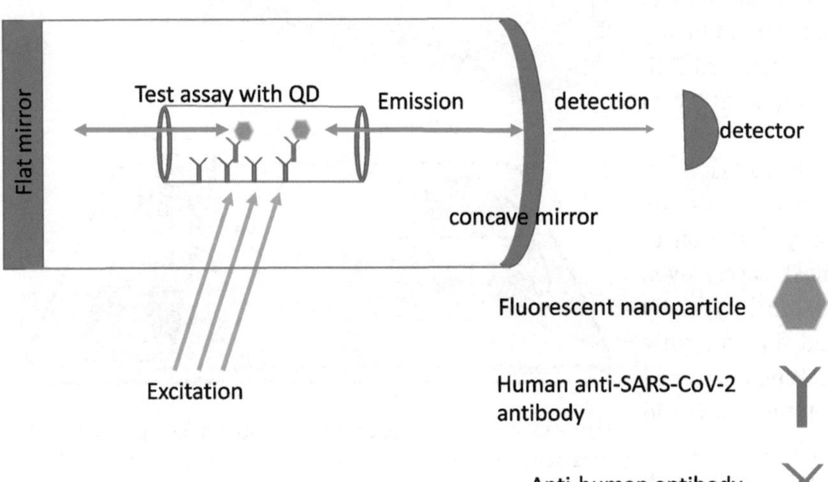

Figure 4: Excitation in one-sided cavity. The green arrows show the excitation light. The red double-arrows show the fluorescence from quantum dots inside the capillary tube. The red arrow on the right shows the output of the cavity which is collected by the detector. Details of the test capillary inside the cavity are shown in Figure 3.

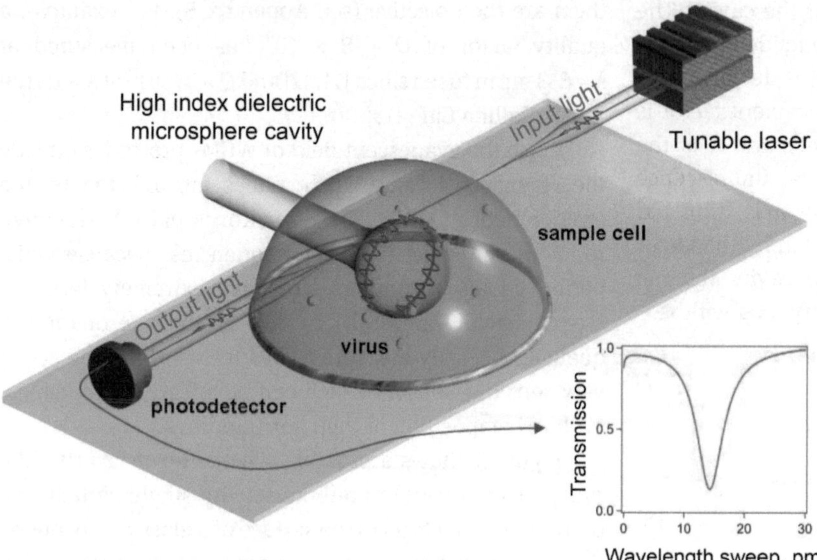

Figure 5: Measuring the presence of a virus using a whispering gallery resonator. As virus molecules bind to the surface, the resonant wavelength position shifts, which is measured by the photodetector. Adapted from F. Vollmer et al. "Single virus detection from the reactive shift of a whispering-gallery mode" ©2008 by The National Academy of Sciences of the USA [14].

transcription qPCR (qRT-PCR) that integrates a rapid-cycle fiber optic PCR chamber with feedback-controlled laser heating and online fluorometric detection of PCR products [22–24] (Refer to Appendix A for definition of PCR, qPCR, RT-PCR). Central to this platform is a suitably tailored optical fiber, whose hollow core provides accommodation for all the PCR components and all the steps of the PCR cycles [25, 26]. Infrared laser radiation is used to provide rapid heating of the fiber optic PCR chamber up to the DNA melting point. The process of laser-induced heating is monitored in real-time by means of all-optical thermometry based on color centers of diamond nanoparticles, after the annealing and extension steps have been performed at lower temperatures as required, the PCR cycles are repeated, providing an exponential amplification of targeted DNA sequences. This process is monitored in real-time using DNA-bound fluorescent reporters driven by visible laser radiation, which is coupled to the hollow fiber core. With careful optimization of laser-induced heating and heat-removal geometry, the photonic version of PCR can radically reduce the PCR cycle duration relative to standard PCR machines, enabling faster testing as a key strategy to mitigate spread of infectious disease.

Our photonic PCR will enable rapid (i.e., as fast as 10 PCR cycles, each cycle being 1 min as described in a study by Li et al. [25]) and highly specific early detection of SARS-CoV-2 soon after infection—in most cases, even before the onset of disease's symptoms [27–31] (typically 10–15 days), providing much-needed lead time for early treatment and disease spread prevention. When designed specifically for SARS-CoV-2 detection, photonic PCR would be run in the qRT-PCR mode. In this mode, a small amount of SARS-CoV-2 RNA present in a biomaterial sample (such as saliva) is first converted into complementary DNA (cDNA) using reverse transcriptase and deoxynucleoside triphosphates (dNTPs) [32]. The cDNA produced through this process is then used as a template for exponential amplification via PCR.

Our photonic platform (Figure 6) is designed to accommodate all of the steps necessary for qPCR/qRT-PCR—denaturation, annealing, and extension. At the denaturation step, the fiber optic thermal cycler is heated, using IR laser radiation, up to 94–98 °C, leading to denaturation of DNA/cDNA by breaking the hydrogen bonds between its bases, thus producing two single-

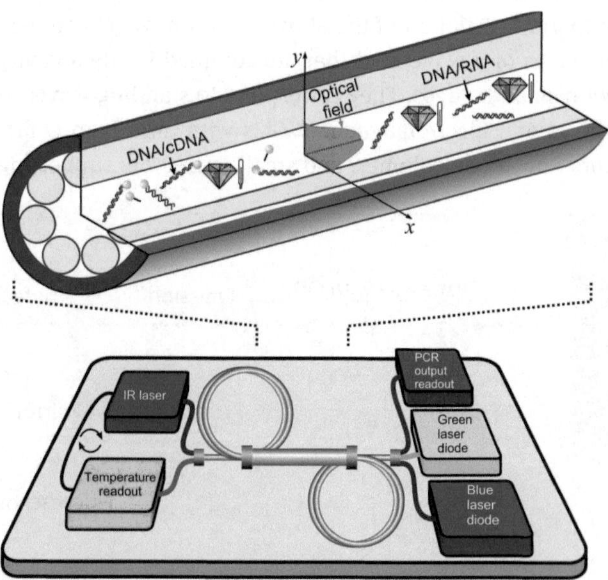

Figure 6: A photonic credit-card-size platform for rapid-cycle real-time polymerase chain reaction (qPCR) and reverse-transcription qPCR.

stranded DNA molecules. At the annealing step, the temperature is lowered to 50–65 °C, allowing left and right primers to base pair to their complementary sequences, thus bracketing the DNA region to be amplified. At the third step, the temperature of the fiber optic thermal cycler is maintained in a typical range of 72–80 °C, as the DNA polymerase synthesizes a new DNA strand complementary to the targeted DNA/cDNA strand by assembling free dNTPs from the reaction mixture to build new strands complementary to the template in the 5′-to-3′ direction. Experimental and theoretical studies of photonic PCR suggest that a PCR master mix of volume Φ will require a nuclease-free water mixture of 0.005Φ of Taq DNA polymerase, 0.1Φ of 10× standard Taq reaction buffer, 0.02Φ of 10 mM dNTPs, 0.02Φ of 10 μM forward primer, and 0.02Φ of 10 μM backward primer.

3 Bioconjugate techniques

This section addresses biological and biochemical terms for readers from other fields, as well as experimentalist who intend to implement these ideas as practical apparatus.

An essential component of the devices we discussed in Section 2 is the preparation of surfaces that are conjugated to bio molecules that we want to detect. Many techniques have been discussed in the literature and compiled in the classic book, Bioconjugate Techniques [33]. Since we are working with various optical fibers, it is natural for us to work with silane coupling agents, while other generic agents, e.g., protein A, G, A/G can be considered (see below).

Silanization is the process of adding silane groups (R3Si–) to a surface. The most useful compounds would be those containing a functional organic component, e.g., 3-glycid-oxy-propyl-trimethoxy-silane (GOPTS) and 3-glycidoxypropyltriethoxysilane that are very commonly used in fiber treatment. These compounds can be used to conjugate thiol-, amine-, or hydroxyl-containing ligands, depending on the pH of the reaction [33]. GOPTS is often used in bioconjugate applications, forming a polymer matrix linked by –Si-O-Si– bonds and binding to glass surfaces joining Si-O– bonds. Biomolecules, antibodies in particular in the context of this paper, can bond to fiber surfaces after reactions with the epoxy group in GOPTS.

Another method for attaching antihuman antibodies to the surface of an optical fiber optic is the use of proteins A, G, and L which are widely used in antibody purification procedures, where a specific antibody needs to be separated from the solution and purified. In some optical detection devices [34, 35], proteins A and G are used to attach the antibody to a substrate that later can be used to detect specific viruses using optical techniques or the naked eye. Using polydopamine and protein G solution [35], any substrate can be functionalized such that the antibody can attach to the substrate using its Fc region. The protein-coated substrate (see Figure 7c), in our case, the hollow-core optical fiber, is primed to attach to the desired antihuman antibody using the Fc region (base, see Figure 7c). This strategy will allow will allow the Fab region of the antibody to be available to interact with the target human antibody (in this case, anti-SARS-CoV-2 IgG and IgM).

In addition, using a fluorescent molecule or, even better, an FNP conjugated to the spike protein of SARS-CoV-2, we can tag the specific antibodies against SARS-CoV-2. In this manner, we will be able to see the signal from very few FNPs or fluorescent molecules. We promote the use of FNPs because they do not bleach under excitation in contrast to fluorescent dyes. This will give us the opportunity to increase the acquisition time for the Charge-Coupled Device (CCD) camera and, as a result, the excitation time. Hence, we will be able to observe very low amounts of FNPs in a few minutes.

A variant of these methods could be coating the optical fiber with the spike protein from SARS-CoV-2. The sample will contain anti-SARS-CoV-2 IgGs and IgMs tagged with FNPs conjugated to antihuman antibodies. As the sample is introduced into the optical fiber, the anti-SARS-CoV-2 antibodies will attach to a substrate. Based on the same principle, we can see the signal from FNPs that are attached to the fiber through linkage between the spike protein and antispike protein antibody tagged with FNPs conjugated with antihuman antibodies. Similar assays have been used to detect water pollutants [36].

Another method that can be used to attach antihuman antibodies to the substrate or optical fiber is to use sulfhydryl (–SH) groups that are naturally existing or added to the antihuman antibodies. In this method, the surface is activated by maleimide. Maleimide is a popular reagent for crosslinking sulfhydryl (–SH) groups, and it is used to form covalent links between the cysteine residues of proteins [37, 38]. Maleimide surfaces are obtained by amino-silanizing a surface with primary amines (–NH2) and reacting them with the heterobifunctional corsslinker, Sulfo-SMCC [39]. The crosslinker sulfo-SMCC is added in order to activate the amino-modified surface. Finally, the sulfhydryl (–SH) groups on the antibody can attach to this substrate by covalent bonds.

The adaptability of our optic fiber approach to existing technologies will make the mass production of this system relatively straight forward and affordable. However,

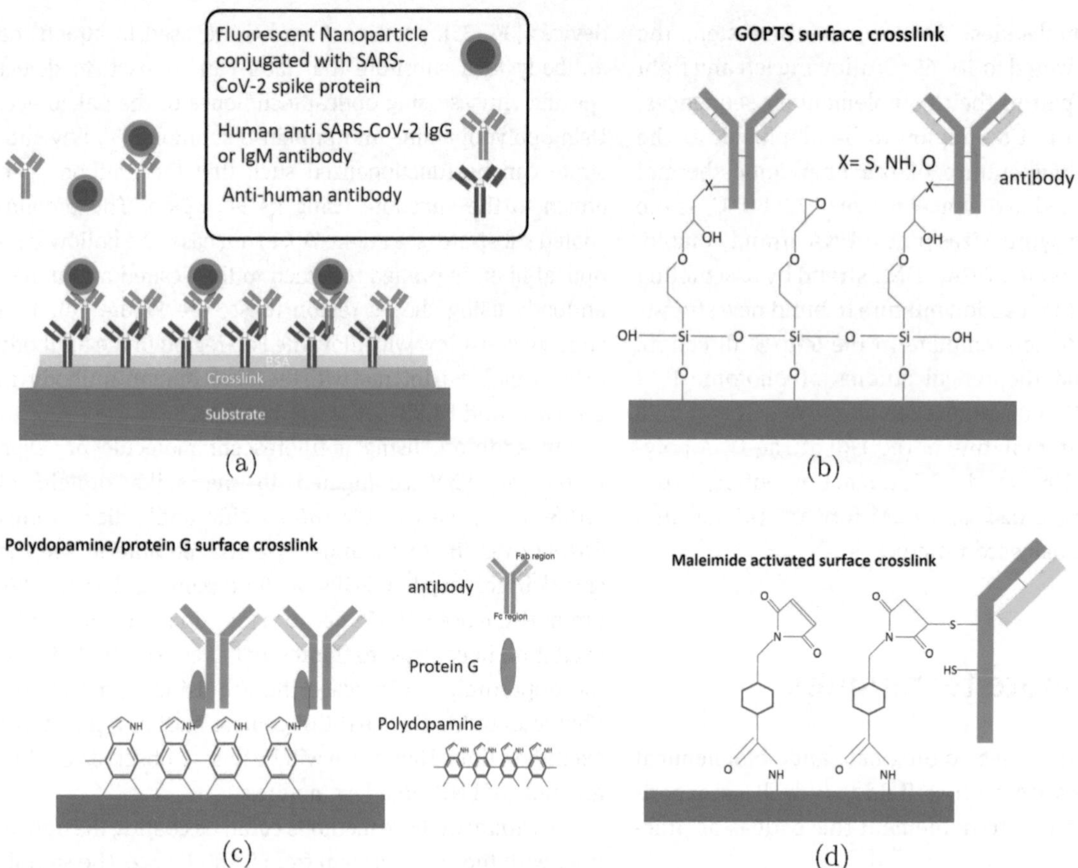

Figure 7: Schematic of substrate assay for detecting human anti-Severe Acute Respiratory Syndrome-Corona Virus-2 (SARS-CoV-2) IgG and IgM. The substrate is the inner surface of a hollow-core fiber optic. The cross-links we can use are 3-glycid-oxy-propyl-trimethoxy-silane (GOPTS), protein A or polydopamine/protein G, or amine-maleimide. The Bovine Serum Albumin (BSA) layer is used to block any other active part of the substrate. The human IgG and IgM bind to antihuman antibody by their Fc region, while the Fab region binds to spike proteins immobilized on the surface of a FNP.

depending on the type of test (PCR, antibody detection, etc.) one may need to choose the proper fiber optic setup and proper sample preparation. For instance, if original blood samples need to be tested, the hollow-core fiber optics chosen must have hollow regions large enough to accommodate for large blood cells. Perhaps, for certain detection applications such as antibodies or other proteins, one will need to filter and separate blood cells from the sample and only use blood plasma (or serum). This will reduce the interference caused by large objects such as red and white blood cells. This filtration can be integrated into the system like the sample pads in lateral flow assays (LFAs), which filter large particles and cells. Sample preparation can also be done before the test to promote the best result.

4 Discussion

One popular configuration for a COVID-19 test is the gold nanoparticle (AuNP) LFA configuration as shown in Figure 8. In this type of setup, a drop of blood mixed with a buffer solution is placed on the sample pad and wicked by capillary action into the conjugation pad which contains gold nanoparticles coated with spiked protein from the SARS-CoV-2. In such a case, the spike protein-conjugated gold nanoparticle looks to an antibody like a virus and the antibody binds to the coated AuNP. This bound, conjugated, gold particle antibody, e.g., SARS-CoV-2–specific IgG, is then wicked downstream to a strip which contains anti-antibodies (i.e., secondary antibodies) which will now bind to the SARS-CoV-2 antibody-gold complex. When there are enough gold particles, the line looks red, and this signals the presence of COVID-19 disease. If there are no antibodies, the gold is swept downstream to the collection pad and a negative result (no infection) is recorded.

The COVID-19 “test setup” discussed in Section 2.1 has points in common with LFA and points which are quite different. For example, the fiber optic platform of Figure 2 replaces the wicking membrane of Figure 8. The gold NPs

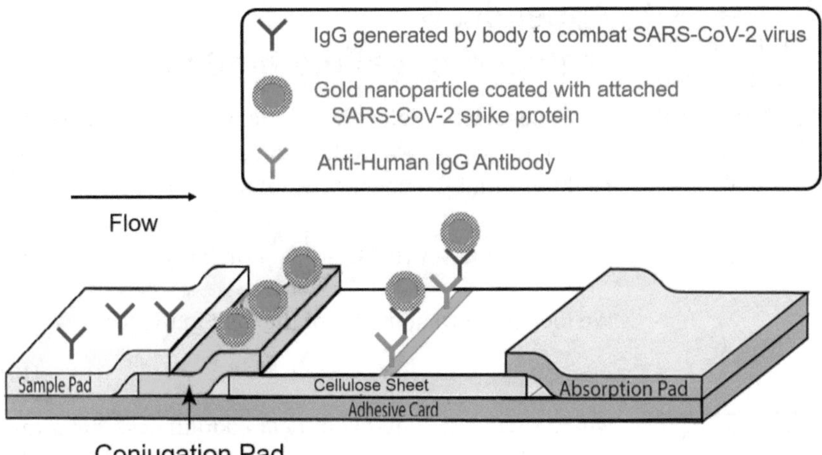

Figure 8: Schematic illustration for the wicking membrane with gold nanoparticle (NP) coated with Severe Acute Respiratory Syndrome-Corona Virus-2 (SARS-CoV-2) (see Table A1) spike protein. When the patient's blood with Corona Virus Disease-2019 (COVID-19) IgG (see Table A1) is flowed, assemblages of IgG and gold NP will be attached to the anti-antibody, and it can produce fluorescent light.

are replaced by the fluorescent NPs as in Figure 3. The fiber optical scheme is easily enclosed within a fiber optic cavity as in Figure 3a.

While antibody sensors provide a measure of the immune response to the virus, PCR is the technology of choice for early diagnostics, before any virus-specific antibodies can be detected in the bloodstream. Real-time state-of-the-art technique, qRT-PCR, which relies on optically excited fluorescent reporters, has become a gold standard in virus detection. The greatest downside of this well-established technology is the time it takes to run a test, limited by the PCR thermal cycle duration; it usually takes time to uniformly and precisely heat the sample volume and then cool it down and then heat again (1–5 days to complete traditional PCR and 45 min to complete qPCR [40]). The number of these cycles determines the DNA multiplication factor. With our fiber-based approach, we expect to have a significant scale-down of the reaction volume, and a corresponding reduction of characteristic cooling and heating time intervals. As a result, we expect to reduce the total time required to obtain test outcomes, to about 10 min. Within our approach, the heating and the temperature control will be conveniently provided through fiber-coupled laser-based methods. Naturally, our photonic qRT-PCR platform will incorporate an online fluorometric detection of PCR products. With careful optimization, this approach will lead to a greatly improved sensitivity, while reducing the test turnaround times, relative to standard PCR systems. In addition, the fiber optic approach, combined with microfluidics engineered for sample and reagent delivery, will allow compact design and scalability, resulting in affordable mass production and point-of-care implementation. Early detection of SARS-CoV-2 virions, enabled by the photonic fiber platform for qRT-PCR, will provide much-needed lead time for early treatment and prevention of disease spread.

Acknowledgments: M. O. S. thanks Chancellor John Sharp for many stimulating discussions and he thanks the Robert A. Welch Foundation, the Air Force Office of Scientific Research, the National Science Foundation (NSF), the Office of Naval Research, and King Abdulaziz City for Science and Technology (KACST) for support. The authors thank Jane Pryor, and Maria Bermudez Cruz for helpful discussions. N. R., S. D., C. L., S. E., M. L. are supported by the Herman F. Heep and Minnie Belle Heep Texas A&M University Endowed Fund held/administered by the Texas A&M Foundation. J. D. C. is supported in part from funds provided by the Texas A&M University System and National Institutes of Health Grant AI104960. P. H. acknowledges financial support from the Government of the Russian Federation (Mega-grant No. 14.W03.31.0028). G. S. A. thanks the Robert A. Welch Foundation grant no A-1943 and the AFOSR award No. FA9550-18-1-0141 for support. A. S. acknowledges the support from the Robert A. Welch Foundation grant no A-1547. A. Z. acknowledges support from the Welch Foundation (Grant No. A-1801-20180324). V. B. and A. Z. acknowledge support from the Russian Foundation for Basic Research (project Nos. 17-00-00212 and 17-00-00214). V. D. acknowledges support from German Research Foundation (CRC 1375 - NOA - C2). V.P. acknowledges the support of the U.S. Army Research Office under Grant W911NF-20-1-0204, the U.S. National Science Foundation under RAPID Grant IIS-2026982, and a grant from the C3.ai Digital Transformation Institute.

Author contribution: All the authors have accepted responsibility for the entire content of this submitted manuscript and approved submission.

Research funding: The research was supported by the Robert A. Welch Foundation (Grant No. A-1943, A-1547, and A-1261), the Air Force Office of Scientific Research (Award No. FA9550-20-1-0366 DEF), National Science Foundation (Grant No. PHY-2013771), Office of Naval

Research (Grant No. N00014-20-1-2184), Texas A&M Foundation, National Institutes of Health (Grant No. AI104960), and the Government of the Russian Federation (14.W03.31.0028). This research is also supported by King Abdulaziz City for Science and Technology (KACST).
Conflict of interest statement: The authors declare no conflicts of interest regarding this article.

Appendix A

Table A1: Summary of abbreviations.

CARS	Coherent anti-Stokes Raman scattering, a technique for using multiple photons for measuring molecular vibration with much greater sensitivity than ordinary (spontaneous) Raman emissions
COVID-19	The corona virus disease that appeared in late 2019
dNTP	Dinucleoside triphosphate, a precursor molecule for DNA
GOPTS	(3-Glycidyloxypropyl)trimethoxysilane is used as a coupling agent for many molecules with glass or mineral surfaces
IgG	The most abundant immunoglobulin (antibody) in human blood; SARS-CoV-2-specific IgG is made by B-cells of the immune system following detection of SARS-CoV-2; attaches to the virus spike protein
IgM	The first immunoglobulin (antibody) to form following detection of an antigen by the human immune system; short-lived relative to IgG
LF4 TAMRA (Figure 2f)	A fluorescent oligonucleotide primer molecule in the DNA polymerase system
NV diamond	Nanosize diamond crystals containing defects in the form of a vacant space next to a nitrogen atom; Such structures fluoresce upon impact of a laser beam of specified wavelength.
PCR/qPCR	Polymerase chain reaction, the process of making new copies of an original sample of DNA (q indicates the quantitative version of PCR)
RT-PCR/qRT-PCR	Quantitative reverse transcriptase PCR, the process of using the enzyme, reverse transcriptase, to read a molecule of RNA (or a fragment, thereof) to form the complimentary DNA
SARS-CoV-2	Severe Acute Respiratory Syndrome-Corona Virus-2, The causative viral agent of COVID-19
Taq DNA polymerase	A thermal-stable DNA polymerase (enzyme)
Sulfo-SMCC	A protein containing an amine-sulfo-NSH-ester on one end and melamide on the other. The former increases water solubility; the latter reacts with SH groups.

Appendix B Whispering gallery modes

Let us consider a dielectric cylinder with a refractive index n and radius R. Maxwell's equation for the dielectric medium with inhomogeneous refractive index $n(\mathbf{r})$ reads

$$\nabla^2\mathbf{E} - \nabla(\operatorname{div}\mathbf{E}) - \frac{1}{c^2}\frac{\partial^2}{\partial t^2}(n^2(\mathbf{r})\mathbf{E}) = 0. \tag{9}$$

We look for normal modes of the field in the form

$$\mathbf{E}(\mathbf{r}) = E(r)e^{-i\omega t + im\phi}\hat{z}, \tag{10}$$

where r, ϕ, and z are cylindrical coordinates, $\hat{z}$ is a unit vector along the cylinder's axis z, and m is the angular quantum number. The mode function (10) is transverse, that is div $(\mathbf{E}) = 0$. In addition, the mode function (10) is independent of z, that is electromagnetic wave does not propagate along the axis of the cylinder.

The Maxwell's Eq. (9) yields that radial part of the mode function (10) inside the cylinder ($r < R$) is given by the cylindrical Bessel function

$$E(r) = C_1 J_m\left(\frac{\omega n}{c}r\right), \tag{11}$$

and by the Hankel function of the first kind outside the cylinder ($r > R$)

$$E(r) = C_2 H_m^{(1)}\left(\frac{\omega}{c}r\right). \tag{12}$$

Solution outside the cylinder describes an outgoing cylindrical wave, and at $r \gg R$ the asymptotic of the electric field is

$$E(t,\mathbf{r}) \sim \frac{1}{\sqrt{r}}e^{i\omega\left(\frac{r}{c}-t\right)+im\phi}.$$

The outgoing cylindrical wave carries energy out of the system and the modes decay with time. The decay rate Γ is given by the imaginary part of the mode frequency

$$\Gamma = -Im(\omega).$$

The normal mode frequencies can be found by imposing the boundary conditions at the cylinder surface, namely, $E(r)$ and $\partial E(r)/\partial r$ must be continuous at $r = R$. This gives a transcendental characteristic equation for ω

$$n\frac{J_m'\left(\frac{\omega n}{c}R\right)}{J_m\left(\frac{\omega n}{c}R\right)} = \frac{H_m^{(1)\prime}\left(\frac{\omega}{c}R\right)}{H_m^{(1)}\left(\frac{\omega}{c}R\right)}, \tag{13}$$

where the prime denotes differentiation with respect to the argument of the Bessel function.

For fixed m, Eq. (13) has an infinite number of solutions which are labeled by the radial quantum

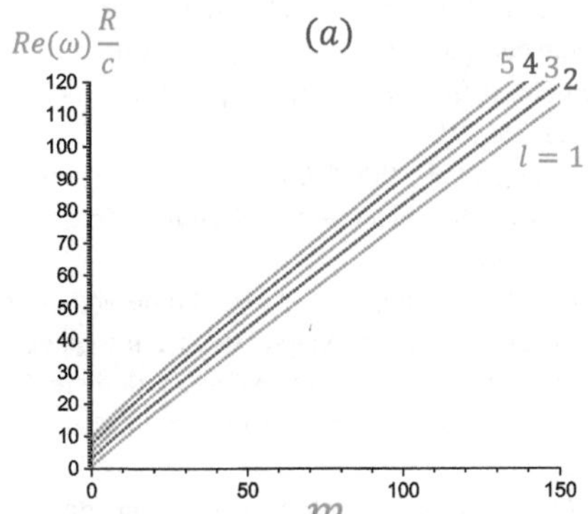

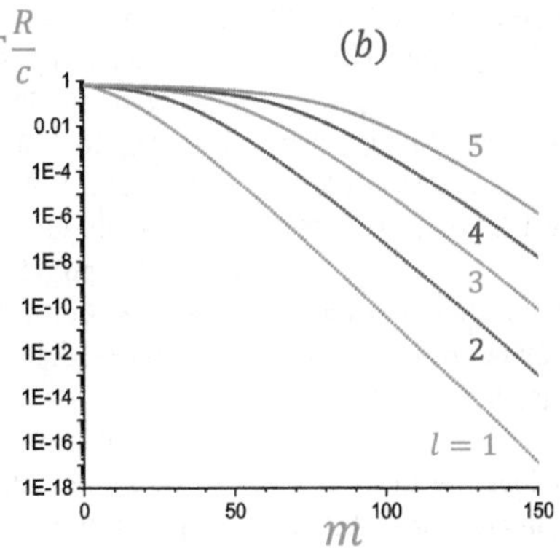

Figure 9: Normal mode frequencies (a) and their decay rates Γ (b) as a function of the angular quantum number m for the radial quantum number l = 1, 2, 3, 4, and 5, obtained by numerical solution of Eq. (13) with n = 1.4.

number l = 1, 2, 3, …. The radial quantum number l determines the number of field oscillations inside the cylinder. In Figure 9, we plot the normal mode frequencies (a) and their decay rates (b) as functions of the angular quantum number m for l = 1, 2, 3, 4, and 5. In these calculations, we assume that cylinder's refractive index is n = 1.4. The vertical axis in Figure 9b has a logarithmic scale.

Figure 9b shows that the decay rate of the modes with large m is exponentially small. The decay rate of the mode is smallest for l = 1. The modes with l = 1 and large m are known as the WGMs. They are a wave analog to a ray reflecting m times off the edge of the cavity at a grazing angle by total internal reflection forming a standing wave.

For large m, the quality factor, Q, of the dielectric microresonators could be only limited by the intrinsic material absorption. A typical value of Q for such microresonators is of the order of 10^8.

High Q implies a narrow resonance width $\Delta\omega = \omega/Q$ or $\Delta\lambda = \lambda/Q$. For example, if the resonance wavelength is $\lambda = 1$ μm then the width of the resonance is about 10 fm. Such small value of the resonance width allows us to measure the resonance frequency shift produced by a single virus molecule attached to the surface of the microresonator. As molecules bind to the surface, the resonant wavelength position, λ, jumps, creating steps in the time dependence $\lambda(t)$. Such steps have been observed, e.g., in the experiment of Armani et al. with interleukin-2 molecules [41].

References

[1] J. S. Ellis and M. C. Zambon, "Molecular diagnosis of influenza," *Rev. Med. Virol.*, vol. 12, pp. 375–389, 2002.

[2] M. O. Scully, G. W. Kattawar, R. P. Lucht, et al., "FAST CARS: engineering a laser spectroscopic technique for rapid identification of bacterial spores," *Proc. Natl. Acad. Sci. U. S. A.*, vol. 99, no. 17, pp. 10994–11001, 2002.

[3] S. O. Konorov, A. M. Zheltikov, and M. Scalora, "Photonic-crystal fiber as a multifunctional optical sensor and sample collector," *Opt. Express*, vol. 13, pp. 3454–3459, 2005.

[4] I. A. Bufetov, A. F. Kosolapov, A. D. Pryamikov, et al., "Revolver hollow core optical fibers," *Fibers*, vol. 6, p. 39, 2018.

[5] P. St and J. Russell, "Photonic crystal fibers," *Science*, vol. 299, pp. 358–362, 2003.

[6] A. M. Zheltikov, "Holey fibers," *Phys. Usp.*, vol. 43, pp. 1125–1136, 2000.

[7] A. B. Fedotov, S. O. Konorov, V. P. Mitrokhin, E. E. Serebryannikov, and A. M. Zheltikov, "Coherent anti-Stokes Raman scattering in isolated air-guided modes of a hollow-core photonic-crystal fiber," *Phys. Rev. A*, vol. 70, p. 045802, 2004.

[8] A. M. Zheltikov, "Nonlinear optics of microstructure fibers," *Phys. Usp.*, vol. 47, pp. 69–98, 2004.

[9] M. Scully and S. Zubairy, *Quantum Optics*, Cambridge, Cambridge University Press, 1996, https://doi.org/10.1017/CBO9780511813993.

[10] G. S. Agarwal, *Quantum Optics*, Cambridge, Cambridge University Press, 2013, p. 175, https://doi.org/10.1017/CBO9781139035170.

[11] K. J. Vahala, "Optical microcavities," *Nature*, vol. 424, pp. 839–846, 2003.

[12] M. L. Gorodetsky, A. A. Savchenkov, and V. S. Ilchenko, "Ultimate Q of optical microsphere resonators," *Opt. Lett.*, vol. 21, pp. 453–455, 1996.

[13] A. A. Savchenkov, A. B. Matsko, V. S. Ilchenko, and L. Maleki, "Optical resonators with ten million finesse," *Opt. Express*, vol. 15, pp. 6768–6773, 2007.

[14] F. Vollmer, S. Arnold, and D. Keng, "Single virus detection from the reactive shift of a whispering-gallery mode," *Proc. Natl. Acad. Sci. U. S. A.*, vol. 105, no. 52, pp. 20701–20704, 30 Dec. 2008.

[15] J. Zhu, S. K. Ozdemir, Y. Xiao, et al., "On-chip single nanoparticle detection and sizing by mode- splitting in an ultra-high-Q microresonator," *Nat. Photonics*, vol. 4, p. 46, 2010.

[16] F. Giovanardi, A. Cucinotta, A. Rozzi, et al., "Hollow core inhibited coupling fibers for biological optical sensing," *J. Lightwave Technol.*, vol. 37, no. 11, 1 Jun. 12019, https://doi.org/10.1109/JLT.2019.2892077.

[17] V. Ahsani, F. Ahmed, M. B. G. Jun, and C. Bradley, "Tapered fiber-optic Mach-Zehnder interferometer for ultra-high sensitivity measurement of refractive index," *Sensors*, vol. 19, p. 1652, 2019.

[18] Y. Liu, P. Hering, and M. O. Scully, "An integrated optical sensor for measuring glucose concentration," *Appl. Phys.*, vol. B54, pp. 18–23, 1992.

[19] J. L. Dominguez-Juarez, G. Kozyreff, J. Martorell, "Whispering gallery microresonators for second harmonic light generation from a low number of small molecules", *Nat. Commun.*, vol. 2, 2011, Art no. 254. https://doi.org/10.1038/ncomms1253.

[20] R. Gao, D. Lu, M. Zhang, and Z. Qi, "Optofluidic immunosensor based on resonant wavelength shift of a hollow core fiber for ultratrace detection of carcinogenic benzo[a]pyrene," *ACS Photonics*, vol. 5, pp. 1273–1280, 2018.

[21] S. Dutta Gupta and G. S. Agarwal, "Strong coupling cavity physics in microspheres with whispering gallery modes," *Opt. Commun.*, vol. 115, p. 597, 1995.

[22] I. V. Fedotov, S. Blakley, E. E. Serebryannikov, et al., "Fiber-based thermometry using optically detected magnetic resonance," *Appl. Phys. Lett.*, vol. 105, p. 261109, 2014.

[23] I. V. Fedotov, L. V. Doronina-Amitonova, D. A. Sidorov-Biryukov, et al., "Fiber-optic magnetic-field imaging," *Opt. Lett.*, vol. 39, pp. 6954–6957, 2014.

[24] S. Blakley, X. Liu, I. Fedotov, et al., "Fiber-optic quantum thermometry with germanium-vacancy centers in diamond," *ACS Photonics*, vol. 6, pp. 1690–1693, 2019.

[25] X. Li, L. V. Nguyen, K. Hill, et al., "Picoliter real-time quantitative polymerase chain reaction (qPCR) in an all-fiber system," in *Asia Communications and Photonics Conference (ACPC) 2019, OSA Technical Digest*, Optical Society of America, 2019. paper S4G.5.

[26] J. M. Mauro, L. K. Cao, L. M. Kondracki, S. E. Walz, and J. R. Campbell, "Fiber-optic fluorometric sensing of polymerase chain reaction-amplified DNA using an immobilized DNA capture protein," *Anal. Biochem.*, vol. 235, pp. 61–72, 1996.

[27] M. M. Arons, K. M. Hatfield, S. C. Reddy, et al., "Presymptomatic SARS-CoV-2 infections and transmission in a skilled nursing facility," *N. Engl. J. Med.*, 2020, https://doi.org/10.1056/NEJMoa2008457, Epub ahead of print.

[28] S. Hoehl, H. Rabenau, A. Berger, et al., "Evidence of SARS-CoV-2 infection in returning travelers from Wuhan, China," *N. Eng. J. Med.*, vol. 283, pp. 1278–1280, 2020.

[29] K. Q. Kam, C. F. Yung, L. Cui, et al., "A well infant with coronavirus disease 2019 with high viral load," *Clin. Infect. Dis.*, 2020, https://doi.org/10.1093/cid/ciaa201.

[30] T. Q. M. Le, T. Takemura, M. L. Moi, et al., "Severe acute respiratory syndrome coronavirus 2 shedding by travelers, Vietnam, 2020," *Emerg. Infect. Dis.*, p. 26, 2020, https://doi.org/10.3201/eid2607.200591.

[31] L. Zou, F. Ruan, M. Huang, et al., "SARS-CoV-2 viral load in upper respiratory specimens of infected patients," *N. Engl. J. Med.*, vol. 382, pp. 1177–1179, 2020.

[32] E. A. Bruce, S. Tighe, J. J. Hoffman, et al., "RT-qPCR detection of Sars-Cov-2 RNA from patient nasopharyngeal swab using Qiagen Rneasy kits or directly via omission of an RNA extraction step", 2020, bioRxiv 2020.03.20.001008, https://doi.org/10.1101/2020.03.20.001008.

[33] G. T. Hermanson, *Bioconjugate Techniques*, London, Academic Press, 2013, https://doi.org/10.1016/C2009-0-64240-9.

[34] A. A. Yanik, M. Huang, O. Kamohara, et al., "An optofluidic nanoplasmonic biosensor for direct detection of live viruses from biological media," *Nano Lett.*, vol. 10, no. 12, pp. 4962–4969, 2010.

[35] J. Moon, J. Byun, H. Kim, et al., "Surface-independent and oriented immobilization of antibody via one-step polydopamine/protein G coating: application to influenza virus immunoassay," *Macromol. Biosci.*, vol. 19, p. 1800486, 2019.

[36] A. Hlaváček, Z. Farka, M. Hübner, et al., "Competitive upconversion-linked immunosorbent assay for the sensitive detection of diclofenac," *Anal. Chem.*, vol. 88, no. 11, pp. 6011–6017, 2016.

[37] http://tools.thermofisher.com/content/sfs/brochures/TR0005-Attach-Ab-glass.pdf.

[38] D. Kim and A. E. Herr, "Protein immobilization techniques for microfluidic assays," *Biomicrofluidics*, vol. 7, no. 4, p. 041501, Jul. 2013.

[39] A. Sen, "Quantification of cell attachment on different materials as candidate electrodes for measurement of quantal exocytosis," PhD thesis, https://doi.org/10.32469/10355/5722.

[40] M. J. Espy, J. R. Uhl, L. M. Sloan, et al., "Real-time PCR in clinical microbiology: application for routine laboratory testing," *Clin. Microbiol. Rev.*, vol. 19, no. 1, pp. 165–256, Jan. 2006.

[41] A. M. Armani, R. P. Kulkarni, S. E. Fraser, R. C. Flagan, and K. J. Vahala, "Label-free, single-molecule detection with optical microcavities," *Science*, vol. 317, p. 783, 2007.

Luca Moretti, Andrea Mazzanti, Arianna Rossetti, Andrea Schirato, Laura Polito, Fabio Pizzetti, Alessandro Sacchetti, Giulio Cerullo, Giuseppe Della Valle*, Filippo Rossi* and Margherita Maiuri*

Plasmonic control of drug release efficiency in agarose gel loaded with gold nanoparticle assemblies

https://doi.org/10.1515/9783110710687-022

Abstract: Plasmonic nanoparticles (NPs) are exploited to concentrate light, provide local heating and enhance drug release when coupled to smart polymers. However, the role of NP assembling in these processes is poorly investigated, although their superior performance as nanoheaters has been theoretically predicted since a decade. Here we report on a compound hydrogel (agarose and carbomer 974P) loaded with gold NPs of different configurations. We investigate the dynamics of light-heat conversion in these hybrid plasmonic nanomaterials via a combination of ultrafast pump-probe spectroscopy and hot-electrons dynamical modeling. The photothermal study ascertains the possibility to control the degree of assembling via surface functionalization of the NPs, thus enabling a tuning of the photothermal response of the plasmon-enhanced gel under continuous wave excitation. We exploit these assemblies to enhance photothermal release of drug mimetics with large steric hindrance loaded in the hydrogel. Using compounds with an effective hydrodynamic diameter bigger than the mesh size of the gel matrix, we find that the nanoheaters assemblies enable a two orders of magnitude faster cumulative drug release toward the surrounding environment compared to isolated NPs, under the same experimental conditions. Our results pave the way for a new paradigm of nanoplasmonic control over drug release.

Keywords: drug delivery; gold nanoparticles; hot electrons; nanoheaters; plasmonics; ultrafast spectroscopy.

***Corresponding authors: Giuseppe Della Valle,** Dipartimento di Fisica, Politecnico di Milano, P.za Leonardo da Vinci 32, 20133, Milan, Italy; Istituto di Fotonica e Nanotecnologie, Consiglio Nazionale delle Ricerche, Piazza Leonardo da Vinci 32, 20133, Milan, Italy, E-mail: giuseppe.dellavalle@polimi.it; **Filippo Rossi,** Dipartimento di Chimica, Materiali e Ingegneria Chimica "Giulio Natta", Politecnico di Milano, via Mancinelli 7, 20131 Milan, Italy, E-mail: filippo.rossi@polimi.it; and **Margherita Maiuri,** Dipartimento di Fisica, Politecnico di Milano, P.za Leonardo da Vinci 32, 20133, Milan, Italy, E-mail: margherita.maiuri@polimi.it
Luca Moretti and Andrea Mazzanti, Dipartimento di Fisica, Politecnico di Milano, P.za Leonardo da Vinci 32, 20133, Milan, Italy
Arianna Rossetti, Fabio Pizzetti and Alessandro Sacchetti, Dipartimento di Chimica, Materiali e Ingegneria Chimica "Giulio Natta", Politecnico di Milano, via Mancinelli 7, 20131 Milan, Italy
Andrea Schirato, Dipartimento di Fisica, Politecnico di Milano, P.za Leonardo da Vinci 32, 20133, Milan, Italy; and Istituto Italiano di Tecnologia, via Morego 30, 16163, Genoa, Italy
Laura Polito, Consiglio Nazionale delle Ricerche, CNR-SCITEC, Via G. Fantoli 16/15, 20138, Milan, Italy
Giulio Cerullo, Dipartimento di Fisica, Politecnico di Milano, P.za Leonardo da Vinci 32, 20133, Milan, Italy; and Istituto di Fotonica e Nanotecnologie, Consiglio Nazionale delle Ricerche, Piazza Leonardo da Vinci 32, 20133, Milan, Italy. https://orcid.org/0000-0002-9534-2702

1 Introduction

The optical properties of metal nanoparticles (NPs), particularly their localized surface plasmon resonances (LSPRs), are well established [1–6]. It is now straightforward to design and fabricate high-quality metal NPs with tailored optical properties (such as optimized absorption, scattering coefficients and narrow LSPR bands) for multiple purposes, ranging from the detection of chemicals and biological molecules [7–11] to light-harvesting enhancement in solar cells [12–15] or applications in nanomedicine [16–19]. By dispersing the NPs in organic compounds (such as polymers) and creating a hybrid material, the robustness, responsiveness and flexibility of the system are enhanced, while preserving the intrinsic properties of the NPs [20]. Specifically, plasmonic nanostructures with high absorption cross sections are desirable for photothermal processes, such as cancer therapy [21] and drug and gene delivery [22].

This article has previously been published in the journal Nanophotonics. Please cite as: L. Moretti, A. Mazzanti, A. Rossetti, A. Schirato, L. Polito, F. Pizzetti, A. Sacchetti, G. Cerullo, G. Della Valle, F. Rossi and M. Maiuri "Plasmonic control of drug release efficiency in agarose gel loaded with gold nanoparticle assemblies" *Nanophotonics* 2021, 10. DOI: 10.1515/nanoph-2020-0418.

The mechanism of heat release upon NP illumination is quite simple: the electromagnetic field of light excites the NPs LSPR (a collective motion of a large number of electrons) which rapidly decays nonradiatively into a distribution of hot electrons, which in turn thermalizes on the picosecond timescale with the phonons of the NP and converts the absorbed light into heat [4, 5]. Finally, heat is transferred from the NP to the environment through phonon-phonon interactions on a timescale of ~100 ps, increasing the temperature of the surrounding medium [7]. Heat generation by metal NPs under optical illumination has attracted much interest [23–34], and it has been demonstrated to enhance performances in the context of drug release [35]. In assemblies of NPs, collective effects can be used to strongly amplify the heating effect and to create local 'hot spots' featuring high temperature [36]. Although the superior performance of NPs assemblies (suprastructures) as nanoheaters has been theoretically predicted since a decade [36, 37], experimental studies of photothermal effects in such assemblies have so far been scarce [38].

The ability of gold (Au) NPs to be embedded in organic networks forming hybrids sensitive to light irradiation is well known in the literature [35, 39] and many research groups described the promising possibility of Au NPs loading within polymeric networks in order to improve cancer treatment [40], antimicrobial activity [41] or bone regeneration [42]; however, no similar studies employing NPs assemblies have been so far reported. Particularly promising biomaterial carriers are hydrogels, hydrophilic biocompatible three-dimensional networks, that found, among others applications in cartilage, central nervous system and bone repair strategies [43]. The hydrogels mild gelling condition and elastic properties allow their use as carriers for drugs and cells at the same time [43]. These features make these hydrogels good candidates for novel strategies in targeted/local delivery of specific drugs and biomolecules, to design advanced therapies for patients with severe or chronic diseases where the body's own response is not enough for the recovery of all functions [44]. By tuning the hydrogel swelling properties, degradation rate and cross-linking density, it is possible to smartly control cell fate and release rates. However, the extremely good results obtained as cell carriers [45] are not matched by similar results in drug delivery applications, both with hydrophobic and hydrophilic drugs [46].

In this paper we investigate photothermal effects in Au NPs assemblies loaded in a hydrogel compound and demonstrate that they enhance drug release performances. We test the photothermal properties of Au NPs assemblies incorporated in a hydrogel library already developed for cell-based therapies [47, 48] to ameliorate its drug delivery performance. In order to compare the drug release performances of isolated versus assembled Au NPs, in one case we decorate the NPs surface with polyethylene glycol (PEG) chains, commonly known as PEGylation, to guarantee that the Au NPs loaded in the gel remain isolated, in contrast with non-PEGylated Au NPs which tend to aggregate in suprastructures. First, we investigate the dynamics of light-heat conversion in these hybrid plasmonic nanomaterials via a combination of ultrafast pump-probe spectroscopy and hot-electrons dynamical modeling. The photothermal study ascertain the possibility to control the degree of assembling via PEGylation of the NPs, thus enabling a tuning of the photothermal response of the plasmon-enhanced gel under continuous wave excitation. Then, the obtained hybrid vehicle material is studied as drug delivery system with and without light irradiation, using a mimic for high steric hindrance therapeutic molecules.

Our results indicate that the presence or absence of PEGylation onto Au NPs can tune the final performance of the drug delivery devices. More specifically, we found that the nanoheaters assemblies enable two orders of magnitude increase of cumulative drug release of the hydrogel toward the surrounding environment compared to isolated NPs, under the same experimental conditions.

2 Materials and methods

2.1 Samples

The chosen hydrogel (HG) was obtained by synthesis from statistical block polycondensation between agarose and carbomer 974P (AC), together with cross-linkers [49]. In previous works, we observed that this HG can remain localized at the site of injection [48], showing high biocompatibility and good ability to provide viability of neural and mesenchymal stem cells together with pathology amelioration after spinal cord injury [50]. However, AC-based HGs are not light-responsive and so, after HG synthesis but before the sol/gel transition takes place, Au NPs were physically entrapped within the AC networks forming the organic-inorganic composite material. Au NPs synthetized with PEG capping (Au-PEG NPs) and in the absence of it were loaded in the HG. Then, the HG-NPs samples were loaded with two fluorescein-based compounds of different sizes, mimicking prototypical drugs: sodium fluorescein (SF) and fluorescein-dextran 70 kDa (DEX). The latter serves as a mimic for high steric hindrance therapeutic molecules like a wide class of proteins, such as anti-Cd11b antibodies, chondroitinase ABC and Anti-Nogo-A antibodies or erythropoietin [51]. We performed drug release experiments on SF-loaded and DEX-loaded HG-NPs in order to compare the efficiency in drug delivery applications of the hydrogels with Au NPs and Au-PEG NPs.

Details on NP synthesis procedures and gel characterization can be found in the Supplementary material.

2.2 Pump-probe measurements

Ultrafast pump-probe experiments were performed starting from a Ti:Sapphire chirped pulse amplified laser (Libra, Coherent), with 1 kHz repetition rate, central wavelength of 800 nm and pulse duration of ≈100 fs. The pump pulses at 400 nm were obtained by frequency doubling the laser output in a 1 mm β-barium borate crystal. Probe pulses in the visible spectral region (450–750 nm) were obtained by supercontinuum generation in a thin sapphire plate. The probe beam transmitted through the sample was detected by a high-speed spectrometer (Entwicklungsbuero EB Stresing). All samples were measured at room temperature in a 1-cm optical path plastic cuvette, filled with water and with the 2-mm-thin HG sample stuck on the cuvette inner wall. Measurements were processed with Matlab™ 2019a and Origin™ software.

2.3 Optical simulations

2.3.1 Steady-state quasi-static model: The steady-state absorbance of the sample is calculated as $A = -\log_{10}(T)$, with $T = \exp(-\sigma_E N_p L)$ the transmission spectrum, retrieved from the extinction cross section σ_E of the considered nanostructures, their concentration, N_p (being 4×10^{11} and 5.6×10^{11} cm^{-3} for the isolated NPs, used to model the Au-PEG NPs, and for NP dimers, used to model the nanoassembled Au NPs samples, respectively), and the gel thickness $L = 2$ mm. The extinction cross section $\sigma_E = \sigma_A + \sigma_S$ is the sum of absorption cross section σ_A and scattering cross section σ_S, calculated under quasi-static approximation from the NP polarizability α. Further details on the model employed for the simulations and on the formulas used to calculate α, both in the case of Au-PEG NPs and non-PEGylated NPs, are reported in the Supplementary material Note 4.1.

2.3.2 The three-temperature model and photoinduced mechanical oscillations: The transient optical model is based on the three-temperature (3TM) model [52, 53], describing the time evolution of the energetic internal degrees of freedom of plasmonic nanostructures upon ultrafast pulse illumination. The model solves for N, the excess energy stored in the nonthermal population of photoexcited electrons, Θ_e, the temperature of the thermalized hot electrons and Θ_l, the Au lattice temperature, and reads:

$$\frac{dN}{dt} = p_a(t) - aN - bN, \tag{1}$$

$$\gamma\Theta_e\frac{d\Theta_e}{dt} = aN - G(\Theta_e - \Theta_l), \tag{2}$$

$$C_l\frac{d\Theta_l}{dt} = bN + G(\Theta_e - \Theta_l) - G_a(\Theta_l - \Theta_m), \tag{3}$$

with Θ_m the surrounding environment (matrix) temperature. In the equations above, a and b are the high energy electron gas heating rate and the nonthermalized electrons-phonons scattering rate, respectively, G is the thermal electron-phonon coupling constant, and G_a a coupling coefficient (to be fitted on dynamical measurements) between Au phonons and the surrounding matrix phonons. Further details on these parameters can be found in the Supplementary material.

The 3TM has been then integrated with a mechanical oscillation model [54] for the expansion of the NP radius, here represented by parameter s, reading:

$$\frac{d^2s}{dt^2} + \frac{2}{\tau}\frac{ds}{dt} + \omega_0^2 s = \omega_0^2\xi(\Theta_l - \Theta_0)R, \tag{4}$$

where $\xi = 1.5 \times 10^{-5}$ K^{-1} is the linear thermal expansion coefficient of gold [54], $\Theta_0 = 300$ K is the room temperature, ω_0 is the resonance frequency of the mechanical mode under consideration and τ is the oscillation damping time. The latter has been taken as a fitting parameter, whose value is set to $\tau = 16$ ps in order to correctly reproduce the damping of the mechanical oscillation observed in the non-PEGylated NPs sample. Regarding the resonance frequency, its expression has been adapted according to the investigated nanostructure, namely in the case of isolated NPs or NPs dimers modeling the assembled sample. Details on the two cases are provided in the Supplementary material.

2.3.3 Nonlinear optical model and transient extinction simulations: We modeled the effect of the dynamical variables appearing in the 3TM on the ultrafast optical response following a segregated approach, reported by some of the authors of this work in the study by Zavelani-Rossi et al. [53], extensively validated and briefly outlined in the Supplementary material.

To model the transient optical response of the experimental samples, a further element affecting their transient permittivity, linked to the NPs thermal expansion and expressed in terms of the sphere stretching mode, needs to be accounted for. In this respect, the harmonic oscillator model outlined above [54] has been considered to act on the NP dimensions, changing the polarizability in a more direct way.

Thus, from the expression of the modified polarizability (refer to Supplementary material Note 4.3), in the quasi-static approximation, it is possible to define the extinction cross section as the sum of absorption and scattering cross section as follows:

$$\sigma_A(\lambda, t) = k\,\mathrm{Im}\{\alpha(\lambda, t)\}, \tag{5}$$

$$\sigma_S(\lambda, t) = \frac{k^4}{6\pi}|\alpha(\lambda, t)|^2, \tag{6}$$

$$\sigma_E(\lambda, t) = \sigma_A(\lambda, t) + \sigma_S(\lambda, t), \tag{7}$$

with k the probe wave-vector in the surrounding environment. Lastly, having defined the differential extinction cross section $\Delta\sigma_E(\lambda, t) = \sigma_E[\alpha(\lambda, t)] - \sigma_E[\alpha_0(\lambda)]$, where $\alpha_0(\lambda)$ is the unperturbed polarizability, the transient transmittance spectra are calculated as follows:

$$\frac{\Delta T}{T}(\lambda, t) = \exp[-\Delta\sigma_E(\lambda, t)N_P L] - 1. \tag{8}$$

2.4 Thermal simulations

To gain insight on the thermal behavior of the samples, a three-dimensional model has been built to solve the heat transfer problem and determine the temperature field (Θ) in the structure both in the time domain (for the pulsed optical excitation) and in the stationary regime (for CW illumination). The temperature Θ, solved for in the heat diffusion problem, is defined across the whole simulation domain. Therefore, $\Theta = \Theta_l$ within the Au NP, whereas $\Theta = \Theta_m$ in the surrounding environment (Θ_l and Θ_m having the same meaning as in Eqs. (1)–(3) above). The cases of a single isolated nanosphere and a $\mathcal{N} \times \mathcal{N}$ square array of sphere dimers have been investigated, with water as the surrounding environment in all simulations, its volume being varied so as to conserve the ratio of water and Au volumes. The general form of the heat transfer equation reads as follows:

$$\rho C_p \frac{\partial \Theta}{\partial t} - \kappa \nabla^2 \Theta = Q, \quad (9)$$

where ρ, C_p and κ are the density, heat capacity and thermal conductivity of the considered material, respectively, and Q the heat source, which has been expressed according to the considered illumination conditions, being either in the pulsed (Q_p) or the CW (Q_c) regime. When solved in the steady state, the equations are formally analogous, with the temporal derivative vanishing. Supplementary material Note 4.4 provides details on the parameters used for the materials properties and thoroughly discusses case-by-case the expression used for the heat source in the four scenarios investigated, i.e., either pulsed or continuous illumination of either isolated or assembled NPs. The numerical analysis was performed using a commercial finite element method–based software (COMSOL Multiphysics 5.4).

2.5 Drug delivery experiment

The two compounds used for the drug delivery experiments (SF and DEX) were chosen mainly because of their steric hindrances, similar to many low steric hindrance drugs (SF) and high steric hindrance biomolecules (DEX), and because of their absorbance, that makes them easily detectable by UV spectroscopy. To perform the measurements, the compounds were mixed with HG-NPs gelling solution, above the sol/gel transition to allow good solute dispersion within the polymeric network. Gelation took place in steel cylinders (0.5 mL, d = 1.1 cm). Three samples for each condition (combination of compound and NPs) were put in excess of PBS and aliquots were collected at defined time points, while the sample volume was replaced by fresh PBS, in order to avoid mass-transfer equilibrium between the gel and the surrounding solution. The percentage of drug released was measured through UV spectroscopy [49]. In parallel, the same drug release studies were conducted on samples irradiated with a continuous wave UV laser diode centered at 375 nm (Omicron-Laserage Laserprodukte GmbH). The obtained release data were used to evaluate diffusion coefficients of the drug mimetic, following a previously described procedure [49] which models the delivery phenomenon with a Fickian behavior (further details are reported in the Supplementary material Note 2.4).

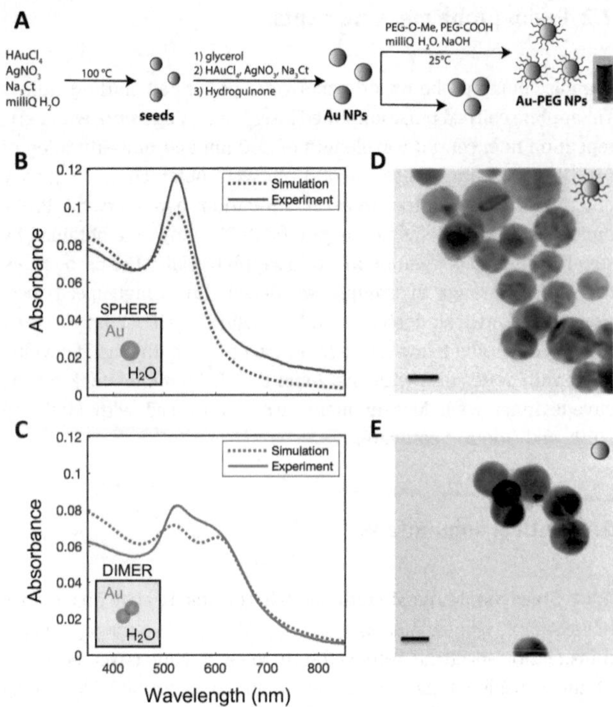

Figure 1: Schematic of the synthesis of gold nanoparticles (NPs) with and without polyethylene glycol (PEG) ligands (A). Absorption spectra of the gold NPs with (B) and without (C) PEG ligand, together with their respective simulated spectra (dashed lines). Corresponding transmission electron microscopy images of the two NPs (from a dried dispersion) with (D) and without (E) the ligand: the scale bar indicates a distance of 10 nm.

3 Results and discussions

3.1 Gold nanoparticles characterization

Figure 1A shows the schematic representation of the synthesis of Au NPs and PEGylated Au NPs (Au-PEG NPs). The procedure followed the Turkevich method, with the seeded-growth one-pot strategy (for a detailed description of the synthesis see Supplementary material) [55].

Figure 1B and C shows the measured absorption spectra of the two types of NPs. Au-PEG NPs (Figure 1B, continuous line) exhibit an absorption peak at 523 nm attributed to the typical LSPR of gold NPs of similar diameters (≈20 nm) [1]. The absorption spectrum of the non-PEGylated NPs (Au NPs) (Figure 1C, continuous line) reveals a more complex structure, with a main peak at 527 nm and a red-shifted broad shoulder. The presence of a second red-shifted band can be ascribed to a resonance caused by the aggregation of multiple NPs, forming NPs assemblies of larger dimensions whose interaction generates a collective plasmonic oscillation [2]. These assumptions are confirmed by the transmission electron microscopy (TEM) images (details in the Supplementary material) of the two samples of Au NPs (from dried dispersion): Figure 1D shows TEM image of the Au-PEG NPs where the NPs are well separated with an average diameter of 22 ± 2.8 nm, obtained from dynamic light scattering (DLS), while in Figure 1E the Au NPs (with an average diameter of 15 ± 1.6 nm, obtained from DLS) are in contact to each other, indicating a more pronounce tendency to assembling.

The absorption spectra were simulated by means of a quasi-static model [56], as detailed in the Materials and Methods section. The static absorption of Au-PEG NPs is correctly reproduced by a single gold sphere model (Figure 1B, dashed line) with 20 nm diameter free standing in water (the few nanometer thin PEG layer was disregarded because field penetration depth in the dielectric is of the order of the optical wavelength). An increased Drude damping coefficient of $5\Gamma_0$ (Γ_0 being the Drude damping in bulk gold) was assumed in order to mimic the inhomogeneous broadening caused by size dispersion in the sample. The aggregated Au NPs were simulated by considering the simplest assembly configuration, a dimer of nanospheres with 15 nm diameter and a center-to-center distance of 11.7 nm (Figure 1C, dashed green line). Again, an increased Drude damping coefficient ($6.2\ \Gamma_0$) was assumed. The agreement with the experimental data is rather good, considering that a perfect match with the data should in fact include a distribution of different types of NPs assemblies (trimers or even bigger aggregates). Even though a precise determination of such distribution is beyond the scope of our study, the fact that the optical response is dominated by a dimeric contribution is a clear-cut indication of assembling in the non-PEGylated sample.

3.2 Tracking photothermal dynamics in plasmonic gels

We loaded the HG network with the two differently capped NPs (isolated and assembled NPs) and demonstrated that the NPs are trapped inside the organic matrix. This evidence was obtained by means of Fourier transform infrared and UV-Vis spectra of the loaded and nonloaded HGs (see Figure S2 of Supplementary material). The presence of Au NPs affects the swelling ratio and the mesh size of the HG depending on its temperature. However, the mechanical properties are preserved, remaining similar to the ones of its nonloaded counterpart (see Supplementary material Note 3 for further details). Then we performed ultrafast optical characterization of the two HG samples, with the aim of tracking the energy flow following photoexcitation and comparing their photothermal dynamics.

Figure 2 shows the 2D differential transmission ($\Delta T/T$) maps, as a function of probe wavelength and delay, acquired by exciting the NPs-HG samples with $\approx$0.17 mJ/cm^2 incident fluence at 400 nm pump wavelength. This wavelength, far from the plasmonic resonance of the samples, was selected in order to guarantee the same level of excitation (under the same pump fluence). Moreover, the two samples exhibit the same extinction coefficient at the chosen excitation wavelength, leading to a similar number of photogenerated excited carriers. This assumption is confirmed by the similar intensity of the $\Delta T/T$ peak signal (at 523 nm) at short time delays for the isolated NPs (Au-PEG NPs) and for the assembled ones (Au NPs), reported in Figure 2A and B, respectively. The $\Delta T/T$ signals up to 5 ps time delay for isolated NPs loaded in the HG

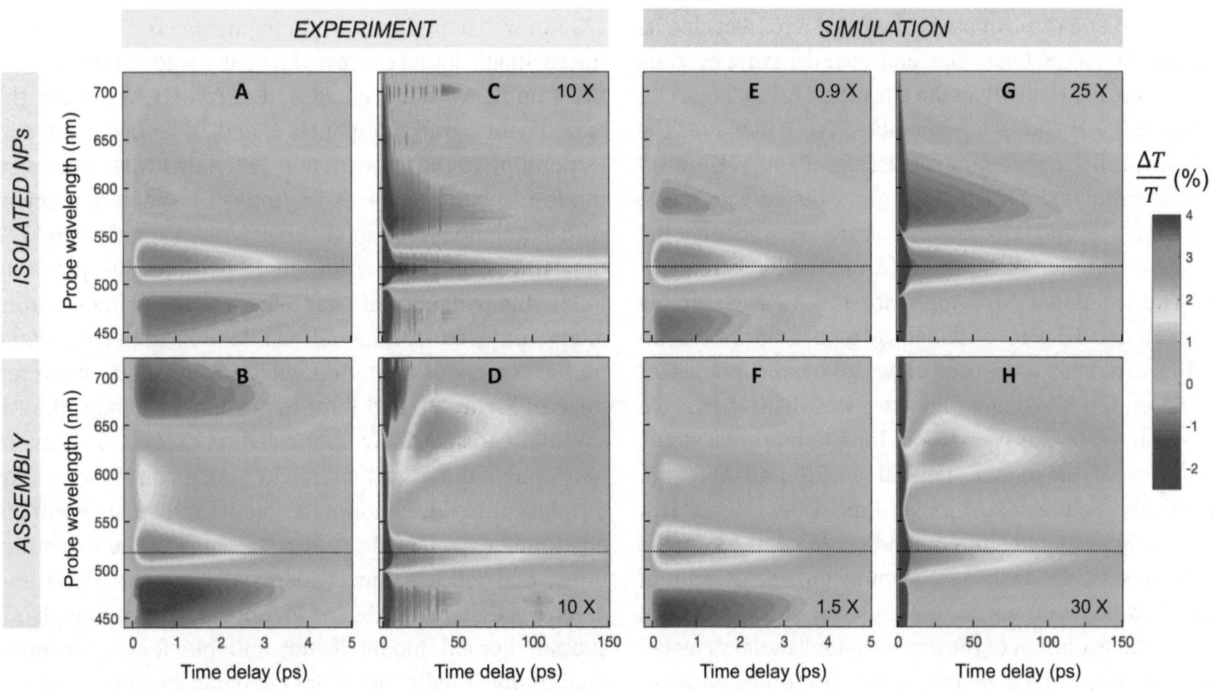

Figure 2: Experimental and simulated 2D $\Delta T/T$ maps of isolated and assembled nanostructures.
Experimental $\Delta T/T$ maps are obtained by pumping at 400 nm wavelength the HG containing Au-polyethylene glycol nanoparticles (PEG NPs) (A and C) and Au NPs (B and D). Panels E and G show simulated $\Delta T/T$ maps for isolated Au NPs, while panels F and H show simulated $\Delta T/T$ maps for a dimeric structure of Au NPs. Panels A, B, E and F show the 2D maps on the short time scale of the dynamics (up to 5 ps) with the same color scale. Panels C, D, G and H show the maps until 150 ps with a magnified $\Delta T/T$ scale in order to better distinguish the long time delay signals. Agreement between experimental and simulated data is remarkable both at short and long time delays.

(Figure 2A) is in perfect agreement with previous numerous works on the ultrafast transient response in similar plasmonic Au NPs [3–6]. The positive band peaked at 520 nm, and the two negative sidebands at longer and shorter wavelengths are due to the transient broadening of the quasi-static extinction peak of the nanosphere, caused by the pump-induced permittivity modulation that, in the considered range of wavelengths, is dominated by an increase of the imaginary part (see e.g., a study by Dal Conte et al. [57]). The $\Delta T/T$ map of the aggregated NPs (Figure 2B) still shows the strong band at 520 nm, with the same signal level, which is a confirmation of the hypothesis that the two samples have been loaded with approximately the same amount of plasmonic structures. However, in the HG with non-PEGylated NPs a second band, peaked at ≈590 nm, is clearly visible. This novel band is assigned to the fingerprint of the assembling, and in particular to the longitudinal plasmonic resonance of the NP dimer. To confirm this assignment, we also investigated the longer time scale of the dynamics. Figure 2C and D shows an expansion of the $\Delta T/T$ maps up to 150 ps (with a magnified signal scale). While Au-PEG NPs (Figure 2C) show a simple monotonic decay of the main plasmonic resonance, the non-PEGylated NPs (Figure 2D) show multiple spectral features. The signal of the longitudinal plasmonic resonance peaked at 590 nm shows a decay similar to the quasi-static resonance of the isolated nanosphere. However, a red-shifted positive band peaked at 647 nm appears at around 5 ps, reaching its maximum at ~30 ps time delay and then decaying over the 150 ps timescale, similarly to the plasmonic resonances.

Figure 2E–H shows the simulated $\Delta T/T$ maps for the two types of NPs, on the same timescales. Panels 2E and 2G show the simulated $\Delta T/T$ maps for an isolated gold nanosphere on a 5 and 150 ps time scale respectively: the agreement with Figure 2A and C (Au-PEG NPs) is remarkable. The simulated $\Delta T/T$ maps for the Au NPs sample (Figure 2F and H) were obtained by modeling the assembled NPs as a dimer composed of two gold nanospheres. At early times, the $\Delta T/T$ maps show two well distinct positive bands, related to transversal and longitudinal plasmonic resonances of the dimer, peaked at 521 and 598 nm, respectively (Figure 2F). At longer time delay (Figure 2H), the formation of a red-shifted band centered at 636 nm is clearly observable, with a maximum growth at around 35 ps. This behavior accurately reproduces the temporal and spectral evolution of the experimental signal detected at 647 nm. This band is related to the spectral response of the longitudinal resonance to the surrounding environment, which is modulated by the mechanical oscillation of the dimer on its stretching mode. This oscillation is coherently triggered in the ensemble of nanostructures by ultrafast heat transfer from the hot electrons to the gold lattice via electron-phonon coupling, taking place on the few-ps time scale. Therefore, the temporal evolution of the signal at 647 nm is an indirect signature of the assembling configuration of the plasmonic structures, dominated by the dimer photothermal dynamics and subsequent modulation of the sample transmittance. The minor discrepancies in the spectral band position are justified by the approximation of the non-PEGylated Au NPs sample to an ensemble of dimers, which is likely not to account for the diverse combination of possible types of aggregates (with their own specific spectral and dynamical features) constituting in fact the experimental sample.

However, the dramatic contrast between the $\Delta T/T$ maps of PEGylated and non-PEGylated samples and the quantitative matching with numerical simulations demonstrate the suitability of ultrafast spectroscopy combined with nonlinear optical modeling for noninvasive analysis of the degree of assembling in plasmon loaded composite nanomaterials.

Interestingly, despite the different spectral features of the two samples in the $\Delta T/T$ maps, the relaxation dynamics both on the short timescale corresponding to electron-phonon coupling and on the longer time scale of phonon-phonon coupling (presiding over heat release to the water environment) looks rather similar. This is ascertained by direct comparison between the $\Delta T/T$ temporal traces at 520 nm wavelength, shown in Figure 3A–D. Experimental traces (black dotted curves) of panels A and C, recorded for the sample with PEGylated NPs, closely resemble the experimental traces of panels B and D, respectively, corresponding to the HG with non-PEGylated NPs. Moreover, the four measured traces are indeed in good agreement with simulations (red solid curves). Interestingly, the $\Delta T/T$ time traces on the few hundreds ps timescale precisely follow the dynamics of heat release to the water environment, retrieved by thermal simulations under the same pulsed regime of excitation, which is detailed by the red trace in Figure 3E and F for the isolated NPs and the NP assembly, respectively, (modeled as a 6×6 array of nanosphere dimers for simplicity). As a consequence, for ultrafast pulsed excitation, the temperature increase in the environment of the NPs is also the same in the two structures, and follows almost identical dynamics (blue and green traces in Figure 3E and F). Note that, even though our simple thermal model is not suitable for an accurate quantitative prediction of the temperature increase in our samples, the calculated temperature dynamics turned out to be very robust against variations of the internal arrangement of NPs within the assembly (see Figure S6 in the Supplementary material). However, as pointed out by Govorov et al. [36] in a seminal paper in 2006, plasmonic

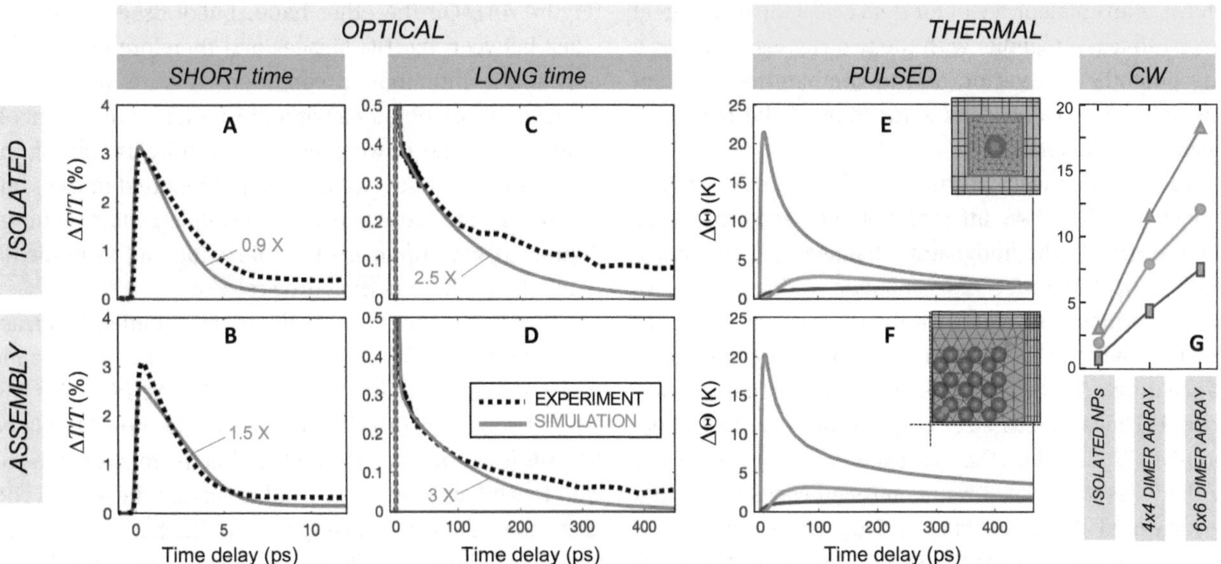

Figure 3: Experimental and simulated $\Delta T/T$ dynamics at 520 nm wavelength in HG samples containing Au-polyethylene glycol nanoparticles (PEG NPs) (A and C) and non-PEGylated Au NPs (B and D).
Panels A and B show the comparison till early times (12 ps) while panels C and D show the dynamics till 450 ps. Panels E, F and G report the results of thermal simulations. Panel E and F show the evolution of the temperatures due to impulsive excitation (as in ultrafast pump-probe experiments) in isolated Au NPs and in a 6 × 6 array of Au NP dimers. Red and green traces are the temperature increase in two selected points of the domain, highlighted with the same color coding in the insets (showing the geometry and mesh used in the simulations, with the first quadrant only, in panel F, for better reading), while blue trace is the average temperature increase of the water domain (10 times magnified for better reading). Panel G shows the temperature increase in the same spots but under continuous wave excitation corresponding to an input laser intensity $I_0 = 10^5$ W/cm^2, which is a typical excitation level assumed when modeling plasmonic nanoheaters in an open domain of water environment [36].

assemblies can behave as very efficient nanoheaters, outperforming isolated nanostructures under illumination with longer pulses. More precisely, one needs a pulse duration τ exceeding the heat diffusion time T across the whole size of the assembly. According to a study by Govorov et al. [36], for an assembly of size l, the heat diffusion time can be estimated as $T = l^2/K_m$, with $K_m = \kappa/(\rho C_p)$ the thermal diffusivity of the matrix embedding the NPs of the assembly. In our case of a water matrix, we obtain $K_m = 1.43 \times 10^{-7}$ m^2/s, and thus $T \simeq 0.3$ µs, for a nanoassembly with $l \simeq 200$ nm. This increased efficiency is mostly related to the cumulative effect deriving from the addition of more heat fluxes with the increasing number of NPs (plus, possibly, the Coulomb interaction between the NPs which is driven by plasmon-enhanced electric fields and depends on NPs distances and arrangements [37]). To ascertain this point, we performed thermal simulations under continuous wave excitation. The results are reported in Figure 3G for three different configurations of NPs: isolated nanospheres, a 4 × 4 and a 6 × 6 array of nanosphere dimers. Our simulations confirm the dramatic improvement of local heating performances for the assembly if compared to isolated structures, in agreement with the scaling law for peak temperature increase reported in a study by Govorov et al. [36]. For a further validation of this picture, we also performed thermal simulations for two systems having exactly the same number of NPs (25 dimers) distributed in the same volume of water, but with two very different degrees of assembling. We found that, even though the average temperature in the water volume is basically the same (because of the same thermal loading of the individual NPs), the temperature in the center of the metallic dimer and close to the edge of the assembly dramatically increases with the degree of assembling (see Figure S7 in the Supplementary material).

3.3 Photothermal drug release study

The outcomes of the optical and photothermal studies detailed above indicate that the degree of plasmonic assembling in the two HGs loaded with PEG or non-PEG NPs is substantially different, and that the non-PEGylated sample is expected to be much more efficient in generating a photothermal effect under CW illumination, compared to the PEGylated one. Since this HG has been demonstrated to behave as an efficient vector for stem cells [48], it is of great interest to test it together with the two classes of NPs as a drug photothermal vector, and to compare the different drug release efficiencies as a function of the degree of plasmonic assembling.

To test the efficacy of the NPs-HG as vehicle for drug delivery, we chose two different types of compounds to be released: (i) SF, (hydrodynamic diameter = 0.07 nm), a small steric hindrance drug mimetic with an effective diameter which is smaller than the HG mesh size, and (ii) DEX (hydrodynamic diameter = 8 nm), a big steric hindrance drug mimetic, which is instead of comparable size to the HG mean mesh size. To perform drug release experiments under photothermal excitation, we used a CW UV diode laser source (375 nm emission wavelength) and we measured the percentage of drug release after fixed temporal intervals under illumination with 10 W/cm^2 intensity. Figure 4A and B schematize the results of our experiments. Under illumination, NPs absorb light and release heat to the surrounding HG matrix, producing an expansion in the mesh size and thus creating more space for the drug to freely move and to be released outside the HG. Generally, for drugs with a size smaller than the HG mesh size, we expect no difference in the release performances between irradiated and non-irradiated samples, since the drug can always escape the HG matrix (Figure 4A). On the other hand, bulky drugs should be stuck between the HG meshes and therefore be released only under illumination conditions (Figure 4B). Similar results to those obtained without irradiation (red symbols in Figure 4C and E) were also observed by irradiating the HG network without Au NPs and not reported in the same Figure for clarity reason (trends in Supplementary material Figure S5), confirming the key role of the Au NPs in the HG network for photothermal drug release.

Figure 4C shows the results of the cumulative release experiments for SF in three conditions: (i) no illumination (red squares); (ii) illumination of the Au-PEG NPs HG (empty squares); (iii) illumination of the non-PEGylated NPs HG (black squares). It is clear that the release of the SF is independent of both the irradiation and the type of NPs used, since data are showing very similar trends. Due to its small size, the release is always allowed, since the drug is not retained by the meshes of the HG even in the non-irradiated (cold and tight) condition. The situation drastically changes if we analyze the cumulative release experiments of the DEX case under the same conditions (Figure 4E). The release of DEX is mostly not allowed in absence of irradiation, keeping a value below 5% all the time. This result is due to the trapping of the large drug in the meshes of the gel. On the other hand, when the NPs-HG is illuminated, a striking distinction between the two types

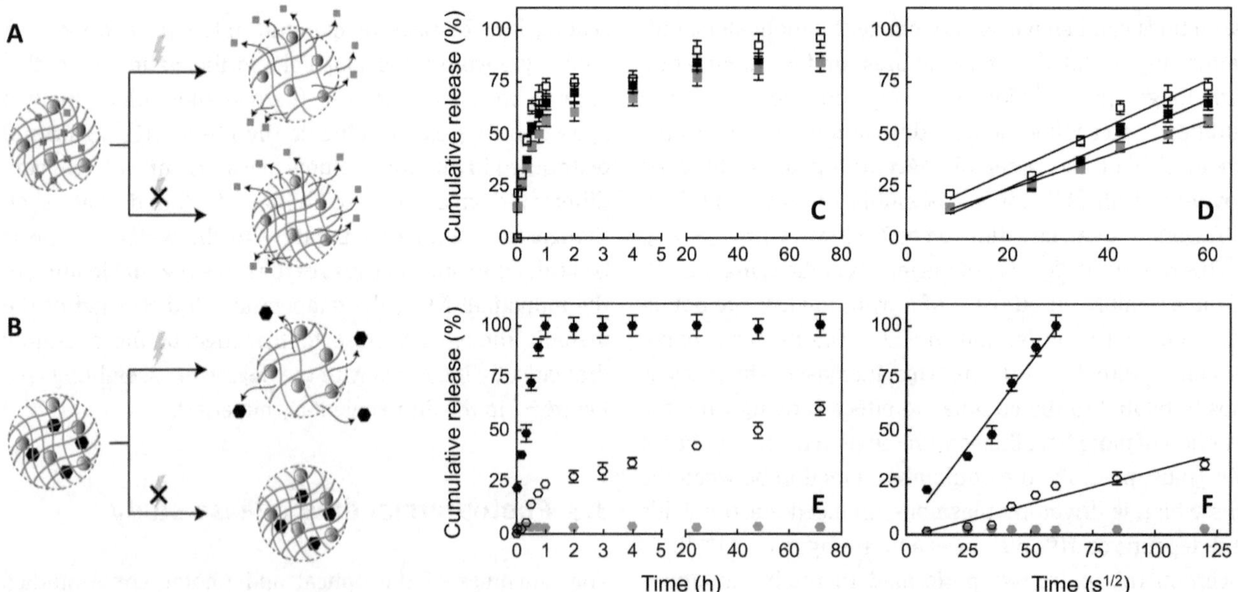

Figure 4: Comparison of NPs-HG with insertion of sodium fluorescein (SF) (A) and fluorescein-dextran 70 kDa (DEX) (B), with and without heating by a continuous UV laser.
Cumulative release of the drug in three different conditions: no irradiation of the pristine HG (red symbols), irradiation of the HG with Au-polyethylene glycol nanoparticles (PEG NPs) (empty symbols), irradiation of the HG with non-PEGylated NPs (black symbols). The cumulative release is shown as function of continuous illumination temporal interval (hours) for SF (C) and DEX (E). Panels (D) and (F) show the corresponding magnified short time interval in square root of seconds in order to extrapolate the diffusion coefficient [49].

of NPs used is observed. The release is clearly present in both cases, in agreement with a mesh size modification due to a local temperature increase (data not shown); however the amount of compound released and the release rate are quite different. In Au-PEG NPs HG, the release reaches 25% within 1 h of illumination and then slowly grows in the whole investigated time range without reaching 100% (Figure 4E, empty hexagons). For the Au NPs HG, the release reaches ~100% value in the first hour of illumination (Figure 4E, black hexagons), after which it saturates. These results confirm the enhancement in nanoheating for plasmonic NP assemblies: aggregated Au NPs contribute to a faster heating of the HG matrix by controlling the drug release in terms of speed and amount. To better investigate and understand the influence of illumination, we plotted the percentage of release against the square root of time (see Figure 4D and F), where a linear relationship is indicative of Fickian diffusion. Figure 4D shows that Fickian diffusion for SF release lasts for about 1 h and then reaches a plateau. The initial burst release is around 20% and almost no differences are visible between the three cases. Diffusion coefficients were calculated using a Fick-based model with cylindrical geometry already validated in literature [58] (details in Supplementary material) and their values are around 1.5×10^{-8} m^2/s, for both neat HG and Au NPs loaded ones. Considering DEX release (Figure 4F), three different cases in terms of Fickian diffusion duration are observed: no Fickian diffusion for neat HG, 1 h for Au NPs (corresponding to a timescale of 60 $s^{1/2}$ in the figure) and about 4 h for Au-PEG NPs (corresponding to 120 $s^{1/2}$). Also burst release presents differences: around 18% for Au NPs and almost negligible for Au-PEG NPs. In accordance, the calculated diffusion coefficients are 2.5×10^{-8} m^2/s for Au NPs and 1.7×10^{-10} m^2/s for Au-PEG NPs.

This set of experiments confirms that the presence of Au NPs influences the release of compounds with effective hydrodynamic diameters bigger than the mesh size of the host matrix. Moreover, we found that the aggregated Au NPs show a much more efficient heat release towards the surrounding environment.

4 Conclusion

In this study, we have investigated the properties of gold NPs, incorporated in a hydrogel matrix, as photothermal agents promoting light-assisted release of drugs with size bigger than the hydrogel mesh size. We have compared NPs capped with PEG chains to uncapped NPs. We have found that PEGylation prevents aggregation of the NPs, which thus behave as isolated nanostructures, while the absence of surface functionalization leads to their aggregation, forming suprastructures which, to a first approximation, can be described as assemblies of dimers. Ultrafast transient absorption spectroscopy, in combination with simulations of the transient optical response, was used to compare isolated and aggregated nanostructures. Experiments, in excellent agreement with the simulations, revealed significant differences between PEGylated and uncapped NPs, allowing to retrieve the characteristic spectral signatures of dimerization. The difference between capped and uncapped NPs was confirmed in photothermal drug release studies, in which the hydrogel, loaded with the NPs as well as with a molecule (fluorescein-dextran 70 kDa) mimicking drugs with big steric hindrance, was subjected to continuous wave UV illumination. A dramatic difference was observed between the photothermal performance of the aggregated and the isolated NPs, with a measured two order of magnitude higher diffusion coefficient for the former with respect to the latter.

Taken together, these results confirm the theoretical prediction that NP assemblies act as efficient nanoheaters, greatly increasing the efficiency of the photothermal effect with respect to isolated NPs because of the cumulative effect of heat fluxes. This is particularly effective in hydrogels loaded with plasmonic nanoassemblies because the breaking of the polymeric network of the gel is mostly sensitive to thermal spots driven at higher temperatures, rather than to a more uniform but moderate heating of the whole matrix, achievable with isolated NPs. The importance of these results is twofold. On the one hand, they open up a new paradigm for nanoplasmonic control over drug release, where the aggregation of the nanostructures can be used as additional degree of freedom to control the efficiency of the process. On the other hand, since the specific agarose-carbomer hydrogel used in these studies has proven to be an effective vector for stem cells, it will be interesting in the future to investigate its performance as a photothermal drug vector.

Acknowledgments: M.M. acknowledges support by the Balzan Foundation for the project Q-EX.

Author contribution: All the authors have accepted responsibility for the entire content of this submitted manuscript and approved submission.

Research funding: This research was funded by the Balzan Foundation for the project Q-EX.

Conflict of interest statement: The authors declare no conflicts of interest regarding this article.

References

[1] P. K. Jain, K. S. Lee, I. H. El-Sayed, and M. A. El-Sayed, "Calculated absorption and scattering properties of gold nanoparticles of different size, shape, and composition: applications in biological imaging and biomedicine," *J. Phys. Chem. B*, vol. 110, pp. 7238–7248, 2006.

[2] S. K. Ghosh and T. Pal, "Interparticle coupling effect on the surface plasmon resonance of gold nanoparticles: from theory to applications," *Chem. Rev.*, vol. 107, pp. 4797–4862, 2007.

[3] J. H. Hodak, A. Henglein, and G. V. Hartland, "Size dependent properties of Au particles: coherent excitation and dephasing of acoustic vibrational modes," *J. Chem. Phys.*, vol. 111, pp. 8613–8621, 1999.

[4] O. L. Muskens, N. Del Fatti, and F. Vallée, "Femtosecond response of a single metal nanoparticle," *Nano Lett.*, vol. 6, pp. 552–556, 2006.

[5] T. S. Ahmadi, S. L. Logunov, and M. A. El-Sayed, "Picosecond dynamics of colloidal gold nanoparticles," *J. Phys. Chem.*, vol. 100, pp. 8053–8056, 1996.

[6] G. V. Hartland, "Coherent excitation of vibrational modes in metallic nanoparticles," *Annu. Rev. Phys. Chem.*, vol. 57, pp. 403–430, 2006.

[7] P. K. Jain, X. Huang, I. H. El-Sayed, and M. A. El-Sayed, "Noble metals on the nanoscale: optical and photothermal properties and some applications in imaging, sensing, biology, and medicine," *Acc. Chem. Res.*, vol. 41, pp. 1578–1586, 2008.

[8] H. Jans and Q. Huo, "Gold nanoparticle-enabled biological and chemical detection and analysis," *Chem. Soc. Rev.*, vol. 41, pp. 2849–2866, 2012.

[9] P. Da, W. Li, X. Lin, Y. Wang, J. Tang, and G. Zheng, "Surface plasmon resonance enhanced real-time photoelectrochemical protein sensing by gold nanoparticle-decorated TiO_2 nanowires," *Anal. Chem.*, vol. 86, pp. 6633–6639, 2014.

[10] C. Wang, and C. Yu, "Detection of chemical pollutants in water using gold nanoparticles as sensors: a review," *Rev. Anal. Chem.*, vol. 32, pp. 1–14, 2013.

[11] I. Ament, J. Prasad, A. Henkel, S. Schmachtel, and C. Sönnichsen, "Single unlabeled protein detection on individual plasmonic nanoparticles," *Nano Lett.*, vol. 12, pp. 1092–1095, 2012.

[12] S. Carretero-Palacios, A. Jiménez-Solano, and H. Míguez, "Plasmonic nanoparticles as light-harvesting enhancers in perovskite solar cells: a user's guide," *ACS Energy Lett.*, vol. 1, pp. 323–331, 2016.

[13] S. Muduli, O. Game, V. Dhas, et al., "TiO2–Au plasmonic nanocomposite for enhanced dye-sensitized solar cell (DSSC) performance," *Sol. Energy*, vol. 86, pp. 1428–1434, 2012.

[14] X. Dang, J. Qi, M. T. Klug, et al., "Tunable localized surface plasmon-enabled broadband light-harvesting enhancement for high-efficiency panchromatic dye-sensitized solar cells," *Nano Lett.*, vol. 13, pp. 637–642, 2013.

[15] S. Chang, Q. Li, X. Xiao, K. Y. Wong, and T. Chen, "Enhancement of low energy sunlight harvesting in dye-sensitized solar cells using plasmonic gold nanorods," *Energy Environ. Sci.*, vol. 5, p. 9444, 2012.

[16] I. Willner, "Stimuli-controlled hydrogels and their applications," *Acc. Chem. Res.*, vol. 50, pp. 657–658, 2017.

[17] M. Wei, Y. Gao, X. Li, and M. J. Serpe, "Stimuli-responsive polymers and their applications," *Polym. Chem.*, vol. 8, pp. 127–143, 2017.

[18] D. A. Urban, L. Rodriguez-Lorenzo, S. Balog, C. Kinnear, B. Rothen-Rutishauser, and A. Petri-Fink, "Plasmonic nanoparticles and their characterization in physiological fluids," *Colloids Surf. B Biointerfaces*, vol. 137, pp. 39–49, 2016.

[19] V. Hirsch, C. Kinnear, M. Moniatte, B. Rothen-Rutishauser, M. J. D. Clift, and A. Fink, "Surface charge of polymer coated SPIONs influences the serum protein adsorption, colloidal stability and subsequent cell interaction in vitro," *Nanoscale*, vol. 5, p. 3723, 2013.

[20] I. Pastoriza-Santos, C. Kinnear, J. Pérez-Juste, P. Mulvaney, and L. M. Liz-Marzán, "Plasmonic polymer nanocomposites," *Nat. Rev. Mater.*, vol. 3, pp. 375–391, 2018.

[21] N. S. Abadeer, and C. J. Murphy, "Recent progress in cancer thermal therapy using gold nanoparticles," *J. Phys. Chem. C*, vol. 120, pp. 4691–4716, 2016.

[22] Y. Wang and D. S. Kohane, "External triggering and triggered targeting strategies for drug delivery," *Nat. Rev. Mater.*, vol. 2, p. 17020, 2017.

[23] H. H. Richardson, Z. N. Hickman, A. O. Govorov, A. C. Thomas, W. Zhang, and M. E. Kordesch, "Thermooptical properties of gold nanoparticles embedded in ice: characterization of heat generation and melting," *Nano Lett.*, vol. 6, pp. 783–788, 2006.

[24] T. Teranishi, S. Hasegawa, T. Shimizu, and M. Miyake, "Heat-induced size evolution of gold nanoparticles in the solid state," *Adv. Mater.*, vol. 13, pp. 1699–1701, 2001.

[25] D. Pissuwan, S. M. Valenzuela, and M. B. Cortie, "Therapeutic possibilities of plasmonically heated gold nanoparticles," *Trends Biotechnol.*, vol. 24, pp. 62–67, 2006.

[26] K. Akamatsu and S. Deki, "TEM investigation and electron diffraction study on dispersion of gold nanoparticles into a nylon 11 thin film during heat treatment," *J. Colloid Interface Sci.*, vol. 214, pp. 353–361, 1999.

[27] K. Akamatsu and S. Deki, "Dispersion of gold nanoparticles into a nylon 11 thin film during heat treatment: in situ optical transmission study," *J. Mater. Chem.*, vol. 8, pp. 637–640, 1998.

[28] T. Shimizu, T. Teranishi, S. Hasegawa, and M. Miyake, "Size evolution of alkanethiol-protected gold nanoparticles by heat treatment in the solid state," *J. Phys. Chem. B*, vol. 107, pp. 2719–2724, 2003.

[29] B. El Roustom, G. Fóti, and C. Comninellis, "Preparation of gold nanoparticles by heat treatment of sputter deposited gold on boron-doped diamond film electrode," *Electrochem. Commun.*, vol. 7, pp. 398–405, 2005.

[30] C. Y. Tsai, H. T. Chien, P. P. Ding, B. Chan, T. Y. Luh, and P. H. Chen, "Effect of structural character of gold nanoparticles in nanofluid on heat pipe thermal performance," *Mater. Lett.*, vol. 58, pp. 1461–1465, 2004.

[31] X. Sun, X. Jiang, S. Dong, and E. Wang, "One-step synthesis and size control of dendrimer-protected gold nanoparticles: a heat-treatment-based strategy," *Macromol. Rapid Commun.*, vol. 24, pp. 1024–1028, 2003.

[32] A. Plech, V. Kotaidis, S. Grésillon, C. Dahmen, and G. von Plessen, "Laser-induced heating and melting of gold nanoparticles studied by time-resolved x-ray scattering," *Phys. Rev. B*, vol. 70, p. 195423, 2004.

[33] M. Hu, X. Wang, G. V. Hartland, V. Salgueiriño-Maceira, and L. M. Liz-Marzán, "Heat dissipation in gold–silica core-shell nanoparticles," *Chem. Phys. Lett.*, vol. 372, pp. 767–772, 2003.

[34] M. M. Maye, W. Zheng, F. L. Leibowitz, N. K. Ly, and C-J. Zhong, "Heating-induced evolution of thiolate-encapsulated gold nanoparticles: a strategy for size and shape manipulations," *Langmuir*, vol. 16, pp. 490–497, 2000.

[35] A. J. Gormley, N. Larson, S. Sadekar, R. Robinson, A. Ray, and H. Ghandehari, "Guided delivery of polymer therapeutics using plasmonic photothermal therapy," *Nano Today*, vol. 7, pp. 158–167, 2012.

[36] A. O. Govorov, W. Zhang, T. Skeini, H. Richardson, J. Lee, and N. A. Kotov, "Gold nanoparticle ensembles as heaters and actuators: melting and collective plasmon resonances," *Nanoscale Res. Lett.*, vol. 1, pp. 84–90, 2006.

[37] A. O. Govorov and H. H. Richardson, "Generating heat with metal nanoparticles," *Nano Today*, vol. 2, pp. 30–38, 2007.

[38] A. Mazzanti, Z. Yang, M. G. Silva, et al., "Light–heat conversion dynamics in highly diversified water-dispersed hydrophobic nanocrystal assemblies," *Proc. Natl. Acad. Sci. U. S. A.*, vol. 116, pp. 8161–8166, 2019.

[39] E. R. Ruskowitz and C. A. DeForest, "Photoresponsive biomaterials for targeted drug delivery and 4D cell culture," *Nat. Rev. Mater.*, vol. 3, p. 17087, 2018.

[40] S.-W. Lv, Y. Liu, M. Xie, et al., "Near-infrared light-responsive hydrogel for specific recognition and photothermal site-release of circulating tumor cells," *ACS Nano*, vol. 10, pp. 6201–6210, 2016.

[41] S. C. T. Moorcroft, L. Roach, D. G. Jayne, Z. Y. Ong, and S. D. Evans, "Nanoparticle-loaded hydrogel for the light-activated release and photothermal enhancement of antimicrobial peptides," *ACS Appl. Mater. Interfaces*, vol. 12, pp. 24544–24554, 2020.

[42] D. N. Heo, W.-K. Ko, M. S. Bae, et al., "Enhanced bone regeneration with a gold nanoparticle–hydrogel complex," *J. Mater. Chem. B*, vol. 2, pp. 1584–1593, 2014.

[43] X. Fu, L. Hosta-Rigau, R. Chandrawati, and J. Cui, "Multi-stimuli-responsive polymer particles, films, and hydrogels for drug delivery," *Inside Chem.*, vol. 4, pp. 2084–2107, 2018.

[44] B. V. Slaughter, S. S. Khurshid, O. Z. Fisher, A. Khademhosseini, and N. A. Peppas, "Hydrogels in regenerative medicine," *Adv. Mater.*, vol. 21, pp. 3307–3329, 2009.

[45] P. M. Kharkar, K. L. Kiick, and A. M. Kloxin, "Designing degradable hydrogels for orthogonal control of cell microenvironments," *Chem. Soc. Rev.*, vol. 42, pp. 7335–7372, 2013.

[46] E. Mauri, F. Rossi, and A. Sacchetti, "Tunable drug delivery using chemoselective functionalization of hydrogels," *Mater. Sci. Eng. C*, vol. 61, pp. 851–857, 2016.

[47] I. Caron, F. Rossi, S. Papa, et al., "A new three dimensional biomimetic hydrogel to deliver factors secreted by human mesenchymal stem cells in spinal cord injury," *Biomaterials*, vol. 75, pp. 135–147, 2016.

[48] G. Perale, F. Rossi, M. Santoro, et al., "Multiple drug delivery hydrogel system for spinal cord injury repair strategies," *J. Contr. Release*, vol. 159, pp. 271–280, 2012.

[49] M. Santoro, P. Marchetti, F. Rossi, et al., "Smart approach to evaluate drug diffusivity in injectable Agar–Carbomer hydrogels for drug delivery," *J. Phys. Chem. B*, vol. 115, pp. 2503–2510, 2011.

[50] S. Papa, I. Vismara, A. Mariani, et al., "Mesenchymal stem cells encapsulated into biomimetic hydrogel scaffold gradually release CCL2 chemokine in situ preserving cytoarchitecture and promoting functional recovery in spinal cord injury," *J. Contr. Release*, vol. 278, pp. 49–56, 2018.

[51] B. K. Kwon, E. B. Okon, W. Plunet, et al., "A systematic review of directly applied biologic therapies for acute spinal cord injury," *J. Neurotrauma*, vol. 28, pp. 1589–1610, 2011.

[52] C.-K. Sun, F. Vallée, L. H. Acioli, E. P. Ippen, and J. G. Fujimoto, "Femtosecond-tunable measurement of electron thermalization in gold," *Phys. Rev. B*, vol. 50, pp. 15337–15348, 1994.

[53] M. Zavelani-Rossi, D. Polli, S. Kochtcheev, et al., "Transient optical response of a single gold nanoantenna: the role of plasmon detuning," *ACS Photonics*, vol. 2, pp. 521–529, 2015.

[54] G. V. Hartland, "Optical studies of dynamics in noble metal nanostructures," *Chem. Rev.*, vol. 111, pp. 3858–3887, 2011.

[55] A. Silvestri, V. Zambelli, A. M. Ferretti, D. Salerno, G. Bellani, and L. Polito, "Design of functionalized gold nanoparticle probes for computed tomography imaging," *Contrast Media Mol. Imaging*, vol. 11, pp. 405–414, 2016.

[56] S. A. Maier, *Plasmonics: Fundamentals and Applications*, New York, Springer Science, 2007.

[57] S. Dal Conte, M. Conforti, D. Petti, et al., "Disentangling electrons and lattice nonlinear optical response in metal-dielectric Bragg filters," *Phys. Rev. B*, vol. 89, p. 125122, 2014.

[58] F. Rossi, R. Ferrari, S. Papa, et al., "Tunable hydrogel—nanoparticles release system for sustained combination therapies in the spinal cord," *Colloids Surf. B Biointerfaces*, vol. 108, pp. 169–177, 2013.

Supplementary Material: The online version of this article offers supplementary material (https://doi.org/10.1515/nanoph-2020-0418).

Shuyan Zhang, Chi Lok Wong, Shuwen Zeng, Renzhe Bi, Kolvyn Tai, Kishan Dholakia and Malini Olivo*

Metasurfaces for biomedical applications: imaging and sensing from a nanophotonics perspective

https://doi.org/10.1515/9783110710687-023

Abstract: Metasurface is a recently developed nanophotonics concept to manipulate the properties of light by replacing conventional bulky optical components with ultrathin (more than 10^4 times thinner) flat optical components. Since the first demonstration of metasurfaces in 2011, they have attracted tremendous interest in the consumer optics and electronics industries. Recently, metasurface-empowered novel bioimaging and biosensing tools have emerged and been reported. Given the recent advances in metasurfaces in biomedical engineering, this review article covers the state of the art for this technology and provides a comprehensive interdisciplinary perspective on this field. The topics that we have covered include metasurfaces for chiral imaging, endoscopic optical coherence tomography, fluorescent imaging, super-resolution imaging, magnetic resonance imaging, quantitative phase imaging, sensing of antibodies, proteins, DNAs, cells, and cancer biomarkers. Future directions are discussed in twofold: application-specific biomedical metasurfaces and bioinspired metasurface devices. Perspectives on challenges and opportunities of metasurfaces, biophotonics, and translational biomedical devices are also provided. The objective of this review article is to inform and stimulate interdisciplinary research: firstly, by introducing the metasurface concept to the biomedical community; and secondly by assisting the metasurface community to understand the needs and realize the opportunities in the medical fields. In addition, this article provides two knowledge boxes describing the design process of a metasurface lens and the performance matrix of a biosensor, which serve as a "crash-course" introduction to those new to both fields.

Keywords: bioimaging; biophotonics; biosensing; metasurface; nanophotonics.

***Corresponding author: Malini Olivo,** Laboratory of Bio-Optical Imaging, Singapore Bioimaging Consortium, A*STAR, Singapore, Singapore, E-mail: malini_olivo@sbic.a-star.edu.sg
Shuyan Zhang, Chi Lok Wong, Renzhe Bi and Kolvyn Tai, Laboratory of Bio-Optical Imaging, Singapore Bioimaging Consortium, A*STAR, Singapore, Singapore. https://orcid.org/0000-0002-1286-4856 (S. Zhang). https://orcid.org/0000-0001-7173-064X (R. Bi)
Shuwen Zeng, XLIM Research Institute, UMR CNRS, Paris, France. https://orcid.org/0000-0003-2188-7213
Kishan Dholakia, SUPA, School of Physics & Astronomy, University of St Andrews, St Andrews, UK; and Department of Physics, Yonsei University, Seoul, South Korea

1 Introduction

From auroras and rainbows to the human eye – light has fascinated scientists for centuries. Today, optical technologies – from lasers to solar cells to cameras – harness light to advance physics and serve societal needs. The understanding and use of the fundamental properties of light are important parts of the progress of human history. Recently, the development of photonic materials, circuitry, devices, and probes on the nanoscale has opened up new opportunities for controlling light in the subwavelength regime.

In 1968, Veselago theoretically predicted the generation of artificial materials [1] by engineering their permittivity and permeability. This was eventually realized by the studies of Pendry et al. [2] and Smith et al. [3] that helped realize Veselago's prediction and have revolutionized our approach to electromagnetics. Such so-termed metamaterials are artificial composite nanostructures that possess unique properties for controlling light and demonstrate numerous exciting new optical effects and applications that cannot be achieved by natural materials [4–7]. Metasurfaces are structures that benefit from the reduced dimensionality of metamaterials. They are relatively easy to fabricate and possess a smaller footprint (thickness of less than a millimeter) than conventional optical components. They consist of optical components (or meta-atoms) patterned on a surface with subwavelength dimensions,

This article has previously been published in the journal Nanophotonics. Please cite as: S. Zhang, C. L. Wong, S. Zeng, R. Bi, K. Tai, K. Dholakia and M. Olivo "Metasurfaces for biomedical applications: imaging and sensing from a nanophotonics perspective" *Nanophotonics* 2021, 10. DOI: 10.1515/nanoph-2020-0373.

which shape the wavefront of light to introduce a desired spatial profile of the optical phase. Since their discovery [8], their exceptional optical properties have led to the development of ultrathin optical devices with various functionalities outperforming their conventional bulky counterparts and also demonstrating new optical phenomena covering a wide range of electromagnetic spectrum from the visible to the terahertz (THz) region. Such metasurface-based flat devices represent a new class of optical components that are compact, flat, and lightweight. Numerous devices based on metasurfaces have been developed, including metasurface lenses (metalenses) [9–19], waveplates [20–23], polarimeters [24–26], and holograms [27–29]. For metalenses, in particular, numerical aperture (NA) as high as 0.97 was demonstrated [15]. Metalenses with NA > 1 was realized by immersion metalenses [30]. There are review papers dedicated to the fundamentals, progress, and applications of metasurfaces. Some representative examples can be found in the references [31–51].

Metasurfaces have generated tremendous interest in the consumer optical and electronics industries. In addition, metasurface-empowered novel bioimaging and biosensing devices have also emerged and reported recently. For many optically based bioimaging devices, their bulk footprint and heavy physical weight have limited their usage in clinical settings. One of the bottlenecks is the large footprint of conventional optical components, such as lens assemblies and spectrometers. There are also limitations in the performance of optical components, for example, spherical lenses suffer from spherical aberrations which greatly impact the imaging quality and require often additional corrective optics. Besides imaging, sensing is another important research field in biomedical engineering. There has been a crucial need for a rapid and reliable sensing method for micro-organisms such as bacteria and viruses. Current methods are tedious and time-consuming, as the growth of these micro-organisms can take up to days and weeks. While methods such as polymerase chain reaction (PCR) and mass spectrophotometry have managed to reduce the time required for the measurements, the steps involved are exceptionally intricate making it challenging to be used for swift detection. As for cancer detection, current detection methods are pricy and time-consuming. Metasurface biosensors can provide a solution for rapid, label-free detection of various micro-organisms and cancer cells.

Given the recent advances and progress of metasurfaces in biomedical engineering, reviews published on the merging of these research areas are gaining attention [52–54]. However, their scopes are limited to either bioimaging or biosensing application. This review article aims to cover the state-of-the-art technology development for metasurfaces and a comprehensive and detailed account of both biomedical applications. The objective of this review is to stimulate interdisciplinary research by introducing the metasurface concept to the biomedical community and informing the metasurface community about the needs and opportunities in the biomedical research area.

Figure 1 gives an overview of the structure of this review paper. Firstly, the general concepts of metasurfaces will be introduced including the governing principles and fabrication techniques. Secondly, recent advances in metasurfaces for various bioapplications will be reviewed. This includes optical chiral imaging, endoscopic optical coherence tomography (OCT), fluorescent imaging, super-resolution imaging, magnetic resonance imaging (MRI), and quantitative phase imaging (QPI). Thirdly, applications of metasurface in biosensing are discussed, including the detection of antibodies and proteins, DNAs, cells, and cancer biomarkers. Lastly, future directions are described including various metasurface functions demonstrated for non-biomedical applications that could be potentially useful for biomedical applications and the development of biomimetic metasurface devices. We conclude with perspectives on present challenges to make metasurface-based biomedical devices toward commercialization and for translational research.

2 Fundamentals of metasurfaces

In contrast to conventional optical components that achieve wavefront engineering by phase accumulation, metasurfaces control the wavefront of light using arrays of optical resonators with subwavelength dimensions. By tailoring the optical properties of each meta-atom, one can spatially control the phase, amplitude, and polarization of the light and consequently shape the wavefront. There has been an increasing interest in the past decade in the field of metasurface as it promises remarkable capabilities for light molding beyond that offered by conventional planar interfaces. Several variations of metasurfaces have arisen to improve their performance, changes such as the transition from plasmonic materials to the use of dielectric materials to improve the efficiency and using different meta-atoms for additional optical functionalities. The fundamentals of metasurfaces are reviewed below from the physical concept, the design methodology, to the fabrication techniques.

2.1 Governing equations

When an electromagnetic wave hits the boundary between two media with different refractive indices, the wave splits

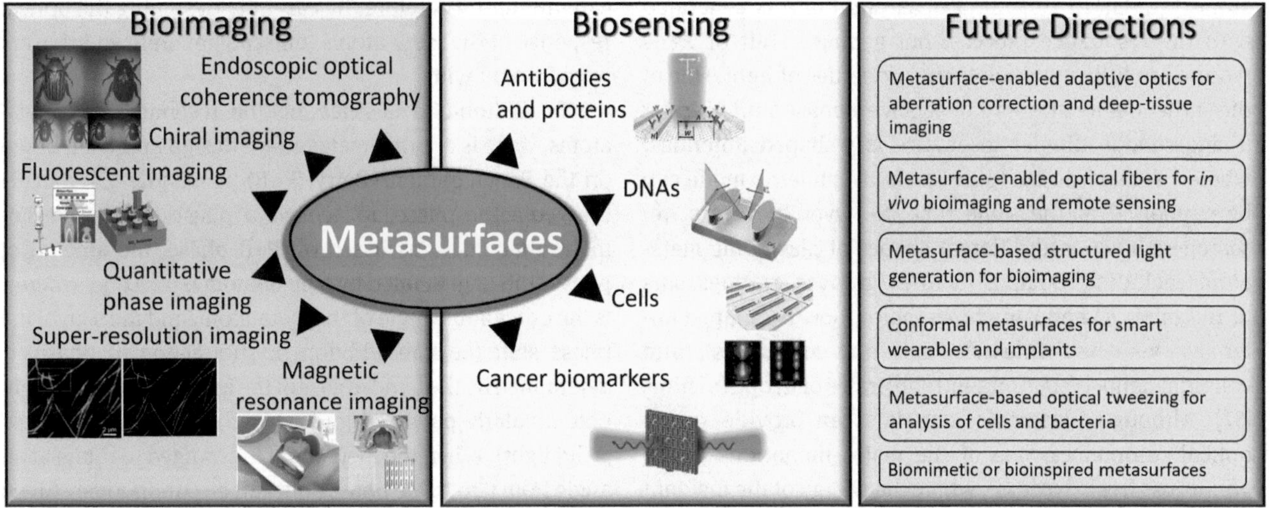

Figure 1: Overview of metasurfaces for bioimaging and biosensing applications and future directions. Chiral imaging: Reproduced with permission from Khorasaninejad et al., Nano Lett. **16**, 4595–4600 (2016) [96]. Copyright 2016 American Chemical Society. Direct link: https://pubs.acs.org/doi/abs/10.1021/acs.nanolett.6b01897. Further permissions related to the material excerpted should be directed to the ACS. Fluorescent imaging: Adapted from Lee et al., Opt. Mater. Express **9**, 3248 (2019) [99]. Copyright 2019 Optical Society of America. Super-resolution imaging: Reproduced with permission from Bezryadina et al., ACS Nano **12**, 8248–8254 (2018) [109]. Copyright 2018 American Chemical Society. Magnetic resonance imaging: Adapted from Schmidt et al., Sci. Rep. **7**, 1678 (2017) [113]. Licensed under Creative Commons Attribution 4.0 International License. Antibodies and proteins: Reproduced with permission from Guo et al., ACS Photonics **1**, 221–227 (2014) [135]. Copyright 2014 American Chemical Society. DNAs: Adapted from Leitis et al., Sci. Adv. **5**, eaaw2871 (2019) [138]. Distributed under a Creative Commons Attribution NonCommercial License 4.0 (CC BY-NC). Cells: Reproduced with permission from Rodrigo et al., Nat. Commun. **9**, 2160 (2018) [145]. Licensed under a Creative Commons Attribution 4.0 International License. Cancer biomarkers: Reproduced with permission from Yan et al., Biosens. Bioelectron. **126**, 485–492 (2019) [148]. Copyright 2019 Elsevier.

into two different waves, the reflected wave is rebounded back into the first medium and the transmitted wave enters the second medium. The angle at which these waves travel is determined by Snell's law. However, Snell's law does not take into account the abrupt phase discontinuities introduced by the metasurface.

The governing equations of the interaction of light and the metasurface for transmission (Eq. 1) and reflection (Eq. 2) are summarized as the generalized Snell's laws in three-dimensional (3D) derived from Fermat's principle [8, 55, 56]:

$$\begin{cases} n_t \sin(\theta_t) - n_i \sin(\theta_i) = \dfrac{\lambda}{2\pi}\dfrac{d\phi}{dx} \\ \cos(\theta_t)\sin(\varphi_t) = \dfrac{1}{n_t}\dfrac{\lambda}{2\pi}\dfrac{d\phi}{dy} \end{cases} \tag{1}$$

$$\begin{cases} \sin(\theta_r) - \sin(\theta_i) = \dfrac{1}{n_i}\dfrac{\lambda}{2\pi}\dfrac{d\phi}{dx} \\ \cos(\theta_r)\sin(\varphi_r) = \dfrac{1}{n_r}\dfrac{\lambda}{2\pi}\dfrac{d\phi}{dy} \end{cases} \tag{2}$$

where the x–z plane is the plane of incidence and the metasurface is located on the x–y plane. n_t, n_i and n_r are the refractive index of the mediums, θ_t, θ_r, and θ_i represent the angle of transmission, reflection, and incidence in the x–z plane, respectively. φ_t and φ_r represent the angle of transmission and reflection in the y–z plane, respectively. $\frac{d\phi}{dx}$ and $\frac{d\phi}{dy}$ is the phase gradient parallel and perpendicular to the plane of incidence, respectively. Note that the lack of a phase gradient will result in a normal form of Snell's law. These equations imply that reflected and transmitted waves can be arbitrarily bent in their respective half space, by engineering the phase gradient's direction and magnitude.

2.2 Design principles

The building blocks of metasurfaces are the meta-atoms on a subwavelength scale. These meta-atoms can be of different materials, such as plasmonic materials, dielectric materials, or a combination of the two. Typical plasmonic materials include metals, transparent conducting oxides (e.g. indium tin oxide), and two-dimensional (2D) materials. Strong light–matter interactions occur in these materials through plasmonic resonances which originate from a process called polarization: the dynamic response of electrons inside the material to the external electromagnetic field. The resonant wavelength depends on the sizes, shapes, materials, and surrounding media of the meta-atoms [31, 33]. In general, a phase shift of π is generated with the resonance process, but a phase shift of 2π is required to fully manipulate the properties of light. One of the earlier demonstrations to achieve a phase shift of 2π for a single-mode dipolar resonance is v-shaped antennas, where a symmetric mode and an antisymmetric mode can be supported at the same time [8]. Over the years, researchers have used different shapes of plasmonic meta-atoms including 1D nanostructures (grooves, gratings, slits or ribbons), 2D nanostructures (holes, bow-tie shaped antennas, v-shaped antennas, split-ring resonators), and colloidal nanocrystals on solid substrates or in optical films [57]. Although plasmonic materials can provide strong optical resonances, one of the major limitations is the ohmic loss (resistive loss), where the energy of the incident light is transferred to heat in the material. For applications that require high efficiency of the device, this is nonideal because the ohmic loss results associated with heat dissipation in the plasmonic materials represent a reduction of the overall efficiency.

To overcome this limitation on reduced efficiency, a class of metasurface devices based on dielectric materials were reported. Representative review papers on this research field can be found in [50, 51]. Typical dielectric materials used include titanium dioxide, silicon, germanium, and tellurium. The real parts of the refractive index of these materials are high so that they can manipulate the properties of light through Mie resonance [58] and achieve a phase shift of 2π with an enhancement of both electric and magnetic fields. The imaginary part of the refractive index is almost negligible so that there is minimal loss due to resistive heating and hence the overall device efficiency remains high.

Besides the type of materials chosen for the meta-atoms, the efficiency of a metasurface device in controlling the transmitted light is also dependent on the reflection at the light–matter interface, i.e. to boost the efficiency for transmission, complete elimination of reflection is essential. This can be achieved by matching the impedance of a metasurface with that of the free space [59]. For example, Huygen's metasurfaces may comprise nonperiodic nanostructures or multilayered structures to engineer the surface impedance locally [46]. According to the Huygen–Fresnel principle [60, 61], every point on a wavefront is itself the source of spherical wavelets and propagates with the same speed as the source wave. The secondary wavelets emanating from different points interfere and the sum of these wavelets forms the new wavefront which is represented as the line tangent to the wavelets. For a Huygen's metasurface, secondary waves transmitted through or reflected from the metasurface gain different phase shifts from the light interaction with the meta-atoms, and they can interfere. Therefore, by carefully designing the optical response of the meta-atoms, one can generate an arbitrary wavefront at will.

In addition to engineering the resonances of meta-atoms, there is another metasurface design approach based on the Pancharatnam–Berry (P–B) phase [62, 63], namely the geometric phase, to achieve a phase shift of 2π. For metasurface designs based on P–B phase, the amount of phase shift of generated by a meta-atom is $\pm 2\theta(x, y)$, where θ is the orientation angle of the meta-atom, and the sign of the phase shift (i.e. the addition or subtraction of phase) is determined by the handedness of the incident polarized light (left circularly polarized [LCP] or right circularly polarized [RCP] light). When the meta-atom is arranged to rotate at an angle from 0 to π, the phase shift can be continuously tuned from 0 to 2π. Since the phase shift is determined by the geometry, P–B phase metasurfaces typically exhibit ultra-broadband performance, however, they only operate on circularly polarized incident light.

So far, we have discussed the overall building mechanisms of a metasurface for the manipulation of light properties. To add further functionalities, recently, a variety of tuning mechanisms have been proposed [41]. For instance, flexible substrates were used for conformal metasurfaces, mechanically, and electrically tunable metasurfaces [4, 64–66]. The high flexibility and stretchability of soft materials such as polydimethylsiloxane (PDMS) and poly(methyl methacrylate) (PMMA) have made it possible for the substrates to be wrapped and bent [67, 68]. In this way, the spacing between meta-atoms is changed and hence the properties of the metasurface become tunable. Phase transition materials have been widely used in optical data-storage systems and nanophotonic systems based on their tunable optical properties. Commonly used phase change materials include vanadium dioxide (VO_2) [69–72] and GeSbTe alloys [73], which show a dramatic change in optical properties when undergoing phase transitions triggered by thermal energy or chemical doping. For example, the change in the refractive index of VO_2 in the mid-infrared wavelength region can reach $\Delta \boldsymbol{n} \approx 4$ and $\Delta \boldsymbol{k} \approx 8$ [74]. This translates to a change in the properties of the metasurface made of these materials. Liquid crystals have also been extensively used for the dynamic control of the optical properties because of their broadband optical nonlinearity and birefringence. Their change of refractive index can be externally tuned by temperature, light, and electric or magnetic fields [75, 76]. Microfluidic technology has been applied to tunable metalenses by controlling the air pressure of the pneumatic valves to change the distribution of liquid metals within a unit cell [77]. In addition, two-dimensional materials with extraordinary optical and electrical properties such as graphene were integrated into electrically tunable metasurfaces. Their optical properties were made tunable through a change in the charge carrier density which was controlled by the gate voltage [78, 79]. For example, in [79], a gate voltage change of 80 V resulted in a modulation depth of up to 100% and a wavelength shift of 1.5 μm in the mid-infrared region.

A brief summary of the design steps of a metalens is given in Box 1 with simulation illustrations. A more general flow chart of designing a general metasurface device not only a lens can be found in [44].

2.3 Fabrication techniques

The fabrication techniques of metasurfaces are microfabrication and nanofabrication techniques depending on the feature size of the metasurface which typically depending on the operating wavelength. For applications with a shorter wavelength range such as in the visible wavelength range, a small feature size is required, so nanofabrication techniques such as electron-beam lithography, focused ion beam (FIB) lithography, and

Box 1 Designing a metalens

Step 1: Choice of unit cell

Metasurfaces consist of carefully arranged "unit cells" or "meta-atoms" with subwavelength structures. The optical response (phase, amplitude, and polarization) of the unit cell to the incident wave changes with its geometry (height, length, width, shape, material, etc.). For a transmission-based metalens, a simple unit cell could be a periodic 2D array of nanopillars. It is the simplest design and the optical response is independent of the incident polarization.

Step 2: Generation of phase map

Normally, the incident wavelength of a metasurface device is known, i.e., it is not a variable. The variables are geometric parameters of the unit cell, e.g. height, radius, and period. By scanning these parameters, one can generate a lookup table of the phase, i.e., a phase map. The height and period that give desired transmission (uniform and high transmission values) and phase (phase coverage from 0 to 2π) properties is chosen. Then the phase of the nanopillar array can be controlled by changing its radius. The phase map describes the relationship between the phase and the radius of the nanopillar for a specific height and period.

Step 3: Mapping of phase profile

With the knowledge of the phase map, it is possible to create a metasurface with an arbitrary phase profile. Note that a phase profile is different from the phase map described in Step 2. A phase profile is the target phase of a metasurface device as a function of the spatial position, i.e., placement of unit cells with different radii based on the phase map. For the case of a metalens, the phase profile is a hyperbola (for 2D focusing into a focal line): $\varphi(x) = -2\pi/\lambda \times (\sqrt{f^2 + x^2} - f)$ or a hyperboloid (for 3D focusing into a focal point): $\varphi(x, y) = -2\pi/\lambda \times (\sqrt{f^2 + x^2 + y^2} - f)$, where λ is the wavelength, f is the focal length of the metalens, x, y are the spatial coordinates. At a given spatial position, it is possible to calculate the required nanopillar radius that can produce a specific phase value (closest to the target phase value) based on the phase map.

Note that for a large metalens (lens diameter > mm) that may consist of billions of such nanopillars, the file size that stores the radius and position information is huge beyond the fabrication capability. In such cases, radial symmetry with fixed edge-to-edge distance or other non-periodic arrangement (i.e. variable period) of the nanopillars can be utilized to reduce the file size.

Step 4: Modeling of full lens

The full metalens is constructed based on the phase map data in Step 2 and the target phase profile in Step 3. The metalens is modeled using finite-difference time-domain (FDTD) technique. Commercial FDTD software is available. The incident waves pass through the metalens when propagating and eventually interfere constructively at the designed focal length. By studying the focal profile (intensity vs position), it is possible to determine the full width at half maximum (FWHM) of the focused spot size and compare it with the theoretical diffraction limit.

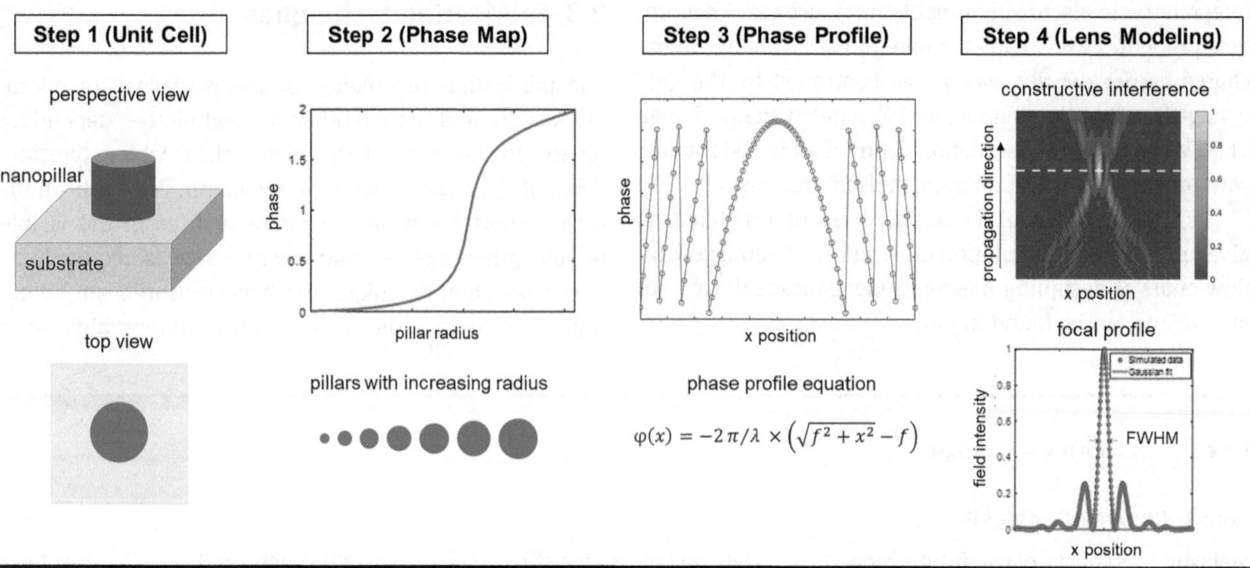

nanoimprint lithography are normally used. For applications with a longer wavelength range such as in the infrared wavelength range, the feature size requirement is relaxed, so microfabrication techniques may be sufficient, such as photolithography. These are top-down–based lithography techniques, and bottom-up-based lithography technique such as self-assembly lithography has also been reported for the fabrication of metasurfaces. A detailed review of various fabrication techniques including the advantages and disadvantages can be found in a study by Su et al. [47].

Photolithography is a high-throughput fabrication method at the microscale and nanoscale. It is the most commonly used fabrication method in the semiconductor industry due to its advantages, such as high reproducibility, high yield, low cost, and mass production availability. Photolithography uses a photomask to cover the photoresist which is a light-sensitive substrate. There are two types of photoresist: positive and negative resist. The photomask is used to transfer the pattern into the photoresist after exposure to light. The photoresist is then used to deposit materials in the desired pattern or used as a photomask for etching patterns onto a substrate placed under the photoresist. After the pattern is transferred to the wafer, the photoresist is removed. The resolution of a photolithography system depends on the wavelength of the light source and the reduction lens system. Traditional UV photolithography based on Hg lamps has a resolution of 1 µm which is suitable for metasurfaces with a large feature size at longer wavelengths (mid-infrared wavelength range and beyond). For example, in a study by Zhang et al. [9, 80, 81], metasurface devices in the mid-infrared wavelength range were fabricated using photolithography, and in a study by Chen et al. [32] and Hu et al. [82], metasurface devices were designed in the terahertz and microwave frequencies. Recently, there has been a trend in using deep-UV photolithography with an excimer laser source to fabricate metasurfaces at the visible [14] and the near-infrared wavelength [10, 83] in an attempt for scalability and mass production.

Electron beam lithography, commonly known as e-beam lithography is a form of fabrication that produces patterns with high resolution down to sub 10 nm. It does not require a mask and uses an electron beam to etch the pattern directly onto the resist. The electron beam changes the solubility of the resist, and the resist is then selectively removed using a solvent. As with photolithography, the purpose of e-beam lithography is to etch patterns into the resist and deposit other materials in the desired pattern or for further etching. Metasurfaces fabricated using e-beam lithography are normally sub-mm [12, 15, 48]. This form of lithography has low throughput, limiting it to low volume production and research purposes only.

FIB lithography is another technique for fabricating patterns with high resolution. It uses a finely focused beam of ions (usually gallium, Ga^+) operating with high beam currents for localized sputtering or milling. The sputtering process is a deposition of Ga atoms on the surface. Milling occurs by the physical bombardment of high energy Ga ions to remove the surface materials. The resolution can reach 10–15 nm. One advantage of using FIB for micromachining or nanomachining is that it is a direct writing process, i.e. no additional etching step is necessary. Metasurfaces fabricated using FIB have been reported in a study by Lin et al. [84] and Liu et al. [85]. Especially it has been widely used to pattern metasurfaces in fiber [24] or on the fiber facets [86, 87]. Compared to electron beam lithography, FIB is also low throughput, but the material selection of the sample is relatively limited. The milling rate varies and greatly depends on the sample material. FIB can be used to pattern nanostructures on fiber tips, but electron beam lithography cannot.

Nanoimprint lithography [88] is a high throughput method that has a high resolution, is low cost, and able to imprint a large area and can be done in a single step. Nanoimprint lithography uses a mold (fabricated using photolithography or e-beam lithography) to imprint the pattern onto a substrate coated with a UV-curable resist layer, the substrate is then processed by heat or UV light followed by a release step and etching process to remove the residual. Nanoimprint lithography is an ideal choice for low cost and large volume metasurface fabrication with a resolution down to 100 nm [89–91], which is sufficient enough for most metasurfaces in the visible wavelength range.

Self-assembly lithography [92], also known as directed self-assembly, is a form of fabrication that can be used to mass produce microparticles and nanoparticles for devices of the high complexity. This method can achieve spontaneous assembly of the pattern by intermolecular attraction and repulsion. Direct assembly lithography uses block copolymer which can be engineered to have different morphology. Surface interactions and thermodynamics between the block copolymers allow the formation of shapes and patterns. However, the interactions will assemble randomly; hence the block copolymer needs to be directed to form the desired pattern. This can be achieved through epitaxy, using a chemical pattern on the surface to direct, or graph epitaxy, using surface topology to align the block copolymers. Examples of metasurfaces fabricated using this technique are included in the studies by Kim et al. [93], Mayer et al. [94] and Wu et al. [95]. This form of lithography is high throughput and low cost but faces challenges in achieving the uniformity and may only generate limited patterns.

3 Bioimaging

Optical bioimaging usually obtains information based on the transmitted, reflected, and scattered light. The information obtained reflects the intensity, phase, and polarization properties of light. Metasurfaces can control these properties through the interaction of light with the meta-atoms and hence improve the imaging quality. In this section, metasurface-based imaging modalities such as chiral imaging study the polarization of light; endoscopic OCT, fluorescent imaging, super-resolution imaging, and MRI study the intensity of light; QPI study the phase of light. Features of metalenses for imaging include diffraction-limited focusing, high-quality imaging, and multifunctionalities [48]. As the phase profile of a metalens can be designed accurately, it can closely approximate the ideal phase of a perfect lens to focus light at the theoretical fundamental diffraction limit without spherical aberrations. The nanostructure arrangement to realize the designed phase profile also gives the flexibility to add functionalities to the metalens while preserving its thinness (hundreds of microns) to make it a complex metasurface device (metadevice). Note that only metasurfaces with demonstrated biological experiments are covered in this section. Metasurfaces with a potential for bio-applications are covered in Section 5. Future Directions.

3.1 Chiral imaging

Biological analytes, such as amino acids, enzymes, glucose, collagen, etc. possess intrinsic chirality (or handedness) which means they cannot be superimposed on their mirror images like the human hands. Chirality also plays a fundamental role in drug discovery, without the correct chirality, many biomolecules cannot function. Hence, optical chiral imaging based on light polarization is an important research topic in the biology community. Conventionally, it is relying on cascading of multiple optical components in a sophisticated setup which is costly and bulky. Khorasaninejad et al. demonstrated using a single metalens (termed "multispectral chiral lens") to replace the bulky setup and imaged a chiral beetle as shown in Figure 2a, b [96]. The metalens consisted of rectangular-shaped nanofins made of TiO_2 with P–B phase shifts. It acted on light with different polarizations and separately focused LCP and RCP light at different locations displaying two images of an object with different handedness within the same field of view. Because the beetle's exoskeleton strongly reflects LCP light and absorbs more RCP light, the skeleton structures are more clearly visualized with the LCP image, and the RCP image appears darker. Another metasurface-based polarization imaging technique that includes a compact polarization camera was developed by Rubin et al. [97]. Based on this approach, they were able to acquire images with the full polarization state information, while no traditional polarization optics and moving parts were used. One of the measurement objects was a human face whose 3D features were discernible only from the polarization image taken by the camera but not from the intensity image. This study

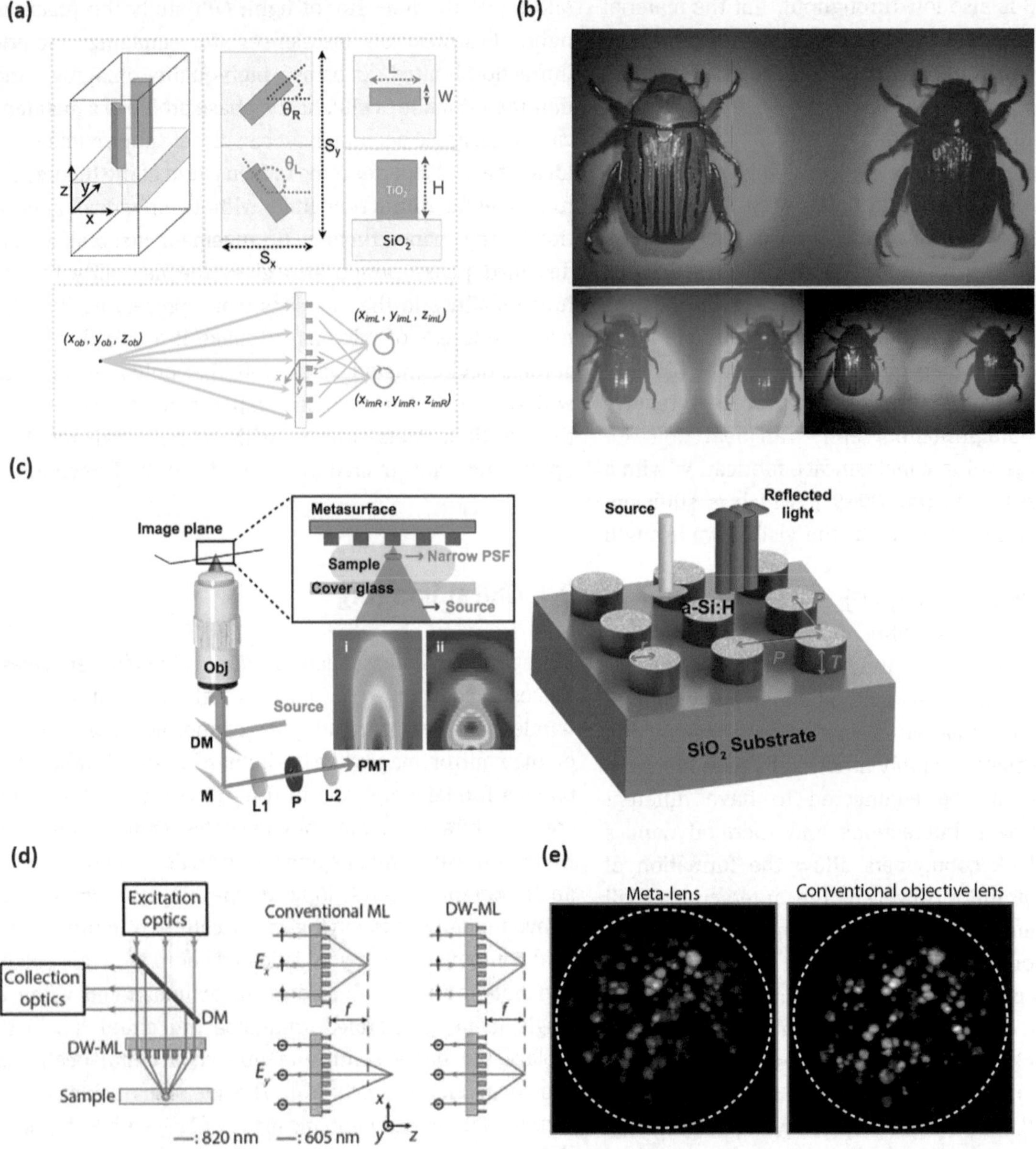

Figure 2: (a) and (b): Chiral imaging; (c)–(e) Fluorescent imaging. (a) (Top row) Perspective view, top view, and side view of the metasurface meta-atoms. (Bottom row) Schematic showing an object is imaged by a metalens which generates two images according to the polarization of the light. The green arrow represents left circularly polarized (LCP) light and blue represents right circularly polarized (RCP) light. (b) A beetle sample imaged by LCP light (left) and RCP light (right) showing different features illuminated with a green, blue, and red LED. (a) and (b) Reproduced with permission from Khorasaninejad et al., Nano Lett. **16**, 4595–4600 (2016) [96]. Copyright 2016 American Chemical Society. Direct link: https://pubs.acs.org/doi/abs/10.1021/acs.nanolett.6b01897. Further permissions related to the material excerpted should be directed to the ACS. (c) (Left) Schematic of the metasurface-enhanced axial-narrowing imaging system. Obj: objective lens, DM: dichroic mirror, M: mirror, L1–L2: lenses, P: pinhole. (Right) Schematic of the metasurface consisting of circular-shaped nanodiscs made of a-Si:H. P: period, r: radius, T: thickness. The metasurface reflects a narrow range of wavelengths. (c) Reproduced with permission from Lee et al., Opt. Mater. Express **9**, 3248 (2019) [99]. Copyright 2019 Optical Society of America. (d) Schematic showing a two-photon microscope setup with a double-wavelength metalens (DW-ML). Light excitation and collection are from the same position from the sample, i.e. from the same focal plane which is different from a conventional metalens. ML: metalens. (e) Two-photon fluorescent images of microspheres captured by metalens and conventional objective lens. (d) and (e) Reproduced with permission from Arbabi et al., Nano Lett. **18**, 4943–4948 (2018) [101]. Copyright 2018 American Chemical Society.

suggests that the metasurface-based polarization technique could be used in 3D facial recognition and reconstruction.

3.2 Endoscopic OCT

High-resolution endoscopic optical imaging is known to be an important tool in biological imaging for studying internal organs. OCT is an imaging technique capable of obtaining tissue microstructures at millimeter depths within the tissue and has been used in ophthalmology, head and neck cancer, cardiovascular diseases, and dermatological clinical diagnosis. OCT uses low-coherence light interferometry configuration to detect the change in the refractive index of the sample and thus offers morphological information. A clinical endoscopic optical imaging device based on OCT integrating a metalens (termed "nano-optics endoscope") was developed [98]. It addressed the spherical aberration and chromatic aberration that are undesirable. The metalens consisted of circularly shaped nanopillars made of amorphous silicon (a-Si) sitting on a SiO_2 substrate. It was attached to a prism at a fiber tip encased in an OCT catheter. By studying the optical path of light traveling through the fiber, the prism, the metalens and the catheter sheath reaching the tissue, the authors designed a metalens that achieved both diffraction-limited focusing resolution of (8.4 μm) and large depth of focus (293 μm) with superior imaging performance than its graded-index lens and ball lens counterparts. Endoscopic OCT imaging in resected human lung specimens and in sheep airways *in vivo* was demonstrated using the probe.

3.3 Fluorescent imaging

Fluorescent imaging of NIH3T3 cells with an enhanced axial resolution was enabled by a metasurface in a laser scanning confocal microscopy configuration [99]. As shown in Figure 2C, the metasurface consisted of an array of a-Si:H nanodiscs working in a reflection mode in the visible wavelength range. The peak reflectance wavelength was tuned by having different radius and period values of the nanodisc array. The enhancement of axial resolution was based on the interference of the excitation and reflection Gaussian beams which was able to confine light beyond the diffraction limit along the axial direction and increase the local electromagnetic field above the metasurface. In order to observe intracellular structures, such as actin filaments, the NIH3T3 cells were stained with red and green fluorescent dyes. Compared to normal confocal images, the metasurface-enhanced confocal images showed a clear distinction between actin filaments (signal) and elliptical shadows of the nuclei (background) and clearer margins of the cells. The axial resolution enhancement was two-fold (around 217 nm), and the signal-to-background enhancement was 1.75 fold.

High-resolution and wide field-of-view (FOV) fluorescent imaging of a *Giardia lamblia* cyst sample was demonstrated with a disorder-engineered metasurface and a spatial light modulator [100]. The goal of the metasurface design as a disordered medium (i.e. diffuser) is to give a far-field pattern that is isotopically scattered over the target angular ranges. One of the advantages of using metasurface for wavefront shaping compared to other types of diffusers is that the disorder of metasurface can be designed, so the input–output characteristics are known as a priori which saves time for initial characterization. It also has excellent stability and a wide angular scattering angle (up to 90°). In particular, it was demonstrated that the proposed system (NA > 0.5, FOV ~ 8 mm) was able to provide a quality image with the same resolution of a 20× objective but with a FOV of a 4× objective and the image quality was maintained for over 75 days.

Two-photon (or multiple-photon) excitation microscopy is a fluorescence imaging technique that utilizes a nonlinear process whereby a fluorophore is excited by two or multiple infrared photons and emits a single photon of higher photon energy in the visible spectrum. Because infrared light can penetrate deeper into the tissue and cause less photodamage, it has advantages over the confocal microscopy that uses single-photon excitation. Recently, a two-photon microscope based on metasurfaces was reported [101]. The novelty is that the metalens was able to focus two wavelengths at the same focal plane (termed "double-wavelength metasurface lens"). The two wavelengths correspond to the excitation and emission wavelength, i.e., 820 and 605 nm, respectively. As shown in Figure 2d, conventional lenses including metalenses typically focus different wavelengths at different focal planes due to the chromatic aberration which may compromise imaging quality. The metalens consisted of rectangular-shaped polycrystalline silicon (p-Si) nanoposts on a fused silica substrate embedded in a SU-8 layer. As a proof of concept, the authors used fluorophore-coated polyethylene microspheres as imaging objects and achieved comparable imaging performance as the conventional objective lens counterpart, as shown in Figure 2e. Almost all fluorescent clusters are visible in both cases, but the metalens is much more compact and easy to integrate with other optical components, such as a dichroic mirror.

Areas of improvement and exploration were also discussed such as using a metasurface with a double layer to correct for off-axis aberration and the effect of using high power pulsed laser sources. Recently, an endoscopic version of the double-wavelength metalens was proposed [102]. The metalens was designed to focus an excitation wavelength at 915 nm and collect with an emission wavelength at 510 nm at the same focal spot with an NA = 0.895 in a mouse brain environment. Based on theoretical calculations, the lateral resolution can reach a value of 0.42 μm.

3.4 Super-resolution imaging

Super-resolution imaging beyond the diffraction limit can greatly improve image resolution and allows tiny biological features to be seen optically without the need for electron microscopy. Techniques such as near-field scanning optical microscopy [103], stimulated emission depletion microscopy (STED) [104], structured illumination microscopy (SIM) [105], and others have been developed over the years. In particular, SIM utilizes light interference patterns to improve the resolution and is uniquely suitable for wide-field biological imaging with high speed. The recently developed plasmonic structured illumination microscopy (PSIM) explores the interference of the counter-propagating surface plasmon waves of adjacent metallic slits of subwavelength dimensions [106, 107]. The tuning of the interference pattern was achieved by changing the light incident angle. As the angle changes, the interference pattern shifts laterally. This expands the detectable spatial frequency range in the Fourier domain which translates to a resolution improvement of 2.6-fold. An improved version of the PSIM system was later developed by utilizing localized plasmonic structured illumination microscopy (LPSIM) [108, 109]. Instead of using propagating wave interference in PSIM, LPSIM makes use of the tight confinement of localized plasmonic fields at the surface. The structure was a metasurface consisting of a 2D hexagonal array of silver nanodiscs embedded in glass (Figure 3a). Near-field patterns generated by the metasurface under different illumination conditions were studied to optimize the design. Imaging of neuron cells expressing Fzd3-TdTomato was demonstrated [108]. With a high-speed camera, video imaging at 4 Hz of microtubules over a 28 × 28 μm FOV under a low laser illumination power was demonstrated. The diffraction-limited images and super-resolution images are shown in Figure 3b and c, respectively. The LPSIM system could image subcellular features with a resolution on the order of $\sim\lambda/11$ [109]. The full width at half maximum (FWHM) of the line profiles show an improvement from 275 to 75 nm (Figure 3d). With a high spatial–temporal resolution, LPSIM has proved the potential for monitoring the dynamics of proteins in the cell membranes of neurons and other tissues. Due to the evanescent nature of localized surface plasmons, LPSIM is best applied to image very thin objects or at the surface of a thick object. An alternative is to use a hyperbolic metamaterial assisted illumination which uses a metallic/dielectric alternating layered structure to achieve super-resolution imaging at the far-field [110]. A resolution of 80 nm which is sixfold resolution enhancement of the diffraction limit has been achieved with an Ag/SiO_2 multilayer structure.

3.5 Magnetic resonance imaging

Metasurfaces have also been used in other modalities of biomedical imaging. MRI is a medical imaging technique that uses strong magnets and radio waves to generate images of the anatomy and the physiological processes of the body or an animal subject. It has been widely used to study different organs especially the brain for neuroimaging. One of the drawbacks of the MRI scan process is the long scanning time which makes the patient uncomfortable. The time required can be reduced by improving the signal-to-noise ratio (SNR) which is an important parameter for characterizing the image quality. Many approaches have been proposed to improve the MRI characteristics including receive coil optimization, the use of special contrast agents, and the use of high dielectric materials. Recently, metasurfaces have been explored to enhance MRI quality. In contrast to other techniques, metasurfaces can controllably redistribute the radiofrequency electromagnetic field at the subwavelength level and hence increase the SNR. By studying the electromagnetic field distribution at different excitation frequencies, one can locally enhance only the magnetic field strength and suppress the electric field. Slobozhanyuk et al. [111] demonstrated that by placing a metasurface pad underneath a fish sample with a 1.5 Tesla (T) MRI system, the SNR was increased by more than two times and the scanning time was reduced by 10 times. The metasurface consisted of an array of metallic wires in a water base holder. A tunable metasurface version was proposed which can tune the resonance frequency of the metasurface pad to match that of the MRI system [112]. This was achieved by changing the water level and thus change the effective permittivity. An SNR improvement of seven times was reported and imaging experiment with grapefruit was conducted with a 1.5 T MRI system. The homogeneity of the enhancement region was improved as well. Not only plant samples, but an *in vivo* experiment with the human brain with a 7 T MRI

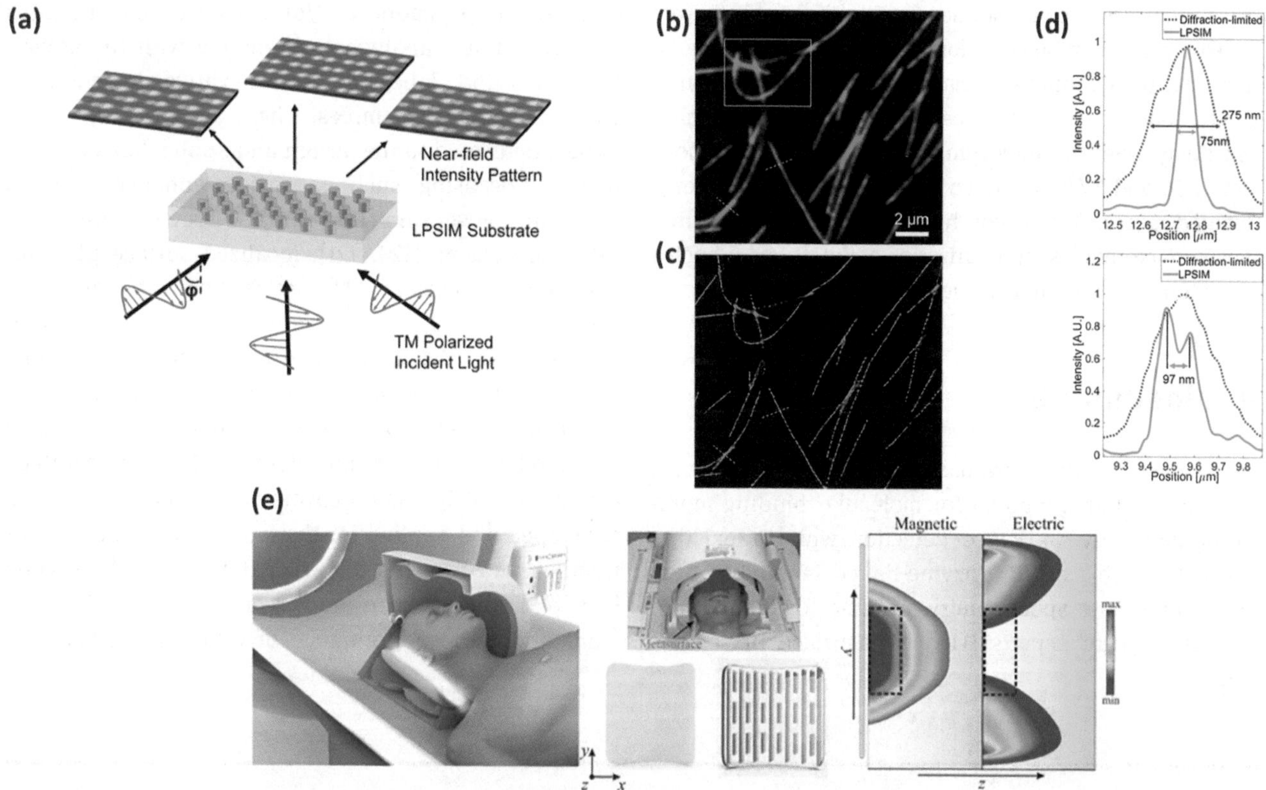

Figure 3: (a)–(d): Super-resolution imaging; (e) magnetic resonance imaging (MRI) imaging. (a) Near-field intensity distributions generated by the metasurface under the illumination of trasnverse-magnetic (TM)-polarized light at an angle of −60°, 0°, and 60°, along the symmetry axes of the localized plasmonic structured illumination microscopy (LPSIM) substrate. (b) One frame of the video recording of green microtubules with diffraction-limited imaging and (c) with super-resolution imaging. (d) Reduced full width at half maximum (FWHM) of the super-resolution imaging compared to diffraction-limited one along the two dashed yellow lines in (b) and (c). (a)–(d) Reproduced with permission from Bezryadina et al., ACS Nano **12**, 8248–8254 (2018) [109]. Copyright 2018 American Chemical Society. (e) Illustration and photograph of the experiment showing an MRI scan for a human brain with a metasurface consisting of a high permittivity dielectric pad and metallic stripes. Numerical simulation showing that metasurface can locally redistribute the magnetic field and electric field. (e) Reproduced with permission from Schmidt et al., Sci. Rep. **7**, 1678 (2017) [113]. Licensed under Creative Commons Attribution 4.0 International License.

system has also been reported, as shown in Figure 3e [113]. The metasurface pad consisting of carefully designed conductor strips and a high permittivity pad was made flexible and ultrathin such that it was able to fit underneath a person's head and a close-fitting receive coil array. An average enhancement factor was about two times. Other experiments include SNR improvement on *in vivo* imaging of a human wrist with a 1.5 T MRI system [114, 115].

3.6 Quantitative phase imaging

QPI is a label-free technique that measures the phase shift quantitatively when light passes through a more optically dense object. It is especially useful to image translucent objects, such as a living human cell which absorbs and scatters little light and difficult to observe in an ordinary optical microscope. Another advantage of QPI is that it is complementary to fluorescence microscopy that uses labels, thus exhibiting lower phototoxicity without photobleaching [116]. The QPI generates phase images instead of intensity images. The phase information obtained provides quantitative information about the morphology and dynamics of individual cells and is sensitive to their thickness and relative refractive index. A typical QPI system is built on a conventional microscope with additional optical components to obtain and analyze the interference patterns, hence it is fundamentally difficult to miniaturize. Recently, a metasurface version of the QPI system with a footprint in the order of 1 mm^3 was demonstrated [117]. It was based on two cascaded metasurface layers made of dielectric materials. The first metasurface layer captured two images for transverse electric and transverse magnetic polarizations and redirected them in three different directions to the second metasurface layer, where three birefringent off-axis metalenses were located. After the second metasurface layer, three interference contrast images with phase offsets were measured and combined to form the final phase image. In this way, the quantitative phase image was obtained in a single shot. A sea urchin cell was imaged. The phase sensitivity was 92.3 mrad/μm and the lateral resolutions were 2.76 and 3.48 μm which made it possible to observe detailed morphology inside the cell. It is impressive that metasurface enables miniaturization of the system to such a small footprint yet can achieve single-cell resolution.

4 Biosensing

In the past decade, there has been a crucial need for a rapid and reliable method for molecular binding monitoring, cell study, and cancer detection, while the existing methods, such as PCR, enzyme-linked immunosorbent assay, and mass spectrometry are time-consuming and require delicate experts [118]. Metasurface biosensors consisting of plasmonic and/or dielectric nanostructures provide a novel analytical alternative with the advantages of rapid, label-free, and sensitive detection for various biological samples. There are existing review papers dedicated to the design and applications of plasmonic biosensing such as surface-enhanced Raman scattering biosensors [119], surface plasmon resonance (SPR) biosensors [120, 121], localized surface plasmon resonance biosensors [122], optical fiber integrated biosensors [123–125], and microfluidics integrated biosensors [126]. Hence we have made the scope of this review paper to metasurface biosensing at the molecular and cellular levels with a focus on results that have appeared after the first introduction of the metasurface concept [8]. This section covers metasurface biosensors developed and applied for the targeted detection of antibodies and proteins [127–136], DNAs [138–143], cells [141–145], and cancer biomarkers [146–149]. Table 1 summarizes the performance of different metasurface

Box 2 Performance factors of metasurface biosensors

1. Bulk RI sensitivity

As mentioned earlier, the main principle of the metasurface biosensor is its ability to detect changes in the refractive index (RI). This is determined by the bulk RI sensitivity (S_B), which is given by the following equation [196]:

$$S_B = \frac{d\lambda_r}{dn_B}$$

where λ_r represents the wavelength where the SPR occurs and n_B represents the RI of the medium that is in contact with the sensor. The S_B is dependent on several factors such as the type of electromagnetic (EM) modes (decay length), resonant wavelength, excitation geometry, and the individual properties of the substrate.

The localization of the EM mode is one of the biggest contributing factor affecting S_B. Based on the following papers [197, 198], delocalized modes yields higher S_B values as opposed to localized modes [197, 199, 200]. By using gold films structured with propagating surface plasmon (PSP), it can produce S_B values far superior than those which are produced by localized surface plasmon (LSP) when excited on gold nanoparticles [197] and on a gold film [198].

Another factor that affects S_B is the resonant wavelength (λ_r), as the resonant wavelength increases, so does the S_B. Research has shown that the shape of the nanoparticle will not affect the S_B [201], but the size [161] and sharpness of the edges [162] of the nanoparticles will. For example, the increasing size of the bipyramid nanoparticles increased the sensitivity, and the less sharp of the edges on polyhedral nanoparticles, the poorer the sensitivity.

2. Figure of Merit (FOM)

The S_B is used to determine the change in RI, it is proportionate to the ability of the sensor to detect minute changes in RI and it is also inversely proportionate to the width (w) of the spectral dips and peaks that are being measured. The features when combined yield the FOM. It allows the quantification of the sensing potential of different nanostructures, it is given by the following equation [196]:

$$\mathrm{FOM}_B = \frac{S_B}{w}$$

PSP sensors can achieve a higher FOM compared to LSP sensors based on results from these papers [197, 199, 200] because of its narrower spectral features. However, FOM does not entirely translate to better performance as seen in [197], when the FOM for LSP excited on gold nanoparticles reached a maximum value at a wavelength of 700 nm. Its performance outperformed that of PSP sensors by 15% with the potential to increase performance further by three fold if the access to the sensing surface for the target molecules is improved.

3. **Limit of Detection (LOD)**

The surface coverage resolution, σ_T, determines the sensitivity of the metasurface biosensor to detect the analyte captured on the surface. It can also be seen as the minimum change that occurs when molecules are bound on to the surface, this can be expressed as [196]:

$$\sigma_T = \frac{\sigma_{SO}}{S_T}$$

where σ_{SO} is the standard deviation and S_T is the sensitivity of the molecular mass bound to the sensor surface. However, this equation does not take into account the analyte transport of the molecules from the solution. Hence, it is necessary to define the LOD. It is the best method to represent the sensitivity of the biosensor. LOD represents the smallest concentration that can be detected by the metasurface biosensor. This is usually determined by how sensitive a sensor is based on three standard deviations of blank measurements, it is expressed as [196]:

$$\mathrm{LOD} = \frac{3\sigma_b}{S_c}$$

where S_c represents the change in the sensor output divided by the change in concentration that results in the change in sensor output. It is dependent on the S_T, which is the rate that the analyte is being transferred from the solution to the surface of the sensor and the required energy for the interaction between the analyte and the binding substrate.

---

biosensors mentioned in this section. Box 2 serves as a reference guide for performance factors of evaluating a biosensor.

4.1 Antibodies and proteins

Biosensing with metasurfaces is closely related to surface plasmon polaritons (SPPs), which has been widely used for such applications [150–159]. However, the 2D periodic structures of metasurfaces can allow lower radiative damping loss at resonance and enable high quality factor detection with Fano resonance and plasmonic induced transparency [160]. Metasurfaces are also capable to provide multiresonance responses that are difficult to achieve with conventional SPP sensors [160]. Kabashin et al. [127] reported a pioneer work on metasurface biosensing using gold nanorods with a prism coupler. Both guided and longitudinal plasmonic modes were excited at the nanorods sensing layer, which allowed the sensor to provide enhanced sensitivity and narrow figure of merit (FOM) performance. The nanorods also offered a larger surface area for biomolecular sensing, which significantly improved the detection sensitivity. The sensor has been applied for the detection of streptavidin-biotin binding and the detection limit was found to be 300 nM. The performance of metasurface-based sensors was further improved by using an asymmetric split-ring array [128]. It excited the low-loss quadrupole and Fano resonance with a narrow linewidth, which showed an order of magnitude enhancement in the quality factor over planar terahertz metamaterials. The sensor sensitivity was found to be 7.75×10^3 and 5.7×10^4 nm/RIU for the quadrupole and Fano resonance, respectively.

A tunable metasurface biosensor consisting of a gold nanoantenna array and a graphene layer has recently been reported by Yu's research group [129, 130]. This operates in the mid-infrared spectral range and provides both quantitative measurement and fingerprint structure information for biomolecules. The structure of the active metasurface is described in Figure 4a. Gold nanorod antennas were fabricated on a graphene layer and a gate voltage was applied to tune the Fermi level of the graphene. A 400 nm SiO_2 layer and a Pt back mirror were deposited below the nanostructure. It supports two hybrid plasmon–phonon modes shown as dips in the reflectance spectra (Figure 4b

Table 1: Summary of metasurface biosensors.

Detection sample	Sample type	Nanostructure	Performance	Reference.
Streptavidin-biotin binding	Small molecule	Random gold nanorod	300 nM	[127]
Human antibody immunoglobulin G (IgG)	Antibodies	Gold nano-antenna array	30 pM	[129]
Mouse IgG (M-IgG)	Protein	Elliptical zigzag array	3 molecules/μm^2	[131]
Protein A/G monolayer	Protein	Elliptical zigzag array	2130 molecules/μm^2	[132]
Insulin, vascular endothelial growth factor (VEGF), and thrombin biomarkers	Protein	Al nanodisks-in-cavities array	1 fmol in a 10 µL droplet	[133]
HIV envelope glycoprotein	Protein	Nanograting	Nil	[134]
Immunoglobulin G (IgG)	Protein	Nanowire array	300 pM	[135]
Streptavidin-biotin binding	Small molecule	Metallic nanogroove array	1 fM	[136]
DNA aptamer-human odontogenic ameloblast-associated protein (ODAM) binding	Protein	Elliptical zigzag array	0.2 pg/mm^2	[138]
DNA aptamer-fructose, chlorpyrifos methyl and thrombin binding	Molecule	Gold patch array	0.2 ng	[139]
DNA–DNA molecular binding	DNA	Three-arm nanoantennas	Nil	[166]
ssDNA	DNA	Graphene–gold metasurface	1 aM	[167]
Epithelium derived oral cancer cell (HSC3)	Oral cancer cell	Concentric gold nanorings	Nil	[144]
GABA neurotransmitter	Cell	Gold nanodipoles	Nil	[145]
Lipid membrane-streptavidin	Cell lipid membrane	Gold nanoantenna array	Nil	[141]
Fungi, *Penicillium chrysogenum*, (penicillum), and Bacteria, *Escherchi coli* (*E. coli*)	Bacteria	Gold square rings	10^7 units/ml (*E. coli*)	[143]
Epidermal growth factor receptor 2 (ErbB2)	Breast cancer biomarker	Silicon nanopost array	0.7 ng/ml	[146]
Alpha-fetoprotein (AFP) and Glutamine transferase isozymes II (GGT-II)	Liver cancer biomarker	Gold split ring	0.02524 µg/ml (AFP) and 5 mu/ml (GGT-II)	[147]
Epithelium-derived oral cancer cell (HSC3)	Oral cancer cell	Gold double split ring array	900 kHz/cell ml^{-1}	[148]

GABA, gamma-aminobutyric acid; AFP, alpha-fetoprotein.

and c). The biosensor can also operate in a passive mode without the graphene layer (Figure 4d). Human antibody immunoglobulin G (IgG) was used to evaluate the performance of the metasurface biosensor. Protein A/G was attached to the gold nanoantennas as the capture agent because it contains both protein A and protein G binding domains which have a high affinity with the fragment crystallizable region of the IgG, this would reduce nonspecific biding allowing better sensitivity and specificity. The results show that as the IgG concentration increased, the resonance shifted to a lower wavelength. Two other peaks were also observed when the intensity increased with the increase in IgG concentration. These two peaks should be assigned as amide I and amide II (Figure 4e). The shift in resonance and the change in intensity of the amide peaks show a strong correlation that the change was due to IgG binding rather than nonspecific binding. The detection limit was 30 pM. Physiologically, it means the device can detect biomarkers for colon cancer which is 10 nM [161]. However, the value still needs to be improved as the detection limits of biomarkers for prostate cancer and ovarian cancer are 1 [162] and 4 pM [163], respectively. The metasurface has further been applied for surface-enhanced Fourier transform infrared (FTIR) measurement for IgG molecules. The measurement results show that it can provide four orders of magnitude improvement in the detection sensitivity compared to conventional attenuated total reflection FTIR method (Figure 4f).

Metasurfaces has also been combined with hyperspectral imaging for ultrasensitive biosensing by Altug's research group [131]. In this case, the metasurface consisted of an array of zigzag structures arranged in pairs. The pairs of nanobars were tilted along the *y*-axis in an asymmetrical manner, which increased the resonance for strong light trapping making it optimal for biosensing and

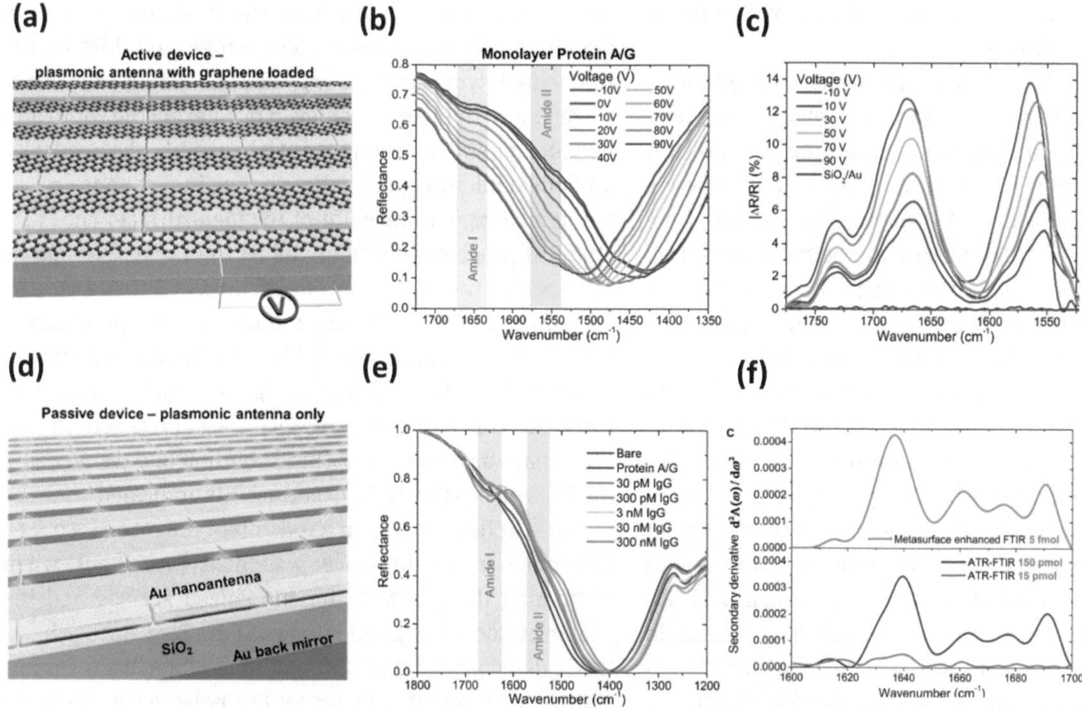

Figure 4: A tunable metasurface biosensor consisting of a gold nanoantenna array and a graphene layer for IgG antibodies detection. (a) Schematic of the active metasurface sensor with a graphene tunable layer. The Fermi level is tuned by an applied gate voltage. (b) Reflectance spectra for the measurement of the protein A/G monolayer at different gate voltages. (c) Relative absorption spectra of the amide I and II bands of the protein A/G monolayer at various voltages. (d) Schematic of the passive metasurface sensor. Gold nanorod antennas were fabricated on the 400 nm SiO_2 thin film and a gold back mirror, while the gap between each nanorod antennas is 30 nm. (e) Measured reflectance spectra for the protein A/G monolayer and the molecular binding with different concentrations of IgG antibodies. (f) Surface-enhanced Fourier transform infrared (FTIR) measurement for IgG molecules. The measurement results show that it can provide four orders of magnitude improvement in the detection sensitivity compared to conventional attenuated total reflection FTIR method. (a)–(f) Reproduced with permission from Li et al., ACS Photonics **6**, 501–509 (2019) [129]. Copyright 2019 American Chemical Society.

allowed spectral tunability. In the optical set-up, a continuously tunable bandpass filter was coupled to a supercontinuum laser source to excite the metasurface at different wavelengths and the spectral images were captured by a CMOS camera. The imaging sensor was demonstrated for the molecular binding detection of rabbit antimouse IgG (R-IgG) and mouse IgG (M-IgG). By comparing the shift maps of 66 different concentrations of samples, the results showed that three molecules per μm^2 binding would result in ~0.4 nm shift in resonance. This is due to the extremely localized fields generated by the tilting of the nanobars. The zigzag array structure was also used in a 2D pixel format to achieve imaging-based spectrometerless molecular fingerprint detection [132]. The resonance position of each pixel was tuned by changing the sizes of the meta-atoms. The 2D spectral barcode system covered the mid-IR range between 1300 and 1800 cm^{-1}, which were located in the spectral absorption regions of protein amide I and amide II functional groups. The system was then demonstrated for protein A/G monolayer detection, where the spectral peaks of amide I (1650 cm^{-1}) and amide II (1540 cm^{-1}) were measured in the transmission imaging measurement and correlated with the results of the FTIR spectroscopy. The detection capability of the system has further been tested with PMMA, PE, and glyphosate pesticide in order to show its potential for various application fields in biosensing, material science, and environmental monitoring.

Aluminum normally provides weak plasmonic response due to the relatively low electron density compared to noble metals. However, Siddique et al. [133] recently presented a metasurface structure comprising aluminum-based nanodisks-in-cavities array which could achieve 1000-fold fluorescence enhancement in the visible wavelength range. This designed structure generated hybrid multipolar lossless plasmonic modes, which enabled strong electromagnetic field confinement. Through these signal enhancement effects, the metasurface sensor was used for the detection of insulin, vascular endothelial, and thrombin biomarkers. These biomolecules with a low concentration of 1 fmol could be detected in a 10 µl droplet.

Recently, a DVD-based metasurface has been reported [134], which provides the advantages of cost-effective, large detection area, and multimodal sensing. A multilayer structure of titanium (Ti), silver (Ag), and gold (Au) layers were deposited on a clean DVD surface. The period of the grating structure was 750 nm and the height was 60 nm. Refractive index sensing of different concentrations of glycerol (1.336–1.428 RIU) has been conducted and the sensor sensitivity was found to be 377.69 nm/RIU, while the asymmetric Fano resonance peak red-shifted from 524.3 to 553.2 nm. The metasurface was functionalized with an anti-gp120 HIV antibody and applied for the detection of HIV envelope glycoprotein (gp120, 200 µg/ml) located at the HIV-1 virus surface. After that, the sensor was used for detecting HIV-1 virus, cumulative resonance wavelength shift was demonstrated at the Fano resonance peak for HIV-1 virus in the concentration of 40 and 80 copies/µl. In addition, the measurement was correlated with SEM images of the HIV-1 virus on the nanograting structures.

A leaky cavity mode resonance (LCMR)–based detection has been demonstrated with a metasurface biosensor consisting of a silicon nanowire array and a graphene monolayer (Figure 5a) [135]. Silicon was used because of its high refractive index, low loss, and its ability to be integrated with current sensing technology. In addition, the silicon nanowire structure was also able to alter the LCMR wavelength; this can be achieved by altering the width and height. This shift in resonance can be readily observed with an optical microscope, i.e. the change in resonance resulted in a change in the color of the metasurface. The color changed progressively from yellow to orange then red as the spacing between the silicon arrays increased. The metasurface was tested with the IgG protein, protein A/G was used as a binding agent. The functionalized graphene allowed for better binding of IgG on the metasurface. Different concentrations of IgG were prepared in the phosphate-buffered saline (PBS) solution and tested using the biosensor (Figure 5b). The results indicated that the increase in IgG resulted in a redshift of the resonance. The difference in performance was also prominent when comparing the graphene functionalized surface and the silicon surface. The graphene functionalized surface has a lower detection limit of 300 pM, while the silicon surface showed a resonant shift below the noise level when the concentration of the IgG was around 30 and 100 nM. This shows that silicon binds poorly with biomolecules as the number of the binding site that was present was low after washing with PBS after the incubation phase.

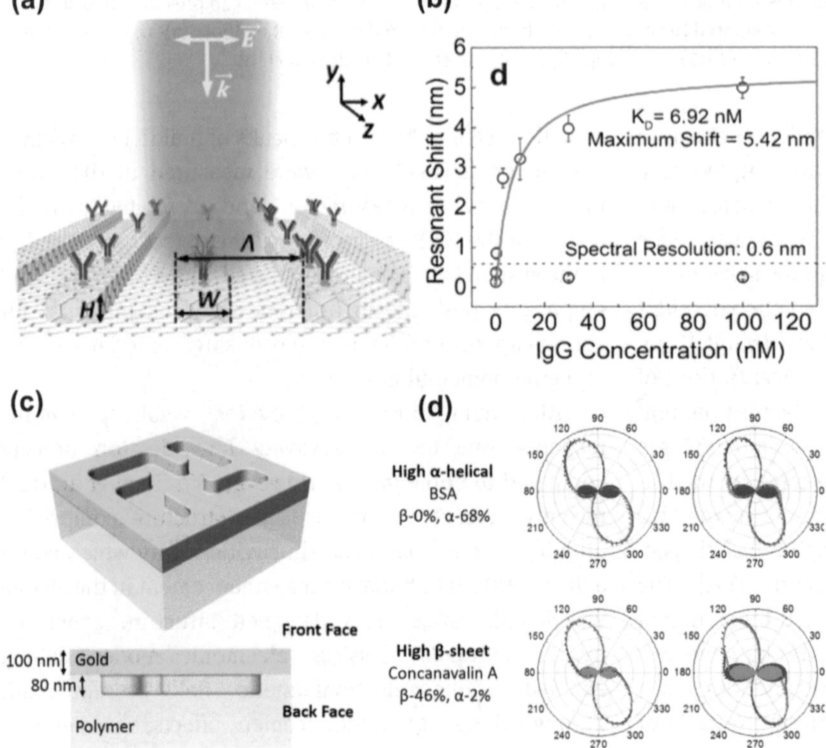

Figure 5: Antibodies and protein detection. (a) Schematic of the leaky cavity mode resonance biosensor with periodic silicon nanowires (SiNWs). A graphene monolayer is overlaid on the SiNW array for the adsorption of biomolecules. (b) The resonant wavelength shift for the molecular bindings between A/G protein and different concentrations of IgG antibody. The detection limit was found to be 300 pM. (a) and (b) Reproduced with permission from Guo et al., ACS Photonics **1**, 221–227 (2014) [135]. Copyright 2014 American Chemical Society. (c) Schematic of the chiral solid-inverse metasurface structure. (d) Polar plots for bovine serum albumin (BSA) and concanavalin A proteins detection with the chiral metasurface. (c) and (d) Reproduced with permission from Jack et al., Nano Lett. **16**, 5806–5814 (2016) [137]. Copyright 2016 American Chemical Society. Direct link: https://pubs.acs.org/doi/10.1021/acs.nanolett.6b02549. Further permissions related to the material excerpted should be directed to the ACS.

In addition, a hyperbolic nanogroove metasurface was developed for bovine serum albumin (BSA)–biotin binding detection [136]. The metasurface used a metallic nanogroove structure on metal. Six patches of nanogroove arrays were carved into the metal with each patch containing 5 × 5 nanogrooves. The metasurface was tunable by altering the width and the period of the groove; in this experiment, a period of 150 nm and a width of 30 nm were used. Calibration of the sensing platform was performed using glycerol of different concentrations. The change of the differential phase was up to 36.448°, this corresponded with the refractive index of 1.3326–1.3338 RIU. This means that the differential phase sensitivity goes up to 30,373 deg/RIU that is five times higher than conventional gold thin-film sensors. The metasurface was then used to monitor the interaction of BSA in mild to extreme dilutions. The introduction of the BSA led to a change in the phase shift which increased exponentially signifying the increase in binding of the BSA. A washing step with distilled water was performed to remove the BSA resulting in a slight decrease in the phase shift. The higher concentrations of BSA would yield a larger differential change in the phase as there was more BSA present. The detection limit of BSA was found to be 1×10^{-19} M. To further investigate the detection ability of the sensing platform, biotin was used. A standard streptavidin-biotin model was adopted, and the streptavidin was functionalized on the metasurface by physisorption forces [164, 165]. The detection limit for biotin was found to be 1×10^{-15} M.

Furthermore, the chiroptical properties of chiral biomaterials have been used for the detection of helical biopolymers and proteins [137]. As shown in Figure 5c, a chiral plasmonic nanostructure with an inverse hybrid gold metafilm was fabricated to sense the reflectivity and chiroptical properties changes of biomacromolecular secondary structures, and it provided specific fingerprints of the α-helical, β-sheet, and disordered motifs. They further demonstrated the measurement of BSA and concanavalin A proteins with polarimetry Figure 5d. It was a new detection approach that did not rely on refractive index sensing and the detection limit could reach the picogram range.

4.2 DNA detections

An angle-multiplexed dielectric metasurface has been introduced for human odontogenic ameloblast-associated protein (ODAM) detection by using single-stranded DNA aptamer for periodontal diseases diagnosis [138]. A zigzag array of elliptical germanium based metasurface was designed for molecular binding detection at the mid-infrared region (Figure 6a). In each unit cell, the plasmonic resonance spectral position can be tuned by different incidence angles. During ODAM detection, polylysine was first bound to the metasurface to allow the immobilization of DNA probes and polylysine interacted with DNA via electrostatic interactions. Then, the ODAM molecules were detected by DNA aptamer. The corresponding signal increased at the amide I and amide II bands as shown in Figure 6b. Experimental results showed that the detection limit of the metasurface was 3000 molecules/μm^2 that would yield a surface mass density of 0.2 pg/mm^2. This value was comparable with the detection limit of the gold standard BIAcore based measurement in molecules detection.

Furthermore, graphene-based metasurface with gold patch resonator array (Figure 6C) were designed for the detection of fructose, chlorpyrifos-methyl, and thrombin with a probe DNA [139]. The graphene layer enhanced the binding of analytes to the metasurface and elicited a stronger response to external molecules with the π–π stacking when molecules interact with delocalized π-electrons of graphene (Figure 6d). Chlorpyrifos-methyl has a benzene ring that interacts with the π-electrons on the graphene, while fructose lacks π-electrons. Measurement results showed that the reflectance shift of chlorpyrifos-methyl is much larger than that of fructose and the detection limit of chlorpyrifos was 0.2 ng (Figure 6e, f). DNA aptamers were further utilized for thrombin detection. The molecular bindings between DNA aptamers and thrombin produced a 4% reflectance intensity decrease.

Surface-enhanced fluorescence spectroscopy has been demonstrated with metasurface for DNA–DNA molecular binding detection [166]. The metasurface was designed with gold log-periodic nanoantennas on a silicon dioxide substrate with a three-arm design which allowed the manipulation of circularly polarized waves (Figure 6g) [139]. A thin film of titanium (3 nm) was used to ensure the adhesion of the meta-atoms to the silicon dioxide substrate. DNA probes labeled with fluorescence tags were then functionalized onto the metasurface with a self-assembled monolayer (Figure 6h). The surface-enhanced fluorescence was excited with a CW diode at 640 nm. Figure 6i shows the image of the fluorescence intensity of the fluorescence molecules on the meta-atoms. Comparing to the fluorescence intensity of the fluorescence molecules on a gold film, it was shown that meta-atoms enhanced the fluorescent signal.

The graphene–gold metasurface architecture was demonstrated by Zeng et al. [167] to achieve ultrasensitive biosensing. At the metasurface interface, a single (or multilayer) layer of graphene was deposited on a gold

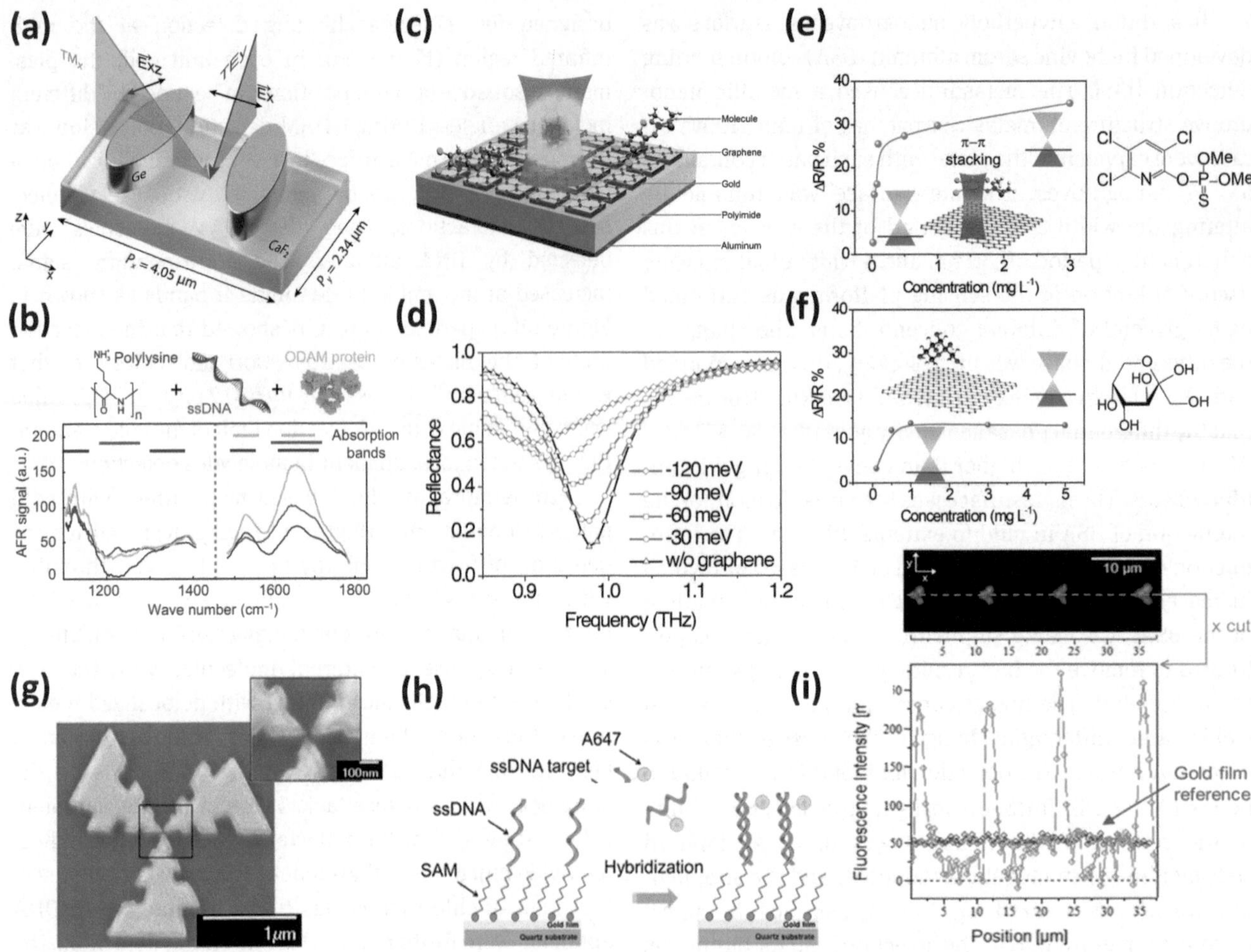

Figure 6: DNA molecules detection.
(a) Design of elliptical zigzag array, while the trasnverse-magnetic (TM) and TE modes vary at different light incident angles. Simulation results show strong near field enhancement surrounding the resonators, which is suitable for signal amplification and molecular vibration detection for analytes. (b) At human odontogenic ameloblast-associated protein (ODAM) detection, the ODAM molecules were detected by DNA aptamer. The corresponding signal increase at the amide I and amide II bands. Experimental results showed that the detection limit of the metasurface was 3000 molecules/μm^2 that would yield a surface mass density of 0.2 pg/mm^2. (a) and (b) Reproduced with permission from Leitis et al., Sci. Adv. **5**, eaaw2871 (2019) [138]. Distributed under a Creative Commons Attribution NonCommercial License 4.0 (CC BY-NC). (c) Gold patch resonator array in 85–95 μm width was fabricated on the polyimide layer (4 μm thickness) and aluminum back-ground mirror (200 nm). The metasurface was covered by a CVD-grown single graphene layer to form graphene-metamaterial heterostructure for biosensing applications. (d) Numerical modeling of different Fermi levels of the graphene layer ranged from 30 to 120 meV. (e) Reflectance signal for different concentrations of chlorpyrifos-methyl molecules (0.02–3.0 mg/l). (f) Reflectance signal for fructose molecules (0.1–5.0 mg/ml). Measurement results showed that the reflectance shift of chlorpyrifos-methyl is much larger than that of fructose and the detection limit of chlorpyrifos was 0.2 ng. (c)–(f) Reproduced with permission from Xu et al., Carbon N. Y. **141**, 247–252 (2019) [139]. Copyright 2019 Elsevier. (g) SEM image of the three-arm log-periodic gold nanoantenna structure. The three-arm design allows polarization independence and it manipulates on circularly polarized waves. (h) DNA probes were attached to the sensor surface to capture Alexa Fluor 647 labeled ssDNA targets. (i) The fluorescence intensity across the nanoantenna array indicates the binding between the probe and DNA target molecules. 4.8 times signal enhancement was recorded with the three-arm nanoantenna structure comparing to a flat gold film. (g)–(i) Reproduced with permission from Aouani et al., J. Phys. Chem. C **117**, 18620–18626 (2013) [166]. Copyright 2013 American Chemical Society.

sensing layer, which allowed a strong localization of the electromagnetic field at the interface. In another approach, gold nanoparticles were used to further enhance the sensor response. The detection limit of ssDNA was 1×10^{-18} M. The authors also reported simulation results on a graphene–MoS_2 hybrid structure [168], which can provide a phase sensor response enhancement factor of greater than 500.

4.3 Cells

Cell apoptosis study is another recent research area with metasurface biosensors. A metasurface biosensor in terahertz was used for *in situ* cell apoptosis measurement [144]. The design contained a planar array of five concentric gold nanorings (20, 28, 36, 44, and 52 µm) on a polyimide substrate (Figure 7a). The distance between each group of concentric gold nanorings was 4 µm and the transmitted terahertz spectrum in Figure 7b showed four resonance peaks. In the biosensing experiment, single layer epithelium-derived oral cancer cells (HSC3) and normal epithelial cells were detected using. Cells of different concentrations were prepared and seeded on the sensor surface. As shown in Figure 7C, the cells caused a change in frequency which increased with the increasing concentration. In the cell apoptosis detection experiment, an oral cancer cell (SCC4) was treated with a targeted cancer drug, cisplatin, which induced cell death to the cancer cell. Measurement results showed that the signal of the drug-treated SCC4 cells was slightly decreased against time, while the signals of the control cells were increased by ~50%. It was predicted that the cell multiplication process in SCC4 cell have been suppressed by the cancer drug, which caused a decrease in sensor signal (Figure 7d). While the cells that were not treated with drugs showed an increase in frequency, this can be attributed to the increasing cell concentration as the cells multiply (Figure 7e).

Recently, a multiresonant metasurface that was able to detect lipid membrane and protein interaction via the methylene and amide band was reported [145]. The metasurface contained an array of gold nanoantennas with two nanodipoles as the meta-atom. The resonance frequency can be altered by adjusting the length of the individual dipoles. Altering the periodicity can also result in a change in the metasurface, in this case, the reflectance is altered. The dipole arrays consisted of two arrays, the second array was made slightly different than the first array by having different lengths and periodicity as shown in Figure 7f. Several tests were performed to demonstrate the ability of the metasurface for multiple analytes detection. The first test was done with streptavidin and a phospholipid membrane. A 10 nm silicon dioxide layer was added to create a hydrophilic surface to allow the membrane to form [140]. The biotinylated lipids were used to facilitate the binding of the streptavidin to the membrane. The authors further investigated the interaction between the lipid membrane and melittin, which is a toxic peptide. Melittin has a strong affinity with the lipid membrane which caused the displacement of lipid membranes resulting in the formation of pores. As shown in Figure 7g, the lipid signal decreased as the melittin signal increased. The metasurface was then to detect small analytes, lipids, and peptides. Cholesterol enriched lipid vesicles with neurotransmitters were used. This was used to simulate synaptic vesicles found in neurons. The vesicles are loaded with gamma-aminobutyric acid (GABA), a neurotransmitter for the inhibition of synaptic transmission. As shown in Figure 7h, the injection of the GABA vesicles resulted in an increase in the sensor signal.

Metasurface with a gold nanoantenna array on top of a calcium fluoride substrate [141] was proposed for lipid and protein fingerprints detection (Figure 7i). To demonstrate the detection specificity of the metasurface, a lipid membrane-protein assay has been tested. Lipid vesicles containing 1,2-dioleoyl-sn-glycerol-3-phosphocholine and biotinylated 1,2-Bis(diphenylphosphino)ethane (DPPE) was injected to the metasurface to form a layer of the lipid membrane. Then, streptavidin was injected to the metasurface, and the SEIRA signal was used to detect the molecular binding between streptavidin and the biotin group on DPPE lipids (Figure 7j).

Muhammad et al. [142] demonstrated a voltage-tunable metasurface for the detection of cells. The design was with parallel gold nanorods arranged in a cross manner. By altering the length, spacing, density of nanorods, and the applied voltage on the lithium niobate film, the metasurface could be tuned to accommodate different resonant modes. To test the performance of the sensor, solvents (water, acetone, 2-propanol, and chloroform) with different refractive indices were measured. The sensor was further applied for cells and biomaterials detection with different applied voltages and resonance wavelengths.

A metasurface biosensor has further been demonstrated for microorganisms detection [143]. This comprised an array of gold square rings with a nanogap. The plasmonic resonance wavelength of the sensor was affected by the capacitance of the nanogap and also the inductance of the outer ring structure. The metasurface was applied for the detection of *Penicillium chrysogenum* fungus and *Escherichia coli* (*E. coli*) bacteria. The injection of penicillium resulted in a 9 GHz shift in frequency and the sensor

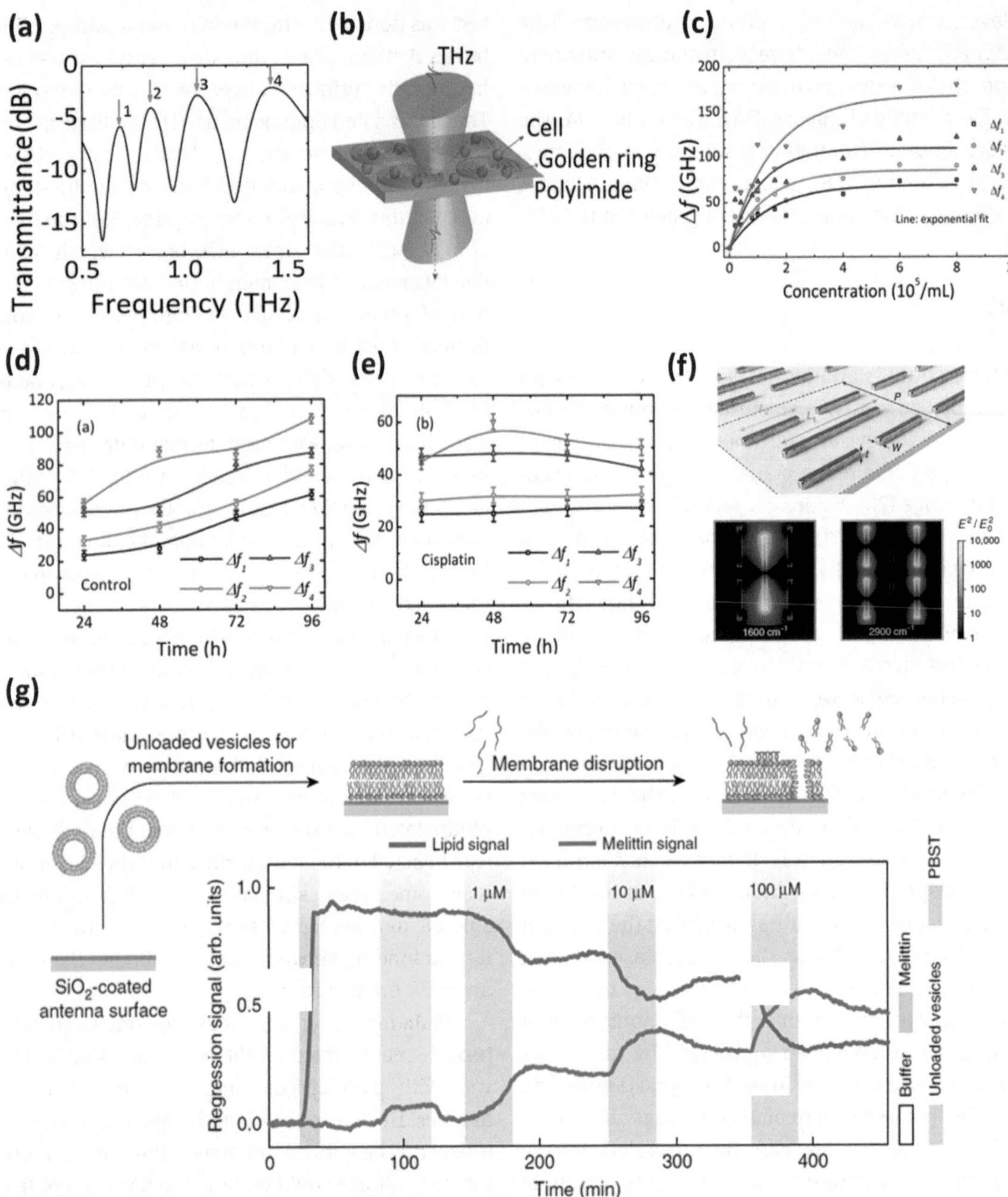

Figure 7: Cell studies and detection.
(a) The transmission spectra of the metasurface showing four peaks, which is produced by the electromagnetically induced transparency (EIT) behavior between adjacent ring structures. (b) Schematic of the metasurface. Each meta-atom consisted of five rings with diameters of 20, 28, 36, 44, and 52 μm and the gap between each ring is 4 μm. (c) The frequency shift for different concentrations of epithelium-derived oral cancer cells (HSC3). (d) The frequency shift of the control oral cancer cells SCC4. (e) The frequency shift of SCC4 cancer cells treated with the chemotherapy drug, Cisplatin, within 24–96 h. (a)–(e) Reproduced with permission from Zhang et al., Appl. Phys. Lett. **108**, 241105 (2016) [144]. Copyright 2016 AIP Publishing. (f) The metasurface structure of gold nanodipoles and simulated reflectance spectrum for the CH_2 (green) and amide (red) band. (g) The presence of melittin (purple) results in a corresponding decrease in lipid (blue) signal, this trend becomes more prominent as the concentration in of melittin increases. (h) Linear regression signals of gamma-aminobutyric acid (GABA), melittin, and lipid over time. (f)–(h) Reproduced with permission from Rodrigo et al., Nat. Commun. **9**, 2160 (2018) [145]. Licensed under a Creative Commons Attribution 4.0 International License. (i) An infrared metasurface biosensor for label-free, real-time lipid membranes monitoring. The metasurface was designed with a gold nanoantenna array on top of a calcium fluoride substrate. The size of the nanoantennas array was 1–1.75 μm in length. (j) Streptavidin was injected into the metasurface and the surface-enhanced infrared absorption (SEIRA) signal was used to detect the molecular binding between streptavidin and the biotin group on DPPE lipids. (i) and (j) Reproduced with permission from Limaj et al., Nano Lett. **16**, 1502–1508 (2016) [141]. Copyright 2016 American Chemical Society.

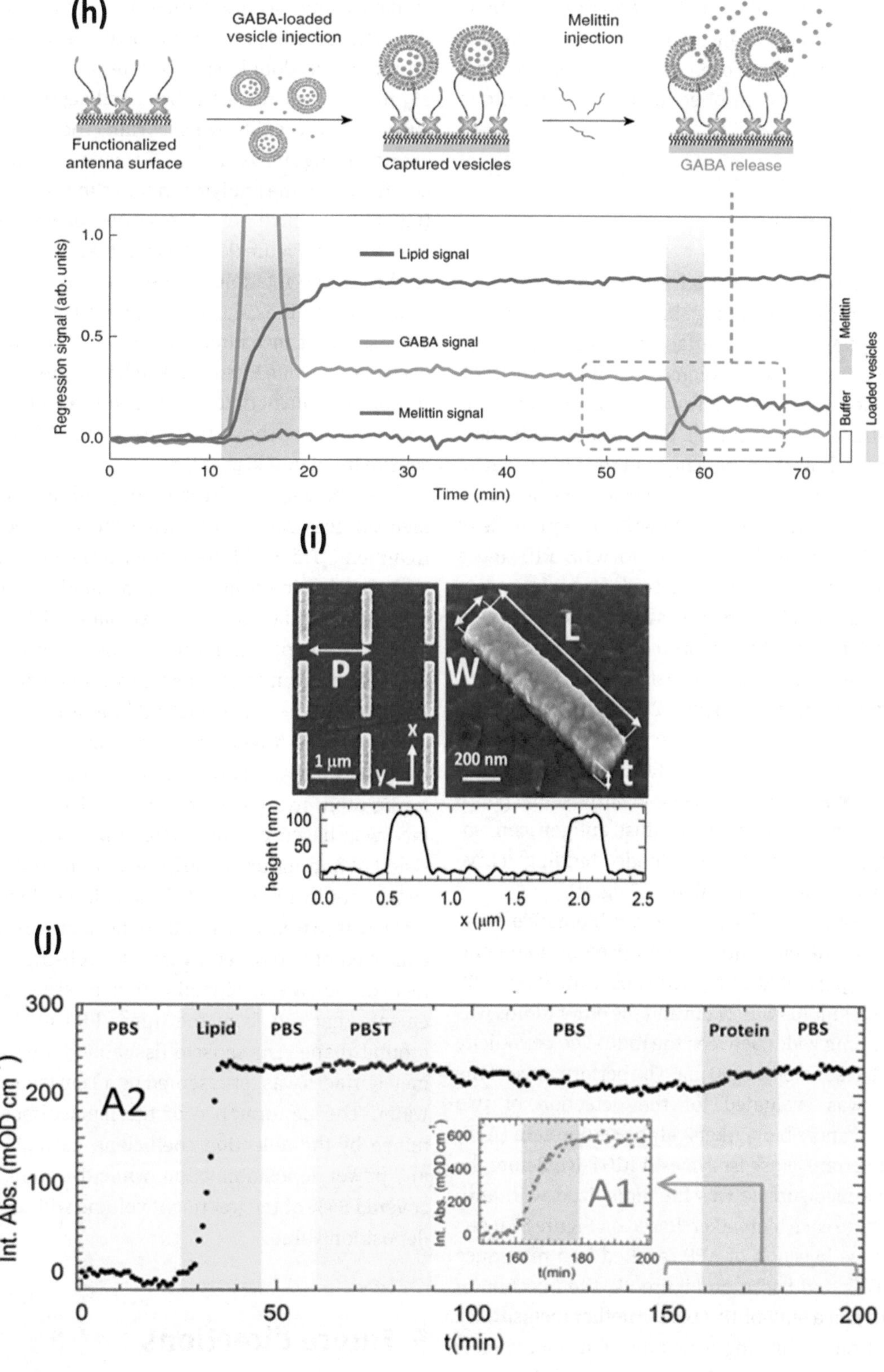

Figure 7: Continued.

response shifted back to the initial reading after the treatment with a fungicide. This signal shift was due to refractive index induced capacitance change at the nanogap. Also, a specific antibody has been used for *E. coli* detection. The injection of *E. coli* resulted in a 23 GHz shift in frequency.

4.4 Cancer biomarkers

Metasurface biosensors also find applications in cancer biomarker detection, including breast, liver, and oral cancer. An optofluidic metasurface biosensor has been developed for breast cancer biomarkers detection [146]. The metasurface was made of an array of silicon nanoposts that is constructed on a silicon on insulator substrate. Figure 8a and b show the schematic and the SEM image of the metasurface structure. The metasurface sensor performance was characterized with different concentrations of ethanol solutions ranged from 1.340 to 1.318 RIU and a sensitivity of 720 nm/RIU was derived from the sensor response curve in refractometric sensing measurement (Figure 8c and d). It has further been demonstrated for the detection of an early-stage breast cancer biomarker, epidermal growth factor receptor 2 (ErbB2). Anti-ErbB2 antibodies were immobilized on the metasurface and the molecular binding with targeted ErbB2 molecules was recorded (Figure 8e). One can see a clear dose–response curve of resonance wavelength against antigen concentrations (Figure 8f) and the kinetic binding curve (Figure 8g) with a detection limit of 0.7 ng/ml.

Another example is for liver cancer biomarker measurement [147]. The metasurface comprised of gold split-ring resonators deposited on a silicon substrate (Figure 8h and i). The inner radius was 24 μm and the outer radius was 30 nm with a 6 nm width between the radii. The periodicity between each resonator was 90 μm. The performance of the metasurface was evaluated for the detection of two different liver cancer biomarkers, alpha-fetoprotein (AFP) and glutamine transferase isozymes II (GGT-II). Before the detection, the metasurface was functionalized with antibodies specific to each biomarker. Based on Figure 8j, it can be seen that the injection of AFP resulted in a resonance shift of 8.6 GHz and based on Figure 8k, the injection of GGT-II resulted in a shift of 18.7 GHz. Another metasurface design based on a split ring resonator with double gaps that supports two resonance dips was proposed. The same experiment was repeated. Based on the results from Figure 8l and m, the addition of the AFP resulted in a shift in resonance for both the dips and the same was observed with GGT-II injection.

Oral cancer cells (HSC3) cell apoptosis process was studied using electromagnetically induced transparency like metasurface [148]. The metasurface was designed with an array gold double split-ring resonator supported on a 25 μm polyimide dielectric layer and deposited on a silicon substrate. The splits were asymmetric of each other as shown in Figure 9a and b. Transmission modeling was conducted for the analyte refractive index from 1 to 1.6 RIU (Figure 9c), which gives the sensor response curve of the metasurface (Figure 9d). It has further been applied for the studies of the relationship between an anti-cancer drug, cisplatin, and the cancer cell apoptosis process. Figure 9e and f show the measurement results for the different doses of cisplatin, while Figure 9g and h indicate the cancer cells' response against time for the anticancer drug. Experimental results show that the metasurface provides a sensitivity of 900 kHz/cell ml^{-1}.

Beyond sensing, Vrba et al. [149] demonstrated that metasurfaces can also be used for hyperthermia cancer treatment. Three different metasurfaces were designed and tested using a muscle tissue model. The first metasurface consisted of four metamaterial units and was based on microcoplanar technology. It was made on an FR4 substrate and used a ground plane beneath the substrate to suppress the backside radiation. The capacitor consisted of 10 fingers of dimensions of 18, 1, and 1 mm in length, width, and gap. The same dimension was applied to the other two metasurfaces too. The second metasurface was based on the microstrip technology and consisted of four inductive strips extending from four fingers capacitors to the grounded capacitive plates. The third metasurface was also based on the microstrip; however, it consisted of two other layers, FR4 substrate and air. The metasurface was designed with five vertical inductive elements that run from the upper FR4 to the lower FR4 ground plane. The muscle tissue model used to test the metasurfaces was represented by a 1 cm layer of deionized water. The performance of the metasurface was determined by the reflection coefficient, current distribution, and power deposition value, while the optimized pattern covered 84% of the treatment volume with 25% the power deposition value.

5 Future directions

Metasurfaces have been demonstrated with applications in nonbiomedical fields with extraordinary features that may in the future be applied for bioimaging and biosensing areas as well. Firstly, metasurfaces-enabled adaptive

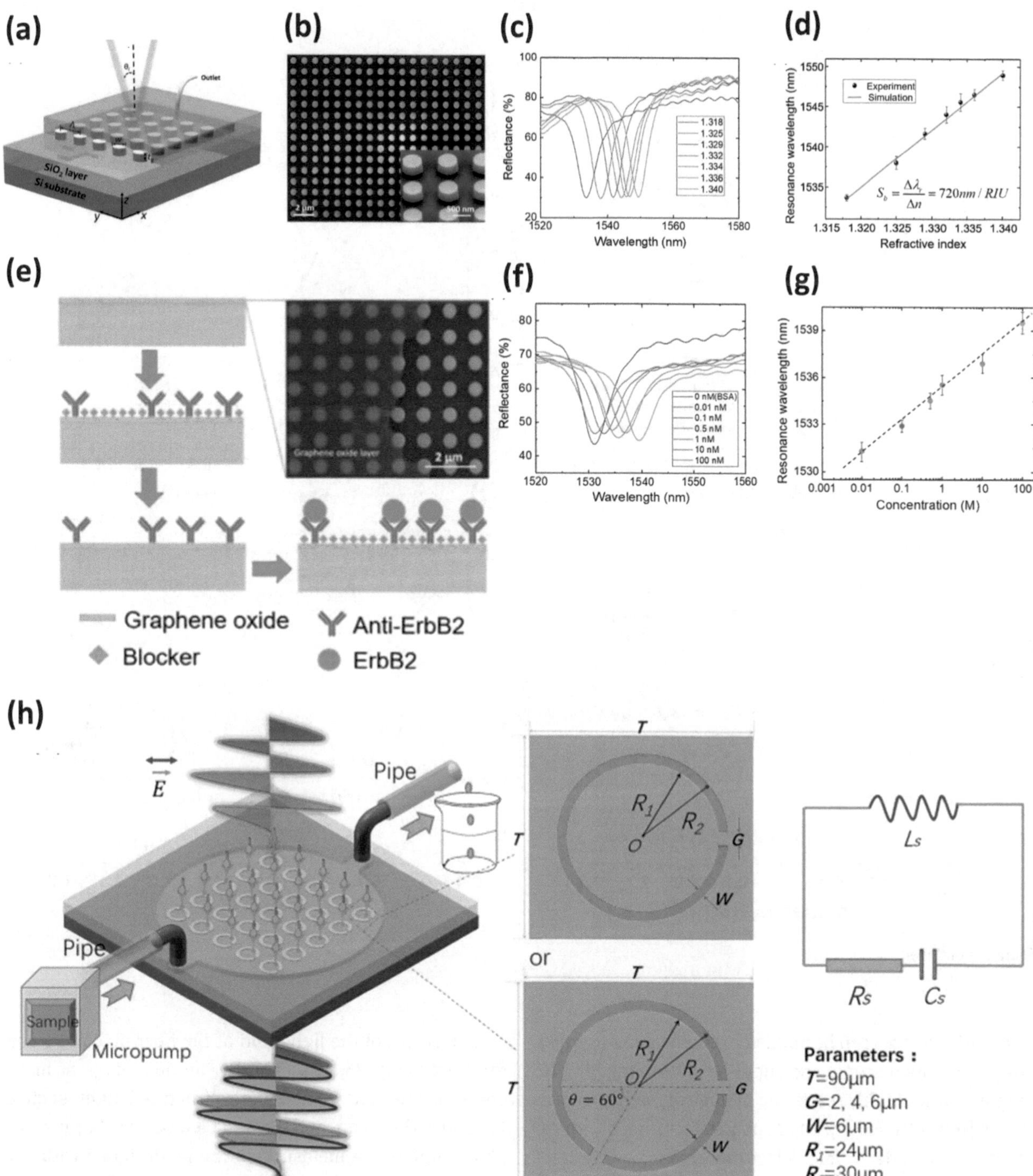

Figure 8: Cancer biomarkers detection.
(a) Schematic of the metasurface with a silicon nanopost array. The sensor surface was integrated with a PDMS based optofluidic chamber for cancer biomarker detection. (b) SEM image of the nanopost array. (c) Measurement results for ethanol solutions with different refractive index values (1.318–1.340 RIU). (d) Experimental and simulated sensor response curve of the metasurface sensor. (e) The sensor surface was functionalized with anti-ErbB2 molecules for the detection of early-stage breast cancer protein biomarker, epidermal growth factor receptor 2 (ErbB2), by monitoring the plasmonic resonance wavelength shift. (f) The reflection spectra for different concentrations of ErbB2 protein (0.01–100 nM). (g) The dose–response curve for the ErbB2 protein. The detection limit was found to be 0.7 ng/ml. (a)–(g) Reproduced with permission from Wang et al., Biosens. Bioelectron. **107**, 224–229 (2018) [146]. Copyright 2018 Elsevier. (h) (Left) Schematic of the THz metasurface integrated with a microfluidic chip. (Right) Design of the meta-atom. (i) Transmission spectra of alpha-fetoprotein (AFP) with the single split-ring resonator. The black curve represents the metasurface without samples and the red curve represents the metasurface with biomolecules. (j) Transmission of GGT-II with the single split-ring resonator. (k) Transmission of AFP with the double split-ring resonator. (l) Transmission of GGT-II with the double split-ring resonator. (h)–(l) Reproduced with permission from Geng et al., Sci. Rep. **7**, 16378 (2017) [147]. Licensed under a Creative Commons Attribution 4.0 International License.

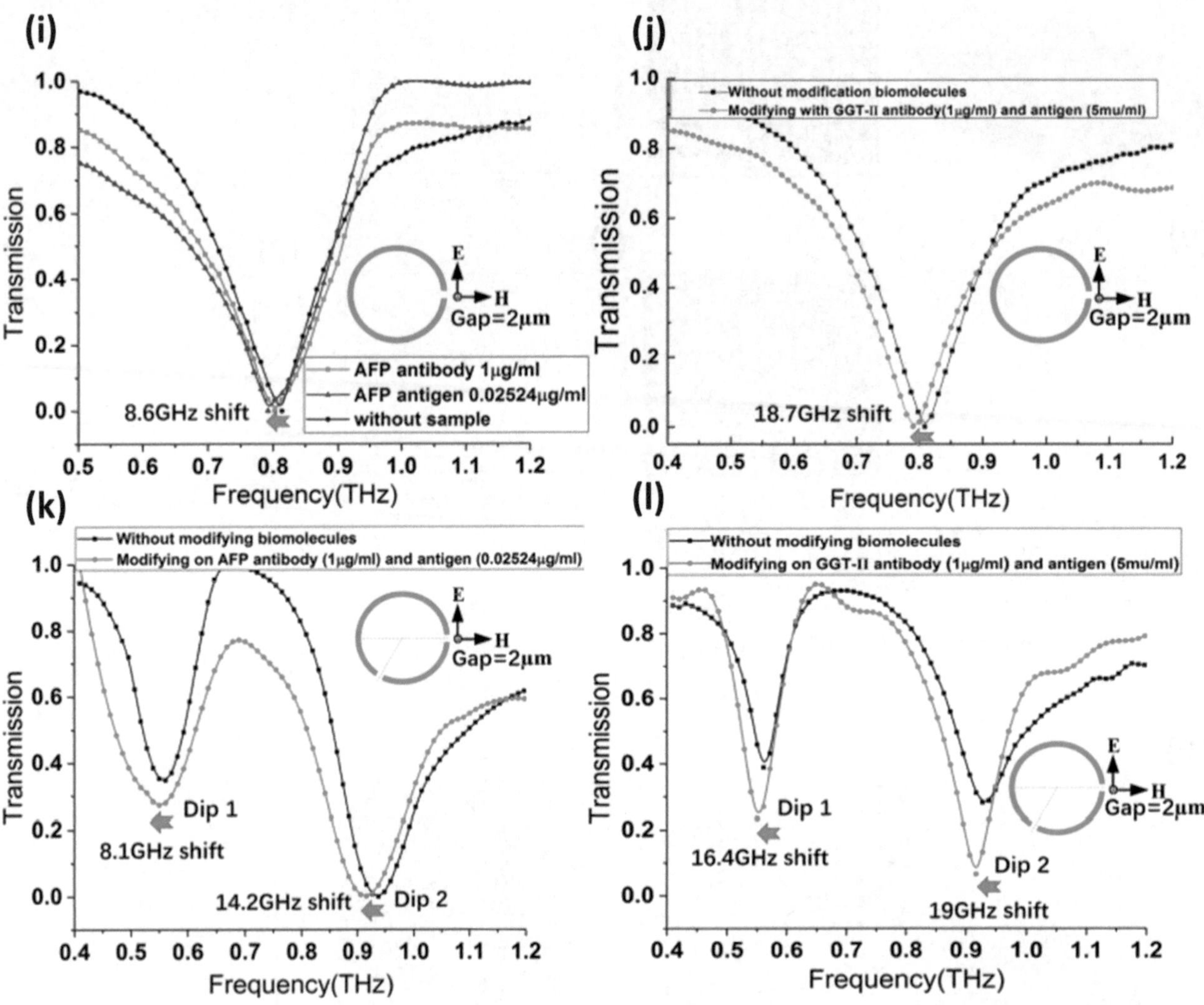

Figure 8: Continued.

optics (AO) devices can be applied in confocal microscopy, two-photon microscopy, and super-resolution imaging for imaging within highly scattering biological media, such as tissues [169–171]. Examples of such metasurfaces with AO features include: microelectromechanical systems (MEMS)–based focal length tuning [172] and coma (off-axis) aberration correction (Figure 10a) [80], soft material–based focal length tuning by mechanical stretching [65, 67] and elastomer actuator–based focal length tuning, astigmatism and shift corrections by electrical deformation (Figure 10b) [66]. Secondly, metasurfaces-enabled optical fibers, which combine the light manipulation capabilities of the metasurface and the flexibility of fibers, can be applied for *in vivo* bioimaging and remote biosensing [125, 173]. For example, light focusing [87] at the fiber tip can be used for high-resolution *in vivo* imaging and surface enhancement of the light field at the fiber tip can be used for biosensing [86, 174]. Thirdly, an advantage of metalenses is the ease to generate structured light such as Laguerre–Gaussian and Bessel beams, etc. because the phase profile of a metasurface can be designed with subwavelength accuracy. Metasurface-based structured beams such as vortex beams [175, 176], Bessel beams (see Figure 10c) [177] and Bessel beams on a fiber tip [178]. This is important for STED microscopy [179], light-sheet microscopy [180], and two-photon microscopy [181]. Fourthly, conformal metasurfaces [64, 182] are another intriguing area of exploration. Metasurface devices can be fabricated on soft substrates to allow for flexibility, stretchability, foldability ("origami metasurfaces" [183]), and actuability, enabling novel form factors and tunable features. Conformal metasurfaces will be useful in smart wearable

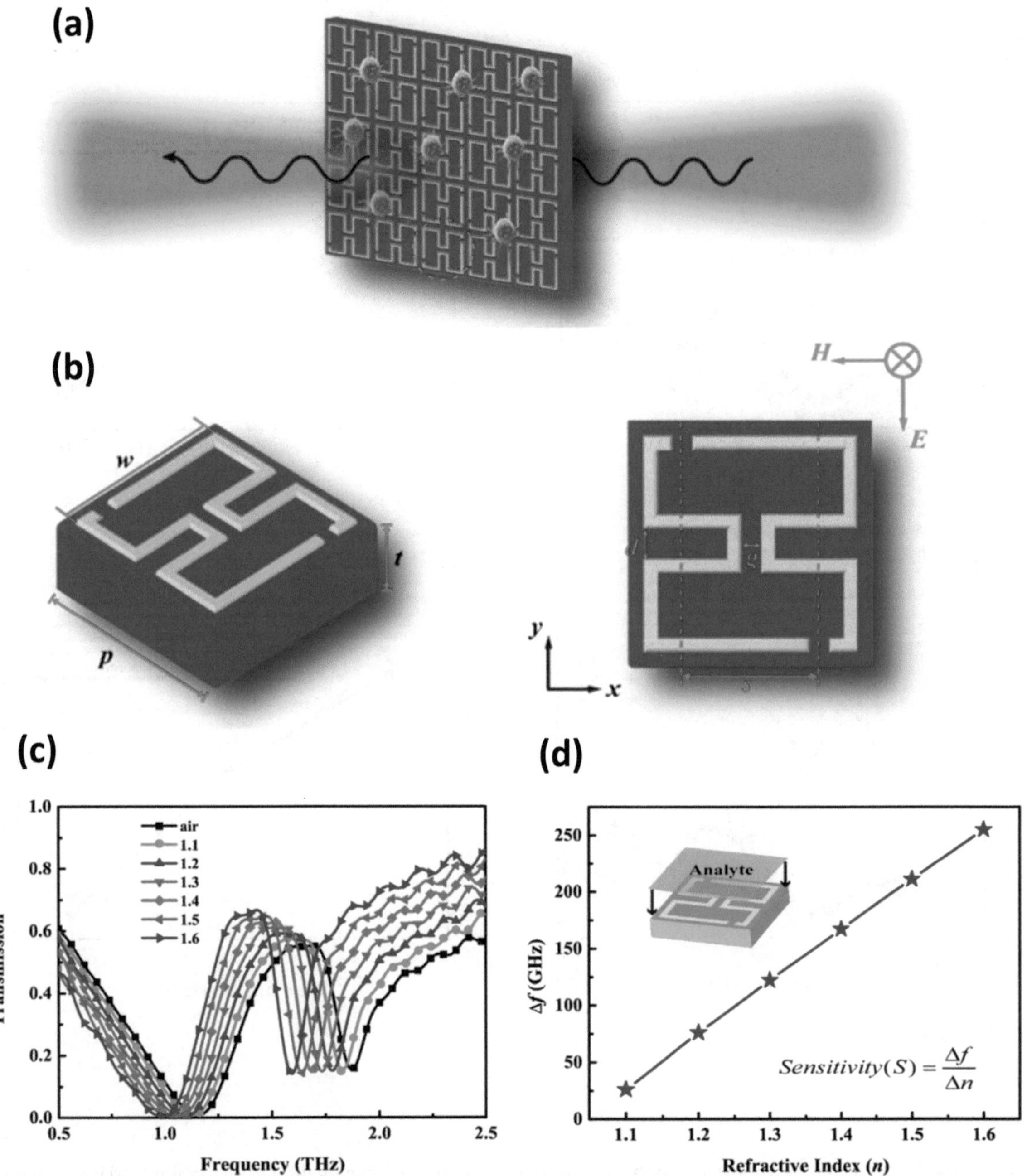

Figure 9: Oral cancer cells (HSC3) apoptosis detection with electromagnetically induced transparency like metasurface. (a) Schematic of the electromagnetically induced transparency like (EIT-like) based metasurface. (b) Metasurface structure with symmetry-breaking double-splits ring resonators. (c) Modeling the transmission curve for different refractive index (1–1.6 RIU). (d) Frequency shift response curve for different refractive indices. (e) and (f) Measurement results for the different dosage of cisplatin. (g) and (h) Cancer cells response against time for the anti-cancer drug. Experimental results show that the metasurface provides a sensitivity of 900 kHz/cell ml^{-1}. (a)–(h) Reproduced with permission from Yan et al., Biosens. Bioelectron. **126**, 485–492 (2019) [148]. Copyright 2019 Elsevier.

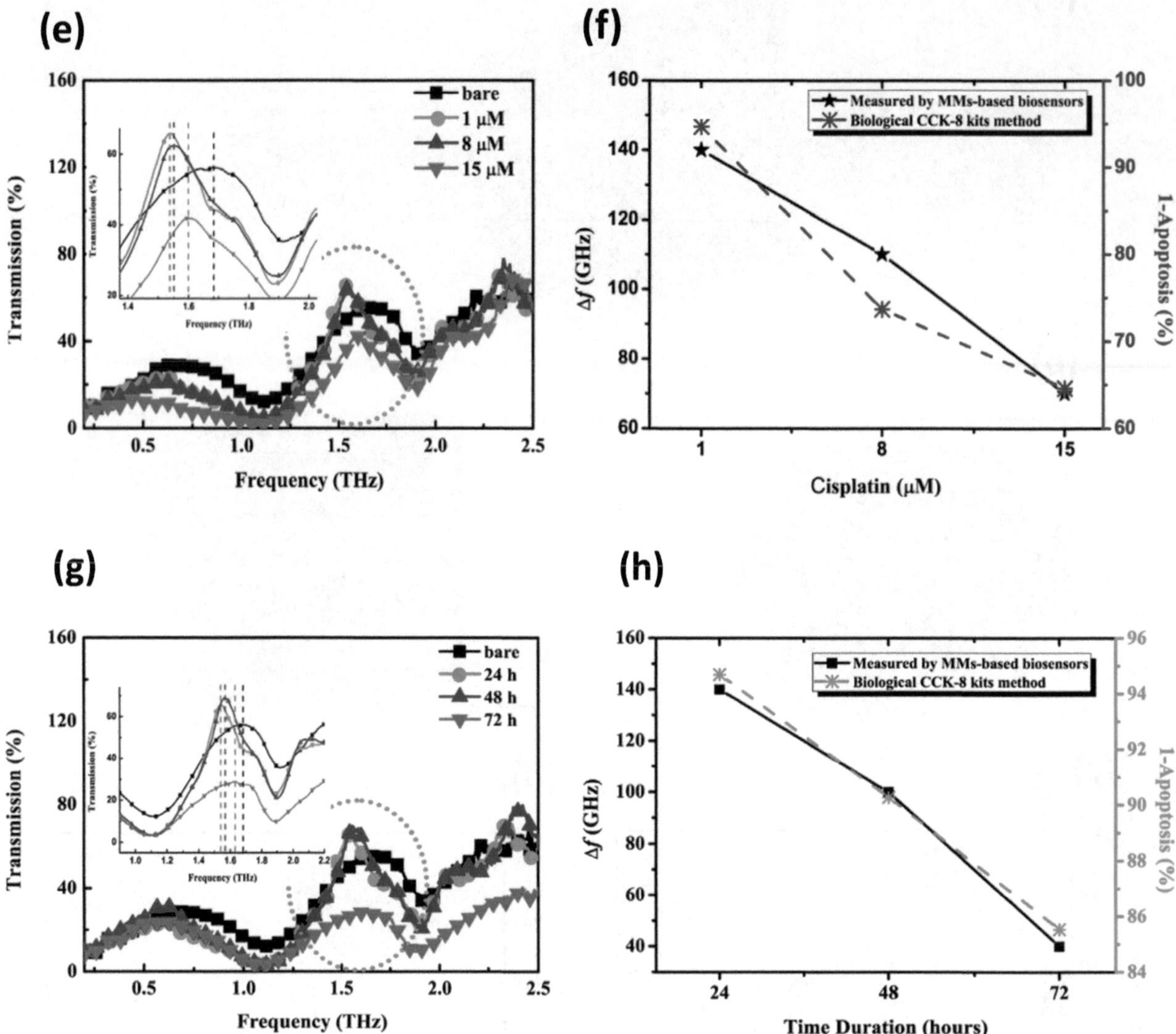

Figure 9: Continued.

and implantable health-care products such as optical bioimagers and biosensors that can be worn on the skin (see Figure 10d) or origami-folded structures directly injected and automatically unfurled inside the body. Lastly, metasurfaces-based optical tweezing and trapping can be applied in studies such as single-molecule biophysics, and cell–cell interactions. Tweezers create a 3D trap that requires tight focusing of light. Traditional microscope objectives are typically used to achieve such a high NA and tightly focused beam, but they are bulky and are often not compatible with lab-on-a-chip systems, where optical can be used to manipulate and analysis of cells and bacteria. Metasurfaces may create an entirely planar structure capable of focusing light from a collimated incident beam into a tight spot, and yield generating traps with structured beams. In recent work [184], planar metalenses created optical traps in a microfluidic environment. The trap stiffness was between 13 and 33 pN μm^{-1} W^{-1} which is modest compared to standard traps but can be improved with better metasurface design. An important facet of the work showed metalenses offer flexibility in the phase profile design and, thus, may lead far beyond the simple focusing of light. The authors then demonstrated the creation of a conical beam for enhanced trapping of objects.

Not only can metasurface devices be used for bio-applications but conversely the vast biological world itself can be an inspiration for the nanophotonics community.

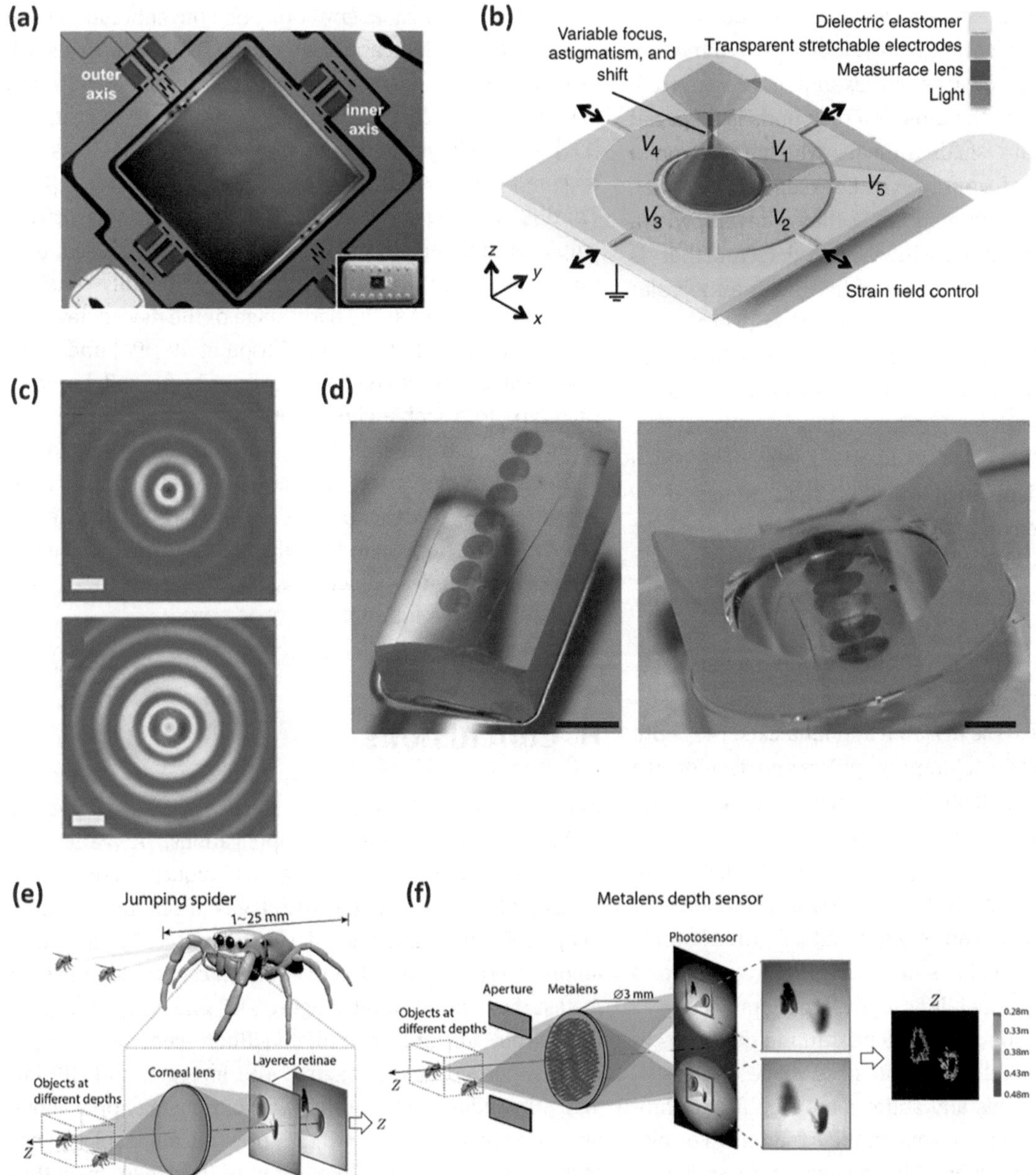

Figure 10: Future directions.
(a) Optical microscope image showing a metalens on a two-axis rotational microelectromechanical system (MEMS) which changes the relative angle between the incident light and the metalens. The inset is the integrated device with a dual in-line package. (a) Reproduced with permission from Roy et al., APL Photonics **3**, 021302 (2018) [80]. Copyright 2018 Author(s). This article is distributed under a Creative Commons Attribution (CC BY) license. (b) Schematic showing an electrically tunable metasurface device with five controllable electrode configurations to correct for defocus, astigmatism, and shift aberrations. (b) Reproduced with permission from She et al., Sci. Adv. **4**, eaap9957 (2018) [66]. Distributed under a Creative Commons Attribution-NonCommercial license. (c) Measured intensity profiles of two Bessel beams with numerical aperture (NA) = 0.7 at $\lambda = 575$ nm. Scale bar, 500 nm. (c) Adapted with permission from Chen et al., Light Sci. Appl. **6**, e16259–e16259 (2017) [177]. Licensed under a Creative Commons Attribution-NonCommercial-ShareAlike 4.0 International License. (d) Conformal metasurface on a convex (left) and a concave glass cylinder (right). In both cases, the metasurfaces make cylinders behave like converging aspherical lenses. Scale bar, 2 mm. Reproduced with permission from Kamali et al., Nat. Commun. **7**, 11618 (2016) [64]. Licensed under a Creative Commons Attribution 4.0 International License. (e) Depth perception of a jumping spider is achieved with the multitiered retina structure of each principal eye which captures multiple images of the same scene with different amounts of defocus. (f) Depth perception of a metalens depth sensor is achieved by a metalens that captures two images of the same scene with different amounts of defocus. The metalens was constructed by spatially multiplexing two phase profiles that focus on the photosensor with a lateral offset and different focal lengths. The depth map was generated by a depth reconstruction algorithm. (e) and (f) Reproduced with permission from Guo et al., Proc. Natl. Acad. Sci. **116**, 22959–22965 (2019) [188]. Copyright 2019 the Author(s). This article is distributed under Creative Commons Attribution License 4.0 (CC BY).

Learning from animals with evolutionary advantages gifted by Nature has led to the development of many biomimetic metasurface devices. For example, Yi et al. [185] observed that sensing the direction of sounds is a special capability of certain small animals which relies on a coherent coupling of soundwaves between the two ears. Inspired by this observation, they developed a metasurface-based ultrasensitive photodetector that is able to detect the incident angle of light with an angular sensitivity of 0.32°. The second example is by Shi et al. [186] who studied the silver ants living in the Saharan desert and developed an idea of a biomimetic metasurface-based coating for the passive effective cooling of objects. The third example is by Basiri et al. [187] who designed a metasurface-based polarization filter device inspired by the polarization sensing of the compound eyes of mantis shrimps. The peak extinction ratio of the metasurface device could reach a value of 35, and transmission efficiency was close to 80% in the near-infrared wavelength region. Another example is the development of a metasurface-based depth sensor inspired by the eyes of jumping spiders. Guo and Shi led the research and mimicked the depth perception mechanism of jumping spiders and developed a metalens equivalent depth sensor which was compact (3 mm in diameter for a depth measurement range of 10 cm), single-shot and required a small amount of computation (fewer than 700 floating-point operations per output pixel with microwatts energy) [188]. Figure 10e and f show the working principles of the depth perception of a jumping spider and a metalens equivalent depth sensor.

Although there are many potential future directions to explore, challenges still exist that need to be addressed. Firstly, metalenses generally suffer more significant chromatic aberration than a conventional curved lens. For bio-applications that require broadband wavelength excitation and focusing or illumination at one wavelength and collection at another wavelength far from that (e.g. multiphoton imaging), large-area metalenses (a lens diameter on the order of centimeter) with achromatic focusing is a technical challenge. In fact, this is a hot research topic in the entire metasurface community. Recently, demonstrations of large-area metalenses with single wavelength operation in the visible and near-infrared wavelength ranges [10, 14, 83, 189] and achromatic focusing of small area metalenses [190, 191] have been reported. Hence, it is promising that this challenge can be solved soon with the advancement in both metasurface design methodologies and fabrication techniques [44]. Secondly, most metasurfaces operate with an incidence light that is normal to the interface, which translates to a limited FOV. Incident light that comes at an angle will result in coma aberration. For conventional optical components, an increased FOV can be realized with a stack of lenses, i.e., a lens assembly. Similarly, for metalenses, FOV may be increased by adopting a multilayered structure [192] with advanced design methodologies such as the inverse design with topological optimization [193]. Thirdly, challenges on clinical translation of metasurface-empowered biomedical devices need to be addressed. Factors such as cost, reliability, and reproducibility must be taken into account at the early stage of the device development journey. The studies by Wilson et al. [194] and Popp et al. [195] review the challenges of transforming a laboratory prototype to a viable clinical product such as conducting preclinical and clinical trials and obtaining regulatory approval. Among these, the most important lesson is to constantly get feedback from the clinicians on the device design and performance to make sure that the device addresses the unmet clinical needs for a smooth clinical translation in the future.

6 Conclusions

In this review, we have covered the use of metasurfaces for bioimaging and biosensing applications. A variety of metasurface-based bioimaging techniques have been developed, spurred by the advantages of uniquely tailored imaging features and small device footprints. In particular, among these, we have described the latest developments in metasurface-based chiral imaging, OCT endoscopy imaging, fluorescent imaging, two-photon fluorescent microscopy, PSIM super-resolution imaging, MRI imaging, and QPI imaging. In addition, metasurfaces for biosensing applications are also reviewed, in which fully customizable nano-patterning and intimate interaction of these extremely thin devices are leveraged to measure biological signals in novel ways. For example, these include the detection of antibody and protein binding using metasurface-based hyperspectral imaging; detection of DNAs using graphene metasurface–based surface-enhanced fluorescence spectroscopy; detection of *in situ* cell apoptosis using a metasurface; detection of cancer biomarkers in the liver and breast using metasurfaces combined with optofluidics and graphene oxide.

It is the authors' view that going forward, the future of the metasurface-based bioimaging and biosensing provides several different exciting paths of exploration. *In vivo* optical imaging is a potential area, where metalenses hold great future promises. In this area, two- or multiphoton imaging and endoscopy are areas of promise for the application of metalenses. Typically, the two-photon imaging systems are complex and bulky, which are major impediments to their practical use, especially in the consideration of animal geometries and animal motion. Another *in vivo* imaging technique is endoscopy which must be engineered while subject to severe space constraints. The conventional micro-optics necessitated by these constraints provide for limited capabilities, which pale in comparison to that which can be built by large, external systems. Metalenses offer interesting engineering possibilities to enable system complexity and extreme compactification, opening up possibilities for working smoothly with living bodies instead of around them. Last but not least, an emerging trend in biophotonics is multimodality: the collection of more than one form of information by light or other methods. For example, this may involve combining morphological information with molecular information. Such a combination of data from a variety of imaging techniques leads to better sensitivity and specificity which in turn means a more informed decision may be made that can directly improve medical diagnosis. The inherent compactness of metasurface devices lends itself to the possibility of integrating multiple devices into an ultrasmall size.

As metasurfaces biomedical devices become better explored, we envision a future where imaging, sensing, and perhaps even treatment in health care can be done with extremely small and flat metasurface devices that close the gap in the interface with living systems while at the same time be significantly less intrusive and noticeable.

Acknowledgments: The research is supported by the Agency of Science, Technology and Research (A*STAR), under its Industry alignment fund prepositioning program, Award H19H6a0025. K. D. thanks the UK Engineering and Physical Sciences Research Council through grant EP/P030017/1. S. Z. thanks Dr. Alan J. She for the helpful discussions and editing of the manuscript.
Author contribution: All the authors have accepted responsibility for the entire content of this submitted manuscript and approved submission.
Research funding: The research is supported by the Agency of Science, Technology and Research (A*STAR), under its Industry alignment fund prepositioning program, Award H19H6a0025. K. D. thanks the UK Engineering and Physical Sciences Research Council through grant EP/P030017/1.
Conflict of interest statement: The authors declare no conflicts of interest regarding this article.

References

[1] V. G. Veselago, "Experimental demonstration of negative index of refraction," *Sov. Phys. Usp.*, vol. 10, p. 509, 1968.
[2] J. B. Pendry, A. J. Holden, W. J. Stewart, and I. Youngs, "Extremely low frequency plasmons in metallic mesostructures," *Phys. Rev. Lett.*, vol. 76, pp. 4773–4776, 1996.
[3] D. R. Smith, W. J. Padilla, D. C. Vier, S. C. Nemat-Nasser, and S. Schultz, "Composite medium with simultaneously negative permeability and permittivity," *Phys. Rev. Lett.*, vol. 84, pp. 4184–4187, 2000.
[4] N. I. Zheludev and Y. S. Kivshar, "From metamaterials to metadevices," *Nat. Mater.*, vol. 11, pp. 917–924, 2012.
[5] W. Cai, U. K. Chettiar, A. V. Kildishev, and V. M. Shalaev, "Optical cloaking with metamaterials," *Nat. Photonics*, vol. 1, p. 224, 2007.
[6] J. B. Pendry, D. Schurig, and D. R. Smith, "Controlling electromagnetic fields," *Science (80-)*, vol. 312, pp. 1780–1782, 2006.
[7] C. M. Soukoulis and M. Wegener, "Optical metamaterials—more bulky and less lossy," *Science (80-)*, vol. 330, pp. 1633–1634, 2010.
[8] N. Yu, P. Genevet, M. a Kats, et al., "Light propagation with phase discontinuities: generalized laws of reflection and refraction," *Science (80-)*, vol. 334, pp. 333–337, 2011.
[9] S. Zhang, M.-H. Kim, F. Aieta, et al., "High efficiency near diffraction-limited mid-infrared flat lenses based on metasurface reflectarrays," *Opt. Express*, vol. 24, p. 18024, 2016.
[10] A. She, S. Zhang, S. Shian, D. R. Clarke, and F. Capasso, "Large area metalenses: design, characterization, and mass manufacturing," *Opt. Express*, vol. 26, p. 1573, 2018.
[11] B. Groever, W. T. Chen, and F. Capasso, "Meta-lens doublet in the visible region," *Nano Lett.*, vol. 17, pp. 4902–4907, 2017.
[12] M. Khorasaninejad, W. T. Chen, R. C. Devlin, J. Oh, A. Y. Zhu, and F. Capasso, "Metalenses at visible wavelengths: diffraction-limited focusing and subwavelength resolution imaging," *Science (80-)*, vol. 352, pp. 1190–1194, 2016.
[13] M. Khorasaninejad, F. Aieta, P. Kanhaiya, et al., "Achromatic metasurface lens at telecommunication wavelengths," *Nano Lett.*, vol. 15, pp. 5358–5362, 2015.
[14] J.-S. Park, S. Zhang, A. She, et al., "All-glass, large metalens at visible wavelength using deep-ultraviolet projection lithography," *Nano Lett.*, vol. 19, pp. 8673–8682, 2019.
[15] A. Arbabi, Y. Horie, A. J. Ball, M. Bagheri, and A. Faraon, "Subwavelength-thick lenses with high numerical apertures and large efficiency based on high-contrast transmitarrays," *Nat. Commun.*, vol. 6, p. 7069, 2015.
[16] F. Aieta, P. Genevet, M. A. Kats, et al., "Aberration-free ultrathin flat lenses and axicons at telecom wavelengths based on plasmonic metasurfaces," *Nano Lett.*, vol. 12, pp. 4932–4936, 2012.
[17] S. J. Byrnes, A. Lenef, F. Aieta, and F. Capasso, "Designing large, high-efficiency, high-numerical-aperture, transmissive meta-lenses for visible light," *Opt. Express*, vol. 24, pp. 1–16, 2015.
[18] M. Khorasaninejad, A. Y. Zhu, C. Roques-Carmes, et al., "Polarization-insensitive metalenses at visible wavelengths," *Nano Lett.*, vol. 16, pp. 7229–7234, 2016.
[19] M. Khorasaninejad, Z. Shi, A. Y. Zhu, et al., "Achromatic metalens over 60 nm bandwidth in the visible and metalens with reverse chromatic dispersion," *Nano Lett.*, vol. 17, pp. 1819–1824, 2017.
[20] Z. Ma, S. M. Hanham, Y. Gong, and M. Hong, "All-dielectric reflective half-wave plate metasurface based on the anisotropic

excitation of electric and magnetic dipole resonances," *Opt. Lett.*, vol. 43, p. 911, 2018.

[21] J. Hu, X. Zhao, Y. Lin, et al., "All-dielectric metasurface circular dichroism waveplate," *Sci. Rep.*, vol. 7, p. 41893, 2017.

[22] E. O. Owiti, H. Yang, P. Liu, C. F. Ominde, and X. Sun, "Highly efficient birefringent quarter-wave plate based on all-dielectric metasurface and graphene," *Opt. Commun.*, vol. 419, pp. 114–119, 2018.

[23] Y. Dong, Z. Xu, N. Li, et al., "Si metasurface half-wave plates demonstrated on a 12-inch CMOS platform," *Nanophotonics*, vol. 9, pp. 149–157, 2020.

[24] J. P. B. Mueller, K. Leosson, and F. Capasso, "Ultracompact metasurface in-line polarimeter," *Optica*, vol. 3, pp. 42–47, 2016.

[25] Q. Jiang, B. Du, M. Jiang, et al., "Ultrathin circular polarimeter based on chiral plasmonic metasurface and monolayer MoSe2," *Nanoscale*, vol. 12, pp. 5906–5913, 2020.

[26] E. Arbabi, S. M. Kamali, A. Arbabi, and A. Faraon, "Full-Stokes imaging polarimetry using dielectric metasurfaces," *ACS Photonics*, vol. 5, pp. 3132–3140, 2018.

[27] X. Ni, A. V Kildishev, and V. M. Shalaev, "Metasurface holograms for visible light," *Nat. Commun.*, vol. 4, p. 2807, 2013.

[28] G. Zheng, H. Mühlenbernd, M. Kenney, G. Li, T. Zentgraf, and S. Zhang, "Metasurface holograms reaching 80% efficiency," *Nat. Nanotechnol.*, vol. 10, pp. 308–312, 2015.

[29] X. Zhang, S. Yang, W. Yue, et al., "Direct polarization measurement using a multiplexed Pancharatnam–Berry metahologram," *Optica*, vol. 6, p. 1190, 2019.

[30] W. T. Chen, A. Y. Zhu, M. Khorasaninejad, Z. Shi, V. Sanjeev, and F. Capasso, "Immersion meta-lenses at visible wavelengths for nanoscale imaging," *Nano Lett.*, vol. 17, pp. 3188–3194, 2017.

[31] S. Chang, X. Guo, and X. Ni, "Optical metasurfaces: progress and applications," *Annu. Rev. Mater. Res.*, vol. 48, pp. 279–302, 2018.

[32] H.-T. Chen, A. J. Taylor, and N. Yu, "A review of metasurfaces: physics and applications," *Rep. Prog. Phys.*, vol. 79, p. 076401, 2016.

[33] N. Meinzer, W. L. Barnes, and I. R. Hooper, "Plasmonic meta-atoms and metasurfaces," *Nat. Photonics*, vol. 8, pp. 889–898, 2014.

[34] Nanfang Yu, P. Genevet, F. Aieta, et al., "Flat optics: controlling wavefronts with optical antenna metasurfaces," *IEEE J. Sel. Top. Quantum Electron.*, vol. 19, p. 4700423, 2013.

[35] A. Y. Zhu, A. I. Kuznetsov, B. Luk'yanchuk, N. Engheta, and P. Genevet, "Traditional and emerging materials for optical metasurfaces," *Nanophotonics*, vol. 6, pp. 452–471, 2017.

[36] G. Li, S. Zhang, and T. Zentgraf, "Nonlinear photonic metasurfaces," *Nat. Rev. Mater.*, vol. 2, p. 17010, 2017.

[37] A. E. Minovich, A. E. Miroshnichenko, A. Y. Bykov, T. V. Murzina, D. N. Neshev, and Y. S. Kivshar, "Functional and nonlinear optical metasurfaces," *Laser Photon. Rev.*, vol. 9, pp. 195–213, 2015.

[38] A. V. Kildishev, A. Boltasseva, and V. M. Shalaev, "Planar photonics with metasurfaces," *Science (80-)*, vol. 339, p. 1232009, 2013.

[39] S. Walia, C. M. Shah, P. Gutruf, et al., "Flexible metasurfaces and metamaterials: a review of materials and fabrication processes at micro- and nano-scales," *Appl. Phys. Rev.*, vol. 2, p. 011303, 2015.

[40] M. Song, D. Wang, S. Peana, et al., "Colors with plasmonic nanostructures: a full-spectrum review," *Appl. Phys. Rev.*, vol. 6, p. 041308, 2019.

[41] C. U. Hail, A. U. Michel, D. Poulikakos, and H. Eghlidi, "Optical metasurfaces: evolving from passive to adaptive," *Adv. Opt. Mater.*, vol. 7, p. 1801786, 2019.

[42] H. Liang, A. Martins, B.-H. V. Borges, et al., "High performance metalenses: numerical aperture, aberrations, chromaticity, and trade-offs," *Optica*, vol. 6, p. 1461, 2019.

[43] F. Ding, Y. Yang, R. A. Deshpande, and S. I. Bozhevolnyi, "A review of gap-surface plasmon metasurfaces: fundamentals and applications," *Nanophotonics*, vol. 7, pp. 1129–1156, 2018.

[44] W. T. Chen, A. Y. Zhu, and F. Capasso, "Flat optics with dispersion-engineered metasurfaces," *Nat. Rev. Mater.*, vol. 5, pp. 604–620, 2020.

[45] F. Ding, A. Pors, and S. I. Bozhevolnyi, "Gradient metasurfaces: a review of fundamentals and applications," *Rep. Prog. Phys.*, vol. 81, p. 026401, 2018.

[46] H.-H. Hsiao, C. H. Chu, and D. P. Tsai, "Fundamentals and applications of metasurfaces," *Small Methods*, vol. 1, p. 1600064, 2017.

[47] V.-C. Su, C. H. Chu, G. Sun, and D. P. Tsai, "Advances in optical metasurfaces: fabrication and applications [Invited]," *Opt. Express*, vol. 26, p. 13148, 2018.

[48] M. Khorasaninejad and F. Capasso, "Metalenses: versatile multifunctional photonic components," *Science (80-)*, vol. 358, p. eaam8100, 2017.

[49] P. Genevet, F. Capasso, F. Aieta, M. Khorasaninejad, and R. Devlin, "Recent advances in planar optics: from plasmonic to dielectric metasurfaces," *Optica*, vol. 4, p. 139, 2017.

[50] S. Jahani, and Z. Jacob, "All-dielectric metamaterials," *Nat. Nanotechnol.*, vol. 11, pp. 23–36, 2016.

[51] A. I. Kuznetsov, A. E. Miroshnichenko, M. L. Brongersma, Y. S. Kivshar, and B. Luk'yanchuk, "Optically resonant dielectric nanostructures," *Science (80-)*, vol. 354, p. aag2472, 2016.

[52] A. Tittl, A. John-Herpin, A. Leitis, E. R. Arvelo, and H. Altug, "Metasurface-based molecular biosensing aided by artificial intelligence," *Angew. Chem. Int. Ed.*, vol. 58, pp. 14810–14822, 2019.

[53] D. Lee, J. Gwak, T. Badloe, S. Palomba, and J. Rho, "Metasurfaces-based imaging and applications: from miniaturized optical components to functional imaging platforms," *Nanoscale Adv.*, vol. 2, pp. 605–625, 2020.

[54] B. Li, W. Piyawattanametha, and Z. Qiu, "Metalens-based miniaturized optical systems," *Micromachines*, vol. 10, p. 310, 2019.

[55] X. Ni, N. K. Emani, A. V. Kildishev, A. Boltasseva, and V. M. Shalaev, "Broadband light bending with plasmonic nanoantennas," *Science (80-)*, vol. 335, p. 427, 2012.

[56] F. Aieta, P. Genevet, N. Yu, M. A. Kats, Z. Gaburro, and F. Capasso, "Out-of-plane reflection and refraction of light by anisotropic optical antenna metasurfaces with phase discontinuities," *Nano Lett.*, vol. 12, pp. 1702–1706, 2012.

[57] S. Gwo, C.-Y. Wang, H.-Y. Chen, et al., "Plasmonic metasurfaces for nonlinear optics and quantitative SERS," *ACS Photonics*, vol. 3, pp. 1371–1384, 2016.

[58] G. Mie, "Beiträge zur Optik trüber Medien, speziell kolloidaler Metallösungen," *Ann. Phys.*, vol. 330, pp. 377–445, 1908.

[59] N. M. Estakhri and A. Alù, "Recent progress in gradient metasurfaces," *J. Opt. Soc. Am. B*, vol. 33, p. A21, 2016.

[60] C. Huygens, *Traité de la Lumière* (drafted 1678; published in Leyden by Van der Aa, 1690), translated by Silvanus P. Thompson as *Treatise on Light (London: Macmillan, 1912; Project Gutenberg edition, 2005)*. Leyden: Van der Aa, p. 19, 1690.

[61] A. Fresnel, Mémoire sur la diffraction de la lumière" (deposited 1818, "crowned" 1819), in *Oeuvres complètes* (Paris: Imprimerie impériale, 1866–1870. vol.1, pp. 247–363; partly translated as "Fresnel's prize memoir on the diffraction of light", in H. Crew (ed.), *The Wave Theory of Light: Memoirs by Huygens*, Young and Fresnel, American Book Co., 1900, pp. 81–144.

[62] M. V. Berry, "The adiabatic phase and pancharatnam's phase for polarized light," *J. Mod. Opt.*, vol. 34, pp. 1401–1407, 1987.

[63] S. Pancharatnam, "Generalized theory of interference and its applications," *Proc. Indian Acad. Sci. Sect. A*, vol. 44, pp. 398–417, 1956.

[64] S. M. Kamali, A. Arbabi, E. Arbabi, Y. Horie, and A. Faraon, "Decoupling optical function and geometrical form using conformal flexible dielectric metasurfaces," *Nat. Commun.*, vol. 7, p. 11618, 2016.

[65] S. M. Kamali, E. Arbabi, A. Arbabi, Y. Horie, and A. Faraon, "Highly tunable elastic dielectric metasurface lenses," *Laser Photon. Rev.*, vol. 10, pp. 1002–1008, 2016.

[66] A. She, S. Zhang, S. Shian, D. R. Clarke, and F. Capasso, "Adaptive metalenses with simultaneous electrical control of focal length, astigmatism, and shift," *Sci. Adv.*, vol. 4, p. eaap9957, 2018.

[67] H.-S. Ee and R. Agarwal, "Tunable metasurface and flat optical zoom lens on a stretchable substrate," *Nano Lett.*, vol. 16, pp. 2818–2823, 2016.

[68] T. Tumkur, G. Zhu, P. Black, Y. A. Barnakov, C. E. Bonner, and M. A. Noginov, "Control of spontaneous emission in a volume of functionalized hyperbolic metamaterial," *Appl. Phys. Lett.*, vol. 99, p. 151115, 2011.

[69] M. R. M. Hashemi, S.-H. Yang, T. Wang, N. Sepúlveda, and M. Jarrahi, "Electronically-controlled beam-steering through vanadium dioxide metasurfaces," *Sci. Rep.*, vol. 6, p. 35439, 2016.

[70] D. Wang, L. Zhang, Y. Gong, et al., "Multiband switchable terahertz quarter-wave plates via phase-change metasurfaces," *IEEE Photonics J*, vol. 8, pp. 1–8, 2016.

[71] D. Wang, L. Zhang, Y. Gu, et al., "Switchable ultrathin quarter-wave plate in terahertz using active phase-change metasurface," *Sci. Rep.*, vol. 5, p. 15020, 2015.

[72] J. Rensberg, S. Zhang, Y. Zhou, et al., "Active optical metasurfaces based on defect-engineered phase-transition materials," *Nano Lett.*, vol. 16, pp. 1050–1055, 2016.

[73] Q. Wang, E. T. F. Rogers, B. Gholipour, et al., "Optically reconfigurable metasurfaces and photonic devices based on phase change materials," *Nat. Photonics*, vol. 10, pp. 60–65, 2016.

[74] M. A. Kats, R. Blanchard, P. Genevet, et al., "Thermal tuning of mid-infrared plasmonic antenna arrays using a phase change material," *Opt. Lett.*, vol. 38, p. 368, 2013.

[75] J. Sautter, I. Staude, M. Decker, et al., "Active tuning of all-dielectric metasurfaces," *ACS Nano*, vol. 9, pp. 4308–4315, 2015.

[76] O. Buchnev, N. Podoliak, M. Kaczmarek, N. I. Zheludev, and V. A. Fedotov, "Electrically controlled nanostructured metasurface loaded with liquid crystal: toward multifunctional photonic switch," *Adv. Opt. Mater.*, vol. 3, pp. 674–679, 2015.

[77] W. Zhu, Q. Song, L. Yan, et al., "A flat lens with tunable phase gradient by using random access reconfigurable metamaterial," *Adv. Mater.*, vol. 27, pp. 4739–4743, 2015.

[78] N. Dabidian, I. Kholmanov, A. B. Khanikaev, et al., "Electrical switching of infrared light using graphene integration with plasmonic Fano resonant metasurfaces," *ACS Photonics*, vol. 2, pp. 216–227, 2015.

[79] Y. Yao, R. Shankar, M. A. Kats, et al., "Electrically tunable metasurface perfect absorbers for ultrathin mid-infrared optical modulators," *Nano Lett.*, vol. 14, pp. 6526–6532, 2014.

[80] T. Roy, S. Zhang, I. W. Jung, M. Troccoli, F. Capasso, and D. Lopez, "Dynamic metasurface lens based on MEMS technology," *APL Photonics*, vol. 3, p. 021302, 2018.

[81] S. Zhang, A. Soibel, S. A. Keo, et al., "Solid-immersion metalenses for infrared focal plane arrays," *Appl. Phys. Lett.*, vol. 113, p. 111104, 2018.

[82] D. Hu, X. Wang, S. Feng, et al., "Ultrathin terahertz planar elements," *Adv. Opt. Mater.*, vol. 1, pp. 186–191, 2013.

[83] T. Hu, Q. Zhong, N. Li, et al., "CMOS-compatible a-Si metalenses on a 12-inch glass wafer for fingerprint imaging," *Nanophotonics*, vol. 9, pp. 823–830, 2020.

[84] J. Lin, J. P. B. Mueller, Q. Wang, et al., "Polarization-controlled tunable directional coupling of surface plasmon polaritons," *Science (80-)*, vol. 340, pp. 331–334, 2013.

[85] Z. Liu, H. Du, Z.-Y. Li, N. X. Fang, and J. Li, "Invited Article: nano-kirigami metasurfaces by focused-ion-beam induced close-loop transformation," *APL Photonics*, vol. 3, p. 100803, 2018.

[86] M. Principe, M. Consales, A. Micco, et al., "Optical fiber meta-tips," *Light Sci. Appl.*, vol. 6, p. e16226, 2017.

[87] J. Yang, I. Ghimire, P. C. Wu, et al., "Photonic crystal fiber metalens," *Nanophotonics*, vol. 8, pp. 443–449, 2019.

[88] G. Barbillon, "Soft UV nanoimprint lithography: a tool to design plasmonic nanobiosensors," *Adv. Unconv. Lithogr.*, pp. 3–14, 2011. https://doi.org/10.5772/20572.

[89] Y. Yao, H. Liu, Y. Wang, et al., "Nanoimprint-defined, large-area meta-surfaces for unidirectional optical transmission with superior extinction in the visible-to-infrared range," *Opt. Express*, vol. 24, p. 15362, 2016.

[90] K. Kim, G. Yoon, S. Baek, J. Rho, and H. Lee, "Facile nanocasting of dielectric metasurfaces with sub-100 nm resolution," *ACS Appl. Mater. Interfaces*, vol. 11, pp. 26109–26115, 2019.

[91] C. A. Dirdal, G. U. Jensen, H. Angelskår, P. C. Vaagen Thrane, J. Gjessing, and D. A. Ordnung, "Towards high-throughput large-area metalens fabrication using UV-nanoimprint lithography and Bosch deep reactive ion etching," *Opt. Express*, vol. 28, p. 15542, 2020.

[92] H.-M. Jin, S.-J. Jeong, H.-S. Moon, et al., "Ultralarge-area block copolymer lithography using self-assembly assisted photoresist pre-pattern," in *IEEE Nanotechnology Materials and Devices Conference (IEEE, 2011)*, 2011, pp. 527–533.

[93] J. Y. Kim, H. Kim, B. H. Kim, et al., "Highly tunable refractive index visible-light metasurface from block copolymer self-assembly," *Nat. Commun.*, vol. 7, p. 12911, 2016.

[94] M. Mayer, M. J. Schnepf, T. A. F. König, and A. Fery, "Colloidal self-assembly concepts for plasmonic metasurfaces," *Adv. Opt. Mater.*, vol. 7, p. 1800564, 2019.

[95] Z. Wu, W. Li, M. N. Yogeesh, et al., "Tunable graphene metasurfaces with gradient features by self-assembly-based moiré nanosphere lithography," *Adv. Opt. Mater.*, vol. 4, pp. 2035–2043, 2016.
[96] M. Khorasaninejad, W. T. Chen, A. Y. Zhu, et al., "Multispectral chiral imaging with a metalens," *Nano Lett.*, vol. 16, pp. 4595–4600, 2016.
[97] N. A. Rubin, G. D'Aversa, P. Chevalier, Z. Shi, W. T. Chen, and F. Capasso, "Matrix Fourier optics enables a compact full-Stokes polarization camera," *Science (80-)*, vol. 365, p. eaax1839, 2019.
[98] H. Pahlevaninezhad, M. Khorasaninejad, Y.-W. Huang, et al., "Nano-optic endoscope for high-resolution optical coherence tomography in vivo," *Nat. Photonics*, vol. 12, pp. 540–547, 2018.
[99] D. Lee, M. Kim, J. Kim, et al., "All-dielectric metasurface imaging platform applicable to laser scanning microscopy with enhanced axial resolution and wavelength selection," *Opt. Mater. Express*, vol. 9, p. 3248, 2019.
[100] M. Jang, Y. Horie, A. Shibukawa, et al., "Wavefront shaping with disorder-engineered metasurfaces," *Nat. Photonics*, vol. 12, pp. 84–90, 2018.
[101] E. Arbabi, J. Li, R. J. Hutchins, et al., "Two-photon microscopy with a double-wavelength metasurface objective lens," *Nano Lett.*, vol. 18, pp. 4943–4948, 2018.
[102] D. Sun, Y. Yang, S. Liu, et al., "Excitation and emission dual-wavelength confocal metalens designed directly in the biological tissue environment for two-photon micro-endoscopy," *Biomed. Opt. Express*, vol. 11, p. 4408, 2020.
[103] J. S. Paiva, P. A. S. Jorge, C. C. Rosa, and J. P. S. Cunha, "Optical fiber tips for biological applications: from light confinement, biosensing to bioparticles manipulation," *Biochim. Biophys. Acta Gen. Subj.*, vol. 1862, pp. 1209–1246, 2018.
[104] V. Westphal, S. O. Rizzoli, M. A. Lauterbach, D. Kamin, R. Jahn, and S. W. Hell, "Video-rate far-field optical nanoscopy dissects synaptic vesicle movement," *Science (80-)*, vol. 320, pp. 246–249, 2008.
[105] J. M. Pullman, J. Nylk, E. C. Campbell, F. J. Gunn-Moore, M. B. Prystowsky, and K. Dholakia, "Visualization of podocyte substructure with structured illumination microscopy (SIM): a new approach to nephrotic disease," *Biomed. Opt. Express*, vol. 7, p. 302, 2016.
[106] F. Wei, D. Lu, H. Shen, et al., "Wide field super-resolution surface imaging through plasmonic structured illumination microscopy," *Nano Lett.*, vol. 14, pp. 4634–4639, 2014.
[107] S. Cao, T. Wang, Q. Sun, B. Hu, U. Levy, and W. Yu, "Graphene on meta-surface for super-resolution optical imaging with a sub-10 nm resolution," *Opt. Express*, vol. 25, p. 14494, 2017.
[108] J. L. Ponsetto, A. Bezryadina, F. Wei, et al., "Experimental demonstration of localized plasmonic structured illumination microscopy," *ACS Nano*, vol. 11, pp. 5344–5350, 2017.
[109] A. Bezryadina, J. Zhao, Y. Xia, X. Zhang, and Z. Liu, "High spatiotemporal resolution imaging with localized plasmonic structured illumination microscopy," *ACS Nano*, vol. 12, pp. 8248–8254, 2018.
[110] Q. Ma, H. Qian, S. Montoya, et al., "Experimental demonstration of hyperbolic metamaterial assisted illumination nanoscopy," *ACS Nano*, vol. 12, pp. 11316–11322, 2018.
[111] A. P. Slobozhanyuk, A. N. Poddubny, A. J. E. E. Raaijmakers, et al., "Enhancement of magnetic resonance imaging with metasurfaces," *Adv. Mater.*, vol. 28, pp. 1832–1838, 2016.
[112] A. V Shchelokova, A. P. Slobozhanyuk, I. V. Melchakova, et al., "Locally enhanced image quality with tunable hybrid metasurfaces," *Phys. Rev. Appl.*, vol. 9, p. 014020, 2018.
[113] R. Schmidt, A. Slobozhanyuk, P. Belov, and A. Webb, "Flexible and compact hybrid metasurfaces for enhanced ultra high field in vivo magnetic resonance imaging," *Sci. Rep.*, vol. 7, p. 1678, 2017.
[114] E. A. Brui, A. V Shchelokova, M. Zubkov, I. V. Melchakova, S. B. Glybovski, and A. P. Slobozhanyuk, "Adjustable subwavelength metasurface-inspired resonator for magnetic resonance imaging," *Phys. Status Solidi Appl. Mater. Sci.*, vol. 215, p. 1700788, 2018.
[115] A. V Shchelokova, A. P. Slobozhanyuk, P. de Bruin, et al., "Experimental investigation of a metasurface resonator for in vivo imaging at 1.5 T," *J. Magn. Reson.*, vol. 286, pp. 78–81, 2018.
[116] Y. Park, C. Depeursinge, and G. Popescu, "Quantitative phase imaging in biomedicine," *Nat. Photonics*, vol. 12, pp. 578–589, 2018.
[117] H. Kwon, E. Arbabi, S. M. Kamali, M. Faraji-Dana, and A. Faraon, "Single-shot quantitative phase gradient microscopy using a system of multifunctional metasurfaces," *Nat. Photonics*, vol. 14, pp. 109–114, 2020.
[118] A. Jayasingh, V. Rompicherla, R. K. N. Radha, and P. Shanmugam, "Comparative study of peripheral blood smear, rapid antigen detection, ELISA and PCR methods for diagnosis of malaria in a tertiary care centre," *J. Clin. Diagn. Res.*, vol. 13, 2019. https:/doi.org/10.7860/JCDR/2019/38213.12483.
[119] T. Moore, A. Moody, T. Payne, G. Sarabia, A. Daniel, and B. Sharma, "In vitro and in vivo SERS biosensing for disease diagnosis," *Biosensors*, vol. 8, p. 46, 2018.
[120] C. L. Wong and M. Olivo, "Surface plasmon resonance imaging sensors: a review," *Plasmonics*, vol. 9, pp. 809–824, 2014.
[121] V. Bochenkov and T. Shabatina, "Chiral plasmonic biosensors," *Biosensors*, vol. 8, p. 120, 2018.
[122] J. R. Mejía-Salazar and O. N. Oliveira, "Plasmonic biosensing," *Chem. Rev.*, vol. 118, pp. 10617–10625, 2018.
[123] Y. Esfahani Monfared, "Overview of recent advances in the design of plasmonic fiber-optic biosensors," *Biosensors*, vol. 10, p. 77, 2020.
[124] A. K. Sharma and C. Marques, "Design and performance perspectives on fiber optic sensors with plasmonic nanostructures and gratings: a review," *IEEE Sens. J.*, vol. 19, pp. 7168–7178, 2019.
[125] X. Yu, S. Zhang, M. Olivo, and N. Li, "Micro- and nano-fiber probes for optical sensing, imaging, and stimulation in biomedical applications," *Photonics Res.* Early Post, 2020. https://doi.org/10.1364/prj.387076.
[126] A. Salim and S. Lim, "Review of recent metamaterial microfluidic sensors," *Sensors*, vol. 18, p. 232, 2018.
[127] A. V. Kabashin, P. Evans, S. Pastkovsky, et al., "Plasmonic nanorod metamaterials for biosensing," *Nat. Mater.*, vol. 8, pp. 867–871, 2009.
[128] R. Singh, W. Cao, I. Al-Naib, L. Cong, W. Withayachumnankul, and W. Zhang, "Ultrasensitive terahertz sensing with high-Q Fano resonances in metasurfaces," *Appl. Phys. Lett.*, vol. 105, p. 171101, 2014.
[129] Z. Li, Y. Zhu, Y. Hao, et al., "Hybrid metasurface-based mid-infrared biosensor for simultaneous quantification and

identification of monolayer protein," *ACS Photonics*, vol. 6, pp. 501–509, 2019.
[130] Y. Zhu, Z. Li, Z. Hao, et al., "Optical conductivity-based ultrasensitive mid-infrared biosensing on a hybrid metasurface," *Light Sci. Appl.*, vol. 7, pp. 1–11, 2018.
[131] F. Yesilkoy, E. R. Arvelo, Y. Jahani, et al., "Ultrasensitive hyperspectral imaging and biodetection enabled by dielectric metasurfaces," *Nat. Photonics*, vol. 13, pp. 390–396, 2019.
[132] A. Tittl, A. Leitis, M. Liu, et al., "Imaging-based molecular barcoding with pixelated dielectric metasurfaces," *Science (80-)*, vol. 360, pp. 1105–1109, 2018.
[133] R. H. Siddique, S. Kumar, V. Narasimhan, H. Kwon, and H. Choo, "Aluminum metasurface with hybrid multipolar plasmons for 1000-fold broadband visible fluorescence enhancement and multiplexed biosensing," *ACS Nano*, vol. 13, pp. 13775–13783, 2019.
[134] R. Ahmed, M. O. Ozen, M. G. Karaaslan, et al., "Tunable fano-resonant metasurfaces on a disposable plastic-template for multimodal and multiplex biosensing," *Adv. Mater.*, vol. 32, p. 1907160, 2020.
[135] Q. Guo, H. Zhu, F. Liu, et al., "Silicon-on-Glass graphene-functionalized leaky cavity mode nanophotonic biosensor," *ACS Photonics*, vol. 1, pp. 221–227, 2014.
[136] L. Jiang, S. Zeng, Z. Xu, et al., "Multifunctional hyperbolic nanogroove metasurface for submolecular detection," *Small*, vol. 13, p. 1700600, 2017.
[137] C. Jack, A. S. Karimullah, R. Leyman, et al., "Biomacromolecular stereostructure mediates mode hybridization in chiral plasmonic nanostructures," *Nano Lett.*, vol. 16, pp. 5806–5814, 2016.
[138] A. Leitis, A. Tittl, M. Liu, et al., "Angle-multiplexed all-dielectric metasurfaces for broadband molecular fingerprint retrieval," *Sci. Adv.*, vol. 5, p. eaaw2871, 2019.
[139] W. Xu, L. Xie, J. Zhu, et al., "Terahertz biosensing with a graphene-metamaterial heterostructure platform," *Carbon N. Y.*, vol. 141, pp. 247–252, 2019.
[140] H.-K. Lee, S. J. Kim, Y. H. Kim, Y. Ko, S. Ji, and J.-C. Park, "Odontogenic ameloblast-associated protein (ODAM) in gingival crevicular fluid for site-specific diagnostic value of periodontitis: a pilot study," *BMC Oral Health*, vol. 18, p. 148, 2018.
[141] O. Limaj, D. Etezadi, N. J. Wittenberg, et al., "Infrared plasmonic biosensor for real-time and label-free monitoring of lipid membranes," *Nano Lett.*, vol. 16, pp. 1502–1508, 2016.
[142] N. Muhammad, Q. Liu, X. Tang, T. Fu, A. Daud Khan, and Z. Ouyang, "Highly flexible and voltage based wavelength tunable biosensor," *Phys. Status Solidi*, vol. 216, p. 1800633, 2019.
[143] S. J. Park, J. T. Hong, S. J. Choi, et al., "Detection of microorganisms using terahertz metamaterials," *Sci. Rep.*, vol. 4, p. 4988, 2015.
[144] C. Zhang, L. Liang, L. Ding, et al., "Label-free measurements on cell apoptosis using a terahertz metamaterial-based biosensor," *Appl. Phys. Lett.*, vol. 108, p. 241105, 2016.
[145] D. Rodrigo, A. Tittl, N. Ait-Bouziad, et al., "Resolving molecule-specific information in dynamic lipid membrane processes with multi-resonant infrared metasurfaces," *Nat. Commun.*, vol. 9, p. 2160, 2018.
[146] Y. Wang, M. A. Ali, E. K. C. Chow, L. Dong, and M. Lu, "An optofluidic metasurface for lateral flow-through detection of breast cancer biomarker," *Biosens. Bioelectron.*, vol. 107, pp. 224–229, 2018.
[147] Z. Geng, X. Zhang, Z. Fan, X. Lv, and H. Chen, "A route to terahertz metamaterial biosensor integrated with microfluidics for liver cancer biomarker testing in early stage," *Sci. Rep.*, vol. 7, p. 16378, 2017.
[148] X. Yan, M. Yang, Z. Zhang, et al., "The terahertz electromagnetically induced transparency-like metamaterials for sensitive biosensors in the detection of cancer cells," *Biosens. Bioelectron.*, vol. 126, pp. 485–492, 2019.
[149] D. Vrba, J. Vrba, D. B. Rodrigues, and P. Stauffer, "Numerical investigation of novel microwave applicators based on zero-order mode resonance for hyperthermia treatment of cancer," *J. Franklin Inst.*, vol. 354, pp. 8734–8746, 2017.
[150] J. N. Anker, W. P. Hall, O. Lyandres, N. C. Shah, J. Zhao, and R. P. Van Duyne, "Biosensing with plasmonic nanosensors," in *Nanoscience and Technology*, UK, Co-Published with Macmillan Publishers Ltd, 2009, pp. 308–319.
[151] K. A. Willets and R. P. Van Duyne, "Localized surface plasmon resonance spectroscopy and sensing," *Annu. Rev. Phys. Chem.*, vol. 58, pp. 267–297, 2007.
[152] P. L. Stiles, J. A. Dieringer, N. C. Shah, and R. P. Van Duyne, "Surface-enhanced Raman spectroscopy," *Annu. Rev. Anal. Chem.*, vol. 1, pp. 601–626, 2008.
[153] C. L. Wong, J. Y. Chan, L. X. Choo, H. Q. Lim, H. Mittman, and M. Olivo, "Plasmonic contrast imaging biosensor for the detection of H3N2 influenza protein-antibody and DNA–DNA molecular binding," *IEEE Sens. J.*, vol. 19, pp. 11828–11833, 2019.
[154] C. L. Wong, M. Chua, H. Mittman, L. X. Choo, H. Q. Lim, and M. Olivo, "A phase-intensity surface plasmon resonance biosensor for avian influenza A (H5N1) detection," *Sensors*, vol. 17, p. 2363, 2017.
[155] C. L. Wong, U. S. Dinish, K. D. Buddharaju, M. S. Schmidt, and M. Olivo, "Surface-enhanced Raman scattering (SERS)-based volatile organic compounds (VOCs) detection using plasmonic bimetallic nanogap substrate," *Appl. Phys. A*, vol. 117, pp. 687–692, 2014.
[156] C. L. Wong, G. C. K. Chen, X. Li, et al., "Colorimetric surface plasmon resonance imaging (SPRI) biosensor array based on polarization orientation," *Biosens. Bioelectron.*, vol. 47, pp. 545–552, 2013.
[157] C. L. Wong, H. P. Ho, K. S. Chan, and S. Y. Wu, "Application of surface plasmon resonance sensing to studying elastohydrodynamic lubricant films," *Appl. Opt.*, vol. 44, p. 4830, 2005.
[158] C. L. Wong, H. P. Ho, K. S. Chan, P. L. Wong, S. Y. Wu, and C. Lin, "Optical characterization of elastohydrodynamic lubricated (EHL) contacts using surface plasmon resonance (SPR) effect," *Tribol. Int.*, vol. 41, pp. 356–366, 2008.
[159] W.-I. K. Chio, W. J. Peveler, K. I. Assaf, et al., "Selective detection of nitroexplosives using molecular recognition within self-assembled plasmonic nanojunctions," *J. Phys. Chem. C*, vol. 123, pp. 15769–15776, 2019.
[160] Y. Lee, S.-J. Kim, H. Park, and B. Lee, "Metamaterials and metasurfaces for sensor applications," *Sensors*, vol. 17, p. 1726, 2017.

[161] S. L. Dodson, C. Cao, H. Zaribafzadeh, S. Li, and Q. Xiong, "Engineering plasmonic nanorod arrays for colon cancer marker detection," *Biosens. Bioelectron.*, vol. 63, pp. 472–477, 2015.
[162] S. Chen, M. Svedendahl, M. Käll, L. Gunnarsson, and A. Dmitriev, "Ultrahigh sensitivity made simple: nanoplasmonic label-free biosensing with an extremely low limit-of-detection for bacterial and cancer diagnostics," *Nanotechnology*, vol. 20, p. 434015, 2009.
[163] J. Yuan, Duan, Luo Yang, and M. Xi, "Detection of serum human epididymis secretory protein 4 in patients with ovarian cancer using a label-free biosensor based on localized surface plasmon resonance," *Int. J. Nanomed.*, vol. 7, pp. 2921–2928, 2012.
[164] E. Ouellet, C. Lausted, T. Lin, C. W. T. Yang, L. Hood, and E. T. Lagally, "Parallel microfluidic surface plasmon resonance imaging arrays," *Lab Chip*, vol. 10, p. 581, 2010.
[165] D. Kim and A. E. Herr, "Protein immobilization techniques for microfluidic assays," *Biomicrofluidics*, vol. 7, p. 041501, 2013.
[166] H. Aouani, M. Rahmani, H. Šípová, et al., "Plasmonic nanoantennas for multispectral surface-enhanced spectroscopies," *J. Phys. Chem. C*, vol. 117, pp. 18620–18626, 2013.
[167] S. Zeng, K. V. Sreekanth, J. Shang, et al., "Graphene–gold metasurface architectures for ultrasensitive plasmonic biosensing," *Adv. Mater.*, vol. 27, pp. 6163–6169, 2015.
[168] S. Zeng, S. Hu, J. Xia, et al., "Graphene–MoS2 hybrid nanostructures enhanced surface plasmon resonance biosensors," *Sens. Actuators B Chem.*, vol. 207, pp. 801–810, 2015.
[169] M. Booth, D. Andrade, D. Burke, B. Patton, and M. Zurauskas, "Aberrations and adaptive optics in super-resolution microscopy," *Microscopy*, vol. 64, pp. 251–261, 2015.
[170] N. Ji, "Adaptive optical fluorescence microscopy," *Nat. Methods*, vol. 14, pp. 374–380, 2017.
[171] M. J. Booth, M. A. A. Neil, R. Juskaitis, and T. Wilson, "Adaptive aberration correction in a confocal microscope," *Proc. Natl. Acad. Sci.*, vol. 99, pp. 5788–5792, 2002.
[172] E. Arbabi, A. Arbabi, S. M. Kamali, Y. Horie, M. Faraji-Dana, and A. Faraon, "MEMS-tunable dielectric metasurface lens," *Nat. Commun.*, vol. 9, p. 812, 2018.
[173] P. Vaiano, B. Carotenuto, M. Pisco, et al., "Lab on fiber technology for biological sensing applications," *Laser Photon. Rev.*, vol. 10, pp. 922–961, 2016.
[174] E. J. Smythe, M. D. Dickey, J. Bao, G. M. Whitesides, and F. Capasso, "Optical antenna arrays on a fiber facet for in situ surface-enhanced Raman scattering detection," *Nano Lett.*, vol. 9, pp. 1132–1138, 2009.
[175] Y.-W. Huang, N. A. Rubin, A. Ambrosio, et al., "Versatile total angular momentum generation using cascaded J-plates," *Opt. Express*, vol. 27, p. 7469, 2019.
[176] R. C. Devlin, A. Ambrosio, N. A. Rubin, J. P. B. Mueller, and F. Capasso, "Arbitrary spin-to-orbital angular momentum conversion of light," *Science (80-)*, vol. 358, pp. 896–901, 2017.
[177] W. T. Chen, M. Khorasaninejad, A. Y. Zhu, et al., "Generation of wavelength-independent subwavelength Bessel beams using metasurfaces," *Light Sci. Appl.*, vol. 6, p. e16259, 2017.
[178] S. Kang, H. E. Joe, J. Kim, Y. Jeong, B. K. Min, and K. Oh, "Subwavelength plasmonic lens patterned on a composite optical fiber facet for quasi-one-dimensional Bessel beam generation," *Appl. Phys. Lett.*, vol. 98, pp. 1–4, 2011.
[179] W. Yu, Z. Ji, D. Dong, et al., "Super-resolution deep imaging with hollow Bessel beam STED microscopy," *Laser Photon. Rev.*, vol. 10, pp. 147–152, 2016.
[180] T. Meinert and A. Rohrbach, "Light-sheet microscopy with length-adaptive Bessel beams," *Biomed. Opt. Express*, vol. 10, p. 670, 2019.
[181] R. Lu, W. Sun, Y. Liang, et al., "Video-rate volumetric functional imaging of the brain at synaptic resolution," *Nat. Neurosci.*, vol. 20, pp. 620–628, 2017.
[182] K. Wu, P. Coquet, Q. J. Wang, and P. Genevet, "Modelling of free-form conformal metasurfaces," *Nat. Commun.*, vol. 9, p. 3494, 2018.
[183] K. Liu, T. Tachi, and G. H. Paulino, "Invariant and smooth limit of discrete geometry folded from bistable origami leading to multistable metasurfaces," *Nat. Commun.*, vol. 10, p. 4238, 2019.
[184] G. Tkachenko, D. Stellinga, A. Ruskuc, M. Chen, K. Dholakia, and T. F. Krauss, "Optical trapping with planar silicon metalenses," *Opt. Lett.*, vol. 43, p. 3224, 2018.
[185] S. Yi, M. Zhou, Z. Yu, et al., "Subwavelength angle-sensing photodetectors inspired by directional hearing in small animals," *Nat. Nanotechnol.*, vol. 13, pp. 1143–1147, 2018.
[186] N. N. Shi, C. C. Tsai, F. Camino, G. D. Bernard, N. Yu, and R. Wehner, "Keeping cool: enhanced optical reflection and radiative heat dissipation in Saharan silver ants," *Science (80-)*, vol. 349, pp. 298–301, 2015.
[187] A. Basiri, X. Chen, J. Bai, et al., "Nature-inspired chiral metasurfaces for circular polarization detection and full-Stokes polarimetric measurements," *Light Sci. Appl.*, vol. 8, p. 78, 2019.
[188] Q. Guo, Z. Shi, Y.-W. Huang, et al., "Compact single-shot metalens depth sensors inspired by eyes of jumping spiders," *Proc. Natl. Acad. Sci. USA*, vol. 116, pp. 22959–22965, 2019.
[189] Z.-B. Fan, Z.-K. Shao, M.-Y. Xie, et al., "Silicon nitride metalenses for close-to-one numerical aperture and wide-angle visible imaging," *Phys. Rev. Appl.*, vol. 10, p. 014005, 2018.
[190] S. Wang, P. C. Wu, V.-C. Su, et al., "A broadband achromatic metalens in the visible," *Nat. Nanotechnol.*, vol. 13, pp. 227–232, 2018.
[191] W. T. Chen, A. Y. Zhu, V. Sanjeev, et al., "A broadband achromatic metalens for focusing and imaging in the visible," *Nat. Nanotechnol.*, vol. 13, pp. 220–226, 2018.
[192] A. Arbabi, E. Arbabi, S. M. Kamali, Y. Horie, S. Han, and A. Faraon, "Miniature optical planar camera based on a wide-angle metasurface doublet corrected for monochromatic aberrations," *Nat. Commun.*, vol. 7, p. 13682, 2016.
[193] Z. Lin, B. Groever, F. Capasso, A. W. Rodriguez, and M. Lončar, "Topology-optimized multilayered metaoptics," *Phys. Rev. Appl.*, vol. 9, p. 044030, 2018.
[194] B. C. Wilson, M. Jermyn, and F. Leblond, "Challenges and opportunities in clinical translation of biomedical optical spectroscopy and imaging," *J. Biomed. Opt.*, vol. 23, p. 1, 2018.
[195] J. Popp, D. Matthews, A. Martinez-Coll, T. Mayerhöfer, and B. C. Wilson, "Challenges in translation: models to

promote translation," *J. Biomed. Opt.*, vol. 23, p. 1, 2017.

[196] B. Spackova, P. Wrobel, M. Bockova, and J. Homola, "Optical biosensors based on plasmonic nanostructures: a review," *Proc. IEEE*, vol. 104, pp. 2380–2408, 2016.

[197] M. A. Otte, B. Sepúlveda, W. Ni, J. P. Juste, L. M. Liz-Marzán, and L. M. Lechuga, "Identification of the optimal spectral region for plasmonic and nanoplasmonic sensing," *ACS Nano*, vol. 4, pp. 349–357, 2010.

[198] J. Homola, Ed., *Surface Plasmon Resonance Based Sensors, Springer Series on Chemical Sensors and Biosensors*, vol. 4, Springer Berlin Heidelberg, 2006.

[199] F. J. Rodríguez-Fortuño, M. Martínez-Marco, B. Tomás-Navarro, et al., "Highly-sensitive chemical detection in the infrared regime using plasmonic gold nanocrosses," *Appl. Phys. Lett.*, vol. 98, p. 133118, 2011.

[200] S. K. Dondapati, T. K. Sau, C. Hrelescu, T. A. Klar, F. D. Stefani, and J. Feldmann, "Label-free biosensing based on single gold nanostars as plasmonic transducers," *ACS Nano*, vol. 4, pp. 6318–6322, 2010.

[201] P. Kvasnička and J. Homola, "Optical sensors based on spectroscopy of localized surface plasmons on metallic nanoparticles: sensitivity considerations," *Biointerphases*, vol. 3, pp. FD4–FD11, 2008.

Giovanna Palermo, Kandammathe Valiyaveedu Sreekanth, Nicolò Maccaferri, Giuseppe Emanuele Lio, Giuseppe Nicoletta, Francesco De Angelis, Michael Hinczewski and Giuseppe Strangi*

Hyperbolic dispersion metasurfaces for molecular biosensing

https://doi.org/10.1515/9783110710687-024

Abstract: Sensor technology has become increasingly crucial in medical research and clinical diagnostics to directly detect small numbers of low-molecular-weight biomolecules relevant for lethal diseases. In recent years, various technologies have been developed, a number of them becoming core label-free technologies for detection of cancer biomarkers and viruses. However, to radically improve early disease diagnostics, tracking of disease progression and evaluation of treatments, today's biosensing techniques still require a radical innovation to deliver high sensitivity, specificity, diffusion-limited transport, and accuracy for both nucleic acids and proteins. In this review, we discuss both scientific and technological aspects of hyperbolic dispersion metasurfaces for molecular biosensing. Optical metasurfaces have offered the tantalizing opportunity to engineer wavefronts while its intrinsic nanoscale patterns promote tremendous molecular interactions and selective binding. Hyperbolic dispersion metasurfaces support high-k modes that proved to be extremely sensitive to minute concentrations of ultralow-molecular-weight proteins and nucleic acids.

Keywords: biosensing; hyperbolic dispersion; metamaterials; metasurfaces.

Giovanna Palermo and Kandammathe Valiyaveedu Sreekanth contributed equally to this work.

***Corresponding author: Giuseppe Strangi,** Department of Physics, Case Western Reserve University, 10600 Euclid Avenue, Cleveland, OH 44106, USA; Department of Physics, University of Calabria, Via P. Bucci, 87036 Rende, CS, Italy; CNR NANOTEC-Istituto di Nanotecnologia, UOS Cosenza, 87036 Rende, CS, Italy, E-mail: Giuseppe.strangi@case.edu
Giovanna Palermo and Giuseppe Emanuele Lio, Department of Physics, University of Calabria, Via P. Bucci, 87036 Rende, CS, Italy; and CNR NANOTEC-Istituto di Nanotecnologia, UOS Cosenza, 87036 Rende, CS, Italy, E-mail: giovanna.palermo@fis.unical.it (G. Palermo). https://orcid.org/0000-0001-5649-735X (G. Palermo)
Kandammathe Valiyaveedu Sreekanth, Centre for Disruptive Photonic Technologies, The Photonic Institute, Nanyang Technological University, Singapore 637371, Singapore
Nicolò Maccaferri, Department of Physics and Materials Science, University of Luxembourg, L-1511 Luxembourg, Luxembourg. https://orcid.org/0000-0002-0143-1510
Giuseppe Nicoletta, Department of Physics, University of Calabria, Via P. Bucci, 87036 Rende, CS, Italy
Francesco De Angelis, Istituto Italiano di Tecnologia, I-16163 Genova, Italy
Michael Hinczewski, Department of Physics, Case Western Reserve University, 10600 Euclid Avenue, Cleveland, OH 44106, USA

1 Introduction

In recent years, a great interest has been spurred by the tremendous technological potential of optically thin nanopatterned surfaces, better known as metasurfaces [1–12]. Metasurfaces allow wavefront engineering, local phase and amplitude control of light along the surface by using dielectric or plasmonic resonators [13–22]. Plasmonic materials are next-generation nanomaterials with enormous potential to transform health care by providing advanced sensors [23–26], imaging devices [27, 28], and therapies [29–32], as well as to advance energy-relevant materials, such as bio-antennae and light-harvesting systems [33–35]. One promising target for plasmonic sensing technologies is the isolation and detection of circulating tumor cells (CTCs), which have received much attention as novel biomarkers in clinical trials of translational cancer research [36–38]. Because CTCs and other cancer-related biomarkers are present in the blood of many patients with cancer, the detection of CTCs can be considered as a real-time liquid biopsy. It has recently been reported that CTCs can leave primary tumors and enter the circulation at a relatively early stage of tumor growth. Therefore, the detection of CTCs is particularly important to study the mechanism of cancer metastasis [39, 40]. The quantification

This article has previously been published in the journal Nanophotonics. Please cite as: G. Palermo, K. V. Sreekanth, N. Maccaferri, G. E. Lio, G. Nicoletta, F. De Angelis, M. Hinczewski and G. Strangi "Hyperbolic dispersion metasurfaces for molecular biosensing" *Nanophotonics* 2021, 10. DOI: 10.1515/nanoph-2020-0466.

and analysis of CTCs and other cancer-related biomarkers in clinical specimens provide a foundation for the development of future noninvasive liquid biopsy techniques [41–44].

Another class of promising biomarkers is exosomes, which could be useful in early diagnosis of various diseases including cancer, and also for the monitoring of cancer progression through noninvasive or minimally invasive procedures [45–47]. Exosomes are nanoscale (30–150 nm) vesicles released by most cells, including cancer cells, and are found in bodily fluids such as blood, urine, and saliva. Exosomes can transfer their cargo containing proteins, lipids, RNA, and DNA to recipient cells and play a key role in different biological process including intercellular signaling, coagulation, inflammation, and cellular homeostasis [48, 49]. With respect to exosomes, the main requirements for next-generation sensors are (i) high levels of specificity to isolate exosomes, (ii) the ability to detect low counts of exosomes (100 target cells per 10^9 blood cells), and (iii) the ability to detect in patient whole blood samples. In this direction, a plasmonic-based, label-free, high-throughput exosome detection approach has been proposed, allowing detection of low counts of exosomes [50–53]. However, label-free detection and quantification of such nanosized objects remains challenging and cumbersome due to the lack of highly sensitive, reproducible, and cost-effective techniques. In particular, the sensitivities of existing optical and plasmonic sensors are not high enough to detect dilute analytes of low molecular weights or low surface coverage of exceedingly small bound molecules [54, 55].

New sensing platforms are needed to address these concerns. To circumvent the limits of plasmonic sensing with traditional materials, research in metamaterials has intensified in the past decade. Metamaterials are a class of engineered materials that do not exist in nature and exhibit exotic and unusual electromagnetic properties that make them attractive for applications in bioengineering and biosensing [56–58]. In particular, metamaterials that show hyperbolic dispersion such as 3D hyperbolic metamaterials (HMMs) and 2D hyperbolic metasurfaces (HMs) have shown extreme sensitivity for low concentrations of smaller bioanalytes [59–62].

HMMs are a class of artificial anisotropic materials, which originates from the concept of optics of crystals. HMM shows hyperbolic dispersion [63, 64] because the out-of-plane dielectric component $\varepsilon_{zz} = \varepsilon_{\perp}$ has an opposite sign to the in-plane dielectric components $\varepsilon_{xx} = \varepsilon_{yy} = \varepsilon_{||}$. As is well known, homogenous isotropic materials exhibit elliptical dispersion, with the dispersion relation $k_x^2 + k_y^2 + k_z^2 = \omega^2/c^2$. However, uniaxial anisotropic materials such as HMMs have a hyperbolic dispersion relation $(k_x^2 + k_y^2)/\varepsilon_{zz} + (k_z^2)/\varepsilon_{xx} = \omega^2/c^2$, where the dielectric response is given by the tensor $\overline{\varepsilon} = [\varepsilon_{xx}, \varepsilon_{yy}, \varepsilon_{zz}]$ [65–67]. This hyperbolic dispersion allows these materials to support waves with infinitely large momentum (bulk plasmon polaritons, BPPs) in the effective medium limit. These waves can propagate inside HMMs, but are evanescent and decay away exponentially in the superstrate [68]. Because the BPP modes are nonradiative with high momentum, they can only be excited using a momentum coupler such as a prism [69] or grating [70], just like in the case of surface plasmon polaritons. It has been shown that the BPP modes of both type I ($\varepsilon_{||} > 0$, $\varepsilon_{\perp} < 0$) and type II ($\varepsilon_{||} < 0$, $\varepsilon_{\perp} > 0$) HMMs can be excited using either coupling technique and the excited modes showed high quality (Q) factor resonances [70]. Because HMMs support high Q-factor multimode BPP resonances, they can be used to develop ultrasensitive multimode biosensors [71]. In recent times, various HMM-based ultrasensitive biosensors using different interrogation schemes have been proposed for noninvasive liquid biopsies [59–61, 72, 73].

Kabashin et al. first proposed to use type I HMM, such as gold nanorods electrochemically grown into a substrate, in biosensing [59]. This plasmonic nanorod HMM excited a bulk guided mode in the near-infrared (NIR) wavelength range using prism coupling and exhibited a record bulk refractive index (RI) sensitivity of 32,000 nm per refractive index unit (RIU). They also demonstrated the real-time binding of small biomolecules such as biotin at a concentration as low as 1 μM. Sreekanth et al. developed a compact multimode biosensor by exciting the BPP modes of an Au/Al_2O_3 HMM using a grating coupling approach [60]. This miniaturized multimode sensor exhibited extreme bulk RI sensitivity with a record figure of merit (FOM) and detected biotin concentrations as low as 10 pM. Since then different hyperbolic dispersion metamaterials have been proposed for sensitivity enhancement purposes [72–82].

One avenue for exploration is using alternative techniques for interrogating the sensor, such as the Goos–Hanchen shift. This has shown to improve the detection limit for small molecules on a 2D hyperbolic metasurface [61]. Another avenue is the development of HMMs free of noble metals, which would also improve the small molecule detection limit because the performance of plasmonic sensors in the visible frequency range is significantly limited by inherent losses in the metallic components. In this direction, a TiN/Sb_2S_3 multilayered HMM-based reconfigurable biosensor has been recently proposed [72]. Plasmonic sensors are also hampered by the limitations of the biomolecular receptors used to target analytes. The receptors can bind to parasitic molecules that cannot be easily distinguished from the targets, which results in a false positive signal. One way around this issue is the use of multiplex detection, realized by developing HMM-based multimode sensors [60] and reconfigurable sensors [72, 83]. In particular, the sensitivity of these sensors can be tuned from maximum to minimum, so that different molecular weight biomolecules including parasitic molecules can be recognized. However, algorithms to analyze proper logic functions based on this concept must be implemented through deep learning or artificial intelligence [9].

In short, significant progress has been made in real-time label-free biosensing of nanoscale targets – such as small molecules and exosomes – at lower concentrations using HMM-based plasmonic biosensor platforms. These in turn can be used to develop cost-effective, noninvasive liquid biopsies for point-of-care (POC) clinical evaluation, early cancer screening and real-time diagnosis of diseases. This review highlights the recent advances in hyperbolic dispersion metasurfaces–based biosensor technology, and in particular discusses biosensors involving metal/dielectric multilayered HMMs. This includes biosensors using different coupling mechanisms and interrogation schemes. We also discuss reconfigurable sensors based on tunable HMMs, localized surface plasmon resonance (LSPR) sensors based on type I HMMs, and biosensing applications of multifunctional hyperbolic nanocavities.

2 Enabling highly tunable engineering of the optical density of states in multifunctional hyperbolic nanocavities for biosensing applications

HMMs of either type I or type II have been investigated in confined in-plane geometries for a variety of applications. In particular, miniaturized optical cavities made of metal-dielectric multilayers have been shown to enable radical increases in the photon density of states. They thus represent an ideal platform to enhance light–matter interactions for applications in optical nanotechnologies [84]. In 2012, Yang et al. reported on nanostructured optical cavities made of indefinite metal-dielectric multilayers that confine the electromagnetic field down to $\lambda/12$ and with ultrahigh (up to 17.4) optical refractive indices [85] (Figure 1).

Their experiments also revealed that these types of cavities display anomalous scaling laws. For instance, cavities with different sizes can be resonant at the same frequency, and higher-order resonant modes oscillate at lower frequencies. These archetypical structures were also proposed to enhance the quantum yield and spontaneous emission of quantum emitters. In 2014, Guclu et al. proposed a radiative emission enhancement mechanism based on the use of hyperbolic nanostructures [86]. In particular, they showed that hyperbolic metamaterial resonators behave like nanoantennas and, owing to their very large density of states, can lead to a 100-fold enhancement of the radiative emission of quantum emitters placed in their vicinity. The same year, Lu et al. demonstrated experimentally enhanced spontaneous emission rates of molecules by using nanopatterned multilayered hyperbolic metamaterials [87] (Figure 2).

They showed that by nanopatterning an HMM of type II made of Ag and Si multilayers, the spontaneous emission rate of rhodamine dye molecules is enhanced 76-fold at tunable frequencies and the emission intensity of the dye increases by a factor 80 compared with the same HMM without nanostructuring. The same group has also

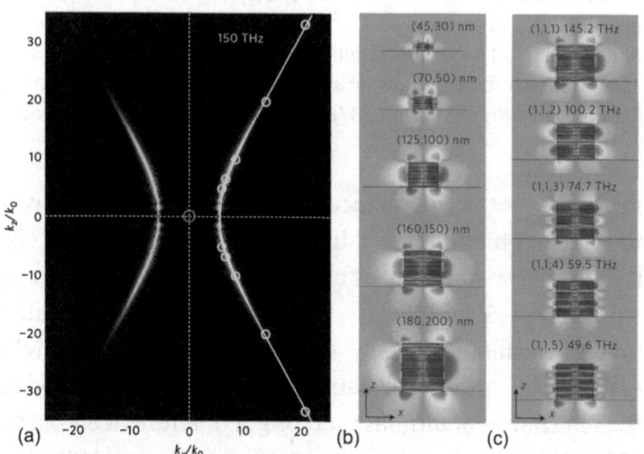

Figure 1: Finite-difference time-domain (FDTD)-calculated iso-frequency contour (IFC) of the multilayer metamaterial and mode profiles of indefinite optical cavities. (a) Cross-sectional view of the hyperbolic IFC for 4 nm silver and 6 nm germanium multilayer metamaterial at 150 THz (bronze curve), which matches the effective medium calculation (white line). The yellow circles represent the resonating wave vectors of the cavity modes shown in b, and the green circle represents the light cone of air. (b) FDTD-calculated electric field (E_z) distributions of the (1, 1, 1) mode for cavities made of 4 nm silver and 6 nm germanium multilayer metamaterial with different size (width, height) combinations but at the same resonant frequency of 150 THz. (c) FDTD-calculated (E_z) distributions of the first five cavity modes along the *z*-direction for the (160, 150) nm cavity. Reproduced with permission [85]. Copyright 2012, Nature Publication Group.

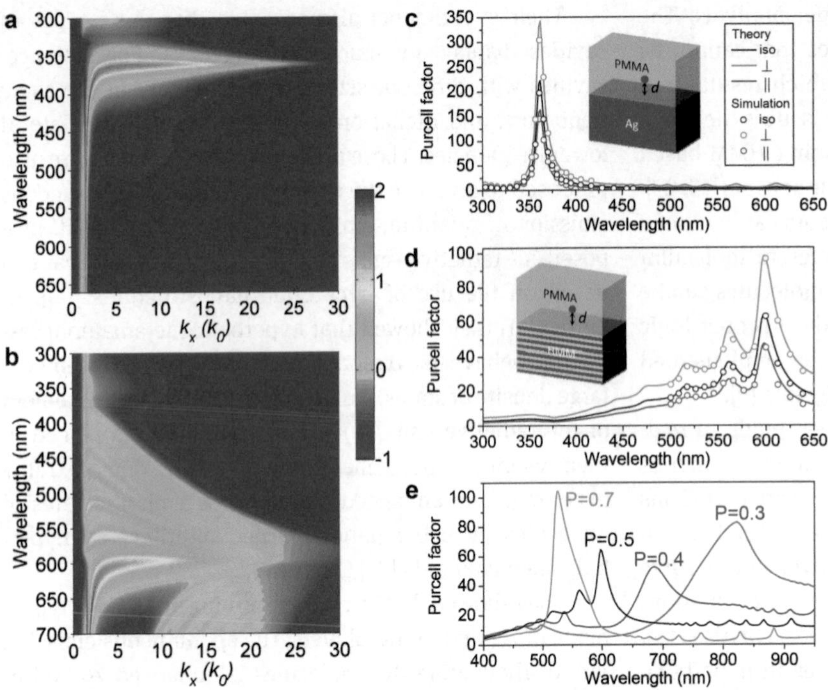

Figure 2: Comparison of Purcell factors for Ag–Si multilayer hyperbolic metamaterials (HMMs) and a pure Ag single layer. a, b, Normalized dissipated power spectra (intensity on a logarithmic scale) for a dipole perpendicular to and at a distance of d = 10 nm above a uniform Ag single layer (a) and a Ag–Si multilayer HMM (b), each with the same total thickness of 305 nm. The multilayer has 15 pairs of Ag and Si layers (each layer thickness is 10 nm). The color scales indicate normalized dissipated power. (c, d) Purcell factor for a dipole located d = 10 nm above the uniform Ag single layer (c) and the Ag–Si multilayer HMM (d), as depicted in the insets. The Purcell factor for isotropic dipoles (iso, black lines) is averaged from that of the dipoles perpendicular (⊥, red lines) and parallel (||, blue lines) to the surface. Corresponding three-dimensional full-wave simulations (open circles) agree with theoretical calculations. (e) Tunable Purcell enhancement across the visible spectra for isotropic dipoles located d = 10 nm above the uniform Ag–Si HMMs by adjusting the volumetric filling ratio of the metal, P. Reproduced with permission [87]. Copyright 2014, Nature Publication Group.

reported a 160-fold enhancement of the spontaneous carrier recombination rates in InGaN/GaN quantum wells thanks to the excitation of hyperbolic modes supported by stacking Ag–Si multilayers, which enable a high tunability in the plasmonic density of states for enhancing light emission at various wavelengths [88]. This approach led to the realization of ultrafast and bright quantum well-based LEDs with a 3-dB modulation bandwidth beyond 100 GHz. More recently, Indukuri et al. proposed nanostructured hyperbolic antennas with an engineered optical density of states and a very large quality-factor/modal volume ratio. Their architecture can enable applications in cavity quantum electrodynamics, nonlinear optics, and biosensors [89]. From a more fundamental point of view, research have been recently focused on layered metal-dielectric HMMs, which support a wide landscape of surface plasmon polaritons and Bloch-like gap-plasmon polaritons with high modal confinement. Light can excite only a subset of these modes, and typically within a limited energy/momentum range if compared with the large set of high-k modes supported by hyperbolic dispersion media, and coupling with gratings [90–92] or local excitation [93] is necessary. Isoniemi et al. recently used electron energy loss spectroscopy (EELS) to achieve a nm-scale local excitation and mapping of the spatial field distribution of bright and dark modes in multilayered type II HMM nanostructures, such as HMM pillars and HMM slot cavities [94] (Figure 3). In particular, as recently demonstrated by Maccaferri et al., HMM nanoantennas can enable a full control of absorption and scattering channels [95].

By varying the geometry of the nanoantennas, the ratio of scattering and absorption and their relative enhancement/quenching can be tuned over a broad spectral range from visible to NIR. Notably, both radiative and nonradiative modes supported by this type of architecture can be excited directly with far-field radiation, even when the radiative channels are almost totally suppressed. Similarly, Song and Zhou showed that multiresonant composite nanoantennas made of vertically stacked building blocks of metal-insulator loop nanoantennas can support multiple nanolocalized modes at different resonant wavelengths, and are thus suitable for multiband operations,

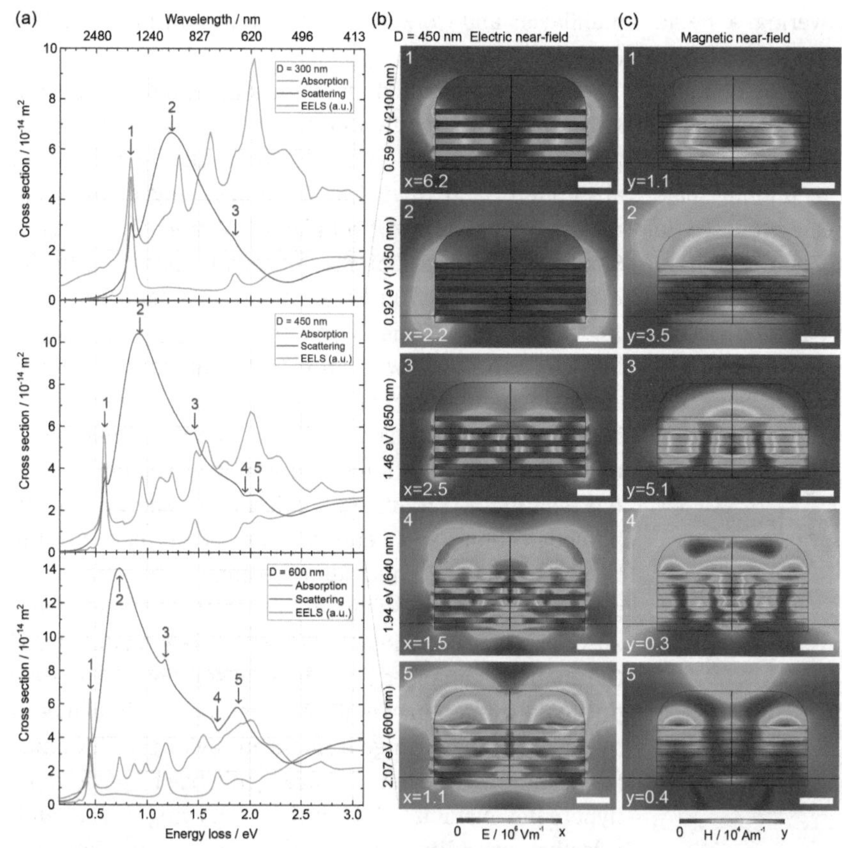

Figure 3: (a) Simulated optical absorption and scattering spectra of hyperbolic metamaterial (HMM) pillars with three different diameters compared with Electron Energy Loss Spectroscopy (EELS) simulations using a vertical beam. Absorption and scattering cross sections are comparable between each other and between pillar sizes, whereas the EEL spectra are comparable only between pillar sizes, using arbitrary units. (b) Electric and (c) magnetic near-field intensities in the 450 nm pillar using plane wave excitation with marked energies. The plane wave propagates from the top, along the axis of symmetry of the pillar. The corresponding resonances are marked with numbered arrows in a, and resonances with similar field profiles for other pillar diameters are marked with the same number. The maximum value of the color scale is marked at lower right in each plot. Scale bars = 100 nm. Reproduced with permission [94]. Copyright 2020, Wiley-VCH.

such as multiphoton processes, broadband solar energy conversion, and wavelength multiplexing [96] (Figure 4).

Furthermore, these hyperbolic nanoantennas have proven to possess both angular and polarization-independent structural integrity, unlocking promising applications as solvable nanostructures. In this framework, Wang et al. experimentally synthetized and measured the optical properties of hyperbolic nanoparticles made with either Au or dielectric nanoobjects coated with alternating SiO_2 and Au multishells [97]. These soluble nanoparticles

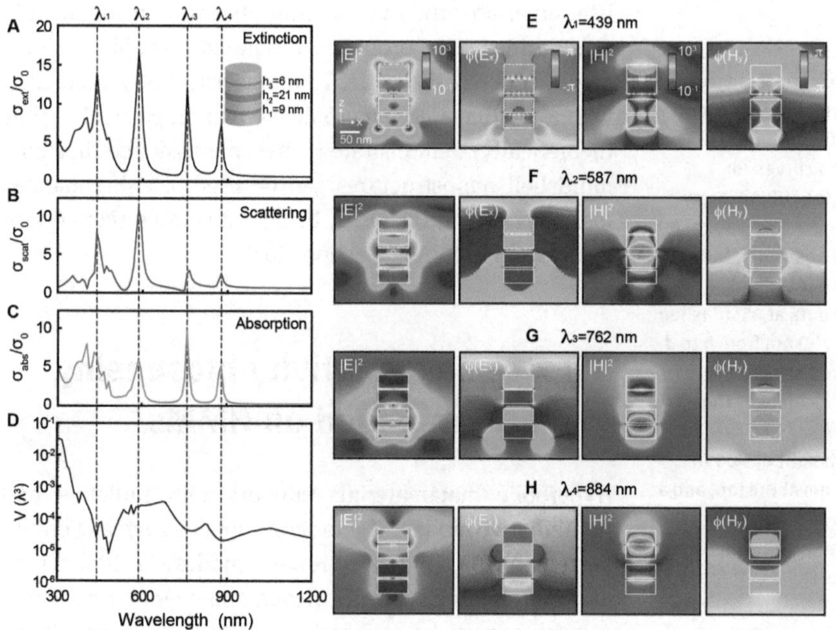

Figure 4: Far-field and near-field optical properties of multiresonant composite nanoantennas. FDTD-calculated spectra of (a) normalized extinction cross-section σ_{ext}/σ_0, (b) normalized scattering cross-section σ_{sca}/σ_0, (c) normalized absorption cross-section σ_{abs}/σ_0, and (d) normalized mode volume V_m/λ^3 for a metal–insulator–metal–insulator–metal–insulator–metal (MIMIMIM) composite nanoantenna with three dielectric layers under linear polarized plane wave excitation at normal incidence angle. (e–h) FDTD-calculated distribution maps of electric field intensity $|E|^2$, phase of in-plane electric field $\Phi(E_x)$, magnetic field intensity $|H|^2$, and phase of magnetic field $\Phi(H_y)$ for MIMIMIM composite nanoantenna at resonant wavelengths of (e) 439 nm, (f) 587 nm, (g) 762 nm, and (h) 884 nm. Reproduced with permission [96]. Copyright 2018, American Chemical Society.

enable highly tunable optical modes covering a broad wavelength range. Moreover, they exhibit high local field intensity enhancement, thus opening excellent opportunities in plasmon-enhanced spectroscopy, nanolasers, design of nonlinear phenomena, photothermal conversions, and hot-electron generation. It is worth mentioning here that nanostructured HMMs have also been proposed to achieve a near-perfect absorption of light. Zhou et al. have proposed a broadband absorber based on a tapered a hyperbolic metamaterial, made of alternating metal-dielectric multilayers and working in the visible and NIR ranges [98] (Figure 5).

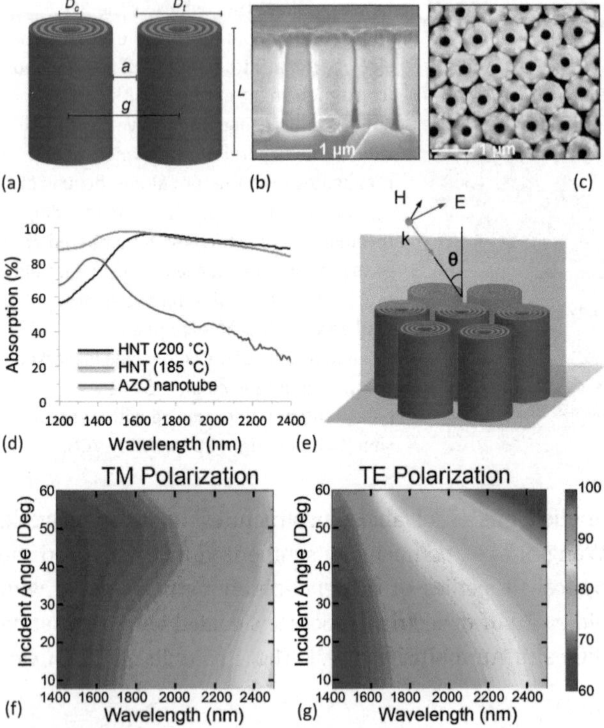

Figure 5: Experimental realization of the hyperbolic metamaterial (HMM) broadband absorber targeting the visible and IR range. (a) Sketch of an array of metal-dielectric multilayered tapers under a transverse magnetic (TM) wave incidence from the top. (b–d) Measured (black) and simulated (red) absorption curves for broadband absorbers with SEM images (inset): (b) 700 nm array period, 9-stack structure (Au–Al_2O_3) absorbing from 1.5 to 3 μm; (c) 700 nm array period, 11-stack structure (Au–Ge) absorbing from 2.5 to 6 μm; (d) 220 nm period, three-stack structure (Al–SiO_2) absorbing from ≈0.4 to 1.2 μm, including angled data at 45° (dashed blue curve). Scale bars represent 500, 500, and 200 nm from b to d. (e) Simulated absorption spectra of a multilayered structure and its effective medium counterpart under a TM wave incidence. The multilayered structure is an array of Ag (20 nm)/Ge (30 nm) multilayered, tapered HMM waveguides with a height of 900 nm, a tapering width from 500 nm at the bottom to 182 nm at the top, and a period of 700 nm. (f) Sketch of the isofrequency contour (IFC) of the multilayered structure (blue) compared with that of the effective medium (red). The dashed blue curve represents the IFC of the multilayered structure at a smaller wavelength. Reproduced with permission [101]. Copyright 2017, National Academy of Sciences.

They demonstrated that light couples into the tapered structure most strongly due to the hyperbolic dispersion phase-matching conditions. Moreover, they obtain a broad absorption band thanks to the broadening of the resonances from an array of coupled HMM tapered structures. Similarly, Sakhdari et al. theoretically propose a vertically integrated hot-electron device based on a nanostructured HMMs that can efficiently couple plasmonic excitations into electron flows, with an external quantum efficiency approaching the physical limit [99]. Their metamaterial-based architecture displays a broadband and omnidirectional response at infrared and visible wavelengths, thus representing a promising platform for energy-efficient photodetection and energy harvesting beyond the bandgap spectral limit. More recently, Abdelatif et al. proposed a funnel-shaped anisotropic metamaterial absorber made of periodic array of nickel–germanium (Ni/Ge) enabling enhanced broadband absorption (up to 96%) due to the excitation of multiple orders of slow-light modes over a wavelength range from the ultraviolet to NIR range [100]. In 2017, Riley et al. have synthetized transferrable hyperbolic metamaterial particles showing broadband, selective, omnidirectional, perfect absorption. Their system is made of hyperbolic nanotubes (HNTs) on a silicon substrate that exhibit near-perfect absorption at telecommunication wavelengths even after being transferred to a mechanically flexible, visibly transparent polymer, which is a more desirable substrate in view of mechanically flexible and low-cost applications [101] (Figure 6).

Finally, Caligiuri et al. presented an approach to achieve super-absorption capabilities based on resonant gain singularities. The proposed mechanism enables a huge amplification of the emitted photons resonantly interacting with a multilayered hyperbolic system. In particular, they theoretically demonstrated that metal/doped-dielectric multishell nanostructures can be used as self-enhanced loss compensated devices, being a favorable scenario for low-threshold SPASER action [102].

3 Extreme sensitivity biosensing platform based on HMMs

Hyperbolic metamaterials support both radiative and nonradiative modes. The radiative modes such as Ferrell–Berreman [103, 104] and Brewster modes [75, 105] can be excited from free space. However, a momentum-matching condition must be satisfied to excite HMM nonradiative

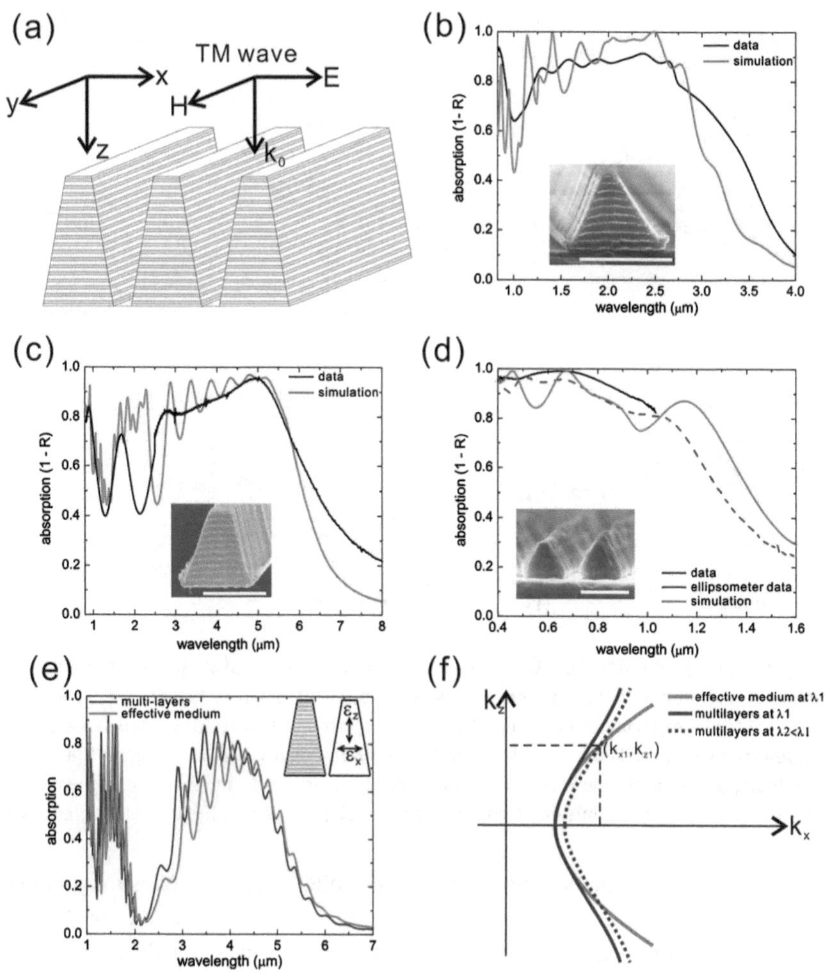

Figure 6: (a) Scanning electron microscopy (SEM) of coupled hyperbolic nanotubes (HNTs). SEM images of the HNT arrays as viewed in the plane (b) parallel and (c) perpendicular to the nanotube axis. (d) Absorption spectra of HNT arrays deposited at a temperature of 185 and 200 °C along with a spectrum of a pure aluminum-doped zinc oxide (AZO) nanotube array deposited at 200 °C. (e) Schematic of the incident radiation at angle θ showing TM polarization. Wide-angle absorption spectra for (f) TM and (g) TE polarizations of an HNT array deposited at 200 °C. The color corresponds to the percentage absorption. Reproduced with permission [98]. Copyright 2014, American Chemical Society.

modes, such as SPP and BPP modes. A phase-sensitive HMM-based biosensor can be developed by exciting the Brewster mode of the HMM because a sudden phase jump occurs at the Brewster angle. In this scheme, the phase shift at the Brewster angle due to the change in RI arising from the presence of the analyte can be recorded as the sensing parameter [75]. However, extreme sensitivity biosensing is only possible by exciting the BPP modes of the HMM. A type II HMM supports only high-k BPP modes, whereas a type I HMM supports both high-k and low-k BPP modes. In the following sections, we discuss the development of multilayered HMM-based biosensors based on metasurface grating and prism coupling excitation schemes.

3.1 Grating-coupled HMM-based multimode biosensor

In this section, the excitation of BPP modes of type II HMMs using grating coupling and its application for the development of biosensors with differential sensitivity is reported. Wavelength and angular interrogation schemes to demonstrate the extreme sensitivity of the biosensor for small molecule detection at low analyte concentrations have been investigated [73, 106]. The HMM-based plasmonic biosensor platform, with integrated microfluidics, is illustrated in Figure 7a and b. As can be seen, it is a combination of a plasmonic metasurface and an HMM. The HMM consists of 16 alternating thin layers of gold and aluminum dioxide (Al_2O_3) with thickness 16 and 30 nm, respectively. The fabricated multilayer is a type II HMM, which shows hyperbolic dispersion at $\lambda \geq 520$ nm, where the real parts of parallel and perpendicular permittivity components are negative and positive, respectively (Figure 7c). To excite the BPP modes of Au–Al_2O_3 HMM, the grating coupling technique has been used [70, 107, 108]. For this purpose, a 2D subwavelength Au diffraction grating with period 500 nm and hole dimeter 160 nm was integrated with the HMM (inset of Figure 7a). Au has been considered because it is the most popular plasmonic

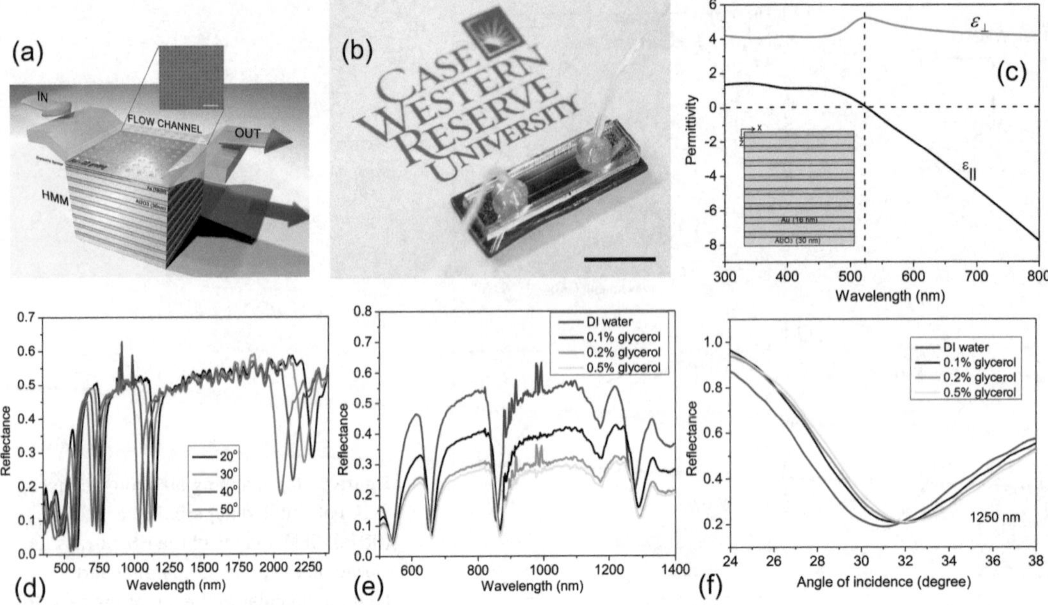

Figure 7: (a) A schematic diagram of the fabricated grating-coupled HMM (GC-HMM)-based sensor device and an SEM image of the 2D subwavelength Au diffraction grating on top the HMM. (b) A photograph of the sensor device. Scale bar = 10 mm. (c) Calculated real parts of effective permittivity for an HMM consisting of eight pairs of gold/Al_2O_3 layers determined using effective media theory. (d) Excited BPP modes of the GC-HMM at different angles of incidence. Reproduced with permission from Sreekanth et al. [60], Copyright 2016, Springer Nature Publication Group. (e, f) Standard sensor calibration tests, injecting different weight percentages of glycerol (0.1–0.5% w/v), using wavelength interrogation (e) and angular interrogation (f). Reproduced with permission from Sreekanth et al. [73], Copyright 2017, EDP Sciences.

material for biosensing applications because of its low oxidation rate and high biocompatibility.

The principle behind grating coupling is that the surface plasmons supported by a metasurface grating can be excited when the parallel wavevectors of the surface plasmons are comparable with the wavevector of the light. Grating diffraction orders are no longer propagating waves when this condition is satisfied, but they are evanescent fields. The enhanced wavevector of the evanescent field is responsible for the coupling of incident light to the surface plasmon modes [109, 110]. The proposed metasurface grating-coupled HMM (GC-HMM)-based sensor works based on the coupling condition between the metasurface grating modes and BPP modes. The coupling condition alters when the RI of the surrounding medium changes, which allows measuring a change in resonance wavelength and resonance angle. Explicitly, the grating coupling condition is given by $k_{SPP} = n_0 k_0 \sin\theta \pm m k_g$, where θ is the incident grazing angle, n_0 is the RI of surrounding medium, $k_0 = 2\pi/\lambda$ is the vacuum wavevector, m is the grating diffraction order, and $k_g = 2\pi/\Lambda$ is the grating wavevector with Λ being the grating period.

To show the BPP modes of the Au–Al_2O_3 HMM, the reflectance spectra of the GC-HMM with varying incidence angle, have been acquired using a variable angle high-resolution spectroscopic ellipsometer. The Q factors of these modes are remarkably high because of strong mode confinement and large modal indices. In Figure 7d, the spectral response of the fabricated sensor at different angles of incidence has been shown. The narrow modes recorded at wavelengths above 500 nm represent highly confined fundamental and higher-order BPP modes of the HMM. The Q-factors of the modes measured at resonance wavelengths 1120, 755, and 580 nm are 29.5, 26, and 23, respectively. The sensor supports many modes with different Q-factors, which increases with increasing mode resonance wavelength.

The performance of the HMM-based sensor using spectral and angular interrogation schemes has been analyzed. The detection limit of the sensor has been first determined by injecting different weight ratios of aqueous solutions of glycerol into the sensor microchannel (sample volume = 1.4 μl). In Figure 7e and f, the reflectance spectra of the sensor with varying concentrations of glycerol in distilled water (0.1–0.5% w/v) have been shown, using spectral and angular scans, respectively. It is evident from Figure 7e that the resonance wavelength corresponding to each BPP mode red shifts and the quality factor of each mode declines with increasing glycerol concentration. In the angular scan, a positive angular shift is obtained when the glycerol weight ratio is increased (Figure 7f). The sensor can record extremely small RI changes of glycerol concentrations in both scans. For example, a significant shift of 12 nm is obtained at 1300 nm even with 0.1% w/v glycerol concentration. It should be noted that the shift increases when the spectral position of the BPP mode varies from visible to NIR wavelengths because the transverse decay of the field in the superstrate strongly varies from one mode to another. Thus the sensor provides different sensitivities as the resonance wavelength increases from visible to NIR wavelengths due to the differential response of the BPP modes. The measured maximum bulk RI sensitivity for the longest wavelength BPP mode in spectral and angular scans is around 30,000 nm/RIU and 2500°/RIU, respectively. The minimum sensitivity is recorded at the shortest wavelength BPP mode, which is 13,333 nm/RIU and 2333°/RIU for spectral and angular scans, respectively. In addition, the sensor exhibits different FOM for each BPP mode, which are 206, 357, 535 and 590 at 550, 660, 880, and 1300 nm, respectively. It is worth noting that the recorded FOMs of the proposed multilayered HMM-based sensor are much higher than existing plasmonic biosensors. An interesting characteristic of the proposed sensor is the flexibility in the selection of a mode for the identification of specific biomolecules since it provides differential sensitivities.

To demonstrate the capabilities of the proposed sensor device for the detection of smaller biomolecules, biotin (molecular weight, 244 Da) has been selected as the analyte because it is a model system for small molecule compounds such as vitamins, cancer-specific proteins, hormones, therapeutics, or contaminants such as pesticides or toxins. To selectively detect biotin, the sensor surface has been immobilized with streptavidin biomolecules and the sensor monitors the wavelength or angular shifts due to the RI change caused by the capture of biotin at the streptavidin sites. The sensor performance is monitored by injecting different concentrations (10 pM–10 μM) of biotin prepared in PBS into the sensor microchannel. The reflectance spectra of the sensor have been recorded after a reaction time of 40 min for each concentration of biotin. PBS was introduced into the microchannel to remove the unbound and weakly attached biotin molecules before each injection of a new concentration of biotin. In Figure 8a and b, the responses of the device during the detection of different concentrations of biotin in spectral and angular scans have been plotted respectively. One can see that an increase in spectral and angular shift is obtained with increasing biotin concentrations. It reflects the increase in RI change due to the capture of biotin molecules on the streptavidin sites. Moreover, a nonlinear variation in shift with increase in biotin concentration is observed in both scans. In Figure 8c, the real-time binding kinetics by

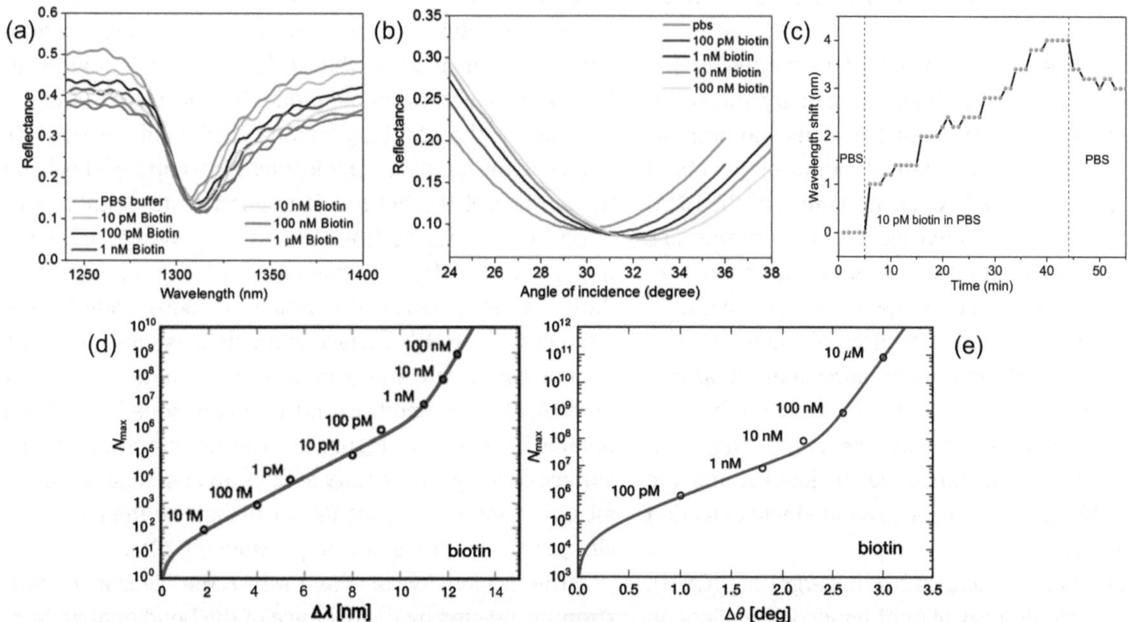

Figure 8: Detection of biotin using the grating-coupled HMM (GC-HMM)-based sensor. (a) Sensor reflectance spectra as a function of wavelength with different concentrations of biotin in PBS at 30° angle of incidence. (b) Sensor reflectance spectra as a function of incident angle with different concentrations of biotin in PBS at wavelength 1250 nm. (c) Demonstration of real-time binding of biotin by injecting 10 pM biotin in PBS (in the time interval between the two dotted lines). (d, e) For the mode located at 1280 nm, the maximum number of biotin molecules adsorbed in the illuminated sensor area versus the corresponding wavelength shift $\Delta\lambda$ (d) and versus the corresponding angular shift $\Delta\theta$ (e) at different biotin concentrations. Reproduced with permission from Sreekanth et al. [73], Copyright 2017, EDP Sciences.

injecting 10 pM biotin has been shown, where a red shift has been observed and discrete steps in resonance wavelength over time, which is due to the 0.2 nm discreteness in the wavelength sensitivity. Variability in the step size was observed, which is due to statistical fluctuations in which larger or smaller numbers of binding events can occur. Finally, the sensitivity of the wavelength and angular shift to the number of adsorbed biotin molecules on the sensor surface have been investigated. The saturation values of the wavelength ($\Delta\lambda$) and angular ($\Delta\theta$) shift for each concentration c of biotin in PBS were considered. Because the shift of the resonance wavelength and angle depends on the number of bound molecules, it is possible to reliably estimate an upper bound N_{max} (c) based on the sensor parameters. In Figure 8d and e, the number of adsorbed biotin molecules with corresponding shifts in wavelength and angle has been plotted, respectively. This behavior is accurately reproduced using a phenomenological double-exponential fitting function [60]. In contrast to type I HMM-based sensor [59], the proposed multilayered type II HMM-based sensor demonstrates the detection of 10 pM biotin in PBS, which shows that sensitivity is increased by six orders of magnitude.

3.2 Biomolecular sensing at the interface between chiral metasurfaces and HMMs

The interface between a chiral metasurface and hyperbolic metamaterials can enable both high sensitivity and specificity for low-molecular-weight nucleic acids and proteins [111–115]. Interestingly, an adapted out-of-plane chiral metasurface enables three key functionalities of the HMM sensor: (i) an efficient diffractive element to excite surface and bulk plasmon polaritons [76, 109]; (ii) an increase in the total sensing surface enabled via out-of-plane binding, improving diffusion-limited detection of small analyte concentrations; and (iii) additional biorecognition assays via circular dichroism (CD) and chiral selectivity that can be optimally tailored to amplify the chiral–chiral interactions between the metamaterial inclusions and the molecules, enabling high-sensitivity handedness detection of enantiomers [116–120].

A sketch of a chiral metasurface hypergrating (CMH), consisting of a periodic array of right-handed Au helices on a type II HMM composed by indium tin oxide (ITO – 20 nm) and silver (Ag – 20 nm) is shown in Figure 9a. From top to bottom, the geometry consists of a superstrate containing the Au helices, the multilayer system (gray stack of ITO/Ag), and the glass substrate. The reflectance and transmittance curves are calculated for a transverse magnetic (TM) wave and an incident angle $\theta_i = 50°$ with respect to the helix axes. This is done by solving the frequency-domain partial differential equation that governs the **E** and **H** fields associated with the electromagnetic wave propagating through the structure [121]. As shown in Figure 9b, for the calculated reflectance for the CMH–HMM, it is possible to distinguish three minima at 635 (mode A), 710 (mode B), and 890 nm (mode C), corresponding to different BPP modes of the underlying HMM, whereas the transmittance remains zero in the entire spectral region.

The sensitivity of the sensing platform can be evaluated as a function of the analyte concentration binding at the chiral metasurface. To this end, different molar fractions of an aqueous solution of 1,2,3-propantriol, characterized by an ultralow molecular weight ($C_3H_8O_3 \approx 60$ Da) [122], have been considered. Figure 9b shows the calculated reflection spectra of the CMH–HMM sensor in measuring 1,2,3-propantriol solutions with different concentrations. As expected, the calculated reflectance minima (mode dips) of the coupled system linearly shift toward longer wavelengths with the increase of RI – from 1.333 (water) to 1.401 (molar fraction of $C_3H_8O_3$ of about 17%). The corresponding limit of detection (LOD) is equal to 1.5×10^{-4} RIU.

An important aspect of the 3D chiral metasurface is the significant increase of the out-of-plane sensing surface because the specific binding of the analytes can occur on the entire helical surface. At the same time, a chiral structure can modify the fluid dynamics around it, inducing an increase of the probability of specific binding.

Clearly, the wavelength shift of BPP modes is strongly related to the quantity of molecules that bind selectively on the surface of the helices. It is possible to quantify this effect by considering different surface coverage of the helix, by considering the maximum RI change equals to 0.068. Coverage was varied from 0 to 100%, where in the latter case the whole surface of the helix is totally covered by molecules; here a maximum spectral shift of 15 nm for mode A, 21 nm for mode B, and 31 nm for mode C have been obtained, as seen in Figure 9c. The minimum detectable surface coverage, necessary to have an appreciable shift of all three modes, is about 16%, whereas for the most sensitive mode (C) alone, it is approximately 12%.

The sensitivity of the CMH–HMM sensor is also strongly affected by the distance of the bond analyte from the HMM surface. In particular, the local change of RI in a small disk surrounding the helix produces appreciable shifts even when the binding is confined exclusively to the upper region of the helix, which represents the maximum distance from the HMM. In Figure 9d a small disk, with $n = 1.401$ and corresponding to an adsorbed surface of 20%,

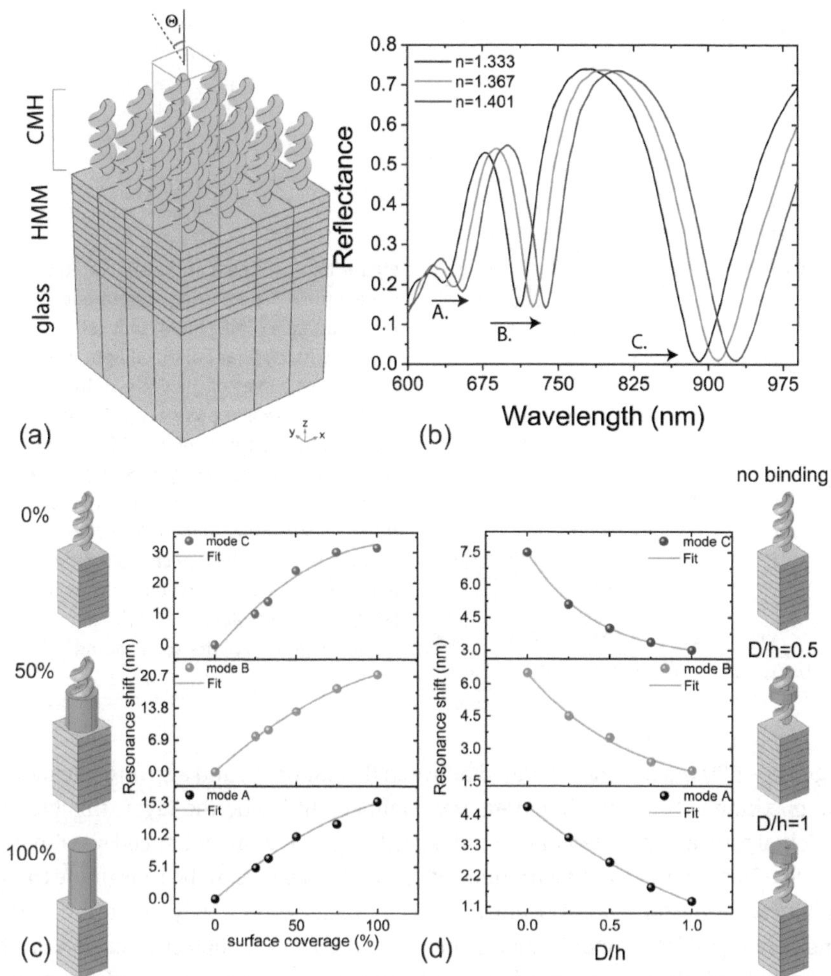

Figure 9: (a) Unit cell of the chiral metasurface hypergrating–hyperbolic metamaterials (CMH–HMM) simulated geometry. (b) Calculated TM-polarization reflectance spectra of an Au helix array on the HMM, with water as the surrounding medium, and angle of incidence $\theta_i = 50°$. The spectrum for pure water is shown in black, whereas red and blue curves correspond to two different mole fractions of 1,2,3-propantriol in distilled water. (c) Sketches of the simulated geometry with different percentages of the helix surfaces covered by bound analytes, and the corresponding resonance wavelength shift for the BPP modes as a function of the surface coverage. (d) Sketch of the geometry and resonance wavelength shift for mode A, B, and C with surface coverage of 20%, but with all analytes bound within a narrow disk at different distances D away from the HMM surface. D is normalized relative to the helix height h.

is positioned at different distances (D) with respect to the surface of the HMM. By plotting the resonance shift of the three modes A, B, and C as a function of the distance D normalized to the helix height (h) it is possible to see that the proposed biosensing platform is able to detect a shift of the considered modes even in the worst case ($D/h = 1$). For this case, the shift is about 1.3 nm for the mode A, 4.0 nm for the mode B, and 4.5 nm for the mode C, as reported in Figure 9d. These results highlight the advantages of having a metasurface to promote the detection of target analytes away from the surface of the HMM, exploiting the increased surface/volume ratio of the 3D CMH exposed to the analytes.

On the other hand, considering the intrinsic chirality of the nanohelices, CMH could excite new BPP modes of the HMM by coupling with their circular polarization-dependent plasmon modes. Indeed, as reported in Figure 10b, different reflectance dips are obtained for left-handed circular polarized (LCP) and right-handed circular polarized (RCP) light. In particular, the reflectance dips obtained for LCP light, indicated in the figure as BPP_3 to BPP_6, are strongly related to a coupling between the plasmonic modes of the gold helix array and the HMM as reported in Figure 10c.

From the chiroptical response of the CMH, it is possible to calculate the CD signal, such as bipolar peaks and crossing points, as seen in Figure 10d. These signatures allow for increased sensitivity and accuracy when monitoring RI changes due to the analyte absorption. This sensing modality offers strong optical contrast even in the presence of highly achiral absorbing media, increasing the signal-to-noise CD measurements of a chiral analyte, relevant for complex biological media with limited transmission [123]. For this purpose, the reflectance curves obtained for LCP and RCP light at $\theta_i = 75°$ are used to calculate the reflectance circular dichroism (RCD) spectra, which characterizes the reflectance difference between LCP and RCP light, leading to an RCD amplitude ($RCD = R_{LCP} - R_{RCP}$). As expected, the RCD spectra for the CMH–HMM exhibit multiple features: different maxima, minima, and crossing points (Figure 10d) that are strongly affected by the RI variation (from 1.333 to 1.401).

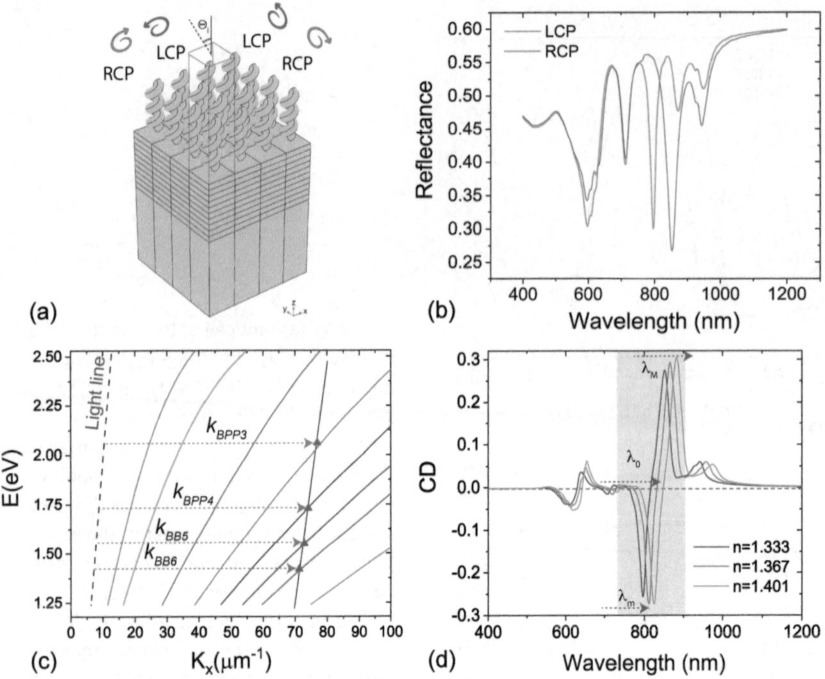

Figure 10: (a) Sketch of the simulated chiral metasurface hypergrating–hyperbolic metamaterials (CMH–HMM) unit cell probed with circular polarized light. (b) Reflectance curves of the CMH–HMM for left-handed circular polarized (LCP) and right-handed circular polarized (RCP) light, with angle of incidence $\theta_i = 75°$. (c) Modal dispersion curves of the HMM, with blue triangles indicating BPP modes (BPP_3 through BPP_6) excited by circular polarized light. (d) Reflectance circular dichroism (RCD) versus wavelength at different refractive indices of the surrounding medium.

It is possible to distinguish in the range 700–900 nm a minimum (λ_m), a crossing point (λ_0) and a maximum (λ_M), respectively, at 797, 816, and 852 nm which significantly shift and modify their intensity as the RI changes. These signals show a chiral plasmon (CP) sensitivity $S_{CP} = \Delta\lambda/\Delta n$ of 412 nm/RIU for λ_m, 485 nm/RIU for λ_0 and 471 nm/RIU for λ_M, respectively. After extracting the classical full-width at half-maximum (FWHM) for the λ_m and λ_M modes and the FWHM in the |RCD| spectrum for λ_0, the FOM, defined as FOM = S_{CP}/FWHM [124], has been calculated. The corresponding FOM values are 18, 20, and 30. The scientific significance of chiral metasurfaces coupled with HMM nanostructures for biosensing lies in the synergistic functionalities which arise from chiral geometries and optical selectivity. This will have important implication in developing next-generation biosensors for gene–protein recognition.

4 Type I HMM-based biosensor

4.1 Type I HMMs: gold nanopillars for high-sensitivity LSPR sensors

Kabashin et al. demonstrated an improvement in biosensing technology using a plasmonic metamaterial that is capable of supporting a guided mode in a porous nanorod layer [59]. Benefiting from a substantial overlap between the probing field and the active biological substance incorporated between the nanorods and a strong plasmon-mediated energy confinement inside the layer, this type I metamaterial (Figure 11c) provides an enhanced sensitivity to RI variations of the medium between the rods (more than 30,000 nm/RIU). They focused on newly emerging plasmonic metamaterials, composite structures consisting of subwavelength-size components, and designed a sensor-oriented metamaterial that is capable of supporting similar or more sensitive guiding modes than SPPs. In particular, they reported on the experimental realization of such a metamaterial-based transducer using an array of parallel gold nanorods oriented normal to a glass substrate (Figure 11a and b). When the distance between the nanorods is smaller than the wavelength, this metamaterial layer supports a guided mode with the field distribution inside the layer determined by plasmon-mediated interaction between the nanorods. This anisotropic guided mode has resonant excitation conditions similar to the SPP mode of a smooth metal film and a large probe depth (500 nm). The structural parameters can be controlled by altering the fabrication conditions – typical ranges span rod lengths of 20–700 nm, rod diameters 10–50 nm, and separations 40–70 nm – thus achieving a nanorod areal density of approximately 10^{10}–10^{11} cm^{-2}. The lateral size and separations between the nanorods are much smaller than the wavelength of light used in the experiments, so only average values of nanorod assembly parameters are important, and individual nanorod size deviations have no influence on the optical properties that are well described

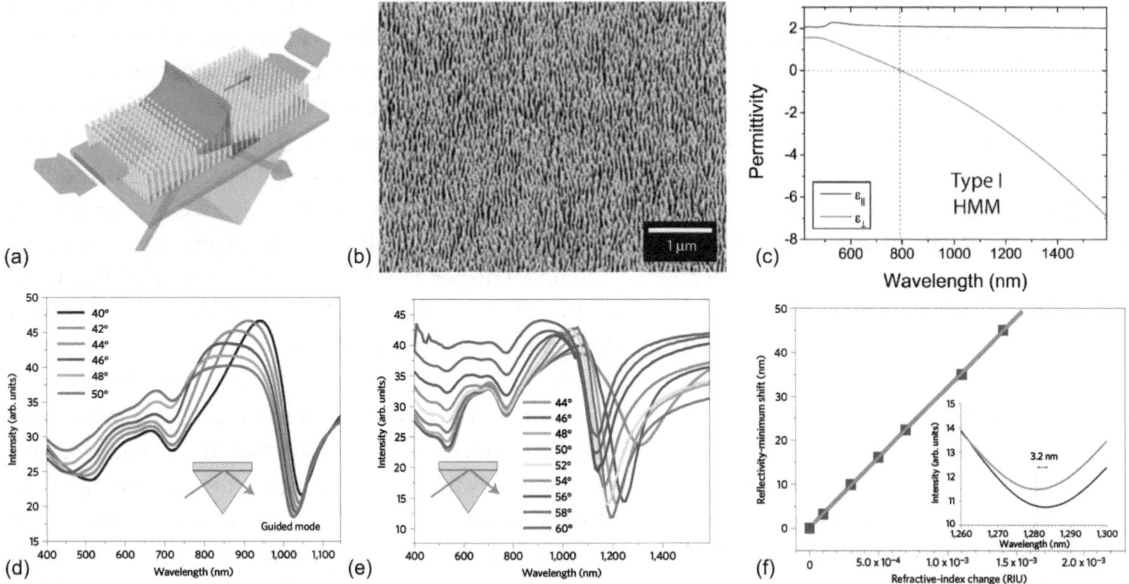

Figure 11: (a) Schematic of the gold nanopillars array and (b) scanning electron micrograph of the nanorod assembly. (c) Components of the effective permittivity tensor of the nanorod array in a water environment calculated using effective-medium theory components of the permittivity along the nanorods ($\varepsilon_{||}$) and perpendicular to nanorods ($\varepsilon_{\perp}$). (d) Reflection spectra of the nanorod array in an air ($n = 1$) (e) and water environment ($n = 1.333$), obtained in the attenuated total internal reflection (ATR) geometry for different angles of incidence. (f) Calibration curve for the metamaterial-based sensor under the step-like changes of the refractive index of the environment using different glycerine–water solutions. The measurements were carried out at a wavelength of 1230 nm. The size of the squares represents error bars. Inset: Reflectivity spectrum modifications with the changes of the refractive index by 10^{-4} RIU. Reproduced with permission [59]. Copyright 2009, Nature Publication Group.

by effective medium theory [125]. The nanorod structures' sensing properties were characterized in both direct transmission and attenuated total internal reflection (ATR) geometries using spectrometric and ellipsometric platforms.

In the direct-transmission geometry, the extinction spectra of the nanorod array in an air environment show two pronounced peaks at 520 and 720 nm. These resonances correspond to the transverse and longitudinal modes of plasmonic excitations in the assembly [126–128]. The transverse mode is related to the electron motion perpendicular to the nanorod long axes and can be excited with light having an electric-field component perpendicular to it, whereas the longitudinal mode is related to the electron oscillations along the nanorod axes and requires p-polarized light to be excited with a component of the incident electric field along the nanorods. The electric-field distribution associated with the longitudinal mode has an intensity maximum in the middle of the nanorods and results from a strong dipole-dipole interaction between the plasmons of individual nanorods in the array and is, therefore, not sensitive to the presence of the superstrate [126–129]. Under oblique incidence and with p-polarized light, the electric field couples to both the transverse and longitudinal modes with the metamaterial slab showing extreme anisotropy of dielectric permittivity similar to that of a uniaxial crystal [130]. The illumination of the same nanorod structure in the ATR geometry reveals the new guided mode in the NIR spectral range which is excited only with p-polarized light and dominates the optical response of the assembly.

The metamaterial works similar to a conventional SPP-based sensor, showing a red shift of the resonance in response to an increase in the RI (Figure 11d and e). Furthermore, a change of the RI by 10^{-4} RIU causes a shift of the resonance by 3.2 nm even without any optimization of the structure (Figure 11f). The corresponding minimum estimation of sensitivity of 32,000 nm/RIU exceeds the sensitivity of localized plasmon-based schemes by two orders of magnitude [23, 56, 131, 132] and an FOM value of 330 is achieved.

4.2 Prism-coupled HMM-based tunable biosensor

In this section, Sreekanth et al. demonstrated the development of a tunable type I HMM and excitation of BPP modes via prism coupling. By using the Goos–Hänchen

(G–H) shift interrogation scheme, reconfigurable sensing with extreme sensitivity biosensing for small molecules has been shown [72]. The G–H shift describes the lateral displacement of the reflected beam from the interface of two media when the angles of incidence are close to the coupling angle [133].

As shown in Figure 12a, to develop a low-loss tunable HMM, 10 alternating thin layers of a low-loss plasmonic material such as TiN (16 nm) and a low-loss phase change material such as Sb_2S_3 (25 nm) are deposited on a cleaned glass substrate. The optical properties of the developed HMM can be tuned by switching the structural phase of Sb_2S_3 from amorphous to crystalline. In Figure 12b, the EMT-derived uniaxial permittivity components of Sb_2S_3–TiN HMM have been shown when Sb_2S_3 is in amorphous and crystalline phases. As can be seen, Sb_2S_3–TiN HMM exhibits type I hyperbolic dispersion at $\lambda > 580$ nm when Sb_2S_3 is in amorphous phase, where $\varepsilon_\perp = \varepsilon_{zz} < 0$ and $\varepsilon_\parallel = \varepsilon_{xx} = \varepsilon_{yy} > 0$. However, the operating wavelength of type I region is slightly blue shifted to 564 nm after the crystallization of Sb_2S_{33} layers in the HMM.

To excite the BPP modes of Sb_2S_3–TiN HMM through prism coupling, a He–Ne laser (632.8 nm) with *p*-polarized light was used as the excitation source. This wavelength belongs to the hyperbolic region of the HMM and the effective index of HMM is less than that of the used BK7 prism index (1.5), so that the momentum matching condition can be satisfied. In Figure 12c, the excited BPP mode of Sb_2S_3–TiN HMM have been plotted, when Sb_2S_3 is in both phases. After switching the phase of Sb_2S_3 from amorphous to crystalline state, the effective index of the HMM decreased, as a result (i) the minimum reflected intensity at the resonance angle declined, (ii) the linewidth of the reflection spectrum decreased, and (iii) the coupling angle slightly shifted. Figure 12d shows the calculated dispersion diagram of fundamental BPP mode of the Sb_2S_3–TiN HMM. Note that the experimentally determined parallel wavevector at 632.8 nm is exactly on the BPP dispersion curve of HMM. Then the Poynting vector (S_x) of the HMM guided mode at 632.8 nm has been calculated (Figure 12e). It shows that the excited mode is a propagating wave inside the HMM, decays exponentially at HMM–water (air) interface, and is leaky in the prism. These are the basic characteristics of a typical BPP mode of an HMM. At resonance, the estimated propagating mode nearest in momentum is 13.08 μm^{-1} with a propagation length of 177 nm. It appears that the excited mode is the fundamental BPP mode of Sb_2S_3–TiN HMM, but it is a low-*k* mode. It is known that the phase difference between TM- and TE-polarized light experiences a sharp singularity at the coupling angle [134]. Thus, as shown in Figure 12f, this phase difference can be actively tuned by switching the phase of Sb_2S_3 from amorphous to crystalline.

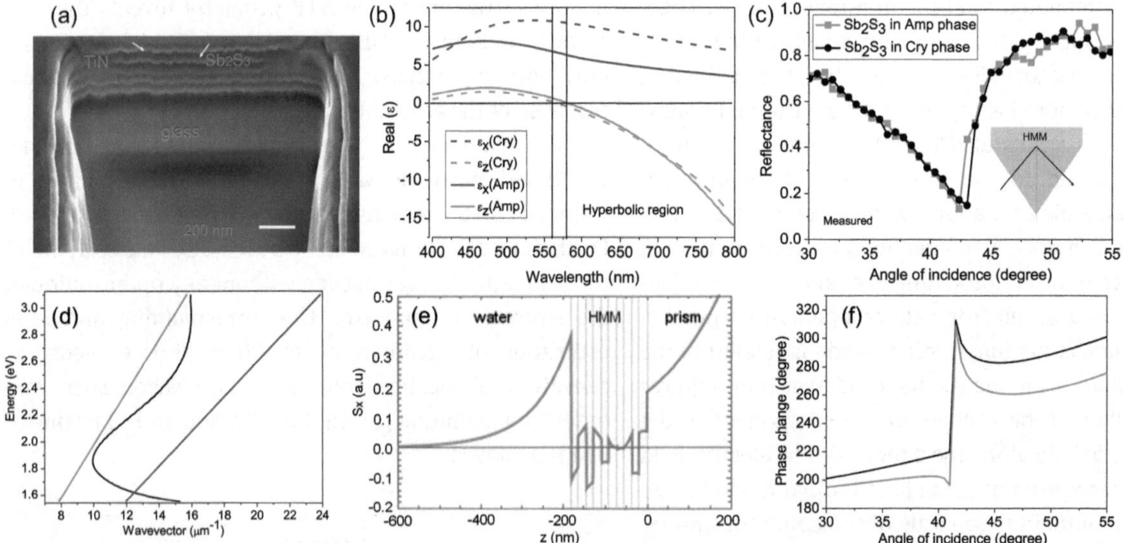

Figure 12: (a) An SEM image of a tunable Sb_2S_3–TiN HMM consisting of five pairs of Sb_2S_3 and TiN, with layer thicknesses of 25 nm for Sb_2S_3 and 16 nm for TiN. (b) EMT-derived real parts of uniaxial permittivity components of Sb_2S_3–TiN HMM when Sb_2S_3 is in the amorphous and crystalline phases. (c) Excited BPP mode of Sb_2S_3–TiN HMM for both phases of Sb_2S_3. (d) Calculated dispersion diagram of fundamental BPP mode of Sb_2S_3–TiN HMM. (e) Poynting vector of the guided mode of HMM. (f) Calculated phase difference between TM and TE polarization for both phases of Sb_2S_3; Reproduced with permission from Sreekanth et al. [72]. Copyright 2019, John Wiley and Sons.

Because the phase derivative at the coupling angle determines the magnitude of the G–H shift, an enhanced and tunable G–H shift at the BPP mode excitation angle can be realized using the Sb_2S_3–TiN HMM. As shown in Figure 13a, the maximum G–H shift is obtained at the coupling angle, where there is a sharp change of phase difference. In addition, a tunable G–H shift is possible by switching the phase of Sb_2S_3 in the HMM from amorphous to crystalline. To validate this tunable behavior experimentally, RI sensing has been performed because the G–H shift strongly depends on the superstrate RI. A differential phase-sensitive setup to record the GH shifts has been used [133–135]. In the experiments, the G–H shifts have been monitored, by injecting different weight ratios (1–10% w/v) of aqueous solutions of glycerol with known refractive indices into sensor channel. The real-time RI sensing data are presented in Figure 13b, where measured G–H shift change with time due to the RI change of glycerol solutions has been report. It is evident that a clear step function in G–H shift by varying the glycerol concentration is obtained. More importantly, a tunable G–H shift is obtained by switching the Sb_2S_3 structural phase from amorphous to crystalline. It is clear from both calculations and experimental data that the maximum G–H shift is possible when Sb_2S_3 in the HMM is in the crystalline phase because the crystalline phase provides a higher phase change at the coupling angle. Therefore, the obtained maximum RI sensitivity for the crystalline phase of HMM results to be 13.4×10^{-7} RIU/nm. The minimum sensitivity obtained for the amorphous phase of HMM is 16.3×10^{-7} RIU/nm. Because the sensor shows tunable sensitivity, a reconfigurable sensor device can be developed for future intelligent sensing applications. Although the obtained tunable range is small, the G–H shift tunability can be further improved using longer wavelength sources and higher RI prisms.

Because enhanced RI sensitivity is possible with the G–H shift interrogation scheme, it can be used to detect small biomolecules at extremely low concentrations. To demonstrate this, it was first sputtered a 10-nm film of gold on top of the crystalline HMM sample. Then the streptavidin–biotin affinity model protocol has been followed for the capture of biotin and the RI change caused by the capture of biotin on the sensor surface was recorded by measuring the G–H shift. By injecting different concentrations (10 fM to 1 μM) of biotin prepared in PBS into the sensor channel with a sample volume of 98 μl and the corresponding G–H shift with increasing concentration was recorded. For biotin concentrations from 10 fM to 1 μM, the response of the sensor with time was analyzed by calculating the marginal G–H shifts ($\delta GH|GH_{PBS} - GH_{biotin}|$). Figure 13c shows the recorded marginal G–H shifts after a reaction time of 40 min and an increase in marginal G–H shifts with increasing biotin concentration is obtained. To confirm the specific binding of biotin on streptavidin sites, the marginal G–H shift change of a single measurement as the biotin molecules accumulate on the sensor surface over time has been recorded, as shown

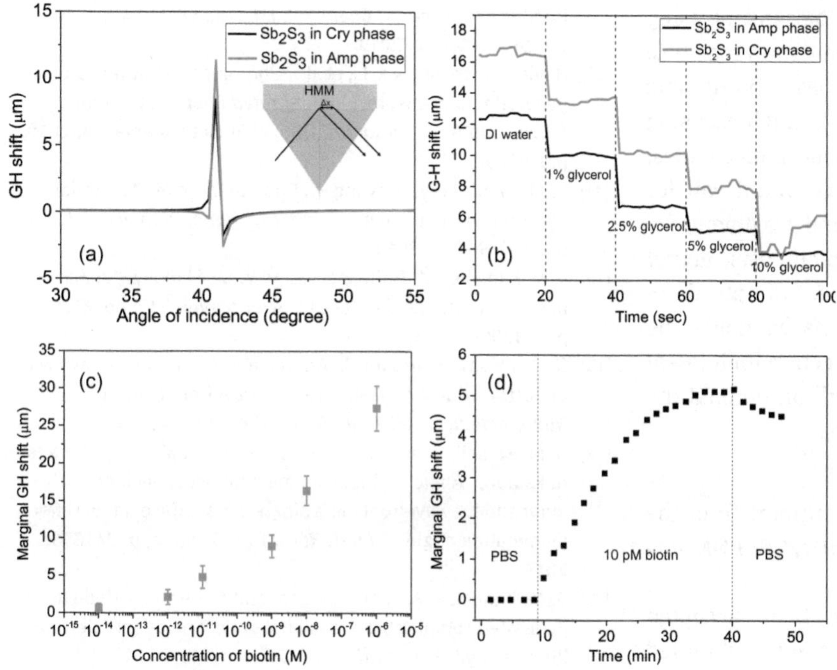

Figure 13: Demonstration of reconfigurable sensing using the Goos–Hänchen (G–H) shift interrogation scheme. (a) Calculated tunable GH shift of the HMM. (b) Real-time tunable refractive index sensing by injecting different weight percentage concentrations of glycerol in distilled water. (c, d) Demonstration of small molecule detection at low concentrations. (c) Measured marginal GH shift for different concentrations of biotin in PBS (10 fM to 1 μM) and (d) Variation of marginal GH shift over time with 10 pM biotin. Reproduced with permission from Sreekanth et al. [72]. Copyright 2019, John Wiley and Sons.

in Figure 13d. A clear step in G–H shift due to the binding of biotin molecules to streptavidin sites is observed. By considering the experimental noise level, it was possible to quantify that the detection limit of the sensor device is less than 1 pM. The detection of ultralow-molecular-weight biomolecules such as biotin at a low concentration of 1 pM is occurred due to the extreme RI sensitivity of the sensor, which is achieved through the G–H shift interrogation scheme. To conclude, it is possible to envision that small molecules such as exosomes can be detected even from bodily fluids using the proposed HMM-based plasmonic platform.

5 Conclusions

In this review, it has been reported the state of the art of hyperbolic dispersion metasurfaces and metamaterials and their applications in the field of molecular biosensing. We started by highlighting the physical aspects characterizing hyperbolic dispersion materials, followed by a review of some relevant designs and properties of HMM that have been proposed by different groups worldwide. HMMs have demonstrated to support extremely sensitive optical modes that can be used to develop cost-effective, noninvasive liquid biopsies for POC clinical evaluation, early cancer screening and real-time diagnosis of diseases. This review highlights the recent advances in hyperbolic dispersion metasurface–based biosensor technology, and in particular discusses biosensors involving metal/dielectric multilayered HMMs. This includes biosensors using different coupling mechanisms and interrogation schemes. We also discuss reconfigurable sensors based on tunable HMMs, LSPR sensors based on type I HMMs and biosensing applications of multifunctional hyperbolic nanocavities. The diagnostic platforms discussed here would call for radical changes in clinical diagnosis and treatment decisions, with the outcomes of lower capital costs, higher accuracy, and specificity. The development of optical devices based on radically new metasurfaces holds the promise to deliver an unprecedented detection limit (areal mass sensitivity at a level of fg/mm) with an optimal dynamic range for proteomic-genomic tests.

Acknowledgments: G.P. acknowledges support from the *"AIM: Attraction and International Mobility"* – PON R&I 2014-2020 Calabria.

Author contribution: All the authors have accepted responsibility for the entire content of this submitted manuscript and approved submission.

Research funding: N.M. acknowledges support from the Luxembourg National Research Fund (CORE Grant No. C19/MS/13624497 "ULTRON") and from the FEDER Program (Grant No. 2017-03-022-19 "Lux-Ultra-Fast").

Conflict of interest statement: The authors declare no conflicts of interest regarding this article.

References

[1] W. T. Chen, A. Y. Zhu, and F. Capasso, "Flat optics with dispersion-engineered metasurfaces," *Nat. Rev. Mater.*, vol. 5, pp. 604–620, 2020.

[2] N. Yu, P. Genevet, M. A. Kats, et al., "Light propagation with phase discontinuities: generalized laws of reflection and refraction," *Science*, vol. 334, no. 6054, pp. 333–337, 2011.

[3] M. Khorasaninejad and F. Capasso, "Metalenses: versatile multifunctional photonic components," *Science*, vol. 358, no. 6367, pp. 1–8, 2017.

[4] F. Capasso, "The future and promise of flat optics: a personal perspective," *Nanophotonics*, vol. 7, no. 6, pp. 953–957, 2018.

[5] N. Yu and F. Capasso, "Flat optics with designer metasurfaces," *Nat. Mater.*, vol. 13, no. 2, pp. 139–150, 2014.

[6] S. M. Kamali, E. Arbabi, A. Arbabi, Y. Horie, and A. Faraon, "Highly tunable elastic dielectric metasurface lenses," *Laser Photonics Rev.*, vol. 10, no. 6, pp. 1002–1008, 2016.

[7] E. Arbabi, A. Arbabi, S. M. Kamali, Y. Horie, M. Faraji-Dana, and A. Faraon, "Mems-tunable dielectric metasurface lens," *Nat. Commun.*, vol. 9, no. 1, pp. 1–9, 2018.

[8] F. Ding, A. Pors, and S. I. Bozhevolnyi, "Gradient metasurfaces: a review of fundamentals and applications," *Rep. Prog. Phys.*, vol. 81, no. 2, p. 026401, 2017.

[9] A. Tittl, A. John-Herpin, A. Leitis, E. R. Arvelo, and H. Altug, "Metasurface-based molecular biosensing aided by artificial intelligence," *Angew. Chem. Int. Ed.*, vol. 58, no. 42, pp. 14810–14822, 2019.

[10] H. Altug, F. Yesilkoy, X. Li, et al., "Photonic metasurfaces for next-generation biosensors," in *Integrated Photonics Research, Silicon and Nanophotonics*, Optical Society of America, 2018, p. ITh3J-5.

[11] S. M. Choudhury, D. Wang, K. Chaudhuri, et al., "Material platforms for optical metasurfaces," *Nanophotonics*, vol. 7, no. 6, pp. 959–987, 2018.

[12] A. V. Kildishev, A. Boltasseva, and V. M. Shalaev, "Planar photonics with metasurfaces," *Science*, vol. 339, no. 6125, p. 1232009, 2013.

[13] S. M. Kamali, E. Arbabi, A. Arbabi, and A. Faraon, "A review of dielectric optical metasurfaces for wavefront control," *Nanophotonics*, vol. 7, no. 6, pp. 1041–1068, 2018.

[14] S. M. Kamali, E. Arbabi, A. Arbabi, Y. Horie, M. Faraji-Dana, and A. Faraon, "Angle-multiplexed metasurfaces: encoding independent wavefronts in a single metasurface under different illumination angles," *Phys. Rev. X*, vol. 7, no. 4, p. 041056, 2017.

[15] S. M. Kamali, E. Arbabi, H. Kwon, and A. Faraon, "Metasurface-generated complex 3-dimensional optical fields for interference lithography," *Proc. Natl. Acad. Sci.*, vol. 116, no. 43, pp. 21379–21384, 2019.

[16] A. Leitis, A. Heßler, S. Wahl, et al., “Huygens’ metasurfaces: all-dielectric programmable huygens’ metasurfaces (adv. Funct. Mater. 19/2020),” *Adv. Funct. Mater.*, vol. 30, no. 19, p. 2070122, 2020.
[17] F. Yesilkoy, E. R. Arvelo, Y. Jahani, et al., “Ultrasensitive hyperspectral imaging and biodetection enabled by dielectric metasurfaces,” *Nat. Photonics*, vol. 13, no. 6, pp. 390–396, 2019.
[18] A. Tittl, A. Leitis, M. Liu, et al., “Imaging-based molecular barcoding with pixelated dielectric metasurfaces,” *Science*, vol. 360, no. 6393, pp. 1105–1109, 2018.
[19] T. D. Gupta, L. Martin-Monier, W. Yan, et al., “Self-assembly of nanostructured glass metasurfaces via templated fluid instabilities,” *Nat. Nanotechnol.*, vol. 14, no. 4, pp. 320–327, 2019.
[20] H. Kwon, E. Arbabi, S. M. Kamali, M. Faraji-Dana, and A. Faraon, “Computational complex optical field imaging using a designed metasurface diffuser,” *Optica*, vol. 5, no. 8, pp. 924–931, 2018.
[21] A. M. Shaltout, J. Kim, A. Boltasseva, V. M. Shalaev, and A. V. Kildishev, “Ultrathin and multicolour optical cavities with embedded metasurfaces,” *Nat. Commun.*, vol. 9, no. 1, pp. 1–7, 2018.
[22] J. Kim, S. Choudhury, C. DeVault, et al., “Controlling the polarization state of light with plasmonic metal oxide metasurface,” *ACS Nano*, vol. 10, no. 10, pp. 9326–9333, 2016.
[23] B. Sepúlveda, P. C. Angelomé, L. M. Lechuga, and L. M. Liz-Marzán, “LSPR-based nanobiosensors,” *Nano Today*, vol. 4, no. 3, pp. 244–251, 2009.
[24] A. Boltasseva and H. A. Atwater, “Low-loss plasmonic metamaterials,” *Science*, vol. 331, no. 6015, pp. 290–291, 2011.
[25] M. C. Estevez, M. Alvarez, and L. M. Lechuga, “Integrated optical devices for lab-on-a-chip biosensing applications,” *Laser Photonics Rev.*, vol. 6, no. 4, pp. 463–487, 2012.
[26] M.-C. Estevez, M. A. Otte, B. Sepulveda, and L. M. Lechuga, “Trends and challenges of refractometric nanoplasmonic biosensors: a review,” *Anal. Chim. Acta*, vol. 806, pp. 55–73, 2014.
[27] X. Chen, L. Huang, H. Mühlenbernd, et al., “Dual-polarity plasmonic metalens for visible light,” *Nat. Commun.*, vol. 3, no. 1, pp. 1–6, 2012.
[28] K. A. Willets, A. J. Wilson, V. Sundaresan, and P. B. Joshi, “Super-resolution imaging and plasmonics,” *Chem. Rev.*, vol. 117, no. 11, pp. 7538–7582, 2017.
[29] X. Huang, P. K. Jain, I. H. El-Sayed, and M. A. El-Sayed, “Plasmonic photothermal therapy (pptt) using gold nanoparticles,” *Laser Med. Sci.*, vol. 23, no. 3, p. 217, 2008.
[30] B. Nasseri, M. Turk, K. Kosemehmetoglu, et al., “The pimpled gold nanosphere: a superior candidate for plasmonic photothermal therapy,” *Int. J. Nanomed.*, vol. 15, p. 2903, 2020.
[31] L. Ricciardi, L. Sancey, G. Palermo, et al., “Plasmon-mediated cancer phototherapy: the combined effect of thermal and photodynamic processes,” *Nanoscale*, vol. 9, no. 48, p. 19279–19289, 2017.
[32] M. R. Ali, M. A. Rahman, Y. Wu, et al., “Efficacy, long-term toxicity, and mechanistic studies of gold nanorods photothermal therapy of cancer in xenograft mice,” *Proc. Natl. Acad. Sci. U.S.A.*, vol. 114, no. 15, pp. E3110–E3118, 2017.
[33] M. Khan and H. Idriss, “Advances in plasmon-enhanced upconversion luminescence phenomena and their possible effect on light harvesting for energy applications,” *Wiley Interdiscip. Rev. Energy Environ.*, vol. 6, no. 6, p. e254, 2017.
[34] S. Manchala, L. R. Nagappagari, S. M. Venkatakrishnan, and V. Shanker, “Solar-light harvesting bimetallic Ag/Au decorated graphene plasmonic system with efficient photoelectrochemical performance for the enhanced water reduction process,” *ACS Appl. Nano Mater.*, vol. 2, no. 8, pp. 4782–4792, 2019.
[35] M. Lee, J. U. Kim, J. S. Lee, B. I. Lee, J. Shin, and C. B. Park, “Mussel-inspired plasmonic nanohybrids for light harvesting,” *Adv. Mater.*, vol. 26, no. 26, pp. 4463–4468, 2014.
[36] M. Poudineh, E. H. Sargent, K. Pantel, and S. O. Kelley, “Profiling circulating tumour cells and other biomarkers of invasive cancers,” *Nat. Biomed. Eng.*, vol. 2, no. 2, pp. 72–84, 2018.
[37] V. Plaks, C. D. Koopman, and Z. Werb, “Circulating tumor cells,” *Science*, vol. 341, no. 6151, pp. 1186–1188, 2013.
[38] M. G. Krebs, R. L. Metcalf, L. Carter, G. Brady, F. H. Blackhall, and C. Dive, “Molecular analysis of circulating tumour cells—biology and biomarkers,” *Nat. Rev. Clin. Oncol.*, vol. 11, no. 3, p. 129, 2014.
[39] M. Cristofanilli, G. T. Budd, M. J. Ellis, et al., “Circulating tumor cells, disease progression, and survival in metastatic breast cancer,” *N. Engl. J. Med.*, vol. 351, no. 8, pp. 781–791, 2004.
[40] S. Maheswaran and D. A. Haber, “Circulating tumor cells: a window into cancer biology and metastasis,” *Curr. Opin. Genet. Dev.*, vol. 20, no. 1, pp. 96–99, 2010.
[41] S. Alimirzaie, M. Bagherzadeh, and M. R. Akbari, “Liquid biopsy in breast cancer: a comprehensive review,” *Clin. Genet.*, vol. 95, no. 6, pp. 643–660, 2019.
[42] C. Alix-Panabières and K. Pantel, “Circulating tumor cells: liquid biopsy of cancer,” *Clin. Chem.*, vol. 59, no. 1, pp. 110–118, 2013.
[43] E. I. Galanzha, Y. A. Menyaev, A. C. Yadem, et al., “In vivo liquid biopsy using cytophone platform for photoacoustic detection of circulating tumor cells in patients with melanoma,” *Sci. Transl. Med.*, vol. 11, no. 496, p. eaat5857, 2019.
[44] G. Rossi and M. Ignatiadis, “Promises and pitfalls of using liquid biopsy for precision medicine,” *Canc. Res.*, vol. 79, no. 11, pp. 2798–2804, 2019.
[45] K. Boriachek, M. N. Islam, A. Möller, et al., “Biological functions and current advances in isolation and detection strategies for exosome nanovesicles,” *Small*, vol. 14, no. 6, p. 1702153, 2018.
[46] T. Huang and C.-X. Deng, “Current progresses of exosomes as cancer diagnostic and prognostic biomarkers,” *Int. J. Biol. Sci.*, vol. 15, no. 1, p. 1, 2019.
[47] J.-H. Kim, E. Kim, and M. Y. Lee, “Exosomes as diagnostic biomarkers in cancer,” *Mol. Cell. Toxicol.*, vol. 14, no. 2, pp. 113–122, 2018.
[48] Y. H. Soung, T. Nguyen, H. Cao, J. Lee, and J. Chung, “Emerging roles of exosomes in cancer invasion and metastasis,” *BMB Rep.*, vol. 49, no. 1, p. 18, 2016.
[49] I. Ament, J. Prasad, A. Henkel, S. Schmachtel, and C. Sonnichsen, “Single unlabeled protein detection on individual plasmonic nanoparticles,” *Nano Lett.*, vol. 12, no. 2, pp. 1092–1095, 2012.
[50] H. Im, H. Shao, Y. I. Park, et al., “Label-free detection and molecular profiling of exosomes with a nano-plasmonic sensor,” *Nat. Biotechnol.*, vol. 32, no. 5, pp. 490–495, 2014.
[51] A. A. I. Sina, R. Vaidyanathan, A. Wuethrich, L. G. Carrascosa, and M. Trau, “Label-free detection of exosomes using a surface plasmon resonance biosensor,” *Anal. Bioanal. Chem.*, vol. 411, no. 7, pp. 1311–1318, 2019.

[52] S. S. Acimovic, M. A. Ortega, V. Sanz, et al., "Lspr chip for parallel, rapid, and sensitive detection of cancer markers in serum," *Nano Lett.*, vol. 14, no. 5, pp. 2636–2641, 2014.
[53] V. Kravets, F. Schedin, R. Jalil, et al., "Singular phase nano-optics in plasmonic metamaterials for label-free single-molecule detection," *Nat. Mater.*, vol. 12, no. 4, pp. 304–309, 2013.
[54] F. De Angelis, F. Gentile, F. Mecarini, et al., "Breaking the diffusion limit with super-hydrophobic delivery of molecules to plasmonic nanofocusing sers structures," *Nat. Photonics*, vol. 5, no. 11, pp. 682–687, 2011.
[55] P. Zijlstra, P. M. Paulo, and M. Orrit, "Optical detection of single non-absorbing molecules using the surface plasmon resonance of a gold nanorod," *Nat. Nanotechnol.*, vol. 7, no. 6, pp. 379–382, 2012.
[56] J. N. Anker, W. P. Hall, O. Lyandres, N. C. Shah, J. Zhao, and R. P. Van Duyne, "Biosensing with plasmonic nanosensors," in *Nanoscience and Technology: A Collection of Reviews from Nature Journals*, Singapore, World Scientific, 2010, pp. 308–319.
[57] K. Q. Le, Q. M. Ngo, and T. K. Nguyen, "Nanostructured metal–insulator–metal metamaterials for refractive index biosensing applications: design, fabrication, and characterization," *IEEE J. Sel. Top. Quant. Electron.*, vol. 23, no. 2, pp. 388–393, 2016.
[58] M. Svedendahl, R. Verre, and M. Käll, "Refractometric biosensing based on optical phase flips in sparse and short-range-ordered nanoplasmonic layers," *Light Sci. Appl.*, vol. 3, no. 11, p. e220, 2014.
[59] A. Kabashin, P. Evans, S. Pastkovsky, et al., "Plasmonic nanorod metamaterials for biosensing," *Nat. Mater.*, vol. 8, no. 11, pp. 867–871, 2009.
[60] K. V. Sreekanth, Y. Alapan, M. ElKabbash, et al., "Extreme sensitivity biosensing platform based on hyperbolic metamaterials," *Nat. Mater.*, vol. 15, no. 6, pp. 621–627, 2016.
[61] L. Jiang, S. Zeng, Z. Xu, et al., "Multifunctional hyperbolic nanogroove metasurface for submolecular detection," *Small*, vol. 13, no. 30, p. 1700600, 2017.
[62] G. Strangi, K. Sreekanth, and M. Elkabbash, "Hyperbolic metamaterial-based ultrasensitive plasmonic biosensors for early-stage cancer detection," in *Next Generation Point-of-Care Biomedical Sensors Technologies for Cancer Diagnosis*, Berlin, Springer, 2017, pp. 155–172.
[63] A. Poddubny, I. Iorsh, P. Belov, and Y. Kivshar, "Hyperbolic metamaterials," *Nat. Photonics*, vol. 7, no. 12, pp. 948–957, 2013.
[64] L. Ferrari, C. Wu, D. Lepage, X. Zhang, and Z. Liu, "Hyperbolic metamaterials and their applications," *Prog. Quant. Electron.*, vol. 40, pp. 1–40, 2015.
[65] I. I. Smolyaninov and V. N. Smolyaninova, "Hyperbolic metamaterials: novel physics and applications," *Solid State Electron.*, vol. 136, pp. 102–112, 2017.
[66] Z. Jacob and E. E. Narimanov, "Optical hyperspace for plasmons: Dyakonov states in metamaterials," *Appl. Phys. Lett.*, vol. 93, no. 22, p. 221109, 2008.
[67] P. Shekhar, J. Atkinson, and Z. Jacob, "Hyperbolic metamaterials: fundamentals and applications," *Nano Convergence*, vol. 1, no. 1, p. 14, 2014.
[68] C. Cortes, W. Newman, S. Molesky, and Z. Jacob, "Quantum nanophotonics using hyperbolic metamaterials," *J. Optics*, vol. 14, no. 6, p. 063001, 2012.
[69] I. Avrutsky, I. Salakhutdinov, J. Elser, and V. Podolskiy, "Highly confined optical modes in nanoscale metal-dielectric multilayers," *Phys. Rev. B*, vol. 75, no. 24, p. 241402, 2007.
[70] K. V. Sreekanth, A. De Luca, and G. Strangi, "Experimental demonstration of surface and bulk plasmon polaritons in hypergratings," *Sci. Rep.*, vol. 3, p. 3291, 2013.
[71] A. G. Brolo, "Plasmonics for future biosensors," *Nat. Photonics*, vol. 6, no. 11, pp. 709–713, 2012.
[72] K. V. Sreekanth, Q. Ouyang, S. Sreejith, et al., "Phase-change-material-based low-loss visible-frequency hyperbolic metamaterials for ultrasensitive label-free biosensing," *Adv. Opt. Mater.*, vol. 7, no. 12, p. 1900081, 2019.
[73] K. V. Sreekanth, M. ElKabbash, Y. Alapan, et al., "Hyperbolic metamaterials-based plasmonic biosensor for fluid biopsy with single molecule sensitivity," *EPJ Appl. Metamater.*, vol. 4, p. 1, 2017.
[74] K. V. Sreekanth, Y. Alapan, M. ElKabbash, et al., "Enhancing the angular sensitivity of plasmonic sensors using hyperbolic metamaterials," *Adv. Opt. Mater.*, vol. 4, no. 11, pp. 1767–1772, 2016.
[75] K. V. Sreekanth, P. Mahalakshmi, S. Han, M. S. Mani Rajan, P. K. Choudhury, and R. Singh, "Brewster mode-enhanced sensing with hyperbolic metamaterial," *Adv. Opt. Mater.*, vol. 7, no. 21, p. 1900680, 2019.
[76] G. Palermo, G. E. Lio, M. Esposito, et al., "Biomolecular sensing at the interface between chiral metasurfaces and hyperbolic metamaterials," *ACS Appl. Mater. Interfaces*, vol. 12, no. 27, pp. 30181–30188, 2020.
[77] F. Abbas and M. Faryad, "A highly sensitive multiplasmonic sensor using hyperbolic chiral sculptured thin films," *J. Appl. Phys.*, vol. 122, no. 17, p. 173104, 2017.
[78] J. Wu, F. Wu, C. Xue, et al., "Wide-angle ultrasensitive biosensors based on edge states in heterostructures containing hyperbolic metamaterials," *Opt. Express*, vol. 27, no. 17, pp. 24835–24846, 2019.
[79] W. He, Y. Feng, Z.-D. Hu, et al., "Sensors with multifold nanorod metasurfaces array based on hyperbolic metamaterials," *IEEE Sens. J.*, vol. 20, no. 4, pp. 1801–1806, 2019.
[80] Y. Yoshida, Y. Kashiwai, E. Murakami, S. Ishida, and N. Hashiguchi, "Development of the monitoring system for slope deformations with fiber Bragg grating arrays," in *Smart Structures and Materials 2002: Smart Sensor Technology and Measurement Systems*, vol. 4694, International Society for Optics and Photonics, 2002, pp. 296–303.
[81] M. Baqir, A. Farmani, T. Fatima, M. Raza, S. Shaukat, and A. Mir, "Nanoscale, tunable, and highly sensitive biosensor utilizing hyperbolic metamaterials in the near-infrared range," *Appl. Opt.*, vol. 57, no. 31, pp. 9447–9454, 2018.
[82] N. Vasilantonakis, G. Wurtz, V. Podolskiy, and A. Zayats, "Refractive index sensing with hyperbolic metamaterials: strategies for biosensing and nonlinearity enhancement," *Opt. Express*, vol. 23, no. 11, pp. 14329–14343, 2015.
[83] D. Rodrigo, O. Limaj, D. Janner, et al., "Mid-infrared plasmonic biosensing with graphene," *Science*, vol. 349, no. 6244, pp. 165–168, 2015.
[84] G. E. Lio, A. Ferraro, M. Giocondo, R. Caputo, and A. De Luca, "Color Gamut behavior in epsilon near-zero nanocavities during propagation of gap surface plasmons," *Adv. Opt. Mater.*, vol. 18, no. 17, p. 2000487, 2020.

[85] X. Yang, J. Yao, J. Rho, X. Yin, and X. Zhang, "Experimental realization of three-dimensional indefinite cavities at the nanoscale with anomalous scaling laws," *Nat. Photonics*, vol. 6, no. 7, pp. 450–454, 2012.

[86] C. Guclu, T. S. Luk, G. T. Wang, and F. Capolino, "Radiative emission enhancement using nano-antennas made of hyperbolic metamaterial resonators," *Appl. Phys. Lett.*, vol. 105, no. 12, p. 123101, 2014.

[87] D. Lu, J. J. Kan, E. E. Fullerton, and Z. Liu, "Enhancing spontaneous emission rates of molecules using nanopatterned multilayer hyperbolic metamaterials," *Nat. Nanotechnol.*, vol. 9, no. 1, pp. 48–53, 2014.

[88] D. Lu, H. Qian, K. Wang, et al., "Nanostructuring multilayer hyperbolic metamaterials for ultrafast and bright green ingan quantum wells," *Adv. Mater.*, vol. 30, no. 15, p. 1706411, 2018.

[89] S. C. Indukuri, J. Bar-David, N. Mazurski, and U. Levy, "Ultrasmall mode volume hyperbolic nanocavities for enhanced light–matter interaction at the nanoscale," *ACS Nano*, vol. 13, no. 10, pp. 11770–11780, 2019.

[90] P. R. West, N. Kinsey, M. Ferrera, A. V. Kildishev, V. M. Shalaev, and A. Boltasseva, "Adiabatically tapered hyperbolic metamaterials for dispersion control of high-*k* waves," *Nano Lett.*, vol. 15, no. 1, pp. 498–505, 2015.

[91] T. Galfsky, H. Krishnamoorthy, W. Newman, E. Narimanov, Z. Jacob, and V. Menon, "Active hyperbolic metamaterials: enhanced spontaneous emission and light extraction," *Optica*, vol. 2, no. 1, pp. 62–65, 2015.

[92] N. Maccaferri, T. Isoniemi, M. Hinczewski, M. Iarossi, G. Strangi, and F. De Angelis, "Designer bloch plasmon polariton dispersion in grating-coupled hyperbolic metamaterials," *APL Photonics*, vol. 5, no. 7, p. 076109, 2020.

[93] F. Peragut, L. Cerutti, A. Baranov, et al., "Hyperbolic metamaterials and surface plasmon polaritons," *Optica*, vol. 4, no. 11, pp. 1409–1415, 2017.

[94] T. Isoniemi, N. Maccaferri, Q. M. Ramasse, G. Strangi, and F. De Angelis, "Electron energy loss spectroscopy of bright and dark modes in hyperbolic metamaterial nanostructures," *Adv. Opt. Mater.*, vol. 8, no. 13, p. 2000277, 2020.

[95] N. Maccaferri, Y. Zhao, T. Isoniemi, et al., "Hyperbolic meta-antennas enable full control of scattering and absorption of light," *Nano Lett.*, vol. 19, no. 3, pp. 1851–1859, 2019.

[96] J. Song and W. Zhou, "Multiresonant composite optical nanoantennas by out-of-plane plasmonic engineering," *Nano Lett.*, vol. 18, no. 7, pp. 4409–4416, 2018.

[97] P. Wang, A. V. Krasavin, F. N. Viscomi, et al., "Metaparticles: Dressing nano-objects with a hyperbolic coating," *Laser Photonics Rev.*, vol. 12, no. 11, p. 1800179, 2018.

[98] J. Zhou, A. F. Kaplan, L. Chen, and L. J. Guo, "Experiment and theory of the broadband absorption by a tapered hyperbolic metamaterial array," *ACS Photonics*, vol. 1, no. 7, pp. 618–624, 2014.

[99] M. Sakhdari, M. Hajizadegan, M. Farhat, and P.-Y. Chen, "Efficient, broadband and wide-angle hot-electron transduction using metal-semiconductor hyperbolic metamaterials," *Nano Energy*, vol. 26, pp. 371–381, 2016.

[100] G. Abdelatif, M. F. O. Hameed, S. Obayya, and M. Hussein, "Ultrabroadband absorber based on a funnel-shaped anisotropic metamaterial," *JOSA B*, vol. 36, no. 10, pp. 2889–2895, 2019.

[101] C. T. Riley, J. S. Smalley, J. R. Brodie, Y. Fainman, D. J. Sirbuly, and Z. Liu, "Near-perfect broadband absorption from hyperbolic metamaterial nanoparticles," *Proc. Natl. Acad. Sci. U.S.A.*, vol. 114, no. 6, pp. 1264–1268, 2017.

[102] V. Caligiuri, L. Pezzi, A. Veltri, and A. De Luca, "Resonant gain singularities in 1D and 3D metal/dielectric multilayered nanostructures," *ACS Nano*, vol. 11, no. 1, pp. 1012–1025, 2017.

[103] W. D. Newman, C. L. Cortes, J. Atkinson, S. Pramanik, R. G. DeCorby, and Z. Jacob, "Ferrell–Berreman modes in plasmonic epsilon-near-zero media," *ACS Photonics*, vol. 2, no. 1, pp. 2–7, 2015.

[104] R. A. Ferrell, "Predicted radiation of plasma oscillations in metal films," *Phys. Rev.*, vol. 111, no. 5, p. 1214, 1958.

[105] C. Xu, J. Xu, G. Song, C. Zhu, Y. Yang, and G. S. Agarwal, "Enhanced displacements in reflected beams at hyperbolic metamaterials," *Opt. Express*, vol. 24, no. 19, pp. 21767–21776, 2016.

[106] G. Luka, A. Ahmadi, H. Najjaran, et al., "Microfluidics integrated biosensors: a leading technology towards lab-on-a-chip and sensing applications," *Sensors*, vol. 15, no. 12, pp. 30011–30031, 2015.

[107] K. Sreekanth, A. De Luca, and G. Strangi, "Excitation of volume plasmon polaritons in metal-dielectric metamaterials using 1D and 2D diffraction gratings," *J. Opt.*, vol. 16, no. 10, p. 105103, 2014.

[108] K. V. Sreekanth, K. H. Krishna, A. De Luca, and G. Strangi, "Large spontaneous emission rate enhancement in grating coupled hyperbolic metamaterials," *Sci. Rep.*, vol. 4, p. 6340, 2014.

[109] S. Thongrattanasiri and V. A. Podolskiy, "Hypergratings: nanophotonics in planar anisotropic metamaterials," *Opt. Lett.*, vol. 34, no. 7, pp. 890–892, 2009.

[110] S. Zeng, D. Baillargeat, H.-P. Ho, and K.-T. Yong, "Nanomaterials enhanced surface plasmon resonance for biological and chemical sensing applications," *Chem. Soc. Rev.*, vol. 43, no. 10, pp. 3426–3452, 2014.

[111] M. Esposito, V. Tasco, M. Cuscuna, et al., "Nanoscale 3D chiral plasmonic helices with circular dichroism at visible frequencies," *ACS Photonics*, vol. 2, no. 1, pp. 105–114, 2015.

[112] M. Esposito, V. Tasco, F. Todisco, et al., "Programmable extreme chirality in the visible by helix-shaped metamaterial platform," *Nano Lett.*, vol. 16, no. 9, pp. 5823–5828, 2016.

[113] J. Rho, and X. Zhang, "Recent progress in hyperbolic, chiral metamaterials and metasurfaces," in *Workshop on Optical Plasmonic Materials*, Optical Society of America, 2014, p. OW4D-1.

[114] F. Li, T. Tang, J. Li, et al., "Chiral coding metasurfaces with integrated vanadium dioxide for thermo-optic modulation of terahertz waves," *J. Alloys Compd.*, vol. 826, p. 154174, 2020.

[115] O. Kotov and Y. E. Lozovik, "Enhanced optical activity in hyperbolic metasurfaces," *Phys. Rev. B*, vol. 96, no. 23, p. 235403, 2017.

[116] Y. Zhao, A. N. Askarpour, L. Sun, J. Shi, X. Li, and A. Alù, "Chirality detection of enantiomers using twisted optical metamaterials," *Nat. Commun.*, vol. 8, no. 1, pp. 1–8, 2017.

[117] Z. Wang, Y. Wang, G. Adamo, et al., "A novel chiral metasurface with controllable circular dichroism induced by coupling localized and propagating modes," *Adv. Opt. Mater.*, vol. 4, no. 6, pp. 883–888, 2016.

[118] M. L. Solomon, J. Hu, M. Lawrence, A. García-Etxarri, and J. A. Dionne, "Enantiospecific optical enhancement of chiral

sensing and separation with dielectric metasurfaces," *ACS Photonics*, vol. 6, no. 1, pp. 43–49, 2018.
[119] E. Yashima, K. Maeda, and T. Nishimura, "Detection and amplification of chirality by helical polymers," *Chem. Eur. J.*, vol. 10, no. 1, pp. 42–51, 2004.
[120] J. T. Collins, C. Kuppe, D. C. Hooper, C. Sibilia, M. Centini, and V. K. Valev, "Chirality and chiroptical effects in metal nanostructures: fundamentals and current trends," *Adv. Opt. Mater.*, vol. 5, no. 16, p. 1700182, 2017.
[121] G. E. Lio, G. Palermo, R. Caputo, and A. De Luca, "A comprehensive optical analysis of nanoscale structures: from thin films to asymmetric nanocavities," *RSC Adv.*, vol. 9, no. 37, pp. 21429–21437, 2019.
[122] F. Koohyar, F. Kiani, S. Sharifi, M. Sharifirad, and S. H. Rahmanpour, "Study on the change of refractive index on mixing, excess molar volume and viscosity deviation for aqueous solution of methanol, ethanol, ethylene glycol, 1-propanol and 1,2,3-propantriol at $t = 292.15$ K and atmospheric pressure," *Res. J. Appl. Sci. Eng. Technol.*, vol. 4, no. 17, pp. 3095–3101, 2012.
[123] H.-H. Jeong, A. G. Mark, M. Alarcón-Correa, et al., "Dispersion and shape engineered plasmonic nanosensors," *Nat. Commun.*, vol. 7, no. 1, pp. 1–7, 2016.
[124] K. M. Mayer and J. H. Hafner, "Localized surface plasmon resonance sensors," *Chem. Rev.*, vol. 111, no. 6, pp. 3828–3857, 2011.
[125] T. C. Choy, *Effective Medium Theory: Principles and Applications*, vol. 165, Oxford, Oxford University Press, 2015.
[126] P. R. Evans, G. A. Wurtz, R. Atkinson, et al., "Plasmonic core/shell nanorod arrays: subattoliter controlled geometry and tunable optical properties," *J. Phys. Chem. C*, vol. 111, no. 34, pp. 12522–12527, 2007.
[127] G. Wurtz, W. Dickson, D. O'connor, et al., "Guided plasmonic modes in nanorod assemblies: strong electromagnetic coupling regime," *Opt. Express*, vol. 16, no. 10, pp. 7460–7470, 2008.
[128] W. Dickson, G. Wurtz, P. Evans, et al., "Dielectric-loaded plasmonic nanoantenna arrays: a metamaterial with tuneable optical properties," *Phys. Rev. B*, vol. 76, no. 11, p. 115411, 2007.
[129] G. A. Wurtz, P. R. Evans, W. Hendren, et al., "Molecular plasmonics with tunable exciton–plasmon coupling strength in *j*-aggregate hybridized au nanorod assemblies," *Nano Lett.*, vol. 7, no. 5, pp. 1297–1303, 2007.
[130] J. Elser, R. Wangberg, V. A. Podolskiy, and E. E. Narimanov, "Nanowire metamaterials with extreme optical anisotropy," *Appl. Phys. Lett.*, vol. 89, no. 26, p. 261102, 2006.
[131] B. Sepúlveda, A. Calle, L. M. Lechuga, and G. Armelles, "Highly sensitive detection of biomolecules with the magneto-optic surface-plasmon-resonance sensor," *Opt. Lett.*, vol. 31, no. 8, pp. 1085–1087, 2006.
[132] M. E. Stewart, C. R. Anderton, L. B. Thompson, et al., "Nanostructured plasmonic sensors," *Chem. Rev.*, vol. 108, no. 2, pp. 494–521, 2008.
[133] K. V. Sreekanth, Q. Ouyang, S. Han, K.-T. Yong, and R. Singh, "Giant enhancement in Goos–Hänchen shift at the singular phase of a nanophotonic cavity," *Appl. Phys. Lett.*, vol. 112, no. 16, p. 161109, 2018.
[134] K. V. Sreekanth, S. Sreejith, S. Han, et al., "Biosensing with the singular phase of an ultrathin metal-dielectric nanophotonic cavity," *Nat. Commun.*, vol. 9, no. 1, pp. 1–8, 2018.
[135] K. V. Sreekanth, W. Dong, Q. Ouyang, et al., "Large-area silver–stibnite nanoporous plasmonic films for label-free biosensing," *ACS Appl. Mater. Interfaces*, vol. 10, no. 41, pp. 34991–34999, 2018.

Part V: **Fundamentals of Optics**

A Tutorial on the Classical Theories of Electromagnetic Scattering and Diffraction

Masud Mansuripur
College of Optical Sciences, The University of Arizona, Tucson

https://doi.org/10.1515/9783110710687-025

Abstract. Starting with Maxwell's equations, we derive the fundamental results of the Huygens-Fresnel-Kirchhoff and Rayleigh-Sommerfeld theories of scalar diffraction and scattering. These results are then extended to cover the case of vector electromagnetic fields. The famous Sommerfeld solution to the problem of diffraction from a perfectly conducting half-plane is elaborated. Far-field scattering of plane waves from obstacles is treated in some detail, and the well-known optical cross-section theorem, which relates the scattering cross-section of an obstacle to its forward scattering amplitude, is derived. Also examined is the case of scattering from mild inhomogeneities within an otherwise homogeneous medium, where, in the first Born approximation, a fairly simple formula is found to relate the far-field scattering amplitude to the host medium's optical properties. The related problem of neutron scattering from ferromagnetic materials is treated in the final section of the paper.

1. Introduction. The classical theories of electromagnetic (EM) scattering and diffraction were developed throughout the nineteenth century by the likes of Augustine Jean Fresnel (1788-1827), Gustav Kirchhoff (1824-1887), John William Strutt (Lord Rayleigh, 1842-1919), and Arnold Sommerfeld (1868-1951).[1-3] A thorough appreciation of these theories requires an understanding of the Maxwell-Lorentz electrodynamics[4-11] and a working knowledge of vector calculus, differential equations, Fourier transformation, and complex-plane integration techniques.[12] The relevant physical and mathematical arguments have been covered (to varying degrees of clarity and completeness) in numerous textbooks, monographs, and research papers.[13-23] The goal of this tutorial is to present the core concepts of the classical theories of scattering and diffraction by starting with Maxwell's equations and deriving the fundamental results using mathematical arguments that should be accessible to students of optical sciences as well as practitioners of modern optical engineering and photonics technologies. A consistent notation and uniform terminology is used throughout the paper. To maintain the focus on the main results and reduce the potential for distraction, some of the longer derivations and secondary arguments have been relegated to the appendices.

The organization of the paper is as follows. After a brief review of Maxwell's equations in Sec.2, we provide a detailed analysis of an all-important Green function in Sec.3. The Huygens-Fresnel-Kirchhoff scalar theory of diffraction is the subject of Sec.4, followed by the Rayleigh-Sommerfeld modification and enhancement of that theory in Sec.5. These scalar theories are subsequently generalized in Sec.6 to arrive at a number of formulas for vector scattering and vector diffraction of EM waves under various settings and circumstances. Section 6 also contains a few examples that demonstrate the application of vector diffraction formulas in situations of practical interest. The famous Sommerfeld solution to the problem of diffraction from a perfectly electrically conducting half-plane is presented in some detail in Sec.7.

Applying the vector formulas of Sec.6 to far-field scattering, we show in Sec.8 how the forward scattering amplitude for a plane-wave that illuminates an arbitrary object relates to the scattering cross-section of that object. This important result in the classical theory of scattering is formally known as the optical cross-section theorem (or the optical theorem).[1,9,24]

An alternative approach to the problem of EM scattering when the host medium contains a region of weak inhomogeneities is described in Sec.9. Here, we use Maxwell's macroscopic equations in conjunction with the Green function of Sec.3 to derive a fairly simple formula for the far-field scattering amplitude in the first Born approximation. The related problem of slow neutron scattering from ferromagnetic media is treated in Sec.10. The paper closes with a few conclusions and final remarks in Sec.11.

This article has previously been published in the journal Nanophotonics. Please cite as: M. Mansuripur "A Tutorial on the Classical Theories of Electromagnetic Scattering and Diffraction" *Nanophotonics* 2021, 10. DOI: 10.1515/nanoph-2020-0348.

2. Maxwell's equations. The standard equations of the classical Maxwell-Lorenz theory of electrodynamics relate four material sources to four EM fields in the Minkowski spacetime $(\boldsymbol{r},t)$.[4-11] The sources are the free charge density ρ_{free}, free current density $\boldsymbol{J}_{\text{free}}$, polarization $\boldsymbol{P}$, and magnetization $\boldsymbol{M}$, while the fields are the electric field $\boldsymbol{E}$, magnetic field $\boldsymbol{H}$, displacement $\boldsymbol{D}$, and magnetic induction $\boldsymbol{B}$. In the *SI* system of units, where the free space (or vacuum) has permittivity ε_0 and permeability μ_0, the displacement is defined as $\boldsymbol{D}=\varepsilon_0\boldsymbol{E}+\boldsymbol{P}$, and the magnetic induction as $\boldsymbol{B}=\mu_0\boldsymbol{H}+\boldsymbol{M}$. Let the total charge-density $\rho(\boldsymbol{r},t)$ and total current-density $\boldsymbol{J}(\boldsymbol{r},t)$ be defined as

$$\rho(\boldsymbol{r},t)=\rho_{\text{free}}(\boldsymbol{r},t)-\boldsymbol{\nabla}\cdot\boldsymbol{P}(\boldsymbol{r},t). \tag{1}$$

$$\boldsymbol{J}(\boldsymbol{r},t)=\boldsymbol{J}_{\text{free}}(\boldsymbol{r},t)+\partial_t\boldsymbol{P}(\boldsymbol{r},t)+\mu_0^{-1}\boldsymbol{\nabla}\times\boldsymbol{M}(\boldsymbol{r},t). \tag{2}$$

The charge-current continuity equation, $\boldsymbol{\nabla}\cdot\boldsymbol{J}+\partial_t\rho=0$, is generally satisfied by the above densities, irrespective of whether their corresponding sources are free (i.e., ρ_{free} and $\boldsymbol{J}_{\text{free}}$), or bound electric charges within electric dipoles (i.e., $-\boldsymbol{\nabla}\cdot\boldsymbol{P}$ and $\partial_t\boldsymbol{P}$), or bound electric currents within magnetic dipoles (i.e., $\mu_0^{-1}\boldsymbol{\nabla}\times\boldsymbol{M}$). Invoking Eqs.(1) and (2), Maxwell's macroscopic equations are written as follows:

$$\varepsilon_0\boldsymbol{\nabla}\cdot\boldsymbol{E}(\boldsymbol{r},t)=\rho(\boldsymbol{r},t). \tag{3}$$

$$\boldsymbol{\nabla}\times\boldsymbol{B}(\boldsymbol{r},t)=\mu_0\boldsymbol{J}(\boldsymbol{r},t)+\mu_0\varepsilon_0\partial_t\boldsymbol{E}(\boldsymbol{r},t). \tag{4}$$

$$\boldsymbol{\nabla}\times\boldsymbol{E}(\boldsymbol{r},t)=-\partial_t\boldsymbol{B}(\boldsymbol{r},t). \tag{5}$$

$$\boldsymbol{\nabla}\cdot\boldsymbol{B}(\boldsymbol{r},t)=0. \tag{6}$$

Taking the curl of both sides of the third equation, using the vector identity $\boldsymbol{\nabla}\times\boldsymbol{\nabla}\times\boldsymbol{V}=\boldsymbol{\nabla}(\boldsymbol{\nabla}\cdot\boldsymbol{V})-\boldsymbol{\nabla}^2\boldsymbol{V}$, and substituting from the first and second equations, we find

$$\boldsymbol{\nabla}\times\boldsymbol{\nabla}\times\boldsymbol{E}=-\mu_0\partial_t\boldsymbol{J}-\mu_0\varepsilon_0\partial_t^2\boldsymbol{E} \quad\rightarrow\quad (\boldsymbol{\nabla}^2-c^{-2}\partial_t^2)\boldsymbol{E}=\mu_0\partial_t\boldsymbol{J}+\varepsilon_0^{-1}\boldsymbol{\nabla}\rho. \tag{7}$$

Here, $c=(\mu_0\varepsilon_0)^{-½}$ is the speed of light in vacuum. Similarly, taking the curl of both sides of Eq.(4) and substituting from Eqs.(5) and (6), we find

$$\boldsymbol{\nabla}\times\boldsymbol{\nabla}\times\boldsymbol{B}=\mu_0\boldsymbol{\nabla}\times\boldsymbol{J}-\mu_0\varepsilon_0\partial_t^2\boldsymbol{B} \quad\rightarrow\quad (\boldsymbol{\nabla}^2-c^{-2}\partial_t^2)\boldsymbol{B}=-\mu_0\boldsymbol{\nabla}\times\boldsymbol{J}. \tag{8}$$

The scalar and vector potentials, $\psi(\boldsymbol{r},t)$ and $\boldsymbol{A}(\boldsymbol{r},t)$, are defined such that $\boldsymbol{B}=\boldsymbol{\nabla}\times\boldsymbol{A}$ and $\boldsymbol{E}=-\boldsymbol{\nabla}\psi-\partial_t\boldsymbol{A}$. With these definitions, Maxwell's third and fourth equations are automatically satisfied. In the Lorenz gauge, where $\boldsymbol{\nabla}\cdot\boldsymbol{A}+c^{-2}\partial_t\psi=0$, Maxwell's second equation yields

$$\boldsymbol{\nabla}\times\boldsymbol{\nabla}\times\boldsymbol{A}=\mu_0\boldsymbol{J}-\mu_0\varepsilon_0\boldsymbol{\nabla}(\partial_t\psi)-\mu_0\varepsilon_0\partial_t^2\boldsymbol{A} \quad\rightarrow\quad (\boldsymbol{\nabla}^2-c^{-2}\partial_t^2)\boldsymbol{A}=-\mu_0\boldsymbol{J}. \tag{9}$$

Similarly, substitution into Maxwell's first equation for the E-field in terms of the potentials yields

$$-\varepsilon_0\boldsymbol{\nabla}\cdot(\boldsymbol{\nabla}\psi+\partial_t\boldsymbol{A})=\rho \quad\rightarrow\quad (\nabla^2-c^{-2}\partial_t^2)\psi=-\rho/\varepsilon_0. \tag{10}$$

Since monochromatic fields oscillate at a single frequency ω, their time-dependence factor is generally written as $\exp(-i\omega t)$. Consequently, the spatiotemporal dependence of all the fields and all the sources can be separated into a space part and a time part. For example, the E-field may now be written as $\boldsymbol{E}(\boldsymbol{r})e^{-i\omega t}$, the total electric charge-density as $\rho(\boldsymbol{r})e^{-i\omega t}$, and so on. Defining the free-space wavenumber $k_0=\omega/c$, the Helmholtz equations (7)-(10) now become

$$(\nabla^2 + k_0^2)\boldsymbol{E}(\boldsymbol{r}) = -\mathrm{i}\omega\mu_0 \boldsymbol{J}(\boldsymbol{r}) + \varepsilon_0^{-1}\nabla\rho(\boldsymbol{r}). \tag{11}$$

$$(\nabla^2 + k_0^2)\boldsymbol{B}(\boldsymbol{r}) = -\mu_0\nabla\times\boldsymbol{J}(\boldsymbol{r}). \tag{12}$$

$$(\nabla^2 + k_0^2)\boldsymbol{A}(\boldsymbol{r}) = -\mu_0 \boldsymbol{J}(\boldsymbol{r}). \tag{13}$$

$$(\nabla^2 + k_0^2)\psi(\boldsymbol{r}) = -\rho(\boldsymbol{r})/\varepsilon_0. \tag{14}$$

In regions of free space, where $\rho(\boldsymbol{r}) = 0$ and $\boldsymbol{J}(\boldsymbol{r}) = 0$, the right-hand sides of Eqs.(11)-(14) vanish, thus allowing one to replace $k_0^2\boldsymbol{E}(\boldsymbol{r})$ with $-\nabla^2\boldsymbol{E}(\boldsymbol{r})$, and similarly for $\boldsymbol{B}(\boldsymbol{r})$, $\boldsymbol{A}(\boldsymbol{r})$, and $\psi(\boldsymbol{r})$, whenever the need arises. These substitutions will be used in the following sections.

3. The Green function. In the spherical coordinate system (r,θ,φ), the Laplacian of the spherically symmetric function $G(\boldsymbol{r}) = e^{\mathrm{i}k_0 r}/r$ equals $\partial^2(rG)/r\partial r^2 = -k_0^2 G(\boldsymbol{r})$ everywhere except at the origin $r = 0$, where the function has a singularity. Thus, $(\nabla^2 + k_0^2)G(\boldsymbol{r}) = 0$ at all points $\boldsymbol{r}$ except at the origin. A good way to handle the singularity at $r = 0$ is to treat $G(\boldsymbol{r})$ as the limiting form of another function that has no such singularity, namely,

$$G(\boldsymbol{r}) = lim_{\varepsilon\to 0}\left(e^{\mathrm{i}k_0 r}/\sqrt{r^2+\varepsilon}\right). \tag{15}$$

The Laplacian of our well-behaved, non-singular function is readily found to be

$$r^{-2}\partial_r\left[r^2\partial_r\left(e^{\mathrm{i}k_0 r}/\sqrt{r^2+\varepsilon}\,\right)\right] = -\left[\frac{3\varepsilon}{(r^2+\varepsilon)^{5/2}} - \frac{2\mathrm{i}k_0\varepsilon}{r(r^2+\varepsilon)^{3/2}} + \frac{k_0^2}{(r^2+\varepsilon)^{1/2}}\right]e^{\mathrm{i}k_0 r}. \tag{16}$$

The first two functions appearing inside the square brackets on the right-hand side of Eq.(16) are confined to the vicinity of the origin at $r = 0$; they are tall, narrow, symmetric, and have the following volume integrals (see Appendix A for details):

$$\int_0^\infty 4\pi r^2(r^2+\varepsilon)^{-5/2}\mathrm{d}r = 4\pi/3\varepsilon. \tag{17}$$

$$\int_0^\infty 4\pi r(r^2+\varepsilon)^{-3/2}\mathrm{d}r = 4\pi/\sqrt{\varepsilon}\cdot \tag{18}$$

Thus, in the limit of sufficiently small ε, the first two functions appearing on the right-hand side of Eq.(16) can be represented by δ-functions,[†] and the entire equation may be written as

$$(\nabla^2 + k_0^2)(e^{\mathrm{i}k_0 r}/\sqrt{r^2+\varepsilon}) = -4\pi(1 - \mathrm{i}2k_0\sqrt{\varepsilon})\delta(\boldsymbol{r}). \tag{19}$$

This is the sense in which we can now state that $(\nabla^2 + k_0^2)G(\boldsymbol{r}) = -4\pi\delta(\boldsymbol{r})$ in the limit when $\varepsilon \to 0$. Shifting the center of the function to an arbitrary point $\boldsymbol{r}_0$, we will have

$$(\nabla^2 + k_0^2)G(\boldsymbol{r},\boldsymbol{r}_0) = -4\pi\delta(\boldsymbol{r}-\boldsymbol{r}_0). \tag{20}$$

The gradient of the Green function $G(\boldsymbol{r},\boldsymbol{r}_0)$, which plays an important role in our discussions of the following sections, is now found to be

$$\nabla G(\boldsymbol{r},\boldsymbol{r}_0) = \nabla\left(\frac{e^{\mathrm{i}k_0|\boldsymbol{r}-\boldsymbol{r}_0|}}{|\boldsymbol{r}-\boldsymbol{r}_0|}\right) = (\mathrm{i}k_0 - |\boldsymbol{r}-\boldsymbol{r}_0|^{-1})\left(\frac{e^{\mathrm{i}k_0|\boldsymbol{r}-\boldsymbol{r}_0|}}{|\boldsymbol{r}-\boldsymbol{r}_0|}\right)\frac{\boldsymbol{r}-\boldsymbol{r}_0}{|\boldsymbol{r}-\boldsymbol{r}_0|}\cdot \tag{21}$$

Appendix B provides an analysis of Eq.(20), an inhomogeneous Helmholtz equation, via Fourier transformation.

[†] When multiplied by ε, as required by Eq.(16), the integrand in Eq.(17) peaks at $\sim\varepsilon^{-1/2}$ at $r = \sqrt{2/3\,\varepsilon}$, then drops steadily to $\sim\sqrt[4]{\varepsilon}$ at $r = \sqrt[4]{\varepsilon}$. Similarly, the integrand in Eq.(18), again multiplied by ε, peaks at ~ 1 at $r = \sqrt{1/2\,\varepsilon}$, then drops steadily to $\sim\sqrt{\varepsilon}$ at $r = \sqrt[4]{\varepsilon}$.

4. The Huygens-Fresnel-Kirchhoff theory of diffraction.[1-3,9] Consider a scalar function $\psi(\boldsymbol{r})$ that satisfies the homogeneous Helmholtz equation $(\nabla^2 + k_o^2)\psi(\boldsymbol{r}) = 0$ everywhere within a volume V of free space enclosed by a surface S. Two examples of the geometry under consideration are depicted in Fig.1.[9] In general, $\psi(\boldsymbol{r})$ can represent the scalar potential or any Cartesian component of the monochromatic E-field, B-field, or A-field associated with an EM wave propagating in free space with frequency ω and wave-number $k_o = \omega/c$. In Fig.1(a), the EM wave arrives at the surface S_1 from sources located on the left-hand side of S_1, and enters the volume V contained within the closed surface $S = S_1 + S_2$. In Fig.1(b), the EM waves emanate from inside the closed surface S_1 and permeate the volume V enclosed by S_1 on one side and by a second closed surface S_2 that defines the outer boundary of V. In both figures, the point $\boldsymbol{r}_0$, where the field is being observed, is located inside the volume V, and the surface normals $\hat{\boldsymbol{n}}$ everywhere on the closed surface S are unit-vectors that point inward (i.e., into the volume V).

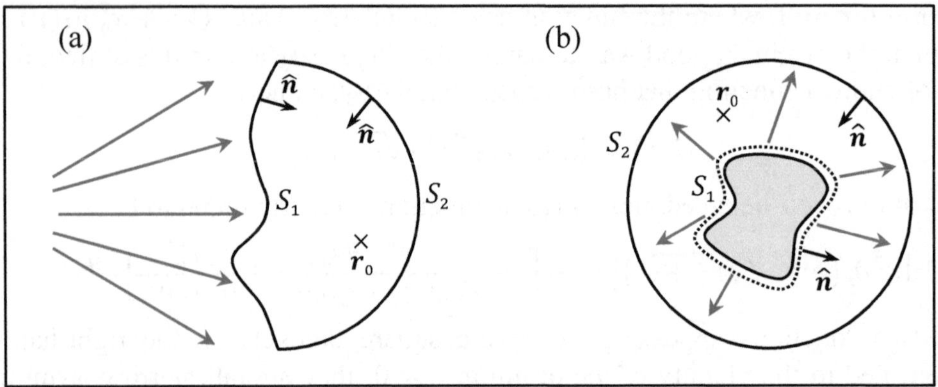

Fig.1. Two surfaces S_1 and S_2 bound the scattering region, which is assumed to be free of sources and material bodies. All the radiation within the scattering region comes from the outside. In (a) the sources of radiation are on the left-hand side of S_1, while in (b) the radiation emanates from the sources inside the closed surface S_1. The observation point $\boldsymbol{r}_o$ is an arbitrary point within the scattering region. All the points located on S_2 are assumed to be far away from $\boldsymbol{r}_0$, so that the fields that reach S_2 do not contribute to the fields observed at $\boldsymbol{r}_0$. At each and every point on S_1 and S_2, the surface normals $\hat{\boldsymbol{n}}$ point *into* the scattering region.

The essence of the theory developed by G. Kirchhoff in 1882 (building upon the original ideas of Huygens and Fresnel)[1] is an exact mathematical relation between the observed field $\psi(\boldsymbol{r}_o)$ and the field $\psi(\boldsymbol{r})$ that exists everywhere on the closed surface S. This relation is derived below using the sifting property of Dirac's δ-function, the relation between the δ-function and the Green function given in Eq.(20), some well-known identities in standard vector calculus, and Gauss's famous theorem of vector calculus, according to which $\int_V \nabla \cdot \boldsymbol{V} \mathrm{d}v = \oint_S \boldsymbol{V} \cdot \mathrm{d}\boldsymbol{s}$. We have

$$\psi(\boldsymbol{r}_o) = \int_V \psi(\boldsymbol{r})\delta(\boldsymbol{r} - \boldsymbol{r}_o)\mathrm{d}\boldsymbol{r} = -(4\pi)^{-1}\int_V \psi(\boldsymbol{r})(\nabla^2 + k_o^2)G(\boldsymbol{r},\boldsymbol{r}_o)\mathrm{d}\boldsymbol{r} \quad \leftarrow \boxed{\text{See Eq.(20)}}$$

$$= -(4\pi)^{-1}\int_V [\psi(\boldsymbol{r})\nabla \cdot \nabla G(\boldsymbol{r},\boldsymbol{r}_o) - G(\boldsymbol{r},\boldsymbol{r}_o)\nabla \cdot \nabla\psi(\boldsymbol{r})]\mathrm{d}\boldsymbol{r} \quad \leftarrow \boxed{\text{Replace } k_o^2\psi \text{ with } -\nabla^2\psi.}$$

$$= -(4\pi)^{-1}\int_V [\nabla \cdot (\psi\nabla G) - \cancel{\nabla G \cdot \nabla\psi} - \nabla \cdot (G\nabla\psi) + \cancel{\nabla\psi \cdot \nabla G}]\mathrm{d}\boldsymbol{r} \quad \leftarrow \boxed{\nabla \cdot (\phi\boldsymbol{V}) = \boldsymbol{V} \cdot \nabla\phi + \phi\nabla \cdot \boldsymbol{V}}$$

$$= (4\pi)^{-1}\int_S [\psi(\hat{\boldsymbol{n}} \cdot \nabla G) - G(\hat{\boldsymbol{n}} \cdot \nabla\psi)]\mathrm{d}s \quad \leftarrow \boxed{\text{Gauss' theorem; } \hat{\boldsymbol{n}} \text{ points into the volume } V.}$$

$$= (4\pi)^{-1}\int_S [\psi(\boldsymbol{r})\partial_n G(\boldsymbol{r},\boldsymbol{r}_o) - G(\boldsymbol{r},\boldsymbol{r}_o)\partial_n\psi(\boldsymbol{r})]\mathrm{d}s. \tag{22}$$

Example. In the extreme situation where S is a small sphere of radius ε centered at $\boldsymbol{r}_0$, we will have $\psi(\boldsymbol{r}) \cong \psi(\boldsymbol{r}_0)$, $G(\boldsymbol{r},\boldsymbol{r}_0) = e^{\mathrm{i}k_0\varepsilon}/\varepsilon$, and $\partial_n G(\boldsymbol{r},\boldsymbol{r}_0) = -(\mathrm{i}k_0 - \varepsilon^{-1})e^{\mathrm{i}k_0\varepsilon}/\varepsilon$. Given that the surface area of the sphere is $4\pi\varepsilon^2$, the second term in Eq.(22) makes a negligible contribution to the overall integral when $\varepsilon \to 0$. The first term, however, contains $e^{\mathrm{i}k_0\varepsilon}/\varepsilon^2$, which integrates to 4π in the limit of $\varepsilon \to 0$, yielding $\psi(\boldsymbol{r}_0)$ as the final result.

Taking the spherical (or hemi-spherical) surface S_2 in Fig.1 to be far away from the region of interest, the field $\psi(\boldsymbol{r})$ everywhere on S_2 should have the general form of $f(\theta,\varphi)e^{\mathrm{i}k_0 r}/r$ and, consequently, $\partial_n\psi(\boldsymbol{r}) \sim (\mathrm{i}k_0 - r^{-1})\psi(\boldsymbol{r})$. Similarly, across the surface S_2, the Green function has the asymptotic form $G(\boldsymbol{r},\boldsymbol{r}_0) \sim e^{\mathrm{i}k_0 r}/r$ and, therefore, $\partial_n G(\boldsymbol{r},\boldsymbol{r}_0) \sim (\mathrm{i}k_0 - r^{-1})G(\boldsymbol{r},\boldsymbol{r}_0)$. Thus, on the faraway surface S_2, the integrand in Eq.(22) must decline faster than $1/r^2$, which means that the contribution of S_2 to $\psi(\boldsymbol{r}_0)$ as given by Eq.(22) should be negligible. Kirchhoff's diffraction integral now relates the field at the observation point $\boldsymbol{r}_0$ to the field distribution on S_1, as follows:

$$\psi(\boldsymbol{r}_0) = (4\pi)^{-1}\int_{S_1}[\psi(\boldsymbol{r})\partial_n G(\boldsymbol{r},\boldsymbol{r}_0) - G(\boldsymbol{r},\boldsymbol{r}_0)\partial_n\psi(\boldsymbol{r})]\mathrm{d}s. \tag{23}$$

To make contact with the Huygens-Fresnel theory of diffraction, Kirchhoff suggested that both $\psi(\boldsymbol{r})$ and $\partial_n\psi(\boldsymbol{r})$ vanish on the opaque areas of the screen S_1, whereas in the open (or transparent, or unobstructed) regions, they retain the profiles they would have had in the absence of the screen. These suggestions, while reasonable from a practical standpoint and resulting in good agreement with experimental observations under many circumstances, are subject to criticism for their mathematical inconsistency, as will be elaborated in the next section.

5. The Rayleigh-Sommerfeld theory. In the important special case where the surface S_1 coincides with the xy-plane at $z = 0$, one can adjust the Green function in such a way as to eliminate either the first or the second term in the integrand of Eq.(23). These situations would then correspond, respectively, to the so-called Neumann and Dirichlet boundary conditions.[1,9] Let the observation point, which we assume to reside on the right-hand side of the planar surface S_1, be denoted by $\boldsymbol{r}_0^+ = (x_0, y_0, z_0)$, while its mirror image in S_1 (located on the left-hand side of S_1) is denoted by $\boldsymbol{r}_0^- = (x_0, y_0, -z_0)$. If we use $G(\boldsymbol{r},\boldsymbol{r}_0^+)$ in Eq.(23), we obtain $\psi(\boldsymbol{r}_0)$ on the left-hand side of the equation, but if we use $G(\boldsymbol{r},\boldsymbol{r}_0^-)$ instead, the integral will yield zero — simply because the peak of the corresponding δ-function now resides outside the integration volume. Thus, we can replace $G(\boldsymbol{r},\boldsymbol{r}_0)$ in Eq.(23) with either of the two functions $G(\boldsymbol{r},\boldsymbol{r}_0^+) \pm G(\boldsymbol{r},\boldsymbol{r}_0^-)$. On the S_1 plane, where $z = 0$, we have $G(\boldsymbol{r},\boldsymbol{r}_0^-) = G(\boldsymbol{r},\boldsymbol{r}_0^+)$ and $\partial_n G(\boldsymbol{r},\boldsymbol{r}_0^-) = -\partial_n G(\boldsymbol{r},\boldsymbol{r}_0^+)$. The resulting diffraction integrals, respectively satisfying the Neumann and Dirichlet boundary conditions, will then be

$$\psi(\boldsymbol{r}_0) = -\frac{1}{2\pi}\int_{S_1}\frac{\exp(\mathrm{i}k_0|\boldsymbol{r}-\boldsymbol{r}_0|)}{|\boldsymbol{r}-\boldsymbol{r}_0|}\partial_n\psi(\boldsymbol{r})\mathrm{d}s. \tag{24}$$

$$\psi(\boldsymbol{r}_0) = \frac{1}{2\pi}\int_{S_1}\frac{(\mathrm{i}k_0 - |\boldsymbol{r}-\boldsymbol{r}_0|^{-1})\exp(\mathrm{i}k_0|\boldsymbol{r}-\boldsymbol{r}_0|)}{|\boldsymbol{r}-\boldsymbol{r}_0|^2}[(\boldsymbol{r}-\boldsymbol{r}_0)\cdot\hat{\boldsymbol{n}}]\,\psi(\boldsymbol{r})\mathrm{d}s. \tag{25}$$

Note that, while Eq.(23) applies to any arbitrary surface S_1, the Rayleigh-Sommerfeld equations (24) and (25) are restricted to distributions that are specified on a flat plane. Given the scalar field profile $\psi(\boldsymbol{r})$ and/or its gradient on a flat plane, all three equations are exact consequences of Maxwell's equations. To apply these equations in practice, one must resort to some form of approximation to estimate the field distribution on S_1. The conventional approximation is that, in the opaque regions of the screen S_1, either $\psi(\boldsymbol{r})$ or $\partial_n\psi(\boldsymbol{r})$ or both are vanishingly small and, therefore, negligible, whereas in the transparent (or unobstructed) regions of S_1, the field $\psi(\boldsymbol{r})$ and/or its gradient $\partial_n\psi(\boldsymbol{r})$ (along the surface-normal) retain the profile they would have had in the absence of the screen. In this way, one can proceed to evaluate the integral on the open (or transparent, or unobstructed) apertures of S_1 in order to arrive at a reasonable estimate of $\psi(\boldsymbol{r}_0)$ at the desired observation location.

As a formula for computing diffraction patterns from one or more apertures in an otherwise opaque screen, the problem with Eq.(23) is that, when combined with Kirchhoff's assumption that *both* ψ and $\partial_n\psi$ vanish on the opaque regions of the physical screen at S_1, it becomes mathematically inconsistent. This is because an analytic function such as $\psi(\boldsymbol{r})$ vanishes everywhere if both ψ and $\partial_n\psi$ happen to be zero on any patch of the surface S_1. In contrast, Eq.(24), when applied to a physical screen, requires only the assumption that $\partial_n\psi$ be zero on the opaque regions of the screen. While still an approximation, this is a much more mathematically palatable condition than the Kirchhoff requirement.[9] Similarly, Eq.(25) requires only the approximation that ψ be zero on the opaque regions. Thus, on the grounds of mathematical consistency, there is a preference for either Eq.(24) or Eq.(25) over Eq.(23). However, given the aforementioned approximate nature of the values chosen for ψ and $\partial_n\psi$ across the screen at S_1, it turns out that scalar diffraction calculations based on these three formulas yield nearly identical results, rendering them equally useful in practical applications.[1,9]

An auxiliary consequence of Eqs.(24) and (25) is that, upon subtracting one from the other, the integral of $\partial_n(G\psi)$ over the entire flat plane S_1 is found to vanish; that is,

$$\int_{S_1} \partial_n[G(\boldsymbol{r},\boldsymbol{r}_0)\psi(\boldsymbol{r})]\mathrm{d}s = 0. \tag{26}$$

We will have occasion to use this important identity in the following section.

6. Vector diffraction. Applying the Kirchhoff formula in Eq.(22), where the integral is over the closed surface $S = S_1 + S_2$, and the function $\psi(\boldsymbol{r})$ is any scalar field that satisfies the Helmholtz equation, to a Cartesian component of the E-field, say, E_x, we write

$$\begin{aligned} 4\pi E_x(\boldsymbol{r}_0) &= \int_S [E_x(\boldsymbol{r})\partial_n G(\boldsymbol{r},\boldsymbol{r}_0) - G(\boldsymbol{r},\boldsymbol{r}_0)\partial_n E_x(\boldsymbol{r})]\mathrm{d}s \qquad \boxed{\boldsymbol{\nabla}(\phi\psi) = \phi\boldsymbol{\nabla}\psi + \psi\boldsymbol{\nabla}\phi} \\ &= \int_S [(\hat{\boldsymbol{n}}\cdot\boldsymbol{\nabla}G)E_x - (\hat{\boldsymbol{n}}\cdot\boldsymbol{\nabla}E_x)G]\mathrm{d}s = \int_S [2(\hat{\boldsymbol{n}}\cdot\boldsymbol{\nabla}G)E_x - \hat{\boldsymbol{n}}\cdot\boldsymbol{\nabla}(GE_x)]\mathrm{d}s \\ &= 2\int_S (\hat{\boldsymbol{n}}\cdot\boldsymbol{\nabla}G)E_x\mathrm{d}s + \int_V \boldsymbol{\nabla}\cdot\boldsymbol{\nabla}(GE_x)\mathrm{d}\boldsymbol{r}. \leftarrow \boxed{\text{Gauss' theorem; } \hat{\boldsymbol{n}} \text{ points into the volume } V} \end{aligned} \tag{27}$$

Given that Eq.(27) is similarly satisfied by the remaining components E_y, E_z of the E-field, the vectorial version of Kirchhoff's formula may be written down straightforwardly. Algebraic manipulations (using standard vector calculus identities described in Appendix C) simplify the final result, yielding the following expression for the E-field at the observation point:[9]

$$\boldsymbol{E}(\boldsymbol{r}_0) = (4\pi)^{-1}\int_S [(\hat{\boldsymbol{n}}\times\boldsymbol{E})\times\boldsymbol{\nabla}G + (\hat{\boldsymbol{n}}\cdot\boldsymbol{E})\boldsymbol{\nabla}G + \mathrm{i}\omega(\hat{\boldsymbol{n}}\times\boldsymbol{B})G]\mathrm{d}s. \tag{28}$$

Once again, it is easy to show that the contribution of the spherical (or hemi-spherical) surface S_2 to the overall integral in Eq.(28) is negligible. This is because, in the far field, $\hat{\boldsymbol{n}}\cdot\boldsymbol{E} \to 0$ and $\hat{\boldsymbol{n}}\times\boldsymbol{B} \to \boldsymbol{E}/c$, while $|\boldsymbol{E}| \sim e^{\mathrm{i}k_0 r}/r$, $G \sim e^{\mathrm{i}k_0 r}/r$ and $\boldsymbol{\nabla}G \sim -(\mathrm{i}k_0 - r^{-1})e^{\mathrm{i}k_0 r}\hat{\boldsymbol{n}}/r$. Consequently,

$$\boldsymbol{E}(\boldsymbol{r}_0) = (4\pi)^{-1}\int_{S_1} [(\hat{\boldsymbol{n}}\times\boldsymbol{E})\times\boldsymbol{\nabla}G + (\hat{\boldsymbol{n}}\cdot\boldsymbol{E})\boldsymbol{\nabla}G + \mathrm{i}\omega(\hat{\boldsymbol{n}}\times\boldsymbol{B})G]\mathrm{d}s. \tag{29}$$

A similar argument can be used to arrive at the vector Kirchhoff formula for the B-field, namely,

$$\boldsymbol{B}(\boldsymbol{r}_0) = (4\pi)^{-1} \int_{S_1} [(\hat{\boldsymbol{n}} \times \boldsymbol{B}) \times \boldsymbol{\nabla} G + (\hat{\boldsymbol{n}} \cdot \boldsymbol{B}) \boldsymbol{\nabla} G - \mathrm{i}(\omega/c^2)(\hat{\boldsymbol{n}} \times \boldsymbol{E}) G] \mathrm{d}s. \tag{30}$$

Needless to say, Eq.(30) could also be derived directly from Eq.(29), or vice versa, although the algebra becomes tedious at times; see Appendix D for one such derivation.

We mention in passing that the arguments that led to Eqs.(29) and (30) could *not* be repeated for the vector potential $\boldsymbol{A}(\boldsymbol{r})$, since in arriving at Eq.(28), in the step where $\boldsymbol{\nabla} \cdot \boldsymbol{E}$ or $\boldsymbol{\nabla} \cdot \boldsymbol{B}$ are set to zero (see Appendix C), we now have $\boldsymbol{\nabla} \cdot \boldsymbol{A} = \mathrm{i}\omega\psi/c^2$.[‡]

In parallel with the arguments advanced previously in conjunction with the Rayleigh-Sommerfeld formulation for scalar fields, one may also modify Eqs.(29) and (30) by setting $G(\boldsymbol{r}, \boldsymbol{r}_0) = 0$ and $\boldsymbol{\nabla} G = 2\partial_n G(\boldsymbol{r}, \boldsymbol{r}_0)\hat{\boldsymbol{n}}$, provided that S_1 is a planar surface. This is equivalent to applying Eq.(25) directly to the x, y, z components of the E-field (or the B-field). It is *not* permissible, however, to retain G and remove $\boldsymbol{\nabla} G$ (again, in the case of a planar S_1), because the gradient of $G(\boldsymbol{r}, \boldsymbol{r}_0^+) + G(\boldsymbol{r}, \boldsymbol{r}_0^-)$ has a nonzero projection onto the xy-plane.

In those special (yet important) cases where S_1 coincides with the xy-plane at $z = 0$, one could begin by applying either Eq.(24) or Eq.(25) to the x, y, z components of the field under consideration. Manipulating the resulting equation with the aid of vector-algebraic identities in conjunction with the fact that $\boldsymbol{\nabla} G = -\boldsymbol{\nabla}_0 G$, leads to vector diffraction formulas that could be useful under special circumstances. For instance, the vectorial equivalent of Eq.(25) yields

$$\begin{aligned} 2\pi \boldsymbol{E}(\boldsymbol{r}_0) &= \int_{S_1} (\hat{\boldsymbol{n}} \cdot \boldsymbol{\nabla} G) \boldsymbol{E}(\boldsymbol{r}) \mathrm{d}s = \int_{S_1} [(\hat{\boldsymbol{n}} \times \boldsymbol{E}) \times \boldsymbol{\nabla} G + (\boldsymbol{E} \cdot \boldsymbol{\nabla} G)\hat{\boldsymbol{n}}] \mathrm{d}s \quad \leftarrow \boxed{(\boldsymbol{a} \times \boldsymbol{b}) \times \boldsymbol{c} = (\boldsymbol{a} \cdot \boldsymbol{c})\boldsymbol{b} - (\boldsymbol{b} \cdot \boldsymbol{c})\boldsymbol{a}} \\ &= \int_{S_1} \boldsymbol{\nabla}_0 G \times (\hat{\boldsymbol{n}} \times \boldsymbol{E}) \mathrm{d}s + \{\textstyle\int_{S_1} [\boldsymbol{\nabla} \cdot (G\boldsymbol{E}) - G\overbrace{\boldsymbol{\nabla} \cdot \boldsymbol{E}}^{0}] \mathrm{d}s\} \hat{\boldsymbol{n}} \quad \leftarrow \boxed{\boldsymbol{\nabla} \cdot (\psi \boldsymbol{V}) = \psi \boldsymbol{\nabla} \cdot \boldsymbol{V} + \boldsymbol{V} \cdot \boldsymbol{\nabla}\psi} \\ &= \int_{S_1} \boldsymbol{\nabla}_0 \times [(\hat{\boldsymbol{n}} \times \boldsymbol{E}) G] \mathrm{d}s + [\textstyle\int_{S_1} \boldsymbol{\nabla} \cdot (G\boldsymbol{E}) \mathrm{d}s] \hat{\boldsymbol{n}} \quad \leftarrow \boxed{\boldsymbol{\nabla} \times (\psi \boldsymbol{V}) = \psi \boldsymbol{\nabla} \times \boldsymbol{V} + \boldsymbol{\nabla}\psi \times \boldsymbol{V}} \\ &= \boldsymbol{\nabla}_0 \times \int_{S_1} [\hat{\boldsymbol{n}} \times \boldsymbol{E}(\boldsymbol{r})] G(\boldsymbol{r}, \boldsymbol{r}_0) \mathrm{d}s + \big[\textstyle\iint_{-\infty}^{\infty} [\overbrace{\partial_x(GE_x)}^{0} + \overbrace{\partial_y(GE_y)}^{0} + \overbrace{\partial_z(GE_z)}^{0}] \mathrm{d}x\mathrm{d}y\big] \hat{\boldsymbol{n}}. \end{aligned} \tag{31}$$

It is easy to see that the first two terms of the second integral on the right-hand side of Eq.(31) vanish since both GE_x and GE_y go to zero in the far out regions of the xy-plane. The vanishing of the third term, however, requires invoking Eq.(26) as applied to the z-component of the E-field. Note, as a matter of consistency, that on the left-hand side of Eq.(31), $\boldsymbol{\nabla}_0 \cdot \boldsymbol{E}(\boldsymbol{r}_0) = 0$ and that, on the right-hand side, the divergence of the curl is always zero.

A similar argument can be advanced for the B-field and, therefore, the following vector diffraction equations are generally valid for a planar surface S_1:

$$\boldsymbol{E}(\boldsymbol{r}_0) = (2\pi)^{-1} \boldsymbol{\nabla}_0 \times \int_{S_1} [\hat{\boldsymbol{n}} \times \boldsymbol{E}(\boldsymbol{r})] G(\boldsymbol{r}, \boldsymbol{r}_0) \mathrm{d}s. \tag{32}$$

$$\boldsymbol{B}(\boldsymbol{r}_0) = (2\pi)^{-1} \boldsymbol{\nabla}_0 \times \int_{S_1} [\hat{\boldsymbol{n}} \times \boldsymbol{B}(\boldsymbol{r})] G(\boldsymbol{r}, \boldsymbol{r}_0) \mathrm{d}s. \tag{33}$$

---

‡ In setting $\boldsymbol{\nabla} \cdot \boldsymbol{E} = 0$, Maxwell's first equation, $\varepsilon_0 \boldsymbol{\nabla} \cdot \boldsymbol{E} = \rho_{\text{total}} = \rho_{\text{free}} + \rho_{\text{bound}}$, has been invoked, with the caveat that the surface S is slightly detached from material bodies where electric charges of one kind or another may reside. No such caveat is needed, however, when setting $\boldsymbol{\nabla} \cdot \boldsymbol{B} = 0$, which is simply Maxwell's fourth equation. The situation is quite different with the vector potential $\boldsymbol{A}$, since setting $\boldsymbol{\nabla} \cdot \boldsymbol{A} = 0$ implies working in the Coulomb gauge. While the standard relations $\boldsymbol{B} = \boldsymbol{\nabla} \times \boldsymbol{A}$ and $\boldsymbol{E} = -\boldsymbol{\nabla}\psi - \partial_t \boldsymbol{A}$ remain valid in all gauges, the equations that relate ψ and $\boldsymbol{A}$ to the charge and current densities are gauge dependent. In particular, $\boldsymbol{A}$ in the Coulomb gauge depends not only on the current-density $\boldsymbol{J}_{\text{total}}$, but also on the charge-density ρ_{total}. While the charge-current continuity equation $\boldsymbol{\nabla} \cdot \boldsymbol{J} + \partial_t \rho = 0$ can be used to arrive at a Helmholtz equation for $\boldsymbol{A}(\boldsymbol{r}, t)$ in the Coulomb gauge, the term appearing on the right-hand side of the equation will be the transverse current-density, which does not necessarily vanish in the free-space regions of the system under consideration.

In similar fashion, the vectorial equivalent of Eq.(24) yields

$$2\pi \boldsymbol{E}(\boldsymbol{r}_0) = -\int_{S_1} G\partial_z \boldsymbol{E} \mathrm{d}s = -\int_{S_1} G[\partial_z E_x \hat{\boldsymbol{x}} + \partial_z E_y \hat{\boldsymbol{y}} - (\partial_x E_x + \partial_y E_y)\hat{\boldsymbol{z}}]\mathrm{d}s$$

$$\boldsymbol{\nabla} \cdot \boldsymbol{E} = 0 \rightarrow \partial_z E_z = -(\partial_x E_x + \partial_y E_y)$$

$$= \int_{S_1} [E_x(\partial_z G\hat{\boldsymbol{x}} - \partial_x G\hat{\boldsymbol{z}}) + E_y(\partial_z G\hat{\boldsymbol{y}} - \partial_y G\hat{\boldsymbol{z}})$$

$$-\partial_z(GE_x)\hat{\boldsymbol{x}} - \partial_z(GE_y)\hat{\boldsymbol{y}} + \partial_x(GE_x)\hat{\boldsymbol{z}} + \partial_y(GE_y)\hat{\boldsymbol{z}}]\mathrm{d}s$$

$$= \int_{S_1} (\hat{\boldsymbol{n}} \times \boldsymbol{E}) \times \boldsymbol{\nabla} G \mathrm{d}s - \iint_{-\infty}^{\infty} [\underbrace{\partial_z(GE_x)}_{0}\hat{\boldsymbol{x}} + \underbrace{\partial_z(GE_y)}_{0}\hat{\boldsymbol{y}} - \underbrace{\partial_x(GE_x)}_{0}\hat{\boldsymbol{z}} - \underbrace{\partial_y(GE_y)}_{0}\hat{\boldsymbol{z}}]\mathrm{d}x\mathrm{d}y$$

$$= \int_{S_1} \boldsymbol{\nabla}_0 G \times (\hat{\boldsymbol{n}} \times \boldsymbol{E})\mathrm{d}s = \boldsymbol{\nabla}_0 \times \int_{S_1} [\hat{\boldsymbol{n}} \times \boldsymbol{E}(\boldsymbol{r})] G(\boldsymbol{r}, \boldsymbol{r}_0)\mathrm{d}s. \quad (34)$$

On the penultimate line of Eq.(34), the first two terms in the integral are seen to vanish when Eq.(26) is applied to the x and y components of the E-field; the 3rd and 4th terms go to zero due to the vanishing of GE_x and GE_y in the far out regions of the xy-plane. Equation (34) thus yields the same expression for $\boldsymbol{E}(\boldsymbol{r}_0)$ as the one reached via Eq.(31).

Example 1. A plane-wave $\boldsymbol{E}(\boldsymbol{r}, t) = (E_{xo}\hat{\boldsymbol{x}} + E_{yo}\hat{\boldsymbol{y}} + E_{zo}\hat{\boldsymbol{z}})e^{\mathrm{i}(k_x x + k_y y + k_z z - \omega t)}$ arriving from the region $z < 0$ is reflected from a perfectly conducting plane mirror located in the xy-plane at $z = 0$. The exact cancellation of the tangential components of the E-field at the mirror surface means that the scattered E-field in the xy-plane at $z = 0^+$ is given by[§]

$$\boldsymbol{E}_s(x, y) = -(E_{xo}\hat{\boldsymbol{x}} + E_{yo}\hat{\boldsymbol{y}} + E_{zo}\hat{\boldsymbol{z}})e^{\mathrm{i}(k_x x + k_y y)}. \quad (35)$$

Beyond the mirror in the region $z > 0$, the scattered E-field is obtained from Eq.(32), as follows:

$$2\pi \boldsymbol{E}_s(\boldsymbol{r}_0) = \boldsymbol{\nabla}_0 \times \iint_{-\infty}^{\infty} \hat{\boldsymbol{n}} \times (-E_{xo}\hat{\boldsymbol{x}} - E_{yo}\hat{\boldsymbol{y}} - E_{zo}\hat{\boldsymbol{z}})e^{\mathrm{i}(k_x x + k_y y)} \frac{\exp\{\mathrm{i}k_0[(x-x_0)^2 + (y-y_0)^2 + z_0^2]^{1/2}\}}{[(x-x_0)^2 + (y-y_0)^2 + z_0^2]^{1/2}} \mathrm{d}x\mathrm{d}y$$

$$= \boldsymbol{\nabla}_0 \times (E_{yo}\hat{\boldsymbol{x}} - E_{xo}\hat{\boldsymbol{y}})e^{\mathrm{i}(k_x x_0 + k_y y_0)} \iint_{-\infty}^{\infty} e^{\mathrm{i}(k_x x + k_y y)} \frac{\exp[\mathrm{i}k_0(x^2 + y^2 + z_0^2)^{1/2}]}{(x^2 + y^2 + z_0^2)^{1/2}} \mathrm{d}x\mathrm{d}y. \quad (36)$$

The 2D Fourier transform appearing in Eq.(36) is readily found to be (see Appendix E):

$$\iint_{-\infty}^{\infty} \frac{\exp\left(\mathrm{i}k_0\sqrt{x^2 + y^2 + z^2}\right)}{\sqrt{x^2 + y^2 + z^2}} e^{\mathrm{i}(k_x x + k_y y)} \mathrm{d}x\mathrm{d}y = \mathrm{i}(2\pi/k_z)e^{\mathrm{i}k_z z}. \quad (37)$$

This is a valid equation whether the incident beam is of the propagating type (i.e., homogeneous plane-wave, with real-valued k_z), or of the evanescent type (i.e., inhomogeneous, with imaginary k_z). Substitution into Eq.(36) now yields

$$\boldsymbol{E}_s(\boldsymbol{r}_0) = \boldsymbol{\nabla}_0 \times \mathrm{i}(E_{yo}\hat{\boldsymbol{x}} - E_{xo}\hat{\boldsymbol{y}})e^{\mathrm{i}(k_x x_0 + k_y y_0 + k_z z_0)}/k_z$$

$$= [-E_{xo}\hat{\boldsymbol{x}} - E_{yo}\hat{\boldsymbol{y}} + (k_x E_{xo} + k_{yo} E_y)\hat{\boldsymbol{z}}/k_z]e^{\mathrm{i}(k_x x_0 + k_y y_0 + k_z z_0)}$$

$$= -(E_{xo}\hat{\boldsymbol{x}} + E_{yo}\hat{\boldsymbol{y}} + E_{zo}\hat{\boldsymbol{z}})e^{\mathrm{i}(k_x x_0 + k_y y_0 + k_z z_0)}. \quad (38)$$

[§] Whereas on opposite facets of the mirror the scattered E_x (as well as the scattered E_y) are identical, the presence of surface charges requires that the sign of the scattered E_z flip between $z = 0^-$ and $z = 0^+$. The scattered E-field amplitude is thus $(-E_{xo}, -E_{yo}, E_{zo})$ at $z = 0^-$ and $(-E_{xo}, -E_{yo}, -E_{zo})$ at $z = 0^+$. Similarly, the existence of surface currents requires the scattered B-field amplitude to be $(B_{xo}, B_{yo}, -B_{zo})$ at $z = 0^-$ and $(-B_{xo}, -B_{yo}, -B_{zo})$ at $z = 0^+$.

As expected, in the half-space $z > 0$, the scattered field precisely cancels out the incident field. This result is quite general and applies to any profile for the incident beam, not just plane-waves, for the simple reason that any incident beam can be expressed as a superposition of plane-waves. It should also be clear that, in the absence of the perfectly conducting reflector in the xy-plane, the distribution of the tangential E-field (or the tangential B-field) throughout the xy-plane at $z = 0$ can be used to reconstruct the entire distribution in the $z > 0$ half-space via either Eq.(32) or Eq.(33).

Example 2. Consider a screen in the xy-plane at $z = 0$ consisting of obstructing segment(s) in the form of thin sheet(s) of perfect conductors and open regions otherwise. A monochromatic incident beam creates surface charges and surface currents on the metallic segment(s) of this planar screen. The scattered fields produced by the induced surface charges and currents can be described in terms of the scattered scalar and vector potentials $\psi_s(\boldsymbol{r},t)$ and $\boldsymbol{A}_s(\boldsymbol{r},t)$. Given that the induced surface current has no component along the z-axis, the vector potential will likewise have only x and y components. Application of Eq.(24) to $A_{sx}(\boldsymbol{r})$ and $A_{sy}(\boldsymbol{r})$ yields

$$\boldsymbol{A}_s(\boldsymbol{r}_0) = -\frac{1}{2\pi}\int_{S_1}\frac{\exp(\mathrm{i}k_0|\boldsymbol{r}-\boldsymbol{r}_0|)}{|\boldsymbol{r}-\boldsymbol{r}_0|}(\partial_z A_{sx}\hat{\boldsymbol{x}} + \partial_z A_{sy}\hat{\boldsymbol{y}})\mathrm{d}s. \tag{39}$$

Now, $\boldsymbol{B}_s(\boldsymbol{r}) = \boldsymbol{\nabla}\times\boldsymbol{A}_s(\boldsymbol{r}) = -\partial_z A_{sy}\hat{\boldsymbol{x}} + \partial_z A_{sx}\hat{\boldsymbol{y}} + (\partial_x A_{sy} - \partial_y A_{sx})\hat{\boldsymbol{z}}$. Consequently,

$$\boldsymbol{A}_s(\boldsymbol{r}_0) = -\frac{1}{2\pi}\int_{S_1}\frac{\exp(\mathrm{i}k_0|\boldsymbol{r}-\boldsymbol{r}_0|)}{|\boldsymbol{r}-\boldsymbol{r}_0|}(B_{sy}\hat{\boldsymbol{x}} - B_{sx}\hat{\boldsymbol{y}})\mathrm{d}s. \tag{40}$$

The symmetry of the scattered field ensures that B_{sx} and B_{sy} are zero in the open areas of the screen, so the integral in Eq.(40) need be evaluated only on the metallic surfaces of the screen. We will have

$$\boldsymbol{B}_s(\boldsymbol{r}_0) = \boldsymbol{\nabla}_0\times\boldsymbol{A}_s(\boldsymbol{r}_0) = (2\pi)^{-1}\boldsymbol{\nabla}_0\times\int_{\text{metal}}[\hat{\boldsymbol{n}}\times\boldsymbol{B}_s(\boldsymbol{r})]\frac{\exp(\mathrm{i}k_0|\boldsymbol{r}-\boldsymbol{r}_0|)}{|\boldsymbol{r}-\boldsymbol{r}_0|}\mathrm{d}s. \tag{41}$$

In situations where a thin, flat metallic object acts as a scatterer, Eq.(41) provides a simple way to compute the scattered field provided, of course, that an estimate of the magnetic field at the metal surface (or, equivalently, an estimate of the induced surface current-density) can be obtained. Needless to say, considering that, in the absence of the scatterer, the continued propagation of the incident beam into the $z > 0$ half-space can be reconstructed from the tangential component of the incident B-field in the $z = 0$ plane (see Example 1), one may add the incident B-field to the scattered field of Eq.(41) and obtain, unsurprisingly, the general vector diffraction formula of Eq.(33).

Example 3. Figure 2 shows a perfectly conducting thin sheet residing in the upper half of the xy-plane at $z = 0$. The incident beam is a monochromatic plane-wave of frequency ω, wavelength $\lambda_0 = 2\pi c/\omega$, linear-polarization aligned with $\hat{\boldsymbol{y}}$, and propagation direction along the unit-vector $\boldsymbol{\sigma}_{\text{inc}} = (\sin\theta_{\text{inc}})\hat{\boldsymbol{x}} + (\cos\theta_{\text{inc}})\hat{\boldsymbol{z}}$, as follows:

$$\boldsymbol{E}(\boldsymbol{r},t) = E_{\text{inc}}\hat{\boldsymbol{y}}\exp[\mathrm{i}k_0(\sin\theta_{\text{inc}}\,x + \cos\theta_{\text{inc}}\,z - ct)]. \tag{42}$$

Using Eq.(32), we derive the diffracted E-field at the observation point $\boldsymbol{r}_0 = x_0\hat{\boldsymbol{x}} + z_0\hat{\boldsymbol{z}}$. (The symmetry of the problem ensures that the field profile is independent of the y-coordinate of the observation point.) In what follows, we use the 0th and 1st order Hankel functions of type 1, $H_0^{(1)}(\zeta)$ and $H_1^{(1)}(\zeta)$, as well as the (complex) Fresnel integral $F(\zeta) = \int_\zeta^\infty \exp(ix^2)\,dx$.[26,27] For a graphical representation of the Fresnel integral via the so-called Cornu spiral,[28] see Appendix F. The following identities will be needed further below:

$$\int_{-\infty}^{\infty} \frac{\exp(i\sqrt{x^2+\zeta^2})}{\sqrt{x^2+\zeta^2}}\,dx = i\pi H_0^{(1)}(\zeta). \quad \text{G\&R}^{27}\ \mathbf{8.421\text{-}11} \tag{43}$$

$$\frac{d}{d\zeta} H_0^{(1)}(\zeta) = -H_1^{(1)}(\zeta). \quad \text{G\&R}^{27}\ \mathbf{8.473\text{-}6} \tag{44}$$

$$H_1^{(1)}(\zeta) \sim \sqrt{\frac{2}{\pi\zeta}}\,e^{i(\zeta - 3\pi/4)}, \quad (\zeta \gg 1). \quad \text{G\&R}^{27}\ \mathbf{8.451\text{-}3} \tag{45}$$

$$\begin{aligned} F(\zeta) &= \int_0^\infty \exp(ix^2)\,dx - \left[\int_0^\zeta \cos(x^2)\,dx + i\int_0^\zeta \sin(x^2)\,dx\right] \\ &= \sqrt{\pi/4}\,e^{i\pi/4} - \sqrt{\pi/2}\,[C(\zeta) + iS(\zeta)]. \quad \text{G\&R}^{27}\ \mathbf{8.250\text{-}2,3} \end{aligned} \tag{46}$$

The E-field at the observation point $\boldsymbol{r}_0$ is found to be

$$\begin{aligned} \boldsymbol{E}(\boldsymbol{r}_0) &\cong (2\pi)^{-1}\boldsymbol{\nabla}_0 \times \int_{x=-\infty}^{0}\int_{y=-\infty}^{\infty} (\hat{\boldsymbol{z}} \times E_{\text{inc}}\hat{\boldsymbol{y}}) e^{ik_0 \sin\theta_{\text{inc}} x} \times \frac{\exp\{ik_0[(x-x_0)^2+y^2+z_0^2]^{1/2}\}}{[(x-x_0)^2+y^2+z_0^2]^{1/2}}\,dy\,dx \\ &= -\tfrac{1}{2}i\boldsymbol{\nabla}_0 \times E_{\text{inc}}\,\hat{\boldsymbol{x}} \int_{-\infty}^{0} e^{ik_0 \sin\theta_{\text{inc}} x} H_0^{(1)}\left[k_0\sqrt{(x-x_0)^2 + z_0^2}\right]dx \\ &= \tfrac{1}{2}iE_{\text{inc}}\,\hat{\boldsymbol{y}} \int_{-\infty}^{0} \frac{k_0 z_0}{[(x-x_0)^2+z_0^2]^{1/2}} e^{ik_0 \sin\theta_{\text{inc}} x} H_1^{(1)}\left[k_0\sqrt{(x-x_0)^2 + z_0^2}\right]dx \\ &\cong \frac{i e^{-i3\pi/4} E_{\text{inc}}\,\hat{\boldsymbol{y}}}{\sqrt{2\pi}} \int_{-\infty}^{0} \frac{\sqrt{k_0 z_0}}{[(x-x_0)^2+z_0^2]^{3/4}} \exp(ik_0 \sin\theta_{\text{inc}}\,x) \exp\left[ik_0\sqrt{(x-x_0)^2 + z_0^2}\right] dx. \end{aligned} \tag{47}$$

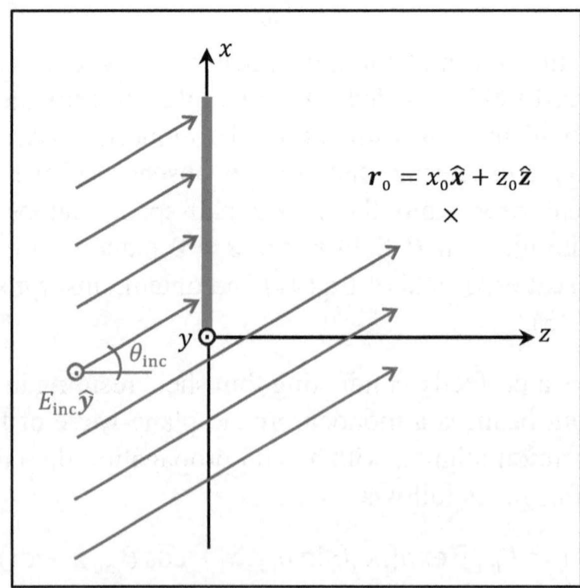

Fig.2. A perfectly conducting thin sheet sits in the upper half of the xy-plane at $z = 0$. The incident plane-wave, which is linearly polarized along the y-axis, has amplitude E_{inc}, frequency ω, wave-number $k_0 = \omega/c$, and propagation direction $\boldsymbol{\sigma}_{\text{inc}} = \sin\theta_{\text{inc}}\,\hat{\boldsymbol{x}} + \cos\theta_{\text{inc}}\,\hat{\boldsymbol{z}}$. The observation point $\boldsymbol{r}_0$ is in the xz-plane.

Here, we have used the large-argument approximate form of $H_1^{(1)}(\zeta)$ given by Eq.(45). Assuming that $z_o \gg |x - x_o|$, we proceed by invoking the following approximation:

$$\exp\left[ik_o\sqrt{(x-x_o)^2+z_o^2}\right]/[(x-x_o)^2+z_o^2]^{3/4} \cong z_o^{-3/2}e^{ik_o[z_o+(x-x_o)^2/2z_o]}. \tag{48}$$

Note that this approximation is not accurate when $|x - x_o|$ acquires large values; however, the rapid phase variations of the integrand in Eq.(47) ensure that the contributions to the integral at points x that are far from x_o are insignificant. Substitution from Eq.(48) into Eq.(47) yields

$$\begin{aligned}\boldsymbol{E}(\boldsymbol{r}_o) &\cong \frac{\sqrt{k_0}\,E_{\text{inc}}\,\hat{\boldsymbol{y}}}{\sqrt{2\pi z_0}}e^{i(k_0z_0-\pi/4)}\int_{-\infty}^{0}\exp(ik_o\sin\theta_{\text{inc}}\,x)\exp[ik_o(x-x_o)^2/(2z_o)]\,dx\\ &= \frac{\sqrt{k_0}\,E_{\text{inc}}\,\hat{\boldsymbol{y}}}{\sqrt{2\pi z_0}}e^{i(k_0z_0-\pi/4)}\exp[ik_o(x_o\sin\theta_{\text{inc}}-½z_o\sin^2\theta_{\text{inc}})]\\ &\quad\times\int_{-\infty}^{0}\exp[ik_o(x-x_o+z_o\sin\theta_{\text{inc}})^2/(2z_o)]\,dx\\ &= \frac{E_{\text{inc}}\,\hat{\boldsymbol{y}}}{\sqrt{\pi}e^{i\pi/4}}e^{ik_0(x_0\sin\theta_{\text{inc}}+z_0-½z_0\sin^2\theta_{\text{inc}})}\int_{\sqrt{k_0/2z_0}(x_0-z_0\sin\theta_{\text{inc}})}^{\infty}\exp(ix^2)\,dx\\ &= \frac{F\left[\sqrt{\pi/(\lambda_0z_0)}\,(x_0-z_0\sin\theta_{\text{inc}})\right]}{\sqrt{\pi}\exp(i\pi/4)}E_{\text{inc}}\,\hat{\boldsymbol{y}}\,e^{ik_0[x_0\sin\theta_{\text{inc}}+z_0(1-½\sin^2\theta_{\text{inc}})]}.\end{aligned} \tag{49}$$

At the edge of the geometric shadow, i.e., the straight line where $x_o = z_o \sin\theta_{\text{inc}}$, we have $F(0) = ½\sqrt{\pi}e^{i\pi/4}$. On this line, the E-field amplitude is $½E_{\text{inc}}$. Above the shadow's edge, the field amplitude steadily declines, whereas below the edge, there occur a large number of oscillations before the field settles into what is essentially the incident plane-wave.[1] Figure 3 shows a typical plot of the far-field intensity (i.e., the square of the E-field amplitude) versus the distance (along the x-axis at a fixed value of z_o) from the edge of the geometric shadow.

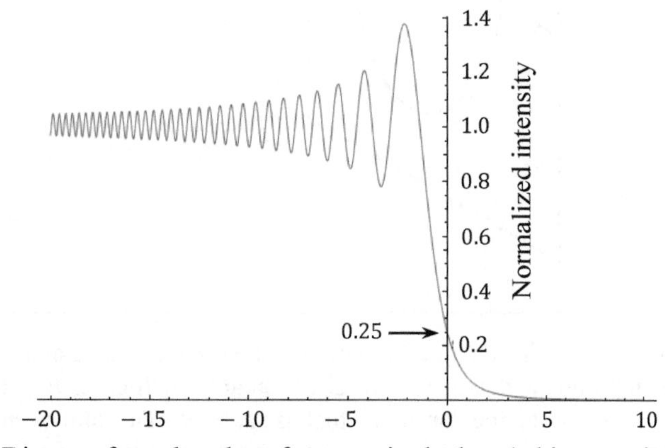

Fig.3. Normalized intensity in the far field of a sharp edge as a function of distance along the x-axis (at fixed z_o) from the edge of the geometric shadow, where $x_o = z_o \sin\theta_{\text{inc}}$.

Example 4. Shown in Fig.4 is a circular aperture of radius R within an otherwise opaque screen located in the xy-plane at $z = 0$. A plane-wave $\boldsymbol{E}(\boldsymbol{r},t) = \boldsymbol{E}_{\text{inc}}\exp[i(k_o\boldsymbol{\sigma}_{\text{inc}}\cdot\boldsymbol{r} - \omega t)]$, where the unit-vector $\boldsymbol{\sigma}_{\text{inc}}$ specifies the direction of incidence, arrives at the aperture from the left-hand side. Maxwell's 3rd equation identifies the incident B-field as $\boldsymbol{B} = \boldsymbol{\sigma}_{\text{inc}} \times \boldsymbol{E}/c$. The observation point $\boldsymbol{r}_o = r_o\boldsymbol{\sigma}_o$ is sufficiently far from the aperture for the following approximation to apply:

$$G(\boldsymbol{r},\boldsymbol{r}_0) = \frac{\exp(\mathrm{i}k_0|\boldsymbol{r}-\boldsymbol{r}_0|)}{|\boldsymbol{r}-\boldsymbol{r}_0|} = \exp\left(\mathrm{i}k_0\sqrt{r^2 + r_0^2 - 2\boldsymbol{r}\cdot\boldsymbol{r}_0}\right)/|\boldsymbol{r}-\boldsymbol{r}_0| \cong \frac{\exp[\mathrm{i}k_0(r_0 - \boldsymbol{\sigma}_0\cdot\boldsymbol{r})]}{r_0}. \quad (50)$$

Let us assume that the screen is a thin sheet of a perfect conductor on whose surface the tangential E-field necessarily vanishes. The appropriate diffraction equation will then be Eq.(32), with the tangential E-field outside the aperture allowed to vanish (i.e., $\hat{\boldsymbol{n}} \times \boldsymbol{E} = 0$ on the opaque parts of the screen). Approximating the E-field within the aperture with that of the incident plane-wave, we will have

$$\begin{aligned}\int_{S_1}[\hat{\boldsymbol{n}} \times \boldsymbol{E}(\boldsymbol{r})]G(\boldsymbol{r},\boldsymbol{r}_0)\mathrm{d}s &\cong (\hat{\boldsymbol{z}} \times \boldsymbol{E}_{\mathrm{inc}}) \int_{\text{aperture}} \exp(\mathrm{i}k_0\boldsymbol{\sigma}_{\mathrm{inc}}\cdot\boldsymbol{r}) \times \frac{\exp[\mathrm{i}k_0(r_0 - \boldsymbol{\sigma}_0\cdot\boldsymbol{r})]}{r_0}\mathrm{d}s \\ &= \frac{\exp(\mathrm{i}k_0 r_0)}{r_0}(\hat{\boldsymbol{z}} \times \boldsymbol{E}_{\mathrm{inc}}) \int_{\text{aperture}} \exp[\mathrm{i}k_0(\boldsymbol{\sigma}_{\mathrm{inc}} - \boldsymbol{\sigma}_0)\cdot\boldsymbol{r}]\,\mathrm{d}s \\ &= \frac{\exp(\mathrm{i}k_0 r_0)}{r_0}(\hat{\boldsymbol{z}} \times \boldsymbol{E}_{\mathrm{inc}}) \int_{r=0}^{R}\int_{\varphi=0}^{2\pi} \exp(\mathrm{i}k_0\zeta r\cos\varphi)\, r\mathrm{d}\varphi\mathrm{d}r. \quad (51)\end{aligned}$$

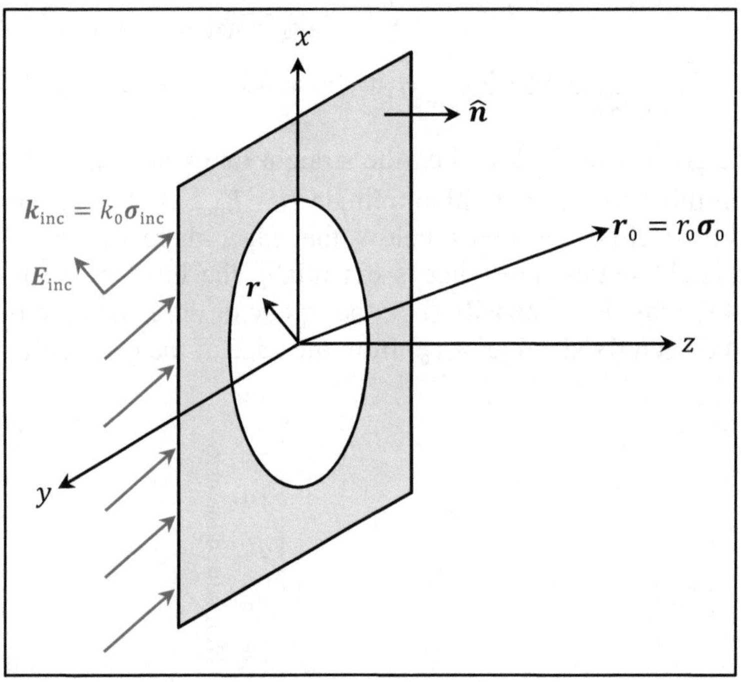

Fig.4. A circular aperture of radius R inside an otherwise opaque screen located at $z = 0$ is illuminated by a plane-wave whose E-field amplitude and propagation direction are specified as $\boldsymbol{E}_{\mathrm{inc}}$ and $\boldsymbol{\sigma}_{\mathrm{inc}}$, respectively. The observation point $\boldsymbol{r}_0 = r_0\boldsymbol{\sigma}_0$ is in the far field; that is, $r_0 \gg R$. The unit-vectors $\boldsymbol{\sigma}_{\mathrm{inc}}$ and $\boldsymbol{\sigma}_0$ have respective polar coordinates $(\theta_{\mathrm{inc}}, \varphi_{\mathrm{inc}})$ and (θ_0, φ_0). The projection of the vector $\boldsymbol{\sigma}_{\mathrm{inc}} - \boldsymbol{\sigma}_0$ onto the xy-plane of the aperture is a vector of length ζ that makes an angle φ with the position vector $\boldsymbol{r} = x\hat{\boldsymbol{x}} + y\hat{\boldsymbol{y}}$.

Here, ζ is the magnitude of the projection of $\boldsymbol{\sigma}_{\mathrm{inc}} - \boldsymbol{\sigma}_0$ onto the xy-plane of the aperture, while φ is the angle between that projection and the position vector $\boldsymbol{r} = x\hat{\boldsymbol{x}} + y\hat{\boldsymbol{y}}$; that is,

$$\boldsymbol{\sigma}_{\mathrm{inc}} - \boldsymbol{\sigma}_0 = (\sin\theta_{\mathrm{inc}}\cos\varphi_{\mathrm{inc}} - \sin\theta_0\cos\varphi_0)\hat{\boldsymbol{x}} + (\sin\theta_{\mathrm{inc}}\sin\varphi_{\mathrm{inc}} - \sin\theta_0\sin\varphi_0)\hat{\boldsymbol{y}} + (\cos\theta_{\mathrm{inc}} - \cos\theta_0)\hat{\boldsymbol{z}}. \quad (52)$$

$$\zeta = \sqrt{\sin^2\theta_{\mathrm{inc}} + \sin^2\theta_0 - 2\sin\theta_{\mathrm{inc}}\sin\theta_0\cos(\varphi_{\mathrm{inc}} - \varphi_0)}. \quad (53)$$

Consequently,

$$\int_{S_1}[\hat{\boldsymbol{n}}\times\boldsymbol{E}(\boldsymbol{r})]G(\boldsymbol{r},\boldsymbol{r}_0)\mathrm{d}s \cong \frac{2\pi\exp(\mathrm{i}k_0r_0)}{r_0}(\hat{\boldsymbol{z}}\times\boldsymbol{E}_{\text{inc}})\int_0^R rJ_0(k_0\zeta r)\mathrm{d}r$$

$J_0(\cdot)$ and $J_1(\cdot)$ are Bessel functions of the first kind, orders 0 and 1.

$$= \frac{2\pi R\exp(\mathrm{i}k_0r_0)}{k_0r_0\zeta}J_1(k_0\zeta R)\,\hat{\boldsymbol{z}}\times\boldsymbol{E}_{\text{inc}}. \quad (54)$$

The E-field at the observation point $\boldsymbol{r}_0 = r_0\boldsymbol{\sigma}_0$ (which is in the far field of the aperture, i.e., $r_0 \gg R$) is now found by computing the curl of the expression on the right-hand side of Eq.(54). Considering that $\hat{\boldsymbol{z}}\times\boldsymbol{E}_{\text{inc}}$ is independent of $\boldsymbol{r}_0$, we use the vector identity $\boldsymbol{\nabla}\times(\psi\boldsymbol{V}_0) = \boldsymbol{\nabla}\psi\times\boldsymbol{V}_0$, then ignore the small (far field) contributions to $\boldsymbol{\nabla}\psi$ due to the dependence of ζ on (θ_0,φ_0), to arrive at

$$\begin{aligned}
\boldsymbol{E}(\boldsymbol{r}_0) &= (2\pi)^{-1}\boldsymbol{\nabla}_0\times\int_{S_1}[\hat{\boldsymbol{n}}\times\boldsymbol{E}(\boldsymbol{r})]G(\boldsymbol{r},\boldsymbol{r}_0)\mathrm{d}s\\
&\cong (R/k_0\zeta)J_1(k_0\zeta R)\boldsymbol{\nabla}_0[\exp(\mathrm{i}k_0r_0)/r_0]\times(\hat{\boldsymbol{z}}\times\boldsymbol{E}_{\text{inc}})\\
&= (R/k_0\zeta)J_1(k_0\zeta R)(\mathrm{i}k_0 - r_0^{-1})[\exp(\mathrm{i}k_0r_0)/r_0]\boldsymbol{\sigma}_0\times(\hat{\boldsymbol{z}}\times\boldsymbol{E}_{\text{inc}})\\
&\cong \frac{\mathrm{i}RJ_1(k_0\zeta R)e^{\mathrm{i}k_0r_0}}{r_0\zeta}[(\boldsymbol{E}_{\text{inc}}\cdot\boldsymbol{\sigma}_0)\hat{\boldsymbol{z}} - \cos(\theta_0)\,\boldsymbol{E}_{\text{inc}}]. \quad (55)
\end{aligned}$$

The approximate nature of these calculations should be borne in mind when comparing the various estimates of an observed field obtained via different routes.[9] For instance, had we started with Eq.(33) and proceeded by setting $\hat{\boldsymbol{n}}\times\boldsymbol{B} = 0$ on the opaque areas of the screen, we would have arrived at the following estimate of the observed E-field:

$$\boldsymbol{E}(\boldsymbol{r}_0) \cong \frac{\mathrm{i}RJ_1(k_0\zeta R)e^{\mathrm{i}k_0r_0}}{r_0\zeta}\{(\boldsymbol{E}_{\text{inc}}\cdot\hat{\boldsymbol{z}})[\boldsymbol{\sigma}_{\text{inc}} - (\boldsymbol{\sigma}_{\text{inc}}\cdot\boldsymbol{\sigma}_0)\boldsymbol{\sigma}_0] - \cos\theta_{\text{inc}}\,[\boldsymbol{E}_{\text{inc}} - (\boldsymbol{E}_{\text{inc}}\cdot\boldsymbol{\sigma}_0)\boldsymbol{\sigma}_0]\}. \quad (56)$$

The differences between Eqs.(55) and (56), which, in general, are *not* insignificant, can be traced to the assumptions regarding the nature of the opaque screen and the approximations involved in equating the $\boldsymbol{E}$ and $\boldsymbol{B}$ fields within the aperture to those of the incident plane-wave.

7. Sommerfeld's analysis of diffraction from a perfectly conducting half-plane. A rare example of an exact solution of Maxwell's equations as applied to EM diffraction was published by A. Sommerfeld in 1896.[25] The simplified version of Sommerfeld's original analysis presented in this section closely parallels that of Ref.[1], Chapter 11. Consider an EM plane-wave propagating in free space and illuminating a thin, perfectly conducting screen that sits in the upper half of the xy-plane at $z = 0$. The geometry of the system is shown in Fig.5, where the incident k-vector is denoted by $\boldsymbol{k}_{\text{inc}}$, the oscillation frequency of the monochromatic wave is ω, the speed of light in vacuum is c, the wave-number is $k_0 = \omega/c$, and the unit-vector along the direction of incidence is $\boldsymbol{\sigma}_{\text{inc}}$. The plane-wave is linearly polarized with its E-field amplitude E_0 along the y-axis and H-field amplitude $H_0 = E_0/Z_0$ in the xz-plane of incidence. ($Z_0 = \sqrt{\mu_0/\varepsilon_0}$ is the impedance of free space.) The electric current density induced in the semi-infinite screen is denoted by $\boldsymbol{J}(\boldsymbol{r},t)$. In the chosen geometry, all the fields (incident as well as scattered) are uniform along the y-axis; consequently, it suffices to specify the position of an arbitrary point in the Cartesian xyz space by its x and z coordinates only; that is,

$$\boldsymbol{r} = x\hat{\boldsymbol{x}} + z\hat{\boldsymbol{z}} = r(\cos\theta\,\hat{\boldsymbol{x}} + \sin\theta\,\hat{\boldsymbol{z}}). \quad (57)$$

Here, $r = \sqrt{x^2+z^2}$, and the angle θ is measured clockwise from the positive x-axis. The range of θ is $(0,\pi)$ for points $\boldsymbol{r}$ on the right-hand side of the screen, and $(-\pi,0)$ for the points on the left-hand side. Figure 5 shows that $\boldsymbol{k}_{\text{inc}}$ makes an angle $\theta_0 \in (0,\pi)$ with the positive x-axis. The incident plane-wave is, thus, fully specified by the following equations:

$$\boldsymbol{k}_{\text{inc}} = k_x\hat{\boldsymbol{x}} + k_z\hat{\boldsymbol{z}} = k_0\boldsymbol{\sigma}_{\text{inc}} = k_0(\cos\theta_0\,\hat{\boldsymbol{x}} + \sin\theta_0\,\hat{\boldsymbol{z}}). \tag{58}$$

$$\begin{aligned}\boldsymbol{E}_{\text{inc}}(\boldsymbol{r},t) &= E_0\hat{\boldsymbol{y}}\exp[\mathrm{i}(\boldsymbol{k}_{\text{inc}}\cdot\boldsymbol{r} - \omega t)] = E_0\hat{\boldsymbol{y}}\exp[\mathrm{i}k_0(\boldsymbol{\sigma}_{\text{inc}}\cdot\boldsymbol{r} - ct)]\\ &= E_0\hat{\boldsymbol{y}}e^{\mathrm{i}k_0 r\cos(\theta-\theta_0)}e^{-\mathrm{i}\omega t}.\end{aligned} \tag{59}$$

$$\boldsymbol{H}_{\text{inc}}(\boldsymbol{r},t) = -(E_0/Z_0)(\sin\theta_0\,\hat{\boldsymbol{x}} - \cos\theta_0\,\hat{\boldsymbol{z}})e^{\mathrm{i}k_0 r\cos(\theta-\theta_0)}e^{-\mathrm{i}\omega t}. \tag{60}$$

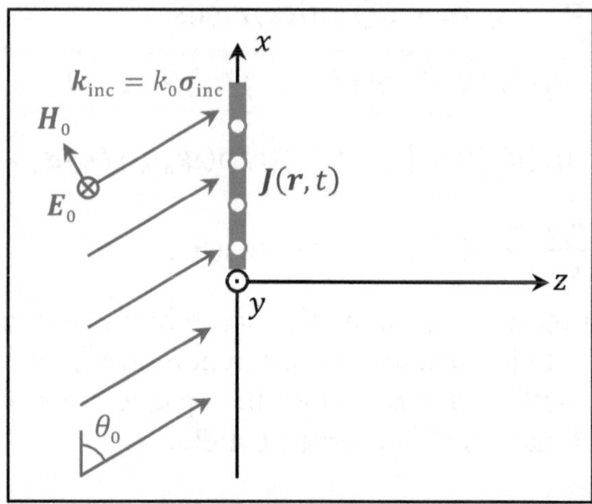

Fig.5. A plane, monochromatic EM wave propagating along the unit-vector $\boldsymbol{\sigma}_{\text{inc}} = \cos\theta_0\,\hat{\boldsymbol{x}} + \sin\theta_0\,\hat{\boldsymbol{z}}$ arrives at a thin, semi-infinite, perfectly electrically conducting screen located in the upper half of the xy-plane. The plane-wave is linearly polarized, with its E-field aligned with the y-axis, while its H-field has components along the x and z directions. The induced current sheet, denoted by $\boldsymbol{J}(\boldsymbol{r},t)$, oscillates parallel to the incident E-field along the direction of the y-axis. (The system depicted here is essentially the same as that in Fig.2, with the exception of the angle of incidence θ_0 being the complement of θ_{inc} of Fig.2.)

With the aid of the step-function $\text{step}(x)$ and Dirac's delta-function $\delta(z)$, we now express the electric current density $\boldsymbol{J}(\boldsymbol{r},t)$ excited on the thin-sheet conductor as a one-dimensional Fourier transform, namely,

$$\boldsymbol{J}(\boldsymbol{r},t) = \text{step}(x)J_s(x)\delta(z)e^{-\mathrm{i}\omega t}\hat{\boldsymbol{y}} = \left[\int_{-\infty}^{\infty}\mathcal{J}(\sigma_x)e^{\mathrm{i}k_0\sigma_x x}\mathrm{d}\sigma_x\right]\delta(z)e^{-\mathrm{i}\omega t}\hat{\boldsymbol{y}}. \tag{61}$$

In the above equation, σ_x represents spatial frequency along the x-axis, and $\mathcal{J}(\sigma_x)$ is the (complex) amplitude of the induced surface-current-density along $\hat{\boldsymbol{y}}$, having spatial and temporal frequencies σ_x and ω, respectively. The first goal of our analysis is to find the function $\mathcal{J}(\sigma_x)$ such that its Fourier transform vanishes along the negative x-axis—as demanded by the step-function in Eq.(61)—while its radiated E-field cancels the incident plane-wave's E-field at the surface of the screen along the positive x-axis. Now, a current sheet $\mathcal{J}(\sigma_x)e^{\mathrm{i}(k_0\sigma_x x-\omega t)}\delta(z)\hat{\boldsymbol{y}}$ that fills the entire xy-plane at $z = 0$ radiates EM fields into the surrounding free space that can easily be shown to have the following structure:

$$\boldsymbol{E}(\boldsymbol{r},t) = -\tfrac{1}{2}(Z_0k_0\hat{\boldsymbol{y}}/k_z)\mathcal{J}(\sigma_x)\exp[\mathrm{i}(k_x x + k_z|z| - \omega t)]. \tag{62}$$

$$\boldsymbol{H}(\boldsymbol{r},t) = \tfrac{1}{2}\mathcal{J}(\sigma_x)[\pm\hat{\boldsymbol{x}} - (k_x/k_z)\hat{\boldsymbol{z}}]\exp[\mathrm{i}(k_x x + k_z|z| - \omega t)]. \tag{63}$$

Here, $k_o = \omega/c$ is the wave-number in free space, while $k_x = k_o\sigma_x$, and $k_z = \sqrt{k_o^2 - k_x^2}$ are the k-vector components along the x and z axes. In general, k_z must be real and positive when $|k_x| \le k_o$, and imaginary and positive otherwise. The $\pm$ signs associated with H_x indicate that the plus sign must be used for the half-space on the right-hand side of the sheet, where $z > 0$, while the minus sign is reserved for the left-hand side, where $z < 0$. The discontinuity of H_x at $z = 0$ thus equals the surface-current-density $\mathcal{J}(\sigma_x)$ along the y-direction, in compliance with Maxwell's requisite boundary conditions. Needless to say, Eqs.(62) and (63) represent a single EM plane-wave on either side of the screen's xy-plane, which satisfy the symmetry requirement of radiation from the current sheet, and with the tangential component H_x of the radiated H-field chosen to satisfy the requisite boundary condition at the plane of the surface current. The plane-waves emanating to the right and left of the xy-plane of the current sheet are homogeneous when k_z is real, and inhomogeneous (or evanescent) when k_z is imaginary.

When working in the complex σ_x-plane, we must ensure that k_z has the correct sign for all values of the real parameter $k_x = k_o\sigma_x$ from $-\infty$ to ∞. Considering that k_z/k_o is the square root of the product of $(1 - \sigma_x)$ and $(1 + \sigma_x)$, we choose for both of these complex numbers the range of phase angles $(-\pi, \pi]$, as depicted in Fig.6(a). The corresponding branch-cuts thus appear as the semi-infinite line segments $(-\infty, -1)$ and $(1, \infty)$, and the integration path along the real axis within the σ_x-plane, shown in Fig.6(b), will consist of semi-infinite line segments slightly above and slightly below the real axis, as well as a short segment of the real-axis connecting -1 to 1. This choice of the integration path ensures that $k_z = k_o\sqrt{1 - \sigma_x^2}$ acquires the correct sign for all values of σ_x.

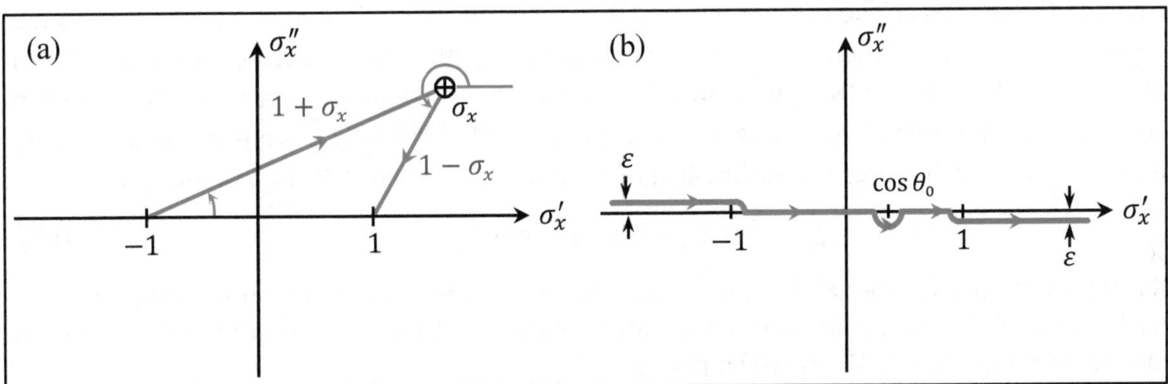

Fig.6. (a) In the complex σ_x-plane, where the phase of the complex numbers $1 \pm \sigma_x$ is measured counterclockwise from the positive σ_x' axis, the range of both angles is confined to the $(-\pi, \pi]$ interval. (b) The integration path along the σ_x' axis is adjusted by shifting the part from $-\infty$ to -1 slightly upward, and the part from 1 to ∞ slightly downward, so that $k_z/k_o = \sqrt{(1 - \sigma_x)(1 + \sigma_x)}$ has the correct sign everywhere. The pole at $\sigma_x = \cos\theta_o$ is handled by locally deforming the contour into a semi-circular path below the real axis.

For reasons that will become clear as we proceed, Sommerfeld suggested the following mathematical form for the surface current-density $\mathcal{J}$ as a function of σ_x:

$$\mathcal{J}(\sigma_x) = \mathcal{J}_o\sqrt{1 - \sigma_x}/(\sigma_x - \cos\theta_o). \tag{64}$$

The domain of this function is the slightly deformed real axis of the σ_x-plane depicted in Fig.6(b). The proposed function has a simple pole at $\sigma_x = \cos\theta_o$, where θ_o is the orientation angle of $\boldsymbol{k}_{\text{inc}}$ shown in Fig.5, and a (complex) constant coefficient $\mathcal{J}_o$ that will be determined shortly. In addition, $\mathcal{J}(\sigma_x)$ contains the term $\sqrt{1-\sigma_x}$, whose branch-cut in the system of Fig.6(a) is the semi-infinite line-segment $(1,\infty)$ along the real axis of the σ_x-plane. It is now possible to demonstrate that the proposed $\mathcal{J}(\sigma_x)$ satisfies its first required property, namely,

$$\int_{-\infty}^{\infty}\mathcal{J}(\sigma_x)e^{\mathrm{i}k_0\sigma_x x}\mathrm{d}\sigma_x = \mathcal{J}_0\int_{-\infty}^{\infty}\frac{\sqrt{1-\sigma_x}}{\sigma_x-\cos\theta_0}e^{\mathrm{i}k_0\sigma_x x}\mathrm{d}\sigma_x = 0, \qquad (x<0). \tag{65}$$

For $x<0$, the integration contour of Fig.6(b) can be closed with a large semi-circular path in the lower half of the σ_x-plane. The only part of the integrand in Eq.(65) that requires a branch-cut is $\sqrt{1-\sigma_x}$, whose branch-cut is the line segment $(1,\infty)$. The integration contour of Fig.6(b), when closed in the lower-half of the σ_x-plane, does not contain this branch-cut. Moreover, the pole at $\sigma_x=\cos\theta_0$ is outside the closed loop of integration, and the contributions to the integral by the singular points $\sigma_x=\pm1$ are zero. Consequently, the current-density $\boldsymbol{J}(\boldsymbol{r},t)$ in the lower half of the xy-plane turns out to be zero, exactly as required.

The radiated E-field, a superposition of contributions from the various $\mathcal{J}(\sigma_x)$ in accordance with Eqs.(62) and (64), must cancel out the incident E-field at the surface of the screen; that is,

$$E_y(x,y,z=0)=\int_{-\infty}^{\infty}E_y(\sigma_x)e^{\mathrm{i}k_0\sigma_x x}\mathrm{d}\sigma_x=-\tfrac{1}{2}Z_0k_0\int_{-\infty}^{\infty}[\mathcal{J}(\sigma_x)/k_z]e^{\mathrm{i}k_0\sigma_x x}\mathrm{d}\sigma_x$$

$$=-\tfrac{1}{2}Z_0\mathcal{J}_0\int_{-\infty}^{\infty}\frac{e^{\mathrm{i}k_0\sigma_x x}}{\sqrt{1+\sigma_x}\,(\sigma_x-\cos\theta_0)}\mathrm{d}\sigma_x=-E_0e^{\mathrm{i}k_0\cos\theta_0 x}, \qquad (x>0). \tag{66}$$

For $x>0$, the contour of integration in Eq.(66) can be closed with a large semi-circle in the upper-half of the σ_x-plane. The branch-cut for $\sqrt{1+\sigma_x}$ in the denominator of the integrand is the line-segment $(-\infty,-1)$, which is below the integration path and, therefore, irrelevant for a contour that closes in the upper-half-plane. For the singularities at $\sigma_x=\pm1$, the residues are zero, whereas for the pole at $\sigma_x=\cos\theta_0$, the residue is $e^{\mathrm{i}k_0\cos\theta_0 x}/\sqrt{1+\cos\theta_0}$. The requisite boundary condition at the screen's surface is thus seen to be satisfied if $\mathcal{J}_0$ is specified as

$$\mathcal{J}_0=-\mathrm{i}(E_0/\pi Z_0)\sqrt{1+\cos\theta_0}. \tag{67}$$

Having found the functional form of $\mathcal{J}(\sigma_x)$, we now turn to the problem of computing the scattered E-field $\boldsymbol{E}_s(\boldsymbol{r})$ at the arbitrary observation point $\boldsymbol{r}=x\hat{\boldsymbol{x}}+z\hat{\boldsymbol{z}}=r(\cos\theta\,\hat{\boldsymbol{x}}+\sin\theta\,\hat{\boldsymbol{z}})$ in accordance with Eqs.(62), (64), and (67); that is,

$$\boldsymbol{E}_s(\boldsymbol{r})=\mathrm{i}(E_0\hat{\boldsymbol{y}}/2\pi)\int_{-\infty}^{\infty}\frac{\sqrt{1+\cos\theta_0}}{\sqrt{1+\sigma_x}\,(\sigma_x-\cos\theta_0)}\exp[\mathrm{i}k_0(\sigma_x x+\sigma_z|z|)]\,\mathrm{d}\sigma_x. \tag{68}$$

The integral in Eq.(68) must be evaluated at positive as well as negative values of x, and also for values of z on both sides of the screen. The presence of k_z in the exponent of the integrand requires that the branch-cuts on both line-segments $(-\infty,-1)$ and $(1,\infty)$ be taken into account. For these reasons, the integration path of Fig.6(b) ceases to be convenient and we need to change the variable σ_x to something that avoids the need for branch-cuts. We now switch the variable from σ_x to φ, where $\sigma_x=\cos\varphi$, with the integration path in the complex φ-plane shown in Fig.7(a). Considering that

$$\cos\varphi=\cos(\varphi'+\mathrm{i}\varphi'')=\cos\varphi'\cosh\varphi''-\mathrm{i}\sin\varphi'\sinh\varphi'', \tag{69}$$

the depicted integration path represents the continuous variation of σ_x from $-\infty$ to ∞. Similarly,

$$\sigma_z = \sqrt{1 - \cos^2 \varphi} = \sin \varphi = \sin \varphi' \cosh \varphi'' + \mathrm{i} \cos \varphi' \sinh \varphi'' \tag{70}$$

is positive real on the horizontal branch, and positive imaginary on both vertical branches of the depicted contour. It is also easy to verify that $\sqrt{1 - \sigma_x} = \sqrt{2} \sin(\varphi/2)$ has identical values at corresponding points on the contours of Figs.(6b) and (7a), as does $\sqrt{1 + \sigma_x} = \sqrt{2} \cos(\varphi/2)$. Recalling that $\mathrm{d}\sigma_x = -\sin \varphi \, \mathrm{d}\varphi = -\sqrt{1 - \cos^2 \varphi} \, \mathrm{d}\varphi$, we can rewrite Eq.(68) as

$$\boldsymbol{E}_s(\boldsymbol{r}) = -\mathrm{i}(E_o \hat{\boldsymbol{y}}/2\pi) \int_{\pi - \mathrm{i}\infty}^{0 + \mathrm{i}\infty} \frac{\sqrt{1 + \cos \theta_0} \sqrt{1 - \cos \varphi}}{\cos \varphi - \cos \theta_0} \exp[\mathrm{i}k_o r \cos(\varphi \mp \theta)] \, \mathrm{d}\varphi. \tag{71}$$

The above equation yields the scattered E-field on both sides of the screen, with the minus sign in the exponent corresponding to $z > 0$, and the plus sign to $z < 0$. In what follows, noting the natural symmetry of the scattered field on the opposite sides of the xy-plane, we confine our attention to Eq.(71) with only the minus sign in the exponent; the scattered E-field on the left hand side of the screen is subsequently obtained by a simple change of the sign of z.

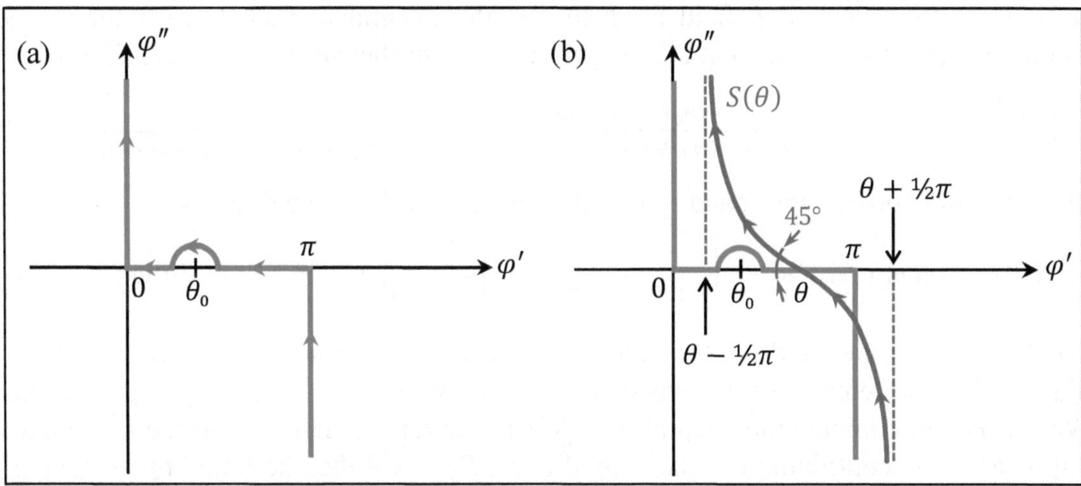

Fig.7. (a) Contour of integration in the complex φ-plane corresponding to the integration path in the σ_x-plane shown in Fig.6(b). By definition, $\sigma_x = \cos \varphi$, which results in $k_z = k_o \sigma_z = k_o \sin \varphi$, a well-defined function everywhere in the φ-plane that does not need branch-cuts. The small bump in the integration path around $\varphi = \theta_0$ corresponds to the small semi-circular part of the contour in Fig.6(b). The negative values of σ_x'' on the semi-circle translate, in accordance with Eq.(69), into positive value of φ'' on the corresponding bump. (b) Along the steepest-descent contour $S(\theta)$ that passes through the saddle-point $\varphi = \theta$, $\mathrm{Re}[\cos(\varphi - \theta)]$ is constant. At the saddle-point, $S(\theta)$ makes a 45° angle with the horizontal and vertical lines, which represent contours along which $\mathrm{Im}[\cos(\varphi - \theta)]$ is constant. $S(\theta)$ also has the property that $\mathrm{Im}[\cos(\varphi - \theta)]$, which is zero at the saddle-point, rises toward ∞ (continuously and symmetrically on both sides of the saddle) as φ moves away from the saddle-point along the contour.

The integration path of Fig.7(a) is now replaced with the steepest-descent contour $S(\theta)$ depicted in Fig.7(b). Passing through the saddle-point of $\exp[\mathrm{i}kr \cos(\varphi - \theta)]$ at $\varphi = \theta$, this contour has the property that $\mathrm{Re}[\cos(\varphi - \theta)]$ everywhere on the contour equals 1. In contrast, $\mathrm{Im}[\cos(\varphi - \theta)]$ starts at zero at the saddle, then rises toward infinity (continuously and symmetrically on opposite sides of the saddle) as φ moves away from the saddle-point at $\varphi = \theta$. It is easy to show that the original integration path of Fig.7(a) can be joined to $S(\theta)$ to form a closed loop with negligible contributions to the overall loop integral by the segments that connect the two contours at infinity.**

The only time when the single pole of the integrand at $\varphi = \theta_0$ needs to be accounted for is when $\theta < \theta_0$, at which point the pole is inside the closed contour, as Fig.7(b) clearly indicates. In the vicinity of the pole, the denominator of the integrand in Eq.(71) can be approximated by the first two terms of the Taylor series expansion of $\cos\varphi$, as follows:

$$\cos\varphi - \cos\theta_0 = [\cos\theta_0 - \sin\theta_0\,(\varphi - \theta_0) + \cdots] - \cos\theta_0 \cong -\sin\theta_0\,(\varphi - \theta_0). \qquad (72)$$

The residue at the pole is thus seen to be $-\exp[ik_0 r\cos(\theta_0 - \theta)]$, with a corresponding contribution of $-E_0\hat{\boldsymbol{y}}\exp[ik_0 r\cos(\theta - \theta_0)]$ to the scattered E-field. It must be emphasized that the scattered field contributed by the pole at $\varphi = \theta_0$ is relevant only when $\theta < \theta_0$, in which case it cancels the contribution of the incident plane-wave of Eq.(59) to the overall EM field on the right-hand side of the screen, where the screen casts its geometric shadow. Outside this geometric shadow, where $\theta_0 < \theta \le \pi$, the incident beam spills into the $z \ge 0$ region—without the counter-balancing effect of the scattered field produced by the pole at $\varphi = \theta_0$.

Returning to the scattered E-field produced by the integral in Eq.(71) over the steepest-descent contour $S(\theta)$, the first term in the integrand can be further streamlined, as follows:

$$\frac{\sqrt{1+\cos\theta_0}\,\sqrt{1-\cos\varphi}}{\cos\varphi - \cos\theta_0} = \frac{2\cos(\theta_0/2)\sin(\varphi/2)}{-2\sin[\tfrac{1}{2}(\varphi+\theta_0)]\sin[\tfrac{1}{2}(\varphi-\theta_0)]} = -\frac{\tfrac{1}{2}}{\sin[\tfrac{1}{2}(\varphi+\theta_0)]} - \frac{\tfrac{1}{2}}{\sin[\tfrac{1}{2}(\varphi-\theta_0)]}. \qquad (73)$$

The scattered E-field on the right-hand side of the screen may thus be written as

$$\boldsymbol{E}_s(\boldsymbol{r}) = \mathrm{i}(E_0\hat{\boldsymbol{y}}/4\pi)e^{\mathrm{i}k_0 r}\int_{S(\theta)}\left(\frac{1}{\sin[\tfrac{1}{2}(\varphi+\theta_0)]} + \frac{1}{\sin[\tfrac{1}{2}(\varphi-\theta_0)]}\right)e^{\mathrm{i}k_0 r[\cos(\varphi-\theta)-1]}\mathrm{d}\varphi\,. \qquad (74)$$

Note that we have factored out the imaginary part of the exponent and taken it outside the integral as $e^{\mathrm{i}k_0 r}$; what remains of the exponent, namely, $\mathrm{i}k_0 r[\cos(\varphi - \theta) - 1]$, is purely real on $S(\theta)$. We proceed to compute the integral in Eq.(74) only for the first term of the integrand and refer to it as $\boldsymbol{E}_{s,1}$; the contribution of the second term, $\boldsymbol{E}_{s,2}$, will then be found by switching the sign of θ_0.

A change of the variable from φ to $\varphi - \theta$ would cause the steepest-descent contour to go through the origin of the φ-plane; this shifted contour will now be denoted by S_0. Taking advantage of the symmetry of S_0 with respect to the origin, we express the final result of integration in terms of the integral on the upper half of S_0, which is denoted by S_0^+. We will have

$$\begin{aligned}\boldsymbol{E}_{s,1}(\boldsymbol{r}) &= \mathrm{i}(E_0\hat{\boldsymbol{y}}/4\pi)e^{\mathrm{i}k_0 r}\int_{S_0^+}\left\{\frac{1}{\sin[\tfrac{1}{2}(\varphi+\theta+\theta_0)]} + \frac{1}{\sin[\tfrac{1}{2}(-\varphi+\theta+\theta_0)]}\right\}e^{\mathrm{i}k_0 r(\cos\varphi-1)}\mathrm{d}\varphi\\ &= \mathrm{i}(E_0\hat{\boldsymbol{y}}/\pi)e^{\mathrm{i}k_0 r}\int_{S_0^+}\frac{\sin[(\theta+\theta_0)/2]\cos(\varphi/2)}{\cos(\varphi) - \cos(\theta+\theta_0)}e^{-\mathrm{i}2k_0 r\sin^2(\varphi/2)}\mathrm{d}\varphi\\ &= -\mathrm{i}(E_0\hat{\boldsymbol{y}}/2\pi)\sin[(\theta+\theta_0)/2]\,e^{\mathrm{i}k_0 r}\int_{S_0^+}\frac{\cos(\varphi/2)\exp[-\mathrm{i}2k_0 r\sin^2(\varphi/2)]}{\sin^2(\varphi/2) - \sin^2[(\theta+\theta_0)/2]}\mathrm{d}\varphi. \end{aligned} \qquad (75)$$

** On the short line-segments that connect the two contours at $\varphi'' = \pm\infty$, the magnitude of the exponential factor in the integrand in Eq.(71) is $\exp[k_0 r\sin(\varphi' - \theta)\sinh(\varphi'')]$. In the upper half-plane, $-1 \le \sin(\varphi' - \theta) \le 0$ and $\sinh(\varphi'') \to \infty$, whereas in the lower half-plane, $0 \le \sin(\varphi' - \theta) \le 1$ and $\sinh(\varphi'') \to -\infty$. Thus, in both cases, the integrand vanishes.

Another change of variable, this time from φ to the real-valued $\zeta = -\exp(\mathrm{i}\pi/4)\sin(\varphi/2)$, where ζ ranges from 0 to ∞ along the steepest-descent contour S_0^+, now yields[††]

$$\boldsymbol{E}_{s,1}(\boldsymbol{r}) = -(E_0\hat{\boldsymbol{y}}/\pi)\sin[(\theta+\theta_0)/2]\,e^{\mathrm{i}k_0 r}e^{-\mathrm{i}\pi/4}\int_0^{\infty}\frac{\exp(-2k_0 r\zeta^2)}{\zeta^2-\mathrm{i}\sin^2[(\theta+\theta_0)/2]}\mathrm{d}\zeta. \tag{76}$$

The integral appearing in the above equation is evaluated in Appendix G, where it is shown that

$$\int_0^{\infty}\frac{\exp(-\lambda\zeta^2)}{\zeta^2-\mathrm{i}\eta^2}\mathrm{d}\zeta = \sqrt{\pi}|\eta|^{-1}e^{-\mathrm{i}\lambda\eta^2}F(|\eta|\sqrt{\lambda}). \tag{77}$$

Here, $F(\alpha) = \int_\alpha^\infty \exp(\mathrm{i}x^2)\,\mathrm{d}x$ is the complex Fresnel integral defined in Eq.(46). We thus find

$$\boldsymbol{E}_{s,1}(\boldsymbol{r}) = -\left(E_0\hat{\boldsymbol{y}}/\sqrt{\pi}\right)e^{-\mathrm{i}\pi/4}e^{\mathrm{i}k_0 r}e^{-\mathrm{i}2k_0 r\sin^2[(\theta+\theta_0)/2]}F\left(\sqrt{2k_0 r}\sin[(\theta+\theta_0)/2]\right). \tag{78}$$

The expression for $\boldsymbol{E}_{s,2}(\boldsymbol{r})$ is derived from Eq.(78) by switching the sign of θ_0. One has to be careful in this case, since $\sin[(\theta-\theta_0)/2]$ may be negative. Given that both θ and θ_0 are in the $(0,\pi)$ interval, the sign of $\sin[(\theta-\theta_0)/2]$ will be positive if $\theta > \theta_0$, and negative if $\theta < \theta_0$. The scattered E-field of Eq.(74) is thus given by

$$\boldsymbol{E}_s(\boldsymbol{r}) = \boldsymbol{E}_{s,1} + \boldsymbol{E}_{s,2} = -\left(E_0\hat{\boldsymbol{y}}/\sqrt{\pi}\right)e^{-\mathrm{i}\pi/4}\left\{e^{\mathrm{i}k_0 r\cos(\theta+\theta_0)}F\left(\sqrt{2k_0 r}\sin[(\theta+\theta_0)/2]\right)\right.$$
$$\left.\pm e^{\mathrm{i}k_0 r\cos(\theta-\theta_0)}F\left(\pm\sqrt{2k_0 r}\sin[(\theta-\theta_0)/2]\right)\right\}. \tag{79}$$

To this result we must add the contribution of the pole, namely, $-E_0\hat{\boldsymbol{y}}\exp[\mathrm{i}k_0 r\cos(\theta-\theta_0)]$ when $\theta < \theta_0$, and the incident beam $E_0\hat{\boldsymbol{y}}\exp[\mathrm{i}k_0 r\cos(\theta-\theta_0)]$ for the entire range $0 \le \theta \le \pi$. Recalling that

$$F(\alpha) + F(-\alpha) = \int_{-\infty}^{\infty}\exp(\mathrm{i}x^2)\,\mathrm{d}x = \sqrt{\pi}e^{\mathrm{i}\pi/4}, \tag{80}$$

the total E-field on the right-hand side of the xy-plane, where $0 \le \theta \le \pi$, becomes

$$\boldsymbol{E}_{\text{total}}(\boldsymbol{r}) = \left(e^{-\mathrm{i}\pi/4}E_0\hat{\boldsymbol{y}}/\sqrt{\pi}\right)\left\{e^{\mathrm{i}k_0 r\cos(\theta-\theta_0)}F\left(-\sqrt{2k_0 r}\sin[(\theta-\theta_0)/2]\right)\right.$$
$$\left.-e^{\mathrm{i}k_0 r\cos(\theta+\theta_0)}F\left(\sqrt{2k_0 r}\sin[(\theta+\theta_0)/2]\right)\right\}. \tag{81}$$

On the left-hand side of the screen, where $z < 0$ and, therefore, $-\pi \le \theta \le 0$, the scattered E-field is obtained by replacing θ with $|\theta|$ in Eq.(79) as well as in the contribution by the pole at $\varphi = \theta_0$. (Appendix H shows that the scattered E-field in the $z < 0$ region can also be evaluated by direct integration over a modified contour in the complex φ-plane.) Once again, adding the incident E-field and invoking Eq.(80), we find

$$\boldsymbol{E}_{\text{total}}(\boldsymbol{r}) = \left(e^{-\mathrm{i}\pi/4}E_0\hat{\boldsymbol{y}}/\sqrt{\pi}\right)\left\{e^{\mathrm{i}k_0 r\cos(\theta-\theta_0)}F\left(\sqrt{2k_0 r}\sin[(\theta-\theta_0)/2]\right)\right.$$
$$\left.-e^{\mathrm{i}k_0 r\cos(\theta+\theta_0)}F\left(-\sqrt{2k_0 r}\sin[(\theta+\theta_0)/2]\right)\right\}. \tag{82}$$

One obtains Eq.(82) by replacing θ in Eq.(81) with $2\pi+\theta$, the latter θ being in the $(-\pi,0)$ interval. Thus, Eq.(81) with $0 \le \theta \le 2\pi$ covers the entire range of observation points $\boldsymbol{r}$. A clear

[††] Upon setting $\zeta^2 = \mathrm{i}\sin^2(\varphi/2)$, we find two possible choices for the new variable, namely, $\zeta = \pm e^{\mathrm{i}\pi/4}\sin(\varphi/2)$. Of these, the one with the minus sign represents the upper-half S_0^+ of the steepest-descent trajectory. To see this, note that as S_0^+ approaches its extremity when $\varphi \to -\tfrac{1}{2}\pi + \mathrm{i}\infty$, we have $\sin(\varphi/2) \to (-1+\mathrm{i})\exp(\varphi''/2)/2\sqrt{2}$, which must be multiplied by $-e^{\mathrm{i}\pi/4}$ for ζ to approach $+\infty$.

understanding of Eq.(81) requires familiarity with the general behavior of the Fresnel integral $F(\alpha)$; Appendix F contains a detailed explanation in terms of the Cornu spiral representation of $F(\alpha)$.‡‡ The general expression for the total (i.e., incident plus scattered) E-field given in Eq.(81) is somewhat simplified in terms of the new function $\Phi(\alpha) = e^{-\mathrm{i}\alpha^2}F(\alpha)$, as follows:

$$\boldsymbol{E}_{\text{total}}(\boldsymbol{r}) = \frac{E_0\hat{\boldsymbol{y}}e^{\mathrm{i}k_0 r}}{\sqrt{\pi}e^{\mathrm{i}\pi/4}}\Phi(-\sqrt{2k_0 r}\sin[(\theta-\theta_0)/2]) - \Phi(\sqrt{2k_0 r}\sin[(\theta+\theta_0)/2]); \quad (0 \le \theta \le 2\pi). \tag{83}$$

It is worth mentioning that, the contribution to the scattered E-field by the pole at $\varphi = \theta_0$, namely, $\boldsymbol{E}_{s,3}(\boldsymbol{r}) = -E_0\hat{\boldsymbol{y}}\exp[\mathrm{i}k_0 r\cos(\theta-\theta_0)]$, which exists only in the intervals $0 \le \theta < \theta_0$ and $2\pi - \theta_0 < \theta \le 2\pi$, cancels the incident E-field in the shadow region behind the screen, while acting as the reflected field in front of the perfectly conducting half-mirror.

7.1. The magnetic field. The total H-field is computed from the E-field of Eq.(83) with the aid of Maxwell's equation $\mathrm{i}k_0 Z_0\boldsymbol{H}(\boldsymbol{r}) = \boldsymbol{\nabla}\times\boldsymbol{E}(\boldsymbol{r}) = -(r^{-1}\partial_\theta E_y)\hat{\boldsymbol{r}} + (\partial_r E_y)\hat{\boldsymbol{\theta}}$. The identity $\Phi'(\alpha) = -1 - 2\mathrm{i}\alpha\Phi(\alpha)$ will be used in this calculation. To simplify the notation, we introduce the new variables $\xi_1 = -\sqrt{2k_0 r}\sin[(\theta-\theta_0)/2]$ and $\xi_2 = \sqrt{2k_0 r}\sin[(\theta+\theta_0)/2]$. We find

$$H_r(\boldsymbol{r}) = \frac{E_0 e^{\mathrm{i}k_0 r}}{\sqrt{\pi}\, e^{\mathrm{i}\pi/4} Z_0}\left[\Phi(\xi_1)\sin(\theta-\theta_0) - \Phi(\xi_2)\sin(\theta+\theta_0) + \frac{\mathrm{i}\sqrt{2}\cos(\theta/2)\cos(\theta_0/2)}{\sqrt{k_0 r}}\right]. \tag{84}$$

$$H_\theta(\boldsymbol{r}) = \frac{E_0 e^{\mathrm{i}k_0 r}}{\sqrt{\pi}\, e^{\mathrm{i}\pi/4} Z_0}\left[\Phi(\xi_1)\cos(\theta-\theta_0) - \Phi(\xi_2)\cos(\theta+\theta_0) - \frac{\mathrm{i}\sqrt{2}\sin(\theta/2)\cos(\theta_0/2)}{\sqrt{k_0 r}}\right]. \tag{85}$$

The Cartesian components $H_x = H_r\cos\theta - H_\theta\sin\theta$ and $H_z = H_r\sin\theta + H_\theta\cos\theta$ of the magnetic field may now be obtained from the polar components H_r and H_θ, as follows:

$$H_x(\boldsymbol{r}) = -\frac{E_0 e^{\mathrm{i}k_0 r}}{\sqrt{\pi}\, e^{\mathrm{i}\pi/4} Z_0}\left\{[\Phi(\xi_1) + \Phi(\xi_2)]\sin\theta_0 - \frac{\mathrm{i}\sqrt{2}\cos(\theta/2)\cos(\theta_0/2)}{\sqrt{k_0 r}}\right\}. \tag{86}$$

$$H_z(\boldsymbol{r}) = +\frac{E_0 e^{\mathrm{i}k_0 r}}{\sqrt{\pi}\, e^{\mathrm{i}\pi/4} Z_0}\left\{[\Phi(\xi_1) - \Phi(\xi_2)]\cos\theta_0 + \frac{\mathrm{i}\sqrt{2}\sin(\theta/2)\cos(\theta_0/2)}{\sqrt{k_0 r}}\right\}. \tag{87}$$

It is readily verified that $H_z = 0$ at the surface of the conductor, where $\theta = 0$, and that H_x equals the x component of the incident H-field in the open half of the xy-plane, where $\theta = \pi$.

8. Far field scattering and the optical theorem. In the system of Fig.1(b), let the object inside the closed surface S_1 be illuminated from the outside, and let the observation point $\boldsymbol{r}_0 = r_0\boldsymbol{\sigma}_0$ be far away from the object, so that the approximate form of $G(\boldsymbol{r},\boldsymbol{r}_0)$ given in Eq.(50) along with its corresponding gradient $\boldsymbol{\nabla}G \cong -\mathrm{i}k_0\boldsymbol{\sigma}_0 G(\boldsymbol{r},\boldsymbol{r}_0)$ would be applicable. Denoting by $\boldsymbol{E}_s(\boldsymbol{r})$ and $\boldsymbol{B}_s(\boldsymbol{r})$ the scattered fields appearing on S_1, Eq.(28) yields the E-field at the observation point as

$$\boldsymbol{E}_s(\boldsymbol{r}_0) \cong \frac{\mathrm{i}k_0\exp(\mathrm{i}k_0 r_0)}{4\pi r_0}\int_{S_1}[c(\hat{\boldsymbol{n}}\times\boldsymbol{B}_s) - (\hat{\boldsymbol{n}}\times\boldsymbol{E}_s)\times\boldsymbol{\sigma}_0 - (\hat{\boldsymbol{n}}\cdot\boldsymbol{E}_s)\boldsymbol{\sigma}_0]\mathrm{e}^{-\mathrm{i}k_0\boldsymbol{\sigma}_0\cdot\boldsymbol{r}}\mathrm{d}s. \tag{88}$$

Considering that the local field in the vicinity of $\boldsymbol{r}_0$ has the character of a plane-wave, the last term in the above integrand, which represents a contribution to $\boldsymbol{E}_s(\boldsymbol{r}_0)$ that is aligned with the local k-vector, is expected to be cancelled out by an equal but opposite contribution from the first term.[9] We thus arrive at the following simplified version of Eq.(88):

‡‡ Our Eq.(81) agrees with the corresponding result in Born & Wolf's *Principles of Optics*,[1] provided that the angles θ and α_0 in their Eq.(22) of Chapter 11, Section 5, are recognized as $2\pi - \theta$ and $\pi - \theta_0$ in our notation.

$$\boldsymbol{E}_s(\boldsymbol{r}_0) \cong -\frac{\mathrm{i}k_0 \exp(\mathrm{i}k_0 r_0)}{4\pi r_0}\boldsymbol{\sigma}_0 \times \int_{S_1}[c\boldsymbol{\sigma}_0 \times (\hat{\boldsymbol{n}} \times \boldsymbol{B}_s) - \hat{\boldsymbol{n}} \times \boldsymbol{E}_s]\mathrm{e}^{-\mathrm{i}k_0\boldsymbol{\sigma}_0\cdot\boldsymbol{r}}\mathrm{d}s. \quad (89)$$

With reference to Fig.8, suppose now that the object is illuminated (and thus excited) by a plane-wave arriving along the unit-vector $\boldsymbol{\sigma}_{\text{inc}}$, whose $\boldsymbol{E}$ and $\boldsymbol{B}$ fields are

$$\boldsymbol{E}_{\text{inc}}(\boldsymbol{r},t) = \boldsymbol{E}_{\text{i}} \exp[\mathrm{i}k_0(\boldsymbol{\sigma}_{\text{inc}}\cdot\boldsymbol{r} - ct)]. \quad (90)$$

$$\boldsymbol{B}_{\text{inc}}(\boldsymbol{r},t) = c^{-1}\boldsymbol{\sigma}_{\text{inc}} \times \boldsymbol{E}_{\text{i}} \exp[\mathrm{i}k_0(\boldsymbol{\sigma}_{\text{inc}}\cdot\boldsymbol{r} - ct)]. \quad (91)$$

The time-averaged total Poynting vector on the surface S_1 is readily evaluated, as follows:

$$\begin{aligned}\langle \boldsymbol{S}_{\text{total}}(\boldsymbol{r})\rangle &= ½\mathrm{Re}\big[(\boldsymbol{E}_{\text{i}}e^{\mathrm{i}k_0\boldsymbol{\sigma}_{\text{inc}}\cdot\boldsymbol{r}} + \boldsymbol{E}_s) \times \mu_0^{-1}(\boldsymbol{B}_{\text{i}}e^{\mathrm{i}k_0\boldsymbol{\sigma}_{\text{inc}}\cdot\boldsymbol{r}} + \boldsymbol{B}_s)^*\big]\\
&= ½\mu_0^{-1}\mathrm{Re}\big(\boldsymbol{E}_{\text{i}} \times \boldsymbol{B}_{\text{i}}^* + \boldsymbol{E}_{\text{s}} \times \boldsymbol{B}_{\text{s}}^* + \boldsymbol{E}_{\text{i}}e^{\mathrm{i}k_0\boldsymbol{\sigma}_{\text{inc}}\cdot\boldsymbol{r}} \times \boldsymbol{B}_{\text{s}}^* + \boldsymbol{E}_{\text{s}} \times \boldsymbol{B}_{\text{i}}^* e^{-\mathrm{i}k_0\boldsymbol{\sigma}_{\text{inc}}\cdot\boldsymbol{r}}\big)\\
&= ½Z_0^{-1}(\boldsymbol{E}_{\text{i}}\cdot\boldsymbol{E}_{\text{i}}^*)\boldsymbol{\sigma}_{\text{inc}} + ½\mu_0^{-1}\mathrm{Re}(\boldsymbol{E}_{\text{s}} \times \boldsymbol{B}_{\text{s}}^*) \leftarrow \boxed{Z_0 = \sqrt{\mu_0/\varepsilon_0} \cong 377\,\Omega}\\
&\quad + ½Z_0^{-1}\mathrm{Re}\{[c\boldsymbol{E}_{\text{i}}^* \times \boldsymbol{B}_{\text{s}} + \boldsymbol{E}_{\text{s}} \times (\boldsymbol{\sigma}_{\text{inc}} \times \boldsymbol{E}_{\text{i}}^*)]e^{-\mathrm{i}k_0\boldsymbol{\sigma}_{\text{inc}}\cdot\boldsymbol{r}}\}.\end{aligned} \quad (92)$$

If we dot-multiply both sides of Eq.(92) by $-\hat{\boldsymbol{n}}$, then integrate over the closed surface S_1, we obtain, on the left-hand side, the total rate of the inward flow of EM energy, which is the absorbed EM power by the object. On the right-hand side, the first term integrates to zero, because $(\boldsymbol{E}_{\text{i}}\cdot\boldsymbol{E}_{\text{i}}^*)\boldsymbol{\sigma}_{\text{inc}}$ is a constant and $\oint_{S_1}\hat{\boldsymbol{n}}\mathrm{d}s = 0$. The integral of the second term will be the negative time-rate of the energy departure from the object via scattering, which can be moved to the left-hand side of the equation. The combination of the two terms on the left-hand side now yields the total EM power that is taken away from the incident beam — either by absorption or due to scattering. We will have

$$\begin{aligned}\textit{Absorbed + Scattered Power} &= -½Z_0^{-1}\mathrm{Re}\oint_{S_1}\hat{\boldsymbol{n}}\cdot[c\boldsymbol{E}_{\text{i}}^* \times \boldsymbol{B}_{\text{s}} + \boldsymbol{E}_{\text{s}} \times (\boldsymbol{\sigma}_{\text{inc}} \times \boldsymbol{E}_{\text{i}}^*)]e^{-\mathrm{i}k_0\boldsymbol{\sigma}_{\text{inc}}\cdot\boldsymbol{r}}\mathrm{d}s\\
\boxed{\boldsymbol{a}\cdot(\boldsymbol{b}\times\boldsymbol{c}) = (\boldsymbol{a}\times\boldsymbol{b})\cdot\boldsymbol{c}} \rightarrow &= ½Z_0^{-1}\mathrm{Re}\oint_{S_1}[c(\hat{\boldsymbol{n}} \times \boldsymbol{B}_{\text{s}})\cdot\boldsymbol{E}_{\text{i}}^* - (\hat{\boldsymbol{n}} \times \boldsymbol{E}_{\text{s}})\cdot(\boldsymbol{\sigma}_{\text{inc}} \times \boldsymbol{E}_{\text{i}}^*)]e^{-\mathrm{i}k_0\boldsymbol{\sigma}_{\text{inc}}\cdot\boldsymbol{r}}\mathrm{d}s\\
&= ½Z_0^{-1}\mathrm{Re}\Big\{\boldsymbol{E}_{\text{i}}^*\cdot\oint_{S_1}[c(\hat{\boldsymbol{n}} \times \boldsymbol{B}_{\text{s}}) - (\hat{\boldsymbol{n}} \times \boldsymbol{E}_{\text{s}}) \times \boldsymbol{\sigma}_{\text{inc}}]e^{-\mathrm{i}k_0\boldsymbol{\sigma}_{\text{inc}}\cdot\boldsymbol{r}}\mathrm{d}s\Big\}.\end{aligned} \quad (93)$$

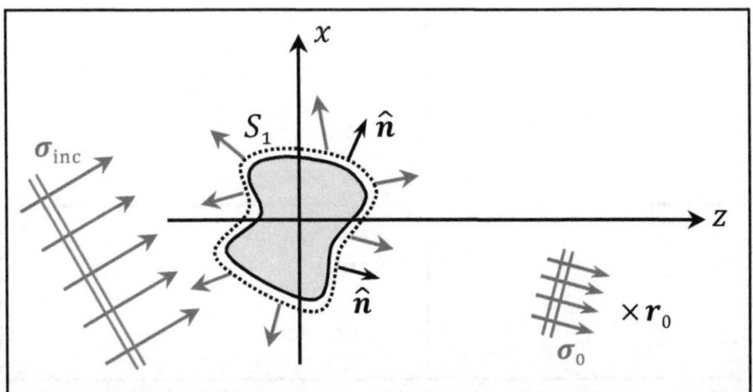

Fig.8. A monochromatic plane-wave propagating along the unit-vector $\boldsymbol{\sigma}_{\text{inc}}$ is scattered from a small object in the vicinity of the origin of the coordinate system. The scattered electric and magnetic fields on the closed surface S_1 surrounding the object are denoted by $\boldsymbol{E}_s(\boldsymbol{r})$ and $\boldsymbol{B}_s(\boldsymbol{r})$. The surface normals $\hat{\boldsymbol{n}}$ at every point on S_1 are outward directed. The scattered light reaching the far away observation point $\boldsymbol{r}_0 = r_0\boldsymbol{\sigma}_0$ has the k-vector $k_0\boldsymbol{\sigma}_0$ and the EM fields $\boldsymbol{E}_s(\boldsymbol{r}_0)$ and $\boldsymbol{B}_s(\boldsymbol{r}_0)$.

Comparison with Eq.(88) reveals that the integral in Eq.(93) is proportional to the scattered E-field in the direction of $\boldsymbol{\sigma}_0 = \boldsymbol{\sigma}_{\text{inc}}$ as observed in the far field. (Recall that the term $(\hat{\boldsymbol{n}} \cdot \boldsymbol{E}_s)\boldsymbol{\sigma}_0$ in the integrand of Eq.(88) has been deemed inconsequential.) If Eq.(93) is normalized by the incident EM power per unit area, namely, $P_{\text{inc}} = \frac{1}{2}Z_0^{-1}\text{Re}(\boldsymbol{E}_{\text{i}} \cdot \boldsymbol{E}_{\text{i}}^*)$, the left-hand side will become the scattering cross-section of the object, while the right-hand side, aside from the coefficient $\text{i}k_0 e^{\text{i}k_0 r_0}/(4\pi r_0)$, will be the projection of the forward-scattered E-field (i.e., $\boldsymbol{\sigma}_0 = \boldsymbol{\sigma}_{\text{inc}}$) on the incident E-field. This important result in the theory of scattering has come to be known as the *optical theorem* (or the optical cross-section theorem).[1,9,24]

9. Scattering from weak inhomogeneities.[9] Figure 9 shows a monochromatic plane-wave of frequency ω passing through a transparent, linear, isotropic medium that has a region of weak inhomogeneities in the vicinity of the origin of coordinates. The host medium is described by its relative permittivity $\varepsilon(\boldsymbol{r},\omega)$ and permeability $\mu(\boldsymbol{r},\omega)$, which consist of a spatially homogeneous background plus slight variations (localized in the vicinity of $\boldsymbol{r} = 0$) on this background; that is,

$$\varepsilon(\boldsymbol{r},\omega) = \varepsilon_h(\omega) + \delta\varepsilon(\boldsymbol{r},\omega). \tag{94}$$

$$\mu(\boldsymbol{r},\omega) = \mu_h(\omega) + \delta\mu(\boldsymbol{r},\omega). \tag{95}$$

The displacement field is thus written as $\boldsymbol{D}(\boldsymbol{r},\omega) = \varepsilon_0\varepsilon(\boldsymbol{r},\omega)\boldsymbol{E}(\boldsymbol{r},\omega)$ and the magnetic induction is given by $\boldsymbol{B}(\boldsymbol{r},\omega) = \mu_0\mu(\boldsymbol{r},\omega)\boldsymbol{H}(\boldsymbol{r},\omega)$. In the absence of free charges and currents, i.e., when $\rho_{\text{free}}(\boldsymbol{r},\omega) = 0$ and $\boldsymbol{J}_{\text{free}}(\boldsymbol{r},\omega) = 0$, Maxwell's macroscopic equations will be[4-11]

$$\boldsymbol{\nabla} \cdot \boldsymbol{D} = 0; \qquad \boldsymbol{\nabla} \times \boldsymbol{H} = -\text{i}\omega\boldsymbol{D}; \qquad \boldsymbol{\nabla} \times \boldsymbol{E} = \text{i}\omega\boldsymbol{B}; \qquad \boldsymbol{\nabla} \cdot \boldsymbol{B} = 0. \tag{96}$$

Noting that the $\boldsymbol{D}$ and $\boldsymbol{B}$ fields depart only slightly from the respective values $\varepsilon_0\varepsilon_h\boldsymbol{E}$ and $\mu_0\mu_h\boldsymbol{H}$ that they would have had in the absence of the $\delta\varepsilon$ and $\delta\mu$ perturbations, we write

$$\begin{aligned}
\boldsymbol{\nabla} \times \boldsymbol{\nabla} \times (\boldsymbol{D} - \varepsilon_0\varepsilon_h\boldsymbol{E}) &= \cancelto{0}{\boldsymbol{\nabla}(\boldsymbol{\nabla} \cdot \boldsymbol{D})} - \boldsymbol{\nabla}^2\boldsymbol{D} - \text{i}\omega\varepsilon_0\varepsilon_h\boldsymbol{\nabla} \times \boldsymbol{B} \\
&= -\boldsymbol{\nabla}^2\boldsymbol{D} - \text{i}\omega\varepsilon_0\varepsilon_h[\boldsymbol{\nabla} \times (\boldsymbol{B} - \mu_0\mu_h\boldsymbol{H}) + \mu_0\mu_h\boldsymbol{\nabla} \times \boldsymbol{H}] \\
&= -\boldsymbol{\nabla}^2\boldsymbol{D} - \text{i}\omega\varepsilon_0\varepsilon_h\boldsymbol{\nabla} \times (\boldsymbol{B} - \mu_0\mu_h\boldsymbol{H}) - (\omega/c)^2\mu_h\varepsilon_h\boldsymbol{D}.
\end{aligned} \tag{97}$$

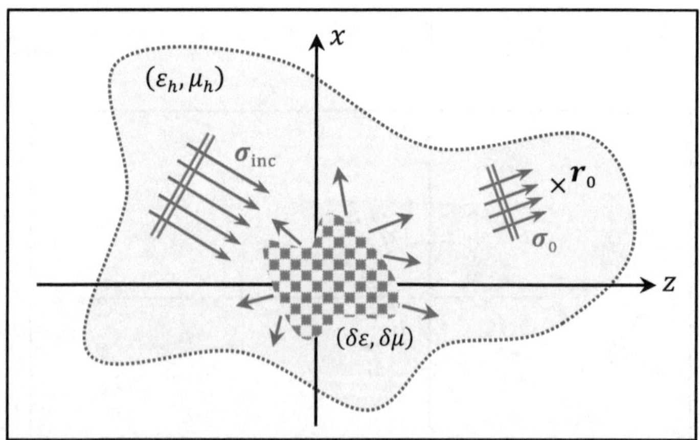

Fig.9. A monochromatic plane-wave of frequency ω and E-field amplitude $\boldsymbol{E}_{\text{inc}}$ propagates along the unit-vector $\boldsymbol{\sigma}_{\text{inc}}$ within a mildly inhomogeneous host medium of refractive index $n_h(\omega) = (\mu_h\varepsilon_h)^{½}$. The inhomogeneous region of the host, a small patch in the vicinity of the origin of coordinates, is specified by its relative permittivity $\varepsilon_h(\omega) + \delta\varepsilon(\boldsymbol{r},\omega)$ and relative permeability $\mu_h(\omega) + \delta\mu(\boldsymbol{r},\omega)$. The scattered field is observed at the faraway point $\boldsymbol{r}_0 = r_0\boldsymbol{\sigma}_0$.

Recalling that the refractive index of the homogeneous (background) material is defined as $n_h = (\mu_h \varepsilon_h)^{1/2}$, and that the free-space wave-number is $k_0 = \omega/c$, Eq.(97) may be rewritten as

$$\boldsymbol{\nabla}^2 \boldsymbol{D} + (k_0 n_h)^2 \boldsymbol{D} = -\mathrm{i}(k_0 \varepsilon_h / c)\boldsymbol{\nabla} \times (\delta\mu \boldsymbol{H}) - \varepsilon_0 \boldsymbol{\nabla} \times \boldsymbol{\nabla} \times (\delta\varepsilon \boldsymbol{E}). \tag{98}$$

This Helmholtz equation has a homogeneous solution, which we denote by $\boldsymbol{D}_h(\boldsymbol{r}, \omega)$, and an inhomogeneous solution, which arises from the local deviations $\delta\varepsilon$ and $\delta\mu$ of the host medium. Recalling that the Green function $G(\boldsymbol{r}, \boldsymbol{r}_0) = \exp(\mathrm{i}k_0 n_h |\boldsymbol{r} - \boldsymbol{r}_0|)/|\boldsymbol{r} - \boldsymbol{r}_0|$ is a solution of the Helmholtz equation $\boldsymbol{\nabla}^2 G + (k_0 n_h)^2 G = -4\pi\delta(\boldsymbol{r} - \boldsymbol{r}_0)$, an integral relation for the scattered field solution $\boldsymbol{D}_s(\boldsymbol{r}, \omega)$ of Eq.(98) in terms of the total fields $\boldsymbol{E}(\boldsymbol{r}, \omega)$ and $\boldsymbol{H}(\boldsymbol{r}, \omega)$ will be

dr stands for $dxdydz$

$$\boldsymbol{D}_s(\boldsymbol{r}_0, \omega) = (4\pi)^{-1} \int_{\text{volume}} [\mathrm{i}(k_0 \varepsilon_h / c)\boldsymbol{\nabla} \times (\delta\mu \boldsymbol{H}) + \varepsilon_0 \boldsymbol{\nabla} \times \boldsymbol{\nabla} \times (\delta\varepsilon \boldsymbol{E})]\, G(\boldsymbol{r}, \boldsymbol{r}_0) \mathrm{d}\boldsymbol{r}. \tag{99}$$

Using a far-field approximation to $G(\boldsymbol{r}, \boldsymbol{r}_0)$ similar to that in Eq.(50), we will have

$$\boldsymbol{D}_s(\boldsymbol{r}_0, \omega) \cong \frac{\exp(\mathrm{i}k_0 n_h r_0)}{4\pi r_0} \int_{\text{volume}} [\mathrm{i}(k_0 \varepsilon_h / c)\boldsymbol{\nabla} \times (\delta\mu \boldsymbol{H}) + \varepsilon_0 \boldsymbol{\nabla} \times \boldsymbol{\nabla} \times (\delta\varepsilon \boldsymbol{E})] e^{-\mathrm{i}k_0 n_h \boldsymbol{\sigma}_0 \cdot \boldsymbol{r}} \mathrm{d}\boldsymbol{r}. \tag{100}$$

The vector identity $(\boldsymbol{\nabla} \times \boldsymbol{V}) e^{-\mathrm{i}\boldsymbol{k} \cdot \boldsymbol{r}} = \mathrm{i}\boldsymbol{k} \times \boldsymbol{V} e^{-\mathrm{i}\boldsymbol{k} \cdot \boldsymbol{r}} + \boldsymbol{\nabla} \times (\boldsymbol{V} e^{-\mathrm{i}\boldsymbol{k} \cdot \boldsymbol{r}})$ can be used to replace the first term in the integrand of Eq.(100) with $-(k_0^2 \varepsilon_h n_h / c) e^{-\mathrm{i}k_0 n_h \boldsymbol{\sigma}_0 \cdot \boldsymbol{r}} \boldsymbol{\sigma}_0 \times \delta\mu \boldsymbol{H}$. The volume integral of $\boldsymbol{\nabla} \times (\boldsymbol{V} e^{-\mathrm{i}\boldsymbol{k} \cdot \boldsymbol{r}})$ becomes the surface integral of $e^{-\mathrm{i}\boldsymbol{k} \cdot \boldsymbol{r}} \boldsymbol{V} \times \mathrm{d}\boldsymbol{s}$, which subsequently vanishes because, for away from the inhomogeneous region, $\delta\mu \to 0$. Similarly, the second term of the integrand is replaced by $\mathrm{i}\varepsilon_0 k_0 n_h \boldsymbol{\sigma}_0 \times [\boldsymbol{\nabla} \times (\delta\varepsilon \boldsymbol{E})] e^{-\mathrm{i}k_0 n_h \boldsymbol{\sigma}_0 \cdot \boldsymbol{r}}$. Another application of the vector identity then replaces the remaining term with $-\varepsilon_0 k_0^2 n_h^2 \boldsymbol{\sigma}_0 \times (\boldsymbol{\sigma}_0 \times \delta\varepsilon \boldsymbol{E}) e^{-\mathrm{i}k_0 n_h \boldsymbol{\sigma}_0 \cdot \boldsymbol{r}}$. We thus arrive at

$$\boldsymbol{D}_s(\boldsymbol{r}_0, \omega) \cong \frac{(k_0 n_h)^2 \exp(\mathrm{i}k_0 n_h r_0)}{4\pi r_0} \int_{\text{volume}} [(\varepsilon_h / n_h c)\delta\mu \boldsymbol{H} + \varepsilon_0 \boldsymbol{\sigma}_0 \times \delta\varepsilon \boldsymbol{E}] \times \boldsymbol{\sigma}_0 e^{-\mathrm{i}k_0 n_h \boldsymbol{\sigma}_0 \cdot \boldsymbol{r}} \mathrm{d}\boldsymbol{r}. \tag{101}$$

In the first Born approximation, the $\boldsymbol{E}(\boldsymbol{r})$ and $\boldsymbol{H}(\boldsymbol{r})$ fields in the integrand of Eq.(101) are replaced with the solutions $\boldsymbol{E}_h(\boldsymbol{r})$ and $\boldsymbol{H}_h(\boldsymbol{r})$ of the homogeneous Helmholtz equation. When the homogeneous background wave is a plane-wave, we will have

$$\boldsymbol{E}_h(\boldsymbol{r}) = \boldsymbol{E}_{\text{inc}} \exp(\mathrm{i}k_0 n_h \boldsymbol{\sigma}_{\text{inc}} \cdot \boldsymbol{r}). \tag{102}$$

$$\boldsymbol{H}_h(\boldsymbol{r}) = \sqrt{\frac{\varepsilon_0 \varepsilon_h}{\mu_0 \mu_h}}\, \boldsymbol{\sigma}_{\text{inc}} \times \boldsymbol{E}_{\text{inc}} \exp(\mathrm{i}k_0 n_h \boldsymbol{\sigma}_{\text{inc}} \cdot \boldsymbol{r}). \tag{103}$$

A final substitution from Eqs.(102) and (103) into Eq.(101) yields

$$\boldsymbol{D}_s(\boldsymbol{r}_0, \omega) \cong \frac{\varepsilon_0 \varepsilon_h k_0^2 \exp(\mathrm{i}k_0 n_h r_0)}{4\pi r_0} \int_{\text{volume}} [(\varepsilon_h \delta\mu \boldsymbol{\sigma}_{\text{inc}} + \mu_h \delta\varepsilon \boldsymbol{\sigma}_0) \times \boldsymbol{E}_{\text{inc}}] \times \boldsymbol{\sigma}_0 e^{\mathrm{i}k_0 n_h (\boldsymbol{\sigma}_{\text{inc}} - \boldsymbol{\sigma}_0) \cdot \boldsymbol{r}} \mathrm{d}\boldsymbol{r}. \tag{104}$$

Thus, in the first Born approximation, the scattered field $\boldsymbol{D}_s(\boldsymbol{r}_0, \omega) = \varepsilon_0 \varepsilon_h \boldsymbol{E}_s(\boldsymbol{r}_0, \omega)$ is directly related to the host medium perturbations $\delta\varepsilon(\boldsymbol{r}, \omega)$ and $\delta\mu(\boldsymbol{r}, \omega)$ via the volume integral in Eq.(104). Here, $\boldsymbol{E}_{\text{inc}}$ embodies not only the strength but also the polarization state of the incident plane-wave, the unit-vector $\boldsymbol{\sigma}_{\text{inc}}$ is the direction of incidence, $\boldsymbol{\sigma}_0 = \boldsymbol{r}_0 / r_0$ is a unit-vector pointing from the origin of coordinates to the observation point $\boldsymbol{r}_0$, and $\boldsymbol{q} = k_0 n_h (\boldsymbol{\sigma}_{\text{inc}} - \boldsymbol{\sigma}_0)$ is the difference between the incident and scattered k-vectors.

10. Neutron scattering from magnetic electrons in Born's first approximation. The scattering of slow neutrons from ferromagnetic materials can be treated in ways that are similar to our analysis of EM scattering from mild inhomogeneities discussed in the preceding section. The wave function $\psi(\boldsymbol{r},t)$ of a particle of mass m in the scalar potential field $V(\boldsymbol{r},t)$ satisfies the following Schrödinger equation:[6-8, 20]

$$i\hbar\partial_t\psi(\boldsymbol{r},t) = -(\hbar^2/2m)\nabla^2\psi(\boldsymbol{r},t) + V(\boldsymbol{r},t)\psi(\boldsymbol{r},t). \tag{105}$$

When the potential is time-independent and the particle is in an eigenstate of energy $\mathcal{E}_n$, we will have the time-independent Schrödinger equation, namely,

$$[(\hbar^2/2m)\nabla^2 + \mathcal{E}_n]\psi(\boldsymbol{r}) = V(\boldsymbol{r})\psi(\boldsymbol{r}). \tag{106}$$

The Green function for Eq.(106) is $G(\boldsymbol{r},\boldsymbol{r}') = e^{ik|\boldsymbol{r}-\boldsymbol{r}'|}/|\boldsymbol{r}-\boldsymbol{r}'|$, where $\hbar k = (2m\mathcal{E}_n)^{½}$. If, in the absence of the potential $V(\boldsymbol{r})$, the solution of the homogeneous equation is found to be $\psi_0(\boldsymbol{r})$, then, when the potential is introduced, we will have

dr′ stands for dx′dy′dz′

$$\psi(\boldsymbol{r}) = \psi_0(\boldsymbol{r}) - (m/2\pi\hbar^2)\iiint_{-\infty}^{\infty} G(\boldsymbol{r},\boldsymbol{r}')V(\boldsymbol{r}')\psi(\boldsymbol{r}')\mathrm{d}\boldsymbol{r}'. \tag{107}$$

Note that Eq.(107) is *not* an actual solution of Eq.(106); rather, considering that the desired wave-function $\psi(\boldsymbol{r})$ appears in the integrand on the right hand-side, Eq.(107) is an integral form of the differential equation (106). In the case of Born's first approximation, one assumes that $V(\boldsymbol{r})$ is a fairly weak potential, in which case $\psi_0(\boldsymbol{r}')$ can be substituted for $\psi(\boldsymbol{r}')$, yielding

$$\psi(\boldsymbol{r}) \cong \psi_0(\boldsymbol{r}) - (m/2\pi\hbar^2)\iiint_{-\infty}^{\infty} G(\boldsymbol{r},\boldsymbol{r}')V(\boldsymbol{r}')\psi_0(\boldsymbol{r}')\mathrm{d}\boldsymbol{r}'. \tag{108}$$

In a typical scattering problem, an incoming particle of mass m and well-defined momentum $\boldsymbol{p} = \hbar\boldsymbol{k}$ has the initial wave-function $\psi_0(\boldsymbol{r}) = e^{i\boldsymbol{k}\cdot\boldsymbol{r}}$. Upon interacting with a weak scattering potential $V(\boldsymbol{r})$, the wave-function will change in accordance with Eq.(108). Let $V(\boldsymbol{r})$ have significant values only in the vicinity of the origin, $\boldsymbol{r} = 0$, and assume that the scattering process is elastic, so that the momentum $\tilde{\boldsymbol{p}}$ of the particle after scattering will have the same magnitude $\hbar k$ as before, but the direction of propagation changes from that of $\boldsymbol{k}$ to that of $\tilde{\boldsymbol{k}}$. At an observation point $\boldsymbol{r}$ far from the origin, that is, $|\boldsymbol{r}| \gg |\boldsymbol{r}'|$, the scattered particle's momentum is expected to be $\tilde{\boldsymbol{p}} = \hbar\tilde{\boldsymbol{k}} = \hbar k\hat{\boldsymbol{r}}$, and the Green function may be approximated as

$$\frac{\exp(ik|\boldsymbol{r}-\boldsymbol{r}'|)}{|\boldsymbol{r}-\boldsymbol{r}'|} \cong \frac{\exp[ik\sqrt{(\boldsymbol{r}-\boldsymbol{r}')\cdot(\boldsymbol{r}-\boldsymbol{r}')}]}{r} \cong \frac{\exp[ik(r-\boldsymbol{r}'\cdot\hat{\boldsymbol{r}})]}{r} = \frac{\exp(ikr)}{r}e^{-i\tilde{\boldsymbol{k}}\cdot\boldsymbol{r}'}. \tag{109}$$

Substitution into Eq.(108) now yields

$$\psi(\boldsymbol{r}) \cong e^{i\boldsymbol{k}\cdot\boldsymbol{r}} - \left(\frac{m}{2\pi\hbar^2}\right)\frac{e^{ikr}}{r}\int_{-\infty}^{\infty} V(\boldsymbol{r}')e^{i(\boldsymbol{k}-\tilde{\boldsymbol{k}})\cdot\boldsymbol{r}'}\mathrm{d}\boldsymbol{r}'. \tag{110}$$

Denoting the change in the direction of the particle's momentum by $\boldsymbol{q} = \boldsymbol{p} - \tilde{\boldsymbol{p}}$, and noting that e^{ikr}/r is simply a spherical wave emanating from the origin, the scattering amplitude $f(\boldsymbol{q})$ is readily seen to be

$$f(\boldsymbol{q}) \cong -\frac{m}{2\pi\hbar^2}\int_{-\infty}^{\infty} V(\boldsymbol{r}')e^{i\boldsymbol{q}\cdot\boldsymbol{r}'/\hbar}\,\mathrm{d}\boldsymbol{r}'. \tag{111}$$

Here, $f(\boldsymbol{q})$ has the dimensions of length (meter in *SI*). Note that the presence of e^{ikr}/r in Eq.(110) makes the wave-function $\psi(\boldsymbol{r})$ dimensionless. Let $\mathrm{d}\Omega = \sin\theta\,\mathrm{d}\theta\mathrm{d}\varphi$ be the differential element of the solid angle viewed from the origin of the coordinates. The differential scattering cross-section will then be $\mathrm{d}\sigma/\mathrm{d}\Omega = |f(\boldsymbol{q})|^2$, with the total cross-section being $\sigma = \int |f(\boldsymbol{q})|^2 \mathrm{d}\Omega$.

Example 1. A particle of mass m is scattered by the spherically symmetric potential $V(r)$ corresponding to a fixed particle located at $\boldsymbol{r} = 0$. The scattering amplitude, computed from Eq.(111), will be

$$f(\boldsymbol{q}) = -\frac{m}{2\pi\hbar^2}\int_{r=0}^{\infty}\int_{\theta=0}^{\pi} V(r)\exp(iqr\cos\theta/\hbar)\,(2\pi r^2\sin\theta)\mathrm{d}\theta\mathrm{d}r = -\frac{2m}{\hbar q}\int_0^{\infty} rV(r)\sin(qr/\hbar)\,\mathrm{d}r. \quad (112)$$

Considering that $\boldsymbol{q} = \boldsymbol{p} - \tilde{\boldsymbol{p}}$, and that $\boldsymbol{p}$ and $\tilde{\boldsymbol{p}}$ have the same magnitude p, we denote by θ the angle between $\boldsymbol{p}$ and $\tilde{\boldsymbol{p}}$, and proceed to write $q = 2p\sin(\theta/2)$. The scattering amplitude thus has circular symmetry around the direction of the incident momentum $\boldsymbol{p}$. The ambiguity of Eq.(112) with regard to forward scattering at $\theta = 0$ is resolved if the forward amplitude $f(0)$ is directly computed from Eq.(110)—with the destructive interference between the incident and scattered amplitudes properly taken into account.

For the Yukawa potential $V(\boldsymbol{r}) = v_0 e^{-\alpha r}/r$, with $\alpha > 0$ being the range parameter, the scattering amplitude is obtained upon integrating Eq.(112), as follows:

$$f(\boldsymbol{q}) = -\frac{2mv_0}{\hbar q}\int_0^{\infty} e^{-\alpha r}\sin(qr/\hbar)\,\mathrm{d}r = -\frac{2mv_0}{q^2 + (\alpha\hbar)^2}. \quad (113)$$

In the limit when $\alpha \to 0$, the Yukawa potential approaches the Coulomb potential $V(\boldsymbol{r}) = v_0/r$. When a particle having electric charge $\pm e$ and energy $\mathcal{E} = p^2/2m$ is scattered from another particle of charge $\pm e$, the scattering cross-section will be

$$\frac{\mathrm{d}\sigma}{\mathrm{d}\Omega} = |f(\boldsymbol{q})|^2 = \left(\frac{e^2}{16\pi\varepsilon_0\mathcal{E}}\right)^2 \frac{1}{\sin^4(\theta/2)}. \quad (114)$$

This is the famous Rutherford scattering cross-section.[9]

Example 2. In low-energy scattering, $k \cong 0$ and the scattering amplitude in all directions becomes $f(\theta,\varphi) \cong -(m/2\pi\hbar^2)\int_{-\infty}^{\infty} V(\boldsymbol{r})\mathrm{d}\boldsymbol{r}$. In the case of low-energy, soft-sphere scattering, where $V(\boldsymbol{r}) = V_0$ when $r \le r_0$ and zero otherwise, we find

$$f(\theta,\varphi) = -(m/2\pi\hbar^2)(4\pi r_0^3/3)V_0. \quad (115a)$$

$$\mathrm{d}\sigma/\mathrm{d}\Omega = |f(\theta,\varphi)|^2 = (2mr_0^3 V_0/3\hbar^2)^2. \quad (115b)$$

$$\sigma = 4\pi(\mathrm{d}\sigma/\mathrm{d}\Omega) = 16\pi m^2 r_0^6 V_0^2/9\hbar^4. \quad (115c)$$

Let us now consider the case of a polarized neutron entering a ferromagnetic medium and getting scattered from the host's magnetic electrons.[29] To obtain an estimate of the corresponding scattering potential $V(\boldsymbol{r})$, we begin by noting that the magnetic field surrounding a point-dipole $m\delta(\boldsymbol{r})\hat{\boldsymbol{z}}$ in free space is

The magnetic dipole moment m should not be confused with the particle's mass m.

$$\boldsymbol{H}(\boldsymbol{r}) = m(2\cos\theta\,\hat{\boldsymbol{r}} + \sin\theta\,\hat{\boldsymbol{\theta}})/(4\pi\mu_0 r^3). \quad (116)$$

A second magnetic point-dipole $\boldsymbol{m}'$, located at $\boldsymbol{r} \ne 0$, will have the energy $\mathcal{E} = -\boldsymbol{m}'\cdot\boldsymbol{H}(\boldsymbol{r})$, which may be written as follows:

$$\begin{aligned}\mathcal{E} &= -\boldsymbol{m}'\cdot m[3\cos\theta\,\hat{\boldsymbol{r}} - (\cos\theta\,\hat{\boldsymbol{r}} - \sin\theta\,\hat{\boldsymbol{\theta}})]/(4\pi\mu_0 r^3)\\ &= \boldsymbol{m}'\cdot m(\hat{\boldsymbol{z}} - 3\cos\theta\,\hat{\boldsymbol{r}})/(4\pi\mu_0 r^3) = [\boldsymbol{m}\cdot\boldsymbol{m}' - 3(\boldsymbol{m}\cdot\hat{\boldsymbol{r}})(\boldsymbol{m}'\cdot\hat{\boldsymbol{r}})]/(4\pi\mu_0 r^3).\end{aligned} \quad (117)$$

These results are consistent with the Einstein-Laub formula $\boldsymbol{F} = (\boldsymbol{m} \cdot \boldsymbol{\nabla})\boldsymbol{H}$ for the force as well as $\boldsymbol{T} = \boldsymbol{m} \times \boldsymbol{H}$ for the torque experienced by a point-dipole in a magnetic field.[30,31] Recall that, in contrast to the standard formula for the dipole moment, our definition of $\boldsymbol{B}$ as $\mu_0\boldsymbol{H} + \boldsymbol{M}$ maintains that the magnitude of $\boldsymbol{m}$ equals μ_0 times the electrical current times the area of a small current loop. Consequently, the aforementioned expression for $\mathcal{E}$ coincides with the well-known expression $\mathcal{E} = -\boldsymbol{m} \cdot \boldsymbol{B}(\boldsymbol{r})$ of the energy of the dipole $\boldsymbol{m}$ in the external field $\boldsymbol{B}(\boldsymbol{r}) = \mu_0\boldsymbol{H}(\boldsymbol{r})$.[6,9]

Equation (117) must be augmented by the contact term $-2\boldsymbol{m} \cdot \boldsymbol{m}'\delta(\boldsymbol{r})/3\mu_0$ to account for the energy of the dipole pair when $\boldsymbol{m}$ and $\boldsymbol{m}'$ overlap in space.[29] We will have

$$\mathcal{E}(\boldsymbol{r}) = \frac{\boldsymbol{m} \cdot \boldsymbol{m}' - 3(\boldsymbol{m} \cdot \hat{\boldsymbol{r}})(\boldsymbol{m}' \cdot \hat{\boldsymbol{r}})}{4\pi\mu_0 r^3} - \frac{2\boldsymbol{m} \cdot \boldsymbol{m}'\delta(\boldsymbol{r})}{3\mu_0}. \tag{118}$$

Suppose the electron has wave-function $\psi(\boldsymbol{r}_e)$ and magnetic dipole moment $\boldsymbol{\mu}_e$, while the incoming neutron has wave-function $\exp(i\boldsymbol{k} \cdot \boldsymbol{r}_n)$, magnetic dipole moment $\boldsymbol{\mu}_n$, and mass m_n. We assume the scattering process does not involve a spin flip, so that both $\boldsymbol{\mu}_e$ and $\boldsymbol{\mu}_n$ retain their orientations after the collision. Moreover, we assume the electron — being bound to its host lattice — does not get dislocated or otherwise distorted, so that $\psi(\boldsymbol{r}_e)$ is the same before and after the collision. Thus, the potential energy distribution across the landscape of the incoming neutron is the integral over $\boldsymbol{r}_e$ of the product of the electron's probability-density function $|\psi(\boldsymbol{r}_e)|^2$ and the dipole-dipole interaction energy $\mathcal{E}(\boldsymbol{r}_e - \boldsymbol{r}_n)$ between the neutron and the electron. Consequently, in the first Born approximation, the scattering amplitude from an initial neutron momentum $\boldsymbol{p} = \hbar\boldsymbol{k}$ to a final momentum $\boldsymbol{p}' = \hbar\boldsymbol{k}'$, is given by

$$f(\boldsymbol{p}', \boldsymbol{p}) = -\left(\frac{m_n}{2\pi\hbar^2}\right) \int_{-\infty}^{\infty} |\psi(\boldsymbol{r}_e)|^2 \mathcal{E}(\boldsymbol{r}_n - \boldsymbol{r}_e) \exp[i(\boldsymbol{k} - \boldsymbol{k}') \cdot \boldsymbol{r}_n] \, d\boldsymbol{r}_e d\boldsymbol{r}_n. \tag{119}$$

Defining the electronic magnetization (i.e., magnetic moment density of the electron) by $\boldsymbol{M}(\boldsymbol{r}_e) = |\psi(\boldsymbol{r}_e)|^2\boldsymbol{\mu}_e$ and its Fourier transform by $\widetilde{\boldsymbol{M}}(\boldsymbol{k}) = \int_{-\infty}^{\infty} \boldsymbol{M}(\boldsymbol{r}_e) \exp(i\boldsymbol{k} \cdot \boldsymbol{r}_e) \, d\boldsymbol{r}_e$, upon substitution from Eq.(118) into Eq.(119) and setting $\boldsymbol{r} = \boldsymbol{r}_n - \boldsymbol{r}_e$, we find

$$f(\boldsymbol{p}', \boldsymbol{p}) = -\frac{m_n}{8\pi^2\mu_0\hbar^2} \int_{-\infty}^{\infty} \left[\frac{\boldsymbol{\mu}_n - 3(\boldsymbol{\mu}_n \cdot \hat{\boldsymbol{r}})\hat{\boldsymbol{r}}}{r^3} - \frac{8\pi\delta(\boldsymbol{r})\boldsymbol{\mu}_n}{3}\right] \cdot \boldsymbol{M}(\boldsymbol{r}_e) \exp[i(\boldsymbol{p} - \boldsymbol{p}') \cdot \boldsymbol{r}_n/\hbar] \, d\boldsymbol{r}_e d\boldsymbol{r}_n. \tag{120}$$

Defining $\boldsymbol{q} = \boldsymbol{p} - \boldsymbol{p}'$ and changing the variables from $(\boldsymbol{r}_e, \boldsymbol{r}_n)$ to $(\boldsymbol{r}_e, \boldsymbol{r} = \boldsymbol{r}_n - \boldsymbol{r}_e)$ — whose transformation Jacobian is 1.0 — substantially simplifies the above integral, yielding

$$f(\boldsymbol{q}) = -\frac{m_n}{8\pi^2\mu_0\hbar^2} \int_{-\infty}^{\infty} \left[\frac{\boldsymbol{\mu}_n - 3(\boldsymbol{\mu}_n \cdot \hat{\boldsymbol{r}})\hat{\boldsymbol{r}}}{r^3} - \frac{8\pi\delta(\boldsymbol{r})\boldsymbol{\mu}_n}{3}\right] \cdot \widetilde{\boldsymbol{M}}(\boldsymbol{q}/\hbar) \exp(i\boldsymbol{q} \cdot \boldsymbol{r}/\hbar) \, d\boldsymbol{r}. \tag{121}$$

Appendix I shows that the exact evaluation of the integral in Eq.(121) leads to

$$f(\boldsymbol{q}) = \left(\frac{m_n}{2\pi\mu_0\hbar^2}\right) \boldsymbol{\mu}_n \cdot \{\widetilde{\boldsymbol{M}}(\boldsymbol{q}/\hbar) - [\widetilde{\boldsymbol{M}}(\boldsymbol{q}/\hbar) \cdot \hat{\boldsymbol{q}}]\hat{\boldsymbol{q}}\}. \tag{122}$$

This is the same result as given in Ref. [29], Eq.(23), in the case of $\lambda = 1$. The coefficient $4\pi\mu_0$ appears here because we have worked in the *SI* system of units with $\boldsymbol{B} = \mu_0\boldsymbol{H} + \boldsymbol{M}$.

11. Concluding remarks. In this paper, we described some of the fundamental theories of EM scattering and diffraction using the electrodynamics of Maxwell and Lorentz in conjunction with standard mathematical methods of the vector calculus, complex analysis, differential equations, and Fourier transform theory. The scalar Huygens-Fresnel-Kirchhoff and Rayleigh-Sommerfeld theories were presented at first, followed by their extensions that cover the case of vector diffraction of EM waves. Examples were provided to showcase the application of these vector diffraction and scattering formulas to certain problems of practical interest. We did not discuss the alternate method of diffraction calculations by means of Fourier transformation, which involves an expansion of the initial field profile in the xy-plane at $z = 0$ into its plane-wave constituents. In fact, with the aid of the two-dimensional Fourier transform of $G(\boldsymbol{r}, 0)$ given in Eq.(37), it is rather easy to establish the equivalence of the Fourier expansion method with the Rayleigh-Sommerfeld formula in Eq.(25), and also with the related vector formulas in Eqs.(32) and (33). Appendix J outlines the mathematical steps needed to establish these equivalencies.

The Sommerfeld solution to the problem of diffraction from a thin, perfectly conducting half-plane described in Sec.7 is one of the few problems in the EM theory of diffraction for which an exact analytical solution has been found; for a discussion of related problems of this type, see Ref.[1], Chapter 11. Scattering of plane-waves from spherical particles of known relative permittivity $\varepsilon(\omega)$ and permeability $\mu(\omega)$, the so-called Mie scattering, is another problem for which an exact solution (albeit in the form of an infinite series) exists; for a discussion of this and related problems the reader is referred to the vast literature of Mie scattering.[1,4,9-11,22]

In our analysis of neutron scattering from ferromagnets in Sec.10, we used the contact term $-2\boldsymbol{m} \cdot \boldsymbol{m}'\delta(\boldsymbol{r})/3\mu_0$ to account for the interaction energy of the dipole pair $\boldsymbol{m}, \boldsymbol{m}'$ when they happen to overlap at the same location in space. This is tantamount to assuming that the dipolar magnetic moments are produced by circulating electrical currents. The contact term would have been $\boldsymbol{m} \cdot \boldsymbol{m}'\delta(\boldsymbol{r})/3\mu_0$ had each magnetic moment been produced by a pair of equal and opposite magnetic monopoles residing within the corresponding particle. Since the Amperian current loop model has been found to agree most closely with experimental findings, we used the former expression for the contact term in Eq.(118); see Ref. [9], Sec.5.7, and Ref. [29] for a pedagogical discussion of the experimental evidence — from neutron scattering as well as the existence of the famous 21 cm astrophysical spectral line of atomic hydrogen — in favor of the Amperian current loop model of the intrinsic magnetic dipole moments of subatomic particles.

Although we did not discuss the Babinet principle of complementary screens that is well known in classical optics, it is worth mentioning here that a rigorous version of this principle has been proven in Maxwellian electrodynamics.[1,9] The original version of Babinet's principle is based on the Kirchhoff diffraction integral of Eq.(23), and the notion that, if S_1 consists of apertures in an opaque screen, then the complement of S_1 would be opaque where S_1 is transmissive, and transmissive where S_1 is opaque. Considering that, in Kirchhoff's approximation, $\psi(\boldsymbol{r})$ and $\partial_n\psi(\boldsymbol{r})$ in Eq.(23) retain the values of the incident beam in the open aperture(s) but vanish in the opaque regions, it is a reasonable conjecture that the observed field in the presence of S_1 and that in the presence of S_1's complement would add up to the observed field when all screens are removed — i.e., when the unobstructed beam reaches the observation point. Similar arguments can be based on either of the Rayleigh-Sommerfeld diffraction integrals in Eqs.(24) and (25), provided, of course, that the Kirchhoff approximation remains applicable. Appendix K describes the rigorous version of the Babinet principle and provides a simple proof that relies on symmetry arguments similar to those used in Example 2 of Sec.6.

Finally, to keep the size and scope of this tutorial within reasonable boundaries, we did not broach the important problem of EM scattering from small dielectric spheres, nor that of EM scattering from small perfectly conducting spheres. The interested reader can find a detailed discussion of these problems in Appendices L and M, respectively.

References

1. M. Born and E. Wolf, *Principles of Optics*, 7th edition, Cambridge University Press, Cambridge, U.K. (1999).
2. C. J. Bouwkamp, "Diffraction theory," in *Reports on Progress in Physics*, edited by A. C. Stickland, Vol. XVII, pp. 35-100, The Physical Society, London (1954).
3. B. B. Baker and E. T. Copson, *The Mathematical Theory of Huygens' Principle*," 2nd edition, Clarendon Press, Oxford (1950).
4. J. A. Stratton, *Electromagnetic Theory*, Wiley, New York (1941).
5. W. K. H. Panofsky and M. Phillips, *Classical Electricity and Magnetism*, Addison-Wesley, Reading, Massachusetts (1956).
6. R. P. Feynman, R. B. Leighton, and M. Sands, *The Feynman Lectures on Physics*, Addison-Wesley, Reading, Massachusetts (1964).
7. L. D. Landau and E. M Lifshitz, *Electrodynamics of Continuous Media*, 2nd edition, Addison-Wesley, Reading, Massachusetts (1984).
8. L. D. Landau and E. M Lifshitz, *The Classical Theory of Fields*, 4th revised English edition, translated by M. Hamermesh, Pergamon Press, Oxford, and Addison-Wesley, Reading, Massachusetts (1987).
9. J. D. Jackson, *Classical Electrodynamics*, 3rd edition, Wiley, New York (1999).
10. A. Zangwill, *Modern Electrodynamics*, Cambridge University Press, Cambridge, United Kingdom (2012).
11. M. Mansuripur, *Field, Force, Energy and Momentum in Classical Electrodynamics*, revised edition, Bentham Science Publishers, Sharjah, U.A.E. (2017).
12. R. Courant and D. Hilbert, *Methods of Mathematical Physics*, Vol.1, Interscience Publishers, New York (1953).
13. J. W. Strutt (Lord Rayleigh), "On the electromagnetic theory of light," *Philosophical Magazine* **12**, 81-101 (1881).
14. J. W. Strutt (Lord Rayleigh), "Wave Theory of Light," *Encyclopedia Britannica*, vol. XXIV, 47-189 (1888).
15. J. W. Strutt (Lord Rayleigh), "On the passage of waves through apertures in plane screens, and allied problems," *Philosophical Magazine* **43**, 259-271 (1897); see also *Scientific Papers by John William Strutt, Baron Rayleigh*, Volumes 1-6, Cambridge University Press, Cambridge, United Kingdom (2011).
16. H. Poincaré, *Théorie Mathématique de la Lumière II*, Georges Carré, Paris (1892).
17. J. Walker, *The Analytical Theory of Light*, Cambridge University Press, Cambridge, U.K. (1904).
18. W. R. Smythe, "The double current sheet in diffraction," *Physical Review* **72**, 1066-70 (1947).
19. M. J. Lax, "Multiple Scattering of Waves," *Reviews of Modern Physics* **23**, 287-310 (1951).
20. P. M. Morse and H. Feshbach, *Methods of Theoretical Physics*, McGraw-Hill, New York (1953).
21. A. Sommerfeld, *Optics*, Academic Press, New York (1954).
22. H. C. van de Hulst, *Light Scattering by Small Particles*, Wiley, New York (1957).
23. J. W. Goodman, *Introduction to Fourier Optics*, 4th edition, W. H. Freeman, New York (2017).
24. M. Mansuripur, "New perspective on the optical theorem of classical electrodynamics," *American Journal of Physics* **80**, 329-333 (2012).
25. A. Sommerfeld, "Mathematische Theorie der Diffraction," *Mathematische Annalen* **47**, 317-374 (1896); translated into English by R. J. Nagem, M. Zampolli, and G. Sandri as *Mathematical Theory of Diffraction*, Springer, New York (2004).
26. E. Jahnke and F. Emde, *Tables of Functions with Formulae and Curves*, Teubner, Leipzig and Berlin; reprinted by Dover Publications, 4th edition, New York (1945).
27. I. S. Gradshteyn and I. M. Ryzhik, *Table of Integrals, Series, and Products*, 7th edition, Academic Press, Burlington, Massachusetts (2007).
28. A. Cornu, "Méthode nouvelle pour la discussion des problèmes de diffraction," *Journal de Physique Théorique et Appliqueé* **3**, 5-15 and 44-52 (1874).
29. J. D. Jackson, "The nature of intrinsic magnetic dipole moments," lecture given in the CERN 1977 Summer Student Lecture Programme under the title: "What can the famous 21 cm astrophysical spectral line of atomic hydrogen tell us about the nature of magnetic dipoles?" CERN Report No.77-17, CERN, Geneva (1977); reprinted in *The International Community of Physicists: Essays on Physics and Society in Honor of Victor Frederick Weisskopf*, edited by V. Stefan, AIP Press/Springer-Verlag, New York (1997).
30. A. Einstein and J. Laub, "Über die im elektromagnetischen Felde auf ruhende Körper ausgeübten ponderomotorischen Kräfte," *Annalen der Physik* **331**, 541–550 (1908) ; the English translation of this paper appears in Einstein's Collected Papers, Vol.2, Princeton University Press (1989).
31. M. Mansuripur, "Force, torque, linear momentum, and angular momentum in classical electrodynamics," *Applied Physics A* **123:653**, pp1-11 (2017).

A. Douglas Stone*, William R. Sweeney, Chia Wei Hsu, Kabish Wisal and Zeyu Wang

Reflectionless excitation of arbitrary photonic structures: a general theory

https://doi.org/10.1515/9783110710687-026

Abstract: We outline and interpret a recently developed theory of impedance matching or reflectionless excitation of arbitrary finite photonic structures in any dimension. The theory includes both the case of guided wave and free-space excitation. It describes the necessary and sufficient conditions for perfectly reflectionless excitation to be possible and specifies how many physical parameters must be tuned to achieve this. In the absence of geometric symmetries, such as parity and time-reversal, the product of parity and time-reversal, or rotational symmetry, the tuning of at least one structural parameter will be necessary to achieve reflectionless excitation. The theory employs a recently identified set of complex frequency solutions of the Maxwell equations as a starting point, which are defined by having zero reflection into a chosen set of input channels, and which are referred to as R-zeros. Tuning is generically necessary in order to move an R-zero to the real frequency axis, where it becomes a physical steady-state impedance-matched solution, which we refer to as a reflectionless scattering mode (RSM). In addition, except in single-channel systems, the RSM corresponds to a particular input wavefront, and any other wavefront will generally not be reflectionless. It is useful to consider the theory as representing a generalization of the concept of critical coupling of a resonator, but it holds in arbitrary dimension, for arbitrary number of channels, and even when resonances are not spectrally isolated. In a structure with parity and time-reversal symmetry (a real dielectric function) or with parity–time symmetry, generically a subset of the R-zeros has real frequencies, and reflectionless states exist at discrete frequencies without tuning. However, they do not exist within every spectral range, as they do in the special case of the Fabry–Pérot or two-mirror resonator, due to a spontaneous symmetry-breaking phenomenon when two RSMs meet. Such symmetry-breaking transitions correspond to a new kind of exceptional point, only recently identified, at which the shape of the reflection and transmission resonance lineshape is flattened. Numerical examples of RSMs are given for one-dimensional multimirror cavities, a two-dimensional multiwaveguide junction, and a multimode waveguide functioning as a perfect mode converter. Two solution methods to find R-zeros and RSMs are discussed. The first one is a straightforward generalization of the complex scaling or perfectly matched layer method and is applicable in a number of important cases; the second one involves a mode-specific boundary matching method that has only recently been demonstrated and can be applied to all geometries for which the theory is valid, including free space and multimode waveguide problems of the type solved here.

Keywords: exceptional point; impedance matching; non-hermitian optics; photonics; photonic design.

***Corresponding author: A. Douglas Stone,** Department of Applied Physics, Yale University, New Haven, CT, 06520, USA; and Yale Quantum Institute, Yale University, New Haven, CT, 06520, USA, E-mail: douglas.stone@yale.edu
William R. Sweeney and Kabish Wisal, Department of Physics, Yale University, New Haven, CT, 06520, USA. https://orcid.org/0000-0002-0487-5571 (W.R. Sweeney)
Chia Wei Hsu, Department of Applied Physics, Yale University, New Haven, CT, 06520, USA; Ming Hsieh Department of Electrical and Computer Engineering, University of Southern California, Los Angeles, CA, 90089, USA
Zeyu Wang, Ming Hsieh Department of Electrical and Computer Engineering, University of Southern California, Los Angeles, CA, 90089, USA

1 Introduction

1.1 Reflectionless excitation of resonant structures

Reflectionless excitation or transmission of waves is a central aspect of harnessing waves for distribution or transduction of energy and information in many fields of applied science and engineering. In the context of radio-frequency and microwave electronics and in acoustics, this is typically referred to as "impedance matching", whereas in optics and photonics, the terms "index matching" and

This article has previously been published in the journal Nanophotonics. Please cite as: A. D. Stone, W. R. Sweeney, C. W. Hsu, K. Wisal and Z. Wang "Reflectionless excitation of arbitrary photonic structures: a general theory" *Nanophotonics* 2021, 10. DOI: 10.1515/nanoph-2020-0403.

"critical coupling" (CC) are more frequently used, as well as "perfect absorption" when the goal is energy transfer or transduction. In the first fields listed, it is typical to represent the response of the media or circuits, which are typically lossy, via a complex impedance, and the simple principle of matching the input impedance to the output impedance is often employed to achieve reflectionless excitation. In optics and photonics, it is more typical to represent the response of the medium by a complex dielectric function or susceptibility, and nearly lossless excitation of dielectric media, as well as free-space excitation, is quite common, so the term impedance matching is less often used. In this article, we will focus on optical and photonic structures/devices and will use the term reflectionless excitation. We will define below the concept of reflectionless scattering modes (RSMs), referring to input wavefronts at specific, discrete *real* frequencies that can be shown to excite a given structure with zero reflection (in a sense to be clearly defined below). Impedance matching or index matching across boundaries between effectively semi-infinite media is well known from textbooks and is not the topic of interest here. Rather here we focus on *finite structures* in any dimension, which are excited by a wave with wavelength smaller than the relevant dimensions of the structure. In this case, reflectionless excitation may be possible but only at discrete frequencies due to the necessity of taking into account multiple internal reflections within the structure. Thus, we are speaking of resonant reflectionless excitation of the structure.

The concept of CC to a resonator, to be discussed in more detail below, is reasonably well known in optics and photonics: a high-Q resonator, when excited by a single electromagnetic channel, either guided or radiative, will generate no reflected waves when it is excited at the resonance frequency, and the input coupling rate to the resonator equals the sum of all *other* loss rates from or within the resonator. The total loss of the resonator is defined as the imaginary part of the complex frequency of the specific quasi-normal mode being excited, which includes the loss through the input channel. The quasi-normal modes (or simply resonances) are rigorously defined as the purely outgoing solutions of the relevant electromagnetic wave equation. These resonances generically have frequencies in the lower half-plane, $\omega = \omega_r - i\gamma$, where $\gamma = 1/2\tau > 0$, τ is the dwell time or intensity decay rate, and $Q = \omega_r\tau$ is the quality factor of the resonance [1–4]. In general, the resonances are not physically realizable steady-state solutions, due to their exponential growth at infinity, but they determine the scattering behavior under steady-state (real frequency) harmonic excitation. However, in electromagnetic scattering with gain, the resonances *can* be realized physically and correspond to the onset of laser emission [5–7]. Thus, in the terminology we use in this work, having a resonance on the real axis does not correspond to the existence of a reflectionless state (in some other contexts the term "resonance" is *used* to refer to a reflectionless state). In the current work, we will define a different set of complex frequency solutions which *do* correspond to the existence of a reflectionless state and which do not in general require the addition of gain or loss to the system to make them accessible via steady-state excitation.

The current theoretical/computational tools available in optics and photonics for determining when and if reflectionless excitation of a structure/resonator is possible consists of analytic calculations in certain one-dimensional structures and transfer matrix computations for more complicated one-dimensional or quasi-one-dimensional structures, along with the principle of CC, which rarely is applied in higher dimensions.

1.2 Limitations of critical coupling concept

The terminology "critical coupling"appears to have been used in microwave/radio-frequency electronics at least 60 years ago but does not appear to have been used extensively in optics until the nineties [8–11]. It always is applied to a structure in which a relatively high-Q resonator with well-separated resonances is effectively excited by a single spatial input channel. The resonator will have some effective coupling-in rate at the surface where the input channel comes in, determined, e.g., by a mirror or facet reflectivity, and it will have some coupling-out/absorption rate within the resonator, due either to other radiative channels or to internal absorption loss or both [12, 13]. Examples in photonics include the asymmetric Fabry–Pérot (FP) semiconductor devices developed in the eighties and nineties, which used the electrooptic effect to switch on and off absorption in the cavity, so as to toggle between a critically coupled condition and a weakly coupled condition [9–11]. In this case, the loss is primarily absorption and represents an irreversible transduction of the energy. Another common, more recent set of examples is the ring resonators side-coupled to silicon waveguides which can be toggled by free-carrier injection between a critically coupled and a weakly coupled state, which turns off and on the transmission through the waveguide [14]. In this case, the loss in the critically coupled state is primarily radiative, and the reflectionless "on" state can be thought of as perfect transmission into the radiative channels, whereas the "off" state corresponds to zero reflection from the ring (only) and hence continued propagation/transmission along the guided waveguide channel. These examples indicate that the source of the loss in the resonator is not important to be able to achieve CC, although it does determine the effect of CC on the exciting wave, i.e., either irreversible transduction or radiative transmission (to a receiver or just to an effective beam sink).

The power of the CC concept is that, if rigorously correct, it implies that it is possible to excite a resonator of arbitrary complexity through a single input channel and have zero reflection, if the total loss from either absorption or radiation into other channels equals the input coupling. To our knowledge, the CC concept is the only general principle relating to reflectionless excitation of a resonator which applies beyond parity-symmetric one-dimensional examples, where it is possible to calculate analytically the reflectionless input frequencies.

However there are obvious questions raised about the meaning and generality of the concept.

- In all cases of which we are aware, the CC condition is derived within a simplified coupled mode theory and not from a first-principles analysis, which would be exact within Maxwell electrodynamics. What principle, if any, underlies its validity in the case of a complex resonator with no symmetries?
- Is there a generalization of the CC concept to the situation in which one is exciting the resonator with more than one input channel? A simple example of this would be a multimode waveguide exciting a resonator or multiple waveguide junction. Similarly, when one is exciting a structure larger than the excitation wavelength in free space, the typical radiation will involve higher multipoles and hence multiple input/output channels. Are there reflectionless solutions in either of these cases?
- CC assumes that the excitation is only of a single resonance, but in any relatively open structure, e.g., a waveguide junction, multiple resonances will often be overlapping and relevant to the scattering process. In this case, the CC concept becomes ill defined. There is no obvious scalar meaning to coupling in and coupling out, so there is no obvious CC condition to apply. In fact, the CC condition has never, to our knowledge, been applied to such situations and would generally be considered irrelevant because one does not have isolated resonances.

This paper outlines and interprets recent results from our group which answers all of these questions and proves that reflectionless states are an exact property of Maxwell electrodynamics in any dimension and for arbitrarily complex structures (larger than the exciting wavelength). Moreover, these states exist even when multiple resonances overlap, and there are no isolated resonances. In general, a single continuous parameter of the resonator/structure needs to be tuned appropriately, and then, the reflectionless state exists only at a single frequency. The reflectionless states can be computed by numerical methods which are closely related to standard techniques in photonics and are tractable for realistic structures. Hence, we believe that the theory presented here provides a powerful tool for the design of photonic structures with controlled excitation which implement perfect coupling, as well as clarifying what sorts of solutions exist generically and what sorts do not. We will present here only the main results of our analysis with illustrative examples; the detailed derivations are given in the study by Sweeney et al. [15].

2 The generalized reflection matrix, R-zeros, and RSMs

2.1 The scattering matrix

To define reflectionless states in electromagnetic scattering, we must first define the scattering matrix (S-matrix) of a finite photonic structure. We consider here the most general system of interest, which consists of an inhomogeneous scattering region or structure, outside of which are asymptotic regions that extend to infinity. To support resonance effects, the scattering region needs to be larger than the wavelength of the excitations created by the input waves within the scatterer. For dielectric systems, this is typically of order but smaller than the input wavelength. However, the theory will also apply to metallic/plasmonic systems (within the Maxwellian framework), where the plasmonic excitations can have orders of magnitude smaller wavelengths. An example of an application of coherent perfect absorption, a special case of the theory, to nanoparticles is cited in the next section. The asymptotic regions are assumed to be time-reversal invariant (so that they support incoming and outgoing asymptotic channels that are related by complex conjugation) and to have some form of translational invariance, e.g., vacuum or uniform dielectric, or a finite set of waveguides, or an infinite periodic photonic crystal. The theory will apply to both free and guided waves. We also focus on media in which the scattering forces are short range, i.e., net neutral media, which is typical for most photonic structures.

A linear and static photonic structure is described by its dielectric function $\varepsilon(\mathbf{r}, \omega)$, which is generically complex valued, with its imaginary part describing absorption and/or gain. The assumed linearity allows the theory to concentrate on scattering at a single real frequency, ω; time-dependent scattering can be studied by superposing solutions. The translational symmetry of the asymptotic regions allows one to define $2N$ power orthogonal propagating "channel states" at each ω. Based on the direction of their fluxes, the $2N$ channels can be unambiguously grouped into N incoming and N outgoing channels, which, as noted, are related by time reversal. Familiar examples of channels include the guided transverse modes of a waveguide and orbital angular momentum waves in free space, with one channel per polarization. In the waveguides, the finite number and width of the waveguides lead to a finite N for a given ω, whereas for the case of a finite scatterer/cavity in free space, the number of propagating angular momentum channels is countably infinite. However, a finite scatterer of linear scale R, with no long-range potential outside, will interact with only a finite number of angular momentum states, such that $l_{\max} \sim \sqrt{\bar{\varepsilon}} R\omega/c$, where $\bar{\varepsilon}$ is the spatially averaged dielectric function in the scattering region, and c is the speed of light. In the general case of polarization mixing in scattering including the polarization index will simply double the size of the S-matrix. Hence, for each ω, we can reasonably truncate the infinite dimensional channel space to a finite, N-dimensional subspace of relevant channels.

A general scattering process then consists of incident radiation, propagating along the N incoming channels, interacting with the scatterer and then propagating out to infinity along the N outgoing channels, as illustrated in Figure 1(a). In a general geometry, which is not partitioned into spatially distinct asymptotic channels, there is no natural definition of reflection and transmission coefficients between different channels, only interchannel scattering vs. backscattering into the same channel; for certain geometries, such as a scattering region within a single or multimode fiber, it is natural to segment the S-matrix into reflection and transmission matrices depending on whether the scattering maintains the sign of propagation (i.e., transmits flux) or reverses it (reflects flux). However, we will define the reflection matrices for a general geometry in a more general way, which need not reduce to this standard definition even in a waveguide geometry. In the channel basis, the wavefronts of the incoming and outgoing fields are given by length-N column vectors $\boldsymbol{\alpha}$ and $\boldsymbol{\beta}$, normalized such that $\boldsymbol{\alpha}^\dagger\boldsymbol{\alpha}$ and $\boldsymbol{\beta}^\dagger\boldsymbol{\beta}$ are proportional to the total incoming and total outgoing energy flux, respectively. The N-by-N scattering matrix $\mathbf{S}(\omega)$, which relates $\boldsymbol{\alpha}$ and $\boldsymbol{\beta}$ at frequency ω is defined as follows:

$$\boldsymbol{\beta} = \mathbf{S}(\omega)\boldsymbol{\alpha}. \tag{1}$$

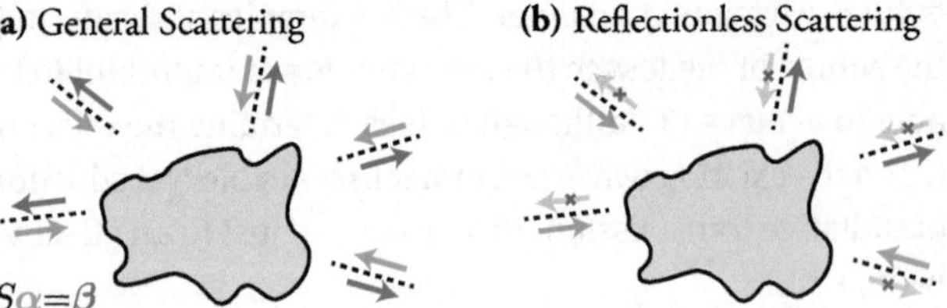

Figure 1: Schematic depicting a (a) general scattering process and (b) reflectionless process. A finite scatterer/cavity interacts with a finite set of asymptotic incoming and outgoing channels, indicated by the red and blue arrows, respectively, related by time reversal. These channels may be localized in space (e.g., waveguide channels) or in momentum space (e.g., angular momentum channels). (a) In the general case without symmetry, all incoming channels will scatter into all outgoing channels. (b) There exist reflectionless states for which there is no reflection back into a chosen set of incoming channels (the inputs), which in general occur at discrete complex frequencies and do not correspond to a steady-state harmonic solution of the wave equation. However, with variation of the cavity parameters, a solution can be tuned to have a real frequency, giving rise to a steady-state reflectionless scattering process for a specific coherent input state, referred to as a reflectionless scattering mode (RSM).

In reciprocal systems, the S-matrix is symmetric, $\mathbf{S} = \mathbf{S}^T$ [16]. If the scatterer is lossless (i.e., ε is real everywhere), then any incoming state leads to a nonzero flux-conserving output, and the S-matrix is unitary. However, the theory outlined below is developed for arbitrary linear S-matrices and complex ε, which then includes the effects of linear absorption or amplification inside the scattering region. Engineering reflectionless states will generally be possible for the case of both unitary and nonunitary S-matrices.

2.2 Coherent perfect absorption

The S-matrix, being well defined at all real frequencies, can be extended to complex frequencies via analytic continuation. As noted in the Introduction section, the resonances or quasi-normal modes of the system are solutions of the wave equation for the structure which are purely outgoing in the asymptotic regions and have discrete complex frequencies in the lower half-plane; those frequencies correspond to poles of the S-matrix. Since purely incoming solutions can be obtained by complex conjugation, this implies that the zeros (frequencies of solutions at which zero flux is outgoing) are simply the complex conjugate of the pole frequencies when the structure is lossless, and these zeros occur in the upper half-plane. The zeros then correspond to a certain type of reflectionless solution to the wave equation but at an unphysical complex frequency. These states can be tuned to a real frequency by adding loss, in just the same manner as poles can be tuned to the real axis by adding gain to initiate lasing [17–20]. A system so tuned is known as a coherent perfect absorber (CPA) and functions as the time reverse of a laser at threshold.

This type of reflectionless state was first pointed out by one of the authors (and coworkers) a decade ago. It is a special case of the reflectionless states we define below, with the additional property that impedance matching is achieved by perfect absorption within the scattering region, and hence irreversible transduction of the incident energy into degrees of freedom within the resonator/scatterer. With the introduction of the CPA concept, it was appreciated that this kind of reflectionless state always exists within a family of arbitrarily complex resonators with tunable loss, for just the same reason that any complex resonator can be made to lase with sufficient gain added. And just as the laser is "perfectly emitting" only for a specific spatial mode of the electromagnetic field, the CPA is perfectly absorbing only for a specific mode, which is the time reverse of the lasing mode, and may have a very complicated spatial structure that is challenging to synthesize. For example, just as there exists random lasers that emit a pseudo-random lasing field when sufficient gain is added, a geometrically similar random structure with an absorbing medium added of similarly strong loss, can, in steady state, perfectly absorb the complex conjugate of this lasing field (an example is shown in Figure 2). As this example shows, in order to be perfectly absorbed, the input "beam" must be focused to roughly the size of the scattering structure so that CPA cannot be achieved in free space for an input beam (plane wave or structured) that is spread out over an area much larger than the transverse size of the structure. Similar constraints apply to the more general reflectionless modes defined below, which will also need to be focused so as to strongly interact with the structure in free space (this constraint is typically automatically imposed by the geometry in the case of guided wave systems).

Figure 2: Reflectionless state of a coherent perfect absorber (CPA), consisting of a random aggregate of lossless glass scattering rods (gray) of index $n = 1.5$ and radius equal to the incident wavelength ($r = \lambda$), surrounding a highly lossy subwavelength central rod (black) of radius $r = 0.15\lambda$ and dielectric constant $\varepsilon = 1.28 + 1.75i$. The color scale indicates the field amplitude for the specific input mode which is perfectly absorbed. This is a steady-state solution in which all of the incident power is dissipated in the central rod, which acts as a perfect sink and is assumed to be a linear absorber. The incident field pattern is found by calculating the complex conjugate of the threshold lasing mode of the analogous random laser and consists of the appropriate coherent superposition of converging cylindrical waves (Hankel functions). The field penetrating into the glass rods is not shown, for clarity. In this very open structure, the resonances strongly overlap, and the concept of critical coupling to a single resonance does not apply; nonetheless, reflectionless states exist.
Figure adapted from animation at http://www1.spms.ntu.edu.sg/~ydchong/research.html, courtesy of Y-D. Chong.

However, as already noted, it can be quite difficult to synthesize the wavefront needed to achieve a CPA, and this limits the application of the CPA concept to complex structures in experiments or devices. In the more common situation in which one seeks to excite a structure in a reflectionless manner, one is not aiming for perfect absorption but is simply seeking to avoid energy flow back into the chosen input channels; in many cases, one does not wish to have any absorption at all. Our theory below includes the CPA as a limiting case but is focused instead on this more common situation of prime importance for the design of photonic devices.

2.3 Generalized reflection matrix and R-zeros

Returning to the S-matrix for an arbitrary finite scattering structure/resonator, we now define reflectionless states in the most general manner possible. The full S-matrix encodes the information about all possible linear excitations of the resonator. Assuming one has access to all N of the possible input channels in the asymptotic region, one can define a particular impedance-matching problem as shown in Figure 1(b) by specifying N_{in} (with $0 < N_{\text{in}} \leq N$) of the incoming channels as the controlled input channels, which, for the appropriate input state, will carry incident flux but no outgoing flux. Conversely, the complementary set of $N_{\text{out}} = N - N_{\text{in}}$ outgoing channels will carry any outgoing flux. This flux can be less than, equal to, or greater than the incident flux, depending on whether the resonator is attenuating, lossless, or amplifying.

Let us for convenience redefine the S-matrix so that for each choice of input channels, the first N_{in} columns of the S-matrix represent the scattering of the chosen N_{in} input channels. This implies that we will only consider scattering input vectors, $\boldsymbol{\alpha}$, which are nonzero for their first N_{in} components (henceforth we will refer to this as the input wavefront). Conversely, for the input wavefront to be reflectionless, the output vector, $\boldsymbol{\beta}$, must have its first N_{in} components be equal to zero. Thus, we can define the upper left $N_{\text{in}} \times N_{\text{in}}$ block of this S-matrix to be a generalized reflection matrix, $\mathbf{R}_{\text{in}}(\omega)$. The condition then for the existence of a reflectionless input state at some frequency $\omega = \omega_{\text{RZ}}$ is that this matrix has an eigenvector with eigenvalue zero; this eigenvector is the N_{in}-component vector consisting of the nonzero components of $\boldsymbol{\alpha}$. The frequencies at which a reflectionless state exists are thus determined by the complex scalar equation.

$$\det \mathbf{R}_{\text{in}}(\omega_{\text{RZ}}) = 0, \tag{2}$$

and in principle the frequencies and input wavefronts for reflectionless states could be found by searching for the zeros of this determinant in the complex frequency plane.

However, another approach is more fruitful. That is to regard the reflectionless boundary conditions as defining a nonlinear eigenvalue problem on the Maxwell wave operator for which $\{\omega_{\text{RZ}}\}$ are the eigenvalues, and use standard methods for solving general nonlinear eigenproblems [21–25]. To impose the appropriate boundary conditions, a familiar method in photonics is the use of perfectly matched layers (PMLs), normally used for finding purely outgoing solutions (resonances) but also applicable here in some cases. For the most general cases, the PML approach is not applicable, but a modification of previous boundary matching methods can be used, as will be discussed briefly below.

Anticipating results which will be demonstrated below, similar to the resonances, we find that the reflectionless input wavefronts will only exist at discrete complex values of the frequency, ω_{RZ}; however, these frequencies will differ from the resonance frequencies, and they represent the complex-valued spectrum of a wave operator with a different physical meaning. We will refer to these frequencies and the associated wave solutions as R-zeros (reflection zeros). Since complex frequency solutions do not represent physically realizable steady-state solutions, a critical step in solving the impedance-matching problem in generic cases will be to tune parameters of the dielectric function in the wave operator in order to move an R-zero to the real frequency axis. When an R-zero is tuned to the real axis, we will refer to the reflectionless physical state which results as an RSM, in analogy to the term "lasing modes" or "CPA mode" which is used for these related but distinct electromagnetic eigenvalue problems. We note that, like a resonance, an R-zero can be transiently realized with a time-dependent input, even when it occurs at a complex frequency [26].

A previous work, [27], has introduced the notion of the R-zero spectrum in the more limited context of waveguides, focusing on single-mode cases, and pointed out its relevance to impedance matching. These authors did not discuss the possibility of tuning the wave operator to create a physical steady state but studied a one-dimensional parity-symmetric case for which the R-zeros can have real frequency due to symmetry.

2.4 The RSM concept

As noted, an RSM is a steady-state wave solution at a real frequency that excites a structure/resonator through specific input channels such that there is zero reflection back into the chosen channels. It can be specified asymptotically by its frequency and wavefront $\boldsymbol{\alpha}$ (where here we refer to the N_{in} nonzero components of $\boldsymbol{\alpha}$). The absence of reflection for the RSM incident wavefront is due to interference: the reflection amplitude of each input channel i destructively interferes with the interchannel scattering from all the other input channels, $(R_{\text{in}})_{ii}\alpha_i + \sum_{j\neq i}(R_{\text{in}})_{ij}\alpha_j = 0$ (which is just a restatement of the fact that $\boldsymbol{\alpha}$ is an eigenvector of $\mathbf{R}_{\text{in}}$ with eigenvalue zero). However, the scattering ("transmission") into the chosen output channels is not obtained from solving this equation alone and must be determined by solving the full scattering problem at ω_{RZ}, i.e., by calculating the full S-matrix at ω_{RZ}.

The case of $N_{\text{in}} = N$ corresponds to the CPA [17], ($\mathbf{R}_{\text{in}} = \mathbf{S}$), which, as noted, has been identified and studied for some time. In this case, the R-zeros are the zeros of the full S-matrix, which are relatively familiar objects. CPA, as an example of an RSM, is rather special because it requires that all the asymptotic channels be controlled in order to have the possibility of achieving a CPA, which is often not practical in complex geometries. Also, only in the case of the CPA is it necessary to violate flux conservation to create an RSM (i.e., by adding an absorbing term to the dielectric function). Thus, the CPA is mainly of interest in cases where the goal is not impedance matching but rather transduction or sinking of energy.

2.5 Wave operator theory of zeros of the S-matrix and generalized reflection matrix

It is both mathematically convenient and helpful for physical insight to consider the R-zero/RSM problem from the point of view of the underlying wave operator with boundary conditions. While Eq. (2) defines the frequencies and wavefronts for which reflectionless states exist for a fixed structure/resonator, solving it will not simultaneously yield the field everywhere within the structure. For the latter, one will need to solve the full Maxwell wave equations subject to the boundary conditions at infinity which follow from the frequency and incident wavefront.

The R-zero/RSM boundary conditions lead to a more constrained scattering problem than the standard scattering boundary conditions in which only the input wavefront is specified, but no constraint is placed on the output wavefront. For the standard problem, a solution exists at every real ω, but it generically involves all of the outgoing channels in the asymptotic region. In the RSM/R-zero problem only, $N - N_{\text{in}}$ of the asymptotic outgoing channels are allowed to appear, and N_{in} of the input channels generically appear. Thus, solutions are constructed from only N of the $2N$ possible asymptotic free solutions and are not guaranteed to exist at all frequencies. As the previous analysis leading to Eq. (2) has shown, having an R-zero requires a certain scattering operator to be noninvertible, which we do not expect to happen generically.

A more familiar situation, in which we impose similar boundary conditions, is in calculating the resonances, using only the N outgoing channel functions at infinity. As already noted, this calculation can be posed as an electromagnetic eigenvalue problem for which the eigenvalues are the frequencies at which solutions exist. For the resonance problem, it is well known that for finite structures of the type considered here, there are an infinite number of solutions at discrete frequencies and that these frequencies correspond to the poles of various response functions including the S-matrix that we have introduced above. In the absence of gain, by causality these poles are restricted to the lower half-plane. We will now discuss a wave operator representation of the S-matrix and the implications for its zeros in preparation for the general theory of R-zeros in the next section. The results we present in the next two sections are based on derivations from the study by Sweeney et al. [15], and here, we simply present the main results without proof, introducing the minimum number of mathematical details necessary to make these results comprehensible. Readers interested only in the basic results and examples may skip to the summary in Section 3.1.

To introduce our notation, consider a wave operator $\hat{A}(\omega)$ acting on state $|\omega\rangle$, which satisfies $\hat{A}(\omega)|\omega\rangle = 0$. For electromagnetic scattering, we may choose the quantity $\langle \mathbf{r}|\omega\rangle$ as the magnetic field, $\mathbf{H}(\mathbf{r})$, under harmonic excitation at ω. The Maxwell operator at frequency ω is given as follows:

$$\langle \mathbf{r}'|\hat{A}(\omega)|\mathbf{r}\rangle = \delta(\mathbf{r}-\mathbf{r}')\left\{\left(\frac{\omega}{c}\right)^2 - \nabla\times\left(\frac{1}{\varepsilon(\mathbf{r},\omega)}\nabla\times\right)\right\}. \tag{3}$$

Here, we have given the full vector Maxwell operator for which all of our results are valid, but in the more detailed analysis below, we only apply the theory to effectively one- and two-dimensional cases in which a scalar Helmholtz-type equation describes the solutions for the appropriate polarization.

In order to express the scattering matrix in terms of its resonances, we divide the system into two regions: the finite, inhomogeneous scattering region Ω in the interior, and the exterior asymptotic region $\overline{\Omega}$ that extends to infinity, which possesses a translational invariance broken only by the boundary, $\partial\Omega$, between Ω and $\overline{\Omega}$. We separate operator $\hat{A}(\omega)$ into three pieces,

$$\hat{A}(\omega) = \left[\hat{A}_0(\omega) \oplus \hat{A}_c(\omega)\right] + \hat{V}(\omega), \tag{4}$$

with $\hat{A}_0(\omega)$ identical to $\hat{A}(\omega)$ on its domain Ω and $\hat{A}_c(\omega)$ is identical to $\hat{A}(\omega)$ on $\overline{\Omega}$. $\hat{V}(\omega)$ represents the residual coupling between the two regions.

The closed-cavity wave operator $\hat{A}_0(\omega)$ on Ω, which we do not assume to be hermitian, has a discrete spectrum of the form $\hat{A}_0(\omega_\mu)|\mu\rangle = 0$ with eigenvalues $\{\omega_\mu\}$. The boundary conditions on $\hat{A}_0(\omega)$ can be chosen to be Neumann, but the effect of coupling terms will introduce a self-energy which will account for the actual continuity conditions at the boundary of Ω. The matrix $\mathbf{A}_0(\omega)$ is naturally defined by its matrix elements as follows:

$$\mathbf{A}_0(\omega)_{\mu\nu} = \langle \mu|\hat{A}_0(\omega)|\nu\rangle\,. \tag{5}$$

The asymptotic wave operator $\hat{A}_c(\omega)$ on $\overline{\Omega}$ has a countable number of eigenfunctions at every real value of ω; these are the propagating channel functions which satisfy $\widehat{A}_c(\omega)|\omega, n\rangle = 0$. The operator $\widehat{V}$ connects these closed and continuum states, and its off-diagonal block is represented by the matrix $\mathbf{W}(\omega)$ as follows:

$$W(\omega)_{\mu n} = \langle \mu | \widehat{V}(\omega) | n, \omega \rangle . \tag{6}$$

While $\mathbf{W}(\omega)$ in general also has a contribution from evanescent channels in $\overline{\Omega}$, we will neglect the effect of such channels henceforth in the current discussion, as they do not change the central results qualitatively [15].

With these definitions, one can derive a general relation [28–31] between the matrices $\mathbf{S}$, $\mathbf{A}_0$, and $\mathbf{W}$, originally developed in nuclear physics, which allows us to find the poles and zeros of $\mathbf{S}$ through its determinant:

$$\det \mathbf{S}(\omega) = \frac{\det(\mathbf{A}_0(\omega) - \mathbf{\Delta}(\omega) - i\mathbf{\Gamma}(\omega))}{\det(\mathbf{A}_0(\omega) - \mathbf{\Delta}(\omega) + i\mathbf{\Gamma}(\omega))}. \tag{7}$$

The two hermitian operators which appear here are $\mathbf{\Delta}$ and $\mathbf{\Gamma} \equiv \pi \mathbf{W}\mathbf{W}^\dagger$, the latter being positive, semidefinite, arise from the coupling operator and, roughly speaking, induce a real and imaginary shift of the eigenvalues of the "closed" system to account for the openness of the system. The operator $\mathbf{\Delta}$ can be expressed in terms of an integral over the $\mathbf{W}\mathbf{W}^\dagger$ matrix [15] and is of less interest in the current context; both operators are infinite dimensional matrices in the space of resonances of the system. The S-matrix however is a finite dimensional matrix in the truncated channel space (as noted above), and hence, Eq. (7) is not a simple identity of linear algebra: the left-hand side is the standard determinant of an N-by-N square matrix, while the right-hand side is a ratio of functional determinants of differential operators on an infinite dimensional Hilbert space (see the study by Sweeney et al. [15] for more details and relevant references). Since the right-hand side is a ratio of determinants of infinite dimensional differential operators upon a finite domain (which will have a countably infinite set of complex eigenvalues which depend on ω), this implies that $\det \mathbf{S}$ indeed has a countably infinite set of zeros and poles (corresponding to the vanishing of the numerator and denominator). When $\mathbf{A}_0$ is hermitian (the scatterer is lossless), then the operators in the numerator and denominator are hermitian conjugates and the poles and zeros come in complex conjugate pairs.

Since the operators in each determinant of the ratio do not commute, one cannot simply say that the eigenvalues of the full operators are the sum of the eigenvalues of each individual operator. However, the more isolated the resonances of the systems are, the more useful is this heuristic interpretation of Eq. (7). Thus, crudely speaking, the scattering resonances will occur at complex frequencies where the real part is given by the real part of the eigenvalue of the closed system containing the scatterer, shifted by the contribution from $\mathbf{\Delta}$, while its imaginary part must be negative, with a value $i(-\gamma_{\mu,\text{rad}} - \gamma_{\mu,\text{int}})$, where $\gamma_{\mu,\text{rad}} > 0$ is an eigenvalue of $\mathbf{\Gamma}$ and represents radiative loss, and $\gamma_{\mu,\text{int}}$ comes from gain or loss in the resonator and can have either sign. With the standard convention we have chosen here, $\gamma_{\mu,\text{int}} > 0$ corresponds to absorption loss so that adding absorption pushes the resonance away from the real axis and hence broadens it, as is familiar. Conversely, the zeros of $\mathbf{S}$ have imaginary part $i(\gamma_{\mu,\text{rad}} - \gamma_{\mu,\text{int}})$, from which we see that there will be a real zero at some frequency when $\gamma_{\mu,\text{rad}} = \gamma_{\mu,\text{int}}$. When the radiative loss equals the absorption loss, the S-matrix has an eigenvalue equal to zero at a real frequency: one can send in a specific wavefront and it will be indefinitely trapped and hence absorbed. The multichannel case of this is the CPA, and the single-channel case, for which a specific wavefront is not required, is the usual CC to a resonator. However, as noted, this only corresponds to CC when all of the loss is due to absorption, whereas we will present the full generalization of reflectionless coupling in the next section.

Up to this point, we have just reviewed the known properties of the S-matrix in this operator representation. Henceforth, we are focusing on the zeros of the matrix $\mathbf{R}_{\text{in}}$, when it differs from the full S-matrix, and seeking the condition for it to have zeros.

We now present results from adapting this formalism to treat R-zeros/RSMs; unlike the results presented in the previous section, these results were not known previous to the derivations in the study by Sweeney et al. [15]. The basic approach is to represent the matrix $\mathbf{R}_{\text{in}}$ through appropriate projection operators applied to the S-matrix and then to perform a similar but more involved set of manipulations to obtain a relationship between the $\det \mathbf{R}_{\text{in}}$ and a ratio of determinants of wave operators to similar to but distinct from those that determine $\det \mathbf{S}$.

The result is as follows [15]:

$$\det \mathbf{R}_{\text{in}}(\omega) = \frac{\det(\mathbf{A}_0(\omega) - \mathbf{\Delta}(\omega) - i[\mathbf{\Gamma}_{\text{in}}(\omega) - \mathbf{\Gamma}_{\text{out}}(\omega)])}{\det(\mathbf{A}_0(\omega) - \mathbf{\Delta}(\omega) + i\mathbf{\Gamma}(\omega))}, \tag{8}$$

where the only (but crucial) difference from Eq. (7) is the replacement of the operator $\mathbf{\Gamma}$ by the difference of two operators associated with the input and output channels, respectively: $\mathbf{\Gamma}_{\text{in}} - \mathbf{\Gamma}_{\text{out}} \equiv \mathbf{W}_{\text{in}}\mathbf{W}_{\text{in}}^\dagger - \mathbf{W}_{\text{out}}\mathbf{W}_{\text{out}}^\dagger$. Here, the subscripts refer to the sectors of the operator $\mathbf{W}$ introduced previously that connect the discrete states of the scatterer/resonator to the asymptotic incoming channels and outgoing channels (respectively), which were specified in the definition of $\mathbf{R}_{\text{in}}$. Eq. (8) is the central mathematical result of our theory of reflectionless states. Similar to Eq. (7), Eq. (8) relates the determinant of the N_{in}-by-N_{in} matrix $\mathbf{R}_{\text{in}}$ to a ratio of wave operator (functional) determinants which describe the discrete but infinite space of eigenvalues of the scatterer/resonator.

3 Properties of R-zeros and RSMs

3.1 General properties

From Eq. (8), we can draw a number of critical conclusions:

- For reasons analogous to those arising from Eq. (7), we can conclude that the matrix $\mathbf{R}_{\text{in}}$ also has a countably infinite set of zeros and poles at discrete complex frequencies.
- Because the denominator is the same as in Eq. (7), the poles of $\mathbf{R}_{\text{in}}$ are identical to those of $\mathbf{S}$ (excluding certain nongeneric cases).
- However, the *zeros* of $\mathbf{R}_{\text{in}}$ (R-zeros) are generically at distinct frequencies in the complex plane from those of $\mathbf{S}$. As noted above, the R-zero spectrum is a new complex spectrum with a distinct physical meaning from S-matrix zeros or poles [15, 27].
- Because the operator in the numerator of Eq. (8) is not the hermitian conjugate of that in the denominator, even when the scatterer is passive (no loss or gain), the R-zeros are not the complex conjugate of the poles and can appear in either the upper or lower half-plane without the addition of loss or gain. In particular, a lossless scatterer can have a real R-zero (RSM), although generically this requires parameter tuning.
- The fact that the positive semidefinite coupling operator $\mathbf{W}_{\text{in}}\mathbf{W}_{\text{in}}^{\dagger}$ gives a contribution to the numerator of Eq. (8) of opposite sign to that of the outgoing channels (which has the usual sign for the S-matrix) implies that qualitatively the incoming channels function as a kind of "radiative gain" for the R-zeros, whereas the outgoing channels function qualitatively in the usual manner as radiative loss. Heuristically, we can expect that an R-zero can become an RSM when the total coupling in from the input channels balances the total coupling out from the radiative channels, or, if there is loss or gain in the resonator itself, when all of these terms are balanced to cancel. If we are in the regime of isolated resonances, there will be only one set of relevant couplings to balance, and this can be regarded as a multichannel generalization of CC. However, to excite this RSM, one will still need to send in the correct coherent wavefront obtained from the eigenvalue equation for $\mathbf{R}_{\text{in}}$.
- Our result shows that the full operator in the numerator of Eq. (8) will have an infinite number of zeros in general, even if we are not in the regime of isolated resonances. In this case, there is no single scalar condition for achieving RSMs and no meaningful generalization of the CC concept. Nonetheless, R-zeros can be calculated, and by varying parameters of the scatterer/resonator, these can be tuned to RSMs. So, the existence of reflectionless states, with tuning, is a robust property of electromagnetic scattering and does not require a high-Q resonator. An example of tuning to RSMs in a low-Q multiwaveguide junction is given in Figure 5 below.
- **Tuning R-zeros to RSMs**: Although there can be complications from interference between resonances, qualitatively speaking, adding loss to the scatterer will cause the R-zeros to flow downward in the complex plane and gain will cause them to flow upward. Similarly, altering the geometry of the scatterer/resonator so as to enhance the coupling of the output channels will cause the R-zeros to move downwards and decreasing the coupling will cause them to move upwards (and vice versa for the input channels). In certain cases, geometric tuning may affect both input and output channels at once and these couplings may not be separately controlled, making it difficult to tune to RSMs. However, in most cases, tuning a single structural parameter will be sufficient to allow perfectly reflectionless excitation of the structure, although the correct relative amplitudes and phases of the input channels will be required to access it.
- When the scatterer/resonator has discrete symmetries, in some cases, no tuning will be required to find RSMs (real R-zeros). Textbook examples are different types of balanced two-mirror resonators, which we will refer to collectively as "Fabry–Pérot" resonators. Such resonators have both parity ($\mathcal{P}$) and time-reversal ($\mathcal{T}$) symmetry. Other more recent examples are one-dimensional photonic structures with balanced gain and loss such that the product of parity and time ($\mathcal{PT}$) is preserved. Here, unlike the FP resonators, the real RSMs are unidirectional and can only be accessed from one side or the other. We will see in the next section that, in both cases, the RSMs can be lost due to spontaneous symmetry breaking at an RSM exceptional point (EP).

3.2 Symmetry properties of R-zeros and RSMs

The general formulas analyzed above only prove the existence of R-zeros in the complex plane, and therefore, since the real axis has zero measure in the plane, without parameter tuning or symmetry constraints, a generic system will have no RSMs. However, well-known examples, such as the FP resonator, have an infinite number of RSMs, apparently due to symmetry. In the study by Sweeney et al. [15], a detailed analysis is given of the implications of various discrete symmetries on the R-zero spectrum. Here, we will only present the main conclusions and illustrate them with simple one-dimensional examples.

We will focus on discrete symmetries and their effect on the R-zero spectrum. The specific symmetries we will analyze here are time reversal, parity, and the product of the two, as well as the very important case of systems with both symmetries ($\mathcal{P}+\mathcal{T}$), exemplified by the FP resonator. In the context of one-dimensional electromagnetic scattering, $\mathcal{T}$ symmetry requires that the resonator has a real dielectric function, $\varepsilon(x)$; $\mathcal{P}$ symmetry requires $\varepsilon(x)=\varepsilon(-x)$ but need not be real, and $\mathcal{PT}$ symmetry requires $\varepsilon(x)=\varepsilon^{*}(-x)$ but again it need not be real.

3.2.1 Time-reversal symmetry ($\mathscr{T}$) and parity symmetry ($\mathscr{P}$)

The time-reversal operator ($\mathcal{T}$) complex conjugates the wave equation in the frequency domain. If the system has $\mathcal{T}$ symmetry, then it maps a left-incident R-zero to a right-incident R-zero of the same cavity but at frequency ω^{*}. Hence the following conclusions are made:

- If a cavity with $\mathcal{T}$ symmetry (real ε) has a left R-zero at ω, then it will have a right R-zero at ω^{*}. If it is tuned to RSMs without breaking $\mathcal{T}$ symmetry, then it will be bidirectional, i.e., it will have a left RSM and right RSM at the same frequency.

The parity operator $\mathcal{P}$ maps $x\to-x$ in the wave equation; if the dielectric function has parity symmetry, then it maps a left R-zero at ω to a right R-zero of the same cavity at the same frequency. Hence the following conclusions are made:

- All R-zeros of parity-symmetric one-dimensional systems are bidirectional, whether or not they are real.

Both $\mathcal{P}$ symmetry and $\mathcal{T}$ symmetry map from left R-zeros to right R-zeros and imply relationships between these spectra but neither alone implies R-zeros are real. Hence, in systems with only one of these symmetries, parameter

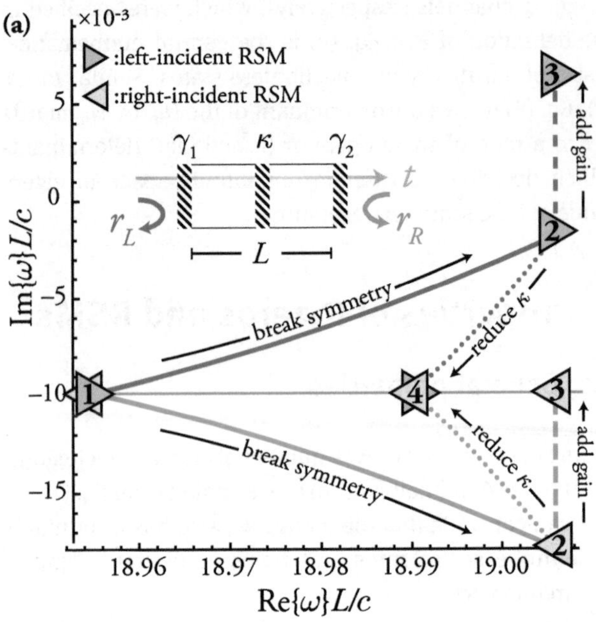

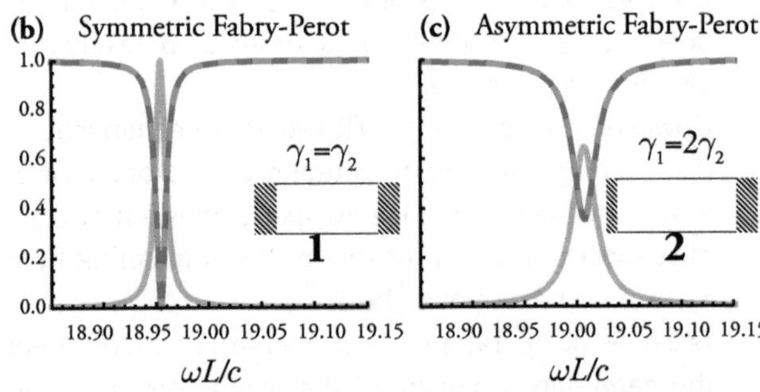

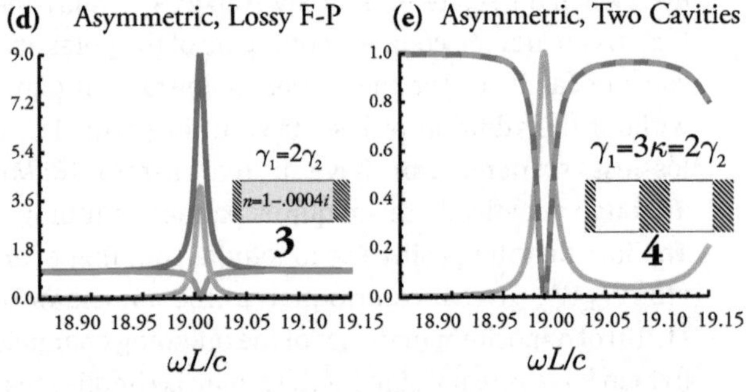

Figure 3: Illustration of reflectionless scattering modes (RSMs) and R-zero spectrum for simple two- and three-mirror resonators of length L in 1D, consisting of δ-function mirrors of strengths $\gamma_1^{-1},\gamma_2^{-1}$, and κ^{-1}, as indicated in the schematic in (a). Throughout, we fix $\gamma_2\equiv c/L$. Blue and red lines $1\to2$ indicate the effect of breaking symmetry by varying γ_1 from $\gamma_2\to2\gamma_2$. A bidirectional RSM [as in (b)] splits into two complex conjugate R-zeros off the real axis and a reflectionless steady-state (RSM) no longer exists, as in (c). Adding gain to the cavity, indicated by blue and red lines $2\to3$, brings the lower R-zero to the real axis (but not the upper one), creating a right-incident amplifying RSM, as in (d). Alternatively, adding a middle mirror and reducing its κ from $\infty\to2\gamma_2/3$ is sufficient to bring both R-zeros back to the real axis ($2\to4$ in (a)), creating simultaneous left and right RSMs at a different frequency from the symmetric Fabry–Pérot (FP) resonator (see (e)), without restoring parity symmetry.

tuning will be required to create RSMs. A simple example illustrating this with an asymmetric FP resonator is shown in Figure 3(a–e). Here, we illustrate tuning to RSMs with both $\mathcal{T}$-preserving (geometric) tuning and $\mathcal{T}$-breaking (gain/loss) tuning.

Starting with a symmetric FP cavity, with $\mathcal{P}+\mathcal{T}$ symmetry, we first break parity symmetry but maintain $\mathcal{T}$ symmetry by simply making the mirror reflectivities unequal (Figure 3(a) solid lines). The left-incident and right-incident R-zeros leave the real axis as complex conjugate pairs, as required by $\mathcal{T}$ symmetry, and there are no remaining RSMs (there are resonances but without zero reflection). To create a right-incident RSM, we add gain to the system, breaking $\mathcal{T}$ symmetry, with the correct value to bring the right R-zero in the lower half-plane back to the real axis; at the same time, the left R-zero is pushed further away, and the RSM created is unidirectional. The left R-zero could also be tuned to RSMs by adding an equivalent amount of loss. To create a bidirectional RSM, we instead add a third lossless mirror to the resonator, and we find that tuning the reflectivity of the new (middle) mirror can bring the system back to real axis. Since $\mathcal{T}$ symmetry has been maintained, the RSM must be bidirectional. Note that in both cases, the tuning has not restored parity symmetry. All such RSMs are in this sense "accidental", achieved by parameter tuning and without an underlying symmetry. Geometric tuning is of particular interest for reflectionless states of complex structures where adding loss or gain may not be practical or desirable (see the examples in Figures 4–6 below).

3.2.2 $\mathcal{P}$ and $\mathcal{T}$ symmetry and symmetry-breaking transition

The most prominent example of a system with an infinite number of RSMs was already mentioned above, two-mirror resonators (where "mirror" includes the many types of reflecting structures used in photonics) which will be referred to collectively as FP resonators. In fact one can easily check that all R-zeros of an FP resonator are real and in one–one correspondence with the resonances of the structure. This example might suggest that any structure with both $\mathcal{P}$ and $\mathcal{T}$ symmetry should have only real R-zeros (RSMs). Previous work on multimirror resonators [32, 33] however has found cases which violate this expectation, but to our knowledge, no general reason for this fact or qualitative understanding of it has been given. The symmetry analysis we present here provides such a framework, with additional implications which are new. A more detailed study of this case is in

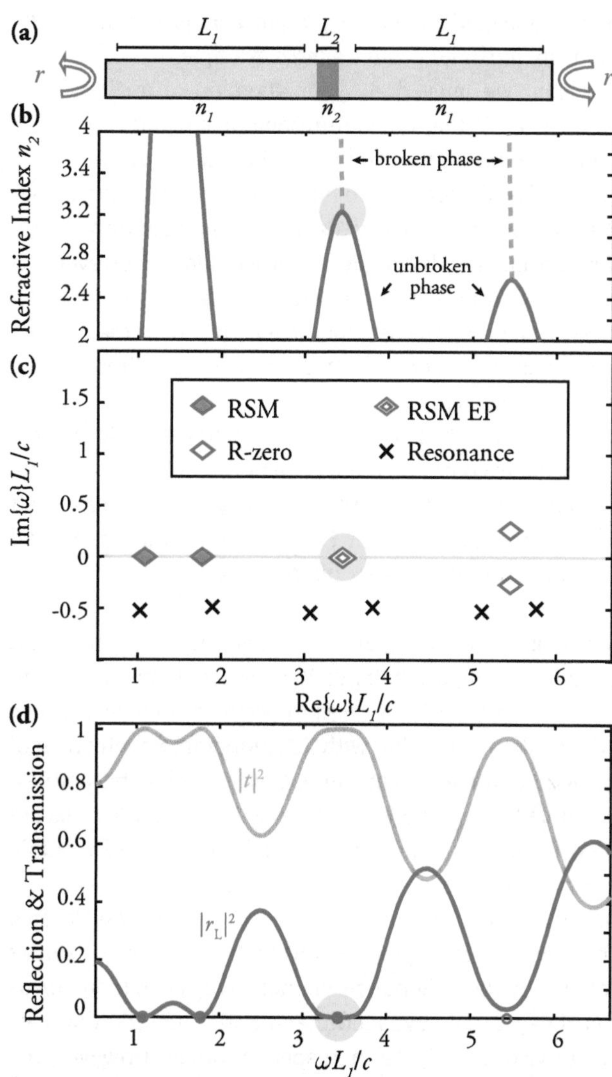

Figure 4: Reflectionless scattering modes (RSMs) in a $\mathcal{P}+\mathcal{T}$ symmetric structure.
(a) Symmetric three-slab heterostructure in air, with refractive index n_1 in the outer sections, which are of length L_1, and variable index n_2 in the middle, of length L_2; here $n_1 = 1.5$ and $L_2 = 0.15L_1/n_2$. (b) Real part of the R-zero frequencies as the central index n_2 is increased. For small n_2, the R-zeros are real-valued RSMs and in the unbroken phase (solid blue lines), while for large n_2, some R-zeros have entered the broken phase (red dashed). After two RSMs meet at an RSM exceptional point (EP), they split into two R-zeros at complex conjugate frequencies as n_2 is further increased. (c) Spectra of the R-zeros/RSMs and resonances in the complex frequency plane at $n_2 = 3.23$ where two (bidirectional) RSMs meet at two degenerate EPs (one for left RSMs and one for right RSMs). (d) Reflection and transmission spectra for the same n_2 as in (c); blue filled dots mark the RSM frequencies, open blue dot is real part of complex R-zero, which has already entered the broken phase. Yellow highlight in (b–d) indicates the same RSM EP, which exhibits quartically flat reflection and transmission.

preparation [34]; here, we present a brief outline of the problem, illustrated with a simple example.

First, we must draw your attention to an important point. Even if the structure/resonator has both $\mathcal{P}$ and $\mathcal{T}$ symmetry, the wave operator for the R-zero spectrum does not. As already noted, both the $\mathcal{P}$ and $\mathcal{T}$ operations map left R-zeros back to right R-zeros; hence, the boundary conditions are not invariant under these operations. However, it is easy to confirm that if a structure has both $\mathcal{P}$ and $\mathcal{T}$ symmetry (including the asymptotic regions), then the product $\mathcal{PT}$ maps the R-zero spectrum for a single directionality back to itself as follows:

- If a cavity has both $\mathcal{P}$ and $\mathcal{T}$ symmetry, its R-zeros occur in complex conjugate pairs or are nondegenerate and real (i.e. RSMs). The frequency spectrum of the left R-zeros and the right R-zeros is the same so that all RSMs and R-zeros are bidirectional.

Since both the cases of real RSMs without tuning and complex conjugate pairs of R-zeros are allowed by symmetry, we expect both cases to occur in some structures. We are already familiar with FP resonators for which all of the R-zeros are RSMs; in Figure 4, we examine the simplest example beyond the FP case for which not all R-zeros are RSMs. Similarly to the example of Figure 3, we add a middle mirror to the FP resonator, but now the initial FP resonator has balanced mirrors and $\mathcal{P} + \mathcal{T}$ symmetry. It is well known that such a coupling mirror will create two coupled cavities for which the original resonances will be paired up as quasi-degenerate symmetric and antisymmetric doublets with twice the original free spectral range. However, the doublets cannot become fully degenerate until the internal mirror becomes totally opaque, and we simply have two separate identical one-sided cavities. In contrast, there are continuity arguments which we omit here that imply that the RSMs must disappear at a finite coupling. In other words, as the coupling mirror becomes more opaque, there is a finite coupling at which a pair of RSMs associated with a resonant doublet meet and then leave the real axis as complex conjugate R-zeros. This is an example of a spontaneous $\mathcal{PT}$ symmetry-breaking transition [35–38] driven by the increase of the coupling mirror opacity. *Thus we have shown the existence of a $\mathcal{PT}$ transition in a lossless system.* This is the first such example to our knowledge, and it is possible because the R-zero boundary conditions themselves are nonhermitian, even if the differential operator in the wave equation does not have a complex dielectric function. This behavior of the three-mirror lossless resonator is illustrated in Figure 4.

Finally, when the two RSMs meet on the real axis, we have a degeneracy of the nonhermitian R-zero eigenvalue problem, so this must correspond to an EP of the underlying wave operator, which in this case happens on the real axis, and not in the complex plane as happens for many other studied cases for which two resonances become degenerate. More precisely, there are two degenerate EPs when bidirectional RSMs meet; one for the left RSM spectra and one for the right RSM spectra. For a second order EP such as this, on the real axis, previous work has shown [39] that the associated lineshape is altered to a quartic flat-top shape; this behavior is visible in Figure 4. The existence of this specific tuned behavior of the three-mirror resonators was predicted long ago [32, 33] and is used in designing "ripple-free" filters. Such filters are the analogs of Butterworth filters in electronics [40]. However, the previous work does not seem to have identified this as an EP, associated with a $\mathcal{PT}$ transition. A much more in-depth analysis of the physics of this transition will be given elsewhere [34].

3.2.3 Parity–time symmetry ($\mathcal{PT}$) and symmetry-breaking transition

Reflectionless states have been previously studied extensively [41, 42] for one-dimensional resonators which have only $\mathcal{PT}$ symmetry but not $\mathcal{P}$ and $\mathcal{T}$ separately; in this case, the condition $\varepsilon(x) = \varepsilon^*(-x)$ holds, but ε is not real (balanced gain and loss). This case is discussed in detail in the study by Sweeney et al. [15]; it is straightforward to show that, in this case, R-zeros are either real or come in complex conjugate pairs, but there is no connection between the left and right R-zero spectra. Hence, all R-zeros and RSMs in this case are unidirectional. If one starts with a system with $\mathcal{P}$ and $\mathcal{T}$ symmetry such as the standard FP resonator and now adds balanced gain and loss to maintain $\mathcal{PT}$ symmetry, again one will see pairs of RSMs move toward each other on the real axis, pass through an EP, before emerging as complex conjugate pairs of R-zeros. This example, given in the study by Sweeney et al. [15], then differs from the example shown in Figure 4 only in that the left R-zeros and right R-zeros no longer move together.

4 Applications of R-zero/RSM theory

4.1 Relationship to coupled mode theory

The preceding results were derived directly from Maxwell's equations and involve no approximation. In many circumstances, an approximate analytic model will be adequate and desirable for simplicity. In photonics, a standard tool is the temporal coupled-mode theory (TCMT) [43–48], which is a phenomenological model widely used in the design and analysis of optical devices [49–52]. The TCMT formalism is derived from symmetry constraints [43–47] rather than from first principles, yet it leads to an analytic relation between the determinant of the scattering matrix and the underlying Hamiltonian that is similar to Eq. (7) and is reasonably accurate in many cases. The appropriate comparison between the TCMT and the exact RSM theory presented here is given in the study by Sweeney et al. [15]. Here, we will only quote one relevant result of that analysis. Not surprisingly, it is possible to adapt the TCMT analysis which leads to an expression for the S-matrix in terms of an "effective Hamiltonian", so as to find an expression for the $\mathbf{R}_{\text{in}}$ matrix, and a condition for it to have an R-zero. In this expression, as in our Eq. (8), the coupling coefficients for the input channels to the resonator appear as an effective gain term and coefficients for the output channels appear as an effective loss term. Very often, in the TCMT formalism, the single-resonance approximation is used, in which the effective Hamiltonian is replaced by a number equal to the complex energy of the resonance. When a similar approximation is used for $\mathbf{R}_{\text{in}}$ to determine the R-zeros, one finds the following [15]:

$$\omega_{\text{RZ}} = (\omega_0 - i\gamma_{\text{nr}}) + i(\gamma_{\text{in}} - \gamma_{\text{out}}),$$

$$\gamma_{\text{in}} \equiv \sum_{n\in F} |d_n|^2/2, \quad \gamma_{\text{out}} \equiv \sum_{n\notin F} |d_n|^2/2, \tag{9}$$

here, ω_{RZ} is the R-zero frequency, ω_0 is the real part of the resonance frequency, d_n is the coupling coefficient (partial width) of the mode to the nth radiative channel, $\gamma_{\text{in}}, \gamma_{\text{out}}$ are then the total radiative rates in and out, respectively, for the resonance, and γ_{nr} is the nonradiative rate associated with loss or gain in the resonator. An RSM arises when the various imaginary terms for ω_{RZ} cancel, and the structure is illuminated with the corresponding wavefront, determined by the eigenvector of $\mathbf{R}_{\text{in}}$. This then corresponds to the CC condition, generalized to multichannel inputs and outputs. Here, we have used the standard notation in the TCMT for the partial coupling rates into or out of a given channel. In our theory, evaluating the matrices $\mathbf{W}_{\text{in}}\mathbf{W}_{\text{in}}^{\dagger}, \mathbf{W}_{\text{out}}\mathbf{W}_{\text{out}}^{\dagger}$ in the limit of a single resonance yields a similar CC relationship.

Hence, the TCMT within the single (high-Q)-resonance approximation gives an analytic basis for the concept of generalized CC introduced above and also shows its limitations. One implication of this result is that R-zeros with

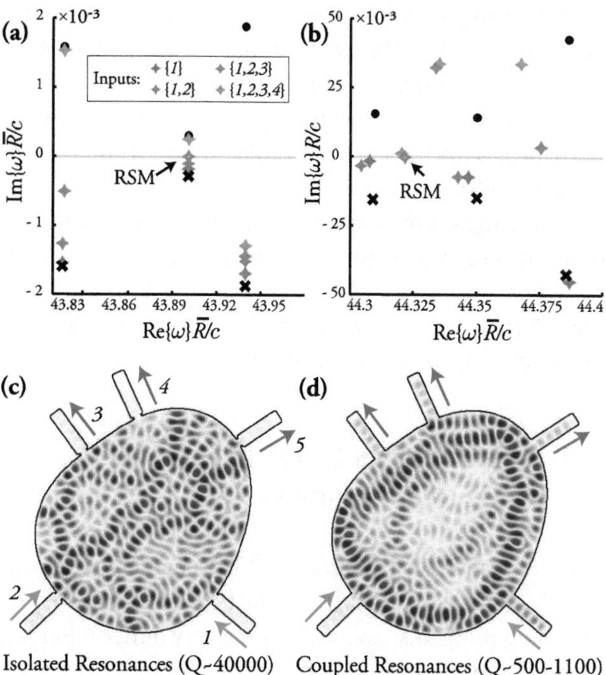

Figure 5: Asymmetric lossless waveguide junction/resonator (mean radius $\bar{R}$) coupled to five single-mode waveguides, with constrictions at the ports to the junction. (a) Numerically calculated R-zero spectrum for a weakly coupled, high-Q junction with well-isolated resonances. Black **x** and dot are purely outgoing (resonance) and incoming (S-matrix zero) frequencies, which are complex conjugates. Colored stars are R-zeros for various choices of input channels; the legend indicates which channels are inputs, with the channel labels given in (c). The R-zeros cluster vertically above the resonance frequency and below the S-matrix zero frequency, as predicted by single-resonance temporal coupled-mode theory (TCMT) approximation in Eq. (9) below. The common width of the constrictions for waveguides {4, 5, 6} is slightly tuned to make a 2-in/3-out R-zero real, creating a reflectionless scattering mode (RSM). (b) R-zero spectrum for the same junction but with the constrictions opened, which lowers the Q of the resonances (note change in vertical scale). The linewidths of the resonances are now comparable to their spacing. Due to multiresonance effects, the R-zeros are spread out along the real and complex frequency axis and are no longer associated with a single resonance. Nonetheless, by slightly tuning the constriction width as before, a 2-in/3-out R-zero is again made real (RSM), as in the high-Q case. (c, d) The mode profiles of the RSMs for the high-Q (c) and low-Q (d) cases.

different numbers of input and output channels will have the same Re{ω} as the underlying resonance and will be simply shifted vertically in the complex plane along a line between the S-matrix pole (resonance) and its zero. The appropriate input wavefront for the RSM will just be that of the outgoing resonance, phase conjugated for the chosen input channels. This behavior is found in the example shown in Figure 5a for the case of a high-Q resonator. Conversely, the TCMT single resonance result fails for a more open resonator (Figure 5b), where multiple resonances mediate the scattering within the resonator. Nonetheless, the exact R-zero/RSM approach can be used to tune to RSMs numerically.

The single-resonance scenario is the simplest example of an R-zero, yet it already is sufficient to explain the impedance-matching conditions previously found using the TCMT in waveguide branches [53], antireflection surfaces [54], and polarization-converting surfaces [55].

4.2 Reflectionless states in complex structures: examples

Here, we show two examples of RSMs engineered in complex photonic structures; these are the kinds of impedance-matching problems, and it would be very difficult to solve without our theory and associated computational approaches. As noted above, R-zero spectra can be calculated by a modified PML method in many cases and by a boundary matching method in all cases. The first of the examples below was done by the PML method, which is somewhat simpler. The second was done by the matching method and could not be done with PMLs because the input and output channels are not separated in the asymptotic regions. We will discuss the solution methods very briefly in the next section.

In Figure 5(a), we show results for an asymmetric cavity much larger than the wavelength of the exciting radiation, with a smooth boundary connected to five single-mode waveguides, without any internal gain or loss. Here and in the next example, we are only going to use geometric/index parameter tuning to achieve a flux-conserving RSM. The structure has no discrete symmetries, so there is no reason that there should exist any RSMs for such structures without parameter tuning. In addition, such cavities are well known to have many pseudorandom wave-chaotic states so that the resonances do not have any simple spatial structure or ray orbit interpretation, making intuitive design approaches to generating the appropriate interference behavior impossible. We consider this structure in two limits: (a) the limit in which the waveguide ports are pinched off by constrictions to create a high-Q cavity and large nonresonant reflection at the ports and (b) the limit of essentially open ports with slight width tuning for which the cavity has much lower Q, and multiple resonances participate in scattering.

In the case (a), the behavior is as predicted by the single-resonance approximation, discussed just above. R-zeros are lined up vertically in the complex plane between the pole and the zero, and one of them (a two-in, three-out case) has been tuned to the real axis. The internal field (real part shown in (c)) is chaotic looking and coincides well with the single resonance associated with these R-zeros.

In the case (b), where we impose the same R-zero boundary conditions, we see very different behavior of the

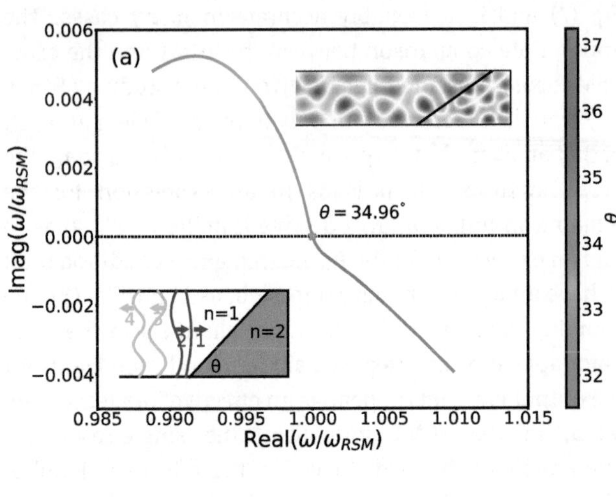

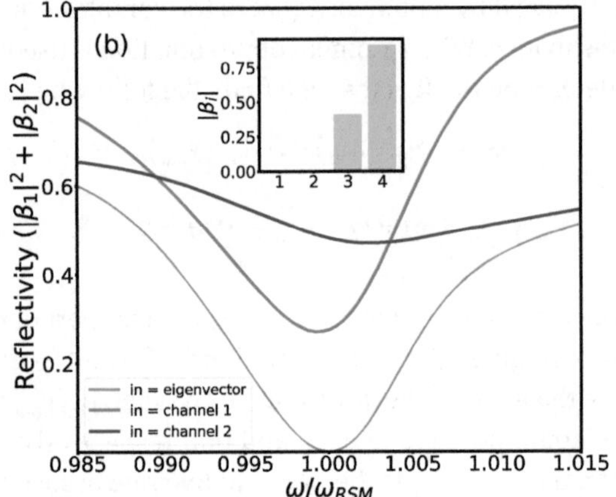

Figure 6: Illustration of a reflectionless scattering mode (RSM) in a four-mode waveguide acting as a mode converter from a superposition of input waveguide modes 1 and 2 into output modes 3 and 4, as the wedge angle of the boundary between the index 1 and index 2 region is varied.

(a) Trajectory of R-zero in the complex frequency plane as wedge-angle θ is tuned. The R-zero crosses the real axis at $\theta = 34.96°$ (red star), becoming an RSM. Insets show a schematic of the structure and the real part of the RSM field profile. (b) Reflectivity into modes 1 and 2 with different incident wavefronts for $\theta = 34.96°$. Red curve has input $\alpha_0(\omega)$, defined to be the eigenvector of $\mathbf{R}_{\text{in}}^{\dagger}(\omega)\mathbf{R}_{\text{in}}(\omega)$ with the smallest eigenvalue. $\mathbf{R}_{\text{in}}(\omega)$ is the 2×2 upper left block of the scattering matrix. The incident wave $\alpha_0(\omega_{\text{RSM}}) = [0.7982, -0.5642 + 0.2158i, 0, 0]$ generates the reflectionless output $\beta(\omega_{\text{RSM}}) = [0, 0, 0.1767 - 0.3689i, 0.7682 + 0.4925i]$. Meanwhile, the inputs from only mode 1 or only mode 2 (green and blue curves) have nonzero reflectivity for all frequencies. The inset shows the output amplitude $|\beta(\omega_{\text{RSM}})|$ for the eigenvector input $\alpha_0(\omega_{\text{RSM}})$.

R-zero spectrum, characteristic of transmission through multiple resonances. The R-zeros are spread out in the complex plane and do not lie on a line coincident with any one resonance, nor is the input wavefront or internal field associated with a single resonance. Tuning to RSMs is achieved by a very slight variation in the width of one of the outgoing waveguides at the port. Further details are given in the figure caption.

The second example, shown in Figure 6, is a structure which functions as a lossless mode converter in reflection. It is a multimode empty waveguide terminated by an angled wedge of purely real index $n = 2$ material and then by a perfectly reflecting wall. In the case where the waveguide has only two modes, it is relatively simple to find a wedge angle which converts, e.g., mode one into mode two perfectly after reflection, and a simple Fresnel scattering analysis could be used to find the necessary angle to a good approximation. Here, however the waveguide has four propagating modes, and the R-zero problem is to use modes one and two as inputs and modes three and four as outputs. Our general theory implies that such a solution should exist at some ω as one tunes a single parameter such as the wedge angle. Indeed, Figure 6(a) shows that as the wedge angle is tuned, one of the R-zeros crosses the real axis and becomes an RSM.

As there are multiple input channels, zero reflection is achieved only when the correct superposition of modes one and two is used. The smallest eigenvalue of $\mathbf{R}_{\text{in}}^{\dagger}(\omega)\mathbf{R}_{\text{in}}(\omega)$ gives the smallest possible reflectivity, and Figure 6(b) shows that this vanishes at ω_{RSM}, while the reflected intensities for single-channel inputs do not anywhere in the vicinity of ω_{RSM}.

Finally, in the study by Sweeney et al. [15], an example was presented of an RSM solution to a different and challenging problem in free-space scattering: designing a dielectric antenna much larger than the input wavelength such that it perfectly reflects a monopole input signal (assuming 2D scalar waves) into higher multipoles in the scattered field. Although here we considered scalar waves, the method can be straightforwardly generalized to vector electromagnetic waves. This additional example, not involving waveguides or mirror resonators, illustrates the versatility of the R-zero/RSM theory for electromagnetic design.

Finding these types of impedance-matched solutions for open multichannel structures would be difficult without our theory, nor were there, to our knowledge, previous formulations which guarantee a solution exists, for the appropriate input wavefront with single parameter tuning. As already noted, the usual CC concept does not extend to these types of low-Q structures.

4.3 Solution method for R-zero/RSM problems

We make a few brief remarks on solving the R-zero/RSM problem numerically, which will be necessary for essentially all cases of interest. In principle, one could find R-zeros by constructing the full S-matrix of the problem and then the relevant $\mathbf{R}_{\text{in}}(\omega)$ and search for the zeros of its determinant in the complex plane. Doing so, however, has a numerical disadvantage because $\mathbf{R}_{\text{in}}(\omega)$ changes rapidly with frequency near resonances, which generally makes it harder for such root finding and for other nonlinear eigenproblem solvers. Our theory demonstrates that the R-zeros can be found directly from imposing boundary conditions of the wave operator on the boundary of a computational cell, without constructing the scattering matrix in the complex plane. This typically more efficient procedure then yields the R-zero spectrum in a given frequency range for an initial structure, which typically contains no RSMs. However, the fact that the R-zeros are eigenvalues of a well-behaved wave operator means that its eigenvalues will move continuously with small changes in the real or imaginary part of the dielectric function. Moreover, the general form of Eq. (8) has given us the intuition to know what kinds of geometric or loss/gain perturbations will move the R-zeros up or down in the complex plane (e.g., increasing the coupling of an outgoing channel will tend to move all R-zeros down and vice versa for an incoming channel). While the simple CC picture is not always valid, the tuning of an R-zero to the real axis is by no means a random search in a parameter space. In addition, obviously, many different tunings will work if any real frequency in a given range is acceptable. The upshot is that, numerically, engineering a single RSM can be achieved by a small number of iterations of the initial calculation. Hence, finding the RSMs in many cases is computationally no more difficult than iterating a resonance calculation of the type available in packages such as COMSOL Multiphysics over a number of weakly perturbed structures.

However, there is one important difference between resonance calculations and RSM calculations; all resonance calculations can be done using the PML method since all the asymptotic channels satisfy the same outgoing boundary conditions. When the system has negligible dispersion, the PML method turns a nonlinear eigenvalue problem into a linear one, for which the boundary condition is independent of ω, and this makes the solution easier [56]. There also exist conjugated PMLs [27], though they are less well known, which can implement purely incoming boundary conditions. However, there is (as yet) no PML method to solve problems, such as the mode converter example above, for which the incoming and outgoing channels overlap in space. Here, we need to impose matching conditions outside the surface of last scattering based on exactly the set of incoming and outgoing channels chosen for the R-zero problem (this is what we mean by R-zero boundary conditions). Those conditions do depend on the frequency, leading to the more complicated nonlinear eigenvalue problem one avoids with PMLs. However, such matching has been done successfully for complex structures in the wave-chaotic regime [57] and in *ab initio* laser theory [58], and we have used similar methods here and in the study by Sweeney et al. [15] to solve the mode converter and multipole converting antenna examples. Even for these more challenging cases, the computations remained quite tractable. More details about the two methods and a derivation of the matching method are given in the study by Sweeney et al. [15].

5 Summary and outlook

This paper outlines a general theory of reflectionless excitation or impedance matching of linear waves to finite structures of arbitrary geometry in any dimension, focusing on the case of classical electromagnetic waves and building on the full theoretical framework presented in the study by Sweeney et al. [15]. The basic framework applies as well to acoustic and other linear classical waves and to some quantum scattering problems as well. Because every impedance-matching problem can be posed as the solution of an electromagnetic eigenvalue problem with certain overdetermined boundary conditions at infinity, similar to the problem of finding resonances, it is possible to completely specify necessary and sufficient conditions for solutions to exist. It is shown that an infinite number of unphysical solutions always exist at discrete complex frequencies, and any single solution can be tuned to the real axis, typically by varying a single parameter of the structure, so as to become a physically realizable steady-state harmonic solution. These solutions are accessible if it is possible to generate the appropriate input wavefront, which can be determined from the solutions. We refer to wave solutions of this type as RSMs because they are steady-state solutions adapted to the particular structure and which specify the behavior of each scattering channel at infinity, similar to lasing modes.

While we have an analytic framework for determining the RSMs, the resulting equations will usually need to be solved numerically, and we present two methods for doing so which are computationally tractable by adapting standard tools of computational photonics. Since reflectionless excitation of fairly complex structures is often a goal in photonics, we believe our theory and approach shows promise for microphotonic and nanophotonic design. The theory may clarify which design goals are guaranteed an exact solution and which ones are not. For example, if one has a three-waveguide junction of some geometry, one is guaranteed to be able to find a design for which excitation from waveguide one into waveguides two and three is reflectionless, typically by tuning a single geometric parameter. However, there is no guarantee that further tuning parameters will find a solution which scatters only into waveguide two in some frequency ranges. However, since there will be many ways to tune to the reflectionless state of waveguide one, it is interesting to propose a search in this parameter space for the way which minimizes the output into waveguide three, which could be a much more efficient search than an *ab initio* combinatoric or machine learning–based search of a huge space of structures, most of which are not close to reflectionless. There are indications in our current results that such an RSM calculation followed by an optimization can succeed. In this manner, we hope that our theory of RSMs can be combined with modern optimization methods to achieve efficiently important design goals in photonic structures.

Acknowledgments: W.R.S. and A.D.S. acknowledge the support of the NSF CMMT program under grant DMR-1743235. The authors thank Yidong Chong for allowing us to adapt Figure 2 from his random CPA animation and Abhinava Chatterjee for initial explorations of the $\mathcal{P}+\mathcal{T}$ symmetry-breaking transition.
Author contribution: All the authors have accepted responsibility for the entire content of this submitted manuscript and approved submission.
Research funding: This research was supported by the NSF Condensed Matter and Materials Theory (CMMT) program under grant DMR-1743235.
Conflict of interest statement: The authors declare no conflicts of interest regarding this article.

References

[1] G. Gamow, “Zur Quantentheorie des Atomkernes,” *Z. Phys.*, vol. 51, pp. 204–212, 1928.
[2] A. Böhm, “Resonance poles and Gamow vectors in the rigged Hilbert space formulation of quantum mechanics,” *J. Math. Phys.*, vol. 22, pp. 2813–2823, 1981.
[3] E. S. C. Ching, P. T. Leung, A. M. van den Brink, W. M. Suen, S. S. Tong, and K. Young, “Quasinormal-mode expansion for waves in open systems,” *Rev. Mod. Phys.*, vol. 70, pp. 1545–1554, 1998.

[4] P. Lalanne, W. Yan, K. Vynck, C. Sauvan, and J.-P. Hugonin, "Light interaction with photonic and plasmonic resonances," *Laser Photon. Rev.*, vol. 12, p. 1700113, 2018.
[5] R. Lang, M. O. Scully, and W. E. Lamb, Jr., "Why is the laser line so narrow? A theory of single-quasimode laser operation," *Phys. Rev. A*, vol. 7, pp. 1788–1797, 1973.
[6] Li Ge, Y. Chong, and A. D. Stone, "Steady-state ab initio laser theory: Generalizations and analytic results," *Phys. Rev. A.*, vol. 82, p. 063824, 2010.
[7] S. Esterhazy, D. Liu, M. Liertzer, et al., "Scalable numerical approach for the steady-state ab initio laser theory," *Phys. Rev. A*, vol. 90, p. 023816, 2014.
[8] R. Adler, L. J. Chu, and R. M. Fano, *Electromagnetic Energy Transmission and Radiation*, New York, Wiley and Sons, 1960.
[9] R. H. Yan, R. J. Simes, and L. A. Coldren, "Electro-absorptive Fabry-Perot reflection modulators with asymmetric mirrors," *IEEE Photon. Technol. Lett.*, vol. 1, pp. 273–275, 1989.
[10] K.-K. Law, R. H. Yan, J. L. Merz, and L. A. Coldren, "Normally-off high-contrast asymmetric Fabry–Perot reflection modulator using Wannier–Stark localization in a superlattice," *Appl. Phys. Lett.*, vol. 56, pp. 1886–1888, 1990.
[11] K.-K. Law, R. H. Yan, L. A. Coldren, and J. L. Merz, "Self-electro-optic device based on a superlattice asymmetric Fabry–Perot modulator with an on/off ratio ≥100:1," *Appl. Phys. Lett.*, vol. 57, pp. 1345–1347, 1990.
[12] M. Cai, O. Painter, and K. J. Vahala, "Observation of critical coupling in a fiber taper to a silica-microsphere whispering-gallery mode system," *Phys. Rev. Lett.*, vol. 85, pp. 74–77, 2000.
[13] A. Yariv, "Critical coupling and its control in optical waveguide-ring resonator systems," *IEEE Photon. Technol. Lett.*, vol. 14, pp. 483–485, 2002.
[14] M. Lipson, "Guiding, modulating, and emitting light on silicon-challenges and opportunities," *J. Lightwave Technol.*, vol. 23, p. 4222, 2005.
[15] W. R. Sweeney, C. W. Hsu, and A. D. Stone, "Theory of reflectionless scattering modes," 2019, arXiv: 1909.04017.
[16] D. Jalas, A. Petrov, M. Eich, et al., "What is—and what is not—an optical isolator," *Nat. Photonics*, vol. 7, p. 579, 2013.
[17] Y. D. Chong, Li Ge, H. Cao, and A. D. Stone, "Coherent perfect absorbers: Time-reversed lasers," *Phys. Rev. Lett.*, vol. 105, p. 053901, 2010.
[18] W. Wan, Y. Chong, L. Ge, H. Noh, A. D. Stone, and H. Cao, "Time-reversed lasing and interferometric control of absorption," *Science*, vol. 331, pp. 889–892, 2011.
[19] H. Noh, Y. D. Chong, A. D. Stone, and H. Cao, "Perfect coupling of light to surface plasmons by coherent absorption," *Phys. Rev. Lett.*, vol. 108, p. 186805, 2012.
[20] D. G. Baranov, A. Krasnok, T. Shegai, A. Alù, and Y. Chong, "Coherent perfect absorbers: Linear control of light with light," *Nat. Rev. Mater.*, vol. 2, p. 17064, 2017.
[21] A. Friedman and M. Shinbrot, "Nonlinear eigenvalue problems," *Acta Math.*, vol. 121, pp. 77–125, 1968.
[22] G. H. Golub and H. A. van der Vorst, "Eigenvalue computation in the 20th century," *J. Comput. Appl. Math.*, vol. 123, pp. 35–65, 2000.
[23] J. Asakura, T. Sakurai, H. Tadano, I. Tsutomu, and K. Kimura, "A numerical method for polynomial eigenvalue problems using contour integral," *Japan J. Indust. Appl. Math.*, vol. 27, pp. 73–90, 2010.
[24] Y. Su and Z. Bai, "Solving rational eigenvalue problems via linearization," *SIAM J. Matrix Anal. Appl.*, vol. 32, pp. 201–216, 2011.
[25] W.-J. Beyn, "An integral method for solving nonlinear eigenvalue problems," *Lin. Algebra Appl.*, vol. 436, pp. 3839–3863, 2012.
[26] G. B. Denis, A. Krasnok, and A. Alù, "Coherent virtual absorption based on complex zero excitation for ideal light capturing," *Optica*, vol. 4, pp. 1457–1461, 2017.
[27] A. S. B. B. Dhia, L. Chesnel, and V. Pagneux, "Trapped modes and reflectionless modes as eigenfunctions of the same spectral problem," *Proc. R. Soc. A*, vol. 474, p. 20180050, 2018.
[28] C. Mahaux and H. A. Weidenmüller, *Shell-model Approach to Nuclear Reactions*, Amsterdam, North-Holland, 1969.
[29] C. W. J. Beenakker, "Random-matrix theory of quantum transport," *Rev. Mod. Phys.*, vol. 69, pp. 731–808, 1997.
[30] C. Viviescas and G. Hackenbroich, "Field quantization for open optical cavities," *Phys. Rev. A*, vol. 67, p. 013805, 2003.
[31] S. Rotter and S. Gigan, "Light fields in complex media: Mesoscopic scattering meets wave control," *Rev. Mod. Phys.*, vol. 89, p. 015005, 2017.
[32] H. van de Stadt and J. M. Muller, "Multimirror Fabry–Perot interferometers," *J. Opt. Soc. Am. A*, vol. 2, pp. 1363–1370, 1985.
[33] J. Stone, L. W. Stultz, and A. A. M. Saleh, "Three-mirror fibre Fabry–Pérot filters of optimal design," *Electron. Lett.*, vol. 26, p. 10731074, 1990.
[34] W. R. Sweeney, C. W. Hsu, K. Wisal, and A. D.Stone, In preparation.
[35] L. Feng, R. El-Ganainy, and L. Ge, "Non-Hermitian photonics based on parity–time symmetry," *Nat. Photonics*, vol. 11, pp. 752–762, 2017.
[36] R. El-Ganainy, K. G. Makris, M. Khajavikhan, Z. H. Musslimani, S. Rotter, and D. N. Christodoulides, "Non-Hermitian physics and symmetry," *Nat. Phys.*, vol. 14, pp. 11–19, 2018.
[37] M. A. Miri and A. Alù, "Exceptional points in optics and photonics," *Science*, vol. 363, p. eaar7709, 2019.
[38] Ş. K. Özdemir, S. Rotter, F. Nori, and L. Yang, "Parity–time symmetry and exceptional points in photonics," *Nat. Mater.*, vol. 18, pp. 783–798, 2019,.
[39] W. R. Sweeney, C. W. Hsu, S. Rotter, and A. D. Stone, "Perfectly absorbing exceptional points and chiral absorbers," *Phys. Rev. Lett.*, vol. 122, 2019, p. 093901.
[40] S. Butterworth, "On the theory of filter amplifiers," *Exp. Wirel. Wirel. Eng.*, vol. 7, pp. 536–541, 1930.
[41] Z. Lin, H. Ramezani, T. Eichelkraut, T. Kottos, H. Cao, and D. N. Christodoulides, "Unidirectional invisibility induced by *PT*-symmetric periodic structures," *Phys. Rev. Lett.*, vol. 106, p. 213901, 2011.
[42] L. Ge, Y. D. Chong, and A. D. Stone, "Conservation relations and anisotropic transmission resonances in one-dimensional *PT*-symmetric photonic heterostructures," *Phys. Rev. A*, vol. 85, p. 023802, 2012.
[43] H. A. Haus, *Waves and Fields in Optoelectronics*, Englewood Cliffs, NJ, Prentice-Hall, 1984, Chap. 7.
[44] S. Fan, W. Suh, and J. D. Joannopoulos, "Temporal coupled-mode theory for the Fano resonance in optical resonators," *J. Opt. Soc. Am. A*, vol. 20, pp. 569–572, 2003.
[45] W. Suh, Z. Wang, and S. Fan, "Temporal coupled-mode theory and the presence of non-orthogonal modes in lossless multimode cavities," *IEEE J. Quant. Electron.*, vol. 40, pp. 1511–1518, 2004.

[46] K. X. Wang, "Time-reversal symmetry in temporal coupled-mode theory and nonreciprocal device applications," *Opt. Lett.*, vol. 43, pp. 5623–5626, 2018.

[47] Z. Zhao, C. Guo, and S. Fan, "Connection of temporal coupled-mode-theory formalisms for a resonant optical system and its time-reversal conjugate," *Phys. Rev. A*, vol. 99, p. 033839, 2019.

[48] F. Alpeggiani, N. Parappurath, E. Verhagen, and L. Kuipers, "Quasinormal-mode expansion of the scattering matrix," *Phys. Rev. X*, vol. 7, p. 021035, 2017.

[49] L. Verslegers, Z. Yu, Z. Ruan, P. B. Catrysse, and S. Fan, "From electromagnetically induced transparency to superscattering with a single structure: A coupled-mode theory for doubly resonant structures," *Phys. Rev. Lett.*, vol. 108, p. 083902, 2012.

[50] B. Peng, Ş. K. Özdemir, F. Lei, et al., "Parity-time-symmetric whispering-gallery microcavities," *Nat. Phys.*, vol. 10, pp. 394–398, 2014,.

[51] C. W. Hsu, B. G. DeLacy, S. G. Johnson, J. D. Joannopoulos, and M. Soljačić, "Theoretical criteria for scattering dark states in nanostructured particles," *Nano Lett.*, vol. 14, pp. 2783–2788, 2014.

[52] B. Zhen, C. W. Hsu, Y. Igarashi, et al., "Spawning rings of exceptional points out of Dirac cones," *Nature*, vol. 525, pp. 354–358, 2015.

[53] S. Fan, S. G. Johnson, J. D. Joannopoulos, C. Manolatou, and H. A. Haus, "Waveguide branches in photonic crystals," *J. Opt. Soc. Am. B*, vol. 18, pp. 162–165, 2001.

[54] K. X. Wang, Z. Yu, S. Sandhu, V. Liu, and S. Fan, "Condition for perfect antireflection by optical resonance at material interface," *Optica*, vol. 1, pp. 388–395, 2014.

[55] Y. Guo, M. Xiao, and S. Fan, "Topologically protected complete polarization conversion," *Phys. Rev. Lett.*, vol. 119, p. 167401, 2017.

[56] P. Lalanne, W. Yan, K. Vynck, C. Sauvan, and J.-P. Hugonin, "Light interaction with photonic and plasmonic resonances," *Laser Photon. Rev.*, vol. 12, p. 1700113, 2018.

[57] H. E. Türeci, H. G. L. Schwefel, P. Jacquod, and A. D. Stone, *Progress in Optics*, Vol. 47, E. Wolf, Ed., Elsevier, 2005, pp. 75–135. Chap. 2.

[58] H. E. Türeci, L. Ge, S. Rotter, and A. D. Stone, "Strong interactions in multimode random lasers," *Science*, vol. 320, pp. 643–646, 2008.

Part VI: **Optimization Methods**

Jiaqi Jiang and Jonathan A. Fan*

Multiobjective and categorical global optimization of photonic structures based on ResNet generative neural networks

https://doi.org/10.1515/9783110710687-027

Abstract: We show that deep generative neural networks, based on global optimization networks (GLOnets), can be configured to perform the multiobjective and categorical global optimization of photonic devices. A residual network scheme enables GLOnets to evolve from a deep architecture, which is required to properly search the full design space early in the optimization process, to a shallow network that generates a narrow distribution of globally optimal devices. As a proof-of-concept demonstration, we adapt our method to design thin-film stacks consisting of multiple material types. Benchmarks with known globally optimized antireflection structures indicate that GLOnets can find the global optimum with orders of magnitude faster speeds compared to conventional algorithms. We also demonstrate the utility of our method in complex design tasks with its application to incandescent light filters. These results indicate that advanced concepts in deep learning can push the capabilities of inverse design algorithms for photonics.

Keywords: categorical optimization; global optimization; multiobjective optimization; neural networks; thin-film stack.

1 Introduction

Inverse algorithms are among the most effective methods for designing efficient, multifunctional photonic devices [1–3]. It remains an open question how to select and implement a design algorithm, and over the last few years, much research has been focused on deep neural networks as inverse design tools [4–6]. Many of these demonstrations are based on the generation of a training set, consisting of device geometries and their optical responses, and modeling these data using discriminative [7, 8] or generative [9–12] neural networks. These methods have proven to be capable of producing high-speed surrogate solvers and can perform inference-type tasks with training data. When the training data are curated using advanced gradient-based optimization methods, such as the adjoint variables [13–17] or objective-first methods [18], the networks can learn to generate high-performing, free-form photonic structures.

To perform global optimization, alternative approaches are required that do not depend on interpolation from a training set. The reason is because the design space is nonconvex and contains multiple local optima, and even devices based on advanced gradient-based optimization methods cannot help a neural network search for the global optimum. In this vein, global optimization networks (GLOnets) have been developed to perform the nonconvex global optimization of free-form photonic devices [19, 20]. GLOnets are gradient-based optimizers that do not use a training set but instead combine a generative neural network with an electromagnetic simulator to perform population-based optimization. The evolution of the generated device distribution is driven by both figure-of-merit values (i.e., efficiencies) and gradients for devices sampled from the generative network. Initial implementations of GLOnets were configured for single-objective problems with binary design variables, such as the maximization of deflection efficiency for a normally incident beam in a metagrating comprising silicon nanostructures. "Single-objective" refers to the optimization of a system operating with one conditional parameter, in this case a system with fixed incidence beam angle, and "binary" refers to silicon and air as our design materials.

A more general formulation of the problem that captures the design space of many photonic technologies is

***Corresponding author: Jonathan A. Fan,** Department of Electrical Engineering, Stanford University, 348 Via Pueblo, Stanford, CA 94305, USA, E-mail: jonfan@stanford.edu
Jiaqi Jiang, Department of Electrical Engineering, Stanford University, 348 Via Pueblo, Stanford, CA 94305, USA. https://orcid.org/0000-0001-7502-0872

This article has previously been published in the journal Nanophotonics. Please cite as: J. Jiang and J. A. Fan "Multiobjective and categorical global optimization of photonic structures based on ResNet generative neural networks" *Nanophotonics* 2021, 10. DOI: 10.1515/nanoph-2020-0407.

multiobjective, categorical optimization with more than two design materials. "Multiobjective" refers to the optimization of a system operating involving more than one objective function to be optimized simultaneously, such as a metagrating operating over a range of incident beam angles, and "categorical" refers to design variables that have two or more categories without intrinsic ordering, such as multiple material types. In this study, we show that GLOnets can be configured as a multiobjective, categorical global optimizer, and we adapt GLOnets to optimize thin-film stacks to demonstrate the capabilities of our algorithms. Thin-film stacks are an ideal model system for multiple reasons. First, the design problem is multiobjective as devices are typically configured for a range of incident wavelengths, angles, and polarizations. Second, the design problem is categorical as individual layer materials are chosen from a library of materials. Third, thin-film stacks are a well-established technology, and there are a number of pre-existing studies that enable proper benchmarking of algorithm performance [21–23].

Thin-film stacks have been widely used in many optical systems including passive radiative coolers [24], efficient solar cells [25, 26], broadband spectral filtering [27, 28], thermal emitters [29], and spatial multiplexing filters [30]. The materials and thicknesses of thin-film layers have to be carefully optimized to achieve the desired transmission and reflection proprieties across a broad wavelength and angular bandwidth. Design methods based on physical intuition result in limited performance, and they are generally difficult to scale to aperiodic thin-film stacks comprising many layers. To address these limitations, various global optimization approaches have been explored, including the Monte Carlo approach [31], particle swarm optimization [32], needle optimization [33–35], and the memetic algorithm [21]. These methods are all derivative-free global optimization algorithms that search the design space through the evaluation of a batch of samples without any gradient calculations, limiting their ability to reliably solve for the global optimum.

2 Method

We consider the design of N-layer thin-film stacks each comprising an isotropic material specified from a material library (Figure 1). The refractive indices of the total stack are denoted as a vector $\mathbf{n}(\lambda) = (n_1(\lambda), n_2(\lambda), \cdots, n_N(\lambda))^T$, where each index term is a function of wavelength to account for dispersion, and the values can be real or complex valued without loss of generality. The thin-film stack thicknesses are $\mathbf{t} = (t_1, t_2, \cdots, t_N)^T$. The material library consists of M material types, and their refractive indices are represented as $\{m_1(\lambda), m_2(\lambda), \cdots, m_M(\lambda)\}$.

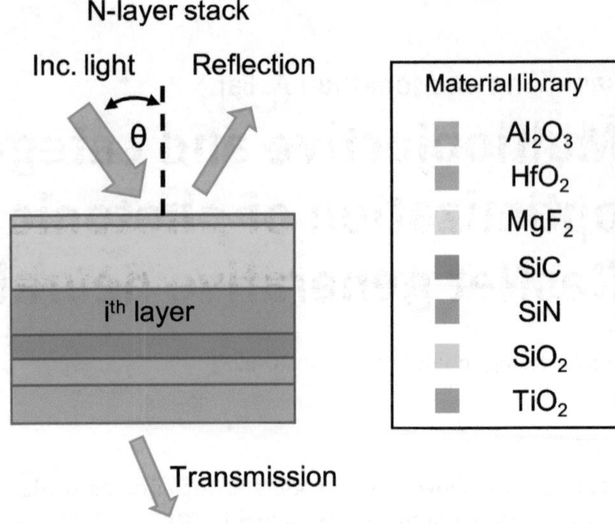

Figure 1: Schematic of the N-layer thin-film stack system. The refractive index and thickness of each layer are optimized to produce a desired reflection profile, and the composition of each layer is constrained to index values specified in a material library.

The optimization problem is posed as finding the proper $\mathbf{n}$ and $\mathbf{t}$ that produces the desired reflection characteristics over a given wavelength bandwidth, incident angle range, and incident polarization:

$$\{\mathbf{n}^*, \mathbf{t}^*\} = \arg\min_{\{\mathbf{n},\mathbf{t}\}} \sum_{\lambda,\theta,\text{pol}} (\mathcal{R}(\mathbf{n}, \mathbf{t}|\, \lambda, \theta, \text{pol}) - \mathcal{R}^*(\lambda, \theta, \text{pol}))^2 \quad (1)$$

The desired reflection spectrum is denoted as $\mathcal{R}^*(\lambda, \theta, \text{pol})$, and $\{\mathbf{n}^*, \mathbf{t}^*\}$ are the corresponding global optimal refractive indices and thicknesses. This optimization problem can be readily cast as the minimization of the objective function: $O(\mathbf{n}, \mathbf{t}) = \sum_{\lambda,\theta,\text{pol}} (\mathcal{R}(\mathbf{n}, \mathbf{t}|\, \lambda, \theta, \text{pol}) - \mathcal{R}^*(\lambda, \theta, \text{pol}))^2$. $\mathbf{n}$ are categorical variables because the index values are chosen from a material database, while $\mathbf{t}$ can span a continuous set of values and is a continuous variable.

2.1 Transfer matrix method solver

A principle requirement of any gradient-based optimizer is a method to calculate local gradients. For thin-film stacks, these gradients indicate how perturbations to the refractive indices and thicknesses of the device can best reduce the objective function. In prior implementations of GLOnets, local gradients were calculated using the adjoint variable method, in which forward and adjoint simulations are calculated using a conventional electromagnetic solver [19, 20].

While the adjoint variables method provides a general formalism to calculating local gradients using any conventional solver, we pursue an alternative approach based on the transfer matrix method (TMM), which is a fully analytic and high-speed solver for thin-film systems. In particular, we program a TMM solver within the automatic differentiation framework in PyTorch [36], which allows gradients to be directly calculated using the chain rule. Automatic differentiation is the basis for calculating gradients during backpropagation in neural network training, and it generally applies to any algorithm that can be described by a differentiable computational graph. Recently, it was implemented in finite-difference time domain (FDTD) and finite-difference frequency domain (FDFD) simulators [37, 38]. Compared to generalized differentiable electromagnetic solvers, such as these FDTD and FDFD implementations, our analytic TMM-based algorithms are faster without loss of accuracy because the thin films are described as layers instead of voxels.

2.2 Res-GLOnet algorithm

A schematic of GLOnets configured for our thin-film stack system is outlined in Figure 2a. We term this GLOnet variant as Res-GLOnets because the generator has a residual network architecture that includes skip connections between layers (blue box inset), which will be discussed in a later section. First, a generative neural network G with trainable weights ϕ produces a distribution of thin-film stack configurations. The input to the generator is a uniformly distributed random vector $\mathbf{z} \sim U(0,1)$, so that the generator can be regarded as a function that maps the uniform distribution to a complex distribution of thin-film stack configurations, $G_\phi : U(0,1) \rightarrow P_\phi(\mathbf{n}, \mathbf{t})$. Different samplings of the input random variable $\mathbf{z}^{(k)}$ map onto different device refractive index and thickness configurations within $P_\phi(\mathbf{n}, \mathbf{t})$, denoted as $\{\mathbf{n}^{(k)}, \mathbf{t}^{(k)}\} = G_\phi(\mathbf{z}^{(k)})$. The generated $\mathbf{n}$ from the network do not take categorical values from the material library but are relaxed to be continuous variables, to stabilize the optimization process. These $\mathbf{n}$ are further processed using a probability matrix to enforce the categorical value constraint, which is discussed in the next section. After processing, the reflection spectra of the generated devices, $\mathcal{R}(\mathbf{n}^{(k)}, \mathbf{t}^{(k)} | \lambda, \theta, \text{pol})$, are calculated using the TMM solver.

The optimization objective, or the loss function, for GLOnet is defined as:

$$L = \mathbb{E}\left[\exp\left(-\frac{O(\mathbf{n},\mathbf{t})}{\sigma}\right)\right] \tag{2}$$

$$= \int \exp\left(-\frac{O(\mathbf{n},\mathbf{t})}{\sigma}\right) P_\phi(\mathbf{n},\mathbf{t})\, d\mathbf{n}d\mathbf{t} \tag{3}$$

$$= \int \exp\left(-\frac{O(G_\phi(\mathbf{z}))}{\sigma}\right) P(\mathbf{z})\, d\mathbf{z} \tag{4}$$

$$\approx \sum_{k=1}^{K} \exp\left(-\frac{O(\mathbf{n}^{(k)},\mathbf{t}^{(k)})}{\sigma}\right) \tag{5}$$

σ is a hyperparameter. These equations follow the derivation of the GLOnet formalism described in the study by Jiang and Fan [20]. To train the generative network and update its weights in a manner that improves the mapping of $\mathbf{z}$ to devices, the gradient of the loss function with respect to the neuron weights, $\nabla_\phi L$, is calculated by backpropagation.

A schematic of the evolution of the generative network over the course of network training is outlined in Figure 2b. Initially, the generator has no knowledge about the design space and outputs a broad distribution of devices spanning the full design space. Over the course of network training, the distribution of generated devices narrows and gets biased toward design space regions that feature relatively small objective function values. Upon the completion of network training, the distribution of generated thin-film stack configurations converges to a narrow distribution centered around the global optimum.

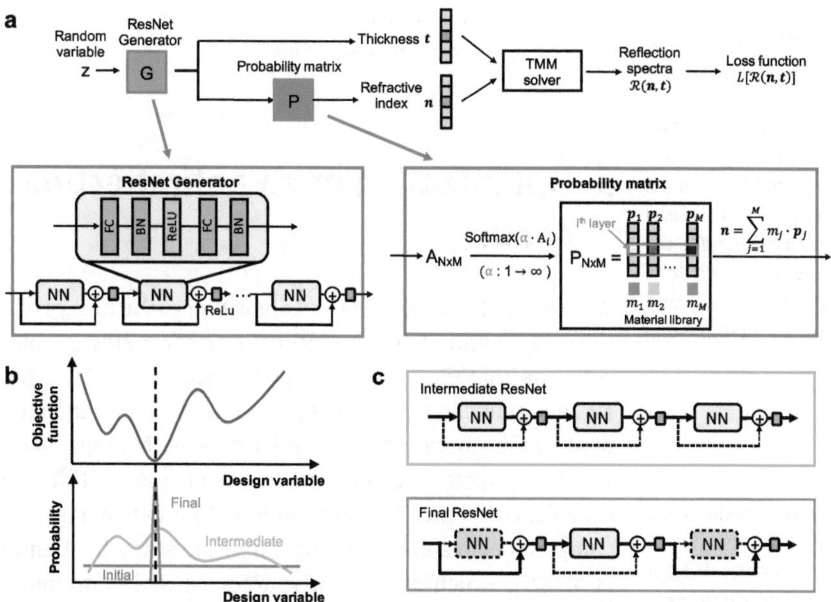

Figure 2: Thin-film global optimization with the Res-GLOnet. (a) Schematic of the Res-GLOnet. A ResNet generator maps a uniformly distributed random variable to a distribution of devices, which are then evaluated using a transfer matrix method solver and used to evaluate the loss function. A probability matrix pushes the continuous generated device indices **n** to discrete values. (b) Evolution of the generated device distribution over the course of network training. The network initially samples the full design space and converges to a narrow distribution centered around the global minimum of the objective function. (c) During training, the network operates as a deep architecture with little impact from the skip connections (Intermediate ResNet). Near training completion, the network evolves to a shallow architecture with large impact from the skip connections (Final ResNet). Bold and dashed lines indicate large and small contributions to the network architecture, respectively. TMM, transfer matrix method; GLOnet, global optimization network.

2.3 Enforcing categorical constraints

To update the weights in the generative network during backpropagation, the chain rule is applied to the entire computation graph of the Res-GLOnet algorithm. One required step is the calculation of the gradient of the reflection spectrum with respect to the refractive indices, $\frac{dR}{d\mathbf{n}}$. If the refractive indices of thin-film stacks outputted by the generator are directly treated as categorical variables, $\mathbf{n}$ is not a continuous function and the gradient term above cannot be calculated.

To overcome this difficulty, we propose a reparameterization scheme in which the generated $\mathbf{n}$ are relaxed to take continuous values and are then processed in a manner that supports convergence to categorical variable values. The concept is outlined in the green box inset in Figure 2a. The network first maps the random vector $\mathbf{z}$ onto an N-by-M matrix A. These values can vary continuously and take any real number value. A softmax function is then applied to each row of A to generate a probability matrix P:

$$P_{ij} = \frac{\exp\left(\alpha \cdot A_{ij}\right)}{\sum_{j=1}^{M} \exp\left(\alpha \cdot A_{ij}\right)} \tag{6}$$

The ith row of matrix P is a $1 \times M$ vector and represents the probability distribution that the ith thin-film layer takes on a particular material choice within the material library. We use the softmax function because it produces a properly normalized probability distribution and is commonly used in other related tasks, such as classification tasks [39]. The expected refractive index of the ith layer given by this distribution, calculated as $n_i(\lambda) = \sum_{j=1}^{M} m_j(\lambda) \cdot P_{ij}$, is used to define the thin-film stack in subsequent TMM calculations in Res-GLOnet. All functions in this algorithm can be expanded into a differentiable computational graph, meaning that the loss function gradient with respect to the refractive index is able to backpropagate through the probability matrix P and to the network weights ϕ.

α is a hyperparameter that tunes the sharpness of the softmax function. Initially, α is set to be one, and the expected refractive index of the ith layer has contributions from many different materials in the material library. Over the course of network training, α is linearly increased as a function of the training iteration number until the probability distribution of the ith thin-film layer is effectively a delta function that has converged to a single material. These concepts build on a similar scheme previously used for image sensor multiplexing design [40].

2.4 ResNet generator

Our optimization problem involves searching within a highly complex, nonconvex design space and is made particularly challenging by device requirements spanning a wide range of incident wavelengths and angles. In the early and intermediate stages of network training, a deep neural network is required to properly generate a complex distribution of devices spanning large regions of the design space. However, toward the latter stages of network training, the distribution of the generated devices should ideally converge to a simple and narrow distribution centered around the global optimum, which is more ideally modeled using a shallow network. GLOnet schemes that train using a fixed network architecture do not have the flexibility to capture these trends: deep architectures have general difficulty in training owing to the well-known vanishing gradient problem, while shallow architectures have the issue of underfitting the design space and are ineffective during the early and intermediate stages of network training [41].

To address these issues, we utilize deep residual networks for the generator architecture, which reformulates our algorithm as Res-GLOnets. Residual networks [41] were developed in the computer vision community to stably process images in very deep networks and overcome the vanishing gradient problem, with the insight that the use of skip connections can enable the depth of the network to be effectively and implicitly tuned over the course of training. A schematic of our Res-GLOnet architecture is shown in the blue box inset in Figure 2a and comprises a series of 16 residual blocks. Each block contains a fully connected layer, a batch normalization layer, and a leaky ReLU nonlinear activation layer. The input x_{in} and output x_{out} of each residual block have the same dimension, and the output of each block contains contributions from both the residual block $f(x_{\text{in}})$ and skip connection: $x_{\text{out}} = f(x_{\text{in}}) + x_{\text{in}}$.

The evolution of the Res-GLOnet architecture over the course of network training is sketched in Figure 2c. When the network is training in the early and intermediate stages of the optimization process, each residual block outputs terms that are typically larger than the skip connection contributions. As a result, the network architecture functions as a deep network, which is required during these stages of Res-GLOnets training. As network training progresses, some of the residual blocks start to output relatively small contributions and $x_{\text{out}} \approx x_{\text{in}}$ owing to the emergence of vanishing gradients. The network architecture now functions as a shallow architecture, having effectively skipped over some of the residual blocks. Note that the increasing contribution of skip connections and reduction of network complexity is not explicitly and externally controlled but evolves over the course of network training, as the loss function guides the network output distribution to a relatively simple form.

3 Optimization of an antireflection coating

We first apply our Res-GLOnet algorithm to the design of a three-layer antireflection (AR) coating for a silicon solar cell. The thin-film AR stack is designed to minimize the average reflection at an air-silicon interface over the incident angle range [0°, 60°] and wavelength range [400, 1100] nanometer for both transverse magnetic (TM) and transverse electric (TE) polarization. As a benchmark, we compare our results with those from the study by Azunre et al. [22], which provides a guaranteed global optimum solution using a parallel branch-and-bound method. The algorithm requires extensive searching through the full design space and utilizes over 19 days of CPU computation to solve for the global optimum. To be consistent with the study by Azunre et al. [22], the refractive indices of the layers in our design implementation do not take discrete categorical values from a material library but are dispersionless and continuously varying in the interval [1.09,

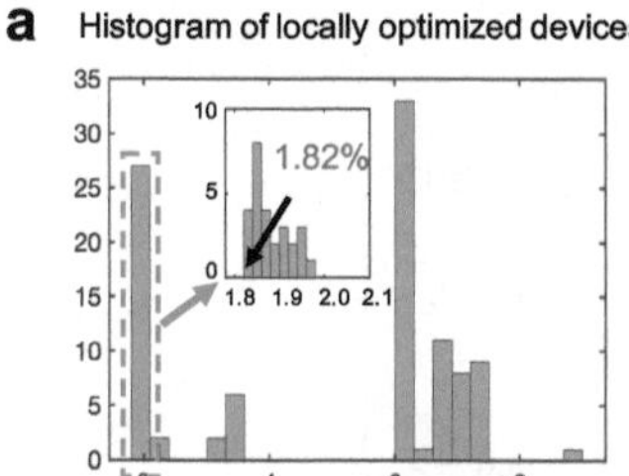

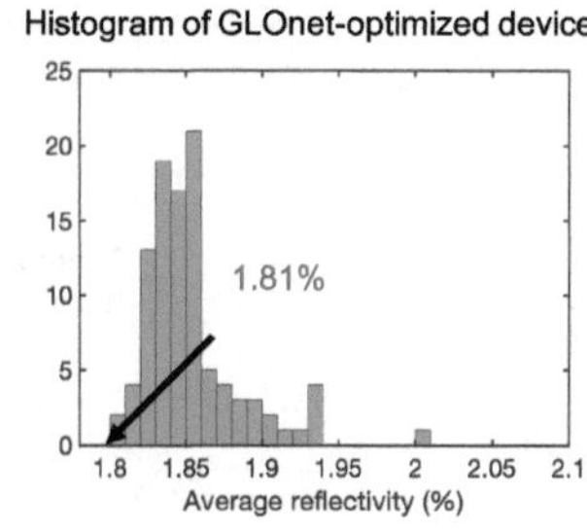

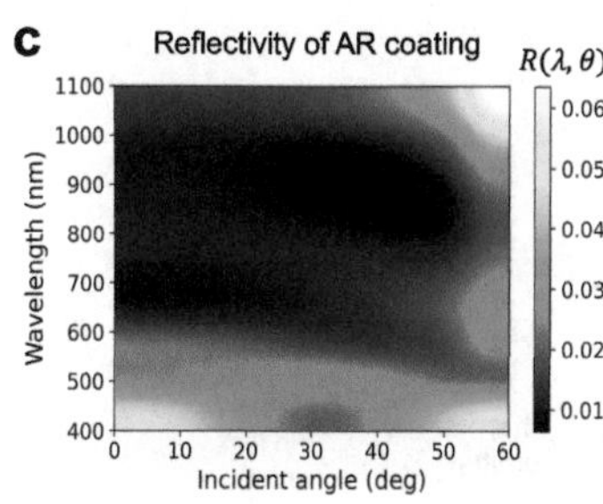

Figure 3: Optimization of a third-layer thin film antireflection (AR) coating on silicon. (a) Histogram of the average reflectivity from 100 AR coatings designed using local gradient-based optimization. The best device has an average reflectivity of 1.82%. (b) Histogram of the average reflectivity from 100 AR coatings designed using a single Res-GLOnet. The best device has an average reflectivity of 1.81%. (c) Contour plot of reflectivity from the best Res-GLOnet–designed AR coating in (b) as a function of the incidence angle and wavelength, averaged for both TE- and TM-polarized waves. GLOnet, global optimization network.

2.60]. The thicknesses of each layer are also continuous variables within the interval [5, 200] nm.

To accommodate the continuous variable nature of the refractive index values in this problem, we modify our categorical optimization scheme by setting the hyperparameter $\alpha = 1$ as a constant and specifying the material library to contain only two materials with constant refractive indices $\{m^L, m^U\}$. $m^L = 1.09$ is the lower bound of the refractive index, while $m^U = 2.60$ is the upper bound. The constraint on thickness can be satisfied by a transformation: $\mathbf{t} = t^L + \text{Sigmoid}(\tilde{\mathbf{t}}) \cdot (t^U - t^L)$. Here, the thickness directly outputted by the generator, $\tilde{\mathbf{t}}$, is normalized to [0, 1] and then linearly transformed to the interval $[t^L, t^U]$, where $t^L = 5$ and $t^U = 200$ are the lower and upper thickness bound, respectively.

As a reference, we first optimize devices using local gradient-based optimization, by replacing the ResNet generator in our Res-GLOnet algorithm with an individual device layout. The optimizations are performed with 100 different devices, initialized using random thickness and refractive index values within the limits of [1.09, 2.60] and [5, 200] nm, respectively. Each optimization is performed over 200 iterations, so that a total of 20,000 sets of calculations are performed for the entire set of optimizations. A histogram of the results (Figure 3a) shows that the optimized devices have average reflectivities that span a wide range of values, from approximately 2 to 10%, demonstrating the highly nonconvex nature of the design space. Average reflectivity is calculated as the reflectivity averaged over the wavelengths, incident angles, and polarizations covered in the design specifications. A fraction of devices are near the global optimum, and the best device has an efficiency of 1.82%.

A histogram of devices sampled from a single trained Res-GLOnet is summarized in Figure 3a. A total of 200 iterations is used together with a batch size of 20 devices, so that a total of 4000 sets of calculations are performed. The total time that Res-GLOnet requires for training is 7 s with a single GPU. All of the devices sampled from the Res-GLOnet are near the global optimum, showing the ability for the generative network to produce a narrow distribution of devices centered at the global optimum. The best device has an efficiency of 1.81%, and its reflectivity for differing incident wavelengths and angles is plotted in Figure 3c. The design of this best device is summarized in Table 1 and is consistent with the result reported in the study by Azunre et al. [22].

4 Optimization of the incandescent light bulb filter

To explore the applicability of Res-GLOnets to more complex problems, we apply our algorithm to optimize incandescent light bulb filters that transmit visible light and reflect infrared light (Figure 4a). In this scheme, the emitter filament heats to a relatively higher temperature using recycled infrared light, thereby enhancing the emission efficiency in the visible range [29].

Table 1: Optimized structure for the AR coating of Si.

Layer #	Refractive index	Thickness (nm)
	Air	Superstrate
1	2.60	54.2
2	1.68	93.6
3	1.17	149.2
	Si	Substrate

AR, antireflection.

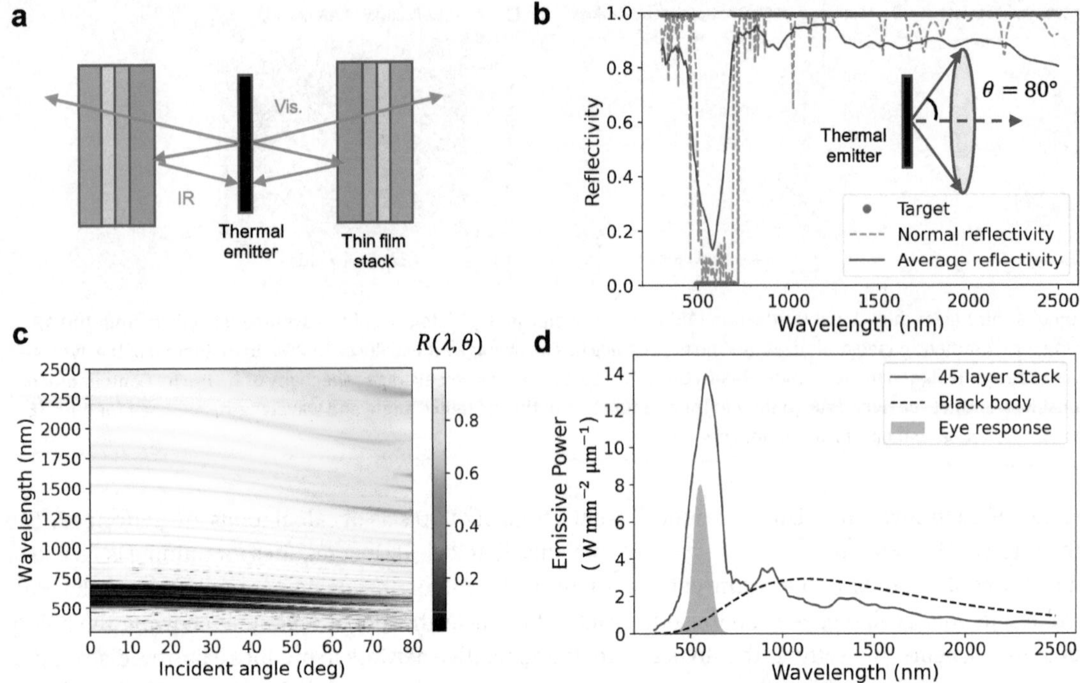

Figure 4: Thin-film stacks for incandescent light bulb filtering. (a) Schematic of an incandescent light bulb filter that transmits visible light and reflects infrared and ultraviolet light. (b) Reflection spectra of a 45-layer Res-GLOnet–optimized device, for normally incidence waves and waves averaged over a large incident solid angle, shown in the inset. (c) Reflection spectra of the device featured in (b) as a function of the incident angle, averaged for TE- and TM-polarized incident waves. (d) Emissive power of a blackbody incandescent source and an equivalent source sandwiched by the filter featured in (b). Also shown is the spectral response of the eye. GLOnet, global topology optimization network.

A range of design methods have been previously applied to this problem. In the initial demonstration of the concept, binary thin-film stacks were designed using a combination of local gradient-based optimization, used to tune the thickness of each layer, and needle optimization, which determined whether an existing layer should be removed or a new layer should be introduced [29]. A memetic algorithm was subsequently applied in which crossover, mutation, and downselecting operations were iteratively performed on a population of thin-film stacks to evolve the quality of devices [21]. Gradient-based local optimizations of device thicknesses were also periodically performed to refine the structures and accelerate algorithm convergence. In a third study, reinforcement learning (RL) was used in which an autoregressive recurrent neural network generated thin-film stacks layer by layer as a sequence [23]. Unlike the GLOnet generator, the probability distribution of the thin-film stack was explicitly outputted by the autoregressive generator. The distribution evolved by optimizing a reward function, and the gradient of the reward function with respect to the neural network weights was calculated using proximal policy optimization.

In our demonstration, we benchmark Res-GLOnets with the memetic and RL studies, which consider a material library comprising seven dielectric material types: Al_2O_3, HfO_2, MgF_2, SiC, SiN, SiO_2, and TiO_2. The superstrate and substrate are both set to be air. The complete wavelength range under consideration is [300, 2500] nm, and the target reflection is set to be 0% for the wavelength range [500, 700] nm and 100% for all other wavelengths. The incident angles span [0, 72] degrees, and both TE and TM polarizations are considered.

We train a Res-GLOnet comprising 16 residue blocks for 1000 iterations with a batch size of 1000. The network is optimized using gradient decent with the momentum algorithm ADAM [42], and a learning rate of 1×10^{-3} is used. The broadband reflection characteristics of a 45-layer device show that the device operates with nearly ideal transmission in the [500, 700] nm interval and nearly ideal reflection at ultraviolet and near-infrared wavelengths, for both normal incidence and for incidence angles averaged over all solid angles within [0, 80] degrees (Figures 4b and 4c). The emission intensity spectrum of the light bulb with and without the thin-film filter is shown in Figure 4d. The input power is fixed at 100 W, and the surface area of the emitter is 20 mm^2.

To evaluate the enhancement of visible light emission due to the filter, we compute the emissivity enhancement factor, χ, as a function of the number of thin-film layers:

$$\chi = \frac{\int_0^\infty E_{\text{emitter+stack}}(P_0, \lambda) V(\lambda) d\lambda}{\int_0^\infty E_{\text{emitter}}(P_0, \lambda) V(\lambda) d\lambda} \tag{7}$$

$E_{\text{emitter+stack}}(P_0, \lambda)$ and $E_{\text{emitter}}(P_0, \lambda)$ are the intensity emission spectrum given the input power P_0. $V(\lambda)$ is the eye's sensitivity spectrum and is shown as the shaded region in Figure 4d. The view factor is the proportion of emitted light from the light bulb filament that can reach the light bulb filter. We use the view factor of 0.95 as was the case for the memetic study [21]. For a 45-layer device, the Res-GLOnet–optimized device achieved a χ of 17.2, and devices with as few as 30 layers still achieved a χ above 15 (Figure 5). The ability to realize high-performance devices with relatively few layers is practically important from a manufacturing and cost point of view. The 45-layer memetic algorithm and RL-optimized device have χ values of 14.8 and 16.6, respectively. We also benchmark Res-GLOnet with GLOnet based on a fixed architecture of four fully connected layers (FC-GLOnet). The benchmark, also plotted in Figure 5, shows that Res-GLOnet performs better in searching for proper devices in this nonconvex optimization problem, particularly for systems with larger numbers of thin films. The points in the plot each corresponds to the results of a single GLOnet run. In terms of computational cost, the memetic algorithm uses 600K simulations (a population size of 3000 and 200 iterations), the RL algorithm uses 30M simulations (a batch size of 3000 and 10,000 iterations), and GLOnets uses 500K simulations (a batch size of 500 and 1000 iterations). As such, GLOnets is demonstrated to be a computationally efficient global optimization algorithm for this problem.

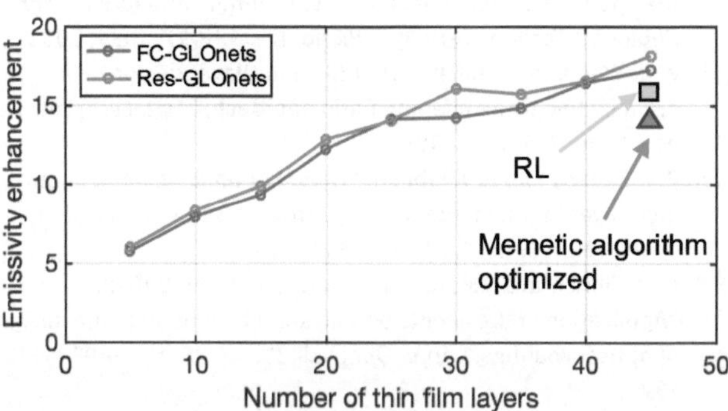

Figure 5: Plot of emissivity enhancement as a function of the number of thin-film layers, for devices optimized using Res-GLOnets and FC-GLOnets. Reference points are also plotted for devices designed using the reinforcement learning (RL) [23] and memetic [21] algorithm. GLOnet, global optimization network.

5 Conclusion

In summary, we show that Res-GLOnets are effective and efficient global optimizers for the multiobjective, categorical design of thin-film stacks. Categorical optimization is performed through the use of a probability matrix, which is fully differentiable and compatible with our neural network training framework. The incorporation of skip connections in our generative neural network helps it evolve from a deep to shallow architecture, which fits with our training objective and improves our search for the global optimum. Benchmarks of our algorithm with a known AR coating and incandescent light filter systems indicate that the Res-GLOnet is effective at searching for global optima, is computationally efficient, and outperforms a number of alternative design algorithms.

We anticipate that concepts developed within Res-GLOnets, particularly those in categorical optimization, can directly apply to the design of other photonics systems, such as lens design where the material type is selected from a material database. We also expect that the implementation of application-specific electromagnetic solvers, in conjunction with automatic differentiation packages, will serve as a foundational concept for many high-speed optimization algorithms beyond those for thin-film stacks. Generalizing the GLOnet algorithm to 3D photonic structures is challenging owing to the requirement of computational expensive simulations. We envision that this roadblock can be overcome by using neural networks as fast surrogate solvers, for which much progress has been made [43, 44]. Looking ahead, we see opportunities for Res-GLOnets to apply to other fields in the physical science, ranging from material science and chemistry to mechanical engineering, where devices and systems are designed using combinations of discrete material types.

Acknowledgments: The simulations were performed in the Sherlock computing cluster at Stanford University.

Author contribution: All the authors have accepted responsibility for the entire content of this submitted manuscript and approved submission.

Research funding: This work was supported by ARPA-E with Agreement Number DE-AR0001212, ONR with Agreement Number N00014-20-1-2105, and the Packard Foundation with Agreement Number 2016-65132

Conflict of interest statement: The authors declare no conflicts of interest regarding this article.

References

[1] S. Molesky, Z. Lin, A. Y. Piggott, W. Jin, J. Vucković, and A. W. Rodriguez, "Inverse design in nanophotonics," *Nat. Photonics*, vol. 12, no. 11, pp. 659–670, 2018.

[2] S. D. Campbell, D. Sell, R. P. Jenkins, E. B. Whiting, J. A. Fan, and D. H. Werner, "Review of numerical optimization techniques for meta-device design," *Opt. Mater. Express*, vol. 9, no. 4, pp. 1842–1863, 2019.

[3] J. A. Fan, "Freeform metasurface design based on topology optimization," *MRS Bull.*, vol. 45, no. 3, pp. 196–201, 2020.

[4] J. Jiang, M. Chen, and J. A. Fan. Deep neural networks for the evaluation and design of photonic devices," arXiv preprint arXiv: 2007.00084, 2020.

[5] K. Yao, R. Unni, and Y. Zheng, "Intelligent nanophotonics: merging photonics and artificial intelligence at the nanoscale," *Nanophotonics*, vol. 8, no. 3, pp. 339–366, 2019.

[6] S. So, T. Badloe, J. Noh, J. Rho, and J. Bravo-Abad, "Deep learning enabled inverse design in nanophotonics," *Nanophotonics*, vol. 9, no. 5, pp. 1041–1057, 2020.

[7] J. Peurifoy, Y. Shen, L. Jing, et al., "Nanophotonic particle simulation and inverse design using artificial neural networks," *Sci. Adv.*, vol. 4, no. 6, p. eaar4206, 2018.

[8] D. Liu, Y. Tan, E. Khoram, and Z. Yu, "Training deep neural networks for the inverse design of nanophotonic structures," *ACS Photonics*, vol. 5, no. 4, pp. 1365–1369, 2018.

[9] W. Ma, F. Cheng, Y. Xu, Q. Wen, and Y. Liu, "Probabilistic representation and inverse design of metamaterials based on a deep generative model with semi-supervised learning strategy," *Adv. Mater.*, vol. 31, no. 35, p. 1901111, 2019.

[10] J. Jiang, D. Sell, S. Hoyer, J. Hickey, J. Yang, and J. A. Fan, "Free-form diffractive metagrating design based on generative adversarial networks," *ACS Nano*, vol. 13, no. 8, pp. 8872–8878, 2019.

[11] Z. Liu, D. Zhu, S. P. Rodrigues, K.-T. Lee, and W. Cai, "Generative model for the inverse design of metasurfaces," *Nano Lett.*, vol. 18, no. 10, pp. 6570–6576, 2018.

[12] F. Wen, J. Jiang, and J. A. Fan, "Robust freeform metasurface design based on progressively growing generative networks," *ACS Photonics*, vol. 7, no. 8, pp. 2098–2104, 2020.

[13] T. W. Hughes, M. Minkov, I. A. D. Williamson, and S. Fan, "Adjoint method and inverse design for nonlinear nanophotonic devices," *ACS Photonics*, vol. 5, no. 12, pp. 4781–4787, 2018.

[14] D. Sell, J. Yang, S. Doshay, R. Yang, and J. A. Fan, "Large-angle, multifunctional metagratings based on freeform multimode geometries," *Nano Lett.*, vol. 17, no. 6, pp. 3752–3757, 2017.

[15] A. Y. Piggott, J. Lu, K. G. Lagoudakis, J. Petykiewicz, T. M. Babinec, and J. Vučković, "Inverse design and demonstration of a compact and broadband on-chip wavelength demultiplexer," *Nat. Photonics*, vol. 9, no. 6, pp. 374–377, 2015.

[16] T. Phan, D. Sell, E. W. Wang, et al., "High-efficiency, large-area, topology-optimized metasurfaces," *Light Sci. Appl.*, vol. 8, no. 1, pp. 1–9, 2019.

[17] J. Yang, D. Sell, and J. A. Fan, "Freeform metagratings based on complex light scattering dynamics for extreme, high efficiency beam steering," *Ann. Phys.*, vol. 530, no. 1, p. 1700302, 2018.

[18] J. Lu and J. Vučković, "Objective-first design of high-efficiency, small-footprint couplers between arbitrary nanophotonic waveguide modes," *Optic Express*, vol. 20, no. 7, pp. 7221–7236, 2012.

[19] J. Jiang and J. A. Fan, "Global optimization of dielectric metasurfaces using a physics-driven neural network," *Nano Lett.*, vol. 19, no. 8, pp. 5366–5372, 2019.

[20] J. Jiang and J. A. Fan, "Simulator-based training of generative neural networks for the inverse design of metasurfaces," *Nanophotonics*, vol. 9, no. 5, pp. 1059–1069, 2019.

[21] Y. Shi, W. Li, A. Raman, and S. Fan, "Optimization of multilayer optical films with a memetic algorithm and mixed integer programming," *ACS Photonics*, vol. 5, no. 3, pp. 684–691, 2017.

[22] P. Azunre, J. Jean, C. Rotschild, V. Bulovic, S. G. Johnson, and M. A. Baldo, "Guaranteed global optimization of thin-film optical systems," *New J. Phys.*, vol. 21, no. 7, p. 073050, 2019.

[23] H. Wang, Z. Zheng, C. Ji, and L. J. Guo. Automated optical multi-layer design via deep reinforcement learning," arXiv preprint arXiv:2006.11940, 2020.

[24] A. P. Raman, M. Abou Anoma, L. Zhu, E. Rephaeli, and S. Fan, "Passive radiative cooling below ambient air temperature under direct sunlight," *Nature*, vol. 515, no. 7528, pp. 540–544, 2014.

[25] W. Li, Y. Shi, K. Chen, L. Zhu, and S. Fan, "A comprehensive photonic approach for solar cell cooling," *ACS Photonics*, vol. 4, no. 4, pp. 774–782, 2017.

[26] A. Lenert, D. M. Bierman, Y. Nam, et al., "A nanophotonic solar thermophotovoltaic device," *Nat. Nanotechnol.*, vol. 9, no. 2, pp. 126–130, 2014.

[27] Y. Shen, D. Ye, I. Celanovic, S. G. Johnson, J. D. Joannopoulos, and M. Soljačić, "Optical broadband angular selectivity," *Science*, vol. 343, no. 6178, pp. 1499–1501, 2014.

[28] F. Cao, Y. Huang, L. Tang, et al., "Toward a high-efficient utilization of solar radiation by quad-band solar spectral splitting," *Adv. Mater.*, vol. 28, no. 48, pp. 10659–10663, 2016.

[29] O. Ilic, P. Bermel, G. Chen, J. D. Joannopoulos, I. Celanovic, and M. Soljačić, "Tailoring high-temperature radiation and the resurrection of the incandescent source," *Nat. Nanotechnol.*, vol. 11, no. 4, pp. 320–324, 2016.

[30] M. Gerken and D. A. B. Miller, "Wavelength demultiplexer using the spatial dispersion of multilayer thin-film structures," *IEEE Photonics Technol. Lett.*, vol. 15, no. 8, pp. 1097–1099, 2003.

[31] W. J. Wild and H. Buhay, "Thin-film multilayer design optimization using a Monte Carlo approach," *Opt. Lett.*, vol. 11, no. 11, pp. 745–747, 1986.

[32] R. I. Rabady and A. Ababneh, "Global optimal design of optical multilayer thin-film filters using particle swarm optimization," *Optik*, vol. 125, no. 1, pp. 548–553, 2014.

[33] A. V. Tikhonravov, M. K. Trubetskov, and G. W. DeBell, "Application of the needle optimization technique to the design of optical coatings," *Appl. Opt.*, vol. 35, no. 28, pp. 5493–5508, 1996.

[34] V. Pervak, A. V. Tikhonravov, M. K. Trubetskov, S. Naumov, F. Krausz, and A. Apolonski, "1.5-octave chirped mirror for pulse compression down to sub-3 fs," *Appl. Phys. B*, vol. 87, no. 1, pp. 5–12, 2007.

[35] A. V. Tikhonravov, M. K. Trubetskov, and G. W. DeBell, "Optical coating design approaches based on the needle

optimization technique," *Appl. Opt.*, vol. 46, no. 5, pp. 704–710, 2007.
[36] A. Paszke, S. Gross, F. Massa, et al., "PyTorch: An imperative style, high-performance deep learning library," in *Advances in Neural Information Processing Systems*, Vancouver, Canada, Curran Associates Inc, 2019, pp. 8026–8037.
[37] T. W. Hughes, I. A. D. Williamson, M. Minkov, and S. Fan, "Forward-mode differentiation of Maxwell's equations," *ACS Photonics*, vol. 6, no. 11, pp. 3010–3016, 2019.
[38] M. Minkov, I. A. D. Williamson, L. C. Andreani, et al., "Inverse design of photonic crystals through automatic differentiation," *ACS Photonics*, vol. 7, no. 7, pp. 1729–1741, 2020.
[39] C. M. Bishop, *Pattern Recognition and Machine Learning*, New Delhi, India, Springer, 2006.
[40] A. Chakrabarti, "Learning sensor multiplexing design through back-propagation," in *Advances in Neural Information Processing Systems*, Barcelona, Spain, Curran Associates Inc, 2016, pp. 3081–3089.
[41] K. He, X. Zhang, S. Ren, and J. Sun, "Deep residual learning for image recognition," in *Proceedings of the IEEE Conference on Computer Vision and Pattern Recognition*, Las Vegas, NV, IEEE Computer Society, 2016, pp. 770–778.
[42] D. P. Kingma and J. Ba, "Adam: A method for stochastic optimization," arXiv preprint arXiv:1412.6980, 2015.
[43] R. Pestourie, Y. Mroueh, T. V. Nguyen, P. Das, and S. G. Johnson, "Active learning of deep surrogates for PDEs: Application to metasurface design," arXiv preprint arXiv:2008.12649, 2020.
[44] M. V. Zhelyeznyakov, S. L. Brunton, and A. Majumdar, "Deep learning to accelerate Maxwell's equations for inverse design of dielectric metasurfaces," arXiv preprint arXiv:2008.10632, 2020.

Supplementary Material: The online version of this article offers supplementary material (https://doi.org/10.1515/nanoph-2020-0407).

Zhaxylyk A. Kudyshev*, Alexander V. Kildishev, Vladimir M. Shalaev and Alexandra Boltasseva*

Machine learning–assisted global optimization of photonic devices

https://doi.org/10.1515/9783110710687-028

Abstract: Over the past decade, artificially engineered optical materials and nanostructured thin films have revolutionized the area of photonics by employing novel concepts of metamaterials and metasurfaces where spatially varying structures yield tailorable "by design" effective electromagnetic properties. The current state-of-the-art approach to designing and optimizing such structures relies heavily on simplistic, intuitive shapes for their unit cells or metaatoms. Such an approach cannot provide the global solution to a complex optimization problem where metaatom shape, in-plane geometry, out-of-plane architecture, and constituent materials have to be properly chosen to yield the maximum performance. In this work, we present a novel machine learning–assisted global optimization framework for photonic metadevice design. We demonstrate that using an adversarial autoencoder (AAE) coupled with a metaheuristic optimization framework significantly enhances the optimization search efficiency of the metadevice configurations with complex topologies. We showcase the concept of physics-driven compressed design space engineering that introduces advanced regularization into the compressed space of an AAE based on the optical responses of the devices. Beyond the significant advancement of the global optimization schemes, our approach can assist in gaining comprehensive design "intuition" by revealing the underlying physics of the optical performance of metadevices with complex topologies and material compositions.

Keywords: machine learning; metasurface; optimization; thermal emitter.

***Corresponding authors: Zhaxylyk A. Kudyshev and Alexandra Boltasseva,** School of Electrical and Computer Engineering, Birck Nanotechnology Center and Purdue Quantum Science and Engineering Institute, Purdue University, West Lafayette, IN, 47906, USA; and The Quantum Science Center (QSC), A National Quantum Information Science Research Center of the U.S. Department of Energy (DOE), Oak Ridge, USA, E-mail: zkudyshev@purdue.edu (Z.A. Kudyshev), aeb@purdue.edu (A. Boltasseva). https://orcid.org/0000-0002-6955-0890 (Z.A. Kudyshev)
Alexander V. Kildishev, School of Electrical and Computer Engineering, Birck Nanotechnology Center and Purdue Quantum Science and Engineering Institute, Purdue University, West Lafayette, IN, 47906, USA
Vladimir M. Shalaev, School of Electrical and Computer Engineering, Birck Nanotechnology Center and Purdue Quantum Science and Engineering Institute, Purdue University, West Lafayette, IN, 47906, USA; and The Quantum Science Center (QSC), A National Quantum Information Science Research Center of the U.S. Department of Energy (DOE), Oak Ridge, USA

1 Introduction

Multiconstrained optimization of metamaterials [1] and metasurfaces [2–5] requires intensive computational efforts. The main goal of such optimization is to determine the distribution of constituent materials within the computational domain, which assures the best performance of the metadevice while satisfying all the constraints of the problem. Recently, various gradient-based [6–10] and metaheuristic algorithms [11, 12] (evolutionary, swarm based) have been adapted to advance nanophotonic design problems. However, even the simplest realizations of these optimization frameworks depend heavily on computationally expensive three-dimensional (3D) full-wave direct electromagnetic solvers, thus making the proposed frameworks very time-consuming and inefficient.

Moreover, the computational costs of conventional optimization methods increase with the number of additional constraints, thus making these methods less practical for highly constrained problems. On the other hand, with the development of novel material platforms and advances in nanofabrication techniques, there is a growing interest in the multiconstrained optimization of such metastructures, which can be decisive in addressing critical problems in the fields of space exploration [13], quantum technology [14], energy [15], and communication [16]. There is a critical demand for efficient optimization frameworks capable of performing global optimization searches within highly dimensional parametric domains with complex optimization landscapes.

This article has previously been published in the journal Nanophotonics. Please cite as: Z. A. Kudyshev, A. V. Kildishev, V. M. Shalaev and A. Boltasseva "Machine learning–assisted global optimization of photonic devices" *Nanophotonics* 2021, 10. DOI: 10.1515/nanoph-2020-0376.

Due to its versatility and efficiency, machine learning (ML) algorithms have been successfully applied to different areas of photonics and optoelectronics. Various ML techniques have demonstrated their potential to address the bottlenecks of the conventional methods. For example, machine and deep learning models have been used in microscopy [17], quantum optics [18–20], and laser physics [21]. Recently, discriminative networks have been applied to various direct and inverse electrodynamics problems in nanophotonics [22–33]. The main advantage of the data-driven frameworks over the conventional electrodynamic simulation methods is the ability of the neural networks (NNs) to identify hidden correlations in the large data sets during the training phase and utilizing the retrieved "knowledge" to provide instantaneous solution searches, without costly computations [34, 35].

Along with the optimization of geometrical parameters and prediction of the optical response of metastructures with simplistic shapes, the advanced deep learning algorithms have been used to perform optimization of the metadevices with complex topologies. Thus, specific classes of generative networks, such as generative adversarial networks (GANs) [36, 37] and autoencoders [38–40], have been applied to nanoantenna design optimization. Recently, GANs and AAEs have been coupled with adjoint topology optimization (TO) technique for optimizing the diffractive dielectric gratings [41, 42], as well as thermal emitters [43]. It has previously been demonstrated that by coupling the AAE network with a conventional adjoint TO formalism, it is possible to get ~4900-time speedup in thermal emitter optimization compared with conventional TO [34].

Within this work, we extended the AAE-based optimization framework beyond the random sampling of the designs from the AAE network. Specifically, we have developed a methodology to perform the multiparametric global optimization (GO) directly within the compressed design space, via coupling the conditional AAE (c-AAE) network with a differential evolution (DE) optimizer. The proposed approach allows us to not only determine the antenna shape/topology but also optimize the geometrical parameters of the unit cell (e.g., the thickness of the dielectric spacer and the array periodicity). Moreover, we demonstrated that supervised training of the c-AAE network allows adding physics-driven regularization to compressed design space during the training phase, which in turn leads to better GO searches. To showcase the performance of the proposed AAE-based GO technique, we optimized the thermal emitter design with two different methods: (i) the c-AAE network coupled with a DE optimizer (c-AAE + DE) and (ii) DE optimization utilizing the compressed design space with physics-driven regularization (c-AAE + rDE).

Section 2 describes the main optimization problem under consideration, while Section 3 introduces the main data generation framework. Specifically, Section 3 describes the structure, training, and design generation process based on conditional AAEs, as well as rapid efficiency estimation process via pretrained conditional convolution NNs. Within Section 4, we focus on the GO scheme based on the developed AAE framework. Section 5 showcases the concept of physics-driven compressed space engineering via supervised training of the conditional AAE network, as well as demonstrates the GO within the regularized compressed design space. Section 6 concludes the work.

2 Optimization problem

Without loss of generality, we focus on the optimization of thermal emitters for thermophotovoltaics (TPVs) utilizing GaSb photovoltaic (PV) cells with a working band ranging from $\lambda_{\min}$ = 0.5 μm to $\lambda_{\max}$ = 1.7 μm (shaded area in Figure 1a). To maximize the generation of electric power from the TPV system, the emissivity of the thermal emitter should maximize in-band radiation (the red shaded area in Figure 1a) and minimize the out-of-band radiation (the blue shaded area in Figure 1a). Hence, the emissivity of the ideal thermal emitter is a step function with $\varepsilon(\lambda_{\min} \leq \lambda \leq \lambda_{\max}) = 1$ and zero elsewhere (depicted as the dashed blue contour in Figure 1a). Recently, gap-plasmon structures have been proposed as a viable solution to the thermal emission reshaping problem [44–47]. However, due to the simplistic, nonoptimal antenna designs, the efficiency of such thermal emitters has been limited.

Within this work, we consider a thermal emitter comprising the titanium nitride (TiN) back reflector [48], a silicon nitride (Si_3N_4) spacer, and a TiN plasmonic antenna in the top layer (Figure 1b). The main goal of the optimization is to determine the shape/topology of the top antenna, as well as the optimal configuration of the entire device, i.e., its 2D periodicity and spacer thickness that would drive the spectral emissivity to the step-like emissivity of the ideal emitter. Here, we do not focus on the details of the TO technique; instead, the paper is centered around the AAE-assisted global optimization framework. More details on the TO used for training set generation can be found in our prior work [43].

For assessing the performance of our designs, we define the efficiency of the thermal emitter as a product of in-band (eff^{in}) and out-of-band (eff^{out}) efficiencies, which is given as follows:

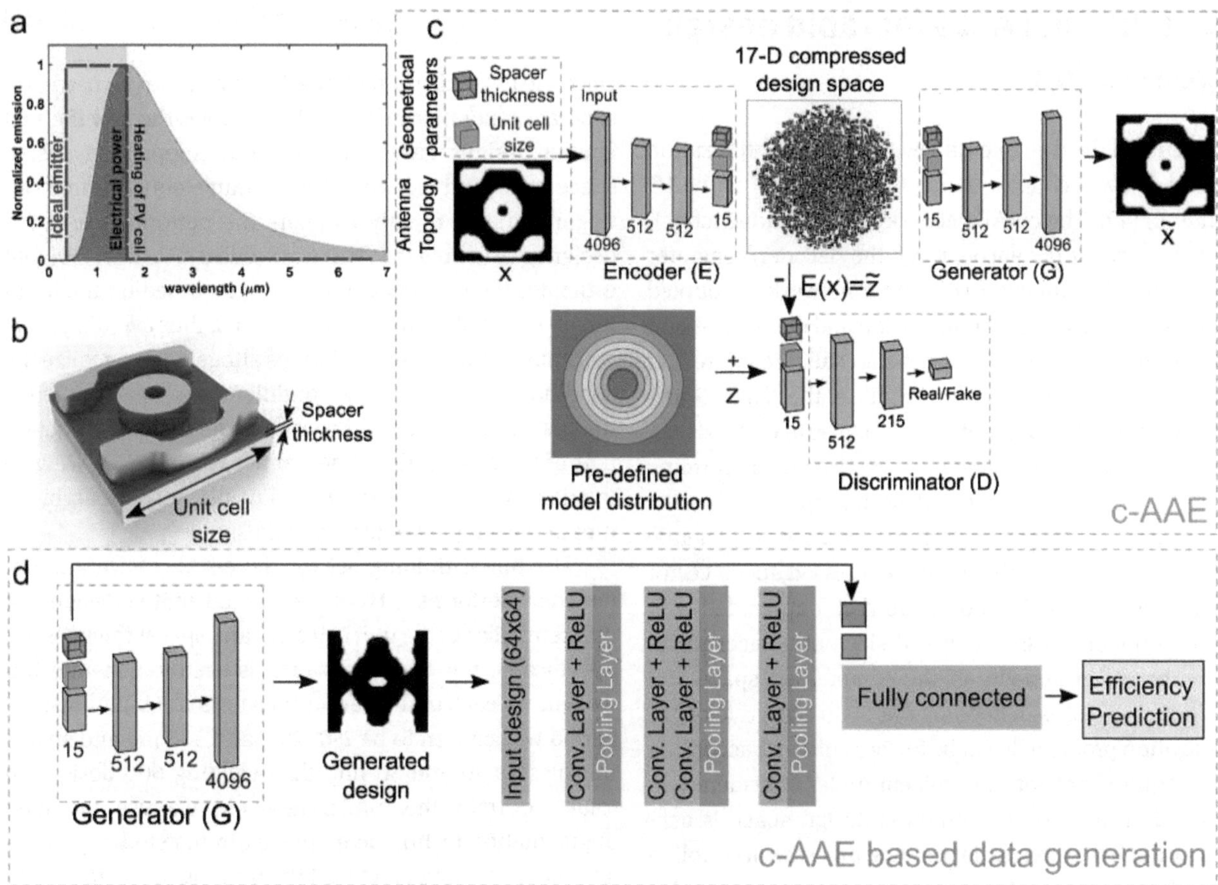

Figure 1: Conditional adversarial autoencoder-based data generation.
(a) Blackbody radiation of a bare heater (solid black curve) corresponding to the emission of a blackbody at 1800 C. The gray rectangular region highlights the GaSb photovoltaic (PV) cell working band. Only in-band radiation is converted into electrical power (red area), while out-of-band radiation causes unwanted heating of the PV cell (blue area). The dark blue dashed contour corresponds to an ideal thermal emitter's emissivity/absorption spectrum. (b) The gap-plasmon thermal emitter design with a metallic TiN back reflector, a Si_3N_4 dielectric spacer, and a top TiN plasmonic antenna. The topology optimization (TO) aims to optimize the top layer with a TiN/air mixer to match the step function emissivity pattern of an ideal emitter. (c) Training of the c-AAE network: Along with the antenna topology, the c-AAE is trained on a conditional vector with geometrical parameters of the unit cell (unit cell size, spacer thickness). (d) c-AAE–assisted rapid design generation: The trained *G* network is coupled with a conditional Visual Geometry Groupnet (VGGnet) type network (c-VGGnet) for rapid efficiency estimation. The geometrical parameters of the unit cell are used as constrained labels for generation of the designs, as well as estimation of the efficiencies. c-AAE, conditional AAE; AAE, adversarial autoencoder.

$$eff = eff^{in} \cdot eff^{out}, \tag{1}$$

here

$$eff^{in} = \int_{\lambda_{min}}^{\lambda_{max}} \varepsilon(\lambda) B_\omega(\lambda, T) d\lambda \Big/ \int_{\lambda_{min}}^{\lambda_{max}} B_\omega(\lambda, T) d\lambda,$$

$$eff^{out} = \int_{\lambda_{max}}^{\infty} \varepsilon_{TiN}(\lambda) B_\omega(\lambda, T) d\lambda \Big/ \int_{\lambda_{max}}^{\infty} \varepsilon(\lambda) B_\omega(\lambda, T) d\lambda.$$

where the Plank law, $B_\omega(\lambda, T) = 2hc\lambda^{-3}(e^{hc/(\lambda k_B T)} - 1)^{-1}$, gives the spectral radiance of the blackbody at a given temperature T and wavelength λ; the fundamental constants include the Planck constant h, the Boltzmann constant k_B, and the speed of light in free space c. In (1) $\varepsilon(\lambda)$ and $\varepsilon_{TiN}(\lambda)$ denote the spectral emissivities of the optimized emitter and a bare TiN back reflector, respectively, T is the working temperature of the emitter, wavelengths λ_{min}, λ_{max} are, respectively, the lower and upper bounds of the PV cell operation band.

eff^{in} is an in-band radiance of the emitter normalized to the in-band radiance of an ideal emitter at 1800°C, while out-of-band efficiency eff^{out} is defined as a ratio of the out-of-band radiance of a back reflector and radiance of the TO design. Such definition of the out-of-band efficiency is dictated by the fact that the response of the gap-plasmon structures in the long-wavelength limit is fully determined by the material properties of the back reflector, and out-of-band emissivity is fundamentally limited by the optical losses of TiN.

3 Conditional AAEs for rapid design generation

To include all the design parameters into the optimization framework, we couple the c-AAE network with TO (Figure 1c) [49]. The c-AAE network is a generative model, which consists of the encoder (E), the decoder/generator (G), and the discriminator (D). The E network is coupled with G network aiming at compression and decompression of the input design (x) through the so-called compressed design space. During the training phase, the E and G networks are trained to minimize the reconstruction loss between input (x) and the generated ($\tilde{x}$) designs by forming the 17-dimensional compressed design space. After the training process, the G network can be used to generate new designs based on the input compressed space vector $\tilde{z} = E(x)$, which is a 15D coordinate vector appended with two conditional labels l (unit cell size and spacer thickness). The dimensionality of the compressed space and conditional vector is defined by the main objective of the optimization problem. It can be further enlarged according to the requirements of the problem under consideration. The regularization of the compress design space is achieved through the adversarial training process via coupling to the D network. The D network is trained to distinguish between samples from the compressed design space distribution $q(\tilde{z})$ and the predefined model $p(z)$. During the training process, the E network is trained to reshape the compressed design space such that the D network cannot distinguish between samples generated from the compressed design space and predefined model. The adversarial training forces the compressed design space to have the same data distribution as the user-defined model $p(z)$. So, the main goal of the E and G networks is to learn the main geometrical features of the antenna designs in the training set, while the D network assures the regularization of the formed compressed design space defined by the predefined model. This is achieved via the optimization of the AAE network weights (E, G, and D networks) during the training phase by minimizing the following loss function:

$$L = L_{\text{adv}} + L_{\text{rec}} \tag{2}$$

The regularization of the compressed space is achieved via a minmax game between E and D networks which aims to minimize the adversarial loss term L_{adv} [49]:

$$L_{\text{adv}} = \min_{E} \max_{D} [\log(D(E(x), l)) + \log(1 - D(\tilde{z}, l))] \tag{3}$$

while the training of the E and G networks is done through to minimize the reconstruction loss L_{rec} [49]:

$$L_{\text{rec}} = -\min_{E,G} [\log p(x|G(E(x), l))] . \tag{4}$$

Once the c-AAE is trained, the G network can be used as a separate generative network that samples the new thermal emitter designs based on the input compressed design space vector and the geometrical parameters of the unit cell (Figure 1d). To rapidly estimate the performance of the designs, we couple the G network with a c-VGGnet [50] that estimates the efficiency of the designs based on the input binary image of the antenna (top view), thus avoiding time-consuming full-wave simulations altogether. To realize the conditional estimation, the conditional vector is coupled to the first fully connected layer after the feature extraction part of the convolutional neural networks (CNN). We give the details on the c-AAE and c-VGGnet network architecture in the supplementary materials.

The initial training set for the c-AAE network is obtained by performing TO of the thermal emitter designs for different cases of the unit cell sizes and spacer thicknesses. Specifically, the TO optimization is used to generate 100 designs for each of the period-thickness combinations: the period was chosen to be 250, 280, and 300 nm and spacer thicknesses 30 and 50 nm, thus yielding 600 designs in total. To train the c-AAE network, we use the data augmentation technique employed in the study by Kudyshev et al. [43] and increased the training set up to 24,000 designs. The periodic nature of the thermal emitters allows us to expand the initial data set by applying lateral and rotational perturbations to the original design. Here, we expand the data set by applying 20 lateral shifts and 90° rotation to each design of the original set. The enlarged design set, as well as corresponding periodicity and spacer thickness labels, has been used to train the c-AAE network. The predefined model distribution is set to be Gaussian that is centered at the origin of 15D compressed space.

Figure 2a shows the adversarial (blue) and reconstruction (red) losses evolution of the c-AAE network as a function of training epochs. The reconstruction loss of the E and G coupled network decreases with the training and saturates at <0.1 value, which indicates that the G network can reconstruct input design correctly from the compressed design space. The adversarial loss decreases at the beginning of the training, which corresponds to the fact that the E network fails to generate the desired compressed space distribution at the early steps of the training. However, with the increasing number of the training epochs, adversarial loss saturates at relatively high values (~0.8), highlighting the ability of the E network to "fool" the D network and passing the sample from the compressed space as "real" through it. This fact indicates that the constructed compressed design space has data distribution close to the

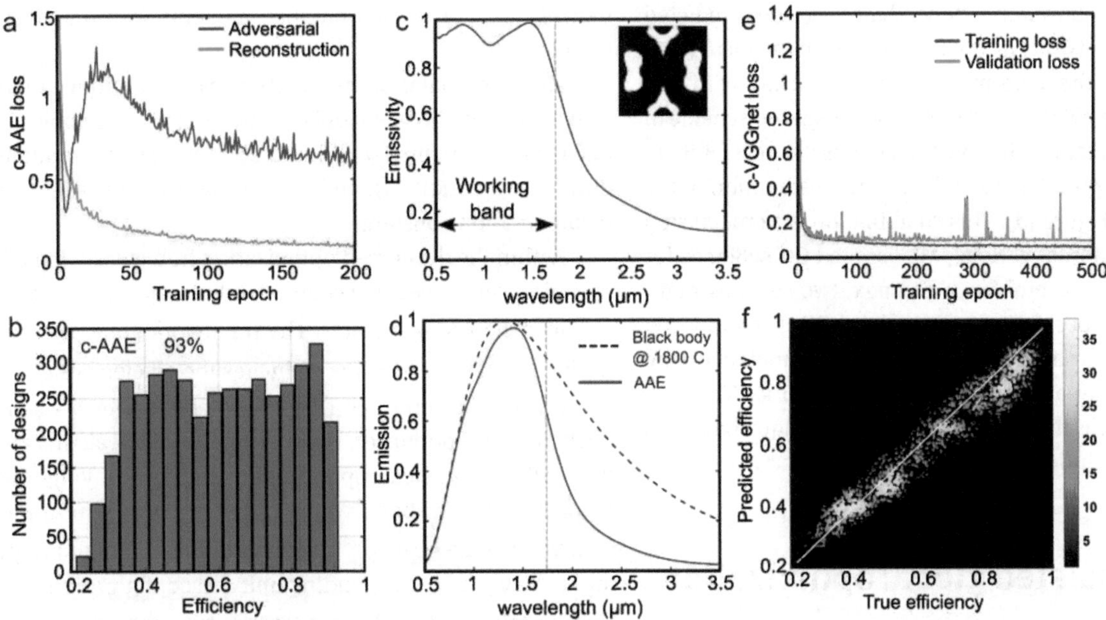

Figure 2: Training of c-AAE and c-VGGnet networks.
(a) Evolution of the adversarial (blue line) and reconstruction (red line) losses during the c-AAE network training. (b) The efficiency distributions of the design set generated by the c-AAE network via random sampling. (c) Spectral emissivity of the best design in the generated set. The dashed vertical black line shows the upper bound of the PV cell's working band. The inset shows the unit cell of the best design in the set (white color: TiN, black: air, unit cell period (*x* and *y*) 280 nm; spacer thickness, 30 nm). (d) Emission spectrum of the blackbody (dashed black line) and thermal emitter (solid blue line) at 1800 C. (e) Evolution of the training (blue line) and validation (red line) losses of the c-VGG predictive network during the training process. (f) The regression results performed by c-VGGnet on the testing data set. White line shows the regression line; the colormap shows the number of designs tested. c-AAE, conditional AAE; AAE, adversarial autoencoder; PV, photovoltaic; c-VGGnet, conditional VGGnet type network.

predefined one. Once the training of the c-AAE network is done, we have generated 4500 thermal emitter designs via random sampling of the compressed space coordinates. Specifically, the 300 designs for each of the combinations of the periodicity (from 200 to 280 nm with 20 nm step) and spacer thicknesses (30, 40, and 50 nm) have been generated. The efficiency of each design has been assessed by Finite-difference time-domain (FDTD) simulation (Lumerical FDTD solver). Figure 2b shows the statistics of the generated data set. The best design in the set has 93% efficiency of thermal emission reshaping and corresponds to 280-nm unit cell size and the 30-nm spacer thickness. Figures 2c and 2d show the corresponding absorption/emissivity spectra, as well as the gray body emission at 1800°C from the best design in the set (blue curves). The inset in Figure 2c shows the antenna design. The complex topology of the c-AAE antenna design enables 94% mean in-band absorption and substantially suppresses the out-of-band absorption spectrum (23% mean out-of-band absorption). This absorption behavior of the thermal emitter leads to its high in-band emission and the rapid decay of the out-of-band emission.

To perform the filtering of highly efficient design, as well as to avoid time-consuming full-wave analysis, the c-AAE network is coupled to the c-VGGnet regression network, which estimates the performance of the design based on the input design and unit cell parameters (Figure 1d). The c-VGGnet regression network is trained on the design set, which is generated by the c-AAE network (Figure 2b). This training is done due to the high designs' variance and a broader range of the unit cell parameters vs. the TO generated set that allows for more efficient training of the c-VGG network. Employing the generated design set, unit cell parameters, and corresponding efficiency values as ground truth, the c-VGGnet is trained by using mean absolute percentage error loss function. Eighty percentage of the designs are used for training, while 20% is used for the validation loss estimation. The loss evolution during the training process is shown in Figure 2e. The figure demonstrates that the regression loss decreases and saturates at 9%, indicating that c-VGGnet is capable of retrieving the efficiency values based on the binary image of the design with high accuracy. For assessing the performance of the regression network, we calculate the coefficient of determination r^2 that quantifies the ability of the network to predict the variance of the true data. While in the ideal case r^2 should be equal to 100%, a sufficiently high value of r^2 (87%) is achieved.

Figure 2f shows the dependence of the predicted values by the network vs. the true values. In the ideal case, the point on the colormap should coincide with the regression line (white, solid line). The integrated scheme of the c-AAE generator with c-VGGnet opens the possibility to perform rapid prototyping and efficiency estimation of the metadevices (Figure 1d). This combination is a crucial step for the realization of various, ML-assisted GO schemes for highly constrained problems. The next two sections highlight one of the possible c-AAE–assisted GO. The proposed approach can be integrated with the other metaheuristic and/or gradient-based optimization frameworks. The next section describes the c-AAE–based GO technique based on differential evolution optimization.

4 AAE-assisted global optimization

Due to the nonconvex nature of many optimization problems, the design spaces usually correspond to the rapidly changing figure of merit (FOM) landscapes that make the brute-force approach inefficient in the quest for the global solution. This issue becomes even more significant for multiconstrained problems since the exhaustive search within highly dimensional parametric landscapes would be extremely resource heavy with no guarantee of retrieving the most optimal solution. Hence, it is crucial to develop a global optimization framework that is capable of using the "best of both worlds": (i) efficiency and scalability of the c-AAE–based optimization framework and (ii) the ability of metaheuristic algorithms to perform global optimization searches most efficiently. To address the issues mentioned above, we develop a c-AAE–assisted GO scheme, which is capable of performing global search directly within the compressed design space.

Within this work, the DE algorithm [51] has been coupled to the c-AAE–based data generation approach for retrieving the global maximum inside the compressed design space. The DE framework is a population-based metaheuristic algorithm, which uses multiple agents to probe the solution space and evolutionally converge to global extremum. At optimization step, the positions of each of the agents in the population are updated by adding the weighted difference between two randomly selected agents from a given population to the agent with the best efficiency at the current iteration:

$$\tilde{z}_n^{i+1} = \tilde{z}_{\text{best}}^i + F\left(\tilde{z}_{r1}^i - \tilde{z}_{r2}^i\right) \tag{5}$$

Here, $\tilde{z}_{\text{best}}^i$ is a compressed design space coordinate of the agent with the best performance at ith iteration, F is a mutation parameter, $r_{1,2}$ = rand(1, N) are the random indices, and N is a total number of agents within a given population.

The coordinates of the agents at the next optimization step are updated with (5) or kept unchanged. This choice is made with a binomial distribution by generating a random number, r = rand(0, 1), and comparing it vs. a predefined recombination constant.

Within the developed optimization framework, the DE optimizer sends the set of compressed design space vectors $\tilde{z}_{1,N}^i$ to the c-AAE generator. The G network generates the designs, and the c-VGG-net estimates the efficiencies $eff_{1,N}^i$ of each design in the set. Once these efficiencies are sent back to the DE optimizer, the coordinates of the agents are updated at the next optimization step employing the algorithm shown above. At the end of the DE optimization run, we use the best $\tilde{z}_{\text{best}}^{\text{final}}$ to generate the antenna design and retrieve the corresponding unit cell configuration. For this, we take two last elements of $\tilde{z}_{\text{best}}^{\text{final}}$ corresponding to the unit cell size and spacer thickness, encoded during the constrained training of the c-AAE network. The c-AAE–based GO framework assures an extremely flexible framework that addresses highly constrained optimization problems by enlarging the compressed design space with a larger number of conditional labels during the c-AAE training. Most importantly, the c-VGGnet regression network removes the need for time-consuming full-wave simulations at the optimization search stage. Additionally, since the proposed approach uses the global optimizer as a black box, the developed c-AAE–assisted framework can be coupled to any global optimization techniques.

The DE optimization is implemented using the SciPy library [52]. The total population size of the DE optimizer is set to 20 agents, with a maximum iteration number of 80. The mutation and recombination coefficients are set to 0.5 and 0.7, respectively. The optimization objective function aims at minimizing the value of 1 – *eff*. Figure 3b shows the typical conversancy curve of the c-AAE–assisted DE optimization (cAAE + DE). The DE optimization starts with the relatively high value of the objective function. However, with the evolution of the optimization, the DE algorithm converges to better efficiencies. The optimization stops when the objective function gets saturated or the maximum iteration is reached. We have performed 60 c-AAE + DE optimization runs. Figure 3c depicts the statistics of the efficiencies obtained from the runs. The analysis shows that the c-AAE + DE approach assures much better performance in comparison with the set generated directly from the c-AAE network. The mean efficiency of the distribution is 85%, while the best obtained design has 95.9% efficiency vs. 93% of the c-AAE set (Figure 2b). Figure 3d

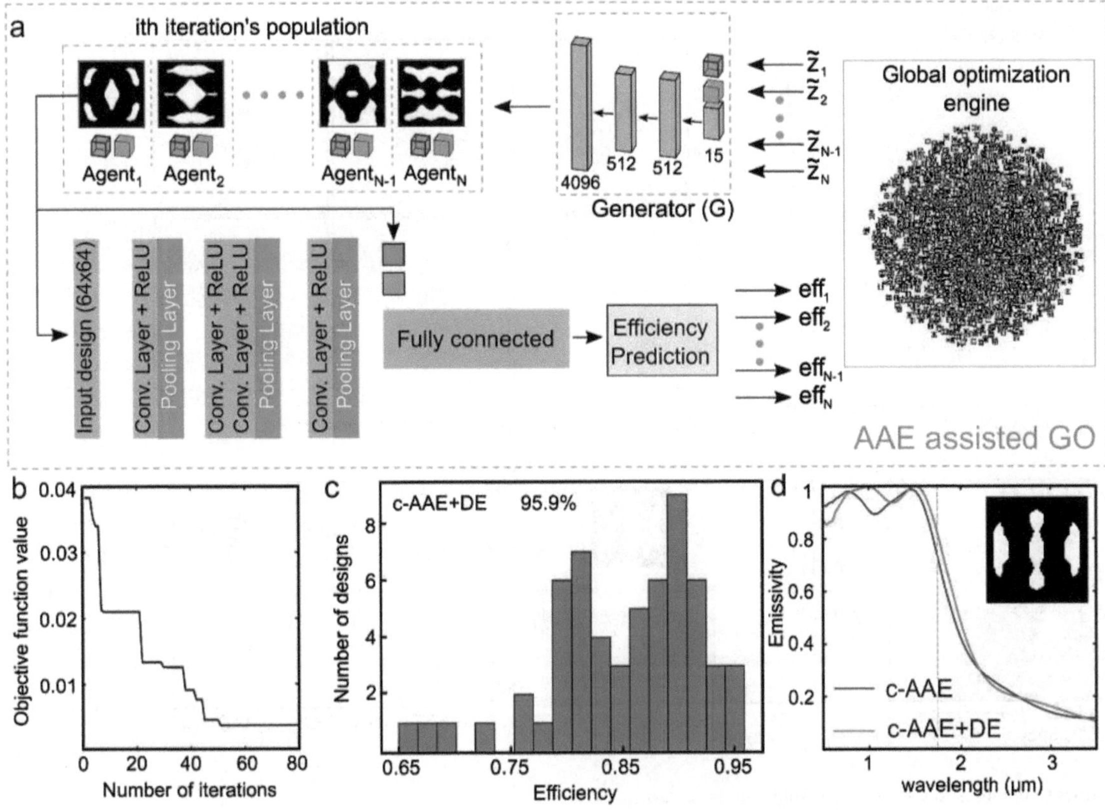

Figure 3: c-AAE–assisted global optimization.
(a) Scheme of the c-AAE based GO algorithm. The GO engine is used as a black box that generates the agents' coordinates within the compressed design space and pass them to the generator of the c-AAE network. The generator samples the design, and the corresponding efficiencies are rapidly assessed via the c-VGGnet network. These efficiencies are returned to the GO engine to update the positions of the agents at the current iteration. (b) Conversancy plot of the c-AAE + DE optimization framework. (c) Efficiency distribution for 60 designs generated by the c-AAE + DE. The legend indicates the maximum efficiencies obtained via random search within the compressed design space (93%, gray) and by GO (95.9%, black). (d) Spectral emissivity/absorption of the best designs generated by random search (blue) and GO (red). Thin vertical dashed line shows the upper bound of the PV cell's working band. The inset depicts the best unit cell design in the set (white color: TiN, black: air; unit cell period (x and y), 190 nm; spacer thickness, 45 nm). c-AAE, conditional AAE; AAE, adversarial autoencoder; PV, photovoltaic; DE, differential evolution; GO, global optimization.

shows the absorption spectra of the best design in the set (red line), while the blue curve indicates the efficiency of the best design generated directly from the c-AAE network (93%). The inset illustrates the best design generated with the c-AAE + DE approach. The c-AAE + DE optimization framework leverageds on the connection of the shape/topology of the metadevices and their optical responses. However, it is highly desirable to construct the physics-driven compressed design space by incorporating available knowledge on the performance of the device into the training process. By doing so, it is possible to regularize the compressed design space for more efficient GO searches. The next section describes the physics-driven compressed space engineering framework and demonstrates the performance of the GO within such design domains.

5 Physics-driven compressed space engineering

The training of the c-AAE network on the TO data set leads to the compressed design spaces, constructed based on geometrical features of the metadevices, omitting the available information regarding the optical responses of the designs. However, using available information about the essential physics of the metadevices, it is possible to preengineer the compressed design space for improved performance of the GO search. Such regularization of the compressed design spaces can be introduced by choosing the physics-driven predefined model of the c-AAE network connected with the FOMs of the metadevices.

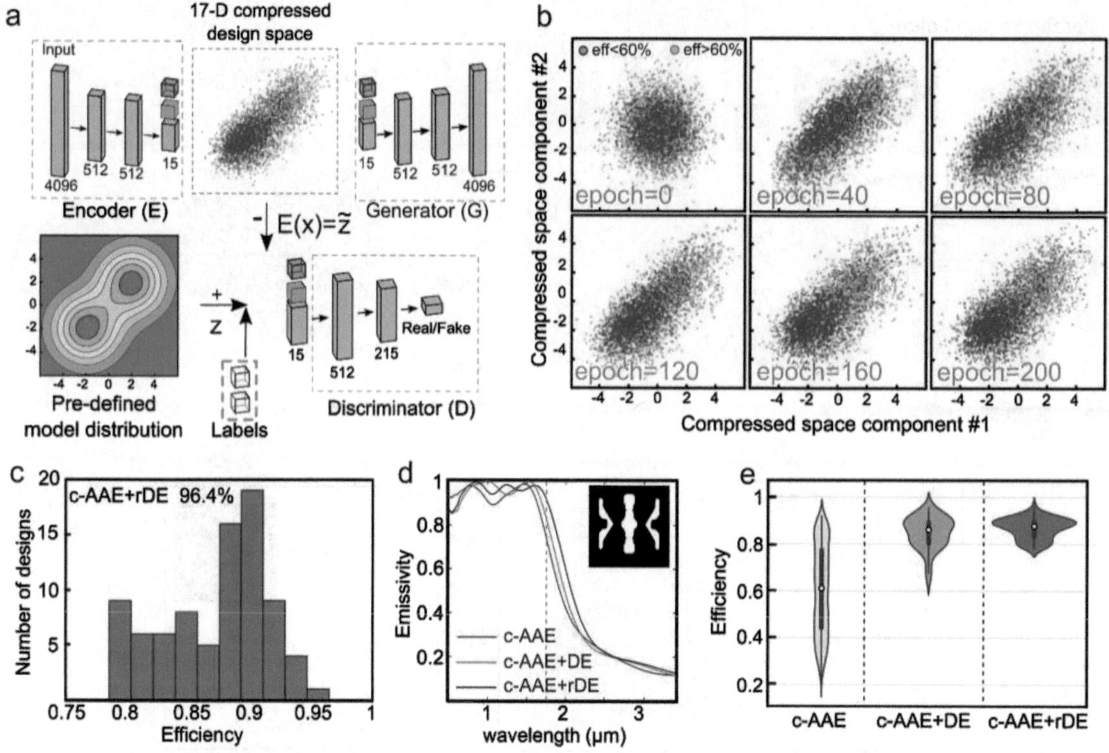

Figure 4: Physics-driven compressed design space engineering for GO.
(a) Scheme of the supervised training of the c-AAE network for physics-driven compressed space construction. The predefined model is set to be a combination of two 15-D Gaussian distributions symmetrically shifted from the origin. The additional "hot" binary vectors are used during the training for regularization of the compressed space during training. (b) Evolution of the compressed design space within the training process. The scatter plots show the distribution of the designs used in the training within 2D plane cut along the first two coordinates of the 17D compressed space. The blue markers depict the LE class data; red markers show the distribution of the HE class. (c) Efficiency distribution of 80 designs globally optimized with the c-AAE + rDE. (d) Spectral emissivity/absorption of the best designs generated by random search (blue dashed), c-AAE + DE (solid red), and c-AAE + rDE (black solid). The thin dashed vertical line shows the upper bound of the PV cell's working band. The inset shows the unit cell of the best design in the set (white color: TiN, black: air, period (*x* and *y*) 290 nm, spacer thickness 45 nm). (e) The "violin plot" of three design sets generated by, (i) random search within unregularized compressed design space (left gray pattern, c-AAE), (ii) DE optimization within the unregularized compressed design space (center light blue pattern, c-AAE + DE), and (iii) DE optimization within the regularized compressed design space (right dark blue pattern, c-AAE + rDE). LE, low efficiency; HE, high efficiency; c-AAE, conditional AAE; AAE, adversarial autoencoder; PV, photovoltaic; DE, differential evolution; GO, global optimization.

This technique can be realized based on the supervised training of the c-AAE network by adjusting the predefined model and passing an additional binary regularization vector into the *D* network at the training stage (see, Figure4a). For demonstrating the physics-driven compressed space engineering, we use two 15D Gaussian distributions in a predefined model and introduce an additional 2D hot vector as a label to the *D* network (Figure 4a). In more detail, the designs in the training data set are divided into two classes based on their efficiencies: (i) high efficiency (HE, *eff* > 60%) and (ii) low efficiency (LE, *eff* < 60%) classes. The predefined model distribution of the c-AAE model has been defined as follows:

$$z_i = \begin{cases} N(\mu = 2, \sigma^2 = 1), eff_i > 60\% \\ N(\mu = -2, \sigma^2 = 1), eff_i < 60\% \end{cases}, \quad (6)$$

here, z_i is a sampling of the predefined model for *ith* design in the set and $N(\mu, \sigma^2)$ is the random normal distribution with mean μ and variance σ^2.

With this approach, we construct the compressed design space, with two clusterization regions separated according to the two efficiency-level classes. By applying the DE optimization within the HE region, we can obtain the thermal emitters with better efficiencies. Figure 4b shows the evolution of the compressed design space throughout the training process. The figure shows two first coordinates of the 17-D compressed design space. The red markers correspond to the HE class, while the blue ones represent the LE designs. Initially, all designs are sampled as a mixture of HE and LE classes. With the training, the clusterization of both classes progressively appears. The final state of the compressed design space is shown in Figure 4b (epoch = 200). Once the training is done, the DE optimization is applied to the HE designs' region of the compressed space. We perform 80 runs with c-AAE + rDE optimization technique. Figure 4c shows the efficiency statistics of the optimized designs. The additional regularization leads to better performance of the DE optimizer vs. the unregularized case (c-AAE + DE). Thus, the c-AAE + rDE ensures 87% mean efficiency, with the best design in the set providing 96.4% efficiency of thermal emission reshaping. Figure 4d shows the emissivity spectrum of the best design in the set. The optimized design delivers 96% of in-band emissivity and substantially suppresses the out-of-band emission. The inset shows the design of the best thermal emitter in the set.

Figure 4e shows the back-to-back comparison of the efficiency distributions of three c-AAE–based optimization frameworks, (i) the set generated from c-AAE by random search (gray), (ii) c-AAE + DE (light blue) case, as well as (iii) c-AAE + rDE (dark blue). A sampling of the emitter designs from the random search (the c-AAE network with no postselection) leads to a broad range of the efficiencies with a mean efficiency of 61% and a maximum efficiency of 93%. It can also be seen that the efficiencies are uniformly distributed along the entire range (almost uniform width of the shadowed area, left subplot in Figure 4e). In contrast, the GO performed inside the unregularized compressed design space enables the design generation with much better efficiencies distribution with data concentrated within [65%, 95.9%] efficiency range and the maximum designs sampled around 90% efficiency (light blue pattern, central subplot in Figure 4e). The regularization of the compressed space coupled with the GO search leads to even better efficiencies distribution within [79%, 96.4%] (dark blue pattern, right subplot in Figure 4e). This analysis clearly shows that the physical regularization of the compressed design space allows adapting the design space configuration to perform a better GO search.

It is important to compare the proposed global optimization vs. previously demonstrated AAE-based "local" optimization frameworks (AAE + TO and AAE + VGGnet) [43]. The first approach (AAE + TO) utilizes a random sampling of the designs from the AAE network and then uses them as the initial condition for the TO refinement. This approach generated a thermal emitter design with the highest efficiency (97.9%) while taking ~31 min per design [43]. Such a relatively high computational time was the result of time-consuming design refinements performed via additional TO iterations. We note that the computation time would be even higher in the case of multiobjective optimization. This leads to poor scalability of the approach for multiobjective optimization. The second approach (AAE + VGGnet) relies on the VGGnet network-based filtering of the highly efficient, robust designs within the randomly sampled AAE design set. While the second approach generated the thermal emitter designs with an efficiency of 95.5%, its computational cost was extremely low (1.2 s per design) [43].

The AAE + DE and AAE + rDE approaches generate the designs with the efficiencies of 95.9 and 96.4%, respectively, while the computational cost for both of these new GO methods is 14 min per design. Thus, both of the new methods provide higher efficiencies in comparison to the random sampling approach based on the AAE + VGGnet method and generate the designs 2.2 times faster than AAE + TO. The optimization efficiency and the speedup of the AAE + rDE method could be further increased by choosing the different types of global optimizers, as well as through more comprehensive physics-driven regularization of the compressed design space. Such flexibility of the proposed c-AAE–based global optimization framework might be instrumental for addressing multiobjective optimization problems with a high number of constraints. All the information on the computation environment is provided in Appendix 3. Detailed information regarding the AAE-assisted local optimization frameworks (AAE + TO, AAE + VGGnet) can be found in the study by Kudyshev et al. [43].

6 Conclusion

In conclusion, we have developed a global optimization framework utilizing a c-AAE network that can be applied to a wide range of highly constrained optimization problems in nanophotonics and plasmonics, as well as in biology, chemistry, and quantum optics. We show that by applying the differential evolution optimization directly to the compressed design space, it is possible to achieve efficient optimization of the metadevices with complex topology. We numerically demonstrate advanced compressed space engineering by utilizing physics-driven regularization of the compressed design space via supervised training of the c-AAE network. The proposed physics-driven design space compression leads to significant improvement in the GO search. We also show that physics-driven regularization of the compressed design space leads to a more intuitive way of performing the GO search within the compressed space, which, in turn, leads to the almost perfect performance of the optimized metadevices.

Preengineering of the compressed design spaces of metastructures can be used in combination with diverse ML algorithms such as principal component analysis [53] and cluster analysis [28] both to retrieve the best possible solution of the problem and to gain hidden knowledge about the physics of the metastructure with complex topologies. For example, analyzing the eigenmodes of the structures sampled from the high-efficiency cluster, it is possible to gain additional intuition regarding the electrodynamic mode components that lead to the optimal metadevices. This technique would allow us to generalize the physics requirements to the device design for achieving the best possible performance and reconstruct the antenna designs based on the first principles approach to the problem.

Author contribution: All the authors have accepted responsibility for the entire content of this submitted manuscript and approved submission.
Research funding: The authors acknowledge partial support from the NSF ECCS award "Machine-Learning-Optimized Refractory Metasurfaces for Thermophotovoltaic Energy Conversion", Army Research Office MURI award no. W911NF-19-1-0279 and DARPA/DSO Extreme Optics and Imaging (EXTREME) Program award no. HR00111720032.
Conflict of interest statement: The authors declare no conflicts of interest regarding this article.

Appendix 1: c-AAE for design production

An adversarial autoencoder (AAE) consists of three coupled NNs: the encoder, decoder/generator, and discriminator. Figure A1 shows a detailed description of the neural networks.

Encoder

The encoder takes a 4096-dimensional vector (that corresponds to a 64 × 64 binary design pattern) as an input. We use two fully connected layers as the hidden layers of the encoder and a 17-neuron layer as an output layer of the encoder so that each of the hidden layers has 512 neurons. The last two elements are the geometrical parameters of the unit cell used as a conditional label, while the rest is a coordinate of the 15D compressed design space. For hidden layers, the rectified linear unit (ReLU) activation function is used, and one batch normalization layer is coupled to the second linear layer.

Decoder

The decoder has the same architecture as the encoder but with the reversed sequence. The decoder generates a 4096-element output vector based on 17-dimensional input (15D coordinate vector + 2 conditional labels). For the output layer, we use the tanh activation function.

Discriminator

The discriminator takes a 17-dimensional latent vector as an input and performs binary classification (fake/real). Hence the output of the discriminator is one neuron. Here, we have used 2 hidden liner layers with 512 and 256 neurons. The activation function for two hidden layers is the ReLU and for the output layer is the sigmoid function.

The compressed space vector consists of 15D latent coordinate and additional 2 labels with geometrical parameters of the unit cell. The training of the conditional AAE network has been realized according to the previously proposed joint disentanglement technique [54].

Appendix 2: c-VGGnet structure

CNN takes 64 × 64 image of the design as an input and passes it through three hidden layers, which consist of convolutional layers with ReLU activation functions. Each hidden layer is followed by the max. pooling layer, which ensures the downsampling of the feature maps. The stack of convolutional layers is followed by one fully connected layer, which is paired with 2 conditional labels corresponding to the unit cell geometrical parameters. The base VGGnet architecture is followed by "linear" activation function with "mean squared error" loss function for efficiency prediction (regression). A detailed description of the VGGnet is shown in Figure A2.

Appendix 3: Time requirement for training set generation, training of the networks

The generation of the original 600 TO designs (training set) was performed on three cluster nodes in parallel,

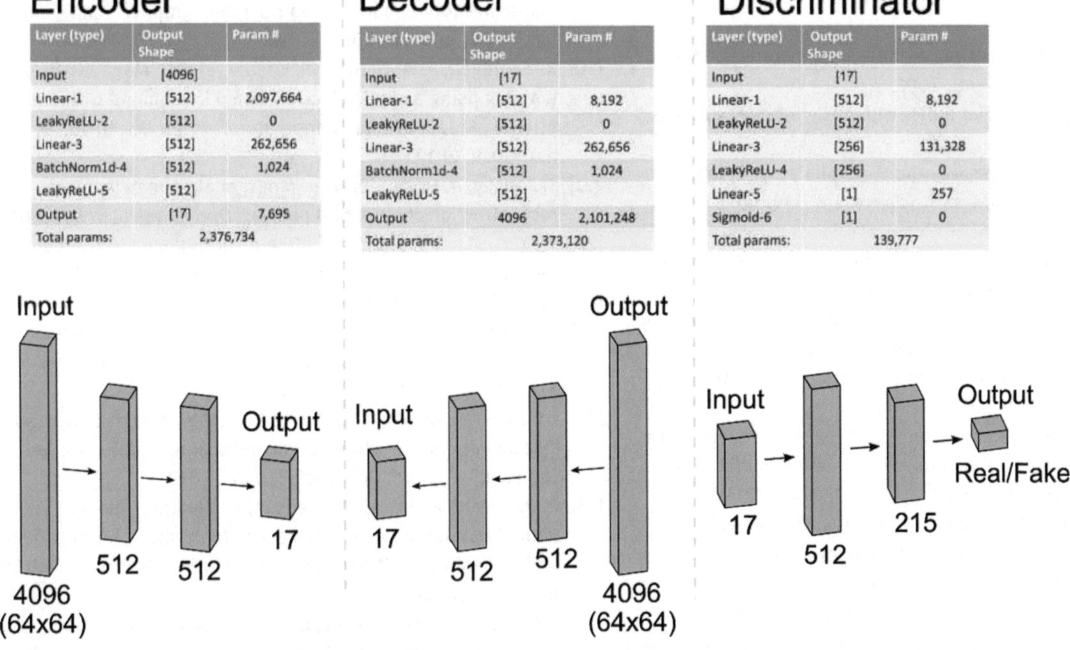

Encoder

Layer (type)	Output Shape	Param #
Input	[4096]	
Linear-1	[512]	2,097,664
LeakyReLU-2	[512]	0
Linear-3	[512]	262,656
BatchNorm1d-4	[512]	1,024
LeakyReLU-5	[512]	
Output	[17]	7,695
Total params:	2,376,734	

Decoder

Layer (type)	Output Shape	Param #
Input	[17]	
Linear-1	[512]	8,192
LeakyReLU-2	[512]	0
Linear-3	[512]	262,656
BatchNorm1d-4	[512]	1,024
LeakyReLU-5	[512]	
Output	4096	2,101,248
Total params:	2,373,120	

Discriminator

Layer (type)	Output Shape	Param #
Input	[17]	
Linear-1	[512]	8,192
LeakyReLU-2	[512]	0
Linear-3	[256]	131,328
LeakyReLU-4	[256]	0
Linear-5	[1]	257
Sigmoid-6	[1]	0
Total params:	139,777	

Figure A1: Structure of the c-AAE network.
c-AAE, conditional AAE; AAE, adversarial autoencoder.

Layer (type)	Output Shape	Param #
Input	[64,64,3]	
conv2d_1	[64, 64, 32]	896
activation_1	[64, 64, 32]	0
batch_normalization_1	[64, 64, 32]	128
max_pooling2d_1	[21, 21, 32]	0
dropout_1	[21, 21, 32]	0
conv2d_2	[21, 21, 64]	18496
activation_2	[21, 21, 64]	0
batch_normalization_2	[21, 21, 64]	256
conv2d_3	[21, 21, 64]	36928
activation_3	[21, 21, 64]	0
batch_normalization_3	[21, 21, 64]	256
max_pooling2d_2	[10, 10, 64]	0
dropout_2	[10, 10, 64]	0
conv2d_4	[10, 10, 128]	73856
activation_4	[10, 10, 128]	0
batch_normalization_4	[10, 10, 128]	512
conv2d_5	[10, 10, 128]	147584
activation_5	[10, 10, 128]	0
batch_normalization_5	[10, 10, 128]	512
max_pooling2d_3	[5, 5, 128]	0
dropout_3	[5, 5, 128]	0
flatten_1	[3200]+[2]	0
dense_1	[1024]	3277824
activation_6	[1024]	0
batch_normalization_6	[1024]	4096
dropout_4	[1024]	0
dense_2	[4]	4100
dense_3	[1]	5
Total params:	3,565,449	

Figure A2: C-VGGnet structure.

which took 328 h. The training of the c-AAE network on topology optimized designs took 35 min, while the training of the c-VGGnet for efficiency regression took 48 min. All the training of the networks, as well as AAE + DE and AAE + rDE optimizations, were done on a cluster node with two 12-core Intel Xeon Gold "Sky Lake" processors @ 2.60 GHz (24 cores per node) and 96 GB of RAM. The initial TO was performed on three cluster nodes: (i) two 12-core Intel Xeon Gold "Sky Lake" processors @ 2.60 GHz with 96 GB of RAM; (ii) two Haswell CPUs @ 2.60 GHz (20 cores per node) and 128 GB of RAM, and (iii) two Haswell CPUs @ 2.60 GHz (20 cores per node) and 256 GB of RAM. Direct full-wave simulation at each iteration was done in parallel, while the filtering, calculation of gradients, and material distribution updates were performed sequentially.

References

[1] V. M. Shalaev, "Optical negative-index metamaterials," *Nat. Photonics*, vol. 1, pp. 41–48, 2007.

[2] N. Yu and F. Capasso, "Flat optics with designer metasurfaces," *Nat. Mat.*, vol. 13, p. 139, 2014.

[3] A. V. Kildishev, A. Boltasseva, and V. M. Shalaev, "Planar photonics with metasurfaces," *Science*, vol. 339, no. 6125, pp. 12320091–12320096, 2013.

[4] O. Quevedo-Teruel, H. Chen, A. Díaz-Rubio, et al., "Roadmap on metasurfaces," *J. Opt.*, vol. 21, p. 073002, 2019.

[5] M. Song, Z. A. Kudyshev, H. Yu, A. Boltasseva, V. M. Shalaev, and A. V. Kildishev, "Achieving full-color generation with polarization-tunable perfect light absorption," *Opt. Mater. Express*, vol. 9, p. 779, 2019.

[6] D. Sell, J. Yang, S. Doshay, R. Yang, and J. A. Fan, "Large-angle, multifunctional metagratings based on freeform multimode geometries," *Nano Lett.*, vol. 17, pp. 3752–3757, 2017.

[7] S. Molesky, Z. Lin, A. Y. Piggott, W. Jin, J. Vucković, and A. W. Rodriguez, "Inverse design in nanophotonics," *Nat. Photonics*, vol. 12, pp. 659–670, 2018.

[8] Z. Lin, V. Liu, R. Pestourie, and S. G. Johnson, "Topology optimization of freeform large-area metasurfaces," *Opt. Express*, vol. 27, p. 15765, 2019.

[9] L. F. Frellsen, Y. Ding, O. Sigmund, and L. H. Frandsen, "Topology optimized mode multiplexing in silicon-on-insulator photonic wire waveguides," *Opt. Express*, vol. 24, p. 16866, 2016.

[10] R. E. Christiansen, J. Michon, M. Benzaouia, O. Sigmund, and S. G. Johnson, "Inverse design of nanoparticles for enhanced Raman scattering," *Opt. Express*, vol. 28, p. 4444, 2020.

[11] S. Jafar-Zanjani, S. Inampudi, and H. Mosallaei, "Adaptive genetic algorithm for optical metasurfaces design," *Sci. Rep.*, vol. 8, p. 11040, 2018.

[12] D. Z. Zhu, E. B. Whiting, S. D. Campbell, D. B. Burckel, and D. H. Werner, "Optimal high efficiency 3D plasmonic metasurface elements revealed by lazy ants," *ACS Photonics*, vol. 6, pp. 2741–2748, 2019.

[13] O. Ilic and H. A. Atwater, "Self-stabilizing photonic levitation and propulsion of nanostructured macroscopic objects," *Nat. Photonics*, vol. 13, pp. 289–295, 2019.

[14] S. I. Bogdanov, A. Boltasseva, and V. M. Shalaev, "Overcoming quantum decoherence with plasmonics," *Science*, vol. 364, pp. 532–533, 2019.

[15] A. Lenert, D. M. Bierman, Y. Nam, et al., "A nanophotonic solar thermophotovoltaic device," *Nat. Nanotechnol.*, vol. 9, pp. 126–130, 2014.

[16] R. C. Devlin, A. Ambrosio, N. A. Rubin, J. P. B. Mueller, and F. Capasso, "Arbitrary spin-to-orbital angular momentum conversion of light," *Science*, vol. 358, pp. 896–901, 2017.

[17] H. Wang, Y. Rivenson, Y. Jin, et al., "Deep learning enables cross-modality super-resolution in fluorescence microscopy," *Nat. Methods*, vol. 16, pp. 103–110, 2019.

[18] Z. A. Kudyshev, S. I. Bogdanov, T. Isacsson, A. V. Kildishev, A. Boltasseva, and V. M. Shalaev, "Rapid classification of quantum sources enabled by machine learning," *Adv. Quantum Technol.*, vol. 3, no. 2000067, 2020, https://doi.org/10.1002/qute.202000067.

[19] A. M. Palmieri, E. Kovlakov, F. Bianchi, et al., "Experimental neural network enhanced quantum tomography," *npj Quant. Inf.*, vol. 6, p. 20, 2020.

[20] R. Santagati, A. A. Gentile, S. Knauer, et al., "Magnetic-field learning using a single electronic spin in diamond with one-photon readout at room temperature," *Phys. Rev. X*, vol. 9, p. 021019, 2019.

[21] T. Zahavy, A. Dikopoltsev, D. Moss, et al., "Deep learning reconstruction of ultrashort pulses," *Optica*, vol. 5, p. 666, 2018.

[22] W. Ma, F. Cheng, and Y. Liu, "Deep-learning-enabled on-demand design of chiral metamaterials," *ACS Nano*, vol. 12, pp. 6326–6334, 2018.

[23] J. Peurifoy, Y. Shen, L. Jing, et al., "Nanophotonic particle simulation and inverse design using artificial neural networks," *Sci. Adv.*, vol. 4, p. eaar4206, 2018.

[24] I. Malkiel, M. Mrejen, A. Nagler, U. Arieli, L. Wolf, and H. Suchowski, "Plasmonic nanostructure design and characterization via deep Learning," *Light Sci. Appl.*, vol. 7, p. 60, 2018.

[25] P. R. Wiecha and O. L. Muskens, "Deep learning meets nanophotonics: a generalized accurate predictor for near fields and far fields of arbitrary 3D nanostructures," *Nano Lett.*, vol. 20, pp. 329–338, 2020.

[26] L. Jin, Y.-W. Huang, Z. Jin, et al., "Dielectric multi-momentum meta-transformer in the visible," *Nat. Commun.*, vol. 10, p. 4789, 2019.

[27] C. C. Nadell, B. Huang, J. M. Malof, and W. J. Padilla, "Deep learning for accelerated all-dielectric metasurface design," *Opt. Express*, vol. 27, p. 27523, 2019.

[28] Y. Kiarashinejad, M. Zandehshahvar, S. Abdollahramezani, O. Hemmatyar, R. Pourabolghasem, and A. Adibi, "Knowledge discovery in nanophotonics using geometric deep learning," *Adv. Intell. Syst.*, vol. 2, p. 1900132, 2020.

[29] Y. Kiarashinejad, S. Abdollahramezani, M. Zandehshahvar, O. Hemmatyar, and A. Adibi, "Deep learning reveals underlying physics of light–matter interactions in nanophotonic devices," *Adv. Theor. Simulat.*, vol. 2, p. 1900088, 2019.

[30] Y. Kiarashinejad, S. Abdollahramezani, and A. Adibi, "Deep learning approach based on dimensionality reduction for designing electromagnetic nanostructures," *npj Comput. Mater.*, vol. 6, p. 12, 2020.

[31] I. Sajedian, J. Kim, and J. Rho, "Finding the optical properties of plasmonic structures by image processing using a combination of convolutional neural networks and recurrent neural networks," *Microsyst. Nanoeng.*, vol. 5, p. 27, 2019.

[32] Z. Liu, D. Zhu, K. Lee, A. S. Kim, L. Raju, and W. Cai, "Compounding meta-atoms into metamolecules with hybrid artificial intelligence techniques," *Adv. Mater.*, vol. 32, p. 1904790, 2020.

[33] H. Ren, W. Shao, Y. Li, F. Salim, and M. Gu, "Three-dimensional vectorial holography based on machine learning inverse design," *Sci. Adv.*, vol. 6, no. 16, eaaz4261, 2020. https://doi.org/10.1126/sciadv.aaz4261.

[34] R. S. Hegde, "Deep learning: a new tool for photonic nanostructure design," *Nanoscale Adv.*, vol. 2, pp. 1007–1023, 2020.

[35] K. Yao, R. Unni, and Y. Zheng, "Intelligent nanophotonics: merging photonics and artificial intelligence at the nanoscale," *Nanophotonics*, vol. 8, pp. 339–366, 2019.

[36] Z. Liu, D. Zhu, S. P. Rodrigues, K. T. Lee, and W. Cai, "Generative model for the inverse design of metasurfaces," *Nano Lett.*, vol. 18, pp. 6570–6576, 2018.

[37] S. So and J. Rho, "Designing nanophotonic structures using conditional deep convolutional generative adversarial networks," *Nanophotonics*, vol. 8, pp. 1255–1261, 2019.

[38] W. Ma, F. Cheng, Y. Xu, Q. Wen, and Y. Liu, "Probabilistic representation and inverse design of metamaterials based on a deep generative model with semi-supervised learning strategy," *Adv. Mater.*, vol. 31, p. 1901111, 2019.

[39] Z. Liu, Z. Zhu, and W. Cai, "Topological encoding method for data-driven photonics inverse design," *Opt. Express*, vol. 28, p. 4825, 2020.

[40] W. Ma and Y. Liu, "A data-efficient self-supervised deep learning model for design and characterization of nanophotonic structures," *Sci. China Phys. Mech. Astron.*, vol. 63, p. 284212, 2020.

[41] J. Jiang, D. Sell, S. Hoyer, J. Hickey, J. Yang, and J. A. Fan, "Free-form diffractive metagrating design based on generative adversarial networks," *ACS Nano*, vol. 13, pp. 8872–8878, 2019.

[42] J. Jiang and J. A. Fan, "Global optimization of dielectric metasurfaces using a physics-driven neural network," *Nano Lett.*, vol. 19, pp. 5366–5372, 2019.

[43] Z. A. Kudyshev, A. V. Kildishev, V. M. Shalaev, and A. Boltasseva, "Machine-learning-assisted metasurface design for high-efficiency thermal emitter optimization," *Appl. Phys. Rev.*, vol. 7, p. 021407, 2020.

[44] M. G. Nielsen, A. Pors, O. Albrektsen, and S. I. Bozhevolnyi, "Efficient absorption of visible radiation by gap plasmon resonators," *Opt. Express*, vol. 20, p. 13311, 2012.

[45] F. Ding, Y. Yang, R. A. Deshpande, and S. I. Bozhevolnyi, "A review of gap-surface plasmon metasurfaces: fundamentals and applications," *Nanophotonics*, vol. 7, pp. 1129–1156, 2018.

[46] J. T. Heiden, F. Ding, J. Linnet, Y. Yang, J. Beermann, and S. I. Bozhevolnyi, "Gap-Surface plasmon metasurfaces for broadband circular-to-linear polarization conversion and vector vortex beam generation," *Adv. Opt. Mater.*, vol. 7, p. 1801414, 2019.

[47] W.-Y. Tsai, C.-M. Wang, C.-F. Chen, et al., "Material-assisted metamaterial: A new dimension to create functional metamaterial," *Sci. Rep.*, vol. 7, p. 42076, 2017.

[48] H. Reddy, U. Guler, Z. Kudyshev, A. V. Kildishev, V. M. Shalaev, and A. Boltasseva, "Temperature-dependent optical properties of plasmonic titanium nitride thin films," *ACS Photonics*, vol. 4, pp. 1413–1420, 2017.

[49] A. Makhzani, J. Shlens, N. Jaitly, I. Goodfellow, and B. Frey, *Adversarial Autoencoders*, preprint at arXiv.org, arXiv: 1511.05644, 2015.

[50] K. Simonyan and A. Zisserman, "Very deep convolutional networks for large-scale image recognition," preprint at arXiv.org, arXiv:1409.1556, 2015.

[51] R. Storn and K. Price, "Differential evolution – A simple and efficient heuristic for global optimization over continuous spaces," *J. Global Optim.*, vol. 11, pp. 341–359, 1997.

[52] P. Virtanen, R. Gommers, T. E. Oliphant, et al., "SciPy 1.0: fundamental algorithms for scientific computing in Python," *Nat. Methods*, vol. 17, pp. 261–272, 2020.

[53] D. Melati, Y. Grinberg, M. Kamandar Dezfouli, et al., "Mapping the global design space of nanophotonic components using machine learning pattern recognition," *Nat. Commun.*, vol. 10, p. 4775, 2019.

[54] D. Polykovskiy, A. Zhebrak, D. Vetrov, et al., "Entangled conditional adversarial autoencoder for de Novo drug discovery," *Mol. Pharm.*, vol. 15, pp. 4398–4405, 2018.

Joeri Lenaerts, Hannah Pinson and Vincent Ginis*

Artificial neural networks for inverse design of resonant nanophotonic components with oscillatory loss landscapes

https://doi.org/10.1515/9783110710687-029

Abstract: Machine learning offers the potential to revolutionize the inverse design of complex nanophotonic components. Here, we propose a novel variant of this formalism specifically suited for the design of resonant nanophotonic components. Typically, the first step of an inverse design process based on machine learning is training a neural network to approximate the non-linear mapping from a set of input parameters to a given optical system's features. The second step starts from the desired features, e.g. a transmission spectrum, and propagates back through the trained network to find the optimal input parameters. For resonant systems, this second step corresponds to a gradient descent in a highly oscillatory loss landscape. As a result, the algorithm often converges into a local minimum. We significantly improve this method's efficiency by adding the Fourier transform of the desired spectrum to the optimization procedure. We demonstrate our method by retrieving the optimal design parameters for desired transmission and reflection spectra of Fabry–Pérot resonators and Bragg reflectors, two canonical optical components whose functionality is based on wave interference. Our results can be extended to the optimization of more complex nanophotonic components interacting with structured incident fields.

Keywords: artificial neural networks; inverse design; optical resonators.

***Corresponding author: Vincent Ginis,** Harvard John A. Paulson School of Engineering and Applied Sciences, Harvard University, 29 Oxford Street, Cambridge, Massachusetts 02138, USA; and Data Lab/Applied Physics, Vrije Universiteit Brussel, Pleinlaan 2, 1050 Brussel, Belgium, E-mail: vincent.ginis@vub.be. https://orcid.org/0000-0003-0063-9608
Joeri Lenaerts and Hannah Pinson, Data Lab/Applied Physics, Vrije Universiteit Brussel, Pleinlaan 2, 1050 Brussel, Belgium

1 Introduction

Inverse design of optical components is the process of calculating the material properties that yield a specific optical response. Popular computational methods for inverse design are evolutionary algorithms [1] and density topology optimization [2–4].

A prevalent method in density topology optimization is the adjoint method. This technique provides a way to efficiently compute gradients of a loss function with respect to design parameters, as was demonstrated in the design of demultiplexers that separate light of different wavelengths [5, 6]. A similar method underlies the design of tunable metasurfaces [7] and nonlinear nanophotonic devices [8]. Recently, Angeris et al. [9] found a technique to calculate the bounds on the performance of these optical structures.

An interesting alternative formalism was introduced by Peurifoy et al. in 2018 [10]. The authors applied deep learning to the inverse design of multi-layered nanoparticles in two different steps. First, a neural network is trained to predict the particles' scattering cross-section as a function of each shell's thicknesses. This network is trained using a supervised learning algorithm. Second, this trained network is used to perform gradient descent on the design parameters. In other words, after the first training step, all weights and biases of the network are fixed, and only the input parameters can be updated to reduce the distance between the predicted and the desired scattering cross-section. So et al. improved this technique of inverse design of nanoparticles in 2019 [11] by adapting the loss function of the neural network to learn the thickness as well as the material of each layer. More advanced neural network architectures, such as a tandem network [12], Generative Adversial Networks (GANs) [13–16] and autoencoders [17, 18] have also been successfully used in the field of inverse design for optical components. In addition, deep learning has found use in various vibrant areas of nanophotonics, including integrated photonics [19], holography [20], diamond photonics [21], plasmonic nanoparticles [22]

This article has previously been published in the journal Nanophotonics. Please cite as: J. Lenaerts, H. Pinson and V. Ginis "Artificial neural networks for inverse design of resonant nanophotonic components with oscillatory loss landscapes" *Nanophotonics* 2021, 10. DOI: 10.1515/nanoph-2020-0379.

and topological photonics [23]. More examples can be found in recent reviews [24, 25].

A problem arises when applying these inverse design methods to optical systems with strong resonances. These resonances are often based on the interference of many partial waves. A highly oscillatory loss function originates because of the periodicity of the phase difference between the interfering waves. The gradient-based inverse design thus often gets stuck in local minima, far away from the global optimum. In this contribution, we propose a solution to this problem, based on Fourier analysis.

2 Predicting the transmission spectra of simple resonators

We first illustrate our technique using a simple one-dimensional Fabry–Pérot resonator: a thin slice of a dielectric medium. When light hits the resonator, it is reflected several times inside the medium. The interference of these partial waves heavily influences the reflected and transmitted light intensities. This leads to a non-trivial dependence of the transmission and reflection spectra on the wavelength of the incident light and the physical parameters of the resonator, as illustrated in Figure 1(A).

Four parameters determine the interference of the partial waves: the wavelength of the incoming light λ, the angle of incidence θ, the index of refraction n and the width of the resonator l. The wavelength-dependent transmission is then given by

$$T(\lambda, \theta, n, l) = \frac{1}{1 + F \sin^2(\delta_0 / (2\lambda))}, \tag{1}$$

where $F(\theta, n) = 4R/(1-R)^2$ is the coefficient of Finesse, and $\delta_0(\theta, n, l) = 4\pi n l \cos(\theta_{mat})$ is a parameter that encodes the phase difference between the different partial waves that are transmitted. Finally, the parameter θ_{mat} is the transmission angle inside the resonator, determined by the incident angle and the index of refraction, in agreement with Snell's law.

We train a neural network to predict the spectral transmission $T(\lambda)$ from given values for F and δ_0. The network consists of an input layer with two nodes, a number of hidden layers with adjustable weights and non-linear activation functions, and an output layer with 200 nodes, representing the transmission spectrum sampled at different wavelengths. During this training procedure, a batch of values of F and δ_0, all normalized between −1 and 1, are

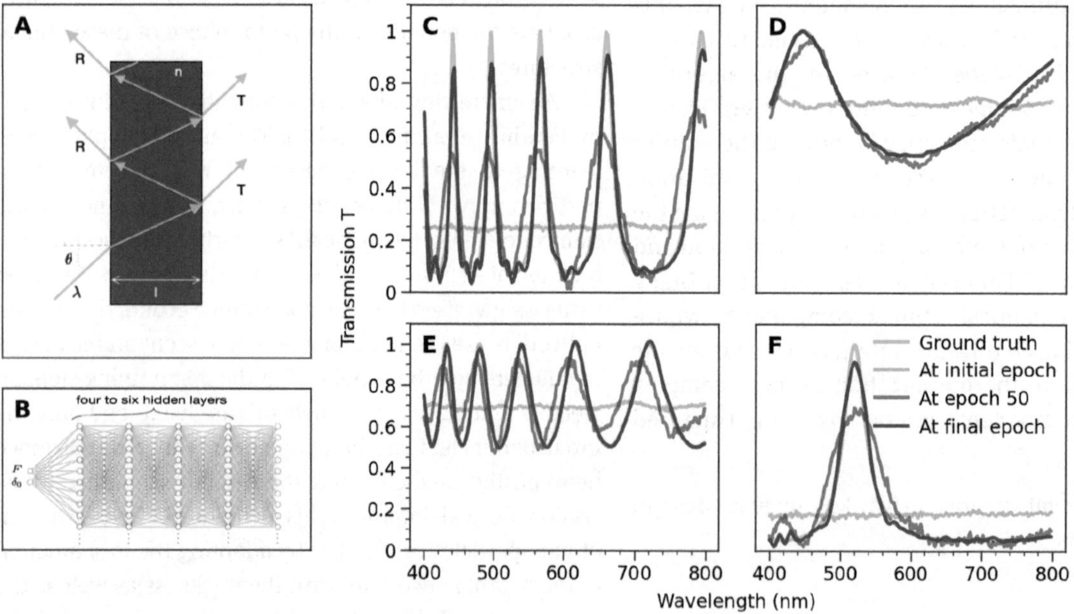

Figure 1: The first step of inverse design using deep learning.
(A) A Fabry–Pérot resonator with refractive index n and width l. Multiple reflections inside the resonator cause interference between different outgoing waves, leading to non-trivial reflection and transmission spectra. (B) A neural network that maps the resonator parameters F and δ_0 to the wavelength-dependent transmission. The networks have four to six hidden layers. The networks we trained had 100 or 200 neurons in the hidden layers and 200 nodes in the output layer, each node corresponding to one wavelength between 400 and 800 nm. The parameters that are updated during training are highlighted in orange. (C)–(F) Predictions of transmissions $T(\lambda)$ for different sets of material parameters. The orange lines show the analytically computed transmissions or ground truth. The blue lines show the predictions during training of the neural network.

provided as input to the network. The network performs a non-linear mapping from this input to an output in the form of a transmission spectrum; the mean squared error (MSE) between this output and the actual transmission spectrum is the loss function. Subsequently, the gradient of this loss function is propagated back through the network to update the values of the weights. This procedure gradually improves the spectral predictions of the network. After a number of epochs—where in each epoch, the entire training data set is used once—predictions of the network closely match the actual spectra. This is visualized in Figure 1.

We performed a search over network architectures and activation functions. The best performing neural network had six hidden layers, 200 nodes in each hidden layer, and the swish activation function. The full details of the meta-search and the resulting network are given in the Supplementary material. The final network model obtains a test mean absolute error (MAE) of 0.485 ± 0.060%. Some examples of predicted spectra $T(\lambda)$ during training are shown in Figure 1(C)–(F). The target spectrum or ground truth, calculated using Eq. (1), is shown in orange; the predictions during different phases of the training are shown in blue. The spectra with sharp peaks are somewhat harder to predict. This can be expected, since we sample the transmission at 200 points. A denser sampling of the transmission should lead to better predictions for transmissions with sharper peaks.

3 Inverse design in an oscillatory loss landscape

In the second step of the inverse design procedure, we harness the fully trained network and the fact that its structure allows us to perform a gradient descent from output to input in a straightforward way. Our goal is to find the material parameters that yield the desired transmission spectrum $T(\lambda)$. The input values are initialized with random values, representing a first guess of the material parameters F and δ_0. Unlike before, only these input values are updated at this point. The weights of the network remain fixed throughout this step. The neural network is used to compute the transmission spectrum from its input values, and subsequently the MSE between the obtained and desired spectrum yields the loss function given by

$$\text{MSE}_{\text{original}} = \sum_{\lambda}\left|\widehat{T}(\lambda) - T(\lambda)\right|^2. \tag{2}$$

The gradient of this loss function is propagated back to the input nodes, where the values are adjusted in a direction that minimizes this loss function. Over time they thus converge to the material parameters that yield a transmission spectrum $\widehat{T}(\lambda)$ that is most similar to the desired transmission $T(\lambda)$.

We repeatedly encountered one specific problem in our experiments: convergence to one of the local minima in the oscillatory loss landscape. As a result, without carefully setting the initial values for F and δ_0, the global minimum is often not found. This is illustrated in Figure 2 and in the left panel of Figure 3.

Figure 2(A) depicts the trained network, while Figure 2(B) shows the desired transmission in orange along with two obtained transmission spectra for two different sets of initial values of F and δ_0 in blue and red. Figure 2(C) shows the mean squared error (MSE) between the desired and obtained transmissions during training. The orange line indicates the loss for predictions with the correct parameters F and δ_0. The other panels represent the evolution of the parameters F and δ_0. Here, the orange line shows the true parameters from which the desired transmission $T(\lambda)$ was computed. We distinguish two cases. A first case where inverse design works, converging to the global minimum, and a second case where the gradient descent gets stuck in a local minimum. In this second case, the obtained and the desired transmissions do not look alike. We also obtain a considerably higher loss than the one obtained for the correct parameters for F and δ_0. In the other case, we choose an initial value sufficiently close to the true δ_0, such that the gradient descent does converge to a global optimum. We see that the spectra are indeed very similar (orange and blue lines). We notice that the loss is as low as it would be for the true parameters.

In Figure 3(A), we see that the loss landscape appears to have strong oscillations along the direction of δ_0, which is causing the local minima. We can illustrate this further by looking at cross-sections of the loss landscape. For a section at constant δ_0, the MSE loss is perfectly suited for gradient descent in the direction of F. There is only one minimum to which we can converge. The story is different for the parameter δ_0 (Figure 3(B)–(C)). The MSE loss shows sharp local minima, from which it is difficult to escape.

In the specific case of the oscillatory loss landscapes, as is found in the inverse design of resonant components, we propose a more efficient solution—one grounded in physical arguments: the key idea is to take the Fourier transform of the desired spectrum $T(\lambda)$ into account. Since δ_0 is linked to the number of oscillations in the transmission spectrum, meaningful information about δ_0 is adequately captured by the Fourier transform of the transmission spectrum. Therefore, we introduce a second

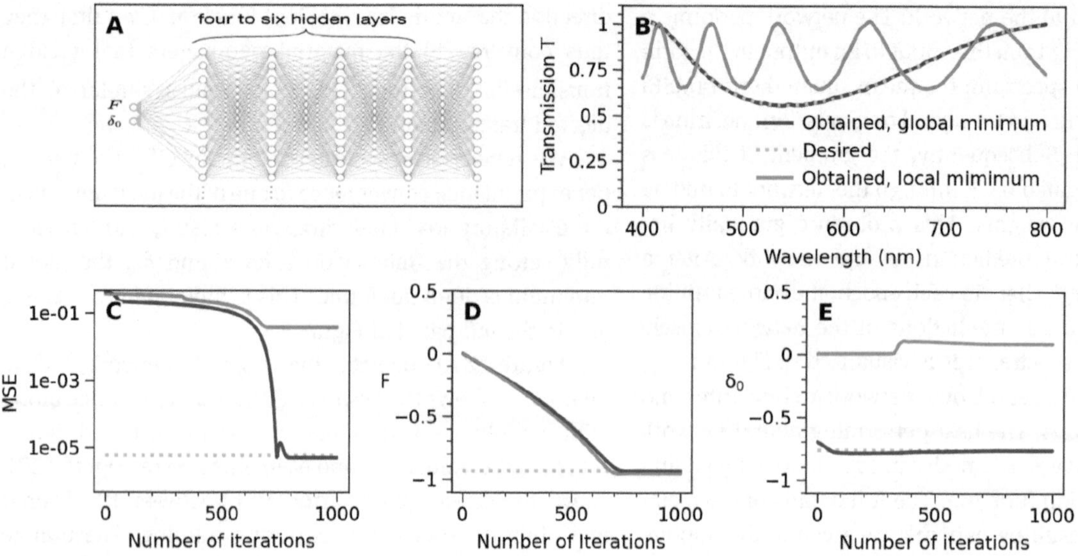

Figure 2: The second step of inverse design using deep learning.
(A) A trained neural network mapping the resonator parameters F and δ_0 to the wavelength-dependent transmission. Only the resonator parameters (the input nodes, in orange) are updated. (B) Example of a desired spectrum and spectra obtained through inverse design. The desired transmission is shown in orange and the transmission predicted by the trained neural network for the obtained resonator parameters is shown in blue and red. Two sets of resonator parameters are obtained for different initializations of the input parameters F and δ_0. (C) Evolution of the loss function during training. (D) Evolution of F during training. The orange line shows the value of F used to create the desired transmission. Values are normalized between −1 and 1. (E) Evolution of δ_0 during training. The orange line shows the value of δ_0 to create the desired transmission.

loss function, the mean squared error (MSE) on the Fourier power spectrum of the target function $T(\lambda)$, given by

$$\mathrm{MSE}_{\mathrm{Fourier}} = \sum_{f=1}^{10} \left|\mathcal{F}^2\{\hat{T}(\lambda)\}(f) - \mathcal{F}^2\{T(\lambda)\}(f)\right|^2 \tag{3}$$

We show the loss landscape of this function in Figure 3(D)–(F). Here, we only consider the first 10 frequency bins, since the transmission spectra do not oscillate more than 10 times in the region of interest. To extract the information on the relative power in the different frequencies, we normalize this total power spectrum to one. The relative power in the different frequencies minimally depends on the oscillation amplitude. This implies that the normalized power spectrum is insensitive to the parameter F, which controls the oscillation amplitude. It also means that every cross-section of the loss landscape along δ_0 with constant F will approximately be the same. We see in Figure 3(D) that this is indeed the case.

We can use this fact to initialize δ_0 sufficiently close to the true value to find the global minimum more quickly in the gradient descent algorithm. Two sections, at $F = 0$ and at the true parameter value $F = -0.932$, are plotted in Figure 3(E)–(F). We can see that these sections reach a global minimum at similar values of δ_0. Thus, we can compute this value for δ_0 without knowing the true value for F beforehand. Simply computing the minimum of the obtained section of the Fourier power spectrum at a chosen value of F, e.g. $F = 0$, yields the desired result.

Unfortunately, the assumption that the sections of the MSE on the Fourier power spectra are equal for all values of F is not entirely valid, as we already see in Figure 3. To mediate this problem, we re-initialize δ_0 after a few iterations of gradient descent, when we are closer to the true parameter for F. In our algorithm, we make this update every 100 epochs. For values of F that are closer to the true parameter, the sections will be more similar to the global minimum section. This way, the initial value for δ_0 gets closer to the true parameter every time we re-initialize.

In the end, we are not interested in matching the Fourier transforms of the spectra, but merely in matching the spectra themselves. The Fourier power spectrum we used so far also does not take information on the phase of different frequencies into account. Therefore, our goal is better achieved by a combined loss function that periodically re-initialises δ_0: the sum of the MSE on the spectra and the MSE on the Fourier power spectra of the spectra. We thus take the MSE loss on the spectra into account in this periodic update:

$$L = \mathrm{MSE}_{\mathrm{original}} + \mathrm{MSE}_{\mathrm{Fourier}}. \tag{4}$$

Our algorithm is thus alternatively minimizing two loss functions. It minimizes the MSE on the spectrum through

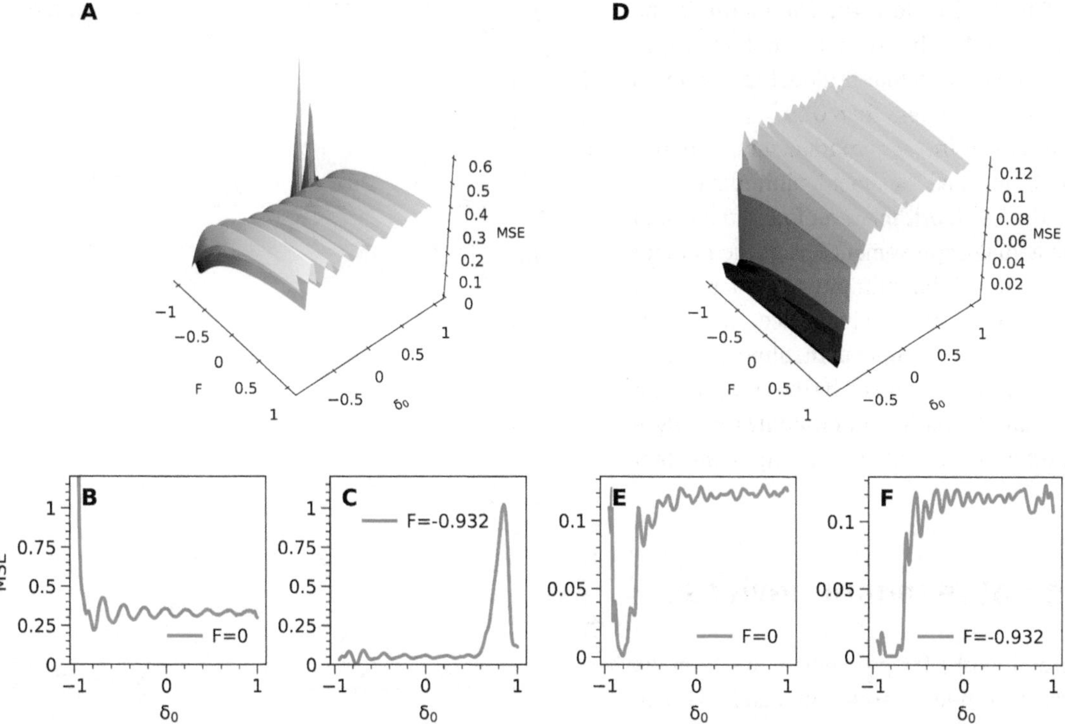

Figure 3: Finding the global minimum in an oscillatory loss landscape.
(A) Mean squared error (MSE) between the desired and predicted transmissions in function of F and δ_0. The desired transmission is an example $T(\lambda)$ from the test set. The loss function shows many local minima. (B) Cross-section of the MSE on the transmissions as a function of δ_0 for $F = 0$. (C) Cross-section of the MSE as a function of δ_0 for $F = 0.932$. (D) MSE between the Fourier power spectra of the desired and predicted transmissions versus F and δ_0. Note the different scale on the y-axis with respect to (A–C). (E) Cross-section of the MSE on the Fourier spectra as a function of δ_0 for $F = 0$. (F) Cross-section of the MSE on the Fourier spectra as a function of δ_0 for $F = -0.932$.

gradient descent, yielding better values for F and δ_0. We avoid getting stuck in local mimima during this process by periodically minimizing a second, combined loss function L given above. The latter minimization procedure consists of a straightforward computation of the value for δ_0 corresponding to a minimum of the loss at the current value for F. The other device parameter F is initialized at 0.0. To verify whether a different initialization would influence the results, we repeated the inverse design of one transmission from the test set for 100 different initializations of F. We demonstrate that all simulations converged to the optimal solution. In the following we explain how this method considerably improves the efficiency of the search in the parameter space.

4 Comparison with other search algorithms

Several solutions exist to overcome the premature convergence on local minima. One solution consists of increasing the stochastic features of the gradient descent. In this way, updates to higher loss regions can occur, allowing the search algorithm to escape from the local minima. Another solution consists of abandoning the neural network gradient descent altogether: we could use the trained network only in the forward direction, to compute the loss for any given set of input parameters, and then combine this with a stochastic local search algorithm such as simulated annealing. Both these approaches potentially visit a vast area of the parameter space before finding the global minimum, thus decreasing the inverse design procedure's efficiency.

To validate our algorithm, we compare it to a classic local search algorithm in the form of simulated annealing (SA) [26]. We use SA to minimize the loss function obtained by the trained neural network's predictions. For a fair comparison, we use both the classic SA as the more advanced dual annealing (DA) algorithm [27]. This DA algorithm combines a larger neighbourhood function with an additional local search and a restart mechanism, resulting in a highly performant search process.

We evaluate our inverse design algorithm and the two simulated annealing algorithms on 1000 transmission spectra from the Fabry–Pérot test set. The metric is the mean absolute error (MAE) between the obtained and desired spectra. We determine that a global minimum is found when we have an MAE less than 0.01.

Without fine tuning of the hyperparameters, both our algorithm and the DA reached the global minimum in 95% of the cases, while the SA algorithm never found the global minimum. But the actual improvement lies in the number of steps taken before the global minimum is reached: DA covers a large part of the parameter space, while our approach by design follows a simple path, almost straight to the global minimum. The path taken by the DA algorithm is 84 times longer than the path of our optimization algorithm. This is further illustrated in the Supplementary material.

5 More general resonant devices

To test if our method works for problems more complex than the Fabry–Pérot resonator, we also applied it to Bragg reflectors. The Bragg reflector consists of two alternating layers of materials with different refractive indices. The interference of many internal reflections again leads to a non-trivial wavelength-dependent reflection spectrum $R(\lambda)$. Now four parameters determine the spectra: the difference Δn between the two refractive indices of the materials, the mean refractive index $\overline{n}$ of the whole structure, the total length Λ of a unit of two alternating layers and the number of these units that are stacked p. The underlying analytical formula for $R(\lambda)$ is given in the Supplemental material.

We trained a neural network to learn the mapping from the four resonator parameters to the reflection $R(\lambda)$. Then we evaluated our inverse design approach on three spectra in the test set. We used the second loss function to initialize the four parameters. We re-initialized them every 300 epochs for Λ and Δn and every 600 epochs for $\overline{n}$ and p. The results of these experiments are shown in Figure 4(A)–(C). These figures show that our inverse design is capable of scaling to a problem with four parameters. In the Supplemental Material, we also show that the performance of the inverse design is not influenced by the number of stacked layers p. The dual annealing algorithm is also able to tackle this problem well. It however takes five times longer to converge, illustrating our algorithm's superior efficiency.

To show that we can also approximate arbitrary reflection spectra with our inverse design method, we borrowed a famous example of a one-dimensional curve from the novel "Le petit prince" by Antoine de Saint-Exupéry. We made a piecewise approximation of an elephant's shape and performed inverse design on this arbitrary "reflection" spectrum. The result, shown in Figure 4(D), illustrates that our inverse design method is capable of approximating a broad range of reflection spectra.

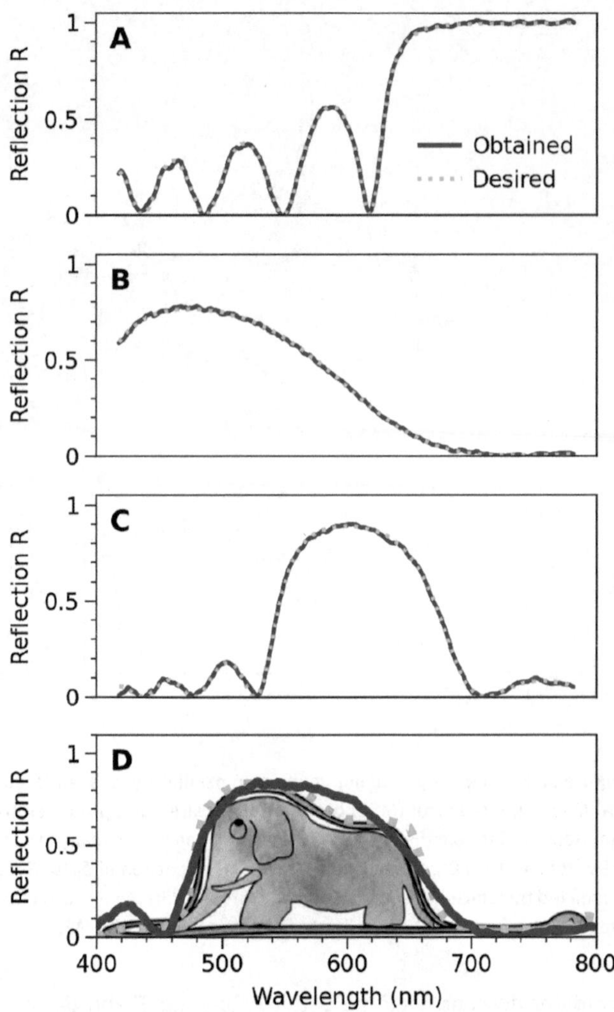

Figure 4: Inverse design of a general Bragg reflector. (A–C) examples of desired reflection spectra from the test set in combination with the spectra obtained through inverse design. The desired spectrum is shown in orange and the spectrum corresponding to the obtained resonator parameters is shown in blue. (D) Fitting an elephant. The spectrum, shown in orange, approximates the shape of the elephant. The spectrum that corresponds with the obtained resonator parameters is shown in blue.

6 Conclusion

Finding the global minimum in a parameter space where there are many local minima is a significant problem in inverse design. In this work, we show that, in the context of resonant photonic components, this problem can be addressed by constructing a combined loss function that also fits the Fourier representation of the target function. Physically, this idea is grounded in the fact that the Fourier representation contains information about the target function's oscillatory behaviour. For many resonant photonic components, these oscillations straightforwardly correspond to specific parameters in the input space. We demonstrated our technique with Fabry–Pérot and Bragg resonators and quantified the superior search efficiency compared to other optimization methods. Our technique will be particularly valuable for inverse design of optical components based on many nanophotonic resonators, such as coupled or multi-layered resonators [28–30], lasers [31], dispersion-engineered metasurfaces [32] and (nonlinear) resonators in the context of integrated optics [33, 34]. Finally, because this method can incorporate several parameters of the incident light, we think it will be interesting to study and optimize complex materials interacting with structured light beams [35–39].

Acknowledgement: H.P. and J.L. acknowledge fellowships from the Research Foundation Flanders (FWO-Vlaanderen) under Grant No. 11A6819N and Grant No. 11G1621N. Work at VUB was partially supported by the Research Council of the VUB.

Author contribution: All the authors have accepted responsibility for the entire content of this submitted manuscript and approved submission.

Research funding: H.P. and J.L. acknowledge fellowships from the Research Foundation Flanders (FWO-Vlaanderen) under Grant No. 11A6819N and Grant No. 11G1621N. Work at VUB was partially supported by the Research Council of the VUB.

Conflict of interest statement: The authors declare no conflicts of interest regarding this article.

References

[1] P. R. Wiecha, C. Majorel, C. Girard, et al., "Design of plasmonic directional antennas via evolutionary optimization," *Opt. Express*, vol. 27, no. 20, p. 29069, 2019.

[2] S. Molesky, Z. Lin, A. Y. Piggott, W. Jin, J. Vucković, and A. W. Rodriguez, "Inverse design in nanophotonics," *Nat. Photonics*, vol. 12, pp. 659–670, 2018, arXiv: 1801. 06715.

[3] Z. Lin, B. Groever, F. Capasso, A. W. Rodriguez, and M. Lončar, "Topology-optimized multilayered metaoptics," *Phys. Rev. Appl.*, vol. 9, no. 4, p. 044030, 2018.

[4] R. Pestourie, C. Pérez-Arancibia, Z. Lin, W. Shin, F. Capasso, and S. G. Johnson, "Inverse design of large-area metasurfaces," *Opt. Express*, vol. 26, no. 26, pp. 33732–33747, 2018.

[5] A. Y. Piggott, J. Lu, K. G. Lagoudakis, J. Petykiewicz, T. M. Babinec, and J. Vučković, "Inverse design and demonstration of a compact and broadband on-chip wavelength demultiplexer," *Nat. Photonics*, vol. 9, no. 6, pp. 374–377, 2015.

[6] L. Su, A. Y. Piggott, N. V. Sapra, J. Petykiewicz, and J. Vuckovic, "Inverse design and demonstration of a compact on-chip narrowband three-channel wavelength demultiplexer," *ACS Photonics*, vol. 5, no. 2, pp. 301–305, 2018.

[7] H. Chung and O. Miller, "Tunable metasurface inverse design for 80% switching efficiencies and 144° angular steering," *ACS Photonics*, vol. 7, no. 8, pp. 2236–2243, 2020, arXiv: 1910.03132 [physics.optics].

[8] T. W. Hughes, M. Minkov, I. A. Williamson, and S. Fan, "Adjoint method and inverse design for nonlinear nanophotonic devices," *ACS Photonics*, vol. 5, no. 12, pp. 4781–4787, 2018.

[9] G. Angeris, J. Vuckovic, and S. P. Boyd, "Computational bounds for photonic design," *ACS Photonics*, vol. 6, no. 5, pp. 1232–1239, 2019.

[10] J. Peurifoy, Y. Shen, L. Jing, et al., "Nanophotonic particle simulation and inverse design using artificial neural networks," *Sci. Adv.*, vol. 4, no. 6, 2017.

[11] S. So, J. Mun, and J. Rho, "Simultaneous inverse design of materials and parameters of core-shell nanoparticle via deep-learning: demonstration of dipole resonance engineering," 2019, arXiv: 1904.02848 [physics.optics].

[12] D. Liu, Y. Tan, E. Khoram, and Z. Yu, "Training deep neural networks for the inverse design of nanophotonic structures," *ACS Photonics*, vol. 5, no. 4, pp. 1365–1369, 2018.

[13] J. Jiang, D. Sell, S. Hoyer, J. Hickey, J. Yang, and J. A. Fan, "Free-form diffractive metagrating design based on generative adversarial networks," *ACS Nano*, vol. 13, no. 8, pp. 8872–8878, 2019.

[14] J. Jiang and J. A. Fan, "Global optimization of dielectric metasurfaces using a physics-driven neural network," *Nano Lett.*, vol. 19, no. 8, pp. 5366–5372, 2019.

[15] J. Jiang and J. A. Fan, "Simulator-based training of generative neural networks for the inverse design of metasurfaces," *Nanophotonics*, vol. 9, no. 5, pp. 1059–1069, 2019.

[16] S. So and J. Rho, "Designing nanophotonic structures using conditional deep convolutional generative adversarial networks," *Nanophotonics*, vol. 8, no. 7, pp. 1255–1261, 2019.

[17] Y. Kiarashinejad, S. Abdollahramezani, and A. Adibi, "Deep learning approach based on dimensionality reduction for designing electromagnetic nanostructures," *Npj Comput. Mater.*, vol. 6, no. 1, 2020, https://doi.org/10.1038/s41524-020-0276-y.

[18] W. Ma, F. Cheng, Y. Xu, Q. Wen, and Y. Liu, "Probabilistic representation and inverse design of metamaterials based on a deep generative model with semi-supervised learning strategy," *Adv. Mater.*, vol. 31, no. 35, 2019, https://doi.org/10.1002/adma.201901111.

[19] M. H. Tahersima, K. Kojima, T. Koike-Akino, et al., "Deep neural network inverse design of integrated photonic power splitters," *Sci. Rep.*, vol. 9, no. 1, pp. 1–9, 2019.

[20] H. Ren, W. Shao, Y. Li, F. Salim, and M. Gu, "Three-dimensional vectorial holography based on machine learning inverse design," *Sci. Adv.*, vol. 6, no. 16, p. eaaz4261, 2020.

[21] C. Dory, D. Vercruysse, K. Y. Yang, et al., "Inverse-designed diamond photonics," *Nat. Commun.*, vol. 10, no. 1, pp. 1–7, 2019.

[22] I. Malkiel, M. Mrejen, A. Nagler, U. Arieli, L. Wolf, and H. Suchowski, "Plasmonic nanostructure design and

characterization via deep learning," *Light Sci. Appl.*, vol. 7, no. 1, pp. 2047–7538, 2018.
[23] Y. Long, J. Ren, Y. Li, and H. Chen, "Inverse design of photonic topological state via machine learning," *Appl. Phys. Lett.*, vol. 114, no. 18, p. 181105, 2019.
[24] S. So, T. Badloe, J. Noh, J. Rho, and J. Bravo-Abad, "Deep learning enabled inverse design in nanophotonics," *Nanophotonics*, vol. 9, no. 5, pp. 1041–1057, 2020.
[25] K. Yao, R. Unni, and Y. Zheng, "Intelligent nanophotonics: merging photonics and artificial intelligence at the nanoscale," *Nanophotonics*, vol. 8, no. 3, pp. 339–366, 2019.
[26] K. Scott, C. D. Gelatt, and M. P. Vecchi, "Optimization by simulated annealing," *Science*, vol. 220, no. 4598, pp. 671–680, 1983.
[27] Y. Xiang, D. Y. Sun, W. Fan, and X. G. Gong, "Generalized simulated annealing algorithm and its application to the Thomson model," *Phys. Lett.*, vol. 233, no. 3, pp. 216–220, 1997.
[28] V. Ginis, M. Piccardo, M. Tamagnone, et al., "Remote structuring of near-field landscapes," *Science*, vol. 369, no. 6502, pp. 436–440, 2020.
[29] A. Zubair, M. Zubair, A. Danner, and M. Q. Mehmood, "Engineering multimodal spectrum of Cayley tree fractal meta-resonator supercells for ultrabroadband terahertz light absorption," *Nanophotonics*, vol. 9, no. 3, pp. 633–644, 2020.
[30] V. Pacheco-Peña and N. Engheta, "Effective medium concept in temporal metamaterials," *Nanophotonics*, vol. 9, no. 2, pp. 379–391, 2020.
[31] H. Sroor, Y. W. Huang, B. Sephton, et al., "High-purity orbital angular momentum states from a visible metasurface laser," *Nat. Photonics*, pp. 1–6, 2020.
[32] W. T. Chen, A. Y. Zhu, and F. Capasso, "Flat optics with dispersion-engineered metasurfaces," *Nat. Rev. Mater.*, pp. 1–17, 2020.
[33] V. Ginis, P. Tassin, C. M. Soukoulis, and I. Veretennicoff, "Enhancing optical gradient forces with metamaterials," *Phys. Rev. Lett.*, vol. 110, no. 5, p. 057401, 2013.
[34] M. Piccardo, B. Schwarz, D. Kazakov, et al., "Frequency combs induced by phase turbulence," *Nature*, vol. 582, no. 7812, pp. 360–364, 2020.
[35] H. Rubinsztein-Dunlop, A. Forbes, M. V. Berry, et al., "Roadmap on structured light," *J. Optic.*, vol. 19, no. 1, p. 013001, 2016.
[36] L. Liu, A. Di Donato, V. Ginis, S. Kheifets, A. Amirzhan, and F. Capasso, "Three-dimensional measurement of the helicity-dependent forces on a mie particle," *Phys. Rev. Lett.*, vol. 120, no. 22, p. 223901, 2018.
[37] A. Forbes, "Structured light from lasers," *Laser Photon. Rev.*, vol. 13, no. 11, p. 1900140, 2019.
[38] B. Bhaduri, M. Yessenov, and A. F. Abouraddy, "Anomalous refraction of optical spacetime wave packets," *Nat. Photonics*, pp. 1–6, 2020.
[39] V. Ginis, "Refracting spacetime wave packets," *Nat. Photonics*, vol. 14, no. 7, pp. 405–407, 2020.

Supplementary Material: The online version of this article offers supplementary material (https://doi.org/10.1515/nanoph-2020-0379).

Raymond A. Wambold, Zhaoning Yu, Yuzhe Xiao, Benjamin Bachman, Gabriel Jaffe, Shimon Kolkowitz, Jennifer T. Choy, Mark A. Eriksson, Robert J. Hamers and Mikhail A. Kats*

Adjoint-optimized nanoscale light extractor for nitrogen-vacancy centers in diamond

https://doi.org/10.1515/9783110710687-030

Abstract: We designed a nanoscale light extractor (NLE) for the efficient outcoupling and beaming of broadband light emitted by shallow, negatively charged nitrogen-vacancy (NV) centers in bulk diamond. The NLE consists of a patterned silicon layer on diamond and requires no etching of the diamond surface. Our design process is based on adjoint optimization using broadband time-domain simulations and yields structures that are inherently robust to positioning and fabrication errors. Our NLE functions like a transmission antenna for the NV center, enhancing the optical power extracted from an NV center positioned 10 nm below the diamond surface by a factor of more than 35, and beaming the light into a ±30° cone in the far field. This approach to light extraction can be readily adapted to other solid-state color centers.

Keywords: adjoint optimization; color center; inverse design; Purcell enhancement; quantum sensing; topological optimization.

***Corresponding author: Mikhail A. Kats,** Department of Electrical and Computer Engineering, University of Wisconsin, Madison, USA; Department of Physics, University of Wisconsin, Madison, USA; and Department of Materials Science and Engineering, University of Wisconsin, Madison, USA, E-mail: mkats@wisc.edu. https://orcid.org/0000-0003-4897-4720

Raymond A. Wambold and Yuzhe Xiao, Department of Electrical and Computer Engineering, University of Wisconsin, Madison, USA. https://orcid.org/0000-0002-3536-7940 (R.A. Wambold). https://orcid.org/0000-0002-0971-2480 (Y. Xiao)

Zhaoning Yu, Department of Electrical and Computer Engineering, University of Wisconsin, Madison, USA; and Department of Physics, University of Wisconsin, Madison, USA. https://orcid.org/0000-0002-8041-0004

Benjamin Bachman and Robert J. Hamers, Department of Chemistry, University of Wisconsin, Madison, USA. https://orcid.org/0000-0002-1356-0387 (B. Bachman). https://orcid.org/0000-0003-3821-9625 (R.J. Hamers)

Gabriel Jaffe, Shimon Kolkowitz and Mark A. Eriksson, Department of Physics, University of Wisconsin, Madison, USA. https://orcid.org/0000-0003-2672-0375 (G. Jaffe). https://orcid.org/0000-0001-7095-1547 (S. Kolkowitz). https://orcid.org/0000-0002-5306-2753 (M.A. Eriksson)

Jennifer T. Choy, Department of Electrical and Computer Engineering, University of Wisconsin, Madison, USA; Department of Engineering Physics, University of Wisconsin, Madison, USA; and Department of Materials Science and Engineering, University of Wisconsin, Madison, USA. https://orcid.org/0000-0002-8689-3801

1 Introduction

Negatively charged nitrogen-vacancy (NV) centers in diamond are optical emitters whose level structure is highly sensitive to external perturbations, which makes them excellent sensors of highly localized electric and magnetic fields, temperature, and strain [1–5]. NV centers are of great interest for quantum computing and communication [6–10] and the study of quantum phenomena such as quantum entanglement and superposition [11, 12]. However, efficiently extracting NV fluorescence is often challenging due to the high index of refraction in diamond (~2.4), which results in high reflectance at the diamond–air interfaces and total internal reflection for emission angles larger than the critical angle. Previous attempts to extract more light from bulk diamond primarily involved the etching of the diamond itself (a complicated fabrication process that can adversely affect NV properties such as spin coherence) [13–19] or fabricating structures that still required a high-numerical-aperture oil-immersion objective to efficiently collect the emission (which adds system complexity and is detrimental to sensing applications) [20–23]. Furthermore, precision etching of diamond around NV centers can be a substantial challenge and can damage the surface of diamond, resulting in roughness and modification of the chemical termination [24], which can degrade the quantum properties of NV centers [25, 26].

Here, we design a silicon-based nanoscale light extractor (NLE) that sits on the top of a flat, unpatterned diamond surface and can enhance the optical output of near-surface NV emitters by more than 35× compared with the unpatterned case, directing the light into a narrow cone

This article has previously been published in the journal Nanophotonics. Please cite as: R. A. Wambold, Z. Yu, Y. Xiao, B. Bachman, G. Jaffe, S. Kolkowitz, J. T. Choy, M. A. Eriksson, R. J. Hamers and M. A. Kats "Adjoint-optimized nanoscale light extractor for nitrogen-vacancy centers in diamond" *Nanophotonics* 2021, 10. DOI: 10.1515/nanoph-2020-0387.

that can be easily collected with low-NA optical systems. Our NLE consists of a patterned silicon structure on top of the diamond surface (Figure 1), directly above a shallow NV center (<300 nm below the surface). The proximity of a resonant high-index dielectric structure close to the emitter enables near-field coupling to and broadband Purcell enhancement of the emitter [23, 27]. We designed the silicon NLE using an adjoint-optimization method, optimizing for NV emission funneled into a narrow cone into the far field. This approach both increases collection efficiency and enhances the radiative emission rate of the NV center.

Our approach focuses on broadband emission enhancement of negatively charged NV centers close to the diamond surface, which is especially useful for sensing applications in which measurement sensitivity is limited by the number of collected photons across the entire NV emission spectrum.

2 Figure of merit for broadband NV emission extraction

We targeted our design toward negatively charged NV centers in [100] diamond. The NV axis is at an angle of 54.7° with respect to the normal to the diamond surface in this orientation [28], and the optical dipole moments are orthogonal to the NV axis [29]. At room temperature, the emission from an NV center is expected to be unpolarized [30] and can therefore be simulated by the incoherent sum of emitted intensities from any two orthogonal linear dipoles that are also orthogonal to the NV axis. In bulk diamond, negatively charged NV centers have optical transitions at ~637 nm (the location of the zero-phonon line in the spectrum), but fluoresce over a bandwidth of >150 nm due to vibrational side bands in the diamond [28, 31]. Therefore, we define a spectrum-averaged figure of merit (FoM) to quantify the degree of light extraction:

$$\text{FoM} = \frac{\int I_{NV}(\lambda) \cdot \eta(\lambda) d\lambda}{\int I_{NV}(\lambda) d\lambda} \quad (1)$$

Here, $I_{NV}(\lambda)$ is the normalized emission spectrum of the NV centers in bulk diamond taken from Ref. [32] and $\eta(\lambda)$ is the extraction efficiency of our NLE, which is defined as the number of photons emitted into free-space in the presence of the NLE divided by the number of photons emitted into free-space by that same NV (i.e., same depth and orientation) with no NLE present, at a wavelength λ. Note that $\eta(\lambda)$ includes both collection and Purcell enhancements. For example, an NLE that results in 10 times as many photons emitted into free space at every wavelength would have $\eta(\lambda) = 10$ and FoM = 10.

For all plots in this paper, the bounds of the integral in Eq. (1) are set to 635–800 nm to cover the NV zero-phonon line and phonon sideband emission spectrum. Our NLE achieves broadband NV fluorescence enhancement, which is particularly useful for sensing applications using shallow NVs, but the design approach can readily be adapted to focus on more-narrow spectral ranges (e.g., the zero-phonon line alone).

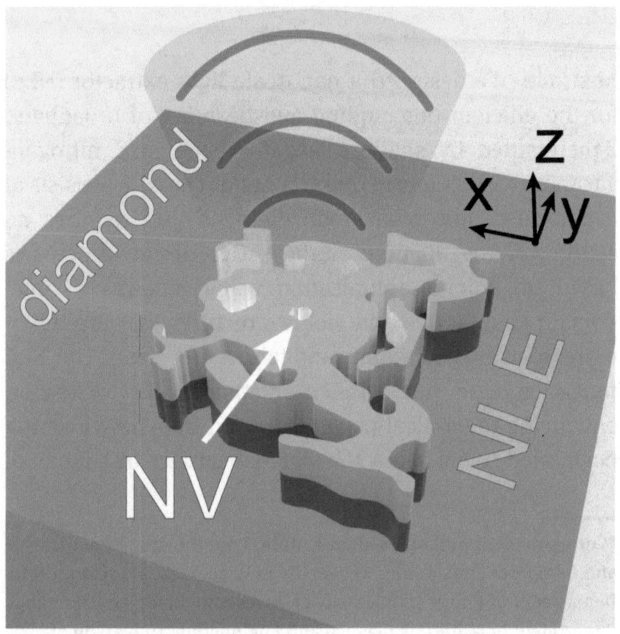

Figure 1: Schematic of the nanoscale light extractor (NLE), which sits on the diamond surface above a nitrogen-vacancy (NV) center and directs fluorescence out of the diamond and into a narrow cone in the far field.
The schematic is a render of the actual optimized structure reported below.

3 Adjoint-optimization method

To design a high-performance NLE, we used adjoint optimization—a design technique that has been used extensively in mechanical engineering and has recently been applied to the design of optical metasurfaces and other photonic structures [33–35]. For free-form structures that are allowed to evolve in three dimensions, the optimization process consists of the evolution of the structure, defined by a refractive-index profile $n(\boldsymbol{r}, \lambda)$, toward maximizing the overlap between a forward simulation field $[E_{\text{fwd}}(\boldsymbol{r}, \lambda)]$ and a specific adjoint simulation field $[E_{\text{adj}}(\boldsymbol{r}, \lambda)]$, where $\boldsymbol{r}$ is a position within the structure.

Owing to the incoherent and unpolarized emission pattern of the NV center at room temperature, our approach was to maximize extraction from two orthogonal optical-dipole orientations [29]. We used two pairs of forward/adjoint simulations for these two orientations. The first of our forward simulations is sourced by an electric dipole 10 nm below the diamond surface, with the dipole moment oriented in the *XZ* plane and tilted at an angle of 35.3° off the *Z* axis. The second was sourced by a dipole placed at the same position and oriented parallel to the *Y* axis. Note that NV centers in [100] diamond can be oriented in one of four directions along the ⟨111⟩ crystal axes. Thus to maximize performance, prior characterization can be done to align the NLE in the proper direction during fabrication [36]. Two adjoint Gaussian beams were used with orthogonal polarizations; one polarized along the *X* axis and the other polarized along the *Y* axis to pair with the forward dipoles in the *XZ* plane and along the *Y* axis, respectively.

We used finite-difference time-domain (FDTD) simulations (implemented in Lumerical FDTD [37]) to determine the forward and adjoint fields in the optimization region across the spectrum of interest for both optimization pairs. At each optimization generation, we calculated a figure-of-merit gradient similar to that of Ref. [38], but averaged across the forward/adjoint simulation pairs:

$$G(\boldsymbol{r},\lambda) \propto \langle n(\boldsymbol{r},\lambda) \cdot Re[E_{\text{fwd}}(\boldsymbol{r},\lambda) \cdot E_{adj}(\boldsymbol{r},\lambda)]\rangle_{pairs} \quad (2)$$

For the case of only one forward/adjoint pair, a positive value of $G(\boldsymbol{r},\lambda)$ indicates that a small increase in the refractive index $n(\boldsymbol{r},\lambda)$ at position $\boldsymbol{r}$ will result in a stronger overlap between the forward and adjoint fields. For more than one pair, like here, $G(\boldsymbol{r},\lambda)$ is a compromise quantity balancing the performance between all pairs. This causes the optimization algorithm to move in the direction that maximizes the gradient with respect to both pairs, and, while not necessarily reaching the same performance levels as with a single pair, is required in our case due to the nature of the NV emission. Furthermore, because it is typically not possible to separately engineer the index at different wavelengths, we define a wavelength-weighted gradient:

$$G(\boldsymbol{r}) = \frac{1}{\Delta\lambda} \int I_{NV}(\lambda) \cdot G(\boldsymbol{r},\lambda) d\lambda \quad (3)$$

To design a structure that can realistically be fabricated using top-down techniques, there should ideally be no material variance in the vertical (Z) direction. We impose this constraint in our optimization by 1) forcing the index in a single column along Z to be constant, and 2) averaging $G(\boldsymbol{r})$ along each column during the update step such that a single G is applied to each column given by

$$G(\bar{\boldsymbol{r}}) = \int_{z_{\min}}^{z_{\max}} \frac{G(\boldsymbol{r})}{z_{\max} - z_{\min}} dz \quad (4)$$

(similar to, e.g., Ref. [38]), where $\bar{\boldsymbol{r}}$ is the two-dimensional position vector in the (x, y) plane.

Figure 2(A) shows the design process for our NLE. To start, the optimization is seeded with a random continuous distribution of complex refractive index, taking values between that of air ($n_{\text{air}} = 1$) and crystalline silicon ($n_{\text{Si}}(\lambda)$, from Ref. [39]), calculated by $n(\bar{\boldsymbol{r}},\lambda) = p(\bar{\boldsymbol{r}}) \cdot n_{\text{si}}(\lambda) + [1 - p(\bar{\boldsymbol{r}})] \cdot n_{\text{air}}$ where $p(\bar{\boldsymbol{r}})$ ranges between 0 and 1, such that if $p = 1$, then $n = n_{\text{si}}$. $G(\bar{\boldsymbol{r}})$ is then used to update the position-dependent refractive-index profile in accordance with $p_{\text{new}}(\bar{\boldsymbol{r}}) = p_{\text{old}}(\bar{\boldsymbol{r}}) + c \cdot G(\bar{\boldsymbol{r}})$, where c is a normalization factor. In principle, this process can be iterated (while evolving c), until $p(\bar{\boldsymbol{r}})$ converges to yield an optimized index profile.

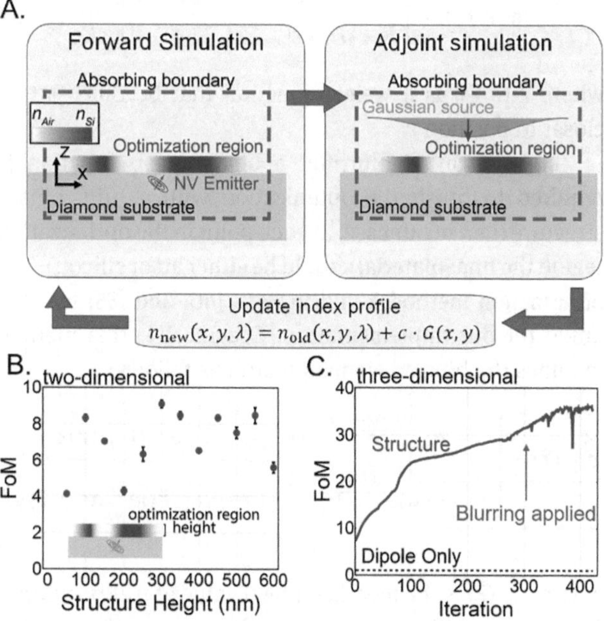

Figure 2: (A) Visualization of the optimization routine, which evolves an index profile situated above an NV emitter. In the forward simulation, the light source is a dipole at the location of the NV. The sources in the adjoint simulations are two orthogonally polarized Gaussian beams injected from free space toward the structure. The sensitivity gradient, $G(\bar{r}) = G(x, y)$, is calculated and then used to evolve the index profile. The updated profile is then used in the next iteration. (B) A sweep of the structure height (i.e., the thickness of the Si membrane), running five full 2D optimization cycles for each height. We found a height of approximately 300 nm to be optimal over the range of the sweep. The error bars represent the variance at each height. The variance is very small for some heights, so the error bars are not visible. (C) FoM vs. the iteration number for a full 3D optimization run to generate our NLE. The dips are due to the application of the secondary blurring later in the optimization. FoM, figure of merit; NLE, nanoscale light extractor; NV, nitrogen-vacancy.

However, the optimization method described above yields an optimized profile $p(\bar{r})$ corresponding to a continuum of index values $n(\bar{r}, \lambda)$. To evolve the index distribution into a binary structure of silicon and air that can be made by lithography and etching, and also ensuring that the feature sizes are not too small, we followed the methods of Sigmund in Ref. [40]. This approach applies a conically shaped blurring function to $p(\bar{r})$ at each iteration to smooth the index distribution to remove features that are smaller than the cone radius, R. For our optimization, we used a conical blurring function with $R = 40$ nm to result in structures that can be readily made with most electron-beam lithography systems. The new blurred profile $p_{\text{blur}}(\bar{r})$, is created according to [40]:

$$p_{\text{blur}}(\bar{r}) = \int \frac{w(\bar{r}, \bar{r}') \cdot p(\bar{r})}{w(\bar{r}, \bar{r}')} d\bar{r}' \tag{5}$$

$$w(\bar{r}, \bar{r}') = \begin{cases} 0, & \| \bar{r}' - \bar{r} \| > R \\ R - \|\bar{r}' - \bar{r}\|, & \| \bar{r}' - \bar{r} \| \leq R \end{cases} \tag{6}$$

where $w(\bar{r}, \bar{r}')$ is a weight function that becomes larger closer to position $\bar{r}$.

In addition to blurring, a binary push function is required to finish the optimization with a fully binary structure (i.e., in our case, at each point in the optimization region the final material should be either air or silicon). The binarization method found in Refs. [40] and [35] was the guide for our implementation. Specifically, this method modifies the blurred element matrix as follows:

$$p_{\text{bin}}(\bar{r}) = \begin{cases} \alpha e^{-\frac{\beta(\alpha - p_{\text{blur}}(\bar{r}))}{\alpha}} - (\alpha - p_{\text{blur}}(\bar{r}))e^{-\beta}, 0 \leq p(\bar{r}) \leq \alpha \\ 1 - (1-\alpha)e^{-\frac{\beta(p_{\text{blur}}(\bar{r}) - \alpha)}{1-\alpha}} - (\alpha - p_{\text{blur}}(\bar{r}))e^{-\beta}, \alpha < p(\bar{r}) \leq 1 \end{cases} \tag{7}$$

where $p_{\text{bin}}(\bar{r})$ is the modified (new) value of p at position $\bar{r}$. Here,α is a cutoff parameter that allows us to define dilated and eroded edges (to account for fabrication errors), but is set to the midpoint value of 0.5 during the optimization process. The parameter β controls the strength of binarization applied to the structure. A β-value of 0 gives no increased binarization to the structure leading to $p_{\text{bin}} = p$. As β increases, values of $p < \alpha$ are pushed toward zero whereas values of $p > \alpha$ are pushed toward one. Like the blurring function, the binarization function is applied at every step of the optimization process, with β starting at 0 and increasing, until a binary structure is achieved at the end of the optimization [~ iteration 400 in Figure 2(C)]. We found that to further remove small features and move the structure away from sharp local maxima, it was beneficial to apply a second blurring function to $p(\bar{r})$ once every 35 iterations late in the optimization process (after iteration 305 in Figure 2(C)).

We leveraged the inherently broadband nature of the FDTD method to simulate and optimize the performance of our device across the entire NV emission spectrum, simultaneously achieving a design that is robust to errors in fabrication. Previous works using adjoint optimization with a frequency-domain electromagnetic solver achieve fabrication robustness through simulating dilated and eroded devices each iteration, which corresponds to a three-fold increase in the number of simulations required [34, 35, 41, 42]. In our optimizations, fabrication robustness was built-in automatically due to our requirement for broadband performance together with the scale invariance of Maxwell's equations (i.e., larger structures at longer wavelengths behave similarly to smaller structures at shorter wavelengths). This correspondence between broadband performance and shape robustness has been observed previously for other inverse-design approaches [43, 44].

Note that our adjoint optimization setup that uses $G(\bar{r})$ to evolve the structure does not directly maximize the FoM in Eq. (1), because the formalism described above requires the selection of specific optical waveforms for the forward and adjoint fields, whereas our FoM does not depend on the particular spatial mode(s) in which light escapes to free space, or the phase of that mode relative to the dipole source. Here, we selected our adjoint source to be a Gaussian beam with a diffraction angle of ±30° injected toward the diamond surface along the Z axis. To relax the phase degree of freedom, we performed multiple optimization runs in which we varied the relative phases of the forward and adjoint sources from 0 to 2π in steps of $\pi/2$. Owing to having two forward and two adjoint sources, this led to 16 total combinations of phases for a given seed. To reduce computation time, only the best 4 phase combinations were selected to finish after 30 generations. While not pursued here, the same type of approach can be taken for accounting for the polarization of the sources.

4 Designing the nanoscale light extractor

To find the ideal height, we first ran a series of a series of 2D optimizations, sweeping over a range of heights of the NLE [Figure 2(A and B)]. Like in the later 3D simulations, the 2D simulations were set up with the dipole embedded 10 nm into diamond, and we used two orthogonal dipole orientations in a plane perpendicular to the NV axis. We swept the NLE height from 50 to 600 nm in steps of 50 nm and ran five optimization runs for each height. Figure 2(B) shows the average of our FoM for each height, with the oscillations reminiscent of Fabry–Perot fringes. Based on these results, we decided to run the full 3D optimization of the NLE for a height of 300 nm. This thickness is a common device-layer thickness in silicon-on-insulator (SOI) technology [45] and we note that SOI lends itself well to fabricating this structure as crystalline silicon device layers can be transferred to diamond using membrane-transfer techniques [46].

The full 3D optimization run can be seen in Figure 2(C). The FoM increases as the optimization progresses, with occasional dips due to the implementation of the secondary blurring function. The optimization terminates when the device is sufficiently binarized, such that the structure is entirely comprised of air and silicon, and the FoM is no longer improving significantly.

5 Results

The structure of our optimized NLE is shown in Figure 3(A), with the extraction efficiency in Figure 3(B). Although the NLE was optimized for an NV depth of 10 nm, we also show the results for depths of 5 and 15 nm but with the same NLE structure. The extraction efficiency increases for NVs closer to the structure due to increased coupling with the NLE, which leads to larger Purcell enhancement [47]. For an NV depth of 10 nm, the Purcell enhancement of our device is ~3, averaged across the emission spectrum. Here, we calculate the Purcell enhancement by dividing the power emitted by the dipole source in the presence of the interface and the NLE by the power emitted by the dipole in homogeneous diamond (no interface).

Figure 3(C and D) shows the near- and far-field radiation patterns. The NLE can shape the emitted fields from the dipole into a beam that is approximately Gaussian. The beam angle is slightly offset from the normal (~7.5°) but the bulk of the beamed power fits within a ±30° cone. The fields in Figure 3(C and D) are for a wavelength of 675 nm, but the beaming persists across the NV spectrum and the peak emission angle does not deviate by more than ±5° (see Supplementary Material for details). For an objective with numerical aperture of 0.75, the collection efficiency, defined as the fraction of emitted optical power that can be collected, is around 40% in the 635–670 nm range, and above 25% across the entire 635–800 nm range (Supplementary Material S4).

Owing to the broadband spectrum incorporated into the optimization, our devices display considerable tolerance to various fabrication defects as well as robustness to alignment errors (Figure 4). Although our optimized NLE structure was based on a fixed NV depth of 10 nm, we also simulated its performance for a variety of depths [Figure 4(A)]. The NLE performs better as the dipole gets closer to the surface because of enhanced near-field coupling. The performance falls off by a factor of 2 at a depth of 40 nm, yet still maintains enhancements of about 15 times the emission of an NV with no NLE. Even down to a dipole depth of 300 nm, the NLE is able to increase the output of the NV by a factor of 3. Note that for NVs at depths substantially different than 10 nm, a more-effective design can very likely be found using the optimization method described previously.

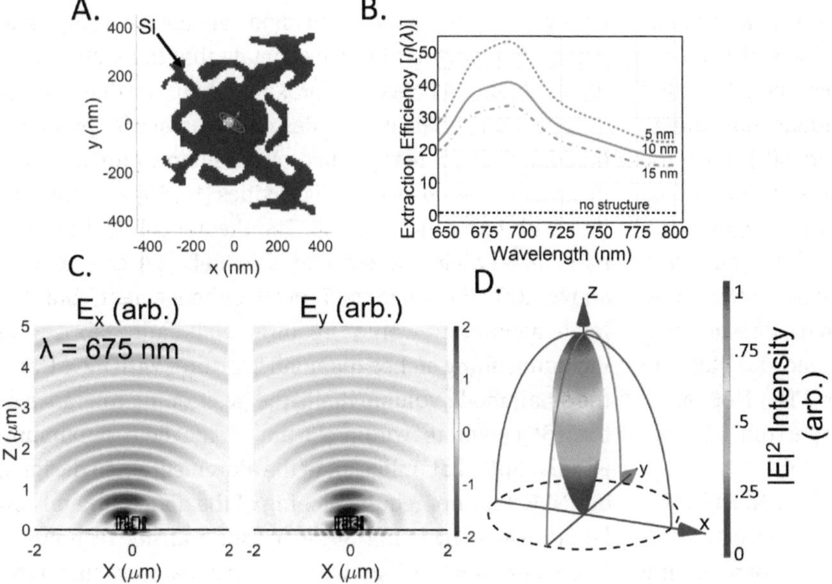

Figure 3: (A) Top-down view of the final NLE optimized for an NV depth of 10 nm. (B) Extraction efficiency [$\eta(\lambda)$] of the NLE for NV depths of 5, 10, and 15 nm below the diamond/air interface. (C and D) Snapshot of the near and meso- fields of the emitted electric field from (left) a dipole in the X–Z plane, and (right) a dipole in the Y direction. (D) Intensity far-field averaged over the two dipole orientations. In C and D, the plots are at a wavelength of 675 nm. The bulk of the beamed power fits within a 60° cone in the far field. NLE, nanoscale light extractor; NV, nitrogen-vacancy.

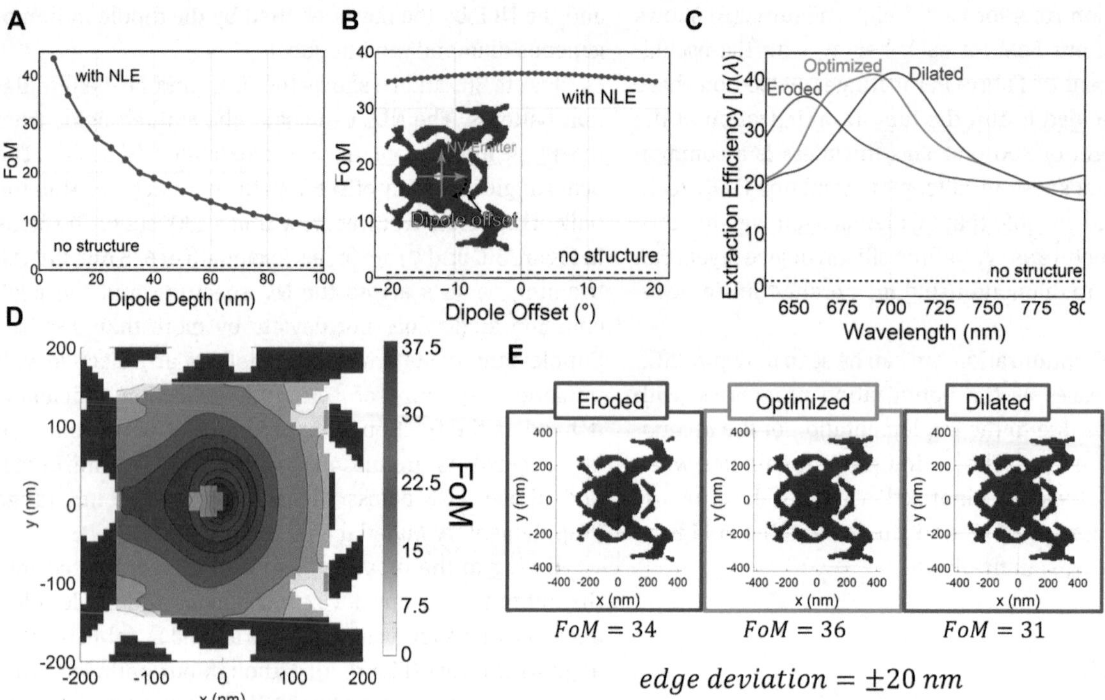

Figure 4: (A) The NLE maintains good performance through a range of depths, with increasing FoM for NVs closer to the surface. (B) FoM dependence on the NV emitter angle emulating angular alignment errors of the NLE. (C) Demonstration of the fabrication robustness of the optimized device for an NV depth of 10 nm. Eroded and dilated structures are based on the optimized structure with edge deviation of ±20 nm to represent fabrication under/over-etching, respectively. Owing to the broadband nature of our optimization, the NLE shows strong tolerance to fabrication errors. (D) Tolerance of the NLE to lateral offsets of the NV center. The FoM remains above 25 for X offsets of ±30 nm and Y offsets of ±40 nm. (e) The geometries of the eroded, optimized, and dilated devices simulated in c. The edge deviation refers to how far the edges shifted inward for the eroded or outward for the dilated cases. FoM, figure of merit; NLE, nanoscale light extractor; NV, nitrogen-vacancy.

The NLE shows minimal performance loss from errors in rotational alignment in the range of −20° to +20° [Figure 4(B)]. The full 360° rotational plot can be found in the Supplementary Material. We also tested the case where the NLE was offset by some amount from the central dipole position [Figure 4(D)], and found that the FoM remains above 25 for X offsets of ±30 nm and Y offsets of ±40 nm.

In practice, one can reasonably expect the fabrication process to cause deviations in edge locations, e.g., due to proximity effects in lithography [48]. Figure 4(C) shows the extraction efficiency of our eroded, optimized, and dilated structures with their index profiles shown in Figure 4(E). We calculated the deviated structures by applying a gaussian blurring filter across the optimized pattern, and then selecting cutoff points of the blurred edges to yield new binarized structures. Here, we selected the blur and cutoff to yield edge deviations of ±20 nm. The NLE maintains good performance across the spectrum despite erosion or dilation [Figure 2(B)].

It is instructive to compare the FoM of our NLE (~35 for NVs at a depth of 10 nm) with some existing structures in the literature designed for broadband vertical outcoupling of light from a diamond slab, with the understanding that the FoM combines the distinct collection-enhancement and Purcell mechanisms. Dielectric structures comprising etched diamond typically do not provide much Purcell enhancement and include bullseye gratings [21], vertical nanowires [49], solid-immersion metalenses [19], and parabolic reflectors [13]; we estimate that these structures have calculated FoMs of approximately 10–20 due entirely to collection enhancement. Resonant geometries proposed thus far have primarily relied on plasmonic enhancement in structures such as metallic cavities [14, 50] and gratings [51], with FoMs of up to ~35. The theoretical FoM of resonant metal-dielectric structures in Ref. [52] can be well above 100 due to large Purcell enhancement, but the implementation would require elaborate fabrication including filling etched diamond apertures with metal, and the small mode volume limits the positioning tolerance of the NV center to within ~5 nm of the field maximum. Finally, note that unlike all of the aforementioned designs, our NLE does not require etching of the diamond, and can be fabricated by using Si membrane transfer techniques [46, 53] or direct CVD growth of Si on a diamond substrate.

We do note that the exact effects of the NLE on the NV-center spin characteristics (e.g., coherence) due to the proximity of the NV to the NLE surface and material are unknown at this time. These effects are difficult to predict and are beyond the scope of this article, warranting future experimental investigations. However, our technique is likely to be less invasive than approaches that require etching the diamond near the NV, which inevitably introduces damage and additional defects.

We note that while the present article was undergoing peer review, a preprint by Chakravarthi et al. describing a similar approach to extracting light for NV centers, but for quantum information applications, was posted on arXiv [54].

6 Conclusion

We presented an NLE designed using adjoint-optimization methods with time-domain simulations to enhance the broadband emission of NV centers in diamond. Our design not only enhances the fluorescence of the NV centers but also demonstrates exceptional beam-shaping qualities, enabling efficient collection with a low-NA lens in free space. Even given reasonable uncertainty in NV center localization and errors in device fabrication, simulations show the NLE maintains a high extraction efficiency. Our results suggest that such robustness to positioning and fabrication errors can be automatically achieved when optimizing for a broadband figure of merit. The NLE can be fabricated with conventional electron-beam lithography techniques without etching of the diamond surface. Our approach can easily be extended to other color centers in diamond, as well as material systems where increased light extraction from defect centers is desired, such as silicon carbide or hexagonal boron nitride [55, 56].

Acknowledgments: The authors thank Jad Salman for useful discussions, and also an anonymous reviewer for identifying an error in the orientation of the NV optical axis in a previous version of the manuscript.
Author contribution: All the authors have accepted responsibility for the entire content of this submitted manuscript and approved submission.
Research funding: This material is based upon work supported by the National Science Foundation under Grant No. CHE-1839174. RW acknowledges support through the DoD SMART program and a scholarship from the Directed Energy Professional Society. Additional contributions by JTC and SK were supported by the U.S. Department of Energy (DOE), Office of Science, Basic Energy Sciences (BES) under Award #DE-SC0020313.
Conflict of interest statement: The authors declare no conflicts of interest regarding this article.

References

[1] F. Jelezko and J. Wrachtrup, “Single defect centres in diamond: a review,” *Phys. Status Solidi A*, vol. 203, no. 13, pp. 3207–3225, 2006.
[2] R. Schirhagl, K. Chang, M. Loretz, and C. L. Degen, “Nitrogen-vacancy centers in diamond: nanoscale sensors for physics and biology,” *Annu. Rev. Phys. Chem.*, vol. 65, no. 1, pp. 83–105, 2014.
[3] I. Aharonovich and E. Neu, “Diamond nanophotonics,” *Adv. Opt. Mater.*, vol. 2, no. 10, pp. 911–928, 2014.
[4] L. Rondin, J.-P. Tetienne, T. Hingant, J.-F. Roch, P. Maletinsky, and V. Jacques, “Magnetometry with nitrogen-vacancy defects in diamond,” *Rep. Prog. Phys.*, vol. 77, no. 5, p. 056503, 2014.
[5] J. F. Barry, J. M. Schloss, E. Bauch, et al., “Sensitivity optimization for NV-diamond magnetometry,” *Rev. Mod. Phys.*, vol. 92, no. 1, p. 015004, 2020.
[6] L. Childress and R. Hanson, “Diamond NV centers for quantum computing and quantum networks,” *MRS Bull.*, vol. 38, no. 2. Cambridge University Press, pp. 134–138, 2013.
[7] A. Russo, E. Barnes, and S. E. Economou, “Photonic graph state generation from quantum dots and color centers for quantum communications,” *Phys. Rev. B*, vol. 98, no. 8, p. 085303, 2018.
[8] A. Beveratos, R. Brouri, T. Gacoin, A. Villing, J. P. Poizat, and P. Grangier, “Single photon quantum cryptography,” *Phys. Rev. Lett.*, vol. 89, no. 18, p. 187901, 2002.
[9] F. Jelezko, T. Gaebel, I. Popa, M. Domhan, A. Gruber, and J. Wrachtrup, “Observation of coherent oscillation of a single nuclear spin and realization of a two-qubit conditional quantum gate,” *Phys. Rev. Lett.*, vol. 93, no. 13, p. 130501, 2004.
[10] M. V. Gurudev Dutt, L. Childress, L. Jiang, et al., “Quantum register based on individual electronic and nuclear spin qubits in diamond,” *Science*, vol. 316, no. 5829, pp. 1312–1316, 2007.
[11] B.-C. Ren and F.-G. Deng, “Hyperentanglement purification and concentration assisted by diamond NV centers inside photonic crystal cavities,” *Laser Phys. Lett.*, vol. 10, no. 11, p. 115201, 2013.
[12] K. Fang, V. M. Acosta, C. Santori, et al., “High-sensitivity magnetometry based on quantum beats in diamond nitrogen-vacancy centers,” *Phys. Rev. Lett.*, vol. 110, no. 13, p. 130802, 2013.
[13] N. H. Wan, B. J. Shields, D. Kim, et al., “Efficient extraction of light from a nitrogen-vacancy center in a diamond parabolic reflector,” *Nano Lett.*, vol. 18, no. 5, pp. 2787–2793, 2018.
[14] J. T. Choy, B. J. M. Hausmann, T. M. Babinec, et al., “Enhanced single-photon emission from a diamond–silver aperture,” *Nat. Photonics*, vol. 5, no. 12, pp. 738–743, 2011.
[15] T. Schröder, F. Gädeke, M. J. Banholzer, and O. Benson, “Ultrabright and efficient single-photon generation based on nitrogen-vacancy centres in nanodiamonds on a solid immersion lens,” *New J. Phys.*, vol. 13, 2011, https://doi.org/10.1088/1367-2630/13/5/055017.

[16] P. Siyushev, F. Kaiser, V. Jacques, et al., "Monolithic diamond optics for single photon detection," *Appl. Phys. Lett.*, vol. 97, no. 24, p. 241902, 2010.

[17] A. Faraon, P. E. Barclay, C. Santori, K.-M. C. Fu, and R. G. Beausoleil, "Resonant enhancement of the zero-phonon emission from a colour centre in a diamond cavity," *Nat. Photonics*, vol. 5, no. 5, pp. 301–305, 2011.

[18] J. P. Hadden, J. P. Harrison, A. C. Stanley-Clarke, et al., "Strongly enhanced photon collection from diamond defect centers under microfabricated integrated solid immersion lenses," *Appl. Phys. Lett.*, vol. 97, no. 24, p. 241901, 2010.

[19] T.-Y. Huang, R. R. Grote, S. A. Mann, et al., "A monolithic immersion metalens for imaging solid-state quantum emitters," *Nat. Commun.*, vol. 10, no. 1, p. 2392, 2019.

[20] J. Zheng, A. C. Liapis, E. H. Chen, C. T. Black, and D. Englund, "Chirped circular dielectric gratings for near-unity collection efficiency from quantum emitters in bulk diamond," *Opt. Express*, vol. 25, no. 26, p. 32420, 2017.

[21] L. Li, E. H. Chen, J. Zheng, et al., "Efficient photon collection from a nitrogen vacancy center in a circular bullseye grating," *Nano Lett.*, vol. 15, no. 3, pp. 1493–1497, 2015.

[22] S. Schietinger, M. Barth, T. Aichele, and O. Benson, "Plasmon-enhanced single photon emission from a nanoassembled metal–diamond hybrid structure at room temperature," *Nano Lett.*, vol. 9, no. 4, pp. 1694–1698, 2009.

[23] M. Y. Shalaginov, V. V. Vorobyov, J. Liu, et al., "Enhancement of single-photon emission from nitrogen-vacancy centers with TiN/(Al,Sc)N hyperbolic metamaterial," *Laser Photonics Rev.*, vol. 9, no. 1, pp. 120–127, 2015.

[24] P. Latawiec, M. J. Burek, Y.-I. Sohn, and M. Lončar, "Faraday cage angled-etching of nanostructures in bulk dielectrics," *J. Vac. Sci. Technol. B Nanotechnol. Microelectron. Mater. Process. Meas. Phenom.*, vol. 34, no. 4, p. 041801, 2016.

[25] S. Sangtawesin, B. L. Dwyer, S. Srinivasan, et al., "Origins of diamond surface noise probed by correlating single-spin measurements with surface spectroscopy," *Phys. Rev. X*, vol. 9, no. 3, p. 031052, 2019.

[26] M. Kaviani, P. Deák, B. Aradi, T. Frauenheim, J. P. Chou, and A. Gali, "Proper surface termination for luminescent near-surface NV centers in diamond," *Nano Lett.*, vol. 14, no. 8, pp. 4772–4777, 2014.

[27] J. Riedrich-Möller, S. Pezzagna, J. Meijer, et al., "Nanoimplantation and Purcell enhancement of single nitrogen-vacancy centers in photonic crystal cavities in diamond," *Appl. Phys. Lett.*, vol. 106, no. 22, p. 221103, 2015.

[28] M. W. Doherty, N. B. Manson, P. Delaney, F. Jelezko, J. Wrachtrup, and L. C. L. Hollenberg, "The nitrogen-vacancy colour centre in diamond," *Phys. Rep.*, vol. 528, no. 1. North-Holland, pp. 1–45, 2013.

[29] R. J. Epstein, F. M. Mendoza, Y. K. Kato, and D. D. Awschalom, "Anisotropic interactions of a single spin and dark-spin spectroscopy in diamond," *Nat. Phys.*, vol. 1, no. 2, pp. 94–98, 2005.

[30] N. Abe, Y. Mitsumori, M. Sadgrove, and K. Edamatsu, "Dynamically unpolarized single-photon source in diamond with intrinsic randomness," *Sci. Rep.*, vol. 7, no. 1, p. 1, 2017.

[31] N. B. Manson and J. P. Harrison, "Photo-ionization of the nitrogen-vacancy center in diamond," *Diam. Relat. Mater.*, vol. 14, no. 10, pp. 1705–1710, 2005.

[32] R. Albrecht, A. Bommer, C. Deutsch, J. Reichel, and C. Becher, "Coupling of a single nitrogen-vacancy center in diamond to a fiber-based microcavity," *Phys. Rev. Lett.*, vol. 110, no. 24, p. 243602, 2013.

[33] J. S. Jensen and O. Sigmund, "Topology optimization for nano-photonics," *Laser Photonics Rev.*, vol. 5, no. 2, pp. 308–321, 2011.

[34] D. Sell, J. Yang, S. Doshay, R. Yang, and J. A. Fan, "Large-angle, multifunctional metagratings based on freeform multimode geometries," *Nano Lett.*, vol. 17, no. 6, pp. 3752–3757, 2017.

[35] E. W. Wang, D. Sell, T. Phan, and J. A. Fan. "Robust design of topology-optimized metasurfaces," *Opt. Mater. Express*, vol. 9, no. 2, pp. 469–482, 2019.

[36] P. R. Dolan, X. Li, J. Storteboom, and M. Gu, "Complete determination of the orientation of NV centers with radially polarized beams," *Opt. Express*, vol. 22, no. 4, pp. 4379–4387, 2014.

[37] Lumerical Inc. Available at: https://www.lumerical.com/products/.

[38] C. M. Lalau-Keraly, S. Bhargava, O. D. Miller, and E. Yablonovitch, "Adjoint shape optimization applied to electromagnetic design," *Opt. Express*, vol. 21, no. 18, p. 21693, 2013.

[39] D. E. Aspnes and A. A. Studna, "Dielectric functions and optical parameters of Si, Ge, GaP, GaAs, GaSb, InP, InAs, and InSb from 1.5 to 6.0 eV," *Phys. Rev. B*, vol. 27, no. 2, pp. 985–1009, 1983.

[40] O. Sigmund, "Manufacturing tolerant topology optimization," *Acta Mech. Sin.*, vol. 25, no. 2, pp. 227–239, 2009.

[41] T. W. Hughes, M. Minkov, I. A. D. Williamson, and S. Fan, "Adjoint method and inverse design for nonlinear nanophotonic devices," *ACS Photonics*, vol. 5, no. 12, pp. 4781–4787, 2018.

[42] J. Andkjær, V. E. Johansen, K. S. Friis, and O. Sigmund, "Inverse design of nanostructured surfaces for color effects," *J. Opt. Soc. Am. B*, vol. 31, no. 1, p. 164, 2014.

[43] J. Lu and J. Vučković, "Nanophotonic computational design," *Opt. Express*, vol. 21, no. 11, p. 13351, 2013.

[44] A. Y. Piggott, J. Lu, K. G. Lagoudakis, J. Petykiewicz, T. M. Babinec, and J. Vucković, "Inverse design and demonstration of a compact and broadband on-chip wavelength demultiplexer," *Nat. Photonics*, vol. 9, no. 6, pp. 374–377, 2015.

[45] D. X. Xu, J. H. Schmid, G. T. Reed, et al., "Silicon photonic integration platform-Have we found the sweet spot?," *IEEE J. Sel. Top. Quantum Electron.*, vol. 20, no. 4, 2014, https://doi.org/10.1109/JSTQE.2014.2299634.

[46] H. Yang, D. Zhao, S. Chuwongin, et al., "Transfer-printed stacked nanomembrane lasers on silicon," *Nat. Photonics*, vol. 6, no. 9, pp. 615–620, 2012.

[47] P. Goy, J. M. Raimond, M. Gross, and S. Haroche, "Observation of cavity-enhanced single-atom spontaneous emission," *Phys. Rev. Lett.*, vol. 50, no. 24, pp. 1903–1906, 1983.

[48] M. Parikh, "Corrections to proximity effects in electron beam lithography. I. Theory," *J. Appl. Phys.*, vol. 50, no. 6, pp. 4371–4377, 1979.

[49] T. M. Babinec, B. J. M. Hausmann, M. Khan, et al., "A diamond nanowire single-photon source," *Nat. Nanotechnol.*, vol. 5, no. 3, pp. 195–199, 2010.

[50] I. Bulu, T. Babinec, B. Hausmann, J. T. Choy, and M. Loncar, "Plasmonic resonators for enhanced diamond NV- center single photon sources," *Opt. Express*, vol. 19, no. 6, p. 5268, 2011,.

[51] J. T. Choy, I. Bulu, B. J. M. Hausmann, E. Janitz, I.-C. Huang, and M. Lončar, "Spontaneous emission and collection efficiency enhancement of single emitters in diamond via plasmonic cavities and gratings," *Appl. Phys. Lett.*, vol. 103, no. 16, p. 161101, 2013.

[52] A. Karamlou, M. E. Trusheim, and D. Englund, “Metal-dielectric antennas for efficient photon collection from diamond color centers,” *Opt. Express*, vol. 26, no. 3, p. 3341, 2018.
[53] C. W. Cheng, K. T. Shiu, N. Li, S. J. Han, L. Shi, and D. K. Sadana, “Epitaxial lift-off process for gallium arsenide substrate reuse and flexible electronics,” *Nat. Commun.*, vol. 4, no. 1, pp. 1–7, 2013.
[54] S. Chakravarthi, P. Chao, C. Pederson, et al., *Inverse-designed Photon Extractors for Optically Addressable Defect Qubits*, ArXiv200712344 Phys. Physicsquant-Ph, 2020 [Online]. Available at: http://arxiv.org/abs/2007.12344 [accessed: Oct. 05, 2020].
[55] M. Atatüre, D. Englund, N. Vamivakas, S. Y. Lee, and J. Wrachtrup, “Material platforms for spin-based photonic quantum technologies,” *Nat. Rev. Mater.*, vol. 3, no. 5. Nature Publishing Group, pp. 38–51, 2018.
[56] M. K. Boll, I. P. Radko, A. Huck, and U. L. Andersen, “Photophysics of quantum emitters in hexagonal boron-nitride nano-flakes,” *Opt. Express*, vol. 28, no. 5, p. 7475, 2020.

Supplementary Material: The online version of this article offers supplementary material (https://doi.org/10.1515/nanoph-2020-0387).

Part VII: **Topological Photonics**

Midya Parto, Yuzhou G. N. Liu, Babak Bahari, Mercedeh Khajavikhan and Demetrios N. Christodoulides*

Non-Hermitian and topological photonics: optics at an exceptional point

https://doi.org/10.1515/9783110710687-031

Abstract: In the past few years, concepts from non-Hermitian (NH) physics, originally developed within the context of quantum field theories, have been successfully deployed over a wide range of physical settings where wave dynamics are known to play a key role. In optics, a special class of NH Hamiltonians – which respects parity-time symmetry – has been intensely pursued along several fronts. What makes this family of systems so intriguing is the prospect of phase transitions and NH singularities that can in turn lead to a plethora of counterintuitive phenomena. Quite recently, these ideas have permeated several other fields of science and technology in a quest to achieve new behaviors and functionalities in nonconservative environments that would have otherwise been impossible in standard Hermitian arrangements. Here, we provide an overview of recent advancements in these emerging fields, with emphasis on photonic NH platforms, exceptional point dynamics, and the very promising interplay between non-Hermiticity and topological physics.

Keywords: exceptional points; non-Hermitian physics; PT symmetry; topological photonics.

***Corresponding author: Demetrios N. Christodoulides,** CREOL, College of Optics and Photonics, University of Central Florida, Orando, 4304 Scorpius St, Orlando, USA, E-mail: demetri@creol.ucf.edu. https://orcid.org/0000-0003-3630-7234
Midya Parto, CREOL, College of Optics and Photonics, University of Central Florida, 4304 Scorpius St, Orlando, Florida, USA. https://orcid.org/0000-0003-2100-5671
Yuzhou G. N. Liu and Babak Bahari, Viterbi School of Engineering, University of Southern California, 3737 Watt Way, Los Angeles, Los Angeles, California, USA
Mercedeh Khajavikhan, Viterbi School of Engineering, University of Southern California, 3737 Watt Way, Los Angeles, Los Angeles, California, USA; CREOL, College of Optics and Photonics, University of Central Florida, 4304 Scorpius St, Orlando, Florida, USA. https://orcid.org/0000-0002-7091-1470

1 Introduction

Quantum mechanics dictates that every observable should be described by means of a self-adjoint or Hermitian operator. In this respect, the Hamiltonian of a system – being no exception to this rule – must exhibit real eigenenergies and orthogonal eigenstates, attributes necessary for unitary evolution and conservation of probability. However, while this universal conservation principle does apply for a closed system as a whole, there is nothing to exclude the possibility of energy exchange among its subsystems. When considered individually, each subsystem can see an overall growth or decay in energy or probability norm – aspects that can phenomenologically be accounted for through the adoption of complex energy eigenvalues. Indeed, approaching quantum mechanical phenomena from such a "nonconservative" standpoint can be traced back to the early studies of Gamow [1] on particle decay or other contributions on neutron scattering [2]. On the other hand, in many classical settings like optics, non-Hermiticity is not always welcome. In particular, energy/power dissipation has been traditionally considered something undesirable, an aspect to be mitigated at all costs (typically through the use of amplification) in order to maintain the performance metrics of a device or a system. At this point, one may naturally ask as to whether loss is always a problem. If not, is there a strategy to use it judiciously in order to attain new degrees of freedom?

A radical change in the way we perceive many of the aforementioned aspects occurred when Bender and Boettcher [3] realized that a large class of non-Hermitian (NH) Hamiltonians can exhibit entirely real spectra, provided that they commute with the parity-time (PT) operator. Here, the parity operator P represents a reflection in the coordinate space with respect to the origin, while T signifies the time-reversal operator. This rather counterintuitive result suggests that a PT-symmetric Hamiltonian can display altogether real eigenvalues whenever its pertinent NH parameters lie in the PT symmetry unbroken phase. On the other hand, once the non-Hermiticity

This article has previously been published in the journal Nanophotonics. Please cite as: M. Parto, Y. G. N. Liu, B. Bahari, M. Khajavikhan and D. N. Christodoulides "Non-Hermitian and topological photonics: optics at an exceptional point" *Nanophotonics* 2021, 10. DOI: 10.1515/nanoph-2020-0434.

parameter exceeds a certain critical threshold, the eigenstates spontaneously break PT symmetry, thus entering into a PT broken phase. This marks the onset of a phase transition that entails NH degeneracies, better known as exceptional points (EPs) [4]. In addition, the eigenfunctions associated with these NH operators are no longer orthogonal with each other; instead, they are now skewed. Starting from these premises, one can then directly show that a necessary (albeit not sufficient) condition for a NH Hamiltonian to be PT symmetric is that its complex potential should satisfy $V(\vec{r}) = V^*(-\vec{r})$ [3]. From here, one can conclude that the real part of the potential must be symmetric with respect to the origin, while the imaginary component must be antisymmetric.

While the physical ramifications of the aforementioned mathematical findings remained for several years a matter of debate, a series of subsequent studies indicated that optics could provide instead an ideal test bed to realize and experimentally investigate the implications of PT symmetry and NH physics in actual settings (see Figure 1A–D) [5–10]. After all, in photonic arrangements, the refractive index profile plays the role of the potential in quantum mechanics. Consequently, optical PT symmetry can be readily established by judiciously distributing the gain and loss in such a way that the refractive index profile is an even function of position while the optical gain/loss emerges as an odd function in the spatial coordinates. These early studies incited a flurry of research activities in many and diverse fields such as microwaves [11], electronics [12], mechanics [13], optomechanics [14, 15], acoustics [16, 17], atomic lattices [18–20], etc., all aiming to harness the very characteristics of PT symmetry and EPs.

A distinctive feature of optical arrangements is the possibility of controlling both the real and imaginary parts of the electromagnetic permittivity in an independent manner, without being over-restricted by the Kramers–Kronig relations. In this regard, photonic platforms can host a multitude of fascinating wave phenomena that solely arise owing to a synergy between PT symmetry and EPs on the one hand and Hermitian symmetries on the other hand. A profound example of such fruitful interactions is the newly emerging field of NH topological photonics. Topological notions originally arose within the context of condensed matter physics after the discovery of topological insulators (TIs), where electron conduction was found to be prohibited in the bulk while it can take place in the periphery of a material via topologically protected unidirectional edge states [21–23]. These developments in turn inspired further research endeavors in employing topological notions in optical arrangements, which led to the observation of unidirectional transport and robust topological edge modes in coupled resonators and waveguide lattices [24–26]. Unlike in early efforts where the emphasis was on conservative optical systems, quite recently, there has been an ever-growing interest in expanding these concepts into NH photonic structures.

In this review article, we focus on novel phenomena in photonics that are enabled by the synergies among

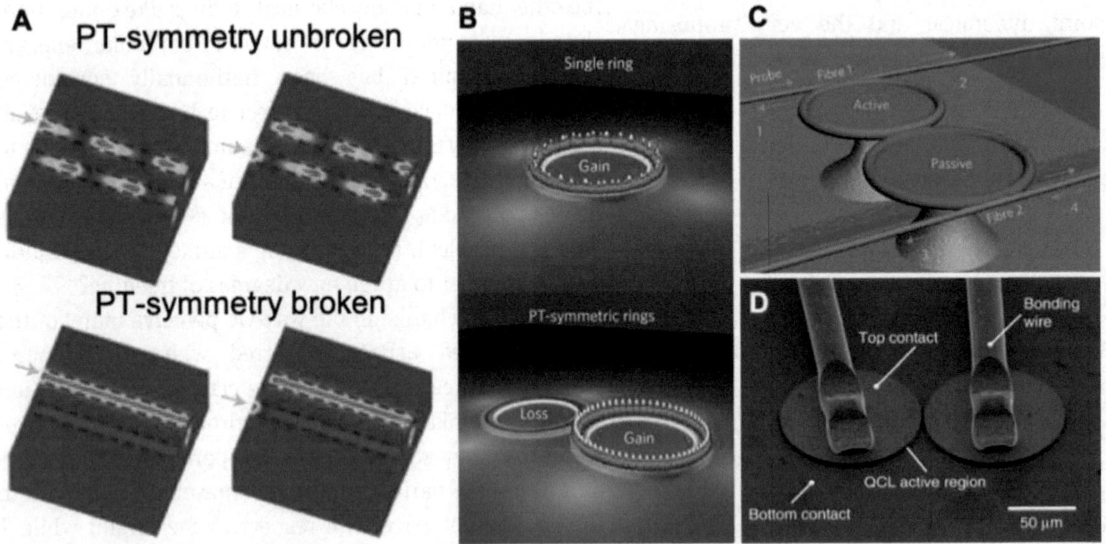

Figure 1: PT symmetry in optics.
(A) Different regimes associated with parity-time (PT)-symmetry breaking manifested in light propagation dynamics within a non-Hermitian (NH) waveguide coupler [9]. (B) Enforcing single-mode lasing in a pair of PT-symmetric microring resonators, each supporting a multitude of lasing states in isolation [55]. (C) A schematic representation of whispering-gallery-mode PT-symmetric microtoroid cavities [48]. (D) Coupled quantum-cascade-laser (QCL) arrangement used to observe pump-induced suppression and revival of lasing [50].

non-Hermiticity, topology, and various other types of symmetries. In Section 2, we discuss PT symmetry and its various realizations in nonconservative optical structures. Section 3 focuses on the physics of EPs and the exotic effects displayed by photonic systems that happen to operate in the vicinity of such NH degeneracies. Section 4 is devoted to topological concepts and their manifestations in connection with non-Hermiticity in the presence of certain classes of symmetries. Finally, Section 5 provides a summary along with an outlook in this general area of research.

2 Non-Hermitian photonics and PT symmetry

As mentioned earlier, an important class of NH Hamiltonians that are capable of exhibiting entirely real eigenvalues is that respecting PT symmetry. In optics, PT symmetry can be established by judiciously distributing gain and loss, processes that are readily accessible in a wide variety of optical platforms ranging from bulk cavities and optical fiber amplifiers to on-chip photonic circuits (Figure 1A). In this regard, optics provides a fertile ground for exploring NH phenomena and their ensuing effects when considered in conjunction with other conservative effects associated with light dynamics. This, in turn, has led to a paradigm shift in molding the flow of light, which was thus far traditionally limited to only shaping the refractive index distribution, starting from the early development of the first lenses and reaching the modern era that witnessed the advent of more sophisticated optical systems like photonic crystal fibers and metamaterials [27–30]. In this section, we focus on some of the exotic behaviors resulting from the introduction of non-Hermiticity and PT symmetry in optical arrangements.

It should be noted that NH effects in photonic settings are not limited to systems exhibiting optical absorption and/or amplification [31–37]. Such phenomena can also arise, for instance, owing to an energy exchange between metastable guided states and leaky modes in optical fibers or whenever a conservative subsystem is open [38–41].

2.1 Lasers and non-Hermitian symmetry breaking

Lasers provide an ideal test bed for studying some of the ramifications of non-Hermiticity. Both gain and loss are indispensable components of any light system, especially when there is a need to overcome optical absorption through the use of amplification. It therefore comes as no surprise that signatures of non-Hermiticity had already been considered in early efforts on lasers, manifested, for example, in mode nonorthogonalities that are known to lead to the Petermann K-factor enhancement [42–44] in the fundamental Schawlow–Townes linewidth [45] of a laser cavity. Nevertheless, until very recently, almost all efforts in designing laser resonators were aimed at lowering the dissipation, something that was traditionally deemed detrimental to the performance of these devices.

The recent developments in NH optics and PT symmetry have provided a systematic framework to explore exotic regimes of behavior in laser systems. Perhaps, the most archetypical example is a pair of two coupled identical cavities, with one being subjected to gain while the other being subjected to an equal amount of loss. Following the previous discussion, this arrangement is PT symmetric, given that the gain/loss distribution is antisymmetric while index-wise is even [46–48]. In the vicinity of the PT phase transition, i.e., at the EP, this structure was shown to exhibit a counterintuitive pump-induced suppression and revival of lasing (Figure 1D) [49–52]. Variations of this behavior have also been demonstrated in a phonon laser arrangement [14], where substantial linewidth enhancement at the EP was found to occur [53]. This latter effect is a by-product of the collapsing supermodes at the degeneracy point, an extreme case of the mode nonorthogonality mentioned earlier.

On many occasions, a primary goal in laser design is to enforce single-mode operation to achieve a coherent, high-quality output. A general trend in fulfilling this requirement is to use miniature semiconductor lasers. However, the inhomogeneously broadened gain bandwidth in such semiconductor-based active materials can easily span a wavelength range that is many times larger than the free spectral range associated with a typical microring cavity, hence eluding the aforementioned goal of exclusively lasing in a single longitudinal mode. One way to mitigate this issue is to use dispersive elements similar to those used, for example, in distributed feedback lasers. An alternative route to accomplish this goal is to exploit the PT symmetry in the underlying cavity structure. In this scenario, by adjusting the pump levels in the system, one can ensure that PT symmetry is broken for only one longitudinal mode, while the other supported states are kept below the PT threshold and hence exhibit real eigenvalues with no amplification. This in turn ensures single-mode operation of the device while all the other principal attributes of the laser cavity remain intact. This scheme has been lately demonstrated in microring lasers (involving whispering gallery modes) without any compromise in terms of slope efficiency or threshold pump intensities (see Figure 1B, C) [54–57]. PT-symmetric lasers have also been demonstrated in various other platforms such as electrically pumped integrated arrangements [58, 59] and optical fibers [60]. A similar approach can be applied to achieve single-frequency lasing through the use of EPs, an aspect demonstrated in dark-state lasers [61, 62].

Non-Hermiticity and loss management in active lattices is shown to be an effective tool for implementing various spin Hamiltonians in an optical platform [63–67]. In these schemes, the loss that vectorial electromagnetic modes experience on the interface of metallic nanocavities provides an effective way to establish the ferromagnetic and antiferromagnetic type of exchanges. Based on such arrangements, large arrays of nanolasers have been realized, which emit in a single mode and with a desired topological singularity [66, 67]. In laser arrays, the interplay between supersymmetry (SUSY) and non-Hermiticity has also been shown to result in single spatial mode operation. In this regard, a waveguide array subject to gain is coupled to a lossy superpartner. With the exception of the fundamental mode that has no counterpart in the partner array, all other modes of the main array share the same eigenfrequencies with those in the superpartner. By adjusting the gain–loss contrast between the two arrays, one can keep all the higher order modes of the array below the PT symmetry breaking point, thus allowing the fundamental mode to experience gain and subsequently lase [68, 69].

On-chip single-mode microlasers based on EPs have so far found applications in generating states with prespecified orbital angular momentum (OAM) on a chip. Typically, the counterpropagating modes within a microring laser cavity form a degenerate pair which tend to simultaneously lase together once the gain exceeds threshold. This precludes a direct generation of a vortex beam with nonzero topological charge since the two opposite azimuthal mode numbers tend to cancel each other. NH schemes based on EPs have recently provided an elegant method to selectively extract only one chiral mode in such lasers [52, 60–72]. This class of devices can emit in a tunable OAM order while operating in a broadband fashion. Another interesting effect that is closely tied to the aforementioned NH aspects is the so-called coherent perfect absorption, where a coherent monochromatic light input is entirely absorbed by a lossy medium [73, 74]. In this respect, a coherent perfect absorber (CPA) acts as a time-reversed version of a laser. Interestingly, it has been shown that a PT-symmetric cavity can simultaneously behave as a laser and a CPA at the same frequency [75, 76].

2.2 PT-symmetric metamaterials and non-Hermitian cloaking

The past two decades have witnessed considerable research efforts that are geared to develop artificial materials, tailored to display properties that are not found in nature. Yet, until recently, such electromagnetic metamaterials have almost exclusively relied only on modifying the real permittivities and permeabilities associated with the material elements [30]. The recent developments in NH photonics have opened up new avenues in exploring light–matter interactions in the entire complex plane of the material constitutive parameters.

In the linear Hermitian domain, a one-dimensional (1D) grating exhibits transmission and reflection properties that are typically independent from the direction the light impinges upon the structure. On the other hand, this feature can be violated in a NH grating. In particular, one can establish gain and loss regions within the unit cell of a grating in such a way that the reflection coefficients are direction dependent (Figure 2A) [77]. This asymmetric wave propagation becomes mostly pronounced at the EP, where light can propagate completely without reflection from one side while exhibiting strong reflection in the opposite direction, a process leading to unidirectional invisibility. Interestingly, because the system is not conservative in this case, the reflection values can greatly exceed unity [77]. Unidirectional invisibility was recently observed in various platforms including photonic mesh lattices (see Figure 2B) [78, 79], passive silicon periodic nanowires [80], multilayer Si/SiO_2 structures [81], organic films [82], and electroacoustic resonators [17].

The peculiar wave transport in NH systems is known to result in other unconventional effects such as unidirectional cloaking [83, 84]. This could be achieved via a PT-symmetric surface that surrounds an object with an arbitrary size. In such a scenario, a lossy part in the surface is designed to entirely absorb the incoming wave while a corresponding active segment can reemit the same amount of power impinging on the object. The result is a broadband cloaking device which benefits from relaxed design constraints owing to its active architecture. Another relevant aspect to this discussion are the so-called "constant intensity waves". Typically, in a lossless medium, an electromagnetic wave (like a plane wave) can remain invariant during propagation when the propagation space is homogeneous. Any inhomogeneity such as an obstacle would inevitably cause reflections and scattering, which disturb the original uniform wavefront. In sharp contrast to this Hermitian picture, by introducing gain and loss in a

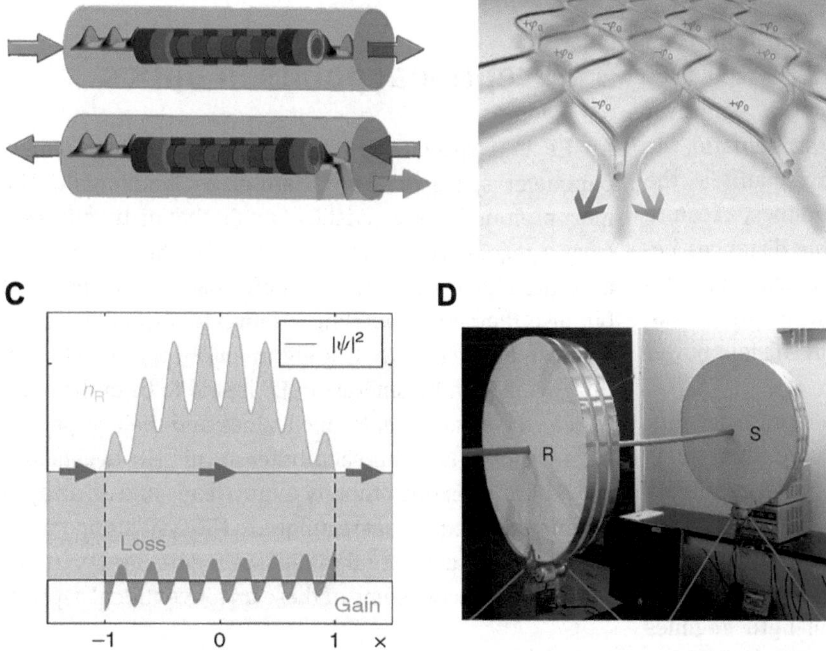

Figure 2: Unidirectional invisibility and parity-time (PT)-symmetry.
(A) A schematic of a PT-symmetric grating that exhibits unidirectional invisibility [77]. (B) Photonic mesh lattices utilized to observe exceptional points [78]. (C) Plane wave propagation in a scattering medium with appropriately patterned gain and loss resulting in a constant intensity wave profile throughout the entire structure [86]. (D) Experimental setup employed to demonstrate efficient wireless power transfer in a nonlinear PT-symmetric arrangement [99].

general class of nonuniform potentials, one can construct a constant-intensity wave solution, even systems with random index distributions (Figure 2C) [85, 86].

Other types of PT-symmetric and complex lattices have been studied and shown to behave quite differently from their Hermitian counterparts. For instance, beam dynamics in a PT-symmetric optical lattice can display band merging effects in the associated complex band structure with EPs emerging in the Bloch momentum space [6]. In addition, even though the material parameters in such lattices are isotropic, the light propagation through the array could exhibit double refraction. This latter unusual behavior stems from the skewedness in the associated Floquet–Bloch modes of the structure. Nonreciprocal Bloch oscillations is yet another peculiar phenomenon that can occur in such complex crystals with no analog in Hermitian arrays [87]. Other works have studied NH symmetry breaking in graphene-like lattices [88], PT Talbot revivals [89], and Anderson localization in disordered PT-symmetric arrangements [90].

Extensions of these ideas have been lately pursued in PT-symmetric metasurfaces. It has been shown that negative refraction and planar focusing could be achieved in such NH sheets without engaging negative-index metamaterials or phase-conjugating surfaces [91–93]. Such structures could enable loss-free all-angle negative refraction and planar lenses in free space. PT-symmetric phase transitions have also been studied in the polarization space of a complex metasurface, in which the eigenstates eventually collapse on each other on the Poincaré sphere when the system reaches an EP [94]. Similar ideas have been exploited to diffract light with asymmetric diffraction orders using a deformed honeycomb metasurface with a diatomic Bravais-lattice topology [95].

2.3 Nonlinear effects in non-Hermitian systems

Nonlinear effects are an integral part of many optical platforms used for implementing NH Hamiltonians. For instance, gain saturation nonlinearities in active materials are known to be responsible for stabilizing laser oscillators. It is therefore natural to investigate the interplay between nonlinearity and non-Hermiticity and the prospect for NH phases such as PT symmetry under such conditions. In this regard, it has been theoretically predicted that optical Kerr nonlinearities can reverse PT-symmetric phases, i.e., transforming a linear system in the PT-unbroken phase to a nonlinear one with a broken symmetry and vice versa [96]. Such nonlinearly induced phase transitions were later shown to also hold in the case of gain saturation nonlinearities and were successfully observed in coupled semiconductor microring lasers [97, 98]. Nevertheless, even in the presence of these nonlinear processes, the eigenmodes of the system still retain their corresponding forms in the linear regime [97]. Interestingly, nonlinear processes can have practical implications in active environments that go beyond lasers. For instance, it has been shown that robust wireless power transfer could be accomplished through the realization of a nonlinear PT-symmetric circuit (Figure 2D) [99]. In this scenario, the nonlinearity guarantees that the system remains in the PT-unbroken phase for strong enough couplings, a feature that allows for a wide range of accessible distances between the source and the receiver. This in turn eliminates the need for constant tunings of the corresponding resonators that is otherwise necessary to attain efficient power transfer.

The impact of non-Hermiticity on nonlinear processes such as that of three- and four-wave mixing has also been investigated. In this vein, non-Hermiticity could assist phase-matching in optical parametric amplifiers when a Hermitian system cannot satisfy these conditions [100, 101]. Such techniques could facilitate parametric amplification in long-wavelength regimes using on-chip semiconductor arrangements with large nonlinearities. In addition, the interplay between nonlinearity and non-Hermiticity in a PT-symmetric optical coupler can lead to interesting functionalities such as optical switching [102] and selective parametric amplification in the spectral domain [103]. Similar concepts have also been considered in optomechanical settings, where the nonlinearity is induced by mechanical vibrations [15]. Along different lines, it has been shown that PT symmetry can be established solely by optical nonlinear processes, without requiring active elements [104]. Such nonlinearity-induced PT effects are instead enabled by parametric gain in a three-wave mixing scenario.

Another interesting function emerging from the combined effects of optical gain and loss is providing additional tools to control light transport in active nonlinear environments [7, 105]. An important class of such NH systems satisfies PT-symmetric conditions with balanced amplification and decay [106–108]. In this vein, PT-symmetry plays a crucial role in determining the stability regimes of nonlinear excitations such as optical solitons in such settings. For instance, it has been predicted and experimentally observed that contrary to other NH nonlinear systems wherein self-trapped states emerge as fixed points in the parameter space, discrete PT solitons can form a continuous family of solutions [79, 108]. Moreover, the synergy between non-Hermiticity and nonlinearity has been deployed to demonstrate unidirectional light transport [109, 110]. Such structures could lead to new approaches in developing all-dielectric on-chip optical components such as circulators.

3 Exceptional points in optics

An EP is a special type of degeneracy in the complex parameter space of a non-Hermitian Hamiltonian. The most profound characteristic of an EP that distinguishes it from a regular Hermitian degeneracy is the fact that not only the eigenvalues degenerately coalesce at this point but also their corresponding eigenvectors simultaneously collapse on each other-leading to an abrupt reduction in dimensionality. In general, an EP is said to be of the order N, if at the same time, N eigenvalues and their respective eigenvectors fuse with each other at this NH degeneracy [111, 112]. This exotic property in turn leads into an array of fascinating effects that are unique to NH systems operating at or close to such singularities (see Figure 3A–D). In this section, we discuss some of these unconventional aspects.

3.1 Enhancement effects around EPs

Over the years, optics has provided some of the most accurate metrology tools for a variety of sensing applications. These devices range from optical gyroscopes and tachometers to chemical and biomedical sensors, to name a few. In this regard, optical resonators have been widely utilized for such purposes mainly due to their ability to provide a strong interaction between the light field and the sensing target. In recent years, the development of ultralow-loss microtoroids [113] and low loss silicon microresonators [114] has sparked a great deal of interest in implementing photonic sensors on a chip.

In spite of their remarkable performance, standard microcavity resonators (characterized by a set of orthogonal modes) are still limited within the bounds imposed by their Hermitian nature. This aspect can be better understood from the perspective of standard perturbation theory. As is well known, if a Hermitian system is perturbed to order ε, then its eigenvalues λ can be obtained from the familiar power series $\lambda = \lambda_0 + \lambda_1 \varepsilon + \lambda_2 \varepsilon^2 + \cdots$, from where one can quickly conclude that the response of any Hermitian arrangement is at best linear with respect to the disturbance ε. In contrast, if a NH system is biased at an EP of order N, once perturbed to order ε, its associated eigenvalues instead follow a Newton–Puiseux series $\lambda = \lambda_0 + \lambda_1 \varepsilon^{1/N} + \lambda_2 \varepsilon^{1/N} + \cdots$ In other words, when an NH configuration is placed at an EP of order N, its first order response is now expected to vary according to $\varepsilon^{1/N}$. Given

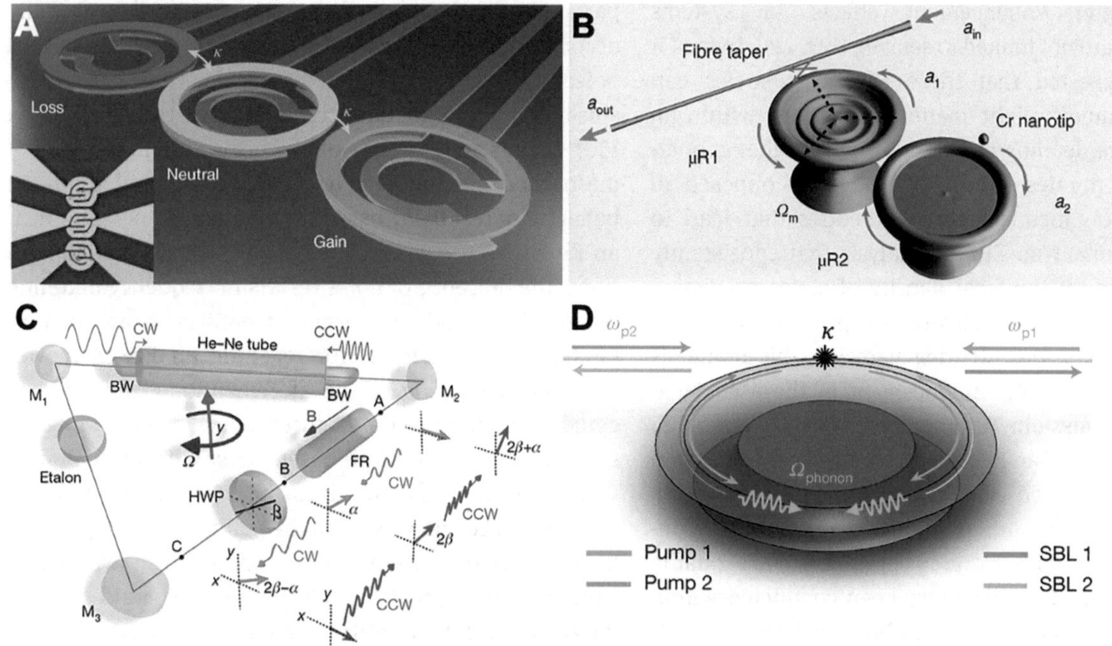

Figure 3: Enhanced sensitivity using exceptional points (EPs).
(A) Coupled microring lasers equipped with microheaters used to realize a third-order EP [118]. (B) A coupled microtoroid phonon laser operating at an exceptional point [53]. (C, D) Enhanced Sagnac effect in EP-based gyroscopes implemented in a ring laser gyroscope [121] and a Brillouin laser cavity [122], respectively.

that for small ε, $\varepsilon^{1/N} \gg \varepsilon$, then it is straightforward to deduce that an NH arrangement (at an EP) can react considerably more drastically than its Hermitian counterpart. Intuitively, this distinctive behavior stems from the abrupt phase transition associated with the reduction in the dimensionality of the corresponding eigenspace. As a result, when dealing with small input signals, the response of EP-based structures can be boosted by orders of magnitude.

Lately, this magnified response provided by EPs has been employed to realize optical sensors with enhanced performance. This can be achieved for example by incorporating two or more scatterers around a passive microcavity which establishes an EP involving the two counterpropagating whispering-gallery modes within the structure [115, 116]. Theoretical results suggest that such an arrangement could enhance single-particle detection sensitivity by a factor of seven as compared to an isolated microcavity [117]. This concept was later demonstrated experimentally both in active and passive photonic platforms. In an active scenario, coupled microring lasers in binary and ternary PT symmetric photonic molecule arrangements have been used to demonstrate second- and third-order EPs, respectively (Figure 3A) [118]. In this case, more than an order of magnitude sensitivity enhancement has been reported which can be boosted even further by the amplitude of the gain present in the system. An alternative realization involves a passive single whispering-gallery optical microresonator which is brought to a second-order EP by using two lossy Rayleigh scatterers [119]. A third scatterer is then used as the target object to be detected which causes a frequency splitting in the complex plane.

From a practical perspective, the augmented response enabled via NH degeneracies has found direct applications in Sagnac-based gyroscopes [120], which can be realized by retrofitting a helium–neon ring laser gyroscope (He–Ne RLG) with a Faraday rotator and a half-wave plate (see Figure 3C) [121]. Combined with Brewster windows surrounding the gain tube, these components can then introduce a differential gain/loss contrast in the system, necessary for establishing an EP. Experimental measurements on such devices indicate a square-root dependence of the frequency response on the gyration speed, in contrast to the linear behavior in a standard arrangement, leading to a twenty-fold enhancement in sensitivity. A different implementation of an EP-enhanced Sagnac effect has been experimentally demonstrated using a microring Brillouin laser (Figure 3D) [122]. By incorporating a fiber taper in the vicinity of the microring, a dissipative coupling takes place between the two counterpropagating modes involved. This in turn induces a second-order EP at a critical pump-detuning frequency, resulting in a four-fold increase in the Sagnac scale factor while allowing for measuring rotations of approximately one revolution per hour.

Interestingly, enhancement effects in systems involving EPs are not limited to sensing [123, 124]. In fact, it has been suggested that these NH degeneracies can considerably modify light–matter interactions within an active structure, leading into substantially higher spontaneous emission rates (Figure 3B) [125]. As opposed to traditional spontaneous emission theories that lead to "infinite" values at an EP, an analysis that consistently takes into account the local density of states predicts a bounded response. Specifically, in passive structures involving second-order EP2 degeneracies, this bound is found to be a four-fold enhancement in the associated spontaneous emission. Another consequence emerging from the presence of an EP in such settings is the emission lineshape itself which can deviate from the Lorentzian profile, thus resulting in a nonlinear scaling of the spontaneous emission with the associated resonance quality factors. Similar studies have also been conducted within the context of NH photonic crystals hosting third-order EP3 degeneracies where an eight-fold enhancement is expected [126]. These results can also be generalized to EPs of higher-order N inducing larger emissions by a factor of $\sqrt{N^3}$ [126].

The fact that EPs are highly sensitive to changes in their environment is not always desirable. For instance, an EP sensor can also be excessively vulnerable to fabrication errors and imperfections that are inevitable in experiments. In this respect, exceptional surfaces have been suggested as a possible avenue to combine the robustness required for practical applications with the characteristic sensitivity offered by EPs [127]. One way to achieve this is to introduce a unidirectional coupling between the counterpropagating modes of a microring cavity. The resulting NH Hamiltonian describing the system features an exceptional surface in the parameter space. In this case, undesired perturbations such as random variations in the coupling coefficients or the resonant frequency of the cavity cause the system to move across this exceptional surface, thus maintaining the useful properties of an EP. On the other hand, when external scatterers start to perturb the cavity, the structure is promptly pushed out of the exceptional surface as a result of the bidirectional coupling that is now induced between the counterpropagating modes. In this latter case, an amplified response could be measured in the spectral splitting of the device. The exceptional surfaces mentioned here are also known to arise in other photonic arrangements such as three-dimensional (3D) PT-symmetric photonic crystals [71, 128].

The prospect of using EPs for optical sensing has recently prompted an investigation of practical aspects associated with these applications such as noise figures. In particular, because boosting the input signal is typically accompanied by an unwanted enhancement of various noise sources, it is not immediately clear if EP-based sensors could offer a superior signal-to-noise ratio (SNR) [129, 130]. To this end, the role of classical noise in the form of mesoscopic fluctuations on the spectral and temporal behavior of resonator-based arrangements operating near an EP has recently been studied [131]. In these configurations, the presence of noise results in frequency detuning among the constituent resonant entities, which in turn modifies the conditions for reaching an EP. Moreover, statistical averaging of the aforementioned fluctuations could smear the spectral features, hence downgrading the effective sensitivity of EP-based sensors to noise-limited values. Along different lines, the performance of EP sensing can be analyzed from the point of view of quantum noise theory [132]. In this vein, it has been shown that by using the quantum Fisher information one could obtain a lower bound for the SNR associated with an EP sensor. These theoretical results predict that by implementing an EP amplifier near the lasing threshold in conjunction with a heterodyne detection scheme, an improved SNR performance as compared with Hermitian sensors can be achieved.

3.2 Encircling EPs and mode conversion

A remarkable behavior of NH degeneracies is related to the dynamical behavior of their associated Hamiltonian. In Hermitian settings, a cyclic evolution that occurs in an adiabatic manner tends to preserve an eigenstate of the system, apart from a geometric phase factor [133]. This picture could completely break down in the case of NH structures involving an EP. A possible scenario in this regard is when the system undergoes a cyclic evolution in a quasi-static fashion. In this case, the instantaneous eigenstates will swap with each other at the end of a single cycle, apart from acquiring a geometric phase [134, 135]. This peculiar behavior can be attributed to the geometry of the intersecting complex Riemann sheets unique to nonconservative systems and has been experimentally observed in microwave [136] and optical cavities [137] as well as exciton–polaritonic arrangements [138]. Alternatively, the NH Hamiltonian may change in such a way that the EP encirclement can no longer be considered adiabatic [139]. Such dynamical evolutions are known to give rise to chiral mode conversions, where the final state of the system is determined by the direction of EP encirclement [140]. This exotic behavior has been experimentally demonstrated in a coupled optical waveguide arrangement [141]. By properly designing the boundaries and losses of each waveguide, a dynamical EP encirclement could be effectively realized in the parameter domain. Owing to the chiral behavior, the output of the system toggles between the even and odd supermodes depending on the direction of propagation of light, regardless of the input beams. A parallel experiment on this chiral mode conversion was carried out in an optomechanical system, where a nonreciprocal transfer was observed between two vibrational modes of a silicon membrane embedded in a high-finesse optical cavity [142]. Similar effects have also been observed in silicon-based photonic architectures [143]. More recently, an analytical explanation of this chiral and robust state conversion mechanism was provided through the asymptotics of exact solutions, along with the fact that this effect can persist even in the presence of nonlinearities [144, 145]. In addition, the possibility of a single-channel optical omni-polarizer was proposed that benefits from this chiral response [144]. Finally, some of the peculiar features arising from the process of winding around multiple NH singularities or EPs have also been explored by invoking the topological notion of homotopy (Figure 4A) [146].

3.3 Symmetries and topology meet EPs

The recent advances in the field of NH physics has incited a flurry of research activities aimed at understanding the interplay between symmetries and topology on the one hand and EPs on the other hand. Such studies can provide a guideline to achieving new symmetry-protected NH phases that have no counterpart whatsoever in the Hermitian domain. For instance, it has been shown that a Dirac point with a nontrivial Berry phase can split into isolated pairs of EPs in the presence of non-Hermiticity [147]. The ensuing double Riemann sheet associated with these EP pairs in turn leads to a bulk Fermi arc which bridges the two EPs in the complex band structure (Figure 4B). This latter effect is a direct by-product of non-Hermiticity and is different from surface Fermi arcs that arise from Weyl points in 3D Hermitian systems. In addition, the EPs obtained in this fashion exhibit half-integer topological

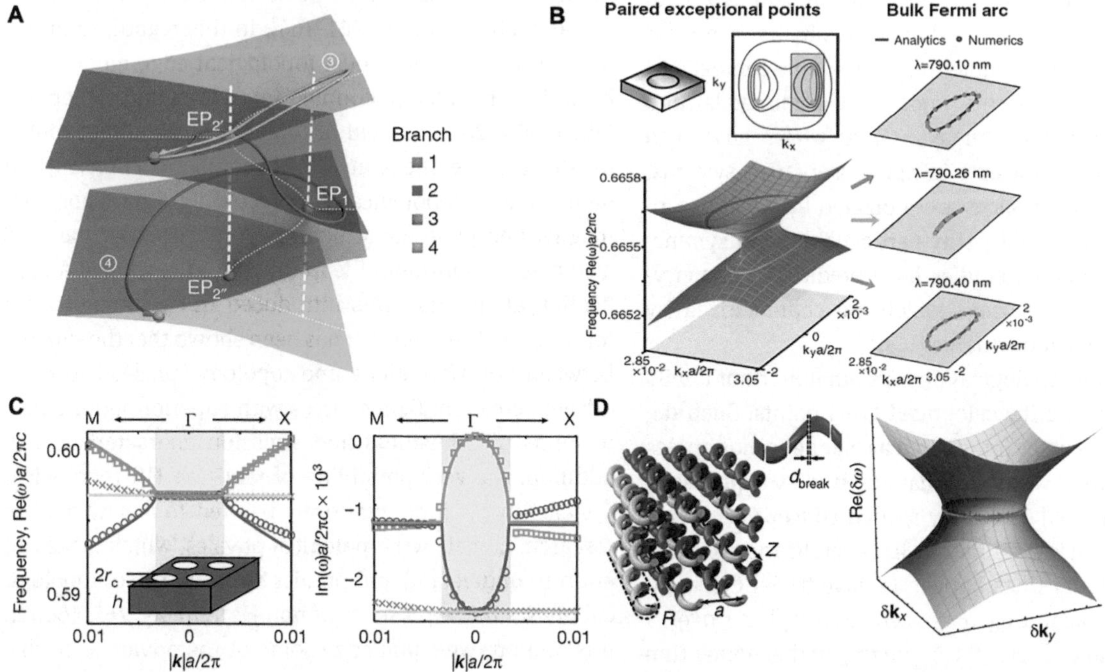

Figure 4: Exceptional rings and higher-order effects.
(A) Exceptional point (EP) encirclement in the parameter space of a system supporting two EPs. In this case, nonhomotopic loops encircling EP1 indicated in the diagram result in different dynamic and stroboscopic evolution behaviors [146]. (B) A bulk Fermi arc that connects a pair of EPs arising from a single Dirac point in the presence of radiation losses in a rhombic lattice having elliptical air holes embedded in a dielectric substrate [147]. (C) Real and imaginary parts of the complex band structure associated with a non-Hermitian square lattice photonic crystal exhibiting a ring of EPs [150]. (D) Helical waveguide lattices with controllable losses utilized to observe a Weyl exceptional ring in their corresponding band structure [156].

invariants, which could manifest themselves in the far-field polarization of the light scattered by the implemented photonic crystal [147]. Along similar lines, there have been proposals to observe bulk Fermi arcs in solid state, where the required non-Hermiticity could be induced via ferromagnetic leads attached to a TI [148].

A Dirac Hamiltonian that deviates from the Hermitian regime is also known to exhibit other interesting effects. To this end, a circuit realization of NH Dirac and Weyl Hamiltonians under the influence of a pseudomagnetic field which is artificially induced by a judicious spatial variation of the circuit elements has been proposed [149]. In this scenario, the combined effect of non-Hermiticity along with the pseudomagnetic field lifts the degeneracy and leads to the emergence of Landau-level–like flat bands in the band structure of the system. Another example involves an accidental degeneracy in the form of a Dirac cone in the Hermitian band structure of a square-lattice photonic crystal [150]. Once non-Hermiticity is at play in the form of radiation losses, this Dirac point transforms into a ring of EPs (Figure 4C) which manifest themselves in the angle-resolved reflection measurements.

Another topic of interest is the synergy between symmetry and EPs. In this vein, symmetry-protected exceptional surfaces are known to generically arise in NH band structures with an increased dimensionality as compared with the case with no symmetries [151]. In this context, NH symmetries can transport nodal NH semimetals into symmetry-protected NH metals. Other works have also investigated these concepts in nonconservative systems, where exceptional surfaces are protected by various constraints such as PT and parity-particle-hole (CP) symmetries [128, 152]. Similar studies have predicted symmetry-protected exceptional rings which are characterized by nonzero topological invariants [153].

In Hermitian topological physics, an important family of degeneracies are the celebrated Weyl points. Such degeneracy points can be interpreted as magnetic monopoles in reciprocal space which are characterized by a quantized Chern number [154]. A closely related concept is a Weyl nodal ring, which is essentially a 3D generalization of Dirac nodes. These latter rings happen to have trivial Chern invariants while acquiring a nonzero Berry phase over a closed path that encircles the entire ring in the momentum space. Quite recently, a set of NH degeneracies termed Weyl exceptional rings have been theoretically predicted which feature both a nontrivial Chern number and a quantized nonzero Berry phase [155]. This intriguing behavior stems from the topology of the Riemann surface which is unique to NH arrangements. Weyl exceptional rings were later experimentally demonstrated in a Floquet system realized in a 3D photonic bipartite lattice comprising evanescently coupled helical waveguides written in silica (Figure 4D) [156]. The required non-Hermiticity in this platform is obtained by incorporating equidistant breaks within the waveguides in one of the sublattices to impose a controlled amount of loss. Experimental measurements confirm a topological transition where Fermi arc surface states emerge in the system after increasing the dissipation levels.

4 Non-Hermitian topological physics

Topological physics is an emerging field that aims to understand and harness a set of new properties arising in a recently discovered phase of matter – properties that tend to remain invariant during a continuous deformation of the system [21–23]. A prime example of such material systems is that of TIs, where electron conduction occurs along the edges while it is prohibited in the bulk [157–160]. In recent years, the prospect of using topological notions in optics to utilize the unique attributes offered by topologically nontrivial structures has been the subject of intense research efforts [25, 26, 161–164]. In this regard, unidirectional transport and robust topological edge modes have been demonstrated in coupled resonators and waveguide lattices [24–26]. The field of topological photonics took a drastic turn after the pioneering experiment of Ref. [25] that demonstrated topologically protected light transport in a magnet-free photonic structure. Quite recently, the field also made a substantial leap forward after optical amplification/attenuation was introduced in conjunction with topology. In this regard, it has been shown that the synergy between non-Hermiticity and topology can lead to more efficient coherent light sources with superior performance in terms of robustness and emission characteristics. In addition, the very possibility of realizing NH topological systems in photonic platforms has led to a new field of research, namely NH topological physics, which makes an effort to understand and predict the response of topological phases in the presence of non-Hermiticity [165, 166]. In this section, we summarize some of the advances in this exciting field.

4.1 Topological lasers

The fact that lasers are susceptible to defects and disorder has always posed a significant challenge in the performance of these devices. Such imperfections are in general inevitable during the fabrication process or may develop in time due to operational degradation and malfunction. These in turn could lead to spatial localization of light within the cavity, eventually leading to lower output powers or even a sudden shutdown of the laser itself. In addition, on many occasions, an array of coupled lasers is used to boost the total emitted power via coherent constructive interference among individual elements. In such scenarios, the system would be even more prone to random deficiencies and failure. It is in this vein that topological features have been sought as a means to develop laser systems that could be immune to perturbations.

In one dimension, an archetypical topological structure is that offered by the Su–Schrieffer–Heeger (SSH) model [167]. When terminated properly, such an array hosts topologically protected defect states with eigenvalues that reside in the middle of the bandgap and are robust against structural disorders. Lasing in this defect state has been demonstrated in systems involving polaritons (Figure 5A) [168], microring cavities (Figure 5B, C) [169, 170], and photonic crystals [171]. Depending on the pumping pattern used, this NH lattice exhibits different regimes of behavior, as dictated by its associated symmetries [169, 172]. In this regard, by appropriately distributing optical gain and loss among the constituent elements, the defect state can be induced to lase in a single-mode fashion while maintaining its topological features in the NH domain. In particular, the lasing edge mode in the SSH array was found to be resilient to both on-site and tunneling disorders [168, 170]. Moreover, unlike the bulk modes, this edge state lases at a wavelength that tends to remain unperturbed even at high pump power levels where non-Hermiticity plays an important role [169].

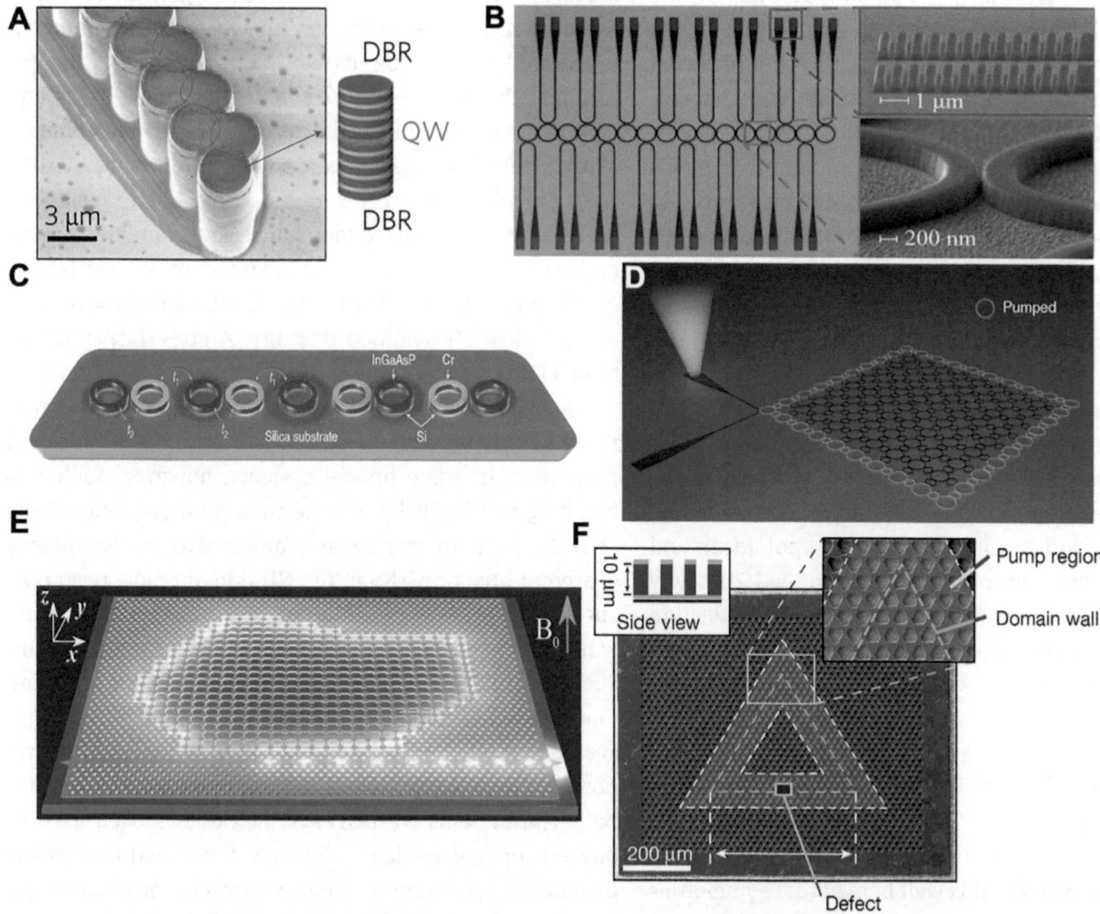

Figure 5: Topological lasers.
(A) Lasing in the topological defect state of a one-dimensional Su–Schrieffer–Heeger (SSH) array of polaritonic micropillar cavities [168]. (B, C) Similar SSH structures implemented based on microring resonators [169, 170]. (D) All-dielectric two-dimensional (2D) topological laser array using microrings coupled via intermediary links to induce an artificial gauge field [174]. (E) Topological lasing in a photonic crystal fabricated on an yttrium iron garnet (YIG) substrate [175]. (F) Electrically pumped topological laser demonstrated in a valley photonic crystal based on quantum cascade lasers (QCLs) [178].

The concept of topological edge transport in a two-dimensional (2D) laser array is yet another avenue that has been pursued in a number of platforms. In this respect, various methods have been used to induce a topologically nontrivial response in these active settings. The prospect of realizing 2D topological lasers has also been pursued in all-dielectric photonic platforms. To achieve this, arrays of coupled microcavities with asymmetric intermediary rings have been used to emulate the presence of an artificial magnetic field (see Figure 5D) [173, 174]. By pumping the boundaries of such a lattice, topological transport in the lasing edge mode has been reported. As compared with a similar but trivial array, higher slope efficiencies and robust single-mode operation is observed even for pump intensities high above threshold. Along different lines, a topologically nontrivial square lattice has been implemented by depositing magneto-optic materials in the substrate of a photonic crystal, where time-reversal symmetry (TRS) can be broken when applying a magnetic field (Figure 5E) [175]. The resulting topological structure is then optically pumped, thus promoting unidirectional lasing along the interface of this square lattice with its surrounding topologically trivial triangular crystal. Similar schemes have been utilized in exciton–polaritonic systems [176]. Alternatively, topological features can be induced in a crystal via the valley degree of freedom [177]. Using this technique, electrically pumped topological lasers have been experimentally realized in the THz regime in a valley photonic crystal inscribed in a QCL wafer (Figure 5F) [178, 179]. The lasing edge mode in such a device can be immune to backscattering due to defects that do not cause intervalley scattering. Interestingly, the role of topology in lasers is not limited to the edge states confined at the boundaries of a photonic system. Rather, it has been suggested that by judiciously interfacing topological and trivial crystals, one can form a highly confined 2D cavity enabled by band inversion reflections [180]. This reflection mechanism can then lead to single-mode lasing with high vertical directionality.

4.2 Non-Hermitian symmetries and topology

In the Hermitian domain, it is well known that symmetries play a pivotal role in topological arrangements. In 2D systems, for example, a Chern insulator could be obtained when the TRS is broken [181]. In fact, the concept of topological protection is often closely intertwined with certain types of symmetries associated with the system [22, 158]. When generalizing topological notions to NH systems, it is therefore natural to ask as to how NH symmetries would interact with topology. Do topologically nontrivial phases exist in certain open systems in various dimensions that satisfy a specific type of symmetry? Can NH symmetries protect a topological edge state?

Answering some of the aforementioned questions has been the subject of numerous recent studies. The existence of topologically protected defect states in PT-symmetric 1D SSH lattices has been predicted and observed in passive coupled waveguides inscribed in silica (Figure 6A) [182, 183]. Various regimes of PT-broken and unbroken were examined as a function of the dimerization in this same structure, and the onset of topological phase transition was shown to be linked to the mean displacement of the light traveling in this discrete lattice. The role of NH symmetries in protecting the defect state of an SSH topological array that respects PT symmetry has also been studied [184]. In such cases, unlike the trivial eigenstates, the topological defect modes preserve their associated eigenvalues even in the presence of perturbations that respect PT symmetry. Remarkably, in some cases, NH symmetries like for example that of NH charge-conjugation or PT symmetry can be the origin of topological edge states, even when a similar Hermitian system is topologically trivial [172, 185, 186]. Therefore, the corresponding defect state resides at the boundary of two regions that are characterized by different NH parameters, and is known to emerge from a continuum in the band structure. Similar effects have been studied in a PT-symmetric Aubry–André–Harper (AAH) model [187].

As mentioned earlier, breaking TRS is yet another way to endow a physical system with nontrivial topological properties. In static optical systems, however, such TRS breaking can typically emerge from gyromagnetic effects [24, 161, 162]. In this regard, alternative methodologies geared at breaking TRS in the NH domain have been pursued, like for example, Floquet systems in 1D quantum walks [188] and 2D lattices for implementing Chern insulators [189, 190]. Alternative techniques for realizing Chern insulators in active platforms have also been suggested that rely on the interplay between nonlinearity and non-Hermiticity [191]. On a different front, endowing topological systems with NH symmetries has been used to develop optical devices with new functionalities. These include PT-symmetric resonator arrays in microwave systems (see Figure 6B) [192], optical isolators in waveguide arrays [193], optical limiters [194], and microring laser arrays capable of light steering [195].

There are currently ongoing efforts aiming to find a unified classification of topological NH systems with different types of symmetries [196, 197]. In this respect,

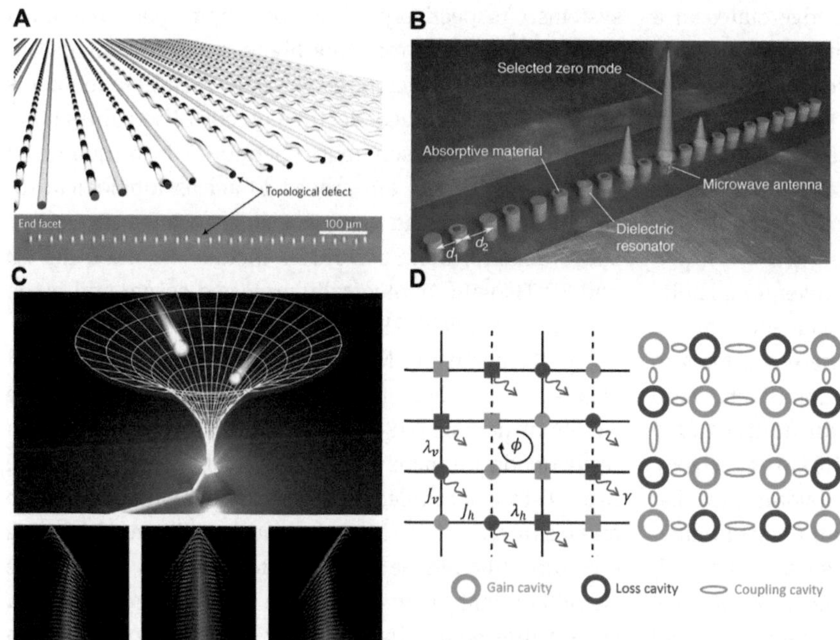

Figure 6: Symmetries and higher-order effects in non-Hermitian (NH) topological systems.
(A) Parity-time (PT)-symmetric Su–Schrieffer–Heeger (SSH) Hamiltonian realized in a waveguide array inscribed in fused silica [182]. (B) Selective enhancement of a topological defect state in a one-dimensional (1D) array of dielectric microwave resonators by utilizing PT-symmetry [192]. (C) Topological funneling of light via the NH skin effect in a 1D chain with anisotropic couplings (top). The input beam is always funneled through the topological interface localized in the middle of the structure, irrespective of its initial launching position (bottom) [209]. (D) A two-dimensional (2D) NH model which supports second-order topological corner states (left). Such a system could be physically realized using microring resonators with intermediary links (right) [219].

even though similar studies in Hermitian TIs [198, 199] can provide a baseline, studies suggest stark distinctions between such arrangements and their NH analogs. For instance, it has been found that the NH counterparts of some of the fundamental symmetries, which are distinct in the Hermitian regime, are in fact equivalent to each other and can be unified [200]. This could lead to nonequilibrium states that are unique to NH topological settings, and implies that in general, the topological classifications of NH symmetries are expected to be quite different from the ones known in the Hermitian domain.

4.3 Non-Hermitian bulk-edge correspondence and non-Hermitian skin effects

The hallmark of topological phases is the presence of edge states that emerge at the boundaries between structures with different topological invariants. In the Hermitian regime, this exotic behavior is a by-product of the bulk-boundary correspondence which relates the topological properties of bulk media to their boundary states [22]. Nevertheless, despite a growing interest in NH topological systems and their applications [46], it is not immediately clear how this correspondence could be translated to such settings [201–204]. Perhaps, a prominent example of how conventional bulk-boundary correspondence could no longer hold in the NH regime is the so-called "non-Hermitian skin effect" [202, 205–208]. In accordance with this, an NH structure with extended bulk states under periodic boundary conditions could behave in a completely different way when terminated with open boundaries. In particular, for certain regions of the NH parameter space, such bulk modes all collapse into localized edge modes – a clear violation of the standard bulk-boundary correspondence. This interesting NH phenomenon has recently been observed in photonic mesh lattices (Figure 6C) [209] and is currently the subject of further studies that aim to shed light on how different NH symmetries could modify the skin effect [196, 207].

Attempts to explain the NH skin effect have been at the core of developing an appropriately modified formalism that could successfully describe the bulk-boundary correspondence in the NH domain. In this regard, two main methods have been implemented so far, each focusing on either the complex spectra of the corresponding NH Hamiltonian and its associated point/line degeneracies [196, 207, 210], or the non-Bloch nature of the eigenstates in an open boundary geometry [205, 206, 211]. In the first approach, topological invariants are interpreted in terms of dynamical phases which depend not only on the eigenstates of the Hamiltonian, but also the associated complex eigenspectrum. In this context, similar to the concept of a topological bandgap in the Hermitian domain, an NH system can be considered as topologically nontrivial if its complex energy spectrum encircles a prespecified base point in the complex plane [210]. Using this, a new kind of bulk-boundary correspondence is then established, wherein a winding number is defined as the degeneracy at the chosen base point. This latter topological invariant represents the number of independent edge states in a semi-infinite arrangement. In this formalism, one finds that unlike Hermitian topological lattices which require at least two bands in their band structure, an NH system could exhibit topological behavior in the absence of any symmetry constraints even within a single band. Other results also related to the role of NH degeneracies in this matter have been reported [212].

Alternatively, the breakdown of the conventional bulk-boundary correspondence in NH arrangements can be attributed to the non-Bloch nature of the eigenstates in such systems. To address this issue, a non-Bloch Chern number could be defined in a 2D lattice which successfully takes into account the aforementioned NH skin effect [206]. Considering a modified NH version of the Haldane model, it has been shown that while the Hermitian Chern number fails to capture topological phase transitions, the new NH Chern number successfully predicts such topological phases and is a faithful indicator of the number of chiral edge modes in this scenario. In contrast to the first approach mentioned previously, this formalism explicitly relies on the non-Bloch eigenstates to redefine the bulk-edge correspondence. Similar findings have been also presented in 1D NH settings [205, 211].

We would like to mention that there are several modifications of the non-Bloch formalism that intend to introduce a self-consistent framework for NH topological systems. One such result exploits the biorthogonal basis for NH Hamiltonians to redefine the bulk as well as edge states in terms of the left and right eigenstates [213]. Based on these, a biorthogonal polarization parameter is then introduced that is shown to be quantized and capable of describing the topological properties of the corresponding NH structure. In addition, it has been suggested that by defining a modified version of the periodic boundary conditions, one could restore the Hermitian bulk-boundary correspondence even in the presence of non-Hermiticity [214]. This could be achieved by introducing a new parameter in this type of boundary conditions. The resulting generalized parameter space can then be used to bridge the open boundary skin effects with the bulk Hamiltonian in a periodic geometry.

4.4 Higher-order non-Hermitian topological effects

In a conventional N-dimensional TI, a nontrivial topological invariant guarantees the existence of N–1–dimensional gapless edge states. Examples include topological transport of surface states and edge modes in 3D and 2D systems, respectively. This paradigm was extended recently [215] by introducing higher-order topological insulators (HOTI), wherein generalized multipole moments are posed as quantized electromagnetic observables [215–217]. By using Wilson loop operators, these quantized multipole moments are proved to act as topological invariants that can lead to topologically protected boundary states. Examples of such HOTIs include second-order 2D and 3D TIs with topologically protected corner and hinge states, respectively [217].

In the context of NH systems, higher-order topological effects could lead to interesting effects that are absent in the Hermitian domain [218]. For instance, HOTIs could arise as a result of non-Hermiticity [219]. This can happen for example in a 2D array of coupled microring cavities with on-site gain and loss arranged in a staggered manner (Figure 6D). In this scenario, although the structure is topologically trivial in the Hermitian limit, by increasing the gain/loss levels above a certain threshold value, a higher-order topological phase transition occurs. The resulting second-order TI is a host to four degenerate corner states in the boundaries of the structure. This behavior can be justified by using a biorthogonal form of nested Wilson loops to establish a higher-order bulk-boundary correspondence [219]. Along different lines, higher-order corner modes within the skin states of NH 2D and 3D lattices have been studied [220] both in reciprocal and nonreciprocal regimes. These NH behaviors have been shown to be governed by a biorthogonal bulk-boundary correspondence [213].

As mentioned before, the usual bulk-boundary correspondence defined for Hermitian TIs can break down in the presence of non-Hermiticity. Interestingly, the higher-order bulk-boundary correspondence could also be modified when translated to the NH domain [221]. In this context, symmetry protected second-order TIs with corner states localized asymmetrically in one boundary have been predicted in 2D NH lattices [221, 222]. In 3D, the breakdown of Hermitian bulk-boundary correspondence can lead to the emergence of anomalous second-order corner modes instead of hinge states. This discrepancy can be rectified via proper use of a non-Bloch eigenstate formalism [221].

5 Summary

In this article, we provided an overview of the recent advances in the field of NH optics. We discussed how various optical platforms exhibiting gain and loss can be utilized to investigate different aspects associated with PT symmetry and NH phenomena. In addition to enabling new functionalities within the discipline of photonics that have no counterparts in the Hermitian domain, such optical realizations have also played a key role in the emerging field of NH topological physics that aims to understand and utilize the synergy between topological notions and non-Hermiticity. More importantly, although PT symmetry, EPs, and the interplay of topology and non-Hermiticity have been extensively explored in the classical and semi-classical regimes, still little is known about the ramification of these developments in a fully quantum domain. In addition, despite intensive recent efforts, a universal theory of NH bulk-boundary correspondence that self-consistently describes topologically nontrivial behavior in all dimensionalities is still elusive. Given that many of these aspects remain unexplored, we believe that further activities in this field will not only yield results that are fundamental in nature, but they could also introduce new tools in photonics and other fields for a new generation of devices and systems.

Acknowledgments: The authors gratefully acknowledge the financial support from DARPA (D18AP00058, HR00111820042, HR00111820038), Army Research Office (ARO; W911NF-16-1-0013, W911NF-17-1-0481), National Science Foundation (ECCS 1454531, DMR 1420620, ECCS 1757025, CBET 1805200, ECCS 2000538, ECCS 2011171), Office of Naval Research (N0001416-1-2640, N00014-18-1-2347, N00014-19-1-2052), Air Force Office of Scientific Research (FA9550-14-1-0037) and US-Israel Binational Science Foundation (BSF; 2016381).

Author contribution: All the authors have accepted responsibility for the entire content of this submitted manuscript and approved submission.

Research funding: We gratefully acknowledge the financial support from DARPA (D18AP00058, HR00111820042, HR00111820038), Army Research Office (ARO; W911NF-16-1-0013, W911NF-17-1-0481), National Science Foundation (ECCS 1454531, DMR 1420620, ECCS 1757025, CBET 1805200, ECCS 2000538, ECCS 2011171), Office of Naval Research (N0001416-1-2640, N00014-18-1-2347, N00014-19-1-2052), Air Force Office of Scientific Research (FA9550-14-1-0037) and US–Israel Binational Science Foundation (BSF; 2016381).

Conflict of interest statement: The authors declare no conflicts of interest regarding this article.

References

[1] G. Gamow, "Zur Quantentheorie des Atomkernes," *Z. Phys.*, vol. 51, pp. 204–212, 1928.

[2] H. Feshbach, C. E. Porter, and V. F. Weisskopf, "Model for nuclear reactions with neutrons," *Phys. Rev.*, vol. 96, pp. 448–464, 1954.

[3] C. M. Bender and S. Boettcher, "Real spectra in non-Hermitian Hamiltonians having PT symmetry," *Phys. Rev. Lett.*, vol. 80, pp. 5243–5246, 1998.

[4] C. M. Bender, S. Boettcher, and P. N. Meisinger, "PT-symmetric quantum mechanics," *J. Math. Phys.*, vol. 40, pp. 2201–2229, 1999.

[5] R. El-Ganainy, K. G. Makris, D. N. Christodoulides, and Z. H. Musslimani, "Theory of coupled optical PT-symmetric structures," *Opt. Lett.*, vol. 32, pp. 2632–2634, 2007.

[6] K. G. Makris, R. El-Ganainy, D. N. Christodoulides, and Z. H. Musslimani, "Beam dynamics in PT symmetric optical lattices," *Phys. Rev. Lett.*, vol. 100, p. 103904, 2008.

[7] Z. H. Musslimani, K. G. Makris, R. El-Ganainy, and D. N. Christodoulides, "Optical solitons in PT periodic potentials," *Phys. Rev. Lett.*, vol. 100, p. 030402, 2008.

[8] A. Guo, G. J. Salamo, D. Duchesne, et al., "Observation of PT-symmetry breaking in complex optical potentials," *Phys. Rev. Lett.*, vol. 103, p. 093902, 2009.

[9] C. E. Rüter, K. G. Makris, R. El-Ganainy, D. N. Christodoulides, M. Segev, and D. Kip, "Observation of parity–time symmetry in optics," *Nat. Phys.*, vol. 6, pp. 192–195, 2010.

[10] S. Klaiman, U. Günther, and N. Moiseyev, "Visualization of branch points in PT-symmetric waveguides," *Phys. Rev. Lett.*, vol. 101, p. 080402, 2008.

[11] S. Bittner, B. Dietz, U. Günther, et al., "PT symmetry and spontaneous symmetry breaking in a microwave billiard," *Phys. Rev. Lett.*, vol. 108, p. 024101, 2012.

[12] J. Schindler, A. Li, M. C. Zheng, F. M. Ellis, and T. Kottos, "Experimental study of active LRC circuits with PT symmetries," *Phys. Rev. A*, vol. 84, p. 040101, 2011.

[13] C. M. Bender, B. K. Berntson, D. Parker, and E. Samuel, "Observation of PT phase transition in a simple mechanical system," *Am. J. Phys.*, vol. 81, pp. 173–179, 2013.

[14] H. Jing, S. K. Özdemir, X.-Y. Lü, J. Zhang, L. Yang, and F. Nori, "PT-symmetric phonon laser," *Phys. Rev. Lett.*, vol. 113, p. 053604, 2014.

[15] H. Jing, Ş. K. Özdemir, Z. Geng, et al., "Optomechanically-induced transparency in parity-time-symmetric microresonators," *Sci. Rep.*, vol. 5, p. 9663, 2015.

[16] X. Zhu, H. Ramezani, C. Shi, J. Zhu, and X. Zhang, "PT-symmetric acoustics," *Phys. Rev. X*, vol. 4, p. 031042, 2014.

[17] R. Fleury, D. Sounas, and A. Alù, "An invisible acoustic sensor based on parity-time symmetry," *Nat. Commun.*, vol. 6, p. 5905, 2015.

[18] C. Hang, G. Huang, and V. V. Konotop, "PT symmetry with a system of three-level atoms," *Phys. Rev. Lett.*, vol. 110, p. 083604, 2013.

[19] Z. Zhang, Y. Zhang, J. Sheng, et al., "Observation of parity-time symmetry in optically induced atomic lattices," *Phys. Rev. Lett.*, vol. 117, p. 123601, 2016.

[20] P. Peng, W. Cao, C. Shen, et al., "Anti-parity–time symmetry with flying atoms," *Nat. Phys.*, vol. 12, pp. 1139–1145, 2016.

[21] D. J. Thouless, M. Kohmoto, M. P. Nightingale, and M. den Nijs, "Quantized Hall conductance in a two-dimensional periodic potential," *Phys. Rev. Lett.*, vol. 49, pp. 405–408, 1982.

[22] M. Z. Hasan and C. L. Kane, "Colloquium: topological insulators," *Rev. Mod. Phys.*, vol. 82, pp. 3045–3067, 2010.

[23] X.-L. Qi and S.-C. Zhang, "Topological insulators and superconductors," *Rev. Mod. Phys.*, vol. 83, pp. 1057–1110, 2011.
[24] Z. Wang, Y. D. Chong, J. D. Joannopoulos, and M. Soljačić, "Reflection-free one-way edge modes in a gyromagnetic photonic crystal," *Phys. Rev. Lett.*, vol. 100, p. 013905, 2008.
[25] M. C. Rechtsman, J. M. Zeuner, Y. Plotnik, et al., "Photonic Floquet topological insulators," *Nature*, vol. 496, pp. 196–200, 2013.
[26] M. Hafezi, S. Mittal, J. Fan, A. Migdall, and J. M. Taylor, "Imaging topological edge states in silicon photonics," *Nat. Photonics*, vol. 7, pp. 1001–1005, 2013.
[27] E. Yablonovitch, "Inhibited spontaneous emission in solid-state physics and electronics," *Phys. Rev. Lett.*, vol. 58, pp. 2059–2062, 1987.
[28] S. John, "Strong localization of photons in certain disordered dielectric superlattices," *Phys. Rev. Lett.*, vol. 58, pp. 2486–2489, 1987.
[29] J. C. Knight, J. Broeng, T. A. Birks, and P. S. J. Russell, "Photonic band gap guidance in optical fibers," *Science*, vol. 282, pp. 1476–1478, 1998.
[30] V. M. Shalaev, "Optical negative-index metamaterials," *Nat. Photonics*, vol. 1, pp. 41–48, 2007.
[31] I. Vorobeichik, U. Peskin, and N. Moiseyev, "Modal losses and design of modal irradiance patterns in an optical fiber by the complex scaled (t, t′) method," *J. Opt. Soc. Am. B*, vol. 12, pp. 1133–1141, 1995.
[32] I. Vorobeichik, N. Moiseyev, D. Neuhauser, M. Orenstein, and U. Peskin, "Calculation of light distribution in optical devices by a global solution of an inhomogeneous scalar wave equation," *IEEE J. Quant. Electron.*, vol. 33, pp. 1236–1244, 1997.
[33] I. Vorobeichik, N. Moiseyev, and D. Neuhauser, "Effect of the second-derivative paraxial term in the scalar Maxwell's equation on amplitude losses and reflections in optical fibers," *J. Opt. Soc. Am. B*, vol. 14, pp. 1207–1212, 1997.
[34] A. M. Kennis, I. Vorobeichik, and N. Moiseyev, "Analysis of an intermediate-mode-assisted directional coupler using Bloch theory," *IEEE J. Quant. Electron.*, vol. 36, pp. 563–573, 2000.
[35] K. Gokhberg, I. Vorobeichik, E. Narevicius, and N. Moiseyev, "Solution of the vector wave equation by the separable effective adiabatic basis set method," *J. Opt. Soc. Am. B*, vol. 21, pp. 1809–1817, 2004.
[36] B. Alfassi, O. Peleg, N. Moiseyev, and M. Segev, "Diverging Rabi oscillations in subwavelength photonic lattices," *Phys. Rev. Lett.*, vol. 106, p. 073901, 2011.
[37] A. Pick and N. Moiseyev, "Polarization dependence of the propagation constant of leaky guided modes," *Phys. Rev. A*, vol. 97, p. 043854, 2018.
[38] N. Moiseyev, L. D. Carr, B. A. Malomed, and Y. B. Band, "Transition from resonances to bound states in nonlinear systems: application to Bose–Einstein condensates," *J. Phys. B: At. Mol. Opt. Phys.*, vol. 37, pp. L193–L200, 2004.
[39] N. Moiseyev and L. S. Cederbaum, "Resonance solutions of the nonlinear Schrödinger equation: tunneling lifetime and fragmentation of trapped condensates," *Phys. Rev. A*, vol. 72, p. 033605, 2005.
[40] O. Peleg, M. Segev, G. Bartal, D. N. Christodoulides, and N. Moiseyev, "Nonlinear waves in subwavelength waveguide arrays: evanescent bands and the "Phoenix Soliton"," *Phys. Rev. Lett.*, vol. 102, p. 163902, 2009.
[41] O. Peleg, Y. Plotnik, N. Moiseyev, O. Cohen, and M. Segev, "Self-trapped leaky waves and their interactions," *Phys. Rev. A*, vol. 80, p. 041801, 2009.
[42] K. Petermann, "Calculated spontaneous emission factor for double-heterostructure injection lasers with gain-induced waveguiding," *IEEE J. Quant. Electron.*, vol. 15, pp. 566–570, 1979.
[43] A. E. Siegman, "Excess spontaneous emission in non-Hermitian optical systems. I. Laser amplifiers," *Phys. Rev. A*, vol. 39, pp. 1253–1263, 1989.
[44] W. A. Hamel and J. P. Woerdman, "Observation of enhanced fundamental linewidth of a laser due to nonorthogonality of its longitudinal eigenmodes," *Phys. Rev. Lett.*, vol. 64, pp. 1506–1509, 1990.
[45] A. L. Schawlow and C. H. Townes, "Infrared and optical masers," *Phys. Rev.*, vol. 112, pp. 1940–1949, 1958.
[46] G. Yoo, H.-S. Sim, and H. Schomerus, "Quantum noise and mode nonorthogonality in non-Hermitian PT-symmetric optical resonators," *Phys. Rev. A*, vol. 84, p. 063833, 2011.
[47] B. Peng, Ş. K. Özdemir, F. Lei, et al., "Parity–time-symmetric whispering-gallery microcavities," *Nat. Phys.*, vol. 10, pp. 394–398, 2014.
[48] L. Chang, X. Jiang, S. Hua, et al., "Parity–time symmetry and variable optical isolation in active–passive-coupled microresonators," *Nat. Photonics*, vol. 8, pp. 524–529, 2014.
[49] M. Liertzer, L. Ge, A. Cerjan, A. D. Stone, H. E. Türeci, and S. Rotter, "Pump-induced exceptional points in lasers," *Phys. Rev. Lett.*, vol. 108, p. 173901, 2012.
[50] M. Brandstetter, M. Liertzer, C. Deutsch, et al., "Reversing the pump dependence of a laser at an exceptional point," *Nat. Commun.*, vol. 5, p. 4034, 2014.
[51] B. Peng, Ş. K. Özdemir, S. Rotter, et al., "Loss-induced suppression and revival of lasing," *Science*, vol. 346, pp. 328–332, 2014.
[52] B. Peng, Ş. K. Özdemir, M. Liertzer, et al., "Chiral modes and directional lasing at exceptional points," *Proc. Natl. Acad. Sci. U. S. A.*, vol. 113, pp. 6845–6850, 2016.
[53] J. Zhang, B. Peng, Ş. K. Özdemir, et al., "A phonon laser operating at an exceptional point," *Nat. Photonics*, vol. 12, pp. 479–484, 2018.
[54] M.-A. Miri, P. LiKamWa, and D. N. Christodoulides, "Large area single-mode parity–time-symmetric laser amplifiers," *Opt. Lett.*, vol. 37, p. 764, 2012.
[55] H. Hodaei, M.-A. Miri, M. Heinrich, D. N. Christodoulides, and M. Khajavikhan, "Parity-time–symmetric microring lasers," *Science*, vol. 346, pp. 975–978, 2014.
[56] L. Feng, Z. J. Wong, R.-M. Ma, Y. Wang, and X. Zhang, "Single-mode laser by parity-time symmetry breaking," *Science*, vol. 346, pp. 972–975, 2014.
[57] H. Hodaei, M.-A. Miri, A. U. Hassan, et al., "Single mode lasing in transversely multi-moded PT-symmetric microring resonators," *Laser Photonics Rev.*, vol. 10, pp. 494–499, 2016.
[58] W. Liu, M. Li, R. S. Guzzon, et al., "An integrated parity-time symmetric wavelength-tunable single-mode microring laser," *Nat. Commun.*, vol. 8, p. 15389, 2017.
[59] W. E. Hayenga, H. Garcia-Gracia, E. Sanchez Cristobal, et al., "Electrically pumped microring parity-time-symmetric lasers," *Proc. IEEE*, vol. 108, pp. 827–836, 2020.

[60] A. K. Jahromi, A. U. Hassan, D. N. Christodoulides, and A. F. Abouraddy, "Statistical parity-time-symmetric lasing in an optical fibre network," *Nat. Commun.*, vol. 8, p. 1359, 2017.
[61] C. M. Gentry and M. A. Popović, "Dark state lasers," *Opt. Lett.*, vol. 39, pp. 4136–4139, 2014.
[62] H. Hodaei, A. U. Hassan, W. E. Hayenga, M. A. Miri, D. N. Christodoulides, and M. Khajavikhan, "Dark-state lasers: mode management using exceptional points," *Opt. Lett.*, vol. 41, pp. 3049–3052, 2016.
[63] A. Marandi, Z. Wang, K. Takata, R. L. Byer, and Y. Yamamoto, "Network of time-multiplexed optical parametric oscillators as a coherent Ising machine," *Nat. Photonics*, vol. 8, pp. 937–942, 2014.
[64] N. G. Berloff, M. Silva, K. Kalinin, et al., "Realizing the classical XY Hamiltonian in polariton simulators," *Nat. Mater.*, vol. 16, pp. 1120–1126, 2017.
[65] M. Nixon, E. Ronen, A. A. Friesem, and N. Davidson, "Observing geometric frustration with thousands of coupled lasers," *Phys. Rev. Lett.*, vol. 110, p. 184102, 2013.
[66] M. Parto, W. Hayenga, A. Marandi, D. N. Christodoulides, and M. Khajavikhan, "Realizing spin Hamiltonians in nanoscale active photonic lattices," *Nat. Mater.*, vol. 19, pp. 725–731, 2020.
[67] M. Parto, W. E. Hayenga, A. Marandi, D. N. Christodoulides, and M. Khajavikhan, "Nanolaser-based emulators of spin Hamiltonians," *Nanophotonics*, vol. 1, 2020. https://doi.org/10.1515/nanoph-2020-0230.
[68] R. El-Ganainy, L. Ge, M. Khajavikhan, and D. N. Christodoulides, "Supersymmetric laser arrays," *Phys. Rev. A*, vol. 92, p. 033818, 2015.
[69] M. P. Hokmabadi, N. S. Nye, R. El-Ganainy, D. N. Christodoulides, and M. Khajavikhan, "Supersymmetric laser arrays," *Science*, vol. 363, pp. 623–626, 2019.
[70] P. Miao, Z. Zhang, J. Sun, et al., "Orbital angular momentum microlaser," *Science*, vol. 353, pp. 464–467, 2016.
[71] J. Ren, Y. G. N. Liu, M. Parto, et al., "Unidirectional light emission in PT-symmetric microring lasers," *Opt. Express*, vol. 26, pp. 27153–27160, 2018.
[72] W. E. Hayenga, M. Parto, J. Ren, et al., "Direct generation of tunable orbital angular momentum beams in microring lasers with broadband exceptional points," *ACS Photonics*, vol. 6, pp. 1895–1901, 2019.
[73] Y. D. Chong, L. Ge, H. Cao, and A. D. Stone, "Coherent perfect absorbers: time-reversed lasers," *Phys. Rev. Lett.*, vol. 105, p. 053901, 2010.
[74] Y. Sun, W. Tan, H. Li, J. Li, and H. Chen, "Experimental demonstration of a coherent perfect absorber with PT phase transition," *Phys. Rev. Lett.*, vol. 112, p. 143903, 2014.
[75] S. Longhi, "PT-symmetric laser absorber," *Phys. Rev. A*, vol. 82, p. 031801, 2010.
[76] Z. J. Wong, Y.-L. Xu, J. Kim, et al., "Lasing and anti-lasing in a single cavity," *Nat. Photonics*, vol. 10, pp. 796–801, 2016.
[77] Z. Lin, H. Ramezani, T. Eichelkraut, T. Kottos, H. Cao, and D. N. Christodoulides, "Unidirectional invisibility induced by PT-symmetric periodic structures," *Phys. Rev. Lett.*, vol. 106, p. 213901, 2011.
[78] A. Regensburger, C. Bersch, M.-A. Miri, G. Onishchukov, D. N. Christodoulides, and U. Peschel, "Parity–time synthetic photonic lattices," *Nature*, vol. 488, pp. 167–171, 2012.
[79] M. Wimmer, M.-A. Miri, D. Christodoulides, and U. Peschel, "Observation of Bloch oscillations in complex PT-symmetric photonic lattices," *Sci. Rep.*, vol. 5, p. 17760, 2015.
[80] L. Feng, Y.-L. Xu, W. S. Fegadolli, et al., "Experimental demonstration of a unidirectional reflectionless parity-time metamaterial at optical frequencies," *Nat. Mater.*, vol. 12, pp. 108–113, 2013.
[81] L. Feng, X. Zhu, S. Yang, et al., "Demonstration of a large-scale optical exceptional point structure," *Opt. Express*, vol. 22, pp. 1760–1767, 2014.
[82] Y. Yan and N. C. Giebink, "Passive PT symmetry in organic composite films via complex refractive index modulation," *Adv. Opt. Mater.*, vol. 2, pp. 423–427, 2014.
[83] X. Zhu, L. Feng, P. Zhang, X. Yin, and X. Zhang, "One-way invisible cloak using parity-time symmetric transformation optics," *Opt. Lett.*, vol. 38, pp. 2821–2824, 2013.
[84] D. L. Sounas, R. Fleury, and A. Alù, "Unidirectional cloaking based on metasurfaces with balanced loss and gain," *Phys. Rev. Applied*, vol. 4, p. 014005, 2015.
[85] K. G. Makris, Z. H. Musslimani, D. N. Christodoulides, and S. Rotter, "Constant-intensity waves and their modulation instability in non-Hermitian potentials," *Nat. Commun.*, vol. 6, p. 7257, 2015.
[86] K. G. Makris, A. Brandstötter, P. Ambichl, Z. H. Musslimani, and S. Rotter, "Wave propagation through disordered media without backscattering and intensity variations," *Light Sci. Appl.*, vol. 6, p. e17035, 2017.
[87] S. Longhi, "Bloch oscillations in complex crystals with PT symmetry," *Phys. Rev. Lett.*, vol. 103, p. 123601, 2009.
[88] A. Szameit, M. C. Rechtsman, O. Bahat-Treidel, and M. Segev, "PT-symmetry in honeycomb photonic lattices," *Phys. Rev. A*, vol. 84, p. 021806, 2011.
[89] H. Ramezani, D. N. Christodoulides, V. Kovanis, I. Vitebskiy, and T. Kottos, "PT-symmetric Talbot effects," *Phys. Rev. Lett.*, vol. 109, p. 033902, 2012.
[90] D. M. Jović, C. Denz, and M. R. Belić, "Anderson localization of light in PT-symmetric optical lattices," *Opt. Lett.*, vol. 37, pp. 4455–4457, 2012.
[91] R. Fleury, D. L. Sounas, and A. Alù, "Negative refraction and planar focusing based on parity-time symmetric metasurfaces," *Phys. Rev. Lett.*, vol. 113, p. 023903, 2014.
[92] F. Monticone, C. A. Valagiannopoulos, and A. Alù, "Parity-time symmetric nonlocal metasurfaces: all-angle negative refraction and volumetric imaging," *Phys. Rev. X*, vol. 6, p. 041018, 2016.
[93] N. Lazarides and G. P. Tsironis, "Gain-driven discrete breathers in PT-symmetric nonlinear metamaterials," *Phys. Rev. Lett.*, vol. 110, p. 053901, 2013.
[94] M. Lawrence, N. Xu, X. Zhang, et al., "Manifestation of PT symmetry breaking in polarization space with terahertz metasurfaces," *Phys. Rev. Lett.*, vol. 113, p. 093901, 2014.
[95] N. S. Nye, A. E. Halawany, C. Markos, M. Khajavikhan, and D. N. Christodoulides, "Flexible PT-symmetric optical metasurfaces," *Phys. Rev. Applied*, vol. 13, p. 064005, 2020.
[96] Y. Lumer, Y. Plotnik, M. C. Rechtsman, and M. Segev, "Nonlinearly induced PT transition in photonic systems," *Phys. Rev. Lett.*, vol. 111, p. 263901, 2013.
[97] A. U. Hassan, H. Hodaei, M.-A. Miri, M. Khajavikhan, and D. N. Christodoulides, "Nonlinear reversal of the PT-symmetric

phase transition in a system of coupled semiconductor microring resonators," *Phys. Rev. A*, vol. 92, p. 063807, 2015.
[98] L. Ge and R. El-Ganainy, "Nonlinear modal interactions in parity-time (PT) symmetric lasers," *Sci. Rep.*, vol. 6, p. 24889, 2016.
[99] S. Assawaworrarit, X. Yu, and S. Fan, "Robust wireless power transfer using a nonlinear parity–time-symmetric circuit," *Nature*, vol. 546, pp. 387–390, 2017.
[100] T. Wasak, P. Szańkowski, V. V. Konotop, and M. Trippenbach, "Four-wave mixing in a parity-time (PT)-symmetric coupler," *Opt. Lett.*, vol. 40, pp. 5291–5294, 2015.
[101] R. El-Ganainy, J. I. Dadap, and R. M. Osgood, "Optical parametric amplification via non-Hermitian phase matching," *Opt. Lett.*, vol. 40, pp. 5086–5089, 2015.
[102] A. A. Sukhorukov, Z. Xu, and Y. S. Kivshar, "Nonlinear suppression of time reversals in PT-symmetric optical couplers," *Phys. Rev. A*, vol. 82, p. 043818, 2010.
[103] D. A. Antonosyan, A. S. Solntsev, and A. A. Sukhorukov, "Parity-time anti-symmetric parametric amplifier," *Opt. Lett.*, vol. 40, p. 4575, 2015.
[104] M.-A. Miri and A. Alù, "Nonlinearity-induced PT-symmetry without material gain," *New J. Phys.*, vol. 18, p. 065001, 2016.
[105] M. J. Ablowitz and Z. H. Musslimani, "Integrable nonlocal nonlinear schrodinger equation," *Phys. Rev. Lett.*, vol. 110, p. 064105, 2013.
[106] N. V. Alexeeva, I. V. Barashenkov, A. A. Sukhorukov, and Y. S. Kivshar, "Optical solitons in PT-symmetric nonlinear couplers with gain and loss," *Phys. Rev. A*, vol. 85, p. 063837, 2012.
[107] Y. V. Kartashov, B. A. Malomed, and L. Torner, "Unbreakable PT symmetry of solitons supported by inhomogeneous defocusing nonlinearity," *Opt. Lett.*, vol. 39, pp. 5641–5644, 2014.
[108] V. V. Konotop, J. Yang, and D. A. Zezyulin, "Nonlinear waves in PT-symmetric systems," *Rev. Mod. Phys.*, vol. 88, p. 035002, 2016.
[109] H. Ramezani, T. Kottos, R. El-Ganainy, and D. N. Christodoulides, "Unidirectional nonlinear PT-symmetric optical structures," *Phys. Rev. A*, vol. 82, p. 043803, 2010.
[110] P. Aleahmad, M. Khajavikhan, D. Christodoulides, and P. LiKamWa, "Integrated multi-port circulators for unidirectional optical information transport," *Sci. Rep.*, vol. 7, p. 2129, 2017.
[111] N. Moiseyev, *Non-Hermitian Quantum Mechanics*, Cambridge, Cambridge University Press, 2011.
[112] M. H. Teimourpour, R. El-Ganainy, A. Eisfeld, A. Szameit, and D. N. Christodoulides, "Light transport in PT-invariant photonic structures with hidden symmetries," *Phys. Rev. A*, vol. 90, p. 053817, 2014.
[113] D. K. Armani, T. J. Kippenberg, S. M. Spillane, and K. J. Vahala, "Ultra-high-Q toroid microcavity on a chip," *Nature*, vol. 421, pp. 925–928, 2003.
[114] V. R. Almeida, C. A. Barrios, R. R. Panepucci, and M. Lipson, "All-optical control of light on a silicon chip," *Nature*, vol. 431, pp. 1081–1084, 2004.
[115] J. Zhu, Ş. K. Özdemir, L. He, and L. Yang, "Controlled manipulation of mode splitting in an optical microcavity by two Rayleigh scatterers," *Opt. Express*, vol. 18, pp. 23535–23543, 2010.
[116] J. Wiersig, "Structure of whispering-gallery modes in optical microdisks perturbed by nanoparticles," *Phys. Rev. A*, vol. 84, p. 063828, 2011.
[117] J. Wiersig, "Enhancing the sensitivity of frequency and energy splitting detection by using exceptional points: application to microcavity sensors for single-particle detection," *Phys. Rev. Lett.*, vol. 112, p. 203901, 2014.
[118] H. Hodaei, A. U. Hassan, S. Wittek, et al., "Enhanced sensitivity at higher-order exceptional points," *Nature*, vol. 548, pp. 187–191, 2017.
[119] W. Chen, Ş. Kaya Özdemir, G. Zhao, J. Wiersig, and L. Yang, "Exceptional points enhance sensing in an optical microcavity," *Nature*, vol. 548, pp. 192–196, 2017.
[120] J. Ren, H. Hodaei, G. Harari, et al., "Ultrasensitive micro-scale parity-time-symmetric ring laser gyroscope," *Opt. Lett.*, vol. 42, pp. 1556–1559, 2017.
[121] M. P. Hokmabadi, A. Schumer, D. N. Christodoulides, and M. Khajavikhan, "Non-Hermitian ring laser gyroscopes with enhanced Sagnac sensitivity," *Nature*, vol. 576, pp. 70–74, 2019.
[122] Y.-H. Lai, Y.-K. Lu, M.-G. Suh, Z. Yuan, and K. Vahala, "Observation of the exceptional-point-enhanced Sagnac effect," *Nature*, vol. 576, pp. 65–69, 2019.
[123] T. Goldzak, A. A. Mailybaev, and N. Moiseyev, "Light stops at exceptional points," *Phys. Rev. Lett.*, vol. 120, p. 013901, 2018.
[124] Q. Zhong, D. N. Christodoulides, M. Khajavikhan, K. G. Makris, and R. El-Ganainy, "Power-law scaling of extreme dynamics near higher-order exceptional points," *Phys. Rev. A*, vol. 97, p. 020105, 2018.
[125] A. Pick, B. Zhen, O. D. Miller, et al., "General theory of spontaneous emission near exceptional points," *Opt. Express*, vol. 25, pp. 12325–12348, 2017.
[126] Z. Lin, A. Pick, M. Lončar, and A. W. Rodriguez, "Enhanced spontaneous emission at third-order Dirac exceptional points in inverse-designed photonic crystals," *Phys. Rev. Lett.*, vol. 117, p. 107402, 2016.
[127] Q. Zhong, J. Ren, M. Khajavikhan, D. N. Christodoulides, Ş. K. Özdemir, and R. El-Ganainy, "Sensing with exceptional surfaces in order to combine sensitivity with robustness," *Phys. Rev. Lett.*, vol. 122, p. 153902, 2019.
[128] H. Zhou, J. Y. Lee, S. Liu, and B. Zhen, "Exceptional surfaces in PT-symmetric non-Hermitian photonic systems," *Optica*, vol. 6, pp. 190–193, 2019.
[129] W. Langbein, "No exceptional precision of exceptional-point sensors," *Phys. Rev. A*, vol. 98, p. 023805, 2018.
[130] H. Wang, Y.-H. Lai, Z. Yuan, M.-G. Suh, and K. Vahala, "Petermann-factor sensitivity limit near an exceptional point in a Brillouin ring laser gyroscope," *Nat. Commun.*, vol. 11, p. 1610, 2020.
[131] N. A. Mortensen, P. a. D. Gonçalves, M. Khajavikhan, D. N. Christodoulides, C. Tserkezis, and C. Wolff, "Fluctuations and noise-limited sensing near the exceptional point of parity-time-symmetric resonator systems," *Optica*, vol. 5, pp. 1342–1346, 2018.
[132] M. Zhang, W. Sweeney, C. W. Hsu, L. Yang, A. D. Stone, and L. Jiang, "Quantum noise theory of exceptional point amplifying sensors," *Phys. Rev. Lett.*, vol. 123, p. 180501, 2019.
[133] M. V. Berry, "Quantal phase factors accompanying adiabatic changes," *Proc. R. Soc. Lond. A Math. Phys. Sci.*, vol. 392, pp. 45–57, 1984.
[134] W. D. Heiss, "The physics of exceptional points," *J. Phys. A Math. Theor.*, vol. 45, p. 444016, 2012.

[135] A. A. Mailybaev, O. N. Kirillov, and A. P. Seyranian, "Geometric phase around exceptional points," *Phys. Rev. A*, vol. 72, p. 014104, 2005.
[136] C. Dembowski, H.-D. Gräf, H. L. Harney, et al., "Experimental observation of the topological structure of exceptional points," *Phys. Rev. Lett.*, vol. 86, pp. 787–790, 2001.
[137] S.-B. Lee, J. Yang, S. Moon, et al., "Observation of an exceptional point in a chaotic optical microcavity," *Phys. Rev. Lett.*, vol. 103, p. 134101, 2009.
[138] T. Gao, E. Estrecho, K. Y. Bliokh, et al., "Observation of non-Hermitian degeneracies in a chaotic exciton-polariton billiard," *Nature*, vol. 526, pp. 554–558, 2015.
[139] R. Uzdin, A. Mailybaev, and N. Moiseyev, "On the observability and asymmetry of adiabatic state flips generated by exceptional points," *J. Phys. A Math. Theor.*, vol. 44, p. 435302, 2011.
[140] T. J. Milburn, J. Doppler, C. A. Holmes, S. Portolan, S. Rotter, and P. Rabl, "General description of quasiadiabatic dynamical phenomena near exceptional points," *Phys. Rev. A*, vol. 92, p. 052124, 2015.
[141] J. Doppler, A. A. Mailybaev, J. Böhm, et al., "Dynamically encircling an exceptional point for asymmetric mode switching," *Nature*, vol. 537, pp. 76–79, 2016.
[142] H. Xu, D. Mason, L. Jiang, and J. G. E. Harris, "Topological energy transfer in an optomechanical system with exceptional points," *Nature*, vol. 537, pp. 80–83, 2016.
[143] J. W. Yoon, Y. Choi, C. Hahn, et al., "Time-asymmetric loop around an exceptional point over the full optical communications band," *Nature*, vol. 562, pp. 86–90, 2018.
[144] A. U. Hassan, B. Zhen, M. Soljačić, M. Khajavikhan, and D. N. Christodoulides, "Dynamically encircling exceptional points: exact evolution and polarization state conversion," *Phys. Rev. Lett.*, vol. 118, p. 093002, 2017.
[145] A. U. Hassan, G. L. Galmiche, G. Harari, et al., "Chiral state conversion without encircling an exceptional point," *Phys. Rev. A*, vol. 96, p. 052129, 2017.
[146] Q. Zhong, M. Khajavikhan, D. N. Christodoulides, and R. El-Ganainy, "Winding around non-Hermitian singularities," *Nat. Commun.*, vol. 9, p. 4808, 2018.
[147] H. Zhou, C. Peng, Y. Yoon, et al., "Observation of bulk Fermi arc and polarization half charge from paired exceptional points," *Science*, vol. 359, pp. 1009–1012, 2018.
[148] E. J. Bergholtz and J. C. Budich, "Non-Hermitian Weyl physics in topological insulator ferromagnet junctions," *Phys. Rev. Research*, vol. 1, p. 012003, 2019.
[149] X.-X. Zhang and M. Franz, "Non-Hermitian exceptional Landau quantization in electric circuits," *Phys. Rev. Lett.*, vol. 124, p. 046401, 2020.
[150] B. Zhen, C. W. Hsu, Y. Igarashi, et al., "Spawning rings of exceptional points out of Dirac cones," *Nature*, vol. 525, pp. 354–358, 2015.
[151] J. C. Budich, J. Carlström, F. K. Kunst, and E. J. Bergholtz, "Symmetry-protected nodal phases in non-Hermitian systems," *Phys. Rev. B*, vol. 99, p. 041406, 2019.
[152] R. Okugawa and T. Yokoyama, "Topological exceptional surfaces in non-Hermitian systems with parity-time and parity-particle-hole symmetries," *Phys. Rev. B*, vol. 99, p. 041202, 2019.
[153] T. Yoshida, R. Peters, N. Kawakami, and Y. Hatsugai, "Symmetry-protected exceptional rings in two-dimensional correlated systems with chiral symmetry," *Phys. Rev. B*, vol. 99, p. 121101, 2019.
[154] X. Wan, A. M. Turner, A. Vishwanath, and S. Y. Savrasov, "Topological semimetal and Fermi-arc surface states in the electronic structure of pyrochlore iridates," *Phys. Rev. B*, vol. 83, p. 205101, 2011.
[155] Y. Xu, S.-T. Wang, and L.-M. Duan, "Weyl exceptional rings in a three-dimensional dissipative cold atomic gas," *Phys. Rev. Lett.*, vol. 118, p. 045701, 2017.
[156] A. Cerjan, S. Huang, M. Wang, K. P. Chen, Y. Chong, and M. C. Rechtsman, "Experimental realization of a Weyl exceptional ring," *Nat. Photonics*, vol. 13, pp. 623–628, 2019.
[157] K. V. Klitzing, G. Dorda, and M. Pepper, "New method for high-accuracy determination of the fine-structure constant based on quantized Hall resistance," *Phys. Rev. Lett.*, vol. 45, pp. 494–497, 1980.
[158] C. L. Kane and E. J. Mele, "Quantum spin Hall effect in graphene," *Phys. Rev. Lett.*, vol. 95, p. 226801, 2005.
[159] B. A. Bernevig, T. L. Hughes, and S.-C. Zhang, "Quantum spin Hall effect and topological phase transition in HgTe quantum wells," *Science*, vol. 314, pp. 1757–1761, 2006.
[160] M. König, S. Wiedmann, C. Brüne, et al., "Quantum spin Hall insulator state in HgTe quantum wells," *Science*, vol. 318, pp. 766–770, 2007.
[161] F. D. M. Haldane and S. Raghu, "Possible realization of directional optical waveguides in photonic crystals with broken time-reversal symmetry," *Phys. Rev. Lett.*, vol. 100, p. 013904, 2008.
[162] Z. Wang, Y. Chong, J. D. Joannopoulos, and M. Soljačić, "Observation of unidirectional backscattering-immune topological electromagnetic states," *Nature*, vol. 461, pp. 772–775, 2009.
[163] M. Hafezi, E. A. Demler, M. D. Lukin, and J. M. Taylor, "Robust optical delay lines with topological protection," *Nat. Phys.*, vol. 7, pp. 907–912, 2011.
[164] A. B. Khanikaev, S. Hossein Mousavi, W.-K. Tse, M. Kargarian, A. H. MacDonald, and G. Shvets, "Photonic topological insulators," *Nat. Mater.*, vol. 12, pp. 233–239, 2013.
[165] T. Ozawa, H. M. Price, A. Amo, et al., "Topological photonics," *Rev. Mod. Phys.*, vol. 91, p. 015006, 2019.
[166] Y. Ota, K. Takata, T. Ozawa, et al., "Active topological photonics," *Nanophotonics*, vol. 9, pp. 547–567, 2020.
[167] W. P. Su, J. R. Schrieffer, and A. J. Heeger, "Solitons in polyacetylene," *Phys. Rev. Lett.*, vol. 42, pp. 1698–1701, 1979.
[168] P. St-Jean, V. Goblot, E. Galopin, et al., "Lasing in topological edge states of a one-dimensional lattice," *Nat. Photonics*, vol. 11, pp. 651–656, 2017.
[169] M. Parto, S. Wittek, H. Hodaei, et al., "Edge-mode lasing in 1D topological active arrays," *Phys. Rev. Lett.*, vol. 120, p. 113901, 2018.
[170] H. Zhao, P. Miao, M. H. Teimourpour, et al., "Topological hybrid silicon microlasers," *Nat. Commun.*, vol. 9, p. 981, 2018.
[171] Y. Ota, R. Katsumi, K. Watanabe, S. Iwamoto, and Y. Arakawa, "Topological photonic crystal nanocavity laser," *Commun. Phys.*, vol. 1, pp. 1–8, 2018.
[172] S. Malzard, C. Poli, and H. Schomerus, "Topologically protected defect states in open photonic systems with non-Hermitian charge-conjugation and parity-time symmetry," *Phys. Rev. Lett.*, vol. 115, p. 200402, 2015.

[173] G. Harari, M. A. Bandres, Y. Lumer, et al., "Topological insulator laser: theory," *Science*, vol. 359, 2018. https://doi.org/10.1126/science.aar4003.
[174] M. A. Bandres, S. Wittek, G. Harari, et al., "Topological insulator laser: Experiments," *Science*, vol. 359, 2018. https://doi.org/10.1126/science.aar4005.
[175] B. Bahari, A. Ndao, F. Vallini, A. E. Amili, Y. Fainman, and B. Kanté, "Nonreciprocal lasing in topological cavities of arbitrary geometries," *Science*, vol. 358, pp. 636–640, 2017.
[176] S. Klembt, T. H. Harder, O. A. Egorov, et al., "Exciton-polariton topological insulator," *Nature*, vol. 562, pp. 552–556, 2018.
[177] T. Ma and G. Shvets, "All-Si valley-Hall photonic topological insulator," *New J. Phys.*, vol. 18, p. 025012, 2016.
[178] Y. Zeng, U. Chattopadhyay, B. Zhu, et al., "Electrically pumped topological laser with valley edge modes," *Nature*, vol. 578, pp. 246–250, 2020.
[179] H. Zhong, Y. Li, D. Song. et al., "Topological valley hall edge state lasing," arXiv:1912.13003 [nlin, physics:physics], 2019.
[180] Z.-K. Shao, H.-Z. Chen, S. Wang, et al., "A high-performance topological bulk laser based on band-inversion-induced reflection," *Nat. Nanotechnol.*, vol. 15, pp. 67–72, 2020.
[181] F. D. M. Haldane, "Model for a quantum Hall effect without Landau levels: condensed-matter realization of the "parity anomaly,"" *Phys. Rev. Lett.*, vol. 61, pp. 2015–2018, 1988.
[182] S. Weimann, M. Kremer, Y. Plotnik, et al., "Topologically protected bound states in photonic parity–time-symmetric crystals," *Nat. Mater.*, vol. 16, pp. 433–438, 2017.
[183] J. M. Zeuner, M. C. Rechtsman, Y. Plotnik, et al., "Observation of a topological transition in the bulk of a non-Hermitian system," *Phys. Rev. Lett.*, vol. 115, p. 040402, 2015.
[184] L. Jin, P. Wang, and Z. Song, "Su-Schrieffer-Heeger chain with one pair of PT-symmetric defects," *Sci. Rep.*, vol. 7, p. 5903, 2017.
[185] K. Takata and M. Notomi, "Photonic topological insulating phase induced solely by gain and loss," *Phys. Rev. Lett.*, vol. 121, p. 213902, 2018.
[186] S. Liu, S. Ma, C. Yang, et al., "Gain- and loss-induced topological insulating phase in a non-Hermitian electrical circuit," *Phys. Rev. Applied*, vol. 13, p. 014047, 2020.
[187] A. K. Harter, T. E. Lee, and Y. N. Joglekar, "PT-breaking threshold in spatially asymmetric Aubry-André and Harper models: hidden symmetry and topological states," *Phys. Rev. A*, vol. 93, p. 062101, 2016.
[188] L. Xiao, X. Zhan, Z. H. Bian, et al., "Observation of topological edge states in parity–time-symmetric quantum walks," *Nat. Phys.*, vol. 13, pp. 1117–1123, 2017.
[189] L. He, Z. Addison, J. Jin, E. J. Mele, S. G. Johnson, and B. Zhen, "Floquet Chern insulators of light," *Nat. Commun.*, vol. 10, p. 4194, 2019.
[190] M. Li, X. Ni, M. Weiner, A. Alù, and A. B. Khanikaev, "Topological phases and nonreciprocal edge states in non-Hermitian Floquet insulators," *Phys. Rev. B*, vol. 100, p. 045423, 2019.
[191] Y. G. N. Liu, P. Jung, M. Parto, J. Leshin, D. N. Christodoulides, and M. Khajavikhan, "Towards a non-magnetic topological Haldane laser," in *Conference on Lasers and Electro-Optics (2019)*, Paper FW3D.1, Optical Society of America, 2019, p. FW3D.1.
[192] C. Poli, M. Bellec, U. Kuhl, F. Mortessagne, and H. Schomerus, "Selective enhancement of topologically induced interface states in a dielectric resonator chain," *Nat. Commun.*, vol. 6, p. 6710, 2015.
[193] R. El-Ganainy and M. Levy, "Optical isolation in topological-edge-state photonic arrays," *Opt. Lett.*, vol. 40, pp. 5275–5278, 2015.
[194] E. Makri, R. Thomas, and T. Kottos, "Reflective limiters based on self-induced violation of PT symmetry," *Phys. Rev. A*, vol. 97, p. 043864, 2018.
[195] H. Zhao, X. Qiao, T. Wu, B. Midya, S. Longhi, and L. Feng, "Non-Hermitian topological light steering," *Science*, vol. 365, pp. 1163–1166, 2019.
[196] K. Kawabata, K. Shiozaki, M. Ueda, and M. Sato, "Symmetry and topology in non-Hermitian physics," *Phys. Rev. X*, vol. 9, p. 041015, 2019.
[197] H. Zhou and J. Y. Lee, "Periodic table for topological bands with non-Hermitian symmetries," *Phys. Rev. B*, vol. 99, p. 235112, 2019.
[198] A. P. Schnyder, S. Ryu, A. Furusaki, and A. W. W. Ludwig, "Classification of topological insulators and superconductors in three spatial dimensions," *Phys. Rev. B*, vol. 78, p. 195125, 2008.
[199] A. Kitaev, "Periodic table for topological insulators and superconductors," *AIP Conf. Proc.*, vol. 1134, pp. 22–30, 2009.
[200] K. Kawabata, S. Higashikawa, Z. Gong, Y. Ashida, and M. Ueda, "Topological unification of time-reversal and particle-hole symmetries in non-Hermitian physics," *Nat. Commun.*, vol. 10, p. 297, 2019.
[201] Y. Xiong, "Why does bulk boundary correspondence fail in some non-Hermitian topological models," *J. Phys. Commun.*, vol. 2, p. 035043, 2018.
[202] T. E. Lee, "Anomalous edge state in a non-Hermitian lattice," *Phys. Rev. Lett.*, vol. 116, p. 133903, 2016.
[203] H. Shen, B. Zhen, and L. Fu, "Topological band theory for non-Hermitian Hamiltonians," *Phys. Rev. Lett.*, vol. 120, p. 146402, 2018.
[204] L. Jin and Z. Song, "Bulk-boundary correspondence in a non-Hermitian system in one dimension with chiral inversion symmetry," *Phys. Rev. B*, vol. 99, p. 081103, 2019.
[205] S. Yao and Z. Wang, "Edge states and topological invariants of non-Hermitian systems," *Phys. Rev. Lett.*, vol. 121, p. 086803, 2018.
[206] S. Yao, F. Song, and Z. Wang, "Non-Hermitian chern bands," *Phys. Rev. Lett.*, vol. 121, p. 136802, 2018.
[207] N. Okuma, K. Kawabata, K. Shiozaki, and M. Sato, "Topological origin of non-Hermitian skin effects," *Phys. Rev. Lett.*, vol. 124, p. 086801, 2020.
[208] K. Kawabata, K. Shiozaki, and M. Ueda, "Anomalous helical edge states in a non-Hermitian Chern insulator," *Phys. Rev. B*, vol. 98, p. 165148, 2018.
[209] S. Weidemann, M. Kremer, T. Helbig, et al., "Topological funneling of light," *Science*, vol. 368, pp. 311–314, 2020.
[210] Z. Gong, Y. Ashida, K. Kawabata, K. Takasan, S. Higashikawa, and M. Ueda, "Topological phases of non-Hermitian systems," *Phys. Rev. X*, vol. 8, p. 031079, 2018.
[211] K. Yokomizo and S. Murakami, "Non-bloch band theory of non-Hermitian systems," *Phys. Rev. Lett.*, vol. 123, p. 066404, 2019.
[212] D. Leykam, K. Y. Bliokh, C. Huang, Y. D. Chong, and F. Nori, "Edge modes, degeneracies, and topological numbers in non-Hermitian systems," *Phys. Rev. Lett.*, vol. 118, p. 040401, 2017.

[213] F. K. Kunst, E. Edvardsson, J. C. Budich, and E. J. Bergholtz, “Biorthogonal bulk-boundary correspondence in non-Hermitian systems,” *Phys. Rev. Lett.*, vol. 121, p. 026808, 2018.
[214] K.-I. Imura and Y. Takane, “Generalized bulk-edge correspondence for non-Hermitian topological systems,” *Phys. Rev. B*, vol. 100, p. 165430, 2019.
[215] W. A. Benalcazar, B. A. Bernevig, and T. L. Hughes, “Quantized electric multipole insulators,” *Science*, vol. 357, pp. 61–66, 2017.
[216] W. A. Benalcazar, B. A. Bernevig, and T. L. Hughes, “Electric multipole moments, topological multipole moment pumping, and chiral hinge states in crystalline insulators,” *Phys. Rev. B*, vol. 96, p. 245115, 2017.
[217] F. Schindler, A. M. Cook, M. G. Vergniory, et al., “Higher-order topological insulators,” *Sci. Adv.*, vol. 4, p. eaat0346, 2018.
[218] C. H. Lee, L. Li, and J. Gong, “Hybrid higher-order skin-topological modes in nonreciprocal systems,” *Phys. Rev. Lett.*, vol. 123, p. 016805, 2019.
[219] X.-W. Luo and C. Zhang, “Higher-order topological corner states induced by gain and loss,” *Phys. Rev. Lett.*, vol. 123, p. 073601, 2019.
[220] M. Ezawa, “Non-Hermitian higher-order topological states in nonreciprocal and reciprocal systems with their electric-circuit realization,” *Phys. Rev. B*, vol. 99, p. 201411, 2019.
[221] T. Liu, Y.-R. Zhang, Q. Ai, et al., “Second-order topological phases in non-Hermitian systems,” *Phys. Rev. Lett.*, vol. 122, p. 076801, 2019.
[222] Z. Zhang, M. Rosendo López, Y. Cheng, X. Liu, and J. Christensen, “Non-Hermitian sonic second-order topological insulator,” *Phys. Rev. Lett.*, vol. 122, p. 195501, 2019.

Mordechai Segev* and Miguel A. Bandres

Topological photonics: Where do we go from here?

https://doi.org/10.1515/9783110710687-032

Abstract: Topological photonics is currently one of the most active research areas in optics and also one of the spearheads of research in topological physics at large. We are now more than a decade after it started. Topological photonics has already proved itself as an excellent platform for experimenting with concepts imported from condensed matter physics. But more importantly, topological photonics has also triggered new fundamental ideas of its own and has offered exciting applications that could become real technologies in the near future.

Keywords: lasers; photonics; topological insulators; topological photonics.

1 Introduction

Topological photonics has started off in a proposal suggesting how to emulate the concepts of topological insulators in an electromagnetic (EM) system [1]. At that time, these concepts were strictly within the realm of condensed matter system. To understand the underlying principles, it is instructive to briefly review the basics of topological insulators. We shall do it on an intuitive basis and attempt to view these principles through the physically observable quantities. In a single sentence – topological insulators are materials that are insulators in the bulk but are perfect conductors on their edges. The robust conduction on the edge is manifested in the fact that the current there is lossless even in the presence of disorder or defects and does not depend on the shape of the edge. That is – the current continues to flow without being scattered into the bulk or being backscattered by local defects, by disorder in the lattice, or by sharp edges. Due to the lack of backscattering, such topological edge current is often viewed as unidirectional. This property is often called "topologically protected transport", and it is illustrated in Figure 1a. It is this property that made topological insulators so important – because – apart from the fundamental physics involved – having a mechanism that can bring to lossless flow of energy, charges, or information, is extremely important for any applications. The important parameter determining this unique robustness is the strength of the variations of the potential that would have normally caused scattering, not the shape nor the position of a particular defect or disorder in the lattice.

To understand the essence of topological transport, it is instructive to recall the first topological insulator ever discovered: the integer quantum Hall effect [2]. This phenomenon occurs in semiconductors under low temperature and a strong magnetic field (Figure 1b). The Lorentz force opens a bandgap in the dispersion curve, and the edge states are characterized by a single line crossing the gap in diagonal. This oversimplified picture of the quantum Hall effect is sketched in Figure 1c, where the red line marks the dispersion curve of the edge states. Notice that there is only one line, as there is no line crossing the gap in the opposite direction. The slope of this line provides the group velocity of any edge excitation (superposition of states on the red line), and it cannot be zero in any topological edge state. The direction of the magnetic field sets the sign of the slope (which determines the direction of the edge current), and the strength of the magnetic field sets the size of the bandgap. Now, if disorder is introduced into this structure, the bands will be slightly modified, and the slope of the red line will be slightly altered. But as long as the strength of the disorder (random variation in the potential) is smaller than some value, scattering will not couple the edge states to bulk states. The implication is that any disorder weaker than (approximately) the bandgap will not cause scattering into the bulk or backscattering. This is the origin of topologically protected transport in the quantum Hall effect, and its key ingredient is the size of the topological

***Corresponding author: Mordechai Segev,** Physics Department, Electrical Engineering Department, and Solid State Institute, Technion, Haifa, Israel, E-mail: msegev@technion.ac.il. https://orcid.org/0000-0002-9421-2148
Miguel A. Bandres, CREOL - The College of Optics and Photonics, University of Central Florida, Orlando, FL, USA

This article has previously been published in the journal Nanophotonics. Please cite as: M. Segev and M. A. Bandres "Topological photonics: Where do we go from here?" *Nanophotonics* 2021, 10. DOI: 10.1515/nanoph-2020-0441.

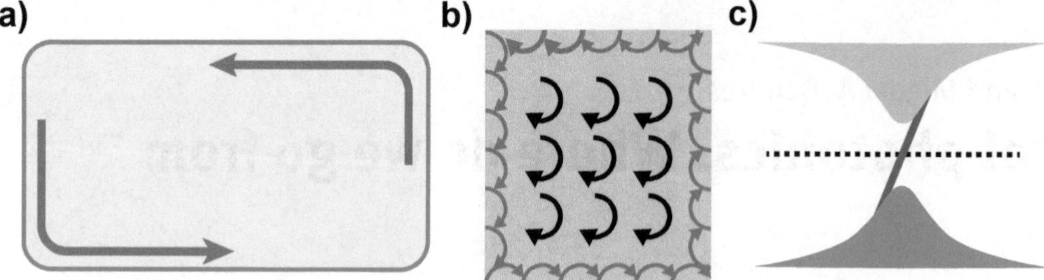

Figure 1: Topological insulators in a nut shell.
(a) Topological insulator: a two-dimensional (2D) material that is insulating in the bulk but exhibits perfect conduction on the edge. (b) Simplified sketch of the integer quantum Hall effect, which was the first topological insulator. (c) Simplified dispersion relation for the quantum Hall effect, with the red line marking the topological edge states.

bandgap, which determines the degree of topological protection of transport. Modern topological insulators rely on the same principles, but the effects can be caused by a variety of other effects, among them fermionic spin–orbit coupling, external modulation, and crystal symmetries, etc. [3–6].

2 The road to topological photonics

Motivated by the growing success on electronic topological insulators, researchers started to ask whether the concept of topological insulators is unique to fermionic systems or is it actually universal. More specifically, can topological protection of transport exist in bosonic systems? Essentially – the challenge was to find a wave system whose dispersion curve resembles Figure 1c. The first suggestion was an EM system that requires breaking time-reversal symmetry to avoid backscattering [1]. Shortly thereafter, a more concrete idea was proposed [7] – based on gyro-optics materials where the application of a magnetic field indeed breaks time-reversal symmetry. This effect is fundamentally weak at all frequencies above THz; hence the resultant topological bandgap would be very small, providing essentially no protection of transport. However, at microwaves frequencies, the effect is strong, opens a large bandgap, and indeed within a year the EM analog of the integer quantum Hall effect was demonstrated [8]. At that point, the challenge was to find a new avenue to bring the concepts of topological insulators into photonics (optical frequencies and near infrared), without relying on the weak gyro-optic effects. Several ideas were proposed – ranging from using polarization as spin in photonic crystals [9] and aperiodic coupled resonators [10] to bianisotropic metamaterials [11]. Eventually, in 2013, the first photonic topological insulators were demonstrated [12], and it indeed displayed topological protection of transport (of light) against defects and disorder. That system relied on periodic modulation, which is the essence of Floquet topological insulators [6, 13]. In electromagnetism, employing temporal modulation in a spatially asymmetric system can lead to optical isolators that block backscattering [14], but still, relating modulation to photonic topological insulators required a lattice structure where the modulation can open a gap. As it turns out, a honeycomb lattice can play this role. Indeed, the first photonic Floquet topological insulators relied on a honeycomb lattice of coupled waveguides, where the modulation is generated by making the waveguides helical [12]. Around the same time, the aperiodic coupled resonator system was also realized in experiment [15] and demonstrated topological protection against disorder in the lattice [16]. Within a few years, numerous other EM topological systems were proposed and demonstrated, among them the topological bianisotropic metamaterials system [11, 17], the so-called "network model" of strongly coupled resonators [18, 19], and the crystalline topological insulator [20, 21]. These photonic topological systems are summarized in Figure 2. A recent comprehensive review on photonic topological insulators provides the details of these systems [23].

3 Photonic realizations of fundamental topological models

We are now more than a decade after the first demonstration of an EM topological insulator [8], and seven years after the first observation of the first photonic topological insulators [12, 15]. These experiments, and the pioneering theoretical papers during those first years, have created a new area from scratch: Topological Photonics. During

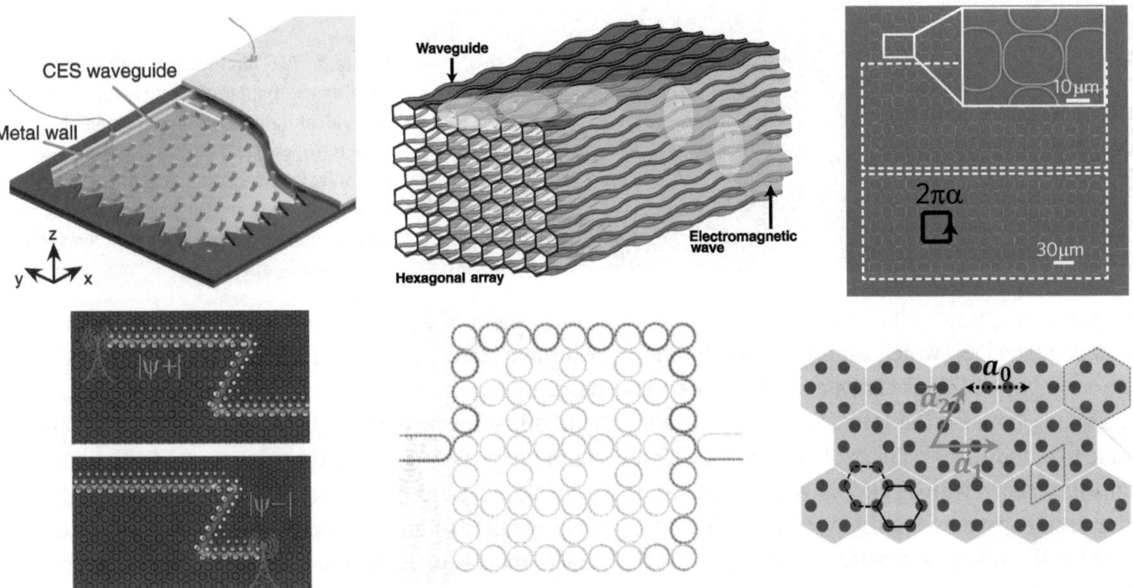

Figure 2: Various schemes for realizing topological insulators for electromagnetic (EM) waves.
Top, left to right: two-dimensional (2D) photonic crystal incorporating gyro-optic materials realizing the quantum Hall effect [8]; honeycomb lattice of helical waveguides realizing Floquet topological insulators [12, 22], aperiodic resonator array realizing the quantum Hall effect [10, 15]. Bottom, left to right: photonic topological insulator based on bianisotropic materials [11, 17], the network model of photonic topological insulators [18, 19], the crystalline photonic topological insulators [20, 21].

these years, this new area has gone a long way in multiple direction. One of the most important directions is ***using photonic systems as platform to experiment with fundamental concepts in physics***, which have been proposed but never realized in experiments. Namely – using photonic platforms to realize models that have been suggested in condensed matter physics. In fact, such photonic realizations were the starting point of topological photonics anyway: both the gyro-optic microwaves system [8] and the aperiodic coupled resonator systems [15] realize the quantum Hall effect, while the helical honeycomb lattice [12] realizes a Floquet topological insulators [6]. Likewise, the valley Hall photonic topological insulator [24] emulates the topological valley Hall transport in bilayer graphene [25]. Interestingly, sometimes the photonic realization of topological insulators came before the analogous experimental observations in solid state, for example, the first Floquet topological insulator ever realized in experiment was the photonic one [12], and it was followed by the solid state realization in Bi_2Se_3 [26]. Another example is the anomalous Floquet topological insulators, which were originally proposed in solid state [27], first observed in photonics [28, 29] and very recently observed with ultracold atoms [30]. In a similar vein, the topological Anderson insulator [31] – a fundamental system which becomes topological only through the introduction of disorder – was first demonstrated in photonics [32], and shortly thereafter with ultracold atoms [33], but realizing this idea in electronic condensed matter systems (in the context it was proposed) seems like a remote possibility. Undoubtedly, the realization of new topological systems, and generally of new phenomena that otherwise cannot be observed in the context they were proposed, has much value. Theoretical work and simulations always idealize the system and assume that the governing equations represent all the relevant physics involved, whereas experiments are never completely isolated from additional effects, and above all, the experiment itself often leads to new ideas and many times offers surprises.

Finally, and most importantly, photonics brought several fundamental and exclusive discoveries on topological physics, by combining non-Hermitian (NH) and topological physics as we describe below. Among those, the most important one (in terms of both fundamentals and applications) is the topological insulator laser [34, 35].

4 Topological photonics in tailored lattices

Another class of phenomena where topological photonics makes a big difference are those that were originally proposed in photonics but have later proved to be universal and could be observed in other fields beyond the domain of EM

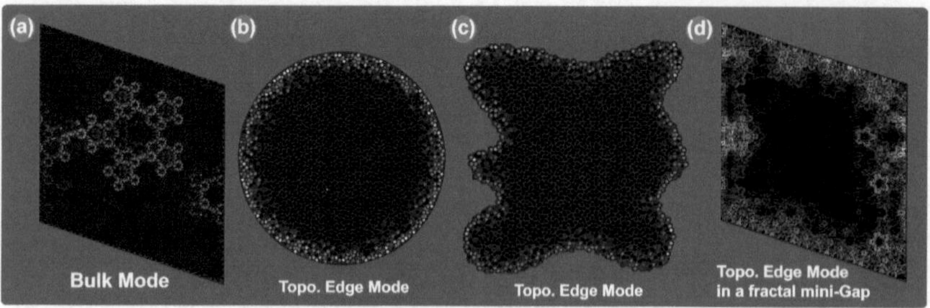

Figure 3: Two-dimensional photonic topological insulator quasicrystals [36]. (a) Typical bulk mode of a quasicrystal lattice. Topological edge states of periodically driven quasicrystals [36], shaped as (b) circle, (c) arbitrary shape, and (d) a topological edge state that lives in a fractal mini-gap.

waves. For example, quasicrystal topological insulators without magnetic fields [36], which were proposed in photonics but the underlying ideas are universal. This system is especially intriguing because the principles of topological insulators seem to require the existence of a bandgap, which is conceptually associated with a periodic system that has a well-defined unit cell, whereas a quasicrystal is not periodic and has no unit cell; rather, its spectrum is fractal. Nevertheless, under periodic modulation quasicrystals can be topological insulators, and its topological edge states exhibit all the features of topological edge states. Figure 3 shows the topological edge states of Floquet quasicrystals of different shapes, highlighting the fact that the shape of the edges is unimportant for having topologically protected transport. In a similar fashion, it was recently suggested that, under periodic modulation, even fractal structures can become topological insulators [37]. For example, the Sierpinski gasket, which has a Hausdorff dimension of 1.585, exhibits topologically protected edge transport even though it has no bulk at all: every site resides on an edge, external or internal. Nevertheless, the periodic modulation of this lattice has topological edge states on its exterior and on its internal edges.

Topological photonics can also offer new ideas that have no counterparts in other physical systems, such as the recent proposal [38] for broadband topological slow light through higher momentum-space winding, which can greatly enhance light–matter interactions.

5 Topological photonics in synthetic space

Another direction where photonics is having an important impact is topological physics in synthetic space. Despite different manifestations in many physical systems, topological insulators usually rely on spatial lattices. The wavepackets propagating in the lattice, whether electrons, photons or phonons, are subjected to gauge fields that give rise to the topological phenomena. However, lattices, periodic structures, do not necessarily have to be a spatial arrangement of sites. Rather, a lattice can also be a ladder of atomic states, or photonic cavity modes, or spin states. Using one (or more) of these ladders in a non-spatial – but synthetic – degree of freedoms, requires the introduction of coupling between the synthetic sites, which can be achieved by external perturbation. In contrast to traditional topological insulators based on a spatial lattice, for topological insulators in synthetic dimensions the transport is not restricted to the spatial edges of the system, but rather transport occurs on the edges of the synthetic space. In this way, it is possible to have topologically protected transport extending over the bulk in real-space. For example, the lowest and highest modes in a system serve as synthetic edges. Based on this concept, topological edge state were observed in cold atoms system, using atomic spin states as a synthetic [39, 40], or the atomic momentum states of a Bose–Einstein condensate [41]. However, using internal degrees of freedom for implementing the synthetic dimensions involves several fundamental problems: the number of these states is small, and the excited states always have a short lifetime. Here is where photonic comes into play and offers numerous ways to realize synthetic dimensions through equally spaced modes of the system. Synthetic dimensions in topological photonics were first introduced for topological pumping [42], where a photonic lattice was mapped onto a corresponding quantum Hall lattice with twice its spatial dimensions. In this vein, photonic topological insulators in synthetic dimensions were proposed with the synthetic space realized through cavity modes [43–45] which offers not only an unlimited number of states but also long lifetime, both marking big advantages for large-scale lattices. More recently, photonic topological insulators in synthetic dimensions were demonstrated in experiments [46]. That system (described in Figure 4) consisted of a two-dimensional (2D) waveguide array, engineered such that it is effectively a 2D lattice, where one of its dimensions is an ordinary spatial dimension, but the second dimension is the mode spectrum of each column of waveguides. This construction enabled observing of the dynamics of topological edge states in synthetic space, highlighting the topologically protected transport [46]. The beauty of this scheme is that it allows for

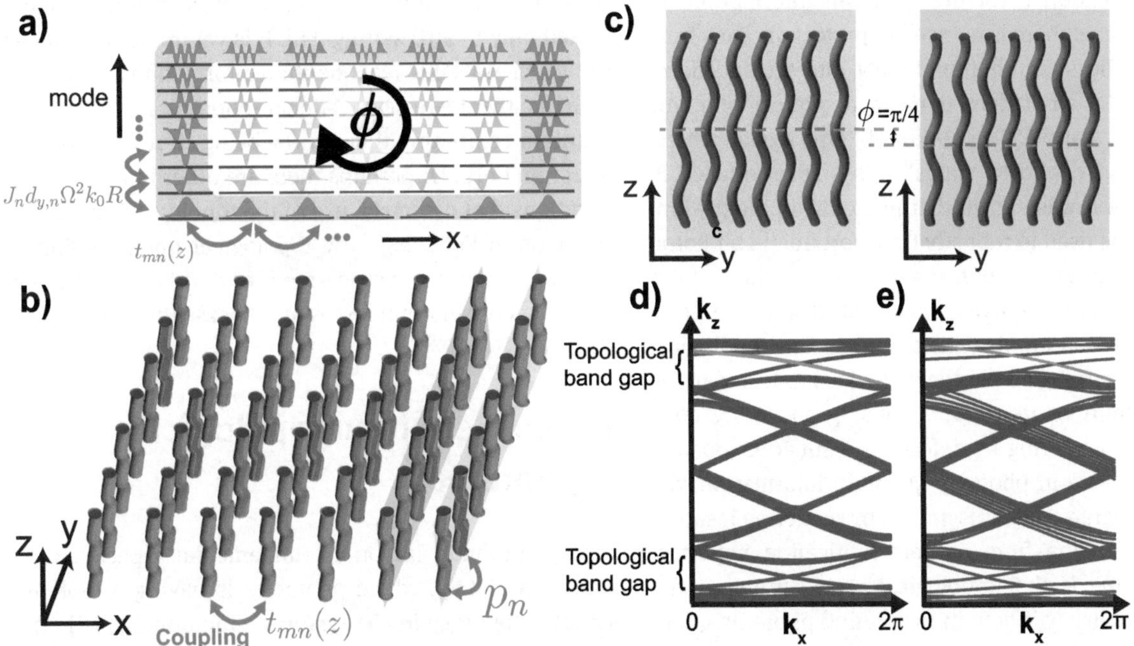

Figure 4: The two-dimensional (2D) synthetic space photonic topological insulator made of an array of judiciously modulated waveguide [46]. (a, b) The synthetic space lattice (a) corresponding to the two-dimensional lattice of waveguides in real-space (b). The basic building block is a one-dimensional (1D) array of *N* evanescently coupled waveguides, with the spacing between waveguides judiciously engineered such that it yields *N* Bloch modes with equally spaced propagation constants [49]. To facilitate transport in the model dimension, the modes are coupled by spatially oscillating the waveguides in the propagation direction, creating a ladder of coupled modes. Arranging *M* such 1D arrays next to one another, with the oscillations phase-shifted from one another, results in the 2D topological insulator in synthetic space. The edge state of the synthetic space (yellow in (a)) resides in the bulk of the waveguide array of (red in (b)). (c) The phase shift between each two adjacent columns of the waveguide array of (b). (d) The Floquet band structure of the lattice with the edge state marked by the red line. (e) The Floquet band structure under random disorder in the coupling between waveguides. Disorder mostly causes shifts and slight deformations in the dispersion curve of the edge state but does not close the topological gap, highlighting the immunity of the topological edge state to disorder.

increasing the dimensionality further to three-dimensional (3D) and even four-dimensional (see Supplementary Material of [46]). Indeed, a closely related system was very recently demonstrated to display topological transport in a 3D synthetic space [47]. In a different photonic realization, dynamics in two synthetic dimensions has been recently demonstrated in a scheme based on a single temporally modulated ring cavity [48]. In this realization, the synthetic dimensions were the frequencies of the cavity modes and the pseudospin states of the clockwise and anticlockwise states. This simple configuration, albeit physically consisting only of a single ring, facilitated demonstrating a variety of effects such as effective spin–orbit coupling, magnetic fields, spin-momentum locking, Meissner-to-vortex phase transition, and signatures of topological chiral one-way edge currents, all completely in synthetic dimensions. This paper demonstrates a new kind of topological protection: transport in synthetic space here means conversion from one mode to another mode, where the topological landscape guarantees robustness to the modal conversion process.

It is now already clear that utilizing the arsenal of photonics to create experimental schemes for topological physics in synthetic dimensions offers a plethora of new possibilities that can hardly be matched outside photonic. For example, it is possible to include additional frequencies to the oscillating 1D columns there, which would induce long-range coupling upon design – leading to new unexplored models of lattice geometries. Finally, adding gain and loss to such lattices in synthetic space would pave the way to party-time (PT) symmetry and exceptional points in synthetic space [45].

6 Topological quantum photonics

The topological protection of transport is in principle a wave phenomenon. However, in condensed matter it was argued that the topological robustness can also protect entanglement, by protecting the entanglement carriers against decoherence [50]. Unlike condensed matter systems where the issue of decoherence is a major obstacle for any application involving quantum information, in photonic systems there is no need to protect photons from decoherence because photons barely interact and decohere slowly. So what does it mean to protect photonic quantum information? The notion of "topological protection of entangled photon states" was first introduced in [51, 52], where it was shown that photonic topological insulators can be used to robustly transport fragile biphoton states. It was shown through simulations that these states maintain their path entanglement despite disorder, in stark contrast with non-topological systems. The topological robustness of entangled photon states is manifested in the robust transport of its unidirectional propagating edge states, where scattering by defects and imperfections are suppressed. Since in photonic quantum information, the scalability to large systems is limited by scattering loss and other errors arising from random fabrication imperfections, the hope is that topological architectures of the photonic circuitry will help in facilitating photonic quantum computing [53]. Proving that topological settings can give rise to robustness of multiphoton quantum states, specifically in architectures relevant to quantum computing, would hold great promise for fault-tolerant quantum logic, which is otherwise very fragile in large-scale settings such as quantum information systems.

Experimentally, transport of quantum edge states using single photons has been demonstrated [54–56], and it is of clear interest for quantum simulation and sensing. These single-photon experiments studied the physics of topologically protected bounds states [55] and topological transitions [56] in photonic quantum walks, as well as demonstrating an interface between a quantum emitter and a photonic topological edge state [54]. However, quantum information systems rely on multiphoton states. Recent experiments in a 1D binary array of waveguides have demonstrated topological protection of biphoton correlations [57] and of entangled photon states [58], which are the key building blocks for robust quantum information systems. These experiments showed that the biphoton states maintain their spatial distribution in the high-dimensional Hilbert space and their propagation constant – as they propagate through a topological nanophotonic lattice with deliberately introduced disorder. In a different experiment [59], the topological edge states of the aperiodic resonator array were used as a platform for generating correlated photon pairs by spontaneous four-wave mixing, and it was shown that they outperform their topologically trivial one-dimensional (1D) counterparts in terms of spectral robustness. Generally, the research on quantum light in topological photonic systems has just started. Its main aspect – studying the topological protection of multiphoton states – was thus far demonstrated only in a 1D setting [58]. It is yet unknown if it can really penetrate into real quantum information technology, such as the large-scale silicon photonics circuitry for quantum computing that is now being developed by PsiQuantum Corp. Or perhaps there are other interesting applications that can make use of the topological protection of quantum light, e.g., the topological quantum-limited traveling wave parametric amplifier that is naturally protected against internal losses and backscattering [60].

7 Non-Hermitian topological photonics

Perhaps the most important fundamental aspect of topological insulators, where photonics is having a profound aspect is ***topology in NH systems***. Traditionally, NH operators have been used in quantum mechanics to describe loss mechanisms, open systems, finite lifetime and dephasing, which would otherwise have to be described by coupling to degrees of freedom outside the system of interest. In this context, the NH version of quantum mechanics is helpful in simplifying calculations, identifying resonances, etc., but the underlying assumption was always that all the observables of a physical system must be real quantities, and consequently the operators must be Hermitian. Twenty years ago, it was found that NH Hamiltonians that obey PT symmetry have a regime of parameters where all their Eigenenergies are real [61]. That discovery implied that, possibly, NH operators can represent physically observable quantities. Still, for another decade that discovery remained with limited physical consequences, until Christodoulides, Makris, El-Ganainy and Musslimani [62], and shortly thereafter independently Moiseyev, Klaiman, and Gunther [63], introduced the concepts of PT-symmetry into optics. The first experiments followed within two years [64, 65]. Actually, introducing NH into optics is very natural, as the NH parts of the operators represent gain and loss, which are present in any laser system. Since then, the field of PT-optics, or in a broader sense – NH optics – has been overwhelmingly flourishing with research activity.

As an important part of the vision for topological photonics was to explore new universal concepts, combining topology with NH physics was a natural but highly challenging goal. Indeed, the first NH topological system was demonstrated in a photonics experiment [66], following an earlier prediction of a topological transition in NH quantum walk [67]. This was a 1D NH lattice, where the presence of loss made it possible to identify the topology of the corresponding passive NH system just from bulk measurements, without the need to investigate edge states. But irrespectively, the combination of topology and NH was bound to create controversies and arguments, at the very least – because topological physics relies on the existence of topological quantities that remain invariant during deformations of the system, and in NH systems it was not clear at all that such quantities can exist. In this spirit – there were theory papers that claimed, for example, that PT-symmetric topological systems cannot exist, casting major doubts on the ability of NH topological systems to exhibit topologically protected transport. Some of the controversy was resolved by the demonstration of 1D topological photonic systems exhibiting broken [68] and full [69] PT-symmetry. However, these 1D systems cannot have topologically protected transport along their edge because the edge in a 1D system is zero-dimensional (0D). The existence of NH topologically protected transport was highlighted by the recent discovery of topological insulator lasers [34, 35]. These are lasers whose cavities are specifically designed to support transport along the cavity edges, while making use of the topological immunity to defects and disorder to enhance the lasing efficiency and maintain single-mode lasing even high above the threshold [34, 35]. One of the models for topological insulator lasers is based on the Haldane model with the addition of gain, loss and nonlinearity [34], where it was shown, unequivocally, that the topological platform give rise to immune transport in this laser system, despite the non-Hermiticity. At that point – it became clear that topological invariants should exist also in NH systems, otherwise – there would be no explanation to the unidirectional transport and the immunity to defects and disorder [34]. This was the goal of several recent papers [70–75]: to present a general framework for classifying topological phases of generic NH systems. This fundamental issue is still one of the outstanding challenges of topological physics at large, where topological photonics is the spearhead of research.

8 Topological insulator laser

Undoubtedly, in the entire field of topological photonics, the research topic that is closest to real technological applications is the topological insulator laser. The vision here is to harness the features of topologically protected transport to force many semiconductor emitters to lock together and behave as a single powerful highly coherent laser source. Technologically, having a high power semiconductor laser has been a challenge of more than four decades, and all attempts to make a "broad-area laser" or a "laser diode array" have generally failed. Laser arrays are currently used only as a strong flashlight to pump solid state lasers (NdYag, etc.), but their coherence is not much better than of a light emitting diode. The vision was therefore to make use of the fundamental features of topological insulators to force injection-locking of many semiconductor laser emitters to act as a single coherent laser [76]. But in the way stood the question of whether NH systems (such as a laser) can support topological protection of any kind. Shortly thereafter, there was a series of works demonstrating lasers emitting from a 0D topological edge state in a 1D chain [77–80], but in those systems there is no edge transport at all, hence no protection to onsite disorder, and the lasing is almost fully confined to a single resonator. Then, there was an attempt to incorporate gyro-optic material in a laser cavity [81], but since the magneto-optic effects at optical frequencies are extremely weak, the bandwidth of the laser was broader than the topological bandgap. The first topological insulator laser was actually demonstrated in experiments a few months earlier [79] and it displayed all the expected features (see Figure 5) [35]. This laser was constructed on a standard optoelectronic platform, as an aperiodic array of 10 × 10 coupled ring-resonators on InGaAsP quantum wells wafer. This 2D setting is comprised a square lattice of ring resonators coupled to each other via auxiliary links. The intermediary links are judiciously spatially shifted to introduce a set of hopping phases, establishing a synthetic magnetic field that yields topological features. To promote lasing of the topologically protected edge modes, only the outer perimeter of the array was pumped, while leaving the interior elements lossy. This topological insulator laser operates in single mode, even considerably above threshold, whereas the corresponding topologically trivial realizations lase in an undesired multimode fashion, see Figure 5. More importantly, the topological laser displays a slope efficiency that is considerably higher than in the corresponding trivial realizations, even in the presence of defects and disorder [35].

Since that visionary work [34, 35] several groups followed with a variety of configurations for realizing topological insulator lasers, e.g., a topological quantum cascade laser with valley edge modes [82], topological bulk laser based on band-inversion-induced reflection [84], and a topological insulator laser with next-nearest-neighbor coupling [85]. Finally, we note very recent experiments on a topological vertical cavity surface emitting laser [86]. Importantly, very recent theoretical work [83] showed that indeed the topological design greatly improves the coherence of a large array of emitters, as envisioned by [34, 76]. From all of this activity, it is now quite clear that currently the topological insulator laser is the most promising application of topological photonics, with many new ideas

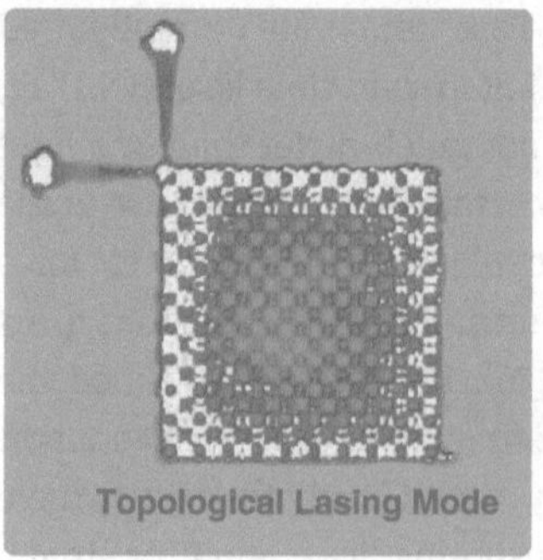

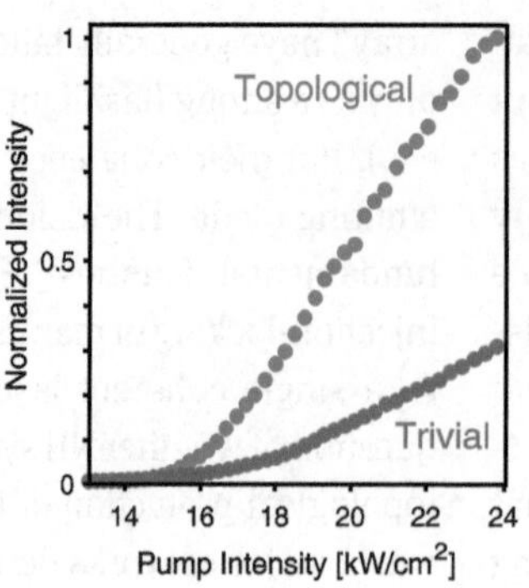

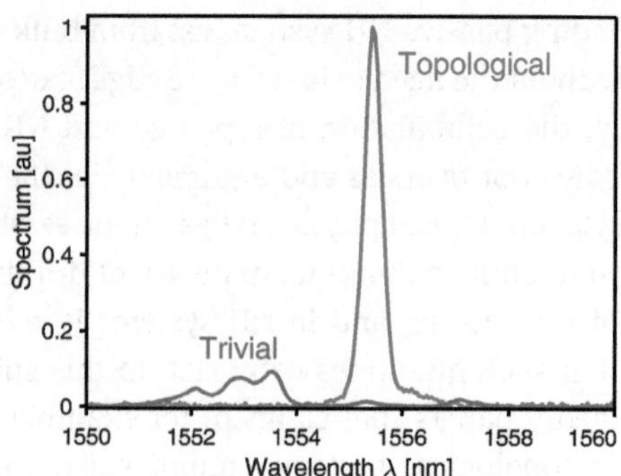

Figure 5: Topological insulator laser [34, 35].
Left to right: Top view photograph of the lasing pattern (topological edge mode) in a 10 × 10 array of topologically connected resonators, and the output ports. Output intensity versus pump intensity for a topological insulator laser and its corresponding trivial counterpart. The enhancement of the slope efficiency is approximately threefold. Emission spectra from a topological insulator laser and its topologically trivial counterpart.

emerging, for example, utilizing topology in synthetic dimensions to force an array of semiconductor lasers to emit mode-locked pulses [87], which could overcome a challenge of three decades.

9 Conclusions

We attempted to provide our perspective on the new field of topological photonics, which is currently at the forefront of photonics research and is also the spearhead topological physics at large. We covered here a small selected list of topics, but in fact there are many more. For example, we did not cover nonlinear topological photonic systems [88], which have started to attract much attention recently with the observation of solitons in a topological bandgap [89]. Likewise, we did not discuss topological exciton–polariton settings [77] and exciton–polariton topological insulators [90], which are extremely interesting because they are a topological symbiosis between light and matter. We did not discuss a plethora of many new ideas that are now frequently emerging in topological photonics. Altogether, it is now clear that within the past decade this new field has gone a long way, and it continues to generate new fundamental ideas and offer exciting applications. Can some of these topological applications become real technology? We can carefully say that the answer seems to be positive, but we will know a more definite answer within five years.

Author contribution: All the authors have accepted responsibility for the entire content of this submitted manuscript and approved submission.

Research funding: The authors gratefully acknowledge the support of the USA–Israel Binational Science Foundation (BSF), the Israel Science Foundation (ISF), Advanced Grant from the European Research Council (ERC), and a grant from the Air Force Office of Scientific Research (AFOSR), USA.

Conflict of interest statement: The authors declare no conflicts of interest regarding this article.

References

[1] F. D. M. Haldane and S. Raghu, “Possible realization of directional optical waveguides in photonic crystals with broken time-reversal symmetry,” *Phys. Rev. Lett.*, vol. 100, no. 1, p. 013904, 2008.

[2] K. V. Klitzing, G. Dorda, and M. Pepper, “New method for high-accuracy determination of the fine-structure constant based on quantized Hall resistance,” *Phys. Rev. Lett.*, vol. 45, no. 6, pp. 494–497, 1980.

[3] C. L. Kane and E. J. Mele, “Quantum spin Hall effect in graphene,” *Phys. Rev. Lett.*, vol. 95, no. 22, p. 226801, 2005.

[4] B. A. Bernevig, T. L. Hughes, and S.-C. Zhang, “Quantum spin Hall effect and topological phase transition in HgTe quantum wells,” *Science*, vol. 314, no. 5806, pp. 1757–1761, 2006.

[5] M. Konig, S. Wiedmann, C. Brune, et al., “Quantum spin Hall insulator state in HgTe quantum wells,” *Science*, vol. 318, no. 5851, pp. 766–770, 2007.

[6] N. H. Lindner, G. Refael, and V. Galitski, “Floquet topological insulator in semiconductor quantum wells,” *Nat. Phys.*, vol. 7, no. 6, pp. 490–495, 2011.

[7] Z. Wang, Y. D. Chong, J. D. Joannopoulos, and M. Soljačić, “Reflection-free one-way edge modes in a gyromagnetic photonic crystal,” *Phys. Rev. Lett.*, vol. 100, no. 1, p. 013905, 2008.

[8] Z. Wang, Y. Chong, J. D. Joannopoulos, and M. Soljačić, “Observation of unidirectional backscattering-immune topological electromagnetic states,” *Nature*, vol. 461, no. 7265, pp. 772–775, 2009.

[9] R. O. Umucalılar and I. Carusotto, “Artificial gauge field for photons in coupled cavity arrays,” *Phys. Rev. A*, vol. 84, no. 4, p. 043804, 2011.

[10] M. Hafezi, E. A. Demler, M. D. Lukin, and J. M. Taylor, “Robust optical delay lines with topological protection,” *Nat. Phys.*, vol. 7, no. 11, pp. 907–912, 2011.

[11] A. B. Khanikaev, S. Hossein Mousavi, W.-K. Tse, M. Kargarian, A. H. MacDonald, and G. Shvets, "Photonic topological insulators," *Nat. Mater.*, vol. 12, no. 3, pp. 233–239, 2013.

[12] M. C. Rechtsman, J. M. Zeuner, Y. Plotnik, et al., "Photonic Floquet topological insulators," *Nature*, vol. 496, no. 7444, pp. 196–200, 2013.

[13] Z. Gu, H. A. Fertig, D. P. Arovas, and A. Auerbach, "Floquet spectrum and transport through an irradiated graphene ribbon," *Phys. Rev. Lett.*, vol. 107, no. 21, p. 216601, 2011.

[14] K. Fang, Z. Yu, and S. Fan, "Realizing effective magnetic field for photons by controlling the phase of dynamic modulation," *Nat. Photonics*, vol. 6, no. 11, pp. 782–787, 2012.

[15] M. Hafezi, S. Mittal, J. Fan, A. Migdall, and J. M. Taylor, "Imaging topological edge states in silicon photonics," *Nat. Photonics*, vol. 7, no. 12, pp. 1001–1005, 2013.

[16] S. Mittal, J. Fan, S. Faez, A. Migdall, J. M. Taylor, and M. Hafezi, "Topologically robust transport of photons in a synthetic gauge field," *Phys. Rev. Lett.*, vol. 113, no. 8, p. 087403, 2014.

[17] W.-J. Chen, S.-J. Jiang, X.-D. Chen, et al., "Experimental realization of photonic topological insulator in a uniaxial metacrystal waveguide," *Nat. Commun.*, vol. 5, no. 1, p. 5782, 2014.

[18] G. Q. Liang and Y. D. Chong, "Optical resonator analog of a two-dimensional topological insulator," *Phys. Rev. Lett.*, vol. 110, no. 20, p. 203904, 2013.

[19] F. Gao, Z. Gao, X. Shi, et al., "Probing topological protection using a designer surface plasmon structure," *Nat. Commun.*, vol. 7, p. 11619, 2016.

[20] L.-H. Wu and X. Hu, "Scheme for achieving a topological photonic crystal by using dielectric material," *Phys. Rev. Lett.*, vol. 114, no. 22, 2015. https://doi.org/10.1103/physrevlett.114.223901.

[21] S. Yves, R. Fleury, T. Berthelot, M. Fink, F. Lemoult, and G. Lerosey, "Crystalline metamaterials for topological properties at subwavelength scales," *Nat. Commun.*, vol. 8, no. 1, p. 16023, 2017.

[22] Y. Chong. "Photonic insulators with a twist." *Nature*, vol. 496, p. 173–174, 2013.

[23] T. Ozawa, H. M. Price, A. Amo, et al., "Topological photonics," *Rev. Mod. Phys.*, vol. 91, no. 1, p. 015006, 2019.

[24] T. Ma and G. Shvets, "All-Si valley-Hall photonic topological insulator," *New J. Phys.*, vol. 18, no. 2, p. 025012, 2016.

[25] L. Ju, Z. Shi, N. Nair, et al., "Topological valley transport at bilayer graphene domain walls," *Nature*, vol. 520, no. 7549, pp. 650–655, 2015.

[26] Y. H. Wang, H. Steinberg, P. Jarillo-Herrero, and N. Gedik, "Observation of Floquet–Bloch states on the surface of a topological insulator," *Science*, vol. 342, no. 6157, pp. 453–457, 2013.

[27] M. S. Rudner, N. H. Lindner, E. Berg, and M. Levin, "Anomalous edge states and the bulk-edge correspondence for periodically driven two-dimensional systems," *Phys. Rev. X*, vol. 3, no. 3, p. 031005, 2013.

[28] L. J. Maczewsky, J. M. Zeuner, S. Nolte, and A. Szameit, "Observation of photonic anomalous Floquet topological insulators," *Nat. Commun.*, vol. 8, no. 1, p. 13756, 2017.

[29] S. Mukherjee, A. Spracklen, M. Valiente, et al., "Experimental observation of anomalous topological edge modes in a slowly driven photonic lattice," *Nat. Commun.*, vol. 8, no. 1, p. 13918, 2017.

[30] K. Wintersperger, C. Braun, F. N. Ünal, et al., "Realization of an anomalous Floquet topological system with ultracold atoms," *Nat. Phys.*, vol. 16, pp. 1058–1063, 2020.

[31] J. Li, R.-L. Chu, J. K. Jain, and S.-Q. Shen, "Topological Anderson insulator," *Phys. Rev. Lett.*, vol. 102, no. 13, p. 136806, 2009.

[32] S. Stützer, Y. Plotnik, Y. Lumer, et al., "Photonic topological Anderson insulators," *Nature*, vol. 560, no. 7719, pp. 461–465, 2018.

[33] E. J. Meier, F. A. An, A. Dauphin, et al., "Observation of the topological Anderson insulator in disordered atomic wires," *Science*, vol. 362, no. 6417, pp. 929–933, 2018.

[34] G. Harari, M. A. Bandres, Y. Lumer, et al., "Topological insulator laser: theory," *Science*, vol. 359, no. 6381, p. eaar4003, 2018.

[35] M. A. Bandres, S. Wittek, G. Harari, et al., "Topological insulator laser: experiments," *Science*, vol. 359, no. 6381, p. eaar4005, 2018.

[36] M. A. Bandres, M. C. Rechtsman, and M. Segev, "Topological photonic quasicrystals: fractal topological spectrum and protected transport," *Phys. Rev. X*, vol. 6, no. 12, p. 011016, 2016.

[37] Z. Yang, E. Lustig, Y. Lumer, and M. Segev, "Photonic Floquet topological insulators in a fractal lattice," *Light Sci. Appl.*, vol. 9, no. 1, p. 128, 2020.

[38] J. Guglielmon and M. C. Rechtsman, "Broadband topological slow light through higher momentum-space winding," *Phys. Rev. Lett.*, vol. 122, no. 15, p. 153904, 2019.

[39] A. Celi, P. Massignan, J. Ruseckas, et al., "Synthetic gauge fields in synthetic dimensions," *Phys. Rev. Lett.*, vol. 112, no. 4, p. 043001, 2014.

[40] B. K. Stuhl, H.-I. Lu, L. M. Aycock, D. Genkina, and I. B. Spielman, "Visualizing edge states with an atomic Bose gas in the quantum Hall regime," *Science*, vol. 349, no. 6255, pp. 1514–1518, 2015.

[41] F. A. An, E. J. Meier, and B. Gadway, "Direct observation of chiral currents and magnetic reflection in atomic flux lattices," *Sci. Adv.*, vol. 3, no. 4, p. e1602685, 2017.

[42] O. Zilberberg, S. Huang, J. Guglielmon, et al., "Photonic topological boundary pumping as a probe of 4D quantum Hall physics," *Nature*, vol. 553, no. 7686, pp. 59–62, 2018.

[43] X.-W. Luo, X. Zhou, J.-S. Xu, et al., "Synthetic-lattice enabled all-optical devices based on orbital angular momentum of light," *Nat. Commun.*, vol. 8, no. 1, p. 16097, 2017.

[44] L. Yuan, Y. Shi, and S. Fan, "Photonic gauge potential in a system with a synthetic frequency dimension," *Opt. Lett.*, vol. 41, no. 4, p. 741, 2016.

[45] T. Ozawa, H. M. Price, N. Goldman, O. Zilberberg, and I. Carusotto, "Synthetic dimensions in integrated photonics: from optical isolation to four-dimensional quantum Hall physics," *Phys. Rev. A*, vol. 93, no. 4, p. 043827, 2016.

[46] E. Lustig, S. Weimann, Y. Plotnik, et al., "Photonic topological insulator in synthetic dimensions," *Nature*, vol. 567, no. 7748, pp. 356–360, 2019.

[47] E. Lustig, Y. Plotnik, Z. Yang, and M. Segev, "3D Parity Time symmetry in 2D photonic lattices utilizing artificial gauge fields in synthetic dimensions," in *Conference on Lasers and Electro-Optics (OSA, 2019)*, p. FTu4B.1.

[48] A. Dutt, Q. Lin, L. Yuan, M. Minkov, M. Xiao, and S. Fan, "A single photonic cavity with two independent physical synthetic dimensions," *Science*, vol. 367, no. 6473, pp. 59–64, 2020.

[49] A. Perez-Leija, R. Keil, A. Kay, et al., "Coherent quantum transport in photonic lattices," *Phys. Rev. A*, vol. 87, no. 1, p. 012309, 2013.

[50] M. Z. Hasan and C. L. Kane, "*Colloquium*: topological insulators," *Rev. Mod. Phys.*, vol. 82, no. 4, pp. 3045–3067, 2010.

[51] M. C. Rechtsman, Y. Lumer, Y. Plotnik, A. Perez-Leija, A. Szameit, and M. Segev, "Topological protection of photonic path entanglement," *Optica*, vol. 3, no. 9, p. 925, 2016.

[52] S. Mittal, V. V. Orre, and M. Hafezi, "Topologically robust transport of entangled photons in a 2D photonic system," *Opt. Express*, vol. 24, no. 14, p. 15631, 2016.
[53] T. Rudolph, "Why I am optimistic about the silicon-photonic route to quantum computing," *APL Photonics*, vol. 2, no. 3, p. 030901, 2017.
[54] S. Barik, A. Karasahin, C. Flower, et al., "A topological quantum optics interface," *Science*, vol. 359, no. 6376, pp. 666–668, 2018.
[55] T. Kitagawa, M. A. Broome, A. Fedrizzi, et al., "Observation of topologically protected bound states in photonic quantum walks," *Nat. Commun.*, vol. 3, no. 1, p. 882, 2012.
[56] F. Cardano, M. Maffei, F. Massa, et al., "Statistical moments of quantum-walk dynamics reveal topological quantum transitions," *Nat. Commun.*, vol. 7, no. 1, p. 11439, 2016.
[57] A. Blanco-Redondo, B. Bell, D. Oren, B. J. Eggleton, and M. Segev, "Topological protection of biphoton states," *Science*, vol. 362, no. 6414, pp. 568–571, 2018.
[58] M. Wang, C. Doyle, B. Bell, et al., "Topologically protected entangled photonic states," *Nanophotonics*, vol. 8, no. 8, pp. 1327–1335, 2019.
[59] S. Mittal, E. A. Goldschmidt, and M. Hafezi, "A topological source of quantum light," *Nature*, vol. 561, no. 7724, pp. 502–506, 2018.
[60] V. Peano, M. Houde, F. Marquardt, and A. A. Clerk, "Topological quantum fluctuations and traveling wave amplifiers," *Phys. Rev. X*, vol. 6, no. 4, p. 041026, 2016.
[61] C. M. Bender, S. Boettcher, and P. N. Meisinger, "PT-symmetric quantum mechanics," *J. Math. Phys.*, vol. 40, no. 5, pp. 2201–2229, 1999.
[62] K. G. Makris, R. El-Ganainy, D. N. Christodoulides, and Z. H. Musslimani, "Beam dynamics in PT-symmetric optical lattices," *Phys. Rev. Lett.*, vol. 100103904, no. 104, pp. 103904(4). 2008.
[63] S. Klaiman, U. Günther, and N. Moiseyev, "Visualization of branch points in PT-symmetric waveguides," *Phys. Rev. Lett.*, vol. 101, no. 8, pp. 080402(4). 2008.
[64] A. Guo, G. J. Salamo, D. Duchesne, et al., "Observation of PT-symmetry breaking in complex optical potentials," *Phys. Rev. Lett.*, vol. 103, no. 9, 2009. https://doi.org/10.1103/physrevlett.103.093902.
[65] C. E. Rüter, K. G. Makris, R. El-Ganainy, D. N. Christodoulides, M. Segev, and D. Kip, "Observation of parity–time symmetry in optics," *Nat. Phys.*, vol. 6, no. 3, pp. 192–195, 2010.
[66] J. M. Zeuner, M. C. Rechtsman, Y. Plotnik, et al., "Observation of a topological transition in the bulk of a non-hermitian system," *Phys. Rev. Lett.*, vol. 115, no. 4, pp. 040402(5). 2015.
[67] M. S. Rudner and L. S. Levitov, "Topological transition in a non-hermitian quantum walk," *Phys. Rev. Lett.*, vol. 102, no. 6, 2009. https://doi.org/10.1103/physrevlett.102.065703.
[68] C. Poli, M. Bellec, U. Kuhl, F. Mortessagne, and H. Schomerus, "Selective enhancement of topologically induced interface states in a dielectric resonator chain," *Nat. Commun.*, vol. 6, no. 1, pp. 6710(5), 2015.
[69] S. Weimann, M. Kremer, Y. Plotnik, et al., "Topologically protected bound states in photonic parity-time-symmetric crystals," *Nat. Mater.*, vol. 16, no. 4, pp. 433–438, 2017.
[70] M. S. Rudner, M. Levin, and L. S. Levitov, "Survival, decay, and topological protection in non-Hermitian quantum transport," ArXiv160507652 Cond-Mat, 2016.
[71] T. E. Lee, "Anomalous edge state in a non-hermitian lattice," *Phys. Rev. Lett.*, vol. 116, no. 13, pp. 133903(5), 2016.
[72] D. Leykam, K. Y. Bliokh, C. Huang, Y. D. Chong, and F. Nori, "Edge modes, degeneracies, and topological numbers in non-hermitian systems," *Phys. Rev. Lett.*, vol. 118, no. 4, pp. 040401(6), 2017.
[73] H. Shen, B. Zhen, and L. Fu, "Topological band theory for non-Hermitian Hamiltonians," *Phys. Rev. Lett.*, vol. 120, no. 14, pp. 146402(6), 2018.
[74] Z. Gong, Y. Ashida, K. Kawabata, K. Takasan, S. Higashikawa, and M. Ueda. "Topological phases of non-Hermitian systems." *Phys. Rev. X 8*, vol. 8, no. 3, p. 031079, 2018.
[75] L. Xiao, T. Deng, K. Wang, et al., "Non-Hermitian bulk–boundary correspondence in quantum dynamics," *Nat. Phys.*, vol. 16, no. 7, pp. 761–766, 2020.
[76] G. Harari, M. A. Bandres, Y. Lumer, Y. Plotnik, D. N. Christodoulides, and M. Segev, "Topological lasers," in *Conference on Lasers and Electro-Optics (OSA, 2016)*, p. FM3A.3.
[77] P. St-Jean, V. Goblot, E. Galopin, et al., "Lasing in topological edge states of a one-dimensional lattice," *Nat. Photonics*, vol. 11, no. 10, pp. 651–656, 2017.
[78] M. Parto, S. Wittek, H. Hodaei, et al., "Edge-Mode Lasing in 1D Topological Active Arrays." *Phys. Rev. Lett*, vol. 120, no. 11, pp. 113901(6), 2018.
[79] S. Wittek, G. Harari, M. Bandres, et al. "Towards the experimental realization of the topological insulator laser." in *Conference on Lasers and Electro-Optics, OSA Technical Digest*, paper FTh1D.3.
[80] H. Zhao, P. Miao, M. H. Teimourpour, et al., "Topological hybrid silicon microlasers," *Nat. Commun.*, vol. 9, no. 1, p. 981, 2018.
[81] B. Bahari, A. Ndao, F. Vallini, A. E. Amili, Y. Fainman, and B. Kanté, "Nonreciprocal lasing in topological cavities of arbitrary geometries," *Science*, vol. 358, no. 6363, pp. 636–640, 2017.
[82] Y. Zeng, U. Chattopadhyay, B. Zhu, et al., "Electrically pumped topological laser with valley edge modes," *Nature*, vol. 578, no. 7794, pp. 246–250, 2020.
[83] I. Amelio, I. Carusotto. "Theory of the coherence of topological lasers." *Phys. Rev.*, 2020. [to appear].
[84] Z.-K. Shao, H.-Z. Chen, S. Wang, et al., "A high-performance topological bulk laser based on band-inversion-induced reflection," *Nat. Nanotechnol.*, vol. 15, no. 1, pp. 67–72, 2020.
[85] Y. G. Liu, P. Jung, M. Parto, W. E. Hayenga, D. N. Christodoulides, and M. Khajavikhan, "Towards the experimental demonstration of topological Haldane lattice in microring laser arrays (Conference Presentation)," in *Novel In-Plane Semiconductor Lasers XIX*, A. A. Belyanin, and P. M. Smowton, Eds., SPIE, 2020, p. 36.
[86] S. Klembt, T. H. Harder, O. A. Egorov, et al., "Exciton–Polariton Topological Insulator (Dataset)", 2018.
[87] Z. Yang, E. Lustig, G. Harari, et al., "Mode-locked topological insulator laser utilizing synthetic dimensions," *Phys. Rev. X*, vol. 10, no. 1, p. 011059, 2020.
[88] Y. Lumer, Y. Plotnik, M. C. Rechtsman, and M. Segev, "Self-localized states in photonic topological insulators," *Phys. Rev. Lett.*, vol. 111, no. 24, p. 243905, 2013.
[89] S. Mukherjee and M. C. Rechtsman, "Observation of Floquet solitons in a topological bandgap," *Science*, vol. 368, no. 6493, pp. 856–859, 2020.
[90] A. Dikopoltsev, T. Harder, E. Lustig, et al., "Topological insulator VCSEL array," in *CLEO 2020* (n.d.).

Aditya Tripathi, Sergey Kruk*, Yunfei Shang, Jiajia Zhou, Ivan Kravchenko, Dayong Jin and Yuri Kivshar

Topological nanophotonics for photoluminescence control

https://doi.org/10.1515/9783110710687-033

Abstract

Objectives: Rare-earth-doped nanocrystals are emerging light sources that can produce tunable emissions in colours and lifetimes, which has been typically achieved in chemistry and material science. However, one important optical challenge – polarization of photoluminescence – remains largely out of control by chemistry methods. Control over photoluminescence polarization can be gained via coupling of emitters to resonant nanostructures such as optical antennas and metasurfaces. However, the resulting polarization is typically sensitive to position disorder of emitters, which is difficult to mitigate.
Methods: Recently, new classes of disorder-immune optical systems have been explored within the framework of topological photonics. Here we explore disorder-robust topological arrays of Mie-resonant nanoparticles for polarization control of photoluminescence of nanocrystals.
Results: We demonstrate polarized emission from rare-earth-doped nanocrystals governed by photonic topological edge states supported by zigzag arrays of dielectric resonators. We verify the topological origin of polarized photoluminescence by comparing emission from nanoparticles coupled to topologically trivial and nontrivial arrays of nanoresonators.
Conclusions: We expect that our results may open a new direction in the study of topology-enable emission properties of topological edge states in many photonic systems.

Keywords: edge states; nanophotonics; polarization control; rare-earth-doped nanocrystals; topological photonics.

***Corresponding author: Sergey Kruk**, Nonlinear Physics Center, Research School of Physics, Australian National University, Canberra, ACT 2601, Australia; and Department of Physics, University of Paderborn, D-33098 Paderborn, Germany,
E-mail: sergey.kruk@anu.edu.au. https://orcid.org/0000-0003-0624-4033
Aditya Tripathi, Nonlinear Physics Center, Research School of Physics, Australian National University, Canberra, ACT 2601, Australia; and Department of Physics, Indian Institute of Technology Delhi, New Delhi 110016, India
Yunfei Shang, Jiajia Zhou and Dayong Jin, Institute for Biomedical Materials and Devices (IBMD), University of Technology Sydney, Sydney, NSW 2007, Australia
Ivan Kravchenko, Center for Nanophase Materials Sciences, Oak Ridge National Laboratory, Oak Ridge, TN 37831, USA
Yuri Kivshar, Nonlinear Physics Center, Research School of Physics, Australian National University, Canberra, ACT 2601, Australia

1 Introduction

Topological phases of light provide unique opportunities to create photonic systems immune to scattering losses and disorder [1]. Motivated by on-chip applications, there have been efforts to bring topological photonics to the nanoscale. Nanostructures made of high-index dielectric materials with judiciously designed subwavelength resonant elements supporting both electric and magnetic Mie resonances [2] show a special promise for implementations of the topological order for light.

Topological structures have recently been studied as powerful tools for harnessing light emission in various systems with topological orders including lasers [3, 4], quantum light sources [5], and nonlinear frequency converters [6]. In addition, topological photonics holds promise for the development of novel types of light emitters, as it provides a systematic way to control the number and degree of localization of spectrally isolated topologically robust edge and corner states.

Here we uncover another class of nontrivial effects in harnessing light emission with topology. We use topological properties of nanoscale photonic systems to control polarization of PL of rare-earth-doped nanocrystals.

Rare-earth-doped nanocrystals are emerging light sources used for many applications of nanotechnology such as bioimaging, sensing, therapy, display, data

This article has previously been published in the journal Nanophotonics. Please cite as: A. Tripathi, S. Kruk, Y. Shang, J. Zhou, I. Kravchenko, D. Jin and Y. Kivshar "Topological nanophotonics for photoluminescence control" *Nanophotonics* 2021, 10. DOI: 10.1515/nanoph-2020-0374.

storage, and photonics devices [7, 8]. The applicability of nanocrystals in the aforementioned aspects is determined by human ability to control their performance in diverse optical dimensions including emission color, spectrum, lifetime, intensity, and polarization.

To date, except for polarization, all other optical parameters of the rare-earth-doped nanocrystals have been extensively engineered by chemistry and material science methods [9–12]. Although polarized emission of rare-earth nanocrystals has been observed [13–16] and demonstrated for flow shear tomography [17], the polarization anisotropy was only detectable through spectroscopic measurements from single or aligned crystals, and most commonly, rod-type nanocrystals. This is because the polarization anisotropy is assigned to each splitting transition of rare-earth ions, and it is determined by the site symmetry in a crystal host. The irregular orientation of the crystalline axis will lead to a neutralization effect. The whole-band-based imaging will also neutralize the polarization information, as narrow peaks from 4f-4f rare-earth transitions may have different dipole orientations. A control of the polarization is even more challenging, as a given nanocrystal always has a fixed crystal phase and site symmetry.

To gain a control over polarization of emission, the emitters can be coupled to resonant nanostructures. Coupling of emitters to individual nanoantennas [18–24] and metasurfaces [25–27] has already been explored for control over the polarization of light [28, 29]. However, the resulting polarization of emission depends typically on the specific positioning of emitters. Precise positioning of emitters on nanostructures is challenging, and typically it requires demanding approaches such as atomic force microscopy [30]. In this regard, polarization control that relies on topologically nontrivial optical modes in nanostructures becomes attractive as topology introduces robustness against disorder in positioning. The topological protection offers to mitigate the dependence of the system on the exact position and orientation of emitters.

In this article, we use topological nanophotonics to control the emission polarization of the rare-earth-doped nanocrystals. Specifically, we use zigzag arrays of dielectric nanoresonators hosting topologically nontrivial optical modes that are robust against perturbations of the system [6]. We couple them with Er^{3+}-doped nanocrystals and observe enhanced polarized PL. We reveal that topological edge states can control polarization of emission in an unusual way. Specifically, in the vicinity of topological edge states, the emission becomes linearly polarized reproducing the polarization of topological edge modes. For the arrays with odd number of nanoresonators, the PL emission from two edges is orthogonally polarized, and for the arrays with even number of nanoresonators, the PL emission becomes copolarized, in accordance with the polarization of the topological states. To test the topological origin of the polarized emission, we study topologically trivial arrays and observe in contrast completely depolarized emission.

2 Results and discussion

We use β-$NaErF_4$@$NaYF_4$ core-shell nanocrystals with average diameter of 25 nm, as shown in the TEM image in Figure 1a and in the schematics in Figure 1b. The nanocrystals are synthesized by using the layer epitaxial growth method [31]. When pumped with 976 nm wavelength laser, the nanocrystals produce PL at around 1532 nm wavelength, which corresponds to the Er^{3+} transition: $^4I_{13/2} \rightarrow {}^4I_{15/2}$, as shown with the relevant energy levels in Figure 1c. The geometry of the zigzag nanostructures is chosen such that the wavelength of the topological edge states matches the wavelength of the nanoparticles' PL. In the topological zigzag array, each disk hosts Mie-resonant

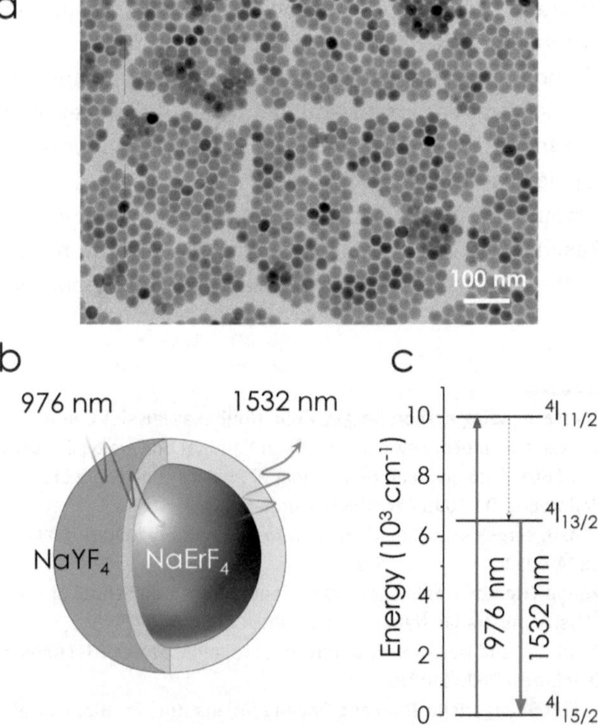

Figure 1: Structure and properties of Er^{3+}-doped core-shell nanoparticles.
(a) Transmission electron microscope image of nanoparticles. (b) Schematic of the core-shell structure of the nanoparticle. (c) Relevant energy levels of Er^{3+}.

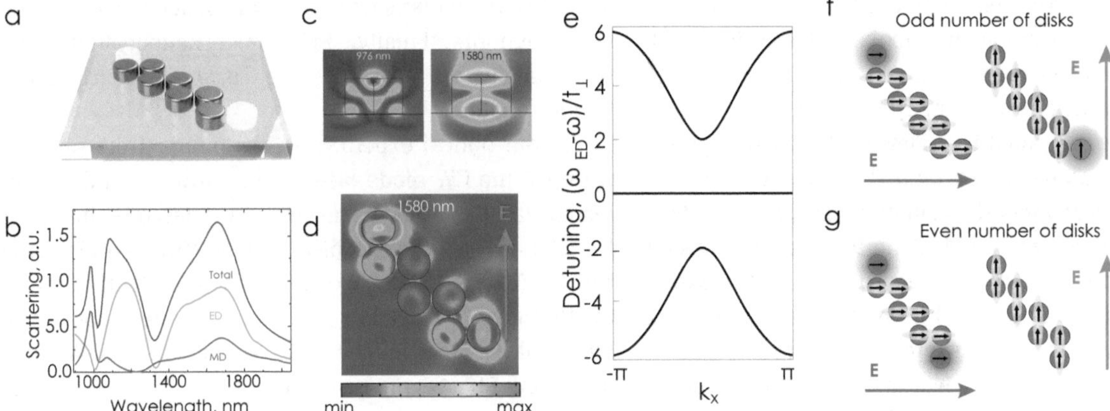

Figure 2: Topological zigzag arrays of dielectric nanoresonators.
(a) Concept image of the zigzag array hosting topological edge states. (b) Multipolar decomposition of a constituent disk nanoresonator. (c) Near-field distributions of a single disk for plane wave excitation at around the pump and the photoluminescence (PL) wavelengths. (d) Full-wave simulation of edge localizations in the zigzag array illuminated with a diagonal polarization. (e) Dispersion curve for Bloch modes in periodic zigzag chain. (f, g) Schematics of the edge states formation in arrays with (f) even and (g) odd number of nanoresonators. The blue and red colors are used to distinguish between two principal polarization of excitation: horizontal and vertical. The yellow joints visualize stronger dipole-dipole coupling between the modes of the near-neighbor disk for the cases of horizontal polarization excitation (left) and vertical polarization excitation (right). Topological states (left side, blue color – horizontally polarized; right side, red color – vertically polarized) form at the edge disks that are weakly coupled to their near-neighbors.

modes [2] (see Figure 2b for details about multipolar expansion and Figure 2c for near-field distributions of the scattered electric field). The electric dipole mode of the disk peaks at around the wavelength of the Er^{3+} electric dipole transition $^4I_{13/2} \rightarrow {}^4I_{15/2}$, thus allowing for coupling between the emitter and the nanostructure.

When such nanoresonators are arranged into zigzag arrays, the geometry of the array introduces altered strong and weak near-neighbor coupling which can be described with a polarization-enriched generalized Su-Schrieffer-Heeger–type model [32–34] with a gauge-independent $\mathbb{Z}(2)$ topological invariant. The formation of topologically nontrivial light localization can be described with the analytical coupled dipoles approximation [32]. The Hamiltonian of the system can be written as

$$\mathcal{H} = \sum_{j,v} \hbar\omega_0 a_{jv}^{\dagger} a_{jv} + \sum_{\langle j,j' \rangle, v, v'} a_{jv}^{\dagger} V_{v,v'}^{\langle j,j' \rangle} a_{jv}$$

where ω_0 denotes the resonance frequency, the indices j and j' label the nanoparticles, $\langle j,j' \rangle$ are the first nearest neighbors in the array, and a_{jv} is the annihilation operator for the multipolar eigenmodes with the polarization v at the jth nanoparticle. We consider the electric dipole modes polarized in the plane of the sample (x, y) in either x- or y-directions. The coupling matrices $V^{\langle j,j' \rangle}$ are then written as

$$V^{\langle j,j' \rangle} = t_{\parallel} e_{\parallel}^{\langle j,j' \rangle} \otimes e_{\parallel}^{\langle j,j' \rangle} + t_{\perp} e_{\perp}^{\langle j,j' \rangle} \otimes e_{\perp}^{\langle j,j' \rangle}$$

where $e_{\parallel}^{\langle j,j' \rangle}$ and $e_{\perp}^{\langle j,j' \rangle}$ are the in-plane unit vectors parallel and perpendicular to the vector linking the near-neighbor particles and $t_{\parallel}$ and $t_{\perp}$ are the coupling constants for the modes, copolarized and cross-polarized with respect to the link vector and the ⊗ sign stands for the direct product. For short-range dipole-dipole interaction, the ratio of the coupling constants can be estimated as $t_{\parallel}/t_{\perp} = -2$ as shown in the study by Slobozhanyuk et al [32]. Figure 2d shows a full-wave simulation of the edge localizations in the zigzag array.

Figure 2e further visualizes the energy spectrum of the zigzag array of coupled resonators. The spectrum features a band gap with two degenerate edge states associated with the near-field distribution shown in Figure 2d. Nonzero longitudinal components of k-vector were accounted in the coupling constants $t \rightarrow t|2 - e^{ik_x}|$ [6]. The edge-state energy corresponds to the energy of a standalone nanoparticle $\hbar\omega_0$. The resulting topological invariant (winding number) equals 1 within the band gap, which corresponds to the emergence of one polarization-dependent state at every edge [32]. Figure 2f and g visualize resonant coupling of the electric dipole resonant modes of the neighboring nanodisks. Depending on the mutual orientation, the neighboring dipoles couple either strongly (visualized in Figure 2f and g with yellow joints) or weakly. In every case, the weakly coupled edge disk hosts an edge mode.

The resulting edge modes are protected by topology against perturbations of the system such as disorder [6]. The modes are linearly polarized. For arrays with odd number of disks the modes at the opposite ends are orthogonally polarized (see Figure 2f), and for the even number of disks, the modes are copolarized (see Figure 2g).

We proceed to experiments and fabricate the zigzag arrays of nanoparticles from amorphous silicon on a fused-silica substrate (500 µm thick). First, a 300-nm amorphous silicon layer was deposited onto the substrate by low-pressure chemical vapor deposition. Subsequently, a thin layer of an electron-resist PMMA A4 950 was spin-coated onto the sample, followed by electron beam lithography and development. A thin Cr film was evaporated onto the sample, followed by a lift-of process to generate a hard mask. Reactive-ion etching was used to transfer the Cr mask pattern into the silicon film. The residual Cr mask was removed via wet etching. This resulted in zigzag arrays of disks 510 nm in diameter with 30 nm spacing between the near-neighbors. Finally, Er^{3+}-doped nanocrystals were spin-coated on top of the arrays of dielectric nanoresonators.

In our optical experiments, we pumped the samples with 976 nm CW diode laser. We narrow down the collimated laser output beam with a telescope made of f = 125 mm lens and an objective lens Mitutoyo Plan Apo NIR HR (X100, 0.7NA). We studied spatial distribution of the PL signal in reflection on a camera Xenics Bobcat-320. The resulting measurement is shown in Figure 3a. We next measured PL spectra in reflection with a spectrometer NIR-Quest from Ocean Optics. Figure 3b shows the comparison of a spectrum measured from nanoparticles on a

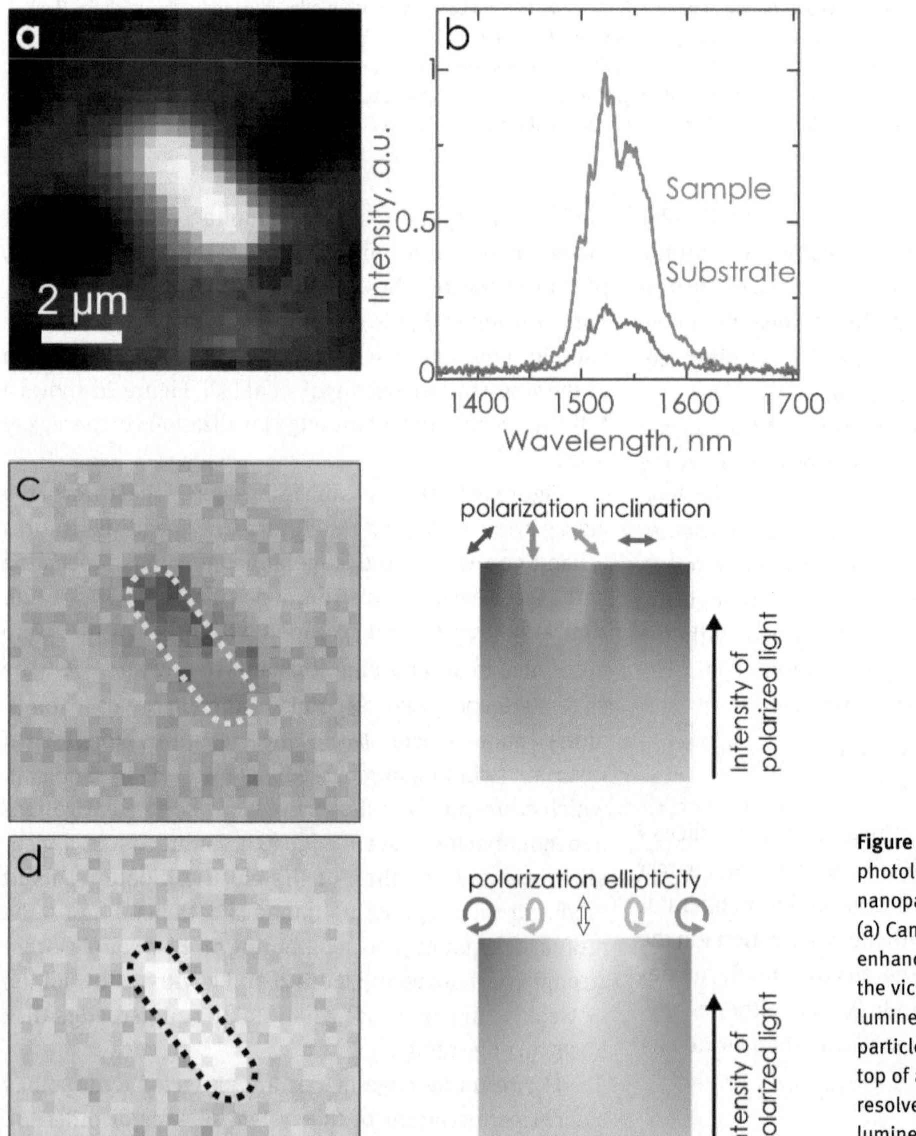

Figure 3: Topology-controlled photoluminescence of Er^{3+} core-shell nanoparticles.
(a) Camera image of the emission enhancement from the Er^{3+} nanoparticles in the vicinity of the zigzag array. (b) Photoluminescence spectra of the Er^{3+} nanoparticles on top of a zigzag array versus on top of a bare substrate. (c, d) Spatially resolved polarization states of photoluminescence showing (c) polarization inclination angles and (d) ellipticity of photoluminescence.

single array with nine nanodisks compared with a spectrum of nanoparticles on a bare glass substrate next to the array. The array provides approximately fivefold enhancement of PL signal.

Next, we experimentally measured spatially resolved polarization states of PL shown in Figure 3c and d. For this, we introduced a quarter-wave plate and a polarizer and performed a Stokes-vector polarimetry. Figure 3c visualizes the polarization inclination angle of PL around the zigzag array. Here, saturation visualizes the intensity of polarized emission (gray – no polarized emission) and the colors visualize the inclination angle of the electric field. Red stands for vertical polarization and blue for horizontal polarization. Figure 3d visualizes ellipticity of emission where similarly saturation reflects the intensity of polarized emission. We notice that ellipticity of polarization is close to zero in all our experiments. We observe that PL from rare-earth-doped nanoparticles becomes polarized in the vicinity of the topological edge states with the polarization state of PL reproducing the polarization of the topological edge mode.

Finally, we compare PL from nanoparticles around the 9-nanodisk zigzag array with 15-nanodisk and 14-nanodisk zigzag arrays as well as with topologically trivial straight array (see Figure 4). The 15-nanodisk zigzag performs qualitatively similar to the 9-nanodisk array as it has same parity of disks (odd). A longer array exhibits sharper localization of the PL as it is expected to provide higher intensity contrast with the bulk [6]. In the 15 disk-long chain PL away from the edge becomes depolarized as the topological modes responsible for polarization control are pinned to the edges. The 14-nanodisk zigzag array in contrast demonstrates copolarized PL in the vicinity of the two edge states in agreement with its parity (even). The bulk of the 15- and 14-nanodisk arrays produces depolarized PL due to lack of polarization sensitivity of zigzag bulk modes. No polarized PL is observed for the topologically trivial array of nanoresonators.

3 Conclusion and outlook

We have studied emission of Er^{3+}-doped nanocrystals deposited on top of topologically nontrivial zigzag arrays of Mie-resonant dielectric nanoparticles. We have observed a novel effect in topological photonics: enhanced and polarized PL emission from the doped nanocrystals located near the topological edge states. We have verified the topological origin of this controlled emission by conducting additional experiments with trivial arrays of the same dielectric nanoresonators, for which we have observed completely depolarized emission.

As the next step in this direction, we expect an efficient control of photon upconversion being crucial for many applications ranging from bioimaging to photovoltaics. By adjusting nanostructure parameters, we expect to achieve a systematical shift of the spectral position of resonances over several hundreds of nanometers, and also observe multifold enhancement of photon upconversion efficiency in nanocrystals which is useful for low-threshold photon upconversion in future solar energy applications.

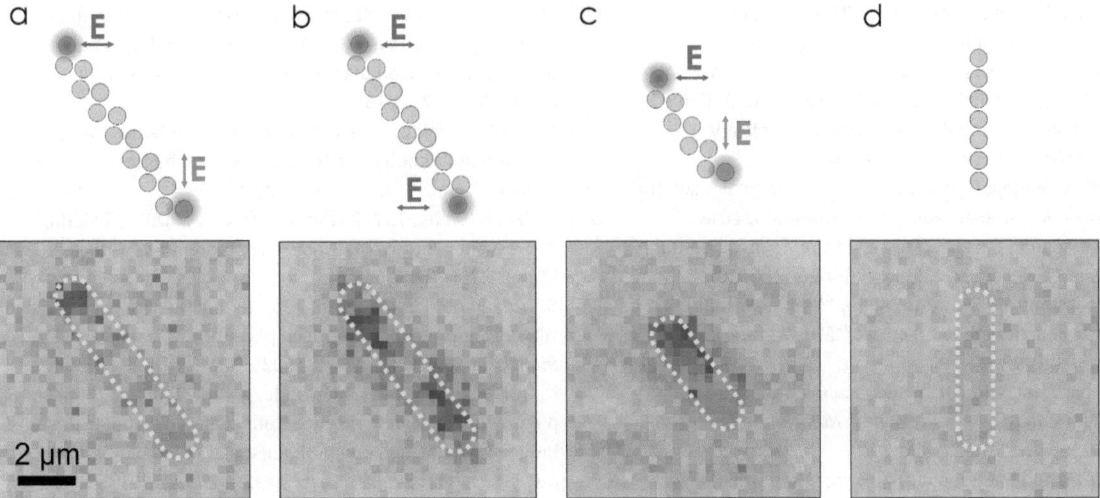

Figure 4: Polarization control of photoluminescence with topological edge states.
Spatially resolved polarization states of photoluminescence for (a) 15-nanodisk zigzag array, (b) 14-nanodisk zigzag array, (c) 9-nanodisk zigzag array, and (d) topologically trivial straight array of resonant nanodisks. The color-map is analogous to that used in Figure 3c.

We expect that our results may open a new direction in the study of topology-enable emission properties of topological edge states in many photonic systems. Indeed, active topological photonics can provide a fertile platform for not only studying interesting fundamental problems involving nontrivial topological phases, but in addition, it may become a route toward the development of novel designs for disorder-immune active photonic device applications, such as robust high-speed routing and switching, nanoscale lasers, and quantum light sources.

Acknowledgments: A part of this research was conducted at the Center for Nanophase Materials Sciences, which is a DOE Office of Science User Facility. A.T. and S.K. thank Barry Luther-Davies for useful discussions and experimental support. S.K. thanks Alexander Poddubny for useful discussions and also acknowledges a financial support from the Alexander von Humboldt Foundation.
Author contribution: All the authors have accepted responsibility for the entire content of this submitted manuscript and approved submission.
Research funding: Australian Research Council (grants DP200101168 and DE180100669), Strategic Fund of the Australian National University, and China Scholarship Council (grant 201706120322).
Conflict of interest statement: The authors declare no conflicts of interest regarding this article.

References

[1] L. Lu, J. D. Joannopoulos, and M. Soljačić, "Topological photonics," *Nat. Photonics*, vol. 8, pp. 821–829, 2014.
[2] Y. Kivshar, "All-dielectric meta-optics and non-linear nanophotonics," *Natl. Sci. Rev.*, vol. 5, pp. 144–158, 2018.
[3] B. Bahari, A. Ndao, F. Vallini, A. El Amili, Y. Fainman, B. Kanté, "Nonreciprocal lasing in topological cavities of arbitrary geometries," *Science*, vol. 358, pp. 636–640, 2017.
[4] M. A. Bandres, S. Wittek, G. Harari, et al., "Topological insulator laser: experiments," *Science*, vol. 359, p. eaar4005, 2018.
[5] S. Barik, A. Karasahin, C. Flower, et al., "A topological quantum optics interface," *Science*, vol. 359, pp. 666–668, 2018.
[6] S. Kruk, A. Poddubny, D. Smirnova, et al., "Nonlinear light generation in topological nanostructures," *Nat. Nanotechnol.*, vol. 14, pp. 126–130, 2019.
[7] B. Zhou, B. Shi, D. Jin, and X. Liu, "Controlling upconversion nanocrystals for emerging applications," *Nat. Nanotechnol.*, vol. 10, pp. 924–936, 2015.
[8] J. Zhou, A. I. Chizhik, S. Chu, and D. Jin, "Single-particle spectroscopy for functional nanomaterials," *Nature*, vol. 579, pp. 41–50, 2020.
[9] F. Wang, R. Deng, J. Wang, et al., "Tuning upconversion through energy migration in core–shell nanoparticles," *Nat. Mater.*, vol. 10, pp. 968–973, 2011.
[10] J. Zhao, D. Jin, E. P. Schartner, et al., "Single-nanocrystal sensitivity achieved by enhanced upconversion luminescence," *Nat. Nanotechnol.*, vol. 8, pp. 729–734, 2013.
[11] Y. Lu, J. Zhao, R. Zhang, et al., "Tunable lifetime multiplexing using luminescent nanocrystals," *Nat. Photonics*, vol. 8, pp. 32–36, 2014.
[12] Y. Fan, P. Wang, Y. Lu, et al., "Lifetime-engineered NIR-II nanoparticles unlock multiplexed in vivo imaging," *Nat. Nanotechnol.*, vol. 13, pp. 941–946, 2018.
[13] J. Zhou, G. Chen, E. Wu, et al., "Ultrasensitive polarized up-conversion of Tm^{3+}-Yb^{3+} doped β-$NaYF_4$ single nanorod," *Nano Lett.*, vol. 13, pp. 2241–2246, 2013.
[14] P. Chen, M. Song, E. Wu, et al., "Polarization modulated upconversion luminescence: single particle vs. few-particle aggregates," *Nanoscale*, vol. 7, pp. 6462–6466, 2015.
[15] P. Rodríguez-Sevilla, L. Labrador-Páez, D. Wawrzyńczyk, et al., "Determining the 3D orientation of optically trapped upconverting nanorods by in situ single-particle polarized spectroscopy," *Nanoscale*, vol. 8, pp. 300–308, 2016.
[16] S. Shi, L. D. Sun, Y. X. Xue, et al., "Scalable direct writing of lanthanide-doped $KMnF_3$ perovskite nanowires into aligned arrays with polarized up-conversion emission," *Nano Lett.*, vol. 18, pp. 2964–2969, 2018.
[17] J. Kim, S. Michelin, M. Hilbers, et al., "Monitoring the orientation of rare-earth-doped nanorods for flow shear tomography," *Nat. Nanotechnol.*, vol. 12, pp. 914–919, 2017.
[18] S. Kühn, U. Håkanson, L. Rogobete, and V. Sandoghdar, "Enhancement of single-molecule fluorescence using a gold nanoparticle as an optical nanoantenna," *Phys. Rev. Lett.*, vol. 97, p. 1, 2006.
[19] A. G. Curto, G. Volpe, T. H. Taminiau, et al., "Unidirectional emission of a quantum dot coupled to a nanoantenna," *Science*, vol. 329, pp. 930–933, 2010.
[20] L. Novotny and N. Van Hulst, "Antennas for light," *Nat. Photonics*, vol. 5, pp. 83–90, 2011.
[21] S. S. Kruk, M. Decker, I. Staude, et al., "Spin-polarized photon emission by resonant multipolar nanoantennas," *ACS Photonics*, vol. 1, pp. 1218–1223, 2014.
[22] M. Cotrufo, C. I. Osorio, and A. F. Koenderink, "Spin-dependent emission from arrays of planar chiral nanoantennas due to lattice and localized plasmon resonances," *ACS Nano*, vol. 10, pp. 3389–3397, 2016.
[23] A. Mohtashami, C. I. Osorio, and A. F. Koenderink, "Angle-resolved polarimetry of antenna-mediated fluorescence," *Phys. Rev. Appl.*, vol. 4, p. 23317019, 2015.
[24] C. Yan, X. Wang, T. V. Raziman, and O. J. Martin, "Twisting fluorescence through extrinsic chiral antennas," *Nano Lett.*, vol. 17, pp. 2265–2272, 2017.
[25] S. Luo, Q. Li, Y. Yang, et al., "Controlling fluorescence emission with split-ring-resonator-based plasmonic metasurfaces," *Laser Photonics Rev.*, vol. 11, p. 1770035, 2017.
[26] K. Q. Le, S. Hashiyada, M. Kondo, and H. Okamoto, "Circularly polarized photoluminescence from achiral dye molecules induced by plasmonic two-dimensional chiral nanostructures," *J. Phys. Chem. C*, vol. 122, pp. 24924–24932, 2018.
[27] A. Vaskin, R. Kolkowski, A. F. Koenderink, and I. Staude, "Light-emitting metasurfaces," *Nanophotonics*, vol. 8, pp. 1151–1198, 2019.
[28] K. J. Russell, T. L. Liu, S. Cui, and E. L. Hu, "Large spontaneous emission enhancement in plasmonic nanocavities," *Nat. Photonics*, vol. 6, pp. 459–462, 2012.

[29] K. Y. Bliokh, F. J. Rodríguez-Fortuño, F. Nori, and A. V. Zayats, "Spin-orbit interactions of light," *Nat. Photonics*, vol. 9, pp. 796–808, 2015.
[30] J. V. Chacko, C. Canale, B. Harke, and A. Diaspro, "Sub-diffraction nano manipulation using STED AFM," *PLoS One*, vol. 8, p. 19326203, 2013.
[31] J. Liao, D. Jin, C. Chen, Y. Li, and J. Zhou, "Helix shape power-dependent properties of single upconversion nanoparticles," *J. Phys. Chem. Lett.*, pp. 2883–2890, 2020. https://doi.org/10.1021/acs.jpclett.9b03838.
[32] A. P. Slobozhanyuk, A. N. Poddubny, A. E. Miroshnichenko, P. A. Belov, and Y. S. Kivshar, "Subwavelength topological edge states in optically resonant dielectric structures," *Phys. Rev. Lett.*, vol. 114, p. 123901, 2015.
[33] Y. Hadad, A. B. Khanikaev, and A. Alù, "Self-induced topological transitions and edge states supported by nonlinear staggered potentials," *Phys. Rev. B*, vol. 93, p. 155112, 2016.
[34] S. Kruk, A. Slobozhanyuk, D. Denkova, et al., "Edge states and topological phase transitions in chains of dielectric nanoparticles," *Small*, vol. 13, p. 16136829, 2017.

Tianshu Jiang, Anan Fang, Zhao-Qing Zhang and Che Ting Chan*

Anomalous Anderson localization behavior in gain-loss balanced non-Hermitian systems

https://doi.org/10.1515/9783110710687-034

Abstract: It has been shown recently that the backscattering of wave propagation in one-dimensional disordered media can be entirely suppressed for normal incidence by adding sample-specific gain and loss components to the medium. Here, we study the Anderson localization behaviors of electromagnetic waves in such gain-loss balanced random non-Hermitian systems when the waves are obliquely incident on the random media. We also study the case of normal incidence when the sample-specific gain-loss profile is slightly altered so that the Anderson localization occurs. Our results show that the Anderson localization in the non-Hermitian system behaves differently from random Hermitian systems in which the backscattering is suppressed.

Keywords: Anderson localization; gain-loss balanced system; non-Hermitian system.

1 Introduction

Anderson localization has been extensively studied for several decades since it was proposed by Anderson in 1958 [1–36]. Classical wave systems are good platforms to study Anderson localization due to the ease of fabrication and characterization and the absence of many-body interactions in photonic or phononic systems [3–7]. It is now well known that the localization of waves is induced by the constructive interference between two counter-propagating backscattering waves, which gives rise to coherent backscattering effects and makes all waves localized in one- and two-dimensional random media [3, 8–11]. For three-dimensional random media, there exist the so-called mobility edges, which separate localized states from extended states [3, 12, 13]. To identify whether the Anderson localization occurs, one needs to check the sample-size dependence of the transmission, which behaves quite differently between the linear decay in the diffusive regime and exponential decay in the localization regime [3]. Since absorption can also lead to an exponential decay of the transmission, it is very difficult to separate the effect of Anderson localization from that of absorption. Therefore, the absorption was excluded in most of the theoretical studies of Anderson localization. Effects of absorption or amplification on the Anderson localization have also been investigated intensively [14–30].

Recently, it has been proposed and shown rigorously that coherent backscattering effects can be totally suppressed in one-dimensional (1D) random systems by adding sample-specific gain and loss balanced profiles into the systems so that total transmission is achieved without reflection when waves are incident from one direction normal to the layers [37]. The experimental demonstration of the total transmission in such non-Hermitian random media has been carried out in an acoustic tube system [38]. Since total transmission can be achieved independent of the sample size, such non-Hermitian random systems possess an infinite localization length. The existence of such a critical point provides us a unique opportunity to study the Anderson localization behaviors in non-Hermitian system in the vicinity of the critical point. Here, we numerically study two situations where Anderson localization can occur. First, the sample-specific gain-loss profile is slightly altered so the total transmission deviates from unity. Second, the incident waves become oblique so that reflections from both sides can occur. In both scenarios, it is expected that the presence of coherent backscattering can lead to the Anderson localization of waves, i.e., the transmission will decay exponentially to zero when the sample is sufficiently large. Our results will be compared with the Anderson localization behaviors found in some special Hermitian random systems, where localization length is also known to be infinite at normal

***Corresponding author: Che Ting Chan,** Department of Physics, Hong Kong University of Science and Technology, Clear Water Bay, Hong Kong, China, E-mail: phchan@ust.hk
Tianshu Jiang, Anan Fang and Zhao-Qing Zhang: Department of Physics, Hong Kong University of Science and Technology, Clear Water Bay, Hong Kong, China.
https://orcid.org/0000-0002-0157-3877 (T. Jiang)

This article has previously been published in the journal Nanophotonics. Please cite as: T. Jiang, A. Fang, Z.-Q. Zhang and C. T. Chan "Anomalous Anderson localization behavior in gain-loss balanced non-Hermitian systems" *Nanophotonics* 2021, 10. DOI: 10.1515/nanoph-2020-0306.

incidence such as random layered media with the same impedance in all layers [31] and pseudospin-1 systems in 1D random potential [32, 33].

It should be pointed out that the gain-loss balanced non-Hermitian random systems considered here are different from the non-Hermitian PT symmetric systems studied extensively recently [39–43]. It has been shown that non-Hermiticity in the PT symmetric systems can give rise to many novel physical phenomena not seen in Hermitian systems due to presence of exceptional points, such as laser absorber [42, 43], impurity immunity [44], unidirectional transmission [45] and negative refraction [46]. Although our systems do not obey the PT symmetry due to the random structures, they do have the property that the spatial integration of the imaginary parts of dielectric constant is zero in every random configuration so that the gain and loss are always balanced.

2 Gain-loss balanced random media

According to a study by Bender and Boettcher [37], the total transmission of a random medium, which is embedded in a homogenous medium, can always be achieved by introducing a sample-specific imaginary part to the relative permittivity of the random medium, i.e., $(-1/k_0)\partial_x W(x)$, where k_0 is the wave vector in the background medium and $W(x)$ denotes the dielectric constant of the random medium. The 1D random system we studied is shown in Figure 1(a). The system has N layers of randomly arranged dielectric media with gain/loss coatings at each interface. The whole system is embedded in a homogeneous medium with an averaged permittivity of the random layers. We assume that the relative permittivity in each layer fluctuates independently with a uniform distribution in the interval $[1-\sigma, 1+\sigma]$, where σ controls the strength of the randomness. In this case, the embedding medium is the vacuum. In this work, we set $\sigma = 0.5$. All layers are nonmagnetic and assumed to have the same thickness $d = 1$. When an interface layer of width d_c is inserted between two adjacent layers $i-1$ and i with dielectric constants n_{i-1} and n_i, respectively, we assume the real part of the dielectric constant of the interface layer $W(x)$ has the form $W(x) = n_{i-1} + (n_i - n_{i-1})(x - x_{i-1})/d_c$, where x_{i-1} denotes the position of the right boundary of the $(i-1)$-th layer. Thus, by using $(-1/k_0)\partial_x W(x)$, the total transmission without reflections can be achieved for waves normally incident from the left if the imaginary parts of the relative permittivity of the interface layer is chosen as $-(n_i - n_{i-1})/d_c k_0$. For the simplicity of calculation, we also assume that the interface layer is much thinner than the wavelength so that the imaginary part of its relative permittivity reduces to the form $(-1/k_0)(n_i - n_{i-1})\delta(x - x_{i-1})$. In this limit, the real part of the relative permittivity of the interface layer becomes irrelevant as there is no phase change occurring in the electric field across the boundary. It should be pointed out that the Anderson localization behavior remains the same if the interface layer has a finite thickness (see Supplementary Note 1). This is expected as the critical behaviors normally do not depend on the details of the system.

In order to study the Anderson localization of the system, we take the following more general form for the relative permittivity of the interface coating between the $(i-1)$-th and i-th layers:

$$\mathrm{Im}(\varepsilon_{i-1,i}/\varepsilon_0) = -\frac{\alpha}{k_0} \cdot (n_i - n_{i-1}) \cdot \delta(x - x_{i-1}), \tag{1}$$

where α is a dimensionless parameter that controls the strength of gain/loss. It is easy to see from Eq. (1) that the sum of $\mathrm{Im}(\varepsilon_{i-1,i})$ for all interfaces is zero. Thus, the system as a whole has no net gain or loss. Since the transmission is always unity for normally incident waves when $\alpha = 1$ independent of random configuration or sample size, the Anderson localization length diverges at this critical point. We will study the divergent behavior of the Anderson localization length when α is close to unity. We will also study the divergent behavior for obliquely incident waves at $\alpha = 1$ when the incident angle is close to zero.

By utilizing Eq. (1) and the Maxwell's equations, we can obtain the boundary conditions at an interface, say, $x = x_{i-1}$ for a harmonic wave with frequency ω (see Supplementary Note 2 for details):

$$E_{i,y}(x = x_{i-1}^+) = E_{i-1,y}(x = x_{i-1}^-), \tag{2}$$

$$E_{i,z}(x = x_{i-1}^+) = E_{i-1,z}(x = x_{i-1}^-), \tag{3}$$

$$H_{i,y}(x = x_{i-1}^+) - H_{i-1,y}(x = x_{i-1}^-) = -\frac{\alpha}{Z_0}(n_i - n_{i-1})E_z(x = x_{i-1}), \tag{4}$$

$$H_{i,z}(x = x_{i-1}^+) - H_{i-1,z}(x = x_{i-1}^-) = \frac{\alpha}{Z_0}(n_i - n_{i-1})E_y(x = x_{i-1}), \tag{5}$$

where $E_{i,y}$ ($H_{i,y}$) and $E_{i,z}$ ($H_{i,z}$) are the y and z components of the electric (magnetic) field in the i-th layer, respectively, and Z_0 is the vacuum impedance. Equations (2) and (3) describe the continuity of the tangential components of the electric field. However, the two tangential components of magnetic field are not continuous due to the presence of the imaginary part of the permittivity at the interface, which acts as a current source in the direction parallel/

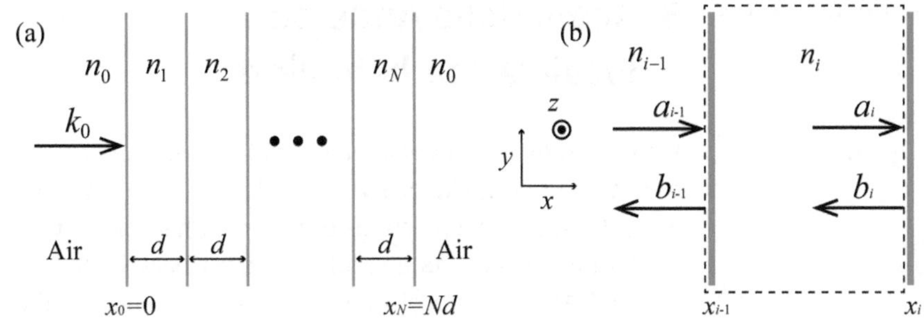

Figure 1: (Color online) (a) The 1D random system composed of random dielectric layers (white) with gain/lossy coatings (blue). The whole system is embedded in air. (b) The schematic of wave propagations inside the 1D random system. The black dashed line indicates the choice of the i-th scattering element. 1D, one dimensional.

antiparallel to the tangential components of the electric field, as shown in Eqs. (4) and (5). Equations (2)–(5) show that due to the vanishing thickness of the coating layer, the real part of the permittivity of the coating does not play a role in the boundary conditions.

It is important to point out that when $\alpha = 1$, Eq. (1) always gives unity transmission with zero reflection when the wave with wavevector k_0 in the background medium is incident from the left, independent of the random arrangement [37]. In this case, the magnitude of electric field is a constant in all layers. To see this explicitly, we consider the case when the electric and magnetic fields are aligned with the y and z axes, respectively, i.e., E_y and H_z. Let $(E_{i-1,y}, H_{i-1,z})$ and $(E_{i,y}, H_{i,z})$ be the electric and magnetic fields of the incident wave in the layer i–1 and transmitted wave in the layer i, respectively. If there is no reflection, the boundary conditions in Eqs. (2) and (5) give the following energy flux change across the interface between layer i–1 and i

$$\begin{aligned}\Delta S &= E_{i,y}(x = x_{i-1}^+)\cdot H_{i,z}(x = x_{i-1}^+) \\ &\quad -E_{i-1,y}(x = x_{i-1}^-)\cdot H_{i-1,z}(x = x_{i-1}^-) \\ &= E_y(x = x_{i-1})\cdot\left[H_{i,z}(x = x_{i-1}^+) - H_{i-1,z}(x = x_{i-1}^-)\right] \\ &= \frac{1}{Z_0}(n_i - n_{i-1})\left|E_y(x = x_{i-1})\right|^2\end{aligned} \tag{6}$$

It is easily seen that the above change of energy flux obeys the Poynting theorem

$\nabla\cdot\mathbf{S} = -\omega\mathrm{Im}(\varepsilon)\cdot|\mathbf{E}(x = x_{i-1})|^2$ if $\mathrm{Im}(\varepsilon)$ has the form prescribed in Eq. (1) with $\alpha = 1$. Thus, the difference in energy flux on each side is totally compensated by the energy flux produced or dissipated at the interface coating and the assumption of no reflection produces the correct physical solution. However, if the wave is incident from the right, the direction of the energy flux is reversed, the energy flux difference changes its sign and reflection will occur in order to make the energy flux conserved.

When $\alpha \neq 1$, reflections occur on both sides of an interface, and Anderson localization will occur as a result of coherent backscattering effects. This is also true for oblique incidence even when $\alpha = 1$.

In order to study the transport properties of the above system, we consider an N-layer sample shown in Figure 1(a) as a successive stack of $N + 1$ scattering elements. In Figure 1(b), we show the i-th scattering element as the region from the right end of the (i–1)-th layer, $x = x_{i-1}$, to that of the i-th layer, $x = x_i$. Using the transfer-matrix method (TMM), one can obtain the transmission and reflection amplitudes across each scattering element. For the i-th scattering element, we let t_{i+} (r_{i+}) be the transmission (reflection) amplitude for waves incident from the left (forward waves), and t_{i-} (r_{i-}) for waves incident from the right (backward waves). Then, for forward waves, the transmission amplitude $t^+(N + 1)$ through the stack of $N + 1$ scattering elements can be obtained from the following recurrence relations [33–36] starting from $i = 1$ till $i = N$,

$$t^+(i+1) = \frac{t^+(i)t_{(i+1)+}}{1 - r_{(i+1)+}r^-(i)}, \tag{7}$$

$$r^-(i) = r_{i-} + \frac{r^-(i-1)t_{i-}t_{i+}}{1 - r^-(i-1)r_{i+}} \quad (i \geq 2), \tag{8}$$

where $t^+(i)$ and $r^-(i)$ denote the forward transmission amplitude and backward reflection amplitude of the first i scattering elements, respectively. Note that when $i = 1$, we have $t^+(1) = t_{1+}$ and $r^-(1) = r_{1-}$. From Eq. (7), we obtain the transmission coefficient $T_N = |t^+(N + 1)|^2$, and

$$\begin{aligned}\ln T_N &= \ln|t^+(N+1)|^2 \\ &= \ln|t^+(N)|^2 + \ln\left|t_{(N+1)+}\right|^2 - 2\ln\left|1 - r_{(N+1)+}r^-(N)\right|,\end{aligned} \tag{9}$$

By applying the recursion equation (9) iteratively, we can express T_N as

$$\ln T_N = \sum_{i=1}^{N+1}\ln|t_{i+}|^2 - 2\sum_{i=2}^{N+1}\ln|1 - r_{i+}r^-(i-1)|. \tag{10}$$

Next, we use the TMM to obtain the transmission and reflection amplitudes of an individual scattering element.

For normal incidence, as shown in Figure 1(b), the electric fields can be expressed as:

$$E_{i-1}(x) = a_{i-1}\exp[in_{i-1}k_0(x - x_{i-1})] + b_{i-1}\exp[-in_{i-1}k_0(x - x_{i-1})], \tag{11}$$

in the $(i-1)$-th layer, and

$$E_i(x) = a_i\exp[in_ik_0(x - x_i)] + b_i\exp[-in_ik_0(x - x_i)], \tag{12}$$

in the i-th layer. Here k_0 is the wave vector in vacuum. The magnetic fields can be obtained through the relation $H_i = E_i/Z_i$, where $Z_i = \sqrt{\mu_0/\varepsilon_i}$ is the impedance of the i-th layer.

Using the boundary conditions in Eqs. (2)–(5), we can obtain the transfer matrix $m^{(i)}$ connecting the electric fields at the interfaces $x = x_{i-1}$ and $x = x_i$,

$$\begin{pmatrix} a_i \\ b_i \end{pmatrix} = \begin{pmatrix} m_{11}^{(i)} & m_{12}^{(i)} \\ m_{21}^{(i)} & m_{22}^{(i)} \end{pmatrix} \cdot \begin{pmatrix} a_{i-1} \\ b_{i-1} \end{pmatrix}, \tag{13}$$

with

$$m_{11}^{(i)} = \frac{1}{2}\left[(1+\alpha) + (1-\alpha)\frac{n_{i-1}}{n_i}\right]\exp(in_ik_0d), \tag{14}$$

$$m_{12}^{(i)} = \frac{1}{2}\left[(1+\alpha) - (1+\alpha)\frac{n_{i-1}}{n_i}\right]\exp(in_ik_0d), \tag{15}$$

$$m_{21}^{(i)} = \frac{1}{2}\left[(1-\alpha) - (1-\alpha)\frac{n_{i-1}}{n_i}\right]\exp(-in_ik_0d), \tag{16}$$

$$m_{22}^{(i)} = \frac{1}{2}\left[(1-\alpha) + (1+\alpha)\frac{n_{i-1}}{n_i}\right]\exp(-in_ik_0d). \tag{17}$$

The transmission and reflection amplitudes of the i-th scattering element can be obtained from the transfer matrix:

$$t_{i+} = \left(m_{11}^{(i)}m_{22}^{(i)} - m_{12}^{(i)}m_{21}^{(i)}\right)/m_{22}^{(i)}, \quad t_{i-} = 1/m_{22}^{(i)}, \tag{18}$$

$$r_{i+} = -m_{21}^{(i)}/m_{22}^{(i)}, \quad r_{i-} = m_{12}^{(i)}/m_{22}^{(i)}. \tag{19}$$

It is easily seen that when $\alpha = 1$ we have $m_{21}^{(i)} = 0$ and $m_{12}^{(i)} \neq 0$, leading to zero reflection for wave incident from left and finite reflection for waves coming from right. Since this behavior is true for all layers, we will have unity transmission and zero reflection for waves entering from the left of the sample. Since the system is reciprocal, unity transmission is also expected for waves coming from right side of the sample even though the reflection is not zero.

3 Anomalous Anderson localization behaviors

We first study the Anderson localization behaviors for the case of normal incidence when α deviates slightly from unity. In this case, reflections occur on both sides of every scattering element. As a result, coherent backscattering-induced Anderson localization can occur, similar to the case of Hermitian random media. The localization length ξ can be obtained from the geometrical mean of the transmission coefficient according to the relation:

$$\xi = -\lim_{L\to\infty}\frac{2L}{\langle \ln T_N\rangle_c}, \tag{20}$$

where $L = Nd$ is the sample length and $\langle\rangle_c$ denotes ensemble averaging.

In our numerical study, we set k_0 and d to be unity for simplicity. As will see in Section 4 that the critical behaviors will not be altered if these parameters are scaled according to $k'd' = k_0d = 1$. We first use Eqs. (13)–(19) to obtain the transmission and reflection amplitudes of each scattering element. Then by applying the recursion equations (7) and (8) iteratively, we obtain the transmission coefficient T_N from Eq. (9). In Figure 2(a) and (b), we plot $\langle \ln T_N\rangle_c$ as a function of the sample length $L = Nd$ (solid circles) at four different values of α for both the cases of $\alpha < 1$ and $\alpha > 1$. Each data point of $\langle \ln T_N\rangle_c$ shown here (as well as in other figures of this work) is obtained from an average of over 200 configurations and the length L used in the calculation of $\langle \ln T_N\rangle_c$ is about five times of the localization length. For each value of α, the plot of $\langle \ln T_N\rangle_c$ vs. L can be well fitted into a straight line with a negative slope, indicating the exponential decay of the transmission T_N with increasing L, i.e., the occurrence of Anderson localization. Following Eq. (20), we can obtain the localization length ξ through the slope of the linear fitting. In the insets of Figure 2(a) and (b), we plot the localization length $\ln\xi$ as a function of $\ln|\alpha - 1|$ for the above two cases, respectively. From the linear fits, we get

$$\ln\xi_- = A_- + B_- \cdot \ln|\alpha - 1|, \quad \alpha < 1, \tag{21}$$

$$\ln\xi_+ = A_+ + B_+ \cdot \ln|\alpha - 1|, \quad \alpha > 1, \tag{22}$$

with $A_- = 4.110$, $B_- = -1.007$, $A_+ = 9.139$ and $B_+ = -1.010$. We can see that for both cases, the log–log plots of ξ vs. $|\alpha - 1|$ follow a straight line with a slope -1 as indicated by the blue lines in the insets, which indicates a $\xi \propto |\alpha - 1|^{-\nu}$ behavior with the exponent $\nu = 1$. Although the critical exponent of ξ is the same for both $\alpha < 1$ and $\alpha > 1$, the magnitude of ξ is about 150 times larger in the region of

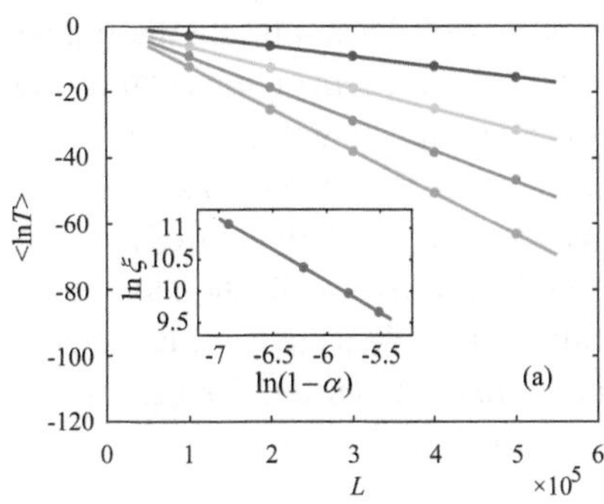

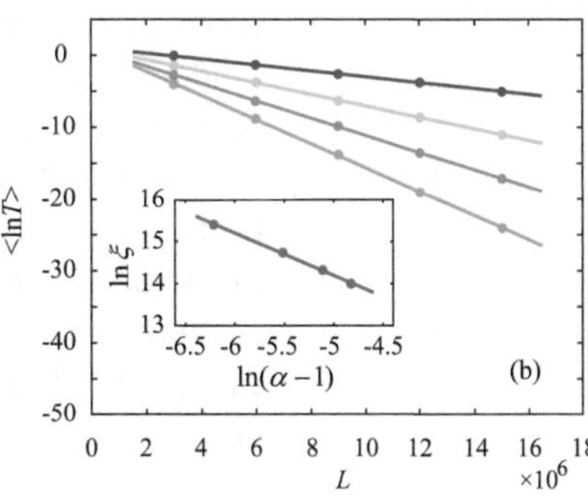

Figure 2: (Color online) (a) The ensemble averaged ln T_N as a function of the sample length L for different values of $\alpha < 1$. (b) Same as (a), but for $\alpha > 1$. The purple, yellow, blue and green solid circles in (a) denote the numerical results for $\alpha = 0.999$, 0.998, 0.997 and 0.996, respectively, while in (b) for $\alpha = 1.002$, 1.004, 1.006 and 1.008. The solid lines are linear fits for different values of α in both (a) and (b). The insets show the localization length ξ retrieved from the slope of the linear fit for different values of $|\alpha - 1|$ (solid circles) for both $\alpha < 1$ [inset in (a)] and $\alpha > 1$ [inset in (b)] cases. All numerical results in the insets are well fitted by a relation $\xi \propto |\alpha - 1|^{\nu}$ with $\nu = 1$ for both $\alpha < 1$ and $\alpha > 1$.

$\alpha > 1$ than in the region of $\alpha < 1$ for a given value of $|\alpha - 1|$. As we will see below, the asymmetry between $\alpha > 1$ and $\alpha < 1$ also leads to very different Anderson localization behaviors in the two regions for obliquely incident waves. The origin of such significant asymmetry will be discussed in Section 4.

In the above simulations, the strength of gain/loss is tuned simultaneously by changing the value of α for all interface coatings. We can also consider the case where the value of α in each coating layer is distributed randomly in an interval $[1-\Delta, 1+\Delta]$. Our simulation results show that the Anderson localization length follows the $\xi \propto \Delta^{-1}$ behavior. Here, the inverse localization length is found to be linearly proportional to the disorder strength Δ of the gain/loss. This is different from the quadratic behavior normally found in random Hermitian systems [3]. The detailed results are shown in Supplementary Note 3.

Now, we study the Anderson localization behaviors when waves are obliquely incident with an incident angle θ_0. We consider S-polarized waves, where the electric field E is perpendicular to the plane of incidence. In this case, the electric field behaves as a scalar wave and can be expressed as (see Figure 3)

$$\begin{aligned} E_{i-1} &= a_{i-1}\exp\{in_{i-1}k_0[(x-x_{i-1})\cos\theta_{i-1} + y\sin\theta_{i-1}]\} \\ &\quad + b_{i-1}\exp\{-in_{i-1}k_0[(x-x_{i-1})\cos\theta_{i-1} - y\sin\theta_{i-1}]\}, \end{aligned} \tag{23}$$

in the $(i-1)$-th layer, and

$$\begin{aligned} E_i &= a_i\exp\{in_ik_0[(x-x_i)\cos\theta_i + y\sin\theta_i]\} \\ &\quad + b_i\exp\{-in_ik_0[(x-x_i)\cos\theta_i - y\sin\theta_i]\}, \end{aligned} \tag{24}$$

in the i-th layer. Using the boundary conditions Eqs. (2)–(5) and the impedance relation $H = \sqrt{\varepsilon/\mu_0}\cdot E$, we obtain the following transfer matrix $m^{(i)}$ connecting the electric fields at the interfaces x_{i-1} and x_i:

$$\begin{aligned} m_{11}^{(i)} &= \frac{1}{2}\left\{\frac{1}{n_i\cos\theta_i}\cdot[\alpha(n_i - n_{i-1}) + n_{i-1}\cos\theta_{i-1}] + 1\right\} \\ &\quad \times \exp(in_ik_0\cos\theta_i d), \end{aligned} \tag{25}$$

$$\begin{aligned} m_{12}^{(i)} &= \frac{1}{2}\left\{\frac{1}{n_i\cos\theta_i}\cdot[\alpha(n_i - n_{i-1}) - n_{i-1}\cos\theta_{i-1}] + 1\right\} \\ &\quad \times \exp(in_ik_0\cos\theta_i d), \end{aligned} \tag{26}$$

$$\begin{aligned} m_{21}^{(i)} &= \frac{1}{2}\left\{-\frac{1}{n_i\cos\theta_i}\cdot[\alpha(n_i - n_{i-1}) + n_{i-1}\cos\theta_{i-1}] + 1\right\} \\ &\quad \times \exp(-in_ik_0\cos\theta_i d), \end{aligned} \tag{27}$$

$$\begin{aligned} m_{22}^{(i)} &= \frac{1}{2}\left\{-\frac{1}{n_i\cos\theta_i}\cdot[\alpha(n_i - n_{i-1}) - n_{i-1}\cos\theta_{i-1}] + 1\right\} \\ &\quad \times \exp(-in_ik_0\cos\theta_i d), \end{aligned} \tag{28}$$

where $\sin\theta_i = n_0\sin\theta_0/n_i$ and θ_0 is the incident angle. It is easy to see that Eqs. (25)–(28) reduce to Eqs. (14)–(17) when

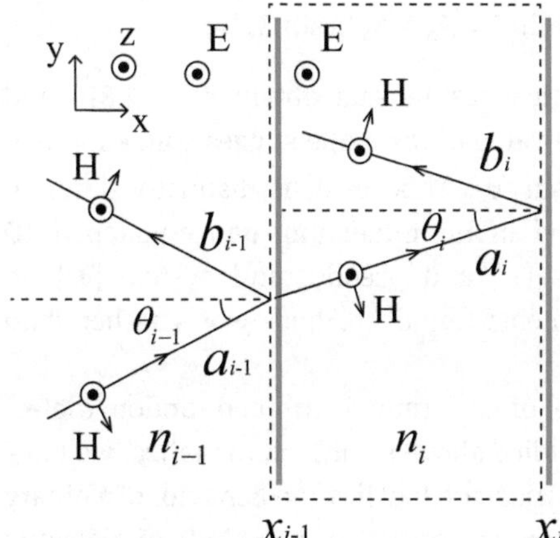

Figure 3: (Color online) The schematic of wave propagations for S-polarized waves.

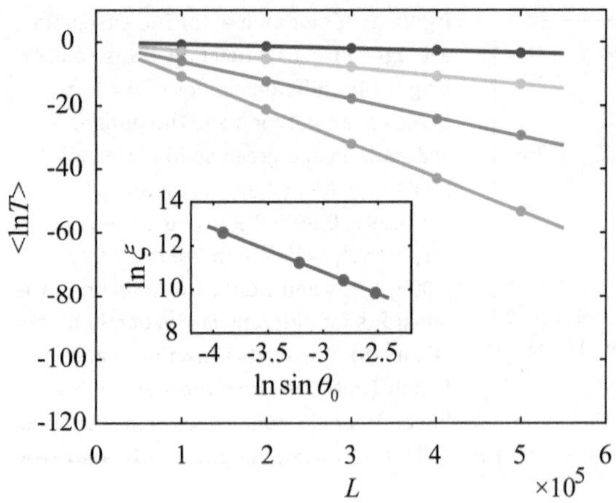

Figure 4: (Color online) The ensemble averaged $\ln T_N$ as a function of the sample length L for different incident angles for S-polarized waves when $\alpha = 1$. The purple, yellow, blue and green solid circles denote the numerical results for $\sin\theta_0 = 0.02, 0.04, 0.06$ and 0.08, respectively. The solid lines are linear fittings. The inset shows the localization length ξ retrieved from the slope of the linear fitting for different $\sin\theta_0$ (solid circles), which are well fitted by a relation $\xi \propto \sin^{-2}\theta_0$(solid line).

$\theta_0 = 0$. By substituting Eqs. (25)–(28) into Eqs. (18) and (19) and using Eqs. (7)–(9), we can study the dependence of $\langle \ln T_N \rangle_c$ on the sample length L. It is expected that the localization length diverges when $\theta_0 = 0$ and $\alpha = 1$. The results of $\alpha = k_0 = 1$ are shown in Figure 4 for different incident angles (solid circles). It can be seen that $\langle \ln T_N \rangle_c$ decays linearly with the sample length L for all incident angles, indicating Anderson localization. In the inset of Figure 4, we plot the localization length ξ (retrieved from the linear fitting of $\langle \ln T_N \rangle_c$ vs. L) as a function of incident angle. We find that the log-log plot of ξ vs. $\sin\theta_0$ shows a straight line, indicating a $\xi \propto \sin^{-\nu'}\theta_0$ behavior at small θ_0. To obtain the critical exponent ν', we use the following linear equation,

$$\ln \xi = A_S + B_S \ln \sin \theta_0, \tag{29}$$

to fit the numerical results and obtain $A_S = 4.810$ and $B_S = -1.996$. The value of the slope suggests an exponent $\nu' = 2$, i.e., $\xi \propto \sin^{-2}\theta_0$. This result is distinctly different from those found in Hermitian impedance-matched 1D random systems [31] and pseudospin-1 system [32], in which the exponents found are both $\nu' = 4$ rather than $\nu' = 2$.

As the model of uniformly distributed random dielectric constants studied above is difficult to realize, we have also studied the Anderson localization behavior of a binary random system, which consists of two kinds of dielectric layers with relative permittivity either 0.5 or 1.5 with equal probability. Our simulation results show the same critical behavior $\xi \propto \sin^{-2}\theta_0$ as that of the uniformly distributed random media, in line with the concept of universality. The detailed results are given in Supplementary Note 4.

In the following, we study the Anderson localization behaviors when α deviates from unity. In Figure 5, we plot the numerical results of the inverse localization length $1/\xi$ as a function of α for different incident angles. These results suggest the following simple expression for the inverse localization length in the region of $\alpha \leq 1$:

$$1/\xi \simeq A(1 - \alpha) + B \sin^2 \theta_0, \tag{30}$$

where A and B are two constants. If we use the fitting results of Eqs. (21) and (29) to set the values of A and B, i.e., $A = \exp(-A_-) = 1.640 \times 10^{-2}$ and $B = \exp(-A_S) = 8.146 \times 10^{-3}$, Eq. (30) gives excellent fitting results of the data at three incident angles in the region of $\alpha \leq 1$ as shown by the solid lines in Figure 5. Eq. (30) indicates that the localization length at any point $(1 - \alpha, \theta_0)$ in the critical region is simply the harmonic mean of the localization length at point $(1 - \alpha, \theta_0 = 0)$ and the localization length at point $(\alpha = 1, \theta_0)$. This also suggests that $1-\alpha$ and θ_0 are two independent parameters in the determination of Anderson localization in the critical region of $\alpha \leq 1$. However, the localization behavior in the region of $\alpha > 1$ is more complicated. As shown in Figure 5, for all the incident angles studied $\sin\theta_0 \neq 0$, the function $1/\xi$ first follows the linear decay of Eq. (30) to a minimum value and then turn around and increases linearly with α. The existence of a linear term $1-\alpha$ instead of a quadratic term in Eq. (30)

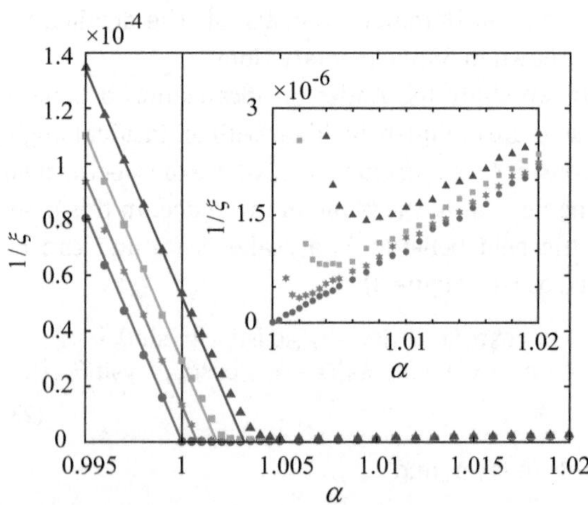

Figure 5: (Color online) The plot of inverse localization length $1/\xi$ as a function of α at different incident angles. The blue circles, red hexagrams, yellow squares and purple triangles denote numerical results for $\sin\theta_0 = 0, 0.04, 0.06$ and 0.08, respectively. The straight lines are the results of Eq. (30). The $\alpha = 1$ is marked by a black solid line. The inset shows the details at small values of $1/\xi$ in the region of $\alpha \geq 1$.

explains the asymmetry in the magnitudes of the localization length in the regions of $\alpha < 1$ and $\alpha > 1$ shown in Figure 2.

4 The origin of anomalous Anderson localization behaviors

In this section, we will start from the stack recursion equation (10) to understand the above anomalous localization behaviors. As can be seen from Eq. (10), the log of the transmission consists of two terms: the first term $\sum_{i=1}^{N+1}\ln|t_{i+}|^2$ involves the transmission coefficients of individual scattering elements only, and we call it noninterference term; the second term $-2\sum_{i=2}^{N+1}\ln|1-r_{i+}r^-(i-1)|$ involves both the modulus and phase of the reflection amplitudes and we will refer to it as the interference term. The stack recursion equation method has been discussed in detail in the literature [33–36].

We first consider the case of normal incidence. In the critical region, we can write $\alpha = 1+\delta$ with δ being a small positive or negative number. By substituting Eqs. (14)–(17) into Eqs. (18) and (19) and taking the small δ limit, we obtain the following expressions for the transmission and reflection amplitudes of the i-th scattering element:

$$t_{i+} \cong \left[1+\frac{1}{2}\left(\frac{n_i}{n_{i-1}}-1\right)\delta\right]\exp(in_ik_0d), \tag{31}$$

$$t_{i-} \cong \left[\frac{n_i}{n_{i-1}}+\left(-\frac{n_i}{2n_{i-1}}+\frac{n_i^2}{2n_{i-1}^2}\right)\delta\right]\exp(in_ik_0d), \tag{32}$$

$$r_{i+} \cong \frac{1}{2}\left(\frac{n_i}{n_{i-1}}-1\right)\delta, \tag{33}$$

$$r_{i-} \cong \left[-1+\frac{n_i}{n_{i-1}}+\left(-\frac{n_i}{2n_{i-1}}+\frac{n_i^2}{2n_{i-1}^2}\right)\delta\right]\exp(2in_ik_0d). \tag{34}$$

It should be noted that k_0 appears only in the phase factors of Eqs. (31)–(34) and scales with $1/d$, In our simulations, we have chosen $k_0 = d = 1$. The choice of another k'_0 will not change the total transmission T_N in Eq. (10) if the layer thickness is scaled to $d' = k_0d/k'_0$. According to Eq. (20), such scaling will only change the localization length of the system by a factor d'/d. It will not change its critical behavior. To the lowest order in δ, the noninterference term in Eq. (10) can be expressed as

$$\sum_{i=1}^{N+1}\ln|t_{i+}|^2 \cong \sum_{i=1}^{N+1}\ln\left[1+\left(\frac{n_i}{n_{i-1}}-1\right)\delta\right] \approx \delta\sum_{i=1}^{N+1}\left(\frac{n_i}{n_{i-1}}-1\right) \tag{35}$$

Thus, the ensemble average of the noninterference term can be written as

$$\left\langle\sum_{i=1}^{N+1}\ln|t_{i+}|^2\right\rangle_c \cong (N+1)\delta\left(\left\langle\frac{n_i}{n_{i-1}}\right\rangle_c-1\right) = (N+1)\cdot C_1(\sigma)\cdot\delta \tag{36}$$

The above linear dependence in δ actually gives rise to both the unity critical exponent and the localization length asymmetry between the regions of $\delta > 0$ and $\delta < 0$. For the uniform distribution of the relative permittivity in the interval $[1-\sigma, 1+\sigma]$, the coefficient $C_1(\sigma)$ can be evaluated as

$$\begin{aligned} C_1(\sigma) &= \frac{1}{4\sigma^2}\int_{1-\sigma}^{1+\sigma}d\varepsilon_i\int_{1-\sigma}^{1+\sigma}d\varepsilon_{i-1}\left(\frac{\sqrt{\varepsilon_i}}{\sqrt{\varepsilon_{i-1}}}-1\right) \\ &= (2/3\sigma^2)\cdot\left(\sigma^2+1-\sqrt{1-\sigma^2}\right)-1, \end{aligned} \tag{37}$$

For our case $\sigma = 0.5$, we can get $C_1(\sigma = 0.5) = 0.0239$. It is important to point out that the linear dependence of the noninterference term shown in Eq. (36) acts as a delocalization effect when $\delta > 0$ as it enhances the transmission. Similarly, using Eq. (33), we can expand the ensemble average of the interference term in Eq. (10) as

$$\begin{aligned} -2\left\langle\sum_{i=2}^{N+1}\ln|1-r_{i+}r^-(i-1)|\right\rangle_c &= -2\left\langle\sum_{i=2}^{N+1}\ln\left|1-\delta\cdot\frac{1}{2}\left(\frac{n_i}{n_{i-1}}-1\right)r^-(i-1)\right|\right\rangle_c \\ &\cong \left\langle\sum_{i=2}^{N+1}\mathrm{Re}\left[\left(\frac{n_i}{n_{i-1}}-1\right)r^-(i-1)\right]\right\rangle_c\cdot\delta \\ &\cong NC_{2,\pm}^{(1)}(\sigma, k_0d)\cdot\delta, \end{aligned} \tag{38}$$

where the subscripts '+' and '−' in $C_{2,\pm}^{(1)}(\sigma, k_0d)$ denote the coefficients calculated in the regions of $\alpha > 1$ and $\alpha < 1$, respectively. The difference is due to the term $r^-(i-1)$ in Eq. (38), which has different behaviors across the critical point at $\alpha = 1$. From Eqs. (10), (20), (36) and (38), the localization length can be expressed as $\xi \cong -\frac{2d}{C_1(\sigma)+\lim_{N\to\infty}C_{2,\pm}^{(1)}(\sigma,k_0d)}\delta^{-1}$ for the normal incidence, which shows explicitly the $\xi \propto |\alpha-1|^{-1}$ behavior found in Figure 2. Since we have already obtained numerically the divergent behavior of ξ in Eqs. (21) and (22) for the case of $k_0 = d = 1$, i.e., $\xi_\pm \cong \exp(A_\pm)\cdot\delta^{-1}$, the values of $\lim_{N\to\infty}C_{2,\pm}^{(1)}(0.5, 1)$ can be obtained from the values of $A_\pm$ and C_1 ($\sigma = 0.5$), from which we can get $\lim_{N\to\infty}C_{2,-}^{(1)}(\sigma = 0.5, 1) = 0.00887$ and $\lim_{N\to\infty}C_{2,+}^{(1)}(\sigma = 0.5, 1) = -0.0241$. Note that $C_{2,-}^{(1)}(0.5, 1)$ and $C_{2,+}^{(1)}(0.5, 1)$ have the opposite signs so that $C_{2,\pm}^{(1)}(0.5, 1)\cdot\delta$ is always negative, which means the interference term always gives positive contribution to the localization effect. From the above analysis, it is clearly seen that the unity critical exponent arising from the linear term δ in the expansion of $\ln T_N$ of Eq. (10) is independent of the choices of k_0d and σ. And the significant asymmetry found in the localization lengths between the regions $\alpha < 1$ and $\alpha > 1$ is due to the delocalization effect of the noninterference term in the region $\alpha > 1$, which significantly enlarge the localization length.

Now we consider the case of oblique incidence with $\alpha = 1$. To obtain the transmission and reflection amplitudes of the i-th scattering element at small θ_0, we expand Eqs. (25)–(28) to the leading order in θ_0 and obtain

$$r_{i+} \cong \left(-\frac{1}{4}+\frac{n_{i-1}}{4n_i}\right)\frac{\rho^2}{n_{i-1}^2}, \tag{39}$$

$$t_{i+} \cong \left[1+\left(-\frac{1}{4}+\frac{n_{i-1}}{4n_i}\right)\frac{\rho^2}{n_{i-1}^2}\right]\exp(in_ik_0\cos\theta_i d), \tag{40}$$

$$t_{i-} \cong \left[\frac{n_i}{n_{i-1}}+\left(\frac{1}{4}-\frac{n_{i-1}}{2n_i}+\frac{n_i}{4n_{i-1}}\right)\frac{\rho^2}{n_{i-1}^2}\right]\exp(in_ik_0\cos\theta_i d), \tag{41}$$

$$r_{i-} = \left[-1+\frac{n_i}{n_{i-1}}+\left(\frac{1}{4}-\frac{n_{i-1}}{2n_i}+\frac{n_i}{4n_{i-1}}\right)\frac{\rho^2}{n_{i-1}^2}\right]\exp(2in_ik_0\cos\theta_i d), \tag{42}$$

where $\rho = n_0\sin\theta_0$. Using Eq. (40), the noninterference term can be written as

$$\begin{aligned}\sum_{i=1}^{N+1}\ln|t_{i+}|^2 &\cong \sum_{i=1}^{N+1}2\ln\left[1+\left(-\frac{1}{4}+\frac{n_{i-1}}{4n_i}\right)\frac{\rho^2}{n_{i-1}^2}\right]\\ &\cong \rho^2\sum_{i=1}^{N+1}\left(-\frac{1}{2}+\frac{n_{i-1}}{2n_i}\right)\frac{1}{n_{i-1}^2}.\end{aligned} \tag{43}$$

The ensemble average of the noninterference term can be expressed as

$$\begin{aligned}\left\langle\sum_{i=1}^{N+1}\ln|t_{i+}|^2\right\rangle_c &\cong (N+1)\rho^2\left\langle\left(-\frac{1}{2}+\frac{n_{i-1}}{2n_i}\right)\frac{1}{n_{i-1}^2}\right\rangle_c\\ &= (N+1)\cdot D_1(\sigma)\cdot\rho^2,\end{aligned} \tag{44}$$

where

$$\begin{aligned}D_1(\sigma) &= \int_{1-\sigma}^{1+\sigma}d\varepsilon_i\int_{1-\sigma}^{1+\sigma}\left(-\frac{1}{2}+\frac{\sqrt{\varepsilon_{i-1}}}{2\sqrt{\varepsilon_i}}\right)\frac{1}{\varepsilon_{i-1}}\cdot d\varepsilon_{i-1}\\ &= -\sigma\cdot\ln\frac{1+\sigma}{1-\sigma}+4\left(1-\sqrt{1-\sigma^2}\right),\end{aligned} \tag{45}$$

Then we can get $D_1(\sigma = 0.5) = -0.0134$ by substituting $\sigma = 0.5$.

Using Eq. (39), the ensemble average of interference term can be expressed as:

$$\begin{aligned}\left\langle -2\sum_{i=2}^{N+1}\ln|1-r_{i+}r^-(i-1)|\right\rangle &= \left\langle -2\sum_{i=2}^{N+1}\ln\left|1-\left(-\frac{1}{4}+\frac{n_{i-1}}{4n_i}\right)\frac{\rho^2}{n_{i-1}^2}\cdot r^-(i-1)\right|\right\rangle\\ &\cong \left\langle\sum_{i=2}^{N+1}\mathrm{Re}\left[\left(-\frac{1}{2}+\frac{n_{i-1}}{2n_i}\right)\frac{1}{n_{i-1}^2}\cdot r^-(i-1)\right]\right\rangle\cdot\rho^2\\ &\cong ND_2^{(1)}\cdot\rho^2,\end{aligned} \tag{46}$$

where $D_2^{(1)}$ denotes the coefficient of the leading term in small ρ^2. We can now express the localization length as $\xi = -\frac{2d}{D_1(\sigma)+\lim_{N\to\infty}D_2^{(1)}(\sigma,k_0d)}\rho^{-2}$, which shows exactly the $\xi \propto \sin^{-2}\theta_0$ behavior found in Figure 4. From the result of $\xi \cong \exp(A_S)\cdot\delta^{-1}$ in Eq. (29) and the value of $D_1(\sigma)$ in Eq. (45), we found $\lim_{N\to\infty}D_2^{(1)}(\sigma = 0.5, k_0d = 1) = -0.00289$. From Eqs. (36), (38), (44) and (46), it is clearly seen that the critical exponents v and v' do not dependent on the choice of σ. Thus, our results are universal for different strengths of randomness.

5 Conclusion

We have numerically studied the Anderson localization behavior of the incident waves of a given wavevector k_0 in 1D random layered non-Hermitian media with sample-specific gain and loss inserted between any two adjacent layers as described in Eq. (1). At normal incidence $\theta_0 = 0$, the systems always possess unity transmission at a specific strength of gain and loss ($\alpha = 1$ in Eq. (1)), independent of the random configuration and the sample size. The existence of such an infinite localization length allows us to study the behavior of Anderson localization in a small critical region surrounding the critical point. We found the following anomalous behaviors. In the case of normal incidence, the localization length behaves like $\xi \approx A|\alpha-1|^{-1}$. While the exponent v is the same in both regions of $\alpha > 1$ and $\alpha < 1$, the prefactor A in the region of $\alpha > 1$ is much greater than that in the $\alpha < 1$ region. In the case of oblique incidence ($\theta_0 \neq 0$), the localization length behaves like $\xi \approx B\sin^{-2}\theta_0$ for S-polarized waves at $\alpha = 1$. The exponent $v' = 2$ is different from the behavior $\xi \propto 1/\sin^4\theta_0$ found in impedance-matched systems and pseudospin-1 Hermitian systems. We have also studied the localization length behavior at oblique incidence when $\alpha \neq 1$. Very different behaviors are found between the region of $\alpha > 1$ and $\alpha < 1$ even though the system is gain/loss balanced on average in both cases. The asymmetry of the localization behaviors in the two regions is due to the presence of a delocalization effect in the noninterference term when $\alpha > 1$, which makes the localization length very large compared to that in the $\alpha < 1$ region.

We note that in ordinary Hermitian random media, the critical point is normally at the ordered media where the localization length diverges. The addition of sample-specific balanced gain and loss in 1D media can move the critical point from ordered to disordered media. This allows us to study the Anderson localization behavior of non-Hermitian disordered systems in the vicinity of its critical point. Although our study was done for some specific choices of disorder strength $\sigma = 0.5$, the critical exponents found here are universal and independent of the choices of σ as can be seen in Section 4. We have also studied a few cases of $k_0 d \neq 1$ and found that different values of $k_0 d$ will only change the magnitude of the localization length. The exponents $\nu = 1$ and $\nu' = 2$ remain unchanged.

Acknowledgments: This work is supported by Hong Kong Research Grants Council (under Grant No. AoE/P-02/12, 16303119 and C6013-18G-A).
Author contribution: All the authors have accepted responsibility for the entire content of this submitted manuscript and approved submission.
Research funding: This work is supported by Hong Kong Research Grants Council (under Grant No. AoE/P-02/12, 16303119 and C6013-18G-A).
Conflict of interest statement: The authors declare no conflicts of interest regarding this article.

References

[1] P. W. Anderson, "Absence of diffusion in certain random lattices," *Phys. Rev.*, vol. 109, no. 5, p. 1492, 1958.
[2] A. Lagendijk, B. Van Tiggelen, and D. S. Wiersma, "Fifty years of Anderson localization," *Phys. Today*, vol. 62, no. 8, pp. 24–29, 2009.
[3] P. Sheng, *Scattering and localization of classical waves in random media*, World Scientific, 1990.
[4] D. S. Wiersma, P. Bartolini, A. Lagendijk, and R. Righini, "Localization of light in a disordered medium," *Nature*, vol. 390, no. 6661, pp. 671–673, 1997.
[5] M. Störzer, P. Gross, C. M. Aegerter, and G. Maret, "Observation of the critical regime near Anderson localization of light," *Phys. Rev. Lett.*, vol. 96, no. 6, 2006, Art no. 063904.
[6] T. Schwartz, G. Bartal, S. Fishman, and M. Segev, "Transport and Anderson localization in disordered two-dimensional photonic lattices," *Nature*, vol. 446, no. 7131, pp. 52–55, 2007.
[7] P. Sheng and Z. Q. Zhang, "Scalar-wave localization in a two-component composite," *Phys. Rev. Lett.*, vol. 57, no. 15, p. 1879, 1986.
[8] S. John, H. Sompolinsky, and M. J. Stephen, "Localization in a disordered elastic medium near two dimensions," *Phys. Rev. B*, vol. 27, no. 9, p. 5592, 1983.
[9] M. Y. Azbel, "Eigenstates and properties of random systems in one dimension at zero temperature," *Phys. Rev. B*, vol. 28, no. 8, p. 4106, 1983.
[10] V. Baluni and J. Willemsen, "Transmission of acoustic waves in a random layered medium," *Phys. Rev.*, vol. 31, no. 5, p. 3358, 1985.
[11] C. M. de Sterke and R. C. McPhedran, "Bragg remnants in stratified random media," *Phys. Rev. B*, vol. 47, no. 13, p. 7780, 1993.
[12] F. M. Izrailev, A. A. Krokhin, and N. M. Makarov, "Anomalous localization in low-dimensional systems with correlated disorder," *Phys. Rep.*, vol. 512, no. 3, pp. 125–254, 2012.
[13] C. Miniatura, L. C. Kwek, M. Ducloy, et al., Eds. *Ultracold Gases and Quantum Information: Lecture Notes of the Les Houches Summer School in Singapore*, vol. 91, July 2009, Oxford University Press; 2011, Chap. 9.
[14] A. Yamilov and B. Payne, "Classification of regimes of wave transport in quasi-one-dimensional non-conservative random media," *J. Mod. Optic.*, vol. 57, no. 19, pp. 1916–1921, 2010.
[15] A. A. Asatryan, N. A. Nicorovici, L. C. Botten, C. M. de Sterke, P. A. Robinson, and R. C. McPhedran, "Electromagnetic localization in dispersive stratified media with random loss and gain," *Phys. Rev. B*, vol. 57, no. 21, p. 13535, 1998.
[16] A. Yamilov, S. H. Chang, A. Burin, A. Taflove, and H. Cao, "Field and intensity correlations in amplifying random media," *Phys. Rev. B*, vol. 71, no. 9, 2005, Art no. 092201.
[17] J. Heinrichs, "Light amplification and absorption in a random medium," *Phys. Rev. B*, vol. 56, no. 14, p. 8674, 1997.
[18] V. Freilikher, M. Pustilnik, and I. Yurkevich, "Effect of absorption on the wave transport in the strong localization regime," *Phys. Rev. Lett.*, vol. 73, no. 6, p. 810, 1994.
[19] Z. Q. Zhang, "Light amplification and localization in randomly layered media with gain," *Phys. Rev. B*, vol. 52, no. 11, p. 7960, 1995.
[20] S. Kalish, Z. Lin, and T. Kottos, "Light transport in random media with PT symmetry," *Phys. Rev.*, vol. 85, no. 5, 2012, Art no. 055802.
[21] T. S. Misirpashaev, J. C. J. Paasschens, and C. W. J. Beenakker, "Localization in a disordered multi-mode waveguide with absorption or amplification," *Phys. Stat. Mech. Appl.*, vol. 236, no. 3-4, pp. 189–201, 1997.
[22] S. A. Ramakrishna, E. K. Das, G. V. Vijayagovindan, and N. Kumar, "Reflection of light from a random amplifying medium with disorder in the complex refractive index: statistics of fluctuations," *Phys. Rev. B*, vol. 62, no. 1, p. 256, 2000.
[23] X. Jiang, Q. Li, and C. M. Soukoulis, "Symmetry between absorption and amplification in disordered media," *Phys. Rev. B*, vol. 59, no. 14, 1999, Art no. R9007.
[24] P. K. Datta, "Transmission and reflection in a perfectly amplifying and absorbing medium," *Phys. Rev. B*, vol. 59, no. 16, 1999, Art no. 10980.
[25] R. Frank, A. Lubatsch, and J. Kroha, "Theory of strong localization effects of light in disordered loss or gain media," *Phys. Rev. B*, vol. 73, no. 24, 2006, Art no. 245107.
[26] L. Y. Zhao, C. S. Tian, Z. Q. Zhang, and X. D. Zhang, "Unconventional diffusion of light in strongly localized open absorbing media," *Phys. Rev. B*, vol. 88, no. 15, 2013, Art no. 155104.
[27] H. Cao, Y. G. Zhao, S. T. Ho, E. W. Seelig, Q. H. Wang, and R. P. H. Chang, "Random laser action in semiconductor powder," *Phys. Rev. Lett.*, vol. 82, no. 11, p. 2278, 1999.
[28] H. Cao, J. Y. Xu, D. Z. Zhang, et al., "Spatial confinement of laser light in active random media," *Phys. Rev. Lett.*, vol. 84, no. 24, p. 5584, 2000.
[29] A. A. Chabanov, M. Stoytchev, and A. Z. Genack, "Statistical signatures of photon localization," *Nature*, vol. 404, no. 6780, pp. 850–853, 2000.
[30] V. Milner and A. Z. Genack, "Photon localization laser: low-threshold lasing in a random amplifying layered medium via wave localization," *Phys. Rev. Lett.*, vol. 94, no. 7, 2005, Art no. 073901.
[31] K. Kim, "Anderson localization of electromagnetic waves in randomly-stratified magnetodielectric media with uniform

impedance," *Optic Express*, vol. 23, no. 11, pp. 14520–14531, 2015.

[32] A. Fang, Z. Q. Zhang, S. G. Louie, et al., "Anomalous Anderson localization behaviors in disordered pseudospin systems," *Proc. Natl. Acad. Sci. Unit. States Am.*, vol. 114, no. 16, pp. 4087–4092, 2017.

[33] A. Fang, Z. Q. Zhang, S. G. Louie, and C. T. Chan, "Nonuniversal critical behavior in disordered pseudospin-1 systems," *Phys. Rev. B*, vol. 99, no. 1, 2019, Art no. 014209.

[34] M. Born and E. Wolf, *Principles of optics*, 7th ed. Cambridge, Cambridge University Press, 2002.

[35] A. A. Asatryan, L. C. Botten, M. A. Byrne, et al., "Suppression of Anderson localization in disordered metamaterials," *Phys. Rev. Lett.*, vol. 99, no. 19, p. 193902, 2007.

[36] M. V. Berry and S. Klein, "Transparent mirrors: rays, waves and localization," *Eur. J. Phys.*, vol. 18, no. 3, p. 222, 1997.

[37] K. G. Makris, A. Brandstötter, P. Ambichl, Z. H. Musslimani, and S. Rotter, "Wave propagation through disordered media without backscattering and intensity variations," *Light Sci. Appl.*, vol. 6, no. 9, 2017, Art no. e17035.

[38] E. Rivet, A. Brandstötter, K. G. Makris, H. Lissek, S. Rotter, and R. Fleury, "Constant-pressure sound waves in non-Hermitian disordered media," *Nat. Phys.*, vol. 14, no. 9, pp. 942–947, 2018.

[39] C. M. Bender and S. Boettcher, "Real spectra in non-Hermitian Hamiltonians having P T symmetry," *Phys. Rev. Lett.*, vol. 80, no. 24, p. 5243, 1998.

[40] C. M. Bender, D. C. Brody, and H. F. Jones, "Complex extension of quantum mechanics," *Phys. Rev. Lett.*, vol. 89, no. 27, 2002, Art no. 270401.

[41] S. Longhi, "Optical realization of relativistic non-Hermitian quantum mechanics," *Phys. Rev. Lett.*, vol. 105, no. 1, 2010, Art no. 013903.

[42] S. Longhi, "PT-symmetric laser absorber," *Phys. Rev.*, vol. 82, no. 3, 2010, Art no. 031801.

[43] Y. D. Chong, L. Ge, and A. D. Stone, "P t-symmetry breaking and laser-absorber modes in optical scattering systems," *Phys. Rev. Lett.*, vol. 106, no. 9, 2011, Art no. 093902.

[44] J. Luo, J. Li, and Y. Lai, "Electromagnetic impurity-immunity induced by parity-time symmetry," *Phys. Rev. X*, vol. 8, no. 3, 2018, Art no. 031035.

[45] L. Ge, Y. D. Chong, and A. D. Stone, "Conservation relations and anisotropic transmission resonances in one-dimensional PT-symmetric photonic heterostructures," *Phys. Rev.*, vol. 85, no. 2, 2012, Art no. 023802.

[46] R. Fleury, D. L. Sounas, and A. Alu, "Negative refraction and planar focusing based on parity-time symmetric metasurfaces," *Phys. Rev. Lett.*, vol. 113, 2014, Art no. 023903.

Supplementary material: The online version of this article offers supplementary material (https://doi.org/10.1515/nanoph-2020-0306).

Part VIII: **Quantum Computing, Quantum Optics, and QED**

Juan Ignacio Cirac*

Quantum computing and simulation

Where we stand and what awaits us

https://doi.org/10.1515/9783110710687-035

Abstract: Quantum computers and simulators can have an extraordinary impact on our society. Despite the extraordinary progress they have made in recent years, there are still great challenges to be met and new opportunities to be discovered.

Keywords: computing; quantum physics; simulation.

1 Introduction

Very recently, Google has announced the construction of a quantum processor based on superconducting qubits that is able to perform certain task much faster than any existing supercomputer [1]. This announcement was preceded by many other announcements by different research groups in which milestones in quantum computing with different platforms have been reported. In fact, in the last few years, the panorama in this field of research abruptly changed when leading technological companies expressed their intention to build such devices and public funding agencies around the globe approved generous support to construct and develop them. Research has gone beyond universities and other research institutions and is now also pursued in companies, which has triggered new dynamics. The field of quantum computing is nowadays attracting the attention of science, media, industry, politics, and the society in general. After many years of research, we are living very agitated times where there is a strong effort worldwide to build such devices, as they promise a variety of applications. We repeatedly read in diverse media how quantum computers are going to impact pharmaceutical industry, medicine, finances, energy production, or climate change. Although there is an obvious exaggeration in all that news and, in fact, in many cases there is no (or very little) evidence of such impact, most scientists working on that subject believe that quantum computers will complement and, in some areas, supersede supercomputers and will strongly impact our society. However, to achieve them, there are still many challenges and obstacles to overcome. In this short article, I want to highlight some of the opportunities quantum computers may offer us, as well as important challenges we are facing. The intention is not to review the state of the art of the different platforms or on quantum algorithms; there exist excellent reviews on those subjects (see, for example, Refs. [2–6]).

To analyze the power and possibilities of existing and future devices, one has to distinguish between different concepts that sometimes lead to confusion and misleading statements. The first is a scalable quantum computer, which can run arbitrary quantum algorithms reliably in the form of a sequence of elementary quantum logic gates. The second is a quantum device which can do the same although the errors accumulate, so that it can only obtain reliable results for limited sizes (or a maximum number of quantum gates) and thus cannot be scaled up to perform arbitrary computations. The third one is an analog quantum device which does not operate based on quantum gates but rather evolves in accordance with some given dynamics which can be engineered to some extent. They cannot solve general problems; however, they are easier to build.

2 Scalable quantum computers

A quantum computer is a device that is able to implement quantum algorithms based on a universal set of gates acting on quantum bits (qubits) with "almost no error". There are a variety of problems, ranging from the simulation of materials or chemistry process, to optimization, for which very efficient quantum algorithms have been develop so that they can be solved by quantum computers way faster than by any other classical one. Those have a variety of applications in drug or material design, industrial processes or data processing, just to name a few. The statement regarding the absence of errors is very important but also very subtle because such a device is based on the laws of quantum

***Corresponding author: Juan Ignacio Cirac**, Max-Planck Institute of Quantum Optics, Hans-Kopfermannstr. 1, D-85748 Garching, Germany, E-mail: ignacio.cirac@mpq.mpg.de. https://orcid.org/0000-0003-3359-1743

This article has previously been published in the journal Nanophotonics. Please cite as: J. I. Cirac "Quantum computing and simulation" *Nanophotonics* 2021, 10. DOI: 10.1515/nanoph-2020-0351.

physics, and thus not deterministic: when we measure at the end of the computation, we can obtain different results with different probabilities. By almost no error, we mean that the probability of obtaining each possible outcome in practice should be very close to the ideal one as dictated by the laws of quantum physics. In a real device, errors will certainly occur, as any interaction with the environment or any tiny imperfection will distort the probabilities of different outcomes. In fact, those errors accumulate during the computation which makes it extraordinarily difficult to build a quantum computer. Furthermore, as we make it bigger, errors will be more likely, so that it seems impossible to scale it to the sizes that are required for most applications. Fortunately, more than 20 years ago, it was discovered that those errors can be mitigated or even corrected so that it should be possible, at least in principle, to scale a quantum computer and yet keeping the condition of having "almost no error". There is, however, a high price to pay: for each qubit, one has to add a number of qubits to correct the errors. In addition, the error procedure can only operate if the error per quantum gate (or time step) is below a threshold, which is of the order 10^{-2}. This leads to an overhead in the number of qubits; that is, the number of required qubits has to be multiplied by a factor of the order 10^3–10^4, depending on the magnitude of the errors produced at each gate. Just to give an order of magnitude, most quantum algorithms that provide speed up with respect to classical ones become useful starting from about 10^3 to 10^4 qubits, so that with the overhead, one will require as many as 10^6–10^8 qubits. The device created by Google, for instance, has 53 qubits that have to operate at extreme conditions of low temperatures and isolation, so that, indeed, building a full-fledged quantum computer imposes a real scientific and technological challenge.

Although there are problems for which a quantum computer can achieve an exponential speed-up (as a function of the number of qubits), there are others in which this advantage is more modest. The first category includes some specific problems, such as factorization or discrete logarithm, and the simulation of quantum many-body systems. The second includes most of the optimization processes. The extra overhead demand will also decrease the advantages of quantum computers in solving some of those problems, especially those in the second category. Indeed, apart from increasing the number of qubits to correct the errors, more operations are required to perform the computation and the corrections so that the size of the problems where the quantum computer offers a speed up with respect to a classical computer may appear at a point where the execution time is extreme large and thus impractical.

Despite the obvious difficulty of scaling up existing technologies, there are different paths in which that task can be simplified. First, key technologies may appear in the way toward the construction of scalable quantum computers, in a similar way that transistors accelerated the development of classical computers. In addition, the combination of technologies may also lead to significant improvements, more compact devices, and smaller errors. There is also room for improvement in the development of error-correcting codes which may be better adapted to the specific errors that appear in different implementations, giving rise to much smaller overheads. For instance, overhead factors of the order of few tens or hundreds may make scaling up a much simpler and doable task in the near future. This will also affect the effectiveness of some of the quantum algorithms for optimization.

3 Noisy intermediate-scale quantum devices

Machines that can implement quantum algorithms but do not correct for errors (in a fault tolerant way) are colloquially called Noisy Intermediate-scale Quantum (NISQ) devices [7]. So far, they have been constructed with different platforms, including trapped ions, superconducting qubits, cold atoms, photons, quantum dots, vacancy centers in diamond, or phosphorous embedded in silicon. The first two are the most advanced, although cold atoms and photons are rapidly catching up. As the errors grow with the system size, they are not scalable and thus unable to solve most of the problems a quantum computer could. However, in the Google experiment, it was clearly shown that, although with 53 qubits and errors per gate of the order of 0.3%, they can still outperform classical computers in a certain task. Despite the fact that this was an academic problem with no practical application, there might be some relevant problems where such noisy devices can help. In particular, it could be expected that NISQ devices with up to few hundred qubits and with errors per gate below 0.1% will be built in the near future, and those may find some specific applications. Although it is hard to envision such applications, the fact that they can be operated in the cloud [8] will certainly open the minds of not only scientist but also of students or entrepreneurs who may find other uses of such devices. Constructing such devices and finding useful applications is a very active field of research and development, and several start-up companies have been created to build both the hardware and software required to operate them.

A very active field of research with NISQ devices is that of variational algorithms which apply to optimization problems in a broad sense, where one wants to find a string of bits (or a state of qubits) that minimizes a cost function. For instance, in the traveling salesman problem, the bits codify different orders in which cities are visited, and the cost function the total distance traveled. Or it can be the expectation value of the energy, so that the problem is to find the ground-state energy of a given Hamiltonian and thus solve problems in physics or chemistry. Variational algorithms create a state in accordance with a quantum circuit, where the quantum gates that are applied depend on some parameters which have to be optimized to minimize the cost function. Although it is not possible to predict the success of this procedure rigorously, it may give good results in practice. However, there are some challenges that need to be better understood and improved. For example, the optimization procedure can also become difficult in practice because of the presence of many local minima, or it may require an enormous number of measurements to compute the cost function, which may reduce their applicability.

This kind of devices can also pave the way toward scalable quantum computers. Although there are obvious limitations to their sizes with current technologies, as well as to the perfection of the quantum gates, one may still be able to introduce, stepwise, some specific error correction (or prevention) schemes that are adapted to the specific platforms, until eventually they become fault tolerant. The whole process has to be accompanied by the development of methods to debug the errors, validate the results, and benchmark different technologies. It is hard to predict whether this can happen progressively or there will be a "quantum winter" in which advance in this direction will be very slow. In view of the media attention and high expectations that quantum computers have raised, this may be very damaging for the field, at least in the short-medium run.

4 Analog quantum devices

This is another class of machines that is called sometimes analog quantum simulators. Those can be viewed as analog quantum computers that cannot implement a universal set of gates, do not have the capability of correcting errors, but yet can solve some specific problems in a more efficient way than classical devices. They are especially suited for problems related to quantum many-body problems that abound condensed matter, high-energy physics, or chemistry. Addressing those problems with classical computers requires resources (computer time and memory) that scale exponentially with the system size; that is, the number of subsystems, or the total volume. The reason is that quantum systems can be in superposition of different configurations, and to specify their quantum state (and thus, be able to compute its properties), one needs to assign a complex number to each configuration. Even in the simplest case where one has two-level systems, the number of configurations scales as 2^N, so that one needs to compute and store such number of complex number, leading to the memory and time requirements. Already for $N = 30$, that figure is so big that even supercomputers cannot cope with it. This obstacle was already pointed out by Feynman [9] about 40 years ago and, in fact, he proposed to use a quantum computer to address such problems. However, for some of them, a quantum computer is not strictly required; one can take a different system which can be controlled to the extent that one can make it behave as the original model one wants to analyze. For instance, to solve the Hubbard model in two dimensions, which describes the motion of electrons in solids and it is a firm candidate to account for high-Tc superconductivity, one may need cold atoms trapped in optical lattices that can hop and interact with each other in accordance with such a model, emulating the electrons in the solid. In a sense, this setup imitates a real solid but with a magnified lattice structure. The larger distances between atoms (or other systems) make quantum simulators more controllable and easier to measure. The number of atoms required equals the number of electrons and thus does not grow exponentially with that number. An experiment with atoms can thus allow us to answer some questions about the Hubbard model which are not reachable with classical computers, like if it features the physical properties found in high-Tc superconductors, thus demonstrating that, indeed, that model can describe such an intriguing phenomenon. One could also use the simulator to address problems in lattice gauge theories, where fermions represent matter and bosons represent the gauge fields. Or in chemistry, where electrons can be represented by fermionic atoms so that one can study the geometric configuration of molecules in equilibrium, their physical properties, or even learn about the chemical reactions that are needed in some drug production.

The main advantage of analog quantum devices with respect to quantum computers is that their operating conditions are easier to reach, as they do not require error correction. However, this means that the results of the simulation will not be perfect and one may wonder why then can they be useful. The reason is related to the fact that in some of those problems, we are interested in learning about specific observables, where the presence of few errors will scarcely affect the result. For instance, in condensed matter physics systems, we are typically interested in observables such as the energy, the magnetization, or superconducting density. If in the final state after the simulation, a small percentage of qubits contain errors, we will still be able to retrieve those properties with sufficient precision. As an example, to know whether the Hubbard model supports d-wave superconductivity, it may suffice to measure the corresponding observable with about 10% precision. However, perhaps the simulation produces more errors than what one could expect because of mismatches in the experimental parameters, or they may accumulate during the dynamics in a way so that the error at the end is much larger than expected. The rigorous formalization of this way of reasoning is still missing, and it goes beyond what computer scientists typically analyze. Furthermore, there are no simple ways of verifying that the simulation is correct because we cannot solve the problem with classical computers. Here, new ways of verification can be thought. For instance, one may attempt to find the solution of the problem with different simulators, or with different algorithms to gain confidence about the result. Besides, there may be some other applications of analog quantum devices beyond quantum simulation because there may be other problems where the final state just has a small percentage of errors which provides the sufficient precision to solve them.

From the experimental side, analog quantum devices are very advanced. Atoms in optical lattices or in tweezers are a leading technology, together with trapped ions. In the first setup, several hundreds of atoms can be well controlled and their interactions can be tailored to mimic specific models in condensed matter physics. Experiments with about 50 trapped ions are also available. Those sizes go beyond what can be simulated with classical supercomputers. Other simulations have been performed with photons, quantum dots, or superconducting devices. In any case, one can expect that in the next few years, we will be able to address some relevant problems in condensed matter physics and, perhaps, in high-energy physics or quantum chemistry as well.

5 Conclusions

Quantum computers have enormous potential to revolutionize many areas of our society. However, this requires building equipment that is scalable, something that needs to leverage existing technologies well beyond current limits or use new ones, as well as improving error correction techniques. All this requires a great deal of funding, as well as close collaboration between industry and research centers. In addition, it is imperative to identify other problems where quantum computers can become a fundamental tool.

During the last few years, there has been a great advance in the construction of NISQ equipment on different platforms and this has culminated in the announcement of the quantum advantage obtained by Google. These computers and those to be developed in the short term will not be able to run most quantum algorithms, as they make mistakes and are not scalable. However, it is very likely that they will give rise to new applications as they have demonstrated that they are capable of performing a specific task more efficiently than a supercomputer. Collaboration between scientists and industry can be key to finding such applications.

One of the most relevant utilities of quantum computers is the possibility of simulating the complex quantum systems that appear in fields such as condensed matter physics, high energies, or chemistry. To do this, it is often not necessary to build a scalable quantum computer, but an analog one is sufficient, called quantum simulators. This equipment is very developed and can help us to solve fundamental problems in physics, or in the design of materials or drugs. For this, apart from the construction of the equipment on different platforms, it is necessary to develop new methods of verification, benchmarking, and debugging.

The field of quantum computing was pushed about 25 years ago by the discovery of quantum algorithms that outperform classical ones, as well as for the identification of several physical systems to build them. After those years, the field has advanced a lot, and now it is already possible to build devices that would be unthinkable a couple of decades ago. However, there is still a very long way to go full of excitement and, probably, many surprises.

Author contribution: All the authors have accepted responsibility for the entire content of this submitted manuscript and approved submission.
Research funding: None declared.
Conflict of interest statement: The author declares no conflicts of interest regarding this article.

References

[1] F. Arute, K. Arya, R. Babbush, et al. "Quantum supremacy using a programmable superconducting processor," *Nature*, vol. 574, pp. 505–510, 2019.
[2] I. Buluta and F. Nori, "Quantum simulators," *Science*, vol. 326, pp. 108–111, 2009.
[3] I. Georgescu, S. Ashhab, and F. Nori, "Quantum simulation," *Rev. Mod. Phys.*, vol. 86, p. 153, 2014.
[4] Special Issue on, "Quantum simulation," *Nat. Phys.*, vol. 8, pp. 263–299, 2012.
[5] Special Issue on, "Quantum simulation," *Ann. Phys.*, vol. 525, no. 10–11 pp. 739–888, 2013.
[6] A. Montanaro, "Quantum algorithms: an overview," *NPJ Quantum Inf.*, vol. 2, pp. 15023, 2016.
[7] J. Preskill, "Quantum computing in the NISQ era and beyond," *Quantum*, vol. 2, pp. 79–99, 2018.
[8] www.ibm.com/quantum-computing. (kein Datum).
[9] R. Feynman, "Simulating physics with computers," *Int. J. Theor. Phys.*, vol. 21, pp. 467–478, 1982.

Andrea Fratalocchi*, Adam Fleming, Claudio Conti and Andrea Di Falco

NIST-certified secure key generation via deep learning of physical unclonable functions in silica aerogels

https://doi.org/10.1515/9783110710687-036

Abstract: Physical unclonable functions (PUFs) are complex physical objects that aim at overcoming the vulnerabilities of traditional cryptographic keys, promising a robust class of security primitives for different applications. Optical PUFs present advantages over traditional electronic realizations, namely, a stronger unclonability, but suffer from problems of reliability and weak unpredictability of the key. We here develop a two-step PUF generation strategy based on deep learning, which associates reliable keys verified against the National Institute of Standards and Technology (NIST) certification standards of true random generators for cryptography. The idea explored in this work is to decouple the design of the PUFs from the key generation and train a neural architecture to learn the mapping algorithm between the key and the PUF. We report experimental results with all-optical PUFs realized in silica aerogels and analyzed a population of 100 generated keys, each of 10,000 bit length. The key generated passed all tests required by the NIST standard, with proportion outcomes well beyond the NIST's recommended threshold. The two-step key generation strategy studied in this work can be generalized to any PUF based on either optical or electronic implementations. It can help the design of robust PUFs for both secure authentications and encrypted communications.

***Corresponding author: Andrea Fratalocchi,** PRIMALIGHT, Faculty of Electrical Engineering, Applied Mathematics and Computational Science, King Abdullah University of Science and Technology, Thuwal 23955-6900, Saudi Arabia, E-mail: andrea.fratalocchi@kaust.edu.sa. https://orcid.org/0000-0001-6769-4439
Adam Fleming and Andrea Di Falco, University of St Andrews, St Andrews, Fife, UK
Claudio Conti, Institute for Complex Systems, National Research Council (ISC-CNR), Via dei Taurini 19, 00185 Rome, Italy; and Department of Physics, University Sapienza, Piazzale Aldo Moro 5, 00185 Rome, Italy

Keywords: artificial intelligence; complex light scattering; physical unclonable functions; random optical nanomaterials; security.

1 Introduction

The modern digital society relies on mobile and ubiquitous optoelectronic devices whose software and hardware security is becoming a global concern owing to the increasing number of disclosed attacks every day [1–4]. The emergence of smart cities, the Internet of things, cloud computing, and big data will generate more challenges in this field [5–8], calling for new opportunities in research. Current cryptography methods for addressing security issues center on the idea of having a digital key, which is safely stored and whose information remains unknown to an adversary. However, implementing this simple concept turns out to be a difficult task: software such as Trojan horses and malware, and side-channel attacks carried out by enemies with single access to the device, can expose the key and lead to security breaches [4, 9–12]. As Tim Cook (Apple CEO) emphasized in a recent interview [13]:

> *"If you put a key under the mat for the cops, a burglar can find it, too. Criminals are using every technology tool at their disposal to hack into people's accounts. If they know there's a key hidden somewhere, they will not stop until they find it."*

These considerations fueled the development of physical unclonable functions (PUFs) [14–16]. A PUF is an object composed of a disordered structure, such as, e.g., a light scatterer, which stores a physical key inside a material layer with no mathematical description. In these systems, a digital key is typically generated by first challenging the PUF with an input signal and then converting into a binary sequence the analog response measured in either time, space, or frequency. The main assumption is that the physical disorder of the PUF cannot be reverse engineered, not even by the original manufacturer. If the PUF is safely stored, an adversary who wants to recreate the key has the

This article has previously been published in the journal Nanophotonics. Please cite as: A. Fratalocchi, A. Fleming, C. Conti and A. Di Falco "NIST-certified secure key generation via deep learning of physical unclonable functions in silica aerogels" *Nanophotonics* 2021, 10. DOI: 10.1515/nanoph-2020-0368.

only possibility of performing a brute force attack, which is practically unfeasible owing to the exponentially large complexity of a PUF [14].

In this field of research, photonics is pioneering technologies for different lines of applications, including authentication [17, 18], secure communications [19–21], and classical equivalent schemes to quantum key distribution with perfect secrecy [22]. The main advantage of photonics PUFs is strong device unclonability: while cloning electronic PUF implementations has been reported [23], no one was ever able to replicate an optical PUF. The main challenge in photonics is the development of general algorithms that transform the response of PUFs into digital keys that appear as unpredictable random sequences. The issue is the local correlations that are present in the PUF response: when transformed into a binary string with conventional techniques, a certain degree of correlation remains in the key and between different keys [24]. To the best of the authors' knowledge, with the exception of the study by Di Falco et al. [22], no optical PUFs has been verified against certification standards that guarantee the genuine unpredictability and uncorrelation of the keys, and no technique has been devised to address this problem controllably for optical PUFs.

Another difficulty originates from the fact that the complex PUFs are strongly sensitive to input conditions. When traditional analog-digital conversion methods are applied to generate the key, such sensibility can generate different keys for apparently identical input conditions [25]. The issue lies in the impossibility of reproducing the same input conditions in different experiments. In a strongly chaotic system such as a PUF, even a small variation in the input parameters can strongly affect the security primitive's reliability. If this problem is addressed, it could also open to new PUFs generated via, e.g., soft-like materials, including gels (e.g., hydrogel, aerogels) and foams. These materials are more input sensitive than solid-state counterparts and are currently not employed as security tokens. However, soft-like structures offer security advantages because their nanoscale disorder can reach a higher entropy than artificial human-made PUFs, which are intrinsically limited by the cost, resolution, and scalability of the present nanofabrication technology [26]. In this article, we propose to address the issues mentioned above by combining PUF with deep learning [27, 28]. We develop a general and versatile two-step key generation strategy, which guarantees the generation of truly random keys verified against the National Institute of Standards and Technology (NIST) standards for cryptographic applications [29], with each key entirely uncorrelated with the others and reliable. We experimentally demonstrate these results with a new class of nonlinear PUFs implemented with silica aerogels (SAs).

2 Results

2.1 All-optical PUFs with aerogels

SA is a material composed of an ultraporous network of sparse silica aggregates. The SA optical response can be adjusted from complete transparency to strongly chaotic scattering by controlling the silica inclusions' size and distribution by either mechanical or optical effects [30]. Owing to a low thermal conductivity, SA exhibits a very strong optothermal nonlinearity [31, 32], which is associated with large and reversible structural deformations, making SA a nonlinear controllable random material that can be employed in different lines of applications [33, 34].

The SA produced in this work is manufactured by a base-catalyzed polymerization process [35, 36], which starts by mixing tetramethyl orthosilicate, methanol, and ammonium hydroxide in a 2:4:1 ratio, producing a gel of good clarity and with minimal defects [37]. The mixture is then poured into a Teflon mold, producing a cuboid-shaped gel of 1 cm side. The gel is subsequently removed from the mold and then washed in a series of acetone baths, each lasting 24 h. The transition from wet gel to aerogel happens by using a low-temperature supercritical CO_2 drying process [38], with a custom setup assembled in our lab. Figure 1a illustrates the setup used to generate all-optical PUFs, acquired as speckle patterns obtained by illuminating the SA sample with a pump-probe configuration. The setup comprises an expanded monochromatic laser probe (wavelength, λ = 632 nm) and a collimated beam with λ = 488 nm waist of approximately 200 μm. The speckle patterns are converted into digital PUFs using a CCD camera placed after the sample (Figure 1a).

In the mapping procedure introduced in this work, it is possible to associate different keys with any class of PUFs that differ in at least a characteristic feature (e.g., distribution or shape of a speckle pattern) that we train the network to resolve. The universal approximation theorem of neural networks [39] guarantees, at least theoretically, that a single neural network that could address this problem exists. In the PUF image acquisition setup illustrated in Figure 1a, it is possible to create PUF images with different speckle features by changing either the laser pump power or the acquisition time. While different pump powers generate diverse characteristic speckles, each pump power triggers a slow dynamical evolution of the speckles over characteristic times of the order of seconds, generating different PUFs in the CCD. Figure 1b shows a typical class of different PUFs that can be acquired at constant pump power (P = 200 mW) and at different times within 1 min of laser illumination. The speckles are observed to be repeatable owing to the good stability properties of SAs [40]. The primary source of entropy that triggers the generation of different speckle patterns in Figure 1b is the spatial fluctuation of the scattering centers of nanoparticles composing the aerogels. These depend on the thermodynamic condition (e.g., temperature, pressure, volume) of the aerogel.

While cloning the soft porous network of SA is hardly imaginable to be feasible now and in the long run owing to the ultradense packing of nanostructured silica components, employing this medium as a PUF generator is also challenging owing to the noticeable spatial fluctuations of the silica nanoparticles, which are visible in the PUF images collected by the CCD (Figure 1b). In the next section, we illustrate a general strategy to address this problem controllably.

2.2 Two-step key generation via deep learning

Figure 2a, b illustrates a high-level schematic of the proposed concept. Ideally, we would like to have at disposal a mapping function $\mathcal{M}$ that, given at the input one experimental PUF generated from the setup of Figure 1a, associates a key k_n with the following properties: i) each key k_n is uncorrelated to the others, ii) each key satisfies the NIST standard to be considered as a real random sequence, and iii) the same key associates with all PUFs experimentally obtained under the same input conditions, controlled with the reproducible accuracy experimentally available.

To address this problem on a general ground, we use a deep neural network (DNN) architecture (Figure 2a, b), which we train to learn the mapping function $\mathcal{M}$ satisfying constraints i)–iii). The DNN used in this work is a 2-layer feedforward neural architecture with a rectified linear unit neural activation function [41]. The network provides a classification of various PUFs into different digital keys k_n, with each key associating a class $c_1, c_2, \ldots, c_N$ of PUF images (Figure 2c, red, green, and blue colors). Each class c_n comprises a series I_{nm} of PUFs (m = 1, 2, ...) that are experimentally obtained under the same input conditions but differ by statistical fluctuations arising in the experimental measurements. The number of PUFs included in each class is not necessarily the same. It can differ according to the fluctuations present for each input challenge considered in the interrogation process of the PUF. The union of all classes $\mathcal{C} \in (c_1, \ldots, c_N)$ constitutes the learning data set, which is fed to the DNN to learn the mapping function $\mathcal{M}$ and predict future keys when we interrogate

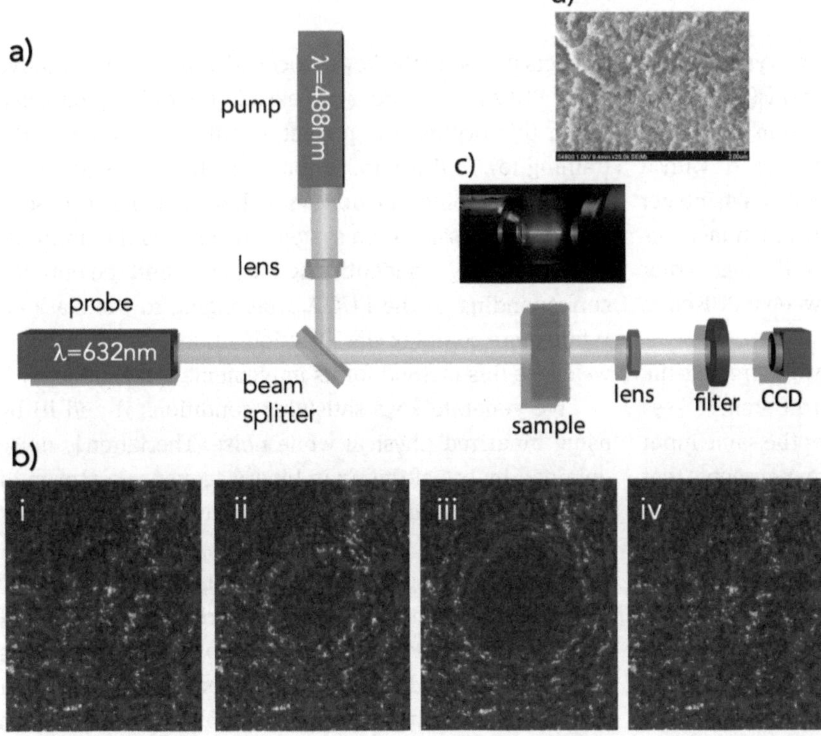

Figure 1: Experimental PUF setup. (a) All-optical PUF aerogel setup and configuration. (b) PUFs collected by the CCD as speckle patterns for t = 0 s (i), t = 20 s (ii), t = 40 s (iii), and t = 60 s (iv). (c) Picture of a real aerogel sample with (d) the corresponding scanning electron microscope (SEM) image. PUF, physical unclonable function.

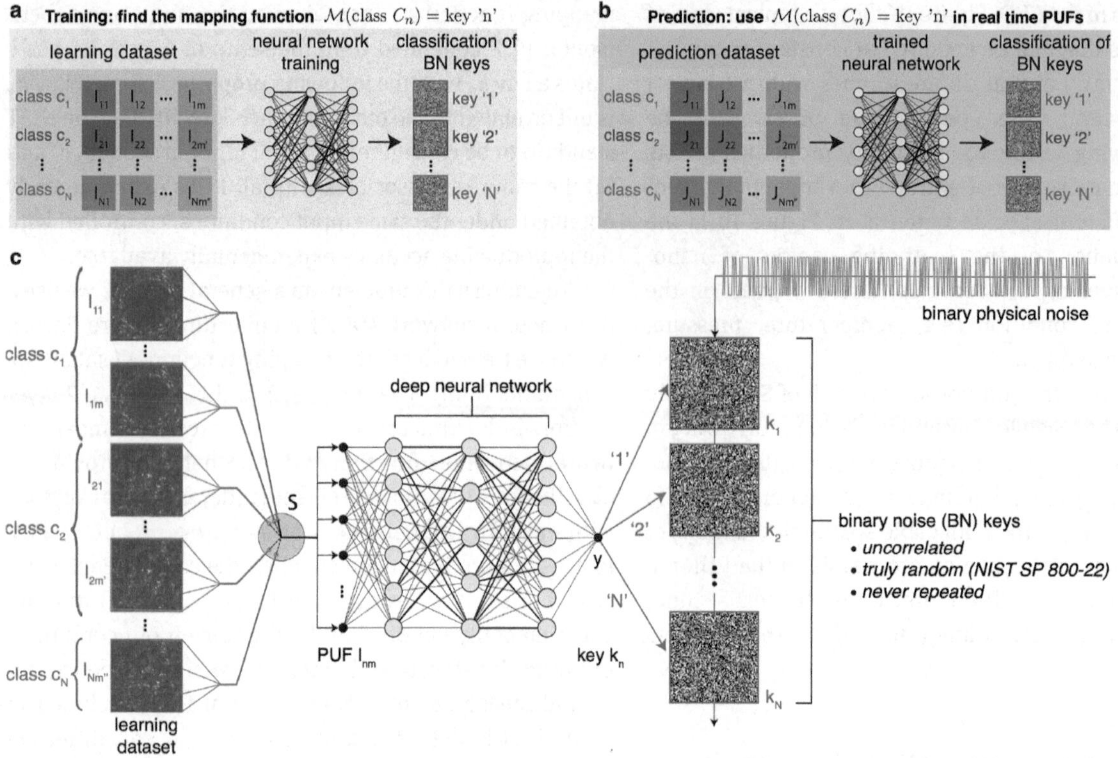

Figure 2: Two-step random key generation via deep learning.
(a–b) Overall process for the training (a) and prediction (b) of key classification and association with PUFs. (c) Detailed schematic workflow of the training procedure: a switch S selects a PUF I_{nm} at the input belonging to single classes c_n (red, green, and blue colors). Each class is mapped by a deep neural network to a different integer number $n = 1, \ldots, N$ at the output y, with each number identifying a binary noise (BN) key k_n. The space of different keys is generated independently by first sampling binary noise (orange solid line) and then splitting the random sequence into consecutive keys $k_1, k_2, \ldots, k_N$ of equal length. The prediction stage (b) employs the trained neural network of (a) in real-time to associate the keys with PUF images J_{mn} acquired in different experiments under the same input conditions of I_{mn} and subjected to experimental fluctuations of input parameters. PUF, physical unclonable function.

the PUF object again. To map PUFs I_{nm} to cryptographic keys k_n, we use a single DNN output channel y (Figure 2c), the latter identifying the output signal from the DNN, and train the network to associate each class c_n with a different integer $n = 1, \ldots, N$ at the output, with each integer n identifying a binary key k_n. Figure 2c illustrates this process visually with different colors, with each color showing the input-output association between a PUF class c_n and a key k_n.

Once we set the DNN weights, the network predicts the key association in future experiments with different classes of PUFs of J_{nm} (Figure 2b), measured under the same input conditions of I_{mn} but acquired in different experiments that differ by uncontrollable fluctuations of the input parameters. The main idea is to include a representative data set comprising a sufficiently large number of PUFs. The DNN learns the features of the experimental fluctuations associated with the different input conditions arising in each class c_n, becoming able to predict the future trends J_{mn} correctly. We increase the data set size until the DNN predicts correctly the key associated with a representative set of PUFs J_{mn} that does not exist in the training data set. When this occurs, the prediction (b) cross-validates the training (a), and it implies that the DNN has learned the required mapping function $\mathcal{M}$ with reasonable accuracy.

In this classification system, the error is the norm between the integer n identifying the key k_n and the output y corresponding to the PUF I_{nm} belonging to the class c_n. While more complex classification strategies are possible, we chose this method for its implementation simplicity.

We generate keys satisfying conditions i) and ii) by using binarized physical white noise. The latter is noise obtained by transforming in binary sequence a stream of white noise generated from a physical object and then split the binary stream into diverse keys $k_1, k_2, \ldots, k_n$ of predefined equal length (Figure 2b, orange binary signal). With the method proposed in this work, the generation of PUFs and cryptographic keys are two different problems that we address independently, overcoming the traditional issues that arise when mapping a complex PUF directly to a binary sequence. The two problems are then combined via machine learning, using a DNN that finds the desired mapping that associates each PUF with a cryptography key.

The mapping function learned by the DNN conserves the security advantages of PUFs: it relies on a PUF object that has no mathematical representation, and it ensures a mapping between an input condition and a random key that cannot be guessed or recreated without the PUF object. From a security perspective, the DNN of Figure 2 acts as an additional, two-step protective layer to the PUF. If the PUF falls in the adversary's hand, the attacker cannot recreate the key without brute forcing all the DNN architecture weights. In a typical integrated electronic system, the space of the combination S_c that the enemy has to explore is $S_c = 2^{64 \cdot N_w}$ possibilities, with 2^{64} the combination required to assess the value of a 64-bit floating point number representing a single weight and N_w the number of network weights. In a DNN with $N_w \geq 4$, the space $S_c = 2^{64 . N_w \geq 256}$ is larger than the space of 2^{256} combinations required to break the 256-bit advanced encryption standard, a NIST-certified cryptography in use by the US government to classify top secret information and presently considered unbreakable by brute force [42].

In the scheme of Figure 2, the DNN operation is typically evaluated by electronic CPU at gigahertz speed. It does not add overhead to the PUF key generation process, which is mainly limited by the camera's acquisition time of the optical PUFs.

2.3 Experimental results on PUF key generation and NIST validation

Figure 3 shows the typical results of cross-validation for two representative classes c_1 and c_2 of PUFs. Experimentally, we observe that classes c_n composed approximately of ≤10 PUFs are sufficient to train the DNN to perform accurate predictions. Figure 3a reports the learning rate obtained by the DNN when training on the learning data set composed of seven PUFs, with I_{11}–I_{13} belonging to c_1 and I_{21}–I_{24} belonging to c_2. These PUFs are acquired in the most fluctuating scenario in the setup of Figure 1a, in which we fix the input power (P = 200 mW) and acquire images at different times. Representative 100 × 100 pixel images of the PUF belonging to each class are shown in Figure 3b.

The results for Figure 3a illustrate that the DNN learns with great accuracy (learning errors below machine precision 10^{-15}) to associate the correct key number n with each PUF in the data set. Figure 3c and d report the resulting performances of the DNN when predicting the key number associated with the prediction data set, composed of six PUFs J_{11}–J_{13} of c_1 and J_{21}–J_{23} of class c_2. Images I_{mn} and J_{mn} are obtained in different experiments and have the same input conditions, i.e., pump power at P = 200 mW and same acquisition time. Although none of the J_{mn} PUFs exist in the learning data set used to train the DNN, the network correctly predicts the right index to each image, with prediction errors below 0.3. These results allow using a simple threshold filter $n \pm 0.5$ to assess correctly the key associated with each PUF, with no error arising from the natural fluctuations present in the experimental measures. The ability of the DNN to learn the feature of each PUF and the required mapping function $\mathcal{M}$ from few images is quite remarkable, especially considering the soft-like nature of the aerogel, whose scattering centers oscillate in time with large spatial fluctuations.

The error, or overfitting, between the DNN prediction and the correct key number in Figure 3c can be reduced by either increasing the length of the training data set and the associated DNN size or by adding a larger number of output channels, with each channel associating the corresponding key with a particular class of PUFs. To create random keys, we sample in time with an analog microphone, the sound arising from the electric engine of a desktop fan (Figure 4a). The sample rate of the microphone is much lower than the fan speed, allowing us to collect a time-varying random stream (Figure 4a, solid blue line). The sequence is then converted to a binary signal (Figure 4a, solid orange line) by using a Gabor transform [14], which associates 1 to all inputs above a threshold value, here chosen as the analog noise mean value. The random binary sequence generated is then partitioned into N different keys $k_1, \ldots, k_N$. We generate a set of N = 100 digital keys in our experiments, each of 10,000 bits.

Figure 4b reports the results of the NIST SP 800–22 test suite on the generated keys. The test comprises a suite of different statistical tests to assess if the key at the output looks like an unpredictable binary sequence in the input's absence of knowledge. The tests analyze the proportion of zeros and ones in the sequence and the existence of harmonic peaks (frequency, block frequency, and FFT), the presence of sequences with identical bits (run, longest run), the occurrence of prespecified target strings (nonoverlapping templates), the rank of disjoint submatrices in the stream (rank), and the sufficient complexity of the sequence to be considered random (serial and linear complexity). Detailed information on each test is available in the NIST reference [43]. Each test results in a proportion, which measures each key's probability to pass the statistical test (Figure 4b, dashed red line). The analysis is

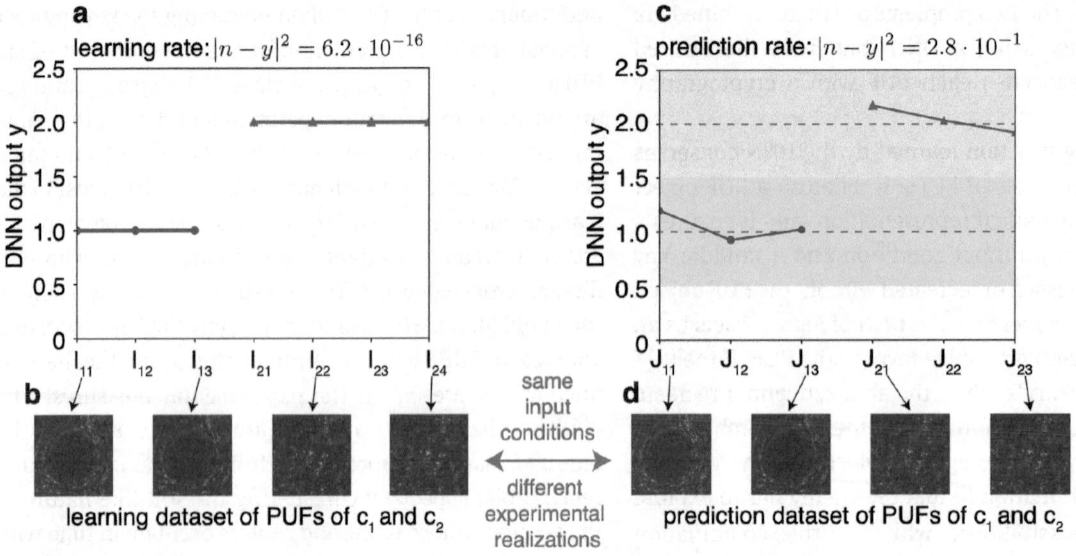

Figure 3: Deep neural network training and key association.
(a) Learning error in the training data set composed of seven PUFs belonging to classes c_1 and c_2, which are obtained with the same pumping power $P = 200$ mW and at different acquisition times. The PUFs I_{11}–I_{13} are associated with the output integer $y = 1$, while I_{21}–I_{24} are associated with $y = 2$. (b) CCD images of the PUFs. Panel (c–d) report the same analysis of (a–b) for the prediction data set, composed of six PUFs J_{11}–J_{13} of class c_1 and J_{21}–J_{23} of c_2. The PUFs in the prediction data set are generated in a different experiment and are not included in the training data set. PUF, physical unclonable function; DNN, deep neural network.

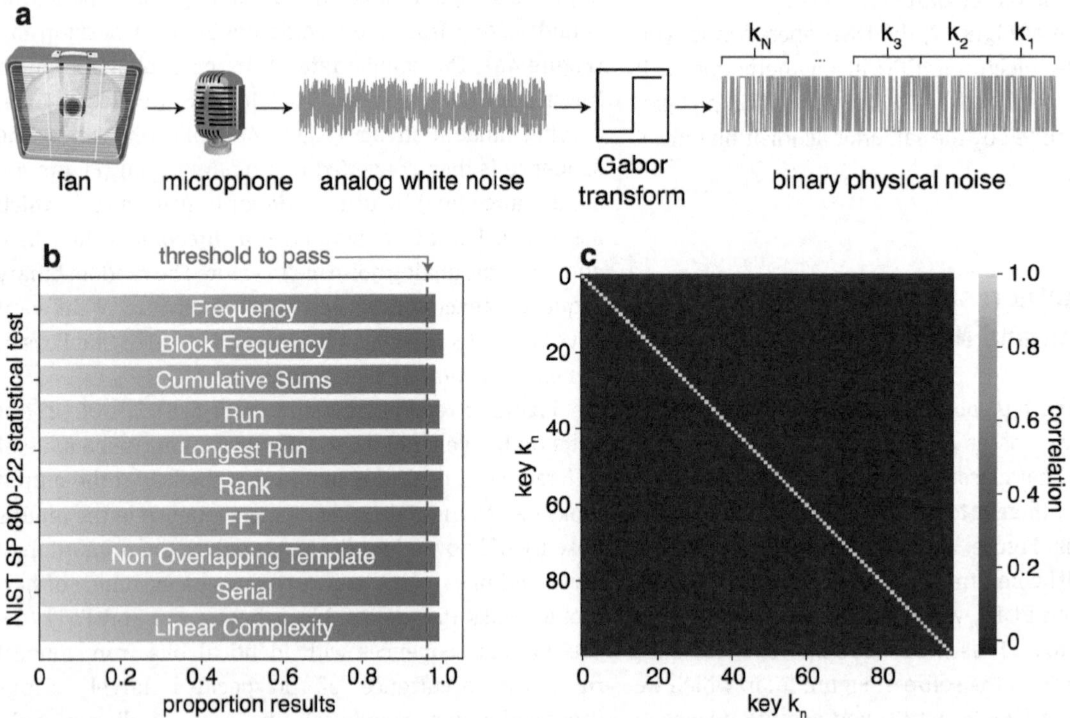

Figure 4: Key generation and NIST certification results.
(a) Generation steps of the binary noise sequence, starting from sampling in time, with a sufficiently low acquisition time, the noise emitted from the electric engine of a desktop fan, and then converting it into a binary sequence with a Gabor transform. The binary stream is then partitioned into $N = 100$ $k_1, \ldots, k_N$ consecutive keys each of 10,000 bits. (b) NIST proportion results on the SP 800-22 statistical test suite applied to the keys generated in (a), with a minimum threshold (dashed red line) recommended by the NIST. (c) Cross-correlation matrix between the keys k_n. NIST, National Institute of Standards and Technology.

performed using the software STS distributed by the NIST [43]. The results of Figure 4b demonstrate that the binary keys pass all the NIST tests well above the minimum threshold, showing that the procedure used to acquire the noise and transform it into a binary sequence generates a truly unpredictable stream of data (Figure 4a). Figure 4c reports the cross-correlation between the keys k_n. The keys generated are completely uncorrelated with each other, with average cross-correlation coefficients between the key k_i and k_j of the order of $\langle C_{ij} \rangle = 10^{-2}$.

Figures 2–4 demonstrate that the technique proposed in this works satisfies requirements i)–iii), with the reliable associations of the same key with experimental PUFs measured after the same challenge with no errors and with each key representing an unpredictable random sequence that is completely uncorrelated to the other.

3 Discussion

We discussed a two-step key generation strategy for PUFs based on deep learning, which can address the shortcomings of unreliability and weak unpredictability of cryptographic keys. The idea explored is to design the PUF independently from the problem of key generation and then use machine learning to train a neural network to find the complex mapping function that can reliably associate the features of PUFs with identical input conditions to a single key. Different binary keys were generated by sampling white noise, representing a physically unpredictable random sequence that passed all validation tests against NIST standards for cryptographic applications. We report experimental results in SAs, exploiting a classification strategy based on integer numbers n, with each number directly identifying a binary key k_n. Despite the high sensitivity of the aerogel to different input conditions, our experiments report that a trained neural network predicts the correct key with no errors. The results of this work can be of help in the development of stronger PUFs for different applications, including authentication and secure communications. The research data supporting this publication can be accessed at https://doi.org/10.17630/50b2f96f-ab3a-4b6e-abcd-c5d14c784de9.

Author contribution: All the authors have accepted responsibility for the entire content of this submitted manuscript and approved submission.

Research funding: C.C. acknowledge funding from Horizon 2020 Framework Programme QuantERA grant QUOMPLEX, by National Research Council (CNR), Grant agreement ID 731473.

Conflict of interest statement: The authors declare no conflicts of interest regarding this article.

References

[1] S. Sakhare and D. Sakhare, "A review—hardware security using puf (physical unclonable function)," in *ICCCE 2019*, A. Kumar, and S. Mozar, Eds., Singapore, Springer Singapore, 2020, pp. 373–377.

[2] D. Adam, "Cryptography on the front line," *Nature*, vol. 413, pp. 766–767, 2001.

[3] S. Chen, "Random number generators go public," *Science*, vol. 360, pp. 1383–1384, 2018.

[4] H. Wang, D. Forte, M. M. Tehranipoor, and Q. Shi, "Probing attacks on integrated circuits: challenges and research opportunities," *IEEE Design Test*, vol. 34, pp. 63–71, 2017.

[5] C. Tankard, "The security issues of the Internet of things," *Comput. Fraud. Secur.*, vol. 2015, pp. 11–14, 2015. Available at: http://www.sciencedirect.com/science/article/pii/S1361372315300841.

[6] F. S. Ferraz and C. A. G. Ferraz, "Smart city security issues: depicting information security issues in the role of an urban environment," in *2014 IEEE/ACM 7th International Conference on Utility and Cloud Computing*, 2014, pp. 842–847.

[7] V. Mayer-Schonberger and K. Cukier, *Big Data: A Revolution That Will Transform How We Live, Work, and Think*, Boston, Houghton Mifflin Harcourt, 2013. Available at: http://www.amazon.com/books/dp/0544002695.

[8] A. AlDairi and L. Tawalbeh, "Cyber security attacks on smart cities and associated mobile technologies," *Procedia Comput. Sci.*, vol. 109, pp. 1086–1091, 2017. Available at: http://www.sciencedirect.com/science/article/pii/S1877050917310669. *8th International Conference on Ambient Systems, Networks and Technologies, ANT-2017 and the 7th International Conference on Sustainable Energy Information Technology*, SEIT 2017, 16–19 May 2017, Madeira, Portugal.

[9] K. Thompson, "Reflections on trusting trust," *Commun. ACM*, vol. 27, pp. 761–763, 1984.

[10] S. Chen, R. Wang, X. Wang, and K. Zhang, "Side-channel leaks in web applications: A reality today, a challenge tomorrow," in *Proceedings of the IEEE Symposium on Security and Privacy (Oakland)*, IEEE Computer Society, 2010. Available at: https://www.microsoft.com/en-us/research/publication/side-channel-leaks-in-web-applications-a-reality-today-a-challenge-tomorrow/.

[11] C. Ashokkumar, R. P. Giri, and B. Menezes, "Highly efficient algorithms for AES key retrieval in cache access attacks," in *2016 IEEE European Symposium on Security and Privacy*, EuroS P, 2016, pp. 261–275.

[12] A. Golder, D. Das, J. Danial, et al., "Practical approaches toward deep-learning-based cross-device power side-channel attack," *IEEE Trans. Very Large Scale Integr. Syst.*, vol. 27, pp. 2720–2733, 2019.

[13] L. Kahney, *Tim Cook: The Genius Who Took Apple to the Next Level*, Penguin Books Limited, 2019. Available at: https://books.google.com.sa/books?id=A5xlDwAAQBAJ.

[14] R. Pappu, B. Recht, J. Taylor, and N. Gershenfeld, "Physical one-way functions," *Science*, vol. 297, pp. 2026–2030, 2002.

[15] C. Herder, M.-D. Yu, F. Koushanfar, and S. Devadas, "Physical unclonable functions and applications: A tutorial," *Proc. IEEE*, vol. 102, pp. 1126–1141, 2014.

[16] B. Škorić, P. Tuyls, and W. Ophey, "Robust key extraction from physical uncloneable functions," in *International Conference on Applied Cryptography and Network Security*, Springer, 2005, pp. 407–422.

[17] S. A. Goorden, M. Horstmann, A. P. Mosk, B. Škorić, and P. W. H. Pinkse, "Quantum-secure authentication of a physical unclonable key," *Optica*, vol. 1, pp. 421–424, 2014.

[18] G. Zhang and Q. Liu, "A novel image encryption method based on total shuffling scheme," *Optic Commun.*, vol. 284, pp. 2775–2780, 2011.

[19] R. Horstmeyer, B. Judkewitz, I. M. Vellekoop, S. Assawaworrarit, and C. Yang, "Physical key-protected one-time pad," *Sci. Rep.*, vol. 3, no. 6, p. 3543, 2013.

[20] M. Leonetti, S. Karbasi, A. Mafi, E. DelRe, and C. Conti, "Secure information transport by transverse localization of light," *Sci. Rep.*, vol. 6, p. 29918, 2016.

[21] B. C. Grubel, B. T. Bosworth, M. Kossey, et al., "Secure communications using nonlinear silicon photonic keys," *Opt. Express*, vol. 26, pp. 4710–4722, 2018.

[22] A. Di Falco, V. Mazzone, A. Cruz, and A. Fratalocchi, "Perfect secrecy cryptography via mixing of chaotic waves in irreversible time-varying silicon chips," *Nat. Commun.*, vol. 10, p. 5827, 2019.

[23] C. Helfmeier, C. Boit, D. Nedospasov, and J.-P. Seifert, "Cloning physically unclonable functions," in *Hardware-Oriented Security and Trust (HOST), 2013 IEEE International Symposium on 1–6*, IEEE, 2013.

[24] U. Rührmair, "Optical pufs reloaded," Eprint.Iacr.Org, 2013, https://doi.org/10.1109/sp.2013.27.

[25] J. Danger, "Physically unclonable functions: principle, advantages and limitations," in *2019 International Conference on Advanced Technologies for Communications (ATC)*, 2019, pp. xxxii–xxxii.

[26] A. Wali, A. Dodda, Y. Wu, et al., "Biological physically unclonable function," *Commun. Phys.*, vol. 2, no. 39, 2019. Available at: https://doi.org/10.1038/s42005-019-0139-3.

[27] I. Goodfellow, Y. Bengio, and A. Courville, *Deep Learning*, Cambridge, The MIT Press, 2016.

[28] G. Marcucci, D. Pierangeli, P. W. H. Pinkse, M. Malik, and C. Conti, "Programming multi-level quantum gates in disordered computing reservoirs via machine learning," *Opt. Express*, vol. 28, pp. 14018–14027, 2020. Available at: http://www.opticsexpress.org/abstract.cfm?URI=oe-28-9-14018.

[29] Bassham, L. E., Andrew, R., Juan, S., et al., "Sp 800-22 rev. 1a. a statistical test suite for random and pseudorandom number generators for cryptographic applications," Tech. Rep., 2010, https://doi.org/10.6028/nist.sp.800-22r1a.

[30] S. M. Jones, "A method for producing gradient density aerogel," *J. Sol. Gel Sci. Technol.*, vol. 44, pp. 255–258, 2007.

[31] S. Gentilini, F. Ghajeri, N. Ghofraniha, A. Di Falco, and C. Conti, "Optical shock waves in silica aerogel," *Opt. Express*, vol. 22, pp. 1667–1672, 2014.

[32] M. C. Braidotti, S. Gentilini, A. Fleming, M. C. Samuels, A. Di Falco, and C. Conti, "Optothermal nonlinearity of silica aerogel," *Appl. Phys. Lett.*, vol. 109, p. 041104, 2016.

[33] A. Fleming, C. Conti, and A. Di Falco, "Perturbation of transmission matrices in nonlinear random media," *Ann. Phys.*, vol. 531, p. 1900091, 2019.

[34] A. Fleming, C. Conti, T. Vettenburg, and A. Di Falco, "Nonlinear optical memory effect," *Opt. Lett.*, vol. 44, pp. 4841–4844, 2019.

[35] G. Nicolaon and S. Teichner, "The preparation of silica aerogels from methylorthosilicate in an alcoholic medium and their properties," 1975.

[36] J. Livage, M. Henry, and C. Sanchez, "Sol-gel chemistry of transition metal oxides," *Prog. Solid State Chem.*, vol. 18, pp. 259–341, 1988.

[37] B. Lin, S. Cui, X. Liu, X. Shen, Y. Liu, and G. Han, "Preparation and characterization of hmds modified hydrophobic silica aerogel," *Curr. Nanosci.*, 2011.

[38] P. H. Tewari, A. J. Hunt, and K. D. Lofftus, "Ambient-temperature supercritical drying of transparent silica aerogels," *Mater. Lett.*, vol. 3, pp. 363–367, 1985.

[39] M. Leshno, V. Y. Lin, A. Pinkus, and S. Schocken, "Multilayer feedforward networks with a nonpolynomial activation function can approximate any function," *Neural Netw.*, vol. 6, pp. 861–867, 1993. Available at: http://www.sciencedirect.com/science/article/pii/S0893608005801315.

[40] E. Strobach, B. Bhatia, S. Yang, L. Zhao, and E. N. Wang, "High temperature stability of transparent silica aerogels for solar thermal applications," *APL Mater.*, vol. 7, p. 081104, 2019.

[41] R. H. R. Hahnloser, R. Sarpeshkar, M. A. Mahowald, R. J. Douglas, and H. S. Seung, "Digital selection and analogue amplification coexist in a cortex-inspired silicon circuit," *Nature*, vol. 405, pp. 947–951, 2000.

[42] J. Schwartz, "U.s. Selects a new encryption technique," 2000. Available at: https://www.nytimes.com/2000/10/03/business/technology-us-selects-a-new-encryption-technique.html.

[43] Computer Security Division, I. T. L, "Nist sp 800-22: documentation and software – random bit generation: Csrc,". Available at: https://csrc.nist.gov/projects/random-bit-generation/documentation-and-software.

Salvatore Savasta, Omar Di Stefano* and Franco Nori

Thomas–Reiche–Kuhn (TRK) sum rule for interacting photons

https://doi.org/10.1515/9783110710687-037

Abstract: The Thomas–Reiche–Kuhn (TRK) sum rule is a fundamental consequence of the position–momentum commutation relation for an atomic electron, and it provides an important constraint on the transition matrix elements for an atom. Here, we propose a TRK sum rule for electromagnetic fields which is valid even in the presence of very strong light–matter interactions and/or optical nonlinearities. While the standard TRK sum rule involves dipole matrix moments calculated between atomic energy levels (in the absence of interaction with the field), the sum rule here proposed involves expectation values of field operators calculated between general eigenstates of the interacting light–matter system. This sum rule provides constraints and guidance for the analysis of strongly interacting light–matter systems and can be used to test the validity of approximate effective Hamiltonians often used in quantum optics.

Keywords: cavity QED; quantum optics; sum rules.

***Corresponding author: Omar Di Stefano,** Dipartimento di Scienze Matematiche e Informatiche, Scienze Fisiche e Scienze della Terra, Università di Messina, I-98166 Messina, Italy; and Theoretical Quantum Physics Laboratory, RIKEN Cluster for Pioneering Research, Wako-shi, Saitama 351-0198, Japan, E-mail: odistefano@unime.it. https://orcid.org/0000-0002-3054-272X
Salvatore Savasta, Dipartimento di Scienze Matematiche e Informatiche, Scienze Fisiche e Scienze della Terra, Università di Messina, I-98166 Messina, Italy, E-mail: ssavasta@unime.it. https://orcid.org/0000-0002-9253-3597
Franco Nori, Theoretical Quantum Physics Laboratory, RIKEN Cluster for Pioneering Research, Wako-shi, Saitama 351-0198, Japan; and Physics Department, The University of Michigan, Ann Arbor, Michigan 48109-1040, USA, E-mail: fnori@riken.jp. https://orcid.org/0000-0003-3682-7432

1 Introduction

1.1 A brief history of sum rules in quantum mechanics

Since the beginning of quantum mechanics, sum rules have proved to be very useful for understanding the general features of difficult problems. These relations, obtained by adding (sum) unknown terms, power tool for the study of physical processes [1]. Historically, the first important sum rule is found in atomic physics and concerns the interaction of electromagnetism with atoms: the Thomas–Reiche–Kuhn (TRK) sum rule [2–4]. It states that the sum of the squares of the dipole matrix moments from any energy level, weighted by the corresponding energy differences, is a constant. The TRK and analogous sum rules, like the Bethe sum rule [5], play an especially important role in the interaction between light and matter. They have widely been applied to the problems of electron excitations in atoms, molecules, and solids [6].

For an atomic electron, the TRK sum rule is a direct consequence, of the canonical commutation relation between position and momentum. It is possible to view it as a necessary condition in order not to violate this commutation relation [7]. Among the many consequences of this sum rule, it constrains the cross sections for absorption and stimulated emission [8]. It has also been shown that useful sum rules can be obtained for nonlinear optical susceptibilities [9–11]. A modified TRK sum rule for the motion of the atomic center of mass and a generalized TRK sum rule to include ions have been also obtained [12]. Extensions of the TRK sum rule to the relativistic case have been studied (see, e.g., [13, 14]). Important sum rules have also been developed in quantum chromodynamics (see, e.g., [15]).

Such sum rules also play a relevant role in the analysis of interacting electron systems [16, 17]. Since they are a direct consequence of particle conservation in the system, their satisfaction is necessary to guarantee a gauge-invariant theory [16, 17] (see, e.g., [18, 19] as two recent examples). In interacting electron systems, the longitudinal version of the TRK sum rule (known as f-sum rule) provides a very useful check on the consistency of any

This article has previously been published in the journal Nanophotonics. Please cite as: S. Savasta, O. Di Stefano and F. Nori "Thomas–Reiche–Kuhn (TRK) sum rule for interacting photons" *Nanophotonics* 2021, 10. DOI: 10.1515/nanoph-2020-0433.

approximate theory and can permit a direct calculation of collective mode frequencies in the long wavelength limit [16]. A striking example of the relevance of sum rules in interacting electron systems is constituted by the apparent gauge invariance difficulty in superconductors (Meissner effect), originating by the violation of the *f*-sum rule of approximate models [20].

Almost all the developed sum rules have been derived for the degrees of freedom of particles. One exception is in the study by Barnett and Loudon [21], where optical sum rules have been derived for polaritons propagating through a linear medium.

1.2 Summary of our main results

Here, we propose a TRK sum rule for electromagnetic fields which is valid even in the presence of very strong light–matter interactions and/or optical nonlinearities [22, 23]. While the standard TRK sum rule involves dipole matrix moments calculated between atomic energy levels (in the absence of interaction with the field), the sum rule here proposed involves the expectation values of the field coordinates or momenta calculated between general eigenstates of the interacting light–matter system (dressed light–matter states) and the corresponding eigenenergies of the interacting system.

In this work, we also present a generalized atomic TRK sum rule for atoms strongly interacting with the electromagnetic field. This sum rule has the same form of the standard TRK sum rule but involves the energy eigenstates and eigenvalues of the interacting system.

The sum rules for interacting light–matter systems proposed here can be useful to analyze general quantum nonlinear optical effects (see, e.g., [24–27]) and many-body physics in photonic systems [28], like analogous sum rules for interacting electron systems, which played a fundamental role for understanding the many-body physics of electron liquids [16, 17, 20]. The proposed sum rules become particularly interesting in the nonperturbative regimes of light–matter inter-actions.

In the last years, several methods to control the strength of the light–matter interaction have been developed, and the ultrastrong coupling (USC) between light and matter has transitioned from theoretical proposals to experimental reality [22, 23]. In this new regime of quantum light–matter interaction, beyond weak and strong coupling, the coupling strength becomes comparable to the transition frequencies in the system or even higher (deep strong coupling [DSC]) [29–32]. In the USC and DSC regimes, approximations widely employed in quantum optics break down [33], allowing processes that do not conserve the number of excitations in the system (see, e.g., [27, 34–37]). The nonconservation of the excitation number gives rise to a wide variety of novel and unexpected physical phenomena in different hybrid quantum systems [35, 38–58]. As a consequence, all the system eigenstates, dressed by the interaction, contain different numbers of excitations. Much research on these systems has dealt with understanding whether these excitations are real or virtual, how they can be probed or extracted, how they make possible higher order processes even at very low excitation densities, and how they affect the description of input and output for the system [22, 23].

The eigenstates of these systems, including the ground state, can display a complex structure involving superposition of several eigenstates of the noninteracting subsystems [22, 23, 59] and can be difficult to calculate. As a consequence, a number of approximation methods have been developed [60, 61]. Moreover, the output field correlation functions, connected to measurements, depend on these eigenstates (see, e.g., [48, 62]). Hence, sum rules providing general guidance and constraints can be very useful to test the validity of the approximations. The general sum rule proposed in this article can also be used to test the validity of effective Hamiltonians often used in quantum optics and cavity optomechanics [58, 63, 64]. In addition, this generalized TRK sum rule applies to the broad emerging field of nonperturbative light–matter interactions, including several settings and subfields, as cavity and circuit quantum electrodynamics (QED) [22], collective excitations in solids [65], optomechanics [63], photochemistry and QED chemistry [59, 66].

2 Sum rule for interacting photons

A key property used for the derivation of the TRK sum rule is that the commutator between the electron coordinate and the electronic Hamiltonian does not depend on the electronic potential, which is a function of the coordinate only, and hence, it is universal. Considering for simplicity, a single electron 1D system, if x is the electron coordinate and $\hat{H}_{\text{at}} = \hat{p}^2/2m + V(x)$ is the electronic Hamiltonian: $[x, \hat{H}_{\text{at}}] = [x, \hat{p}^2/2m] = i(\hbar/m)\hat{p}$.

In the Coulomb gauge, the (transverse) vector potential $\mathbf{A}$ represents the field coordinate, while its conjugate momentum Π is proportional to the transverse electric field:

$$\Pi(\mathbf{x}, t) = -\varepsilon_0 \hat{E}(\mathbf{x}, t) = \varepsilon_0 \dot{\hat{A}}(\mathbf{x}, t). \tag{1}$$

A general feature of the light–matter interaction Hamiltonians derived from the *minimal coupling replacement* (as for the Coulomb gauge) is that the momenta of the matter system are coupled only to the field coordinate. We can express the total light–matter quantum Hamiltonian as $\widehat{H} = \widehat{H}_F + \widehat{H}_M + \widehat{H}_I$, where the first two terms on the r.h.s. are the field- and matter system–free Hamiltonians, and the third describes the light–matter interaction. Using (1) and the Heisenberg equation $i\hbar\dot{\widehat{A}} = [\widehat{A}, \widehat{H}]$, we obtain the following relation:

$$i\hbar\Pi = \varepsilon_0\left[\widehat{A}, \widehat{H}\right] = \varepsilon_0\left[\widehat{A}, \widehat{H}_F\right], \tag{2}$$

where the second equality follows from $[\widehat{A}, \widehat{H}_I] = 0$, which holds, e.g., in the Coulomb gauge. For simplicity, we consider the case of a quasi 1*D* electromagnetic resonator of length L so that the expression for the electric field operator can be simplified to $\widehat{E}(\mathbf{r}, t) \to \tilde{s}\widehat{E}(x, t)$, where $\tilde{s} = \mathrm{y}/|\mathrm{y}|$, where x is the coordinate along the cavity axis, and y is a coordinate along an axis orthogonal to the cavity axis. The vector potential (as well as the electric field operator) can be expanded in terms of photon creation and destruction operators as

$$\widehat{A}(x, t) = \sum_m A_m(x)\widehat{a}_m e^{-i\omega_m t} + \text{h.c.}$$

and

$$\widehat{E}(x, t) = \sum_m E_m(x)\widehat{a}_m e^{-i\omega_m t} + \text{h.c.},$$

where

$$A_m(x) = [\hbar/(2\omega_m\varepsilon_0 S)]^{1/2} u_m(x),$$

and

$$E_m(x) = i\omega_m A_m(x).$$

Here SL is the resonator volume, the subscript m labels a generic mode index with frequency ω_m, and $u_m(x)$ is the normal modes of the field chosen as real functions. For example, imposing the vanishing of the electric field at the two end walls at $x = \pm L/2$ of the cavity,

$$u_m(x) = \left(1/\sqrt{L}\right)\sin k_m(x + L/2),$$

where $k_m = \pi m/L$.

Let us now consider the matrix elements of the operators in (2) between two generic eigenstates $|\psi_i\rangle$ of the total Hamiltonian $\widehat{H}$. We obtain the following:

$$\Pi_{ij} = i\varepsilon_0\omega_{ij}\mathbf{A}_{ij}, \tag{3}$$

where $\omega_{ji} = \omega_j - \omega_i$ and we used the notation $O_{ij} = \langle\psi_i|\widehat{O}|\psi_j\rangle$. Here and in the following, $j = 0$ indicates the system ground state, and the energy levels are ordered according to their energy: $j > i$ if $\omega_j > \omega_i$. We now multiply both sides of (3) by $u_m(x)$ and integrate over x. By defining

$$\widehat{\mathcal{Q}}^{(m)} = (\widehat{a}_m + \widehat{a}_m^\dagger)/\sqrt{2},$$

and

$$\widehat{\mathcal{P}}^{(m)} = i(\widehat{a}_m^\dagger - \widehat{a}_m)/\sqrt{2},$$

we obtain the corresponding relation for the individual modes:

$$\omega_m\mathcal{P}_{ij}^{(m)} = i\omega_{ij}\mathcal{Q}_{ij}^{(m)}. \tag{4}$$

It is worth noticing that, in the limit when the light–matter interaction vanishes, $\left|\mathcal{P}_{ij}^{(m)}\right| = \left|\mathcal{Q}_{ij}^{(m)}\right|$, and (4) can easily be verified analytically. When the interaction becomes relevant, so that the system eigenstates differ from the harmonic spectrum for free fields, the ratio between the two quadratures can be very different from 1 and can be determined by the only knowledge of the energy spectrum, independently on the specific interacting system. Equation (4) is the first result of this work. It shows that the ratio between the two field quadratures is uniquely determined by the energy spectrum. The two quadratures can display very different matrix elements when the interaction with the matter system changes significantly the energy levels of the interacting systems, as it occurs in the USC and DSC regimes.

Let us now consider the commutator between the mode coordinate and its conjugate momentum:

$$i = \left[\widehat{\mathcal{Q}}^{(m)}, \widehat{\mathcal{P}}^{(m)}\right] = \frac{1}{i\hbar\omega_m}\left[\widehat{\mathcal{Q}}^{(m)}, \left[\widehat{\mathcal{Q}}^{(m)}, \widehat{H}_F\right]\right], \tag{5}$$

where we used

$$\omega_m\widehat{\mathcal{P}}^{(m)} = \dot{\widehat{\mathcal{Q}}}^{(m)}, \text{ and } \left[\widehat{\mathcal{Q}}^{(m)}, \widehat{H}\right] = \left[\widehat{\mathcal{Q}}^{(m)}, \widehat{H}_F\right].$$

Developing the double commutator, considering its matrix elements between two generic eigenstates of the total Hamiltonian $\widehat{H}$ and inserting the identity operators $\left(\widehat{I} = \sum_k |\psi_k\rangle\langle\psi_k|\right)$, we obtain the following relation:

$$\sum_k \frac{\omega_{k,i} + \omega_{k,j}}{\omega_m}\mathcal{Q}_{i,k}^{(m)}\mathcal{Q}_{k,j}^{(m)} = \delta_{i,j}, \tag{6}$$

which reduces (choosing $j = i$) to the TRK sum rule for interacting fields:

$$2\sum_k \frac{\omega_{k,i}}{\omega_m}\left|\mathcal{Q}_{i,k}^{(m)}\right|^2 = 1. \tag{7}$$

By using (4), (7) can be also expressed in terms of the momenta matrix elements:

$$2\omega_m\sum_k \left|\mathcal{P}_{i,k}^{(m)}\right|^2 / \omega_{k,i} = 1.$$

Formally, it coincides with the TRK sum rule for atoms; however, in (7) the matrix elements of the field-mode coordinate replace the atomic *electric dipole* matrix elements. An important difference is that the atomic TRK sum rule [67] considers atomic energy eigenstates, calculated in the absence of interaction with the field. On the contrary, this sum rule is very general since it holds in the presence of interactions with *arbitrary* matter systems every time the interaction occurs via the field coordinate (e.g., Coulomb gauge). We also observe that (7) describes a collection of sum rules, one for each field mode m. Actually, following the same reasoning which led us to (7), a generalized atomic TRK sum rule for atoms strongly interacting with the electromagnetic field [analogous to (7)] can be easily obtained, as shown in Section 4.

3 Applications

3.1 Quantum Rabi model

The quantum Rabi Hamiltonian describes the dipolar coupling between a two-level atom and a single mode of the quantized electromagnetic field. Recently, it has been shown [68] that the correct (satisfying the gauge principle) quantum Rabi Hamiltonian in the Coulomb gauge

$$\hat{H}_C = \hbar\omega_c\hat{a}^\dagger\hat{a} + \frac{\hbar\omega_0}{2}\{\hat{\sigma}_z \cos\left[2\eta\left(\hat{a}+\hat{a}^\dagger\right)\right] + \hat{\sigma}_y \sin\left[2\eta\left(\hat{a}+\hat{a}^\dagger\right)\right]\}, \tag{8}$$

strongly differs from the standard model (see also the studies by De Bernardis et al [69], Stokes et al [70], and Settineri et al [71] for gauge issues in the USC regime). Here, ω_c is the resonance frequency of the cavity mode, ω_0 is the transition frequency of a two-level atom, $\hat{a}$ and $\hat{a}^\dagger$ are the destruction and creation operators for the cavity field, respectively, while the qubit degrees of freedom are described by the Pauli operators $\hat{\sigma}_i$. The parameter

$$\eta = A_0 d/\hbar$$

(A_0 is the zero-point fluctuation amplitude of the field potential and d is the atomic dipole moment) in (A) describes the normalized light–matter coupling strength. When the normalized coupling strength is small ($\eta \ll 1$), considering only first-order contributions in η, the standard interaction term $\hbar\omega_0\eta\,(\hat{a}+\hat{a}^\dagger)\hat{\sigma}_y$ is recovered. If the system is prepared in its first excited state, the photodetection rate for cavity photons is proportional to $|\mathcal{P}_{1,0}|^2$ (see [62, 71]). Figure 1(a) displays this quantity (black dashed curve) as well as $|\mathcal{Q}_{1,0}|^2$ (dotted blue) versus the normalized coupling η, calculated after the numerical diagonalization of (8). The two quantities are equal only at negligible coupling. When the coupling strength increases, the two quantities provide very different results. However, in agreement with (4), the numerically calculated $(\omega_{1,0}^2/\omega_c^2)|\mathcal{Q}_{1,0}|^2$ coincides with $|\mathcal{P}_{1,0}|^2$. In contrast, the Jaynes Cummings (JC) model,

$$\hat{H}_{JC} = \hbar\omega_c\hat{a}^\dagger\hat{a} + \hbar\omega_0/2\hat{\sigma}_z + \hbar\eta\omega_c\,(\hat{a}\hat{\sigma}_+^\dagger + \text{h.c.}),$$

violates (4) providing coupling-independent values $|\mathcal{Q}_{1,0}|^2 = |\mathcal{P}_{1,0}|^2$ [the horizontal line in Figure 1(a)].

These findings show that, using the wrong quadrature ($\mathcal{Q}$ instead of $\mathcal{P}$) for the calculation of the photodetection rate for systems in the USC regime can result into significantly wrong results. This is a direct consequence of (3).

In order to understand how the sum rule in (7) applies to the quantum Rabi model, we calculate partial sums with an increasing number of states. Specifically, we calculate

$$\sum_{j=1}^{N} \mathcal{F}_{0j}, \quad \text{where} \quad \mathcal{F}_{0j} = 2(\omega_{j,0}/\omega_c)|\mathcal{Q}_{0,j}|^2.$$

Here and in the following, the eigenstates of the total Hamiltonian, obtained for a given coupling strength η, are labeled so that $i > j$ for $\omega_i > \omega_j$. Differently from the JC model, the quantum Rabi model does not conserve the excitation number. Therefore, expectation values like $\mathcal{Q}_{0,j}$ (and hence $\mathcal{F}_{0,j}$) can be different from zero also for $j > 2$. Figure 1(b) displays such partial sums as a function of the number of levels included, obtained for different values of η. For small values ($\eta = 0.01$), only the two lowest excited levels contribute to the sum with approximately equal weights, in good agreement with the JC model. For $\eta = 0.2$, still only two transitions contribute to the sum rule; however, the second transition provides a larger contribution to the sum. For $\eta = 0.5$, the contribution of the lowest energy transition becomes smaller, while $\mathcal{F}_{02} = 0$ owing to the parity selection rule. Note that, at $\eta = \eta_{cr} \simeq 0.44$, there is a crossing between the levels 2 and 3 [see inset in Figure 1(b)] so that, for $\eta > \eta_{cr}$, state $|2\rangle$ has the same parity of state $|0\rangle$. It is sufficient to include $\mathcal{F}_{03}$ to approximately satisfy the sum rule. For $\eta = 1$, $\mathcal{F}_{0,1}$ is very small and $\mathcal{F}_{0,2} = 0$. In this case, the sum rule is satisfied mainly with the contributions $\mathcal{F}_{0,j}$ with $3 \le j \le 6$. Finally, for very high values of the normalized coupling strength ($\eta = 1.8$), only one contribution ($\mathcal{F}_{0,3}$) becomes relevant. This effect is due to the light–matter decoupling [71] which occurs at very high values of η, where the system ground state $|0\rangle$ is well approximated by $|g, 0\rangle$ [the first entry in the ket labels the photon number, the second labels the qubit state: ground (g) or excited (e)], then $|1\rangle \simeq |e, 0\rangle$, $|2\rangle \simeq |e, 1\rangle$, $|3\rangle \simeq |g, 1\rangle$, and so on: the higher energy levels are of the kind $\simeq |g\,(e), n > 1\rangle$. This explains why for $\eta = 1.8$, the only significant contribution to the sum is $\mathcal{F}_{0,3}$. These behaviors of the partial sums and of the terms

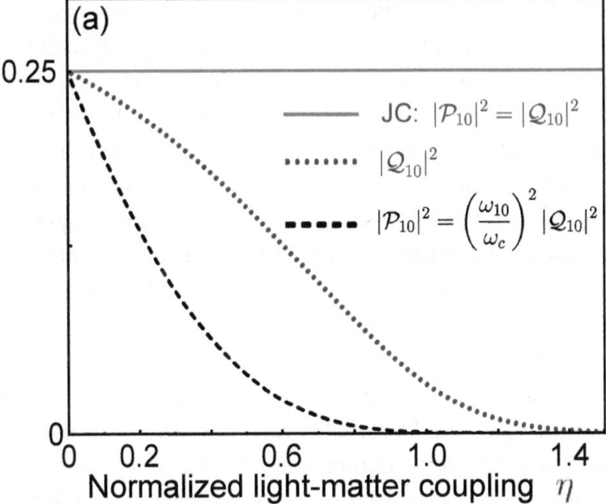

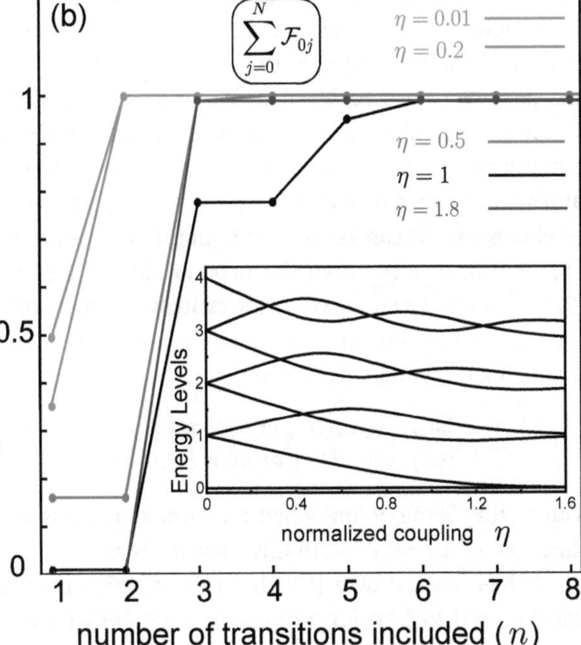

Figure 1: (a) $\mathcal{P} - \mathcal{Q}$ relation: calculation of $|\mathcal{P}_{1,0}|^2$ (proportional to the photodetection rate for cavity photons) (black dashed) and of $|\mathcal{Q}_{1,0}|^2$ (dotted blue) versus the normalized coupling η. (b) Thomas–Reiche–Kuhn (TRK) sum rule for interacting fields: partial sums $\sum_{j=0}^{N} \mathcal{F}_{0j}$ as function of the number N of levels included for different normalized coupling rates η. Inset: energy spectrum for the first energy levels $\omega_{k,0}$ versus the normalized coupling strength.

$\mathcal{F}_{i,j}$ are closely connected to accessible experimental features, as explicitly shown in the example below.

3.2 Nonlinear electromagnetic resonator

As a further test, we analyze a single-mode nonlinear optical system described by the following effective Hamiltonian:

$$\hat{H} = \hbar\omega_c \hat{a}^\dagger \hat{a} + \eta\hbar\omega_c \left(\hat{a} + \hat{a}^\dagger\right)^3 + \frac{\eta}{10}\hbar\omega_c \left(\hat{a} + \hat{a}^\dagger\right)^4. \qquad (9)$$

Here $\hat{H}_F = \hbar\omega_c \hat{a}^\dagger \hat{a}$, while the nonlinear terms are assumed to arise from the dispersive interaction with some material system [72]. Note that the nonlinear terms in (9) commute with the field coordinate $\hat{\mathcal{Q}} = (\hat{a} + \hat{a}^\dagger)/\sqrt{2}$; hence, Eqs. (4) and (7) hold. In contrast, the presence of a standard self-Kerr term $\propto \hat{a}^{\dagger 2}\hat{a}^2$ (see, e.g., [73]) would violate them. The inset in Figure 2 shows the anharmonic energy spectrum $\omega_{k,0}$ as a function of η. Figure (2) displays the partial sums $\sum_{j=1}^{N} \mathcal{F}_{0j}$ as versus the number of included levels, calculated for different values of η. Increasing the anharmonicity coefficient η, the number of contributions in the sum increases at the expense of the contribution $\mathcal{F}_{01}$ of the lowest energy transition. This behavior is closely connected with accessible experimental features which can be observed, e.g., in linear transmission spectra. For a two-port (equally coupled to the external modes) nonlinear resonator, the transmission spectrum (see Appendix A for supporting content) can be written as follows:

$$T(\omega) = \omega^2 \left| \sum_k \frac{\Gamma_{k,0}/\omega_{k,0}}{\omega_{k,0} - \omega - i\Gamma_k} \right|^2, \qquad (10)$$

where the radiative decay rates are

$$\Gamma_{k,j} = 2\pi g^2(\omega_{k,j}) |\mathcal{Q}_{k,j}|^2, \quad \Gamma_k = \sum_{j<k} \Gamma_{k,j},$$

and we assumed an ohmic coupling with the external modes ($g^2(\omega) \propto \omega$). When the anharmonicity is switched off ($\eta = 0$), $\Gamma_{k,0} \propto \mathcal{F}_{0k} = 0$ for $k \neq 1$, and the transmission spectrum presents a single peak at $\omega = \omega_c$ [dashed curve in Figure 2(b)]. When $\eta \neq 0$, $\Gamma_{k,0} \propto \mathcal{F}_{0k} \neq 0$, and the transmission spectrum in Figure 2 evolves accordingly (the blue continuous curve shows the spectrum calculated for $\eta = 0.12$). By integrating the individual spectral lines in (10), we obtain for each line a contribution $\simeq \pi\Gamma_{k,0}^2/\Gamma_k$, which is approximately proportional to $\mathcal{F}_{0k}$ in the sum (notice that $\Gamma_k \sim k\Gamma_1$). The inset in Figure 2 shows the integrated lines for two values of η.

3.3 Frequency conversion in ultrastrong cavity QED

The relations in (4) and (7) are very general. So far, we applied them to single-mode fields; however, they are also valid in the presence of (even interacting) multimode fields (see, e.g., [74, 75]). Here, we analyze the TRK sum rule for interacting photons in a three-component system constituted by two single-mode resonators ultrastrongly coupled to a single superconducting flux qubit. This coupling can

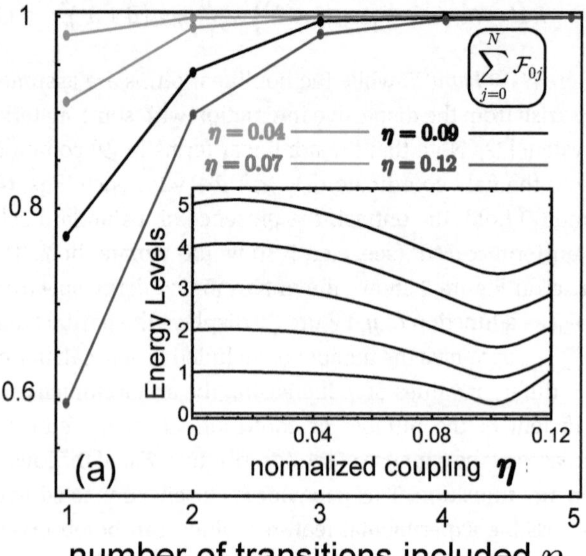

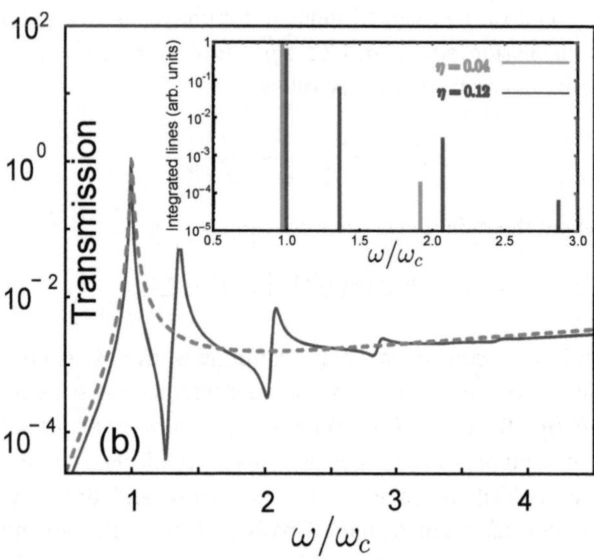

Figure 2: (a) Thomas–Reiche–Kuhn (TRK) sum rule for a single-mode nonlinear system: partial sums $\sum_{j=0}^{N} \mathcal{F}_{0j}$ versus the number (N) of levels included for different normalized coupling strengths η. Inset: anharmonic energy spectrum $\omega_{k,0}$ versus η. (b) Transmission spectrum $T(\omega)$ for a two-port nonlinear resonator for $\eta = 0.12$. The inset shows the integrated lines for two values of η.

induce an effective interaction between the fields of the two resonators. Using suitable parameters for the three components, the system provides a method for frequency conversion of photons which is both versatile and deterministic. It has been shown that it can be used to realize both single and multiphoton frequency conversion processes [52]. The system Hamiltonian is given as follows:

$$\hat{H} = \hbar\omega_a \hat{a}^\dagger \hat{a} + \hbar\omega_b \hat{b}^\dagger \hat{b} + \frac{\hbar\omega_0}{2}\hat{\sigma}_z + \hbar\left[g_a\left(\hat{a} + \hat{a}^\dagger\right) + g_b\left(\hat{b} + \hat{b}^\dagger\right)\right]\left[\cos(\theta)\hat{\sigma}_x + \sin(\theta)\hat{\sigma}_z\right], \tag{11}$$

where $(\hat{a}, \omega_a, g_a)$ and $(\hat{b}, \omega_b, g_b)$ describe the photon operator, the frequency mode, and the coupling with the qubit for the two resonators. The angle θ encodes the qubit flux offset which determines parity symmetry breaking. A zero flux offset implies $\theta = 0$. Figure 3(a) displays the lowest normalized energy levels $(\omega - \omega_g)/\overline{\omega}_0$ (we indicated with $\hbar\omega_g$ the ground state energy) versus the qubit frequency $\omega_0/\overline{\omega}_0$ obtained diagonalizing numerically the Hamiltonian in (11). We used the parameters $\omega_a = 3\overline{\omega}_0$, $\omega_b = 2\overline{\omega}_0$, $\theta = \pi/6$, $g_a = g_b = 0.2\overline{\omega}_0$, where $\overline{\omega}_0$ is a reference point for the qubit frequency. Notice that the two resonators are set in order that their resonance frequencies satisfy the relationship $\omega_a = \omega_b + \overline{\omega}_0$. The first excited level is a line with slope $\simeq 1$, corresponding to the approximate eigenstate $|\psi_1\rangle \simeq |0, 0, e\rangle$, where the first two entries in the ket indicate the number of photons in resonator a and b, respectively, while the third entry indicates the qubit state. The second excited level is a horizontal line corresponding to the eigenstate $|\psi_2\rangle \simeq |0, 1, g\rangle$; the next two lines on the left of the small rectangle in Figure 3(a) (for values of $\omega_0/\overline{\omega}_0$ before the apparent crossing) correspond to the states $|\psi_3\rangle \simeq |0, 1, e\rangle$ and $|\psi_4\rangle \simeq |1, 0, g\rangle$. The apparent crossing in the rectangle is actually an avoided level crossing, as can be inferred from the enlarged view in Figure 3(b). It arises from the hybridization of the states $|0, 1, e\rangle$ and $|1, 0, g\rangle$ induced by the counter-rotating terms in the system Hamiltonian. The resulting eigenstates can be approximately written as follows:

$$\begin{aligned} |\psi_3\rangle &\simeq \cos\theta |0, 1, e\rangle - \sin\theta |1, 0, g\rangle \\ |\psi_4\rangle &\simeq \sin\theta |0, 1, e\rangle + \cos\theta |1, 0, g\rangle. \end{aligned} \tag{12}$$

The mixing is maximum when the level splitting is minimum (at $\omega_0/\overline{\omega}_0 \simeq 1.056$). In this case, $\theta = \pi/4$.

It has been shown [52] that this effective coupling can be used to transfer a quantum state constituted by an arbitrary superposition of zero and one photon in one resonator (e.g., a) to a quantum state corresponding to the same superposition in the resonator at frequency ω_b.

This system represents an interesting example of two interacting optical modes (with the interaction mediated by a qubit). In order to understand how the sum rule in (7) applies to such a system, we investigate its convergence, calculating partial sum rules for the two modes. Figure 4 shows $\sum_{j=0}^{N} \mathcal{F}_{0j}^a$ (a) and $\sum_{j=1}^{N} \mathcal{F}_{1j}^b$ (b) for different values of N. The black line describes the zero detuning case, while the dashed blue line, the case $\delta = (\omega_0 - \overline{\omega}_0)/\overline{\omega}_0 = -6 \times 10^{-3}$. The results in Figure 4(a) can be understood observing that

$$\mathcal{F}_{0j}^a \propto \left|\langle 0|\hat{a} + \hat{a}^\dagger|j\rangle\right|^2.$$

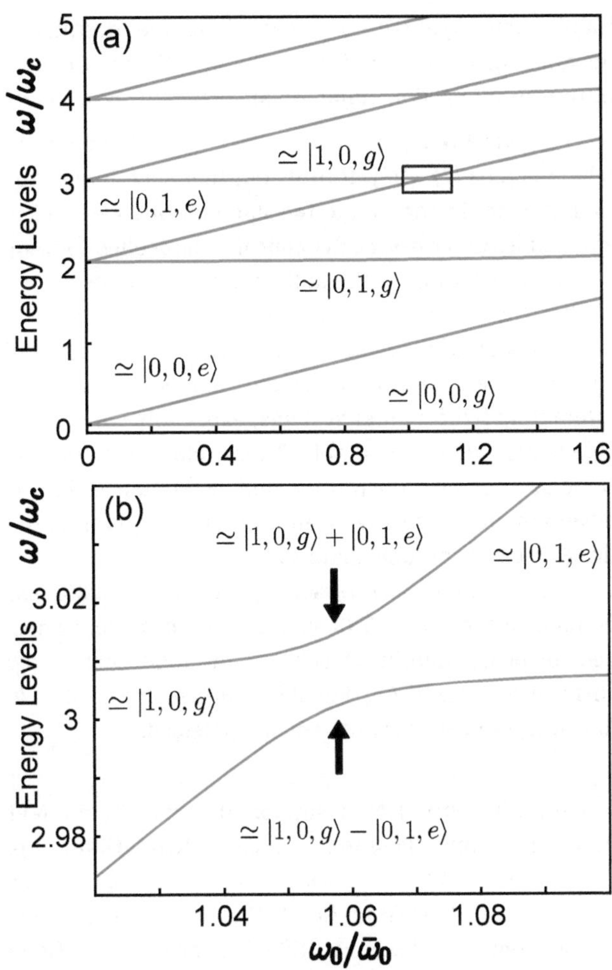

Figure 3: Energy spectum obtained from the numerical diagonalization of (11).
(a) Lowest normalized energy levels versus the qubit frequency. (b) Enlarged view of the spectrum inside the rectangle in (a) showing the presence of an avoided level crossing. Parameters are given in the text.

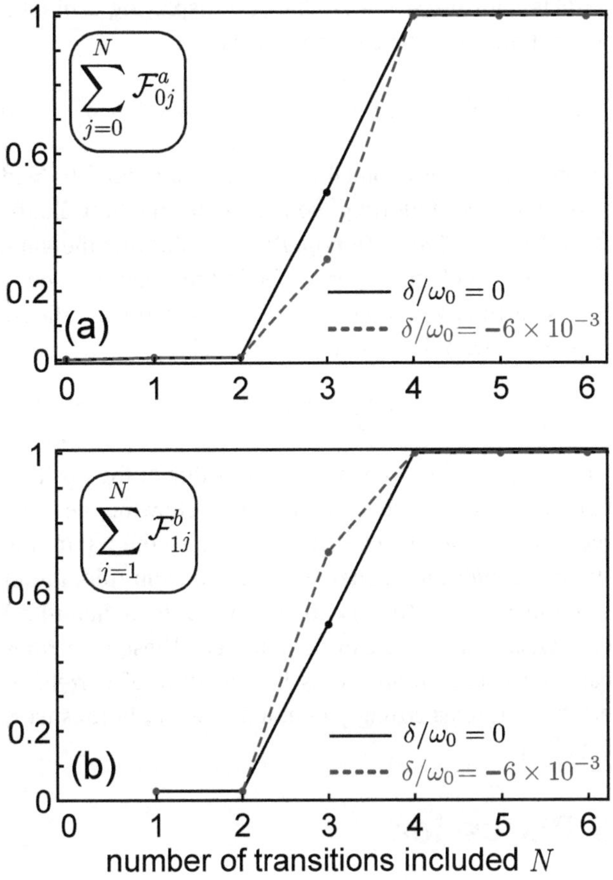

Figure 4: Thomas–Reiche–Kuhn (TRK) sum rule for interacting photons in the three-component system described by the Hamiltonian in (11).
(a) Partial sum rules $\sum_{j=1}^{N} \mathcal{F}^{a}_{0j}$ relative to the first resonator and (b) $\sum_{j=1}^{N} \mathcal{F}^{b}_{1j}$ relative to the second resonator, both for different values of levels N. The black segmented line describes the zero detuning case $\delta = 0$, while the dashed blue segmented lines refer to the case $\delta = (\omega_0 - \overline{\omega}_0)/\overline{\omega}_0 = -6 \times 10^{-3}$. Parameters are given in the text.

since

$$|0\rangle \simeq |0,0,g\rangle, |1\rangle \simeq |0,0,e\rangle, |2\rangle \simeq |0,1,g\rangle, |3\rangle, \text{and } |4\rangle$$

are provided in (C), it is easy to obtain

$$\mathcal{F}^{a}_{01} \simeq \mathcal{F}^{a}_{02} \simeq 0, \mathcal{F}^{a}_{03} \propto \sin^2\theta, \text{and } \mathcal{F}^{a}_{04} \propto \cos^2\theta,$$

in agreement with the results in Figure 4(a). Notice that for $\delta = 0$, it results in $\theta = \pi/4$, and hence, $\mathcal{F}^{a}_{03} \simeq \mathcal{F}^{a}_{04}$. A similar analysis can be carried out for the results in Figure 4(b).

4 TRK sum rule for atoms interacting with photons

The standard atomic TRK sum rule [67] considers atomic energy eigenstates, calculated in the absence of interaction with the transverse electromagnetic field. A recent interesting example of descriptions including the electron–electron interaction can be found in the study by Andolina et al. [18].

Following the same reasoning which led us to (7), a generalized atomic TRK sum rule for atoms strongly interacting with the electromagnetic field [analogous to (7)] can be easily obtained, starting from the dipole gauge. In this gauge (see, e.g., [68]), the light–matter interaction term does not depend on the particle momentum, and the same steps used to obtain (7) can thus be followed. The resulting atomic generalized TRK sum rule formally coincides with the standard one, with the only difference that all the expectation values are calculated using the eigenstates of the *total* light–matter system. For example, we consider a system described by a single effective particle with mass m and charge q displaying a dipolar interaction with a single mode resonator:

$$\hat{H}_D = \frac{1}{2m}\hat{p}^2 + V(x) + \frac{q^2\omega_c A_0^2}{\hbar}x^2 + iq\omega_c A_0 x\left(\hat{a}^\dagger - \hat{a}\right), \quad (13)$$

where A_0 is the zero-point fluctuation amplitude of the field potential. The following commutation relation holds: $[x, \hat{H}_D] = [x, \hat{p}^2/2m] = i(\hbar/m)\hat{p}$. From it, following the same steps used to obtain (7) or to obtain the standard atomic TRK sum rule, we obtain the TRK sum rule for a dipole interacting with the electromagnetic field:

$$2m\sum_k \omega_{k,j}\left|x_{k,j}\right|^2 = 1, \quad (14)$$

where $x_{k,j} \equiv \langle i|x|j\rangle$ is the expectation value of the position operator between two dressed states. Following the same reasoning, it can also be shown that also the f-sum rule [16] (the longitudinal analog of the TRK sum rule) for an electron system strongly interacting with a quantized electromagnetic field can be obtained. These sum rules can find useful applications in the study of correlated electron systems strongly interacting with photons (see, e.g., [76]).

5 Discussion

The TRK sum rule for interacting photons proposed here can be useful for investigating general quantum nonlinear optical effects and many-body physics in photonic systems (see, e.g., [24–28]), like the corresponding sum rules for interacting electron systems, which played a fundamental role for understanding the many-body physics of interacting electron systems [16, 17, 20].

We provided a few examples showing how the light–matter interaction can change significantly the number of excited photonic states exhausting the sum rule. Using the sum rule, one can prove without explicit calculations that other excited states have negligible oscillator strength.

The relations in (4) and (7) are very general. They are also valid in systems including several dipoles (see, e.g., [77, 78]) and modes (see, e.g., [75]). These relations provide a very useful check on the consistency of approximate models in quantum optics. Approximate Hamiltonians and effective models can violate one of them. Such a violation indicates that the model may miss some relevant physics [16]. For example, we have shown that the JC model, a widespread description for the dipolar coupling between a two-level atom and a quantized electromagnetic field, violates the relation (4). An additional example of a model violating this relation is provided by the well-known and widely employed cavity optomechanical interaction Hamiltonian $\hbar g\hat{a}^\dagger\hat{a}(\hat{b} + \hat{b}^\dagger)$ (here $\hat{b}$ is the destruction operator for the mechanical oscillator) [79]. On the contrary, the interaction Hamiltonian obtained by a microscopic model [63] $\hbar g(\hat{a}^\dagger + \hat{a})^2(\hat{b} + \hat{b}^\dagger)$ satisfies both of these relations [Eqs (7), (14)]. It turns out that such interaction Hamiltonian, in addition to the standard optomechanical effects, also describes the dynamical Casimir effect [58, 64].

An interesting feature of the relations proposed here is that they hold in the presence of light–matter interactions of arbitrary strength. Moreover, the obtained sum rule can be useful for the analysis of strongly interacting light–matter systems, especially when exact eigenstates are not available. These relations in (4) and (7) can provide constraints and a guidance in the development of effective Hamiltonians in quantum optics and cavity optomechanics.

Following the same reasoning leading to (7), we also proposed a generalized TRK sum rule for the matter component involving transitions between the *total* light–matter energy eigenstates [(14)], describing particle conservation in the presence of arbitrary light–matter interactions.

Acknowledgments: F.N. is supported in part by the NTT Research, Army Research Office (ARO) (Grant No. W911NF-18-1-0358), Japan Science and Technology Agency (JST) (via the CREST Grant No. JPMJCR1676), Japan Society for the Promotion of Science (JSPS) (via the KAKENHI Grant No. JP20H00134, and the grant JSPS-RFBR Grant No. JPJSBP120194828) and the Grant No. FQXi-IAF19-06 from the Foundational Questions Institute Fund (FQXi), a donor advised fund of the Silicon Valley Community Foundation. S.S. acknowledges the Army Research Office (ARO) (Grant No. W911NF1910065).

Author contribution: All the authors have accepted responsibility for the entire content of this submitted manuscript and approved submission.

Research funding: F.N. is supported in part by the NTT Research, Army Research Office (ARO) (Grant No. W911NF-18-1-0358), Japan Science and Technology Agency (JST) (via the CREST Grant No. JPMJCR1676), Japan Society for the Promotion of Science (JSPS) (via the KAKENHI Grant No. JP20H00134, and the grant JSPS-RFBR Grant No. JPJSBP120194828), the Asian Office of Aerospace Research and Development (AOARD), and the Grant No. FQXi-IAF19-06 from the Foundational Questions Institute Fund (FQXi), a donor advised fund of the Silicon Valley Community Foundation. S.S. acknowledges the Army Research Office (ARO) (Grant No. W911NF1910065).

Conflict of interest statement: The authors declare no conflicts of interest regarding this article.

Appendix: A Linear response theory and transmission of a nonlinear optical system

This section provides a derivation of the transmission coefficient of a nonlinear optical system based on the dressed master equation approach [80, 81].

The dressed master equation in the Schrödinger picture can be written as follows [80, 81]:

$$\dot{\hat{\rho}}(t) = -i[\hat{H}_S, \hat{\rho}(t)] + \mathcal{L}\hat{\rho}(t), \tag{A.1}$$

where $\rho(t)$ is the density matrix operator for the nonlinear optical system,

$$\hat{H}_S = \sum_k \omega_k |k\rangle\langle k|, \tag{A.2}$$

is the system Hamiltonian expressed in the dressed basis, constituted by the energy eigenstates of the nonlinear system. Dissipation is described by the Lindbladian superoperator defined by

$$\mathcal{L}\hat{\rho}(t) = \sum_i \sum_{j,k<j} \{\Gamma_{jk}^{(i)} n(\omega_{jk}, T_i)\mathcal{D}[|j\rangle\langle k|]\hat{\rho}(t)$$

$$+\Gamma_{jk}^{(i)}[1 + n(\omega_{jk}, T_i)]\mathcal{D}[|k\rangle\langle j|]\hat{\rho}(t)\}, \tag{A.3}$$

This equation includes the thermal populations

$$n(\Delta_{jk}, T_i) = [\exp\{\omega_{jk}/k_B T_i\} - 1]^{-1}, \tag{A.4}$$

and the damping rates

$$\Gamma_{jk}^{(i)} = 2\pi g_i^2(\omega_{jk})|X_{jk}|^2. \tag{A.5}$$

Here, $i = \{L, R\}$ indicates the input–output ports, $g(\omega)$ is the system reservoir coupling strength, $\hat{X}$ is the system operator interacting with the external modes, and

$$\mathcal{D}[\hat{O}]\hat{\rho} = \frac{1}{2}(2\hat{O}\hat{\rho}\hat{O}^\dagger - \hat{\rho}\hat{O}^\dagger\hat{O} - \hat{O}^\dagger\hat{O}\hat{\rho}). \tag{A.6}$$

At $T = 0$, being $n(\Delta_{jk}, T_i) = 0$, we obtain the following:

$$\mathcal{L}\hat{\rho} \underset{T=0}{=\!=} \mathcal{L}_0\hat{\rho} = \sum_i \sum_{j,k<j} \{\Gamma_i^{jk}\mathcal{D}[|k\rangle\langle j|]\hat{\rho}\}. \tag{A.7}$$

We also consider a coherent drive entering from the left port, described by the following interaction Hamiltonian:

$$\hat{H}_d(t) = i\hat{X}\int d\omega g_L(\omega)[e^{-i\omega t}\beta_L(\omega) - e^{i\omega t}\beta_L^*(\omega)], \tag{A.8}$$

where $\hat{X}$ is the system operator interacting with the external modes, and

$$\beta_L(\omega) = \langle\hat{b}_L(\omega)\rangle$$

is a c-number corresponding to the mean value of the external (left) field operators, assumed to be in a coherent state. We will also assume

$$\hat{X} = \hat{Q} = (\hat{a} + \hat{a}^\dagger)/\sqrt{2},$$

where $\hat{a}$ is the photon destruction operator for a single-mode electromagnetic resonator. The master equation(A.1) becomes

$$\dot{\hat{\rho}}(t) = -i[\hat{H}_S + \hat{H}_d(t), \hat{\rho}(t)] + \mathcal{L}_0\hat{\rho}(t). \tag{A.9}$$

We assume that the light field from the left port is coherent with driving frequency ω:

$$\langle\hat{b}_\omega\rangle = \beta_L(\omega)\exp[-i\omega t].$$

Retaining only the terms depending linearly from the input field and using Eqs. (A1), (A7), (A8), assuming

$$\rho_{n0}(t) = \rho_{n0}\exp[-i\omega t]$$

(i.e., oscillating resonantly with the driving field), and using the rotating wave approximation, we obtain (to first order in the field)

$$\rho_{n0}^{(1)} = \frac{ig_L(\omega)\beta_L(\omega)X_{n0}}{(\omega - \omega_{n0}) + i\sum_i\sum_{k<n}\Gamma_i^{n,k}}, \tag{A.10}$$

where, being $T = 0$, only the ground state is populated in the absence of interaction ($\rho_{00}^{(0)} = 1$). In order to calculate the transmitted signal that can be experimentally detected, we consider a system constituted by an *LC*-oscillator coupled to a transmission line and use the input–output relations [71] for the positive frequency component of the output (input) vector potential operator defined as follows:

$$\hat{\Phi}^+_{\text{out(in)}}(t) = \Lambda\int_0^\infty \frac{d\omega}{\sqrt{\omega}}\,\hat{b}_\omega^{\text{out(in)}}(t), \tag{A.11}$$

where, for the sake of simplicity, we disregarded the spatial dependence, and $\Lambda = \sqrt{\hbar Z_0/4\pi}$, with Z_0 the impedance of the in-out transmission line(s). In addition, we consider two distinct ports for the input (L) and the output (R) [for simplicity we assume $g_L(\omega) = g_R(\omega) = g(\omega)$], and we have for the output voltage operator [71] $\hat{V}_{\text{out}}^{(R)+}(t) = \dot{\hat{\Phi}}_{\text{out}}^{(R)+}(t)$:

$$\hat{V}_{\text{out}}^{(R)+}(t) = -2\pi\Lambda\sum_j \frac{g(\omega_{j0})}{\sqrt{\omega_{j0}}}X_{0j}\dot{\hat{P}}_{0j}(t), \tag{A.12}$$

which can be expressed as follows:

$$\hat{V}_{\text{out}}^{(R)+}(t) = -K\hat{V}^+(t), \tag{A.13}$$

where

$$\hat{V}^+ = \Phi_{\text{zpf}}\sum_j X_{0j}\dot{\hat{P}}_{0j}(t). \tag{A.14}$$

Assuming $g(\omega) = G\sqrt{\omega}$, the constants K and Φ_{zpf} satisfy the following relation:

$$\frac{K\Phi_{zpf}}{\Lambda} = 2\pi G. \tag{A.15}$$

Using (A.11), we have for the mean value of the input sent through the port (L)

$$\langle \widehat{V}_{in}^{(L)+}(t)\rangle = \langle \dot{\widehat{\phi}}_{in}^{(L)+}(t)\rangle = -i\Lambda\sqrt{\omega}\,\beta_L(\omega), \tag{A.16}$$

where we assumed a coherent drive input at frequency ω:

$$\langle \widehat{b}_{\omega'}^{L}(t)\rangle = \beta_L(\omega)\delta(\omega' - \omega).$$

Considering the linear response only, the projection operator oscillates at the frequency ω of the drive,

$$\dot{\widehat{P}}_{0j}(t) = -i\omega\widehat{P}_{0j}(t),$$

using Eqs. (A13) and (A14), the mean value for the output is given as follows:

$$\langle \widehat{V}_{out}^{(R)+}(t)\rangle = iK\Phi_{zpf}\omega\sum_j X_{0j}\rho_{j0}(t), \tag{A.17}$$

where $\widehat{\rho}$ is the density matrix and we used the following relation:

$$\langle \widehat{P}_{0j}(t)\rangle = \rho_{j0}(t).$$

Using Eqs. (A15)–(A17), we can calculate the transmission coefficient $T(\omega)$ due to the signal detected from the port (R) when a driving field is sent through the port (L) as follows:

$$T(\omega) = \left|\frac{\langle \widehat{V}_{out}^{(R)+}(t)\rangle}{\langle \widehat{V}_{in}^{(L)+}(t)\rangle}\right|^2 = \omega^2\left|\sum_j \frac{\Gamma_{j0}/\omega_{j0}}{(\omega - \omega_{j0}) + i\sum_i\sum_{k<n}\Gamma_j^{nk}}\right|^2, \tag{A.18}$$

where $\Gamma_{j0} = 2\pi|g(\omega_{j0})|^2|X_{j0}|^2$. Recalling that we assumed $\widehat{X} = \widehat{Q}$, (A.18) corresponds to (10).

References

[1] G. Orlandini and M. Traini, "Sum rules for electron-nucleus scattering," *Rep. Prog. Phys.*, vol. 54, p. 257, 1991.

[2] W. Thomas, "Über die Zahl der Dispersionselektronen, die einem stationären Zustande zugeordnet sind. (Vorläufige Mitteilung)," *Naturwissenschaften*, vol. 13, p. 627, 1925.

[3] W. Kuhn, "ber die Gesamtstrke der von einem Zustande ausgehenden Absorptionslinien," *Z. Phys.*, vol. 33, p. 408, 1925.

[4] F. Reiche and W. Thomas, "ber die Zahl der Dispersionselektronen, die einem stationren Zustand zugeordnet," *Z. Phys.*, vol. 34, p. 510, 1925.

[5] H. Bethe, "Zur Theorie des Durchgangs schneller Korpuskularstrahlen durch Materie," *Ann. Phys.*, vol. 397, p. 325, 1930.

[6] S. Wang, "Generalization of the Thomas-Reiche-Kuhn and the Bethe sum rules," *Phys. Rev. A*, vol. 60, p. 262, 1999.

[7] S. M. Barnett and R. Loudon, "Sum rule for modified spontaneous emission rates," *Phys. Rev. Lett.*, vol. 77, p. 2444, 1996.

[8] E. Merzbacher, *Quantum Mechanics*, 2nd ed., New York, Wiley, 1970.

[9] F. Bassani and S. Scandolo, "Dispersion relations and sum rules in nonlinear optics," *Phys. Rev. B*, vol. 44, p. 8446, 1991.

[10] S. Scandolo and F. Bassani, "Nonlinear sum rules: the three-level and the anharmonic-oscillator models," *Phys. Rev. B*, vol. 45, p. 13257, 1992.

[11] S. Scandolo and F. Bassani, "Kramers-Kronig relations and sum rules for the second-harmonic susceptibility," *Phys. Rev. B*, vol. 51, p. 6925, 1995.

[12] C. Baxter, "Center-of-mass motion of anN-particle atom or ion and the Thomas-Reiche-Kuhn sum rule," *Phys. Rev. A*, vol. 50, p. 875, 1994.

[13] J. S. Levinger, M. L. Rustgi, and K. Okamoto, "Relativistic corrections to the dipole sum rule," *Phys. Rev.*, vol. 106, p. 1191, 1957.

[14] J. L. Friar and S. Fallieros, "Relativistic and retardation effects in the Thomas-Reiche-Kuhn sum rule for a bound particle," *Phys. Rev. C*, vol. 11, p. 274, 1975.

[15] M. Nielsen, F. S. Navarra, and S. H. Lee, "New charmonium states in QCD sum rules: A concise review," *Phys. Rep.*, vol. 497, p. 41, 2010.

[16] D. Pines and P. Nozieres, *The Theory of Quantum Liquids*, New York, W. A. Benjamin, 1966.

[17] G. Giuliani and G. Vignale, *Quantum Theory of the Electron Liquid*, Cambridge, Cambridge University Press, 2005.

[18] G. M. Andolina, F. M. D. Pellegrino, V. Giovannetti, A. H. MacDonald, and M. Polini, "Cavity quantum electrodynamics of strongly correlated electron systems: A no-go theorem for photon condensation." *Phys. Rev. B*, vol. 100, p. 121109, 2019.

[19] L. Garziano, A. Settineri, O. Di Stefano, S. Savasta, and F. Nori, "Gauge invariance of the Dicke and Hopfield models," *Phys. Rev. A*. vol. 102, p. 023718, 2020.

[20] P. W. Anderson, "Random-phase approximation in the theory of superconductivity," *Phys. Rev.*, vol. 112, p. 1900, 1958.

[21] S. M. Barnett and R. Loudon, "Optical Thomas-Reiche-Kuhn sum rules," *Phys. Rev. Lett.*, vol. 108, p. 013601, 2012.

[22] A. F. Kockum, A. Miranowicz, S. D. Liberato, S. Savasta, and F. Nori, "Ultrastrong coupling between light and matter," *Nat. Rev. Phys.*, vol. 1, p. 19, 2019.

[23] P. Forn-Díaz, L. Lamata, E. Rico, J. Kono, and E. Solano, "Ultrastrong coupling regimes of light-matter interaction," *Rev. Mod. Phys.*, vol. 91, p. 025005, 2019.

[24] T. Peyronel, O. Firstenberg, L. Qi-Yu, et al., "Quantum nonlinear optics with single photons enabled by strongly interacting atoms," *Nature*, vol. 488, p. 57, 2012.

[25] D. E. Chang, V. Vuletić, and M. D. Lukin, "Quantum nonlinear optics – photon by photon," *Nat. Photonics*, vol. 8, p. 685, 2014.

[26] T. Guerreiro, A. Martin, B. Sanguinetti, et al., "Nonlinear interaction between single photons," *Phys. Rev. Lett.*, vol. 113, p. 173601, 2014.

[27] A. F. Kockum, A. Miranowicz, V. Macrì, S. Savasta, and F. Nori, "Deterministic quantum nonlinear optics with single atoms and virtual photons," *Phys. Rev. A*, vol. 95, p. 063849, 2017a.

[28] I. Carusotto and C. Ciuti, "Quantum fluids of light," *Rev. Mod. Phys.*, vol. 85, p. 299, 2013.
[29] S. De Liberato, "Light-matter decoupling in the deep strong coupling regime: The breakdown of the Purcell effect," *Phys. Rev. Lett.*, vol. 112, p. 016401, 2014.
[30] J. J. García-Ripoll, B. Peropadre, and S. De Liberato, "Light-matter decoupling and A2 term detection in superconducting circuits," *Sci. Rep.*, vol. 5, p. 16055, 2015.
[31] A. Bayer, M. Pozimski, S. Schambeck, et al., "Terahertz light-matter interaction beyond unity coupling strength," *Nano Lett.*, vol. 17, p. 6340, 2017.
[32] F. Yoshihara, T. Fuse, S. Ashhab, K. Kakuyanagi, S. Saito, and K. Semba, "Characteristic spectra of circuit quantum electrodynamics systems from the ultrastrong- to the deep-strong-coupling regime," *Phys. Rev. A*, vol. 95, p. 053824, 2017.
[33] A. Ridolfo, M. Leib, S. Savasta, and M. J. Hartmann, "Photon blockade in the ultrastrong coupling regime," *Phys. Rev. Lett.*, vol. 109, p. 193602, 2012.
[34] T. Niemczyk, F. Deppe, H. Huebl, et al., "Circuit quantum electrodynamics in the ultrastrong-coupling regime," *Nat. Phys.*, vol. 6, p. 772, 2010.
[35] J. Casanova, G. Romero, I. Lizuain, J. J. García-Ripoll, and E. Solano, "Deep strong coupling regime of the Jaynes-Cummings model," *Phys. Rev. Lett.*, vol. 105, p. 263603, 2010.
[36] L. Garziano, V. Macrì, R. Stassi, O. Di Stefano, F. Nori, and S. Savasta, "One photon can simultaneously excite two or more atoms," *Phys. Rev. Lett.*, vol. 117, p. 043601, 2016.
[37] R. Stassi, V. Macrì, A. F. Kockum, et al., "Quantum nonlinear optics without photons," *Phys. Rev. A*, vol. 96, p. 023818, 2017.
[38] S. De Liberato, C. Ciuti, and I. Carusotto, "Quantum vacuum radiation spectra from a semiconductor microcavity with a time-modulated vacuum Rabi frequency," *Phys. Rev. Lett.*, vol. 98, p. 103602, 2007.
[39] S. Ashhab and F. Nori, "Qubit-oscillator systems in the ultrastrong-coupling regime and their potential for preparing nonclassical states," *Phys. Rev. A*, vol. 81, p. 042311, 2010.
[40] I. Carusotto, S. De Liberato, D. Gerace, and C. Ciuti "Back-reaction effects of quantum vacuum in cavity quantum electrodynamics," *Phys. Rev. A*, vol. 85, p. 023805, 2012.
[41] A. Auer and G. Burkard, "Entangled photons from the polariton vacuum in a switchable optical cavity," *Phys. Rev. B*, vol. 85, p. 235140, 2012.
[42] L. Garziano, A. Ridolfo, R. Stassi, O. Di Stefano, and S. Savasta, "Switching on and off of ultrastrong light-matter interaction: Photon statistics of quantum vacuum radiation," *Phys. Rev. A*, vol. 88, p. 063829, 2013.
[43] R. Stassi, A. Ridolfo, O. Di Stefano, M. J. Hartmann, and S. Savasta, "Spontaneous conversion from virtual to real photons in the ultrastrong-coupling regime," *Phys. Rev. Lett.*, vol. 110, p. 243601, 2013.
[44] L. Garziano, R. Stassi, A. Ridolfo, O. Di Stefano, and S. Savasta, "Vacuum-induced symmetry breaking in a superconducting quantum circuit," *Phys. Rev. A*, vol. 90, p. 043817, 2014.
[45] J.-F. Huang and C. K. Law, "Photon emission via vacuum-dressed intermediate states under ultrastrong coupling," *Phys. Rev. A*, vol. 89, p. 033827, 2014.
[46] G. Benenti, A. D'Arrigo, S. Siccardi, and G. Strini, "Dynamical Casimir effect in quantum-information processing," *Phys. Rev. A*, vol. 90, p. 052313, 2014.
[47] L. Garziano, R. Stassi, V. Macrì, A. F. Kockum, S. Savasta, and F. Nori, "Multiphoton quantum Rabi oscillations in ultrastrong cavity QED," *Phys. Rev. A*, vol. 92, p. 063830, 2015.
[48] R. Stassi, S. Savasta, L. Garziano, B. Spagnolo, and F. Nori, "Output field-quadrature measurements and squeezing in ultrastrong cavity-QED," *New J. Phys.*, vol. 18, p. 123005, 2016.
[49] T. Jaako, Z.-L. Xiang, J. J. Garcia-Ripoll, and P. Rabl, "Ultrastrong-coupling phenomena beyond the Dicke model," *Phys. Rev. A*, vol. 94, p. 033850, 2016.
[50] S. De Liberato, "Virtual photons in the ground state of a dissipative system," *Nat. Commun.*, vol. 8, p. 1465, 2017.
[51] M. Cirio, K. Debnath, N. Lambert, and F. Nori, "Amplified optomechanical transduction of virtual radiation pressure," *Phys. Rev. Lett.*, vol. 119, p. 053601, 2017.
[52] A. F. Kockum, V. Macrì, L. Garziano, S. Savasta, and F. Nori, "Frequency conversion in ultrastrong cavity QED," *Sci. Rep.*, vol. 7, p. 5313, 2017b.
[53] F. Albarrán-Arriagada, G. Alvarado Barrios, F. Cárdenas-López, G. Romero, and J. C. Retamal, "Generation of higher dimensional entangled states in quantum Rabi systems," *J. Phys. A: Math. Theor.*, vol. 50, p. 184001, 2017.
[54] O. Di Stefano, R. Stassi, L. Garziano, A. F. Kockum, S. Savasta, and F. Nori, "Feynman-diagrams approach to the quantum Rabi model for ultrastrong cavity QED: stimulated emission and reabsorption of virtual particles dressing a physical excitation," *New J. Phys.*, vol. 19, p. 053010, 2017.
[55] V. Macrì, F. Nori, and A. Kockum, "Simple preparation of Bell and Greenberger-Horne-Zeilinger states using ultrastrong-coupling circuit QED," *Phys. Rev. A*, vol. 98, p. 062327, 2018.
[56] L.-L. Zheng, X.-Y. Lü, Q. Bin, Z.-M. Zhan, S. Li, and Y. Wu, "Switchable dynamics in the deep-strong-coupling regime," *Phys. Rev. A*, vol. 98, p. 023863, 2018.
[57] S. Felicetti, D. Z. Rossatto, E. Rico, E. Solano, and P. Forn-Díaz, "Two-photon quantum Rabi model with superconducting circuits," *Phys. Rev. A*, vol. 97, p. 013851, 2018.
[58] V. Macrì, A. Ridolfo, O. Di Stefano, A. F. Kockum, F. Nori, and S. Savasta, "Nonperturbative dynamical Casimir effect in optomechanical systems: Vacuum Casimir-Rabi splittings," *Phys. Rev. X*, vol. 8, p. 011031, 2018.
[59] J. Flick, C. Schäfer, M. Ruggenthaler, H. Appel, and A. Rubio, "Ab initio optimized effective potentials for real molecules in optical cavities: photon contributions to the molecular ground state," *ACS Photonics*, vol. 5, p. 992, 2018.
[60] E. Sánchez-Burillo, L. Martín-Moreno, J. J. García-Ripoll, and D. Zueco, "Single photons by quenching the vacuum," *Phys. Rev. Lett.*, vol. 123, p. 013601, 2019.
[61] U. Mordovina, C. Bungey, H. Appel, P. J. Knowles, A. Rubio, and F. R. Manby, "Polaritonic coupled-cluster theory," *Phys. Rev. Research*, vol. 2, p. 023262, 2019.
[62] O. Di Stefano, A. F. Kockum, A. Ridolfo, S. Savasta, and F. Nori, "Photodetection probability in quantum systems with arbitrarily strong light-matter interaction," *Sci. Rep.*, vol. 8, p. 17825, 2018.
[63] C. K. Law, "Interaction between a moving mirror and radiation pressure: A Hamiltonian formulation," *Phys. Rev. A*, vol. 51, p. 2537, 1995.
[64] O. Di Stefano, A. Settineri, V. Macrì, et al., "Interaction of mechanical oscillators mediated by the exchange of virtual photon pairs," *Phys. Rev. Lett.*, vol. 122, p. 030402, 2019.

[65] P. Kirton, M. M. Roses, J. Keeling, and E. G. Dalla Torre, "Introduction to the dicke model: from equilibrium to nonequilibrium, and vice versa," *Adv. Quantum Technol.*, vol. 2, p. 1800043, 2019.

[66] J. Flick, M. Ruggenthaler, H. Appel, and A. Rubio, "Atoms and molecules in cavities, from weak to strong coupling in quantum-electrodynamics (QED) chemistry," *Proc. Natl. Acad. Sci. U.S.A.*, vol. 114, p. 3026, 2017.

[67] J. J. Sakurai, *Modern Quantum Mechanics*, Reading, MA, Addison-Wesley Publishing Company, Inc., 1994.

[68] O. Di Stefano, A. Settineri, V. Macrì, et al., "Resolution of gauge ambiguities in ultrastrong-coupling cavity quantum electrodynamics," *Nat. Phys.*, vol. 15, p. 803, 2019.

[69] D. De Bernardis, P. Pilar, T. Jaako, S. De Liberato, and P. Rabl, "Breakdown of gauge invariance in ultrastrong-coupling cavity QED," *Phys. Rev. A*, vol. 98, p. 053819, 2018a.

[70] A. Stokes and A. Nazir, "Gauge ambiguities imply Jaynes-Cummings physics remains valid in ultrastrong coupling QED," *Nat. Commun.*, vol. 10, 2019, https://doi.org/10.1038/s41467-018-08101-0.

[71] A. Settineri, O. Di Stefano, D. Zueco, S. Hughes, S. Savasta, and F. Nori, "Gauge freedom, quantum measurements, and time-dependent interactions in cavity and circuit QED," arXiv: 1912.08548, 2019.

[72] K. Jacobs and A. J. Landahl, "Engineering giant nonlinearities in quantum nanosystems," *Phys. Rev. Lett.*, vol. 103, p. 067201, 2009.

[73] S. Ferretti and D. Gerace, "Single-photon nonlinear optics with Kerr-type nanostructured materials," *Phys. Rev. B*, vol. 85, p. 033303, 2012.

[74] M. Malekakhlagh and H. E. Türeci, "Origin and implications of an A2-like contribution in the quantization of circuit-QED systems" *Phys. Rev. A*, vol. 93, p. 012120, 2016.

[75] C. Sánchez Muñoz, F. Nori, and S. De Liberato, "Resolution of superluminal signalling in non-perturbative cavity quantum electrodynamics" *Nat. Commun.*, vol. 9, p. 1924, 2018, arXiv: 1709.09872.

[76] P. Knüppel, S. Ravets, M. Kroner, S. Fält, W. Wegscheider, and A. Imamoglu, "Nonlinear optics in the fractional quantum Hall regime," *Nature*, vol. 572, p. 91, 2019.

[77] D. De Bernardis, T. Jaako, and P. Rabl, "Cavity quantum electrodynamics in the nonperturbative regime" *Phys. Rev. A*, vol. 97, p. 043820, 2018b.

[78] Q. Bin, X. Lü, T. Yin, Y. Li, and Y. Wu, "Collective radiance effects in the ultrastrong-coupling regime" *Phys. Rev. A*, vol. 99, p. 033809, 2019.

[79] M. Aspelmeyer, T. J. Kippenberg, and F. Marquardt, "Cavity optomechanics," *Rev. Mod. Phys.*, vol. 86, p. 1391, 2014.

[80] F. Beaudoin, J. M. Gambetta, and A. Blais, "Dissipation and ultrastrong coupling in circuit QED" *Phys. Rev. A*, vol. 84, p. 043832, 2011, arXiv:1107.3990.

[81] A. Settineri, V. Macrì, A. Ridolfo, et al., "Dissipation and thermal noise in hybrid quantum systems in the ultrastrong-coupling regime," *Phys. Rev. A*, vol. 98, p. 053834, 2018.

Johannes Feist*, Antonio I. Fernández-Domínguez and Francisco J. García-Vidal*

Macroscopic QED for quantum nanophotonics: emitter-centered modes as a minimal basis for multiemitter problems

https://doi.org/10.1515/9783110710687-038

Abstract: We present an overview of the framework of macroscopic quantum electrodynamics from a quantum nanophotonics perspective. Particularly, we focus our attention on three aspects of the theory that are crucial for the description of quantum optical phenomena in nanophotonic structures. First, we review the light–matter interaction Hamiltonian itself, with special emphasis on its gauge independence and the minimal and multipolar coupling schemes. Second, we discuss the treatment of the external pumping of quantum optical systems by classical electromagnetic fields. Third, we introduce an exact, complete, and minimal basis for the field quantization in multiemitter configurations, which is based on the so-called emitter-centered modes. Finally, we illustrate this quantization approach in a particular hybrid metallodielectric geometry: two quantum emitters placed in the vicinity of a dimer of Ag nanospheres embedded in a SiN microdisk.

Keywords: emitter-centered modes; hybrid cavities; macroscopic quantum electrodynamics; quantum nanophotonics.

***Corresponding authors: Johannes Feist**, Departamento de Física Teórica de la Materia Condensada and Condensed Matter Physics Center (IFIMAC), Universidad Autónoma de Madrid, E-28049 Madrid, Spain, E-mail: johannes.feist@uam.es. https://orcid.org/0000-0002-7972-0646; and **Francisco J. García-Vidal**, Departamento de Física Teórica de la Materia Condensada and Condensed Matter Physics Center (IFIMAC), Universidad Autónoma de Madrid, E-8049 Madrid, Spain; and Donostia International Physics Center (DIPC), E-20018 Donostia/San Sebastian, Spain, E-mail: fj.garcia@uam.es. https://orcid.org/0000-0003-4354-0982
Antonio I. Fernández-Domínguez, Departamento de Física Teórica de la Materia Condensada and Condensed Matter Physics Center (IFIMAC), Universidad Autónoma de Madrid, E-28049 Madrid, Spain, E-mail: a.fernandez-dominguez@uam.es. https://orcid.org/0000-0002-8082-395X

1 Introduction

In principle, quantum electrodynamics (QED) provides an "exact" approach for treating electromagnetic (EM) fields, charged particles, and their interactions, within a full quantum field theory where both matter and light are second quantized (i.e., both photons and matter particles can be created and annihilated). However, this approach is not very useful for the treatment of many effects of interest in fields such as (nano)photonics and quantum optics, which take place at "low" energies (essentially, below the rest mass energy of electrons), where matter constituents are stable and neither created nor destroyed, and additionally, there are often "macroscopic" structures such as mirrors, photonic crystals, metallic nanoparticles etc. involved. Owing to the large number of material particles (on the order of the Avogadro constant, $\approx 6 \times 10^{23}$), it then becomes unthinkable to treat the electrons and nuclei in these structures individually. At the same time, a sufficiently accurate description of these structures is usually given by the macroscopic Maxwell equations, in which the material response is described by the constitutive relations of macroscopic electromagnetism. In many situations, it is then desired to describe the interactions between light and matter in a setup where there are one or a few microscopic "quantum emitters" (such as atoms, molecules, quantum dots, etc.) and additionally a "macroscopic" material structure whose linear response determines the local modes of the EM field interacting with the quantum emitter(s).

The quantization of the EM field in such arbitrary material environments, i.e., the construction of a second quantized basis for the medium-assisted EM field that takes into account the presence of the "macroscopic" material structure, is a longstanding problem in QED. The most immediate strategy is to calculate the (classical) EM modes of a structure and to quantize them by normalizing their stored energy to that of a single photon at the mode frequency, $\hbar\omega$ [1]. However, an important point to remember here is that, even for lossless materials, EM modes always

This article has previously been published in the journal Nanophotonics. Please cite as: Johannes Feist, Antonio I. Fernández-Domínguez and Francisco J. García-Vidal "Macroscopic QED for quantum nanophotonics: emitter-centered modes as a minimal basis for multiemitter problems" *Nanophotonics* 2021, 10. DOI: 10.1515/nanoph-2020-0451.

form a continuum in frequency, i.e., there exist modes at any positive frequency ω. Consequently, there are in general no truly bound EM modes, and what is normally thought of as an isolated "cavity mode" is more correctly described as a resonance embedded in the continuum, i.e., a quasi-bound state that decays over time through emission of radiation. An interesting exception here are guided modes in systems with translational invariance (i.e., where momentum in one or more dimensions is conserved), as modes lying outside the light cone $\omega = ck$ then do not couple to free-space radiation [2]. An additional exception is given by "bound states in the continuum" [3], which arise owing to destructive interference between different resonances.

As a further obstacle to a straightforward quantization strategy as described above, the response functions describing material structures are necessarily dissipative owing to causality (as encoded in, e.g., the Kramers–Kronig relations). When these losses cannot be neglected, quantization is complicated even further by the difficulty to define the energy density of the EM field inside the lossy material [4–6].

Given all the points above, it is not surprising that there are many different approaches to quantizing EM modes in lossy material systems [2, 7–15]. In the following, we give a concise overview of a particularly powerful formal approach that resolves these problems, called macroscopic QED [16–24]. While there are excellent reviews of this general framework available (e.g., [22, 23]), we focus on its application in the context of quantum nanophotonics and strong light–matter coupling. In particular, we discuss the implications and lessons that can be taken from this approach on gauge independence and, in particular, the role of the so-called dipole self-energy term in the light–matter interaction in the Power–Zienau–Woolley (PZW) gauge, which has been the subject of some recent controversy [25–38]. We then review in detail a somewhat nonstandard formulation of macroscopic QED that allows one to construct a minimal quantized basis for the EM field interacting with a collection of multiple quantum emitters. This approach was first introduced by Buhmann and Welsch [39] and subsequently rediscovered independently by several other groups [40–44]. The fact that this very useful approach has been reinvented by different researchers over the past decade or so partially motivates the current article, which intends to give a concise and accessible overview, and presents some explicit relations that have (to our knowledge) not been published before. We also note that with "minimal basis," we are here referring to a minimal *complete* basis for the medium-assisted EM field, i.e., this basis contains all the information about the material structure playing the role of the cavity or antenna, and no approximations are made in obtaining it. This then makes it appropriate to serve as a starting point either for numerical treatments [45, 46] or for deriving simpler models where, e.g., the full EM spectrum is described by a few lossy modes [47].

In the final part of the article, we then present an application of the formalism to a specific problem, a hybrid dielectric-plasmonic structure [14, 48, 49]. In particular, we consider a dimer of metallic nanospheres placed within a dielectric microdisk, a geometry that is similar to that considered by Doeleman et al. [50].

2 Theory

Macroscopic QED provides a recipe for quantizing the EM field in any geometry, including with lossy materials. One particularly appealing aspect is that the full information about the quantized EM field is finally encoded in the (classical) EM dyadic Green's function $\mathbf{G}(\mathbf{r}, \mathbf{r}', \omega)$. While there are several ways to derive the general formulation (see, e.g., the review by Scheel and Buhmann [22] for a discussion of various approaches), a conceptually simple way to understand the framework is to represent the material structures through a collection of fictitious harmonic oscillators coupled to the free-space EM field (which itself corresponds to a collection of harmonic oscillators [51]). Formally diagonalizing this system of coupled harmonic oscillators leads to a form where the linear response of the medium can be expressed through the coupling between the material oscillators and the EM field. The end result is that the fully quantized medium-supported EM field is represented by an infinite set of bosonic modes defined at each point in space and each frequency and labeled with index λ (see below), $\hat{\mathbf{f}}_\lambda(\mathbf{r}, \omega)$, which act as sources for the EM field through the classical Green's function. These modes are called "polaritonic" as they represent mixed light–matter excitations [17]. While this is a completely general approach for quantizing the EM field in arbitrary structures, it cannot be used "directly" in practice owing to the extremely large number of modes that describe the EM field (a vectorial-valued four-dimensional continuum). Most uses of macroscopic QED thus apply this formalism to derive expressions where the explicit operators $\hat{\mathbf{f}}_\lambda(\mathbf{r}, \omega)$ have been eliminated, e.g., through adiabatic elimination, perturbation theory, or the use of Laplace transform techniques [22–24, 52–55]. In particular, macroscopic QED has been widely used in the context of dispersion forces, which are responsible for, e.g., the Casimir effect and Casimir–Polder force [23, 24].

2.1 Minimal coupling

In the following, we represent a short overview of the general theory of light–matter interactions in the framework of macroscopic QED. Since full details can be found in the literature [22, 23], we do not attempt to make this a fully self-contained overview, but rather highlight and discuss some aspects that are not within the traditional focus of the theory, in particular in the context of quantum nanophotonics.

The first step in the application of macroscopic QED is the separation of all matter present in the system to be treated into two distinct groups: one is the macroscopic structure (e.g., a cavity, plasmonic nanoantenna, photonic crystal, ...) that will be described through the constitutive relations of electromagnetism, while the other are the microscopic objects (atoms, molecules, quantum dots, ...) that are described as a collection of charged particles. This separation constitutes the basic approximation inherent in the approach and relies on the assumptions that macroscopic electromagnetism is valid for the material structure (the medium) and its interaction with the charged particles. While this is often an excellent approximation, some care has to be taken for separations in the subnanometer range, where the atomic structure of the material can have a significant influence [56–61]. One significant advantage of this approach is that the microscopic objects are governed by the "standard" Hamiltonian of charged particles interacting through the Coulomb force. They can thus be represented using standard approximations, e.g., using the methods of atomic and molecular physics and quantum chemistry to obtain few-level approximations, or also of solid-state physics to obtain effective descriptions of their band structure, although care has to be taken with gauge invariance when such approximations are performed [62].

For simplicity, we assume that the medium response is local and isotropic in space, such that it can be encoded in the position- and frequency-dependent scalar relative permittivity $\epsilon(\mathbf{r},\omega)$ and relative permeability $\mu(\mathbf{r},\omega)$ that describe the local matter polarization and magnetization induced by external EM fields[1]. The extension of the quantization scheme to nonlocal response functions can be found in reference [22]. We directly give the Hamiltonian $\mathcal{H} = \mathcal{H}_A + \mathcal{H}_F + \mathcal{H}_{AF}$ within the minimal coupling scheme [22, 23]

$$\mathcal{H}_A = \sum_\alpha \frac{\hat{\mathbf{p}}_\alpha^2}{2m_\alpha} + \sum_\alpha \sum_{\beta>\alpha} \frac{q_\alpha q_\beta}{4\pi\varepsilon_0 |\hat{\mathbf{r}}_\alpha - \hat{\mathbf{r}}_\beta|}, \tag{1a}$$

$$\mathcal{H}_F = \sum_\lambda \int_0^\infty \mathrm{d}\omega \int \mathrm{d}^3\mathbf{r}\, \hbar\omega\, \hat{\mathbf{f}}_\lambda^\dagger(\mathbf{r},\omega)\hat{\mathbf{f}}_\lambda(\mathbf{r},\omega), \tag{1b}$$

$$\mathcal{H}_{AF} = \sum_\alpha \left[q_\alpha \hat{\phi}(\hat{\mathbf{r}}_\alpha) - \frac{q_\alpha}{m_\alpha}\hat{\mathbf{p}}_\alpha \cdot \hat{\mathbf{A}}(\hat{\mathbf{r}}_\alpha) + \frac{q_\alpha^2}{2m_\alpha}\hat{\mathbf{A}}^2(\hat{\mathbf{r}}_\alpha)\right]. \tag{1c}$$

Here, the "atomic" Hamiltonian $\mathcal{H}_A$ describes a (nonrelativistic) collection of point particles with position and momentum operators $\hat{\mathbf{r}}_\alpha$ and $\hat{\mathbf{p}}_\alpha$ and charges and masses q_α and m_α. The field Hamiltonian $\mathcal{H}_F$ is expressed in terms of the bosonic operators $\hat{\mathbf{f}}_\lambda(\mathbf{r},\omega)$ discussed above, which obey the commutation relations

$$\left[\hat{\mathbf{f}}_\lambda(\mathbf{r},\omega), \hat{\mathbf{f}}_{\lambda'}^\dagger(\mathbf{r}',\omega')\right] = \delta_{\lambda\lambda'}\boldsymbol{\delta}(\mathbf{r}-\mathbf{r}')\delta(\omega-\omega'), \tag{2a}$$

$$\left[\hat{\mathbf{f}}_\lambda(\mathbf{r},\omega), \hat{\mathbf{f}}_{\lambda'}(\mathbf{r}',\omega')\right] = \left[\hat{\mathbf{f}}_\lambda^\dagger(\mathbf{r},\omega), \hat{\mathbf{f}}_{\lambda'}^\dagger(\mathbf{r}',\omega')\right] = \mathbf{0}, \tag{2b}$$

where $\boldsymbol{\delta}(\mathbf{r}-\mathbf{r}') = \delta(\mathbf{r}-\mathbf{r}')\mathbf{1}$, and $\mathbf{1}$ and $\mathbf{0}$ are the Cartesian (3×3) identity and zero tensors, respectively. The index $\lambda \in \{e, m\}$ labels the electric and magnetic contributions, with the magnetic contribution disappearing if $\mu(\mathbf{r},\omega) = 1$ everywhere in space. The particle-field interaction Hamiltonian $\mathcal{H}_{AF}$ contains the interaction of the charges both with the electrostatic potential $\hat{\phi}(\mathbf{r})$ and the vector potential $\hat{\mathbf{A}}(\mathbf{r})$, both of which can be expressed through the fundamental operators $\hat{\mathbf{f}}_\lambda(\mathbf{r},\omega)$ [22, 23]. Note that in the above, we have neglected magnetic interactions owing to particle spin. We explicitly point out that although we work in Coulomb gauge, $\nabla\cdot\hat{\mathbf{A}}(\mathbf{r}) = 0$, the electrostatic potential $\hat{\phi}(\mathbf{r})$ is in general nonzero owing to the presence of the macroscopic material structure, which also implies that the name "$\mathbf{p}\cdot\mathbf{A}$-gauge," which is sometimes employed for the minimal coupling scheme, is misleading in the presence of material bodies.

The light–matter interaction Hamiltonian can be simplified in the long-wavelength or dipole approximation, i.e., if we assume that the charged particles are sufficiently close to each other compared to the spatial scale of local field variations that a lowest order approximation of the positions of the charges relative to their center of mass position $\hat{\mathbf{r}}_i$ is valid. For an overall neutral collection of charges, this leads to

$$\mathcal{H}_{AF} \approx -\hat{\mathbf{d}}\cdot\hat{\mathbf{E}}^\parallel(\hat{\mathbf{r}}_i) - \sum_\alpha \frac{q_\alpha}{m_\alpha}\hat{\mathbf{p}}_\alpha\cdot\hat{\mathbf{A}}(\hat{\mathbf{r}}_i) + \sum_\alpha \frac{q_\alpha^2}{2m_\alpha}\hat{\mathbf{A}}^2(\hat{\mathbf{r}}_i), \tag{3}$$

1 Note that since the response functions are considered time independent, effects due to the motion of the structure, such as in cavity optomechanics [63], cannot be treated without further modifications.

where $\hat{\mathbf{d}} = \sum_\alpha q_\alpha \hat{\bar{\mathbf{r}}}_\alpha$ is the electric dipole operator of the collection of charges, while $\hat{\bar{\mathbf{r}}}_\alpha$ and $\hat{\bar{\mathbf{p}}}_\alpha$ are the position and momentum operators in the center-of-mass frame of the charge collection[2]. Equation (3) explicitly shows that in the presence of material bodies, the emitter-field interaction has two contributions, one from longitudinal (electrostatic) fields owing to the Coulomb interaction with charges in the macroscopic body and one from transverse fields (described by the vector potential). The relevant fields are given by

$$\hat{\mathbf{E}}^{\parallel}(\hat{\mathbf{r}}) = \sum_\lambda \int_0^\infty \mathrm{d}\omega \int \mathrm{d}^3\mathbf{r}' \,{}^{\parallel}\mathbf{G}_\lambda(\mathbf{r},\mathbf{r}',\omega)\cdot\hat{\mathbf{f}}_\lambda(\mathbf{r}',\omega) + \mathrm{H.c.}, \tag{4a}$$

$$\hat{\mathbf{A}}(\hat{\mathbf{r}}) = \sum_\lambda \int_0^\infty \frac{\mathrm{d}\omega}{i\omega} \int \mathrm{d}^3\mathbf{r}' \,{}^{\perp}\mathbf{G}_\lambda(\mathbf{r},\mathbf{r}',\omega)\cdot\hat{\mathbf{f}}_\lambda(\mathbf{r}',\omega) + \mathrm{H.c.}, \tag{4b}$$

where the longitudinal and transverse components of a tensor $\mathbf{T}(\mathbf{r},\mathbf{r}')$ are given by

$${}^{\parallel/\perp}\mathbf{T}(\mathbf{r},\mathbf{r}') = \int \mathrm{d}^3\mathbf{s}\,\boldsymbol{\delta}^{\parallel/\perp}(\mathbf{r}-\mathbf{s})\mathbf{T}(\mathbf{s},\mathbf{r}'), \tag{5}$$

with $\boldsymbol{\delta}^{\parallel/\perp}(\mathbf{r}-\mathbf{s})$ the standard longitudinal or transverse delta function in 3D space. The functions $\mathbf{G}_\lambda(\mathbf{r},\mathbf{r}',\omega)$ are related to the dyadic Green's function $\mathbf{G}(\mathbf{r},\mathbf{r}',\omega)$ through

$$\mathbf{G}_e(\mathbf{r},\mathbf{r}',\omega) = i\frac{\omega^2}{c^2}\sqrt{\frac{\hbar}{\pi\epsilon_0}\mathrm{Im}\,\epsilon(\mathbf{r}',\omega)}\mathbf{G}(\mathbf{r},\mathbf{r}',\omega), \tag{6a}$$

$$\mathbf{G}_m(\mathbf{r},\mathbf{r}',\omega) = i\frac{\omega}{c}\sqrt{\frac{-\hbar}{\pi\epsilon_0}\mathrm{Im}\,\mu^{-1}(\mathbf{r}',\omega)}\left[\nabla'\times\mathbf{G}(\mathbf{r}',\mathbf{r},\omega)\right]^T. \tag{6b}$$

We note that in the derivation leading to the above expressions, it is assumed that $\mathrm{Im}\,\epsilon(\mathbf{r},\omega) > 0$ and $\mathrm{Im}\,\mu(\mathbf{r},\omega) > 0$ for all $\mathbf{r}$, i.e., that the materials are lossy everywhere in space. The limiting case of zero losses in some regions (e.g., in free space) is only taken at the very end of the calculation. We will see that in the reformulation in terms of emitter-centered modes, subsection 2.4, $\mathbf{G}_\lambda(\mathbf{r},\mathbf{r}',\omega)$ disappears from the formalism relatively early, and only the "normal" Green's function $\mathbf{G}(\mathbf{r},\mathbf{r}',\omega)$ is needed (for which the limit is straightforward).

As mentioned above, the electrostatic contribution is not present in free space, and in the literature, it is often assumed that any abstract "cavity mode" corresponds to a purely transverse EM field. This is a good approximation for emitters that are far enough away from the material, e.g., in "large" (typically dielectric) structures such as Fabry–Perot planar microcavities, photonic crystals, micropillar resonators, etc. [64], but can break down otherwise. In general, this happens for coupling to evanescent fields [65] and in particular when sub-wavelength confinement is used to generate extremely small effective mode volumes, such as in plasmonic [66, 67] or phonon-polaritonic systems [68, 69]. This observation is particularly relevant as such subwavelength confinement is the only possible strategy for obtaining large enough light–matter coupling strengths to approach the single-emitter strong coupling regime at room temperature [70–73]. For subwavelength separations, it is well known that the Green's function is dominated by longitudinal components, while transverse components can be neglected [74]. In this *quasistatic* approximation, we thus have $\hat{\mathbf{A}} \approx 0$, cf. Eq. (4), and the light–matter interactions are all due to electrostatic (or Coulomb) interactions, even within the minimal coupling scheme[3]. Conversely, owing to the strongly subwavelength field confinement, the long-wavelength or dipole approximation is not necessarily appropriate, and an accurate description requires either the direct use of the expression in terms of the electrostatic potential [75] or the inclusion of higher order terms in the interaction [76]. In this context, it should be noted that within an *ab initio* description (i.e., when the emitters are not treated as few-level systems), the term $-\hat{\mathbf{d}}\cdot\hat{\mathbf{E}}^{\parallel}(\hat{\mathbf{r}}_i)$ within the dipole approximation can be problematic as, e.g., the interaction can become arbitrarily large when the computational box is too big, which in particular affects the more extended excited states, but even leads to the lack of a ground state in infinite space [28, 33]. For the electrostatic (longitudinal) interactions considered here, this has to be resolved by going beyond the dipole approximation, such that the potential is accurately represented within the whole computational box and in particular disappears at large distances to the material [35]. It is also important to keep in mind that in any case, the approximations inherent in macroscopic QED, i.e., that the emitter wave functions do not overlap with the material part, break down for large computational boxes for the emitter wave function.

2 While we here assumed a single emitter (i.e., a collection of close-by charged particles), the extension of the above formula to multiple emitters is trivial. One important aspect to note is that the electrostatic Coulomb interaction between charges (second term in Eq. (1a)) is still present, i.e., there are direct instantaneous Coulomb interactions between the charges in different emitters.

3 It could then be discussed what the field modes should be called in the limit when they contain negligible contributions from propagating photon modes. However, since *all* EM modes that are not just freely propagating photons will always have a somewhat mixed light–matter character, and since these modes always solve the macroscopic Maxwell equations, they are conventionally referred to as "light," "EM," or "photon" modes. It is thus important to remember that this does not imply that they are simply modes of the transverse EM field.

2.2 Multipolar coupling

We now discuss the PZW gauge transformation [77–79], which is used to switch from the minimal coupling scheme discussed up to now to the so-called multipolar coupling scheme, which will then in turn form the basis for the emitter-centered modes we discuss later. This scheme has several advantageous properties: it expresses all light–matter interactions through the fields **E** and **B** directly, without needing to distinguish between longitudinal and transverse fields and allows a systematic expansion of the field-emitter interactions in terms of multipole moments. Additionally, in the multiemitter case, it also removes direct Coulomb interactions between charges in different emitters, which instead become mediated through the EM fields. This property makes it easier to explicitly verify and guarantee that causality is not violated through faster-than-light interactions. We only point out and discuss some specific relevant results here and again refer the reader to the literature for full details [22, 23]. The PZW transformation is carried out by the unitary transformation operator

$$\hat{U} = \exp\left[\frac{i}{\hbar}\int d^3\mathbf{r}\sum_i \hat{\mathbf{P}}_i(\mathbf{r})\cdot\hat{\mathbf{A}}(\mathbf{r})\right] \tag{7}$$

where we have explicitly grouped the charges into several emitters, i.e., distinct (nonoverlapping) collections of charges, labeled with index i. The polarization operator $\hat{\mathbf{P}}_i(\mathbf{r})$ of emitter i is usually defined as

$$\hat{\mathbf{P}}_i(\mathbf{r}) = \sum_{\alpha\in i} q_\alpha \hat{\bar{\mathbf{r}}}_\alpha \int_0^1 d\sigma\,\delta\left(\mathbf{r}-\hat{\mathbf{r}}_i-\sigma\hat{\bar{\mathbf{r}}}_\alpha\right), \tag{8}$$

where $\hat{\mathbf{r}}_i$ is the center-of-mass position operator of emitter i and $\hat{\bar{\mathbf{r}}}_\alpha = \hat{\mathbf{r}}_\alpha - \hat{\mathbf{r}}_i$ is the relative position operator of charge α belonging to emitter i. We note that there is considerable freedom in choosing a definition for $\hat{\mathbf{P}}_i(\mathbf{r})$ as only its longitudinal component is physical. There are subtle issues associated with this choice, in particular, owing to the assumption of a point charge model and the lack of a UV cutoff leading to singular expressions. These issues are not specific to macroscopic QED, but general to the PZW transformation and have been discussed in detail in the literature [29, 80].

Since Eq. (7) describes a unitary transformation, physical results are unaffected in principle, although the convergence behavior with respect to different approximations can be quite different [26, 36]. Applying the transformation gives the new operators $\hat{O}' = \hat{U}\hat{O}\hat{U}^\dagger$. Expressing the Hamiltonian Eq. (1) in terms of these new operators then gives the multipolar coupling form. The effect can be summarized by noting that the operators $\hat{\mathbf{A}}$ and $\hat{\mathbf{r}}_\alpha$ are unchanged, while their canonically conjugate momenta $\hat{\mathbf{\Pi}}$ and $\hat{\mathbf{p}}_\alpha$ are not. In light of the discussion of longitudinal (electrostatic) versus transverse interactions, it is interesting to point out that the electrostatic potential $\hat{\phi}(\mathbf{r})$ is also unaffected, i.e., in the quasistatic limit, the PZW transformation has no effect on the emitter-field interaction and discussions of gauge dependence for this specific case become somewhat irrelevant. However, both the bare-emitter and the bare-field Hamiltonian are changed, as $\hat{\mathbf{f}}_\lambda(\mathbf{r},\omega)' \neq \hat{\mathbf{f}}_\lambda(\mathbf{r},\omega)$, i.e., the separation into emitter and field variables is different than in the minimal coupling scheme. We here directly show the multipolar coupling Hamiltonian after additionally neglecting interactions containing the magnetic field. This leads to

$$\mathcal{H}_A = \sum_i\left[\sum_{\alpha\in i}\frac{\hat{\mathbf{p}}_\alpha^2}{2m_\alpha} + \frac{1}{2\varepsilon_0}\int d^3\mathbf{r}\,\hat{\mathbf{P}}_i^2(\mathbf{r})\right], \tag{9a}$$

$$\mathcal{H}_F = \sum_\lambda\int_0^\infty d\omega\int d^3\mathbf{r}\,\hbar\omega\,\hat{\mathbf{f}}_\lambda^\dagger(\mathbf{r},\omega)\hat{\mathbf{f}}_\lambda(\mathbf{r},\omega), \tag{9b}$$

$$\mathcal{H}_{AF} = -\sum_i\int d^3\mathbf{r}\,\hat{\mathbf{P}}_i(\mathbf{r})\cdot\hat{\mathbf{E}}(\mathbf{r}), \tag{9c}$$

where all operators are their PZW-transformed (primed) versions, but we have not included explicit primes for simplicity. In the long-wavelength limit, Eq. (9c) becomes simply

$$\mathcal{H}_{AF} = -\sum_i \hat{\mathbf{d}}_i\cdot\hat{\mathbf{E}}(\hat{\mathbf{r}}_i), \tag{10}$$

i.e., all field-emitter interactions are expressed through the dipolar coupling term, with the electric field operator given explicitly by

$$\hat{\mathbf{E}}(\mathbf{r}) = \sum_\lambda\int_0^\infty d\omega\int d^3\mathbf{r}'\,\mathbf{G}_\lambda(\mathbf{r},\mathbf{r}',\omega)\cdot\hat{\mathbf{f}}_\lambda(\mathbf{r}',\omega) + \text{H.c.} \tag{11}$$

We note that the form of the bare-emitter Hamiltonian $\mathcal{H}_A = \sum_i H_i$ [Eq. (9a)] under multipolar coupling is changed compared to the minimal coupling picture and in particular can be rewritten as

$$H_i = \sum_{\alpha\in i}\frac{\hat{\mathbf{p}}_\alpha^2}{2m_\alpha} + \sum_{\alpha,\beta\in i}\frac{q_\alpha q_\beta}{8\pi\varepsilon_0|\mathbf{r}_\alpha-\mathbf{r}_\beta|} + \frac{1}{2\varepsilon_0}\int d^3\mathbf{r}\left[\hat{\mathbf{P}}_i^\perp(\mathbf{r})\right]^2, \tag{12}$$

which makes explicit the fact that the emitter Hamiltonian in the multipolar gauge is equivalent to the emitter Hamiltonian in minimal coupling plus a term containing the transverse polarization only. In order to arrive at this form, we have used that the Coulomb interaction can be rewritten as an integral over the longitudinal polarization. The transverse part of the polarization in Eq. (12) corresponds to the so-called dipole self-energy term [28]. When a single- or few-mode approximation of the EM field is performed *before* doing the PZW transformation, this term depends on the square of the mode-emitter coupling strength, but not on any photonic operator. However, when all modes of the EM field are included, as implicitly done here and as motivated by the fact that the term is not mode selective (it does not depend on any EM field operator), it is seen directly that this term becomes completely independent of any characteristics of the surrounding material structure, i.e., it cannot be modified by changing the environment that the emitter is located in. Instead, the bare-emitter Hamiltonian in the multipolar approach is simply slightly different than under minimal coupling. This raises the question whether in a few-mode approximation, such a term should be included in simulations of strongly coupled light–matter systems, i.e., whether the few-mode approximation should be performed before the PZW transformation or after [38]. Including the term improves some mathematical properties of the dipole approximation, in particular in large computational boxes and/or for very large coupling strengths [28, 33, 81]. However, it should also be remembered here that in realistic cavities capable of reaching few-emitter strong coupling, the dominant interaction term is due to longitudinal fields, for which this term does not exist (see discussion in subsection 2.1). Furthermore, it should be mentioned that a similar term can arise if the environment-mediated electrostatic interactions are taken into account explicitly instead of through the quantized modes, and one (or some) of the EM modes is additionally treated explicitly. The action of these modes on the emitters then has to be subtracted from the electrostatic interaction to avoid double counting them [27].

2.3 External (classical) fields

Adapting an argument by Sánchez-Barquilla et al. [45], here we show that macroscopic QED also enables a straightforward treatment of external incoming EM fields, in particular for the experimentally most relevant case of a classical laser pulse. Assuming that the incoming laser field at the initial time $t = 0$ has not yet interacted with the emitters (i.e., it describes a pulse localized in space in a region far away from the emitters), it can simply be described by a product of coherent states of the EM modes for the initial wave function, $|\psi(0)\rangle = \prod_n |\alpha_n(0)\rangle = \prod_n e^{\alpha_n(0)a_n^\dagger - \alpha_n^*(0)a_n}|0\rangle$, where the index n here runs over all indices of the EM basis $(\lambda, \mathbf{r}, \omega)$, and the $\alpha_n(0)$ correspond to the classical amplitudes obtained when expressing the laser pulse in the basis defined by these modes. In order to avoid the explicit propagation of this classical field within a quantum calculation, the classical and the quantum field can be split in the Hamiltonian using a time-dependent displacement operator [82] $T(t) = e^{\sum_n \alpha_n^*(t)a_n - \alpha_n(t)a_n^\dagger}$, where $\alpha_n(t) = \alpha_n(0)e^{-i\omega_n t}$. Applying this transformation to the wavefunction, $|\psi'\rangle = T(t)|\psi\rangle$, adds an (time-dependent) energy shift that does not affect the dynamics and splits the electric field term in Eq. (10) into a classical and a quantum part, $\hat{\mathbf{E}}(\mathbf{r})' = \hat{\mathbf{E}}(\mathbf{r}) + \mathbf{E}_{\text{cl}}(\mathbf{r}, t)$.

We note that the above properties imply that within this framework, the action of any incoming laser pulse on the *full* emitter-cavity system can be described purely by the action of the medium-supported classical electric field on the emitters, with no additional explicit driving of any EM modes. This is different to, e.g., standard input–output theory, where the EM field is split into modes inside the cavity and free-space modes outside, and external driving thus affects the cavity modes[4].

Importantly, $\mathbf{E}_{\text{cl}}(\mathbf{r}, t)$ is the field obtained at the position of the emitter upon propagation of the external laser pulse through the material structure, i.e., it contains any field enhancement and temporal distortion. Since the Hamiltonian expressed by the operators $\hat{\mathbf{f}}_\lambda(\mathbf{r}, \omega)$ by construction solves Maxwell's equations in the presence of the material structure, $\mathbf{E}_{\text{cl}}(\mathbf{r}, t)$ can be obtained by simply solving Maxwell's equations using any classical EM solver without ever expressing the pulse in the basis of the modes $\hat{\mathbf{f}}_\lambda(\mathbf{r}, \omega)$. It is important to remember that EM field observables are also transformed according to

$$\langle\psi|O|\psi\rangle = \langle\psi'|T(t)OT^\dagger(t)|\psi'\rangle, \tag{13}$$

such that, e.g., $\langle\psi|a_n|\psi\rangle = \langle\psi'|a_n + \alpha_n(t)|\psi'\rangle$. This takes into account that the "quantum" field generated by the laser-emitter interaction interferes with the classical pulse propagating through the structure and ensures a correct description of absorption, coherent scattering, and similar effects.

2.4 Emitter-centered modes

Following references [39–44], we now look for a linear transformation of the bosonic modes $\hat{\mathbf{f}}_\lambda(\mathbf{r}, \omega)$ at each frequency in such a way that in the new basis, only a minimal

4 It should be noted that in principle, both macroscopic QED and input–output theory can provide for a complete description of the system. In fact, a few-mode description suitable for treatment within input–output theory or master equation formalisms can be obtained from macroscopic QED by, essentially, "reversing" Fano diagonalization and reexpressing the EM modes through a finite number of discrete modes coupled to exterior continua [15, 47].

number of EM modes couples to the emitters. To this end, we start with the macroscopic QED Hamiltonian within the multipolar approach Eq. (9), with the emitter-field interaction treated within the long-wavelength approximation Eq. (10). For simplicity of notation, we assume that the dipole operator of each emitter only couples to a single-field polarization, $\hat{\mathbf{d}}_i = \hat{\mu}_i \mathbf{n}_i$. Alternatively, the sum over i could simply be extended to include up to three separate orientations per emitter. Our goal can then be achieved by defining *emitter-centered* or *bright* (from the emitter perspective) EM modes $\hat{B}_i(\omega)$ associated with each emitter i:

$$\hat{B}_i(\omega) = \sum_\lambda \int \mathrm{d}^3\mathbf{r}\, \boldsymbol{\beta}_{i,\lambda}(\mathbf{r},\omega) \cdot \hat{\mathbf{f}}_\lambda(\mathbf{r},\omega) \tag{14a}$$

$$\boldsymbol{\beta}_{i,\lambda}(\mathbf{r},\omega) = \frac{\mathbf{n}_i \cdot \mathbf{G}_\lambda(\mathbf{r}_i,\mathbf{r},\omega)}{G_i(\omega)}, \tag{14b}$$

where $G_i(\omega)$ is a normalization factor. Using Eq. (2), the commutation relations of the operators $\hat{B}_i(\omega)$ reduce to overlap integrals of their components, $[\hat{B}_i(\omega), \hat{B}_j^\dagger(\omega')] = S_{ij}(\omega)\delta(\omega - \omega')$, with

$$\begin{aligned} S_{ij}(\omega) &= \sum_\lambda \int \mathrm{d}^3\mathbf{r}\, \boldsymbol{\beta}^*_{i,\lambda}(\mathbf{r},\omega) \cdot \boldsymbol{\beta}_{j,\lambda}(\mathbf{r},\omega) \\ &= \frac{\hbar\omega^2}{\pi\epsilon_0 c^2} \frac{\mathbf{n}_i \cdot \operatorname{Im}\mathbf{G}(\mathbf{r}_i,\mathbf{r}_j,\omega)\cdot \mathbf{n}_j}{G_i(\omega)G_j(\omega)}, \end{aligned} \tag{15}$$

such that the overlap matrix $S_{ij}(\omega)$ at each frequency is real and symmetric (since $\mathbf{G}(\mathbf{r},\mathbf{r}',\omega) = \mathbf{G}^T(\mathbf{r}',\mathbf{r},\omega)$). In the above derivation, we have used the Green's function identity

$$\sum_\lambda \int \mathrm{d}^3\mathbf{s}\, \mathbf{G}_\lambda(\mathbf{r},\mathbf{s},\omega) \cdot \mathbf{G}_\lambda^{*T}(\mathbf{r}',\mathbf{s},\omega) = \frac{\hbar\omega^2}{\pi\epsilon_0 c^2} \operatorname{Im}\mathbf{G}(\mathbf{r},\mathbf{r}',\omega) \tag{16}$$

to obtain a compact result [22, 23]. The normalization factor $G_i(\omega)$ is obtained by requiring that $S_{ii}(\omega) = 1$, so

$$G_i(\omega) = \sqrt{\frac{\hbar\omega^2}{\pi\epsilon_0 c^2} \mathbf{n}_i \cdot \operatorname{Im}\mathbf{G}(\mathbf{r}_i,\mathbf{r}_i,\omega)\cdot \mathbf{n}_i}. \tag{17}$$

We note that the coupling strength $G_i(\omega)$ of the emitter-centered mode $\hat{B}_i(\omega)$ to emitter i is directly related to the EM spectral density $J_i(\omega) = [\mu G_i(\omega)/\hbar]^2$ for transition dipole moment μ [83]. In the regime of weak coupling, i.e., when the EM environment can be approximated as a Markovian bath, the spontaneous emission rate at an emitter frequency ω_e is then given by $2\pi J_i(\omega_e)$.

Since the overlap matrix $S_{ij}(\omega)$ of the modes associated with emitters i and j is determined by the imaginary part of the Green's function between the two emitter positions, it follows that the modes $\hat{B}_i(\omega)$, or equivalently, the coefficient functions $\boldsymbol{\beta}_i(\mathbf{r},\omega)$, are not orthogonal in general. This can be resolved by performing an explicit orthogonalization, which is possible as long as the modes are linearly independent. When this is not the case, linearly dependent modes can be dropped from the basis until a minimal set is reached [42]. In the following, we thus assume linear independence. We can then define new orthonormal modes $\hat{C}_i(\omega)$ as a linear superposition of the original modes (and vice versa)

$$\hat{C}_i(\omega) = \sum_{j=1}^N V_{ij}(\omega)\hat{B}_j(\omega), \tag{18a}$$

$$\boldsymbol{\chi}_{i,\lambda}(\mathbf{r},\omega) = \sum_{j=1}^N V_{ij}(\omega)\boldsymbol{\beta}_{j,\lambda}(\mathbf{r},\omega), \tag{18b}$$

which also implies that $\hat{B}_i(\omega) = \sum_{j=1}^N W_{ij}(\omega)\hat{C}_j(\omega)$, where $\mathbf{W}(\omega) = \mathbf{V}(\omega)^{-1}$. In the above, the transformation matrix $\mathbf{V}(\omega)$ is chosen such that $[\hat{C}_i(\omega), \hat{C}_j^\dagger(\omega')] = \delta_{ij}\delta(\omega - \omega')$, which implies $\mathbf{V}(\omega)\mathbf{S}(\omega)\mathbf{V}^\dagger(\omega) = \mathbb{1}$. The coefficient matrices $\mathbf{V}(\omega)$ can be chosen in various ways, corresponding to different unitary transformations of the same orthonormal basis. We mention two common approaches here. One consists in performing a Cholesky decomposition of the overlap matrix, $\mathbf{S}(\omega) = \mathbf{L}(\omega)\mathbf{L}^T(\omega)$, with $\mathbf{V}(\omega) = \mathbf{L}(\omega)^{-1}$, where $\mathbf{L}(\omega)$ and $\mathbf{L}(\omega)^{-1}$ are lower triangular matrices. This is the result obtained from Gram–Schmidt orthogonalization, with the advantage that $\hat{C}_i(\omega)$ only involves $\hat{B}_j(\omega)$ with $j \le i$ and vice versa, such that emitter i only couples to the first i photon continua. Another possibility is given by Löwdin orthogonalization, with $\mathbf{V}(\omega) = \mathbf{S}(\omega)^{-1/2}$, where the matrix power is defined as $\mathbf{S}^a = \mathbf{U}\boldsymbol{\Lambda}^a\mathbf{U}^\dagger$, with $\mathbf{U}(\boldsymbol{\Lambda})$ a unitary (diagonal) matrix containing the eigenvectors (eigenvalues) of $\mathbf{S}$. This approach maximizes the overlap $[\hat{C}_i(\omega), \hat{B}_i^\dagger(\omega)]$ while ensuring orthogonality and can thus be seen as the "minimal" correction required to obtain an orthonormal basis.

Using the orthonormal set of operators $\hat{C}_i(\omega)$, which are themselves linear superpositions of the operators $\hat{\mathbf{f}}_\lambda(\mathbf{r},\omega)$, we can perform a unitary transformation (separately for each frequency ω) of the $\hat{\mathbf{f}}_\lambda(\mathbf{r},\omega)$ into a basis spanned by the emitter-centered (or bright) EM modes and an infinite number of "dark" modes $\hat{D}_i(\omega)$ that span the orthogonal subspace and do not couple to the emitters, such that

$$\hat{\mathbf{f}}_\lambda(\mathbf{r},\omega) = \sum_{i=1}^N \boldsymbol{\chi}^*_{i,\lambda}(\mathbf{r},\omega)\hat{C}_i(\omega) + \sum_j \boldsymbol{d}^*_{j,\lambda}(\mathbf{r},\omega)\hat{D}_j(\omega). \tag{19}$$

where $\int \mathrm{d}^3\mathbf{r}\, \boldsymbol{d}_j^*(\mathbf{r},\omega) \cdot \boldsymbol{\chi}_i(\mathbf{r},\omega) = 0$. The Hamiltonian can then be written as

$$\mathcal{H} = \int_0^\infty \mathrm{d}\omega \Bigg[\sum_{i=1}^N \hbar\omega \widehat{C}_i^\dagger(\omega)\widehat{C}_i(\omega) + \sum_j \hbar\omega \widehat{D}_j^\dagger(\omega)\widehat{D}_j(\omega) - \sum_{i,j=1}^N \widehat{\mu}_i\big(g_{ij}(\omega)\widehat{C}_j(\omega) + \mathrm{H.c.}\big)\Bigg] + \sum_{i=1}^N \widehat{H}_i \tag{20}$$

where $g_{ij}(\omega) = G_i(\omega)W_{ij}(\omega)$. We note here that $\mathbf{V}(\omega)$ and thus $\mathbf{W}(\omega)$ and $g_{ij}(\omega)$ can always be chosen real owing to the reality of $S_{ij}(\omega)$, but we here treat the general case with possibly complex coefficients. Since the dark modes are decoupled from the rest of the system, they do not affect the dynamics and can be dropped, giving

$$\mathcal{H} = \sum_{i=1}^N \widehat{H}_i + \int_0^\infty \mathrm{d}\omega \Bigg[\sum_{i=1}^N \hbar\omega \widehat{C}_i^\dagger(\omega)\widehat{C}_i(\omega) - \sum_{i,j=1}^N \widehat{\mu}_i\big(g_{ij}(\omega)\widehat{C}_j(\omega) + \mathrm{H.c.}\big)\Bigg]. \tag{21}$$

We mention for completeness that if the dark modes are initially excited, including them might be necessary to fully describe the state of the system. We have now explicitly constructed a Hamiltonian with only N independent EM modes $\widehat{C}_i(\omega)$ for each frequency ω [39–44]. This Hamiltonian corresponds to a set of N emitters coupled to N continua and can be treated using, e.g., a wide variety of methods developed in the context of open quantum systems [84–87].

Furthermore, one can obtain an explicit expression for the electric field operator based on the modes $\widehat{C}_i(\omega)$, which to the best of our knowledge has not been presented previously in the literature. This is obtained by inserting Eq. (19) in Eq. (11), again dropping the dark modes $\widehat{D}_i(\omega)$ and again using the integral relation for Green's functions from Eq. (16), leading to

$$\widehat{\mathbf{E}}^{(+)}(\mathbf{r}) = \sum_{i=1}^N \int_0^\infty \mathrm{d}\omega\, \mathbf{E}_i(\mathbf{r},\omega)\widehat{C}_i(\omega) \tag{22a}$$

$$\mathbf{E}_i(\mathbf{r},\omega) = \sum_{j=1}^N V_{ij}^*(\omega)\boldsymbol{\varepsilon}_j(\mathbf{r},\omega) \tag{22b}$$

$$\boldsymbol{\varepsilon}_j(\mathbf{r},\omega) = \frac{\hbar\omega^2}{\pi\epsilon_0 c^2}\frac{\operatorname{Im}\mathbf{G}(\mathbf{r},\mathbf{r}_j,\omega)\cdot\mathbf{n}_j}{G_j(\omega)} \tag{22c}$$

These relations show that we can form explicit photon modes in space at each frequency by using orthonormal superpositions of the emitter-centered EM modes $\boldsymbol{\varepsilon}_j(\mathbf{r},\omega)$. We note that this also provides a formal construction for the emitter-centered EM modes, i.e., the modes created by the operators $\widehat{B}_i(\omega)$, with a field profile corresponding to the imaginary part of the Green's function associated with that emitter. These modes are well behaved: they solve the source-free Maxwell equations, are real everywhere in space, and do not diverge anywhere (here, it should be remembered that the real part of the Green's function diverges for $\mathbf{r} = \mathbf{r}'$, while the imaginary part does not). As a simple example, we can take a single z-oriented emitter at the origin in free space. The procedure used here then gives exactly the $l = 1$, $m = 0$ spherical Bessel waves, i.e., the only modes that couple to the emitter when quantizing the field using spherical coordinates. Furthermore, it should be noted that these modes contain the absorption of the EM field in the material, which is encoded in the Green's function that solves the macroscopic Maxwell equations including losses. This can be understood by remembering that macroscopic QED can be seen as solving the dynamics of an infinite set of coupled harmonic oscillators corresponding to the EM modes and material excitations (including reservoir modes that describe dissipation), cf. subsection 2. Since the coupled system consists only of harmonic oscillators, this problem can be diagonalized without approximations (i.e., without using a master equation or similar formalisms), resulting in a description of the full system in terms of continua of oscillators. The electric field operator as described by Eq. (22) can then be seen as a projection of the full solution onto the field subspace [23]. The end effect of this is that although the Hamiltonian, Eq. (21), is Hermitian, dissipation in the material is fully included. Finally, we note that it can be verified easily that inserting Eq. (22) in Eq. (9) and simplifying leads exactly to Eq. (21).

3 Example

As an example, we now treat a complex metallodielectric structure, as shown in Figure 1. It is composed of a dielectric microdisk resonator supporting whispering gallery modes, with a metallic sphere dimer antenna placed within. The SiN ($\epsilon = 4$) disk is similar to that considered in Ref. [50], with a radius of 2.03 µm and a height of 0.2 µm. Two 40-nm-diameter Ag (with permittivity taken from Ref. [88]) nanospheres separated by a 2-nm gap are placed 1.68 µm away from the disk axis. Two point-dipole emitters modeled as two-level systems and oriented along the dimer axis are placed in the central gap of the dimer antenna and just next to the antenna (labeled as points 1 and 2 in Figure 1). Their transition frequencies are chosen as $\omega_{e,1} = \omega_{e,2} = 2\,\mathrm{eV}$, while the dipole transition moments are $\mu_1 = 0.1$ e nm and $\mu_2 = 3$ e nm (roughly corresponding to typical single organic molecules and J-aggregates, respectively [89]). As shown in the theory section, the emitter dynamics are then fully determined by the Green's function between the emitter positions (as also found in multiple scattering approaches [54]). In Figure 2, we show the relevant values $\mathbf{n}_i \cdot \operatorname{Im} \mathbf{G}(\mathbf{r}_i, \mathbf{r}_j, \omega) \cdot \mathbf{n}_j$, which clearly reveals the relatively sharp Mie resonances of the dielectric disk, hybridized with the short-range plasmonic modes of the metallic nanosphere dimer. In addition, it also shows that the coupling between the two emitters, i.e., the off-diagonal term with $i = 1, j = 2$, has a significant structure and changes sign several times within the frequency interval.

We now study the dynamics for the Wigner–Weisskopf problem of spontaneous emission of emitter 1, i.e., for the case where emitter 1 is initially in the excited state, while emitter 2 is in the ground state and the EM field is in its vacuum state, such that $|\psi(t = 0)\rangle = \sigma_1^+|0\rangle$, where $|0\rangle$ is the global vacuum without any excitations and σ_1^+ is a Pauli matrix acting on emitter 1. We also treat the light–matter coupling within the rotating wave approximation (RWA), i.e., we use $\sum_{i,j=1}^{N} \mu_i (g_{ij}(\omega)\sigma^+ \hat{C}_j(\omega) + \text{H.c.})$ as the light–matter interaction term. The number of excitations is then conserved, and the system can be solved easily within the single-excitation subspace by discretizing the photon continua in frequency [51]. The resulting emitter dynamics, i.e., the population of the excited states of the emitters, are shown in Figure 3. This reveals that the EM field created by emitter 1 due to spontaneous emission is partially reabsorbed by emitter 2. Comparison with the dynamics of emitter 1 when emitter 2 is not present (shown as a dashed

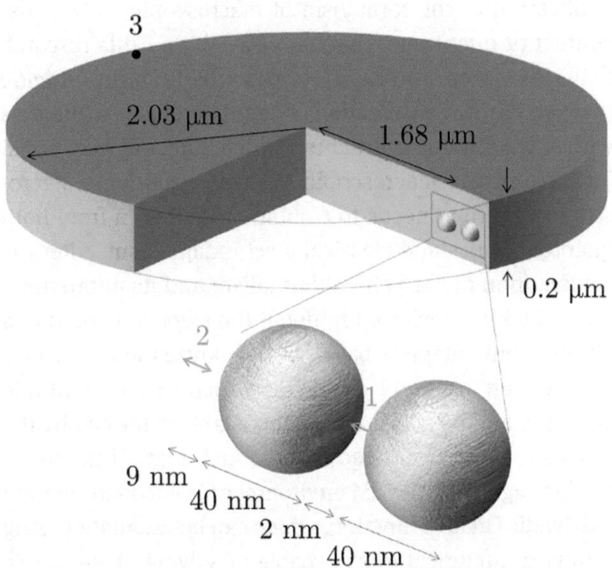

Figure 1: Sketch of the hybrid metallodielectric structure we treat: a dielectric microdisk resonator with a metallic dimer antenna placed on top. The positions of the two emitters are indicated by points 1 and 2, while the field evaluation point used later is indicated as point 3.

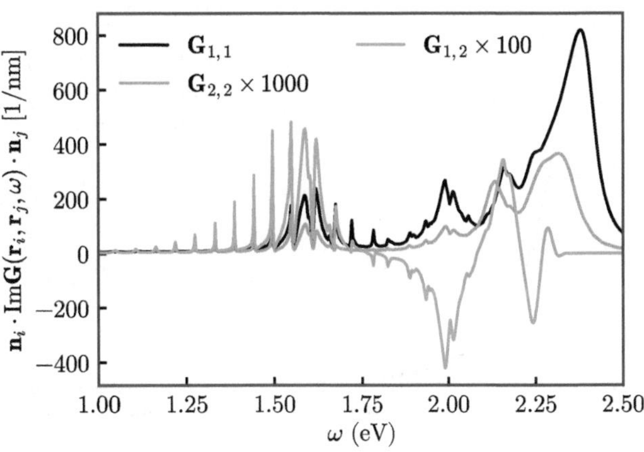

Figure 2: Green's function factor $\mathbf{n}_i \cdot \operatorname{Im} \mathbf{G}(\mathbf{r}_i, \mathbf{r}_j, \omega) \cdot \mathbf{n}_j$ connecting the two emitters in the geometry of Figure 1.

blue line in Figure 3) furthermore reveals that there is also significant transfer of the population back from emitter 2 to emitter 1. We note that while the system treated here is in the weak coupling regime and no Rabi oscillations are present, the method works equally well under strong light–matter coupling [45–47]. Within the RWA employed here, the emitter populations for long times go to zero. Without the RWA, the ground state of the system would be a dressed state with an energy shift corresponding exactly to the Casimir–Polder potential of the emitter in the structure (usually calculated within lowest order perturbation theory) [23]. In our example, the final state at long times would then be the new dressed ground state plus a wave packet of photons emitted into free space. Furthermore, since the number of excitations (in the uncoupled basis) is not conserved without the RWA, the initial state $\sigma_1^+|0\rangle$ would actually have nonzero probability of emitting multiple photons. When the dressing is sufficiently large, the so-called ultrastrong coupling regime is entered [90]. We note for completeness that the theory presented here can treat this regime without problems. In fact, since ultrastrong coupling effects do not depend on (approximate) resonance between modes and levels, taking into account the full spectrum of EM modes is particularly important, and macroscopic QED is thus particularly well suited to treat such effects.

The direct access to the photonic modes in this approach provides interesting insight into, e.g., the photonic mode populations, which are shown in Figure 4 at the final time considered here, $t = 1000$ fs. As mentioned above, there is some freedom in choosing the orthonormalized continuum modes $C_j(\omega)$ as any linear superposition of modes at the same frequency is also an eigenmode of the EM field. We have here chosen the modes obtained through Gram–Schmidt orthogonalization,

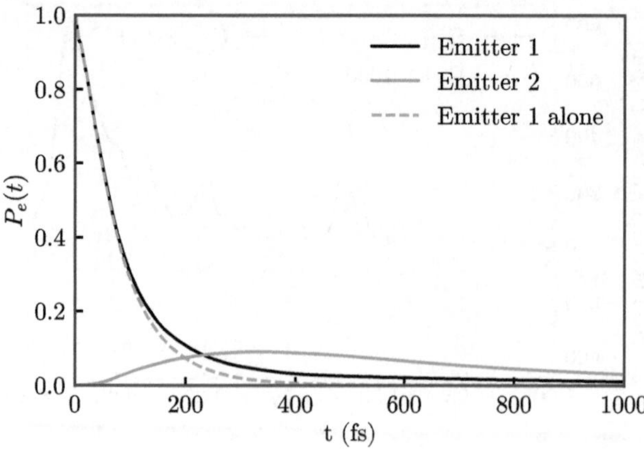

Figure 3: Population of emitters 1 (black line) and 2 (orange line) for the Wigner–Weisskopf problem, with emitter 1 initially excited. The blue dashed line shows the dynamics of emitter 1 if emitter 2 is not present.

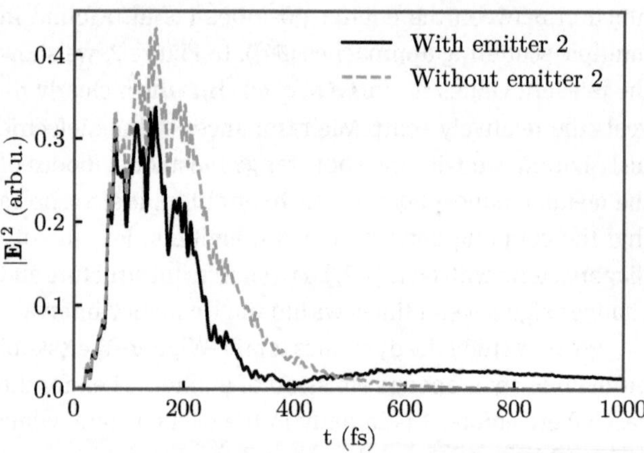

Figure 5: Time-dependent electric field intensity $|\mathbf{E}|^2$ at position 3 in Figure 1, for the case when both emitters are present (black solid line) and when only emitter 1 is present (orange dashed line).

which, as discussed above, have the advantage that emitter i only couples to the first i photon continua. In particular, emitter 1 only couples to a single continuum, $\widehat{C}_1(\omega)$, while emitter 2 couples to the same continuum, and additionally to its "own" continuum $\widehat{C}_2(\omega)$. This makes the comparison between the case of having both emitters present or only including emitter 1 quite direct, as can be observed in Figure 4. In particular, any population in the modes $\widehat{C}_2(\omega)$ must come from emitter 2, after it has in turn been excited by the photons emitted by emitter 1 into continuum 1.

Finally, we also evaluate the electric field in time at a third position (indicated as point 3 in Figure 1), as determined by Eq. (22). This is displayed in Figure 5 and shows a

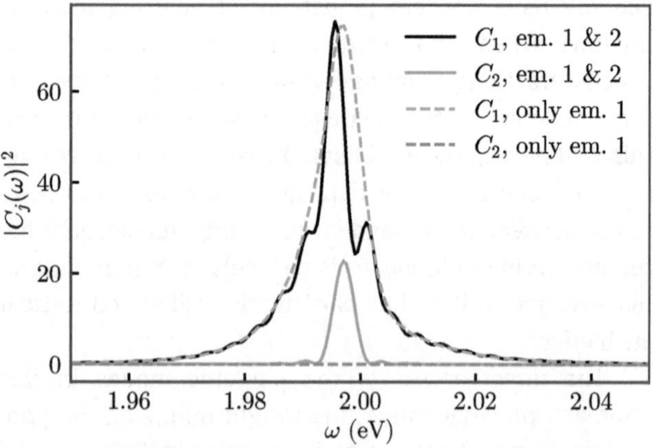

Figure 4: Population $\langle C_j^\dagger(\omega)C_j(\omega)\rangle$ of electromagnetic (EM) modes at time $t = 1000$ fs in the presence of both emitters (solid black and orange lines) and in the presence of only emitter 1 (dashed blue and green lines). Gram–Schmidt orthogonalization has been used here, so that emitter 1 only couples to continuum 1 (i.e., populations for $j = 2$ are identically zero when emitter 2 is not present), while emitter 2 couples to both continua.

broad initial peak due to the fast initial decay of emitter 1 (filtered by propagation through the EM structure, with clear interference effects visible) and then a longer tail due to the longer-lived emission from both emitters, which is mostly due to emitter 2 (which is less strongly coupled to the EM field) and its backfeeding of the population to emitter 1.

4 Conclusions

In this article, we have presented a general overview of the application of the formalism of macroscopic QED in the context of quantum nanophotonics. Within this research field, it is often mandatory to describe from an *ab initio* perspective how a collection of quantum emitters interacts with a nanophotonic structure, which is usually accounted for by utilizing macroscopic Maxwell equations. Macroscopic QED then needs to combine tools taken from both quantum optics and classical electromagnetism. After the presentation of the general formalism and its approximations, we have reviewed in detail the steps to construct a minimal but complete basis set to analyze the interaction between an arbitrary dielectric structure and multiple quantum emitters. This minimal basis set is formed by the so-called emitter-centered modes, such that all the information regarding the EM environment is encoded into the EM dyadic Green's function, which can be calculated using standard numerical tools capable of solving macroscopic Maxwell equations in complex nanophotonic structures. As a way of example and to show its full potential, in the final part of this article, we have applied this formalism to solve both the population dynamics and EM field generation associated with the coupling of two quantum emitters with a hybrid plasmodielectric structure composed of a dielectric microdisk within which a metallic nanosphere dimer is immersed. We emphasize that this formalism can be used not only to provide exact solutions to problems in quantum nanophotonics but also to serve as a starting point for deriving simpler models and/or approximated numerical treatments.

Acknowledgments: We thank M. Ruggenthaler for interesting discussions.
Author contribution: All the authors have accepted responsibility for the entire content of this submitted manuscript and approved submission.
Research funding: This work has been funded by the European Research Council (doi: 10.13039/501100000781) through grant ERC-2016-StG-714870 and by the Spanish Ministry for Science, Innovation, and Universities – Agencia Estatal de Investigación (doi: 10.13039/501100011033) through grants RTI2018-099737-B-I00, PCI2018-093145 (through the QuantERA program of the European Commission), and MDM-2014-0377 (through the María de Maeztu program for Units of Excellence in R&D).
Conflict of interest statement: The authors declare no conflicts of interest regarding this article.

References

[1] R. Loudon, *The Quantum Theory of Light*, 3rd ed., Oxford, New York, Oxford University Press, 2000.
[2] D. R. Abujetas, J. Feist, F. J. García-Vidal, J. Gómez Rivas, and J. A. Sánchez-Gil, "Strong coupling between weakly guided semiconductor nanowire modes and an organic dye," *Phys. Rev. B*, vol. 99, p. 205409, 2019.
[3] C. W. Hsu, B. Zhen, A. D. Stone, J. D. Joannopoulos, and M. Soljačić, "Bound states in the continuum," *Nat. Rev. Mater.*, vol. 1, p. 1, 2016.
[4] R. Loudon, "The propagation of electromagnetic energy through an absorbing dielectric," *J. Phys. Gen. Phys.*, vol. 3, p. 515, 1970.
[5] S. A. Maier, *Plasmonics: Fundamentals and Applications*, New York, Springer, 2007.
[6] J. E. Vázquez-Lozano and A. Martínez, "Optical chirality in dispersive and Lossy media," *Phys. Rev. Lett.*, vol. 121, p. 043901, 2018.
[7] A. Imamoğlu, "Stochastic wave-function approach to non-Markovian systems," *Phys. Rev. A*, vol. 50, p. 3650, 1994.
[8] B. M. Garraway, "Decay of an atom coupled strongly to a reservoir," *Phys. Rev. A*, vol. 55, p. 4636, 1997.
[9] B. M. Garraway, "Nonperturbative decay of an atomic system in a cavity," *Phys. Rev. A*, vol. 55, p. 2290, 1997.
[10] B. J. Dalton, S. M. Barnett, and B. M. Garraway, "Theory of pseudomodes in quantum optical processes," *Phys. Rev. A*, vol. 64, p. 053813, 2001.
[11] E. Waks and D. Sridharan, "Cavity QED treatment of interactions between a metal nanoparticle and a dipole emitter," *Phys. Rev. A*, vol. 82, p. 043845, 2010.
[12] A. González-Tudela, P. A. Huidobro, L. Martín-Moreno, C. Tejedor, and F. J. García-Vidal, "Theory of strong coupling between quantum emitters and propagating surface plasmons," *Phys. Rev. Lett.*, vol. 110, p. 126801, 2013.
[13] F. Alpeggiani and L. C. Andreani, "Quantum theory of surface plasmon polaritons: planar and spherical geometries," *Plasmonics*, vol. 9, p. 965, 2014.
[14] S. Franke, S. Hughes, M. Kamandar Dezfouli, et al., "Quantization of quasinormal modes for open cavities and plasmonic cavity quantum electrodynamics," *Phys. Rev. Lett.*, vol. 122, p. 213901, 2019.
[15] D. Lentrodt and J. Evers, "Ab initio few-mode theory for quantum potential scattering problems," *Phys. Rev. X*, vol. 10, p. 011008, 2020.
[16] U. Fano, "Atomic theory of electromagnetic interactions in dense materials," *Phys. Rev.*, vol. 103, p. 1202, 1956.
[17] J. Hopfield, "Theory of the contribution of excitons to the complex dielectric constant of crystals," *Phys. Rev.*, vol. 112, p. 1555, 1958.
[18] B. Huttner and S. M. Barnett, "Quantization of the electromagnetic field in dielectrics," *Phys. Rev. A*, vol. 46, p. 4306, 1992.
[19] H. T. Dung, L. Knöll, and D.-G. Welsch, "Three-dimensional quantization of the electromagnetic field in dispersive and absorbing inhomogeneous dielectrics," *Phys. Rev. A*, vol. 57, p. 3931, 1998.
[20] S. Scheel, L. Knöll, and D.-G. Welsch, "QED commutation relations for inhomogeneous Kramers–Kronig dielectrics," *Phys. Rev. A*, vol. 58, p. 700, 1998.
[21] S. Y. Buhmann and D.-G. Welsch, "Dispersion forces in macroscopic quantum electrodynamics," *Prog. Quant. Electron.*, vol. 31, p. 51, 2007.
[22] S. Scheel and S. Y. Buhmann, "Macroscopic quantum electrodynamics – concepts and applications," *Acta Phys. Slovaca*, vol. 58, p. 675, 2008.
[23] S. Y. Buhmann, *Dispersion Forces I, Springer Tracts in Modern Physics*, vol. 247, Berlin, Heidelberg, Springer Berlin Heidelberg, 2012.
[24] S. Y. Buhmann, *Dispersion Forces II, Springer Tracts in Modern Physics*, vol. 248, Berlin, Heidelberg, Springer Berlin Heidelberg, 2012.
[25] A. Vukics, T. Grießer, and P. Domokos, "Elimination of the *A*-square problem from cavity QED," *Phys. Rev. Lett.*, vol. 112, p. 073601, 2014.
[26] D. De Bernardis, P. Pilar, T. Jaako, S. De Liberato, and P. Rabl, "Breakdown of gauge invariance in ultrastrong-coupling cavity QED," *Phys. Rev. A*, vol. 98, p. 053819, 2018.
[27] D. De Bernardis, T. Jaako, and P. Rabl, "Cavity quantum electrodynamics in the nonperturbative Regime," *Phys. Rev. A*, vol. 97, p. 043820, 2018.
[28] V. Rokaj, D. M. Welakuh, M. Ruggenthaler, and A. Rubio, "Light–matter interaction in the long-wavelength limit: no ground-state without dipole self-energy," *J. Phys. B*, vol. 51, p. 034005, 2018.
[29] D. L. Andrews, G. A. Jones, A. Salam, and R. G. Woolley, "Perspective: quantum Hamiltonians for optical Interactions," *J. Chem. Phys.*, vol. 148, p. 040901, 2018.

[30] A.Vukics, G.Kónya, and P.Domokos, "The gauge-invariant Lagrangian, the Power–Zienau–Woolley picture, and the choices of field momenta in nonrelativistic quantum electrodynamics," arXiv:1801.05590v2.
[31] E.Rousseau and D.Felbacq, "Reply to "The Equivalence of the Power–Zineau–Woolley Picture and the Poincaré Gauge from the Very First Principles" by G. Kónya, et al.," arXiv:1804.07472.
[32] C. Sánchez Muñoz, F. Nori, and S. De Liberato, "Resolution of superluminal signalling in non-perturbative cavity quantum electrodynamics," *Nat. Commun.*, vol. 9, p. 1924, 2018.
[33] C. Schäfer, M. Ruggenthaler, and A. Rubio, "Ab initio nonrelativistic quantum electrodynamics: bridging quantum chemistry and quantum optics from weak to strong coupling," *Phys. Rev. A*, vol. 98, p. 043801, 2018.
[34] A. Stokes and A. Nazir, "Gauge ambiguities imply Jaynes–Cummings physics remains valid in ultrastrong coupling QED," *Nat. Commun.*, vol. 10, p. 499, 2019.
[35] J. Galego, C. Climent, F. J. Garcia-Vidal, and J. Feist, "Cavity Casimir–Polder forces and their effects in ground-state chemical reactivity," *Phys. Rev. X*, vol. 9, p. 021057, 2019.
[36] O. Di Stefano, A. Settineri, V. Macrì, et al., "Resolution of gauge ambiguities in ultrastrong-coupling cavity Quantum electrodynamics," *Nat. Phys.*, vol. 15, p. 803, 2019.
[37] M. A. D. Taylor, A. Manda, W. Zhou, and P. Huo, "Resolution of gauge ambiguities in molecular cavity quantum electrodynamics," *Phys. Rev. Lett.*, vol. 125, p. 123602, 2020.
[38] A.Stokes and A.Nazir, "Gauge non-invariance due to material truncation in ultrastrong-coupling QED," arXiv:2005.06499.
[39] S. Y. Buhmann and D.-G. Welsch, "Casimir–Polder forces on excited atoms in the strong atom-field coupling regime," *Phys. Rev. A*, vol. 77, p. 012110, 2008.
[40] T. Hümmer, F. J. García-Vidal, L. Martín-Moreno, and D. Zueco, "Weak and strong coupling regimes in plasmonic QED," *Phys. Rev. B*, vol. 87, p. 115419, 2013.
[41] B. Rousseaux, D. Dzsotjan, G. Colas des Francs, H. R. Jauslin, C. Couteau, and S. Guérin, "Adiabatic passage mediated by plasmons: a route towards a decoherence-free quantum plasmonic platform," *Phys. Rev. B*, vol. 93, p. 045422, 2016.
[42] D. Dzsotjan, B. Rousseaux, H. R. Jauslin, G. C. des Francs, C. Couteau, and S. Guérin, "Mode-selective quantization and multimodal effective models for spherically layered systems," *Phys. Rev. A*, vol. 94, p. 023818, 2016.
[43] A. Castellini, H. R. Jauslin, B. Rousseaux, et al., "Quantum plasmonics with multi-emitters: application to stimulated Raman adiabatic passage," *Eur. Phys. J. D*, vol. 72, p. 223, 2018.
[44] H. Varguet, B. Rousseaux, D. Dzsotjan, H. R. Jauslin, S. Guérin, and G. Colas des Francs, "Non-Hermitian Hamiltonian description for quantum plasmonics: from dissipative dressed atom picture to Fano states," *J. Phys. B Atom. Mol. Opt. Phys.*, vol. 52, p. 055404, 2019.
[45] M. Sánchez-Barquilla, R. E. F. Silva, and J. Feist, "Cumulant expansion for the treatment of light–matter interactions in arbitrary material structures," *J. Chem. Phys.*, vol. 152, p. 034108, 2020.
[46] D. Zhao, R. E. F. Silva, C. Climent, J. Feist, A. I. Fernández-Domínguez, and F. J. García-Vidal, "Plasmonic purecell effect in organic molecules," arXiv:2005.05657.
[47] I. Medina, F. J. García-Vidal, A. I. Fernández-Domínguez, and J. Feist, "Few-mode field quantization of arbitrary electromagnetic spectral densities," arXiv:2008.00349.
[48] P. Peng, Y.-C. Liu, D. Xu, et al., "Enhancing coherent light–matter interactions through microcavity-engineered plasmonic resonances," *Phys. Rev. Lett.*, vol. 119, p. 233901, 2017.
[49] B. Gurlek, V. Sandoghdar, and D. Martín-Cano, "Manipulation of quenching in nanoantenna–emitter systems enabled by external detuned cavities: a path to enhance strong-coupling," *ACS Photonics*, vol. 5, p. 456, 2018.
[50] H. M. Doeleman, E. Verhagen, and A. F. Koenderink, "Antenna-cavity hybrids: matching polar opposites for purcell enhancements at any linewidth," *ACS Photonics*, vol. 3, p. 1943, 2016.
[51] G. Grynberg, A. Aspect, C. Fabre, and C. Cohen-Tannoudji, *Introduction to Quantum Optics: From the Semi-classical Approach to Quantized Light*, Cambridge, Cambridge University Press, 2010.
[52] M. Wubs, L. G. Suttorp, and A. Lagendijk, "Multiple-scattering approach to interatomic interactions and superradiance in Inhomogeneous dielectrics," *Phys. Rev. A*, vol. 70, p. 053823, 2004.
[53] P. Yao, C. Van Vlack, A. Reza, M. Patterson, M. Dignam, and S. Hughes, "Ultrahigh purcell factors and lamb shifts in slow-light metamaterial waveguides," *Phys. Rev. B*, vol. 80, p. 195106, 2009.
[54] A. Delga, J. Feist, J. Bravo-Abad, and F. J. Garcia-Vidal, "Quantum emitters near a metal nanoparticle: strong coupling and quenching," *Phys. Rev. Lett.*, vol. 112, p. 253601, 2014.
[55] A. Asenjo-Garcia, J. D. Hood, D. E. Chang, and H. J. Kimble, "Atom–light interactions in quasi-one-dimensional nanostructures: a Green's-function perspective," *Phys. Rev. A*, vol. 95, p. 033818, 2017.
[56] K. J. Savage, M. M. Hawkeye, R. Esteban, A. G. Borisov, J. Aizpurua, and J. J. Baumberg, "Revealing the quantum regime in tunnelling plasmonics," *Nature*, vol. 491, p. 574, 2012.
[57] P. Zhang, J. Feist, A. Rubio, P. García-González, and F. J. García-Vidal, "Ab initio nanoplasmonics: the impact of atomic structure," *Phys. Rev. B*, vol. 90, p. 161407(R), 2014.
[58] G. Aguirregabiria, D. C. Marinica, R. Esteban, A. K. Kazansky, J. Aizpurua, and A. G. Borisov, "Role of electron tunneling in the nonlinear response of plasmonic nanogaps," *Phys. Rev. B*, vol. 97, p. 115430, 2018.
[59] R. Sinha-Roy, P. García-González, H.-C. Weissker, F. Rabilloud, and A. I. Fernández-Domínguez, "Classical and ab Initio plasmonics meet at sub-nanometric noble metal Rods," *ACS Photonics*, vol. 4, p. 1484, 2017.
[60] R. Zhang, L. Bursi, J. D. Cox, et al., "How to identify plasmons from the optical response of nanostructures," *ACS Nano*, vol. 11, p. 7321, 2017.
[61] X. Chen and L. Jensen, "Morphology dependent near-field response in atomistic plasmonic nanocavities," *Nanoscale*, vol. 10, p. 11410, 2018.
[62] Y. Wang, M. Tokman, and A. Belyanin, "Second-order nonlinear optical response of graphene," *Phys. Rev. B*, vol. 94, p. 195442, 2016.
[63] M. Aspelmeyer, T. J. Kippenberg, and F. Marquardt "Cavity optomechanics," *Rev. Mod. Phys.*, vol. 86, p. 1391, 2014.
[64] D. Sanvitto and S. Kéna-Cohen, "The road towards polaritonic devices," *Nat. Mater.*, vol. 15, p. 1061, 2016.
[65] J. Petersen, J. Volz, and A. Rauschenbeutel, "Chiral nanophotonic waveguide interface based on spin–orbit interaction of light," *Science*, vol. 346, p. 67, 2014.

[66] A. I. Fernández-Domínguez, F. J. García-Vidal, and L. Martín-Moreno, "Unrelenting plasmons," *Nat. Photonics*, vol. 11, p. 8, 2017.

[67] J. J. Baumberg, J. Aizpurua, M. H. Mikkelsen, and D. R. Smith, "Extreme nanophotonics from ultrathin metallic gaps," *Nat. Mater.*, vol. 18, p. 668, 2019.

[68] S. Foteinopoulou, G. C. R. Devarapu, G. S. Subramania, S. Krishna, and D. Wasserman, "Phonon-polaritonics: enabling powerful capabilities for infrared photonics," *Nanophotonics*, vol. 8, p. 2129, 2019.

[69] C. R. Gubbin and S. De Liberato, "Optical nonlocality in polar dielectrics," *Phys. Rev. X*, vol. 10, p. 021027, 2020.

[70] K. Santhosh, O. Bitton, L. Chuntonov, and G. Haran, "Vacuum Rabi splitting in a plasmonic cavity at the single quantum emitter limit," *Nat. Commun.*, vol. 7, p. 11823, 2016.

[71] R. Chikkaraddy, B. de Nijs, F. Benz, et al., "Single-molecule strong coupling at room temperature in plasmonic nanocavities," *Nature*, vol. 535, p. 127, 2016.

[72] H. Groß, J. M. Hamm, T. Tufarelli, O. Hess, and B. Hecht, "Near-field strong coupling of single quantum dots," *Sci. Adv.*, vol. 4, p. eaar4906, 2018.

[73] O. S. Ojambati, R. Chikkaraddy, W. D. Deacon, et al., "Quantum electrodynamics at room temperature coupling a single vibrating molecule with a plasmonic nanocavity," *Nat. Commun.*, vol. 10, p. 1049, 2019.

[74] S. Y. Buhmann, "Casimir–Polder forces on atoms in the presence of magnetoelectric bodies", PhD thesis, Friedrich-Schiller-Universität Jena, 2007.

[75] T. Neuman, R. Esteban, D. Casanova, F. J. García-Vidal, and J. Aizpurua, "Coupling of molecular emitters and plasmonic cavities beyond the point-dipole approximation," *Nano Lett.*, vol. 18, p. 2358, 2018.

[76] A. Cuartero-González and A. I. Fernández-Domínguez, "Light-forbidden transitions in plasmon-emitter interactions beyond the weak coupling Regime," *ACS Photonics*, vol. 5, p. 3415, 2018.

[77] E. A. Power and S. Zienau, "Coulomb gauge in non-relativistic quantum electro-dynamics and the shape of spectral lines," *Philos. Trans. R. Soc. Lond. Math. Phys. Sci.*, vol. 251, p. 427, 1959.

[78] R. G. Woolley, "Molecular quantum electrodynamics," *Proc. R. Soc. Lond. A Math. Phys. Sci.*, vol. 321, p. 557, 1971.

[79] R. G. Woolley, "Power–Zienau–Woolley representations of nonrelativistic QED for atoms and molecules," *Phys. Rev. Res.*, vol. 2, p. 013206, 2020.

[80] A. Vukics, T. Grießer, and P. Domokos, "Fundamental limitation of ultrastrong coupling between light and atoms," *Phys. Rev. A*, vol. 92, p. 043835, 2015.

[81] C. Schäfer, M. Ruggenthaler, V. Rokaj, and A. Rubio, "Relevance of the quadratic diamagnetic and self-polarization terms in cavity quantum electrodynamics," *ACS Photonics*, vol. 7, p. 975, 2020.

[82] C. Cohen-Tannoudji, J. Roc, and G. Grynberg, *Photons and Atoms. Introduction to Quantum Electrodynamics*, New York, Wiley-Interscience, 1987.

[83] L. Novotny and B. Hecht, *Principles of Nano-optics*, 2nd ed., Cambridge, Cambridge University Press, 2012.

[84] C. W. Gardiner and P. Zoller, *Quantum Noise: A Handbook of Markovian and Non-Markovian Quantum Stochastic Methods with Applications to Quantum Optics*, Springer Berlin Heidelberg, 2004.

[85] H.-P. Breuer and F. Petruccione, *The Theory of Open Quantum Systems*, Oxford University Press, 2007.

[86] J. Prior, A. W. Chin, S. F. Huelga, and M. B. Plenio, "Efficient simulation of strong system-environment Interactions," *Phys. Rev. Lett.*, vol. 105, p. 050404, 2010.

[87] I. de Vega and D. Alonso, "Dynamics of non-Markovian open quantum systems," *Rev. Mod. Phys.*, vol. 89, p. 015001, 2017.

[88] A. D. Rakić, A. B. Djurišić, J. M. Elazar, and M. L. Majewski, "Optical properties of metallic films for vertical-cavity optoelectronic devices," *Appl. Opt.*, vol. 37, p. 5271, 1998.

[89] J. Moll, S. Daehne, J. R. Durrant, and D. A. Wiersma, "Optical dynamics of excitons in J aggregates of a carbocyanine dye," *J. Chem. Phys.*, vol. 102, p. 6362, 1995.

[90] A. Frisk Kockum, A. Miranowicz, S. D. Liberato, S. Savasta, and F. Nori, "Ultrastrong coupling between light and matter," *Nat. Rev. Phys.*, vol. 1, p. 19, 2019.

Mikhail Tokman, Maria Erukhimova, Yongrui Wang, Qianfan Chen and Alexey Belyanin*

Generation and dynamics of entangled fermion–photon–phonon states in nanocavities

https://doi.org/10.1515/9783110710687-039

Abstract: We develop the analytic theory describing the formation and evolution of entangled quantum states for a fermionic quantum emitter coupled simultaneously to a quantized electromagnetic field in a nanocavity and quantized phonon or mechanical vibrational modes. The theory is applicable to a broad range of cavity quantum optomechanics problems and emerging research on plasmonic nanocavities coupled to single molecules and other quantum emitters. The optimal conditions for a tripartite entanglement are realized near the parametric resonances in a coupled system. The model includes dissipation and decoherence effects due to coupling of the fermion, photon, and phonon subsystems to their dissipative reservoirs within the stochastic evolution approach, which is derived from the Heisenberg–Langevin formalism. Our theory provides analytic expressions for the time evolution of the quantum state and observables and the emission spectra. The limit of a classical acoustic pumping and the interplay between parametric and standard one-photon resonances are analyzed.

Keywords: cavity optomechanics; cavity quantum electrodynamics; entanglement; quantum acoustics; quantum information; quantum optics.

***Corresponding author: Alexey Belyanin,** Department of Physics and Astronomy, Texas A&M University, College Station, TX, 77843, USA, E-mail: belyanin@tamu.edu. https://orcid.org/0000-0001-5233-8685
Mikhail Tokman and Maria Erukhimova, Institute of Applied Physics, Russian Academy of Sciences, Nizhny Novgorod, 603950, Russia, E-mail: tokman@appl.sci-nnov.ru (M. Tokman), maria.erukhimova@gmail.com (M. Erukhimova)
Yongrui Wang and Qianfan Chen, Department of Physics and Astronomy, Texas A&M University, College Station, TX, 77843, USA, E-mail: wangyongrui@physics.tamu.edu (Y. Wang), qxc76@tamu.edu (Q. Chen)

1 Introduction

There is a lot of recent interest in the quantum dynamics of fermion systems coupled to both an electromagnetic (EM) mode in a cavity and quantum or classical mechanical/acoustic oscillations or phonon vibrations. This problem is related to the burgeoning fields of cavity optomechanics [1–3] and quantum acoustics [4–6]. Another example where this situation can be realized is a molecule placed in a plasmonic nanocavity [7, 8]. In this case, the fermion system may comprise two or more electron states forming an optical transition, whereas the phonon field is simply a vibrational mode of a molecule. One can also imagine a situation where a quantum emitter such as a quantum dot (QD) or an optically active defect in a solid matrix is coupled to the quantized phonon modes of a crystal lattice, which would be an extension of an extremely active field of research on phonon–polaritons or plasmon–phonon–polaritons [9, 10] into a fully quantum regime.

Apart from the fundamental interest, the studies of such systems are motivated by quantum information applications. Indeed, the presence of a classical or quantized acoustic mode provides an extra handle to control the quantum state of a coupled fermion–boson quantum system. In the extreme quantum limit in which the fermionic degree of freedom and all bosonic degrees of freedom (both photons and phonons) are quantized, a strong enough coupling between them leads to an entangled fermion–photon–phonon state, which is a complex enough system to implement basic gates for quantum computation or other applications. Such a system has not been realized experimentally. However, many ingredients have been already demonstrated, such as strong coupling between a nanocavity mode and a single molecule [11], numerous examples of strong coupling between nanocavity modes and a single fermionic quantum emitter such as a color center [12] or a QD (see e.g., the studies by Yoshie et al. [13] and Reithmaier et al. [14] for semiconductor cavity–QD systems and the studies by Leng et al. [15], Bitton et al. [16], and Park et al. [17] for plasmonic cavities), strong coupling and entanglement of acoustic phonons [18, 19], resolving the energy levels of a nanomechanical oscillator [6], or

This article has previously been published in the journal Nanophotonics. Please cite as: M. Tokman, M. Erukhimova, Y. Wang, Q. Chen and A. Belyanin "Generation and dynamics of entangled fermion–photon–phonon states in nanocavities" *Nanophotonics* 2021, 10. DOI: 10.1515/nanoph-2020-0353.

cooling a macroscopic system into its motional ground state [20].

Interaction of three or more modes of oscillations, whether they are classical or quantized, is strongly enhanced close to the parametric resonance, which is therefore the most interesting region to study. Fortunately for theorists, the analysis near the parametric resonance is greatly simplified because some form of a slowly varying amplitude method for classical systems [21, 22] or the rotating wave approximation (RWA) for quantum systems [23] can be applied. The use of RWA restricts the coupling strength to the values much lower than the characteristic energies in the system, such as the optical transition or vibrational energy. The emerging studies of the so-called ultra-strong coupling regime [24] have to go beyond the RWA. Nevertheless, for the vast majority of experiments, including nonperturbative strong coupling dynamics and entanglement, the RWA is adequate and provides some crucial simplifications that allow one to obtain analytic solutions.

In particular, within Schrödinger's description, the equations of motion for the components of an infinitely dimensional state vector $|\Psi\rangle$ that describes a coupled fermion–boson system can be split into the blocks of low dimensions if the RWA is applied. This is true even if the dynamics of the fermion subsystem is nonperturbative, e.g., the effects of saturation are important. Note that there is no such simplification in the Heisenberg representation, i.e., when solving the equations of motion

$$\frac{d}{dt}\hat{g} = \frac{i}{\hbar}\left[\hat{H}, \hat{g}\right], \tag{1}$$

where $\hat{g}$ is the Heisenberg operator of a certain physical observable g and $\hat{H}$ is the Hamiltonian of the system. Operator-valued Eq. (1) is generally impossible to split into smaller blocks, even within the RWA. This happens because some matrix elements $g_{AB}(t)$ of the Heisenberg operator are determined by states $|A\rangle$, $|B\rangle$ which belong to different blocks that evolve independently in the Schrödinger picture. The simplification could only be possible for specially selected initial conditions in which the Heisenberg operator is determined on a "truncated" basis belonging to only one of the independent blocks. The Schrödinger's approach also leads to fewer equations for the state vector components than the approach based on the von Neumann master equation for the elements of the density matrix.

Obviously, the Schrödinger equation in its standard form cannot be applied to describe open systems coupled to a dissipative reservoir. In this case, the stochastic versions of the equation of evolution for the state vector have been developed, e.g., the method of quantum jumps [23, 25]. This method is optimal for numerical analysis in the Monte-Carlo type schemes. Here, we formulate the stochastic equation for the state vector derived from the Heisenberg–Langevin approach which is more conducive to the analytic treatment. Its key element is an assumption that there exists the operator of evolution $\hat{U}$, which is determined unambiguously not only by the parameters of the dynamical system but also by the statistical properties of a dissipative reservoir.

There were a number of theoretical studies of tripartite entanglement in open optomechanical systems, either for purely bosonic field modes or involving the atomic degree of freedom (see, e.g., [26–32]). The existing studies are based on either the Heisenberg–Langevin approach or the master equation. In these cases, the analytic solution is possible only after some drastic approximations such as adiabatic elimination of some degrees of freedom or within the linear perturbation theory, when the atomic populations are unperturbed. The work closest to our model is the study by Liao et al. [31], which considers tripartite entanglement in the vicinity of a parametric resonance by numerically solving the Lindblad master equation. The present paper is different in several important aspects. First, our work develops a new formalism based on the stochastic evolution of the state vector, which allowed us to obtain explicit analytic expressions for the evolution of the quantum states and all relevant observables: field and atom energies, emission spectra, etc. We were also able to derive analytic criteria for the separation of resonances, which is important for real experimental situations in which the frequency of the phonons or mechanical oscillations is much lower than the photon or electron transition frequencies, so the parametric and one-photon resonances can easily overlap.

Second, the study by Liao et al. [31] takes into account only the dissipation in the atomic subsystem, whereas our work includes relaxation and fluctuations in all subsystems: electronic, EM, and phonon/mechanical. We were able to write explicit expressions for the relaxation and noise terms which include all relevant relaxation channels and ensure that the system goes into a physically meaningful equilibrium in the absence of external excitation. Moreover, our analysis takes into account both inelastic processes (dissipation) and purely elastic decoherence processes.

The paper is structured as follows. Section II formulates the model and the Hamiltonian for coupled quantized fermion, photon, and phonon fields in a nanocavity. Section III derives the solution for the quantum states of a closed system in the vicinity of a parametric resonance and analyzes its properties. In Section IV, we provide the stochastic equation describing the evolution of quantum states of an open system in contact with a dissipative reservoir and describe the observables. In Section V, we consider the case of a classical acoustic pumping. Section VI describes the interplay of parametric and standard one-photon resonances and provides the conditions under which these resonances can be separated. Section VII gives an example of manipulating entangled electron–photon states by an acoustic pumping. Appendix contains the derivation of the stochastic equation of evolution from the Heisenberg–Langevin approach and compares with Lindblad density matrix formalism.

The focus of the paper is to provide analytic solutions for the quantum dynamics in systems of coupled electron, photon, and phonon excitations including dissipation and decoherence effects. Here, we emphasize "analytic" which means that we provide the expressions for the time evolution of the state vector and observable quantities in the form which shows explicitly the dependence on all experimental parameters: transition energies and frequencies, matrix elements of the optical transitions, the spatial structure of the field modes, relaxation rates for all constituent subsystems, ambient temperatures, etc. That is why we believe that the results obtained in this paper will be useful for the experimentalists working on a broad range of nanophotonic systems.

2 A coupled quantized electron–photon–phonon system: the model

Consider a quantized electron system coupled to the quantum EM field of a nanocavity and classical or quantized vibrational (phonon) modes, see Figure 1 which sketches two out of many possible scenarios.

Here, the electron transition energy is W, the photon and phonon mode frequencies are ω and Ω, respectively. The decay constants γ, μ_{ω}, and μ_{Ω} of the electron, photon, and phonon subsystems due to couplings to their respective dissipative reservoirs are also indicated. Figure 1a sketches a single molecule in a nanocavity formed by a nanotip and a metal substrate. Here, the EM field is coupled to a transition between electron states, and this coupling is modulated by molecular vibrations. Figure 1b shows an electron transition in a QD coupled to the EM field. The figure implies that it is a QD which experiences mechanical or acoustic vibrations, but our treatment below works for any mechanism of relative displacement between the electron system and the field of an EM cavity mode, including the situations where it is the wall of a nanocavity which experiences oscillations.

a

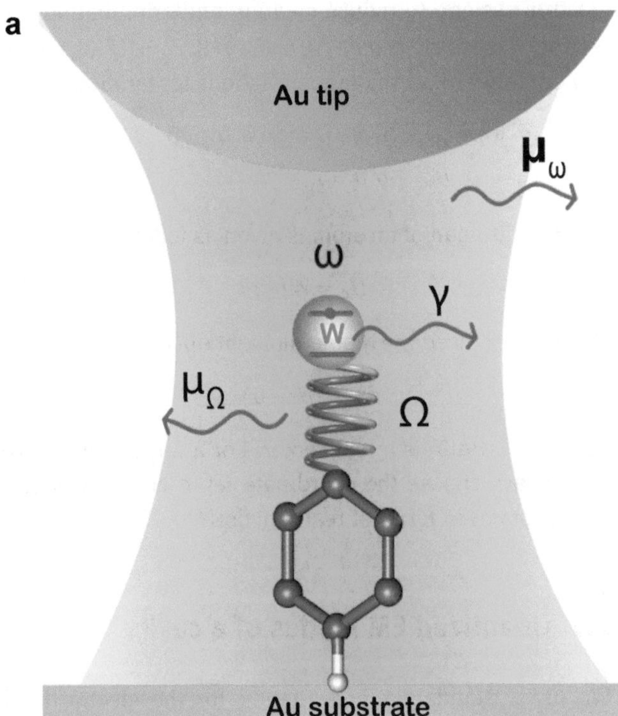

b

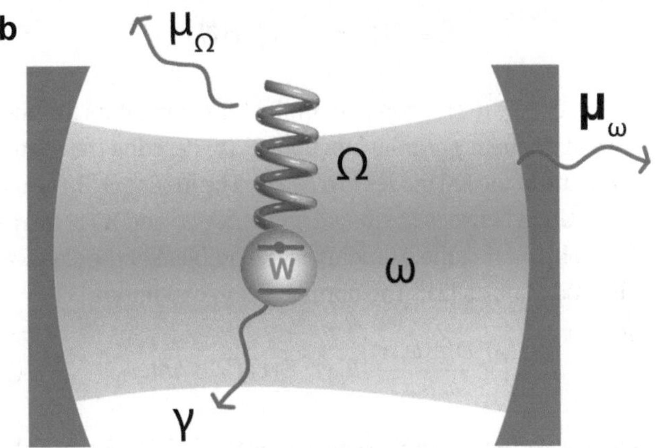

Figure 1: (a) A sketch of a molecule in a nanocavity created by a metallic nanotip and a substrate; (b) A sketch of a quantum dot coupled to optical and mechanical vibrational modes in a nanocavity.

We start from writing down a general Hamiltonian for a coupled quantized electron–photon–phonon system and derive its various approximate forms: the RWA, small-amplitude acoustic oscillations, classical versus quantum phonon mode, etc.

2.1 The fermion subsystem

Consider the simplest version of the fermion subsystem: two electron states $|0\rangle$ and $|1\rangle$ with energies 0 and W, respectively. We will call it an "atom" for brevity, although it can be electron states of a molecule, a QD, or any other electron system. Introduce creation and annihilation operators of the excited state $|1\rangle$, $\widehat{\sigma} = |0\rangle\langle 1|$, $\widehat{\sigma}^\dagger = |1\rangle\langle 0|$, which satisfy standard commutation relations for fermions:

$$\widehat{\sigma}^\dagger|0\rangle = |1\rangle, \widehat{\sigma}|1\rangle = |0\rangle, \widehat{\sigma}\widehat{\sigma} = \widehat{\sigma}^\dagger\widehat{\sigma}^\dagger = 0; \left[\widehat{\sigma},\widehat{\sigma}^\dagger\right]_+ = \widehat{\sigma}\widehat{\sigma}^\dagger + \widehat{\sigma}^\dagger\widehat{\sigma} = 1.$$

The Hamiltonian of an atom is given as follows:

$$\widehat{H}_a = W\widehat{\sigma}^\dagger\widehat{\sigma}. \tag{2}$$

We will also need the dipole moment operator,

$$\widehat{\mathbf{d}} = \mathbf{d}(\widehat{\sigma}^\dagger + \widehat{\sigma}), \tag{3}$$

where $\mathbf{d} = \langle 1|\widehat{\mathbf{d}}|0\rangle$ is a real vector. For a finite motion, we can always choose the coordinate representation of stationary states in terms of real functions.

2.2 Quantized EM modes of a cavity

We use a standard representation for the electric field operator in a cavity:

$$\widehat{\mathbf{E}} = \sum_i \left[\mathbf{E}_i(\mathbf{r})\widehat{c}_i + \mathbf{E}_i^*(\mathbf{r})\widehat{c}_i^\dagger\right], \tag{4}$$

where $\widehat{c}_i{}^\dagger, \widehat{c}_i$ are creation and annihilation operators for photons at frequency ω_i; the functions $\mathbf{E}_i(\mathbf{r})$ describe the spatial structure of the EM modes in a cavity. The functions $\mathbf{E}_i(\mathbf{r})$ and the relation between the modal frequency ω_i and $\mathbf{E}_i(\mathbf{r})$ can be found by solving the boundary value problem of the classical electrodynamics [23]. The normalization conditions [33]

$$\int_V \frac{\partial\left[\omega_i^2\varepsilon(\omega_i,\mathbf{r})\right]}{\omega_i\partial\omega_i}\mathbf{E}_i^*(\mathbf{r})\,\mathbf{E}_i(\mathbf{r})d^3r = 4\pi\hbar\omega_i \tag{5}$$

ensure correct bosonic commutators $[\widehat{c}_i,\widehat{c}_j^\dagger] = \delta_{ij}$ and the field Hamiltonian in the form

$$\widehat{H}_{em} = \hbar\sum_i \omega_i\left(\widehat{c}_i^\dagger\,\widehat{c}_i + \frac{1}{2}\right). \tag{6}$$

here, V is a quantization volume and $\varepsilon(\omega,\mathbf{r})$ is the dielectric function of a dispersive medium that fills the cavity. Eq. (5) is true for any fields satisfying Maxwell's equations as long as intracavity losses can be neglected and the flux of the Poynting vector through the total cavity surface is zero (see e.g., Refs. [33–36]). Of course the photon losses are always important when calculating the decoherence rates and fluctuations. What matters for Eq. (5) is that the effect of losses on the *spatial structure* of the cavity modes is insignificant. The latter is true as long as it makes sense to talk about cavity modes at all, which means in practice that the cavity Q-factor is at least around 10 or greater.

2.3 The quantized phonon field

We assume that our two-level atom is dressed by a phonon field which can be described by the displacement operator:

$$\widehat{\mathbf{q}} = \sum_i \widehat{\mathbf{q}}_i;\quad \widehat{\mathbf{q}}_i = \mathbf{Q}_i(\mathbf{r})\widehat{b}_i + \mathbf{Q}_i^*(\mathbf{r})\widehat{b}_i^\dagger \tag{7}$$

here, $\widehat{b}_i$ and $\widehat{b}_i^\dagger$ are annihilation and creation operators of phonons, and the functions $\mathbf{Q}_i(\mathbf{r})$ determine the spatial structure of oscillations at frequencies Ω_i. Expression (7) can be used when the amplitude of oscillations is small enough. One can always choose the normalization of functions $\mathbf{Q}_i(\mathbf{r})$ corresponding to standard commutation relations for bosons, $[\widehat{b}_i,\widehat{b}_j^\dagger] = \delta_{ij}$ and a standard form for the Hamiltonian of mechanical oscillations:

$$\widehat{H}_p = \hbar\sum_i \Omega_i\left(\widehat{b}_i^\dagger\,\widehat{b}_i + \frac{1}{2}\right). \tag{8}$$

2.4 An atom coupled to quantized EM and phonon fields

Now, we can combine all ingredients into a coupled quantized system. Adding the interaction Hamiltonian with a EM cavity mode in the electric dipole approximation, $-\widehat{\mathbf{d}}\cdot\widehat{\mathbf{E}}$, the Hamiltonian of an atom coupled to a single mode EM field can be written as follows:

$$\widehat{H} = \widehat{H}_{em} + \widehat{H}_a - \mathbf{d}(\widehat{\sigma}^\dagger + \widehat{\sigma})\cdot\left[\mathbf{E}(\mathbf{r})\widehat{c} + \mathbf{E}^*(\mathbf{r})\widehat{c}^\dagger\right]_{\mathbf{r}=\mathbf{r}_a}, \tag{9}$$

where $\mathbf{r} = \mathbf{r}_a$ denotes the position of an atom inside the cavity. The effect of "dressing" of the coupled atom–EM field system by mechanical oscillations in its most general form can be included by adding the Hamiltonian of phonon modes $\widehat{H}_p$ and substituting $\mathbf{r}_a \Rightarrow \mathbf{r}_a + \widehat{\mathbf{q}}$ in Eq. (9). This will work for an arbitrary relative displacement of an atom with respect to the EM cavity mode. Keeping only one phonon mode for simplicity, in which

$$\widehat{\mathbf{q}} = \mathbf{Q}(\mathbf{r})\widehat{b} + \mathbf{Q}^*(\mathbf{r})\widehat{b}^\dagger, \tag{10}$$

and expanding in Taylor series in the vicinity of $\mathbf{r} = \mathbf{r}_a$, we obtain the total Hamiltonian,

$$\begin{aligned}\widehat{H} = \widehat{H}_{em} + \widehat{H}_a + \widehat{H}_p - (\chi\widehat{\sigma}^\dagger\widehat{c} + \chi^*\widehat{\sigma}\widehat{c}^\dagger + \chi\widehat{\sigma}\widehat{c} + \chi^*\widehat{\sigma}^\dagger\widehat{c}^\dagger)\\ -(\eta_1\widehat{\sigma}^\dagger\widehat{c}\widehat{b} + \eta_1^*\widehat{\sigma}\widehat{c}^\dagger\widehat{b}^\dagger + \eta_2\widehat{\sigma}^\dagger\widehat{c}\widehat{b}^\dagger\\ +\eta_2^*\widehat{\sigma}\widehat{c}^\dagger\widehat{b} + \eta_1\widehat{\sigma}\widehat{c}\widehat{b} + \eta_1^*\widehat{\sigma}^\dagger\widehat{c}^\dagger\widehat{b}^\dagger + \eta_2\widehat{\sigma}\widehat{c}\widehat{b}^\dagger + \eta_2^*\widehat{\sigma}^\dagger\widehat{c}^\dagger\widehat{b}\right)\end{aligned} \tag{11}$$

where

$$\chi = (\mathbf{d}\cdot\mathbf{E})_{\mathbf{r}=\mathbf{r}_a},\quad \eta_1 = [\mathbf{d}(\mathbf{Q}\cdot\nabla)\mathbf{E}]_{\mathbf{r}=\mathbf{r}_a},\quad \eta_2 = \left[\mathbf{d}(\mathbf{Q}^*\cdot\nabla)\mathbf{E}\right]_{\mathbf{r}=\mathbf{r}_a}.$$

Our model corresponds to the situation when the amplitude of phonon (acoustic) or mechanical oscillations is much larger than the size of an atom. In this case, an acoustic field shifts the potential well for electrons as a whole, rather than deforming it. It is possible to have an opposite situation when the acoustic field deforms the potential well for electrons, thus modulating the dipole moment of the optical transition. This will change the expression for the effective constant of the parametric coupling but will not change the resulting Hamiltonian.

We can always take the functions $\mathbf{E}(\mathbf{r})$ and $\mathbf{Q}(\mathbf{r})$ to be real *at the position of an atom*, but we cannot keep the derivatives real at the same time if the modal structure $\propto e^{i\mathbf{k}\cdot\mathbf{r}}$. However, for ideal cavity modes, the latter is possible. As we will see below, the best conditions for electron–photon–phonon entanglement are reached in the vicinity of the *parametric resonance*:

$$\frac{W}{\hbar} \approx \omega \pm \Omega. \tag{12}$$

When the upper sign is chosen in Eq. (12), the RWA applied to the Hamiltonian (11) yields the following equation:

$$\hat{H} = \hat{H}_{em} + \hat{H}_a + \hat{H}_p - \left(\eta\hat{\sigma}^{\dagger}\hat{c}\hat{b} + \eta^*\hat{\sigma}\hat{c}^{\dagger}\hat{b}^{\dagger}\right) \tag{13}$$

where $\eta \equiv \eta_1$. For the lower sign in Eq. (12), the RWA Hamiltonian is as follows:

$$\hat{H} = \hat{H}_{em} + \hat{H}_a + \hat{H}_p - \left(\eta\hat{\sigma}^{\dagger}\hat{c}\hat{b}^{\dagger} + \eta^*\hat{\sigma}\hat{c}^{\dagger}\hat{b}\right) \tag{14}$$

where $\eta \equiv \eta_2$.

2.5 An atom coupled to the quantized EM field and dressed by a classical acoustic field

For classical acoustic oscillations, the operator $\hat{\mathbf{q}} = \mathbf{Q}(\mathbf{r})\hat{b} + \mathbf{Q}^*(\mathbf{r})\hat{b}^{\dagger}$ in Eq. (10) becomes a classical function

$$\mathbf{q} = \mathbf{Q}(\mathbf{r})\, e^{-i\Omega t} + \mathbf{Q}^*(\mathbf{r})\, e^{i\Omega t} \tag{15}$$

where $\mathbf{Q}$ is a coordinate-dependent complex amplitude of classical oscillations. Near the parametric resonance $\left(\omega + \Omega \approx \frac{W}{\hbar}\right)$, the RWA Hamiltonian takes the following form:

$$\hat{H} = \hat{H}_{em} + \hat{H}_a - \left(\mathfrak{R}\hat{\sigma}^{\dagger}\hat{c}e^{-i\Omega t} + \mathfrak{R}^*\hat{\sigma}\hat{c}^{\dagger}e^{i\Omega t}\right). \tag{16}$$

where $\mathfrak{R} = [\mathbf{d}(\mathbf{Q}\cdot\nabla)\mathbf{E}]_{\mathbf{r}=\mathbf{r}_a}$. The value of the acoustic frequency Ω in Eq. (16) can be of either sign, corresponding to the choice "±" in the parametric resonance condition Eq. (12); when the sign of Ω changes from positive to negative, one should replace $\mathbf{Q}$ with $\mathbf{Q}^*$ in the above expression for $\mathfrak{R}$.

Qualitatively, Hamiltonian (13) corresponds to the decay of the fermionic excitation into a photon and phonon; Hamiltonian (14) corresponds to the decay of a photon into a phonon and fermionic excitation, whereas Hamiltonian (16) describes parametric decay of a photon into an atomic excitation and back, mediated by classical acoustic oscillations.

3 Parametric resonance in a closed system

When the system is closed and there is no dissipation, the general analytic solution to the dynamics of coupled fermions, photons, and phonons can be obtained in the RWA. We write the state vector as follows:

$$\Psi = \sum_{\alpha,n=0}^{\infty} \left(C_{\alpha n0}|\alpha\rangle\,|n\rangle\,|0\rangle + C_{\alpha n1}|\alpha\rangle\,|n\rangle\,|1\rangle\right). \tag{17}$$

here, Greek letters denote phonon states, Latin letters denote photon states, and numbers 0, 1 describe fermion states. We will keep the same sequence of symbols throughout the paper:

$$C_{\text{phonon photon fermion}}\ |\text{phonon}\rangle\,|\text{photon}\rangle\,|\text{fermion}\rangle.$$

Next, we substitute Eq. (17) into the Schrödinger equation,

$$i\hbar\frac{\partial}{\partial t}|\Psi\rangle = \hat{H}|\Psi\rangle \tag{18}$$

where $\hat{H}$ is the RWA Hamiltonian. For definiteness, we consider the vicinity of the parametric resonance with a plus sign, $\omega + \Omega \approx \frac{W}{\hbar}$, which corresponds to the Hamiltonian (13). In this case, the equations for the coefficients in Eq. (17) can be separated into the pairs of coupled equations

$$\frac{d}{dt}\begin{pmatrix} C_{\alpha n0} \\ C_{(\alpha-1)(n-1)1} \end{pmatrix} + \begin{pmatrix} i\omega_{\alpha,n} & -i\Omega_R^{(\alpha,n)*} \\ -i\Omega_R^{(\alpha,n)} & i\omega_{\alpha,n} - i\Delta \end{pmatrix}\begin{pmatrix} C_{\alpha n0} \\ C_{(\alpha-1)(n-1)1} \end{pmatrix} = 0, \tag{19}$$

and a separate equation for the lowest energy state:

$$\dot{C}_{000} + i\omega_{0,0}C_{000} = 0, \tag{20}$$

where

$$\Omega_R^{(\alpha,n)} = \frac{\eta}{\hbar}\sqrt{\alpha n}, \quad \omega_{\alpha,n} = \Omega\left(\alpha + \frac{1}{2}\right) + \omega\left(n + \frac{1}{2}\right),$$
$$\Delta = \Omega + \omega - \frac{W}{\hbar}.$$

Note that approximate Eqs. (19) and (20) preserve the norm exactly:

$$|C_{000}|^2 + \sum_{\alpha=1,n=1}^{\infty,\infty} \left(|C_{\alpha n0}|^2 + |C_{(\alpha-1)(n-1)1}|^2 \right)$$
$$= \sum_{\alpha=0,n=0}^{\infty,\infty} \left(|C_{\alpha n0}|^2 + |C_{\alpha n1}|^2 \right) = \text{const.}$$

The solution to Eq. (20) is trivial: $C_{000}(t) = C_{000}(0)\exp(-i\omega_{0,0}t)$. The solution to Eq. (19) takes the following form:

$$\begin{pmatrix} C_{\alpha n0} \\ C_{(\alpha-1)(n-1)1} \end{pmatrix} = Ae^{-\Lambda_1^{(\alpha,n)}t} \begin{pmatrix} 1 \\ a_1^{(\alpha,n)} \end{pmatrix} + Be^{-\Lambda_2^{(\alpha,n)}t} \begin{pmatrix} 1 \\ a_2^{(\alpha,n)} \end{pmatrix}, \quad (21)$$

where the constants A and B are determined from initial conditions. Here, the eigenvalues $\Lambda_{1,2}^{(\alpha,n)}$ and eigenvectors $\begin{pmatrix} 1 \\ a_{1,2}^{(\alpha,n)} \end{pmatrix}$ of the matrix of coefficients in Eq. (19) are given as follows:

$$\Lambda_{1,2}^{(\alpha,n)} = i\omega_{\alpha,n} - i\delta_{1,2}^{(\alpha,n)}, \; a_{1,2}^{(\alpha,n)} = \frac{\delta_{1,2}^{(\alpha,n)}}{\Omega_R^{(\alpha,n)*}}, \quad (22)$$

where

$$\delta_{1,2}^{(\alpha,n)} = \frac{\Delta}{2} \pm \sqrt{\frac{\Delta^2}{4} + |\Omega_R^{(\alpha,n)}|^2}. \quad (23)$$

Figure 2 shows the eigenfrequencies of the system given by Eq. (22) with $\alpha = n = 1$, shifted by $\omega_{1,1}|_{\Delta=0}$. One can see the anticrossing with splitting by $2\Omega_R^{(1,1)}$ at the parametric resonance.

As an example, consider an exact parametric resonance $\frac{W}{\hbar} = \Omega + \omega$ and the simplest initial state $\Psi_0 = |0\rangle|0\rangle|1\rangle$ corresponding to the initially excited atom in a cavity. In this case, the only nonzero amplitudes are C_{001} and C_{110}:

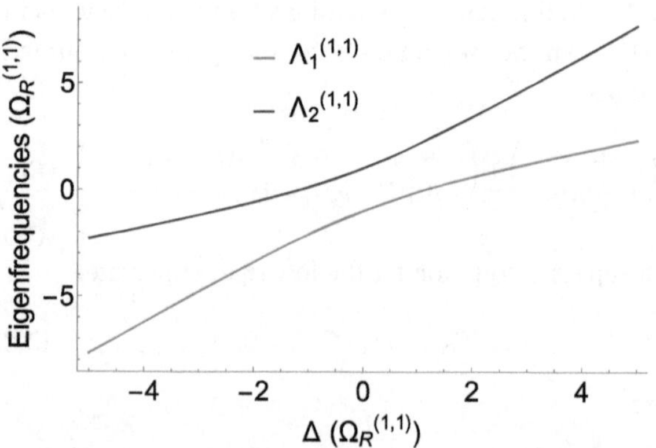

Figure 2: Frequency eigenvalues of the coupled electron–photon–phonon quantum system as a function of detuning from the parametric resonance $\frac{W}{\hbar} = \Omega + \omega$. All frequencies are in units of the generalized Rabi frequency $\Omega_R^{(1,1)}$. The values of eigenfrequencies are shifted vertically by $\omega_{1,1}|_{\Delta=0}$.

$$\begin{pmatrix} C_{110} \\ C_{001} \end{pmatrix} = \frac{1}{2} e^{-i\left(\omega_{1,1} - |\Omega_R^{(1,1)}|\right)t} \begin{pmatrix} e^{-i\theta} \\ 1 \end{pmatrix} + \frac{1}{2} e^{-i\left(\omega_{1,1} + |\Omega_R^{(1,1)}|\right)t} \begin{pmatrix} -e^{-i\theta} \\ 1 \end{pmatrix}, \quad (24)$$

where

$$\omega_{1,1} = \Omega\left(1 + \frac{1}{2}\right) + \omega\left(1 + \frac{1}{2}\right), \Omega_R^{(1,1)} = \frac{\eta}{\hbar} = |\Omega_R^{(1,1)}|e^{i\theta}.$$

The resulting state vector is given as follows:

$$\Psi = e^{-i\omega_{1,1}t}\left[ie^{-i\theta}\sin\left(|\Omega_R^{(1,1)}|t\right)|1\rangle\,|1\rangle\,|0\rangle + \cos\left(|\Omega_R^{(1,1)}|t\right)|0\rangle\,|0\rangle\,|1\rangle\right]. \quad (25)$$

This is clearly an entangled electron–photon–phonon state, which is not surprising. In the absence of dissipation, any coupling between these subsystems leads to entanglement.

State (25) is a tripartite entangled state which belongs to the family of Greenberger–Horne–Zeilinger (GHZ) states. It can be reduced to a standard GHZ state by local operations [37, 38], e.g., by rotations on the Bloch sphere of each qubit. In most cases discussed in the literature, the GHZ states are made of identical subsystems, e.g., photons [39, 40], which determine the way how they can be manipulated and used. In our case, each subsystem is of different nature: a fermionic electron system, a bosonic EM field, and a bosonic phonon field, so we envision at least two interesting applications. One is to determine the statistics of phonons or atomic excitations by measuring the statistics of photons. The latter is relatively easy to do, whereas phonon counting or direct measurement of quantized mechanical oscillations is extremely difficult due to their low energy and the lack of suitable detectors. The second application is control of a bipartite entangled state of two subsystems, say an atom and a photon mode, by using the state of the third subsystem, say phonons or mechanical oscillations, as a control handle. One example is given in Section 7 of the paper. One can come up with other combinations involving more complex qubits consisting of coupled fermion–boson subsystems, for example, an entanglement of the atom–phonon system by a classical EM field, etc.

The dynamics of the corresponding physical observables, such as the energy of the field and the atom, is Rabi oscillations at the frequency which generalizes a standard Rabi frequency to the case of a parametric photon–phonon–atom resonance and which depends on both the spatial structure of the photon and phonon fields and their occupation numbers:

$$\langle\Psi|\hat{\mathbf{E}}^2|\Psi\rangle = |\mathbf{E}(\mathbf{r})|^2\left[2 - \cos\left(2|\Omega_R^{(1,1)}|t\right)\right] \quad (26)$$

$$\langle\Psi|\hat{H}_a|\Psi\rangle = W\frac{1+\cos\left(2\left|\Omega_R^{(1,1)}\right|t\right)}{2} \quad (27)$$

It is illustrated in Figure 3 which shows the normalized EM field energy density and energy of an atom as a function of time. Note that the EM field energy never reaches zero because of the presence of zero-point vacuum energy. With detuning from the parametric resonance, the amplitude of the oscillations will decrease.

4 Dynamics of an open electron–photon–phonon system

4.1 Stochastic evolution equation

Now we include the processes of relaxation and decoherence in an open system, which is (weakly) coupled to a dissipative reservoir. We will use the stochastic equation of evolution for the state vector, which is derived in Appendix. This is basically the Schrödinger equation modified by adding a linear relaxation operator and the noise source term with appropriate correlation properties. The latter are related to the parameters of the relaxation operator, which is a manifestation of the fluctuation–dissipation theorem [41].

Within our approach, the system is described by a state vector which has a fluctuating component: $|\Psi\rangle = \overline{|\Psi\rangle} + \widetilde{|\Psi\rangle}$, where the straight bar means averaging over the statistics of noise, and the wavy bar denotes the fluctuating component. This state vector is of course very different from the state vector obtained by solving a standard Schrödinger equation for a closed system. In fact, coupling to a dissipative reservoir leads to the formation of a mixed state, which can be described by a density matrix $\hat{\rho} = \overline{|\Psi\rangle}\cdot\overline{\langle\Psi|} + \overline{\widetilde{|\Psi\rangle}\widetilde{\langle\Psi|}}$. In Appendix, we derived the general form of the stochastic equation of evolution from the Heisenberg–Langevin equations [23, 34, 42] and showed how physically reasonable constraints on the observables determine the properties of the noise sources. We also demonstrated the relationship between our approach and the Lindblad method of solving the master equation.

One can view the stochastic equation approach as a convenient formalism for calculating physical observables which allows one to obtain analytic solutions for the evolution of a coupled system even in the presence of dissipation and decoherence. In this section, we use the stochastic equation for the state vector given by Eqs. (A9) and (A10). The effective Hamiltonian in Eqs. (A9) and (A10) is determined by an approximation the user wants. If one wants the Markovian approximation, the Hamiltonian is obtained simply by summing up partial Lindbladians for all subsystems, whatever they are (in our case, these are a fermion emitter, an EM cavity mode, and a phonon mode). Then, the noise source term is determined unambiguously by conservation of the norm of the state vector and the requirement that the system should approach thermal equilibrium when the external perturbation is turned off. This immediately gives Eqs. (28) and (29) below.

Following the derivation in Appendix, Eqs. (19) and (20) are modified as follows:

$$\frac{d}{dt}\begin{pmatrix} C_{an0} \\ C_{(a-1)(n-1)1}\end{pmatrix} + \begin{pmatrix} i\omega_{a,n}+\gamma_{an0} & -i\Omega_R^{(a,n)*} \\ -i\Omega_R^{(a,n)} & i\omega_{a,n}-i\Delta+\gamma_{(a-1)(n-1)1}\end{pmatrix}\begin{pmatrix} C_{an0} \\ C_{(a-1)(n-1)1}\end{pmatrix} = -\frac{i}{\hbar}\begin{pmatrix} R_{an0} \\ R_{(a-1)(n-1)1}\end{pmatrix}, \quad (28)$$

$$\dot{C}_{000} + (i\omega_{0,0}+\gamma_{000})C_{000} = -\frac{i}{\hbar}R_{000}. \quad (29)$$

Coupling to a reservoir introduces two main differences to Eqs. (28) and (29) as compared to Eqs. (19) and (20) for a closed system. First, eigenfrequencies acquire imaginary parts which describe relaxation:

$$\omega_{a,n} \Rightarrow \omega_{a,n} - i\gamma_{an0},\ \omega_{a,n} - \Delta \Rightarrow \omega_{a,n} - \Delta - i\gamma_{(a-1)(n-1)1},$$
$$\omega_{0,0} \Rightarrow \omega_{0,0} - i\gamma_{000}.$$

The relaxation constants are determined by the properties of all subsystems. They are derived in Appendix, and their explicit form is given in the end of this section.

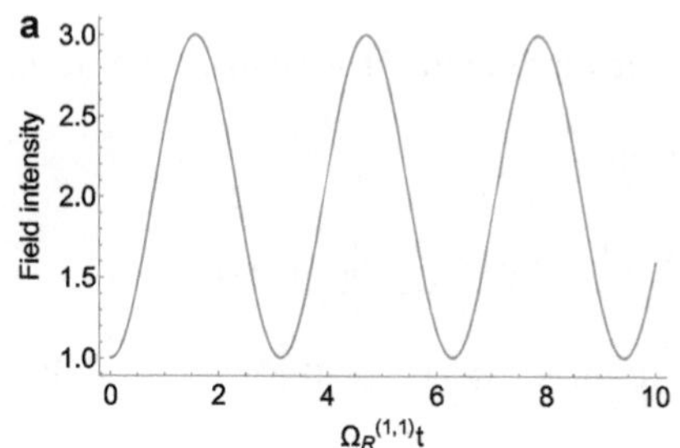

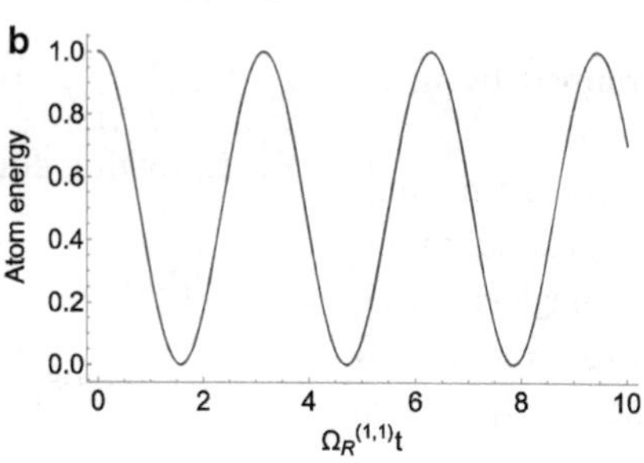

Figure 3: (a) Normalized field intensity, $\langle\Psi|\hat{\mathbf{E}}^2|\Psi\rangle/|\mathbf{E}(\mathbf{r})|^2$, and (b) normalized atom energy $\langle\Psi|\hat{H}_a|\Psi\rangle/W$ as a function of time in units of the generalized Rabi frequency $\Omega_R^{(1,1)}$.

Second, the right-hand side of Eqs. (28) and (29) contain noise sources $-\frac{i}{\hbar}R_{an0}$, $-\frac{i}{\hbar}R_{(\alpha-1)(n-1)1}$, and $-\frac{i}{\hbar}R_{000}$. They are equal to 0 after averaging over the noise statistics: $\overline{R_{an0}} = \overline{R_{(\alpha-1)(n-1)1}} = \overline{R_{000}}$. The averages of the quadratic combinations of noise source terms are nonzero. Including the noise sources is crucial for consistency of the formalism: it ensures the conservation of the norm of the state vector and leads to a physically meaningful equilibrium state. Note that the Weisskopf–Wigner theory does not enforce the conservation of the norm.

4.2 Evolution of the state amplitudes and observables

The solution to Eq. (29) is given as follows:

$$C_{000} = e^{-(i\omega_{0,0}+\gamma_{000})t}\left[C_{000}(0) - \frac{i}{\hbar}\int_0^t e^{(i\omega_{0,0}+\gamma_{000})\tau}R_{000}(\tau)d\tau\right]. \tag{30}$$

The solution to Eq. (28) is determined again by the eigenvalues and eigenvectors of the matrix of coefficients, which are now modified by relaxation rates:

$$\Lambda_{1,2}^{(\alpha,n)} = i\omega_{\alpha,n} - i\delta_{1,2}^{(\alpha,n)},\ a_{1,2}^{(\alpha,n)} = \frac{\delta_{1,2}^{(\alpha,n)} - i\gamma_{an0}}{\Omega_R^{(\alpha,n)*}}, \tag{31}$$

where

$$\delta_{1,2}^{(\alpha,n)} = \frac{\Delta}{2} + i\frac{\gamma_{an0} + \gamma_{(\alpha-1)(n-1)1}}{2} \pm \sqrt{\frac{\left[\Delta + i\left(\gamma_{(\alpha-1)(n-1)1} - \gamma_{an0}\right)\right]^2}{4} + \left|\Omega_R^{(\alpha,n)}\right|^2}. \tag{32}$$

The solution to Eq. (28) takes the following form:

$$\begin{pmatrix} C_{an0} \\ C_{(\alpha-1)(n-1)1} \end{pmatrix} = e^{-\Lambda_1^{(\alpha,n)}t}\begin{pmatrix} 1 \\ a_1^{(\alpha,n)} \end{pmatrix}\left(A - \frac{i}{\hbar}\int_0^t e^{\Lambda_1^{(\alpha,n)}\tau}\frac{R_{an0}(\tau)a_2^{(\alpha,n)} - R_{(\alpha-1)(n-1)1}(\tau)}{a_2^{(\alpha,n)} - a_1^{(\alpha,n)}}d\tau\right) + e^{-\Lambda_2^{(\alpha,n)}t}\begin{pmatrix} 1 \\ a_2^{(\alpha,n)} \end{pmatrix}\left(B - \frac{i}{\hbar}\int_0^t e^{\Lambda_2^{(\alpha,n)}\tau}\frac{R_{(\alpha-1)(n-1)1}(\tau) - R_{an0}(\tau)a_1^{(\alpha,n)}}{a_2^{(\alpha,n)} - a_1^{(\alpha,n)}}d\tau\right) \tag{33}$$

where the constants A and B are determined by initial conditions.

As an example, we consider the reservoir at low temperatures, when the steady-state population should go to the ground state $|0\rangle|0\rangle|0\rangle$. Also, we will neglect purely elastic dephasing processes which lead to atomic decoherence without changing the populations. The elastic processes will be added later. In this case, we can take $\gamma_{000} = 0$, as shown below. We will also assume that the only nonzero correlator of noise is delta-correlated in time:

$$\overline{R_{000}(t+\xi)R_{000}^*(t)} = \hbar^2\delta(\xi)D_{000}. \tag{34}$$

Then, Eqs. (29) and (30) yield the following:

$$\frac{d}{dt}\overline{|C_{000}|^2} = D_{000}, \tag{35}$$

whereas Eq. (28) gives the following:

$$\frac{d}{dt}\left(\overline{|C_{an0}|^2} + \overline{\left|C_{(\alpha-1)(n-1)1}\right|^2}\right) = -2\left(\gamma_{an0}\overline{|C_{an0}|^2} + \gamma_{(\alpha-1)(n-1)1}\overline{\left|C_{(\alpha-1)(n-1)1}\right|^2}\right). \tag{36}$$

This equation guarantees that the system occupies the ground state at $t \to \infty$.

The noise intensity is determined by the condition that the norm of the state vector be conserved. This gives the following equation:

$$D_{000} = 2\sum_{\alpha=1,n=1}^{\infty,\infty}\left(\gamma_{an0}\overline{|C_{an0}|^2} + \gamma_{(\alpha-1)(n-1)1}\overline{\left|C_{(\alpha-1)(n-1)1}\right|^2}\right). \tag{37}$$

In Appendix, we discuss in detail the dependence of the noise correlator on the averaged dyadic components of the state vector. We also show how to find the correlators which ensure that the system approaches thermal distribution at a finite temperature.

The above formalism allows us to obtain analytic solutions to the state vector and observables at any temperatures and detunings from the parametric resonance, while still within the RWA limits. However, the resulting expressions are very cumbersome, and they are better to visualize in the plots. Let us give an example of the solution at zero reservoir temperature and exactly at the parametric resonance $\frac{W}{\hbar} = \Omega + \omega$, when the expressions are more manageable. Consider the initial state $\Psi_0 = |0\rangle|0\rangle|1\rangle$ when an atom is excited and boson modes are in the ground state. In this case, the only nonzero amplitudes are C_{000}, C_{001}, and C_{110}. To make the algebra a bit simpler, we assume that the dissipation is weak enough and its effect on the eigenvectors $\begin{pmatrix} 1 \\ a_{1,2}^{(\alpha,n)} \end{pmatrix}$ can be neglected. As a result, we obtain the following equation:

$$\Psi = e^{-\left(i\omega_{1,1}+\frac{\gamma_{001}+\gamma_{110}}{2}\right)t}\left[ie^{-i\theta}\sin\left(\left|\tilde{\Omega}_R^{(1,1)}\right|t\right)|1\rangle|1\rangle|0\rangle + \cos\left(\left|\tilde{\Omega}_R^{(1,1)}\right|t\right)|0\rangle|0\rangle|1\rangle\right] + C_{000}|0\rangle|0\rangle|0\rangle, \tag{38}$$

where

$$\overline{|C_{000}|^2} = 1 - e^{-(\gamma_{110}+\gamma_{001})t},\ \tilde{\Omega}_R^{(1,1)} = \sqrt{\left|\Omega_R^{(1,1)}\right|^2 - \frac{(\gamma_{001}-\gamma_{110})^2}{4}},$$
$$\theta = \mathrm{Arg}\left[\Omega_R^{(1,1)}\right].$$

As we see, dissipation leads not only to the relaxation of the entangled part of the state vector but also to the frequency shift of the Rabi oscillations. This shift is absent if $\gamma_{001} = \gamma_{110}$.

The resulting expressions for the observables such as the EM field intensity and the energy of the atomic excitation are given as follows:

$$\langle\Psi|\hat{\mathbf{E}}^2|\Psi\rangle = |\mathbf{E}(\mathbf{r})|^2\Big[1 + e^{-(\gamma_{110}+\gamma_{001})t} - \cos\left(2\left|\tilde{\Omega}_R^{(1,1)}\right|t\right)e^{-(\gamma_{110}+\gamma_{001})t}\Big], \quad (39)$$

$$\langle\Psi|\hat{H}_a|\Psi\rangle = W\frac{1+\cos\left(2\left|\tilde{\Omega}_R^{(1,1)}\right|t\right)}{2}e^{-(\gamma_{110}+\gamma_{001})t} \quad (40)$$

Figure 4 illustrates the dynamics of observables in Eqs. (39) and (40).

Note that the Weisskopf–Wigner theory would give the same expression (40) for the atomic energy but a wrong expression for the EM field intensity:

$$\langle\Psi|\hat{\mathbf{E}}^2|\Psi\rangle = |\mathbf{E}(\mathbf{r})|^2\left[2 - \cos\left(2\left|\tilde{\Omega}_R^{(1,1)}\right|t\right)\right]e^{-(\gamma_{110}+\gamma_{001})t},$$

which does not approach the correct vacuum state.

4.3 Emission spectra

According to the study by Scully and Zubairy [23], the power spectrum of the emission is given as follows:

$$S(\mathbf{r},\nu) = \frac{1}{\pi}\mathrm{Re}\int_0^\infty d\tau G^{(1)}(\mathbf{r},\mathbf{r};\tau)e^{i\nu\tau}, \quad (41)$$

where $G^{(1)}(\mathbf{r},\mathbf{r};\tau)$ is the field autocorrelation function at the position $\mathbf{r}$ of the detector:

$$G^{(1)}(\mathbf{r},\mathbf{r};\tau) = |\mathbf{E}(\mathbf{r})|^2\int_0^\infty dt\overline{\langle\hat{c}_d^\dagger(t)\hat{c}_d(t+\tau)\rangle}. \quad (42)$$

where $\hat{c}_d(t), \hat{c}_d^\dagger(t)$ are annihilation and creation operators for the photons which interact with the detector, and the Heisenberg picture is used. We will assume that the coupling between the photons and the detector is weak, so the photon detection does not affect the dynamics of the intracavity photons. According to the study by Madsen and Lodahl [45], $\hat{c}_d(t) \propto \hat{c}(t)$ for a nanocavity, so we can calculate the $G^{(1)}(\mathbf{r},\mathbf{r};\tau)$ using operators for the cavity field $\hat{c}(t), \hat{c}^\dagger(t)$, up to a constant factor in the result. Note that the lower limit of the integral over t is set to be $t = 0$, which requires that no photons exist before $t = 0$.

In the Heisenberg–Langevin approach, an operator in the Heisenberg picture can be expressed through Schrödinger's operators using the effective Hamiltonian $\hat{H}_{\mathrm{eff}}$, which contains the anti-Hermitian part, see the Appendix. At the same time, the inhomogeneous term proportional to the noise sources should be added. Including these noise terms in the solution for the field operators when calculating the emission spectra is equivalent to taking into account the detection of thermal radiation which seeps into the cavity from outside and spontaneous emission resulting from thermal excitation of an atom. We assume that the reservoir temperature in energy units is much lower than W and $\hbar\omega$ so that the contribution of these noise terms to the emission spectra can be neglected (although noise is still needed to preserve the norm).

Then, the average correlator $\overline{\langle\hat{c}^\dagger(t)\hat{c}(t+\tau)\rangle}$ is expressed as follows:

$$\overline{\langle\hat{c}^\dagger(t)\hat{c}(t+\tau)\rangle}$$
$$= \overline{\langle\Psi(t=0)|e^{i\hat{H}_{\mathrm{eff}}^\dagger t/\hbar}\hat{c}^\dagger e^{-i\hat{H}_{\mathrm{eff}}t/\hbar}e^{i\hat{H}_{\mathrm{eff}}^\dagger(t+\tau)/\hbar}\hat{c}e^{-i\hat{H}_{\mathrm{eff}}(t+\tau)/\hbar}|\Psi(t=0)\rangle}$$
$$= \overline{\langle\Psi(t)|\hat{c}^\dagger e^{-i\hat{H}_{\mathrm{eff}}t/\hbar}e^{i\hat{H}_{\mathrm{eff}}^\dagger(t+\tau)/\hbar}\hat{c}|\Psi(t+\tau)\rangle}, \quad (43)$$

where $|\Psi(t)\rangle$ is the state vector of the system which we found in the previous subsection. It can be written as $|\Psi(t)\rangle = \sum_{n=0}^{\infty}C_n(t)|n\rangle\left|\Psi_n^{a,e}(t)\right\rangle$, where $\left|\Psi_n^{a,e}(t)\right\rangle$ is the part describing phonons and electrons. Therefore,

Consider a simple example when the initial state is $|0\rangle|0\rangle|1\rangle$. Within the RWA, the system can only reach states

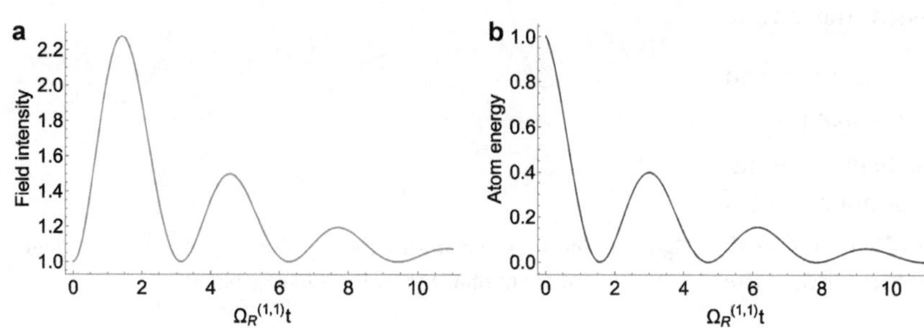

Figure 4: (a) Normalized field intensity, $\langle\Psi|\hat{\mathbf{E}}^2|\Psi\rangle/|\mathbf{E}(\mathbf{r})|^2$, and (b) normalized atom energy $\langle\Psi|\hat{H}_a|\Psi\rangle/W$ a function of time in units of the generalized Rabi frequency $\Omega_R^{(1,1)}$. Here, $\gamma_{110} + \gamma_{001} = 0.3\Omega_R^{(1,1)}$.

$$\begin{aligned}
&\overline{\langle \hat{c}^\dagger(t)\hat{c}(t+\tau)\rangle} \\
&= \left(\sum_{n=0}^{\infty} C_n^*(t)\langle n|\langle \Psi_n^{\alpha,e}(t)|\right)\hat{c}^\dagger e^{-i\widehat{H}_{\text{eff}}t/\hbar} e^{i\widehat{H}_{\text{eff}}^\dagger(t+\tau)/\hbar}\hat{c}\left(\sum_{n=0}^{\infty} C_n(t+\tau)|n\rangle|\Psi_n^{\alpha,e}(t+\tau)\rangle\right) \\
&= \left(\sum_{n=0}^{\infty} \sqrt{n}\,C_n^*(t)\langle n-1|\langle \Psi_n^{\alpha,e}(t)|\right) e^{-i\widehat{H}_{\text{eff}}t/\hbar} e^{i\widehat{H}_{\text{eff}}^\dagger(t+\tau)/\hbar}\left(\sum_{n=0}^{\infty} \sqrt{n}C_n(t+\tau)|n-1\rangle|\Psi_n^{\alpha,e}(t+\tau)\rangle\right)
\end{aligned} \quad (44)$$

$|0\rangle|0\rangle|1\rangle$, $|1\rangle|1\rangle|0\rangle$, and $|0\rangle|0\rangle|0\rangle$. After acting with $\hat{c}$ on a state of the system, a new state $|1\rangle|0\rangle|0\rangle$ can also appear, but it cannot evolve into other states. So, in this case, we have the following:

$$\begin{aligned}
&\overline{\langle \hat{c}^\dagger(t)\hat{c}(t+\tau)\rangle} \\
&= (C_1^*(t)\langle 0|\langle \Psi_1^{\alpha,e}(t)|)e^{-i\widehat{H}_{\text{eff}}t/\hbar}e^{i\widehat{H}_{\text{eff}}^\dagger(t+\tau)/\hbar}(C_1(t+\tau)|0\rangle|\Psi_1^{\alpha,e}(t+\tau)\rangle)) \\
&= (C_{110}^*(t)\langle 1|\langle 0|\langle 0|)e^{-i\widehat{H}_{\text{eff}}t/\hbar}e^{i\widehat{H}_{\text{eff}}^\dagger(t+\tau)/\hbar}(C_{110}(t+\tau)|1\rangle|0\rangle|0\rangle) \\
&= C_{110}^*(t)C_{110}(t+\tau)\exp\left[i\omega_{1,0}\tau - \gamma_{100}(2t+\tau)\right],
\end{aligned} \quad (45)$$

where we used Eqs. (A32) and (A33) and assumed that the noise for state $|1\rangle|0\rangle|0\rangle$ has zero correlator. Since

$$C_{110}(t) = i\sin\left(\left|\tilde{\Omega}_R^{(1,1)}\right|t\right)\exp\left[-i\omega_{1,1}t - \frac{\gamma_{110}+\gamma_{001}}{2}t\right], \quad (46)$$

we obtain

$$\begin{aligned}
\overline{\langle \hat{c}^\dagger(t)\hat{c}(t+\tau)\rangle} &= \sin\left(\left|\tilde{\Omega}_R^{(1,1)}\right|t\right)\sin\left(\left|\tilde{\Omega}_R^{(1,1)}\right|(t+\tau)\right) \\
&\exp\left[-i\omega\tau\right]\exp\left[-\gamma_{\text{ac}}(2t+\tau)\right],
\end{aligned} \quad (47)$$

where we introduced the notation $\gamma_{\text{ac}} \equiv \gamma_{100} + \frac{\gamma_{110}+\gamma_{001}}{2}$. Then, the power spectrum is found to be

$$\begin{aligned}
S(\mathbf{r},\nu) &\propto \frac{1}{\pi}|\boldsymbol{E}(\mathbf{r})|^2\frac{\left|\tilde{\Omega}_R^{(1,1)}\right|^2}{4\gamma_{\text{ac}}\left(\left|\tilde{\Omega}_R^{(1,1)}\right|^2+\gamma_{\text{ac}}^2\right)} \\
&\text{Re}\left[\frac{2\gamma_{\text{ac}} - i(\nu-\omega)}{[\gamma_{\text{ac}} - i(\nu-\omega)]^2 + \left|\tilde{\Omega}_R^{(1,1)}\right|^2}\right].
\end{aligned} \quad (48)$$

The normalized power spectra are shown in Figure 5 for various values of $\left|\tilde{\Omega}_R^{(1,1)}\right|/\gamma_{\text{ac}}$. For $\left|\tilde{\Omega}_R^{(1,1)}\right| < \gamma_{\text{ac}}$, the spectrum has a single maximum at zero detuning $\nu = \omega$. For $\left|\tilde{\Omega}_R^{(1,1)}\right| > \gamma_{\text{ac}}$, the spectra are split and their maxima (same value for all spectra) are reached at detunings given by $(\nu-\omega)^2 = \left|\tilde{\Omega}_R^{(1,1)}\right|^2 - \gamma_{\text{ac}}^2$. Therefore, to reach the strong coupling regime, the Rabi frequency $\left|\tilde{\Omega}_R^{(1,1)}\right|$ has to exceed the combination of the decoherence rates denoted by γ_{ac}.

Note that in a standard Lindblad formalism, γ parameters are inverse decay times for the populations or field energies, i.e., they are related to the quantities which are quadratic with respect to the state vector within the Schrödinger's approach. That is why our γ's in the stochastic equation of evolution, which is linear with respect to the state vector, correspond to the Lindblad γ's divided by 2.

4.4 Relaxation rates

Finally, we give explicit expressions for the relaxation constants γ_{an0} and γ_{an1}. They were derived in Appendix using the Lindblad master equation approach and assuming statistical independence of "partial" dissipative reservoirs for the atomic, EM, and phonon subsystems. Within the model which neglects purely elastic dephasing processes, the result is as follows:

$$\begin{aligned}
\gamma_{an0} &= \frac{\gamma}{2}N_1^{T_a} + \frac{\mu_\omega}{2}\left[\overline{n}_\omega^{T_{em}}(n+1) + \left(\overline{n}_\omega^{T_{em}}+1\right)n\right] \\
&+ \frac{\mu_\Omega}{2}\left[\overline{n}_\Omega^{T_p}(\alpha+1) + \left(\overline{n}_\Omega^{T_p}+1\right)\alpha\right],
\end{aligned} \quad (49)$$

$$\begin{aligned}
\gamma_{an1} &= \frac{\gamma}{2}N_0^{T_a} + \frac{\mu_\omega}{2}\left[\overline{n}_\omega^{T_{em}}(n+1) + \left(\overline{n}_\omega^{T_{em}}+1\right)n\right] \\
&+ \frac{\mu_\Omega}{2}\left[\overline{n}_\Omega^{T_p}(\alpha+1) + \left(\overline{n}_\Omega^{T_p}+1\right)\alpha\right],
\end{aligned} \quad (50)$$

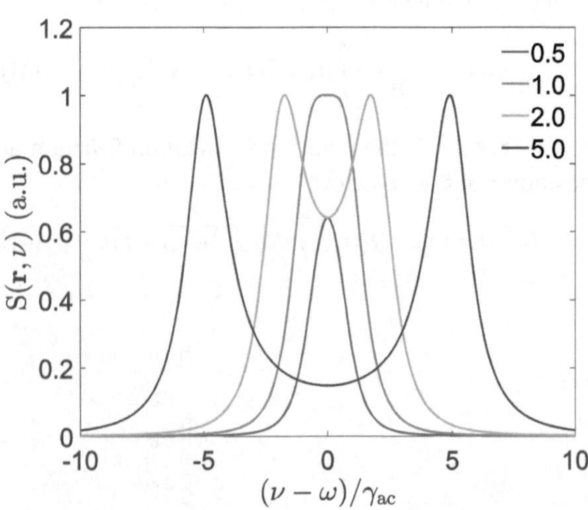

Figure 5: The emission spectra for $\left|\tilde{\Omega}_R^{(1,1)}\right|/\gamma_{\text{ac}}$ equal to 0.5, 1, 2, and 5. All spectra are normalized by the same constant.

where γ, μ_ω, and μ_Ω are partial relaxation rates of the atomic, photon, and phonon subsystems, respectively; $N_0^{T_a} = \frac{1}{1+e^{-\frac{W}{T_a}}}$, $N_1^{T_a} = \frac{e^{-\frac{W}{T_a}}}{1+e^{-\frac{W}{T_a}}}$, $\bar{n}_\omega^{T_{em}} = \frac{1}{e^{\frac{\hbar\omega}{T_{em}}}-1}$, $\bar{n}_\Omega^{T_p} = \frac{1}{e^{\frac{\hbar\Omega}{T_p}}-1}$ are their occupation numbers at thermal equilibrium; $T_{a,em,p}$ are temperatures of partial atom, photon, and phonon reservoirs in energy units, respectively. As a reminder, the atom energy is equal to 0 in state $|0\rangle$ and W in state $|1\rangle$.

If all reservoirs are at zero temperature, we obtain the following:

$$\gamma_{an0} = \frac{\mu_\omega}{2}n + \frac{\mu_\Omega}{2}\alpha, \quad \gamma_{an1} = \frac{\gamma}{2} + \frac{\mu_\omega}{2}n + \frac{\mu_\Omega}{2}\alpha. \tag{51}$$

Eq. (51) shows that $\gamma_{000} = 0$, validating our choice earlier in this section. We also obtain physically intuitive expressions for γ_{110} and γ_{001}: $\gamma_{110} = \frac{\mu_\omega}{2} + \frac{\mu_\Omega}{2}$, $\gamma_{001} = \frac{\gamma}{2}$.

Purely elastic processes which lead to the atomic decoherence with characteristic time $1/\gamma_{el}$ can be taken into account by adding γ_{el} to the relaxation constant γ_{001}. At the same time, one has to modify the noise correlator $\overline{R_{001}(t+\xi)R_{001}^{*}(t)}$ by adding to it the quantity $2\gamma_{el}\hbar^2\overline{|C_{001}|^2}$. This prescription will lead to correct dynamics of the observables; see the last section in the Appendix. Note that the population relaxation times will not depend on γ_{el} as it should be.

The above expressions allow one to quickly estimate the feasibility of reaching strong coupling regime and quantum entanglement when all fields are quantized. In a semiconductor dielectric cavity at near-infrared wavelengths ~1 μm for the refractive index ~3.5 and a typical dipole matrix element of the interband optical transition d/e~0.5 nm, the maximum vacuum Rabi frequency is of the order of 100–200 μeV [13, 14], diffraction limited by the cavity size. This sets the upper limit for the sum of relaxation rates $\gamma_{ac} \equiv \gamma_{100} + \frac{\gamma_{110}+\gamma_{001}}{2}$ introduced above. In the experiments with single QDs [13, 14], the phonons were not involved and the sum of relaxation rates in the electron and cavity subsystems was kept below 100 μeV at low temperatures, allowing them to reach the strong coupling regime.

In plasmonic cavities, a sub-nm field localization can be achieved, leading to Rabi frequencies of the order of 100–200 meV for the same order of the transition dipole moments. However, the combined relaxation rate is much higher, up to ~100 meV, typically dominated by cavity losses. Here, the strong coupling of plasmons to a single quantum emitter has been achieved in multiple experiments, as discussed in the Section 1. However, reaching the quantum regime for the EM and especially the phonon fields remains a challenge. It could be beneficial to consider longer wavelength emitters with the optical transition at the midinfrared and even terahertz wavelengths. Indeed, with increasing wavelength, the plasmon losses go down and the matrix element of a dipole-allowed transition increases, whereas the plasmon localization stays largely the same.

5 Classical acoustic or mechanical oscillations

Quantization of acoustic or mechanical oscillations is very difficult to achieve because of their low energy of the quantum. In most experiments, they remain classical and therefore cease to be an independent degree of freedom. Instead, their amplitude becomes an external time-dependent parameter, like an external pumping. In this case, the RWA Hamiltonian is given by Eq. (16). It depends only on quantum operators $\hat{\sigma}, \hat{\sigma}^\dagger$ and $\hat{c}, \hat{c}^\dagger$; therefore, the state vector has to be expanded over the basis states $|n\rangle\,|0\rangle$ and $|n\rangle\,|1\rangle$:

$$\Psi = \sum_{n=0}^{\infty} (C_{n0}|n\rangle|0\rangle + C_{n1}|n\rangle|1\rangle). \tag{52}$$

Substituting Eq. (52) in the Schrödinger equation with the Hamiltonian (16), we again get separation into a block of two equations,

$$\dot{C}_{n0} = -i\omega_n C_{n0} + i\frac{\mathfrak{R}^*}{\hbar}e^{i\Omega t}C_{(n-1)1}\sqrt{n}, \tag{53}$$

$$\dot{C}_{(n-1)1} = -i\left(\omega_{n-1} + \frac{W}{\hbar}\right)C_{(n-1)1} + i\frac{\mathfrak{R}}{\hbar}e^{-i\Omega t}C_{n0}\sqrt{n}, \tag{54}$$

and a separate equation for the amplitude of the ground state $|0\rangle\,|0\rangle$ of the system:

$$\dot{C}_{00} = -i\omega_0 C_{00}, \tag{55}$$

where $\omega_n = \omega\left(n+\frac{1}{2}\right)$ and $\mathfrak{R} = [\mathbf{d}(\mathbf{Q}\cdot\nabla)\mathbf{E}]_{\mathbf{r}=\mathbf{r}_a}$ (see Eq. (16)). After making the substitution $C_{(n-1)1} = G_{(n-1)1}e^{-i\Omega t}$, Eqs. (53) and (54) give the equations similar in form to Eq. (19):

$$\frac{d}{dt}\begin{pmatrix} C_{n0} \\ G_{(n-1)1} \end{pmatrix} + \begin{pmatrix} i\omega_n & -i\Omega_R^{(n)*} \\ -i\Omega_R^{(n)} & i\omega_n - i\Delta \end{pmatrix}\begin{pmatrix} C_{n0} \\ G_{(n-1)1} \end{pmatrix} = 0, \tag{56}$$

where

$$\Omega_R^{(n)} = \frac{\mathfrak{R}}{\hbar}\sqrt{n}, \; \Delta = \Omega + \omega - \frac{W}{\hbar}, \; \omega_n - \Delta = \omega_{n-1} + \frac{W}{\hbar}.$$

Eqs. (55) and (56) are different from Eqs. (19) and (20) only in one aspect: they do not contain the index of the quantum state of the phonon field, whereas the Rabi frequency depends on the amplitude of classical acoustic oscillations $\mathbf{Q}(\mathbf{r}_a)$, see Section 2.5. Obviously, the solution to Eqs. (55)

and (56) will have the same form and the expressions (26), (27) for the observables will remain the same, after dropping the index of the quantum phonon state and redefining the Rabi frequency.

Dissipation due to coupling to a reservoir can be included using the stochastic equation of evolution of the state vector, see the Appendix. The corresponding equations are again similar to those for a fully quantum problem given by Eqs. (28) and (29):

$$\dot{C}_{00} + i(\omega_0 + \gamma_{00})C_{00} = -\frac{i}{\hbar}R_{00}, \tag{57}$$

$$\frac{d}{dt}\begin{pmatrix} C_{n0} \\ C_{(n-1)1} \end{pmatrix} + \begin{pmatrix} i\omega_n + \gamma_{n0} & -i\Omega_R^{(n)*} \\ -i\Omega_R^{(n)} & i\omega_n - i\Delta + \gamma_{(n-1)1} \end{pmatrix}\begin{pmatrix} C_{n0} \\ C_{(n-1)1} \end{pmatrix} = -\frac{i}{\hbar}\begin{pmatrix} R_{n0} \\ R_{(n-1)1} \end{pmatrix}. \tag{58}$$

Since the acoustic field is now a given external pumping, the relaxation constants should not depend on the parameters of a phonon reservoir. They can be obtained after obvious simplification of Eqs. (49) and (50):

$$\gamma_{n0} = \frac{\gamma}{2}N_1^{T_a} + \frac{\mu_\omega}{2}\left[\bar{n}_\omega^{T_{em}}(n+1) + \left(\bar{n}_\omega^{T_{em}} + 1\right)n\right], \tag{59}$$

$$\gamma_{n1} = \frac{\gamma}{2}N_0^{T_a} + \frac{\mu_\omega}{2}\left[\bar{n}_\omega^{T_{em}}(n+1) + \left(\bar{n}_\omega^{T_{em}} + 1\right)n\right], \tag{60}$$

All expressions for the state vector and observables can be obtained from the corresponding expressions in Section 4 after dropping the index α of the quantum state of the phonon field and redefining the frequency of Rabi oscillations.

6 Separation and interplay of the parametric and one-photon resonance

For an electron system coupled to a EM cavity mode and dressed by a phonon field, the phonon frequency Ω can be much lower than the optical frequency. In this case, the overlap of the parametric (three-wave) resonance $\omega \pm \Omega \approx \frac{W}{\hbar}$ and the one-photon (two-wave) resonance $\omega \approx \frac{W}{\hbar}$ can be an issue.

Here, we derive the analytic criteria for the separation of these resonances and show numerically what happens when they overlap. Throughout this section, we neglect losses. Since the parametric and one-photon resonances are separated by the phonon frequency Ω, the losses can be neglected when the value of Ω exceeds the sum of the spectral widths of the EM cavity mode and the electron transition which originate from dissipation and decoherence. If this condition is violated, resonances overlap strongly and their separation is impossible anyway.

The separation criterion imposes certain restrictions on the Rabi frequencies of the two resonances. To derive these restrictions, we neglect dissipation and retain in the Hamiltonian (11) both the RWA terms near the parametric resonance $\omega \pm \Omega \approx \frac{W}{\hbar}$ and the terms near a one-photon resonance $\omega \approx \frac{W}{\hbar}$. Since the result will be almost the same whether the phonon field is quantized or classical, we will consider the classical phonon field to keep the expressions a bit shorter. The resulting Hamiltonian is given as follows:

$$\hat{H} = \hbar\omega\left(\hat{c}^\dagger\hat{c} + \frac{1}{2}\right) + W\hat{\sigma}^\dagger\hat{\sigma} - (\chi + \Re e^{-i\Omega t})\hat{\sigma}^\dagger\hat{c} - (\chi^* + \Re^* e^{i\Omega t})\hat{\sigma}\hat{c}^\dagger, \tag{61}$$

where $\chi = (\mathbf{d}\cdot\mathbf{E})_{\mathbf{r}=\mathbf{r}_a}$ and $\Re$ was defined in Eq. (16); $\mathbf{Q}$ is now a complex-valued amplitude of classical phonon oscillations. The value of Ω in Eq. (61) can be both positive and negative, corresponding to the choice of an upper or lower sign in the parametric resonance condition $\omega \pm \Omega \approx \frac{W}{\hbar}$. The change of sign in Ω corresponds to replacing $\mathbf{Q}$ with $\mathbf{Q}^*$ in the expression for $\Re$.

The state vector should be sought in the form of Eq. (52). After substituting it into the Schrödinger equation, we obtain coupled equations for the amplitudes of basis states $|n\rangle|0\rangle$, $|n-1\rangle|1\rangle$:

$$\dot{C}_{n0} + i\omega_n C_{n0} - \frac{i}{\hbar}(\chi^* + \Re^* e^{i\Omega t})\sqrt{n}C_{(n-1)1} = 0, \tag{62}$$

$$\dot{C}_{(n-1)1} + i\left(\omega_{n-1} + \frac{W}{\hbar}\right)C_{(n-1)1} - \frac{i}{\hbar}(\chi + \Re e^{-i\Omega t})\sqrt{n}C_{n0} = 0, \tag{63}$$

and

$$\dot{C}_{00} + i\omega_0 C_{00} = 0, \tag{64}$$

where $\omega_n = \omega\left(n + \frac{1}{2}\right)$. To compare these equations with Eqs. (53) and (54), it is convenient to assume that the system is exactly at one of the resonances and study the behavior of the solution with increasing the detuning from another resonance. For example, we assume an exact parametric resonance $\omega + \Omega = \frac{W}{\hbar}$. In this case, the detuning from the two-wave resonance is $\frac{W}{\hbar} - \omega = \Omega$. After the substitution $C_{n0} = G_{n0}e^{-i\omega_n t}$ and $C_{(n-1)1} = G_{(n-1)1}e^{-i\left(\omega_{n-1} + \frac{W}{\hbar}\right)t}$, we obtain from Eqs. (62) and (63) that

$$\dot{G}_{n0} - \frac{i}{\hbar}(\chi^* + \Re^* e^{i\Omega t})\sqrt{n}G_{(n-1)1}e^{-i\left(\frac{W}{\hbar} - \omega\right)t} = 0, \tag{65}$$

$$\dot{G}_{(n-1)1} - \frac{i}{\hbar}\left(\chi + \Re e^{-i\Omega t}\right)\sqrt{n}G_{n0}e^{i\left(\frac{W}{\hbar}-\omega\right)t} = 0. \quad (66)$$

If we neglect at first the perturbation of the system in the vicinity of the two-wave resonance, the solution to Eqs. (65) and (66) at $\chi = 0$ is

$$\begin{pmatrix} G_{n0} \\ G_{(n-1)1} \end{pmatrix} = Ae^{i\Omega_R^{(3)}t}\begin{pmatrix} 1 \\ 1 \end{pmatrix} + Be^{-i\Omega_R^{(3)}t}\begin{pmatrix} 1 \\ -1 \end{pmatrix}, \quad (67)$$

where $\Omega_R^{(3)} = \frac{1}{\hbar}\Re\sqrt{n}$ is the Rabi frequency of the parametric resonance and A and B are arbitrary constants. The state described by Eq. (67) is obviously entangled.

To write the formal solution to Eqs. (65) and (66), we make another substitution of variables: $G_{n0} \pm G_{(n-1)1} = G_\pm$. The result is

$$\dot{G}_\pm \mp i\Omega_R^{(3)}G_\pm = i\Omega_R^{(2)*}e^{-i\Omega t}G_{n0} \pm i\Omega_R^{(2)}e^{i\Omega t}G_{(n-1)1},$$

where $\Omega_R^{(2)} = \frac{1}{\hbar}\chi\sqrt{n}$ is the Rabi frequency corresponding to the one-photon (two-wave) resonance. The solution to the last equation is given as follows:

$$G_\pm = 2(A,B)e^{\pm i\Omega_R^{(3)}t} + ie^{\pm i\Omega_R^{(3)}t}\int_0^t e^{\mp i\Omega_R^{(3)}\tau}\left[\Omega_R^{(2)*}e^{-i\Omega\tau}G_{n0}(\tau) \pm \Omega_R^{(2)}e^{i\Omega\tau}G_{(n-1)1}(\tau)\right]d\tau. \quad (68)$$

Considering the terms proportional to $\Omega_R^{(2)}$ as perturbation, we seek the solution as

$$\begin{pmatrix} G_{n0} \\ G_{(n-1)1} \end{pmatrix} = Ae^{i\Omega_R^{(3)}t}\begin{pmatrix} 1 \\ 1 \end{pmatrix} + Be^{-i\Omega_R^{(3)}t}\begin{pmatrix} 1 \\ -1 \end{pmatrix} + \begin{pmatrix} \delta G_{n0} \\ \delta G_{(n-1)1} \end{pmatrix}.$$

To estimate the magnitude of the perturbation, we substitute Eq. (67) into Eq. (68). After some algebra, we obtain that under the condition $\Omega_R^{(3)} \ll \Omega$, the magnitude of the perturbation is given as follows:

$$\delta G_{n0,\,(n-1)1} \sim \left|\frac{\Omega_R^{(2)}}{\Omega}\right| G_{n0,\,(n-1)1},$$

whereas if $\Omega_R^{(3)} \sim \Omega$, the magnitude of the perturbation is given as follows:

$$\delta G_{n0,\,(n-1)1} \sim \left|\frac{\Omega_R^{(2)}}{\Omega_R^{(3)}}\right| G_{n0,\,(n-1)1}.$$

To summarize this part, if both Rabi frequencies $\Omega_R^{(3)}$, $\Omega_R^{(2)} \ll \Omega$, the two resonances can be treated independently for any relationship between the magnitudes of $\Omega_R^{(3)}$ and $\Omega_R^{(2)}$. If the above inequality is violated, one can neglect one of the resonances only if its associated Rabi frequency is much lower than the Rabi frequency of another resonance. These restrictions are obvious from qualitative physical reasoning: either the magnitudes of the Rabi splittings are much smaller than the distance between resonances or one of the splittings is much weaker than another one.

When the effect of the neighboring resonance is non-negligible, it can still be taken into account in the solution. Indeed, consider the solution to Eqs. (62) and (63), taking into account only the two-wave resonance, i.e., taking $\Re = 0$. After obvious substitutions, we arrive at

$$\begin{pmatrix} C_{n0} \\ C_{(n-1)1} \end{pmatrix} = Ae^{-i\left(\omega_n - \frac{\Omega}{2} + \sqrt{\frac{\Omega^2}{4} + \left|\Omega_R^{(2)}\right|^2}\right)t} \times \begin{pmatrix} 1 \\ \dfrac{-\frac{\Omega}{2} + \sqrt{\frac{\Omega^2}{4} + \left|\Omega_R^{(2)}\right|^2}}{\Omega_R^{(2)}} \end{pmatrix}$$

$$+ Be^{-i\left(\omega_{n-1} + \frac{W}{\hbar} + \frac{\Omega}{2} - \sqrt{\frac{\Omega^2}{4} + \left|\Omega_R^{(2)}\right|^2}\right)t} \times \begin{pmatrix} \dfrac{\Omega_R^{(2)}}{\frac{\Omega}{2} - \sqrt{\frac{\Omega^2}{4} + \left|\Omega_R^{(2)}\right|^2}} \\ 1 \end{pmatrix}; \quad (69)$$

In the limit $\Omega \gg \Omega_R^{(2)}$, we obtain the following:

$$\begin{pmatrix} C_{n0} \\ C_{(n-1)1} \end{pmatrix} \approx Ae^{-i\left(\omega_n + \frac{|\Omega_R^{(2)}|^2}{\Omega}\right)t}\begin{pmatrix} 1 \\ \dfrac{|\Omega_R^{(2)}|}{\Omega} \end{pmatrix} + Be^{-i\left(\omega_{n-1} + \frac{W}{\hbar} - \frac{|\Omega_R^{(2)}|^2}{\Omega}\right)t}\begin{pmatrix} -\dfrac{|\Omega_R^{(2)}|}{\Omega} \\ 1 \end{pmatrix} \quad (70)$$

It is clear from Eq. (70) that the entanglement of states described by C_{n0} and $C_{(n-1)1}$ is determined by a small parameter $\frac{|\Omega_R^{(2)}|}{\Omega}$, whereas at exact resonance, the entanglement is always stronger, see Eq. (67). Therefore, when $\Omega_R^{(2)} \ll \Omega$, we can neglect the contribution of the two-photon resonance to the entanglement of states $|n\rangle|0\rangle$ and $|n-1\rangle\,|1\rangle$. However, it follows from Eq. (69) that the two-wave resonance shifts the eigenfrequencies of the system. Qualitatively, these shifts can be included by putting $\chi = 0$ in Eqs. (62) and (63) but replacing the eigenfrequencies ω_n and ω_{n-1} according to Eq. (70):

$$\omega_n \Rightarrow \omega_n + \frac{\left|\Omega_R^{(2)}\right|^2}{\Omega}, \quad \omega_{n-1} + \frac{W}{\hbar} \Rightarrow \omega_{n-1} + \frac{W}{\hbar} - \frac{\left|\Omega_R^{(2)}\right|^2}{\Omega}. \quad (71)$$

If $\Omega_R^{(3)} \ll \frac{\left|\Omega_R^{(2)}\right|^2}{\Omega}$, these shifts can be significant in order to interpret the spectra near the three-wave parametric resonance.

The same reasoning can be carried out to analyze the effect of a detuned three-wave resonance on the solution near the two-wave resonance.

These results can be verified by an exact numerical solution of Eqs. (62) and (63) for given initial conditions. After that, we can obtain the spectra of C_{n0} and $C_{(n-1)1}$. Since they are oscillating functions, their spectra form discrete lines at frequencies which we denote as ω_{osc}.

As an example, we select the case of $n = 1$, set $|\Omega_R^{(2)}| = |\Omega_R^{(3)}| = 0.1\Omega$, and choose the initial condition as $C_{n0}(0) = 0$ and $C_{(n-1)1}(0) = 1$. The frequencies ω_{osc} of the spectral lines for C_{n0} and $C_{(n-1)1}$ are shown in Figure 6. Their values are shifted by $\omega_{osc,0} = \omega_1|_{\omega=W/\hbar}$. The area of the dot for each spectral line is proportional to the square of its amplitude. If a marker is not visible, it means the corresponding line is very weak and can be neglected. The anticrossing can be seen at both the one-photon resonance and parametric resonance.

As an illustration of the violation of the condition for resonance separation, we show the oscillation frequencies for $|\Omega_R^{(2)}| = |\Omega_R^{(3)}| = 0.5\,\Omega$ in Figure 7. Here, the anticrossing picture of isolated resonances is smeared and cannot be observed.

7 Control of entangled states

In order to control the quantum state of the system, turn the entanglement on/off, read or write information into a qubit, or implement a logic gate, one has to vary the parameters of a system, for example, the detuning from resonance, the field amplitude of the EM mode at the atom position, or the intensity of a classical acoustic pumping. The analytic results obtained in previous sections can be readily generalized when the variation of a parameter is adiabatic, i.e., slower than the optical frequencies ω or $\frac{W}{\hbar}$. Since the space is limited, the time-dependent problem will be considered elsewhere. Here, we consider just one example, namely turning on/off of a classical acoustic pumping $\mathbf{q} = \mathbf{Q}(\mathbf{r})e^{-i\Omega t} + \mathbf{Q}^*(\mathbf{r})e^{i\Omega t}$.

For maximal control, it is beneficial to place an atom at the point where $\mathbf{E}(\mathbf{r} = \mathbf{r}_a) \to 0$, whereas $(\mathbf{Q}\cdot\nabla)\mathbf{E}_{\mathbf{r}=\mathbf{r}_a}$ is maximized. The equations of motion for quantum state amplitudes were derived in Section 5, see Eqs. (53)–(55).

Consider an exact parametric resonance $\omega + \Omega = \frac{W}{\hbar}$ for simplicity, when

$$\omega_n = \omega_{n-1} + \frac{W}{\hbar}.$$

The solution to Eqs. (53)–(55) when the acoustic pumping is turned off is given as follows:

$$\Psi = C_{00}(0)e^{-i\omega_0 t}|0\rangle|0\rangle + \sum_{n=1}^{\infty}\Big(C_{n0}(0)e^{-i\omega_n t}|n\rangle|0\rangle + C_{(n-1)1}(0)e^{-i\left(\omega_{n-1}+\frac{W}{\hbar}\right)t}|n-1\rangle|1\rangle\Big)$$

The solution when the acoustic pumping is turned on is given as follows:

$$\Psi = C_{00}(0)e^{-i\omega_0 t}|0\rangle|0\rangle \sum_{n=1}^{\infty}\Big[\Big(A_n e^{-i|\Omega_R^{(n)}|t} + B_n e^{i|\Omega_R^{(n)}|t}\Big)e^{-i\omega_n t}|n\rangle|0\rangle + \Big(-A_n e^{-i|\Omega_R^{(n)}|t} + B_n e^{i|\Omega_R^{(n)}|t}\Big)e^{i\theta - i\left(\omega_{n-1}+\frac{W}{\hbar}\right)t}|n-1\rangle|1\rangle\Big] \quad (72)$$

Assume that the initial quantum state before the pumping was turned on was not entangled, for example, an atom was in an excited state and there were no photons:

$$\Psi = e^{-i\left(\omega_0+\frac{W}{\hbar}\right)t}|0\rangle|1\rangle.$$

If the acoustic pumping is turned on at $t = 0$, the quantum state becomes entangled:

$$\Psi = ie^{-i\omega_1 t - i\theta}\sin\big(|\Omega_R^{(1)}|t\big)|1\rangle|0\rangle + e^{-i\left(\omega_0+\frac{W}{\hbar}\right)t}\cos\big(|\Omega_R^{(1)}|t\big)|0\rangle|1\rangle, \quad (73)$$

Then the acoustic pumping can be turned off. Depending on the turnoff moment of time, one can obtain various entangled photon–atom states, e.g., Bell states, etc. The above reasoning is valid when the turn-on/off rate is slower

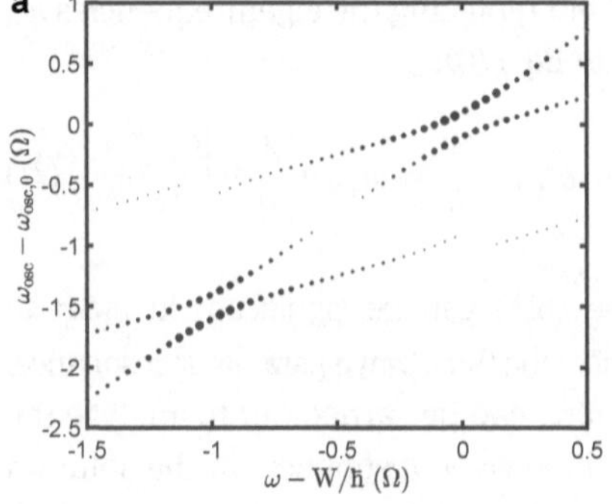

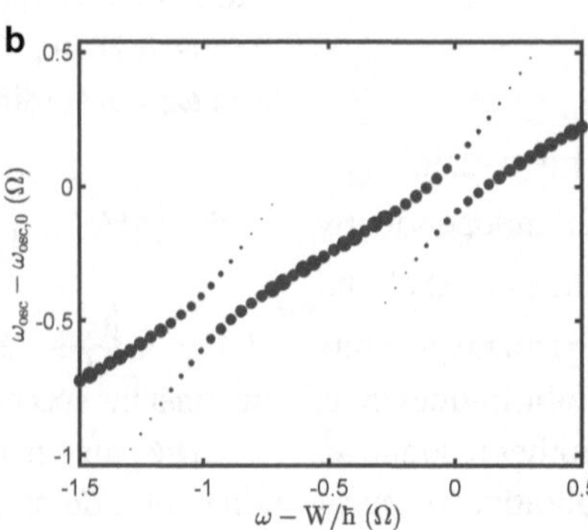

Figure 6: The frequencies ω_{osc} of the spectral lines for C_{n0} (left panel) and $C_{(n-1)1}$ (right panel), with $n = 1$, as functions of the photon frequency ω. The photon frequencies are shifted by $W/\hbar$, and the positions of spectral lines ω_{osc} are shifted by $\omega_{osc,0} = \omega_1|_{\omega=W/\hbar}$. The area of a marker is proportional to the amplitude squared of the spectral line. Both axes are in units of Ω. The parameters are $|\Omega_R^{(2)}| = |\Omega_R^{(3)}| = 0.1\,\Omega$, and the initial condition is $C_{n0}(0) = 0$ and $C_{(n-1)1}(0) = 1$.

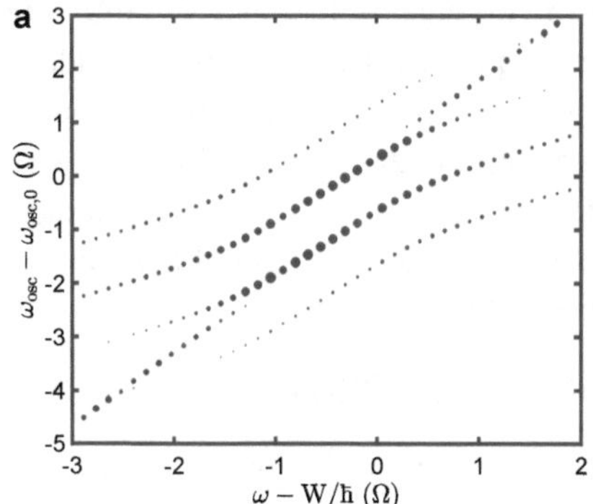

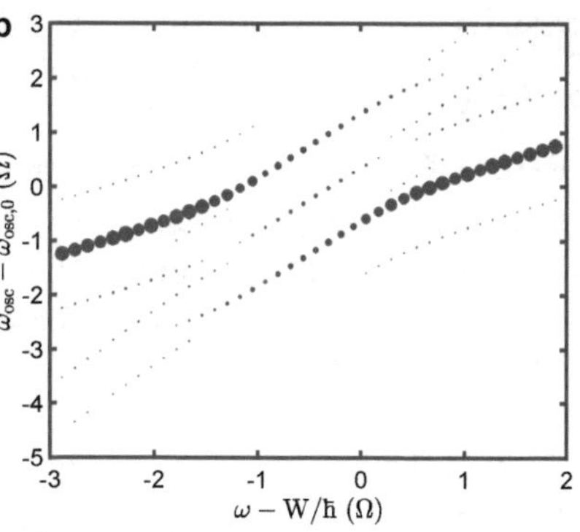

Figure 7: The frequencies ω_{osc} of the spectral lines for C_{n0} (left panel) and $C_{(n-1)1}$ (right panel), with $n = 1$. The notations are the same as in Figure 6. The parameters are $|\Omega_R^{(2)}| = |\Omega_R^{(3)}| = 0.5\,\Omega$, and the initial condition is $C_{n0}(0) = 0$ and $C_{(n-1)1}(0) = 1$.

than the optical frequencies and the detuning from the two-wave resonance $\omega = \frac{W}{\hbar}$.

8 Conclusions

In conclusion, we showed how the entanglement in a system of a fermionic quantum emitter coupled to a quantized EM field in a nanocavity and quantized phonon or mechanical vibrational modes emerges in the vicinity of a parametric resonance in the system. We developed analytic theory describing the formation and evolution of entangled quantum states, which can be applied to a broad range of cavity quantum optomechanics problems and emerging nanocavity strong coupling experiments. The model includes decoherence effects due to coupling of the fermion, photon, and phonon subsystems to their dissipative reservoirs within the stochastic evolution approach, which is derived from the Heisenberg–Langevin formalism. We showed that our approach provided the results for physical observables equivalent to those obtained from the density matrix equations with the relaxation operator in Lindblad form. We derived analytic expressions for the time evolution of the quantum state and observables and the emission spectra. The limit of a classical acoustic pumping, the control of entangled states, and the interplay between parametric and standard two-wave resonances were discussed.

Author contribution: All the authors have accepted responsibility for the entire content of this submitted manuscript and approved submission.

Research funding: This work has been supported in part by the Air Force Office for Scientific Research through Grant No. FA9550-17-1-0341, National Science Foundation Award No. 1936276, and Texas A&M University through STRP, X-grant and T3-grant programs. M.T. acknowledges the support from the Russian Foundation for Basic Research Grant No. 20-02-00100, and M.E. acknowledges the support from the Federal Research Center Institute of Applied Physics of the Russian Academy of Sciences (Project No. 0035-2019-004).

Conflict of interest statement: The authors declare no conflicts of interest regarding this article.

Appendix A
The stochastic equation of evolution for the state vector

The description of open quantum systems within the stochastic equation of evolution for the state vector is usually formulated for a Monte Carlo–type numerical scheme, e.g., the method of quantum jumps [23, 25]. We developed an approach suitable for analytic derivations. Our stochastic equation of evolution is basically the Schrödinger equation modified by adding a linear relaxation operator and the noise source term with appropriate correlation properties. The latter are related to the parameters of the relaxation operator in such a way that the expressions for the statistically averaged quantities satisfy certain physically meaningful conditions.

The protocol of introducing the relaxation operator with a corresponding noise source term to the quantum dynamics is well known in the Heisenberg picture, where it is called the Heisenberg–Langevin method [23, 34, 42]. We develop a conceptually similar approach for the Schrödinger equation. Here, we derive the general form of the stochastic equation of evolution from the Heisenberg–Langevin equations and track how certain physically reasonable constraints on the observables determine the correlation properties of the noise sources.

1. From Heisenberg–Langevin equations to the stochastic equation for the state vector

The Heisenberg–Langevin equation for the operator $\hat{g}$ of a certain observable quantity takes the following form [23, 34, 42]:

$$\frac{d}{dt}\hat{g} = \frac{i}{\hbar}\left[\hat{H},\hat{g}\right] + \hat{R}(\hat{g}) + \hat{L}_g(t), \tag{A1}$$

where $\hat{R}(\hat{g})$ is the relaxation operator, $\hat{L}_g(t)$ is the Langevin noise source satisfying $\overline{\hat{L}_g(t)} = 0$, where the bar means statistical averaging. For given commutation relations of the two operators, $[\hat{g}_1, \hat{g}_2] = C$, where C is a constant, correct Langevin sources should ensure the conservation of commutation relations at any moment of time, despite the presence of the relaxation operator in Eq. (A1), see [34, 43, 44].

The group of terms $\frac{i}{\hbar}[\hat{H},\hat{g}] + \hat{R}(\hat{g})$ can often be written as follows:

$$\frac{i}{\hbar}\left[\hat{H},\hat{g}\right] + \hat{R}(\hat{g}) = \frac{i}{\hbar}\left(\hat{H}_{\text{eff}}^{\dagger}\,\hat{g} - \hat{g}\hat{H}_{\text{eff}}\right), \tag{A2}$$

where $\hat{H}_{\text{eff}}$ is a non-Hermitian operator. For example, if the relaxation operator describes dissipation with relaxation constant γ so that $\overline{\hat{g}} \propto e^{-\gamma t}$, then $\hat{H}_{\text{eff}} = \hat{H} - i\hbar\frac{\gamma}{2}\hat{1}$, where $\hat{1}$ is a unit operator. Note that in the master equation for the density matrix, the relaxation is often introduced in a conceptually similar way [25], $[\hat{H},\hat{\rho}] \Rightarrow \hat{H}_{\text{eff}}\hat{\rho} - \hat{\rho}\hat{H}_{\text{eff}}^{\dagger}$, which is however slightly different from the form used in Eq. (A2): $[\hat{H},\hat{g}] \Rightarrow \hat{H}_{\text{eff}}^{\dagger}\,\hat{g} - \hat{g}\hat{H}_{\text{eff}}$. The difference is because the commutator of an unknown operator with Hamiltonian enters with opposite sign in the master equation as compared to the Heisenberg equation.

Now consider the transition from the Heisenberg–Langevin equation to the stochastic equation for the state vector. The key point is to assume that there exists the operator of evolution $\hat{U}(t)$, which is determined not only by the system parameters but also by the properties of the reservoir. This operator determines the evolution of the state vector:

$$|\Psi(t)\rangle = \hat{U}(t)|\Psi_0\rangle,\ \langle\Psi(t)| = \langle\Psi_0|\hat{U}^{\dagger}(t), \tag{A3}$$

where $\Psi_0 = \Psi(0)$. Hereafter, we will denote the operators in the Schrödinger picture with index "s" to distinguish them from the Heisenberg operators. An observable can be calculated as follows:

$$g(t) = \langle\Psi(t)|\hat{g}_S|\Psi(t)\rangle = \langle\Psi_0|\hat{g}(t)|\Psi_0\rangle$$

which leads to

$$\hat{g}(t) = \hat{U}^{\dagger}(t)\hat{g}_S\hat{U}(t), \tag{A4}$$

Since the substitution of Eqs. (A3) and (A4) into the standard Heisenberg equation leads to the standard Schrödinger equation, it makes sense to apply the same procedure to the Heisenberg–Langevin equation in order to obtain the "stochastic variant" of the Schrödinger equation. The solution of the latter should yield the expression for an observable,

$$g(t) = \overline{\langle\Psi(t)|\hat{g}_S|\Psi(t)\rangle},$$

which is different from the standard expression by additional averaging over the noise statistics.

Note that an open system interacting with a reservoir is generally in a mixed state and should be described by the density matrix. We are describing the state of the system with a state vector which has a fluctuating component. For example, in a certain basis $|\alpha\rangle$, the state vector will be $C_\alpha(t) = \overline{C_\alpha} + \tilde{C}_\alpha$, where the fluctuating component is denoted with a wavy bar. The elements of the density matrix of the corresponding mixed state are $\rho_{\alpha\beta} = \overline{C_\alpha C_\beta^*} = \overline{C_\alpha}\cdot\overline{C_\beta^*} + \overline{\tilde{C}_\alpha\cdot\tilde{C}_\beta^{\,*}}$.

The solution to the Heisenberg–Langevin equation can be expressed through the evolution operator $\hat{U}(t)$ using Eq. (A4). The noise source terms should be chosen to ensure the conservation of commutation relations at any moment of time, despite the presence of the relaxation operator. Since commutation relations between any two operators are conserved if and only if the evolution operator $\hat{U}(t)$ is unitary, a correct noise source in the Heisenberg–Langevin equation will automatically ensure the condition $\hat{U}^{\dagger}\hat{U} = \hat{1}$.

We implement the above protocol. Substituting Eq. (A4) together with $\hat{H}_{\text{eff}} = \hat{U}^{\dagger}\hat{H}_{\text{eff},S}\hat{U}$ and $\hat{H}_{\text{eff}}{}^{\dagger} = \hat{U}^{\dagger}\hat{H}_{\text{eff},S}{}^{\dagger}\,\hat{U}$ into Eqs. (A1) and (A2) and using $\hat{U}^{\dagger}\hat{U} = \hat{1}$, we arrive at the following:

$$\left(\frac{d}{dt}\hat{U}^{\dagger} - \frac{i}{\hbar}\hat{U}^{\dagger}\hat{H}_{\text{eff},S}{}^{\dagger}\right)\hat{g}_S\hat{U} + \hat{U}^{\dagger}\hat{g}_S\left(\frac{d}{dt}\hat{U} + \frac{i}{\hbar}\hat{H}_{\text{eff},S}\hat{U}\right) = \hat{L}_g \tag{A5}$$

Next, we introduce the operator $\hat{F}$, defined as follows:

$$\hat{L}_g = 2\hat{U}^{\dagger}\hat{g}_S\hat{F} \tag{A6}$$

For the operator $\hat{L}_g^{\dagger}$, Eq. (A6) gives $\hat{L}_g^{\dagger} = 2\hat{F}^{\dagger}(\hat{U}^{\dagger}\hat{g}_S)^{\dagger} = 2\hat{F}^{\dagger}\hat{g}_S^{\dagger}\,\hat{U}$. Since $\hat{g}$ and $\hat{g}_S$ are Hermitian operators, $\hat{L}_g$ has to be Hermitian too. (One can develop the Heisenberg–Langevin formalism for non-Hermitian operators too, for example, creation or annihilation operators, but the derivation becomes longer.) Then the operator $\hat{L}_g$ can be "split" between the two terms on the left-hand side of Eq. (A5) using the following relationship:

$$\widehat{L}_g = \widehat{U}^\dagger \widehat{g}_S \widehat{F} + \widehat{F}^\dagger \widehat{g}_S \widehat{U} \tag{A7}$$

Substituting the latter into Eq. (A5), we obtain the following:

$$\left(\frac{d}{dt}\widehat{U}^\dagger - \frac{i}{\hbar}\widehat{U}^\dagger \widehat{H}^\dagger_{\mathrm{eff},S} - \widehat{F}^\dagger\right)\widehat{g}_S \widehat{U} + \widehat{U}^\dagger \widehat{g}_S\left(\frac{d}{dt}\widehat{U} + \frac{i}{\hbar}\widehat{H}_{\mathrm{eff},S}\widehat{U} - \widehat{F}\right) = 0.$$

For simplicity, we will assume operator $\widehat{H}_{\mathrm{eff}}$ to be constant with time, i.e., we will not differentiate between $\widehat{H}_{\mathrm{eff}}$ and $\widehat{H}_{\mathrm{eff},S}$.

The last equation is satisfied for sure if

$$\frac{d}{dt}\widehat{U} = -\frac{i}{\hbar}\widehat{H}_{\mathrm{eff}}\widehat{U} + \widehat{F}, \frac{d}{dt}\widehat{U}^\dagger = \frac{i}{\hbar}\widehat{U}^\dagger \widehat{H}^\dagger_{\mathrm{eff}} + \widehat{F}^\dagger. \tag{A8}$$

Multiplying Eq. (A8) by the initial state vector $|\Psi_0\rangle$ from the right and from the left, we obtain the stochastic equation for the state vector and its Hermitian conjugate:

$$\frac{d}{dt}|\Psi\rangle = -\frac{i}{\hbar}\widehat{H}_{\mathrm{eff}}|\Psi\rangle - \frac{i}{\hbar}|R(t)\rangle \tag{A9}$$

$$\frac{d}{dt}\langle\Psi| = \frac{i}{\hbar}\langle\Psi|\widehat{H}^\dagger_{\mathrm{eff}} + \frac{i}{\hbar}\langle R(t)| \tag{A10}$$

where we introduced the notations $i\hbar\widehat{F}|\Psi_0\rangle \Rightarrow |R(t)\rangle$, $-i\hbar\langle\Psi_0|\widehat{F}^\dagger \Rightarrow \langle R(t)|$. We will also need Eqs. (A9) and (A10) in a particular basis $|\alpha\rangle$:

$$\frac{d}{dt}C_\alpha = -\frac{i}{\hbar}\sum_\nu\left(\widehat{H}_{\mathrm{eff}}\right)_{\alpha\nu}C_\nu - \frac{i}{\hbar}R_\alpha, \tag{A11}$$

$$\frac{d}{dt}C^*_\alpha = \frac{i}{\hbar}\sum_\nu C^*_\nu\left(\widehat{H}^\dagger_{\mathrm{eff}}\right)_{\nu\alpha} + \frac{i}{\hbar}R^*_\alpha, \tag{A12}$$

where $R_\alpha = \langle\alpha|R\rangle$, $(\widehat{H}_{\mathrm{eff}})_{\alpha\beta} = \langle\alpha|\widehat{H}_{\mathrm{eff}}|\beta\rangle$.

Applying the same procedure to the standard Heisenberg Eq. (1), we obtain that in Eqs. (A9) and (A10): $\widehat{H}_{\mathrm{eff}} \equiv \widehat{H}^\dagger_{\mathrm{eff}} = \widehat{H}$ and $|R(t)\rangle \equiv 0$, which corresponds to the standard Schrödinger equation and its Hermitian conjugate.

Note that intermediate relations (A8) for the evolution operator and in particular operator $\widehat{F}$ should not depend on the choice of a particular physical observable g in the original Heisenberg–Langevin Eq. (A1). We assume that the Langevin operators in the original equation do not contradict this physically reasonable requirement.

In general, statistical properties of noise that ensure certain physically meaningful requirements impose certain constraints on the noise source $|R\rangle$ which enters the right-hand side of the stochastic equation for the state vector. In particular, it is natural to require that the statistically averaged quantity $\overline{|R\rangle} = 0$. We will also require that the noise source $|R\rangle$ has the correlation properties that preserve the norm of the state vector averaged over the reservoir statistics:

$$\overline{\langle\Psi(t)|\Psi(t)\rangle} = 1. \tag{A13}$$

2. Noise correlator

The solution to Eqs. (A9) and (A10) can be formally written as follows:

$$|\Psi\rangle = e^{-\frac{i}{\hbar}\widehat{H}_{\mathrm{eff}}t}|\Psi_0\rangle - \frac{i}{\hbar}\int_0^t e^{\frac{i}{\hbar}\widehat{H}_{\mathrm{eff}}(\tau - t)}|R(\tau)\rangle d\tau, \tag{A14}$$

$$\langle\Psi| = \langle\Psi_0|e^{\frac{i}{\hbar}\widehat{H}^\dagger_{\mathrm{eff}}t} + \frac{i}{\hbar}\int_0^t \langle R(\tau)|e^{-\frac{i}{\hbar}\widehat{H}^\dagger_{\mathrm{eff}}(\tau - t)}d\tau, \tag{A15}$$

In the basis $|\alpha\rangle$, Eqs. (A14) and (A15) can be transformed into the following equations:

$$C_\alpha = \langle\alpha|e^{-\frac{i}{\hbar}\widehat{H}_{\mathrm{eff}}t}|\Psi_0\rangle - \frac{i}{\hbar}\int_0^t \langle\alpha|e^{\frac{i}{\hbar}\widehat{H}_{\mathrm{eff}}(\tau - t)}|R(\tau)\rangle\, d\tau, \tag{A16}$$

$$C^*_\alpha = \langle\Psi_0|e^{\frac{i}{\hbar}\widehat{H}^\dagger_{\mathrm{eff}}t}|\alpha\rangle + \frac{i}{\hbar}\int_0^t \langle R(\tau)|e^{-\frac{i}{\hbar}\widehat{H}^\dagger_{\mathrm{eff}}(\tau - t)}|\alpha\rangle\, d\tau. \tag{A17}$$

In order to calculate the observables, we need to know the expressions for the averaged dyadic combinations of the amplitudes. We can find them using Eqs. (A11) and (A12):

$$\begin{aligned}\frac{d}{dt}\overline{C_\alpha C^*_\beta} = &-\frac{i}{\hbar}\sum_\nu\left(H^{(h)}_{\alpha\nu}\overline{C_\nu C^*_\beta} - \overline{C_\alpha C^*_\nu}H^{(h)}_{\nu\beta}\right)\\ &-\frac{i}{\hbar}\sum_\nu\left(H^{(ah)}_{\alpha\nu}\overline{C_\nu C^*_\beta} + \overline{C_\alpha C^*_\nu}H^{(ah)}_{\nu\beta}\right)\\ &+\left(-\frac{i}{\hbar}\overline{C^*_\beta R_\alpha} + \frac{i}{\hbar}\overline{R^*_\beta C_\alpha}\right),\end{aligned} \tag{A18}$$

where we separated the Hermitian and anti-Hermitian components of the effective Hamiltonian: $\langle\alpha|\widehat{H}_{\mathrm{eff}}|\beta\rangle = H^{(h)}_{\alpha\beta} + H^{(ah)}_{\alpha\beta}$. Substituting Eqs. (A16) and (A17) into the last term in Eq. (A18), we obtain the following:

$$-\frac{i}{\hbar}\overline{C^*_\beta R_\alpha} + \frac{i}{\hbar}\overline{C_\alpha R^*_\beta} = \frac{1}{\hbar^2}\int_{-t}^0 \overline{\langle R(t+\xi)|e^{-\frac{i}{\hbar}\widehat{H}^\dagger_{\mathrm{eff}}\xi}|\beta\rangle\langle\alpha|R(t)\rangle}d\xi$$

$$+\frac{1}{\hbar^2}\int_{-t}^0 \overline{\langle R(t)|\beta\rangle\langle\alpha|e^{\frac{i}{\hbar}\widehat{H}_{\mathrm{eff}}\xi}|R(t+\xi)\rangle}d\xi.$$

To proceed further with analytical results, we need to evaluate these integrals. The simplest situation is when the noise source terms are delta-correlated in time (Markovian). In this case, only the point $\xi = 0$ contributes to the integrals. As a result, Eq. (A18) is transformed to the following equation:

$$\frac{d}{dt}\overline{C_\alpha C_\beta^*} = -\frac{i}{\hbar}\sum_\nu\left(H_{\alpha\nu}^{(h)}\overline{C_\nu C_\beta^*} - \overline{C_\alpha C_\nu^*}H_{\nu\beta}^{(h)}\right)$$
$$-\frac{i}{\hbar}\sum_\nu\left(H_{\alpha\nu}^{(ah)}\overline{C_\nu C_\beta^*} + \overline{C_\alpha C_\nu^*}H_{\nu\beta}^{(ah)}\right) + D_{\alpha\beta}, \qquad \text{(A19)}$$

where the correlator $D_{\alpha\beta}$ is defined as follows:

$$\overline{R_\beta^*(t+\xi)R_\alpha(t)} = \overline{R_\beta^*(t)R_\alpha(t+\xi)} = \hbar^2\delta(\xi)D_{\alpha\beta} \qquad \text{(A20)}$$

The time derivative of the norm of the state vector is given as follows:

$$\frac{d}{dt}\sum_\alpha\overline{|C_\alpha|2} = -\sum_\alpha\left[\frac{i}{\hbar}\sum_\nu\left(H_{\alpha\nu}^{(ah)}\overline{C_\nu C_\alpha^*} + \overline{C_\alpha C_\nu^*}H_{\nu\alpha}^{(ah)}\right) - D_{\alpha\alpha}\right] \qquad \text{(A21)}$$

Clearly, the components $D_{\alpha\alpha}$ of the noise correlator need to compensate the decrease in the norm due to the anti-Hermitian component of the effective Hamiltonian. Therefore, the expressions for $H_{\alpha\beta}^{(ah)}$ and $D_{\alpha\alpha}$ have to be mutually consistent. This is the manifestation of the fluctuation–dissipation theorem [41].

Note that the noise correlator could depend on the averaged combinations (e.g., dyadics) of the components of the state vector. This is because the noise source term $|R\rangle$ introduced above depends on the initial state $|\Psi_0\rangle$ and the evolution operator $\widehat{U}$, and these quantities form the state vector components at any given time. Of course, what we call a "state vector" is the solution of the stochastic equation of motion, which is very different from the solution of the conventional Schrödinger equation for a closed system. In particular, we postulated the existence of the evolution operator $\widehat{U}$ determined not only by the parameters of the dynamical system but also by the properties of a dissipative reservoir, although we did not specify any particular expression for $\widehat{U}$.

As an example, consider a simple diagonal anti-Hermitian operator $H_{\alpha\nu}^{(ah)}$:

$$H_{\alpha\nu}^{(ah)} = -i\hbar\gamma_\alpha\delta_{\alpha\nu} \qquad \text{(A22)}$$

and introduce the following models:

(i) Populations relax much slower than coherences (expected for condensed matter systems). In this case, we can choose $D_{\alpha\neq\beta} = 0$, $D_{\alpha\alpha} = 2\gamma_\alpha\overline{|C_\alpha|^2}$; within this model, the population at each state will be preserved.

(ii) The state $\alpha = \alpha_{\text{down}}$ has a minimal energy, while the reservoir temperature $T = 0$. In this case, it is expected that all populations approach zero in equilibrium, whereas the occupation number of the ground state approaches 1, similar to the Weisskopf–Wigner model. The adequate choice of correlators is $D_{\alpha\neq\beta} = 0$, $D_{\alpha\alpha} \propto \delta_{\alpha\alpha_{\text{down}}}$, $\gamma_{\alpha_{\text{down}}} = 0$. The expression for the remaining nonzero correlator,

$$D_{\alpha_{\text{down}}\alpha_{\text{down}}} = \sum_{\alpha\neq\alpha_{\text{down}}} 2\gamma_\alpha\overline{|C_\alpha|^2}, \qquad \text{(A23)}$$

ensures the conservation of the norm:

$$\frac{d}{dt}\sum_{\alpha\neq\alpha_{\text{down}}}\overline{|C_\alpha|^2} = -\sum_{\alpha\neq\alpha_{\text{down}}}2\gamma_\alpha\overline{|C_\alpha|^2} = -\frac{d}{dt}\overline{|C_{\alpha_{\text{down}}}|^2}.$$

This is an example of the correlator's dependence on the state vector that we discussed before.

3. Comparison with the Lindblad method

One can choose the anti-Hermitian Hamiltonian $H_{\alpha\beta}^{(ah)}$ and correlators $D_{\alpha\beta}$ in the stochastic equation of motion in such a way that Eq. (A19) for the dyadics $\overline{C_n C_m^*}$ corresponds exactly to the equations for the density matrix elements in the Lindblad approach. Indeed, the Lindblad form of the master equation has the following form [23, 25]:

$$\frac{d}{dt}\widehat{\rho} = -\frac{i}{\hbar}\left[\widehat{H},\widehat{\rho}\right] + \widehat{L}(\widehat{\rho}) \qquad \text{(A24)}$$

where $\widehat{L}(\widehat{\rho})$ is the Lindbladian:

$$\widehat{L}(\widehat{\rho}) = -\frac{1}{2}\sum_k\gamma_k\left(\widehat{l}_k^\dagger\widehat{l}_k\widehat{\rho} + \widehat{\rho}\widehat{l}_k^\dagger\widehat{l}_k - 2\widehat{l}_k\widehat{\rho}\widehat{l}_k^\dagger\right), \qquad \text{(A25)}$$

Operators $\widehat{l}_k$ in Eq. (A25) and their number are determined by the model which describes the coupling of the dynamical system to the reservoir. The form of the relaxation operator given by Eq. (A25) preserves automatically the conservation of the trace of the density matrix, whereas the specific choice of relaxation constants ensures that the system approaches a proper steady state given by thermal equilibrium or supported by an incoherent pumping.

Eq. (A24) is convenient to represent in a slightly different form:

$$\frac{d}{dt}\widehat{\rho} = -\frac{i}{\hbar}\left(\widehat{H}_{\text{eff}}\widehat{\rho} - \widehat{\rho}\widehat{H}_{\text{eff}}^\dagger\right) + \delta\widehat{L}(\widehat{\rho}) \qquad \text{(A26)}$$

where

$$\widehat{H}_{\text{eff}} = \widehat{H} - i\hbar\sum_k\gamma_k\widehat{l}_k^\dagger\widehat{l}_k,\ \delta\widehat{L}(\widehat{\rho}) = \sum_k\gamma_k\widehat{l}_k\widehat{\rho}\widehat{l}_k^\dagger. \qquad \text{(A27)}$$

Writing the anti-Hermitian component of the Hamiltonian in Eqs. (A11) and (A12) as

$$H_{\alpha\beta}^{(ah)} = -i\hbar\langle\alpha|\sum_k\gamma_k\widehat{l}_k{}^\dagger\widehat{l}_k|\beta\rangle, \qquad \text{(A28)}$$

and defining the corresponding correlator of the noise source as

$$\overline{R^{*}_{\beta}(t+\xi)R_{\alpha}(t)} = \hbar^2\delta(\xi)D_{\alpha\beta}, \quad D_{\alpha\beta} = \langle\alpha|\delta\widehat{L}(\widehat{\rho})|\beta\rangle_{\rho_{mn}=\overline{C_n C^{*}_m}}, \tag{A29}$$

We obtain the solution in which averaged over noise statistics dyadics $\overline{C_n C^{*}_m}$ correspond exactly to the elements of the density matrix within the Lindblad method.

Instead of deriving the stochastic equation of evolution of the state vector from the Heisenberg–Langevin equations, we could postulate it from the very beginning. After that, we could justify the choice of the effective Hamiltonian and noise correlators by ensuring that they lead to the same observables as the solution of the density matrix equations with the relaxation operator in the Lindblad form [25, 46, 47]. However, the demonstration of direct connection between the stochastic equation of evolution of the state vector and the Heisenberg–Langevin equation provides an important physical insight.

4. Relaxation rates for coupled subsystems interacting with a reservoir

Whenever we have several coupled subsystems (such as electrons, photon modes, and phonons in this paper), each coupled to its reservoir, the determination of relaxation rates of the whole system becomes nontrivial. The problem can be solved if we assume that these "partial" reservoirs are statistically independent. In this case, it is possible to add up partial Lindbladians and obtain the total effective Hamiltonian.

Consider the Hamiltonian (11) of the system formed by a two-level electron system coupled to an EM mode field and dressed by a phonon field:

$$\widehat{H} = \widehat{H}_{em} + \widehat{H}_a + \widehat{H}_p + \widehat{V}. \tag{A30}$$

where, $\widehat{H}_{em} = \frac{\hbar\omega}{2}(\widehat{c}^{\dagger}\widehat{c} + \widehat{c}\widehat{c}^{\dagger})$ is the Hamiltonian for a single EM mode field, $\widehat{H}_a = W_1\widehat{\sigma}^{\dagger}\widehat{\sigma} + W_0\widehat{\sigma}\widehat{\sigma}^{\dagger}$ is the Hamiltonian for a two-level "atom" with energy levels $W_{0,1}$, $\widehat{H}_p = \frac{\hbar\Omega}{2}(\widehat{b}^{\dagger}\widehat{b} + \widehat{b}\widehat{b}^{\dagger})$ is the Hamiltonian for a phonon mode, $\widehat{V} = \widehat{V}_1 + \widehat{V}_2$ is the interaction Hamiltonian, where $\widehat{V}_{1,2}$ describe the atom–photon and atom–photon–phonon coupling, respectively:

$$\widehat{V}_1 = -(\chi\widehat{\sigma}^{\dagger}\widehat{c} + \chi^{*}\widehat{\sigma}\widehat{c}^{\dagger} + \chi\widehat{\sigma}\widehat{c} + \chi^{*}\widehat{\sigma}^{\dagger}\widehat{c}^{\dagger}),$$

$$\widehat{V}_2 = -\big(\eta_1\widehat{\sigma}^{\dagger}\widehat{c}\widehat{b} + \eta_1^{*}\widehat{\sigma}\widehat{c}^{\dagger}\widehat{b}^{\dagger} + \eta_2\widehat{\sigma}^{\dagger}\widehat{c}\widehat{b}^{\dagger} + \eta_2^{*}\widehat{\sigma}\widehat{c}^{\dagger}\widehat{b} + \eta_1\widehat{\sigma}\widehat{c}\widehat{b} + \eta_1^{*}\widehat{\sigma}^{\dagger}\widehat{c}^{\dagger}\widehat{b}^{\dagger} + \eta_2\widehat{\sigma}\widehat{c}\widehat{b}^{\dagger} + \eta_2^{*}\widehat{\sigma}^{\dagger}\widehat{c}^{\dagger}\widehat{b}\big),$$

where χ, η_1, η_2 are coupling constants defined before.

Summing up the known (see e.g., [23, 25]) partial Lindbladians of two bosonic (infinite amount of energy levels) and one fermionic (two-level) subsystems, we obtain the following:

$$\begin{aligned} L(\widehat{\rho}) = &-\frac{\gamma}{2}N_1^{T_a}(\widehat{\sigma}\widehat{\sigma}^{\dagger}\widehat{\rho} + \widehat{\rho}\widehat{\sigma}\widehat{\sigma}^{\dagger} - 2\widehat{\sigma}^{\dagger}\widehat{\rho}\widehat{\sigma}) - \frac{\gamma}{2}N_0^{T_a}(\widehat{\sigma}^{\dagger}\widehat{\sigma}\widehat{\rho} + \widehat{\rho}\widehat{\sigma}^{\dagger}\widehat{\sigma} - 2\widehat{\sigma}\widehat{\rho}\widehat{\sigma}^{\dagger}) \\ &-\frac{\mu_\omega}{2}\overline{n}_\omega^{T_{em}}(\widehat{c}\widehat{c}^{\dagger}\widehat{\rho} + \widehat{\rho}\widehat{c}^{\dagger}\widehat{c} - 2\widehat{c}^{\dagger}\widehat{\rho}\widehat{c}) - \frac{\mu_\omega}{2}(\overline{n}_\omega^{T_{em}} + 1)(\widehat{c}^{\dagger}\widehat{c}\widehat{\rho} + \widehat{\rho}\widehat{c}\widehat{c}^{\dagger} - 2\widehat{c}\widehat{\rho}\widehat{c}^{\dagger}) \\ &-\frac{\mu_\Omega}{2}\overline{n}_\Omega^{T_p}(\widehat{b}\widehat{b}^{\dagger}\widehat{\rho} + \widehat{\rho}\widehat{b}^{\dagger}\widehat{b} - 2\widehat{b}^{\dagger}\widehat{\rho}\widehat{b}) - \frac{\mu_\Omega}{2}(\overline{n}_\Omega^{T_p} + 1)(\widehat{b}^{\dagger}\widehat{b}\widehat{\rho} + \widehat{\rho}\widehat{b}\widehat{b}^{\dagger} - 2\widehat{b}\widehat{\rho}\widehat{b}^{\dagger}) \end{aligned} \tag{A31}$$

where γ, μ_ω, and μ_Ω are partial relaxation rates of the systems,

$$N_{0,1}^{T_a} = \left(1 + e^{-\frac{W_1 - W_0}{T_a}}\right)^{-1} e^{-\frac{W_{0,1} - W_0}{T_a}}, \quad \overline{n}_\omega^{T_{em}} = \left(e^{\frac{\hbar\omega}{T_{em}}} - 1\right)^{-1}, \quad \overline{n}_\Omega^{T_p} = \left(e^{\frac{\hbar\Omega}{T_p}} - 1\right)^{-1},$$

$T_{a,em,p}$ are the temperatures of partial reservoirs. For the Lindblad master equation in the form Eq. (A26), we get the following:

$$\widehat{H}_{\text{eff}} = \widehat{H} - i\widehat{\Gamma}, \tag{A32}$$

where

$$\begin{aligned} \widehat{\Gamma} = \frac{\hbar}{2}\Big\{&\gamma\big(N_1^{T_a}\widehat{\sigma}\widehat{\sigma}^{\dagger} + N_0^{T_a}\widehat{\sigma}^{\dagger}\widehat{\sigma}\big) + \mu_\omega\Big[\overline{n}_\omega^{T_{em}}\widehat{c}\widehat{c}^{\dagger} + \big(\overline{n}_\omega^{T_{em}} + 1\big)\widehat{c}^{\dagger}\widehat{c}\Big] \\ &+ \mu_\Omega\Big[\overline{n}_\Omega^{T_p}\widehat{b}\widehat{b}^{\dagger} + \big(\overline{n}_\Omega^{T_p} + 1\big)\widehat{b}^{\dagger}\widehat{b}\Big]\Big\}. \end{aligned} \tag{A33}$$

Using the effective Hamiltonian given by Eqs. (A32) and (A33), we arrive at the stochastic equation for the state vector in the following form:

$$\begin{aligned} \frac{d}{dt}C_{\alpha n0} = &-i\frac{W_0 + \hbar\omega\big(n + \frac{1}{2}\big) + \hbar\Omega\big(\alpha + \frac{1}{2}\big)}{\hbar}C_{\alpha n0} \\ &- \frac{i}{\hbar}\langle\alpha|\langle n|\langle 0|\widehat{V}|\Psi\rangle - \gamma_{\alpha n0}C_{\alpha n0} - \frac{i}{\hbar}R_{\alpha n0}, \end{aligned} \tag{A34}$$

$$\begin{aligned} \frac{d}{dt}C_{\alpha n1} = &-i\frac{W_1 + \hbar\omega\big(n + \frac{1}{2}\big) + \hbar\Omega\big(\alpha + \frac{1}{2}\big)}{\hbar}C_{\alpha n1} - \frac{i}{\hbar}\langle\alpha|\langle n|\langle 1|\widehat{V}|\Psi\rangle \\ &- \gamma_{\alpha n1}C_{\alpha n1} - \frac{i}{\hbar}R_{\alpha n1}, \end{aligned} \tag{A35}$$

where

$$\begin{aligned} \gamma_{\alpha n0} = &\frac{\gamma}{2}N_1^{T_a} + \frac{\mu_\omega}{2}\Big[\overline{n}_\omega^{T_{em}}(n+1) + \big(\overline{n}_\omega^{T_{em}} + 1\big)n\Big] \\ &+ \frac{\mu_\Omega}{2}\Big[\overline{n}_\Omega^{T_p}(\alpha+1) + \big(\overline{n}_\Omega^{T_p} + 1\big)\alpha\Big], \end{aligned} \tag{A36}$$

$$\begin{aligned} \gamma_{\alpha n1} = &\frac{\gamma}{2}N_0^{T_a} + \frac{\mu_\omega}{2}\Big[\overline{n}_\omega^{T_{em}}(n+1) + \big(\overline{n}_\omega^{T_{em}} + 1\big)n\Big] \\ &+ \frac{\mu_\Omega}{2}\Big[\overline{n}_\Omega^{T_p}(\alpha+1) + \big(\overline{n}_\Omega^{T_p} + 1\big)\alpha\Big], \end{aligned} \tag{A37}$$

Eqs. (A36) and (A37) determine the rules of combining the "partial" relaxation rates for several coupled subsystems.

5. Including purely elastic dephasing processes

So far, we used the Lindbladian which includes only the dissipation and does not include purely elastic dephasing processes. In order to take them into account, we need to modify the Lindbladian to include the explicit dependence on the operators of populations. We will follow the prescription which can be found in the study by Fain and Khanin [48]. Using Eq. (A25) for the "partial" Lindbladian $\widehat{L}_{2l}$ of a two-level atom, where $k = 1,2,3$, $\widehat{l}_1 = \widehat{\sigma}$, $\widehat{l}_2 = \widehat{\sigma}^\dagger$, and $\widehat{l}_3 = \widehat{\sigma}\widehat{\sigma}^\dagger - \widehat{\sigma}^\dagger\widehat{\sigma}$, we obtain the following:

$$L_{2l}(\widehat{\rho}) = -\frac{\gamma}{2}N_1^T(\widehat{\sigma}\widehat{\sigma}^\dagger\widehat{\rho} + \widehat{\rho}\widehat{\sigma}\widehat{\sigma}^\dagger - 2\widehat{\sigma}^\dagger\widehat{\rho}\widehat{\sigma}) - \frac{\gamma}{2}N_0^T(\widehat{\sigma}^\dagger\widehat{\sigma}\widehat{\rho} + \widehat{\rho}\widehat{\sigma}^\dagger\widehat{\sigma} - 2\widehat{\sigma}\widehat{\rho}\widehat{\sigma}^\dagger) - \frac{\gamma_{el}}{2}\widehat{\rho} + \delta\widehat{L}_{el}(\widehat{\rho}), \tag{A38}$$

where

$$\delta\widehat{L}_{el}(\widehat{\rho}) = \frac{\gamma_{el}}{2}(\widehat{\sigma}\widehat{\sigma}^\dagger - \widehat{\sigma}^\dagger\widehat{\sigma})\widehat{\rho}(\widehat{\sigma}\widehat{\sigma}^\dagger - \widehat{\sigma}^\dagger\widehat{\sigma}). \tag{A39}$$

For the evolution of a two-level system, using the Lindbladian (A39) leads to standard density matrix equations with inverse relaxation times for the coherence, $\frac{1}{T_2} = \frac{\gamma}{2} + \gamma_{el}$, and populations, $\frac{1}{T_1} = \gamma$.

Furthermore, using the scheme developed in Section 3 of the Appendix, we obtain that adding elastic processes to the stochastic equation of evolution for the state vector leads to the following modifications for the anti-Hermitian part of the effective Hamiltonian,

$$H_{\alpha\beta}^{(ah)} \Rightarrow H_{\alpha\beta}^{(ah)} - i\delta_{\alpha\beta}\frac{\hbar}{4}\gamma_{el},$$

and the correlators of noise sources,

$$D_{\alpha\beta} \Rightarrow D_{\alpha\beta} + \gamma_{el}\left(\delta_{\alpha\beta}\overline{|C_\alpha|^2} - \frac{1}{2}C_\alpha C_\beta^*\right).$$

This is a general prescription. Since, in this work, we are only interested in the RWA dynamics of states $|\alpha\rangle|n\rangle|0\rangle$ and $|\alpha-1\rangle|n-1\rangle|1\rangle$, the same expressions for the observables can be obtained with a much simpler modification of the formalism. One can show that it is sufficient to modify the relaxation constants $\gamma_{(\alpha-1)(n-1)1}$ according to

$$\gamma_{(\alpha-1)(n-1)1} \Rightarrow \gamma_{(\alpha-1)(n-1)1} + \gamma_{el}$$

and correlators $D_{(\alpha-1)(n-1)1;(\alpha-1)(n-1)1}$ as

$$D_{(\alpha-1)(n-1)1;(\alpha-1)(n-1)1} \Rightarrow D_{(\alpha-1)(n-1)1;(\alpha-1)(n-1)1} + 2\gamma_{el}\overline{|C_{(\alpha-1)(n-1)1}|^2}.$$

References

[1] M. Aspelmeyer, T. J. Kippenberg, and F. Marquardt, "Cavity optomechanics," *Rev. Mod. Phys.*, vol. 86, p. 1391, 2014.

[2] P. Meystre, "A short walk through quantum optomechanics," *Ann. Phys.*, vol. 525, p. 215, 2013.

[3] J.-M. Pirkkalainen, S.U. Cho, F. Massel, et al., "Cavity optomechanics mediated by a quantum two-level system," *Nat. Commun.*, vol. 6, p. 6981, 2015.

[4] Y. Chu, P. Kharel, W. H. Renninger, et al., "Quantum acoustics with superconducting qubits," *Science*, vol. 358, p. 199, 2017.

[5] S. Hong, R. Riedinger, I. Marinkovic, et al., "Hanbury Brown and Twiss interferometry of single phonons from an optomechanical resonator," *Science*, vol. 358, p. 203, 2017.

[6] P. Arrangoiz-Arriola, E. A. Wollack, Z. Wang, et al., "Resolving the energy levels of a nanomechanical oscillator," *Nature*, vol. 571, p. 537, 2019.

[7] F. Benz, M. K. Schmidt, A. Dreismann, et al., "Single-molecule optomechanics in "picocavities"," *Science*, vol. 354, p. 726, 2016.

[8] K.-D. Park, E. A. Muller, V. Kravtsov, et al., "Variable-temperature tip-enhanced raman spectroscopy of single-molecule fluctuations and dynamics," *Nano Lett.*, vol. 16, p. 479, 2016.

[9] M. S. Tame, K. R. McEnery, S. K. Ozdemir, J. Lee, S. A. Maier, and M. S. Kim, "Quantum plasmonics," *Nat. Phys.*, vol. 9, p. 329, 2013.

[10] F. C. B. Maia, B. T. OCallahan, A. R. Cadore, et al., "Anisotropic flow control and gate modulation of hybrid phonon-polaritons," *Nano Lett.*, vol. 19, p. 708, 2019.

[11] R. Chikkaraddy, B. de Nijs, F. Benz, et al., "Single-molecule strong coupling at room temperature in plasmonic nanocavities," *Nature*, vol. 535, p. 127, 2016.

[12] A. Sipahigil, R. E. Evans, D. D. Sukachev, et al., "An integrated diamond nanophotonics platform for quantum-optical networks," *Science*, vol. 354, p. 847, 2016.

[13] T. Yoshie, A. Scherer, J. Hendrickson, et al., "Vacuum Rabi splitting with a single quantum dot in a photonic crystal nanocavity," *Nature*, vol. 432, p. 200, 2004.

[14] J. P. Reithmaier, G. Sek, A. Loffler, et al., "Strong coupling in a single quantum dot-semiconductor microcavity system," *Nature*, vol. 432, p. 197, 2004.

[15] H. Leng, B. Szychowski, M.-C. Daniel, and M. Pelton, "Strong coupling and induced transparency at room temperature with single quantum dots and gap plasmons," *Nat. Commun.*, vol. 9, p. 4012, 2018.

[16] O. Bitton, S. N. Gupta, and G. Haran, "Quantum dot plasmonics: from weak to strong coupling," *Nanophotonics*, vol. 8, p. 559, 2019.

[17] K.-D. Park, M. A. May, H. Leng, et al., "Tip-enhanced strong coupling spectroscopy, imaging, and control of a single quantum emitter," *Sci. Adv.*, vol. 5, p. eaav5931, 2019.

[18] K. J. Satzinger, Y. P. Zhong, H. Chang, et al., "Quantum control of surface acoustic-wave phonons," *Nature*, vol. 563, p. 661665, 2018.

[19] A. Bienfait, Y. P. Zhong, H.-S. Chang, et al., "Quantum erasure using entangled surface acoustic phonons," *Phys. Rev. X*, vol. 10, p. 021055, 2020.
[20] U. Delic, M. Reisenbauer, K. Dare, et al., "Cooling of a levitated nanoparticle to the motional quantum ground state," *Science*, vol. 367, p. 892, 2020.
[21] N. Bloembergen, *Nonlinear Optics*, Singapore, World Scientific, 1996.
[22] Y. A. Bogoliubov and N. N. Mitropolsky, *Asymptotic Methods in the Theory of Nonlinear Oscillations*, London, Gordon and Breach, 1961.
[23] M. O. Scully and M. S. Zubairy, *Quantum Optics*, Cambridge, Cambridge University Press, 1997.
[24] P. Forn-Diaz, L. Lamata, E. Rico, J. Kono, and E. Solano, "Ultrastrong coupling regimes of light-matter interaction," *Rev. Mod. Phys.*, vol. 91, p. 025005, 2019.
[25] M. B. Plenio and P. L. Knight, "The quantum-jump approach to dissipative dynamics in quantum optics," *Rev. Mod. Phys.*, vol. 70, p. 101, 1998.
[26] C. Genes, D. Vitali, and P. Tombesi, "Emergence of atom-light-mirror entanglement inside an optical cavity," *Phys. Rev. A*, vol. 77, no. R, p. 050307, 2008.
[27] G. Chiara, M. Paternostro, and G. M. Palma, "Entanglement detection in hybrid optomechanical systems," *Phys. Rev. A*, vol. 83, p. 052324, 2011.
[28] Y. Xiang, F. X. Sun, M. Wang, Q. H. Gong, and Q. Y. He, "Detection of genuine tripartite entanglement and steering in hybrid optomechanics," *Opt. Expr.*, vol. 23, p. 30104, 2015.
[29] Y. Wang, S. Chesi, and A. A. Clerk, "Bipartite and tripartite output entanglement in three-mode optomechanical systems," *Phys. Rev. A*, vol. 91, p. 013807, 2015.
[30] X. Yang, Y. Ling, X. Shao, and M. Xiao, "Generation of robust tripartite entanglement with a single-cavity optomechanical system," *Phys. Rev. A*, vol. 85, p. 052303, 2017.
[31] Q. Liao, Y. Ye, P. Lin, N. Zhou, and W. Nie, "Tripartite entanglement in an atom-cavity-optomechanical system," *Int. J. Theor. Phys.*, vol. 57, p. 1319, 2018.
[32] C. Jiang, S. Tserkis, K. Collins, S. Onoe, Y. Li, and L. Tian, "Switchable bipartite and genuine tripartite entanglement via an optoelectromechanical interface," *Phys. Rev. A*, vol. 101, p. 042320, 2020.
[33] M. Tokman, Y. Wang, I. Oladyshkin, A. R. Kutayiah, and A. Belyanin, "Laser-driven parametric instability and generation of entangled photon-plasmon states in graphene," *Phys. Rev. B*, vol. 93, p. 235422, 2016.
[34] M. Tokman, X. Yao, and A. Belyanin, "Generation of entangled photons in graphene in a strong magnetic field," *Phys. Rev. Lett.*, vol. 110, p. 077404, 2013.
[35] M. D. Tokman, M. A. Erukhimova, and V. V. Vdovin, "The features of a quantum description of radiation in an optically dense medium," *Ann. Phys.*, vol. 360, p. 571, 2015.
[36] M. Tokman, Z. Long, S. AlMutairi, Y. Wang, M. Belkin, and A. Belyanin, "Enhancement of the spontaneous emission in subwavelength quasi-two-dimensional waveguides and resonators," *Phys. Rev. A*, vol. 97, p. 043801, 2018.
[37] W. Dur, G. Vidal, and J. I. Cirac, "Three qubits can be entangled in two inequivalent ways," *Phys. Rev. A*, vol. 62, p. 062314, 2000.
[38] M. M. Cunha, A. Fonseca, and E. O. Silva, arXiv:1909.00862v2.
[39] L. K. Shalm, D. R. Hamel, Z. Yan, C. Simon, K. J. Resch, and T. Jennewein, "Three-photon energy-time entanglement," *Nat. Phys.*, vol. 9, p. 19, 2012.
[40] A. Agusti, C. W. Sandbo Chang, F. Quijandria, G. Johansson, C. M. Wilson, and C. Sabin, "Tripartite genuine non-Gaussian entanglement in three-mode spontaneous parametric down-conversion," *Phys. Rev. Lett.*, vol. 125, p. 020502, 2020.
[41] L. D. Landau and E. M. Lifshitz, *Statistical Physics, Part 1*, Oxford, Pergamon, 1965.
[42] C. Gardiner and P. Zoller, *Quantum Noise*, Berlin, Heidelberg, Springer-Verlag, 2004.
[43] M. Erukhimova and M. Tokman, "Squeezing of thermal fluctuations in four-wave mixing in a scheme," *Phys. Rev A*, vol. 95, p. 013807, 2017.
[44] M. Tokman, Z. Long, S. Almutairi, et al., "Purcell enhancement of the parametric down-conversion in two-dimensional nonlinear materials," *APL Photonics*, vol. 4, p. 034403, 2019.
[45] K. H. Madsen and P. Lodahl, "Quantitative analysis of quantum dot dynamics and emission spectra in cavity quantum electrodynamics," *New J. Phys.*, vol. 15, p. 025013, 2013.
[46] S. Haroche and J.-M. Raymond, *Exploring the Quantum. Atoms, Cavities, and Photons*, Oxford, UK, Oxford University Press, 2006.
[47] K. Blum, *Density Matrix Theory and Applications*, Heidelberg, Springer, 2012.
[48] V. M. Fain and Y. I. Khanin, *Quantum Electronics. Basic Theory*, Cambridge, MA, MIT, 1969.

Charles A. Downing and Luis Martín-Moreno*

Polaritonic Tamm states induced by cavity photons

https://doi.org/10.1515/9783110710687-040

Abstract: We consider a periodic chain of oscillating dipoles, interacting via long-range dipole–dipole interactions, embedded inside a cuboid cavity waveguide. We show that the mixing between the dipolar excitations and cavity photons into polaritons can lead to the appearance of new states localized at the ends of the dipolar chain, which are reminiscent of Tamm surface states found in electronic systems. A crucial requirement for the formation of polaritonic Tamm states is that the cavity cross section is above a critical size. Above this threshold, the degree of localization of the Tamm states is highly dependent on the cavity size since their participation ratio scales linearly with the cavity cross-sectional area. Our findings may be important for quantum confinement effects in one-dimensional systems with strong light–matter coupling.

Keywords: cavity quantum electrodynamics; one-dimensional systems; polaritonics; Tamm states.

1 Introduction

In 1932, Tamm [1] showed the existence of surface states in a one-dimensional (1-D) crystal lattice, due to the abrupt termination of the periodic crystal at an interfacing surface, such as the vacuum [2–4]. This result highlighted a surprising failure of the theory of a periodic potential with cyclic boundary conditions at the elementary level of the electronic bandstructure, despite its great utility in explaining the bulk properties of solids [5, 6]. Tamm surface states have since been shown to have profound consequences for the rich field of surface science, including for photoluminescence in mesoscopic systems [7], photocurrents in superlattices [8], and the absorption spectra of molecular chains [9–11].

Over the last two decades, various theories of Tamm states in the latest condensed matter systems have been developed [12]. For example, with exciton–polaritons in multilayer dielectric structures [13], with plasmons at at the boundary between a metal and a dielectric Bragg mirror [14], and with phonons in graphene nanoribbons [15]. Pioneering experimental work has seen the observance of Tamm states in magnetophotonic structures [16], in organic dye-doped polymer layers [17], and latterly in photonic crystals [18–20].

Due to the rise of topological physics in photonics and plasmonics [21–26], there is an ongoing interest in finding and classifying unconventional light–matter states. Indeed, the latest advances in topological matter have been made in photon-based systems, leading to the rapidly expanding subfield of topological nanophotonics [27–33]. It is therefore crucial to also classify and understand surface states of a nontopological origin, such as Tamm-like edge states, in systems with strong light–matter coupling. Indeed, there are recent experimental studies of polariton micropillars, where the localization of both topologically trivial and topologically nontrivial modes is examined in detail [34, 35].

In this work, we consider a nanophotonic system which exhibits Tamm-like edge states: a 1-D chain of regularly spaced nanoresonators, coupled via dipole–dipole interactions, which are housed inside a cavity waveguide (see Figure 1(a)). The linear dipolar chain is of some importance since it is a simple system where one may study the subwavelength transportation of energy and information [36–40]. In our theory, we place the dipolar chain inside a cuboid cavity in order to study the effect of controllable light–matter interactions. Modulating the cross-sectional area of the cavity allows one to tune both

***Corresponding author: Luis Martín-Moreno,** Instituto de Nanociencia y Materiales de Aragón (INMA), CSIC-Universidad de Zaragoza, Zaragoza 50009, Spain; and Departamento de Física de la Materia Condensada, Universidad de Zaragoza, E-50009, Zaragoza, Spain, E-mail: lmm@unizar.es. https://orcid.org/0000-0001-9273-8165
Charles A. Downing, Instituto de Nanociencia y Materiales de Aragón (INMA), CSIC-Universidad de Zaragoza, Zaragoza 50009, Spain; and Departamento de Física de la Materia Condensada, Universidad de Zaragoza, E-50009, Zaragoza, Spain, E-mail: downing@unizar.es. https://orcid.org/0000-0002-0058-9746

This article has previously been published in the journal Nanophotonics. Please cite as: Charles A. Downing and L. Martín-Moreno "Polaritonic Tamm states induced by cavity photons" *Nanophotonics* 2021, 10. DOI: 10.1515/nanoph-2020-0370.

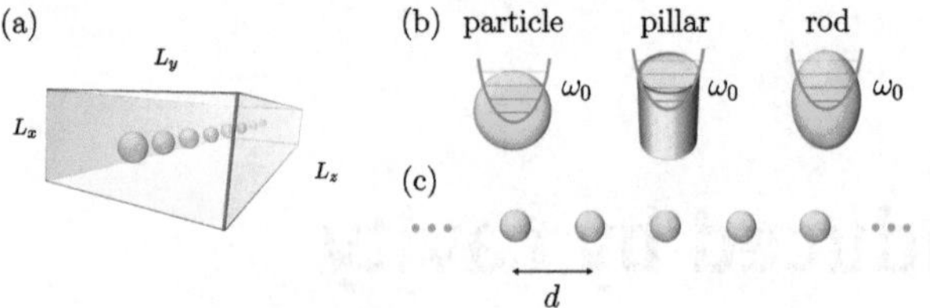

Figure 1: Panel (a): a sketch of our system: a chain of dipoles embedded inside a cuboid cavity waveguide of dimensions $L_x \times L_y \times L_z$. Panel (b): Each dipole is modeled as a harmonic oscillator of resonance frequency ω_0. It can be realized by the Mie resonance in a dielectric nanoparticle, spin waves in a magnetic micropillar, or localized surface plasmons in a metallic nanorod. Panel (c): the long chain of $\mathcal{N} \gg 1$ oscillating dipoles, regularly spaced by the center-to-center separation d.

the light–matter coupling strength and the light–matter detuning. In the strong coupling regime, the dipolar excitations in the resonator chain hybridize with the cavity photons to form polaritonic excitations [41, 42]. The resulting polaritons, which display half-light and half-matter properties, can lead to the emergence of a highly localized edge state of a nontopological origin: a Tamm-like state. Notably, neither the dipolar chain nor the cavity photons display Tamm states when the light–matter interaction is switched off. We discuss the properties of the emergent polaritonic Tamm state, including how its creation requires the cavity cross section to be above a critical size and how its localization properties scale with the cavity cross-sectional area.

The presented theory of a chain of oscillating dipoles embedded inside a cuboid cavity may be realized in a wide range of dipolar systems, as alluded to in the sketches in Figure 1(b). At the subwavelength scale, exploiting the Mie resonances in a chain of dielectric nanoparticles is a promising option since the system does not suffer from high losses and is hence ideal for energy transportation [43–45]. Localized surface plasmons hosted by metallic nanoparticles are another accessible platform [46–48], and there are several recent experimental studies of plasmonic nanoparticles in cavity geometries [49–52]. Exciting spin waves in magnetic microspheres is another appealing possibility [53] since cavity magnons have been well studied experimentally in recent years [54]. Finally, implementations with Rydberg [55, 56] and ultracold atoms [57, 58], as well as helical resonators [59], are also viable settings for the versatile theory presented here.

The rest of the manuscript is organized as follows: we describe our model in Section 2, we unveil the polaritonic Tamm states in Section 3, and we draw some conclusions in Section 4. The Supplementary material contains additional theoretical details.

2 Model

The Hamiltonian of a chain of oscillating dipoles embedded inside a cavity reads as follows [60–62]:

$$H = H_{\text{dp}} + H_{\text{ph}} + H_{\text{dp-ph}}, \tag{1}$$

accounting for the dipolar dynamics, the cavity photons, and the light–matter coupling, respectively. Importantly, the couplings in H_{dp} go beyond the nearest-neighbor approximation [25, 40], which is essential for a proper treatment of the type of Tamm states discussed in this system.

We sketch in Figure 1(c) the model of our system: a 1-D array of dipoles, regularly spaced at the interval d, which is encased inside a cuboid cavity of dimensions $L_x \times L_y \times L_z$ (see panel [a]). Tuning the size of the cavity cross-sectional area ($L_x \times L_y$) modulates both the light–matter coupling strength and the light–matter detuning, such that polariton excitations may be formed by the mixing between the cavity photons and dipolar excitations (which are generally treated as harmonic oscillators, see Figure 1(b) for some physical realizations).

2.1 Dipolar Hamiltonian

The dipolar Hamiltonian (H_{dp} in Eq. (1)) describes a linear chain of $\mathcal{N} \gg 1$ dipoles (cf. Figure 1[c]), oscillating in the $\hat{x}$-direction with transverse polarization ($\uparrow\uparrow\uparrow\cdots$), and coupled to each other via dipole–dipole interactions. Setting $\hbar = 1$ throughout, this Hamiltonian reads (see Refs. [39, 40, 63] for details) as follows:

$$H_{\text{dp}} = \sum_q \left\{ \omega_0 b_q^\dagger b_q + \frac{\Omega}{2} f_q \left[b_q^\dagger \left(b_q + b_{-q}^\dagger \right) + \text{h.c.} \right] \right\}, \tag{2}$$

where the bosonic creation (annihilation) operator $b_q^\dagger$ (b_q) creates (destroys) a dipolar excitation of wavevector q, where $q \in [-\pi/d, +\pi/d]$ spans the first Brillouin zone. The dipolar resonance frequency of a single dipole is ω_0, which is associated with the length scale a [63]. The weak dipolar coupling constant $\Omega \ll \omega_0$ reads $\Omega = (\omega_0/2)(a/d)^3$, exhibiting the inverse-cubic dependence characteristic of dipole–dipole interactions, and d is the center-to-center separation between the dipoles [63]. In Eq. (2), we have introduced the lattice sum $f_q = 2\sum_{n=1}^{\infty} \cos(nqd)/n^3 = 2\text{Cl}_3(qd)$, where $\text{Cl}_s(z)$ is the Clausen function of order s. Crucially, f_q takes into account long-range interactions between all of the resonators, which is known to be important in dipolar systems [40]. Significantly, going beyond the nearest-neighbor approximation also changes the constraints at the edge of the chain from standard hard-wall boundary conditions. Namely, the edge resonators 1 and $\mathcal{N}$ do not just feel the penultimate resonators 2 and $\mathcal{N}-1$ but also those in the bulk. When considering the nearest-neighbor (nn) coupling approximation, one should replace the lattice sum f_q with the standard result $f_q^{\mathrm{nn}} = 2\cos(qd)$.

After ignoring the counter-rotating terms in Eq. (2) (see Ref. [63] for the full treatment), the eigenfrequencies ω_q^{dp} of the collective dipolar modes follow immediately as

$$\omega_q^{\mathrm{dp}} = \omega_0 + \Omega f_q. \tag{3}$$

Equation (3) describes the usual space quantization of eigenfrequencies into a solitary band. Within the nearest-neighbor coupling approximation $f_q \to f_q^{\mathrm{nn}}$, the spectrum of Eq. (3) reduces to the more familiar cosine expression, $\omega_q^{\mathrm{dp,nn}} = \omega_0 + 2\Omega\cos(qd)$. The most noticeable impact of the above approximation, at the level of the continuum bandstructure, is a reduction in the dipolar bandwidth B^{dp} to the nearest-neighbor value $B^{\mathrm{dp,nn}} = 4\Omega$. The higher "all-coupling" value, which follows from Eq. (3), is $B^{\mathrm{dp}} = (7/2)\zeta(3)\Omega \simeq 4.21\Omega$. Here $\zeta(3) = 1.202\ldots$ is Apéry's constant and $\zeta(s)$ is the Riemann zeta function.

2.2 Polaritonic Hamiltonian

The photonic Hamiltonian (H_{ph} in Eq. (1)) describes the cavity photons inside the long cuboid cavity of dimensions $L_z \gg L_y > L_x$ (see Figure 1[a]). In terms of the photonic creation (annihilation) operator $c_q^\dagger$ (c_q), it reads (see Refs. [63, 67, 68] for details) as follows:

$$H_{\mathrm{ph}} = \sum_q \omega_q^{\mathrm{ph}} c_q^\dagger c_q. \tag{4}$$

where the cavity photon dispersion ω_q^{ph} is given as follows:

$$\omega_q^{\mathrm{ph}} = c\sqrt{q^2 + \left(\frac{\pi}{L_y}\right)^2}, \tag{5}$$

where c is the speed of light in vacuum. The cavity width is L_y and the cavity aspect ratio $L_y > L_x$, such that only the photonic band of Eq. (5) is relevant for the problem. The full light–matter coupling Hamiltonian ($H_{\mathrm{dp-ph}}$ in Eq. (1)) reads (see Ref. [63] for the derivation) as follows:

$$H_{\mathrm{dp-ph}} = \sum_q \left\{ \mathrm{i}\xi_q \left[b_q^\dagger \left(c_q + c_{-q}^\dagger\right) - \mathrm{h.c.}\right] + \frac{\xi_q^2}{\omega_0}\left[c_q^\dagger\left(c_q + c_{-q}^\dagger\right) + \mathrm{h.c.}\right]\right\}, \tag{6}$$

where, we have introduced the light–matter coupling constant as follows:

$$\xi_q = \omega_0 \left(\frac{2\pi a^3}{L_x L_y d}\frac{\omega_0}{\omega_q^{\mathrm{ph}}}\right)^{1/2}. \tag{7}$$

The diamagnetic term (on the second line of Eq. (6)) simply leads to a renormalization of the photon dispersion ω_q^{ph}, as defined in Eq. (5), into

$$\widetilde{\omega}_q^{\mathrm{ph}} = \omega_q^{\mathrm{ph}} + \frac{2\xi_q^2}{\omega_0}, \tag{8}$$

a shift which can be safely disregarded throughout this work, since it only leads to small quantitative changes to the results presented here. The paramagnetic term (on the first line of Eq. (6)) is important and gives rise to the formation of polaritonic excitations.

Ignoring counter-rotating terms in the polaritonic Hamiltonian (formed by Eqs. (2), (4), and (6)), we may write the resulting rotating wave approximation polaritonic Hamiltonian as follows:

$$H_{\mathrm{pol}}^{\mathrm{RWA}} = \sum_q \widehat{\psi}^\dagger \mathcal{H}_{\mathrm{pol}}^{\mathrm{RWA}} \widehat{\psi}, \quad \mathcal{H}_{\mathrm{pol}}^{\mathrm{RWA}} = \begin{pmatrix} \omega_q^{\mathrm{dp}} & \mathrm{i}\xi_q \\ -\mathrm{i}\xi_q & \omega_q^{\mathrm{ph}} \end{pmatrix}, \tag{9}$$

where the polaritonic Bloch Hamiltonian is $\mathcal{H}_{\mathrm{pol}}^{\mathrm{RWA}}$ and where we used the basis $\widehat{\psi} = (b_q, c_q)$. We arrive by bosonic Bogoliubov transformation at the diagonal form of Eq. (9)

$$H_{\mathrm{pol}}^{\mathrm{RWA}} = \sum_{q\tau} \omega_{q\tau}^{\mathrm{pol}} \beta_{q\tau}^\dagger \beta_{q\tau}, \tag{10}$$

where the index $\tau = \pm$ labels the upper and lower polariton bands. The polariton dispersion $\omega_{q\tau}^{\mathrm{pol}}$ in Eq. (10) reads as follows:

$$\omega_{q\tau}^{\mathrm{pol}} = \overline{\omega}_q + \tau\Omega_q, \tag{11}$$

where the average frequency of the uncoupled dispersions $\overline{\omega}_q$, the effective coupling constant Ω_q, and the light–matter detuning Δ_q are given as follows:

$$\overline{\omega}_q = \frac{1}{2}\left(\omega_q^{\mathrm{ph}} + \omega_q^{\mathrm{dp}}\right), \quad \Omega_q = \sqrt{\xi_q^2 + \Delta_q^2}, \qquad \Delta_q = \frac{1}{2}\left(\omega_q^{\mathrm{ph}} - \omega_q^{\mathrm{dp}}\right). \tag{12}$$

The bosonic Bogoliubov operators $\beta_{q\tau}$ in Eq. (10) are defined as follows:

$$\beta_{q+} = \sin\theta_q b_q - i\cos\theta_q c_q, \beta_{q-} = \cos\theta_q b_q + i\sin\theta_q c_q, \tag{13}$$

where the Bogoliubov coefficients are as follows:

$$\cos\theta_q = \frac{1}{\sqrt{2}}\left(1 + \frac{\Delta_q}{\Omega_q}\right)^{1/2}, \sin\theta_q = \frac{1}{\sqrt{2}}\left(1 - \frac{\Delta_q}{\Omega_q}\right)^{1/2}, \tag{14}$$

in terms of the quantities defined in Eq. (12).

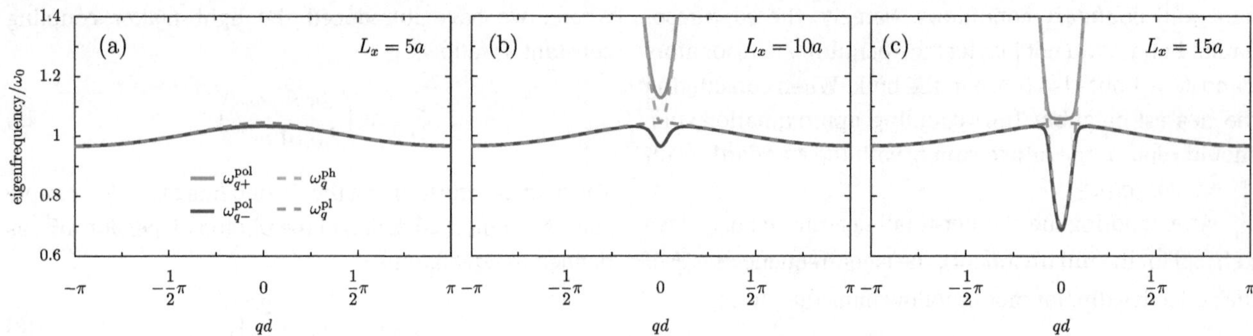

Figure 2: The polariton dispersion $\omega_{q\tau}^{\rm pol}$ in the first Brillouin zone ([cf. Eq. (11)) for the cavity heights (a) $L_x = 5a$, (b) $L_x = 10a$, and (c) $L_x = 15a$. The upper (lower) polaritons with $\tau = +(-)$ are denoted by solid blue (red) lines. The uncoupled photonic (dipolar) dispersions $\omega_q^{\rm ph}$ ($\omega_q^{\rm dp}$) are shown as dashed cyan (orange) lines (cf. Eqs. (5) and (3)). In the figure, the interdipole separation $d = 3a$, the dipole strength $\omega_0 c/a = 1/10$, and the cavity aspect ratio $L_y = 3L_x$ are shown.

We plot in Figure 2 the polariton dispersion of Eq. (11) for the cavity heights $L_x = \{5a, 10a, 15a\}$ in panels (a), (b), and (c) respectively, where the cavity aspect ratio is fixed at $L_y = 3L_x$ and the interdipole separation at $d = 3a$. With increasing cavity cross-sectional area in going from panel (a) to (b) to (c), the light–matter detuning Δ_q is reduced (cf. Eq. (12)), leading to increasingly noticeable deviations of the polariton bands (solid lines) from the uncoupled dispersions (dashed lines). The upper (lower) polariton band is given by the red (blue) lines. The photonic bandstructure is denoted by orange lines, while the dipolar bands are in cyan. Notably, in panel (a), only a single polaritonic band is visible on this scale since the (mostly photonic) upper polariton band lies significantly above the frequency scale of ω_0. Panel (b) demonstrates the strong coupling regime and its associated highly reconstructed polariton dispersion, while panel (c) displays the usual band anticrossing behavior as the detuning is further reduced.

The Bogoliubov operators of Eq. (13) imply the pair of polaritonic Bloch spinors $\psi_{q+} = (\sin\theta_q, -i\cos\theta_q)^{\rm T}$ and $\psi_{q-} = (\cos\theta_q, i\sin\theta_q)^{\rm T}$. Notably, unlike the celebrated spinors describing excitations in some topologically nontrivial systems [21–27], there is not a q-dependent phase factor difference (like $e^{i\delta_q}$) between the upper and lower components of each individual spinor $\psi_{q\tau}$. This suggests the absence of any topological physics, which can be confirmed by analyzing the Bloch Hamiltonian. The Hamiltonian of $\mathcal{H}_{\rm pol}$ in Eq. (9) can be decomposed into a 1-D Dirac-like Hamiltonian as follows:

$$\mathcal{H}_q^{\rm pol} = \bar{\omega}_q I_2 - \Delta_q\sigma_z - \xi_q\sigma_y, \tag{15}$$

where $\{\sigma_x, \sigma_y, \sigma_z\}$ are the Pauli matrices, and I_2 is the two-dimensional identity matrix. Despite this Dirac mapping, the associated spinors $\psi_{q\tau}$ indeed lead to a trivial Zak phase of zero [70, 71]. This triviality follows from the symmetries of the Bloch Hamiltonian of Eq. (15), which displays broken inversion ($\sigma_x\mathcal{H}_{-q}^{\rm pol}\sigma_x \neq \mathcal{H}_q^{\rm pol}$) and chiral ($\sigma_z\mathcal{H}_q^{\rm pol}\sigma_z \neq -\mathcal{H}_q^{\rm pol}$) symmetries [72]. This Zak phase analysis classifies the system as topologically trivial, which hence implies an absence of topologically protected edge states. This fact motivates us to understand the highly localized, and yet nontopological, states which can nevertheless be supported by this system (as is shown in the next section).

Perhaps surprisingly, the mixing between the dipolar and photonic modes into polaritons also gives rise to the formation of Tamm-like edge states. These localized states are missing in Figure 2 since their emergence requires a finite system (which precludes the use of periodic boundary conditions).

3 Polaritonic Tamm states

In order to search for the edge states in our system of a chain of resonators inside a cavity, it is necessary to solve the eigenproblem of Eq. (1) in real space, thus removing the periodic boundary condition assumption in the Fourier space calculation of the previous section. This procedure leads to the eigenfrequencies $\omega_m^{\rm pol}$ (and $\omega_m^{\rm pol,nn}$ in the nearest-neighbor approximation), where each eigenstate is labeled with the index m. Each eigenstate $\psi(m) = (\psi_1, \cdots, \psi_{\mathcal{N}})$ spans every site in the chain of $\mathcal{N}$ dipoles. The localization of the states may be classified by the participation ratio PR(m), as defined as follows [73, 74]:

$$\mathrm{PR}(m) = \frac{\left(\sum_{n=1}^{\mathcal{N}}|\psi_n(m)|^2\right)^2}{\sum_{n=1}^{\mathcal{N}}|\psi_n(m)|^4}, \tag{16}$$

where the summations are over all of the dipole sites n. Extended states residing in the bulk part of the spectrum are characterized by a participation ratio scaling linearly with the system size and in the nearest-neighbor approximation $\mathrm{PR}(m) \simeq (2/3)\mathcal{N}$ [63]. Notably, the participation ratio of edge states does not scale with the system size $\mathcal{N}$.

In Figure 3(a), we plot the polariton bandstructure from Eq. (11) with $L_x = 10a$ (cf. Figure 2(b)), where all-neighbor coupling $\omega_{q\tau}^{\mathrm{pol}}$ (nearest-neighbor coupling $\omega_{q\tau}^{\mathrm{pol,\,nn}}$) is denoted by solid lines (dashed lines). The upper (lower) polariton band is red (blue) for all-neighbor coupling and green (pink) for the nearest-neighbor coupling approximation. The horizontal gray line is a guide for the eye at the eigenfrequency corresponding to the top of the lower polariton band ($\tau = -1$). Clearly, the impact on the continuum bandstructure of going beyond the nearest-neighbor approximation is negligible, perhaps making the appearance of edge states in the corresponding finite system even more surprising.

Using Eq. (16), Figure 3(b) displays the participation ratio $\mathrm{PR}(m)$ for the equivalent problem in real space for a chain of $\mathcal{N} = 1000$ dipoles, and the color scheme is the same as in panel (a). Strikingly, the participation ratio of the polariton states in the nearest-neighbor coupling case (green and pink triangles) is essentially uniform [$\mathrm{PR}^{\mathrm{nn}}(m) \simeq (2/3)1000 \simeq 667$], while for the all-neighbor case, the participation ratio of the lower polariton band (blue circles) is markedly different, especially near to the top of the lower polariton band (TLB). In particular, the state at the very top of the lower polariton band in the nearest-neighbor approximation is associated with $\mathrm{PR}^{\mathrm{nn}}(m_{\mathrm{TLB}}) \simeq 667$, while in the all-coupling case, the Tamm state just above this band (which we ascribe with the state index m_{Tamm}) has $\mathrm{PR}(m_{\mathrm{Tamm}}) \simeq 41$ (see the insert in panel (b) for a zoom in on the Tamm state). This last result suggests a highly localized state, induced by the different boundary conditions in the all-coupling case, as compared to the standard hard-wall conditions in the nearest-neighbor approximation.

We plot in Figure 3(c) the probability density $|\psi_n|^2$ along the dipolar chain, where the sites are labeled by n, for the polariton eigenstate at the top of the lower polariton band, where the nearest (all)-neighbor coupling is given by the thin green (thick red) line and is associated with the index m_{TLB} (m_{Tamm}). This panel clearly displays the emergence of the Tamm-like edge state in the all-coupling case, induced by (i) the strong light–matter coupling in the cavity and (ii) the all-coupling boundary conditions. This state is not associated with a topological invariant (see the discussion after Eq. (15)), and so we term it a polaritonic Tamm state. This is in direct analogy with the non-topological surface states studied in solid state physics, which also typically arise in 1-D tight-binding models.

In Figure 4(a) and (b), we show the dependence of the participation ratio $\mathrm{PR}(m)$ on the number of dipoles in the chain $\mathcal{N}$ for the cavity heights $L_x = 5.64a$ in panel (a) (the reason for this choice will become apparent in what follows) and $L_x = 10a$ in panel (b). The results for the Tamm states are denoted by thin blue lines, while the results for the bulk states are represented by thick red lines, and the equation of the line is labeled nearby. These results confirm that the bulk states behave according to the standard formula $\mathrm{PR}(m_{\mathrm{bulk}}) \simeq (2/3)\mathcal{N}$ (see Ref. [63]) and reveal that the exotic state revealed in Figure 3(b) is indeed a highly

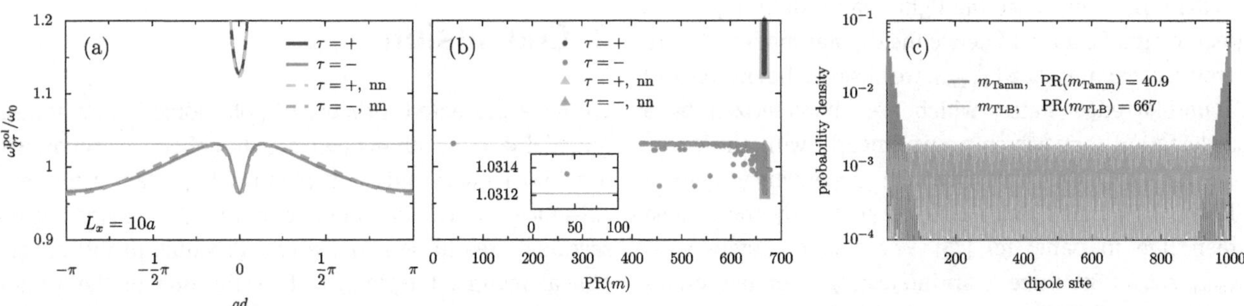

Figure 3: Panel (a): the polariton dispersion in the first Brillouin zone, with all-neighbor coupling $\omega_{q\tau}^{\mathrm{pol}}$ (nearest-neighbor coupling $\omega_{q\tau}^{\mathrm{pol,\,nn}}$) (cf. Eq. (11)). The upper $\tau = +$ polariton band is denoted by solid blue (dashed pink) lines, and the lower $\tau = -$ polariton band is denoted by solid red (dashed green) lines for all (nearest)-neighbor coupling. Horizontal gray line: guide for the eye at the eigenfrequency which corresponds to the top of the bulk band. Panel (b): the polariton eigenfrequencies with all-neighbor coupling ω_m^{pol} (nearest-neighbor coupling $\omega_m^{\mathrm{pol,\,nn}}$), calculated in real space for a chain of $\mathcal{N} = 1000$ dipoles, as a function of the participation ratio PR(m), where m labels the eigenstate (cf. Eq. (16)). The color scheme is the same as in panel (a). Inset: a zoom in of the Tamm state, which lies just above the bulk band. Panel (c): the probability density across the dipolar chain for the polariton eigenstate at the top of the lower polariton band m_{TLB} (for the Tamm state m_{Tamm}), where nearest (all)-neighbor coupling is given by the thin green (thick red) solid line. In the figure, the interdipole separation $d = 3a$, the dipole strength $\omega_0 c/a = 1/10$, the cavity height $L_x = 10a$, and the cavity aspect ratio $L_y = 3L_x$ are shown.

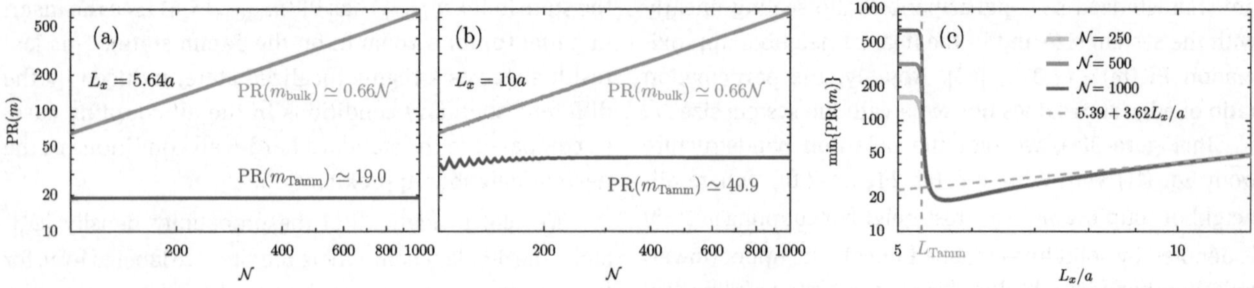

Figure 4: Panels (a) and (b): the participation ratio PR(m) as a function of the number of dipoles $\mathcal{N}$ in the chain, where bulk (Tamm) states are denoted by the thick red (thin blue) lines (cf. Eq. (16)). We show results for the cavity heights $L_x = 5.64a$ (panel [a]) and $L_x = 10a$ (panel [b]). Panel (c): the minimum of the participation ratio min{PR(m)} as a function of the reduced cavity height L_x/a, calculated for $\mathcal{N} = \{250, 500, 1000\}$ dipoles. The linear fitting valid for $L_x \gtrsim 8a$ is given by the dashed green line. The critical cavity size L_{Tamm} is denoted by the vertical dashed gray line. In the figure, the interdipole separation $d = 3a$, the dipole strength $\omega_0 c/a = 1/10$, and the cavity aspect ratio $L_y = 3L_x$ are shown.

localized edge state since it persists with $\mathrm{PR}(m_{\mathrm{Tamm}}) \simeq$ constant with increasing $\mathcal{N}$. Clearly, the increased cavity height L_x in going from panel (a) to (b) in Figure 4 has led to an increased participation ratio of the Tamm states, suggesting weaker localization (explicitly a rise from $\mathrm{PR}(m_{\mathrm{Tamm}}) \simeq 19.0$ in panel [a] to $\mathrm{PR}(m_{\mathrm{Tamm}}) \simeq 40.9$ in panel (b)). This reveals the simple modulation of the cavity cross-sectional area as a tool to control the degree of localization of the edge state (supplementary plots for other cavity sizes are given in Ref. [63]).

We investigate the cavity size Tamm state relationship in Figure 4(c), which shows the minimum of the participation ratio min{PR(m)} as a function of the cavity height L_x, for chains of $\mathcal{N} = \{250, 500, 1000\}$ dipoles. It exposes the identity of the critical length scale $L_{\mathrm{Tamm}} \simeq 5.30a$, the cavity height above which the Tamm states first appear in the system. For subcritical cavities ($L_x < L_{\mathrm{Tamm}}$), no edge states are present, as in a regular dipolar chain uncoupled to cavity photons, since the light–matter detuning is too great to significantly influence the dipolar modes. For supercritical cavities ($L_x \geq L_{\mathrm{Tamm}}$), we observe the presence of Tamm-like edge states, which are characterized by a participation ratio which grows linearly with the cavity size for $L_x \gtrsim 8a$. Explicitly, the dependency here is $\mathrm{PR}(m_{\mathrm{Tamm}}) \simeq 3.62(L_x/a) + 5.39$, as shown by the dashed green line in panel (c) [75]. For smaller cavity sizes $L_{\mathrm{Tamm}} \leq L_x \lesssim 8a$, there is an interesting nonmonotonous behavior, and a global minimum of $\mathrm{PR}(m_{\mathrm{Tamm}}) \simeq 19.0$ occurs at $L_x = 5.64a$ (cf. the results of Figure 4[a]). Of course, these numbers depend on the chosen interdipole separation ratio d/a, dimensionless dipole strength $\omega_0 c/a$, and cavity aspect ratio L_y/L_x.

We have therefore demonstrated an unusual, nontopological (see the discussion after Eq. (15)) phase transition demarcating the absence and presence of Tamm-like edge states, which are induced by cavity interactions and boundary conditions in the chain beyond those used in the nearest-neighbor approximation. The observation of these proposed Tamm states requires careful sweeping in energy, due to their proximity to bulk states. Such careful measurements can be performed using the latest techniques in cathodoluminescence spectroscopy [76] and optical microscopy and spectroscopy [77]. The detection of such states in the strong coupling regime provides perspectives for the fundamental understanding of the interplay between edge states and light–matter coupling and for controlling the localization of polariton states in nanoscale waveguiding structures. Furthermore, while there is a great quest to find topological nanophotonic states [27–33], our findings highlight that after the experimental observation of an edge state, one should also consider possible nontopological mechanisms of generation.

4 Conclusion

We have presented a theory of polaritonic Tamm states, forged due to the mixing between the collective excitations in a dipolar chain and cavity photons. Importantly, the very existence of Tamm states requires the cavity cross-sectional area to be above a critical value. In this supercritical regime, the degree of localization of the Tamm states is highly dependent on the cavity size, with the participation ratio scaling linearly with the cavity cross-sectional area. We have also shown the crucial role played by dipole–dipole interactions beyond the celebrated nearest-neighbor approximation, without which the Tamm edge states do not form. The theory demonstrates the possibility of light trapping in nontopological 1-D structures, which may be important for waveguiding at the nanoscale. Our results also highlight how edge states may be generated via nontopological means, quite distinct from iconic topological models.

Our proposed model can be implemented in a host of systems based upon dipolar resonators, including dielectric and metallic nanoparticles [78]. Our theoretical proposal therefore offers the opportunity to finely control the propagation and localization of collective light–matter excitations at the subwavelength scale and provides perspectives for more complicated and higher dimensional nanophotonic systems [79, 80].

Author contribution: All the authors have accepted responsibility for the entire content of this submitted manuscript and approved submission.
Research funding: This work was supported by the Aragón government through the project Quantum Materials and Devices (Q-MAD) and Ministerio de Economía y Competitividad (MINECO) (Contract No. MAT2017-88358-C3-I-R). CAD acknowledges support from the Juan de la Cierva program (MINECO, Spain).
Conflict of interest statement: The authors declare no conflicts of interest regarding this article.

References

[1] I. E. Tamm, "On the possible bound states of electrons on a crystal surface," *Phys. Z. Sowjetunion*, vol. 1, p. 733, 1932.
[2] R. H. Fowler, "Notes on some electronic properties of conductors and insulators," *Proc. Roy. Soc. A*, vol. 141, p. 56, 1933.
[3] W. Shockley, "On the surface states associated with a periodic potential," *Phys. Rev.*, vol. 56, p. 317, 1939.
[4] F. Forstmann, "The concepts of surface states," *Prog. Surf. Sci.*, vol. 42, p. 21, 1993.
[5] S. G. Davison and M. Steslicka, *Basic Theory of Surface States*, Oxford, Oxford University Press, 1996.
[6] S. Y. Ren, *Electronic States in Crystals of Finite Size: Quantum Confinement of Bloch Waves*, Berlin, Springer, 2005.
[7] H. Ohno, E. E. Mendez, J. A. Brum, et al., "Observation of "Tamm states" in superlattices," *Phys. Rev. Lett.*, vol. 64, p. 2555, 1990.
[8] H. Ohno, E. E. Mendez, A. Alexandrou, and J. M. Hong, "Tamm states in superlattices," *Surf. Sci.*, vol. 267, p. 161, 1990.
[9] V. M. Agranovich, K. Schmidt, and K. Leo, "Surface states in molecular chains with strong mixing of Frenkel and charge-transfer excitons," *Chem. Phys. Lett.*, vol. 325, p. 308, 2000.
[10] K. Schmidt, "Quantum confinement in linear molecular chains with strong mixing of Frenkel and charge-transfer excitons," *Phys. Lett. A*, vol. 293, p. 83, 2002.
[11] M. Hoffmann, "Mixing of Frenkel and charge-transfer excitons and their quantum confinement in thin films," *Thin Films Nanostruct.*, vol. 31, p. 221, 2003.
[12] For a review of Tamm states in photonic crystals, see A. P. Vinogradov, A. V. Dorofeenko, A. M. Merzlikin, and A. A. Lisyansky, "Surface states in photonic crystals," *Phys.-Usp.*, vol. 53, p. 243, 2010.
[13] A. Kavokin, I. Shelykh, and G. Malpuech, "Optical Tamm states for the fabrication of polariton lasers," *Appl. Phys. Lett.*, vol. 87, p. 261105, 2005.
[14] M. Kaliteevski, I. Iorsh, S. Brand, et al., "Tamm plasmon-polaritons: possible electromagnetic states at the interface of a metal and a dielectric Bragg mirror," *Phys. Rev. B*, vol. 76, p. 165415, 2007.
[15] A. V. Savin and Y. S. Kivshar, "Vibrational Tamm states at the edges of graphene nanoribbons," *Phys. Rev. B*, vol. 81, p. 165418, 2010.
[16] T. Goto, A. V. Dorofeenko, A. M. Merzlikin, et al., "Optical Tamm states in one-dimensional magnetophotonic structures," *Phys. Rev. Lett.*, vol. 101, p. 113902, 2008.
[17] S. Núñez-Sánchez, M. Lopez-Garcia, M. M. Murshidy, et al., "Excitonic optical Tamm states: a step toward a full molecular-dielectric photonic integration," *ACS Photonics*, vol. 3, p. 743, 2016.
[18] A. Juneau-Fecteau and L. G. Fréchette, "Tamm plasmon-polaritons in a metal coated porous silicon photonic crystal," *Opt. Mater. Express*, vol. 8, p. 2774, 2018.
[19] A. Juneau-Fecteau, R. Savin, A. Boucherif, and L. G. Fréchette, "Tamm phonon-polaritons: localized states from phonon–light interactions," *Appl. Phys. Lett.*, vol. 114, p. 141101, 2019.
[20] Y. Nakata, Y. Ito, Y. Nakamura, and R. Shindou, "Topological boundary modes from translational deformations," *Phys. Rev. Lett.*, vol. 124, p. 073901, 2020.
[21] A. P. Slobozhanyuk, A. N. Poddubny, A. E. Miroshnichenko, P. A. Belov, and Y. S. Kivshar, "Subwavelength topological edge states in optically resonant dielectric structures," *Phys. Rev. Lett.*, vol. 114, p. 123901, 2015.
[22] C. W. Ling, M. Xiao, C. T. Chan, S. F. Yu, and K. H. Fung, "Topological edge plasmon modes between diatomic chains of plasmonic nanoparticles," *Opt. Express*, vol. 23, p. 2021, 2015.
[23] C. A. Downing and G. Weick, "Topological collective plasmons in bipartite chains of metallic nanoparticles," *Phys. Rev. B*, vol. 95, p. 125426, 2017.
[24] S. R. Pocock, X. Xiao, P. A. Huidobro, and V. Giannini, "Topological plasmonic chain with retardation and radiative effects," *ACS Photonics*, vol. 5, p. 2271, 2018.
[25] C. A. Downing and G. Weick, "Topological plasmons in dimerized chains of nanoparticles: robustness against long-range quasistatic interactions and retardation effects," *Eur. Phys. J. B*, vol. 91, p. 253, 2018.
[26] Á. Gutiérrez-Rubio, L. Chirolli, L. Martín-Moreno, F. J. García-Vidal, and F. Guinea, "Polariton anomalous Hall effect in transition-metal dichalcogenides," *Phys. Rev. Lett.*, vol. 121, p. 137402, 2018.
[27] L. Lu, J. D. Joannopoulos, and M. Soljačić, "Topological photonics," *Nat. Photonics*, vol. 8, p. 821, 2014.
[28] A. B. Khanikaev and G. Shvets, "Two-dimensional topological photonics," *Nat. Photonics*, vol. 11, p. 763, 2017.
[29] X.-C. Sun, C. H. Xiao, P. Liu, M.-H. Lu, S.-N. Zhu, and Y.-F. Chen, "Two-dimensional topological photonic systems," *Prog. Quantum. Electron.*, vol. 55, p. 52, 2017.
[30] V. M. Martinez Alvarez, J. E. Barrios Vargas, M. Berdakin, and L. E. F. Foa Torres, "Topological states of non-Hermitian systems," *Eur. Phys. J. Spec. Top.*, vol. 227, p. 1295, 2018.
[31] T. Ozawa, H. M. Price, A. Amo, et al., "Topological photonics," *Rev. Mod. Phys.*, vol. 91, p. 015006, 2019.

[32] M. S. Rider, S. J. Palmer, S. R. Pocock, X. Xiao, P. A. Huidobro, and V. Giannini, "A perspective on topological nanophotonics," *J. Appl. Phys.*, vol. 125, p. 120901, 2019.
[33] Y. Ota, K. Takaka, T. Ozawa, et al., "Active topological photonics," *Nanophotonics*, vol. 9, p. 547, 2020.
[34] P. St-Jean, V. Goblot, E. Galopin, et al., "Lasing in topological edge states of a one-dimensional lattice," *Nat. Photonics*, vol. 11, p. 651, 2017.
[35] C. E. Whittaker, E. Cancellieri, P. M. Walker, et al., "Effect of photonic spin-orbit coupling on the topological edge modes of a Su-Schrieffer-Heeger chain," *Phys. Rev. B*, vol. 99, p. 081402(R), 2019.
[36] M. L. Brongersma, J. W. Hartman, and H. A. Atwater, "Electromagnetic energy transfer and switching in nanoparticle chain arrays below the diffraction limit," *Phys. Rev. B*, vol. 62, p. R16356(R), 2000.
[37] S. A. Maier, P. G. Kik, and H. A. Atwater, "Optical pulse propagation in metal nanoparticle chain," *Phys. Rev. B*, vol. 67, p. 205402, 2003.
[38] S. Y. Park and D. Stroud, "Surface-plasmon dispersion relations in chains of metallic nanoparticles: an exact quasistatic calculation," *Phys. Rev. B*, vol. 69, p. 125418, 2004.
[39] A. Brandstetter-Kunc, G. Weick, C. A. Downing, D. Weinmann, and R. A. Jalabert, "Nonradiative limitations to plasmon propagation in chains of metallic nanoparticles," *Phys. Rev. B*, vol. 94, p. 205432, 2016.
[40] C. A. Downing, E. Mariani, and G. Weick, "Retardation effects on the dispersion and propagation of plasmons in metallic nanoparticle chains," *J. Phys. Condens. Matter*, vol. 30, p. 025301, 2018.
[41] R. Ameling and H. Giessen, "Microcavity plasmonics: strong coupling of photonic cavities and plasmons," *Laser Photonics Rev.*, vol. 7, p. 141, 2013.
[42] P. Ginzburg, "Cavity quantum electrodynamics in application to plasmonics and metamaterials," *Rev. Phys.*, vol. 1, p. 120, 2016.
[43] R. S. Savelev, A. V. Yulin, A. E. Krasnok, and Y. S. Kivshar, "Solitary waves in chains of high-index dielectric nanoparticles," *ACS Photonics*, vol. 3, p. 1869, 2016.
[44] R. M. Bakker, Y. Feng, Y. R. Paniagua-Domínguez, B. Luk'yanchuk, and A. I. Kuznetsov, "Resonant light guiding along a chain of silicon nanoparticles," *Nano Lett.*, vol. 17, p. 3458, 2017.
[45] K. Koshelev, S. Kruk, E. Melik-Gaykazyan, et al., "Subwavelength dielectric resonators for nonlinear nanophotonics," *Science*, vol. 367, p. 288, 2020.
[46] S. J. Barrow, D. Rossouw, A. M. Funston, G. A. Botton, and P. Mulvaney, "Mapping bright and dark modes in gold nanoparticle chains using electron energy loss spectroscopy," *Nano Lett.*, vol. 14, p. 3799, 2014.
[47] F. N. Gür, C. P. T. McPolin, S. Raza, et al., "DNA-assembled plasmonic waveguides for nanoscale light propagation to a fluorescent nanodiamond," *Nano Lett.*, vol. 18, p. 7323, 2018.
[48] H. T. Rekola, T. K. Hakala, and P. Törmä, "One-dimensional plasmonic nanoparticle chain lasers," *ACS Photonics*, vol. 5, p. 1822, 2018.
[49] M. Barth, S. Schietinger, S. Fischer, et al., "Nanoassembled plasmonic-photonic hybrid cavity for tailored light–matter coupling," *Nano Lett.*, vol. 10, p. 891, 2010.
[50] F. M. Huang, D. Wilding, J. D. Speed, A. E. Russell, P. N. Bartlett, and J. J. Baumberg, "Dressing plasmons in particle-in-cavity architectures," *Nano Lett.*, vol. 11, p. 1221, 2011.
[51] M. A. Schmidt, D. Y. Lei, L. Wondraczek, V. Nazabal, and S. A. Maier, "Hybrid nanoparticle–microcavity–based plasmonic nanosensors with improved detection resolution and extended remote-sensing ability," *Nat. Commun.*, vol. 3, p. 1108, 2012.
[52] Y. Yin, J. Wang, X. Lu, et al., "In situ generation of plasmonic nanoparticles for manipulating photon–plasmon coupling in microtube cavities," *ACS Nano*, vol. 12, p. 3726, 2018.
[53] F. Pirmoradian, B. Z. Rameshti, M. Miri, and S. Saeidian, "Topological magnon modes in a chain of magnetic spheres," *Phys. Rev. B*, vol. 98, p. 224409, 2018.
[54] X. Zhang, C.-L. Zou, L. Jiang, and H. X. Tang, "Strongly coupled magnons and cavity microwave photons," *Phys. Rev. B*, vol. 113, p. 156401, 2014.
[55] A. Browaeys, D. Barredo, and T. Lahaye, "Experimental investigations of dipole–dipole interactions between a few Rydberg atoms," *J. Phys. B At. Mol. Opt. Phys.*, vol. 49, p. 152001, 2016.
[56] S. de Léséleuc, V. Lienhard, P. Scholl, et al., "Observation of a symmetry-protected topological phase of interacting bosons with Rydberg atoms," *Science*, vol. 365, p. 775, 2019.
[57] H. Weimer, N. Y. Yao, C. R. Laumann, and M. D. Lukin, "Long-range quantum gates using dipolar crystals," *Phys. Rev. Lett.*, vol. 108, p. 100501, 2012.
[58] N. R. Cooper, J. Dalibard, and I. B. Spielman, "Topological bands for ultracold atoms," *Rev. Mod. Phys.*, vol. 91, p. 015005, 2019.
[59] C. R. Mann, T. J. Sturges, G. Weick, W. L. Barnes, and E. Mariani, "Manipulating type-I and type-II Dirac polaritons in cavity-embedded honeycomb metasurfaces," *Nat. Commun.*, vol. 9, p. 2194, 2018.
[60] D. P. Craig and T. Thirunamachandran, *Molecular Quantum Electrodynamics: An Introduction to Radiation-Molecule Interactions*, London, Academic Press, 1984.
[61] A. Salam, *Molecular Quantum Electrodynamics: Long-Range Intermolecular Interactions*, New Jersey, Wiley, 2010.
[62] C. A. Downing, T. J. Sturges, G. Weick, M. Stobińska, and L. Martín-Moreno, "Topological phases of polaritons in a cavity waveguide," *Phys. Rev. Lett.*, vol. 123, p. 217401, 2019.
[63] See the Supplemental material, which contains Refs. [64]–[66] for further descriptions, details, and derivations.
[64] C. A. Downing, E. Mariani, and G. Weick, "Radiative frequency shifts in nanoplasmonic dimers," *Phys. Rev. B*, vol. 96, p. 155421, 2017.
[65] J. J. Hopfield, "Theory of the contribution of excitons to the complex dielectric constant of crystals," *Phys. Rev.*, vol. 112, p. 1555, 1958.
[66] C. A. Downing, J. C. López Carreño, A. I. Fernández-Domínguez, and E. del Valle, "Asymmetric coupling between two quantum emitters," *Phys. Rev. A*, vol. 102, p. 013723, 2020.
[67] P. W. Milonni, *The Quantum Vacuum: An Introduction to Quantum Electrodynamics*, London, Academic Press, 1994.
[68] K. Kakazu and Y. S. Kim, "Quantization of electromagnetic fields in cavities and spontaneous emission," *Phys. Rev. A*, vol. 50, p. 1830, 1994.
[69] J. C. G. Henriques, T. G. Rappoport, Y. V. Bludov, M. I. Vasilevskiy, and N. M. R. Peres, "Topological photonic Tamm states and the Su-Schrieffer-Heeger model," *Phys. Rev. A*, vol. 101, p. 043811, 2020.
[70] M. V. Berry, "Quantal phase factors accompanying adiabatic changes," *Proc. R. Soc. A*, vol. 392, p. 45, 1984.
[71] J. Zak, "Berry's phase for energy bands in solids," *Phys. Rev. Lett.*, vol. 62, p. 2747, 1989.

[72] J. K. Asboth, L. Oroszlany, and A. Palyi, *A Short Course on Topological Insulators*, Heidelberg, Springer, 2016.
[73] R. J. Bell and P. Dean, "Atomic vibrations in vitreous silica," *Discuss. Faraday Soc.*, vol. 50, p. 55, 1970.
[74] D. J. Thouless, "Electrons in disordered systems and the theory of localization," *Phys. Rep.*, vol. 13, p. 93, 1974.
[75] Of course, for very large cavities where $L_x \gg a$, any Tamm-like edge states are lost as they merge into the bulk part of the spectrum. This limit recovers the regular dipolar chain (uncoupled to light) which exhibits the complete absence of any highly localized states. However, this weak light-matter coupling limit goes beyond our strong coupling model, as encapsulated by the 2×2 Hamiltonian of Eq. (9). A proper treatment of the problem in the regime $L_x \gtrsim 15a$ requires many photonic bands to be included in the theory, as discussed in Ref. [63].
[76] S. Peng, N. J. Schilder, X. Ni, et al., "Probing the band structure of topological silicon photonic lattices in the visible spectrum," *Phys. Rev. Lett.*, vol. 122, p. 117401, 2019.
[77] N. S. Mueller, Y. Okamura, B. G. M. Vieira, et al., "Deep strong light–matter coupling in plasmonic nanoparticle crystals," *Nature*, vol. 583, p. 780, 2020.
[78] D. E. Chang, J. S. Douglas, A. Gonzalez-Tudela, C.-L. Hung, and H. J. Kimble, "Colloquium: quantum matter built from nanoscopic lattices of atoms and photons," *Rev. Mod. Phys.*, vol. 90, p. 031002, 2018.
[79] I. D'Amico, D. G. Angelakis, F. Bussieres, et al., "Nanoscale quantum optics," *Riv. Nuovo Cimento*, vol. 42, p. 153, 2019.
[80] L. Huang, L. Xu, M. Woolley, and A. E. Miroshnichenko, "Trends in quantum nanophotonics," *Adv. Quantum Technol.*, vol. 42, p. 153, 2020.

Supplementary Material: The online version of this article offers supplementary material (https://doi.org/10.1515/nanoph-2020-0370).

Tao Gong, Matthew R. Corrado, Ahmed R. Mahbub, Calum Shelden and Jeremy N. Munday*

Recent progress in engineering the Casimir effect – applications to nanophotonics, nanomechanics, and chemistry

https://doi.org/10.1515/9783110710687-041

Abstract: Quantum optics combines classical electrodynamics with quantum mechanics to describe how light interacts with material on the nanoscale, and many of the tricks and techniques used in nanophotonics can be extended to this quantum realm. Specifically, quantum vacuum fluctuations of electromagnetic fields experience boundary conditions that can be tailored by the nanoscopic geometry and dielectric properties of the involved materials. These quantum fluctuations give rise to a plethora of phenomena ranging from spontaneous emission to the Casimir effect, which can all be controlled and manipulated by changing the boundary conditions for the fields. Here, we focus on several recent developments in modifying the Casimir effect and related phenomena, including the generation of torques and repulsive forces, creation of photons from vacuum, modified chemistry, and engineered material functionality, as well as future directions and applications for nanotechnology.

Keywords: Casimir effect; Casimir force; Casimir torque; quantum electrodynamics; quantum fluctuations.

***Corresponding author: Jeremy N. Munday,** Department of Electrical and Computer Engineering, University of California, Davis, CA 95616, USA, E-mail: jnmunday@ucdavis.edu. https://orcid.org/0000-0002-0881-9876
Tao Gong, Department of Electrical and Computer Engineering, University of California, Davis, CA 95616, USA; Department of Materials Science and Engineering, University of California, Davis, CA 95616, USA. https://orcid.org/0000-0002-5033-8750
Matthew R. Corrado, Department of Physics, University of California, Davis, CA 95616, USA. https://orcid.org/0000-0002-3141-7322
Ahmed R. Mahbub and Calum Shelden, Department of Electrical and Computer Engineering, University of California, Davis, CA 95616, USA. https://orcid.org/0000-0002-9987-9777 (A.R. Mahbub). https://orcid.org/0000-0001-9404-4247 (C. Shelden)

1 Introduction

Quantum fluctuations of electromagnetic waves are a fascinating consequence of the quantization process that occurs when light is pushed to the quantum regime. These fluctuations are responsible for many everyday phenomena from spontaneous emission and the Lamb shift to the wetting processes at solid–liquid interfaces and the biological grip of geckos [1–4]. Another intriguing phenomenon that results from the confinement of quantum fluctuations is the Casimir effect, which occurs when two or more materials are placed within close proximity and perturb the available vacuum modes. If the vacuum energy density varies with displacement of one of the bodies with respect to the others, a force results (Figure 1). In 1948, Casimir [5] considered how the density of vacuum modes changed when two parallel, uncharged metal plates were brought nearby, and he found an attractive force exists between them, which is given by $F = -\frac{\pi^2 \hbar c}{240 d^4}$, where d is the separation between the plates and $\hbar$ and c are fundamental constants. In the decades that followed, several experiments confirmed Casimir's prediction [6–10], and methods have been developed to improve the measurement by taking into account surface roughness [11–14], electrostatic patch potentials [15–21], and thermal contributions [22, 23]. Recently, experiments have begun to show other novel regimes where the strong deviations in the magnitude of the force or power law can be obtained through modification of the geometry [24–31] or optical properties [32–39].

Our ability to tailor electromagnetic boundary conditions is opening new opportunities to engineer not only the Casimir force but also other related phenomena ranging from the generation of photons from moving cavities to modifications of chemical reactions and material phase transitions. In this review, we will highlight some of the recent advancements in the field of Casimir physics and point to some of the exciting new developments and future directions. Given the number of excellent review articles and books pertaining to the Casimir effect [1–3, 40–51], we will limit our focus to connecting recent works with new

This article has previously been published in the journal Nanophotonics. Please cite as: T. Gong, M. R. Corrado, A. R. Mahbub, C. Shelden and J. N. Munday "Recent progress in engineering the Casimir effect – applications to nanophotonics, nanomechanics, and chemistry" *Nanophotonics* 2021, 10. DOI: 10.1515/nanoph-2020-0425.

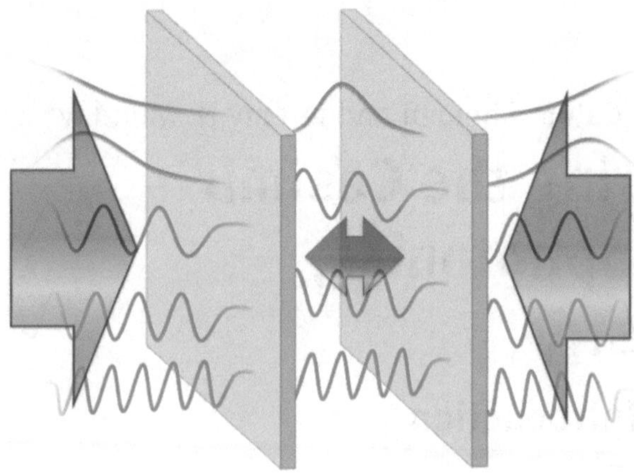

Figure 1: Casimir effect. Quantum fluctuations of electromagnetic fields give rise to a virtual photon pressure on two metal plates. The pressure between the plates is smaller than the pressure from outside, leading to an attractive force between them.

directions for engineering the Casimir effect and related phenomena.

This review article is divided as follows. After a brief introduction to the Casimir force and an historical perspective on its measurement, we discuss how both the optical properties and the geometry of the objects involved affect the magnitude and sign of the force. Next, we describe how optically anisotropic materials can give rise to a new rotational effect, known as the Casimir torque. We follow with a discussion of how systems can be pushed out of equilibrium to further modify the force. Beyond the generation of forces and torques, we discuss how nonadiabatic disturbances of the vacuum field state can result in the generation of photons through the rapid modulation of an electromagnetic boundary and how the Casimir energy can be used to modify molecular states and chemistry. Finally, we conclude with a discussion of applications to nanoscale mechanical devices and other future directions.

Within a decade of Casimir's prediction, initial experiments by Abrikosova and Derjaguin [6] and Sparnaay [7] showed evidence of a force, but limitations in their ability to rule out spurious effects resulted in Sparnaay's claim that his measurements merely "do not contradict Casimir's theoretical prediction." Twenty years passed until van Blokland and Overbeek [8] presented clear evidence of the Casimir force using a chromium layer and compensating voltage to reduce electrostatic interactions. For the next two decades, there were no significant experimental advances until the precision measurement of the Casimir effect by Lamoreaux [9] using modern experimental techniques. This measurement ushered in a new wave of Casimir force experiments using a variety of techniques [10, 23, 24, 35, 52–57]. Figure 2 gives a brief overview of several experimental achievements pertaining to detecting and modifying the Casimir force with a focus on early experimental techniques and advancements [5–10, 55], the effect of the material's optical properties [27, 32, 33, 35–38, 58–60], and the effect of surface texturing and geometry [24, 26, 28–30, 61].

2 Effect of dielectric response

While Casimir's derivation invoked perfect electrical conductors (i.e., ideal mirrors) for the two plates, the force persists when the materials are described by realistic optical properties [62], and its strength depends on the reflectivity of these interacting surfaces (higher reflectivity results in a stronger force). Most metals are excellent broadband reflectors; hence, the interaction between metallic plates is usually much stronger than that between dielectrics. For example, the Casimir force was found to decrease by a factor of two when one of the metallic surfaces was replaced with a transparent conducting oxide [36]. Similarly, there is interest in observing the Casimir effect as a material goes through a phase change, which modifies its optical and electrical properties [60, 63, 64]. However, just because a surface is transparent in the visible region, it does not mean that a significant reduction of the Casimir effect will occur because the force is extremely broadband (see, for example, the case of hydrogen-switchable mirrors, where no significant difference was found between the transparent and reflective states [33]). Further, the individual frequency contributions to the force are highly oscillatory, containing both attractive and repulsive components [65, 66], which adds complexity to both the calculation (e.g., ensuring that the total force converges numerically) and to using intuition from optics to engineer the interaction (e.g., metamaterials involving resonant structures may appear to greatly affect the force [67, 68], but in reality, the results are usually minor [69–71]).

To efficiently calculate the Casimir force between real materials, we often perform a contour integration in the complex plane (Wick rotation) to remove the oscillatory behavior that occurs for real frequencies. This procedure amounts to replacing ω with $i\xi$ (noting that no poles exist in the upper half of the complex plane as a physical consequence of causality) and causes the force integrand to decay rapidly with increasing ξ [62, 66, 72]. Another consequence of the Wick rotation is that the resulting dielectric response $\epsilon(i\xi)$ is a monotonically decreasing function with respect to $i\xi$, which is why the mathematical integration to obtain the force is greatly simplified. From a physical perspective, $\epsilon(i\xi)$ depicts how the charges within a material respond to exponentially increasing fields rather than oscillating ones [73].

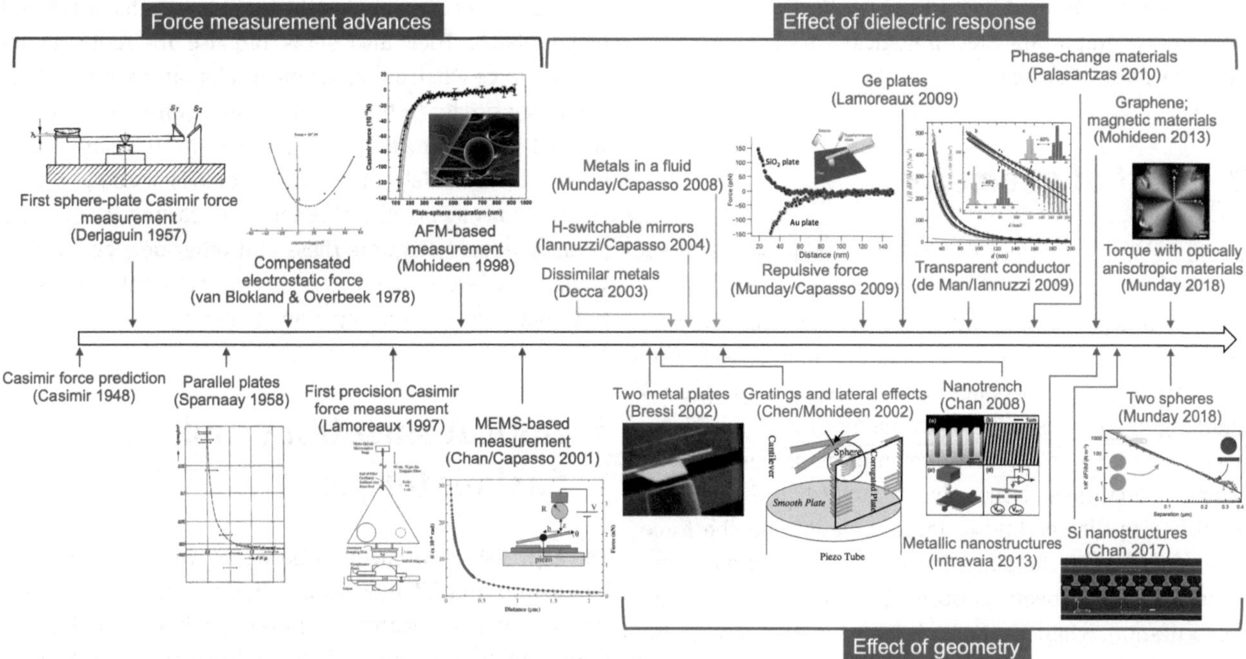

Figure 2: A brief history of Casimir force measurements and the use of dielectric response and geometry to modify the interaction. The images in this chart are adapted with permission from the studies by Abrikosova and Derjaguin [6]. Copyright 1957, MAIK Nauka/Interperiodica. Sparnaay [7]. Copyright 1958, Elsevier B.V. van Blokland and Overbeek [8]. Copyright 1978, Royal Society of Chemistry. Lamoreaux [9]. Copyright 1997, American Physical Society. Mohideen and Roy [10]. Copyright 1998, American Physical Society. Chan et al. [55]. Copyright 2001, The American Association for the Advancement of Science. Munday et al. [59]. Copyright 2009, Nature Publishing Group. de Man et al. [36]. Copyright 2009, American Physical Society. Somers et al. [32]. Copyright 2018, Nature Publishing Group. Bressi et al. [24]. Copyright 2002, American Physical Society. Chen et al. [29]. Copyright 2002, American Physical Society. Chan et al. [61]. Copyright 2008, American Physical Society. Tang et al. [30]. Copyright 2017, Nature Publishing Group. Garrett et al. [26]. Copyright 2018, American Physical Society.

For two mirror symmetric objects made of the same material, the force is always attractive, irrespective of the intermediate medium [74]; however, if the symmetry is broken, a net repulsive force can arise, provided the interacting materials are properly chosen. The condition to achieve repulsion between two semi-infinite half-spaces separated by a third material is $\epsilon_1(i\xi) > \epsilon_3(i\xi) > \epsilon_2(i\xi)$, where the subscript 1, 2, and 3 denote the materials of interacting objects (1 and 2) and intervening medium (3) [75]. It should be noted that this condition has to hold over a sufficiently large frequency range to render the force repulsive; hence, finding materials that satisfy this condition is not easy. In 2009, the first measurement of the long-range repulsive Casimir force was reported using a gold sphere, a silica plate, and a fluid of bromobenzene [59], which suggests a potential way to mitigate stiction problems in microelectromechanical systems (MEMS) or nanoelectromechanical systems (NEMS), and also suggests the possibility of other intriguing phenomena, such as quantum levitation, superlubricity [76], and stable mechanical suspension of objects in fluids [77]. Recently, a teflon-coated gold surface was also used to stably trap a gold nanoplate in ethanol [78], and new theoretical approaches are being developed to analyze other geometries in a fluid [79]. For optically anisotropic materials, it has been pointed out that two identical birefringent materials can lead to repulsion for particular orientations of the objects, provided the optical properties are specifically chosen [80]. Materials with a strong magnetic response have also been proposed to yield a repulsive force when placed near another material with a strong electric response [81, 82]; however, questions have been raised about the ability to tailor such materials [83, 84]. One potential implementation involves superparamagnetic metamaterials [85], although the magnitude of the force appears to be beyond current experimental capabilities. Finally, materials that have $\epsilon(\omega) = 0$ can be used to create cavities that suppress quantum fluctuations over a particular bandwidth [86], which may provide an additional opportunity for modifying the Casimir effect if materials can be engineered over appropriate bandwidths.

These examples show how selecting and engineering the dielectric response over a broad frequency range can lead to significant modifications of the Casimir effect with potentially important technological applications for nanomechanical devices.

3 Effect of geometry

While Casimir's original derivation considered the force between two parallel plates, many alternative geometries have been explored both due to experimental convenience and for the possibility of generating repulsive forces. Experimentally, the plate–plate configuration proved challenging, as it is difficult to maintain parallelism between two plates at submicron separations as a result of the strongly nonlinear force power law (d^{-4}) that would amplify any tilt, leading to large variations in the force magnitude across the plate. For this reason, nearly all experiments have been performed in the sphere–plate configuration, where the Casimir force persists, albeit with a slightly different form: $F_C = -\frac{\pi^3 \hbar c R}{360 d^3}$, where R is the radius of the sphere (valid for $R >> d$).

New experimental setups are enabling researchers to now move beyond the sphere–plate configuration (see Figure 2). The first significant modification was achieved by corrugating one of the surfaces [29]. This configuration allowed for both a modification of the force as well as the observation of a lateral contribution. Advances in silicon microfabrication have also allowed for the use of MEMS devices with interlaced parts (similar to the interlaced comb-drive actuator) to explore more complex, in-plane geometries [30]. Atomic force microscope (AFM)-based techniques have been further expanded to allow for 3D alignment and force detection, opening the door to a variety of more complex configurations, including the recent measurement of the Casimir force between two metal-coated spheres [26].

Experimental advances to study more complex geometries have been guided by the theoretical potential of new geometries to give rise to repulsive forces or situations where analytical approximations begin to break down. One of the earliest geometries that appeared to give rise to a repulsive interaction is the hollow sphere [87]; however, closer inspection suggests that only attractive interactions should exist in this case [74, 82, 88]. More recently, a geometry has been proposed consisting of a needle-like metal object and a pinhole, which is predicted to lead to repulsion [89]. This result has led to the prediction of a class of systems that could yield repulsive forces [90]. 2D materials, such as those in the graphene family, have unique scaling laws for the Casimir energy compared to the traditional behavior of bulk metal plates. They also show promise for reducing the Casimir force when adhered onto a substrate in applications where sensitivity to the Casimir interaction may cause unwanted consequences [91–93]. While the development of new methods to detect the Casimir force is ongoing, significant advancements in technology are starting to allow experimentalists to explore these and other new geometries, which may lead to the measurement of a repulsive Casimir force between vacuum-separated materials.

4 Optical anisotropy and the Casimir torque

The expressions for the force between two surfaces derived by Casimir and Lifshitz assume that the interacting materials are optically isotropic; however, what would happen if these materials were optically *anisotropic* (see Figure 3)? In the early 1970s, Kats [94] and Parsegian and Weiss [95] independently answered this question. The expression derived by Parsegian and Weiss [95] is for the nonretarded interaction energy (which is generally valid for $\lesssim$10 nm and where the interaction is considered instantaneous) between two anisotropic dielectric plates, with in-plane optical axes that are parallel to the surface, immersed in a third medium. They found that the interaction energy depends on the angular orientation of the two plates, resulting in a torque that would cause their optical axes to align. In 1978, Barash [96] analyzed a similar problem using the Helmholtz free energy, which included retardation effects. In the nonretarded regime, both expressions are in agreement.

A number of proposals were made for experiments to detect this torque [97–102], and final experimental confirmation came in 2018 [32]. The difficulty in detecting this torque comes from the need to measure the rotation of one relatively large body in close proximity to another. For the case of two parallel, optically anisotropic plates, one could use a torsional rod to suspend one plate above the other; however, it is difficult to maintain parallelism of the plates at separations of less than 1 μm. To circumvent this obstacle, we made two modifications in our recent experiments [32]. First, we replaced the vacuum gap with an isotropic solid (Al_2O_3), which keeps the two plates separated at a fixed distance and parallel to each other without touching. The addition of this isotropic layer also increases the magnitude of the torque [103]. However, if all three layers are solids, frictional forces will prevent rotation.

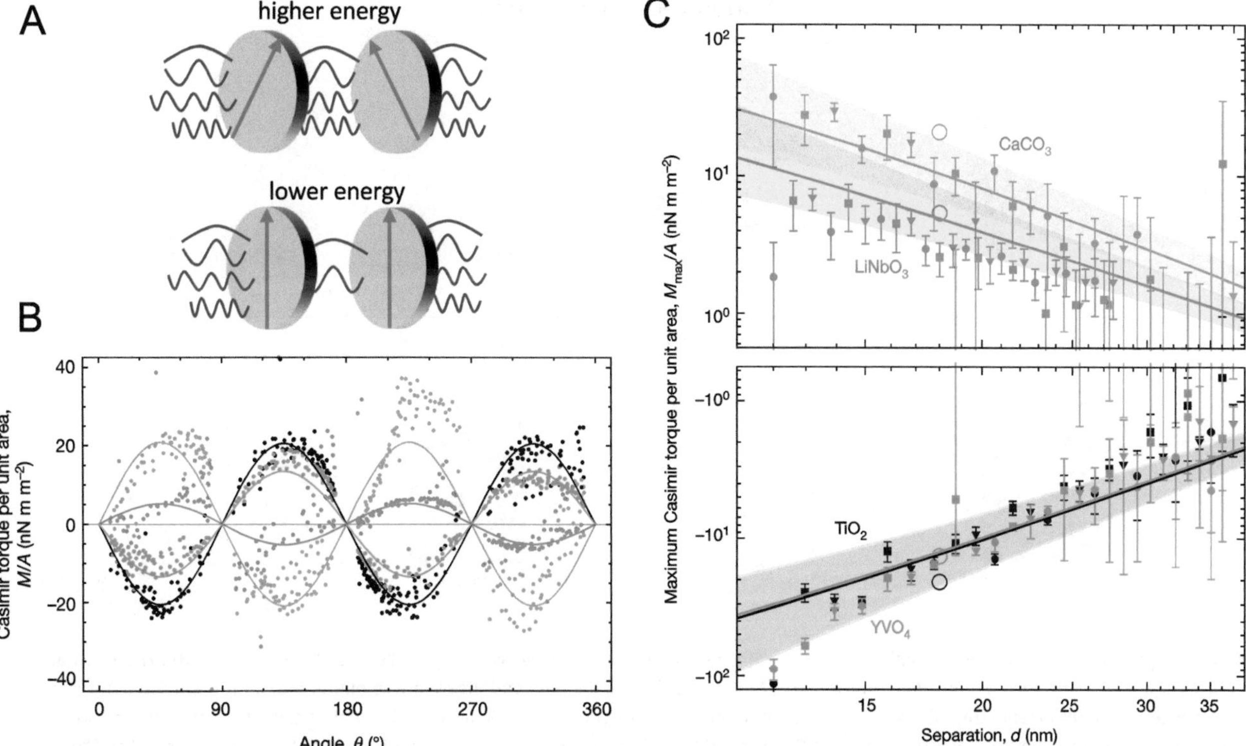

Figure 3: Casimir torque. (A) When materials with optical anisotropy are used instead of isotropic metal plates, the energy associated with the quantum fluctuations depends on their orientation and results in a torque. This Casimir torque is found to (B) have a sin(2θ) dependence and (C) decrease in magnitude with increasing separation between the plates. (B and C) Adapted with permission from the study by Somers et al. [32]. Copyright 2018, Nature Publishing Group.

Second, we replaced one of the solid anisotropic plates with a liquid crystal. The liquid crystal wets the Al_2O_3 layer and is free to rotate. This configuration allows for an all-optical measurement technique to detect the liquid crystal rotation caused by the Casimir torque (Figure 3). Further advances and new techniques to measure ultrasmall torques [104, 105] could potentially be used in future experiments to measure the Casimir torque between two solid objects or other rotational or quantum friction effects [106].

The Casimir torque unveils new engineering possibilities to create nanoscale torsional devices and gears based on quantum fluctuations. Because the torque depends on the optical properties of the interacting objects, there are further opportunities to tune this effect using materials with enhanced anisotropy [107, 108] or one-dimensional gratings [109]. Similarly, for some materials, different frequency components yield positive or negative contributions to the torque, suggesting that there could be a distance dependence in the sign of the torque for some materials or configurations [110], just as there can be a sign change in the force. All of these possibilities are now within experimental reach.

5 Nonequilibrium systems

The theories presented by both Casimir and Lifshitz assume all interacting objects are in thermal equilibrium with their surroundings; however, a variety of interesting phenomena can occur when this condition is relaxed (Figure 4A–C). An early approach by Henkel et al. [111] showed that for a particle–plate system where the plate is held at a finite temperature and the particle is held at 0 K, the particle will experience a temperature-dependent repulsive force attributed to the thermal fluctuations from the plate. Using a Rb Bose–Einstein condensate and a dielectric substrate, Obrecht et al. [112] were able to make the first measurement of these nonequilibrium forces and verify the previous theoretical predictions [113]. Interestingly, they found that by tuning the temperature of the substrate, the force due to the thermal fluctuations could either enhance the attraction or exert a repulsion.

For configurations involving two solid bodies at different temperatures, e.g., two spheres, a sphere and a plate [114], or two cylinders [115], not only has enhancement of the attraction or repulsion been predicted but also

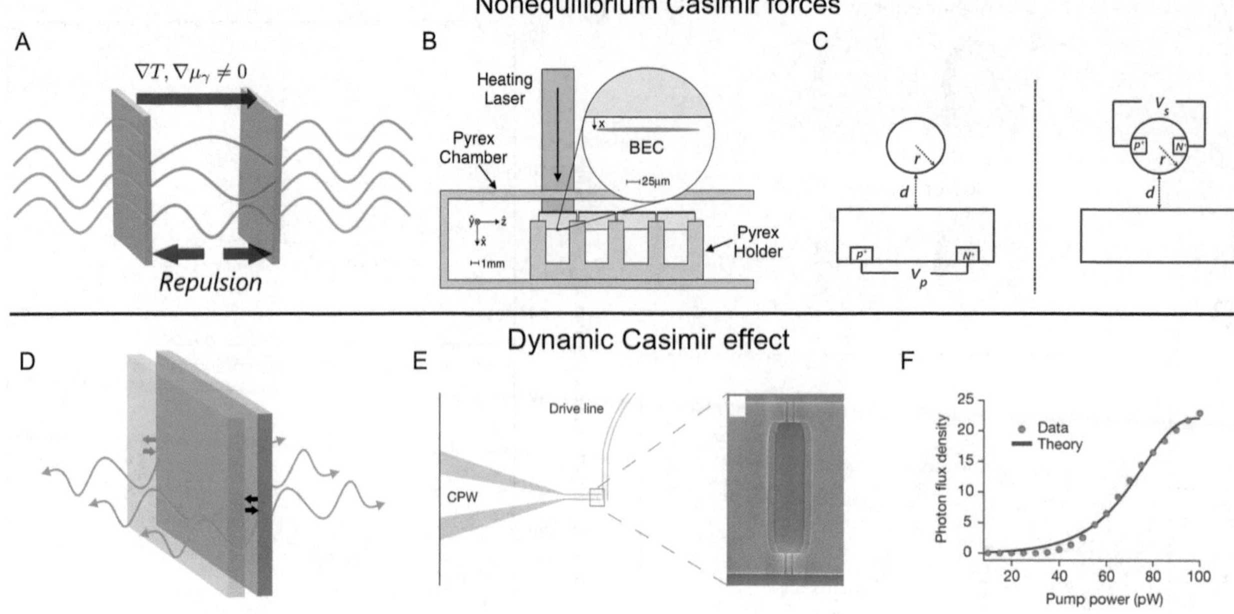

Figure 4: The Casimir effect beyond static equilibrium. (A) Schematic showing the system pushed out of equilibrium. (B) Experimental configuration showing the Casimir–Polder interaction between a Bose–Einstein condensate and a plate held at different temperatures. Reproduced with permission from the study by Obrecht et al. [112]. Copyright 2007, American Physical Society. (C) Configuration where photons are given a chemical potential based on the applied bias. Reproduced with permission from the study by Chen and Fan [116]. Copyright 2016, American Physical Society. (D) Schematic of the dynamic Casimir effect (DCE) showing the generation of photons resulting from an oscillating plate. (E) Schematic of a ~43-mm aluminum coplanar waveguide (CPW) terminated by a SQUID used to detect the DCE. The parametric inductance of the SQUID is tuned by applying a magnetic flux, which allows for a dynamically changing boundary condition. (F) Generated photon flux density due to the DCE at a frequency detuned from the central frequency (10.3 GHz) by 764 MHz with increasing pump power. (E and F) Reproduced with Permission from the study by Wilson et al. [117]. Copyright 2011, Nature Publishing Group.

a new oscillation in the force is expected as a result of scattering and interference of electromagnetic waves in the vicinity of the objects. This apparent oscillation between the attractive and repulsive regimes leads to multiple stable separations where the net force is zero. However, one major drawback of using a temperature difference to push the system out of equilibrium is that within the experimentally accessible (but still difficult) temperature difference of a few hundred Kelvin, the repulsive force is quite small. Further, the attractive equilibrium component typically dominates the interaction, especially at the nanoscale. A far greater temperature difference is needed for the repulsive, nonequilibrium component to dominate at the submicron regime.

One potential solution to the barriers faced using a temperature difference to induce a nonequilibrium state is to instead use an applied bias on one of the bodies to modify the chemical potential of the photons [116]. Calculations show, similar to a temperature difference, an applied bias to the plate modifies the chemical potential of the associated photons and can cause the sphere to experience a repulsion away from the plate. In the far field (beyond 1 μm), the nonequilibrium force shows no dependence on the sphere–plate separation and can simply be interpreted as radiation pressure. In the near field (down to separations as small as 50 nm), calculations predict a repulsion conveying the dominance of the nonequilibrium component well into the submicron regime. The magnitude of the nonequilibrium force calculated in this investigation is well within the measurement capability of current Casimir force techniques, suggesting the possibility of attaining a repulsive nonequilibrium Casimir force at nanoscale separations.

6 Generation of photons from vacuum

One interpretation of the Casimir effect is that it arises from the spatial mismatch of virtual photon modes inside and outside of a cavity. If the boundary of a cavity is oscillating sufficiently fast to render the process nonadiabatic, a temporal mismatch of the virtual photon modes inside the cavity occurs and real photons can be emitted from the cavity (Figure 4D). This process is referred to as the dynamic Casimir effect (DCE) [118].

The simplest system that can produce photons via the DCE is a single mirror in nonuniform motion [119]. The emitted photons in turn exert a radiation/damping force on the mirror, with a magnitude that is proportional to the time derivative of the acceleration of the mirror [120, 121]. If the mirror is oscillating at the frequency Ω with an amplitude of motion a, photons will be generated in pairs, and the emission spectrum is proportional to $\left(\frac{a}{c}\right)^2 \omega(\Omega - \omega)$ based on a simplified 1D calculation and is symmetric about half the oscillation frequency [122]; see also the study by Neto and Machado [123] for the 3D case. The total number of generated photons within time T is found to vary quadratically with the maximum velocity of the oscillating mirror at the rate $\frac{\Omega}{3\pi}\left(\frac{v}{c}\right)^2$ [124].

Notwithstanding its theoretical interest, a single moving mirror can hardly produce experimentally detectable photons due to the limitation of the practically achievable mechanical oscillation speed (~10^{-8} c) and frequency (~3 GHz) [119]. However, it is predicted that by adding another parallel mirror to form a simple cavity, the photon generation rate can be enhanced by approximately the cavity Q-factor to be as high as 10^5 [125], which is attributed to a parametric enhancement effect [124]. Additionally, a number of studies have shown that when the intermode coupling condition is not satisfied, the number of generated photons grows exponentially over time as $N = \sinh^2\left(\Omega t \frac{v}{c}\right)$ under resonant oscillation $\Omega = 2\omega_m$, where ω_m is the mth resonance mode of the cavity [126]. Yet the enhanced photon generation rate (10^{-1}–10 s^{-1}) is still not adequate to be measured practically in this configuration.

To overcome the limitation imposed by mechanical oscillation, a time-varying boundary condition of a cavity can alternatively be induced with electromagnetic modulation, which gives rise to a modulation of the effective cavity length as a function of time. Various approaches along this direction have been proposed, most of which are focused on illuminating the cavity with short laser pulses to alter the dielectric function of the intracavity material (filled completely or partially with a semiconductor) [119, 127, 128]. In the microwave frequency range, semiconductors can behave like a reflective metal mirror under excitation by a laser pulse and become transparent after relaxing from excitation, which can result in oscillation lengths as large as a few millimeters [129–131]. A cavity inserted with nonlinear materials (e.g., $\chi^{(2)}$ or $\chi^{(3)}$ materials) with an effective refractive index that changes with applied light field has been proposed as another promising configuration to achieve measurable DCE photons [132, 133]. These aforementioned modulation methods are theoretically predicted to yield a much larger photon generation rate (10^5–10^{10} s^{-1}), enabling experimental detection. In recent years, other novel cavity modulation methods have drawn increasing attention. In particular, a superconducting quantum interference device (SQUID) modulated by a magnetic field flux can mimic a moving mirror for generating DCE photons, which has experimentally been realized in the microwave regime [117, 134] (see Figure 4E, F), demonstrating great promises of extracting the vacuum photons consistently and reliably. Very recently, an analog DCE has been realized in the near-infrared regime using a dispersion-oscillating photonic crystal fiber [135].

7 Vacuum fluctuation chemistry

Apart from the Casimir effect, vacuum field fluctuations have been attributed to a number of other nonclassical phenomena, among which their significant influence on chemistry and materials science has been gaining attention. As an example, the strong coupling between vacuum fields in an optical cavity and electronic transitions in a molecule can give rise to two hybrid polaritonic states (P+ and P−), separated by the Rabi splitting energy (Figure 5A). In organic molecules, the splitting can be as large as hundreds of meV due to the large transition dipole moments and can hence modify the energy landscape of the reactant molecules. The strong coupling with vacuum fields is found to dramatically influence the kinetics of the photoisomeric conversion from spiropyran into merocyanine (MC) [136]. When the cavity resonance is tuned with the MC absorption at 560 nm, the photoisomeric reaction rate is drastically reduced but the final MC yield is increased by 10% (Figure 5B, C). The modified energy landscape of a molecule can also be expected to alter the electron affinity and work function and may find applications in optoelectronic device design such as light-emitting diodes (LEDs) and photovoltaics.

Molecular vibrational modes are also able to strongly couple with vacuum field fluctuations in a microcavity at room temperature, resulting in Rabi splitting and a change of the Morse potential and the resultant chemical reactivity [138–140]. Vibrational strong coupling (VSC) has drawn increasing interest as the role of the vibrational modes is decisive in determining the kinetics of isomeric and other chemical processes, and the potential impact on bond-selective chemistry is significant. For example, the ground-state deprotection reaction of an alkynylsilane,

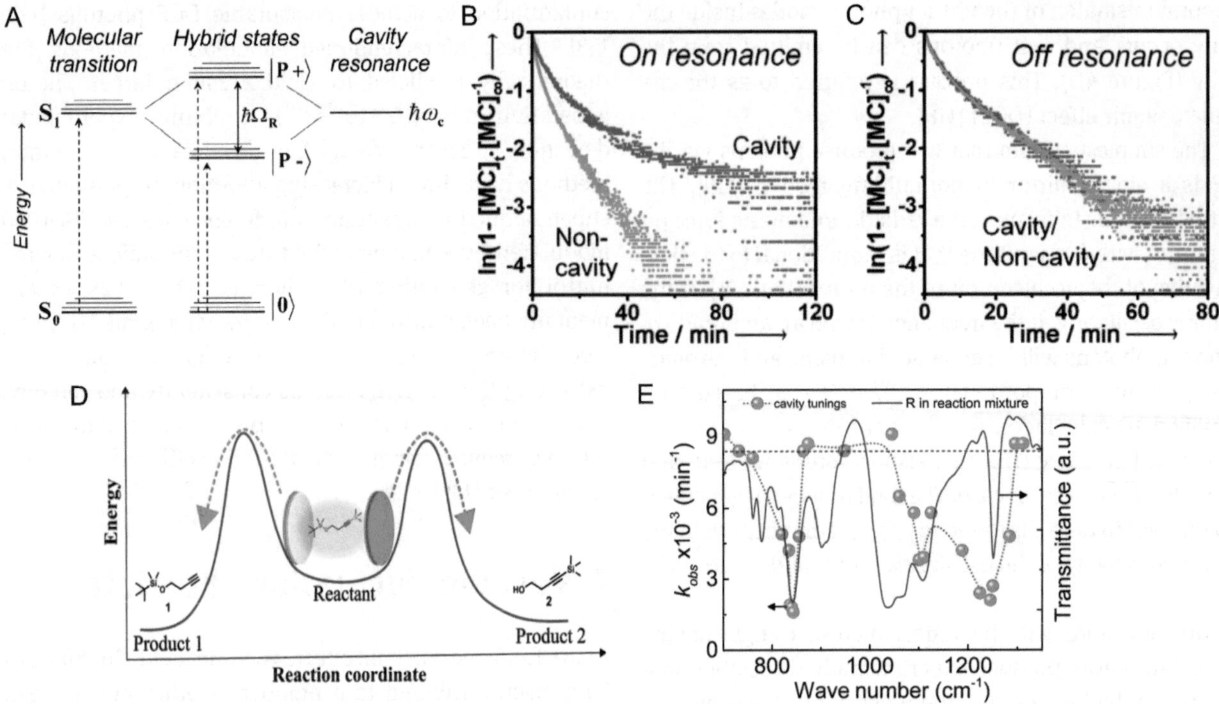

Figure 5: Vacuum fluctuations affecting chemistry. (A) Schematic of the hybridization of energy states when the molecular electron transitions are strongly coupled with vacuum fluctuations in a cavity. The kinetics of the photoisomeric conversion from spiropyran (SPI) into merocyanine (MC) measured for bare molecules (red) and molecules coupled with cavity vacuum fluctuation (green) in both (B) on-resonance (when cavity is tuned on resonance with the MC absorption wavelength at 560 nm) and (C) off-resonance cases. (D) Schematic of two silyl cleavage pathways for a silane derivative with tetrabutylammonium fluoride in a 1:1 mixture of methanol and tetrahydrofuran. (E) The total reaction rate as a function of cavity tuning (red dots) is modified from the rate outside of the cavity (blue dashed line) when the cavity is tuned in resonance with the transmission dips (solid blue line) that correspond to the vibrational modes from the infrared absorption spectrum. (A, B, and C) Reproduced with permission from the study by Hutchison et al. [136]. Copyright 2012, Wiley-VCH. (D and E) Reproduced with permission from the study by Thomas et al. [137]. Copyright 2019, The American Association for the Advancement of Science.

1-phenyl-2-trimethylsilylacetylene (PTA) placed inside a microfluid microcavity is found to be slowed by a factor of 5 when the S–C vibrational stretching mode of the reactant is strongly coupled with the cavity vacuum fields [141]. Further, site selectivity can be modified in reactions with multiple final products so that one product can be favored over the others. As an example, two distinct sites for prospective silyl bond cleavage in a silane derivative (see Figure 5D) can be selectively excited by tuning the cavity (two parallel Au-coated ZnSe mirrors) into resonances corresponding to the two distinct vibrational modes (Si–C stretching bond at 842 cm^{-1} and Si–O stretching bond at 110 cm^{-1}) [137]. When the Si–C stretching bond vibration is strongly coupled with the cavity vacuum field, the total reaction rate is reduced by a factor of 3.5, whereas the overall rate is slowed by a factor of 2.5 when the Si–O stretching vibration is strongly coupled (Figure 5E). More interesting, the ratio of the final products indicates that in either case the VSC modifies the reaction landscape to favor the Si–O cleavage over the Si–C cleavage when off-resonance or outside the cavity, which reverses the selectivity. Vacuum fluctuations can help to control the chemical landscape and aid in a better understanding of the reaction mechanisms for a variety of chemical reactions, including effects in biochemistry and enzyme catalysis [142].

8 Applications to nanotechnology

Because the Casimir force has a strong power law dependence with separation, it becomes the dominant interaction between objects on the nanoscale. For example, the pressure resulting from the Casimir force between two parallel plates separated by 10 nm is approximately 1 atm (or 10^5 N/m^2). Thus, the Casimir force provides both an impediment and an opportunity for nanoscale devices [48, 143].

Advances in silicon processing and integrated circuit technology have allowed for the development of advanced MEMS. These devices can be found in a variety of

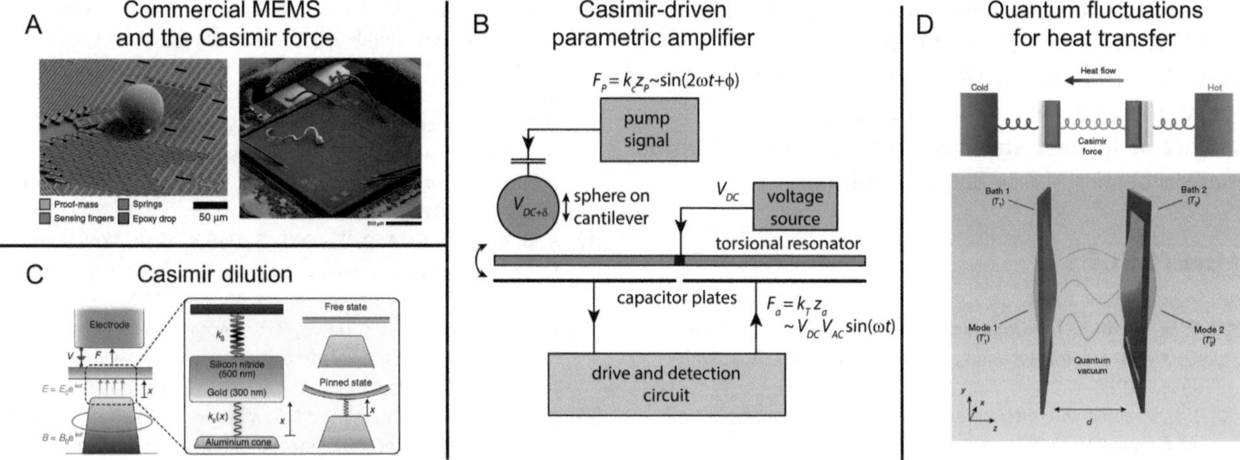

Figure 6: Application of the Casimir force to nanotechnology. (A) Use of a commercial MEMS sensor to measure the Casimir force. Reproduced with permission from the study by Stange et al. [150]. Copyright 2019, Nature Publishing Group. (B) Casimir parametric amplifier. Reproduced with permission from the study by Imboden et al. [151]. Copyright 2014, AIP Publishing. (C) Casimir force incorporated into an optomechanical cavity giving rise to dissipation dilution. Reproduced with permission from the study by Pate et al. [152]. Copyright 2020, Nature Publishing Group. (D) Heat transfer driven by quantum fluctuations. Reproduced with permission from the study by Fong et al. [153]. Copyright 2019, Nature Publishing Group.

technologies ranging from airbag sensors to cell phones. With a push toward further miniaturization (and the transition from MEMS to NEMS), movable parts in these systems begin to experience the Casimir force, which can lead to stiction and adhesion if not properly designed around [48, 144]. Alternatively, the Casimir effect can be advantageous in these devices due to its sensitivity to small changes in distance – examples include actuators [55], devices exploiting bistability and hysteresis [145–147], chaotic behavior [148], virtually frictionless bearings [149], etcetera.

Recently, it has been shown that an off-the-shelf MEMS sensor can be used to measure the Casimir force [150], which shows how commercial technologies and the Casimir effect are beginning to overlap (Figure 6A). Another example is a Casimir-driven parametric amplifier [151], see Figure 6B. This concept involves using the Casimir force to modulate a MEMS parametric amplifier to enable a very sensitive voltage measurement technique, which could be integrated into other sensing devices.

The Casimir force has also found its way into cavity optomechanics and nanoscale heat transfer. One example is the use of the Casimir force to engineer dilution in a macroscopic system involving a SiN membrane and a photonic cavity [152]. In this case, the Casimir force causes an additional strain on the acoustic membrane, allowing the mechanical *Q* factor to be increased (Figure 6C). In another example, it has been found that heat can be transferred through quantum fluctuations, rather than through the conventional routes of conduction, convection, and radiation [153], see Figure 6D. This effect is the result of induced phonon coupling across a vacuum gap due to the electromagnetic fluctuations. This new method of heat transfer could play a significant role in thermal management of future nanoscale technologies.

9 Conclusions

Quantum fluctuations are responsible for a number of phenomena, which we can begin to engineer through material properties, geometry, temperature, etcetera. As we continue to pursue these research directions, the lines between optics, physics, and chemistry blur as these fields coalesce and applications emerge in nanoscale devices, catalytic behavior, and novel chemistry. Herein, we have outlined several of the recent developments in Casimir physics and related fields and point to future directions and developments that we believe will have a significant impact on science and technology in the coming decades through the engineering of quantum vacuum fluctuations.

Acknowledgments: The authors would like to thank Joseph Garrett and Marina Leite for helpful discussions and suggestions pertaining to the manuscript.
Author contribution: All the authors have accepted responsibility for the entire content of this submitted manuscript and approved submission.

Research funding: The authors acknowledge funding from a DARPA YFA Grant No. D18AP00060, a DARPA QUEST grant No. HR00112090084, and the National Science Foundation under grant number PHY-1806768.
Conflict of interest statement: The authors declare no conflicts of interest regarding this article.

References

[1] P. W. Milonni, *The Quantum Vacuum: An Introduction to Quantum Electrodynamics*, 1st ed., Boston, US, Academic Press, 1993.
[2] J. N. Israelachvili, *Intermolecular and Surface Forces*, 3rd ed., Academic Press, 2011.
[3] V. A. Parsegian, *Van der Waals Forces: A Handbook for Biologists, Chemists, Engineers, and Physicists*, New York, Cambridge University Press, 2005.
[4] K. Autumn, Y. A. Liang, S. T. Hsieh, et al., "Adhesive force of a single gecko foot-hair," *Nature*, vol. 405, pp. 681–685, 2000.
[5] H. B. G. Casimir, "On the attraction between two perfectly conducting plates," *Proc. K. Ned. Akad.* vol. 360, pp. 793–795, 1948. https://www.dwc.knaw.nl/DL/publications/PU00018547.pdf.
[6] I. I. Abrikosova, B. V. Deriagin, "Direct measurement of molecular attraction of solid bodies. II. Method for measuring the gap. Results of experiments," *Sov. Phys. JETP* vol. 4, pp. 2–10, 1957. http://www.jetp.ac.ru/cgi-bin/e/index/e/4/1/p2?a=list.
[7] M. J. Sparnaay, "Measurements of attractive forces between flat plates," *Physica*, vol. 24, pp. 751–764, 1958.
[8] P. H. G. M. van Blokland, and J. T. G. Overbeek, "van der Waals forces between objects covered with a chromium layer," *J. Chem. Soc. Faraday Trans.*, vol. 74, pp. 2637–2651, 1978.
[9] S. K. Lamoreaux, "Demonstration of the Casimir force in the 0.6 to 6 μm range," *Phys. Rev. Lett.*, vol. 78, pp. 5–8, 1997.
[10] U. Mohideen and A. Roy, "Precision measurement of the Casimir force from 0.1 to 0.9 μm," *Phys. Rev. Lett.*, vol. 81, pp. 4549–4552, 1998.
[11] A. A. Maradudin and P. Mazur, "Effects of surface roughness on the van der Waals force between macroscopic bodies," *Phys. Rev. B*, vol. 22, pp. 1677–1686, 1980.
[12] V. B. Bezerra, G. L. Klimchitskaya, and C. Romero, "Casimir force between a flat plate and a spherical lens: application to the results of a new experiment," *Mod. Phys. Lett. A*, vol. 12, pp. 2613–2622, 1997.
[13] P. A. Maia Neto, A. Lambrecht, and S. Reynaud, "Roughness correction to the Casimir force: beyond the proximity force approximation," *Europhys. Lett.*, vol. 100, p. 29902, 2012.
[14] P. J. van Zwol, G. Palasantzas, and J. T. M. De Hosson, "Influence of random roughness on the Casimir force at small separations," *Phys. Rev. B*, vol. 77, p. 075412, 2008.
[15] J. L. Garrett, J. Kim, and J. N. Munday, "Measuring the effect of electrostatic patch potentials in Casimir force experiments," *Phys. Rev. Res.*, vol. 2, p. 023355, 2020.
[16] W. J. Kim, M. Brown-Hayes, D. A. R. Dalvit, J. H. Brownell, and R. Onofrio, "Anomalies in electrostatic calibrations for the measurement of the Casimir force in a sphere-plane geometry," *Phys. Rev. A*, vol. 78, p. 020101(R), 2008.
[17] R. O. Behunin, F. Intravaia, D. A. R. Dalvit, P. A. M. Neto, and S. Reynaud, "Modeling electrostatic patch effects in Casimir force measurements," *Phys. Rev. A*, vol. 85, p. 012504, 2012.
[18] D. Garcia-Sanchez, K. Y. Fong, H. Bhaskaran, S. Lamoreaux, and H. X. Tang, "Casimir force and in situ surface potential measurements on nanomembranes," *Phys. Rev. Lett.*, vol. 109, p. 027202, 2012.
[19] R. O. Behunin, D. A. R. Dalvit, R. S. Decca, et al., "Kelvin probe force microscopy of metallic surfaces used in Casimir force measurements," *Phys. Rev. A*, vol. 90, p. 062115, 2014.
[20] R. O. Behunin, D. A. R. Dalvit, R. S. Decca, and C. C. Speake, "Limits on the accuracy of force sensing at short separations due to patch potentials," *Phys. Rev. D*, vol. 89, p. 051301, 2014.
[21] J. L. Garrett, D. Somers, and J. N. Munday, "The effect of patch potentials in Casimir force measurements determined by heterodyne Kelvin probe force microscopy," *J. Phys. Condens. Matter*, vol. 27, p. 214012, 2015.
[22] A. O. Sushkov, W. J. Kim, D. A. R. Dalvit, and S. K. Lamoreaux, "Observation of the thermal Casimir force," *Nat. Phys.*, vol. 7, pp. 230–233, 2011.
[23] G. Bimonte, D. López, and R. S. Decca, "Isoelectronic determination of the thermal Casimir force," *Phys. Rev. B*, vol. 93, p. 184434, 2016.
[24] G. Bressi, G. Carugno, R. Onofrio, and G. Ruoso, "Measurement of the Casimir force between parallel metallic surfaces," *Phys. Rev. Lett.*, vol. 88, p. 041804, 2002.
[25] M. Brown-Hayes, D. A. R. Dalvit, F. D. Mazzitelli, W. J. Kim, and R. Onofrio, "Towards a precision measurement of the Casimir force in a cylinder-plane geometry," *Phys. Rev. A*, vol. 72, pp. 1–11, 2005.
[26] J. L. Garrett, D. A. T. Somers, and J. N. Munday, "Measurement of the Casimir force between two spheres," *Phys. Rev. Lett.*, vol. 120, p. 040401, 2018.
[27] W. J. Kim, A. O. Sushkov, D. A. R. Dalvit, and S. K. Lamoreaux, "Measurement of the short-range attractive force between Ge plates using a torsion balance," *Phys. Rev. Lett.*, vol. 103, p. 060401, 2009.
[28] F. Intravaia, S. Koev, I. W. Jung, et al., "Strong Casimir force reduction through metallic surface nanostructuring," *Nat. Commun.*, vol. 4, p. 2515, 2013.
[29] F. Chen, U. Mohideen, G. L. Klimchitskaya, and V. M. Mostepanenko, "Demonstration of the lateral Casimir force," *Phys. Rev. Lett.*, vol. 88, p. 101801, 2002.
[30] L. Tang, M. Wang, C. Y. Ng, et al., "Measurement of non-monotonic Casimir forces between silicon nanostructures," *Nat. Photonics*, vol. 11, pp. 97–101, 2017.
[31] D. E. Krause, R. S. Decca, D. López, and E. Fischbach, "Experimental investigation of the Casimir force beyond the proximity-force approximation," *Phys. Rev. Lett.*, vol. 98, p. 050403, 2007.
[32] D. A. T. Somers, J. L. Garrett, K. J. Palm, and J. N. Munday, "Measurement of the Casimir torque," *Nature*, vol. 564, pp. 386–389, 2018.
[33] D. Iannuzzi, M. Lisanti, and F. Capasso, "Effect of hydrogen-switchable mirrors on the Casimir force," *Proc. Natl. Acad. Sci. U. S. A.*, vol. 101, p. 4019, 2004.
[34] A. Le Cunuder, A. Petrosyan, G. Palasantzas, V. Svetovoy, and S. Ciliberto, "Measurement of the Casimir force in a gas and in a liquid," *Phys. Rev. B*, vol. 98, p. 201408, 2018.

[35] R. S. Decca, D. Lopez, E. Fischbach, and D. E. Krause, "Measurement of the Casimir force between dissimilar metals," *Phys. Rev. Lett.*, vol. 91, p. 050402, 2003.

[36] S. de Man, K. Heeck, R. J. Wijngaarden, and D. Iannuzzi, "Halving the Casimir force with conductive oxides," *Phys. Rev. Lett.*, vol. 103, p. 040402, 2009.

[37] A. A. Banishev, C. C. Chang, G. L. Klimchitskaya, V. M. Mostepanenko, and U. Mohideen, "Measurement of the gradient of the Casimir force between a nonmagnetic gold sphere and a magnetic nickel plate," *Phys. Rev. B*, vol. 85, p. 195422, 2012.

[38] A. A. Banishev, H. Wen, J. Xu, et al., "Measuring the Casimir force gradient from graphene on a SiO_2 substrate," *Phys. Rev. B*, vol. 87, p. 205433, 2013.

[39] A. A. Banishev, G. L. Klimchitskaya, V. M. Mostepanenko, and U. Mohideen, "Demonstration of the Casimir force between ferromagnetic surfaces of a Ni-coated sphere and a Ni-coated plate," *Phys. Rev. Lett.*, vol. 110, p. 137401, 2013.

[40] P. W. Milonni and M.-L. Shih, "Casimir forces," *Contemp. Phys.*, vol. 33, pp. 313–322, 1992.

[41] L. Spruch, "Long-range (Casimir) interactions," *Science*, vol. 272, p. 1452, 1996.

[42] V. Mostepanenko and N. N. Trunov, *The Casimir Effect and Its Applications*, 1st ed., Oxford, U.K., Clarendon Press, 1997.

[43] K. A. Milton, *The Casimir Effect: Physical Manifestations of Zero-point Energy*, River Edge, USA, World Scientific, 2001.

[44] M. Bordag, U. Mohideen, and V. M. Mostepanenko, "New developments in the Casimir effect," *Phys. Rep.*, vol. 353, pp. 1–205, 2001.

[45] P. A. Martin, P. R. Buenzli, "The Casimir effect," *Acta Phys. Pol. B* vol. 37, pp. 2503–2559, 2006. https://www.actaphys.uj.edu.pl/R/37/9/2503.

[46] A. Lambrecht, "The Casimir effect: a force from nothing," *Phys. World*, vol. 15, pp. 29–32, 2002.

[47] S. K. Lamoreaux, "Resource letter CF-1: Casimir force," *Am. J. Phys.*, vol. 67, pp. 850–861, 1999.

[48] F. Capasso, J. N. Munday, D. Iannuzzi, and H. B. Chan, "Casimir forces and quantum electrodynamical torques: physics and nanomechanics," *IEEE J. Sel. Top. Quant.*, vol. 13, pp. 400–414, 2007.

[49] A. W. Rodriguez, F. Capasso, and S. G. Johnson, "The Casimir effect in microstructured geometries," *Nat. Photonics*, vol. 5, pp. 211–221, 2011.

[50] L. M. Woods, D. A. R. Dalvit, A. Tkatchenko, P. Rodriguez-Lopez, A. W. Rodriguez, and R. Podgornik, "Materials perspective on Casimir and van der Waals interactions," *Rev. Mod. Phys.*, vol. 88, p. 045003, 2016.

[51] D. Iannuzzi, M. Lisanti, J. N. Munday, and F. Capasso, "The design of long range quantum electrodynamical forces and torques between macroscopic bodies," *Solid State Commun.*, vol. 135, pp. 618–626, 2005.

[52] A. Roy, C.-Y. Lin, and U. Mohideen, "Improved precision measurement of the Casimir force," *Phys. Rev. D*, vol. 60, p. 111101, 1999.

[53] B. W. Harris, H. F. Chen, and U. Mohideen, "Precision measurement of the Casimir force using gold surfaces," *Phys. Rev. A*, vol. 62, p. 052109–052101, 2000.

[54] R. S. Decca, D. López, E. Fischbach, G. L. Klimchitskaya, D. E. Krause, and V. M. Mostepanenko, "Precise comparison of theory and new experiment for the Casimir force leads to stronger constraints on thermal quantum effects and long-range interactions," *Ann. Phys.*, vol. 318, pp. 37–80, 2005.

[55] H. B. Chan, V. A. Aksyuk, R. N. Kleiman, D. J. Bishop, and F. Capasso, "Quantum mechanical actuation of microelectromechanical systems by the Casimir force," *Science*, vol. 291, pp. 1941–1944, 2001.

[56] S. de Man, K. Heeck, and D. Iannuzzi, "Halving the Casimir force with conductive oxides: experimental details," *Phys. Rev. A*, vol. 82, p. 062512, 2010.

[57] J. L. Garrett, D. A. T. Somers, K. Sendgikoski, and J. N. Munday, "Sensitivity and accuracy of Casimir force measurements in air," *Phys. Rev. A*, vol. 100, p. 022508, 2019.

[58] J. N. Munday, F. Capasso, V. A. Parsegian, and S. M. Bezrukov, "Measurements of the Casimir-Lifshitz force in fluids: the effect of electrostatic forces and Debye screening," *Phys. Rev. A*, vol. 78, p. 032109, 2008.

[59] J. N. Munday, F. Capasso, and V. A. Parsegian, "Measured long-range repulsive Casimir-Lifshitz forces," *Nature*, vol. 457, pp. 170–173, 2009.

[60] G. Torricelli, P. J. van Zwol, O. Shpak, et al., "Switching Casimir forces with phase-change materials," *Phys. Rev. A*, vol. 82, p. 010101, 2010.

[61] H. B. Chan, Y. Bao, J. Zou, et al., "Measurement of the Casimir force between a gold sphere and a silicon surface with nanoscale trench arrays," *Phys. Rev. Lett.*, vol. 101, pp. 1–4, 2008.

[62] E. M. Lifshitz, "The theory of molecular attractive forces between solids," *Sov. Phys. JETP-USSR* vol. 2, pp. 73–83, 1956. http://www.jetp.ac.ru/cgi-bin/dn/e_002_01_0073.pdf.

[63] G. Bimonte, E. Calloni, G. Esposito, L. Milano, and L. Rosa, "Towards measuring variations of Casimir energy by a superconducting cavity," *Phys. Rev. Lett.*, vol. 94, p. 180402, 2005.

[64] E. G. Galkina, B. A. Ivanov, S. Savel'ev, V. A. Yampol'skii, and F. Nori, "Drastic change of the Casimir force at the metal-insulator transition," *Phys. Rev. B*, vol. 80, p. 125119, 2009.

[65] L. H. Ford, "Spectrum of the Casimir effect," *Phys. Rev. D*, vol. 38, pp. 528–532, 1988.

[66] A. Rodriguez, M. Ibanescu, D. Iannuzzi, J. D. Joannopoulos, and S. G. Johnson, "Virtual photons in imaginary time: computing exact Casimir forces via standard numerical electromagnetism techniques," *Phys. Rev. A*, vol. 76, p. 032106, 2007.

[67] U. Leonhardt and T. G. Philbin, "Quantum levitation by left-handed metamaterials," *New J. Phys.*, vol. 9, pp. 254, 2007.

[68] R. Zhao, J. Zhou, T. Koschny, E. N. Economou, and C. M. Soukoulis, "Repulsive Casimir force in chiral metamaterials," *Phys. Rev. Lett.*, vol. 103, p. 103602, 2009.

[69] A. P. McCauley, R. Zhao, M. T. H. Reid, et al., "Microstructure effects for Casimir forces in chiral metamaterials," *Phys. Rev. B*, vol. 82, 2010, https://doi.org/10.1103/physrevb.82.165108.

[70] F. S. S. Rosa, "On the possibility of Casimir repulsion using Metamaterials," *J. Phys. Conf. Ser.*, vol. 161, p. 012039, 2009.

[71] F. S. S. Rosa, D. A. R. Dalvit, and P. W. Milonni, "Casimir-lifshitz theory and metamaterials," *Phys. Rev. Lett.*, vol. 100, p. 183602, 2008.

[72] A. W. Rodriguez, A. P. McCauley, J. D. Joannopoulos, and S. G. Johnson, "Casimir forces in the time domain: theory," *Phys. Rev. A*, vol. 80, p. 012115, 2009.

[73] L. D. Landau, E. M. Lifshitz, and L. P. Pitaevskii, *Electrodynamics of Continuous Media*, 2nd ed., Oxford, Pergamon Press, 1960.

[74] O. Kenneth and I. Klich, "Opposites attract: a theorem about the Casimir force," *Phys. Rev. Lett.*, vol. 97, p. 160401, 2006.
[75] I. E. Dzyaloshinskii, E. M. Lifshitz, and L. P. Pitaevskii, "The general theory of van der Waals forces," *Adv. Phys.*, vol. 10, pp. 165–209, 1961.
[76] A. A. Feiler, L. Bergström, and M. W. Rutland, "Superlubricity using repulsive van der Waals forces," *Langmuir*, vol. 24, pp. 2274–2276, 2008.
[77] A. W. Rodriguez, J. N. Munday, J. D. Joannopoulos, F. Capasso, D. A. R. Dalvit, and S. G. Johnson, "Stable suspension and dispersion-induced transitions from repulsive Casimir forces between fluid-separated eccentric cylinders," *Phys. Rev. Lett.*, vol. 101, p. 190404, 2008.
[78] R. K. Zhao, L. Li, S. Yang, et al., "Stable Casimir equilibria and quantum trapping," *Science*, vol. 364, pp. 984–987, 2019.
[79] B. Spreng, P. A. Maia Neto, and G.-L. Ingold, "Plane-wave approach to the exact van der Waals interaction between colloid particles," *J. Chem. Phys.*, vol. 153, p. 024115, 2020.
[80] D. A. T. Somers and J. N. Munday, "Conditions for repulsive Casimir forces between identical birefringent materials," *Phys. Rev. A*, vol. 95, p. 022509, 2017.
[81] T. H. Boyer, "Van der Waals forces and zero-point energy for dielectric and permeable materials," *Phys. Rev. A*, vol. 9, pp. 2078–2084, 1974.
[82] O. Kenneth, I. Klich, A. Mann, and M. Revzen, "Repulsive Casimir forces," *Phys. Rev. Lett.*, vol. 89, p. 033001, 2002.
[83] D. Iannuzzi and F. Capasso, "Comment on "Repulsive Casimir forces"," *Phys. Rev. Lett.*, vol. 91, p. 029101, 2003.
[84] O. Kenneth, I. Klich, A. Mann, and M. Revzen, "Kenneth et al. reply," *Phys. Rev. Lett.*, vol. 91, p. 029102, 2003.
[85] V. Yannopapas and N. V. Vitanov, "First-principles study of Casimir repulsion in metamaterials," *Phys. Rev. Lett.*, vol. 103, p. 120401, 2009.
[86] I. Liberal and N. Engheta, "Zero-index structures as an alternative platform for quantum optics," *Proc. Natl. Acad. Sci. U. S. A.*, vol. 114, p. 822, 2017.
[87] T. H. Boyer, "Quantum electromagnetic zero-point energy of a conducting spherical shell and the Casimir model for a charged particle," *Phys. Rev.*, vol. 174, pp. 1764–1776, 1968.
[88] R. L. Jaffe, "Unnatural acts: unphysical consequences of imposing boundary conditions on quantum fields," *AIP Conf. Proc* vol. 687, pp. 3–12, 2003.
[89] M. Levin, A. P. McCauley, A. W. Rodriguez, M. T. H. Reid, and S. G. Johnson, "Casimir repulsion between metallic objects in vacuum," *Phys. Rev. Lett.*, vol. 105, 2010, https://doi.org/10.1103/physrevlett.105.090403.
[90] P. S. Venkataram, S. Molesky, P. Chao, and A. W. Rodriguez, "Fundamental limits to attractive and repulsive Casimir-Polder forces," *Phys. Rev. A*, vol. 101, p. 052115, 2020.
[91] S. Tsoi, P. Dev, A. L. Friedman, et al., "van der Waals screening by single-layer graphene and molybdenum disulfide," *ACS Nano*, vol. 8, pp. 12410–12417, 2014.
[92] P. Rodriguez-Lopez, W. J. M. Kort-Kamp, D. A. R. Dalvit, and L. M. Woods, "Casimir force phase transitions in the graphene family," *Nat. Commun.*, vol. 8, pp. 1–9, 2017.
[93] G. Bimonte, G. L. Klimchitskaya, and V. M. Mostepanenko, "How to observe the giant thermal effect in the Casimir force for graphene systems," *Phys. Rev. A*, vol. 96, p. 012517, 2017.
[94] E. I. Kats, "Van der Waals forces in non-isotropic systems," *Sov. Phys. JETP*, vol. 33, p. 634, 1971.
[95] V. A. Parsegian and G. H. Weiss, "Dielectric anisotropy and the van der Waals interaction between bulk media," *J. Adhes.*, vol. 3, pp. 259–267, 1972.
[96] Y. S. Barash, "Moment of van der Waals forces between anisotropic bodies," *Radiophys. Quantum Electron.*, vol. 21, pp. 1138–1143, 1978.
[97] J. N. Munday, D. Iannuzzi, Y. Barash, and F. Capasso, "Torque on birefringent plates induced by quantum fluctuations," *Phys. Rev. A*, vol. 71, p. 042102, 2005.
[98] J. N. Munday, D. Iannuzzi, and F. Capasso, "Quantum electrodynamical torques in the presence of Brownian motion," *New J. Phys.*, vol. 8, pp. 244, 2006.
[99] R. B. Rodrigues, P. A. M. Neto, A. Lambrecht, and S. Reynaud, "Vacuum-induced torque between corrugated metallic plates," *Europhys. Lett.*, vol. 76, pp. 822–828, 2006.
[100] X. Chen and J. C. H. Spence, "On the measurement of the Casimir torque," *Phys. Status Solidi B*, vol. 248, pp. 2064–2071, 2011.
[101] R. Guérout, C. Genet, A. Lambrecht, and S. Reynaud, "Casimir torque between nanostructured plates," *Europhys. Lett.*, vol. 111, p. 44001, 2015.
[102] D. A. T. Somers and J. N. Munday, "Rotation of a liquid crystal by the Casimir torque," *Phys. Rev. A*, vol. 91, p. 032520, 2015.
[103] D. A. T. Somers and J. N. Munday, "Casimir-lifshitz torque enhancement by retardation and intervening dielectrics," *Phys. Rev. Lett.*, vol. 119, p. 183001, 2017.
[104] J. Ahn, Z. Xu, J. Bang, et al., "Optically levitated nanodumbbell torsion balance and GHz nanomechanical rotor," *Phys. Rev. Lett.*, vol. 121, p. 033603, 2018.
[105] J. Ahn, Z. Xu, J. Bang, P. Ju, X. Gao, and T. Li, "Ultrasensitive torque detection with an optically levitated nanorotor," *Nat. Nanotechnol.*, vol. 15, pp. 89–93, 2020.
[106] R. Zhao, A. Manjavacas, F. J. García de Abajo, and J. B. Pendry, "Rotational quantum friction," *Phys. Rev. Lett.*, vol. 109, p. 123604, 2012.
[107] S. Niu, G. Joe, H. Zhao, et al., "Giant optical anisotropy in a quasi-one-dimensional crystal," *Nat. Photonics*, vol. 12, pp. 392–396, 2018.
[108] A. Segura, L. Artús, R. Cuscó, T. Taniguchi, G. Cassabois, and B. Gil, "Natural optical anisotropy of h-BN: highest giant birefringence in a bulk crystal through the mid-infrared to ultraviolet range," *Phys. Rev. Mater.*, vol. 2, p. 024001, 2018.
[109] M. Antezza, H. B. Chan, B. Guizal, V. N. Marachevsky, R. Messina, and M. Wang, "Giant Casimir torque between rotated gratings and the θ = 0 anomaly," *Phys. Rev. Lett.*, vol. 124, p. 013903, 2020.
[110] P. Thiyam, P. Parashar, K. V. Shajesh, et al., "Distance-dependent sign reversal in the Casimir-lifshitz torque," *Phys. Rev. Lett.*, vol. 120, p. 131601, 2018.
[111] C. Henkel, K. Joulain, J.-P. Mulet, and J.-J. Greffet, "Radiation forces on small particles in thermal near fields," *J. Opt. A*, vol. 4, pp. S109–S114, 2002.
[112] J. M. Obrecht, R. J. Wild, M. Antezza, L. P. Pitaevskii, S. Stringari, and E. A. Cornell, "Measurement of the temperature dependence of the Casimir-polder force," *Phys. Rev. Lett.*, vol. 98, p. 063201, 2007.

[113] M. Antezza, L. P. Pitaevskii, and S. Stringari, "New asymptotic behavior of the surface-atom force out of thermal equilibrium," *Phys. Rev. Lett.*, vol. 95, p. 113202, 2005.

[114] M. Krüger, T. Emig, G. Bimonte, and M. Kardar, "Non-equilibrium Casimir forces: spheres and sphere-plate," *Europhys. Lett.*, vol. 95, p. 21002, 2011.

[115] V. A. Golyk, M. Krüger, M. T. H. Reid, and M. Kardar, "Casimir forces between cylinders at different temperatures," *Phys. Rev. D*, vol. 85, p. 065011, 2012.

[116] K. Chen and S. Fan, "Nonequilibrium Casimir force with a nonzero chemical potential for photons," *Phys. Rev. Lett.*, vol. 117, p. 267401, 2016.

[117] C. M. Wilson, G. Johansson, A. Pourkabirian, et al., "Observation of the dynamical Casimir effect in a superconducting circuit," *Nature*, vol. 479, pp. 376–379, 2011.

[118] G. T. Moore, "Quantum theory of the electromagnetic field in a variable-length one-dimensional cavity," *J. Math. Phys.*, vol. 11, pp. 2679–2691, 1970.

[119] C. Braggio, G. Bressi, G. Carugno, et al., "A novel experimental approach for the detection of the dynamical Casimir effect," *Europhys. Lett.*, vol. 70, pp. 754–760, 2005.

[120] L. H. Ford and A. Vilenkin, "Quantum radiation by moving mirrors," *Phys. Rev. D*, vol. 25, pp. 2569–2575, 1982.

[121] D. T. Alves, E. R. Granhen, and M. G. Lima, "Quantum radiation force on a moving mirror with Dirichlet and Neumann boundary conditions for a vacuum, finite temperature, and a coherent state," *Phys. Rev. D*, vol. 77, p. 125001, 2008.

[122] A. Lambrecht, M.-T. Jaekel, and S. Reynaud, "Motion induced radiation from a vibrating cavity," *Phys. Rev. Lett.*, vol. 77, pp. 615–618, 1996.

[123] P. A. M. Neto and L. A. S. Machado, "Quantum radiation generated by a moving mirror in free space," *Phys. Rev. A*, vol. 54, pp. 3420–3427, 1996.

[124] J. R. Johansson, G. Johansson, C. M. Wilson, and F. Nori, "Dynamical Casimir effect in superconducting microwave circuits," *Phys. Rev. A*, vol. 82, p. 052509, 2010.

[125] D. F. Mundarain and P. A. Maia Neto, "Quantum radiation in a plane cavity with moving mirrors," *Phys. Rev. A*, vol. 57, pp. 1379–1390, 1998.

[126] V. V. Dodonov, A. B. Klimov, and D. E. Nikonov, "Quantum phenomena in nonstationary media," *Phys. Rev. A*, vol. 47, pp. 4422–4429, 1993.

[127] M. Crocce, D. A. R. Dalvit, F. C. Lombardo, and F. D. Mazzitelli, "Model for resonant photon creation in a cavity with time-dependent conductivity," *Phys. Rev. A*, vol. 70, p. 033811, 2004.

[128] T. Kawakubo and K. Yamamoto, "Photon creation in a resonant cavity with a nonstationary plasma mirror and its detection with Rydberg atoms," *Phys. Rev. A*, vol. 83, p. 013819, 2011.

[129] M. Crocce, D. A. R. Dalvit, and F. D. Mazzitelli, "Resonant photon creation in a three-dimensional oscillating cavity," *Phys. Rev. A*, vol. 64, p. 013808, 2001.

[130] H. Johnston and S. Sarkar, "Moving mirrors and time-varying dielectrics," *Phys. Rev. A*, vol. 51, pp. 4109–4115, 1995.

[131] M. Uhlmann, G. Plunien, R. Schutzhold, and G. Soff, "Resonant cavity photon creation via the dynamical Casimir effect," *Phys. Rev. Lett.*, vol. 93, p. 193601, 2004.

[132] F. X. Dezael and A. Lambrecht, "Analogue Casimir radiation using an optical parametric oscillator," *Europhys. Lett.*, vol. 89, 2010, https://doi.org/10.1209/0295-5075/89/14001.

[133] D. Faccio and I. Carusotto, "Dynamical Casimir effect in optically modulated cavities," *Europhys. Lett.*, vol. 96, 2011, https://doi.org/10.1209/0295-5075/96/24006.

[134] P. Lähteenmäki, G. S. Paraoanu, J. Hassel, and P. J. Hakonen, "Dynamical Casimir effect in a Josephson metamaterial," *Proc. Natl. Acad. Sci. U. S. A.*, vol. 110, pp. 4234–4238, 2013.

[135] S. Vezzoli, A. Mussot, N. Westerberg, et al., "Optical analogue of the dynamical Casimir effect in a dispersion-oscillating fibre," *Commun. Phys.*, vol. 2, p. 84, 2019.

[136] J. A. Hutchison, T. Schwartz, C. Genet, E. Devaux, and T. W. Ebbesen, "Modifying chemical landscapes by coupling to vacuum fields," *Angew Chem. Int. Edit.*, vol. 51, pp. 1592–1596, 2012.

[137] A. Thomas, L. Lethuillier-Karl, K. Nagarajan, et al., "Tilting a ground-state reactivity landscape by vibrational strong coupling," *Science*, vol. 363, p. 616, 2019.

[138] J. P. Long and B. S. Simpkins, "Coherent coupling between a molecular vibration and Fabry-Perot optical cavity to give hybridized states in the strong coupling limit," *ACS Photonics*, vol. 2, pp. 130–136, 2015.

[139] A. Shalabney, J. George, J. Hutchison, G. Pupillo, C. Genet, and T. W. Ebbesen, "Coherent coupling of molecular resonators with a microcavity mode," *Nat. Commun.*, vol. 6, p. 5981, 2015.

[140] J. George, T. Chervy, A. Shalabney, et al., "Multiple Rabi splittings under ultrastrong vibrational coupling," *Phys. Rev. Lett.*, vol. 117, p. 153601, 2016.

[141] A. Thomas, J. George, A. Shalabney, et al., "Ground-state chemical reactivity under vibrational coupling to the vacuum electromagnetic field," *Angew Chem. Int. Edit.*, vol. 55, pp. 11462–11466, 2016.

[142] R. M. A. Vergauwe, J. George, T. Chervy, et al., "Quantum strong coupling with protein vibrational modes," *J. Phys. Chem. Lett.*, vol. 7, pp. 4159–4164, 2016.

[143] J. Bárcenas, L. Reyes, and R. Esquivel-Sirvent, "Scaling of micro- and nanodevices actuated by Casimir forces," *Appl. Phys. Lett.*, vol. 87, p. 263106, 2005.

[144] W. Broer, H. Waalkens, V. B. Svetovoy, J. Knoester, and G. Palasantzas, "Nonlinear actuation dynamics of driven Casimir oscillators with rough surfaces," *Phys. Rev. Appl.*, vol. 4, p. 054016, 2015.

[145] F. M. Serry, D. Walliser, and G. J. Maclay, "The anharmonic Casimir oscillator (ACO)-the Casimir effect in a model microelectromechanical system," *J. Microelectromech. Syst.*, vol. 4, pp. 193–205, 1995.

[146] E. Buks and M. L. Roukes, "Metastability and the Casimir effect in micromechanical systems," *Europhys. Lett.*, vol. 54, pp. 220–226, 2001.

[147] H. B. Chan, V. A. Aksyuk, R. N. Kleiman, D. J. Bishop, and F. Capasso, "Nonlinear micromechanical Casimir oscillator," *Phys. Rev. Lett.*, vol. 87, p. 211801, 2001.

[148] F. Tajik, M. Sedighi, A. A. Masoudi, H. Waalkens, and G. Palasantzas, "Sensitivity of chaotic behavior to low optical frequencies of a double-beam torsional actuator," *Phys. Rev. E*, vol. 100, p. 012201, 2019.

[149] D. Iannuzzi, J. N. Munday, and F. Capasso, "*Ultra-low friction configuration*," US Patent Application 20070066494, 2005.

[150] A. Stange, M. Imboden, J. Javor, L. K. Barrett, and D. J. Bishop, "Building a Casimir metrology platform with a commercial MEMS sensor," *Microsyst. Nanoeng.*, vol. 5, p. 14, 2019.

[151] M. Imboden, J. Morrison, D. K. Campbell, and D. J. Bishop, "Design of a Casimir-driven parametric amplifier," *J. Appl. Phys.*, vol. 116, p. 134504, 2014.

[152] J. M. Pate, M. Goryachev, R. Y. Chiao, J. E. Sharping, and M. E. Tobar, "Casimir spring and dilution in macroscopic cavity optomechanics," *Nat. Phys.*, 2020, https://doi.org/10.1038/s41567-020-0975-9.

[153] K. Y. Fong, H.-k. Li, R. Zhao, S. Yang, Y. Wang, and X. Zhang, "Phonon heat transfer across a vacuum through quantum fluctuations," *Nature*, vol. 576, pp. 243–247, 2019.

Zhujing Xu, Zubin Jacob and Tongcang Li*

Enhancement of rotational vacuum friction by surface photon tunneling

https://doi.org/10.1515/9783110710687-042

Keywords: optical levitation; perovskite materials; surface photon tunneling; vacuum friction.

Abstract: When a neutral sphere is rotating near a surface in vacuum, it will experience a frictional torque due to quantum and thermal electromagnetic fluctuations. Such vacuum friction has attracted many interests but has been too weak to be observed. Here we investigate the vacuum frictional torque on a barium strontium titanate (BST) nanosphere near a BST surface. BST is a perovskite ferroelectric ceramic that can have large dielectric responses at GHz frequencies. At resonant rotating frequencies, the mechanical energy of motion can be converted to electromagnetic energy through resonant photon tunneling, leading to a large enhancement of the vacuum friction. The calculated vacuum frictional torques at resonances at sub-GHz and GHz frequencies are several orders larger than the minimum torque measured by an optically levitated nanorotor recently, and are thus promising to be observed experimentally. Moreover, we calculate the vacuum friction on a rotating sphere near a layered surface for the first time. By optimizing the thickness of the thin-film coating, the frictional torque can be further enhanced by several times.

***Corresponding author: Tongcang Li,** Department of Physics and Astronomy, Purdue University, West Lafayette, IN 47907, USA; School of Electrical and Computer Engineering, Purdue University, West Lafayette, IN 47907, USA; Purdue Quantum Science and Engineering Institute, Purdue University, West Lafayette, IN 47907, USA; and Birck Nanotechnology Center, Purdue University, West Lafayette, IN 47907, USA, E-mail: tcli@purdue.edu. https://orcid.org/0000-0003-3308-8718
Zhujing Xu, Department of Physics and Astronomy, Purdue University, West Lafayette, IN 47907, USA. https://orcid.org/0000-0003-1780-9132
Zubin Jacob, School of Electrical and Computer Engineering, Purdue University, West Lafayette, IN 47907, USA; Purdue Quantum Science and Engineering Institute, Purdue University, West Lafayette, IN 47907, USA; and Birck Nanotechnology Center, Purdue University, West Lafayette, IN 47907, USA

1 Introduction

Quantum fluctuations of electromagnetic fields cause an attractive force between two neutral metallic plates, which was first calculated in 1948 and well known as Casimir force [1]. Besides the Casimir force between static macroscopic objects, the fluctuating electromagnetic fields can cause noncontact vacuum friction between two surfaces in relative motion [2, 3]. Many efforts have been made for studying the vacuum friction between various materials and configurations [4–10]. However, former attempts using atomic force microscopes had not successfully detected the vacuum friction [11–13]. The difficulty mainly comes from the small value of vacuum friction. It is thus worthwhile to understand the mechanisms of vacuum friction better and look for enhancements of the vacuum friction. At the same time, it is essential to develop a suitable ultrasensitive detector for the measurement.

In this letter, we investigate the vacuum friction on a rotating nanosphere levitated near a flat plate (Figure 1(a)), inspired by recent breakthroughs in levitated optomechanics. There are growing interests in using optically levitated dielectric particles in vacuum for precision measurements since they are well isolated from the environment and have ultrahigh sensitivity [14–21]. Microspheres and nanospheres have been optically levitated near surfaces [22, 23]. Meanwhile, levitated nanoparticles have been driven to rotate up to 5 GHz by a circularly polarized laser [19]. Remarkably, a rotating nanoparticle levitated at 10^{-5} torr has measured a torque as small as 5×10^{-28} Nm in just 100 s [19].

Similar to the vacuum frictional force between two plates that have relative motions [2, 3], there will be a vacuum frictional torque on a nanosphere rotating at a high speed [6, 10]. Thus a levitated nanorotor provides a promising method to detect the vacuum frictional torque. Former calculations have found that the vacuum frictional torque on a 150 nm-diameter silica nanosphere near a silica

This article has previously been published in the journal Nanophotonics. Please cite as: Z. Xu, Z. Jacob and T. Li "Enhancement of rotational vacuum friction by surface photon tunneling" *Nanophotonics* 2021, 10. DOI: 10.1515/nanoph-2020-0391.

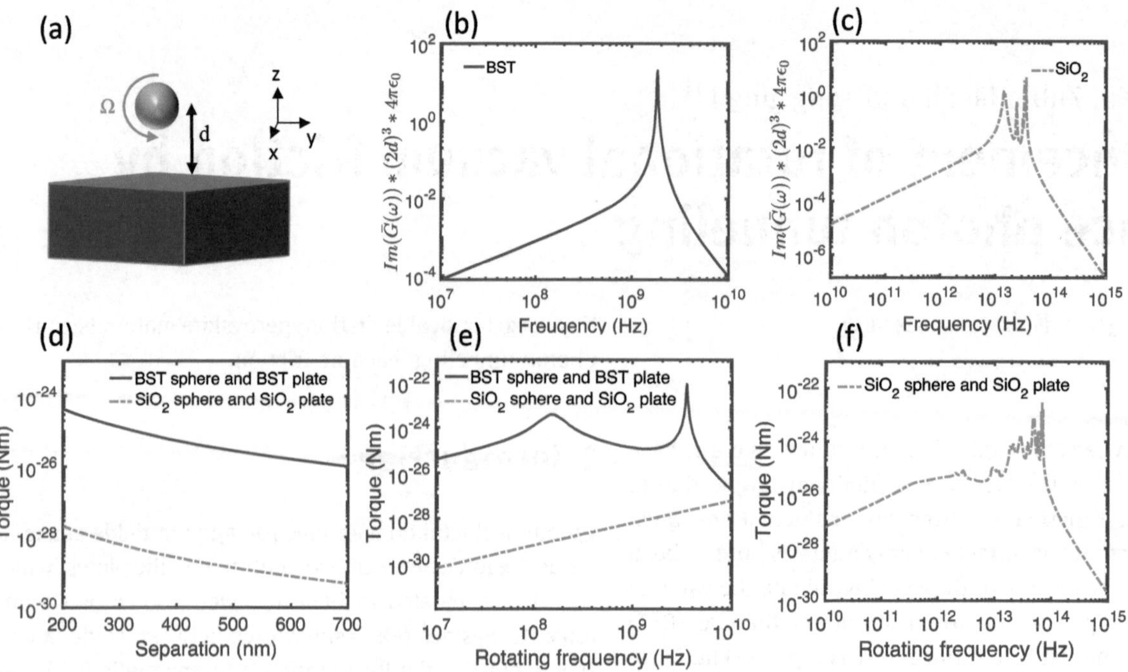

Figure 1: Vacuum friction between a rotating nanosphere and a substrate.
(a) A sphere with a radius of 75 nm rotates around the *x* axis. The *z* axis is normal to the substrate. (b) The imaginary part of the Green function (Eq. (3)) between a barium strontium titanate (BST) substrate and a nearby nanosphere. (c) The imaginary part of the Green function between a SiO_2 substrate and a nearby nanosphere. (d) Calculated vacuum friction torque acting on a BST/silica nanosphere near a BST/silica surface as a function of the separation between the sphere and the substrate. The sphere rotates at a frequency of 150 MHz, which is a resonant frequency of BST. (e) The frictional torque on a BST/silica sphere near a BST/silica surface is shown as a function of the rotating frequency of the nanosphere. The separation between the sphere and the substrate is 300 nm. (f) The friction torque of a SiO_2 sphere near a SiO_2 surface is shown at a higher frequency range. The separation is 300 nm.

surface is about 6×10^{-29} Nm at a separation of 300 nm when the sphere is rotating at 1 GHz and both the surface and the sphere are at a temperature of 300 K [19]. To measure the vacuum friction between a silica sphere and a silica plate, the experiment needs to be conducted under ultrahigh vacuum below 10^{-9} torr [19]. In this work, we propose to use BST, which is a perovskite ferroelectric ceramic that can have a large dielectric resonance at GHz frequency [24], to dramatically enhance the vacuum friction by resonant surface photon tunneling at sub-GHz and GHz frequencies. We show that the vacuum frictional torque on a BST sphere near a BST plate can be significantly enhanced to 3.6×10^{-24} Nm and 8.1×10^{-23} Nm through resonant photon tunneling at a rotating frequency of 150 MHz and 3.5 GHz, respectively. So the torque on a BST sphere near a BST plate can be measured at a pressure of about 10^{-4} torr.

The enhancement of the vacuum friction comes from photon tunneling between two surfaces. At the resonant condition, the conversion of the mechanical energy of motion to the electromagnetic energy is realized efficiently. In the domain of the rotation frequency, there are two resonances that correspond to the normal and anomalous Doppler effect [25–27]. The resonant tunneling occurs when the mechanical rotation frequency matches the sum or the difference of the surface plasmon or phonon polariton frequencies. For metals and semiconductors, the surface plasmon or phonon polariton frequency is usually on the order of 10^{14} Hz and 10^{12} Hz, respectively. Such a high surface polariton frequency makes it extremely difficult to experimentally reach the resonant condition [25–27]. Here we choose a ferroelectric material BST which has low surface polariton frequency at GHz range [24]. The dielectric properties of BST can also be tuned by fabrication and doping. When the mechanical rotating frequency of the sphere is also around GHz, the vacuum frictional torque can be significantly enhanced due to the resonant photon tunneling. In this way, the vacuum frictional torque is several orders higher than the minimum detectable torque. Besides the huge enhancement, for the first time we will show the vacuum friction between a BST sphere and a multilayer coating on top of a BST substrate. The torque can be further increased by several times by optimizing the thickness of the coating layer.

This paper is organized as follows. We show the general method of calculating the vacuum frictional torque on a rotating nanosphere and discuss the condition for the resonant photon tunneling in Section 2. The calculated friction torque acting on a BST sphere near a BST plate is presented in Section 3. In this section, we show a significant enhancement of the friction torque. Section 4 is devoted to showing the torque on a BST sphere near a multilayer coating on top of a BST substrate. The transfer matrix for calculating the optical matrix of the multilayer structure is introduced. Further enhancement of the friction torque is demonstrated.

2 Resonances in a rotating nanosphere near a surface

As a ferroelectric material, BST possesses many exceptional dielectric properties such as high dielectric constant, low loss and large electric-field dielectric tunability over a wide frequency range [24]. Especially, it has low surface polariton frequency which is favorable for enhancing the vacuum friction torque at sub-GHz and GHz frequencies.

We investigate on a case that a sphere with radius r is rotating with frequency Ω in vacuum and is placed at a separation of d from the substrate as shown in Figure 1(a). We work within the nonrelativistic and near-field limit. Thus the separation needs to be far smaller than both c/Ω ($\approx$5 cm at 1 GHz) and $c\hbar/k_B T_j$ ($\approx$8 μm at 300 K). Besides, the dipole approximation requires the radius of the nanosphere to be sufficiently smaller than the separation. The friction torque experienced by the sphere is written as [10].

$$M_p = -\frac{2\hbar}{\pi}\int_{-\infty}^{\infty}[n_1(\omega-\Omega)-n_0(\omega)] \times \mathrm{Im}[\alpha(\omega-\Omega)]\mathrm{Im}[\overline{G}(\omega)]d\omega \tag{1}$$

where $n_j(T) = [\exp(\hbar\omega/k_B T_j) - 1]^{-1}$ is the Bose–Einstein distribution function at temperature T_j and $j = 0, 1$ are for the substrate and the sphere, respectively. $\alpha(\omega)$ is the electric polarizability of the nanosphere. $\overline{G}(\omega)$ is the Green function such that $\overline{G}(\omega) = [G_{yy}(\omega) + G_{zz}(\omega)]/2$, where G_{yy} and G_{zz} are electromagnetic Green tensor components and x is the axis of the rotation. The polarizability and the Green function can be described as

$$\alpha(\omega) = 4\pi\epsilon_0 R^3 \frac{\epsilon_{sp}(\omega)-1}{\epsilon_{sp}(\omega)+2}, \tag{2}$$

$$\overline{G}(\omega) = \frac{3}{8\pi\epsilon_0(2d)^3}\frac{\epsilon_{sub}(\omega)-1}{\epsilon_{sub}(\omega)+1}, \tag{3}$$

where ϵ_{sp} and ϵ_{sub} are the dielectric functions of the sphere and the substrate, respectively. The equations above are for the situation that the rotation axis of the sphere is parallel to the substrate surface. When the rotating axis is normal to the surface, the Green function has a factor of $\frac{2}{3}$ difference and hence the friction torque is [28].

$$M_n = \frac{2}{3}M_p, \tag{4}$$

where M_n and M_p stand for the torque when the sphere is rotating normal and parallel to the surface, respectively. Notice that all the following calculations of the vacuum frictional torque are for the parallel case. As we can see, the friction torque depends on the optical properties of the interacting materials. The dielectric functions of different dielectric materials can be modeled as Lorentz oscillators, which treat each discrete vibrational mode as a classical damped harmonic oscillator:

$$\epsilon(\omega) = \epsilon_\infty\left(1 + \sum\frac{\omega_L^2-\omega_T^2}{\omega_T^2-\omega^2-i\Gamma\omega}\right), \tag{5}$$

where ϵ_∞ is the permittivity in the high-frequency limit, ω_L is the longitudinal polar-optic phonon frequency, ω_T is the transverse polar-optic phonon frequency and γ is the damping coefficient.

As mentioned above, we are looking at the resonant photon tunneling between surface phonon polaritons to realize enhancement of the vacuum friction. Similar to the normal and anomalous Doppler effect for the system of two sliding plates [25–27], rotational friction between a rotating sphere and a flat plate also has such resonances. BST is a good candidate for realizing such resonant photon tunneling since it can support low-frequency surface polariton modes. The optical phonon parameters of BST are $\epsilon_\infty = 2.896$, $\omega_L = 1.3\times10^{10}\ \mathrm{s}^{-1}$, $\omega_T = 5.7\times10^{9}\ \mathrm{s}^{-1}$, $\Gamma = 2.8\times10^{8}\ \mathrm{s}^{-1}$ [24].

The friction increases significantly at the rotating frequency $\Omega = |\omega_{1p} \pm \omega_{2p}|$ which correspond to the resonant generation of surface polaritons. Here Ω is the rotating frequency of the sphere. ω_{1p} and ω_{2p} are surface polariton frequencies such that $\mathrm{Re}(\epsilon(\omega_{1p})) = -2$ and $\mathrm{Re}(\epsilon(\omega_{2p})) = -1$, which lead to large $\alpha(\omega)$ and $\overline{G}(\omega)$. Two resonant frequencies are $\omega_{1p} = 1.06\times10^{10}\ \mathrm{s}^{-1}$ and $\omega_{2p} = 1.15\times10^{10}\ \mathrm{s}^{-1}$. Therefore, when the rotation frequency $\Omega = |\omega_1 - \omega_2| = 2\pi\times150\,\mathrm{MHz}$ or $\Omega = \omega_1 + \omega_2 = 2\pi\times3.52\,\mathrm{GHz}$, the friction will be greatly enhanced. Figure 1(b) shows the Green function that connects the dipole moment fluctuation of the sphere and the induced electromagnetic field on a BST surface. As a comparison, the Green function for a silica surface is presented in Figure 1(c). The frequency of surface polaritons for a silica surface [29] is around 10^{13} Hz, which is several orders higher than the frequency of BST. To meet the condition of resonant photon tunneling between a silica sphere and a silica plate, the rotation frequency of the sphere will need to reach 10^{13} Hz which cannot be achieved experimentally.

3 Vacuum friction between a BST nanosphere and a BST plate

The calculated vacuum friction between a BST sphere and a BST plate is presented in this section. Here, we investigate on a case that the radius of the sphere is 75 nm [19] and the sphere and the plate have the same temperature of 300 K. At a separation of 300 nm, the calculated friction torque is shown as a function of rotating frequency in Figure 1(e). The blue solid curve corresponds to the friction between a BST sphere and a BST plate. It shows two resonant peaks of the torque at 150 MHz and 3.52 GHz as expected before. The amplitude of the friction torque at two resonant frequencies are several orders higher than the nonresonant part. The maximum torque of 8.06×10^{-23} Nm is achieved at 3.52 GHz. The red dashed curve is the friction torque acting on a silica sphere near a silica plate, for comparison. The low-frequency surface polariton mode and resonant photon tunneling significantly enhance the amplitude of the vacuum friction. At 150 MHz and 3.52 GHz, the friction torque between a BST sphere and a BST surface has an enhancement of about 4×10^5 compared to the torque between a silica sphere and a silica surface.

Figure 1(f) shows the calculated torque for silica at much higher rotation frequencies. Similarly, silica also has resonant photon tunneling conditions and at such frequencies, the torque can also be greatly enhanced. At a rotating frequency of around 100 THz, the torque can also be near 10^{-22} Nm. However, this frequency is far beyond the current experimental limit. The calculated friction torque as a function of the separation between the sphere and the substrate is shown in Figure 1(d) at a rotating frequency of 150 MHz, which is the resonant frequency of BST. For each separation, the enhancement is more than five orders.

A recent experiment has detected a torque of 5×10^{-28} Nm with an optically levitated nanosphere at 10^{-5} torr [19]. It is four orders smaller than the vacuum frictional torque on a BST sphere at resonant frequency. Another factor that can affect the detection of vacuum friction is the air damping torque. The air damping torque on a rotating nanosphere due to residual air molecules in the vacuum chamber can be described as [30]:

$$M_{\text{air}} = \frac{\pi p \Omega (2r)^4}{11.976} \sqrt{\frac{2m_{\text{gas}}}{\pi k_B T}}, \tag{6}$$

where p is the air pressure, r is the sphere radius, $m_{\text{gas}} = 4.8 \times 10^{-26}$ kg is the mass of the air molecule and T is the temperature of the surrounding air molecules. At the pressure of 10^{-4} torr, the air damping torque acting on the sphere is 4.5×10^{-24} Nm at a rotating frequency of 150 MHz at room temperature, which is about one orders smaller than the targeted vacuum friction torque. 10^{-4} torr and lower pressures have been achieved in levitation experiments [19, 31]. Compared to silica, the requirement of pressure for measuring vacuum friction on a BST nanosphere is substantially relaxed.

4 Effects of surface coating

We have demonstrated the vacuum friction between a BST sphere and a BST plate and showed that the friction can be enhanced by several orders due to the surface resonant photon tunneling. It will be easier to experimentally realize this condition when the resonant frequency is as low as possible. In this section we show that the resonant frequency can be lowered by thin-film coating.

We investigate on the friction torque on a BST sphere near a surface that has a thin layer of dielectric coating on top of a BST substrate as shown in Figure 2(a). To calculate the vacuum frictional torque, we need to get the reflection coefficients for the layered structure by the transfer matrix method. For a surface that has a thin layer on top of a substrate, the transfer matrix for the $p(s)$ polarization is given as [32, 33]

$$T^{p(s)} = D^{p(s)}_{0\to1} P(L_1) D^{p(s)}_{1\to2}, \tag{7}$$

where L_1 is the thickness of the thin layer. Here $j = 0, 1, 2$ stands for the vacuum between the sphere and the surface, the thin layer, and the BST substrate, respectively. $D^{p(s)}_{j,j+1}$ is the transmission matrix between layer j and j+1 for the $p(s)$ polarization and it can be written as

$$D^{p(s)}_{j\to j+1} = \frac{1}{2}\begin{bmatrix} 1+\eta^{p(s)}_{j,j+1} & 1-\eta^{p(s)}_{j,j+1} \\ 1-\eta^{p(s)}_{j,j+1} & 1+\eta^{p(s)}_{j,j+1} \end{bmatrix}, \tag{8}$$

where $\eta^{p(s)}_{j,j+1}$ is given as

$$\eta^{p}_{j,j+1} = \frac{\epsilon_j k_{j+1,z}}{\epsilon_{j+1} k_{j,z}}, \quad \eta^{s}_{j,j+1} = \frac{k_{j+1,z}}{k_{j,z}}. \tag{9}$$

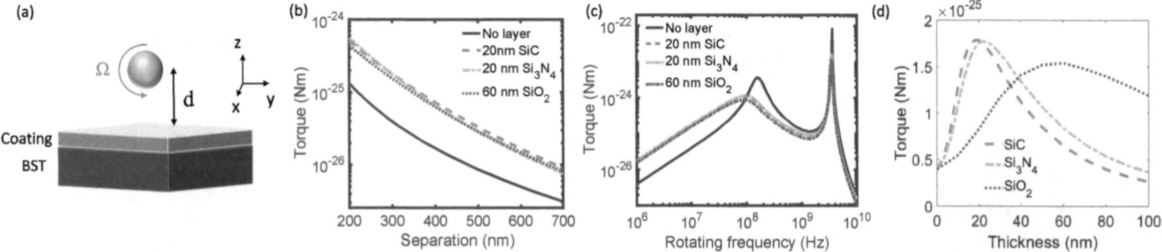

Figure 2: Vacuum friction between a barium strontium titanate (BST) sphere and a BST substrate with a single-layer coating.
(a) A sphere with a radius of 75 nm rotates around the x axis. The z axis is the direction normal to the substrate. The plate consists of a thin layer of dielectric material (shown in yellow) and a BST substrate (shown in blue). (b) Calculated vacuum frictional torque acting on the nanosphere as a function of the separation between the sphere and the substrate. The sphere rotates at a frequency of 10 MHz. The red, yellow, and green dashed curve correspond to the case for a 20-nm SiC, 20-nm Si_3N_4, and a 60-nm SiO_2 thin layer. The blue solid curve is for the case when there is no coating. (c) The frictional torque is shown as a function of the rotating frequency of the nanosphere. The separation is 300 nm. The blue solid curve shows the case when there is no coating on the BST substrate. The red, yellow, and green dashed curve correspond to the case for a 20-nm SiC, a 20-nm Si_3N_4, and a 60-nm SiO_2 thin layer on top of the BST substrate. (d) The frictional torque is shown as a function of the thickness of the thin layer on top of the BST substrate for the case of SiC, Si_3N_4, and SiO_2. Here the separation is 300 nm and the rotating frequency is 10 MHz.

Here $k_{j,z} = \sqrt{\varepsilon\omega^2/c^2 - k_\parallel^2}$ is the vertical wave vector for the jth layer. $k_\parallel$ is the wave vector parallel to the surface. $P(L_j)$ is the propagation matrix in the jth layer for both p and s polarizations and it is given as

$$P(L_j) = \begin{bmatrix} e^{-ik_{j,z}L_j} & 0 \\ 0 & e^{ik_{j,z}L_j} \end{bmatrix}. \tag{10}$$

With the transfer matrix M, the optical properties such as reflection, transmission and absorption for the layered structure can be calculated. The reflection coefficients for the $p(s)$ polarization of the layered structure is [32, 33].

$$r_{p(s)} = T_{21}^{p(s)} / T_{11}^{p(s)}. \tag{11}$$

where $T_{21}^{p(s)}$ and $T_{21}^{p(s)}$ are the components of the transfer matrix T. To calculate the vacuum friction between a sphere and a layered structure, we need to get the electromagnetic Green tensor and it is given as [34]

$$G(\omega) = i\int_0^\infty dk_\parallel k_\parallel e^{2ik_z d}\cdot\left[r_p(\omega,k_\parallel)\begin{pmatrix} -k_z/2 & 0 & 0 \\ 0 & -k_z/2 & 0 \\ 0 & 0 & k_\parallel^2/k_z \end{pmatrix} + r_s(\omega,k_\parallel)\begin{pmatrix} k^2/2k_z & 0 & 0 \\ 0 & k^2/2k_z & 0 \\ 0 & 0 & 0 \end{pmatrix}\right], \tag{12}$$

where d is the separation between the sphere and plate, $k = \omega/c$ is the total wave vector, $k_\parallel$ is the wave vector parallel to the surface, $k_z = \sqrt{k^2 - k_\parallel^2}$ is the wave vector perpendicular to the surface. r_s and r_p are the reflection coefficients of the layered structure which have been calculated by the transfer matrix. Therefore, the Green function that associates the dipole moment fluctuations of the sphere and the induced electromagnetic field on the layered surface is given as

$$\begin{aligned} \overline{G}(\omega) &= \frac{1}{2}(G_{yy} + G_{zz}) \\ &= \frac{i}{2}\int_0^\infty dk_\parallel k_\parallel e^{2ik_z d}\left[-r_p(\omega,k_\parallel)\frac{k_z}{2} \right. \\ &\quad \left. + r_s(\omega,k_\parallel)\frac{k^2}{2k_z} + r_p(\omega,k_\parallel)\frac{k_\parallel^2}{k_z}\right]. \end{aligned} \tag{13}$$

Based on the transfer matrix introduced above, now we can calculate the friction torque experienced by a sphere near a layered structure as shown in Figure 2(a). At a rotating frequency of 10 MHz, the calculated friction as a function of the separation is shown in Figure 2(b). The red, yellow, and green dashed curve is the case for a single layer of 20-nm SiC, 20-nm Si_3N_4, and 60-nm SiO_2 on top of the BST. The blue solid curve is the case for a BST sphere and a BST substrate, for comparison. The optical properties of SiC, Si_3N_4, and SiO_2 can be found in [29, 34]. We can see that the torque for the single-layer coating is enhanced by about four times compared to the no-coating case. We also show the calculated friction as a function of the rotating frequency in Figure 2(c) when the separation is 300 nm. Notice that one of the resonant frequencies (originally at 150 MHz) shifts to the left. This explained the enhancement of the friction torque for the single-layer structure at low-frequency range. This layered structure is favorable for the experiment since it gives a larger torque at lower rotating frequency. At this frequency, the mechanical rotor behaves more stable and hence easier for the measurement to be performed. We also show the dependence of thickness of

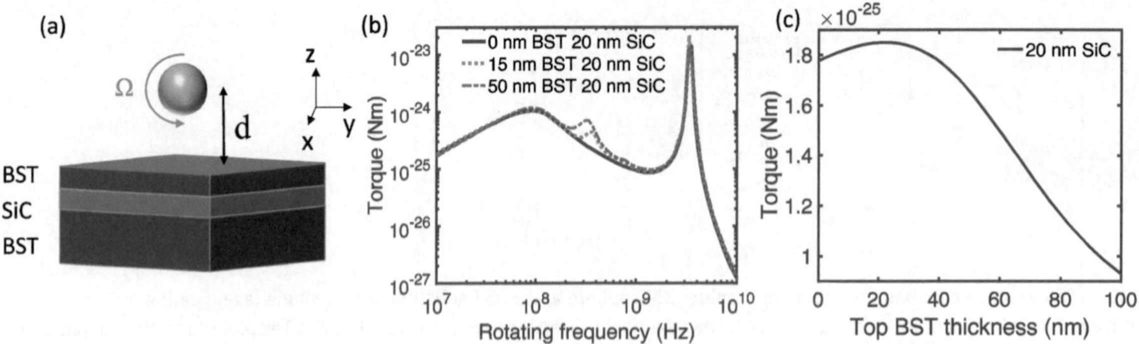

Figure 3: Vacuum friction between a barium strontium titanate (BST) sphere and a BST substrate with a double-layer coating on top. (a) A sphere with a radius of 75 nm rotates around the *x* axis, which is parallel to the plate. The plate consists of a thin layer of BST (shown in blue), a thin layer of SiC (shown in yellow) and a BST substrate. (b). Calculated vacuum frictional torque as a function of the rotating frequency of the nanosphere. The separation is 300 nm. The blue solid, red dashed, and green dashed curves correspond to a thin layer of 0-, 15-, and 50-nm BST on top of a 20-nm SiC-coated BST plate. (c). When the thickness of the second layer (SiC) is fixed to 20 nm and the rotating frequency is 10 MHz, the friction torque is shown as a function of the thickness of the first layer (BST). The separation is 300 nm.

the thin layer on the friction torque in Figure 2(d). It shows that 20-nm SiC, 20-nm Si_3N_4, and 60-nm SiO_2 give the largest friction torque. It is essential to find the optimum thickness of the coating to get the largest torque.

Beyond that, we also calculated the friction torque for the double-layer coating. The configuration is shown in Figure 3(a). The top is a thin layer of BST as shown in blue. The middle one is a 20-nm SiC thin layer similar to the former case and the bottom is a BST substrate. Similar to the single-layer coating, the transfer matrix for the $p(s)$ polarization is given as $T^{p(s)} = D^{p(s)}_{0\to1} P(L_1) D^{p(s)}_{1\to2} P(L_2) D^{p(s)}_{2\to3}$, where L_1 and L_2 is the thickness of the first and second layer coating, respectively. $D^{p(s)}_{j\to j+1}$ is the transmission matrix and $P(L_j)$ is the propagation matrix. At a separation of 300 nm, the calculated friction torque as a function of the rotating frequency is presented in Figure 3(b). The blue solid, red dashed, and green dashed curves corresponds to a thin layer of 0-, 15-, and 50-nm BST on top of a 20-nm SiC--coated BST plate. The torque for the double-layer is only slightly larger than the one with the single-layer coating at low-frequency range. Figure 3(c) shows the friction torque as a function of the thickness of the top BST layer when the rotating frequency is 10 MHz and the separation is 300 nm.

5 Conclusion

In this paper, we have calculated the vacuum friction between a rotating sphere and a close surface. We notice that BST has low surface phonon polariton frequency which can be mechanically excited. At two resonant rotating frequencies, the mechanical energy of the rotating sphere can be converted to the electromagnetic field energy and hence significantly enhance the vacuum friction torque. This provides us a more practical scheme to measure the long-sought vacuum friction torque. Moreover, we find that applying a single-layer coating on top of the BST substrate can further increase the friction torque by four times. The torque not only depends on the rotating frequency and the separation, but also relates to the thickness of the layer. The friction torque for the double-layer coating on the BST substrate is also presented in the content. In the future, it will be interesting to study whether more complex metamaterials [35] and topological materials [36] can further enhance the vacuum friction.

Acknowledgments: The authors are grateful to receive support from the DARPA QUEST program and the Office of Naval Research under Grant No. N00014-18-1-2371.

Author contribution: All the authors have accepted responsibility for the entire content of this submitted manuscript and approved submission.

Research funding: This research was supported by the DARPA QUEST program and the Office of Naval Research under Grant No. N00014-18-1-2371.

Conflict of interest statement: The authors declare no conflicts of interest regarding this article.

References

[1] H. B. G. Casimir, "On the attraction between two perfectly conducting plates," *Proceedings*, vol. 51, pp. 793–795, 1948.

[2] J. B. Pendry, "Shearing the vacuum - quantum friction," *J. Phys. Condens. Matter*, vol. 9, pp. 10301–10320, 1997.

[3] A. I. Volokitin and B. N. J. Persson, "Theory of friction: the contribution from a fluctuating electromagnetic field," *J. Phys. Condens. Matter*, vol. 11, pp. 345–359, 1999.
[4] P. A. Maia Neto and S. Reynaud, "Dissipative force on a sphere moving in vacuum," *Phys. Rev. A*, vol. 47, pp. 1639–1646, 1993.
[5] F. Intravaia, R. O. Behunin and D. A. R. Dalvit, "Quantum friction and fluctuation theorems," *Phys. Rev. A*, vol. 89, p. 050101, 2014.
[6] A. Manjavacas and F. J. García de Abajo, "Vacuum friction in rotating particles," *Phys. Rev. Lett.*, vol. 105, p. 113601, 2010.
[7] M. S. Tomassone and A. Widom, "Electronic friction forces on molecules moving near metals," *Phys. Rev. B*, vol. 56, pp. 4938–4943, 1997.
[8] A. I. Volokitin and B. N. J. Persson, "Resonant photon tunneling enhancement of the van der waals friction," *Phys. Rev. Lett.*, vol. 91, p. 106101, 2003.
[9] A. I. Volokitin and B. N. J. Persson, "Quantum friction," *Phys. Rev. Lett.*, vol. 106, p. 094502, 2011.
[10] R. Zhao, A. Manjavacas, F. J. García de Abajo, and J. B. Pendry, "Rotational quantum friction," *Phys. Rev. Lett.*, vol. 109, p. 123604, 2012.
[11] I. Dorofeyev, H. Fuchs, G. Wenning, and B. Gotsmann, "Brownian motion of microscopic solids under the action of fluctuating electromagnetic fields," *Phys. Rev. Lett.*, vol. 83, pp. 2402–2405, 1999.
[12] B. Gotsmann and H. Fuchs, "Dynamic force spectroscopy of conservative and dissipative forces in an al-au(111) tip-sample system," *Phys. Rev. Lett.*, vol. 86, pp. 2597–2600, 2001.
[13] B. C. Stipe, H. J. Mamin, T. D. Stowe, T. W. Kenny, and D. Rugar, "Noncontact friction and force fluctuations between closely spaced bodies," *Phys. Rev. Lett.*, vol. 87, p. 096801, 2001.
[14] G. Ranjit, M. Cunningham, K. Casey, and A. A. Geraci, "Zeptonewton force sensing with nanospheres in an optical lattice," *Phys. Rev. A*, vol. 93, p. 053801, 2016.
[15] Z. Xu and T. Li, "Detecting casimir torque with an optically levitated nanorod," *Phys. Rev. A*, vol. 96, p. 033843, 2017.
[16] J. Ahn, Z. Xu, J. Bang, et al., "Optically levitated nanodumbbell torsion balance and ghz nanomechanical rotor," *Phys. Rev. Lett.*, vol. 121, p. 033603, 2018.
[17] R. Reimann, M. Doderer, E. Hebestreit, et al., "Ghz rotation of an optically trapped nanoparticle in vacuum," *Phys. Rev. Lett.*, vol. 121, p. 033602, 2018.
[18] C. P. Blakemore, A. D. Rider, S. Roy, Q. Wang, A. Kawasaki, and G. Gratta, "Three-dimensional force-field microscopy with optically levitated microspheres," *Phys. Rev. A*, vol. 99, p. 023816, 2019.
[19] J. Ahn, Z. Xu, J. Bang, P. Ju, X. Gao, and T. Li, "Ultrasensitive torque detection with an optically levitated nanorotor," *Nat. Nanotechnol.*, vol. 15, pp. 89–93, 2020.
[20] Y. Zheng, L.-M. Zhou, Y. Dong, et al., "Robust optical-levitation-based metrology of nanoparticle's position and mass," *Phys. Rev. Lett.*, vol. 124, p. 223603, 2020.
[21] J. Bang, T. Seberson, P. Ju, et al., "5D Cooling and precession-coupled nonlinear dynamics of a levitated nanodumbbell," arXiv:2004.02384, 2020.
[22] R. Diehl, E. Hebestreit, R. Reimann, F. Tebbenjohanns, M. Frimmer, and L. Novotny, "Optical levitation and feedback cooling of a nanoparticle at subwavelength distances from a membrane," *Phys. Rev. A*, vol. 98, p. 013851, 2018.
[23] L. Magrini, R. A. Norte, R. Riedinger, et al., "Near-field coupling of a levitated nanoparticle to a photonic crystal cavity," *Optica*, vol.5, pp. 1597–1602, 2018.
[24] A. O. Turky, M. Mohamed Rashad, A. E.-H. Taha Kandil, and M. Bechelany, "Tuning the optical, electrical and magnetic properties of ba0.5sr0.5tixm1–xo3 (bst) nanopowders," *Phys. Chem. Chem. Phys.*, vol. 17, pp. 12553–12560, 2015.
[25] Y. Guo and Z. Jacob, "Giant non-equilibrium vacuum friction: role of singular evanescent wave resonances in moving media," *J. Opt.*, vol. 16, p. 114023, 2014.
[26] Y. Guo and Z. Jacob, "Singular evanescent wave resonances in moving media," *Opt. Express*, vol. 22, pp. 26193–26202, 2014.
[27] A. Volokitin, "Resonant photon emission during relative sliding of two dielectric plates," *Mod. Phys. Lett.*, vol. 35, p. 2040011, 2020.
[28] A. I. Volokitin, "Singular resonance in fluctuation-induced electromagnetic phenomena at the rotation of a nanoparticle near the surface of a condensed medium," *JETP Lett.*, vol. 108, pp. 147–154, 2018.
[29] J. Kischkat, S. Peters, B. Gruska, et al., "Mid-infrared optical properties of thin films of aluminum oxide, titanium dioxide, silicon dioxide, aluminum nitride, and silicon nitride," *Appl. Opt.*, vol. 51, pp. 6789–6798, 2012.
[30] J. Corson, G. W. Mulholland, and M. R. Zachariah, "Calculating the rotational friction coefficient of fractal aerosol particles in the transition regime using extended kirkwood-riseman theory," *Phys. Rev. E*, vol. 96, p. 013110, 2017.
[31] F. Tebbenjohanns, M. Frimmer, A. Militaru, V. Jain, and L. Novotny, "Cold damping of an optically levitated nanoparticle to microkelvin temperatures," *Phys. Rev. Lett.*, vol. 122, p. 223601, 2019.
[32] L. Ge, X. Shi, Z. Xu, and K. Gong, "Tunable casimir equilibria with phase change materials: from quantum trapping to its release," *Phys. Rev. B*, vol. 101, p. 104107, 2020.
[33] T. Zhan, X. Shi, Y. Dai, X. Liu, and J. Zi, "Transfer matrix method for optics in graphene layers," *J. Phys. Condens. Matter*, vol. 25, p. 215301, 2013.
[34] A. Manjavacas, F. J. Rodríguez-Fortuño, F. J. García de Abajo, and A. V. Zayats, "Lateral casimir force on a rotating particle near a planar surface," *Phys. Rev. Lett.*, vol. 118, p. 133605, 2017.
[35] X. Ni, Z. J. Wong, M. Mrejen, Y. Wang, and X. Zhang, "An ultrathin invisibility skin cloak for visible light," *Science*, vol. 349, p. 1310, 2015.
[36] Q.-D. Jiang and F. Wilczek, "Quantum atmospherics for materials diagnosis," *Phys. Rev. B*, vol. 99, p. 201104, 2020.

Part IX: **Plasmonics and Polaritonics**

Fan Yang, Kun Ding and John Brian Pendry*

Shrinking the surface plasmon

https://doi.org/10.1515/9783110710687-043

Abstract: Surface plasmons at an interface between dielectric and metal regions can in theory be made arbitrarily compact normal to the interface by introducing extreme anisotropy in the material parameters. We propose a metamaterial structure comprising a square array of gold cylinders and tune the filling factor to achieve the material parameters we seek. Theory is compared to a simulation wherein the unit cell dimensions of the metamaterial are shown to be the limiting factor in the degree of localisation achieved.

Keywords: metamaterials; surface plasmons; transformation optics.

Surface plasmons exist at an interface between a metal, $\varepsilon_m < 0$, and dielectric, $\varepsilon_d > 0$ [1–3]. If these surface states have in-plane wave vectors k_x, they are confined normal to the surface by imaginary wave vectors,

$$\begin{aligned} k_z &= +i\sqrt{k_x^2 - \varepsilon_d(\omega_{sp})c_0^{-2}\omega_{sp}^2}, \quad z > 0, \\ k_z &= -i\sqrt{k_x^2 - \varepsilon_m(\omega_{sp})c_0^{-2}\omega_{sp}^2}, \quad z < 0, \end{aligned} \tag{1}$$

where we assume that the dielectric occupies the space $z > 0$, ω_{sp} is the surface plasmon frequency and c_0 is the velocity of light in free space. We have assumed isotropic media: ε_d, ε_m are the permittivities of the dielectric and metal, respectively. They are in general dependent on frequency. At lower values of k_x, the surface plasmon is rather diffuse in extent, but at large values of k, the surface plasmon becomes compact and increasingly electrostatic in nature: $k_z \to \pm ik_x$ and is confined to the surface region. This compact nature results in a high density of states in the immediate vicinity of the surface, which is exploited in many applications. In this letter, we show how by exploiting transformation optics theory, surface plasmons can in principle be made arbitrarily compact, depending only on the availability of suitable materials. We propose a new metamaterial designed to address the latter issue.

Transformation optics [4–7] is a theory that relates distortions of geometry to redefined values of permittivity and permeability. For example, in our case, we seek to compress the surface plasmon normal to the surface. In the study by Kundtz et al. [7], we learn that if we compress the wave fields by a factor β ($\beta < 1$) so that the new imaginary wave vectors increase by a factor β^{-1}, in order that the compressed wave fields continue to obey Maxwell's equations, we must introduce new values of permittivity as follows,

$$\begin{aligned} \varepsilon_{d\parallel} &= \beta^{-1}\varepsilon_d, \quad \varepsilon_{m\parallel} = \beta^{-1}\varepsilon_m, \\ \varepsilon_{dz} &= \beta\varepsilon_d, \quad \varepsilon_{mz} = \beta\varepsilon_m, \end{aligned} \tag{2}$$

with analogous formulas for the permeability. This formula solves our problem at a stroke, always provided of course that we can find suitably anisotropic materials. It is also possible to expand the surface plasmon by choosing $\beta > 1$ [8].

To solve the problem of finding permittivities tunable in the fashion required, we turn to metamaterials [9–13]. These are composite materials structured on a scale much less than the relevant wavelengths in the problem, whose properties owe more to their structure than to their chemical composition. Tuning the magnetic response is more of a problem because even metamaterials struggle with magnetism at optical frequencies. However, here, we appeal to the mainly electrostatic nature of the surface plasmon at higher values of k_x and show that a high degree of compression can be achieved by tuning the electrical response alone.

Our target metamaterial structure is shown in Figure 1. In the first instance, we use a simple approximation to find the effective medium parameters of our structure, which we then check against COMSOL simulations. The Maxwell Garnett theory gives the following formula for the metamaterial parameters [14, 15],

$$\varepsilon_\parallel = \varepsilon_d\frac{(1+f_m)\varepsilon_m + (1-f_m)\varepsilon_d}{(1+f_m)\varepsilon_d + (1-f_m)\varepsilon_m}, \quad \varepsilon_z = f_m\varepsilon_m + (1-f_m)\varepsilon_d, \tag{3}$$

where f_m is the metal volume filling fraction of the cylinders. We model the metal with a Drude permittivity and the dielectric as vacuum,

***Corresponding author: John Brian Pendry,** The Blackett Laboratory, Imperial College London, London, SW7 2AZ, UK,
E-mail: j.pendry@imperial.ac.uk. https://orcid.org/0000-0001-5145-5441
Fan Yang: Department of Electrical and Computer Engineering, University of California, San Diego, 9500 Gilman Dr, La Jolla, CA, 92093, USA
Kun Ding: The Blackett Laboratory, Imperial College London, London, SW7 2AZ, UK. https://orcid.org/0000-0002-0185-2227

This article has previously been published in the journal Nanophotonics. Please cite as: F. Yang, K. Ding and J. B. Pendry "Shrinking the surface plasmon" *Nanophotonics* 2021, 10. DOI: 10.1515/nanoph-2020-0361.

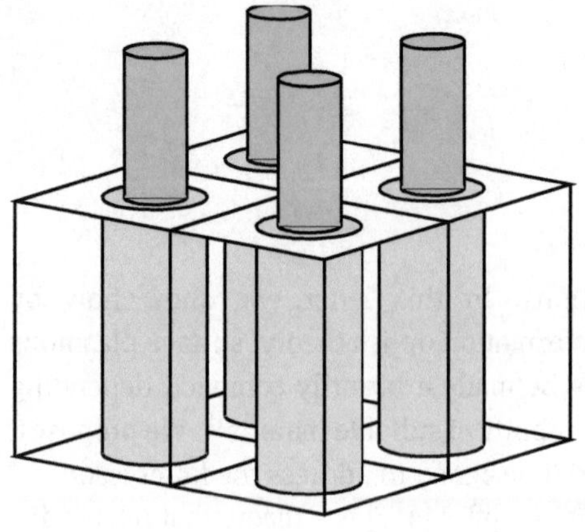

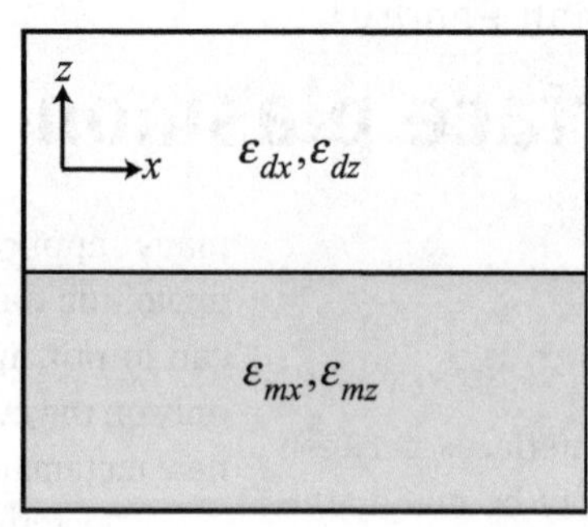

Figure 1: A two-dimensional square array of metallic cylinders much smaller than the relevant wavelengths, embedded in a dielectric. In the plane $z = 0$, there is an interface between two sets of cylinders. This is where the interface plasmon forms. We tune the volume fraction to achieve the desired properties of an effective medium shown on the right.

$$\varepsilon_m = 1 - \frac{\omega_p^2}{\omega(\omega + i\gamma)}, \quad \omega_p = 8.95\,\text{eV}, \quad \gamma = 0.329\,\text{eV}, \quad \varepsilon_d = 1, \tag{4}$$

with the metal parameters chosen to model gold. In the first instance, we shall neglect losses, $\gamma = 0$, but later when comparing to COMSOL simulations, loss is taken into account.

Although (3) is an approximation, it can be shown to be highly accurate [16]. Figure 2 compares a COMSOL simulation of transmission and reflection coefficients for the structure shown in Figure 1, with an effective medium calculation using the parameters given by the Maxwell Garnet formula.

The challenge is to design two metamaterials each with huge anisotropies, but one pair taking negative values and the other taking positive values. This will realize our requirements for compression of a surface plasmon at the interface between the two.

Recognizing that $\varepsilon_d, \varepsilon_m$ have opposite signs, inspection of (3) shows that we can make the real part of ε_z very small by choice of $f_m = \varepsilon_d/(\varepsilon_d + |\varepsilon_m|)$. The imaginary part of ε_m will be a limiting factor in how close we can come to our ideal. Having chosen f_m, we can solve for,

$$\varepsilon_\parallel = \frac{-\varepsilon_d|\varepsilon_m|(\varepsilon_d + |\varepsilon_m|)}{2|\varepsilon_d|^2 + |\varepsilon_m|\varepsilon_d - |\varepsilon_m|^2}, \tag{5}$$

where we have recognized that $\varepsilon_m < 0$. We can arrange that the denominator takes a very small value by adjusting $|\varepsilon_m|/\varepsilon_d = 2$ and hence $f_m = 1/3$. Thus, by exploiting the

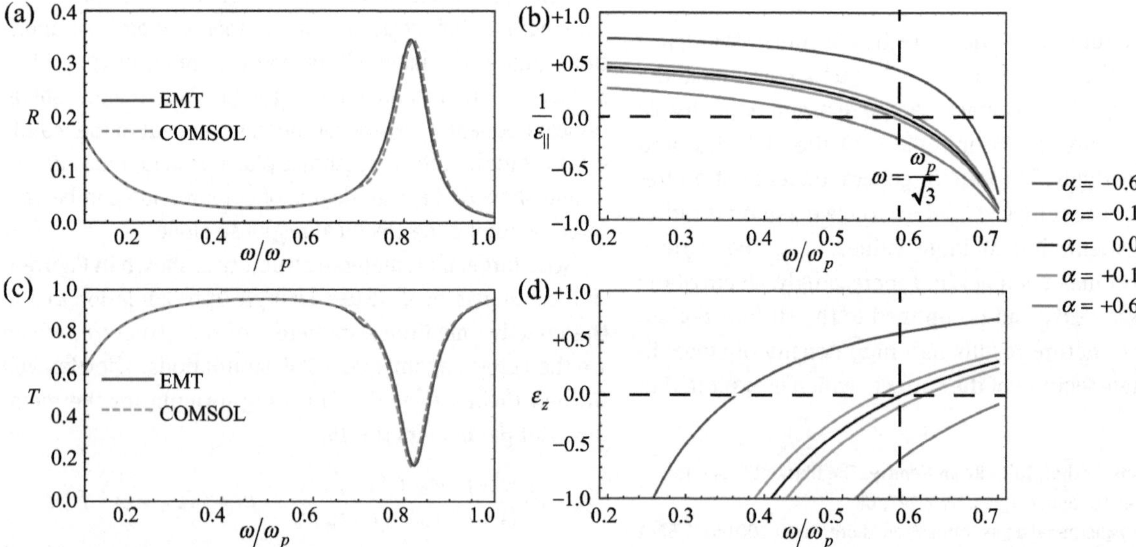

Figure 2: (a) Reflection from and (c) transmission through one unit cell of a nanowire metamaterial with period 10 nm and filling ratio $f_m = 1/3$, calculated with effective medium theory (EMT) theory and COMSOL. The incident angle is at 45° to the z-axis, and the electric field has both E_z and E_x components. The magnetic field has only a H_y component. (b) $1/\varepsilon_\parallel$ for various $-0.6 < \alpha < 0.6$, plotted against ω/ω_p. (d) ε_z for various $-0.6 < \alpha < 0.6$.

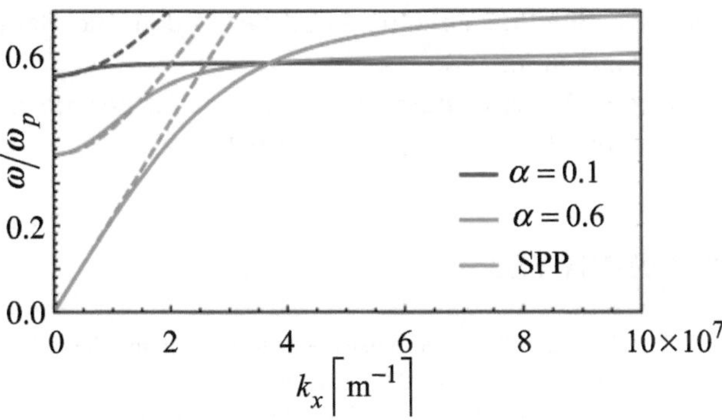

Figure 3: Dispersion of the surface plasmons for $\alpha = 0.1$ and $\alpha = 0.6$ together with dispersion of the pure metal-vacuum surface plasmon plotted against k_x in units of m^{-1}. The dotted lines show the associated light lines in the dielectric.

properties of a plasmonic material, we can achieve our goal of extreme anisotropy. Furthermore, if we vary f_m about the singular point,

$$f_m = (1 + \alpha)/3, \tag{6}$$

we find that for $\alpha > 0$, we have an extremely anisotropic metal, and for $\alpha < 0$, we have an extremely anisotropic dielectric.

In Figure 2b and d, we plot the Maxwell Garnett formula for the metallic and dielectric metamaterial anisotropic permittivities for several values of α. When $\alpha = 0$, the curves intersect at zero and a frequency of $\omega/\omega_p = 1/\sqrt{3} = 0.5774$. Somewhere in the range where the $\alpha > 0$ parameters are both negative and the $\alpha < 0$ parameters are both positive, we expect to find a surface plasmon.

Next, we present some calculations to demonstrate the feasibility of our theory.

We also need to recognize that the metamaterial concept only holds good on length scales greater than the metamaterial structure, which we take to be 10 nm.

Figure 3 shows dispersion of the surface plasmon trapped between the metametal and the metadielectric calculated for an effective medium corresponding to two media with filling factors defined by $\pm\alpha$. Shown on the same plot is the light line for the metadielectric. Dispersion curves to the right of this line represent surface plasmons trapped at the surface; to the left of this line, dispersion curves represent waves that are perfectly transmitted across the interface in the manner of a Brewster condition. This is a typical behaviour when a surface plasmon dispersion curve appears to cross the light line. For example, the $\alpha = \pm 0.1$ surface plasmon exists between $0.56268 < \omega/\omega_p < 0.57765$. The pure metal-vacuum surface plasmon is shown for comparison. It disperses much more rapidly with frequency than the compressed surface plasmon and therefore has a much lower density of states. All dispersion curves are degenerate at $\omega = \omega_p/\sqrt{3}$; increasing α lowers the frequency at $k_x = 0$ while increasing the limiting frequency at $k_x \to \infty$.

Figure 4a shows an interface surface state plotted as a function of distance from the interface calculated in the effective medium approximation. Compared to the surface

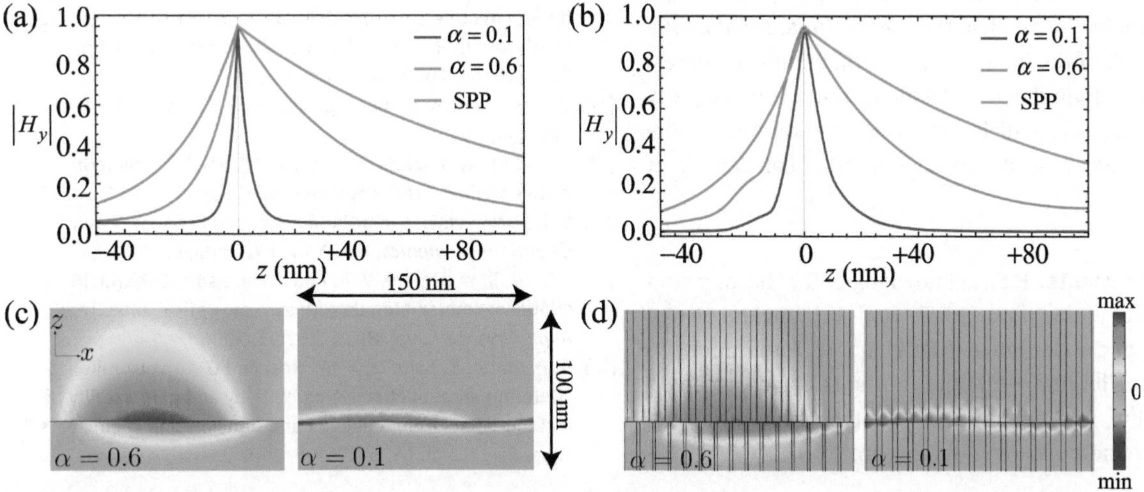

Figure 4: (a) Modulus of the magnetic field for the interface plasmon calculated in the effective medium approximation at $k_x = 2.26 \times 10^7\ m^{-1}$ compared to a surface plasmon that exists between a pure metal and pure vacuum at the same frequency. (b) The same calculation but now deploying COMSOL on the metamaterial structure, lattice spacing $a = 10$ nm. (c) An effective medium calculation of the magnetic field distribution plotted in the vicinity of the interface. (d) The same calculation but now deploying COMSOL on the metamaterial structure and plotted in a plane taken through the centre of the cylinders. The surface mode is excited by a surface current along the x-direction at the interface, which makes the magnetic field discontinuous.

plasmon existing between a pure metal and pure vacuum, we can see very large compression by a factor of about 20 when $\alpha = \pm 0.1$. Furthermore, as noted in Figure 3, the density of states is greatly enhanced by the flattened dispersion of the interface surface plasmon.

So far, we have worked in the effective medium approximation, but now, we make a more realistic test by including the microstructure of the metamaterial and the loss parameter $\gamma = 0.329$ eV, which so far, we have taken to be zero. Loss is also included in the effective medium calculation in Figure 4. Figure 4b presents a COMSOL simulation of a metamaterial structure in which the metal permittivity includes loss as described in (4), and the lattice period is 10 nm. At the interface, the two sets of cylinders on either side are coaxial with one other and touch at the interface. The pure metal/dielectric SPP is unchanged of course, but we see spreading of the metamaterial interface plasmon. This is mainly due to the finite dimensions of the metamaterial unit cell: on length scales <10 nm, the effective medium approximation breaks down. Nevertheless, our model metamaterial still shows substantial compression of the surface plasmon by about a factor of 7.

In conclusion, we have shown that in theory, interface surface plasmons can be arbitrarily compressed provided that the specified anisotropic material parameters can be realized. We proposed a metamaterial structure based on a square array of gold cylinders and showed how the design parameters can be tuned to approach the ideal anisotropic parameters for the metamaterials on each side of the interface. In practice, although substantial compression of the interface surface plasmon can be achieved, the extreme values predicted by an ideal theory are limited first by the finite unit cell of the metamaterial, which limits compression to no less than the unit cell dimensions, and secondly by metallic losses, which limit the compression of the density of states and hence also of the local density of states.

Acknowledgements: F.Y. acknowledges the Gordon and Betty Moore Foundation. J.B.P. and K.D. acknowledge funding from the Gordon and Betty Moore Foundation.

Author contribution: All the authors have accepted responsibility for the entire content of this submitted manuscript and approved submission.

Research funding: Funding from the Gordon and Betty Moore Foundation.

Conflict of interest statement: The authors declare no conflicts of interest regarding this article.

References

[1] R. H. Ritchie, "Plasma losses by fast electrons in thin films," *Phys. Rev.*, vol. 106, pp. 874–881, 1957.

[2] W. L. Barnes, A. Dereux, and T. W. Ebbesen, "Surface plasmon subwavelength optics," *Nature*, vol. 424, pp. 824–830, 2003.

[3] S. A. Maier, *Plasmonics: Fundamentals and Applications*, New York, Springer Science & Business Media, 2007.

[4] A. Ward and J. B. Pendry, "Refraction and geometry in Maxwell's equations," *J. Mod. Opt.*, vol. 43, pp. 773–793, 1996.

[5] J. B. Pendry, D. Schurig, and D. R. Smith, "Controlling electromagnetic fields," *Science*, vol. 312, pp. 1780–1782, 2006.

[6] U. Leonhardt, "Optical conformal mapping," *Science*, vol. 312, pp. 1777–1780, 2006.

[7] N. B. Kundtz, D. R. Smith, and J. B. Pendry, "Electromagnetic design with transformation optics," *Proc. IEEE*, vol. 99, pp. 1622–1633, 2011.

[8] F. Yang, S. Ma, K. Ding, S. Zhang, and J. B. Pendry, "Continuous topological transition from metal to dielectric," *Proc. Natl. Acad. Sci.*, 2020, to appear. https://doi.org/10.1073/pnas.2003171117.

[9] J. B. Pendry, A. J. Holden, D. J. Robbins, and W. J. Stewart, "Magnetism from conductors and enhanced non-linear phenomena," *IEEE Trans. Microw. Theor. Tech.*, vol. 47, pp. 2075–2084, 1999.

[10] D. R. Smith, J. B. Pendry, and M. C. K. Wiltshire, "Metamaterials and negative refractive index," *Science*, vol. 305, pp. 788–792, 2004.

[11] J. B. Pendry, "Metamaterials and the control of electromagnetic fields", in *Coherence and Quantum Optics IX*, N. P. Bigelow, J. H. Eberly and C. R. Stroud, Jr., Eds., Washington, DC, OSA Publications, 2009, 2008, pp. 42–52.

[12] W. Cai and V. M. Shalaev, *Optical Metamaterials*, New York, Springer, 2010.

[13] A. Poddubny, I. Iorsh, P. Belov, and Y. Kivshar, "Hyperbolic metamaterials," *Nat. Photonics*, vol. 7, pp. 948–957, 2013.

[14] S. I. Bozhevolnyi, L. Martin-Moreno, and F. Garcia-Vidal, *Quantum Plasmonics*, New York, NY, Springer, 2017.

[15] J. Elser, R. Wangberg, V. A. Podolskiy, and E. E. Narimanov, "Nanowire metamaterials with extreme optical anisotropy," *Appl. Phys. Lett.*, vol. 89, p. 261102, 2006.

[16] D. R. Smith, S. Schultz, P. Markos, and C. M. Soukoulis, "Determination of effective permittivity and permeability of metamaterials from reflection and transmission coefficients," *Phys. Rev. B*, vol. 65, p. 195104, 2002.

D. N. Basov*, Ana Asenjo-Garcia, P. James Schuck, Xiaoyang Zhu and Angel Rubio

Polariton panorama

https://doi.org/10.1515/9783110710687-044

Abstract: In this brief review, we summarize and elaborate on some of the nomenclature of polaritonic phenomena and systems as they appear in the literature on quantum materials and quantum optics. Our summary includes at least 70 different types of polaritonic light–matter dressing effects. This summary also unravels a broad panorama of the physics and applications of polaritons. A constantly updated version of this review is available at https://infrared.cni.columbia.edu.

Keywords: portions; quantum electrodynamics; quantum materials; quantum optics.

Polaritons are commonly described as light–matter hybrid quasiparticles. Polaritons inherit their attributes from both their light and matter constituents. More rigorously, a polariton is a quantum mechanical superposition of a photon with a matter excitation, the latter being a collective mode in solids and superconducting circuits or an electron in atoms, molecules or even superconducting qubits. As such, the notion of polaritons is a unifying universal concept between the fields of quantum materials (QMs) and quantum optics/electrodynamics. Until fairly recently, these subfields of contemporary physics evolved largely independently of each other. Among the unintended consequences of these divisions is the ambiguity in polaritonic terminology with the same terms used markedly differently in QMs and cavity quantum electrodynamics (QED) with atomic systems. Here, we attempt to summarize (in alphabetical order) some of the polaritonic nomenclature in the two subfields. We hope this summary will help readers to navigate through the vast literature in both of these fields [1–520]. Apart from its utilitarian role, this summary presents a broad panorama of the physics and technology of polaritons transcending the specifics of particular polaritonic platforms (Boxes 1 and 2). We invite readers to consult with reviews covering many important aspects of the physics of polaritons in QMs [1–3], atomic and molecular systems [4], and in circuit QED [5, 6], as well as general reviews of the closely related topic of strong light–matter interaction [7–11, 394]. A constantly updated version is available at https://infrared.cni.columbia.edu.

Anderson–Higgs polaritons [12, 13]. The matter constituent of these polaritons originates from the amplitude mode in superconductors [14] (Figure 1). Anderson–Higgs polaritons are yet to be experimentally observed.

Bardasis–Schrieffer polaritons. The matter constituent of Bardasis–Schrieffer (BaSh) polaritons is associated with the fluctuations of subdominant order parameter in superconductors [15, 16], charge density wave systems [13], and excitonic insulators [17]. This novel theoretical concept still awaits experimental confirmation. The requisite experiments include nanospectroscopy and nanoimaging of polaritonic dispersion in the terahertz (THz) frequency range below the energy gap of superconductors. These are challenging scanning probe measurements, as they have to be carried out at cryogenic temperatures. Nano-THz imaging at cryogenic temperatures have been recently fulfilled [18], paving the wave to the exploration of polaritonic phenomena in superconductors (see also *Cooper pair plasmon polaritons* and *Josephson plasmon polaritons*).

Berreman polaritons: Phonon polaritons in anisotropic materials and multilayer structures are also referred to as epsilon-near-zero or *ENZ polaritons* [29–31]. ENZ materials, artificial structures, and nanocavities reveal exotic electromagnetic responses with a broad range of technological applications [31–35]. For example, ENZ nanocavities facilitate ultrastrong coupling between plasmonic and phononic modes [36], as well as the so-called photonic doping [37].

Berry plasmon polaritons: chiral plasmonic modes whose dispersion is explicitly impacted by the Berry curvature and

***Corresponding author: D. N. Basov**, Department of Physics, Columbia University, New York, NY 10027, USA, E-mail: db3056@columbia.edu
Ana Asenjo-Garcia, Department of Physics, Columbia University, New York, NY 10027, USA
P. James Schuck, Department of Mechanical Engineering, Columbia University, New York, NY 10027, USA
Xiaoyang Zhu, Department of Chemistry, Columbia University, New York, NY 10027, USA
Angel Rubio, Max Planck Institute for the Structure and Dynamics of Matter, Luruper Chaussee 149, 22761 Hamburg, Germany; and Center for Computational Quantum Physics (CCQ), Flatiron Institute, 162 Fifth Avenue, New York, NY 10010, USA

This article has previously been published in the journal Nanophotonics. Please cite as: D. N. Basov, A. Asenjo-Garcia, P. J. Schuck, X. Zhu and A. Rubio "Polariton panorama" *Nanophotonics* 2021, 10. DOI: 10.1515/nanoph-2020-0449.

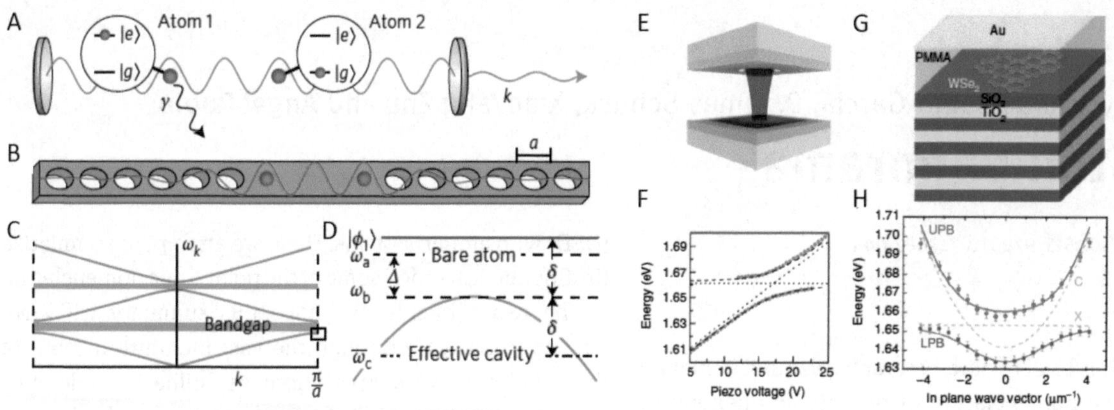

Box 1: Cavity quantum electrodynamics and cavity polaritons. In cavity quantum electrodynamics (QED), the spontaneous emission of atoms, molecules, and solids is governed not only by the properties of the emitter *per se* but is also controlled by its local electromagnetic environment. Optical cavities assembled from two parallel mirrors have long been used to confine light, to enhance light–matter interaction and to promote lasing [19]. The probability of interaction between light and matter is enhanced by the number of bounces the photon makes between the mirrors before leaving the cavity, which is conventionally quantified by the cavity finesse *F*. Cavities with high quality factors promote extremely efficient light matter couplings. In the strong-coupling regime (where the coherent interactions between the matter excitation and the cavity mode overcome the dissipation, i.e., when the vacuum Rabi splitting is much larger than the linewidth), the atomic or material excitation hybridizes with the photonic mode and produces a cavity polariton. The minimum separation upper polariton branch and lower polariton branch $E_{UPB}–E_{LPB}$ in Panel H is commonly referred to the normal-mode splitting in analogy to the Rabi splitting of a single-atom cavity system [20] (also Figure 4). Rabi splitting can reach fractions of eV in QMs and can exceed 1 eV in molecules [21, 22]. Strong coupling leads to photon blockade, where the presence of a photon in a cavity blocks a second one from coming in the study by Tian and Carmichael [23] and Imamoğlu et al. [24]. See also *microcavity polaritons*.
Panel A: cavity-mediated coherent interactions between two atoms in a Fabry–Perot resonator. Two atoms are coupled with strength g_c to a single mode of a Fabry–Perot cavity, enabling an excited atom (atom 1) to transfer its excitation to atom 2 and back. The coherence of this process is reduced by dissipation in the form of the cavity decay at a rate κ and atomic spontaneous emission into free space at a rate γ (adapted from a study by Douglas et al. [25]). Panel B is the photonic crystals, dielectric materials with a periodic modulation of their refractive index, which provide a rich playground for realizing tailored atom–atom interactions. Photonic crystals act as cavities that localize photonic modes (red) at defect sites, created by altering the periodicity (here, by removing certain holes). Atoms coupled to such a system may then interact via this mode in a manner analogous to that in A. Panel C is a typical band structure of a one-dimensional photonic crystal, illustrating the guided mode frequency ωk versus the Bloch wavevector k in the first Brillouin zone. Photonic crystals allow for the exploration of waveguide QED, where atoms are coupled to a propagating photon. Atoms coupled to the crystal have resonance frequency ωa close to the band edge frequency ωb, with $\Delta \equiv \omega a - \omega b$ (adapted from a study by Douglas et al. [25]). Panel D presents the effective cavity mode properties and energy level diagram for the photonic crystal dressed state $|\phi 1\rangle$ (blue), provided the atomic resonance lies inside the bandgap (a frequency region that does not support photon propagation). An excited atom hybridizes with the photonic mode giving rise to an atom–photon bound state, where the photon is localized around the atom, effectively forming a cavity. The dressed state energy ω is detuned by δ from the band edge into the bandgap (band shown in red). The atom is coupled to an effective cavity mode with frequency $\omega c = \omega b - \delta$ formed by superposition of modes in the band (adapted from a study by Douglas et al. [25]). Panel E is an open cavity based on two separated distributed Baragg reflector (DBR) mirrors (shaded blue). The monolayer of active semiconductor material (dark gray) is located on top of the bottom mirror [26]. Panel F is the distance between the mirrors in panel E which can be controlled by a piezo actuator, enabling the tuning of the optical cavity mode into resonance with the excitonic transition. The net effect is the observation of the anticrossing at resonance between the excitonic band and the cavity mode. Adapted from a study by Dufferwiel et al. [26]. Panels G and H are hybrid DBR microcavity with thin semitransparent metallic mirror on top [27]. The lower and upper polariton branches are observed. Trace C displays the cavity resonance C and line X marks the exciton resonance in the absence of coupling. Similar results for strong light–matter coupling in MoS_2 semiconductor integrated in DBR cavity were originally reported in a study by Liu et al. [28].

anomalous velocity in chiral media [38–40]. Berry plasmon polaritons are yet to be experimentally observed.

Bose–Hubbard polaritons: cavity QED polaritons with matter component associated with transitions across the Mott gap in the system of interacting atoms [41] (see also *Mott polaritons*).

Bragg polaritons. Bragg reflectors (Box 1 panel G, Figures 2 and 4) are routinely utilized to implement polaritonic cavities. Bragg polaritons pertain to systems in which multiple excitonic layers and/or quantum wells are periodically integrated in a DBR cavity [47, 48] (see also *polaritonic lattices*). The inherent anisotropy of Bragg

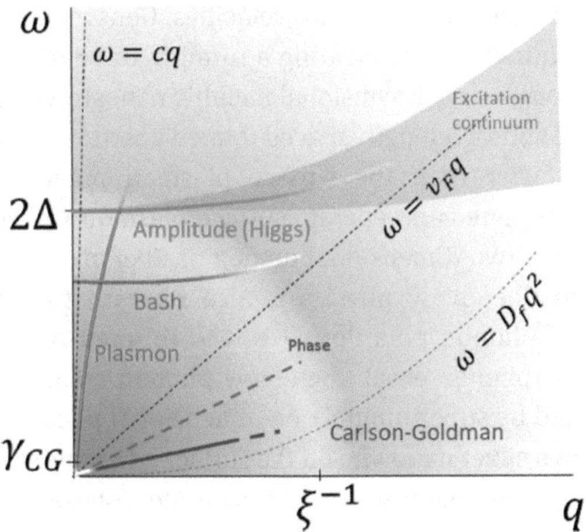

Figure 1: Schematic representation in the frequency–momentum plane of the collective modes that may appear in the electrodynamical response of a two-dimensional (2D) superconductor. The blue area shows the low-energy and long-wavelength region, where weakly damped collective modes may be observed. Anticrossing between the plasmon and Higgs mode and the Bardasis–Schrieffer (BaSh) mode is not shown here. Here, c is the speed of light, v_F is the Fermi velocity, and Df is the normal-state diffusion coefficient. Adapted from a study by Sun et al. [13].

multilayer structures may enable hyperbolic electrodynamics [49] (see *hyperbolic polaritons*).

Cavity (microcavity) polaritons. Weisbuch et al. [142] devised and implemented the first semiconductor (micro) cavity device revealing Rabi splitting of exciton polaritons (Boxes 1 and 2). Semiconductor microcavites emerged as a powerful platform for the investigation of strong light–matter interaction in semiconductors [50, 51]. Microcavity structures reveal intriguing phenomena including polariton parametric amplification [52] and its spontaneous counterpart, the parametric photoluminescence [53]. Parametric photoluminescence is a purely quantum process. An appealing attribute of polariton parametric photoluminescence is that signal-idler polariton pairs are produced in nonclassical states with quantum correlations. The quest for Bose–Einstein condensation of microcavity polaritons has produced a stream of breakthrough results [54, 55] (see also *exciton polaritons and their condensates*). Microcavity exciton polaritons display quantum effects including entanglement [56] and polariton blockade [57, 58] and may serve as a platform for the implementation of qubits [59].

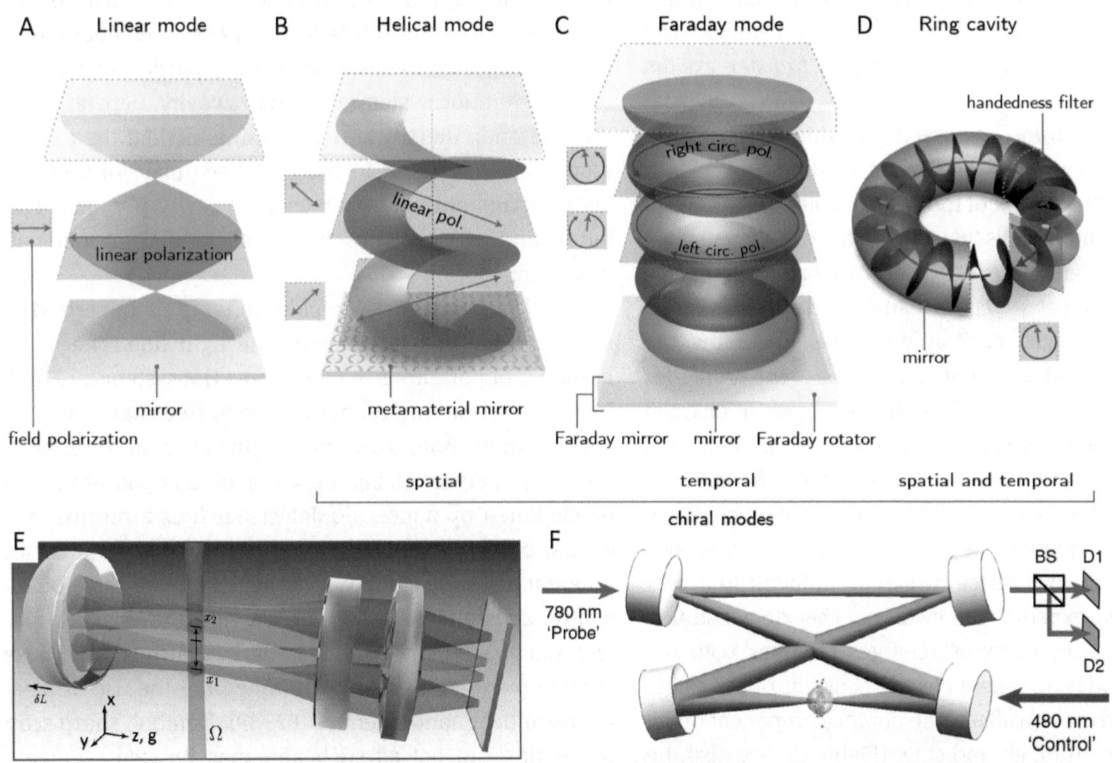

Box 2: Panorama of cavities and cavity modes. A common Fabry–Perot cavity (panel A) formed by two parallel mirrors supports linear modes and maintains time reversal symmetry. Cavities employing chiral metasurfaces support helical modes (panel B). A possible realization of time reversal symmetry breaking is offered by the use of Faraday mirrors in panel (panel C). Ring mode cavities (panel D) sustain running waves of a chosen circular polarization and break time reversal symmetry by means of a handedness filter realizable with a combination of a Faraday rotator and polarization optics. Advanced cavities are well suited for the exploration of the physics of spin vortices and skyrmion spin textures in exciton polariton condensates originating from the optical spin Hall effect [42, 43]. Panels A–D from a study by Hubener et al. [44]. Panel E is a multimode cavity quantum electrodynamics (QED) enabling local light–matter coupling. The schematic displays two ^{87}Rb Bose–Einstein condensates trapped at locations x_1 and x_2 on opposite sides of the cavity center [45]. Panel F is the schematic of a strongly interacting *polaritonic quantum dot* formed by 150 Rydberg-dressed Rubidium atoms in a single-mode optical resonator [46]. BS, beamsplitter; D1 and D2, single-photon detectors.

Channel polaritons are supported by materials and structures with a straight channel cut in polaritonic medium [60]. Channel polaritons were utilized for the implementation of waveguide components including interferometers and ring resonators [61]. Polaritons guided along the nanoslit are predicted to form *hybrid polaritons,* giving rise to both bonding and antibonding modes [62].

Charge transfer polaritons. The formation of plasmon polaritons in graphene or semiconductors relies on the high carrier density that can be introduced by electrostatic gating [63, 64], ferroelectric polarization [65], chemical doping [66], or photoexcitation [67]. Alternatively, the requisite carrier density can be introduced by charge transfer across the interface between proximal materials with dissimilar work functions. Such charge transfer plasmon polaritons have been demonstrated for graphene residing on another van der Waals material $RuCl_3$ [68]. Experiments on metallic nanoparticles show that charge transmitted between the pair of nanoparticles through a conducting pathway leads to a characteristic plasmonic response [69] termed *charge transfer plasmons.* Interlayer exciton in transition metal dichalcogenide (TMDC) heterostructures (e.g., $MoSe_2/WSe_2$) also involves charge transfer from one layer to another; the relevant microcavity polaritons [70] are classified as *charge transfer exciton polaritons.*

Charged polariton. Charged polaritons posess a nonvanishing electric charge. This interesting concept was introduced in the context of the cavity exciton polaritons in GaAs/AlAs quantum wells that also hosted two-dimensional electron gas with the density n_e. Spectroscopic experiments in a study by Forg et al. [71] have identified several distinct properties of charged exciton polaritons, including the scaling of the coupling strength analogous to the properties of atomic QED system [72]. The effective mass of charged polaritons exceeds the band mass of a GaAs quantum well by a factor of 200. Tiene et al. [73] have theoretically demonstrated the unique utility of charged microcavity polaritons for exploring the physics of electron–hole systems with charge imbalance, which are difficult to access with alternative experimental methods. They demonstrated how the Fermi sea of excess charges modifies both the exciton properties and the dielectric constant of the cavity medium, which in turn affects the photon component of the many-body polariton ground state (Figure 2). See also the closely related entries of *Fermi-edge exciton polaritons* and *trion polaritons.*

Cherenkov polaritons. In the Cherenkov effect [74], a charged particle moving with a velocity faster than the phase velocity of light in the medium radiates light. The emitted radiation forms a cone with a half angle determined by the ratio of the two velocities. Genevet et al. [75] demonstrated that by creating a running wave of polarization along a one-dimensional metallic nanostructure consisting of subwavelength-spaced rotated apertures that propagates faster than the surface plasmon polariton phase velocity, one can generate surface plasmon wakes that serve as a two-dimensional analog of Cherenkov radiation. The Cherenkov physics is also relevant to the properties of phonon polaritons [76, 77]. Infrared nano-imaging experiments reveal Cherenkov phonon polariton wakes emitted by superluminal one-dimensional plasmon polaritons in a silver nanowire on the surface of hexagonal boron nitride [78]. See also *Exciton polariton X-waves* on superluminal properties in the system of exciton polaritons.

Cooper pairs polaritons (in QMs and cold fermionic cavity systems). *Cooper pair plasmon polaritons* emerge in superconductors. The matter component of these polaritons is associated with the superfluid density (from a study by Basov et al. [1]). The dispersion of Cooper pairs plasmon polaritons in layered cuprate high-Tc superconductors has been investigated theoretically [13, 79] but is yet to be explored in experiments. Recently, the formalism of the Bardeen Cooper and Schrieffer theory of superconductivity has been applied to describe the quasiparticle excitations of a cold fermion system coupled to a cavity. Depending on the excitation density and atomic interaction, the excited atoms and holes and in the Fermi sea may form bound Cooper pairs strongly coupled with cavity photons. This latter kind of polaritons were also termed Cooper pair polaritons [80].

Dark polaritons in QMs: polaritons are characterized by a wavevector that lies beyond the light line. The lower branches of polaritons in many/most QM systems are dark by this criterion and do not couple to free space photons because of the notorious "momentum mismatch" problem (Box 1 F, H, Figure 1). Light excition of dark polaritons can be mediated by nanoscale defects such as a protrusions, divots, or cracks, exploiting the high spatial frequencies inherent to these deeply subwavelength objects. Better controlled strategies can also provide the missing momentum needed for coupling to dark polartons [81]. These include prism and grating coupling, and the use of plasmonic optical nanoantennas [82–89]. Notably, sharp scan probe tips can act as such antennas [90–95], allowing polaritonic waves to be launched and visualized. Scanning probe antenna-based nano-optics has emerged as an indespensible research tool enabling spectroscopy and visualization of polaritons in QMs [1, 88, 96].

Dark-state polaritons in atomic ensembles: typically, this refers to polaritons in atomic ensembles that propagate in the

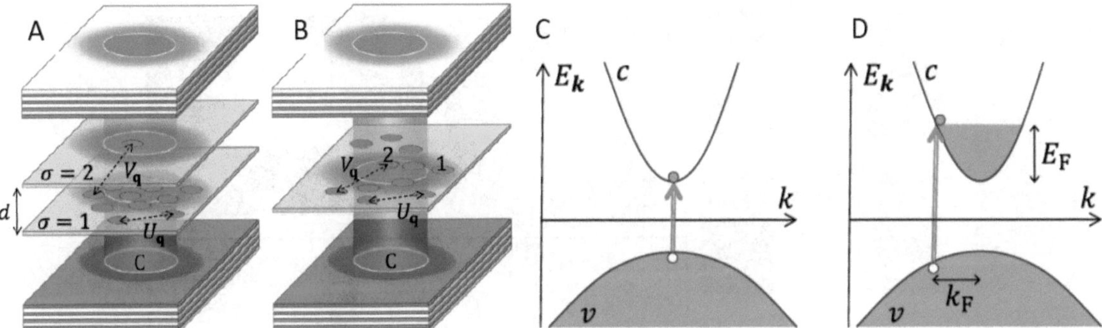

Figure 2: Charged exciton polaritons.
Panel A: two quantum wells, labeled with the indicies $\sigma = 1, 2$ and separated by a distance d, form an electron–hole bilayer in the extremely imbalanced limit. The minority species belongs to the $\sigma = 2$ layer, while the majority species at $\sigma = 1$ forms an interacting Fermi sea. U_q and V_q are, respectively, intraspecies and interspecies Coulomb interactions. The bilayer is located inside a planar cavity that confines the cavity photon mode C. The (blue) shaded area represents the finite-size external laser pump spot. Panel B: the same setup in a single quantum well geometry. Here, the majority $\sigma = 1$ and minority $\sigma = 2$ species belong to the same well. Panels C and D: the particle–hole excitation process via a photon without and with Fermi sea, respectively. All photon-mediated transitions are approximately vertical in a cavity. Adapted from a study by Tiene et al. [73].

regime of electromagnetically induced transparency (EIT) [97–100]. The darkness arises from the photon mixing strongly with a collective atomic excitation, resulting in a state with only a minute photonic component. See also *EIT polaritons* below. In ordered atomic arrays, dark (also often referred to as subradiant) states emerge due to interference in photon emission and absorption. At the single photon level, these darks states are collective spin excitations with a wave vector that lies beyond the light line, preventing the coupling with radiation modes (exactly the same phenomenon of "momentum mismatch" described above for QMs) [101–104]. Polaritons arising in atomic lattices have applications in quantum information storage and processing [103].

Demons: or density modes were introduced by David Pines [105], an early protagonist of plasmons research. *Demons* are particularly relevant to the response of the Dirac fluid in graphene in hydrodynamic regime [106] and adiabatic plasmon amplification [107].

Dirac plasmon polaritons are formed by hybrids of infrared photons with Dirac electrons in graphene [63, 64, 108, 109]. Direct nanoimaging experiments uncovered extraordinarily long propagation lengths of highly confined Dirac polaritons and have established fundamental limits underlying their decoherence and losses [110].

Dyakonov surface polaritons: the surface modes that propagate along the interface between isotropic and uniaxial materials is known as Dyakonov surface polaritons [111–113]. A special case of Dyakonov polaritons is realized in anisotropic crystals of layered van der Waals materials. One example is that of the *hyperbolic surface phonon polaritons* propagating along the edges of slabs prepared from hexagonal boron nitride [114–116].

Edge magneto plasmons. Two-dimensional (2D) electron gas subjected to the magnetic field normal to the plane of the 2D conductors reveals two distinct field-dependent resonances: the cyclotron resonance mode with frequency increasing with the magnetic field and another mode that redshifts with the applied field. The latter mode has been linked to the edge plasmons of the charged sheet and can be viewed as the 2D analog of surface plasmons in three-dimensional (3D) systems [117]. Specifically, edge magneto plasmons can propagate along the physical boundary of the 2D conductors [118, 119]. Edge magneto plasmons constitute a spectacular manifestation of the dynamical Hall effect. Edge magneto plasmons are chiral. Their chirality is a direct implication of the applied Lorentz force [120]. Graphene reveals rich plasmonic phenomena in the presence of magnetic fields [121–125].

Edge plasmon polaritons: one-dimensional plasmonic modes propagating along the physical boundaries of two-dimensional materials (Figure 3) is called edge plasmon polaritons. They reveal an approximately 10% shorter wavelength compared to the interior of the plasmonic medium [128]. Qualitatively, the shorter wavelength can be ascribed to the effective reduction of the Drude weight since free carriers exist only on one side of the physical boundary. *Dyakonov hyperbolic phonon polaritons* are a lattice analog of edge plasmon polaritons. Berini reported on an in-depth numerical analysis of edge and corner plasmon polariton modes in thin conducting slabs [129]. *Whispering-galley polaritons* is a special example of an edge polaritons that loops around the ridge of polartonic medium [130] or along the circumference of nanoholes [131, 132].

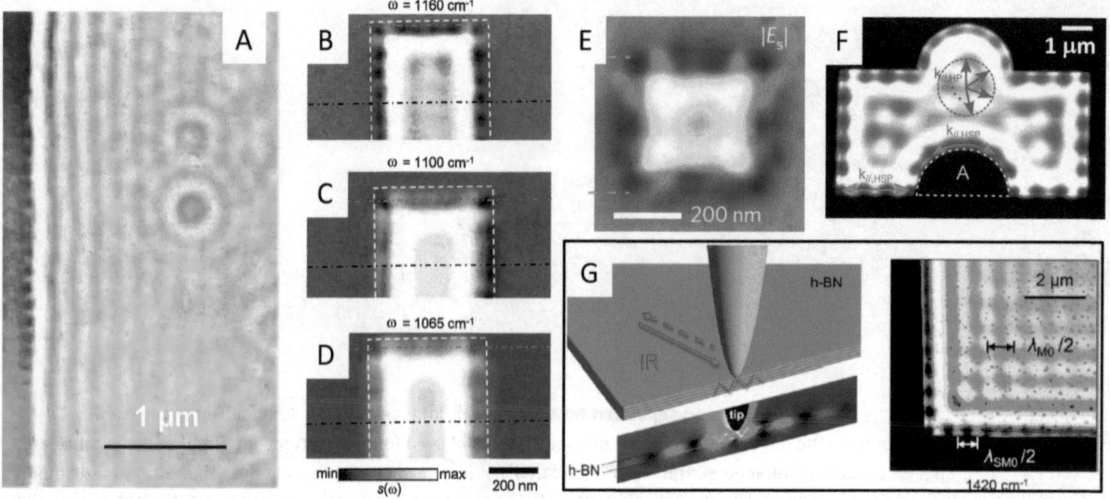

Figure 3: Interior and edge polaritons in van der Waals quantum materials.
Panel A: charge transfer plasmon polaritons at the interface of graphene and a-RuCl3 visualized by means of nanoinfrared methods ($\omega = 898\ cm^{-1}$, $T = 60$ K). Three types of plasmonic fringes are observed: (i) edge plasmon polaritons (dark spots at along the physical boundary of graphene crystal), (ii) interior plasmon polaritons (oscillating wave pattern emanating from the boundary of graphene on the left), and (iii) defect-launched plasmon polaritons forming circular patterns in the interior of the sample. Adapted from a study by Rizzo et al. [68]. Panels B–D: nano-IR imaging of edge plasmons on graphene nanoribbons. White dashed lines mark the boundaries of the crosscut GNR. Adapted from a study by Fei et al. [126]. Panel E: nanoinfrared image of edge plasmons in a square sample of graphene. Adapted from a study by Nikitin et al. [128]. Panels F: nanoinfrared images of edge phonon polaritons in the 25-nm-thick slab of hBN. Adapted from a study by Dai et al. [114]. Panel G: edge and interior phonon polaritons in a 40-nm-thick slab of hBN [127].

EIT in nanoplasmonic structures [133, 134], EIT with plasmon polaritons in graphene [135, 136] and EIT with exciton polaritons in microcavities [137].

EIT polaritons propagate in atomic systems under conditions of EIT. A remarkable aspect of EIT polaritons is that they can be slowed down to 10s of meters per second [176] or even brought to a standstill [177, 178]. EIT polaritons can be dark (decoupled from radiation, more "atom-like") or bright (coupled to radiation, more "photon-like"). The darkness/brightness of the polaritons is controlled by an external laser beam. EIT polaritons can be strongly interacting, if coupled to Rydberg states (see *Rydberg polaritons* below). The EIT phenomenon is also observed in materials and nanostructures. Examples include:

ENZ polaritons: epsilon-near-zero or ENZ polaritons are equivalent to *Berreman polaritons* above.

Exciton polaritons and their condensates. Exciton polaritons are bosonic quasiparticles originating from photons hybridized with hydrogen-like bound electron–hole pairs. Semiconductor microcavities (Box 1 and Figure 4A) offer an outstanding platform for the investigation of exciton polaritons and the attendant strong light–matter coupling, provided a high-quality microcavity is nearly resonant with an excitonic transition. Trapped photons may be emitted and reabsorbed multiple times before being lost to dissipation or cavity leakage. Absorption and re-emission of photons in the cavity give rise to light–matter mixed eigenstates [138]. When sufficiently long-lived, exciton polaritons may form coherent quantum states [139–145]. Bose–Einstein condensates (BECs) of exciton polaritons are appealing quantum liquids in part because their coherent state is created and controlled by light [146–148]. The binding energies of excitons in organic molecules [149], TMDCs, and lead halide perovskites can be as high as 0.75 eV [150–158]; these extraordinary high binding energies underlie the theoretical predictions of condensation and superfluidity at $T = 300$ K [159–161]. BECs of exciton polaritons were predicted to form spatially and temporarily ordered states: time crystals [162]. Exciton polariton condensates may also enable energy-efficient lasers [163].

Exciton polariton X-waves: wavepackets of exciton polaritons that sustain their shape without spreading, even in the linear regime. In a study by Gianfrate et al. [164]. Self-generation of an X-wave out of a Gaussian excitation spot is obtained via a weakly nonlinear asymmetric process with respect to two directions of the nonparabolic polariton dispersion. Notably, X-waves were found to propage with supluminal peak speed with respect to the group velocity of the polaritonic system.

Fermi edge exciton polaritons [165, 166] are observed in microcavities where the active semiconductor is

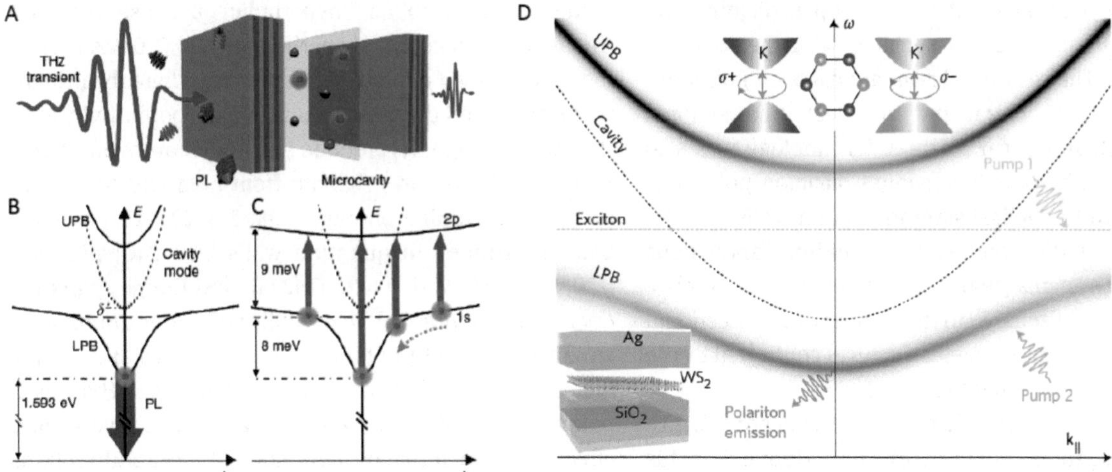

Figure 4: Cavity exciton polaritons.
Panel A: polaritons (pink spheres with blue halo) emerge from strong coupling between the excitonic resonance in a quantum well (transparent sheet) and the photonic mode of a GaAs/AlGaAs microcavity. THz probing (blue curve) maps out the matter component of the polaritons, while photoluminesce (PL, red arrows) leaking through a Bragg mirror reveals the photonic component. Panel B: normal-mode splitting. The heavy hole 1*s* exciton resonance (dashed curve) and the photonic mode (dotted curve) are replaced by the upper polariton branch and lower polariton branch (UPB and LPB, respectively; solid curves). PL (thick red arrow) originates from the radiative decay of polaritons at small in-plane momenta $k_{||}$. Panel C: THz absorption probes hydrogen-like intraexcitonic transitions. While the 1*s* state is spectrally shifted by strong light–matter coupling, the optically dark 2*p* exciton is not affected by the cavity. The resulting momentum dependence of the THz transition energy allows us to map out the momentum distribution of the polaritons as they relax toward $k_{||} = 0$ (green dotted arrow). From a study by Ménard et al. [174]. Panel D: schematic of the valley exciton polariton phenomena. The solid (gray) curves indicate LPB and UPB. The bare cavity and the exciton dispersion are shown by the black and orange dashed curves, respectively. Pump 1 is used to excite directly the exciton reservoir, whereas pump 2 excites the lower polariton branch at specific $k_{||}$ and ω. The emission is collected at smaller angles. The top inset shows the valley polarization phenomena in 2D transition metal dichalcogenide (TMDC) semiconductors caused by the broken inversion symmetry. In these materials, the *K* and *K′* points correspond to the band edges separated in momentum space but energetically degenerate. The bottom inset is a schematic of the microcavity structure with silver and a SiO_2 cavity layer embedded with prototypical TMDC materials WS_2. From a study by Sun et al. [175].

heavily doped to form the Fermi edge. Fermi edge exciton polaritons are formed of electron hole pair excitations involving electron and hole states with in-plane wave vectors around the Fermi edge: $k_{IIe} = k_{IIh} \sim k_F$, where k_F is the Fermi wavevector. In some literature, this latter form of polaritonic states are referred to Mahan exciton polaritons [167], recognizing a prediction of excitonic bound states in doped semiconductors beyond the critical density of the insulator to metal transition states by Mahan [168, 169]. See also *Quantum Hall polaritons* below.

Floquet polaritons. The concept of Floquet engineering refers to the control of a system using a time periodic optical field and is being broadly applied in atomic physics, as well as in the field of QMs [170]. The notion of Floquet polaritons pertains to polaritons in a system of Floquet-engineered atomic states [171] or electronic states in solids [172]. The concept of Floquet engineering by time period optical fields has been extended to coherent phonons in QMs [173]. *Chiral Floquet polaritons* are predicted [44] to form in chiral cavities, in which fundamental matter symmetries are broken (Box 2).

Frenkel exciton polaritons. The matter constituent of these polaritons originates from Frenkel excitons characterized by the Bohr radii of the same order as the size of the unit cell. Frenkel exciton polaritons are common in organic semiconductors [179]. The high exciton binding energy (~eV) and large oscillator strength may lead to room temperature exciton polariton condensates [180–182].

Fuchs–Kliewer interface polaritons: phonon polaritons occurring at surfaces and interfaces [183] with the matter part are originating from Fuchs–Kliewer surface phonons [184]. Huber et al. [185] employed nanoinfrared methods to visualize propagating Fuchs–Kliewer surface phonon polaritons in SiC. Surface phonon polaritons are observed in insulating and semiconducting materials including hBN [97, 98], SiC [186–189], GaAs [190], and many others.

Helical plasmon polaritons: were predicted to form in topologically nontrivial Weyl semimetals [191]. Plasmon polariton dispersion may enable the detection of a chiral anomaly: a charge imbalance between the Weyl nodes in the presence of electric and magnetic fields [192]. The Fermi surface of Weyl semimental features open disjoint segments – the Fermi arcs – associated with the topolical surface states. The resulting *Fermi arc plasmon polaritons* are predicted to be chiral and to reveal unidirectional propagation [193]. Helical plasmon terminology was also applied to describe one-dimensional plasmon polaritons associated with the helical state in domain walls of topologically nontrivial conductors including anomalous quantum Hall systems [194]. Helicity dependence of plasmon polaritons is discussed in the context of unidirectional propagation in plasmonic metastructures controlled by the circular polarization of light [195, 196].

Hopfield polaritons: a bold theoretical concept of light–matter hybridization proposed by John Hopfield in his doctoral thesis back in 1958 (in a study by Hopfield [197]). Hopfield also coauthored the first experimental paper on polaritons devoted to the study of phonon polariton dispersion in GaP by means of Raman scattering [198]. Other early contributions to the theory of polaritons (short of introducing this term) were made by Fano [199], Huang [200], and Tolpygo [201].

Hybrid polaritons. Different types of polaritons hosted by the same material are prone to hybridization [202]. For example, *intersubband polaritons* and *phonon polaritons* hybridize in semiconductor quantum wells [203–205]. Hybridization can also occur in multilayered structures. In all-dielectric layered structures, phonon polaritons associated with the neighboring layers couple to form hybrid modes [87, 206, 207]. Semiconductor heterostructures [208, 209] and especially van der Waals heterostructures offer a fertile platform for the implementation of hybrid polaritons [210, 211]. One such example (Figure 5B and C) is graphene surrounded by insulating layers of hexagonal boron nitride hBN or silicon dioxide. Plasmons associated with graphene layers hybridize with phonon polaritons in proximal SiO_2 or hBN layers to form *plasmon–phonon polaritons* [63, 64, 212, 213]. Hybrid polaritons at the interface of graphene with high-T_c superconductors were proposed as a tool to probe Anderson–Higgs electrodynamics [214]. Hybrid polariton at the interface of graphene with a charge density wave materials were theoretically proposed to "melt" the density wave order [215]. Hybrid modes produced by plasmons in graphene and molecular vibrations of absorbates on the graphene surface may enable high-selectivity sensing mechanisms [216, 217]. A special case of hybrid modes is *hybrid longitudinal–transverse phonon polaritons* [218]. Polaritonic heterostructures with phase change materials enable persistent switching of polaritonic response under thermal and optical stimuli [219].

Hyperbolic polaritons. Anisotropic media are predicted to support an interesting class of polaritonic light–matter modes referred to as "hyperbolic" because their isofrequency surface is a hyperboloid [213, 220–227]. These modes exist over a range of frequencies where the in-plane permittivity and the out-of-plane (c-axis) permittivity are of the opposite sign. Hyperbolic electrodynamics and hyperbolic polaritons can originate from a variety of physical processes including phonons [219, 223, 228–237] intersubband transitions in quantum wells [238–240] plasmons [220, 226, 241–244], excitons [245], and Cooper pairs (see *Cooper pair polaritons*). Hyperbolic polaritons dramatically enhance the local photonic density of states and are predicted to give rise to strong nonlinearities [246]. Hyperbolic polaritons enable canalization imaging [247] with image effectively transferred by high-momentum subdiffractional polaritonic rays from back to front surface of the polaritonic medium [248–251].

Image polaritons: virtual polariton modes produced by image charges at the interface of a polaritonic medium and a metal are called image polaritons. Lee et al. [252] have experimentally demonstrated low loss response of image polaritons at the interface of hBN separated with a thin spacer from a metallic substrate (Figure 5D).

Interband polaritons. The matter constituent of these polaritons originates from contributions of the optical response associated with transitions across the energy gap in the electronic spectrum of a material. These include transitions across the energy gap in semiconductors [253] and superconductors or transitions involving minibands/flat bands in moire superlattices of van der Waals materials [254–257] (see also *Moire polaritons*). The frequency dependence of $\sigma_2(\omega)x\omega$, where $\sigma_2(\omega)$ is the imaginary part of the complex conductivity, is informative for the analysis of interband polaritons [255, 258]. Spectra of $\sigma_2(\omega)x\omega$ reveal a series of steps separated by plateaus, with each step uncovering the energy scale associated with separate interband contributions. In the limit of $\omega \to 0$, the product $\sigma_2(\omega)x\omega$ quantifies the spectral weight of intraband processes to the plasmon polaritons. Interband effects play a central role in theoretical proposals for the implementation of population inversion [259], gain and *superluminal plasmon polaritons* [260].

Intersubband polaritons. Dini et al. [261, 262] reported the first experimental observation of the vacuum-field Rabi splitting of an intersubband transition inside a planar microcavity hosting two-dimensional electron gas. Nonlinearities associated with intesubband transitions in semiconductors can be dramatically enhanced by in hybrid structure with plasmonic metasurfaces [263] (see also *hybrid polaritons*).

Josephson plasmon polariton: an inherent attribute of strongly anisotropic layered superconductors is the

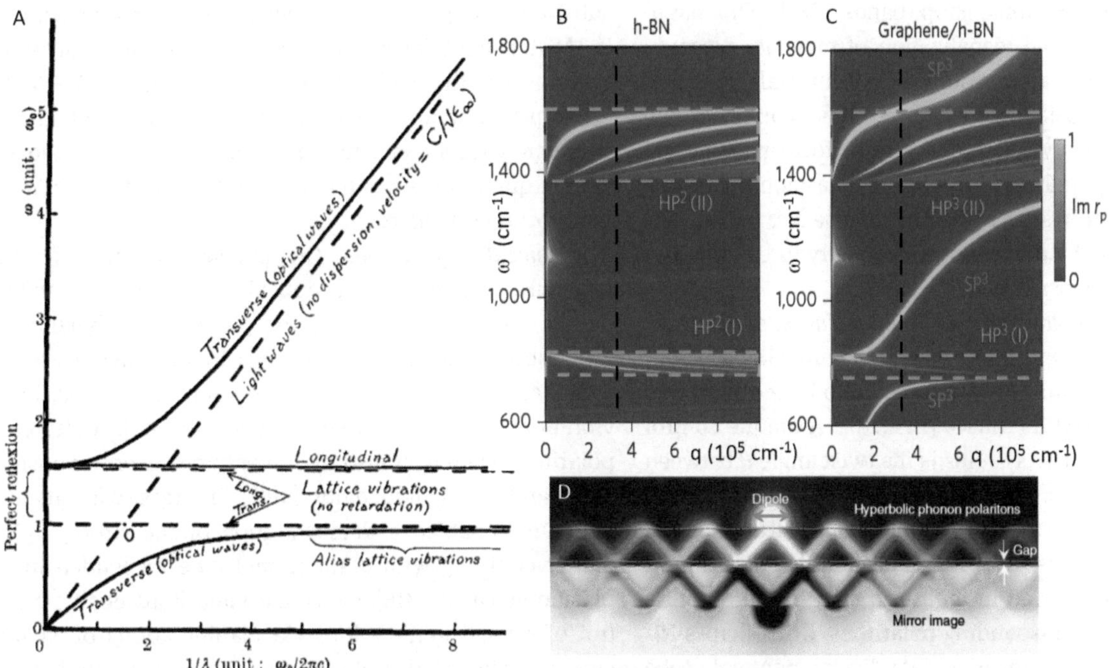

Figure 5: Phonon polaritons, hybrid plasmon–phonon polaritons, and image polaritons.
Panel A: dispersion of phonon polaritons in ionic crystals predicted by Huang (in a study by Sun et al. [175]). In the original publication, Huang did not use the term polariton. Panel B: calculated dispersion of the hyperbolic phonon polaritons in hBN (HP2). Panel C: calculated dispersion of the hyperbolic phonon polaritons in h-BN coupled to plasmon polaritons in the graphene layer and forming hyperbolic plasmon–phonon polaritons (HP3) and surface plasmon–phonon polaritons (SP3). Adapted from a study by Bezares et al. [212]. Panel D: concept of image polaritons at the interface of hBN and a metal. From a study by Yoo et al. [36].

Josephson plasmon polariton. The matter constituent of Josephson plasmon polaritons originates from interlayer Josephson plasmon in layered superconducting materials such as cuprates [79, 264]. Josephson plasmons are the electromagnetic signature of three-dimensional superconductivity in highly anisotropic layered high-T_c superconductors [265]. Josephson plasma waves can be parametrically amplified under illumination with pulsed THz fields [266], paving the way for active Josephson polaritonics.

Kane polaritons: surface plasmon polaritons formed with Kane quasiparticles is the Kane polaritons. Kane polaritons were recently observed in pump–probe experiments on narrow gap II–VI semiconductors [267].

Landau polaritons. The matter component of Landau polaritons originates from cyclotron resonances and transitions between quantized Landau levels relevant in low-dimensional electron gases subjected to high magnetic fields [268, 269]. See also *magneto plasmon polariton*.

Luttinger liquid polaritons: plasmon polaritons in one-dimensional conductors recently revealed by infrared nanoimaging of single-wall and multiple-wall carbon nanotubes [270]. Interacting electrons confined in one dimension are generally described by the Luttinger liquid formalism [271, 272]. Anomalous dependence of the plasmonic quality factor on gate voltage was interpreted in terms of plasmon–plasmon interaction in carbon nanotubes [273].

Magnon polaritons. The matter constituent of these polaritons originate from antiferromagnetic [274, 275] and ferromagnetic resonances [15]. In weak magnetic fields, *surface magnon polaritons* are predicted to acquire nonreciprocal properties. Macedo and Camley [276] analyzed the propagation of surface magnon polaritons in anisotropic antiferromagnets. Sloan et al. [277] predicted that surface magnon polaritons will strongly enhance the spin relaxation of quantum emitters in the proximity of antiferromagnetic materials such as MnF_2 or FeF_2. Kruk et al. [278] developed artificial structures with hyperbolic magnetic response with principal components of the magnetic permeability tensor having the opposite signs. Magnetic materials also support *hybrid polaritons*, including hybrid *magnon–phonon polaritons* recently observed in $ErFeO_3$/$LiNbO_3$ multilayers [279].

Magneto plasmon polaritons: coupled modes of magneto plasmons and THz/infrared photons [280, 281]. Theoretically predicted unconventional properties of magneto polaritons in Weyl semimetals include hyperbolic dispersion and photonic stop bands [282]. The nano-infrared imaging and visualization of magneto plasmon polaritons remains an unresolved experimental challenge. Once technical obstacles are circumvented, it may become possible to directly explore both the focusing and the nonreciprocity predicted for magneto plasmon polaritons [283]. Plasmonic system driven by intense a.c. field is predicted to reveal spontaneous symmetry breaking and nonlinear magnetism [284].

Microcavity polaritons: see cavity polaritons.

Moire polaritons. Atomic layers comprising van der Waals materials can be reassembled into heterostructures with nearly perfect interfaces [285–287]. A unique control knob specific to vdW systems is the twist angle θ between the adjacent layers. Varying θ forms moiré superlattices that can radically modify the electronic structure and attendant properties [288–302]. Plasmons, phonons, and excitons are all altered in moire superlattices prompting changes of the corresponding polaritons. G/hBN[cross-Ni-Moire]. Infrared nanoimaging data display rich real space patterns of polaritonis with selected examples of moire polaritons displayed in Figure 6. Morie design principle can be applied to epitaxially grown thin films on dielectric substrates [303]. Recent experiments on interlayer excitons in TMDC heterobilayers have revealed the trapping of these excitons on the moire potential landscape [304–307]. When placed in an optical cavity, such moire trapped excitons may form an exciton polariton lattice and serve as analog quantum simulators (QSs) (see *polaritonic lattices and quantum simulators*).

Molecular polaritons. Organic semiconductors and molecules embedded in optical (nano)cavities under strong and ultrastrong coupling promote the dynamical formation of molecular polaritons: hybrid energy eigenstates composed of entangled photonic, electronic, and vibrational degrees of freedom [34, 312, 313]. Molecular polaritons were demonstrated to enhance energy transfer [314] and DC conductivity [315]. Progress with nanostructures enabled a demonstration of the strong–light matter coupling with a single molecule embedded in a plasmonic cavity [316]. Molecular molaritons enable control of optical nonlinearities via manipulations of cavity characteristics [317]. Molecular polaritons can form *hybrid polaritons* by coupling to surface plasmons [318], for example. We remark that molecular polaritons are commonly referred to as *vibrational polaritons*.

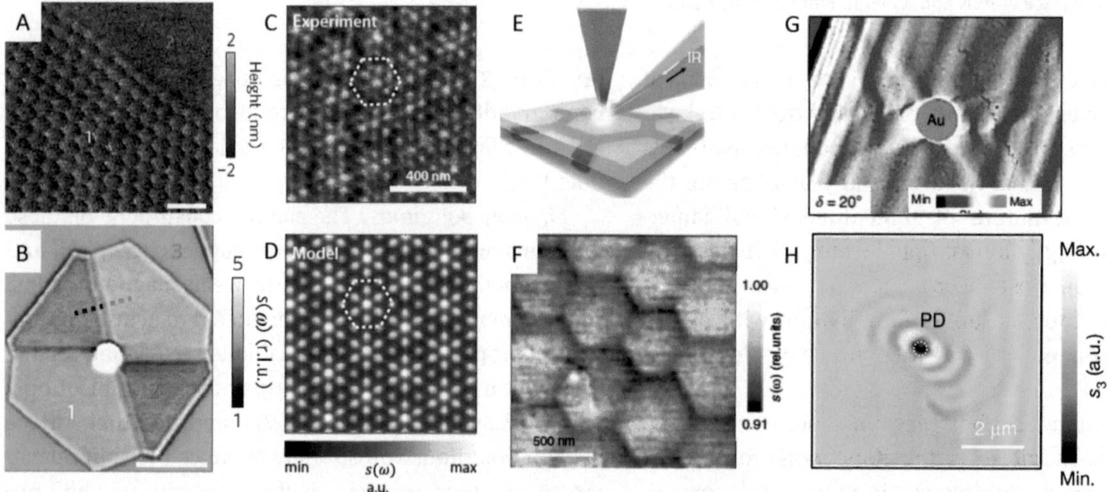

Figure 6: Moire polaritons and topological phonon polaritons in twisted van der Waals materials.
Panel A: atomic force friction image of the graphene/hBN structure at the boundary between the moiré-superlattice and plain graphene (marked in Panel B). Moire reconstruction leads to a periodic pattern with the periodicity of 14 nm. Scale bar 1 μm. Panel B: nanoinfrared image of the graphene/hBN structure. Darker contrast occurs in the moire region. The analysis of plasmon polariton fringes along the boundary between moire superlattice and plain graphene allows one to reconstruct the gross feature of the altered electronic structure in the moire superlattice region. Adapted from a study by Ni et al. [255]. Panel C: nanoinfrared image of plasmon polaritons interference patterns in a moiré superlattice formed by twisted layers of graphene. The dashed hexagons represent the boundaries of a single unit cell. From a study by Sunku et al. [308]. Panel D: plasmon polariton superposition model, which accounts for the gross features of the image in C. Panel E: schematic of the nano-IR imaging showing an AFM tip illuminated by a focused IR beam. Panel F: nanoinfrared image of moire superlattice pattern in hBN. The contrast is formed by the shift and broadening of the phonon polariton resonance. Adapted from a study by Ni et al. [309]. Panel G: nanoinfrared image of phonon polaritons in a twisted structure of MoO_3 slabs rotated by $\theta = 20°$, revealing complex wavefront geometry. adapted from a study by Chen et al. [310]. Panel H: topological phonon polaritons in twisted MoO_3 slabs rotated by $\theta = 77°$. From a study by Hu et al. [311].

Mott polaritons (QED): nonequilibrium driven states in an array of circuit QED cavities or optical resonators [319, 320] is the Mott polaritons. See also *polaritonic lattices*.

Mott polaritons (QM) were also introduced in context of the resonant coupling between strongly correlated electrons in solid Mott insulators integrated in a single-mode cavity [321].

Phonon polaritons: is a collective excitation comprised (infrared) light coupled with a polar lattice vibration. Like other polaritons, phonon polaritons can be understood in terms of an anticrossing of the dispersion curves of light and matter constituents (Figure 5). Early observations of phonon polaritons (see *Hopfield polaritons*) in bulk crystals and films were made using a variety of spectroscopic methods [322, 323]. More recent work [324] has focused on the generation, detection, and on picosecond polaritons dynamics [325–329]. By matching the phonon polariton velocity in $LiNbO_3$ crystal to the group velocity of the *fs* punp pulse Yeh et al. [330] have been able to generate intense THz fields of the order of 10 μJ energy. Advanced nanoimaging/spectroscopy methods [331–333] were employed for the real-space visualization of phonon polariton standing ways. Phonon polaritons play a major role in nanoscale thermal transport at nanoscale and mesoscale [86, 334–339]. Phonon polaritons in the anisotropic oxide material MoO_3 reveal both elliptical and hyperbolic dispersions [339–341]. The dispersion and propagation of phonon polaritons can be controlled by nanostructuring [342] and twist-angle (moire) engineering (Figure 6). The recent discovery of parametric phonon amplification in SiC paves the way for the exploration of nonlinear and active phonon polariton phenomena [343]. *Surface phonon polaritons* (see also *Fuchs–Kliewer* interface polaritons) reveal a dispersion branch located between longitudinal and transverse vibrational modes (see *hybrid polaritons*). Dai et al. [344] detected surface phonon polaritons in monolayers of hBN.

Plasmon polaritons: probably the most thoroughly studied class of polaritons. A surface plasmon polariton is a transverse magnetic (TM)-polarized optical surface wave that, for example, propagates along a flat metal–dielectric interface, typically at visible or infrared wavelengths [345–347]. Plasmon polaritons have rich implications for technology [348–351]. Nonlinear [352–354] and quantum [355–359] properties of plasmonic structures are in the vanguard of current research. Plasmon polaritons can be controlled at femto-second timescales [67, 267, 360–363] enabling access to novel physics and applications [364, 365]. Plasmonic waveguides have been incorporated with light-emitting materials, paving the way for integrated plasmonic and photonic structures [366]. Plasmon polaritons have been harnessed to implement high-quality factors such as whispering gallery microcavities [367]. In parallel, many research groups are searching for new plasmonic media with the properties optimized for different classes of plasmonic effects [368–371]. Van der Waals materials, and especially graphene, are emerging as outstanding plasmonic media in light of their inherent tunability with different stimuli (see *Dirac plasmons*). *Acoustic plasmon polaritons* are a special example of *hybrid polaritons* whose frequency-momentum $\omega(q)$ dispersion is predicted to be linear [372–376]. Acoustic plasmon plaritons have been demonstrated [377–379] in structures, where graphene resides in close proximity to metallic surfaces. *Spoof surface plasmons polaritons* were introduced describe plasmon polaritons on the surface of artificial metallic structures and metamaterials [380]. *Airy surface plasmon polaritons* are the surface counterparts of nondiffracting airy waves [381] and have been demonstrated by direct nanoimaging [382]. *Chiral plasmon polaritons* [383] were predicted to occur in twisted bilayer graphene [384] (Figure 7).

Plexcitons are a specific example of hybrid polaritons. The matter constituent of plexcitons originates from plasmon exciton coupled modes [386–392] Historically, plexciton studies have focused primarily on localized states [387, 393]. Propagating plexciton states also exist and offer potential for compact quantum information carriers as well as opportunities for mediating emitter–emitter coupling [394–396]. Composite structures and multilayers can feature plexcitons. An interesting recent example of plexciton study has been conducted in the setting of scanning probe nano-optical imaging and spectroscopy (Figure 8). This work by May et al. [398], along with a study by Groß et al. [397], implemented the scanning optical cavities formed between a nano-optical antenna and the substrate. The authors investigated CdSe/ZnS quantum dots using this scanning cavity approach and observed plexitonic Rabi splitting of 163 meV.

Polaritons parametric amplification, gain, and lasing have been demonstrated for exciton polaritons in microcavities [52, 399–401]. Resonant coupling between photons and excitons in microcavities can efficiently generate significant single-pass optical gains [399]. Polaritonic lasing has been implemented and analyzed in different material systems hosting *plasmon polartions* and *exciton polaritons* [402]. Amplification of *demons* [107] has been predicted as well but is yet to be experimentally demonstrated.

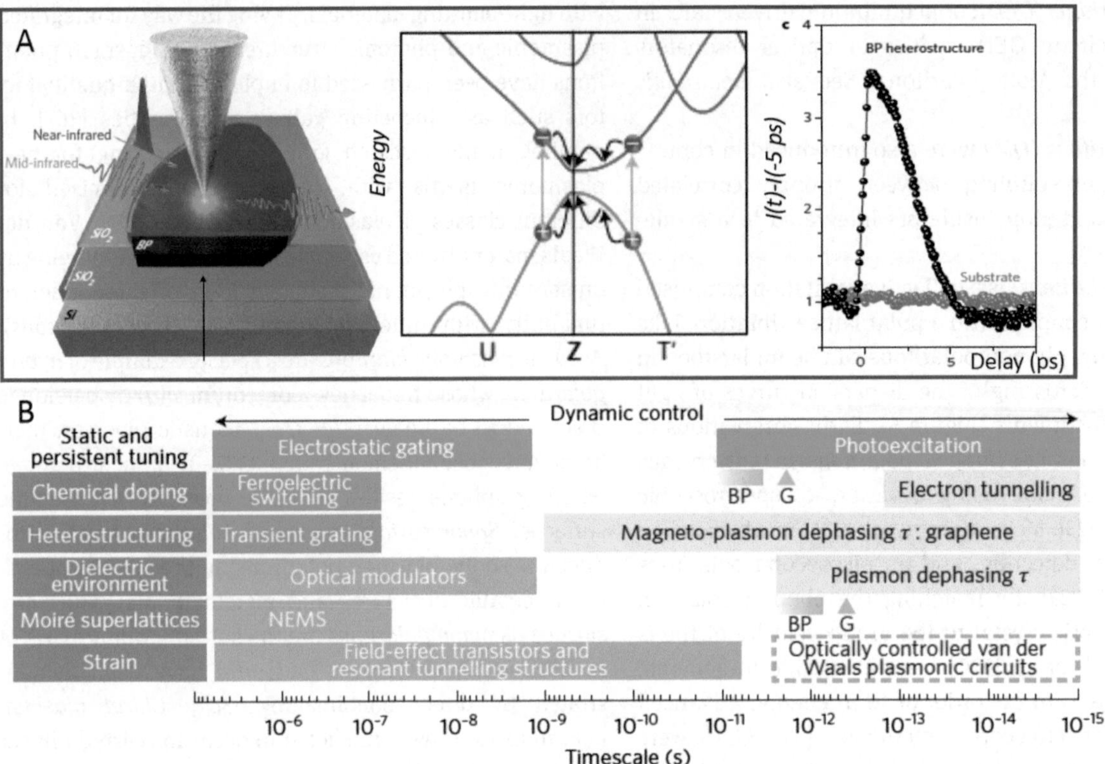

Figure 7: Ultrafast plasmonic effects in van der Waals materials.
Panel A: nanoinfrared spectroscopy and imaging of switchable plasmon polaritons in black phosphorous (bP) semiconductor. Left: experimental schematics. Middle: band structure of bP. Orange arrows indicate electron–hole pairs excited by a near-infrared pulse centered at a wavelength of 1560 nm. The curved black arrows indicate carrier cooling toward the band extrema. Right: Ultrafast pump–probe dynamics of the scattered near-field intensity normalized to the signal at the negative delay time (equilibrium). The SiO_2 substrate (blue points) shows no dynamics, whereas the SiO_2/bP/SiO_2 heterostructure (black points) features a strong pump–probe signal. Adapted from a study by Eisele et al. [362]. Panel B: methods for controlling plasmons in van der Waals materials and the corresponding timescales. Static and persistent tuning methods are displayed in the blue boxes; dynamical control methods are displayed in the orange ones. The yellow boxes show the dephasing times (τ) of plasmons and magneto plasmons in van der Waals materials along with characteristic timescales of electron tunneling in these systems. The green boxes represent timescales pertinent for various photonics technologies. The box with the dashed green outline indicates the desired timescales for future ultrafast plasmonic circuits. NEMS, nanoelectromechanical systems; G, graphene. Adapted from a study by Basov et al. [385].

Polaritonic chemistry: an emerging field focused on modifying pathways of chemical reactions in molecular systems coupled to photonic cavities [403–407].

Polaritonic circuits, devices, arrays, and systems. Both light and matter constituents of polaritons are amenable to controls with external stimuli [408]. The use of *exciton polaritons* as building blocks for future information processing such as spin switches [409], spin memory [410], transistors [411], logic gates [412], resonant tunneling diodes [413], routers [414], and lasers [415] has recently been demonstrated. The first polaritonic systems are also emerging and include QSs and networks for neuromorphic computers [416]. TMDC material WSe_2 integrated into microcavity devices acts as efficient light emitting device [417].

Polaritonic lattices, and QSs. A variety of experimental approaches have been utilized to implement one- and two-dimensional arrays of interacting polaritons. In the field of *microcavity exciton polaritons* gate arrays, spatially dependent optical potential as well as surface acoustic waves [418], have been utilized to generate arrays/lattices [419]. One-dimensional *exciton polariton* superlattices reveal weak lasing assigned to a novel type of a phase transition in this interacting system [420]. Arrays of evanescently coupled cavities hosting neutral atoms [421] have been proposed as QSs, where the photon blockade provided by the atom limits the occupancy of each cavity to one, allowing for the implementation of the Bose–Hubbard model. QSs require controllable quantum systems that efficiently simulate a Hamiltonian of interest, which may encode phases with a significant degree of entanglement and is not amenable to calculations by classical computer [422–427]. Lattices of *exciton polaritons* [422, 428–432] have emerged as a promising platform for QS, along with ultracold atoms [425, 433], trapped ions [434–436], and superconducting circuits

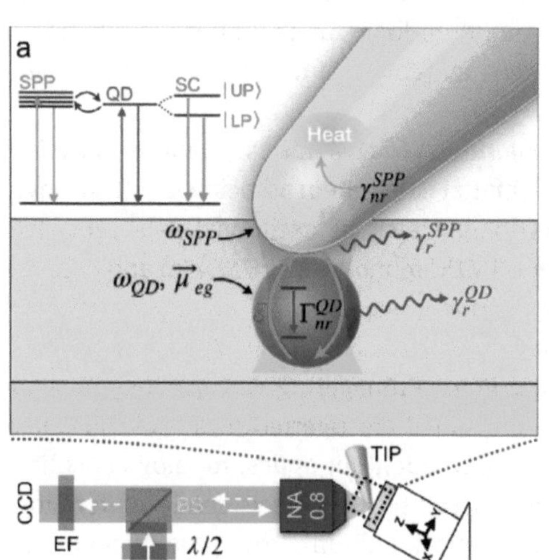

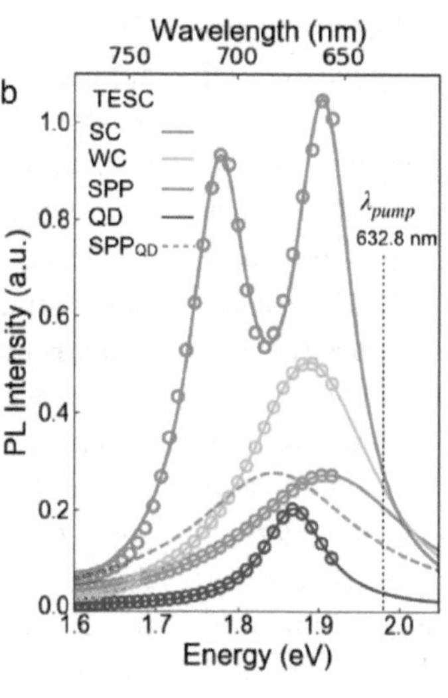

Figure 8: Tip-enhanced spectroscopy of plexcitons.
Panel A: the strongly confined $|E_z|$ field in a plasmonic nanogap cavity surrounding a single isolated CdSe/ZnS quantum dot (QD) and a tilted Au tip induce coupling between the plasmon and exciton. Panels B: Measured PL spectra for the QD, cavity plasmon polariton, weakly coupled system (WC) and strongly coupled states (SC) with coupling strength $g = 141$ meV. A Lorentzian lineshape representing the redshifted plasmon resonance in the presence of the QD is calculated from the fitted values (SPP_{QD}) [398].

[437, 438]. Moire superlattices of *plasmon polaritons* (Figure 6) present yet another example of polaritonic lattices. Moire superlattices were realized in graphene devices with nanostructured gate electrodes [439], as well as in moire superlattices of twisted graphene layers [308].

Polaritonic interference, refraction, collimation, front shaping, and waveguiding. All these common wave phenomena are relevant to polaritons (Figure 9). In van der Waals materials, domain wall boundaries can act a polaritonc reflectors [440–442], or conductors [443]. Zia and Brongersma [444] demonstrated Young's double-slit experiments with surface *plasmon polaritonss*. Beyond analogs of geometrical optics effects, polaritons offer at least two novel routes for image formation. First, *hyperbolic polaritons* enable canalization imaging [247], with images effectively transferred by high-momentum subdiffractional polaritonic rays from the back to the front surface of the polaritonic medium [248, 250, 311, 445] (Figure 9C). Second, polaritons are amenable to guiding and steering using methods of *transformation optics*. Polaritonic waveguides have been implemented over a broad range of frequencies from THz [446] and infrared regions to visible light. Peier et al. observed phonon–polariton tunneling across the airgap [447]. Advanced polaritonic launchers and metalenses (Figure 9D) are well suited for defining the trajectories of polaritonic surface "beams" [448, 449]. In highly nonlinear regime polaritons are predicted to display self-focusing effects and to form solitons [450].

Polariton–polariton interactions. The interaction of polaritons stems from their underlying matter constituents. In close analogy with other interacting systems, polariton–polariton interactions renormalize the dispersion and also prompt a blue shift of the emission energy as the polariton density increases [451, 452]. Polariton–polariton interaction

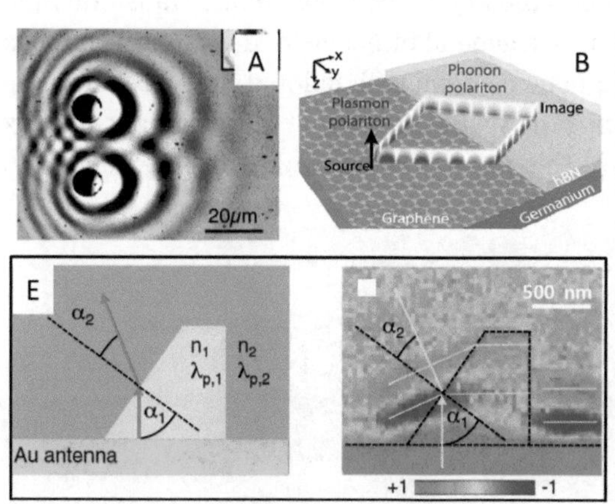

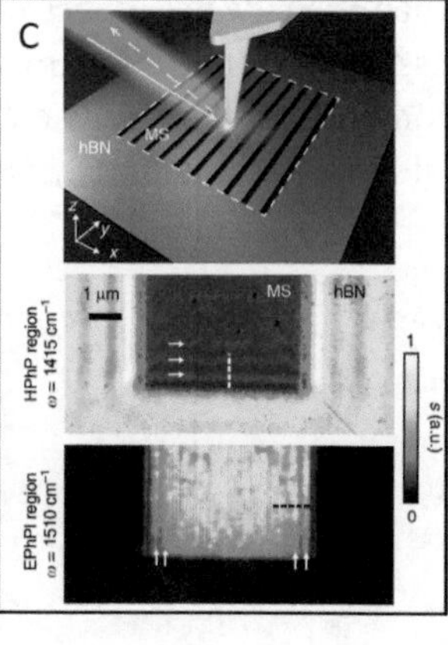

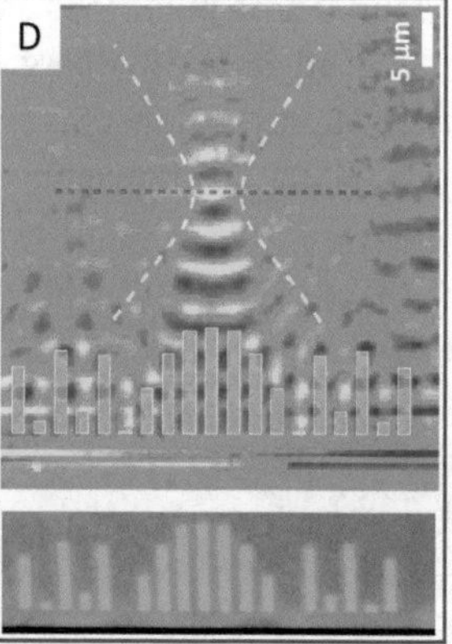

Figure 9: Infrared nanoimaging of polaritonic waves.
Panel A: nano-IR image of the interference pattern of surface *phonon polaritons* on a SiC launched by circular Au discs [331]. Panel B: prediction of in-plane negative refraction between *plasmon polaritons* in graphene and *phonon polaritons* in an hBN slab [455]. Panel C: nano-IR imaging of polariton evolution and canalization in an hBN metasurface [248]. Panel D: optical images of the laser-written metalense (bottom). Nano-IR image of revealing focusing of *phonon polaritons* at 1452 cm^{-1} [456]. Panel E: refraction of graphene *plasmon polaritons* at the prism formed by bilayer graphene [457].

effects have been recently demonstrated for *microcavity exciton polaritons* [453, 454]. See also *quantum Hall polaritons.*

Polaron polaritons. In TMDC monolayers, the itinerant electrons dynamically screen exciton to form new quasiparticle branches – the attractive and repulsive polaron – each with a renormalized mass and energy [458, 459]. *Microcavity polaritons* with the matter constituent linked to these polaron branches are referred to as polaron polaritons [458].

Quantum Hall polaritons are a product of coupling cavity photons to the cyclotron resonance excitations of electron liquids in high-mobility semiconductor quantum wells or graphene sheets [460, 461]. The edge channels of the quantum Hall effect offer a platform for probing interference and entanglement effects in the setting of a condensed matter system since the edge states propagation is ballistic, one-dimensional, and chiral. This platform enables experimental implementation of electron quantum optics [462–465] and may be suitable for the realization of flying qubits. In a parallel development, Smolka et al. [466]. investigated cavity *exciton polaritons* in the presence of high-mobility 2D electron gas subjected to external magnetic field and discovered novel correlated electron phases. Knuppel et al. [467] reported on strong *polariton–polariton interactions* in the fractional quantum Hall regime.

Rydberg polaritons (QED): photons dressed by highly excited atomic Rydberg states under conditions of electromagnetic induced transparency. These polaritons can either reside in a cavity or propagate throughout an atomic ensemble. In a cavity, Rydberg dressing bestows an atomic ensemble with the character of a two-level system: the excitation of a single Rydberg polariton prevents the creation of a second one, in the so-called "Rydberg blockade" regime. Under conditions of *electromagnetic induced transparency*, polaritons can propagate within an optically dense atomic cloud. These polaritons can then be made to interact with each other via Rydberg dressing: the first Rydberg polariton alters the transparency condition for the second one, preventing its propagation within a certain "blockade radius" [470–473]. Rydberg polaritons are appealing for quantum logic functionalities [474] and for realizing synthetic materials via many-body states of light [140, 171].

Rydberg polaritons (QM): a special example of *exciton polaritons* with matter constituent associated with strongly interacting Rydberg states of excitons [137]. Candidate systems include TMDC monolayers [475, 476] and cuprous oxide, where Rydberg states with principal quantum numbers of up to $n = 25$ are feasible [477].

Soliton polaritons. Propagating wavepackets in semiconductor micorcavities are referred to as soliton polaritons (Figure 10C). In quantum optics, topological soliton polaritons refer to composite objects made of fermions trapped in an optical soliton. The prototypical one-dimensional (1D) model of solitons posessing nontrivial topology is the model of Su–Schrieffer–Heeger (SSH) chains [478]. Variants of the SSH Hamiltonian have been emulated in the 1D lattices of *microcavity exciton polaritons* [479] and also in the system of quantum emitters coupled to a photonic waveguide [480]. Topological phases of polritons in cavity waveguides were analyzed in a study by Downing et al. [481].

Spin polaritons: this term was coined in the context of polariton microcavity diode lasers operating via injection of spin polarized currents [482].

Spin plasmon polaritons are relevant to the plasmonic response of spin-polarized electron gas [483]. Alternatively, spin–orbit interaction may lift the degeneracy between the spin states and give rise to transitions responsible for peculiar dispersion features of spin plasmon polaritons [484]. The surface plasmon of a helical electron liquid is predicted to carry spin and is also referred to as a spin plasmon polariton [485].

Transformation optics with polaritons. Transformation optics refers to a general principle for designing a complex electromagnetic medium with tailored properties by carefully crafting the spatial patterns of the local optical index [486, 487]. This general principle has been extended to

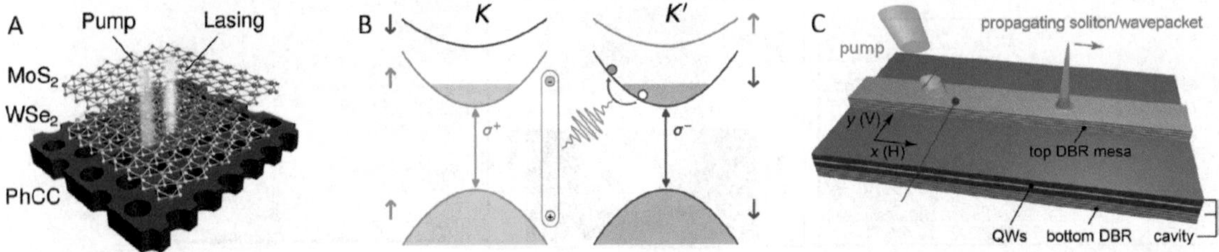

Figure 10: Panel A: schematic of MoS_2/WSe_2 heterobilayer nanolaser integrated in photonic crystal cavity [402]. Panel B: polaron–polaritons in TMDC semiconductors. Schematic to illustrate the conduction and valence band structure and optical selection rules of monolayer $MoSe_2$ close to the K and K' points. An exciton in the K valley interacts with conduction band electron–hole pairs in the Fermi sea of the K' valley to form an intervalley polaron. From a study by Bing Tan et al. [468]. Panel C: experimental setup for the exploration of propagating solitons in the system of microcavity exciton polaritons [469].

polaritons [488] and polartonic cavities [489], and specifically to *plasmon polaritons* in graphene [490]. Losses present the most significant experimental roadblock for practical transformational polaritonics. Recent advances with highly confined but low-loss plasmon polaritons [110] and phonon polaritons [223] fulfill important experimental preconditions for the realization of transformation optics ideas in polaritonic systems.

Tamm surface plasmon polaritons are associated with Tamm states at metallic surfaces [491]. Common surface *plasmon polaritons* are formed with a TM polarization at the boundary of metallic and dielectric surfaces and lie to the right of the light cone. Tamm polaritons are found with both TM and transverse electric polarizations, and their dispersion can be within the light cone [492, 493].

Trion polaritons. The matter constituent of these polaritons is formed by charged excitons or trions (see also *charged polaritons*). Trion polaritons are commonly found in the response of TMDC semiconductors [494, 495] and also in carbon nanotubes [496].

Tunneling plasmon polaritons were predicted [497] and observed [498] in an atomically thick tunable quantum tunneling devices consisting of two layers of graphene separated by 1 nm of h-BN. By applying a bias voltage between the graphene layers, one creates an electron gas coupled to a hole gas. Even though the total charge of the devices is zero, this system supports propagating graphene plasmons.

Valley polaritons. The matter constituent of these polaritons originates from valley polarized excitons in TMDC semiconductors (Figure 4B and *exciton polaritons*). The electronic structure of two-dimensional TMDC semiconductors endows this class of materials with the spin–valley degree of freedom that provides an optically accessible route for the control and manipulation of electron spin [499–501].

Vibrational polaritons: see molecular polaritons.

Wannier or *Wannier–Mott polaritons* borrow their matter part from Wannier excitons in semiconductors [502].

Waveguides and photonic crystals for polaritons. Waveguides and photonic crystals allow one to design and control the properties of photons, and thus of polaritons, both in quantum optics and QMs. In waveguide QED, different type of emitters (neutral atoms, quantum dots, color centers, superconducting qubits) are coupled to a

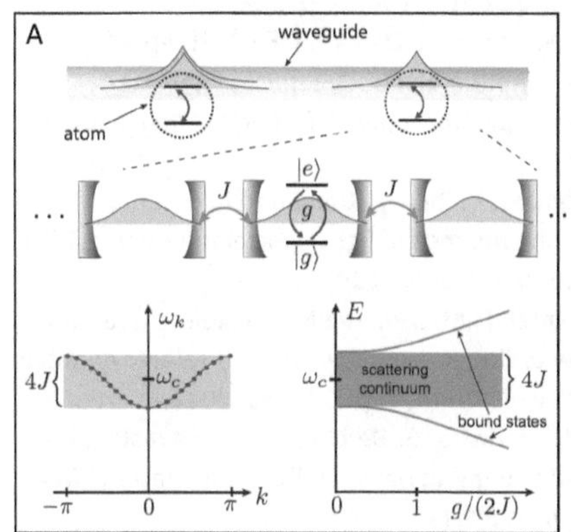

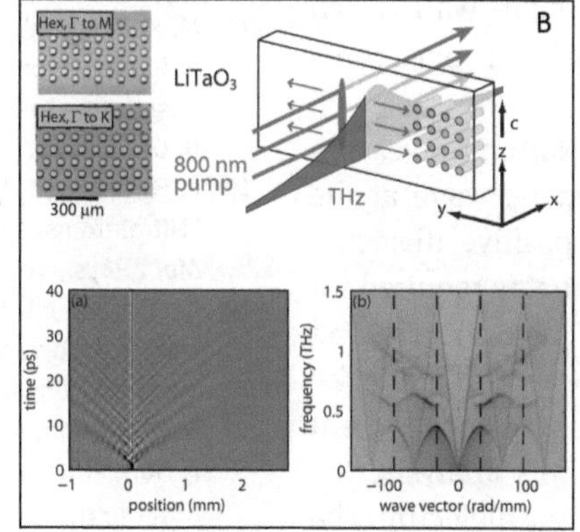

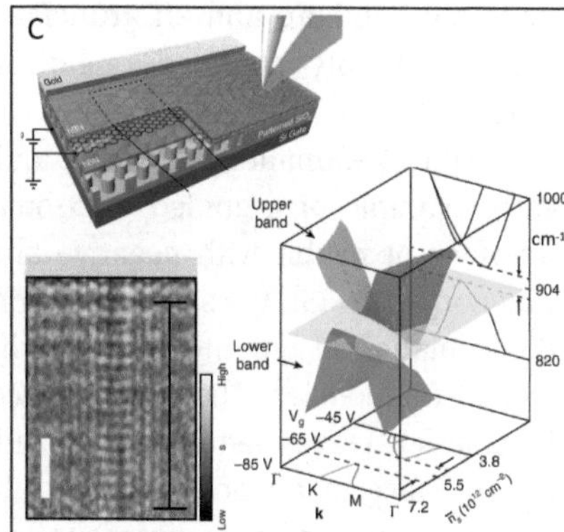

Figure 11: Polariton waveguide QED.
Panel A: emergence of bound atom–photon dressed states in 1D waveguides with finite bandwidth. The slow-light waveguide can be modeled as a large array of coupled optical resonators with nearest-neighbor coupling J. Lower left: band structure of the waveguide without atoms. Lower right: single-photon spectrum as a function of the atom–photon coupling g in the case of a single atom (with $\omega a = \omega c$) coupled to the waveguide, showing the emergence of bound states. Reproduced from a study by Calajo et al. [503]. Panel B: Photonic crystal for phonon polaritons in $LaTaO_3$. Top left: optical microscope images of the photonic crystal patterns. Top right: schematic of pump–probe experiments. Bottom left: space–time plot of THz waves generated directly inside a square photonic crystal. The edges of the image are the edges of the photonic crystal. Bottom right: dispersion diagram obtained from a 2D Fourier transform of the space–time plot in bottom left panel. The region highlighted in yellow represents the light cone. The regions highlighted in orange show the locations of the leaky modes. Adapted from a study by Ofori-Okai et al. [504]. Panel C: tunable and switchable photonic crystal for surface plasmon polaritons in graphene. Top: Schematic of a photonic crystal comprised of a graphene monolayer fully encapsulated by hexagonal boron nitride on top of an array of SiO_2 pillars. Pixelated gate insulator implemented in the form of nanopillars enables the local modulation of the carrier density and therefore of the plasmonic density of states. Bottom left: near-field nano-IR image of plasmonic standing waves for a structure in the top panel. Scale bar 400 nm. Bottom right: calculated plasmonic band structure as a function of wave vector k and average carrier density ns. A vertical cut parallel to the ω–k plane (back panel) generates the plasmonic band structure at fixed carrier density $n_s = 5.5 \times 10^{12}$ cm^2. The dashed lines mark the range of a complete plasmonic bandgap. A horizontal cut parallel to n_s–k plane (bottom panel) generates the plasmonic dispersion as a function of average carrier density ns and wave vector k, at laser frequency $\omega = 904$ cm^{-1}; a complete bandgap is evident for carrier density around $n_s = 5.5 \times 10^{12}$ cm^2.

one-dimensional (1D) optical channels [505], such as fibers [506, 507], photonic crystals [508, 509], and transmission lines [510, 511] (Box 1 and Figure 11A). Channel with a bandgap give rise to atom–photon bound states (i.e., polaritonic bound states), provided the atomic resonance frequency is close to the band edge. Beyond the band-edge, photons are bound to the atoms, forming localized polaritonic cavities that can be harnessed for realizing quantum simulation and quantum information processing (Box 1). If the coupling between photons and atoms is strong enough, bound states emerge even if the atomic resonance frequency lies inside the band (i.e., as a "bound states in the continuum") due to multiple scattering [503, 512]. In the field of QMs, photonic crystal structures were fabricated using common phonon–polariton oxide systems $LiTaO_3$ and $LiNbO_3$ (Figure 11B). Pump–probe experiments in Figure 11B revealed the key attributes of the dispersion control by these periodic structures. A significant deficiency of conventional photonic crystals is that they do not allow for dynamical dispersion engineering. Xiong et al. circumvented this limitation and demonstrated a broadly tunable two-dimensional photonic crystal for *surface plasmon polaritons* [cross-ref-xiond]. Infrared nanoimaging revealed the formation of a photonic bandgap and an artificial domain wall which supports highly confined one-dimensional plasmonic modes.

Zenneck–Sommerfeld waves and Norton waves: an early example of a guided electromagnetic wave at the interface of media with negative and positive dielectric function [513–515], the same condition that is required for the formation of polaritonic modes in THz, infrared, and optical frequencies. The original prediction of Zenneck–Sommerfeld waves pertained to the radiofrequency wave at the interface of air and the Earth. In this analysis, the surface of the Earth was regarded as a lossy dielectric. The concept of Zenneck–Sommerfeld waves and closely related Norton waves has been applied to a broad class of wave patterns on the surface of metallic [516–519] and dielectric materials [520].

Acknowlegements: Research at Columbia is supported as part of Programmable Quantum Materials, an Energy Frontier Research Center funded by the U.S. Department of Energy (DOE), Office of Science, Basic Energy Sciences (BES), under award DE-SC0019443. D.N.B. is Vannevar Bush Faculty Fellow ONR-VB: N00014-19-1-2630 and Moore Investigator in Quantum Materials #9455.

Author contribution: All the authors have accepted responsibility for the entire content of this submitted manuscript and approved submission.

Research funding: Research at Columbia is supported as part of Programmable Quantum Materials, an Energy Frontier Research Center funded by the U.S. Department of Energy (DOE), Office of Science, Basic Energy Sciences (BES), under award DE-SC0019443.

Conflict of interest statement: The authors declare no conflicts of interest regarding this article.

References

[1] D. N. Basov, M. M. Fogler, and F. J. Garcia de Abajo, "Polaritons in van der Waals materials," *Science*, vol. 354, p. 195, 2016.

[2] D. N. Basov, R. D. Averitt and D. Hsieh, "Towards properties on demand in quantum materials," *Nat. Mater.*, vol. 16, p. 1077, 2017.

[3] T. Low, A. Chaves, J. D. Caldwell, et al., "Polaritons in layered two-dimensional materials," *Nat. Mater.*, vol. 16, p. 182, 2017.

[4] D. E. Chang, J. S. Douglas, A. Gonzalez-Tudela, C.-L. Hung, and H. J. Kimble, "Colloquium: Quantum matter built from a nanoscopic lattices of atoms and photons," *Rev. Mod. Phys.*, vol. 90, p. 031002, 2018.

[5] A. A. Clerk, K. W. Lehnert, P. Bertet, J. R. Petta and Y. Nakamura, "Hybrid quantum systems with circuit quantum electrodynamics," *Nat. Phys.*, vol. 16, p. 257, 2020.

[6] I. Carusotto, A. A. Houck, P. Roushan, D. I. Schuster, and J. Simon, "Photonic materials in circuit quantum electrodynamics," *Nat. Phys.*, vol. 16, p. 268, 2020.

[7] M. Ruggenthaler, N. Tancogne-Dejean, J. Flick, H. Appel, and A. Rubio, "From a quantum-electrodynamical light–matter description to novel spectroscopies," *Nat. Rev. Chem.*, vol. 2, p. 0118, 2018.

[8] P. Forn-Díaz, L. Lamata, E. Rico, J. Kono, and E. Solano, "Ultrastrong coupling regimes of light–matter interaction," *Rev. Mod. Phys.*, vol. 91, p. 025005, 2019.

[9] A. Michael, J. L. Sentef, F. Künzel, and M. Eckstein, "Quantum to classical crossover of Floquet engineering in correlated quantum system,s" *Phys. Rev. Res.*, vol. 2, p. 033033, 2020.

[10] A. F. Kockum, A. Miranowicz, S. De Liberato, S. Savasta, and F. Nori, "Ultrastrong coupling between light and matter," *Nat. Rev. Phys.*, vol. 1, p. 19, 2019.

[11] R. Jestädt, M. Ruggenthaler, M. J. T. Oliveira, A. Rubio, and H. Appel "Light–matter interactions within the Ehrenfest–Maxwell–Pauli–Kohn–Sham framework: fundamentals, implementation, and nano-optical applications," *Adv. Phys.*, vol. 68, p. 225, 2020.

[12] Z. M. Raines, A. A. Allocca, M. Hafezi, and V. M. Galitski, "Cavity Higgs polaritons," *Phys. Rev. Res.*, vol. 2, p. 013143, 2020.

[13] Z. Sun, M. M. Fogler, D. N. Basov, and A. J. Millis, "Collective modes and THz near field response of superconductors," *Phys. Rev. Res.*, vol. 2, p. 023413, 2020.

[14] P. B. Littlewood and C. M. Varma, "Gauge-invariant theory of the dynamical interaction of charge density waves and superconductivity" *Phys. Rev. Lett.*, vol. 47, p. 811, 1981.

[15] L. R. Walker, "Magnetostatic modes in ferromagnetic resonance," *Phys. Rev.*, vol. 105, p. 390, 1957.

[16] A. A. Allocca, Z. M. Raines, J. B. Curtis, and V. M. Galitski, "Cavity superconductor-polaritons" *Phys. Rev. B*, vol. 99, p. 020504(R), 2019.

[17] Z. Sun and A. Millis, "Bardasis–Schrieffer polaritons in excitonic insulators," *Phys. Rev. B*, vol. 102, p. 041110, 2020.

[18] H. T. Stinson, A. Sternbach, O. Najera, et al., "Imaging the nanoscale phase separation in vanadium dioxide thin films at terahertz frequencies," *Nat. Commun.*, vol. 9, p. 3604, 2018.

[19] N. G. Basov, "Semiconductor lasers," *Science*, vol. 149, p. 821, 1965.

[20] H. Deng, "Exciton-polariton Bose–Einstein condensation," *Rev. Mod. Phys.*, vol. 82, p. 1489, 2010.

[21] T. Schwartz, J. A. Hutchison, C. Genet, and T. W. Ebbesen, "Reversible switching of ultra-strong coupling," *Phys. Rev. Lett.*, vol. 106, p. 196405, 2011.

[22] S. Kéna-Cohen, S. A. Maier, and D. D. C. Bradley, "Ultrastrongly coupled exciton-polaritons in metal-clad organic semiconductor microcavities," *Adv. Opt. Mater.*, vol. 1, p. 827, 2013.

[23] L. Tian and H. J. Carmichael, "Quantum trajectory simulations of the two-state Behavior of an optical cavity containing one atom," *Phys. Rev. A*, vol. 46, p. R6801, 1992.

[24] A. Imamoğlu, H. Schmidt, G. Woods, and M. Deutsch, "Strongly interacting Photons in a nonlinear cavity," *Phys. Rev. Lett.*, vol. 79, p. 1467, 1997.

[25] J. S. Douglas, H. Habibian, C.-L. Hung, A. V. Gorshkov, H. J. Kimble and D. E. Chang, "Quantum many-body models with cold atoms coupled to photonic crystals," *Nat. Photonics*, vol. 9, p. 326, 2015.

[26] S. Dufferwiel, S. Schwarz, F. Withers, et al., "Exciton-polaritons in van der waals heterostructures embedded in tunable microcavities," *Nat. Commun.*, vol. 6, p. 8579, 2015.

[27] N. Lundt, S. Klembt, E. Cherotchenko, et al., "Room-temperature Tamm-plasmon exciton-polaritons with a WSe_2 monolayer," *Nat. Commun.*, vol. 7, p. 13328, 2016.

[28] X. Liu, T. Galfsky, Z. Sun, et al., "Strong light–matter coupling in two-dimensional atomic crystals," *Nat. Photonics*, vol. 9, p. 30, 2015.

[29] J. Sik, M. Schubert, T. Hofman and V. Gottschalch, Free-Carrier Effects and Optical Phonons in GaNAs/GaAs Superlattice Heterostructures Measured by Infrared Spectroscopic Ellipsometry, vol. 5, Cambridge University Press, 2014.

[30] E. L. Runnerstrom, K. P. Kelley, E. Sachet, C. T. Shelton, and J. P. Maria, "Epsilon-near-zero modes and surface plasmon resonance in fluorine-doped cadmium oxide thin films," *ACS Photonics*, vol. 4, p. 188, 2017.

[31] K. P. Kelley, E. L. Runnerstrom, E. Sachet, et al., "Multiple epsilon-near-zero Resonances in multilayered cadmium oxide: designing metamaterial-like optical Properties in monolithic materials," *ACS Photonics*, vol. 6, p. 1139, 2019.

[32] A. Alù, M. G. Silveirinha, A. Salandrino, and N. Engheta, "Epsilon-near-zero metamaterials and electromagnetic sources: tailoring the radiation phase pattern,"*Phys. Rev. B*, vol. 75, p. 155410, 2007.

[33] V. Bruno, C. DeVault, S. Vezzoli, et al., "Negative refraction in time-varying strongly coupled plasmonic-antenna–epsilon-near-zero systems," *Phys. Rev. Lett.*, vol. 124, p. 043902, 2020.

[34] T. G. Folland, G. Lu, A. Bruncz, J. R. Nolen, M. Tadjer, and J. D. Caldwell, "Vibrational coupling to epsilon-near-zero waveguide modes," *ACS Photonics*, vol. 7, p. 614, 2020.

[35] M. H. Javani and M. I. Stockman, "Real and imaginary Properties of epsilon-near-zero materials," *Phys. Rev. Lett.*, vol. 117, p. 107404, 2016.

[36] D. Yoo, F. de Leon-Perez, I.-H. Lee, et al., "Ultrastrong plasmon-phonon coupling via epsilon-near-zero nanocavities," arXiv: 2003.00136, 2020.

[37] A. M. Mahmoud, Y. Li, B. Edwards, and N. Engheta, "Photonic doping of epsilon-near-zero media Iñigo Liberal," *Science*, vol. 355, p. 10568, 2017.

[38] J. C. W. Song and M. S. Rudner, "Chiral plasmons without magnetic field," *Proc. Natl. Acad. Sci. U.S.A.*, vol. 113, no. 17, p. 4658, 2016.

[39] A. Kumar, A. Nemilentsau, K. Hung Fung, G. Hanson, N. X. Fang, and T. Low, "Chiral plasmon in gapped Dirac systems," *Phys. Rev. B*, vol. 93, p. 041413(R), 2016.

[40] Li-kun Shi and Justin C. W. Song, "Plasmon geometric phase and plasmon Hall shift," *Phys. Rev. X*, vol. 8, p. 021020, 2018.

[41] P. M. J. Bhaseen, M. Hohenadler, A. O. Silver, and B. D. Simons, "Polaritons and pairing Phenomena in bose-hubbard mixtures," *Phys. Rev. Lett.*, vol. 102, p. 135301, 2009.

[42] P. Cilibrizzi, H. Sigurdsson, T. C. H. Liew, et al., "Half-skyrmion spin textures in polariton microcavities," *Phys. Rev. B*, vol. 94, p. 045315, 2016.

[43] S. Donati, L. Dominici, G. Dagvadorj, et al., "Twist of generalized skyrmions and spin vortices in a polariton superfluid," *Proc. Natl. Acad. Sci. U.S.A.*, vol. 113, p. 14926, 2016.

[44] H. Hubener, U. De Giovannini, C. Schafer, et al., "Quantum cavities and Floquet materials engineering: the power of chirality," *Nat. Mater.* 2020 (in press).

[45] V. D. Vaidya, Y. Guo, R. M. Kroeze, et al., "Tunable-range, photon-mediated atomic interactions in multimode cavity QED," *Phys. Rev. X*, vol. 8, p. 011002, 2018.

[46] N. Jia, N. Schine, A. Georgakopoulos, et al., "A strongly interacting polaritonic quantum dot," *Nat. Phys.*, vol. 14, p. 550, 2018.

[47] S. Faure, C. Brimont, T. Guillet, et al., "Relaxation and emission of Bragg-mode and cavity-mode polaritons in a ZnO microcavity at room temperature," *Appl. Phys. Lett.*, vol. 95, p. 121102, 2009.

[48] D. Goldberg, L. I. Deych, A. A. Lisyansky, et al., "Exciton-lattice polaritons in multiple-quantum-well-based photonic crystals," *Nat. Photonics*, vol. 3, p. 662, 2009.

[49] E. S. Sedov, I. V. Iorsh, S. M. Arakelian, A. P. Alodjants, and A. Kavokin, "Hyperbolic metamaterials with Bragg polaritons," *Phys. Rev. Lett.*, vol. 114, p. 237402, 2015.

[50] C. Weisbuch and H. Benisty, "Microcavities in ecole polytechnique federale de lausanne, ecole polytechnique (France) and elsewhere: past, present and future," *Phys. Stat. Sol.(b)*, vol. 242, p. 2345, 2005.

[51] A. Kavokin and G. Malpuech, *Cavity Polaritons*, vol. 32, p. 246, Academic Press, 2003.

[52] P. G. Savvidis, J. J. Baumberg, R. M. Stevenson, M. S. Skolnick, D. M. Whittaker, and J. S. Roberts, "Angle-resonant stimulated polariton amplifier," *Phys. Rev. Lett.*, vol. 84, p. 1547, 2000.

[53] R. M. Stevenson, V. N. Astratov, M. S. Skolnick, et al., "Continuous wave observation of massive polariton redistribution by stimulated scattering in semiconductor microcavities," *Phys. Rev. Lett.*, vol. 85, p. 3680, 2000.

[54] H. Deng, G. Weihs, C. Santori, J. Bloch, and Y. Yamamoto, "Condensation of semiconductor microcavity exciton polaritons" *Science*, vol. 298, p. 1999, 2002.

[55] T. Guillet and C. Brimont, "Polariton condensates at room temperature," *Compt. Rendus Phys.*, vol. 17, p. 946, 2016.

[56] Á. Cuevas, J. C. López Carreño, B. Silva, et al., "First observation of the quantized exciton-polariton field and effect of interactions on a single polariton," *Sci. Adv.*, vol. 4, p. eaao6814, 2018.
[57] A. Delteil, T. Fink, A. Schade, S. Höfling, C. Schneider, and A. İmamoğlu, "Towards polariton blockade of confined exciton-polaritons," *Nat. Mater.*, vol. 18, p. 219, 2019.
[58] G. Muñoz-Matutano, A. Wood, M. Johnsson, et al., "Emergence of quantum correlations from interacting fibre-cavity polaritons," *Nat. Mater.*, vol. 18, p. 213, 2019.
[59] S. S. Demirchyan, Y. Chestnov, A. P. Alodjants, M. M. Glazov, and A. V. Kavokin, "Qubits Based on polariton Rabi oscillators," *Phys. Rev. Lett.*, vol. 112, p. 196403, 2014.
[60] I. V. Novikov and A. A. Maradudin, "Channel polaritons" *Phys. Rev. B*, vol. 66, p. 035403, 2002.
[61] S. I. Bozhevolnyi, V. S. Volkov, E. Devaux, J-Y. Laluet, and T. W. Ebbesen, "Channel plasmon subwavelength waveguide components including interferometers and ring resonators," *Nature*, vol. 440, p. 508, 2006.
[62] P. A. D. Gonçalves, S. Xiao, N. M. R. Peres, and N. Asger Mortensen, "Hybridized plasmons in 2D nanoslits: from Graphene to anisotropic 2D materials," *ACS Photonics*, vol. 4, p. 3045, 2017.
[63] Z. Fei, A. S. Rodin, G. O. Andreev, et al., "Gate-tuning of graphene plasmons revealed by infrared nano-imaging," *Nature*, vol. 487, p. 82, 2012.
[64] J. Chen, M. Badioli, P. Alonso-Gonzalez, et al., "Optical nano-imaging of gate-tunable graphene plasmons," *Nature*, vol. 487, p. 77, 2012.
[65] M. D. Goldflam, G.-X. Ni, K. W. Post, et al., "Tuning and persistent switching of graphene plasmons on a ferroelectric substrate," *Nano Lett.*, vol. 15, p. 4859, 2015.
[66] S. Xiao, X. Zhu, B.-H. Li, and N. Asger Mortensen, "Graphene-plasmon polaritons: from fundamental properties to potential applications," *Front. Phys.*, vol. 11, p. 117801, 2016.
[67] M. Wagner, Z. Fei, A. S. McLeod, et al., "Ultrafast and nanoscale plasmonic phenomena in exfoliated graphene revealed by infrared pump–probe nanoscopy," *Nano Lett.*, vol. 14, p. 894, 2014.
[68] D. J. Rizzo, B. S. Jessen, Z. Sun, et al., "Graphene/α-RuCl3: an emergent 2D plasmonic interface," arXiv:2007.07147, 2020.
[69] F. Wen, Y. Zhang, S. Gottheim, et al., "Charge transfer plasmons: optical frequency Conductances and tunable infrared resonances," *ACS Nano*, vol. 9, p. 6428, 2015.
[70] M. Forg, L. Colombier, R. K. Patel, et al., "Cavity-control of interlayer excitons in van der Waals heterostructures," *Nat. Commun.*, vol. 10, p. 3697, 2019.
[71] R. Rapaport, R. Harel, E. Cohen, A. Ron, E. Linder, and L. N. Pfeiffer, "Negatively charged quantum well Polaritons in a GaAs/AlAs microcavity: an analog of atoms in a cavity," *Phys. Rev. Lett.*, vol. 84, p. 1607, 2000.
[72] J. M. Raimond and S. Haroche, "Confined electrons and photons," edited by E. Burstein and C. Weisbuch, *Springer- New Physics and Applications*, vol. 383, 1995.
[73] A. Tiene, J. Levinsen, M. M. Parish, A. H. MacDonald, J. Keeling, and F. M. Marchetti, "Extremely imbalanced two-dimensional electron–hole–photon systems," *Phys. Rev. Res.*, vol. 2, p. 023089, 2020.
[74] P. A. Cherenkov. "Visible emission of clean liquids by action of γ radiation," *Dokl. Akad. Nauk SSSR*, vol. 2, p. 451, 1934.
[75] P. Genevet, D. Wintz, A. Ambrosio, A. She, R. Blanchard and F. Capasso, "Controlled steering of Cherenkov surface plasmon wakes with a one-dimensional metamaterial," *Nat. Nanotechnol.*, vol. 10, p. 804, 2015.
[76] D. H. Auston, K. P. Cheung, J. A. Valdmanis, and D. A. Kleinman, "Cherenkov radiation from femtosecond optical pulses in electro-optic media," *Phys. Rev. Lett.*, vol. 53, p. 1555, 1984.
[77] C. Maciel-Escudero, A. Konečńa, R. Hillenbrand, and J. Aizpurua, "Probing and steering bulk and surface phonon polaritons in uniaxial materials using fast electrons: hexagonal boron nitride," arXiv:2006.05359v1, 2020.
[78] Y. Zhang, C. Hu, B. Lyu, et al., "Tunable Cherenkov radiation of phonon Polaritons in silver nanowire/hexagonal boron nitride heterostructures," *Nano Lett.*, vol. 20, p. 2770, 2020.
[79] H. T. Stinson, J. S. Wu, B. Y. Jiang, et al., "Infrared nanospectroscopy and imaging of collective superfluid excitations in anisotropic superconductors," *Phys. Rev. B*, vol. 90, p. 014502, 2014.
[80] A. Dodel, A. Pikovski, I. Ermakov, et al., "Cooper pair polaritons in cold fermionic atoms within a cavity," *Phys. Rev. Res.*, vol. 2, p. 013184, 2020.
[81] W. L. Barnes, A. Dereux, D T. W. Ebbesen, "Surface plasmon subwavelength optics," *Nature*, vol. 424, p. 824, 2003.
[82] P. J. Schuck, D. P. Fromm, A. Sundaramurthy, G. S. Kino, and W. E. Moerner, "Improving the mismatch between light and nanoscale objects with gold bowtie nanoantennas," *Phys. Rev. Lett.*, vol. 94, p. 017402, 2005.
[83] P. Biagioni, J. S. Huang, and B. Hecht, "Nanoantennas for visible and infrared radiation," *Rep. Prog. Phys.*, vol. 75, p. 024402, 2012.
[84] L. Novotny and N. Van Hulst, "Antennas for light," *Nat. Photonics*, vol. 5, p. 83, 2011.
[85] J. J. Baumberg, J. Aizpurua, M. H. Mikkelsen, and D. R. Smith, "Extreme nanophotonics from ultrathin metallic gaps," *Nat. Mater.*, vol. 18, p. 668, 2019.
[86] M. Tamagnone, A. Ambrosio, K. Chaudhary, et al., "Ultra-confined mid-infrared resonant phonon polaritons in van der Waals nanostructures," *Sci. Adv.*, vol. 4, p. eaat7189, 2018.
[87] A. M. Dubrovkin, B. Qiang, T. Salim, D. Nam, N. I. Zheludey and Q. J. Wang, "Resonant nanostructures for highly confined and ultra-sensitive surface phonon-laritons," *Nat. Commun.*, vol. 11, p. 1863, 2020.
[88] L. V. Brown, M. Davanco, Z. Sun, et al., "Nanoscale mapping and spectroscopy of nonradiative hyperbolic modes in hexagonal boron nitride nanostructures," *Nano Lett.*, vol. 18, p. 1628, 2018.
[89] X. Chen, D. Hu, R. Mescall, et al., "Modern scattering-type scanning near-field optical microscope for advanced material research," *Adv. Mater.*, vol. 31, p. 1804774, 2019.
[90] Z. Fei, G. O. Andreev, W. Bao, et al., "Infrared nanoscopy of Dirac plasmons at the graphene–SiO_2 interface," *Nano Lett.*, vol. 11, p. 4701, 2011.
[91] P. J. Schuck, A. Weber-Bargioni, P. D. Ashby, D. F. Ogletree, A. Schwartzberg, and S. Cabrini, "Life beyond diffraction: opening new routes to materials characterization with next-generation optical near-field approaches," *Adv. Funct. Mater.*, vol. 23, p. 2539, 2013.
[92] S. Berweger, J. M. Atkin, R. L. Olmon, and M. B. Raschke, "Light on the tip of a needle: plasmonic nanofocusing for spectroscopy on the nanoscale," *J. Phys. Chem. Lett.*, vol. 3, p. 945, 2012.

[93] K. D. Park, T. Jiang, G. Clark, X. Xu, M. B. Raschke, "Radiative control of dark excitons at room temperature by nano-optical antenna-tip Purcell effect," *Nat. Nanotechnol.*, vol. 13, p. 59, 2017.
[94] B.-Y. Jiang, L. M. Zhang, A. H. Castro Neto, D. N. Basov, and M. M. Fogler, "Generalized spectral method for near-field optical microscopy," *J. Appl. Phys.*, vol. 119, p. 054305, 2016.
[95] F. Keilmann and R. Hillenbrand, "Near-field microscopy by elastic light scattering from a tip," *J. Philos. Trans. R. Soc. Lond. Ser. A Math. Phys. Eng. Sci.*, vol. 362, p. 787, 2014.
[96] J. M. Atkin, S. Berweger, A. C. Jones and M. B. Raschke, "Nano-optical imaging and spectroscopy of order, phases, and domains in complex solids," *Adv. Phys.*, vol. 61, p. 745, 2012.
[97] M. Fleischhauer and M. D. Lukin, "Dark-state polaritons in electromagnetically induced transparency," *Phys. Rev. Lett.*, vol. 84, p. 5094, 2000.
[98] S. E. Harris, J. E. Field, and A. Imamoglu, "Nonlinear optical processes using electromagnetically induced transparency," *Phys. Rev. Lett.*, vol. 64, p. 1107, 1990.
[99] M. D. Lukin, "Colloquium: trapping and manipulating photon states in atomic ensembles," *Rev. Mod. Phys.*, vol. 75, p. 457, 2003.
[100] M. Fleischhauer, A. Imamoglu and J. P. Marangos, "Electromagnetically induced transparency: optics in coherent media," *Rev. Mod. Phys.*, vol. 77, p. 633, 2005.
[101] H. Zoubi and H. Ritsch, "Lifetime and emission characteristics of collective electronic excitations in two-dimensional optical lattices," *Phys. Rev. A*, vol. 83, p. 063831, 2011.
[102] R. J. Bettles, S. A. Gardiner, and C. S. Adams, "Cooperative eigenmodes and scattering in one-dimensional atomic arrays," *Phys. Rev. A*, vol. 94, p. 043844, 2016.
[103] A. Asenjo-Garcia, M. Moreno-Cardoner, A. Albrecht, H. J. Kimble, and D. E. Chang, "Exponential improvement in photon storage fidelities using subradiance and "selective radiance," in atomic arrays" *Phys. Rev. X*, vol. 7, p. 031024, 2017.
[104] E. Shahmoon, D. S. Wild, M. D. Lukin, and S. F. Yelin, "Cooperative resonances in light scattering from two-dimensional atomic arrays," *Phys. Rev. Lett.*, vol. 118, p. 113601, 2017.
[105] D. Pines, "Electron Interaction in solids," *Can. J. Phys.*, vol. 34, p. 1379, 1956.
[106] Z. Sun, D. N. Basov, and M. M. Fogler, "Universal linear and nonlinear electrodynamics of a Dirac fluid," *Proc. Natl. Acad. Sci. U.S.A.*, vol. 115, p. 3285, 2018.
[107] Z. Sun, D. N. Basov, and M. M. Fogler, "Adiabatic amplification of plasmons and demons in 2D systems," *Phys. Rev. Lett.*, vol. 117, p. 076805, 2016.
[108] M. Jablan, H. Buljan, and M. Soljačić, "Plasmonics in graphene at infrared frequencies," *Phys. Rev. B*, vol. 80, p. 245435, 2009.
[109] A. Woessner, M. B. Lundeberg, Y. Gao, et al., "Highly confined low-loss plasmons in graphene-boron nitride heterostructures," *Nat. Mater.*, vol. 14, p. 421, 2015.
[110] G. X. Ni, A. S. McLeod, Z. Sun, et al., "Fundamental limits to graphene plasmonics," *Nature*, vol. 557, p. 530, 2018.
[111] M. I. Dyakonov, "New type of electromagnetic wave propagating at an interface," *Sov. Phys. JETP*, vol. 67, p. 714, 1988.
[112] O. Takayama, L. Crasovan, D. Artigas, and L. Torner, "Observation of Dyakonov surface waves," *Phys. Rev. Lett.*, vol. 102, p. 043903, 2009.
[113] J. Zubin and E. E. Narimanov, "Optical hyperspace for plasmons: Dyakonov states in metamaterials," *Appl. Phys. Lett.*, vol. 93, p. 221109, 2008.
[114] S. Dai, M. Tymchenko, Y. Yang, et al., "Manipulation and steering of hyperbolic surface polaritons in hexagonal boron nitride," *Adv. Mater.*, vol. 30, p. 1706358, 2018.
[115] F. J. Alfaro-Mozaz, P. Alonso-Gonzalez, S. Velez, et al., "Nanoimaging of resonating hyperbolic polaritons in linear boron nitride antennas," *Nat. Commun.*, vol. 8, p. 15624, 2017.
[116] I. Dolado, F. J. Alfaro-Mozaz, P. Li, et al., "Nanoscale guiding of infrared light with hyperbolic volume and surface polaritons in van der Waals material ribbons," *Adv. Mater.*, vol. 32, p. 1906530, 2020.
[117] D. B. Mast, A. J. Dahm, and A. L. Fetter, "Observation of bulk and edge magnetoylasmons in a two-dimensional electron fluid," *Phys. Rev. Lett.*, vol. 54, p. 1706, 1985.
[118] V. A. Volkov and S. A. Mikhailov, "Edge magnetoplasmons: low frequency weakly damped excitations in inhomogeneous two-dimensional electron systems," *Radio Eng. Electron. Inst. USSR Acad. Sci. Zh. Eksp. Teor. Fiz.*, vol. 94, p. 217, 1988.
[119] N. Kumada, P. Roulleau, B. Roche, et al., "Resonant edge magnetoplasmons and their decay in graphene," *Phys. Rev. Lett.*, vol. 113, p. 266601, 2014.
[120] I. Petkovic, F. I. B. Williams and D. Christian Glattli, "Edge magnetoplasmons in graphene," *J. Phys. D Appl. Phys.*, vol. 47, 2014.
[121] M. L. Sadowski, G. Martinez, M. Potemski, C. Berger, and W. A. de Heer, "Landau level spectroscopy of ultrathin graphite layers," *Phys. Rev. Lett.*, vol. 97, p. 266405, 2006.
[122] Z. Jiang, E. A. Henriksen, L. C. Tung, et al., "Infrared spectroscopy of Landau levels in graphene," *Phys. Rev. Lett.*, vol. 98, p. 197403, 2007.
[123] M. Orlita, C. Faugeras, R. Grill, et al., "Carrier scattering from dynamical magneto-conductivity in quasineutral epitaxial graphene," *Phys. Rev. Lett.*, vol. 107, p. 216603, 2011.
[124] Z. G. Chen, Z. Shi, W. Yang, et al., "Observation of an intrinsic bandgap and Landau level renormalization in graphene/boron-nitride heterostructures," *Nat. Commun.*, vol. 5, p. 4461, 2014.
[125] I. O. Nedoliuk, S. Hu, A. K. Geim, and A. B. Kuzmenko, "Colossal infrared and terahertz magneto-optical activity in a two-dimensional Dirac material,"*Nat. Nano*, vol. 14, p. 756, 2019.
[126] Z. Fei, M. D. Goldflam, J.-S. Wu, et al., "Edge and surface plasmons in graphene nanoribbons," *Nano Lett.*, vol. 15, pp. 8271–8276, 2015.
[127] P. Li, I. Dolado, F. J. Alfaro-Mozaz, et al., "Optical nanoimaging of hyperbolic surface polaritons at the Edges of van der Waals materials," *Nano Lett.*, vol. 17, p. 228, 2017.
[128] A. Y. Nikitin, P. Alonso-González, S. Vélez, et al., "Real-space mapping of tailored sheet and edge plasmons in graphene nanoresonators," *Nat. Photonics*, vol. 10, p. 239, 2016.
[129] P. Berini, "Plasmon-polariton waves guided by thin lossy metal films of finite width: Bound modes of symmetric structures," *Phys. Rev. B*, vol. 61, p. 10484, 2000.
[130] M. Lorente-Crespo, G. C. Ballesteros, C. Mateo-Segura, and C. García-Meca, "Edge-plasmon whispering-gallery modes in nanoholes" *Phys. Rev. Appl.*, vol. 13, p. 024050, 2020.
[131] T. Rindzevicius, Y. Alaverdyan, B. Sepulveda, et al., "Nanohole plasmons in optically thin gold films," *J. Phys. Chem. C*, vol. 111, p. 1207, 2007.

[132] A. Degiron, H. Lezec, N. Yamamoto, and T. Ebbe-sen, "Optical transmission properties of a single subwave-length aperture in a real metal," *Opt. Commun*, vol. 239, p. 61, 2004.

[133] N. Liu, L. Langguth, T. Weiss, et al., "Plasmonic analogue of electromagnetically induced transparency at the Drude damping limit," *Nat. Mater.*, vol. 8, p. 758, 2009.

[134] R. Taubert, M. Hentschel, J. Kastel, and H. Gissen, "Classical analog of electromagnetically induced absorption in plasmonics," *Nano Lett.*, vol. 12, p. 1367, 2012.

[135] H. Yan, T. Low, F. Guinea, F. Xia, and P. Avouris, "Tunable phonon-induced transparency in bilayer graphene nanoribbons," *Nano Lett.*, vol. 14, p. 4581, 2014.

[136] S. Xia, X. Zhai, L. Wang, and S. Wen, "Plasmonically induced transparency in in-plane isotropic and anisotropic 2D materials," *Opt. Express*, vol. 28, p. 7980, 2020.

[137] V. Walther, R. Johne and T. Pohl, "Giant optical nonlinearities from Rydberg excitons in semiconductor microcavities," *Nat. Commun.*, vol. 9, p. 1309, 2018.

[138] C. Weisbuch, M. Nishioka, A. Ishikawa, and Y. Arakawa, "Observation of the coupled exciton-photon mode splitting in a semiconductor quantum microcavity," *Phys. Rev. Lett.*, vol. 69, p. 3314, 1992.

[139] J. Kasprzak, M. Richard, S. Kundermann, et al., "Bose–Einstein condensation of exciton polaritons" *Nature*, vol. 443, p. 409, 2006.

[140] I. Carusotto and C. Ciuti, "Quantum fluids of light," *Rev. Mod. Phys*, vol. 85, p. 29,2013.

[141] T. Byrnes, N. Y. Kim, and Y. Yamamoto, "Exciton-polariton condensates," *Nat. Phys.*, vol. 10, p. 803, 2014.

[142] A. Griffin, D. W. Snoke, and S. Stringari, eds., *Bose–Einstein Condensation*, Cambridge University Press, 1996.

[143] A. A. High, J. R. Leonard, A. T. Hammack, et al., "Spontaneous coherence in a cold exciton gas," *Nature*, vol. 483, p. 7391, 2012.

[144] D. Snoke, "Spontaneous Bose coherence of excitons and polaritons," *Science*, vol. 298, p. 1368, 2002.

[145] J. P. Eisenstein and A. H. MacDonald, "Bose–Einstein condensation of excitons in bilayer electron systems," *Nature*, vol. 432, p. 691, 2004.

[146] H. Deng, H. Haug, and Y. Yamamoto, "Exciton-polariton Bose-Einstein condensation," *Rev. Mod. Phys.*, vol. 82, p. 1489, 2010.

[147] D. W. Snoke and J. Keeling, "The new era of polariton condensates," *Phys. Today*, vol. 70, p. 54, 2017.

[148] Y. Sun, P. Wen, Y. Yoon, et al., "Bose–Einstein condensation of long-lifetime polaritons in thermal equilibrium," *Phys. Rev. Lett.*, vol. 118, p. 016602, 2017.

[149] G. Lerario, A. Fieramosca, F. Barachati, et al., "Room-temperature superfluidity in a polariton condensate," *Nat. Phys.*, vol. 13, p. 837, 2013.

[150] Y. Fu, H. Zhu, J. Chen, M. P. Hautzinger, X.-Y. Zhu, and S. Jin, "Metal halide perovskite nanostructures for optoelectronic applications and the study of physical properties," *Nat. Rev. Mater.*, vol. 4, p. 169, 2019.

[151] A. T. Hanbicki, M. Currie, G. Kioseoglou, A. L. Friedman, and B. T. Jonker, "Measurement of high exciton binding energy in the monolayer transition-metal dichalcogenides WS_2 and WSe_2," *Sol. State. Commun.*, vol. 203, p. 16, 2015.

[152] T. Cheiwchanchamnangij and W. R. L. Lambrecht, "Quasiparticle band structure calculation of monolayer, bilayer, and bulk MoS_2," *Phys. Rev. B*, vol. 85, p. 205302, 2012.

[153] A. Ramasubramaniam, "Large excitonic effects in monolayers of molybdenum and tungsten dichalcogenides," *Phys. Rev. B*, vol. 86, p. 115409, 2012.

[154] H.-P. Komsa and A. V. Krasheninnikov, "Effects of confinement and environment on the electronic structure and exciton binding energy of MoS_2 from first principles," *Phys. Rev. B*, vol. 86, p. 241201, 2012.

[155] Y. Liang, S. Huang, R. Soklaski, and L. Yang, "Quasiparticle band-edge energy and band offsets of monolayer of molybdenum and tungsten chalcogenides," *Appl. Phys. Lett.*, vol. 103, p. 042106, 2013.

[156] K. He, N. Kumar, L. Zhao, et al., "Tightly bound excitons in monolayer WSe_2," *Phys. Rev. Lett.*, vol. 113, p. 026803, 2014.

[157] Z. Ye, T. Cao, K. O'Brien, et al., "Probing excitonic dark states in single-layer tungsten disulphide," *Nature*, vol. 513, p. 214, 2014.

[158] K. Yao, A. Yan, S. Kahn, et al., "Optically discriminating carrier-induced quasiparticle band gap and exciton energy renormalization in monolayer MoS_2," *Phys. Rev. Lett.*, vol. 119, p. 087401, 2017.

[159] H. Min, R. Bistritzer, J. Su and A. H. MacDonald, "Room-temperature superfluidity in graphene bilayers" *Phys. Rev. B*, vol. 78, p. 121401, 2008.

[160] M. M. Fogler, L. V. Butov, and K. S. Novoselov, "High-temperature superfluidity with indirect excitons in van der Waals heterostructures" *Nat. Commun.*, vol. 5, p. 4555, 2014.

[161] F.-C. Wu, F. Xue, and A. H. MacDonald, "Theory of two-dimensional spatially indirect equilibrium exciton condensates," *Phys. Rev. B*, vol. 92, p. 165121, 2015.

[162] A. V. Nalitov, H. Sigurdsson, S. Morina, et al., "Optically trapped polariton condensates as semiclassical time crystals," *Phys. Rev. A*, vol. 99, p. 033830, 2019.

[163] M. D. Fraser, S. Hofling, and Y. Yamamoto, "Physics and applications of exciton-polariton lasers," *Nat. Mater.*, vol. 15, p. 1049, 2016.

[164] A. Gianfrate, L. Dominici, O. Voronych, et al., "Superluminal X-waves in a polariton quantum fluid," *Light Sci. Appl.*, vol. 7, p. 17119, 2018.

[165] A. Gabbay, Y. Preezant, E. Cohen, B. M. Ashkinadze and L. N. Pfeffier, "Fermi edge Polaritons in a microcavity containing a high-density two-dimensional electron gas,"*Phys. Rev. Lett.*, vol. 99, p. 157402, 2007.

[166] D. Pimenov, J. von Delft, L. Glazman, and M. Goldstein, "Fermi-edge exciton-polaritons in doped semiconductor microcavities with finite hole mass," *Phys. Rev. B*, vol. 96, p. 155310, 2017.

[167] M. Baetena and M. Wouters, "Mahan polaritons and their lifetime due to hole recoil," *Eur. Phys. J. D*, vol. 69, p. 243, 2015.

[168] G. D. Mahan, "Excitons in degenerate semiconductors," *Phys. Rev.*, vol. 153, p. 882, 1967.

[169] G. D. Mahan, "Excitons in metals" *Phys. Rev. Lett.*, vol. 18, p. 448, 1967.

[170] T. Oka and S. Kitamura, "Floquet engineering of quantum materials," *Ann. Rev. Condensed Matter Phys.*, vol. 10, p. 387, 2019.

[171] L. W. Clark, N. Jia, N. Schine, C. Baum, A. Georgakopoulos, and J. Simon, "Interacting Floquet polaritons," *Nature*, vol. 571, p. 532, 2019.

[172] F. Mahmood, C.-K. Chan, Z. Alpichshev, et al., "Selective scattering between Floquet-Bloch and Volkov states in a topological insulator," *Nat. Phys.*, vol. 12, p. 306, 2016.

[173] H. Hübener, U. D. Giovannini, and A. Rubio, "Phonon driven Floquet matter," *Nano Lett.*, vol. 18, p. 1535, 2018.

[174] J.-M. Ménard, C. Poellmann, M. Porer, et al., "Revealing the dark side of a bright exciton-polariton condensate," *Nat. Commun.*, vol. 5, p. 4648, 2018.

[175] Z. Sun, J. Gu, A. Ghazaryan, et al., "Optical control of room-temperature valley polaritons," *Nat. Photonics*, vol. 11, p. 491, 2017.

[176] L. Vestergaard Hau, S. E. Harris, Z. Dutton, and C. H. Behroozi, "Light speed reduction to 17 metres per second in an ultracold atomic gas," *Nature*, vol. 397, p. 594, 1999.

[177] O. Kocharovskaya, Y. Rostovtsev, and M. O. Scully, "Stopping light via hot atoms," *Phys. Rev. Lett.*, vol. 86, p. 628, 2001.

[178] M. Bajcsy, A. S. Zibrov, and M. D. Lukin, "Stationary pulses of light in an atomic medium," *Nature*, vol. 426, p. 638, 2003.

[179] S. Betzold, M. Dusel, O. Kyriienko, et al., "Coherence and Interaction in confined room-temperature polariton condensates with Frenkel excitons," *ACS Photonics*, vol. 7, p. 384, 2020.

[180] D. G. Lidzey, D. D. C. Bradley, T. Virgili, S. Walker, and D. M. Walker, "Strong exciton–photon coupling in an organic semiconductor microcavity," *Nature*, vol. 395, p. 53, 1998.

[181] G. Lerario, A. Fieramosca, F. Barachati, et al., "Room-temperature superfluidity in a polariton condensate," *Nat. Phys.*, vol. 13, p. 837, 2017.

[182] J. Keeling and S. Kena-Cohen, "Bose–Einstein condensation of exciton-polariton in organic microcavities," *Annu. Rev. Phys. Chem.*, vol. 71, p. 435, 2020.

[183] S. Foteinopoulou, G. Chinna Rao Devarapu, G. S. Subramania, S. Krishna and D. Wasserman, "Phonon-polaritonics: enabling powerful capabilities for infrared photonics," *Nanophotonics*, vol. 8, p. 2129, 2019.

[184] K. I. Kliewer and R. Fuchs, "Optical modes of vibration in an ionic crystal slab including retardation. I. Nonradiative region," *Phys. Rev.*, vol. 144, p. 495, 1966.

[185] A. Huber, N. Ocelic, D. V. Kazantsev, and R. Hillenbrand, "Near-field imaging of mid-infrared surface phonon polariton propagation," *Appl. Phys. Lett.*, vol. 87, p. 081103, 2005.

[186] R. Hillenbrand, T. Taubner and F. Keilmann, "Phonon-enhanced light-matter interaction at the nanometre scale," *Nature*, vol. 418, p. 159, 2002.

[187] J. D. Caldwell, O. J. Glembocki, N. Sharac, et al., "Low-loss, extreme sub-diffraction photon confinement via silicon carbide surface phonon polariton nanopillar resonators," *Nano Lett.*, vol. 13, p. 3690, 2013.

[188] T. E. Tiwald, J. A. Woolam, S. Zollner, et al., "Carrier concentration and lattice absorption in bulk and epitaxial silicon carbide determined using infrared ellipsometry," *Phys. Rev. B*, vol. 60, p. 11464, 1999.

[189] T. Taubner, D. Korobkin, Y. Urzhumov, G. Shvets, and R. Hillenbrand, "Near-field microscopy through a SiC superlens," *Science*, vol. 313, p. 1595, 2006.

[190] W. J. Moore and R. T. Holm. "Infrared dielectric constant of GaAs," *J. Appl. Phys.*, vol. 80, p. 6939, 1996.

[191] M. Francesco, D. Pellegrino, M. I. Katsnelson, and M. Polini, "Helicons in Weyl semimetals," *Phys. Rev. B*, vol. 92, p. 201407, 2015.

[192] J. Zhou, H.-R. Chang, and D. Xiao, "Plasmon mode as a detection of the chiral anomaly in Weyl semimetals," *Phys. Rev. B*, vol. 91, p. 035114, 2015.

[193] C. Justin W. Song, and M. S. Rudner, "Fermi arc plasmons in Weyl semimetals," *Phys. Rev. B*, vol. 96, p. 205443, 2017.

[194] I. Iorsh, G. Rahmanova, and M. Titov, "Plasmon-polariton from a helical state in a Dirac magnet," *ACS Photonics*, vol. 6, p. 2450, 2019.

[195] J. Lin, J. P. Bathasar Mueller, Q. Wang, et al., "Polarization-controlled tunable directional coupling of surface plasmon polaritons," *Science*, vol. 340, p. 331, 2013.

[196] L. Huang, X. Chen, B. Bai, et al., "Helicity dependent directional surface plasmon polariton excitation using a metasurface with interfacial phase discontinuity," *Light Sci. Appl.*, vol. 2, p. 70, 2013.

[197] J. J. Hopfield, "Theory of the contribution of excitons to the complex dielectric constant of crystals," *Phys. Rev.*, vol. 112, p. 1555, 1958.

[198] C. H. Henry and J. J. Hopfield, "Raman scattering by polaritons," *Phys. Rev. Lett.*, vol. 15, p. 964, 1965.

[199] U. Fano, "Atomic theory of electromagnetic interactions in dense materials," *Phys. Rev.*, vol. 103, p. 1202, 1956.

[200] K. Huang, "Lattice vibrations and optical waves in ionic crystals," *Nature*, vol. 167, p. 779, 1951.

[201] K. B. Tolpygo, "Physical properties of a rock salt lattice made up of deformeable ions," Translated and reprinted from *Zh. Eksp. Teor. Fiz.*, vol. 20, no. 6, p. 497, 1950.

[202] J. D. Caldwell, I. Vurgaftman, J. G. Tischler, O. J. Glembocki, J. C. Owrutsky, and T. L. Reinecke, "Atomic-scale photonic hybrids for mid-infrared and terahertz nanophotonics," *Nat. Nanotechnol.*, vol. 11, p. 9, 2016.

[203] L. Wendler and R. Haupt, "Long-range surface plasmon-phonon-polaritons," *J. Phys. C Solid State Phys.*, vol. 19, p. 1871, 1986.

[204] B. Askenazi, A. Vasanelli, A. Delteil, et al., "Ultra-strong light–matter coupling for designer Reststrahlen band," *New J. Phys.*, vol. 16, p. 043029, 2014.

[205] C. Franckié, K. Ndebeka-Bandou, J. Ohtani, and M. Faist, "Quantum model of gain in phonon-polariton lasers," *Phys. Rev. B*, vol. 97, p. 075402, 2018.

[206] N. C. Passler, C. R. Gubbin, T. G. Folland, et al., "Strong coupling of epsilon-near-zero phonon Polaritons in polar dielectric heterostructures," *Nano Lett.*, vol. 18, p. 428, 2018.

[207] A. A. Strashko and V. M. Agranovich, "To the theory of surface plasmon-polaritons on metals covered with resonant thin films," *Opt. Commun.*, vol. 332, p. 201, 2014.

[208] D. C. Ratchford, C. J. Winta, I. Chatzakis, et al., "Controlling the infrared dielectric function through atomic-scale heterostructures," *ACS Nano*, vol. 13, p. 6730, 2019.

[209] E. L. Runnerstrom, K. P. Kelley, T. G. Folland, et al., "Polaritonic hybrid-epsilon-near-zero modes: beating the plasmonic confinement vs propagation-length trade-off with doped cadmium oxide bilayers," Nano Lett., vol. 19, p. 948, 2019.

[210] K. Chaudhary, M. Tamagnone, M. Resaee, et al., "Engineering phonon polaritons in van der Waals heterostructures to enhance in-plane optical anisotropy," *Sci. Adv.*, vol. 5, p. eaau7171, 2019.

[211] S. Dai, Q. Ma, M. K. Liu, et al., "Graphene on hexagonal boron nitride as a tunable hyperbolic metamaterial," *Nat. Nanotechnol.*, vol. 10, p. 682, 2015.
[212] F. J. Bezares, A. De Sanctis, J. R. M. Saavedra, et al., "Intrinsic plasmon-phonon interactions in highly doped graphene: a near-field imaging study," *Nano Lett.*, vol. 17, p. 5908, 2017.
[213] A. Kumar, T. Low, K. Hung Fung, P. Avouris, and N. X. Fang, "Tunable light–matter interaction and the role of hyperbolicity in graphene-hBN system," *Nano Lett.*, vol. 15, p. 3172, 2015.
[214] A. T. Costa, P. A. D. Gonçalves, H. Frank, et al., "Harnessing ultra-confined graphene plasmons to probe the electrodynamics of superconductors," arXiv:2006.00748, 2020.
[215] H. Dehghani, Z. M. Raines, V. M. Galitski, and M. Hafezi, "Optical enhancement of superconductivity via targeted destruction of charge density waves," *Phys. Rev. B*, vol. 101. p. 195106, 2020.
[216] D. Rodrigo, O. Limaj, D. Janner, et al., "Mid-infrared plasmonic biosensing with graphene," *Science*, vol. 349, p. 165, 2015.
[217] M. Autore, P. Li, I. Dolado, et al., "Boron nitride nanoresonators for phonon-enhanced molecular vibrational spectroscopy at the strong coupling limit," *Light Sci. Appl.*, vol. 7, p. 17172, 2018.
[218] C. R. Gubbin, R. Berte, M. A. Meeker, et al., "Hybrid longitudinal-transverse phonon polaritons," *Nat. Commun.*, vol. 10, p. 1682, 2019.
[219] T. G. Folland, A. Fali, S. T. White, et al., "Reconfigurable infrared hyperbolic metasurfaces using phase change materials," *Nat. Commun.*, vol. 9, p. 4371, 2018.
[220] A. Poddubny, I. Iorsh, P. Belov, and Y. Kivshar, "Hyperbolic metamaterials," *Nat. Photonics*, vol. 7, p. 948, 2013.
[221] J. S. Gomez-Diaz and A. Alu, "Flatland optics with hyperbolic metasurfaces," *ACS Photonics*, vol. 3, p. 2211, 2016.
[222] J. S. Gomez-Diaz, M. Tymchenko, and A. Alù, "Hyperbolic plasmons and topological transitions over uniaxial metasurfaces," *Phys. Rev. Lett.*, vol. 114, p. 233901, 2015.
[223] J. D. Caldwell, A. V. Kretinin, Y. Chen, et al., "Sub-diffractional volume-confined polaritons in the natural hyperbolic material hexagonal boron nitride," *Nat. Commun.*, vol. 5, p. 5221, 2014.
[224] Y. Guo, W. Newman, C. L. Cortes, and Z. Jacob, "Applications of hyperbolic metamaterial substrates," *Adv. Optoelectron.* 452502, (2012).
[225] C. L. Cortes, W. Newman, S. Molesky and Z. Jacob "Quantum nanophotonics using hyperbolic metamaterials," *J. Opt.*, vol. 14, p. 063001, 2012.
[226] J. Sun, N. M. Litchinitser, and J. Zhou, "Indefinite by nature: from ultraviolet to terahertz," *ACS Photonics*, vol. 1, p. 293, 2014.
[227] E. E. Narimanov and A. V. Kildishev, "Metamaterials naturally hyperbolic," *Nat. Photonics*, vol. 9, p. 214, 2015.
[228] S. Dai, Z. Fei, Q. Ma, et al., "Tunable phonon polaritons in atomically thin van der Waals crystals of boron nitride," *Science*, vol. 343, p. 1125, 2014.
[229] A. J. Giles, S. Dai, I. Vurgaftman, et al., "Ultralow-loss polaritons in isotopically pure boron nitride," *Nat. Mater.*, vol. 17, p. 134, 2017.
[230] J. Taboada-Gutiérrez, G. Álvarez-Pérez, J. Duan, et al., "Broad spectral tuning of ultra-low-loss polaritons in a van der Waals crystal by intercalation," *Nat. Mater.*, vol. 19, p. 964, 2020.
[231] J. D. Caldwell, I. Aharonovich, G. Cassabois, J. H. Edgar, B. Gil and D. N. Basov, "Photonics with hexagonal boron nitride" *Nat. Mater. Rev.*, vol. 4, no. 8, p. 552, 2019.
[232] A. Fali, S. T. White, T. G. Folland, et al., "Refractive index-based Control of hyperbolic phonon-polariton propagation," *Nano Lett.*, vol. 9, no. 11, 7725–7730, 2019.
[233] A. Ambrosio, L. A. Jauregui, S. Dai, et al., "Mechanical detection and imaging of hyperbolic phonon polaritons in hexagonal boron nitride," *ACS Nano*, vol. 11, p. 8741, 2017.
[234] A. Ambrosio, M. Tamagnone, K. Chaudhary, et al., "Selective excitation and imaging of ultraslow phonon polaritons in thin hexagonal boron nitride crystals," *Light Sci. Appl.*, vol. 7, p. 27, 2018.
[235] G. Hu, J. Shen, C. W. Qiu, A. Alù, and S. Dai, "Phonon Polaritons and hyperbolic Response in van der waals materials," *Adv. Opt. Mater. Spec. Issue Polarit. Nanomater.*, vol. 8, p. 1901393, 2020.
[236] S. Dai, J. Quan, G. Hu, et al., "Hyperbolic phonon polaritons in suspended hexagonal boron nitride," *Nano Lett.*, vol. 19, p. 100, 2019.
[237] S. Dai, M. Tymchenko, Z. Q. Xu, et al., "Internal nanostructure diagnosis with hyperbolic phonon polaritons in hexagonal boron nitride," *Nano Lett.*, vol. 18, p. 5205, 2018.
[238] A. J. Hoffman, A. Sridhar, P. X. Braun, et al., "Midinfrared semiconductor optical metamaterials," *J. Appl. Phys.*, vol. 105, p. 122411, 2009.
[239] K. Feng, G. Harden, D. L. Sivco, and A. J. Hoffman, "Subdiffraction confinement in all-semiconductor hyperbolic metamaterial resonators," *ACS Photonics*, vol. 4, p. 1621, 2017.
[240] D. Lu, H. Qian, K. Wang, et al., "Nanostructuring multilayer hyperbolic metamaterials for ultrafast and bright green InGaN quantum wells," *Adv. Mater.*, vol. 30, p. 15, 2018.
[241] I. V. Iorsh, I. S. Mukhin, I. V. Shadrivov, P. A. Belov, and Y. S. Kivshar, "Hyperbolic metamaterials based on multilayer graphene structures," *Phys. Rev. B*, vol. 87, p. 075416, 2013.
[242] E. E. Narimanov and A. V. Kildishev, "Naturally hyperbolic," *Nat. Photonics*,", vol. 9, p. 214, 2015.
[243] G. Hu, A. Krasnok, Y. Mazor, C.-W. Qiu, and A. Alù, "Moiré hyperbolic metasurfaces," *Nano Lett.*, vol. 20, p. 3217, 2020.
[244] C. Wang, S. Huang, Q. Xing, et al., "Vander Waals thin films of WTe_2 for natural hyperbolic plasmonic surfaces," *Nat. Commun.*, vol. 11, p. 1158, 2020.
[245] E. Itai, A. J. Chaves, D. A. Rhodes, et al., "Highly confined In-plane propagating exciton-polaritons on monolayer semiconductors," *2D Mater.*, vol. 7, p. 3, 2020.
[246] N. Riveraa, G. Rosolen, J. D. Joannopoulosa, I. Kaminera, and M. Soljacic, "Making two-photon processes dominate one-photon processes using mid-IR phonon polaritons," *Proc. Natl. Acad. Sci. U.S.A.*, vol. 114, p. 3607, 2017.
[247] P. A. Belov and Y. Hao, "Subwavelength imaging at optical frequencies using a transmission device formed by a periodic layered metal-dielectric structure operating in the canalization regime," *Phys. Rev. B*, vol. 73, p. 113110, 2006.
[248] P. Li, G. Hu, I. Dolado, et al., "Collective near-field coupling and nonlocal phenomena in infrared-phononic metasurfaces for nano-light canalization," *Nat. Commun.*, vol. 11, p. 3663, 2020.
[249] S. Dai, Q. Ma, T. Andersen, et al., "Subdiffractional focusing and guiding of polaritonic rays in a natural hyperbolic material," *Nat. Commun.*, vol. 6, p. 6963, 2015.
[250] D. Correas-Serrano, A. Alù, and J. Sebastian Gomez-Diaz, "Plasmon canalization and tunneling over anisotropic metasurfaces," *Phys. Rev. B*, vol. 96, p. 075436, 2017.

[251] P. Li, M. Lewin, A. V. Kretinin, et al., "Hyperbolic phonon-polaritons in boron nitride for near-field optical imaging and focusing," *Nat. Commun.*, vol. 6, p. 7507, 2015.
[252] I.-H. Lee, M. He, X. Zhang, et al., "Image polaritons in boron nitride for extreme polariton confinement with low losses," *Nat. Commun.*, vol. 11, p. 3649, 2020.
[253] A. Stahl, "Polariton Structure of interband Transitions in semiconductors," *Phys. Stat. Sol.(b)*, vol. 94, p.221, 1979.
[254] A. Tomadin, F. Guinea, and M. Polini, "Generation and morphing of plasmons in graphene superlattices," *Phys. Rev. B*, vol. 90, p. 161406, 2014.
[255] G. X. Ni, H. Wang, J. S. Wu, et al., "Plasmons in graphene moiré superlattices," *Nat. Mater.*, vol. 14, p. 1217, 2015.
[256] N. C. H. Hesp, I. Torre, D. R-Legrian, et al., "Collective excitations in twisted bilayer graphene close to the magic angle," arXiv:1910.07893, 2019.
[257] P. Novelli, I. Torre, F. H. L. Koppens, F. Taddei, and M. Polini, "Optical and plasmonic properties of twisted bilayer graphene: Impact of interlayer tunneling asymmetry and ground-state charge inhomogeneity," arXiv:2005.09529, 2020.
[258] D. N. Basov, R. Liang, D. A. Bonn, et al., "In-plane anisotropy of the penetration depth in $YBa_2Cu_3O_{7-x}$ and $YBa_2Cu_4O_8$ superconductors," *Phys. Rev. Lett.*, vol. 74, p. 598, 1995.
[259] P. A. Wolff, "Plasma-wave instability in narrow-gap semiconductors," *Phys. Rev. Lett.*, vol. 24, p. 266, 1970.
[260] T. Low, P.-Y. Chen, and D. N. Basov "Superluminal plasmons with resonant gain in population inverted bilayer graphene" *Phys. Rev. B*, vol. 98, p. 041403(R), 2018.
[261] D. Dini, R. Kohler, A. Tredicucci, G. Biasiol, and L. Sorba, "Microcavity polariton splitting of intersubband transitions," *Phys. Rev. Lett.*, vol. 90, p. 116401, 2003.
[262] D. Ballarini and S. D. Liberato, "Polaritonics: from microcavities to sub-wavelength confinement," *Nanophotonics*, vol. 8, p. 641, 2019.
[263] J. Lee, M. Tymchenko, C. Argyropoulos, et al., "Giant nonlinear response from plasmonic metasurfaces Coupled to intersubband transitions," *Nature*, vol. 511, p. 65, 2014.
[264] Y. Laplace, S. Fernandez-Pena, S. Gariglio, J. M. Triscone, and A. Cavalleri, "Proposed cavity Josephson plasmonics with complex-oxide heterostructures," *Phys. Rev. B*, vol. 93, p. 075152, 2016.
[265] D. N. Basov and T. Timusk, "Electrodynamics of high-Tc superconductors," *Rev. Mod. Phys.*, vol. 77, p. 721, 2005.
[266] S. Rajasekaran, E. Casandruc, Y. Laplace, et al., "Parametric amplification of a superconducting plasma wave," *Nat. Phys.*, vol. 12, p. 1012, 2016.
[267] A. Charnukha, A. Sternbach, H. T. Stinson, et al., "Ultrafast nonlocal collective dynamics of Kane plasmon-polaritons in a narrow-gap semiconductor," *Sci. Adv.*, vol. 5, p. eaau9956, 2019.
[268] X. Li, M. Bamba, Q. Zhang, et al., "Vacuum Bloch–Siegert shift in Landau polaritons with ultra-high cooperativity," *Nat. Photonics*, vol. 12, p. 324, 2018.
[269] G. L. Paravicini-Bagliani, F. Appugliese, E. Richter, et al., "Magneto-transport controlled by Landau polariton states," *Nat. Phys.*, vol. 15, p. 186, 2019.
[270] Z. Shi, X. Hong, H. A. Bechtel, et al., "Observation of a Luttinger-liquid plasmon in metallic single-walled carbon nanotubes," *Nat. Photonics*, vol. 9, p. 515, 2015.
[271] J. M. Luttinger, "An exactly soluble model of a many-fermion system," *J. Math. Phys.*, vol. 4, p. 1154, 1963.
[272] F. D. M. Haldane, "Luttinger liquid theory, of one-dimensional quantum fluids. I. Properties of the Luttinger model and their extension to the general 1D interacting spinless Fermi gas," *J. Phys. C Solid State Phys.*, vol. 14, p. 2585, 1981.
[273] S. Wang, S. Zhao, Z. Shi, et al., "Nonlinear Luttinger liquid plasmons in semiconducting single-walled carbon nanotubes," *Nat. Mater.*, vol. 19, p. 986, 2020.
[274] E. Camley, "Long-wavelength surface spin waves on antiferromagnets," *Phys. Rev. Lett.*, vol. 45, p. 283, 1980.
[275] R. E. Camley and D. L. Millis, "Surface polaritons on uniaxial antiferromagnets," *Phys. Rev. B*, vol. 26, p. 1280, 1982.
[276] R. Macêdo and R. E. Camley, "Engineering terahertz surface magnon-polaritons in hyperbolic antiferromagnets," *Phys. Rev. B*, vol. 99, p. 014437, 2019.
[277] J. Sloan, N. Rivera, J. D. Joannopoulos, I. Kaminer, and M. Soljăci, "Controlling spins with surface magnon polaritons," *Phys. Rev. B*, vol. 100, p. 235453, 2019.
[278] S. S. Kruk, Z. J. Wong, E. Pshenay-Severin, et al., "Magnetic hyperbolic optical metamaterials," *Nat. Commun.*, vol. 7, p. 11329, 2016.
[279] P. Sivarajah, A. Steinbacher, B. Dastrup, et al., "THz-frequency magnon-phonon-polaritons in the collective strong-coupling regime," *J. Appl. Phys.*, vol. 125, p. 213103, 2019.
[280] J. J. Brion, R. F. Wallis, A. Hartstein and E. Burstein, "Theory of surface magnetoplasmons in semiconductors," *Phys. Rev. Lett.*, vol. 28, p. 1455, 1972.
[281] B. Hu, Y. Zhang and Q. J. Wang, "Surface magneto plasmons and their applications in the infrared frequencies," *Nanophotonics*, vol. 4, p. 4, 2015.
[282] Z. Long, Y. Wang, M. Erukhimova, M. Tokman, and A. Belyanin, "Magnetopolaritons in Weyl semimetals in a strong magnetic field," *Phys. Rev. Lett.*, vol. 120, p. 037403, 2018.
[283] R. L. Stamps and R. E. Camley, "Focusing of magnetoplasmon polaritons," *Phys. Rev. B*, vol. 31, p. 4924, 1985.
[284] M. S. Rudner and J. C. W. Song, "Self-induced Berry flux and spontaneous non-equilibrium magnetism," *Nat. Phys.*, vol. 15, p. 1017, 2019.
[285] A. K. Geim and I. V. Grigorieva, "Van der Waals heterostructures" *Nature*, vol. 499, p. 419, 2013.
[286] K. S. Novoselov, A. Mishchenko, A. Carvalho and A. H. Castro Neto, "2D materials and van der Waals heterostructures" *Science*, vol. 353, p. 6298, 2016.
[287] A. Castellanos-Gomez, "Why all the fuss about 2D semiconductors?" *Nat. Photonics*, vol. 10, p. 202, 2016.
[288] H. Schmidt, T. Ludtke, P. Barthold, E. McCann, V. I. Fal'ko and R. J. Haug, "Tunable graphene system with two decoupled monolayers," *Appl. Phys. Lett.*, vol. 93, p. 172108, 2008.
[289] G. Li, A. Luican, J. M. B. Lopes dos Santos, et al., "Observation of Van Hove singularities in twisted graphene layers," *Nat. Phys.*, vol. 6, p. 109, 2010.
[290] J. D. Sanchez-Yamagishi, T. Taychatanapat, K. Watanabe, T. Taniguchi, A. Yacoby and P. Jarillo-Herrero, "Quantum Hall effect, screening, and layer-polarized insulating States in twisted bilayer graphene," *Phys. Rev. Lett.*, vol. 108, p. 076601, 2012.
[291] D. S. Lee, C. Riedl, T. Beringer, et al., "Quantum Hall effect in twisted bilayer graphene" *Phys. Rev. Lett.*, vol. 107, p. 216602, 2011.
[292] J. D. Sanchez-Yamagishi, J. Y. Luo, A. F. Young, et al., "Helical edge states and fractional quantum Hall effect in a graphene electron–hole bilayer," *Nat. Nanotechnol.*, vol. 12, p. 118, 2017.

[293] Y. Cao, J. Y. Luo, V. Fatemi, et al., "Superlattice-induced insulating states and valley-protected orbits in twisted bilayer graphene" *Phys. Rev. Lett.*, vol. 117, p. 116804, 2016.
[294] K. Liu, L. M. Zhang, T. Cao, et al., "Evolution of interlayer coupling in twisted molybdenum disulfide bilayers," *Nat. Commun.*, vol. 5, p. 4966, 2014.
[295] M. Barbier, P. Vasilopoulos and F. M. Peeters, "Extra Dirac points in the energy spectrum for superlattices on single-layer graphene," *Phys. Rev. B*, vol. 81, p. 075438, 2010.
[296] C. R. Woods, L. Britnell, A. Eckmann, et al., "Commensurate-incommensurate transition in graphene on hexagonal boron nitride," *Nat. Phys.*, vol. 10, p. 451, 2014.
[297] B. Hunt, J. D. Sanchez-Yamagishi, A. F. Young, et al., "Massive Dirac fermions and hofstadter butterfly in a van der Waals heterostructure," *Science*, vol. 340, p. 6139, 2013.
[298] M. Yankowitz, J. Xue, D. Cormode, et al., "Emergence of superlattice Dirac points in graphene on hexagonal boron nitride," *Nat. Phys.*, vol. 8, p. 382, 2012.
[299] Y. Cao, V. Fatemi, S. Fang, et al., "Unconventional superconductivity in magic-angle graphene superlattices" *Nature*, vol. 556, p. 43, 2018.
[300] D. M. Kennes, L. Xian, M. Claassen, and A. Rubio, "One-dimensional flat bands in twisted bilayer germanium selenide," *Nat. Commun.*, vol. 11, p. 1124, 2020.
[301] L. Xian, D. M. Kennes, N. Tancogne-Dejean, M. Altarelli, and A. Rubio, "Multiflat bands and strong correlations in twisted bilayer boron nitride: doping-induced correlated insulator and superconductor," *Nano Lett.*, vol. 19, p. 4934, 2019.
[302] D. M. Kennes, M. Claassen, L. Xian, A. Georges, et al.," Moiré heterostructures: a condensed matter quantum simulator", to appear in *Nat. Phys.*, 2020.
[303] X. Chen, X. Fan, L. Li, et al., "Moiré engineering of electronic phenomena in correlated oxides," *Nat. Phys.*, vol. 16, p. 631, 2020.
[304] K. L. Seyler, P. Rivera, H. Yu, et al., "Signatures of moire-trapped valley excitons in $MoSe_2/WSe_2$ heterobilayers," *Nature*, vol. 567, p. 66, 2019.
[305] Yuan, B. Zheng, J. Kuntsmann, et al., "Twist-angle-dependent interlayer exciton diffusion in WS_2–WSe_2 heterobilayers," *Nat. Mater.*, vol. 19, p. 617, 2020.
[306] W. Li, X. Lu, S. Dubey, L. Devenica and A. Srivastava, "Dipolar interactions between localized interlayer excitons in van der Waals heterostructures," *Nat. Mater.*, vol. 19, p. 624, 2020.
[307] Y. Bai, L. Zhou, J. Wang, et al., "Excitons in strain-induced one-dimensional moire potentials at transition metal dichalcogenide heterojunctions," *Nat. Mater.*, vol. 19, p. 1068, 2020.
[308] S. S. Sunku, G. X. Ni, B. Y. Jiang, et al., "Photonic crystals for nano-light in moiré graphene superlattices" *Science*, vol. 362, p. 1153, 2018.
[309] G. X. Ni, H. Wang, B.-Y. Jiang, et al., "Soliton superlattices in twisted hexagonal boron nitride," *Nat. Commun.*, vol. 10, p. 4360, 2019.
[310] M. Chen, X. Lin, T. H. Dinh, et al., "Configurable phonon polaritons in twisted α-MoO_3," *Nat. Mater.*, 2020. https://doi.org/10.1038/s41563-020-0781-x.
[311] G. Hu, Q. Ou, G. Si, et al., "Topological polaritons and photonic magic angles in twisted α-MoO3 bilayers," *Nature*, vol. 582, p. 209, 2020.
[312] F. Herrera and J. Owrutsky, "Molecular polaritons for controlling chemistry with quantum optics," *J. Chem. Phys.*, vol. 152, p. 100902, 2020.
[313] J. P. Long and B. K. Simpkins, "Coherent coupling between a molecular vibration and Fabry–Perot optical cavity to give hybridized states in the strong coupling limit," *ACS Photonics*, vol. 2, p. 130, 2015.
[314] B. Xiang, R. F. Ribeiro, M. Du, et al., "Intermolecular vibrational energy transfer enabled by microcavity strong light–matter coupling," *Science*, vol. 368, p. 665, 2020.
[315] E. Orgiu, J. George, J. A. Hutchison, et al., "Conductivity in organic semiconductors hybridized with the vacuum field," *Nat. Mater.*, vol. 14, p. 1123, 2015.
[316] R. Chikkaraddy, B. de Nijs, F. Benz, et al., "Single-molecule strong coupling at room temperature in plasmonic nanocavities," *Nature*, vol. 535, p. 127, 2016.
[317] B. Xiang, R. F. Ribeiro, Y. Li, et al., "Manipulating optical nonlinearities of molecular polaritons by delocalization," *Sci. Adv.*, vol. 5, p. eaax5196, 2019.
[318] H. Memmi, O. Benson, S. Sadofev, and S. Kalusniak, "Strong coupling between surface plasmon polaritons and molecular vibrations," *Phys. Rev. Lett.*, vol. 118, p. 126802, 2017.
[319] L. Henriet, Z. Ristivojevic, P.P. Orth, and K. Le Hur, "Quantum dynamics of the driven and dissipative Rabi model," *Phys. Rev. A*, vol. 90, p. 023820, 2014.
[320] S. Schmidt and J. Koch, "Circuit QED lattices towards quantum simulation with superconducting circuits," *Ann. Phys. (Berlin)*, vol. 525, p. 395, 2013.
[321] M. Kiffner, J. Coulthard, F. Schlawin, A. Ardavan and D. Jaksch, "Mott polaritons in cavity-coupled quantum materials," *N. J. Phys.*, vol. 21, p. 073066, 2019.
[322] W. L. Faust and C. H. Henry, "Mixing of visible and near-resonance infrared light in GaP," *Phys. Rev. Lett.*, vol. 17, p. 1265, 1966.
[323] S. A. Holmstrom, T. Stievater, M. W. Pruessner, et al., "Guided-mode phonon-polaritons in suspended waveguides," *Phys. Rev. B*, vol. 86, p. 165120, 2012.
[324] J. D. Caldwell, L. Lindsay, V. Giannini, et al., "Low-loss, infrared and terahertz nanophotonics using surface phonon polaritons," *Nanophotonics*, vol. 4, 2015. https://doi.org/10.1515/nanoph-2014-0003.
[325] S. Vassant, F. Marquier, J. J. Greffet, F. Pardo, and J. L. Pelouard, "Tailoring GaAs terahertz radiative properties with surface phonons polaritons," *Appl. Phys. Lett.*, vol. 97, p. 161101, 2010.
[326] P. C. M. Planken, L. D. Noordam, T. M. Kermis, and A. Lagendijk, "Femtosecond time-resolved study of the generation and propagation of phonon polaritons in LiNbo," *Phys. Rev. B*, vol. 45, p.13, 1992.
[327] T. Feurer, N. S. Stoyanov, D. W. Ward, J. C. Vaughan, E. R. Statz, and K. A. Nelson, "Terahertz polaritonics," *Annu. Rev. Mater. Res.*, vol. 37, p. 317, 2007.
[328] S. Kojima, N. Tsumura and M. W. Takeda, "Far-infrared phonon-polariton dispersion probed by terahertz time-domain spectroscopy," *Phys. Rev. B*, vol. 67, p. 035102, 2003.
[329] H. J. Bakker, S. Hunsche, and H. Kurz, "Coherent phonon polaritons as probes of anharmonic phonons in ferroelectrics," *Rev. Mod. Phys.*, vol. 70, p. 2, 1998.
[330] K.-L. Yeh, M. C. Hoffman, J. Hebling, and K. A. Nelson, "Generation of 10 μJ ultrashort terahertz pulses by optical rectification," *Appl. Phys. Lett.*, vol. 90, p. 171121, 2007.
[331] A. J. Huber, N. Ocelic, and R. Hillenbrand, "Local excitation and interference of surface phonon polaritons studied by near-field infrared microscopy," *J. Microsc.*, vol. 229, p. 389, 2008.

[332] A. J. Huber, R. Hillenbrand, B. Deutsch, and L. Novotny, "Focusing of surface phonon polaritons," *Appl. Phys. Lett.*, vol. 92, p. 203104, 2008.

[333] A. A. Goyyadinov, A. Konecna, A. Chuvilin, et al., "Probing low-energy hyperbolic polaritons in van der Waals crystals with an electron microscope," *Nat. Commun.*, vol. 8, p. 95, 2017.

[334] D.-Z. A. Chen, A. Narayanaswamy, and G. Chen, "Surface phonon-polariton mediated thermal conductivity enhancement of amorphous thin films," *Phys. Rev. B*, vol. 72, p. 155435, 2005.

[335] P. S. Venkataram, J. Hermann, A. Tkatchenko, and A. W. Rodriguez, "Phonon-polariton mediated thermal radiation and heat transfer among molecules and macroscopic bodies: nonlocal electromagnetic response at mesoscopic scales," *Phys. Rev. Lett.*, vol. 121, p. 045901, 2018.

[336] D. G. Cahill, P. V. Braun, G. Chen, et al., "Nanoscale thermal transport," *Appl. Phys. Rev.*, vol. 1, p. 011305, 2014.

[337] K. Kim, B. Song, V. Fernández-Hurtado, et al., "Radiative heat transfer in the extreme near field," Nature, vol. 528, p. 387, 2015.

[338] D. Thompson, L. Zhu, R. Mittapally, et al., "Hundred-fold enhancement in far-field radiative heat transfer over the blackbody limit," *Nature*, vol. 561, p. 216, 2018.

[339] Z. Zheng, J. Chen, Y. Wang, et al., "Highly confined and tunable hyperbolic phonon Polaritons in van der waals semiconducting transition metal oxides," *Adv. Mater.*, vol. 30, p. 1705318, 2018.

[340] W. Ma, P. Alonso-Gonzalez, S. Li, et al., "In-plane anisotropic and ultra-low-loss polaritons in a natural van der Waals crystal," *Nature*, vol. 562, p. 557, 2018.

[341] Z. Zheng, N. Xu, S. L. Oscurato, et al., "A mid-infrared biaxial hyperbolic van der Waals crystal," *Sci. Adv.*, vol. 5, p. eaav8690, 2019.

[342] P. Li, I. Dolado, F. J. Alfaro-Mozaz, et al., "Infrared hyperbolic metasurface based on nanostructured van der Waals materials," *Science*, vol. 359, p. 892, 2018.

[343] A. Cartella, T. F. Nova, M. Fechner, R. Merlin, and A. Cavalleri, "Parametric amplification of optical phonons," *Proc. Natl. Acad. Sci. U.S.A.*, vol. 115, p. 12148, 2018.

[344] S. Dai, F. Wenjing, N. Rivera, et al., "Phonon polaritons in monolayers of hexagonal boron nitride" *Adv. Mater.*, vol. 31, p. 1806603, 2019.

[345] A. V. Zayats, I. I. Smolyaninov and A. A. Maradudin, "Nano-optics of surface plasmon polaritons," *Phys. Rep.*, vol. 408, p. 131, 2005.

[346] S. A. Maier, *Plasmonics: Fundamentals and Applications*, Berlin, Springer, 2007.

[347] D. K. Gramotney and S. I. Bozhevolnvi, "Plasmonics beyond the diffraction limit," *Nat. Photonics*, vol. 4, p. 83, 2010.

[348] H. A. Atwater and A. Polman, "Plasmonics for improved photovoltaic devices," *Nat. Mater.*, vol. 9, p. 205, 2010.

[349] P. Berini and I. De Leon, "Surface plasmon-polariton amplifiers and lasers," *Nat. Photonics*, vol. 6, p. 16, 2012.

[350] S. I. Bogdanov, A. Boltasseva, and V. M. Shalev, "Overcoming quantum decoherence with plasmonics," *Science*, vol. 364, p. 532, 2019.

[351] J. Lee, S. Jung, P. Y. Chen, et al., "Ultrafast electrically-tunable polaritonic metasurfaces," *Adv. Opt. Mater.*, vol. 2, p. 1057, 2014.

[352] M. Kauranen and A. V. Zayats, "Nonlinear plasmonics," *Nat. Photonics*, vol. 6, p. 73, 2012.

[353] J. Lee, N. Nookala, J. S. Gomez-Diaz, et al., "Ultrathin gradient nonlinear metasurfaces with giant nonlinear response," *Optica*, vol. 3, p. 283, 2016.

[354] M. Tymchenko, J. S. Gomez-Diaz, J. Lee, M. A. Belkin, and A. Alù, "Gradient nonlinear Pancharatnam–Berry metasurfaces," *Phys. Rev. Lett.*, vol. 115, p. 207403, 2015.

[355] Z. Jacob, and V. M. Shalev, "Plasmonics goes quantum," *Science*, vol. 334, p. 463, 2011.

[356] M. S. Tame, K. R. McEnery, S. K. Ozdemir, J. Lee, S. A. Maier, and M. S. Kim, "Quantum plasmonics," *Nat. Phys.*, vol. 9, p. 329, 2013.

[357] S. I. Bozhevolnvi and J. B. Khurgin, "The case for quantum plasmonics," *Nat. Photonics*, vol. 11, p. 398, 2017.

[358] J. S. Fakonas, A. Mitskovets, and H. A. Atwater, "Path entanglement of surface plasmons," *New J. Phys.*, vol. 17, p. 023002, 2015.

[359] M.-C. Dheur, F. Devaux, T. W. Ebbesen, et al., "Single-plasmon interferences," *Sci. Adv.*, vol. 2, p. e1501574, 2016.

[360] M. Wagner, A. S. McLeod, S. J. Maddox, et al., "Ultrafast dynamics of surface plasmons in InAs by time-resolved infrared nanospectroscopy," *Nano Lett.*, vol. 14, p. 4529, 2014.

[361] K. F. MacDonald, Z. L. Samson, M. I. Stockman and N. I. Zheludev, "Ultrafast active plasmonics," *Nat. Photonics*, vol. 3, p. 55, 2009.

[362] M. Eisele, T. L. Cocker, M. A. Huber, et al., "Ultrafast multi-terahertz nano-spectroscopy with sub-cycle temporal resolution," *Nat. Photonics*, vol. 8, p. 841, 2014.

[363] M. A. Huber, F. Mooshammer, M. Plankl, et al., "Femtosecond photo-switching of interface polaritons in black phosphorus heterostructures," *Nat. Nanotechnol.*, vol. 12, p. 207, 2017.

[364] Z. Yao, S. Xu, D. Hu, X. Chen, Q. Dai, and M. Liu, "Nanoimaging and nanospectroscopy of polaritons with time resolved s-SNOM," *Adv. Opt. Mater.*, vol. 8, p. 1901042, 2020.

[365] A V Krasavin, A. V. Zayats and N. I. Zheludev, "Active control of surface plasmon-polariton waves," *J. Opt. Pure Appl. Opt.*, vol. 7, p. S85, 2005.

[366] J. Shi, M. H. Lin, I. T. Chen, et al., "Cascaded exciton energy transfer in a monolayer semiconductor lateral heterostructure assisted by surface plasmon polariton," *Nat. Commun.*, vol. 8, p. 35, 2017.

[367] B. Min, E. Ostby, V. Sorger, et al., "High-Q surface-plasmon-polariton whispering-gallery microcavity," *Nature*, vol. 457, p. 455, 2009.

[368] A. Boltasseva and H. A. Atwater, "Low-loss plasmonic metamaterials," *Science*, vol. 331, p. 290, 2011.

[369] G. V. Naik and V. M. Shalaev, "Alternative plasmonic materials: beyond gold and silver," *Adv. Mater.*, vol. 25, p. 3264, 2013.

[370] F. Xia, H. Wang, D. Xiao, M. Dubey, and A. Ramasubramaniam, "Two-dimensional material nanophotonics," *Nat. Photonics*, vol. 8, p. 899, 2014.

[371] F. H. da Jornada, L. Xian, A. Rubio, and S. G. Louie, "Universal slow plasmons and giant field enhancement in atomically thin quasi-twodimensional metals," *Nat. Commun.*, vol. 11, p. 1013, 2020.

[372] E. H. Hwang and S. Das Sarma, "Plasmon modes of spatially separated doublelayer graphene," *Phys. Rev. B*, vol. 80, p. 205405, 2009.

[373] A. Principi, R. Asgari, and M. Polini, "Acoustic plasmons and composite hole-acoustic plasmon satellite bands in graphene on a metal gate," *Solid State Commun.*, vol. 151, p. 1627, 2011.

[374] R. E. V. Profumo, R. Asgari, M. Polini, and A. H. MacDonald, "Double-layer graphene and topological insulator thin-film plasmons," *Phys. Rev. B*, vol. 85, p. 085443, 2012.

[375] S. Chen, M. Autore, J. Li, P. Li, et al., "Acoustic graphene plasmon nanoresonators for field-enhanced infrared molecular spectroscopy," *ACS Photonics*, vol. 4, p. 3089, 2017.
[376] T. Stauber and G. Gomez-Santos, "Plasmons in layered structures including graphene," *New J. Phys.*, vol. 14, p. 105018, 2012.
[377] I.-H. Lee, D. Yoo, P. Avouris, T. Low and S.-H. Oh, "Graphene acoustic plasmon resonator for ultrasensitive infrared spectroscopy, "*Nat. Nanotechnol.*, vol. 14, p. 313, 2019.
[378] P. Alonso-Gonzalez, A. Y. Nikitin, Y. Gao, et al., "Acoustic terahertz graphene plasmons revealed by photocurrent nanoscopy," *Nat. Nanotechnol.*, vol. 12, p. 31, 2017.
[379] M. B. Lundeberg, Y. Gao, R. Asgari, et al., "Tuning quantum nonlocal effects in graphene plasmonics," *Science*, vol. 347, p. 187, 2017.
[380] J. B. Pendry, L. Martin-Moreno, and F. J. Garcia-Vidal, "Mimicking surface plasmons with structured surfaces," *Science*, vol. 305, p. 847, 2014.
[381] M. V. Berry and N. L. Balazs, "Nonspreading wave packets," *Am. J. Phys.*, vol. 47, p. 264, 1979.
[382] A. Minovich, A. E. Klein, N. Janunts, T. Pertsch, D. N. Neshev, and Y. S. Kivshar, "Generation and near-field Imaging of airy surface plasmons," *Phys. Rev. Lett.*, vol. 107, p. 116802, 2011.
[383] M. Hentschel, M. Schäferling, X. Duan, H. Giessen, N. Liu, "Chiral plasmonics," *Sci. Adv.*, vol. 3, p. e1602735, 2017.
[384] T. Stauber, T. Low, and G. Gómez-Santos, "Chiral response of twisted bilayer graphene,"*Phys. Rev. Lett.*, vol. 120, p. 046801, 2018.
[385] D. N. Basov and M. M. Fogler, "Quantum Materials: The quest for ultrafast plasmonics," *Nat. Nanotechnol.*, vol. 12, p. 187, 2017.
[386] P. Vasa, W. Wang, R. Pomraenke, et al., "Real-time observation of ultrafast Rabi oscillations between excitons and plasmons in metal nanostructures with J-aggregates," *Nat. Photonics*, vol. 7, p. 128, 2013.
[387] N. T. Fofang, T. H. Park, O. Neumann, N. A. Mirin, P. Nordlander and N. J. Halas, "Plexcitonic nanoparticles: plasmon-exciton coupling in nanoshell-J-aggregate complexes," *Nano Lett.*, vol. 8, p. 3481, 2008.
[388] N. T. Fofang, N. K. Grady, Z. Fan, A. O. Govorov, and N. J. Halas, "Plexciton dynamics: exciton-plasmon coupling in a J-aggregate–Au nanoshell complex provides a mechanism for nonlinearity," *Nano Lett.*, vol. 11, p. 1556, 2011.
[389] A. Manjavacas, F. J. Garcia de Abajo and P. Nordlander, "Quantum plexcitonics: strongly interacting plasmons and excitons," *Nano Lett.*, vol. 11, p. 2318, 2011.
[390] J. Yuen-Zhou, S. K. Saikin, T. Zhu, et al., "Plexciton Dirac points and topological modes," *Nat. Commun.*, vol. 7, p. 11783, 2016.
[391] A. P. Manuel, A. Kirkey, N. Mahdi and K. Shankar, "Plexcitonics – fundamental principles and optoelectronic applications," *J. Mater. Chem. C*, vol. 7, 2019. https://doi.org/10.1039/c8tc05054f.
[392] K. Wu, W. E. Rodriguez-Cordoba, Y. Yang, and T. Lian, "Plasmon-induced hot electron transfer from the Au tip to CdS rod in CdS-Au nanoheterostructures," *Nano Lett.*, vol. 13, p. 5255, 2013.
[393] P. Torma and W. L. Barnes, "Strong coupling between surface plasmon polaritons and emitters: a review," *Rep. Prog. Phys.*, vol. 78, p. 013901, 2015.
[394] P. A. D. Goncalves, L. P. Bertelsen, S. Xiao and N. Mortensen, "Plasmon-exciton polaritons in two-dimensional semiconductor/metal interfaces," *Phys. Rev. B*, vol. 97, p. 041402, 2018.
[395] V. Karanikolas, I. Thanopulos, and E. Paspalakis, "Strong interaction of quantum emitters with a WS_2 layer enhanced by a gold substrate," *Opt. Lett.*, vol. 44, p. 2049, 2019.
[396] T. Chervy, S. Azzini, E. Lorchat, et al., "Room temperature chiral coupling of valley excitons with spin-momentum locked surface plasmons" *ACS Photonics*, vol. 5, p. 1281, 2018.
[397] H. Groß, J. M. Hamm, T. Tufarelli, O. Hess, and B. Hecht, "Near-field strong coupling of single quantum dots," *Sci. Adv.*, vol. 4, p. eaar4906, 2018.
[398] M. A. May, D. Fialkow, T. Wu, et al., "Nano-cavity QED with tunable nano-tip interaction," *Adv. Quantum Technol.*, vol. 3, p. 190087, 2020.
[399] M. Saba, C. Ciuti, J. Bloch, et al., "High-temperature ultrafast polariton parametric amplification in semiconductor microcavities," *Nature*, vol. 414, p. 731, 2001.
[400] S. Kéna-Cohen and S. R. Forrest, "Room-temperature polariton lasing in an organic single-crystal microcavity," *Nat. Photonics*, vol. 4, p. 371, 2010.
[401] H. Deng, G. Weihs, D. Snoke, J. Bloch, and Y. Yamamoto, "Polariton lasing vs. photon lasing in a semiconductor microcavity," *Proc. Natl. Acad. Sci. U.S.A.*, vol. 23, p. 5318, 2003.
[402] Y. Liu, H. Fang, A. Rasmita, et al., "Room temperature nanocavity laser with interlayer excitons in 2D heterostructures," *Sci. Adv.*, vol. 5, p. eaav4506, 2019.
[403] J. Flick, N. Rivera and P. Narang, "Strong light–matter coupling in quantum chemistry and quantum photonics," *Nanophotonics*, vol. 7, p. 1479, 2018.
[404] J. A. Hutchison, T. Schwartz, C. Genet, E. Devaux, and T. W. Ebbesen, "Modifying chemical landscapes by coupling to vacuum fields," *Angew. Chem. Int. Ed.*, vol. 51, p. 1592, 2012.
[405] J. Yuen-Zhou and V. M. Menon, "Polariton chemistry: thinking inside the (photon) box," *Proc. Natl. Acad. Sci. U.S.A.*, vol. 116, p. 5214, 2019.
[406] J. Flick, M. Ruggenthaler, H. Appel, and A. Rubio "Atoms and molecules in cavities: from weak to strong coupling in QED chemistry," *Proc. Natl. Acad. Sci. U.S.A.*, vol. 114, p. 3026, 2017.
[407] C. Schäfer, M. Ruggenthaler, V. Rokaj, and A. Rubio, "Relevance of the quadratic diamagnetic and self-polarization terms in cavity quantum electrodynamics," *ACS Photonics*, vol. 7, p. 975, 2020.
[408] D. Sanvitto and S. Kéna-Cohen, "The road towards polaritonic devices," *Nat. Mater.*, vol. 15, p. 1061, 2016.
[409] A. Amo, T. C. H. Liew, C. Adrados, et al., "Exciton-polariton spin switches," *Nat. Photonics*, vol. 4, p. 361, 2010.
[410] T. Gao, P. Eldridge, T. Liew, et al., "Polariton condensate transistor switch," *Phys. Rev. B*, vol. 85, p. 235102, 2012.
[411] D. Ballarini, M. De Giorgi, E. Cancellieri, et al., "All-optiocal polariton transistor," *Nat. Commun.*, vol. 4, p. 1778, 2013.
[412] C. Antón, T. C. H. Liew, J. Cuadra, et al., "Quantum refelctions and shunting of polariton condensate wave trains: Implementation of a logic AND gate," *Phys. Rev. B*, vol. 88, p. 245307, 2013.

[413] H. S. Nguyen, D. Vishnevsky, C. Sturm, et al., "Realization of a double-barrier resonant tunneling Diode for cavity polaritons," *Phys. Rev. Lett.*, vol. 110, p. 236601, 2013.

[414] F. Marsault, H. S. Nguuyen, D. Tanese, et al., "Realization of an all optical exciton-polariton router," *Appl. Phys. Lett.*, vol. 107, p. 201115, 2015.

[415] A. Kavokin, T. C. H. Liew, C. Schneider, S. Hofling "Bosonic lasers" *Low Temp. Phys.*, vol. 42, p. 323, 2016.

[416] D. Ballarini, A. Gianfrate, R. Panico, et al., "Polaritonic neuromorphic computing outperforms linear classifiers," *Nano Lett.*, vol. 20, p. 3506, 2020.

[417] J. Gu, B. Chakraborty, M. Khatoniar and V. M. Menon, "A room-temperature polariton light-emitting diode based on monolayer WS2," *Nat. Nanotechnol.*, vol. 14, p. 1024, 2019.

[418] E. A. Cerda-Méndez, D. N. Krizhanovskii, M. Wouters, et al., "Polariton Condensation in dynamic acoustic lattices," *Phys. Rev. Lett.*, vol. 105, p. 116402, 2010.

[419] A. Amo and J. Bloch, "Exciton-polaritons in lattices: a non-linear photonic simulator," *Compt. Rendus Phys.*, vol. 17, p. 934, 2016.

[420] L. Zhang, W. Xie, J. Wang, et al., "Weak lasing in one-dimensional polariton superlattices," *Proc. Natl. Acad. Sci. U.S.A.*, vol. 112, p. E1516, 2015.

[421] M. J. Hartmann, F. G. S. L. Brandao, and M. B. Plenio, "Quantum many-body phenomena in coupled cavity arrays," *Laser Photonics*, vol. 2, p. 6, 2008.

[422] N. Y. Kim and Y. Yamamoto, "*Exciton-Polariton Quantum Simulators in Quantum Simulations with Photons and Polaritons*," vol. 91, D. Angelakis, ed., Springer, 2017.

[423] I. M. Georgescu, S. Ashhab and F. Nori, "Quantum simulation," *Rev. Mod. Phys.*, vol. 86, p. 153, 2014.

[424] J. I. Cirac and P. Zoller, "Golas and opportunities in quantum simulation," *Nat. Phys.*, vol. 8, p. 264, 2012.

[425] I. Bloch, J. Dalibard, and S. Nascimbene, "Quantum simulations with ultracold quantum gases," *Nat. Phys.*, vol. 8, p. 267, 2012.

[426] N. C. Harris, G. R. Steinbrecher, M. Prabhu, et al., "Quantum transport simulations in a programmable nanophotonic processor," *Nat. Photonics*, vol. 11, p. 447, 2017.

[427] A. Aspuru-Guzik and P. Walther, "Photonic quantum simulators," *Nat. Phys.*, vol. 8, p. 285, 2012.

[428] M. J. Hartmann, F. G. S. L. Brandao, and M. B. Plenio, "Strongly interacting polaritons in coupled arrays of cavities," *Nat. Phys.*, vol. 2, p. 849, 2006.

[429] A. D. Greentree, C. Tahan, J. H. Cole, and L. C. L. Hollenberg, "Quantum phase transitions of light," *Nat. Phys.*, vol. 2, p. 856, 2006.

[430] T. Byrnes, P. Recher, and Y. Yamamoto, "Mott transitions of excitons polaritons and indirect excitons in a periodic potential," *Phys. Rev. B*, vol. 81, p. 205312, 2010.

[431] N. Na and Y. Yamamoto, "Massive parallel generation of indistinguishable single photons iva the polaritonic superfluid to Mott-insulator quantum phase transition," *New J. Phys.*, vol. 12, p. 123001, 2010.

[432] N. G. Berlo, M. Silva, K. Kalinin, et al., "Realizing the classical XY Hamiltonian in polariton simulators," *Nat. Mater.*, vol. 16, p. 1120, 2017.

[433] T. Esslinger, "Fermi-hubbard Physics with Atoms in an optical lattice," *Ann. Rev. Condensed Matter Phys.*, vol. 1, p. 129, 2010.

[434] R. Blatt and C. F. Roos, "Quantum simulations with trapped ions," *Nat. Phys.*, vol. 8, p. 277, 2012.

[435] K. Kim, S. Korenblit, R. Islam, et al., "Quantum simulation of the transverse Ising model with trapped ions," *New J. Phys.*, vol. 13, p. 105003, 2011.

[436] J. W. Britton, B. C. Sawyer, A. C. Keith, et al., "Engineered two-dimensional Ising interactions in a trapped-ion quantum simulator with hundreds of spins," *Nature*, vol. 484, p. 489, 2012.

[437] A. A. Houck, H. Tureci, and J. Koch, "On-chip quantum simulation with superconducting circuits," *Nat. Phys.*, vol. 8, p. 292, 2012.

[438] J. Koch, A. A. Houck, K. Le Hur, and S. M. Girvin, "Time-reversal-symmetry breaking in circuit-QED-based photon lattices," *Phys. Rev. A*, vol. 82, p. 043811, 2010.

[439] L. Xiong, C. Forsythe, M. Jung, et al., "Photonic crystal for graphene plasmons" *Nat. Commun.*, vol. 10, p. 4780, 2019.

[440] B.-Y. Jiang, G.-X. Ni, Z. Addison, et al., "Plasmon reflections by topological electronic boundaries in bilayer graphene," *Nano Lett.*, vol. 17, p. 7080, 2017.

[441] Z. Fei, G. X. Ni, B. Y. Jiang, M. M. Fogler, and D. N. Basov, "Nanoplasmonic phenomena at electronic boundaries in graphene," *ACS Photonics*, vol. 4, no. 12, p. 2971, 2017.

[442] J. Chen, M. L. Nesterov, A. Y. Nikitin, et al., "Strong plasmon reflection at nanometer-size gaps in monolayer graphene on SiC," *Nano Lett.*, vol. 13, p. 6210, 2013.

[443] E. H. Hasdeo and J. C. W. Song, "Long-lived domain wall plasmons in gapped bilayer graphene," *Nano Lett.*, vol. 17, p. 7252, 2017.

[444] R. Zia and M. L. Brongersma, "Surface plasmon polariton analogue to Young's double-slit experiment," *Nat. Nanotechnol.*, vol. 2, p. 426, 2007.

[445] P. Li, M. Lewin, A. V. Kretinin, et al., "Hyperbolic phonon-polaritons in boron nitride for near-field optical imaging and focusing," *Nat. Commun.*, vol. 7, p. 7507, 2015.

[446] C. Yang, Q. Wu, J. Xu, K. A. Nelson, and C. A. Werley, "Experimental and theoretical analysis of THz-frequency, direction-dependent, phonon polariton modes in a subwavelength, anisotropic slab waveguide," *Opt. Express*, vol. 18, p. 26351, 2010.

[447] P. Peier, K. A. Nelson and T. Feurer, "Coherent phase contrast imaging of THz phonon-polariton tunneling," *Appl. Phys. B*, vol. 99, p. 433, 2010.

[448] J. Lin, J. Dellinger, P. Genevet, B. Cluzel, F. Fornel, and F. Capasso, "Cosine-gauss plasmon beam: a localized long-range nondiffracting surface wave," *Phys. Rev. Lett.*, vol. 109, p. 093904, 2012.

[449] I. Epstein and A. Arie, "Arbitrary bending plasmonic light waves," *Phys. Rev. Lett.*, vol. 112, p. 023903, 2014.

[450] A. R. Davoyan, I. V. Shadrivov, and Y. S. Kivshar, "Self-focusing and spatial plasmon-polariton solitons," *Opt. Express*, vol. 17, p. 21732, 2009.

[451] J. Levinsen, G. Li, and M. M. Parish, "Microscopic description of exciton-polaritons in microcavities," *Phys. Rev. Res.*, vol. 1, p. 033120, 2019.

[452] L. Nguyen-th^e, S. De Liberato, M. Bamba, and C. Ciuti, "Effective polariton-polariton interactions of cavity-embedded two-dimensional electron gases," *Phys. Rev. B*, vol. 87, p. 235322, 2013.

[453] Y. S. Yoseob Yoon, M. Steger, G. Liu, et al., "Direct measurement of polariton–polariton interaction strength," *Nat. Phys.*, vol. 13, p. 870, 2017.

[454] E. Estrecho, T. Gao, N. Bobrovska, et al., "Direct measurement of polariton-polariton interaction strength in the Thomas–Fermi regime of exciton-polariton condensation," *Phys. Rev. B*, p. 035306, 2019.

[455] X. Lina, Y. Yangc, N. Riverab, et al., "All-angle negative refraction of highly squeezed plasmon and phonon polaritons in graphene–boron nitride heterostructures," *Proc. Natl. Acad. Sci. U.S.A.*, vol. 114, p. 6717, 2017.

[456] K. Chaudhary, M. Tamagnone, X. Yin, et al., "Polariton nanophotonics using phase-change materials," *Nat. Commun.*, vol. 10, p. 4487, 2019.

[457] P. Alonso-González, A. Y. Nikitin, F. Golmar, et al., "Controlling graphene plasmons with resonant metal antennas and spatial conductivity patterns," *Science*, vol. 344, p. 1369, 2014.

[458] M. Sidler, P. Back, O. Cotlet, et al., "Fermi polaron-polaritons in charge-tunable atomically thin semiconductors," *Nat. Phys.*, vol. 13, p. 255, 2017.

[459] D. K. Efimkin and A. H. MacDonald, "Many-body theory of trion absorption Features in two-dimensional semiconductors," *Phys. Rev. B*, vol. 95, p. 035417, 2017.

[460] G. Scalari, C. Maissen, D. Turcinkova, et al., "Ultrastrong coupling of the cyclotron transition of a 2D electron gas to a THz metamaterial," *Science*, vol. 335, p. 1323, 2012.

[461] F. M. D. Pellegrino, V. Giocannetti, A. H. MacDonald, and M. Polini, "Modulated phases of graphene quantum Hall polariton fluids," *Nat. Commun.*, vol. 7, p. 13355, 2016.

[462] I. Neder, N. Ofek, Y. Chung, M. Heiblum, D. Mahalu, and V. Umansky, "Interference between two indistinguishable electrons from independent sources," *Nature (London)*, vol. 448, p. 333, 2007.

[463] C. Bäuerle, D. C. Glattli, T. Meunier, et al., "Coherent control of single electrons: a review of current progress," *Rep. Prog. Phys.*, vol. 81, p. 056503, 2018.

[464] J. Splettstoesser and R. J. Haug, "Single-electron control in solid state devices," *Phys. Status Solidi*, vol. B254, p. 1770217, 2017.

[465] D. C. Glattli and P. S. Roulleau, "Levitons for electron quantum optics," *Phys. Status Solidi*, vol. B254, p. 1600650, 2016.

[466] S. Smolka, W. Wuester, F. Haupt, S. Faelt, W. Wegscheider, and A. Imamoglu, "Cavity quantum electrodynamics with many-body states of a two-dimensional electron gas," *Science*, vol. 346, p. 332, 2014.

[467] P. Knuppel, S. Ravets, M. Kroner, S. Falt, W. Wegscheider, and A. Imamoglu, "Nonlinear optics in the fractional quantum Hall regime," *Nature*, vol. 572, p. 91, 2019.

[468] L. Bing Tan, O. Cotlet, A. Bergschneider, et al., "Interacting polaron-polaritons," *Phys. Rev. X*, vol. 10, p. 021011, 2020.

[469] M. Sich, L. E. Tapia-Rodriguez, H. Sigurdsson, et al., "Spin domains in one-dimensional conservative polariton solitons," *ACS Photonics*, vol. 5, p. 5095, 2018.

[470] M. D. Lukin, M. Fleischhauer, R. Cote, et al., "Dipole Blockade and quantum information Processing in mesoscopic atomic ensembles," *Phys. Rev. Lett.*, vol. 87, p. 037901, 2001.

[471] D. Comparat and P. Pillet, "Dipole blockade in a cold Rydberg atomic sample," *J. Opt. Soc. Am. B*, vol. 27, p. A208, 2010.

[472] A. V. Gorshkov, J. Otterbach, M. Fleischhauer, T. Pohl, and M. D. Lukin, "Photon–photon interactions via Rydberg blockade," *Phys. Rev. Lett.*, vol. 107, p. 133602, 2011.

[473] E. Shahmoon, G. Kurizki, M. Fleischhauer, and D. Petrosyan, "Strongly interacting photons in hollow-core waveguides," *Phys. Rev. A*, vol. 83, p. 033806, 2011.

[474] T. Peyronel, O. Firstenberg, Q.-Y. Liang, et al., "Quantum nonlinear optics with single photons enabled by strongly interacting atoms," *Nature*, vol. 488, p. 57, 2012.

[475] A. Chernikov, T. C. Berkelbach, H. M. Hill, et al., "Exciton binding energy and nonhydrogenic Rydberg series in monolayer WS_2," *Phys. Rev. Lett.*, vol. 113, p. 076802, 2014.

[476] P. Merkl, F. Mooshammer, S. Brem, et al., "Twist-tailoring Coulomb correlations in van der Waals homobilayers," *Nat. Commun.*, vol. 11, p. 2167, 2020.

[477] T. Kazimierczuk, D. Fröhlich, S. Scheel, H. Stolz, and M. Bayer, "Giant Rydberg excitons in the copper oxide Cu_2O," *Nature*, vol. 514, p. 343, 2014.

[478] W. P. Su, J. R. Schrieffer, and A. J. Heeger, "Solitons in polyacetylene," *Phys. Rev. Lett.*, vol. 42, p. 1698, 1979.

[479] P. St-Jean, V. Goblot, E. Galopin, et al., "Lasing in topological edge states of a one-dimensional lattice," *Nat. Photonics*, vol. 11, p. 651, 2017.

[480] M. Bello, G. Platero, J. I. Cirac, and A. González-Tudela Bello, "Unconventional quantum optics in topological waveguide QED," *Sci. Adv.*, vol. 5, p. eaaw0297, 2019.

[481] C. A. Downing, T. J. Sturges, G. Weick, M. Stobinska, and L. Martin-Moreno, "Topological phases of polaritons in a cavity waveguide," *Phys. Rev. Lett.*, vol. 123, p. 217401, 2019.

[482] A. Bhattacharya, M. Zunaid Baten, I. Iorsh, T. Frost, A. Kavokin, and P. Bhattacharya, "Room-temperature spin polariton diode laser," *Phys. Rev. Lett.*, vol. 119, p. 067701, 2017.

[483] A. Agarwal, M. Polini, G. Vignale, and M. E. Flatt, "Long-lived spin plasmons in a spin-polarized two-dimensional electron gas," *Phys. Rev. B*, vol. 90, p. 155409, 2014.

[484] L. I. Magarill, A. V. Chaplik, and M. V. Éntin, "Spin-plasmon oscillations of the two-dimensional electron gas," *J. Exp. Theor. Phys.*, vol. 92, p. 15, 2001.

[485] S. Raghu, S. Bum Chung, X. L. Qi, and S.-C. Zhang, "Collective Modes of a helical liquid," *Phys. Rev. Lett.*, vol. 104, p. 116401, 2010.

[486] J. B. Pendry, D. Schurig, and D. R. Smith, "Controlling electromagnetic electromagnetic fields," *Science*, vol. 312, p. 1780, 2006.

[487] Ulf Leonhardt, "Optical conformal mapping," *Science*, vol. 312, p. 1777, 2006.

[488] P. A. Huidobro, M. L. Nesterov, L. Martın-Moreno, and F. J. Garcıa, "Vidal transformation optics for plasmonics," *Nano Lett.*, vol. 10, p. 1985, 2010.

[489] V. Ginis, P. Tassin, J. Danckaert, C. M. Soukoulis and I. Veretennicoff, "Creating electromagnetic cavities using transformation optics," *New J. Phys.*, vol. 14, p. 03300, 2012.

[490] A. Vakil and N. Engheta, "Transformation optics using graphene," *Science*, vol. 332, p. 1291, 2011.

[491] I. Tamm, "Über eine mögliche Art der Elektronenbindung an Kristalloberflächen," *Z. Phys.*, vol. 76, p. 849, 1932.

[492] M. Kaliteevski, I. Iorsh, S. Brand, et al., "Tamm plasmon-polaritons: possible electromagnetic states at the interface of a metal and a dielectric Bragg mirror," *Phys. Rev. B*, vol. 76, p. 165415, 2007.

[493] B. Liu, R. Wu, and V. M. Menon, "Propagating hybrid Tamm exciton Polaritons in organic microcavity," *J. Phys. Chem. C*, vol. 123, no. 43, p. 26509, 2019.

[494] S. Dhara, C. Chakraborty, K. M. Goodfellow, et al., "Anomalous dispersion of microcavity trion-polaritons," *Nat. Phys.*, vol. 14, p. 130, 2017.
[495] R. P. A. Emmanuele, M. Sich, O. Kyriienko, et al., "Highly nonlinear trion-polaritons in a monolayer semiconductor," *Nat. Commun.*, vol. 11, p. 3589, 2020.
[496] C. Möhl, A. Graf, F. J. Berger, et al., "Trion-polariton formation in single-walled carbon nanotube microcavities," *ACS Photonics*, vol. 5, p. 2074, 2018.
[497] S. de Vega and F. Javier García de Abajo, "Plasmon generation through electron tunneling in graphene," *ACS Photonics*, vol. 4, p. 2367, 2017.
[498] A. Woessner, A. Misra, Y. Cao, et al., "Propagating plasmons in a charge-neutral quantum tunneling transistor," *ACS Photonics*, vol. 4, p. 3012, 2017.
[499] X. Xu, W. Yao, D. Xiao, and T. Heinz, "Spin and pseudospins in layered transition metal dichalcogenides," *Nat. Phys.*, vol. 10, p. 343, 2014.
[500] R. Peng, C. Wu, H. Li, X. Xu, and M. Li, "Separation of the valley exciton-polariton in two-dimensional semiconductors with an anisotropic photonic crystal," *Phys. Rev. B*, vol. 101, p. 245418, 2020.
[501] S. Guddala, R. Bushati, M. Li, A. B. Khanikaev, and V. M. Menon, "Valley selective optical control of excitons in 2D semiconductors using a chiral metasurface," *Opt. Mater. Express*, vol. 9, p. 536, 2019.
[502] I. Egri and A. Stahl, "Real space wave equation for exciton-polaritons pf wannier type," *Phys. Status Solidi*, vol. 96, p. K83, 1979.
[503] G. Calajo, F. Ciccarello, D. Chang, and P. Rabl, "Atom-field dressed states in slow-light waveguide QED," *Phys. Rev. A*, vol. 93, p. 033833, 2016.
[504] B. K. Ofori-Okai, P. Sivarajah, C. A. Werley, S. M. Teo, and K. A. Nelson, "Direct experimental visualization of waves and band structure in 2D photonic crystal slabs," *New J. Phys.*, vol. 16, p. 053003, 2014.
[505] A. Asenjo-Garcia, J. D. Hood, D. E. Chang, and H. J. Kimble, "Atom-light interactions in quasi-one-dimensional nanostructures: a Green's-function perspective," *Phys. Rev. A*, vol. 95, p. 033818, 2017.
[506] E. Vetsch, D. Reitz, G. Sagué, R. Schmidt, S. T. Dawkins, and A. Rauschenbeutel, "Optical interface created by laser-cooled atoms trapped in the evanescent field surrounding an optical nanofiber," *Phys. Rev. Lett.*, vol. 104, p. 203603, 2010.
[507] A. Goban, K. S. Choi, D. J. Alton, et al., "Demonstration of a state-insensitive, compensated nanofiber trap," *Phys. Rev. Lett.*, vol. 109, p. 033603, 2012.
[508] A. Goban, C.-L. Hung, J. D. Hood, et al., "Superradiance for atoms trapped along a photonic crystal waveguide," *Phys. Rev. Lett.*, vol. 115, p. 063601, 2015.
[509] J. D. Hood, A. Goban, A. Asenjo-Garcia, et al., "Atom–atom interactions around the band edge of a photonic crystal waveguide," *Proc. Natl. Acad. Sci. U.S.A.*, vol. 113, p. 10507, 2016.
[510] M. Mirhosseini, E. Kim, X. Zhang, et al., "Cavity quantum electrodynamics with atom-like mirrors," *Nature*, vol. 569, p. 692, 2019.
[511] Y. Liu and A. A. Houck, "Quantum electrodynamics near a photonic bandgap," *Nat. Phys.*, vol. 13, p. 48, 2017.
[512] T. Shi, Y.-H. Wu, A. González-Tudela, and J. I. Cirac, "Bound states in boson impurity models," *Phys. Rev. X*, vol. 6, p. 021027, 2016.
[513] J. Zenneck, "Fortplfanzung ebener elektromagnetischer Wellenlängs einer ebenen Leiterfläche," *Ann. Phys.*, vol. 328, p. 846, 1907.
[514] K. A. Norton, "Propagation of radio waves over the surface of the earth and in the upper atmosphere," *Proc. IRE*, vol. 24, p. 1367, 1936.
[515] A. Sommerfeld, "Uber die Ausbreitung der Wellen in derdrahtlosen Telegraphie,"*Ann. Phys.*, vol. 333, p. 665, 1909.
[516] M. Sarrazin and J.-P. Vigneron, "Light transmission assisted by Brewster–Zennek modes in chromium films carrying a subwavelength hole array," *Phys. Rev. B*, vol. 71, p. 075404, 2005.
[517] A. Shivola, J. Qi, and I. V. Lindell, "Bridging the gap between plasmonics and Zenneck waves," *IEEE Antenn. Propag. Mag.*, vol. 52, p. 124, 2010.
[518] K. A. Michalski and J. R. Mosig, "The sommerfeld halfspace problem redux: alternative field representations, Role of Zenneck and surface plasmon waves," *IEEE Trans. Antenn. Propag.*, vol. 63, 2015. https://doi.org/10.1109/tap.2015.2489680.
[519] A. Yu Nikitin, S. G. Rodrigo, F. J. García-Vidal and L. Martín-Moreno, "In the diffraction shadow: Norton waves versus surface plasmon polaritons in the optical region," *New J. Phys.*, vol. 11, p. 123020, 2009.
[520] V. E. Babicheva, S. Gamage, L. Zhen, S. B. Cronin, V. S. Yakovlev, and Y. Abate, "Near-field surface waves in few-layer MoS_2," *ACS Photonics*, vol. 5, p. 2106, 2018.

Rituraj*, Meir Orenstein and Shanhui Fan*

Scattering of a single plasmon polariton by multiple atoms for in-plane control of light

https://doi.org/10.1515/9783110710687-045

Abstract: We study the interaction of a single photon in a surface plasmon polariton mode with multiple atoms. We propose a system of two atoms to achieve a tunable scattering from subscattering to superscattering regimes by changing the angle of the incident photon. We also demonstrate a perfect electromagnetically-induced transparency using two atoms with two-level structures. The proposed framework is efficiently scalable to a system with a large number of atoms and opens up the possibility of designing novel atom-based optical devices. We design an atomically thin parabolic mirror to focus single photons and form a quantum mirage in a cavity built from atoms.

Keywords: electromagnetically induced transparency; nanophotonics; plasmonics; quantum electrodynamics; scattering; waveguide QED.

1 Introduction

Photon–atom interaction is an important subject with considerable theoretical and practical interests [1–3]. With the development of nanotechnology it has now become possible to tailor this interaction by designing nanophotonic structures with unique optical properties as well as artificial atoms like a superconducting qubit, quantum dot or a Rydberg atom in a highly excited state [4–6]. In recent years there have been numerous studies investigating the coherent scattering of a few photon Fock states by an atom [7–16]. Most of these studies, however, are concerned with a single or a few atoms coupled to one dimensional (1D) continuum of photonic modes of the waveguide [7–14, 17]. It has been shown that a two-level atom coupled to a waveguide acts as a perfect reflector near the resonant frequency, despite its subwavelength size. Similarly, it has been shown that an atom exhibits a cross section much larger than its physical dimensions for single photon scattering in free space near the resonant frequency [16]. Unlike a 1D waveguide, higher dimensionality provides much richer opportunities of manipulating photons through careful geometric arrangement of atoms [18]. Still, very few works have been done regarding scattering of single photons in two (2D) or three dimensions (3D) by multiple atoms.

In a recent work, we presented a general model for the scattering of surface plasmon polariton (SPP) mode by a single atom (in general any two-level quantum system) without making the usual dipole approximation [15]. Since the coupling of the atom to the slow surface modes is much stronger than its coupling to the free space modes, the system essentially represents an atom interacting with a 2D photonic environment. In the current work, we further develop the formalism to compute the scattering properties for a more complicated scenario of multiple atoms coupled to a single photon in the SPP mode. The proposed model is general and includes all the multiple scattering events. The 2D setting with multiple atoms allows us to implement complex photon based quantum circuitry, and here we exemplify it by a few basic examples. We show that a system of two atoms can be tuned to exhibit either subscattering or superscattering by simply changing the photon angle of incidence. We also achieve a perfect atom cloaking with zero scattering at a certain frequency between the resonant frequency of the two atoms. This is different from the usual electromagnetically-induced transparency (EIT) which is based on interference between the transition paths in an atom with at least a three-level structure [1, 19–21]. We further explore the possibilities of designing novel atom-based optical devices to manipulate single photons and demonstrate multiple atoms based single photon focusing and the formation of quantum mirage in a 2D cavity like system.

***Corresponding authors: Rituraj and Shanhui Fan,** Electrical Engineering, Stanford University, Stanford, USA,
E-mail: rituraj@stanford.edu (Rituraj), shanhui@stanford.edu (S. Fan). https://orcid.org/0000-0002-7842-6808 (Rituraj)
Meir Orenstein, Electrical Engineering, Technion Israel Institute of Technology, Haifa, Israel

This article has previously been published in the journal Nanophotonics. Please cite as: Rituraj, M. Orenstein and S. Fan "Scattering of a single plasmon polariton by multiple atoms for in-plane control of light" *Nanophotonics* 2021, 10. DOI: 10.1515/nanoph-2020-0340.

2 Mathematical formulation

Here, we develop the formalism for a system of N two-level atoms coupled to a single photon in 2D SPP mode. An atom–SPP system is shown schematically in Figure 1 for two atoms. The infinite 2D surface that supports the SPP mode (shown in green) is taken to be the $z = 0$ plane and the nth atom (represented by red cylinder), is placed at the coordinates $(\mathbf{r}_n \equiv (x_n, y_n), h_n)$. We assume that the atom is separated from the surface by vacuum, and the SPP mode is ideal without any propagation losses. In the Coulomb gauge, the SPP vector potential operator $\mathbf{A}(\mathbf{r}, z)$ in the upper half space $(z > 0)$ is given by:

$$\mathbf{A}(\mathbf{r},z) = \frac{1}{2\pi} \underbrace{\iint dk_x\, dk_y \sqrt{\frac{\hbar}{2L_{\mathbf{k}}\epsilon_0\omega_{\mathbf{k}}}}\left(i\hat{k} - \frac{k}{\kappa}\hat{z}\right)e^{-\kappa z}e^{i\mathbf{k}\cdot\mathbf{r}}}_{\mathbf{A}_{\mathbf{k}}(\mathbf{r},z)}\, a_{\mathbf{k}} + H.c, \quad (1)$$

where, $a_k, a_k^\dagger$ are the Bosonic annihilation and creation operators for the SPP mode satisfying the commutation $[a(\mathbf{k}), a^\dagger(\mathbf{k}')] = \delta^2(\mathbf{k} - \mathbf{k}')$, $\hbar$ is the reduced Planck's constant, ϵ_0 is the vacuum permittivity, $\kappa = \sqrt{k^2 - \omega_{\mathbf{k}}^2/c^2}$ is the spatial decay rate of the mode along z, $L_{\mathbf{k}}$ is the characteristic modal length given by $L_{\mathbf{k}} = (\kappa^2 + k^2)/\kappa^3$ and is derived through normalization consideration [22–25], $H.c$ stands for Hermitian conjugate. $\omega_{\mathbf{k}}$ is the SPP mode frequency at in-plane wavevector $\mathbf{k} \equiv (k_x, k_y)$, and could be well approximated by Eq. (2) for frequency close to the surface plasmon frequency ω_{sp}:

$$\omega_{\mathbf{k}} = \sqrt{\frac{1+\varepsilon_m}{\varepsilon_m}}ck \approx \frac{\omega_p}{\sqrt{2}}\left(1 - \frac{k_p^2}{8k^2}\right) = \omega_{sp} - \frac{\beta}{k^2}, \quad (2)$$

where, ϵ_m is the real dielectric constant of the metal given by the Drude model, ω_p is its plasma frequency, k_p is the free space wave-vector magnitude at frequency $\omega_p (= ck_p)$, c is the speed of light in vacuum [26–28].

We assume that the atoms couple only to the SPP mode and ignore coupling to the free-space electromagnetic modes.

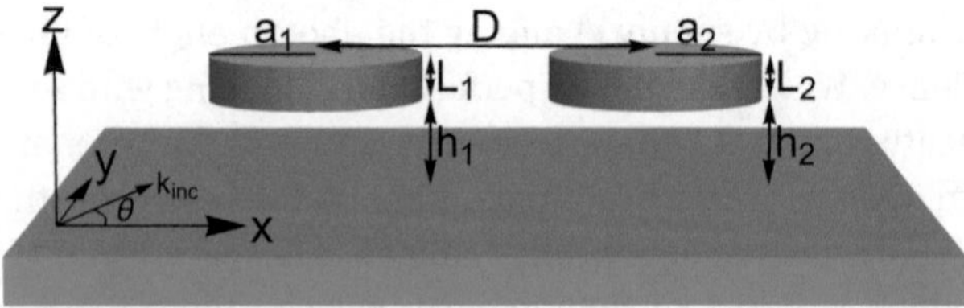

Figure 1: Two atoms (shown in red) coupled to the surface plasmon polariton (SPP) mode of an infinite two dimensional (2D) surface (shown in green). The surface is the $(z = 0)$ plane and each of the atom-like system confines the electron wave functions in a cylinder with radius a_n and height L_n, located at a distance h_n above the surface.

This assumption is well justified since the near field coupling to the SPP modes is much stronger than coupling to the free-space modes. It is also assumed that the atoms do not directly interact with each other. Starting with the standard minimal coupling Hamiltonian, for resonant coupling near the surface plasmon frequency, we can show that the light–matter interaction in the atom–SPP system can be described by the following spatial domain Hamiltonian [15]:

$$\begin{aligned} H = {} & \hbar\omega_{sp} \iint dx\, dy c^\dagger(\mathbf{r})c(\mathbf{r}) \\ & + \hbar\beta \iint dx\, dy \iint dx'\, dy' \frac{\ln|\mathbf{r}-\mathbf{r}'|}{2\pi} c^\dagger(\mathbf{r})c(\mathbf{r}') \\ & + \sum_{n=1}^{N} E_g^n b_g^{n\dagger} b_g^n + E_e^n b_e^{n\dagger} b_e^n + \sum_{n=1}^{N} \iint dx\, dy \Big(V_n(\mathbf{r}) c^\dagger(\mathbf{r}) b_g^{n\dagger} b_e^n \\ & + V_n^*(\mathbf{r}) c(\mathbf{r}) b_e^{n\dagger} b_g^n \Big) \end{aligned} \quad (3)$$

where, E_g^n and E_e^n are the ground and excited state energies for the electron in the nth atom, and $b_g^n, b_g^{n\dagger}, b_e^n, b_e^{n\dagger}$ are the respective Fermionic annihilation and creation operators. $c(\mathbf{r}), c^\dagger(\mathbf{r})$ are the spatial Bosonic annihilation and creation operators respectively as defined by Eq. (4), and satisfy the commutation $[c(\mathbf{r}), c^\dagger(\mathbf{r}')] = \delta^2(\mathbf{r} - \mathbf{r}')$. $V_n(\mathbf{r})$ is the Fourier transform of the atom–field coupling strength $V_{\mathbf{k}}^n$ given by Eq. (5):

$$c(\mathbf{r}) = \frac{1}{2\pi} \iint dk_x\, dk_y\, e^{i\mathbf{k}\cdot\mathbf{r}} a_{\mathbf{k}}\ ;\ c^\dagger(\mathbf{r}) = \frac{1}{2\pi} \iint dk_x\, dk_y\, e^{-i\mathbf{k}\cdot\mathbf{r}} a_{\mathbf{k}}^\dagger \quad (4)$$

$$V_{\mathbf{k}}^{n*} = -\frac{e}{m}\langle e_n|\mathbf{A}_{\mathbf{k}} \cdot \mathbf{p}_n|g_n\rangle;\ V_n(\mathbf{r}) = \frac{1}{2\pi} \iint dk_x\, dk_y\, e^{i\mathbf{k}\cdot\mathbf{r}} V_{\mathbf{k}}^n \quad (5)$$

where, e and m are respectively the charge and the rest mass of the electron, $\mathbf{p}_n = -i\hbar\vec{\nabla}_n$ is the canonical momentum operator for the electron in the nth atom. In the Hamiltonian of Eq. (3), we have ignored the terms related to intrinsic spin angular momentum and the $\mathbf{A}^2$ term. We have also made the usual rotating wave approximation in the interaction Hamiltonian [1, 2, 29]. These approximations are justified in the weak coupling regime (small $V_{\mathbf{k}}$), which is the case here and are also validated from the results shown later where the linewidths in scattering spectrum are much smaller than the resonant frequency. The logarithmic form in the second term of the Hamiltonian of Eq. (3) arises from the SPP dispersion relation in the short wavelength limit (Eq. (2)).

Consider an incident SPP photon with a wavevector $\mathbf{q}$ interacting with the atoms, the resulting stationary state can then be written as:

$$\begin{aligned} |\psi_{\mathbf{q}}\rangle = {} & \iint dx\, dy\, \phi(\mathbf{r}) c^\dagger(\mathbf{r}) |g, g, \ldots g, 0\rangle \\ & + \sum_{n=1}^{N} e_{\mathbf{q}}^n b_e^{n\dagger} b_g^n |g, g, \ldots g, 0\rangle \end{aligned} \quad (6)$$

where, $|g, g, \ldots g, 0\rangle$ is the state with all atoms in the ground state and zero photons in the SPP mode, $\phi(\mathbf{r})$ is the photon field amplitude, and $e^n_{\mathbf{q}}$ is the excited state amplitude for the nth atom. Eq. (6) represents a complete basis for the system [7, 30]. Now, using Eq. (6) in the eigenvalue problem $H|\psi_{\mathbf{q}}\rangle = E_{\mathbf{q}}|\psi_{\mathbf{q}}\rangle$ gives the following equations:

$$\left(\hbar\omega_{sp} + \sum_{n=1}^{N} E^n_g - E_{\mathbf{q}}\right)\phi(\mathbf{r}) + \hbar\beta \iint dx'\, dy' \frac{\ln|\mathbf{r}-\mathbf{r}'|}{2\pi}\phi(\mathbf{r}') = -\sum_{n=1}^{N} e^n_{\mathbf{q}} V_n(\mathbf{r}) \tag{7}$$

$$(E_{\mathbf{q}} - E^n_e)e^n_{\mathbf{q}} = \iint dx\, dy\, V^*_n(\mathbf{r})\phi(\mathbf{r}). \tag{8}$$

Taking 2D Laplacian $(\widehat{\Delta} \equiv \partial^2/\partial x^2 + \partial^2/\partial y^2)$ of both sides of integral equation (7) and substituting $E_{\mathbf{q}} = \sum_n E^n_g + \hbar\omega_{sp} - \hbar\beta/q^2$, we get the 2D Helmholtz equation with N distributed source terms

$$\widehat{\Delta}\phi(\mathbf{r}) + q^2\phi(\mathbf{r}) = -\frac{q^2}{\hbar\beta}\sum_{n=1}^{N} e^n_{\mathbf{q}}\widehat{\Delta}V_n(\mathbf{r}). \tag{9}$$

Writing in dimensionless units by defining $\mathbf{r}' = k_{sp}\mathbf{r}$ and $\mathbf{q}' = \mathbf{q}/k_{sp}$, where $k_{sp} = \omega_{sp}/c$, we obtain

$$\widehat{\Delta}'\left(\frac{\phi(\mathbf{r}')}{k_{sp}}\right) + q'^2\left(\frac{\phi(\mathbf{r}')}{k_{sp}}\right) = -q'^2\left(\frac{\omega_{sp}k_{sp}}{\beta}\right)\sum_{n=1}^{N} e^n_{\mathbf{q}'}\widehat{\Delta}'\left(\frac{V_n(\mathbf{r}')}{\hbar\omega_{sp}k_{sp}}\right) \tag{10}$$

$$\frac{(E_{\mathbf{q}'} - E^n_e)}{\hbar\omega_{sp}}e_{\mathbf{q}'} = \iint dx'\, dy' \left(\frac{V^*_n(\mathbf{r}')}{\hbar\omega_{sp}k_{sp}}\right)\left(\frac{\phi(\mathbf{r}')}{k_{sp}}\right). \tag{11}$$

Following the standard procedure of computing scattering eigenstates, far away from the atoms where $V_n(\mathbf{r})$ approaches 0 fast enough as r increases, $\phi(\mathbf{r})$ could be expressed as a sum of incident and scattered waves [31]:

$$\phi(\mathbf{r}) = \phi_{inc} + \phi_{sca} = e^{i\mathbf{q}\cdot\mathbf{r}} + \phi_{sca}. \tag{12}$$

In Eq. (12), the incident wave ϕ_{inc} can be any SPP wave in the absence of coupling to the atoms $(V_n(\mathbf{r}) = 0)$ and is taken here as a 2D SPP plane wave. The scattered wave ϕ_{sca} of the ensemble of atoms can be computed from the knowledge of the scattered fields for all the individual (isolated) atoms ϕ^n_{sca} $(1 \le n \le N)$ [32–34]. The isolated atom scattering is computed by the procedure outlined in the previous work [15]. We model the two-level atom as an infinite cylindrical potential well as shown in Figure 1 with the following wavefunctions for the ground $(l_g = 1)$ and excited $(l_e = 2)$ states [35]:

$$\psi^n_{g,e}(\mathbf{r}, z) = \sqrt{\frac{2}{L_n}}\sin\frac{l_{g,e}\pi(z-h_n)}{L_n} \times \frac{\sqrt{2}}{aJ_1(j_0)}J_0(j_0|\mathbf{r}-\mathbf{r}_n|/a_n)\Theta_n(r,z) \tag{13}$$

where, J_n is the nth order Bessel function, j_0 is the first zero of the 0th order Bessel function, $\Theta_n(r,z)$ is a scalar function which is unity inside the cylinder $(|\mathbf{r}-\mathbf{r}_n| < a_n, h_n < z < h_n + L_n)$ and zero outside. In this case, $V_n(\mathbf{r})$ has azimuthal symmetry (Eq. (5)) and the scattered field could be expressed in terms of the 0th order Hankel function of the first kind $H^{(1)}_0(q|\mathbf{r}-\mathbf{r}_n|)$.

$$\begin{aligned}\phi_n(\mathbf{r}) &= e^{i\mathbf{q}\cdot\mathbf{r}} + \phi^n_{sca} = e^{i\mathbf{q}\cdot\mathbf{r}_n}\left(e^{i\mathbf{q}\cdot(\mathbf{r}-\mathbf{r}_n)} + b_{0n}H^{(1)}_0(q|\mathbf{r}-\mathbf{r}_n|)\right)\\ &= e^{i\mathbf{q}\cdot\mathbf{r}_n}\left(\sum_{\substack{m=-\infty\\ m\neq 0}}^{\infty} i^m J_m(q|\mathbf{r}-\mathbf{r}_n|)e^{im\theta} + J_0(q|\mathbf{r}-\mathbf{r}_n|) + b_{0n}H^{(1)}_0(q|\mathbf{r}-\mathbf{r}_n|)\right)\end{aligned} \tag{14}$$

In the above expression for the total field $\phi_n(\mathbf{r})$, only the $m = 0$ angular momentum component of the incident wave is scattered and the scattering coefficient for the nth isolated atom b_{0n} is computed using appropriate boundary conditions as done in the previous work [15]. This procedure is repeated for all the atoms to determine $b_{0n}, \forall n \in \{1, \ldots, N\}$. It follows from the linearity of Eq. (10) that the scattered wave for the system of N atoms can be expressed as a superposition of the scattered waves by individual atoms as:

$$\phi(\mathbf{r}) = \phi_{inc} + \phi_{sca} = e^{i\mathbf{q}\cdot\mathbf{r}} + \sum_{n=1}^{N} b_n H^{(1)}_0(q|\mathbf{r}-\mathbf{r}_n|), \tag{15}$$

where, b_n is now the scattering coefficient for the nth atom, and is in general different from the isolated atom scattering coefficient b_{0n} due to multiple scattering between the different atoms. It can be shown that the above expression for the scattered field satisfies Eq. (10) in the far field where $V_n(\mathbf{r})$ goes to 0. To compute the scattering coefficients b_n, we use the Graf's addition theorem for the Hankel function (Eq. (16)) and express all the scattered waves $H^{(1)}_0(q|\mathbf{r}-\mathbf{r}_n|)$ in Eq. (15) in terms of Bessel functions centered at a particular jth atom.

$$H^{(1)}_0(q|\mathbf{r}-\mathbf{r}_n|) = \sum_{m=-\infty}^{\infty} H^{(1)}_m(q|\mathbf{r}_n-\mathbf{r}_j|)J_m(q|\mathbf{r}-\mathbf{r}_j|)e^{im\theta} \tag{16}$$

$$\begin{aligned}\phi(\mathbf{r}) &= e^{i\mathbf{q}\cdot\mathbf{r}} + b_j H^{(1)}_0(q|\mathbf{r}-\mathbf{r}_j|) + \sum_{\substack{n=1\\ n\neq j}}^{N}\sum_{m=-\infty}^{\infty} b_n H^{(1)}_m(q|\mathbf{r}_n-\mathbf{r}_j|)J_m(q|\mathbf{r}-\mathbf{r}_j|)e^{im\theta}\\ &= \sum_{m=-\infty}^{\infty} e^{im\theta}\left(i^m e^{i\mathbf{q}\cdot\mathbf{r}_j}J_m(q|\mathbf{r}-\mathbf{r}_j|) + \sum_{\substack{n=1\\ n\neq j}}^{N} b_n H^{(1)}_m(q|\mathbf{r}_n-\mathbf{r}_j|)J_m(q|\mathbf{r}-\mathbf{r}_j|)\right)\\ &\quad + b_j H^{(1)}_0(q|\mathbf{r}-\mathbf{r}_j|)\end{aligned} \tag{17}$$

The above expression can be interpreted as follows. The jth atom sees the combination of the incident plane wave and the waves scattered by all the other atoms as its total incident wave and only scatters the $m = 0$ angular momentum component. From linearity, the scattering coefficient b_j is proportional to the net amplitude of the $J_0(q|\mathbf{r} - \mathbf{r}_j|)$ term in Eq. (17) and the proportionality constant is the same as for that for the isolated atom in Eq. (14):

$$b_j = b_{0j}\left(e^{i\mathbf{q}\cdot\mathbf{r}_j} + \sum_{\substack{n=1\\n\neq j}}^{N} b_n H_m^{(1)}(q|\mathbf{r}_n - \mathbf{r}_j|)\right). \tag{18}$$

Repeating this for all the atoms ($j \in \{1,\ldots,N\}$), we obtain N linear equations which can be expressed as $\mathbf{Mb} = \mathbf{in}$, where $\mathbf{b} = (b_1, b_2, \ldots, b_N)^T$, $\mathbf{in} = (e^{i\mathbf{q}\cdot\mathbf{r}_1}, e^{i\mathbf{q}\cdot\mathbf{r}_2}, \ldots, e^{i\mathbf{q}\cdot\mathbf{r}_N})^T$ and $\mathbf{M}$ is a $N \times N$ matrix with elements $M_{m,n} = \delta_{m,n}/b_{0m} + (\delta_{m,n} - 1)H_0^{(1)}(q|\mathbf{r}_m - \mathbf{r}_n|)$. $\delta_{m,n}$ is the Kronecker delta function. Solving this system of linear equations then allows to treat the scattering from N atoms.

3 Results

Having presented a general framework to compute the scattering eigenstates of N atoms coupled to the 2D SPP mode, in this section we study a few interesting applications involving light manipulation using atoms.

3.1 Subscattering and superscattering

As the first case, we examine the interaction of a single SPP with two quantum dots with subwavelength horizontal separation. We choose the following parameters for the quantum dots: $a = 10$ nm, $L = 3.35$ nm, and they are separated by a distance $D = 260$ nm and placed at a height $h = 50$ nm from the surface as shown in Figure 1. For these values the atomic transition energy $\hbar\Omega_n = (E_e^n - E_g^n)$ lies close to the surface plasmon energy $\hbar\omega_{sp} = 0.1$ eV. We choose for each atom a slightly different transition frequency amounting to different detuning values ($\Delta\omega_n = \Omega_n - \omega_{sp}$, $\hbar\Delta\omega_1 = -9.5$ meV, $\hbar\Delta\omega_2 = -10$ meV) which could be achieved either by choosing slightly different dimensions for the quantum dots or by applying static external fields. As shown in Figure 1, the angle of incidence is defined as the angle between the incident SPP wave vector $\mathbf{q}$ and the line joining the center of the two atoms (x axis). Figure 2 plots the total scattering cross section σ_T, given by Eq. (19), as a function of frequency for different angles of incidence θ [34].

$$\sigma_T = \frac{\lambda}{\pi^2}\int_0^{2\pi} d\phi |T(\phi)|^2 = \frac{2\lambda}{\pi}\left(|b_1|^2 + |b_2|^2 + 2J_0(qD)\mathrm{Re}(b_1 b_2^*)\right) \tag{19}$$

The dotted blue curves in Figure 2a show the individual atomic scattering cross sections ($|b_{01}|^2, |b_{02}|^2$), while the red curves correspond to the total scattering cross section for the system of two atoms. The two detuning values are chosen such that their difference ($\Delta\omega_1 - \Delta\omega_2$) is smaller than the individual atomic linewidths resulting in a significant overlap between the two spectra (Figure 2a). As expected, we observe a highly anisotropic scattering behavior. For incidence angles close to the normal ($\theta = \pi/2$), the spectrum shows two peaks (Figure 2a and b). In a small frequency range between the peaks, we observe EIT-like behavior (subscattering) where the total scattering cross section is smaller than the individual scattering cross sections. At small angles of incidence, the spectrum shows only a single prominent resonant peak (Figure 2c and d). This is the typical signature of superscattering behavior where the total scattering cross section is larger than the individual ones. A classical analog of the current setup was considered in a previous work [36] where the authors used coupled mode theory formalism to analyze the transmission cross section of a metal film with two slits each supporting a localized resonance. The spectral response is characterized by two resonant modes: superradiant mode (the broad resonance in Figure 2a) and the subradiant mode (narrow resonance in Figure 2a). The two modes can also be computed as eigenmodes of the Hamiltonian of the

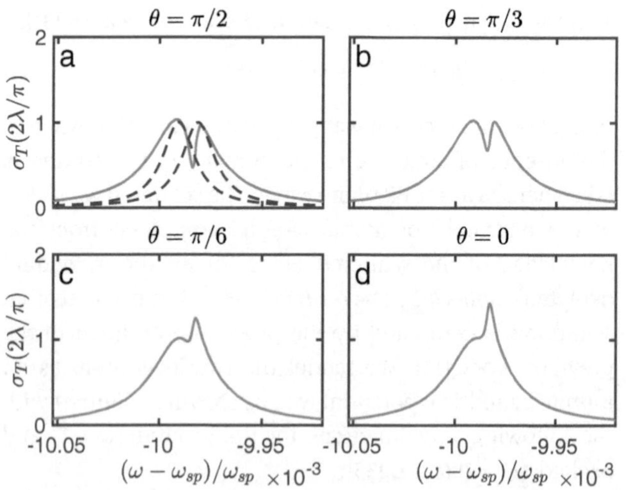

Figure 2: Scattering cross section of the system of two atoms coupled to the surface plasmon polariton (SPP) mode as shown in Figure 1, plotted as a function of frequency for different angles of incidence (plotted in red). The dotted blue curves show the scattering amplitudes ($|b_{0,1}|^2, |b_{0,2}|^2$) for the two individual atoms.

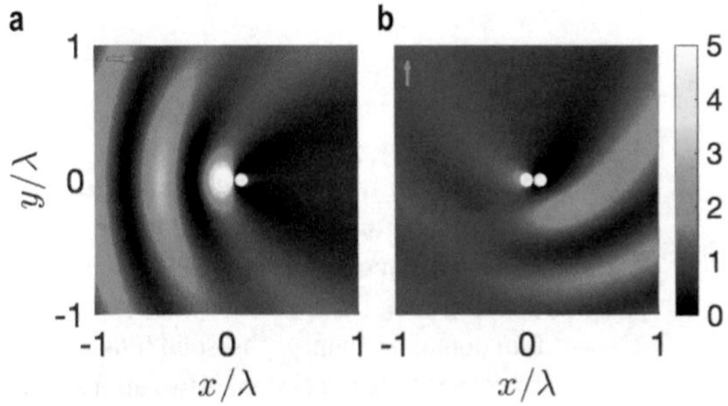

Figure 3: Total field intensity ($|\phi(\mathbf{r})|^2$) for an incident wave with frequency ($\omega - \omega_{sp} = -9.985 \times 10^{-3}\omega_{sp}$) and angle of incidence θ. (a) superscattering ($\theta = 0$), (b) subscattering ($\theta = \pi/2$). The green arrow indicates the direction of the incident plane wave.

system of two atoms after integrating out the photonic degree of freedom which introduces a coupling between the atoms. The EIT-like dip(superscattering) arises because of destructive(constructive) interference of radiation/scattering from the two resonances depending on the angle of incidence [16, 36–38]. The lateral inter-atom distance D is crucial for the required characteristic of the 2D scattering, since the radiation/scattering patterns from the two resonant modes will be very distinct for a large separation. Thus, EIT-like response is only observed for subwavelength separation where the overlap between the emission patterns can yield efficient destructive interference. The vertical separation $|h_1 - h_2|$ is less crucial, since it does not affect the overlap between the emission patterns and primarily controls the relative linewidth of the spectrum from the two atoms. Figure 3 shows the squared field amplitude $|\phi(\mathbf{r})|^2$ plot for two orthogonal directions of incidence at a frequency corresponding to the wavevector magnitude $q = 5.004\,k_{sp}$ or, wavelength $\lambda = 2.465\,\mu\text{m}$. We observe subscattering behavior at normal incidence and a superscattering behavior for grazing incidence. There is a high field amplitude concentrated near the two quantum dots which is more pronounced for the superscattering case as compared to the subscattering case.

3.2 Perfect atom cloaking

In the previous section, even though we observe an EIT-like behavior, the scattering cross section does not go to zero. One of the conditions required to obtain a perfect scattering cancellation for a system of two resonators as pointed out in Ref. [36] is to have identical radiation profile for the subradiant and superradiant eigenmodes or equivalently for the two resonators. It is possible to achieve this condition in our system, where the two atoms predominantly couple to the same 2D surface mode, by aligning them horizontally and displacing vertically as shown in Figure 4a. The two quantum dots are identical to those of the previous section ($a = 10\,\text{nm}, L = 3.35\,\text{nm}$) but with different detuning values ($\hbar\Delta\omega_1 = -0.15\,\text{meV}, \hbar\Delta\omega_2 = -0.16\,\text{meV}$) and ($h_1 = 200\,\text{nm}, h_2 = 250\,\text{nm}$). Similar to the previous case, these parameters are chosen such that the two individual atomic spectra have a significant overlap with slightly different resonant frequencies and linewidths. Here, since the system has azimuthal symmetry, only $m = 0$ angular momentum mode is scattered and the scattering is independent of angle of incidence. Figure 4b plots the scattering cross section as a function of frequency. The solid blue curve corresponds to the system of two atoms, whereas the dotted black curves correspond to the scattering spectrum of the individual atoms. We observe two peaks in the total scattering plot close to the two individual atomic resonances. In between the two peaks, at a certain frequency the scattering cross section goes to zero showing a perfect cloaking of the atoms. Furthermore we also see a broadening in the scattering linewidth evident from the broader tail in the blue spectrum where the net scattering cross section is larger than that of the individual atoms. The EIT-like spectrum is qualitatively similar to the one discussed in the previous subsection (Figure 2a) and arises from Fano interference between the two resonant pathways corresponding to the superradiant (broader peak in Figure 4b) and subradiant (narrow peak in Figure 4b) modes.

Figure 4c and d show the squared field amplitude ($|\phi(\mathbf{r})|^2$) plot for two different frequencies corresponding to perfect cloaking or EIT ($\sigma_T = 0$) and the higher frequency resonance ($\sigma_T = 2\lambda/\pi$) respectively. In the EIT case even though the scattered field is zero and field amplitude is constant far away from the atom, the near field amplitude is non-uniform. This is expected as the excited state amplitudes $e^n_{\mathbf{q}}$ for both the atoms is non-zero resulting in a net non-zero distributed source term in Eq. (10). At the EIT frequency, the emitted/scattered fields from the two source terms (atoms) cancel each other in the far field limit, and there is some energy stored in the near field and the excited state population. For the resonant scattering case shown in Figure 4d, we observe a large shadow behind the atoms and a large near field amplitude. This cloaking scheme is scalable and could be used to generate omnidirectional cloak for any distribution of atoms near a desired EIT frequency. It is also interesting to note that one can tune from an EIT-like response to Autler-Townes splitting (ATS) like behavior, which is characterized by a well separated doublet peak in spectral response, by introducing a direct (electronic) coupling between the two atoms [39–42].

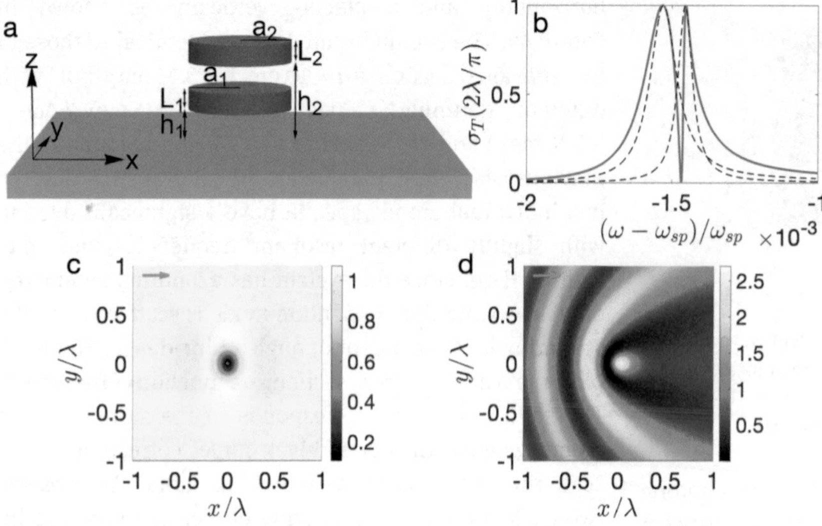

Figure 4: (a) The system of two horizontally aligned quantum dots coupled to the surface plasmon polariton (SPP) mode. (b) Scattering cross section ($|b_0|^2$) as a function of frequency. The solid blue curve corresponds to the system of two atoms and the dotted black curves are for the individual atoms. Total field intensity ($|\phi(\mathbf{r})|^2$) plot for (c) electromagnetically-induced transparency (EIT) ($\omega - \omega_{sp} = -1.5 \times 10^{-3}\omega_{sp}$), and (d) resonance ($\omega - \omega_{sp} = -1.4 \times 10^{-3}\omega_{sp}$). The green arrow indicates the direction of the incident wave.

3.3 Large structures

Having looked at the system of two atoms with sub-wavelength separation to achieve control and tuning of the scattering in the previous sections, now we design larger atomic optical devices to manipulate single photons in useful ways. In particular, we demonstrate an atomically thin parabolic mirror which concentrates light at its focus [43–46] and quantum mirage formation in an elliptical cavity like structure [47]. For the following computation, all the quantum dots are assumed to be identical with the same parameter values used in Section 3.1 and detuning $\hbar\Delta\omega = -9.5\,\text{meV}$.

Figure 5 shows the results for the parabolic mirror, with N identical atoms uniformly spaced along the parabola $y = -x^2/(4f)$, with $f = 50\lambda, \lambda = 2.46\,\mu\text{m}$. The incident plane wave is propagating along the axis of symmetry of the parabola (y axis). Figure 5a and b show the results for $N = 100$ atoms whereas Figure 5c and d correspond to $N = 1200$ atoms. Figure 5a and c show the total field amplitude squared $|\phi(\mathbf{r})|^2$ plot for the frequency corresponding to the atomic resonant frequency. The white dots represent the position of the atoms. We clearly see a high field amplitude concentrated near the focus of the parabola $\mathbf{f} = (0, -f)$. Furthermore, we observe a higher field amplitude in front of the mirror ($y < -x^2/(4f)$) and a much lower field behind it. The field amplitude at the focus and the contrast between the fields in front of and behind the mirror is greater for larger N as seen from Figure 5c. Thus the system acts as a parabolic mirror and the performance improves as the number of atoms N is increased. Also, from the inset in Figure 5c one can see that the focal spot size is smaller than a wavelength and is diffraction-limited [46]. When the atoms are more than a wavelength apart, the field amplitude at the focus increases almost linearly with N (intensity increases quadratically). Figure 5b and d show the bandwidth of the mirror and plot the maximum scattered field amplitude (normalized) near the focus of the parabola as a function of frequency (solid red curve). The dotted black curves correspond to the scattering coefficient for the individual atoms. For small N (Figure 5b), the separation between the atoms is large and the net interaction resulting from the multiple scattering between the atoms, is relatively weak resulting in a Lorentzian spectrum for the mirror which is quite similar to the scattering spectrum for individual atoms. As the number of atoms is increased (Figure 5d), the separation between the atoms decreases resulting in a stronger interaction. This leads to a much broader and frequency-shifted spectrum as observed in Figure 5d.

Finally, we discuss an interesting case of quantum mirage formation [47] for a system of 150 identical atoms uniformly spaced along the circumference of an ellipse ($x^2/a^2 + y^2/b^2 = 1$) with ($a = 5\lambda, b = 4\lambda$) and the wavelength corresponding to the individual atomic resonance ($\lambda = 2.46\,\mu\text{m}$). The atom to be imaged is placed at the positive focus of the ellipse ($\mathbf{f}_+ = (\sqrt{a^2 - b^2}, 0)$). Figure 6 shows the total field amplitude profile ($|\phi(\mathbf{r})|^2$) for an incident cylindrical wave of angular momentum $m = 0$, i.e., $\phi_{inc} = J_0(q|\mathbf{r} - \mathbf{f}_+|)$. The white dots show the position of the atoms. Besides observing a high field amplitude near the atom placed at the positive focus $\mathbf{f}_+$, we also observe a high field concentrated near the negative focus $-\mathbf{f}_+$ of the ellipse which is a mirage or optical image of the real atom. This results from a coherent superposition of the waves originally scattered by the atom at $\mathbf{f}_+$ and subsequently

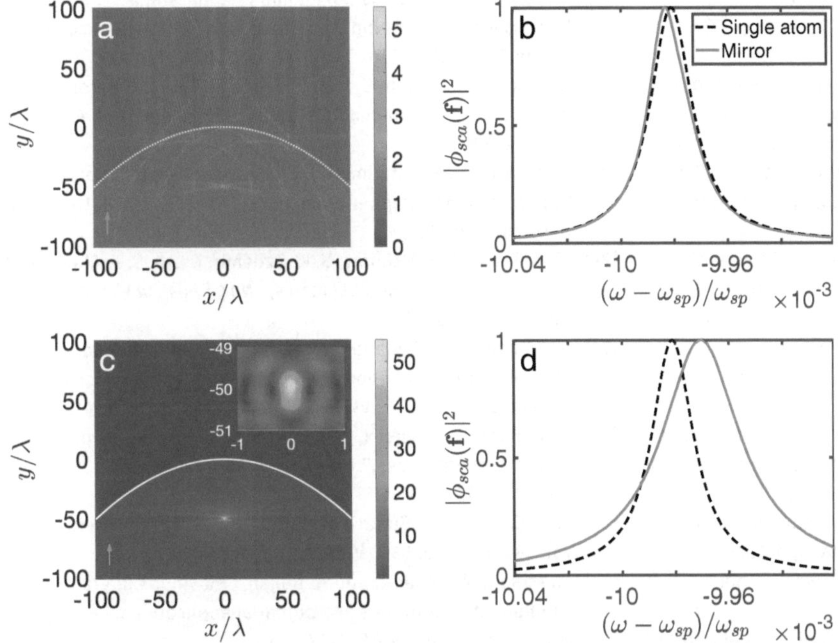

Figure 5: Total field intensity ($|\phi(\mathbf{r})|^2$) plots for a system with (a) 100 atoms (marked by white dots), and (c) 1200 atoms, placed along a parabola. The green arrow indicates the direction of the incident wave. Maximum scattered field amplitude normalized and plotted in solid red as a function of frequency for (b) 100 atoms, and (d) 1200 atoms. The dotted black curve corresponds to the scattered field amplitude for a single atom.

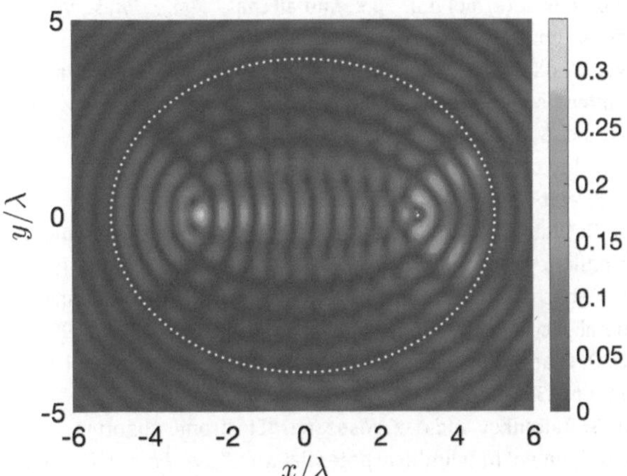

Figure 6: Total field intensity ($|\phi(\mathbf{r})|^2$) plot for a system with 150 atoms (marked by white dots) placed along the circumference of an ellipse, and one atom at the positive focus of the ellipse. The incident wave is taken to be $\phi_{inc} = J_0(q|\mathbf{r} - \mathbf{f}_+|)$.

reflected by the atoms on the ellipse. This is because of the geometric property of the ellipse where the sum of distances of the two foci from any point on the ellipse is a constant ($2a$). Note that we get a mirage only for a specific wavelength where the path difference ($\delta = 2a - 2|\mathbf{f}_+|$) is an integral multiple of wavelength i.e., $q\delta = 2n\pi$ for some integer n.

4 Conclusion

We have demonstrated the possibility of realizing optical devices using multiple atoms interacting with a surface plasmon polariton. Such atomic structures are not only the smallest possible realization of optical structures but also offer unprecedented control over the optical response as compared to the conventional dielectric structures. Among various examples that we have shown, we have realized, for the first time, a perfect EIT for 2D SPP mode using only two-level atoms unlike the usual EIT which requires at least three-level quantum system. The proposed EIT scheme is general and could also be applied to cloak a large number of atoms.

Acknowledgment: This work is supported by a Vannevar Bush Faculty Fellowship from the U. S. Department of Defense (Grant No. N00014-17-1-3030). Rituraj acknowledges the support from a Stanford Graduate Fellowship.

Author contribution: All the authors have accepted responsibility for the entire content of this submitted manuscript and approved submission.

Research funding: This work is supported by a Vannevar Bush Faculty Fellowship from the U. S. Department of Defense (Grant No. N00014-17-1-3030).

Conflict of interest statement: The authors declare no conflicts of interest regarding this article.

References

[1] M. O. Scully and M. S. Zubairy, *Quantum Optics*, 1999.

[2] D. F. Walls and G. J. Milburn, *Quantum Optics*, Springer Science & Business Media, 2007.

[3] J. L. O'brien, A. Furusawa, and J. Vučković, "Photonic quantum technologies," *Nat. Photonics*, vol. 3, no. 12, p. 687, 2009.

[4] A. Goban, C. L. Hung, S. P. Yu, et al., "Atom–light interactions in photonic crystals," *Nat. Commun.*, vol. 5, p. 3808, 2014.

[5] A. F. Van Loo, A. Fedorov, K. Lalumiere, B. C. Sanders, A. Blais, and A. Wallraff, "Photon-mediated interactions between distant artificial atoms," *Science*, vol. 342, no. 6165, pp. 1494–1496, 2013.

[6] A. V. Gorshkov, J. Otterbach, M. Fleischhauer, T. Pohl, and M. D. Lukin, "Photon-photon interactions via Rydberg blockade," *Phys. Rev. Lett.*, vol. 107, no. 13, p. 133602, 2011.

[7] J. T. Shen and S. Fan, "Coherent photon transport from spontaneous emission in one-dimensional waveguides," *Opt. Lett.*, vol. 30, no. 15, pp. 2001–2003, 2005.

[8] J. T. Shen and S. Fan, "Coherent single photon transport in a one-dimensional waveguide coupled with superconducting quantum bits," *Phys. Rev. Lett.*, vol. 95, no. 21, p. 213001, 2005.

[9] S. Fan, Ş. E. Kocabaş, and J. T. Shen, "Input-output formalism for few-photon transport in one-dimensional nanophotonic waveguides coupled to a qubit," *Phys. Rev.*, vol. 82, no. 6, p. 063821, 2010,.

[10] P. Longo, P. Schmitteckert, and K. Busch, "Dynamics of photon transport through quantum impurities in dispersion-engineered one-dimensional systems," *J. Opt. Pure Appl. Opt.*, vol. 11, no. 11, p. 114009, 2009.

[11] L. Zhou, Z. R. Gong, Y. X. Liu, C. P. Sun, and F. Nori, "Controllable scattering of a single photon inside a one-dimensional resonator waveguide," *Phys. Rev. Lett.*, vol. 101, no. 10, p. 100501, 2008.

[12] J. Q. Liao, Z. R. Gong, L. Zhou, Y. X. Liu, C. P. Sun, and F. Nori, "Controlling the transport of single photons by tuning the frequency of either one or two cavities in an array of coupled cavities," *Phys. Rev.*, vol. 81, no. 4, p. 042304, 2010.

[13] D. Witthaut and A. S. Sørensen, "Photon scattering by a three-level emitter in a one-dimensional waveguide," *New J. Phys.*, vol. 12, no. 4, p. 043052, 2010.

[14] H. Zheng, D. J. Gauthier, and H. U. Baranger, "Waveguide-QED-based photonic quantum computation," *Phys. Rev. Lett.*, vol. 111, no. 9, p. 090502, 2013.

[15] M. O. Rituraj and S. Fan, "Two-level quantum system as a macroscopic scatterer for ultraconfined two-dimensional photonic modes," *Phys. Rev.*, vol. 102, no. 1, p. 013717, 2020.

[16] J. Liu, M. Zhou, and Z. Yu, "Quantum scattering theory of a single-photon Fock state in three-dimensional spaces," *Opt. Lett.*, vol. 41, no. 18, pp. 4166–4169, 2016.

[17] L. Zhou, H. Dong, Y. X. Liu, C. P. Sun, and F. Nori, "Quantum supercavity with atomic mirrors," *Phys. Rev.*, vol. 78, no. 6, p. 063827, 2008.

[18] E. Shahmoon, D. S. Wild, M. D. Lukin, and S. F. Yelin, "Cooperative resonances in light scattering from two-dimensional atomic arrays," *Phys. Rev. Lett.*, vol. 118, no. 11, p. 113601, 2017.

[19] S. E. Harris, J. E. Field, and A. Imamoğlu, "Nonlinear optical processes using electromagnetically induced transparency," *Phys. Rev. Lett.*, vol. 64, no. 10, p. 1107, 1990.

[20] K. J. Boller, A. Imamoğlu, and S. E. Harris, "Observation of electromagnetically induced transparency," *Phys. Rev. Lett.*, vol. 66, no. 20, p. 2593, 1991.

[21] M. Fleischhauer, A. Imamoğlu, and J. P. Marangos, "Electromagnetically induced transparency: Optics in coherent media," *Rev. Mod. Phys.*, vol. 77, no. 2, p. 633, 2005.

[22] K. J. Blow, R. Loudon, S. J. D. Phoenix, and T. J. Shepherd, "Continuum fields in quantum optics," *Phys. Rev.*, vol. 42, no. 7, p. 4102, 1990.

[23] B. A. Ferreira, B. Amorim, A. J. Chaves, and N. M. R. Peres, "Quantization of graphene plasmons," *Phys. Rev.*, vol. 101, no. 3, p. 033817, 2020.

[24] M. S. Tame, K. R. McEnery, Ş. K. Özdemir, J. Lee, S. A. Maier, and M. S. Kim, "Quantum plasmonics," *Nat. Phys.*, vol. 9, no. 6, pp. 329–340, 2013,.

[25] A. Archambault, F. Marquier, J. J. Greffet, and C. Arnold, "Quantum theory of spontaneous and stimulated emission of surface plasmons," *Phys. Rev. B*, vol. 82, no. 3, p. 035411, 2010.

[26] W. L. Barnes, A. Dereux, and T. W. Ebbesen, "Surface plasmon subwavelength optics," *Nature*, vol. 424, no. 6950, pp. 824–830, 2003.

[27] E. N. Economou, "Surface plasmons in thin films," *Phys. Rev.*, vol. 182, no. 2, p. 539, 1969.

[28] M. Jablan, M. Soljačić, and H. Buljan, "Plasmons in graphene: Fundamental properties and potential applications," *Proc. IEEE*, vol. 101, no. 7, pp. 1689–1704, 2013.

[29] E. A. Power and T. Thirunamachandran, "On the nature of the Hamiltonian for the interaction of radiation with atoms and molecules: (e/mc) p.a, - μ.e, and all that," *Am. J. Phys.*, vol. 46, no. 4, pp. 370–378, 1978.

[30] V. I. Yudson and P. Reineker, "Multiphoton scattering in a one-dimensional waveguide with resonant atoms," *Phys. Rev.*, vol. 78, no. 5, p. 052713, 2008.

[31] J. R. Taylor, *Scattering Theory: The Quantum Theory of Nonrelativistic Collisions*, Courier Corporation, 2006.

[32] G. O. Olaofe, "Scattering by an arbitrary configuration of parallel circular cylinders," *JOSA*, vol. 60, no. 9, pp. 1233–1236, 1970.

[33] D. Felbacq, G. Tayeb, and D. Maystre, "Scattering by a random set of parallel cylinders," *JOSA A*, vol. 11, no. 9, pp. 2526–2538, 1994.

[34] G. O. Olaofe, "Scattering by two cylinders," *Radio Sci.*, vol. 5, no. 11, pp. 1351–1360, 1970.

[35] A. S. Baltenkov and A. Z. Msezane, "Electronic quantum confinement in cylindrical potential well," *Eur. Phys. J. D*, vol. 70, no. 4, p. 81, 2016.

[36] L. Verslegers, Z. Yu, Z. Ruan, P. B. Catrysse, and S. Fan, "From electromagnetically induced transparency to superscattering with a single structure: A coupled-mode theory for doubly resonant structures," *Phys. Rev. Lett.*, vol. 108, no. 8, p. 083902, 2012.

[37] R. G. DeVoe and R. G. Brewer, "Observation of superradiant and subradiant spontaneous emission of two trapped ions," *Phys. Rev. Lett.*, vol. 76, no. 12, p. 2049, 1996.

[38] R. H. Dicke, "Coherence in spontaneous radiation processes," *Phys. Rev.*, vol. 93, no. 1, p. 99, 1954.

[39] S. H. Autler and C. H. Townes, "Stark effect in rapidly varying fields," *Phys. Rev.*, vol. 100, no. 2, p. 703, 1955.

[40] D. D. Smith, H. Chang, K. A. Fuller, A. T. Rosenberger, and R. W. Boyd, "Coupled-resonator-induced transparency," *Phys. Rev.*, vol. 69, no. 6, p. 063804, 2004.

[41] B. Peng, Ş. K. Özdemir, W. Chen, F. Nori, and L. Yang, "What is and what is not electromagnetically induced transparency in whispering-gallery microcavities," *Nat. Commun.*, vol. 5, no. 1, pp. 1–9, 2014,.

[42] P. Ginzburg and M. Orenstein, "Slow light and voltage control of group velocity in resonantly coupled quantum wells," *Opt. Express*, vol. 14, no. 25, pp. 12467–12472, 2006.
[43] A. Vakil and N. Engheta, "One-atom-thick reflectors for surface plasmon polariton surface waves on graphene," *Opt. Commun.*, vol. 285, no. 16, pp. 3428–3430, 2012.
[44] G. Alber, J. Z. Bernád, M. Stobińska, L. L. Sánchez-Soto, and G. Leuchs, "QED with a parabolic mirror," *Phys. Rev.*, vol. 88, no. 2, p. 023825, 2013.
[45] L. H. Ford and N. F. Svaiter, "Focusing vacuum fluctuations," *Phys. Rev.*, vol. 62, no. 6, p. 062105, 2000.
[46] P. N. Melentiev, A. A. Kuzin, D. V. Negrov, and V. I. Balykin, "Diffraction-limited focusing of plasmonic wave by a parabolic mirror," *Plasmonics*, vol. 13, no. 6, pp. 2361–2367, 2018.
[47] H. C. Manoharan, C. P. Lutz, and D. M. Eigler, "Quantum mirages formed by coherent projection of electronic structure," *Nature*, vol. 403, no. 6769, pp. 512–515, 2000.

Qixin Shen, Amirhassan Shams-Ansari, Andrew M. Boyce, Nathaniel C. Wilson, Tao Cai, Marko Loncar and Maiken H. Mikkelsen*

A metasurface-based diamond frequency converter using plasmonic nanogap resonators

https://doi.org/10.1515/9783110710687-046

Abstract: Diamond has attracted great interest as an appealing material for various applications ranging from classical to quantum optics. To date, Raman lasers, single photon sources, quantum sensing and quantum communication have been demonstrated with integrated diamond devices. However, studies of the nonlinear optical properties of diamond have been limited, especially at the nanoscale. Here, a metasurface consisting of plasmonic nanogap cavities is used to enhance both $\chi^{(2)}$ and $\chi^{(3)}$ nonlinear optical processes in a wedge-shaped diamond slab with a thickness down to 12 nm. Multiple nonlinear processes were enhanced simultaneously due to the relaxation of phase-matching conditions in subwavelength plasmonic structures by matching two excitation wavelengths with the fundamental and second-order modes of the nanogap cavities. Specifically, third-harmonic generation (THG) and second-harmonic generation (SHG) are both enhanced 1.6×10^7-fold, while four-wave mixing is enhanced 3.0×10^5-fold compared to diamond without the metasurface. Even though diamond lacks a bulk $\chi^{(2)}$ due to centrosymmetry, the observed SHG arises from the surface $\chi^{(2)}$ of the diamond slab and is enhanced by the metasurface elements. The efficient, deeply subwavelength diamond frequency converter demonstrated in this work suggests an approach for conversion of color center emission to telecom wavelengths directly in diamond.

***Corresponding author: Maiken H. Mikkelsen,** Department of Electrical and Computer Engineering, Duke University, Durham, NC, 27708, USA, E-mail: m.mikkelsen@duke.edu. https://orcid.org/0000-0002-0487-7585

Qixin Shen and Nathaniel C. Wilson, Department of Physics, Duke University, Durham, NC, 27708, USA

Amirhassan Shams-Ansari and Marko Loncar, John A. Paulson School of Engineering and Applied Sciences, Harvard University, Cambridge, MA, 02138, USA

Andrew M. Boyce and Tao Cai, Department of Electrical and Computer Engineering, Duke University, Durham, NC, 27708, USA

Keywords: diamond; frequency conversion; nanogap cavity; nonlinear generation; plasmonics.

The unique properties of diamond such as a superior thermal conductivity, a high index of refraction, an ultrawide transparency window and negligible birefringence [1–3] have made this material a promising platform for nanophotonics both in the classical [4–7] and quantum regimes [8–14]. Additionally, with its relatively high third-order nonlinear susceptibility ($\chi^{(3)}$), diamond is an appealing candidate for integrated nonlinear devices. To date, Raman lasing [15, 16], supercontinuum generation [17] and frequency combs [18] have been demonstrated in an integrated, single-crystal diamond platform. One of the distinctive features of diamond is its ability to host defects in its crystal lattice known as color centers, which are key components for quantum communication. These optically addressable spin qubits are unexplored in terms of their interactions with nonlinear processes such as frequency conversion to shift their emission wavelength [19]. However, this task has remained an outstanding challenge due to the weak intrinsic response of nonlinear processes and the phase-matching requirements. Plasmonic structures have proven to be a well-suited platform to investigate and enhance nonlinear optical processes [20–24]. The deeply subwavelength scale of these devices simultaneously allows for large confinement and enhancement of electric fields [25–29], as well as a relaxation of phase-matching conditions [30–32].

Here, film-coupled, plasmonic nanogap cavities created by a nondisruptive transfer method are utilized to enhance the nonlinear response in nanoscale diamond films without patterning the diamond itself. When the excitation wavelength overlaps with the cavity resonance, both third-harmonic generation (THG) and second-harmonic generation (SHG) are dramatically enhanced compared to a thin diamond slab reference. Furthermore, THG, sum frequency generation (SFG) and four-wave mixing (FWM) were enhanced simultaneously by leveraging two different cavity modes, further highlighting the versatility of this platform as a frequency converter.

This article has previously been published in the journal Nanophotonics. Please cite as: Q. Shen, A. Shams-Ansari, A. M. Boyce, N. C. Wilson, T. Cai, M. Loncar and M. H. Mikkelsen "A metasurface-based diamond frequency converter using plasmonic nanogap resonators" *Nanophotonics* 2021, 10. DOI: 10.1515/nanoph-2020-0392.

The diamond slab is sandwiched in between a 75 nm gold film and gold nanoparticles, forming plasmonic nanogap cavities. Arrays of nanoparticles were fabricated on a silicon substrate by electron beam lithography (EBL) and then transferred onto the diamond slab using a polydimethylsiloxane (PDMS) stamp (see Supplementary material for fabrication details). As shown in Figure 1a, there are three key areas on the sample: (A) transferred nanoparticles on a 12-nm-thick diamond film, forming nanogap cavities; (B) transferred nanoparticles on a ~200 nm diamond film, which behave as decoupled nanoparticles due to the increased diamond thickness and (C) diamond (12 nm thick) on a gold substrate, serving as a reference. An additional reference consists of a bare diamond slab on PDMS with a similar thickness and gradient as the diamond slab on gold. The thicknesses of the different areas of the diamond slab are confirmed by atomic force microscopy as shown in Figure 1b. The height profile in Figure 1c demonstrates that the thinnest section of the diamond is 12 nm thick. A schematic of the sample is shown in Figure 1d and consists of arrays of nanoparticles (220 nm particle side length) with a pitch of 440 nm on a wedge-shaped diamond thin film (gradient of ~2.6 nm/µm).

The simulated field distribution for a nanocavity with a 12 nm diamond gap illustrates that the highly confined electric fields in the cavity are enhanced up to 40-fold in comparison with the original incident field, facilitating enhanced nonlinear generation (Figure 1e). A reflection spectrum is measured from 700 to 1600 nm to determine the cavities' resonance wavelengths. For the nanogap cavities, the fundamental resonance mode is at 1455 nm and the second-order mode is at 840 nm (Figure 1f). The fundamental mode is blue shifted to 1130 nm for the decoupled nanoparticles, and no obvious second-order mode is observed. We attribute the fundamental mode from the decoupled nanoparticles to a localized surface plasmon resonance (LSPR) mode from the nanoparticle itself. This mode is distinct from the gap mode in nanogap cavities because the ~200 nm separation between the gold nanoparticles and substrate is too thick to support gap plasmons.

To leverage the field enhancement in the nanogap cavities for nonlinear generation, a pump wavelength of 1455 nm, matching the fundamental cavity resonance, is used to excite THG. The presence of THG is confirmed via observation of a third-order power law during power dependence measurements on all three regions of the sample and on the control as shown in Figure 2a. The THG response from the nanogap cavities is enhanced 1.6×10^7-fold compared to bare diamond on PDMS (details on calculation of enhancement factor in Supplementary material). Next, to investigate if the large enhancement is arising from the nanogap mode or simply the presence of the gold nanoparticles, THG is also measured from the decoupled nanoparticles. Measurements are performed using both 1455 nm excitation, which overlaps with the fundamental resonance of the nanogap mode (Figure 2a), and 1130 nm excitation, which overlaps with the LSPR mode of the decoupled nanoparticles (Figure 2b). For the decoupled nanoparticles, nearly three orders of magnitude less enhancement is observed for both excitation wavelengths, even though a larger amount of nonlinear medium (i.e. a thicker diamond layer) contributes to the THG response. The electric field intensity is much less in the diamond for the decoupled nanoparticles, resulting in less THG enhancement.

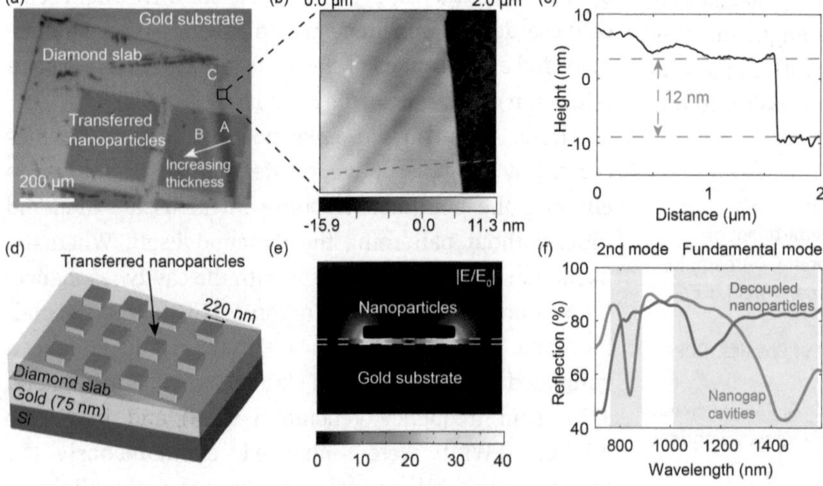

Figure 1: (a) Optical image of the sample. Area A: transferred nanoparticles on ~12 nm diamond, forming nanogap cavities; area B: transferred nanoparticles on ~200 nm diamond, forming decoupled nanoparticles; area C: diamond on gold. (b) Atomic force microscopy (AFM) image of the area in (a) delineated by the square. (c) Height profile along the red, dashed line in (b). (d) Schematic of the sample structure: a thin diamond wedge is sandwiched in between a 75 nm evaporated gold film and electron beam lithography (EBL)–fabricated gold nanoparticles with a height of 30 nm. (e) Simulated electric field enhancement distribution of nanogap cavities at 1455 nm. (f) Measured reflection spectra from nanogap cavities (red) and decoupled nanoparticles (blue).

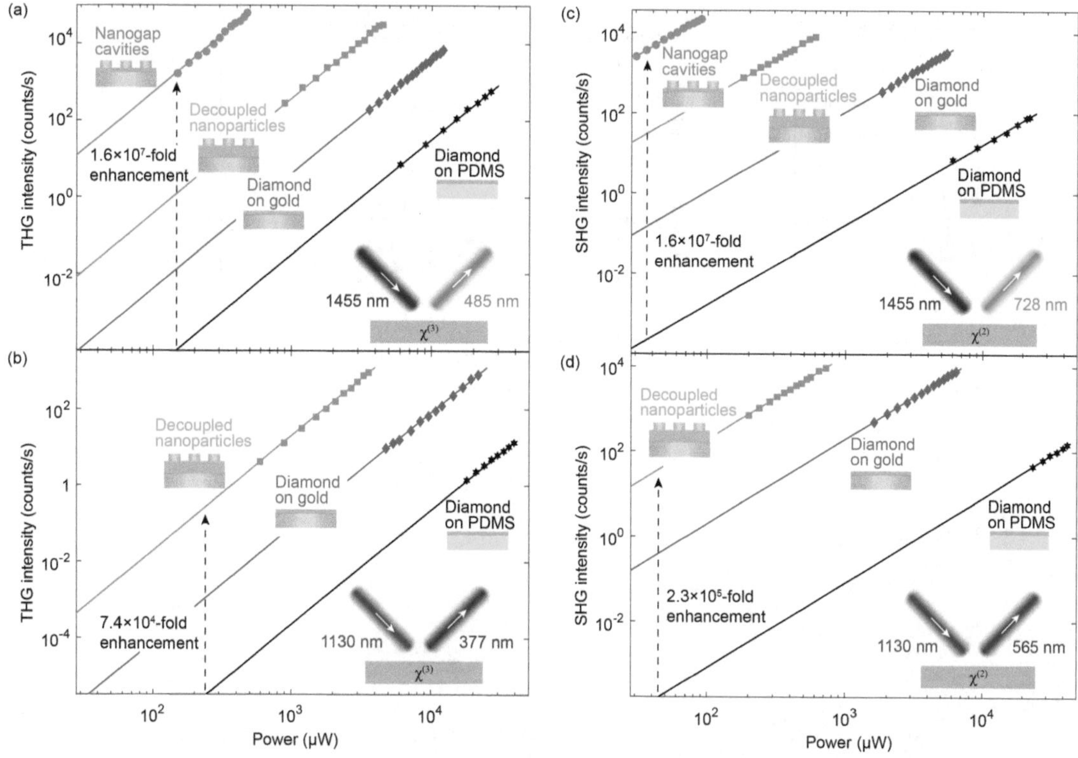

Figure 2: (a and b) Third-harmonic generation (THG) intensity as a function of excitation power shown on a log scale. Data points are experimental results, and lines are cubic polynomial fits. (a) 1455 nm excitation. (b) 1130 nm excitation. (c and d) Second-harmonic generation (SHG) intensity as a function of excitation power shown on a log scale. Data points are experimental results, and lines are quadratic polynomial fits. (c) 1455 nm excitation. (d) 1130 nm excitation. Nanogap cavities: transferred nanoparticles on ~12 nm diamond with an underlying gold film; decoupled nanoparticles: transferred nanoparticles on ~200 nm diamond with an underlying gold film.

As seen in Figure 2a and b, the nonlinear response from the diamond on gold is also enhanced compared with the diamond on PDMS. This may be explained by a contribution from the nonlinearity of gold or surface effects, as the nonlinear response from PDMS itself is negligible. In addition to the large enhancement, the nanogap cavities have relatively high nonlinear conversion efficiencies considering their nanoscale dimensions. The efficiency is defined as the power of the generated nonlinear intensity from embedded diamond nanocavities divided by the power of the incident excitation. The power of the nonlinear signal is derived from the photon counts and a calibrated light source (Labsphere) positioned at the focal plane of the objective lens (details on determining the efficiency in Supplementary material). With an observed damage threshold of 5 mW of excitation power, the maximum THG conversion efficiency is estimated to be 2.33×10^{-5}%.

Similarly, power dependence measurements are performed for SHG, and a quadratic power dependence is observed for both 1455 nm excitation and 1130 nm excitation (Figure 2c and d). SHG is enhanced 1.6×10^{7}-fold for diamond in the nanogap cavities compared with bare diamond on PDMS for 1455 nm excitation. This enhancement is comparable to the observed THG enhancement even though it is from a lower order nonlinear process. This can be explained by the modified $\chi^{(2)}$ profile within diamond due to the presence of plasmonic structures, giving rise to a higher SHG intensity than simply the enhancement resulting from the electric field confinement in the cavity. The maximum SHG conversion efficiency from the nanogap cavities for a 5 mW excitation power is estimated to be 7.59×10^{-6}%.

Next, the excitation wavelength was varied to further investigate the importance of the plasmonic nanogap mode for enhanced nonlinear responses. Specifically, laser excitation at wavelengths ranging from 1430 to 1480 nm with constant power is utilized for nanogap cavities, and the resulting spectra are shown in Figure 3a for THG and Figure 3b for SHG (additional data in Supplementary material). As expected, the maximum nonlinear intensity occurs at the cavity resonance wavelength (1455 nm) for both the THG and SHG signals. The nonlinear intensity decreases dramatically when the excitation is detuned from 1455 nm due to the reduced electric field intensity

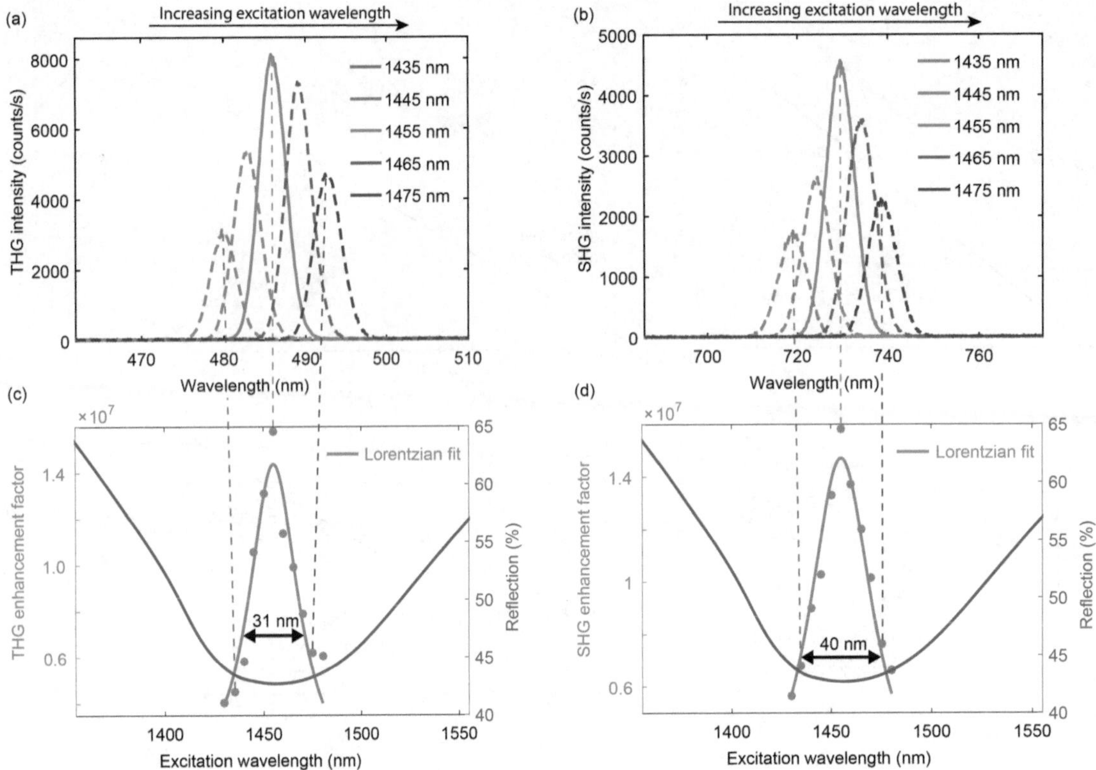

Figure 3: (a) THG response spectra as a function of excitation wavelength. (b) SHG response spectra as a function of excitation wavelength. (c) THG enhancement factor as a function of excitation wavelength (red) compared to the reflection spectrum (blue) which shows the plasmon resonance. The full widths at half maximum (FWHM) of the THG enhancement is 31 nm. (d) SHG enhancement factor as a function of excitation wavelength (red) compared to the reflection spectrum (blue) which shows the plasmon resonance. The FWHM of the SHG enhancement is 40 nm, while the FWHM of the reflection spectrum is 204 nm in both (c) and (d). THG, third-harmonic generation; SHG, second-harmonic generation.

within the diamond layer. Thus, the nonlinear intensity is strongly related to the enhanced electric field intensity confined within the diamond, illustrating that the observed nonlinear response mainly originates from the diamond. The full widths at half maximum (FWHM) of the THG and SHG enhancement as a function of excitation wavelength are noted to be much narrower than the cavity reflection spectrum as shown in Figure 3c and 3d, which is a consequence of the nonlinear dependence of THG and SHG intensity on the electric field intensity. It is observed that THG decreases more rapidly than SHG when the excitation wavelength is detuned from the cavity resonance as the THG depends on the third power of the electric field in the cavity, whereas SHG has a second-order power dependence. THG is expected to have a third-order Lorentzian lineshape and SHG a second-order Lorentzian lineshape. A coupled mode theory analysis gives a ratio between the FWHM of the THG and SHG enhancement of $\frac{\sqrt{\sqrt[3]{2}-1}}{\sqrt{\sqrt{2}-1}} = 0.79$ which agrees very well with the experimentally observed ratio of $\frac{31}{40} = 0.78$ (further details in Supplementary material).

Next, we investigate nonlinear light generation arising from multiple nonlinear optical processes occurring simultaneously within a single nanocavity. Here, SFG and FWM are selected to demonstrate this capability along with the previously characterized THG process. Excitation wavelengths of 840 nm (ω_1) and 1455 nm (ω_2), matching the cavities' fundamental and second-order modes, were selected. The output SFG frequency equals $\omega_{SFG} = \omega_1 + \omega_2$ while the degenerate FWM frequency equals $\omega_{FWM} = 2\omega_1 - \omega_2$ (indicated in the inset schematic in Figure 4c–f).

A variable delay stage is employed to control the relative time delay between the two excitation pulses. The SFG (532 nm) and FWM (590 nm) responses start to appear when the time delay between the two excitation pulses is below approximately 150 fs, which is the pulse duration of the excitation laser. The THG intensity remains constant as it only depends on the 1455 nm excitation (Figure 4a). The largest SFG and FWM intensities are observed at zero time delay when the two excitation pulses are perfectly overlapped (Figure 4b). The emission peak observed at 485 nm is from THG as confirmed above. Similar power dependence measurements are performed to demonstrate that

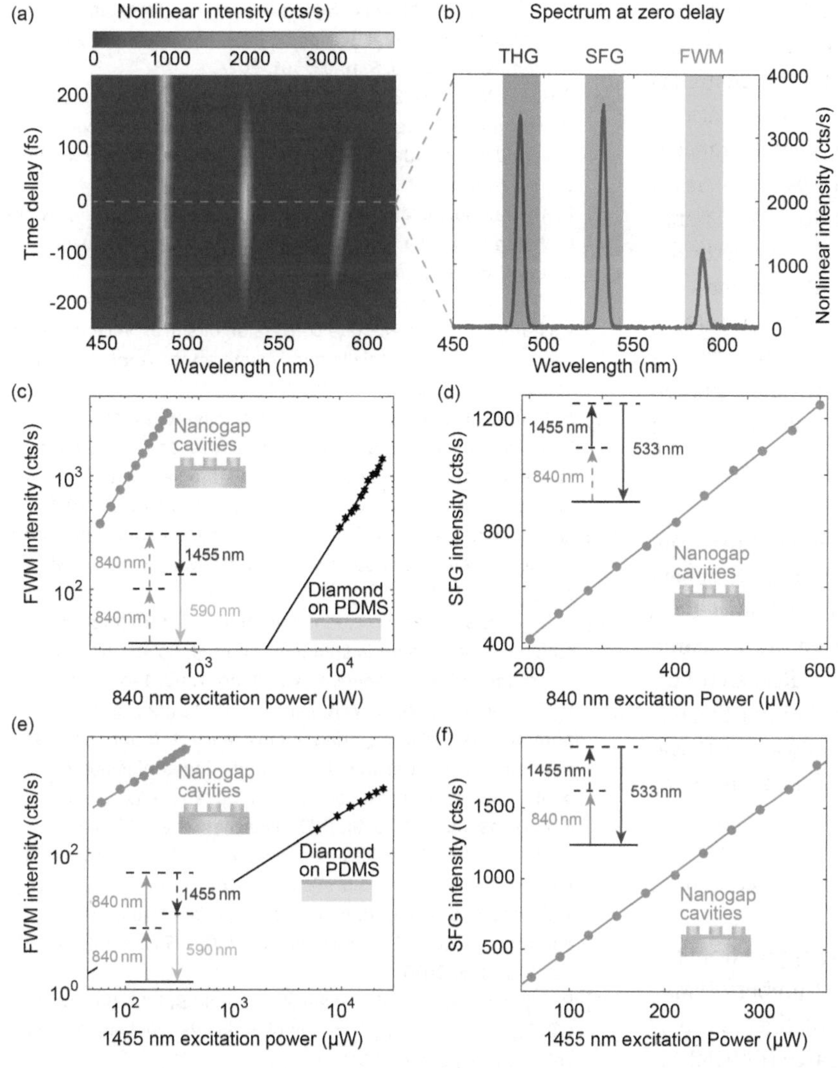

Figure 4: (a) The spectra of nonlinear responses as a function of the time delay between the two excitation pulses at 840 and 1455 nm. (b) Nonlinear response spectra at zero time delay. The three peaks correspond to THG, SFG and FWM as indicated in the figure. (c and d) Power dependence measurements for variable 840 nm excitation power. (c) FWM intensity. The 1455 nm excitation powers for nanogap cavities and the diamond reference are 180 μW and 18 mW, respectively. (d) SFG intensity. The 1455 nm excitation power for nanogap cavities is 180 μW, and SFG intensity from the diamond reference is too small to observe. (e and f) Power dependence measurements for variable 1455 nm excitation power. (e) FWM intensity. The 840 nm excitation powers for nanogap cavities and the diamond reference are 400 μW and 14 mW, respectively. (f) SFG intensity. The 840 nm excitation power for nanogap cavities is 400 μW, and the SFG intensity from the diamond reference is too small to observe. The insets in (c–f) show the energy diagram for the corresponding frequency conversion process. THG, third-harmonic generation; SHG, second-harmonic generation; SFG, sum frequency generation; FWM, four-wave mixing.

the other two peaks are from SFG and FWM processes. Specifically, the signal at the longest wavelength (590 nm) scales quadratically with the 840 nm excitation power and scales linearly with the 1455 nm excitation power. Therefore, it is confirmed that this signal is from FWM as it obeys the expected power law and, furthermore, occurs at the wavelength given by the frequency conversion relation. Similarly, the signal in the middle at 532 nm scales linearly with both the 840 and 1455 nm excitation powers, indicating that it is indeed from SFG.

Large enhancement is expected for SFG and FWM from diamond embedded in the nanogap cavities due to the strong electric field at both excitation wavelengths. Experimentally, we find that FWM is enhanced 3.0×10^5-fold compared to bare diamond on PDMS as extracted from Figures 4c and e. For SFG, the signal from bare diamond is too weak to detect since only the surface of diamond contributes SFG due to the inversion symmetry in its crystal lattice. Thus, we cannot determine an enhancement factor for this process.

In summary, we embed a diamond slab into plasmonic nanogap cavities formed by a gold ground plane and EBL-fabricated nanoparticles transferred using a PDMS stamp. By overlapping the excitation wavelength with the nanocavities' resonance wavelength, 1.6×10^7-fold enhancement is observed for both THG and SHG. This large enhancement is accompanied by relatively high nonlinear conversion efficiencies, approaching $2.33 \times 10^{-5}\%$ and $7.59 \times 10^{-6}\%$ for THG and SHG, respectively. These efficiencies are comparable with other reported results using plasmonic structures [20, 22, 24, 32–34] but achieved with a unique material – diamond – which have not previously been experimentally demonstrated. Furthermore, simultaneous enhancement of multiple nonlinear processes was demonstrated, specifically THG, SFG and FWM, enabled by the relaxed phase-matching conditions in the deeply subwavelength cavities. To further enhance nonlinear generation, the damage threshold of the structure could be increased by utilizing refractory materials [35–37] such as TiN or gold nanoparticles coated with an ultrathin atomic layer deposition (ALD) layer to prevent deformation [38]. The PDMS transfer process provides a convenient and nondisruptive method to place nanoparticles. This technique, along with the cavity's vertically oriented gap, offers the potential to embed diamond containing color centers. These studies demonstrate a metasurface-based diamond frequency converter that is promising for on-chip nonlinear devices and single-photon frequency conversion of the emission of color centers in diamond from visible to telecommunication wavelengths.

Acknowledgments: M.H.M. acknowledges support from the National Science Foundation (NSF) Grant Numbers EFMA-1640986 and DMR-1454523. M.L. acknowledges support from the Air Force Office of Scientific Research (FA9550-19-1-0376), the Defense Advanced Research Projects Agency (W31P4Q-15-1-0013), and the National Science Foundation (DMR-1231319). Authors acknowledge Pawel Latawiec for developing the diamond thin-film on insulator process.

Author contribution: All the authors have accepted responsibility for the entire content of this submitted manuscript and approved submission.

Research funding: M.H.M. acknowledges support from the National Science Foundation (NSF) Grant Numbers EFMA-1640986 and DMR-1454523. M.L. acknowledges support from the Air Force Office of Scientific Research (FA9550-19-1-0376), the Defense Advanced Research Projects Agency (W31P4Q-15-1-0013), and the National Science Foundation (DMR-1231319).

Conflict of interest statement: The authors declare no conflicts of interest regarding this article.

References

[1] I. Aharonovich, A. D. Greentree, and S. Prawer, "Diamond photonics," *Nat. Photonics*, vol. 5, no. 7, pp. 397–405, 2011.

[2] R. P. Mildren and J. R. Rabeau, *Optical Engineering of Diamond*, John Wiley & Sons, 2013.

[3] A. A. Kaminskii, V. G. Ralchenko, and V. I. Konov, "CVD-diamond – a novel $\chi^{(3)}$-nonlinear active crystalline material for SRS generation in very wide spectral range," *Laser Phys. Lett.*, vol. 3, no. 4, pp. 171–177, 2006.

[4] M. J. Burek, Y. Chu, M. S. Z. Liddy, et al., "High quality-factor optical nanocavities in bulk single-crystal diamond," *Nat. Commun.*, vol. 5, no. 1, pp. 1–7, 2014.

[5] H. A. Atikian, P. Latawiec, M. J. Burek, et al., "Freestanding nanostructures via reactive ion beam angled etching," *APL Photonics*, vol. 2, no. 5, p. 051301, 2017.

[6] F. Gao, J. Van Erps, Z. Huang, H. Thienpont, R. G. Beausoleil, and N. Vermeulen, "Directional coupler based on single-crystal diamond waveguides," *IEEE J. Sel. Top. Quantum Electron.*, vol. 24, no. 6, p. 6100909, 2018.

[7] B. Khanaliloo, M. Mitchell, A. C. Hryciw, and P. E. Barclay, "High-Q/V monolithic diamond microdisks fabricated with quasi-isotropic etching," *Nano Lett.*, vol. 15, no. 8, pp. 5131–5136, 2015.

[8] K. Bray, D. Y. Fedyanin, I. A. Khramtsov, et al., "Electrical excitation and charge-state conversion of silicon vacancy color centers in single-crystal diamond membranes," *Appl. Phys. Lett.*, vol. 116, no. 10, p. 101103, 2020.

[9] S. I. Bogdanov, M. Y. Shalaginov, A. S. Lagutchev, et al., "Ultrabright room-temperature sub-nanosecond emission from single nitrogen-vacancy centers coupled to nanopatch antennas," *Nano Lett.*, vol. 18, no. 8, pp. 4837–4844, 2018.

[10] L. Li, T. Schröder, E. H. Chen, et al., "Coherent spin control of a nanocavity-enhanced qubit in diamond," *Nat. Commun.*, vol. 6, no. 1, pp. 1–7, 2015.

[11] Y. I. Sohn, S. Meesala, B. Pingault, et al., "Controlling the coherence of a diamond spin qubit through its strain environment," *Nat. Commun.*, vol. 9, no. 1, pp. 1–6, 2018.

[12] H. Siampour, O. Wang, V. A. Zenin, et al., "Ultrabright single-photon emission from germanium-vacancy zero-phonon lines: deterministic emitter-waveguide interfacing at plasmonic hot spots," *Nanophotonics*, vol. 9, no. 4, pp. 953–962, 2020.

[13] R. E. Evans, M. K. Bhaskar, D. D. Sukachev, et al., "Photon-mediated interactions between quantum emitters in a diamond nanocavity," *Science (80-)*, vol. 362, no. 6415, pp. 662–665, 2018.

[14] T. Schröder, S. L. Mouradian, J. Zheng, et al., "Quantum nanophotonics in diamond [invited]," *J. Opt. Soc. Am. B*, vol. 33, no. 4, p. B65, 2016.

[15] P. Latawiec, V. Venkataraman, M. J. Burek, B. J. M. Hausmann, I. Bulu, and M. Lončar, "On-chip diamond Raman laser," *Optica*, vol. 2, no. 11, p. 924, 2015.

[16] P. Latawiec, V. Venkataraman, A. Shams-Ansari, M. Markham, and M. Loncar, "An integrated diamond Raman laser pumped in the near-visible," *Opt. Lett.*, vol. 43, no. 2, pp. 318–321, 2017.

[17] A. Shams-Ansari, P. Latawiec, Y. Okawachi, et al., "Supercontinuum generation in angle-etched diamond waveguides," *Opt. Lett.*, vol. 44, no. 16, pp. 4056–4059, 2019.

[18] B. J. M. Hausmann, I. Bulu, V. Venkataraman, P. Deotare, and M. Loncar, "Diamond nonlinear photonics," *Nat. Photonics*, vol. 8, no. 5, pp. 369–374, 2014.

[19] Z. Lin, S. G. Johnson, A. W. Rodriguez, and M. Loncar, "Design of diamond microcavities for single photon frequency down-conversion," *Opt. Express*, vol. 23, no. 19, p. 25279, 2015.

[20] H. Aouani, M. Rahmani, M. Navarro-Cía, and S. A. Maier, "Third-harmonic-upconversion enhancement from a single semiconductor nanoparticle coupled to a plasmonic antenna," *Nat. Nanotechnol.*, vol. 9, no. 4, pp. 290–294, 2014.

[21] P. Genevet, J. P. Tetienne, E. Gatzogiannis, et al., "Large enhancement of nonlinear optical phenomena by plasmonic nanocavity gratings," *Nano Lett.*, vol. 10, no. 12, pp. 4880–4883, 2010.

[22] Q. Shen, T. B. Hoang, G. Yang, V. D. Wheeler, and M. H. Mikkelsen, "Probing the origin of highly-efficient third-

harmonic generation in plasmonic nanogaps," *Opt. Express*, vol. 26, no. 16, pp. 20718–20725, 2018.
[23] Y. Zhang, F. Wen, Y.-R. Zhen, P. Nordlander, and N. J. Halas, "Coherent fano resonances in a plasmonic nanocluster enhance optical four-wave mixing," *Proc. Natl. Acad. Sci.*, vol. 110, no. 23, pp. 9215–9219, 2013.
[24] M. Celebrano, X. Wu, M. Baselli, et al., "Mode matching in multiresonant plasmonic nanoantennas for enhanced second harmonic generation," *Nat. Nanotechnol.*, vol. 10, no. 5, pp. 412–417, 2015.
[25] J. B. Lassiter, F. McGuire, J. J. Mock, et al., "Plasmonic waveguide modes of film-coupled metallic nanocubes," *Nano Lett.*, vol. 13, no. 12, pp. 5866–5872, 2013.
[26] G. M. Akselrod, C. Argyropoulos, T. B. Hoang, et al., "Probing the mechanisms of large purcell enhancement in plasmonic nanoantennas," *Nat. Photonics*, vol. 8, no. 11, pp. 835–840, 2014.
[27] Q. Shen, A. M. Boyce, G. Yang, and M. H. Mikkelsen, "Polarization-controlled nanogap cavity with dual-band and spatially overlapped resonances," *ACS Photonics*, vol. 6, no. 8, pp. 1916–1921, 2019.
[28] J. J. Baumberg, J. Aizpurua, M. H. Mikkelsen, and D. R. Smith, "Extreme nanophotonics from ultrathin metallic gaps," *Nat. Mater.*, vol. 18, no. 7, pp. 668–678, 2019.
[29] A. Rose, T. B. Hoang, F. McGuire, et al., "Control of radiative processes using tunable plasmonic nanopatch antennas," *Nano Lett.*, vol. 14, no. 8, pp. 4797–4802, 2014.
[30] G. Sartorello, N. Olivier, J. Zhang, et al., "Ultrafast optical modulation of second- and third-harmonic generation from cut-disk-based metasurfaces," *ACS Photonics*, vol. 3, no. 8, pp. 1517–1522, 2016.
[31] S. Liu, P. P. Vabishchevich, A. Vaskin, et al., "An all-dielectric metasurface as a broadband optical frequency mixer," *Nat. Commun.*, vol. 9, no. 1, pp. 1–6, 2018.
[32] Q. Shen, W. Jin, G. Yang, A. W. Rodriguez, and M. H. Mikkelsen, "Active control of multiple, simultaneous nonlinear optical processes in plasmonic nanogap cavities," *ACS Photonics*, vol. 7, no. 4, pp. 901–907, 2020.
[33] S. Park, J. W. Hahn, and J. Y. Lee, "Doubly resonant metallic nanostructure for high conversion efficiency of second harmonic generation," *Opt. Express*, vol. 20, no. 5, p. 4856, 2012.
[34] Y. Zhang, N. K. Grady, C. Ayala-Orozco, and N. J. Halas, "Three-dimensional nanostructures as highly efficient generators of second harmonic light," *Nano Lett.*, vol. 11, no. 12, pp. 5519–5523, 2011.
[35] G. Albrecht, M. Ubl, S. Kaiser, H. Giessen, and M. Hentschel, "Comprehensive study of plasmonic materials in the visible and near-infrared: linear, refractory, and nonlinear optical properties," *APL Photonics*, vol. 5, no. 3, pp. 1058–1067, 2018.
[36] U. Guler, A. Boltasseva, and V. M. Shalaev, "Refractory plasmonics," *Science (80-.)*, vol. 344, no. 6181, pp. 263–264, 2014.
[37] M. P. Wells, R. Bower, R. Kilmurray, et al., "Temperature stability of thin film refractory plasmonic materials," *Opt. Express*, vol. 26, no. 12, p. 15726, 2018.
[38] G. Albrecht, S. Kaiser, H. Giessen, and M. Hentschel, "Refractory plasmonics without refractory materials," *Nano Lett.*, vol. 17, no. 10, pp. 6402–6408, 2017.

Supplementary Material: The online version of this article offers supplementary material (https://doi.org/10.1515/nanoph-2020-0392).

Nicolò Accanto, Pablo M. de Roque, Marcial Galvan-Sosa, Ion M. Hancu and Niek F. van Hulst*

Selective excitation of individual nanoantennas by pure spectral phase control in the ultrafast coherent regime

https://doi.org/10.1515/9783110710687-047

Abstract: Coherent control is an ingenious tactic to steer a system to a desired optimal state by tailoring the phase of an incident ultrashort laser pulse. A relevant process is the two-photon–induced photoluminescence (TPPL) of nanoantennas, as it constitutes a convenient route to map plasmonic fields, and has important applications in biological imaging and sensing. Unfortunately, coherent control of metallic nanoantennas is impeded by their ultrafast femtosecond dephasing times so far limiting control to polarization and spectral optimization. Here, we report that phase control of the TPPL in resonant gold nanoantennas is possible. We show that, by compressing pulses shorter than the localized surface plasmon dephasing time (<20 fs), a very fast coherent regime develops, in which the two-photon excitation is sensitive to the phase of the electric field and can therefore be controlled. Instead, any phase control is gone when using longer pulses. Finally, we demonstrate pure phase control by resorting to a highly sensitive closed-loop strategy, which exploits the phase differences in the ultrafast coherent response of different nanoantennas, to selectively excite a chosen antenna. These results underline the direct and intimate relation between TPPL and coherence in gold nanoantennas, which makes them interesting systems for nanoscale nonlinear coherent control.

Keywords: closed-loop control; coherent control; hot spot; nanoantenna; spectral phase control; ultrafast.

Present address: Pablo M. de Roque, KPMG Lighthouse – Data Analytics & AI, 08908 L'Hospitalet de Llobregat, Barcelona, Spain; and **Ion M. Hancu,** Gauss & Neumann, Search Engine Marketing Technology.

***Corresponding author: Niek F. van Hulst,** ICFO—Institut de Ciences Fotoniques, The Barcelona Institute of Science & Technology, 08860 Castelldefels, Barcelona, Spain; and ICREA—Institució Catalana de Recerca i Estudis Avançats, 08010 Barcelona, Spain, E-mail: Niek.vanHulst@icfo.eu. https://orcid.org/0000-0003-4630-1776https://orcid.org/0000-0003-4630-1776
Nicolò Accanto, ICFO—Institut de Ciences Fotoniques, The Barcelona Institute of Science & Technology, 08860 Castelldefels, Barcelona, Spain; and Institut de la Vision, Sorbonne Université, Inserm S968, CNRS UMR7210, 17 Rue Moreau, 75012 Paris, France, E-mail: nicolo.accanto@inserm.fr. https://orcid.org/0000-0003-2491-7190
Pablo M. de Roque and Ion M. Hancu, ICFO—Institut de Ciences Fotoniques, The Barcelona Institute of Science & Technology, 08860 Castelldefels, Barcelona, Spain, E-mail: pablom.roque@gmail.com (P. M. de Roque), im.hancu@gmail.com (I. M. Hancu). https://orcid.org/0000-0002-0751-9126 (P. M. de Roque)
Marcial Galvan-Sosa, Neurochemistry and Neuroimaging Laboratory, University of La Laguna, Tenerife, Spain, E-mail: igalvans@ull.es

1 Introduction

In the field of optics, coherent control refers to the capability of precisely tailored laser fields to actively manipulate the outcome of certain light–matter interactions by exploiting the coherent properties of the system under study. One common experimental implementation of coherent control makes use of broadband phase-controlled laser pulses to generate interference phenomena [1–6]. By inducing constructive interference between the pathways that connect the initial state to the desired final state and destructive interference among those leading to unwanted final states, the laser pulse actively steers the investigated system towards the desired target. However, because interference is strictly related to phase, this approach to coherent control is only applicable to systems that retain the phase information for a sufficiently long time, i.e. for a time longer than the pulse duration [3, 4, 6]. In the coherent control formalism, the characteristic time after which the system loses its phase memory is called dephasing or coherence time and is indicated with T_2. The condition for coherent control to be effective is thus that $T_2 > \delta$, where δ represents the pulse duration. Systems with T_2 shorter than the pulse duration are elusive to coherent control.

Optical nanoantennas are metallic nanostructures that support localized surface plasmon resonances (LSPRs) in the optical or near-infrared spectral region. Light couples

This article has previously been published in the journal Nanophotonics. Please cite as: N. Accanto, P. M. de Roque, M. Galvan-Sosa, I. M. Hancu, N. F. van Hulst "Selective excitation of individual nanoantennas by pure spectral phase control in the ultrafast coherent regime" *Nanophotonics* 2021, 10. DOI: 10.1515/nanoph-2020-0406.

to the nanoantennas driving a LSPR confined to the nanoantenna surface, providing control of the near fields at subwavelength spatial scales [7]. Optical nanoantennas are used to direct light emission [8–10], to change excitation and emission rates in molecules [11–13], to improve nonlinear optical effects [14] and as active elements for sensing [15, 16]. Two-photon–induced photoluminescence (TPPL) microscopy, in which an ultrashort laser pulse induces two-photon excitation (TPE), is one of the most commonly used techniques to map surface plasmons in optical nanoantennas [17–19]. In recent years, the TPPL from gold nanoantennas emerged as a promising contrast mechanism for biological imaging and even for cancer therapy [20–22]. The perspective of coherent control strategies to manipulate TPPL in gold nanoantennas by tuning the phase of the laser pulse is attractive for optimal contrast control.

Stockman et al. [23] and Stockman [24] first linked the field of coherent control to the study of optical nanoantennas with the objective of simultaneously achieving nanometre and femtosecond control of optical fields. Experimental demonstrations of such control principle include: the phase and polarization manipulation of nanooptical fields in the incoherent regime [25, 26]; the simultaneous high space and time resolution imaging of surface plasmon dynamics in the coherent and incoherent regimes [27–33]; and the phase-dependent control of the propagation of LSPRs through nanoparticle arrays caused by different amounts of dispersion acquired in different propagation directions [34, 35].

The TPPL in gold nanoantennas was studied by Biagioni et al. [18] as a function of increasing chirp applied to an ultrashort pulse. The observed lack of change in the TPPL for pulses shorter than 100 fs was interpreted as stemming from a two-step single photon absorption involving an intermediate state, as also shown in other works [36, 37]. In contrast, phase dependence of the TPPL on a rough gold film was reported [38], and such dependence was used to deduce the LSPRs of different gold nanoantennas [39]. These early works point towards the necessity of a unique model to clear out the origin of the phase dependence of the TPPL in gold nanoantennas and to find under which conditions coherent phase control in these systems can be made to use.

Here, we show that the intrinsic nature of the TPE in gold nanoantennas, in which an intermediate state (the LSPR) mediates the process, leads to a very fast initial coherent regime. This regime is only accessible by ultrashort laser pulses, shorter than the coherence time T_2 (<50 fs), for which the TPE is sensitive to the spectral phase of the electric field. We then demonstrate that phase-only coherent control is capable to manipulate the TPE in different nanoantennas and obtain contrast in the emitted TPPL. To achieve this, we used a newly developed closed-loop phase control scheme with single-molecule sensitivity [40] that adapts to the subtle differences in the coherent response of different nanoantennas with detuned LSPRs to selectively excite them.

2 The initial coherent regime

In order to unveil the coherent response of gold nanoantennas, we first investigated the ultrafast dynamics of the TPE by varying the excitation pulse duration in the range of 15 fs–1 ps and detecting the TPPL with a high-resolution confocal microscope (see Methods). As sketched in Figure 1a, we used a pulse shaper to actively control the spectral phase $\varphi(\omega)$ of the laser field, which allowed us to change the pulse duration by applying different amounts of linear chirp.

The time duration dependence of the TPPL from a single resonant 90 nm gold nanorod antenna is plotted in Figure 1b, together with that of the second harmonic (SH) produced by a single nonresonant barium–titanate ($BaTiO_3$) nanoparticle that provides the reference for a pure TPE process. As expected from theory [41], the SH intensity scales as the inverse of the pulse duration δ, being maximal for the shortest laser pulse. In contrast, the TPPL from the gold nanoantenna presents a very different behaviour, and three regimes can be identified: (i) for $\delta \lesssim 50$ fs, the TPPL increases as the pulse duration decreases; (ii) for $100 \lesssim \delta \lesssim 300$ fs, the TPPL is almost constant; (iii) for $\delta \gtrsim 300$ fs, the dependence on the inverse of δ is recovered. Regimes (ii) and (iii) were already studied in the study by Biagioni et al. [18] and are consistent with the established model in which the TPE process in gold involves two successive single-photon absorptions, mediated by a real transition with lifetime T_1 in the few hundreds to thousands femtosecond range [18, 19, 32, 37]. For nanoantennas resonant in the spectral region of the excitation spectrum, the LSPR acts as the intermediate state, enhancing the TPE process [32, 37]. In this model, a second photon can induce TPE if it impinges on the system before the intermediate state population excited by a first photon has decayed, which explains the observed TPPL dependence in regimes (ii) and (iii). This behaviour can be described mathematically as follows [18]:

$$\mathrm{TPE}_{\delta>100} \propto \int E_\delta(t)dt \int E_\delta(t-t_1)e^{-t_1/T_{LF}}dt_1 \tag{1}$$

where $E_\delta(t)$ represents the time-varying electric field as a function of the pulse duration δ. A fit of the experimental

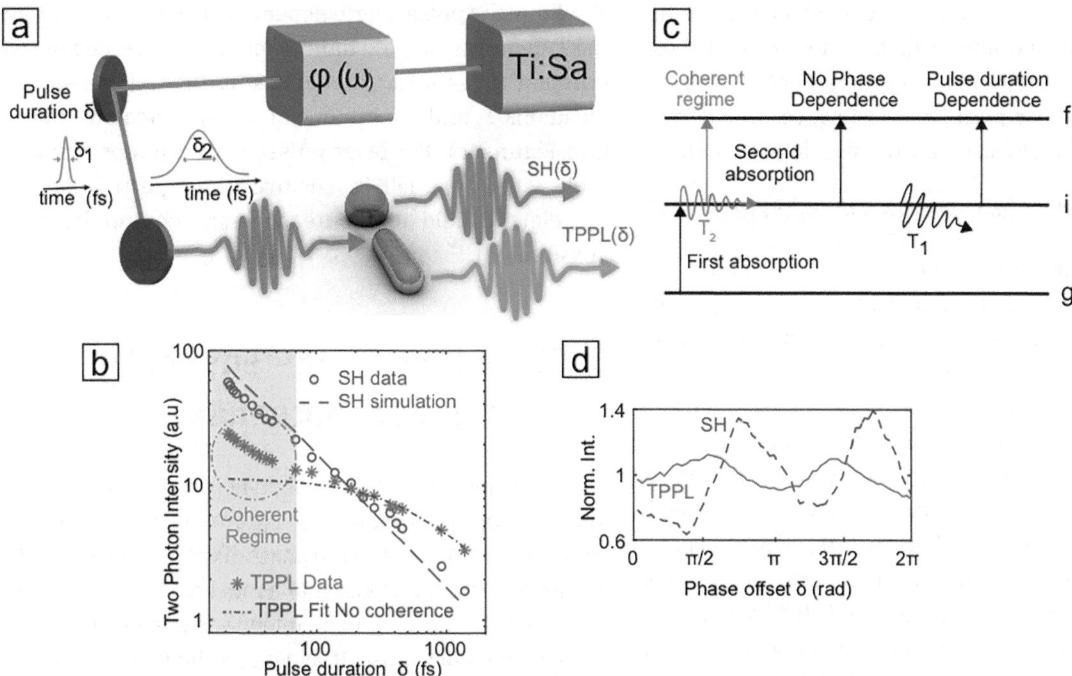

Figure 1: TPPL as a function of the pulse duration.
(a) Schematic of the experiment: a pulse shaper changes the time duration of a laser pulse in the range of 15 fs–1 ps. The TPPL from a resonant nanoantenna and the SH from a nonresonant $BaTiO_3$ are detected as a function of the pulse duration. (b) Experimental dependence of the SH (blue circles) and the TPPL from a resonant 90-nm long gold nanoantenna (green stars) on the pulse length δ. The blue dashed curve is the predicted behaviour of the SH, which agrees well with the experiment. The dark green dot-dashed curve corresponds to a fit to the TPPL for $\delta > 100$ fs using Eq. (1) from which we infer a lifetime of the intermediate state $T_1 \cong 350 \pm 50$ fs. For pulses shorter than ~50 fs, a coherent regime develops. (c) Sketch of the three different regimes involved in TPE as a function of the pulse length. (d) Effect of a spectral phase scan on the TPPL from a resonant gold nanoantenna (green curve) and the SH from a $BaTiO_3$ nanoparticle (blue curve). Note that the modulation depth is substantially reduced for the TPPL response. TPPL, two-photon–induced photoluminescence; TPE, two-photon excitation; SH, second harmonic.

TPPL data in Figure 1b using Eq. (1) produces the dark green dot-dashed curve, which well reproduces the experimental points for $\delta \gtrsim 100$ fs, yielding a lifetime $T_1 = 350 \pm 50$ fs, a bit shorter than the value of 650 fs reported by Biagioni et al. [19]. The lifetime should be compared to typical relaxation rates of the hot electron gas in noble metals [42, 43]. The TPE in gold can be understood on the basis of the gold band structure as follows [18, 36]: the first photon induces an intraband transition that excites an electron to the *sp* conduction band above the Fermi level, leaving a hole in the *sp* conduction band below the Fermi level. The second photon successively induces an interband transition, exciting an electron from the *d* band to the *sp* band, which can recombine with the previously created *sp* hole. The first photon in this picture creates a nonequilibrium situation (hot electrons) that has to decay to the thermal equilibrium. Typical internal thermalization times of the hot electron gas in noble metals were determined to be in the range of 350 fs–3 ps [42–44]. Biagioni et al. [19] reported on crystalline gold gap-antennas, while we study amorphous gold nanorod antennas, which will explain our shorter value of T_1.

Equation (1), however, completely fails to describe the behaviour of the TPPL for $\delta \lesssim 50$ fs, exactly where our experiment is most sensitive. Equation (1) was derived under the explicit assumption that coherent (and thus phase) effects can be neglected for sufficiently long laser pulses [18]. The dephasing time T_2 associated with LSPRs is in the order $T_2 \sim 20$ fs and thus within the time resolution achievable with our experiment [27–32, 45]. We therefore attribute the deviation of the experimental points for $\delta \lesssim 50$ fs from the prediction of Eq. (1) to the development of a coherent regime in which the interaction of the two successive single photons with the nanoantenna depends on the relative phase of the two photons. In this regime, Eq. (1) needs to be modified to account for phase effects.

The phase dependence of resonance-mediated TPE processes was studied for the case of very narrow absorption lines in atoms [46] and for relatively broader absorptions in rare Earth systems [47], in both cases under the assumption of pulse durations much shorter than the dephasing time of the intermediate state. None of those cases is therefore directly applicable here. For $\delta \lesssim 50$ fs and not considering other plasmonic resonances at the two-photon energy, the TPE can be written as follows [47]:

$$\mathrm{TPE}_{\delta<50} \propto \int d\omega_f \left| \int d\omega_i \; A(\omega_i) E_A(\omega_f - \omega_i) E_A(\omega_i) \right|^2 \quad (2)$$

where $A(\omega)$ is a real function corresponding to the LSPR that mediates the excitation, and the term $E_A(\omega)$ is the complex electric field at the nanoantenna, consisting of an amplitude and a spectral phase component $\varphi(\omega)$. Following the formalism we used in a recent work [48] and neglecting the small contribution from the intrinsic phase response of the nanoantenna, $E_A(\omega)$ can be written as $E_A(\omega) = E(\omega)A(\omega)$, where $E(\omega)$ is the complex laser field. In this limit, the phase dependence in the TPE is recovered. Moreover, considering that $A(\omega)$ is a real function, the dependence on the spectral phase predicted by Eq. (2) is similar to that of the SH from a broadband pulse (see, for example, the study by Lozovoy et al. [41]). Indeed, for very short pulse durations, the TPPL in Figure 1b almost recovers a $\sim\delta^{-1}$ dependence, similar to the SH curve, which strongly supports our interpretation.

To verify the phase dependence of the TPPL, we performed phase scans on a resonant gold nanoantenna (detecting the TPPL) and compared it directly to the SH trace measured from a $BaTiO_3$ nanoparticle. Specifically, we carried out a multiphoton intrapulse interference phase scan (MIIPS), ramping a sinusoidal phase modulation through the laser spectrum [49, 50]. The effect of the spectral phase scanning on TPPL and SH is shown in Figure 1d. Clearly, the two traces show an oscillatory character, with maxima and minima located at different positions. The TPPL does show phase response, yet the total modulation depth is higher in the case of the SH nanoparticle. This difference can be explained based on the three-level model discussed above. As the pulse in the MIIPS is in the range of 20–100 fs, the green curve in Figure 1d is probing a regime in which the intermediate state dynamics starts to kick in, decreasing the achievable modulation depth. In fact, MIIPSs on gold nanoantennas using pulses longer than 100 fs do not show any modulation at all. Our current findings also explain earlier MIIPSs on gold nanoantennas, detecting TPPL to obtain information on their LSPRs [39]. In those scans, as the phase was changed, the time duration of the pulse also varied; as a result, the TPPL showed dependence on the MIIPS, yet mixing phase dependence with time-dependent signature of the intermediate state [39].

From the pulse length dependence, we conclude that, the TPE process in gold nanoantennas is governed by two different timescales: the dephasing time of the surface plasmons T_2 and the lifetime of the intermediate state T_1 (see Figure 1c). For laser pulses with durations comparable with T_2, the TPE is sensitive to the spectral phase of the electric field and pure coherent control becomes possible [51].

3 Closed-loop coherent control of pairs of nanoantennas

We then demonstrated such coherent control by resorting to a closed-loop optimization strategy, in which a deterministic algorithm [40, 52] adjusts the phase of the laser field in order to maximize the targeted signal. From Eq. (2), one can verify that the TPE in a gold nanoantenna is maximized by the shortest possible pulse (the Fourier limited pulse) [47]. Performing a blind maximization of the TPPL on a single nanoantenna would therefore produce Fourier limited pulses, which is a trivial control experiment. To test the control efficiency on the ultrafast TPE, we simultaneously excited two nanorod antennas of different but precisely defined lengths, characterized by different LSPRs and $A(\omega)$ terms, and used the ratio of their emitted TPPL as the feedback variable for the maximization. The concept of the experiment is sketched in Figure 2a (detail in Methods).

First, we acquired a two-dimensional TPPL image of an array of nanoantennas in the sample (Figure 2b). The nanoantennas were fabricated in lines of paired antennas. The length of the lower nanoantennas in the pair was kept constant at 90 nm, whereas the upper nanoantenna length was swept from 100–190 nm. The TPPL image in Figure 2b was taken using a Fourier limited excitation pulse. Based on Figure 2b, nanoantennas of length in the range of 90–120 nm produced the highest signals and therefore were resonant with the excitation laser. Moreover, since the spectral position of LSPRs depends on the length of the nanoantennas (for a fixed width), from Figure 2b we expect 90 nm (120 nm) – long antennas to be resonant with the blue (red) side of the laser spectrum (simulations shown in Figure 3 confirm this). In order to obtain the best control, understood as the highest obtainable contrast compatible with our laser spectrum, we implemented the closed-loop optimization on a pair of 90–120 nm nanoantenna. The feedback variable for the search algorithm was the quotient $A_{120\mathrm{nm}}/A_{90\mathrm{nm}}$, which represents the ratio between the TPPL from the 120 nm and the 90 nm nanoantenna in the pair.

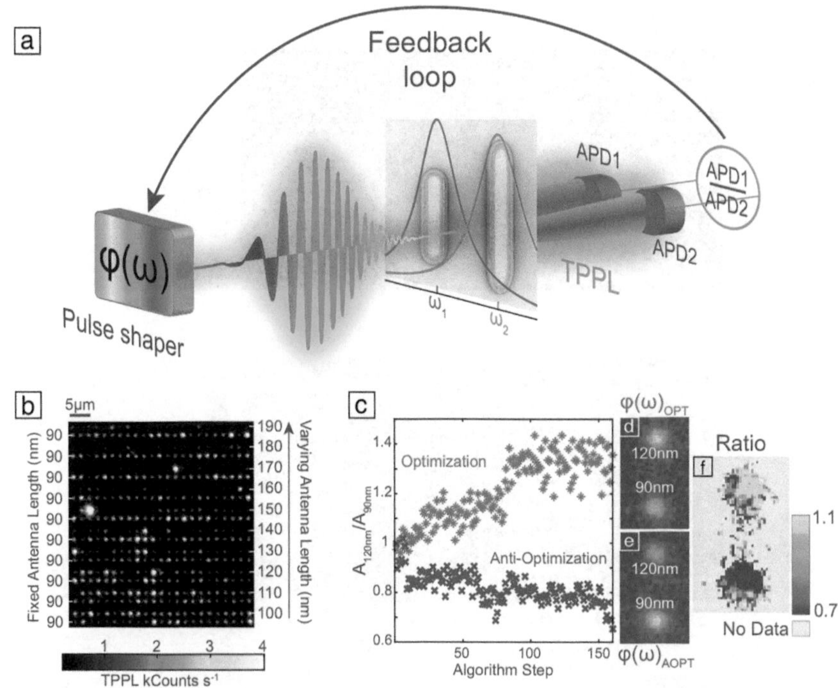

Figure 2: Closed-loop control experiment. (a) The pulse shaper controls the spectral phase of the laser pulse, which simultaneously excites two resonant nanoantennas, characterized by detuned plasmon resonances ω_1 and ω_2, inducing TPE. The TPPL generated by the nanoantennas is sent to two different single-photon avalanche photodiodes (APD1 and APD2), and the ratio between the two signals (APD1/APD2) is maximized in a feedback loop optimization. The search algorithm determines the spectral phase that enhances the ratio of the two-photon excitation in the nanoantennas. (b) Two-photon–induced photoluminescence image of an array of nanoantennas showing the arrangement of the nanoantennas in the sample: fixed length of 90 nm alternating with increasing length (100–190 nm) from bottom to top and the same length for each row. Note that the nanoantennas giving the highest signal (90–120 nm) are the ones that are resonant with the laser spectrum. (c) Plot of the $A_{120\text{nm}}/A_{90\text{nm}}$ ratio as a function of the algorithm step for both the optimization and antioptimization case. (d, e) TPPL images of the two nanoantennas taken with the optimization ($\varphi(\omega)_{\text{OPT}}$) and antioptimization phase ($\varphi(\omega)_{\text{AOPT}}$), respectively. Note that the maximum of the two-photon–induced photoluminescence switches from the 120 nm to the 90 nm antenna. (f) Ratio between images d and e emphasizing the contrast between the two nanoantennas. TPPL, two-photon–induced photoluminescence; TPE, two-photon excitation.

We performed two distinct coherent control experiments: in the first one (called optimization experiment), the algorithm maximized $A_{120\text{nm}}/A_{90\text{nm}}$ and in the second one (antioptimization) instead, minimized it. The results of the two experiments are plotted in Figure 2c as a function of the algorithm step. Experimental data are normalized such that the initial $A_{120\text{nm}}/A_{90\text{nm}}$ value (corresponding to the Fourier limited pulses) is one. As Figure 2c clearly shows, in the optimization and the antioptimization experiment, the ratio $A_{120\text{nm}}/A_{90\text{nm}}$ changed, with a final contrast being approximately 2. The spectral phases obtained at the end of the coherent control experiment are called $\varphi(\omega)_{\text{OPT}}$ and $\varphi(\omega)_{\text{AOPT}}$ and correspond to the optimization and antioptimization case, respectively. Imaging the two nanoantennas using these phases produced the images shown in Figure 2d and e from which one can see that the TPPL maximum switched from one nanoantenna to the other. To make the contrast even more apparent, in Figure 2f, we plot the ratio between Figure 2d and e. To produce this image (and equivalently Figure 3a), we discarded any pixel in which the signal was lower than 200 counts $s{-}1$, as they correspond to areas in which no nanoantenna was present, and the signal was only generated by random dark counts in the avalanche photodiodes (APDs).

These results prove that closed-loop coherent control of the TPE in single nanoantennas is possible. To emphasize the role of coherence, we wish to stress that the algorithm acted purely on the spectral phase of the laser pulse, while the laser intensities over the full spectrum were kept constant. Therefore, during an optimization, the laser always resonantly excited the LSPRs of each nanoantennas in the same way, keeping the populations of the intermediate states constant. The control derives from an interference process occurring purely among photons at the energy of the final state, which results in the modulation of the final state populations. This is in contrast to experiments in which the delay between two pulses is varied, as, for instance, in the studies by Kubo et al. [27], Sun et al. [28] and Nishiyama et al. [32]. Scanning the delay between two pulses effectively changes the excitation spectrum and therefore modulates the population of the intermediate state, which is reflected in the population of the final state. At the same time, the results shown here could not be obtained by using longer pulses [25, 26], as these would not interact coherently with the nanoantennas.

We next checked the robustness of the found solutions for nanoantennas of the same type. The upper image in Figure 3a is a contrast image (obtained with the same

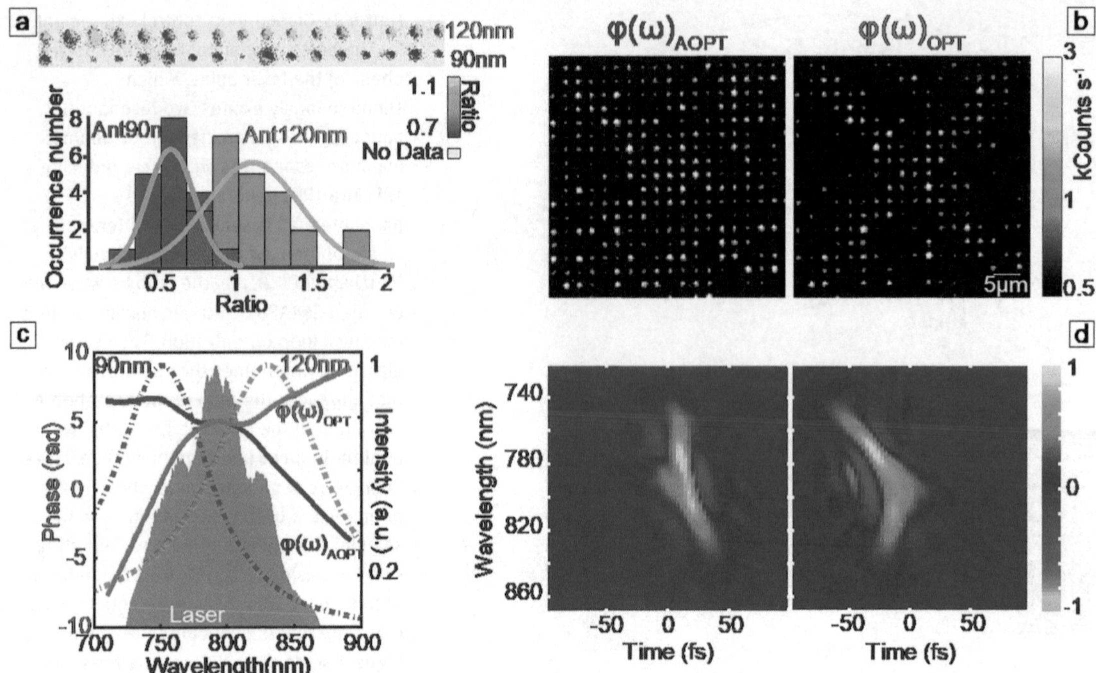

Figure 3: Coherent control of nanoantenna arrays.
(a) The upper image is the ratio between the TPPL images of an array of nanoantenna pairs obtained with $\varphi(\omega)_{OPT}$ and $\varphi(\omega)_{AOPT}$, respectively. Nanoantennas of 120 nm are brighter when using $\varphi(\omega)_{OPT}$, whereas the 90 nm ones are brighter when excited with $\varphi(\omega)_{AOPT}$. The histogram on the bottom quantitatively analyses the ratio of all the 90 nm antennas and 120 nm nanoantennas with the two different phases. The cyan and orange curves superimposed on the histogram are fits to the data assuming a normal distribution. Note that, in the histogram, a full array of nanoantennas is considered for the statistics (23 pairs), whereas in the upper image, a zoom on 17 pairs of nanoantennas is shown. We also note that for some of the 90 nm antennas, the TPPL intensity emitted with $\varphi(\omega)_{OPT}$ is so low that it is not taken into account in the histogram graph. (b) Two-photon–induced photoluminescence images of an array of mixed 120 and 90 nm antennas using $\varphi(\omega)_{AOPT}$ or $\varphi(\omega)_{OPT}$. The encoded message clearly appears when using $\varphi(\omega)_{OPT}$ and is erased by $\varphi(\omega)_{AOPT}$. (c) Laser spectrum (shaded region), $\varphi(\omega)_{OPT}$ (green curve) and $\varphi(\omega)_{AOPT}$ (violet curve) as measured with the multiphoton intrapulse interference phase scan (see Methods). The red and blue dashed curves are the resonance response of the 90 nm and the 120 nm nanoantennas as obtained by Finite-Difference Time-Domain (FDTD) simulations. (d) Time–frequency Wigner representation of the pulses corresponding to the optimization (right) and antioptimization (left) case. TPPL, two-photon–induced photoluminescence.

procedure as for Figure 2f) of one array of paired 90 and 120 nm nanoantennas, obtained under excitation with $\varphi(\omega)_{OPT}$ and $\varphi(\omega)_{AOPT}$. From this image, one can see that changing from $\varphi(\omega)_{OPT}$ to $\varphi(\omega)_{AOPT}$ produced the same overall effect in most of the 90–120 nm nanoantenna pairs. Statistical analysis is given in the histogram of Figure 3a, which helps visualizing this trend. The distribution of the ratios is clearly shifted, as confirmed by fits to the data presented in the histogram assuming a normal distribution function. The width of the distributions is caused by the intrinsic variability of fabrication, which is always present for e-beam nanofabricated antennas. It must also be noted that a higher level of control could be obtained for the 90 nm nanoantennas. By changing the spectral phase from $\varphi(\omega)_{AOPT}$ to $\varphi(\omega)_{OPT}$, the TPPL from the 120 nm nanoantennas increased on average by ~1.12, whereas that from 90 nm nanoantennas decreased by ~1.76, with a total contrast of ~2. This difference could be due to distinct coherence time for the two antenna species, as explained below.

Now, we can use this coherent selectivity in nonlinear microscopy, as shown for ensembles of molecules and quantum dots [53–56], and extend the control to the field of nanoplasmonics. Figure 3b demonstrates this concept at the level of individual nanoantennas. We fabricated an array of mixed 120 and 90 nm nanoantennas. The 120 nm nanoantennas were arranged such to form an encoded message, in this case the letter 'X', on a background of 90 nm ones. We acquired two different two-dimensional TPPL images of such array using $\varphi(\omega)_{OPT}$ and $\varphi(\omega)_{AOPT}$, respectively. As Figure 3b demonstrates, by using pure coherent control, we could deterministically reveal or completely hide the encoded message. In this case, complete switch of the TPPL maximum from 120 to 90 nm nanoantennas did not occur (as in the case of Figure 2d and f) as the 120 nm ones always emitted more TPPL. Still, clear contrast between the two images is visible, confirming that coherent control can extract sensitive microscopic information about different resonant species.

We are now in the position to give a solid interpretation to the coherent control experiment, as illustrated in the bottom panel of Figure 3. In Figure 3c, we plot the excitation laser spectrum, together with the spectral phases $\varphi(\omega)_{OPT}$ and $\varphi(\omega)_{AOPT}$ obtained after the optimization. The blue and red dashed curves correspond to the LSPR profiles for the 90 and 120 nm nanoantennas, respectively, as obtained from Finite-Difference Time-Domain (FDTD) simulations (see Methods). As anticipated above, the LSPR corresponding to the 90 nm antennas is resonant on the blue side and that of the 120 nm one on the red side of the laser spectrum. These curves represent the intermediate states in the TPE process, namely the terms $A(\omega)$ in Eq. (1). From this plot, it is apparent that the two spectral phases $\varphi(\omega)_{OPT}$ and $\varphi(\omega)_{AOPT}$ have symmetric behaviours with respect to the positions of the LSPRs.

As both the time and the frequency information is relevant for the interpretation of the current experiment, in Figure 3d, we report the time–frequency representation (details in the study by Paye [57]) for the two electric fields corresponding to the optimization (right) and antioptimization (left) case, respectively. In the spectral domain (vertical axis of the plots), one can see that in the optimization case, the red part of the laser spectrum is only slightly distorted (it is almost a straight line at time zero), whereas the blue part of it is chirped. A flat spectral phase drives the 120 nm antenna, whereas a dispersed spectral phase drives the 90 nm antenna. The opposite situation is found for the antioptimization case, and the respective time–frequency Wigner representations are almost inverted along their time development. The effect of coherent control is therefore to distort the pulse in the spectral region of the unwanted LSPR and make TPE through the desired plasmon resonance more favourable. This is consistent with our findings that $\varphi(\omega)_{OPT}$ optimizes the TPPL ratio between 120 and 90 nm antennas and vice versa. However, as explained above, the phase control is only effective for short enough pulses (within the coherent regime of Figure 1b), whereas for longer pulses, the spectral phase has no effect. Indeed, the pulses represented in Figure 3d are still very short, with a total duration of less than 50 fs and therefore can interact coherently with the nanoantennas. Note that our algorithm optimizes on the differences between the nanoantennas, without any assumption on coherent or incoherent regimes. The algorithm, in the configuration used [40], has the capability to stretch the pulse beyond 200 fs and could have ended up in pulses longer than the coherence time. Instead, the algorithm choses to exploit the phase and chirp the pulse in a selected spectral regime. This again confirms the validity of the three-level model discussed above.

As a final confirmation that the interpretation given here is correct, we calculated the contrast in the TPE probability in the two nanoantennas using Eq. (2), which is approximately valid in the very short pulse regime. To do so, we considered the resonant curves of Figure 3c as the excitation lineshapes (the $A(\omega)$ terms) of the nanoantennas and calculated the laser electric field based on the measurements of the laser spectrum and the spectral phases $\varphi(\omega)_{OPT}$ and $\varphi(\omega)_{AOPT}$. The calculated TPE probabilities are in good agreement with the experiment, yielding a contrast of ~2.0 for the 90 nm nanoantenna and of ~1.3 for the 120 nm one, with a total contrast of ~2.6. The spectral width of the LSPR is wider for the 120 nm nanoantenna, i.e. the longer antenna is more lossy and the coherence time T_2 is shorter. The algorithm changes the phase and therefore the time duration of the laser pulse and reaches first the incoherent regime of the 120 nm before the 90 nm nanoantenna. As a result, the algorithm has a larger time–frequency parameter space to distort the pulse in the case of the shorter 90 nm nanoantenna, as in fact shown by the time–frequency representations in Figure 3d, and higher contrast is obtainable for the 90 nm nanoantennas.

4 Conclusions

In this work, we have presented a detailed and comprehensive study of the TPE process and its dependence on the phase of an ultrashort laser pulse in gold nanoantennas. We have shown that two relevant timescales govern the dynamics of the two-step excitation process: the dephasing time of the localized surface plasmons and the lifetime of the intermediate state involved. For very short pulses with duration shorter than or comparable to the plasmon dephasing time, the TPE in resonant gold nanoantennas is sensitive to the spectral phase of the laser pulse and coherent control becomes possible. For longer pulses (longer than 100 fs), no phase control of the gold nanoantennas is possible. The less lossy high-index semiconductor antennas might be interesting for control at longer timescales [58].

We have subsequently demonstrated pure coherent phase control by using a closed-loop optimization strategy that actively manipulates the coherence of surface plasmons to selectively excite two different nanoantennas solely by shaping the laser spectral phase. This is a conclusive demonstration of highly sensitive closed-loop coherent control on the nanoscale.

5 Methods

The laser source used in the experiment was a broadband titanium sapphire laser (Octavius 85 M, Menlo System, Menlo Systems GmbH, Am Klopferspitz 19a, D-82152 Martinsried, Germany) with a spectrum centred at 800 nm and with bandwidth of ~150 nm. The pulse shaper was arranged in a 4-f configuration, and the active element was a spatial light modulator composed by two 640-pixel liquid crystal masks. The pulse shaper was used exclusively for phase shaping, and no amplitude shaping was performed.

In the pulse duration dependence experiment, the laser beam was sent to a confocal microscope and focused on the sample from the bottom by a 1.4 numerical aperture (NA) objective. The sample was mounted on a nanometre precision piezoelectric scanner for two-dimensional imaging. The nonlinear signals generated at the sample were collected through the same objective, spectrally filtered to cut the reflected laser light and sent to the photodetector: a Si APD in counting mode. The pulse shaper was used to change the pulse duration by applying different amounts of linear chirp.

The $BaTiO_3$ nanoparticles were purchased from Sigma-Aldrich: Sigma-Aldrich, Inc., nanopowder (cubic crystalline phase), <100 nm particle size, ≥99% trace metals basis, dispersed in ethanol, and deposited onto a microscope coverslip, following a previously developed procedure [49].

In the closed-loop coherent control experiment, the laser was focused from the top of the sample by a 0.15 NA objective on the sample containing the gold nanoantennas. The lateral size of the excitation spot at the sample position was about 8 μm and, together with the arrangement of the nanoantennas in the sample, was chosen such that two different nanoantennas could be excited at the same time. The excitation polarization was kept circular in the experiment. The TPPL was collected in transmission through the 1.4 NA objective and sent to two different APDs after being separated by a beam splitter. The combination of the 1.4 NA objective, the system of lenses that focused the light on the APDs and the physical size of the active area of the APDs was such that the APDs receive light form a much smaller area than the excitation spot, being roughly of 1 μm. This allowed us to spatially distinguish between one and the other nanoantenna and to change the actual detection area by spatially displacing the APDs. By slightly moving the two APDs in different directions, we could align them on the TPPL emitted by the two different nanoantennas.

The nanoantennas studied were gold nanorods of length varying between 90 and 190 nm, fabricated using electron beam lithography on a glass coverslip coated with a 10 nm-thin layer of indium tin oxide (ITO). The width and height of the nanoantennas were both fixed at 50 nm, and they were fabricated to be resonant with different portions of the laser spectrum. In order to study two nanoantennas characterized by different LSPRs at the same time, they were arranged in pairs of different lengths separated by 2 μm.

The spectral phases shown in Figure 3c were measured using the MIIPS described in the studies by Xu et al. [50] and Galvan-Sosa et al. [52] combined with SH nanoparticles, following a method we previously developed [49].

The extinction spectra of the nanoantennas were calculated using FDTD Solutions (Lumerical Inc.): www.Lumerical.com, Lumerical Inc., 1700–1095 West Pender Street, Vancouver, BCV6E 2M6, Canada. The nanoantennas themselves were 130 × 40 × 40 and 100 × 40 × 40 nm^3, with the top corners rounded (radius = 10 nm) to approximate the fabricated nanoantennas. The nanoantenna lied on a 10 nm layer of ITO, which was itself on silicon dioxide. The permittivity of Au was taken using a fit of Palik's data, and the ITO data were taken from in-house measurements and also fitted, in both cases using their multicoefficient model of the experimental data. The mesh size was 2 nm and encompassed the entire nanoantenna. The source was a total field scattered field (TFSF) plane wave at normal incidence to the glass surface, injected from within the air side. To measure the absorbed and scattered powers, plane monitors were arranged to form two boxes around the nanoantenna, one inside and one outside of the TFSF source volume, and each monitor recorded the power transmitted through its planar surface. Given the nature of the TFSF source (i.e. it subtracts the source fields at the edge of its defined volume), the power box outside was used to calculate the scattering spectrum and the one inside to calculate the absorption. An identical simulation without nanoantenna was also realized to calculate the ITO absorption, which is then subtracted from each nanoantenna's absorption spectrum.

Author contribution: All the authors have accepted responsibility for the entire content of this submitted manuscript and approved submission.

Research funding: This research was funded by the European Commission (ERC Adv. Grant 670949-LightNet); Spanish Severo Ochoa Programme for Centres of Excellence in R&D (SEV-2015-0522); Plan National Projects MICINN FIS2012-35527, MINECO FIS2015-69258-P and MCIU PGC2018-096875-B-I00 and cofunded by the Fondo Europeo de Desarrollo Regional (FEDER); the Catalan AGAUR (2014 SGR01540 and 2017 SGR1369); Fundació Privada Cellex; Fundació Privada Mir-Puig and Generalitat de Catalunya through the Centres de Recerca de Catalunya (CERCA) program. PMdR acknowledges financial support from the Spanish Government MINECO-FPI grant and European Science Foundation under the PLASMON-BIONANOSENSE Exchange Grant program. MGS acknowledges financial support from grants MICINN TEC2011-22422 and MINECO TEC2014-52642-C2-1-R.

Conflict of interest statement: The authors declare no conflicts of interest regarding this article.

References

[1] D. J. Tannor, R. Kosloff, and S. A. Rice, "Coherent pulse sequence induced control of selectivity of reactions: exact quantum mechanical calculations," *J. Chem. Phys.*, vol. 85, pp. 5805–5820, 1986.

[2] M. Shapiro and P. Brumer, *Quantum Control of Molecular Processes*, Wiley-VCH Verlag GmbH & Co. KGaA, 2011, https://doi.org/10.1002/9783527639700.

[3] M. Dantus and V. V. Lozovoy, "Experimental coherent laser control of physicochemical processes," *Chem. Rev.*, vol. 104, pp. 1813–1859, 2004.

[4] P. Nuernberger, G. Vogt, T. Brixner, and G. Gerber, "Femtosecond quantum control of molecular dynamics in the condensed phase," *Phys. Chem. Chem. Phys.*, vol. 9, p. 2470, 2007.

[5] C. Brif, R. Chakrabarti, and H. Rabitz, “Control of quantum phenomena: past, present and future,” *New J. Phys.*, vol. 12, 2010. https://doi.org/10.1088/1367-2630/12/7/075008.
[6] Y. Silberberg, “Quantum coherent control for nonlinear spectroscopy and microscopy,” *Annu. Rev. Phys. Chem.*, vol. 60, pp. 277–292, 2009.
[7] L. Novotny and N. F. van Hulst, “Antennas for light,” *Nat. Photon.*, vol. 5, pp. 83–90, 2011.
[8] A. G. Curto, G. Volpe, T. H. Taminiau, et al., “Unidirectional emission of a quantum dot coupled to a nanoantenna,” *Science*, vol. 329, pp. 930–933, 2010.
[9] I. M. Hancu, A. G. Curto, M. Castro-López, M. Kuttge, and N. F. van Hulst, “Multipolar interference for directed light emission,” *Nano Lett.*, vol. 14, pp. 166–171, 2014.
[10] D. Vercruysse, X. Zheng, Y. Sonnefraud, et al., “Directional fluorescence emission by individual V-antennas explained by mode expansion,” *ACS Nano*, vol. 8, pp. 8232–8241, 2014.
[11] G. Zengin, M. Wersäll, S. Nilsson, et al., “Realizing strong light-matter interactions between single-nanoparticle plasmons and molecular excitons at ambient conditions,” *Phys. Rev. Lett.*, vol. 114, p. 157401, 2015.
[12] A. Kinkhabwala, Z. Yu, S. Fan, et al., “Large single-molecule fluorescence enhancements produced by a bowtie nanoantenna,” *Nat. Photon.*, vol. 3, pp. 654–657, 2009.
[13] E. Wientjes, J. Renger, A. G. Curto, R. Cogdell, and N. F. van Hulst, “Strong antenna-enhanced fluorescence of a single light-harvesting complex shows photon antibunching,” *Nat. Commun.*, vol. 5, pp. 1516–1519, 2014.
[14] M. Kauranen and A. V. Zayats, “Nonlinear plasmonics,” *Nat. Publ. Gr.*, vol. 6, pp. 737–748, 2012.
[15] S. S. Aćimović, M. A. Ortega, V. Sanz, et al., “LSPR chip for parallel, rapid, and sensitive detection of cancer markers in serum,” *Nano Lett.*, vol. 14, pp. 2636–2641, 2014.
[16] P. Zijlstra, P. M. R. Paulo, and M. Orrit, “Optical detection of single non-absorbing molecules using the surface plasmon resonance of a gold nanorod,” *Nat. Nanotechnol.*, vol. 7, pp. 379–382, 2012.
[17] P. Ghenuche, S. Cherukulappurath, T. H. Taminiau, N. F. van Hulst, and R. Quidant, “Spectroscopic mode mapping of resonant plasmon nanoantennas,” *Phys. Rev. Lett.*, vol. 101, p. 116805, 2008.
[18] P. Biagioni, M. Celebrano, M. Savoini, et al., “Dependence of the two-photon photoluminescence yield of gold nanostructures on the laser pulse duration,” *Phys. Rev. B Condens. Matter*, vol. 80, pp. 1–5, 2009.
[19] P. Biagioni, M. H. Lee, N. J. Halas, et al., “Dynamics of four-photon photoluminescence in gold nanoantennas,” *Nano Lett.*, vol. 12, pp. 2941–2947, 2012.
[20] A.M. Gobin, M. H. Lee, N. J. Halas, et al., “Near-infrared resonant nanoshells for combined optical imaging and photothermal cancer therapy,” *Nano Lett.*, vol. 7, pp. 1929–1934, 2007.
[21] N. J. Durr, T. Larson, D. K. Smith, et al., “Two-photon luminescence imaging of cancer cells using molecularly targeted gold nanorods,” *Nano Lett.*, vol. 7, pp. 941–945, 2007.
[22] N. Gao, Y. Chen, L. Li, et al., “Shape-dependent two-photon photoluminescence of single gold nanoparticles,” *J. Phys. Chem. C*, vol. 118, pp. 13904–13911, 2014.
[23] M. I. Stockman, S. V. Faleev, and D. J. Bergman, “Coherent control of femtosecond energy localization in nanosystems,” *Phys. Rev. Lett.*, vol. 88, pp. 67402/1–67402/4, 2002.
[24] M. I. Stockman, “Ultrafast nanoplasmonics under coherent control,” *New J. Phys.*, vol. 10, 2008. https://doi.org/10.1088/1367-2630/10/2/025031.
[25] M. Aeschlimann, M. Bauer, D. Bayer, et al., “Spatiotemporal control of nanooptical excitations,” *Proc. Natl. Acad. Sci. USA*, vol. 107, pp. 5329–5333, 2010.
[26] M. Aeschlimann, M. Bauer, D. Bayer, et al., “Adaptive subwavelength control of nano-optical fields,” *Nature*, vol. 446, pp. 301–304, 2007.
[27] A. Kubo, K. Onda, H. Petek, et al., “Femtosecond imaging of surface plasmon femtosecond imaging of surface plasmon dynamics in a nanostructured silver film,” *Nano*, vol. 5, pp. 1123–1127, 2005.
[28] Q. Sun, K. Ueno, H. Yu, et al., “Direct imaging of the near field and dynamics of surface plasmon resonance on gold nanostructures using photoemission electron microscopy,” *Light Sci. Appl.*, vol. 74, p. e118, 2013.
[29] T. Hanke, J. Cesar, V. Knittel, et al., “Tailoring spatiotemporal light confinement in single plasmonic nanoantennas,” *Nano Lett.*, vol. 12, pp. 992–996, 2012.
[30] R. Mittal, R. Glenn, I. Saytashev, V. V. Lozovoy, and M. Dantus, “Femtosecond nanoplasmonic dephasing of individual silver nanoparticles and small clusters,” *J. Phys. Chem. Lett.*, vol. 6, pp. 1638–1644, 2015.
[31] E. Mårsell, A. Losquin, R. Svärd, et al., “Nanoscale imaging of local few-femtosecond near-field dynamics within a single plasmonic nanoantenna,” *Nano Lett.*, vol. 15, pp. 6601–6608, 2015.
[32] Y. Nishiyama, K. Imaeda, K. Imura, and H. Okamoto, “Plasmon Dephasing in Single Gold Nanorods Observed by Ultrafast Time-Resolved Near-Field Optical Microscopy,” *J. Phys. Chem. C*, vol. 119, pp. 16215–16222, 2015.
[33] Y. Nishiyama, K. Imura, and H. Okamoto, “Observation of plasmon wave packet motions via femtosecond time-resolved near-field imaging techniques,” *Nano Lett.*, vol. 15, pp. 7657–7665, 2015.
[34] J. M. Gunn, S. H. High, V. V. Lozovoy, and M. Dantus, “Measurement and control of ultrashort optical pulse propagation in metal nanoparticle-covered dielectric surfaces,” *J. Phys. Chem. C*, vol. 114, pp. 12375–12381, 2010.
[35] J. M. Gunn, M. Ewald, and M. Dantus, “Polarization and phase control of remote surface-plasmon-mediated two-photon-induced emission and waveguiding,” *Nano Lett.*, vol. 6, pp. 2804–2809, 2006.
[36] K. Imura, T. Nagahara, and H. Okamoto, “Near-field two-photon-induced photoluminescence from single gold nanorods and imaging of plasmon modes,” *J. Phys. Chem. B*, vol. 109, pp. 13214–13220, 2005.
[37] X.-F. Jiang, Y. Pan, C. Jiang, et al., “Excitation nature of two-photon photoluminescence of gold nanorods and coupled gold nanoparticles studied by two-pulse emission modulation spectroscopy,” *J. Phys. Chem. Lett.*, vol. 4, pp. 1634–1638, 2013.
[38] K. Imaeda and K. Imura, “Optical control of plasmonic fields by phase-modulated pulse excitations,” *Opt. Express*, vol. 21, p. 27481, 2013.
[39] D. Brinks, M. Castro-Lopez, R. Hildner, and N. F. van Hulst, “Plasmonic antennas as design elements for coherent ultrafast nanophotonics,” *Proc. Natl. Acad. Sci. USA*, vol. 110, pp. 18386–18390, 2013.

[40] N. Accanto, P. M. de Roque, M. Galvan-Sosa, S. Christodoulou, I. Moreels and N. F. van Hulst, "Rapid and robust control of single quantum dots," *Light Sci. Appl.*, vol. 6, p. e16239, 2016.
[41] V. V. Lozovoy, I. Pastirk, K. A. Walowicz, and M. Dantus, "Multiphoton intrapulse interference. II. Control of two- and three-photon laser induced fluorescence with shaped pulses," *J. Chem. Phys.*, vol. 118, pp. 3187–3196, 2003.
[42] N. Del Fatti, C. Voisin, M. Achermann, et al., "Nonequilibrium electron dynamics in noble metals," *Phys. Rev. B*, vol. 61, pp. 16956–16966, 2000.
[43] C.-K. Sun, F. Vallée, L. H. Acioli, E. P. Ippen, and J. G. Fujimoto, "Femtosecond-tunable measurement of electron thermalization in gold," *Phys. Rev. B*, vol. 50, pp. 15337–15348, 1994.
[44] A. Block, M. Liebel, R. Yu, F.J. García de Abajo, Y. Sivan, and N.F. van Hulst, "Tracking ultrafast hot-electron diffusion in space and time by ultrafast thermo-modulation microscopy," *Sci. Adv.*, vol. 5, p. eaav8965, 2019.
[45] A. Anderson, K. S. Deryckx, X. G. Xu, G. Steinmeyer, and M. B. Raschke, "Few-plasmon dephasing of a single metallic nanostructure from optical response function reconstruction by interferometric frequency resolved optical gating," *Nano Lett.*, vol. 10, pp. 2519–2524, 2010.
[46] N. Dudovich, B. Dayan, S. M. Gallagher Faeder, and Y. Silberberg, "Transform-limited pulses are not optimal for resonant multiphoton transitions," *Phys. Rev. Lett.*, vol. 86, pp. 47–50, 2001.
[47] S. Zhang, C. Lu, T. Jia, J. Qiu, and Z. Sun, "Coherent phase control of resonance-mediated two-photon absorption in r femtosecond are-earth ions," *Appl. Phys. Lett.*, vol. 103, pp. 1–5, 2013.
[48] N. Accanto, L. Piatkowski, J. Renger, and N. F. van Hulst, "Capturing the optical phase response of nanoantennas by coherent second-harmonic microscopy," *Nano Lett.*, vol. 14, pp. 4078–4082, 2014.
[49] Y. Coello, V. V. Lozovoy, T. C. Gunaratne, et al., "Interference without an interferometer: a different approach to measuring, compressing, and shaping ultrashort laser pulses," *J. Opt. Soc. Am. B*, vol. 25, p. A140, 2008.
[50] B. Xu, J. M. Gunn, J. M. Dela Cruz, V. V. Lozovoy, and M. Dantus, "Quantitative investigation of the multiphoton intrapulse interference phase scan method for simultaneous phase measurement and compensation of femtosecond laser pulses," *J. Opt. Soc. Am. B*, vol. 23, p. 750, 2006.
[51] V. Remesh, M. Stührenberg, L. Saemisch, N. Accanto, and N.F. van Hulst, "Phase control of plasmon enhanced two-photon photoluminescence in resonant gold nanoantennas," *Appl. Phys. Lett.*, vol. 113, p. 211101, 2018.
[52] M. Galvan-Sosa, J. Portilla, J. Hernandez-Rueda, et al., "Optimization of ultra-fast interactions using laser pulse temporal shaping controlled by a deterministic algorithm," *Appl. Phys. Mater. Sci. Process*, vol. 114, pp. 477–484, 2014.
[53] T. Brixner, N. H. Damrauer, P. Niklaus, and G. Gerber, "Photoselective adaptive femtosecond quantum control in the liquid phase," *Nature*, vol. 414, pp. 57–60, 2001.
[54] M. Ruge, R. Wilcken, M. Wollenhaupt, A. Horn, and T. Baumert, "Coherent control of colloidal semiconductor nanocrystals," *J. Phys. Chem. C*, vol. 117, pp. 11780–11790, 2013.
[55] J. P. Ogilvie, D. Débarre, X. Solinas, et al., "Use of coherent control for selective two-photon fluorescence microscopy in live organisms," *Opt. Express*, vol. 14, pp. 759–766, 2006.
[56] I. Pastirk, J. Dela Cruz, K. Walowicz, V. Lozovoy, and M. Dantus, "Selective two-photon microscopy with shaped femtosecond pulses," *Opt. Express*, vol. 11, pp. 1695–1701, 2003.
[57] J. Paye, "The chronocyclic representation of ultrashort light pulses," *IEEE J. Quant. Electron.*, vol. 28, pp. 2262–2273, 1992.
[58] V. Remesh, G. Grinblat, Y. Li, S. A. Maier, and N. F. van Hulst, "Coherent multiphoton control of gallium phosphide nanodisk resonances," *ACS Photon.*, vol. 6, pp. 2487–2491, 2019.

Angela Vasanelli, Yanko Todorov, Baptiste Dailly, Sébastien Cosme, Djamal Gacemi, Andrew Haky, Isabelle Sagnes and Carlo Sirtori*

Semiconductor quantum plasmons for high frequency thermal emission

https://doi.org/10.1515/9783110710687-048

Abstract: Plasmons in heavily doped semiconductor layers are optically active excitations with sharp resonances in the 5–15 μm wavelength region set by the doping level and the effective mass. Here, we demonstrate that volume plasmons can form in doped layers of widths of hundreds of nanometers, without the need of potential barrier for electronic confinement. Their strong interaction with light makes them perfect absorbers and therefore suitable for incandescent emission. Moreover, by injecting microwave current in the doped layer, we can modulate the temperature of the electron gas. We have fabricated devices for high frequency thermal emission and measured incandescent emission up to 50 MHz, limited by the cutoff of our detector. The frequency-dependent thermal emission is very well reproduced by our theoretical model that let us envision a frequency cutoff in the tens of GHz.

Keywords: mid-infrared; plasmons; thermal emission.

***Corresponding author: Carlo Sirtori,** Laboratoire de Physique de l'Ecole Normale Supérieure, ENS, Université PSL, CNRS, Sorbonne Université, Université de Paris, Paris, France,
E-mail: carlo.sirtori@ens.fr. https://orcid.org/0000-0003-1817-4554
Angela Vasanelli, Yanko Todorov, Djamal Gacemi and Andrew Haky, Laboratoire de Physique de l'Ecole Normale Supérieure, ENS, Université PSL, CNRS, Sorbonne Université, Université de Paris, Paris, France, E-mail: angela.vasanelli@ens.fr (A. Vasanelli), yanko.todorov@ens.fr (Y. Todorov), djamal.gacemi@ens.fr (D. Gacemi), andrew.haky@phys.ens.fr (A. Haky). https://orcid.org/0000-0003-1945-2261 (A. Vasanelli). https://orcid.org/0000-0002-2359-1611 (Y. Todorov)
Baptiste Dailly and Sébastien Cosme, Laboratoire Matériaux et Phénomènes Quantiques, CNRS, Université de Paris, Paris, France, E-mail: baptiste.dailly@gmail.com (B. Dailly), cosme.sebastiensc@gmail.com (S. Cosme)
Isabelle Sagnes, Centre for Nanosciences and Nanotechnology, CNRS, Universite Paris-Saclay, UMR 9001, 10 Boulevard Thomas Gobert, 91120 Palaiseau, France,
E-mail: isabelle.sagnes@c2n.upsaclay.fr. https://orcid.org/0000-0001-8068-6599

1 Introduction

Plasmons are collective excitations of electron gas that concentrate the interaction with light. In metals, they have been studied for more than a century as volume and surface plasmons and have sparked fundamental semiclassical theories that we are still using to determine the high frequency conductivity and the upper limit of transparency (the plasma frequency) of many metals and highly doped semiconductors. The Drude theory, revisited then by Sommerfeld, is the most successful story. Originally conceived to describe the thermal properties of metals, it contains also the essential features on how light couples to an ensemble of electrons and is still largely used for explaining the dielectric constant of conductive solids.

Since the 70s, plasmons have also been studied in very thin metallic films of few tens of Angstroms, where size confinement gives rise to sizable shifts of their energy [1–3]. Moreover, at the edges of the potential well, plasmons develop a huge optically active dipole and become easily observable by spectroscopy techniques [4]. The resulting optical resonance is called Berreman or Ferrell–Berreman mode. The advent of nanotechnology has brought this field much farther by making available extremely small metallic or semiconductor particles in which plasmonic optical resonances can be tuned by size confinement [5, 6]. Remarkably, plasmonics with objects of nanometric dimensions has permitted to concentrate the electromagnetic radiation well below the diffraction limit and has open the field of nanoantennas [7, 8]. The study and the understanding of these properties are of major interest today and go under the name of quantum plasmonics [9].

Another challenge that is actively studied in this field is the generation of infrared radiation by thermally excited plasmons [10–12]. Due to their superradiant nature, excited plasmons radiatively decay by spontaneous emission [13]. Thermal radiation is of major importance for several applications, such as lighting, energy harvesting and management, sensing, tagging, and imaging [14]. There are

This article has previously been published in the journal Nanophotonics. Please cite as: A. Vasanelli, Y. Todorov, B. Dailly, S. Cosme, D. Gacemi, A. Haky, I. Sagnes and C. Sirtori "Semiconductor quantum plasmons for high frequency thermal emission" *Nanophotonics* 2021, 10. DOI: 10.1515/nanoph-2020-0413.

several ways of engineering thermal emission by exploiting a material resonance, such as phonon modes [15–17], optical transitions in a quantum well [18], or plasmonic resonances [19]. The interest of using plasmons is that they have huge dipoles and produce sharp optical resonances that can be as dark as a perfect blackbody on a narrow frequency band, ideal for a thermal emitter. Moreover, in a semiconductor, the energy of the plasmonic resonance can be adjusted by selecting the level of doping. Dynamic control of incandescence [20] can be achieved through time modulation of the electronic temperature. This can be obtained by transferring energy through electrical injection either in the electron gas or in an adjacent resistance. The combination of electrical injection and the very low thermal capacitance of the electron gas allows a high frequency modulation of the temperature and therefore of the thermal emission.

In this work, we show that mid-infrared optically active plasmons in highly doped GaInAs layers embedded between two sharp AlInAs barriers are substantially identical to those arising in highly doped GaInAs layers without the confining barrier. In the following of the paper, we will concentrate on the realization of incandescent devices in which the emission spectrum is controlled by a Berreman mode. The modulation of the current injected in the doped layer induces fast temperature variations of the electron gas. In this configuration, we were able to modulate the thermal emission up to 50 MHz, the highest frequency response of our detector.

2 Volume plasmons in highly doped semiconductor layers

Bulk plasmons are collective excitations of a three-dimensional electron gas existing in metals and semiconductors. In thin films with thickness smaller than their wavelength, these excitations develop dipoles at the edges, due to instantaneous charge separation, and they can therefore emit or absorb photons from free space radiation [1, 2, 21]. These optically active confined plasmon modes are called Berreman modes. In highly doped semiconductor layers, their optical resonance, at the plasma frequency Ω_p, is in the mid-infrared range that corresponds to optical wavelengths between 5 and 10 µm, depending on their effective mass and doping:

$$\Omega_p = \sqrt{\frac{e^2}{\epsilon_0 \epsilon_s} \frac{N_v}{m^*}} \tag{1}$$

with e the electronic charge, ϵ_0 the vacuum electric constant, ϵ_s the background dielectric constant, and m^* the effective mass, and N_v the electron density per unit volume.

Figure 1a illustrates the plasmons that we are investigating. They develop an optical dipole in the direction perpendicular to the layer and they are spatially modulated in the plane, thus possess a wavevector, $k_{//}$. This quantity is the quantum number that is preserved in emission or absorption and can be selected by changing the angle of emission/detection, θ, measured from the normal to the surface.

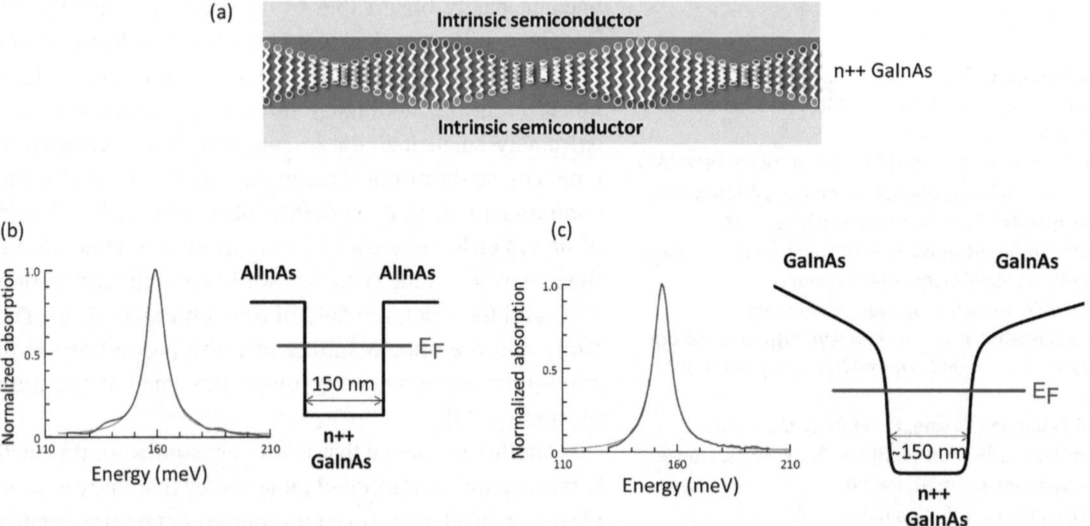

Figure 1: (a) Sketch of the system: a highly doped GaInAs layer embedded between two intrinsic semiconductor layers sustains a Berreman mode characterized by a dipole along the growth direction. (b) and (c) Conduction band profile and normalized absorption spectrum measured at 10° internal angle (blue line) for a sample in which the two intrinsic layers are (panel b) AlInAs barriers or (c) intrinsic GaInAs layers. Red lines are Lorentzian fit of the absorption spectra.

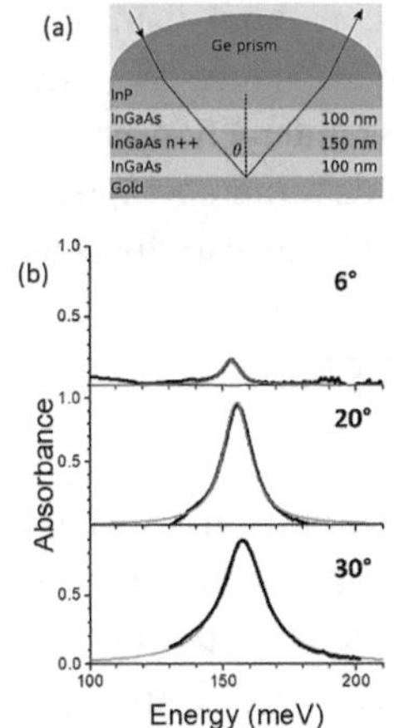

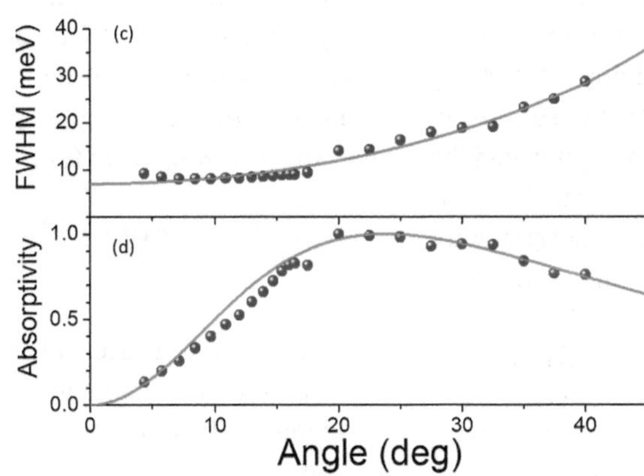

Figure 2: (a) Sketch of the sample and of the geometry used for reflectivity experiments. (b) Absorption spectra at three different internal angles (black lines) and Lorentzian fit of the resonances (red lines). (c) Full width at half the maximum (FWHM) of the absorption peak (blue symbols) as a function of the internal angle. Red line presents the calculated broadening, including radiative and nonradiative contribution. (d) Peak absorptivity extracted from the absorption spectra as a function of the angle, compared with the calculated one (red line) by using Eq. (3).

Plasmons appear in layers embedded between two confining barriers in a wide well configuration (Figure 1b). AlInAs barriers however are not necessary to confine a plasmon, as a simple heavily doped layer in an original "flat potential" bends the conduction band profile due to charge separation and form a potential well (Figure 1c). In both cases, the confining potential forms a set of discrete electronic energy levels which are the basis that we use for the theoretical description of these collective excitations. Given the wide dimension of the potential wells, 150 nm, it is remarkable that the electronic wavefunctions maintain their coherence and still form a set of energy states separated of few meV. The role of electronic confinement on plasmonic resonances with different potential geometries is therefore crucial to bring quantum engineering to the realm of plasmonics [22]. Moreover, a semiconductor platform instead of metals brings clear evidence on the quantum nature of the electrons building the plasmonic response. Figure 1b and c show measured normalized absorption spectra from two 150 nm doped layers that correspond to the potentials sketched in the figure. In both cases, the absorption has a Lorentzian shape and peaks at ~160 meV, in excellent agreement with our theory reported in the study by Pegolotti et al [23]. Surprisingly, in spite of the disorder introduced by the donors, the spectra measured at small angles are characterized by sharp resonances with a quality factor on the order of 20. At higher angles, the collective excitations are dominated by spontaneous superradiant emission [13] and become radiatively broadened (Figure 2).

3 Superradiance, perfect absorption

The very strong interaction that plasmons have with light is intimately related to their collective nature, i.e., they are a symmetric superposition of all the optically active electronic transitions of the system. The optical properties of the plasmon mode have been calculated by solving quantum Langevin equations in the input–output formalism [24]. The bright plasmon is coupled with two bosonic baths: a bath of electronic excitations and a bath of free space photons. The coupling with the electronic bath is considered phenomenologically, and it gives rise to a nonradiative broadening γ of the plasmon mode, which is typically 20 times smaller than the plasmon frequency Ω_P. The coupling between the plasmon and the free space photons is calculated thanks to our quantum model. It results in a radiative decay rate Γ, which is a function of the angle of emission, θ, and of the photon frequency ω through:

$$\Gamma(\theta, \omega) = \Gamma_0 \frac{\omega}{\Omega_P} \frac{\sin^2\theta}{\cos\theta} \tag{2}$$

Γ_0 depends on the oscillator strength of the plasmon, and its complete expression can be found in the study by Huppert et al [24]. It can be shown that Γ_0 is proportional to the density of the electron gas, an indication of the superradiant nature of the plasmon emission [13]. The $\sin^2\theta$ dependence in Eq. (2) comes from the fact that the plasmon has a dipole oriented along the growth direction, while the $1/\cos\theta$ dependence is associated with the photonic density of states. The angular dependence of the radiative rate allows us to vary the interaction between the plasmon and the electromagnetic field by simply changing the light propagation angle. As it will be discussed later, this permits to experimentally investigate different regimes of the light–matter interaction. It is also important to note that in our quantum model the spontaneous emission of the plasmons into the free space is calculated without employing a perturbative approach, and it results in a dependence of Γ on the photon frequency which is characteristics of non-Markovian dynamics.

The angular behavior of the plasmon spectra from the sample without the confining barriers (Figure 1c) has been experimentally investigated by performing reflectivity spectra, after evaporating a gold mirror on the surface. In order to span a wide angular range, we have illuminated the sample through a Germanium hemispherical lens, as sketched in Figure 2a. In this way, absorption spectra at internal angles between 6° and 40° have been measured. Few representative spectra are presented in Figure 2b, showing the Berreman mode and its evolution with the angle in terms of peak absorptivity and linewidth. In Figure 2c, the full width at half maximum of the absorption peaks, Γ_{tot}, is plotted as a function of the angle, θ. The results are well reproduced by Eq. (2) assuming that $\Gamma_{tot} = \gamma + \Gamma(\theta, \Omega_P)$ is the sum of the radiative and nonradiative contribution. By changing the propagation angle of the light, the radiative rate, Γ, varies from 0 to a value much bigger than the nonradiative rate, γ, as it can be seen in Figure 2c. At 6° internal angle, the radiative contribution to the plasmon linewidth is negligible, and therefore, we can extract the value of the nonradiative broadening, γ = 8 meV. The quality factor of the plasmon mode, centered at 153 meV, is thus 19. This value is quite remarkable as the plasmon is simply defined by the presence of a highly doped GaInAs region between two intrinsic GaInAs layers.

Figure 2d presents the peak absorptivity as a function of the internal angle extracted from the measured absorptivity spectra (symbols) and calculated (solid line) by using the result of our model which is expressed below, in Eq. (3). As expected from our quantum model, the peak absorptivity increases with increasing the angle up to θ_c = 20° where the critical coupling is reached. At this angle, the radiative loss rate equals nonradiative ones, which is the condition for perfect absorption where responsivity reaches 1. For greater angles, the peak absorptivity decreases while the linewidth increases. We also observe a blue shift of the plasmon peak, which can be well described as a cooperative Lamb shift [25]. Notice that the perfect absorption gives rise to a perfect blackbody emitter at the frequency of the Berreman mode.

The plasmon absorptivity is calculated by considering an optical input in our quantum model. In the presence of a single bright plasmon mode, close to a perfect gold mirror, the absorptivity is given by the following formula [24]:

$$\alpha(\theta,\omega) = \frac{\frac{4\Omega_P^2}{(\Omega_P+\omega)^2}\gamma\Gamma(\theta,\omega)}{(\omega-\Omega_P)^2 + \frac{4\Omega_P^2}{(\Omega_P+\omega)^2}\left(\frac{\gamma}{2}+\frac{\Gamma(\theta,\omega)}{2}\right)^2} \quad (3)$$

This expression is obtained beyond the rotating wave approximation, and it takes into account the antiresonant terms of the light–matter interaction, responsible of the ultrastrong light–matter coupling [26]. Our model perfectly reproduces the experimental results, without using any fitting parameter, as it can be seen in Figure 2d. Three different regimes of the light–matter interaction can be identified. For small angles, $\gamma \gg \Gamma(\theta,\omega)$, the absorptivity takes the following form for frequencies close to the peak:

$$\alpha(\theta,\omega) \approx \frac{2}{\gamma}\frac{2\Gamma}{1+\left(\frac{\omega-\Omega_P}{\gamma/2}\right)^2} \quad (4)$$

The absorption spectrum is thus Lorentzian, with a full width at half the maximum given by the nonradiative broadening. The amplitude of the peak is proportional to twice the radiative rate $\Gamma \approx \Gamma_0\frac{\sin^2\theta}{\cos\theta} \approx \Gamma_0\theta^2$ due to the presence of the gold mirror. This is exactly the result expected in a perturbative description of the plasmon absorption. In this regime, the absorptivity peak increases with the angle, while the spectrum progressively broadens.

The second regime occurs at the angle θ_c for which the radiative rate equals the nonradiative one: $\gamma = \Gamma(\theta_c, \Omega_P)$. In this case, the absorption spectrum is still a Lorentzian function, with unitary peak absorptivity and with full width at half the maximum 2γ:

$$\alpha(\theta_c,\omega) \approx \frac{1}{1+\left(\frac{\omega-\Omega_P}{\gamma}\right)^2} \quad (5)$$

This is the critical coupling regime, giving rise to perfect absorption at the plasmon frequency: $\alpha(\theta_c, \Omega_P) = 1$. For angles greater than the critical coupling angle θ_c, the plasmon is overdamped: the absorption spectrum is radiatively broadened and the absorptivity peak decreases.

Our quantum model allows the description of incandescent emission induced by injecting an electrical current in the highly doped layer sustaining the Berreman mode [13, 27]. The application of an in-plane current induces the Joule heating of the electron gas and thus increases the temperature of the electronic reservoir T_e with respect to that of the photonic one, T_{ph}. In this out-of-equilibrium situation, the incandescent emission process can be seen as a plasmon-mediated exchange between the electronic and the photonic reservoir. The number of output photons, calculated by considering that both baths are in an incoherent thermal input state, is equal to the product of the absorptivity times the thermal occupancy at temperature T_e [24]. This is equivalent to Kirchhoff's law of thermal emission, stating the equivalence between the emissivity and the absorptivity at a given frequency and emission angle. The incandescent power $P(\theta, \Omega_P)$ emitted at the plasma energy $\hbar\Omega_P$ and angle θ is thus proportional to $\alpha(\theta,\omega)/(\exp(\frac{\hbar\Omega_P}{k_B T}) - 1)$. In this configuration, by modulating the injected current, we can modulate the temperature of the electron gas and therefore the incandescent emission. We will show in the following that a modulation frequency as high as 50 MHz has been measured, due to the small heat capacity of the electron gas.

4 High frequency incandescent emission

Figure 3a presents a sketch of our device, which is fabricated exploiting a highly doped 45-nm GaInAs layer embedded between two undoped AlInAs barriers. This highly doped layer sustains a Berreman mode at 166 meV, which is inserted in a microcavity of total thickness 1.125 μm, delimited on the bottom by a highly doped 2-μm-thick GaInAs mirror. The sample has been grown by metal organic chemical vapor deposition and processed into a field-effect transistor-like structure, consisting of two ohmic contacts made of TiAu. The gate size is $L \times W = 50 \times 50$ μm^2, where L and W stand for the length and width of the channel, respectively. On the top of the gate, we have realized a two-dimensional metallic grating by e-beam lithography and Au evaporation. The top part of Figure 3a presents a scanning electron microscope image of the grating, on which we have also indicated the dimensions of the square gold patches. The thickness of the cavity, defined as the distance between the bottom InGaAs mirror and the top grating, determines the energy of the cavity modes, while the ratio between the size of the patches and their periodicity determines the mode contrast, i.e., their coupling with free space. The right top part of Figure 3a presents the calculated absorption spectrum of the device. Due to the huge dipole of the collective electronic excitation, the cavity mode strongly couples with the Berreman mode and gives rise to polariton modes, as indicated in the spectrum. The grating parameters and the cavity thickness have been designed in order to have two polaritonic modes with peak absorptivity, and hence peak emissivity, close to 1. Figure 3b (top) shows the setup that has been used for our high frequency spectral resolved measurements. The plasmonic emitter is mounted on a high frequency holder to inject microwave into the electron gas that can also be biased with a

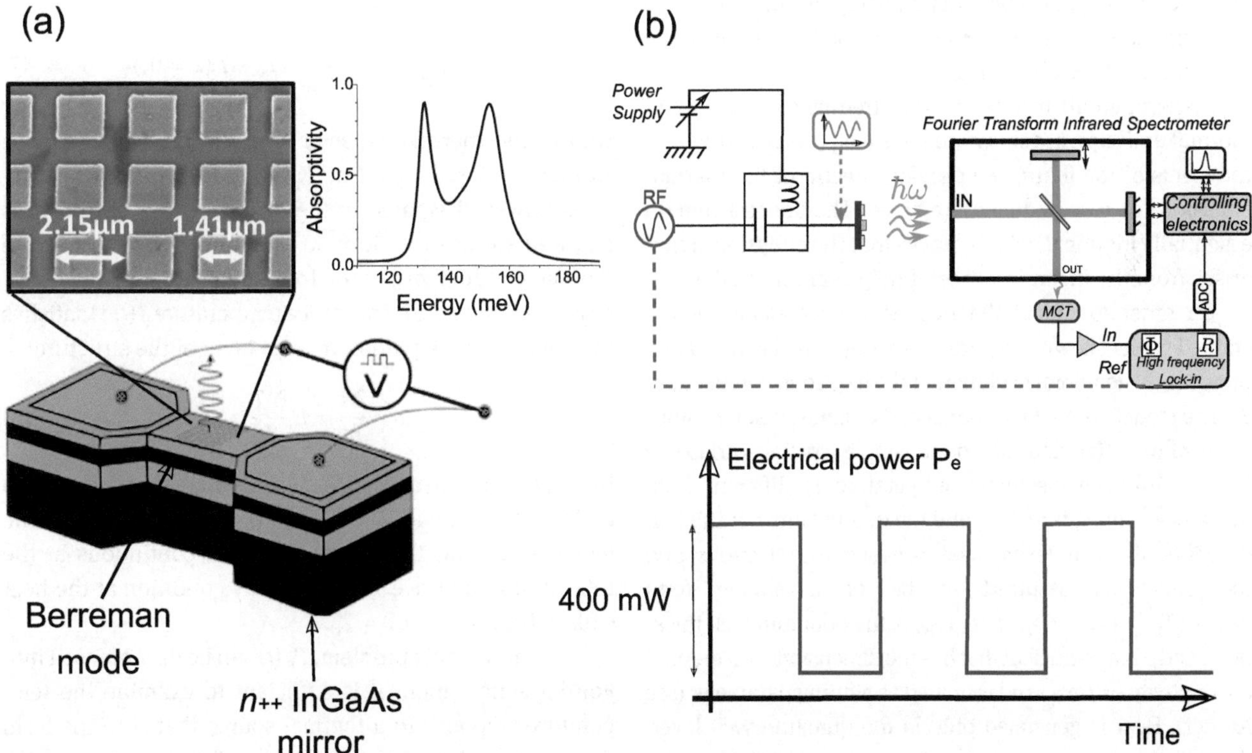

Figure 3: (a) Sketch of the device fabricated for thermal emission experiments. It consists of a 50 × 50 μm^2 mesa with a top grating for the extraction of incandescent emission. In order to electrically excite the thermal emission, two lateral Ti/Au contacts (125 × 100 μm^2) have been evaporated on the GaInAs layer sustaining the Berreman mode, after inductively coupled plasma reactive ion etching. The top part of the panel presents a scanning electron microscope image of the grating, with its characteristic dimensions, and the calculated absorption spectrum, displaying two polariton modes with almost unity peak absorptivity. (b) Sketch of the setup used for high frequency modulation of thermal emission. An electrical current is injected between the two lateral contacts. Its time modulation results in a modulation of the emitted power.

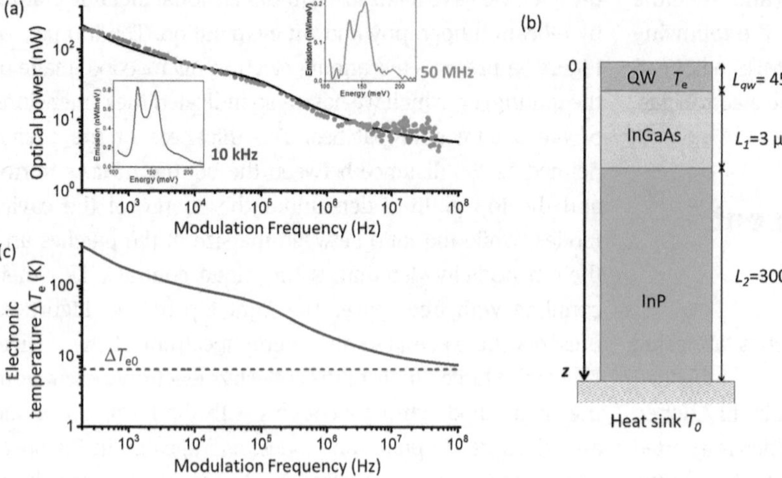

Figure 4: (a) Emitted power as a function of the modulation frequency (dots). The solid line is a result of modeling as described in the main text. The insets show two typical spectra for modulation frequency of 10 kHz and 50 MHz. (b) Schematic of semiconductor layers from the quantum well down to the heat sink used for our thermal model. (c) Estimations of the frequency-modulated part of the excess electronic temperature $\Delta T_e(\omega)$ resulting from Eq. (9). The dashed line is the result of a simplified model where the effect of the semiconductor substrate has been neglected.

direct current (dc) power supplier. The emitted power is collected into a Fourier transform infrared spectrometer and then measured using a fast Mercury Cadmium Telluride (MCT) detector with a frequency cutoff at 50 MHz. The spectra are measured in a step-scan mode using a high frequency lock-in amplifier. In the bottom part of Figure 3b, we indicate the temporal form of the input current used to modulate the temperature of the electron gas.

Experimental results of the frequency modulated thermal emission are shown in Figure 4a (dotted curve). In the insets of the figure, we provide spectra of the thermal emission for two frequency regimes. The spectra remain essentially identical, which proves that the emission signal arises from the incandescence of the polariton modes.

In order to model the frequency dependence of the emitted power shown in Figure 4a, we developed a model that provides the temporal evolution of the electronic temperature T_e. The model is one dimensional and assumes that the lattice temperature $T(z)$ depends on a single spatial variable z (Figure 4b). The electronic temperature is different from the lattice temperature $T(z)$, and the thermal flux density that describes the heat exchanged between the electrons and the lattice is assumed to be of a simple form $\beta/L_{QW}(T_e - T(z=0))$. Here, L_{QW} is the quantum well thickness, and β is a coefficient that has the dimensions of a thermal conductivity (W/m K) and is treated as a fitting parameter (see further). Heat is generated only in the quantum well layer, through the dissipation of electrical power, and then propagates toward the heat sink at the bottom of the semiconductor substrate maintained at a temperature T_O = 300 K. In Figure 4b, we indicate the different semiconductor layers between the quantum well and the heat sink.

Using energy conservation, we can write the following equations that couple the electronic and lattice temperature at the position of the quantum well $z = 0$:

$$c_e \frac{\partial T_e}{\partial t} = -\frac{\beta}{L_{QW}^2}(T_e - T(z=0)) + \frac{P_e(t)}{\sigma L_{QW}} \tag{6}$$

$$-\lambda_1 \left[\frac{\partial T}{\partial z}\right]_{z=0} = \frac{\beta}{L_{QW}}(T_e - T(z=0)) \tag{7}$$

with c_e the thermal capacity of electrons, which can be assumed to be $c_e \approx 10^3$ J/m^3 K at 300 K [28], $P_e(t)$ the electrical power dissipated in the quantum well of thickness L_{QW} and surface $\sigma = 50 \times 50$ μm^2, and $\lambda_1 = 5$ W/m K the thermal conductivity of the InGaAs layer below the quantum well (Figure 4b). The lattice temperature $T(t, z)$ satisfies the Fourier heat equation in each layer of the structure:

$$\frac{\partial T}{\partial t} = D_i \frac{\partial^2 T}{\partial z^2} \tag{8}$$

Here, D_i is the thermal diffusion coefficient of the InGaAs and InP layers, with (i = 1) and (i = 2), respectively. The temperature and the thermal flux are continuous at the InGaAs/InP interface. The boundary condition at the heat sink is $T(t, z = L_1 + L_2) = T_O$.

To analyze this problem, $P_e(t)$ can be decomposed into Fourier series; then, it is sufficient to examine the temperature response to a thermal source that is of the form $P_e(t) = \langle P_e \rangle + P_1 \cos(\omega t)$. In the following, we will consider only the time-varying part of the temperature that is due to the source term $P_1 \cos(\omega t)$. The stationary part $\langle P_e \rangle$ provides a static temperature response that is not relevant for the frequency dependence of the emitted power. In the experiment, the electrical power is modulated using a square wave with a typical amplitude P_0 = 400 mW (Figure 3b). The signal is detected using a lock-in amplifier, locked on the fundamental harmonic of the square wave, and therefore, in Eq. (6), the time-varying source can be considered of the form $\left(\frac{2}{\pi}\right)P_0 \cos(\omega t)$. The temperature response thus splits into a static and a modulated contribution using a complex notation, $T = T_{st} + \Delta T(\omega)e^{i\omega t}$. By solving the above system with the appropriate boundary conditions, we obtain the following expression for the modulated part of the complex electronic temperature:

$$\Delta T_e(\omega) = \frac{\Delta T_{e0}}{\frac{i\omega}{\omega_0} + \frac{1}{1+\frac{\beta L_1}{\lambda_1 L_{QW}} S(\omega)}},$$

$$S(\omega) = \frac{1}{k_1 L_1} \frac{\tanh(2k_1L_1) + \frac{\lambda_1 k_1}{\lambda_2 k_2}\tanh(2k_2L_2)}{1 + \frac{\lambda_1 k_1}{\lambda_2 k_2}\tanh(2k_1L_1)\tanh(2k_2L_2)} \quad (9)$$

Here, $\Delta T_{e0} = \left(\frac{2P_0}{\pi}\right)(L_{QW}/(\beta\sigma))$ and $\omega_0 = \beta/(c_e L_{QW}^2)$ is a cutoff frequency that describes the thermal dynamics of the electron gas alone. The frequency-dependent coefficient $S(\omega)$ takes into account the temperature distribution in the InGaAs and InP layers. The quantities, k_i, are defined as $k_i = (1 + i)\sqrt{\omega/2D_i}$, and therefore, their reciprocal values, $1/k_i$, have the meaning of frequency-dependent thermal diffusion lengths across the different layers.

Once the temperature $\Delta T_e(\omega)$ is known, the optical power can be deduced from the Bose–Einstein occupation factor of the plasmon modes $n_B(\hbar\Omega_P, T) = 1/(\exp(\frac{\hbar\Omega_P}{k_B T}) - 1)$ [13]: $P_{opt}(t) \propto n_B(\hbar\Omega_P, T_{st} + \mathrm{Re}\{\Delta T_e(\omega)\exp(i\omega t)\})$, where $\hbar\Omega_P$ = 155 meV is the energy of the plasmon. The signal amplitude detected by the lock-in amplifier by our setup described in Figure 3b can therefore be modeled as follows:

$$P_{signal}(\omega) = P_0[n_B(\hbar\Omega_P, T_{st} + |\Delta T_e(\omega)|) - n_B(\hbar\Omega_P, T_{st} - |\Delta T_e(\omega)|)] \quad (10)$$

P_0 is a constant adjusted so that the theoretical curve $P_{signal}(\omega)$ can be directly compared to the experimental data (Figure 4a).

All the physical parameters of the model are known except the constant β. The material parameters for InGaAs and InP are λ_1 = 5 W/m K and $D_1 = 3.33 \times 10^{-6}$ m²/s (the effect of the doping in the 2-μm-thick part has been neglected) and λ_2 = 68 W/m K and $D_2 = 41 \times 10^{-6}$ m²/s, respectively. The parameter β is determined from the best fit of the experimental data which provides β = 0.7 W/m K. As shown in Figure 4a, a very good fit of the optical power is obtained in the entire frequency range, which covers almost six decades. In Figure 4a, the bump observed at ~10^5 Hz is due to the InGaAs/InP interface. Therefore, up to 1 MHz, the thermal dynamics of the system is dominated by the substrate underlying the electron gas. At higher frequencies, the curve flattens as only the electronic dynamics is sufficiently fast to provide a temperature change. Indeed, from the value of β = 0.7 W/m K, we determine the thermal cutoff frequency of the electron gas: $\omega_0/(2\pi)$ = 55 GHz. Thermal sources based on electronic heating enable therefore an ultrafast modulation of thermal radiation. Notice that the value of the cutoff ω_0 is determined fundamentally by the thermal coefficient β that sets the temperature difference between the electrons gas and the lattice. The reciprocal of the cutoff gives a time constant in the order of 18 ps, not far from the lifetime of the optical phonons (8 ps) which is, in our interpretation, the physical limit for the electron cooling at room temperature. Indeed, for shorter time than $1/\omega_0$, the electron gas and the phonon bath are in thermal equilibrium via absorption and emission of optical phonons and have no temperature difference.

In Figure 4c, we plot the electronic temperature $|\Delta T_e(\omega)|$ predicted from Eq. (9). Typical value observed at 10 kHz is 100 K, which corresponds to a thermal resistance $R_{th} = |\Delta T_e(\omega = 2\pi\ 10^4)|/P_0$ ~250 K/W, a value consistent with previous experiments [27]. Furthermore, the values of the excess electronic temperature are comparable with those reported in the literature for the case of quantum cascade lasers below threshold [29]. Note that in Eq. (9), the effect of the substrate on the temperature dynamics of electrons is described by the function $S(\omega)$. It can be shown that this function vanishes at very high frequencies: $S(\omega \to \infty) \to 0$. In this limit, the temperature of the substrate can be considered constant and does not affect the electron dynamics. Equation (9) becomes a low-pass filter, $\Delta T_e(\omega) = \Delta T_{e0}/(1 + i\omega/\omega_0)$ making evident the cutoff $\omega_0 = \beta/(c_e L_{QW}^2)$. In the current frequency range of Figure 4c, this simplified form appears as the flat part of the filter with ΔT_{e0} = 6.5 K. This model confirms that the effects of the substrate become negligible in the MHz range and the emission observed at 50 MHz is already essentially due to the heating of the electron gas alone.

5 Conclusions and perspectives

We have exploited the peculiar quantum properties of plasmons confined in semiconductor layers to demonstrate perfect absorption and dynamic modulation of thermal emission up to 50 MHz in the mid-infrared frequency range. When they are confined in a layer with thickness smaller than the plasma wavelength, confined plasmons are optically active and display absorption resonances, known as Berreman modes, with a quality factor in the order of 20 at low internal angle of light propagation. Plasmon radiative rate increases as a function of the angle of emission and largely exceeds the nonradiative rate, a demonstration of the superradiant character of this collective electronic excitation. We have demonstrated that when the radiative rate equals the nonradiative one, a critical coupling regime occurs, corresponding to perfect absorption of the plasmon mode. All our experimental results are supported by a quantum model of the plasmon optical properties, which goes beyond Markov and rotating wave approximation.

The optical properties of the plasmon have been exploited to realize a perfect blackbody, in which incandescence is induced by Joule heating of the electron gas. We have shown that a time modulation of the injected current results in a modulation of the incandescent emission, up to 50 MHz, limited by the detector bandwidth. Our experimental results have been analyzed through an analytical model which takes into account the different dynamic behavior of electronic and lattice temperature under Joule heating. This model gives an excellent agreement with the experimental results in the entire frequency range, which covers almost six decades.

Berreman modes in semiconductors offer a huge possibility to control emissivity, wavelength, directionality, and dynamic behavior of mid-infrared thermal sources in technologically mature material platforms. We believe that the versatility of this system can be beneficial to different fields of applications of thermal emitters, particularly whenever a high frequency response is necessary, like for thermal camouflage, friend-or-foe identification, or trace detection. Finally, the superradiant character of these excitations can be exploited also for radiative cooling in ultradense electronic circuits [30].

Acknowledgments: The authors acknowledge financial support from the Agence Nationale de la Recherche (Grant ANR-14-CE26-0023-01). The authors thank Stephan Suffit for help with the device fabrication, Eugenio Maggiolini for preliminary experiments on high modulation of thermal emission, Jean- Jacques Greffet for the loan of the 50 MHz-bandwidth MCT detector, and Emilie Sakat for discussions on the cavity design.

Author contribution: All the authors have accepted responsibility for the entire content of this submitted manuscript and approved submission.

Research funding: The authors acknowledge financial support from the Agence Nationale de la Recherche (Grant ANR-14-CE26-0023-01).

Conflict of interest statement: The authors declare no conflicts of interest regarding this article.

References

[1] R. A. Ferrell, "Predicted radiation of plasma oscillations in metal films," *Phys. Rev.*, vol. 111, pp. 1214–1222, 1958.

[2] A. R. Melnyk and M. J. Harrison, "Resonant excitation of plasmons in thin films by elecromagnetic waves," *Phys. Rev. Lett.*, vol. 21, pp. 85–88, 1968.

[3] I. Lindau and P. O. Nilsson, "Experimental verification of optically excited longitudinal plasmons," *Phys. Scr.*, vol. 3, pp. 87–92, 1971.

[4] A. R. Melnyk and M. J. Harrison, "Theory of optical excitation of plasmons in metals," *Phys. Rev. B*, vol. 2, pp. 835–850, 1970.

[5] J. A. Scholl, A. L. Koh, and J. A. Dionne, "Quantum plasmon resonances of individual metallic nanoparticles," *Nature*, vol. 483, pp. 421–427, 2012.

[6] W. P. Halperin, "Quantum size effects in metal particles," *Rev. Mod. Phys.*, vol. 58, pp. 533–606, 1986.

[7] P. Mühlschlegel, H.-J. Eisler, O. J. F. Martin, B. Hecht, and D. W. Pohl, "Resonant optical antennas," *Science*, vol. 308, p. 1607, 2005.

[8] P. J. Schuck, D. P. Fromm, A. Sundaramurthy, G. S. Kino, and W. E. Moerner, "Improving the mismatch between light and nanoscale objects with gold bowtie nanoantennas," *Phys. Rev. Lett.*, vol. 94, p. 017402, 2005.

[9] M. S. Tame, K. R. McEnery, Ş. K. Özdemir, J. Lee, S. A. Maier, and M. S. Kim, "Quantum plasmonics," *Nat. Phys.*, vol. 9, pp. 329–340, 2013.

[10] J. A. Mason, S. Smith, and D. Wasserman, "Strong absorption and selective thermal emission from a midinfrared metamaterial," *Appl. Phys. Lett.*, vol. 98, p. 241105, 2011.

[11] S. Huppert, A. Vasanelli, T. Laurent, et al., "Radiatively broadened incandescent sources," *ACS Photonics*, vol. 2, pp. 1663–1668, 2015.

[12] S. Campione, F. Marquier, J. P. Hugonin, et al., "Directional and monochromatic thermal emitter from epsilon-near-zero conditions in semiconductor hyperbolic metamaterials," *Sci. Rep.*, vol. 6, p. 34746, 2016.

[13] T. Laurent, Y. Todorov, A. Vasanelli, et al., "Superradiant emission from a collective excitation in a semiconductor," *Phys. Rev. Lett.*, vol. 115, p. 187402, 2015.

[14] D. G. Baranov, Y. Xiao, I. A. Nechepurenko, A. Krasnok, A. Alù, and M. A. Kats, "Nanophotonic engineering of far-field thermal emitters," *Nat. Mater.*, vol. 18, pp. 920–930, 2019.

[15] J.-J. Greffet, R. Carminati, K. Joulain, J. P. Mulet, S. Mainguy, and Y. Chen, "Coherent emission of light by thermal sources," *Nature*, vol. 416, pp. 61–64, 2002.

[16] G. Lu, J. R. Nolen, T. G. Folland, M. J. Tadjer, D. G. Walker, & J. D. Caldwell, "Narrowband polaritonic thermal emitters driven by waste heat," *ACS Omega*, vol. 5, pp. 10900–10908, 2020.

[17] T. Wang, P. Li, D. N. Chigrin, et al., "Phonon-polaritonic bowtie nanoantennas: controlling infrared thermal radiation at the nanoscale," *ACS Photonics*, vol. 4, pp. 1753–1760, 2017.

[18] M. De Zoysa, T. Asano, K. Mochizuki, A. Oskooi, T. Inoue, & S. Noda, "Conversion of broadband to narrowband thermal emission through energy recycling," *Nat. Photonics*, vol. 6, pp. 535–539, 2012.
[19] Z.-Y. Yang, S. Ishii, T. Yokoyama, et al., "Narrowband wavelength selective thermal emitters by confined tamm plasmon polaritons," *ACS Photonics*, vol. 4, pp. 2212–2219, 2017.
[20] T. Inoue, M. D. Zoysa, T. Asano, and S. Noda, "Realization of dynamic thermal emission control," *Nat. Mater.*, vol. 13, pp. 928–931, 2014.
[21] B. Askenazi, A. Vasanelli, A. Delteil, et al., "Ultra-strong light–matter coupling for designer Reststrahlen band," *New J. Phys.*, vol. 16, p. 043029, 2014.
[22] A. Vasanelli, S. Huppert, A. Haky, T. Laurent, Y. Todorov, and C. Sirtori, *Semiconductor quantum plasmonics*, Available at: https://arxiv.org/abs/2003.01543v2.
[23] G. Pegolotti, A. Vasanelli, Y. Todorov, and C. Sirtori, "Quantum model of coupled intersubband plasmons," *Phys. Rev. B*, vol. 90, p. 035305, 2014.
[24] S. Huppert, A. Vasanelli, G. Pegolotti, Y. Todorov, and C. Sirtori, "Strong and ultrastrong coupling with free-space radiation," *Phys. Rev. B*, vol. 94, p. 155418, 2016.
[25] G. Frucci, S. Huppert, A. Vasanelli, et al., "Cooperative Lamb shift and superradiance in an optoelectronic device," *New J. Phys.*, vol. 19, p. 043006, 2017.
[26] C. Ciuti, G. Bastard, and I. Carusotto, "Quantum vacuum properties of the intersubband cavity polariton field," *Phys. Rev. B*, vol. 72, p. 115303, 2005.
[27] T. Laurent, Y. Todorov, A. Vasanelli, I. Sagnes, G. Beaudoin, & C. Sirtori, "Electrical excitation of superradiant intersubband plasmons," *Appl. Phys. Lett.*, vol. 107, p. 241112, 2015.
[28] A. N. Chakravarti, K. P. Ghatak, S. Ghosh, and A. K. Chowdhury, "Effect of heavy doping on the electronic heat capacity in semiconductors," *Phys. Status Solidi (b)*, vol. 109, pp. 705–710, 1982.
[29] V. Spagnolo, G. Scamarcio, H. Page, and C. Sirtori, "Simultaneous measurement of the electronic and lattice temperatures in $GaAs/Al_{0.45}Ga_{0.55}As$ quantum-cascade lasers: influence on the optical performance," *Appl. Phys. Lett.*, vol. 84, pp. 3690–3692, 2004.
[30] S. Buddhiraju, W. Li, and S. Fan, "Photonic refrigeration from time-modulated thermal emission," *Phys. Rev. Lett.*, vol. 124, p. 077402, 2020.

Cheng Zong and Ji-Xin Cheng*

Origin of dispersive line shapes in plasmon-enhanced stimulated Raman scattering microscopy

https://doi.org/10.1515/9783110710687-049

Abstract: Plasmon-enhanced stimulated Raman scattering (PESRS) microscopy has been recently developed to reach single-molecule detection limit. Unlike conventional stimulated Raman spectra, dispersive-like vibrational line shapes were observed in PESRS. Here, we propose a theoretical model together with a phasor diagram to explain the observed dispersive-like line shapes reported in our previous study. We show that the local enhanced electromagnetic field induced by the plasmonic nanostructure interferes with the molecular dipole-induced field, resulting in the dispersive profiles of PESRS. The exact shape of the profile depends on the phase difference between the plasmonic field and the molecular dipole field. We compared plasmon-enhanced stimulated Raman loss (PESRL) and plasmon-enhanced stimulated Raman gain (PESRG) signals under the same pump and Stokes laser wavelength. The PESRL and PESRG signals exhibit similar signal magnitudes, whereas their spectral line shapes show reversed dispersive profiles, which is in an excellent agreement with our theoretical prediction. Meanwhile, we verify that the nonresonant background in PESRS mainly originates from the photothermal effect. These new insights help the proper use of PESRS for nanoscale bioimaging and ultrasensitive detection.

Keywords: plasmon; plasmon-enhanced spectroscopy; stimulated Raman scattering.

***Corresponding author: Ji-Xin Cheng**, Department of Electrical and Computer Engineering, Boston University, Boston, MA 02215, USA; Department of Biomedical Engineering, Boston University, Boston, MA 02215, USA; Photonics Center, Boston University, Boston, MA 02215, USA; and Department of Chemistry, Boston University, Boston, MA 02215, USA, E-mail: jxcheng@bu.edu. https://orcid.org/0000-0003-2048-6207

Cheng Zong: Department of Electrical and Computer Engineering, Boston University, Boston, MA 02215, USA, E-mail: czongcz@bu.edu

1 Introduction

Raman spectroscopy offers label-free contrasts for bioimaging based on the vibrational information of biomolecules such as lipids, protein, DNA, and small metabolites [1]. However, spontaneous Raman spectroscopy suffers from its poor sensitivity due to inherent small cross-sections for most endogenous biomolecules. It needs a long exposure time to obtain a high-quality Raman spectrum. Owing to the generation of a coherent light field, stimulated Raman scattering (SRS) produces a signal that is many orders of magnitude larger than that of spontaneous Raman spectroscopy [2]. Because of its fast imaging speed comparable with fluorescence microscopy, SRS microscopy has been extensively utilized to capture chemical images of live cells, large-area tissues, living animals, and even human subjects [3].

Despite these attractive attributions, the imaging sensitivity of SRS is not sufficient for performing measurements on the single-molecule level without other enhancement processes [2]. To detect SRS signals from a single molecule, one approach is the involvement of electronic resonance which increases the nonlinear susceptibility of a molecule. Min and coworkers reported electronic pre-resonance SRS to achieve sub-micromolar-sensitivity detection for chromophores [4, 5]. Later, Min et al. reported a stimulated Raman excited fluorescence method that encoded the vibrational resonance into the fluorescence emission. This approach can obtain the Raman signal of the single chromophore by detecting its fluorescence signal [6]. Another route to achieving single-molecule detection is to enlarge the driving light field. Due to the localized surface plasmon resonance (LSPR) effect, the plasmonic nanostructure dramatically increases the intensity of the local electromagnetic field on and near the metal surface [7, 8]. The local enhanced electromagnetic field interacts with molecules close to the nanostructure surface to amplify the SRS signals. The Van Duyne's group reported surface-enhanced femtosecond SRS (SE-FSRS) spectra, including stimulated Raman loss (SRL) and stimulated Raman gain (SRG) from molecules embedded in a SiO_2-coated gold

This article has previously been published in the journal Nanophotonics. Please cite as: C. Zong and J.-X. Cheng "Origin of dispersive line shapes in plasmon-enhanced stimulated Raman scattering microscopy" *Nanophotonics* 2021, 10. DOI: 10.1515/nanoph-2020-0313.

nanodumbbell colloid [9–12]. Very recently, Cheng and coworkers demonstrated a label-free plasmon-enhanced SRS (PESRS) microscopy which is able to detect single molecules having extremely small Raman cross-sections (~10^{-30} $cm^2\,sr^{-1}$) with the aid of gold nanoparticles (Au NPs) and the picoJoule high-repetition-rate laser pulses [13]. The reported PESRS microscopy opened a new window for ultrasensitive and ultrafast bio-imaging.

Unlike the Lorentzian-line shape in traditional SRS spectroscopy, SE-FSRS spectroscopy and PESRS hyperspectral imaging both produced dispersive line shape spectra [9–15]. The dispersive line shape was attributed to Fano interference of the discrete molecular vibrational states on the continuum-like plasmon resonance [9, 11]. Ziegler et al. indicated that the cross term between the heterodyned SE-FSRS signal and the plasmon light emission resulted in the dispersive-like line shape [16]. Schatz et al. proposed that the asymmetric line shape resulted from the combination of two Fano contributions arising from the interference between both the real and imaginary components of the Raman local electromagnetic field and the plasmon field [17]. Both theoretical descriptions were based on reported SE-FSRS results and only discussed the SRG process. Yet, there are significant differences between the SE-FSRS measurement and PESRS imaging in experiment conditions, including the laser pulse duration, the modulation frequency, the laser excitation, collection approach, and the data processing. Besides, the recent wavelength-dependent SE-FSRS result did not match predictions of any proposed SE-FSRS theory, indicating an incomplete understanding of the plasmon-enhanced SRS process [12, 18].

In this article, we employ the classic light-matter interaction theory and the phasor diagram to elucidate the origin of dispersive shapes in plasmon-enhanced stimulated Raman loss (PESRL) and plasmon-enhanced stimulated Raman gain (PESRG) measurements. The phasor graphs provide an intuitive view of the origin of dispersion, which derives from an interaction between the plasmon-enhanced electric field and the molecular dipole field. We also develop a theoretical model to simulate both PESRL and PESRG spectra. Furthermore, PESRL and PESRG measurements using the same pump and Stokes laser wavelength are conducted to confirm our theoretical prediction.

2 Theory

We start with basic SRS in the absence of a plasmonic field. For details on the classical model of the SRS process, readers can read references [19, 20]. Assuming a pump field (E_P) at the frequency of ω_P and a Stokes field (E_S) at the frequency of ω_S. Based on heterodyne detection, the SRL and SRG intensities can be written as

$$I_{\mathrm{SRL}} = 2\mathrm{Re}\left[E_P^* E_{\mathrm{SRL}}\right] \tag{1a}$$

$$I_{\mathrm{SRG}} = 2\mathrm{Re}\left[E_S^* E_{\mathrm{SRG}}\right] \tag{1b}$$

where

$$E_{\mathrm{SRL}} = i\chi_{\mathrm{SRL}} E_S E_P E_S^* \tag{2a}$$

$$E_{\mathrm{SRG}} = i\chi_{\mathrm{SRG}} E_P E_S E_P^* \tag{2b}$$

and

$$\chi_{\mathrm{SRL}} = \chi_{\mathrm{NR}} + \frac{\alpha}{\delta - i\Gamma} \tag{3a}$$

$$\chi_{\mathrm{SRG}} = \chi_{\mathrm{NR}} + \frac{\alpha}{\delta + i\Gamma} \tag{3b}$$

Here $\delta = \Omega_v - (\omega_P - \omega_S)$, Γ is the half width at height maximum (HWHM). α is the oscillator strength of the molecular vibration. Ω_v is the Raman vibrational frequency. As shown in Figure S1, the imaginary part of χ_R shows a Lorentzian form, whereas the real part features a Fano dispersive profile [20]. We note that χ_{NR}, a constant, is contributed by the cross-phase modulation (XPM), the photothermal effect (PT), or the transient absorption (TA), which could be minimized after optimizing the optical design [21].

For SRL, the first term and second term in χ_{SRL} produce a non-Raman resonant field ($E_{\mathrm{SRL}}^{\mathrm{NR}}$) and a Raman resonant field (E_{SRL}^{R}), respectively. In a phasor representation for SRL, we set the local oscillator E_P as a reference field with zero phases ($E_P = 1$). The SRL intensity is a projection of E_{SRL} on the x-axis. As shown in Figure 1a, $E_{\mathrm{SRL}}^{\mathrm{NR}}$ has a $\pi/2$ phase shift from E_P and therefore is decoupled from E_P. E_{SRL}^{R} has an additional phase shift θ from $E_{\mathrm{SRL}}^{\mathrm{NR}}$, with $\theta = tg^{-1}(\Gamma/\delta)$. Thus, only E_{SRL}^{R} is coupled to E_P and the x-axis projection of E_{SRL}^{R} (the imaginary part of χ_{SRL}) produces an intensity loss in the pump beam. For SRG, E_S is set as a reference field with zero phase ($E_s = 1$). $E_{\mathrm{SRG}}^{\mathrm{NR}}$ is decoupled from E_S, and E_{SRG}^{R} is coupled to E_S as shown in Figure 1b. The x-axis projection of E_{SRG}^{R} (the imaginary part of χ_{SRG}) generates an intensity gain in the Stokes beam due to a phase shift $\theta = tg^{-1}(-\Gamma/\delta)$ between E_{SRG}^{R} and $E_{\mathrm{SRG}}^{\mathrm{NR}}$. Both SRL and SRG show a Lorentzian line shape.

In the case of PESRS, the signal is from the interaction of a molecule with the local enhanced electromagnetic field. Thus, the PESRS intensity is rewritten as:

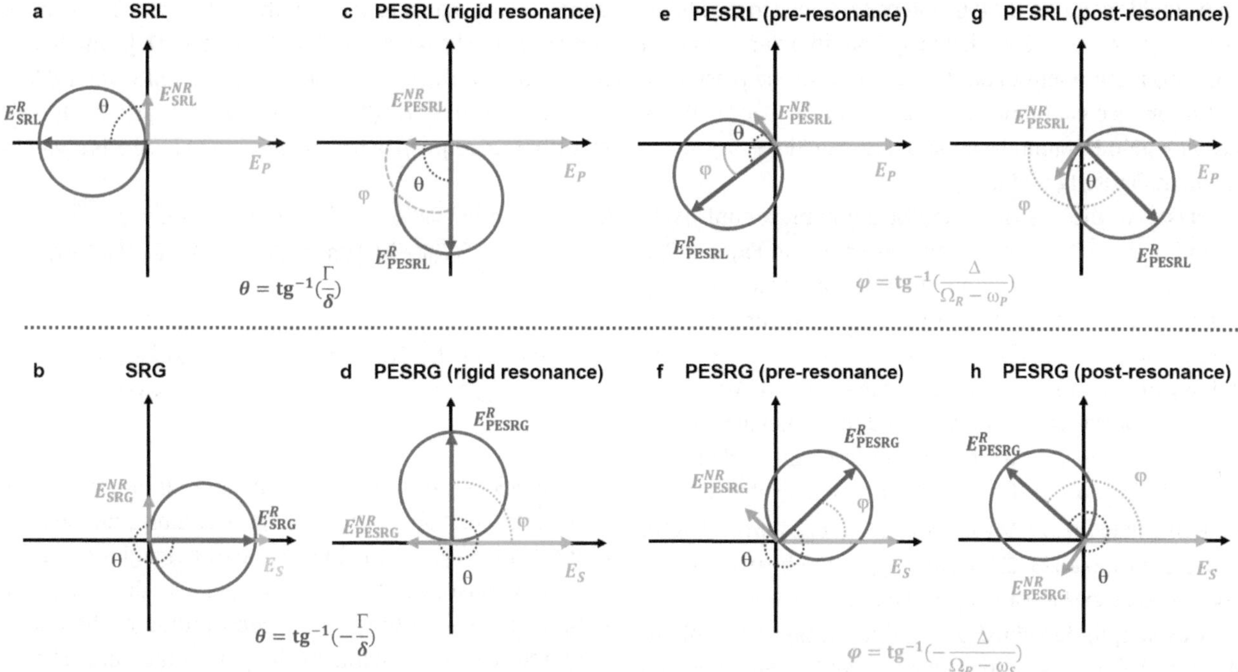

Figure 1: Phasor graphs of stimulated Raman scattering (SRS) and plasmon-enhanced stimulated Raman scattering (PESRS). Phasor representation of stimulated Raman loss (SRL) (a) and stimulated Raman gain (SRG) (b) without plasmon field, plasmon-enhanced stimulated Raman loss (PESRL) (c) and plasmon-enhanced stimulated Raman gain (PESRG) (d) at rigid plasmon resonance, PESRL (e) and PESRG (f) at plasmon pre-resonance, and PESRL (g) and PESRG (h) at plasmon post-resonance. E_P and E_S are set as a reference field with zero phase.

$$I_{\mathrm{SRL}}^{\mathrm{PE}} = 2\mathrm{Re}\left[E_P^* E_{\mathrm{PESRL}}\right] \tag{4a}$$
$$I_{\mathrm{SRG}}^{\mathrm{PE}} = 2\mathrm{Re}\left[E_S^* E_{\mathrm{PESRG}}\right] \tag{4b}$$

where

$$E_{\mathrm{PESRL}} = i\chi_{\mathrm{SRL}} E_S^{\mathrm{PE}} E_P^{\mathrm{PE}} E_S^{\mathrm{PE}*} \tag{5a}$$

$$E_{\mathrm{PESRG}} = i\chi_{\mathrm{SRG}} E_P^{\mathrm{PE}} E_S^{\mathrm{PE}} E_P^{\mathrm{PE}*} \tag{5b}$$

Here, the local enhancement fields are described as $E_P^{\mathrm{PE}} = E_P F_P$ and $E_S^{\mathrm{PE}} = E_S F_S$. The plasmon enhancement coefficient is given by [22],

$$F_P = 1 + \frac{A\Delta}{\Omega_R - \omega_P - i\Delta} \tag{6a}$$

$$F_S = 1 + \frac{A\Delta}{\Omega_R - \omega_S - i\Delta} \tag{6b}$$

where Ω_R is the LSPR frequency, Δ is the HWHM of plasmon peak with the Lorentzian shape, and A is the local field enhancement factor. When the pump and Stokes fields are tuned to near plasmon resonance and A is large enough, the first term in eq. (6) can be ignored. Thus, E_P^{PE} and E_S^{PE} express a phase shift of $\varphi_P = tg^{-1}(\Delta/\Omega_R - \omega_P)$ and $\varphi_S = tg^{-1}(\Delta/\Omega_R - \omega_S)$ from the E_P and the E_S field, respectively, in the phasor representation. The φ_P and the φ_S are transferred to the signal field E_{PESRL} and E_{PESRG}, respectively. The cases of plasmon rigid resonance ($\varphi_P = \varphi_S = \pi/2$) are shown in Figure 1c and d. The phasor representations of E_{PESRL} and E_{PESRG} both show a $\pi/2$ anticlockwise rotation about the origin E_{SRL} and the E_{SRG} field, respectively. For PESRL, both E_{PESRL}^R and $E_{\mathrm{PESRL}}^{\mathrm{NR}}$ are coupled to the local oscillator E_P. The $E_{\mathrm{PESRL}}^{\mathrm{NR}}$ generates an intensity loss in the pump beam. The x-axis projection of E_{PESRL}^R (the real part of χ_{SRL}) produces an intensity loss in the pump beam at the lower wavenumber region and an intensity gain in the pump beam at the higher wavenumber region. Similarly, for PESRG, E_{PESRG}^R and $E_{\mathrm{PESRG}}^{\mathrm{NR}}$ are coupled to the local oscillator E_S. The $E_{\mathrm{PESRG}}^{\mathrm{NR}}$ generates an intensity loss in the Stokes beam. The x-axis projection of E_{PESRG}^R (the real part of χ_{SRG}) generates an intensity gain in the Stokes beam at the lower wavenumber region and generates an intensity loss at the higher wavenumber region. Notably, the x-axis projections of E_{PESRL}^R and E_{PESRG}^R are zero at the Raman resonance ($\omega_P - \omega_S = \Omega_v$) condition.

Figure 1e and f presents the pre-resonance conditions ($\Omega_R - \omega_P > 0$ or $\Omega_R - \omega_S > 0$). For PESRL, since the nonresonant background and the SRL signal are largely on the left side of y axis, the total intensity change in the pump beam should appear as a sum of the stimulated Raman signal and the nonresonant background. For PESRG, since the nonresonant background and the SRL signal are mostly on the opposite side of y axis, the total intensity change in the Stokes beam appears as a subtraction of the stimulated Raman signal from the nonresonant background. At the plasmonic post-resonance condition ($\Omega_R - \omega_P < 0$ or $\Omega_R - \omega_S < 0$), we predict a gain in pump beam and a loss in Stokes beam, as shown in Figure 1g and h.

Next, we discuss the origin of the nonresonant background in the PESRS microscopy. As shown in Figure S2, a strong nonresonant background, in the form of an intensity loss, is generated in the Au NPs substrates with and without adenine adsorption. The background also shows a broad and asymmetric profile in a delay time domain. Moreover, the strong backgrounds are observed in the negative delay time region. Thus, we assign the nonresonant background to the PT effect of Au NPs. The PT effect can be viewed as a scattering field ($E_{PT} \propto \Delta nVE_{pr} \propto \sigma_{abs}E^2_{heat}E_{pr}$) induced by an equivalent dipole [23–25]. Here, E_{heat} is the heat laser (the modulated laser) field and E_{pr} is the probe laser field. σ_{abs} is the wavelength-depended absorption cross-section of Au NPs, which depends on the plasmon profiles [26]. Thus, the PT signal intensity (I_{PT}) corresponds to the interference of the scattering field with the incident probe beam (E_{pr}) as following [23]:

$$I_{PT} = 2\mathrm{Re}\left[E^*_{pr}E_{PT}\right] = C\sigma_{abs}|E_{heat}|^2|E_{pr}|^2 \quad (7a)$$

$$\sigma_{abs} = \frac{A\Delta^2}{(\Omega_R - \omega_{heat})^2 + \Delta^2} \quad (7b)$$

Here, C is a constant which reflect the magnitude of PT contribution. We note that other pump-probe processes including XPM and TA may also contribute a small part of nonresonant background. Both XPM and TA can be considered to be wavelength independent [21], which does not impact the dispersive line shapes of PESRL and PESRG.

By inserting eq. (5) into eq. (4) and adding the PT contribution (eq. (7)), a full expression of PESRS is given by

$$\begin{aligned} I^{PE}_{SRL} &= I_{PT} - 2\mathrm{Re}\left(E^*_P E^{PE}_{SRL}\right) = I_{PT} - 2\mathrm{Re}\left(E^*_P i\chi_{SRL}E^{PE}_P\right)|E^{PE}_S|^2 \\ &= I_{PT} - 2|E_S F_S|^2|E_P|^2\left[\mathrm{Im}(F_P)\mathrm{Re}(\chi_{SRL}) + \mathrm{Re}(F_P)\mathrm{Im}(\chi_{SRL})\right] \end{aligned} \quad (8a)$$

$$\begin{aligned} I^{PE}_{SRG} &= I_{PT} + 2\mathrm{Re}\left(E^*_S E^{PE}_{SRG}\right) = I_{PT} + 2\mathrm{Re}\left(E^*_S i\chi_{SRG}E^{PE}_S\right)|E^{PE}_P|^2 \\ &= I_{PT} + 2|E_P F_P|^2|E_S|^2\left[\mathrm{Im}(F_S)\mathrm{Re}(\chi_{SRG}) + \mathrm{Re}(F_S)\mathrm{Im}(\chi_{SRG})\right] \end{aligned} \quad (8b)$$

The PESRS can be viewed as a linear combination of an asymmetric line shape (the real part of molecular susceptibility) and a symmetric line shape (the imaginary part of molecular susceptibility). The dispersive line shape of PESRS depends on the relative contribution of the imaginary and real components of the plasmonic field. The PT term normally is a negative constant value in a short spectral window. As shown in eq. (8), the interference between the molecular susceptibility and the plasmon enhancement coefficient leads to the dispersive line shapes in PESRS. When $I_{PT} = 0$ and $F_P = F_S = 1$ (no plasmon resonance), eq. (8) is reduced to the normal SRG and SRL cases.

We performed simulations of PESRS based on eq. (8). The Raman peak is centered at 731 cm^{-1} with a 7 cm^{-1} HWHM. The pump and Stokes beam wavelengths are centered at 969 nm and 1040 nm, respectively. The plasmon peak is kept at 800 nm. The HWHM of the plasmonic peak is set as 5000 cm^{-1}. The PT constant C is set

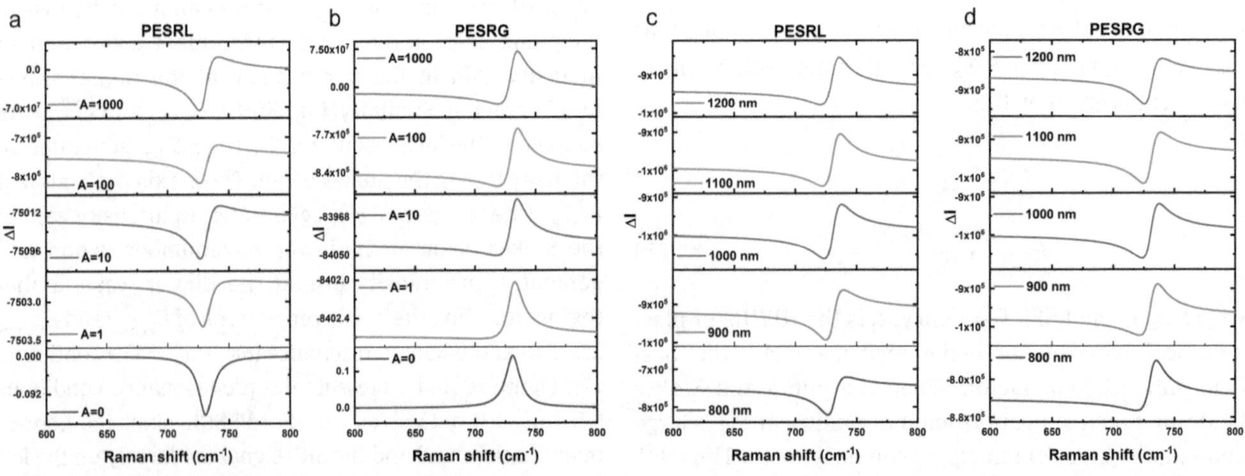

Figure 2: Simulation results. (a–b) Simulated plasmon-enhanced stimulated Raman loss (PESRL) (a) and plasmon-enhanced stimulated Raman gain (PESRG) (b) as a function of the plasmonic enhancement factor, A. The plasmon peak is kept at 800 nm. (c–d) Simulated PESRL (c) and PESRG (d) as a function of the plasmon peak Ω_R and enhancement factor is kept at 100. For all simulations, the pump and Stokes beam wavelengths are centered at 969 nm and 1040 nm, respectively. The Raman peak is centered at 731 cm^{-1} with an half width at height maximum (HWHM) of 7 cm^{-1}. The photothermal effect (PT) constant C is set as −10,000.

as −10,000, indicating an intensity loss in the probe beam. The simulated PESRL and PESRG results are shown in Figure 2. Figure 2a and b presents the PESRL and PESRG spectra as a function of the enhancement factor A. As we expected, without plasmon enhancement ($A = 0$), the SRL and SRG spectral profiles are determined by the imaginary part of χ, both showing symmetric Lorentzian-line shapes with a negative and a positive peak, respectively. With the plasmonic effect, SRL and SRG both show dispersive line shapes. All PESRL spectra show a strong negative peak at lower wavenumber and a weak positive peak at higher wavenumber. All PESRG spectra show a strong positive peak at higher wavenumber and a weak dip at lower wavenumber. As we predicted, the asymmetric profiles of PESRL and PESRG are opposite. In addition, PESRL and PESRG acquire more prominent dispersive characters as the enhancement factor increases. Similar results were reported by Ziegler [16]. While we find that the dispersive characters do not increase when the enhancement factor reaches a point that is large enough. As we discussed above, the dispersive line shape depends on the plasmon resonance frequency relative to the pump/Stokes frequency. As is evident in Figure 2c and d, the increasing dispersive line shapes and more negative PT backgrounds are shown when the plasmonic peak approaches the pump or Stokes wavelength. Intriguingly, under the plasmon post-resonance condition (e.g., first rows in the panel c and d), as predicted by our phasor diagram, PESRL appears as a gain at post-resonance condition, i.e., the positive peak becomes stronger than the negative peak. Similarly, PESRG appears as a loss, i.e., the negative peak dominates.

3 Result and discussion

We have experimentally investigated the line shapes of PESRL and PESRG predicted by our theoretical model above. The hyperspectral SRS data are collected by using a spectral-focusing SRS microscope which has been previously described [13]. Figure 3 presents the scheme of the hyperspectral SRS microscope. In brief, an 80 MHz tunable femtosecond laser (InSight DS+, Spectra-Physics) provides the pump (969 nm, ~120 fs) and Stokes (1040 nm, ~200 fs) lasers. Two acousto-optic modulators are installed in the pump and the Stokes beams for implementing SRL and SRG, respectively. The pump and the Stokes beams are collinearly overlapped and sent to an upright microscope with a 2D galvo mirror system for the laser scanning. To obtain spectral-domain information, the pump and the Stokes pulses are stretched by 5 and 6 glass rods, respectively. In this way, about 7 cm^{-1} spectral resolution is achieved. The laser powers (the pump ~ 150 μW and the Stokes ~ 150 μW at the sample) are sufficiently low to ensure no damage of molecules or nanostructures during the experiments. A water immersion objective (Olympus, NA = 0.8, 40×) is used to focus the pump and the Stokes lasers on a sample. An oil condenser (Olympus, NA = 1.4) is used to collect the laser light in the forward direction. To obtain SRL signals, the Stokes beam is modulated at 2.2 MHz. Two 1000 nm short-pass filters block the Stokes laser before a photodiode with a lab-built resonant amplifier. To obtain SRG signals, the pump beam is modulated at 2.2 MHz. Two 1050/50 nm band-pass filters and two 1000 nm long-pass filters block the pump beam before the photodiode. A lock-in amplifier demodulates the SRS signals and the R channel is used to read out SRS signals. The hyperspectral PESRS datacube contains 200 × 200 pixels with 120 Raman channels. Each datacube took about 2 min with a 10 μs dwell time per pixel. The image area (30 μm × 30 μm) and pixel size (150 nm) are the same in all experiments. We note that Van Duyne's group used pump/Stokes and pump/anti-Stokes pulses to obtain SE-FSRG and SE-FSRL signals to sample different portions of LSPR of nanodumbbell colloids [12]. Here, we used the same pump and Stokes pulses and exchanged the modulated beam to obtain PESRL and PESRG to probe the same LSPR portions of the nanoparticle aggregation.

First, we compared the SRL and SRG responses of non-plasmonic enhanced SRS. Figure S3a shows SRL and SRG images of a solid adenine sample at the same location. The SRL and SRG spectra of adenine that are equal in the bandwidth and the Raman peak position are shown in Figure S3b. In principle, the SRL and the SRG signal are equal in magnitude. However, due to the wavelength-dependent quantum efficiency of the photodiode and different transmittances of filter sets, the SRL response in our microscope is about three times larger than the SRG response. The SRL and SRG response correction will be considered in the following results to ensure the PESRS spectra which could be compared accurately in magnitudes.

Next, we examined the PESRL and PESRG signals of the adenine adsorbed on Au NPs. 5 μL of a 1 mM aqueous adenine solution was added to 2–4 μL of concentrated 60 nm Au NPs colloid sol. Then, the Au NPs were dropped on a cover glass followed by vacuum drying. Crucially, the analytes induced the aggregation of Au NPs before drying, creating hot spots occupied by the analyte itself. We randomly measured three locations on the aggregated Au NPs substrate. Both PESRL and PESRG signals were collected in every location to avoid the interference of aggregation states. To decrease the intensity variation, the

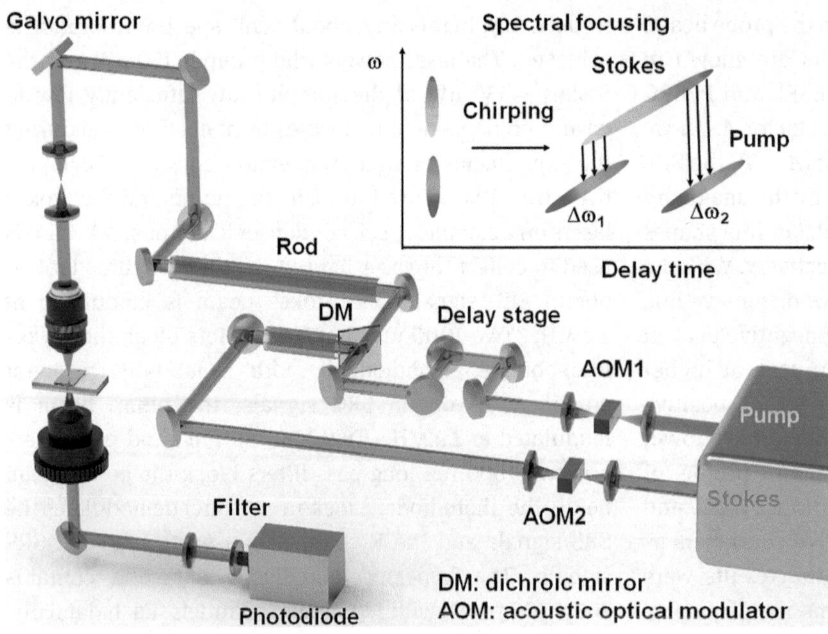

Figure 3: Schematic of spectral-focusing hyperspectral stimulated Raman scattering (SRS) microscope for stimulated Raman loss (SRL) and stimulated Raman gain (SRG) measurement.

PESRS spectra from all pixels in the whole image were averaged. The raw PESRL and PESRG spectra of adenine are displayed in Figure 4. The resulting PESRL spectra (Figure 4a) from the adenine-adsorbed Au NPs aggregates consists of a narrower feature at 727 cm^{-1} on the top of a strong and broad nonresonant background with a high end on the left. A penalized least squares fitting was used to subtract the broad spectral background [27]. The fitted backgrounds of PESRS are shown in Figure 4 as the same color dash lines. All PESRG spectra (Figure 4b) showed a broad background with a high end on the right and a sharp negative peak at 735 cm^{-1}. The end-for-end reversed backgrounds in PESRL and PESRG can be explained as following. For the PESRL measurement, the 969 nm laser is the "probe" beam and the modulated 1040 nm laser is the "heat" beam for the PT process. The "heat" and "probe" beam are exchanged with each other for the PESRG measurement, and relative delay time turns end for end. In this way, the PT decay profiles under two different conditions are just reversed. Additionally, the SRL and the PT effect both produce an intensity loss in the "probe" beam. Thus, the raw data of PESRL are the sum of the PT signal and the SRL signal. While the SRG is a "probe" beam gain process, the total signal is the subtraction of the SRG signal from the PT signal. Consequently, the PESRG appears as a negative peak on a broad PT background.

Figure 5 shows the vibrational resonant components of the PESRL and PESRG spectra obtained from the subtraction of the fitted backgrounds from the observed PESRS signals. For a better comparison, we set the strong peak of PESRG data as a positive peak. Due to our experimental conditions, the intensity of PESRL is about three times larger than that of PESRG. After the correction of the system response, the signal intensities are nearly identical between PESRL and PESRG. Remarkably, the PESRG result shows a blue shift in peak positions from the PESRL result and with an oppositely phased dispersion. In the PESRL result, the vibrational feature is characterized by a dip on the high wavenumber side and a peak on the low wavenumber side. While, in the PESRG result, the dip is on the low wavenumber side and the peak is on the high wavenumber side. This is an excellent agreement with our theoretical prediction of opposite dispersive line shapes between PESRL and PESRG.

Then, we simulated the spectra of adsorbed adenine on Au NPs based on eq. (8). The vibrational peak of adenine is at 731 cm^{-1} with a HWHM of 7 cm^{-1}. The PESRS spectra were simulated by using a plasmon response with $\Delta = 5000\ cm^{-1}$, $A = 100$, and $\Omega_R = 800$ nm. Experimentally, the enhancement factor of PESRS signal is about 10^6–10^7 in previous results [9, 13]. The PESRS signal is proportional to the cubic of the plasmonic field, as shown in our theory. Thus, here the value of A is selected as 100 for the simulation. The normalized simulated PESRL and PESRG are shown as green dot lines in Figure 5. For a better comparison, we set stimulated PESRL profile as a positive peak and show normalized PESRL and PESRG spectra. We find a good agreement in peak/dip positions and dispersive profiles between the experimental and simulated results. Simulated results also show a reversal in the sign of the dispersive PESRL and PESRG line shape. The excellent agreement between the experimental and simulated

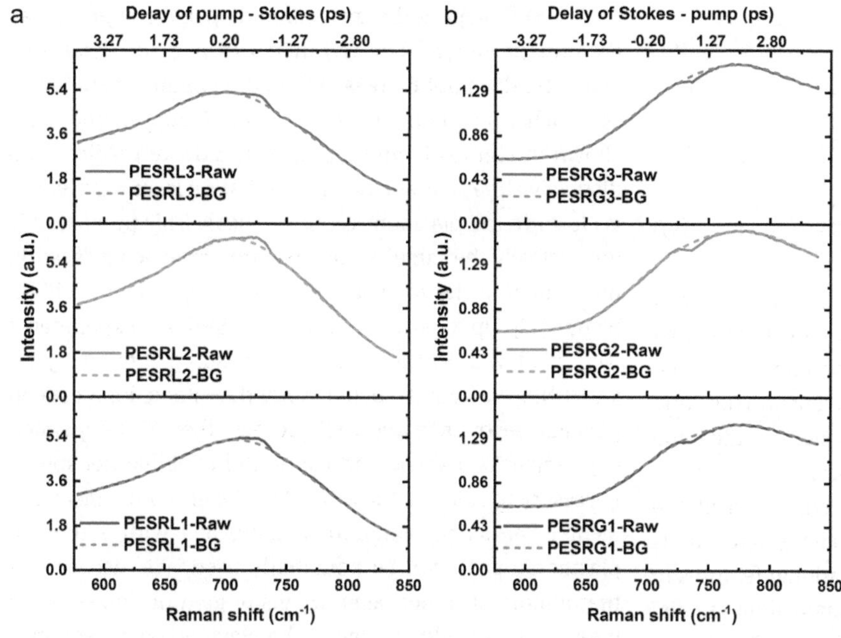

Figure 4: Three raw plasmon-enhanced stimulated Raman loss (PESRL) (a) and plasmon-enhanced stimulated Raman gain (PESRG) (b) spectra of adenine adsorbed on aggregated Au NPs which are obtained from different locations. Fitting backgrounds are displayed as dash lines.

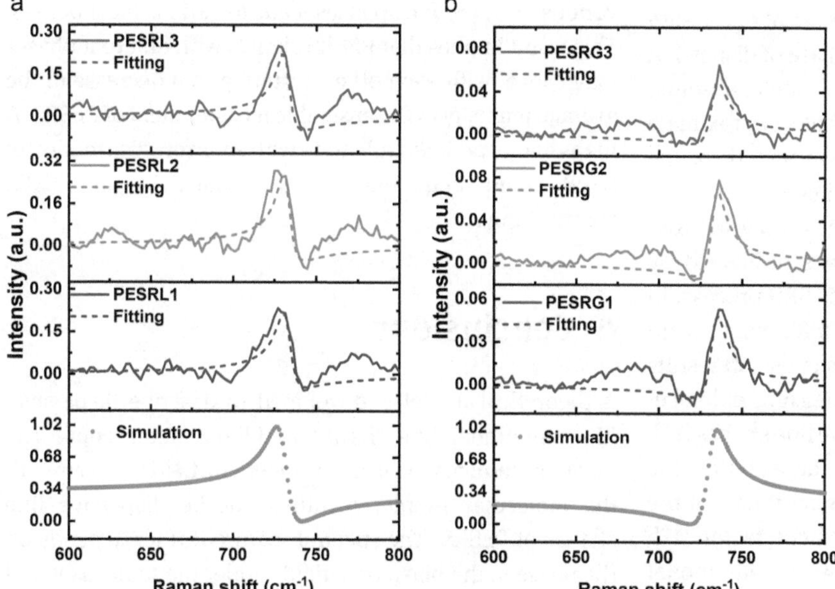

Figure 5: The experimental and simulated plasmon-enhanced stimulated Raman loss (PESRL) (a) and plasmon-enhanced stimulated Raman gain (PESRG) (b) spectra of adenine. The background-corrected PESRL and PESRG spectra are presented as solid lines (black, red, and blue). The Fano-like fitting spectral profiles based on eq. (9) are presented as the same color dash lines. Simulated PESRL and PESRG spectra based on eq. (8) are presented as green dot lines.

results collectedly confirms the reliability of our theoretical model.

In the literature, the Fano function has been used to describe a dispersive line shape of SE-FSRS [11]. As shown in eq. (8), the PESRS can be viewed as a linear combination of a Fano-line shape (the real part of the molecular susceptibility in which Fano parameter value is one as shown in eq. S1) and a Lorentzian line shape (the imaginary of the molecular susceptibility). For a better comparison with the Fano line shape, we insert eq. (3) into eq. (8) and the eq. (8) are reformatted into a Fano-like line shape as following:

$$I_{\mathrm{SRL}}^{\mathrm{PE}} = I_{\mathrm{PT}} - A_{\mathrm{SRL}}\left[\frac{\left(q_{\mathrm{SRL}} + \frac{\delta}{\Gamma}\right)^2}{1 + \left(\frac{\delta}{\Gamma}\right)^2} - \frac{\left(\frac{\delta}{\Gamma}\right)^2}{1 + \left(\frac{\delta}{\Gamma}\right)^2}\right] \tag{9a}$$

$$I_{\mathrm{SRG}}^{\mathrm{PE}} = I_{\mathrm{PT}} + A_{\mathrm{SRG}}\left[\frac{\left(q_{\mathrm{SRG}} + \frac{\delta}{\Gamma}\right)^2}{1 + \left(\frac{\delta}{\Gamma}\right)^2} - \frac{\left(\frac{\delta}{\Gamma}\right)^2}{1 + \left(\frac{\delta}{\Gamma}\right)^2}\right] \tag{9b}$$

where

$$q_{\mathrm{SRL}} = \frac{2\mathrm{Re}(F_P)}{\mathrm{Im}(F_P)} \tag{9c}$$

$$q_{\mathrm{SRG}} = -\frac{2\mathrm{Re}(F_S)}{\mathrm{Im}(F_S)} \tag{9d}$$

$$A_{\mathrm{SRL}} = |E_S F_S|^2 |E_P|^2 \frac{\mathrm{Im}(F_P)^2}{2\mathrm{Re}(F_P)\Gamma} \tag{9e}$$

$$A_{\mathrm{SRG}} = |E_P F_P|^2 |E_S|^2 \frac{\mathrm{Im}(F_S)^2}{2\mathrm{Re}(F_S)\Gamma} \tag{9f}$$

As shown in eq. (9), the PESRS profiles are the subtraction of two Fano profiles where the Fano parameter (q) of the first Fano function term depends on the real part and the imaginary part of plasmon enhancement coefficient and the Fano parameter of the second term is zero. The signs of the Fano parameters of PESRL and PESRG are opposite.

To valid this relation, we used the eq. (9) to fit the experimental spectra. We can obtain excellent agreements using $q = -1.7$ and $q = 2$ for all of the vibrational features in PESRL and PESRG, respectively. The fitting results are displayed in Figure 5 as the same color dash lines. The fitting results of the peak frequency, the bandwidth, and the amplitude are reported in Table S1. The reversal of the sign of q indicates an opposite change in the phase of dispersion between the PESRL and the PESRG process. Similar absolute values of q for PESRL and PESRG indicate approximate plasmonic responses due to the close wavelength of the pump and the Stokes beams in our experiments.

It is interesting to compare our phasor diagram and theoretical model to the SE-FSRS theories published by Ziegler and Schatz who only discussed the SRG process. In Ziegler's theory, the interference between Raman susceptibility and a stimulated surface plasmon emission results in asymmetric line shapes, and the increasing enhancement factors result in increasing dispersive line shapes [16]. In Schatz's theory, the Fano parameter depends on the ratio between the real part and the imaginary part of the square of the plasmonic field enhancement factor [17]. Different from Schatz's and Ziegler's work, our model makes an important prediction about the plasmonic enhancement factor which is given by $|F_S|^2 F_P$ and $|F_P|^2 F_S$ for PESRL and PESRG, respectively. The PESRS can be viewed as the result of the incoming propagating field interferences with the local plasmon-induced molecular dipole field having the same frequency as the incoming field. The incoming light plays the leading role in the far-field detection. This heterodyne detection is significantly different from surface-enhanced Raman scattering and surface-enhanced coherent anti-Stokes scattering, where the dipole re-radiation effect dominates the far-field contribution of the generated new frequency fields [28–30]. Different from Ziegler's work, we predict that PESRL and PESRG acquire the increasing dispersive character as the enhancement factor (A) increases and the dispersive character does not increase when the enhancement factor (A) reaches a certain level as shown in Figure 2a and b. As shown in Figures 1c and d and 2c and d, we predict more dispersive line shapes when the excitation laser wavelength is closer to the plasmon resonance peak. This prediction is qualitatively the same as the deduction from Schatz's work, not matching the previous frequency-dependent SE-FSRS results [12]. Up to now, all PESRS and SE-FSRS experiments were achieved on the ensemble measurements on aggregated NPs clusters. To better study the interaction between plasmon resonance and SRS process, the LSPR-depended experiment of a single particle or an individual hot spot is necessary to rule out the artifact of the ensemble measurements. The single nanorod or nanoshell with a tunable plasmonic peak can be potentially used. We also predict that abnormal pump beam intensity gain or Stokes beam intensity loss could happen in the plasmon post-resonance conditions. More importantly, a random substrate with various local LSPR responses leads to various local phases. This result implies that PESRS signals with different phases may cancel with each other, resulting in a decrease of the average intensity in the ensemble measurement of PESRS. A high-quality periodic substrate with a narrow distribution of LSPR could help improve the sensitivity of PESRS spectroscopy.

4 Conclusions

A theoretical model is developed to describe dispersive-like vibrational line shapes in PESRS microscopy. The plasmon-induced local electromagnetic field interacts with the molecular dipole, resulting in the dispersive line shapes of PESRS. The spectral feature further depends on the phase in the plasmonic field. Under the same laser and molecular conditions, an intensity loss in pre-plasmon resonance could become an intensity gain in post-plasmon resonance. At the same pump and Stokes wavelength setting, we experimentally observed PESRL and PESRG signals with reversed dispersive profiles, which fall into line with our theoretical prediction. We also verified that the nonresonant background in PESRS mainly originated from the photothermal effect. Together, we provided new insight into the nature of PESRS profiles and the mechanism of plasmonic enhancement of coherent Raman scattering. Results reported here can be used to improve signal enhancement in nanoscale coherent Raman imaging towards ultrasensitive detection of molecules.

Acknowledgments: We thank Lawrence D. Ziegler for his valuable suggestions in theory. We also thank Yurun Xie for his assistance with PESRS measurement.
Author contribution: All the authors have accepted responsibility for the entire content of this submitted manuscript and approved submission.
Research funding: This work was supported by NIH R01 grants GM118471, AI141439, and a Keck Foundation grant to J.X.C.
Conflict of interest statement: The authors declare no conflicts of interest regarding this article.

References

[1] Y. Shen, F. Hu, and W. Min, "Raman imaging of small biomolecules," *Annu. Rev. Biophys.*, vol. 48, pp. 347–369, 2019.
[2] R. C. Prince, R. R. Frontiera, and E. O. Potma, "Stimulated Raman scattering: from bulk to nano," *Chem. Rev.*, vol. 117, pp. 5070–5094, 2017.
[3] J.-X. Cheng and X. S. Xie, "Vibrational spectroscopic imaging of living systems: an emerging platform for biology and medicine," *Science*, vol. 350, p. aaa8870, 2015.
[4] L. Wei and W. Min, "Electronic preresonance stimulated Raman scattering microscopy," *J. Phys. Chem. Lett.*, vol. 9, pp. 4294–4301, 2018.
[5] L. Wei, Z. Chen, L. Shi, et al., "Super-multiplex vibrational imaging," *Nature*, vol. 544, p. 465, 2017.
[6] H. Xiong, L. Shi, L. Wei, et al., "Stimulated Raman excited fluorescence spectroscopy and imaging," *Nat. Photonics*, vol. 13, pp. 412–417, 2019.
[7] K. A. Willets and R. P. Van Duyne, "Localized surface plasmon resonance spectroscopy and sensing," *Annu. Rev. Phys. Chem.*, vol. 58, pp. 267–297, 2007.
[8] C. Zong, M. Xu, L.-J. Xu, et al., "Surface-enhanced Raman spectroscopy for bioanalysis: reliability and challenges," *Chem. Rev.*, vol. 118, pp. 4946–4980, 2018.
[9] R. R. Frontiera, A.-I. Henry, N. L. Gruenke, and R. P. Van Duyne, "Surface-enhanced femtosecond stimulated Raman spectroscopy," *J. Phys. Chem. Lett.*, vol. 2, pp. 1199–1203, 2011.
[10] L. E. Buchanan, N. L. Gruenke, M. O. McAnally, et al., "Surface-enhanced femtosecond stimulated Raman spectroscopy at 1 MHz repetition rates," *J. Phys. Chem. Lett.*, vol. 7, pp. 4629–4634, 2016.
[11] R. R. Frontiera, N. L. Gruenke, and R. P. Van Duyne, "Fano-like resonances arising from long-lived molecule-plasmon interactions in colloidal nanoantennas," *Nano Lett.*, vol. 12, pp. 5989–5994, 2012.
[12] L. E. Buchanan, M. O. McAnally, N. L. Gruenke, G. C. Schatz, and R. P. Van Duyne, "Studying stimulated Raman activity in surface-enhanced femtosecond stimulated Raman spectroscopy by varying the excitation wavelength," *J. Phys. Chem. Lett.*, vol. 8, pp. 3328–3333, 2017.
[13] C. Zong, R. Premasiri, H. Lin, et al., "Plasmon-enhanced stimulated Raman scattering microscopy with single-molecule detection sensitivity," *Nat. Commun.*, vol. 10, p. 5318, 2019.
[14] N. L. Gruenke, M. O. McAnally, G. C. Schatz, and R. P. Van Duyne, Balancing the effects of extinction and enhancement for optimal signal in surface-enhanced femtosecond stimulated Raman spectroscopy," *J. Phys. Chem. C*, vol. 120, pp. 29449–29454, 2016.
[15] B. Negru, M. O. McAnally, H. E. Mayhew, et al., "Fabrication of gold nanosphere oligomers for surface-enhanced femtosecond stimulated Raman spectroscopy," *J. Phys. Chem. C*, vol. 121, pp. 27004–27008, 2017.
[16] A. Mandal, S. Erramilli, and L. Ziegler, "Origin of dispersive line shapes in plasmonically enhanced femtosecond stimulated Raman spectra," *J. Phys. Chem. C*, vol. 120, pp. 20998–21006, 2016.
[17] M. O. McAnally, J. M. McMahon, R. P. Van Duyne, and G. C. Schatz, "Coupled wave equations theory of surface-enhanced femtosecond stimulated Raman scattering," *J. Chem. Phys.*, vol. 145, 2016, Art no. 094106.
[18] A. Fast and E. O. Potma, "Coherent Raman scattering with plasmonic antennas," *Nanophotonics*, vol. 8, pp. 991–1021, 2019.
[19] H. Rigneault and P. Berto, "Tutorial: Coherent Raman light matter interaction processes," *APL Photonics*, vol. 3, 2018, Art no. 091101.
[20] J.-X. Cheng and X. S. Xie, *Coherent Raman Scattering Microscopy*, Boca Raton, CRC Press, 2012.
[21] D. Zhang, M. N. Slipchenko, D. E. Leaird, A. M. Weiner, and J.-X. Cheng, "Spectrally modulated stimulated Raman scattering imaging with an angle-to-wavelength pulse shaper," *Opt. Express*, vol. 21, pp. 13864–13874, 2013.
[22] X. Hua, D. V. Voronine, C. W. Ballmann, A. M. Sinyukov, A. V. Sokolov, and M. O. Scully, "Nature of surface-enhanced coherent Raman scattering," *Phys. Rev. A*, vol. 89, 2014, Art no. 043841.
[23] A. Gaiduk, P. V. Ruijgrok, M. Yorulmaz, and M. Orrit, "Detection limits in photothermal microscopy," *Chem. Sci.*, vol. 1, pp. 343–350, 2010.
[24] M. Selmke, M. Braun, and F. Cichos, "Photothermal single-particle microscopy: detection of a nanolens," *ACS Nano*, vol. 6, pp. 2741–2749, 2012.
[25] S. Berciaud, D. Lasne, G. A. Blab, L. Cognet, and B. Lounis, "Photothermal heterodyne imaging of individual metallic nanoparticles: theory versus experiment," *Phys. Rev. B*, vol. 73, 2006, Art no. 045424.
[26] J. Olson, S. Dominguez-Medina, A. Hoggard, L.-Y. Wang, W.-S. Chang, and S. Link, "Optical characterization of single plasmonic nanoparticles," *Chem. Soc. Rev.*, vol. 44, pp. 40–57, 2015.
[27] Z.-M. Zhang, S. Chen, and Y.-Z. Liang, "Baseline correction using adaptive iteratively reweighted penalized least squares," *Analyst*, vol. 135, pp. 1138–1146, 2010.
[28] S.-Y. Ding, E.-M. You, Z.-Q. Tian, and M. Moskovits, "Electromagnetic theories of surface-enhanced Raman spectroscopy," *Chem. Soc. Rev.*, vol. 46, pp. 4042–4076, 2017.
[29] M. Kerker, D.-S. Wang, and H. Chew, "Surface enhanced Raman scattering (SERS) by molecules adsorbed at spherical particles: errata," *Appl. Opt.*, vol. 19, pp. 4159–4174, 1980.
[30] C. Steuwe, C. F. Kaminski, J. J. Baumberg, and S. Mahajan, "Surface enhanced coherent anti-stokes Raman scattering on nanostructured gold surfaces," *Nano Lett.*, vol. 11, pp. 5339–5343, 2011.

Supplementary Material: The online version of this article offers supplementary material (https://doi.org/10.1515/nanoph-2020-0313).

Chang-Wei Cheng, Soniya S. Raja, Ching-Wen Chang, Xin-Quan Zhang, Po-Yen Liu, Yi-Hsien Lee, Chih-Kang Shih and Shangjr Gwo*

Epitaxial aluminum plasmonics covering full visible spectrum

https://doi.org/10.1515/9783110710687-050

Abstract: Aluminum has attracted a great deal of attention as an alternative plasmonic material to silver and gold because of its natural abundance on Earth, material stability, unique spectral capability in the ultraviolet spectral region, and complementary metal-oxide-semiconductor compatibility. Surprisingly, in some recent studies, aluminum has been reported to outperform silver in the visible range due to its superior surface and interface properties. Here, we demonstrate excellent structural and optical properties measured for aluminum epitaxial films grown on sapphire substrates by molecular-beam epitaxy under ultrahigh vacuum growth conditions. Using the epitaxial growth technique, distinct advantages can be achieved for plasmonic applications, including high-fidelity nanofabrication and wafer-scale system integration. Moreover, the aluminum film thickness is controllable down to a few atomic monolayers, allowing for plasmonic ultrathin layer devices. Two kinds of aluminum plasmonic applications are reported here, including precisely engineered plasmonic substrates for surface-enhanced Raman spectroscopy and high-quality-factor plasmonic surface lattices based on standing localized surface plasmons and propagating surface plasmon polaritons, respectively, in the entire visible spectrum (400–700 nm).

Keywords: aluminum epitaxial film; molecular-beam epitaxy; monolayer transition metal dichalcogenide; plasmonic surface lattice; surface-enhanced Raman spectroscopy; surface plasmon interferometry.

***Corresponding author: Shangjr Gwo,** Department of Physics, National Tsing-Hua University, Hsinchu 30013, Taiwan; Institute of NanoEngineering and Microsystems, National Tsing-Hua University, Hsinchu 30013, Taiwan; and Research Center for Applied Sciences, Academia Sinica, Nankang, Taipei 11529, Taiwan,
E-mail: gwo@phys.nthu.edu.tw. https://orcid.org/0000-0002-3013-0477
Chang-Wei Cheng, Department of Physics, National Tsing-Hua University, Hsinchu 30013, Taiwan. https://orcid.org/0000-0001-8937-5084
Soniya S. Raja, Institute of NanoEngineering and Microsystems, National Tsing-Hua University, Hsinchu 30013, Taiwan
Ching-Wen Chang, Research Center for Applied Sciences, Academia Sinica, Nankang, Taipei 11529, Taiwan
Xin-Quan Zhang, Po-Yen Liu and Yi-Hsien Lee, Department of Materials Science and Engineering, National Tsing-Hua University, Hsinchu 30013, Taiwan
Chih-Kang Shih, Department of Physics, The University of Texas at Austin, Austin, TX 78712, USA

1 Introduction

Plasmonics is a rapidly evolving field that takes advantage of strong light confinement and drastically enhanced light–matter interactions beyond the diffraction limit near the surfaces and interfaces of metal nanostructures. In the past few decades, remarkable advances based on plasmonic nanostructures, metamaterials, and metasurfaces have been made for surface-enhanced spectroscopies, sensors, photovoltaics, super-resolution microscopy and lithography, metalenses, biomedical therapeutics, nonlinear optics, and integrated nanophotonics [1–3]. However, there are still significant material issues to be resolved such that plasmonics can be elevated to a transformative technology for general applications.

One critical issue of plasmonics is related to the intrinsic properties of available plasmonic materials. Nearly all of the research results adopt noble metals (silver [Ag] and gold [Au] are the most popular choices) as the plasmonic materials because they exhibit negative real and small imaginary parts of dielectric function in the visible and near-infrared spectral regions. However, noble metals suffer from high material cost (Au), low material stability (Ag), and incompatibility with existing semiconducting technology (Ag, Au). Furthermore, owing to the interband transitions in noble metals, spectral responses of plasmonic devices are limited in some specific ranges. To overcome these difficulties, alternative plasmonic materials, such as aluminum (Al), copper (Cu), transition metal nitrides, conducting metal oxides, and graphene have

This article has previously been published in the journal Nanophotonics. Please cite as: C.-W. Cheng, S. S. Raja, C.-W. Chang, X.-Q. Zhang, P.-Y. Liu, Y.-H. Lee, C.-K. Shih and S. Gwo "Epitaxial aluminum plasmonics covering full visible spectrum" *Nanophotonics* 2021, 10. DOI: 10.1515/nanoph-2020-0402.

been extensively pursued in recent years [4–6]. Among them, aluminum is particularly interesting because it acts as an ideal Drude metal except a narrow interband transition window in the near-infrared (at 800 nm) [7]. In particular, for the ultraviolet (UV) and deep-UV plasmonic applications, aluminum is the best plasmonic material due to the negative real and relatively small imaginary parts of aluminum dielectric function in the UV region.

Considering practical applications, aluminum is also a sustainable plasmonic material since it is naturally abundant in the Earth's crust and has a native oxide protection layer (~3–5 nm Al_2O_3) [8, 9]. However, aluminum was not previously considered as a good candidate for alternative plasmonic material [5] before the advent of high-quality aluminum nanocrystals [10, 11] and epitaxial films [9, 12, 13] with greatly improved material properties. During the past few years, the fast development of aluminum plasmonics has attract a great deal attention not only for the expected good performance of aluminum for UV plasmonics, such as UV surface-enhanced Raman spectroscopy (UV-SERS) [14, 15], plasmonic lasers [16–22], and deep-UV resonances [8, 23], but also for its unexpected excellent performance in the visible region, including complementary metal-oxide-semiconductor (CMOS)-compatible color filters [24–28], photocatalysis [29], nonlinear optics [30–32], and SERS [15, 33]. Very recently, aluminum has even been found to outperform silver in some important plasmonic applications [15, 34].

The key to understand these finding is that the performance of plasmonic materials depends not only on their intrinsic optical properties but also their material properties, such as crystallinity, surface and interface quality, as well as stability. In the literature, aluminum nanostructures and metasurface are typically fabricated by lithographic methods using thermally evaporated aluminum films. In such cases, amorphous or polycrystalline film growth, as well as residual oxygen in the growth chamber, will eventually affect the performance of aluminum-based plasmonic devices. Recently, epitaxial aluminum film growth on commercially available substrates (silicon [13, 35, 36], GaAs [12, 37, 38], sapphire [9, 39, 40]) has been developed by using the molecular-beam epitaxy (MBE) technique under ultrahigh vacuum conditions. The availability of high-quality aluminum epitaxial films opens the way to explore aluminum plasmonics for real-world applications [8, 41, 42].

Here, we report on aluminum epitaxial films grown on sapphire substrates by MBE. This heteroepitaxial system is possible since the hexagonal lattice of Al (111) plane is close to that of the *c*-plane sapphire substrate (lattice mismatch is about 3.9%). Using these aluminum epitaxial films, we can have some distinct advantages for plasmonic applications, including high-fidelity nanofabrication for precise control of surface plasmon resonances owing to the single-crystalline material structure and large-scale, highly uniform plasmonic structures required for SERS substrates and high-quality-factor (high-Q) plasmonic surface lattices. Moreover, the aluminum film thickness is controllable down to a few atomic monolayers, allowing for ultrathin metal layer plasmonic applications [43–45]. It is worth noting that, since aluminum is considered as the "silicon" of superconductivity [46], aluminum epitaxial films can also be used as the building material for quantum computers requiring high-performance superconducting qubits.

2 Epitaxial growth and structural properties

Previous works on silver epitaxial films and nanostructures have demonstrated that crystalline properties and surface morphologies play an important role in plasmonic applications [47–53]. Especially, uniform and controllable plasmonic hot spots can realized by high-fidelity top-down nanofabrication on ultrasmooth, single-crystalline Ag colloidal crystals [53, 54]. In this work, aluminum epitaxial growth was conducted by using a MBE system under ultrahigh vacuum conditions. Two-inch double-side-polished *c*-plane sapphire (0001) wafers were used as the substrates, and the base vacuum pressure was kept about 1×10^{-10} Torr during growth. Before growth, the *c*-plane sapphire substrate was thermally cleaned at 950 °C for 2 h. A streaky reflection high-energy electron diffraction (RHEED) pattern of the *c*-plane sapphire surface can be obtained after this cleaning step (Figure 1A). Then, aluminum was evaporated by using a Kundsen cell. The deposition rate (about 200 nm/h) was controlled by the cell temperature, and the substrate temperature was maintained at room temperature (~300 K) during growth. A streaky RHEED pattern (Figure 1A) of aluminum film indicates a smooth film morphology during epitaxial growth.

The crystal orientation of epitaxial aluminum film was measured by X-ray diffraction (XRD) using the copper $K_{\alpha 1}$ line at the wavelength of 0.15406 nm. The XRD pattern (Figure 1B) shows the aluminum film is single-crystalline and grown along the (111) direction. The Al (111) and *c*-sapphire (0006) diffraction peaks are at about 38.5° and 42°, respectively. Due to the abrupt aluminum/sapphire interface and smooth film surface morphology, clear X-ray interference fringes of ultrathin aluminum films can be observed by high-resolution XRD (Figure 1B). Thus, we can

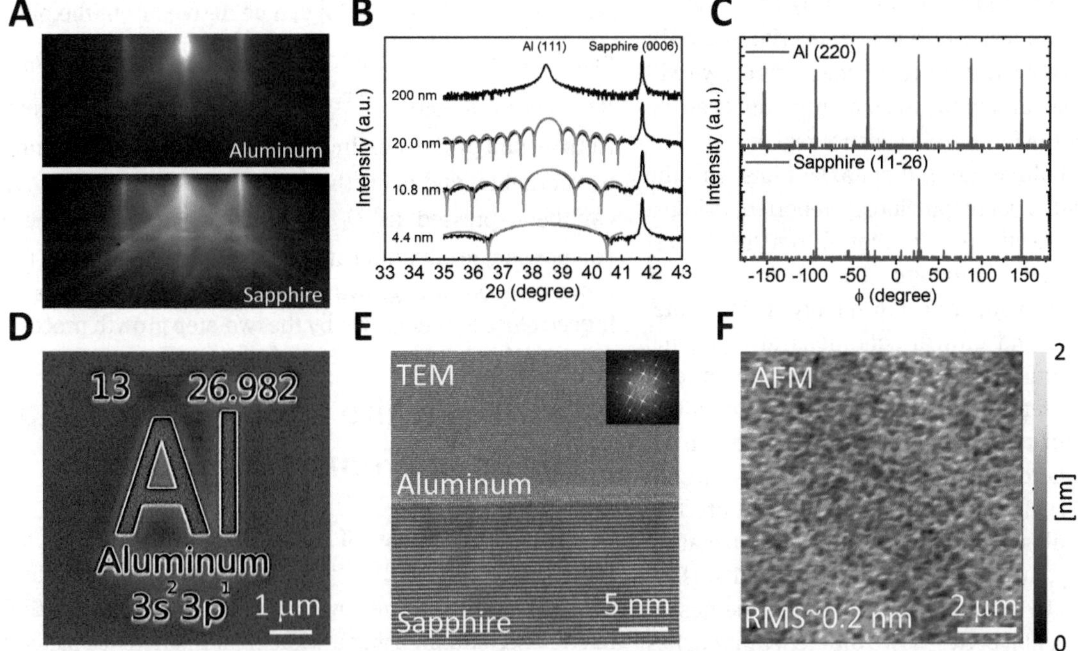

Figure 1: Structural properties of epitaxial aluminum films on sapphire substrates.
(A) In situ reflection high-energy electron diffraction (RHEED) patterns of *c*-plane sapphire and epitaxial aluminum film. (B) The X-ray diffraction patterns (2θ-scan) for 4.4 nm (~19 ML), 10.8 nm (~47 ML), 20.0 nm (~ 87 ML) and 200 nm-thick films, showing clearly the Al (111) peak, X-ray inference fringes, and the *c*-plane sapphire peak. The experimental data and the fitting curve are shown by black and red line, respectively. (C) The in-plane X-ray diffraction patterns (ϕ-scan) of Al (220) and *c*-sapphire (11–26) planes. (D) Scanning electron microscope (SEM) image of a focused ion beam patterned epitaxial aluminum. (E) High-resolution transmission electron microscope (TEM) image of an epitaxial aluminum film showing an abrupt interface between the epitaxial aluminum film and the *c*-sapphire substrate. The inset shows the fast Fourier transform of Al lattice along the [02-2] zone axis. (F) Atomic force microscope (AFM) image (10×10 μm^2) taken on the native oxide (~3 nm)-covered aluminum film. The film roughness is about 0.2 nm.

precisely determine the film thickness, ranging from a few nanometers, 4.4 nm (~19 monolayers [ML]), 10.8 nm (~47 ML), and 20.0 nm (~87 ML) to bulk-like (~200 nm). In Figure 1C, we show the in-plane X-ray diffraction scan performed for the Al (220) and *c*-sapphire (11–26) peaks, confirming the expected six-fold in-plane symmetry for epitaxial growth.

In Figure 1D, the scanning electron microscope image of a patterned aluminum epitaxial film (this pattern is adopted from an element periodic table) demonstrates that high-fidelity nanofabrication can be achieved via focused ion beam (FIB) lithography (FEI Helios NanoLab 600i) at an ion beam current of 7.7 pA due to the single-crystalline film properties. The high-resolution transmission electron microscope image (Figure 1E) shows the abrupt interface between the epitaxial aluminum film and the sapphire substrate. The root-mean-square (RMS) roughness of epitaxial aluminum film surface was measured by atomic force microscope (AFM), showing the RMS roughness of epitaxial aluminum film is atomically smooth (about 0.2 nm, Figure 1F).

3 Optical properties

The wafer scale epitaxial aluminum film is mirror-like (Figure 2A) due to a high optical reflectivity. We can compare the dielectric functions of literature data measured by spectroscopic ellipsometry (SE) for silver [51] and aluminum [9, 13] epitaxial films (Figure 2B and C). Previous studies have shown the Drude–Lorentz model can be used to fit dielectric functions of epitaxial silver and aluminum films [51, 55], which is expressed as

$$\begin{aligned}\varepsilon(\omega) &= \varepsilon_1(\omega) + i\varepsilon_2(\omega)\\ &= \varepsilon_b - \frac{\omega_p^2}{\omega\left(\omega + i\gamma_p\right)} + \sum_{j=1}^{N} \frac{f_j\tilde{\omega}_j^2}{\left(\tilde{\omega}_j^2 - \omega^2 - i\omega\Gamma_j\right)},\end{aligned} \quad (1)$$

where ε_b is the polarization response from the core electrons (background permittivity), ω_p is the bulk plasmon frequency, γ_p is the relaxation rate (electron-electron scattering loss), f_j and $\tilde{\omega}_j$ are the strengths and resonant frequencies of interband transitions (N is the number of interband transitions used for modeling), and Γ_j is the damping rates of interband transitions. In Figure 2C (inset), aluminum is clearly a better plasmonic material in the UV region compared to silver. In the following sections, we will show that epitaxial aluminum plasmonics can even be extended to cover the full visible spectral region.

Silver [51] and aluminum [13] epitaxial films with the best material quality were previously reported using a refined two-step growth process that shows the lowest optical loss. However, although the two-step growth technique can lead to superior film quality, it is a time-consuming process and growth conditions at cryogenic temperatures are difficult to achieve. Instead, room temperature and near-zero-Celsius-degree growth conditions are widely used for aluminum epitaxial films grown on high-quality commercial substrates (silicon [35, 36], gallium arsenide [37, 38], and sapphire [9]). There are two types of surface plasmons on these films: propagating surface plasmon polaritons (SPPs) and localized surface plasmons (LSPs). To compare the optical properties of epitaxial aluminum film between two different approaches [9, 13], we plot SPP (Q_{SPP}) and LSP (Q_{LSP}) quality factors in Figure 2D for both cases using the published data, where $Q_{SPP} = \frac{\mathrm{Re}(k_{SPP})}{\mathrm{Im}(k_{SPP})} \approx 2\frac{\varepsilon_1+\varepsilon_d}{\varepsilon_1\varepsilon_d}\frac{(\varepsilon_1)^2}{\varepsilon_2}$ [5] can be derived from the plasmon dispersion relation $k_{SPP} = \sqrt{\frac{\varepsilon_d\varepsilon(\omega)}{\varepsilon_d+\varepsilon(\omega)}}k_0 = n_{eff}k_0 = \mathrm{Re}(k_{SPP}) + i\mathrm{Im}(k_{SPP}) = \frac{2\pi\,\mathrm{Re}(n_{eff})}{\lambda} + \frac{2\pi\,i\mathrm{Im}(n_{eff})}{\lambda}$ (ε_d is the dielectric constant of surrounding medium, $k_0 = \frac{2\pi}{\lambda}$ is the vacuum wavenumber, and n_{eff} is the SPP effective index) and Q_{LSP} can be expressed as $Q_{LSP} = \omega\left(d\varepsilon_1/d\omega\right)/2\varepsilon_2$ [56]. These comparison results indicate the optical properties of the room temperature–grown aluminum epitaxial films is indeed close to that grown by the two-step growth process.

4 Surface white light interface and plasmon propagation length

To demonstrate long SPP propagation length in the full visible spectral region, we fabricated surface double-groove structures by FIB milling. We measured SPP interference spectra using a white-light interference method [47–49, 52, 53] (WLI, Figure 3A). The incident SPPs are generated by a halogen light source with an oblique incident

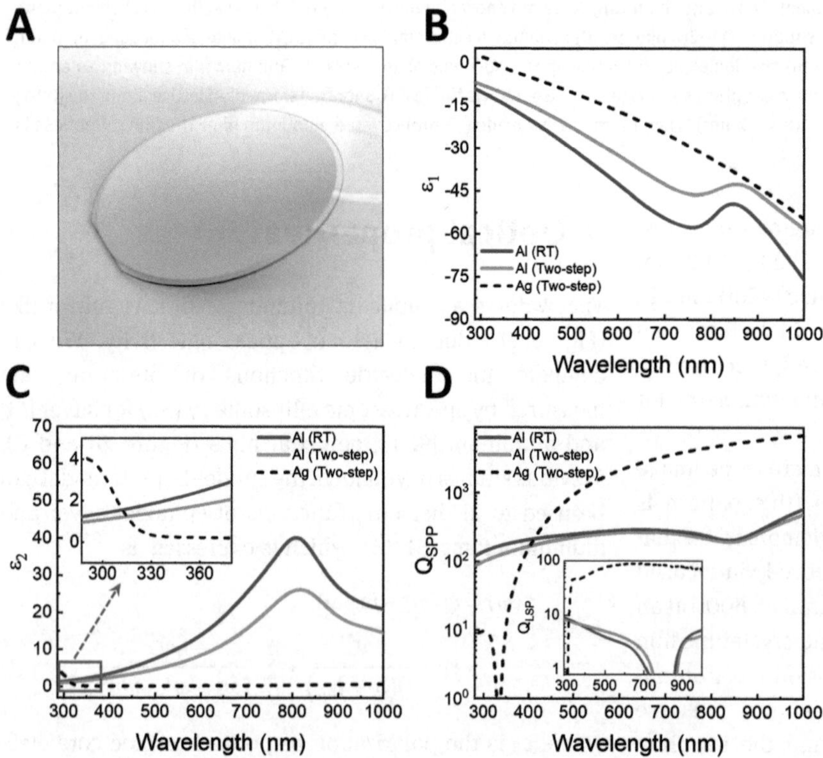

Figure 2: Optical properties of Al and Ag epitaxial films. (A) Optical image of Al epitaxial film on a 2-inch, *c*-plane sapphire substrate. (B) Real part of the dielectric function (ε_1) extracted from the literature data as a function of wavelength (blue solid curve). (C) Imaginary part of the dielectric constant (ε_2). For comparison, literature data of epitaxial Al (blue [9] and red line [13]) and Ag (black dash line [51]). The inset shows the wavelength from 287 to 380 nm. (D) The quality factor of SPP ($Q_{SPP} = \mathrm{Re}(k_{spp})/\mathrm{Im}(k_{spp}) \approx 2\frac{\varepsilon_1+\varepsilon_d}{\varepsilon_1\varepsilon_d}\frac{(\varepsilon_1)^2}{\varepsilon_2}$) comparison for Al and Ag epitaxial films. The inset shows the localized surface plasmons (LSP) quality factor ($Q_{LSP} = \omega\left(d\varepsilon_1/d\omega\right)/2\varepsilon_2$) comparison for Al and Ag epitaxial films.

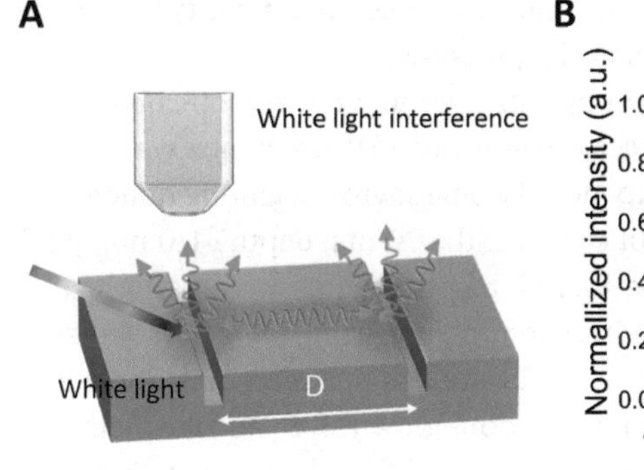

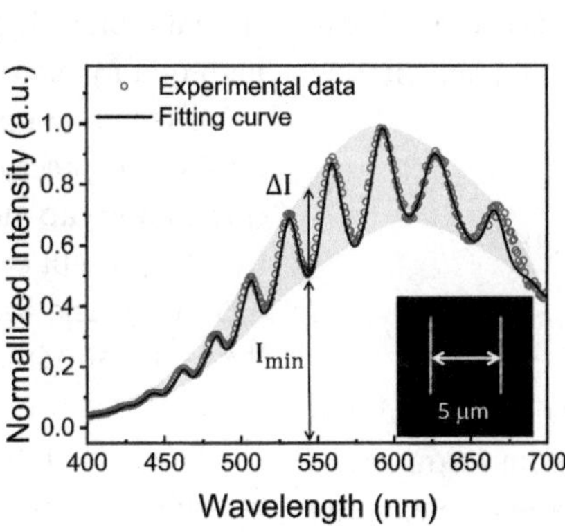

Figure 3: White-light surface plasmon interference. (A) Schematic of white-light interference method for measuring the surface plasmon polaritons (SPP) propagation length. (B) Scattering spectrum collected by the optical microscope objective. The inset is the optical scattering image from the double-groove structure. The groove separation (D) is 5 μm.

angle around 75–80° (see Supplementary material for experimental setup). The incident photons partially coupled to surface plasmons, which propagate along the aluminum surface covered with a 3-nm-thick native oxide (see Supplementary material for more details about the influence of oxide layer on optical properties). In order to reduce grain and surface scattering effects in the WLI measurements, single-crystalline and atomic-smooth aluminum surface are necessary, as reported in the previous works [47–49, 52, 53].

Figure 3B shows a clearly interference spectrum, indicating that propagating SPPs can reflect back and forth multiple times between two grooves. After multiple reflections and decoupling into far-field radiation at the incident groove, a microscope objective (100×, numerical aperture = 0.8) is used to collect the decoupled photons. The standard constructive and destructive interference conditions can be used to find the real part of SPP wave number

$$\begin{aligned} &\text{Peaks:}\ \ 2\mathrm{Re}(k_{SPP})D = 2q\pi \\ &\text{Dips:}\ \ 2\mathrm{Re}(k_{SPP})D = (2q+1)\pi, \end{aligned} \tag{2}$$

where D is the distance between two grooves, and q is an arbitrary integer number (q = 0, 1, 2, 3, ...). In order to determine the exact q value and the real part of n_{eff}, we utilize the extracted dielectric function by SE in the long wavelength region for this purpose. After that, we can apply the Drude–Lorentz model (Eq. 1) to determine the imaginary part of n_{eff}. In general, the dielectric function acquired in this approach can match well with that determined by SE, and it has been confirmed for the present case of aluminum epitaxial film.

By using the conventional theory of Fabry–Pérot interferometry [57], we can further derive the electric field of scattered SPPs at the incident groove ($E_{SPP}(\lambda)$) as the following:

$$\begin{aligned} E_{SPP}(\lambda) &= \frac{\sqrt{I_{inc}(\lambda)}}{1 - R\exp(-2\mathrm{Im}(k_{SPP})D)\exp(2i\mathrm{Re}(k_{SPP})D)} \\ &= \frac{\sqrt{I_{inc}(\lambda)}}{1-(RA)\exp(2i\mathrm{Re}(k_{SPP})D)}, \end{aligned} \tag{3}$$

where $I_{inc}(\lambda)$ is the incident SPP field intensity at the incident groove, R is the SPP reflectivity of both grooves, and $A \equiv \exp(-2\mathrm{Im}(k_{SPP})D) = \exp(-D/L_{spp})$ is the plasmonic propagation loss (absorption) factor between two grooves. Here, $L_{spp} \equiv 1/2\mathrm{Im}(k_{SPP}) = \lambda/4\pi\mathrm{Im}(n_{eff})$ is defined as the surface plasmon propagation length. According to Eq. (3), the ratio of interference maximum (I_{max}) and minimum (I_{min}) intensities can be expressed as the following:

$$\frac{I_{max}}{I_{min}} = \frac{\max\left(|E_{SPP}|^2\right)}{\min\left(|E_{SPP}|^2\right)} = \frac{(1+RA)^2}{(1-RA)^2} = \frac{(1+r)^2}{(1-r)^2}, \tag{4}$$

where $r \equiv RA$ is the round-trip reflectivity between two grooves, taking into account the plasmonic material loss. Furthermore, we can define the relative modulation depth [47]

$$\frac{\Delta I}{I_{min}} = \frac{I_{max} - I_{min}}{I_{min}} = \frac{4r}{(1-r)^2}, \tag{5}$$

where $\Delta I = I_{max} - I_{min}$ is the difference between the envelopes enclosing the intensity maxima and minima of the WLI pattern, and it indicates quantitatively how pronounced the SPP interference effect is.

In our measurements, $I_{max}(\lambda)$, $I_{min}(\lambda)$, as well as $r(\lambda)$ can be determined by experiment, as shown in Figure 3B. Furthermore, we can determine the values of A, L_{spp} and $R = r/A$ by using the dielectric function. Significant advantages of the WLI method are that we can directly correlate with the dielectric function determined by SE and measure the surface plasmon propagation lengths on the actual film surface in the full visible spectral range. As shown in Figure 3B, we can also numerically fit the experimental interference pattern with the following expression

$$I_{interference}(\lambda) = \frac{I_{max}(\lambda)(1-r)^2}{|1 - R\exp(4\pi i(n_{SPP})D/\lambda)|^2}, \tag{6}$$

where the fitting parameters R_{max} is found to be ~0.25 at 550 nm and L_{SPP} ranges from 5 to 13 μm in the spectral range of 400–700 nm. The plasmon propagation length measured for the aluminum epitaxial film is comparable to that measured for single-crystalline silver nanowires at 785 nm (10 μm [47]).

5 Surface-enhanced Raman spectroscopy

Following the discussion of fundamental structural and optical properties, we now turn to the demonstration of epitaxial film–based plasmonic applications, including aluminum SERS substrate [15] and plasmonic surface lattices [22, 34, 58]. For the SERS study (see Supplementary material for experimental setup), we use a vertically stacked molybdenum disulfide (MoS_2)/tungsten diselenide (WSe_2) heterostructures on top of the aluminum SERS substrate (nanogroove grating) as a uniform two-dimensional analyte to evaluate the SERS performance (Figure 4A). The nanogroove array structure fabricated on the aluminum epitaxial film has the benefits of large-area spatial uniformity and wide-spectral tunability because of high-quality material properties.

The nanogroove array structure was fabricated by FIB milling for precise control of LSP resonance wavelength (plasmonic gap mode) via nanogroove grating dimensions (for the present case: width ~70 nm, depth ~110 nm, pitch ~250 nm). The Raman measurements were performed in the backscattering configuration using a 532-nm solid-state laser. Using this SERS substrate design [15], large-area chemical vapor deposition–grown MoS_2 and mechanically exfoliated WSe_2 stack on Al-SERS substrate can be clearly distinguishable (Figure 4B, D, and F) by Raman intensity mapping because of a spatially uniform Raman hot zone and atomically smooth film surface, which prevent the formation of inhomogeneous stochastic hot-spots.

In the Raman scattering spectra shown in Figure 4C and E, three prominent Raman active peaks can be identified with E^1_{2g} (384 cm^{-1}), A_{1g} (403 cm^{-1}), and $2LA(M)$ (447 cm^{-1}) vibrational modes, originating from monolayer MoS_2. On the other hand, a mixed Raman active

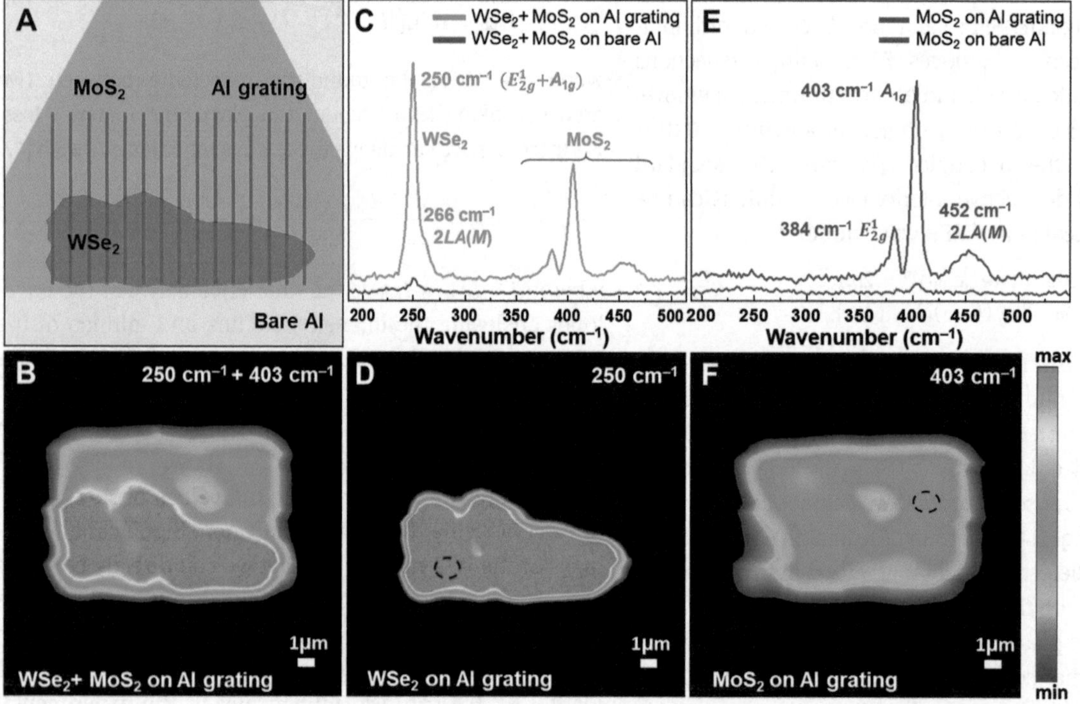

Figure 4: Raman intensity mapping of a vertically stacked monolayer transition metal dichalcogenide (TMDC) heterostucture. (A) Schematic of the surface-enhanced Raman spectroscopy (SERS) substrate structure, consisting of a grating with the size of 12 × 8 μm². The grating is sequentially stacked with MoS_2 and WSe_2 monolayers (WSe_2 on top). (B) Combining Raman intensity mapping of the A_{1g} peak (403 cm^{-1}) of MoS_2 monolayer and the $E^1_{2g} + A_{1g}$ mixed peak (250 cm^{-1}) of WSe_2 monolayer. The 532-nm excitation laser has a spot size ~2 μm, which limits the lateral resolution. (C) Enhanced Raman signals of MoS_2 and WSe_2 monolayers, which are recorded at the circled area in Figure 4D. The Raman signal of MoS_2 monolayer is still present since it is directly below the WSe_2 monolayer. (D) Raman intensity mapping at 250 cm^{-1}, corresponding to the $E^1_{2g} + A_{1g}$ mixed peak of WSe_2 monolayer. (E) Enhanced Raman signal of MoS_2 monolayer. (F) Raman intensity mapping at 403 cm^{-1}, corresponding to the A_{1g} peak of MoS_2 monolayer. The laser power density is ~30 kW/cm² and the exposure time is 1 s per data point. Scanning area is 20 × 20 μm².

peak $E^1_{2g} + A_{1g}$ (250 cm^{-1}) is identified for monolayer WSe_2. The E^1_{2g} and A_{1g} Raman peaks arise from the first-order, in-plane vibration of two S(Se) atoms relative to Mo(W) atom and out-of-plane vibration of S(Se) atoms, respectively. Due to the strong plasmonic field enhancement, we can also clearly observe the $2LA(M)$ peaks of MoS_2 (452 cm^{-1}) and WSe_2 (266 cm^{-1}), originating from a second-order process involving the longitudinal acoustic phonons at the M point of the Brillouin zone.

6 Plasmonic surface lattice

Generally, aluminum nanostructure dipolar resonators can only exhibit a small quality factor ($Q < 5$) due to a high radiative loss in the visible range [8, 11]. Here, we fabricate aluminum nanohole arrays (i.e., "antipartcle" arrays, in contrast to conventional nanoparticle arrays) on the aluminum epitaxial films as the plasmonic surface lattices to demonstrate high-Q plasmonic surface lattice [34, 58] modes based on propagating SPPs (the nanoholes are used

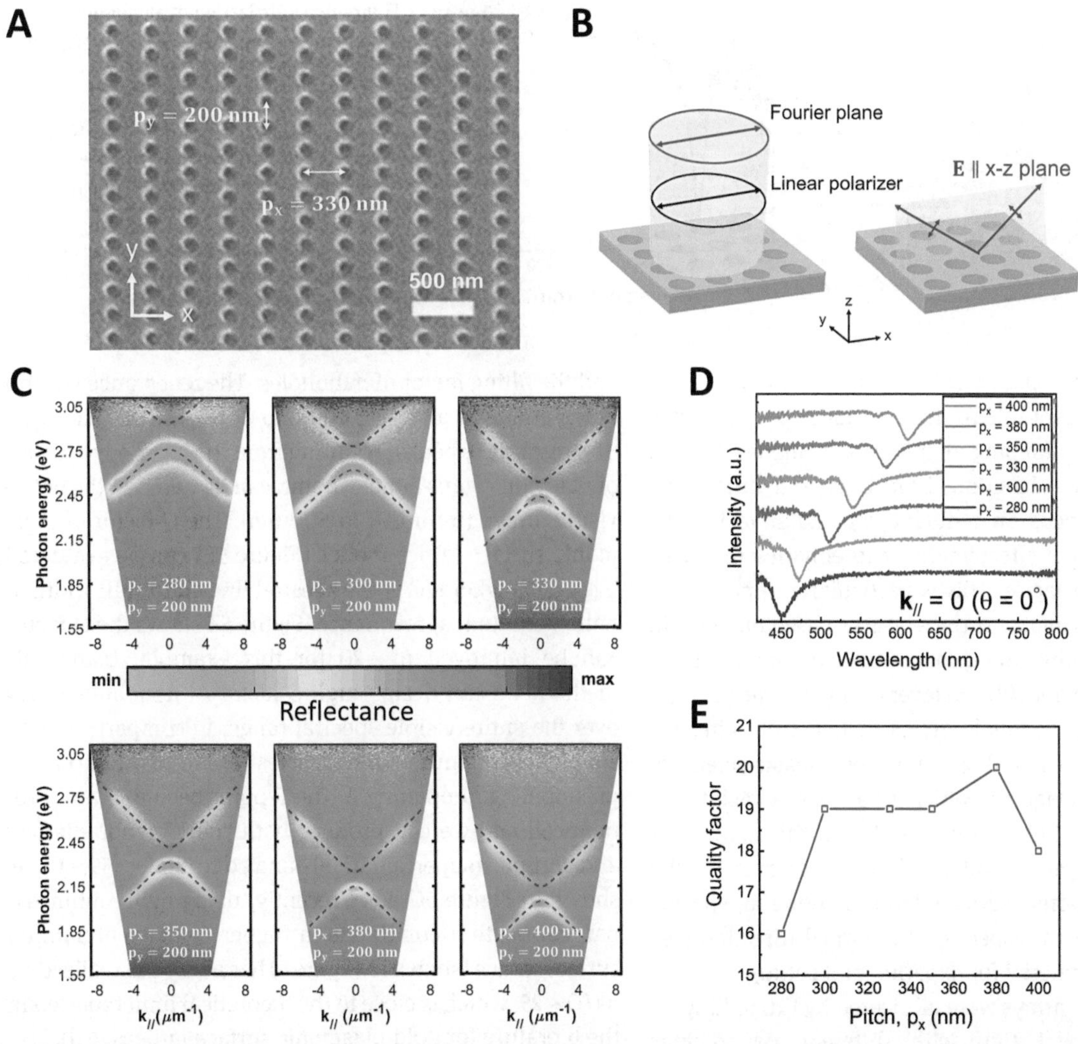

Figure 5: Optical properties of plasmonic surface lattices on epitaxial aluminum film. (A) Sample configuration of surface lattices patterned on epitaxial aluminum film. (B) Schematic of the optical setup for angle-resolved reflectance measurements, illustrating the orientations of spectrometer slit and polarizer with respect to the Al nanohole array capped with a polydimethylsiloxane (PDMS) layer. The Fourier spectra were collected by a 100× objective (N.A. = 0.55). (C) Angle-resolved reflectance mapping of the Al nanohole arrays, showing the evolution of the plasmonic surface lattice modes while the pitch along the x-axis (p_x) is increased from 280 to 400 nm. (D) Reflectance spectra of Al nanohole arrays extracted from the angle-resolved reflectance spectra (Figure 5C). All of resonance peaks are shown at the emission angle (θ) equal to 0°. (E) Measured quality factor of plasmonic surface lattice modes, which is defined as $Q = \lambda_{res}/\Delta\lambda$.

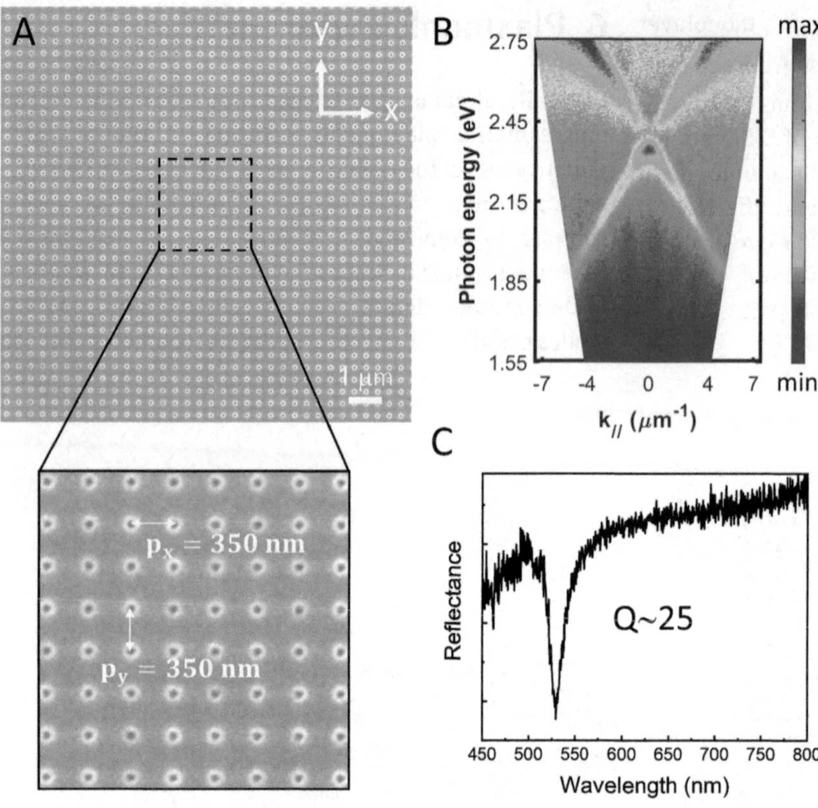

Figure 6: Optimization of aluminum nanohole surface lattice. (A) SEM image of the optimized aluminum plasmonic surface lattice. The inset shows the lattice constants (p_x and p_y) are 350 nm. (B) Angle-resolved reflectance mapping of the Al nanohole surface lattice. (C) Reflectance spectrum extracted from the angle-resolved reflectance mapping. The resonance peak is shown at the emission angle (θ) equal to 0°.

to define the spatial periodicities their LSP resonances do not play a significant role in these surface lattices). Since the nonradiative damping of plasmonic material is an intrinsic property originating from inter and intra band transitions, we focus on reducing the radiative loss by increasing the local effective index to enhance the near-field confinement factor. This is particularly important to realize high-*Q* plasmonic surface lattices based on SPPs. In this work, to improve the *Q*-factor, we cap one-mm-thick polydimethylsiloxane (PDMS) layer (refractive index: ~1.4) on the aluminum nanohole arrays to boost the effective index. The angle-resolved spectroscopic measurement results are shown in Figure 5, which is based on a back-focal-plane imaging technique (Figure 5B). In this setup, the angle-resolved spectra collect photons emitted by the aluminum plasmonic surface lattices along a specific emission angle with respect to the normal direction (see Supplementary material for experimental setup).

The nanohole arrays were fabricated by FIB milling on aluminum epitaxial film with different periodicities p_x = 280, 300, 330, 350, 380, 400 nm along the *x*-direction, while keeping a constant p_y = 200 nm, in order to tune the plasmonic surface lattice resonances at the Γ point of the lattice Brillouin zone. The angle-resolved reflectance spectra shown in Figure 5C illustrate the plasmonic surface lattice modes are well controlled by changing the pitch (p_x) and the filling factor of nanoholes. The reflectance spectra at the emission angle (θ) equal to 0° (Figure 5D) show the SPP resonance peaks, resulting from the band edge modes of plasmonic gaps at the Γ point, can be tuned in the entire visible range (from 450 to 600 nm). The *Q*-factor of plasmonic surface lattice modes (Figure 5E) can be evaluated by $\lambda_{res}/\Delta\lambda$, where $\Delta\lambda$ is the spectral linewidth (full width at half maximum) at resonance. Figure 5E shows the *Q*-factor can be improved to ~20 for this example. Using this method, we can design high *Q*-factor Al nanohole arrays over the entire visible spectral range, in comparison with colloidal aluminum nanoparticles [8, 11]. In addition, the plasmonic gap opening at the Γ point becomes very pronounced and we can produce both "dark" (subradiative) and "bright" (superradiative) band edge modes [59, 60], as shown in Figure 5C. Very recently, using more optimized nanofabrication conditions, a higher *Q*-factor aluminum surface lattice (shown in Figure 6) has also been realized by us ($Q \sim 25$, which is close to the theoretical limit reported in the literature for gold plasmonic surface lattices [61]).

7 Conclusions and outlook

In summary, we have demonstrated that epitaxial aluminum films are a promising material platform for general plasmonic applications for both ultraviolet and visible spectral regions. White-light surface plasmon interferometry measurement in the entire visible range (400–700 nm) shows that long SPP propagation lengths (5–13 μm) can be achieved due to high-quality Al film. As the first example of plasmonic applications, the aluminum epitaxial film is used to fabricate large-area, highly uniform SERS substrates at 532 nm. A vertically stacked monolayer WSe_2/MoS_2 heterostructure is adopted as a uniform analyte to show the chemical mapping capability of SERS for two-dimensional material characterization. In the second application, large-area plasmonic surface lattices (periodic hole arrays) with precisely engineered lattice parameters are fabricated on an aluminum epitaxial film. By tuning the lattice parameter along one axis, we can control dark and bright band edge modes over the full visible spectrum. To improve the Q-factor of surface lattice modes, we cap the surface lattices with a PDMS dielectric layer to increase the local effective index. The Q-factor can be improved to ~25, compared to bare Al nanostructures without capping ($Q < 5$). The epitaxial approach reported here paves the way for widespread applications of aluminum plasmonics, especially for sensing, photonic system integration, and novel devices requiring ultrathin plasmonic layers. Moreover, it also can be applied for quantum information processing requiring high-quality aluminum superconducting qubits.

Acknowledgments: We would like to acknowledge funding support from the Ministry of Science and Technology in Taiwan for this research and Ragini Mishra for her help in XRD measurements. S.G. and Y.-H.L. were partially supported by Frontier Research Center on Fundamental and Applied Sciences of Matters at National Tsing-Hua University, the Featured Areas Research Center Program within the framework of the Higher Education Sprout Project by the Ministry of Education in Taiwan. C.-K.S was partially supported by the Yushan Scholar Program by the Ministry of Education in Taiwan.

Author contribution: All the authors have accepted responsibility for the entire content of this submitted manuscript and approved submission.

Research funding: This research was supported by the Ministry of Science and Technology in Taiwan under grants numbers MOST-108-2119-M-007-008 and MOST-107-2923-M-007-004-MY3.

Conflict of interest statement: The authors declare no conflicts of interest regarding this article.

References

[1] W. L. Barnes, A. Dereux, and T. W. Ebbesen, "Surface plasmon subwavelength optics," *Nature*, vol. 424, pp. 824–830, 2003.

[2] S. A. Maier and H. A. Atwater, "Plasmonics: localization and guiding of electromagnetic energy in metal/dielectric structures," *J. Appl. Phys.*, vol. 98, p. 011101, 2005.

[3] M. Khorasaninejad and F. Capasso, "Metalenses: versatile multifunctional photonic components," *Science*, vol. 358, p. eaam8100, 2017.

[4] I. Zorić, M. Zach, B. Kasemo, and C. Langhammer, "Gold, platinum, and aluminum nanodisk plasmons: material independence, subradiance, and damping mechanisms," *ACS Nano*, vol. 5, pp. 2535–2546, 2011.

[5] G. V. Naik, V. M. Shalaev, and A. Boltasseva, "Alternative plasmonic materials: beyond gold and silver," *Adv. Mater.*, vol. 25, pp. 3264–3294, 2013.

[6] B. Doiron, M. Mota, M. P. Wells, et al., "Quantifying figures of merit for localized surface plasmon resonance applications: a materials survey," *ACS Photonics*, vol. 6, pp. 240–259, 2019.

[7] H. Ehrenreich, H. R. Philipp, and B. Segall, "Optical properties of aluminum," *Phys. Rev.*, vol. 132, pp. 1918–1928, 1963.

[8] M. W. Knight, N. S. King, L. Liu, H. O. Everitt, P. Nordlander, and N. J. Halas, "Aluminum for plasmonics," *ACS Nano*, vol. 8, pp. 834–840, 2014.

[9] C. W. Cheng, Y. J. Liao, C. Y. Liu, et al., "Epitaxial aluminum-on-sapphire films as a plasmonic material platform for ultraviolet and full visible spectral regions," *ACS Photonics*, vol. 5, pp. 2624–2630, 2018.

[10] M. J. McClain, A. E. Schlather, E. Ringe, et al., "Aluminum nanocrystals," *Nano Lett.*, vol. 15, pp. 2751–2755, 2015.

[11] A. Sobhani, A. Manjavacas, Y. Cao, et al., "Pronounced linewidth narrowing of an aluminum nanoparticle plasmon resonance by Interaction with an aluminum metallic film," *Nano Lett.*, vol. 15, pp. 6946–6951, 2015.

[12] H. W. Liu, F. C. Lin, S. W. Lin, et al., "Single-crystalline aluminum nanostructures on a semiconducting GaAs substrate for ultraviolet to near-infrared plasmonics," *ACS Nano*, vol. 9, pp. 3875–3886, 2015.

[13] F. Cheng, P.-H. Su, J. Choi, S. Gwo, X. Li, and C.-K. Shih, "Epitaxial growth of atomically smooth aluminum on silicon and its intrinsic optical properties," *ACS Nano*, vol. 10, pp. 9852–9860, 2016.

[14] S. K. Jha, Z. Ahmed, M. Agio, et al., "Deep-UV surface-enhanced resonance Raman scattering of adenine on aluminum nanoparticle arrays," *J. Am. Chem. Soc.*, vol. 134, pp. 1966–1969, 2012.

[15] S. S. Raja, C.-W. Cheng, Y. Sang, et al., "Epitaxial aluminum surface enhanced Raman spectroscopy substrates for large-scale 2D material characterization," *ACS Nano*, vol. 14, pp. 8838–8845, 2020.

[16] Q. Zhang, G. Y. Li, X. F. Liu, et al., "A room temperature low-threshold ultraviolet plasmonic nanolaser," *Nat. Commun.*, vol. 5, p. 4953, 2014.

[17] Y.-H. Chou, Y.-M. Wu, K.-B. Hong, et al., "High-operation-temperature plasmonic nanolasers on single-crystalline aluminum," *Nano Lett.*, vol. 16, pp. 3179–3186, 2016.

[18] B.-T. Chou, Y.-H. Chou, Y.-M. Wu, et al., "Single-crystalline aluminum film for ultraviolet plasmonic nanolasers," *Sci. Rep.*, vol. 6, p. 19887, 2016.
[19] Y. Chou, K. Hong, Y. Chung, et al., "Metal for plasmonic ultraviolet laser: Al or Ag?," *IEEE J. Sel. Top. Quant. Electron.*, vol. 23, pp. 1–7, 2017.
[20] Y. J. Liao, C. W. Cheng, B. H. Wu, et al., "Low threshold room-temperature UV surface plasmon polariton lasers with ZnO nanowires on single-crystal aluminum films with Al_2O_3 interlayers," *RSC Adv.*, vol. 9, pp. 13600–13607, 2019.
[21] H. Li, J. H. Li, K. B. Hong, et al., "Plasmonic nanolasers enhanced by hybrid graphene-insulator-metal structures," *Nano Lett.*, vol. 19, pp. 5017–5024, 2019.
[22] R. Li, D. Wang, J. Guan, et al., "Plasmon nanolasing with aluminum nanoparticle arrays," *J. Opt. Soc. Am. B*, vol. 36, pp. E104–E111, 2019.
[23] G. Maidecchi, G. Gonella, R. P. Zaccaria, et al., "Deep ultraviolet plasmon resonance in aluminum nanoparticle arrays," *ACS Nano*, vol. 7, pp. 5834–5841, 2013.
[24] T. Xu, Y.-K. Wu, X. Luo, and L. J. Guo, "Plasmonic nanoresonators for high-resolution colour filtering and spectral imaging," *Nat. Commun.*, vol. 1, p. 59, 2010.
[25] S. Yokogawa, S. P. Burgos, and H. A. Atwater, "Plasmonic color filters for CMOS image sensor applications," *Nano Lett.*, vol. 12, pp. 4349–4354, 2012.
[26] J. Olson, A. Manjavacas, L. Liu, et al., "Vivid, full-color aluminum plasmonic pixels," *Proc. Natl. Acad. Sci. U.S.A.*, vol. 111, p. 14348, 2014.
[27] S. J. Tan, L. Zhang, D. Zhu, et al., "Plasmonic color palettes for photorealistic printing with aluminum nanostructures," *Nano Lett.*, vol. 14, pp. 4023–4029, 2014.
[28] D. Fleischman, K. T. Fountaine, C. R. Bukowsky, G. Tagliabue, L. A. Sweatlock, and H. A. Atwater, "High spectral resolution plasmonic color filters with subwavelength dimensions," *ACS Photonics*, vol. 6, pp. 332–338, 2019.
[29] L. Zhou, C. Zhang, M. J. McClain, et al., "Aluminum nanocrystals as a plasmonic photocatalyst for hydrogen dissociation," *Nano Lett.*, vol. 16, pp. 1478–1484, 2016.
[30] D. Krause, C. W. Teplin, and C. T. Rogers, "Optical surface second harmonic measurements of isotropic thin-film metals: gold, silver, copper, aluminium, and tantalum," *J. Appl. Phys.*, vol. 96, pp. 3626–3634, 2004.
[31] M. Castro-Lopez, D. Brinks, R. Sapienza, and N. F. van Hulst, "Aluminum for nonlinear plasmonics: resonance-driven polarized luminescence of Al, Ag, and Au nanoantennas," *Nano Lett.*, vol. 11, pp. 4674–4678, 2011.
[32] W. P. Guo, W. Y. Liang, C. W. Cheng, et al., "Chiral second-harmonic generation from monolayer WS_2/aluminum plasmonic vortex metalens," *Nano Lett.*, vol. 20, pp. 2857–2864, 2020.
[33] S. Tian, O. Neumann, M. J. McClain, et al., "Aluminum nanocrystals: a sustainable substrate for quantitative SERS-based DNA detection," *Nano Lett.*, vol. 17, pp. 5071–5077, 2017.
[34] X. Zhu, G. M. Imran Hossain, M. George, et al., "Beyond noble metals: high Q-factor aluminum nanoplasmonics," *ACS Photonics*, vol. 7, pp. 416–424, 2020.
[35] I. Levine, A. Yoffe, A. Salomon, W. J. Li, Y. Feldman, and A. Vilan, "Epitaxial two dimensional aluminum films on silicon (111) by ultra-fast thermal deposition," *J. Appl. Phys.*, vol. 111, p. 124320, 2012.
[36] Y. H. Tsai, Y. H. Wu, Y. Y. Ting, C. C. Wu, J. S. Wu, and S. D. Lin, "Nano-to atomic-scale epitaxial aluminum films on Si substrate grown by molecular beam epitaxy," *AIP Adv.*, vol. 9, p. 105001, 2019.
[37] S. W. Lin, J. Y. Wu, S. D. Lin, M. C. Lo, M. H. Lin, and C. T. Liang, "Characterization of single-crystalline aluminum thin film on (100) GaAs substrate," *Jpn. J. Appl. Phys.*, vol. 52, p. 045801, 2013.
[38] K. D. Zhang, S. J. Xia, C. Li, et al., "Interface engineering and epitaxial growth of single-crystalline aluminum films on semiconductors," *Adv. Mater. Interfaces*, p. 2000572, 2020. https://doi.org/10.1002/admi.202000572.
[39] Y. N. Zhu, W. L. Wang, W. J. Yang, H. Y. Wang, J. N. Gao, and G. Q. Li, "Nucleation mechanism for epitaxial growth of aluminum films on sapphire substrates by molecular beam epitaxy," *Mater. Sci. Semicond. Process.*, vol. 54, pp. 70–76, 2016.
[40] S. W. Hieke, G. Dehm, and C. Scheu, "Annealing induced void formation in epitaxial Al thin films on sapphire (α-Al_2O_3)," *Acta Mater.*, vol. 140, pp. 355–365, 2017.
[41] G. Davy and K. G. Stephen, "Aluminium plasmonics," *J. Phys. D Appl. Phys.*, vol. 48, p. 184001, 2015.
[42] C. J. DeSantis, M. J. McClain, and N. J. Halas, "Walking the walk: a giant step toward sustainable plasmonics," *ACS Nano*, vol. 10, pp. 9772–9775, 2016.
[43] R. A. Maniyara, D. Rodrigo, R. Yu, et al., "Tunable plasmons in ultrathin metal films," *Nat. Photonics*, vol. 13, pp. 328–333, 2019.
[44] B. Frank, P. Kahl, D. Podbiel, et al., "Short-range surface plasmonics: localized electron emission dynamics from a 60-nm spot on an atomically flat single-crystalline gold surface," *Sci. Adv*, vol. 3, p. e1700721, 2017.
[45] S. Campione, I. Brener, and F. Marquier, "Theory of epsilon-near-zero modes in ultrathin films," *Phys. Rev. B*, vol. 91, p. 121408, 2015.
[46] J. M. Martinis, M. H. Devoret, and J. Clarke, "Quantum Josephson junction circuits and the dawn of artificial atoms," *Nat. Phys.*, vol. 16, pp. 234–237, 2020.
[47] H. Ditlbacher, A. Hohenau, D. Wagner, et al., "Silver nanowires as surface plasmon resonators," *Phys. Rev. Lett.*, vol. 95, p. 257403, 2005.
[48] P. Nagpal, N. C. Lindquist, S.-H. Oh, and D. J. Norris, "Ultrasmooth patterned metals for plasmonics and metamaterials," *Science*, vol. 325, pp. 594–597, 2009.
[49] J. H. Park, P. Ambwani, M. Manno, et al., "Single-crystalline silver films for plasmonics," *Adv. Mater.*, vol. 24, pp. 3988–3992, 2012.
[50] Y.-J. Lu, J. Kim, H.-Y. Chen, et al., "Plasmonic nanolaser using epitaxially grown silver film," *Science*, vol. 337, pp. 450–453, 2012.
[51] Y. Wu, C. Zhang, N. M. Estakhri, et al., "Intrinsic optical properties and enhanced plasmonic response of epitaxial silver," *Adv. Mater.*, vol. 26, pp. 6054–6055, 2014.
[52] F. Cheng, C.-J. Lee, J. Choi, et al., "Epitaxial growth of optically thick, single crystalline silver films for plasmonics," *ACS Appl. Mater. Interfaces*, vol. 11, pp. 3189–3195, 2019.
[53] C. Y. Wang, H. Y. Chen, L. Y. Sun, et al., "Giant colloidal silver crystals for low-loss linear and nonlinear plasmonics," *Nat. Commun.*, vol. 6, p. 7734, 2015.
[54] S. Gwo, H.-Y. Chen, M.-H. Lin, et al., "Nanomanipulation and controlled self-assembly of metal nanoparticles and

nanocrystals for plasmonics," *Chem. Soc. Rev.*, vol. 45, pp. 5672–5716, 2016.
[55] A. D. Rakic, A. B. Djurisic, J. M. Elazar, and M. L. Majewski, "Optical properties of metallic films for vertical-cavity optoelectronic devices," *Appl. Opt.*, vol. 37, pp. 5271–5283, 1998.
[56] F. Wang and Y. R. Shen, "General properties of local plasmons in metal nanostructures," *Phys. Rev. Lett.*, vol. 97, p. 206806, 2006.
[57] B. E. A. Saleh and M. C. Teich, *Fundamentals of Photonics*, New York, John Wiley & Sons, 1991.
[58] V. G. Kravets, A. V. Kabashin, W. L. Barnes, and A. N. Grigorenko, "Plasmonic surface lattice resonances: a review of properties and applications," *Chem. Rev.*, vol. 118, pp. 5912–5951, 2018.
[59] S. R. K. Rodriguez, A. Abass, B. Maes, O. T. A. Janssen, G. Vecchi, and J. Gómez Rivas, "Coupling bright and dark plasmonic lattice resonances," *Phys. Rev. X*, vol. 1, p. 021019, 2011.
[60] T. K. Hakala, H. T. Rekola, A. I. Väkeväinen, et al., "Lasing in dark and bright modes of a finite-sized plasmonic lattice," *Nat. Commun.*, vol. 8, p. 13687, 2017.
[61] S. R. K. Rodriguez, M. C. Schaafsma, A. Berrier, and J. Gómez Rivas, "Collective resonances in plasmonic crystals: size matters," *Phys. B Condens. Matter*, vol. 407, pp. 4081–4085, 2012.

Supplementary Material: The online version of this article offers supplementary material (https://doi.org/10.1515/nanoph-2020-0402).

Part X: **Metaoptics**

Nader Engheta*

Metamaterials with high degrees of freedom: space, time, and more

https://doi.org/10.1515/9783110710687-051

Abstract: In this brief opinionated article, I present a personal perspective on metamterials with high degrees of freedom and dimensionality and discuss their potential roles in enriching light–matter interaction in photonics and related fields.

Keywords: dimensionality; light–matter interaction; metamaterial; photonics.

1 Main text

To control and manipulate photons we often use materials. The light–matter interaction, which is at the heart of optics and photonics, is governed by the laws of electrodynamics, both classical as well as quantum electrodynamics. In dealing with electromagnetic wave and field interaction with materials, one can always start at the quantum level where photons interact with atoms. However, we often prefer to parameterize such interaction and consequently we commonly use the notion of *macroscopic* light–matter interaction, in which we assign certain material parameters to media with which waves and fields interact. Such electromagnetic parameters are well known quantities, such as permittivity (ε), permeability (μ), conductivity (σ), chirality (ξ), nonlinear susceptibilities ($\chi^{(2)}, \chi^{(3)}, \ldots$), etc., which are based on electric and magnetic dipolar terms, an approximation from the formal treatment of multipole expansion in electromagnetic interaction [1]. (The higher multipolar terms, such as quadrupoles, octopoles, etc. can also be added and augmented, but often dipolar terms suffice.) In Nature, there are numerous materials, all derived from the elements of the periodic table, that interact with electromagnetic waves. Such materials are made of atoms and molecules with specific arrangements. Their chemical compositions determine their various properties, ranging from mechanical, thermal, acoustical, optical, and electromagnetic properties, just to name a few. For these materials, the material parameters listed above attain certain ranges of values, which depend on various factors including frequency of operation. However, specially designed material structures, often termed *metamaterials* and *metastructures*, have provided scientists and engineers with the opportunities to go beyond this limitation by considering structural arrangements, instead of (and in addition to) natural chemical arrangements, as collections of tiny structures with man-made patterns embedded in host media [2–5]. Depending on the compositions, arrangements, alignments, densities, shapes and geometries of these inclusions, and their host media, such metamaterials may offer unusual wave–matter features and their effective parameters may attain values beyond what Nature has given us. This has provided unprecedented opportunities for innovations in material-based devices and systems.

To achieve desired functionality from light–matter interaction, we often utilize spatial inhomogeneities in material parameters. For example, a simple convex lens operates based on the light propagation from air to the lens, formed by a material with $\varepsilon(x, y, z)$ different from that of air and with a proper shape. Waveguides are another common example of spatial inhomogeneity, in which the core has higher relative permittivity than that of the cladding, thus confining the guided waves within and along the guide. In general, based on spatial distributions of material parameters, one can categorize the spatial dimensionality of electromagnetic problems as three-dimensional (3D) such as general 3D scattering, two-dimensional (2D) such as parallel-plate waveguides, one-dimensional (1D) such as transmission lines, and 'zero-dimensional' (0D) as lumped circuit elements [6]. This categorization has also been utilized in the field of metamaterials, leading to the notions of 3D volumetric metamaterials [2–5], 2D metasurfaces [7–14], 1D left-handed transmission lines [4, 5], and 0D optical metatronics [15, 16]. However, a question

***Corresponding author: Nader Engheta**, University of Pennsylvania, Philadelphia, PA 19104, USA, Web: www.seas.upenn.edu/~engheta/, E-mail: engheta@ee.upenn.edu. https://orcid.org/0000-0003-3219-9520

This article has previously been published in the journal Nanophotonics. Please cite as: N. Engheta "Metamaterials with high degrees of freedom: space, time, and more" *Nanophotonics* 2021, 10. DOI: 10.1515/nanoph-2020-0414.

may be naturally raised here: Can we have higher dimensional metamaterials? Can we have metastructures with four dimensions? How about higher degrees of freedom? This indeed may be possible. But why should one be interested in such increase in material dimensionality? On the one hand, in most cases the higher the dimensions, the more degrees of freedom we can use to get to our goals of desired functionalities in wave-based devices and systems. On the other hand, frequently the complexity also increases with the dimensionality. As one approach to increasing the dimensionality of metamaterials, one can consider bringing the dimension of "time" into this platform. This means that one would desire to have parameters such as ε also change with time, instead of (or in addition to) its variation in space, i.e., $\varepsilon(t)$ or $\varepsilon(x, y, z, t)$. Such time-varying, or time and space inhomogeneous platforms can certainly increase the number of "knobs" we can "turn" to control and achieve desired functionalities in light–matter interaction. The research on time-varying electromagnetic platforms has a long history dating back to 1950s, when one of the early works, by Morgenthaler [17], explored theoretically what would happen to a monochromatic electromagnetic wave in a medium when the medium's phase velocity is rapidly changed in time. In the past decades, there have been numerous efforts in investigating various phenomena related to the temporal variation of electromagnetic media, providing temporal boundaries [18], temporal holography [19], Doppler cloaking [20], temporal band gaps [21–23], temporal effective medium concept [24], temporal impedance matching [25, 26], Fresnel drag in spatiotemporal metamaterials [27], spacetime cloaks [28, 29], modeling time and causality with metamaterials [30, 31], rapidly growing plasma and accelerating reference frame [32], and exceptional points in time-varying media [33], just to name a few.

Although in the Maxwellian electrodynamics there are symmetries between space (x, y, z) and time (t), such as the time-reversal and space-reversal phenomena, and consequently numerous features have similar and analogous spatial and temporal characteristics, there are of course some fundamental differences. For instance, we have three dimensions of x, y, z in the Euclidean space, whereas in time (t) we have only one. For the spatial dimensions, we have no spatial "causality", i.e., we can have positive and negative values for x, y, z, without any special orders. However, we have causality in time (t), so events always move forward in time, consistent with the second law of thermodynamics. Such differences between the space (x, y, z) and time (t) provide additional richness to wave–matter interaction in spatiotemporal metamaterials. For example, such four-dimensional (4D) structures can be used to break the reciprocity without using any biasing DC magnetic field, as demonstrated by several groups [34–43]. Moreover, temporal media exhibit wave features that are more complex than those of their time-invariant counterparts, such as time-varying polarizabilities [44] and a generalization of the Kramers–Kronigs relations based on the mathematics of linear time-variant systems [45].

Clearly, 4D metamaterials have brought, and continue to bring, various exciting new phenomena into the realm of light–matter interaction. But can we have 4D metamterials with more variables and higher degrees of freedom? Adding additional degrees of freedom can bring more novelty in controlling the wave phenemona using such structures. One possible way to do so is by merging temporal variation of the materials parameters, e.g., $\varepsilon(t)$ or $\varepsilon(x, y, z, t)$ with the notion of anisotropy of these paramteres. In other words, instead of changing isotropic values of ε in time, can we consider a temporal change of ε from an isotropic value to an anisotropic tensoral value, which implies that one may want to change a single scalar value of ε to different values for each element of anisotropic permittivity tensors? In this way, for example, can we change a wave platform temporally from a simple isotropic medium to a unixial or a bixial crystal? This approach, which we can call "temporal anisotropy" can bring a new set of degrees of freedom into the electromagnetic platforms. Since wave propagation in anisotropic media has specific features, distinct from that in isotropic media, such temporal anisotropy can bring novel wave phenomena. For example, it is well known that in anisotopic media waves can experience birefringence, i.e., refractive index is different for different directions of propagation and/or different wave polarizations [1]. Moreover, it is also known that in anisotropic media the direction of phase flow (i.e., the wave vector) is not necessarily the same as the direction of energy flow (i.e., the Poynting vector). These two features have been recently utilized in developing new phenomena of the inverse prism [46], in which after the rapid change of refractive index from isotropic to an anisotropic set of values the waves experience different frequencies in different directions of propagation, and the temporal aiming [47], in which the direction of energy propagation for a wave packet can change midstream by rapid change of permittivity from an isotropic to an anisotropic case. Together with my collaborators, we are currently investigating various aspects of such 4D metamaterials with high degree of freedom.

Although in the above discussion, we mentioned the 4D structures with high degrees of freedom based only on one of the material parameters, namely the permittivity, obviously this methodology can also be applied to other material parameters such as, permeability (μ), conductivity (σ), chirality (ξ), nonlinear susceptibilities ($\chi^{(2)}, \chi^{(3)}, \ldots$), etc. Therefore, one can imagine such high degree-of-freedom media for several of these parameters concurrently. This would bring interesting curiosity-driven research possibilities. For example, what would happen if the permittivity of a structure can be engineered spatiotemporally with given spatial and temporal distributions, while at the same time and in the same place, its permeability and/or its chirality can change with space and time with a different set of spatiotemporal distributions? What if in a layered structure, each layer consists of spatiotemporal structures with a set of space–time modulated parameters different from those of other layers, some isotropic and some anisotropic? What about the applications of mathematical optimization and inverse design tools to such metamaterials? One can of course think about many other possibilities for increasing and merging various degrees of freedom. This will certainly provide far richer (and admittedly more complex) platforms for wave–matter interaction, which may provide novel material-based devices and components.

To summarize, in my opinion it seems that increasing degrees of freedom in spatiotemporal metamaterials, while adds more complexity to the structures, can open new directions in research in light–matter interaction, with the goals towards devices and systems with new functionalities. Such possibilities are limitless.

Acknowledgments: The author acknowledges the partial support from the Vannevar Bush Faculty Fellowship program sponsored by the Basic Research Office of the Assistant Secretary of Defense for Research and Engineering, funded by the Office of Naval Research through grant N00014-16-1-2029, and the partial support from the US Air Force Office of Scientific Research (AFOSR) Multidisciplinary University Research Initiative (MURI) grant number FA9550-17-1-0002.

Author contribution: The author has accepted responsibility for the entire content of this submitted manuscript and approved submission.

Research funding: This research was funded partially by the Office of Naval Research through grant N00014-16-1-2029, and partially by the US Air Force Office of Scientific Research (AFOSR) Multidisciplinary University Research Initiative (MURI) grant number FA9550-17-1-0002.

Conflict of interest statement: The authors declare no conflicts of interest regarding this article.

References

[1] C. H. Papas, *Theory of Electromagnetic Wave Propagation*, Mineola, NY, Dover, 1965.

[2] N. Engheta, R. W. Ziolkowski, and R. W. Metamaterials, *Physics and Engineering Explorations. Metamaterials: Physics and Engineering Explorations*, Hoboken, NJ, IEEE-Wiley, 2006.

[3] W. Cai and V. M. Shalaev, *Optical Metamaterials: Fundamentals and Applications*, New York, NY, Springer, 2010.

[4] G. V. Eleftheriades and K. G. Balmain, *Negative-Refraction Metamaterials: Fundamental Principles and Applications*, Hoboken, NJ, John Wiley & Sons, 2005.

[5] C. Caloz and T. Itoh, *Electromagnetic Metamaterials: Transmission-Line Theory and Microwave Applications*, Hoboken, NJ, John Wiley & Sons, 2006.

[6] F. Yang and Y. Rahmat-Samii, *Surface Electromagnetics – with Applications in Antennas, Microwave and Optical Engineering*, Cambridge University Press, 2019.

[7] N. Yu, P. Genevet, M. A. Kats, et al., "Light propagation with phase discontinuities: Generalized laws of reflection and refraction," *Science*, vol. 334, pp. 333–337, 2011.

[8] X. Ni, N. K. Emani, A. V. Kildishev, A. Boltasseva, and V. Shalaev, "Broadband light bending with plasmonic nanoatennas," *Science*, vol. 335, p. 427, 2012.

[9] P. Lalanne, S. Astilean, P. Chavel, E. Cambril, and H. Launois, "Blazed binary subwavelength gratings with efficiencies larger than those of conventional echelette gratings," *Opt. Lett.*, vol. 23, no. 14, pp. 1081–1083, 1998.

[10] E. Hasman, V. Kleiner, G. Biener, and A. Niv, "Polarization dependent focusing lens by use of quantized Pancharatnam–Berry phase diffractive optics," *Appl. Phys. Lett.*, vol. 82, no. 3, pp. 328–330, 2003.

[11] E. F. Kuester, M. A. Mohamed, M. Piket-May, and C. L. Holloway, "Averaged transition conditions for electromagnetic fields at a metafilm," *IEEE Trans. Antenna Propag.*, vol. 51, no. 10, pp. 2641–2651, 2003.

[12] A. M. Shaltout, V. M. Shalaev, and M. L. Brongersma, "Spatiotemporal light control with active metasurfaces," *Science*, vol. 364, 2019, Art no. eaat3100.

[13] Y. Hadad, D. L. Sounas, and A. Alu, "Space-time gradient metasurfaces," *Phys. Rev. B*, vol. 92, p. 100304, 2015.

[14] D. Lin, P. Fan, E. Hasman, and M. L. Brongersma, "Dielectric gradient metasurface optical elements," *Science*, vol. 345, pp. 298–302, 2014.

[15] N. Engheta, "Circuits with light at nanoscales: Optical nanocircuits inspired by metamaterials," *Science*, vol. 317, pp. 1698–1702, 2007.

[16] N. Engheta, A. Salandrino, and A. Alu, "Circuit elements at optical frequencies: Nanoinductor, nanocapacitor, and nanoresistor," *Phys. Rev. Lett.*, vol. 95, 2005, Art no. 095504.

[17] F. Morgenthaler, "Velocity modulation of electromagnetic waves," *IRE Trans. Microw. Theory Tech.*, vol. 6, no. 2, pp. 167–172, 1958.

[18] Y. Xiao, D. N. Maywar, and G. P. Agrawal, "Reflection and Transmission of electromagnetic waves at temporal boundary," *Opt. Lett.*, vol. 39, p. 577, 2014.

[19] V. Bacot, M. Labousse, A. Eddi, M. Fink, and E. Fort, "Time reversal and holography with spacetime transformations," *Nat. Phys.*, vol. 12, pp. 972–977, 2016.

[20] D. Ramaccia, D. L. Sounas, A. Alu, A. Toscano, and F. Bilotti, "Doppler cloak restores invisibility to objects in relativistic motion," *Phys. Rev. B*, vol. 95, 2017, Art no. 075113.
[21] J. R. Zurita-Sanchez, P. Halevi, and J. C. Cervantes-Gonzalez, "Reflection and transmission of a wave incident on a slab with a time-periodic dielectric function epsilon (t)," *Phys. Rev. A*, vol. 79, 2009, Art no. 053821.
[22] J. S. Martínez-Romero, O. M. Becerra-Fuentes, and P. Halevi, "Temporal photonic crystals with modulations of both permittivity and permeability," *Phys. Rev. A*, vol. 93, 2016, Art no. 063813.
[23] A. M. Shaltout, A. V. Kildishev, and V. M. Shalaev, "Photonic time crystals and momentum band-gaps," in *Conference on Lasers and Electro-Optics (CLEO)*, 2016, paper FM1D.4.
[24] V. Pacheco-Pena and N. Engheta, "Effective-medium concepts in temporal metamaterials," *Nanophotonics*, vol. 9, no. 2, pp. 379–391, 2020.
[25] V. Pacheco-Pena and N. Engheta, "Antireflection temporal coatings," *Optica*, vol. 7, p. 323, 2020.
[26] A. Shlivinski and Y. Hadad, "Beyond the Bode-Fano bound: Wideband impedance matching for short pulses using temporal switching of transmission-line parameters," *Phys. Rev. Lett.*, vol. 121, p. 204301, 2018.
[27] P. A. Huidobro, E. Galiffi, S. Guenneau, R. V. Craster, and J. B. Pendry, "Fresnel drag in space-time-modulated metamaterials," *Proc. Natl. Acad. Sci. U S A.*, vol. 116, no. 50, pp. 24943–24948, 2019.
[28] M. W. McCall, A Favaro, P. Kinsler, and A. Boardman, "A spacetime cloak, or a history editor," *J. Optic.*, vol. 13, 2011, Art no. 024003.
[29] P. Kinsler and M. W. McCall, "Transformation devices: Event carpets in space and space-time," *Phys. Rev. A.*, vol. 89, 2014, Art no. 063818.
[30] I. I. Smolyaninov and Y.-J. Hung, "Modeling time with metamaterials," *J. Opt. Soc. Am. B*, vol. 28, no. 7, pp. 1591–1595, 2011.
[31] I. I. Smolyaninov, "Modeling of causality with metamaterials," *J. Optic.*, vol. 15, 2013, Art no. 025101.
[32] E. Yablonovitch, "Accelerating reference frame for electromagnetic waves in a rapidly growing plasma: Unruh-Davies-Fulling-DeWitt radiation and the nonadiabatic Casimir effect," *Phys. Rev. Lett.*, vol. 62, no. 15, pp. 1742–1745, 1989.
[33] T. T. Koutserimpas and R. Fleury, "Electromagnetic fields in a time-varying medium: exceptional points and operator symmetries," *IEEE Trans. Antenn. Propag.*, 2020, https://doi.org/10.1109/TAP.2020.2996822.
[34] C. Caloz, A. Alu, S. Tretyakov, D. Sounas, K. Achouri, and Z.-L. Deck-Leger, "Electromagnetic nonreciprocity," *Phys. Rev. Appl.*, vol. 10, 2018, Art no. 047001.
[35] Z. Yu and S. Fan, "Complete optical isolation created by indirect interband photonic transitions," *Nat. Photon.*, vol. 3, p. 91, 2009.
[36] R. Fleury, D. L. Sounas, C. F. Sieck, M. R. Haberman, and A. Alu, "Sound isolation and giant linear nonreciprocity in a compact acoustic circulator," *Science*, vol. 343, p. 516, 2014.
[37] Z. Yu, G. Veronis, Z. Wang, and S. Fan, "One-way electromagnetic waveguide formed at the interface between a plasmonic metal under a static magnetic field and a photonic crystal," *Phys. Rev. Lett.*, vol. 100, 2008, Art no. 023902.
[38] A. Mock, D. Sounas, and A. Alu, "Magnet-free circulator based on spatiotemporal modulation of photonic crystal defect cavities," *ACS Photon.*, vol. 6, pp. 2056–2066, 2019.
[39] N. A. Estep, D. L. Sounas, J. Soric, and A. Alu, "Magnetic-free non-reciprocity and isolation based on parametrically modulated coupled-resonator loops," *Nat. Phys.*, vol. 10, p. 923, 2014.
[40] H. Lira, Z. Yu, S. Fan, and M. Lipson, "Electrically driven nonreciprocity induced by interband photonic transition on a silicon chip," *Phys. Rev. Lett.*, vol. 109, 2012, Art no. 033901.
[41] D. L. Sounas, C. Caloz, and A. Alu, "Giant non-reciprocity at the subwavelength scale using angular momentum based metamaterials," *Nat. Commun.*, vol. 4, p. 2407, 2013.
[42] T. Kodera, D. L. Sounas, and C. Caloz, "Artificial faraday rotation using a ring metamaterial structure without static magnetic field," *Appl. Phys. Lett.*, vol. 99, 2011, Art no. 031114.
[43] D. L. Sounas and A. Alu, "Non-reciprocal photonics based on time modulation," *Nat. Photon.*, vol. 11, p. 774, 2017.
[44] M. S. Mirmoosa, T. T. Koutserimpas, G. A. Ptitcyn, S. A. Tretyakov, and R. Fleury, *Dipole Polarizability of Time-Varying Particles*, 2020, arXiv:2002.12297 [physics.app-ph].
[45] D. M. Solis, and N. Engheta, *A Generalization of the Kramers-Kronigs Relations in Linear Time-Varying Media*, 2020, arXiv: 2008.04304 [physics.optics].
[46] A. Akbarzadeh, N. Chamanara, and C. Caloz, "Inverse prism based on temporal discontinuity and spatial dispersion," *Opt. Lett.*, vol. 43, pp. 3297–3300, 2018.
[47] V. Pacheco-Pena and N. Engheta, "Temporal aiming," *Light Sci. Appl.*, vol. 9, p. 129, 2020.

Mark L. Brongersma*

The road to atomically thin metasurface optics

https://doi.org/10.1515/9783110710687-052

Abstract: The development of flat optics has taken the world by storm. The initial mission was to try and replace conventional optical elements by thinner, lightweight equivalents. However, while developing this technology and learning about its strengths and limitations, researchers have identified a myriad of exciting new opportunities. It is therefore a great moment to explore where flat optics can really make a difference and what materials and building blocks are needed to make further progress. Building on its strengths, flat optics is bound to impact computational imaging, active wavefront manipulation, ultrafast spatiotemporal control of light, quantum communications, thermal emission management, novel display technologies, and sensing. In parallel with the development of flat optics, we have witnessed an incredible progress in the large-area synthesis and physical understanding of atomically thin, two-dimensional (2D) quantum materials. Given that these materials bring a wealth of unique physical properties and feature the same dimensionality as planar optical elements, they appear to have exactly what it takes to develop the next generation of high-performance flat optics.

Keywords: 2D materials; flat optics; metasurfaces.

1 A brief historical perspective of passive metasurfaces

In optics, we traditionally control and measure the behavior of light using bulky optical components. The field of flat optics aims to manipulate the flow of light with more compact elements and we are currently seeing an explosion of research on this topic. To appreciate the somewhat daunting number of directions this field is currently moving into, it is important to briefly look back at its historical development. Many reviews already provide valuable, different perspectives on the topic [1–6] and in this opinionated article I just highlight some of the key developments. During the second world war, early work by Kock already aimed to create compact diffractive lenses in the microwave range [7, 8]. However, only in the last five decades have researchers attempted to replace bulky refractive optical elements operating in the visible and infrared spectral ranges by razor-thin, planar components. This journey started with the development of diffractive optical elements (DOEs), which came with the exciting promise to reduce the size and weight of complex optical systems. Early work includes research on metallic artificial index gratings that could serve as beam deflectors and polarizers [9]. The first dielectric échelette-type DOEs, such as the Fresnel lens, shaped the phase front of light by manipulating its propagation phase. They were created by judiciously structuring the surface of transparent materials to create wavelength-scale height variations. These designer topographic features directly translate into spatially variant changes of the local phase pickup as light propagates through the element. Binary blazed gratings evolved from échelette components and they manipulate the propagation phase by nanostructuring thin films into dense forests of pillars that feature a binary height-profile [10]. If the refractive index of the pillar material is sufficiently high, they can serve as tiny, truncated waveguides whose physical dimensions determine the locally incurred phase lag upon propagation from one end to the other. By carefully choosing all of the pillar cross-sections in an array, it is again possible to imprint spatially varying phase changes onto an incident light wave. These elements are easier to fabricate and offer higher diffraction efficiencies for significant deviation angles (high-NA) by avoiding undesired shadowing effects. For very dense arrays of high-index pillars or ridges, optical coupling cannot be ignored and a physical picture involving coupled Bloch-modes [11] becomes the most insightful to explain some of the unique properties that are achievable at higher areal densities, such as broadband reflectance ($\Delta\lambda/\lambda \approx 0.3$) and resonances with very narrow linewidths ($\Delta\lambda/\lambda < 10^{-3}$). This kicked off the field of high-contrast gratings with many integrated optics applications [12]. It was shown that similar types of

***Corresponding author: Mark L. Brongersma**, Geballe Laboratory for Advanced Materials, Stanford University, Stanford, CA 94305, USA, E-mail: brongersma@stanford.edu. https://orcid.org/0000-0003-1777-8970

This article has previously been published in the journal Nanophotonics. Please cite as: M. L. Brongersma "The road to atomically thin metasurface optics" *Nanophotonics* 2021, 10. DOI: 10.1515/nanoph-2020-0444.

wavefront manipulation can also be attained by engineering the light transmission through nanoscale holes [13, 14] or engineered gap plasmonic waveguides [15] carved into thin metallic films.

In 2001 geometric-phase, flat optical elements entered the scene and provided a fundamentally new way to shape wavefronts by varying the orientation as opposed to the size and shape of subwavelength structures [16–19]. They offer complete 0 to 2π phase control by simply rotating structures, whose geometry and spacing is optimized to achieve high diffraction efficiencies [20]. Geometric phase elements have fabrication advantages as all scattering elements can be the same size and shape. By having the structure's orientation control the phase, it was for the first time possible to realize high-negative-permittivity metallic optical elements that are deep-subwavelength in thickness [21]. A unique and sometimes challenging feature, is that the local phase pickup depends on the polarization handedness of the incident light. This can be used to one's advantage in e.g. polarimetry applications [22, 23] or avoided through clever design [24].

Another important conceptual step in the development of flat optics was the realization that optical resonances in first metallic [25, 26] and later high-index semiconductor [27, 28] nanostructures could also be harnessed to create flat optical elements. Resonances can notably enhance light scattering cross sections of small particles and effectively control the phase [29]. Their design closely follows that of radiofrequency antennas and were consequently termed optical antennas [30]. As these structures resonate guided optical modes with strongly reduced wavelengths, their size can be much smaller than the free-space wavelength [31–34]. For this reason, optical antennas can be used to further thin optical elements that rely on manipulating propagation phase, which includes dielectric geometric-phase-elements [27]. The resulting flat optical element can be treated as 2D and appear to cause a discontinuity in the phase of transmitted and reflected light waves. This brought about the notion that the optical properties of a nanopatterned surface of a material can be as important as their bulk optical properties (i.e. the refractive index) and led to the formulation of generalized laws of refraction [25, 35] and the term metasurface [1, 5]. Taxonomy is a dynamic science and nowadays, the concept of metasurface-based optics and flat optics are often used interchangeably. Independent of the nomenclature, it is important to realize that the transmission, reflection, absorption, and polarization conversion coefficients of ultrathin metasurfaces satisfy a set of well defined fundamental relations [36, 37] that can assist in metasurface design thinking. One challenge with simple resonant antennas is that their Lorentzian polarizability only offers a limited 0 to π phase pickup. This spurred new ideas on Kerker resonators with two degenerate resonances [28, 38] and overcoupled gap plasmon resonators on a reflective substrate [39–41] capable of producing phase pickups across a full 2π range. Another challenge with antennas is that the light scattering properties (amplitude and phase) are strongly wavelength dependent. However, with new computational inverse and topological design techniques [42–44] the dispersive nature of nanoscale antennas can also be an invaluable asset in creating high-efficiency and virtually aberration-free metasurfaces [45].

The various concepts above have been instrumental in realizing small-form-factor optics delivering multifunctionality [46–49], very high-numerical apertures [50, 51], meaningful integration with conventional optics [52], minimal aberrations [53–56], nonlinear optics [57, 58], 3D and holograms [59–62], and control over the light field [63–65]. Given the extreme complexity of the current elements, it is good to note that there are still some very basic needs and challenges with even the simplest flat optical elements such as a lens. For example, while it is possible to supply a lens with a broadband antireflection coating, it is very challenging to avoid reflections from high numerical (N.A.) flat optics. This makes it hard to cascade them into multicomponent systems while avoiding ghost images and most likely a blend of flat and conventional optics can provide the highest performance.

2 Metasurfaces in optoelectronic devices

Concurrent with the development of flat optics for passive wavefront manipulation, it has also been realized that the layers in thin film optoelectronic devices can be nanostructured to enhance, manipulate, and control light absorption and emission functions. In such devices, metasurfaces and metafilms are serving as sources or terminals for light and are not used to transform light waves by manipulating scattering and interference phenomena. The typical questions about imaging aberrations are shifted to new questions pertaining to the best ways to facilitate and control light emission or absorption. In current semiconductor devices, the critical dimensions are already at the nanoscale and it is important to co-optimize the size and shape of the nanostructures to achieve the best electronic and (resonant) optical properties. For example, in today's optoelectronic devices the metallic electrodes typically only serve as electrical leads. However, with a simple patterning step they can also increase light absorption in solar cells and detectors [66, 67], enable polarimetry [22, 23, 68] and spectral filtering functions [69–71], or prevent radiative decay of quantum emitters into lossy plasmonic modes in solid state light sources [72]. The semiconductor layers in photodetection systems can also be structured at a subwavelength scale to create metafilms with reengineered absorption spectra [73–77], structural color [78–80], valuable color sorting functions at the nanoscale [81], or achieve an angle-sensitive response [82]. Metasurfaces of Mie and plasmonic resonators also enabled the design of conceptually new types of high-performance antireflection coatings [66, 83–85]. New research further shows that active light emitting layers can be patterned at the subwavelength scale to enhance and direct light emission [86–89]. This is an area where commercialization of metasurface concepts can be possibly fastest as the device infrastructure for realizing nanostructured optoelectronic devices is already very mature. Existing, commercially viable technologies may be re-envisioned by exploring whether minor, extra patterning steps in a process can yield tremendous optical performance benefits.

3 Open challenges and opportunities for metasurfaces

Despite the many impressive advances, there remain a great number of exciting open challenges and opportunities for further development and research. To a large extent, these opportunities are brought to us by emerging application areas for metasurfaces. These include computational imaging, dynamic wavefront manipulation, ultrafast spatiotemporal control of light, quantum communication, thermal emission management, novel display technologies, and sensing. It is of value to analyze these areas and pinpoint the specific new functionalities that they require.

It is the focus of the computational imaging field to extract high-dimensional data from images and transform it into useful numerical or symbolic information that can be further processed, interpreted and used in decision-making [90]. With advances in deep learning with artificial neural networks, digital computers are now able to analyze images with a logic structure that is similar to how humans think [91, 92]. However, there are many imaging applications that require high-throughput, real-time, and low power image processing for which digital electronics is not ideally suited. To enable such applications, it is possible to off-load certain critical and computationally intensive tasks to passive, nonenergy consuming, and ultrafast flat and deep optics [93–101]. Next-generation computational imaging systems may therefore include a series of deep and flat optical layers that need to be easily stackable and have a series of very demanding linear, nonlocal and nonlinear optical properties. When properly designed, basic image processing tasks can be performed simply as light flows through these stacks.

A number of emerging metasurface applications need to deliver dynamic wavefront control, such as light detection and ranging (LIDAR) and dynamic holography [102, 103]. This requires the design of optical antennas capable producing very large changes in their resonant light scattering response. Effective ways to manipulate resonances are currently applying mechanical motion [104–109], electrical gating to modify carrier concentrations [110–116] or the Stark effect [117], electrochemistry [118–122], liquid crystals [123–125], and phase change materials [126–130].

Metasurfaces with ultrafast responses also show great promise as it is becoming eminently clear that such surfaces are not bound by the fundamental limits of static elements [131–133] and novel incarnations of effective medium theories are required to describe their behavior [134]. They can be applied to break Lorentz reciprocity [135], attain frequency conversion [136, 137], and new opportunities for wavefront manipulation [138]. For this reason, new materials need to be identified that are easily incorporated in metasurfaces and can offer very strong, tunable, and ultrafast light-matter interactions.

Nanophotonics and quantum have been a natural match as one of the key strengths of nanophotonic elements is to concentrate light and enhance light–matter interaction for quantum objects [139–141] and overcome quantum decoherence [142]. Now, new types of metasurfaces are being developed that can control quantum properties of emitted, transmitted, and reflected light [143–148]. Further development requires emitting materials with large oscillator strengths and related radiative decay rates that well exceed nonradiative and pure dephasing decay rates.

Thermal emission from objects tends to be spectrally broadband, unpolarized, and temporally invariant. These common notions are now challenged with the emergence of new nanophotonic structures and metasurface concepts that afford on-demand, active manipulation of the thermal emission process [149–156]. The thermal emission spectra of metasurfaces, which are directly connected to their spectral absorption properties through Kirchhoff's law [157], can be engineered with tremendous flexibility through nanostructuring. This ability can impact diverse application areas, including solid state lighting, sensing, thermal imaging, thermophotovoltaics, and personal/commercial heat management [158]. However, further progress will be reliant on the availability of ultrathin elements that can easily be patterned and exhibit strong, electrically tunable absorption resonances.

For display applications, especially in wearables we need to identify lightweight materials that can be grown over large areas and easily be incorporated/patterned into metasurfaces. Materials and structures need to be identified that can enhance and control emission for displays with extreme pixel densities exceeding 10^4 pixels per square inch [159]. Wavefront manipulating optics needs to be developed that can handle the light field at wavelengths across the visible range and can assist in reducing aberrations in imaging systems comprised of miniature projectors and multiple cascaded optical components.

For application in chemical, biological, and environmental sensing, it would be very helpful to identify flat optical elements that deliver high-quality-factor (high-Q) optical resonances that can enhance light–matter interaction to e.g., increase sensitivity to subtle changes in refractive index or produce strong Raman signals. Several structures have recently been pursued in this direction that rely on the creation of bound states in the continuum, breaking symmetries in highly symmetric resonators, or the excitation of quasi-guided modes [11, 160–164].

4 2D, or not 2D, that is the question

In the design and optimization of next generation metasurfaces, it is important to realize that their ultimate physical and practical limitations can always be traced back to the properties of the materials and building blocks that the metasurfaces are constructed from. The current metasurface scene is dominated by truncated waveguides and plasmonic- or Mie-resonant antennas. The latter two can afford strong scattering and absorption with the relevant cross-sections often exceeding the geometric cross sections [165, 166]. The emerging metasurface applications, however, demand much more. From the aforementioned discussion, it is clear that we need metasurface materials and building blocks that are easily grown and patterned over large areas, straightforward to stack, and offer strong, tunable, nonlinear, ultrafast optical responses. The current building blocks can be hard to accurately place and stack over larger areas and into complex 3D architectures. Their tunability also tends to be very limited as the magnitude of most electroabsorption and electrorefraction effects in bulk (3D) metals and semiconductors are very weak [167]. In thinking what the ideal tunable, compact, stackable metasurfaces may look like, it is worth investigating the possibility of creating metasurfaces from atomically thin materials.

Concurrently with the prolific developments on metasurfaces, there has been incredible progress in the synthesis, handling, and understanding of 2D van der Waals (vdW) materials as is detailed in several comprehensive review articles [168–170]. The ability to realize these atomically-thin layers with wafer-scale uniformity is demonstrating their commercial potential [171, 172]. These materials also display a fascinating diversity of quantum, collective, topological, nonlinear, and ultrafast behaviors. It is exciting to think how such materials may open up new functions for metasurfaces. They come in the form of atomically thin sheets of insulators (e.g., hexagonal boron nitride), semiconductors (transition metal dichalcogenides [TMDCs], such as molybdenum disulphide) and semimetals (e.g., graphene). Some hard-to-realize properties in metamaterials also appear naturally in the 2D materials. For example, hyperbolic dispersion [173] and negative refraction [174] can be attained in graphite materials. Weak vdW forces bind these sheets together and this facilitates facile separation and stacking [175] to create heterojunctions and complex multilayer systems [176]. As their surfaces are naturally passivated without dangling bonds, they can also be easily integrated with other electronic and photonic elements. For these reasons, the opportunities for incorporating them in nanophotonic devices seem virtually limitless [177–179].

The suitability of 2D materials for incorporation into flat optics is in part founded on their natural ability to provide strong light–matter interaction. For example, due to its unique band structure, the absorption of graphene can be linked to the fine structure constant $\alpha_0 \approx 1/137$ as $\pi\alpha_0 = 2.3\%$ across a wide spectral range [180]. As doped graphene supports long-lived plasmons [181, 182], the absorption can be further increased to near-unity values by patterning a graphene layer [183]. In 2D semiconductors, the resonant excitation of excitons, electron-hole pairs bound by the Coulomb force, can also give rise to sharply peaked absorption features, just below the bandgap of a semiconductor [184]. A number of 2D semiconductors exhibit very high (~10%) absorption by a single layer near excitonic resonances, well exceeding that of many 3D semiconductors [185, 186]. The reduced screening in these 2D materials leads to exciton binding energies of hundreds of meV and they can consequently be harnessed in nanophotonic devices operating at room temperature [187–190]. Polar 2D dielectric materials, such as hBN, support strong optical phonon resonances with long lifetimes. As a result, they can support low-loss and highly confined phonon-polariton modes [191, 192] and can also be patterned to create low-loss metasurface building blocks.

The 2D materials also offer impressive tunability with various external stimuli. Graphene's optical properties can easily be altered by changing the Fermi energy through chemical or electrostatic gating, which is significantly more challenging in conventional metals and semiconductors. Black phosphorus, an attractive material for mid-infrared optoelectronics, also offers notable tunability of its optical properties with electrical gating [193–195] and affords ultrafast, dynamic polarization-dependent optical responses [196]. Many years of research on electro-optical modulation, also indicate that the strongest, high-speed modulation of materials optical properties is achieved by manipulating excitons [197]. Exciton resonances can also effectively be tuned over several 100 meV with changes in the materials composition, environmental index [198–200], electric/magnetic fields [198, 201], and strain [202, 203]. The suppression of exciton states through carrier injection [204–206] can have an even stronger impact on the material optical properties. Effective electrical modulation in graphene [207, 208] and 2D semiconductors materials [209] has already been demonstrated at speeds exceeding tens of GHz. Given the very high carrier mobility of graphene at 200,000 $cm^2/(V{\cdot}s)$ at room temperature and picosecond photocarrier relaxation processes, there is significant room for much higher speeds and ultrafast optical modulation [210]. The rapid ongoing improvements in the growth, processing, and encapsulation of high-quality 2D materials is certain to also further improve the tuning performance.

The noted superior electronic and highly tunable optical properties of 2D materials raise the intriguing question whether they can be used to create dynamic and atomically thin metasurface components. The creation of metasurfaces from graphene first arose from the question what good conductors are for plasmonics [211]. Since then a wide range of metasurfaces with graphene strips, discs, split-ring-resonators, and cones have been developed for the mid-IR and THz ranges [212]. They have demonstrated possibilities to achieve unity absorption, cloaking, polarization control, wavefront manipulation, and been applied to quantum communication, nonlinear optics, and sensing [179, 212]. Based on recent measurements showing near-unity reflectivity from monolayers of $MoSe_2$ [213, 214], it is clear that the interaction with single layer 2D semiconductors can also be extremely strong. Given the strong light–matter interaction, a variety of passive optical elements has already been realized by patterning multilayer 2D TMDCs [215, 216], multilayer graphene [217], 200-nm-thick graphene oxide [218], and even monolayer TMDCs [219]. Complex patterning and integration of graphene with subwavelength cavities have also been pursued to create metasurface devices capable of performing dynamic functions, such as beamsteering [220]. Most recently, active modulation of atomically thin WS_2 lenses has been demonstrated by turning "on" and "off" excitonic resonances in the visible spectral range by electrical gating [148]. This work is now opening the door to create electrically-tunable flat optics for the shorter wavelengths, where phonon resonances in polar dielectrics and plasmon resonances in graphene are not accessible.

Higher diffraction efficiencies and new functions for flat optics based on 2D materials may become achievable by stacking. The 2D materials are easily stacked to create metamaterials with new engineered properties that go beyond those of the individual building blocks. The atomically thin layers are certainly much smaller than the relevant wavelength of light, suggesting that standard homogenization should be allowed to guide the design. However, one should keep in mind the possible strong interactions (some not electromagnetic) between the different layers that can modify the optical properties of the constituent layers. The transdimensional properties of few-layers systems could prove to be a rich ground for discovery by itself [221, 222]. Stacking has already been used to create negative index materials [223], one of the major milestones in metamaterials. Heterostructures can also support highly confined plasmons that facilitate creation of metasurface elements that offer single photon nonlinearities due to the extraordinarily strong light–matter interaction [224].

5 The imminent fusion of 2D metasurface optics and 2D vdW metasurfaces

Achieving high efficiencies for manipulating light with 2D materials is challenging, as the interaction length/time with free-space optical beams is short. Smart geometries, such attenuated total internal reflection setups [225] and integration with cavities [226–228] has been pursued to access strong light–matter coupling physics [229–231]. Plasmonic [232–237] and Mie-resonant [238–240] antennas as well as metasurfaces [241, 242] can also naturally be integrated with 2D materials of the same dimensionality to enhance light–matter interactions without notably altering the properties of the 2D materials. This has been successfully pursued to enhance light absorption, scattering, and emission phenomena [234], whereas conventional metasurfaces can help improve the performance and add functionality to 2D materials, and the reverse has also proven to be extremely valuable. For example, the electrically tunable properties of graphene have been used to activate otherwise-passive plasmonic antennas and metasurfaces [116, 243, 244]. It is clear that there will be many beneficial synergies in bringing together the two classes of 2D materials.

6 Conclusions and outlook

A brief look at the history and tremendous opportunities for flat optics, paint a bright future for the field. The value of combining the rich physics of 2D materials with the design flexibility of 2D metasurfaces is already clearly evident. Much research will be needed on how to best integrate these distinct material classes to allow easy fabrication and maximum synergy. It is exciting to watch the growing importance and diversity in optical resonances that are being used in flat optics, starting from the geometry-controlled plasmon and Mie resonances to materials-based exciton and phonon resonances. As flat optics continues to get flatter, it appears that there is still plenty of room at the bottom.

Acknowledgments: The author acknowledges valuable discussions with and help from Philippe Lalanne, Jorik van de Groep, Qitong Li, and Pieter Kik in the preparation of this manuscript.
Author contribution: The author has accepted responsibility for the entire content of this submitted manuscript and approved submission.
Research funding: For this article, we would like to acknowledge funding from the US Air Force Office of Scientific Research (AFOSR) Multidisciplinary University Research Initiative (MURI) under grant number FA9550-17-1-0002 and individual investigator grants from the Department of Energy (Grant DE-FG07-ER46426) and AFSOR (Grants FA9550-18-1-0323 and FA9550-18-1-0208).
Conflict of interest statement: The authors declare no conflicts of interest regarding this article.

References

[1] N. Yu and F. Capasso, "Flat optics: controlling wavefronts with optical antenna metasurfaces," *IEEE J. Sel. Top. Quant. Electron.*, vol. 19, p. 4700423, 2013.

[2] H. T. Chen, A. J. Taylor, and N. Yu, "A review of metasurfaces: physics and applications," *Rep. Prog. Phys.*, vol. 79, p. 076401, 2016.

[3] A. Y. Zhu, A. I. Kuznetsov, B. Luk'Yanchuk, N. Engheta, and P. Genevet, "Traditional and emerging materials for optical metasurfaces," *Nanophotonics*, vol. 6, pp. 452–471, 2017.

[4] P. Lalanne and P. Chavel, "Metalenses at visible wavelengths: past, present, perspectives," *Laser Photonics Rev.*, vol. 11, p. 1600295, 2017.

[5] A. V. Kildishev, A. Boltasseva, and V. M. Shalaev, "Planar photonics with metasurfaces," *Science*, vol. 339, p. 1232009, 2013.

[6] S. B. Glybovski, S. A. Tretyakov, P. A. Belov, Y. S. Kivshar, and C. R. Simovski, "Metasurfaces: from microwaves to visible," *Phys. Rep.*, vol. 634, pp. 1–72, 2016.

[7] W. E. Kock, "Metal-lens antennas," *Proc. IRE*, vol. 34, pp. 828–836, 1946.

[8] W. E. Kock, "Metallic delay lenses," *Bell Syst. Technol. J.*, vol. 27, pp. 58–82, 1948.

[9] P. Kipfer, M. Collischon, H. Haidner, et al., "Infrared optical components based on a microrelief structure," *Opt. Eng.*, vol. 33, pp. 79–84, 1994.

[10] P. Lalanne, S. Astilean, P. Chavel, E. Cambril, and H. Launois, "Blazed binary subwavelength gratings with efficiencies larger than those of conventional échelette gratings," *Opt. Lett.*, vol. 23, p. 1081, 1998.

[11] P. Lalanne, J. P. Hugonin, and P. Chavel, "Optical properties of deep lamellar gratings: a coupled Bloch-mode insight," *J. Light. Technol.*, vol. 24, pp. 2442–2449, 2006.

[12] C. J. Chang-Hasnain, "High-contrast gratings as a new platform for integrated optoelectronics," *Semicond. Sci. Technol.*, vol. 26, p. 014043, 2011.

[13] F. M. Huang, N. Zheludev, Y. Chen, and F. Javier Garcia De Abajo, "Focusing of light by a nanohole array," *Appl. Phys. Lett.*, vol. 90, p. 091119, 2007.

[14] X. Ni, S. Ishii, A. V Kildishev, and V. M. Shalaev, "Ultra-thin, planar, Babinet-inverted plasmonic metalenses," *Light Sci. Appl.*, vol. 2, p. e72, 2013.

[15] P. B. Catrysse, P. B. Catrysse, Z. Yu, et al., "Planar lenses based on nanoscale slit arrays in a metallic film," *Nano Lett.*, vol. 9, pp. 235–238, 2009.

[16] S. Pancharatnam, "Generalized theory of interference and its applications," *Proc. Indian Acad. Sci.*, vol. A44, pp. 247–262, 1956.

[17] M. V. Berry, "The adiabatic phase and Pancharatnam's phase for polarized light," *J. Mod. Opt.*, vol. 34, pp. 1401–1407, 1987.

[18] Z. Bomzon, V. Kleiner, and E. Hasman, "Formation of radially and azimuthally polarized light using space-variant subwavelength metal stripe gratings," *Appl. Phys. Lett.*, vol. 79, p. 1587, 2001.

[19] Z. Bomzon, G. Biener, V. Kleiner, and E. Hasman, "Space-variant Pancharatnam–Berry phase optical elements with computer-generated subwavelength gratings," *Opt. Lett.*, vol. 27, pp. 1141–1143, 2002.

[20] E. Hasman, V. Kleiner, G. Biener, and A. Niv, "Polarization dependent focusing lens by use of quantized Pancharatnam–Berry phase diffractive optics," *Appl. Phys. Lett.*, vol. 82, pp. 328–330, 2003.

[21] Z. Bomzon, V. Kleiner, and E. Hasman, "Pancharatnam–Berry phase in space-variant polarization-state manipulations with subwavelength gratings," *Opt. Lett.*, vol. 26, pp. 1424–1426, 2001.

[22] A. Pors, M. G. Nielsen, and S. I. Bozhevolnyi, "Plasmonic metagratings for simultaneous determination of Stokes parameters," *Optica*, vol. 2, pp. 716–723, 2015.

[23] J. P. Balthasar Mueller, K. Leosson, and F. Capasso, "Ultracompact metasurface in-line polarimeter," *Optica*, vol. 3, pp. 42–47, 2016.
[24] D. Lin, A. L. Holsteen, E. Maguid, et al., "Polarization-independent metasurface lens employing the Pancharatnam–Berry phase," *Opt. Express*, vol. 26, pp. 24835–24842, 2018.
[25] N. Yu, P. Genevet, M. A. Kats, et al., "Light propagation with phase discontinuities: generalized laws of reflection and refraction," *Science*, vol. 334, pp. 333–337, 2011.
[26] X. Ni, N. K. Emani, A. V Kildishev, A. Boltasseva, and V. M. Shalaev, "Broadband light bending with plasmonic nanoantennas," *Science*, vol. 335, p. 427, 2012.
[27] D. Lin, P. Fan, E. Hasman, and M. L. Brongersma, "Dielectric gradient metasurface optical elements," *Science*, vol. 345, pp. 298–302, 2014.
[28] Y. F. Yu, A. Y. Zhu, R. Paniagua-Domínguez, et al., "High-transmission dielectric metasurface with 2π phase control at visible wavelengths," *Laser Photonics Rev*, vol. 9, pp. 412–418, 2015.
[29] C. F. Bohren and D. R. Huffman, *Absorption and Scattering of Light by Small Particles*, New York, Wiley, 1983.
[30] P. Bharadwaj, B. Deutsch, and L. Novotny, "Optical antennas," *Adv. Opt. Photonics*, vol. 1, pp. 438–483, 2009.
[31] S. I. Bozhevolnyi and T. Søndergaard, "General properties of slow-plasmon resonant nanostructures: nano-antennas and resonators," *Opt. Express*, vol. 15, pp. 10869–10877, 2007.
[32] P. E. Landreman, H. Chalabi, J. Park, and M. L. Brongersma, "Fabry-Perot description for Mie resonances of rectangular dielectric nanowire optical resonators," *Opt. Express*, vol. 24, pp. 29760–29772, 2016.
[33] T. H. Taminiau, F. D. Stefani, and N. F. Van Hulst, "Optical nanorod antennas modeled as cavities for dipolar emitters: evolution of sub- and super-radiant modes," *Nano Lett.*, vol. 11, pp. 1020–1024, 2011.
[34] L. Novotny, "Effective wavelength scaling for optical antennas," *Phys. Rev. Lett.*, vol. 98, p. 266802, 2007.
[35] S. Larouche and D. R. Smith, "Reconciliation of generalized refraction with diffraction theory," *Opt. Lett.*, vol. 37, p. 2391, 2012.
[36] Y. Ra'di, C. R. Simovski, and S. A. Tretyakov, "Thin perfect absorbers for electromagnetic waves: theory, design, and realizations," *Phys. Rev. Appl.*, vol. 3, p. 037001, 2015.
[37] A. Arbabi and A. Faraon, "Fundamental limits of ultrathin metasurfaces," *Sci. Rep.*, vol. 7, p. 43722, 2017.
[38] M. Decker, I. Staude, M. Falkner, et al., "High-efficiency dielectric Huygens' surfaces," *Adv. Opt. Mater.*, vol. 3, pp. 813–820, 2015.
[39] S. Sun, Q. He, S. Xiao, Q. Xu, X. Li, and L. Zhou, "Gradient-index meta-surfaces as a bridge linking propagating waves and surface waves," *Nat. Mater.*, vol. 11, pp. 426–431, 2012.
[40] S. Sun, K. Y. Yang, C. M. Wang, et al., "High-efficiency broadband anomalous reflection by gradient metasurfaces," *Nano Lett.*, vol. 12, pp. 6223–6229, 2012.
[41] A. Pors, O. Albrektsen, I. P. Radko, and S. I. Bozhevolnyi, "Gap plasmon-based metasurfaces for total control of reflected light," *Sci. Rep.*, vol. 3, p. 2155, 2013.
[42] S. Molesky, Z. Lin, A. Y. Piggott, W. Jin, J. Vucković, and A. W. Rodriguez, "Inverse design in nanophotonics," *Nat. Photonics*, vol. 12, pp. 659–670, 2018.
[43] J. A. Fan, "Freeform metasurface design based on topology optimization," *MRS Bull.*, vol. 45, pp. 196–201, 2020.
[44] J. S. Jensen and O. Sigmund, "Topology optimization for nano-photonics," *Laser Photonics Rev*, vol. 5, pp. 308–321, 2011.
[45] W. T. Chen, A. Y. Zhu, and F. Capasso, "Flat optics with dispersion-engineered metasurfaces," *Nat. Rev. Mater.*, vol. 5, pp. 604–620, 2020.
[46] E. Maguid, I. Yulevich, D. Veksler, V. Kleiner, M. L. Brongersma, and E. Hasman, "Photonic spin-controlled multifunctional shared-aperture antenna array," *Science*, vol. 352, pp. 1202–1206, 2016.
[47] S. M. Kamali, A. Arbabi, Y. Horie, M. S. Faraji-Dana, and A. Faraon, "Angle-multiplexed metasurfaces: encoding independent wavefronts in a single metasurface under different illumination angles," *Phys. Rev. X*, vol. 7, p. 041056, 2017.
[48] E. Maguid, I. Yulevich, M. Yannai, V. Kleiner, M. L. Brongersma, and E. Hasman, "Multifunctional interleaved geometric-phase dielectric metasurfaces," *Light Sci. Appl.*, vol. 6, p. e17027-7, 2017.
[49] M. Khorasaninejad and F. M. Capasso, "Versatile multifunctional photonic components," *Science*, vol. 358, p. eaam8100, 2017.
[50] R. Paniagua-Domínguez, Y. F. Yu, E. Khaidarov, et al., "A metalens with a near-unity numerical aperture," *Nano Lett.*, vol. 18, pp. 2124–2132, 2018.
[51] H. Liang, Q. Lin, X. Xie, et al., "Ultrahigh numerical aperture metalens at visible wavelengths," *Nano Lett.*, vol. 18, pp. 4460–4466, 2018.
[52] W. T. Chen, Y. Ibrahim, A. Y. Zhu, and F. Capasso, "Hybrid metasurface-refractive lenses," in *Design and Fabrication Congress OT2A*, vol. 4, 2019.
[53] V. Sanjeev, A. Y. Zhu, V. Sanjeev, et al., "A broadband achromatic metalens for focusing and imaging in the visible," *Nat. Nanotechnol.*, vol. 13, pp. 220–226, 2017.
[54] M. Khorasaninejad, et al., "Metalenses at visible wavelengths: diffraction-limited focusing and subwavelength resolution imaging," *Science*, vol. 352, pp. 1190–1194, 2016.
[55] S. Shrestha, A. C. Overvig, M. Lu, A. Stein, and N. Yu, "Broadband achromatic dielectric metalenses," *Light Sci. Appl.*, vol. 7, p. 85, 2018.
[56] S. Wang, P. C. Wu, V. C. Su, et al., "A broadband achromatic metalens in the visible," *Nat. Nanotechnol.*, vol. 13, pp. 227–232, 2018.
[57] G. Li, S. Zhang, and T. Zentgraf, "Nonlinear photonic metasurfaces," *Nat. Rev. Mater.*, vol. 2, p. 17010, 2017.
[58] A. Krasnok, M. Tymchenko, and A. Alù, "Nonlinear metasurfaces: a paradigm shift in nonlinear optics," *Mater. Today*, vol. 21, pp. 8–21, 2018.
[59] X. Ni, A. V Kildishev, and V. M. Shalaev, "Metasurface holograms for visible light," *Nat. Commun.*, vol. 4, p. 2807, 2013.
[60] G. Zheng, H. Mühlenbernd, M. Kenney, G. Li, T. Zentgraf, S. Zhang, "Metasurface holograms reaching 80% efficiency," *Nat. Nanotechnol.*, vol. 10, pp. 308–312, 2015.
[61] Y. W. Huang, W. T. Chen, W.-Y. Tsai, et al., "Aluminum plasmonic multicolor meta-hologram," *Nano Lett.*, vol. 15, pp. 3122–3127, 2015.
[62] H. Feng, Q. Li, W. Wan, et al., "Spin-switched three-dimensional full-color scenes based on a dielectric meta-hologram," *ACS Photonics*, vol. 6, pp. 2910–2916, 2019.
[63] R. J. Lin, V.-C. Su, S. Wang, et al., "Achromatic metalens array for full-colour light-field imaging," *Nat. Nanotechnol.*, vol. 14, pp. 227–231, 2019.

[64] A. L. Holsteen, D. Lin, I. Kauvar, G. Wetzstein, and M. L. Brongersma, "A light-field metasurface for high-resolution single-particle tracking," *Nano Lett.*, vol. 19, pp. 2267–2271, 2019.

[65] N. Engheta, A. Salandrino, and A. Alù, "Circuit elements at optical frequencies: nanoinductors, nanocapacitors, and nanoresistors," *Phys. Rev. Lett.*, vol. 95, p. 95504, 2005.

[66] H. A. Atwater and A. Polman, "Plasmonics for improved photovoltaic devices," *Nat. Mater.*, vol. 9, pp. 205–213, 2010.

[67] M. Esfandyarpour, E. Garnett, Y. Cui, M. D. McGehee, M. L. B. Mark, and L. Brongersma, "Metamaterial mirrors in optoelectronic devices," *Nat. Nanotechnol.*, vol. 9, pp. 542–547, 2014.

[68] A. Basiri, X. Chen, J. Bai, et al., "Nature-inspired chiral metasurfaces for circular polarization detection and full-Stokes polarimetric measurements," *Light Sci. Appl.*, vol. 8, p. 78, 2019.

[69] S. Yokogawa, S. P. Burgos, and H. A. Atwater, "Plasmonic color filters for CMOS image sensor applications," *Nano Lett.*, vol. 12, pp. 4349–4354, 2012.

[70] M. L. Brongersma, "Plasmonic photodetectors, photovoltaics, and hot-electron devices," *Proc. IEEE*, vol. 104, pp. 2349–2361, 2016.

[71] A. McClung, S. Samudrala, M. Torfeh, M. Mansouree, and A. Arbabi, "Snapshot spectral imaging with parallel metasystems," *Sci. Adv.*, vol. 6, p. eabc7646, 2020.

[72] M. Esfandyarpour, A. G. Curto, P. G. Kik, N. Engheta, and M. L. Brongersma, "Optical emission near a high-impedance mirror," *Nat. Commun.*, vol. 9, p. 3324, 2018.

[73] L. Cao, P. Fan, A. P. Vasudev, et al., "Semiconductor nanowire optical antenna solar absorbers," *Nano Lett.*, vol. 10, pp. 439–445, 2010.

[74] S. J. Kim, P. Fan, J.-H. Kang, and M. L. Brongersma, "Creating semiconductor metafilms with designer absorption spectra," *Nat. Commun.*, vol. 6, p. 7591, 2015.

[75] S. J. Kim, J. Park, M. Esfandyarpour, et al., "Superabsorbing, artificial metal films constructed from semiconductor nanoantennas," *Nano Lett.*, vol. 16, pp. 3801–3808, 2016.

[76] S. J. Kim, I. Thomann, J. Park, J.-H. Kang, A. P. Vasudev, M. L. Brongersma, "Light trapping for solar fuel generation with Mie resonances," *Nano Lett.*, vol. 14, pp. 1446–1452, 2014.

[77] T. J. Kempa, R. W. Day, S.-K. Kim, H.-G. Park, and C. M. Lieber, "Semiconductor nanowires: a platform for exploring limits and concepts for nano-enabled solar cells," *Energy Environ. Sci.*, vol. 6, p. 719, 2013.

[78] L. Cao, P. Fan, E. S. Barnard, A. M. Brown, and M. L. Brongersma, "Tuning the color of silicon nanostructures," *Nano Lett.*, vol. 10, pp. 2649–2654, 2010.

[79] V. Neder, S. L. Luxembourg, and A. Polman, "Efficient colored silicon solar modules using integrated resonant dielectric nanoscatterers," *Appl. Phys. Lett.*, vol. 111, p. 073902, 2017.

[80] W. Yang, et al., "All-dielectric metasurface for high-performance structural color," *Nat. Commun.*, vol. 11, p. 1864, 2020.

[81] S. J. Kim, J.-H. Kang, M. Mutlu, et al., "Anti-Hermitian photodetector facilitating efficient subwavelength photon sorting," *Nat. Commun.*, vol. 9, p. 316, 2018.

[82] S. Yi, M. Zhou, Z. Yu, et al., "Subwavelength angle-sensing photodetectors inspired by directional hearing in small animals," *Nat. Nanotechnol.*, vol. 13, pp. 1143–1147, 2018.

[83] P. Spinelli, M. A. Verschuuren, and A. Polman, "Broadband omnidirectional antireflection coating based on subwavelength surface Mie resonators," *Nat. Commun.*, vol. 3, p. 692, 2012.

[84] E. F. Pecora, A. Cordaro, P. G. Kik, and M. L. Brongersma, "Broadband Antireflection coatings employing multiresonant dielectric metasurfaces," *ACS Photonics*, vol. 5, pp. 4456–4462, 2018.

[85] A. Cordaro, J. van de Groep, S. Raza, E. F. Pecora, F. Priolo, M. L. Brongersma, "Antireflection high-index metasurfaces combining Mie and fabry-pérot resonances," *ACS Photonics*, vol. 6, pp. 453–459, 2019.

[86] A. Vaskin, R. Kolkowski, A. F. Koenderink, and I. Staude, "Light-emitting metasurfaces," *Nanophotonics*, vol. 8, pp. 1151–1198, 2019.

[87] S. Liu, A. Vaskin, S. Addamane, et al., "Light-emitting metasurfaces: simultaneous control of spontaneous emission and far-field radiation," *Nano Lett.*, vol. 18, pp. 6906–6914, 2018.

[88] E. Khaidarov, Z. Liu, R. Paniagua-Domínguez, et al., "Control of LED emission with functional dielectric metasurfaces," *Laser Photonics Rev.*, vol. 14, p. 1900235, 2020.

[89] P. P. Iyer, R. A. DeCrescent, Y. Mohtashami et al., "Unidirectional luminescence from InGaN/GaN quantum-well metasurfaces," *Nat. Photonics*, vol. 14, pp. 543–548, 2020.

[90] R. Klette, *Concise Computer Vision – An Introduction into Theory and Algorithms*, Berlin, Springer, 2014. https://doi.org/10.1007/978-1-4471-6320-6.

[91] S. Ren, K. He, R. Girshick, and J. Sun, "Faster R-CNN : towards real-time object detection with region proposal networks," *IEEE Trans. Pattern Anal. Mach. Intell.*, vol. 39, pp. 1137–1149, 2017.

[92] R. Rithe, P. Raina, N. Ickes, S. V. Tenneti, and A. P. Chandrakasan, "Reconfigurable processor for energy-Efficient computational photography," *IEEE J. Solid State Circuits*, vol. 48, pp. 2908–2919, 2013.

[93] A. Silva, F. Monticone, G. Castaldi, V. Galdi, A. Alu, N. Engheta "Performing mathematical operations with metamaterials," *Science*, vol. 343, pp. 160–164, 2014.

[94] H. Kwon, D. Sounas, A. Cordaro, A. Polman, and A. Alù, "Nonlocal metasurfaces for optical signal processing," *Phys. Rev. Lett.*, vol. 121, p. 173004, 2018.

[95] C. Guo, M. Xiao, M. Minkov, Y. Shi, and S. Fan, "Photonic crystal slab Laplace operator for image differentiation," *Optica*, vol. 5, p. 251, 2018.

[96] A. Pors, M. G. Nielsen, and S. I. Bozhevolnyi, "Analog computing using reflective plasmonic metasurfaces," *Nano Lett.*, vol. 15, pp. 791–797, 2015.

[97] X. Lin, Y. Rivenson, N. T. Yardimci, et al., "All-optical machine learning using diffractive deep neural networks," *Science*, vol. 361, pp. 1004–1008, 2018.

[98] H. Kwon, E. Arbabi, S. M. Kamali, M. S. Faraji-Dana, and A. Faraon, "Single-shot quantitative phase gradient microscopy using a system of multifunctional metasurfaces," *Nat. Photonics*, vol. 14, pp. 109–114, 2020.

[99] Y. Zhou, H. Zheng, I. I. Kravchenko, and J. Valentine, "Flat optics for image differentiation," *Nat. Photonics*, vol. 14, pp. 316–323, 2020.

[100] D. Lee, J. Gwak, T. Badloe, S. Palomba, and J. Rho, "Metasurfaces-based imaging and applications: from

miniaturized optical components to functional imaging platforms," *Nanoscale Adv.*, vol. 2, pp. 605–625, 2020.
[101] Z. Wu, M. Zhou, E. Khoram, B. Liu, and Z. Yu "Neuromorphic metasurface," *Photonics Res.*, vol. 8, pp. 46–50, 2020.
[102] N. I. Zheludev and Y. S. Kivshar, "From metamaterials to metadevices," *Nat. Mater.*, vol. 11, pp. 917–924, 2012.
[103] B. Schwarz, "Lidar: mapping the world in 3D," *Nat. Photonics*, vol. 4, pp. 429–430, 2010.
[104] A. L. Holsteen, S. Raza, P. Fan, P. G. Kik, and M. L. Brongersma, "Purcell effect for active tuning of light scattering from semiconductor optical antennas," *Science*, vol. 358, pp. 1407–1410, 2017.
[105] L. Gao, Y. Zhang, H. Zhang, et al., "Optics and nonlinear buckling mechanics in large-area, highly stretchable arrays of plasmonic nanostructures," *ACS Nano*, vol. 9, pp. 5968–5975, 2015.
[106] N. I. Zheludev and E. Plum, "Reconfigurable nanomechanical photonic metamaterials," *Nat. Nanotechnol.*, vol. 11, pp. 16–22, 2016.
[107] P. Cencillo-abad, J. Ou, E. Plum, and N. I. Zheludev, "Electro-mechanical light modulator based on controlling the interaction of light with a metasurface," *Sci. Rep.*, vol. 7, p. 5404, 2017.
[108] A. Afridi, J. Canet-Ferrer, L. Philippet, J. Osmond, P. Berto, R. Quidant, "Electrically driven varifocal silicon metalens," *ACS Photonics*, vol. 5, pp. 4497–4503, 2018.
[109] E. Arbabi, A. Arbabi, S. M. Kamali, Y. Horie, M. S. Faraji-Dana, A. Faraon, "MEMS-tunable dielectric metasurface lens," *Nat. Commun.*, vol. 9, p. 812, 2018.
[110] Y. Yao, M. A. Kats P. Genevet, et al., "Broad electrical tuning of graphene-loaded plasmonic antennas," *Nano Lett.*, vol. 13, pp. 1257–1264, 2013.
[111] N. K. Emani, T.-F. Chung, X. Ni, A. V. Kildishev Y. P. Chen, A. Boltasseva, "Electrically tunable damping of plasmonic resonances with graphene," *Nano Lett.*, vol. 12, pp. 5202–5206, 2012.
[112] J. Park, J. Kang, S. J. Kim, X. Liu, and M. L. Brongersma, "Dynamic reflection phase and polarization control in metasurfaces," *Nano Lett.*, vol. 17, pp. 407–413, 2017.
[113] Y. Huang, H. W. H. Lee, R. Sokhoyan, et al., "Gate-tunable conducting oxide metasurfaces," *Nano Lett.*, vol. 16, pp. 5319–5325, 2016.
[114] P. P. Iyer, M. Pendharkar, C. J. Palmstrøm, and J. A. Schuller, III–V, "Heterojunction platform for electrically reconfigurable dielectric metasurfaces," *ACS Photonics*, vol. 6, pp. 1345–1350, 2019.
[115] Y. C. Jun, J. Reno, T. Ribaudo, et al., "Epsilon-near-zero strong coupling in metamaterial-semiconductor hybrid structures," *Nano Lett.*, vol. 13, pp. 5391–5396, 2013.
[116] M. C. Sherrott, P. W. C. Hon, K. T. Fountaine, et al., "Experimental demonstration of >230° phase modulation in gate-tunable graphene-gold reconfigurable mid-infrared metasurfaces," *Nano Lett.*, vol. 17, pp. 3027–3034, 2017.
[117] P. C. Wu, R. A. Pala, G. Kafaie Shirmanesh, et al., "Dynamic beam steering with all-dielectric electro-optic III–V multiple-quantum-well metasurfaces," *Nat. Commun.*, vol. 10, p. 3654, 2019.
[118] A. Emboras, J. Niegemann, P. Ma, et al., "Atomic scale plasmonic switch," *Nano Lett.*, vol. 16, pp. 709–714, 2016.
[119] G. Di Martino, S. Tappertzhofen, S. Hofmann, and J. Baumberg, "Nanoscale plasmon-enhanced spectroscopy in memristive switches," *Small*, vol. 12, pp. 1334–1341, 2016.
[120] D. T. Schoen, A. L. Holsteen, and M. L. Brongersma, "Probing the electrical switching of a memristive optical antenna by STEM EELS," *Nat. Commun.*, vol. 7, p. 12162, 2016.
[121] Y. Li, J. Van De Groep, A. A. Talin, and M. L. Brongersma, "Dynamic tuning of gap plasmon resonances using a solid-state electrochromic device," *Nano Lett.*, vol. 19, pp. 7988–7995, 2019.
[122] X. Duan, S. Kamin, and N. Liu, "Dynamic plasmonic colour display," *Nat. Commun.*, vol. 8, p. 14606, 2017.
[123] A. Minovich, J. Farnell, D. N. Neshev, et al., "Liquid crystal based nonlinear fishnet metamaterials," *Appl. Phys. Lett.*, vol. 100, p. 121113, 2012.
[124] O. Buchnev, J. Y. Ou, M. Kaczmarek, N. I. Zheludev, and V. A. Fedotov, "Electro-optical control in a plasmonic metamaterial hybridised with a liquid-crystal cell," *Opt. Express*, vol. 21, p. 1633, 2013.
[125] S. Q. Li, X. Xu, R. Maruthiyodan Veetil, V. Valuckas R. Paniagua-Domínguez, A. I. Kuznetsov, "Phase-only transmissive spatial light modulator based on tunable dielectric metasurface," *Science*, vol. 364, pp. 1087–1090, 2019.
[126] M. Wuttig, H. Bhaskaran, and T. Taubner, "Phase-change materials for non-volatile photonic applications," *Nat. Photonics*, vol. 11, pp. 465–476, 2017.
[127] F. Ding, Y. Yang, and S. I. Bozhevolnyi, "Dynamic metasurfaces using phase-change chalcogenides," *Adv. Opt. Mater.*, vol. 7, p. 1801709, 2019.
[128] B. Gholipour, J. Zhang, K. F. MacDonald, D. W. Hewak, and N. I. Zheludev, "An all-optical, non-volatile, bidirectional, phase-change meta-switch," *Adv. Mater.*, vol. 25, pp. 3050–3054, 2013.
[129] Q. Wang, E. T. F. Rogers, B. Gholipour, et al., "Optically reconfigurable metasurfaces and photonic devices based on phase change materials," *Nat. Photonics*, vol. 10, pp. 60–65, 2015.
[130] J. Park, S. J. Kim, P. Landreman, and M. L. Brongersma, "An over-coupled phase-change metasurface for efficient reflection phase modulation," *Adv. Opt. Mater.*, vol. 8, p. 2000745, 2020.
[131] C. Caloz and Z.-L. Deck-Léger, "Spacetime metamaterials – Part I : general concepts," *IEEE Trans. Antennas Propag.*, vol. 68, pp. 1569–1582, 2020.
[132] C. Caloz, "Spacetime metamaterials – Part II : theory and applications," *IEEE Trans. Antennas Propag.*, vol. 68, pp. 1583–1598, 2020.
[133] A. M. Shaltout, V. M. Shalaev, and M. L. Brongersma, "Spatiotemporal light control with active metasurfaces," *Science*, vol. 364, pp. 374–377, 2019.
[134] V. Pacheco-Peña and N. Engheta, "Effective medium concept in temporal metamaterials," *Nanophotonics*, vol. 9, pp. 379–391, 2020.
[135] Z. Yu and S. Fan, "Complete optical isolation created by indirect interband photonic transitions," *Nat. Photonics*, vol. 3, pp. 91–94, 2009.
[136] M. Notomi and S. Mitsugi, "Wavelength conversion via dynamic refractive index tuning of a cavity," *Phys. Rev. A At. Mol. Opt. Phys.*, vol. 73, p. 051803(R), 2006.

[137] N. Karl, P. P. Vabishchevich, M. R. Shcherbakov, et al., "Frequency conversion in a time-variant dielectric metasurface," *Nano Lett.*, vol. 20, pp. 7052–7058, 2020.

[138] A. M. Shaltout, K. G. Lagoudakis, J. van de Groep, et al., "Spatiotemporal light control with frequency-gradient metasurfaces," *Science*, vol. 365, pp. 374–377, 2019.

[139] N. P. de Leon, M. D. Lukin, and H. Park, "Quantum plasmonic circuits," *IEEE J. Sel. Top. Quantum Electron.*, vol. 18, pp. 1781–1791, 2012.

[140] S. I. Bozhevolnyi and N. A. Mortensen, "Plasmonics for emerging quantum technologies," *Nanophotonics*, vol. 6, pp. 1185–1188, 2017.

[141] R.-J. Shiue, D. K. Efetov G. Grosso, C. Peng, K. C. Fong, D. Englund, "Active 2D materials for on-chip nanophotonics and quantum optics," *Nanophotonics*, vol. 6, pp. 1329–1342, 2017.

[142] S. I. Bogdanov, A. Boltasseva, and V. M. Shalaev, "Overcoming quantum decoherence with plasmonics," *Science*, vol. 364, pp. 532–533, 2019.

[143] P. K. Jha, X. Ni, C. Wu, Y. Wang, and X. Zhang, "Metasurface-enabled remote quantum interference," *Phys. Rev. Lett.*, vol. 115, pp. 1–5, 2015.

[144] T. Stav, A. Faerman, E. Maguid, et al., "Quantum entanglement of the spin and orbital angular momentum of photons using metamaterials," *Science*, vol. 361, pp. 1101–1104, 2018.

[145] R. Bekenstein, I. Pikovski H. Pichler E. Shahmoon, S. F. Yelin, M. D. Lukin, "Quantum metasurfaces with atom arrays," *Nat. Phys.*, vol. 16, pp. 676–681, 2020.

[146] L. Li, Z. Liu X. Ren, et al., "Metalens-array-based high-dimensional and multiphoton quantum source," *Science*, vol. 368, pp. 1487–1490, 2020.

[147] E. Shahmoon, D. S. Wild, M. D. Lukin, and S. F. Yelin, "Cooperative resonances in light scattering from two-dimensional atomic arrays," *Phys. Rev. Lett.*, vol. 118, p. 113601, 2017.

[148] J. van de Groep, J.-H. Song, U. Celano, Q. Li, P. G. Kik M. L. Brongersma "Exciton resonance tuning of an atomically thin lens," *Nat. Photonics*, vol. 14, pp. 426–430, 2020.

[149] J.-J. Greffet, R. Carminati, K. Joulain, J.-P. Mulet, S. Mainguy, Y. Chen, "Coherent emission of light by thermal sources," *Nature*, vol. 416, pp. 61–64, 2002.

[150] L. Cao, J. S. White J.-S. Park et al., "Engineering light absorption in semiconductor nanowire devices," *Nat. Mater.*, vol. 8, pp. 643–647, 2009.

[151] D. Costantini, A. Lefebvre, A. L. Coutrot, et al., "Plasmonic metasurface for directional and frequency-selective thermal emission," *Phys. Rev. Appl.*, vol. 4, pp. 2–7, 2015.

[152] A. Gopinath, S. V Boriskina, B. M. Reinhard, and L. Dal Negro, "Deterministic aperiodic arrays of metal nanoparticles for surface-enhanced Raman scattering (SERS)," *Opt. Express*, vol. 17, pp. 3741–3753, 2009.

[153] S. V. Boriskina, J. K. Tong W.-C. Hsu, et al., "Heat meets light on the nanoscale," *Nanophotonics*, vol. 5, pp. 134–160, 2016.

[154] S. Vassant, A. Archambault, F. Marquier, et al., "Epsilon-near-zero mode for active optoelectronic devices," *Phys. Rev. Lett.*, vol. 109, p. 237401, 2012.

[155] J. Park, J.-H. Kang, X. Liu, et al., "Dynamic thermal emission control with InAs-based plasmonic metasurfaces," *Sci. Adv.*, vol. 4, p. eaat3163, 2018.

[156] H. Chalabi, A. Alù, and M. L. Brongersma, "Focused thermal emission from a nanostructured SiC surface," *Phys. Rev. B Condens. Matter Mater. Phys.*, vol. 94, p. 094307, 2016.

[157] J.-J. Greffet and M. Nieto-Vesperinas, "Field theory for generalized bidirectional reflectivity: derivation of Helmholtz's reciprocity principle and Kirchhoff's law," *J. Opt. Soc. Am. A*, vol. 15, pp. 2735–2743, 1998.

[158] A. P. Raman, M. A. Anoma, L. Zhu, E. Rephaeli, and S. Fan, "Passive radiative cooling below ambient air temperature under direct sunlight," *Nature*, vol. 515, pp. 540–544, 2014.

[159] Y. Kwon, S. H. Song, J. C. Bae, A. Jo, M. Kwon, S. Han, "Metasurface-driven OLED displays beyond 10,000 pixels per inch," *Science*, vol. 370, pp. 459–463, 2020.

[160] M. V. Rybin, K. L. Koshelev, Z. F. Sadrieva, et al., "High-Q supercavity modes in subwavelength dielectric resonators," *Phys. Rev. Lett.*, vol. 119, p. 243901, 2017.

[161] F. Yesilkoy, E. R. Arvelo Y. Jahani, et al., "Ultrasensitive hyperspectral imaging and biodetection enabled by dielectric metasurfaces," *Nat. Photonics*, vol. 13, pp. 390–396, 2019.

[162] S. Campione, S. Liu, L. I. Basilio, et al., "Broken symmetry dielectric resonators for high quality factor fano metasurfaces," *ACS Photonics*, vol. 3, pp. 2362–2367, 2016.

[163] D. H. Lien, S. Z. Uddin M. Yeh, et al., "Electrical suppression of all nonradiative recombination pathways in monolayer semiconductors," *Science*, vol. 364, pp. 468–471, 2019.

[164] A. A. Basharin, V. Chuguevsky, N. Volsky, M. Kafesaki, and E. N. Economou, "Extremely high Q-factor metamaterials due to anapole excitation," *Phys. Rev. B*, vol. 95, p. 035104, 2017.

[165] J. A. Schuller, E. S. Barnard, W. Cai, Y. C. Jun, J. S. White, M. L. Brongersma, "Plasmonics for extreme light concentration and manipulation," *Nat. Mater.*, vol. 9, pp. 193–204, 2010.

[166] A. I. Kuznetsov, A. E. Miroshnichenko, M. L. Brongersma, Y. S. Kivshar, and B. Luk'yanchuk, "Optically resonant dielectric nanostructures," *Science*, vol. 354, p. 2472, 2016.

[167] D. A. B. Miller, "Attojoule optoelectronics for low-energy information processing and communications," *J. Light. Technol.*, vol. 35, pp. 346–396, 2017.

[168] A. K. Geim and K. S. Novoselov, "The rise of graphene," *Nat. Mater.*, vol. 6, pp. 183–191, 2007.

[169] S. Manzeli, D. Ovchinnikov, D. Pasquier, O. V. Yazyev, and A. Kis, "2D transition metal dichalcogenides," *Nat. Rev. Mater.*, vol. 2, p. 17033, 2017.

[170] M. Naguib, V. N. Mochalin, M. W. Barsoum, and Y. Gogotsi, "25th anniversary article: MXenes: a new family of two-dimensional materials," *Adv. Mater.*, vol. 26, pp. 992–1005, 2014.

[171] R. Dong, T. Zhang, and X. Feng, "Interface-assisted synthesis of 2D materials: trend and challenges," *Chem. Rev.*, vol. 118, pp. 6189–6325, 2018.

[172] K. Kang, S. Xie, L. Huang, et al., "High-mobility three-atom-thick semiconducting films with wafer-scale homogeneity," *Nature*, vol. 520, pp. 656–660, 2015.

[173] J. Sun, J. Zhou, B. Li, and F. Kang, "Indefinite permittivity and negative refraction in natural material: graphite," *Appl. Phys. Lett.*, vol. 98, pp. 2009–2012, 2011.

[174] H. Harutyunyan, R. Beams, and L. Novotny, "Controllable optical negative refraction and phase conjugation in graphite thin films," *Nat. Phys.*, vol. 9, pp. 423–425, 2013.

[175] K. S. Novoselov, D. Jiang, F. Schedin, et al., "Two-dimensional atomic crystals," *Proc. Natl. Acad. Sci. U.S.A.*, vol. 102, pp. 10451–10453, 2005.
[176] K. S. Novoselov and A. H. Castro Neto, "Two-dimensional crystals-based heterostructures: materials with tailored properties," *Phys. Scr.*, vol. T146, p. 014006, 2012.
[177] F. Xia, H. Wang, D. Xiao, M. Dubey, and A. Ramasubramaniam, "Two-dimensional material nanophotonics," *Nat. Photonics*, vol. 8, pp. 899–907, 2014.
[178] Q. H. Wang, K. Kalantar-Zadeh, A. Kis, J. N. Coleman, and M. S. Strano, "Electronics and optoelectronics of two-dimensional transition metal dichalcogenides," *Nat. Nanotechnol.*, vol. 7, pp. 699–712, 2012.
[179] Z. Dai, G. Hu, Q. Ou, et al., "Artificial metaphotonics born naturally in two dimensions," *Chem. Rev.*, vol. 120, pp. 6167–6246, 2020.
[180] R. R. Nair, P. Blake, A. N. Grigorenko et al., "Fine structure constant defines visual transparency of graphene," *Science*, vol. 320, p. 1308, 2008.
[181] M. Jablan, H. Buljan, and M. Soljačić, "Plasmonics in graphene at infrared frequencies," *Phys. Rev. B Condens. Matter Mater. Phys.*, vol. 80, p. 245435, 2009.
[182] F. H. L. Koppens, D. E. Chang, and F. J. García De Abajo, "Graphene plasmonics: a platform for strong light-matter interactions," *Nano Lett.*, vol. 11, pp. 3370–3377, 2011.
[183] S. Thongrattanasiri, F. H. L. Koppens, and F. J. García De Abajo, "Complete optical absorption in periodically patterned graphene," *Phys. Rev. Lett.*, vol. 108, p. 047401, 2012.
[184] Y. Li, A. Chernikov, X. Zhang, et al., "Measurement of the optical dielectric function of monolayer transition-metal dichalcogenides: MoS_2, $MoSe_2$, WS_2, and WSe_2," *Phys. Rev. B*, vol. 90, p. 205422, 2014.
[185] G. Eda and S. A. Maier, "Two-dimensional crystals: managing light for optoelectronics," *ACS Nano*, vol. 7, pp. 5660–5665, 2013.
[186] S. Gupta, S. N. Shirodkar, A. Kutana, and B. I. Yakobson, "Pursuit of 2D materials for maximum optical response," *ACS Nano*, vol. 12, pp. 10880–10889, 2018.
[187] K. F. Mak, C. Lee, J. Hone, J. Shan, and T. F. Heinz, "Atomically thin MoS_2: a new direct-gap semiconductor," *Phys. Rev. Lett.*, vol. 105, p. 136805, 2010.
[188] L. Cao, "Two-dimensional transition-metal dichalcogenide materials: toward an age of atomic-scale photonics," *MRS Bull.*, vol. 40, pp. 592–599, 2015.
[189] A. Splendiani, L. Sun, Y. Zhang et al., "Emerging photoluminescence in monolayer MoS_2," *Nano Lett.*, vol. 10, pp. 1271–1275, 2010.
[190] T. Mueller and E. Malic, "Exciton physics and device application of two-dimensional transition metal dichalcogenide semiconductors," *NPJ 2D Mater. Appl.*, vol. 2, p. 29, 2018.
[191] J. D. Caldwell, A. V. Kretinin, Y. Chen, et al., "Sub-diffractional volume-confined polaritons in the natural hyperbolic material hexagonal boron nitride," *Nat. Commun.*, vol. 5, p. 5221, 2014.
[192] S. Dai, Z. Fei, Q. Ma et al., "Tunable phonon polaritions in atomically thin van der Waals crystals of boron nitride," *Science*, vol. 343, p. 1125, 2014.
[193] B. Deng, V. Tran, Y. Xie, et al., "Efficient electrical control of thin-film black phosphorus bandgap," *Nat. Commun.*, vol. 8, p. 14474, 2017.
[194] C. Chen, X. Lu, B. Deng, et al., "Widely tunable mid-infrared light emission in thin-film black phosphorus," *Sci. Adv.*, vol. 6, p. eaay6134, 2020.
[195] W. S. Whitney, M. C. Sherrott, D. Jariwala, et al., "Field effect optoelectronic modulation of quantum-confined carriers in black phosphorus," *Nano Lett.*, vol. 17, pp. 78–84, 2020.
[196] B. Liao, H. Zhao, E. Najafi, et al., "Spatial-Temporal imaging of anisotropic photocarrier dynamics in black phosphorus," *Nano Lett.*, vol. 17, pp. 3675–3680, 2017.
[197] D. A. B. Miller, "Attojoule optoelectronics for low-energy information processing and communications," *J. Light. Technol.*, vol. 35, pp. 346–396, 2017.
[198] A. V. Stier, N. P. Wilson, G. Clark, X. Xu, and S. A. Crooker, "Probing the lifluence of dielectric environment on excitons in monolayer WSe_2: insight from high magnetic fields," *Nano Lett.*, vol. 16, pp. 7054–7060, 2016.
[199] A. Raja, A. Chaves, J. Yu, et al., "Coulomb engineering of the bandgap and excitons in two-dimensional materials," *Nat. Commun.*, vol. 8, p. 15251, 2017.
[200] G. Gupta, S. Kallatt, and K. Majumdar, "Direct observation of giant binding energy modulation of exciton complexes in monolayer $MoSe_2$," *Phys. Rev. B*, vol. 96, p. 081403, 2017.
[201] A. V. Stier, N. P. Wilson, K. A. Velizhanin, J. Kono, X. Xu, S. A. Crooker, "Magnetooptics of exciton Rydberg states in a monolayer semiconductor," *Phys. Rev. Lett.*, vol. 120, p. 057405, 2018.
[202] D. Lloyd, X. Liu, J. W. Christopher et al., "Band gap engineering with ultralarge biaxial strains in suspended monolayer MoS_2," *Nano Lett.*, vol. 16, pp. 5836–5841, 2016.
[203] O. B. Aslan, M. Deng, and T. F. Heinz, "Strain tuning of excitons in monolayer WSe_2," *Phys. Rev. B*, vol. 98, p. 115308, 2018.
[204] J. S. Ross, S. Wu, H. Yu, et al. "Electrical control of neutral and charged excitons in a monolayer semiconductor," *Nat. Commun.*, vol. 4, p. 1474, 2013.
[205] A. Chernikov, A. M. van der Zande, H. M. Hill, et al., "Electrical tuning of exciton binding energies in monolayer WS_2," *Phys. Rev. Lett.*, vol. 115, p. 126802, 2015.
[206] Y. Yu, L. Huang, H. Peng, L. Xiong, L. Cao, "Giant gating tunability of optical refractive index in transition metal dichalcogenide monolayers," *Nano Lett.*, vol. 17, pp. 3613–3618, 2017.
[207] M. Liu, X. Yin, E. Ulin-Avila, et al., "A graphene-based broadband optical modulator," *Nature*, vol. 474, pp. 64–7, 2011.
[208] C. Qiu, W. Gao R. Vajtai, P. M. Ajayan J. Kono Q. Xu, "Efficient modulation of 1.55 μm radiation with gated graphene on a silicon microring resonator," *Nano Lett.*, vol. 14, pp. 6811–6815, 2014.
[209] I. Datta, S. H. Chae G. R. Bhatt et al., "Low-loss composite photonic platform based on 2D semiconductor monolayers," *Nat. Photonics*, vol. 14, pp. 256–262, 2020.
[210] S. Luo, Y. Wang, X. Tong, and Z. Wang, "Graphene-based optical modulators," *Nanoscale Res. Lett.*, vol. 10, p. 199, 2015.
[211] C. M. Soukoulis, T. Koschny, P. Tassin, N. H. Shen, and B. Dastmalchi, "What is a good conductor for metamaterials or plasmonics," *Nanophotonics*, vol. 4, pp. 69–74, 2015.
[212] Y. Fan, N. H. Shen, F. Zhang, et al., "Graphene plasmonics: a platform for 2D optics," *Adv. Opt. Mater.*, vol. 7, pp. 1–14, 2019.
[213] G. Scuri, Y. Zhou, A. A. High, et al., "Large excitonic reflectivity of monolayer $MoSe_2$ encapsulated in hexagonal boron nitride," *Phys. Rev. Lett.*, vol. 120, p. 37402, 2018.

[214] P. Back, S. Zeytinoglu, A. Ijaz, M. Kroner, and A. Imamoğlu, "Realization of an electrically tunable narrow-bandwidth atomically thin MirrorUsing monolayer $MoSe_2$," *Phys. Rev. Lett.*, vol. 120, p. 037401, 2018.
[215] J. Yang, Z. Wang, F. Wang, et al., "Atomically thin optical lenses and gratings," *Light Sci. Appl.*, vol. 5, p. e16046, 2016.
[216] C.-H. Liu, J. Zheng S. Colburn et al., "Ultrathin van der Waals metalenses," *Nano Lett.*, vol. 18, pp. 6961–6966, 2018.
[217] X. T. Kong, A. A. Khan, P. R. Kidambi, et al., "Graphene-based ultrathin flat lenses," *ACS Photonics*, vol. 2, pp. 200–207, 2015.
[218] X. Zheng, B. Jia, H. Lin, L. Qiu, D. Li, M. Gu, "Highly efficient and ultra-broadband graphene oxide ultrathin lenses with three-dimensional subwavelength focusing," *Nat. Commun.*, vol. 6, p. 8433, 2015.
[219] H. Lin, Z. Q. Xu, G. Cao, et al., "Diffraction-limited imaging with monolayer 2D material-based ultrathin flat lenses," *Light Sci. Appl.*, vol. 9, p. 137, 2020.
[220] F. Lu, B. Liu, and S. Shen, "Infrared wavefront control based on graphene metasurfaces," *Adv. Opt. Mater.*, vol. 2, pp. 794–799, 2014.
[221] D. Shah, Z. A. Kudyshev, S. Saha, V. M. Shalaev, and A. Boltasseva, "Transdimensional material platforms for tunable metasurface design," *MRS Bull.*, vol. 45, pp. 188–195, 2020.
[222] C. Hsu, R. Frisenda, R. Schmidt et al., "Thickness-dependent refractive index of 1L, 2L, and 3L MoS_2, $MoSe_2$, WS_2, and WSe_2," *Adv. Opt. Mater.*, vol. 7, p. 1900239, 2019.
[223] A. Al Sayem, M. M. Rahman, M. R. C. Mahdy, I. Jahangir, and M. S. Rahman, "Negative refraction with superior transmission in graphene-hexagonal boron nitride (hBN) multilayer hyper crystal," *Sci. Rep.*, vol. 6, p. 25442, 2016.
[224] A. Woessner, M. B. Lundeberg, Y. Gao, et al., "Highly confined low-loss plasmons in graphene-boron nitride heterostructures," *Nat. Mater.*, vol. 14, pp. 421–425, 2015.
[225] G. Pirruccio, L. Martín Moreno, G. Lozano, and J. Gómez Rivas, "Coherent and broadband enhanced optical absorption in graphene," *ACS Nano*, vol. 7, pp. 4810–4817, 2013.
[226] V. Thareja, J.-H. Kang H. Yuan, et al., "Electrically tunable coherent optical absorption in graphene with ion gel," *Nano Lett.*, vol. 15, pp. 1570–1576, 2015.
[227] L. Zhang, R. Gogna, W. Burg, E. Tutuc, and H. Deng, "Photonic-crystal exciton-polaritons in monolayer semiconductors," *Nat. Commun.*, vol. 9, p. 713, 2018.
[228] S. Wu, S. Buckley, J. R. Schaibley et al., "Monolayer semiconductor nanocavity lasers with ultralow thresholds," *Nature*, vol. 520, pp. 69–72, 2015.
[229] C. Schneider, M. M. Glazov, T. Korn, S. Höfling, and B. Urbaszek, "Two-dimensional semiconductors in the regime of strong light–matter coupling," *Nat. Commun.*, vol. 9, p. 2695, 2018.
[230] D. Ballarini and S. De Liberato, "Polaritonics: from microcavities to sub-wavelength confinement," *Nanophotonics*, vol. 8, pp. 641–654, 2019.
[231] A. Krasnok, S. Lepeshov, and A. Alú, "*Nanophotonics with 2D transition metal dichalcogenides*," *Optic Express*, vol. 26, pp. 2443–2447, 2018.
[232] T. J. Echtermeyer, L. Britnell, P. K. Jasnos, et al., "Strong plasmonic enhancement of photovoltage in graphene," *Nat. Commun.*, vol. 2, p. 458, 2011.
[233] Z. Fang, Z. Liu, Y. Wang, P. M. Ajayan, P. Nordlander, N. J. Halas, "Graphene-antenna sandwich photodetector," *Nano Lett.*, vol. 12, pp. 3808–13, 2012.
[234] Z. Li, Y. Li, T. Han, et al., "Tailoring MoS_2 exciton–plasmon interaction by optical spin–orbit coupling," *ACS Nano*, vol. 11, pp. 1165–1171, 2017.
[235] M. Wang, A. Krasnok, T. Zhang, et al., "Tunable Fano resonance and plasmon–exciton coupling in single Au nanotriangles on monolayer WS_2 at room temperature," *Adv. Mater.*, vol. 30, p. 1705779, 2018.
[236] Z. Wang, Z. Dong, Y. Gu, et al., "Giant photoluminescence enhancement in tungsten-diselenide-gold plasmonic hybrid structures," *Nat. Commun.*, vol. 7, p. 11283, 2016.
[237] J. Fang, D. Wang, C. T. DeVault, et al., "Enhanced graphene photodetector with fractal metasurface," *Nano Lett.*, vol. 17, pp. 57–62, 2017.
[238] A. F. Cihan, A. G. Curto, S. Raza, P. G. Kik, and M. L. Brongersma, "Silicon Mie resonators for highly directional light emission from monolayer MoS_2," *Nat. Photonics*, vol. 12, pp. 284–291, 2018.
[239] D. G. Baranov, D. A. Zuev, S. I. Lepeshov, et al., "All-dielectric nanophotonics: the quest for better materials and fabrication techniques," *Optica*, vol. 4, pp. 814–824, 2017.
[240] S. Lepeshov, A. Krasnok, and A. Alu, "Enhanced excitation and emission from 2D transition metal dichalcogenides with all-dielectric nanoantennas," *Nanotechnology*, vol. 30, p. 254004, 2019.
[241] V. Thareja, M. Esfandyarpour, P. G. Kik, and M. L. Brongersma, "Anisotropic metasurfaces as tunable SERS substrates for 2D materials," *ACS Photonics*, vol. 6, pp. 1996–2004, 2019.
[242] R. Mupparapu, T. Bucher, and I. Staude, "Integration of two-dimensional transition metal dichalcogenides with Mie-resonant dielectric nanostructures," *Adv. Phys. X*, vol. 5, p. 1734083, 2020.
[243] Y. Yao, R. Shankar, M. A. Kats, et al., "Electrically tunable metasurface perfect absorbers for ultrathin mid-infrared optical modulators," *Nano Lett.*, vol. 14, pp. 6526–6532, 2014.
[244] N. K. Emani, T.-F. Chung, X. Ni, A. V. Kildishev, Y. P. Chen A. Boltasseva "Electrically tunable damping of plasmonic resonances with graphene," *Nano Lett.*, vol. 12, pp. 5202–5206, 2012.

Stephanie C. Malek, Adam C. Overvig, Sajan Shrestha and Nanfang Yu*

Active nonlocal metasurfaces

https://doi.org/10.1515/9783110710687-053

Abstract: Actively tunable and reconfigurable wavefront shaping by optical metasurfaces poses a significant technical challenge often requiring unconventional materials engineering and nanofabrication. Most wavefront-shaping metasurfaces can be considered "local" in that their operation depends on the responses of individual meta-units. In contrast, "nonlocal" metasurfaces function based on the modes supported by many adjacent meta-units, resulting in sharp spectral features but typically no spatial control of the outgoing wavefront. Recently, nonlocal metasurfaces based on quasi-bound states in the continuum have been shown to produce designer wavefronts only across the narrow bandwidth of the supported Fano resonance. Here, we leverage the enhanced light-matter interactions associated with sharp Fano resonances to explore the active modulation of optical spectra and wavefronts by refractive-index tuning and mechanical stretching. We experimentally demonstrate proof-of-principle thermo-optically tuned nonlocal metasurfaces made of silicon and numerically demonstrate nonlocal metasurfaces that thermo-optically switch between distinct wavefront shapes. This meta-optics platform for thermally reconfigurable wavefront shaping requires neither unusual materials and fabrication nor active control of individual meta-units.

Keywords: metasurface; nonlocal; optical modulator; quasi-bound states in the continuum.

***Corresponding author: Nanfang Yu,** Department of Applied Physics and Applied Mathematics, Columbia University, New York, NY 10027, USA, E-mail: ny2214@columbia.edu. https://orcid.org/0000-0002-9462-4724
Stephanie C. Malek and Sajan Shrestha, Department of Applied Physics and Applied Mathematics, Columbia University, New York, NY 10027, USA
Adam C. Overvig, Department of Applied Physics and Applied Mathematics, Columbia University, New York, NY 10027, USA; Photonics Initiative, Advanced Science Research Center, City University of New York, New York, NY 10031, USA

1 Introduction

Metasurfaces are optically thin planar structured photonic devices [1, 2]. We can consider metasurfaces to be local if the independent scattering events of individual meta-units dictate device behavior, or nonlocal if many adjacent meta-units support a collective mode that governs the response of the device [3]. Generically, local metasurfaces shape the wavefront across a broad spectrum and may achieve functionalities such as lensing and holography. Nonlocal metasurfaces, in contrast, have sharp spectral control but typically without wavefront shaping capabilities, with prototypical examples including guided mode resonance filters [4] and photonic crystal slabs [5].

A significant ongoing challenge in metasurfaces is active tunability and reconfigurability of device functionality to realize devices such as varifocal metalenses or switchable metasurface holograms [6, 7]. A few emerging mechanisms of active tunability in local metasurfaces include mechanical strain, thermal or electrical control with designer materials, and complex electrical actuation of individual meta-units. By modifying phase profiles via mechanical strain, local metasurfaces on stretchable substrates have yielded mechanically tunable zoom metalenses [8, 9] and switchable metasurface holograms [10]. Electrical methods for actuating these stretchable metasurfaces have been introduced to simplify the mechanical system, including dielectric elastomeric actuators [11]. Phase change materials have been employed in local metasurfaces to realize thermally tunable devices such as a varifocal metalens with Ge-Sb-Se-Te meta-units [12] and dynamic meta-holograms based on vanadium dioxide [13]. Two-dimensional materials have also shown promise for active optics, including a tunable Fresnel zone plate based on the electrically tunable excitonic resonance in monolayer WS_2 [14]. More functionally versatile but electrically cumbersome approaches to tunable metasurfaces require electrical gating of individual meta-units, realizing, for example, one-dimensional beam steering by electrically control of liquid crystals [15] or InSb carrier concentration [16], and two-dimensional beam steering with a network of indium tin oxide–based electrodes [17].

In parallel, recent efforts toward active tuning of nonlocal metasurfaces have demonstrated tuning of the placement and linewidth of spectral features. Mechanical

This article has previously been published in the journal Nanophotonics. Please cite as: S. C. Malek, A. C. Overvig, S. Shrestha and N. Yu "Active nonlocal metasurfaces" *Nanophotonics* 2021, 10. DOI: 10.1515/nanoph-2020-0375.

tuning of the resonant wavelength has been demonstrated in stretchable dielectric photonic crystals [18] and plasmonic lattices [19, 20]. Other approaches include thermal [21] or electrical [22] tuning of photonic crystals. Our previous works computationally demonstrated tunable nonlocal metasurfaces based on quasi-bound states in the continuum (q-BICs), using electro-optic tuning of silicon to tune resonant frequencies [23] and mechanical tuning to control quality factors by active symmetry breaking [24]. Nonlocal metasurfaces have the notable advantage over local metasurfaces of enhanced light-matter interactions due to the long optical lifetime states, increasing the efficacy of minute refractive index changes. However, they have so far been limited to modulating optical spectra, with no active control over the shape of the optical wavefronts.

In our previous works, we have demonstrated theoretically [24, 25] and experimentally [26] a generalization of nonlocal metasurfaces capable of shaping an optical wavefront by spatially varying the polarization properties of q-BICs. In particular, by using a geometric phase associated with the linear [25] or circular [27] dichroism of Fano resonances supported by suitably perturbed photonic crystal slabs, nonlocal metasurfaces have been demonstrated that shape light only across the narrow spectral bandwidth of the resonance. This platform inherits both the enhanced light-matter interactions of the arbitrarily narrow linewidths of q-BICs [28] and the metasurface physics encapsulated by the generalized Snell's law [29].

In this work, we experimentally demonstrate thermo-optic tuning of resonant frequencies in silicon nonlocal metasurfaces based on q-BICs, a simple platform that can be extended to realize more complex nonlocal metasurfaces. We then explore the spatial and spectral reconfigurability of nonlocal metasurfaces. We show that wavefront shaping of narrowband incident light can be turned on and off by shifting the resonant wavelength through refractive-index tuning (Figure 1A), and that both wavefront shape and resonant wavelength can be tuned by

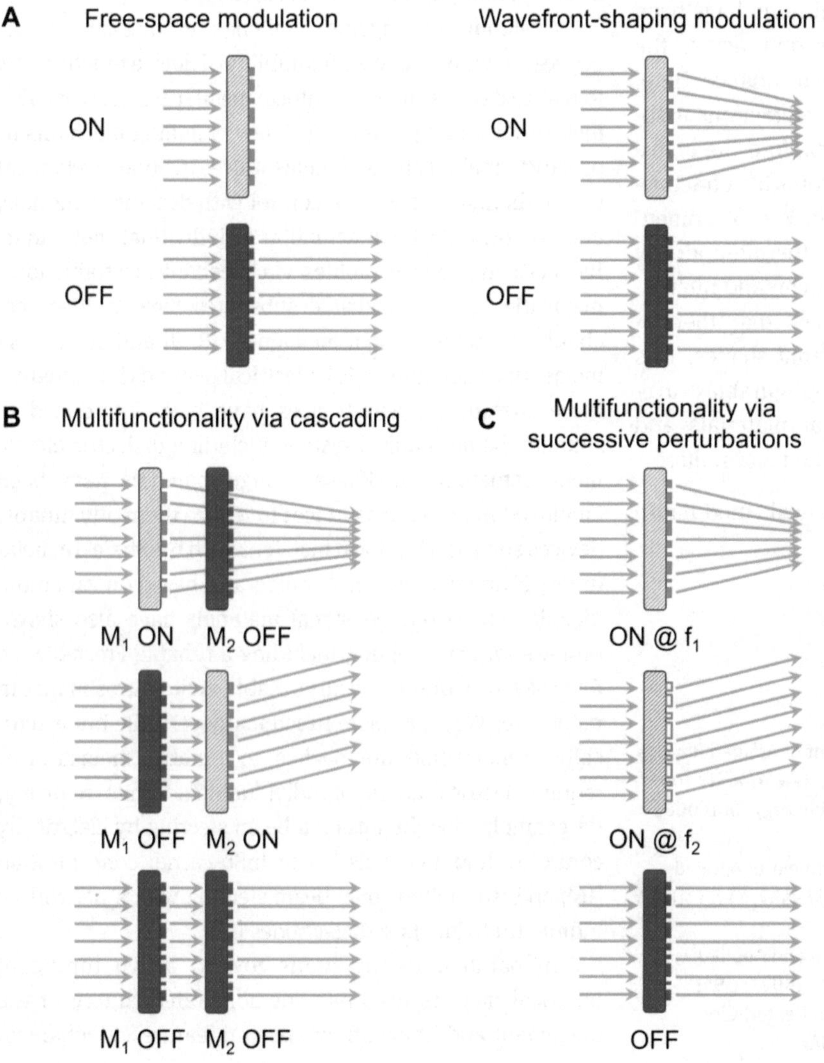

Figure 1: Schematics illustrating several functionalities of nonlocal metasurface modulators. (A) Free-space modulation (left) and wavefront-shaping modulation (right) based on single nonlocal metasurfaces that can be switched on or off when its resonance is aligned with or detuned from the frequency of the incident light. (B) Multifunctional switchable modulation based on cascading nonlocal metasurfaces with distinct wavefront-shaping capabilities at distinct resonant frequencies. (C) Multifunctional switchable modulation based on a single metasurface containing orthogonal perturbations to implement distinct wavefront-shaping capabilities at distinct resonant frequencies.

mechanical strain. Our independently tunable metasurfaces can be cascaded to achieve reconfigurable or tunable multifunctional devices (Figure 1B). Additionally, we demonstrate a multifunctional metasurface exhibiting two distinct functionalities while switching between its two resonances (Figure 1C).

2 Metasurfaces supporting quasi-bound states in the continuum

Bound states in the continuum (BICs) are states that are momentum-matched to free space but are nevertheless bound to a device [30]. BICs can be either accidental in that they incidentally have a coupling coefficient to free space of zero or symmetry protected in that a symmetry incompatibility forbids a mode from coupling to free space. We consider here symmetry-protected BICs that are controlled through a symmetry-breaking perturbation, yielding states (i.e., q-BICs) that radiate to free space with designer polarization dependence and quality factors (Q-factors), which vary inversely with the perturbation strength (δ) as $Q \sim 1/\delta^2$ [23, 28]. We focus on period-doubling perturbations that double the period along a real-space dimension and therefore halve the period in k-space. This effectively folds the bandstructure such that modes that were previously bound and under the light line are folded to the Γ-point, where they can be excited by free-space light at normal incidence. These q-BICs obey selection rules [24] governing whether excitation of a given mode is forbidden or allowed according to the symmetries of the mode, perturbation, and polarization of incoming light. Briefly, light at normal incidence will couple from free space to a dimerized structure only if the out-of-plane component of the mode's E or H field gives an effective net in-plane dipole moment in the direction of the incident polarization. The variety of possible optical responses has been cataloged based on symmetry degeneration from square or hexagonal lattices [24], a process which presents three key insights.

First is a rational design scheme for dimerized photonic crystal slabs with the following steps: (1) choose a real-space mode profile (e.g., large modal overlap with tunable materials) by choice of a photonic crystal structure and one of its high-symmetry modes, (2) adjust geometrical parameters to minimize photonic band curvature, as flat bands are associated with small group velocities and thus reduced device footprints, and (3) choose a type of dimerizing perturbation according to the selection rules [24] to target a specific free-space polarization state, and tune the perturbation strength to achieve desired Q-factors. The second key insight is that metasurfaces with local $p2$ plane group symmetry impart a geometric phase to incident circularly polarized light. The third insight is that we can add successive independent perturbations to a single nonlocal metasurface for realizing multifunctional device operation [25].

The first insight allows us to design free-space metasurface modulators with small footprints, desired Q-factors, and most importantly, deliberate field overlap with the active material. The second insight allows for resonant, wavefront-shaping metasurfaces by tiling meta-units with distinct geometric phase responses. Taken together, these two insights uniquely enable a wavefront-shaping modulator that shapes the wavefront when narrowband incident light is aligned with the resonance but leaves the wavefront shape unaltered when the incident light and resonance are intentionally misaligned by modulation. Finally, adding successive perturbations controlling two q-BICs with a small spectral separation allows us to demonstrate a multifunctional metasurface modulator that can be modulated to switch between distinct wavefront-shaping functionalities.

3 Free-space thermo-optic modulators

The rational design process described above can be leveraged to design free-space modulators based on metasurfaces that provide a large field overlap with the active material. We begin by developing thermo-optic nonlocal metasurface modulators in silicon and then extend this principle to wavefront-shaping nonlocal metasurfaces. Silicon is a common choice of active material with a thermo-optic coefficient of $\sim 2 \times 10^{-4}\ \mathrm{K}^{-1}$ near telecommunications wavelengths [31]. The simplest case of a dimerized nonlocal metasurface is a one-dimensional (1D) grating. In 1D dimerizing perturbation, we can either create a "gap perturbation", where the gap size between grating fingers alternates between two values for every other finger, or a "width perturbation" where the width of every other finger alternates between two values. In both cases, the perturbation doubles the period of the structure in real space, which halves the period in k-space. This effectively folds the first Brillouin zone, resulting in 0th-order diffractive modes excitable from free space (above the light line) in the perturbed structure that were bound modes (below the light line) in the unperturbed structure.

For a silicon metasurface on an insulator substrate, we choose a gap perturbation as shown in Figure 2A and consider the perturbation strength δ to be the difference in the widths of adjacent gaps. The structure supports a flat band (Figure 2B) due to mode hybridization as a result of out-of-plane symmetry breaking by the substrate [32]. The electric-field mode profile (Figure 2A) for excitation polarized parallel to the fingers has a good overlap with the silicon fingers. We fabricate these devices on a silicon-on-insulator substrate with electron beam lithography and dry etch with an alumina hard mask. We demonstrate a set of devices with a footprint of 500 μm × 500 μm and varying perturbation strengths (Figure 2C). The latter control the optical lifetime with smaller perturbations producing higher measured Q-factors of up to $Q \sim 300$ (Figure 2D). Measurements of the reflection spectra of these devices over a 100 °C temperature range show a 4.6-nm shift in the resonant wavelength and an extinction ratio of 2.4 at $\lambda = 1549$ nm (Figure 2E). We note that this device can also be used to enhance third-harmonic generation in argon gas in the gaps between the silicon fingers due to enhanced optical fields there [33].

We also demonstrate polarization-insensitive thermo-optic modulators in silicon. According to the selection-rule catalog [24], polarization-insensitive behavior requires degenerate E-type modes that are preserved by four-fold rotational symmetry. We choose a two-dimensional (2D) structure belonging to the *p4g* plane group as depicted in Figure 3A and adjust the period and fill factor to minimize the band curvature for a mode at the telecommunications wavelengths (Figure 3B). The out-of-plane electric-field profiles on resonance of the degenerate modes for x- and y-polarized incident light show a large modal overlap with the metasurface (Figure 3A). We fabricate 2D devices with a range of perturbation strengths (Figure 2C) and experimentally obtain Q-factors as high as $Q \sim 600$ for a device consisting of rectangular silicon pillars with in-plane dimensions of 505 nm × 425 nm (i.e., $\delta = 80$ nm) (Figure 2D). $Q \sim 600$ is comparable to the highest experimentally realized Q-factors in q-BIC metasurfaces [34–36]. For a device with $Q \sim 290$, measured reflection spectra show a 3.2-nm shift in resonant wavelengths over a 100 °C temperature range and an extinction ratio of 1.18 at $\lambda = 1529$ nm (Figure 3E).

4 Wavefront-shaping modulators

Next, we design thermally tunable wavefront-shaping metasurfaces by choosing a device geometry that provides a large modal overlap with the active material and a spatially tailorable geometric phase [24, 25] using the

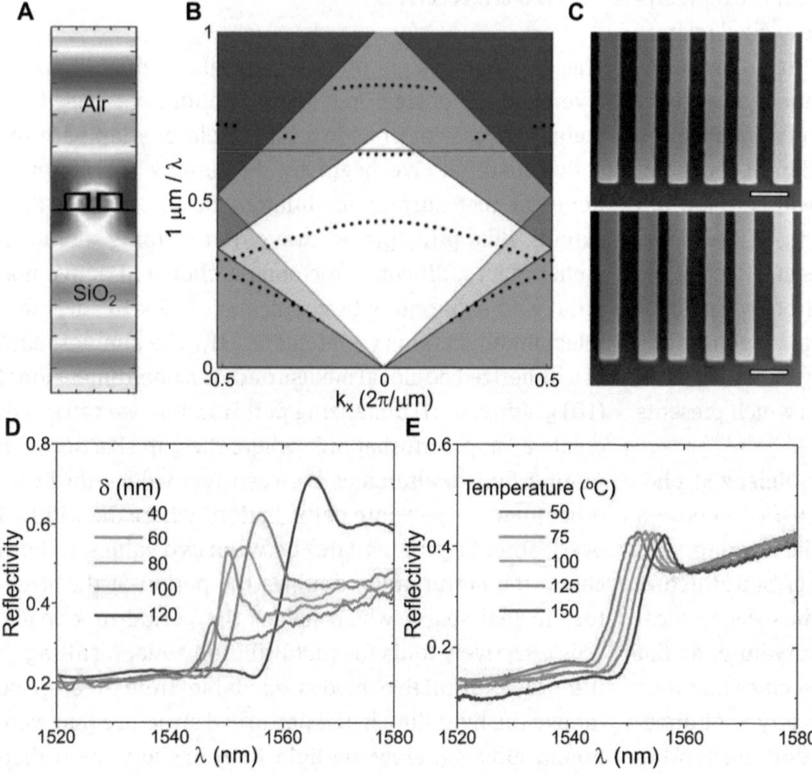

Figure 2: Design and experimental results of one-dimensional (1D) free-space metasurface modulators.
(A) Electric-field profile of a quasi-bound states in the continuum (q-BIC) mode at $\lambda = 1.55$ μm over one period of a "gap perturbed" 1D metasurface. The cross sections of silicon grating fingers are outlined in black and the incident polarization is along the fingers. Device dimensions: period of 950 nm, finger width of 270 nm, finger height of 250 nm, and gap perturbation of $\delta = 20$ nm (i.e., difference in size of adjacent gaps of 20 nm). (B) Band diagram of the 1D metasurface. Red line indicates $\lambda = 1.55$ μm. (C) Scanning electron microscope (SEM) images of fabricated structures with $\delta = 40$ nm (top) and $\delta = 100$ nm (bottom). Scale bar: 500 nm. (D) Measured reflection spectra of devices with different δ. (E) Measured reflection spectra of a device with $\delta = 60$ nm at five different temperatures.

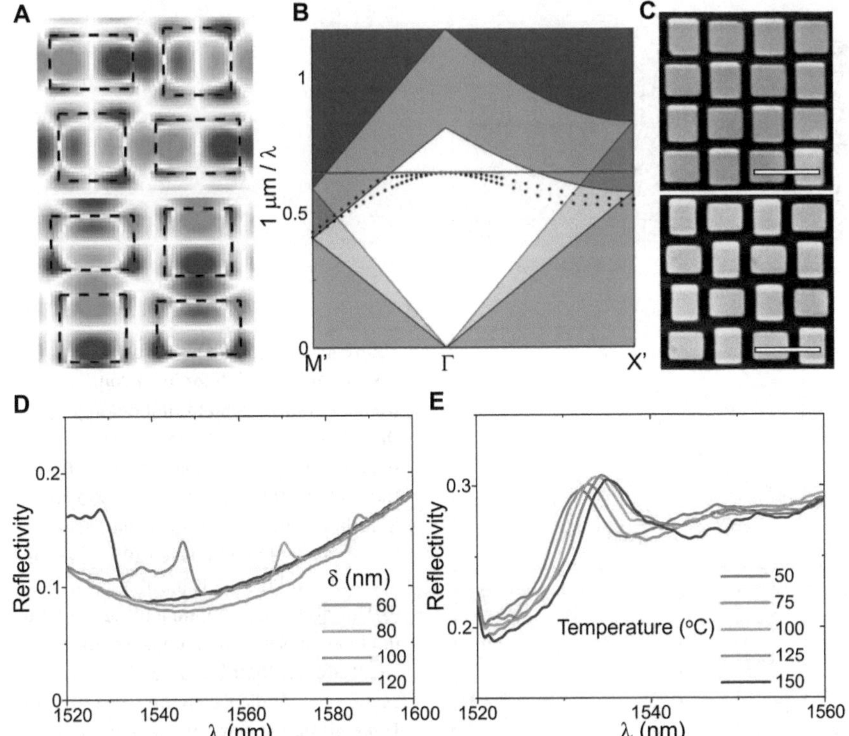

Figure 3: Design and experimental results of two-dimensional (2D) polarization-insensitive metasurface modulators. (A) Mode profiles (i.e., out-of-plane electric-field component of degenerate E-modes) of a 2D structure belonging to the *p4g* plane group excited by *y*-polarized (top) and *x*-polarized (bottom) incident light. (B) Folded band diagram of an unperturbed 2D device consisting of a square lattice of square silicon pillars. Red line indicates $\lambda = 1.55$ µm. Device dimensions: unperturbed period of 650 nm, square width of 450 nm, and silicon thickness of 250 nm. (C) Scanning electron microscope (SEM) images of fabricated 2D devices with $\delta = 60$ nm (505 nm × 445 nm rectangles) (top) and $\delta = 120$ nm (505 nm × 385 nm rectangles) (bottom). Scale bar: 1 µm. (D) Measured reflection spectra of devices with different δ. (E) Measured reflection spectra at five different temperatures of a device with a period of 1300 nm and consisting of rectangular pillars with cross-sectional dimensions of 525 nm × 440 nm.

proof-of-principle platform demonstrated in Figures 2 and 3. Our selection-rule catalog [24] shows that in a meta-unit belonging to the *p2* plane group, for example, two rectangular apertures defined in a thin dielectric film (Figure 4A), the in-plane rotation angle α of the apertures controls the far-field polarization angle ϕ that can couple to a q-BIC mode, such that $\phi \sim 2\alpha$. This relationship between ϕ and α derives from the "parent-child" symmetry relation between higher-symmetry plane groups *pmm* and *pmg* as "parent" groups and *p2* as the "child" group. Comparing the geometry and selection rules of the *pmm* and *pmg* plane groups reveals that a change of the in-plane rotation angle α of 45° gives a 90° change in couplable polarization angle ϕ, leading to the $\phi \sim 2\alpha$ relationship. This linear dichroism introduces a geometric phase twice as much as the conventional geometric phase of local metasurfaces: when circularly polarized light is incident onto the meta-unit, one factor of the geometric phase, $\Phi_{in} = \phi \sim 2\alpha$, is produced from coupling into the linearly polarized q-BIC mode; subsequently, another factor of the geometric phase, $\Phi_{out} = \phi \sim 2\alpha$, is produced when light couples out into the free space and is decomposed into circularly polarized light of opposite handedness. As such, there is a total geometric phase of $\Phi = \Phi_{in}+\Phi_{out} \sim 4\alpha$. However, there is no geometric phase for transmitted light with the same handedness of circular polarization as that of the incident light because $\Phi_{in} = \phi$ and $\Phi_{out} = -\phi$ cancel each other. In other words, we get a geometric phase of $\Phi \sim 4\alpha$ because dichroic elements impart a geometric phase of $\Phi \sim 2\phi$, and these nonlocal metasurfaces have $\phi \sim 2\alpha$. Typical local metasurfaces have a $\phi \sim \alpha$ relation and therefore give a geometric phase $\Phi \sim 2\alpha$. It is also relevant that the maximum transmission efficiency of converted light on resonance is ¼ as this nonlocal metasurface is a four-port system with ½ of the incident light transmitted and ½ of the incident light reflected on resonance where at most ½ of the transmitted light has converted circular polarization.

We begin designing the wavefront-shaping modulator by choosing *p2* structure with rectangular apertures in silicon and selecting a q-BIC transverse magnetic (TM) mode with A_1 symmetry as shown in the transverse and longitudinal cross sections of the mode profile (Figure 4B, C) to ensure a large modal overlap with the active material, silicon, for efficient thermo-optic modulation. For this mode, there is a Lorentzian transmission peak with $Q \sim 150$ for light of converted circular polarization and a dip for light of unconverted circular polarization (Figure 4D). We then confirm that geometric phase follows the $\Phi \sim 4\alpha$ relationship (Figure 4E). Meta-units with different values of α and thus different phase responses can be tiled to form spatially varying phase profiles, creating devices such as lenses and beam deflectors. The resulting devices shape the wavefront only on resonance and only for transmitted light of converted handedness of circular polarization.

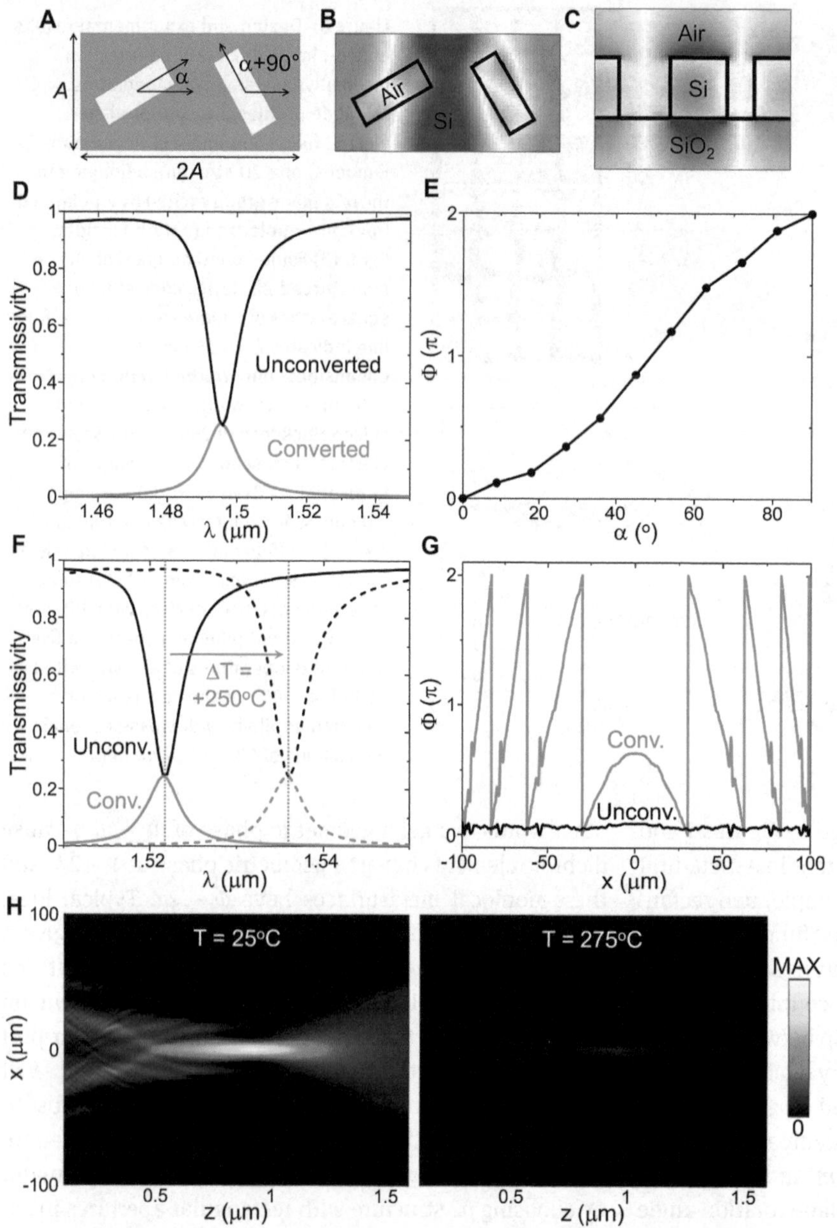

Figure 4: Design of a wavefront-shaping thermo-optic modulator in silicon. (A) Schematic of a meta-unit belonging to the *p*2 plane group. It consists of rectangular apertures defined in a 250-nm-thick silicon thin film. Here, $A = 400$ nm, α is variable, and the aperture dimensions vary linearly with α between 100 × 275 nm at $\alpha = 45°$ and 120 × 235 nm at $\alpha = 0°$ or 90°. (B) Out-of-plane component of electric field (E_z) in a cross section parallel to the substrate. (C) Distribution of E_z in a cross section perpendicular to the substrate. (D) Transmission spectra of one meta-unit (with periodic boundary condition) for light with converted circular polarization (red) and unconverted circular polarization (black) at 25 °C. (E) Geometric phase, Φ, as a function of in-plane rotation angle α, showing a relation of $\Phi \sim 4\alpha$. (F) Transmission spectra of light with converted circular polarization (red) and unconverted circular polarization (black) for a numerical aperture (NA) = 0.1 nonlocal metalens at 25 and 275 °C. (G) Phase profile of the metalens for light with converted circular polarization (red) and unconverted circular polarization (black). (H) Far-field intensity distributions at $\lambda = 1.521$ μm of the metalens at 25 °C (left) and 275 °C (right).

With this meta-unit library, we create a cylindrical metalens with a numerical aperture (NA) of 0.1 and a dimension along the phase profile direction of 200 μm. Simulated transmission spectra of light with converted handedness of the device at 25 °C (n = 3.45) and 275 °C (n = 3.50) show a shift in the resonant wavelength of 14.0 nm and an extinction ratio of 37.9 at λ = 1521 nm (Figure 4F). Simulated far-field distributions of the metalens at λ = 1521 nm (Figure 4H) demonstrate that the device acts as a lens at 25 °C but not at 275 °C, where little light of converted handedness is transmitted. Hence for narrowband incident light, the device exhibits thermally switchable functionalities between that of a lens and that of an unpatterned substrate. We have previously demonstrated experimentally that nonlocal metasurfaces with distinct resonant wavelengths can be cascaded to achieve multifunctional behavior [26]. Cascaded and independently tunable nonlocal metasurface modulators could enable multifunctional switchable behavior by tuning the resonance of each metasurface to be aligned or misaligned with the narrowband incident light (Figure 1B). We caution that aligning the resonances of multiple cascaded metasurfaces to the narrowband incident light does not provide a meaningful use case on account of its low efficiency as each metasurface can transmit only ~25% of the incident light on resonance for each circular polarization.

The pathway for switchable multifunctional thermo-optic modulators on a single metasurface requires successive orthogonal perturbations to control distinct q-BIC modes (Figure 1C). We have previously proposed a scheme of successive orthogonal perturbations that produce independent geometric phases for up to four wavelengths: a single-layered metasurface can generate distinct wavefront shapes at spectrally separated resonances that are associated with orthogonal q-BIC modes [25]. Here, in order to leverage successive perturbations to design a thermally switchable multifunctional nonlocal metasurfaces, we must consider one new design constraint: the spectral spacing between the adjacent orthogonal modes must be small enough for thermo-optic modulation to redshift the "blue" resonance to align with the initial "red" resonance. We satisfy this constraint by beginning with two modes that are degenerate in the unperturbed lattice and apply a set of two perturbations simultaneously to lift the degeneracy and to supply the two modes with distinct selection rules [24]. Figure 5A shows a schematic of our chosen meta-unit consisting of apertures defined in a silicon thin film: gray circular apertures represent the unperturbed structure, and red and blue apertures represent the two orthogonal perturbations. Near telecommunications wavelengths, this meta-unit supports two TM modes (Figure 5B, right panels) each controlled by a separate perturbation such that rotating the red apertures controls the geometric phase of the redshifted mode following an

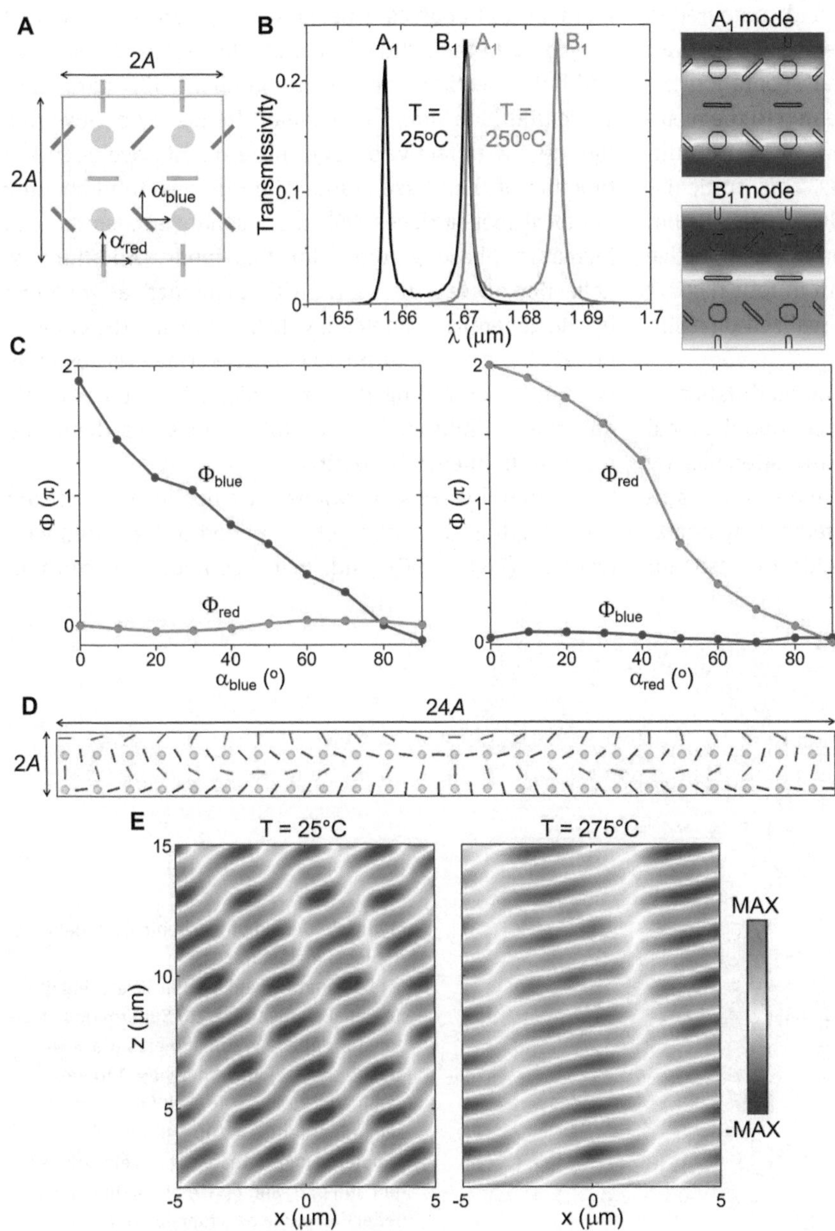

Figure 5: Design of a multifunctional wavefront shaping thermo-optic modulator in silicon.
(A) Composite meta-unit with two orthogonal perturbations (red and blue). Features denote apertures etched in a 250-nm-thick silicon slab. The unperturbed lattice consists of a square array of circular apertures with a diameter of 125 nm and the lattice constant is $A = 900$ nm. The colored rectangular apertures represent two types of perturbations and have a dimension of 175 nm × 25 nm. (B) Left panel: transmission spectra of the meta-unit for light of converted circular polarization at 25 °C (black curve) and 250 °C (magenta curve). Right panels: out-of-plane electric-field components of the "blue" (A_1 type) and "red" (B_1 type) modes. (C) Geometric phases control by the red (right) and blue (left) apertures for the red and blue resonances, respectively. (D) One superperiod of a multifunctional beam deflector where red and blue apertures impart distinct phase gradients to the two modes. (E) Far field electric-field (real part) profiles at $\lambda = 1.649$ μm of the multifunctional beam deflector at 25 and 275 °C.

approximately $\Phi_{red} = 4\alpha_{red}$ relation but has negligible impact on the blueshifted mode (Figure 5C, right panel). Conversely, the blue apertures impart a geometric phase $\Phi_{blue} \sim 4\alpha_{blue}$ to the blueshifted mode but not the redshifted one (Figure 5C, left panel). With this meta-unit library, we devise a device such that each of the orthogonal perturbations is tiled to create a distinct phase profile, leading to anomalous refraction of light to a distinct angle. A schematic superperiod of this device is shown in Figure 5D. The far-field electric-field profiles at $\lambda = 1649$ nm of light with converted circular polarization confirm that light is refracted to a 35-degree angle at 25 °C and a 16.7-degree angle at 275 °C (Figure 5E). In this way, for narrowband incident light, we can thermally switch the function of the nonlocal metasurface. We note that each of the resonances of the superperiod are spectrally blueshifted compared to the meta-unit design by the in-plane *k*-vector of the phase gradient [26]. Consequently, the spectral separation between the modes increases slightly compared to the meta-unit design therefore requiring a larger temperature differential to spectrally overlap the modes. This particular design considered only two distinct q-BIC resonances but with more advanced optimization and potentially an active material affording stronger tunability, up to four distinct resonances and functionalities could be realized on a single metasurface [25].

Finally, we explore wavefront-shaping modulation in a mechanically stretchable system. Most commonly, local metasurfaces on stretchable substrates are stretched by uniform biaxial tensile stress to control the phase distribution and nonlocal metasurfaces are stretched by uniaxial tensile stress to control the *Q*-factor or resonant wavelength. For our nonlocal metasurfaces, we consider nonuniform biaxial strain (Figure 6A). We begin by devising a periodic 1D linear phase profile constructed with a *p*2 meta-unit library with $Q \sim 150$ and consisting of silicon pillars embedded in polydimethylsiloxane, serving as a stretchable polymer substrate and superstrate. We then study the optical response of the embedded device as a function of stretching both along the phase gradient direction (*x*-direction) and orthogonal to it (*y*-direction), as shown in Figure 6B. We exclude shear from our analysis and assume that the silicon pillars are simply displaced in the direction of the applied strain. In this case, *p*2 symmetry is maintained upon stretching and the selection rules governing the geometric phase are unaffected. Stretching along the *x*-direction decreases the magnitude of the phase gradient and enables a beam steering functionality, and stretching in the *y*-direction alters the resonant frequency without affecting the phase gradient (in particular, stretching blueshifts the resonant frequency by lowering the effective refractive index). The resonant wavelength is a function of the phase gradient in our wavefront-shaping nonlocal metasurfaces [25]: the resonance redshifts with increased phase gradient for this mode. Additionally, deflection angle is dispersive with wavelength as governed by the generalized Snell's law [29], following the conventional dispersion of diffractive devices. However, by independently controlling the strain along both the *x* and *y* directions, simultaneous control of phase gradient and resonant frequency is possible.

With full-wave simulations, we calculate the resonant wavelength as a function of the period in the *x* and *y* directions (Figure 6C) and mark contours of constant

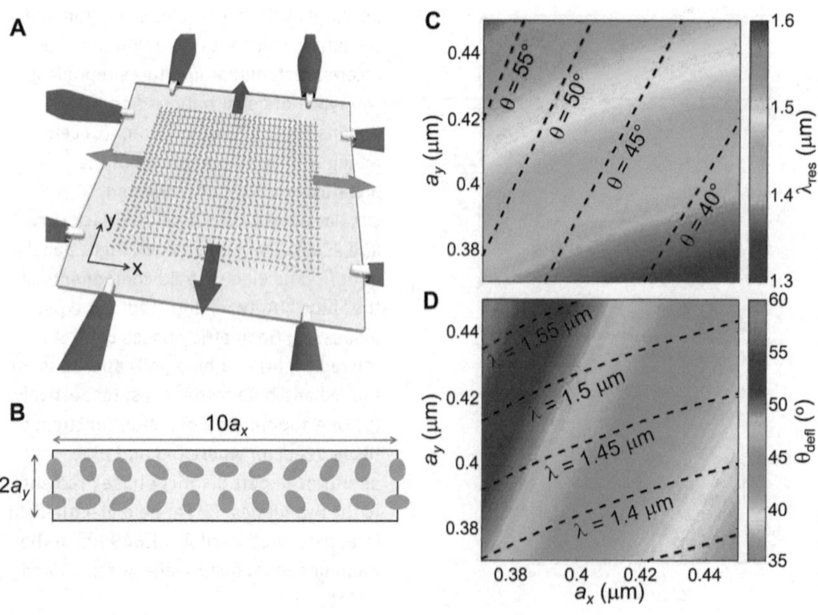

Figure 6: Design of mechanically tunable nonlocal metasurfaces.
(A) Schematic of anisotropic stretching of a nonlocal metasurface. (B) Superperiod of a constant-phase-gradient metasurface with ellipse dimensions of height = 300 nm, semimajor radius = 160–183 nm, and semiminor radius = 100–110 nm. (C) Simulated map of resonant wavelength as a function of a_x and a_y. (D) Simulated map of deflection angle as a function of a_x and a_y.

deflection angle calculated by the generalized Snell's law. We also map deflection angle as a function of period and mark contours of constant resonant wavelength (Figure 6D). These results confirm the prediction of simultaneous control of resonant wavelength and deflection angle, and suggest two use cases for a stretchable nonlocal device: (1) a device that is stretched to deflect different wavelengths of light to a constant angle by following a contour lines in Figure 6C, and (2) a device that is stretched to deflect the same wavelength of light to different angles by following a contour line in Figure 6D. The first case enables devices with reconfigurable operating wavelength but without chromatic dispersion, while the latter case is active beam steering devices for specific operating wavelengths. More generally, these results demonstrate that judiciously stretched nonlocal metasurfaces based on flexible substrates may deflect light to a wide range of angles over a wide range of operating wavelengths. We have therefore demonstrated that nonlocal metasurface platforms may achieve simultaneous spectral and spatial control of light.

5 Conclusion

We have experimentally demonstrated thermo-optic free-space modulators based on nonlocal metasurfaces and designed thermally and mechanically tunable wavefront-shaping nonlocal metasurfaces. Past this proof-of-principle demonstration, multifunctional behavior can be realized on a single metasurface incorporating orthogonal perturbations and cascaded tunable metasurfaces can achieve more complex multifunctionality. We note that these modulators behave differently than conventional tunable local and nonlocal metasurfaces both in the context of refractive-index tuning (Table 1) and mechanical tuning (Table 2), and that the simultaneous control of spectra and wavefront afforded by nonlocal metasurfaces offers unique opportunities for active meta-optics. In particular, thermally tunable nonlocal metasurfaces provide a simpler platform compared with previous tunable local metasurface approaches by eliminating the need for designer materials (e.g., phase change materials), novel nanofabrication techniques, or electrical control of individual meta-units. Future work may incorporate electro-optic tuning or electrically controlled thermo-optic tuning of nonlocal metasurfaces to provide a simple functional pathway for coveted electrically reconfigurable optical metasurfaces modulating both the spectrum and wavefront of light.

Last, we note that the primary limitations of our wavefront-shaping nonlocal metasurfaces are as follows: (1) a maximum efficiency of ~25% that is associated with two instances of conversion of polarization states (i.e., circularly polarized incident light coupling into linearly polarized q-BIC modes, and coupling of the latter to circularly polarized output), and (2) angular dispersion of the resonant wavelength (i.e., a relation between resonant frequency and the magnitude of phase gradient, which is dependent upon the flatness of the band structure),

Table 1: Comparison of properties tuned by refractive-index tuning.

Property	Local geometric phase metasurface	Local truncated waveguide metasurface	Local Huygens metasurfaces [12, 37, 38]	Nonlocal metasurface	Wavefront-shaping nonlocal metasurface
Resonant wavelength (λ_{res})	N/A	N/A	Engineerable	Engineerable	Engineerable
Quality factor (Q)	N/A	N/A	N/A	Trivial	Trivial
Phase distribution ($\phi(x,y)$)	Not engineerable	Limited control (using material dispersion)	Engineerable	N/A	Engineerable (distinct $\phi(x,y)$ at distinct resonances)
Band structure	N/A	N/A	N/A	Trivial	Trivial

Table 2: Comparison of properties tuned by mechanical strain.

Property	Local metasurface	Nonlocal metasurface	Wavefront-shaping nonlocal metasurface
Resonant wavelength (λ_{res})	N/A [8] or trivial [9]	Engineerable [20, 39]	Engineerable
Quality factor (Q)	N/A [8] or trivial [9]	Engineerable [24]	Engineerable
Phase distribution ($\phi(x,y)$)	Engineerable [8–10]	N/A	Engineerable
Band structure	N/A	Engineerable	Engineerable

resulting in a tradeoff between NA and *Q*-factor of devices [24, 25]. The first challenge can be addressed by introducing chirality into nonlocal metasurfaces to achieve near unity efficiency in reflection mode [27], while the second is solvable by sophisticated bandstructure engineering or judicious adjustment of the local effective refractive index to maintain a constant resonant frequency across the device. Overcoming these challenges will enable high efficiency switchable meta-optics such as holograms and high NA lenses for use in display and imaging applications such as augmented reality.

Acknowledgments: The authors thank the Lipson Nanophotonics Group for assistance with optical measurements. The work was supported by the National Science Foundation (grant nos. ECCS-1307948 and QII-TAQS-1936359), the Defense Advanced Research Projects Agency (grant nos. D15AP00111 and HR0011-17-2-0017), and the Air Force Office of Scientific Research (grant no. FA9550-14-1-0389). A.C.O. acknowledges support from the NSF IGERT program (grant no. DGE-1069240). S.C.M acknowledges support from the NSF Graduate Research Fellowship Program (grant no. DGE-1644869).
Author contribution: All the authors have accepted responsibility for the entire content of this submitted manuscript and approved submission.
Research funding: The work was supported by the National Science Foundation (grant nos. ECCS-1307948 and QII-TAQS-1936359), the Defense Advanced Research Projects Agency (grant nos. D15AP00111 and HR0011-17-2-0017), and the Air Force Office of Scientific Research (grant no. FA9550-14-1-0389). A.C.O. acknowledges support from the NSF IGERT program (grant no. DGE-1069240). S.C.M acknowledges support from the NSF Graduate Research Fellowship Program (grant no. DGE-1644869).
Conflict of interest statement: The authors declare no conflicts of interest regarding this article.

References

[1] N. Yu and F. Capasso, "Flat optics with designer metasurfaces," *Nat. Mater.*, vol. 13, pp. 139–150, 2014.
[2] P. Genevet, F. Capasso, F. Aieta, M. Khorasaninejad, and R. Devlin, "Recent advances in planar optics: from plasmonic to dielectric metasurfaces," *Optica*, vol. 4, pp. 139–152, 2017.
[3] H. Kwon, D. Sounas, A. Cordaro, A. Polman, and A. Alù, "Nonlocal metasurfaces for optical signal processing," *Phys. Rev. Lett.*, vol. 121, p. 173004, 2018.
[4] S. Tibuleac and R. Magnusson, "Reflection and transmission guided-mode resonance filters," *J. Opt. Soc. Am. A*, vol. 14, pp. 1617–1626, 1997.
[5] S. G. Johnson, S. Fan, P. R. Villeneuve, J. D. Joannopoulos, and L. A. Kolodziejski, "Guided modes in photonic crystal slabs," *Phys. Rev. B*, vol. 60, pp. 5751–5758, 1999.
[6] I. Kim, G. Yoon, J. Jang, P. Genevet, K. T. Nam, and J. Rho, "Outfitting next generation displays with optical metasurfaces," *ACS Photonics*, vol. 5, pp. 3876–3895, 2018.
[7] Q. He, S. Sun, and L. Zhou, "Tunable/reconfigurable metasurfaces: physics and applications," *Research*, no. 2, pp. 1–16, 2019, https://doi.org/10.34133/2019/1849272.
[8] S. M. Kamali, E. Arbabi, A. Arbabi, Y. Horie, and A. Faraon, "Highly tunable elastic dielectric metasurface lenses," *Laser Photonics Rev.*, vol. 10, pp. 1002–1008, 2016.
[9] H.-S. Ee and R. Agarwal, "Tunable metasurface and flat optical zoom lens on a stretchable substrate," *Nano Lett.*, 2016, https://doi.org/10.1021/acs.nanolett.6b00618.
[10] S. C. Malek, H.-S. Ee, and R. Agarwal, "Strain multiplexed metasurface holograms on a stretchable substrate," *Nano Lett.*, vol. 17, pp. 3641–3645, 2017.
[11] A. She, S. Zhang, S. Shian, D. R. Clarke, and F. Capasso, "Adaptive metalenses with simultaneous electrical control of focal length, astigmatism, and shift," *Sci. Adv.*, vol. 4, p. eaap9957, 2018.
[12] M. Y. Shalaginov, S. An, Y. Zhang, et al., "Reconfigurable all-dielectric metalens with diffraction limited performance," arXiv: 1911.12970 [physics] (2019), https://arxiv.org/abs/1911.12970.
[13] X. Liu, Q. Wang, X. Zhang, et al., "Thermally dependent dynamic meta-holography using a vanadium dioxide integrated metasurface," *Adv. Opt. Mater.*, vol. 7, p. 1900175, 2019.
[14] J. van de Groep, J.-H. Song, U. Celano, Q. Li, P. G. Kik, and M. L. Brongersma, "Exciton resonance tuning of an atomically thin lens," *Nat. Photonics*, vol. 14, pp. 1–5, 2020.
[15] A. I. Kuznetsov, A. E. Miroshnichenko, M. L. Brongersma, Y. S. Kivshar, and B. Luk'yanchuk, "Optically resonant dielectric nanostructures," *Science*, vol. 354, 2016, https://doi.org/10.1126/science.aag2472.
[16] P. P. Iyer, M. Pendharkar, and J. A. Schuller, "Electrically reconfigurable metasurfaces using heterojunction resonators," *Adv. Opt. Mater.*, vol. 4, pp. 1582–1588, 2016.
[17] G. K. Shirmanesh, R. Sokhoyan, P. C. Wu, and H. A. Atwater, "Electro-optically tunable multifunctional metasurfaces," *ACS Nano*, 2020, https://doi.org/10.1021/acsnano.0c01269.
[18] C. L. Yu, H. Kim, N. de Leon, et al., "Stretchable photonic crystal cavity with wide frequency tunability," *Nano Lett.*, vol. 13, pp. 248–252, 2013.
[19] M. L. Tseng, J. Yang, M. Semmlinger, C. Zhang, P. Nordlander, and N. J. Halas, "Two-dimensional active tuning of an aluminum plasmonic array for full-spectrum response," *Nano Lett.*, vol. 17, pp. 6034–6039, 2017.
[20] Y. Cui, J. Zhou, V. A. Tamma, and W. Park, "Dynamic tuning and symmetry lowering of Fano resonance in plasmonic nanostructure," *ACS Nano*, vol. 6, pp. 2385–2393, 2012.
[21] T. Lewi, N. A. Butakov and H. A. Evans, "Thermally reconfigurable meta-optics," *IEEE Photonics J.*, vol. 11, pp. 1–16, 2019.
[22] C. Qiu, J. Chen, Y. Xia, and Q. Xu, "Active dielectric antenna on chip for spatial light modulation," *Sci. Rep.*, vol. 2, p. 855, 2012.
[23] A. C. Overvig, S. Shrestha, and N. Yu, "Dimerized high contrast gratings," *Nanophotonics*, vol. 7, pp. 1157–1168, 2018.
[24] A. C. Overvig, S. C. Malek, M. J. Carter, S. Shrestha, and N. Yu, "Selection rules for quasibound states in the continuum," *Phys. Rev. B*, vol. 102, p. 035434, 2020.

[25] A. C. Overvig, S. C. Malek, and N. Yu, "Multifunctional nonlocal metasurfaces," *Phys. Rev. Lett.*, vol. 125, p. 11125, 2020.

[26] S. C. Malek, A. C. Overvig, S. Shrestha, and N. Yu, "Resonant wavefront-shaping metasurfaces based on quasi-bound states in the continuum," in *Conference on Lasers and Electro-Optics*, 2020.

[27] A. Overvig, N. Yu, and A. Alu, "Chiral quasi-bound states in the continuum," arXiv:2006.05484 [physics] (2020), http://arxiv.org/abs/2006.05484.

[28] K. Koshelev, S. Lepeshov, M. Liu, A. Bogdanov, and Y. Kivshar, "Asymmetric metasurfaces with high-Q resonances governed by bound states in the continuum," *Phys. Rev. Lett.*, vol. 121, p. 193903, 2018.

[29] N. Yu, P. Genevet, M. A. Kats, et al., "Light propagation with phase discontinuities: generalized laws of reflection and refraction," *Science*, vol. 334, pp. 333–337, 2011.

[30] C. W. Hsu, B. Zhen, A. D. Stone, J. D. Joannopoulos, and M. Soljačić, "Bound states in the continuum," *Nat. Rev. Mater.*, vol. 1, pp. 1–13, 2016.

[31] G. Cocorullo, F. G. Della Corte, and I. Rendina, "Temperature dependence of the thermo-optic coefficient in crystalline silicon between room temperature and 550 K at the wavelength of 1523 nm," *Appl. Phys. Lett.*, vol. 74, pp. 3338–3340, 1999.

[32] S. Cueff, F. Dubois, M. S. R. Huang, et al., "Tailoring the local density of optical states and directionality of light emission by symmetry-breaking," *IEEE J. Sel. Top. Quantum Electron*, vol. 25, pp. 1–7, 2019.

[33] J. S. Ginsberg, A. C. Overvig, M. M. Jadidi, et al., "Enhancement of harmonic generation in gases using an all-dielectric metasurface," in *Conference on Lasers and Electro-Optics (2019), Paper FM4M.7*, Optical Society of America, 2019, p. FM4M.7.

[34] S. Campione, S. Liu, L. I. Basilio, et al., "Broken symmetry dielectric resonators for high quality factor Fano metasurfaces," *ACS Photonics*, vol. 3, pp. 2362–2367, 2016.

[35] F. Yesilkoy, E. R. Arvelo, Y. Jahani, et al., "Ultrasensitive hyperspectral imaging and biodetection enabled by dielectric metasurfaces," *Nat. Photonics*, vol. 13, pp. 390–396, 2019.

[36] Y. Yang, I. I. Kravchenko, D. P. Briggs, and J. Valentine, "All-dielectric metasurface analogue of electromagnetically induced transparency," *Nat. Commun.*, vol. 5, p. 5753, 2014.

[37] A. Afridi, J. Canet-Ferrer, L. Philippet, J. Osmond, P. Berto, and R. Quidant, "Electrically driven varifocal silicon metalens," *ACS Photonics*, vol. 5, pp. 4497–4503, 2018.

[38] P. P. Iyer, R. A. DeCrescent, T. Lewi, N. Antonellis, and J. A. Schuller, "Uniform thermo-optic tunability of dielectric metalenses," *Phys. Rev. Appl.*, vol. 10, p. 044029, 2018.

[39] J.-H. Choi, Y.-S. No, J.-P. So, et al., "A high-resolution strain-gauge nanolaser," *Nat. Commun.*, vol. 7, p. 11569, 2016.

Ahmed Mekawy and Andrea Alù*

Giant midinfrared nonlinearity based on multiple quantum well polaritonic metasurfaces

https://doi.org/10.1515/9783110710687-054

Abstract: Ultrathin engineered metasurfaces loaded with multiple quantum wells (MQWs) form a highly efficient platform for nonlinear optics. Here we discuss different approaches to realize mid infrared metasurfaces with localized second-harmonic generation based on optimal metasurface designs integrating engineered MQWs. We first explore the combination of surface lattice resonances and localized electromagnetic resonances in nanoresonators to achieve very large field concentrations. However, when we consider finite size effects, the field enhancement drops significantly together with the conversion efficiency. To overcome this shortcoming, we explore nonetched L-shaped dielectric nanocylinders and etched arrow-shaped nanoresonators that locally support multiple overlapped resonances maximizing the conversion efficiency. In particular, we show the realistic possibility to achieve up to 4.5% efficiency for a normal incident pump intensity of 50 kW/cm^2, stemming from inherently local phenomena, including saturation effects in the MQW. Finally, we present a comparison between pros and cons of each approach. We believe that our study provides new opportunities for designing highly efficient nonlinear responses from metasurfaces (MSs) coupled to MQW and to maximize their impact on technology.

Keywords: metasurface; multiple quantum wells; nonlinear optics.

***Corresponding author: Andrea Alù**, Photonics Initiative, Advanced Science Research Center, City University of New York, New York, NY 10031, USA; Department of Electrical Engineering, City College of The City University of New York, New York, NY 10031, USA; and Physics Program, Graduate Center, City University of New York, New York, NY 10016, USA, E-mail: aalu@gc.cuny.edu. https://orcid.org/0000-0002-4297-5274
Ahmed Mekawy, Photonics Initiative, Advanced Science Research Center, City University of New York, New York, NY 10031, USA; and Department of Electrical Engineering, City College of The City University of New York, New York, NY 10031, USA

1 Introduction

The quest for novel applications in the midinfrared spectral range is rapidly increasing, drawing a lot of attention in recent years. This spectral range is of particular interest for chemical sensing, since the natural absorption lines of various gases and liquids lie in this region [1, 2]. Moreover, atmospheric transmission windows lie within this same frequency range, offering interesting opportunities for enhanced data transmission and a door for outer space communications [3]. Recently, there have been significant interest in exploring midinfrared metasurfaces for different functionalities, like far-field engineering [4, 5] and thermal emission control [6, 7]. Of particular interest in some of these implementations has been the integration of suitably engineered multiple quantum well (MQW) materials with tailored intersubband (ISB) midinfrared transitions efficiently coupled to photons through electromagnetic engineering of metasurfaces, leading to strong polaritonic responses over an ultrathin platform. These surfaces support strongly nonlinear phenomena, ideal for frequency conversion and generation, and addressing the important need for midinfrared sources with broad tunability operating at room temperature, necessary to explore the applications mentioned above [8].

MQWs have been shown to possess high intrinsic nonlinear susceptibility associated with their ISB transitions, several orders of magnitude larger than conventional nonlinear crystals [9–12]. However, the electronic ISB nonlinearity can be excited only by electric fields polarized normal to the MQW barriers [9], which shows the necessity for an intermediate stage whose main function is to tailor the fields of a normal incident pump to the desired direction to leverage this large nonlinearity. Ultrathin metasurfaces (MSs) made of subwavelength plasmonic inclusions have been shown to provide an ideal bridge between the normal incident pump and the ISB dipole moment in two ways. The MSs can be fabricated on the top of the MQW thin layer, and not only they provide polarization conversion but also they can be designed to support multiply-resonant fields at the pump frequency and the desired converted frequencies, enabling large field

This article has previously been published in the journal Nanophotonics. Please cite as: A. Mekawy and A. Alù “Giant midinfrared nonlinearity based on multiple quantum well polaritonic metasurface” *Nanophotonics* 2021, 10. DOI: 10.1515/nanoph-2020-0408.

enhancement in the MQW that strongly and efficiently excites the nonlinear transitions. Compared to conventional nonlinear processes, here the response is highly localized around each individual resonator, and sufficiently strong to relax the need for phase matching and enable the control of nonlinear processes at the subwavelength scale, a property that is truly unprecedented in nonlinear optics. MSs can be hence designed to control and steer at will the nonlinear wavefront. This approach has led to a plethora of interesting applications in the midinfrared range, including record-high second harmonic generation (SHG) [13–15], difference frequency generation [16], third harmonic generation (THG) [17], and more. Nonlinear wavefront manipulation based on controlling the local phase and amplitude, such as in the case of Pancharatnam–Berry MSs coupled to MQW has been also shown for wave steering and focusing [18], and for spin-controlled wave mixing [19].

It should be notes that plasmonic MSs on their own can contribute to enhanced nonlinear processes, due to their intrinsic nonlinear response [20]. However, plasmonic nanoparticles showcase a strong nonlinearity at higher frequencies than the midinfrared range of interest for MQW materials, typically around the frequency range where plasmonic resonances arise [21]. At these frequencies, the field enhancement can reach significant values [22], and many nonlinear applications can be explored, including frequency conversion [23–25]. In general specifically tailored plasmonic inclusions can be considered for different nonlinear applications, for example, split-ring resonators have been studied for SHG because they are not centrosymmetric [26, 27], while dolmen-type structures have been used for THG as they provide high field enhancement with narrow spectral linewidth [28]. In addition, due to the ability of controlling the phase response of the frequency converted wave, a single inclusion can be geometrically tuned for more involved nonlinear applications beyond just frequency conversion, for instance for nonlinear wavefront manipulation. Twisted nanodimer antennas have been used for nonlinear chiral imaging [29], and dipole nanoantennas loaded with nonlinear materials for phase conjugation in reflection and transmission modes [30]. Moreover, MSs based on more complex plasmonic inclusions, such as nano-Kirigami inclusions, were explored for unitary optical circular dichroism of SH conversion [31], V-antennas for nonlinear holography with dual polarizations [32], and star-like antennas for nonlinear optical image encoding [33]. In the following, we also use metallic inclusions loaded by MQWs, but we stress that here the MQW substrate provides the core nonlinear response, while the metallic antennas are used to engineer the optical response, enhance the fields and control the polarization, amplitude and phase of the nonlinearly generated light for different nonlinear applications.

While there have been various approaches to MQW polaritonic metasurfaces, based on T-shaped, gamma-shaped, V-shaped and split-ring resonators [14–19, 34], the optimal approach to achieve the largest generation efficiency remains elusive. This work is an attempt to present and compare a few approaches to polaritonic MS designs leveraging MQWs to achieve large SHG. We propose four different MS designs that we believe span the available design space for SHG. They consist of surface lattice resonances (SLRs) [35], embedded eigenstates [36], nonetched and etched plasmonic resonators supporting two overlapped modes at pump and SH frequency, and dielectric resonators.

2 Discussion

A typical SHG metasurface operating in reflection mode, as considered in the rest of this work, is shown in Figure 1a. The nonlinear semiconductor heterostructure is made of repetitions of MQW layers [11] of total thickness h_{MQW}, as shown in the figure. We aim to synthesize a subwavelength MS unit cell (cartoon shown in the inset) that maximizes the second harnmonic generation (SHG) conversion efficiency. Such inclusion, integrated with the MQW and the back-metal reflector, forms our basic “nanocavity” element. We aim at coupling the electronic ISB transitions in MQWs, defined by the conduction band diagram (see inset) with the electromagnetic resonances provided by the nanocavity to maximize the nonlinear response of the system.

SHG is a nonlinear conversion process achieved when an impinging pump beam at wavelength (frequency) 2λ (ω) excites strong normalized electric fields $\zeta_z^{(\omega)}$ inside the MQW cavity polarized in the z direction. When the nanocavity resonance matches the ISB energy, the MQW layers trigger an ISB transition [9, 11] between two quantized energy levels in the conduction band of the quantum wells, as shown in the inset. This transition induces nonlinear polarization currents that oscillate at 2ω, generating normalized fields $\zeta_z^{(2\omega)}$ radiating with wavelength λ. The efficiency of the radiated SH depends on the spatial field overlap at the pump and SH wavelengths, given by

$$\eta = \frac{\omega^2 h_{\mathrm{MQW}}^2}{2\epsilon_0 c_0^3}\left|\chi_{\mathrm{eff}}^{(2)}\right|^2 I_{\mathrm{inc}}\,; \quad \chi_{\mathrm{eff}}^{(2)} = \frac{1}{V_{\mathrm{MQW}}}\int_{V_{\mathrm{MQW}}} \chi_{zzz}^{(2)}(I_z)\zeta_{\omega,z}^2\zeta_{2\omega,z}dV \tag{1}$$

where V_{MQW} is the MQW volume, h_{MQW} is the MQW height, c_0 is the speed of light, $\zeta_{i,z}$ is the normalized local field at

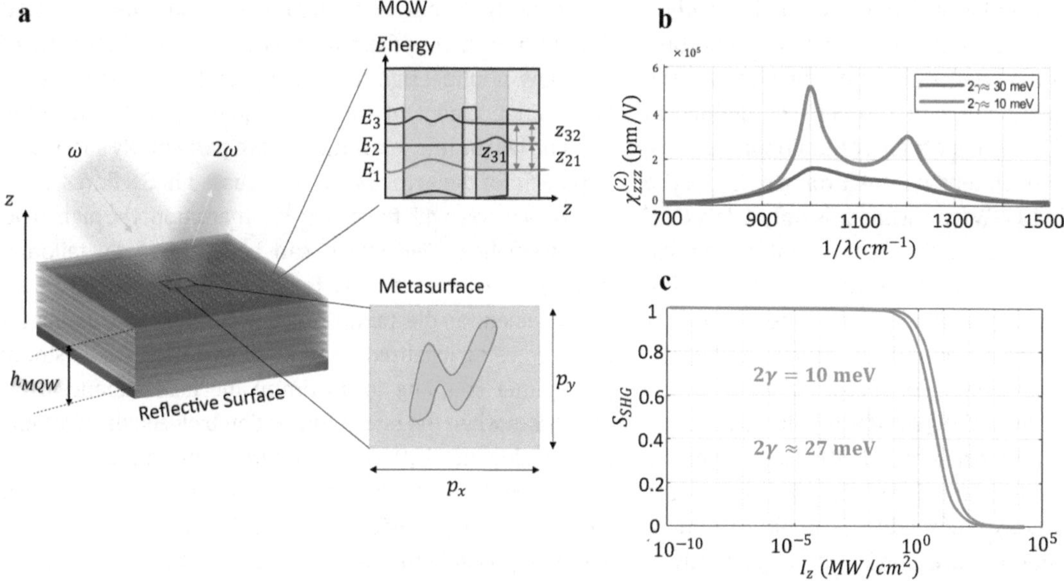

Figure 1: Multiple quantum wells (MQW)-loaded metasurface for second harmonic generation (SHG).
(a) 3D view of the MQW integrated with a metasurface and backed by a metallic reflector for second-harmonic generation in reflection mode. The top panel inset shows the band diagram of the MQW. The bottom panel shows one unit-cell of the metasurface. (b) Nonlinear susceptibility $\chi^{(2)}_{zzz}$ for the bare MQW, and (c) saturation factor S_{SHG} *for* MQW parameters $E_{21} = 150$ meV, $E_{31} = 248$ meV, $z_{12} = -1.6$ nm, $z_{23} = -2.3$ nm, and $z_{13} = -97$ nm for $2\gamma_{12,13,23} = 10$, and 27 meV.

frequency i polarized along z, $\chi^{(2)}_{zzz}$ is the intrinsic tensorial component of the nonlinear susceptibility of the MQW.

Inspecting Eq. (1), we find two main sets of parameters that can control the SHG efficiency. One set corresponds to the MQW material design, identified through the nonlinear susceptibility $\chi^{(2)}_{zzz}$, which can be controlled through material quantum engineering. The second set of parameters that can maximize the nonlinear response is given by the nanocavity geometrical and electromagnetic parameters, which can be aimed at maximizing the overlap integral (1). We start by examining the MQW design parameters that can be optimized for improved SHG efficiency.

2.1 MQW parameters for high SHG efficiency

The nonlinear susceptibility tensor element for SHG at pump frequencies close to the ISB resonances may be approximately written as:

$$\chi^{(2)}_{zzz} = S_{\mathrm{SHG}}\left(I_z\right)\frac{N_e e^3}{\hbar^2\epsilon_0}\frac{z_{12}z_{23}z_{31}}{\left(E_{31} - 2E - i\gamma_{31}\right)\left(E_{21} - E - i\gamma_{21}\right)} \tag{2}$$

where $S_{\mathrm{SHG}}\ (I_z)$ is the factor that accounts for saturation effects in the MQW, I_z is the intensity normal to the MQW barriers, N_e is the average bulk doping density, $\hbar\omega_{ij}$, $2\hbar\gamma_{ij}$ and ez_{ij} are the energy, linewidth and dipole moment, respectively. Saturation in MQWs occurs when the input intensity is so large that the ISB absorption empties the quantum well ground state of energy E_1 [38, 39]. While this effect can be beneficial for other applications involving limiters [40] or ultrafast optical switches [41], it is the limiting factor for the overall nonlinear frequency conversion achievable with this approach. Equation (1) indeed shows that the SHG efficiency scales with the input power, hence as we increase the incident power we would expect increasing conversion efficiency. However, as we crank up the power, saturation effects emerge, and the efficiency starts to decrease. Therefore, there is an optimal intensity for each metasurface design that guarantees maximum efficiency. Saturation effects in MQWs for ultrathin MS can be predicted analytically [34], and the aim of this paper is to design optimal MS, after considering saturation effects, that achieve maximum generation efficiency complaint with practical available midinfrared laser power levels (50–100 kW/cm^2).

Equation (1) shows that, in addition to designing an MS that maximizes the overlap integral, the intrinsic nonlinear susceptibility of the MQW $\chi^{(2)}_{zzz}$ can be increased by narrowing the linewidth. While the transition energies may be fixed given a spectral window of interest, the ISB linewidth is a prominent factor in determining the nonlinear response. For instance, Figure 1b shows the nonlinear susceptibility for two different linewidths keeping the other parameters constants. We find that it reaches 200 nm/V for linewidths of 30 meV and is scaled up by five times when the resonance linewidth is 10 meV. Another important feature of reducing the linewidth of ISB transitions is the corresponding relaxation of saturation effects, defined in the factor S_{SHG} (I). Figure 1c plots the saturation factor S_{SHG} (I) as a function of input intensity. The turning point for the narrow linewidth is 10 MW/cm^2 while it is only 5 MW/cm^2 for 30 meV linewidth. This implies that the saturation will kick in faster in case of a wider linewidth, thereby limiting the overall efficiency. We conclude that it is better to work with smaller linewidths of the ISB transitions, and correspondingly sharper material resonances.

For SHG applications of MQWs, it was believed that the ideal scenario for the most efficient SHG was given by the condition that the 1–2 transition energy E_{12} equals to the 2–3 transition energy E_{23}, referred to as doubly resonant MQW. Such transition corresponds to a permittivity dispersion following a Drude–Lorentzian model with large imaginary part at the transition energy, which increases the losses at the pump [13] and causes faster saturation. It was recently shown, however, that, different from the doubly resonant MQW designs ($E_{21} = E_{32} = E_{31}/2$) used in MQW-based MSs [7, 13], enhanced efficiency can be achieved if the transition energy between states 1 and 2 is purposefully detuned from the optimal pump energy to reduce optical losses in the nanoresonator cavity and delay saturation effects. In this work, we employ MQWs with engineered transition parameters slightly detuned from the pump photon energy to avoid quick saturation and achieve better conversion efficiency.

2.2 Metasurface parameters for high SHG efficiency

Building on previous explorations for MQW parameters that improve SHG efficiency according to (1), there are still several unexplored venues for the MS design that may enhance the efficiency in (1). While a simple inspection of (1) indicates that the MS design should support large field intensity enhancement, recent attempts have been focused on two enhancement strategies: nanoparticle arrays associated with narrow resonances and high-quality factors (Q-factors), for instance, supporting SLRs [37, 42, 43]; multiply-resonant nanostructures for which the resonance enhancement occurs both at pump wavelength and at the SH wavelength [13–15]. However, it is important to realize that high field enhancements at both frequencies will not necessarily result in improved efficiency. For instance, very high field intensities at the pump frequency can cause saturation to kick in sooner degrading the efficiency significantly. It may be preferable to ensure uniform field distributions across the nanocavity to maximize the use of the MQW nonlinearity. If we then resort of maximizing the field enhancement at the SH frequency, saturation phenomena will be less relevant. This is a particularly successful approach for dielectric MSs, as discussed in Section 3.3.

Hence, we will target an MS that supports high field enhancement at 2ω, which can be achieved by tailoring high Q resonances at this frequency, and moderate field enhancement at the pump frequency, which can be easily achieved using localized resonances. We also explore strong light-matter coupling to maximize the field enhancement. This arises when the rate of interaction between the photonic mode inside the MQW cavity and the transition dipole (electronic mode) is faster than the dissipation rate. Therefore, it requires narrow linewidths of the ISB transitions, and it results in high field intensity enhancement [44, 45]. The spatial distribution of these resonances should still be engineered so that the integral in (1) is maximized, i.e., there should be good overlap between the resonant fields at ω and at 2ω.

3 Results and discussion

We explore in this section different MS designs, and characterize their nonlinear response using full-wave simulations. To model the MQW parameters, the ISTs are characterized by an anisotropic Lorentzian oscillator model, whose dielectric tensor is given by $\epsilon = \epsilon_t(\hat{\mathbf{x}}\hat{\mathbf{x}} + \hat{\mathbf{y}}\hat{\mathbf{y}}) + \epsilon_z\hat{\mathbf{z}}\hat{\mathbf{z}}$, where $\epsilon_t \approx 10.24$ and

$$\epsilon_z = \epsilon_t + \frac{N_e e^2}{\epsilon_0 \hbar}\left(\frac{z_{12}^2}{E_{21} - E - i\gamma_{21}} + \frac{z_{13}^2}{E_{31} - E - i\gamma_{31}}\right) \quad (3)$$

Similarly, in our plasmonic designs, gold is used as the material of choice, and its permittivity is taken from experimental data by Olmon [46]. We provide the details on our full-wave simulations, including how we consider saturation effects, in Section 5.

3.1 Surface lattice resonance

As a first example of high Q resonances for MQW metasurfaces, we explore the use of SLRs. They are collective resonances that arise when individual nanostructures couple to in-plane diffraction orders [47, 48]. The most basic SLR arises when the MS periodicity equals the wavelength at the cutoff of the ±1st diffraction orders, forming standing waves whose quality factor depends on the localized surface plasmon resonance (LSPR) of the inclusion [43, 49]. SLRs are typically characterized by high Q

factors, and they have been explored to enhance the weak nonlinear optical responses of conventional MS [43]. The inclusions themselves can support any kind of additional resonance, as desired, which is particularly useful to enhance the overlap integral between fundamental frequency and SH. Here we explore the role of SLRs in enhancing the nonlinear conversion efficiency in MQW MSs by comparing the SHG efficiency between a conventional MS supporting localized resonances at the two frequencies, and one in which we space the same nanocavities by one wavelength at the second-harmonic frequency to sustain an SLR.

In particular, we consider an etched L-shaped nanocavity as the unit cell of our MS, as shown in Figure 2a, top panel, with the MQW sandwiched between a gold reflector and a 50 nm top gold nanoantenna. This geometry was optimized to support a resonance at the SH wavelength $\lambda = 4$ μm and one at the pump wavelength $2\lambda = 8$ μm, as shown in the absorption spectra in the bottom panel. These simulations are assuming a densely populated array, in which the unit cell periods are much smaller than the generated wavelength, as indicated in the figure. We calculated the mode overlap in the integrand of (1) in the middle panel, showing that the nonlinear polarization currents responsible for nonlinear radiation are localized in a small area near the edges of the nanoantenna.

We now modify the dimensions and the period of the nanocavities, as shown in Figure 2b, such that it support an additional SLR with high Q at λ and another resonance with moderate Q at 2λ. This is achieved by simply making the structure in Figure 2a less dense. Since the nanocavities support a localized resonance at 4 μm, the resonances of the individual nanocavities strongly couple with the lattice resonance resulting in a narrower resonance at 5 μm, and a high $Q \approx 100$ SLR at $\lambda = 4$ μm, as shown in the absorption spectra in the bottom panel. The high Q resonance at λ has two advantages: the spatial distribution of this resonance is a standing wave, as shown in Figure 2b top panel, and this leads to a good mode overlap that extends over a larger area, as shown in the middle panel, when compared to the

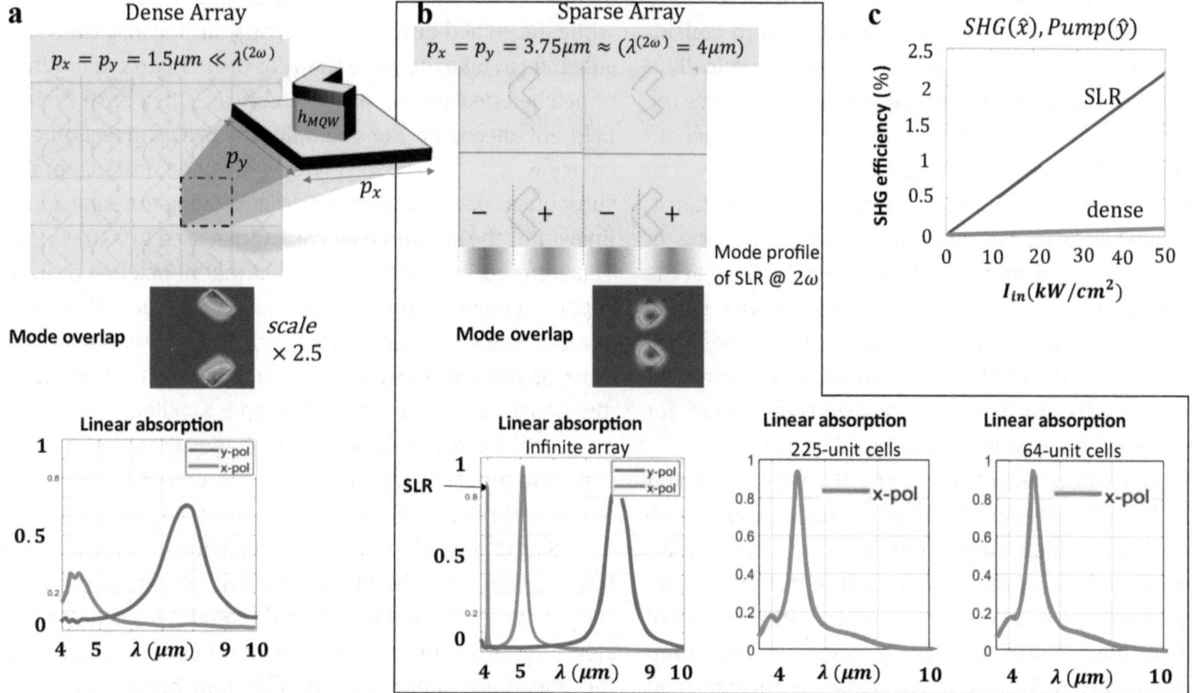

Figure 2: Second harmonic generation (SHG) efficiency from dense metasurfaces supporting a dual resonance, and sparse metasurfaces with additional surface lattice resonance.
(a) Dense array (top panel) supporting dual orthogonal resonances, x-polarized resonance at ω and y-polarized resonance at 2ω (bottom panel), and mode overlap at ω and 2ω at $h_{MQW}/2$ (middle panel). The arm length of the L-shaped nanocavity is 0.6 μm and its width is 0.4 μm, $h_{MQW} = 0.4$ μm, top gold thickness is 50 nm. (b) Sparse array with electric filed distribution at 2ω plotted in the bottom of the figure (top panel) supporting x-polarized resonance at ω, and surface lattice resonance at 2ω for infinite array (bottom panel), and mode overlap between modes at ω and 2ω at $h_{MQW}/2$ (middle panel). The linear absorption for finite array size is shown in the bottom panel for 225 unit cells, and 64 unit cells illuminated. The arm length of the L-shaped nanocavity is 0.755 μm, its width is 0.4 μm, and $h_{MQW} = 0.4$ μm, top gold thickness is 50 nm. (c) $\hat{y}$ polarized second-harmonic generation efficiency for the sparse and dense array at 4 μm and $\hat{x}$-polarized pump at 8 μm with (bottom panel). MQW parameters same as in a.

dense array. In addition, due to its high Q nature, this resonance supports high field intensity enhancement. The drawback is of course that the number of resonators per unit surface is reduced in this second example.

In order to show the difference between the dense and sparse array, we plot the SHG conversion efficiency, neglecting saturation effects for the moment, in Figure 2c for pump wavelength $\lambda = 8$ µm polarized along $\hat{y}$ and SH at $\lambda = 4$ µm polarized along $\hat{x}$ (notice that inversion symmetry is broken for the tensorial vector xyy). The conversion efficiency of the sparse array is 26 times larger than the dense array, opening interesting perspectives for nonlinear generation. These results are in line with the results in a study by Czaplicki et al. [37], where it was shown that the two enhancement properties mentioned above, enhanced overlap integral and enhanced fields, improve the nonlinear properties even if at the cost of making the array less dense. As an additional advantage, when saturation phenomena are considered, we expect a reduced sensitivity to saturation in the SLR scenario because of the more spread overlap integral of the resonant fields.

There are however a couple of drawbacks in considering SLR MSs: first, the pump laser power is typically a finite Gaussian beam focused on the surface to enhance the local intensity. Hence, especially for midinfrared excitations, for which power levels are limited, we can expect to illuminate only a finite number of unit cells. For instance, the available data for the pump focal spot on the MSs in similarly coupled plasmonic MQW structure is 35 µm at the laser wavelength of 9 µm [13, 14]. Second, the lattice resonance inherently relies on a collective, highly nonlocal effect, and hence does not allow controlling the nonlinear wavefront at will as in the case of localized resonances, for instance using the approach introduced in [18].

We study the effect of truncating the number of unit cells in the array on the SLR quality factor in Figure 2b bottom panel, which shows the linear absorption for infinite arrays, 225 unit cells and 64 unit cells, therefore we effectively expect a focal spot illumination with diameter of 56 and 30 µm, respectively. These values of the pump illumination beam focal spot are close to the reported values used in experiments [13, 14]. The high-Q resonance of the infinite array rapidly fades away as we decrease the number of unit cells illuminated in the array. We believe this is a fundamental limiting factor for SLR MS in our applications. The typical experiments reported by our group and others for MQW MSs have been typically excited by focused beams with significantly smaller regions than those required to observe the advantages of an SLR.

To overcome this shortcoming, we consider next MS designs based on localized resonances at λ and at 2λ. In order to enhance the local density of states and hence the nonlinear response, we consider engineered MQW responses with ISB transitions with a linewidth of 10 meV, in line with recent experimental realizations of high-quality MQWs [44, 50]. In these designs, we target SHG in the wavelength range $\lambda \approx 5.65$–5.9 µm and consequently pump wavelength range $2\lambda \approx 11.3$–11.8 µm.

3.2 Nonetched L-shaped nanocavities

The development of MQW-loaded MSs for nonlinear harmonic generation has gone through several stages. The first examples were based on nonetched designs, as in [13], in order to maximize the nonlinear material within the substrate of the metasurface. In later stages, etched designs were considered [14–19], noticing that the reduced material volume was largely compensated by enhanced field confinement and overlapped resonances. The nonetched designs typically rely on metallic nanoantennas patterned on the top of an MQW layer, as in the geometry of Figure 2b, while the etched designs consider patterning also the MQW substrate to take the same shape as the nanoantenna. While nonetched designs are easy to fabricate, they have limited field enhancement. For example, the SHG conversion efficiency was 2×10^{-6}% in [13] for nonetched designs, but in subsequent optimized etched designs using the same MQW linewidths the measured efficiency increased to 0.075% [14]. This difference is rooted in the fact that nonetched designs support a mode volume larger than etched ones. This leads to reduced filed intensity close to the nanoantennas, where most of the nonlinear generation occurs. In addition, as mentioned above, the mode overlap is smaller.

Consider for instance an L-shape unetched MS, as shown in the inset of Figure 3a. The structure is optimized to support plasmonic resonances at $2\lambda = 11.61$ µm and at $\lambda = 5.8$ µm for an MQW with ISB at 5.9 µm as shown in the linear absorption in Figure 3a. The mode splitting at 5.9 µm (the ISB wavelength is denoted by red dots) is a clear signature of strong polaritonic coupling between the plasmonic mode and the ISB, and hence high field intensity enhancement. We calculate the conversion efficiency for this structure, considering also saturation effects in the MQW, for pump wavelength $\lambda = 11.8$ µm polarized along $\hat{x}$ and SH at $\lambda = 5.9$ µm polarized also along $\hat{x}$ (notice that inversion symmetry is broken for the tensorial vector xxx), as shown in Figure 3b. The conversion efficiency is 0.6% at pump power 20 kW/cm^2, much larger than the one reported in a study by Lee et al. [14] because of the improved design and the sharper transition linewidth of 10 meV.

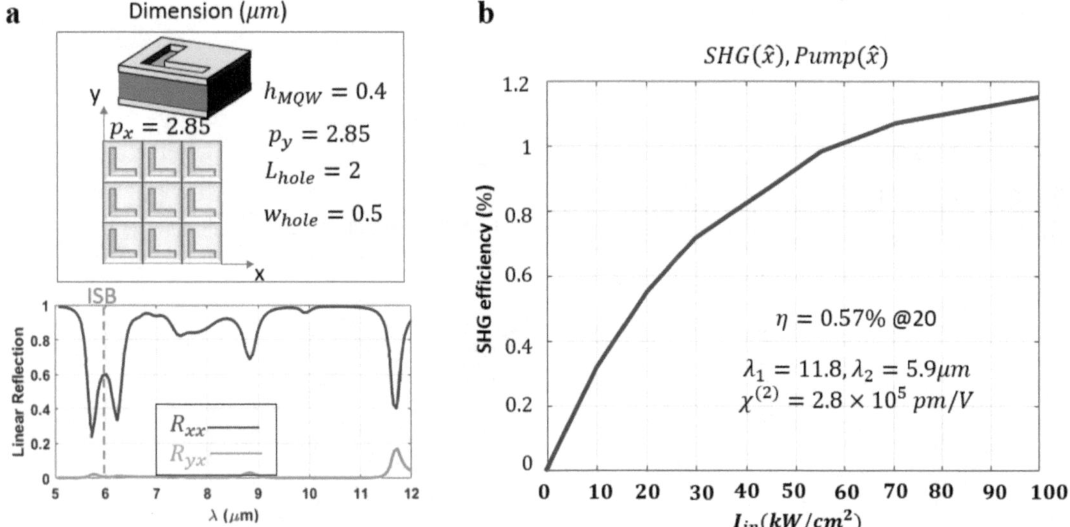

Figure 3: Second harmonic generation (SHG) efficiency for nonetched plasmonic metasurface.
(a) Linear copolymerization and crosspolarization reflection for an x-polarized input wave. The inset shows a 3D view of one unit cell, and 2D view of the metasurface looking top down with dimensions in μm. (b) SHG efficiency for $\hat{x}$-polarized pump wavelength 11.8 μm and $\hat{x}$-polarized second-harmonic wavelength 5.9 μm for the metasurface in (a) with inset showing the multiple quantum well (MQW) parameter, and the efficiency at 20 kW/cm^2. MQW parameters are $E_{13} = 207.8$ meV, $E_{21} = 157.3$ meV, $2\gamma = 10$ meV, $z_{12} = -1.76$ nm, $z_{23} = -4.37$ nm and $z_{13} = -1.22$ nm.

3.3 Etched MQWs without metallic nanoantennas

Recently, all-dielectric MSs have emerged as a promising platform to realize exotic light-matter interactions avoiding the inevitable Ohmic losses in metals. As an additional advantage, all-dielectric structures can stand higher power intensities before their breakdown compared to the melting of metallic nanostructures [51]. MQWs are particularly well suited in this regard, since they naturally support a large index of refraction, ideal for field confinement without metal based on Mie resonances. We explore here the possibility of SHG from an all-dielectric MQW metasurface obtained by etching the substrate but avoiding the metallic nanoantenna on top of each etched region. The resulting optimized nanocavity is shown in the inset of Figure 4a. The geometry is optimized to support resonances at fundamental and second-harmonic frequency, and enhancement of the z-polarized field in the MQW volume.

We study the response of these all-dielectric metasurfaces in two scenarios. First, we design the structure to support resonances at λ and 2λ. Second, we slightly detune the resonance at 2λ (the dimensions for this geometry are shown in the inset). In general, the MQW is modeled as an anisotropic dielectric, as given by Eq. (3). The dimensions are optimized to give the desired resonance at λ and 2λ. To easily quantify the resonances of such a material we first assume that there is no ISB transition, i.e., $N_e = 0$ in Eq. (3). In this case, the MQW becomes a uniaxial lossless material with refractive index $n_e \approx n_o \approx n_{\mathrm{MQW}} = 3.2$. From there we can easily choose the dimensions of a disk that support a dielectric Mie resonance at the desired frequency. Once we chose the geometry, we introduce in the model the ISB transition in Eq. (3) and perform full-wave simulations. After optimizing the structure dimension, we see that the first geometry leads to larger field enhancement at the pump wavelength 2λ, as expected (see inset of Figure 4b, for the field enhancement for the first geometry on the right, and for the second design on the left). The linear absorption for the detuned geometry is shown in Figure 4a, and it can be readily seen that, due to the broken symmetry of the dielectric, it supports many resonances that can be accessed by a normally incident wave. The reason for the multiple resonances observed in the spectrum is that we need to integrate two resonances at λ and around 2λ in one dielectric inclusion. Hence, the dimensions will be determined by the larger wavelength and, as a result, there are multiple dielectric modes supported at shorter wavelengths. A clear evidence of the supported resonance at $\lambda = 6$ μm is provided by the Rabi splitting associated with the strong polaritonic coupling of electromagnetic and material resonances (the ISB wavelength is denoted by red dots). In addition, it shows a detuned resonance at 10.5 μm. In Figure 4b, we plot the SHG conversion efficiency for this geometry for pump wavelength $\lambda = 11.61$ μm polarized in the direction $\hat{r} = 1/\sqrt{2}(\hat{y} + \hat{x})$ and SH at $\lambda = 5.8$ μm polarized in the $\hat{x}$ direction (notice that inversion symmetry is also here broken for the tensorial vector xrr), and we compare the result with the case in which the fundamental resonance is tuned with the pump, leading to higher field

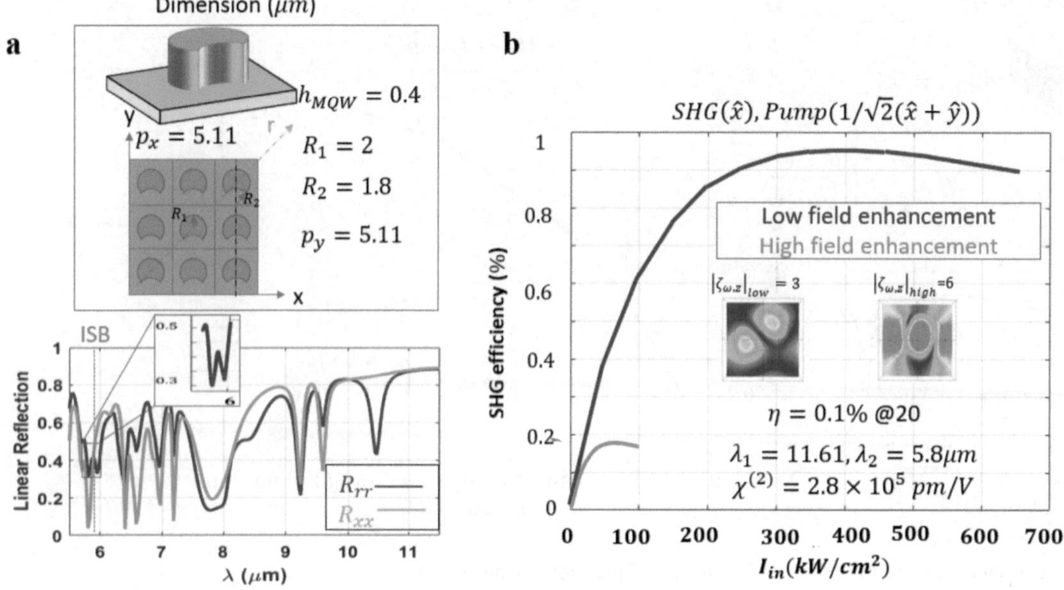

Figure 4: Second harmonic generation (SHG) efficiency for dielectric metasurface.
(a) Linear copolarized reflection for *x*-polarized, and *r*-polarized input wave. The *r* vector is drawn in the inset such that, $\sqrt{2}\hat{r} = \hat{x} - \hat{y}$. The inset shows a 3D view of one-unit cell, and 2D view of the metasurface looking top down with dimensions in μm. (b) SHG efficiency for $\hat{r}$-polarized pump with wavelength 11.61 μm and $\hat{x}$-polarized second harmonic with wavelength 5.8 μm for the metasurface in (a); the inset shows the field enhancement at the pump wavelength for the detuned and tuned designs at the pump wavelength. The efficiency at 20 kW/cm^2 is 0.1%. The multiple quantum well (MQW) parameters are the same as in Figure 3.

enhancement. It is clear that the detuned MS offers much more resilience to saturation effects, due to the lower field profile, enhancing the achievable level of SHG efficiency. Moreover, the dielectric inclusion achieves high efficiency at much larger power compared to the metallic MS considered in Figure 3. For example, the detuned MS achieves a 0.6% SHG efficiency at a pump power of 100 kW/cm^2, compared to only 20 kW/cm^2 for the nonetched MS in Figure 3. While it is true that dielectric MS have lower loss compared to those considering metallic inclusions, it should be noted that the dimensions of the dielectric MS compared to the metallic ones are much larger because of the larger size of the nanoantennas to support Mie resonances. For this reason, both the lateral size and the height of each MQW nanocavity is significantly larger than the metallic MS, limiting the field confinement, overlap integrals and also making more challenging the fabrication.

3.4 Etched arrow-shape MS

To better improve the SHG efficiency, we finally explored etched nanocavities with metallic nanoantennas. We found optimal geometries with an arrow shape, as shown in the inset of Figure 5a. This nanoantenna is somewhat similar to the T-shaped nanoantenna considered in a study by Lee et al. [14], but with tilted arms, and added small vertical arm that improves the spatial overlap integral. Figure 5b shows the calculated efficiency of this design, for pump polarized in the $\hat{x}$ direction at wavelength $2\lambda = 11.3$ μm and SH polarized to the $\hat{y}$ direction at wavelength $\lambda = 5.65$ μm. The SHG efficiency is 3% at 20 kW/cm^2, which we believe represents the best theoretical value for efficiency reported so far for these MQW material parameters. We stress that we have been considering the same material features and linewidth in the calculations in Figures 3–5, showing that the etched substrates capped by metallic nanoantennas appear to provide the ideal response for nonlinear SHG. Field confinement and overlap integral are maximized in these designs, and open to the possibility of achieving close to 5% at moderate input intensities and in a deeply subwavelength volume relying on purely local phenomena, hence ideally suited for nonlinear wavefront shaping and nonlinear metasurface manipulation.

4 Conclusions

In summary, we have presented here a comprehensive overview of the available opportunities in the context of midinfrared metasurface design for SHG. We have shown that SLR approaches can provide very large conversion efficiency, even when large ISB linewidths are considered, but

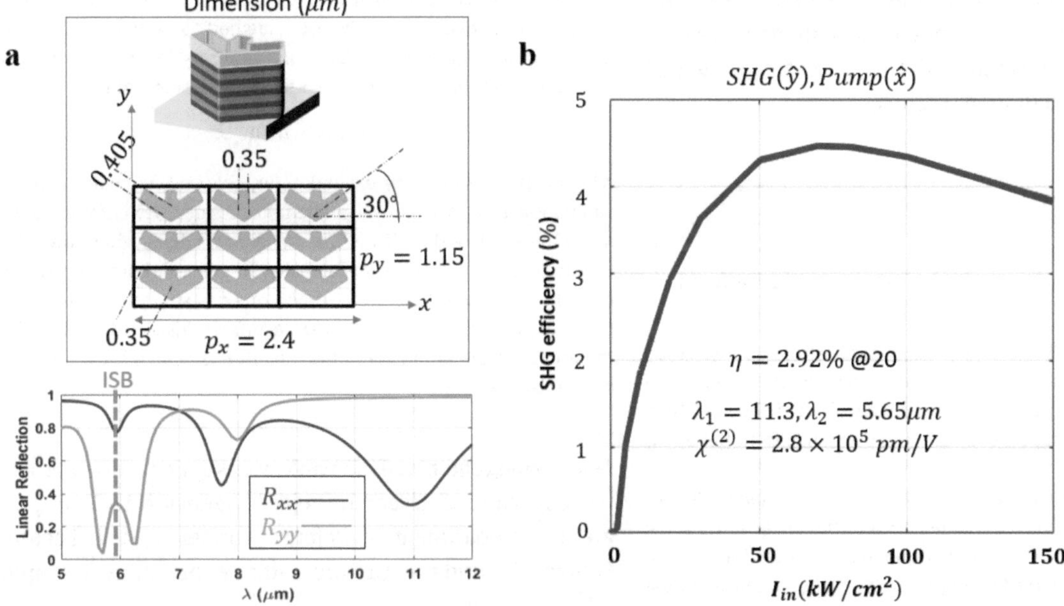

Figure 5: Second harmonic generation (SHG) efficiency for etched metallic metasurface.
(a) Linear copolarized reflection for x-polarized and y-polarized input waves. The inset shows a 3D view of one unit cell, and 2D view of the metasurface looking top down with dimensions in µm. (b) SHG efficiency for $\hat{x}$-polarized multiple quantum well (MQW) pump wavelength 11.3 µm and $\hat{y}$-polarized second harmonic at 5.65 µm for the metasurface in (a) with the inset showing the MQW parameters, and the efficiency at 20 kW/cm^2. MQW parameters are the same as in Figure 3.

only provided that we can excite the structures uniformly over many unit cells. These requirements prevent the use of this technique for nonlinear metasurface designs. Alternatively, we have explored a few other MS designs based on local overlapped resonances, including non-etched, etched and all-dielectric designs. Our results show the realistic possibility of achieving close to 5% efficiencies with modest pump powers, with the most promising approach relying on etched metallic metasurfaces.

Table 1 summarizes the various approaches described in this paper and their pros and cons. We envision a promising future for midinfrared nonlinear metasurfaces based on MQWs, as we improve the growth and design of these materials and their transition properties, and at the same time we optimize the fabrication and design of the resulting

Table 1: Comparison between different MSs presented in this work and their pros and cons.

Design	η@15 kW/cm^2	Advantages	Challenges
Surface lattice resonance $2\gamma = 30$ meV	0.5%	Large η even with larger ISB linewidth $2\gamma = 27$ meV. High efficiency at low pump power.	Requires very large uniform illumination, as it relies to a highly nonlocal resonance. It does not allow local manipulation of the nonlinear wavefront.
Dielectric inclusions on gold substrate. (ISB linewidth $2\gamma = 10$ meV)	0.05%	It does not require cooling. No metal oxidation or damage. Lower loss, lower heat. Larger saturation thresholds.	Need larger nanoantennas, in the order of $\frac{\lambda}{2}$. It requires high power to reach large efficiencies. Height of MQW is large.
Metallic nonetched designs (ISB linewidth $2\gamma = 10$ meV)	0.3%	Good efficiency even at low power. Small periods $\ll\lambda$. Easy fabrication as the MQW is not patterned.	Requires cooling. Highest efficiency at relatively high power. Metal losses, oxidation, and possible damage at high power levels.
Metallic etched designs (ISB linewidth $2\gamma = 10$ meV)	1.8%	Largest efficiency at low power. Small period $\ll\lambda$. Deeply subwavelength thickness.	Requires cooling. Metal losses, oxidation, and possible damage at high power levels.

ISB, intersubband.

metasurfaces. The reported features open a truly new paradigm for flatland nonlinear optics in the midinfrared region, with applications spanning defense, sensing, thermal manipulation and midinfrared sources.

5 Methods

5.1 Nonlinear simulations including saturation effects

We calculated the SHG efficiency using COMSOL Multiphysics by initially doing two linear frequency domain simulations at ω and at 2ω corresponding to the pump and SH angular frequencies and then doing some post processing over the resulting fields from the linear simulations.

Initially, we assume that all the involved materials are linear, including the MQW, which is modeled using the linear anisotropic permittivity as given in Eq. (3). We perform the simulation for one-unit cell of the MS with periodic boundary conditions excited by a normal incident plane wave with small intensity $I_{inc,0}$ (in simulation we can choose $I_{inc,0}$ such that it is much less than the saturation intensity). This excitation will induce an electric field inside the MQW, with normalized field given as, $\zeta_{\alpha,z}(\mathbf{r}) = \frac{E_{\alpha,z}(\mathbf{r})}{E_0}$, where E_0 is the electric field value of the incident plane wave and $\alpha = \omega$, *or* 2ω is the input frequency.

To calculate the SHG efficiency as defined in Eq. (1), knowing the normalized z field components at ω ($\zeta_{\omega,z}$) and at 2ω ($\zeta_{2\omega,z}$) from the previous step, we perform a spatial integration over the MQW volume as:

$$P(I_{inc,0}) = \int_{MQW} S_{SHG}\left(I_z(\mathbf{r}, I_{inc,0})\right)\left(\zeta_{\omega,z}(\mathbf{r})\right)^2 \zeta_{2\omega,z}(\mathbf{r})dv \quad (4)$$

Notice that $S_{SHG}(I_z(\mathbf{r}, I_{inc,0})) = 1$ since $I_{inc,0} \ll I_{sat}$, where I_{sat} is the saturation intensity corresponding to the intensity turning point in Figure 1c, and $I_z = 2n_{MQW}\epsilon_0 c|E_{\omega,z}(\mathbf{r}, I_{inc,0})|^2$, where $\mathbf{r}$ is the spatial position inside the MQW, and n_{MQW} is the refractive index of the MQW, ϵ_0 is the vacuum permittivity, and c is the speed of light.

We can plug the previous integral value in Eq. (1) to calculate the low power SHG efficiency, i.e., at $I_{inc,0}$. As we increase the incident power to higher value $I_{inc} > I_{inc,0}$, we do not need to perform further simulations to calculate the saturation effect. Since we assume that the material parameters do not change significantly, this is true for the given MQW structures which have low saturation intensities, therefore increasing the incident power affects only the term $S_{SHG}(I_z(\mathbf{r}, I_{inc,0}))$ in Eq. (4) which can be written as a function of the saturation factor S_{SHG} at the low incident power $I_{inc,0}$ as,

$$S_{SHG}(I_z(\mathbf{r}, I_{inc})) = S(I_z(\mathbf{r}, I_{inc,0})I_{inc}/I_{inc,0})$$

Therefore, we recalculate the spatial integral at all intensities, given as,

$$P(I_{inc}) = \int_{MQW} S_{SHG}(I_z(\mathbf{r}, I_{inc,0})I_{inc}/I_{inc,0})\left(\zeta_{\omega,z}(\mathbf{r})\right)^2 \zeta_{2\omega,z}(\mathbf{r})dv$$

It is readily seen from Eq. (1) that the efficiency η is proportional to the multiplication $P(I_{inc})\,I_{inc}$, so there is an optimal power level for the highest efficiency as shown in Figures 3b, 4b, 5b as a result of the competition between two functions, $P(I_{inc})[I_{inc}]$ that decreases [increases] with increasing the input intensity.

While this is the method we used to calculate the SHG efficiency in this work, one commonly used method is based on coupled nonlinear simulations. In this case, one performs linear simulations at frequency ω producing a nonlinear polarization current P_{NL}

$$P_{NL,z} = i2\omega_0 S(\mathbf{r}, I_{inc})\chi^{(2)}_{zzz}E^2_{\omega,z},$$

which will be used as the excitation for a second frequency domain simulation at $2f_0$. We then calculated the radiated power from the MS at 2ω to calculate the SHG efficiency by normalizing that power intensity to the incident power intensity. In general, both approaches provide similar results but the former one has shortest computation time, yet it is valid for very thin metasurfaces $h_{MQW} \ll \lambda$, and for SH radiation without additional diffraction orders, which is the scenario of interest in this work.

Acknowledgements: This work was supported by the Defense Advanced Research Projects Agency NASCENT program, a Department of Defense Vannevar Bush Faculty Fellowship, and the Air Force Office of Scientific Research MURI program.

Author contribution: All the authors have accepted responsibility for the entire content of this submitted manuscript and approved submission.

Research funding: This work was supported by the Defense Advanced Research Projects Agency NASCENT program, a Department of Defense Vannevar Bush Faculty Fellowship, and the Air Force Office of Scientific Research MURI program.

Conflict of interest statement: The authors declare no conflicts of interest regarding this article.

References

[1] B. Van Eerdenbrugh and L. S. Taylor, "Application of mid-IR spectroscopy for the characterization of pharmaceutical systems," *Int. J. Pharm.*, vol. 417, nos. 1–2, pp. 3–16, 2011.

[2] G. V. Naik, V. M. Shalaev, and A. Boltasseva, "Alternative plasmonic materials: beyond gold and silver," *Adv. Mater.*, vol. 25, no. 24, pp. 3264–3294, 2013.

[3] M. Sieger and B. Mizaikoff, "Toward on-chip mid-infrared sensors," *Anal. Chem.*, vol. 88, no. 11, pp. 5562–5573, 2016.

[4] D. G. Baranov, Y. Xiao, I. A. Nechepurenko, A. Krasnok, A. Alù, and M. A. Kats, "Nanophotonic engineering of far-field thermal emitters," *Nat. Mater.*, vol. 18, pp. 920–930, 2019.

[5] R. Duggan, Y. Ra'di, and A. Alù, "Temporally and spatially coherent emission from thermal embedded eigenstates," *ACS Photonics*, vol. 6, no. 11, pp. 2949–2956, 2019.

[6] M. De Zoysa, T. Asano, K. Mochizuki, A. Oskooi, T. Inoue, and S. Noda, "Conversion of broadband to narrowband thermal emission through energy recycling," *Nat. Photonics*, vol. 6, no. 8, pp. 535–539, 2012.

[7] T. Inoue, M. De Zoysa, T. Asano, and S. Noda, "High-*Q* mid-infrared thermal emitters operating with high power-utilization efficiency," *Opt. Express*, vol. 24, no. 13, pp. 15101–15109, 2016.

[8] J. Faist, F. Capasso, C. Sirtori, et al., "High power mid-infrared ($\lambda \sim 5$ μm) quantum cascade lasers operating above room temperature," *Appl. Phys. Lett.*, vol. 68, no. 26, pp. 3680–3682, 1996.
[9] E. Rosencher, A. Fiore, B. Vinter, V. Berger, P. Bois, and J. Nagle, "Quantum engineering of optical nonlinearities," *Science*, vol. 271, pp. 168–173, 1996.
[10] E. Rosencher, P. Bois, J. Nagle, and S. Delattre, "Second harmonic generation by intersub-band transitions in compositionally asymmetrical MQWs," *Electron. Lett.*, vol. 25, p. 1063, 1989.
[11] F. Capasso, C. Sirtori, and A. Y. Cho, "Coupled quantum well semiconductors with giant electric field tunable nonlinear optical properties in the infrared," *IEEE J. Quantum Electron.*, vol. 30, pp. 1313–1326, 1994.
[12] M. M. Fejer, S. J. B. Yoo, R. L. Byer, A. Harwit, and J. S. Harris Jr., "Observation of extremely large quadratic susceptibility at 9.6–10.8 μm in electric-field-biased AlGaAs quantum wells," *Phys. Rev. Lett.*, vol. 62, pp. 1041–1044, 1989.
[13] J. Lee, M. Tymchenko, C. Argyropoulos, et al., "Giant nonlinear response from plasmonic metasurfaces coupled to intersubband transitions," *Nature*, vol. 511, no. 7507, pp. 65–69, 2014.
[14] J. Lee, N. Nookala, J. S. Gomez-Diaz, et al., "Ultrathin second-harmonic metasurfaces with record-high nonlinear optical response," *Adv. Opt. Mater.*, vol. 4, no. 5, pp. 664–670, 2016.
[15] R. Sarma, D. de Ceglia, N. Nookala, et al., "Broadband and efficient second-harmonic generation from a hybrid dielectric metasurface/semiconductor quantum-well structure," *ACS Photonics*, vol. 6, no. 6, pp. 1458–1465, 2019.
[16] M. Tymchenko, J. S. Gomez-Diaz, J. Lee, M. A. Belkin, and A. Alù, "Highly-efficient THz generation using nonlinear plasmonic metasurfaces," *J. Opt.*, vol. 19, no. 10, p. 104001, 2017.
[17] J. Yu, S. Park, I. Hwang, D. Kim, J. Y. Jung, and J. Lee, "Third-harmonic generation from plasmonic metasurfaces coupled to intersubband transitions," *Adv. Opt. Mater.*, vol. 7, no. 9, p. 1801510, 2019.
[18] M. Tymchenko, J. S. Gomez-Diaz, J. Lee, N. Nookala, M. A. Belkin, and A. Alù, "Gradient nonlinear pancharatnam-berry metasurfaces," *Phys. Rev. Lett.*, vol. 115, no. 20, p. 207403, 2015.
[19] D. Kim, H. Chung, J. Yu, et al., "Spin-controlled nonlinear harmonic generations from plasmonic metasurfaces coupled to intersubband transitions," *Adv. Opt. Mater.*, vol. 8, p. 2000004, 2020.
[20] E. Rahimi and R. Gordon, "Nonlinear plasmonic metasurfaces," *Adv. Opt. Mater.*, vol. 6, no. 18, p. 1800274, 2018.
[21] S. A. Maier, *Plasmonics: Fundamentals and Applications*, New York, NY, Springer Science & Business Media, 2007.
[22] H. Yu, Y. Peng, Y. Yang, and Z. Y. Li, "Plasmon-enhanced light–matter interactions and applications," *NPJ Comput. Mater.*, vol. 5, no. 1, pp. 1–14, 2019.
[23] M. Kauranen and A. V. Zayats, "Nonlinear plasmonics," *Nat. Photonics*, vol. 6, no. 11, pp. 737–748, 2012.
[24] G. Li, S. Chen, N. Pholchai, et al., "Continuous control of the nonlinearity phase for harmonic generations," *Nat. Mater.*, vol. 14, no. 6, pp. 607–612, 2015.
[25] N. C. Panoiu, W. E. I. Sha, D. Y. Lei, and G. C. Li, "Nonlinear optics in plasmonic nanostructures," *J. Opt.*, vol. 20, no. 8, p. 083001, 2018.
[26] C. Ciracì, E. Poutrina, M. Scalora, and D. R. Smith, "Origin of second-harmonic generation enhancement in optical split-ring resonators," *Phys. Rev. B*, vol. 85, no. 20, p. 201403, 2012.
[27] S. Linden, F. B. P. Niesler, J. Förstner, Y. Grynko, T. Meier, and M. Wegener, "Collective effects in second-harmonic generation from split-ring-resonator arrays," *Phys. Rev. Lett.*, vol. 109, no. 1, p. 015502, 2012.
[28] B. Metzger, T. Schumacher, M. Hentschel, M. Lippitz, and H. Giessen, "Third harmonic mechanism in complex plasmonic Fano structures," *ACS Photonics*, vol. 1, no. 6, pp. 471–476, 2014.
[29] M. J. Huttunen, G. Bautista, M. Decker, S. Linden, M. Wegener, and M. Kauranen, "Nonlinear chiral imaging of subwavelength-sized twisted-cross gold nanodimers," *Opt. Mater. Express*, vol. 1, no. 1, pp. 46–56, 2011.
[30] P. Y. Chen and A. Alù, "Subwavelength imaging using phase-conjugating nonlinear nanoantenna arrays," *Nano Lett.*, vol. 11, no. 12, pp. 5514–5518, 2011.
[31] Y. Tang, Z. Liu, J. Deng, K. Li, J. Li, and G. Li, "Nano-Kirigami metasurface with giant nonlinear optical circular dichroism," *Laser Photonics Rev.*, vol. 14, p. 2000085, 2020.
[32] E. Almeida, O. Bitton, and Y. Prior, "Nonlinear metamaterials for holography," *Nat. Commun.*, vol. 7, no. 1, pp. 1–7, 2016.
[33] F. Walter, G. Li, C. Meier, S. Zhang, and T. Zentgraf, "Ultrathin nonlinear metasurface for optical image encoding," *Nano Lett.*, vol. 17, no. 5, pp. 3171–3175, 2017.
[34] J. S. Gomez-Diaz, M. Tymchenko, J. Lee, M. A. Belkin, and A. Alù, "Nonlinear processes in multi-quantum-well plasmonic metasurfaces: electromagnetic response, saturation effects, limits, and potentials," *Phys. Rev. B*, vol. 92, no. 12, p. 125429, 2015.
[35] M. Minkov, D. Gerace, and S. Fan, "Doubly resonant χ (2) nonlinear photonic crystal cavity based on a bound state in the continuum," *Optica*, vol. 6, no. 8, pp. 1039–1045, 2019.
[36] Z. Sakotic, A. Krasnok, N. Cselyuszka, N. Jankovic, and A. Alú, "Berreman embedded eigenstates for narrowband absorption and thermal emission," arXiv preprint arXiv:2003.07897, 2020.
[37] R. Czaplicki, A. Kiviniemi, M. J. Huttunen, et al., "Less is more: enhancement of second-harmonic generation from metasurfaces by reduced nanoparticle density," *Nano Lett.*, vol. 18, no. 12, pp. 7709–7714, 2018.
[38] K. L. Vodopyanov, V. Chazapis, C. C. Phillips, B. Sung, and J. S. Harris Jr., "Intersubband absorption saturation study of narrow III–V multiple quantum wells in the $\lambda = 2.8–9$ μm spectral range," *Semicond. Sci. Technol.*, vol. 12, p. 708, 1997.
[39] R. W. Boyd, *Nonlinear Optics*, 3rd ed., New York, Academic, 2008, p. 640.
[40] H. Qian, S. Li, Y. Li, et al., "Nanoscale optical pulse limiter enabled by refractory metallic quantum wells," *Sci. Adv.*, vol. 6, no. 20, p. eaay3456, 2020.
[41] G. Günter, A.A. Anappara, J. Hees, et al., "Sub-cycle switch-on of ultrastrong light–matter interaction," *Nature*, vol. 458, no. 7235, pp. 178–181, 2009.
[42] M. J. Huttunen, P. Rasekh, R. W. Boyd, and K. Dolgaleva, "Using surface lattice resonances to engineer nonlinear optical processes in metal nanoparticle arrays," *Phys. Rev. A*, vol. 97, no. 5, p. 053817, 2018.
[43] M. J. Huttunen, O. Reshef, T. Stolt, K. Dolgaleva, R. W. Boyd, and M. Kauranen, "Efficient nonlinear metasurfaces by using multiresonant high-*Q* plasmonic arrays," *JOSA B*, vol. 36, no. 7, pp. E30–E35, 2019.
[44] A. Benz, S. Campione, S. Liu, et al., "Strong coupling in the sub-wavelength limit using metamaterial nanocavities," *Nat. Commun.*, vol. 4, no. 1, pp. 1–8, 2013.
[45] O. Wolf, A. A. Allerman, X. Ma, et al., "Enhanced optical nonlinearities in the near-infrared using III-nitride

heterostructures coupled to metamaterials," *Appl. Phys. Lett.*, vol. 107, no. 15, p. 151108, 2015.

[46] R. L. Olmon, B. Slovick, T. W. Johnson, et al., "Optical dielectric function of gold," *Phys. Rev. B*, vol. 86, no. 23, p. 235147, 2012.

[47] V. G. Kravets, A. V. Kabashin, W. L. Barnes, and A. N. Grigorenko, "Plasmonic surface lattice resonances: a review of properties and applications," *Chem. Rev.*, vol. 118, no. 12, pp. 5912–5951, 2018.

[48] D. Khlopin, F. Laux, W. P. Wardley, et al., "Lattice modes and plasmonic linewidth engineering in gold and aluminum nanoparticle arrays," *JOSA B*, vol. 34, no. 3, pp. 691–700, 2017.

[49] M. Saad Bin-Alam, O. Reshef, Y. Mamchur, et al., Ultra-high-*Q* resonances in plasmonic metasurfaces," arXiv preprint arXiv: 2004.05202v1, 2020, https://doi.org/10.1109/pn50013.2020.9167033.

[50] S. Campione, A. Benz, J. F. Klem, M. B. Sinclair, I. Brener, and F. Capolino, "Electrodynamic modeling of strong coupling between a metasurface and intersubband transitions in quantum wells," *Phys. Rev. B*, vol. 89, no. 16, p. 165133, 2014.

[51] P. R. West, S. Ishii, G. V. Naik, N. K. Emani, V. M. Shalaev, and A. Boltasseva, "Searching for better plasmonic materials," *Laser Photonics Rev.*, vol. 4, no. 6, pp. 795–808, 2010.

Roberto Merlin*

Near-field plates and the near zone of metasurfaces

https://doi.org/10.1515/9783110710687-055

Abstract: A brief, tutorial account is given of the differences between the near and far regions of the electromagnetic field emphasizing the source-dependent behavior of the former and the universal properties of the latter. Field patterns of near-field plates, that is, metasurfaces used for sub-wavelength applications, are discussed in some detail. Examples are given of fields that decay away from the plates in an exponential manner, a ubiquitous feature of many interface problems, and metasurfaces for which the decay is not exponential, but algebraic. It is also shown that a properly designed system of two parallel near-field plates can produce fields that exhibit pseudo minima, which are potentially useful for near-field tweezer-like applications.

Keywords: electromagnetic near field; metasurfaces; near-field plates; subwavelength focusing; wireless power transfer.

1 Introduction

Metasurfaces are artificial two-dimensional structures used to generate a desired electromagnetic (EM) field pattern or modify an incoming wave to obtain a predetermined result [1]. Most of the EM applications of metasurfaces, from microwave receivers and transmitters [2] to reconfigurable devices [3] and flat optical lenses [4], pertain to the radiation (or far) zone. Metasurfaces whose primary function involves the near field are known as near-field plates (NFPs) [5, 6]. Since the near zone encodes detailed information about the sources, ignoring restrictions imposed by the standard diffraction limit [7] and, moreover, because it allows one to separate the electric from the magnetic field, the interest in NFPs centers primarily on subwavelength focusing [8] and wireless power transfer [9].

In this work, we present an abridged review of the near field properties of metasurfaces, emphasizing the forms of decay of the EM field (although the elastic field is not considered here, many of the results apply also to acoustic metasurfaces [10]). Other than the ever present exponential decay, commonly associated with interface phenomena [11], we show cases where the decay of the near field is algebraic in nature. We also introduce an arrangement of a pair of metasurfaces and describe its potential use as EM tweezers.

2 The near field and evanescent waves

2.1 Localized charges and currents

Textbooks tell us that the EM field of a confined distribution of moving charges behaves very differently in regions that are close to and far from the charges, with the length scale determined by the wavelength of the radiation, λ [12, 13]. Differentiation between the near and far behaviors appears already in the expressions for the fields resulting from the motion of a point charge. The corresponding Liénard–Wiechert potentials involve the sum of two terms: (i) the near-field contribution, which is associated with the static fields and does not contain the acceleration of the sources, and (ii) the radiation or far-field term, which vanishes as the acceleration goes to zero. The separation is also apparent in the expressions for the EM field of a sinusoidally, time-varying electric (magnetic) dipole where the electric (magnetic) field dominates in the near zone whereas, far away from the dipole, the electric and magnetic field are of the same magnitude (Gaussian units), and both decay inversely proportional to the distance. For various reasons, and leaving aside the question of the sources needed to generate a particular field, it is useful to frame the broad near-zone versus far-zone discussion around the behavior of the EM field in vacuum. For fields with time dependence given by

***Corresponding author: Roberto Merlin,** The Harrison M. Randall Laboratory of Physics, University of Michigan, Ann Arbor, Michigan 48109-1040, USA, E-mail: merlin@umich.edu. https://orcid.org/0000-0002-5584-0248

This article has previously been published in the journal Nanophotonics. Please cite as: R. Merlin "Near-field plates and the near zone of metasurfaces" *Nanophotonics* 2021, 10. DOI: 10.1515/nanoph-2020-0307.

$e^{-i\omega t}$ (ω is the angular frequency), the empty-space potentials in the Lorenz gauge satisfy the homogeneous Helmholtz equation

$$(\nabla^2 + \omega^2/c^2)F = 0 \tag{1}$$

where F is the electrostatic potential Φ or a Cartesian component of the potential vector $\mathbf{A}$, and c is the speed of light, with $\nabla \cdot \mathbf{A} - i\omega c^{-2}\Phi = 0$. For localized sources, it is convenient to expand F in terms of outgoing spherical waves,

$$F = \sum_{lm} A_{lm} h_l^{(1)}(kr) Y_{lm}(\theta, \varphi), \tag{2}$$

themselves solutions of Eq. (1). Here, r is the radial distance, θ and φ are the polar and azimuthal angles, $k = \omega/c$ is the wave-vector and A_{lm} are expansion coefficients; $h_l^{(1)}$ and Y_{lm} are, respectively, a spherical Hankel function of the first kind and order l and a spherical harmonic. Since $h_l^{(1)} \to e^{ikr}/r$ for $kr \gg 1$, regardless of l, the far-field $1/r$ decay behavior is universal. Instead, the near-field properties depend on the particulars of the distribution. Consider sources confined to a region of space of dimensions $d \ll \lambda$. Then, for $d \ll r \ll \lambda$, $h_l^{(1)} \sim r^{-(l+1)}$ and, since $\nabla^2[r^{-(l+1)} Y_{lm}(\theta, \varphi)] = 0$, we get $\nabla^2 F \approx 0$. This is to be expected given that the Helmholtz equation becomes Laplace's equation in the limit $c \to \infty$, when retardation can be ignored (since magnetism is a relativistic effect, care must be exerted when taking this limit for $\mathbf{A}$). To lowest order, we have $\Phi \sim 1/r$ and $A_i \sim 1/r^2$, which correspond, respectively, to the fields of a static electric monopole and magnetic dipole. We note that, unlike the Liénard–Wiechert expressions, the spherical-wave expansion does not split into a separate sum of near- and far-zone terms. Rather, each term in Eq. (2) gives the corresponding near- and far-field expressions in two separate limits.

2.2 Sources behind a plane: cylindrical and Cartesian waves, and exponential decay

The above considerations are not applicable to extended sources of dimensions $\gtrsim \lambda$ or, for that matter, to metasurfaces, which divide space into two halves. Let us assume that all the sources are in the half-space defined by $z < 0$ and expand the empty half-space potentials in terms of the complete set of solutions of Helmholtz equation in cylindrical coordinates

$$F(\rho, \varphi, z) = \sum_n e^{in\varphi} \int B_n(q) J_n(q\rho) e^{i\kappa(q)z}\, dq \tag{3}$$

where $\rho = \sqrt{r^2 - z^2}$, $\kappa^2 = k^2 - q^2$, J_n is a Bessel function of order n and B_n are the parameters of the expansion. Notice that $B_n = B_0\delta_{n0}$ for azimuthally symmetric fields like, *e.g.*, the axicon [14]. Eq. (3) divides into exponentially-decaying components for which $q > k$, and traveling waves, commonly known as Bessel beams [15], with real $\kappa < k$. This allows for a clear separation between the near and far fields, associated primarily with the evanescent ($\kappa = i|\kappa|$) and traveling components, respectively. Using the orthogonality condition for Bessel functions, we get

$$B_m(q) = \frac{q}{2\pi} \int F(\rho, \varphi, 0) e^{-im\varphi} J_m(q\rho) \rho\, d\rho\, d\varphi, \tag{4}$$

which, together with Eq. (3), establishes an exact relationship between values of the field at two parallel planes (two values of z).

Consider now the equivalent Cartesian coordinate approach [16, 17]. Let $F(x, y, z_0)$ be the potential field in the plane $z = z_0$. The angular spectrum in this plane is defined as the Fourier transform

$$\alpha(q_x, q_y, z_0) = \frac{1}{4\pi^2} \iint F(x, y, z_0) e^{i(q_x x + q_y y)}\, dx dy. \tag{5}$$

As before, we assume that all the sources are in the half-space $z < 0$. Then, the field at any point in the source-free half-space is given by

$$F(x, y, z) = \int \alpha(q_x, q_y, z_0) e^{i[q_x x + q_y y + \kappa(z - z_0)]}\, dq_x dq_y \tag{6}$$

where

$$\kappa = \begin{cases} i\left|\left(q_x^2 + q_y^2 - k^2\right)^{1/2}\right| & q_x^2 + q_y^2 \geq k^2 \\ \left|\left(k^2 - q_x^2 - q_y^2\right)^{1/2}\right| & q_x^2 + q_y^2 < k^2 \end{cases}. \tag{7}$$

Note that, by construction, $(\nabla^2 + \omega^2/c^2)F = 0$. Like its cylindrical counterpart, Eqs. (3) and (4), this expression can be used (i) to calculate the field propagation in the forward direction or, by back propagation, (ii) to infer the field at $z = z_0$ that will produce a desired field pattern further ahead. The latter approach was applied to the design of a NFP [5] mimicking the behavior of a negative-index slab [18] to attain focusing beyond the standard diffraction limit; one of the first realizations of a metasurface [6].

As for the cylindrical-wave representation, the (Cartesian) angular decomposition allows for a sharp separation between the near- and the far-field depending on whether κ is purely imaginary or real [19]. This, however, should not be construed to imply that the decay of the near-field component of F is always exponential, for planar geometries or otherwise (recall that the near field of a localized charge distribution decays algebraically). Several examples of non-evanescent decay of the metasurface near field are given below. Eq. (6) shows that a sufficient condition for exponential decay to occur is to have a sharp peak in the angular spectrum at a wave-vector $\mathbf{q}_0$ of modulus $q_0 > k$ so that $F \sim e^{-(q_0^2-k^2)^{1/2}z}$. This is precisely the condition met by important interface phenomena, such as total internal reflection, surface polaritons [20] and focusing by a negative-index slab [16]. Similar considerations apply to cylindrical fields for which $B_m(q) \propto \delta(q-q_0)$ with $q_0 > k$; see Eqs. (3) and (4).

3 One-dimensional metasurfaces and near-field plates

From now on, and for simplicity, we focus the attention on problems for which $\partial F/\partial x \equiv 0$. In this case, the most general outgoing-wave solution to Helmholtz' equation is of the form

$$F = \sum_n C_n H_n^{(1)}(k\eta) e^{in\beta} \tag{8}$$

where $\eta^2 = y^2 + z^2$ and $\tan\beta = y/z$. $H_n^{(1)}$ is a Hankel function of the first kind and order n, and C_n are constants. Since $\eta = 0$ is the only singularity of the Hankel functions, this expansion accounts for the field on the source-free side of a one-dimensional metasurface placed at arbitrary $z > 0$. Once again, the behavior of the (two-dimensional) far field is universal since $H_n^{(1)}(k\eta) \to e^{ik\eta}/\eta^{1/2}$ for $k\eta \gg 1$, while the near field properties depend on the specifics of the source distribution. Replacing the Hankel functions by their small argument limit, we get

$$\lim_{k\eta\to 0} F = \frac{2iC_0}{\pi}\ln(k\eta/2) - \frac{i}{\pi}\sum_n C_n \Gamma(n)(2/k\eta)^n e^{in\beta}. \tag{9}$$

This expression is formally identical to the general solution of the two-dimensional Laplace's equation in polar coordinates that is consistent with $F \to 0$ at infinity. With F becoming a harmonic function (that is, a function that satisfies $\nabla^2 F = 0$ in two dimensions) and, after some rearrangement of terms, the limit $k\eta \to 0$ leads to the multipole expansion of the static electric (Φ) and magnetic ($\mathbf{A}$) potentials (we observe once more that caution must be applied when calculating $\mathbf{A}$ since the magnetic field vanishes for $c \to \infty$).

Figure 1 shows results for

$$F = \sum_{n=0}^{N} \frac{H_n^{(1)}(k\eta)}{H_n^{(1)}(kR)} \cos n\beta. \tag{10}$$

At large N, this expression gives a field that is sharply peaked at $\beta = 0$ in the circle $\eta = R$. The contour plot, Figure 1A, and Figure 1B (y = 0) reveal three distinctive regions with the subwavelength scale R setting the boundary between the two that belong to the near field. For $y = 0$ and $z \ll R \ll \lambda$, the behavior of F is determined by $H_N^{(1)}$, the leading term of which is z^{-N} while, in the intermediate near-zone region $R \ll z \ll \lambda$, the dominant term is $H_0^{(1)} \sim \ln kz$. As expected, $F \propto z^{-1/2}$ in the far field ($kz \gg 1$). Also important is that, for $z \ll R$, F shows a peak of width $\approx z/N$ centered at $y = 0$ (Figure 1A covers too large a range to notice the z-dependence of the width).

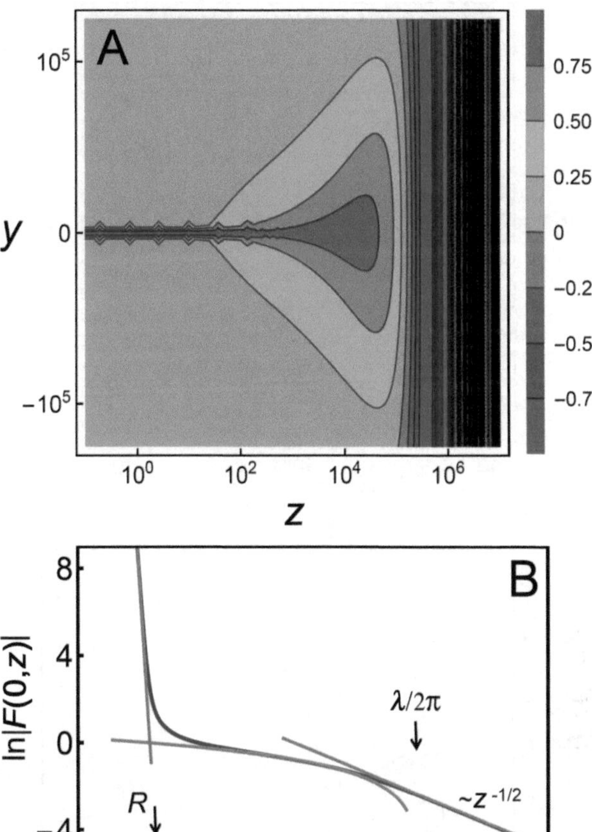

Figure 1: Near and far field of a one-dimensional metasurface; see Eq. (10). (A) Contour plot of the normalized field, $\mathrm{Re}[F(y,z)]/|F(0,z)|$, for $N = 20$, $R = 10$ and $\lambda = 10^6$. (B) $\ln|F(0,z)|$ versus z (blue curve). Gray curves are asymptotes. For small and intermediate distances from the metasurface, $F \propto z^{-N}$ and $F \propto \ln z$ ($10^2 \lesssim z \lesssim 10^5$). These near-field forms correspond, respectively, to the small argument limit of $H_N^{(1)}$ and $H_0^{(1)}$. In the far field, $z \gtrsim 10^5$, $F \propto 1/\sqrt{z}$.

Returning to the angular spectrum representation, consider a situation where $F(y, 0)$ is localized mainly in a segment of length $\ll\lambda$, as in the above example and, more generally, for a generic NFP. We can then ignore contributions to $\alpha(q, 0)$ from all, but spatial frequencies $\gg k$, so that the near field can be approximated by the harmonic expression

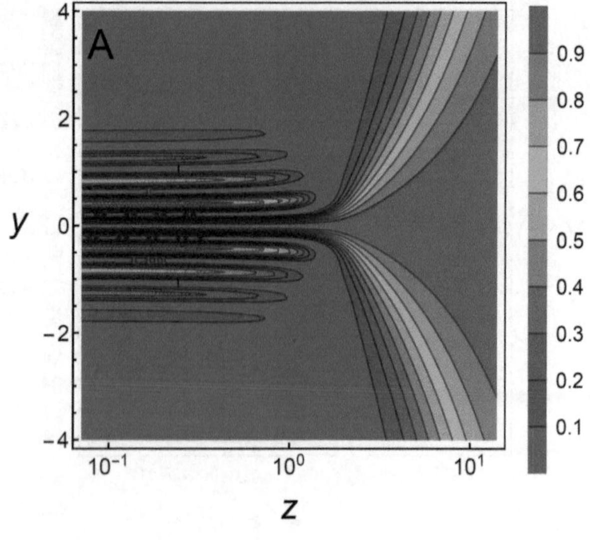

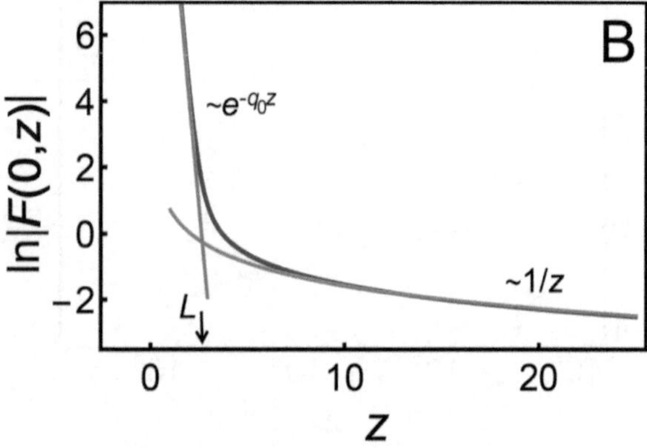

Figure 2: Near field of the modulated-grating metasurface; see Eq. (12). (A) Contour plot of the normalized intensity, $\mathrm{Re}^2[F(y, z)]/\mathrm{Re}^2[F(0, z)]$, for $q_0 = 7$ and $L = 2.5$. (B) $\ln|F(0, z)|$ versus z (blue curve). Gray curves are asymptotes. For small (large) z, F decays exponentially ($F \propto 1/z$).

$$F(y, z) \approx \int \alpha(q, 0) e^{iqy} e^{-|q|z} dq. \tag{11}$$

It is well known that arbitrary functions of the form $f(iy \pm z)$ are solutions to Laplace's equation in two dimensions. Thus, the near field associated with a subwavelength-localized $F(y, 0)$ is a harmonic function that is analytic in the half-space $z > 0$ and decays with increasing z.

4 Modulated grating and single-aperture near-field plates

Figure 2 shows calculations using Eq. (11) for

$$F(y, 0) = \frac{e^{iq_0 y}}{y^2 + L^2}; \tag{12}$$

the corresponding angular spectrum is $\alpha(q, 0) = e^{-L|q-q_0|}/2L$. The modulated-grating profile mirrors the behavior of the EM field at the exit side of a negative-index slab [5, 21]. If $q_0 \gg k$, near identical results are obtained using the exact expression, Eq. (6). We emphasize that the plots show only the near field. Because of the dominant peak of $\alpha(q, 0)$ at $q = q_0$, F decays first exponentially, turning later into a $1/z$ dependence for $z \gg q_0^{-1}$. Not shown in Figure 2 is the far field, which manifests itself for $z \gg k^{-1}$ and decays $\sim 1/z^{1/2}$ for $y = 0$, as expected. For completeness, we give below the explicit expression of the near field in terms of harmonic functions:

$$F(y, z) = \left[\frac{1}{(L + iy + z)} - \frac{1}{(L + iy - z)}\right] \frac{e^{-q_0 L}}{2L} + \frac{e^{q_0 (iy - z)}}{(L + iy - z)(L - iy + z)}. \tag{13}$$

From here, one can show that, as the distance z from the metasurface increases, the width of $|F(y, z)|$ decreases reaching a minimum value $\sim q_0^{-1}$ at $z = L$. This closely

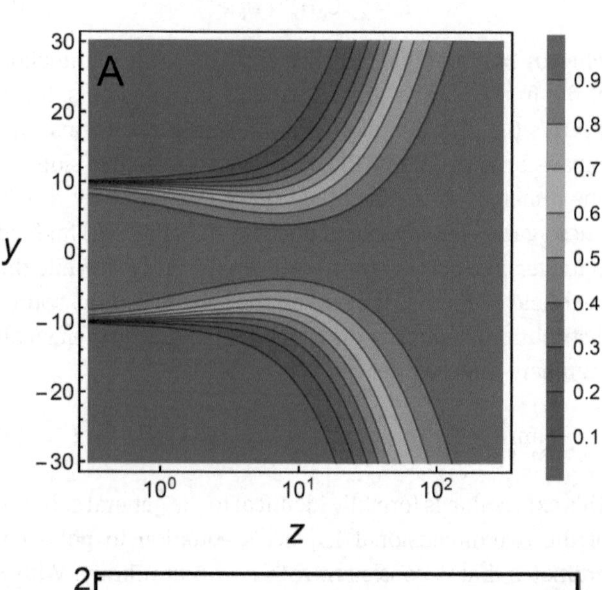

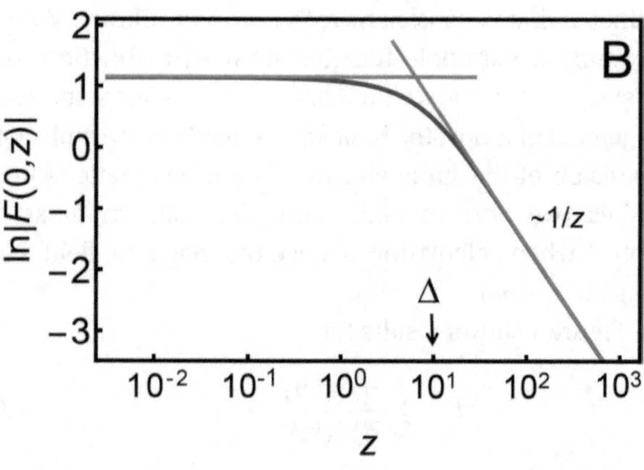

Figure 3: Near field of the single-aperture metasurface; see Eq. (14). (A) Contour plot of $\mathrm{Re}^2[F(y, z)]/\mathrm{Re}^2[F(0, z)]$ for $\Delta = 10$ and $L = 1$. (B) $\ln|F(0, z)|$ versus z (blue curve). Gray curves are asymptotes. For small (large) z, F is nearly constant ($F \propto 1/z$).

simulates the behavior of subwavelength focusing by a negative-index slab [18]. Because the field magnitude decreases with increasing z, the focus is a saddle point, something that could have been anticipated since harmonic functions do not have absolute maxima or minima [22].

Figure 3 shows an example of near-field focusing without exponential decay. The calculations are for a single-aperture NFP with

$$F(y, 0) = \pi/2 - \arctan\frac{y^2 - \Delta^2}{L^2}. \quad (14)$$

This expression gives a Lorentzian-like peak of half-width L for $L \gg \Delta$ whereas, for $L \ll \Delta$, $F \approx 0$ except for $|y| < \Delta$ where it is nearly constant. The corresponding angular spectra are roughly of the form $e^{-|q|L}$ and $\sin(q\Delta)/q$, respectively. Using these approximations and Eq. (11), we get

$$F(y, z) \approx \begin{cases} \dfrac{L+z}{y^2 + (L+z)^2} & (L \gg \Delta) \\ \dfrac{1}{2}\left(\tan^{-1}\dfrac{\Delta + y}{z} + \tan^{-1}\dfrac{\Delta - y}{z}\right) & (L \ll \Delta) \end{cases} \quad (15)$$

In the limit $L \gg \Delta$ $(L \ll \Delta)$, and for $z \ll L$ (Δ), $F(0, z)$ is nearly constant while $F(y, z)$ exhibits a peak at $y = 0$ of half-width $\approx L$ (Δ). At large values of z (but $z \ll \lambda$), $F \propto 1/z$. The case $\Delta = 10$, $L = 1$ is illustrated in Figure 3. As for the calculations of Figure 2, we underline the fact that these plots depict only the near field. The results show that neither the amplitude (Figure 3B) nor the width of the field vary much in the range $z \ll \Delta$. This makes the single-aperture NFP a promising candidate for subwavelength focusing and power-transfer applications, with the drawback that the high resolution is restricted to distances from the NFP on the order of the resolution itself (the same limitation as in evanescent-wave focusing [23]).

5 Pair of parallel near-field plates

As discussed earlier, at distances $z \ll \lambda$, the near-field of a metasurface is well described by a harmonic function that decays away from the plate. For a pair of parallel plates, however, the decay requirement does not longer apply and, thus, an arbitrary singularity-free harmonic function represents a physically realizable near field. This offers a path for the development of near-field devices that could meet the needs of particular applications. Of interest here is the design of near-field structures that can be used to manipulate subwavelength objects by means of radiation pressure, like optical tweezers. Even though harmonic functions such as F do not possess maxima or minima [22], the saddle points make it possible to devise an intensity pattern that comes close to providing a confinement potential for the intensity. Figure 4 shows one such an example. The three-dimensional plot gives $\mathrm{Re}^2[F(y, z)]$, which is proportional to the intensity, for

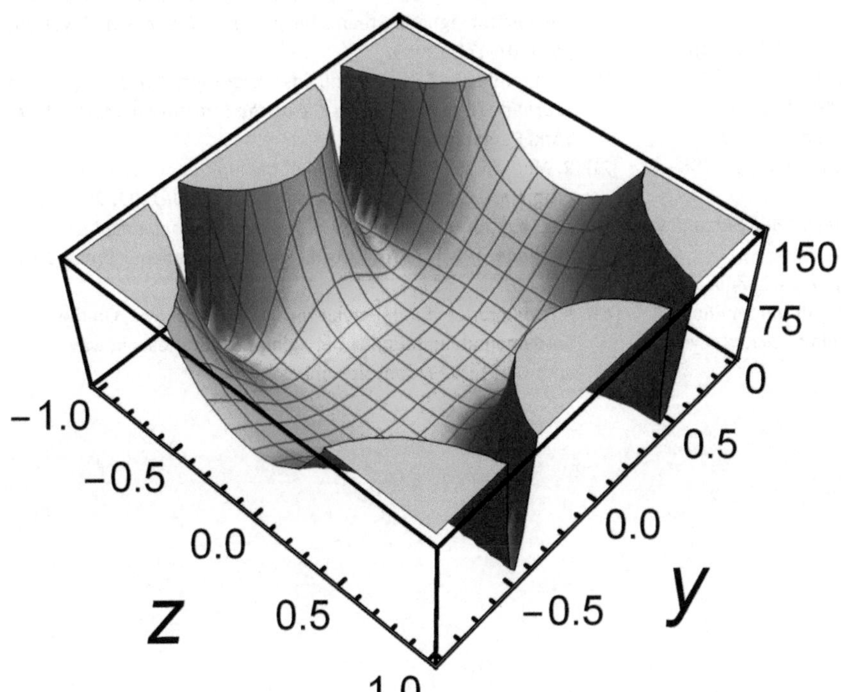

Figure 4: Near field of a system of two-metasurfaces; see Eq. (16). The 3D plot shows the square of the field, $\mathrm{Re}^2[F(y, z)]$.

$$F = \frac{e^{q_0(iy-z)} - 1}{iy - z} + \frac{e^{q_0(iy+z)} - 1}{iy + z} - V_0, \tag{16}$$

with parameters $q_0 = 4$ and $V_0 = 8$. The latter value was chosen so that there is a saddle point at $F \approx 0$. The depression around the origin is not a true minimum for it is possible to descend towards regions of lower intensity using the canyon-like features along the planes $x = \pm 1$. Nevertheless, the dip can be used to trap a particle with a refractive index smaller than that of the surrounding medium provided its size is large enough to prevent escape through the canyons.

6 Conclusions

We showed that, unlike the far field, the EM field close to a metasurface exhibits a variety of dependencies with distance, from that of evanescent waves, often (and wrongly) viewed as the hallmark of the near field, to various forms of algebraic decay. We also demonstrated that fields between two parallel near-field plates can have pseudo minima, a property that is potentially useful for applications as near-field EM tweezers.

Research funding: None declared.
Conflict of interest statement: The author declare no conflicts of interest regarding this article.

References

[1] O. Quevedo-Teruel, H. Chen, A. Díaz-Rubio, et al., "Roadmap on metasurfaces," *J. Opt.*, vol. 1, p. 073002, 2019.
[2] G. M. Minatti, E. Martini, F. Caminita, et al., "Modulated metasurface antennas for space: synthesis, analysis and realizations," *IEEE Trans. Antennas Propag.*, vol. 63, pp. 1288–1300, 2015.
[3] Y. Radi and A. Alù, "Reconfigurable metagratings," *ACS Photonics*, vol. 5, pp. 1779–1785, 2018.
[4] M. Khorasaninejad, W. T. Chen, R. C. Devlin, J. Oh, A. Y. Zhu, and F. Capasso, "Metalenses at visible wavelengths: diffraction-limited focusing and subwavelength resolution imaging," *Science*, vol. 352, pp. 1190–1194, 2016.
[5] R. Merlin, "Radiationless electromagnetic interference: evanescent-field lenses and perfect focusing," *Science*, vol. 317, pp. 927–929, 2007.
[6] A. Grbic, L. Jiang, and R. Merlin, "Near-field plates: subdiffraction focusing with patterned surfaces," *Science*, vol. 320, pp. 511–513, 2009.
[7] E. Abbe, "Beiträge zur Theorie des Mikroskops und der mikroskopischen Wahrnehmung," *Arch. Mikrosk. Anat.*, vol. 9, pp. 413–468, 1873.
[8] A. Grbic, R. Merlin, E. M. Thomas, and M. F. Imani, "Near-field plates: metamaterial surfaces/arrays for subwavelength focusing and probing," *Proc. IEEE*, vol. 99, pp. 1806–1815, 2011.
[9] M. F. Imani and A. Grbic, "Unidirectional wireless power transfer using near-field plates," *J. Appl. Phys.*, vol. 117, p. 184903, 2015.
[10] Y. Xie, W. Wang, H. Chen, A. Konneker, B. I. Popa, and S. A. Cummer, "Wavefront modulation and subwavelength diffractive acoustics with an acoustic metasurface," *Nat. Commun.*, vol. 5, p. 5553, 2014.
[11] C. Girard, C. Joachim, and S. Gauthier, "The physics of the near field," *Rep. Prog. Phys.*, vol. 63, pp. 893–938, 2000.
[12] J. D. Jackson, *Classical Electrodynamics*, 3rd ed. Hoboken, Wiley, 1999.
[13] J. A. Stratton, *Electromagnetic Theory*, New York, McGraw-Hill, 1941.
[14] J. H. McLeod, "The Axicon: a new type of optical element," *J. Opt. Soc. Am.*, vol. 44, p. 592, 1954.
[15] J. Durnin, J. J. Miceli, Jr., and J. H. Eberly, "Diffraction-free beams," *Phys. Rev. Lett.*, vol. 58, p. 1499, 1987.
[16] P. C. Clemmow, *The Plane Wave Spectrum Representation of Electromagnetic Fields*, Oxford, Pergamon, 1966.
[17] J. W. Goodman, *Introduction to Fourier Optics*. New York, McGraw-Hill, 1968, p. 48 ff.
[18] J. B. Pendry, "Negative refraction makes a perfect lens," *Phys. Rev. Lett.*, vol. 85, pp. 3966–3969, 2000.
[19] H. F. Arnoldus and J. T. Foley, "Travelling and evanescent parts of the electromagnetic Green's tensor," *J. Opt. Soc. Am. A*, vol. 19, pp. 1701–1711, 2005.
[20] E. Burstein and F. De Martini, Eds. *Polaritons, Proceedings of the Taormina Research Conference on the Structure of Matter*, New York, Pergamon, 1974.
[21] R. Merlin, "Analytical solution of the almost-perfect-lens problem," *Appl. Phys. Lett.*, vol. 84, pp. 1290–1292, 2004.
[22] See, e.g., G. B. Arfken, and H. J. Weber, *Mathematical Methods for Physicists*, 7th ed., Waltham, Academic Press, 2012, ch. 6.
[23] A. Y. Piggott, L. Su, J. Petykiewicz, and J. Vučković, On the fundamental limitations of imaging with evanescent waves, arXiv:2001.00237 [physics.optics].

Dongyi Wang, Tong Liu, Yuejiao Zhou, Xiaoying Zheng, Shulin Sun, Qiong He* and Lei Zhou*

High-efficiency metadevices for bifunctional generations of vectorial optical fields

https://doi.org/10.1515/9783110710687-056

Abstract: Vectorial optical fields (VOFs) exhibiting tailored wave fronts and spatially inhomogeneous polarization distributions are particularly useful in photonic applications. However, devices to generate them, made by natural materials or recently proposed metasurfaces, are either bulky in size or less efficient, or exhibit restricted performances. Here, we propose a general approach to design metadevices that can efficiently generate two distinct VOFs under illuminations of circularly polarized lights with different helicity. After illustrating our scheme via both Jones matrix analyses and analytical model calculations, we experimentally demonstrate two metadevices in the near-infrared regime, which can generate vortex beams carrying different orbital angular momenta yet with distinct inhomogeneous polarization distributions. Our results provide an ultracompact platform for bifunctional generations of VOFs, which may stimulate future works on VOF-related applications in integration photonics.

Keywords: bifunctional metasurfaces; local polarization distributions; Pancharatnam–Berry phase; resonance phase; spin-decoupled functionalities; vectorial optical fields.

Dongyi Wang, Tong Liu, and Yuejiao Zhou contributed equally to this work.

***Corresponding authors: Qiong He and Lei Zhou,** State Key Laboratory of Surface Physics, Key Laboratory of Micro and Nano Photonic Structures (Ministry of Education) and Department of Physics, Fudan University, Shanghai 200438, China; Academy for Engineering and Technology, Fudan University, Shanghai 200433, China; and Collaborative Innovation Centre of Advanced Microstructures, Nanjing 210093, China, E-mail: qionghe@fudan.edu.cn (Q. He), phzhou@fudan.edu.cn (L. Zhou) https://orcid.org/0000-0002-4966-0873 (Q. He)
Dongyi Wang, Tong Liu, Yuejiao Zhou and Xiaoying Zheng, State Key Laboratory of Surface Physics, Key Laboratory of Micro and Nano Photonic Structures (Ministry of Education) and Department of Physics, Fudan University, Shanghai 200438, China, E-mail: dywang17@fudan.edu.cn (D. Wang), tongliu16@fudan.edu.cn (T. Liu), 18110190015@fudan.edu.cn (Y. Zhou), 18110190014@fudan.edu.cn (X. Zheng). https://orcid.org/0000-0002-8947-707X (T. Liu)
Shulin Sun, Shanghai Engineering Research Centre of Ultra-Precision Optical Manufacturing, Green Photonics and Department of Optical Science and Engineering, Fudan University, Shanghai 200433, China, E-mail: sls@fudan.edu.cn. https://orcid.org/0000-0003-3046-1142

1 Introduction

Vectorial optical fields (VOFs) are special solutions of Maxwell's equations, which exhibit well-defined wave fronts and tailored inhomogeneous distributions of polarization (also called "spin") state [1, 2]. The latter, unique to electromagnetic (EM) waves being vectorial in nature, make VOFs particularly useful in many applications such as optical communications, biosensing and chemical sensing, particle trapping and high-resolution imaging [2, 3]. However, conventional approaches to generate VOFs require separate devices to control wave fronts (e.g., spatial light modulators [4]) and inhomogeneous polarization distributions (e.g., Q-plate [5] or spiral phase elements [6]) of light, which are bulky and complicated. Moreover, usually a single system can only generate a particular VOF. All these limitations make conventional devices unfavorable for integration photonics applications.

Metasurfaces, ultrathin metamaterials composed by planar subwavelength microstructures (e.g., meta-atoms) exhibiting tailored optical properties, attracted immense interests recently owing to their unprecedented capabilities to control EM waves [7–9]. Designing metasurfaces to exhibit certain anisotropic or spatially inhomogeneous phase distributions for reflected/transmitted waves, researchers have demonstrated separate manipulations on polarization [10–12] or wave front [13–18] properties of EM waves, leading to many practical applications (e.g.,

This article has previously been published in the journal Nanophotonics. Please cite as: D. Wang, T. Liu, Y. Zhou, X. Zheng, S. Sun, Q. He and L. Zhou "High-efficiency metadevices for bifunctional generations of vectorial optical field" *Nanophotonics* 2021, 10. DOI: 10.1515/nanoph-2020-0465.

polarization control [10–12], light-bender [13–17], metalens [19–23], and metahologram [24–27]). More recently, metadevices exhibiting combined functionalities of polarization and wave front manipulations were proposed [18, 28–39], some of which could already generate particular VOFs as desired [29–31, 37–43] These devices are generally flat and ultracompact, being very promising for on-chip photonic applications.

Despite of the fast developments along this research direction, several limitations still hinder the practical applications of these VOF metagenerators. First of all, metadevices so far proposed can usually generate only one particular VOF [18, 30, 31, 38, 40–42], while multi-functional VOF generators are highly desired in future applications. Secondly, most reported metadevices can only generate VOFs with restricted polarization distributions (e.g., standard cylindrical polarization distributions [18, 31, 39, 44–46]), while VOFs with *arbitrary* polarization distributions are rarely seen. The inherent physics behind such issues are that previous approaches only utilized a single mechanism (either resonance or geometric mechanism) to design meta-atoms in constructing metadevices and only explored certain polarization-control capabilities of the constitutional meta-atoms [28–32, 38, 45–49]. Although some sporadic works have reported the VOF generations employing multiple mechanisms [46, 50], few of those designs was demonstrated experimentally in optical regime.

In this paper, we propose a generic approach to design metadevices that can efficiently generate two distinct VOFs with *designable* polarization distributions, upon excitations of circularly polarized (CP) lights with two opposite helicity. We achieve this end by choosing meta-atoms possessing reflection phases governed by two different mechanisms (resonance and geometric ones) and taking *full* use of the polarization-control capabilities of constitutional meta-atoms. After explaining the basic concept with Jones matrix analyses, we explicitly illustrate our strategy based on analytical Green's function (GF) calculations. As the proof of our concept, we experimentally demonstrate two metadevices working at telecom wavelengths, which can achieve bifunctional generations of distinct VOFs exhibiting vortex wave fronts with different orbital angular momenta (OAM) and inhomogeneous polarization distributions including standard and more general ones. We finally discuss potential applications of our strategy and present our own perspectives on possible future works, before concluding this paper.

2 Results and discussions

2.1 Basic concept

We start from discussing the phase distributions required by our metadevices, which are supposed to be able to generate two distinct VOFs under illuminations of CP lights with different helicity (see Figure 1b). In this paper, we study reflective metasurfaces as an example to illustrate our key idea and extensions to the transmissive case are straightforward. To realize such spin-delinked dual functionalities, our meta-atoms should exhibit both spin-*dependent* geometric phases and spin-*independent* resonant phases, as already pointed out in recent literature [33, 35, 39, 46, 51, 52]. Moreover, to generate a predesigned polarization distribution in the VOF, our meta-atoms should further possess desired local polarization-control capabilities, corresponding to certain paths on Poincare's sphere (see Figure 1a).

Based on the above two requirements, we consider a generic reflective meta-atom exhibiting mirror symmetries with respect to two principle axes denoted as u and v, which is then rotated by an angle ξ with respect to the laboratory coordinate system (see inset to Figure 1a). Jones matrix of such a meta-atom can be written as $\mathbf{R} = \mathbf{M}(\xi)\begin{pmatrix} r_{uu} & 0 \\ 0 & r_{vv} \end{pmatrix}\mathbf{M}^{-1}(\xi)$ in linear polarization (LP) bases in laboratory coordinate system, where $\mathbf{M}(\xi) = \begin{pmatrix} \cos\xi & -\sin\xi \\ \sin\xi & \cos\xi \end{pmatrix}$, $r_{uu} = e^{i\Phi_u}$ and $r_{vv} = e^{i\Phi_v}$. Here, we neglect material losses as we establish the theoretical framework, and later we show that adding losses back does not significantly change the established picture. Transform the bases from LP ones to CP ones, the resulting Jones matrix reads $\tilde{\mathbf{R}} = \mathbf{S}\mathbf{R}\mathbf{S}^{-1}$ with $\mathbf{S} = \frac{\sqrt{2}}{2}\begin{pmatrix} 1 & -i \\ 1 & i \end{pmatrix}$ [48, 49]. Shine such a meta-atom by CP light exhibiting different spin $|\sigma\rangle$ with $|+\rangle = \begin{pmatrix} 1 \\ 0 \end{pmatrix}$ and $|-\rangle = \begin{pmatrix} 0 \\ 1 \end{pmatrix}$ denoting left circular polarization (LCP) and right circular polarization (RCP), respectively, we find that the reflected wave becomes

$$\tilde{\mathbf{R}}|\sigma\rangle = e^{i\Phi_{\text{ini}}^{\sigma}}|f^{\sigma}\rangle. \tag{1}$$

Here, $|f^{\sigma}\rangle = \begin{pmatrix} e^{-i\Psi^{\sigma}/2}\cos(\Theta^{\sigma}/2) \\ e^{i\Psi^{\sigma}/2}\sin(\Theta^{\sigma}/2) \end{pmatrix}$ denotes the polarization state of the reflected wave represented by a point $(\Theta^{\sigma}, \Psi^{\sigma})$ on Poincare's sphere and $\Phi_{\text{ini}}^{\sigma}$ is the initial phase carried by

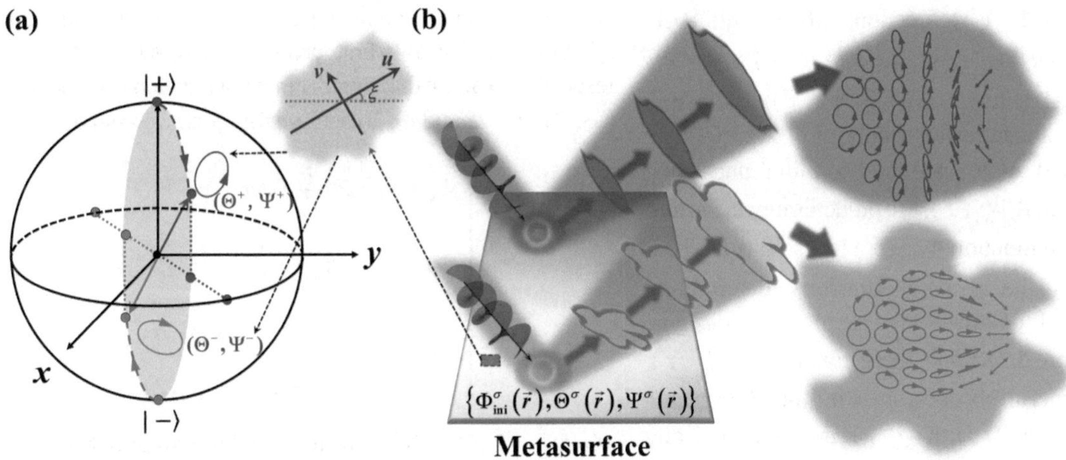

Figure 1: Schematics of spin-delinked bifunctional vectorial optical field (VOF) generations.
(a) Dashed lines in blue and purple colors illustrate how the polarization change from initial left circular polarization (LCP) and right circular polarization (RCP) states to two final states denoted by two points (Θ^+, Ψ^+) and (Θ^-, Ψ^-) on Poincare's sphere, as our meta-atom is shined by incident CP light with different helicity. Inset to (a) illustrates the local coordinate of a meta-atom. (b) Schematics of light scatterings at a metasurface exhibiting pre-designed distributions of $\{\Phi^\sigma_{ini}(\vec{r}), \Theta^\sigma(\vec{r}), \Psi^\sigma(\vec{r})\}$ under LCP and RCP illuminations, generating two distinct reflected beams with different wave fronts and inhomogeneous polarization distributions.

the reflected wave. It is worth noting that two angles $(\Theta^\sigma, \Psi^\sigma)$ unambiguously determine the polarization state, with $\cos\Theta^\sigma$ representing the ellipticity and $\Psi^\sigma/2$ dedicating the polar angle of the polarization.

This set of parameters, $\{\Theta^\sigma, \Psi^\sigma, \Phi^\sigma_{ini}\}$, are determined by the properties of our meta-atoms $\{\Phi_u, \Phi_v, \xi\}$. For the cases of $\Delta\Phi = \Phi_v - \Phi_u \neq \pm\pi$, we find

$$\begin{cases} \Theta^+ = \Delta\Phi, & \Psi^+ = 2\xi - \dfrac{\pi}{2} \\ \Theta^- = \pi - \Delta\Phi, & \Psi^- = 2\xi + \dfrac{\pi}{2} \end{cases} \tag{2}$$

and

$$\Phi^\sigma_{ini} = \Phi_{res} + \sigma \cdot \Phi_{geo} \tag{3}$$

where $\Phi_{res} = \overline{\Phi} - \pi/4$ (with $\overline{\Phi} = (\Phi_u + \Phi_v)/2$) is a spin-*independent* term originated from structural resonances and $\Phi_{geo} = \xi$ is a spin-*dependent* term originated from the geometric effect (e.g., structural rotation). The case of $\Delta\Phi = \pm\pi$ corresponds to a singular point in the gauge that we choose here, and $\Psi^\sigma, \Phi_{res}, \Phi_{geo}$ take different expressions:

$$\Psi^\sigma = 0\,; \quad \Phi_{res} = \arg\left(e^{i\Phi_u} - e^{i\Phi_v}\right); \quad \Phi_{geo} = 2\xi. \tag{4}$$

Equations (1)–(4) contain the crucial physics presented in this paper. After reflections by such a meta-atom, the polarization of light can change from the initial LCP or RCP to different final states represented by $(\Theta^\sigma, \Psi^\sigma)$ (see Figure 1a). Meanwhile, initial phases Φ^σ_{ini} carried by waves reflected by different meta-atoms under different CP excitations are useful for constructing VOFs exhibiting predesigned wave fronts (see Figure 1b). Therefore, if we choose a set of meta-atoms with appropriate scattering properties to form a metasurface exhibiting certain $\Theta^\sigma(\vec{r})$, $\Psi^\sigma(\vec{r})$ and $\Phi^\sigma_{ini}(\vec{r})$ distributions, we can thus realize bifunctional generations of VOFs with different wave fronts (dictated by $\Phi^\sigma_{ini}(\vec{r})$) and polarization distributions (dictated by $\Theta^\sigma(\vec{r})$, $\Psi^\sigma(\vec{r})$) as shining the metasurface by CP lights with different spin.

Before closing this sub-section, we discuss several important physics. First of all, the presence of both spin-*independent* Φ_{res} and spin-*dependent* Φ_{geo} in Eq. (3) is crucial to construct two *discorrelated* phase distributions (i.e., $\Phi^+_{ini}(\vec{r})$ and $\Phi^-_{ini}(\vec{r})$), making spin-delinked bifunctional controls possible. Secondly, varying $\Delta\Phi$ and ξ can *continuously* change the polarization state of the reflected wave (see Eq. (2)), which is crucial to realize desired inhomogeneous polarization distributions. Thirdly, we note that for the cases of $\Delta\Phi \neq \pm\pi$, the geometric phase $\Phi_{geo} = \xi$ does not exhibits the standard Pancharatnam–Berry (PB) form (i.e., $\Phi_{PB} = 2\xi$ [48, 53–56]). The underlying physics is that the final spin states in such cases carry a phase factor $e^{i\xi}$, which is different from the special case of $\Delta\Phi = \pm\pi$. More discussions on this point are presented in Sec. 1 of Supplementary material (SM).

Finally, we discuss the limitations of our proposed bifunctional VOF generations. Since the meta-atoms we choose here only exhibit three independent parameters (i.e., Φ_u, Φ_v and ξ, or equivelantly, $\Delta\Phi$, Φ_{res} and ξ), they can *not* be used to realize metasurfaces with *freely* designable

$\Theta^\sigma(\vec{r})$, $\Psi^\sigma(\vec{r})$ and $\Phi^\sigma_{ini}(\vec{r})$ functions which require six free parameters. In fact, Eq. (2) shows that $(\Theta^+(\vec{r}), \Psi^+(\vec{r}))$ are inherently correlated with $(\Theta^-(\vec{r}), \Psi^-(\vec{r}))$, indicating that the resulting polarization distributions of two VOFs are ultimately connected. Also, once the two initial phase distributions ($\Phi^+_{ini}(\vec{r})$ and $\Phi^-_{ini}(\vec{r})$) are predetermined, we find from Eq. (3) that the function $\Phi_{geo}(\vec{r})$ is also fixed, which subsquently locks the $\Psi^\pm(\vec{r})$ functions determining the polar angles of polarizations. In practical designs, we usually only pre-determine three functions $\{\Phi^+_{ini}(\vec{r}), \Phi^-_{ini}(\vec{r}), \Theta^+(\vec{r})\}$ from all six ones, since the remaining three ones (i.e., $\{\Theta^-(r^\rightarrow), \Psi^+(r^\rightarrow), \Psi^-(r^\rightarrow)\}$) are automatically determined by $\{\Phi^+_{ini}(\vec{r}), \Phi^-_{ini}(\vec{r}), \Theta^+(\vec{r})\}$ as shown in Eqs. (2)–(4). We emphasize that the designable distributions of $\Theta^+(\vec{r})$ enable the full control on both local ellipticity and inhomogeneous helicity.

2.2 Verifications by GF calculations on an ideal model

We now illustrate how to implement our strategy based on GF calculations on an ideal model. As a particular example, we choose to generate two divergent vectorial vortex beams carrying different OAMs (with topological charges $l = 2$ and $l = 0$, respectively) and exhibiting distinct polarization distributions (see Figures 2a and d). To achieve this goal, we assume that:

$$\begin{cases} \Phi^+_{ini}(\vec{r}) = 2\pi\frac{r}{P_s} + 2\varphi \\ \Phi^-_{ini}(\vec{r}) = 2\pi\frac{r}{P_s} - \frac{\pi}{2} \\ \Theta^+(\vec{r}) = \begin{cases} \varphi & 0 \le \varphi < \pi \\ 2\pi - \varphi & \pi \le \varphi < 2\pi \end{cases} \end{cases} \tag{5}$$

where $P_s = 1.55\lambda$ with λ being the working wavelength. φ and r are the polar angle and radius of a vector $\vec{r}$ in cylindrical coordinate system. Based on the correlations set up in Eqs. (2) and (3), we can readily retrieve the remaining three functions

$$\begin{cases} \Theta^-(\vec{r}) = \begin{cases} \pi - \varphi & 0 \le \varphi < \pi \\ \varphi - \pi & \pi \le \varphi < 2\pi \end{cases} \\ \Psi^+(\vec{r}) = 2\varphi \\ \Psi^-(\vec{r}) = 2\varphi + \pi \end{cases} \tag{6}$$

Equations (5) and (6) reveal the properties of two VOFs to be generated. The first two lines in Eq. (5) suggest that two reflected beams exhibit vortex wave fronts carrying

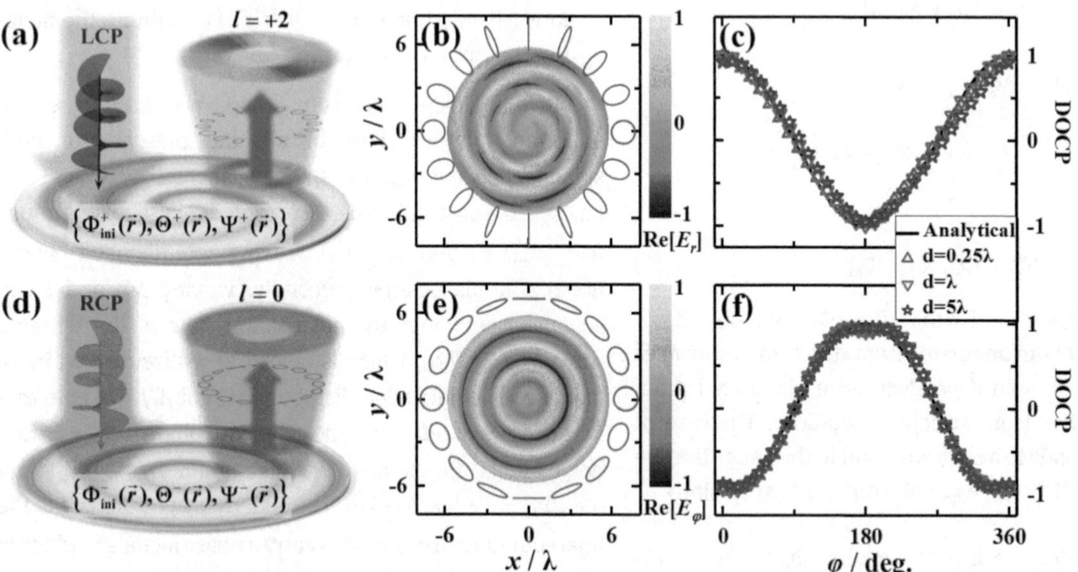

Figure 2: Analytical calculations on a model system.
Schematics of light scatterings by a model metasurface exhibiting $\{\Phi^\sigma_{ini}(\vec{r}), \Theta^\sigma(\vec{r}), \Psi^\sigma(\vec{r})\}$ distributions as given by Eqs. (5) and (6), under illuminations of (a) left circular polarization (LCP) and (d) right circular polarization (RCP) incidences. (b) $Re[E_r]$ distribution on an *xy* plane at a distance $d = \lambda$ above the metasurface for reflected beam under LCP incidence, and (e) $Re[E_\varphi]$ distribution on an *xy* plane at a distance $d = \lambda$ above the metasurface for reflected beam under RCP incidence, calculated by the Green's function approach. Ellipses illustrate the polarization patterns at different angles. Degrees of circular polarization as functions of the azimuthal angle φ, calculated by the Green's function approach for reflected waves on *xy* planes at different distances (denoted by *d*) above the metasurface under (c) LCP and (f) RCP incidences, respectively. Solid lines in (c) and (f) are directly obtained from the final polarizations defined by Eqs. (5) and (6).

different topological charges. Meanwhile, the third line in Eq. (5) and the first line in Eq. (6) indicate that the ellipticity of local polarization changes dramatically inside two reflected beams. Finally, the last two lines in Eq. (6) imply that polar angle of local polarization also changes as varying φ, which are along the radial and tangential directions for the cases of LCP and RCP incidences, respectively.

With Eq. (5) known, we numerically retrieve the scattering properties of all meta-atoms according to Eqs. (2)–(4), represented by the $\Phi_u(\vec{r})$, $\Phi_v(\vec{r})$ and $\xi(\vec{r})$ distributions. Knowing these information, we then employ the standard GF approach to study the beams reflected by our model system, simply replacing each meta-atom by appropriate oscillating surface current yielding the same scattered field (see Sec. 2 in SM for more details). Figures 2b and e depict, respectively, the distributions of scattered $\mathrm{Re}[E_r]$ and $\mathrm{Re}[E_\varphi]$ on a plane $d = \lambda$ above the metasurface, as it is shined by CP lights with different spin. We could clearly see that the reflected beams exhibit OAM features with topological charges $l = 2$ and $l = 0$, respectively. Meanwhile, we also calculated the polarization patterns at different local points inside the reflected beams and then schematically illustrated in Figures 2b and e how the polarization changes along the azimuthal angle (see those ellipses surrounding the patterns). Quantatively, we depict in Figures 2c and f how the local degree of circular polarization (DOCP), defined as $\mathrm{DOCP} = (|E_r - iE_\varphi|^2 - |E_r + iE_\varphi|^2)/(|E_r - iE_\varphi|^2 + |E_r + iE_\varphi|^2)$, vary against the azimuthal angle φ, on xy planes at different distances above the metasurface. These results clearly show that the generated inhomogeneous polarization distribution preserves well as the beams propagate in air, in consistency with our analytical predictions (solid lines in Figure 2c and f) based on the predetermined $(\Theta^\pm(\vec{r}), \Psi^\pm(\vec{r}))$ functions. Details on the GF approach and more analytical results can be found in Sec. 2 of SM.

2.3 Meta-atom designs

We now choose the near infrared (NIR) regime to experimentally demonstrate our idea, starting from designing appropriate meta-atoms. As shown in the inset to Figure 3a, our basic meta-atom is in metal-insulator-metal (MIM) configuration, which consists of a gold (Au) cross-shaped resonator (with principle axes rotated by an angle ξ) and a continuous 125-nm-thick Au film separated by a 100-nm-thick SiO_2 spacer. Such an MIM meta-atom can reflect incident lights polarized along two principle axes efficiently (100% in ideal lossless cases), with reflection phases Φ_u and Φ_v dictated by two bar-lengths L_u and L_v.

We perform finite-difference time-domain (FDTD) simulations to study how Φ_u and Φ_v vary against L_u and L_v at the working wavelength of 1550 nm, with material losses fully taken into account. Figure 3a shows that Φ_u sensitively depends on L_u and varying L_u can drive Φ_u to near cover the whole 2π range, manifesting a typical magnetic resonance behavior [57]. Meanwhile, Φ_v exhibits a similar dependence on L_v (see more simulation results in Fig. S4 in SM). For the benefits of future designs, we further depict in Figures 3b and c how Φ_{res} and $\Delta\Phi$ vary against L_u and L_v, respectively. There results can help us quickly sort out the geometric parameters (L_u, L_v) and orientation angle of meta-atoms located at different points, from the $\{\Phi_{res}(\vec{r}), \Delta\Phi(\vec{r}), \xi(\vec{r})\}$ distributions retrieved from the pre-determined $\{\Phi^+_{ini}(\vec{r}), \Phi^-_{ini}(\vec{r}), \Theta^+(\vec{r})\}$ distributions.

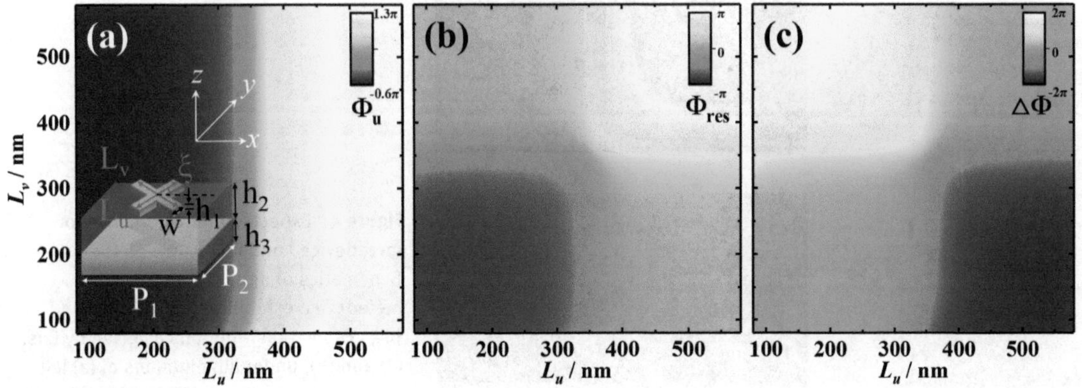

Figure 3: Meta-atom designs at the wavelength of 1550 nm.
(a) finite-difference time-domain (FDTD)–computed reflection phase Φ_u versus L_u and L_v for meta-atoms (see inset) under illuminations of $\hat{u}$-polarized light. Other geometrical parameters of the meta-atoms are fixed as $h_1 = 30$ nm, $h_2 = 100$ nm, $h_3 = 125$ nm, $w = 80$ nm, and $P_1 = P_2 = 600$ nm. (b) Resonance phase Φ_{res} and (c) cross-polarization phase difference $\Delta\Phi$ versus L_u and L_v, computed by FDTD simulations for our meta-atoms. The working wavelength is set as 1550 nm.

2.4 Experimental demonstrations

2.4.1 Metadevice I: bifunctional generations of cylindrically polarized beams

We now experimentally verify our concept, starting from demonstrating the most standard VOFs exhibiting cylindrical polarizations. Without losing generality, we assume our metadevice to exhibit the following distributions:

$$\Phi_{\text{ini}}^{+}(\vec{r}) = 3\varphi; \quad \Phi_{\text{ini}}^{-}(\vec{r}) = \varphi - \frac{\pi}{2}; \quad \Theta^{+}(\vec{r}) = \frac{\pi}{2} \tag{7}$$

which, based on the inherent restrictions set in Eqs. (2) and (3), yield the following explicit forms of the remaining three functions :

$$\Theta^{-}(\vec{r}) = \frac{\pi}{2}; \quad \Psi^{+}(\vec{r}) = 2\varphi; \quad \Psi^{-}(\vec{r}) = 2\varphi + \pi \tag{8}$$

Equations (7) and (8) clearly reveal the expected properties of two VOFs to be generated. Distributions of $(\Phi_{\text{ini}}^{+}(\vec{r}), \Phi_{\text{ini}}^{-}(\vec{r}))$ in Eq. (7) indicate that two reflected beams are of vortex wave fronts carrying different topological charges ($l = 3$ and $l = 1$, see Figures 4a and b). Yet, $\Theta^{+}(\vec{r}) = \Theta^{-}(\vec{r}) = \pi/2$ dictates that all local spin states inside the two reflected beams should be LPs. Finally, $\Psi^{+}(\vec{r})$ and $\Psi^{-}(\vec{r})$ distributions in Eqs. (7) and (8) suggest that the polar directions of these LPs are along the radial or tangential directions, respectively, for two different input spins.

Following the strategy described in last subsection, we first retrieve from Eqs. (7) and (8) the distributions of $\{\Phi_{\text{res}}(\vec{r}), \Delta\Phi(\vec{r}), \xi(\vec{r})\}$, which further assist us to determine the geometric parameters of all meta-atoms $\{L_u(\vec{r}), L_v(\vec{r}), \xi(\vec{r})\}$ based on Figures 3b and c. We finally design and fabricate out the sample (see scanning electron microscopy (SEM) pictures in Figures 4c and d) according to these parameter distributions, and characterize the scattering properties of this sample (Figure 5).

Figures 5a–c illustrate the essential properties of the beam reflected by our metadevice under LCP plane wave excitation. To clearly characterize the vortex properties of the generated VOF, we perform interference measurement with a homemade Michelson interferometer. Interference between the generated VOF and an incident spherical wave yields a 3rd-order spiral shape in the interference pattern (see dashed lines in Figure 5a), which is the clear evidence of an $l = 3$ vortex. Interferences with a plane wave reinforced our conclusion (see Figs. S5 in SM). We now experimentally characterize the polarization distribution of the generated VOF. Placing a linear polarizer in front of our charge-coupled device (CCD), we find that the recorded intensity image changes dramatically as we rotate the polarizer, visualizing the desired inhomogeneous polarization distribution as expected (see Figs. S6

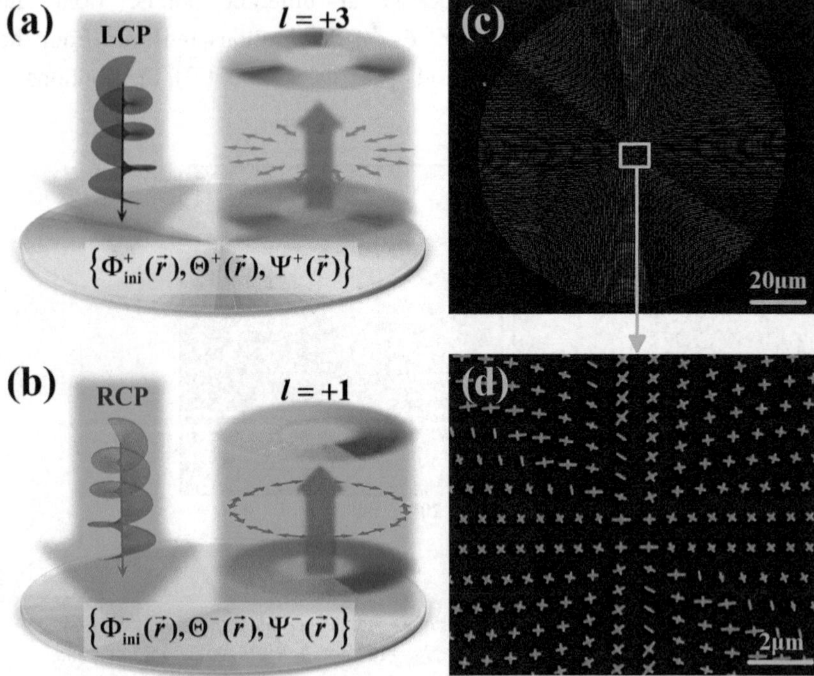

Figure 4: Expected bifunctionalities of metadevice I and its sample picture. Schematics of light scatterings at a metadevice exhibiting $\{\Phi_{\text{ini}}^{+}, \Theta^{+}, \Psi^{+}\}$ and $\{\Phi_{\text{ini}}^{-}, \Theta^{-}, \Psi^{-}\}$ distributions as given by Eqs. (7) and (8), under illuminations of (a) left circular polarization (LCP) and (b) right circular polarization (RCP) incidences. (c) Scanning electron microscopy (SEM) image and (d) a zoom-in view of the fabricated sample.

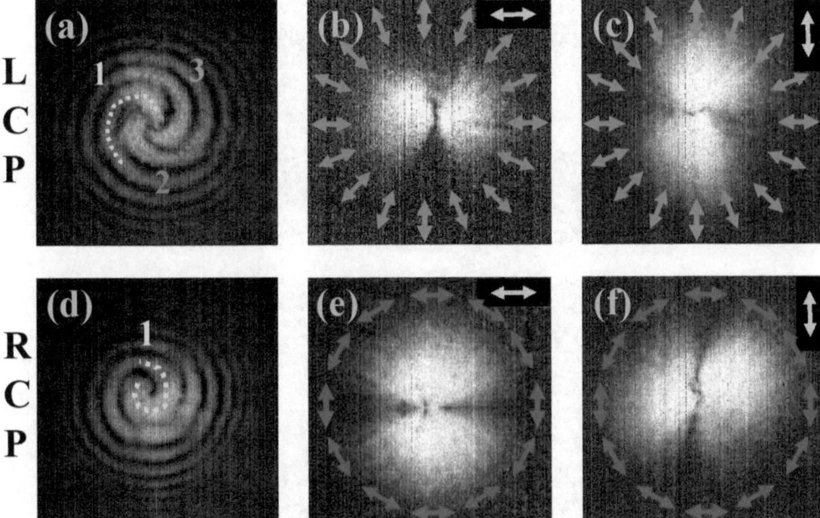

Figure 5: Experimental characterizations on metadevice I.
Measured far-field interference patterns between a spherical reference wave and light beams reflected by metadevice I under (a) left circular polarization (LCP) and (d) right circular polarization (RCP) incidences, respectively. Far-field images recorded by our charge-coupled device (CCD) with a linear polarizer placed in front of it, tilted by an angle of (b, e) 0° and (c, f) 90°, as metadevice I is illuminated by (b, c) LCP and (e, f) RCP incidences, respectively. Arrows in (b, c, e, f) illustrate the expected polarization patterns at different angles.

in SM). In particular, the image profile obtained with our polarizer placed horizontally (Figure 5b) or vertically (Figure 5c) exhibits a nice donut shape with intensity zeros appearing at the angles perpendicular to the polarizer, well illustrating the radial polarization distribution as expected.

Under the RCP incidence, however, both wave front and polarization distribution of the reflected beam change dramatically. Now the interference pattern contains a 1st-order spiral shape indicating that $l = 1$ (Figure 5d), in consistency with our expectation. Meanwhile, repeating the intensity measurements with a rotating polarizer, we find that now the intensity zeros appear at the directions parallel to the polarizer (Figures 5e and f), indicating that now the polarization distribution changes to a tangentially polarized one, as expected. More experimental results can be found in Sec. 4 of SM.

Since it is difficult to experimentally characterize the working efficiency of our metadevice due to the technical limitation of our experimental setup, the highly inhomogeneous polarization distribution of generated VOFs and the presence of undesired stray light, we numerically estimate it as an average of efficiencies of all individual meta-atoms constructing our metadevice. Due to material loss which is the only reason, the efficiency of our metadevices is limited to 55% at working wavelength but still comparable to those of recent PB metadevices at different working frequencies [25, 48, 56]. We emphasize that the working efficiency of our metadevices can be further improved to 100% by constructing our meta-atoms with less lossy materials (e.g. dielectric resonators) (see Section 7 in SM).

2.4.2 Metadevice II: bifunctional generations of vectorial beams beyond cylindrical polarizations

We proceed to experimentally demonstrate another device, which, upon excitations of CP lights with different spin, can generate vectorial beams with polarization distributions *beyond* the standard cylindrical ones. Explicitly, we require metadevice II to exhibit the following expressions for the chosen three functions:

$$\Phi^{+}_{\text{ini}}(\vec{r}) = 3\varphi; \quad \Phi^{-}_{\text{ini}}(\vec{r}) = -\varphi - \frac{\pi}{2}; \quad \Theta^{+}(\vec{r}) = \frac{\pi}{2}. \tag{9}$$

According to Eqs. (2)–(4), we immediately obtain the forms of remaining functions:

$$\Theta^{-}(\vec{r}) = \frac{\pi}{2}; \quad \Psi^{+}(\vec{r}) = 4\varphi; \quad \Psi^{-}(\vec{r}) = 4\varphi + \pi \tag{10}$$

Compared to Eqs. (7) and (8), the most crucial difference is that, for this device, the orientations of local LPs are no longer along $\hat{e}_r$ and $\hat{e}_\varphi$ directions, as revealed by its $\{\Psi^{+}(\vec{r}), \Psi^{-}(\vec{r})\}$ distributions. Meanwhile, two reflected beams still exhibit vortex wave fronts but with topological charges $l = 3$ and $l = -1$, respectively. Figures 6a and e schematically depict the expected wave front and polarization properties of beams reflected by metadevice II under LCP and RCP incidences, respectively. Following the design strategy as discussed in Sec. 2.3, we successfully retrieve from Eqs. (9) and (10) the geometric parameters $\{L_u(\vec{r}), L_v(\vec{r}), \xi(\vec{r})\}$ of all needed meta-atoms, which help us to finally fabricate out the sample (see Fig. S11 in SM for its SEM picture).

We perform experiments similar to those in last subsection to characterize the essential properties of two reflected

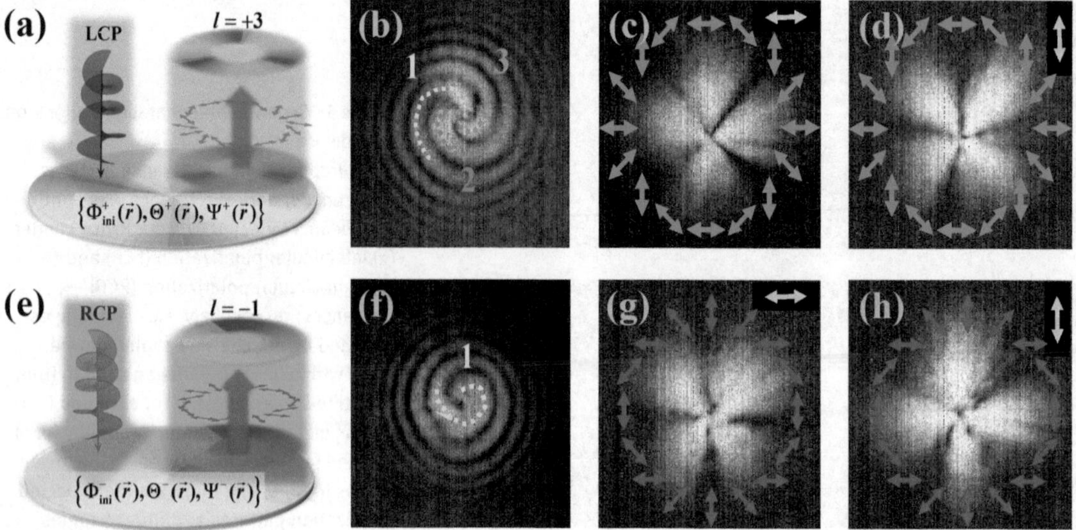

Figure 6: Experimental characterizations of metadevice II.
Schematics of light scatterings at a metadevice exhibiting $\{\Phi^+_{\text{ini}}, \Theta^+, \Psi^+\}$ and $\{\Phi^-_{\text{ini}}, \Theta^-, \Psi^-\}$ distributions as given by Eqs. (9) and (10), under illuminations of (a) left circular polarization (LCP) and (e) right circular polarization (RCP) incidences. Measured far-field interference patterns between a spherical reference wave and light beams reflected by metadevice II under (b) LCP and (f) RCP incidences, respectively. Far-field images recorded by our CCD with a linear polarizer placed in front of it, tilted by an angle of (c, g) 0° and (d, h) 90°, as metadevice II is illuminated by (c, d) LCP and (g, h) RCP incidences, respectively. Arrows in (c, d, g, h) illustrate the expected polarization patterns at different angles.

beams. Inferenced with a spherical wave, the resulting patterns show that now the reflected beam under LCP incidence exhibits an OAM with $l = +3$ (Figure 6b) while that under RCP incidence exhibits an OAM of $l = -1$ (Figure 6f), manifested by the opposite spiral direction as compared to Figure 5d. Meanwhile, Figures 6c, d, g, and h depict the measured intensity patterns of two reflected beams with a linear polarizer placed horizontally or vertically, respectively. Again, compared to those shown in Figure 5, here more intensity zeros appear in the measured patterns, at angles precisely consistent with the expected polarization distributions depicted in the figure. The working efficiency of our metadevice is numerically evaluated as 55% (see more details in Sec. 7 of SM). One can notice some image distortions in Figures 6b–h, which are caused by the material loss–induced stray light with undesired OAM and local polarization properties. Such issue reducing the working efficiency of metadevices could be solved by employing lossless meta-atom designs such as dielectric resonators.

We emphasize that the polarization distributions of the generated VOFs are not confined to those realized in Figure 6, but can in principle be rather general. Equations (2)–(4) reveal that the ellipticity distribution of polarization pattern (dictated by $\Theta^+(\vec{r})$) can be freely chosen, so that the local polarizations are not necessary LPs. Meanwhile, the distribution of local polar direction, dictated by the function $\Psi^+(\vec{r})$, can also change easily, although it is intimately linked with the two expected wave fronts via $\Psi^+(\vec{r}) = \Phi^+_{\text{ini}}(\vec{r}) - \Phi^-_{\text{ini}}(\vec{r}) - \pi/2$. Therefore, through tuning these functions $(\Theta^+(\vec{r}), \Psi^+(\vec{r}))$ we can readily design metadevices to realize highly nontrivial polarization distributions beyond the standard ones. As a particular example, we successfully retrieve the geometric parameters $\{L_u(\vec{r}), L_v(\vec{r}), \xi(\vec{r})\}$ of all needed meta-atoms to form a metadevice exhibiting the functionalities specified by Eqs. (5) and (6). The device layout is shown in Fig. S14 in SM. We have numerically studied the wave front and polarization properties of two VOFs generated by such a device, with realistic material losses fully taken into account. The computed results clearly verify our predictions that two generated VOFs indeed exhibit inhomogeneous distributions of elliptical polarizations dictated by the functions $(\Theta^\pm(\vec{r}), \Psi^\pm(\vec{r}))$ as specified in Eqs. (5) and (6) (see more details in Section 6 of SM). Such numerical calculations also well validate our design strategy established in Section 2 based on ideal meta-atoms.

3 Conclusions and perspectives

In short, we have established a generic strategy to design high-efficiency metadevices to bifunctionally create complex VOFs with desired wave fronts and polarization distributions, through exploring the full capabilities of meta-atoms in controlling light polarizations and combining two different mechanisms (resonance phases and geometric phases) to generate phase shifts for incident light. Based on the established guidelines, we design and fabricate two metadevices working at telecom wavelengths and experimentally demonstrate their bifunctional generations of two vortex vectorial beams possessing different topological charges and distinct polarization distributions, as shined by CP light with different helicity.

Our results pave the road to generate complex VOFs with desired properties, which may inspire many future works on both fundamental and application sides of research. For example, switching the LCP and RCP incidences to two cross-polarized LP ones, the resulting VOFs generated by our devices, obtained by linear combinations of previous two, thus change accordingly. These new patterns not only provide more VOFs for potential applications but also make dynamical tuning of the VOFs possible. Moreover, generating VOFs with truly delinked properties and their optical characterizations are very interesting and challenging future works.

4 Materials and methods

4.1 Numerical simulations

We perform FDTD simulations using numerical software Concerto 7.0. The permittivity of Au is described by the Drude model $\varepsilon_r(\omega) = \varepsilon_\infty - \frac{\omega_p^2}{\omega(\omega+i\gamma)}$ with $\varepsilon_\infty = 9, \omega_p = 1.367 \times 10^{16}\,\mathrm{s}^{-1}, \gamma = 1.224 \times 10^{14}\,\mathrm{s}^{-1}$, obtained by fitting with experimental results. The SiO_2 spacer was considered as a lossless dielectric with permittivity $\varepsilon = 2.085$. Additional losses caused by surface roughness and grain boundary effects in thin films, as well as dielectric losses are effectively considered in fitting the parameter γ.

4.2 Sample fabrications

All samples were fabricated using standard thin-film deposition and electron-beam lithography (EBL) techniques. We firstly deposit 5-nm Cr, 125-nm Au, 5-nm Cr and a 100-nm SiO_2 dielectric layer onto a silicon substrate using magnetron DC-sputtering (Cr and Au) and RF-sputtering (SiO_2). Secondly, we lithographed the cross structures with EBL employing a ~100-nm-thick PMMA2 layer at an acceleration voltage of 20 keV. After development in a 1:3 solution of methyl isobutyl ketone and isopropyl alcohol, a 5-nm Cr adhesion layer and a 30-nm Au layer were deposited subsequently using thermos evaporation. The Au patterns were finally formed on top of the SiO_2 film after a lift-off process using acetone.

4.3 Experimental setup

We use a homemade NIR microimaging system equipped with an NIR CCD (NIRvana: 640-ST from Princeton Instruments) and an additional interference optical path to characterize the performances of our metadevices. A broadband supercontinuum laser source and a fiber-coupled grating spectrometer (ideaoptics NIR2500) were used in far-field measurement. Beam splitter, linear polarizer, and visible CCD are also used to measure the reflectance and analyze the polarization distributions. More details can be found in SM.

Acknowledgements: This work was funded by National Natural Science Foundation of China (No. 11734007, No. 91850101, No. 11674068 and No. 11874118), National Key Research and Development Program of China (No. 2017YFA0303504 and No. 2017YFA0700201), Natural Science Foundation of Shanghai (No. 20JC1414601 and No.18ZR1403400), Fudan University-CIOMP Joint Fund (No. FC2018-006). L. Zhou and Q. He acknowledge technical supports from the Fudan Nanofabrication Laboratory for sample fabrications.

Author contribution: D.W., T.L. and Y.Z. contributed equally to this work. D.W. carried out simulations, fabricated the samples and conducted part of the measurements; T.L. did the theoretical calculations and designed the samples; Y.Z. and X.Z. built the experimental setup and conducted part of measurements; S.S. provided technical supports for simulations and data analyses. L.Z. and Q.H. conceived the idea and supervised the project. All the authors contributed to the preparation of the manuscript, and have accepted responsibility for the entire content of this submitted manuscript and approved submission.

Research funding: This work was funded by National Natural Science Foundation of China (No. 11734007, No. 91850101, No. 11674068, No. 11874118), National Key Research and Development Program of China (No. 2017YFA0303504 and No. 2017YFA0700201), Natural Science Foundation of Shanghai (No. 20JC1414601 and No.18ZR1403400), Fudan University-CIOMP Joint Fund (No. FC2018-006).

Conflict of interest statement: The authors declare no conflict of interest.

References

[1] D. G. Hall, "Vector-beam solutions of Maxwell's wave equation," *Opt. Lett.*, vol. 21, no. 1, p. 9, 1996.

[2] Q. Zhan, "Cylindrical vector beams: from mathematical concepts to applications," *Adv. Opt. Photonics*, vol. 1, no. 1, p. 1, 2009.

[3] J. Chen, C. Wan, and Q. Zhan, "Vectorial optical fields: recent advances and future prospects," *Sci. Bull.*, vol. 63, no. 1, pp. 54–74, 2018.
[4] A. S. Ostrovsky, C. Rickenstorff-Parrao, and V. Arrizón, "Generation of the "perfect" optical vortex using a liquid-crystal spatial light modulator," *Opt. Lett.*, vol. 38, no. 4, p. 534, 2013.
[5] P. Chen, W. Ji, B. Y. Wei, W. Hu, V. Chigrinov, and Y. Q. Lu, "Generation of arbitrary vector beams with liquid crystal polarization converters and vector-photoaligned q-plates," *Appl. Phys. Lett.*, vol. 107, no. 24, 2015. https://doi.org/10.1063/1.4937592.
[6] Z. Liu, Y. Liu, Y. Ke, et al., "Generation of arbitrary vector vortex beams on hybrid-order Poincaré sphere," *Photonics Res.*, vol. 5, no. 1, p. 15, 2017.
[7] N. Yu and F. Capasso, "Flat optics with designer metasurfaces," *Nat. Mater.*, vol. 13, no. 2, pp. 139–150, 2014.
[8] Q. He, S. Sun, S. Xiao, and L. Zhou, "High-efficiency metasurfaces: principles, realizations, and applications," *Adv. Opt. Mater.*, vol. 6, no. 19, p. 1800415, 2018.
[9] S. Sun, Q. He, J. Hao, S. Xiao, and L. Zhou, "Electromagnetic metasurfaces: physics and applications," *Adv. Opt. Photonics*, vol. 11, no. 2, p. 380, 2019.
[10] J. Hao, Y. Yuan, L. Ran, et al., "Manipulating electromagnetic wave polarizations by anisotropic metamaterials," *Phys. Rev. Lett.*, vol. 99, no. 6, p. 063908, 2007.
[11] S. Jiang, X. Xiong, Y.-S. Hu, et al., "Controlling the polarization state of light with a dispersion-free metastructure," *Phys. Rev. X*, vol. 4, no. 2, p. 021026, 2014.
[12] J. P. Balthasar Mueller, N. A. Rubin, R. C. Devlin, B. Groever, and F. Capasso, "Metasurface polarization optics: independent phase control of arbitrary orthogonal states of polarization," *Phys. Rev. Lett.*, vol. 118, no. 11, p. 113901, 2017.
[13] N. Yu, P. Genevet, M. A. Kats, et al., "Light propagation with phase discontinuities: generalized laws of reflection and refraction," *Science*, vol. 334, no. 6054, pp. 333–337, 2011.
[14] X. Ni, N. K. Emani, A. V. Kildishev, A. Boltasseva, and V. M. Shalaev, "Broadband light bending with plasmonic nanoantennas," *Science*, vol. 335, no. 6067, p. 427, 2012.
[15] S. Sun, Q. He, S. Xiao, Q. Xu, X. Li, and L. Zhou, "Gradient-index meta-surfaces as a bridge linking propagating waves and surface waves," *Nat. Mater.*, vol. 11, no. 5, pp. 426–431, 2012.
[16] S. Sun, K. Y. Yang, C. M. Wang, et al., "High-efficiency broadband anomalous reflection by gradient meta-surfaces," *Nano Lett.*, vol. 12, no. 12, pp. 6223–6229, 2012.
[17] L. Huang, X. Chen, H. Mühlenbernd, et al., "Dispersionless phase discontinuities for controlling light propagation", *Nano Lett*, vol. 12, 5750–5755, 2012.
[18] A. Arbabi, Y. Horie, M. Bagheri, and A. Faraon, "Dielectric metasurfaces for complete control of phase and polarization with subwavelength spatial resolution and high transmission," *Nat. Nanotechnol.*, vol. 10, no. 11, pp. 937–943, 2015.
[19] T. Cai, S. W. Tang, G. M. Wang, et al., "High-Performance bifunctional metasurfaces in transmission and reflection geometries," *Adv. Opt. Mater.*, vol. 5, p. 1600506, 2017.
[20] X. Li, S. Xiao, B. Cai, Q. He, T. J. Cui, and L. Zhou, "Flat metasurfaces to focus electromagnetic waves in reflection geometry," *Opt. Lett.*, vol. 37, no. 23, p. 4940, 2012.
[21] M. Khorasaninejad, W. T. Chen, R. C. Devlin, J. Oh, A. Y. Zhu, and F. Capasso, "Metalenses at visible wavelengths: diffraction-limited focusing and subwavelength resolution imaging," *Science*, vol. 352, no. 6290, pp. 1190–1194, 2016.
[22] S. Wang, P. C. Wu, V.-C. Su, et al., "A broadband achromatic metalens in the visible," *Nat. Nanotechnol.*, vol. 13, no. 3, pp. 227–232, 2018.
[23] Z.-B. Fan, Z.-K. Shao, M.-Y. Xie, et al., "Silicon nitride metalenses for close-to-one numerical aperture and wide-angle visible imaging," *Phys. Rev. Appl.*, vol. 10, no. 1, p. 014005, 2018.
[24] W. T. Chen, K.-Y. Y. Yang, C.-M. M. Wang, et al., "High-efficiency broadband meta-hologram with polarization-controlled dual images," *Nano Lett.*, vol. 14, no. 1, pp. 225–230, 2014.
[25] G. Zheng, H. Mühlenbernd, M. Kenney, G. Li, T. Zentgraf, and S. Zhang, "Metasurface holograms reaching 80% efficiency," *Nat. Nanotechnol.*, vol. 10, pp. 308–312, 2015.
[26] B. Wang, F. Dong, Q.-T. T. Li, et al., "Visible-frequency dielectric metasurfaces for multiwavelength Achromatic and highly dispersive holograms," *Nano Lett.*, vol. 16, no. 8, pp. 5235–5240, 2016.
[27] L. Wang, S. Kruk, H. Tang, et al., "Grayscale transparent metasurface holograms," *Optica*, vol. 3, no. 12, p. 1504, 2016.
[28] N. Yu, F. Aieta, P. Genevet, M. a Kats, Z. Gaburro, and F. Capasso, "A broadband, background-free quarter-wave plate based on plasmonic metasurfaces," *Nano Lett.*, vol. 12, no. 12, pp. 6328–6333, 2012.
[29] H. Zhou, B. Sain, Y. Wang, et al., "Polarization-encrypted orbital angular momentum multiplexed metasurface holography," *ACS Nano*, vol. 14, no. 5, pp. 5553–5559, 2020.
[30] J. T. Heiden, F. Ding, J. Linnet, Y. Yang, J. Beermann, and S. I. Bozhevolnyi, "Gap-surface plasmon metasurfaces for broadband circular-to-linear polarization conversion and vector vortex beam generation," *Adv. Opt. Mater.*, vol. 7, no. 9, p. 1801414, 2019.
[31] F. Ding, Y. Chen, and S. I. Bozhevolnyi, "Focused vortex-beam generation using gap-surface plasmon metasurfaces," *Nanophotonics*, vol. 9, no. 2, pp. 371–378, 2020.
[32] N. K. Grady, J. E. Heyes, D. R. Chowdhury, et al., "Terahertz metamaterials for linear polarization conversion and anomalous refraction," *Science*, vol. 340, no. 6138, pp. 1304–7, 2013.
[33] S. Li, Z. Wang, S. Dong, et al., "Helicity-delinked manipulations on surface waves and propagating waves by metasurfaces," *Nanophotonics*, vol. 9, p. 3473, 2020.
[34] E. Wang, J. Niu, Y. Liang, et al., "Complete control of multichannel, angle-multiplexed, and arbitrary spatially varying polarization fields," *Adv. Opt. Mater.*, vol. 8, no. 6, p. 1901674, 2020.
[35] Y. Xu, Q. Li, X. X. Zhang, et al., "Spin-Decoupled multifunctional metasurface for asymmetric polarization generation," *ACS Photonics*, vol. 6, no. 11, pp. 2933–2941, 2019.
[36] Z. Deng, M. Jin, X. Ye, et al., "Full-color complex-amplitude vectorial holograms based on multi-freedom metasurfaces," *Adv. Funct. Mater.*, vol. 30, no. 21, p. 1910610, 2020.
[37] E. Maguid, I. Yulevich, M. Yannai, V. Kleiner, M. L. Brongersma, and E. Hasman, "Multifunctional interleaved geometric-phase dielectric metasurfaces," *Light Sci. Appl.*, vol. 6, no. 8, p. e17027, 2017.
[38] Y. Yang, W. Wang, P. Moitra, I. I. Kravchenko, D. P. Briggs, and J. Valentine, "Dielectric meta-reflectarray for broadband linear polarization conversion and optical vortex generation," *Nano Lett.*, vol. 14, no. 3, pp. 1394–1399, 2014.

[39] G. Ding, K. Chen, X. Luo, J. Zhao, T. Jiang, and Y. Feng, "Dual-helicity decoupled coding metasurface for independent spin-to-orbital angular momentum conversion," *Phys. Rev. Appl.*, vol. 11, no. 4, p. 044043, 2019.

[40] H. Zhao, B. Quan, X. Wang, C. Gu, J. Li, and Y. Zhang, "Demonstration of orbital angular momentum multiplexing and demultiplexing based on a metasurface in the terahertz band," *ACS Photonics*, vol. 5, no. 5, pp. 1726–1732, 2018.

[41] F. Yue, D. Wen, C. Zhang, et al., "Multichannel polarization-controllable superpositions of orbital angular momentum states," *Adv. Mater.*, vol. 29, no. 15, p. 1603838, 2017.

[42] F. Yue, D. Wen, J. Xin, B. D. Gerardot, J. Li, and X. Chen, "Vector vortex beam generation with a single plasmonic metasurface," *ACS Photonics*, vol. 3, no. 9, pp. 1558–1563, 2016.

[43] S. Kruk, B. Hopkins, I. I. Kravchenko, A. Miroshnichenko, D. N. Neshev, and Y. S. Kivshar, "Broadband highly efficient dielectric metadevices for polarization control," *APL Photonics*, vol. 1, no. 3, p. 030801, 2016.

[44] E. Arbabi, S. M. Kamali, A. Arbabi, and A. Faraon, "Vectorial holograms with a dielectric metasurface: ultimate polarization pattern generation," *ACS Photonics*, vol. 6, no. 11, pp. 2712–2718, 2019.

[45] F. Ding, Y. Chen, Y. Yang, and S. I. Bozhevolnyi, "Multifunctional metamirrors for broadband focused vector-beam generation," *Adv. Opt. Mater.*, vol. 7, no. 22, p. 1900724, 2019.

[46] Y. Xu, H. Zhang, Q. Li, et al., "Generation of terahertz vector beams using dielectric metasurfaces via spin-decoupled phase control," *Nanophotonics*, vol. 9, p. 3393, 2020.

[47] M. Jia, Z. Wang, H. Li, et al., "Efficient manipulations of circularly polarized terahertz waves with transmissive metasurfaces," *Light Sci. Appl.*, vol. 8, no. 1, p. 16, 2019.

[48] W. Luo, S. Xiao, Q. He, S. Sun, and L. Zhou, "Photonic spin Hall effect with nearly 100% efficiency," *Adv. Opt. Mater.*, vol. 3, no. 8, pp. 1102–1108, 2015.

[49] S. Ma, X. Wang, W. Luo, et al., "Ultra-wide band reflective metamaterial wave plates for terahertz waves," *Europhys. Lett.*, vol. 117, no. 3, p. 37007, 2017.

[50] S. Wang and Q. Zhan, "Reflection type metasurface designed for high efficiency vectorial field generation," *Sci. Rep.*, vol. 6, p. 29626, 2016.

[51] H. X. Xu, L. Han, Y. Li, et al., "Completely spin-decoupled dual-phase hybrid metasurfaces for arbitrary wavefront control," *ACS Photonics*, vol. 6, p. 211, 2019.

[52] Z. Wang, S. Li, X. Zhang, et al., "Excite spoof surface plasmons with tailored wavefronts using high-efficiency terahertz metasurfaces," *Adv. Sci.*, p. 2000982, 2020. https://doi.org/10.1002/advs.202000982.

[53] S. Pancharatnam, "Generalized theory of interference and its applications," *Proc. Indian Acad. Sci. Sect. A*, vol. 44, no. 6, p. 398, 1956.

[54] M. V. Berry, "Quantal phase factors accompanying adiabatic changes," *Proc. R. Soc. A Math. Phys. Eng. Sci.*, vol. 392, no. 1802, pp. 45–57, 1984.

[55] Z. Bomzon, G. Biener, V. Kleiner, and E. Hasman, "Space-variant Pancharatnam–Berry phase optical elements with computer-generated subwavelength gratings," *Opt. Lett.*, vol. 27, no. 13, p. 1141, 2002.

[56] W. Luo, S. Sun, H.-X. Xu, Q. He, and L. Zhou, "Transmissive ultrathin Pancharatnam–Berry metasurfaces with nearly 100% efficiency," *Phys. Rev. Appl.*, vol. 7, no. 4, p. 044033, 2017.

[57] C. Qu, S. Ma, J. Hao, et al., "Tailor the functionalities of metasurfaces based on a complete phase diagram," *Phys. Rev. Lett.*, vol. 115, no. 23, p. 235503, 2015.

Supplementary Material: The online version of this article offers supplementary material (https://doi.org/10.1515/nanoph-2020-0465).

Qinghua Song, Samira Khadir, Stéphane Vézian, Benjamin Damilano, Philippe de Mierry, Sébastien Chenot, Virginie Brandli, Romain Laberdesque, Benoit Wattellier and Patrice Genevet*

Printing polarization and phase at the optical diffraction limit: near- and far-field optical encryption

https://doi.org/10.1515/9783110710687-057

Abstract: Securing optical information to avoid counterfeiting and manipulation by unauthorized persons and agencies requires innovation and enhancement of security beyond basic intensity encryption. In this paper, we present a new method for polarization-dependent optical encryption that relies on extremely high-resolution near-field phase encoding at metasurfaces, down to the diffraction limit. Unlike previous intensity or color printing methods, which are detectable by the human eye, analog phase decoding requires specific decryption setup to achieve a higher security level. In this work, quadriwave lateral shearing interferometry is used as a phase decryption method, decrypting binary quick response (QR) phase codes and thus forming phase-contrast images, with phase values as low as 15°. Combining near-field phase imaging and far-field holographic imaging under orthogonal polarization illumination, we enhanced the security level for potential applications in the area of biometric recognition, secure ID cards, secure optical data storage, steganography, and communications.

Keywords: dielectric metasurface; information security; meta-hologram; nanoantenna; optical encryption.

***Corresponding author: Patrice Genevet,** CNRS-CRHEA, Université Cote d'Azur, Rue Bernard Gregory, Sophia Antipolis, Valbonne, 06560, France, E-mail: Patrice.Genevet@crhea.cnrs.fr
Qinghua Song, Samira Khadir, Stéphane Vézian, Benjamin Damilano, Philippe de Mierry, Sébastien Chenot and Virginie Brandli: CNRS-CRHEA, Université Cote d'Azur, Rue Bernard Gregory, Sophia Antipolis, Valbonne, 06560, France. https://orcid.org/0000-0002-4622-0418 (Q. Song)
Romain Laberdesque and Benoit Wattellier: Phasics Company, Bâtiment Explorer, Espace Technologique, Route de l'Orme des Merisiers, St. Aubin, 91190, France

1 Introduction

Information security is an important concern in daily life from civil to military applications [1, 2]. Optical encryption for information security has gained enormous attention owing to its ability to provide many degrees of freedom to encode the information relying on various optical channels such as frequency, amplitude, phase, and polarization [3, 4]. These efforts led to various encryption methods, including Lippmann plate [5], spatial correlators [6], and holograms [7]. Recently, an ultrathin artificial material with subwavelength structure called metasurface [8–10] has been developed to control the electromagnetic wave across the entire frequency range for various optical applications such as lenses [11–15], cloaking [16, 17], holograms [18–21], polarizers [22–24], perfect absorbers [25, 26], retroreflectors [27], etc. The compactness and versatile functionality make metasurfaces perfect candidates for optical encryption. The basic idea is to leverage on different optical channels of the metasurface to encode different optical information. Proof-of-concept demonstration of metasurface spectral encoding, displaying sharp and hyper-resolved images observable under a microscope, has been demonstrated [28, 29]. Metasurfaces have also been designed to perform far-field holographic encoding using the concept of meta-holograms. The latter is the most common approach in the area of metasurface encryption using different light sources to decrypt the holographic images. A great deal of attention has been paid on the design of light source–dependent meta-hologram, such as polarization selectivity [30–33], orbital angular momentum selectivity [34, 35], incoming direction of the incident light [36–38], etc. Other attempts have also been made for metasurface encryption, such as the combination of color printing and the holographic image [39, 40], image postprocessing based on spatial frequency [41], and tunable meta-hologram [42]. All of the proposed metasurface encryptions, relying either on the intensity of the color

This article has previously been published in the journal Nanophotonics. Please cite as: Q. Song, S. Khadir, S. Vézian, B. Damilano, P. de Mierry, S. Chenot, V. Brandli, R. Laberdesque, B. Wattellier and P. Genevet "Printing polarization and phase at the optical diffraction limit: near- and far-field optical encryption" *Nanophotonics* 2021, 10. DOI: 10.1515/nanoph-2020-0352.

print or on the holographic image projection, are directly perceived by human eyes, which could limit their applicability for information security.

Here, we introduce a new class of optical encryption that combines the near- and far-field information. On the one hand, the near-field encryption is based on phase imaging at the metasurface plane. The information is encoded in the phase of the transmitted light without any modulation of its intensity, making the information inaccessible with a conventional microscope. In order to resolve the encoded information, one has to obtain the phase map information, which we measured in this article using quadriwave lateral shearing interferometry (QLSI) technique [43–46] (see more details in Section 5). The phase-addressing capability of metasurfaces, i.e., the spatial phase distribution of the phase elements, can be scaled down to the single pixel with a pitch of 300 nm, which is beyond Abbe's classical diffraction limit with an incident wavelength higher than 600 nm. Obviously, as to retrieve the phase information, one is using traditional optical characterization tools generally limited to diffraction limits, and it is not necessary to encode with such offensive resolution. As a proof of concept, we have considered phase pixels of about 750 nm. The proposed phase encryption could complement the class of color printing at the optical diffraction limit [28, 29] in high-density spectrally encoded optical data storage and so on. On the other hand, far-field encryption relying on the holographic image requires a full 2π phase range, which according to our results, appears to be not necessary for near-field encoding mentioned above, allowing more flexibility for realistic applications. The combination of phase and projected intensity, respectively encoded in the near field and far field, improves optical information security.

In the following section, we describe the theoretical design as well as the experimental fabrication and characterization methods. Two different examples of phase encryption are demonstrated in Section 3. The first example shows several functional phase-encoded binary quick response (QR) codes with decreasing phase-encoding values ranging from 180° down to 15° phase difference, below which the experimental realization fails to deliver readable codes. In the second example, near-field encryption, i.e., phase imaging at the metasurface plane, is multiplexed with far-field encryption, i.e., holographic image projection, to display two different "Bat" and "Batman" images.

2 Design and fabrication

The principle of the dual-mode encryption is shown in Figure 1. The metasurface consists of a birefringent dielectric meta-molecule array with Jones matrix of $\begin{pmatrix} J_{xx} & 0 \\ 0 & J_{yy} \end{pmatrix} = \begin{pmatrix} e^{i\varphi_{xx}} & 0 \\ 0 & e^{i\varphi_{yy}} \end{pmatrix}$ (assuming full transmission of the incident wave). When a plane wave with x-polarization

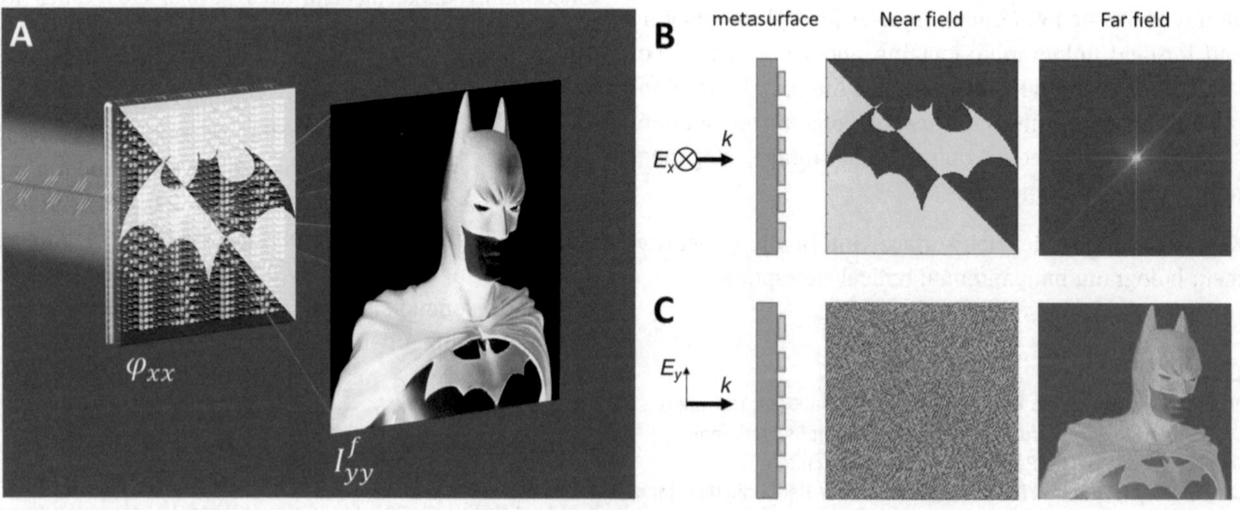

Figure 1: Design principle of dual-mode encryption combining near-field phase imaging and far-field holographic image projection. (A) Schematic of the dual-mode encryption, in which the near field is encrypted by a phase imaging of "Bat" and the far field is encrypted by a holographic image of "Batman." (B) Metasurface under x-polarized incidence (left panel). A phase imaging of "Bat" is retrieved by extracting the metasurface spatial phase distribution (middle panel). A random intensity profile is displayed in the far field (right panel). (C) Metasurface under y-polarized incidence (left panel). Complex holographic phase imaging is observed (middle panel). A holographic image of "Batman" is observed in the far field (right panel). "Bat" and "Batman" images are adapted from Wikemedia.org and Pexels.com, respectively.

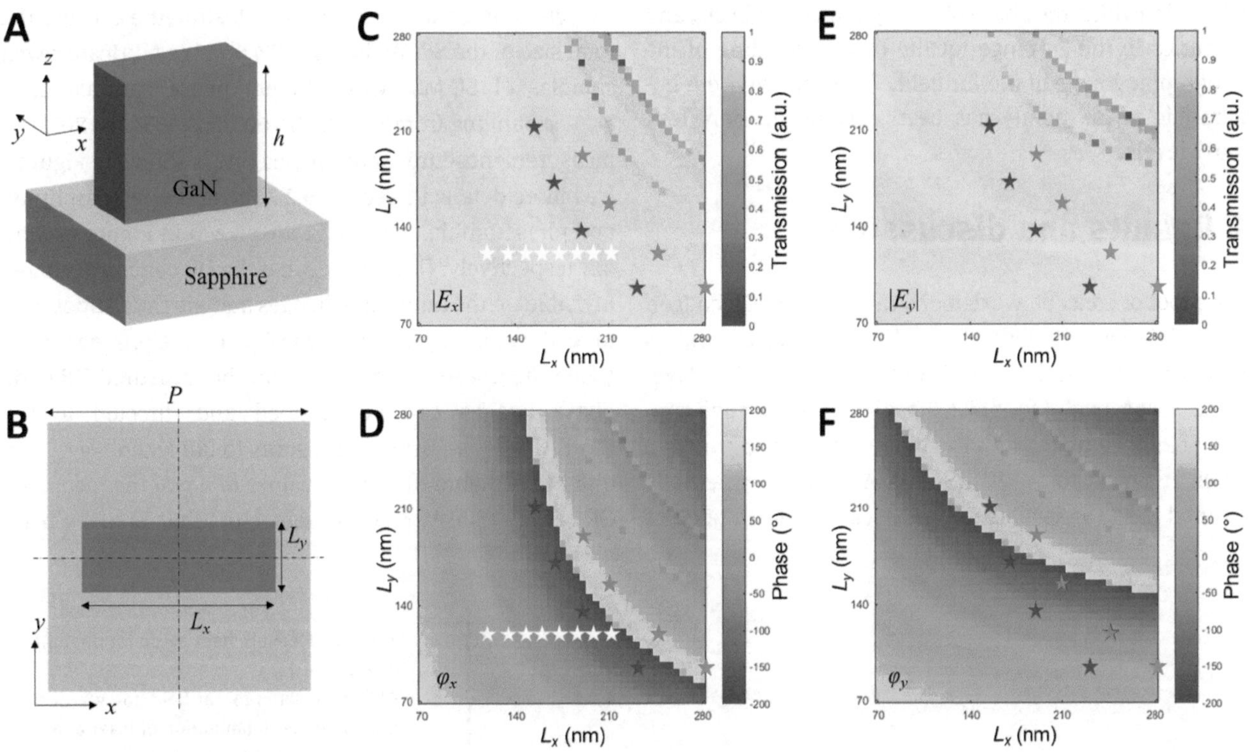

Figure 2: Simulation results of metasurface consisting of GaN nanopillars on sapphire substrate.
(A) Perspective view and (B) top view of the metasurface with one meta-molecule. (C) Simulated amplitude and (D) phase of the transmission with *x*-polarized incidence. (E) Simulated amplitude and (F) phase of the transmission with *y*-polarized incidence by changing the size of the GaN blocks. The white stars are the selected structures for the design of SC-MS. The purple and red stars are selected for the design of DC-MS. DC-MS: dual-channel metasurface; SC-MS: single-channel metasurface.

($\mathbf{E_x}$) illuminates the metasurface, as shown in Figure 1B (left panel), the output near field of the metasurface is $\mathbf{E_x}e^{i\varphi_{xx}}$, which is encoded by a phase information of φ_{xx}. By locally designing different metastructures, we can get a specific phase pattern in the near field (Figure 1B, middle) and a random intensity profile in the far field (Figure 1B, right). However, the *y*-polarized light works in the opposite way, creating a disordered near-field phase image but encrypting a holographic image in the far field, as shown in Figure 1C.

The design of the metasurfaces is realized using full-wave finite-difference time-domain simulations at the wavelength of 617 nm. The unit cell of the metasurface consists of semiconductor-based GaN nanopillars with a thickness of $h = 1$ µm and period of $P = 300$ nm on sapphire substrate, as shown in Figure 2A and 2B. GaN has the merits of (1) low loss in the entire visible range to design the metasurfaces with high efficiency and (2) high refractive index to manipulate the phase in the full 2π range using relatively short nanostructures. Other low loss and high refractive index like SiN and TiO_2 could be alternative choices [47, 48]. The GaN nanopillars support waveguide-like modes for which the accumulated phase of the transmitted light is strongly dependent on the dimensions L_x and L_y [13]. The simulated results of the transmitted amplitude and phase, as shown in Figure 2C–2E, demonstrate that the transmitted amplitude of the metasurface is almost near unity, while the phase is widely covering a full 2π range.

In this study, we propose two sets of encrypting metasurfaces, (1) a single-channel metasurface (SC-MS) encryption working only for *x*-polarized light and (2) a dual-channel metasurface (DC-MS) encryption that works differently for *x*- and *y*-polarized light. SC-MS keeps $L_y = 120$ nm, and L_x is controlled to achieve different phase information for *x*-polarized light, as shown by the white stars in Figure 2C and 2D. DC-MS controls both phase information of *x*- and *y*-polarized light, as shown in the purple and red stars in Figure 2C–2F. By controlling the size of L_x and L_y of the DC-MS, the transmitted phase of *x*-polarized light φ_{xp} and φ_{xr} (subscript "*p*" indicating purple stars and "*r*" indicating red stars) is constant along the purple and red stars, respectively, with a phase difference $\varphi_{xp} - \varphi_{xr}$ of 90° for binary phase imaging in the near field. Meanwhile, the phase of the y-polarized light φ_{yp} and φ_{yr} varies in the 2π range for the phase encoding of the holographic image in the far field. For simplicity, the holographic phase profile has been encoded in only four phase levels.

3 Results and discussion

As a proof of concept, we demonstrate a binary QR code in the SC-MS with the binary encoded phase of φ_h and φ_l (subscripts "*h*" and "*l*" indicating high- and low-level phase, respectively) for phase imaging. The QR code consists of 29 × 29 pixels with the binary phase of φ_h and φ_l, which redirects to the CNRS-CRHEA website when it is scanned using a portable QR code scanner. Each pixel of the QR code contains 20 × 20 metastructures. Thus, the total size of the SC-MS is 174 × 174 μm. We fabricate seven samples of SC-MS with different phase contrast $\Delta\varphi = \varphi_h - \varphi_l$ ranging from 5°, 15°, 30°, 60°, 90°, 135°, to 180°. The measurement setup for phase imaging is shown in Figure 3 (see more details in Section 5.2). The measured amplitude and phase profile using QLSI are shown in Figure 4A and 4B, respectively. One can see that the QR codes are almost invisible on the amplitude images and are thus impossible to read while they are clearly visible on the phase images. When the phase contrast $\Delta\varphi \geq 15°$, the measured QR code phase images can be scanned and directed to the CNRS-CRHEA website, but it starts to fail when $\Delta\varphi \leq 5°$. In order to estimate the actual values of $\Delta\varphi$ of the measured QR code phase images, we plot the pixel-by-pixel

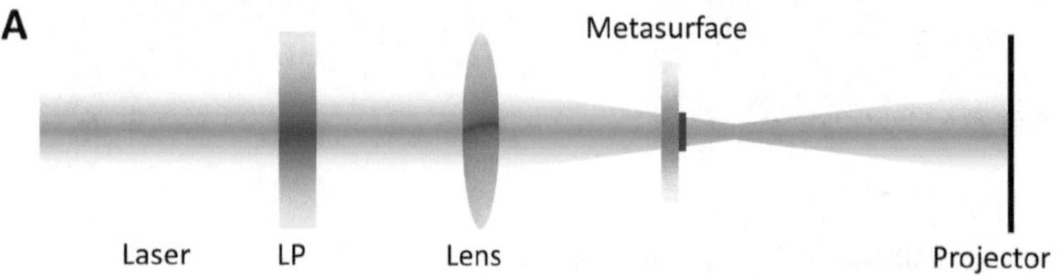

Figure 3: Measurement setup for dual-mode encryption.
(A) Optical setup for far-field holographic imaging under illumination of laser at a wavelength of 617 nm. (B) Phase imaging measurement setup using QLSI. An LED with a wavelength centered at 617 nm integrated in a Köhler configuration is used as an illumination source on the metasurface with a controlled optical plane wave. The transmitted signal is then collected by the wavefront analyzer. D1 and D2 are field and aperture diaphragms, respectively. LED: light-emitting diode; LP: linear polarizer; MHM: modified Hartmann mask; QLSI: quadriwave lateral shearing interferometry; SCMOS: scientific complementary metal-oxide-semiconductor.

Figure 4: Experimental results of binary QR codes using SC-MS.
(A) Amplitude and (B) phase images measured by QLSI. From left to right panels, the expected binary phase contrast values are 5°, 15°, 30°, 60°, 90°, 135°, and 180° (all the phase images have the same scale of 0–180°). The QR code is designed to link the CNRS-CRHEA website (www.crhea.cnrs.fr). The phase images with binary phase contrast ≥15° can be redirected to the CNRS-CRHEA website when scanned by a portable QR code scanner, and it starts to fail for extremely small phase contrast below 5°. The size of the QR code is 174 × 174 μm. QLSI: quadriwave lateral shearing interferometry; QR: quick response; SC-MS: single-channel metasurface.

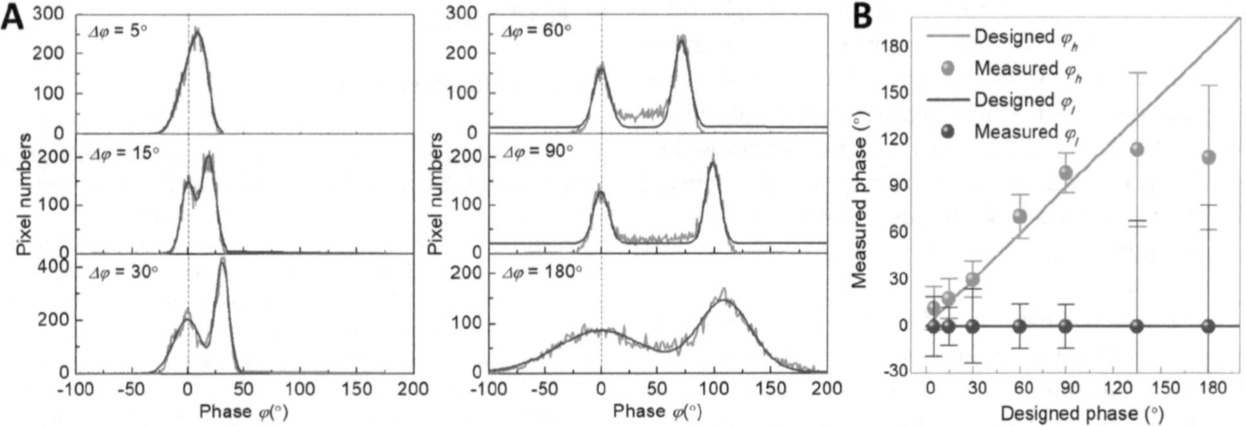

Figure 5: (A) Pixel-by-pixel histogram of the measured phase images for the different designed phase contrast of 5°, 15°, 30°, 60°, 90°, and 180°. Red curve: measured results. Blue curve: fitting curves based on two Gaussian functions. (B) The measured phase and standard deviation of binary phase imaging.

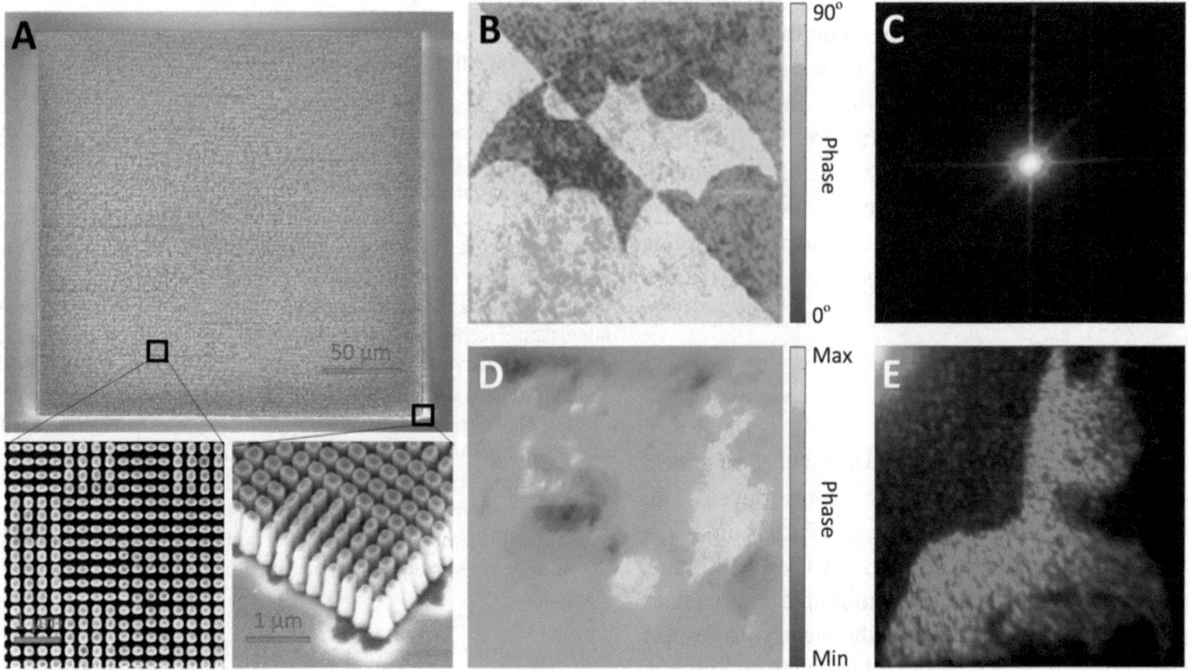

Figure 6: Experimental results of DC-MS.
(A) SEM images of the fabricated metasurface. (B) Measured phase image and (C) far-field image with *x*-polarized incidence. (D) Measured phase image and (E) far-field holographic image with *y*-polarized incidence. DC-MS: dual-channel metasurface; SEM: scanning electron microscopy.

histograms for each image, as shown in Figure 5A. The histogram is fitted with Gaussian functions with two peaks, one corresponding to the low-level phase φ_l and the other corresponding to the high-level phase φ_h. The difference between the two peaks gives the phase contrast $\Delta\varphi$, while the full width at half maximum estimates the error of the extracted phase. The results for different phase contrast from 15° to 180° are shown in Figure 5A. The expected measured phase and standard deviation of φ_h and φ_l are shown in Figure 5B.

It can be seen that when $\Delta\varphi$ is too small, $\Delta\varphi = 5°$, the width of each Gaussian profile is larger than $\Delta\varphi$, which makes the two peaks indistinguishable. The discrepancy of the measured phase mainly comes from the fabrication error because the length difference of the GaN nanopillars corresponding to the binary phase of 5° is only 2.5 nm, which is lower than our electron-beam lithography (EBL) precision. Increasing the designed phase contrast, i.e., such as $\Delta\varphi \geq 15°$, the two peaks become distinguishable, and the extracted phase contrasts agree well with the expected ones. However, when the phase contrast is close to 180°, such as $\Delta\varphi = 135°$ and $\Delta\varphi = 180°$, the standard deviation is increasing owing to the phase wrapping issues from the phase retrieval algorithm used to extract the pixel phase value.

The near-field phase imaging can be further combined with far-field holographic imaging so that a dual-mode optical encryption can be achieved. Here, the phase contrast of near-field phase imaging is chosen as $\Delta\varphi = 90°$, according to its good performance shown in Figure 5. The experimental results of DC-MS for near-field and far-field encryption are shown in Figure 6A. When the metasurface is illuminated by x-polarized light, a phase image of "Bat" measured by QLSI is shown in Figure 6B; however, owing to the designed phase in the near field only, the holographic image in the far field has no useful information, as shown in Figure 6C. Instead, considering a y-polarized light illumination, a phase profile without useful information is shown in Figure 6D, displaying a holographic image of "Batman" in the far field, as shown in Figure 6E, to achieve a dual-mode optical encryption able to combine near-field phase imaging and far-field holographic image projection.

4 Conclusion

In conclusion, we have demonstrated a new class of optical encryption relying on dual-mode metasurfaces. Our method is able to print optical phase information at the nanoscale, which can be extracted using QLSI technique. We encoded several phase QR codes in simple binary phase images, demonstrating significant advantage with respect to conventional full-phase holographic image encryption. We demonstrated that a phase difference as low as 15° is sufficient to successfully scan the QR code and redirect to the desired website page. It is interesting to point out that gradually varying the nanopillar geometry enables analog phase addressing, offering an interesting perspective for multiplexed coding capabilities. Moreover, by combining the holographic display in the orthogonal polarization, a dual-mode optical encryption is demonstrated for near-field and far-field encryption, expanding promising applications in optical information security, data encryption, optical ID tags for authentication and verification, high-density optical data storage, etc.

5 Methods

5.1 Fabrication processes

The fabrication of the metasurface starts from growing a 1-μm GaN thin film on a double-side polished c-plane sapphire substrate using a molecular beam epitaxy RIBER system. Conventional EBL is used to pattern the GaN nanopillars. The GaN thin film is spin-coated by a double layer of 200-nm Polymethyl methacrylate (PMMA) resist (495A4) and baked on a hot plate at a temperature of 125 °C. The resist is exposed to an electron beam of 20 keV (Raith ElphyPlus, Zeiss Supra 40) considering the desired pattern and then developed using IPA:MIBK (3:1) solution. A 50-nm-thick nickel film is deposited on the sample in the e-beam evaporator, and a nickel hard mask is obtained by removing the resist through the lift-off process in acetone solution. The GaN nanopillars are obtained by reactive ion etching (Oxford system) with a plasma composed of Cl_2CH_4Ar gases. Finally, the nickel hard mask on the GaN nanopillars is dissolved in $HCl:HNO_3$ (1:2) solution, revealing only the GaN nanopillar pattern on the transparent sapphire substrate.

5.2 Measurement setup

The measurement setup for far-field holographic imaging is shown in Figure 3A. A laser with a wavelength of 617 nm is used as the light source. The latter passes through a linear polarizer followed by a lens to weakly focus on the metasurface and creates a holographic image in the far field, which is captured by a projector placed after the metasurface at a distance of 10 cm. The setup for phase imaging is shown in Figure 3B. It utilizes a quantitative phase microscopy technique based on QLSI [42–45]. A light-emitting diode with a wavelength centered at 617 nm integrated in a Köhler configuration is used as an illumination source with a controlled optical plane wave (controlled illuminated area and controlled numerical aperture). The light beam passing through the metasurface is collected by a microscope objective and sent to the QLSI wavefront analyzer (SID4 camera from Phasics Company) to measure both the intensity and the optical path difference of the light at the microscope's imaging plane, coinciding with the metasurface position.

Acknowledgments: The authors acknowledge the support from the European Research Council (ERC) under the European Union's Horizon 2020 research and innovation program (grant agreement no. 639109).
Author contribution: All the authors have accepted responsibility for the entire content of this submitted manuscript and approved submission.
Research funding: The study was supported from the European Research Council (ERC) under the European Union's Horizon 2020 research and innovation program (grant agreement no. 639109).
Conflict of interest statement: The authors declare no conflicts of interest regarding this article.

References

[1] R. Anderson and T. Moore, "The economics of information security," *Science*, vol. 314, pp. 610–613, 2006.
[2] M. E. Whitman and H. J. Mattord, *Principles of Information Security*, Boston, Cengage Learning, 2011. ISBN10: 1-337-10206-7; ISBN13: 978-1-337-10206-3.
[3] O. Matoba, T. Nomura, E. Perez-Cabre, M. S. Millan, and B. Javidi, "Optical techniques for information security," *Proc. IEEE*, vol. 97, pp. 1128–1148, 2009.
[4] B. Javidi, Ed. *Optical and Digital Techniques for Information Security*, New York, Springer, 2005.
[5] G. Lippmann, "Sur la théorie de la photographie des couleurs simples et composées par la méthode interférentielle," *J. Phys. Theor. Appl.*, vol. 3, pp. 97–107, 1894.
[6] Y. Li, K. Kreske, and J. Rosen, "Security and encryption optical systems based on a correlator with significant output images," *Appl. Opt, AO*, vol. 39, pp. 5295–5301, 2000.
[7] O. Bryngdahl and F. Wyrowski, "I digital holography – computer-generated holograms. In: Wolf E., Ed., editor. Progress in Optics, 28. Amsterdam, Netherlands: Elsevier; 1990. pp. 1–86.
[8] N. Yu, P. Genevet, M. A. Kats, et al., "Light propagation with phase discontinuities: generalized laws of reflection and refraction," *Science*, vol. 334, pp. 333–337, 2011.
[9] N. Yu and F. Capasso, "Flat optics with designer metasurfaces," *Nat. Mater.*, vol. 13, pp. 139–150, 2014.
[10] P. Genevet, F. Capasso, F. Aieta, M. Khorasaninejad, and R. Devlin, "Recent advances in planar optics: from plasmonic to dielectric metasurfaces," *Optica*, vol. 4, 2017, pp. 139–152.
[11] M. Khorasaninejad and F. Capasso, "Metalenses: Versatile multifunctional photonic components," *Science*, vol. 358, p. eaam8100, 2017.
[12] W. T. Chen, A. Y. Zhu, V. Sanjeev, et al., "A broadband achromatic metalens for focusing and imaging in the visible," *Nat. Nanotechnol.*, vol. 13, pp. 220–226, 2018.
[13] S. Wang, P. C. Wu, V.-C. Su, et al., "A broadband achromatic metalens in the visible," *Nat. Nanotechnol.*, vol. 13, pp. 227–232, 2018.
[14] C. Sun, "Shrinking the camera size: metasurface lens," *Nat. Mater.*, vol. 16, pp. 11–12, 2017.
[15] W. M. Zhu, Q. Song, L. Yan, et al., "A flat lens with tunable phase gradient by using random access reconfigurable metamaterial," *Adv. Mater.*, vol. 27, pp. 4739–4743, 2015.
[16] J. Y. H. Teo, L. J. Wong, C. Molardi, and P. Genevet, "Controlling electromagnetic fields at boundaries of arbitrary geometries," *Phys. Rev. A*, vol. 94, 2016, Art no. 023820.
[17] Y. Y. Xie, P. N. Ni, Q. H. Wang, et al., "Metasurface-integrated vertical cavity surface-emitting lasers for programmable directional lasing emissions," *Nat. Nanotechnol.*, vol. 15, pp. 125–130, 2020.
[18] G. Zheng, H. Mühlenbernd, M. Kenney, G. Li, T. Zentgraf, and S. Zhang, "Metasurface holograms reaching 80% efficiency," *Nat. Nanotechnol.*, vol. 10, pp. 308–312, 2015.
[19] P. Genevet, J. Lin, M. A. Kats, and F. Capasso, "Holographic detection of the orbital angular momentum of light with plasmonic photodiodes," *Nat. Commun.*, vol. 3, p. 1278, 2012.
[20] L. Huang, X. Chen, H. Mühlenbernd, et al., "Three-dimensional optical holography using a plasmonic metasurface," *Nat. Commun.*, vol. 4, p. 2808, 2013.
[21] Q. Song, A. Baroni, R. Sawavt, et al., "Ptychography retrieval of fully polarized holograms from geometric-phase metasurfaces," *Nat. Commun.*, vol. 11, pp. 1–8, 2020.
[22] J. K. Gansel, M. Thiel, M. S. Rill, et al., "Gold helix photonic metamaterial as broadband circular polarizer," *Science*, vol. 325, pp. 1513–1515, 2009.
[23] Q. H. Song, P. C. Wu, W. M. Zhu, et al., "Split Archimedean spiral metasurface for controllable GHz asymmetric transmission," *Appl. Phys. Lett.*, vol. 114, p. 151105, 2019.
[24] Y. Z. Shi, T. Zhu, T. Zhang, et al., "Chirality-assisted lateral momentum transfer for bidirectional enantioselective separation," *Light Sci. Appl.*, vol. 9, pp. 1–12, 2020.
[25] G. M. Akselrod, J. Huang, T. B. Hoang, et al., "Large-area metasurface perfect absorbers from visible to near-infrared," *Adv. Mater.*, vol. 27, pp. 8028–8034, 2015.
[26] Q. Song, W. M. Zhu, P. C. Wu, et al., "Liquid-metal-based metasurface for terahertz absorption material: frequency-agile and wide-angle," *APL Mater.*, vol. 5, p. 066103, 2017.
[27] L. Yan, W. M. Zhu, M. F. Karim, et al., "0.2 λ_0 thick adaptive retroreflector made of spin-locked metasurface," *Adv. Mater.*, vol. 30, p. 1802721, 2018.
[28] K. Kumar, H. Duan, R. S. Hegde, S. C. Koh, J. N. Wei, and J. K. Yang, "Printing colour at the optical diffraction limit," *Nat. Nanotechnol.*, vol. 7, pp. 557–561, 2012.
[29] Y. Gu, L. Zhang, J. K. Yang, S. P. Yeo, and C. W. Qiu, "Color generation via subwavelength plasmonic nanostructures," *Nanoscale*, vol. 7, pp. 6409–6419, 2015.
[30] X. Luo, Y. Hu, X. Li, et al., "Integrated metasurfaces with microprints and helicity-multiplexed holograms for real-time optical encryption," *Adv. Opt. Mater.*, vol. 8, p. 1902020, 2020.
[31] R. Zhao, L. Huang, C. Tang, et al., "Nanoscale polarization manipulation and encryption based on dielectric metasurfaces," *Adv. Opt. Mater.*, vol. 6, p. 1800490, 2018.
[32] J. Deng, L. Deng, Z. Guan, et al., "Multiplexed anticounterfeiting meta-image displays with single-sized nanostructures," *Nano Lett.*, vol. 20, pp. 1830–1838, 2020.
[33] L. Deng, J. Deng, Z. Guan, et al., "Malus-metasurface-assisted polarization multiplexing," *Light Sci. Appl.*, vol. 9, p. 101, 2020.
[34] H. Ren, G. Briere, X. Fang, et al., "Metasurface orbital angular momentum holography," *Nat. Commun.*, vol. 10, p. 2986, 2019.
[35] X. Fang, H. Ren, and M. Gu, "Orbital angular momentum holography for high-security encryption," *Nat. Photon.*, vol. 14, pp. 102–108, 2020.
[36] K. Chen, G. Ding, G. Hu, et al., "Directional Janus metasurface," *Adv. Mater.*, vol. 32, p. 1906352, 2020.
[37] Y. Chen, X. Yang, and J. Gao, "3D Janus plasmonic helical nanoapertures for polarization-encrypted data storage," *Light Sci. Appl.*, vol. 8, p. 45, 2019.
[38] D. Frese, Q. Wei, Y. Wang, L. Huang, and T. Zentgraf, "Nonreciprocal asymmetric polarization encryption by layered plasmonic metasurfaces," *Nano Lett.*, vol. 19, pp. 3976–3980, 2019.
[39] G. Yoon, D. Lee, K. T. Nam, and J. Rho, ""Crypto-display" in dual-mode metasurfaces by simultaneous control of phase and spectral responses," *ACS Nano*, vol. 12, pp. 6421–6428, 2018.
[40] K. T. P. Lim, H. Liu, Y. Liu, and J. K. W. Yang, "Holographic colour prints for enhanced optical security by combined phase and amplitude control," *Nat. Commun.*, vol. 10, p. 25, 2019.
[41] J. Deng, Y. Yang, J. Tao, et al., "Spatial frequency multiplexed meta-holography and meta-nanoprinting," *ACS Nano*, vol. 13, pp. 9237–9246, 2019.

[42] J. Li, S. Kamin, G. Zheng, F. Neubrech, S. Zhang, and N. Liu, "Addressable metasurfaces for dynamic holography and optical information encryption," *Sci. Adv.*, vol. 4, p. eaar6768, 2018.
[43] P. Bon, G. Maucort, B. Wattellier, and S. Monneret, "Quadriwave lateral shearing interferometry for quantitative phase microscopy of living cells," *Opt. Express*, vol. 17, p. 13080, 2009.
[44] S. Khadir, D. Andrén, P. C. Chaumet, et al., "Full optical characterization of single nanoparticles using quantitative phase imaging," *Optica*, vol. 7, p. 243, 2020.
[45] S. Khadir, D. Andrén, D. Verre, et al., "optical imaging and characterization of graphene and other 2D materials using quantitative phase microscopy," *ACS Photonics*, vol. 4, no. 12, pp. 3130–3139, 2017.
[46] J. Primot and N. Guérineau, "Extended Hartmann test based on the pseudoguiding property of a Hartmann mask completed by a phase chessboard," *Appl. Optic.*, vol. 39, pp. 5715–5720, 2000.
[47] S. Colburn, A. Zhan, E. Bayati, et al., "Broadband transparent and CMOS-compatible flat optics with silicon nitride metasurfaces," *Opt. Mater. Express*, vol. 8, pp. 2330–2343, 2018.
[48] M. Khorasaninejad, W. T. Chen, R. C. Devlin, J. Oh, A. Y. Zhu, and F. Capasso, "Metalenses at visible wavelengths: diffraction-limited focusing and subwavelength resolution imaging," *Science*, vol. 352, pp. 1190–1194, 2016.

Mutasem Odeh, Matthieu Dupré, Kevin Kim and Boubacar Kanté*

Optical response of jammed rectangular nanostructures

https://doi.org/10.1515/9783110710687-058

Abstract: Random jammed dipole scatterers are natural composite and common byproducts of various chemical synthesis techniques. They often form complex aggregates with nontrivial correlations that influence the effective dielectric description of the medium. In this work, we investigate the packing dynamic of rectangular nanostructure under a close packing protocol and study its influence on the optical response of the medium. We show that the maximum packing densities, maximum scattering densities, and percolation threshold densities are all interconnected concepts that can be understood through the lens of Onsager's exclusion area principle. The emerging positional and orientational correlations between the rectangular dipoles are studied, and various geometrical connections are drawn. The effective dielectric constants of the generated ensembles are then computed through the strong contrast expansion method, leading to several unintuitive results such as scattering suppression at maximum packing densities, as well as densities below the percolation threshold, and maximum scattering in between.

Keywords: disorder; metamaterials; metasurfaces; nanostructures.

***Corresponding author: Boubacar Kanté,** Department of Electrical Engineering and Computer Sciences, University of California, Berkeley, CA 94720, USA; and Department of Electrical and Computer Engineering, University of California San Diego, La Jolla, CA 92093, USA; and Materials Sciences Division, Lawrence Berkeley National Laboratory, 1 Cyclotron Road, Berkeley, California 94720, USA, E-mail: bkante@berkeley.edu. https://orcid.org/0000-0001-5633-4163
Mutasem Odeh, Department of Electrical Engineering and Computer Sciences, University of California, Berkeley, CA 94720, USA
Matthieu Dupré and Kevin Kim, Department of Electrical and Computer Engineering, University of California San Diego, La Jolla, CA 92093, USA

1 Introduction

Random packing persists to be an alluring topic, pertinent to fundamental questions in physics, chemistry, and biology [1–3]. Within the field of optics and photonics, in particular, understanding light–matter interactions in random packed media is crucial and urged by the growing usage of optical sensors and imaging systems in probing complex living cells, liquids, and granular media. In addition, the thriving genre of disordered photonics domesticates randomness toward various applications in light trapping [4], radiative cooling [5], and random lasing [6].

In this work, we investigate the optical response of packed rectangular nanostructures on a surface, as they are commonly employed as dipole scatterers in optical devices [7–9] for various applications including light harvesting [10] and biosensing [11]. However, a detailed electromagnetic simulation of such an ensemble is a computationally expensive task to perform and one rather seeks the effective medium description as an approximation. Many homogenization theories with varying degrees of applicability and complexity have been developed toward this aim [12, 13]. Bruggeman's theory models aggregate structure with constituents that are treated on an equal footing and therefore cannot be applied in this case [14]. Maxwell–Garnett approximation, on the other hand, models inclusions dispersed in a continuous host medium. Its analytic simplicity arises from the consideration of only the one-point probability function (density) where convergence is assured under the dilute and long wavelength limit. However, at large packing density, the positional and orientational correlations between the dipoles are not negligible anymore and can drastically alter the effective dielectric constant of the ensemble. The *strong contrast expansion method* presented in the study by Rechtsmanand and Torquato [13] is rather a generic and exact approach that includes the contribution of high-order point probability functions and thus captures the correction due to the emerged correlations.

In this article, we investigate the influence of the packing dynamic on the optical response of jammed rectangular

This article has previously been published in the journal Nanophotonics. Please cite as: M. Odeh, M. Dupré, K. Kim and B. Kanté "Optical response of jammed rectangular nanostructures" *Nanophotonics* 2021, 10. DOI: 10.1515/nanoph-2020-0431.

nanostructures. A comprehensive workflow chart can be found in the supplementary material (section S1). In section 2, we define the packing protocol used in the study and compute the maximum achieved packing densities at various aspect ratios. We proceed in section 3 with point process statistical analysis to unravel the short-range ordering and spontaneous alignment between the packed rectangles. In section 4, we model the ensemble as a two-phase isotropic medium and estimate the effective dielectric constant (ε_{eff}) using the strong contrast expansion method. The last section concludes and summarizes the work.

2 Random close packing

We consider the packing of hard rectangles of length l and width w on a square substrate, where the interaction potential $\phi(x)$ is infinite inside the rectangle region and zero otherwise. Therefore, a stable state is a state with no overlapping elements. The maximum packing density and the state characteristics are protocol dependent. In this work, we implement a collective rearrangement packing protocol which models an abrupt surge or sedimentation of the rectangle concentration near an adsorbing edge which is common in many chemical synthesis techniques [15, 16]. The algorithm starts by placing N identical rectangles of aspect ratio $\alpha = l/w$ at random positions and orientations on a flat substrate of area A_s with an initial packing fraction $\phi_i = Nlw/A_s$, as shown in Figure 1A. In an iterative procedure, each overlapping rectangle is individually displaced from its initial position with a random radial distance and orientation that are uniformly distributed on the range of $[0, 2l]$ and $[0, \pi]$, respectively. The procedure persists until the rectangle avoids overlapping or n attempts has been reached after which the process is terminated, and the rectangle is removed from the packing process. To calculate the maximum packing density ($\phi_{\max}$), we configured the initial density to $\phi_i = 0.6$, number of attempts to $n = 100{,}000$, and particle number to $N = 10{,}000$. Periodic boundaries were used to avoid finite size effects, and the results were averaged over multiple realizations to ensure convergence. The maximum packing density achieved for different aspect ratios is shown in Figure 1B. We note that $\phi_{\max}$ is aspect ratio dependent with a unique cusp around $\alpha = 1.5$. The decrease of $\phi_{\max}$ to the right of the cusp can be explained through the lens of Onsager's exclusion area principle [17]. The principle states that each rectangle has an excluding area $A(l, w, \theta)$ within which other similar rectangles with a relative angel θ are forbidden to occupy if overlapping is to be avoided, as illustrated by the inset in Figure 1B. The ratio of the rectangle area lw to what it excludes on average $\langle A \rangle$ across all probable angels $[0, \pi]$ has been shown to give an accurate estimation of the percolation threshold [18] and is given by the following equation:

$$\phi_{\max}(\alpha) = \frac{lw}{\langle A \rangle} = \frac{\alpha}{c_1(1+\alpha^2) + c_2\alpha} \tag{1}$$

where $c_1 = 2/\pi$ and $c_2 = (2 + 8/\pi)$ in the case of penetrable rectangles. However, our protocol induces a nontrivial positional and orientational correlation that makes the excluded area principle difficult to be derived for $\phi_{\max}$ estimation. In fact, this is the parking problem in 2D which is still an open question to be answered. Nonetheless, we found that the behavior can still be well captured by Eq. (1), using a fitted value of the c_1 and c_2 coefficients, as shown in Figure 1B. However, the model fails to fit the cusp and $\phi_{\max}$ at low aspect ratios. Similar behavior has been observed for

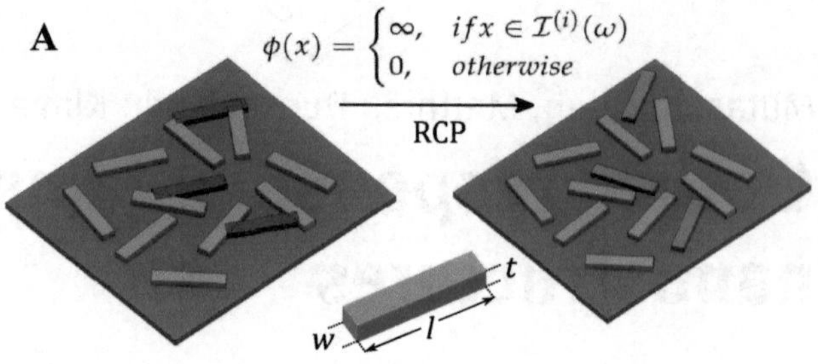

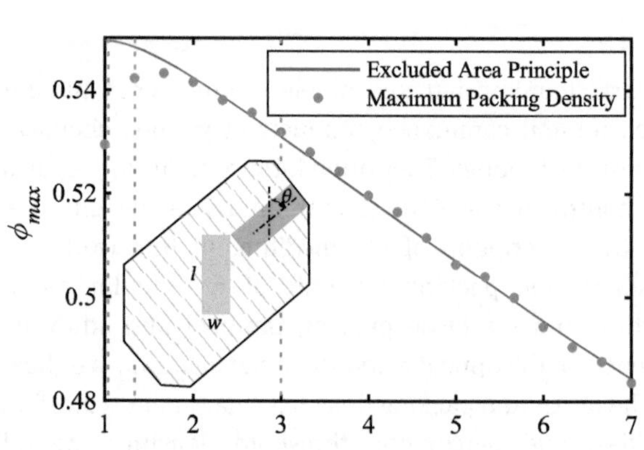

Figure 1: Random close packing of hard rectangles: (A) illustration of the collective rearrangement packing protocol with $\phi(x)$ as the interaction potential and $I^{(i)}(\omega)$ is the indicator function of phase i in realization ω and (B) is the computed maximum packing density ($\phi_{\max}$) as a function of the aspect ratios (α). The blue dots represent the simulated data, and the red curve is a fitting line based on the excluded area principle given by Eq. (1) with $c_1 = 0.047$ and $c_2 = 1.725$. The inset illustrates the excluded area of a rectangle of length l and width w with another identical rectangle with a relative tilt θ.

packing hard ellipsoids, explained through the isostatic conjecture, and verified through the famous *M&M's* experiment [19, 20]. The conjecture states that the mean contact number between packed elements is on average twice the number of degrees of freedom in a jammed configuration. Consequently, deviating from squares to rectangles introduces an additional degree of freedom (orientation) to the packing process, which increases the average contact numbers and consequently ϕ_{max}. Strictly speaking, our packing protocol does not lead to a jammed state rather than a saturated state with an average contact number of zero. Nonetheless, the state can still be treated as a jammed one since the element cannot move once they settled on a nonoverlapping position [20]. Therefore, there are two competing effects occurring in the close packing protocol. On the one hand, adding an extra degree of freedom to squares gives extra flexibility to pack at higher densities (isostatic conjecture), yet on the other hand, increasing the degree of anisotropy prevents short-range ordering and limits the maximum packing density instead (exclusion area principle). These behaviors have a direct influence on the effective dielectric constant of the random jammed media. For example, engineering a random medium with high dielectric constant requires, in general, a high density of packed rectangles. Unintuitively, this can be optimally approached by choosing rectangles of 1.5 aspect ratio, as can be deduced from Figure 1B.

3 Point process statistical analysis

The collective rearrangement packing protocol produces a statistically homogeneous medium that we assume ergodic (any single realization of the ensemble is representative of the ensemble in the infinite area-limit). We start our investigation by performing a stochastic point process analysis. Each rectangle is represented by its two midpoint coordinates (x, y) and the angle (θ) that the longer axis makes with respect to a fixed global reference. We calculate two important statistical descriptors. The first is the radial *pair correlation* function, which is defined as follows:

$$g_2(r) = \frac{\rho_2(r)}{\rho^2}, \quad (2)$$

where ρ is the number density (number of points per unit area in the infinite area limit) and $\rho_2(r)$ is the number of points within a distance of r and $r + dr$ from a reference rectangle. Thus, a deviation of $g_2(r)$ from the unity provides a measure of positional correlation or anticorrelation between the rectangles. The second statistical descriptor is the orientational correlation function $g_\theta(r)$ defined in the study by Ma and Torquato [21] as follows:

$$g_\theta(r) = <\cos(2[\theta(0) - \theta(r)])>, \quad (3)$$

which is an average measure of the degree of alignment between two rectangles within a distance of r and $r + dr$. Thus, $g_\theta > 0$ suggests statistically parallel rectangles, $g_\theta < 0$ suggests statistically perpendicular rectangles, and $g_\theta = 0$ suggests the lack of any preferential orientation. The two statistical descriptors were applied to the states of maximum packed density that were presented in Figure 1B. The width is normalized $(w = 1)$ without loss of generality, and the descriptors are plotted as shown in Figure 2A and B. We observe the emergence of three distinct features on both surfaces that can be fitted by three lines (p_1, p_2, p_3) that intersect at $(\alpha = 1, r = 1)$. Transitioning from squares to rectangles causes an interesting trifurcation of the first appearing peak at $\alpha = 1$, as clarified by the top view insets of Figure 2A and B. The trifurcation is a clear sign of the isostatic conjecture. The conjecture states that in a jammed configuration, the mean contact number between packed elements is on average twice the number of degrees of freedom (DOFs) [19, 20]. For simplicity, we can approximate squares as circles and rectangles as ellipsoids by ignoring the sharp edges that have a negligible influence on the packing dynamic. Building on this approximation, there are two DOFs for squares (x, y) and three DOFs for rectangles (x, y, θ). According to the isostatic conjecture, we can conclude that the average contact number increases from 4 to 6 when deforming squares to rectangles. Consequently, the ensemble contains two extra possible configurations, that is, in addition to the original configuration, and forms the trifurcation observed in Figure 2A and B. The low g_2 correlation in the triangular region between p_1 and p_2 is a direct consequence of the excluding area principle. In other words, as the aspect ratio increases, it becomes statistically difficult to pack rectangles within close proximity. Figure 2C shows a cross-sectional plot of both g_2 and g_θ for two different aspect ratios. For the square case $(\alpha = 1)$, we note a sharp increase in g_2 at $r > 1$ indicating a high probability of occupancy. On contrary to the case of circles packing where the increase is abrupt, rectangle packing has a finite slope that is attributed to their radial asymmetry. In addition, the high probability of occupancy of the first neighbor square induces a negative correlation on the next adjacent regions, a repeated process that explains the oscillatory behavior of $g_2(r)$ that is damped with distance. We also note that g_θ is approximately zero, suggesting the lack of any preferential alignment between the packed squares. On the other hand, for $\alpha = 3$, g_θ is positively correlated in the region between p_2 and p_3, whereas g_r is negatively correlated. The statistical

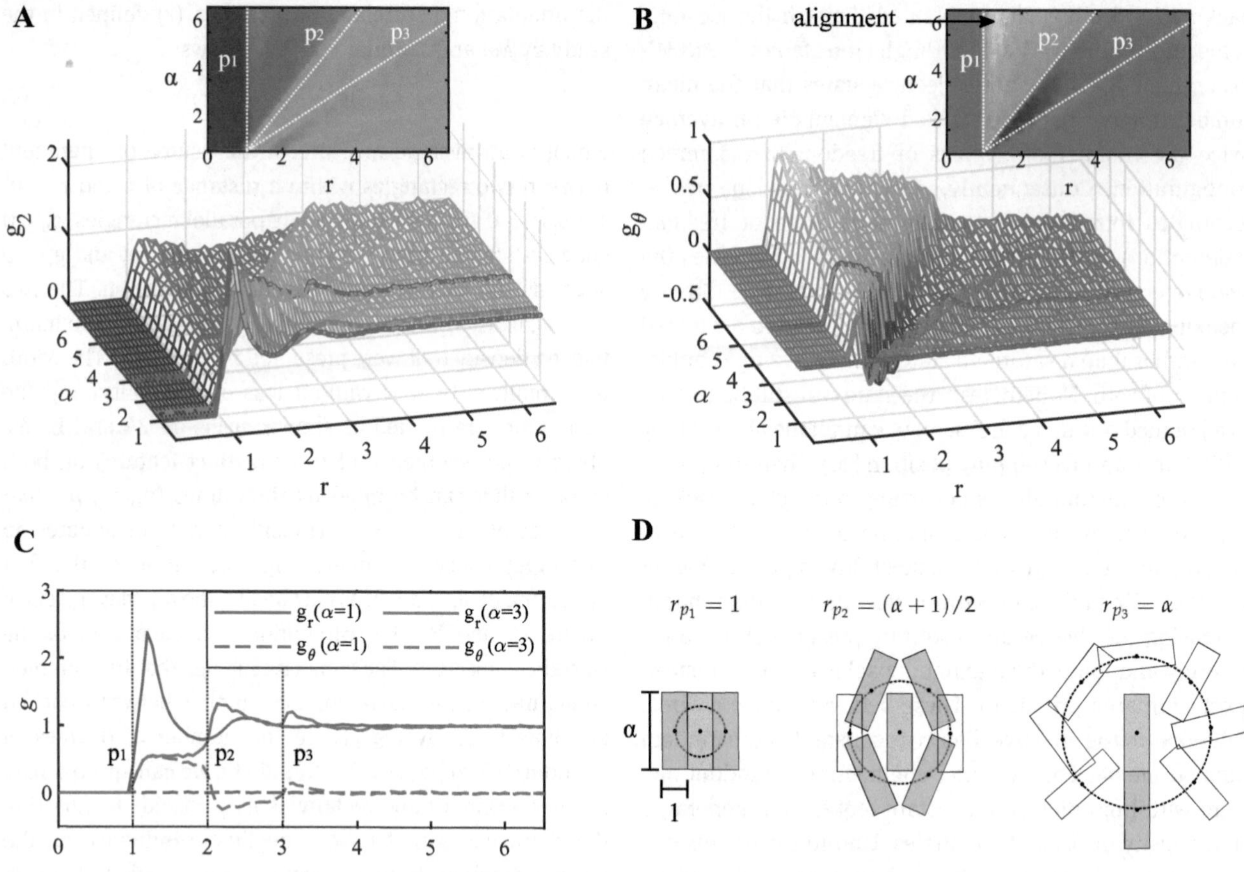

Figure 2: Point process analysis of maximally random packed rectangles: (A) radial pair correlation functions $g_2(r, \alpha)$ given by Eq. (2) and (B) orientational correlation function $g_\theta(r, \alpha)$ given by Eq. (3). Three different aspect ratios are highlighted on both surfaces with different colors; the blue line corresponds to $\alpha = 1$, green to $\alpha = 1.3$, and red to $\alpha = 3$. The insets on both surfaces correspond to a top view perspective with three important features highlighted by p_1, p_2, p_3 lines. (C) A cross sectional plot of g_2 and g_θ surfaces at $\alpha = 1$ and $\alpha = 3$ and (D) geometrical illustration of the origin of p_1, p_2, p_3 lines, respectively. For each configuration, the dashed circle with radius r_{pi} represents the shortest distance between rectangles' center for overlap to be avoided.

interpretation indicates that it is highly constrained to place two rectangles in proximity, yet if it is deemed necessary, they must be well aligned. However, such constraint is lifted at p_2 and further relaxed at p_3. The three constraints can be traced geometrically, as illustrated in Figure 2D. The p_2 line equals to the minimum distance when two perpendicular rectangles are not overlapping, that is, when $r_{p_2} = (\alpha + 1)/2$. The p_3 line equals to the minimum distance of two stacked rectangles on their longer axes, that is, when $r_{p_3} = \alpha$. We also note that at each relaxation point (p_1, p_2, p_3), a transition in the sign of g_θ occurs.

We conclude from this analysis the lack of long-range translational or nematic order in the ensemble. The effective permittivity in the 2D plane is thus macroscopically isotropic and polarization independent at all aspect ratios. In addition, high aspect ratios have a destructive behavior on short-range positional order, and therefore, their scattering features will be weaker. Furthermore, the average rectangle orientation after lifting the p_2 constrain is approximately $\theta \approx \pi/4$. In the simple dipole picture, this suggests a spectral redshift for the resonance mode supported along the longer axis (h), as illustrated in the middle configuration of Figure 2D.

4 Strong contrast expansion of the effective dielectric constant

The statistical properties of phase i in two-phase heterogeneous media can be specified by an infinite set of n-point probability functions S_n^i [22]. In a homogeneous and isotropic medium, the first term reduces simply to the density of phase i ($S_1^{(i)} = \phi_i$), whereas the second term $S_2^{(i)}(r)$ is interpreted as the probability of finding both endpoints of a line segment of length r in phase i. In the

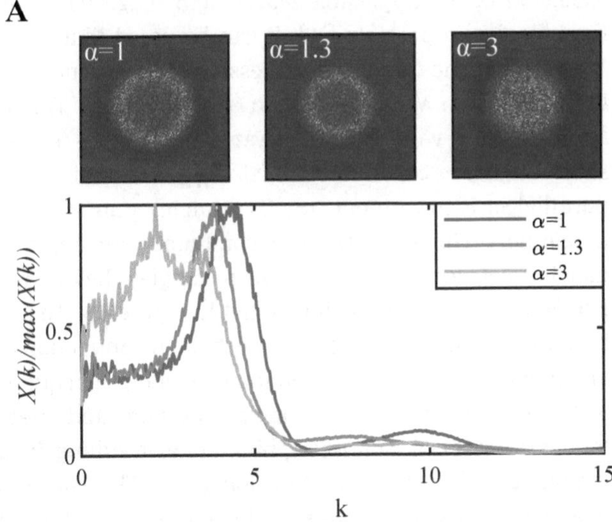

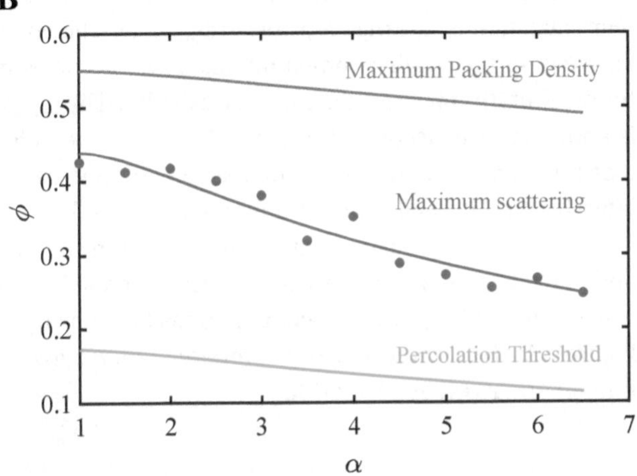

Figure 3: Two-phase heterogeneous analysis: (A) spectral density images for $\alpha = 1, 1.3, 3$ at ϕ_{max} followed by a radial average plot and (B) density plot as a function of the aspect ratio, showing three curves fitted using the excluded are principle given by Eq. (1): (i) percolation threshold curve (yellow) fitted with $c_1 = 2/\pi$ and $c_2 = (2 + 8/\pi)$ where ensembles below exhibit weak correlation; (ii) maximum scattering density curve (blue) fitted with $c_1 = 0.3780$ and $c_2 = 1.5255$ represents the ensembles at which the zeroth wavevector spectral density function $\chi_v(k = 0)$ exhibits a maximum; and (iii) the maximum packing density curve (red) fitted with $c_1 = 0.047$ and $c_2 = 1.725$.

following discussion, we will drop the superscript (i) and implicitly refer to the rectangle phase. The scattering behavior of the ensemble is captured by the spectral density function $\chi_v(k)$ which is the Fourier transform of the autocovariance function $\chi_v(r) = S_2(r) - S_1^2$. Figure 3A shows the spectral density for three different aspect ratios $(\alpha = 1, 1.3, 3)$ at their maximum packing density limit. Structures with low aspect ratios exhibit a clear attenuation in their scattering behavior at small k wave vectors. This is attributed to the suppression of long-range density fluctuations due to the positional ordering of the packed elements. The attenuation becomes weaker as α increases due to the destructive role of the addition of an extra DOF to the packing process, as discussed previously. The scattering behavior is reminiscent of hyperuniformity where $\chi_v(k) \to 0$ as $k \to 0$ [23]. Hyperuniform structures have been shown to exhibit unique optical properties including the formation of isotropic photonic bandgap that can be used for light guiding and confinement [24, 25]. In addition, the suppressed scattering can lead to transparency in a dielectric medium [26] and enhanced optical absorption in a lossy medium [27]. Although, strictly speaking, our packing dynamic does not lead to hyperuniform structure ($\chi_v(k) \neq 0$ as $k \to 0$), shared properties are expected. Given that the scattering is weak below the percolation threshold due to the low number of scatterers and similarly weak at maximum packing densities due to the suppression of long-range fluctuations, we expect that there should be an intermediate regime where the scattering events are maximum. We evaluated the zeroth wavevector spectrum density function $\chi_v(k = 0)$ for different aspect ratios (α) and located the densities at which the function is maximum, as shown in Figure 3B. Indeed, such an intermediate regime exists between the percolation and maximum packing density limits. In addition, its aspect ratio dependence can also be fitted by the exclusion area principle using Eq. (1).

From the calculated autocovariance function, we can proceed in calculating the effective-dielectric constant by the strong contrast expansion method. The expressions presented in the study by Rechtsman and Torquato [13] were formulated for 3D random structures. We rederive the method for two-phase medium in 2D and truncate the expansions up to the second order to include the 2-point probability function $S_2^{(i)}(r)$. In addition, we focus on the effective dielectric constant experienced by a transverse magnetic plane wave propagating parallel to the 2D medium plane $(k_z = 0, E_z = 0)$. The effective permittivity can be obtained from the following expression (see section S2 for full discussion):

$$\beta_{pq}^2 \phi_p^2 \beta_{eq}^{-1} = \phi_p \beta_{pq} - \underbrace{A_2^{(p)} \beta_{pq}^2}_{\text{2nd order}}, \tag{4}$$

where ϕ_p is the area density of the packed rectangles, $\beta_{pq} = (\varepsilon_p - \varepsilon_q)/(\varepsilon_p + \varepsilon_q)$ is the rectangles polarizability (phase p) with respect to the environment (phase q), and $\beta_{eq} = (\varepsilon_{\text{eff}} - \varepsilon_q)/(\varepsilon_{\text{eff}} + \varepsilon_q)$ is the effective polarizability with respect to the environment. A_2^p is the coefficient of the second-order correction for a 2D isotropic medium

$$A_2^{(p)} = \frac{i\pi k_q^2}{2} \int_0^{r_{\max}} dr\, r H_0^{(1)}(k_q r)\, \chi_V(r), \tag{5}$$

where $H_0^{(1)}$ is the Hankel function of the first kind which acts as propagator function for $\chi_V(r)$. It is important to emphasize that Eq. (5) is approximately valid when $\lambda_q \gg r_{\max}$, where λ_q is the wavelength in the background medium and $r_{\max}$ is the radius at which the positional correlation is negligible $(\chi_V(r > r_{\max}) \approx 0)$. In other words, all positionally correlated dipoles are assumed to be excited in phase by an external plane wave of wavelength λ_q. However, the condition is not stringent, and an extension of the applicable wavelength range has been recently shown possible [28]. It can be noted that when $A_2^p\beta_{pq}^2$ approaches zero, the expression reduces simply to the well-known Maxwell–Garnett approximation. This is approximately valid for large wavelength $(k_q \to 0)$ and short-range autocovariance function $\chi_V(r)$. In this limit, pure dielectric constituents $(\varepsilon_p, \varepsilon_q)$ result in a pure effective permittivity $(\mathrm{Im}[\varepsilon_{\mathrm{eff}}] = 0)$. The conditions are approximately met for subwavelength rectangles below the percolation threshold, as illustrated in Figure 3B. This is because, above the percolation threshold, the formation of an infinite sized cluster of overlapping rectangles requires many adjustments of the rectangles' positions and orientations to reach a nonoverlapping state, resulting in a long-range autocovariance function $\chi_V(r)$. The real and imaginary parts of $\varepsilon_{\mathrm{eff}}$ were evaluated as a function of wavelength for three different aspect ratios $(\alpha = 1, 1.3, 3)$ at their maximum packing densities $(\phi_{\max})$, as shown in Figure 4. In the studied configurations, $\chi_V(r)$ becomes in the order of 10^{-6} for $r > 5$. Therefore, $\varepsilon_{\mathrm{eff}}$ calculated by Eq. (4) is approximately accurate for $\lambda_q \gg 5$. It can also be noted that the complex permittivity approaches the one calculated by the Maxwell–Garnett approximation as $\lambda_q \to \infty$. Furthermore, large aspect ratios damp the resonance and spread the scattering for larger wavelengths. This is consistent with power spectrum density shown in Figure 3A since there exists a proportionality between $\mathrm{Im}[\varepsilon_{\mathrm{eff}}]$ and $\chi_V(k)$ as $k \to 0$ [13].

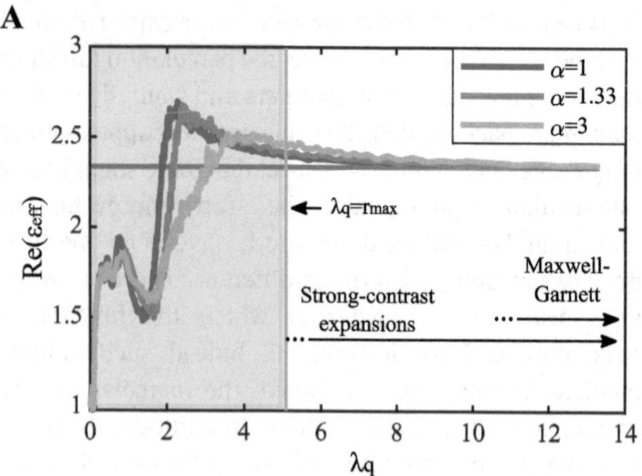

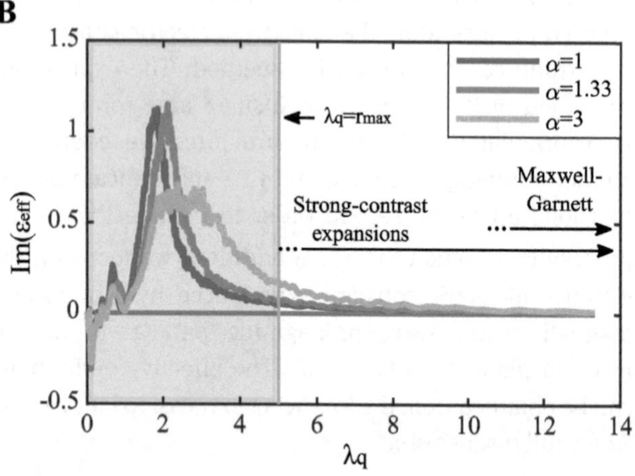

Figure 4: The complex effective dielectric constant $(\varepsilon_{\mathrm{eff}})$ for jammed rectangular nanostructures calculated by the strong contrast expansion method and plotted as a function of wavelength (λ_q). The 2D rectangles have width $w = 1$, height $h = \alpha$, permittivity $\varepsilon_p = 9$, and are maximally packed $(\phi_{\max})$ in a background medium with permittivity $\epsilon_q = 1$. (A) The real part of $\varepsilon_{\mathrm{eff}}$ showing an explicit redshift as α increases, and (B) the imaginary part of $\varepsilon_{\mathrm{eff}}$ showing larger scattering as α increases for $\lambda_q \gg r_{\max}$. In the long wavelength limit, both the real and imaginary parts of $\varepsilon_{\mathrm{eff}}$ approach the Maxwell–Garnett approximation.

5 Conclusion

Random packed media are a ubiquitous and natural outcome of various chemical synthesis techniques. In the subwavelength limit, the complex inhomogeneous medium can be described by an effective homogeneous one with great accuracy. In this work, we statistically analyze jammed rectangular dipoles under the random close packing protocol for various densities and aspect ratios. The arising microscopic correlations were traced and shown to have direct and indirect consequences on the effective dielectric constant of the medium. Statistical tools and concepts such as Onsager's excluded area principle, the positional correlation function, and the orientational correlation function, are of great utilities in describing the state of the ensemble and deduce some of the optical characteristics such as polarization dependence and spectral shifts. To study the influence of structural correlations on the macroscopic optical response, we accommodate the strong contrast expansion method to two-dimensional structure and use it to estimate the effective dielectric constant for the generated ensembles. This allows us to capture various effects beyond what Maxwell–Garnett approximation can, such as scattering enhancement and suppression as well as correlation-induced spectral shift. This work paves a systematic path toward engineering random medium with tailored optical properties.

Author contribution: All the authors have accepted responsibility for the entire content of this submitted manuscript and approved submission.

Research funding: This work was supported by the National Science Foundation Career Award (ECCS-1554021), the Office of Naval Research Young Investigator Award (N00014-17-1-2671), the ONR JTO MRI Award (N00014-17-1-2442), and the DARPA DSO-NLM Program no. HR00111820038.

Conflict of interest statement: The authors declare no conflicts of interest regarding this article.

References

[1] J. D. Bernal, "A geometrical approach to the structure of liquids," *Nature*, vol. 183, pp. 141–147, 1959.

[2] P. Schaaf and J. Talbot, "Kinetics of random sequential adsorption," *Phys. Rev. Lett.*, vol. 62, pp. 175–178, 1989.

[3] J. Feder, "Random sequential adsorption," *J. Theor. Biol.*, vol. 87, pp. 237–254, 1980.

[4] H. Atwater and A. Polman, "Plasmonics for improved photovoltaic devices," *Nat. Mater.*, vol. 9, pp. 205–213, 2010.

[5] Y. Zhail, Y. Ma, S. David, et al., "Scalable-manufactured randomized glasspolymer hybrid metamaterial for daytime radiative cooling," *Science*, vol. 355, pp. 1062–1066, 2017.

[6] N. Lawandy, R. Balachandran, A. Gomes, et al., "Laser action in strongly scattering media," *Nature*, vol. 368, pp. 436–438, 1994.

[7] M. Dupre, L. Hsu, and B. Kante, "On the design of random metasurface based devices," *Sci. Rep.*, vol. 8, p. 7162, 2018.

[8] H. Nasari, M. Dupré, and B. Kanté, "Efficient design of random metasurfaces," *Opt. Lett.*, vol. 43, pp. 5829–5832, 2018.

[9] J. Park, A. Ndao, W. Cai, et al., "Symmetry-breaking-induced plasmonic exceptional points and nanoscale sensing," *Nat. Phys.*, vol. 16, pp. 462–468, 2020.

[10] Y.-Z. Zheng, X. Tao, J.-W. Zhang, et al., "Plasmonic enhancement of light-harvesting efficiency in tandem dye-sensitized solar cells using multiplexed gold core/silica shell nanorods," *J. Power Sources*, vol. 376, pp. 26–32, 2018.

[11] A. Abbas, L. Tian, J. J. Morrissey, et al., "Hot spot-localized artificial antibodies for label-free plasmonic biosensing," *Adv. Funct. Mater.*, vol. 23, pp. 1789–1797, 2013.

[12] M. Safdari, M. Baniassadi, H. Garmestani, et al., "A modified strong-contrast expansion for estimating the effective thermal conductivity of multiphase heterogeneous materials," *J. Appl. Phys.*, vol. 112, p. 114318, 2012.

[13] M. Rechtsman and S. Torquato, "Effective dielectric tensor for electromagnetic wave propagation in random media," *J. Appl. Phys.*, vol. 103, pp. 1–15, 2008.

[14] G. A. Niklasson, C. G. Granqvist, and O. Hunderi, "Effective medium models for the optical properties of inhomogeneous materials," *Appl. Opt.*, vol. 20, pp. 26–30, 1981.

[15] A. Bertei, C. C. Chueh, J. G. Pharoah, et al., "Modified collective rearrangement sphere-assembly algorithm for random packings of nonspherical particles: towards engineering applications," *Powder Technol.*, vol. 253, pp. 311–324, 2014.

[16] J. Perez-Justea, I. Pastoriza-Santosa, L. Liz-Marzana, et al., "Gold nanorods: synthesis, characterization and applications," *Coord. Chem. Rev.*, vol. 249, pp. 1870–1901, 2005.

[17] L. Onsager, "The effects of shape on the interaction of colloidal particles," *Ann. N. Y. Acad. Sci.*, vol. 51, pp. 627–659, 1949.

[18] I. Balberg, C. H. Anderson, S. Alexander, et al., "Excluded volume and its relation to the onset of percolation," *Phys. Rev. B*, vol. 30, pp. 3933–3943, 1984.

[19] A. Donev, I. Cisse, D. Sachs, et al., "Improving the density of jammed disordered packings using ellipsoids," *Science*, vol. 303, pp. 990–993, 2004.

[20] P. Chaikin, A. Donev, W. Man, et al., "Some observations on the random packing of hard ellipsoids," *Ind. Eng. Chem. Res.*, vol. 45, pp. 6960–6965, 2006.

[21] Z. Ma and S. Torquato, "Hyperuniformity of generalized random organization models," *Phys. Rev. E*, vol. 99, p. 022115, 2019.

[22] S. Torquato, *Random Heterogeneous Materials*, Berlin, Springer, 2002.

[23] S. Torquato, "Hyperuniform states of matter," *Phys. Rep.*, vol. 745, pp. 1–95, 2018.

[24] W. Man, M. Florescu, K. Matsuyama, et al., "Photonic band gap in isotropic hyperuniform disordered solids with low dielectric contrast," *Opt. Express*, vol. 21, pp. 19972–19981, 2013.

[25] W. Man, M. Florescu, E. P. Williamson, et al., "Isotropic band gaps and freeform waveguides observed in hyperuniform disordered photonic solids," *Proc. Natl. Acad. Sci. U.S.A.*, vol. 40, pp. 15886–15891, 2013.

[26] O. Leseur, R. Pierrat, and R. Carminati, "High-density hyperuniform materials can be transparent," *Optica*, vol. 3, pp. 763–767, 2016.

[27] F. Bigourdan, R. Pierrat, and R. Carminati, "Enhanced absorption of waves in stealth hyperuniform disordered media," *Opt. Express*, vol. 27, pp. 8666–8682, 2019.

[28] S. Torquato and J. Kim, "Nonlocal effective electromagnetic wave characteristics of composite media: beyond the quasistatic regime," arXiv preprint arXiv:2007.00701.

Supplementary Material: The online version of this article offers supplementary material (https://doi.org/10.1515/nanoph-2020-0431).

Sun-Je Kim, Hansik Yun, Sungwook Choi, Jeong-Geun Yun, Kyungsoo Park, Sun Jae Jeong, Seung-Yeol Lee, Yohan Lee, Jangwoon Sung, Chulsoo Choi, Jongwoo Hong, Yong Wook Lee and Byoungho Lee*

Dynamic phase-change metafilm absorber for strong designer modulation of visible light

https://doi.org/10.1515/9783110710687-059

Abstract: Effective dynamic modulation of visible light properties has been significantly desired for advanced imaging and sensing technologies. In particular, phase-change materials have attracted much attention as active material platforms owing to their broadband tunability of optical dielectric functions induced by the temperature-dependent phase-changes. However, their uses for visible light modulators are still limited to meet multi-objective high performance owing to the low material quality factor and active tunability in the visible regime. Here, a design strategy of phase-change metafilm absorber is demonstrated by making the use of the material drawbacks and extending design degree of freedom. By engineering tunability of effective anisotropic permittivity tensor of VO_2-Ag metafilm around near-unity absorption conditions, strong dynamic modulation of reflection wave is achieved with near-unity modulation depth at desired wavelength regions without sacrificing bandwidth and efficiency. By leveraging effective medium theory of metamaterial and coupled mode theory, the intuitive design rules and theoretical backgrounds are suggested. It is also noteworthy that the dynamic optical applications of intensity modulation, coloring, and polarization rotation are enabled in a single device. By virtue of ultrathin flat configuration of a metafilm absorber, design extensibility of reflection spectrum is also verified. It is envisioned that our simple and powerful strategy would play a robust role in development of miniaturized light modulating pixels and a variety of photonic and optoelectronic applications.

Keywords: coupled mode theory; effective medium theory; metafilm; phase-change material; vanadium dioxide; visible light modulation.

***Corresponding author: Byoungho Lee,** Inter-University Semiconductor Research Center and School of Electrical and Computer Engineering, Seoul National University, Gwanakro 1, Gwanak-Gu, Seoul, 08826, Republic of Korea, E-mail: byoungho@snu.ac.kr
Sun-Je Kim, Hansik Yun, Jeong-Geun Yun, Yohan Lee, Jangwoon Sung, Chulsoo Choi and Jongwoo Hong: Inter-University Semiconductor Research Center and School of Electrical and Computer Engineering, Seoul National University, Gwanakro 1, Gwanak-Gu, Seoul, 08826, Republic of Korea
Sungwook Choi, Kyungsoo Park, Sun Jae Jeong and Yong Wook Lee: Interdisciplinary Program of Biomedical Mechanical & Electrical Engineering and School of Electrical Engineering, Pukyong National University, Yongso-ro 45, Nam-Gu, Busan, 48513, Republic of Korea
Seung-Yeol Lee: School of Electronics Engineering, College of IT Engineering, Kyungpook National University, Daehakro 80, Buk-gu, Daegu, 702-701, Republic of Korea

1 Introduction

The field of dynamic nanophotonics [1–3] suggests the ultimate goal of high performance integrated optical modulation with compact volume and improved functionality via combination of active optical materials and advanced metamaterial technologies [1–3]. It is implied that general light properties, such as optical amplitude, phase, spectrum, and polarization, can be designed with large degree of freedom and tuned by the application of external stimuli to active optical materials. In the context, there have been intensive efforts on the goal of dynamic light modulation based on numerous active optical materials [4–6].

Among those, nanophotonics platform using phase-change materials (PCMs) has risen as powerful candidates in the field of dynamic nanophotonics [1, 4, 5, 6]. As conventional PCMs exhibit much larger tunability of complex refractive index in the near-infrared and visible range compared to other active material platforms such as elastic nanoparticle assemblies [7, 8], low-dimensional graphene [9], transparent conducting oxides [10–14], or thermally tunable semiconductors [15, 16], PCMs are highly advantageous to design photonic nanostuructures considering large modulation depth, broad bandwidth,

This article has previously been published in the journal Nanophotonics. Please cite as: S.-J. Kim, H. Yun, S. Choi, J.-G. Yun, K. Park, S. J. Jeong, S.-Y. Lee, Y. Lee, J. Sung, C. Choi, J. Hong, Y. W. Lee and B. Lee "Dynamic phase-change metafilm absorber for strong designer modulation of visible light" *Nanophotonics* 2021, 10. DOI: 10.1515/nanoph-2020-0264.

and miniaturization of device over the broad optical regime.

Representative PCMs such as VO_2 [17–29], $Ge_2Sb_2Te_5$ (GST) [30–41], $SmNiO_3$ [42] have been thoroughly studied with the application of external thermal, electrical, and optical signal for the last decade. These PCMs were integrated to metamaterial and metasurface structures for dynamically tunable extraordinary responses of optical wavefront, spectrum, and polarization within ultrathin thickness, particularly, in the mid [28, 29, 39–42] and near-infrared [18–20, 26, 27, 30–34] and visible range [35–38].

However, even exploiting PCMs, when it comes to the light modulation in the visible regime, it has been quite difficult to achieve good performance exhibiting high contrast and efficiency, and broad operation bandwidth, simultaneously. The main obstacle originates from the large extinction coefficients and moderate tunability of dielectric functions of PCMs in the visible spectrum (Figure S1 of Supplementary Materials) compared to those in the infrared regime. Since these properties limit efficient oscillation of guided photonic mode and quality factor of PCMs, modulation depth and efficiency are hardly increased, simultaneously. To overcome these intrinsic problems in the visible regime, there have been numerous studies introducing various geometric nano-antenna resonances to strongly couple incident optical energy with active PCM nano-structures. Gap plasmon [26, 39], dielectric [32, 35], magnetic [18], localized surface plasmon [27], and ultrathin Gires-Tournois absorber type [24, 37, 38] resonances have been utilized for VO_2 and GST based phase-change metasurface devices. Particularly, Gires-Tournois absorber has proved to be the simplest useful design with large tunability. By actively exploiting absorptive properties of PCMs rather than suppressing them, tunable near-unity absorptions are achieved and shifted according to the phase-changes [24, 37, 38, 43, 44].

In 2014 and 2017, H. Bhaskaran group [37] and S. Y. Lee et al. [38] proposed impressive demonstrations of reconfigurable switching of visible color using ultrathin Gires-Tournois absorber configurations that consist of indium tin oxide-capped GST thin film on metallic mirrors, respectively. Nevertheless, their devices exhibit switching between two differently encoded visible spectra for certain fixed thicknesses of absorbing PCMs. Thus, it is hard to integrate those PCM based Gires-Tournois absorber pixels with different thicknesses for multiple spectral operations on a single substrate. To improve design capability of operation bandwidth in a single active nanophotonic device, without sacrificing modulation depth and efficiency, more versatile and extensible thin film structure with enlarged design degree of freedom is necessary. The improved Gires-Tournois absorber, where lateral encoding of desired active dielectric functions in an active material is available, would be desirable rather than changing the film thickness.

Here, we propose a simple and powerful strategy of novel Gires-Tournois absorbers using phase-change VO_2-Ag metafilm for effective designer modulation of visible light. The dual bandwidth of strong active modulation is designed by metamaterial-assisted phase-change effects based on judicious embedding of subwavelength noble metal nanobeams into an ultrathin VO_2 layer. The polarization-controlled metafilm approach exactly meets the abovementioned fundamental demands to design and encode multiple dielectric functions in a single active layer, and extends design degree of freedom in a single chip. The proposed phase-change metafilm absorber (PCMA) is made by stacking metallic mirror, a noble metal-embedded VO_2 layer, and a transparent dielectric substrate. Based on the recent successes in high-contrast near-unity absorption tunings in Gires-Tournois absorber configurations [24, 37, 38], dynamic insulator-metal transition (IMT) of VO_2 is tuned in anisotropic manner for strong designer modulation. The rest parts of the paper are organized as follows. As the ground work, firstly, a dynamic Gires-Tournois VO_2 absorber is investigated to explore tunable near-unity absorptions in the visible range. Secondly, effects of embedded nanobeams in VO_2 film based on the effective medium approximation are designed and experimentally verified. Then, theoretical analysis on the tunable resonance mechanisms is suggested by help of temporal coupled mode theory with effective medium approximation. For the next, as versatile optical applications of a PCMA device, reconfigurable high-contrast modulation of reflected intensity, color, and polarization direction is presented. Finally, design extensibility of dynamic PCMA is also studied numerically.

2 Results

2.1 Design of near-unity absorptions in a dynamic phase-change metafilm absorber

We start from investigating tunable complex refractive index of a baer 40 nm-thick VO_2 film via the reconfigurable gradual IMT phenomenon. Figure 1A describes measured thermally-driven reconfigurable evolutions of (n, k) at the four representative wavelengths (blue: 473 nm, green: 532 nm, red: 633 nm, and deep red: 700 nm). n and k denote real and imaginary parts of complex refractive index of VO_2, respectively. Interestingly, increase of temperature leads to continuously decreasing tendencies of both n and k values regardless of wavelength while they show simultaneously large values [45]. The gradual IMT between the saturated insulating and metallic phases depends on the intermediate temperatures during heating and cooling processes. It implies that heating and cooling would induce gradual blue and red shifts of resonances in a VO_2-included nanophotonic resonator.

A baer VO_2 film shows low modulation performance owing to theoretically limited maximal absorption of 0.5 [46, 47]. On the other hand, it is well known that a Gires-Tournois VO_2 absorber configuration (Figure 1B and Figure S2 of Supplementary Materials) shows large modulation depth (η_m) of light intensity by excitation and modulation of near-unity absorption. Here, η_m is defined as $\eta_m = |I_i - I_m|/\max(I_i, I_m)$ and I_i and I_m correspond to reflection intensities at the insulating and metallic phases, respectively. The two distinct near-unity absorptions in the insulating and metallic phases at the wavelengths of 655 and 550 nm are verified through the theoretical calculation and measurement (Inset reflectance spectra of Figure 1B and Figure S2 of Supplementary Materials). It means the gradual blue and red shifts of absorption dip with change of the resonance wavelength about 100 nm, respectively. Yet, it is still hard to design tunable near-unity absorption with high reflection tunability by only adjusting thickness of VO_2 film to meet complex interference conditions of reflection phasors in highly dispersive

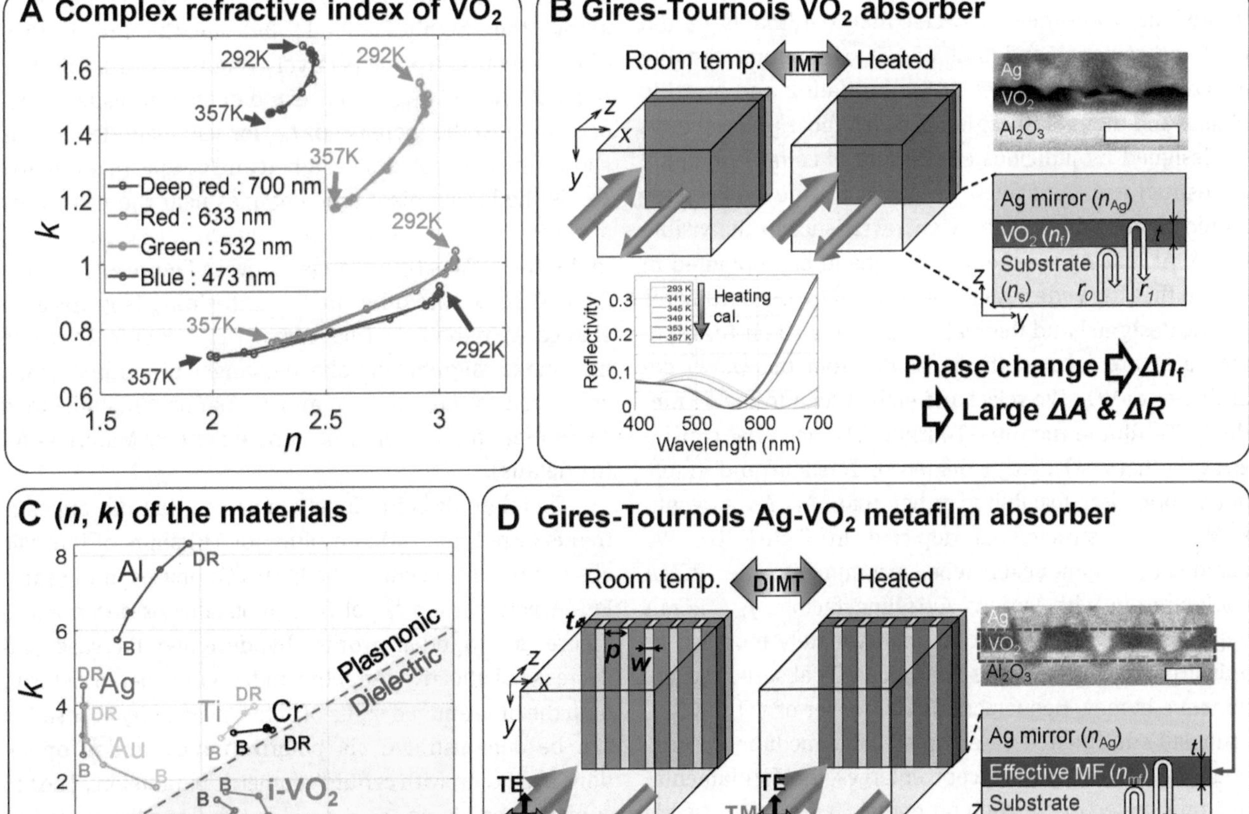

Figure 1: Concept of dynamically tunable phase-change VO_2 metafilm absorber.
(A) Measured thermally driven cyclic evolutions of (n, k) coordinates of VO_2 at the representative blue (473 nm), green (532 nm), red (633 nm), and deep red (700 nm) colors. (B) Schematic illustration, cross-sectional scanning electron microscopy (SEM) image of the sapphire substrate-VO_2-Ag Gires-Tournois absorber ($t = 40$ nm), and calculated results of tunable near-unity absorption spectra of the absorber during heating. (C) Complex refractive index map for various materials and VO_2 at the representative four visible wavelengths (B: blue (473 nm), DR: deep red (700 nm)). (D) Schematic description of phase-change metafilm based dynamic Gires-Tournois absorber. SEM image shows cross-sectional image of the fabricated dynamic Gires-Tournois metafilm absorber with Ag filling factor about 0.2. The scale bars in SEM images of (B) and (D) denote 100 nm.

and absorptive PCM such as VO_2 (the Section 3 of the Supplementary Materials).

Here, we introduce a metamaterial-based solution into the Gires-Tournois VO_2 absorber scheme to expand range of possible effective refractive index of VO_2 film by geometric implanting of subwavelength-spaced periodic nanostructures in an intrinsic PCM VO_2, which is called metafilm strategy [48–51]. We would regard an anisotropic Ag-VO_2 film as a tunable effective metafilm with designed effective refractive index and the designed IMT (DIMT). The first step of designing phase-change VO_2 metafilm is to choose a proper material to embed geometrically in a VO_2 film. Figure 1C describes (n, k) coordinates of the several metals and the two distinct VO_2 phases in the visible range. Once a phase-change metafilm is designed as an effective mixture of VO_2 and another photonic nanostructure, effective refractive index coordinate of a metafilm would be located at somewhere between the coordinates of the photonic material and that of VO_2. Therefore, tunable effective refractive indices at the insulating and metallic phases and thermal transition between those phases could be designed by judicious embedding of certain photonic nanostructures. As shown in Figure 1C, the noble plasmonic metals with less lossy characteristics in the visible range, Al and Ag are the best options to be implanted in VO_2 so that the widest range of artificial (n, k) coordinates can be designed and thermally tuned since their (n, k) coordinates are located the most far from the (n, k) coordinates of VO_2. We select and embed Ag into the 40 nm-thick VO_2 film in the Gires-Tournois VO_2 absorber considering both (n, k) design degree of freedom and stable nanofabrication feasibility rather than Al. As a result, PCMA is constructed as depicted in Figure 1D. We demonstrate nanobeam type metafilm in the PCMA configuration with various Ag filling factors, f_{Ag}. PCMA depicted in Figure 1D can be approximately modeled as anisotropic effective Gires-Tournois VO_2 absorber where dynamic homogeneous permittivity tensor of VO_2 layer is artificially designated based on effective medium approximation. This simple concept enlarges the fundamental dynamic dielectric function properties in terms of possible range of spatial encoding as well as polarization-controlled characteristics of it in a single certain device while large dynamic tunability is guaranteed.

For systematic design of VO_2 metafilm layer in PCMA configuration, we propose the design rule based on anisotropic counter-intuitive properties of the Wiener's bounds of effective medium approximation visualized in a complex (n, k) plane [52–54]. The two bound curves correspond to the normal and parallel effective dielectric functions, ε_{TM} and ε_{TE}, depending on f_{Ag}, respectively. If deep subwavelength resonances around Ag nanobeams are neglected, the ideal analytic first-order approximations of ε_{TM} and ε_{TE} are harmonic and arithmetic means of constitutive dielectric functions, respectively [52–54]. In Figure 2A, B, transverse magnetic (TM) and transverse electric (TE) polarization cases are numerically calculated for f_{Ag} varying from 0.05 to 0.95 based on complex reflection and transmission coefficients [55, 56].

The effective refractive index retrieval results regarding subwavelength resonant behaviors show similar trend of the zeroth order effective medium approximation (Figure S4 of Supplementary Materials). As f_{Ag} increases, effective refractive indices for TM and TE polarizations, $\varepsilon_{TM} = (f_{Ag}\varepsilon_{Ag}^{-1} + (1-f)\varepsilon_{VO2}^{-1})^{-1}$ and $\varepsilon_{TE} = f_{Ag}\varepsilon_{Ag} + (1-f_{Ag})\varepsilon_{VO2}$, start to deviate from coordinates of metallic (m-VO_2) and insulting phases (i-VO_2) along the highly curved and nearly linear two-dimensional paths, respectively (Figure 2A, B). The real parts of ε_{TM} and ε_{TE}, increase and decrease, respectively, according to the increase of f_{Ag} for values under 0.5. As shown in Figure 2A, B a designed anisotropic (n, k) clearly yields the large anisotropic changes near the near-unity absorptions from the IMT (the gray arrow in Figure 2A, B) to the two DIMTs (the orange arrow in Figure 2A, B when f_{Ag} is 0.2). On the other hand, embedding isotropically-shaped Ag nanodisks (marked as ND in Figure 2A, B) do not make significant phase-change tunability when compared to intrinsic one and Ag nanobeam embedded cases (See the Section 4 of Supplementary Materials for the details).

Therefore, it is intuitively expected that the opposite trends of red (for TM) and blue (for TE) shifts of intrinsic absorption dips occur at the both VO_2 phases in the same PCMA with certain f_{Ag} of Ag nanobeams as described in Figure 2C. In other words, by designed increase and decrease of effective refractive indices of a metafilm layer, both the blue and red shifts of active near-unity absorption can be demonstrated via polarization-controlled operations in a PCMA with certain geometric parameters. The top subfigure of Figure 2D describes the periodic (period of 250 nm) unit cell of PCMA with Ag nanobeams considering tapering effect of ion beam milling fabrication. The four images below the subfigure are top view scanning electron microscopy (SEM) images of the fabricated nanobeam type PCMAs with the four different f_{Ag} values (Figure S5 of Supplementary Materials). The fabricated samples would be called as the NB1 (f_{Ag} = 0.17), NB2 (f_{Ag} = 0.23), NB3 (f_{Ag} = 0.34), and NB4 (f_{Ag} = 0.40). The corresponding parameter sets of $(\underline{w_1}, w_2)$ of the NB1, NB2, NB3, and NB4 are (18 nm, 51 nm), (20 nm, 70 nm), (40 nm, 94 nm), (40 nm,

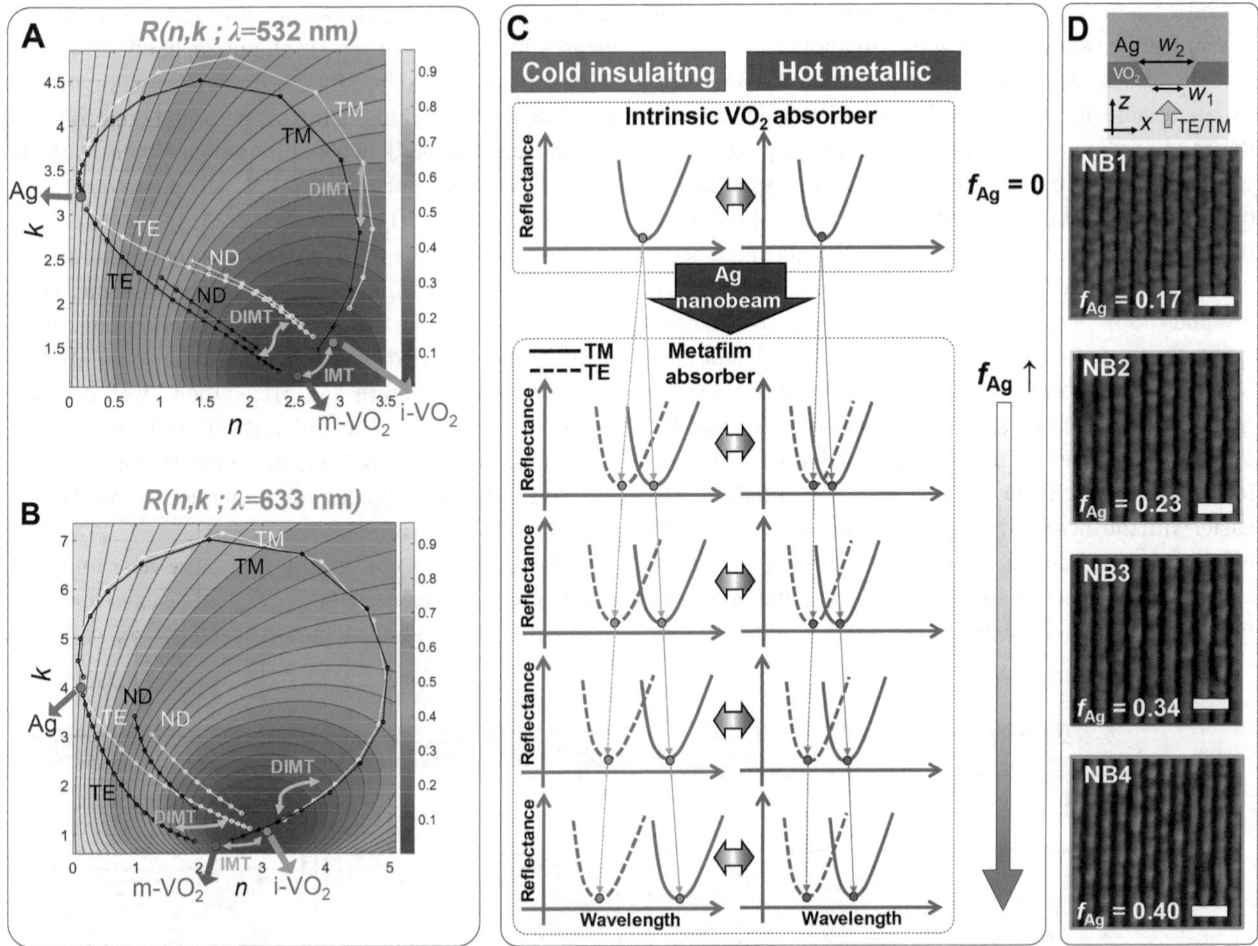

Figure 2: Design principles of anisotropic effective complex refractive indices of VO_2 metafilm and designed reflection tunability. The numerically retrieved Wiener bounds of effective refractive indices with corresponding reflectance maps at the (A) green (532 nm) and (B) red (633 nm) colors. The points marked as Ag, i-VO_2, and m-VO_2 correspond to (n, k) coordinates of Ag, high index VO_2 at the insulating phase, and low index VO_2 at the metallic phase at the certain wavelengths, respectively. The white and black dotted lines denote effective refractive index coordinates with VO_2 at the insulating and metallic phases, respectively. ND in (A) and (B) implies nanodisk. The ND curves in (A) and (B) imply the VO_2 metafilm cases where isotropic circular Ag nanodisks are embedded in the VO_2 layer with same f_{Ag} ($=\pi r^2/p^2$ where r and p are radius and period of Ag nanodisks) values compared to the Ag nanobeam embedded metafilm absorbers. f_{Ag} values of nanobeam and nanodisk type metafilms are varied from 0.05 to 0.95, and from 0.05 to 0.45, respectively. (C) Schematic illustration of intended mechanism and working principle of dynamic Gires-Tornois metafilm absorbers according to periodic photonic doping level of Ag nanobeams. (D) Schematic unitcell description with sidewall tapering effect and top view SEM images of the phase-change metafilm absorbers (PCMAs). Scale bars denote 500 nm.

120 nm), respectively. Here, NB stands for nanobeam. The difference between w_1 and w_2 accounts for modeling of abovementioned realistic tapering effect in ion beam milling fabrication.

2.2 Underlying physics of tunable phase-change effects using Ag nanobeams

In this section, the physical mechanisms of tunable effective refractive indices and their thermal DIMTs are theoretically studied with respect to effective medium approximation and Fano-like resonance interpretation of the near-unity absorptions. The theoretical concern starts by re-considering validity and assumption of effective medium theory discussed in the previous section. Based on the literature [54–56] and our numerical data of effective refractive index retrieval (Figure 2A, B and Figure S4 of Supplementary Materials), it is clear that the zeroth order approximation of effective medium approximation is only useful for prediction of anisotropic trend rather than calculation of exact effective refractive index values as the period of nanobeams (250 nm) is not small enough to guarantee the validity of the approximation.

The nanoscale physical origins of such difference between numerically retrieved effective refractive index and the zeroth order analytic effective medium approximation can be explained in terms of plasmonic and Mie-like resonances around Ag nanobeams. In case of TM polarized illumination, Ag nanobeams excite surface plasmons at the both VO_2-Ag and Al_2O_3–Ag interfaces (Figure 3A, B). If f_{Ag} is low and widths of nanobeams (w_1 and w_2) are narrow, near-unity absorption occurs mainly at the Ag-VO_2 interface between Ag nanobeams as shown in Figure 3A. However, if f_{Ag}, w_1, and w_2 increases, incident field funneling into the VO_2 region between Ag nanobeams gets harder while surface plasmon oscillation at the Al_2O_3–Ag interface gets easier with the increased interface length and SP excitation efficiency at the boundary of Al_2O_3–Ag. As refractive index of Al_2O_3 and effective SP mode index at the Al_2O_3–Ag interface is smaller than that of VO_2 and that at the VO_2-Ag interface, resultantly, overall effective resonance area increases by increase of f_{Ag}. On the other hand, as TE illumination cannot excite SPs, the dominant region of light trapping and absorbing is the inside of the dielectric VO_2 nanobeams capped between the Ag nanobeams via leaky Mie-like TE resonance (Figure 3C, D). As a result, the decreased size of VO_2 nanobeams owing to increased f_{Ag} induces decrease of effective resonator size and blue shift of resonance. In Figure 4A–H, anisotropic designer shifts and thermal tunings of near-unity absorption dips in the devices are verified numerically (Figure 4A–D) and experimentally (Figure 4E–H). When we compare Figure 4A–D, the red (TM) and blue (TE) shifts of the near-unity absorption dips owing to the opposite changes of the effective refractive indices are numerically verified according to the increase of f_{Ag}.

Moreover, it is noteworthy that large near-unity absorption is achieved even though f_{Ag} increases so that large modulation depth is still achieved over the two wavelength

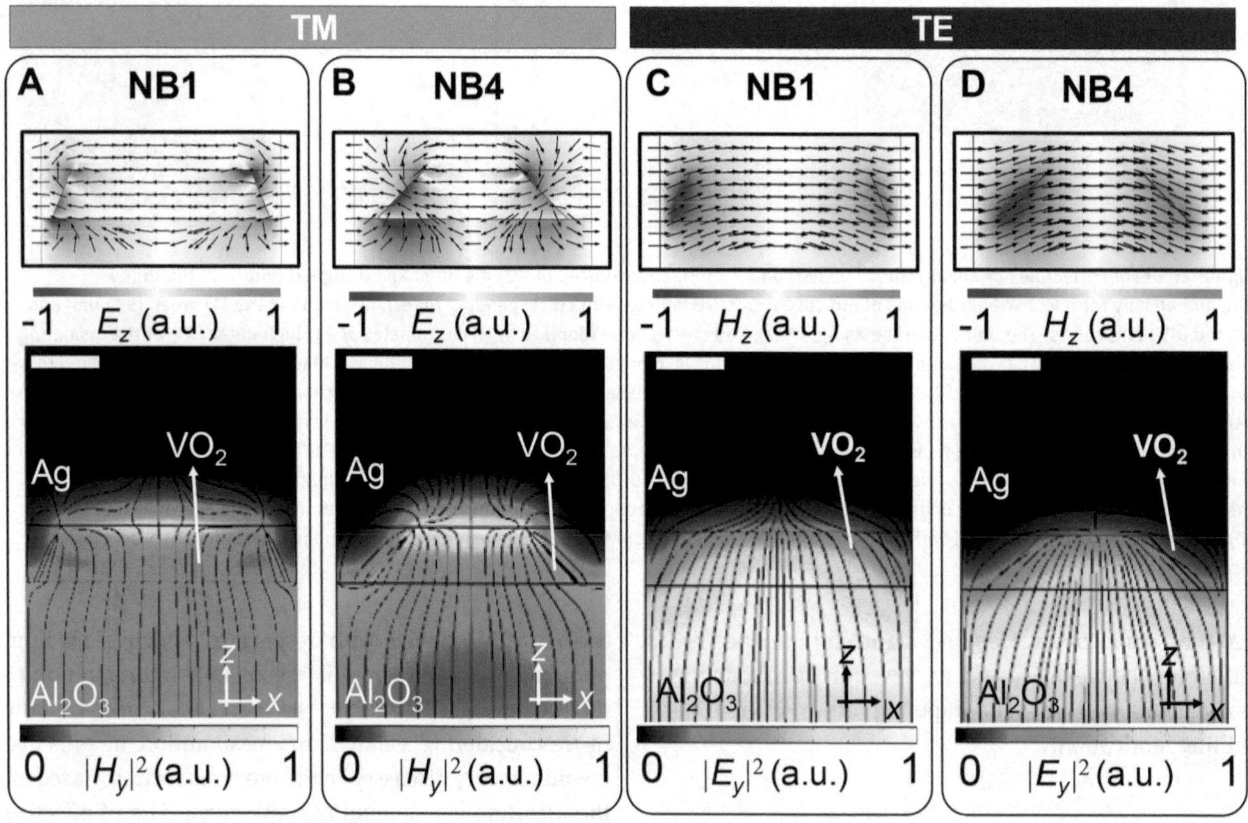

Figure 3: Nanoscopic mechanisms of the resonant near-unity absorptions: subwavelength light funneling into the metafilms. Electromagnetic field profiles of (A, C) NB1 and (B, D) NB4 devices are numerically investigated at the resonances in the insulating phase for (A, B) TM and (C, D) TE illumination, respectively. Upper figures show the (A, B) E_z and (C, D) H_z, with the normalized black vector arrows of (a, b) displacement current and (c, d) magnetic field, respectively. Lower figures show normalized electromagnetic field profiles of (A, B) $|H_y|^2$ and (C, D) $|E_y|^2$. Black lines in the lower figures depict the flow of Poynting vectors coming from the free space of Al_2O_3 substrate to the PCMA structures. The scale bars correspond to 60 nm.

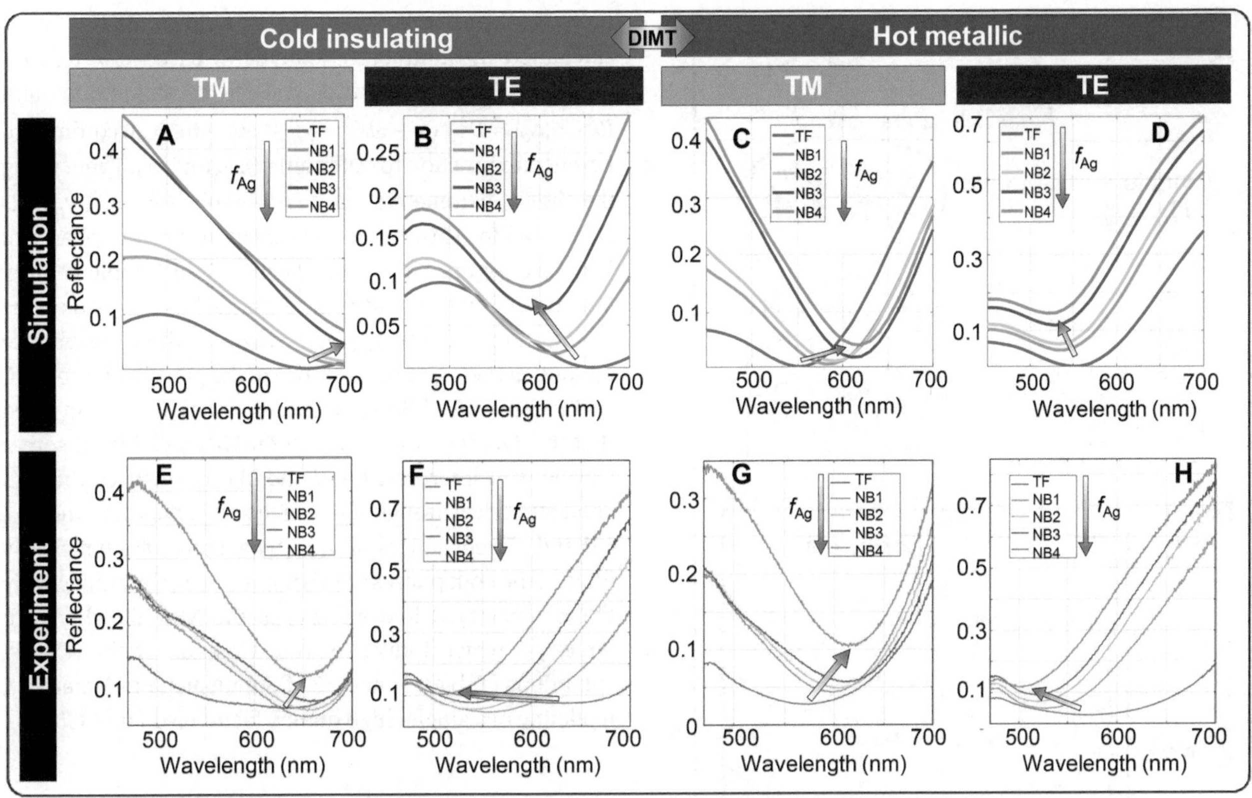

Figure 4: Effects of Ag nanobeams on the dynamic modulation range.
Polarization and temperature dependent reflectance spectra of the PCMAs with varying f_{Ag} (0, 0.17, 0.23, 0.34, 0.4) obtained from (A–D) simulations and (E–H) experiments. (A, B, E, F) present polarization dependent spectra at the cold insulating phase of VO_2 while (C, D, G, H) present those at the hot metallic phase of VO_2. (A, C, E, G) and (B, D, F, H) show reflectance under TM and TE polarized normal illumination cases, respectively. The legend, TF, refers to the case of intrinsic thin film VO_2 without Ag embedding (the Gires-Tournois VO_2 abosrber case when f_{Ag} is zero). The direction of gradually colored arrows in (A-H) imply the increase of f_{Ag}.

bands for a certain PCMA device under the two polarizations (TM and TE). Figure 4E–H successfully verifies the numerical designs of extraordinary modulations via experimental results. We fabricated the micron scale PCMAs via four steps with pulsed laser deposition, e-beam evaporation, and focused ion beam milling. We measured the temperature-dependent backscattering spectra from the PCMA samples (See Experimental sections for the details.).

The f_{Ag}-controlled design of reflection spectra in Figure 4 involves the fascinating interference phenomena through the two distinct reflection mechanisms owing to both effective medium approximation and the sub-wavelength absorption resonances (Figure 3). Here, we introduce Fano-like resonance modeling based on temporal coupled mode theory for quantitative analysis on the two effects to understand the underlying mechanisms of the high-contrast reflection wave modulations. Over the decade, temporal coupled mode theory has been widely utilized to interpret Fano and Fano-like resonances stemming from interference between direct and in-direct scatterings of transmission and reflection waves [57–61]. In this case, the proposed PCMA can be modeled as a resonator with a single port and a single resonance at the each VO_2 phase (Figure 5A). Here, a direct reflection wave can be explained by the Fresnel-Airy reflection formula based on the first order analytic effective medium approximation (Section 6 of Supplementary Materials) and in-direct reflection wave can be modeled as scattering from TM (plasmonic) and TE (Mie-like) resonances discussed above [58, 61].

Assuming $e^{i\omega t}$ time dependence of modes, total complex reflectivity (r_{tot}), a sum of direct (r_d) and in-direct (r_{id}) ones, can be derived as Eq. (1) (See Section 6 of Supplementary Materials for derivation.).

$$r_{tot} = r_d(\omega) + r_{id}(\omega) = r_d(\omega) + \frac{De^{i\varphi}}{i(\omega - \omega_0) + \gamma_{tot}}. \quad (1)$$

Here, D, γ, ω_0, and φ are real valued constants and D, γ, and ω_0 are positive real values. The fitting to extract proper

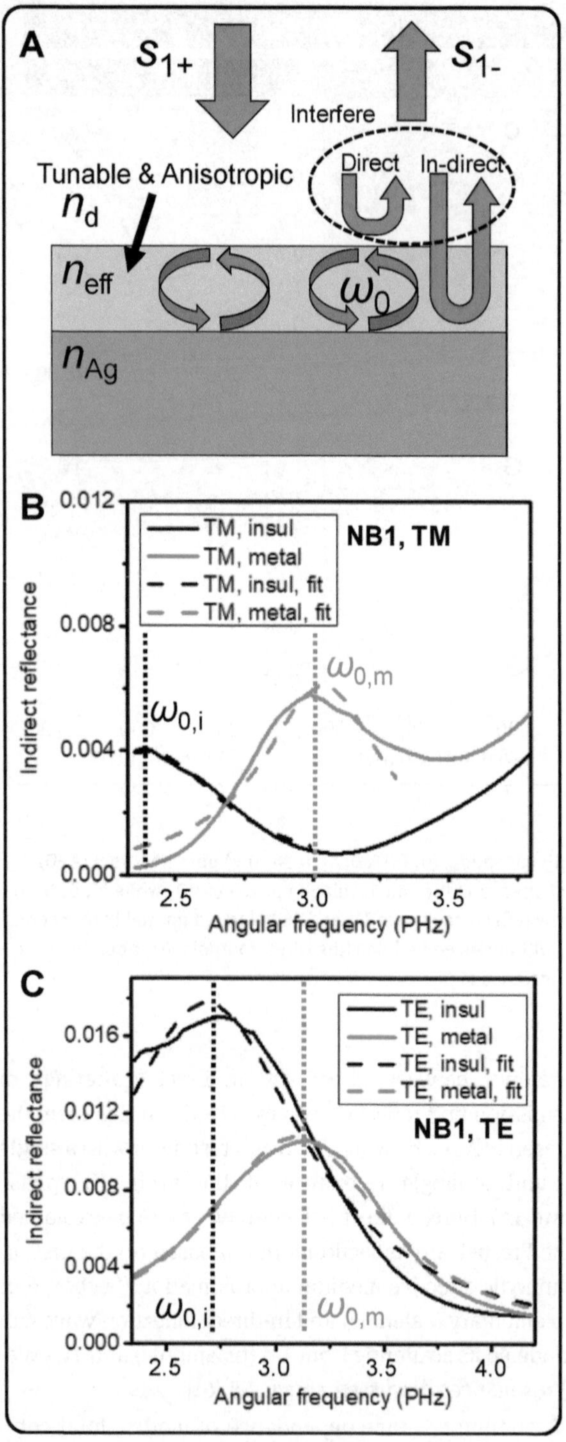

Figure 5: Fano-like resonance modeling.
(A) Schematic illustration of the modeling of the dynamic dual resonances as Fano-like interferences between direct and in-direct reflection waves. (B, C) Indirect reflectance spectra of the NB1 metafilm when f_{Ag} is 0.17 neglecting tapering walls and their Lorentzian fitting results. The Indirect reflectance spectra and their fitting results for (B) TM and (C) TE polarizations, respectively. The legends account for polarization direction and phase of VO_2. The dashed lines in (B) and (C) are the results of Lorentzian fitting near resonances.

D, γ_{tot}, ω_0, and φ is conducted as follows. At first, r_{id} is calculated by subtracting analytic r_d based n_{eff} from r_{tot} obtained from numerical full field simulation and $R_{id}\,(= |r_{id}| = D^2/\{(\omega - \omega_0)^2 + \gamma_{ror}^2\})$ is fitted according to Lorentzian function to determine D, γ, and ω_0. Then, using the fitted parameters, $\varphi = \angle r_{id} + \tan^{-1}((\omega - \omega_0)/\gamma_{tot})$ is calculated for TM and TE resonances at the both phases of VO_2 (See Figure 5B, C, and Section 7 of Supplementary Materials). But the validity of this modeling is based on the assumptions of low γ_{tot} and single in-direct resonance mode at ω_0 [54, 55, 57]. As shown in the plots in Figure 5B, C and Figure S6A of Supplementary Materials, it is revealed that the dynamic resonances of the NB1 metafilm absorber can be well interpreted as Fano-like resonances through good numerical fittings of r_{id}. On the other hand, the indirect TM and TE reosnances of the NB4 absorber are not clearly fitted at the both phases of VO_2 (Figure S6 B, C). It is basically due to enalarged loss and γ_{tot} of the NB4 absorber with larger f_{Ag} which induces severe violation of the two assumptions (Tables S1 and S2 of Supplemenatry Materials), modeling of a single, high quality factor resonance [58, 60].

2.3 Versatile optical applications and design extensibility

In this section, we prove excellent multiple functionalities of a single PCMA device as dynamic optical applications based on high-contrast gradual phase-change capabilities and polarization-controlled responses. As the NB1 sample with f_{Ag} about 0.17 shows the best (high-contrast and largely anisotropic) dynamic performances of near-unity absorptions among the demonstrated NB-type samples, it is considered to investigate gradual thermal modulation of both large anisotropy and near-unity absorptions. Figure 6A, B exhibit gradual blue shifts of resonance in the NB1 sample in cases of TM and TE-polarized illuminations via thermal heating in the different wavelength regions. It implies that the NB1 sample with a fixed geometry can continuously shift a near-unity absorption dip in a broad bandwidth of wavelength ranging from about 530 nm to about 670 nm by help of polarization control. Under TM and TE polarizations, blue shifts of absorption dips are verified from about 670 to 590 nm and from about 570 to 530 nm, respectively. The gradual measurement results of the NB1 sample show good agreement with numerical simulation results both in heating and cooling processes (Figure 3A, B and Section 8 of Supplementary Materials).

These largely anisotropic and gradually tunable phase-change characteristics of a single PCMA device can

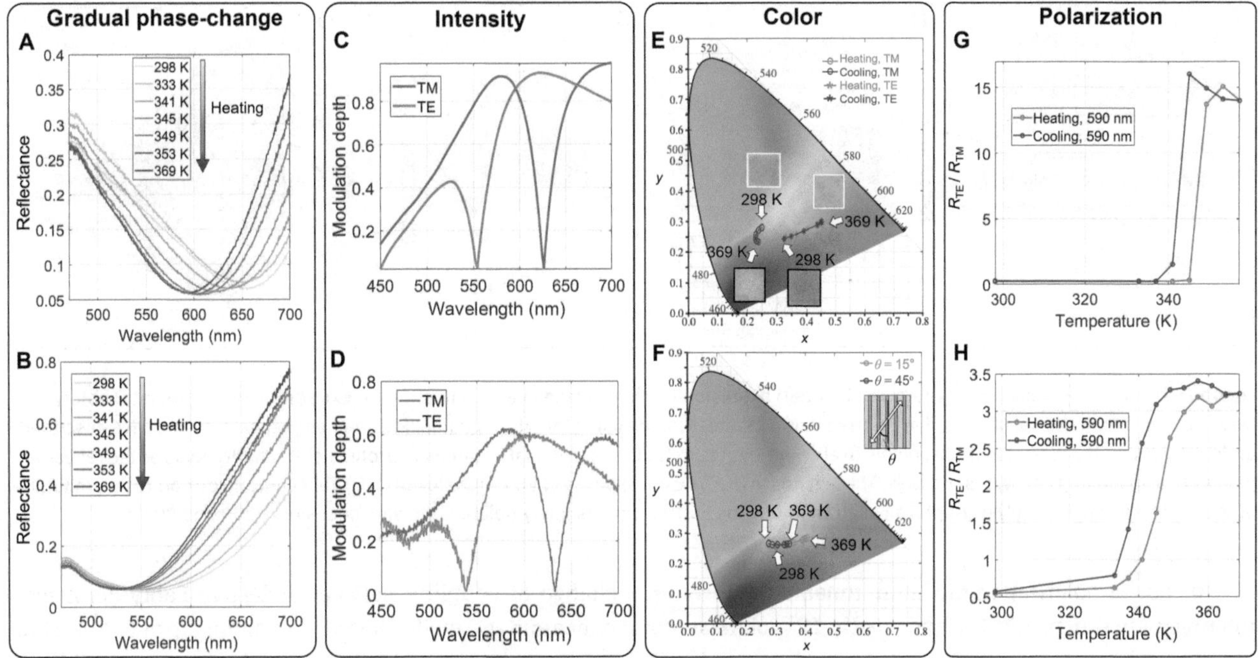

Figure 6: Optical applications of an anisotropic PCMA: modulation of intensity, color, and polarization.
(A, B) Heating induced gradual tuning of near-unity absorption spectra of the NB1 sample under illumination of (A) TM- and (B) TE-polarized light, respectively. Anisotropic broadband modulation depth spectra of reflection intensity from the NB1 sample obtained from (C) full-field simulation and (D) experiment. (E) Polarization-dependent gradual multi-level coloring of the NB1 sample described in CIE space. (F) Polarization rotation controlled intermediate color generations with the NB1 sample between TE and TM curves during the heating processes. The inset figure depicts top view scheme when linearly polarized light normally illuminates the sample with rotation angle, θ. (G, H) Thermally switchable R_{TE}/R_{TM} of the NB1 device, at the wavelength of 590 nm derived from (G) simulation and (H) experiment, respectively.

be applied to multiple high-contrast modulations of broadband intensity, reflected color, and polarization direction. Numerical and experimental results of Figure 6C, D shows that intensity of reflected wave is thermally modulated with large modulation depth over broad bandwidths both for TM and TE polarizations. In terms of the locations of modulation depth minimum and maximum, the results of the simulation and measurement show good agreement. However, the maximum value of measured modulation depth (~0.6) is a bit lower than that of simulation (~0.9). We guess that a bit larger reflectance of measurement rather than simulation is mainly due to offset baseline reflection from the air-substrate side (seen in measurement configuration in Figure S11) not considered in simulations, and sidewall imperfection of fabrication (seen in Figure S5). In particular, the baseline reflection from air-sapphire substrate for normal illumination is about 0.077 (7.7%) at the wavelength of 600 nm, which is a considerable value when calculating modulation depth.

Moreover, color spectrum and polarization direction of reflected light are also gradually modulated with high contrast in the certain device. Figure 6E shows thermally-driven gradual color generation along largely separated Commission internationale de l'éclairage (CIE) curves according to polarization direction. Blue to light violet and violet to orange thermal colorings are verified experimentally under illumination of TM and TE polarizations, respectively. Here, reflection Jones matrix of a PCMA, $r(\lambda, T)$, can be written as the Eq. (2).

$$r(\lambda, T) = \begin{pmatrix} r_{TE}(\lambda, T) & 0 \\ 0 & r_{TM}(\lambda, T) \end{pmatrix}. \tag{2}$$

In this formalism, $(1\ 0)^T$ and $(0\ 1)^T$ correspond to TE and TM polarization Jones vectors while r_{TE} and r_{TM} are complex reflectivity values dependent on operation wavelength and temperature, respectively. The intermediate colors in a CIE space between the curves for two orthogonal polarizations can be formulated as $R_\theta(\lambda, T) = R_{TE}(\lambda, T)\cos^2\theta + R_{TM}(\lambda, T)\sin^2$ when $R_{TE} = |r_{TE}|^2$ and $R_{TM} = |r_{TM}|^2$. Such dynamic colorings are demonstrated with intermediate polarization direction rotated (rotation angle of θ) from TM or TE (illustrated in the sub-figure of Figure 6F). In Figure 6F, the red and blue dotted-curves correspond to the cases when polarization rotation angle, θ, is 15 ° and 45 °, respectively.

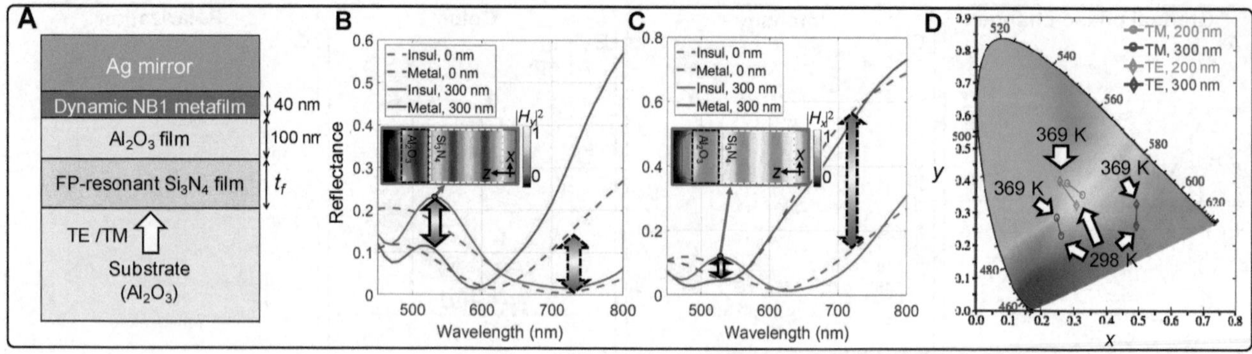

Figure 7: Dynamic metafilm heterostructure absorber: extensibility of the phase-change metafilm for extraordinary spectrum tunability. (A) Schematic illustration of of Si_3N_4 film inserted PCMA heterostructure using the NB1 metafilm. Phase-change tunability of device described under (B) TM and (C) TE illuminations, respectively. Sub-figures of (B, C) present spatial intensity profiles of H_y and H_x fields at the reflection peaks at the insulating phase, respectively. The legends in (B, C) describe phase of VO_2 and value of t_f. (D) Reflected anisotropic color tunability of the device described in (A) is graphically shown in CIE space. The legends imply polarization and values of t_f (200 and 300 nm).

In case of polarization direction, tunable polarization filtering between nearly-TM and nearly-TE polarizations can be achieved at certain resonant wavelength. In Figure 6A, B, the near-unity absorption dip of the NB1 sample at the insulating phase for TE polarization and that at the metallic phase for TM polarization are closely met (The dip wavelengths are about 610 and 590 nm, respectively. Also see Section 9 of Supplementary Materials.). At the wavelengths near those dip positions, 590 and 610 nm, reflectance is highly anisotropic. Particularly at the wavelength of 590 nm, numerically calculated reflectance ratio defined as $R_{TE}/R_{TM} = |r_{TE}/r_{TM}|^2$ is modulated between about 0 and 14 via heating and cooling (Figure 6G and the Section 8 of Supplementary Materials). As described in Figure 6H, experimental results at the same wavelength exhibit similar trend that reflectance ratio is modulated between about 0.5 and 3.3 in the reconfigurable manner with hysteresis.

As discussed above, it has proved that multi-functional optical applications with the capability of high contrast active thermal modulation are achieved with the PCMAs. Lastly, we would address the design extensibility of the concept for more design degree of freedom in a simple method. Once effective refractive indices of a certain metafilm layer is numerically characterized, this layer can be coupled with other thin film resonators vertically as an example described in Figure 7A. Then, it is possible to calculate reflectance and its tunability only by transfer matrix method without numerical full-field simulations. In case of an example of PCMA heterostructure in Figure 7A, it can exhibit not only near-unity absorptions but also resonant reflection peaks mainly stemming from a Si_3N_4 etalon. As the transparent dielectric film can induce reflection peaks of Fabry-Pérot resonance (See field profiles in subfigures of Figure 7B, C), a change of VO_2 phase can induce change of coupling between reflective Fabry-Pérot resonance and thermally shifted absorption dip of PCMA. Thus, as shown in Figure 7B, C, thermally-driven blue shifts of absorption dips induce weakened Fabry-Pérot resonance strengths with polarization-dependent manners. Variation of t_f would enable shift of a spectral location of a Fabry–Pérot peak and more diverse spectral tunability in the visible spectrum, potentially. In the perspective of dynamic color generation, this novel and simple strategy can dynamically enhance and tune purities of reflection color (Figure 7D) without significant fabrication problems compared to the original PCMA devices. Such PCMA heterostructures also hold a potential of vertical integration with other functional metasurfaces [62–64].

3 Conclusion

Inspired by near-unity absorption in ultrathin Gires-Tournois absorber and effective medium approximation, we have shown the simple and powerful framework, PCMA, for effective and versatile modulations of visible light with a single chip in a reconfigurable manner. We expect that the proposed concept would be fruitful for improvement of various integrated optoelectronics technologies for compact spatial light modulator, photodetector, compact reflective display, and multiplexed image sensors that essentially require compact and dynamic visible range operation. Moreover, it seems to be promising that the proposed strategy based on noble metal-embedding could be a milestone research for sub-micron active modulator pixels, if pixel-by-pixel electro-thermal control is achieved in the future, which is one of the ultimate goals of dynamic nanophotonics.

4 Experimental section

4.1 Sample fabrication

The device was fabricated through the four steps. Firstly, we use a pulsed laser deposition method (LAMBDA PHYSIK, COMPEX 205) with a KrF excimer laser at 248 nm for deposition of 40 nm-thick VO_2 on a sapphire substrate. Secondly, 150 nm-thick Ag was deposited for a hard mask on a VO_2 film by e-beam evaporation (Korea Vacuum Tech, KVE-3004). Then, 10 μm by 10 μm sized nanobeam-type patterns were defined by focused ion beam milling machine (FEI, Quanta 200 3D). Ag hard mask enables accurate high resolution focused ion beam patterning in large area and protects VO_2 film from direct stoichiometric Ga^+ ion contamination which would deteriorate molecular bonding structure as defects [23]. At last, we deposited a 200 nm-thick Ag film by e-beam evaporation to form metallic inclusions and mirrors.

4.2 Material property measurement

Temperature-dependent cyclic evolution of dielectric functions of 40 nm-thick VO_2 film is measured by using a variable angle spectroscopic ellipsometer system (J. A. Woollam, V-VASE) and temperature-controlled Peltier stage (Linkam, PE120). The complex refractive indices and effective exact thickness of VO_2 film have been fitted by Kramers-Kroning relation and general oscillator model with combination of Tauc-Lorentz, Lorentz, and Drude oscillators (Section 10 of Supplementary Materials). We used the measured permittivity data of metals and Al_2O_3 measured by E. D. Palik [65] and I. H. Malitson [66]. Temperature-dependent resistivity of a VO_2 film is measured for the verification of IMT phenomena using a source meter, a hot plate, and a thermocouple (Figure S12 of Supplementary Materials). Atomic force micrograph (Park Systems, XE150) is scanned with non-contact type tip for surface roughness and grain measurement of VO_2 film (Figure S10 of Supplementary Materials).

4.3 Temperature-controlled bright field backscattering spectroscopy and imaging

Bright field backscattering spectra from microscopic metafilm samples are measured using our custom-built setup with an optical microscope and spectrometer (Princeton Instruments, SpectraPro 2300) (Figure S11 of Supplementary Materials). Temperature of the samples were controlled by Peltier stage (Linkam, PE120) in our thermostatic laboratory of about 20 °C. Spectral data is scanned over about 6 μm by 6 μm region of samples through an aperture of spectrometer. Broadband illumination of white light (Thorlabs, MNWHL4 LED) is focused on the sample through the iris, broadband polarizer, and 50X objective lens. The spectra are normalized by the reflection spectrum from a sapphire substrate coated by optically thick silver film (deposited with the thickness over 500 nm by e-beam evaporation). The 100X magnified charge coupled device images are captured under the incidence of our neutral white light emitting diode.

4.4 Numerical simulation

Electromagnetic full field simulations are conducted with commercial finite element method tool (COMSOL Multiphysics 5.3, RF module, Frequency domain solver).

Acknowledgments: This work was supported by the National Research Foundation of Korea (NRF) grant funded by the Korea government (MSIT) (No. 2020R1A2B5B02002730). We deeply thank Dr. So Yeun Kim of IBS Center for Correlated Electron Systems for their help of the VO_2 ellipsometry measurement. We also thank to Dr. Joonsoo Kim for fruitful discussions about theoretical analysis. Part of this study has been performed using facilities at IBS Center for Correlated Electron Systems of Seoul National University, Electronics and Telecommunications Research Institute, and National Center for Inter-University Research Facilities of Seoul National University.

Author contribution: S.-J. K. conceived the idea, conducted material investigation, analytical modeling, numerical design, fabrication and measurement of the devices, and preparation of the draft. H. Y., J.-G. Y., and S.-Y. L. helped measurement and data analysis. Y. L., J. S., and C. C. helped measurement. S. C., K. P., S. J. J., and Y. W. L. deposited vanadium dioxide thin films. J. H. helped theoretical discussions. B. L. initiated and supervised the project. All of the authors participated in discussions and approved the submission of the manuscript.

Research funding: None declared.

Conflict of interest statement: The authors declare no conflict of interest.

References

[1] M. Ferrera, N. Kinsey, A. Shaltout, C. DeVault, V. Shalaev, and A. Boltasseva, "Dynamic nanophotonics," *J. Opt. Soc. Am. B*, vol. 34, no. 1, pp. 95–103, 2017.

[2] S. V. Makarov, A. S. Zalogina, M. Tajik, et al., "Light-induced tuning and reconfiguration of nanophotonic structures," *Laser Photon. Rev.*, vol. 11, no. 5, p. 1700108, 2017.

[3] N. I. Zheludev and Y. Kivshar, "From metamaterials to metadevices. *Nat. Mater.*, vol. 11, no. 11, pp. 917–924, 2012.

[4] S. M. Choudhury, D. Wang, K. Chaudhuri, et al., "Material platforms for optical metasurfaces," *Nanophotonics*, vol. 7, no. 6, pp. 959–987, 2018.

[5] M. Wuttig, H. Bhaskaran, and T. Taubner, "Phase-change materials for non-volatile photonic applications," *Nat. Photon.*, vol. 11, no. 8, p. 465, 2017.

[6] N. R. Hosseini and J. Rho, "Metasurfaces based on phase-change material as a reconfigurable platform for multifunctional devices," *Materials*, vol. 10, no. 9, p. 1046, 2017.

[7] D. M. Wu, M. L. Solomon, G. V. Naik, et al., "Chemically responsive elastomers exhibiting unity-order refractive index modulation," *Adv. Mater.*, vol. 30, no. 7, p. 1703912, 2018.

[8] J. Y. Kim, H. Kim, B. H. Kim, et al., "Highly tunable refractive index visible-light metasurface from block copolymer self-assembly," *Nat. Commun.*, vol. 7, no. 1, pp. 1–9, 2016.

[9] M. S. Jang, V. W. Brar, M. C. Sherrott, et al., "Tunable large resonant absorption in a midinfrared graphene Salisbury screen," *Phys. Rev. B*, vol. 90, no. 16, p. 165409, 2014.

[10] V. E. Babicheva, A. Boltasseva, and V. Lavrinenko, "Transparent conducting oxides for electro-optical plasmonic modulators," *Nanophotonics*, vol. 1, pp. 165–185, 2015.

[11] A. Howes, W. Wang, I. Kravchenko, and J. Valentine, "Dynamic transmission control based on all-dielectric Huygens metasurfaces," *Optica*, vol. 5, no. 7, pp. 787–792, 2018.

[12] J. Park and S. J. Kim, "Subwavelength-spaced transmissive metallic slits for 360-degree phase control by using transparent conducting oxides," *Appl. Opt.*, vol. 57, no. 21, pp. 6027–6031, 2018.

[13] G. K. Shirmanesh, R. Sokhoyan, R. A. Pala, and H. A. Atwater, "Dual-gated active metasurface at 1550 nm with wide (> 300°) phase tunability," *Nano Lett.*, vol. 18, no. 5, pp. 2957–2963, 2018.

[14] E. Li, Q. Gao, R. T. Chen, and A. X. Wang, "Ultracompact silicon-conductive oxide nanocavity modulator with 0.02 lambda-cubic active volume," *Nano Lett.*, vol. 18, no. 2, pp. 1075–1081, 2018.

[15] Y. Horie, A. Arbabi, E. Arbabi, S. M. Kamali, and A. Faraon, "High-speed, phase-dominant spatial light modulation with silicon-based active resonant antennas," *ACS Photon.*, vol. 5, no. 5, pp. 1711–1717, 2017.

[16] R. Bruck, K. Vynck, P. Lalanne, et al., "All-optical spatial light modulator for reconfigurable silicon photonic circuits," *Optica*, vol. 3, no. 4, pp. 396–402, 2016.

[17] T. Driscoll, H. T. Kim, B. G. Chae, et al., "Memory metamaterials," *Science*, vol. 325, no. 5947, pp. 1518–1521, 2009.

[18] M. J. Dicken, K. Aydin, I. M. Pryce, et al., "Frequency tunable near-infrared metamaterials based on VO_2 phase transition," *Opt. Express*, vol. 17, no. 20, pp. 18330–18339, 2009.

[19] S. J. Kim, H. Yun, K. Park, et al., "Active directional switching of surface plasmon polaritons using a phase transition material," *Sci. Rep.*, vol. 7, p. 43723, 2017.

[20] S. J. Kim, S. Choi, C. Choi, et al., "Broadband efficient modulation of light transmission with high contrast using reconfigurable VO_2 diffraction grating," *Opt. Express*, vol. 26, no. 26, pp. 34641–34654, 2018.

[21] M. Seo, J. Kyoung, H. Park, et al., "Active terahertz nanoantennas based on VO_2 phase transition," *Nano Lett.*, vol. 10, no. 6, pp. 2064–2068, 2010.

[22] M. A. Kats, R. Blanchard, S. Zhang, et al., "Vanadium dioxide as a natural disordered metamaterial: perfect thermal emission and large broadband negative differential thermal emittance," *Phys. Rev. X*, vol. 3, no. 4, p. 041004, 2013.

[23] J. Rensberg, S. Zhang, Y. Zhou, et al., "Active optical metasurfaces based on defect-engineered phase-transition materials," *Nano Lett.*, vol. 16, no. 2, p. 1050, 2016.

[24] M. A. Kats, D. Sharma, J. Lin, et al., "Ultra-thin perfect absorber employing a tunable phase change material," *Appl. Phys. Lett.*, vol. 101, no. 22, p. 1050, 2012.

[25] T. Driscoll, S. Palit, M. M. Qazilbash, et al., "Dynamic tuning of an infrared hybrid-metamaterial resonance using vanadium dioxide," *Appl. Phys. Lett.*, vol. 93, no. 2, p. 024101, 2008.

[26] Z. Zhu, P. G. Evans, R. F. Haglund, Jr., and J. G. Valentine, "Dynamically reconfigurable metadevice employing nanostructured phase-change materials," *Nano Lett.*, vol. 17, no. 8, pp. 4881–4885, 2017.

[27] Y. Ke, X. Wen, D. Zhao, R. Che, Q. Xiong, and Y. Long, "Controllable fabrication of two-dimensional patterned VO_2 nanoparticle, nanodome, and nanonet arrays with tunable temperature-dependent localized surface plasmon resonance," *ACS Nano*, vol. 11, no. 7, pp. 7542–7551, 2017.

[28] N. A. Butakov, I. Valmianski, T. Lewi, et al., "Switchable plasmonic–dielectric resonators with metal–insulator transitions," *ACS Photon.*, vol. 5, no. 2, pp. 371–377, 2018.

[29] Z. Xu, Q. Li, K. K. Du, et al., "Spatially resolved dynamically reconfigurable multilevel control of thermal emission," *Laser Photon. Rev.*, vol. 14, no. 1, p. 1900162, 2020.

[30] B. Gholipour, J. Zhang, K. F. MacDonald, D. W. Hewak, and N. I. Zheludev, "An all-optical, non-volatile, bidirectional, phase-change meta-switch," *Adv. Mater.*, vol. 25, no. 22, pp. 3050-4, 2013.

[31] Q. Wang, E. T. Rogers, B. Gholipour, et al., "Optically reconfigurable metasurfaces and photonic devices based on phase change materials," *Nat. Photon.*, vol. 10, no. 1, p. 60, 2016.

[32] A. Karvounis, B. Gholipour, K. F. MacDonald, and N. I. Zheludev, "All-dielectric phase-change reconfigurable metasurface," *Appl. Phys. Lett.*, vol. 109, no. 5, p. 051103, 2016.

[33] X. Yin, T. Steinle, L. Huang, et al., "Beam switching and bifocal zoom lensing using active plasmonic metasurfaces," *Light Sci. Appl.*, vol. 6, no. 7, p. e17016, 2017.

[34] C. Choi, S. Y. Lee, S. E. Mun, et al., "Metasurface with nanostructured $Ge_2Sb_2Te_5$ as a platform for broadband-operating wavefront switch," *Adv. Opt. Mater.*, vol. 7, no. 12, p. 1900171, 2019.

[35] C. Y. Hwang, G. H. Kim, J. H. Yang, et al., "Rewritable full-color computer-generated holograms based on color-selective diffractive optical components including phase-change materials," *Nanoscale*, vol. 10, no. 46, pp. 21648–21655, 2018.

[36] B. Gholipour, A. Karvounis, J. Yin, C. Soci, K. F. MacDonald, and N. I. Zheludev, "Phase-change-driven dielectric-plasmonic transitions in chalcogenide metasurfaces," *NPG Asia Mater.*, vol. 10, no. 6, pp. 533–539, 2018.

[37] P. Hosseini, C. D. Wright, and H. Bhaskaran, "An optoelectronic framework enabled by low-dimensional phase-change films," *Nature*, vol. 511, no. 7508, pp. 206–211, 2014.
[38] S. Y. Lee, Y. H. Kim, S. M. Cho, et al., "Holographic image generation with a thin-film resonance caused by chalcogenide phase-change material," *Sci. Rep.*, vol. 7, no. 1, p. 41152, 2017.
[39] A. Tittl, A. K. U. Michel, M. Schäferling, et al., "A switchable mid-infrared plasmonic perfect absorber with multispectral thermal imaging capability," *Adv. Mater.*, vol. 27, no. 31, pp. 4597–4603, 2015.
[40] K. K. Du, Q. Li, Y. B. Lyu, et al., "Control over emissivity of zero-static-power thermal emitters based on phase-changing material GST," *Light Sci. Appl.*, vol. 6, no. 1, p. e16194, 2017.
[41] Y. Qu, Q. Li, L. Cai, et al., "Thermal camouflage based on the phase-changing material GST," *Light Sci. Appl.*, vol. 7, no. 1, p. 1, 2018.
[42] Z. Li, Y. Zhou, H. Qi, et al., "Correlated perovskites as a new platform for super-broadband-tunable photonics," *Adv. Mater.*, vol. 28, no. 41, pp. 9117–9125, 2016.
[43] M. A. Kats, R. Blanchard, P. Genevet, and F. Capasso, "Nanometre optical coatings based on strong interference effects in highly absorbing media," *Nat. Mater.*, vol. 12, no. 1, pp. 20–24, 2013.
[44] J. Park, S. J. Kim, and M. L. Brongersma, "Condition for unity absorption in an ultrathin and highly lossy film in a Gires–Tournois interferometer configuration," *Opt. Lett.*, vol. 40, no. 9, pp. 1960–1963, 2015.
[45] M. M. Qazilbash, A. A. Schafgans, K. S. Burch, et al., "Electrodynamics of the vanadium oxides VO_2 and V_2O_3," *Phys. Rev. B*, vol. 77, no. 11, p. 115121, 2008.
[46] W. W. Salisbury. "Absorbent body for electromagnetic waves," U.S. 2,599,944, 1952.
[47] R. L. Fante and M. T. Mccormack, "Reflection properties of the Salisbury screen," *IEEE Trans. Antennas Propag.*, vol. 36, no. 10, pp. 1443–1454, 1988.
[48] I. Liberal, A. M. Mahmoud, Y. Li, B. Edwards, and N. Engheta, "Photonic doping of epsilon-near-zero media," *Science*, vol. 355, no. 6329, pp. 1058–1062, 2017.
[49] S. J. Kim, P. Fan, J. H. Kang, and M. L. Brongersma, "Creating semiconductor metafilms with designer absorption spectra," *Nat. Commun.*, vol. 6, no. 1, pp. 1–8, 2015.
[50] S. J. Kim, J. Park, M. Esfandyarpour, E. F. Pecora, P. G. Kik, and M. L. Brongersma, "Superabsorbing, artificial metal films constructed from semiconductor nanoantennas," *Nano Lett.*, vol. 16, no. 6, pp. 3801–3808, 2016.
[51] M. Esfandyarpour, E. C. Garnett, Y. Cui, M. D. McGehee, and M. L. Brongersma, "Metamaterial mirrors in optoelectronic devices," *Nat. Nanotechnol.*, vol. 9, no. 7, p. 542, 2014.
[52] O. Wiener. "Die Theorie des Mischkorpers fur das Feld der stationaren Stromung." *Abh Math-Phys Klasse Koniglich Sachsischen Des Wiss*, vol. 32, pp. 507–604, 1912.
[53] W. Cai and V. Shalaev, Optical metamaterials: fundamentals and applications, New York, NY, Springer Science & Business Media, 2009.
[54] P. Lalanne and J. P. Hugonin, "High-order effective-medium theory of subwavelength gratings in classical mounting: application to volume holograms," *J. Opt. Soc. Am. A*, vol. 15, no. 7, pp. 1843–1851, 1998.
[55] A. Fang, T. Koschny, and C. M. Soukoulis, "Optical anisotropic metamaterials: negative refraction and focusing," *Phys. Rev. B*, vol. 79, no. 24, p. 245127, 2009.
[56] D. R. Smith, D. C. Vier, T. Koschny, and C. M. Soukoulis, "Electromagnetic parameter retrieval from inhomogeneous metamaterials," *Phys. Rev. E*, vol. 71, no. 13, p. 036617, 2005.
[57] H. A. Haus, *Waves and fields in optoelectronics*, New Jersey, United States, Prentice-Hall, 1984.
[58] S. Fan, W. Suh, and J. D. Joannopoulos, "Temporal coupled-mode theory for the Fano resonance in optical resonators," *J. Opt. Soc. Am. A*, vol. 20, no. 3, pp. 569–572, 2003.
[59] K. X. Wang, Z. Yu, S. Sandhu, and S. Fan, "Fundamental bounds on decay rates in asymmetric single-mode optical resonators," *Opt. Lett.*, vol. 38, no. 2, pp. 100–102, 2013.
[60] J. W. Yoon and R. Magnusson, "Fano resonance formula for lossy two-port systems," *Opt. Exp.*, vol. 21, no. 15, pp. 17751–17759, 2013.
[61] A. Cordaro, J. Van de Groep, S. Raza, E. F. Pecora, F. Priolo, and M. L. Brongersma, "Antireflection high-index metasurfaces combining Mie and Fabry-Pérot resonances," *ACS Photon.*, vol. 6, no. 2, pp. 453–459, 2019.
[62] J. Van de Groep and M. L. Brongersma, "Metasurface mirrors for external control of Mie resonances," *Nano Lett.*, vol. 18, no. 6, pp. 3857–3864, 2018.
[63] J. G. Yun, J. Sung, S. J. Kim, H. Yun, C. Choi, and B. Lee, "Ultracompact meta-pixels for high colour depth generation using a bi-layered hybrid metasurface," *Sci. Rep.*, vol. 9, no. 1, pp. 1–9, 2019.
[64] Y. Zhou, I. I. Kravchenko, H. Wang, H. Zheng, G. Gu, and J. Valentine, "Multifunctional metaoptics based on bilayer metasurfaces," *Light Sci. Appl.*, vol. 8, no. 1, pp. 1–9, 2019.
[65] E. D. Palik, *Handbook of Optical Constants of Solids*, Orlando, Academic Press, 1985.
[66] I. H Malitson and M. J. Dodge, "Refractive Index and Birefringence of Synthetic Sapphire," *J. Opt. Soc. Am.*, vol. 62, no. 1405, pp. 11797–2999, 1972.

Supplementary material: The online version of this article offers supplementary material https://doi.org/10.1515/nanoph-2020-0264.

Marco Piccardo* and Antonio Ambrosio*

Arbitrary polarization conversion for pure vortex generation with a single metasurface

https://doi.org/10.1515/9783110710687-060

Abstract: The purity of an optical vortex beam depends on the spread of its energy among different azimuthal and radial modes, also known as ℓ- and *p*-modes. The smaller the spread, the higher the vortex purity and more efficient its creation and detection. There are several methods to generate vortex beams with well-defined orbital angular momentum, but only few exist allowing selection of a pure radial mode. These typically consist of many optical elements with rather complex arrangements, including active cavity resonators. Here, we show that it is possible to generate pure vortex beams using a single metasurface plate—called *p*-plate as it controls radial modes—in combination with a polarizer. We generalize an existing theory of independent phase and amplitude control with birefringent nanopillars considering arbitrary input polarization states. The high purity, sizeable creation efficiency, and impassable compactness make the presented approach a powerful complex amplitude modulation tool for pure vortex generation, even in the case of large topological charges.

Keywords: dielectric metasurfaces; modal purity; optical vortex generation; polarization conversion.

1 Introduction

The characterizing feature of an optical vortex is a zero of intensity, which coincides with a phase singularity of the field. The phase circulates around this point of null intensity endowing the beam of light with orbital angular momentum (OAM) [1]. The OAM and the specific ring-shaped intensity distribution of these beams make them attractive for a number of applications [2], ranging from quantum information [3, 4] to super-resolution microscopy [5, 6], and have motivated the development of several techniques of optical vortex generation [7]. However, most of these methods, such as spiral phase plates [8], computer-generated holograms [9, 10], spatial light modulators [11], *q*-plates [12, 13], and *J*-plates [14, 15], rely on phase-only (PO) transformations. The azimuthal phase modulation imparted by these optical elements allows creation of an optical vortex, but the lack of amplitude modulation prevents the output beam from being a solution of the paraxial wave equation. The missing amplitude term is compensated by the spreading of the beam energy during propagation on high-order radial modes, thus leading to impure states consisting of a superposition of vortex modes [16].

There exist a few methods that can provide both the required phase and amplitude (PA) modulation for pure vortex generation, such as mode conversion in active resonators [17–19], but these require either specific input beams or rather complex cavity set-ups. Here, we introduce a method that allows conversion of an arbitrary input beam into a pure vortex mode using a single metasurface plate—called *p*-plate as it controls radial modes—which is easy to implement in practical optics experiments and could be of use in any vortex application, requiring that the beam power is contained within a specific *p*-mode.

2 Methods

After presenting our metasurface theory, we will prove the approach by applying it to the design of dielectric metasurfaces operating in the near-infrared region. This spectral range is only chosen for the sake of illustration, without loss of generality. We consider amorphous silicon nanopillars with a rectangular cross section lying on a silica substrate. The pillars have a height of 600 nm and are arranged in a hexagonal close packed lattice (600-nm pillar-to-pillar separation). The wavelength of the source is 1064 nm. A library of the transmission coefficients and phases imparted by the pillars as a function of their size $L_{x,y}$ (in the 100- to 480-nm range) is constructed using the finite-difference time-domain (FDTD) module of Lumerical and selecting the

***Corresponding authors: Marco Piccardo and Antonio Ambrosio,** Center for Nano Science and Technology, Fondazione Istituto Italiano di Tecnologia, Milan 20133, Italy, E-mail: marco.piccardo@iit.it (M. Piccardo); antonio.ambrosio@iit.it (A. Ambrosio). https://orcid.org/0000-0001-8851-1638 (M. Piccardo)

This article has previously been published in the journal Nanophotonics. Please cite as: M. Piccardo and A. Ambrosio "Arbitrary polarization conversion for pure vortex generation with a single metasurface" *Nanophotonics* 2021, 10. DOI: 10.1515/nanoph-2020-0332.

elements with the best performance in terms of amplitude transmission and phase accuracy [20]. The average transmittance of the pillars chosen from the library to design the metasurfaces considered in this work is typically around 95%. The design code sets the sizes and orientation angles of the pillars according to our theory to produce a desired Laguerre–Gaussian (LG) beam. We validate the method by carrying out FDTD simulations of p-plates with 30-μm diameter, using plane wave or Gaussian sources with linear or elliptical polarizations. The beam waist of the designed LG modes is set to $\omega_0 = 5\,\mu\text{m}$. The simulated near field is propagated to a hemispherical surface in the far field at 1 m from the metasurface and then projected to a plane for visual representation using direction cosines. Finally, the propagated field is decomposed on an LG basis to evaluate its modal purity. The basis is truncated at the 20th radial mode. The complex beam parameter of the LG basis is obtained by an optimization algorithm that maximizes the overlap probability of the far-field beam with its decomposition over the finite basis [21].

3 Phase and amplitude control

The operation of a single birefringent nanopillar on an incident field can be represented in Jones calculus in the pillar frame of reference as

$$M = e^{i\psi}\begin{pmatrix} 1 & 0 \\ 0 & e^{i\Delta\phi} \end{pmatrix} \tag{1}$$

where the birefringence arises from the form factor of the pillar. We assume unity transmission; thus, M is unitary, and the pillar cannot modulate the amplitude of the field. Meta-atom geometries capable of controlling directly both the phase and amplitude of the field have been demonstrated [22, 23] but suffer in general from fabrication complexities as they rely on resonances, may be limited in efficiency, or may not allow full modulation ranges, i.e. from 0 to 1 in amplitude and from 0 to 2π in phase. However, a simple birefringent pillar represented by the operator M allows conversion of the polarization of the incident field, thanks to the phase retardance $\Delta\phi$. Thus, by adding a linear polarizer after the metasurface, one can in principle use the projection of the converted polarization state to modulate the transmitted amplitude of the field, as well as control its phase via the global $e^{i\psi}$ factor. This simple but powerful scheme was demonstrated with metasurfaces designed for the specific cases of linearly polarized [24] and circularly polarized [25] light inputs. Here, we generalize this approach to input states of arbitrary polarization.

The operation principle of the proposed scheme of PA control is shown in Figure 1A. Given an elliptically polarized input, the pillar dimensions and orientation angle α are designed so that the polarization-converted beam after passing through a polarizer has the desired phase and amplitude. The phase is determined by one of the two dimensions of the pillar, which sets ψ, while the amplitude can be controlled via a combination of $\Delta\phi$ and α, as shown in Figure 1B.

Now, we concentrate on the specific problem of finding analytically the exact conditions to obtain the extinction. This is much more important than determining the conditions for unity transmission as the extinction is critical to mask certain parts of the input beam. On the other hand, the conditions for unity transmission only influence the efficiency of the device and will be discussed later. The problem is illustrated in Figure 1C. We consider an incident elliptical state represented by the Jones vector

$$\begin{pmatrix} \cos\chi \\ e^{i\delta}\sin\chi \end{pmatrix} \equiv |\chi, \delta\rangle \tag{2}$$

and a linear polarizer with angle $\theta_{pol} = \pi/2$ with respect to the x-axis. Our goal is to find the pillar parameters that convert the input state into a linear state oriented perpendicularly to the polarizer axis. In the following, the subscript "0" refers to parameters and operators corresponding to the extinction condition. First of all, we use the rotation matrix R_0 to rotate the field on the pillar basis as $R_0|\chi,\delta\rangle \equiv |\chi',\delta'\rangle$, which turns the ellipse angle from θ to $\theta - \alpha_0$. Then, the new state can be linearized via the pillar birefringence operator as $M_0|\chi',\delta'\rangle$ with the condition $\Delta\phi_0 = -\delta'$. After linearization, the orientation of the state is further rotated to an angle $\theta_{lin} = \chi'$. We impose that after returning back to the original reference frame using $R_0^{-1}M_0R_0|\chi,\delta\rangle$, the output state lies parallel to the x-axis, which gives the condition $\theta_{lin} + \alpha_0 = 0$. By combining the previous conditions, we obtain $\alpha_0 = -\chi'$. Thus, all that remains is to express the rotated state $|\chi',\delta'\rangle$ in terms of the input one $|\chi,\delta\rangle$. This finally leads to the general extinction condition for the orientation angle

$$\alpha_0 = \frac{1}{2}\tan^{-1}(\tan\chi\sec\delta). \tag{3}$$

valid for an arbitrary polarization state. A closed-form expression can also be deduced for the phase retardance $\Delta\phi_0$, though not as compact (Supplementary Material).

Figure 1D and E represent the solutions for the extinction condition for all the possible input polarization states shown on the (χ,δ) plane. In the case of circularly polarized light (i.e. $|\pi/4,\pi/2\rangle$), one obtains the intuitive result of $\Delta\phi_0 = \pi/2$, corresponding to a quarter-waveplate, and $\alpha_0 = \pi/4$. Clearly, for any polarization state, it is always possible to achieve a full dynamic range of amplitude modulation by following a suitable path in the parameter space connecting the zero of amplitude with the global maximum (cf. Figure 1B). However, such trajectory cannot be described with a simple analytical form, and it may be

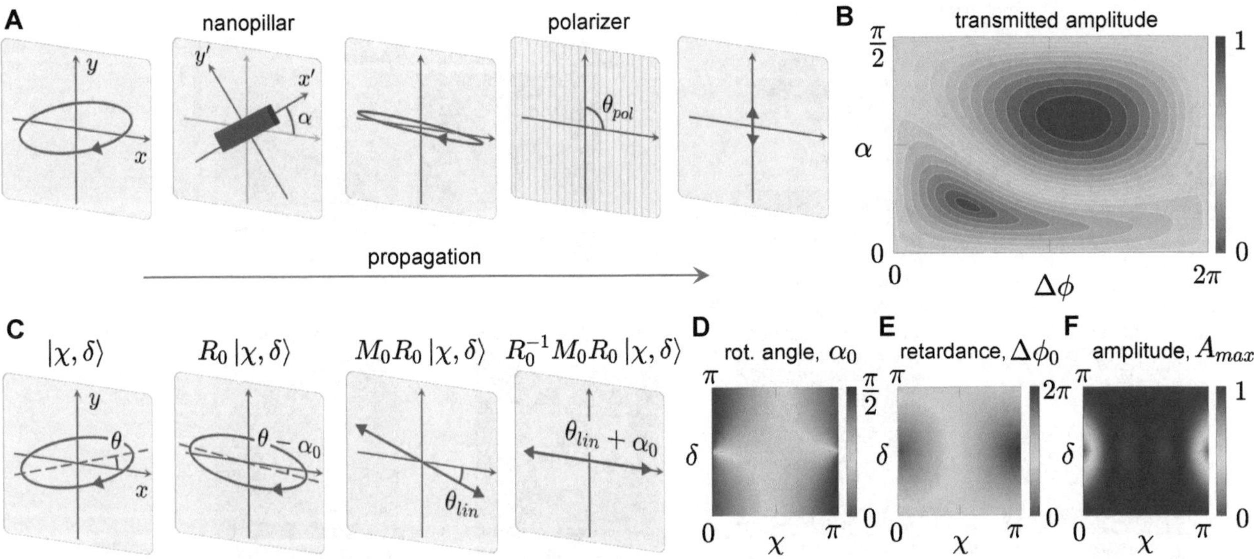

Figure 1: Mapping polarization to amplitude.
(A) An arbitrary input polarization state is converted by a birefringent nanopillar into a new polarization state that, after passing through a polarizer, is reduced in amplitude. This approach still allows the phase of the output state to be controlled independently. The nanopillar orientation angle and polarizer axis are α and θ_{pol}, respectively. (B) Contour plot showing the transmitted amplitude of the elliptical state shown in A after propagation through the nanopillar–polarizer system as a function of the orientation angle α and phase retardance $\Delta\phi$ of the pillar. (C) Conditions to achieve amplitude extinction through the nanopillar–polarizer system when the polarizer axis is aligned along the y-axis. The input is an arbitrary polarization state $|\chi, \delta\rangle$ that is rotated by an angle α_0 in the nanopillar frame of reference (operator R_0), linearized via birefringence (operator M_0), and finally brought back to the original frame of reference (operator R_0^{-1}) resulting orthogonal to the polarizer axis. (D and E) Design parameter maps corresponding to the pillar angle and phase retardance needed to obtain extinction for any input polarization state, represented in the (χ, δ) plane. (F) Maximum amplitude transmission through the nanopillar–polarization system for any input state when the phase retardance is fixed to $\Delta\phi_0$ and only α is adjusted. If instead both parameters are allowed to vary a full modulation range from 0 to 1 can be achieved, as in B.

convenient in practical metasurface designs to set one of the two parameters, e.g. $\Delta\phi_0$, and vary the other one to modulate the transmitted amplitude according to an analytical formula (Supplementary Material). We show in Figure 1F that with this restriction, only a small fraction of polarization states remains limited in modulation, while for the majority, it is possible to achieve a large modulation range, from 0 to almost 1.

4 Pure vortex generation

The principle of arbitrary polarization conversion illustrated above for the single nanopillar is now applied to the design of optical plates for pure vortex generation. The operation scheme is shown in Figure 2A. The input beam can be any wave of known PA distribution, such as a plane or Gaussian wave, and arbitrary polarization. In the case of a plane wave, the p-plate is designed to impart the PA profile of an LG mode [26]

$$\mathrm{LG}_{\ell,p} \propto \left(\frac{r\sqrt{2}}{\omega_0}\right)^{|\ell|} L_p^{|\ell|}\left(\frac{2r^2}{\omega_0^2}\right) e^{-r^2/\omega_0^2} e^{-i\ell\varphi} \equiv A(r)e^{i\psi(\varphi)} \quad (4)$$

where ℓ and p are indices denoting the OAM and radial mode of the LG beam, respectively; r and φ are the radial and azimuthal coordinates, respectively; ω_0 is the beam waist; and $L_p^{|\ell|}(x)$ is the generalized Laguerre polynomial of argument x. The amplitude mask $A(r)$ is assimilated in the angles of the pillars $\alpha(r)$, which vary along the radial direction of the plate (inset of Figure 2A), while the azimuthal phase profile $\psi(\varphi)$ defines the pillar dimension $L_x(\varphi)$. The other pillar dimension $L_y(\varphi)$ is determined by $\psi(\varphi) + \Delta\phi_0$, where $\Delta\phi_0$ is a fixed parameter in the design corresponding to the phase retardance for the extinction condition, as previously described. The output of the metasurface is a vector vortex beam as it carries OAM and exhibits a nonuniform polarization distribution. After the beam is filtered by a linear polarizer, which could be a separate element or directly integrated in the metasurface substrate as a wire-grid polarizer [24], its amplitude distribution corresponds to that of a pure LG beam.

The near-field intensity maps of an $\mathrm{LG}_{5,0}$ vortex generator obtained by FDTD simulations are shown in Figure 2B. The uniform intensity of the input plane wave is separated by the p-plate into a vertically polarized component, which is aligned with the polarizer axis and

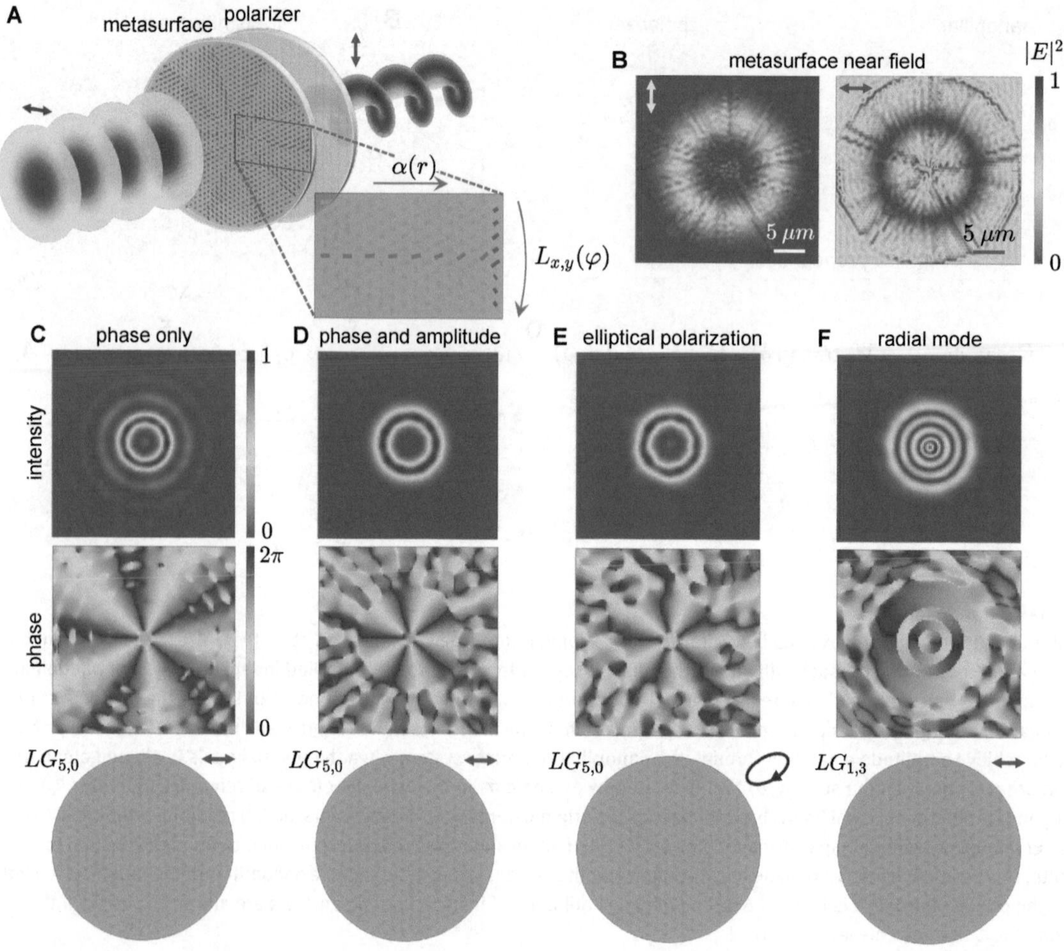

Figure 2: Pure vortex generation from a single metasurface.
(A) Schematic of the system implementation for pure vortex generation. An input wave of known amplitude, phase, and polarization, here represented as a linearly polarized Gaussian beam, propagates through the metasurface and polarizer and exits as a pure vortex beam. The angles of the nanopillars along the radial directions $\alpha(r)$ define the amplitude mask of the metasurface, while their rectangular dimensions set the azimuthal phase profile $L_{x,y}(\varphi)$ of the beam. (B) Near-field intensity maps obtained by finite-difference time-domain (FDTD) simulations for a metasurface generating an $LG_{5,0}$ mode. Only the vertically polarized component (left) is transmitted through the polarizer. (C–F) Far-field intensity and phase distributions obtained by FDTD simulations for different structures: (C) a phase-only metasurface generating a superposition of radial modes; (D–F) phase and amplitude metasurfaces generating pure vortex beams for different input and output states. In all cases, the bottom row shows the central area (10-μm diameter) of the metasurface structures, the target Laguerre–Gaussian beam, and the input polarization state (either linear or elliptical).

exhibits the characteristic ring-shaped distribution of a vortex mode, and a horizontally polarized component, which contains essentially the complementary intensity distribution and will be filtered by the polarizer. (A small fraction of light goes also into the longitudinal component, which is not shown here.) After propagating the component filtered by the polarizer to the far field, we obtain the intensity and phase distributions shown in Figure 2D, which can be compared with those obtained from a PO metasurface operating without a polarizer (Figure 2C). While in both cases, the phase shows an azimuthal modulation consisting of five sectors ($\ell = 5$), only the PA metasurface produces a single ring of intensity as one would expect for the fundamental radial mode ($p = 0$). The PO metasurface instead shows multiple rings corresponding to the superposition of different p-modes [16]. A modal purity analysis shows that 97% of the generated beam power is in the $p = 0$ mode for the PA metasurface, while this value drops to 23% for the PO metasurface. The efficiency in generating pure beams with the two types of metasurfaces will be discussed in Section 5.

The purity of the intensity profile generated by the PA metasurface all originates from a proper use of the pillar angle degree of freedom, as highlighted by the comparison of the device designs (cf. Figure 2C and D). As described in our theory, the PA control works for any input polarization state. In Figure 2E, we show that also for elliptically polarized light, where we chose as an example $|\chi = \pi/5, \delta = \pi/3\rangle$, one can generate a high-purity vortex mode with a single intensity ring. While the generation of such beams is of interest in OAM applications, where the energy needs to be confined in the fundamental radial mode, the control of higher order p-modes may be appealing in other applications, such as super-resolution microscopy. In Figure 2F, we show that this is possible using a PA metasurface, demonstrating as an example the generation of an $LG_{1,3}$ mode. Also, in this case, the purity of the beam is high as a very large fraction of the beam power transmitted through the polarizer (92%) is retained in the target mode.

5 Discussion

After having demonstrated that PA metasurfaces perform better than PO metasurfaces in terms of pure vortex generation, we wish to compare their efficiencies. We define the efficiency as the fraction of power of the beam incident on the metasurface that is converted into the target $LG_{\ell,0}$ mode. For this comparison, it is important to consider an input wave that carries finite power and that can be readily available as a practical source in a laboratory; thus, we choose a Gaussian beam. In the case of PO metasurfaces, the efficiency is known to drop rapidly with the OAM charge [16]. As shown in Figure 3A, the converted beam power spreads towards higher radial modes as $|\ell|$ grows owing to the lack of the amplitude modulation term, and nearly, no energy is contained in the $p = 0$ mode for $|\ell| > 10$. In PA metasurfaces instead, all the power transmitted through the polarizer is already in the $p = 0$ mode; thus, no modal filtering is needed. What limits the efficiency in this case is the fraction of input power that needs to be absorbed by the polarizer. To maximize the generation efficiency, there is an optimum choice for the beam waist ω_S of the Gaussian source, which is in general larger than the beam waist ω_0 of the Gaussian embedded in the target LG mode (Eq. 4) and is calculated as $\omega_S/\omega_0 = \sqrt{|\ell| + 1}$ based on the overlap of the source intensity profile with the transmittance mask of the metasurface (Supplementary Material). The efficiency of the PA metasurface for the optimum ω_S/ω_0 ratio is shown in Figure 3B. It shows a clear benefit for the generation of beams with large OAM charge as it remains sizeable even for very large values of $|\ell|$ (e.g. 5% efficiency at $|\ell| = 100$). Another advantage of PA metasurfaces in this respect lies in the sampling of the azimuthal phase profile. While PO metasurfaces suffer from poor phase resolution close to the singularity point owing to the strong phase gradient of large OAM beams and finite size of the meta-atoms, in the case of PA metasurfaces, the phase sampling only matters in the ring-shaped intensity region of the target $LG_{\ell,0}$ mode. Considering that the peak intensity of the ring occurs at a radial position $r_{max} = \omega_0\sqrt{|\ell|/2}$ from the singularity and that the length of an arc spanning a phase period from 0 to 2π is $p = 2\pi r_{max}/|\ell|$, we can estimate the number of pillars covering a phase period in a PA metasurface as

$$\frac{p}{U} \sim \sqrt{\frac{2}{|\ell|}}\frac{\pi\omega_0}{U} \tag{5}$$

where U is the size of the unit cell of the metasurface. Thus, for any OAM charge and for a unit cell size fixed by the phase library, the phase sampling of the PA metasurface can be chosen arbitrarily large just by adjusting ω_0.

The PA metasurfaces called p-plates introduced here also present some limitations with respect to the existing approaches for optical vortex generation. Differently from J-plates [14, 15], which implement PO transformations, p-plates are not spin–orbit converters. In particular, they can operate only on a single input polarization state. The degree of freedom that is used in J-plates to achieve independent OAM conversion of two orthogonal input polarization states here is used to apply a desired amplitude modulation. In comparison with q-plate lasers [17] and J-plate lasers [19], which incorporate PO transformation

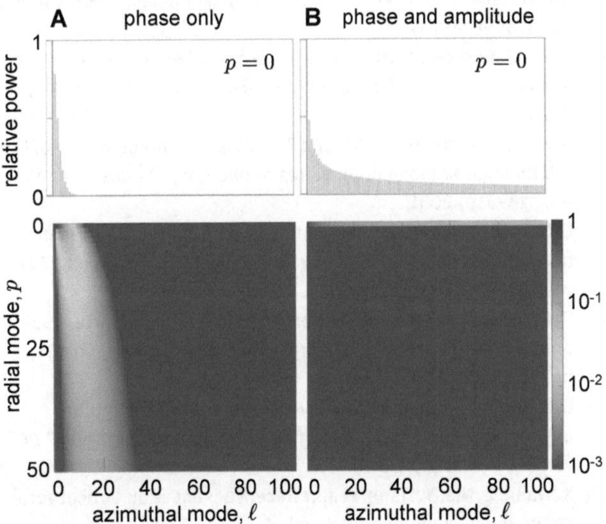

Figure 3: Purity efficiency of phase-only and phase-amplitude transformations.
Fraction of the input beam power that is contained in the fundamental radial mode $p = 0$ (top) and distributed among all p-modes (bottom) for different values of orbital angular momentum ℓ imparted by a metasurface plate. Both the case of phase-only metasurfaces (A) and phase-amplitude metasurfaces (B) are shown.

optics in active resonator cavities, p-plates can neither convert all the input energy into a pure LG state nor generate a continuously variable superposition of OAM states [19], only a fixed one. In view of these differences, we do not expect that p-plates will replace the existing technologies for OAM beam generation but may represent a very convenient and powerful approach in several applications, thanks to their simple and compact implementation scheme, sizeable efficiency, and high purity.

Acknowledgements: This work has been financially supported by the European Research Council (ERC) under the European Union's Horizon 2020 research and innovation programme "METAmorphoses", grant agreement no. 817794.
Author contribution: All the authors have accepted responsibility for the entire content of this submitted manuscript and approved submission.
Research funding: This work has been financially supported by the European Research Council (ERC) under the European Union's Horizon 2020 research and innovation programme "METAmorphoses", grant agreement no. 817794.
Conflict of interest statement: The authors declare no conflicts of interest regarding this article.

References

[1] L. Allen, M. W. Beijersbergen, R. J. C. Spreeuw, and J. P. Woerdman, "Orbital angular momentum of light and the transformation of Laguerre–Gaussian laser modes," *PRA*, vol. 45, pp. 8185–8189, 1992.
[2] S. Franke-Arnold, L. Allen, and M. Padgett, "Advances in optical angular momentum," *Laser Photon. Rev.*, vol. 2, pp. 299–313, 2020.
[3] A. Mair, A. Vaziri, G. Weihs, and A. Zeilinger, "Entanglement of the orbital angular momentum states of photons," *Nature*, vol. 412, pp. 313–316, 2001.
[4] E. Nagali, F. Sciarrino, F. De Martini, et al., "Quantum information transfer from spin to orbital angular momentum of photons," *PRL*, vol. 103, p. 013601, 2009.
[5] S. Fürhapter, A. Jesacher, S. Bernet, and M. Ritsch-Marte, "Spiral phase contrast imaging in microscopy," *Opt. Express*, vol. 13, pp. 689–694, 2005.
[6] L. Li and F. Li, "Beating the Rayleigh limit: orbital-angular-momentum-based superresolution diffraction tomography," *PRE*, vol. 88, p. 033205, 2013.
[7] X. Wang, Z. Nie, Y. Liang, et al., "Recent advances on optical vortex generation," *Nanophotonics*, vol. 7, pp. 1533–1556, 2018.
[8] M. W. Beijersbergen, R. P. C. Coerwinkel, M. Kristensen, and J. P. Woerdman, "Helicalwavefront laser beams produced with a spiral phaseplate," *Opt. Commun.*, vol. 112, pp. 321–327, 1994.
[9] N. R. Heckenberg, R. McDuff, C. P. Smith, and A. G. White, "Generation of optical phase singularities by computer-generated holograms," *Opt. Lett.*, vol. 17, pp. 221–223, 1992.
[10] A. Ambrosio, L. Marrucci, F. Borbone, A. Roviello, and P. Maddalena, "Light-induced spiral mass transport in azo-polymer films under vortex-beam illumination," *Nat. Commun.*, vol. 3, p. 989, 2012.
[11] S. Ngcobo, I. Litvin, L. Burger, and A. Forbes, "A digital laser for on-demand laser modes," *Nat. Commun.*, vol. 4, p. 2289, 2013.
[12] R. C. Devlin, A. Ambrosio, D. Wintz, et al., "Spin-to-orbital angular momentum conversion in dielectric metasurfaces," *Opt. Express*, vol. 25, pp. 377–393, 2017.
[13] A. Rubano, F. Cardano, B. Piccirillo, and L. Marrucci, "Q-plate technology: a progress review [Invited]," *J. Opt. Soc. Am. B*, vol. 36, pp. 70–87, 2019.
[14] R. C. Devlin, A. Ambrosio, N. A. Rubin, J. P. B. Mueller, and F. Capasso, "Arbitrary spin-to-orbital angular momentum conversion of light," *Science*, vol. 358, pp. 896–901, 2017.
[15] Y.-W. Huang, et al., "Versatile total angular momentum generation using cascaded," *J Plates. Opt. Express*, vol. 27, pp. 7469–7484, 2019.
[16] B. Sephton, A. Dudley, and A. Forbes, "Revealing the radial modes in vortex beams," *Appl. Optic.*, vol. 55, pp. 7830–7835, 2016.
[17] E. Maguid, et al., "Topologically controlled intracavity laser modes based on pancharatnam-berry phase," *ACS Photon.*, vol. 5, pp. 1817–1821, 2018.
[18] R. Uren, S. Beecher, C. R. Smith, and W. A. Clarkson, "Method for generating high purity Laguerre–Gaussian vortex modes," *IEEE J. Quant. Electron.*, vol. 55, pp. 1–9, 2019.
[19] H. Sroor, Y.-W. Huang, B. Sephton, et al., "High-purity orbital angular momentum states from a visible metasurface laser," *Nat. Photon.*, vol. 14, pp. 498–503, 2020.
[20] A. Arbabi, Y. Horie, M. Bagheri, and A. Faraon, "Dielectric metasurfaces for complete control of phase and polarization with subwavelength spatial resolution and high transmission," *Nat. Nanotechnol.*, vol. 10, pp. 937–943, 2015.
[21] G. Vallone, "Role of beam waist in Laguerre–Gauss expansion of vortex beams," *Opt. Lett.*, vol. 42, pp. 1097–1100, 2017.
[22] Q. Wang, et al., "Broadband metasurface holograms: toward complete phase and amplitude engineering," *Sci. Rep.*, vol. 6, p. 32867, 2016.
[23] S. L. Jia, X. Wan, P. Su, Y. J. Zhao, and T. J. Cui, "Broadband metasurface for independent control of reflected amplitude and phase," *AIP Adv.*, vol. 6, p. 045024, 2020.
[24] S. Divitt, W. Zhu, C. Zhang, H. J. Lezec, and A. Agrawal, "Ultrafast optical pulse shaping using dielectric metasurfaces," *Science*, vol. 364, pp. 890–894, 2019.
[25] A. C. Overvig, S. Shrestha, S. C. Malek, et al., "Dielectric metasurfaces for complete and independent control of the optical amplitude and phase," *Light Sci. Appl.*, vol. 8, p. 92, 2019.
[26] B. E. Saleh and M. Teich, *Fundamentals of Photonics*, Hoboken, United States, Wiley, 2007.

Supplementary material: The online version of this article offers supplementary material (https://doi.org/10.1515/nanoph-2020-0332).

Jared S. Ginsberg*, Adam C. Overvig, M. Mehdi Jadidi, Stephanie C. Malek, Gauri N. Patwardhan, Nicolas Swenson, Nanfang Yu and Alexander L. Gaeta

Enhanced harmonic generation in gases using an all-dielectric metasurface

https://doi.org/10.1515/9783110710687-061

Abstract: Strong field confinement, long-lifetime resonances, and slow-light effects suggest that metasurfaces are a promising tool for nonlinear optical applications. These nanostructured devices have been utilized for relatively high efficiency solid-state high-harmonic generation platforms, four-wave mixing, and Raman scattering experiments, among others. Here, we report the first all-dielectric metasurface to enhance harmonic generation from a surrounding gas, achieving as much as a factor of 45 increase in the overall yield for Argon atoms. When compared to metal nanostructures, dielectrics are more robust against damage for high power applications such as those using atomic gases. We employ dimerized high-contrast gratings fabricated in silicon-on-insulator that support bound states in the continuum, a resonance feature accessible in broken-symmetry planar devices. Our 1D gratings maintain large mode volumes, overcoming one of the more severe limitations of earlier device designs and greatly contributing to enhanced third- and fifth-harmonic generation. The interaction lengths that can be achieved are also significantly greater than the 10's of nm to which earlier solid-state designs were restricted. We perform finite-difference time-domain simulations to fully characterize the wavelength, linewidth, mode profile, and polarization dependence of the resonances. Our experiments confirm these predictions and are consistent with other nonlinear optical properties. The tunable wavelength dependence and quality factor control we demonstrate in these devices make them an attractive tool for the next generation of high-harmonic sources, which are anticipated to be pumped at longer wavelengths and with lower peak power, higher repetition rate lasers.

Keywords: bound state in the continuum; harmonic generation; metasurface; nonlinearity.

***Corresponding author: Jared S. Ginsberg,** Department of Applied Physics and Applied Mathematics, Columbia University, 500 West 120th Street, New York, New York, 10027-6902, USA, E-mail: jsg2208@columbia.edu. https://orcid.org/0000-0002-0071-2169

Adam C. Overvig, M. Mehdi Jadidi, Stephanie C. Malek, Nanfang Yu and Alexander L. Gaeta, Department of Applied Physics and Applied Mathematics, Columbia University, 500 West 120th Street, New York, New York, 10027-6902, USA. https://orcid.org/0000-0002-9462-4724 (N. Yu)

Gauri N. Patwardhan, Department of Applied Physics and Applied Mathematics, Columbia University, 500 West 120th Street, New York, New York, 10027-6902, USA; Department of Applied and Engineering Physics, Cornell University, Ithaca, New York, USA

Nicolas Swenson, Department of Applied and Engineering Physics, Cornell University, Ithaca, New York, USA

1 Introduction

Strong laser field excitation of rare gas atoms has been the preferred method of generating short wavelength coherent radiation down into the soft X-ray range [1–5]. High-order harmonic generation (HHG) has been a pillar of ultrashort pulse generation for decades [6, 7] and has been the catalyst for significant advances in nonlinear optics and attosecond science. Next-generation studies of ultrafast atomic phenomena demand higher repetition rates and shorter pulses with greater bandwidths. Thus, the quadratic dependence of the maximum HHG bandwidth on pump wavelength [8] provides strong motivation to push the wavelength of the pump field deeper into the mid-infrared. Strict intensity requirements of nearly 100 terawatt/cm^2 in order to access the nonperturbative regime of HHG in gases have, however, mostly restricted the field to the use of chirped pulse, multipass, and regeneratively amplified lasers. These large-footprint systems have low repetition rates, high average power, and are limited to the near-visible regime, hindering progress toward mid-infrared pumping. Our work is motivated by a simultaneous pursuit of efficient HHG at longer wavelengths and lower pump powers.

The lower intensities required to observe HHG from solids [9] have generated significant interest and led to

This article has previously been published in the journal Nanophotonics. Please cite as: J. S. Ginsberg, A. C. Overvig, M. M. Jadidi, S. C. Malek, G. N. Patwardhan, N. Swenson, N. Yu and A. L. Gaeta "Enhanced harmonic generation in gases using an all-dielectric metasurface" *Nanophotonics* 2021, 10. DOI: 10.1515/nanoph-2020-0362.

rapid innovation in the decade since its initial discovery. A promising direction has been the merging of nonlinear optics with metasurfaces to create an efficient platform for HHG [10–15]. These nanostructured devices allow for the engineering of far-field emission profiles, selective wavelength enhancement, and exceptionally strong field confinement. Subwavelength scale plasmonics and dielectrics have both been demonstrated for these purposes. While the strong field confinement permitted by these two types of systems further reduces the pump power requirements for harmonic generation in solid-state systems, enhancing intensities within the device material poses its own challenges. Plasmonic devices are severely hampered by ohmic losses and are susceptible to melting, even at modest intensities [16]. Both plasmonics and dielectrics must also contend with the reabsorption of harmonics within an opaque generation material [17], reducing the effective interaction length of the nonlinear medium to only a few tens of nanometers. Alternatively, spatially selective confinement of pump energy within the regions just outside the metasurface can be accomplished with bound states in the continuum (BIC), a class of optical resonance with broadly engineerable mode profiles supported by all-dielectric metasurfaces [18–21]. A variety of BIC-enhanced light–matter interactions including optical absorption [22], third-harmonic generation (THG) in solids [23], and Raman scattering [24] have been reported.

In this work, we demonstrate an all-dielectric metasurface to significantly enhance the harmonic yield from gas-filled gaps by utilizing dimerized high-contrast gratings (DHCGs) supporting strong BIC resonances near 1550 or 2900 nm. DHCG metasurfaces offer independent parameters for band structure design, wavelength selection, and quality factor tuning [25], making them well suited for HHG. We perform experiments that confirm the mode location, polarization dependence, and resonance wavelength predicted by our device simulations and show that harmonic emission from Argon atoms can be enhanced by more than an order of magnitude via resonant excitation of our devices.

2 Device principles

Bound states in the continuum were first proposed as a mathematical curiosity of quantum mechanics [26] but have since been realized in a number of experimental environments beyond quantum systems. In periodic planar optical devices, BICs manifest as infinitely strong resonances of zero linewidth [27] and are therefore not experimentally accessible. One such example is a normally incident plane wave or Gaussian beam and a high-contrast grating with an antisymmetric mode at its gamma point, that is, wavevector $k = 0$. The inability of an even symmetry incident wave to couple to an odd symmetry device mode is the working principle of a so-called symmetry-protected BIC [28]. Coupling to such a resonance therefore requires either the breaking of the even symmetry of the incident beam or of the odd symmetry of the mode, a process that results in a state that is quasi-bound in the continuum (a quasi-BIC). The former requires nonnormal incidence of the pump beam, while the latter can be achieved by a periodic perturbation to the grating unit cell. The type and magnitude of the perturbation offer significant flexibility when selecting mode profiles and device bandwidth [29].

We designed and fabricated silicon-on-insulator DHCGs on a standard wafer with a 250-nm-thick silicon device layer, 1-μm of oxide, and a 500-μm silicon substrate. Full-wave device simulations were performed with Lumerical, a commercial finite-difference time-domain (FDTD) software package. Figure 1A shows a top view SEM image of the finalized device, fabricated by e-beam lithography, which consists of a series of silicon fingers 270 nm wide with a total grating period of 920 nm. The large grating period is a result of the dimerization of an unperturbed high-contrast grating with a period of 460 nm. We also fabricated a second grating with approximately twice the width and spacing, allowing us to test two very different operating wavelengths of 1550 and 2900 nm. By scaling the period and duty cycle, the system can be made to operate at any wavelength and is limited only by fabrication sensitivity.

The dimerization of a high-contrast grating alters the physics of the system in ways beyond the perturbation to the mode symmetry described above. In the momentum space picture, the doubling of the spatial period of the structure implies a halving in k-space, known as Brillouin zone folding [30]. This "folds" modes at the edges of the first Brillouin zone (FBZ) to the gamma point, making them accessible to normal incidence excitation. Such modes had previously only existed in an unusable state below the light line. The Brillouin zone folding of a high-contrast grating is illustrated in Figure 1B. In a 1D DHCG, such as the ones described in this work, two types of periodic perturbations can be implemented to dimerize the unit cell. The finger width or spacing is alternatingly increased and decreased by a perturbation δ, while the other parameter is held constant. We choose to modulate the spacing (i.e., the gap perturbation) since this folds the unique mode depicted in Figure 1C to the gamma point. The mode is centered in the gap between adjacent grating fingers, where it provides a uniform two orders of magnitude enhancement in the local

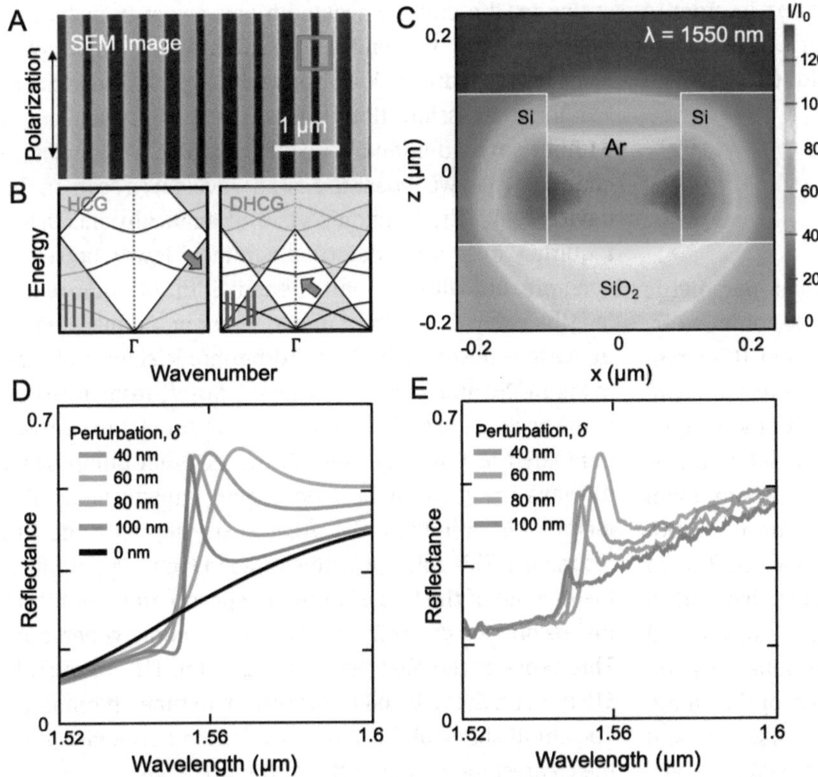

Figure 1: Device fabrication and linear characterization.
(A) Scanning electron microscope (SEM) image of the fabricated dimerized high-contrast grating (DHCG). The height of the grating is 250 nm, the finger width is 270 nm, and the period is 920 nm. The dimerizing perturbation to the grating finger spacing results in alternating smaller gaps of 150 nm and larger gaps of 230 nm. (B) Typical band structures of unperturbed high-contrast grating (HCG) (left) and perturbed DHCG (right). Red arrows indicate the mode shown in (C) which is brought to the gamma point by Brillouin zone folding. Shaded blue regions are below the light line. (C) Finite-difference time-domain (FDTD) simulation showing the relative power enhancement in the region marked by a red square in (A), now a side view. The strong field confinement mode is centered in the Argon gas region, with leakage into the walls of the grating fingers and substrate, outlined in white. (D) Simulated unpolarized reflectance spectrum of DHCGs with alternating gaps of 190 nm $\pm\delta$. The reflectance peaks at the bound states in the continuum (BIC) Fano resonance design wavelength near 1550 nm. Smaller perturbations correspond to larger Q-factors and narrower linewidths. (E) Fourier transform infrared spectroscopy (FTIR) measured reflectance of unpolarized near-infrared light. Losses lead to the observed variation in reflectance magnitude.

field intensity for a modest quality factor Q = 1500. This mode affords us the ability to inject a gaseous nonlinear medium (e.g., Argon atoms) into the gaps. Only when the polarization of the pump is oriented along the finger direction will this particular Transverse Electric (TE) mode be efficiently excited. The total number of modes in the system is also a function of the grating depth. By restricting the device to 250 nm in height, we thereby reduce the likelihood of a parasitic resonance being located near our gap-centered mode. Alternatively, a mode centered purely within the silicon could have been achieved with a finger width perturbation and fixed gaps.

The modes that exist at the Brillouin zone boundaries are typically much flatter than modes found at the gamma point, and the slope of the band is directly proportional to the group velocity of light in the mode [31, 32]. Therefore, high degrees of confinement in small devices benefit from the flattest possible bands. Furthermore, the combination of small slopes and a reduced FBZ means DHCGs can exhibit sharper spectral features with narrower linewidths than either unperturbed or low-contrast gratings. However, for all DHCG modes, the quality factor of the resonance can be fully tuned by the magnitude of the dimerizing perturbation according to $Q \propto \delta^{-2}$ for small perturbations δ. The inverse quadratic scaling of the Q-factor is a general feature of all quasi-BIC supporting metasurfaces [33]. Figure 1D and E show the simulated and measured unpolarized reflectance spectra of a set of DHCGs designed with a central wavelength of 1550 nm and a range of Q-factors as high as 1000. We also include the symmetry-protected case of unperturbed gap widths of 190 nm and confirm that no resonance is observed. In our experiments, we choose to increase the size of the perturbation in order to accommodate the bandwidth of our laser pulses, while simultaneously making it easier for the field to couple into the mode.

3 Experiments

Our experimental setup utilizes an optical parametric amplifier (Light Conversion HE Topas Prime) pumped by an amplified Titanium–sapphire laser system (Coherent Legend Elite) operating at a 1-kHz repetition rate with 6 mJ of pulse energy to generate tunable, 60-fs duration signal pulses with center wavelengths from 1485 to 1580 nm. This parametric amplifier output is used to pump an additional difference frequency generation module for our mid-infrared measurements at 2.9 μm with a pulse duration of 70 fs. The pump beam polarization is controlled with a broadband zero-order half wave plate before being focused on the sample by a 5-cm focal length CaF_2 lens. Any unconverted seed light is removed with the appropriate long-pass filters. Reflected harmonics are collected and measured on a fiber-coupled ultraviolet–visible spectrometer (Ocean Optics QE65+ custom built spectrometer) with a 350–1100 nm detection range. The devices are mounted in a high-pressure stainless steel gas cell with a 1-inch diameter sapphire window. For sufficiently large beam diameters, the intensity on the sapphire window is low enough that it does not contribute to the nonlinear signal in our experiments. Argon pressures in excess of 200 psi could be held on the devices within the cell.

The THG signals generated on and off the 1550 nm-resonant DHCG are shown in Figure 2. Off the metasurface structure, the pump wavelength was swept from 1485 to 1580 nm, and the emitted third-harmonic signal was recorded in Figure 2A. We observe the expected behavior of harmonic emission that tracks the pump wavelength without any additional spectral features. This picture changes when we repeat the same wavelength scan on the device itself. The harmonic yield dramatically increases, requiring over an order of magnitude lower power to generate, but within a limited bandwidth. At pump wavelengths below 1530 nm or above 1580 nm, no measurable increase is observed in the third-harmonic signals plotted in Figure 2B, and only a small signal, far-off-resonance due to the peak of the 1485-nm pump, is measured (blue arrow). Compared to this small signal, the 1550-nm pump beam displays the greatest harmonic yield enhancement of a factor of 45, implying the greatest overlap with the BIC resonance. This concept is illustrated in Figure 2C, in which the overlap of the Gaussian pump spectra and Lorentzian BIC resonance determines efficient harmonic generation. This explains the observed pinning of the THG to roughly 516 nm and the reduced linewidths of detuned harmonics. The cutoff shape of the short wavelength harmonics gives the clearest indication of the resonance edge.

We also examine the polarization dependence of our fabricated near-infrared (IR) metasurfaces. As described in the previous section, the grating mode under study is excited by TE polarized light parallel to the long direction of the grating fingers. We hereafter refer to this polarization direction as 0°. The theoretical dependence of the peak reflectance from the grating is studied in a series of FDTD simulations and plotted as a blue line in Figure 3B. For clarity, we subtract any contribution to the reflectance due to etalon effects of the thin device and oxide layers, hence

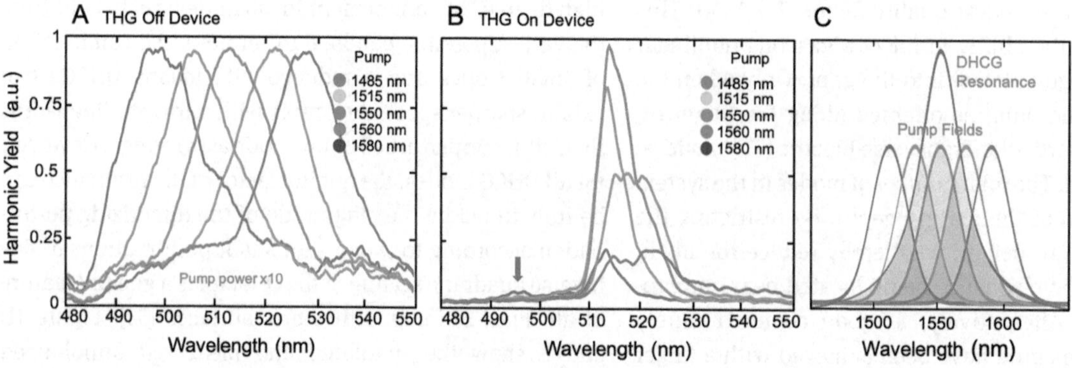

Figure 2: Near-infrared device wavelength characterization.
(A) Normalized third-harmonic signal generated from the bare substrate for pump wavelengths in the range of 1485 to 1580 nm. Pump powers are 10 × greater than those in B. (B) Third-harmonic spectra for the same five pump wavelengths shown in (A), now on the dimerized high-contrast gratings (DHCG). The large field enhancement on resonance pins the third harmonic to 513 nm. The small signal at 495 nm (blue arrow) corresponds to the third-harmonic signal from the peak of the 1485-nm pump that is generated off-resonance. (C) Schematic of the pump field and resonance overlap resulting in the harmonics measured in B. The shaded overlap regions contribute to the nonlinear signal.

the reflectance has a minimum of zero. As the polarization is rotated by an angle θ beginning from 0°, the peak power reflectance due to the BIC follows precisely the expected $\cos^2\theta$ behavior. The component of the field oriented along 90° therefore makes no contribution, and the $\cos^2\theta$ behavior is a result of the projection of the pump beam onto the 0° axis. We compare to this reflectance curve the integrated (measured) THG yield of 1550-nm pulses on resonance, shown as red dots in Figure 3B, corresponding to the spectra in Figure 3A. The experimental data follow a similar trend but with some important differences. Were it a perturbative third-order nonlinearity with $I \propto \cos^2\theta$, a THG efficiency proportional to $\cos^6\theta$ may be expected, given by the red line in Figure 3B. The measured harmonics follow neither the exact $\cos^2\theta$ behavior of the resonance or the perturbative $\cos^6\theta$ dependence. The observed saturation of the intensity dependence of the Nth order harmonic below I^N is a signature of nonperturbative harmonic generation. Thus, we conclude that in these experiments the generated signal corresponds to the nonperturbative regime with more than 32 times larger emission at 0° than at 80°. Within the nonperturbative regime for THG pumped in the mid-IR, a conversion efficiency upper bound of about 10^{-7} is found for low-pressure Argon gas [34]. Fifth-harmonic conversion efficiencies saturate about one order of magnitude lower at 10^{-8}. While the intensity enhancement provided by the DHCGs demonstrated here does not raise this efficiency limit, it does successfully lower the input power at which the maximum conversion efficiency is reached. At 90° polarization, the harmonic yield converges back to the signal off-resonance.

We also characterize the performance using the scaled up mid-infrared devices, allowing us to explore the third and fifth harmonics, both of which fall within our detection range for a 2.9-μm pump wavelength. Figure 4 shows the third- and fifth-harmonic spectra of these structures for a range of Argon gas pressures. For both harmonic orders, the harmonic yields scale quadratically with gas pressure (see Figure 4B) confirming that the emission is due to the atoms in the gap regions. At a pump intensity of 2×10^{12} W/cm^2, this emission occurs at field strengths where efficient harmonic generation in Argon is not expected to take place. In the absence of Argon, we still observe a nonzero third- and fifth-harmonic signal, which we attribute to the silicon finger surfaces. This is further supported by the expected overlap of the mode and grating predicted in our simulation (Figure 1C). At 200 psi of Argon pressure, the harmonic yields from Argon significantly exceed that of the device material.

4 Discussion and outlook

We have observed a greater than tenfold enhancement of the third- and fifth-harmonic yields from Argon atoms in the presence of a BIC-supporting DHCG, with the yield enhancement for higher order harmonics being greater due to the higher order dependence on pump intensity. We have evidence based on the polarization dependence of THG depicted in Figure 3 that we have achieved the strong field, nonperturbative regime, as the integrated THG does not follow a $\cos^6\theta$ dependence on the polarization angle. From

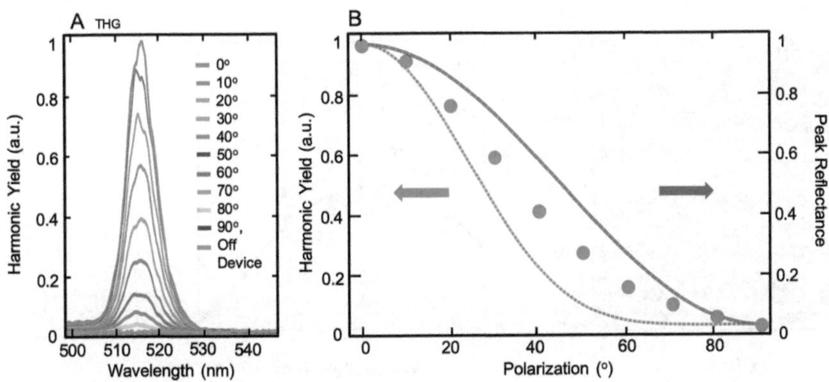

Figure 3: Grating polarization characterization.
(A) Normalized third-harmonic spectra of a 1550-nm pump, corresponding to the red points in B. Also included is the signal generated off the device, which is unmeasurable. One atmosphere of Argon pressure was flowed over the sample. Contributions to the third harmonic from both gas and silicon are present. (B) Simulated peak reflectance of the ideal grating (blue line) and integrated experimental third-harmonic yield (red dots), as a function of the pump polarization. 0° is the polarization parallel to the grating fingers. The red dashed line indicates $\cos^6\theta$, corresponding to the expected scaling of perturbative third-harmonic generation (THG). The mismatch between the red theoretical curve and experimental dots indicates that the measured THG is instead following a nonperturbative scaling typical of higher pump intensities.

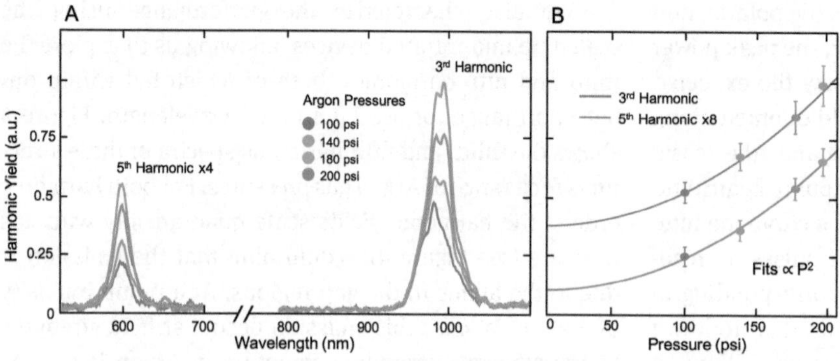

Figure 4: Pressure dependence of mid-infrared device.
(A) Third and fifth harmonics of a 2.9-μm pump beam on a 2.9-μm resonant dimerized high-contrast grating (DHCG). Argon pressures are varied from 100 to 200 psi in a high-pressure gas cell. The fifth harmonic has been scaled up for visibility. (B) Integrated harmonic yields from the spectra in (A), matched by color. Pressure dependence of both harmonic orders is fit to the square of the pressure added to a pressure independent contribution that we assume is due to the silicon fingers. The fifth harmonic is similarly scaled up.

this observation, we note that the hybrid gas–DHCG technique lowers the required pump input intensity to saturate the THG conversion efficiency at about 10^{-7}. Furthermore, we have determined that for pressures between 100 and 200 psi, the contributions to our total signal from enhanced harmonics within the grating material and Argon are of the same order of magnitude. With this, we confirm the mode profile produced by our FDTD simulations which predicts a gap-centered mode with leakage into the silicon fingers.

Further improvement of our hybrid metasurface–gas harmonic scheme requires the ability to restrict the strong field confinement mode to only the gas-filled region. Ideally, this would be accomplished while maintaining relatively large mode areas and all of the tunable wavelength and quality factor properties afforded to us by the DHCG platform. We propose a device in Figure 5, based on a 2-dimensional realization of a DHCG, that meets all of these stated requirements. Figure 5A shows a representation of the device, which consists of silicon dimers containing one circular and one elliptical nanopillar. The perturbation of the unit cell is to the ellipse minor axis. For a target resonance wavelength in the mid-infrared, the disk major axes should be 500 nm, and the grating period should be 2.1 μm. Figure 5B shows the BIC mode of interest in the system, chosen at a wavelength of 3.15 μm. In this system, the strongest field enhancement no longer spatially overlaps with the silicon. This offers tremendous flexibility to the proposed system in two forms. First, the contribution to the harmonic signal from any nongas sources would be greatly diminished. Second, the quality

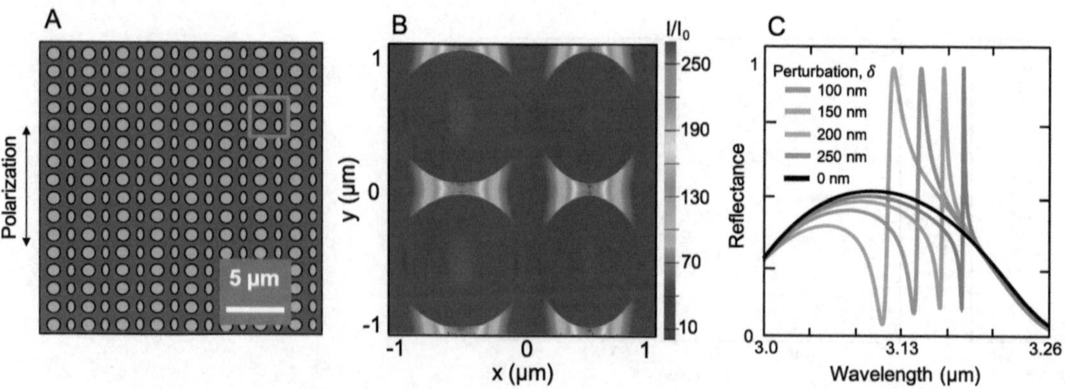

Figure 5: 2D dimerized high-contrast grating (DHCG) simulation.
(A) Proposed 2D DHCG. The height of the grating is 250 nm, disk radii is 500 nm, and the period is 2.1 μm. The dimerizing perturbation is to every other disk's minor axis. (B) FDTD simulation showing the relative power enhancement in the region marked by a red square in (A) (top view). The strong field confinement mode is entirely contained in the Argon gas region, with no leakage into the walls of the grating or substrate (C) Simulated polarized reflectance spectrum of 2D DHCGs with alternating minor axes of 500 nm − δ for four magnitudes of the perturbation as well as the unperturbed case. The reflectance peaks at the BIC Fano resonance wavelength near 3.15 μm. Smaller perturbations correspond to larger Q-factors and narrower linewidths. FDTD, finite-difference time-domain; BIC, bound states in the continuum.

factors could be raised to their fabrication limits, greatly increasing the local powers experienced by the atomic targets without increasing the likelihood of device damage. Quality factor tuning of the two-dimensional (2D) DHCG observes the required δ^{-2} dependence, but now the lower-Q peaks in the reflectance curve tend toward the short wavelength side as in Figure 5C. Smaller ellipse minor axes corresponding to larger perturbations result in less silicon in each unit cell and hence a lower effective index, which leads to a blueshift in the reflectance peak.

The potential for greatly amplified field strengths within gas-filled gaps gives renewed relevance to high-harmonic generation from gases in the presence of metasurfaces. Since a critical factor in the generation of high energy harmonics is, among other things, a sufficiently large intensity, the approach we have developed in this work can be extended readily to higher order harmonic generation. The possibility to tailor mode profiles to exclude leakage into the grating and therefore eliminate the damage that has so far limited plasmonic and dielectric devices alike deserves further study. Using a transparent nonlinear medium such as Argon extends the interaction length of the harmonic generation process well beyond the 10's of nanometers available in either metal- or silicon-based sources, while dramatically reducing harmonic absorption when compared with solid-state media. The 250-nm tall gratings used in our experiments, for example, are tall enough to allow greater coherent buildup of the harmonic signal, while being sufficiently subwavelength in extent to sidestep phasematching restrictions. Lower power requirements open up the possibility to use higher repetition rate sources to initiate HHG in gases and have the potential to make smaller footprint table-top attosecond sources more viable. The ease with which Q-factors can be adjusted in the fabrication process means the DHCG devices shown here can be made to match the bandwidth of a given pump laser. Alternatively, the resonance linewidth can be chosen to select for the desired linewidth of the generated harmonics. Moreover, the control provided by DHCGs can be further extended to simultaneously include both wavefront and polarization shaping.

Looking to the future, our hybrid gas–DHCG scheme could be applied to the entire suite of nonlinear optical studies being conducted in atomic systems. Beyond the gas phase, two-dimensional materials are emerging as a promising nonlinear optical platform moving forward [35–37]. The single- to few-atom thickness of graphene and transition metal dichalcogenides limits their nonlinear interaction length, making them prime candidates for integration with field-confining metasurfaces. Furthermore, the all-dielectric metasurface is a leading platform for metalenses, capable of providing the same level of phase control and integrability with 2D materials as seen in previous plasmonic nonlinear metalenses [38]. This is due to the high degree of engineerability of modes provided by all-dielectric DHCGs, but they also bring the added benefit of eliminating potential field hotspots in sensitive 2D materials, especially at higher harmonic orders and greater field intensities.

Author contribution: All the authors have accepted responsibility for the entire content of this submitted manuscript and approved submission.

Research funding: This study is funded by National Science Foundation (ECCS-1307948QII-TAQS-1936359), NSF IGERT program (DGE-1069240), NSF Graduate Research Fellowship Program (DGE-1644869), AFOSR Multidisciplinary University Research Initiative (FA9550-16-1-0121).

Conflict of interest statement: The authors declare no conflicts of interest regarding this article.

References

[1] T. Pfeifer, C. Spielmann, and G. Gerber, “Femtosecond X-ray science,” *Rep. Prog. Phys.*, vol. 69, pp. 443–505, 2006.

[2] P. B. Corkum, “Plasma perspective on strong field multiphoton ionization,” *Phys. Rev. Lett.*, vol. 71, pp. 1994–1997, 1993.

[3] C. Gohle, T. Udem, M. Herrmann, et al., “A frequency comb in the extreme ultraviolet,” *Nature*, vol. 436, pp. 234–237, 2005.

[4] G. Porat, C. M. Heyl, S. B. Schoun, et al., “Phase-matched extreme-ultraviolet frequency-comb generation,” *Nat. Photonics*, vol. 12, pp. 387–391, 2018.

[5] T. Popmintchev, M.-C. Chen, D. Popmintchev, et al., “Bright coherent ultrahigh harmonics in the keV X-ray regime from mid-infrared femtosecond lasers,” *Science*, vol. 336, pp. 1287–1291, 2012.

[6] X. F. Li, A. L'Huillier, M. Ferray, L. A. Lompré, and G. Mainfray, “Multiple-harmonic generation in rare gases at high laser intensity,” *Phys. Rev. A*, vol. 39, pp. 5751–5761, 1989.

[7] P. B. Corkum and F. Krausz, “Attosecond science,” *Nat. Phys.*, vol. 3, pp. 381–387, 2007.

[8] J. L. Krause, K. J. Schafer, and K. C. Kulander, “High-order harmonic generation from atoms and ions in the high intensity regime,” *Phys. Rev. Lett.*, vol. 68, pp. 3535–3538, 1992.

[9] S. Ghimire, A. D. DiChiara, E. Sistrunk, P. Agostini, L. F. DiMauro, and D. A. Reis, “Observation of high-order harmonic generation in a bulk crystal,” *Nat. Phys.*, vol. 7, pp. 138–141, 2011.

[10] H. Liu, C. Guo, G. Vampa, et al., “Enhanced high-harmonic generation from an all-dielectric metasurface,” *Nat. Phys.*, vol. 14, pp. 1006–1010, 2018.

[11] G. Vampa, B. G. Ghamsari, S. Siadat Mousavi, et al., “Plasmon-enhanced high-harmonic generation from silicon,” *Nat. Phys.*, vol. 13, pp. 659–662, 2017.

[12] S. Kim, J. Jin, Y.-J. Kim, I.-Y. Park, Y. Kim, and S.-W. Kim, “High-harmonic generation by resonant plasmon field enhancement,” *Nature*, vol. 453, pp. 757–760, 2008.

[13] S. Han, H. Kim, Y. W. Kim, et al., "High-harmonic generation by field enhanced femtosecond pulses in metal-sapphire nanostructure," *Nat. Commun.*, vol. 7, p. 13105, 2016.
[14] M. Sivis, M. Taucer, G. Vampa, et al., "Tailored semiconductors for high-harmonic optoelectronics," *Science*, vol. 357, pp. 303–306, 2017.
[15] I.-Y. Park, S. Kim, J. Choi, et al., "Plasmonic generation of ultrashort extreme-ultraviolet light pulses," *Nat. Photonics*, vol. 5, pp. 677–681, 2011.
[16] N. Pfullmann, C. Waltermann, M. Kovačev, et al., "Nano-antenna-assisted harmonic generation," *Appl. Phys. B*, vol. 113, pp. 75–79, 2013.
[17] S. Ghimire, A. D. DiChiara, E. Sistrunk, et al., "Generation and propagation of high-order harmonics in crystals," *Phys. Rev. A*, vol. 85, p. 043836, 2012.
[18] C. W. Hsu, B. Zhen, J. Lee, et al., "Observation of trapped light within the radiation continuum," *Nature*, vol. 499, pp. 188–191, 2013.
[19] C. W. Hsu, B. Zhen, A. D. Stone, J. D. Joannopoulos, and M. Soljačić, "Bound states in the continuum," *Nat. Rev. Mater.*, vol. 1, p. 16048, 2016.
[20] D. C. Marinica, A. G. Borisov, and S. V. Shabanov, "Bound states in the continuum in photonics," *Phys. Rev. Lett.*, vol. 100, p. 183902, 2008.
[21] Y. Plotnik, O. Peleg, F. Dreisow, et al., "Experimental observation of optical bound states in the continuum," *Phys. Rev. Lett.*, vol. 107, p. 183901, 2011.
[22] M. Zhang and X. Zhang, "Ultrasensitive optical absorption in graphene based on bound states in the continuum," *Sci. Rep.*, vol. 5, p. 8266, 2015.
[23] L. Carletti, S. S. Kruk, A. A. Bogdanov, C. De Angelis, and Y. Kivshar, "High-harmonic generation at the nanoscale boosted by bound states in the continuum," *Phys. Rev. Res.*, vol. 1, p. 023016, 2019.
[24] S. Romano, G. Zito, S. Managò, et al., "Surface-enhanced Raman and fluorescence spectroscopy with an all-dielectric metasurface," *J. Phys. Chem. C*, vol. 122, pp. 19738–19745, 2018.
[25] A. C. Overvig, S. Shrestha, and N. Yu, "Dimerized high contrast gratings," *Nanophotonics*, vol. 7, pp. 1157–1168, 2018.
[26] J. von Neumann and E. P. Wigner, "Über merkwürdige diskrete Eigenwerte," in *The Collected Works of Eugene Paul Wigner*, A. S. Wightman, Ed., Springer Berlin Heidelberg, 1993, pp. 291–293.
[27] J. W. Yoon, S. H. Song, and R. Magnusson, "Critical field enhancement of asymptotic optical bound states in the continuum," *Sci. Rep.*, vol. 5, p. 18301, 2015.
[28] S. Li, C. Zhou, T. Liu, and S. Xiao, "Symmetry-protected bound states in the continuum supported by all-dielectric metasurfaces," *Phys. Rev. A*, vol. 100, p. 063803, 2019.
[29] A. C. Overvig, S. C. Malek, M. J. Carter, S. Shrestha, and N. Yu, "Selection Rules for Symmetry-Protected Bound States in the Continuum," *Arxiv*, 2019. https://doi.org/10.1364/cleo_qels.2019.ff2a.5.
[30] C. W. Neff, T. Yamashita, and C. J. Summers, "Observation of Brillouin zone folding in photonic crystal slab waveguides possessing a superlattice pattern," *Appl. Phys. Lett.*, vol. 90, p. 021102, 2007.
[31] M. S. Bigelow, N. N. Lepeshkin, and R. W. Boyd, "Observation of ultraslow light propagation in a ruby crystal at room temperature," *Phys. Rev. Lett.*, vol. 90, p. 113903, 2003.
[32] T. Baba, "Slow light in photonic crystals," *Nat. Photonics*, vol. 2, pp. 465–473, 2008.
[33] K. Koshelev, S. Lepeshov, M. Liu, A. Bogdanov, and Y. Kivshar, "Asymmetric metasurfaces with high-Q resonances governed by bound states in the continuum," *Phys. Rev. Lett.*, vol. 121, p. 193903, 2018.
[34] J. Ni, J. Yao, B. Zeng, et al., "Comparative investigation of third- and fifth-harmonic generation in atomic and molecular gases driven by midinfrared ultrafast laser pulses," *Phys. Rev. A*, vol. 84, p. 063846, 2011.
[35] S.-Y. Hong, J. I. Dadap, N. Petrone, P.-C. Yeh, J. Hone, and R. M. Osgood, "Optical third-harmonic generation in graphene," *Phys. Rev. X*, vol. 3, p. 021014, 2013.
[36] Y. Song, R. Tian, J. Yang, R. Yin, J. Zhao, and X. Gan, "Second harmonic generation in atomically thin $MoTe_2$," *Adv. Opt. Mater.*, vol. 6, p. 1701334, 2018.
[37] J. W. You, S. R. Bongu, Q. Bao, and N. C. Panoiu, "Nonlinear optical properties and applications of 2D materials: theoretical and experimental aspects," *Nanophotonics*, vol. 8, pp. 63–97, 2018.
[38] J. Chen, K. Wang, H. Long, et al., "Tungsten disulfide–gold nanohole hybrid metasurfaces for nonlinear metalenses in the visible region," *Nano Lett.*, vol. 18, pp. 1344–1350, 2018.

Qiong He, Fei Zhang, MingBo Pu, XiaoLiang Ma, Xiong Li, JinJin Jin, YingHui Guo and XianGang Luo*

Monolithic metasurface spatial differentiator enabled by asymmetric photonic spin-orbit interactions

https://doi.org/10.1515/9783110710687-062

Abstract: Spatial differentiator is the key element for edge detection, which is indispensable in image processing, computer vision involving image recognition, image restoration, image compression, and so on. Spatial differentiators based on metasurfaces are simpler and more compact compared with traditional bulky optical analog differentiators. However, most of them still rely on complex optical systems, leading to the degraded compactness and efficiency of the edge detection systems. To further reduce the complexity of the edge detection system, a monolithic metasurface spatial differentiator is demonstrated based on asymmetric photonic spin-orbit interactions. Edge detection can be accomplished via such a monolithic metasurface using the polarization degree. Experimental results show that the designed monolithic spatial differentiator works in a broadband range. Moreover, 2D edge detection is experimentally demonstrated by the proposed monolithic metasurface. The proposed design can be applied at visible and near-infrared wavelengths by proper dielectric materials and designs. We envision this approach may find potential applications in optical analog computing on compact optical platforms.

Keywords: asymmetric photonic spin-orbit interactions; edge detection; metasurface; spatial differentiator.

1 Introduction

Edge detection captures the edge information and provides the most significant outlines of an image or an object, which has been widely applied in image processing and computer vision owning to the high processing speed and low data volume [1–4]. The most important process for edge detection is differentiation, a process generally operated by a spatial differentiator, either in a digital computation or an analog computation way [5–7]. Compared with digital computing, optical analog computing can process parallel information with high efficient and low power consumption, holding great potential in real-time detections [8–11]. However, in the traditional optical computing system, a 4-F optical system containing at least two lenses and a spatial filter is required for Fourier transform and inverse Fourier transform, which is bulky and complicated, hindering the applications in modern optoelectronics with high integration level.

Metasurfaces [12–15], single or few-layer subwavelength structures which are capable of modulating the phase, amplitude, and polarization of electromagnetic waves [16–19], have provided new strategies for various optical elements and systems including flat lenses [20–22], holograms [23–25], electromagnetic stealth [26–28], vortex beam generator [29], tunable optical components [30, 31], flat displays [32, 33], and so on. Notably, optical spatial computing based on metamaterials including differentiators, integrators, and equation solvers is proposed by Silva et al in 2014 [34], opening new avenues for optical analog computing. Since then, various differentiator metasurfaces are developed based on photonic crystals [5, 35],

Qiong He and Fei Zhang: These authors contributed equally to this work.

***Corresponding author: XianGang Luo**, State Key Laboratory of Optical Technologies on Nano-Fabrication and Micro-Engineering, Institute of Optics and Electronics Chinese Academy of Sciences, Chengdu, 610209, China; and School of Optoelectronics, University of Chinese Academy of Sciences, Beijing, 100049, China, E-mail: lxg@ioe.ac.cn. https://orcid.org/0000-0002-1401-1670
Qiong He, Fei Zhang and JinJin Jin, State Key Laboratory of Optical Technologies on Nano-Fabrication and Micro-Engineering, Institute of Optics and Electronics Chinese Academy of Sciences, Chengdu, 610209, China
MingBo Pu, XiaoLiang Ma, Xiong Li and YingHui Guo, State Key Laboratory of Optical Technologies on Nano-Fabrication and Micro-Engineering, Institute of Optics and Electronics Chinese Academy of Sciences, Chengdu, 610209, China; and School of Optoelectronics, University of Chinese Academy of Sciences, Beijing, 100049, China

This article has previously been published in the journal Nanophotonics. Please cite as: Q. He, F. Zhang, M. Pu, X. Ma, X. Li, J. Jin, Y. Guo and X. Luo "Monolithic metasurface spatial differentiator enabled by asymmetric photonic spin-orbit interactions" *Nanophotonics* 2021, 10. DOI: 10.1515/nanoph-2020-0366.

spin Hall effect [36], surfaces plasmons [37], high-contrast gratings [8], and Pancharatnam-Berry (PB) phase [38, 39]. However, additional prisms or lenses are still required for plasmon coupling or Fourier transform in those applications [37, 40], which is incompatible with the flat and compact optical systems. Besides, many metasurface spatial differentiators work only in 1D edge detection [8, 9, 35, 36], restricting the practical applications in image recognition and processing.

To achieve a monolithic metasurface for spatial differentiation, the focusing and differentiation abilities should be integrated into one metasurface, which can be realized by asymmetric photonic spin-orbit interactions (PSOIs) of light. Here, we propose a monolithic metasurface spatial differentiator for edge detection without any additional lenses. Spin-dependent PB phase and spin-independent propagation phase are merged to arbitrarily and independently manipulate wavefronts of two converted spins, leading to asymmetric PSOIs of light [41, 42]. It is demonstrated that the monolithic metasurface is capable of forming both left-handed circularly polarized (LCP) and right-handed circularly polarized (RCP) images when illuminated by a linear polarized (LP) light and the edge image is acquired by using an orthogonal linear polarizer. The independent manipulation on LCP and RCP lights of the proposed monolithic metasurface enables edge detection without the 4F system, leading to edge detection systems with higher integration level and compactness. Four samples (three for 1D differentiation with different edge resolutions and one for 2D differentiation) are fabricated to demonstrate the tunable resolution at different orientations and 2D edge detection, respectively. The monolithic metasurface spatial differentiator works at a broadband with an edge resolution of about 49.4 μm. Notably, by integrating the linear polarizer into the metasurface, the overall compactness can be further improved. This work is helpful for the applications of monolithic metasurface in edge detection and the multifunctionality of other monolithic metasurfaces based on PSOIs.

2 Theoretical analyses

The principle of the monolithic metasurface spatial differentiator is based on asymmetric PSOIs, as shown in Figure 1. When the LP light illuminates on the object and passes through the metasurface, the LCP and RCP images are formed at different positions with a slight distance due to the independent manipulation on the wavefronts of both LCP and RCP light. By inserting a linear polarizer, the overlapped LP light is eliminated, leaving the edge image. It is worth noting that traditional metasurfaces edge detection performs differentiation operations in the spatial Fourier domain by using the 4-F optical system, while the metasurface in this manuscript can manipulate the image directly, enabling a monolithic metasurface spatial differentiator.

For simplicity, we first analyze the mechanism of 1D edge imaging. Under the illumination of the LP light along the x-axis, the proposed metasurface can project two images of the object with a transverse shift of $\pm\Delta$, corresponding to LCP and RCP images, respectively. Thus, the image can be expressed as

$$E_{\text{image}}(x,y) = E_0[(x-\Delta),y]\begin{bmatrix}1\\-i\end{bmatrix} + E_0[(x+\Delta),y]\begin{bmatrix}1\\i\end{bmatrix} \quad (1)$$

When Δ is far smaller than the imaging distance, the phase difference between the LCP and RCP components at the same point can be ignored. As a result, the overlapped area of the

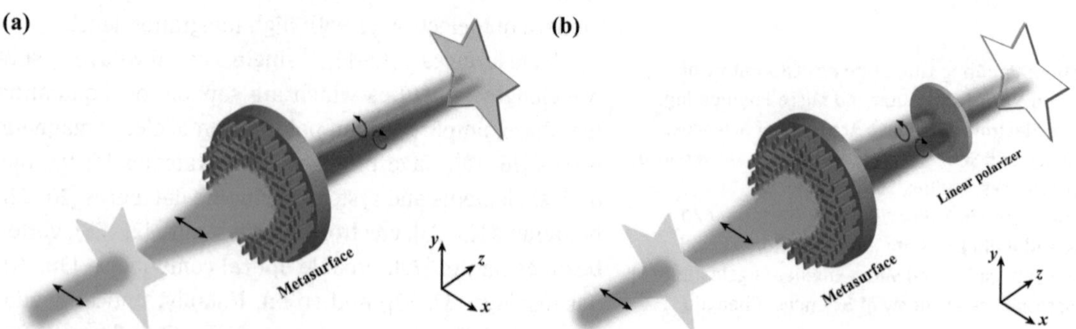

Figure 1: Schematic of the proposed monolithic metasurface.
(a) The monolithic metasurface for left-handed circularly polarized (LCP) and right-handed circularly polarized (RCP) imaging with an opposite shift along the phase gradient direction (x-axis in this figure). (b) A linear polarizer is applied to eliminate the LP image of the overlapped LCP and RCP images for edge detection.

LCP and RCP images retains the original incident polarization state and thus can be filtered with an orthogonal linear polarizer. Then, the final edge can be expressed as [38]

$$E_{\text{edge}}(x,y) = \{E_0[(x+\Delta),y] - E_0[(x-\Delta),y]\}\begin{bmatrix} 0 \\ i \end{bmatrix} \quad (2)$$

If the Δ is small enough, the edge information along the x-axis is recorded, which is tunable by adjusting the shift distances between LCP and RCP images. This description suggests that the proposed metasurface can generate two different phase distributions for LCP and RCP light. In accordance with the aforementioned analysis, the two-phase distribution can be written as

$$\begin{aligned} \phi_\sigma(x,y) &= -k_0\sqrt{x^2+y^2+f^2} + k_\sigma x \\ k_\sigma &= \sigma k_0 \sin\zeta \end{aligned} \quad (3)$$

where f is the focal length; $k_0 = 2\pi/\lambda$ is the wavenumber in free space; $\sigma = \pm 1$ indicates RCP and LCP light; k_σ corresponds to the additional horizontal momentum [43], ζ determines the degree of separation between LCP and RCP images. When ζ is small enough, the shift value Δ along the x-axis can be approximately equal to ζv, where v indicates the imaging distance and ζ should be converted to radians. Thus, the edge width depends on the separation and the resolution of the obtained edge images is limited by the diffraction limit of the designed metasurface due to the nearly perfect focusing phase distribution as indicated by Eq. (3). For 2D edge detection, the aforementioned phase distribution should be modified as

$$\phi_\sigma(x,y) = -k_0\sqrt{x^2+y^2+f^2} + k_\sigma\sqrt{x^2+y^2} \quad (4)$$

The independent control of LCP and RCP light is the key for edge detection with a monolithic device, which can be realized by asymmetric PSOIs. For simplicity, assuming a lossless anisotropic subwavelength structure with the phase shift of $\beta \pm \delta/2$ along its major and minor axes, the Jones matrix can be described as diag[exp($i\beta + i\delta/2$), exp($i\beta - i\delta/2$)]. After being rotated by θ along the optical axis and under the normal illumination of circularly polarized (CP) light of $[1\ -i\sigma]^T$, the output optical field can be given as [41, 44]

$$\cos\frac{\delta}{2}e^{i\beta}\begin{bmatrix} 1 \\ -i\sigma \end{bmatrix} - i\sin\frac{\delta}{2}e^{i(-2\sigma\theta+\beta)}\begin{bmatrix} 1 \\ i\sigma \end{bmatrix} \quad (5)$$

From the Eq. (5), the flipped spin component carries not only spin-independent propagation phase β but also spin-dependent PB phase $-2\sigma\theta$. The propagation phase can break the conjugation symmetry of optical fields generated by PB-based metasurfaces, leading to asymmetric PSOIs. Such two phases can be independently controlled by changing the size and orientation of the subwavelength structure, respectively. Then, the propagation phase and orientation at the coordinate (x, y) can be given as

$$\beta(x,y) = \frac{1}{2}[\phi_1(x,y) + \phi_{-1}(x,y)] \quad (6)$$

$$\theta(x,y) = \frac{1}{4}[\phi_{-1}(x,y) - \phi_1(x,y)] \quad (7)$$

where $\phi_1(x, y)$ and $\phi_{-1}(x, y)$ are independent phase distributions for $\sigma = \pm 1$.

3 Simulation and sample fabrication

As a proof-of-concept demonstration, a metasurface spatial differentiator that works at 10.6 μm has been designed and fabricated. The unit cell is a chamfered pillar on the substrate and silicon is chosen as the dielectric material because of the straightforward fabrication processes and negligible loss at the infrared band, as illustrated in Figure 2a. The pillars with the same height of H = 7 μm but different transverse sizes (width W and length L) and orientations θ are mounted on the hexagonal unit cells with a lattice constant of P. Because such high-contrast pillars can work as weakly coupled low-quality-factor Fabry-Pérot resonators, propagation and PB phases can be independently controlled by changing the size and orientation [45]. Eight unit cells with an incremental propagation phase of ~$\pi/4$ are designed at the wavelength of 10.6 μm, with their propagation phase and transmittances shown in Figure 2b. As can be seen, a 0–2π phase coverage, high cross-polarized transmittance, and low co-polarized transmittance are realized by these eight unit cells, which indicate that all the unit cells approximately work as a local half-wave plate, that is, $\delta = \pi$. In fact, the unit cells of the metasurfaces cannot behave like a perfect half-wave plate, indicating the existence of the co-polarized lights. However, the co-polarized LCP and RCP lights will not be separated and the polarization states remain unchanged, and thus, the merged light is still LP, which will be filtered by the orthogonal linear polarizer. Consequently, the remaining co-polarized transmittance lights only decrease the efficiency of the metasurface spatial differentiator and cannot contribute to the differential functionality. In accordance with Eqs. (6) and (7), four edge detectors are designed with the same diameter of 1 cm and a focal length of 2 cm working at 10.6 μm. These four samples correspond three 1D edge detectors (denoted by S1, S2, and S3 with respective ζ = 0.1, 0.075 and 0.05°) and one 2D edge detector (denoted by S4 with ζ = 0.05°).

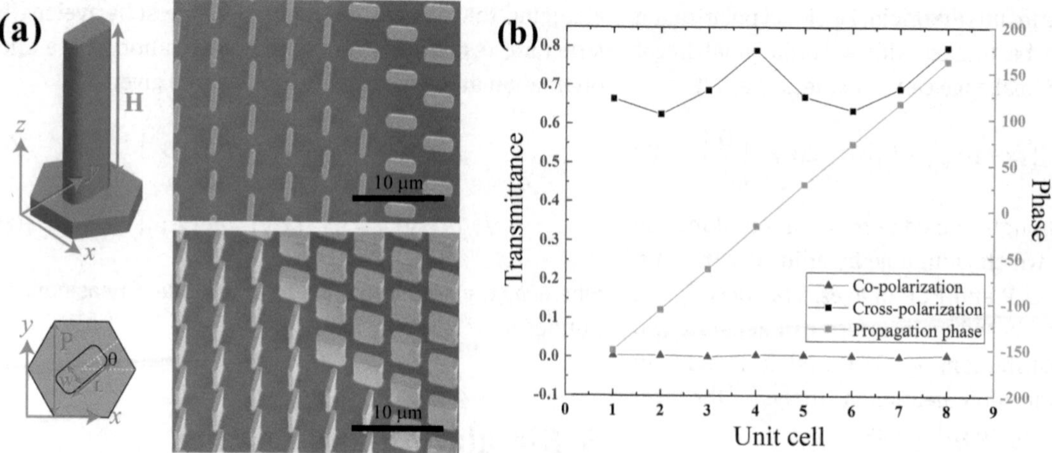

Figure 2: Design and simulation results of the unit cells.
(a) Schematic images of the unit cell and scanning electron microscope (SEM) images of the samples. A hexagonal shape is designed because the circular symmetry of the hexagonal shape is higher than the square, leading to more uniform responses of the unit cells at different rotation angles for Pancharatnam-Berry (PB) phase. The period P of the unit cell and the height H of the pillar are 4.8 and 7 μm, respectively. The edges of the pillar in this design are bent by a radius of $W/4$, where W is the width of the pillar. (b) Simulation results of eight unit cells including co-polarization, cross-polarization transmittance and propagation phase at 10.6 μm. The sizes of unit cells 1–4 are L = 3.8, 3.45, 3.1, 3.5 μm and W = 1.75, 1.6, 1.43, 1 μm, respectively. The results of unit cells 5–8 are obtained by rotating unit cells 1–4 by 90°.

The metasurfaces are fabricated by the laser direct writing and inductively coupled plasma (ICP) etching processes. Specifically, 1-μm thick positive photoresist (AZ1500) was spin-coated onto the clean silicon substrates and prebaked at 150 °C for 5 min, followed by the laser direct writing and corresponding developing processes. Next, ICP etching was used to fabricate the silicon pillars where the positive photoresist acts as the mask. The fabrication processes are simple and low cost, holding great potential for the applications of such monolithic metasurface in edge detection systems. Morphologies of the samples were characterized by scanning electron microscope, as shown in the right part of Figure 2a. More photographs of the samples and spectra of the unit cells are shown in the Supplementary materials.

4 Results and discussion

We first demonstrate the 1D edge detection at different orientations by the experimental setup shown in Figure 3a. A CO_2 laser was utilized as the light source. After passing through a linear polarizer and a 1/4 wave plate, the beam was sent through the object and then transmitted to the sample. The transmitted beam was filtered by another linear polarizer and recorded by an infrared detector (800 × 600 pixels with a pixel size of 17 × 17 μm) with a video capture card (TC-UB625). The 1D edge images and 2D edge images were magnified through the built-in interpolation algorithm to four and two times, respectively, indicating the effective pixel sizes of respective 4.25 × 4.25 μm and 8.5 × 8.5 μm for 1D and 2D images throughout this experiment. It is worth noting that the first linear polarizer and the 1/4 wave plate are required just for CP imaging and the edge detection can be realized through only the metasurface and a linear polarizer. With the development of micro-/nano-fabrication, the linear polarization can be fabricated on the backside of the metasurface, further reducing the complexity and improving the integration level of the spatial differentiation system. Therefore, compared with other optical edge detection systems, our experimental setup is free of the 4F system, which means that the system volume along the optical axis can be reduced by 50%, suggesting higher integration level and compactness in practical applications.

From the first row of Figure 3b, one can see clear LCP and RCP images, indicating excellent independent control on the wavefronts of both LCP and RCP light enabled by the high-efficiency asymmetric PSOIs. When the overlapped area of the LCP and RCP image is filtered by the orthogonal linear polarizer, a clear edge along the x-axis appears. Besides, the phase gradient of the metasurfaces can be designed at arbitrary directions, implying edge detection at different orientations. By rotating the metasurface, CP images and the corresponding edge images at different orientations were obtained as shown in Figure 3b. Similar to the first row of Figure 3b, LCP and RCP images for other orientations shift to opposite directions along the designed phase gradient, leaving the clear edge images. However, the fabrication error can lead

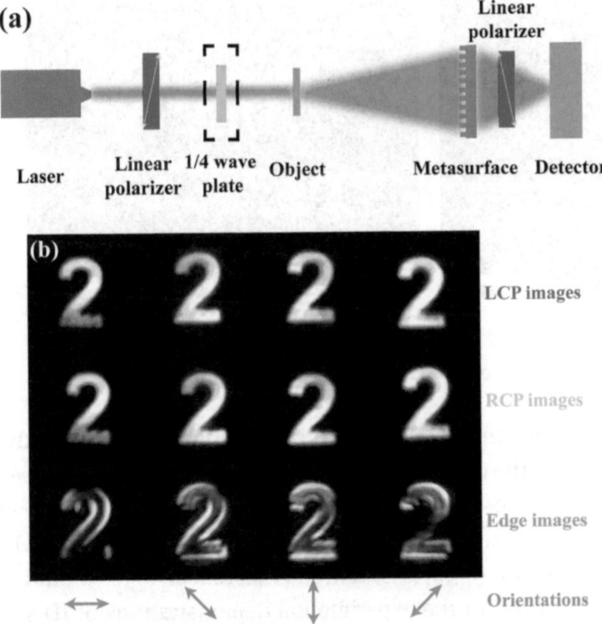

Figure 3: 1D edge detection setup and the corresponding images. (a) Schematic illustration of the measurement setup for edge detection. (b) LCP images, RCP images, and 1D edge images of the number '2' at different orientations. The rows from top to bottom are LCP images, RCP images, and 1D edge images. The columns from left to right are images at 0, 45, 90, and 135°.

to the imbalance of intensities between LCP and RCP lights, and thus the merged light will be elliptically polarized. Consequently, the linear polarizer is not able to eliminate the overlapped light completely, leaving the inner fields in the edge images.

Then, the tunable resolution of the edge detection is further demonstrated. The edge images of S1, S2, and S3 are shown in Figure 4a–c. It is clear that when ζ is decreased, the edge of the image turns thinner, which is attributed to the proportional relation between the edge width and the ζ, as described by $w = 2\Delta = 2\zeta v$. The three samples share the same f (2.0 cm) and object distance u (6.8 cm) during the measurement, so the imaging distance v can be calculated to be 2.8 cm in accordance with the Gauss formula. Accordingly, the theoretical edge widths for S1, S2, and S3 are, respectively, 98.8, 74.1, and 49.4 μm, which can be easily distinguished by the detector with a resolution of 34 μm. Here, the edge widths along the x-axis of the three images are measured to be 101.6, 78.1, and 54.7 μm, which are close to the theoretical values. In accordance with the relation between the edge width and ζ, the edge width can be further reduced by decreasing the ζ. However, considering that the final edge image is also constrained by the resolution of the detector, which means edge width smaller than the resolution of the detector will not be

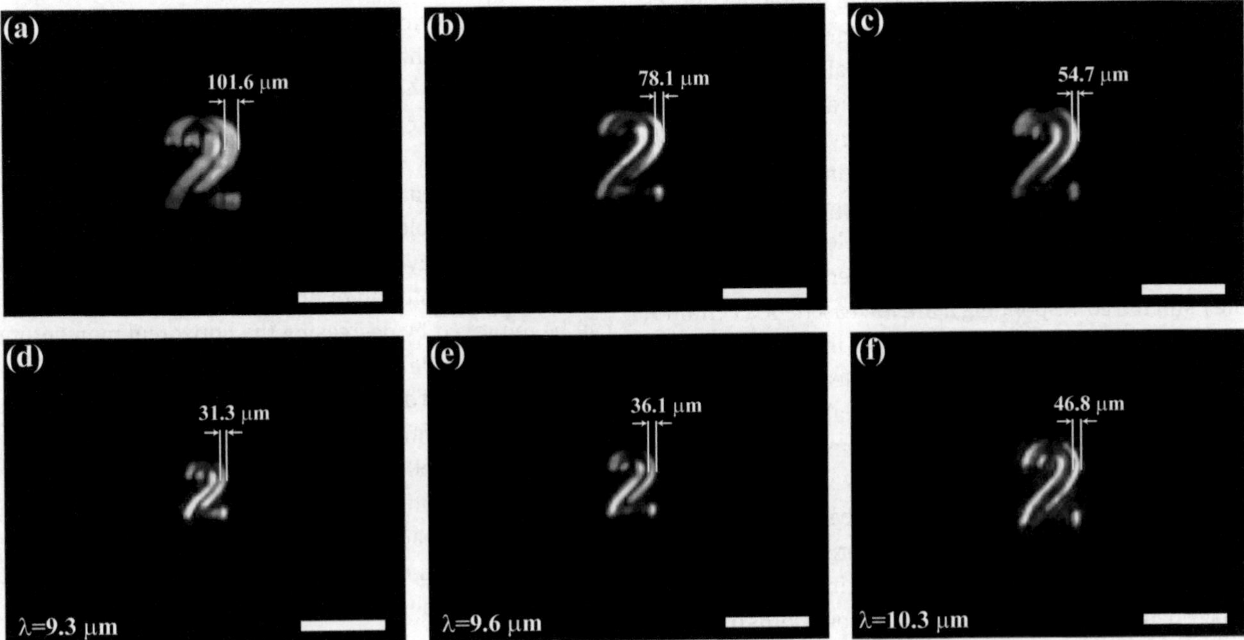

Figure 4: Demonstration of tunable edge resolution 1D edge detection and broadband 1D edge detection. (a), (b), and (c) 1D edge images of the number "2" by using S1, S2, and S3 at 10.6 μm. (d), (e), and (f) Broadband 1D edge detection demonstrated by S3 at the wavelengths of 9.3, 9.6, and 10.3 μm, respectively. The edge detection can work at broadband from 9.3 to 10.6 μm (scale bar is 500 μm).

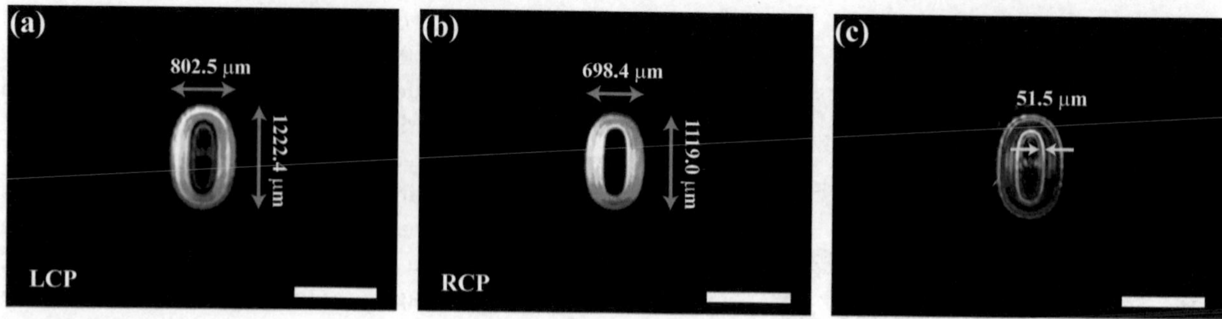

Figure 5: Experimental demonstration of 2D edge detection by S4.
(a) The expanded LCP image of "0". (b) The shrunk RCP image of "0". (c) Edge image of "0". (Scale bar is 1000 μm).

distinguished, the metasurface should be carefully designed in accordance with the specific application and detector to ensure an optimal edge image.

The work bandwidth is further tested by changing the wavelength of a CO_2 laser. Different from the continuous laser, the wavelength of the CO_2 laser cannot be tuned to an arbitrary value, so we chose some specific wavelengths of the CO_2 laser, 9.3, 9.6, and 10.3 μm to verify the edge detection bandwidth of S3. By changing the wavelength and object distance, the corresponding edge images are obtained with fixed image distance, as shown in Figure 4d–f. The clear edge images for different wavelengths indicate that the metasurface differentiator works well from the designed working wavelength at 10.6–9.3 μm with a slight degradation of edge image quality. Although the wavelength beyond 10.6 μm is not measured limited by the CO_2 laser, it could be inferred from the high image quality at 10.6 μm that such differentiator can also work at wavelength beyond 10.6 μm to a certain range, indicating a broadband (over 2.3 μm) working ability. Besides, one can see from Figure 4c–f that the size of the image "2" increases with the increasing working wavelength, which origins from the wavelength-dependent focal length of the metasurface. Specifically, when the wavelength is decreased, the f and the corresponding u are increased for a certain v, leading to a smaller magnification, and thus, the image size is smaller at shorter wavelength. Meanwhile, reduced magnification implies decreased equivalent ζ, so the edge width at a shorter wavelength is thinner, as indicated by Figure 4c–f.

Finally, the 2D edge detection capability of S4 is demonstrated with the same experimental setup in Figure 3a. Figure 5 shows the LCP/RCP image and the edge image of the number "0". The heights and widths of the "0" for LCP and RCP are, respectively, approximately 1222.4, 802.5, 1119.0, and 698.4 μm, which results from that the LCP image expands while the RCP image shrinks along the radial direction. If the overlapped LP image is eliminated by a linear polarizer, the 2D edge image appears, suggesting the capability of 2D edge detection of the proposed monolithic metasurface. Moreover, the edge width in Figure 5c is ~51.5 μm, which is in accordance with that of 1D edge width (Figure 4c) and the theoretical edge resolution S4. Except for the experimental demonstrations of 1D and 2D edge detection shown here, some simulation results of 2D edge detection for more complex objects are provided in the Supplementary materials, which further demonstrate the ability of edge detection for the proposed metasurface spatial differentiator.

5 Conclusions

In conclusion, owing to the independent manipulation on LCP and RCP lights by a monolithic metasurface, edge detection without 4F system is demonstrated, suggesting a simpler and more compact edge detection system in practical applications. The proposed monolithic metasurface spatial differentiator works at a wide bandwidth over 2.3 μm with tunable resolutions. Besides, both 1D and 2D differentiation have been experimentally demonstrated with an edge resolution of about 49.4 μm (~4.7 λ), which can be improved by decreasing the horizontal momentum carried on circularly polarized light. Notably, such a monolithic spatial differentiator can be designed to work at visible and near-infrared wavelengths, which would satisfy more possible applications. We believe that the monolithic spatial differentiation metasurface in this article can be applied in compact optical imaging systems and the asymmetric PSOIs can be utilized to build more kinds of multifunctional metasurfaces.

Acknowledgments: The authors acknowledge the financial support by the National Natural Science Foundation of China under contract Nos. 611975210 and 61822511.

Author contribution: All the authors have accepted responsibility for the entire content of this submitted manuscript and approved submission.

Research funding: This research was supported the National Natural Science Foundation of China under contract Nos. 611975210 and 61822511.

Conflict of interest statement: The authors declare no conflicts of interest regarding this article.

References

[1] D. Ziou and S. Tabbone, "Edge detection techniques – an overview," *Pattern Recogn. Image Anal.*, vol. 8, pp. 537–559, 1998.

[2] D. R. Solli and B. Jalali, "Analog optical computing," *Nat. Photonics*, vol. 9, pp. 704–706, 2015.

[3] H. S. Hsu, and W. H. Tsai, "Moment-preserving edge detection and its application to image data compression," *Opt. Eng.*, vol. 32, pp. 1596–1609, 1993.

[4] S. Fürhapter, A. Jesacher, S. Bernet, and M. Ritsch-Marte, "Spiral phase contrast imaging in microscopy," *Opt. Express*, vol. 13, pp. 689–694, 2005.

[5] Y. Zhou, H. Zheng, I. I. Kravchenko, and J. Valentine, "Flat optics for image differentiation," *Nat. Photonics*, vol. 14, pp. 316–323, 2020.

[6] A. Pors, M. G. Nielsen, and S. I. Bozhevolnyi, "Analog computing using reflective plasmonic metasurfaces," *Nano Lett.*, vol. 15, pp. 791–797, 2015.

[7] W. Wu, W. Jiang, J. Yang, S. Gong, and Y. Ma, "Multilayered analog optical differentiating device: performance analysis on structural parameters," *Opt. Lett.*, vol. 42, pp. 5270–5273, 2017.

[8] Z. Dong, J. Si, X. Yu, and X. Deng, "Optical spatial differentiator based on subwavelength high-contrast gratings," *Appl. Phys. Lett.*, vol. 112, p. 181102, 2018.

[9] Y. Zhou, W. Wu, R. Chen, W. Chen, R. Chen, and Y. Ma, "Analog optical spatial differentiators based on dielectric metasurfaces," *Adv. Opt. Mater.*, vol. 8, p. 1901523, 2019.

[10] M. Farmahini-Farahani, J. Cheng, and H. Mosallaei, "Metasurfaces nanoantennas for light processing," *J. Opt. Soc. Am. B*, vol. 30, pp. 2365–2370, 2013.

[11] H. Kwon, D. Sounas, A. Cordaro, A. Polman, and A. Alu, "Nonlocal metasurfaces for optical signal processing," *Phys. Rev. Lett.*, vol. 121, p. 173004, 2018.

[12] N. Yu and F. Capasso, "Flat optics with designer metasurfaces," *Nat. Mater.*, vol. 13, pp. 139–150, 2014.

[13] X. Luo, *Engineering Optics 2.0: A Revolution in Optical T heories, Materials, Devices and Systems*, Singapore, Springer, 2019.

[14] A. V. Kildishev, A. Boltasseva, and V. M. Shalaev, "Planar photonics with metasurfaces," *Science*, vol. 339, p. 1232009, 2013.

[15] X. Luo, "Principles of electromagnetic waves in metasurfaces," *Sci. China Phys. Mech.*, vol. 58, p. 594201, 2015.

[16] F. Zhang, Q. Zeng, M. Pu, et al., "Broadband and high-efficiency accelerating beam generation by dielectric catenary metasurfaces," *Nanophotonics*, vol. 9, pp. 2829–2837, 2020.

[17] J. P. Balthasar Mueller, N. A. Rubin, R. C. Devlin, B. Groever, and F. Capasso, "Metasurface polarization optics: independent phase control of arbitrary orthogonal states of polarization," *Phys. Rev. Lett.*, vol. 118, p. 113901, 2017.

[18] W. T. Chen, A. Y. Zhu, and F. Capasso, "Flat optics with dispersion-engineered metasurfaces," *Nat. Rev. Mater.*, vol. 5, pp. 607–620, 2020.

[19] J. Guo, T. Wang, B. Quan, et al., "Polarization multiplexing for double images display," *Opto-Electron. Adv.*, vol. 2, p. 180029, 2019.

[20] M. Khorasaninejad, W. T. Chen, R. C. Devlin, J. Oh, A. Y. Zhu, and F. Capasso, "Metalenses at visible wavelengths: Diffraction-limited focusing and subwavelength resolution imaging," *Science*, vol. 352, pp. 1190–1194, 2016.

[21] S. Wang, P. C. Wu, V. C. Su, et al., "A broadband achromatic metalens in the visible," *Nat. Nanotechnol.*, vol. 13, pp. 227–232, 2018.

[22] K. Dou, X. Xie, M. Pu, et al., "Off-axis multi-wavelength dispersion controlling metalens for multi-color imaging," *Opto-Electron. Adv.*, vol. 3, p. 190005, 2020.

[23] G. Zheng, H. Muhlenbernd, M. Kenney, G. Li, T. Zentgraf, and S. Zhang, "Metasurface holograms reaching 80% efficiency," *Nat. Nanotechnol.*, vol. 10, pp. 308–312, 2015.

[24] X. Li, L. Chen, Y. Li, et al., "Multicolor 3D meta-holography by broadband plasmonic modulation," *Sci. Adv.*, vol. 2, p. e1601102, 2016.

[25] X. Xie, K. Liu, M. Pu, et al., "All-metallic geometric metasurfaces for broadband and high-efficiency wavefront manipulation," *Nanophotonics*, vol. 9, pp. 3209–3215, 2019.

[26] W. Cai, U. K. Chettiar, A. V. Kildishev, and V. M. Shalaev, "Optical cloaking with metamaterials," *Nat. Photonics*, vol. 1, pp. 224–227, 2007.

[27] X. Xie, X. Li, M. Pu, et al., "Plasmonic metasurfaces for simultaneous thermal infrared invisibility and holographic illusion," *Adv. Funct. Mater.*, vol. 28, p. 14, 2018.

[28] X. Ma, M. Pu, X. Li, Y. Guo, and X. Luo, "All-metallic wide-angle metasurfaces for multifunctional polarization manipulation," *Opto-Electron. Adv.*, vol. 2, p. 180023, 2019.

[29] P. Genevet, N. Yu, F. Aieta, et al., "Ultra-thin plasmonic optical vortex plate based on phase discontinuities," *Appl. Phys. Lett.*, vol. 100, p. 013101, 2012.

[30] P. C. Wu, N. Papasimakis, and D. P. Tsai, "Self-affine graphene metasurfaces for tunable broadband absorption," *Phy. Rev.Appl.*, vol. 6, p. 044019, 2016.

[31] Q. Wang, E. T. F. Rogers, B. Gholipour, et al., "Optically reconfigurable metasurfaces and photonic devices based on phase change materials," *Nat. Photonics*, vol. 10, pp. 60–65, 2016.

[32] M. A. Kats, R. Blanchard, P. Genevet, and F. Capasso, "Nanometre optical coatings based on strong interference effects in highly absorbing media," *Nat. Mater.*, vol. 12, pp. 20–24, 2013.

[33] M. A. Kats, D. Sharma, J. Lin, et al., "Ultra-thin perfect absorber employing a tunable phase change material," *Appl. Phys. Lett.*, vol. 101, p. 221101, 2012.

[34] A. Silva, F. Monticone, G. Castaldi, V. Galdi, A. Alu, and N. Engheta, "Performing mathematical operations with metamaterials," *Science*, vol. 343, pp. 160–163, 2014.

[35] A. Cordaro, H. Kwon, D. Sounas, A. F. Koenderink, A. Alù, and A. Polman, "High-index dielectric metasurfaces performing

mathematical operations," *Nano Lett.*, vol. 19, pp. 8418–8423, 2019.

[36] T. Zhu, Y. Lou, Y. Zhou, et al., "Generalized spatial differentiation from the spin hall effect of light and its application in image processing of edge detection," *Phys. Rev. Appl.*, vol. 11, p. 3, 2019.

[37] T. Zhu, Y. Zhou, Y. Lou, et al., "Plasmonic computing of spatial differentiation," *Nat. Commun.*, vol. 8, p. 15391, 2017.

[38] J. Zhou, H. Qian, C. F. Chen, et al., "Optical edge detection based on high-efficiency dielectric metasurface," *Proc. Natl. Acad. Sci. U. S. A.*, vol. 116, pp. 11137–11140, 2019.

[39] J. Zhou, H. Qian, J. Zhao, et al. Two-dimensional optical spatial differentiation and high-contrast imaging. *Natl. Sci. Rev.* 2020, https://doi.org/10.1093/nsr/nwaa176.

[40] P. Huo, C. Zhang, W. Zhu, et al., "Photonic spin-multiplexing metasurface for switchable spiral phase contrast imaging," *Nano Lett.*, vol. 20, pp. 2791–2798, 2020.

[41] F. Zhang, M. Pu, X. Li, et al., "All-dielectric metasurfaces for simultaneous giant circular asymmetric transmission and wavefront shaping based on asymmetric photonic spin-orbit interactions," *Adv. Funct. Mater.*, vol. 27, p. 47, 2017.

[42] F. Zhang, M. Pu, J. Luo, H. Yu, and X. Luo, "Symmetry breaking of photonic spin-orbit interactions in metasurfaces," *Opto-Electro. Eng.*, vol. 44, pp. 319–325, 2017.

[43] X. Luo, M. Pu, X. Li, and X. Ma, "Applications. Broadband spin hall effect of light in single nanoapertures," *Light-Sci. Appl.*, vol. 6, p. e16276, 2017.

[44] M. Pu, X. Li, X. Ma, et al., "Catenary optics for achromatic generation of perfect optical angular momentum," *Sci. Adv.*, vol. 1, p. e1500396, 2015.

[45] A. Arbabi, Y. Horie, M. Bagheri, and A. Faraon, "Dielectric metasurfaces for complete control of phase and polarization with subwavelength spatial resolution and high transmission," *Nat. Nanotechnol.*, vol. 10, pp. 937–943, 2015.

Supplementary material: The online version of this article offers supplementary material (https://doi.org/10.1515/nanoph-2020-0366).